ENVIRONMENTAL ENGINEERING AND SCIENCE

AN INTRODUCTION

Ram S. Gupta, PhD

Government Institutes
Rockville, MD

621866

Government Institutes, Inc., 4 Research Place, Rockville, Maryland 20850, USA.

Copyright ©1997 by Government Institutes. All rights reserved.

01 00 99 98 97   5 4 3 2

No part of this work may be reproduced or transmitted in any form or by any means, electronic or mechanical, including photocopying, recording, or any information storage and retrieval system, without permission in writing from the publisher. All requests for permission to reproduce material from this work should be directed to Government Institutes, Inc., 4 Research Place, Rockville, Maryland 20850, USA.

The reader should not rely on this publication to address specific questions that apply to a particular set of facts. The authors and publisher make no representation or warranty, express or implied, as to the completeness, correctness, or utility of the information in this publication. In addition, the authors and publisher assume no liability of any kind whatsoever resulting from the use of or reliance upon the contents of this book.

Library of Congress Cataloging-in-Publication Data

Gupta, Ram S.
    Environmental engineering and science: an introduction / by Ram S. Gupta
        p.   cm.
    Includes index.
    ISBN: 0-86587-548-0
    1. Environmental sciences. 2. Environmental engineering.
    I. Title.
    GE105.G86   1997
    628--dc21
                    96-27082
                    CIP

Printed in the United States of America

TABLE OF CONTENTS

Tables and Figures xiii
Preface xvii
About the Author xix

Chapter 1: AN ENVIRONMENTAL MODEL 1

Delineation of Parameters 1
The Three E's Connection 4
A Basic Environmental Quality Model 6
Model Components 8
Extension of the Basic Model 9
Characteristics of the Model Components 12
    Human Population Trends 13
    Arithmetic Growth Model 14
    Exponential Growth Model 15
    Declining Rate of Growth 15
Resources Consumption 17
Pollution of the Environment 21
Summary 22
Problems 24
Study Questions 27
References 29

Chapter 2: MATTER AND MATERIALS BALANCE 31

Organization of Matter 31
Atomic Theory of Matter 34
    Atomic Dimensions 35
Radioactive Matter 36
Properties of Matter 38
First Property of Matter: Mass 38
    The Mass-Force Relation: Newton's Three Basic Laws 38
    Force of Attraction Between Masses: Newton's Law of Universal Gravitation 40
The Law of Conservation of Mass— The Continuity Equation 42
Materials Balance and Pollution 43

A Steady-State Equilibrium Conserved System 44
A Steady-State Stored Conserved System 46
A Steady-State Equilibrium Nonconserved System 49
Problems 53
Study Questions 56
References 58

Chapter 3: PRINCIPLES OF ENERGY AND ENERGY ALTERNATIVES 59

Work and Energy 59
Units of Energy 60
Forms of Energy 61
The Laws of Energy 63
The Concept of Conservation of Mass-Energy 65
The Source of Energy 66
The Outward Flow of Energy from the Earth 70
A Simple Inflow-Outflow Blackbody Model 72
The Role of Meteorology in Atmospheric Pollution 74
Matter and Energy in Society Infrastructure 77
Energy for the 21st Century 78
Alternative Energy Strategy 79
    Conservation of Energy 80
    Renewable Energy Resources 81
Solar Energy 83
    Passive Solar Energy Conversion for Heating and Cooling 83
    Active Solar Energy for Heating Space and Water 84
    Solar-Thermal-Electric Engines 84
    Converting Solar Energy Directly to Electricity by Photovoltaic (PV) Cells 85
    Advantages and Disadvantages of Solar Energy 85
Nuclear Energy 86
Problems 87
Study Questions 89
References 91

Chapter 4: PRINCIPLES OF ENVIRONMENTAL CHEMISTRY 95

Chemical Reactions 95
Solutions 98
    Liquid Solutions 98
Solubility 99

Compound's Degree of Solubility	100
Ions	100
Acids and Bases	100
Equilibrium Constants	101
The pH Concept	103
Gases	105
Expression of Gaseous Pollutants	107
Partial Pressures of Gases	108
Solubility of Gases	108
Chemistry of Organic Chemicals	109
Types of Organic Compounds	109
Organic Formulas and Names	112
Alkane Group of Hydrocarbons	114
Alkene Group of Hydrocarbons	114
Alkyne Group of Hydrocarbons	114
Aromatic Group of Hydrocarbons	115
Derivatives of Hydrocarbons	116
Other Organic Compounds	116
The Chemistry of Energy Generation	117
Petroleum	117
Coal	118
Natural Gas	118
Methanol and Ethanol	118
Hydrogen Gas	118
Fuel Recovery from Garbage	119
Photovoltaic Cells	119
Nuclear Energy	119
Problems	121
Study Questions	125
References	127

Chapter 5: PRINCIPLES OF ECOLOGY AND MICROBIOLOGY 129

Ecosystem Concepts	129
Biotic Component Organization in an Ecosystem	130
Principles Pertaining to the Abiotic Component	136
Role of the Abiotic Component in an Ecosystem	136
Inorganic Substances	137
Organic Substances	137
Temperature	137
Humidity	138

Winds, Currents, and Waves 138
Pressure 138
Light 139
Fire 139
Soil 139
The Flow of Energy Through an Ecosystem 141
Energy Flow in a Modern City 143
Matter and Energy Transfer Through Trophic Levels 146
Microorganisms and Their Role in the Environment 148
Microbial Kinetics 149
Biological Parameters of Waste Quality 153
Indicator Organisms for Pollution 153
Biochemical Oxygen Demand (BOD) 156
Problems 158
Study Questions 162
References 164

Chapter 6: THE FUNCTIONING ECOSYSTEM 167

Materials Cycling in an Ecosystem 167
A Steady-State Biogeochemical Cycle 169
The Water Cycle 169
The Carbon Cycle 171
The Nitrogen Cycle 173
The Oxygen Cycle 176
Stresses on an Ecosystem 179
Homeostasis 182
Inertia 182
Resilience 182
Adaptability 182
Succession 184
Biodiversity and its Significance 185
Ethical 185
Aesthetic 186
Sustainability 186
Existence 186
Ecological 186
Causes of Biodiversity Loss 186
Population Growth 187
Increased Resource Consumption 187
Ignorance About the Ecosystem 187

    Government Policies 187
    Trading Systems 187
    Inequitable Resources Distribution 188
    Wetland Encroachment 188
    Deforestation 188
Strategy to Conserve Biodiversity 188
    Policy Changes 188
    Integrated Land Use 189
    Species Protection 189
    Habitat Protection 189
    Ex Situ Conservation 190
    Pollution Control 190
Ecosystem Diversity 190
Freshwater Ecosystems 191
Marine Ecosystems 193
Terrestrial Ecosystems 195
Problems 199
Study Questions 202
References 204

Chapter 7: PROCESS ENGINEERING 205

Environmental Processes 205
Transport Processes 205
    Convective Transport 206
    Diffusive or Fickian Transport 206
    Combined Convection-Diffusion Transport 209
Settling Processes 209
    Sedimentation Process 210
    The Concept of Terminal Velocity 212
    Sediment Basin Design Considerations 214
Filtration Processes: Liquid-Solid Separation 216
    Sequence of Filtration Operation 217
    Hydraulics of Filtration and Backwash 218
Chemical Processes 220
    Reactors and Resident Time 221
    Types of Reactors 221
Reaction Kinetics 221
Chemical Processes in Dissolved Material Removal Trains 223
    Chemical Precipitation Process 223
    Ion Exchange Process 225

    Gas Transfer Process 226
    Adsorption Process 226
    Membrane-Separation Process 227
    Disinfection Process 228
Biological Processes in Suspended Material Removal Trains 229
Types of Biological Processes 230
    Activated Sludge Process 230
    Trickling Filter Process 232
    Anaerobic Digestion 232
Problems 235
Study Questions 239
References 241

Chapter 8: THE WATER ENVIRONMENT 243

Quantitative Assessment of Water Resources 243
Quality of Water Resources 244
    Quality of Natural Waters 247
    Effluent Discharge into Natural Waters 247
    Quality of Processed Water 249
Pollution of Water Resources 252
Fate of Pollutants in the Water Environment 254
    Bioconcentration Concept 255
Measurements of Quality Characteristics 257
Analytical Procedures 258
    Gravimetric Analysis 258
    Volumetric Analysis 258
    Colorimetric Analysis 259
    Direct Instrumental Measurements 261
Basic Examination of Water and Wastewater 263
    pH 263
    Acidity 263
    Alkalinity 263
    Dissolved Oxygen 264
    Total Solids 265
    Biochemical Oxygen Demand 266
    Coliform Bacteria 266
Problems 267
Study Questions 270
References 272

Chapter 9: POLLUTION AND TREATMENT OF THE WATER ENVIRONMENT 275

The Problems of Aquatic Pollution 275
Pollution of Atmospheric Water: Acid Rain 275
Stream Pollution 275
    Toxins 277
    Sediments 277
    Thermal Regime 277
    Organic Matter 277
    The Oxygen Sag Curve 279
Lake Pollution 282
    Eutrophication 282
    Cultural Eutrophication Factors 283
    Stratification 283
    Control of Eutrophication 284
Groundwater Contamination 285
    An Overview of Groundwater 285
    Mechanism of Groundwater Contamination 288
    Fate of Contaminants in Groundwater 290
    Fate of Soluble Contaminants 291
    Immiscible or Nonaqueous Phase Liquids (NAPL) 292
    Groundwater Pollution Control 296
    Groundwater Treatment Techniques 297
Ocean Pollution 298
    Oil Spills into the Ocean 299
    Control of Spills 300
Treatment of Water and Wastewater 301
Water Treatment 302
    Physical Unit Operations 302
    Chemical Unit Processes 302
Wastewater Treatment 303
    Primary Treatment 303
    Secondary Treatment 304
Design of Wastewater Units 305
    Primary Sedimentation Tank 305
    Secondary Treatment Units 306
    Activated Sludge Aeration Tank 306
    Trickling Filter 307
    Final Sedimentation Tank (Clarifier) 309
Efficiency of Treatment 310
Advanced or Tertiary Treatment 311

Sludge Treatment 311
Sludge Weight-Volume Relation 312
Problems 314
Study Questions 317
References 319

Chapter 10: THE ATMOSPHERIC ENVIRONMENT 321

Introduction to the Atmosphere 321
    Troposphere 323
    Stratosphere 323
    Mesosphere 323
    Upper Atmosphere 324
    Pressure Variation in the Atmosphere 324
Quality of the Atmosphere 325
    Units of Pollutant Measurement 326
    Quality Standards of Pollutants 327
    Sources, Types, and Effects of Pollutants 329
Fate of Chemicals in the Atmosphere 330
    Removal of Pollutants by Deposition 331
    Dispersion of Pollutants in the Atmosphere 333
    Atmospheric Chemical Reactions 334
Problems of Atmospheric Pollution 334
Acid Rain 335
    Transport, Transformation and Deposition of Acid Rain 335
    Acid Rain Damage 337
    Acid Rain Control Measures 337
Smog Formation 337
    Thermal Inversion: The Mechanism of Smog Formation 338
Greenhouse Effect 341
    Enhanced Greenhouse Effect or Radioactive Forcing 343
    Impacts of Greenhouse Warming 345
Ozone Depletion 347
    Basic Theory of Depletion 347
    Ozone-Depleting Chemicals 348
    Impacts of Ozone Depletion 349
Air Pollution Control 350
Particulate Control 351
    Gravity Settlers 351
    Cyclone Separators 352
    Fabric Filters 353

         Electrostatic Precipitators (ESPs) 353
         Wet Scrubbers 353
         Collection Efficiency of Control Equipment 353
Control of Gases and Vapors 354
Problems 358
Study Questions 360
References 362

Chapter 11: THE TERRESTRIAL ENVIRONMENT 365

The Terrestrial System and Solid Waste 365
Sources and Classification of Solid Wastes 366
Unit Waste Factors 368
Physical Properties of Waste Components 369
Determining Waste Generation 370
Fate of Pollutants in Soil Matrix 371
Municipal Solid Waste (MSW) Management 372
         Handling and Separating of MSW at the Source 373
         Collection and Transfer Operations 373
         Processing Operations 377
         Resource Conservation and Recovery 377
         Treatment and Disposal 378
MSW Landfill Design Considerations 379
         Landfill Configurations 379
         Landfill Size 381
Characteristics of Hazardous Wastes 382
         Ignitability 382
         Corrosivity 382
         Reactivity 382
         Toxicity 383
Regulation of Hazardous Waste 383
The Cradle-to-Grave Concept 384
Hazardous Waste Management 385
         Waste Minimization Methods 385
         Hazardous Waste Treatment Technology 386
         Disposal Technology 387
Problems of the Terrestrial Environment 387
Deforestation 387
         Importance of Forests 388
Preservation of Forests 388
         Fuelwood and Agroforestry 389

Upland Watershed Land Use 389
Forest Management for Industrial Uses 389
Conservation of Tropical Forest Ecosystems 389
Research, Training, and Extension Programs 390
Wetlands Encroachment 390
Wildlife Habitats 390
Water Quality 391
Groundwater Recharge 391
Flood Attenuation 391
Wetland Destruction 391
Wetland Protection Program 392
Underground Storage Tank Technology 392
RCRA UST Program 394
New UST System 394
Upgrading the Existing UST System 396
Closure of UST Systems 396
Aboveground Storage Tank Technology 397
Cleaning of Contaminated Soils 400
Appropriate Tests for Soil Contaminants 400
Soil Treatment Technologies 401
Problems 403
Study Questions 406
References 407

Appendix A – Equivalency Tables **411**
Appendix B – Useful Conversion Factors **415**
Appendix C – Physical Properties of Water **417**
Appendix D – Physical Properties of Air **419**
Appendix E – Saturation Values of Dissolved Oxygen in Water **421**
Appendix F – Quantity of Air Pollutant Emissions **423**
Appendix G – Population/GDP and Energy Consumption in the US **425**
Appendix H – Quantity of Municipal Solid Waste (MSW) Generated **427**

Answers to Chapter Problems **429**

Glossary **439**

Index **491**

TABLES AND FIGURES

INSIDE COVER Prefixes, The Greek Alphabet, and Numerical Designations . Front Cover
Periodic Table of the Elements .... Back Cover

CHAPTER 1	**AN ENVIRONMENTAL MODEL**	1
Figure 1-1	Vertical Structure of the Earth	2
Figure 1-2	System Representation of the Three E's	4
Table 1-1	Historic Data for the U.S. CO_2 Emissions and Contributing Factors ..	10
Table 1-2	Percent Growth Rates Based on the Historic Data	12
Figure 1-3	Population Growth Patterns	13
Figure 1-4	Classification of Natural Resources	18
Figure 1-5	Resource Base Divisions	18
Figure 1-6	Levels of Pollution Control	22
CHAPTER 2	**MATTER AND MATERIAL BALANCE**	31
Figure 2-1	Classification of Matter	32
Figure 2-2	Hierarchical Organization of Matter	33
Figure 2-3	A Simplified Atomic Structure	34
Figure 2-4	Closed Volume System of Stream-Sewage Discharge	45
Figure 2-5	Stored Water Volume System of a Tank	47
Figure 2-6	Stored System with Outflow	48
Figure 2-7	A Nonconserved Sewage Outflow System	50
Figure 2-8	Box Model of a Town System	51
CHAPTER 3	**PRINCIPLES OF ENERGY AND ENERGY ALTERNATIVES** .	59
Table 3-1	Units of Energy	60
Table 3-2	Important Forms of Energy	62
Figure 3-1	Energy Transformations and Energy Conservation	63
Figure 3-2	The Mass-Energy Link	66
Table 3-3	Order of Merit of Different Energy Forms in the Universe	67
Figure 3-3	Flow of Energy to and from the Earth	68
Figure 3-4	Spectrum of Electromagnetic Waves	71
Figure 3-5	A Simple Blackbody Model	73
Figure 3-6	Possible Temperature Profiles	76
Figure 3-7	A Simple One-Way or Throwaway Society Model	77

| Figure 3-8 | A Model of Sustainable Society | 78 |
| Table 3-4 | Evaluation of Energy Alternatives | 81 |

CHAPTER 4 PRINCIPLES OF ENVIRONMENTAL CHEMISTRY 95

Table 4-1	Periodic Table of the Elements	97
Figure 4-1	pH of Fluids	104
Figure 4-2 (a)	Continuous Chain	110
Figure 4-2 (b)	Chain with Branches	110
Figure 4-2 (c)	Ring Bond	110
Figure 4-2 (d)	Double Ring Bond	110
Figure 4-3 (a)	Alkane Compound	111
Figure 4-3 (b)	Alkene Compound	111
Figure 4-3 (c)	Alkyne Compound	111
Figure 4-4 (a)	Alkane (Saturated) Hydrocarbon	112
Figure 4-4 (b)	Different Alkane of Similar Molecular Formula	112
Table 4-2	Hydrocarbon Series	113
Figure 4-5	Example 4-10	113
Figure 4-6	Aromatic Compound	115
Table 4-3	Benzene Series Hydrocarbons	116
Table 4-4	Energy Liberated by Various Fuels (in kcal/g of fuel)	120
Figure 4-7	Problem 4-39	124
Figure 4-8	Problem 4-40	124

CHAPTER 5 PRINCIPLES OF ECOLOGY AND MICROBIOLOGY 129

Figure 5-1	Hierarchical Organization of an Ecosystem	130
Figure 5-2	Producer-Consumer Levels and Food Chain	132
Figure 5-3	Food Web for an Aquatic Ecosystem	133
Table 5-1	Soil Profile	140
Figure 5-4	Flow of Energy and Circulation of Matter Through an Ecosystem	142
Figure 5-5	Lindeman's Model for the Transfer of Energy in an Ecosystem	144
Figure 5-6	A Linear Model Showing the Flow of Energy Through a Food Chain	145
Figure 5-7	Pyramids of Biomass and Energy	147
Figure 5-8	Microbial Growth in Pure Culture	150
Figure 5-9	Successive Dilution of Samples for MPN Testing	154
Table 5-2	Sample Table for Most Probable Number of Coliforms per 100mL	155
Figure 5-10	Biochemical Oxygen Demand Curve	156

CHAPTER 6 THE FUNCTIONING ECOSYSTEM 167

Figure 6-1	A Generalized Geochemical Model	168
Figure 6-2	The Water or Hydrologic System	170
Figure 6-3	The Carbon Cycle	172
Figure 6-4	Varieties and Processes in the Nitrogen Cycle	175

Tables and Figures xv

Figure 6-5	Quantities in the Nitrogen Cycle	177
Figure 6-6	The Sphere of Oxygen Cycles	178
Figure 6-7	Range of Species Population Growth	180
Figure 6-8	Dynamic Balance of an Ecosystem	180
Figure 6-9	Evolution of a Species	183
Figure 6-10	Energy Penetration Zones in Lakes	191
Figure 6-11	Zonation of the Sea	193
Figure 6-12	Triangle of Biomes Showing Broad Distribution of World's Plant Formations	196
CHAPTER 7	**PROCESS ENGINEERING**	**205**
Figure 7-1	Simple Diffusion Device	207
Figure 7-2	Categories of Sedimentation	211
Figure 7-3	Settlement of a Discrete Particle	212
Figure 7-4	An Ideal Sedimentation Basin	214
Figure 7-5	Rapid Sand Filter	217
Figure 7-6	Reactors and Their Features	222
Figure 7-7	Double-Boundary Layer Theory	224
Table 7-1	Common Chemical Precipitants	225
Table 7-2	Principal Biological Processes	230
Figure 7-8	Flow Diagram of an Activated Sludge Process	231
Figure 7-9	Flow Diagram of a Trickling Filter	232
Figure 7-10	Anaerobic Digester	233
CHAPTER 8	**THE WATER ENVIRONMENT**	**243**
Table 8-1	Common Characteristics of Water Quality Assessment	245
Table 8-2	Major Sources and Types of Pollutants	253
Figure 8-1	Pathways of Pollutants to Natural Waters	254
Figure 8-2	Schematic Diagram of a Photoelectric Device	260
Figure 8-3	Schematic Diagram of a Gas Chromatograph	262
Figure 8-4	Steps of Laboratory Procedures to Determine Solids Concentration	265
CHAPTER 9	**POLLUTION AND TREATMENT OF THE WATER ENVIRONMENT**	**275**
Figure 9-1	Discharge from a Point Source	276
Table 9-1	Saturation Dissolved Oxygen Concentration	278
Figure 9-2	Oxygen Sag Curve	279
Figure 9-3	Plot of Dissolved Oxygen vs Time	282
Figure 9-4	Summer Stratification Ending with Fall Overturn	284
Figure 9-5	Meinzer's Groundwater Classification	286
Figure 9-6	Model of Channel and Lake	288
Figure 9-7	Pathways of Groundwater Contamination	289

Figure 9-8a	Typical Behavior of Light Nonaqueous Phase Liquid (LNAPL) Underground	294
Figure 9-8b	Typical Behavior of Dense Nonaqueous Phase Liquid (DNAPL) Underground	295
Figure 9-9	Schematic Diagram of a Water Treatment System	301
Figure 9-10	Wastewater Treatment System	304

CHAPTER 10 THE ATMOSPHERIC ENVIRONMENT ... 321

Table 10-1	Composition of Dry Air	321
Figure 10-1	Stratification of the Atmosphere	322
Table 10-2	Values of Gas Constant, R	325
Table 10-3	National Ambient Air Quality Standards	328
Table 10-4	Types, Sources, and Effects of Common Pollutants	329
Figure 10-2	Terminal Velocity of Spherical Particles in Air at 25 °C	332
Figure 10-3	Fate of Acidifying gases after Emission	336
Figure 10-4	Atmospheric Stability	339
Figure 10-5	Strongly Stable Atmosphere and Smog Formation	340
Figure 10-6	Temperature Profile of Example 10-4	341
Figure 10-7	Atmospheric Absorption of Solar Radiation	342
Figure 10-8	A Simplified Diagram of the Greenhouse Effect	342
Table 10-5	Characteristics of Greenhouse Gases	344
Table 10-6	Profile of Ozone-Depleting Chemicals	349
Figure 10-9	Contributory Factor Based Classification	350
Figure 10-10	Pollutant Based Classification	351
Figure 10-11	Cyclone Separator	352
Table 10-7	Example 10-6 Computations	354

CHAPTER 11 THE TERRESTRIAL ENVIRONMENT ... 365

Table 11-1	Average Consumption of the Residential Municipal Solid Waste	367
Table 11-2	Unit Waste Factors	368
Table 11-3	Physical Properties of Wastes	369
Figure 11-1	Route Collection Sequence	374
Figure 11-2	Cross-Section of a Double-Liner Containment System	380
Table 11-4	Requirements for UST Systems	395
Table 11-5	Tests for Contaminated Soils and Groundwaters	398
Table 11-6	Soil Treatment Technologies	398
Figure 11-3	In Situ Bioremediation System	402

PREFACE

The study of the environment is a multidisciplinary field. It combines principles of chemistry, physics, biology, ecology, energy, engineering, and management. Although most environmental professionals use aspects of these specialties in their everyday design, management, and compliance activities, few have received training or have extensive experience in all of them.

When I taught *ENVR101: Introduction to Environmental Engineering* at the Roger Williams University to the first freshman class of the Environmental Engineering Science Program, I recognized the lack of a suitable textbook for the course. The books available on the subject were either too technically intense for an overview course, too mathematically advanced for students just finishing high school, or essentially nontechnical in nature, suitable only for a general education course. I noted that professionals also did not have a book they could refer to for referencing topics they could not recall from their college days, unless they were prepared to brush up on calculus first.

Environmental Science and Engineering: An Introduction provides comprehensive coverage, presenting a blend of scientific principles and engineering theory and application in quantitative terms at a basic level. It is appropriate for a science-oriented or an engineering-oriented course, and is perfect for professional reference. The fundamentals of the scientific disciplines of physics, chemistry, and biology, as they relate to the environment, have been presented in separate chapters without assuming that readers have had any prerequisite coursework. Depending upon the preparation of the reader, some of these basic science chapters can be skipped. The mathematics has been kept at the high school algebra level throughout the book.

Engineering aspects commence with an introduction to the concept of modeling and a simple environmental model. Important engineering processes involved in environmental unit operations and designs have been presented, followed up by the theory and applications with respect to each of the three physical environmental resources—air, water, and land. Mathematical formulations have been simplified to the extent possible in order to gain quantitative understanding of the phenomena at a very basic level. All environmentally related issues (such as acid rain, smog and thermal inversion, ozone erosion, greenhouse effect, groundwater contamination, ocean oil spills, and waste discharge into streams and lakes) are fully covered, including details on the current state of the art in control measures.

Two of my former students were very helpful in the preparation of the book: Neal Personeus created all the artwork on the AUTOCAD, and Stacey Kurbiec arranged the references for various chapters. Raina Robins, an Environmental Engineering senior student helped in the preparation of the solutions manual. To the reviewers, I would like to gratefully acknowledge the valuable assistance provided toward the finalization of the text: specifically, Dr. Gary Shook of Boise State University, Mr. Jack Daugherty, Environmental Safety Engineer, Jackson, Mississippi, and Dr. Igor Runge, Dean of School of Engineering, Roger Williams University, Bristol, RI. The staff of the School of Engineering extended a helping hand in numerous tasks involved in manuscript preparation. James Devlin, Director of Laboratories, always welcomed me in North Campus Lab and also helped in resolving many computer problems. My sincere thanks to all of them.

It is my hope that as a basic text and desk reference in one volume, *Environmental Science and Engineering: An Introduction* will be referred to repeatedly by environmental professionals throughout their careers..

Ram S. Gupta

ABOUT THE AUTHOR

Ram S. Gupta is a professor of environmental engineering at Roger Williams University and president of Delta Engineers, an engineering consulting firm specializing in on-site disposal systems, drainage systems, landfill design, coastal structures, and containment remediation. Dr. Gupta has over 30 years of field and teaching experience in civil and environmental engineering.

He has worked on water projects in the United States, Australia, India, and Liberia, and was instrumental in developing the highly successful Environmental Science and Engineering Program at Roger Williams University. He is a member of the American Society of Civil Engineers and Water Environment Federation and is the author of *Hydrology and Hydraulic Systems*. A Registered Professional Engineer on Massachusetts and Rhode Island, Dr. Gupta holds doctoral and masters degrees in civil engineering from Polytechnic University of New York and University of Roorkee, India, respectively.

Chapter 1

AN ENVIRONMENTAL MODEL

Delineation of Parameters

Of the billions upon billions of entities that exist in the immense span of the universe, only one, the planet Earth, is known to be capable of supporting an intelligent form of life. The vertical structure of the planet Earth is illustrated in Figure 1-1. The planet comprises the earth's interior, the upper part of which is known as the *lithosphere*; the envelope of air mass, known as the *atmosphere*; and the surrounding body of water, known as the *hydrosphere*. Only a small fraction of the planet is favorable to life in some form. The part in which life exists in any form—plants, animals, and protist—is referred to as the *biosphere*. The biosphere is remarkably limited in depth, being concentrated close to the surface of the earth; it is equal to the skin if the earth is compared to an apple. The biosphere penetrates the earth's crust to about 4 mi. which is only 1/1000 of the earth's radius; it envelopes 4–5 mi. into the atmosphere, and extends to a depth of 5–6 mi. in the hydrosphere. However, in the lateral direction, life has great mobility, extending throughout the periphery of the planet.

Along with the living organisms that are grouped together into the *biotic component*, lifeless objects exist side by side in the biosphere. These include natural factors such as sunlight, wind, temperature, and topography, as well as material objects such as soil, minerals, and water. These are grouped into the *abiotic component*. The biotic and abiotic components function together as a single unit called an *ecological system,* or *ecosystem*. The study of the relationships and interactions between the biotic and abiotic components is *ecology*. Thus, an ecosystem is a functional unit in ecology. An ecosystem could be as complex as the entire earth system or as simple as a small aquarium in a household, as long as the biotic and the abiotic components are present and operate together to achieve a functional stability for an appreciable period of time.

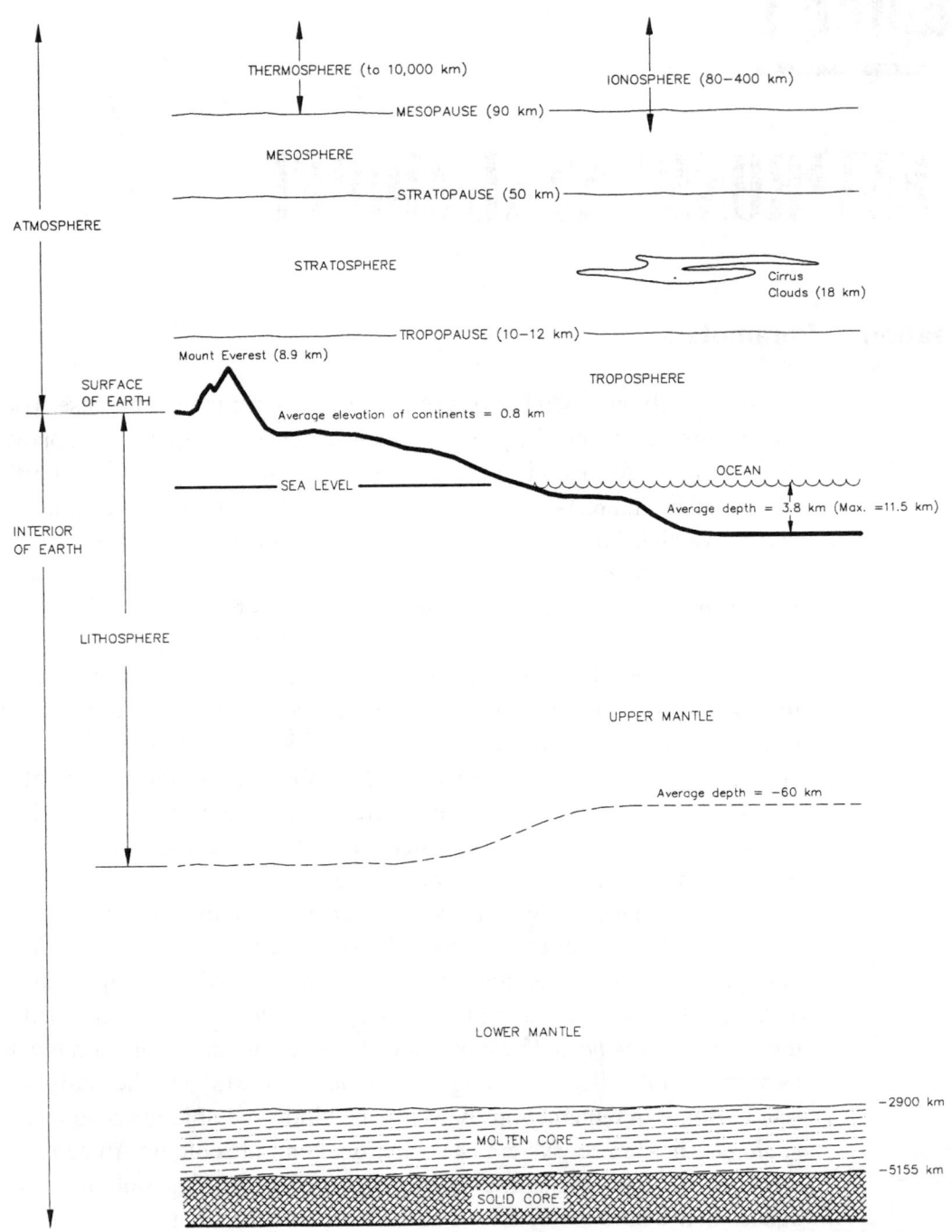

Figure 1-1 Vertical Structure of the Earth
From Ehlich, Ehlich, and Holden; © 1977 by W.H. Freeman & Co.
Used with permission.

The abiotic surroundings of living organisms, that interact with them and affect them continually, collectively constitute the *environment* of the living organism. It includes the natural factors that are responsible for seasons, weather, wind, waves, sunshine, and rain, and the material objects that control the activities of and provide necessary resources for the survival of living beings. The atmosphere serves as both protector and preserver; it transmits the sunlight to plants and animals for their growth and yet also shields them from the harmful radiation from outer space. The sun directly and indirectly nourishes life on the earth. Likewise, living organisms influence the abiotic environment abundantly.

Within the ecosystem any physical entity that occupies space, can be seen, smelled or felt, is known as *matter*. The content of matter is its *mass*. The gravitational pull exerted by the earth on the mass gives *weight* to matter.

Energy, which is defined as "the ability to perform work," is a mover of matter. Energy is essential for all vital processes at every ecosystem level. Without it, the ecosystem will collapse and life will cease to exist. The sun is a prime mover of the earth as well as the other planets of the solar system. Thus, the sun is the prime energy source on the earth in a direct or a derivative form.

For energy and material objectives on the earth, the law of the balance of nature applies, according to which new energy and matter are not created; the same energy and matter appear in different forms. This means that the misuse of energy and matter may render them unfit for further consumption, thus affecting the useable supply.

The foremost among the biotic components, human beings, maintained a balance with nature or with our abiotic component in earlier times. However, this position changed gradually. The present technological society is disturbing this balance with massive use of fossil fuel in the form of coal, oil, gas, and heavy pollution of resources.

Nature created a unique planet with a favorable climate and fertile land on which life was able to grow and advance. The process of evolution led to development of the intelligent form of life that has been bestowed with immense faculty of reasoning. A natural calamity that is capable of making this planet unfit for habitation is not expected on the human time scale for millions of years to come. A real danger, on the other hand, lies from the action of human beings, ourselves.

Humans, through advancement of technology, are currently very powerful. With our expanding ability we have been directly or indirectly interfering, both knowingly and unknowingly, with the functioning of various ecosystems that have existed for billions of years. Our power to change and control seems to be increasing faster than our understanding of the results of the profound changes of which we are now capable. This is a potentially dangerous situation.

The Three E's Connection

Energy, ecosystem, and environment are three inseparable phases in the human-to-nature relationship. The living organisms that constitute an ecosystem require energy for their survival and growth. As such, energy imposes a direct restriction on the evolution of the ecological system. The environment controls the activities of the living organisms that exist within it. At the same time, living organisms constantly influence the environment and distribution of energy in many ways; they return new compounds and energy sources to the environment. This forward and backward action-reaction chain continues until an equilibrium is achieved among the energy, environment, and ecosystem. A fixed set of formulas cannot adequately describe the relationship among these elements because the entire process is dynamic and too complex.

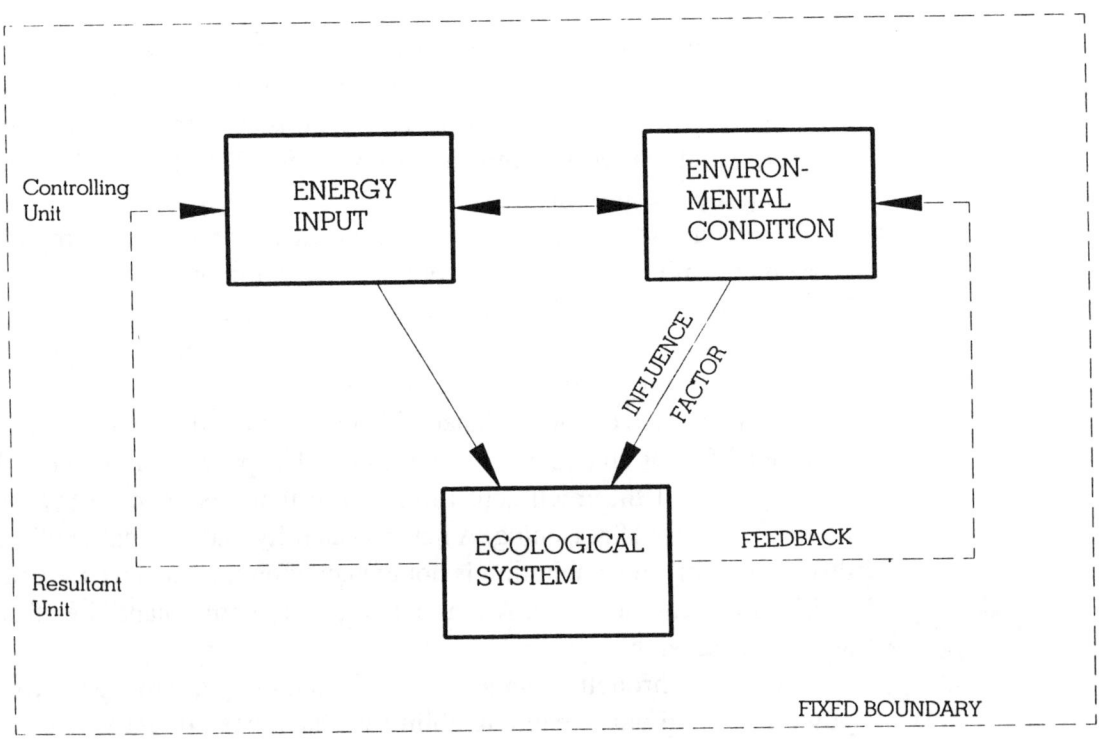

Figure 1-2 System Representation of the Three E's

A generalized model of the natural system is shown in Figure 1-2. The boxes represent three key elements of the system. The solid connecting arrows are the influencing factors. The intricate interactions among the elements are self evident. From the ecological system's viewpoint, energy input and the environmental conditions are the two controlling units; the structural form and features of the ecosystem that ultimately develop are the cumulated effect of these two controlling units. A similar assessment can be made from the viewpoint of the energy input or the environmental condition as the ultimate resultant unit, with the other two components acting as the controlling units. An important feature of the system is the presence of a feedback loop, represented by the broken arrows, which indicate that the resultant unit transfers its stresses back to the controlling units, affecting them consequently.

For example, consider the impact of automobiles on Los Angeles. The source of energy is gasoline. The region's climatic conditions and topographic surroundings set up the environmental condition. In the presence of sunlight, automobile emissions trigger *photochemical smog*, to which that setting is vulnerable. The effect on the ecosystem is in terms of damage to plants and animals, disappearance of certain species, and health-related problems among humans. The changing weather patterns, emission control regulations, use of car pools, and the clean air legislation can be viewed as responses of the ecosystem that will, in turn, cast their effects on the environment and energy distribution in the region, and a chain reaction continues.

The engineering approach to studying any problem is to set limits of space and time around the problem. An analysis is performed within a defined boundary over a fixed period of time so that all input and output parameters can be traced and accounted for. These boundaries may be confined to a single nucleus or extend to the entire universe, within a fraction of a second or to a span of a light-year. This is the *controlled-volume* or *closed-boundary* approach.

A model is a convenient tool for studying an engineering problem. It can be used to study how a particular process will perform or to directly derive the outcome of a system without ascertaining the factual explanation of the process itself. There are three types of models:

1. A *physical* model depicts the system through an actual scale replica of the original. Experiments performed on the model are interpreted for the system's performance,

2. A *simulated* or *analogue* model draws an analogy between two different systems. For example, a model of groundwater is made using an electrical circuit in which the flow of current represents the flow of water. The correlation of the two systems helps to convert the results.

3. A *mathematical* model consists of a set of equations representing the various phenomena involved in the environmental system. The equations are solved for the conditions similar to those existing in the system. The statistical relations can be used in the modeling without a theoretical understanding. However, unless the system's features and conditions are duplicated, the equations themselves do not represent the model. This depiction, known as setting of the boundary conditions, is achieved through the application of the finite difference or finite element methods. Mathematical models are almost always computer based and, in essence, they are implemented by the computer programs.

All types of models are invalid until they are *calibrated*, which means they should be run using the historic or known values to verify that the results derived are similar to the actual values obtained from the real system. The calibration provides a means to adjust the model.

A Basic Environmental Quality Model

Environmental pollution can be expressed by the following simple relation in accordance with the statements by Ehrlich and Holdren (1971):

$$Q = IFP \tag{1-1}$$

where Q = level of pollution
P = size of population
F = amount of material resource or energy resource consumed per person
I = impact index (pollution contributed) per unit of resource consumed.

To represent a model, the above relation has to be considered within the defined limits of space and time and with respect to a particular type of pollutant. The relation will also need to be calibrated with a known set of data with respect to all parameters of the model. The relative weights can be assigned to the components during calibration while the simple structure of the model is retained.

Ehrlich and Holdren (1971) argued that the size of the population is a major component in environmental degradation; the other two components F and I are not independent, but themselves are functions of the population size, P. In other words, the effect of population increase is not linear on pollution contribution.

It has been shown that the cost of reducing contaminants to progressively lower levels increases very rapidly. For example, sewage treatment at a secondary treatment plant removes about 80% of contaminants (in the form of biochemical oxygen demand [BOD]). Advanced treatment to remove a higher percentage of contamination may cost two to four times as much. Consider a case in which the situation demands that the pollution level Q should be held constant as the population doubles. If the per capita consumption F does not change in Equation 1-1, the contaminant contribution per resource unit I has to be halved; i.e., the per unit effectiveness of the pollution control measures has to be doubled, at two to four times the usual price. Thus, for a doubled population, the total cost would be four- to- eight times of the original cost.

Conversely, according to Commoner and others (1971) the major component of the model is the increased environmental impact per unit of production/consumption I. Commoner examined the rise in pollution in the U.S. as a whole during the post-war period, 1946-68. The change in the pollution level was estimated at between 200 and 1000%. The increase in the U.S. population for the period amounted to about 43%. The gross national product (GNP), an average indicator of the per capita resource production, increased about 59% during the period. It was, accordingly, concluded that two components, P and F, together are not enough to account for the much higher level of pollution recorded. Commoner and others (1971) accordingly attributed the major problem to the increased environmental impact per unit of production (rise in the rate of pollution) I caused by the high degree of technological changes and large number of product replacements.

Example 1-1

The town of Bristol, RI has a population of 25,000. The average wastewater (sewage) generated is 100 gal/person/day (GPCD). The pollution strength of the wastewater is 200 mg/L (in terms of BOD). Determine the level of pollution of the wastewater.

Solution:

Time duration: daily
Space limits: Bristol Town
Type of pollutant: BOD
P = 25,000 persons
F = 100 gal/person/day or 379 L/person/day
I = 200 mg/L.

From Equation 1-1

$$P = \left(200 \frac{mg}{L}\right)\left(379 \frac{L}{(persons)(day)}\right)(25{,}000 \text{ persons})$$

$$= 1895 \times 10^6 \text{ mg/day or } 1895 \text{ kg/day}$$
$$\text{or } 4170 \text{ lbs/day (of BOD).}$$

Model Components

While opinions may vary on the relative effectiveness of the degradation factors in the U.S. context, on a global basis, certain features of the model components are well recognized.

The world's population in 1990 was 5.3 billion and is expected to reach 8.5 billion by 2025. If the countries of the world are divided in two groups based on the average annual per capita GNP, the *more developed countries* (MDC), (comprising all countries in Europe, North America, Japan, Australia, New Zealand, and the former USSR) had a total population of 1.2 billion in 1990, which is 23% of the world's population. This is projected to reach 1.4 billion in 2025, registering an annual growth rate of 0.5%. All other countries (including Asia, Africa, and Central and South America) grouped under the *less developed countries* (LDC), had 4.1 billion persons or 77% of the world's population in 1990. This is expected to rise to 7.1 billion by 2025, indicating an annual growth rate of 2.1%. Thus, the component of population P in LDCs has a dominating effect that needs to be addressed.

The second component of the model relates to the rate of resource use. A *resource* is anything needed by a living organism, the increased availability of which leads to a rise in energy flow through the organisms. Matter and energy sources fit the definition of resources. MDCs, with a meager population, consume about 80% of the world's mineral and energy resources, leaving only 20% for the vast majority of population of the LDCs. MDCs bear a large extent of responsibility for pollution due to resource use. The third component of the model, relating to the degree of pollution of the resource used, has been greatly impacted by technology. Scientific advancements have played both positive and negative roles.

The manufacturing of new resources such as plastics, pesticides, certain chlorinated fluorocarbons (CFCs), nuclear materials and increased use of coal, oil, and natural gas have increased the pollution level. Scientists and engineers have also developed substitutes for scarce resources, devised more efficient systems, and invented methods for controlling and cleaning up the pollution, thus reducing the impact of pollution. The MDC's research efforts should be directed toward earth-friendly advancements. Appropriate technology transfers can then be made to LDCs to sustain the earth.

Extension of the Basic Model

For energy-related environmental degradation, the resource term F can be broken into two identities to relate with GNP. Then Equation 1-1, can be expressed as follows:

$$Environmental\ degredation = \left(\frac{pollution}{energy}\right)\left(\frac{energy}{GNP}\right)\left(\frac{GNP}{population}\right)(population) \quad (1-2)$$

The second term relates to the amount of energy required to produce a unit of GNP. This is a measure of efficiency. The efforts towards reducing degradation have mostly concentrated on the first two elements of the model. The amount of pollution per unit of energy is controlled with technologies such as scrubbing and precipitation. The amount of energy required to produce a unit of GNP is reduced with application of technologies that improve the efficiency of conversion and energy utilization.

The third element of per capita GNP is a common economic measure of the standard of living or economic well-being of a society. It is interesting to note that in the model, a higher per capita GNP means a negative contribution to the environment. Thus, an important question: What is an adequate sustainable standard of living? The importance of integrating energy and environmental policy with economic policy becomes evident.

In Equation 1-2, the environmental degradation may be related to some specific emission; say CO_2 for example. Then, the relationship will be of the form:

$$CO_2 = \left(\frac{CO_2}{energy}\right)\left(\frac{energy}{GNP}\right)\left(\frac{GNP}{population}\right)(population) \quad (1-3)$$

The application of the above relation for analysis of environmental effects has been made by Kaya et al. (1989), Kaya (1990), and Gray et al. (1990). Gray proposed the use of Gross Domestic Product (GDP) in place of GNP in the context of historic and prospective analysis for the U.S.

Example 1-2

From the data given in Appendices F-1 and G, yearly information have been compiled in Table 1-1 for carbon intensity (CO_2/energy ratio), energy efficiency (energy/GDP ratio), per capita GDP, and population size for the U.S. Determine the environmental degradation in terms of the emissions of CO_2.

Solution:

From Equation 1-2, for 1980:

$$CO_2 = \left(16.61 \times 10^{-6} \frac{mg}{Btu}\right) \times \left(28.26 \times 10^3 \frac{Btu}{\$}\right) \times \left(11.8 \times 10^3 \frac{\$}{persons}\right)$$
$$\times (228 \times 10^6 \; persons)$$
$$= 1{,}262{,}870 \times 10^6 \; mg \;\; or \;\; 1{,}263 \times 10^3 \; kg$$

Similarly, the values for other years are computed as shown in the last column of Table 1-1.

Table 1-1 Historic Data for the U.S. CO_2 Emissions and Contributing Factors

Year	CO_2/Energy (mg/Btu × 10^{-6})	Energy/GDP (Btu/$ × 10^3)	GDP/Person ($/person × 10^3)	Population (person × 10^6)	CO_2 emission (kg × 10^3)
1980	16.61	28.26	11.80	228	1263
1981	16.40	24.61	13.08	230	1214
1982	16.26	22.72	13.44	232	1152
1983	16.42	21.06	14.30	234	1157
1984	16.10	19.94	15.73	236	1192
1985	16.46	18.67	16.61	238	1215
1986	16.65	17.79	17.34	241	1238
1987	16.69	17.26	18.32	243	1282
1988	16.92	16.68	19.60	245	1355
1989	16.84	15.85	20.73	248	1372
1990	16.57	15.07	21.58	250	1347

The annual growth rate (yearly difference) of each identity in Table 1-1 can be determined by successively subtracting the values of each year from the following year. The percent growth rate of each identity can be obtained by making a ratio of the annual growth rate with the yearly value for that identity in Table 1-1. The percent growth rates are arranged in Table 1-2. As a first-order approximation, the cumulated percent growth rate of CO_2 or any other pollutant can be ascertained by adding the percent growth rate of each identity. Thus:

$$\{c\} = \left\{\frac{c}{e}\right\} + \left\{\frac{e}{o}\right\} + \left\{\frac{o}{p}\right\} + \{p\} \tag{1-4}$$

where $\{c\}$ = percent growth rate of pollutant

$\left\{\frac{c}{e}\right\}$ = percent growth rate of pollutant to energy ratio

$\left\{\frac{e}{o}\right\}$ = percent growth rate in energy efficiency

$\left\{\frac{o}{p}\right\}$ = percent growth rate in economic well-being

$\{p\}$ = percent growth rate in population

Example 1-3
From the annual growth rate data in Table 1-2, determine the yearly growth rate of CO_2.

Solution:
From Equation 1-4, for 1980–81:

$\{c\} = \{-1.26\} + \{-12.92\} + \{10.85\} + \{0.88\}$
   $= -2.45$ (a decline by 2.45%.)

Similarly, the values for other years are computed in the last column of Table 1-2.

Table 1-2 Percent Growth Rates Based on the Historic Data

Period	$\left\{\dfrac{c}{e}\right\}$	$\left\{\dfrac{e}{o}\right\}$	$\left\{\dfrac{o}{p}\right\}$	$\{p\}$	$\{c\}$
1980–81	-1.26	-12.92	10.85	0.88	-2.45
1981–82	-0.85	-7.68	1.22	0.87	-6.44
1982–83	0.98	-7.31	6.40	0.86	0.93
1983–84	-1.95	-5.32	10.00	0.85	3.58
1984–85	2.20	-6.37	5.59	0.85	2.27
1985-86	1.18	-4.71	4.39	1.27	2.13
1986-87	0.24	-3.00	5.65	0.83	3.72
1987-88	1.38	-3.36	7.00	0.82	5.84
1988-89	-0.47	-5.00	5.77	1.22	1.52
1989-90	-1.60	-4.92	4.10	0.81	-1.61

From the analysis, the role of efficiency improvements in reducing CO_2 emissions is clear. For example, during 1980–81, efficiency improvement (12.92% per year) is many times more important than emission control measures (1.26% per year). Obviously, economic well-being and population growth are major factors contributing to pollution. Overall, a decrease was achieved because of energy improvements and control measures. Without these, the CO_2 emissions would have risen by 11.73% during 1980–81. It might be noted from Table 1-2, that economic well-being continues to be an important factor contributing to enhanced CO_2 but with a generally diminishing effect.

Characteristics of the Model Components

The basic model of environmental degradation described earlier on page 6 incorporated three elements: size of population, use of resources, and pollution of the resources used. The basic concepts as they relate to the classification, trend, and distribution of each of these elements are discussed below.

Human Population Trends

In the early days, the world population grew very slowly. It took millions of years during the Stone, Bronze, Iron, and Middle Ages to accumulate the first billion people by the year 1800. It took another 130 years to add the second billion. In the Modern Age, the trend shifted to rapid growth. The next billion were added in 30 years by 1960. The population of 4 billion was reached in the next 14 years by 1974, and only 13 years later it touched the level of 5 billion in 1987. If the trend continues, one billion people will be added each decade. The pattern of growth is very uneven throughout the world; the more-developed countries registered only a 15% rise from 1970 to 1990, as compared to 55% among the less-developed countries.

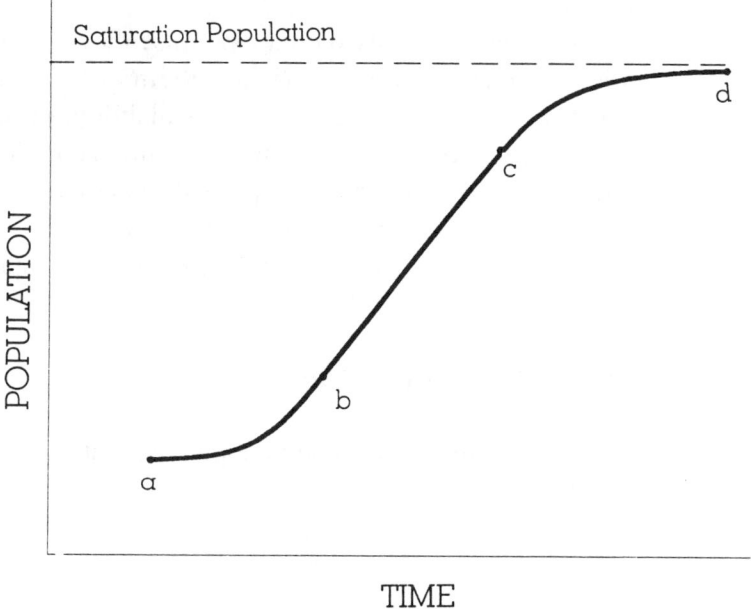

Figure 1-3 Population Growth Patterns

A population model is an appropriate tool to trace the population growth in an individual country or a smaller unit, and also to project future population. For long-term projections of more than 10 years, sophisticated models are used that account for births, deaths, and migrations within the population unit. The mathematical models assume one of the three forms of growth pattern of population as shown by the three segments in Figure 1-3. For each segment, there is a distinct equation applied. Segment '*bc*' indicates arithmetic growth. This is typical of a place where the population grows by some fixed number during each unit of time. For example, a city of 10,000 that records a growth of 1000 every year. After 5 years, the population will rise by 5,000 to 15,000 persons.

Segment '*ab*' refers to exponential growth, where the rise is a fixed percent of the then existing population. For example, if the population of a city of 10,000 increases by 10% per year, the first-year increase will be 1000 with an end of the year population of 11,000. The second year rise will be 10% of 11,000 (or 1100) and the end of the second year population will be 12,100. The subsequent three year additions will be 1210, 1331, and 1464, producing at the end of five years a population of 16,105. This is similar to the compounding of interest on money in a bank. With exponential growth a quantity grows faster and faster with the passage of time. A quantity that is growing exponentially will take a fixed period of time to double regardless of its size at the start. Similarly, a quantity decaying (decreasing) exponentially will have a fixed *half-life*[1] or, the time needed to reach to one-half of the starting level.

Segment '*cd*' is based on the premise that any place has a certain fixed holding capacity referred to as the *saturation population,* which is dictated by the physical constraints, resource availability, and infrastructure facilities. As the population of a place approaches this saturation level, the growth rate slows, i.e., the rate of growth is proportional to the deficit in population with respect to the saturation population ($P_{sat}-P$).

The mathematical framework of the models representing each of these segments, follows.

Arithmetic Growth Model

The growth rate (change in population with time) is a fixed quantity, or:

$$\frac{dP}{dt} = K$$

where P = population
t = time
K = growth constant.

Integrating, we obtain:

$$P_t = P_o + K(t) \tag{1-5}$$

---

[1] Used in the context of radioactivity. It is the time during which one-half of the radioactive nuclei in a given sample will decay.

where P_0 = initial population or population at time, $t = 0$
P_t = projected population t years after P_0
t = period of projection

and
$$K = \frac{P_2 - P_1}{\Delta t} \qquad (1\text{-}6)$$

where P_1 and P_2 are recorded populations (two levels of the resource) in the past at some Δt interval apart.

Exponential Growth Model

For exponential or geometric rate of growth, the rate of population rise is proportional to the level of current population, that is

$$\frac{dP}{dt} = KP$$

Integration of the above equation leads to the following, similar to Equations 1-5 and 1-6:

$$\ln P_t = \ln P_0 + K(t) \qquad (1\text{-}7)$$

and
$$K = \frac{\ln P_2 - \ln P_1}{\Delta t} \qquad (1\text{-}8)$$

Declining Rate of Growth

In Equations 1-7 and 1-8, the population parameters P are substituted by the values of population deficit, i.e., $(P_{sat} - P)$. The same two relations are valid for declining rate of growth, as given:

$$\ln (P_{sat} - P_t) = \ln (P_{sat} - P_0) + K(t) \qquad (1\text{-}9)$$

and
$$K = \frac{\ln (P_{sat} - P_2) - \ln (P_{sat} - P_1)}{\Delta t} \qquad (1\text{-}10)$$

Example 1-4
The population of a town has been recorded in 1970 and 1990 as 15,000 and 22,000, respectively. Estimate the year 2000 population, assuming arithmetic growth.

Solution:
From Equation 1-6:

$$K = \frac{22,000 - 15,000}{20} = 350$$

From Equation 1-5: $P_{2000} = P_{1990} + K(t)$
$$= 22,000 + 350 (10)$$
$$= 25,500 \text{ persons}$$

Example 1-5
In Example 1-4, estimate the year 2000 population, assuming geometric growth.

Solution:
From Equation 1-8:
$$K = \frac{\ln 22,000 - \ln 15,000}{20}$$
$$= \frac{9.9999 - 9.616}{20} = 0.0192$$

From Equation 1-7:

$$\ln P_{2000} = \ln 22,000 + 0.0192(10)$$
$$= 10.191$$

$$P_{2000} = \text{inv } \ln 10.191 \text{ } or \text{ } e^{10.191}$$
$$= 26,640 \text{ persons}$$

Example 1-6
In Example 1-4, if the saturation of the town is 50,000, estimate the year 2000 population, assuming declining rate of growth.

Solution:
From Equation 1-10:
$$K = \frac{\ln(50,000 - 22,000) - \ln(50,000 - 15,000)}{20}$$
$$= \frac{10.240 - 10.463}{20} = -0.011$$

From Equation 1-9:

$$\ln(50{,}000 - P_{2000}) = \ln(50{,}000 - 22{,}000) + (-0.011)(10) = 10.13$$

$$P_{2000} = 50{,}000 - e^{10.13}$$
$$= 24{,}920 \text{ persons}$$

In accordance with model requirements, a population unit with a distinct boundary such as a town, city, or region, has to be modeled. The projection has to be made for a defined time interval. The model has to be calibrated from historic data for the same population unit. The historic population data should be plotted with respect to time. The match of the plot to one of the three segments of Figure 1-3 indicates which model type is applicable for the population projection in question.

For large population centers such as large cities, states, or nations, the *Logistic Curve Method* is used. In this method, a set of mathematical equations representing an empirical S shaped curve, similar to Figure 1-3, is used. The coefficients of these equations are evaluated (calibrated) from historic data. As stated earlier, for long term projections of more than 10 years, sophisticated models are used that simulate trends of the population units through various factors.

Resources Consumption

All environmental problems arise because of resource consumption. Broadly, all resources belong to two groups: *stock resources* that are fixed in quantity and, therefore, nonrenewable, and *flow resources* which are continually available and as such, are renewable. Further subdivisions of the two categories are shown in Figure 1-4, with examples.

The finite quantity of a given stock resource on the earth is known as the *resource base*. The further classification of the resource base is shown in Figure 1-5. It is the proven reserves that can be extracted at economic prices (for economic gain at current prices) with existing technology. The boundary between proven and conditional reserves is subject to continuous revision in the light of available technology. Certain nonrenewable resources that are discarded after use can be recycled. However, recycling is limited by the level of technology and cost considerations. Thus, many potentially recoverable resources end up within the "consumed by use" category. The renewable resources, though consumed, are capable of natural replacement on a human time scale.

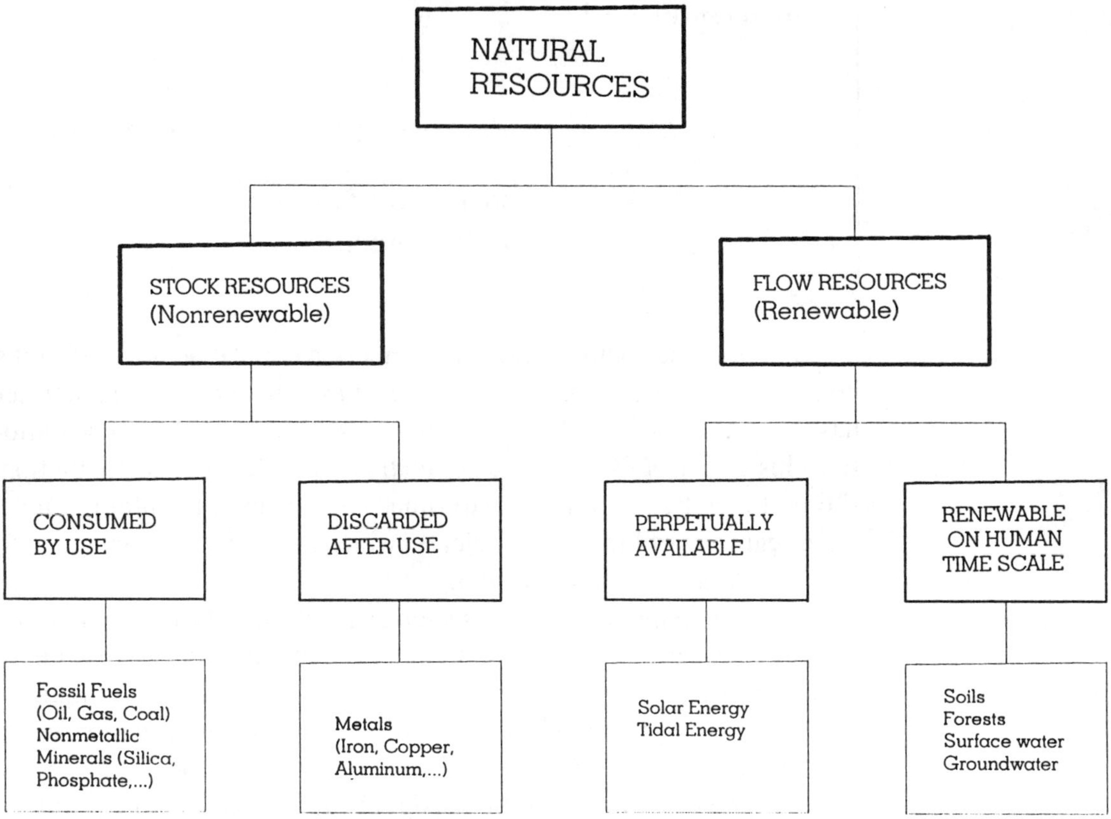

NATURAL RESOURCES
STOCK RESOURCES (Nonrenewable)
FLOW RESOURCES (Renewable)
CONSUMED BY USE
DISCARDED AFTER USE
PERPETUALLY AVAILABLE
RENEWABLE ON HUMAN TIME SCALE
Fossil Fuels (Oil, Gas, Coal) Nonmetallic Minerals (Silica, Phosphate,...)
Metals (Iron, Copper, Aluminum,...)
Solar Energy Tidal Energy
Soils Forests Surface water Groundwater

Figure 1-4 Classification of Natural Resources

USED RESOURCE
PROVEN
RESERVES
CONDITIONAL
RESERVES
HYPOTHETICAL RESOURCES
UNCERTAIN
SPECULATIVE RESOURCES
RESERVES
RESOURCE BASE
RESOURCE BASE

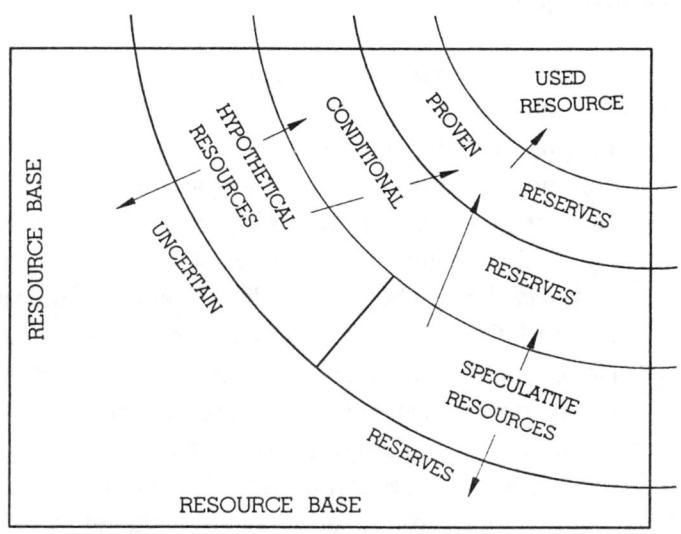

Figure 1-5 Resource Base Divisions
(As modified from Owens and Owens, 1991)

An Environmental Model

A renewable resource will last indefinitely if used with care within the limits of the *sustained yield,* which is the highest usage rate without bringing any adverse effect on the renewable potential. When over-exploited, the renewable resources may become nonrenewable or even nonexistent. There are many common-property renewable resources—the air we breathe, water in rivers and lakes, or natural forests— that belong to everyone or to no one. The unrestricted exploitation of the common-property renewable resources can cause many problems. The stock of many non renewable resources deplete completely with over exploitation.

The consumption of a resource may grow arithmetically, exponentially, or by a declining growth rate similar to human population growth (the saturation population will be equivalent to the reserves of the resource, in this case). In such a case, the relevant equations referred to in the previous section would be applicable to predict the resource consumption. When a quantity decreases exponentially, Equations 1-7 and 1-8 are relevant but the rate constant has a negative value. However, the consumption pattern during the entire cycle of resource use is quite complicated. It might follow one of the defined patterns such as arithmetic, exponential, and declining rate of increase (or a combination thereof) during a phase of resource development; at other times the consumption might not fit into any defined mode. Hubbert (1969) suggested a bell-shaped curve for the complete consumption cycle of a resource has been suggested. It begins with exponential growth while the resource is plentiful, eventually consumption peaks out, and then a downward trend takes place as the resource becomes scarce and substitutions take effect. The *normal distribution function* might be used to define this model.

Example 1-7
The worldwide growth rate of of natural gas consumption is 3.5% per year. If growth continues at the same rate, how long will it take for gas consumption to double?

Solution:
Let the present consumption level, P_0 = X units (trillion ft^3)
Future consumption level, P_t = 2X units
Rate of growth = 0.035 per year
Period = doubling time = t_p

From Equation 1-7:

$$\ln 2X = \ln X + K(t_p)$$

$$\ln \frac{2X}{X} = K(t_p)$$

$$\ln 2 = K(t_p)$$

hence... $t_p = \dfrac{0.693}{K}$

$t_p = 20$ years, by substituting $K = 0.035$.

Note that in the mathematics of exponential growth, the doubling period is independent of the level of consumption. It depends only on the rate of growth (decay). Equation 1-11 gives the doubling time or the half-life.

Example 1-8

A form of carbon (^{14}C) is radioactive. It has a half-life of 5568 years. If 0.5 g of this isotope is contained in a matter, how much radioactivity will be left after 100 years?

Solution:

Half-life, $t_p = 5568$ years
Present radioactivity $= 0.5$ g
Projection time, $t = 100$ years

From Equation 1-11:

$$5568 = \frac{0.693}{-K}$$

since the quantity is decreasing, K has a negative sign
hence, $K = -1.245 \times 10^{-4}$ per year.

From Equation 1-7:
$$\ln P_t = \ln 0.5 + (-1.245 \times 10^{-4})(100)$$
$$= -0.7056$$
$$P_t = 0.494 \text{ g.}$$

Pollution of the Environment

The pollution is defined in many ways. A widely accepted definition of pollution is as follows (Owens, 1991):

Pollution is the introduction by human action, directly or indirectly, of substances or energy into the environment, resulting in deleterious effects of such a nature as to endanger human health, harm living resources or ecosystems, and impair or interfere with amenities and other legitimate uses of the environment.

Two important points that emerge from this definition are that pollution is manmade; sulfur emissions from an industry constitute pollution, but emissions from a volcano don't, and pollution occurs only when damage is imminent; an equal amount of a substance may pollute one ecosystem and not the other if the second ecosystem has a capacity to absorb and neutralize the substance.

The term *contamination* is used to indicate the presence of a substance in the environment that has the potential for damage. This is a fine line of distinction from the definition of pollution that involves direct harmful effects as a criterion. Sometimes, a time lag is involved between accumulation of substances (such as DDT or mercury) and damage to the environment; i.e. there is a delay between contamination and pollution. However, the distinction between definitions is often ignored and both pollution and contamination are used interchangeably to include all substances that cause damage or have potential for damage.

Environmental resources that are polluted are public goods, or common property resources that are considered practically free inputs to the production process. Thus, the pollution is an external cost of production. These externalities can be internalized to some extent by making the polluters install particular types of pollution control equipment, pay penalties for noncompliance, pay taxes according to the amount of pollution produced, and buy the pollution rights. It would be ideal to eliminate emission of pollutants altogether or to seek *zero pollution*. This is, obviously, impractical. An alternative approach is to seek a balance between cost of pollution control and cost paid without pollution control. The cost of pollution control escalates steeply as the level of pollution approaches zero, as shown by the pollution control curve in Figure 1-6.

Conversely, the damage potential decreases sharply as the pollution level goes down. The intersection of the two curves indicates the *socially optimum level of pollution*. However, it is usually not feasible to identify the socially optimum level because it is difficult to quantify all the damages in monetary terms, such as effects on diversity in the ecosystem, cultural heritage, and long-term human health. Since the above-mentioned two goals of zero pollution and socially optimum level of pollution are not possible in practice, the pollution control policy is designed to achieve the *acceptable level of pollution control*. This level is based on *best available technology* (BAT), which changes over time and space and varies between different societies.

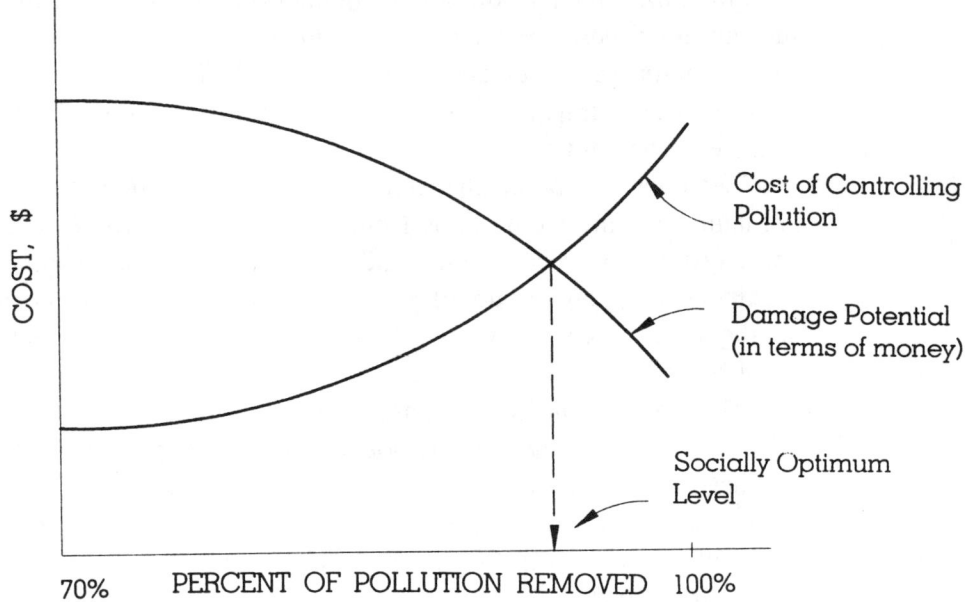

Figure 1-6 Levels of Pollution Control

There are two sources of pollution. *Point pollution* is when the source can be identified such as a sewage plant effluent pipe or factory exhaust chimney. *Nonpoint pollution* refers to dispersed pollution from one or more sources, such as pollutants leaching from an area.

There are two approaches to pollution control. *Input control* directs efforts at the source of pollution, whereas *output control* concentrates on the receiving environment (for example, a lake, a forest, or the air). Input control is more favorable because it provides for efficient treatment, allows for a stringent control at the source, and implies setting of the uniform emission standards. A pollution problem is often transferred from one sector to another rather than solved; for example, when wastewater is treated, solid waste is formed and must be disposed. This *cross-media* effect should be avoided by considering the overall impact.

Summary

Everything that occupies space is matter. Everything we do, uses energy. The thinking we do, the movie we watch, the food we eat, the fire we burn, and the automobile in which we ride, are all examples of energy consumption. A person's intelligence is the same kind of energy as that of water stored behind a hydropower dam. The birth of the universe began with energy and the prime mover of the grand system is also energy. Life on the earth owes its existence to energy from the sun. The rise of civilization is a story of the development of energy resources. The key word to advancement is energy. The flow and use of energy on the earth, the subsistence of life in the biosphere, and the state of our physical environment, are three vital elements that are connected intimately with each other. When we misuse any energy resource, we disturb the accompanied environment and degrade the related ecosystem, and vice versa.

Environmental degradation can be quantified with a simple three-element model comprised of population size, per capita resource consumption, and pollution of each unit of resource used. In an extension of this model, the energy-related environmental degradation has been correlated with four elements comprising the pollution produced per unit of energy, efficiency improvements to energy conversions, economic well-being, and the population size. The environmental, energy, and economic policies are, thus, interwoven.

Population growth and resource consumption follow similar mathematical formulations. Three models make short-term projections of population sizes and resource development. Technology impacts all elements of the environmental model, but it plays a vital role in energy improvements, and pollution control. Improved economic welfare and growth of population make a negative contribution to environmental quality, and, as such, emphasize the need for a sustained standard of living and controlled population.

PROBLEMS

1. In Example 1-1, for the town of Bristol (population: 25,000, wastewater flow 100 GPCD), the pollution (BOD) level is reduced to 40 mg/L at the secondary treatment plant before the effluent (treated wastewater) is discharged into Mount Hope Bay. What is the total quantity of pollution in the effluent? If the price to treat to this stage is 25¢/1000 gal, what is the cost of the treatment?

2. The drinking water supply for a city is from a river that has a dissolved oxygen concentration of 6.2 mg/L. The city population is 15,000 and the rate of supply is 80 GPCD. For the water to be supplied at the saturated dissolved oxygen level of 8 mg/L, how much total oxygen supply needs to be added in a week?

3. In a bar, the air exhaust system refreshes the air every hour. The following statistics have been noted on the average: number of customers per day: 500; number of cigarettes each customer smokes: 5; nicotine added to air per cigarette: 0.002 mg. Determine the level of the nicotine in the stale air at the bar.

4. A town supplies well water to its 2,000 residents. The well water has a hardness of 400 mg/L (as $CaCO_3$). The water demand is 60 GPCD for domestic use. The water is first treated to lower its hardness to 150 mg/L before it is supplied. Determine the total level of the hardness in the water that is supplied to the residents, and the amount of hardness removed.

5. The population of the town in Problem 1 is expected to increase to 35,000. It is stipulated that the total amount of pollution due to effluent discharged in the bay should not increase from the level in Problem 1. Determine the pollutant (BOD) level to which the effluent should be treated.

6. If the following price schedule applies to sewage treatment of 200 mg/L strength, determine the cost of treatment in Problem 5. Compare the cost with Problem 1: 80% treatment = 25¢/1000 gal; 85% treatment = 40¢/1000 gal; and 90% treatment = 70¢/1000 gal.

7. Carbon monoxide emissions per unit of energy are given below. Using other variables from Table 1-1, determine the yearly emissions of carbon monoxide in the U.S.

Year	1980	1981	1982	1983	1984	1985	1986	1987	1988	1989	1990
CO/energy (mg/Btu × 10^{-6})	1.54	1.51	1.49	1.49	1.38	1.32	1.28	1.17	1.12	1.04	1.03

An Environmental Model

8. For Problem 7, prepare a percent growth rate table similar to Table 1-2. Determine, the yearly growth rate of carbon monoxide, and the relative contribution to pollution by various factors.

9. The prospective annual growth rates of CO_2 emissions in the U.S. are given below along with certain other percent growth factors. Determine the contribution of energy efficiency in reducing the CO_2 emissions. Without energy efficiency, project the CO_2 growth pattern for the future.

| | Percent Growth Rates | | | | |
Period	$\{c/e\}$	$\{e/o\}$	$\{o/p\}$	$\{p\}$	$\{c\}$
1985–2000	-0.28	?	2.26	0.74	-0.53
2000–2025	-1.62	?	2.18	0.32	-2.16
2025–2050	-1.59	?	2.06	-0.06	-1.35
2050–2075	-0.85	?	1.53	-0.03	-0.89
2075–2100	-0.72	?	0.99	0.01	-1.15

10. The population of a city has been recorded in 1975 and 1990 as 100,000 and 150,000, respectively. Estimate the 1998 population, assuming arithmetic growth.

11. Solve Problem 10, assuming geometric growth.

12. Solve Problem 10, assuming a declining rate of growth. The saturation population is 220,000.

13. Using the following census figures, estimate the population for 1997 by the appropriate model.

Year	Population (in thousands)
1960	35.8
1970	38.2
1980	40.7
1990	43.3

14. From the following census data, estimate the 1975 and 1998 population by the appropriate model.

Year	Population
1960	25,000
1970	28,190
1980	31,780
1990	35,830

15. For Problem 11, determine how many years it will take the population to double to 200,000.

16. The worldwide production of coal in 1990 was 5.3 billion tons/year. The recoverable reserves are estimated at 4 trillion tons. The growth has been exponential in the past at a rate of 2.7%/year. How long will the coal supply last? At the zero growth rate (production at the current level), how long will it take to exhaust the reserves?

17. Data show that there was an exponential growth at 4.1% per year for oil energy until the 1973 oil embargo, after which time it has leveled off to near zero growth. In 1973, the energy input was about 80 exaJoules (80×10^{18} Joules) or 675.9×10^{15} Btu. Determine the level of energy input in 1960.

18. The worldwide annual aluminum production in 1975 and 1990 was 74.9 million tons and 109.1 million tons, respectively. The bauxite (source of aluminum) world reserves are 21.8 billion tons. Determine the life (years the resources will last), and the time required to double the production of aluminum.

19. The concentration of the carbon dioxide in the atmosphere rose exponentially from 316 parts per million in 1959 to 355 parts per million in 1991. If this trend continues, determine the concentration level by the turn of the century.

20. A radioisotope has a half-life of 1 year. If 100g of the isotope is present now, how long it will take until 10g (radioactivity) is left?

STUDY QUESTIONS

1. Treating the environmental condition of an industrial town as the resultant unit, describe the relationships among energy, ecosystem, and environment.

2. Among environmentalists, there are two opinions about the impact of population on the environment. Some believe that population increase is the main factor of environmental degradation. Others contend that technological changes are the principal causes of degraded environmental quality. Present your arguments in favor of each of the two schools of thought.

3. Based on human population trends, draw a graph of population vs. time showing how population has grown from early time to the present. Make an approximate forecast of the likely population in 2010 by extending the graph.

4. Based on population growth rates from 1970 to 1990, make simple extrapolations to ascertain and compare the population doubling time between the more developed countries (MDC) and the less developed countries (LDC). Based on resource consumption trends, will the early population doubling time of LDC impose a serious constraint on the resources?

5. Describe the simple environmental degradation model. Discuss the characteristics of each component. What role does technology play on each component?

6. A higher per capita GNP is a measure of the economic well-being of a society. However, it makes a negative contribution to the environment in light of the extension model. Discuss the pros and cons of this position.

7. What are the major types of resources? How are they degraded? From environmental consideration, which is the best type of resource to be used? Which one is most abundantly used? Is the use of resources consistent with environmental priority?

8. Write down the definition of pollution from a dictionary. Compare it to the textbook definition. List the distinguishing criteria. Discuss the pollution control measures with respect to zero pollution, socially optimum level of pollution, and acceptable level of pollution.

9. What are the two major sources of pollution? Identify five items under each category.

10. Define the following:
    a. Atmosphere, lithosphere, hydrosphere, and biosphere
    b. Biotic and abiotic components
    c. Mathematical models
    d. Less developed countries (LDC) and more developed countries (MDC)
    e. Arithmetic, exponential, and declining rate of growth
    f. Stock resources and flow resources

REFERENCES

Commoner, B., Corr, M., and Stamler, P.J., "The Causes of Pollution," *Environment,* April 1971, Reprinted Goldfarb, T.D., ed., *Taking Sides*, 4th ed., Dushkin Publishing Group, Guilford, CT, 1991.

Ehrlich, P. R., Ehrlich, A. H., and Holdren, J. P., *Ecoscience, Population, Resources, Environment,* W. H. Freeman and Company, San Francisco, CA, 1977.

Ehrlich, P. R., and Holdren, J. P., "Impact of Population Growth," *Science*, March 26, 1971.

Gibbons, J. H., "The Interface of Environmental Science, Technology and Policy," Tester, J. W., Wood, D. O., and Ferrari, N. A., ed., *Energy and the Environment in the 21st Century,* The MIT Press, Cambridge, MA, 1991.

Gray, P. E., Tester, J. W., and Wood, D. O., "Energy Technology: Problems and Solutions," Tester, J. W., Wood, D. O., and Ferrari, N. A., ed., *Energy and the Environment in the 21st Century,* The MIT Press, Cambridge, MA, 1991.

Kaya, Y., Kenjl, Y., and Matsuhoaki, R., " A Grand Strategy for Global Warming," A paper at the Tokyo Conference on the Global Environment and Human Response Toward Sustainable Development, September, 11-13, Tokyo, Japan, 1989.

Kaya, Y., Contribution to "Policy Strategies for Managing the Global Environment," Tester, J. W., Wood, D. O., and Ferrari, N. A., ed., *Energy and the Environment in the 21st Century,* Cambridge, MA, 1991.

Liu, P. I., *Introduction to Energy and the Environment*, Van Nostrand Reinhold, New York, 1993.

Marden, P. G., and Hodgson, D., ed., *Population Environment and the Quality of Life*, AMS Press, New York, 1975.

Michael, B. C., and Stamer, P. J., "The Causes of Pollution," *Environment*, April, 1971.

Miller, G. T., Jr., *Environmental Science*, 5th ed., Wadsworth Publishing Company, Belmont, CA, 1995.

Odum, E. P., *Fundamentals of Ecology*, 3rd. ed., W. B. Saunders Company, Philadelphia, PA, 1971.

Owens, S., and Owens, P. L., *Environment, Resources and Conservation*, Cambridge University Press, Cambridge, U.K., 1991.

Rees, J., *Natural Resources,* Methuen and Company, London, UK, 1985.

White, I. D., Mottershead, D. N., and Harrison, S. J., *Environmental Systems: An Introductory Text*, 2nd ed., Chapman and Hall, London, U.K., 1984.

Chapter 2

MATTER AND MATERIALS BALANCE

Organization of Matter

Chapter 1 described matter as substance and energy as the mover of the substance, together forming the cosmos. Matter is the building block of the universe. A resource that is at the root of all environmental problems consists of either a matter or an energy source. Matter consists of basic substances called *elements* such as oxygen, carbon, and sodium. If we take a piece of carbon (an element), and continue dividing it into smaller and smaller pieces, we would eventually come to the smallest particle, which if further divided would not show the properties of carbon. The smallest particle of an element that still has all the properties of that element is called an *atom*. Atoms are so minute that it takes hundreds of millions of atoms to draw a line one inch long. Thus, the simplest form of a substance is an element that cannot be broken down into a simpler substance because the atom of an element cannot be split except through nuclear transformation, and in that case, it would lose the characteristics of that element. The atoms of elements unite to form *molecules*. Two or more atoms of the same element can combine to form a molecule, which may be a more stable form of that element. A *compound* is formed by the combination of two or more atoms of different elements held together through a chemical bond. A molecule is the smallest particle in a compound that shows all the properties of that compound. A substance formed by mixing different elements or compounds that are not chemically united is called a *mixture*. Air and soil are examples of mixtures. A *solution* is a special kind of mixture, for example, table salt dissolved in water. The three categories of matter are shown in Figure 2-1.

Eighty-eight natural and 18 manmade elements exist, accounting for a total of 106 known elements. Each element has a distinct symbol made up of the principal letter or letters in the name of the element; Latin names are used for antique elements, such as potassium (kalium) K, copper (cuprum) Cu; and gold (aurum) Au. In the universe, matter is distributed in many material forms comprising the basic 106 elements and over a million compounds formed by the combination of these elements and their mixtures. The hierarchical organization of matter, including living organisms, ranges from minute subatomic particles to gigantic celestial bodies, as shown in Figure 2-2. In between lies the realm of ecology that is discussed in Chapter 5. The flow of energy is involved in each step of the organization of matter.

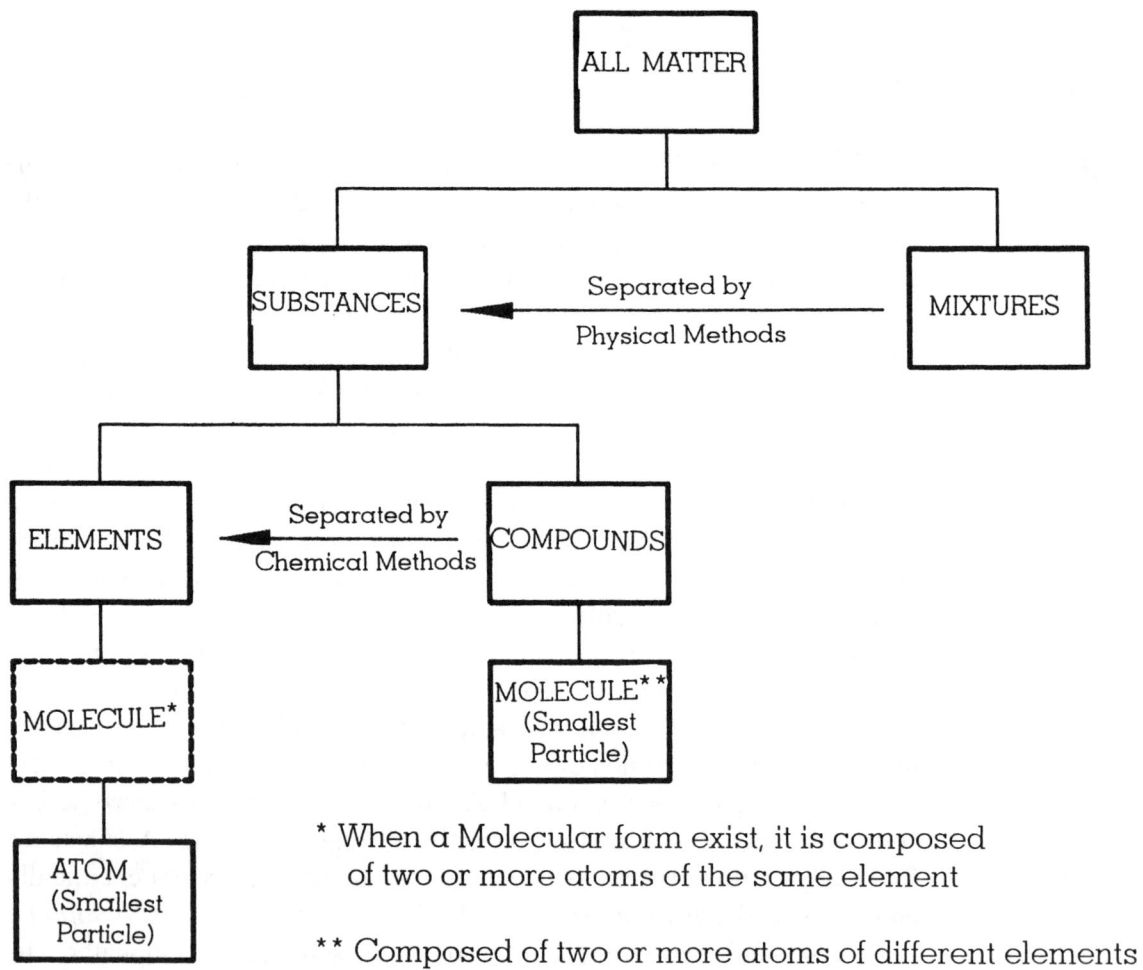

Figure 2-1 Classification of Matter

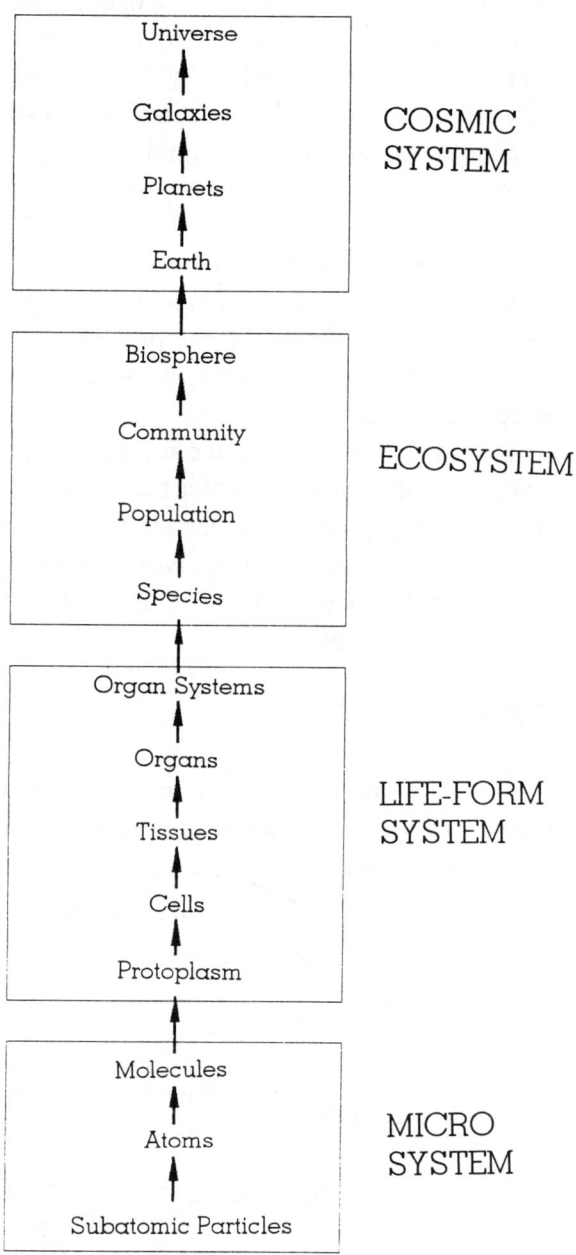

Figure 2-2 Heirarchical Organization of Matter

Matter appears in three states: gaseous, liquid, and solid. Matter in the gaseous state has a tendency to occupy all of the available space. The volume, pressure, and temperature of gaseous matter are dependent on each other and bear a set of well defined relationships. If the temperature of a gas is cooled below a certain value, and the pressure is gradually increased, the gas condenses into a liquid. Liquid flows into the shape of the container but its volume does not change appreciably with pressure and temperature. Matter in the liquid state has greater density. When liquids are cooled to form crystals, they are said to have passed into the solid state. A solid retains its shape and does not change appreciably with pressure and temperature. Solid matter resists shearing and has properties, like hardness, brittleness, and ductility, that may differ in various spatial directions in a solid mass, unlike gaseous and liquid states.

The distribution of matter and transformation of energy within living organisms and nonliving entities that guide our day-to-day life are controlled by the laws of physics. Understanding these laws is an important step in learning how things around us work and how they affect our environment. Many useful concepts are developed in this chapter that have relevance throughout the book.

Atomic Theory of Matter

The modern concept of atomic structure is based upon the presence of three subatomic particles within an atom. An atom is depicted as a tiny sphere in Figure 2-3.

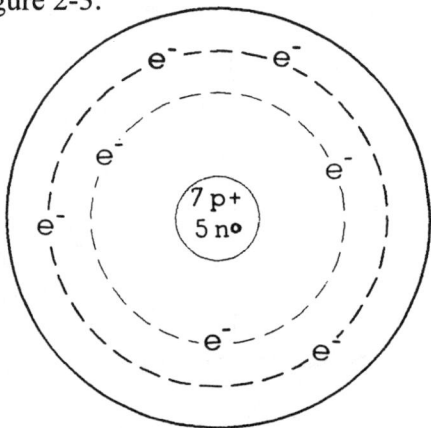

e- Electron, p+ Proton, no Neutron

Atomic Number 7

Atomic Weight 12

Figure 2-3 A Simplified Atomic Structure

An *electron* has a negative electrical charge, a *proton* has a positive electrical charge, and the *neutron*, has no electric charge. The positive charge on a proton exactly equals the negative charge on the electron. The atom as a whole is neutral in charge. According to Bohr's 1913 model of the atom, every atom has a nucleus that contains all of the protons and neutrons. At a distance from the nucleus, there are as many electrons as there are protons, thus balancing the charge to make the atom neutral. The electrons are in continuous motion around the nucleus at a speed in the range of the speed of light. The space between the electrons and the nucleus is empty. Each electron and the nucleus also spin on their own axis and the atom as a whole spins about or vibrates around one spot. The configuration of an atom is similar to the planetary system. Nearly all the mass of an atom is in the nucleus; each proton and neutron weighs about 1837 times as much as an electron; but the electron has a volume 2.5 times that of a proton.

The number of protons in the atom of an element is called the *atomic number* of the element. The total number of protons and neutrons in the atom of an element is called its *mass number*. The weight of an atom is determined by the mass number since electrons have practically zero weight. A weight of one unit is assigned to each proton and neutron. Thus, the *atomic weight* of an atom is equal to the mass number in relative terms of atomic mass unit (AMU). Refer to the following section for a description of units. The atoms of two different elements cannot be similar, and all atoms of a particular element are identical in terms of the number of protons and electrons but they may have a different number of neutrons in the nucleus, and thus could have different mass numbers. Atoms with different mass numbers are called *isotopes*. Isotopes of the same element are identified by appending the mass number to the symbol of the element or using the mass number as a superscript: ^1H and ^2H are two isotopes of hydrogen. Some 8,000 isotopes are believed to exist. Since almost all elements have isotopes (gold is an exception with only one form), the atomic weight listed for an element is the average of the atomic weights of the isotopes of the element, taking into consideration the relative abundance of each isotope in a natural sample. Because of the relative size of subatomic particles, it is difficult to draw a scale model of an atom on a sheet of paper.

Atomic Dimensions

A hydrogen atom is the simplest and smallest atom, with only one proton and one electron. The atom of element cesium is thought to be one of the largest atoms. The dimension of these two atoms are as follows:

	Diameter (cm)	Mass(g)
Hydrogen	2×10^{-8}	1.7×10^{-24}
Cesium	6×10^{-8}	2.2×10^{-22}

The variation in diameter is threefold, but the variation in mass is 1,000 fold. However, these sizes are so minute that they are invisible even under the most powerful microscopes.

An atom with a radius of 1×10^{-8} cm has a nucleus whose radius is only of the order of 10^{-12} cm. Thus, a model of an atom with a dot representing the nucleus would have to represent electrons drawn many yards away. A scale model is, therefore, impractical.

For specifying the size of atoms, physicists have adopted a smaller unit of length called the *Angstrom* (Å); $1 \text{Å} = 1 \times 10^{-8}$ cm. Similarly, for mass of atoms, a relative unit is selected known as the *atomic mass unit* (AMU). This is also referred to as the *atomic weight unit*. One AMU is taken as equal to 1/12 of the atomic mass of the most common form of carbon, which is arbitrarily set at 12.000. The actual mass of 1/12 of carbon, i.e., 1 AMU = 1.66×10^{-24} g. Since this unit is only a relative number, it is dimensionless.

The proton and neutron each have a mass of approximately 1 AMU. Therefore, the mass number of an atom is approximately equal to it's AMU.

Radioactive Matter

Most elements have stable nuclei that do not change or decay with time. However, nuclei of some elements or some of their isotopes, especially those having neutron-to-proton ratio of greater than 1.5, are unstable. These nuclei change spontaneously into different kinds of nuclei by emitting highly energetic nuclear particles. Such elements are said to possess *radioactivity* and their emissions are known as *nuclear radiation*. Some 30 natural elements have been found to be radioactive. The particles emitted by a radioactive nucleus fall in three distinct classes: some are positively charged, some are negatively charged, and some have no charge at all. In earlier days when the nature of these particles was not clear, they were differentiated by the first three letters of the Greek alphabet as alpha (α), beta (β), and gamma (γ) rays; these names are still used. The emission of positively charged alpha rays (particles) by a nucleus necessarily changes the number of protons in an atom and, both the atomic and mass number of the atom is changed; that is, a new element is formed by the alpha decay. For example, when radium-226 emits alpha rays, radon-222 is formed. Beta decay involves emission of negatively charged particles from the nucleus of an atom. A beta (negatively charged) particle does not exist in the nucleus before ejection; it is generated by the conversion of a neutron to a proton. Thus, in beta decay the number of protons increases and the number of neutrons decreases which changes the atomic number while the mass number remains unchanged. Thus, both alpha and beta rays (particles) are emitted by disintegrating or breaking the atomic nuclei. Conversely, gamma rays (which are not particles but electromagnetic rays of very high frequency) result without the disintegration of the atom, due to a rearrangement of neutrons and protons within a nucleus spontaneously. Since high-energy gamma rays have no charge and no particle mass, they are highly penetrating and the most difficult of all radioactive emissions to shield against. Some radioactive nuclei emit only one type of rays (radiation); others emit two or three types.

Alpha, beta, and gamma rays possess more than enough energy to dislodge one or more electrons from atoms of any matter they hit. When an electron is removed from a neutral atom, it takes a positive charge, and is called a *positive ion*. Similarly, an atom may pick up an added electron to become a *negatively charged ion*. This ionization property of radioactive rays is very harmful to living tissues. Ionization disrupts the normal chemical processes in a living cell and causes the cell to grow abnormally or to die. Nuclear radiation may cause *somatic* damage to an organism, which is bodily damage within the life span of an individual. It also causes *genetic* mutation, when affected DNA molecules are passed on to future generations.

The previous section described the nuclear reactions that occur naturally in atoms of unstable nuclei. However, it is possible to take an extremely stable nucleus and make it unstable by bombarding it with highly energetic particles such as alpha particles. These particles can force their way inside the stable nucleus and produce a new unstable nucleus which then decays into yet another nucleus and other smaller particles. This process is called *induced or artificial radioactivity*. This is the basis of *nuclear fission* in which the nucleus of a heavy atom such as uranium-235 is split into lighter nuclei along with high-energy particles known as fission fragments. This produces not only highly radioactive fission products, but also releases a tremendous amount of energy[1]

A reverse process known as *fusion*, in which two light nuclei are combined or fused to form a heavier nucleus, is also accompanied by the release of a large amount of energy. Fusion occurs in the sun constantly and is the source of energy to the Earth. The problem with the fusion process is that it occurs at very high temperatures of hundreds of millions of degrees Celsius. Therefore, it is much more difficult to achieve fusion than fission. To sum up, the three possible sources of nuclear radiation comprise natural radioactivity, fission, and fusion.

---

[1] A small change in mass produces enormous energy. See Chapter 3.

Properties of Matter

One form of matter is distinguished from the other by its properties. There are two types of properties: *physical properties* that describe the nature of matter as it exists, and *chemical properties* that describe its ability to react chemically with other matter. Mass, hardness, color, and odor are physical properties; the first is the most distinct property, as discussed subsequently. Chemical properties are discussed in Chapter 4.

First Property of Matter: Mass

Mass is the first property of matter. It is a fundamental quantity in a body. The quantity of matter present in any object is its mass. A big body has a larger mass than a small body. Mass does not change when the shape of the body is changed. If we dissolve a metal block in acid, the combined mass of the acid and the metal will remain unchanged by the chemical reaction. This fact concerning mass constitutes the first of several statements called the *conservation laws* of physics. French scientist Lavoisier (1743–1794) conclusively demonstrated through his experiments that in chemical reactions no mass is gained or lost; in other words, matter is neither created nor destroyed. This is the first law of conservation known as the *law of conservation of mass*[2]. Thus, except for a small gain or loss, the total supply of matter on the earth is essentially constant. This principle appears in the form of a balance or cycle of elementary materials. However, the underlying problems result from the conversion of a material into its harmful physical or chemical state through the indiscriminate use of the useful form of a compound.

The mass of matter is a measure of its inertia. Inertia is the property that causes matter to resist any change in its condition. For example, the inertia of an automobile has to be overcome to set it into motion or to stop it if it is moving.

The following laws apply to the mass of a body:

The Mass–Force Relation: Newton's Three Basic Laws

Although Italian physicist Galileo (1564–1642) was the first to recognize the inertial property of matter, it was English scientist Newton (1642–1727)

---

[2]This is true for all every day processes that we encounter in practice. However, Einstein's relativity theory predicts that mass is interchangeable with energy. The quantity that is actually conserved is the combination of mass and energy. See Chapter 3 for further explanation.

who incorporated it with force in his three laws of motion, which laid the foundation of mechanics on which the entire structure of physics is based[3].

According to Newton's *first law of motion*, known as the *principle of inertia*, a body will tend to stay at rest or, if it is moving, it will tend to continue to move in a straight line. A force, which is any type of push or pull, is required to overcome the inertia of a body, that is, to stop a moving body or to move a body that is at rest.

Newton's second law expands on his first law. It is sometimes proposed that his first law is contained within his second law. According to the *second law of motion*, the force required to accelerate an object is proportional to the mass of the object and to the acceleration given it. Acceleration is defined as the rate of change of velocity. In mathematical terms, the law is expressed as:

$$F = ma \tag{2-1}$$

where F = force in pounds Newtons or dynes[4]
m = mass in slugs[5], kilograms or grams
a = acceleration or rate of change of velocity in ft/sec^2, m/sec^2, or cm/sec^2

The larger a mass is, the greater is the force needed to produce a given acceleration; or, an object of a given mass will accelerate faster as the force is increased. The acceleration is zero when the velocity of an object does not change with time. From Equation (2-1) the force is zero when acceleration of the object is zero, which means the object will continue to move at a constant speed or stay at rest without any external force.[6] This is the first law stated above.

When an object is falling under the influence of gravity, its downward acceleration, due to the attraction exerted by the earth, is always the same.

---

[3]Newton is also known to be the inventor of calculus. He did ingenious work in the field of optics as well.

[4]In English or FPS units, the measure of *pound* is a weight unit. The corresponding less known mass unit is *slug*, which is obtained when a pound weight is divided by acceleration due to gravity (g = 32.2 ft/sec^2). Likewise, in CGS (centimeter-gram-second) units, the common measure of *gram* is a mass unit. The corresponding weight unit is *dyne* which equals the gram mass multiplied by acceleration due to gravity (g = 981 cm/sec^2). See Equation 2-2 for mass-weight relation.

[5]See Footnote 4.

[6]This does not happen because of friction force.

This downward acceleration is given by a constant, g, and is known as the acceleration due to gravity. The magnitude of the gravitational force is termed as the *weight* of the object. Equation 2-1 takes the form:

$$W = mg \qquad (2\text{-}2)$$

Where m = mass in slugs, kilograms, or grams
W = weight in pounds, Newtons, or dynes
g = acceleration due to gravity
= 32.2 ft/sec^2
= 9.81 m/sec^2
= 981 cm/sec^2

Since the acceleration due to gravity is constant on the earth, weight is proportional to mass and the two terms are often used interchangeably. However, with reference to other planets, the mass of an object is the same everywhere, but acceleration due to gravity is different; hence, weight varies on different planets. For example, the acceleration due to gravity on the moon is only about one-sixth that of the earth. Hence, the weight of an object of the same mass on the moon's surface would be only one-sixth of that on the earth.

Newton's third law of motion states that whenever an object A exerts a force on another object B, the object B exerts a force back on A; these two forces are equal in magnitude and opposite in direction. The force exerted by A on B is called the *action* and the back force of B on A is called the *reaction*. The law is often referred to as the *action-reaction law*. The action and reaction occurs on two different objects. This is a very important principle used to solve many engineering problems.

Force of Attraction Between Masses: Newton's Law of Universal Gravitation

Newton's insight was not restricted to the motion of objects in a straight line over the earth (expressed through the three laws explained in the previous section). He recognized the universal nature of the gravitational force that exists between any two masses, including celestial objects millions of miles apart such as the earth and the moon. Newton stated that every body in the universe, small or large, attracts every other body, small or large, close or far, by a force that is directly proportional to the product of the masses of two bodies and inversely proportional to the square of the distance between the centers of two bodies. This is known as *Newton's Law of Universal Gravitation*,[7] or:

$$F_G = \frac{G(M_1)(M_2)}{R^2} \tag{2-3}$$

where M_1 = mass of object 1
M_2 = mass of object 2
R = distance between center to center of two objects
F_G = Force of attraction between two bodies
G = Universal Gravitational constant
= 6.672×10^{-11} m³/(kg-s²)

This is sometimes considered the greatest single discovery in the history of science. This gravitational force of mutual attraction holds the universe together; it anchors things to the earth so that they do not float into space. The occurrence of tides due to the moon can be explained by Newton's law of gravitation. Between ordinary objects on the earth, this force is actually extremely small.

Example 2-1
What is the magnitude of force of the attraction of the sun to the earth?

Solution:
Mass of the Sun = 1.99×10^{30} kg
Mass of the Earth = 5.98×10^{24} kg
Distance between Sun and Earth = 1.49×10^{11} m

From Equation (2-3):

$$F_G = \frac{(6.672 \times 10^{-11})(1.99 \times 10^{30})(5.98 \times 10^{24})}{(1.49 \times 10^{11})^2}$$

$$= 3.58 \times 10^{22} \text{ Newtons } or\ 8.0 \times 10^{21} \text{ lb.}$$

---

[7]Newton's simple law of universal gravitation has been refined by Einstein's more complicated law of gravitation, which is part of his general theory of relativity proposed in 1916. However, the difference between the two laws are very minute. Newton's law needs modification only in certain special cases.

This enormous force keeps the Earth in its orbit around the Sun. Without this force, the Earth would fly off in a straight line into space.

The Law of Conservation of Mass–The Continuity Equation

Let us consider a flow system, such as a pipe, of cross section A through which a fluid having a density, ρ, flows at uniform velocity, V. If t is the time required to fill in a segment, then the length of the filled segment will be $L=Vt$, the volume of the fluid in the segment will be $\forall = AVt$, and its mass will be $m = \rho AVt$. The mass rate of fluid flow per unit time (the flux) can be expressed as

$$\dot{m} = \frac{m}{t} = \rho A V$$

For fluids that are incompressible (their density does not vary) such as most liquids (not gases), the mass rate can be substituted by the volumetric flow rate by dividing the above equation by density, thus expressed as follows

$$Q = AV \tag{2-4}$$

where Q = volume flow rate or discharge
A = cross-sectional area of flow system
V = velocity of flow.

In Equation 2-4, if area, A or volume, $\forall$ changes in any section or at any time without involving any change of flow, Equation 2-4 will be adjusted as follows:

$$Q = A_1 V_1 = A_2 V_2 = ... \tag{2-5}$$

This is known as the *equation of continuity*, which appears frequently in fluid flow applications.

Example 2-2
Five hundred gallons per minute of water flows through a pipe of 12 in. diameter. Determine the velocity of flow. If the diameter tapers down to 9 in., what will be the velocity of flow?

Solution:

$$Q = \left(500 \frac{\text{gal}}{\text{min}}\right)\left[\frac{1}{7.48}\frac{\text{ft}^3}{\text{min}}\right]\left[\frac{1}{60}\frac{\text{min}}{\text{sec}}\right] \quad \text{(See footnote)}[8]$$

$$= 1.11 \text{ ft}^3/\text{s}$$

For $d_1 = 12$ in. or 1 ft

$$A_1 = \frac{\pi}{4}(1)^2 = 0.785 \text{ ft}^2$$

From Equation 2-4:
$1.11 = 0.785 \, V_1$
$V_1 = 1.41 \text{ ft/sec}$

For $d_2 = 9$ in. or 0.75 ft

$$A_2 = \frac{\pi}{4}(0.75)^2 = 0.442 \text{ ft}^2$$

From Equation 2-5:
$1.11 = 0.442 \, V_2$
or $V_2 = 2.51 \text{ ft/sec}$.

Materials Balance and Pollution

The law of conservation is an important concept in science and engineering applications. Scientific studies are performed within a fixed space. Either a realistic or an imaginary boundary is drawn around the system to keep a track of the material balance. In a chemical process, a mixing tank might represent the boundary and in a flow analysis, the boundary might be formed by the stretch of a stream. These are referred to as *closed-boundary* or *closed-volume* studies, as described on page 5 in the context of modeling. Within a fixed space, if conditions for a system change with time, it is referred to as the *unsteady-state* system. For example, a stream carries different flows from time to time. Conversely, if things do not change within a given time period, the system is known to be in the *steady-state*. In mathematical terms, a steady-state system can be expressed as:

---

[8] The term in [ ] converts units throughout the book.

(input quantity) = (output quantity) + (storage change within system) + (decay) (2-6)

The input and output can be from more than one source. The last term relates to some quantity getting out of the system or converting to another form due to chemical, biological, or nuclear reactions. If a system is in an unsteady-state, the above relation is applicable at any instant of time.

Equation 2-6 can be simplified in three ways. The first simplification results when the system acquires a steady-state, i.e., the time parameter is not a factor. In the second simplification there is no decay, i.e., the system is *conserved*. The third simplification considers an *equilibrium* state wherein there is no change in the storage term. The combination of these conditions can lead to many stages in which a system can exist. The mass balance relations for simplified systems are discussed below:

A Steady-State Equilibrium Conserved System

This is the simplest stage, which reduces to the following form:

(input mass quantity or rate) = (output mass quantity or rate) (2-7)

In the context of pollution of a resource, if C is the concentration of a pollutant (substance) in mass per unit volume of a resource and Q is the volume or flow rate of the resource, then the total mass (or mass rate) of the pollutant will be CQ, which should remain constant according to equation. (2-7). In addition to the balance of the pollutant, the mass relation applies to the resource quantity as well. There could be more than one source of inflow or outflow in the relation. Mathematically, this can be expressed by the following set of equations.

$$\Sigma Q_i = \Sigma Q_o \quad (2\text{-}8)$$

$$\Sigma C_i Q_i = \Sigma C_o Q_o \quad (2\text{-}9)$$

where C = pollutant (substance) concentration, mass/volume
Q = quantity of a resource, volume or volume/time
i = subscript for inflow
o = subscript for outflow
Σ = summation of inflows or outflows from all sources.

The relation is $\Sigma \rho_i Q_i = \Sigma \rho_o Q_o$. However, density ρ_i and ρ_o for inflow and outflow are the same, hence, dropped out.

Example 2-3

A stream has a flow of 10 million gal/day (MGD). Its dissolved oxygen concentration is 8 mg/L. An effluent of 1 MGD from a sewage treatment plant is discharged into the stream that has a dissolved oxygen concentration of 2 mg/L. What will be the oxygen concentration downstream?

Solution:

The system is shown in Figure 2-4.

$$\Sigma Q_i = Q_1 + Q_2 = 10 + 1 = 11 \text{ MGD}$$

From Equation 2-8:

$$\Sigma Q_o = \Sigma Q_i = 11 \text{ MGD}$$
$$C_i Q_i = C_1 Q_1 + C_2 Q_2$$
$$= 8(10) + 2(1) = 82$$

From Equation 2-9:

$$\Sigma q_o C_o = \Sigma Q_i C_i$$
$$\text{hence, } C_o(11) = 82$$
$$\text{or } C_o = 81/11 = 7.45 \text{ mg/L}$$

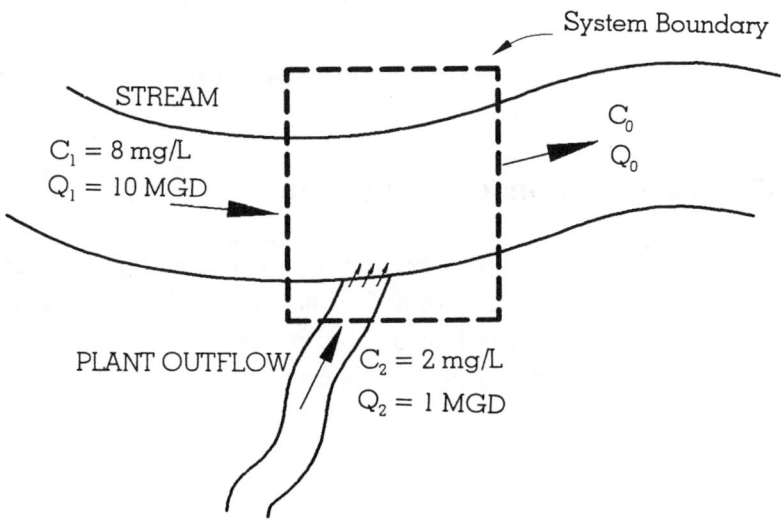

Figure 2-4 Closed Volume System of Stream-Sewage Discharge

Example 2-4

A stream has a flow rate of 10 m³/s. The oxygen content of the stream is 8 mg/L. A town discharges sewage into the stream at a rate of 2 m³/sec with no oxygen content. Determine the oxygen content after complete mixing of flows from stream and sewage.

Solution:

The system is similar to that shown in Figure 2-4.
Assume the oxygen concentration of combined flow to be C mg/L:
1 m³ = 1000 L
Stream flow rate = 10 × 1000 = 10,000 L/sec
Sewage flow rate = 2 × 1000 = 2000 L/sec
Combined flow rate = 12,000 L/sec

$$\text{Total quantity of oxygen in stream} = \left(10,000 \frac{L}{s}\right)\left(8 \frac{mg}{L}\right)$$

$$= 80,000 \text{ mg/s}$$

$$\text{Total quantity of oxygen in sewage} = 2000 \times 0$$
$$= 0 \text{ mg/s}$$
$$\Sigma c_i Q_i = 80,000 + 0 = 80,000$$

Total quantity of oxygen in combined flow = 12,000 × C

From Equation 2-9: For oxygen quantity

$$[80,000 + 0] = [12,000 \times C]$$
or, $C = 80,000/12,000 = 6.67$ mg/L

Thus, the stream is polluted by sewage discharge that reduces the oxygen level to 6.67 mg/L.

A Steady-State Stored Conserved System

In this and the next section, we assume that the stored volume provides a complete mixing of the substance, and that prior to the inflow in question the storage was free of concentration of the substance. For dimensional homogeneity, the time interval appears in the balance equation. The following relation corresponds to Equation 2-8.

$$\Sigma Q_i = \Sigma Q_o + \frac{\Delta S}{\Delta t} \tag{2-10}$$

where ΔS = change in storage in interval, Δt

Equation 2-10 is commonly used to describe the hydrologic cycle and other materials cycles. The pollutant mass balance relation can be obtained by multiplying the two sides of the equation by their respective concentrations. However, it has a limited practical application because of the assumption made about the storage volume being free of the substance prior to inflow.

Example 2-5
Runoff into a tank is 1 ft³/sec (cfs). How much water will be collected in the tank in 2 hrs?

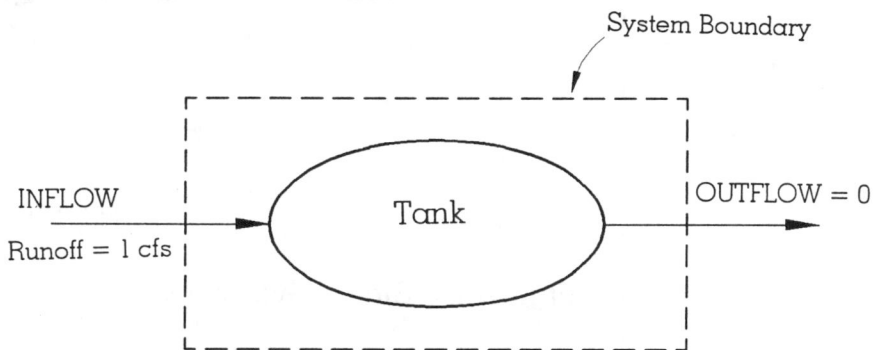

Figure 2-5 Stored Water Volume System of a Tank

Solution:
The system diagram is given in Figure 2-5.

Q_i = 1 ft³/sec
Q_o = 0
t = 2 hr or 7200 sec

From Equation 2-10:

$$1 = 0 + \frac{\Delta S}{7200}$$

hence, $\Delta S = 7200 \, ft^3$

Example 2-6

The annual rainfall in a basin of 2000 acres of drainage area is 14 in. It is estimated that 60% of this rainfall is evaporated. Annual river flow at the outlet of the basin is 1 cfs. Determine the annual volume of storage in the basin. The river inflow into the basin is zero.

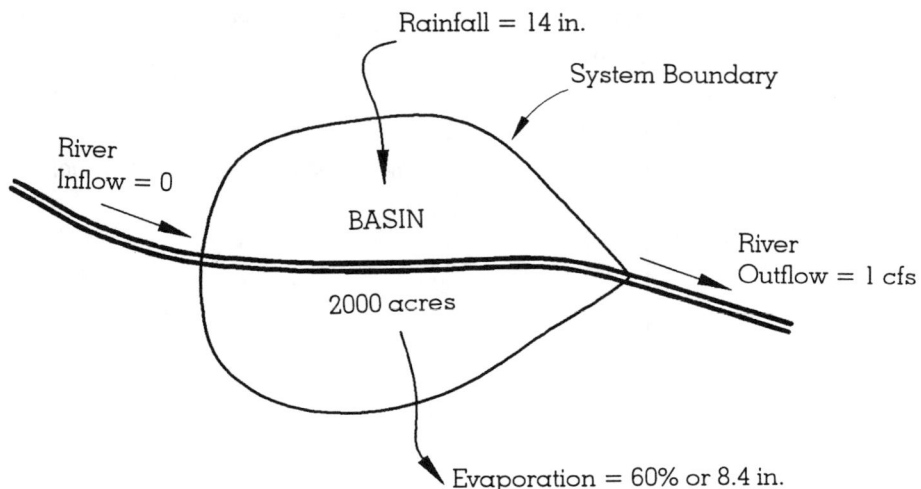

Figure 2-6 Stored System with Outflow

Solution:

The diagram of the system is shown in Figure 2-6.

Inflow: Rain of 14 in. deep over 2000 acres

Q_i = volume of rain per year

$$= (2000 \text{ acres})\left(14 \frac{\text{in}}{\text{yr}}\right)\left[\frac{43{,}650 \text{ ft}^2}{1 \text{ acre}}\right]\left[\frac{1 \text{ ft}}{12 \text{ in}}\right]$$

Outflows:
a) Evaporation: = 60% of rainfall

$$= \frac{60}{100}(101.64 \times 10^6) = 61 \times 10^6 \text{ ft}^3/\text{year}$$

b) River flow

$$= \left(1 \frac{\text{ft}^3}{\text{yr}}\right)\left(\frac{60 \times 60 \times 24 \times 365 \text{ sec}}{1 \text{ year}}\right)$$

$$= 31.54 \times 10^6 \text{ ft}^3/\text{year}$$

$$\Sigma Q_o = 61 \times 10^6 + 31.54 \times 10^6 = 92.54 \times 10^6 \text{ ft}^3/\text{year}$$

From Equation 2-10:

$\Delta t = 1$ year

$$101.64 \times 10^6 = 92.54 \times 10^6 + \frac{\Delta S}{1}$$

hence, $\Delta S = 9.1 \times 10^6$ ft^3/year

A Steady-State Equilibrium Nonconserved System

Since the storage does not change (system is in equilibrium), the relevant terms in this system are the input rate, output rate, and the decay. As stated earlier, this assumes that the pollutant does not remain in the storage. Some of it is decayed and the balance passes through outflow. Frequently, the decay is considered to be a first-order reaction in which the rate of loss of substance is proportional to the amount of the substance present in the stored volume. The decay term can be expressed as

$$\text{Decay rate} = kCV \tag{2-11}$$

where k = reaction rate coefficient
C = concentration at any time t, $C = C_i \, e^{-kt}$, C_i being the initial concentration of a substance
V = stored volume.

For the balance of the resource volume or flow rate, Equation 2-8 is applicable.

For the balance of the substance concentration, including the decay term at the output and dropping the storage term from Equation 2-6, the following relation holds:

$$\Sigma C_i Q_i = \Sigma C_o Q_o + k C_o V \tag{2-12}$$

Example 2-7

A sewage treatment plant is discharging its waste into a holding pit of 30 million ft^3 volume. The sewage discharge is 20 cfs and has a BOD (an indicator of pollutant) concentration of 80 mg/L. The pit overflows into a river. The rate of coefficient of decay of BOD in the pit is 0.15/day. Find the concentration of BOD in the outflow to the river.

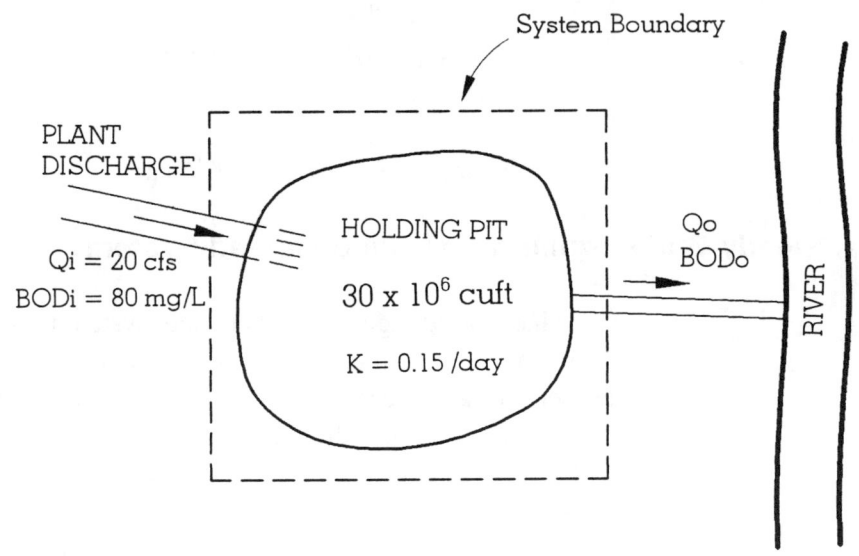

Figure 2-7 A Nonconserved Sewage Outflow System

Solution:
The diagram of the system is shown in Figure 2-7.

Time unit seconds
Volume of pit: $V = 30 \times 10^6$ ft^3
Suppose the outflow BOD is C_o mg/L

From Equation 2-8:
$$\Sigma Q_i = \Sigma Q_o,$$
hence $Q_o = 20$ ft^3/sec

Input concentration rate = $C_i Q_i$: $(80)(20) = 1600$ (mixed units)
Decay rate: $k = 0.15$/day or 1.74×10^{-6} per sec
$kCV = (1.74 \times 10^{-6})(C_o)(30 \times 10^6) = 52.2\ C_o$
Output concentration rate: $C_o Q_o = 20\ Q_o$

From Equation 2-12:
$1600 = 20 C_o + 52.2 C_o$
hence, $C_o = 22.2$ mg/L.

Example 2-8

A fairly rectangular town of 10 mi² area is affected by smog. An inversion layer (the concept will be explained in another section) at about 1000 ft above the ground acts as a blanket that prevents the vertical dispersion of air pollutants. Hills on the two sides restrict horizontal dispersion, practically forcing the pollutant to move in the direction of the wind that flows at an average rate of 100 million cfs. The addition of carbon monoxide due to car emissions in the town is at a rate of 80 kg/sec. If the decay (conversion of carbon monoxide to another form) coefficient is 0.2 /hr, estimate the concentration of carbon monoxide in the town.

Solution:

Refer to Figure 2-8.
Time unit seconds
Assume the pollutant concentration in atmosphere is C_o mg/L

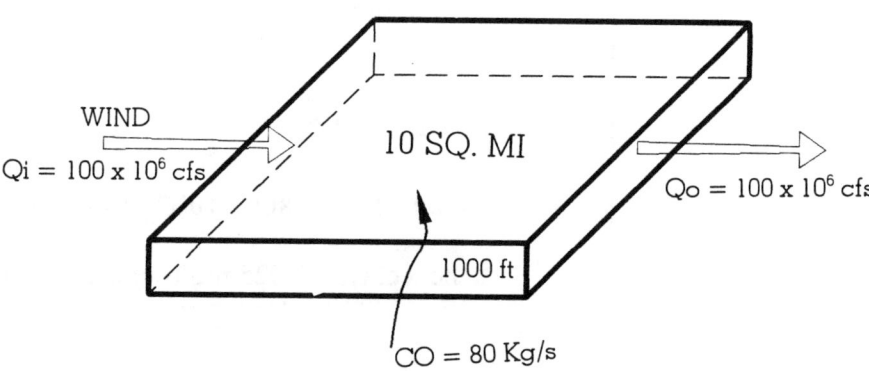

Figure 2-8 Box Model of a Town System

Concentration in g/ft³

$$= \left(C_o \frac{mg}{L}\right) \left[\frac{g}{mg}\right] \left[\frac{L}{ft^3}\right]$$

$$= (C_o) \left(\frac{1}{100}\right) \left(\frac{28.32}{1}\right)$$

$$= 0.028 \, C_o \, (g/ft^3)$$

Input Rate of CO = 80×10^3 /sec
Output Rate of $C_o Q_o = \left(0.028 \, C_o \frac{g}{ft^3}\right) \left(100 \times 10^6 \frac{ft^3}{sec}\right)$

$$= 2800 \times 10^3 \, C_o$$

For decay rate:

$$k = 0.2/hr \text{ or } 55.6 \times 10^{-6} \text{ per sec}$$

$$V = (1000 \text{ ft}) \, (10 \text{ mi}^2) \left[\frac{5280^2 \text{ ft}^2}{1 \text{ mile}^2}\right]$$

$$= 278.8 \times 10^9 \text{ ft}^3$$

$$kVC_o = \left(55.6 \times 10^{-6} \frac{1}{sec}\right) (278.8 \times 10^9 ft^3) \left(0.028 C_o \frac{g}{ft^3}\right)$$

$$= 434 \times 10^3 \text{ g/sec}$$

From Equation 2-12:

$$80 \times 10^3 = 2800 \times 10^3 C_o + 434 \times 10^3 C_o$$

hence, $C_o = 0.025$ mg/L of carbon monoxide.

Matter and Materials Balance 53

PROBLEMS

1. The atomic number and the mass number of the following elements are given. (1) Sketch their atomic structure, (2) determine their AMU, and (3) determine the masses of elements in grams

Element	Atomic no.	Mass no.
Carbon (C)	6	12
Chlorine (Cl)	17	35
Gold (Au)	79	197
Mercury (Hg)	80	200
Uranium (U)	92	238

2. The density of water is measured to be 1000 kg/m³ on the earth where the acceleration due to gravity is g = 9.81 m/s². Determine the density of water on a planet A, where g = 4 m/sec², and the weight of 1m³ of water on the planet A. (Hint: mass = volume × density; for any volume V find mass on earth which is the same on planet A. The density on planet A is equal to the mass divided by the volume. Determine weight using Equation 2-2.)

3. An empty container weighs 3.2 lb. Filled with water at 4°C, the mass of the container (with water) is 1.95 slugs. Find the weight of water in the container.

4. What is the gravitational force exerted by the earth on an object the mass of which is 1 kg? Solve the problem by Newton's basic laws of motion, and Newton's Law of Universal Gravitation. The mass of the earth is 5.98×10^{24} kg. The radius of the earth is 6.37×10^6 m.

5. Calculate the gravitational force exerted by the earth on the moon. The mass of the moon is 7.34×10^{22} kg. The distance between moon and earth is 3.84×10^8 m.

6. A 2-kg glider moves at constant velocity of 10 m/sec along a horizontal air track (friction is negligible). What horizontal force acts on the glider?

7. A 2-kg glider moves horizontally at a uniform acceleration of 10 m/s². What horizontal force acts on the glider? Discuss difference between Problem 6 and Problem 7.

8. In Problem 6, what is the gravitational force acting on the glider?

9. A 2-kg book is pushed across a horizontal table by a force of 5 Newtons. The friction between table and book that retards the motion of the book has a magnitude of 3 Newtons. What is the acceleration of the book? What is the gravitational force? What is the upward force exerted by the table on the book?

10. If the earth were three times farther from the sun than it is now, how would the gravitational force compare with the present value?

11. Compute the force of attraction between two 20,000-ton ships 100 feet apart.

12. Water flows through a 14-in. diameter pipe at a velocity of 6.4 ft/sec. Determine the rate of flow (discharge) in gallons per minute.

13. One thousand Liters per minute of water flows through a 400-mm diameter pipe. Determine the velocity of flow.

14. A pipe tapers from 3.5 in. diameter to 2 in. diameter. The rate of flow through the pipe is 0.7 cfs. Determine the velocity of flow at the two ends of the pipe.

15. A rectangular stream has been measured at two sections. At Section 1, it is 200 ft wide and 5.5 ft deep. At Section 2, it is 150 ft wide and 6.5 ft deep respectively. The velocity of flow at Section 1 is 2.5 ft/sec. Determine the stream flow rate, and the velocity of flow at Section 2.

16. In Example 3, the BOD is 2 mg/L for the stream and 50 mg/L for the effluent. Find the BOD at the downstream end of the stream.

17. One stream flows in a northeast direction. Above this, another stream flows toward southeast. After meeting the combined stream flows east. The northeast stream has a flow rate of 100 cfs and a total dissolved solids (TDS) concentration of 200 mg/L. The southeast stream has a flow of 60 cfs and TDS of 300 mg/L. Determine the TDS in the water after the confluence.

18. In Problem 17, a third stream free of TDS joins from the north further downstream of the confluence of the first two streams. If the final TDS downstream of three streams is 100 mg/L, determine the flow rate of the third stream.

19. Exhausts from two factories are continuously pumping stale air to a common chamber in which the stale air is mixed with the fresh air before it is released to the atmosphere. The exhaust rate of the first factory is 2000 m^3/hr with a carbon dioxide (CO_2) concentration of 0.6 mg/m^3. The rate and concentration from the second factory are 1200 m^3/hr and 0.8 mg/m^3, respectively. The concentration of CO_2 in the atmosphere should not exceed 0.2 mg/m^3. What should be the rate of the fresh air to the pool exhaust? Assume no decay.

Matter and Materials Balance 55

20. The initial storage in a river reach is 55.3 acre-ft. The inflow into the reach is 375 cfs and the outflow is 563 cfs. Determine the change in storage and the storage volume after 2 hr. (Hint: Acre-ft is a volume term that can be converted to ft^3.)

21. From a reservoir of 500 acres surface area, the evaporation losses are 6.5 million ft^3/day. The outflow rate from the reservoir is 50 cfs. There is no inflow into the reservoir. Determine the change in water level of the reservoir in a day.

22. A filter chamber is being used to remove dust from air. The rate of flow of dust particles into the chamber is 1500 g/min. The dust removed from the chamber in 1 hr is 80 kg. What is the content of dust per minute in the filtered air from the chamber?

23. In a coal-fired electric power plant, 1 million kg of coal is burned per day. The coal has an ash content (portion of ash produced from coal) of 10% by mass. From the bottom of the furnace where coal is burned, 35% ash falls to the bottom. The remaining ash from the furnace is transferred to an electrostatic precipitator (ESP). The ESP removes 97% of the ash that comes into it and emits 3% into the atmosphere. Calculate the mass emission rate per day of ash into the atmosphere per day.

24. A lagoon receives effluent from a factory at a rate of 1m^3/s with a pollutant concentration of 25 mg/L. The pollutant decays in the lagoon at a rate of 0.01/hr. The lagoon has a holding capacity of 12 million m^3 beyond which it discharges into a stream. Determine the pollutant concentration in the overflow from lagoon.

25. In Problem 24, beside the factory effluent, a stream that is free of the pollutant also discharges into the lagoon at a rate of 10·m^3/s. What will be the effect on the pollutant concentration as compared to Problem 24?

26. A lake is fed by a stream at a rate of 200 cfs. The organic solids concentration of the stream is 12 mg/L. Domestic wastes are also discharged into the lake at a rate of 25 cfs, with organic solids concentration of 200 mg/L. The lake has a reaction rate coefficient of 0.01/hr. The outflow from the lake must have an organic solids concentration of not more than 5 mg/L. What should be the capacity of the lake?

27. In a restaurant, the smoking section has a size of 50 ft long, 40 ft wide, 10 ft high. It has a capacity of 50 persons. On an average, each person smokes two cigarettes in 1 hour that add to a pollution of 2 mg/cigarette (This is the input cone.). The pollutant is mixed with fresh air (that enters the section) at a rate of 2 ft^3/sec. The stale air also leaves the section at the same rate. If the decay rate (conversion to a different form) is 1.4×10^{-4}/sec, what is the average concentration of the pollutant in the room?

STUDY QUESTIONS

1. What is matter? What is its importance in the environmental context? Make a hierarchical presentation of all matter available in the universe.

2. Outline an atomic structure labeling the relative size/amount of its constituents. Why it is not possible to make a scale model of an atom?

3. Discuss three nuclear changes (radiations) that can take place in atoms. What are their applications in nature and by humans? What are their disadvantages? Which one is the most difficult to achieve in practice?

4. What is the scientific law that governs the distribution of matter on the earth? How do you interpret the saying, "There is nowhere we can discard the things that we do not want any more"? What is the significance of this statement in the environmental context?

5. State in simple words Newton's three laws of motion and the law of universal gravitation. Give a practical example of each.

6. Differentiate between mass and weight. A rock is buried deep beneath the surface of the earth. Will its mass be the same as on the surface? Will its weight be the same as on the surface? If not, will it be more or less?

7. Demonstrate how the scientific law of matter controls the natural flow of a river. If the river widens from one place to another, discuss how the flow will be affected?

8. Demonstrate how the scientific law of matter controls the quality of river water. How is it used to enforce discharge from an outfall into the river?

9. Differentiate between a steady-state equilibrium conserved system, a steady-state stored conserved system, and a steady-state equilibrium nonconserved system. Draw a sketch of each system.

10. Define the following:
    a. Atom and molecule
    b. Proton, electron, and neutron
    c. Element and compound
    d. Mixture and solution
    e. Gas, liquid, and solid
    f. Mass number
    g. Atomic number
    h. Radioactivity
    i. Alpha, beta, and gamma rays
    j. Positive and negative ions
    k. Isotopes
    l. Mass, weight, and force
    m. Universal gravitation constant
    n. Steady and unsteady state
    o. Closed-volume system

REFERENCES

Asimov, I., *The History of Physics*, Walker and Company, New York, 1966.

Freeman, I. M., Revised by Durden, W. J., *Physics Made Simple*, Doubleday, New York, 1990.

Hewitt, P. G., *Conceptual Physics: A New Introduction to Our Environment*, Little, Brown and Company, Boston, MA, 1971.

Inhaber, H., *Physics of the Environment*, Ann Arbor Science, MI, 1978.

Life Science Library, *Matter*, Time Inc., New York, 1963.

Marion, J. B., *Physics and the Physical Universe*, John Wiley and Sons, New York, 1971.

Masters, G. M., *Introduction to Environmental Engineering and Science*, Prentice Hall, NJ, 1991.

Mulligan, J. F., *Introductory College Physics*, 2nd. ed., McGraw-Hill Book Company, New York, 1991.

Romer, R. H., *Energy: An Introduction to Physics*, W. H. Freeman and Company, San Francisco, CA, 1976.

Wanielista, M. P., et al., *Engineering and the Environment*, Robert E. Krieger Publishing Company, Malabar, FL, 1990.

Chapter 3

PRINCIPLES OF ENERGY AND ENERGY ALTERNATIVES

Work and Energy

Newton's three laws of motion related matter to force. Force is linked with energy through the concept of work. *Work* is defined as the product of the applied force on an object and the distance that object moves in the direction of the force. Force and distance are two essential requirements for work. Both can appear in many forms. Distance may involve movement of an entire object as machines do or it may be on the scale of individual molecules as in the case of electricity. The former is referred to as *macroscopic work* as compared to *microscopic work* for the latter. Force may be in a form of direct pull or push, attraction between two objects, chemical bonds that hold atoms together, nucleus strength of protons and neutrons, or motion of electric and magnetic waves.

According to this definition, since motion is an essential part of work, if a person holds a book in a stationary position for a long time, no work is accomplished. For such a situation, two categories of work are defined: *processing* work, where direct motion is involved during the work, and *storage work*, where input of work has been made to achieve the current position.

Energy is a capacity to do work. In quantitative terms, work is a measure of energy. As work has many forms, so does energy that directly relates to work.

Power is the rate at which energy flows or is used, that is, it is energy divided by the time of application of energy.

Example 3-1
Determine the work done when (1) a person walks on a flat surface holding a 2 lb book, and when (2) a horizontal force of 5 Newtons is applied to a box on a shelf to move it with a constant speed of 2 m/s.

Solution:
1. Since the weight of the book is acting vertically down, there is no force acting horizontally. Hence, no work is being done on the book in the sense of the physical definition of work.
2. Since there is no acceleration of the box, the force is zero according to Newton's first law of motion. Whatever force is applied just balances the retarding frictional force so that net force = 0. However, since the external force of 5 Newtons is being applied and there is a displacement of the box, the work is being done at the following rate:

Distance moved per second = 2 meter
Work done per second: $5 \times 2 = 10$ Newton–meter or Joules
Thus, whether an object accelerates or not, work is done.

Units of Energy

In a most direct form, work or energy is given by:

(work or energy) = (force) × (displacement of force)

$$\text{or } E = Fd \qquad (3\text{-}1)$$

The two measurement systems commonly used in the U.S. for units of energy, are listed in Table 3-1.

Table 3-1 Units of Energy

Units /System[1]	Force	Displacement	Energy
FPS or English	Pound	Foot	Foot-pound
SI or Metric	Newton	Meter	Joule

The calorie unit of energy is commonly used in the thermal field, and the Calorie unit is used (with the capital letter C rather than lower case c as in the former case) in nutrition; one kilocalorie (thermal unit) is equal to 1 Calorie and represents the energy needed to raise the temperature of 1 kg of water by 1°C.

The unit for power in metric units is Joule per second, also known as a Watt (W). The English unit for power is horsepower (hp); 1 hp = 746 W.

---

[1]The conversion is as follows: 1 Newton = 0.225 lb, 1 Joule = 0.738 ft-lb, 1 kilocalorie (or food Calorie) = 4184 Joules.

Forms of Energy

There are two basic forms of energy: the energy associated with motion and the energy associated with position. The energy possessed by an object by virtue of its motion is known as *kinetic energy*. Derived from Newton's second law of motion, it is expressed as

(kinetic energy) = ½ (mass) (velocity)²

$$\text{or} \quad KE = \frac{1}{2}mv^2 \tag{3-2}$$

The unit is the same as discussed above.

When an object is lifted above the ground, work is done against the gravitational force. Energy acquired by the object is equal to this work. This energy possessed by the body by virtue of its position above the ground is referred to as *potential energy*. It is expressed as:

(potential energy) = (mass)(acceleration due to gravity)(height above ground)

$$\text{or} \quad PE = mgH \tag{3-3}$$

When the object falls from a certain height, its potential energy is converted into kinetic energy. Thus, energy can change form; several transformations of energy take place in the universe.

Under each of the two basic forms, energy can assume many special forms. Table 3-2 lists some common forms of energy:

Currently, chemical energy is the most common source of energy; the energy in coal, gas, and oil is used to run cars, heat houses, and generate electricity, and the energy in food sustains life on the earth.

Heat is another important form of energy.[2] All chemical processes are either accompanied by the release of heat (exothermic processes) or require an input of heat (endothermic processes). Heat is a part of other energy conversion processes as well. The branch of physics that relates heat energy to mechanical work is called *thermodynamics*. Accordingly, the principles of energy are also recognized as the laws of thermodynamics.

---

[2]Leibnitz, a German scientist, first theorized in 1695 that heat was a form of energy.

According to modern physics theory consequent to Einstein's Special Theory of Relativity, mass is considered to be a form of energy and even a small mass conversion can lead to enormous energy, as demonstrated later in this chapter.

Table 3-2 Important Forms of Energy

Type of Energy	Basic Form	Example
Gravitational	PE	Object's weight
Mechanical	KE	Rotating wheel
	PE	Water behind a dam
Chemical	PE	Fossil fuel
	PE	Food
Sound	KE	Earthquake
Electrical	PE	Battery
	KE	Electric current
Electromagnetic	PE + KE[3]	Sunlight
Radiation	PE + KE	Radio waves
Heat	KE	Temperature increase
		Molecular agitation
Mass or nuclear	PE + KE	Sun's energy
		Nuclear fusion

Legend:
PE= potential energy; KE= kinetic energy

---

[3]Electromagnetic and nuclear energy are basically potential energies that are released as kinetic energy.

Principles of Energy and Energy Alternatives 63

The Laws of Energy

The first law of energy is a very important concept that forms the basis of many theories of modern physics. It also controls day-to-day life of human beings. According to this law energy can be neither created nor destroyed by any means; it simply changes from one form to another. This statement is called the *principle of conservation of energy*, also known as the *First Law of Thermodynamics*. Thus, for a closed physical system

$$\text{Total energy} = KE + PE = \text{constant} \qquad (3\text{-}4)$$

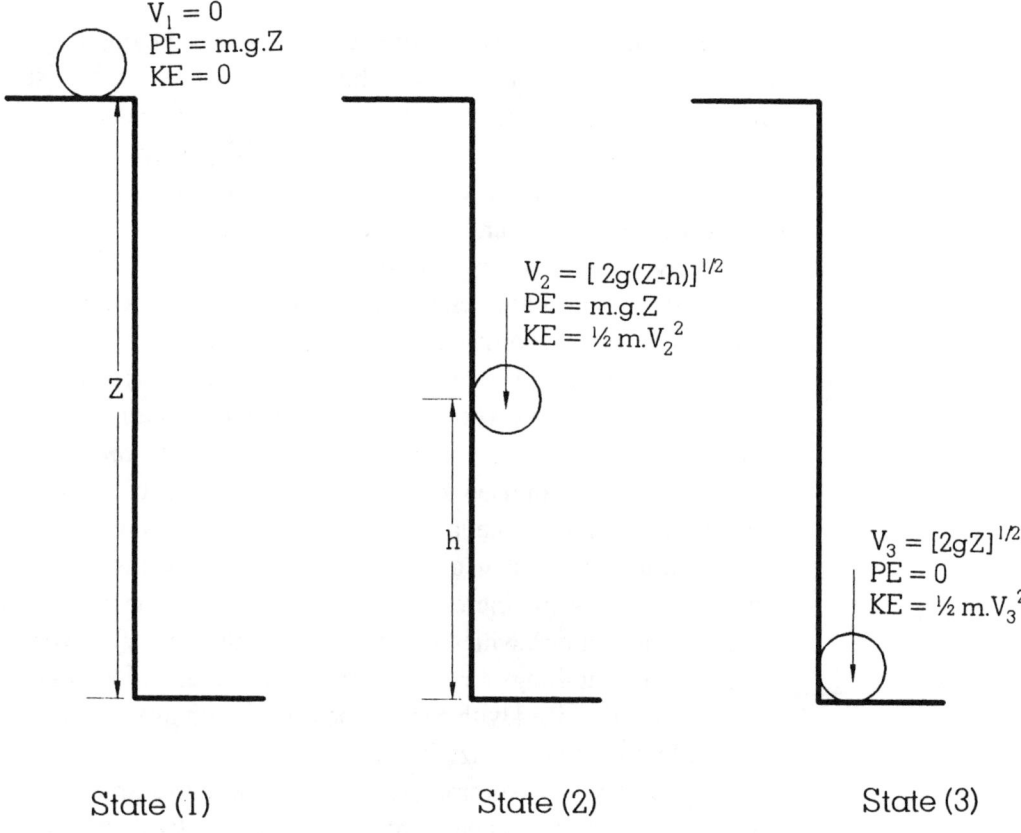

PE at State (1) = (PE + KE) at State (2) = KE at State (3)

Figure 3-1 Energy Transformations and Energy Conservation

Figure 3-1 shows a simple system demonstrating the law of conservation of energy. Many formulas have been derived from this relation and a large number of energy problems have been solved using this principle.

Example 3-2
If 2 m³/s of water flows over a 20 m-high dam, what is the power produced by the hydropower station at the base? Assume no conversion loss.

Solution:
Unit weight of water = 9.81 kN/m^3
Weight of water flowing/sec = 9.81 × 2 = 19.62 kN/sec
Energy/sec or Power = WH = 19.62 × 20
   = 392.4 kNm/sec or kW.

In as much as the first law of energy concerns the quantity of energy, the *Second Law of Energy* relates to the quality of energy. It is stated in many different ways. Put simply, as the temperature of a heat source decreases its available energy for doing work decreases. Dispersed (low grade) heat may not have any ability to do work. Thus, energy can be viewed as concentrated or high-quality energy and dispersed or low-quality energy. The second law essentially stipulates three conditions: (1) low quality energy cannot be converted to high-quality energy without expending external work; (2) in all energy transactions, some energy is degraded in quality, that is, no system is 100% efficient; and (3) energy tends to flow (convert) from high-quality to low-quality energy and a continual increase in low quality energy takes place. This is also known as the *Second Law of Thermodynamics*.

Entropy is a term used to measure disorder of energy. It is a measure of the amount of energy unavailable for conversion into work. Thus, according to condition 3(a) above, in any energy conversion within a closed system, the entropy of the system increases. The second law of thermodynamics is often stated in this manner, which is recognized as the *law of entropy*. The essence of this law is that things tend to go from order to disorder. An organized desk becomes disorganized with papers and books with time, and it requires effort (external work) to organize it again.

The two laws are summarized with a simple example: Suppose we take 2 kg of water at 50° C and want to separate it in two parts; 1 kg 100°C and 1 kg at 0°C. The first law applies since we are neither creating nor destroying any energy, but simply redistributing it. However, the second law states that we cannot do it, because energy cannot be converted from 50°C to 100°C without doing external work (energy flow cannot take place under increasing temperature).

While it is true that energy is never destroyed, it can be degraded. Thus, the aim to conserve energy actually is an attempt to retain the useful forms of energy.

Example 3-3
Explain the application of energy law in the human system.

Solution:
The human body is not a closed system since we get a continuous supply of energy from food. In a balanced condition, the energy taken in as food equals the energy lost in the form of waste products, plus the work done by the body plus the heat lost to the surroundings. If the food energy exceeds this, the excess energy is stored in the body as muscle tissue or fat depending on the body's level of activity.

The Concept of Conservation of Mass-Energy

Einstein's theory slightly modified the concept of mass remaining as mass and energy remaining as energy. According to Einstein's equation, the total energy (E) locked into a mass (m) is equal to the mass times the square of the velocity of light, which is 186,272 mi/sec, or: $E = mc^2$.

The results are astonishing. For example, 1 lb. of any substance, if totally converted into energy, would yield:

Mass of substance = 1/32.2 = 0.031 lb. mass or slug

$$\text{Speed of light} = \left(186{,}272 \ \frac{\text{miles}}{\text{sec}}\right)\left[\frac{5{,}280 \text{ ft}}{1 \text{ mile}}\right]$$

$$= 983.5 \times 10^6 \text{ ft/sec.}$$
$$\text{hence, } E = (0.031)(983.5 \times 10^6)^2$$
$$= 30 \times 10^{15} \text{ ft/lb, or}$$
$$11 \text{ billion kW-hr.}$$

Thus, energy from 1 lb of a substance is enough to drive a car 180,000 times around the Earth. However, there is not the slightest indication that this energy can ever be fully realized.

Einstein's theory that a tiny amount of matter could yield enormous energy opened the door to the atomic age when scientists learned to pry open the heart of matter.

According to Einstein, any change in energy, such as the heating of an object, will affect its mass. However, this change of mass is imperceptible within the practical range of energy changes.

Because of the possibility of converting mass into energy (although under special conditions only), the principles of conservation of mass and conservation of energy can be absorbed into one broad principle of conservation of mass-energy. This is shown in Figure 3-2.

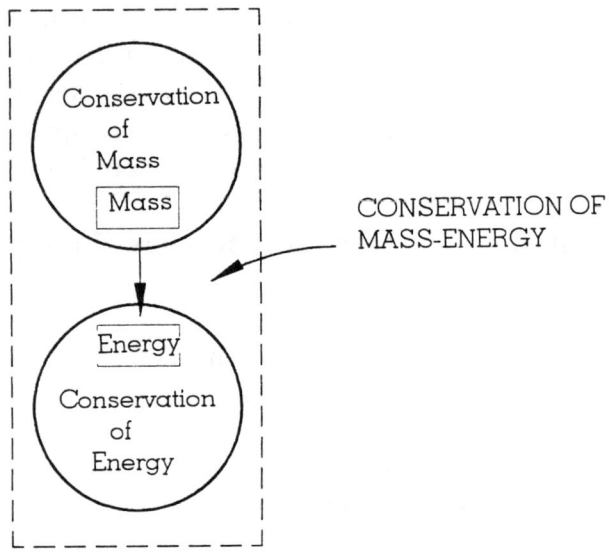

Figure 3-2 The Mass-Energy Link

The Source of Energy

There is no beginning of energy; it has always existed in the universe. The primordial "fireball," from which the universe was theoretically born, possessed the potential energy of immense pressure and extreme high temperature. This was converted into kinetic energy of explosion as the mass spread out into space. Then, the potential energy of gravitation, acting like an invisible spring, came into play to prevent all the bit and pieces of masses that constituted the celestial bodies from flying out. Most of the energy in the universe resides within the gravity field. All celestial bodies in space (billions in number) contain gravitational energy. When celestial masses contract during the evolution of stars, some of this energy is converted into energy of motion, heat, and light. Extremely high temperatures, which cause fusion of hydrogen elements to helium, result in release of nuclear energy.

Although it is only a small fraction in the universe, nuclear energy opposes the gravitational collapse of stars beyond a certain point; thus the stars remain in their position and continue to shine. Energy adopts many expressions between the two basic forms of potential and kinetic energy through the universe. Chemical energy, which plays a vital role in sustaining life on Earth, counts for very little in the universe.

According to the second law, the energy flows from a "higher quality" to a "lower quality" form. Gravitational energy is of the highest form and it has no entropy. Table 3-3 indicates the entropy associated with different forms of energy, arranged in order of merit.

On the Earth, the source of energy is directly and indirectly from the sun. The sun emits energy at an enormous rate. Every second, 4.6 million tons of the sun's mass is converted into energy by fusion. The energy from the sun radiates into space in all directions at a rate of $380,000 \times 10^{21}$W. Still, it will take over 5 billion years to fully use up the sun's energy. The earth, located at a radius of 93 million miles from the sun, receives only 173×10^{15} W, which is five ten-billionth's ($5/10^{10}$) of the sun's energy[4].

Table 3-3 Order of Merit of Different Energy Forms in the Universe

Form of Energy	Entropy per Unit of Energy*
Gravitation	0
Energy of rotation	0
Energy of orbital motion	0
Nuclear reactions	10^{-6}
Internal heat of stars	10^{-3}
Sunlight	1
Chemical reactions	1 – 10
Terrestrial waste heat	10 – 100
Cosmic microwave radiation	10,000

*Expressed in units of inverse electron volts (1×10^6 electron volts = 1.6×10^{-13} Joules)

---

[4]The mean value for the solar constant is 1.395 kW/m² or 2 cal/cm²/min, which is intercepted by the earth's diametric plane of 1.275×10^{14} square meter.

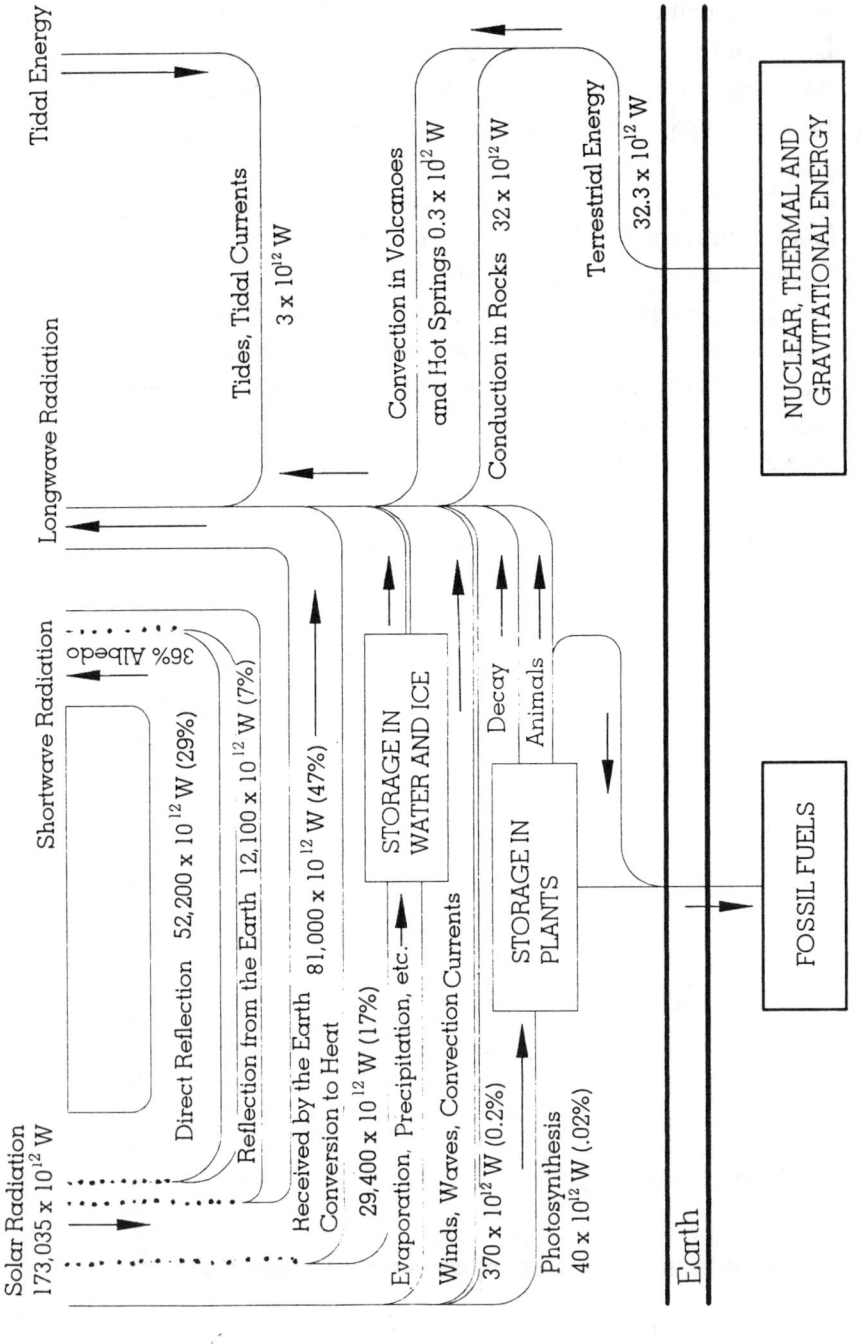

Figure 3-3 Flow of Energy to and from the Earth
(As Modified from Energy and Power, Scientific American, 1971)

The three sources of energy flow toward the earth's surface are: solar radiation of $173{,}035 \times 10^{12}\text{W}$[5] (99.98%) on the top of the earth's atmosphere, heat from the earth's interior of $32.3 \times 10^{12}\text{W}$ (0.02%); and tidal energy from the earth–moon–sun driven system of $3 \times 10^{12}\text{W}$ (0.002%). The solar energy cycle is shown in Figure 3-3.

Of the solar radiation:

- about 29% ($50{,}200 \times 10^{12}$ W) is directly reflected back into space as short–wavelength[6] radiation energy without reaching the earth;
- another 7% ($12{,}100 \times 10^{12}$ W) is reflected from the earth surface, making up a total of 36% albedo[7];
- the atmosphere absorbs 17% ($29{,}400 \times 10^{12}\text{W}$) which is consumed in the hydrologic cycle of precipitation, evaporation, surface water flow, and groundwater infiltration;
- the earth's surface receives 47% ($81{,}000 \times 10^{12}$ W) which is converted into heat at the ambient surface temperature and then reradiated to the atmosphere;
- a small fraction, 0.2% (370×10^{12} W), drives atmospheric and oceanic convections and circulations, and the ocean waves, and dissipates as frictional heat; and
- a still smaller fraction, 0.02% ($40 \times 10^{12}\text{W}$) is converted into chemical potential energy through *photosynthesis*, which is a natural chemical reaction performed by plants to produce food for themselves and for animals.

Triggered by photosynthesis, the energy flow in the food chain follows its own cycle as discussed subsequently. A small fraction of plant and animal matter is buried under conditions of incomplete oxidation and decay. This process produces *fossil fuels* in the forms of coal, oil, and natural gas over a period of millions of years. Thus, with the help of the sun's energy, carbohydrates are formed that serve as food or fossil fuel, depending on the use to which they are put.

---

[5]The conventional unit for energy is foot-pound which is in reference to mechanical energy. For electricity or electromagnetic radiation energy, the unit is Joule. However, the commonly used unit is Watt, which represents power, the rate of energy. 1 Watt = 1 Joule/second.

Heat energy is often expressed in kilocalorie (kcal) which is the amount of heat needed to raise 1 kilogram of water by 1°C. 1 food Calorie is equivalent to 1 kcal. The corresponding English Unit, the British thermal unit (Btu), is the heat required to raise 1 pound of water by 1 °F.

[6]This term is defined in the next section.

[7]Albedo is a measure of the fraction of light that is simply reflected, as from a mirror surface.

The amount of radiation (rate of energy) from the sun falling on the upper atmosphere perpendicular to the direction of the suns rays is 1400W/m². Determine the total radiation absorbed by the earth's atmosphere.

Solution:

Assuming the earth to be a sphere of diameter 7900 mi (12.7 million m), the cross-sectional area perpendicular to the sun's rays is the target of the incidental radiation. Because the earth's albedo is 36%, only 64% of the radiation is absorbed by the earth.

$$\text{Earth's area intercepting radiation} = \frac{\pi}{4}(12.7 \times 10^6)^2$$
$$= 127 \times 10^{12} \, m^2$$

$$\text{Total radiation absorbed} = 0.64 \times 1400 \times 127 \times 10^{12}$$
$$= 114 \times 10^{15} \, W$$

This is a simplified presentation of energy flow. The problem is complicated with exchange of energy between the atmosphere and the earth's surface. Only a small quantity of matter is embodied in living organisms and their requirement of energy in terms of food is only a tiny fraction of the total solar energy falling to the earth. If the entire amount of solar radiation falling on the earth could be utilized, within one- half hour, it would satisfy the energy demand of the entire human race for one full year. The vast resources of solar radiation, tidally generated energy, and the earth's interior heat are, however, returned to space unharnessed, whereas humans are currently exploiting the small storage of fossil fuels.

The Outward Flow of Energy from the Earth

Energy may be transferred by the displacement of matter from one place to another or by means of wave motion. All of radiation energy, including that from the sun, travels as electromagnetic waves. Electromagnetic waves consist of two vibrating electric and magnetic fields traveling together, in the same phase but at right angles to each other. The waves propagate as ripples forming crests and troughs. The distance between two successive crests of a wave is known as the *wavelength*. The number of times that a wave travels from one crest to the next crest in 1 second is known as its *frequency*. The product of the frequency and wavelength is the *speed* of the wave. All electromagnetic waves have a speed equal to the velocity of light.

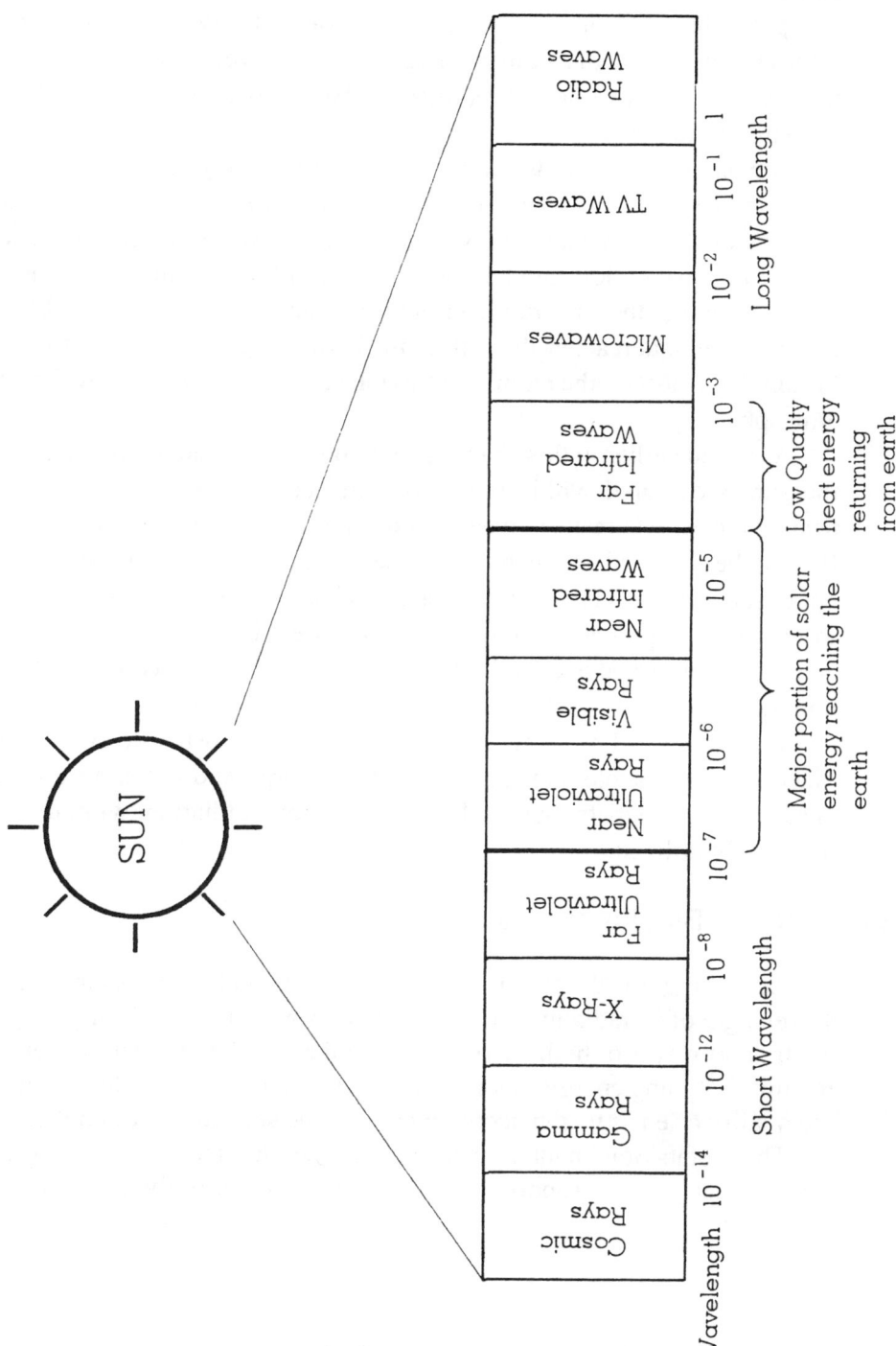

Figure 3-4 Spectrum of Electromagnetic Waves

All electromagnetic waves belong to a single family. Thus, light, heat, electric, radio waves, X-rays, microwaves, and ultraviolet rays are all similar except for their differing wavelengths. The longer the wavelength, the lower is the energy content of a radiant energy wave. Electromagnetic waves range from a very small value of less than 1×10^{-15} m to a large value of 1 m, as shown in Figure 3-4.

The representation of the waves that constitute a ray of the sun, is known as the *solar spectrum*, as shown in Figure 3-4. The shorter wavelengths such as ultraviolet, X-rays, gamma rays, and cosmic rays, have a very high energy content due to their ionization state and are harmful to living organisms. The major portion of the spectrum consisting of ultraviolet rays, and visible and infrared light that reaches the earth, falls in the range from 10^{-7} m to 10^{-5} m. Fortunately, most of the harmful ultraviolet rays are filtered out by the upper atmosphere.

As stated earlier in this chapter, about 64% of the radiation energy from the sun is captured within the earth-atmospheric system; this goes into heating the earth to make it liveable and to replenish the earth's water supply. If all of the absorbed energy remained here, the earth would soon become too hot to support life. The amount of energy absorbed inevitably radiates back to space. This process maintains a thermal equilibrium because if more energy was absorbed than emitted back, the earth would become continually hotter.

The outward flow of energy from the earth-atmospheric system takes place as long wavelength radiation. Tidal energy and the earth's interior energy also dissipate out as long wavelength radiation without being practically utilized.

A Simple Inflow-Outflow Blackbody Model

According to Wein's Law, the hotter a body, the shorter is the wavelength of its radiating electromagnetic waves. The sun, being very hot, emits short waves of high energy. On the other hand, the much cooler earth re-emits outgoing energy as long wavelength radiation. This degradation of the quality of energy is in accordance with the second law of energy.

The Stefan-Boltzmann Law also states that the rate of energy radiating from a blackbody is proportional to the fourth power of the temperature of the body:

$$S = \sigma T^4 \qquad (3\text{-}5)$$

where S = energy radiated per second per sq meter
T = temperature in Kelvin, (C + 273)
σ = Stefan-Boltzmann constant
   = 5.67×10^{-8} W/m^2/K^4.

A simple model uses the Wein and Stefean-Boltzmann laws, treating earth as a blackbody but accounting for the solar energy reflected back into space (the earth's albedo), as shown in Figure 3-5. This model differs from reality since it does not take into account the meteorological factors.

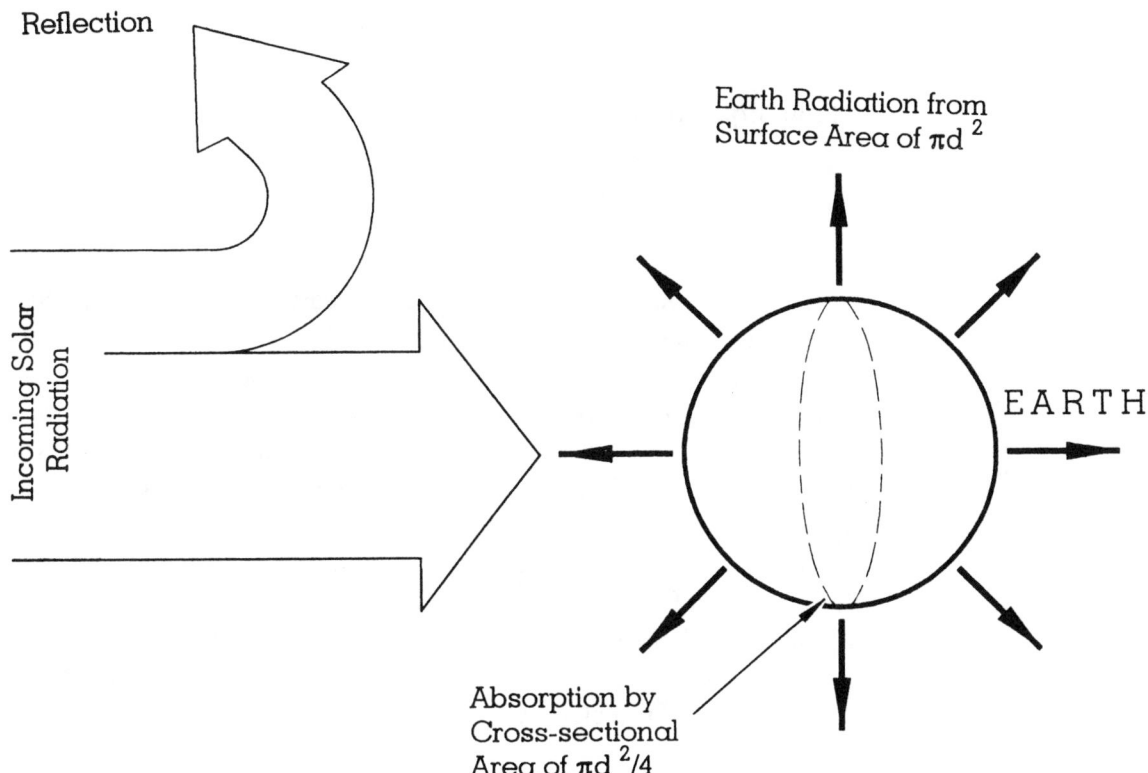

Figure 3-5 A Simple Blackbody Model

Example 3-5
Using the blackbody concept, find the earth's temperature (at the upper atmosphere).

Solution:
Because the earth is at an equilibrium condition for temperature (ignoring the greenhouse effect), the amount of energy emitted is equal to the amount of energy absorbed, which is 114×10^{15}W (Joules/s) as per example 3-4. This energy is emitted from the surface area of the earth.

Assume the earth to be a sphere of diameter, d:

Emission area = surface area of sphere = πd^2
$= \pi (12.7 \times 10^6)^2$
$= 506 \times 10^{12} \text{ m}^2$

Energy radiated/surface area = $\dfrac{114 \times 10^{15}}{506 \times 10^{12}}$

$= 225 \text{ W/m}^2$

From Equation 3-5:
$225 = 5.67 \times 10^{-8} T^4$
or $T = 251 \text{ K} = -22 \text{ °C}$.

This is the temperature at the edge of the earth's upper atmosphere that is modified by the atmospheric blanket.

As discussed earlier in this chapter, the flow of energy from the earth is at a fairly defined rate. This tends to stabilize the earth's surface temperature. The presence of the atmosphere, which reradiates a portion of energy back toward the earth's surface, helps in making the temperature more even and hospitable.

The addition of radiation energy by human activities, such as excessive burning of fossil fuels causes a rise in the equilibrium temperature according to the Stefan-Boltzmann relation. For a known amount of heat added through human activities, the effect on the earth's temperature or thermal pollution could be broadly projected by applying the Stefan-Boltzmann equation.

The Role of Meteorology in Atmospheric Pollution

The energy conversions produce a large amount of gaseous wastes. These are disposed of in the atmosphere in much the same manner as liquid and solid wastes are dumped on land or in water. The solar-powered mechanisms responsible for meteorologic and geographic features perform natural cleansing. The atmospheric air serves two functions: dilution and dispersion. The enormous ocean of air mixes the waste thoroughly. The air currents and winds carry the waste away and spread it throughout the atmosphere. A lot of atmosphere exists on our planet, estimated at 5.6 trillion tons. Compared to this, the waste disposal is very small. Thus, the concentration never rises above small fraction of a percent and the composition of atmosphere essentially remains unchanged. However, there are certain meteorological conditions under which this disposal mechanism fails, giving rise to pollution distress.

A parcel of air warmed by contact with the ground becomes less dense and rises. As the air rises, the atmospheric pressure around it decreases, making the air expand and cool. This decrease in temperature with altitude is known as *adiabatic lapse rate* (ALR.)[8] The ALR for dry air is -9.8°C per 1000 m rise. For moist air it could be as low as -3.6°C per 1000 m. The average ALR is -7°C per 1000 m.

The temperature of the atmosphere normally decreases as the altitude increases. The average temperature at ground level is 10°C, which decreases to -60°C at 10,000 m, i.e., a rate of 7°C/1000 m. This temperature profile of atmosphere depends on many factors such as the wind, sunshine intensity, and humidity.

The following three conditions could exist for the air parcel and the atmospheric temperature profile:

1. If the temperature of the atmosphere decreases more rapidly than the ALR of the air parcel, then the air parcel containing the pollutant released at ground level would be warmer than the surrounding air at all levels, as shown in Figure 3-6a. Thus, the parcel continues to move upward indefinitely. This leads to good vertical mixing and a pollution-free environment. However, this is a condition of meteorological instability that rarely occurs

2. In a stable condition, the temperature of the atmosphere drops less rapidly with altitude than does the temperature of the air parcel. The warmed polluted air released near the ground surface rises in the ambient cooler atmosphere until at some level it is no longer warmer than the surroundings. Then it ceases to rise as shown in Figure 3-6b.

3. The conditions could exist when the atmospheric temperature increases with altitude as shown in Figure 3-6c. In such a case any polluted warm air released near the ground does not disperse vertically but 'hangs over' a city. This condition is known as *thermal inversion*. This concept will be described in detail with an example in Chapter 10.

---

[8]In thermodynamics, the word adiabatic means without changing heat with the neighboring air.

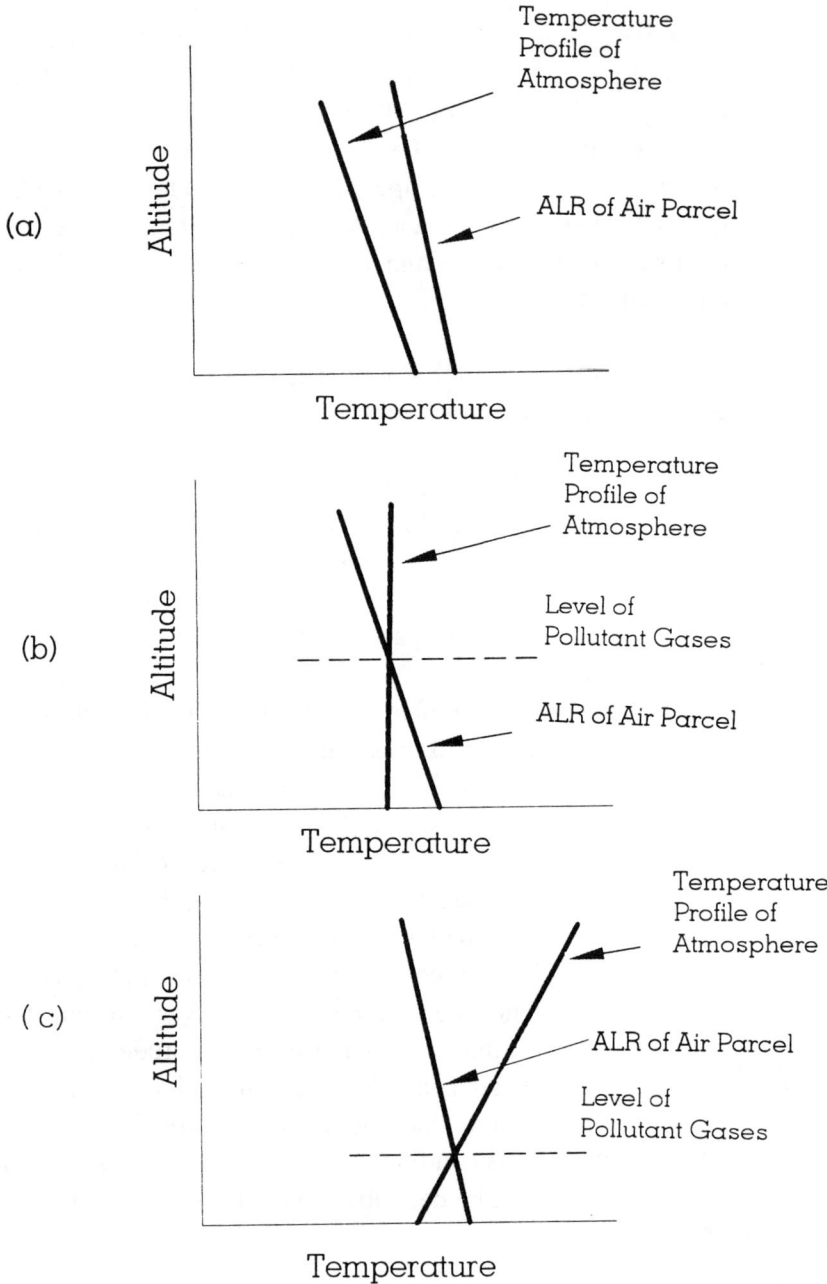

Figure 3-6 Possible Temperature Profiles

Principles of Energy and Energy Alternatives 77

Matter and Energy in Society Infrastructure

On the basis of matter circulation and energy flow on the earth, two models of the societies can be presented. Both of these models are applicable to advanced industrialized countries such as the U.S.

The first model given in Figure 3-7 represents a present-day modern society:

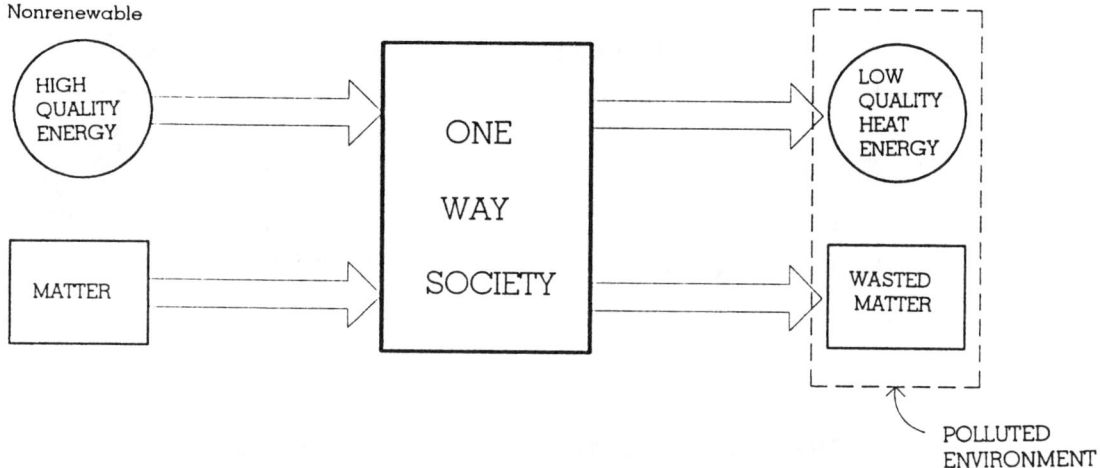

Figure 3-7 A Simple One-Way or Throwaway Society Model

This society is based on the premise of using fresh material and energy resources, discarding them after usage, and then looking for fresh resources again. This is a model of a throwaway society in which only a one-way flow of matter and energy takes place. The society believes that new matter and energy resources will be available forever, which is contrary to the principles of matter and energy. This kind of society is causing harm in three ways:

- Rapidly rendering the useful resources of the world to a degraded form;
- Adding to the problem of trash at a fast rate; and
- Producing excessive pollution of the environment.

As opposed to the example above, a model of a sustainable-earth society is presented in Figure 3-8.

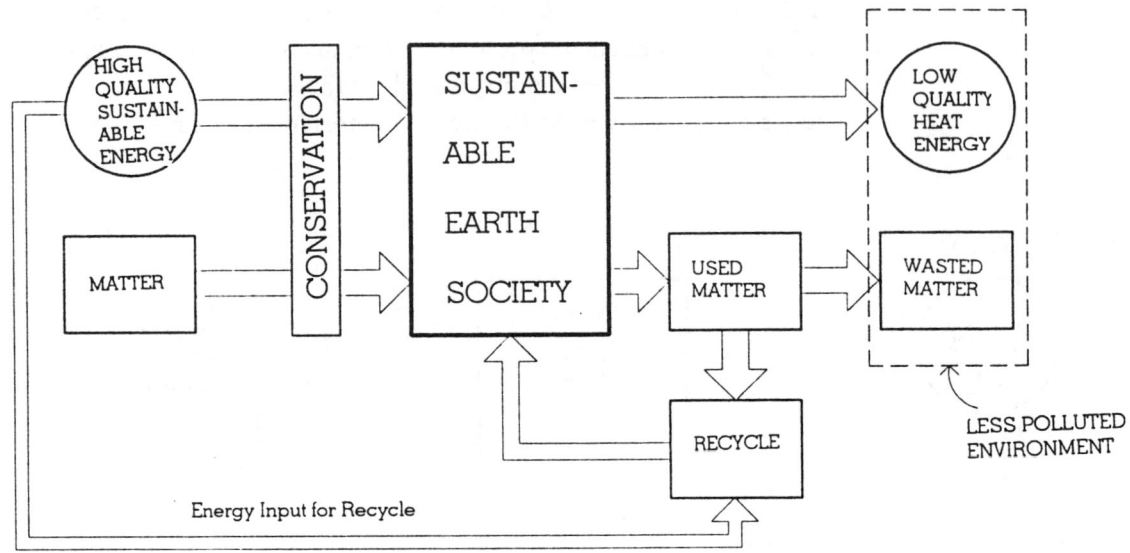

Figure 3-8 A Model of Sustainable Society

This society has three distinct characteristics:

1. It conserves both matter and energy resources, controls population, and exercises input control on pollution;
2. It recycles and reuses matter to a maximum extent; and
3. It explores and uses alternative energy resources of a more sustainable nature.

In this model, an additional input of high-quality energy is needed for recycling the matter. However, this requirement is far less in quantity than the requirement for processing of virgin resources. This form of society minimizes the three problems associated with the throwaway society. However, it demands a fundamental change in lifestyle and attitude of the population. The laws of matter and energy dictate that ultimately a form of society in least conflict with nature has to be adopted.

Energy for the 21st Century

Each phase of growth of human civilization had been directly associated with improvements of energy resources. Wood was the primary fuel before the industrial revolution. With the discovery of coal as a fuel source, the

English paved the way for the industrial society. By 1900, other forms of fossil fuel (oil and natural gas) which burned much cleaner, appeared along with coal. In 1950, the use of oil increased to be at par with coal at 40% and natural gas contributed another 15%. Since the 1980s, 85% or more of the energy needs in the U.S. have been met from three nonrenewable fossil fuels–oil 40%, natural gas 25%, and coal 20%.

Two-thirds of the world's oil reserves are located in just five Middle-East countries. At a best estimate, most of the oil will be exhausted by 2070. Natural gas is a mixture of 50–90% methane and remaining amounts of propane and butane. The former Soviet Union has more than 40% of the total reserves of natural gas. The U.S. reserves account for less than 10%. The world's supply of natural gas is projected to last until 2045. More than two-thirds of world's coal reserves are in three countries: the U.S., the former Soviet Union, and China. The identified coal reserves are expected to last more than 200 years and the unidentified at least for another 200 years. By far, coal is the most abundant fossil fuel but it is also the dirtiest. The use of coal, accordingly, has declined in last 100 years.

The burning of fossil fuels produces carbon dioxide, sulfur dioxide, nitrogen oxides, carbon monoxide, and particulate matter. This has contributed to the greenhouse effect; acid deposition; smog formation; destruction of trees, crops, and wild animals; and corrosion of metals. The depletion of fossil fuel resources and widespread negative environmental effects caused by them are convincing reasons to seek alternative energy policies.

Alternative Energy Strategy

A two-point strategy on energy needs to focus on the measures to conserve energy, and a switch to alternative forms of energy. Energy conservation is the most effective strategy; it has the highest energy yield, has no adverse environmental impact, and promotes economy by saving money. According to one estimate, at least 40% of all energy used in the U.S. is unnecessarily wasted. Improvements in energy efficiency can be achieved in all areas–residential, commercial, industrial, and transportation.

There are many choices of energy alternatives, although not all of them have equally proven technologies. These can be divided in three categories:

1. Energy-based on renewable and unharnessed sources. Leading the list is solar energy. Others are hydropower, windpower, tidal energy, ocean thermal energy, geothermal (earth's interior) energy, biofuels, and hydrogen gas.

2. Clean coal and synthetic liquid and gaseous fuels made from coal, which is nonrenewable but relatively abundant in supply
3. Nuclear power, preferably by fusion or by fission process. This is a nonrenewable resource but practically unharnessed in the U.S.

Solar and nuclear energy are two projected alternatives for the 21st century.

Conservation of Energy

Some experts believe that the U.S. can cut down its current energy demand by as much as two-thirds. The savings can be effected by two ways: (1) From reduced consumption by adjusting the habits of energy use, and (2) by using a smaller amount of energy to perform the same work. The savings can be made at all levels, as follows:

- Improving residential and commercial efficiency: More than 50% of energy used in buildings can be saved with existing technology. Some energy-saving measures are:
    1. getting as much heat and cooling as possible from natural sources– sun, wind, shade, etc.,
    2. arranging to heat and cool only the space as necessary,
    3. insulating, caulking, and weather-stripping homes and using energy-efficient windows, and
    4. using energy-efficient lights and appliances. Fluorescent light bulbs use only one-fourth the electricity of conventional incandescent bulbs.

- Improving industrial efficiency: Industrial processes consume 36% of the energy in the U.S. Installation of cogeneration units can recover a substantial portion of waste energy during the generation process. A large portion of electricity in industry is used to drive motors that are very inefficient. High-efficiency motors can reduce this wastage. Better management of energy through computer-controlled systems that shut off equipment not in operation can save energy. Refuse-derived fuel (RDF) and recycling of products are significant steps to reduce energy waste.

- Improving transportation energy efficiency: Transportation accounts for more than 50% of the daily oil use in the U.S. The methods to achieve transportation-fuel savings comprise:
    1. to improving the fuel efficiency of vehicles;
    2. encouraging commuter car pooling;

3. making greater use of mass transit;
4. hauling freight by the most efficient modes; and
5. increasing the tax on gasoline.

Renewable Energy Resources

Alternative technologies are desirable to develop inexhaustible energy sources that will eliminate concerns over depletion of resources. The pollution and environmental degradation from renewable sources would be far less as compared to fossil fuel sources. Except for a general agreement in solar energy, the opinions about the best energy option are much divided. The main deterrent factors are the high cost of changeover and the fact that new technologies are less well-defined and not fully developed. Table 3-4 summarizes the advantages and disadvantages of renewable and unharnessed energy alternatives and states the theory of energy development under each alternative. This information is based on the current state of knowledge; future innovations will change the information in the table. The two promising alternatives of the 21st century are discussed in the subsequent sections.

Table 3-4 Evaluation of Energy Alternatives

Alternative	Principle of Energy Conversion	Advantages	Disadvantages
Solar Energy	Refer to next section		
Hydropower	Water stored behind a dam falls from a height to spin turbines and generate electricity.	1. No emissions of air pollutants during operation 2. Long life span of dams 3. Low operating and maintenance cost 4. Multipurpose uses possible 5. Small sites are available that have fewer environmental impacts	1. Few suitable sites are left for large dams 2. Initial cost is high 3. Environmental impact is high: submerges areas, destroys wildlife, uproots people, reduces aquatic life and reduces downstream flow
Wind Power	The blades of the windmill rotate when wind strikes them. The blades turn a central shaft which operates an electric generator.	1. No emissions of air pollutants 2. Wind farms can be built in a few months 3. Land can be used for grazing and other purposes	1. Suitable only in areas with sufficient winds 2. Power generation only during windy conditions- supplemental source necessary 3. Visual pollution 4. Danger to flying birds 5. Noise pollution

(Table 3-4 continued)

Alternative	Principle of Energy Conversion	Advantages	Disadvantages
Tidal Energy	Technology is similar to low-head hydropower. Tides cause water levels to fluctuate twice daily; water flows in and out of banks along the shoreline. The water flow is directed through turbines connected to generators. The turbines are designed so that the water flow both into and out of the banks produce electricity	1. Low operating costs 2. No emissions of pollutants 3. Little disturbance to land	1. Only a few suitable sites with high tides 2. Construction cost high 3. A backup system is necessary 4. Power plant can be damaged by storm surges 5. Corrosion by sea water 6. Aquatic life in estuaries can be disrupted
Ocean Thermal Energy	A large temperature difference exists between the sun-warmed surface and the cold deep waters of the tropical oceans. An ocean thermal energy conversion (OTEC) plant anchored to the bottom of the ocean uses the temperature gradient to produce electricity	1. Except for the release of carbon dioxide, no other air pollution 2. Requires no land 3. No backup system needed	1. Technology in research and development stage only 2. Sites limited in number 3. Construction cost high 4. Operation and maintenance cost high 5. Plant could be damaged by severe weather 6. Algae causes corrosion of meals and fouling of heat exchanges 7. Water pumping disrupts aquatic life 8. Means to transport energy to the shore must be developed
Geothermal Energy	Earth's molten interior possesses an immense amount of thermal energy. At some locations this geothermal energy has moved up close to the crust underground, over millions of years. This reservoir of energy exists in the form of dry steam, wet steam (with water droplets), and hot water. Wells are drilled to tap this energy. This can be used as heat or to generate electricity	1. Sites are easily accessible close to the earth's surface 2. Far less emission of carbon dioxide than burning of fossil fuels 3. Production cost low	1. Number of sites scarce 2. Can deplete the nonrenewable resource 3. Destroys or degrades forests 4. Moderate-to-high air and water pollution 5. Poses danger for soil salination 6. Land subsidence risk 7. Disrupts the aquifer 8. Seismicity is induced 9. Technology to be improved
Biofuels	Woods, dry plant matter, agricultural wastes, and garbage can be burned as a fuel. This can be burned as solid fuel or converted by bacterial and chemical processes into gaseous or liquid biofuels	1. If the rate of burning is matched with the rate of replenishment, no increase in carbon dioxide 2. Much less sulfur dioxide and nitrogen oxides than caused by the burning of coal 3. Use of methane will reduce greenhouse effect	1. Energy yield low 2. Lots of land needed to grow biomass fuel; compete with food crops 3. Collecting and hauling expensive 4. Environmental impacts: soil erosion, nutrient depletion, loss of wildlife, reduced biodiversity, and water pollution by pesticides and fertilizers 5. Accidents from house fires 6. Indoor emission of particulate matter and carbon monoxide 7. Gasoline form of biomass has less energy, is more flammable, and is more costly

(Table 3-4 continued)

Alternative	Principle of Energy Conversion	Advantages	Disadvantages
Hydrogen	Decomposition of water by passing electric current produces hydrogen and oxygen.	1. Can be provided from water 2. No carbon dioxide emission, only a small amount of nitrogen oxides emitted 3. High energy yield 4. Can be distributed in pipeline under pressure and stored in tanks 5. Can be converted into solid form to release hydrogen on heating	1. Requires another energy source to produce electric current to decompose water; more energy needed than release 2. Technology to be improved to use solar energy to produce hydrogen fuel 3. Well suited to the role of secondary fuel 4. Hydrogen gas is more dangerous, more flammable and ignitable; solid form will not explode 5. Electrolyzers now in operation emit asbestos particles

Solar Energy

All of the energy alternatives presented in Table 3-4, except for tidal and geothermal, are derivatives of solar energy. More than 99.9% of the energy available on the earth's surface is from incoming solar radiation. Any single day, the sunlight that falls on the earth surface has 15,000-times more energy than the total energy consumption of the world in one day. In one month, we receive as much energy from the sun as that combined in all known reserves of coal, oil, gas, and uranium. However, the direct conversion of the energy from the sun at present is negligible. In view of the facts that the convenience of fossil fuel is diminishing in the wake of adverse environmental impacts and the projected 30% growth for energy requirements by 2025, solar technology is expected to revolutionize the energy field. This will bring beneficial effects to the environment and to the global climate. In addition to direct solar energy conversion, other forms of energy developments described in Table 3-4 will emerge in the 21st century; the regional factors will favor some alternatives over others.

There are several forms of direct solar energy conversions.

Passive Solar Energy Conversion for Heating and Cooling

The sun's rays falling on a structure increase the structure's temperature. Passive systems make use of this natural solar radiation without using any motor or pump devices. It has become increasingly popular to effectively use sunlight to heat residences and commercial buildings. The passive solar design approach consists of measures such as considering the orientation of buildings, (designing south-facing windows, and paying attention to shadow lines from adjoining structures and trees) and preventing drafts and leaks.

The choice of building materials can contribute to passive energy retention. Products of modern technology can play an important part in passive systems. Passive approaches are also well suited to retrofitting existing structures. A performance study of residential buildings has indicated that up to 70% of heating requirements could be met from passive solar designs.

Passive cooling can be provided by blocking the sun's rays by trees or window overhangs and by reflecting the rays on foil sheets. Earth tubes buried at about 10 ft deep can relay cool air into a home, with a simple installation.

Active Solar Energy for Heating Space and Water

Heating of space in residential and commercial buildings and heating of water for domestic use with the help of the sun are common applications of solar energy. A collector system often comprises a number of flat-plate units mounted on the roof. A unit, facing the sun, is covered by one or two glass or plastic panels. The heat absorption surface, underneath the cover, comprises a metal sheet painted black for maximum solar absorptivity. The tubes or pipes carrying water or antifreeze solution are directly attached to the absorption sheet for efficient heat conduction. The fluid inside the tube carries heat to a heat exchanger system, which could be a water storage tank. The hot water from the tank is used directly or circulated to heat space.

Solar-Thermal-Electric Engines

These devices have four components: a collector to focus the sunlight, a receiver to absorb the heat, a heat exchanger to produce steam, and a generator to produce electricity. The huge reflectors known as collectors are used to focus sunlight on a central heat collection element. Reflectors are either long cylindrical trough sections or spherical dishes made of mirrors or other shiny metal. Their parabolic shapes cause the reflected rays to converge at a common point. It is at this point that a receiver is located. A power tower absorbs sunlight from hundreds of reflectors to achieve high temperatures between 1000 and 2000°C. The cold feedwater is pumped to the receiver to make high-pressure steam. The steam can be either delivered directly to the turbine and generator or to the storage system. With an energy storage system, electricity could continue to be generated for several hours without sunlight. It is assessed that the complete electric needs of Los Angeles could be met with coverage of a very small area in the Mojave Desert.

Converting Solar Energy Directly to Electricity by Photovoltaic (PV) Cells

Photovoltaic or solar cells represent the fundamental power conversion units of the sun's energy. They are made from semiconductor material. Many different cells of different materials and different structures are available. Currently, crystalline silicon cells are the most reliable and most technologically developed. Solar cells are produced from sand in four stages:

1. Sand is converted to pure silicon by chemical transformation;
2. Silicon is grown by a crystallization process into a single-crystal silicon or silicon ingot. At this stage, a controlled amount of another material (boron or phosphorus) is added in what is known as doping to make the silicon conduct electricity,
3. Crystals are sliced into wafers by a sawing machine; and
4. Cells are fabricated by subjecting wafers to several chemical, thermal, and deposition treatments.

A cell consists of several layers including front and back metal contacts and an antireflection coat. When the sun's rays strike a cell, the movement of charged particles produces electric current. About 36 cells are interconnected and encapsulated to form a module that can last 20 years or more. Modules are mounted on a support structure to make an array. The amount of electricity produced by a single module is small; hence, many modules must be wired together and mounted on a roof.

The use of thin-film solar cells can reduce the cost and provide an integrally interconnected module rather than individual cells requiring separate interconnections. The thin-film technology holds promise for future solar conversion. Four types of thin-film cells on the market include the amorphous silicon cell, thin multicrystalline silicon, copper indium diselenide, and cadmium telluride.

Advantages and Disadvantages of Solar Energy

Advantages of solar energy are as follows:

- The energy source is perpetual, free, and naturally available.
- Technology for passive and active uses is fairly well-developed.
- The fuel source produces no environmental pollution. Any contamination occurs in the conversion process only.
- There is very little or no land disturbance involved.

- It is a quiet process, have no moving parts, and requires little maintenance.
- It can be easily installed and expanded.

Disadvantages are as follows:

- Some people believe that the collectors on rooftops are unsightly.
- A large area is needed for collectors/solar cells.
- Cost of the system is high.
- Technology needs to be improved, especially for direct electric conversion.

Nuclear Energy

There are two mechanisms by which nuclear energy is obtained: the fission process and the fusion process. In a fission process, a fuel such as enriched uranium (^{235}U) is used. The ^{235}U is contained in fuel rods placed in the core of a reactor. When a ^{235}U atom is bombarded by neutrons, the nucleus of the ^{235}U atom captures a neutron, becomes unstable, and splits into smaller atoms. The resulting fission products (atoms) have slightly less mass. The loss of small mass represents conversion into tremendous energy according to Einstein's principle. When an atom splits, several free electrons are released. These are available to strike other ^{235}U atoms causing them to split too thus, creating a chain reaction. If this chain reaction is to continue, enough ^{235}U must be packed together to achieve the critical mass. There are many types of reactors; the breeder-type reactors are significant because they can convert nonfissionable uranium-238 into fissionable plutonium-239.

The heat generated in a fission reaction is transferred to a liquid that turns a turbine. The spent fuel rods (that can no longer sustain nuclear fission) contain fragments (smaller atoms and neutrons) that are intensely radioactive. It is the radioactivity of the fission products that makes nuclear energy undesirable.

In the other nuclear process of fusion, lighter atoms such as hydrogen isotopes are combined (fused) to form a heavier element. Extremely high temperatures–hotter than the sun–are required to achieve the fusion. Fusion produces more energy and has less environmental impact since atoms released by this process have a much shorter half-life (12.3 years for common atoms formed compared to thousands of years in fission). The process currently is not feasible on an industrial scale. However, it is predicted that the fusion reactor will be an important source of energy by the middle of the 21st century.

Principles of Energy and Energy Alternatives 87

PROBLEMS

1. A stone falls from a hill. How fast will it be falling 2 sec after it has dropped? If the acceleration due to gravity at the moon is 1/6 the value on the earth, what will be its speed on the moon?

2. A 3-lb paint bucket falls from a 10-ft ladder. What will be its speed when it hits the floor? How much energy will it exert at the floor?

3. A flower pot falls from a balcony of a building. It reaches the ground in 6 sec. How high is the balcony? If a man ten times heavier falls from the balcony, how long will he take to hit the ground?

4. At a 200-ft waterfall, water flow (unit weight 62.4 lb/ft^3) of 1000 ft^3 is recorded over a period of time. What is the energy potential at the site? Also express the energy in Joules and kilowatt-hours, using a conversion table.

5. A 10 kg object falls from a height in 20 sec under its own weight. How much work (energy) and power are being expended?

6. An 80 kg girl starts at the top of a 10-meter high water slide at an amusement park. What is her total potential energy? When she slides down to a point 4 m above the ground, what will be her potential energy? What will be her velocity at that point? Assume no losses from friction.

7. A man pushes a large box with a force of 100 Newtons for 30 sec but cannot move it. How much energy has he spent?

8. How much energy is required to increase the speed of a 1200 kg car of from 10 m/s to 25 m/s in 3 sec?

9. The earth receives radiation energy from the sun at a rate of 2 cal/min/cm^2. Incidental radiation is perpendicular to the earth's diameter. How much energy is received during the life time (10 billion years) of the earth?

10. The food intake of a laborer is 4000 Calories per day. He performs 1 million J of work every day. What fraction of his food energy has been converted to work?

11. The complete combustion of 1 kg of coal releases 3 x 10^7 Joules of heat. The world's coal reserves are 8.4 × 10^{12} tons (1 ton = 909 kg). How much is the total energy content of the coal?

12. The sun's energy shines on the earth is at a rate of 173×10^{15} Watts or J/sec. How many year's worth of supply are represented by the entire coal reserves in terms of solar radiation?

13. Suppose the sun is made of coal and it continues to emit radiation at the present rate of 4×10^{26} Watts (J/s). The combustion of 1 kg of coal releases about 30 million J of heat energy. If the sun's mass is 2×10^{30} kg, how long will the sun radiate the energy?

14. The sun has a surface temperature of 5750 K. It has a radius of 696×10^6 m. Determine the sun's luminosity (rate of radiation of energy into the space).

15. The rate of radiation from the sun falling on the earth is 1400 W/m². If all of this is absorbed by the earth, what will be the temperature of the earth?

16. On an average summer day, suppose the city of Boston intercepts 45% of the radiation constant of 1400 W/m². How many Joules of energy is received by Boston (area 4000 mi²) during the daylight time of 8 hr?

17. From the information presented on pp. 79-82, prepare a table similar to Table 3-4 for solar energy, listing various forms of solar energy conversions separately.

STUDY QUESTIONS

1. How are matter and energy related? What are the units of measurement of energy? What are the major forms of energy?

2. What are the two scientific laws of energy? Why are they also known as the laws of thermodynamics? What is the significance of these laws to the functioning of the ecosystem and the environment?

3. The first law of thermodynamics is sometimes stated as "you can not get something for nothing", and the second law is stated as "you can not even break even". Clarify these statements in the light of the two laws explained in the chapter. Trace the history of thermal energy of our body to show that the ultimate source is the sun. Does the sun lose any energy by doing work on the earth?

4. You drive your car to school one morning. Upon your return, you park your car in the same spot you were in that morning. There is no change in net energy. Explain the gasoline energy that has been used up. Identify all forms of energy conversions.

5. Give at least five examples of energy conversions that are taking place in your classroom. Why can't the high quality energy of electricity used to heat a room be converted back into useful work at some later time? What is the environmental implication of this energy consumption?

6. The earth receives a tremendous amount of energy from the sun, then why the temperature of the earth does not rise continuously? What effects will clouds of dust have on the average temperature of the earth?

7. Outline and differentiate between a throwaway society model and a sustainable society model. Can a modern society adopt the sustainable model? Explain.

8. What means of energy conservation can you suggest for your home, town, and nation?

9. Evaluate the alternative energy sources with their pros and cons. Which of these appear promising for the future?

10. Evaluate solar energy as the energy of the future. What supplemental sources, if any, would you recommend?

11. It has been decided to locate a nuclear power plant in your town. What are your opinions about this decision?

12. Define the following:
    a. Work, energy, and power
    b. Quality of energy
    c. Entropy
    d. British thermal unit (Btu)
    e. Wavelength and frequency
    f. Shortwave and longwave radiations
    g. Wein's law
    h. Stefan-Boltzmann law
    i. Adiabatic lapse rate (ALR)
    j. Hydropower
    k. Geothermal energy
    l. Hydrogen energy
    m. Ocean thermal energy

REFERENCES

Allen, J. L., ed., *Annual Editions Environment 92/93*, The Dushkin Publishing Group, Guilford, CT, 1992.

Berger, J. J., *Environmental Restoration: Science and Strategies for Restoring the Earth*, Island Press, Washington, D.C., 1990.

Cartledge, B., ed., *Energy and the Environment: The Linacre Lectures 1991-2*, Oxford University Press, Oxford, U.K., 1993.

Cassedy, E. S., and Grossman, P. Z., *Introduction to Energy: Resources, Technology and Society*, Cambridge University Press, Cambridge, U.K., 1990.

Chiras, D. D., ed., *Voices for the Earth*, Johnson Books, Boulder, CO, 1995.

Christensen, J. W., *Global Science: Energy, Resources and Environment*, 3rd ed., Kendall/Hunt, Dubuque, IA, 1990.

Colorado Energy Research Institute, *Net Energy Analysis: An Energy Balance Study of Fossil Fuel Resources*, Colorado Energy Research Institute, Golden, CO, 1976.

Crawley, G. M., *Energy*, Macmillan, New York, 1975.

Deudney, D., *Rivers of Energy: The Hydropower Potential*, Paper 44, Worldwatch Institute, Washington, D.C., 1981.

Flavin, C., *Wind/Power: A Turning Point*, Paper 45, Worldwatch Institute, Washington, D.C., 1981.

Fowler, J. M., *Energy and the Environment*, 2nd ed., McGraw-Hill Book Company, New York, 1984.

Gibbons, J. H., and Chandler, W. U., *Energy: The Conservation Revolution*, Plenum Press, New York, 1981.

Hewitt, P. G., *Conceptual Physics*, Little, Brown, Boston, MA, 1977.

Hoffman, E. J., *The Concept of Energy: An Inquiry Into Origins and Application*, Ann Arbor Science, Ann Arbor, MI, 1977.

Hollander, J. M., (Ed.), *The Energy-Environment Connection*, Island Press, Washington, D.C., 1992.

Hubbert, M. K., "The Energy Sources of the Earth," *Scientific American*, 224, 3, pp. 60-70, September, 1971.

Kraushaar, J. J., and Ristinen, R. A., *Energy and Problems of a Technical Society*, John Wiley and Sons, New York, 1988.

Landsberg, H. H., Chairman Study Group, *Energy: The Next Twenty Five Years*, Ballinger Publishing Company, Cambridge, MA, 1979.

Lenssen, N., *Empowering Development: The New Energy Equation*, Paper 111, Worldwatch Institute, Washington, D.C., November 1992.

Life Science Library, *Energy*, Time-Life Books, New York, 1968.

Marion, J. B., *Energy in Perspective*, Academic Press, New York, 1974.

Markvart, T., ed., *Solar Electricity*, John Wiley and Sons, New York, 1994.

Mathews, J. T., ed., *Preserving the Global Environment*, W. W., Norton and Company, New York, 1991.

Miller, G. T., Jr., *Environmental Science*, 5th ed., Wadsworth Publishing Company, Belmont, CA, 1995.

Morrison, W. E., and Readling, C. L., *An Energy Model for the U. S.*, U. S. Department of the Interior, Bureau of Mines, No. 8384, Washington, D.C., 1968.

Murdoch, W. W., ed., *Environment: Resources, Pollution and Society*, Sinauer Assoc., Stamford, CT, 1971.

National Geographic, *Energy: Facing Up to the Problems, Getting Down to the Solutions*, National Geographic Society, Washington, D.C., Feb. 1981.

National Research Council, Committee on Nuclear and Alternative Energy Systems, *Energy Transition 1985-2010*, Final Report, Published for National Academy of Sciences by W. H. Freeman and Co., San Francisco, CA, 1980

Odum, H.T., and Odum, E.C., *Energy Basis for Man and Nature*, McGraw Hill Book Company, New York, 1976.

Ogden, J. M., and Williams, R. H., *Solar Hydrogen: Moving Beyond Fossil Fuels*, World Resources Institute, Washington, D.C., 1989.

Olszewski, B, J., and Shiavo F. R., *Readings in Environmental Studies*, Kendall/Hunt Publishing Company, Dubuque, IA, 1992.

Priest, J., *Problems of Our Physical Environment*, Addison-Wesley, Reading, MA, 1976.

Repetto, R., *World Enough and Time: Successful Strategies for Resource Management*, Yale University Press, New Haven, CT, 1986.

Romer, R. H., *Energy — An Introduction to Physics*, W. H. Freeman, San Francisco, CA, 1976.

Saperstein, A. M., *Physics: Energy in the Environment*, Little, Brown, Boston, MA, 1975.

Scientific American, *Energy and Power*, W. H. Freeman and Company, San Francisco, CA, 1971.

Scientific American, *New Technologies for the 21st Century*, 150th Anniversary Issue, v. 273, n. 3, September 1995.

Sterrett, F. S., ed., *Alternative Fuels and the Environment*, Lewis Publishers, Boca Raton, FL, 1995.

Tester, T.W., Wood, D.O., and Ferrari, N.A., *Energy and the environment in the 21st Century*, MIT Press, Cambridge, MA, 1991.

White, I. D., Mottershead, D. N., and Harrison, S. J., *Environmental Systems: An Introductory Text*, 2nd ed., Chapman and Hall, London, U.K., 1992.

Wilson, R., and Jones, W. J., *Energy, Ecology, and Environment*, Academic Press, New York, 1974.

Work Group on Energy Products, *Man's Impact on the Global Environment: Assessment and Recommendations for Action*, The MIT Press, Boston, MA, 1970.

World Commission on Environment and Development, "Energy Choices for Environment and Development," *Our Common Future*, Oxford University Press, Oxford, U.K., 1987.

Young, J. J., *Discarding the Throwaway Society*, Paper 101, Worldwatch Institute, Washington, D.C., 1991.

Chapter 4

PRINCIPLES OF ENVIRONMENTAL CHEMISTRY

Chemical Reactions

When hydrogen gas is burned in air (oxygen), it forms water. This can be expressed as:

$$\text{Hydrogen gas} + \text{oxygen gas} \xrightarrow{\Delta} \text{water} \qquad (4\text{-}1)$$

where Δ denotes high temperature

This is an equation of reaction between the two chemicals hydrogen and oxygen, which are called the reactants. The chemical(s) on the right side of the arrow that is formed by the reaction is called the product. Replacing words with chemical symbols, the above relation can be written as

$$H_2 + O_2 \xrightarrow{\Delta} H_2O \qquad (4\text{-}2)$$

In accordance with the law of conservation of mass, an atom should not be lost in a reaction. The same number of atoms of each element should appear on both sides of the equation, making it balanced. The equation is balanced below:

$$2H_2 + O_2 \xrightarrow{\Delta} 2H_2O \qquad (4\text{-}3)$$

It is read: Two molecules of hydrogen react with one molecule of oxygen to produce two molecules of water. The words high temperature (denoted by Δ) indicate the necessary condition for the reaction to occur. Almost all chemical reactions either release heat or absorb heat. Any reaction that gives off heat is called an *exothermic process*. An *endothermic process* is the one in which heat has to be supplied for reaction to take place, as in Equation 4-3. This condition may not be indicated on the equation.

In pollution problems, it is more useful to know the quantities in grams or pounds of the various substances involved in the reactions. As stated in Chapter 2, the weight of an atom is expressed by a relative number referred to as the atomic mass unit (AMU). Table 4-1, called the *periodic table*, gives the name, symbol, atomic number, and atomic weight of each element. The vertical columns of the table are called *groups*. Each group of elements has the same electron structure in the outer shell, and thus demonstrates similar chemical behavior. The horizontal rows are called *periods*. Physical properties of elements progress in an orderly fashion up and down a period. For example, in ordinary conditions, fluorine (F) is a gas, bromine (Br) is a liquid, and iodine (I) is a solid.

The molecular weight of a molecule is the summation of the atomic weights of all atoms in the molecular formula. To express mass or weight of a substance, a relative term, *mole,* is used as follows:

$$\text{Moles} = \frac{\text{mass of substance}}{\text{molecular weight in AMU}} \qquad (4\text{-}4)$$

Moles are designated (g)-mol or (lb)-mol, for mass in grams or pounds respectively.

According to Avogadro's Hypothesis, made in 1811, one (g)-mol of any substance always contains 6.022×10^{23} molecules. In other words, if we take 6.022×10^{23} molecules (a fixed number) or one mole of any substance, its mass will be equal to the molecular weight of the substance in grams. Accordingly, in place of molecules, a chemical reaction might be considered to take place on a much larger scale, in terms of stacks of 6.022×10^{23} molecules or moles of each chemical. Thus, Equation 4-3 can be restated. Two moles of hydrogen react with one mole of oxygen to produce two moles of water. This provides a basis for the quantitative analysis of chemical reactions known as *stoichiometry*.

Example 4-1
How much water would be produced if 100 g of hydrogen gas is fully burned?

Table 4-1 Periodic Table of the Elements

Solution:
Use Table 4-1 for atomic weights, molecular weights and, hence, the grams per mole of each chemical in the reaction:

$H_2 = 2 \times 1 = 2$ g/mole
$O_2 = 2 \times 16 = 32$ g/mole
$H_2O = (2 \times 1) + 16 = 18$ g/mole.

From Equation 4-3:

$$2H_2 + O_2 \rightarrow 2H_2O$$

2 moles	1 mole	2 moles
2×2 = 4 g	32 g	2×18 = 36 g

Since 4 g hydrogen produces 36 g of water.
Hence, 100 g of hydrogen will produce 36/4 ×100 = 900 g of water.

Note that the mass is conserved in the above equation with 36 g on either side.

Solutions

A solution is a homogeneous mixture of two or more substances. The substance in the smaller amount is the *solute* and the other is the *solvent*. A solution can be gaseous, solid or liquid. The quantity of solute in a solution is known as its *concentration*. A standard solution is any solution of an accurately known concentration.

Liquid Solutions

Water is a common solvent. In the environmental field, many important chemical reactions and virtually all biological reactions take place in a water medium.

The common methods of expressing concentration of a liquid solution follow. Gaseous concentrations are discussed later in this chapter.

1. parts per million (ppm)[1] = $\dfrac{\text{amount of substance in mg}}{\text{weight of solution in kg}}$ (4-5)

---

[1] ppb (parts per billion) when the amount of substance is measured in μg (microgram). This unit is for very small concentrations.

2. $\text{mg/L} = \dfrac{\text{amount of substance dissolved in mg}}{\text{liters}^2 \text{ of solution}}$ (4-6)

The two terms (1) and (2) are related shown by:

3. $\text{mg/L} = (\text{ppm})(\text{specific gravity of solution})$ (4-7)

Since the concentration of most pollutants is quite small in a water solution, the solution has a specific gravity of 1, that of water. Thus, 1 mg/L = 1 ppm, practically.

Molarity is the number of moles of solute per liter of solution.

4. $\text{Molarity (in M or mol/L)} = \dfrac{\text{number of moles in dissolved substance}}{\text{Liters of solution}}$ (4-8)

Since the number of moles of a substance is given by the mass of substance divided by the molecular weight, Equation 4-8 reduces to the following form connected with the first term:

5. $\text{mg/L} = (\text{molarity})(\text{molecular weight})(1000)$ (4-9)

Example 4-2
Twenty grams of sodium chloride (NaCl) is dissolved in water to make a 500 mL solution. Find the molarity of the solution.

Solution:
Molecular weight of NaCl = 58.4

$$\text{No. of moles} = \dfrac{20}{58.4} = .034$$

$$\text{Molarity} = \dfrac{0.34}{0.5} = 0.68 \text{ M or mol/L.}$$

Solubility

The solubility of a substance is the maximum amount of it that can be dissolved in a given amount of solvent at a specified temperature and pressure. For gases, as the temperature increases, the solubility decreases. For solids, as the temperature increases, the solubility usually increases. For liquids, the change in solubility with temperature follows no general rule.

---

[2] 1 Liter = 1000 cm^3 (cubic centimeter)

The following terms relate to solubility characteristics of substances in water.

Compound's Degree of Solubility

Based on the solubility, compounds are divided into three categories:
1. *Soluble*: those having solubility of 1 g or more per 100 mL of water,
2. *Slightly soluble*: those having solubility of less than 1 g but more than 0.1 g per 100 mL, and
3. *Insoluble*: those having solubility of less than 0.1 g per 100 mL.

Ions

In ordinary chemical changes, the positively charged protons in the nucleus of an atom remain intact but the electrons may be gained or lost. When electrons are removed or added to a neutral atom, the charged atom is called an *ion*. An ion that loses electrons, i.e., bears a positive charge is a *cation* and an ion that gains electrons to have negative charge is an *anion*.

All solutes in water solutions fall into two groups: *electrolytes* are substances that when dissolved in water can conduct electricity through water. *Nonelectrolytes* do not conduct electricity. Electrolytes dissociate into positively and negatively charged ions in solution, such as:

$$HCl = [H^+] + [Cl^-] \tag{4-10}$$

Conversely, nonelectrolytes dissolve in water as neutral molecules rather than ions.

When all dissolved molecules of a substance are fully ionized, a strong electrolyte, such as hydrochloric acid (HCl), is formed. A weak electrolyte is ionized only to a slight extent, such as acetic acid (CH_3COOH).

Based on the electrolytes concept, chemical compounds are divided into two categories. *Ionic compounds* in which cations and anions are held together with the neutral charge as reverse of Equation 4-10, and the dissolved portion gets ionized, and *molecular compounds* in which the smallest dissolved units are individual molecules. Thus, a compound can be insoluble but a strong electrolyte, which means that whatever small quantity dissolves in water gets fully dissociated into ions, such as AgCl.

Acids and Bases

Electrolytes can be divided into three types of substances:

1. *Acids* are defined as substances that yield hydrogen ions [H^+] when dissolved in water.
2. *Bases* are substances that yield hydroxide ions [OH^-] when dissolved in water. It is not necessary for a base to contain hydroxide ions in its structure but it must produce [OH^-] ions, such as in ammonia (NH_3)
3. *Salts* are substances that ionize to produce neither hydrogen or hydroxide ions.

Equilibrium Constants

There are four types of reactions:

1. *Combination* reactions in which two or more substances react to form one product: A + B → AB.
2. *Decomposition* reactions in which one substance undergoes a reaction to produce two or more substances: AB → A + B.
3. *Displacement* reactions in which an atom or an ion in a compound is replaced by another element: A + BC → AC + B.
4. *Metathesis* reactions which are doubled displacement reactions :
   AD + BC → AC + BD.

After reaching a stage of chemical equilibrium, a reaction stabilizes, which means that the process in one direction (the reactants making the products) is balanced by the process in the opposite direction (the products combining to form the reactants). A system (reaction) tends to maintain the state of equilibrium. If any stress is imposed by change in temperature, pressure, or concentration, the system adjusts the position of equilibrium to relieve that stress. This is the principle of *Le Chatelier,* contained in the *law of mass action,* as follows:

For an equation expressed by:

$$mA + nB \rightleftarrows pC + qD, \qquad (4\text{-}11)$$

at equilibrium condition

$$\frac{[C]^p [D]^q}{[A]^m [B]^n} = K \qquad (4\text{-}12)$$

K is known as the *equilibrium constant.* [ ] denotes molar concentration of the substance contained within the bracket, expressed in moles per liter and not milligrams per Liter. The equilibrium relationship helps to compute the concentration of substances entering into a reaction.

When applied to a solution of an electrolyte, K is known as the *ionization constant*.

Example 4-3

A 0.15 M solution of acetic acid (CH_3COOH) is ionized 1% at equilibrium. Find the ionization constant.

Solution:

Reaction of acetic acid:

$$CH_3COOH \rightleftharpoons [H^+] + [CH_3COO^-] \tag{4-13}$$

Acetic acid conc. = 0.15 M

$$\text{Acid ionization} = \frac{1(0.15)}{100} = 0.0015 \text{ M}$$

One mole of CH_3COOH ionizes to give one $[H^+]$ and one $[CH_3COO^-]$ ion, i.e., the amount is in equilibrium:

	CH_3COOH	$[H^+]$	$[CH_3COO^-]$
Initial	0.15 M	0	0
Change	(-) 0.0015 M	(+) 0.0015 M	(+) 0.0015 M
Equilibrium	0.1485 M	0.0015 M	0.0015 M

From Equation 4-12:

$$K = \frac{(0.0015)(0.0015)}{0.1485} = 1.5 \times 10^{-5}$$

In addition to the electrolyte ionization analysis, Equation 4-12 is applicable to analysis of solubility of compounds as well. It can be applied to a saturated solution of a substance at the condition of solubility equilibrium. The denominator in that case relates to the substance being dissolved. When a solid substance is dissolved, its concentration is regarded as a constant. Thus, the equilibrium constant for dissolution of solids can be written, from Equation 4-12:

$$\frac{[C]^p[D]^q}{\text{constant}} = K$$

or $[C]^p [D]^q = (\text{constant})K = K_{sp}$ (4-14)

K_{sp} is another constant. It is called the *solubility product constant*.

The smaller the K_{sp}, the less soluble is the compound.

Example 4-4
The solubility of silver chloride is 0.19 g per 100 L. Find the solubility product constant.

Solution:
Molecular weight of AgCl = 143.35

Molar solubility = $\dfrac{0.19}{100\,(143.35)}$ = 1.33×10^{-5} M

Reaction	AgCl ⇌	[Ag$^+$] +	[Cl$^-$]
Initial		0 M	0 M
Change		1.33×10^{-5} M	1.33×10^{-5} M
Equilibrium		1.33×10^{-5} M	1.33×10^{-5} M

From Equation 4-14:

$K_{sp} = (1.33 \times 10^{-5})(1.33 \times 10^{-5}) = 1.77 \times 10^{-10}$

The pH Concept

As stated earlier, the hydrogen ion concentration [H$^+$] of a substance in aqueous solution is a measure of its acidity. But the expression of hydrogen ion in terms of molar concentration is cumbersome. In practice[3], the hydrogen ion concentration is expressed as its negative logarithm and designated the pH. Thus:

$$\text{pH} = -\log [H^+]$$

---

[3]Proposed by Sorensen in 1909.

$$\text{or pH} = \log \frac{1}{[H^+]} \qquad (4\text{-}15)$$

The pH of water is an important characteristic. The quality of water for dissolving chemicals, its effectiveness in coagulation, suitability to support life, and potential to corrosiveness all depend on the pH value.

Example 4-5
Determine the pH of pure water. The solubility product constant for water at 25°C, $K_w = 10^{-14}$

Solution:
By itself, water is a very weak electrolyte. Its minute ionization is as follows:
$$H_2O \rightleftharpoons [H^+] + [OH^-]$$

From Equation 4-12:
$$\frac{[H^+][OH^-]}{[H_2O]} = K_w$$

By convention $[H_2O]$ is unity, and for neutral condition $[H^+] = [OH^-]$, thus:
$$[H^+][OH^-] = K_w = 1 \times 10^{-14}$$
$$[H^+] = [OH^-] = 1 \times 10^{-7}$$
$$pH = \log \frac{1}{(1 \times 10^{-7})} = 7$$

This is considered the neutral value of pH.

The pH scale is represented by a range from 0 to 14. A value of 7 is neutral, a value lower than 7 indicates that the hydrogen ion concentration is more than that of the hydroxide ions, thus the medium is acidic. A medium is basic when the pH is greater than 7. Dissolving a substance in pure water changes its pH. The pH of several fluids is indicated in Figure 4-1.

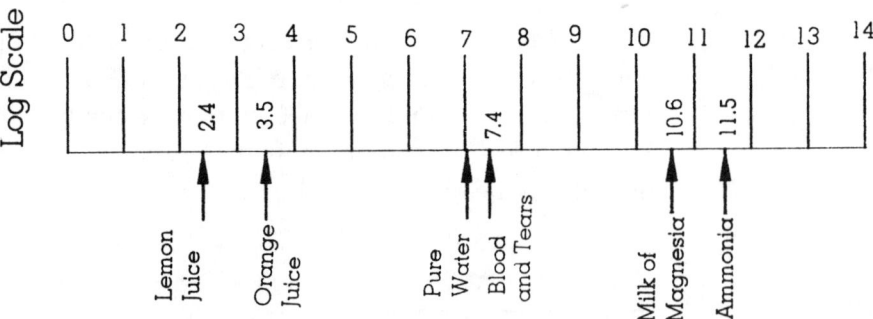

Figure 4-1 pH of Fluids

Example 4-6
The pH of rainwater is 5.25. What are the hydrogen and hydroxide concentrations in the rainwater?

Solution:
$$pH = 5.25 = \log \frac{1}{[H^+]}$$

$[H^+] = 5.6 \times 10^{-6}$ M
$[H^+][OH^-] = K_w = 1 \times 10^{-14}$

$$[OH^-] = \frac{1 \times 10^{-14}}{5.6 \times 10^{-6}} = 1.8 \times 10^{-9} \text{ M}$$

Gases

The study of air pollution deals with gaseous concentrations. A gas has no internal boundary; it expands to fill the volume of a container. The volume occupied by a gas is highly variable. Two factors have direct influence upon the gaseous volume and on its density: pressure and temperature. Boyle provided the relationship between the pressure and volume according to which the volume of a fixed amount of gas is inversely proportional to the gas pressure at a constant temperature. Charles (and Gay-Lussac) correlated the other set of parameters by a law stating that the volume occupied by a gas is directly proportional to the absolute temperature[4] of the gas at a given pressure. Combining Boyle's Law and Charles's Law, and expressing the constant per mole of the gas :

$$\frac{pV}{T} = nR \qquad (4\text{-}16)$$

where p = pressure of gas in units of above atmosphere
V = volume of gas in Liters
T = absolute temperature in °K
n = number of moles of gas = mass/molecular weight
R = gas constant (0.082 L-atm/mole-°K).

Avogadro's Hypothesis, which specified a fixed 6.022×10^{23} molecules in 1 (g)-mole of any substance, also stated that equal volumes of all gases contain an equal number of molecules under similar temperature and pressure conditions. At standard temperature (0°C) and pressure (1 atm), 1 (g)-mole

---

[4] On Kelvin scale, °K = °C + 273. On Rankine scale, °R = °F + 460

of any gas will occupy 22.4 L of space. This is a very useful relation for volumetric analysis. Substituting these values in Equation 4-16, R is equal to 0.082 L-atm/mole/°K.

An ideal gas is a hypothetical gas that follows the relationship of Equation 4-16. The ideal gas equation is used for real gases as well since discrepancy over reasonable temperature and pressure conditions is not significant.

Example 4-7
What tank volume is required to hold 1000 kg of methane gas at 20°C and 44.1 psi pressure?

Solution:
20°C = 293°K
44.1 psi = 3 atm

$$1000 \text{ kg of } CH_4 = \frac{1000(10^3)}{16} = 62,500 \text{ (g)-moles}$$

From Equation 4-16:

$$V = \frac{nRT}{p}$$

$$= \frac{(62,500)(0.082)(293)}{3}$$

$$= 500,542 \text{ L}.$$

Example 4-8
A 3×10^{-3} M solution of acetic acid is aerobically digested. Determine the oxygen demand in the solution. How much volume of the carbon dioxide will be released per liter of solution at standard temperature and pressure (STP)?

Solution:
Balanced equation for digestion:

CH_3COOH +	$2O_2$	→	$2CO_2$ +	$2H_2O$
1 mole	2 mole		2 mole	2 mole
60 g	2×32 =64 g		2×44 =88 g	2×18 =36 g

From Equation 4-9:
milligrams per Liter of acetic acid = $(3 \times 10^{-3})(60)(1000) = 180$ mg/L

Principles of Environmental Chemistry

60 g of acetic acid requires 64 g of oxygen

180 mg/L of acidic acid will require $\frac{64}{60} \times 180 = 192$ mg/L O_2

60 g of acetic acid releases 88 g of CO_2

180 mg/L of acidic acid will release $\frac{88}{60} \times 180 = 264$ mg/L CO_2

No. of moles of CO_2 per L of solution. $= \frac{(264 \times 10^{-3})}{44} = 0.006$

1 mole occupies 22.4 L
0.006 will occupy $22.4 \times 0.006 = 0.13$ L (at STP).

Expression of Gaseous Pollutants

For gaseous pollutants in air, the parts-per-million concentration is expressed in volumetric terms instead of mass ratio. The following three relations apply to gaseous concentration, in place of Equations (4-5), (4-6), and (4-7):

1. Parts per million (ppm) $= \dfrac{\text{volume of gaseous pollutant}}{\text{volume of air + pollutant}} \times 10^6$ (4-17)

2. $\mu g/m^3 = \dfrac{\text{micrograms of pollutant}}{\text{cubic meter of air}}$ (4-18)[5]

The volumetric relation of Equation 4-17 is converted to the mass relation of Equation 4-18 by using the gas law of Equation 4-16. The following expression emerges:

3. $\mu g/m^3 = \dfrac{(ppm)pMW}{TR} \times 10^3$ (4-19)

where p = pressure in atm. (101.3 kPa or 14.7 psi = 1 atm)
T = temperature in K (0°C = 273 K)
MW = molecular weight of pollutant
R = gas law constant in atm-L/(g)-mol-K (R=0.08206)

Example 4-9
If the carbon monoxide concentration in a room of 2000 m³ capacity is 200 L what is the concentration, in parts per million, and micrograms per cubic meter at standard temperature (0°C) and pressure (1 atm)?

---

[5] $\mu g/m^3 = $ mg/L $\times 10^6$

Solution:
MW of CO = 28
Volume of CO = 200 L or 0.2 m³

From Equation 4-17:
$$\text{ppm} = \frac{0.2}{2000 + 0.2} \times 10^6$$
$$= 100 \text{ ppm}$$

From Equation 4-19:
$$\mu g/m^3 = \frac{(100)(1)(28)}{(273)(0.08206)} \times 10^3$$
$$= 0.125 \times 10^6 \, \mu g/m^3 \text{ or } 0.125 \text{ g/m}^3$$

Partial Pressures of Gases

In a mixture of gases A, B, C, etc., the total pressure p_t exerted is the summation of the pressure by each gas:

$$p_T = p_A + p_B + p_C + \ldots \quad (4\text{-}20)$$

Individual pressures can be determined from Equation 4-16 from known values of V, T, and n. This is *Dalton's Law*. If water vapor is also present in the same space, the vapor pressure obtained from a handbook should be added to p_T value.

Solubility of Gases

External pressure acting over a solvent surface has practically no effect on solubility of solids and liquids. However, in the case of gases, the solubility is controlled by the pressure. According to *Henry's Law* the solubility of a gas in a liquid is proportional to the partial pressure exerted by that gas over the solution:

$$C_a = (K)(p) \quad (4\text{-}21)$$

where C_a = concentration of dissolved gas (mol/L)
p = pressure of dissolved gas over the solution (atm)
K = Henry's constant (mol/L-atm)

The values of K are listed in standard chemistry handbooks and textbooks.

Chemistry of Organic Chemicals

Carbon is a key element present in organic chemicals. Organic chemicals include biological matter, synthetic polymers, industrial compounds, and agricultural chemicals all of which have a significant impact on the environment. More than 4 million organic compounds are known to exist as compared to approximately 1000 inorganic compounds. This is because of the ability of carbon atoms to form bonds with each other and with other elements in a virtually limitless variety of forms. Certain salient features of organic compounds, as distinct from inorganic materials, are

1. They are usually less soluble in water.
2. Their melting and boiling points are lower.
3. They are usually combustible.
4. They have high level of energy. They are formed from inorganic matter of low energy level through an input of energy:

$$\text{inorganics} + \text{energy} \rightarrow \text{organics} \tag{4-22}$$

5. They react very slowly and have a tendency to form mixtures rather than to go to the completion of a reaction; hence, they do not break down easily.
6. Their molecular weight is very high. Even simple compounds may contain 60 carbon atoms. Complex molecules such as starch contain a staggering number of carbon atoms.
7. Usually, the compounds with single bonds among carbon atoms are least active and those with triple bonds are most active.

Types of Organic Compounds

The key organic element, carbon, has four covalent bonds because it has four electrons in its outer cell. These bonds of a carbon atom are represented as four arms with which it grabs other atoms. The linking together of the carbon atoms by covalent bonding is possible in many forms, as follows:

1. As a continuous chain, as in Figure 4-2a
2. As a chain with branches, as in Figure 4-2b
3. In a ring, as in Figure 4-2c
4. In a double bond ring, as in Figure 4-2d

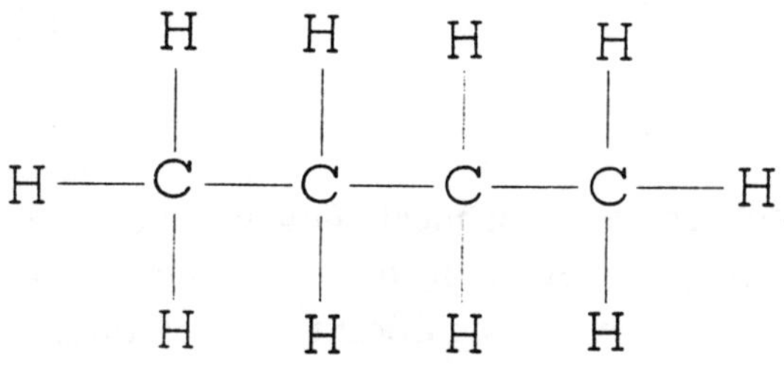

Figure 4-2a Continuous Chain

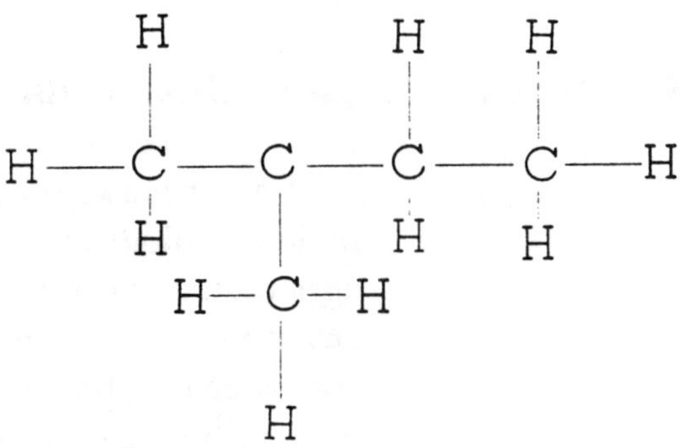

Figure 4-2b Chain with Branches

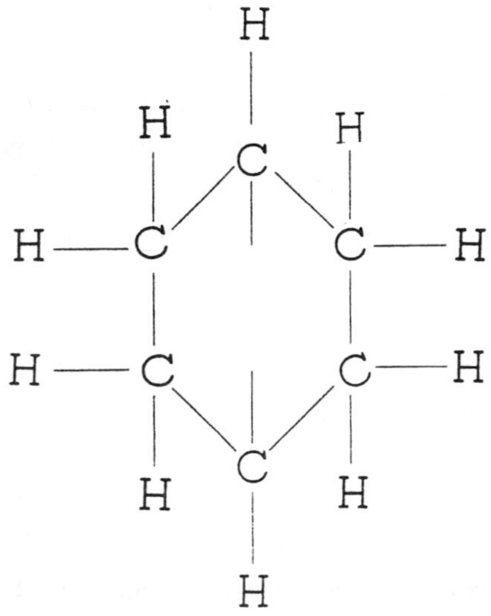

Figure 4-2c Ring Bond

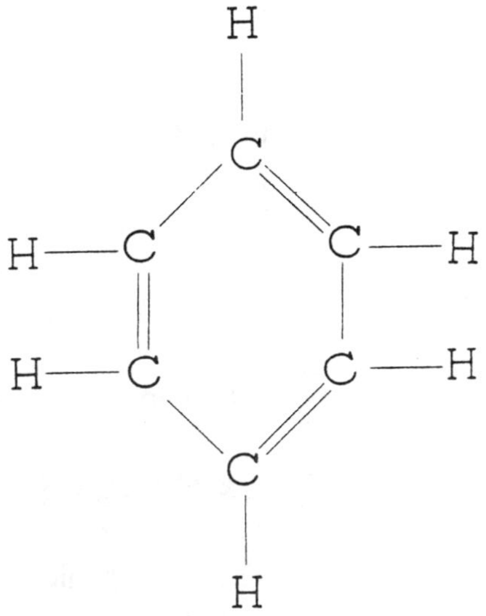

Figure 4-2d Double Ring Bond

Hydrogen or another atom attaches to the end of the open arm. Based on the linking pattern, the organic compounds are classified in three major types: aliphatic, aromatic, and heterocyclic. The *aliphatic* compounds are those in which the formation is a straight or branched carbon chain, as shown in Figure 4-2a and b or is a ring form, shown in Figure 4-2c. The *aromatic* compounds have a ring form with three double bonds as shown in Figure 4-2d. The *heterocyclic* compounds have a ring structure in which one member is an element other than carbon; they could be aliphatic or aromatic.

When only hydrogen atoms attach to carbon atoms to form compounds, i.e., the compounds are made of only two elements, carbon and hydrogen, they are known as hydrocarbons. Hydrocarbons are the most common types of organic compounds. Hydrocarbons have many families, with many compounds in each family. The aliphatic category of hydrocarbons is further subdivided into three groups:

1. *Alkanes*: The straight or branched chain consists of a single bond among carbon atoms, as shown in Figure 4-3a.
2. *Alkenes*: At least one double bond exists between carbon to carbon atoms, as shown in Figure 4-3b.
3. *Alkynes*: Compounds that have at least one triple carbon to carbon bond, as shown in Figure 4-3c.

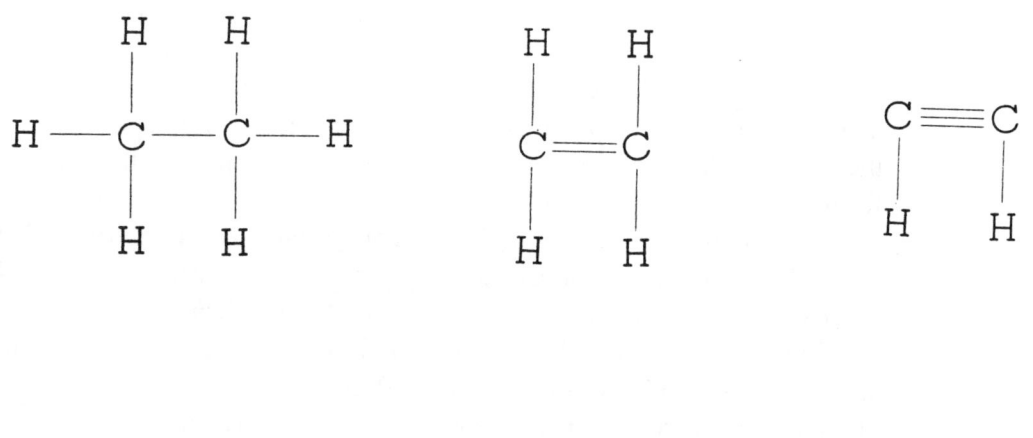

a) Alkane   b) Alkene   c) Alkyne

Figure 4-3 Hydrocarbon Compounds

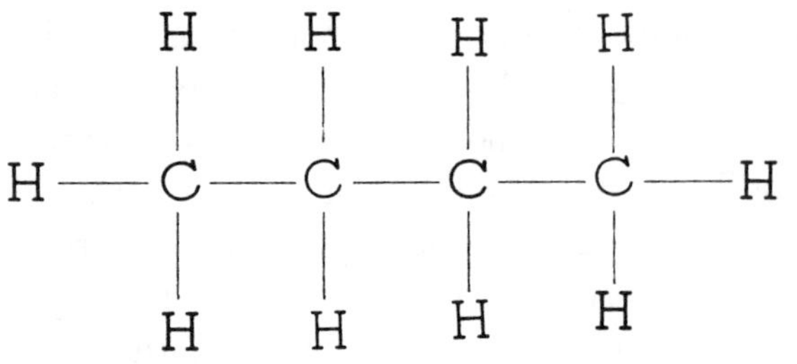

Figure 4-4a Alkane (Saturated) Hydrocarbons

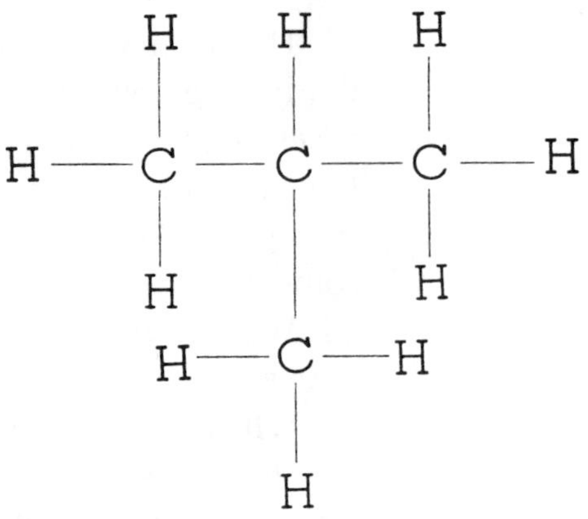

Figure 4-4b Different Alkane of Similar Molecular Formula

Organic Formulas and Names

A molecular formula gives the number of each kind of atom constituting a compound. Figure 4-4a and Figure 4-4b both have a molecular formula of C_4H_{10}. However, they represent two different compounds. Accordingly, it becomes essential in organic chemistry to show the arrangement of atoms through the structural formula. Compounds that have the same molecular formula but different structural formula are known as the isomers. Each isomer has distinct chemical and physical properties. To save space, the structural formula could be abbreviated to a single condensed structural formula. For example, for Figure 4-4a and b, the structural formulas could be written as:

$$CH_3\text{-}CH_2\text{-}CH_2\text{-}CH_3 \quad \text{and} \quad CH_3\text{-}CH\text{-}(CH_3)\text{-}CH_3.$$

The group (CH_3) is placed in parenthesis to show that it is a branch attached to the long continuous chain of carbon atoms. It is understood that the group is attached to the carbon immediately preceding it in the condensed formula.

The naming system of organic compounds is based on the rules (1960) of the International Union of Pure and Applied Chemistry (IUPAC). The names are called the *Systematic Names,* or *IUPAC Names.*

The first rule of the IUPAC system is to look for the longest chain of carbon atoms, eliminating all branches off the longest continuous chain. The group names of the branches are prefixed sequentially, indicating the point of attachment by the number assigned to the carbon atom on the longest chain. The longest carbon chain is numbered beginning from the left or right end, whichever causes the point of attachment to be the smallest number.

Table 4-2 Hydrocarbon series

Name	Formula	Important Members
Alkane	C_nH_{2n+2}	methane, CH_4; ethane C^2H_6; propane, C_3H_8; butane C_4H_{10}; pentane, C_5H_{12}; octane, C_8H_{18}
Alkene	C_nH_{2n}	ethylene, C_2H_4; propene, C_3H_6; butene, C_4H_8; pentene, C_5H_{10}
Alkyne	C_nH_{2n-2}	acetylene, C_2H_2; propyne, C_3H_4
Aromatic	C_nH_{2n-6}	benzene, C_6H_6; toluene, C_7H_8

The parent names derived from the longest straight chains are listed in Table 4-2. The prefixes to a parent name are added on the basis of the group names of the branches attached; i.e., methyl (CH_3), ethyl (CH_2-CH_3), *n*-propyl (CH_2-CH_2-CH_3), *n*-butyl (CH_2-CH_2-CH_2-CH_3), chloro (Cl), bromo (Br), amino (NH_2), etc.

Example 4-10
Give the systematic name to the alkane compound shown in Figure 4-5.

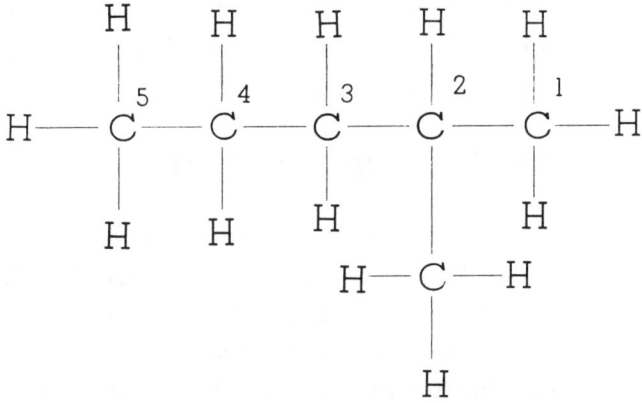

Figure 4-5 Example 4-10

Solution:
The name is 2-methylpentane.
The pentane part of the name is applied since there are five carbon atoms in the longest continuous chain. Pent stands for five and Ane is due to the alkane compound. A methyl group (CH_3) is attached at the second carbon atom, hence the prefix of 2-methyl is added to the name. Note that the numbering is done from the right because this will provide the smallest value to the carbon at the point of attachment.

Alkane Group of Hydrocarbons

The alkanes, also called saturated hydrocarbons or paraffins, form a series of compounds starting with methane, CH_4 and then each successive member increasing by CH_2. When the formulas of a series of compounds differ by a common increment, such as CH_2, the series is referred to as a *homologous series*. Different categories of hydrocarbons have their own series that fit the general formula of that series. Alkanes, with single bonds, have a maximum hydrogen-to-carbon ratio (completely occupied by hydrogen atoms, hence the name saturated), and bear a general formula C_nH_{2n+2}.

The compounds of the series have the -ane ending to their names (for alkane), such as methane, ethane, propane. When one hydrogen is replaced from a molecule of the series, the *-ane* ending is dropped and *-yl* is added, ie., CH_3 is a methyl and C_2H_5 is an ethyl group. That series itself is known as alkyl. Principal members of the alkane and other hydrocarbon series are listed in Table 4-2.

Alkanes are generally not very reactive substances. At room temperature members C_1 through C_5 are gases, those from C_6 to C_{17} are liquids, and those above C_{17} are solids. They are colorless, practically odorless, and quite insoluble in water.

The principal source of alkanes is petroleum. Gasoline and diesel are mixtures that contain several alkanes. Methane (the first alkane) is a chief ingredient of natural gas.

Alkene Group of Hydrocarbons

Alkenes are unsaturated hydrocarbons with at least one carbon-to-carbon double bond. The names of this series end with *-ene*. This homologous series has a general formula, C_nH_{2n}. There have to be two carbon atoms, at least, to make a double bond. Therefore, ethane of the alkane series is the first member that can loose two hydrogen atoms to form an alkene member. The member takes its name from the alkane member as ethylene or ethene. In the names of alkenes, a number indicates where the double carbon bond appears. For example, $CH_2=CH-CH_2-CH_3$ is 1-butene. Alkenes are used in the manufacture of polymers.

Alkyne Group of Hydrocarbons

Alkynes are unsaturated hydrocarbons with a triple bond. They have a general formula of C_nH_{2n-2}. Their names end with, *-yne*. The position of the triple bond is indicated by a number, as in the case of alkenes. Unsaturated hydrocarbons can be converted to saturated hydrocarbons by the addition of hydrogen atoms and breaking the double or triple bonds. This process, known as the hydrogenation, is used to convert vegetable oils to solid shortenings. Unsaturated hydrocarbons are very reactive. They are prone to combine with each other to form polymers that serve as the basis for many synthetic products.

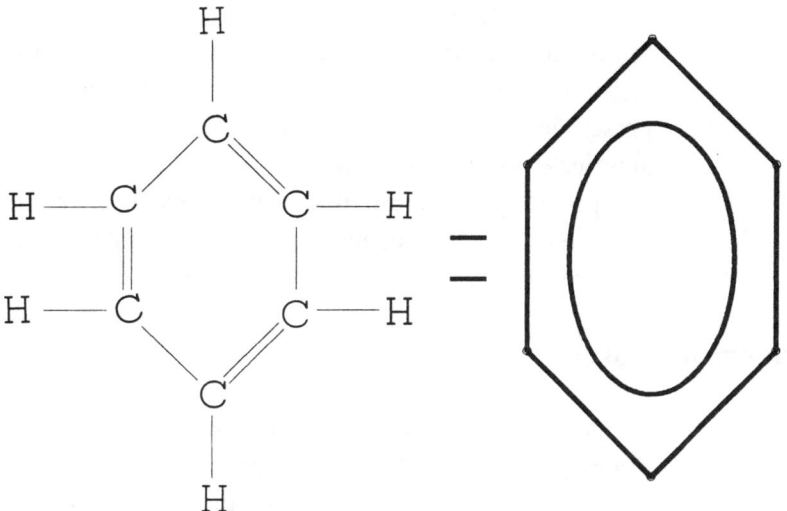

Figure 4-6 Aromatic Compound

Aromatic Group of Hydrocarbons

The simplest aromatic compound is made up of six carbon atoms as shown in Figure 4-6. This is known as the benzene ring (C_6H_6), also represented by a simplified drawing on the right side of Figure 4-6. The specific bonding makes benzene stable and inert. It does not undergo addition reactions. Instead, the common reaction is substitution in which a new group replaces a hydrogen or the carbon skeleton of the ring. There are two series of aromatic hydrocarbons: those with only one benzene ring, and those with multiple rings. In single-ring benzene series, the homologous compounds are made up of the substitution of hydrogen by an alkyl group (C_nH_{2n+1}). The compounds of the series are shown in Table 4-3. They are found in coal tar and many crude petroleum products. They are important constituents of lead-free gasoline.

Table 4-3 Benzene Series Hydrocarbons

Benzene	C_6H_6
Toluene	$C_6H_5CH_3$
Xylene	$C_6H_4(CH_3)_2$
Ethylbenzene	$C_6H_4C_2H_5$

The multi-ring aromatic molecules known as the polycyclic aromatic hydrocarbons (PAH's) are produced by the combustion of hydrocarbons under oxygen deficient conditions. They may be formed from higher alkanes present in fuel. They are highly stable compounds. Napthalene ($C_{10}H_8$) is the simplest polyring compound. Benzopyrene is a most cited polyring compound. Some compounds with many rings are carcinogens. PAH's may be present in the atmosphere as particulate material. Any group that contains one or more benzene rings is called an *aryl* group.

Derivatives of Hydrocarbons

A large number of compounds are formed when one or more of the hydrogen atoms of hydrocarbons are replaced by one of many functional groups, such as an OH (hydroxyl) group. The functional groups consist of a specific bonding configuration of atoms in organic molecules. Since these functional groups are responsible for most of the reactions of the parent compounds, the functional groups offer a very convenient basis with which to classify the derivatives of organic compounds. Since hydrocarbons are highly combustible substances, their successive oxidation leads to a series of compounds. Alcohols are primary oxidation products of hydrocarbons. All alcohols contain the hydroxyl group. Oxidation products of alcohols are aldehydes and ketones. They have a carbonyl group (C=O). Organic acids represent the highest organic oxidation state of hydrocarbons; all organic acids contain a carboxyl group (COOH). Further oxidation of organic acids leads to conversion into the inorganic state, with the formation of carbon dioxide and water.

Other Organic Compounds

The presence of elements other than hydrogen and carbon vastly diversifies the characteristics and applications of organic chemicals, many of which are formed synthetically. An important series of compounds formed naturally by plants from carbon dioxide and water through the process of photosynthesis are carbohydrates. These are compounds of carbon, hydrogen, and oxygen that usually have a hydrogen-to-oxygen ratio of 2 to 1, the same as in water. Our bodies have enzymes to convert carbohydrates to glucose, which is our energy source.

A variety of organic compounds contain nitrogen. They are formed when ammonia or nitric acid reacts with the alkyl group or aromatic benzene compounds. Their significance is as disinfectants, and agriculture fertilizers. The halogen-substituted hydrocarbons that contain at least one atom of halogen (F, Cl,Br, and I) are known as organohalides. Some of these, such as chlorofluorocarbons (CFCs), are a major source of pollution. Similarly, a variety of organosulfur compounds are formed by the substitution of alkyl or aryl hydrocarbon groups on hydrogen sulfide.

A wide variety of synthetic polymers are produced when small molecules called monomers bond together to form large molecules. Alkenes and aromatic hydrocarbons are raw materials for the manufacture of these monomers. Polyvinylchloride (PVC), polyethylene (plastic bags), polypropylene (impact-resistant plastics), polystryrene (foam insulation), polytetrafluoroethylene (nonstick coatings) and polyacrylonitrile (carpets) are polymers in common use. Many hazardous pollutants arise from these.

The Chemistry of Energy Generation

The energy used by us is produced by physical or chemical means. Hydroelectric energy from storage behind a dam, wind energy from rotating vanes, and geothermal energy from tapping underground stream beds are examples of physical energy. However, most forms of energy are chemical. The chemical reactions involving oxygen, known as oxidation, take place and heat energy is released. The common sources of chemical energy are petroleum, coal, and natural gas. Methanol, ethanol, and hydrogen are other chemical energy sources. Food is also a chemical source that releases energy by the respiration process.

Petroleum

A mixture of hydrocarbons largely from the Alkane family, petroluem reacts with oxygen as follows:

$$2C_nH_{2n+2} + (3n+1)O_2 \rightarrow 2_nCO_2 + 2(n+1)H_2O + [\text{heat energy}] \quad (4\text{-}23)$$

where $2n+2 = 1, 2$, etc.

Coal

The carbon in coal is burned directly:

$$C + O_2 \rightarrow CO_2 + [\text{heat energy}] \tag{4-24}$$

Also, carbon can be reacted with steam to produce water gas:

$$C + H_2O \rightarrow CO + H_2 + [\text{heat energy}] \tag{4-25}$$

The mixture of CO and H_2, known as water gas, can be adjusted to make liquid fuel comparable to gasoline:

$$6CO + 13H_2 \rightarrow C_6H_{14} + 6H_2O \tag{4-26}$$

where C_6H_{14} is a form of gasoline in the alkane family.

Natural Gas

For the most part, natural gas is methane (CH_4) with some ethane (C_2H_6), propane (C_3H_8) and trace of higher alkanes. The reaction is similar to Equation 4-23 for n=1.

Methanol and Ethanol

Methanol (CH_3OH) can be made by destructive distillation heating of wood in the absence of air, or from a mixture of CO and H_2 from coal, Equation 4-25. It burns as follows:

$$2CH_3OH + 3O_2 \rightarrow 2CO_2 + 4H_2O + [\text{heat energy}] \tag{4-27}$$

Ethanol (CH_3CH_2OH) can be made by the fermentation of carbohydrates from grain, corn, sugarcane, cassava roots, and other high-starch-producing plants. It burns as follows:

$$CH_3CH_2OH + 3O_2 \rightarrow 2CO_2 + 3H_2O + [\text{heat energy}] \tag{4-28}$$

Hydrogen Gas

Electrolysis of water can provide hydrogen and oxygen. Hydrogen could then be used as a fuel:

$$2H_2 + O_2 \rightarrow 2H_2O + [\text{heat energy}] \tag{4-29}$$

Fuel Recovery from Garbage

Garbage contains about 40% organic wastes and a large percentage of water. The decay of the organic compounds inside the landfills in the absence of air (known as anaerobic digestion or anaerobic fermentation) leads to the production of methane and carbon dioxide gases, as follows:

$$8C_aH_bO_cN_d + 2mH_2O \xrightarrow{\text{bacterial conversion}} nCH_4 + sCO_2 + 8dNH_3 \tag{4-30}$$

where m = 4a-b-2c+3d
n = 4a+b-2c-3d
s = 4a-b+2c+3d.

This is a biological process. A chemical process has been developed that converts garbage into a petroleum-like material. The garbage is placed in a thick-walled steel reactor. The temperature in the reactor is raised to 380°C. Carbon monoxide is introduced at 1000 psi pressure. The steam generated from the moisture in the garbage reacts with carbon monoxide to generate nascent or atomic hydrogen which in turn combines with the organic material in the garbage to form petroleum-like combustible hydrocarbons.

Photovoltaic Cells

As discussed in Chapter 3, an ultrapure silicon cell is used to convert solar energy to electrical energy. When the sun's rays strike an atom of silicon, they knock an electron from the outer shell, giving rise to electric current. This is being used on a smaller scale to power calculators, watches, and to heat water for homes.

Nuclear Energy

As discussed in Chapter 3, nuclear fission involves splitting apart the nucleus of heavy atoms by bombardment with neutrons. Nuclear fusion is the opposite, wherein nuclei of light atoms are fused together under extraordinarily high temperatures. Both of these processes release enormous amounts of energy. These are not chemical reactions in the ordinary sense in which atoms are retained, but these are chemical reactions involving nuclear transmutations. In ordinary chemical reactions, the law of conservation of mass holds. In nuclear transmutation, however, some mass transforms into energy. The combined law applies: that the total of mass-energy is neither created nor destroyed.

Einstein's relation between mass and energy, as defined in Chapter 3, is reproduced as:

$$E = mc^2 \qquad (4\text{-}31)$$

where E = energy in Ergs (4·184 × 10^7 Ergs = 1 calorie)
m = mass in grams
c = speed of light (3 × 10^{10} cm/sec).

Example 4-11
If 4 moles of hydrogen are fused to make 1 mole of helium, how much energy is released in the process?

Solution:
4 moles of hydrogen = 4 × 1.008 = 4.032 g
1 mole of helium = 4.000 g
Loss of mass (transformed into energy) = 0.032 g
From Equation 4-31: $E = 0.032 \times (3 \times 10^{10})^2$
$= 0.288 \times 10^{20}$ Ergs
or 7 × 10^8 kcal.
This is 14 million times the energy produced by burning 4 g of petroleum or natural gas.

The energy content of various fuel sources is given in Table 4-4.

Table 4-4 Energy Liberated by Various Fuels (in kcal/g of fuel)

Fuel	Energy Release	Fuel	Energy Release
Natural Gas	11.6	Wood	4.5
Petroleum or Oil	11.3	Hydrogen	34.2
Coal (anthracite)	7.3	Carbon monoxide (CO)	2.4
Coal (bituminous)	7.0	Methane (CH_4)	13.3
Ethanol (C_2H_5OH)	7.1	Fusion 2 $^2H = {}^4He$	1 × 10^8
Methanol (CH_3OH)	5.4	Fission ^{235}U	2 × 10^7

1 Btu = 0.252 kcal

PROBLEMS

1. Find the percentage of copper (Cu) in hydrated copper sulfate ($CuSO_4 \cdot 5H_2O$).

2. How many moles are contained in 50 g of calcium bicarbonate, $Ca(HCO_3)_2$?

3. Methane gas (CH_4) is completely oxidized (burned with oxygen) to produce carbon dioxide and water. Write the balanced equation for the reaction.

4. What mass of carbon dioxide would be given off if 100 g of methane is fully oxidized?

5. What mass of water would be produced if 100 g of methane is fully oxidized?

6. How much oxygen would be needed to fully oxidize 100 g of methane gas?

7. Balance the following equations:

   (a) C_4H_{10} (butane gas) $+ O_2 = CO_2 + H_2O$
   (b) $NaCl + H_2SO_4 = Na_2SO_4 + HCl$
   (c) $Na_2CO_3 + Ca(OH)_2 = NaOH + CaCO_3$
   (d) $Ca_3(PO_4)_2 + SiO_2 + C = CaSiO_3 + CO + P$

8. By anaerobic decomposition (without oxygen) of a waste (CH_3COOH), carbon dioxide and methane gas are produced. If 2000 kg of waste is decomposed, how much methane gas will be formed?

9. TNT ($C_7H_5N_3O_6$) explodes in the presence of oxygen to produce water, carbon dioxide, and nitrogen.

   (a) Write a balanced equation of the reaction.
   (b) How many moles of oxygen are required to explode 2 moles of TNT?
   (c) How many grams of oxygen are required to produce 100 g of CO_2?

10. Aluminum sulfate (alum) is a common coagulant used in water treatment according to the following reaction:

    $Al_2(SO_4)_3 \cdot 14H_2O + Ca(HCO_3)_2 = Al(OH)_3 + CaSO_4 + H_2O + CO_2$

    (a) Balance the equation.
    (b) How many moles of carbon dioxide will be formed by 5 moles of aluminum sulfate?
    (c) What mass of aluminum hydroxide floc will be formed by 2 kg of alum?

11. A quantity of 25 g of Na_2CO_3 is dissolved in water to make up a 400 mL of solution. Determine the concentration in milligrams per liter and parts per million and molarity of the solution.

12. The total dissolved solids content of 50 ft³ of water is 0.5 lb. Determine the concentration in parts per million and milligrams per liter.

13. Find the molarity of the following:

    (a) 0.05 moles of NaCl in an 80 mL solution.
    (b) 250 g of NaOH in a 4 L solution.
    (c) 300 g of $CaCl_2$ in 5000 g of water.

14. A solution contains 300 mg/L of glucose ($C_6H_{12}O_6$). Find the oxygen required to fully oxidize the solution to CO_2 and H_2O.

15. A glucose solution has a molarity of 0.002 M. Determine the amount of carbon dioxide produced in complete oxidation.

16. A 0.1 M solution of Formic acid (HCOOH) is 0.41% ionized at equilibrium. Find the ionization constant.

17. A 0.12 M solution of acetic acid (CH_3COOH) is 1.4% ionized at equilibrium. Find the ionization constant.

18. The ionization constant of acetic acid is 1.17×10^{-5}. Find the ionic concentration of $[H^+]$ and $[CH_3COO^-]$ in a 0.01 M solution of the acid.

19. What is the percent ionization in Problem 18?

20. The solubility of calcium fluoride (CaF_2) is 0.0015 gram per 100 mL. Find its solubility product constant. (Hint: one molecule of CaF_2 yields one ion of Ca and two ions of F. Thus the molecular concentration of F will be twice that of Ca and CaF_2].

21. The solubility product constant of silver sulfate (Ag_2SO_4) is 1.4×10^{-5}. What is the molar solubility of silver sulfate?

22. In a saturated solution of magnesium hydroxide, $Mg(OH)_2$, the concentration of hydroxide ion is 8.3×10^{-4} M. What is the concentration of magnesium ions? ($K_{sp} = 1.2 \times 10^{-11}$)

23. A solution of acetic acid has a hydrogen ion concentration of 1.4×10^{-3} M. What is the pH of the solution?

24. Hydrochloric acid has a pH of 2.5. What is the concentration of hydrogen ions in the acid?

25. In a NaOH solution, the hydroxide ion concentration is 3.2×10^{-4} M. What is the pH of the solution? (Hint: Find pOH; pH + pOH = 14)?

26. In a water analysis of a lake, the concentration of hydrogen was 3.6×10^{-5} mg/L. What is the pH of water?

27. Calculate the volume occupied by 9 g of CO_2 gas at standard temperature and pressure.

28. Determine the volume of a tank to store 10 kg of ethane (C_2H_6) gas at 20°C and 29.4 psi (2 atm) pressure.

29. A gas occupies 500 L at 30°C and 2 atm. What will be its volume at 0°C and 3 atm?

30. At the bottom of a tank, where the temperature and pressure are 10°C and 10 atm respectively, the volume of an air bubble is 2 mL. What will be the volume of the bubble when it rises to the surface at 20°C and 1 atm?

31. The amount of carbon monoxide present in 1000 L of air is 800 mL. Determine the concentration of CO in parts per million.

32. For a pressure of 1 atm and temperature of 20°C, what is the concentration in micrograms per cubic meter for Problem 31?

33. The concentration of sulfur dioxide is measured to be 650 µg/m³ at 25°C and 1.6 atm pressure in a place. What is the concentration in parts per million?

34. Exhaust from an automobile contains 0.5% by volume of hydrocarbons ($MW = 100$). What is the concentration in parts per million and milligrams per liter at 4°C and 20 psi?

35. One mole of air contains 0.78 mole of nitrogen and 0.22 mole of oxygen. Find the partial pressure of nitrogen if the total pressure is 2 atm.

36. On heating, potassium chloride ($KClO_3$) breaks down to KCl and oxygen gas. What volume of O_2 at standard conditions would be produced by a complete decomposition of 10.2 g of potassium chloride?

37. Nitrogen gas and hydrogen gas are combined to produce ammonia gas (NH$_3$). Seventy Liters of hydrogen at standard temperature and pressure are used for this reaction. What volume of nitrogen would be required?

38. What volume of ammonia gas would be produced in Problem 37?

39. Systematically name the organic compound shown in Figure 4-7, according to IUPAC.

40. Systematically name the compound shown in Figure 4-8, according to IUPAC.

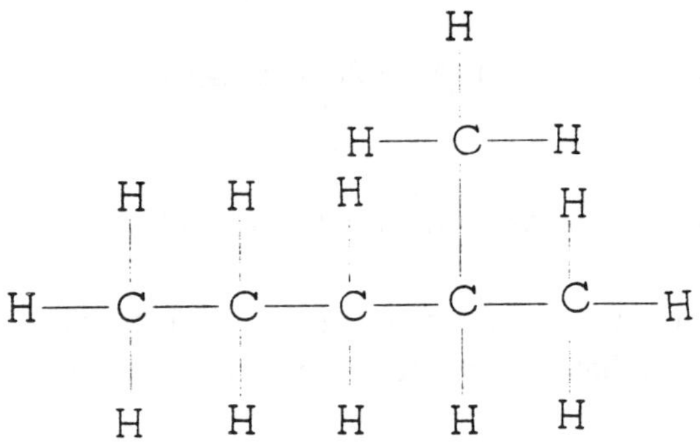

Figure 4-7

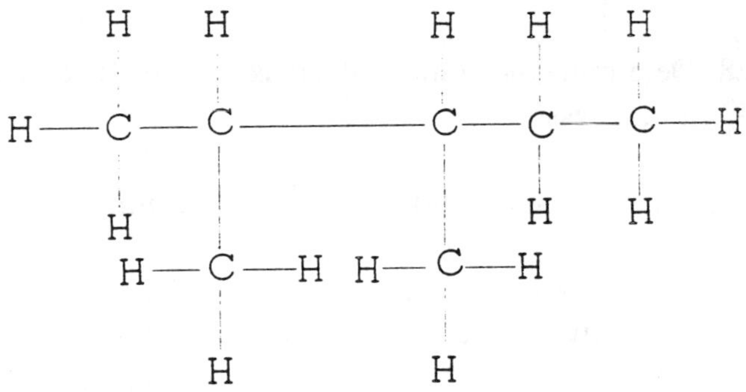

Figure 4-8

41. The products obtained by fission of 1 mole of uranium (atomic weight 235) were 1 mole of strontium, 1 mole of xenon, and other elements amounting to 16 g. How much energy is released by the nuclear explosion?

42. How much energy is released in British thermal units per mole of hydrogen burned? Use Table 4-4.

STUDY QUESTIONS

1. What is a chemical reaction? Describe, in words, the meaning of the following equation. Express the equation in symbolic form.

   Calcium Carbide + Water $\xrightarrow{\Delta}$ Calcium Hydroxide + Acetylene

2. What are the four types of reactions? What is the equilibrium reaction? Define and differentiate among the equilibrium constant, the ionization constant, and the solubility product constant.

3. Describe the solution process in water. What are the units measuring concentration in a liquid solution? How are they related?

4. The pressure-volume relations of gases are related by Boyle's law. The temperature-volume relations are described by Charles law. The mass-volume relations are stated by Avogadro's hypothesis. The ideal gas law combines the three. Describe each law of these four laws. Can these be applied to real gases?

5. Describe the solution process of gases in liquid. What are the units of measuring gaseous pollutants in air? How are they related?

6. What is the significance of the pH of a solution? Can a solution have a negative pH? Explain your answer. Prepare a pH scale showing relative locations of highly acidic, highly alkaline, and practically neutral solutions.

7. Differentiate between inorganic and organic chemicals. What are the three major types of organic compounds? What are the subdivisions of aliphatic and aromatic hydrocarbons?

8. What is a structural formula? Describe the naming convention of organic compounds.

9. Evaluate the energy generation potential of various organic chemicals. Summarize the energy released by them per unit mass.

10. Define the following:
    a. Reactants and products
    b. Periodic table
    c. Parts per million
    d. Molarity
    e. Solubility
    f. Electrolytes
    g. Ions- cations and anions
    h. Acid and base
    i. Law of mass action
    j. Ideal gas law
    k. Dalton's law
    l. pH
    m. Avogadro's hypothesis (law)
    n. Henry's law
    o. Alkane, alkene, and alkyne
    p. Isomers
    q. Aromatic hydrocarbons
    r. Photovoltaic cells

REFERENCES

Bailey, R. A., et al., *Chemistry of the Environment*, Academic Press, New York, 1978.

Benfey, O. T., *The Names and Structures of Organic Compounds*, John Wiley and Sons, New York, 1966.

Butler, J. D., *Air Pollution Chemistry*, Academic Press, London, U.K., 1979.

Chang, R., *General Chemistry*, Random House, New York, 1986.

Harrison, R. M., deMora, S. J., Rapsomanikis, S., Johnston, W. R., *Introductory Chemistry for the Environmental Sciences*, Cambridge University Press, Cambridge, U.K., 1991.

Hess, F. C., *Chemistry Made Simple*, Revised ed., Doubleday, New York, 1984.

Horne, R. A., *The Chemistry of Our Environment*, John Wiley and Sons, New York, 1978.

Manahan, S. E., *Environmental Chemistry*, 5th ed., Lewis Publishers, MI, 1991.

Moore, J. W., and Moore, E. A., *Environmental Chemistry*, Academic Press, New York, 1976.

O'Neill, P., *Environmental Chemistry*, George, Allen and Urwin, London, U.K., 1985.

Sawyer, C. N., and McCarty, P. L., *Chemistry of Environmental Engineering*, 3rd ed., McGraw-Hill Book Company, New York, 1978.

Scientific American, *Chemistry in the Environment*, W. H. Freeman and Company, San Francisco, CA, 1973.

Seinfeld, J. H., *Lectures in Atmospheric Chemistry*, American Institute of Chemical Engineers, New York, 1980.

Wayne, R. P., ed., *Chemistry of Atmosphere*, Clarendon Press, Oxford, U.K., 1991.

White, E. H., *Chemical Background for the Biological Sciences*, Prentice Hall, NJ, 1964.

Williams, A. L., et al., *General Chemistry*, Addison-Wesley Publishing Company, MA, 1970.

Chapter 5
PRINCIPLES OF ECOLOGY AND MICROBIOLOGY

Ecosystem Concepts

As stated in Chapter 1, ecology[1] is concerned with the relationships between plants and animals and the environment in which they live. If we extend this idea with reference to a fixed space, then there are many kinds of organisms that live in a given space. Together they are known as a *community*. Within a community, the total number of individuals of any one kind of organisms is called a *population*. Thus, in an ecological sense, a community includes all of the populations occupying a given area. Each community is surrounded by certain nonliving entities. A dictionary meaning of a system is the interacting and interdependent components that make a unified whole. Thus, the community and its nonliving environment make a system known as an *ecological system*, or *ecosystem*, as follows:

$$\text{(biotic component)} + \text{(abiotic component)} = \text{(ecosystem)} \qquad (5\text{-}1)$$
(community)   *(nonliving environment)*

An ecosystem is a basic functional unit of living entities on the earth. The circulation of matter and flow of energy are essential for sustaining life, and, as such, are indispensable ingredients of an ecosystem. Another important attribute of the ecosystem is that it is self-sufficient, which means it is independent of external (outside of the system) sources of matter and energy except for the light from the sun. An ecosystem is a dynamic entity that is responsive to changing conditions of the environment.

Ecosystems occur in various forms and sizes. A pond is an example of a small ecosystem. The largest ecosystem is the biosphere, which includes all living organisms on the earth (extending to the lithosphere, hydrosphere, and

---

[1] The literal meaning is the study of living organisms in their homes, i.e., environment.

atmosphere) that interact with the entire physical environment of the planet. The same ecosystem can be viewed from several levels. A biological unit may act as a self-sufficient ecosystem and it may be a part of a bigger system as well; a small system can exist within a large system. A hierarchical organization of an ecosystem is shown in Figure 5-1. Every lower level from bottom to top and from left to right is contained within the next higher level. The double arrows show interaction at each level.

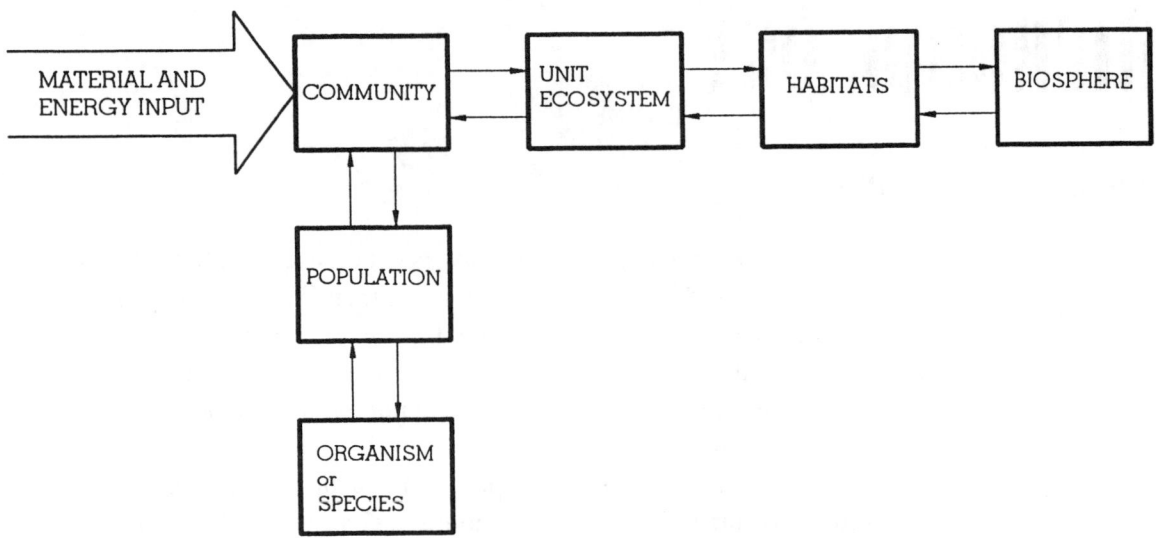

Figure 5-1 Hierarchical Organization of an Ecosystem

Biotic Component Organization in an Ecosystem

All living organisms constitute the biotic component of an ecosystem. Organics (organic compounds) are the constituent part of living organisms; carbohydrates, fats, and proteins are major compounds. Carbohydrates and fats consist of carbon, hydrogen, and oxygen. Proteins include nitrogen and sulfur as well. Thus, the chemical composition of the matter within a living organism (the protoplasm) can be expressed as $C_aH_bO_cN_dP_eS_f$. Organics are formed from inorganic substances with input of energy. Once a living state is ended, organics decompose and convert back to low-energy-level inorganic products.

All living organisms are grouped in three categories: producers, consumers, and decomposers. *Producers* are the first and most important members of the ecosystem. The role played by an individual in the ecosystem is called its *niche*. Green plants occupy the niche of producers because they make their own food (chemical energy). Plants contain an abundance of energy. When we burn a piece of wood, energy becomes evident in the forms of heat and light. Plants derive energy in chemical form from the sun by a process known as *photosynthesis*, as follows:

$$\begin{pmatrix} \text{carbon dioxide} \\ \text{from the air} \end{pmatrix} + (\text{moisture}) + \begin{pmatrix} \text{suns radiant} \\ \text{energy} \end{pmatrix} = \begin{pmatrix} \text{glucose chemical} \\ \text{energy} \end{pmatrix} + (\text{oxygen}) \quad (5\text{-}2a)$$

This chemical equation[2] can be expressed as:

$$xCO_2 + yH_2O \xrightarrow{\text{sun's energy}} C_x(H_2O)y + xO_2 \quad (5\text{-}2b)$$

$C_x(H_2O)y$ represents carbohydrates. The simplest carbohydrate involved in photosynthesis is glucose, $C_6H_{12}O_6$. Photosynthesis is an endogenic process that requires energy.

For each mole of glucose (180 g) formed, 675 kcal[3] of the sun's energy is needed. On a typical summer day, about 1300 cal/cm²-day of solar energy reaches the earth. Of this, only 2 to 5% is converted into biomass (net potential plant production of 7×10^{-3} g/cm²-day) by photosynthesis.

Example 5-1
How much of the light energy absorbed by plants converts into biomass?

Solution:
Assume that glucose is the material produced.
One mole of glucose has a mass of 180 g/mole.
675 kcal of energy per mole is needed.
Energy needed per gram = $675 \times 10^3/180$ = 3750 cal/g.
With net production of 7×10^{-3} g/cm²-day.
energy needed for biomass produced = $3750 \times (7 \times 10^{-3})$
   = 26.3 cal/cm²-day.
If the sun's incidence is 1300 cal/cm²-day.
photosynthesis efficiency = 26.3/1300 = 5%.

---

[2] Some of the carbohydrates are converted to fats, proteins, and other organics as required for cell growth. An expanded relation can be expressed as:

$$CO_2 + H_2O + PO_4 + NH_3 + SO_4 \xrightarrow{\text{suns energy}} C_aH_bO_cN_dP_eS_f + O_2$$
(This equation is not balanced) (5-3)

[3] A food calorie, Calorie (written with capital letter C is a kilocalorie (kcal) in physics.

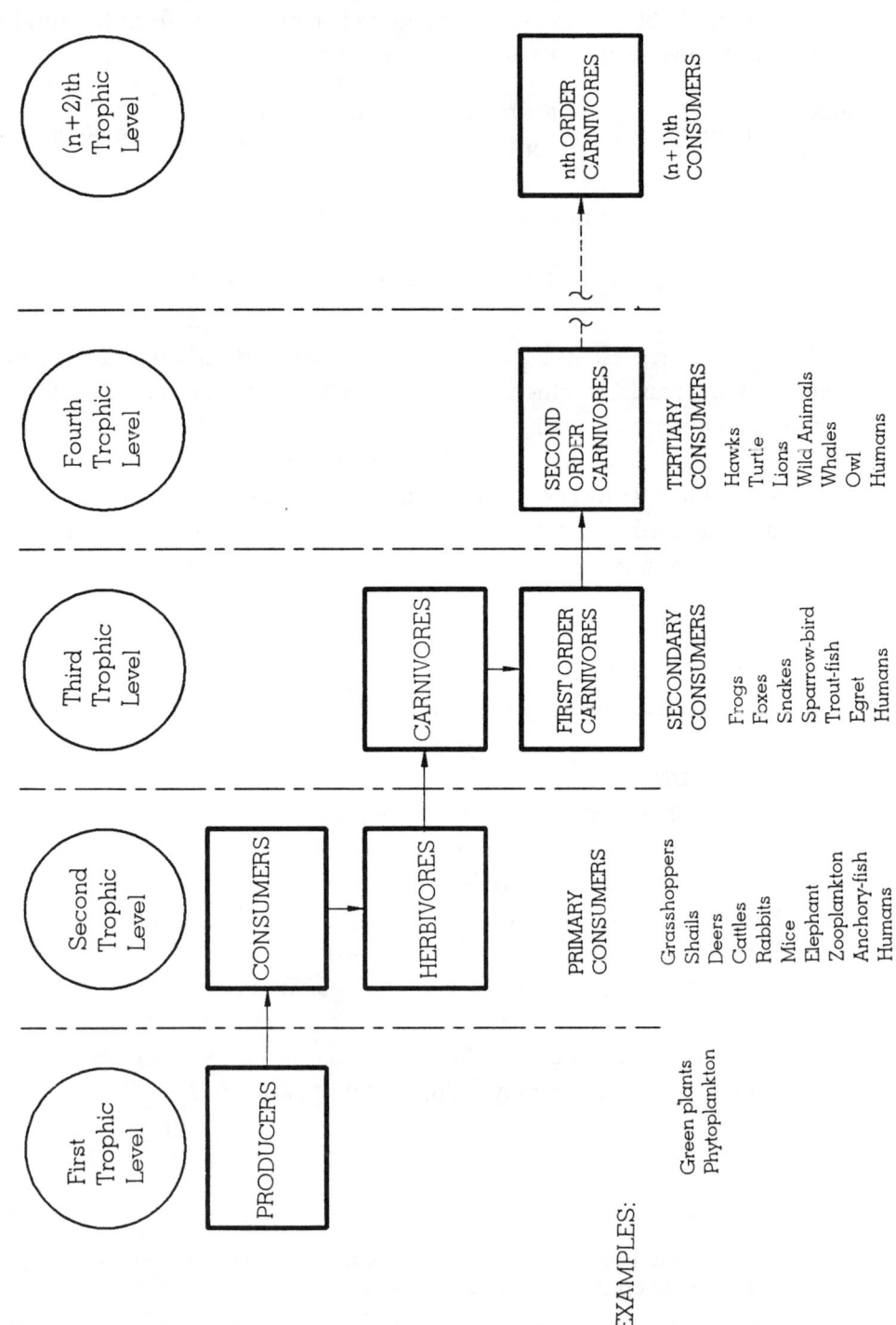

Figure 5-2 Producer-Consumer Levels and Food Chain

Principles of Ecology and Microbiology

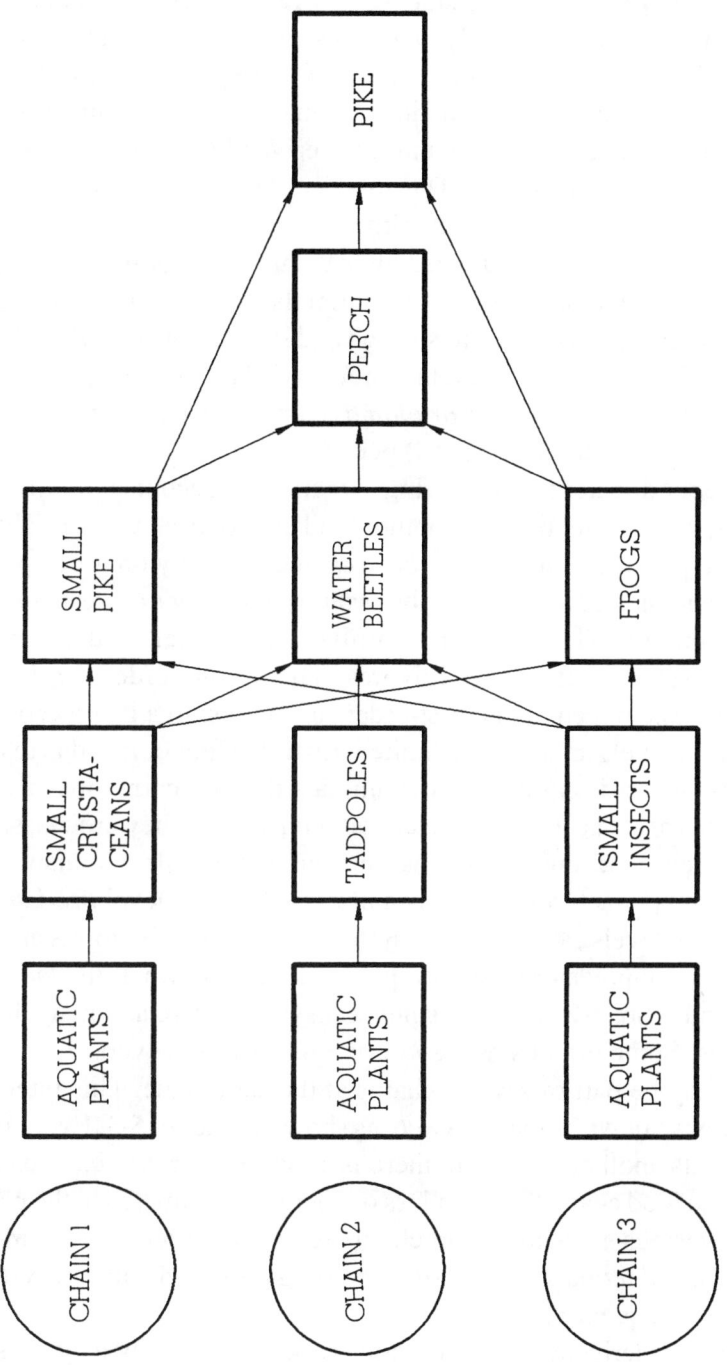

Figure 5-3 Food Web for an Aquatic Ecosystem

Photosynthesis is a natural process that has not been replicated in laboratories and is unique to plants (and a few bacteria). This process of energy conversion can occur only in the presence of certain pigments that are located in plant cells. The most important of these pigments is *chlorophyll*, which gives most plants their distinct green color. Some mineral salts are also needed by plants for growth and are derived by them from the abiotic component of soil and water. The total energy produced during photosynthesis by a plant in a given time is known as the *gross primary productivity*. Some of this is utilized by the plant itself for life processes and the remaining is available as food for animals, which is known as the *net primary productivity*. This remains stored and is released only when the plant is consumed by an animal. The total amount of plant material present at any time is called the *standing crop or biomass*, and is usually expressed as the dry weight (after water is removed) per unit area.

The second category of living organisms occupies the niche of *consumers*, who cannot directly capture food energy from the sun. There are two major types of consumers. *Herbivores*, feed directly on green plants or producers and are also known as the *primary consumers*. *Carnivores* feed on other animals. The first-order carnivores, also called the *secondary consumers*, feed directly on herbivores. The second-order carnivores or *tertiary consumers* feed on the first-order carnivores. In a given ecosystem, the number of levels involved are limited usually to four or five due to energy constraints (to be discussed). Some animals that are carnivores at certain times and herbivores at others are called *omnivores*. They are classified at different levels depending upon what they are eating. Most humans belong to this category. Each prior level is a *prey* to the later-level *predator*. The above order of levels exists in an ecosystem wherein the life forms are linked through the prey-predator relationship. This link is known as the *food chain*, as shown in Figure 5-2. Within a single ecosystem, there are many different food chains. Some organisms may derive foods from several food chains; i.e., they may be part of several chains at the same time. This interwoven relationship is known as the *food web*, as shown in Figure 5-3. Every time an organism eats another organism, there is a loss of energy; i.e., at each step (level) of the food chain, there is a loss of energy as compared to the previous step. The various steps in a food chain are called *trophic* levels and have significance in terms of the flow of energy through an ecosystem as discussed subsequently.

The energy derivation or conversion process from the food by consumers is called *respiration*. A previous-level organism serves as food for the subsequent-level organism. The respiration process as indicated below is the reverse of photosynthesis. It is an exogenic process that releases energy.

$$\begin{pmatrix} \text{glucose} \\ \text{from food} \end{pmatrix} + \begin{pmatrix} \text{oxygen from} \\ \text{air and water} \end{pmatrix} = \begin{pmatrix} \text{carbon dioxide} \\ \text{into atmosphere} \end{pmatrix} + (\text{water}) + (\text{energy}) \quad (5\text{-}4a)$$

This chemical equation[4] can be written:

$$C_6H_{12}O_6 + 6O_2 \rightarrow 6CO_2 + 6H_2O + (\text{energy}) \quad (5\text{-}4b)$$

All processes–physical, chemical and biological–involved in the above reaction, whereby organic matter is utilized by living beings to obtain energy, are known as *metabolism*, which is a distinct characteristic of living things. The rate of reaction that takes place within the living body is called the *metabolic rate*. For a person at complete rest, the rate of reaction in the form of heat production is the basic metabolic rate, which is between 1340 and 1750 kcal/day (65–85 Watts).

Dead consumers become a food for another group of organisms. These are mostly micro-sized organisms. However, a special category of macro-sized organisms directly consume dead animals and plants. These are known as *detritus feeders, and scavengers*. Examples are turtles, vultures, crayfish, termites, earthworms, ants, and wood beetles. Termites that eat wood (plants) are at the same trophic level as primary consumers but differ from them in an important way: they feed on dead rather than living vegetation. Secondary, tertiary, and higher-order detritus feeders feed on dead organisms of the orders parallel to the consumers category.

Microorganisms, however, feed by a different process. They decompose or break down the dead organisms into simpler inorganic matter so that these substances become available once again to be used by producers. Organisms such as bacteria, protozoa, and fungi belonging to this category and occupy the niche of *decomposers*. Fungi include molds, mushrooms, and puffballs. Bacteria are present everywhere. They appear even in a sterilized environment in time. Most bacteria are harmless; they even help in various routines, such as the fermentation process, making of yogurt, and breadmaking. A few that cause diseases are recognized as pathogenic bacteria.

From the broad consideration of trophe or nourishment, an ecosystem can be divided into two groups: The *autotrophic* or self-nourishing group to which producers belong, and the *heterotrophic* or other nourishing group that includes consumers and decomposers.

---

[4]The expanded form is:

$$C_aH_bO_cN_dP_eS_f + O_2 \rightarrow CO_2 + H_2O + NO_3 + PO_4 + \text{energy}$$
(This equation is not balanced) (5-5)

Principles Pertaining to the Abiotic Component

The abiotic component bears important relations to the biotic component, (similar to the interrelations among the biotic elements themselves) in controlling, and distributing the abundance of organisms in an ecosystem. Each organism requires certain kinds and quantities of nutrients and energy resources on which the presence and success of that organism depends. There are three basic principles that relate resources to organisms. These are: the law of the minimum, the law of tolerance, and the law of limiting factors.

According to the first law proposed by German chemist Liebig in 1840, among the basic required resources, the one that is present in the minimum quantity controls the growth of the organism and not the material available in abundance. For example, carbon dioxide, phosphorus, nitrogen, and sunlight are necessary for any plant. If phosphorus is depleted, growth of the plant will stop even though other resources may be available. This *law of the minimum* is applicable under steady-state conditions.

Studies have also shown that too much of a resource can limit growth as well as too little. For example, excessive heat, light, and water can be harmful to plants. The concept of the limiting effect of minimum as well as maximum was incorporated in the *law of tolerance* by Shelford in 1913, according to which organisms have a range of tolerance for different factors, between minimum and maximum limits.

Both of these laws have been united in a broad *law of limiting factors*. The presence or success of any organism depends upon a number of factors (resources). With respect to each factor, a range of magnitude is tolerated by any given organism. For example, most forms of life can exist within a temperature range of -200°C to 100°C. If the accepted range of tolerance of any factor for an organism is exceeded, that factor is said to the limiting factor for that organism, which causes retardation of the organisms growth. The important items that can play the role of limiting factor in organism growth are discussed below.

Role of the Abiotic Component in an Ecosystem

The abiotic component of an ecosystem can be grouped into four categories:
1. *Inorganic substances* (carbon, oxygen, nitrogen, water, minerals, etc.) that are involved in material cycles;
2. *Oganic substances* (protein, carbohydrates, fats, humic acids, etc.) that are found within or outside of living organisms;
3. *Climatic regime* (temperature, humidity, pressure, etc.) that regulates the ecosystem; and

4. *Physical factors* (soil, fire, light, etc.) that serve as energy resources.

Inorganic Substances

Living organisms require some 40 chemical elements for growth and development. Water and oxygen are two very basic substances that are essential to support life. In addition, organisms need nitrogen to make protein and plants require carbon dioxide to synthesize food. Other nutrients considered vital for life include phosphorus, potassium, sulfur, calcium, magnesium, iron, boron, zinc, chlorine, molybdenum, cobalt, iodine, and sodium. All of these are available naturally and cycle through the ecosystem. Their cycling is recognized as *nutrient cycles* or *biogeochemical cycles*, as described in Chapter 6. Many nonessential inorganic elements also cycle through ecosystem, including radioactive substances.

Organic Substances

The composition of organic compounds consists of carbon atoms combining to strings of atoms of hydrogen, oxygen, nitrogen and sulfur. Organic compounds mostly originate from living sources such as plants and animals. Compounds such as carbohydrates, proteins, and lipids that serve as human food are forms of stored chemical energy. While some substances such as ATP (adenosine triphosphate), DNA (deoxyribonucleic acid)[5] and chlorophyll have significance only within living cells, some other substances such as humic matter[6] appear only outside of living cells.

Some organic nutrients also cycle through the ecosystem in the same general manner as inorganic elements.

Temperature

Growth of life took place on the earth because of favorable climatic conditions. Organisms adapt to any substantive changes in climate only within certain limits, beyond which they perish. The law of limiting factors holds. A *temperature* range of thousands of degrees occurs in the universe but life forms can exist only within a narrow range. Fluctuation of temperature is also important. It has been noted that a constant temperature tends to slow growth and makes organisms depressed as compared to those

---

[5]ATP is a molecule that acts as the primary energy carrier in the living cell. DNA molecules, located within cell nuclei, contain genes. Genes are responsible for similarities among species and for characteristics of an individual member of a species.

[6]These are organic wastes that resist decomposition and remain in the soil for a long time.

exposed to variable temperatures. Similar to thermal chemical processes, animals are also either *exothermic* (they exploit outside energy in addition to their own metabolism to maintain their body heat), or they are endothermic, (or, *homeothermic*) when their own metabolism is sufficiently high to maintain their body temperature. Most aquatic life and reptiles belong to the first category, and birds and mammals belong to the second category.

Humidity

Humidity represents the amount of water in vapor form in the air at a given place. When this amount is expressed as a ratio to the amount of water vapor that air can hold at a given place at an existing temperature and pressure, it is known as *relative humidity*. As air warms, its moisture holding capacity rises, thus lowering the relative humidity for the same moisture content in the air. The relative humidity varies from time to time and place to place. The moisture in air accumulates due to *evaporation* of water by solar radiation from land and water bodies, and due to *transpiration*, which is evaporation from plant leaves and grass. Evaporation and transpiration or (evapotranspiration) increase as the relative humidity decreases. Temperature and relative humidity together, to a large extent, determine the type and distribution of organisms. Low or high temperatures exert a more severe effect when combined with conditions of moisture extremes of very low or very high relative humidity. High temperatures are more uncomfortable when relative humidity is also high.

Winds, Currents, and Waves

Winds, currents, and waves are instrumental in changing the climate and controlling the distribution of various nutrients and concentration of gases in the atmosphere and hydrosphere. For example, the currents in a stream are principally responsible for the differences in the types of species in a stream and a pond. Waves are responsible for differences in marine and freshwater ecosystems. Winds exert influence on the land regime in the same manner. They directly limit the growth of plants, and cause dispersal of seeds, insects, and small animals over a wide area, thus offering the opportunity to establish diverse species.

Pressure

Barometric pressure is linked to weather and climate, which are direct limiting factors to organisms. In water, the pressure increases by 1 atm for every 10 m depth. In general, the higher pressures exerted in the deep areas of oceans and lakes slow the pace of aquatic life.

Light

Light, fire, and soil are key natural resources of immense consequence to the ecosystem. Sunlight triggers the entire process of photosynthesis and sustains the food chain. The entire biosphere is powered by the sun, which makes the earth liveable. Sunlight is the ultimate source of energy on the planet. Energy is the key to the evolution of the biosphere and growth of civilization. Thus, light is a vital limiting factor. The quality of light (color), the intensity (rate of energy), and the duration, are all ecologically important.

Fire

Fire serves both as a friend and a foe. Humans started their journey to civilization with the use of fire. In addition to being a source of energy, fires were used to clear grounds for hunting and agriculture, to develop grass and vegetation, and to open up countryside. Two extreme types of fires are *crown fires* that destroy all the vegetation, and *surface fires* that destroy only the undergrowth. A controlled moderate surface fire has many advantages. It reduces complex organic substances to their elemental state for recycling. It allows waste (particularly trash and litter) to be reduced to ash instead of accumulating. It increases the nitrogen content of topsoil by converting vegetation litter into ash which contains some nitrogen. Further increases in nitrogen come from a marked increase in nitrogen-fixing legumes following a fire, which is good for stimulating new plant growth. Fire also exposes the minerals and stimulates the germination of certain seeds. Grasslands respond to fire by releasing new shoots and displaying increased root activity, thus raising their net productivity. It can improve forage stands and seed beds for regeneration of certain forest types and tree species and, it can be used to improve wildlife habitats. Some studies advocate the use of controlled burning to increase timber and game production. Fires often combine with plant-produced antibiotics to bring about rhythmic changes in vegetation that result in alternate stabilization and rejuvenation of primary production and species diversity (Odum, 1971). Conversely, uncontrolled crown and severe surface fires destroy the entire ecosystem. The biotic community must start developing again from the beginning, a process that can take many decades.

Soil

Soil supports the terrestrial community. A stratum of soil consists of solid grains, interconnected pores or voids, and air and water within the voids. The soil grains consist of inorganic minerals, most of which originate from the parent rock materials that form the earth's crust. The grains also include inorganic minerals and simple organic molecules decomposed from dead animals and animal excreta. The organic matter should finally decompose into inorganic substances; however, this does not happen easily due to the lack of soil aeration, and 1-5% of organic material by weight remains in the soil. This organic material gives soil its black or brown color. Soil as a constituent of land mass participates in the nutrient cycles and acts as a reservoir.

If a cut edge of a soil bank or a trench is examined, the soil is found to be composed of distinct layers of color and texture, known as *soil horizons*. The sequence of horizons constitutes the *soil profile*. The possible horizons are listed in Table 5-1. Most soils display horizons to a varied extent but some new soil formations might not have any distinct horizons.

Table 5-1 Soil Profile

Designation	Name	Types of Soil
A-Horizon		
A-0	Litter	Fresh fallen leaves and organic debris
A-1	Top Soil	Humus, plant roots, living organisms, and some inorganic minerals; dark colored
A-2	Leached zone	Dissolved or suspended materials leached by downward flow of water; light colored
B-Horizon	Mineral soil	Organic compounds converted to inorganic by decomposers
C-Horizon	Parent material	Original mineral formation

Particle size varies widely in soils. Depending on the size, particles are called clay (less than 0.002 mm), silt (0.002–0.06 mm), sand (0.06–2 mm), and gravel (larger than 2 mm). A soil sample contains a combination of different-sized particles that can vary infinitely; no two soils are identical in particle distribution. Accordingly, the soil classification system is based on general characteristics of a mass of the soil and not on the exact soil particle makeup. In engineering applications, a common classification is the Unified Classification System proposed by the U.S. Army Corps of Engineers.

Soil is known to be *saturated* when its voids are fully filled with water. This happens when a certain amount of water infiltrates deeper below the ground surface after each rain and because of seepage from stream beds. The enormous amount of stored water produced in this way is called *groundwater*, as discussed in Chapter 8. However, the portion of rainwater that seeps to a shallow depth only partially fills the soil voids. This water, known as *soil water*, contains dissolved minerals. Plant roots spread within this stratum and derive the water and minerals as needed for photosynthesis.

In summary, an ecosystem has three distinct features:

- The two members, the biotic and the abiotic, form an integral part of an ecosystem.
- The system is self-maintained. All essential materials are continuously cycled through the system. The cycles of important materials are described in the next chapter.
- The system is self-regulated through the one-way flow of energy that passes from one trophic level to the next in the food chain. A certain amount is lost at each level, as discussed subsequently.

These three features are depicted in Figure 5-4. The interconnection of the biotic elements is provided through the circulation of material and the flow of energy. The solid arrows represent movement of material among producers, consumers, and decomposers, forming a closed loop. The dashed arrows show exchange of energy, some of which is lost at each biotic element level. The entire biotic part operates within an abiotic environment shown by the surrounding rectangle. From the viewpoint of the functioning of an ecosystem, the system can be studied in terms of

- the food chain and the trophic levels,
- the cycling of materials,
- the flow of energy, and
- the stresses on the system and diversity of species.

The Flow of Energy Through an Ecosystem

Matter and energy are as indispensable for an ecosystem as for the universe. Progression and self-duplication are the essence of life. Life is only possible through synthesis of matter and exchange of energy. The same laws of physics that govern nonliving systems as described in Chapter 3 are applicable to ecological systems. The principle of conservation is an essential attribute of both matter and energy and it dictates a closed loop in the distribution of these two items. In nature, energy cycles at the level of the solar system. Energy that enters the earth's surface as light is balanced by what leaves the earth into space as invisible heat radiation. In the process energy flows through the ecosystem to keep it functioning. Matter cycles at the planetary level within the earth's system and in smaller cycles through ecological subsystems.

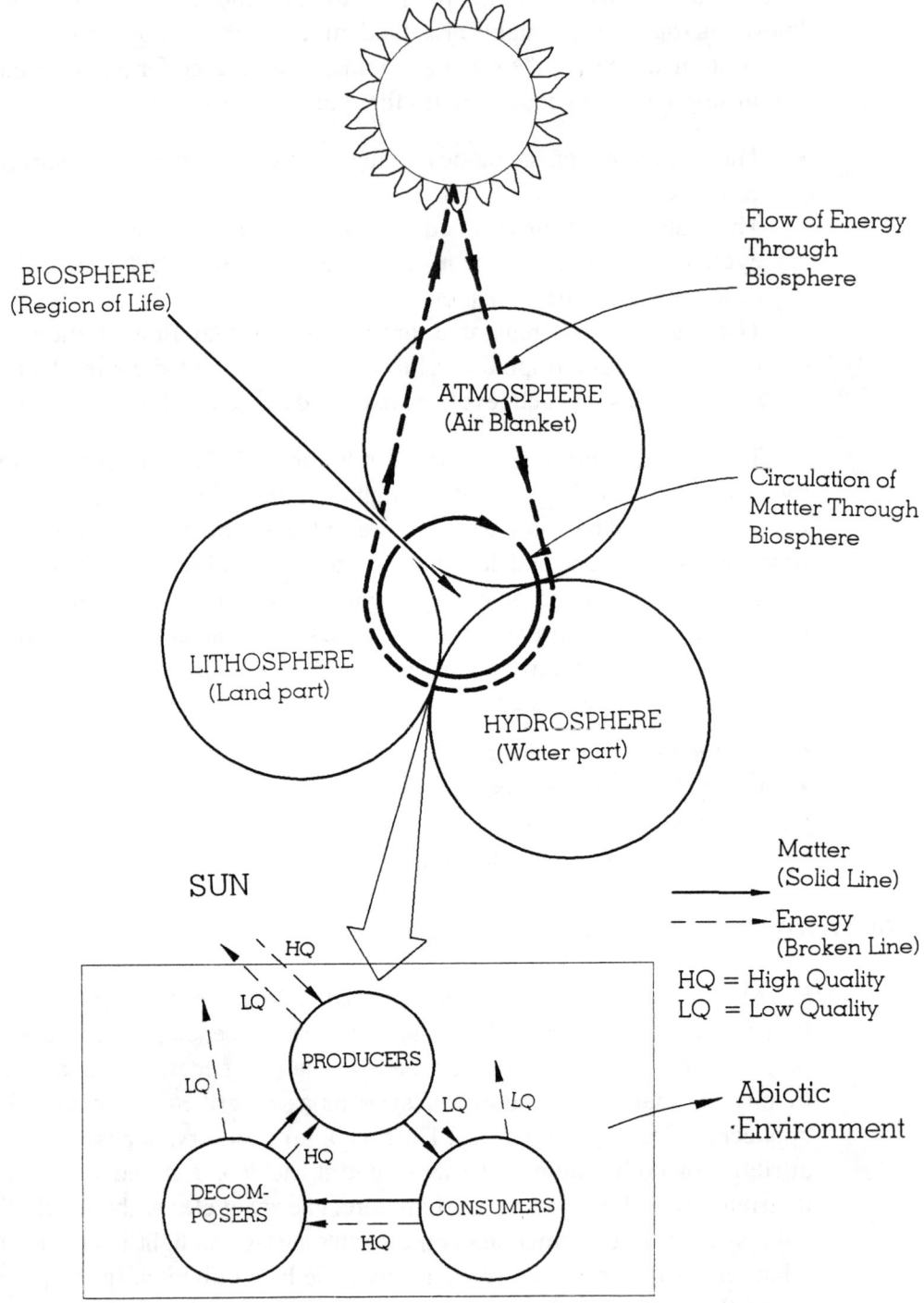

Figure 5-4 Flow of Energy and Circulation of Matter Through an Ecosystem

Energy and matter flow together through the ecosystem. The flow of energy is one way, i.e., it is not recycled because at each transfer it undergoes a change of form which cannot be reused by organisms again. Conversely, material recirculates through the ecosystem between nonliving and living beings. Energy is expended to recycle the material.

Although we recognize that the ultimate source of all energy on the planet is the sun, there are many other forms of energy in terms of direct sources: fossil fuel, tidal action, heavy rain, irrigation, fertilizer, transportation, labor, and so on. Certain large ecosystems such as big lakes and oceans effectively use only the solar energy. These are known as the *solar-powered* ecosystems. However, many systems augment energy from other sources. These are known as the *energy-subsidized* ecosystems; in such systems also, most energy needs, including that captured by plants, are met from solar radiation.

Energy captured at the earth's surface serves two purposes: a major part of it goes to ecosystem upkeep including maintaining the structure between abiotic and biotic components, cycling of materials, driving up of the water cycle, and retaining adequate temperature on the earth, and a small fraction of energy of the solar origin is absorbed by photosynthetic plants that maintain all forms of life in the ecosystem. The path of the fraction of energy entering the ecosystem by way of photosynthesis (sustaining the remarkable diversity of living organisms) is traced in Figure 5-5 through a simple model suggested by R.L. Lindeman in 1942. Energy distribution has been considered within the closed boundary of a system. Producers get energy from the sun. Consumers derive their energy from the producers. The nth order refers to tertiary or higher-order consumers. Decomposers receive energy from producers and from all trophic-level consumers. This model does not separate the decomposers themselves into different trophic levels. Many modified models do this to show the possibility of the cross-pathways between consumers and decomposers; for example, a secondary consumer (a bird) feeding on a primary decomposer (a termite). As shown in the model, at each level there is an energy loss through heat generated by work. Eventually all energy entering an ecosystem by the process of photosynthesis is lost in the form of heat; none is recycled. The position of transfer of matter and energy from one trophic level to other is discussed subsequently.

Energy Flow in a Modern City

Natural energy subsidies such as tides in estuaries, rains on cropland, dead leaves falling on ponds, winds, and flowing streams to produce power, are helpful in increasing energy inflow substantially. However, humans have been instrumental in transporting energy from one ecosystem to another on a scale that never happens in nature. In a modern city, the flow of energy is tremendous. There is only small primary production. Most of the energy needs, in the form of fossil fuels, are obtained from somewhere else. Thus, a modern city is a subsidized system. Also, it is heavily dependent on other ecosystems and, as such, it is not self-sustaining. One of the requirements of an ecosystem is that it should be self-sustaining. Accordingly, even though it possesses many attributes of an ecosystem, a modern city is not an ecosystem in the opinion of many scientists.

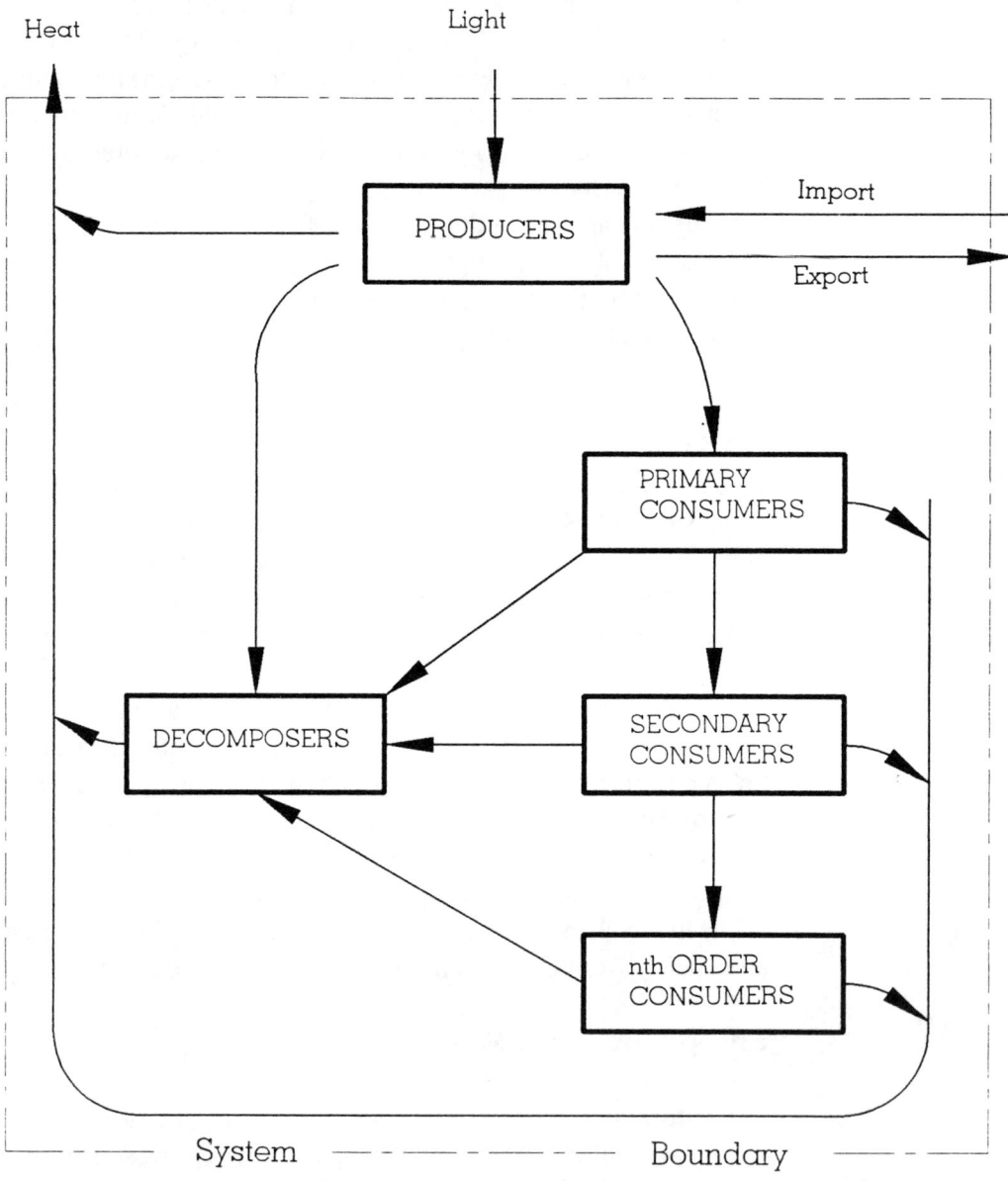

Figure 5-5 Lindeman's Model for the Transfer of Energy in an Ecosystem

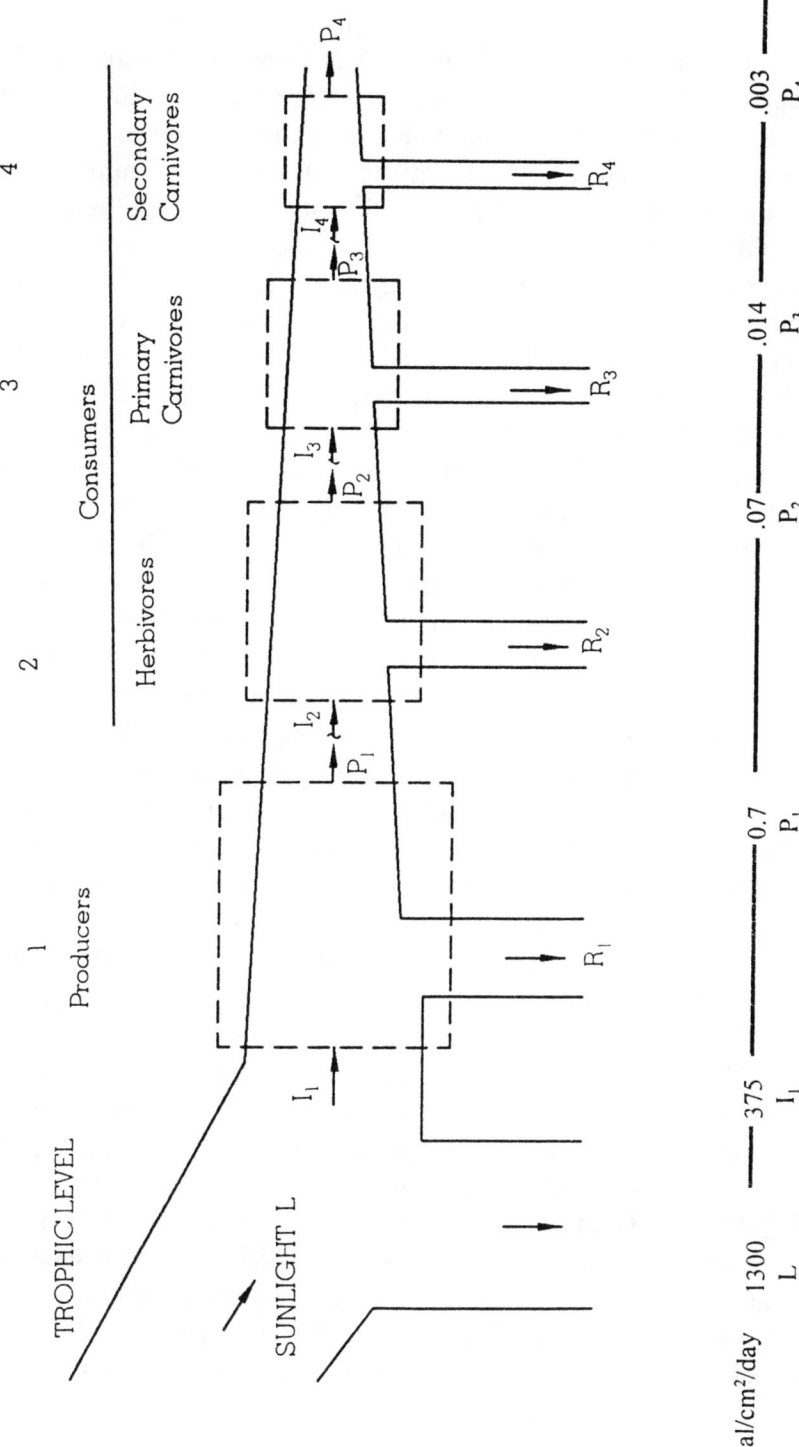

Figure 5-6 A Linear Model Showing the Flow of Energy Through a Food Chain

Matter and Energy Transfer Through Trophic Levels

A linear model of a food chain is shown in Figure 5-6. The boxes represent trophic levels and the channels depict the flow of energy. The plants produce energy in a form that can be used by animals. Every time an animal eats plants, some energy is converted into animal tissue or the production of biomass (P) and some is lost as heat or respiration (R). At each trophic level, energy inflows (I) balance energy outflows (P plus R) according to the first law of energy, and each transfer is accompanied by dispersion of energy as heat in accordance with the second law of energy. Plants and animals are highly inefficient in energy assimilation. Of the 81×10^{15} W or 1300 cal/cm^2-day of energy reaching the earth's surface, only 25–50% (375 to 750 cal/cm^2-day) is absorbed by plant cover to stimulate photosynthesis, and only a fraction of it (0.7 cal/cm^2-day or 40×10^{12} W)[7] is actually converted to chemical energy by plants at the first trophic level. At the second trophic level comprising herbivores, the efficiency is about 10%, and at successive levels (carnivores) the efficiency is about 20%. The production energy becomes very small at the fourth trophic level. Thus, the levels are limited to four or five in a food chain. The nearer an organism is to the beginning of the chain, the greater is the available energy. It is energy efficient to be a vegetarian. The characteristics of energy loss at each transfer necessitates that plants should be more abundant than animals (many plants are required to support a rabbit) in an ecosystem and that herbivorous animals should be more abundant than carnivorous animals (many rabbits will support one fox), and so on. The abundance can be measured either in terms of the rate of energy, as discussed above, or in terms of the dry unit weight of the living organism existing at each trophic level at any one time, known as the *standing crop biomass*. Each ecosystem has a definite trophic structure that can be shown graphically by means of an ecological pyramid with the producer level at the base and the highest consumer level at the apex. When indicated by weight of the biomass, the trophic structure is the *pyramid of biomass* as shown in Figure 5-7a, and when expressed in terms of energy flow at each trophic level, it is the *pyramid of energy* as shown in Figure 5-7b. If an organism at a lower level is small and short-lived, its biomass at any particular time may be less than that of the successive-level organism, even though the metabolism rate and energy flow at the lower level will be high. The biomass pyramid in such a case can be inverted with a narrow base. The energy pyramid by far gives the best picture of the functional aspect of the ecosystem.

---

[7]This is the amount based on the global energy budget. The radiation energy varies considerably from place to place and time to time.

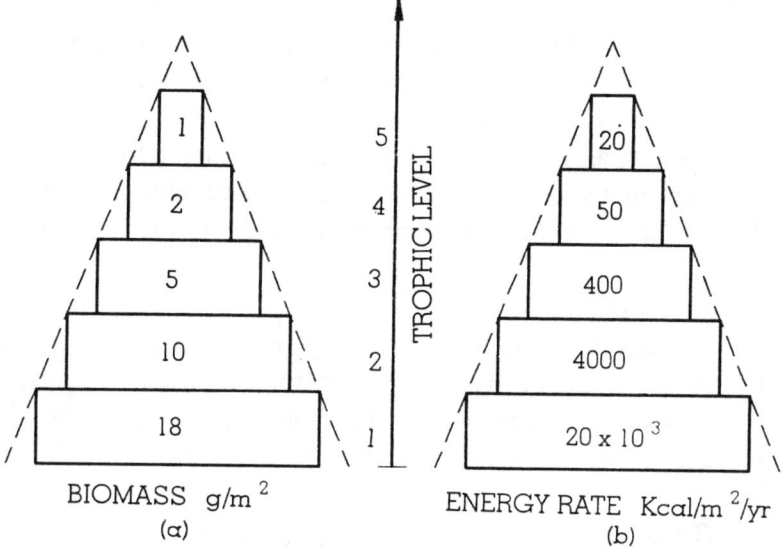

Figure 5-7 Pyramids of Biomass and Energy

For energy balance at each trophic level:

$$I_i = P_i + R_i \tag{5-6}$$

where I = trophic level
I = inflow of energy at trophic level I
P = energy available for growth and maintenance at level I
R = energy lost through respiration at level I.

The P value becomes I for the next trophic level; inversely, I for a level is P of the previous level. When energy changes take place under a constant pressure condition, as is usually the case with a living system open to the atmosphere (pressure of 1 atm), Equation 5-6 can be expressed in terms of the thermal energy relation of chemistry:

$$\Delta H_i = \Delta G_i + T\Delta S_i \tag{5-7}$$

where ΔH_i is the change of enthalpy, which is a measure of heat content of a system at level I, expressed in energy units.
ΔG_i is the change in free energy of a system at level I. The adjective *free* indicates the usable energy, i.e., in a form that could be used to do work.
ΔS_i is the change of entropy of a system at level I, which is a measure of disorder of the system. Multiplied by the absolute

temperature, T (in Kelvin or Rankine), it indicates the energy form that is not available for work.

The enthalpy change for a reaction is equal to the sum of enthalpies of each product subtracted by the sum of enthalpies of all reactants. This is *Hess's Law*. A similar principle applies for entropy of a reaction.

Example 5-2
At a trophic level, the total energy consumed by an ecosystem is 7×10^6 cal/day. Assume that 75% of this is lost in respiration, leakage, and other losses. The average temperature for the process is 40°C. Determine the amount of free energy and entropy of the system.

Solution:
Enthalpy, $\Delta H = \left(7 \times 10^6 \dfrac{\text{cal}}{\text{day}}\right)\left(\dfrac{4.184 \text{ J}}{1 \text{ cal}}\right)$

$\qquad = 29.3 \times 10^6$ J/day
$\qquad$ or 29.3×10^3 kJ/day

$T\Delta S = \dfrac{75}{100} \times 29.3 \times 10^3 = 22 \times 10^3$ kJ/day

$T = 40 + 273 = 313$ K

$\Delta S = \dfrac{22 \times 10^3 (1000)}{313} = 70.3 \times 10^3$ J/ K-day

$\Delta G = \Delta H - T\Delta S$
$\qquad = 29.3 \times 10^3 - 22 \times 10^3$
$\qquad = 7.3 \times 10^3$ kJ/day.

Microorganisms and Their Role in the Environment

Microorganisms, as decomposers, convert organic material into simpler inorganic matter, in accordance with the respiration process, thus making those substances available for use once again in a cycle. In a sample of wastewater, aerobic bacteria (those requiring oxygen) and facultative bacteria (those that can survive with or without free oxygen) start decomposing the organics using the oxygen. On depletion of the oxygen, the facultative bacteria use the bound oxygen in nitrate. Next, the oxygen of sulfate is used, producing the rotten-egg smell of H_2S. If still more organic matter is available and methane-forming bacteria are present, the process will further proceed by formation of methane gas and CO_2. There is a variety of microorganisms whose basic composition can be represented by a generalized protoplasm relationship of the form, $C_aH_bO_cN_dP_eS_f$.

Bacteria are simple, colorless, one-celled organisms that range in size from 0.5 to 5 µm, and therefore, are visible only through a microscope. Their reproduction is by binary fission; that is, a cell divides into two new cells, each of which matures and divides into two again, and so on. The fission occurs every 15–30 min. Protozoa are also single-celled organisms that multiply by binary fission, too. They range in size from 10 to 300 µm. These two microorganisms play a key role in wastewater treatment systems.

Food is required by microorganisms both for growth and energy; they can not utilize the sun as a source of energy since they do not contain chlorophyll. The essential elements of food for microorganisms are those that are found in their cells and consist of carbon, hydrogen, oxygen, nitrogen, sulfur, phosphorus, manganese, copper, zinc, and cobalt. Micronutrients include magnesium, potassium, calcium, and iron. Microorganisms, principally bacteria, are classified based on the food requirements. There are two major groups. Those that use CO_2 as a sole source of carbon are called *autotrophs*, whereas those that derive carbon from organic sources are called *heterotrophs* or *saprophytes*. The term *facultative* is applied to those that can use both CO_2 and organic compounds as carbon sources. Heterotrophic microorganisms are further subdivided into three categories based on their oxygen requirement for metabolism. *Aerobes* require free dissolved oxygen, and *anaerobes* live in complete absence of dissolved oxygen by using oxygen bound in other compounds. *Facultative* microorganisms use free oxygen when available but can also live in the absence of dissolved oxygen.

Microbial Kinetics

The knowledge of microbial kinetics is used to develop relationships that are applied in biologic process designs. The following discussion relates to the growth of bacteria, the microorganism of primary importance in biological treatment. Two underlying assumptions about the classic growth pattern are that:

1. All essential ingredients for growth, (i.e., food, micronutrients, moisture, temperature and pH) are present at the start; and
2. It is a batch-culture situation, wherein the culture is isolated; there is no inflow into to or outflow from the system (tank) in which a population of a single species of bacteria is introduced. The growth-time graph is plotted with time in hours as the abscissa and either bacterial count or biomass growth on a log scale (since the range is very high) as ordinate, as shown in Figure 5-8.

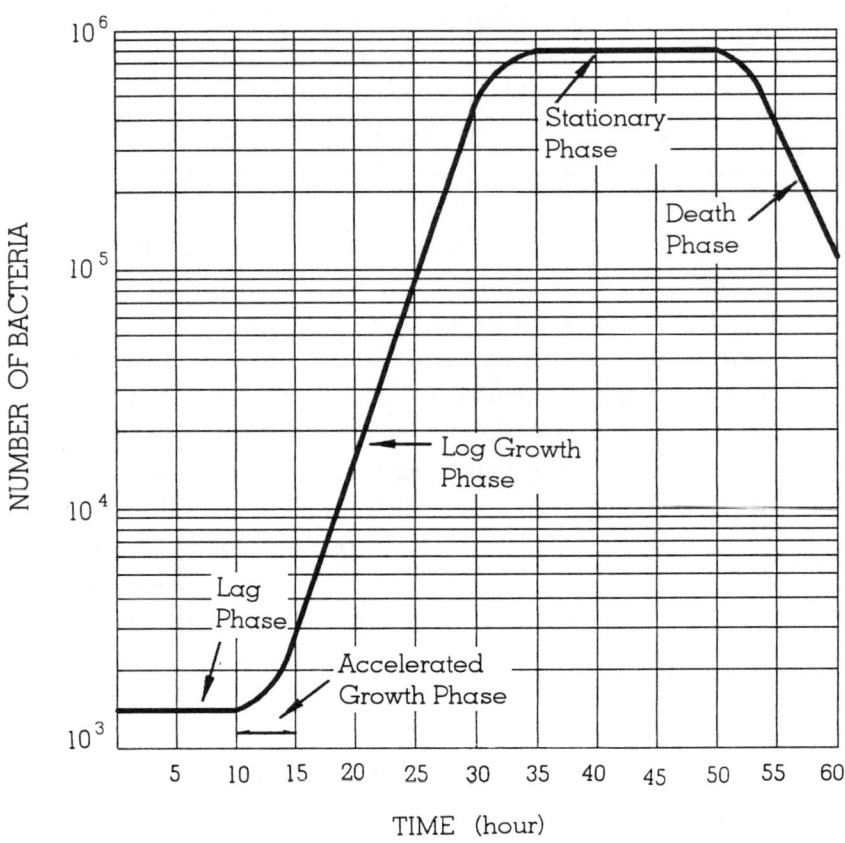

Figure 5-8 Microbial Growth in Pure Culture

At first, no growth is recorded since bacteria must get used to their new environment. This is the *lag phase*. Soon bacteria start multiplying by binary fission, registering a steady rise of population as the *exponential,* or, *log growth phase*. As food[8] becomes limited, the population becomes constant by balancing death and reproduction rates, as shown by the *stationary phase*. Following this, as food is exhausted, the bacteria begin to die faster than they reproduce in the *death*, or, *endogenous phase*.

In the exponential phase[9]

$$\frac{dX}{dt} = \mu X \tag{5-8}$$

---

[8]Usually referred to as substrate: a substance or group of substances that is utilized as a nutrient or micronutrient by the microorganisms.

[9]This equation is similar to the population growth model seen in Chapter 1.

where $\frac{dX}{dt}$ = growth rate of biomass (bacteria) mg/L-t

μ = specific growth rate (1/t)
X = Concentration of biomass (mg/L).

In the endogenous phase, including the decay of bacteria:

$$\frac{dX}{dt} = \mu x - K_d X \qquad (5\text{-}9)$$

where K_d = specific decay rate or endogenous decay coefficient (1/t).

The biomass production (of bacteria) is directly dependent upon the rate of food (substrate) utilization:

$$\frac{dS}{dt} = -\frac{1}{Y}\frac{dX}{dt} \qquad (5\text{-}10)$$

where $\frac{dS}{dt}$ = rate of food utilization (mg/L-t)

$\frac{dX}{dt}$ = growth rate of biomass

Y = fraction of food converted to biomass or yield coefficient.

The specific growth rate, μ in Equations 5-8 and 5-9, is also dependent on the substrate concentration, S. The Monod model[10] adequately expresses this relationship.

Example 5-3
The initial concentration of organic waste (substrate) and cells in a reactor are 800 mg/L and 200 mg/L, respectively. After 1 hr, the cell concentration increased to 500 mg/L. If the endogenous decay coefficient is 0.04/hr, determine the specific growth rate in the (a) log phase, and (b) death phase.

Solution:
From Equation 5-9:

$$\frac{dX}{X} = (\mu - K_d) dt$$

---

[10] Not included here at the elementary level of discussion.

Integrating:

$$\ln \frac{X_1}{X_0} = (\mu - K_d)(t_1 - t_0)$$

$$\ln \frac{500}{200} = (\mu - K_d)1$$

or $0.92 = (\mu - K_d)$

(a) For log phase, when $k_d = 0$, $\mu = 0.92$/hr
(b) For death phase, when $k_d = 0.04$, $\mu = 0.96$/hr.

Example 5-4

Determine the amount of waste remaining after 1 hr in Example 5-3. $Y = 0.4$ mg/mg.

Solution:

From Equation 5-10:

$$\frac{dS}{dt} = -\frac{1}{Y}\frac{dX}{dt}$$

Expressed numerically:

$$\frac{800 - S^2}{1} = -\frac{1}{0.4}\frac{(200-500)}{1}$$

$$S^2 = 50 \text{ mg/L}.$$

A good treatment of waste is achieved when the microbial kinetics is held in the endogenous phase, because bacteria starve for food and bring about a high degree of organic matter removal. This is controlled by maintaining a low food-to-microorganism (F/M) ratio.

The above discussion is based on a pure culture consisting of a single species. When organic matter is fed to a mixed population of microorganisms, competition arises for food. Bacteria are very competitive feeders that dominate both in aerobic and anaerobic cases. Protozoa that consume bacteria are a dominating predator. The presence of protozoa produces more rigorous growth of both species for efficient waste treatment.

Biological Parameters of Waste Quality

Coliform bacteria and biochemical oxygen demands are two convenient parameters used to ascertain pollution from organic wastes. These are described below:

Indicator Organisms for Pollution

It is not practical or necessary to analyze samples for all disease-causing (pathogenic) microorganisms. It is far more beneficial if a few tests can indicate the presence of pathogenic bacteria. The coliform group of bacteria reside in the intestinal tract of humans and warm-blooded animals and are extracted in a very large number in their feces, averaging about 50 million per gram. Pathogenic bacteria also originate from the same source, i.e., the fecal discharges of diseased persons or animals. Thus, any water that has been polluted by a fecal discharge is considered to be potentially dangerous. Water can be tested for the presence of coliform bacteria that is so abundantly present in feces. However, just the presence of coliform bacteria is not a sure sign of human fecal pollution because these bacteria can originate from animal feces and soil as well. An elevated temperature test separates organisms of fecal and nonfecal origin, because high temperature is lethal to nonfecal coliforms.

The criteria in terms of number of coliform organisms per 100 mL of water sample have been established for drinking water and water used for other purposes. These criteria are supported by the fact that the presence of pathogens is significantly low relative to coliforms, and the die-off rates of pathogens are far greater outside the intestinal tract.

A three-step procedure for coliform analysis consists of:

1. The presumptive test based on gas production from lactose,
2. The confirmed test on a positive presumptive test sample by growing cultures on a medium that suppresses other organisms, and
3. The completed test based on the ability of the cultures grown in the confirmed test sample to again produce gas from lactose. Normally only the first two tests are performed.

The results of the tests are expressed as the *most probable number* (MPN) per 100 mL. The procedure includes serial dilutions and multiple-tube fermentation. It involves the use of three 100-mL bottles and six sets of five 10-mL fermentation tubes (30 tubes in all). The tubes contain a suitable lactose culture medium. A 100 mL sample of water to be tested is collected in bottle #1. Two successive 100 times dilutions of this are obtained as shown in bottles #2 and #3 in Figure 5-9.

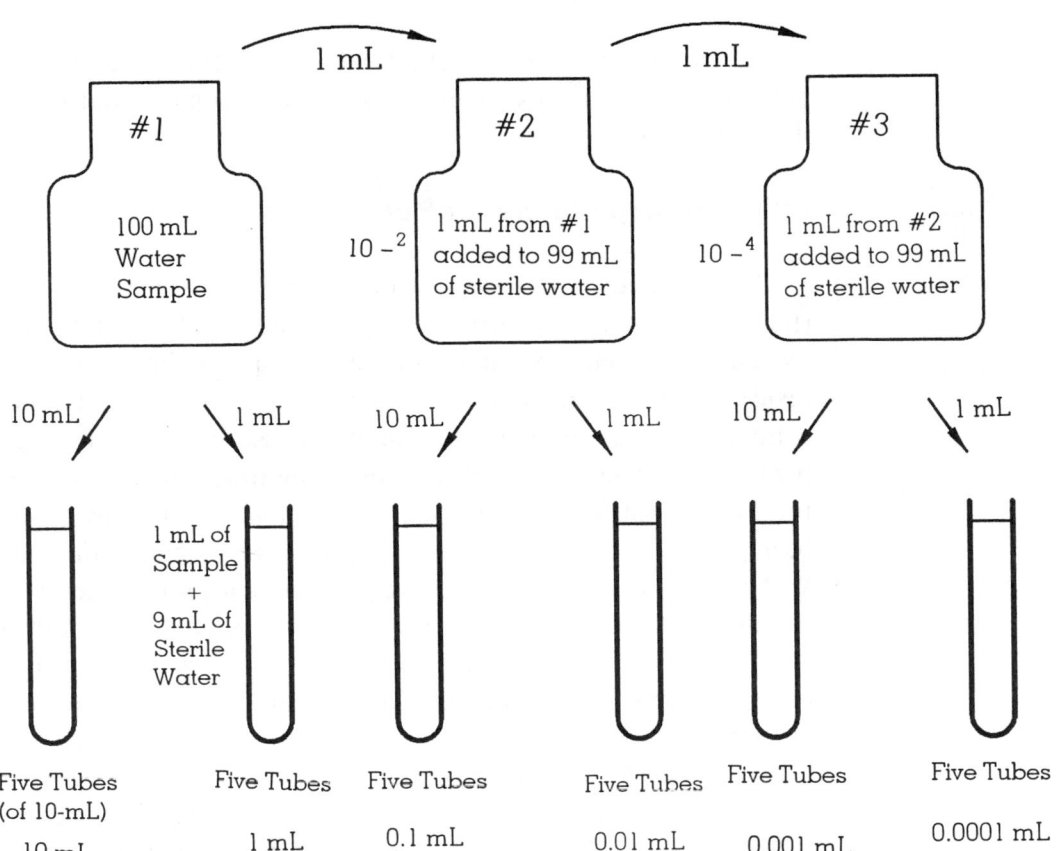

Figure 5-9 Successive Dilution of Samples for MPN Testing

Every bottle dispenses to two sets of tubes, each set consisting of five tubes each; from each bottle, first a 10-mL portion of the sample is transferred to one set of five tubes, and then a 1-mL of sample is transferred (and 9 mL of pure water is mixed) to the second set of five tubes. The tubes are incubated for 24 hours at 35°C. The tubes are examined for positive reactions. The number of positive tubes, out of five in each diluted series are recorded. For various combinations of positive tube results from three successive test runs, the MPN Index is read from a Standard Table or computed from a formula.[11] A sample table for MPN is given in Table 5-2. The MPN is based on the statistical (Poisson's) distribution. It is not a precise count of microorganisms, but only a statistical estimate of the concentration.

The MPN table values correspond to tubes containing 10-mL, 1-mL, and 0.1-mL sample portions. If smaller sample amounts (more diluted sample

---

[11]Not included here; readers should refer to a microbiology textbook.

runs) were referred to in the table, the value should be adjusted by a factor of 10/maximum sample portion used.

Example 5-5
The following are the results from a presumptive-stage coliform analysis. Determine the MPN. Use Table 5-2.

Sample Portion (mL)	Number of Positive Reactions (out of five tubes)
0.01	5
0.001	2
0.0001	0

Table 5-2 Sample Table for Most Probable Number of Coliforms per 100 mL

Number Positive Tubes			MPN	Number Positive Tubes			MPN
10 mL	1 mL	0.1 mL		10 mL	1 mL	0.1 mL	
0	0	0	0	3	0	0	7.8
0	2	5	13	3	2	5	31
0	5	5	19	3	5	5	45
1	0	0	2	4	0	0	13
1	2	5	17	4	2	5	50
1	5	5	24	4	5	5	81
2	0	0	3	5	0	0	23
2	2	5	22	5	2	0	49
2	5	5	32	5	2	5	180
				5	5	4	1600

Solution:
From the MPN table for 5,2,0 positive, the MPN index = 49

Adjustment for sample size = $\dfrac{10}{0.01} = 1000$

MPN for presumptive test = 49 × 1000 = 49,000.

Biochemical Oxygen Demand (BOD)

Bacteria growing within wastewater utilize organic matter as food, as discussed above. To carry out this respiration process, oxygen is derived from the dissolved oxygen in wastewater. The amount of oxygen used in this process is called the *biochemical oxygen demand* (BOD). Since the requirement of oxygen is directly dependent upon the organic matter to be digested, the BOD is an indicator of the pollutant strength of the wastewater although by itself it is not a pollutant. The oxygen used in the reaction is in the dissolved state in water and is inversely proportional to water temperature. The maximum (saturation) values of dissolved oxygen in water at various temperatures are given Appendix E.

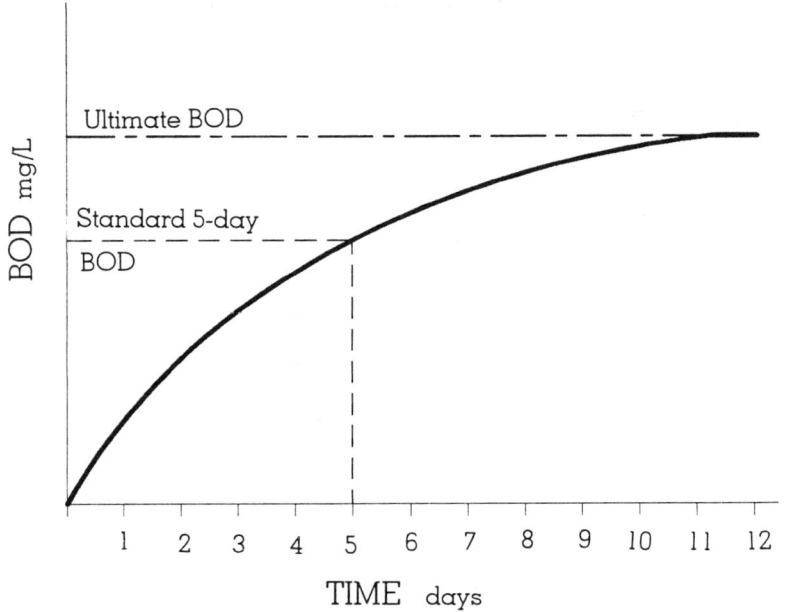

Figure 5-10 Biochemical Oxygen Demand Curve

The oxygen demand with respect to time can be expressed by a curve as shown in Figure 5-10. This curve has the following mathematical form:

$$BOD_t = L(1 - e^{-kt}) \qquad (5\text{-}11)$$

where BOD_t = Biochemical oxygen demand in mg/L after *t* days
L = ultimate BOD in mg/L
k = deoxygenation rate constant in per day.

The BOD testing for 5 days at 20°C has become a standard practice. This 5-day value for domestic wastewater represents approximately two-thirds of the demand that would be exerted to oxidize all biologically oxidizable waste. Because the solubility (availability) of oxygen in water is limited, a strong waste capable of exerting an oxygen demand in excess of the saturation oxygen limit is tested only after it is diluted with water. A measured volume of wastewater diluted with water is poured into a 300-mL BOD bottle. Seed microorganisms are supplied if not already present in the wastewater sample. The dissolved oxygen level is measured after 15 min. The bottle is incubated at 20°C for 5 days and the dissolved oxygen level is again determined. The following equation is used to calculate the BOD.

$$BOD_5 = \text{Dilution}(DO_0 - DO_5) \tag{5-12}$$

where DO_0 = dissolved oxygen (mg/L) of diluted sample after 15 min
DO_5 = dissolved oxygen (mg/L) after 5 days
$$\text{Dilution} = \frac{\text{volume of dilution water plus wastewater}}{\text{volume of wastewater sample used}}$$

Example 5-6

The data from a domestic wastewater BOD test are:
5 mL of waste sample diluted to fill a 300-mL bottle, initial (after 15 min) DO of 8.0 mg/L; after 5 days, DO of 4.0 mg/L. Compute the BOD_5 and the ultimate BOD, assuming the rate constant = 0.3 per day.

Solution:

$$\text{Dilution} = \frac{300}{5} = 60$$

From Equation 5-12:

$$BOD_5 = 60\,(8.0 - 4.0) = 240 \text{ mg/L}$$

From Equation 5-11:

$$240 = L[1 - e^{-(.3)(5)}]$$

$$\text{or } L = \frac{240}{1 - 0.223} = 309 \text{ mg/L}$$

PROBLEMS

1. In a photosynthesis process, assume that 600 kcal of energy is needed for formation of one mole of glycerol ($C_3H_8O_3$). One kilogram of gycerol is generated from each square meter. How much energy is needed to produce glycerol from 15 acres?

2. If only 10 kcal/m²/hr of the solar energy contributes to the biomass creation, how much biomass of glycerol is produced over 10 acres of land every day? Assume the same level of energy is needed per mole as in Problem 1.

3. With 674 kcal of light energy needed per mole of glucose ($C_6H_{12}O_6$) produced in photosynthesis, what is the energy content of each gram of glucose? For the glucose biomass production of 7×10^{-3} g/cm²-day, how much of the solar energy is consumed over 50 acres in a week?

4. In photosynthesis, 40×10^{12} W of solar power is consumed. If the energy content of the glucose produced is 3744 cal/g, calculate the total plant matter produced (biomass as glucose in grams) each year on the planet. (Hint: convert Watts to calories per year and divide by the glucose content.)

5. Calculate the food production as glucose per acre per year based on the entire planet area from the information presented in Problem 4.

6. A person has a metabolic rate of 160 W. How many kcal a day does that person consume? If the human body is only 15% efficient, how much is the useful energy?

7. Balance Equation 5-3. For 1 kg cell growth of the microorganism ($C_5H_{7N}O_2$), how much carbon is required?

8. Balance Equation 5-5. For complete metabolism of 5 lb of glucose ($C_6H_{12}O_6$), how much oxygen is required?

9. Two pounds of organic material was fully decomposed microbiologically under aerobic conditions. Determine the oxygen demand. The material was $C_{11}H_{14}O_4N$.

10. The energy needs of an ecosystem are as follows:

Trophic level	Energy (kcal/month)
1	52,000
2	10,400
3	1,100
4	200
5	50

Draw a diagram similar to Figure 5-6 showing the flow of energy through the food chain. Determine the efficiency at each trophic level which is the ratio of energy available for growth at a level P_i, to inflow of energy I_i.

11. Prepare a pyramid of energy on a yearly basis for Problem 10. Make a logarithm of energy values and plot one side of the pyramid (one-half of the values).

12. The rate of loss of energy is 85% at each step in an ecosystem. Prepare a food chain showing available energy at each trophic level up to four trophic levels. If 1000 kcal of energy is needed at the fourth level, how much energy input is needed at the start of the chain in kcal? Develop a suitable mathematical expression for energy input in terms of trophic levels.

13. From the energy levels given at the bottom of Figure 5-6, fill in the values of I, R, and P for each trophic level based on Equation 5-6.

14. The Silver Springs, Florida, ecosystem consists of a clean spring-fed stream with vegetation covering the bottom and numerous species of animals living in or around the water. Energy flow in the ecosystem consists of 1,700,000 units of solar radiation reaching the earth, of which 410,000 units is absorbed by plants. During photosynthesis, 20,800 units result in gross primary productivity and the balance converts to heat. Net plant productivity is 8,830 units. The efficiency of energy production by herbivores and carnivores is 10% each. Draw the diagram of the ecosystem. All energy units are in kilocalories per square meter per year.

15. Metabolism of glucose ($C_6H_{12}O_6$) in the human body results in CO_2 and water. The enthalpy of this reaction is 2800 kJ/mole (energy is given up). The corresponding entropy is 260 J/K-mole. The metabolism takes place at 40°C. Determine the energy that becomes available to synthesize biological molecules and sustained activities.

16. At a trophic level in an ecosystem, the entropy is 20,000 J/K. The energy losses in the system are 80%. The average process temperature is 50°C. Determine the amount of enthalpy and free energy in calories.

17. A wastewater sample contains organics, sulfate, nitrate, and dissolved oxygen. If bacteria are present in the sample, in what order will the breakdown take place? At what point will the odor appear?

18. The initial concentration of food (substrate) and cells in a reactor is 200 and 100 mg/L respectively. After 2 hrs, the cell concentration increased to 150 mg/L. Determine:
    a. The specific growth rate for the log phase;
    b. The specific growth rate for the death phase; and
    c. The amount of substrate after 1 hr. $K_d = 0.08$ per hr, $Y = 0.5$ mg/mg.

19. The initial concentration of substrate and bacterial biomass in a treatment unit is 8000 mg/L and 50 mg/L respectively. If the specific growth rate is 0.2/hr, the decay rate is 0.05/hr, and $Y = 0.3$, what will be the microbial biomass after one day? What is the final amount of substrate after one day?

20. In Problem 19, determine the waste (substrate) remaining after 2 hr in the log or exponential phase.

21. The results from a coliform analysis for the presumptive and confirmed phases are as follows. Determine the most probable number (MPN) for the two tests.

Sample Portion	Number of positives out of five tubes	
	Presumptive Test	Confirmed Test
1	–	4
0.1	5	3
0.01	2	1
0.001	1	1–n

The relevant values from the MPN table are extracted below:

Number of tubes giving a positive reaction out of:			MPN Index
5 tubes of 10 mL	5 tubes of 1 mL	5 tubes of 0.1 mL	
4	3	1	33
5	2	1	70

22. A BOD test was conducted by pouring 100 mL of polluted surface water to a 300-mL bottle and filling it up with dilution water. The initial dissolved oxygen measured 7.9 mg/L, and after 5 days of incubation, dissolved oxygen measured 2.5 mg/L. Calculate the BO_5.

23. A BOD test was conducted on municipal waste. Five samples were prepared. In each sample, the wastewater portion added to a 300-mL bottle was 30 mL and the bottles were filled with water. The dissolved oxygen values measured at different times are listed below. Plot a BOD vs. time curve and determine the 5-day BOD.

Time, days	0	2	4	6	10
D.O. mg/L	8.7	6.7	5.7	4.9	3.9

24. Determine the 1-day BOD and Ultimate BOD for wastewater that has a 5-day BOD of 150 mg/L. $k = 0.23$/day

25. If the ultimate (first stage) BOD is 295 mg/L and the 5-day BOD is 200 mg/L, what is the reaction constant?

26. Determine the size of sample (volume of wastewater) if the 5-day BOD is 380 mg/L and the total oxygen used in a 300-mL bottle is 2 mg/L.

STUDY QUESTIONS

1. Describe the three major categories of biotic components. Discuss their roles within an ecosystem.

2. List various abiotic factors. Describe their roles within an ecosystem.

3. Distinguish clearly between photosynthesis and respiration. Outline their roles within an ecosystem.

4. Describe and distinguish among food chain, food web, and trophic levels. Graphically show the trophic structure of the biomass in terms of weight of biomass at different trophic levels.

5. What are the three laws related to abiotic components? Give an example of how they can affect an ecosystem?

6. Describe with illustration, the flow of energy through an ecosystem.

7. Sketch a labeled diagram of the microbial kinetics in a pure culture. Why is a wastewater processing plant operated in the endogenous phase?

8. Explain the rationale of using coliform bacteria to evaluate drinking water quality. Describe the procedure of analysis. How are the most probable numbers (MPN) interpreted?

9. How is biochemical oxygen demand related to water pollution? How is it tested?

10. Which of the following are considered ecosystems? Why or why not?
    a. Boston, Massachusetts
    b. A 100-acre corn farm
    c. Lake Michigan
    d. Your classroom
    e. The Connecticut river

Principles of Ecology and Microbiology 163

11. Define the following:
    a. System
    b. Ecosystem
    c. Gross and net primary productivity
    d. Herbivores, carnivores, and omnivores
    e. Primary consumers, secondary consumers
    f. Exothermic and endothermic
    g. Autotrophic and heterotrophic
    h. Detritus feeders
    i. Relative humidity
    j. Soil horizons
    k. Energy-subsidized ecosystem
    l. Hess's law
    m. Autotroph, heterotroph, and facultative organisms
    o. Aerobe, anaerobe, and facultative organisms
    p. BOD_5

REFERENCES

Boughey, A. S., *Man and the Environment: An Introduction to Human Ecology and Evolution*, The Macmillan Company, New York, 1971.

Davis, M.L., and Cornwell, D.A., *Introduction to Environmental Engineering,* 2nd ed., McGraw-Hill Book Company, New York, 1991.

Ehrlich, P. R., Ehrlich, A. H., and Holdren, J. P., *Ecoscience: Population, Resources, Environment*, W. H. Freeman and Company, San Francisco, CA, 1977.

Enger, E. D., et al., *Concepts in Biology*, 5th ed., Wm. C. Brown Publishers, Dubuque, IA, 1988.

Foin, T. C., *Ecological Systems and the Environment*, Houghton Mifflin Company, Boston, MA, 1976.

Freedman, B., *Environmental Ecology*, Academic Press, San Diego, CA, 1989.

Gates, D. M., *Energy Exchange in the Biosphere*, Harper and Row, New York, 1962.

Kormondy, E., *Concepts of Ecology*, Prentice Hall, New Jersey, 1969.

Life Science Library, *Ecology*, Time-Life Books, New York, 1963.

National Research Council, Ecosystem Impact Resource Group, Risk and Impact Panel of the Committee on Nuclear and Alternative Energy Systems, *Energy and the Fate of Ecosystems*, National Academy Press, Washington, D.C., 1980.

Nebel, B. J., *Environmental Science*, 2nd ed., Prentice Hall, New Jersey, 1987.

Odum, E. P., *Ecology and Our Endangered Life-Support Systems*, 2nd ed., Sinauer Assoc. Publishers, Sunderland, MA, 1993.

Odum, E. P., *Fundamentals of Ecology*, 3rd ed., W. B. Saunders Company, Philadelphia, PA, 1971.

Odum, H. T., *Environment, Power, and Society*, Wiley-Interscience, New York, 1971.

Odum, H. T., and Odum, E. C., *Energy Basis for Man and Nature*, 3rd ed. McGraw-Hill Book Company, New York, 1981.

Owen, D. F., *What is Ecology?*, 2nd ed., Oxford University Press, U.K., 1980.

Phillipson, J., *Ecological Energetics*, Edward Arnold Publishers, London, U.K., 1966.

Ramade, F., *Ecology of Natural Resources*, John Wiley and Sons, Paris, 1981.

Sarnoff, P., *The New York Times Encyclopedic Dictionary of the Environment*, Quadrangle Books, New York, 1971.

Scientific American, *Ecology, Evolution and Population Biology*, W. H. Freeman and Company, San Francisco, CA, 1974.

Smith, R. L., *Elements of Ecology and Field Biology*, Harper and Row, New York, 1977.

Sutton, D. B., and Harmon, N. P., *Ecology: Selected Concepts*, John Wiley and Sons, New York, 1973.

Turk, A., Turk, J., Wittes, J. T., and Wittes, R., *Environmental Science*, W. B. Saunders Company, Philadelphia, PA, 1974.

Chapter 6

THE FUNCTIONING ECOSYSTEM

Materials Cycling in an Ecosystem

Because of the conserving nature of matter, chemical elements tend to move repeatedly within the earth's crust, the ocean, and the atmosphere via living organisms. There is very little loss to outer space or to the earth's mantle. This continuous and repeated movement of chemical elements or species is known as a *biogeochemical cycle*. *Bio* refers to living organisms, *geo* to the rocks, air, and water through which movement of chemicals takes place. All elements that occur in nature exhibit biogeochemical cycling. Some forty of these elements are required by living beings. The biogeochemical cycles pertaining to these nutrients are also referred to as the *nutrient cycles*. Carbon, oxygen, and nitrogen are the three elements that are needed in relatively large quantities. These are considered to be the energy elements. The cycles of matter are driven by one-way natural flow of energy as described in Chapter 5. Nutrients follow the same pathways as energy, passing on from one trophic level to the other, the only difference being that while energy is dissipated at every level, nutrients cycle through the chain.

A general model of a biogeochemical cycle is shown in Figure 6-1; the blocks are reservoirs that designate storage units such as land and atmosphere, and the lines are transportation paths of matter exchange from one reservoir (division of biosphere) to another. The mechanisms of material transfer are also given in the figure by the numerals. Two categories of cycles exist: the gaseous/liquid type in which the reservoir lies in the atmosphere or ocean, and the sedimentary type in which the reservoir is in the earth's crust. For gaseous cycles, the cycling period is relatively short, i.e,. material stays for a short time in a reservoir. In sedimentary cycles, elements generally are retained for a long time in the ocean and the land mass before release to the organisms. Water is a common resource distributed throughout land, ocean, and atmosphere that provides a medium of circulation for biogeochemical cycles. The solvent property of water helps in the transfer process. Thus, the *water cycle,* or *hydrologic cycle* is a parent cycle without which bio-geochemical cycles could not exist. All cycles are inseparably linked to the water cycle.

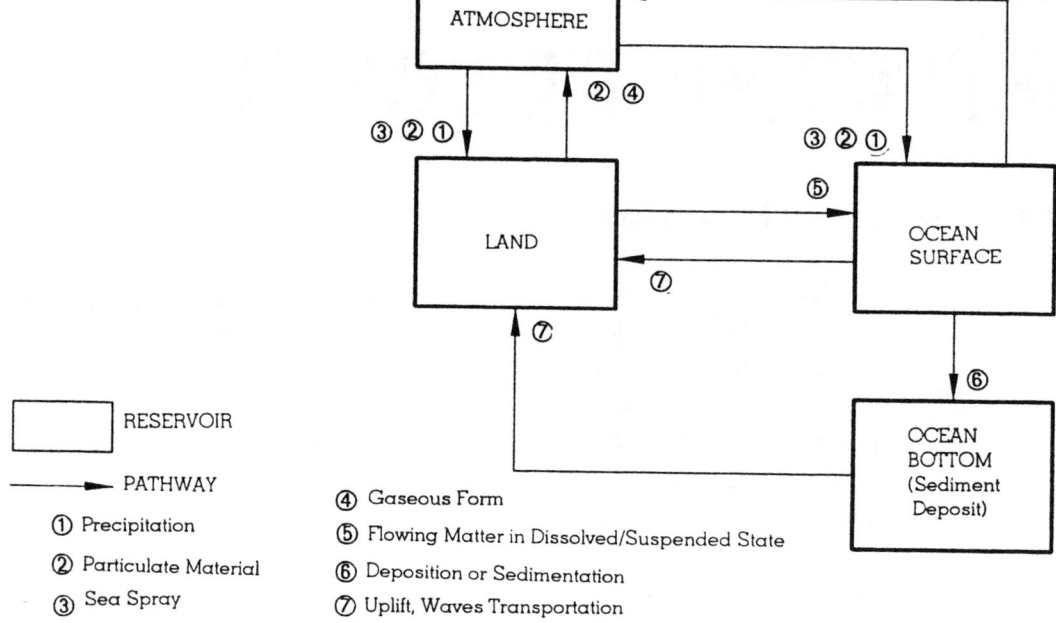

Figure 6-1 A Generalized Geochemical Model

In a cycle, a reservoir can be divided into many subunits. The number of reservoir subunits used in a cycle depends on the degree of detail with which a system is to be studied. In a quantitative study of a biogeochemical cycle, estimates are made both for the amount of matter present in a reservoir and the flux between the reservoirs (the rate of material passing along a particular transport pathway). If the amount of a chemical remains constant in a reservoir for a period of time, then the amount entering the reservoir is the same as the amount going out of the reservoir. As recognized in Chapter 2, such a system is known to be in a *steady-state,* or *stable condition.* A cycle may be stable globally but it may have wide fluctuations on a local level. A quantitative model should be able to predict what happens at the local or regional level.

A Steady-State Biogeochemical Cycle

In a steady-state cycle, the *residence time, renewal period,* or *turnover time* of a chemical in a reservoir is given by:

$$\text{residence time} = \frac{\text{amount of chemical in a reservoir}}{\text{rate of addition or removal of chemical element.}} \quad (6\text{-}1)$$

Example 6-1
The total amount of salt dissolved in the ocean is approximately 42×10^{18} lb. The annual addition of salt is 200×10^9 lb/yr. What is the residence time of salt in the ocean?

Solution:
From Equation 6-1:

$$\text{Residence time} = \frac{42 \times 10^{18}}{200 \times 10^9}$$

$$= 210 \text{ million years.}$$

Biogeochemical cycles are naturally stable. But human activities have caused many changes. Not only have the cycles been overburdened by the increased rate of movement of many natural chemicals (such as carbon dioxide), but completely new chemicals are being added (such as chlorofluorocarbons), to which the cycles are unable to adjust immediately. The materials injected into the biosphere that affect functioning of the ecosystems and produce adverse effects on living organisms are collectively called *pollutants*. Pollutants spread through the disturbed biogeochemical cycles and move through the food chain and include many toxic substances, e.g., lead, mercury, and chlorinated hydrocarbons.

The Water Cycle

The water cycle is shown in Figure 6-2. Solar energy causes the evaporation of water from the oceans and land and also the transpiration from plant leaves. The evaporated water vapor in the atmosphere forms clouds and eventually falls as some form of precipitation. Some precipitation falls directly over the oceans. Another portion that reaches the land mass divides into three parts. Some of it appears as surface runoff, draining into the rivers and streams. Some of it moves into the ground by infiltration and eventually reaches the rivers and streams as groundwater flow. A fraction is retained by vegetation for growth. Rivers and streams empty into the oceans.

The evaporation and transpiration process, from ocean and land, commences again, restarting the cycle.

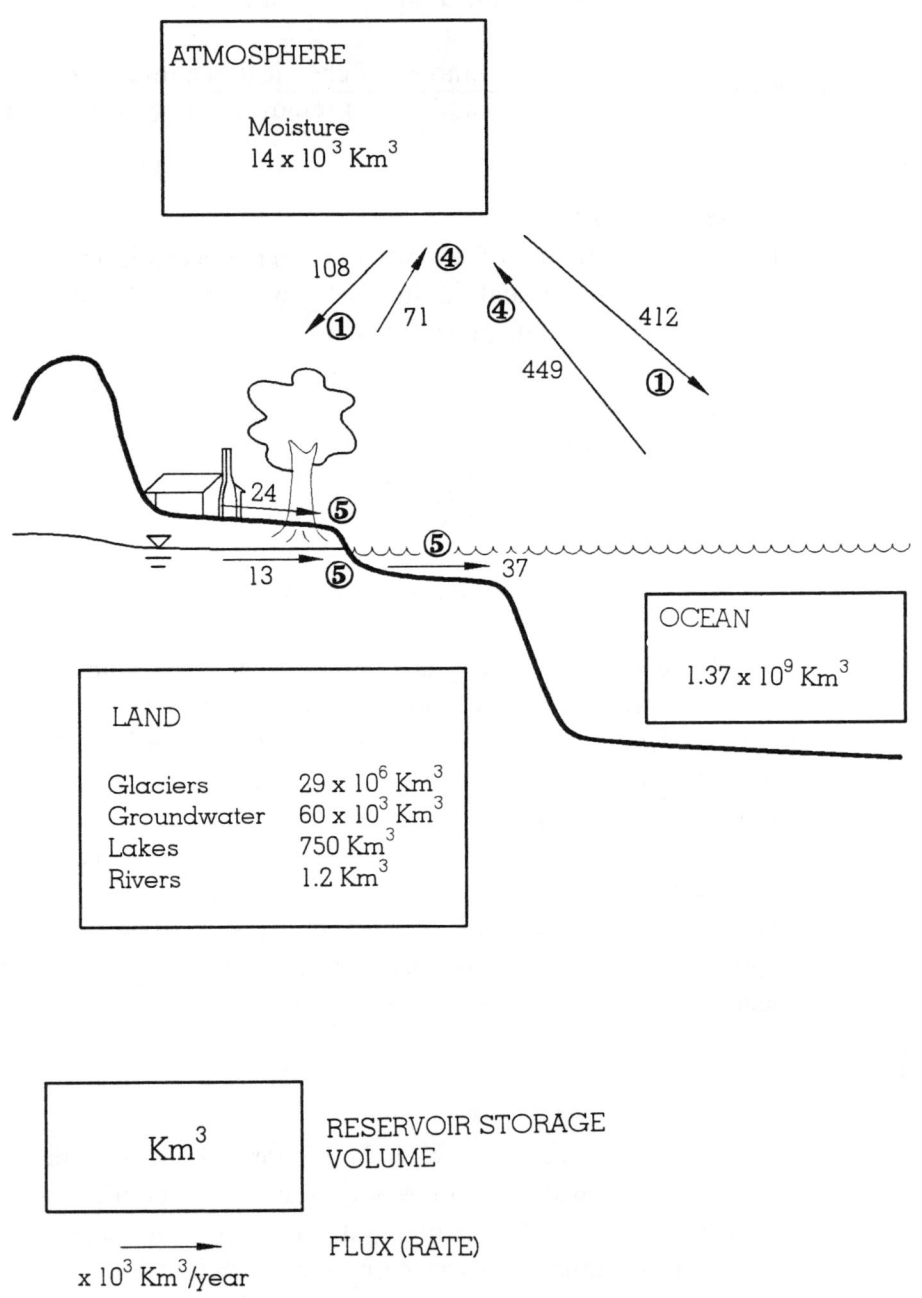

Figure 6-2 The Water or Hydrologic System

Since there are no geographic boundaries in the atmosphere, the rain that falls in our neighborhood may have been water from Amazon Basin in the past. The water cycle is global in nature. The annual fluxes (the rates of movement) and the reservoir storages for the global cycle are shown in Figure 6-2. Oceans, which cover about 71% of the earth's surface, contain 97% or 1.37×10^9 km^3 of world's water supply. Thus, only the remaining 3% represents the fresh water on the planet. Of this, more than 2% is in the polar ice caps and glaciers that cover about 10% of the land surface. This leaves only 1% available as fresh water for direct use by humans.

In terms of fluxes, a balance exists between the water evaporating (including transpiration) and precipitating, both are 520×10^3 km^3/yr. The precipitation on land is 108×10^3 km^3/yr (21% of the total), whereas the evaporation from the land is 71×10^3 km^3/yr (14% of total). The extra supply of 37×10^3 km^3/yr (7% of the total), comprising of 24×10^3 km^3/yr of runoff and 13×10^3 km^3/yr of groundwater flow, is used for growth by plants, animals, and humans, and finally returns to the oceans. In addition to the global scale, the water balance is achieved in the large basins on a long-term basis.

Water that evaporates from oceans or land leaves behind dissolved solids. Ocean spray sends fine salt particles into suspension in the atmosphere. Rain also dissolves various impurities existing in the air; where there is sufficient accumulation of oxides of sulfur, nitrogen, and carbon in the atmosphere, the problem of acid rain becomes apparent, as discussed later in Chapter 10.

The Carbon Cycle

Water and carbon compounds are two main components in living organisms. Their circulation is of primary importance to the support of life. Just as most of the water resources are contained in oceans, most of the carbon supply is located in the land mass. Carbon deposits in the earth's crust of 16×10^{18} kg, or 18×10^{15} tons, amount to 99.75% of the total carbon supply, mostly in the form of carbonate sedimentary rocks. The oceans contain 29×10^{15} kg, or 0.17% of the total, mostly in the form of bicarbonate and carbonate ions. Only 580×10^{12} kg, or 0.003%, exists in the atmosphere, mostly in the form of carbon dioxide. The deposits of fossil fuel that are so important to modern society amount to only 8×10^{15} kg, or 0.05% of the total supply.

The sedimentary exchange of carbon between ocean and land is practically a closed loop. The circulation of carbon through the atmosphere also involves two distinct cycles; one between atmosphere and ocean and one between atmosphere and land. This phase of the cycle is shown in Figure 6-3.

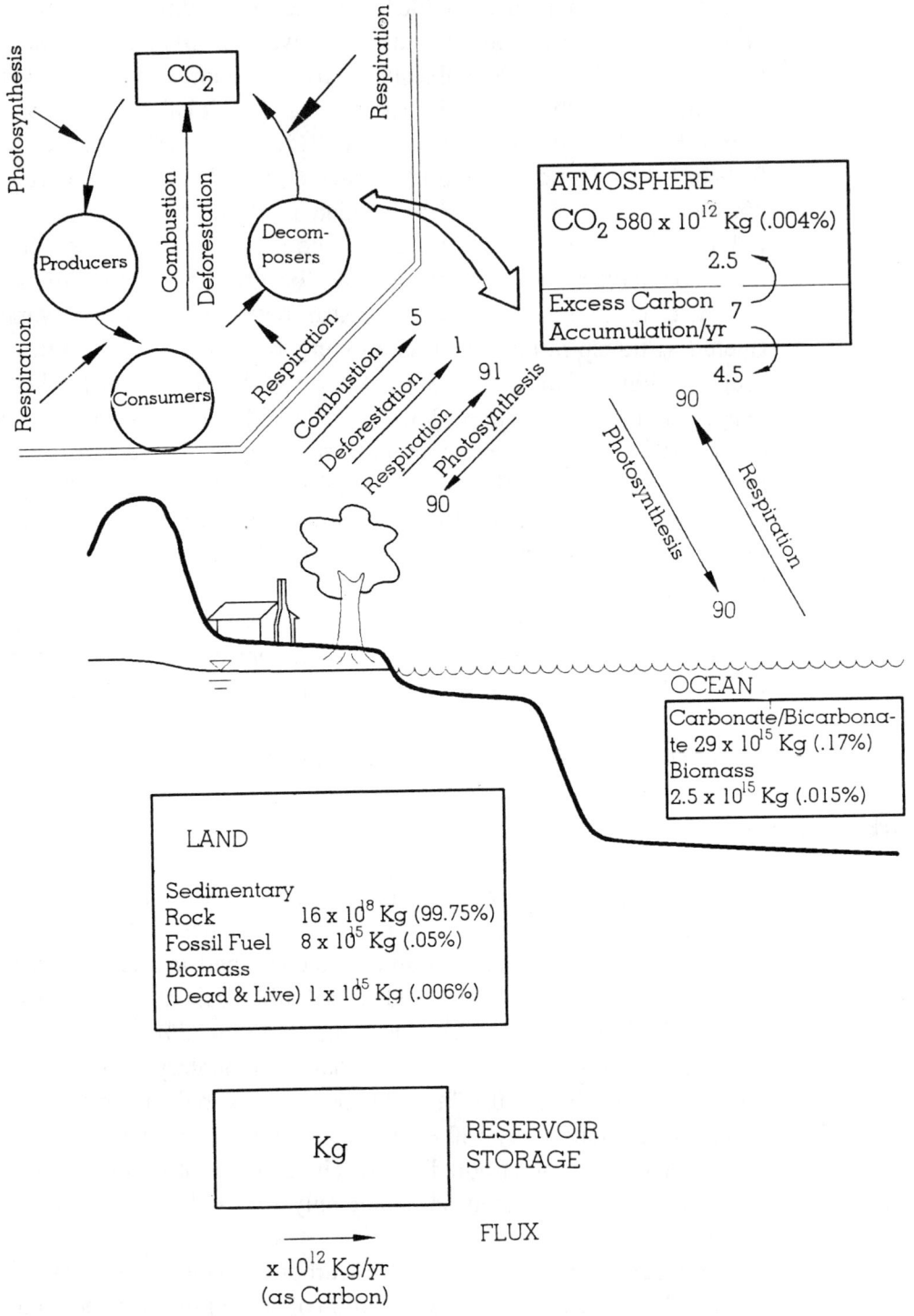

Figure 6-3 The Carbon Cycle

Atmospheric carbon dioxide is fixed by plants by the process of photosynthesis. The carbohydrates and oxygen produced are consumed by animals. Animals respire to release some carbon dioxide. When plants and animals die, they are decomposed by microorganisms. The carbon in their tissues is oxidized to carbon dioxide and returns to the atmosphere, completing the cycle. A similar carbon cycle takes place in the ocean, started by phytoplankton. The photosynthesis by terrestrial vegetation removes about 90×10^{12} kg of carbon dioxide (measured as the carbon content) per year from the atmosphere. The marine biomass removes another 90×10^{12} kg per year of carbon, making up a total of 180×10^{12} kg/yr of carbon removal. In a naturally balanced cycle, the same amount of carbon should be returned by the respiration and decay. The marine ecosystem is in equilibrium. However, on land human activities are changing the environment in many ways. Currently, about 5×10^{12} kg of fossil fuel carbon is released every year. Large areas once covered with forests have been cleared. Rapid deforestation is resulting in a surge of release of 1×10^{12} kg of carbon annually. Respiration and decay releases are slightly in excess of photosynthesis. The total amount of carbon reaching the atmosphere is 187×10^{12} kg per year, i.e,. 7×10^{12} kg per year in excess. About one-third of this excess carbon stays in the atmosphere, while two-thirds is absorbed by the oceans or added to the land mass. Thus, the dynamic equilibrium has been disturbed and the cycle could be considered in a transition state. The balancing process could take decades. This addition has caused an increase in atmospheric carbon dioxide concentration from 290 mg/L to 350 mg/L over past 100 years and it is projected to rise to 375–400 mg/L by the year 2010. The resultant effect is the warming of the earth at a rate of $0.5 - 1\,°C$ per decade. Further, the burning of fossil fuels produces sulfur dioxide and nitrogen oxides as well, which are the main causes of the acid rain. These phenomena are discussed in detail in Chapter 10.

The Nitrogen Cycle

Nitrogen is needed to constitute protein, the building block of all living things. We breath in the ocean of nitrogen, since the earth's atmosphere consists of 78% nitrogen by volume (76% by weight). However, this atmospheric nitrogen is in the form of dinitrogen (N_2) which is an inert gas. In order for it to be utilized, nitrogen gas needs to be converted to a chemical form that can be accepted by plants, a process known as *fixing of nitrogen*. This conversion constitutes a major part of the nitrogen cycle. As a result, the production of food is limited more by the availability of fixed nitrogen than by any other nutrient. Nitrogen has many oxidation levels, i.e., it can form many compounds in combination with oxygen, hydrogen, and other atoms.

A variety of nitrogen species appear in the cycle. Specific terminology is used to describe the chain of reactions involved in the cycle. The formulation of species, the reactions involved, and the energy yield of each step of the cycle are shown in Figure 6-4. Inert nitrogen gas is fixed to ammonia by one of the processes described subsequently. Before it is fixed it is first activated from the molecular form to free atoms of nitrogen. The fixation of nitrogen requires an investment of energy that is subsequently released in other steps of the cycle. The cycle has several loops. In one loop ammonia is precipitated and absorbed by the soil. From there it is taken by the plant roots, which convert it into amino acids and then into proteins. When plants are eaten by animals, new types of proteins are formed. Some proteins are recycled to be again used by plants and animals, thus forming a small loop within the loop. Eventually, proteins return to the soil as plant and animal wastes and dead products. They are decomposed to ammonia again by bacteria and fungi and the loop is completed. In the other loop, ammonia is converted into nitrite ion (NO_2) in the presence of oxygen by a specialized group of microorganisms. Another group of microorganisms oxidize nitrite ions to nitrate ions (NO_3) Nitrates are taken in by plants from the soil and amino acids and proteins are formed. Then the cycle becomes part of the first loop, as shown in Figure 6-4. The preferred source of nitrogen by plants, whether ammonia or nitrate ion, is a subject of investigation. In the soil, numerous kinds of bacteria reduce nitrite and nitrate ions to nitrogen gas, completing the major loop. Specific terminology describes the processes in the cycle, as summarized below.

1. *Activation.* Molecular nitrogen (N_2) is split into two atoms of free nitrogen.
2. *Fixation.* Conversion of inert nitrogen (N_2) to a chemical compound that can be utilized by plants and animals.
3. *Ammonification.* Amino acids and proteins are converted (oxidized) by microorganisms to ammonia.
4. *Nitrification.* Ammonia is converted (oxidized) by microorganisms to nitrite and nitrate ions.
5. *Denitrification.* Nitrite and nitrate ions are converted (reduced) by microorganisms to gaseous nitrogen.

Nitrogen is fixed in four ways:

1. The largest source of fixed nitrogen is by nitrogen-fixing bacteria associated with plants.
2. A smaller but significant amount of fixation occurs through the natural phenomena of cosmic radiation, meteor trails, or lightning which provide high energy for reactions.

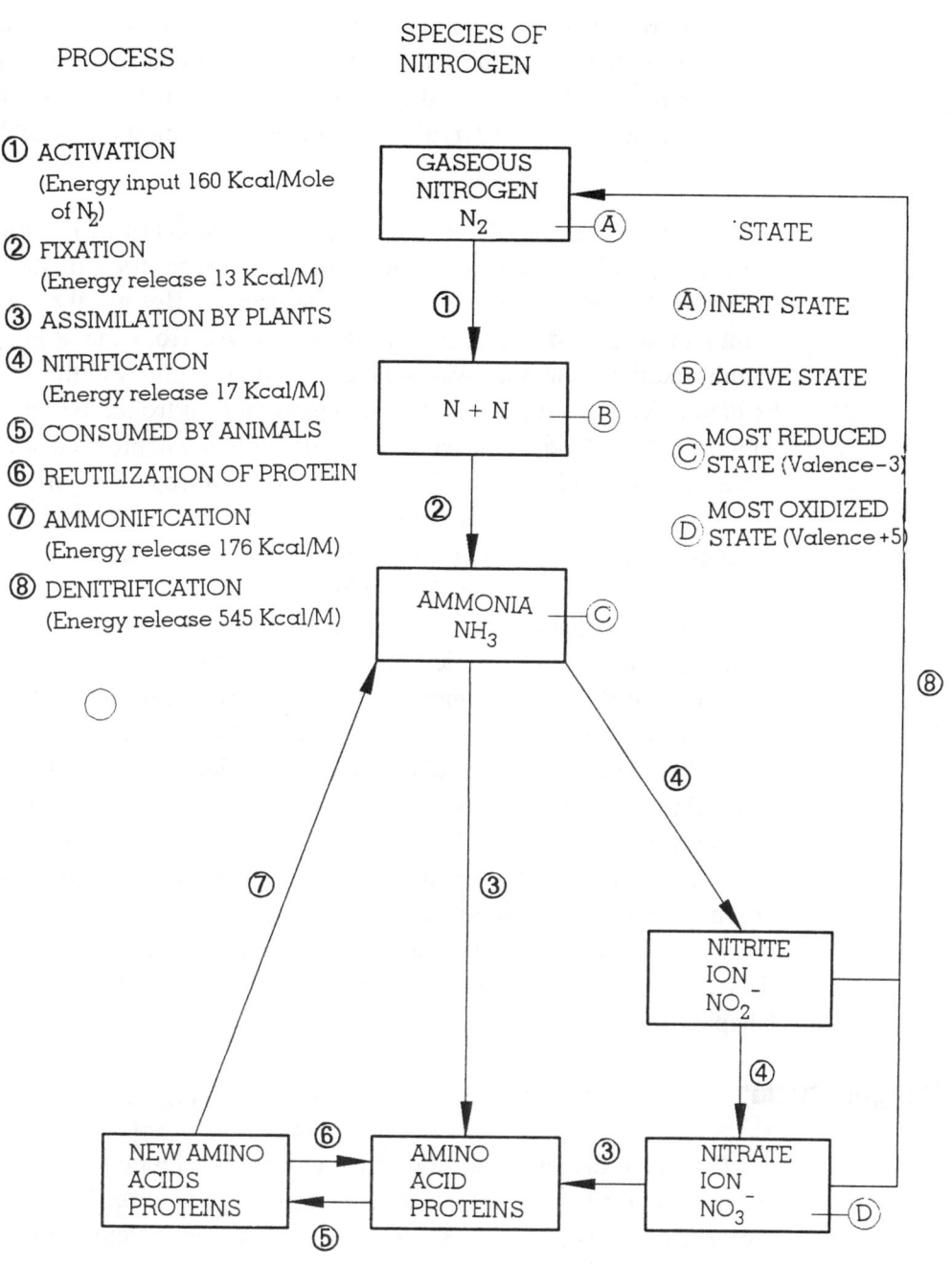

Figure 6-4 Varieties and Processes in the Nitrogen Cycle

3. Volcanic eruptions and weathering of igneous rocks add to the fixed nitrogen by a small amount.
4. The production of fertilizer is the human-initiated effort toward nitrogen fixing by combining hundreds of degrees of temperature and thousands of pounds of force within a synthetic ammonia reactor. This industrial fixation by the fertilizer industry, currently about 73×10^9 kg/yr, almost matches the natural biological fixation of 82×10^9 kg/yr on land.

The quantities involved in the cycle are indicated in Figure 6-5 and represent rough estimates, since there are uncertainties in the amounts involved in the various processes described. Before the large-scale manufacture of fertilizers, the nitrogen removed from the atmosphere by natural fixation processes was balanced by that returned to the atmosphere by denitrification. At present, the total amount of nitrogen being fixed is at a rate of 189×10^9 kg/yr, whereas the amount being denitrified and returned to the atmosphere is 181×10^9 kg/yr; that is, the nitrogen is building up at a rate of some 8×10^9 kg/yr in the soil, rivers, lakes, and oceans. The consequences of this over an extended period are not known. According to *Scientific American* (1974) "We do know that excessive runoff of nitrogen compounds in streams and rivers can result in blooms of algae and intensified biological activity that deplete the available oxygen and destroy fish and other oxygen-dependent organisms." Lake eutrophication is a direct result.

In addition to the main cycle, a number of other reactions occur in the atmosphere that involve the subcycling of the oxides of nitrogen, nitric oxide, NO and nitrogen dioxide, NO_2. Although, the concentration of these oxides is low on the global basis (3×10^{-3} mg/L), localized high concentrations (of 1–2 mg/L) occur due to burning of fossil fuels in factories and automobile engines. When combined with atmospheric moisture, these oxides of nitrogen appear as acid rain. When associated with hydrocarbon and sunlight, they contribute to *photochemical smog*. These problems are discussed in Chapter 10.

The Oxygen Cycle

Oxygen is the only truly abundant element. It is found in the universe, the earth, the earth's crust, the ocean, the atmosphere, and the biosphere. In the atmosphere, oxygen is the second largest constituent, after nitrogen, accounting for 21% by volume (or 23% by weight) of the air, mainly in the form of dioxygen gas (O_2). Most of the oxygen has been produced by the process of photosynthesis described earlier in Chapter 5. A small amount of oxygen is produced when a water molecule is broken by the action of ultraviolet rays in sunlight, known as the photodissociation.

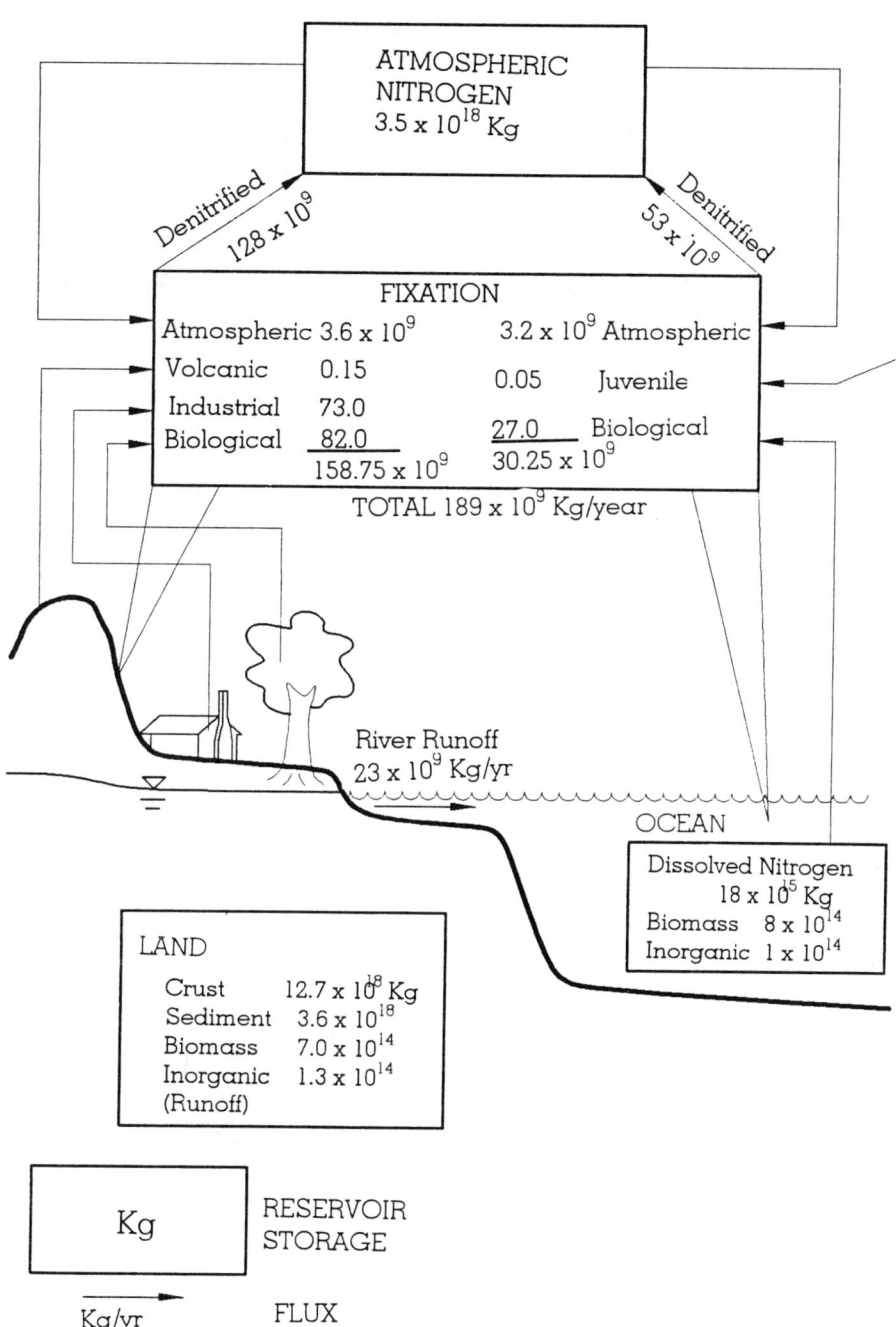

Figure 6-5 Quantities in the Nitrogen Cycle

Photosynthesis and respiration are cyclic as shown in the carbon cycle, and would balance out with each other. Thus, the oxygen pool of 10×10^{17} kg in the atmosphere represents the portion that was not utilized because some photosynthesis products remained unoxidized forming the reserves of coal, oil, and gas.

In contrast to inert nitrogen gas, the atmospheric oxygen is highly reactive and combines with other elements such as hydrogen to form water, carbon and carbohydrates to form carbon dioxide, nitrogen to form nitrate, sulfur to for sulfur dioxide, and iron to form oxides of iron. All of these reactions are parts of the geochemical cycles of the respective elements. Thus, the oxygen cycle merges with other cycles and has a controlling influence on them.

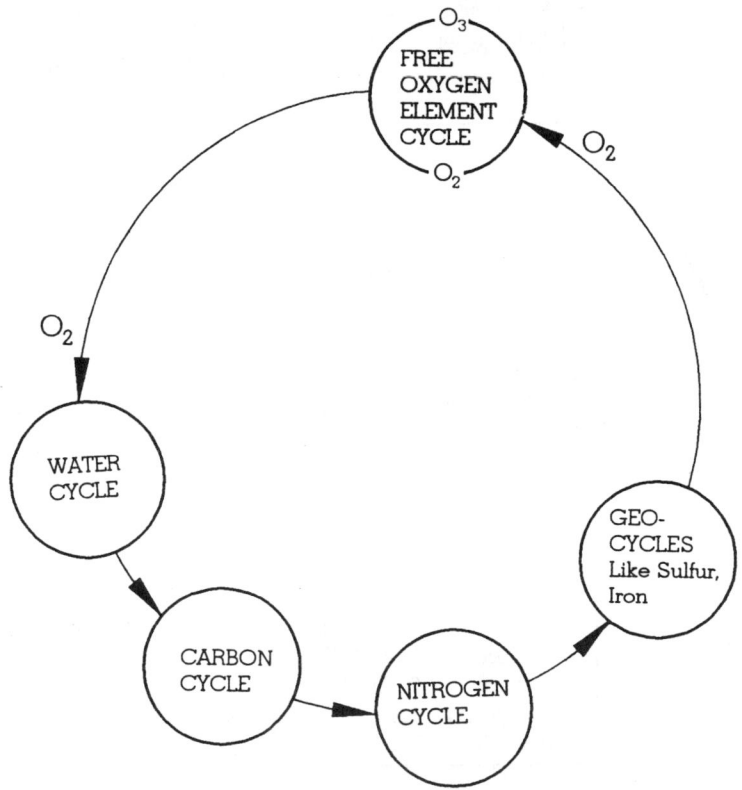

Figure 6-6 The Sphere of Oxygen Cycles

Figure 6-6 shows the sphere of the oxygen cycles. The cycling of oxygen is quite complex because of its association with a number of other cycles. The cycle for the free oxygen element itself comprises the continuous back-and-forth conversion between triatomic oxygen and dioxygen molecules in the upper atmosphere (the stratosphere). The ultraviolet light responsible for splitting water molecules into oxygen also breaks the bond of some dioxygen molecules to form two free oxygen atoms. These atoms join with the molecular dioxygen to form triatomic oxygen, called ozone, or O_3. The process is known as *photolysis*. Ozone has the property of absorbing ultraviolet light and is very unstable. Some ozone converts back to dioxygen after exposure to ultraviolet rays by the process of photodissociation. This pattern repeats and the relative amount of dioxygen and ozone in the stratosphere remains in balance, although there are daily variations from the mean by a few percent. The disturbing of the ozone balance by man-made mechanisms is a cause of serious concern at present, as discussed later in Chapter 10.

Stresses on an Ecosystem

An early section of Chapter 5 (Principles Pertaining to the Abiotic Component) stated that the success of an organism depends upon a number of factors, such as temperature, pH level, salinity, water, and nutrients, and another section (Role of Abiotic Component in an Ecosystem) discussed these limiting factors. With respect to each of these factors, there is a range of magnitude that is tolerated by a given organism. Figure 6-7 shows a plot of the species population vs. environmental factors. The lowest and highest ends on the environmental factor scale mark the limits of tolerance. The peak is the point of maximum growth. A band on two sides of the peak to the points of inflection is the optimal range. Beyond the optimal range are the zones of increasing stresses in the direction of the limits of tolerance. Any one factor being outside the optimal range will cause the stress and limit the growth of organisms. Outside the limits of tolerance, with respect to any factor, the stress conditions are too severe for the survival of organisms. The shape, magnitude and range of the curve in Figure 6-7 varies for different species.

If all factors are within the optimum range, the population of a species will continue to grow as shown on the left side in the beginning part of Figure 6-8. The optimum growth of a species under all favorable conditions is referred to as the *biotic potential*. If this trend continues, the species population will be immense. This does not happen, however, (except for humans who can manipulate the environment to their advantage) because all factors are seldom favorable for an extended period of time.

Abiotic factors outside the optimal range (such as temperature, pH, salinity, water, food shortage, and adverse climate) and biotic factors (such as predators, parasites, diseases, and lack of habitat) are stresses that act to reduce population growth. The combined effect of all the factors that inhibit the growth of a species is known as the *environmental resistance*.

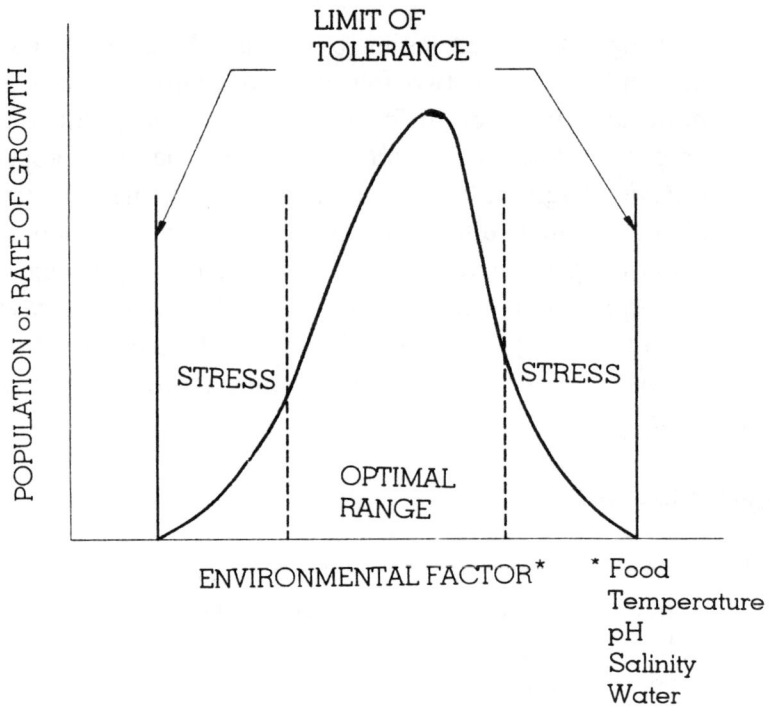

Figure 6-7 Range of Species Population Growth

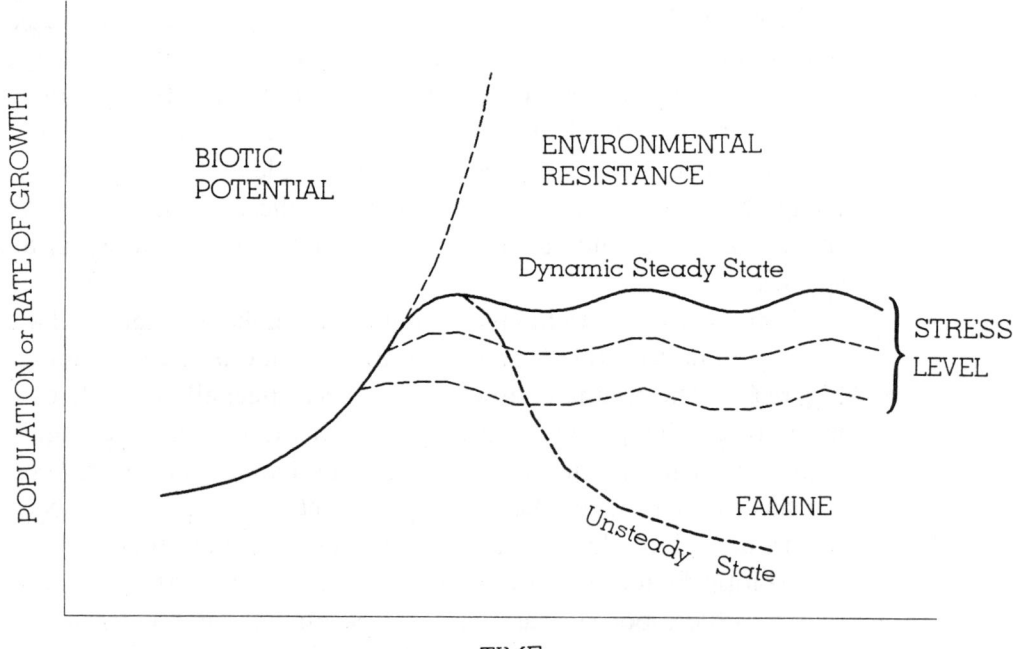

Figure 6-8 Dynamic Balance of an Ecosystem

The population level of a species is a direct result of the dynamic balance between the biotic potential and environmental resistance. The size at which the population is stabilized depends on the level of stress exercised through the environmental resistance. A steady-state system maintains a flat curve, with only slight fluctuations with time above and below the carrying capacity of the ecosystem, which is the saturation level of a species population that the ecosystem is able to support indefinitely under the existing set of conditions. The fluctuations occur because the factors of environmental resistance are density dependent. For example, when the rabbit density (number of rabbits per unit area) in an ecosystem is low, they have ample food and plenty of hiding places. The environmental resistance is low and the rabbit population rises. When their density becomes high, food gets scarce and shelters are limited. The rabbits become easy prey fox and the population declines. As the parallel lines on Figure 6-8 indicate, the balanced systems are not stable at one level all the time. The environmental factors such as climate, drought, floods, fire, and frost, change every year and so do the stress conditions. Depending on the severity of stress changes, the ecosystem may stabilize once again at a different level or may suffer long-term damage as in the case of a famine.

Within the stress levels, beyond the optimal range, to the limits of tolerance shown in Figure 6-7, an ecosystem is usually able to regenerate itself, although it might suffer substantial damage. However, the severe stress imbalances beyond the limits of tolerance may lead to the extinction of a species. Often the stresses that act on certain species within an ecosystem do not equally affect the entire ecosystem.

Two attributes that provide stability to an ecosystem:

1. *Biological diversity,* or *biodiversity* (covered in the following section); and
2. Keystone species.

Many kinds of species living together in a complex ecosystem can counteract stresses more effectively than can fewer species in a simpler system. A single or monoculture ecosystem is vulnerable to extinction as a result of a certain disease or from the impact of an acute stress of any other kind that may affect the entire system. In a diversified system, the entire ecosystem is not likely to be destroyed because a varied gene pool will contain more options with which to respond to changing conditions, and all species have less likelihood of becoming extinct. Each system contains certain key species when they are removed, the food chain is broken and the ecosystem collapses.

An ecosystem acquires certain built-in characteristics to sustain impact of stresses. These characteristics are: homeostasis, inertia, resilience, adaptability, and succession.

Homeostasis

Homeostasis is a tendency of an ecosystem to maintain its equilibrium. When a system is disturbed, the natural regulatory mechanism of the system opposes the disruption to keep the system in existence. For example, in the case of a short drought, the grass, plants, and animals show a protective behavioral response to lack of food to safeguard their own population as well as others.

Inertia

Inertia is a tendency to resist change. Broadly, it is a part of homeostasis. For example, the character of a tropical rain forest is not easily altered when subjected to disturbances unless a tract is completely cleared.

Resilience

Resilience is another property that indicates the ability of a damaged system to restore itself once the disturbed conditions are removed. Nature shows remarkable resilience. A burned grassland recovers quickly unless all its roots are destroyed.

These three properties demonstrate the resistant nature of the ecosystem. The next properties relate to the process of accommodation under stresses of long duration.

Adaptability

Consider that a species is introduced into a new ecosystem. Either that species will survive and flourish, or will suffer and die out. In the former case, the species has to:

1. Adapt to cope with the climate of the ecosystem.
2. Adapt to available food and energy sources.
3. Adapt to live with other species; to escape from predators and live on prey.
4. Adapt to being resistant to diseases and parasites.
5. Adapt to reproduce in the existing conditions.

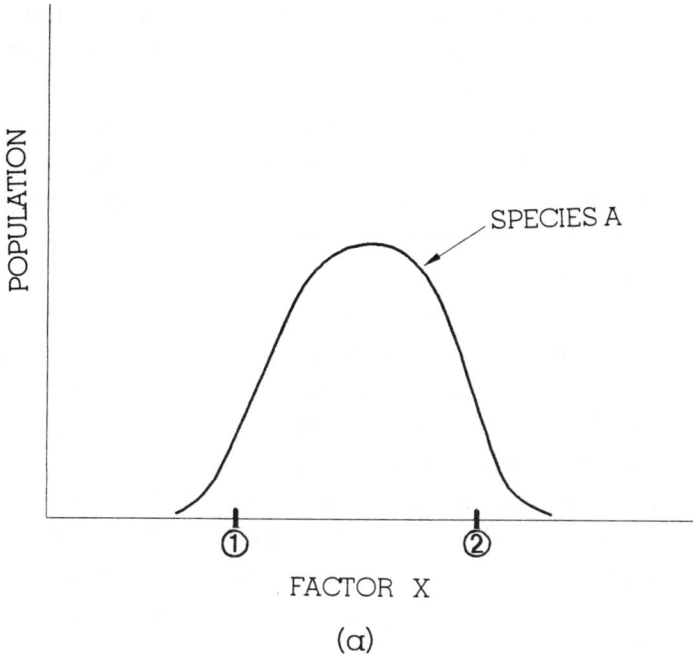

(a)

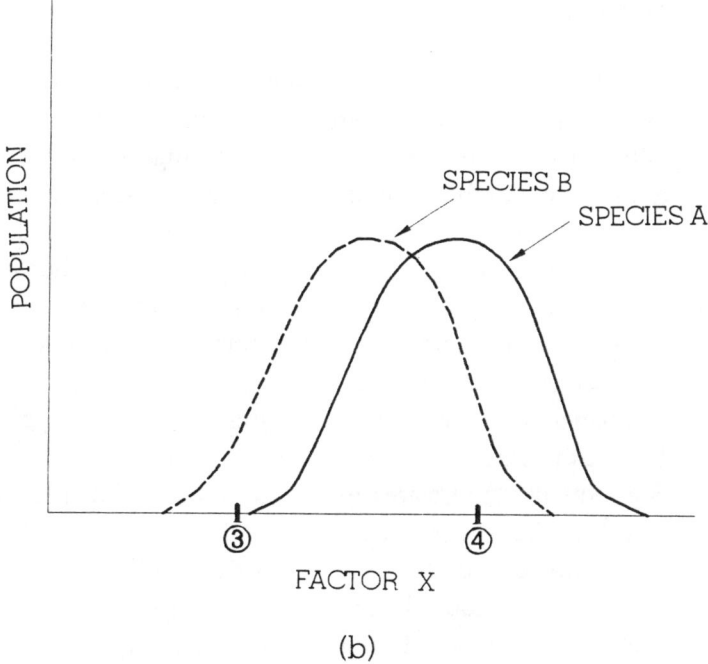

(b)

Figure 6-9 Evolution of a Species

The gene pool of the species makes sufficient variations to accommodate new conditions. This is the process of *adaptation*. If the conditions are outside the limits of tolerance of the gene pool, the species will become extinct.

Through natural selection, the species will gradually become more adapted to new conditions until the members are well adapted to the existing environment. However, the environmental conditions do not remain the same indefinitely. In Figure 6-9a, the range 1–2 represents the existing condition of a certain factor X. Since this is within the tolerance limits of species A, the population grows. Suppose that the condition of factor X is changed to 3–4 as shown in Figure 6-9b. Since this condition falls outside the tolerance limit of species A, the population will die out. Some individuals of the species could survive this stress if their gene pools adapt to new conditions. The changes will continue to provide better adaptation until the population becomes significantly different from the original, and is recognized as a different species B, for which the range 3–4 is within the tolerance limits. This process of evolution is called *speciation*. Indirectly this idea was presented by Charles Darwin in 1859 in his well known "survival of the fittest" concept.

Succession

The above discussion centered around one species within an ecosystem at one time. An ecosystem includes numerous species of different kinds. When a factor of the ecosystem is changed, some species as a whole are disfavored and their population declines or is sometimes eliminated, while other species are favored and register growth. As a result, a change occurs in which one or more species are displaced by other kinds of species. The process is known as *succession*, and is a natural phenomenon through which an ecosystem develops with time. Succession reaches the *climax* when all of the species involved reach a balance with each other in the environment. In an abandoned agricultural field, grasses and plants grow. These are invaded by brush which is then followed by fire cherry, pine, and aspen trees. Subsequently, these are replaced by a forest of maple, oak, and hickory trees. Each of these steps represents a developmental stage in the process of succession that ends in the final stage or climax of a stable forest.

Stresses induced by nature are slow, the degree of imbalance imposed on the ecosystem is not significant, and the changes produced are gradual. Conversely, human induced stresses through pollution, toxic waste and diversion of waterways are sudden and drastic; these can cause ecological upsets. Human activities could overly simplify a complex diverse ecosystem, leading to its collapse. For example, when a forest is cleared for a highway or a parking lot, a great variety of species are wiped out, some of which could be key species in the context of the ecosystem in question.

Biodiversity and Its Significance

Biodiversity is a very broad term used to signify the total variety of life on the earth. It encompasses all species of living beings (plants, animals, and microorganisms) and the ecosystems of which they are parts. Biodiversity is considered at three levels: genetic diversity, species diversity, and ecosystem diversity. Genetic diversity considers genetic variations within an individual species, species diversity refers to the variety of living organisms on the earth, and ecosystem diversity relates to different types of biotic communities and their habitats.

Biologists believe that there are between 5 million and 30 million kinds of different living organisms on the earth. Of these, only 1.4 million have been named by scientists. These consist of 750,000 (54%) insects, 265,000 (19%) plants, 41,000 (3%) vertebrates, and the remaining 344,000 (24%) microorganisms, fungi, algae, and invertebrates. In general, cold and temperate regions hold less variety of species than do tropical regions. It is estimated that the rain forests that make up only 7% of the earth's surface possess more than 50% of the world's total species. With unprecedented destruction of forests at a rate of 80 acres/min, the species are disappearing very fast. Other factors responsible for the loss of species are pollution of soil, air and water; draining and filling of wetlands; and the conversion of wild lands to urban and agricultural use. Because all species have not been cataloged, it is difficult to calculate the exact number of species that are becoming extinct. It is estimated that up to 15% of the earth's species may vanish by the year 2025; a rate of 50,000 species per year.

The following factors outline the significance of biodiversity:

Ethical

The World Charter for Nature of the United Nations (October, 1982) recognizes that humankind is part of nature; that every form of life is unique and warrants respect regardless of its worth to human beings; and that lasting benefits from nature depend upon the maintenance of essential ecological processes and life-support systems, and upon the diversity of life forms.

It is a moral responsibility of humans to maintain the equity and dignity of the entire world community. Human culture must be built upon a profound respect for nature.

Aesthetic

A variety of species adds to richness and beauty of life on the earth. Enjoyment of nature offers a relief to humans from stressful urban life. Once a species is extinct, it is lost forever. Future generations will pay a high price for any irresponsible acts of the current generation.

Sustainability

Other species are essential for human survival. Humans depend on other species for all of their food needs. Medicines and products such as timber, game meat, fish, and ivory are biological resources. Approximately one-fourth of all prescribed medicines in the U.S. are derived from plants. Worldwide, more than 75% of people depend upon medicines extracted from plants. Pharmaceutical companies invest heavily in research on plants. Once a promising substance is found, chemists try to synthesize it.

Intricate functioning of other species provide essential aspects such as photosynthesis, distribution of water on the earth, purification of water, cycling of nutrients, regulation of climate, and breakdown of pollutants.

Existence

A single crop or species may become extinct from a specific disease. Diversity is a check against mass extinction. Moreover, from a pool of diverse genes, new varieties might evolve that will resist the disease and adapt to changing conditions. It is a wholesome world in which the well-being of one part depends upon the well-being and healthy state of the other parts. Diversity adds to the strength of the whole, and is a raw material from which a new life form springs.

Ecological

In an ecosystem, many species are intricately connected. Humans are subject to the same ecological principles as all other species; they are part of the same nutrients cycles and energy supply. Destroying one species leads to extinction of many others and even collapse of the ecosystem with destruction of a key species.

Causes of Biodiversity Loss

The loss of biodiversity can be in any of these forms:

- Reduction of genetic variations within a species;
- Extinction of a species; and
- Loss of unique habitat or ecosystem.

The threat to biodiversity arises from pollution and a shift in land uses. The following are the major contributory factors.

Population Growth

The world's population has more than tripled in the 20th century. More space is needed for living and agriculture. There is an increasing demand on timber for everything from housing and furniture to paper. Wood is still a major source of energy in developing countries. This is causing depletion of forests, and along with it, the loss of species.

Increased Resource Consumption

Increased population combined with a throwaway lifestyle have placed a higher demand on resources. These trends produce more wastes which are in turn responsible for pollution.

Ignorance About the Ecosystem

Knowledge about the world's life forms is lacking, with only 1.4 million species out of 5–30 million named so far. Knowledge about the structure and function of ecosystems is also scant. Understanding of the values of biological diversity is limited. A species might be lost because we did not know it existed at a development site. Many species known and used beneficially by native people may die out because of a change in lifestyle.

Government Policies

Government policies directed to encourage sectors such as agriculture or forestry result in the destruction of biodiversity.

Trading Systems

Countries rely on agricultural commodities for export income. This puts pressure on specialization. For example, developing countries concentrate on certain crops such as coffee, cocoa, and bananas. As the number of crop species decreases, the supporting species, such as insects that pollinate the crops, decline as well.

Inequitable Resources Distribution

The poorer countries where natural resources exist often do not control the resources that are exploited by the richer countries with advanced technology and financial support. For example, large U.S. rubber companies maintain and control plantations in Africa.

Wetland Encroachment

Wetlands (areas that lie between land and water) are nursery grounds for fish and shellfish and habitats for waterfowl and valuable wildlife. Wetlands support a prevalence of vegetation. Wetlands have many other values; they moderate floods, buffer the shore against storm damage, assimilate wastes, and provide hardwoods and medicinal herbs. Wetlands will be discussed later in Chapter 11.

Deforestation

Forests have declined significantly throughout the world. Deforestation continues at an alarming pace in developing countries. In addition to the loss of ecology, this has lead to many other environmental problems including the greenhouse effect. This is discussed further in Chapter 11.

The causes described do not operate in isolation but tend to act with one another and contribute to a widespread loss of habitat.

Strategy to Conserve Biodiversity

In general six types of action can be taken at international, national, and local levels to conserve biodiversity:

- Policy changes;
- Integrated land-use management;
- Species protection;
- Habitat protection;
- Ex situ (off-site) conservation; and
- Ppollution control.

Each approach depends on the others for its success.

Policy Changes

Since government policies are one of the root causes for biodiversity depletion, policy amendments are a necessary first step. The Global

Biodiversity Strategy, the result of the combined efforts of the World Resources Institute, the World Conservation Union, and the United Nations Environmental Program, identifies four major areas of policy reform as follows:

- Reforming the existing policies that contribute to the loss of biodiversity, such as wasteful logging practices, and the excessive use of pesticides, fertilizers, and water.
- Adopting new policies and accounting methods that promote conservation. This includes modification of an accounting system to reflect the loss that results when a biological resource is degraded, adoption of fisheries policies based on sustainable harvest, and setting aside marine habitats in coastal developments.
- Reducing demand for biological resources through recycling and conservation.
- Integrating biodiversity conservation into national planning. This will help countries to set biodiversity priorities.

Integrated Land Use

The conservation effort needs to be woven together with agriculture, forestry, fisheries, transport, national defense, and other developmental efforts. Optimum and lasting benefits can be realized by bringing the processes of conservation and development together. In preparing the conservation strategy, government agencies, nongovernmental organizations, private interests, and the community at large should be involved.

Species Protection

Species are the building blocks of an ecosystem. Species and their genetic pool benefit all human beings. Their protection is the most direct approach to biological conservation. Many government and nongovernment organizations are playing active roles in protecting particular species or groups from destructive exploitation.

Habitat Protection

Species are best conserved as part of their community within a large ecosystem where they can continue to adapt to changing conditions. Thus, the most effective mechanism for the conservation of biodiversity is habitat protection. If a habitat is preserved, the animals and plants within it will protect themselves. Certain habitats are protected by governments by the establishment of national parks and reserves that extend into forests, reefs, and wetlands.

Ex Situ Conservation

Measures to promote ex situ conservation include botanical gardens, game farms, zoos, agricultural research centers, and gene banks. These facilities can be critical components of a comprehensive conservation program. They supplement in situ measures such as habitat protection.

Pollution Control

Biological diversity is threatened by various forms of chemical and organic pollution. This is critical because the future projected environmental changes extend beyond the range of variations to which living organisms have been exposed to over the past millions of years. The measures to curb the contamination of the biosphere are expensive and very difficult to implement.

Ecosystem Diversity

Three major habitats exist in the biosphere; freshwater, marine, and terrestrial. Each of these supports many kinds of ecosystems as listed below.

Freshwater Habitat Supports
- Lakes and ponds
- Marshes, swamps, and bogs
- Streams and rivers

Marine Habitat Supports
- Estuaries
- Ocean systems

Terrestrial habitat supports
- Forests
- Grasslands
- Deserts

The organisms living in water fall in the categories of producers, macro and micro consumers, and decomposers, as in terrestrial habitats. However, based on their life forms, they are further classified into the following units:

- *Rooted plants* that generally grow in shallow waters.

- *Plankton* or floating organisms. Floating plants (producers) such as algae, are known as *phytoplankton* and floating animals (consumers) are known as *zooplankton*.
- *Benthic organisms*, or, *benthos* are attached to or live in the bottom sediment, such as a clam or snail.
- *Nekton* are swimming organisms such as fish and insects.
- *Neuston* are organisms that rest or swim on the water surface.
- *Saprotrophic* are aquatic bacteria such as certain fungi and flagellates.

Freshwater Ecosystems

Freshwater habitats may be conveniently considered in two classes: *standing-water habitats* and *running-water habitats*. There is no sharp distinction between two classes. Lakes, which are defined as basins filled with water with no immediate means of flowing to the sea, have relatively still waters. Ponds are small lakes in which rooted plants on the top layer reach to the bottom. In lakes, oxygen, light (the sunlight's intensity), and temperature vary with the depth and also undergo seasonal changes every year, forming distinct layers or stratifications. Sunlight is the energy source to lakes that can penetrate to a limited depth only due to turbidity of water. This has a profound influence on the life forms in a lake. A lake can be divided into three zones based on distribution of the sunlight, as shown in Figure 6-10.

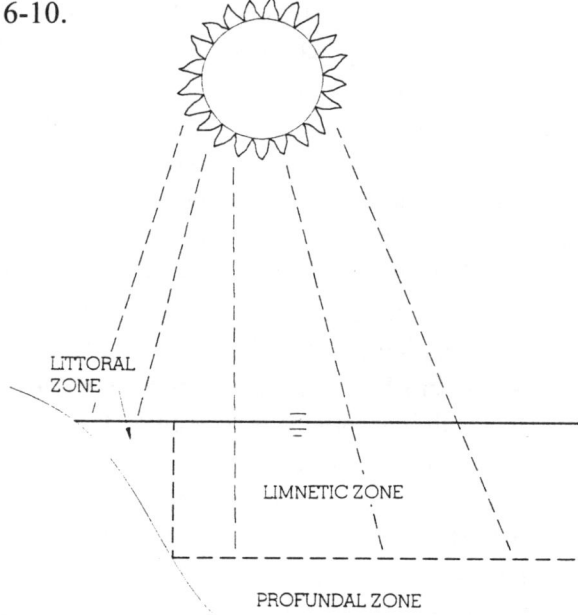

Figure 6-10 Energy Penetration Zones in Lakes

Where the water is shallow, such as around the edges of lakes and throughout a pond, the sunlight can reach to the bottom. This is the *littoral zone*, in which aquatic life occupied by rooted plants is richest. Beyond this, in the open water of the lake, the *limnetic zone* is a layer to the depth of light penetration, which is active in photosynthesis. This zone is inhabited by plankton (floating organisms), nekton (swimming organisms and fish) and neuston (organisms resting on the lake surface). The deep-water area, below the depth of light penetration, is the *profundal zone*. Life in this zone depends on the oxygen level and temperature, and on the raining of organic material as an energy source, from the layer above. The profundal zone supports life, such as plankton, nekton, and benthic organisms living on bottom sediments. The organic matter sinks to the bottom and is washed in by inflowing water to make up the bottom sediment on which the *benthos* organisms live. Prominent among these are certain forms of worms, shellfish, clams, snails, and anaerobic bacteria.

Marshes and swamps are wetlands developed along the shallow margins of lakes and ponds in low, poorly drained lands and on the flood plains of large river systems. In marshes the grass life form is dominant, whereas swamps are wooded wetlands. Both may range from shallow to deep water and both support diverse and abundant animal life. Ducks and a variety of birds add richness to swamps and marshes. Bogs develop in humid regions where drainage is blocked and are most abundant in the cold northern regions. Due to drainage congestion and low temperatures, accumulation of peat, growth of cushion-like vegetation, and formation of semifloating mat of plant growth take place in bogs. Generally, acidic conditions prevail. The variety of plants and species are restricted, but the quantity is abundant. Marshes, swamps, and bogs are subject to neglect by humans and frequently considered places that can be conveniently filled in.

Running water is an outstanding feature of streams and rivers. The character of a stream and the presence of organisms are influenced by the velocity of water flow. The dividing line between slow streams and fast streams is an average velocity of flow of 2 ft/sec. At this velocity, all particles up to a 0.2 in. diameter will be washed off, leaving behind a stony bottom. The fast streams have two segments: the *rapid,* or, *riffle zone* and the *pool zone*. In shallow depths, the velocity of current is above 2 ft/sec and the bottom is cleared of loose silt and other material, providing a firm substrate. The organisms that can cling to the firm substrate against the currents are present in this zone. A different type of algae and water moss and strong swimmers such as darters fish occupy this zone. Many organisms live on the underside of rubble and gravel in the rapid zone where they are protected from the current. The rapid zone insects are the nymphs of mayflies, caddisflies, true flies, stoneflies, and alderflies. Below the rapid zone is a pool zone of deeper water. The reduced velocity allows silt and other loose materials to settle, thus providing a soft bottom. In the pool zone, as in slow streams and rivers, the composition of the aquatic community approaches that of lakes and ponds.

Marine Ecosystems

More than 70% of the planet is covered by sea, comprising a single connected system to which different names have been given. These include the Pacific, Atlantic, and Indian Oceans with a number of minor seas. The Pacific Ocean is the largest, covering an area of 180 million km². Just like lakes, seas exhibit zonation as shown in Figure 6-11.

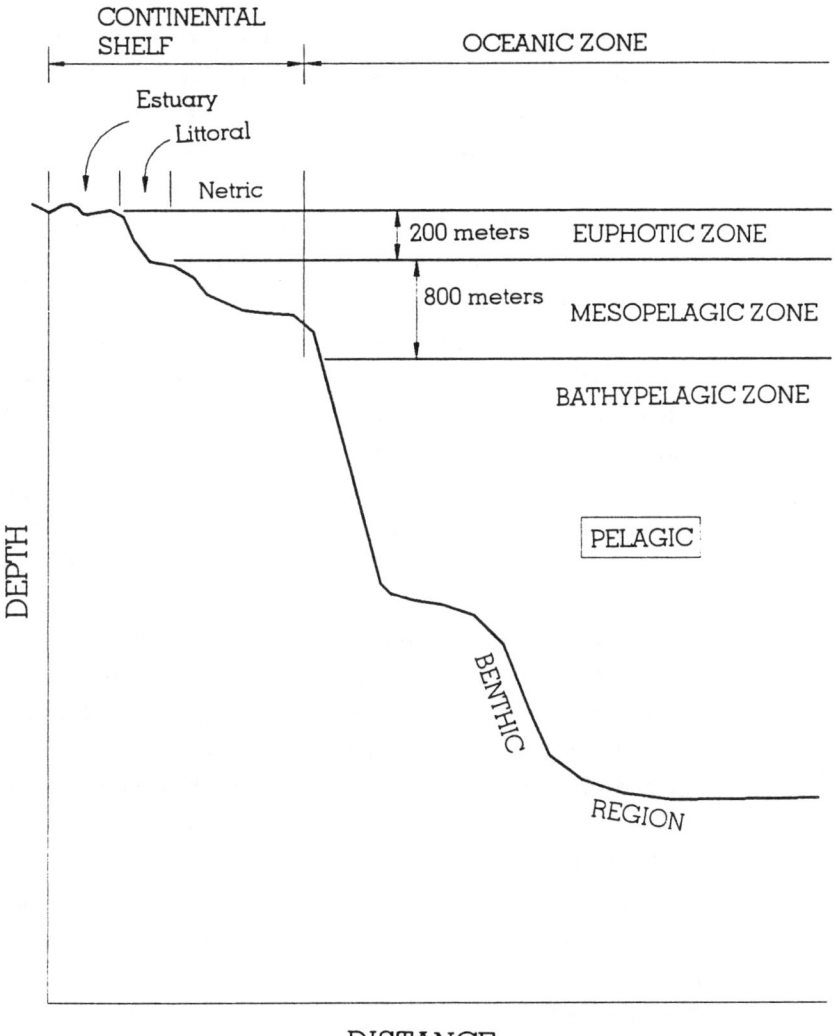

Figure 6-11 Zonation of the Sea

The sea is divided into two portions: the *pelagic*, denoting the whole body of water and the *benthic*, representing the bottom region. The body of water is further subdivided horizontally and vertically. The offshore distance extending up to the point where the bottom suddenly drops steeply is the *continental shelf*. On the shelf, the portion where the river meets the ocean is the *estuary*. The distance measured horizontally between the edges of high and low tides is the *littoral zone*, and the shallow depth of water over the shelf is the neritic zone. Beyond the continental shelf is the *oceanic zone*. Vertically, from the surface to about 200 m where the sunlight can penetrate, is the *euphotic zone*. In this layer, the sharp gradients or stratification of light intensity, temperature, and salinity exist. From 200 to 1000 m, where very little light penetrates and temperature is more even, is the *mesopelagic zone*. This layer contains minimum oxygen and maximum nitrate and phosphate. Below 1000 m to a depth of about 6000 m, where darkness is complete, temperature is low, and pressure is great, is the *bathypelagic zone*.

The marine environment is distinctly different from the freshwater world as a result salinity, waves, tides, currents, and depth. The salinity of the open sea averages about 35×10^3 ppm (or mg/L) which is mainly contributed by sodium and chlorine that constitute more than 85% of the salt in the sea. There is a marked change in salinity with depth that produces density layers in the sea. Salinity and depth are the major barriers to free movement of marine organisms. Seawater is in continuous circulation due to wind-driven surface currents, temperature, and salinity-induced deep currents. As a result of this circulation, the oxygen depletion that occurs in freshwater lakes is rare in seawater. The sea is dominated by waves stirred by wind. As a wave moves, the particles of water remain largely in the same place, following an elliptical orbit. As the waves break on shore, they dissipate energy against the coast, eroding and building beaches. Tides are produced by the pull of the moon and the sun. Tides occur twice daily in a locality with a periodicity of 12.5 hr. being about 50 min later each successive day. Every two weeks, the *spring tide* (with the highest tidal range) occurs due to the combined effects of the sun and moon. Midway into two weeks, the *neap tide* (with the smallest tidal range) occurs when the effects of the sun and the moon tend to cancel each other. In a marine ecosystem, the concentration of nutrients such as nitrates and phosphates is low and, as such, constitutes a limiting factor.

An estuary is a place where freshwater joins seawater. It is a semi-enclosed coastal body of water wherein seawater is diluted with water coming from land. It is strongly affected by tidal action of the sea. Salinity in an estuary varies through the year, being highest during the summer when less freshwater flows into the estuary. Organisms that inhabit the estuary face both the problems of maintaining the position under tidal action and of adjustment to changing salinity conditions. Most estuarine organisms are benthic and are securely attached to the bottom, such as in oyster beds and oyster reefs. The organisms of the estuary are essentially of marine type that are able to sustain full seawater salinity; freshwater organisms are not found there. Species decline when the salinity becomes low. Plankton population depends on the currents; seaward movement transports plankton out to the sea. Movable inhabitants consist mainly of fish and crustaceans. The estuary provides a shelter for larvae and young species protecting them from predators that cannot withstand the lower salinity. Estuarine species move to seawater with higher saline content as they mature. Around the estuary on the alluvial plains exists the tidal marshes where salt-tolerant vegetation grow. Tidal marshes serve as feeding grounds because organic matter is carried into estuaries by tides. An assortment of animal life inhabits a tidal marsh, including crabs, snails, mosquitoes, flies, and birds. An estuarine ecosystem supports more biological productivity than the sea or the freshwater system because both the freshwater cycle and tidal cycle provide energy subsidies.

Because of nutrient deficiency and light penetration limited to 100 meters, biological productivity in the open sea is far less than in the terrestrial system, although the former occupies more than two-thirds of the earth's surface. The dominant plant life is phytoplankton. The main consumers in the sea are zooplankton. At higher trophic levels are small fish, squid, and the large carnivores of the sea such as whales and seals.

Terrestrial Ecosystems

The terrestrial ecosystem has very high biological productivity or biomass as compared to the aquatic system. It is a highly developed system featuring the most complex and specialized forms of all organisms. The following features are important for comparison.

- Temperature variations are more pronounced on land than in water.
- Moisture is a limiting factor for organism growth on land.
- Land offers a firm support to plants and animals resulting in their strong skeletons.
- Soil is a source of highly variable nutrients.
- A complete mixing of air provides a uniform content of oxygen and carbon dioxide.

To divide the terrestrial system, many classification schemes in the past have attempted to combine plant and animal distributions into a single scheme. However, this failed because the distribution of plants and animals in regions did not coincide and too many combination units resulted. In another approach, only plants have been considered as the biotic units because of their dominance as food producers. The terrestrial system has been divided into the blocks of similar plant formations or vegetation. The terrestrial community in each block or biotic unit is called a *biome*. This approach works fairly well since animal life depends on the plant base. Each biome represents a distinct spectrum of life forms.

In terrestrial ecosystem classification, a biome can be recognized as a large land community in which a distinct climatic pattern and animal spectrum exist and the life form of vegetation is uniform throughout in a fully developed or climax stage. Three major biomes exist in the world: forests, grasslands, and deserts; subdivisions of these may vary from scientist to scientist. A close correlation between the biomes and the climatic factors such as temperature and precipitation has been observed, as shown by the triangle in Figure 6-12. The figure provides a broad classification of the biomes of the world in relation to precipitation and temperature.

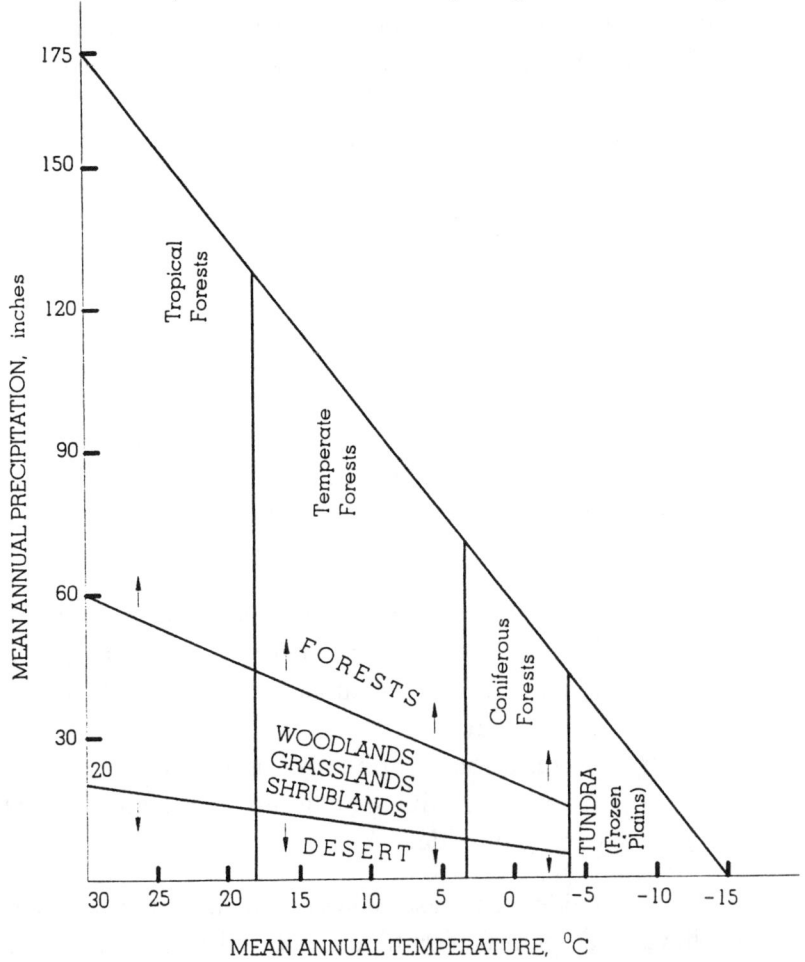

Figure 6-12 Triangle of Biomes Showing Broad Distribution of World's Plant Formations

In North America, if a transect is run east to west, it will pass through many ecosystems based on the moisture gradient resulting from variation of precipitation. Beginning with moist mesophytic forest of the Appalachians, the biome will successively change to oak-hickory forest, to oak woodland, tall-grass prairie, short-grass plains, and desert grassland, ending in desert shrubland. The north-south transect reflects ecosystems that are based on the temperature gradient. It passes through the arctic tundra, boreal coniferous forest, mixed northern hardwood forest, temperate deciduous forest, subtropical forest of Florida, and tropical forest of Mexico. There are no sharp dividing lines between the ecosystems. One system blends into another with common species in the transitional region.

Of all the biomes, the forest ecosystem is most widespread and diverse. Forests are highly stratified from top canopy to shaded ground layer. Diversity of animal life is associated with the stratification. The greatest concentration of life forms exists on or near the ground layer.

Coniferous, deciduous, and tropical rain forest are three dominant types of forests on the earth. Coniferous forests are confined to the northern hemisphere where the winters are long and cold. Pine, hemlock, cedar, spruce, and fir are characteristic of coniferous forest. There are three strata in coniferous forests. The strata below the canopy is poorly developed and the ground layer consists of ferns and mosses. The deciduous forest grows in mild climate and moderate precipitation. This forest type is found in eastern Asia, western Europe, and eastern and central North America. Deciduous forest is dominated by broad-leaved trees. Maple, oak-hickory, oak-chestnut; and oak are characteristic of this type. Four stratifications exist consisting of two canopy layers, a shrub layer, and a field layer of herbs, ferns, and mosses.

The tropical rain forest forms a worldwide belt around the equator. A rain forest grows in a land where the precipitation is heavy, the humidity is high, the annual temperature is about 28°C, and the seasonal climatic changes are minimal. Rain forests occur in three regions:the Amazon and Orinoco basins in South America and Central America, the Congo, Niger, and Zambezi basins in central and western Africa, and the Indo-Malaya-Borneo-New Guinea belt in Asia. Tropical rain forests have the greatest diversity. The stratification of life is most pronounced in tropical rain forests; six distinct feeding groups are recognized between the upper layer above the canopy of trees to the ground-level animals. Tree species are broad-leaved and evergreen and number in the thousands. Where rainfall diminishes, semi-rain forests or monsoon forests grow that differ from rain forests in that they shed their leaves in the dry season. The transitional variations produce other forest biomes that include the temperate evergreen coniferous forest, temperate rain forest and broad-leaved evergreen subtropical forest.

Grasslands occur where precipitation is too low to support the growth of a forest but is more than what can lead to the formation of a desert. All grassland regions have high rates of evaporation, are subjected to periodic droughts, lie in rolling to flat terrain, and are dominated by grazing and burrowing animals. The grassland biome can be subdivided into tall grasses (5–8ft), mid grasses (2–4ft) and short grasses (0.5–1.5 ft). Grasslands with scattered trees, known as *tropical savannas*, are found in warm regions that have prolonged dry seasons and 40– 60 in. of rainfall per year, in Africa, South America, and Australia.

Deserts have very low rainfall (less than 10 in. annually), very high evaporation (practically all rain evaporates), wide daily temperature range (hot by day, cool by night), low humidity (high solar insolation), and stark topography. Deserts extend to a worldwide belt around the Tropic of Cancer and the Tropic of Capricorn, occupying almost one-seventh of the land surface of the earth. Deserts are deprived of rainfall because of factors including movement of moisture-carrying air masses over the earth, mountain ranges causing a rain shadow on the lee side, and remoteness of places from the source of oceanic moisture. North American deserts, on the basis of temperature, are divided into hot deserts and cold deserts. The Great Basin is a northern cold desert dominated by sagebrush plant life. The deserts of southwest (the Mohave, the Sonoran, and the Chihuahua) are hot deserts dominated by creosote bush. There are three life forms of plants in deserts: the annuals that grow when there is adequate moisture, the succulents that store water, such as cactus, and the desert shrubs that can withstand prolonged dry periods. The plants are scattered thinly with large bare areas in between. When small amounts of rain arrive and plants flourish, the desert is swarmed with insects such as crickets, grasshoppers, ants, bees, wasps, butterflies, moths, and beetles. Desert birds, reptiles and other animals such as rodents, kitfox, jackrabbits, kangaroo rats, and pocket mice thrive in deserts. Large mammals include camels, african antelope and oryx. These animals are either drought evaders who go into estivation (the state of summer dormancy) or are drought resisters who have evolved means of storing water and reducing their water requirements.

In semiarid regions where winters are mild and rainy summers are long and hot, a shrub ecosystem is supported. Shrublands are characterized by woody structures, with dense branching and low height. Shrublands have their own distinct wildlife. They are valuable in their own way as grazing lands, wildlife habitats, and sources of food and pharmaceutical materials. Shrublands and deserts are not properly appreciated by humans and are frequently destroyed through development and poor management.

PROBLEMS

1. The preindustrial concentration of carbon dioxide in the atmosphere (amount in reservoir) was 275 ppm. The atmospheric life (residence) time for carbon dioxide is 110 years on an average. What is the steady-state rate of addition of carbon dioxide? The concentration increased to 360 ppm in 1995. If it is to be maintained at that level, what would be the rate of addition to the atmosphere?

2. From the data given in Problem 1, determine the percent increase in a steady rate of addition from the 1995 level that will double the preindustrial concentration of carbon dioxide on a long-term basis.

3. The present atmospheric concentration (not at a steady state) of CFCs (reservoir amount) is 800×10^{-6} ppm. A typical atmospheric life (residence) time is 139 years. It is estimated that a reduction of 75% from the current emission rate will be required to stabilize concentrations at the present level (for a steady state). What is the percent rate of emission of CFCs? If the rate of emission continues at the present level, what will be the new steady state concentration in the atmosphere?

4. The volume of a wastewater tank (reservoir) is 500,000 ft^3. The rate of waste flow into the tank is 2.5 million gal/day What is the residence (detention) time of waste in the tank? If the wastewater has to stay for a minimum of one day in the tank, what is the peak flow rate, in million gal/day, that the tank could handle?

5. In a certain 220 square mile region where the hydrologic cycle holds well, the following fluxes were observed on a yearly basis: precipitation 24 in., evaporation 8 in., transpiration 7 in., and surface runoff 5 in. Determine the yearly rate of infiltration of water.

6. Draw a hydrologic cycle for the region in Problem 5. Infiltration eventually becomes the groundwater flow. Leave the reservoir pools blank. Infiltration eventually becomes the groundwater flow.

7. A region in which the hydrologic cycle is valid has an area of 100 sq. mi. The observed values of fluxes per year were: precipitation 20 in., runoff 1 billion ft^3, and evaporation and transpiration together amounted to 60% of precipitation. Determine the amount of infiltration. (Hint: Divide runoff by area to convert to linear dimension to be consistent with the other data.).

8. From the values of the reservoir storage volumes given in Figure 6-2 for the hydrologic cycle, if the ocean, land, and atmospheric water constitute the total water resources, determine the following:

a. Percent of total supply in the ocean
b. Percent of freshwater (supply other than in the ocean)
c. Of the freshwater supply, percent in glaciers
d. Of the freshwater, percent of available water (freshwater other than in glaciers)
e. Of the available water, percent of surface water in lakes and rivers (including the quantity in atmosphere).

9. Prepare the sulfur cycle from the following information: The reservoir pools comprise a storage of 4×10^9 kg of sulfur in atmosphere, 1.3×10^{18} kg in ocean, and a larger pool of 8×10^{18} kg in sedimentary rocks on land. The circulation comprises (all quantities in terms of sulfur content in billion kg per year) emissions into atmosphere, from trees: H_2S of 23; from factories: SO_2 of 52; from volcanoes: SO_4 of 30; from rivers and lakes: H_2S of 20; and from ocean (in addition to rivers and lakes emptying into the ocean): SO_4 of 60. In the atmosphere, the hydrogen sulfide (H_2S) converts to sulfur dioxide (SO_2), to sulfur trioxide (SO_3)$_s$ to sulfate (SO_4). The deposition of sulfate from the atmosphere comprises 100 on land and 85 into the ocean. Between land and the ocean, a back and forward circulation of 220 takes place. Use Figure 6-2 as a model.

10. The carbon cycle in Figure 6-3 indicates an excess carbon dioxide buildup of 7×10^{12} kg/yr of which about one third stays in the atmosphere. Assume the lower atmospheric volume of $20 \times 10^{15} m^3$. What will be the yearly increase in atmospheric carbon dioxide concentration in milligrams per liter?

11. If the earth is warming at a rate of 0.5°F for each milligram per liter of CO_2 increase, project the rise of temperature after a decade, using information from Problem 10. If the present concentration level of CO_2 is 29 mg/L, what is the percent increase of concentration of CO_2 in a decade?

12. In a region, the following nitrogen fluxes are taking place at yearly intervals: Fixation on land and atmosphere is 41,580 lbs; of this, the runoff into rivers and lakes is 5100 lbs. and denitrification on land is 36,480 lbs. The fixation in rivers, lakes, and atmosphere is 6600 lbs., and denitrification from streams and lakes is 10,660 lbs. Prepare a simplified nitrogen cycle. Determine if there is any buildup of the nitrogen in the streams and lakes. What possible effects will this have in the region?

13. In the nitrogen cycle, the various processes (such as activation, nitrification, etc.) as indicated in Figure 6-4 comprise a closed loop (cycle). Prepare a cycle of the nitrogen processes (processes shown by blocks); indicate the magnitude of energy involved at each stage of the process.

14. Suppose you want to balance the carbon cycle so that there is no excess buildup into the atmosphere. Discuss various alternatives in terms of physical measures to balance out fluxes due to photosynthesis, deforestation, respiration, and combustion.

STUDY QUESTIONS

1. Discuss the statement, "All essential elements are continuously cycled through the ecosystem." List essential nutrients.

2. Draw a boxed diagram of the carbon cycle. How do humans impact the cycle?

3. Sketch the main features of the hydrologic cycle. Explain each of its elements. How do humans impact the cycle?

4. Describe the nitrogen cycle. What is its significance to agriculture? How do humans impact the cycle?

5. What are the biotic potential and the environmental resistance? How do they work in a dynamic system to determine a balanced ecosystem?

6. What is biodiversity? What is its significance to the ecosystem?

7. What characteristics does an ecosystem possess to resist external stresses?

8. Discuss species succession in light of external stresses on the ecosystem. How does it contribute to the emergence of a climax ecosystem?

9. What are the major types of biomes? Draw the biomes triangle showing the classification of plants of the world.

10. Distinguish between freshwater and marine ecosystems. What are the major types of aquatic ecosystems? Outline their salient features.

11. Define the following:
    a. Biogeochemical cycles
    b. Residence or turnover time
    c. Fixing of nitrogen
    d. Nitrification and denitrification
    e. Photodissociation
    f. Adaptation
    g. Speciation
    h. Keystone species
    i. Biomes
    j. Limnetic and profundal zones
    k. Pelagic and benthic
    l. Littoral, neuritic, and continental shelf
    m. Euphotic, mesopelagic, and bathypelagic

REFERENCES

Agsteribbe, M. A., *A Guide to the Study of Freshwater Ecology*, Prentice Hall, NJ, 1972.

Benarde, M. A., *Our Precarious Habitat*, Revised ed., W. W. Norton and Company, New York, 1973.

Benton, M., *Life on Earth*, Warwick Press, New York, 1986.

Davies, N. D., et al., *A Guide to the Study of Soil Ecology*, Prentice Hall, NJ, 1973.

McNeely, J. A., et al, *Conserving the World's Biological Diversity*, Published by International Union for Conservation of Nature and Natural Resources, World Resources Institute; Conservation International World Wildlife Fund-U.S., and the World Bank, Gland, Switzerland and Washington, D.C., 1990.

Mooney, H. A., and Godron, M., ed., *Disturbance, and Ecosystems: Components of Response*, Springer-Verlag, Berlin, 1983.

Odum, E. P., *Ecology and Our Endangered Life-Support Systems*, 2nd ed., Sinauer Assoc. Publishers, Sunderland, MA, 1993.

Odum, E. P., *Fundamentals of Ecology*, 3rd ed., W. B. Saunders Company, Philadelphia, PA, 1971.

O'Neill, P., *Environmental Chemistry*, George, Allen and Urwin, London, U.K., 1985.

Raven, P. H., Berg, L. R., and Johnson, G. B., *Environment*, Saunders College Publishing, Philadelphia, PA, 1993.

Rich, L. G., *Environmental Systems Engineering*, McGraw-Hill Book Company, New York, 1973.

Ryan, J. C., *Life Support: Conserving Biological Diversity*, Paper 108, Worldwatch Institute, Washington, D.C., 1992.

Scientific American, *Ecology, Evolution and Population Biology*, W. H. Freeman and Company, San Francisco, CA, 1974.

Smith, R. L., *Elements of Ecology and Field Biology*, Harper and Row Publishers, New York, 1977.

Turk, A., Turk, J., Wittes, J. T., and Wittes, R., *Environmental Science*, W. B. Saunders Company, Philadelphia, PA, 1974.

Wilson, E. O., *The Diversity of Life*, W. W. Norton and Company, New York, 1992.

World Resources Institute, Chapter on "Wildlife and Habitat," in World Resources, 1992-93, Oxford University Press, New York, 1992.

Chapter 7
PROCESS ENGINEERING

Environmental Processes

Contaminants are not stationary in the environment. They move through air, water, and soil. This unit provides the theoretical framework as it relates to transport of contaminants, their fate in the environment, and their removal from the waste stream. The study of processes leads to the proper design of treatment units. The processes encompass physical, chemical, and biological phenomena. Some important processes of environmental engineering are listed below.

Physical processes include transport processes and settlement processes. Chemical processes include chemical precipitation, gas transfer, ion exchange, adsorption, membrane separation, and disinfection. Biological processes include activated sludge processes, trickling filter processes, and digestion processes.

A single process or combination of several processes might be involved in dealing with contamination. A *treatment train* is defined as application of one or more of the processes to accomplish a particular treatment objective; for example, the removal of discrete solids may be an objective. A system may consist of one or more treatment trains. A water treatment system consists of several trains each designed to meet a specific treatment objective.

Transport Processes

When a crystal of a colored substance is placed in a bottle filled with water, the color slowly spreads out. Eventually, the entire solution acquires the same color. The basic phenomenon related to transportation of a substance (mass) dissolved in a fluid (a gas in air or a chemical in water) is *diffusion*. Diffusion is caused by random molecular motion that takes place over a small molecular distance at a given time. Any inequality in a fluid, such as concentration of a dissolved substance, velocity, or temperature tends to get leveled out over a period of time by diffusion. Transport by such random motion is described as *diffusive*, or, *Fickian* transport. The mass (substance) transfer takes place both in static and moving fluids. When a fluid is in motion, in addition to be moved by diffusion, the mass (substance) is transported in the direction of flow due to bulk movement of fluid. This phenomenon of transport is known as *convection* (called *advection* in water). Convection can be a result of an externally applied pressure difference, known as *forced convection*, as in the case of pipe flow. Alternatively, it can be due to gravity or density changes, referred to as *free convection.*

Convective Transport

Any substance introduced into blowing wind or a flowing stream is transported in the direction of and by the magnitude of the velocity of bulk fluid motion. At the same time, spreading takes place due to Fickian transport. The distance that mass is transported by convection is usually much larger than that by Fickian movement. The convective mass flux, which is the substance transport rate per unit area with the bulk fluid, is equal to the product of the concentration of the substance and the velocity of the medium (air or water).

$$N_c = CV \qquad (7\text{-}1)$$

where $N_c = \dfrac{\text{(mass moved due to convection)}}{\text{(time)(area)}}$

C = chemical concentration of a substance
V = fluid velocity.

Equation 7-1 requires that variables be expressed in consistent units.

Diffusive or Fickian Transport

Two basic models describe the mass transport process arising from random movement of molecules of a substance in fluid. Both are based on the premise that mass flux (rate at which mass moves per unit area) is proportional to the difference in concentration of the mass. According to first model based on Fick's Law:

$$\text{Mass flux, } N_x = D\left(\frac{\Delta C}{\Delta z}\right) \qquad (7\text{-}2)$$

where N_x = diffusive mass flux = $\dfrac{\text{mass moved}}{\text{(time)(area)}}$ or $\dfrac{\text{rate of mass moved}}{\text{area}}$

ΔC = concentration difference
Δz = distance in which ΔC takes place
D = diffusion coefficient.

Example 7-1

Two bulbs are connected by a capillary tube of 1 mm diameter and 300 mm length. One bulb contains 100 mg/L concentration of carbon dioxide and the other is filled with air. The system is at a constant temperature and pressure. Determine how fast the carbon dioxide will mix initially. The diffusion coefficient is 0.15 cm²/sec.

Solution:

Refer to Figure 7-1:

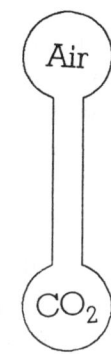

Figure 7-1 Simple Diffusion Device

Concentration in one bulb = 100 mg/L or 0.1 mg/cm³

Capillary tube area = $\frac{\pi}{4}$ $(0.1)^2 = 0.00785$ cm²
Length = 30 cm

From Equation 7-2:

$$\left(\frac{\text{Rate moved}}{\text{area}}\right) = D \left(\frac{\text{concentration difference}}{\text{capillary length}}\right)$$

$$\text{Rate moved} = D \text{ (area)} \left(\frac{\text{concentration difference}}{\text{capillary length}}\right)$$

$$= (0.15 \tfrac{\text{cm}^2}{\text{sec}})(0.0785 \text{ cm}^2)\left(0.1 - 0 \tfrac{\text{mg}}{\text{cm}^3}\right)\left(\frac{1}{30 \text{ cm}}\right)$$

$$= 3.9 \times 10^{-6} \text{ mg/sec.}$$

When a very large change of concentration occurs in a very short distance, as in the case of diffusion across the interface of two different liquids where the solution is well mixed away from the boundary, the term Δz is difficult to assess, and $(D/\Delta z)$ in Equation 7-2 is substituted by a lumped parameter representing the mass transfer rate over the surface. This is the second model, known as the mass transfer model:

$$\text{mass flux,} \quad N_x = k\Delta C \tag{7-3}$$

where k = mass transfer coefficient.

The second lumped parameter model is simpler and useful when only the average concentration is to be considered.

When two different phases are involved, i.e., air (atmosphere) and water, Equation 7-3 provides a very important relation to ascertain water-to-air exchange (or converse) of a chemical which is the most common mechanism of chemical removal from the aquatic environment. For two distinct phases, Equation 7-3 is expressed as:

$$N_x = k(C_s - C) \tag{7-4}$$

where C_s = saturated concentration of a dissolved gas or vapor in water (liquid)
C = actual concentration of a dissolved gas/vapor.

In accordance with Henry's Law,[1] $C_s = C_a/H$, where C_a is the concentration of the gas or vapor in air and H is the dimensionless form of Henry's Law Constant.

When the concentration C is higher than C_s, a gaseous chemical dissolved in liquid, if volatile, will leave the liquid and enter the atmosphere as gas or vapor, and, if not volatile, it will separate into a layer. If C_s is higher than C, the chemical present in the atmosphere may dissolve into liquid.

Example 7-2

The concentration of methane gas in a water tank is 15 mg/L. The closed atmosphere over the tank has a methane concentration of 200 mg/L. Determine the flux of methane gas. The dimensionless Henry's Law Constant is 27 and mass transfer coefficient is 5×10^{-3} cm/sec.

Solution:

$C_s = 200/27 = 7.41$ mg/L
$N_x = k(C_s - C)$
  $= 5 \times 10^{-3} (7.41 - 15)$
  $= \left(5 \times 10^{-5} \dfrac{\text{cm}}{\text{sec}}\right)\left(7.41 - 15 \dfrac{\text{mg}}{\text{L}}\right)\left[\dfrac{1}{1000} \dfrac{\text{L}}{\text{cm}^3}\right]$
  $= -3.8 \times 10^{-5}$ mg/cm²-sec (from tank to atmosphere)

---

[1] Traditionally, the law is expressed in terms of partial pressure of the gas; see Chapter 4.

In addition to molecular diffusion, a random movement of mass takes place due to turbulence or eddies present in the flowing air and water. In groundwater flow, water takes random paths while causing mixing and net transport of a chemical from one point to another. Despite different mechanisms, all random behaviors of mass transport including those due to turbulent diffusion are depicted by Fick's Law, in Equation 7-2; however, the coefficient D of Equation 7-2 is substituted by E, the dispersion coefficient. The dispersion coefficient, which combines all dispersal effects, bears little relationship to the molecular diffusion coefficient; only the form of the equation is the same.

Determination of the dispersion coefficient is cumbersome and inaccurate. The values found from experiments are more reliable but have limited applicability. Based on the premise that the dispersion of airborne pollutants usually follows a Gaussian statistical distribution, the dispersion coefficient has been correlated with standard deviations of the Gaussian profile of the dispersed pollutant. Dispersion in lakes and rivers is still difficult because of bed configuration. Experimental data on dispersion coefficients in pipelines and porous medium are available but these cannot be used to generalize. Certain empirical relations for the dispersion coefficient also exist.

Combined Convection-Diffusion Transport

The summation of Equations 7-1 and 7-2 indicates the total flux. In differential form, the rate of change of mass at any point (dC/dt) is equal to the rate of mass moved by convection (VdC/dz), plus by diffusion,

$\left(\frac{d}{dz}(DdC/dz)\right)$ (without considering reaction taking place during transport).

The differential form of this relation can be solved (integrated) for specific boundary conditions such as a fixed mass of a substance released instantly at $t=0$, and $z=0$. The solution, in terms of $C(z,t)$, is an exponential form equation that gives the concentration of the mass at any time t at any distance z downstream. At any given time, the concentration has a bell-shaped curve with respect to distance. For truly Fickian transport (no convection), it has the Gaussian shape curve (distribution), as stated above.

Settling Processes

Contrary to the transport processes that mix and distribute contaminants throughout the environment, the settlement processes separate and remove contaminants from air and water environments. Particle size plays an important role in the settling process. Particulate materials are divided in three groups based on the particles size range:

- *Discrete particulate matter* particle size greater than 0.1 mm (100 µm)
- *Suspended particulate matter* particle size between 0.001 mm and 0.1 mm
- *Dissolved matter* particle size less than 0.001 mm.

Discrete matter can be settled by sedimentation with or without flocculation. Some very small discrete particles do not settle by themselves. Two basic mechanisms are employed to agglomerate particles together into sufficiently large aggregates to facilitate their settling from suspension. The first, known as *coagulation*, consists of adding chemical agents to induce tendency for particles to floc together. The second mechanism, known as *flocculation*, is agitation of chemically treated water so that particles meet each other to actually agglomerate. Suspended matter is separated by filtration, which is also a physical process. However, the organic material in suspension is removed by biological processes. The dissolved matter is usually removed through chemical processes.

Sedimentation Process

Sedimentation is the removal of solid particles by gravity. In a treatment train, sedimentation is performed three ways:

1. Directly on a waste stream to reduce solids that settle;
2. After chemical treatment to remove flocculated impurities; or
3. After biological treatment to remove biodegraded solids.

Sedimentation processes are classified into four categories on the basis of the concentration of the contaminant particles and their flocculating properties :

- Type I settlement;
- Type II settlement;
- Zone settlement; and
- Compression settlement, as shown in Figure 7-2.

Type I and II are for sedimentation of dilute suspensions, the difference being that Type I is essentially for discrete particles and Type II is for flocculating material. In the latter case, because the heavier particles settle faster, they coagulate with smaller particles to form still larger floc.

The opportunity for contact among the settling solids increases with depth. No mathematical relation exists to simulate flocculation behavior. Accordingly, to evaluate this type of sedimentation, analysis for settlement is made in the laboratory through a column of suspension.

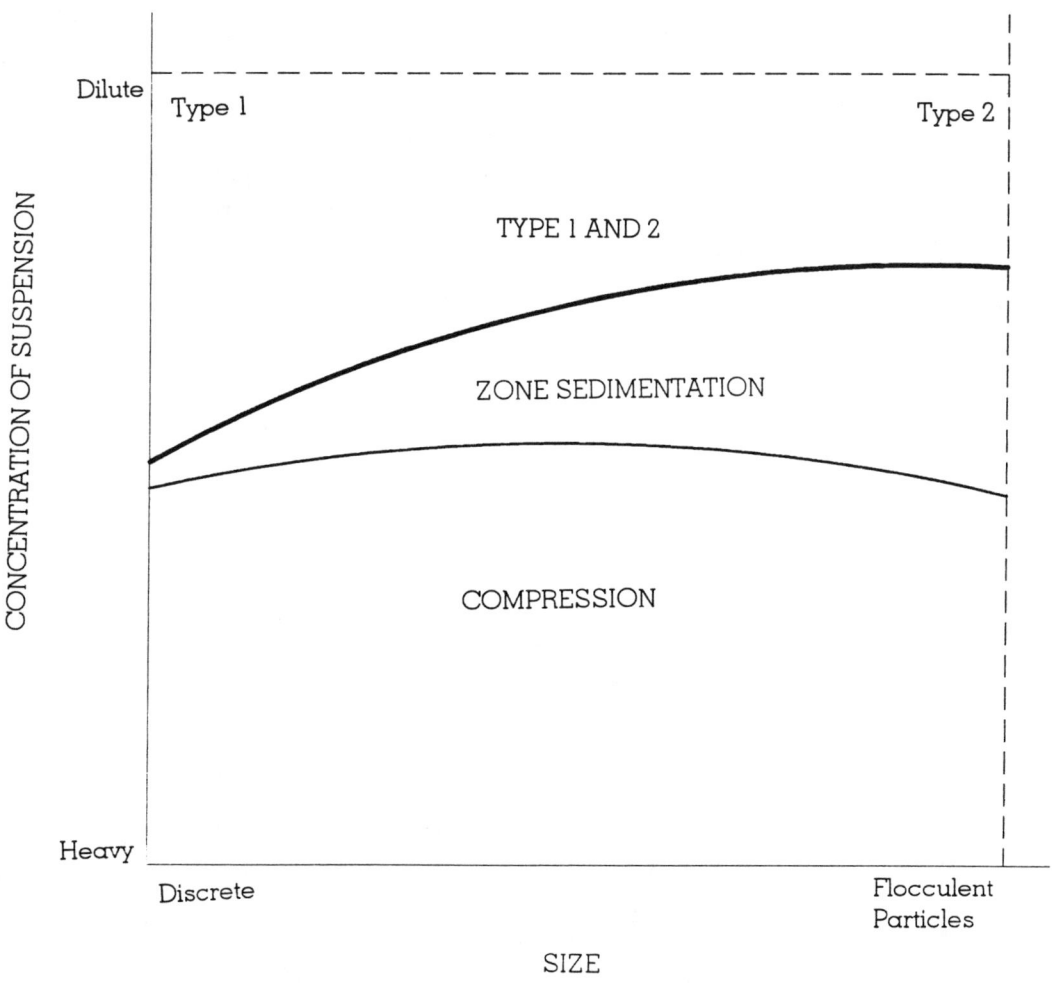

Figure 7-2 Categories of Sedimentation

As the suspension acquires the intermediate concentration, the particles come close enough for the interparticle forces to hold them together in a fixed position relative to each other. The particles settle through the column in their same relative positions. This is the zone settling in which a distinct interface between the clear water and settling solids is formed.

When the concentration becomes too high, the particles come into contact with each other. Settlement then has to occur by compression of the pore space between highly concentrated particles. This is a very time-consuming process.

The Concept of Terminal Velocity

Consider settlement of discrete particles (Type I sedimentation). As a particle falls in a fluid under the influence of gravity only, it accelerates until the downward acting force due to its weight is balanced by the upward-acting force of buoyancy and the force of drag through the fluid. The particle then acquires a fixed velocity called *terminal velocity*.

Figure 7-3 shows a free-body diagram of a spherical particle for which W is the weight of the object, F_b is the buoyant force and, F_d is the viscous drag force.

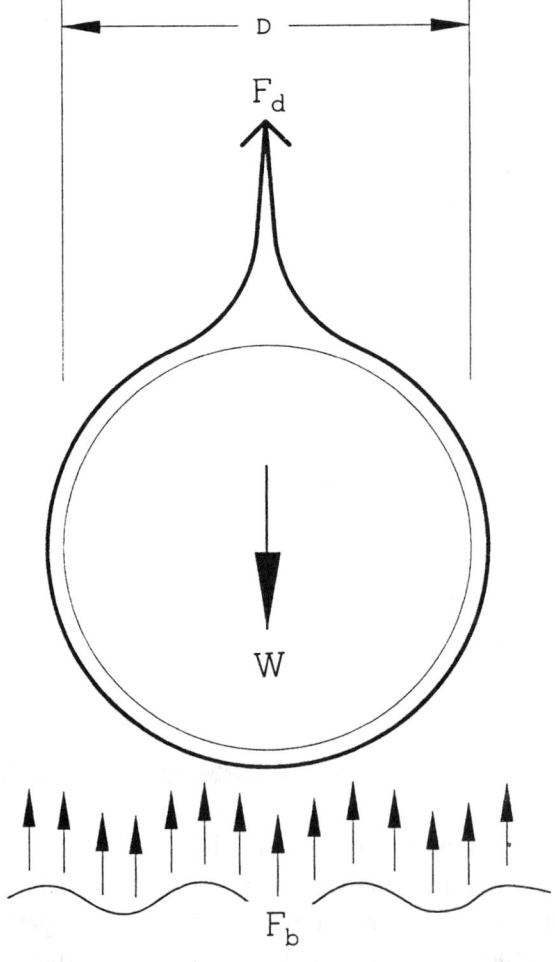

Figure 7-3 Settlement of a Discrete Particle

In equilibrium condition:

$$W - F_b - F_d = 0 \tag{a}$$

If ρ_s is the density of particle, and ρ_f is the density of fluid (medium) and V is the volume of the spherical particle ($V = \frac{\pi}{6} D^3$), then

$$W = \rho_s g \left(\frac{\pi}{6} D^3\right) \tag{b}$$

$$F_b = \rho_f g \left(\frac{\pi}{6} D^3\right) \tag{c}$$

The drag on an object moving through fluid is given by:

$$F_d = \tfrac{1}{2} C_d A \rho_f V^2 \tag{d}$$

where C_d is the drag coefficient, and V is the terminal velocity, and A is the cross-sectional area of the object $A = \frac{\pi}{4} D^2$

For a small velocity in viscous flow, $C_d = 24/R_e$, where Reynolds Number, $R_e = \rho_f V D/\mu$, μ being the dynamic viscosity of fluid. Thus, the drag relation reduces to:

$$F_d = 3\pi\mu V D \tag{e}$$

Substituting (b), (c), and (d) in (a), and solving for V:

$$V = \frac{(\rho_s - \rho_f) g D}{18\mu} \tag{7-5}$$

This is Stoke's Law for terminal velocity.

Example 7-3

The specific gravity of particles settling in water at 20°C is 1.2. The average particle size is 0.1 mm. What is the settling velocity of the particle?

Solution:
For water, at 20°C
$\rho_w = 1$ g/cm³, $\mu = 1.0087 \times 10^{-2}$ poise or g/cm-sec (Appendix C)
$\rho_s = G_s \rho_w = 1.2$ g/cm³
$g = 981$ cm/s²
$D = 0.1$ mm or 0.01 cm

From Equation 7-5:

$$V = \frac{(1.2 - 1.0 \text{ g/cm}^3)(981 \text{ m/sec}^2)(0.01 \text{cm})^2}{18 (1.0087 \times 10^{-2} \text{ g/cm-sec})}$$

$$= 0.108 \text{ cm/sec}.$$

Sediment Basin Design Considerations

An ideal sedimentation basin for discrete particle settlement is shown in Figure 7-4. Q is inflow as well as outflow of the basin, L is basin length, W is basin width, and H is the settling height in the basin. The particles entering the basin have a horizontal velocity equal to the velocity of the fluid.

$$V_H = \frac{Q}{WH}$$

Any particle entering at (a) in Figure 7-4, which has the vertical velocity of at least V_s, will reach (b), i.e., will be retained in the basin. Others will be washed off in the outlet zone.

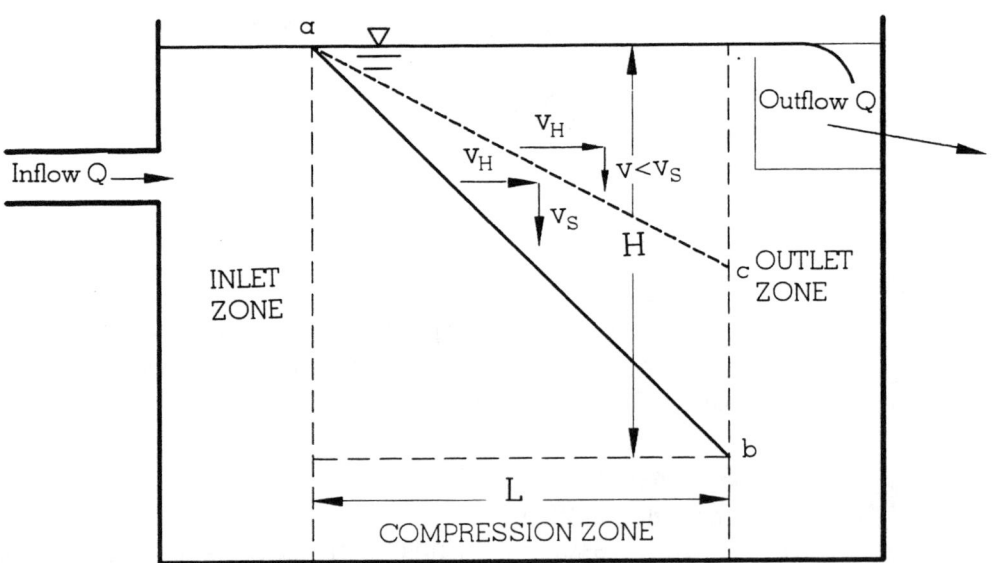

Figure 7-4 An Ideal Sedimentation Basin

From the geometry of the basin:

$$\frac{V_s}{V_H} = \frac{H}{L}$$

or $V_s = V_H H/L$

or $V_s = \left(\frac{Q}{WH}\right)\left(\frac{H}{L}\right)$

or $V_s = \frac{Q}{A}$

where A is the surface area of the settling zone (LW).

Velocity, V_s given by Equation 7-5, is defined as the *surface overflow rate* (SOR). The terminal velocities of the particles can be determined by Equation 7-4, Stoke's Law. Those particles having velocities equal to or greater than the SOR (Equation 7-5) will be entirely removed while those with lower velocities will be only partly removed depending on their location upon entrance into the basin. An equation for the removal efficiency of the basin is derived on this basis.

Example 7-4

The effective (settling zone) dimensions of a settling basin are 40 ft long, 20 ft wide, and 15 ft deep. It has a processing rate of 4 million gal/day. Determine the surface overflow rate in gallons per square foot per day and centimeters per second. Determine whether the particles of Example 7-3 will settle within the basin.

Solution:

$Q = 4 \times 10^6$ gal/day
$A = WL = (20)(40) = 800$ ft^2

From Equation 7-5:

$$V_s = \frac{Q}{A} = \frac{4 \times 10^6}{800} = 5000 \text{ gal/ft2-day}$$

The conversion of units:

$$V_s = \left(5000 \frac{\text{gal}}{\text{ft}^2\text{-day}}\right)\left[\frac{1}{7.48}\frac{\text{ft}^3}{\text{gal}}\right]\left[\frac{1}{24 \times 60 \times 60}\frac{\text{day}}{\text{sec}}\right]$$

$= 0.00774$ ft/sec or 0.236 cm/sec.

Since the terminal velocity of 0.108 cm/sec (Example 7-3) is less than the surface overflow rate, the particles entering at the top will not settle.

Filtration Processes: Liquid-Solid Separation

Filtration is performed by four different means: gravity, vacuum, pressure, and centrifugation. Gravity filtration is very common and is applied to the removal of particles from relatively dilute suspensions by passing the suspension through a porous unconsolidated media. Use of a fabric filter is common in air pollution control. A layered bed of anthracite coal or sand or a combination is commonly used for water and wastewater filtration. The other methods find application in partial separation of liquids from concentrated suspensions such as sludge dewatering and slurry thickening. Certain salient features of gravity filtration are

- The quantity of particles removed by a layer of filter medium is proportional to the concentration of particles entering that layer.
- A layer eventually becomes saturated with removed particles. At this point, the filter is no longer effective in clearing a suspension.
- The saturation condition starts at the filter face and proceeds through the filter in the direction of flow.
- As the successive layers become saturated, the head loss through the filter increases due to constriction of the passage. When the head loss becomes excessive, a partial vacuum (negative pressure) condition might be created.
- Several formulas have been developed to calculate the loss of head through a granular-media filter. The rate of loss is directly proportional to the rate of filtration, and inversely proportional to the square of the grains diameter in the filter media.
- When head loss makes filtration ineffective, the filter media must be backwashed. The backwashing process consists of passing filtered water or air upward through the bed at such a velocity that it causes the bed to expand until its thickness is 40% greater than that during filtering. This causes the filtered matter to wash out of the voids of the filter.
- The rate of backwash is computed from the relationship of the fluidizing velocity, i.e., when the grains are in the stage of flotation in water. This condition exists when the buoyancy force exerted by the washwater equals the particle weight. For most suspensions, the backwash rate must exceed 1 ft/min. Backwash generally continues for 5 min. Hence, the amount of water required for backwash is 5 ft^3/ft^2 area of the filter.

Process Engineering

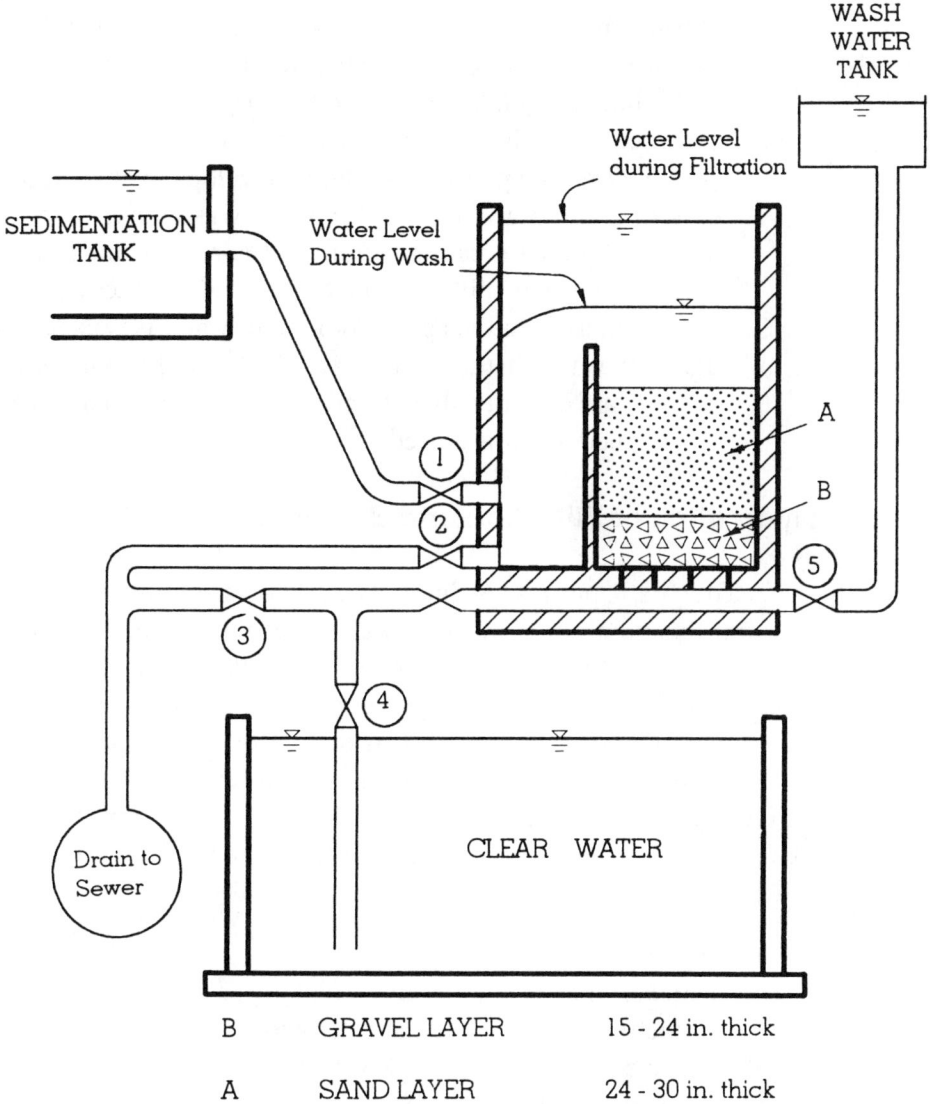

B	GRAVEL LAYER	15 - 24 in. thick
A	SAND LAYER	24 - 30 in. thick

Figure 7-5 Rapid Sand Filter

Sequence of Filtration Operation

Figure 7-5 is a diagram of a rapid sand filter. The distinct characteristics of a rapid filter as compared to those of a slow filter are

- Limit on turbidity loading or pretreatment of water is needed;
- High filtration rate at 2–5 gal/min/ft^2 is achieved (as compared to 0.05-0.1 gal/min/ft^2 for slow filter); and
- Backwashing is required instead of scraping off the top layer.

The sequence of operation is as follows:

1. During filtration, valves 1 and 4 are open; 2, 3, and 5 are closed. Water passes through the filter bed into the clean water tank. The underdrains regulate the rate of discharge.
2. To proceed with backwashing when the head loss becomes excessive (zero hydrostatic pressure condition develops), valves 1, 3, and 4 are closed; 2 and 5 are opened. From the wash tank, water moves up through the bed creating a fluidized condition for sand particles. Dirty wash water spills into the trough and goes to the sewer drain.
3. To switch to filtration again, some filter water is wasted to flush out the wash water that remained in the bed. For this purpose, valves 2, 4, and 5 are closed; valves 1 and 3 are opened. After a while, valve 4 is opened and 3 is closed.

Hydraulics of Filtration and Backwash

A filter is a porous medium. Water flows through the voids of the medium in smooth distinct lines; water does not mix as it is filtered. This kind of flow is known as *laminar* or *streamline* flow. In laminar flow, viscosity of moving fluid plays the dominating role. As liquid moves through the constricted pore openings, a loss of energy takes place, expressed in a linear dimension, as a head loss. Hazen-Poiseuille and Darcy-Weisbach derived the relations for head loss. Kozeny and many other scientists extended this relation, specifically, to filter media. In these relations, the head loss depends on the:

- Viscosity of fluid;
- Unit weight (weight per unit volume) of fluid;
- Velocity of fluid above the filter bed;
- Particle sizes (distribution) of filter medium; and
- Porosity of filter medium.

For head loss through a filter, when water temperature is 50°F, the simplified form of the Kozeny equation is as follows:

$$\frac{h}{l} = 1.2 \times 10^{-4} \frac{(1-p)^2 v}{p^3 d^2} \tag{7-6}$$

where h = head loss through filtration, in ft
l = depth of filter, in ft
p = porosity of filter bed
v = velocity of fluid above the bed, in ft/sec
d = mean grain diameter, in ft

Example 7-5

Calculate the head loss through a uniform sand filter. The rate of flow through the filter is 3600 gal/min. The filter surface area is 45 ft × 40 ft and thickness is 2 ft. The filter has a porosity of 0.42, and a mean grain diameter of 2×10^{-3} ft. Assume the water temperature is 50° F.

Solution:

$Q = 3600$ gal/min or 8.03 ft^3/sec

$$v = \frac{Q}{A} = \frac{8.03}{45 \times 40} = 4.5 \times 10^{-3} \text{ft/sec}$$

From Equation 7-6:

$$\frac{h}{l} = 1.2 \times 10^{-4} \frac{(1-0.42)^2}{(0.42)^3} \frac{(4.5 \times 10^{-3})}{(2 \times 10^{-3})^2}$$

$$= \frac{1.817 \times 10^{-7}}{2.96 \times 10^{-7}}$$

$$= 0.61$$

hence $h = (0.61)(2) = 1.22$ ft.

Sand pores clog with filter operation. Once a day, the filter is backwashed by running clean water upward through the under drainage system at seven-or-eight times the rate of filtration (typically 15 gpm/ft^2 or 2 ft/min). The backwash operation continues for a period of 5 to 10 min, and it consumes about 2 to 4% of the filter water. On completion of backwashing, the sand bed stands expanded by at least 40% more than the thickness prior to backwash.

The backwash process is based on the principle of fluidization whereby the upward force due to backwash velocity just overcomes the weight of a grain so that the grain attains a suspended or fluidized state and the dirt washes out. Several other processes have been developed that achieve backwashing with or without fluidization. These include scrubbing, surface wash, air scour, and air-water backwash at a nonfluidization rate.

From a practical consideration, the backwash rate is controlled by the following:

$$V_b > 7 \text{ gal/min-ft}^2 \tag{7-7a}$$

$$V_b > V_t p^{4.5} \tag{7-7b}$$

where V_b = backwash velocity
V_t = settling (terminal) velocity
p = porosity.

The backwash volume can be computed as follows:

(volume of backwash) = (backwash velocity)(backwash time)(filter area) (7-8)

Example 7-6
Calculate the backwash rate and volume for a uniform sand filter. The settling velocity of sand particles is 8.2 m/min and porosity is 0.45. Backwash time is 5 min and filter size is 50 ft × 45 ft.

Solution:
From Equation (7-7a): $V_b > 7$ gpm/ft^2
From Equation (7-7b): $V_b > (8.2)(.45)^{4.5} = 0.226$ m/min or 5.5 gpm/ft^2

V_b should be more than 7 gpm/ft^2; use the typical value of 15 gpm/ft^2:
Hence, $V_b = 15$ gpm/ft^2 or 2 ft/min.

From Equation 7-8:
Backwash volume = (2 ft/min)(5 min)(50 ft × 45 ft) = 22,500 ft^3.

Chemical Processes

Those processes in which a change of the environmental condition is brought about by means of a chemical reaction are known as *chemical processes*. Chemical processes, unlike physical or biological processes, are usually additive, i.e., something is added to achieve removal of something else. When a reaction takes place in a single phase (solid, liquid, or gas) it is known as a *homogeneous reaction*. The reactions that involve more than one phase are *heterogeneous reactions*. Chemical reactions involve heat transfer—an endothermic reaction requires the addition of heat while an exothermic reaction produces heat. Many reactions will simply not occur if a substance known as a *catalyst* is not present. A catalyst by itself does not enter into a reaction but it influences it by either promoting or retarding the reaction. The unit (tank or container) within which chemical or biological reactions take place is called a *reactor*.

The following factors are crucial in chemical process analyses and designs:

1. The stoichiometric relations expressed by the chemical formulas that describe the mass balance among different materials entering into the reaction.
2. The type of reactor used that controls the process mechanism.
3. The kinetics of the chemical process that describes the rate of progress of the reaction.

The first item is discussed in Chapter 4. The following two items are discussed subsequently. The major chemical processes involved in environmental pollution control are chemical precipitation, ion exchange, gas transfer, adsorption, and disinfection.

Reactors and Resident Time

The hydraulic character (characteristic with respect to flow) of a reactor is defined by the *hydraulic detention time* or the *mean residence time* for which the particles of the liquid stay in the reactor

$$t_R = \frac{V}{Q} \tag{7-9}$$

where t_R = mean residence time
V = volume of reactor
Q = volumetric rate of flow through reactor.

The time needed to complete a reaction to a specific degree will dictate the value of t_R. This time will depend upon the reaction kinetics and the type of reactor. Once the residence time is decided, the required tank size is determined for the design flow from Equation 7-9. Thus, t_R is a design control parameter.

Types of Reactors

The three principal types of reactors are shown in Figure 7-6. The mass balance relations of Chapter 2 are applicable to continuous-flow reactors.

Reaction Kinetics

The rate at which a reaction proceeds is critical since it dictates the residence time, and, hence, the design of a reactor. There are three categories of reaction rates:

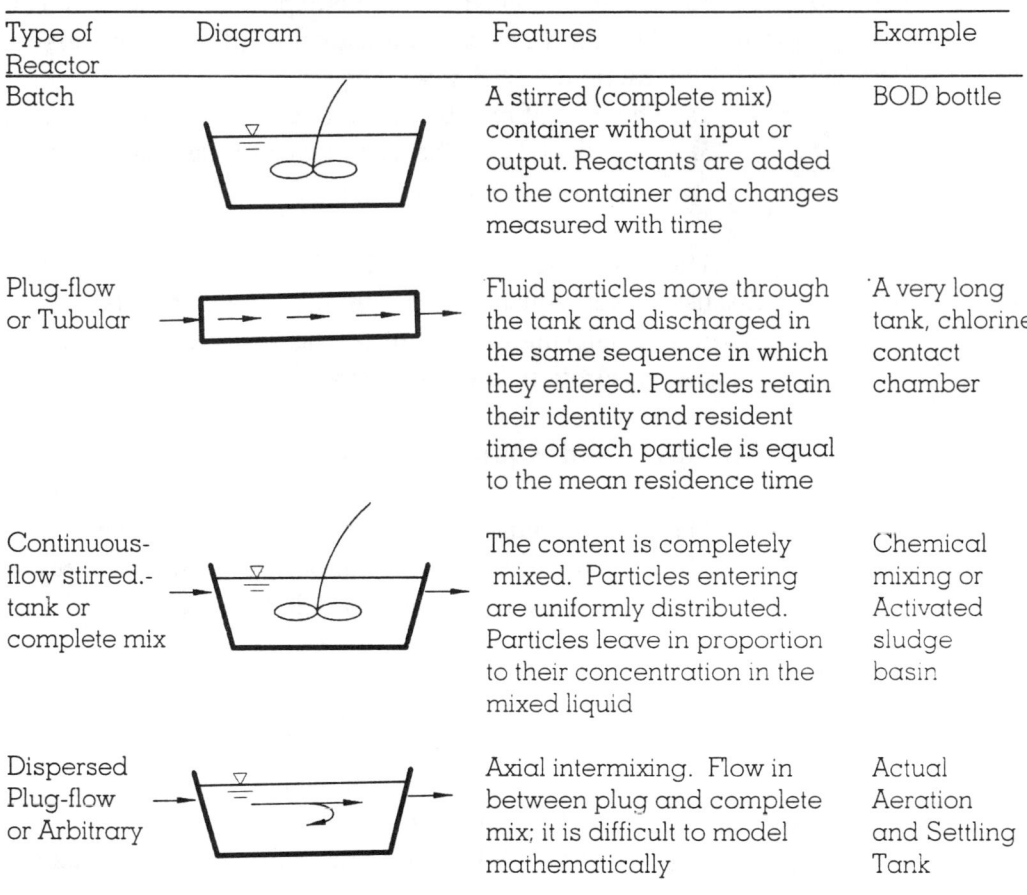

Figure 7-6 Reactors and Their Features

1. Zero-order reactions take place at a pace that does not depend on the concentration of any reactant or product; i.e., $\Delta C/\Delta t = -k$, where $\Delta C/\Delta t$ is the rate of change in concentration of a substance and k is the specific reaction rate constant. This is similar to the arithmetic model of population growth discussed earlier in Chapter 1. The reaction rate constant is quite sensitive to temperature.
2. First-order reactions proceed at a rate that is proportional to the remaining concentration of a reactant at any time, i.e., $\Delta C/\Delta t = -kC$. This is similar to the exponential model in Chapter 1.
3. In second-order reactions, the rate is proportional to the square of concentration of a reactant at any time.

The volume of a reactor is determined by considering the reaction rate and the mass balance. For example, the residence time of a continuous-flow stirred-tank reactor, as defined in Figure 7-6, under first-order reaction is given by:

$$t_R = \frac{1}{k}\left(\frac{C_0}{C}-1\right) \tag{7-10}$$

where C_0 = influent concentration of a reactant
C = final concentration of reactant in the reactor
k = first-order reaction rate
t_R = residence time.

Based on Equation 7-9, flow rate multiplied by residence time gives the volume of the reactor.

Example 7-7
Assuming that a first-order reaction applies, determine the required residence time of a continuous-flow stirred-tank reactor to achieve a 90% reduction in concentration for a reactant having a k value of 1/day. The flow rate is 4 million gal/day.

Solution:
If $C_0 = 1$, $C = 0.1$ for 90% removal

$$t_R = \frac{1}{1}\left(\frac{1}{0.1}-1\right)$$
$$= 9 \text{ days}$$

$\mathcal{V} = (4 \times 10^6 \text{ gal/day})(9 \text{ days}) = 36$ million gal.

Chemical Processes in Dissolved Material Removal Trains

The chemical reaction processes that are used in a treatment train, with or without a combination of physical and biological processes, are known as chemical unit processes. The important processes are described in the following section.

Chemical Precipitation Process

The process involves adding chemicals into a solution that interact with substances present or mixed in the solution. The reactions alter the physical state of the solids dissolved in the solution. Precipitates are produced consisting of fine suspensions of particles that are removed by the sedimentation process. Sometimes, the alteration is only slight and removal is achieved by entrapping the fines within the voluminous precipitate of the coagulant itself.

Extremely small-sized (1–200 nm) particles stay in a solution either because of their chemical combination with water or because of surface electric charge. According to the *double-layer theory*, a fine particle carries a negative charge. It is covered by a fixed layer of positive ions from the solution. This stationary layer, known as the *stern layer*, is then surrounded by a second movable *diffuse layer* of counter ions, as shown in Figure 7-7.

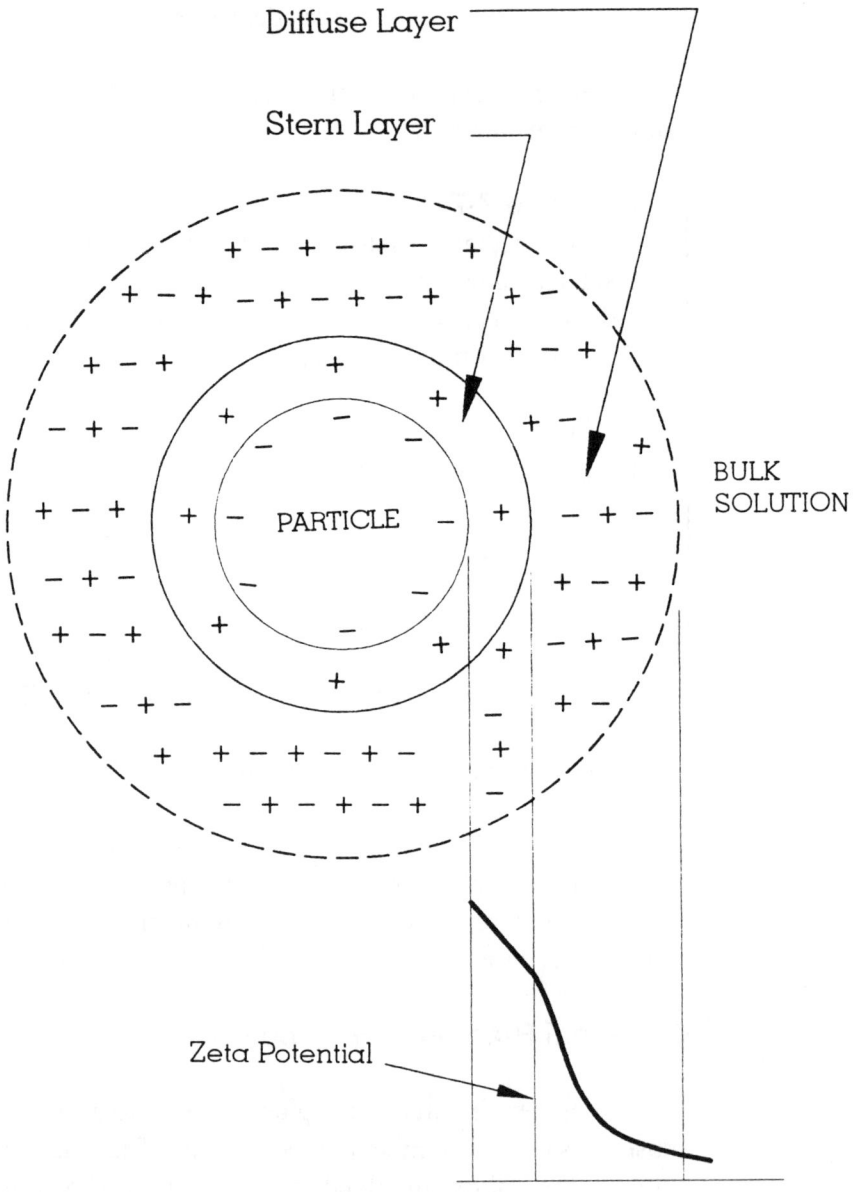

Figure 7-7 Double-Boundary Layer Theory

The charge at the surface where the diffuse layer can shear off is known as the *zeta potential*. This charge produces the repulsive force that does not allow the particles to aggregate. Particles with high zeta potential produce a stable *sol*, or colloidal dispersion.

To bring about aggregation of particles for precipitation, it is necessary to reduce the particle layer charge or to overcome the effect of this charge. This is accomplished by addition of certain chemicals or organic compounds (polymers). Most common precipitants are listed in Table 7-1. Chemical precipitation processes of importance in water treatment are water softening (removal of calcium and magnesium ions) and iron and manganese removal, and the removal of phosphate, heavy metals, and emulsified oil in wastewater treatment.

Table 7-1 Common Chemical Precipitants

Chemical	Formula
Alum	$Al_2(SO_4)_3 \cdot 18H_2O$
Lime	$Ca(OH)_2$
Ferrous sulfate	$FeSO_4 \cdot 7H_2O$
Ferric sulfate	$Fe_2(SO_4)_3$
Ferric chloride	$FeCl_3$

Ion Exchange Process

Ion exchange is used for selective impurity removal. Ion exchangers are insoluble natural or synthetic compounds. They are called *zeolites*. There are two kinds of ion exchange. In cation exchange, a cation of zeolite is displaced by a cation of the material in solution that is to be removed. This can be expressed by the equilibrium reaction:

$$A_1^+\underline{\quad} + A_2^+R^- \leftrightarrows A_2^+\underline{\quad} + A_1^+R^- \qquad (7\text{-}11)$$
Material in solution   Exchanger

where R^- is the exchange resin.

For example, the hardness-producing calcium and magnesium ions are replaced by sodium as follows:

$$Ca^+(HCO_3)_2^- + Na_2^+R^- \leftrightarrows 2Na^+HCO_3^- + Ca^+R^- \qquad (7\text{-}12)$$

This shows that when water containing calcium is passed through an ion exchanger, the calcium cation is taken up by the exchanger. When the sodium of the exchanger is exhausted, it can be regenerated by applying a solution of sodium chloride:

$$Ca^+R^- + 2Na^+Cl^- \rightarrow Na_2^+R^- + Ca^+Cl_2^- \quad (7\text{-}13)$$

This is the reverse reaction of Equation 7-12.

In anion exchange, anions from the solution are replaced by the anion species of the zeolite that contains cation exchange resin.

$$\underline{\quad}B_1^- + R^+B_2^- \rightleftharpoons \underline{\quad}B_2^- + R^+B_1^- \quad (7\text{-}14)$$
Material in solution   Exchanger

The exchange process behaves as a reversible equilibrium reaction to which an equilibrium constant relation similar to Equation 4-12 applies. If a solution that contains equal concentrations of species A_1^+ and A_3^+ is brought into contact with zeolite $A_2^+R^-$, the species having the greater equilibrium constant with respect to the exchange material (zeolite) will be picked up over the other species.

Gas Transfer Process

The two-film theory of gas transfer between two phases, usually air and water, assumes that two thin films exist at the gas-liquid interface—one gas film and one liquid film. These films offer resistance to the passage of gas molecules between the bulk-liquid and bulk-gas phases. The diffusion principle of transport earlier in this chapter applies to this phenomenon. Equation 7-4 is used for gas transfer analysis.

Gas transfer is vital to a number of chemical and biological processes. Some notable applications include oxygen transfer to biological units, chlorine transfer for disinfection, ammonia or other gaseous chemical removal from water, and hydrocarbon plumes movement in groundwater flow.

Adsorption Process

Adsorption is a process of collecting soluble substances from a solution on a suitable interface. The interface could be between a liquid and gas, liquid and another liquid, and liquid and a solid. Activated carbon is a common adsorption element in environmental processes. Activated carbon is prepared by first making a char by heating coal to red-heat in insufficient oxygen to sustain combustion. The char particles are then activated by exposure to an oxidizing gas at a high temperature. This develops a porous structure in the char with tremendous surface area in the range of 1,000 m^2/g. Either the powdered or granular (diameter of more than 0.1 mm) form of the product is used in the adsorption process. In the granular form, a fixed-bed column of activated carbon is used through which the solution is passed from the top and collected in bottom drains. In the powdered form, the activated carbon is added into the liquid that settles to the bottom. The adsorbed molecules of the substance that leave solution are held onto solid carbon by physical and chemical bonding. The molecules are called the *adsorbate* and the solid (carbon) is called the *adsorbent*. The process occurs in three stages:

- Transfer of adsorbate molecules through the film that surrounds the adsorbent;
- Diffusion through the pores of the adsorbent; and
- Uptake and bonding of the adsorbate molecules by the active surface of the adsorbent.

Membrane-Separation Process

Two broad categories of membrane processes exist: pressure-driven and electrodialysis. The pressure-driven processes include hyperfiltration or reverse osmosis (RO), nanofiltration (NF), ultrafiltration (UF), and microfiltration (MF), representing removal of small ionic and organic materials by RO, to removal of large suspended colloidal particles by MF in ascending order. The operating pressure range for RO is from 200 to 2,000 lb/in.^{2} (psi) which gradually decreases to 5–30 psi for MF. The flux (filtration rate) for RO is 3–20 gal/ft-day (gfd), which increases to 100–1000 gfd for MF.

A membrane separating a concentrated solution from a dilute solution will permit flow from dilute solution to concentrated solution in the natural osmosis process. In reverse osmosis, the flow is forced to take place from a solution of high concentration to one of dilute solution by applying the pressure. The membrane is made of cellulose acetate which rejects 99% of ion species. This membrane permeates water from highly concentrated solution to dilute solution but prevents the solute from passing. The process separates the clear water and the high solution concentrate.

Electrodialysis membranes are sheets of high-capacity ion-exchange resins that pass ions but prevent the passage of water. Cation membranes transmit only cations and anion membranes allow only anions to pass. This characteristic is due to the ion-exchange nature of the material of which the membranes are made. In the process, a large number of membranes are arranged in a stack in an alternate sequence of cation and anion membranes.

The alternate cells of clear water and high solution concentrate are formed on completion of the process. High concentrates separated from the solution have to be dealt with separately.

Disinfection Process

Disinfection refers to destruction of disease-causing microorganisms as distinct from sterilization which is the destruction of all microorganisms. Chlorine and its compounds are commonly used disinfectant.

Two important factors in disinfection are the time of contact and the concentration of the disinfecting agent. This is given by a model:

$$Kill = kC^n t \qquad (7\text{-}15)$$

where k = rate constant
n = a factor greater than zero
C = concentration
t = time.

The model is calibrated for k and n values from field test data. Note that with long contact time, a low dosage of disinfectant may suffice, whereas a short contact time requires a high concentration of disinfectant. For similar conditions, factor n is higher for agents that are more effective as disinfectants. For example, the value of n will be higher for a free chlorine dosage than for combined chlorine compounds.

Example 7-8
A contact time of 30 min with a free-chlorine concentration of 10 mg/L achieves a 90% kill of pathogenic bacteria. How long it will take for 99% removal of pathogens at 8 mg/L concentration? Assume $n = 2$.

Solution:
From Equation 7-15:
$$90 = k(10)^2 (30)$$

$$k = \frac{90}{(100)(30)}$$

$$= 0.03$$

For 99% kill:
$$99 = (0.03)(8)^2 t$$

$$t = \frac{99}{(0.03)(64)}$$

$$= 51.6 \text{ min}$$

Biological Processes in Suspended Material Removal Trains

The nonsettling suspended organic matter (mostly of food origin) in wastewater streams is removed by biological means since chemical processes are not highly effective in the removal of organic waste. Living microbial organisms are conveniently used to eat away the organic wastes. Organisms undergo growth kinetics as discussed in Chapter 5. Some useful terms related to biological processes are defined below.

1. *BOD Removal.* Biochemical oxygen demand (BOD) refers to use of the dissolved oxygen by microorganisms in the consumption (by oxidation) of organic matter to produce cell tissues and gaseous end products. BOD removal is a measure of the conversion of the organic matter in wastewater. During this stage of conversion, the nitrogen present is converted to ammonia.
2. *Nitrification.* A two-stage biological process in which ammonia is converted first to nitrite and then to nitrate.
3. *Denitrification.* Biological process in which nitrate is converted to nitrogen.
4. *Stabilization.* Biological process in which organic matter is converted to cell tissues and gases.
5. *Substrate.* Denotes the organic matter or nutrients that are stabilized during a biological treatment.
6. *Aerobic processes.* The processes that occur in the presence of oxygen.
7. *Anaerobic processes.* The processes that occur in the absence of free oxygen. Anaerobes are bacteria that survive in the absence of free oxygen.
8. *Suspended growth processes.* The microorganisms responsible for conversion of organic matter are in a state of suspension within the liquid.
9. *Attached growth processes.* The microorganisms responsible for conversion of organic wastes are attached to some inert media such as rocks.
10. *Sludge.* Solid part with a high water content that is removed from water or wastewater.

Types of Biological Processes

Based upon definitions (6) through (9) above, the biological processes are divided into four main groups, as shown in Table 7-2. Each group is further subclassified into several separate processes based on the type of reactor used. Some processes combine the attached and suspended growth features. Some processes function both under aerobic and anaerobic conditions. These are not listed in the table.

Activated sludge, trickling filters, and anaerobic digestion are three common biological processes; the first two being applied to wastewater or sewage treatment and the last to sludge treatment.

Table 7-2 Principal Biological Processes

Type	Specific Process Name
Aerobic suspended growth	Activated sludge Aerated lagoons Aerobic digestion
Aerobic attached growth	Trickling filters Rotating biological contractors
Anaerobic suspended growth	Anaerobic digestion Anaerobic contact process
Anaerobic attached growth	Anaerobic filters Anaerobic lagoons

Activated Sludge Process

The process is shown schematically in Figure 7-8. Wastewater containing degradable organics is brought into a reactor in which microorganisms exist in suspension. They utilize organic matter for food. The reactor is aerated to maintain an aerobic environment. The content of the reactor is known as the *mixed liquor suspended solids*, (MLSS): a mixture of organic wastes and biological solids (microorganisms). After a specified time, the MLSS is passed into a settling tank where the organisms settle. A portion of the organisms is cycled; the remaining is removed as waste. A term *mixed liquor volatile suspended solids*, (MLVSS) is used to represent the microorganism portion of the MLSS.

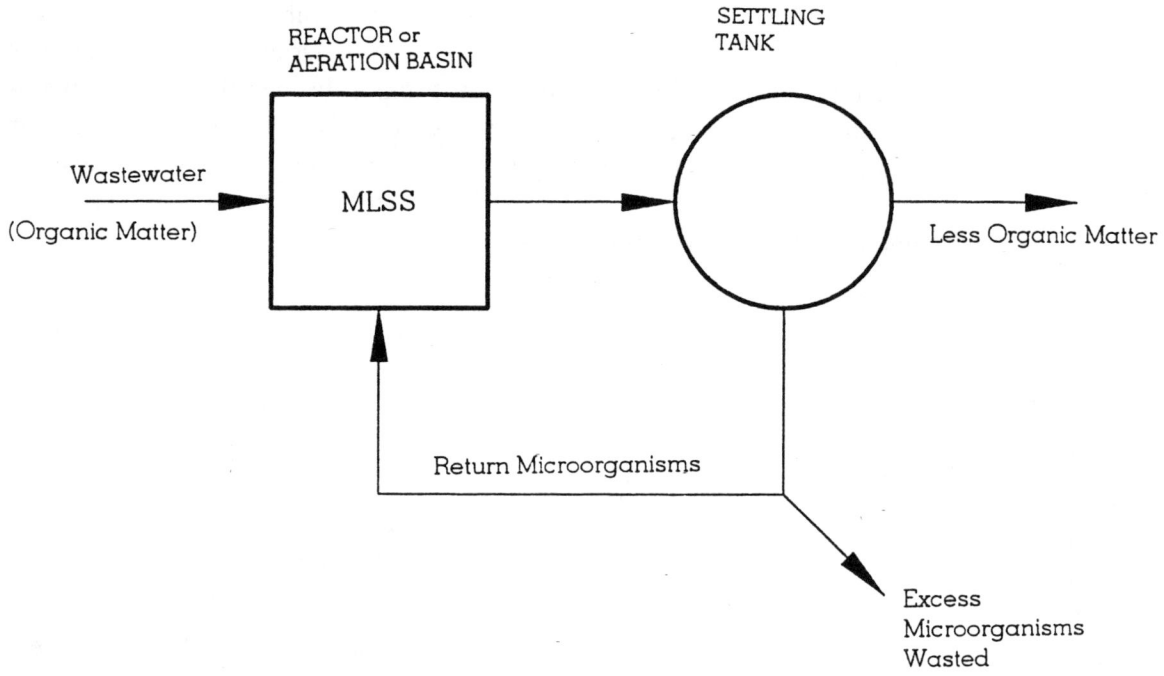

Figure 7-8 Flow Diagram of an Activated Sludge Process

A mass-balance relation is expressed for the microorganisms in the system. This is superimposed by the microbial kinetic relations explained in Chapter 5. The resulting equations are the basis for the design of activated sludge process units.

Two most common parameters for activated sludge process control and design are (1) the food-to-microorganism ratio (F/M), and (2) the mean cell (microorganism) residence time in the system. The two parameters are interrelated. A process is designed for a specified F/M ratio which is defined as follows:

$$F/M = \frac{BOD}{t(MLVSS)} \qquad (7\text{-}16)$$

where BOD = BOD of wastewater (influent), mg/L
 t = hydraulic detention time of aeration tank (V/Q) in days
 MLVSS = Concentration of volatile suspended solids in the aeration tank, mg/L.

Trickling Filter Process

A trickling filter is a bed of permeable media, such as crushed rocks, over which microbial growth takes place. The wastewater containing organics is sprayed over the bed and percolates or trickles through the media flowing over the microorganisms in a thin layer. Oxygen becomes available from air moving through voids in the media. Microorganisms eat away the waste. A trickling filter system comprises a primary settling tank and a final settling tank (clarifier) as shown in Figure 7-9.

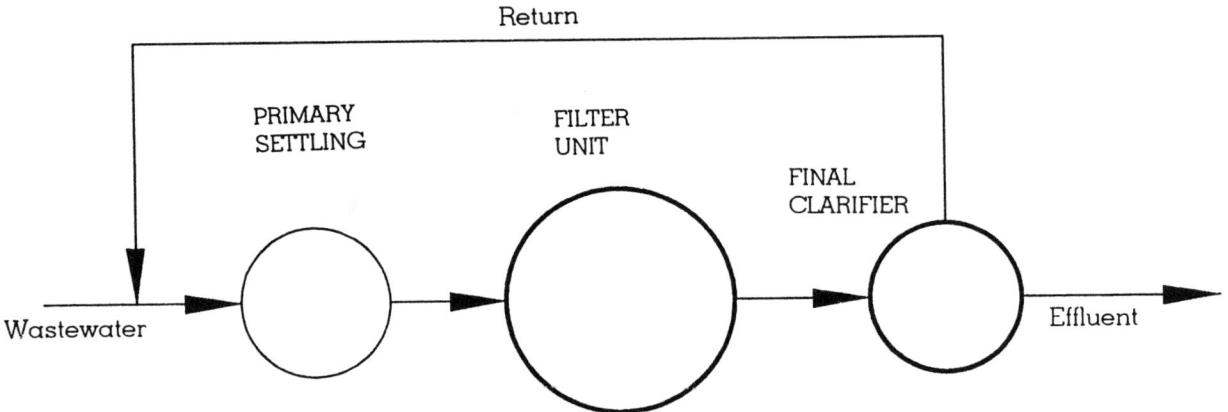

Figure 7-9 Flow Diagram of a Trickling Filter

The treated wastewater and biological solids that have detached from the media are collected in an underdrain and passed on to a final settling tank where the solids are separated from effluent. The solids in the final settling tank are returned to the beginning, mixed with the raw wastewater, and settled in the primary settling tank.

Certain empirical relations for the rate of flux of organic material into the media have been suggested. The flux relation is superimposed on the mass relation of the substrate in the system. This results in equations for the design of a trickling filter.

Anaerobic Digestion

The anaerobic process employs microbes that do not require free oxygen to oxidize organic matter. They derive oxygen from combined sources that may include carbonate, sulfate, nitrate, and phosphate. While the end product of an aerobic process is the growth of cells and carbon dioxide, the anaerobic end products are growth of new cells, methane, carbon dioxide, and unused organics. Anaerobic digestion involves interaction by two groups of bacteria: organic acid-forming bacteria and methane-producing bacteria. The former converts the organic matter into organic fatty acids. The methane-producing bacteria use these organic acids as a substrate and produce methane and carbon dioxide. Anaerobic processes result in waste stabilization producing small amounts of sludge. However, a disadvantage of the process is that the rate is very slow and the digestion process takes a long time (20–60 days).

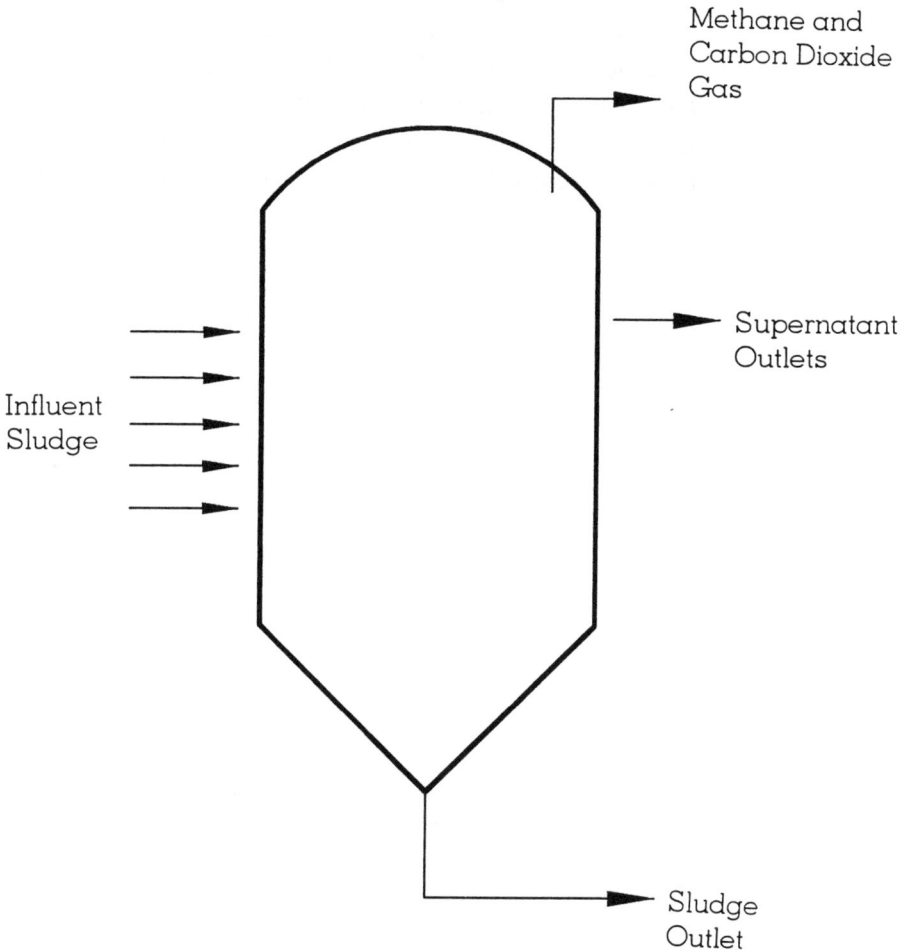

Figure 7-10 Anaerobic Digester

The process consists of introducing sludge into an airtight reactor at temperatures ranging from 85 to 95°F, as shown in Figure 7-10. The reactor is a concrete cylindrical tank 12–36 ft deep and 15–125 ft in diameter. The floor is shaped like an inverted cone. The digester contents are stratified from top bottom to top into

- Digested sludge;
- Digesting sludge;
- Supernatant;
- Scum; and
- Gases.

The digested sludge is withdrawn through the pipe located at the center of the inverted cone. Supernatant liquor is the water released during digestion. It is high in BOD and suspended solids. Usually, it is fed back to the influent to the primary clarifier. The gases consisting of 55–75% methane, 45–25% carbon dioxide, and a trace of hydrogen sulfide, hydrogen, and nitrogen collect in the dome of the floating cover.

PROBLEMS

1. If the dissolved solids in a stream flowing at a velocity of 10 ft/sec, are 500 mg/L, what is the convective mass flux of solids in the stream?

2. In Example 7-1, if the area of the capillary tube connecting two bulbs and its length are doubled, what will be the rate of transport?

3. In Example 7-1, what will be the rate of transport when the concentration of CO_2 in one bulb is 80 mg/L and the other bulb is 20 mg/L?

4. A 1-mm thin film separates two solutions of the same solute. On one side, the concentration of the solute is 1000 mg/L and on the other side it is 100 mg/L. What is the flux (rate of transfer per unit area) across the film? The diffusion coefficient is 0.28 cm^2/sec.

5. There is a leak of gasoline from a gas station to a nearby residence. A gasoline concentration of 4×10^{-6} g/cm^3 has been measured 3 m below the dirt floor of the basement of the residence. What is the rate (flux) of gasoline vapor into the house? The diffusion coefficient of gasoline vapor is 1.1×10^{-2} cm^2/sec. (Hint: Assume zero concentration at basement floor level.)

6. TCE (trichloroethylene) has a concentration of 10 ppb (10 µg/L) in water. If the mass transfer coefficient is 2×10^{-3} cm/sec, what is the flux rate of transfer of TCE from the water? Assume zero saturated concentration of TCE in water.

7. Due to an accident, hydrocarbon is spilled into a bay. The concentration of hydrocarbon is 500 mg/L. What is the flux rate of vaporization of the hydrocarbon from the bay? The transfer coefficient is 60 cm/min.

8. For oxygen in water, the diffusion coefficient at 25°C is 2.1×10^{-5} cm^2/sec and the mass transfer coefficient is 2.1×10^{-3} cm/sec. If the actual concentration of oxygen in seawater (at chloride content 20,000 mg/L) is 2 mg/L, what is the oxygenation (flux rate of transfer of oxygen) of seawater? The saturated value of dissolved oxygen for a chloride content of 20,000 mg/L at 25°C is 6.74 mg/L.

9. Specific gravity of particles settling in water at 20°C is 1.25. The average diameter of the particles is 0.08 mm. What is the settling velocity?

10. A grit particle of 0.2 mm size and 2.6 specific gravity and an organic particle of 8 mm size and 1.08 specific gravity settle through a column of water at 20°C. Which will settle faster?

11. A fine sand particle of specific gravity of 2.65 and a diameter of 0.004 in. settles in water at 60°F. Determine the settling rate.

12. In a water softening process, very fine crystals are formed. These crystals flocculate and settle in clusters possessing a specific gravity of 1.2. Assuming clusters with an average diameter of 0.06 mm, calculate their settling velocity at 25 °C using Stoke's Law.

13. What is the terminal velocity for a spherical particle of 2 g/cm^3 density and 100 μm diameter falling through the air? Neglect the upward buoyancy force of the air. The atmospheric temperature is 20°C. (Hint: Drop the term ρ_f from Equation 7-4.)

14. The settling zone dimensions of a sedimentation tank are: 50 ft long, 25 ft wide, and 10 ft high. The rate of flow in the tank is 3 MGD. Determine the surface overflow rate in gallons per square foot per day and in centimeters per second. Will the particles of Problem 9 settle in this basin?

15. A settling basin is designed for a surface overflow rate of 33 m/day. Determine the smallest size of the particles of specific gravity 1.2 that will be fully removed. The water temperature is 20°C.

16. The surface overflow rate of a basin is 0.09 ft/sec. The basin has the effective dimensions of 40 ft long, 20 ft wide, and 15 ft deep. What is the flow rate through the basin?

17. Calculate the head loss through a sand filter. The filtration velocity is 0.0027 m/sec. The average grain size is 0.65 mm. The bed has a depth of 1 m with a porosity of 0.45. Water temperature is 50°C.

18. Find the head loss through a uniform-grained filter having the following data:

Flow rate	= 5000 gal/min
Surface area of filter	= 50 ft × 50 ft
Porosity	= 0.4
Depth of bed	= 2 ft
Grain size	= 5.3 × 10^{-3} ft
Temperature	= 50°F

19. Calculate the backwash velocity and the volume of water for backwash of a sand filter. The grains have a settling velocity of 20 m/min and a porosity of 0.42. The backwash time is 5 min. The filter has the dimensions of 50 m × 40 m.

20. For a filter having the mean grain size of 0.08 cm and specific gravity of 2.65, determine the backwash rate. The bed porosity is 0.47. The temperature of water is 65°F. (Hint: Determine the settling velocity using Stoke's law.)

21. For backwater velocity, the following relations are valid for a filter with sand of specific gravity of 2.65 and anthracite of specific gravity of 1.55:

V_b (m/min) = $0.59 D_{60}^{1.82}$ for sand, D^*_{60} in mm
V_b (m/min) = $0.21 D_{60}^{1.82}$ for anthracite
*Diameter from which 60% material is finer.

Calculate the backwater velocity for a sand filter with D_{60} of 0.8mm and porosity of 0.46. What is the maximum terminal velocity for not exceeding this backwater rate?

22. Water enters a tank at a rate of 3000 gal/min. The hydraulic detention time is 6 hr. What is the capacity of the tank?

23. The mean residence time for a reaction is 3 hr. The flow into the reactor is 4 MGD. Determine the size of the reactor if the width and height are approximately equal to one-half of the length.

24. The following data are obtained for the reaction A → B. Determine the order of reaction and the value of reaction rate constant, k.

t, min	0	20	40	80	120
A, mg/L	90	72	57	36	23

(Hint: For the zero-order, the plot of t (X-scale) vs. A (Y-scale) on an arithmetic graph paper is a straight line with the slope of −k. For the first-order, the plot of t vs. log A on a semilog paper is a straight line of slope, −k, and for the second-order, the plot of t vs. 1/A on arithmetic graph paper is a straight line of slope, −k.)

25. Assuming a first-order reaction for a process, determine the residence time for a continuous-flow stirred tank to obtain an 80% reaction completion, k=2 per day. If the rate of flow is 1 MGD, what reactor size is required?

26. For Problem 25, compute and plot the relationship between the residence time and the removal efficiency in a reactor.

27. Which is a more effective disinfection: (1) 15 min contact with chlorine at 10 mg/L concentration, or (2) 50 min contact at 5 mg/L concentration? $n = 2$.

28. A 90% disinfection (*kill*) is achieved when a treated wastewater is exposed for 30 min to a free-chlorine concentration of 12 mg/L, when $n = 3$. To achieve the same level of disinfection in 60 min in combined chlorine exposure, what should be the concentration, if $n = 1.5$?

29. The BOD of a domestic waste is 200 mg/L. The aeration (detention) time is 3 hr. The MLSS is 4,000 mg/L and the ratio of MLVSS to MLSS is 0.8. Determine the design F/M ratio for the system.

30. The F/M ratio of an activated sludge process is 0.5/day. The BOD and MLVSS are 300 mg/L and 3,500 mg/L. What is the volume of the aeration tank to handle a wastewater flow of one MGD?

STUDY QUESTIONS

1. How is a transport process different from a settlement process for particle movement? Differentiate between convection and diffusion.

2. How is sedimentation of particles different from filtration? Describe and differentiate various types of sedimentation processes in water treatment.

3. Show the configuration of a typical rapid sand filter. Describe the sequence of its operation.

4. What is the significance of a reactor in the materials removal process? Summarize various types of reactors with their distinct features.

5. What are the major chemical processes involved in pollution control? Describe and differentiate each process.

6. What are the objectives of the following in water treatment?
   a. Addition of alum
   b. Applying of chlorine
   c. Adding of activated carbon
   d. Passing through zeolite (ion-exchanger)

7. Describe the double boundary layer theory of colloidal particles in suspension. Describe the role of coagulant chemicals added into the suspension.

8. Can chemical processes be used in place of biological processes in wastewater treatment? Outline the principal biological processes with their objectives.

9. Differentiate between an aerobic process and an anaerobic process. Describe the configuration of a typical anaerobic digester.

10. Define the following:
    a. Treatment train
    b. Convection or advection
    c. Diffusion
    d. Stoke's law
    e. Surface overflow rate (SOR)
    f. Hydraulic detention time
    g. Homogeneous and heterogeneous reactions
    h. Continuous-flow stirred-tank reactor
    i. Zeta potential
    j. Ion-exchange
    k. Suspended and attached growth processes

REFERENCES

Ayers, K. W., et al., *Environmental Science and Technology Handbook*, Government Institutes, Rockville, MD, 1994.

Balchen, J. G., and Mumme, K. I., *Process Control: Structures and Applications*, Van Nostrand Reinhold Company, New York, 1988.

Cheremisinoff, P. N., *Biomanagement of Wastewater and Wastes*, Prentice Hall, NJ, 1994.

Cussler, E. L., *Diffusion: Mass Transfer in Fluid Systems*, Cambridge University Press, Cambridge, U.K., 1984.

Eckert, E. R. G., and Drake, R. M., *Analysis of Heat and Mass Transfer*, Hemisphere Publishing Corp., Washington, D.C., 1987.

Hemond, H. F., and Fechner, E. J., *Chemical Fate and Transport in the Environment*, Academic Press, San Diego, CA, 1994.

Jackson, A. T., *Process Engineering in Biotechnology*, Prentice Hall, NJ, 1991.

Kay, J. M., and Nedderman, R. M., *Fluid Mechanics and Transfer Processes*, Cambridge University Press, Cambridge, U.K., 1985.

Metcalf and Eddy, Inc., *Wastewater Engineering: Treatment, Disposal and Reuse*, 3rd ed., McGraw-Hill Book Company, New York, 1991.

Ney, R. E., Jr., *Fate and Transport of Organic Chemicals in the Environment: A Practical Guide*, 2nd ed., Government Institute, Rockville, MD, 1995.

Ray, B. T., *Environmental Engineering*, PWS Publishing Company, Boston, MA, 1995.

Sawyer, C. N., and McCarty, P. L., *Chemistry for Environmental Engineering*, 3rd ed., McGraw-Hill Book Company, New York, 1978.

Steel, E. W., and McGhee, T. J., *Water Supply and Sewerage*, 5th ed., McGraw-Hill Book Company, New York, 1979.

Suffet, I. H., ed., *Fate of Pollutants in the Air and Water Environment*, Part 1 and Part 2, Wiley-Interscience, New York, 1977.

Viessman, W., and Hammer, M. J., *Water Supply and Pollution Control*, 4th ed., Harper and Row, New York, 1985.

Warren, M. R., and Harry, C., *Heat, Mass, and Momentum Transfer*, Prentice Hall, NJ, 1961.

Chapter 8

THE WATER ENVIRONMENT

Quantitative Assessment of Water Resources

The earth's total water resources have been assessed at 1.37×10^9 km^3. As Figure 6-2 shows, 97% of this quantity belongs to the oceans of the world. The remaining 3% is stored on land and only 0.001% is retained by the atmosphere. Of the supply on land, 2% is tied up within the ice caps and glaciers, leaving only 1% for freshwater lakes, rivers, and groundwater combined. Groundwater to a depth of 13,000 ft contains much of this supply.

These quantities indicate the permanent pool of water for each category. However, the more significant quantities are the amounts of water in continuous circulation over the planet, given as fluxes in the hydrological cycle in Figure 6-2. The annual budget of water in circulation is 520×10^3 km^3/yr. Precipitation and evaporation, two counter-balancing arcs of the cycle, each represent this quantity. Precipitation comprises 412×10^3 km^3/year on the world's oceans and 108×10^3 km^3/year on the land areas, and evaporation from the oceans and land comprises of 449×10^3 km^3/year and 71×10^3 km^3/year, respectively. The difference between precipitation and evaporation on land of 37×10^3 km^3/year represents the runoff to the ocean from rivers and underground.

The U.S. possesses a total water resources volume of 146×10^3 km^3, which is only about 0.5% of the water on the land area of the world. (It is only 0.01% of the total water resources of the planet.) Of the resources available in the U.S., 86% belongs to shallow and deep groundwater at 43% each, 13% is stored in freshwater lakes, and the remaining 1% is the total stored volume of other sources including streams. The components of the hydrologic cycle of the U.S. comprising a budget of annual water circulation of 5807 km^3/year, are indicated below:

	km^3/yr	Percentage
Precipitation	5807	100
Surface and groundwater runoff	1936	33
Evaporation	3871	67

The total consumption of water in the U.S. for all purposes in 1980 was 443 billion gal/day or 612 km^3/yr. This amounts to an average overall use of about 2000 gal/person/day – the highest in the world. The largest usage is by irrigation: 136 billion gal/day (30% of the total).

Runoff is an indicator of water available for use. The total consumption in the U.S. is 32% of the annual runoff. On this basis, the water demands are fully met. However, there are spatial (place-to-place), and temporal (year-to-year, month-to-month, day-to-day) variabilities in the runoff (stream flows) that necessitate adoption of resources development and management measures, consisting of the following:

- Resource conservation;
- Interbasin transfer;
- Dams and reservoirs;
- Integrated surface and groundwater usage; and
- Desalination.

Among various natural drainage regions, in the continental U.S. (excluding Hawaii and Alaska), the Columbia–North Pacific region has the highest mean annual runoff of 290 km^3/yr. The lowest mean annual runoff of 7 km^3/yr is observed in the Great Basin and Rio Grande regions.

Quality of Water Resources

The water source on land is from precipitation as part of the hydrologic cycle. Precipitation in its mode of reaching the earth acquires impurities from dissolving gases and particulate matter in the atmosphere. Upon reaching the land, some of it runs off as surface water into rivulets, streams, and rivers; some infiltrates the soil to become groundwater; and some evaporates back into the atmosphere. Both surface water and groundwater dissolve many chemicals and organic compounds while flowing through the soil. Thus, the natural water distributed on the earth possesses many impurities. These impurities impart to it a character that is collectively recognized as the *quality of water*. To these natural impurities, many impurities are routinely added by humans by way of their lifestyle. This constitutes humans' qualitative interference with the hydrologic cycle. When water from a source cannot be used for an intended purpose due to an excessive quantity of an unwanted substance, it is considered to be *polluted*.[1] In this sense, any substance in the wrong place at wrong time acquires the status of a pollutant. However, the natural pollutants are usually convenient to control. Certain human-induced

---

[1] An accurate definition of pollution is given in Chapter 1.

pollutants, on the other hand, make a significant adverse impact on the environment. This is elaborated later in this chapter.

The impurity indicators of water quality belong to physical, chemical, and biological categories. Some common indicators are not distinguished by physical, chemical, and biological attributes. These are known as the *gross measures* of quality. The examples of such measures are suspended solids, odor, alkalinity, hardness, and BOD. They are very common descriptors of water quality. The second type, the *specific measures* of water quality, are the indicators of individual quality characteristic from the category of physical, chemical, or biological attributes such as heavy metals, or toxic organic compounds. The common parameters used to assess water quality are summarized in Table 8-1.

Table 8-1 Common Characteristics of Water Quality Assessment

A. Gross Measures of Quality
 1. Odor
 2. Solids: total solids; suspended solids;
 3. Alkalinity
 4. Hardness
 5. Biochemical oxygen demand
 6. pH
 7. Temperature

B. Specific Measures of Quality:
 I. *Physical characteristics*:
  1. Turbidity
  2. Color
 II. *Chemical characteristics*:
  1. Inorganic chemical parameters:
   i. Toxic chemicals frequently found in water: arsenic, asbestos, barium, cadmium, chromium, copper, fluoride, lead, mercury, nitrate, and selenium
   ii. High level of occurrence but safe at concentration expected *or* their occurrence is extremely rare, at a level at which toxicity is a concern: calcium, carbon dioxide, chloride, iron, lithium, magnesium, manganese, dissolved oxygen, phosphate, potassium, silica, bromide, chorine, iodine, and ozone
   iii. Either occurrence is rare *or* concentration at a level at which toxicity is not a concern: aluminum, cyanide, molybdenum, nickel, silver, sulfate, zinc, and sodium
   iv. Frequently found in low concentration at which toxicity is not a concern: antimony, beryllium, cobalt, tin, thorium, and vanadium

Table 8-1 continued...

> 2. Organic chemical parameters:
>   i. Detergents or surfactants
>   ii. Phenols
>   iii. Halogenated chloro-organic compounds (THMS), such as chloroform
>   iv. Volatile organic chemicals (VOC): benzene, vinyl chloride, carbon tetrachloride, 1,2-dichloroethane, trichloroethylene, 1,1-dichloroethylene, trachloroethylene,1,1,1-trichloroethane, p-dichlorobenzene
>   v. Synthetic organic chemicals (SOC): solvents, pesticides, insecticides, herbicides, fumigants, dyes, PCB, wood preservatives, plastics, resins, insulators, synthetic rubber, perfumes, gasoline, pharmaceutical products, motor oil additives, and refrigerants
>
> III. *Biological characteristics:*
>   1. Total coliform
>   2. Heterotrophic bacteria
>   3. Viruses
>   4. Pathogenic protozoa

Water is a resource of primary importance to humans and other living things. It has multiple uses, ranging from drinking to recreation. The quality requirements in terms of physical, chemical, and biological characteristics are, obviously, not the same for all intended uses, being most stringent for drinking purposes. Because each type of use sets its own limitations, a uniform quality requirement cannot be imposed on all water sources. This aspect has been recognized in the Clean Water Act (CWA) of 1987[2] which requires that each state should:

- Classify the waters[3] within the state boundary according to intended use, e.g., public drinking water supply, propagation of fish and wildlife, recreation, industrial, agriculture, and other uses; and
- Establish numerical quality standards for physical, chemical, and biological characteristics of waters as necessary to support the designated use. U.S. Environmental Protection Agency (U.S. EPA) has published water quality criteria for inorganic and organic chemicals including toxic substances.

These criteria are continuously revised and expanded. State's quality criteria are normally based on and must be at least as stringent as the Federal criteria.

---

[2]The Act can trace its roots to the Federal Water Pollution Control Act, 1972.

[3]Groundwater is excluded from the scope of the Act.

The first step defines the Environmental Quality Objective (EQO) and the second step defines the Environmental Quality Standards (EQS) to secure the objective. This EQO/EQS approach is an endeavor to restore and maintain integrity of the nation's water. The Act has set a goal to achieve a level of water quality that provides for the protection and propagation of fish, shellfish, and wildlife and for recreation in and on the water.

The other goal of the CWA is to prohibit the discharge of any pollutant in natural waters except as authorized by the permit so that the quality of receiving waters does not deteriorate.

Quality of Natural Waters

There are two types of water uses: withdrawal or abstractive uses and nonwithdrawal uses. In abstractive uses, such as for drinking water or irrigation, water is removed from a source for consumption and again returned to the same or a different source through the hydrologic cycle. A water body serves as a raw water source, the quality characteristics of which are modified to meet the standards of the intended usage. The nonwithdrawal uses, such as for fish, wildlife, and recreation, seek to utilize the resource within the water body; the natural water is expected to meet the standards of the usage. As stated above, the CWA sets the goal for the nation's water body to achieve a quality that provides for the protection of fish and wildlife. To meet this objective, the U.S. EPA has prepared the quality criteria for protection of aquatic life covering about 100 parameters including inorganic, synthetic organic, and halogenated organic compounds.

Effluent Discharge into Natural Waters

As a part of the goal requirement, the CWA stipulates that from any point source,[4] no discharge can be made to a water body unless a permit is obtained under the National Pollutants Discharge Elimination System (NPDES) Program of the Act. This permit issued by U.S. EPA or an authorized state agency, allows the discharge of specified pollutants in specified quantity from a designated outfall. To calculate how much can be discharged by the permittee, the following procedure is followed:

---

[4]The Clean Water Act, 1987, recognized two sources of pollution:
1. *Point source* is defined to include "any discernible confined and discrete conveyance from which pollutants are or may be discharged." This may include pipes, ditches, channels, or gullies, which need not be man-made.
2. *Nonpoint sources* cover discharges other than point sources. One of the largest nonpoint contributor is agricultural runoff.

1. The state water quality standard for the pollutant to be discharged is determined. This is a numerical limit that cannot be exceeded in the ambient water to protect the designated use. For example, if arsenic is a pollutant to be discharged, its quality criteria for aquatic life is 0.14 mg/L.
2. The dilution factor is determined. This is the ratio of the combined flow of the receiving water and the effluent to the flow of the effluent from the outfall.
3. The ambient quality standard of the item is multiplied by the dilution factor of the item to obtain the permit limitation.

Thus:

$$\text{TMDL} = \text{DF}(I) \tag{8-1}$$

where TMDL = total maximum daily load of pollutant
DF = dilution factor
I = ambient water quality criteria for the pollutant.

In locations where discharges are expected from more than one source, the state will determine the TMDL for a segment of the water body and allocate this amount among individual dischargers. In locations for which the quality standards are not established or it is difficult to set quality-based permit limitations, the discharge is based on effluent toxicity testing. This involves exposing selected species of aquatic life to an effluent in a laboratory to determine the short-term and long-term effects of exposure to the effluent.

Example 8-1

The level of arsenic for trout propagation has been set at 0.15 mg/L by a state. From an outfall discharging 10,000 gal/day into a river carrying a low flow (7-day, 10% flow) of 2 million gal/day, determine the permit limitation for arsenic discharge.

Solution:

$I = 0.15$ mg/L

$$\text{DF} = \frac{2.0 \times 10^6 + 10{,}000}{10{,}000}$$

$= 201$

From Equation 8-1:

$\text{TMDL} = (201)(0.15) = 30.2$ mg/L

Quality of Processed Water

Water quality in withdrawal usages can be treated first before putting it to use. Many international (World Health Organization, UN agencies, European Economic Community), Federal (U.S. EPA, U.S. Geological Survey, Department of Commerce, Department of Agriculture, Public Health Service), state governments, other institutions (American Water Works Association, water control boards, municipalities), and private-sector organizations have set the quality standards for processed waters for various usages such as drinking, industry, power, food and beverage, irrigation, livestock, and poultry purposes. Most comprehensive guidelines and regulatory requirements have been provided for drinking water systems including groundwater under the Safe Drinking Water Act (SDWA), enacted in 1974 and amended in 1986. The SDWA requires U.S. EPA to establish the national standards to be implemented and enforced by the states.

The first step in the process of setting up the standards by U.S. EPA is to propose the *maximum contaminant level goal*, (MCLG) for all substances expected to be present in water. An MCLG is set at the level at which no known or anticipated adverse effect on the health of a person occurs and which allows an adequate margin of safety. Once U.S. EPA promulgates an MCLG for a substance, the next step is for the U.S. EPA is to issue a *maximum contaminant level*, (MCL) for that chemical, that is as close to the MCLG as is feasible to achieve.

The procedure to determine an MCLG is as follows:

For most of the toxic or hazardous substances, scientists have conducted laboratory tests on animals and have determined concentration values below which the substance will not cause any adverse health effects. These are known as *no observable adverse effects levels* (NOAELs), which are expressed in milligrams per kilogram of body weight per day. A *reference dose*, (RfD) is determined by applying a factor of safety to NOAEL. To arrive at an RfD value in milligrams per day it has to be multiplied by the weight of an average person (assumed to be 70 kg for an adult and 10 kg for a child):

$$\text{RfD} = \frac{(\text{NOAEL in mg/kg/day}) (70 \text{ kg})}{\text{FS}} \quad (8\text{-}2)$$

where FS = factor of safety; 10 when good acute or chronic exposure data are available for humans and other species, 100 when good data are available but not for humans, 1,000 when data are limited and incomplete.

The MCLG is computed as follows:

$$\text{MCLG (mg/day)} = (\text{RfD}) - (\text{contribution from food}) - (\text{contribution from air})$$

When insufficient data are available for air and food, a 20% contribution from drinking water (i.e., an MCLG of 20% of RfD) is assumed. To compute the MCLG in milligrams per liter, Equation 8-3 should be divided by the water intake by a person in a day (assumed to be 2 L/day for an adult and 1 L/day for a child). MCLGs have been promulgated by U.S. EPA for non-carcinogenic substances. For carcinogens, U.S. EPA has not been able to determine whether there is an exposure threshold below which a carcinogen will not cause the adverse health effects. Accordingly, U.S. EPA has usually set carcinogens MCLGs=0.

Example 8-2
Based on the research work at the National Academy of Sciences, an NOAEL of 2.8 mg/kg/day has been determined for arsenic. Determine the MCLG, assuming an uncertainty factor of 1000, air intake of 0.12 μg/day, and food intake of 61.5 μg/day.

Solution:
From Equation 8-2:

$$\text{RfD} = \frac{2.8(70)}{1000} = 0.2 \text{ mg/day}$$

$$\text{MCLG} = 0.2 - (61.5 \times 10^{-3}) - (0.12 \times 10^{-3}) = 0.138 \text{ mg/day}$$

$$\text{MCLG (mg/L)} = \frac{0.138}{2} = 0.069 \text{ mg/L.}[5]$$

U.S. EPA has categorized the pollutants to be regulated into several groups based on a common trait such as the properties, treatment techniques, or formation pathways. These groups are:

- Volatile organic chemicals (VOC);
- Synthetic organic chemicals (SOC);
- Inorganic chemicals (IOC);
- Techniques for lead and copper;
- Radionuclides;
- Surface water treatment rules (SWTR) for microbial contaminant; and
- Disinfection and disinfection byproducts (DBP).

---

[5]The EPA recommended MCLG is 0.05 mg/L

These are briefly discussed below:

- Eight VOC's were regulated by U.S. EPA in 1987 through establishment of the MCLGs and MCLs. Another 51 unregulated VOC's were identified for monitoring purposes.
- Thirty selected SOCs, including important gasoline constituents and PCBs, were regulated by fixing MCLs in 1989.
- U.S. EPA also promulgated (in 1989), MCL regulations for nine inorganic chemicals.
- Monitoring program and treatment techniques were set by U.S. EPA (in 1988) for MCLCs and MCLs for lead and copper..
- Radionuclides refer to wastes of radioactive origin. U.S. EPA regulated gross alpha particle activity, beta particle and photon radioactivity, radium 226 and 228, and radon.
- U.S. EPA (in 1989) revised the MCLs for total coliform bacteria and specified the four acceptable analytical methods for determination of total coliform bacteria.
- U.S. EPA (in 1989) promulgated the SWTR. This rule contained disinfection requirements for surface water supplies. The rule prescribed the techniques to protect against pathogens, heterotrophic bacteria, and viruses.
- SDWA (1986) requires that U.S. EPA promulgate disinfection requirements for all public water supplies. In 1992 U.S. EPA proposed the Groundwater Disinfection Rule. A list of disinfection byproducts (DBP) that are expected to occur from addition of commonly used disinfectants comprising chlorine, chloramine, chlorine dioxide, and ozone has been established. One class of DBP, known as trihalomethane (THMs) has been regulated by MCL including four compounds, chloroform being the common one. Limits for other disinfectant byproducts are being developed.

There are certain quality characteristics that, beyond certain limits, tend to make water disagreeable to use, although their MCLs are either high or not regulated and, hence, they do not pose any adverse public health effects at those limits. SDWA provided for the *secondary maximum contaminant levels*, (SMCL) that are advisable levels but not enforced by U.S. EPA for contaminants such as chloride, color, copper,[6] corrosivity, foaming agents, hydrogen sulfide, iron, manganese, odor, pH, sulfate, total dissolved solids, and zinc.[7]

[6,7]For these substances MLCs have been fixed that are higher than SMCLs.

Pollution of Water Resources

The need for the detailed quality measures as discussed above arose because of a widespread degradation of the national water resources by human activities. The pollution, as earlier stated, falls under two categories; point or direct sources and the nonpoint or diffuse sources. The former, as the name suggests, covers an identifiable discharge point. The principal sources and types of pollutants under the two categories are given in Table 8-2.

A major contamination occurs due to gross organic pollution originating from sewage treatment effluents and farm slurries. These organic materials provide food and nutrients for the growth of microorganisms in receiving waters. This contributes to two problems:

1. Microorganisms use the dissolved oxygen of the water body rapidly, thereby killing higher forms of aquatic life; and
2. Waste, particularly domestic sewage, might contain pathogenic organisms that make water unhealthy to use.

Many diseases such as cholera, typhoid, dysentery, gastroenteritis, hepatitis, and various forms of diarrhea are linked to pathogens in water. The gross pollutants are generally measured in concentrations greater than 1 mg/L. The laboratory tests commonly used to measure gross organic pollutants are BOD, chemical oxygen demand (COD), total organic carbon, and total oxygen demand. This type of pollution is well understood and its control is mainly a matter of waste treatment, as discussed subsequently.

On the other hand, the priority pollutants consisting of toxic inorganic and organic chemicals are capable of causing adverse health effects at very minute concentrations because these substances remain unchanged in the aquatic environment for a very long time. The toxicity is measured at two levels: acute, which causes an effect (usually death) within a short period, and chronic, which causes an effect over a prolonged period of time. These substances include inorganic chemicals such as antimony, arsenic, beryllium, cadmium, chromium, cobalt, copper, lead, mercury, nickel, selenium, silver, thallium, vanadium, and zinc. The organic chemicals include synthetic compounds belonging to fuels, detergents, pesticides, solvents, plasticizers, pharmaceuticals, feed additives, brighteners, paint lacquers, rubber chemicals, and fibers. Their carcinogenic effect is a major concern. Their concentration is measured in the range of 10^{-12} to 10^{-3} mg/L.

Among the metals, almost any element of the periodic table can be present in water. It is the presence of higher than desired concentration levels of a particular metal that poses the pollution problem.

Table 8-2 Major Sources and Types of Pollutants

	Point Sources	**Nonpoint Sources**
Sources	Domestic Industrial Storm drains	Atmospheric fallout Agricultural runoff Urban runoff Septic systems Landfills/seepage Spills
Types	Gross organic pollutants Priority chemicals Inorganic minerals Thermal discharges	Acidity (acid rain) Fertilizers (nitrates, phosphates) Pesticides Sediments
Control Measures	NPEDS permit for effluent discharge under the CWA, 1987 Storm water discharge permit under CWA, 1987 Special criteria for discharge of heat under Section 316 of the CWA, 1987 Pretreatment of discharge into public sanitary systems Special criteria for ocean discharge under CWA, 1987 Ocean dumping regulated under the Marine Protection Research and Sanctuaries Act (MPRSA)	Measures required to be undertaken by the states to reduce the pollutant loadings under Section 319 of the CWA, 1987.

Since the quality constraints are different, the polluting potential of a water body has to be assessed in the context of water use. The trace metals in minute concentrations discussed above have similar relevance. Thermal pollution is caused by cooling water, industrial, and municipal discharges. The increased temperature reduces oxygen absorption capacity, which alters the aquatic species composition.

The pollution from nonpoint sources is widespread and, in general, more difficult to control. Nitrogen and phosphorus are essential nutrients for plant and animal growth. These are extensively used as fertilizers in agriculture. Excess amounts of these wash off from agricultural lands and domestic septic fields into surface waters, stimulating algal and plant growth that later dies and settles at the bottom of the water body to demand more oxygen. Erosion and sedimentation of soils from strip mining operations, urban runoff, crop land and deforestation fill in stream channels, reservoirs, and lakes and form a blanket over the aquatic organisms at the bottom, suffocating them.

Fate of Pollutants in the Water Environment

The determination of what happens to pollutants after they enter the water is an important phase. The pathways that pollutants follow on their introduction to water bodies—streams, lakes, estuaries, and the ocean—are depicted in Figure 8-1.

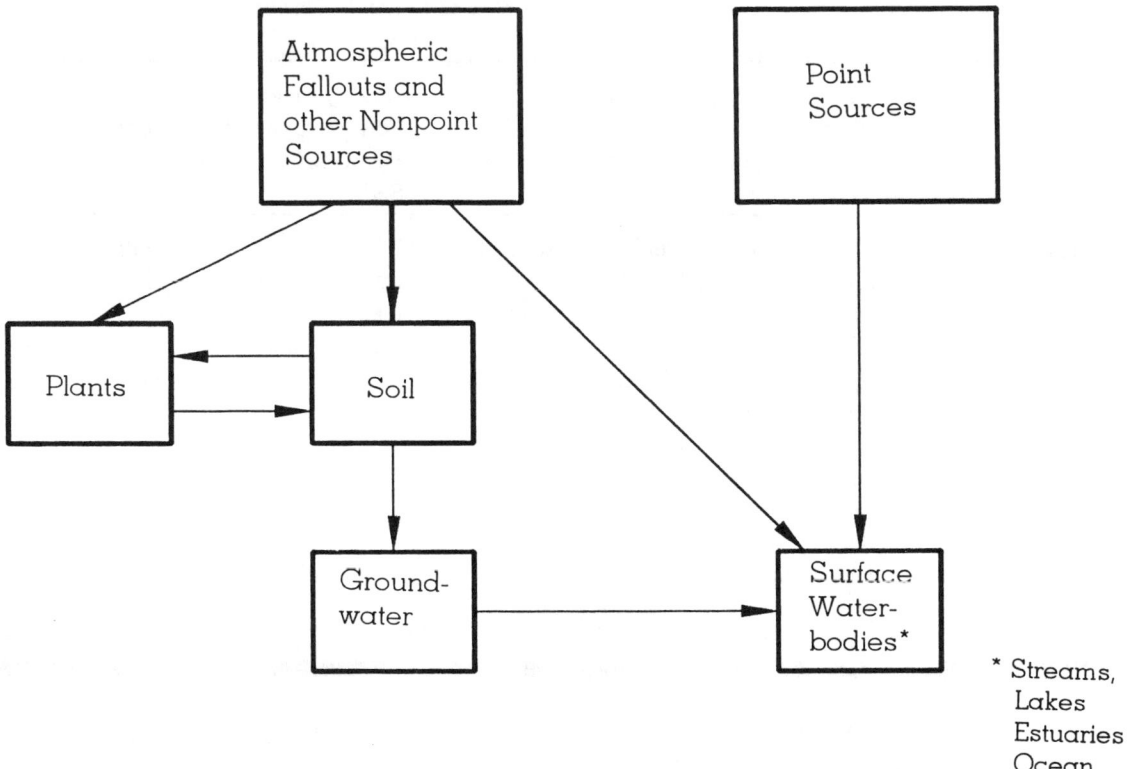

Figure 8-1 Pathways of Pollutants to Natural Waters

Point sources usually discharge directly into receiving waters. Nonpoint atmospheric fallout in the form of acidity and particulate matter return to earth via rainfall. This pollutant is retained by soil, exchanged with plants and soil organisms, and along with other sources such as septic systems, reaches surface waters directly or through groundflow. Other nonpoint sources follow both direct and land routes.

After the pollutants reach a water body, they are subjected to physical, chemical, and biological influences in accordance with environmental processes described in Chapter 7. The pollutants are transported by convective and diffusion modes and some settle to the bottom by sedimentation. Physicochemical reactions cause them to precipitate, coagulate, adsorb with bottom silt, and form many chemical species. Microorganisms commence biochemical oxidation on degradable organics. However, another key factor in the water environment exercises a major influence on the fate of pollutants. Natural waters have many species of aquatic life. Most aquatic organisms, including plants and animals, absorb pollutants that are lipid soluble, either directly from water or indirectly through eating contaminated food-chain organisms. The pollutants are accumulated with time within living organisms in relatively high concentrations. This process is termed as *bioconcentration* or *ecological magnification*. This has lead to serious problems in water pollution. U.S. EPA has proposed to evaluate bio-concentration as a major criterion in setting up of effluent quality standards.

Bioconcentration Concept

Among water pollutants, a class of pollutants consists of synthetic organic compounds, toxic metals, and radionuclides that are usually present in very small quantities. These micropollutants, however, have significant impact because of their properties of :

- Low-water and high-lipid solubility;
- Bioconcentration; and
- Toxicity or carcinogenicity.

When pollutants are consumed by aquatic organisms, the water-soluble compounds are excreted but the water-insoluble and lipid-soluble compounds are absorbed by cell membranes or walls, and stored in tissue lipids. The rate of absorption is a first-order process expressed as a form of Fick's Law (Equation 7-2) wherein it is a function of the concentration of the pollutant, the time of exposure, the area of absorbing surface of the cell, the thickness of membrane or skin and the diffusion coefficient. The concentration of pollutant in tissue, even though some detoxification or clearing process occurs, increases with the age of the aquatic organism. Data from laboratory studies of model ecosystems indicate that for synthetic organic compounds there is an inverse relationship between water solubility and bio-concentration.

In a large pool of an aquatic environment contaminated at a constant level, such as a lake, the concentration of the pollutant within a single homogeneous aquatic organism can be given by the following relation pertaining to drug absorption and metabolism:

$$\frac{dC}{dt} = \frac{C_o K_a}{V} - K_c C \qquad (8\text{-}3)$$

where C = concentration of pollutant within organism at time t; ppm
K_a = absorption rate constant determined from lipid/water partition coefficient of the pollutant (mg per unit time or volume per unit time
V = mass, in mg, or volume of organism
C_o = concentration of pollutant in water; ppm
$K_c t$ = clearance rate constant of the pollutant by degradation, elimination, or other process, per unit time.

By integrating Equation 8-3:

$$C = \frac{C_o K_a}{K_c V} (1 - e^{-K_c t}) \qquad (8\text{-}4)$$

Instead of the direct absorption of a pollutant by an aquatic organism, if a contaminated prey is ingested by another organism in a food chain, Equation 8-4 could be used to compute the accumulated concentration of the pollutant within the consuming organism. The term $C_o K_a$ should then be substituted by $c(C_p W_p)$, where C_p is the concentration of pollutant in the prey, W_p is the mass of daily intake of prey, and c is a suitable factor to measure extraction of the contaminant from the food.

In accordance with Equation 8-4, the concentration within the organism will increase until a steady state is reached when the clearance equals the absorption, then the time will disappear.

Thus, at the steady state;

$$C = \frac{C_o K_a}{K_c V} \qquad (8\text{-}5a)$$

or

$$\frac{C}{C_o} = \frac{K_a}{K_c V} \qquad (8\text{-}5b)$$

where C/C_o is known as the *bioconcentration factor*.

When the rate of clearance K_c is very small as compared to the rate of absorption K_a, the value of C at steady state will be so large that it will not be attained within the life span of the organism, and the bioconcentration will continue to rise.

In a study by the U.S. EPA (1972), on lake trout in Lake Michigan, the bioconcentration factors for DDT and PCB were more than 3×10^6 each.

In the case of fully water-soluble compounds, for which K_a is zero, there is no accumulation of pollutant within tissue lipids. Essentially, the organisms have the ability to convert relatively water insoluble (lipid soluble) compounds into more water soluble compounds that are excreted from body. This detoxification or clearing process is carried out by microsomal oxidase enzymes present in organisms. The process is known as *biotransformation*. However, the level of microsomal oxidase in living beings cannot cope with hundreds of thousand of toxic pollutants injected into the environment by humans, many of which have been synthesized to persist for decades.

Example 8-3
A lake has a DDT concentration of 0.0006 ppm. The population of trout in the lake has an absorption rate of 1 mg/year for each milligram mass of the trout. The clearance rate is 0.005 per year. Determine the bioconcentration in the lake trout at a steady-state.

Solution:
$C_o = 0.0006$ ppm
$K_a/V = 1$ mg per year/mg
$K_c = 0.005$ per year

From Equation 8-5a:

$$C = \frac{(0.0006)(1)}{(0.005)} = 0.12 \text{ ppm or mg/kg.}$$

Measurements of Quality Characteristics

Water quality analysis deals with determination of the exact amount of a substance by weight present in a known weight or volume of water (or wastewater). Two terms used alternatively to express the result are: milligrams per Liter, referring to milligrams of the substance present in 1 L of the sample, and parts per million, indicating how many parts of the substance exist in one million parts of the substance and water combined, measured by weight. These units are discussed in Chapter 4. The quality standards, defined earlier in this chapter, are set in terms of these units. Against those standards, the results of laboratory analyses are compared to ascertain how efficiently a plant is operating and to predict and prevent problems within a process. Many different methods can be proposed to determine the quantity of the same substance. The analytical results obtained by different approaches, however, are often not identical. This leads to the obvious problem of data interpretation in cases of a litigation. To avoid this situation, the American Public Health Association proposed the Standard Methods of Analysis in 1905, which were subsequently elaborated and enlarged by the American Water Works Association and the Water Pollution Control Federation. The 14th edition of the Standards appeared in 1976 and is designed to be used by trained analysts having knowledge of chemistry, biology, and bacteriology. Only basic analytical methods are presented here.

Laboratory results are only as good as the sample collected. Two types of samples are collected for analysis. Grab samples are taken in full at any given time, usually during low flows. Composite samples are collected in a small bottle at regular intervals, usually every hour, pooled over 24 hours, into a larger bottle and kept refrigerated at 4°C to 10°C. The number of samples to be collected depends upon the nature of the project, the type of material to be sampled, and changes that can occur in the character of the material.

Analytical Procedures

One of the following four procedures is applied to analyze a water or wastewater sample.

Gravimetric Analysis

This method involves weighing a substance by means of an analytical balance. Examples are total solids, suspended solids, and volatile solids determination. Gravimetric methods are not commonly used.

Volumetric Analysis

Also known as the *titration process*, this method involves measuring the volume of a standard solution (of a known concentration) that is needed to complete a reaction with a particular substance present in a fixed volume of the sample. It is a commonly used method of analysis. Examples are chlorides, dissolved oxygen, BOD, and chemical oxygen demand. The following relation is used to compute the concentration of the substance:

$$V_1 N_1 = V_2 N_2 \tag{8-6}$$

where V_1 = volume of standard solution
 N_1 = normality of standard solution (known)
 V_2 = volume of sample used
 N_2 = normality of substance in the sample.

Example 8-4

A 1700 mg/L (0.01 Normal[8]) solution of silver nitrate ($AgNO_3$) is used in volumetric analysis of chloride content in water. Ten milliliters of $AgNO_3$ is required to reach the equivalent point with 100 mL of the sample. Determine the concentration of chloride in water.

Solution:
From Equation 8-6:
$(10)(1700) = (100)(N_2)$
$N_2 = 0.001$ Normal

Chloride concentration = (normality)(equivalent weight)(1000)
= (0.001)(35.5)(1000)
= 35.5 mg/L

Colorimetric Analysis

This method uses color of solution as a measure of concentration of a substance. The solution formed must have definite color characteristics that are directly proportional to the concentration of the substance being measured. The procedure is fast, economical, convenient, and is very widely used.

Colorimetric measurements can be made with a wide range of equipment grouped in the following categories: standard color comparison tubes (Nessler tubes), photoelectric calorimeters, and spectrophotometers.

Color comparison tubes are simple to use and were very commonly used at one time. In this method, a series of standard color solutions are used to match the color of the solution containing the substance. Applications of the method are limited because fresh color standards have to be prepared for each test and human observation is inaccurate.

A schematic diagram of a photoelectric device is shown in Figure 8-2. The device works on the following principle:

1. A source of energy, such as an ordinary light bulb, is used.
2. The beam of light is passed through a color filter and monochromatic rays are obtained.
3. Monochromatic light rays are directed through a cell that contains the sample.
4. The light that penetrates the sample hits a photoelectric cell.

---

[8] mg/L = normality × equivalent weight × 1000,
or mg/L = molarity × molecular weight × 1000

5. The current developed by the photoelectric cell is measured by a galvanometer, which is a measure of the absorbance of light by the sample; it is a distinct characteristic of a substance.
6. The instrument is calibrated for a concentration measure. A blank sample containing distilled water is used for initialization.

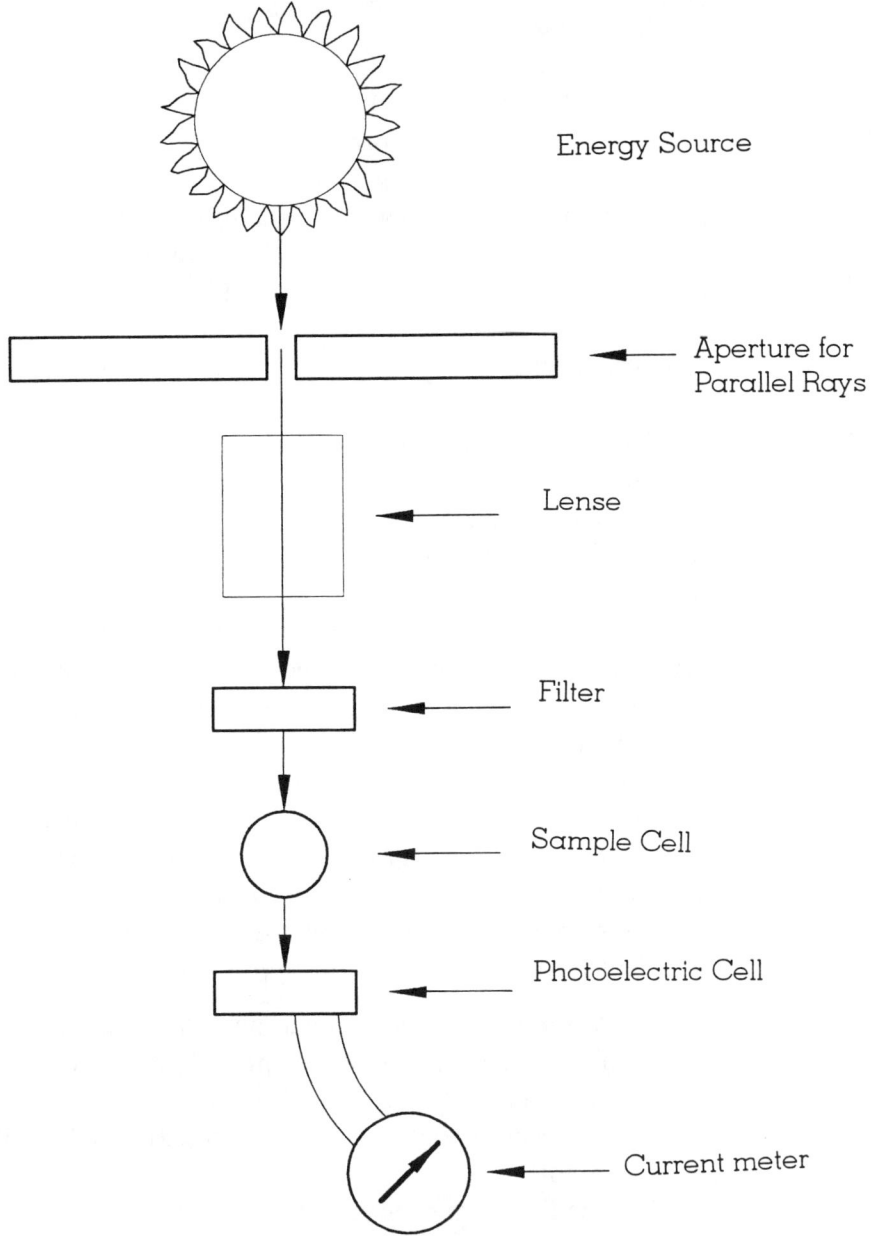

Figure 8-2 Schematic Diagram of a Photoelectric Device

A limitation on the device is the requirement for separate color filter for each determination. Spectrophotometers essentially operate with a similar principle except for the manner in which monochromatic light is obtained. A prism or diffraction grating is used that enables the selection of monochromatic rays of any wavelength in the visible spectrum of light. Spectrophometers have a very wide adaptability.

Direct Instrumental Measurements

A variety of sensitive instruments have been developed that use a distinct property of an element or compound to detect presence and concentration of a substance. For example, the ability of a colored solution of a substance to absorb or transmit light, the capacity of a solution to carry a current, or the travel rate of a substance in a gaseous state, are used as a basis of measurements. The instruments are grouped as follows:

Optical Instruments: These are based on the interaction of light or radiant energy on matter. The most frequently observed parameter is the wavelength. The radiant energy may be visible light, ultraviolet rays, X-rays, and radiowaves.

Optical instruments may be further subclassified based on whether they have been designed to measure the absorption, emission, or dispersion of energy by a substance.

Spectrophometers measure the absorption of the radiant energy by a substance. Instrumentation is extended to ultraviolet spectrophotometers, which are suitable for measurements of trace concentrations of organic compounds, and infrared spectrophotometers, which are suitable for pesticides and complex chemicals.

Flame photometry, *atomic absorption spectrometry*, and *emission spectroscopy* are methods that measure the emission of energy by a substance. In flame photometry, the sample is placed into the flame. The radiated energy is passed through a prism to isolate a portion. A photocell with amplifier is used to measure the intensity of the isolated radiation. Atomic absorption directs a light beam as well, at the sample in the flame, thus combining absorption measurement with radiation measurement. Spectroscopy uses another source instead of a flame, such as a high-voltage spark. Emission methods are suitable for measurements of trace elements in water.

Turbidimeter and *nephelometer* are examples of instruments that measure the dispersion or scatter of energy by a substance in a solution.

Electrical instruments: These instruments are based on the relationship between electrical and chemical phenomena; i.e., voltage developed across electrodes in a solution, flow of current when a small voltage is applied, and equivalence relations between amount of electricity and quantity of chemical changes. The methods based on these principles are known as *potentiometric*, *polarographic*, and *coulometric*, respectively. The pH meter and dissolved oxygen analyzer are the most common instruments of this category.

Gas chromatography: Gas chromatography is a highly versatile instrumental method developed in 1951 that provides a convenient means to perform routine quantitative analysis. A schematic diagram is shown in Figure 8-3.

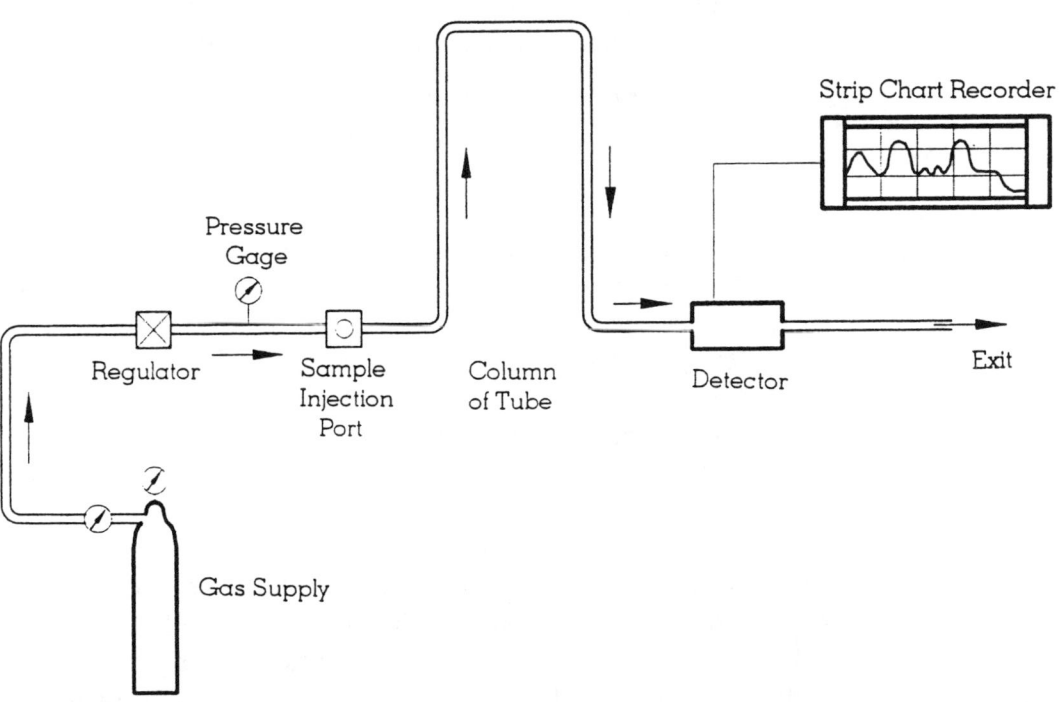

Figure 8-3 Schematic Diagram of a Gas Chromatograph

A gas cylinder continuously supplies a carrier gas such as hydrogen, helium, or nitrogen at a constant rate. A small sample is injected at the sample port where it instantly evaporates into various constituent components. The carrier gas carries these gaseous components through a chromatic column, which is a long tube of a narrow diameter. Because gaseous forms of different constituent substances travel at different rates, they emerge from the column at different times and the peaks are recorded on a strip chart. Each peak represents a specific chemical compound. The area under the peak is proportional to the concentration of the compound in the sample. The time required for each compound to emerge from the column is a characteristic feature of the compound.

Basic Examination of Water and Wastewater

The following laboratory tests are commonly performed to describe water and wastewater quality.

pH

pH is important in all phases of environmental engineering because it effects all chemical and biological processes. It is a numerical expression of the intensity of acidity or basicity; a value of 7.0 is a neutral pH. A number below 7.0 denotes acidity and that above 7.0, basicity. pH can be measured electrometrically. A pH meter employing a gas electrode can be used to make direct measurements of the pH value. The instrument should be standardized against a buffer solution of a known pH. The electrodes should never be allowed to dry and are stored in a solution with high ionic strength.

Acidity

Acidity is significant because it contributes to corrosiveness. Acidity of natural water is caused by carbon dioxide and minerals. Acidity is imparted to rain by burning of fossil fuels. The testing procedure is:

1. Measure 100 mL of a water sample into an Erlenmeyer flask
2. Add three drops of phenolphthalein indicator
3. Add 0.02 N (800 mg/L) sodium hydroxide (NaOH) from a buret and constantly swirl the flask contents above a white surface until a pink color appears and lasts for at least 3 sec.
4. Total acidity (mg/L as $CaCO_3$) = $\dfrac{\text{mL of 0.02 N NaOH} \times 1000}{\text{mL of water sample}}$ (8-7)

Alkalinity

Alkalinity is contributed by hydroxide, carbonate (CO_3), or bicarbonate (HCO_3). It is due to salts of weak acids and strong bases. Alkalinity is a measure of the buffer capacity to resist a drop in pH resulting from acid addition. The test procedure is as follows:

1. Place 100 mL of sample in an Erlenmeyer flask.
2. Add two drops of phenolphthalein which will make the sample color pink; if no pink color appears, there is no hydroxide or carbonate alkalinity.
3. Add 0.02 N (980 mg/L) sulfuric acid (H_2SO_4) from a buret and swirl the flask over a white surface until the pink color disappears.
4. Hydroxide and carbonate alkalinity:

$$(\text{mg/L as } CaCO_3) = \frac{\text{mL of 0.02 N } H_2SO_4 \times 1000}{\text{mL of water sample}} \qquad (8\text{-}8)$$

5. Add two drops of methyl orange indicator to the flask and continue adding H_2SO_4 from the buret until the color of the flask turns faint pink.
6. Total (including bicarbonate) alkalinity:

$$(\text{mg/L as } CaCO_3) = \frac{\text{total mL of 0.02 N, } H_2SO_4 \times 1000}{\text{mL of water sample}} \qquad (8\text{-}9)$$

Dissolved Oxygen

Dissolved oxygen (DO) is one of the most important parameters of water quality in natural waters. The higher the concentration of DO, the better the water quality. There are two methods of testing: the membrane electrode method and titration method.

Membrane electrode meters have probes that can be lowered directly into water. The electrode probe senses a small electric current that is proportional to the DO concentration in water.

The steps of the titration method follow:

1. Completely fill a 300-mL BOD bottle with water, allowing it to overflow.
2. Add 2 mL manganese sulfate (480 g/L solution) and 2 mL alkaline iodide-sodium azide solution[9], below the surface of the sample; replace stopper and shake well.
3. Remove stopper, add 2 mL of concentrated H_2SO_4, mix until no floc is visible, allow to stand for at least 5 min (up to 2 hr).

---

[9]Dissolve 500 g sodium hydroxide in 500- 600 mL distilled water; dissolve 150 g potassium iodide in 200- 300 mL distilled water in a separate container. Mix both solution when they are cool. Dissolve 10 g sodium azide in 40 mL of distilled water; add this to previous solution; then add distilled water to the mixture to make 1 Liter volume.

Exercise caution in the procedure; the products are caustic.

4. Withdraw 200 mL of solution into an Erlenmeyer flask and titrate with sodium thiosulfate (0.025 N or 6.205 g/L) solution until the yellow color disappears.
5. Add 1 mL starch (6 g/L) solution and continue titration with sodium thiosulfate until the blue color disappears.
6. Number of milliliters of sodium thiosulfate used in the titration represents the DO in milligrams per liter.

Total Solids

This refers to the residue left in a drying dish after evaporation of a water or wastewater sample, usually expressed in milligrams per liter. It is an indicator of the combined inorganic and organic content of the wastewater. The volatile fraction of the total solids, which is ignited or burned at 550°C, is a rough measure of the organic content of the waste.

Laboratory steps to determine total solids and total volatile solids are shown in Figure 8-4.

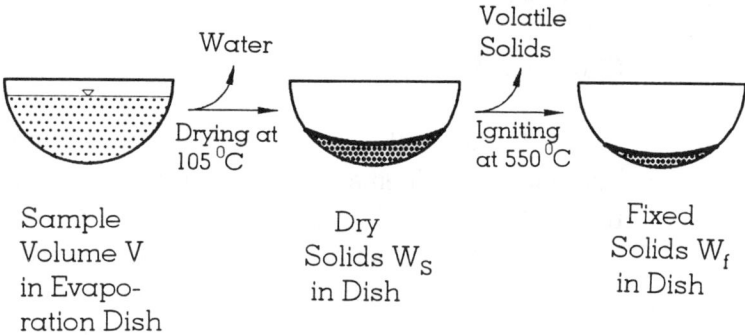

Figure 8-4 Steps of Laboratory Procedures to Determine Solids Concentration

Example 8-5

From a test on a wastewater sample, the following data were recorded. Calculate the total and volatile solids concentration in the wastewater.

Volume of wastewater sample placed in evaporation dish = 100 cm^3
Weight of empty dish (W_1) = 58.942 g
Weight of dish plus dry solids at 105°C (W_2) = 59.049 g
Weight of dish plus ignited solids at 150°C (W_3) = 59.00 g.

Solution:
Dry solids $W_s = W_1 - W_2 = 59.049 - 58.942 = 0.107$ g
Volatile solids $W_v = W_2 - W_3 = 59.049 - 59.000 = 0.049$ g

$$\text{Total dry solids}^{10} = \left(\frac{0.107 \text{ g}}{100 \text{ cm}^3}\right)\left[\frac{1000 \text{ mg}}{1 \text{ g}}\right]\left[\frac{1000 \text{ cm}^3}{1 \text{ L}}\right]$$
$$= 1070 \text{ mg/L}$$

$$\text{Total volatile solids}^{11} = \left(\frac{0.049}{100}\right)(1000)(1000)$$
$$= 490 \text{ mg/L}$$

Biochemical Oxygen Demand

The BOD test determines the amount of organic material in wastewater by measuring the oxygen consumed by microorganisms while they decompose the waste. The test consists of diluting the sample with high-quality distilled water and determining the DO prior to and following a five day incubation of the sample at 20°C. The theory of the procedure has been described in Chapter 5.

Coliform Bacteria

Escherhia coli (E-coli) bacteria from the coliform group inhabit the intestinal tract of humans and animals and are extracted with feces. They can be easily tested for in a laboratory and are considered nonpathogenic. However, their presence in a water sample shows that the water has made contact with feces. Since the feces may contain some other pathogenic bacteria, the presence of coliform bacteria is a useful indication of possible pollution. The procedure is described in Chapter 5.

---

[10] Inorganic and organic content.

[11] Organic content.

PROBLEMS

1. It has been determined that antimony is toxic to freshwater algae at 0.6 mg/L. Determine the permitted load of antimony from an outfall of 50,000 gal/day discharging into a river carrying a low flow of 1 MGD, to preserve the algae growth.

2. For fluorides, a minimum concentration of 0.1 mg/L has been recommended for protection of saltwater species. A concentration exceeding 1.5 mg/L constitutes a hazard to the marine environment. Determine the minimum and maximum daily load from a pipe discharging 5,000 gal/ day in a rivulet carrying a low flow of 100,000 gal/day.

3. In Problem 2, if the low flow is 0.2 ft^3 per sec, determine the maximum load.

4. In a stretch of a stream, four outlets are discharging at a rate of 20,000, 30,000, 40,000, and 50,000 gal/day, respectively. The down-stream low flow of the stream is 5 MGD. A water quality criteria for protection of aquatic life has been set at 1.2 µg/L for silver. Determine the total daily permitted load of silver and its allocation among outlets based upon their respective discharges into the stream.

5. For cadmium, no adverse effects were observed at 0.75 mg/kg/day based on a study on rats. Assuming an uncertainty factor of 1000, a respiratory intake of 0.2 µg/day, and the dietary intake of 50 µg/day, determine the maximum contaminant level goal (MCLG).

6. In laboratory rats supplied with 25 mg/L of chromium for one year, the NOAEL was 2.41 mg/kg/day. With an uncertainty factor of 500, determine the MCLG. Assume air and food intakes of 15 and 100 µg/day, respectively.

7. The MCLG for copper has been fixed at 1.3 mg/L by the U.S. EPA. The daily intake through air and food are 0.005 and 2 mg/day. Compute the NOAEL when the factor of uncertainty is 10.

8. The MCLG is 4 mg/L for fluoride. Assume 20% exposure from drinking water and an uncertainty factor of 10, determine the RfD and NOAEL for fluoride.

9. A lake has a PCB concentration of 0.00082 ppm. The bioconcentration factor in the fish population of the lake is 3×10^4. What is the concentration of PCB in fish tissues? If the rate of clearance is 0.0001 per year, what is the rate of absorption for each unit mass of fish for a steady state?

10. Methylmercury concentration in tuna from sea water is 0.2 ppm. For an average weight (mass) of 5 g, the absorption rate is 10×10^3 mg/year and the clearance rate constant is 0.0002 per year. What is the average concentration of the mercury in the seawater in a steady state?

11. The residue of DDT in trout from Cayuka Lake, NY has been evaluated to be 1 ppm at 1 year of age. This increased to 11 ppm at 20 years of age. Assuming the absorption and clearance rates to be constant and ignore the variation of weights; determine the clearance rate constant. The system is not in a steady state.

12. In Problem 11, if the absorption rate is 1,000 mg/year per mg mass of trout, what is the concentration of DDT in Cayuka Lake?

13. Coho salmon form about 40% of the diet of the mink. The PCB concentration in Coho salmon is 15 ppm. The intake-to-organism mass ratio (W/V) is 0.001. One-tenth of the daily intake is absorbed as contaminant (factor c). Determine the PCB concentration in mink. The coefficient of degradation is 0.005/day. Treat it is as a steady-state system. (Hint: $W_p = 0.4W$ or $W_p/V = (0.4)(0.001)$.)

14. A 100-mL sample containing calcium hydroxide, $Ca(OH)_2$ is titrated with 1,000 mg/L (0.02 N) solution of sulfuric acid, H_2SO_4. A volume of 10 mL of H_2SO_4 is required to reach the end point. Determine the concentration of calcium hydroxide.

15. Twelve milliters of 49 g/L (1 N) solution of H_2SO_4 exactly neutralizes 100 mL of sodium hydroxide, NaOH. What is the concentration of NaOH in the solution?

16. A sample contains 20 g/L of hydrochloric acid solution. Calculate the volume of the sample required if 15.5 mL of 56 g/L solution of KOH fully neutralizes the acid.

17. A 100-mL sample requires 12 mL of 0.02 N solution of sodium hydroxide to end point titration. What is the acidity of the sample as $CaCO_3$?

18. The following data are from a test on a wastewater sample. Calculate the total and volatile solids concentrations in milligrams per liter.

Weight of evaporation dish = 69.502 g
Weight of oven-dried sample with dish = 69.777 g
Weight of furnace-dried sample with dish = 69.658 g
Volume of sample = 200 cm^3

19. Listed below are total solids data from a laboratory test. Calculate the inorganic and organic solids concentration.

    Weight of empty dish = 80.237 g
    Weight of dish plus ovendried solids = 80.372 g
    Weight of dish plus ignited solids = 80.275 g
    Volume of sample = 85 mL

STUDY QUESTIONS

1. Why is it necessary to consider the potential use of a water resource for judging its quality? Which type of use expects the most comprehensive quality requirements? Why?

2. What are water quality criteria? How are they determined?

3. How has the US Environmental Protection Agency grouped water pollutants? Briefly discussed these groups.

4. What are the two broad categories of water pollutants? What are the principal sources of pollutants under these categories?

5. Discuss the types of pollutants found in water supply. Describe the problems expected to result from each type of pollutant.

6. What is bioconcentration? How does it occur? What is its significance?

7. What are the four categories of procedures to analyze a water sample? Outline and differentiate each of them.

8. Summarize the basic laboratory tests performed on water and wastewater samples.

9. Discuss the sequence of reagents used in a laboratory test for dissolved oxygen by the titration method.

10. What are total solids, inorganic solids, and organic solids? How are they determined by a simple laboratory test?

11. What is the principle of the photoelectric colorimeter? Make a schematic diagram of such a device. Differentiate between a photoelectric colorimeter and a spectrophotometer.

12. Define the following:
    a. Gross measures of quality
    b. Specific measures of quality
    c. Withdrawal uses
    d. National pollutants discharge elimination system (NPDES)
    e. Ambient water quality criteria (standards)
    f. Maximum contaminant level goals (MCLG)
    g. No observable adverse effects levels (NOAEL)
    h. Reference Dose (RfD)
    i. Fick's law
    j. Standard methods of analysis
    k. Direct instrument measurements
    l. Optical instruments
    m. Gas chromatography

REFERENCES

American Public Health Association, American Water Works Association, Water Pollution Control Federation, Clesceri, L.S., ed., *Selected Physical and Chemical Standard Methods for Students,* 17th ed., American Public Health Association, Washington, D.C., 1990.

Currie, J. C., and Pepper, A. T., ed., *Water and the Environment*, Ellis Horwood, New York, 1993.

Davis, M. L., and Cornwell, D. A., *Introduction to Environmental Engineering*, 2nd ed., McGraw-Hill Book Company, New York, 1991.

DeZuane, J., *Handbook of Drinking Water Quality*, Van Nostrand Reinhold Company, New York, 1990.

Gallagher, L.A., and Miller, L., *Clean Water Handbook*, 2nd ed., Government Institutes, Rockville, MD, 1996.

Gupta, R. S., *Hydrology and Hydraulic Systems*, Waveland Press, Prospect Heights, IL, 1995.

McGauhey, P. H., *Engineering Management of Water Quality*, McGraw-Hill Book Company, New York, 1968.

McGhee, T. J., *Water Supply and Sewerage*, 6th ed., McGraw-Hill Book Company, New York, 1991.

Pettyjohn, W. A., ed., *Water Quality in a Stressed Environment*, Burgess Publishing Company, MN, 1972.

Sawyer, C. N., and McCarty, P, *Chemistry for Environmental Engineering*, 3rd ed., McGraw-Hill, NY, 1978.

Suffet, I. H., ed., *Fate of Pollutants in the Air and Water Environment*, 1 & 2, John Wiley & Sons, NY, 1977.

Sullivan, T. F. P., ed., *Environmental Law Handbook*, 14th ed., Government Institutes, Rockville, MD, 1997.

Tchobanoglous, G., and Schroeder, E. D., *Water Quality*, Addison-Wesley Publishing Co., Reading, MA, 1987.

U. S. Environmental Protection Agency, *An Evaluation of DDT and Dieldrin in Lake Michigan*, Ecological Research Service, EPA R3-72-003, Washington, D.C., August, 1972.

Vesilind, P. A., Peirce J. J., and Weiner, R. F., *Environmental Pollution and Control*, 3rd ed., Butterworth-Heinemann, Boston, MA, 1990.

Viessman, W., Jr., and Hammer, M. J., *Water Supply and Pollution Control*, 4th ed., Harper and Row Publishers, New York, 1985.

Water Pollution Control Federation, Technical Practice Committee, *Simplified Laboratory Procedures for Wastewater Examination*, Water Control Federation, Washington, DC, 1976.

William R.B., and Culp, G.L., ed., *Handbook of Public Water Systems*, Van Nostrand Reinhold, NY, 1986.

Williams, R. B., and Culp, G. L., ed., *Handbook of Public Water Systems*, Van Nostrand Reinhold Company, New York, 1986.

Chapter 9

POLLUTION AND TREATMENT OF THE WATER ENVIRONMENT

The Problems of Aquatic Pollution

Water resources are foremost in drawing attention toward environmental pollution problems because they are most vulnerable to the polluting effects. Through the hydrologic cycle, the atmosphere and land are intricately connected with hydrosphere in an overall planetary circulation of water. The pollution of the former two contributes to the pollution of the latter as well. The pollution problem extends to all forms of the moisture supply. The nature of problems is, however, different depending upon the character of the moisture distribution. The common problems of pollution relate to distribution of

- Atmospheric or rainwater;
- Streamflow;
- Impounded water;
- Groundwater; and
- Oceans.

Pollution of Atmospheric Water: Acid Rain

This problem is related to air pollution since the chemicals released into the atmosphere contribute to acidification of rainwater. This has been discussed within the atmospheric pollution in Chapter 10.

Stream Pollution

Rivers and streams are common areas in which domestic and industrial wastes are disposed. The dilution capacity of the mass of water, flushing action of the flowing water, assimilating ability of microbial population, and reaeration property of exposed water are the factors that had favored this practice for decades. However, the modern-day quality of pollutants, in terms of their intensity, amount, and nonbiodegradability or persistent behavior, is too offensive and waterways are often harmed beyond their self-purification threshold. Accordingly, the standards for effluent discharges and the permit requirements have been stipulated in the Clean Water Act (CWA).

Dilution is an effective means of dealing with a pollutant. Immediately beyond the point of discharge, the process of mixing and dilution begins. The complete mixing, however, does not take place at the outfall. Instead, a waste plume is formed that gradually widens, as shown in Figure 9-1. The length and dispersion (width) of the plume depend upon river geometry, flow velocity, and flow depth. Beyond the mixing zone, the concentration of the pollutant can be given by a mass balance relation, similar to Equation 2-9, as follows:

$$C_d = \frac{(Q_s C_s) + (Q_w C_w)}{(Q_s + Q_w)} \tag{9-1}$$

The terms are explained in Figure 9-1. Various types of pollution problems associated with streamflows are summarized below.

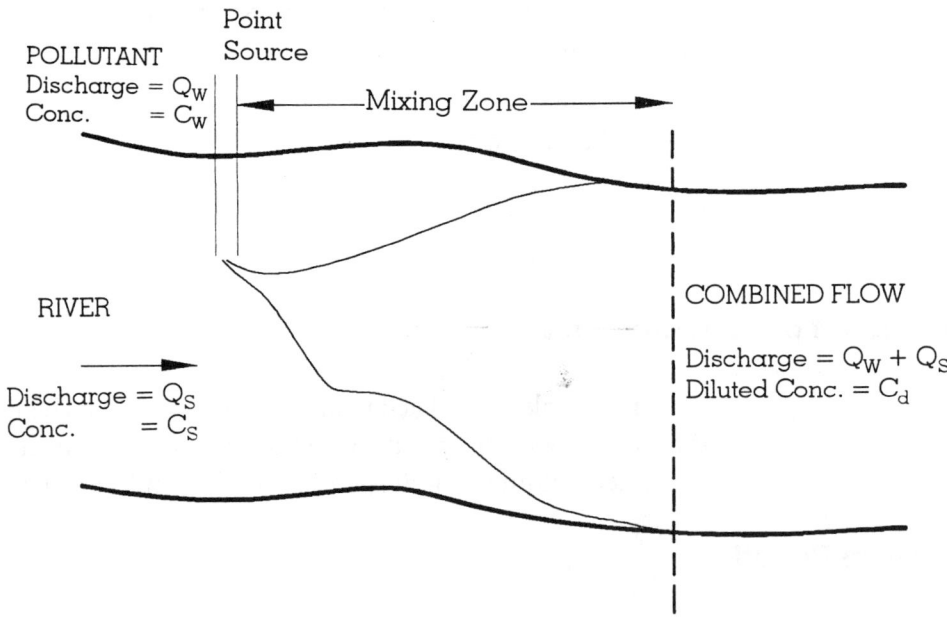

Figure 9-1 Discharge from a Point Source

Toxins

A variety of substances can damage plants and animals. These include:

- Heavy metals such as zinc, copper, cadmium, chromium, mercury, and lead;
- Organic compounds such as cyanide, phenols, detergents, and chlorinated hydrocarbons; and
- Poisonous gases such as ammonia and hydrogen sulfide.

Some chemicals do not dissolve in water such as PCBs and cannot be diluted. They become trapped in river sediments where they persist for decades.

Sediments

Erosion of soils during runoff, washoff from agriculture lands and construction sites, and quarrying and mining operations bring sediment deposits to streams. Besides causing turbidity and filling in the channel section, these deposits destroy vegetation and seriously damage spawning sites for fish.

Thermal Regime

Heated outfalls affect streams in many ways. The heat can change hatching patterns of aquatic life, increase the rate of biological activities, reduce the saturation dissolved oxygen level, and magnify the toxicity of poisonous substances.

Organic Matter

Domestic wastes containing oxygen-demanding organics are a common category of pollutants affecting rivers, estuaries, and near-shore ocean bodies. Natural waters contain some dissolved oxygen, which is essential for aquatic life just as it is essential for us in the atmosphere. The optimum level at which it exists in water exposed to the atmosphere is known as the *saturation dissolved oxygen concentration*, which depends on the water temperature, solids content, and the atmospheric pressure. The values are listed in Table 9-1.

Table 9-1 Saturation Dissolved Oxygen Concentration
(milligrams per Liter at 1 atm)

Temperature (°C)	Fresh Water	Seawater (chloride concentration 20,000 mg/L)
0	14.62	11.32
4	13.13	10.25
20	9.17	7.42
30	7.63	6.13

If any water is deficient in oxygen, it will capture oxygen from the atmosphere gradually. If excess oxygen is injected in water, it will escape into the atmosphere. The saturation concentration is a stable level that water tends to attain. The external factors that bring changes in dissolved oxygen levels (DOs) are described below.

- Biodegradable organic matter, mostly of domestic origin, removes DO. This is because the microorganisms present in water consume the organics as food (substrate) and utilize oxygen to accomplish respiration as per Equation 5-4. The more organics that are present, the larger is the demand on oxygen.
- Aquatic animals, including organisms in sediment remove DO.
- Plants add DO during the day via photosynthesis but remove it at night by respiration. Dying and decaying plants diminish DO.
- In summer, the increased water temperature reduces DO solubility.
- Tributaries draining into or wastewater discharging into a river bring their own oxygen supplies that affect DO of the river on mixing.
- Low river flows slow the rate of oxygen transfer into water from the atmosphere.

The DO level is used as an index of the health of a river; it is one of the important tests used by environmental engineers. A minimum DO concentration of 4 mg/L is necessary for warm-water fish. Cold-water fish require a still higher level. Below 2 mg/L, even the growth of microorganisms is affected. When a stream is devoid of oxygen, ammonia and hydrogen sulfide are formed among other gases; black sludge solids float to the surface. An anaerobic stream can be recognized by its foul smell. It can support only a few specialized types of organisms.

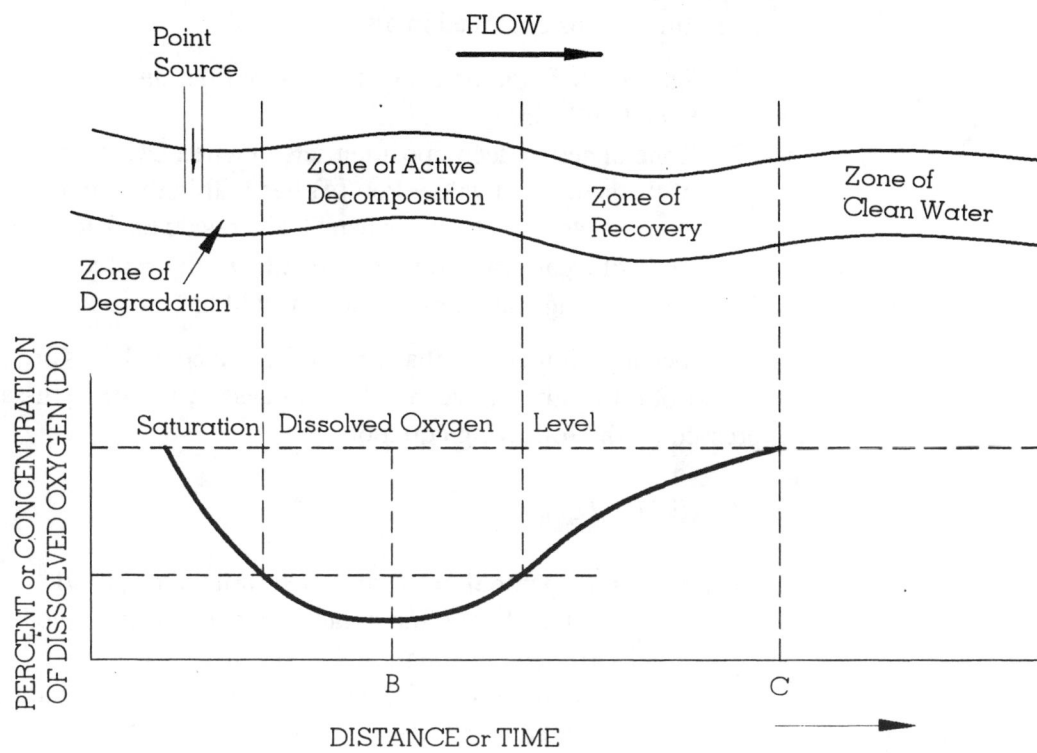

Figure 9-2 Oxygen Sag Curve

The Oxygen Sag Curve

Suppose a town discharges its sewage into a stream at location A in Figure 9-2, above. The various oxygen subtracting and adding processes listed above will commence. A simple model, referred to as the *Streeter-Phelps Model*, considers only two key processes:

- Oxygen depletion by microorganisms to biodegrade the wastes; and
- Oxygen replenishment subsequently through reaeration at the river surface from the atmosphere.

At the end of the first step at B, the oxygen level is lowest. When the sewage loading is excessive, the oxygen level will drop to zero, the anaerobic conditions will set in, and all higher forms of animals will be killed or driven out. Beyond point B, when most of the oxygen-demanding material is consumed, the DO concentration gradually rises again further downstream by atmospheric replenishment. As a result of the shape of this curve, it is called the *oxygen sag curve*. The self-purification process of the stream is completed at location C when water acquires the saturation DO level again. Four distinct zones observed in a stream are

1. Zone of degradation characterized by floating solids, turbidity, and visual pollution;
2. Zone of active decomposition sets in when DO drops to about 40% of the saturation value. Point B lies within this zone;
3. Zone of recovery begins when DO increases back to about 40%; and
4. Zone of clean water follows when DO fully replenished. Any town withdrawing water prior to point C will have a contaminated supply.

Thus, it becomes imperative that streams be loaded with less organic wastes.

Incorporating the above two key processes, the oxygen-sag curve is expressed by the following equation:[1]

$$D = (kLt + D_a) e^{-kt} \qquad (9\text{-}2)$$

where D = oxygen deficit in stream at time t days, mg/L
L = initial BOD after stream and wastewater have been mixed (based on Equation 9-1), mg/L
D_a = initial deficit of DO after stream and wastewater have been mixed, mg/L (based on Equation 9-1)
k = deoxygenation and reaeration rate, per day.

Example 9-1
A city produces sewage at the rate of 24 million gal/day. The sewage plant effluent has a BOD of 28 mg/L and DO of 1.8 mg/L. It discharges into a stream having a flow of 250 ft³/sec (cfs). The temperature of water is 24°C. The stream has a BOD of 3.6 mg/L and is saturated with oxygen. The deoxygenation coefficient is 0.5 per day. Determine the minimum DO in the stream and draw the oxygen sag curve.

Solution:
1 MGD = 1.55 cfs
Sewage flow = 24 × 1.55 = 37.2 cfs
At 24°C, saturation DO = 8.53 mg/L

---

[1] This relation is applicable for the case when the deoxygenation rate is equal to the reaeration rate.

From Equation 9-1:

The combined value at the initial time

$$L = BOD_0 = \frac{Q_s (BOD_s) + Q_w (BOD_w)}{(Q_s + Q_w)}$$

$$= \frac{250(3.6) + 37.2(28)}{250 + 37.2} = 6.8 \text{ mg/L}$$

From Equation 9-1:

$$DO_0 = \frac{Q_s (DO_s) + Q_w (DO_w)}{(Q_s + Q_w)}$$

$$= \frac{250(8.53) + 37.2(1.8)}{250 + 37.2} = 7.66 \text{ mg/L}$$

D_a = saturation DO - DO_0 = 8.53 - 7.66 = 0.87 mg/L.

From Equation 9-2:

$$D = [(0.5)(6.8)t + 0.87] \, e^{-0.5t}$$
$$\text{or } D = (3.4t + 0.87) \, e^{-0.5t}. \tag{9-3}$$

The values of D for selected time t are computed below from Equation 9-3:

t, days select	D, mg/L (from Equation 9-3)	DO = saturation DO - D (in mg/L)
1	2.59	8.53 - 2.59 = 5.94
1.5	2.82	5.71
1.75	2.84	5.69
2	2.82	5.71
3	2.47	6.06
5	1.47	7.06
10	0.2	38.3

DO vs. t are plotted in the form of an oxygen sag curve in Figure 9-3. The minimum DO from the curve = 5.7 mg/L.

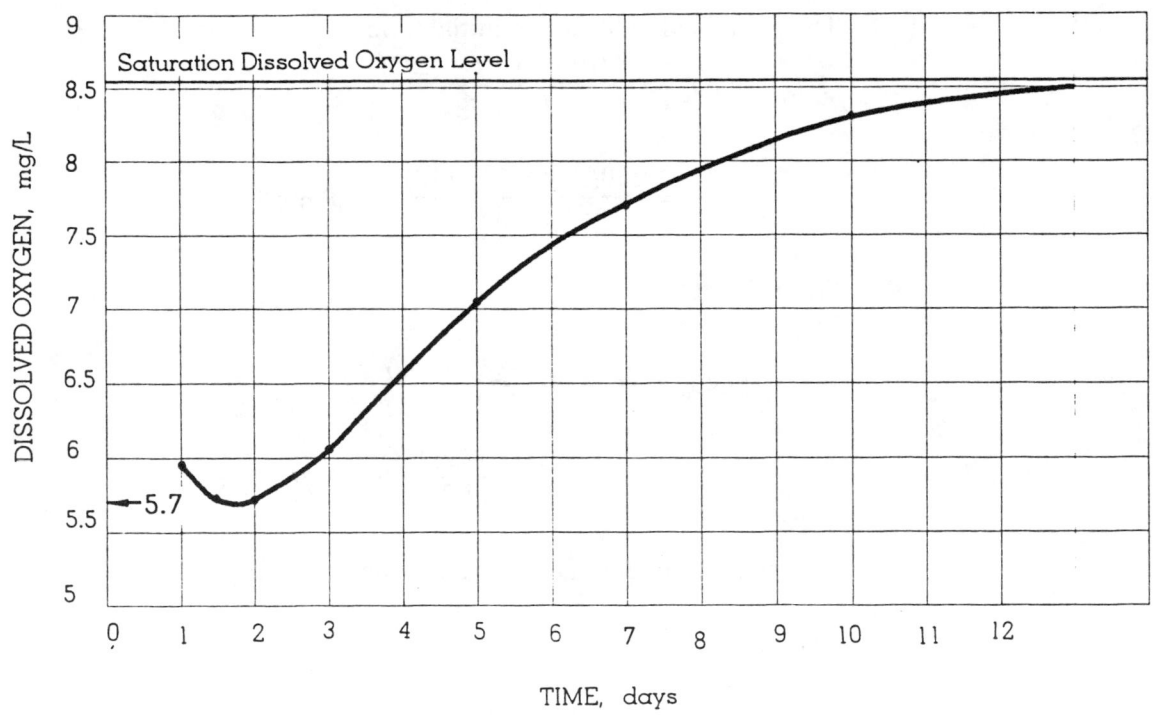

Figure 9-3 Plot of Dissolved Oxygen vs Time

Lake Pollution

Eutrophication

Lakes are subject to natural aging. The stages in a lake's life cycle follow.

- *Oligotrophic stage.* In this stage a lake is a clear body of water. It has very few nutrients, and plant and fish life are scarce.
- *Mesotrophic stage.* In this stage nutrients and sediments begin to accumulate and a variety of aquatic life appears.
- *Eutrophic stage.* As the name eutrophic (meaning well-fed) suggests, the lake accumulates enough nutrients, silt, and organic debris. This causes it to get shallower and warmer, more plants take root along the edges, organisms take over, and the lake gets choked with algae bloom and weed growth.
- *Senescent stage.* With further aging the lake fills to a large extent and vegetation emerges throughout the lake. What once was a lake eventually becomes a marsh or bog.

This natural aging phenomenon, known as *eutrophication*, is exceedingly slow and takes thousands of years to complete. However, human activities can speed this process tremendously. The human-manipulated process is called *cultural eutrophication*. It is one of the most significant current water quality problems: two-thirds of all lakes in the U.S. are reported to be significantly degraded as a result of cultural eutrophication.

The pollution of lakes poses problems that are different from those of streams, essentially because of lack of water flow in lakes. The short residence time for a specific mass of water in rivers reduces the effect of eutrophication in spite of the presence of the nutrients. In lakes, water quality is more dependent on the presence of nutrients than are organic wastes; sewage is usually not discharged into lakes due to lack of flushing action. Together, the chemical, biological, and physical factors cause eutrophication problems in lakes.

Cultural Eutrophication Factors

Chemical factors "wash off" nutrients in lakes. Nitrogen and phosphorus are two critical plant nutrients. They are mostly contributed by agricultural runoff containing excess fertilizers and by human-generated wastes containing detergents and human excrement. It has been reported that for phosphorus and nitrogen, any concentration in excess of 0.015 mg/L and 0.3 mg/L, respectively, can cause bloom of algae and excessive plant growth.

Biological factors in eutrophication consist of biomass overgrowth at all levels in the food chain due to abundant nutrients. The production of algae, which is a primary producer, exceeds the consumptive demand. Algae-bloom increases turbidity. Diatoms and green algae are replaced by blue-green algae which is known to cause taste and odor problems. The decaying algae depletes oxygen. Overall fish population in an eutrophic lake may be higher but consists of fewer favorable species.

Physical factors consist of light penetration and temperature-induced stratification. Turbidity and algal bloom restrict light penetration to only the top few feet; hence, photosynthesis occurs in that zone. Layering or stratification and mixing or overturning of lakes due to temperature differences occurs twice a year in temperate climates and once a year in warm climates.

Stratification

Stratification during summer is shown in Figure 9-4. A layer of warm water, called the *epilimnion*, is formed at the top. The colder and denser water remains at the bottom in the *hypolimnion* layer. A thin layer called the *thermocline* prevents mixing of water at the top and bottom layers. Epilimnion receives sunlight and allows algal growth. In hypolimnion, decaying sediments deplete dissolved oxygen, sometimes causing anaerobic conditions to develop. During fall, as the air temperature decreases, the epilimnion water cools, becomes dense, and sinks towards the bottom causing mixing of the lake. This circulation is called the *fall overturn*.

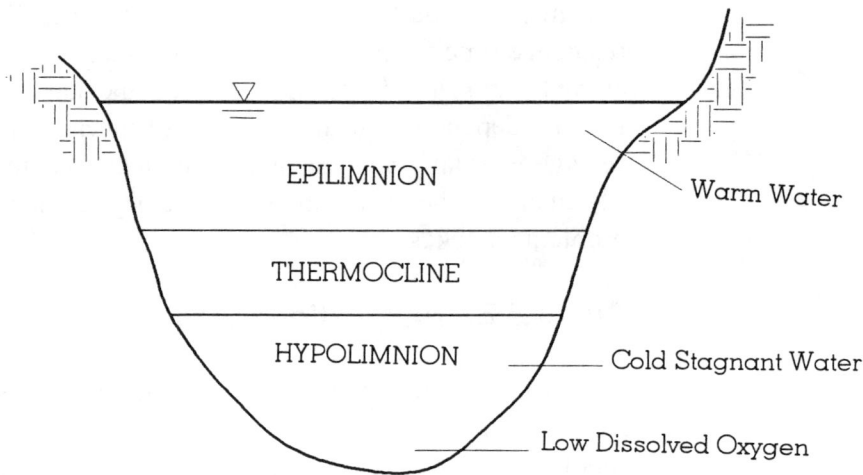

Figure 9-4 Summer Stratification Ending with Fall Overturn

During winter, an ice layer covers the lake surface, leaving the warmer water at the bottom. Reduced light penetration inhibits biological production. As spring approaches, the ice melts, becomes dense, and sinks to the bottom. Aided by wind, this causes mixing of water while the water temperature gradually increases. This is called the *spring overturn*. During fall and spring overturns, the undesirable matter is mixed throughout the lake and water taste and odor problems may be intensified. In warm climates where the temperature never drops below 4°C, lakes stratify during the summer only.

Control of Eutrophication

Two main factors control the rate of plant production: light and concentration of nutrients. Cultural eutrophication can be controlled by minimizing the nutrient input since there is practically no control on the light input. The list of nutrients is long but, fortunately, the choice of limiting nutrients is either phosphorus or nitrogen. According to stoichiometric analysis, it takes about seven times more nitrogen than phosphorus to produce a given amount of algae. However, from practical considerations, when the nitrogen to phosphorus ratio in lake water is more than 20:1, phosphorus is a limiting factor. When the ratio is 5:1 or less, nitrogen is a limiting factor.

The following measures are proposed to reduce nutrient input into lakes.

For nonpoint sources:
Soil erosion control;
More effective use of fertilizers.

For point sources:
Alternative disposal of effluent on land by irrigation;
Diversion of watewater around or away from lakes, such as into streams; and
Advanced waste treatment to remove phosphorus and nitrogen.

Temporary control measures include:
Chemical control (use of algicides and herbicides [e.g. $CuSO_4$] kills algae);
Harvesting of aquatic plants;
Mechanical destruction (boat-mounted underwater weed cutters);
Aeration of hypolimnion layer; and
Dredging to remove sediments.

Groundwater Contamination

An Overview of Groundwater

According to Meinzer's definition (1923) which is very widely accepted today, *subsurface water* designates all waters that occur underground and *groundwater* is meant to be the water within the zone of saturation as shown in Figure 9-5. The formation that can yield a significant quantity of water is known as an *aquifer*. Alluvial deposits are the best form of aquifers. There are two main types of aquifers: unconfined and confined aquifers. In unconfined aquifers, the upper surface of the groundwater or zone of saturation is exposed to the atmosphere and is known as a *water table*. The confined aquifer, also known as a *pressure* or *artesian* aquifer, occurs where groundwater is held under pressure due to an overlying rock. If this impervious layer of rock is punctured, the level to which water will rise is known as a *piezometric head*.

Aquifers are characterized by their two basic properties: the ability to conduct water through void spaces within the aquifer (*hydraulic conductivity,* or *coefficient of permeability*) and the ability to release water when the water table drops (*specific yield,* or *storage coefficient*).

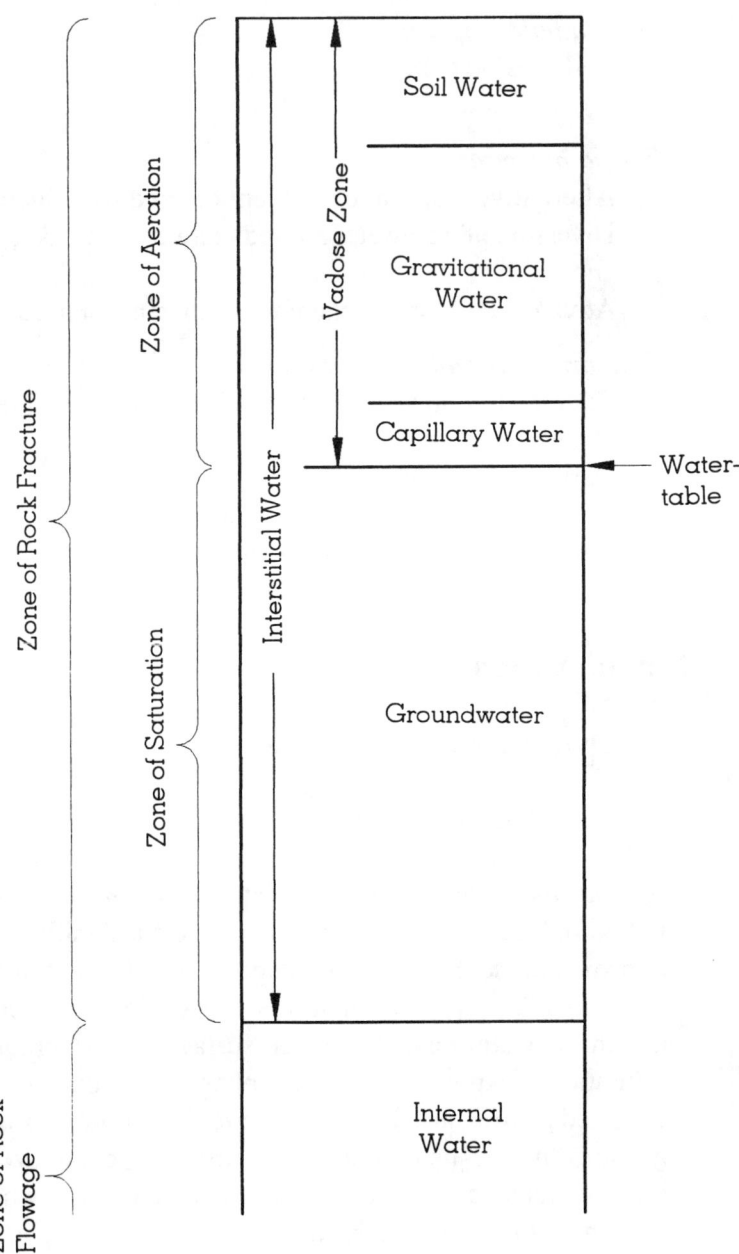

Figure 9-5 Meinzer's Groundwater Classification

For a steady-state condition in which the water table or piezometric head does not change within a specified time, the groundwater movement is expressed by the following relations:

$$\text{Pore velocity or advection, } v = \frac{K(h_1 - h_2)}{nL} \tag{9-4}$$

Pore area of flow, $A_v = nA$
Combining the two, since $Q = Av$,

$$\text{Rate of groundwater flow, } Q = \frac{K(h_1 - h_2)A}{L} \tag{9-5}$$

Where Q = rate of groundwater flow
v = pore velocity or advection
K = hydraulic conductivity
A = aquifer cross-section area through which flow takes place
h_1 = water head at upstream end
h_2 = water head at downstream end
L = distance between h_1 and h_2
n = porosity.

Equation 9-5 is known as *Darcy's Law*. The term $(h_1 - h_2)/L$, which is the slope of the water table or piezometric line, is known as the *hydraulic gradient*.

Example 9-2

An irrigation channel runs almost parallel to a lake. They are 2,000 ft apart. A pervious formation of 30 ft average thickness connects them. Hydraulic conductivity and porosity of the pervious formation are 10 ft/day and 0.56, respectively. The water level in the channel is at an elevation of 110 ft and at 100 ft in the lake. Determine the rate of seepage from the channel to the lake.

Solution:
Refer to Figure 9-6:

$$\text{Hydraulic gradient, } i = \frac{h_1 - h_2}{L} = \frac{110 - 100}{2000} = 0.005$$

For each 1 ft width (perpendicular to cross-section in Figure 9-6)
$A = 1 \times 30 = 30 \text{ ft}^2$

From Equation 9-5:
$$Q = (10 \text{ ft/day})(0.005)(30 \text{ ft}^2) = 1.5 \text{ ft}^3/\text{day/ft width}$$

From Equation 9-4:
$$\text{seepage velocity, } v = \frac{K(h_1 - h_2)}{nL} = \frac{(10)(0.005)}{0.56} = 0.089 \text{ ft/day.}$$

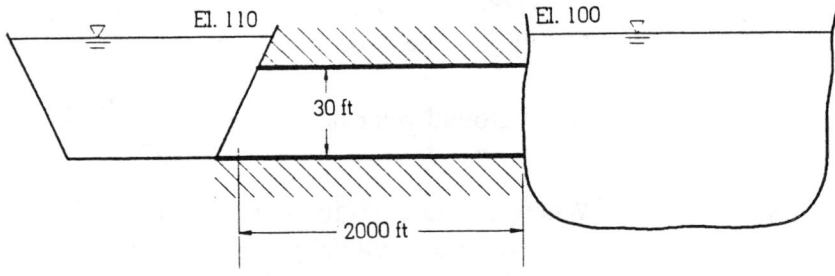

Figure 9-6 Model of Channel and Lake

Mechanism of Groundwater Contamination

Contaminants found in groundwater include the entire range of inorganic chemicals, organic chemicals, pathogenic organisms, and radioactive substances as discussed in the context of water quality. There are three main sources of groundwater contamination. First, the natural processes such as seepage, runoff, and evapotranspiration can contribute to groundwater contamination. Second, the waste disposal practices of humans directly contribute to contamination. The third category also results from human activities, but is unrelated to the waste-disposal practices. This includes accidental spills, highway deicing salts, acid rain, and improper groundwater exploitation.

The sources of contamination that relate to human waste-disposal practices are listed below.

- Wastewater impoundments such as ponds, lagoons, and pits;
- Dumps and landfills;
- Septic tanks and leach fields;
- Underground storage tanks;
- Land-disposed sludges;
- Brine and other fluids disposed from petroleum exploration;
- Disposal of mine wastes;
- Agricultural and animal feedlot wastes;

- Collection, treatment, and disposal of municipal wastewater;
- Leaking sewers;
- Disposal wells;
- Leaks around wells;
- Saltwater intrusion; and
- Overpumping of wells.

The pathways describing how contaminants from these various sources enter groundwater are depicted in Figure 9-7.

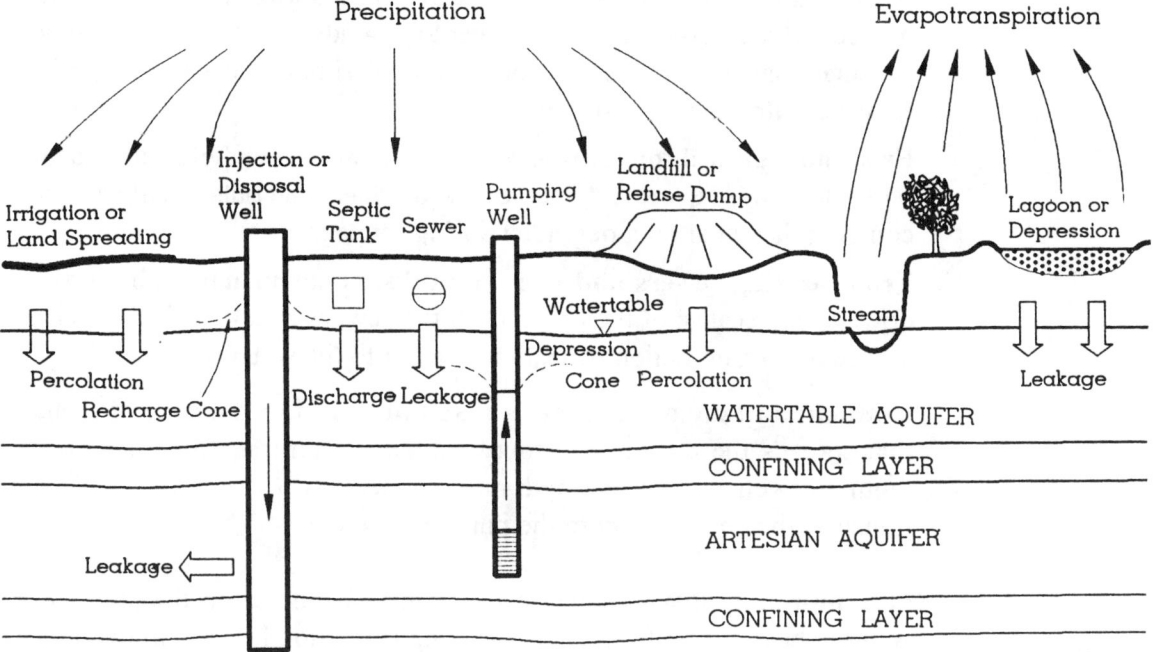

Figure 9-7 Pathways of Groundwater Contamination

The soil stratum from the point of entry at the surface to the groundwater level normally provides adequate filtration and chemical and biological reactions to remove the contaminants. Thus, the formations that are most susceptible to contamination are:

- where the water table can rise too high to make contact with or come in proximity of the above-referred sources of contamination;

- where there are fissures and cracks in the formation that can convey contaminants to groundwater;
- where the aquifer depths are shallow; and
- when the contaminant concentrations are large and chemicals are inert to reactions.

Some mechanisms of contamination are as follows.

From a lagoon or basin: the contaminants flow downward to form a recharge mound on the top of a water table; from there contaminants disperse laterally. Many municipal water supply wells are located in proximity of streams and lakes for augmented flows. The drawdowns in these wells induce recharge from the surface water to groundwater. Where surface water is contaminated, it affects the groundwater. A similar mechanism applies to seawater intrusion into coastal wells.

From stockpiles, dumps, disposal sites, fills, and leach fields: the leachates move downward to the water table. The leachate materials usually are stable compounds that are not degraded during seepage.

From leaking sewers and storage tanks: contaminants drain vertically downward because of the pressure in the sewer or tank. This forms a line source of contamination beneath the sewer to the watertable.

Contaminant introduced into a disposal well: forms a cone of recharge that spreads the contaminant outward away from the well. If there is a pumping well nearby, the hydraulic connection between two wells will transport the contaminant to the pumping well.

Overpumping of wells: may cause saline or other objectionable natural water to be pulled in from the underlying formation into a shallow aquifer. This is another cause of saltwater intrusion in addition to the lateral contamination described previously.

The above mechanisms are a simplistic presentation of the complicated groundwater contamination problem. The final result is that once an aquifer is contaminated, it is extremely difficult to reclaim it.

Fate of Contaminants in Groundwater

Once a contaminant reaches an aquifer by one of the above means, its medium of movement is through groundwater, forming a plume of contamination. Two types of contaminants exist: *soluble contaminants* that, to a large extent, dissolve into the groundwater, and *insoluble* or *immiscible fluids*, such as oil, that do not appreciably mix with water. The first group of substances moves along with groundwater, but the second group of substances may not move with water as they move in a distinct place.

Fate of Soluble Contaminants

Since the contaminants move with groundwater, the basic movement is given by the advection relation of Equation 9-5. In addition, contaminants disperse causing the solute to spread in both a longitudinal and transverse direction. Thus, a plume widens and thickens as it travels downstream. The dispersion is given by Fick's Law as described in Chapter 7. The two terms are combined in what is widely known as the advection-dispersion equation. In an expanded form, the advection-dispersion equation also includes a term for the chemical reaction to depict the contaminant that is lost from the flow as a result of adsorption.

Dispersion causes spreading of a contaminant over a large aquifer volume. Dispersion, together with adsorption, indicate dilution or attenuation of a contaminant as it travels. In addition to these two factors, other chemical reactions contribute to the attenuation of contaminants in an aquifer, including precipitation, dissolution, oxidation, and reduction. These chemical reactions are analyzed separately if the purpose is to assess the concentration of a contaminant on completion of various chemical processes. The purpose of the advection-dispersion analysis is to assess the rate of movement of the contaminant, shape of the plume, and prediction of the arrival time of of the contaminant to a point of groundwater discharge such as a stream, lake, wetland, or tidal water.

According to many studies, the diffusion-type relation (based on Fick's Law), used in the advection-dispersion equation is not fully valid for contaminant transport through a porous medium. This is because in groundwater, dispersion is caused by both microscopic and macroscopic effects. Mechanical dispersion and molecular diffusion are microscopic-scale components. Mechanical dispersion is due to deviations in pore velocity from the average pore velocity. These variations are caused because the water in the center of a pore space moves faster than the water at the walls of the pore, and around the solid grains of a porous material repeated branching of flow paths occurs. Molecular diffusion is due to the molecules' movement from higher to lower concentration, as discussed in Chapter 7. On a macroscopic scale, dispersion is caused by the presence of large-scale heterogeneities in the soil. For example, pockets of high permeability within a medium of essentially low permeability will cause distortion of streamlines. Considerable uncertaintities exist in quantifying macroscopic dispersions in groundwater. It has been found that after a certain time and distance from the source of contamination, the dispersion component of the advection-dispersion equation can be represented by Fick's Law, and therefore, the classic advection-dispersion equation could be applied. However, for some distances from the source, which could be hundreds of meters, uncertainties exist. Using a time or distance dependent factor for dispersion in the classic advection-dispersion equation is one of the suggested approaches. Stochastic methods have been developed based on the variations in hydraulic conductivity values because that characteristic is responsible for macroscopic dispersion.

Immiscible or Nonaqueous Phase Liquids (NAPL)

The movement of liquids that do not mix with water takes place as a separate phase distinct from water flow. Only to the extent that the liquids dissolve in water is their flow controlled by the principle discussed in the previours subsection. Some of these liquids, such as gasoline and diesel fuel, may be lighter than water; these are known as light nonaqueous phase liquids (LNAPL). Others, that are denser than water, such as chlorinated hydrocarbons, are called dense nonaqueous phase liquids (DNAPL). When a liquid makes contact with another substance, whether solid, another immiscible liquid, or gas, a force of attraction acts on the interface as demonstrated by the formation of a curved surface. This phenomenon, related to molecular attraction among substances, is called the *surface tension*. When two liquids compete for a single contact surface for interfacial tension, one dominates and coats the solid surface as a *wetting fluid*. The other acts as the *nonwetting fluid*. In a water-oil system, water is the wetting fluid unless the surface is already coated with oil before water makes contact.

Consider a pore space that is saturated with a wetting-fluid and a nonwetting fluid slowly starts displacing the wetting fluid. At a certain stage, this draining process stops and no more wetting fluid will be displaced. The content of wetting fluid at this stage is referred to as the *residual wetting saturation*.

Now consider that the wetting fluid is introduced again to displace the nonwetting fluid. When this process stops, some nonwetting fluid still remains in the pores. This amount is referred to as the *residual nonwetting saturation*. These phenomena are important to the understanding of the flow of immiscible fluids through porous media.

Water will not flow in a pore space until its content exceeds the residual wetting fluid saturation. Similarly, a nonwetting fluid will not begin to flow until the residual nonwetting fluid saturation is exceeded. In other words, in a two-phase water-oil system, if the water content is less than the residual saturation, the oil can flow but the water will be held within the pores. Similarly, when the oil concentration is less than the residual nonwetting saturation, water can flow but oil will be held within the pores. When their contents exceed the residual saturations, both immiscible fluids will flow through the pore space in distinct phases at different rates. In a steady state (when there is a continuous flow at a uniform rate) saturated condition, a part of the pore space is filled with one fluid and the remainder with the other fluid. The flow of each fluid can be given by the modified Darcy's Law.

For the flow rate of water, in Darcy's Equation 9-5, the term hydraulic conductivity, K, refers to the conductivity of water[2] and the term $(h_1 - h_2/L)$ refers to gradient head of water. For the flow rate of oil, the hydraulic conductivity refers to conductivity of oil and $(h_1 - h_2/L)$ to the gradient head of oil.

When a nonaqueous phase liquid (NAPL) is spilled at the land surface, it travels through the larger pore openings as a nonwetting fluid vertically downward through the vadose or aeration zone, displacing the air. A view of the distribution of NAPL in soil is shown in Figures 9-8a and b. As the NAPL moves forward, a fraction remains behind throughout the thickness of the vadose zone as a residual oil. In moving down, the NAPL may displace some water in the vadose zone, causing a water layer to move in advance of the NAPL front. Once a capillary zone is reached, the NAPL accumulates. Eventually, the capillary fringe is squeezed out and an oil table forms on the top of the water table.

If it is an LNAPL, a core of the LNAPL will remain, slightly depressing the water table by its own weight.

Conversely, a DNAPL can continue to move downward below the water table. For DNAPL to migrate downward, the water in the pores must be expelled. To achieve this, the DNAPL must have sufficient height so that its weight can displace pore water. The critical height of the DNAPL can be determined from the principle of interfacial tension and capillary pressure between water and DNAPL. For well-sorted, well-rounded grains, the critical height can be expressed by:

$$hc = \frac{16.5 \, \sigma \cos\theta}{d(\gamma_w - \gamma_o)} \qquad (9\text{-}6)$$

---

[2] $K = \dfrac{k\gamma}{\mu}$

Where, k is intrinsic permeability of water or oil
γ is specific weight of water or oil
μ is dynamic viscosity of water or oil

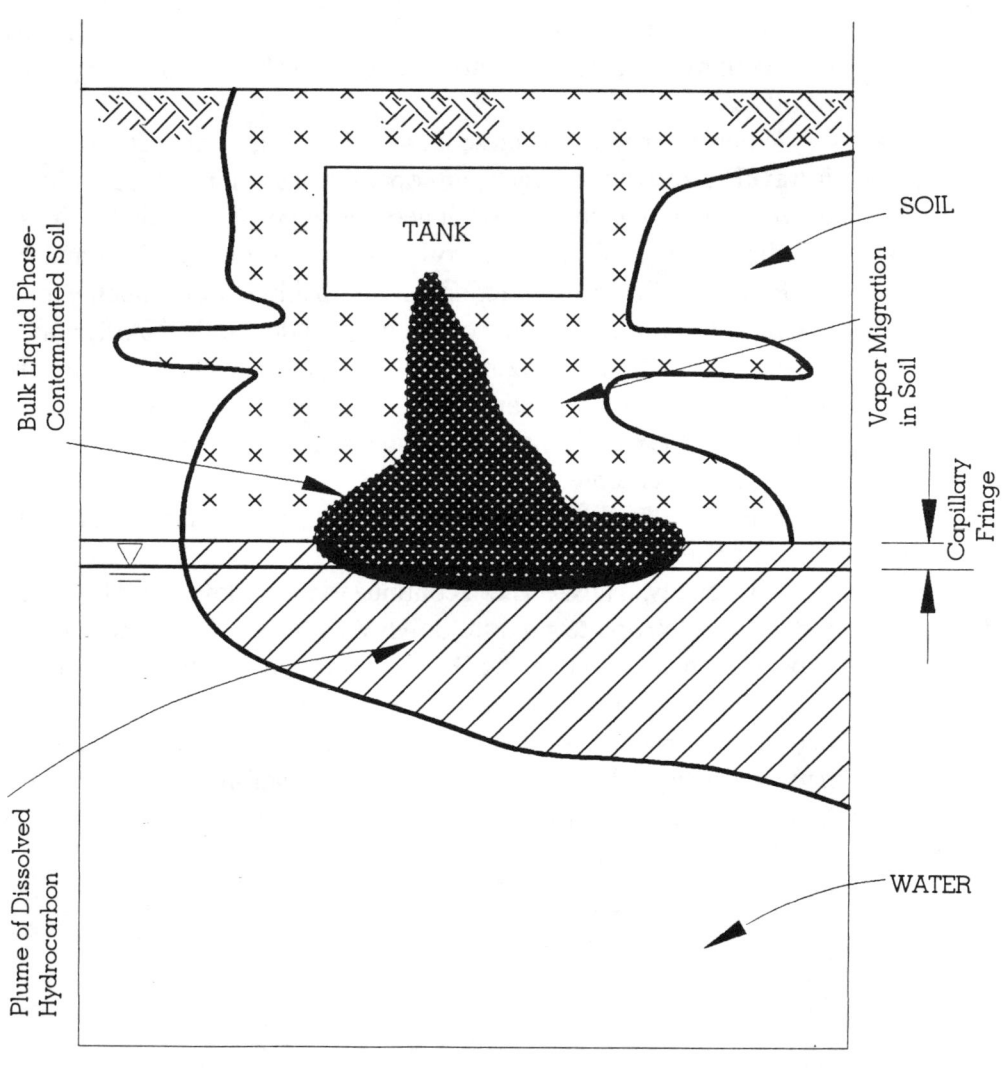

Figure 9-8a Typical Behavior of Light Nonaqueous Phase Liquid (LNAPL) Underground

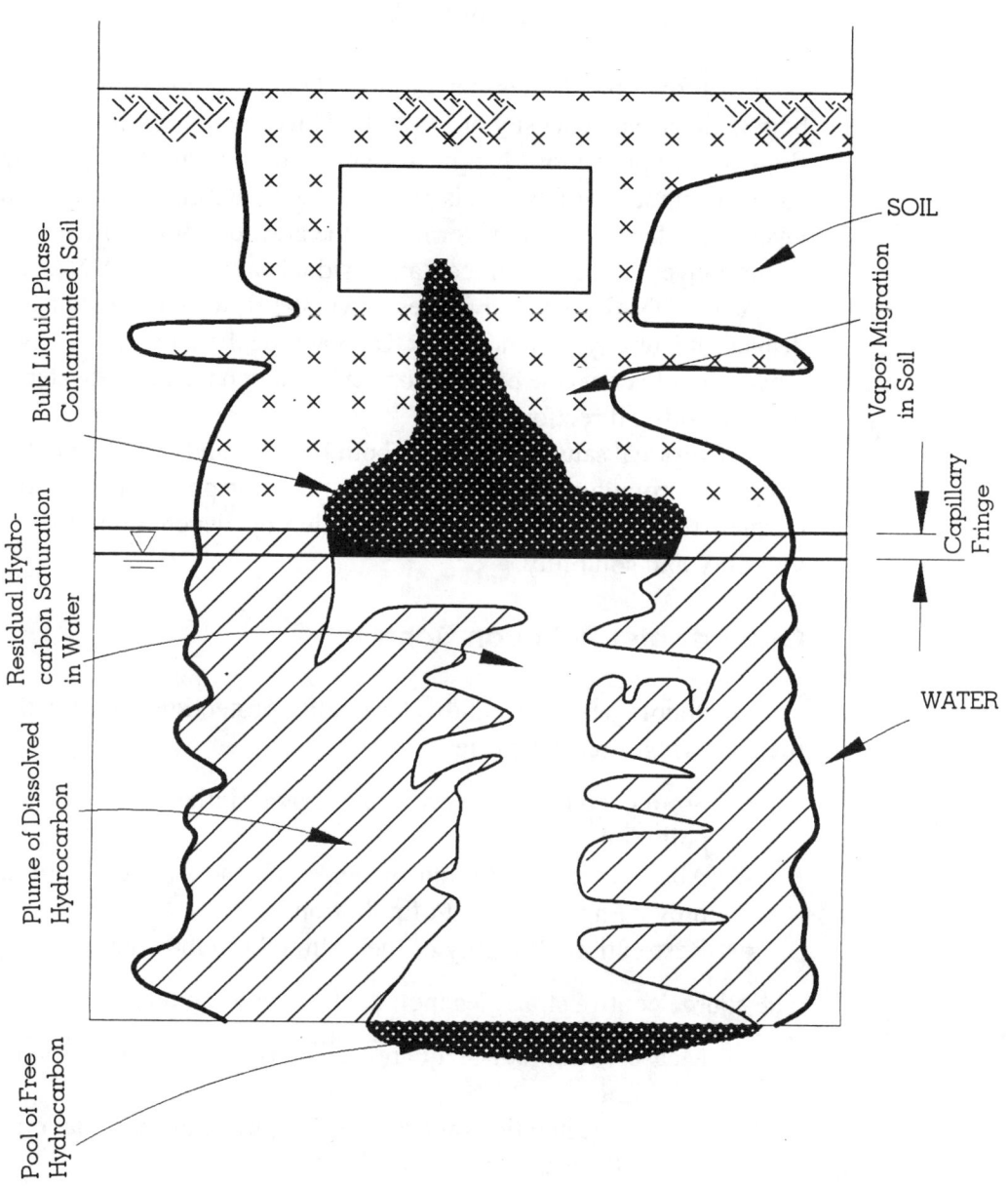

Figure 9-8b Typical Behavior of Dense Nonaqueous Phase Liquid (DNAPL) Underground

where h_c = critical height of DNAPL
σ = surface tension between fluids (water and oil)
θ = wetting angle
d = diameter of grains
γ_w = unit weight of water
γ_o = unit weight of DNAPL.

Equation 9-6 indicates that the critical height is inversely proportional to both grain size and density of DNAPL. Thus, smaller grains and less dense fluid will require more height to overcome the capillary pressure. If an adequate amount of DNAPL is present, it will continue to migrate down. A layer of the DNAPL will be formed at the aquifer bottom. In the column of water above, the pores will contain residual saturation of DNAPL.

As the DNAPL stringer moves down, the flowing groundwater tends to displace it laterally to some extent. On reaching the bottom, the DNAPL will move laterally down the bottom slope even though the groundwater flow may be in the other direction.

The residual saturation NAPL (both LNAPL and DNAPL) in the vadose zone and groundwater can partition into vapor phase and solution phase through the pores, the degree of partitioning depends upon the relative volatility and solubility.

Groundwater Pollution Control

Methodologies for groundwater quality protection can be grouped into the following three categories:

- Source control or preventing or minimizing the occurrence of pollution;
- Abatement of pollution or removing the source of pollution and preventing movement of pollution; and
- Restoration of quality of the polluted groundwater.

The source control strategies include the following measures:

- Measures to reduce quantitative use by resource recovery and recycling;
- Chemical alterations and detoxification and biodegradation of wastes before disposal; and
- Stabilization and solidification including fixing the waste into a solid mass prior to land disposal.

The abatement measures are designed to control the movement of underground water such that pollutants are not transported from the sources into aquifer or the reach of the pollution is isolated. The methods in this category include:

- Using a system of injection and withdrawal wells to manipulate the groundwater table;
- Using an interceptor system comprising trenches and collection drains below the water table;
- Capping and lining of landfills;
- Forming a sheet piling as a form a barrier to flow; and
- Using slurry walls to curtail flow into or out of a site.

The restoration phase relates to treatment of the contaminated groundwater and adjoining soil formations to make them usable. Treatment techniques fall into two categories:

- Ex-situ methods based on the concept of pumping the contaminated water out of an aquifer and taking it to a treatment system above ground; and
- In-situ methods designed to treat contaminated water without withdrawal from the site.

Physical and chemical methods are used for treatment of inorganic and organic substances. Biological methods are also used in the case of organic contaminants. Several bioremediation techniques have been developed and many new ones are being researched.

Groundwater Treatment Techniques

The common techniques of ex-situ and in-situ treatment using physical, chemical, and biological procedures, are summarized below. Each technique may contain several methods of treatment based on the same concept.

Ex situ physical and chemical methods include pure compound recovery of immiscible fluids, air stripping, carbon adsorption, and chemical precipitation.

Pure compound recovery of immiscible fluids: As discussed earlier in this chapter, immiscible fluids either float on the top of an aquifer or collect at the bottom. A well is placed in the accumulated mass. Controlled pumping with an oil-water separation device helps to remove the immiscible fluid.

Air stripping: This is used to remove volatile compounds. The approach is to bring the contaminated water into intimate contact with moving air causinng vaporization of the contaminant fluid. The aeration can be accomplished by several means: aeration tanks, spray basins, column-packed towers, and cascade aerators.

Carbon adsorption: Adsorption occurs when a molecule of a liquid is brought in contact with a solid surface where it is held either by physical attraction caused by the surface tension of the solid or by chemical bonding between the solid and liquid. Adsorption of organics by activated carbon is of a physical kind that is carried out by batch, column, or fluidized-bed operations.

Chemical precipitation: For inorganic contaminants, the removal by precipitation (adding of chemicals), flocculation-sedimentation-filtration, ion exchange, and membrane separation are well-known techniques as discussed in Chapter 7.

Ex situ biological methods are suitable for organic wastes. The techniques are similar to wastewater treatment as described later in this chapter. Microorganisms oxidize the waste, using it as a substrate (food) in their metabolic processes. For groundwater, aerated lagoons or pools are convenient reactors. The contaminated groundwater is pumped into the lagoon; the microorganisms in the lagoon degrade the organics and create new cells. Biological treatment is often used in conjunction with other methods as pre-or post-treatment. For example, contaminated groundwater may undergo metal removal and high-temperature air stripping before biological treatment, or the biological methods may be followed by the carbon adsorption technique.

In situ treatment theoretically can be done to inorganic and organic contaminants through the use of chemical neutralizing agents. However, such methods only demonstrate a distinct advantage when they involve the biological technique. Generally, the methods involve installation of injection wells at the head or within the plume of contaminated groundwater through which the treatment agents are pumped into the aquifer. Essential elements of the treatment are the control of groundwater movement, and spreading of oxygen, nutrients, microorganisms, and any other treatment agent throughout the aquifer. For each of these elements several techniques have been developed. The combination of in-situ and ex-situ methods have also produced very promising results. Between 1987 and 1997, the engineering applications of bioremediation techniques have grown significantly. New technologies are emerging.

Ocean Pollution

The oceans are an ultimate sink for nature's water; all rivers empty into the oceans. Relying on their enormous capacity for dilution, dispersion, and degradation, oceans are being indiscriminately used to dump most of the waste matter we produce. Specifically, the following materials find their way into the oceans:

- Treated or untreated wastewater from communities or industries through submerged outfalls;
- Solid wastes comprising dredged material, residential and industrial wastes, and sewage sludge dumped from barges and ships; and
- Accidental and intentional oil spills from tankers and off shore drilling platforms.

A wastewater outfall consists of a conduit that may or may not lie submerged under the sea-water. Treated wastewater discharged through the conduit is expected to get sufficiently diluted and diffused by seawater and moved away by the currents. Unfortunately, the discharged waste is sometimes carried toward the shore. Moreover, the massive input of untreated or partially treated wastewater has overwhelmed the dilution capacity of the near-shore regions, causing widespread pollution of coastal areas, harbors, and bays. Wastewater that is lighter than ocean water tends to collect on the surface.

Dumping of solid wastes has been permitted by law in deep water because of the assimilating capacity of the ocean. However, this practice has resulted in over-dumping of wastes by big cities onto the nearby ocean shelf. Moreover, about one-third of modern waste consists of plastics, pesticides, and other synthetic materials that do not degrade easily and are known to be highly toxic and carcinogenic. Marine habitats have suffered heavy casualties as a result.

Oil Spills into the Ocean

The sources of oil in the ocean include: oil disposed of on land that finds its way into the ocean; oil that escapes from off-shore-drilled boreholes (blowouts); intentional releases of oil from off-shore wells, transportation barges, and ships approaching a shore; and tanker accidents.

The last item receives a lot of publicity but accounts for only 10–15 % of the total oil in the ocean. When an oil spill occurs, an ocean ecosystem is affected in many ways, as follows:

- A thin film of light oil spreads very quickly on the surface. This cuts off the oxygen supply to the underwater aquatic life and coats the feathers of birds that dwell on the surface.
- Some tar-like globs remain floating on the surface. These stick to birds, sea otters, seals, rocks, and any other objects that come in contact with the oil. Birds adjust their body temperature according to contact with the atmosphere. The oil coating acts as an insulating layer, causing birds to lose their body heat, resulting in their deaths.

- Oil slicks are driven toward land by wind and tides. This interferes with recreational uses of beach areas. Beaches may remain closed for many days to years.
- Heavy oil components sink to the bottom. Bottom-dwelling organisms such as crabs, oysters, clams, and more vulnerable eggs and larvae of sea life are killed. This causes a long-term impact on the ocean environment.

Control of Spills

Measures adopted to control oil spills are 1) input controls directed toward conservation and recycling of oil, and plans to reduce tanker accidents, including improved tankers design, and strict regulations on operation, maintenance, and disposal of oil, and 2) containment and physical removal of spilled oil including the following methods:

Floating booms: These are flexible barriers that contain oil by restricting its surface movement. It is important that a boom be placed downstream from the spill as quickly as possible. After placement, the boom is towed to permit the slick to move along the boom into a spot where it can be removed by one of the following procedures.

Absorption: Hay or straw bundles, chicken-feather pillows, and polyurethane foams can absorb oil many times their weight. The material must have a good capacity to retain oil so that it does not drip out as the absorbent is lifted.

Skimming: In this procedure the oil layer is either made to spill over an adjusted notch into a sump from which the spilled quantity is pumped out, or an endless belt makes the oil adhere to it and is then lifted out of water; or alternatively, a rotating impeller causes formation of a vortex, towards which the oil flows. The oil is then pumped out.

Dispersion: Some detergent chemicals change the property of oil so that it does not stick to solid surfaces. This makes the oil spread out and become diluted. Ferric colloidal dispersion agents impart a magnetic property to oil so that it then be picked up by a magnet.

Bacteria: Oil-eating bacteria efficiently biodegrade compounds in oil.

Ignition: Light crude oil can be ignited on the surface, and a burning agent can be used to support combustion.

Dispersion and bacteria are objected to by many environmentalists because of possible side effects from introducing foreign matters into the ocean ecosystem. Ignition is objected to because it deprives oxygen supply to aquatic life for a period of time while the burning occurs.

Treatment of Water and Wastewater

The quality of natural water, even when unsatisfactory, cannot be modified significantly within the body of water because of huge volume and flow conditions. Accordingly, the quality control approach is directed to the water withdrawn from a source for a specific use. The drawn water is treated before its use, and then the spent or wastewater is treated again after the use, prior to its disposal into a water body. As the name implies, a wastewater is more than 99% water and less than 1% waste materials that are dissolved or suspended in water. The waste materials from water and wastewater are removed by physical, chemical, and/or biological means. Chapter 7 describes various processes involved under each of the three categories. The enclosure or vessel within which each individual process takes place is referred to as an unit. Physical treatments are called *physical unit operations* and chemical and biological treatments are termed *chemical unit processes* and *biological unit processes*, respectively. Thus, an overall treatment may include a series of physical unit operations and chemical unit processes and/or biological unit processes, depending upon the quality of contaminated water, desired quality of effluent, and operational limitations.

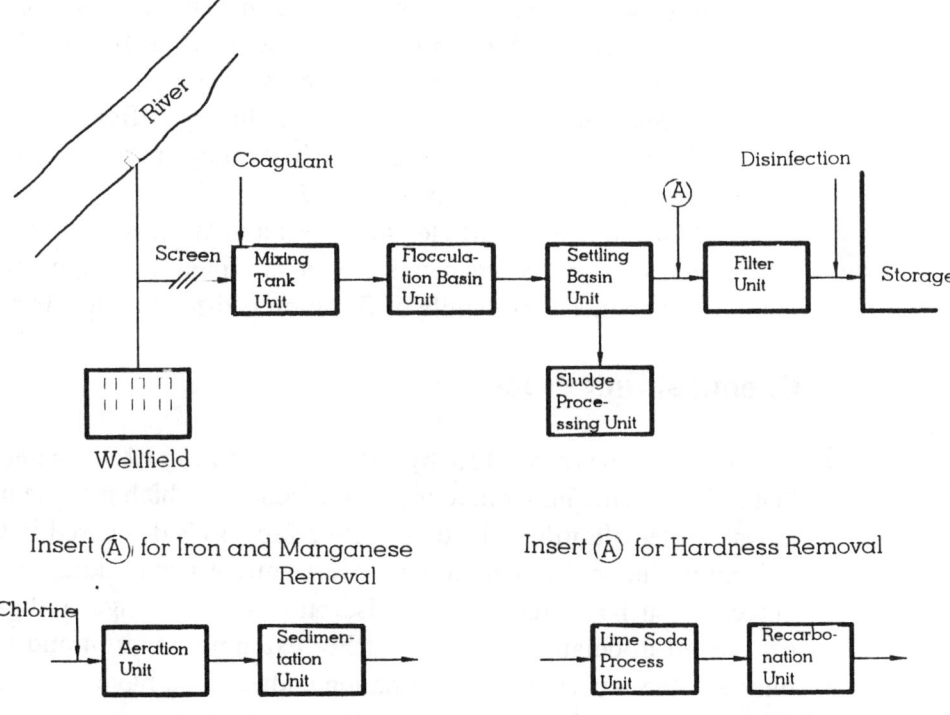

Figure 9-9 Schematic Diagram of a Water Treatment System

Commonly, water treatment methods are physical unit operations and chemical unit processes. As a result of dilution and assimilative capacity of water sources, a need for biological processes does not arise in water treatment. Conversely, organic matter is dominant in wastewater that is conveniently treated by the biological unit processes. Removal of BOD is a major goal of wastewater treatment.

Water Treatment

Many process combinations are possible; the choice is made to suit the quality of raw and finished waters. A flow diagram of a conventional treatment system is shown in Figure 9-9.

Physical Unit Operations

These operations include the following:

1. *Screening*: Used to remove large sized floating and suspended debris.
2. *Mixing*: The colloidal and fine suspensions cause turbidity. The chemicals known as coagulants are mixed to make tiny particles stick together. Alum, for example, listed in Table 7-1, is a coagulant.
3. *Flocculation*: Water mixed with coagulants is given a slow motion to allow particles to meet and floc together.
4. *Sedimentation*: Water is detained for a sufficient time so that flocculated particles settle to the bottom by gravity. The principle of settling is discussed in Chapter 7.
5. *Filtration*: Fine particles still remaining in water after sedimentation and some microorganisms present are filtered through a bed of sand and coal. The principle of filtration is discussed in Chapter 7.

Chemical Unit Processes

Disinfection is carried out by applying chlorine or ozone or another agent that kills the remaining microorganisms, some of which may be pathogenic in water after filtration. The disinfection process is discussed in Chapter 7. Depending upon other pollutants, water might need additional chemical processes, such as precipitation, adsorption, ion-exchange, and gas transfer (discussed in detail in Chapter 7). For example, for iron and manganese removal, the processes of chlorination and aeration have to be carried out. Similarly, for hardness (calcium and magnesium) removal, a unit of excess lime and soda and a unit of recarbonation have to be provided, shown as insert A in Figure 9-9.

The sludge collected from filter backwash might be returned to the intake line prior to mixing in the tank. The sludge from the settling tank is disposed of in lagoons or other places after dewatering.

Wastewater Treatment

There are two categories of wastewater treatment systems. The first is a decentralized or individual sewage disposal system (ISDS), wherein the treatment is carried out for individual houses or clusters of houses or industries near the source of the waste. Usually, this involves a septic tank and an absorption field with their associated devices. Many innovative techniques are being developed at present. The second is a centralized wastewater treatment plant (WWT), in which a sanitary sewer network within a town brings in wastewater to a treatment plant by gravity or by pumping. At the treatment plant, the entire process is divided into two major steps.

The primary treatment involving physical operations and chemical processes similar to those used in water treatment, and the secondary treatment consisting of the biological processes to remove organic waste, together with sludge treatment collected during the two steps. This is shown in Figure 9-10.

Optionally, when nutrients such as nitrogen and phosphorus are to be removed to control eutrophication or when any specific contaminant like a heavy metal has to be removed, additional or tertiary treatments are done.

Primary Treatment

Primary treatment consists of the following physical unit operations performed in sequential order. The first two are also recognized as preliminary treatment.

- Bar screens remove large floating objects from wastewater such as rags, wood, and plastics.
- A grit chamber (a large tank) holds the water just long enough to drop the obvious large particles such as sand, grit, and gravel, to the tank bottom.
- A sedimentation tank, also known as a primary clarifier, holds water for several hours. Solids that can settle by gravity are removed at this stage. During this stage, substances that float, such as grease and oil, rise to the surface where they are skimmed off. The sludge is sucked out of the bottom of the tank and is processed separately.

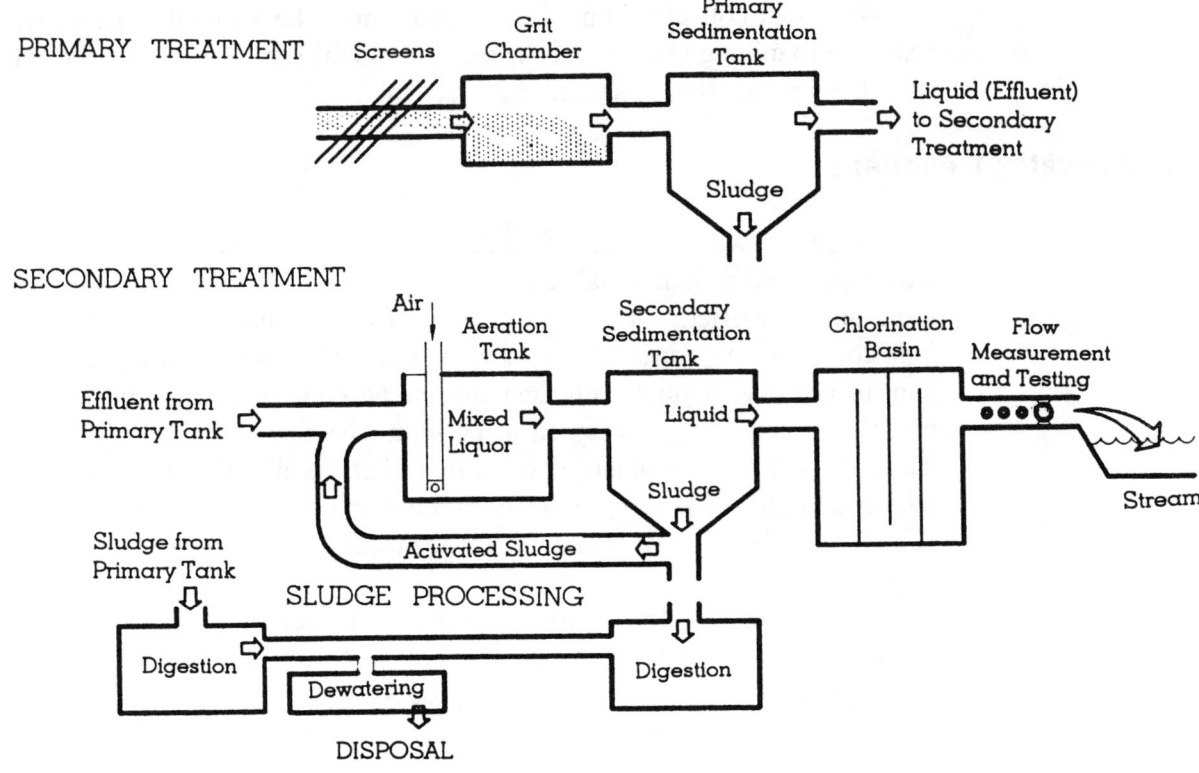

Figure 9-10 Wastewater Treatment System
(As modified from Water Control Federation: Clean Water for Today)

Secondary Treatment

Secondary treatment is carried out using essentially biological processes. Two common techniques are employed: activated sludge process and trickling filter process. These are discussed in Chapter 7. Both techniques use two units. In the first unit, microorganisms are brought in contact with the wastewater on which they feed. The correct temperature, nutrient, and oxygen levels are maintained for microorganisms to encourage optimum waste consumption. The final products are carbon dioxide, water, and new microorganism cells. From the first unit, the mixture of water and microorganisms goes to a second unit: a sedimentation tank similar to the one used in primary treatment. In this unit, also known as the final clarifier, microorganisms and other solids settle to the bottom.

A portion of the solids that settle in the final clarifier, are circulated back to the beginning of the process to serve as seed organisms. The remaining solids are withdrawn for further processing of the sludge. The water portion of the wastewater from the final clarifier, known as effluent, is disinfected

Pollution and Treatment of the Water Environment

before it is discharged to receiving waters. Secondary treatment, at the end of which more than 85% of the BOD is removed, is usually the end of wastewater treatment.

Design of Wastewater Units

Primary Sedimentation Tank

Two design parameters for primary sedimentation tanks are: the detention time, which is the ratio of tank volume to rate of wastewater flow, as given by Equation 7-9; and the surface loading or overflow rate (SOR) expressed as the rate of wastewater flow per square foot of tank surface area. Normally, tanks are designed to provide 1.5 – 2.5 hr of detention, based on the average daily rate of wastewater flow. The surface overflow rates, typically range between 800 and 1200 gal/day-ft^2.

Tanks are either circular or rectangular, with a length-to-width ratio in the range of 2–5. If the plant is not very small, two or more tanks are provided so that the process can continue when one tank is being maintained.

> ### *Example 9-3*
> Design a primary sedimentation tank for an average daily wastewater flow of 1 million gal/day (MGD).
>
> ### *Solution:*
> Adopt a design detention time, $t = 2$ hr.
> Adopt a design surface overflow rate, SOR = 1000 gal/ft-day
>
> Tank volume, $V = Qt$
>
> $$= \left(1 \times 10^6 \frac{\text{gal}}{\text{day}}\right)(2 \text{ hrs})\left[\frac{1}{7.48} \frac{\text{ft}^3}{\text{gal}}\right]\left[\frac{1 \text{ day}}{24 \text{ hrs}}\right]^3$$
>
> $$= 11{,}141 \text{ ft}^3$$
>
> Surface area, $A = \dfrac{Q}{\text{SOR}}$
>
> $$= \frac{1 \times 10^6 \text{ gal/day}}{1000 \text{ gal/day-ft}^2}$$
>
> $$= 1000 \text{ ft}^2$$

---

[3] The term [ ] converts units of flow and time.

Provide two rectangular tanks of length equal to three times the width.

For each tank, $V = \frac{11,141}{2} = 5,570.5 \text{ ft}^3$ and $A = \frac{1000}{2} = 500 \text{ ft}^2$

Hence, $3W(W) = 500$
or $W = 12.9$ or 13 ft
and $L = 3 \times 13 = 39$ or 40 ft
and $H = \frac{\text{volume}}{\text{area}} = \frac{5570.5}{13 \times 40} = 10.7$ or 11 ft.

Secondary Treatment Units

Secondary treatment constitutes a combination of two units; a biological reactor and a clarifying or sedimentation basin. The biological process is carried out either in a suspension mode (*suspended growth process*), wherein the bacteria swim within wastewater liquid, or in an attached mode (*attached growth process*), wherein the bacteria stay in a place on a solid medium and a stream of wastewater passes through them. The activated sludge treatment is the most commonly used suspension mode and the trickling filter treatment is the most commonly used attached mode process. The activated sludge reactor is known as an *aeration tank* since an external supply of air is a critical element of the method.

Activated Sludge Aeration Tank

The contents of the aeration tank is known as *mixed liquor*. Suspended solids in the tank, mostly microorganisms (organics are mostly in dissolved form), are known as *mixed liquor suspended solids* (MLSS). Two factors are important in the process: an adequate population of microorganisms, and the availability of oxygen. Correspondingly, the two design parameters of aeration tanks are food-to-microorganism (F/M) ratio and aeration period.

F/M ratio, also known as BOD loading is expressed as pounds of BOD per day per pound of MLSS, as follows:[4]

$$\frac{F}{M} = \frac{133,690(\text{BOD}_s) Q_o}{(\text{MLSS}) V} \qquad (9\text{-}7)$$

where BOD_s = settled BOD_5 (5-day BOD) from primary tank, mg/L
Q_o = average daily wastewater flow, MGD

---

[4]This equation is the same as Equation 7-16, expressed differently.

MLSS = mixed liquor suspended solids, mg/L
V = volume of tank, ft³
133,690 = conversion of units.

Aeration period (similar to detention time) is expressed by dividing the tank volume by average daily flow. Both of these parameters are computed without regard to return sludge amount. The quantity of return sludge flow is fixed based on the level of MLSS desired in the tank. There are many variations in activated sludge operations with a wide range of prescribed aeration periods and F/M ratios.

A conventional activated sludge process has a BOD5 loading (F/M ratio) range of 0.2–0.5 per day and an aeration period between 6 and 7.5 hrs. Recommended tank depth is 10–15 ft.

Example 9-4

Design a conventional aeration tank for Example 9-3. The BOD from primary clarifier is 120 mg/L. MLSS is 2000 mg/L.

Solution:

Adopt design F/M = 0.5 per day
Adopt design aeration period, t = 6 hr.
From Equation 9-7:

$$0.50 = \frac{133{,}690\,(120)(1)}{(2000)\,V}$$

or V = 16,043 ft³

Aeration tank volume, $V = Qt$

$$= \left(1 \times 10^6 \frac{\text{gal}}{\text{day}}\right)(6\text{ hrs})\left[\frac{1}{7.48}\frac{\text{ft}^3}{\text{gal}}\right]\left[\frac{1\text{ day}}{24\text{ hr}}\right]$$

$$= 33{,}422 \text{ ft}^3 \leftarrow \text{controls}$$

Assume a depth of 10 ft and a length of twice of the width:

$$A = \frac{33{,}422}{10} = 3342 \text{ ft}^2$$

$(2W)(W) = 3342$
$W = 41$ ft
$L = 82$ ft.

Trickling Filter

The design parameters used for trickling filter units are the BOD and hydraulic loading.

BOD loading in this case expressed in terms of pounds per day of BOD_5 applied per 1000 ft³ volume of the filter, is as follows:

$$\text{BOD loading} = \frac{8340\,(BOD_s)(Q_o)}{V} \qquad (9\text{-}8)$$

where BOD_s = settled BOD_5 from primary, mg/L
Q_o = Average wastewater flow rate, MGD
V = filter volume, ft³
8340 = conversion of units.

Hydraulic loading is expressed in terms of rate of flow per acre of surface area of the filter. BOD loading is calculated without regard to the BOD in the returned (recirculated) flow. However, for hydraulic loading, the raw wastewater flow and recirculated flow are added together, as follows:

$$\text{Hydraulic load} = \frac{Q_o + R}{A} \qquad (9\text{-}9)$$

where Q_o = average wastewater flow rate, MGD
R = recirculated flow = Q_o × circulation ratio
A = filter area, in acres.

Low-rate trickling filters have no recirculation. High-rate trickling filters have a circulation ratio of 0.5 to 3.0. For high-rate filters, the standards are: BOD loading of 25 to 45 lb/1000 ft³-day, hydraulic load of 10 to 30 MGD/acre, and depth of 5–7 ft.

For the design of a filter, an empirical relation has also been developed that relates the influent BOD and effluent BOD with filter depth and applied rate of wastewater flow. The following simple design has been based on the loading standards only.

Example 9-5
Design a high-rate trickling filter for Example 9-3.
Recirculation ratio is 0.5.

Solution:
Adopt design BOD loading = 25 lb/1000 ft³-day
Adopt design hydraulic loading = 10 MGD/acre
From Equation 9-8:

$$\text{Volume of filter media, } V = \frac{8340\,(120)(1)}{(25)}$$

$$= 40{,}032 \text{ ft}^3$$

From Equation 9-9:

$$\text{Area of filter, } A = \frac{Q_o + R}{\text{hydraulic loading}}$$

$$Q_o = 1 \text{ MGD}, R = 0.5 \text{ MGD}$$

$$\text{Hence, } A = \frac{1.0 + 0.5}{10} = 0.15 \text{ acre or } 6534 \text{ ft}^2$$

$$\text{Depth, } H = \frac{V}{A} = \frac{40,032}{6,534} = 6.12 \text{ ft (between 5 and 7 ft, OK)}$$

$$\text{Diameter of filter, } D = \left[\frac{4(6534)}{\pi}\right]^{1/2} = 91 \text{ ft}$$

Use a filter of 95 ft diameter with 6 ft depth.

Final Sedimentation Tank (Clarifier)

Zone settling, as discussed Chapter 7, takes place in the final clarifier. Solids handling during peak flows is an important consideration in design. Hence, in addition to the detention time and the surface overflow rate, based on the average daily wastewater flow (without regard to return sludge flow), similar to a primary tank, two other parameters are important in the design of the final clarifier. These are the surface overflow rate for peak wastewater flow condition (without regard to return sludge flow) and the solids loading based on the average daily wastewater flow, including the return sludge flow, expressed as follows:

$$\text{Solids loading (in lbs/day-ft}^2\text{)} = \frac{8.34 \text{ (MLSS)} (Q_o + R)}{A} \quad (9\text{-}10)$$

where MLSS = mixed liquor suspended solids in aeration tank, mg/L
Q_o = average daily wastewater flow, MGD
R = return sludge flow, MGD
A = clarifier surface area, ft².

Typical surface overflow rates based on average daily design flow are 600–800 gal/day-ft². During peak flows, the surface overflow rates should not exceed 1200–1600 gal/day-ft². Minimum detention time is 2–3 hours. The allowable solids loading are 20–30 lb/day-ft². For good settling sludge in deep clarifiers, the solids loading can be as high as 60 lb/day-ft².

Example 9-6
Design the final clarifier for Example 9-3. The MLSS is 2000 mg/L. Recirculation ratio is 0.5. Peak flow is 2.2 times of the average flow.

Solution:
Adopt design SOR for average daily flow = 600 gal/day-ft^2
Adopt design SOR for peak flow = 1200 gal/day-ft^2
Adopt design detention time, t = 3 hr
Adopt design solids load = 20 lb/day-ft^2

Peak flow = 2.2 × 1 = 2.2 MGD
Flow + return flow = $Q_o + R$ = 1.0 + 0.5 = 1.5 MGD

$$\text{Volume of tank} = Q_o t = \left(1 \times 10^6 \frac{\text{gal}}{\text{day}}\right)(3 \text{ hrs})\left[\frac{1}{7.48}\frac{\text{ft}^3}{\text{gal}}\right]\left[\frac{1 \text{ day}}{24 \text{ hr}}\right]$$

$$= 16{,}712 \text{ ft}^3.$$

Based on average flow, surface area, $A = \dfrac{Q_o}{\text{SOR}} = \dfrac{1 \times 10^6}{600} = 1667 \text{ ft}^2$

Based on peak flow, $A = \dfrac{Q_o}{\text{SOR}_{\text{peak}}} = \dfrac{2.2 \times 10^6}{1200} = 1833 \text{ ft}^2$ ← controls

From Equation 9-10:
Based on solids leading, $A = \dfrac{(8.34)(2000)(1.5)}{20} = 1{,}251 \text{ ft}^2$

Use two tanks each with $V = \dfrac{16{,}712}{2} = 8356 \text{ ft}^3$

and $A = \dfrac{1833}{2} = 916.5 \text{ ft}^2$

Depth $= \dfrac{V}{A} = \dfrac{8{,}356}{916.5} = 9.11 \text{ ft}$

Diameter $= \left[\dfrac{4\,(916.5)}{\pi}\right]^{1/2} = 34.2 \text{ ft.}$

Use two tanks of 35 ft diameter and 10 ft depth.

Efficiency of Treatment

Average domestic wastewater has a suspended solids concentration of 240 mg/L and BOD$_5$ of 200 mg/L. In primary sedimentation alone, approximately 40–50% of suspended solids and 24–40% of BOD$_5$ are removed. Including the secondary treatment, but without any advanced treatment, the overall efficiency of a plant is above 85% for both BOD_5 and suspended solids. For efficiency of the secondary-stage of treatment (biological unit and final clarifier combination), the National Research Council has proposed a separate formula set for trickling filters and activated sludge processes. However, the overall plant efficiency can be expressed by the following simple relation:

$$\eta = \frac{BOD_i - BOD_e}{BOD_i} \times 100 \qquad (9\text{-}11)$$

where $BOD_i = BOD_5$ of the influent (raw) wastewater, mg/L
$BOD_e = BOD_5$ of the effluent from final clarifier, mg/L.

Advanced or Tertiary Treatment

In situations for which a higher degree of treatment is desired to remove nutrients such as nitrogen and phosphorus or toxic substances such as heavy metals or when a more complete removal of BOD is required so that the treated water can be reused, extra processing is carried out beyond the secondary stage. This is recognized by a general term *advanced wastewater treatment*, (AWT). Different types of AWT processes are designed to deal with specific constituents. Important AWT processes are summarized below:

- Increased BOD and suspended solids removal by filtration techniques.
- Phosphorus precipitation by coagulation with alum or lime.
- Two-step process for nitrogen removal. First, the converting of ammonia, which is the most common form of nitrogen in wastewater, into nitrate by aerobic bacteria. This is nitrification. Second the conversion of nitrate to nitrogen gas by anaerobic bacteria. This is denitrification.
- Heavy metals removal by lime precipitation along with other metal coagulants.

Sludge Treatment

The settled portion of waste collected from the bottom of primary and final sedimentation tanks is referred to as *sludge* which is a slurry of water and solids, typically containing about 3% solids. It is not practical to dispose of such volumes of wastes. Accordingly, techniques have been designed to increase the solid concentration to as much as 20–50%, bringing the sludge to a consistency ranging from wet mud to chunky solids. As discussed in the next section, this can reduce the sludge volume by as much as a factor of twenty. Large volumes of liquid extracted from sludge are returned to treatment facilities. The various processes involved in handling sludge are as follows:

Conditioning: This includes adding chemicals or providing heat treatment to make the sludge release water more easily. This is done prior to thickening and then prior to dewatering operations.

Thickening: Thickening techniques use gravity, vacuum, flotation (introduction of air bubbles), or centrifugal force to separate water from the solids. A thickened sludge contains 6–10% solids.

Stabilizing: This is an important step in which the organic matter present in the sludge is oxidized so that solids do not decompose further. This is done biologically by anaerobic or by aerobic digestion. These techniques are described in Chapter 7. Sometimes the oxidation is done chemically using lime.

Dewatering: Dewatering involves further removal of water from digested sludge. Mechanical devices include filters, centrifuges, and presses. Other effective dewatering techniques include drying beds and drying lagoons, where such facilities are available.

Disposal: After completion of above treatment, the concentrated solids are placed in landfills, applied to land, spread in the ocean, incinerated, or composted with wood chips, leaves, or shredded papers. These practices are discussed in the context of solid waste disposal in Chapter 11.

Sludge Weight-Volume Relation

Sludge contains a substantial amount of water and, hence, occupies a large volume. Each percent increase of solids content cuts down the volume considerably. The volume of sludge can be related to its weight as follows:

$$V = \frac{W}{sG_s} \qquad (9\text{-}12)$$

where V = volume of sludge, in L
W = Weight of solids in sludge, Kg
s = solids content, fraction
G_s = Specific gravity of sludge.

Before dewatering, the sludge has a specific gravity of 1.01–1.04.

Example 9-7
The sludge from a primary clarifier has a solids content of 3%, which is increased to 6% after the thickening operation. Determine the reduction in volume.

Solution:

Adopt a sludge specific gravity of 1.03 that remains unaffected. Assume any weight of sludge, for example, 1000 kg

$$\text{Volume of primary sludge} = \frac{1000}{(0.03)(1.03)} = 32{,}362 \text{ L}$$

$$\text{Volume of thickened sludge} = \frac{1000}{(0.06)(1.03)} = 16{,}181 \text{ L}$$

$$\text{Percent reduction} = \frac{32{,}362 - 16{,}181}{32{,}362} = 50\%.$$

PROBLEMS

1. The average flow in a river is 100 ft³/sec (cfs). The BOD_5 of natural river water is 5 mg/L. Effluent from a treatment plant, carrying a BOD_5 of 30 mg/L, is discharged at a rate of 5 cfs. What is the BOD_5 concentration in the river after complete mixing?

2. A river having a flow of 100 cfs. has the total suspended solids (TSS) concentration of 10 mg/L. The maximum allowable concentration of TSS from a plant that discharges 5 MGD of effluent is 30 mg/L. What is the concentration of the TSS in the river after plant discharge?

3. A city produces sewage at a rate of 23 MGD andt has a BOD of 20 mg/L and a dissolved oxygen (DO) of 1.5 mg/L. The streamflow is 150 cfs. The stream DO and BOD levels are 8.4 mg/L at saturation and 1.5 mg/L respectively. Graph the oxygen sag curve. $k = 0.45$/day.

4. A city discharges 5 MGD of waste with no dissolved oxygen and 200 mg/L of BOD into a river, the discharge of which is 100 cfs. River water is clean with no BOD and a saturated DO level. The water temperature is 18°C and deoxygenation rate is 0.3/day. What is the critical level of DO? At what time does it occurs?

5. In Problem 4, if the river is flowing at a velocity of 1.5 ft/sec, how far from the discharge point will the critical deficiency occur? Draw the oxygen sag curve.

6. Phosphorus and nitrogen contents of a lake are 0.12 mg/L and 2 mg/L respectively. Determine whether an algae bloom will occur in the lake, and whether phosphorus or nitrogen is the controlling factor. The critical concentration of phosphorus and nitrogen are 0.015 mg/L and 0.3 mg/L respectively.

7. A dug-out trench used for storm water conveyance runs parallel to a fresh-water channel. Both are 1000 ft apart and are connected by a pervious formation 20 ft thick. The hydraulic conductivity and porosity of formation are 0.6 ft/hr and 0.75, respectively. The water level in the trench is 5 ft higher than in the channel. Determine the advection velocity and the rate of seepage of storm water from the trench into the channel.

8. In a locality, two wells are drilled through a pervious formation having a hydraulic conductivity of 400 ft/day. Depth of wells through the formation is 100 ft. The wells are 1500 ft apart. The elevations of water in the wells are 120 ft and 115 ft, respectively. Determine the flow rate of groundwater per foot width of the formation.

9. In a two-phase flow, water and oil steadily flow vertically through a 15-ft-thick soil layer. The drop of head of water and oil through the soil layer are 5 ft and 3 ft, respectively. The hydraulic conductivity of water and oil are 6 ft/day and 4 ft/day, respectively. Determine the rate of flow of two immiscible fluids through a surface area of 500 ft².

10. Oil leaks continuously from an underground tank into the water table through a soil medium. The head difference between the oil in the tank and water table is 18 ft and the depth of the water table to the tank bottom is 10 ft. If the leaking surface area is 100 ft^3., determine the rate of oil flow. Hydraulic conductivity of oil is 0.16 ft/hr.

11. When the average grain size of a medium is 1 mm, the critical height of a DNAPL is 1.2 m to migrate down through water. If the grain size is reduced to 0.075 mm, determine the critical height, if other conditions remain unchanged.

12. A 2-ft-thick layer of DNAPL having a density of 1.2 g/cm^3 can migrate through water in a sand bed with an average grain size of 0.1 mm. Determine how much thickness of the layer of DNAPL has to build up of the similar capillary characteristics but 1.08 g/cm^3 density to migrate through a bed of fine sand of 0.5 mm diameter.

13. Design a rectangular primary sedimentation tank for an average daily wastewater flow rate of 2 MGD. Adopt the detention time and surface overflow rate of 2 hr and 1000 gal/day-ft^2. Design a single tank having a length of twice the width.

14. For Problem 13, design two circular tanks.

15. A tank has dimensions of 100 ft long × 40 ft wide × 15 ft high. It is used to treat wastewater flow of 3 MGD. Determine the detention time and the surface overflow rate.

16. Design a conventional activated sludge aeration tank for Problem 13. The settled BOD$_5$ is 140 mg/L and MLSS is 2500 mg/L. Use design F/M of 0.4/day and an aeration period of 7 hrs.

17. The BOD$_5$ from a primary clarifier is 140 mg/L. The flow rate is 0.12 cfs and the MLSS is 2000 mg/L. For a tank of dimensions, 40 ft long × 20 ft wide × 10 ft high, calculate the F/M ratio.

18. Design a trickling filter for Problem 9.13. The settled BOD$_5$ 140 mg/L and the recirculation ratio is 0.5. Use design of BOD loading of 30 lb/1000 ft^3/day and hydraulic loading of 15 MGD/acre.

19. The raw flow from a municipality is 1.5 MGD. The BOD$_5$ from the primary is 117 mg/L. Determine the diameter and depth of a low-rate trickling filter (without any recirculation). Standard BOD loading is 15 lb/1000 ft^3/day and hydraulic loading is 3 MGD/acre. Depth is between 5 and 7 ft.

20. Two final clarifiers of 100 ft diameter and 15 ft deep are designed to treat 12 MGD of the average daily flow. Calculate the surface overflow rate and the detention time. If the aeration tank is operated at MLSS of 4000 mg/L and the recirculation ratio is 0.4, calculate the solids loading on the clarifier.

21. Design the two clarifiers for Problem 13. The MLSS is 2000 mg/L, and the recirculation ratio is 0.5. Peak flow is twice of the average flow. Use the lowest suggested limits for the design parameters.

22. BOD_5 of a domestic wastewater is 220 mg/L. The effluent from a secondary treatment plant has a BOD_5 of 25 mg/L. What is the efficiency of the treatment plant?

23. A raw wastewater has a BOD of 200 mg/L. After primary sedimentation, the BOD is 130 mg/L and after secondary treatment, the BOD reduces to 30 mg/L. Determine, the efficiency of the primary treatment, the secondary treatment alone, and the overall efficiency.

24. The raw sludge from a treatment plant has solids content of 4% and a specific gravity of 1.02. After treatment, the solids content increases to 50% and the specific gravity to 1.1. Determine the percent reduction in the volume of the sludge.

25. The sludge produced from a treatment is 10,000 lb/day containing 3% solids. How much volume of sludge is produced every day? If the volume is to be reduced to one quarter of the original, to what solids content should the sludge be thickened? Assume specific gravity of 1.03.

STUDY QUESTIONS

1. Explain the differences that may be expected in water quality characteristics of a lake and a stream. What are the typical problems associated with each?

2. Evaluate the significance of oxygen demanding substances. What is the main goal of a pollution control program directed largely to remove BOD?

3. Discuss occurrences in a stream after the discharge of organic matter. What impact can it have on aquatic life? What impact can it have on the domestic water supply?

4. Sketch a labeled diagram of a typical oxygen gas curve. Differentiate between the curves for a stream under low and high rates of organic loadings (pollutants discharge).

5. Define the terms: Oligotrophic, Mesotrophic, and Eutrophic in the context of lakes. What are the appropriate measures to reduce eutrophication of lakes?

6. How do contaminants make contact with groundwater? What happens to them after they have contacted the groundwater?

7. Trace the movement of nonaqueous phase liquids below the ground surface.

8. Describe various techniques of groundwater treatment.

9. How does an oil spill affect the ecosystem? Discuss the measures to control the spill after an accident.

10. Water supply from a river is treated by the following operations and chemical additions. Arrange them in a sequence they will occur. Briefly describe the purpose of each operation.
    a. Filtration
    b. Addition of coagulant
    c. Screening
    d. Chlorination
    e. Flocculation
    f. Sedimentation
    g. Mixing

11. Outline, in proper sequence, the various processes undertaken during the primary treatment and the secondary treatment of a wastewater supply. State the purpose of each operation.

12. Compare trickling filtration and activated sludge operation. Highlight their similarities and differences.

13. The sludge treatment is comprised of the following processes. Arrange them in a proper sequence. State the purpose of each operation.
    a. Dewatering
    b. Conditioning
    c. Stabilizing
    d. Thickening
    e. Disposal

14. Sketch a flow sheet for a complete municipal water and wastewater system starting with withdrawal from a river source and ending with discharging of treated wastewater into the same river. The main components of operations are comprised as follows:(not necessarily in the proper order)
    a. Physical units
    b. Chemical processes
    c. Conveyance system from river to water treatment
    d. Water distribution system
    e. Primary treatment
    f. Secondary treatment
    g. Advanced treatment
    h. Sewer System to waste treatment
    i. Disposal line to river
    j. Sludge processing

15. Define the following:
    a. Thermal regime
    b. Saturation dissolved oxygen concentration
    c. Thermal stratification
    d. Aquifer
    e. Watertable
    f. Hydraulic conductivity
    g. Darcy's law
    h. Light nonaqueous phase liquids (LNAPL)
    i. Ex situ methods
    j. Physical unit operations
    k. Chemical unit processes
    l. Individual sewage disposal system (ISDS)
    m. Aeration tank
    n. Mixed liquor suspended solids
    o. F/M ratio
    p. Advanced treatment

REFERENCES

Ayers, K.W., et al., *Environmental Science and Technology Handbook*, Government Institutes, Inc., Rockville, MD, 1994.

Canter, L. W., and Knox, R. C., *Groundwater Pollution Control*, Lewis Publishers, Boca Raton, FL, 1986.

Carberry, J. B., *Environmental Systems and Engineering*, Saunders College Publishing, Philadelphia, PA, 1990.

Cheremisinoff, P. N., *Biomanagement of Wastewater and Wastes*, Prentice Hall, NJ, 1994.

Currie, J. C., and Pepper, A. T., ed., *Water and the Environment*, Ellis Horwood, New York, 1993.

Fetter, C. W., *Applied Hydrogeology*, Merrill Publishing Company, Columbus, OH, 1988.

Gupta, R. S., *Hydrology and Hydraulic Systems*, Waveland Press, Prospect Heights, IL, 1995.

Guswa, J. H., et al., *Groundwater Contamination and Emergency Response Guide*, Noyes Publications, Park Ridge, NJ, 1984.

Hammer, M. J., and Mackichan, K. A., *Hydrology and Quality of Water Resources*, John Wiley and Sons, New York, 1981.

Institute of Geology and Mines of Spain, *Groundwater Pollution: Technology, Economics and Management*, Food and Agriculture Organization of the UN, Rome, 1979.

Knowles, P.C., ed., Dames and Moore, *Fundamentals of Environmental Science and Technology*, Government Institutes, Inc., Rockville, MD, 1992.

McGauhey, P. H., *Engineering Management of Water Quality*, McGraw-Hill Book Company, New York, 1968.

McGhee, T. J., *Water Supply and Sewerage*, 6th ed., McGraw-Hill Book Company, New York, 1991.

Metcalf and Eddy, Inc., Revised by Tchobanoglous, G, and Burton, F. L., *Wastewater Engineering: Treatment, Disposal and Reuse*, 3rd ed., McGraw-Hill Book Company, New York, 1991.

Miller, D. W., ed., *Waste Disposal Effects on Groundwater*, Premier Press, Berkley, CA, 1980.

National Research Council, Geophysics Study Committee, *Groundwater Contamination*, National Academy Press, Washington, DC, 1984.

Ney, R.E., *Fate and Transport of Organic Chemicals in the Environment: A Practical Guide*, 2nd ed., Government Institutes, Inc., Rockville, MD, 1995.

Nyer, E. K., *Groundwater Treatment Technology*, Van Nostrand Reinhold Company, New York, 1985.

Peterson, S., and Strong, B., "Ecological Engineering for Wastewater Treatment," *US Water News*, P.7, April, 1993.

Reynolds, T. D., *Unit Operations and Processes in Environmental Engineering*, Brooks/Cole Engineering Division, Monterey, CA, 1982.

Shaheen, E. I., *Environmental Pollution: Awareness and Control*, Engineering Technology, IL, 1974.

Sierra Club Legal Defense Fund, Jorgensen, E. P., ed., *The Poisoned Well*, Island Press, Washington, DC, 1989.

Tchobanoglous, G., and Schroeder, E. D., *Water Quality*, Addison-Wesley Publishing Company, Reading, MA, 1987.

Vesilind, P. A., Peirce, J. J., and Weiner, R. F., *Environmental Pollution and Control*, 3rd ed., Butterworth-Heinemann, Boston, MA, 1990.

Viessman, W. Jr., and Hammer, M. J., *Water Supply and Pollution Control*, 5th ed., Harper Collins College Publishers, New York, 1993.

Ward, C. H., ed., *Groundwater Quality*, John Wiley and Sons, New York, 1985.

Water Pollution Control Federation, *Clean Water for Today: What is Wastewater Treatment?*, Water Pollution Control Federation, Alexandria, VA, (no date).

Williams, R. B., and Culp, G. L., ed., *Handbook of Public Water Systems*, Van Nostrand Reinhold Company, New York, 1986.

Chapter 10

THE ATMOSPHERIC ENVIRONMENT

Introduction to the Atmosphere

The envelope of gases surrounding a planet constitutes its atmosphere. For the earth, just as the oceans are unique, the atmosphere is distinct; together they make life possible on this planet. The atmosphere is vast; it extends up to 600 miles over the earth's surface. Beyond it is an infinite expanse of space in which billions upon billions of stars reside. The weight of atmosphere is 5.7×10^{15} tons, about one–millionth of the weight of the earth. The lower portion of the atmosphere is filled with a mixture of the following gases recognized as dry air. On the average, water vapor to 0.7% by volume is also present causing slight changes in the composition given in Table 10-1.

Table 10-1 Composition of Dry Air

Gas	% By Volume	Molecular Weight
Nitrogen (N_2)	78.08	28.02
Oxygen (O_2)	20.98	32.00
Argon (Ar)	0.9	39.88
Carbon dioxide (CO_2)	0.04	44.00
Helium, hydrogen, methane, etc.	trace	

It is believed that the early atmosphere, prior to evolution of life on the planet, did not have the same composition. It was full of carbon dioxide and notably devoid of the oxygen; it gradually changed with the evolution of green plants.

The atmosphere is stratified with altitude with respect to temperature. With increasing altitude, the pressure decreases exponentially as shown later in this chapter. The density of air also decreases but not as drastically. The stratification of the atmosphere is given in Figure 10-1 and summarized below.

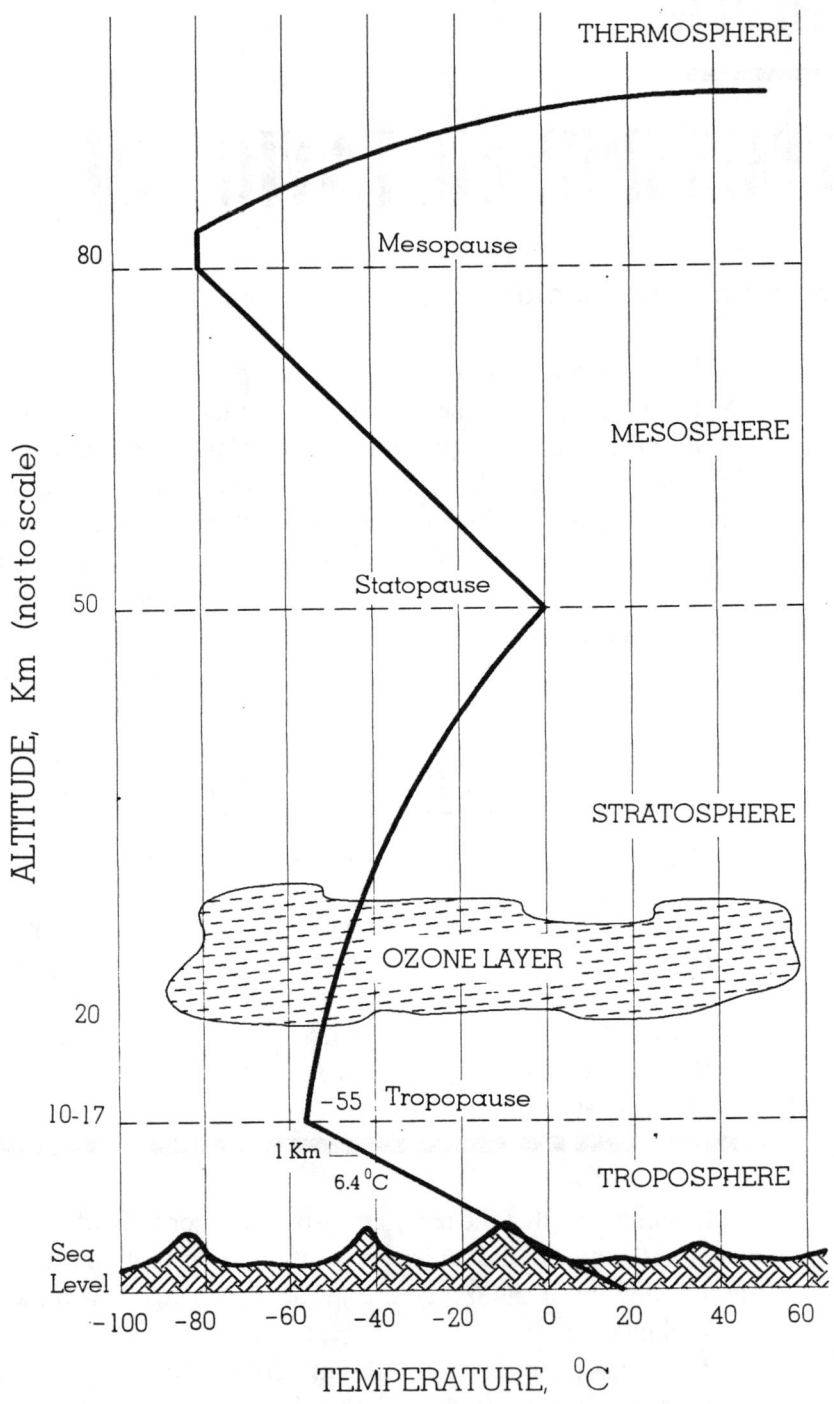

Figure 10-1 Stratification of the Atmosphere

Troposphere

The troposphere is the lowest layer that extends from sea level to a height of about 10 to 17 km (6 to 11 mi). It varies from an average of about 7.5 km near the poles to 17 km near the equator and fluctuates seasonally. This layer contains 75% of the total air mass on the planet and virtually all of the water vapor. All cloud formations, precipitation, and seasonal changes occur in the troposphere. Winds and other turbulence that are prevalent in the troposphere cause Fickian (diffusive) transport of atmospheric chemicals along with advective transport. By comparison, the chemicals that mix in a few weeks in the troposphere might take years to mix in the upper layer. Throughout this layer, there is a general decrease of temperature with height at a mean rate of 6.4°C per km. Typically, the temperature at the top of the layer is -55°C.

At sea level, the standard atmospheric pressure is 1 atm or 14.7 lb/in.^{2}, or psi. This decreases to 1.6 psi at an altitude of 15 km. Regions of transition or pauses exist between the layers. The transition between the troposphere and the next stratification is called the *tropopause*, a thin layer of relatively stable temperature. Because of this stable layer, the troposphere remains self-contained to a large extent. This portion of the atmosphere is commonly referred to as *lower atmosphere*.

Stratosphere

The second major atmospheric layer, the stratosphere, extends up to 50 km. Almost all of the remaining air mass is contained here, and together with troposphere, 99% of the air occurs within the two layers. Most of the atmospheric natural ozone is contained in the stratosphere from 10 to 30 km altitude, being at its highest concentration at an altitude of 22 km. The ozone concentration is higher above the poles than at the equator and also varies seasonally. This plays a significant environmental role as discussed later in this chapter.

The stratospheric temperature is fairly uniform up to 25 to 30 km and then rises steadily. At the top of the layer at 50 km, the temperature reaches to 0°C. A passenger jet plane can fly to within the lower stratosphere.

Mesosphere

Above the stratopause, the temperature drops from 0°C to -90°C at 80 km, the top of the mesosphere. The pressure decreases from 0.0145 psi to 0.000145 psi (practically zero). The stratosphere and mesosphere are commonly referred to as *middle atmosphere*.

Upper Atmosphere

Above the mesosphere occurs the thermosphere or ionosphere, in which temperature continuously rises. In this region solar radiation creates ionized layers. The temperature can reach 1500°C at 350 km. Above that is the exosphere where sparse air molecules and solar particles exist.

Pressure Variation in the Atmosphere

In any fluid medium of specific weight γ, pressure p, at any depth h, is $p = \gamma h$. Instead of depth, the pressure change with height is given with a minus sign.

$$dp = -\gamma dh \qquad (a)$$

According to the Ideal Gas Law:

$$pV = nRT, \text{ where } n = \text{mass/MW} \qquad (b)$$

$$\text{or } p = \frac{\text{mass}}{(MW)V}RT = \frac{\rho}{(MW)}RT = \frac{\gamma}{(MW)g}RT \qquad (c)$$

$$\text{or } \gamma = \frac{p(MW)g}{RT} \qquad (d)$$

where R = gas constant, the values listed in Table 10-2
T = absolute temperature, K = °C + 273 and °R = °F + 460
n = Number of moles = mass/MW
MW = molecular weight of gas, for air MW = 29
g = gravitational constant
ρ = density of air, where applicable
γ = specific weight of air.

Substituting, γ from (d) into (a) and simplifying:

$$\frac{dp}{p} = \frac{-(MW)g}{RT}dh \qquad (e)$$

Integrating between two heights the corresponding two pressures:

$$\ln\frac{p_1}{p_2} = \frac{-(MW)g}{RT}(H_1 - H_o) \qquad (f)$$

where $\dfrac{(MW)g}{R}$ = 0.0188 in FPS units and 0.0342 in SI units

Thus,

$$p_1 = p_o \, e^{-0.0188 \Delta H/T} \quad \text{(for FPS Units)} \tag{10-1a}$$

$$\text{or } p_1 = p_o \, e^{-0.0342 \Delta H T} \quad \text{(for SI Units)} \tag{10-1b}$$

In Equation 10-1, the temperature has been considered constant with altitude. For a large altitude difference, the average temperature might be used with reasonable accuracy.

Example 10-1
Determine the pressure at 600 m above the earth's surface. Assume the standard pressure (1 atm) and 15°C temperature at the earth.

Solution:
$T = 15 + 273 = 288$ K

From Equation 10-1b:
$p_1 = (1) \, e^{-0.0342(600)/288}$
  $= 0.93$ atm.

Table 10-2 Values of Gas Constant, R

Gas Constant, R	Units
49749	psf-ft^3/lb-mol-R
345.5	psi-ft^3/lb-mol-R
8312	Pa-m^3/g-mol-K
0.0821	atm-L/g-mol-K
8312	J/kg-mol-K
82.06	atm-cm^3/g-mol-K

Quality of the Atmosphere

Many physical attributes of the atmosphere such as wind, temperature, precipitation, and relative humidity are regularly observed as weather phenomena. Some chemical aspects such as acid rain, smoke clouds, and smog conditions are readily noticed. However, some other less obvious mechanisms and processes also take place in the atmosphere that cause long-term changes in atmospheric quality.

As apparent from our interest in monitoring day-to-day weather, what happens in the atmosphere affects us directly. In fact, the earth and atmosphere could be considered one earth-atmosphere system.

Human activities contribute to large-scale fluxes of particulate matter and gaseous substances to the atmosphere. A simple action such as burning a fire adds smoke composed of gases and particles. When any of these substances interfere with human or animal life in any way, they are regarded as *pollutants*. The atmosphere is enormous and has an immense dilution capacity. However, some substances show their effects even when present in a minute quantity or when concentrated locally before being dispersed.

The units in which the pollutants are measured, the limits that have been established on quantities of pollutants, and the types, sources, and effects of major pollutants are discussed subsequently.

Units of Pollutant Measurement

For air and gaseous pollutants, two units are used to measure the degree of pollution: the volumetric measure expressed as *parts of pollutant per million parts of air* (ppm), and the mass measure expressed as *microgram of pollutant per cubic meter of air* ($\mu g/m^3$). They are interrelated. Equation 4-17 defines the volumetric measure, 4-18 indicates the pollutant mass, Equation 4-19 interconnects them.

Example 10-2
When a fuel is burned, 800 mL of carbon dioxide is emitted to 100 L of air. Calculate the concentration. Temperature = 20°C, pressure = 1 atm.

Solution:
Volume of CO_2 = 800 mL
Volume of air + CO_2 = 100,000 + 800 = 100,800 mL

From Equation 4-17:
$$\text{ppm} = \frac{800}{100,800} \times 10^6 = 7937 \text{ ppm}$$

$T = 20 + 273 = 293$ K
MW of CO_2 = 44
$R = 0.0821$ from Table 10-2

From Equation 4-19:

$$\mu g/m^3 = \frac{(ppm)\, p\, (MW)}{TR} \times 10^3$$

$$= \frac{(7{,}937)(1)(44)}{(293)(0.0821)} \times 10^3 = 14.5 \times 10^6 \; \mu g/m^3$$

Quality Standards of Pollutants

The Clean Air Act Ammendments of 1990 (CAA, 1990), which can trace their roots to the Air Pollution Control Act of 1955, regulate the air quality in the U.S. Under the act, two types of the standards have been defined:

- *National Ambient Air Quality Standards* (NAAQS) are those that set limits for concentration of the pollutants in the outdoor atmosphere in parts per million or micrograms per cubic meter, and
- *New Source Performance Standards* (NSPS) are those that apply to emissions of pollutants from specific sources.

These are written in terms of mass emitted per unit time (grams per minute) or per unit production (grams per kilogram).

The NAAQS are implemented through NSPS contained in the State Implementation Plans (SIPs) of each state. U.S. EPA sets the minimum criteria for NSPS. NAAQS have been established for six pollutants: particulate matter (PM-10[1]), sulfur dioxide, nitrogen oxides, carbon monoxide, ozone, and lead.

For each of these pollutants, NAAQS are set based on two criteria: primary standards to protect public health, and secondary standards to protect public well-being (which may be non-health related). These standards are presented in Table 10-3. These values are at a temperature of 25°C (room temperature) and 1 atm pressure. Note that the concentrations, as per standards in Table 10-3 cannot be exceeded more than once per calendar year.

Besides the NAAQS and NSPS criteria, the Clean Air Act (CAA), addressed some other key issues as well. The 1977 version included the following:

- In regions cleaner than NAAQS, recognized as attainment areas, a prevention of significant deterioration (PSD) program was set up that allowed only specific limited emissions; thus air that was already clean is not subjected to deterioration.

---

[1]Particles of 10 microns (μm) or smaller

- In nonattainment areas that could not meet NAAQS for one or more pollutants, a policy of emissions offset was adopted that permitted construction of a new major source of pollution in nonattainment area provided it can be offset by an equal or more reduction in emissions from existing sources.

The 1990 version of the Act more directly addressed the global issues. It provided the following:

- Mobile sources were recognized as constituting a high portion of pollution. For mobile sources such as automobiles, emission standards were prescribed: non-methane hydrocarbons 0.25 g/mile, CO 3.4 g/mile, No_x 0.4 g/mile, S in diesel fuel 0.05%, and non-leaded fuel from 1996.
- Air toxins or hazardous air pollutants comprising 189 organics and metals were listed for which U.S. EPA had to promulgate standards and source regulations.
- Two precursors of acid rain, SO_2 and No_x, were controlled.
- Ozone depleting substances were scheduled to be phased out. The use of CFC's, halons, and carbon tetrachloride will be prohibited by the year 2000, and the use of hydrochlorofluorocarbons will be prohibited by the year 2030.

Table 10-3 National Ambient Air Quality Standards
(not to be exceeded more than once per year)

Pollutant	Primary Standard, ($\mu g/m^3$)	Secondary Standard, ($\mu g/m^3$)	Time
PM-10	150	150	24 hr
SO_2	365		24 hr
		1300	3 hr
CO	10,000	—	8 hr
	40,000	—	1 hr
NO_2	100	100	Annual mean
O_3	235	235	1 hr
Pb	1.5	1.5	3 months

Sources, Types, and Effects of Pollutants

There are two broad categories of air pollution sources: point or stationary sources, and diffused or mobile sources. Stationary sources relate to fixed entities such as electric utilities, industries, commercial establishments, and residential units. Four processes contribute to pollution from stationary sources: fuel combustion, industrial manufacturing, waste disposal, and forest fires. Mobile sources constitute various modes of transportation. Because it is more difficult to control emissions from mobile sources, the pollution problem from those sources is growing more severe each year.

The common pollutants contributed by these sources are: particulate matter, sulfur dioxide, nitrogen oxides, carbon monoxide, volatile organic compounds (VOCs), lead, carbon dioxide, CFCs, and methane. The first six of these are criteria pollutants covered under the NAAQS; VOCs are a precursor to O_3. The other three pollutants have global environmental consequences.

The types, sources, and effects of the pollutants are summarized in Table 10-4.

Table 10-4 Types, Sources, and Effects of Common Pollutants

Pollutant	Major Sources	Effects
Particle matter	Industrial processes (40%) Combustion (20%) Mobile (20%) Fire (15%)	Causes upper respiratory infection, cardiac disorder, bronchitis, asthma, pneumonia, and emphysema Reduces visibility Intensifies other chemical reactions
Sulfur	Coal burning power plant (65%) Metal smelting (15) Furnaces (10%)	Causes acid rain Damages plants and animals Irritates eyes, nose, throat, causes chronic bronchitis
Nitrogen oxides	Mobile sources (40%) Power plants (35%) Furnaces (15%)	Causes acid rain Causes smog Damages plants and animals Causes lower respiratory tract illness; irritates nose and eyes, causes pulmonary disorders, bronchitis, pneumonia Reduces visibility Nitrous oxides contributes to the Greenhouse Effect

Table 10-4 *continued*

Pollutant	Major Sources	Effects
Carbon monoxide	Mobile (60%) Forest fires (10%) Residential combustion (10%)	Causes slight headaches to nausea to death, depending on concentration Reacts with hemoglobin in blood to prevent oxygen transfer
VOC (nonmethane)	Industrial prosesses: petrochemical, surface coating, organic solvent (45%) Mobile sources (35%)	Reacts in atmosphere to produce ozone Some forms are carcinogenic Ozone irritates eyes, nose, and throat Ozone reduces lung function Ozone damages plants Ozone causes cracking of rubber, textile, and paints
Lead	Mobile sources Mining, smelting, incineration, steel production, battery manufacturing	Causes anemia, destructive behavior, learning disabilities, seizures, brain damage, and death
Carbon dioxide	Power plants (26%) Mobile sources (25%) Industrial combustion (19%) Residential combustion (10%) Fires and deforestation (20%)	Contributes to the Greenhouse Effect
CFCs (halogenated hydrocarbon)	Industrial origin (100%); aerosol propellant, refrigerant, foam-blowing agents, solvents, and fire retardants	Contributes to ozone depletion Contributes to the Greenhouse Effect
Methane	Natural wastelands (20%) Rice paddies (20%) Animals (15%) Biomass burning (10%) Landfills (10%)	Contributes to the Greenhouse Effect

Fate of Chemicals in the Atmosphere

It is important to understand what happens to the polluting matter after its release from a source before air quality can be determined. The three common modes to which the atmospheric pollutants subjected are summarized below.

- Physical deposition of relatively large-sized particles and liquid droplets occurs to soils, water bodies, and vegetation.

- Light particles and gaseous chemicals are transported downwind by advection while a widening plume causes the chemicals to become more dispersed.
- Significantly reactive chemicals go through a chain of photochemical and oxidizing reactions resulting in production of many chemical species.

Computer models simulate these effects on different scales. Local scale models present a single point source, the urban scale models cover many locations, and global models extend far beyond the points of pollutants' release.

Removal of Pollutants by Deposition

Several deposition mechanisms exist: gravitational settling, impaction, absorption, and wet deposition with precipitation. Particles larger than 10 µm have significant settling velocity. A 10-µm particle of 1 g/cm^3 density (density of water) settles at a rate of 0.3 cm/sec in air. The principle of terminal settling velocity was presented in Chapter 7. A major difficulty in determining the terminal velocity is assessing the drag coefficient, C_d which is highly dependent on the shape of the particle and the Reynolds Number. For viscous or streamflow regime, the terminal velocity is given by Stoke's Law of Equation 7-5. For flow-through air, Stoke's Law is applicable when the Reynolds Number is less than 1. Particles smaller than 50 µm exhibit this behavior. Commonly, Stoke's Law can be used with little error for particles up to 100 µm. Most particulate matter is less than 100 µm. The term ρ_f for density of air in Equation 7-5 is omitted.

When particles are very small (less than about 5 µm), they tend to slip past the air molecules and the settling velocity becomes greater than that produced by Stoke's Law. A Cunningham Correction Factor is applied to Stoke's terminal velocity to account for the slippage.

For particles having the Reynolds Number (R_e) greater than 1, when Stoke's Law is not applicable, the drag coefficient C_d is ascertained from C_d-R_e experimental correlation or by empirical relations available in the technical literature.[2]

Using the appropriate values of C_d, the terminal velocities for particles of various sizes are presented in Figure 10-2 for particles of densities 1.0, 1.5, 2, and 3 g/cm^3, respectively. For particle sizes above 30 µm, the left and the upper axes and the curves with corresponding arrows should be used.

---

[2]Cooper and Alley (1994, pp 117)

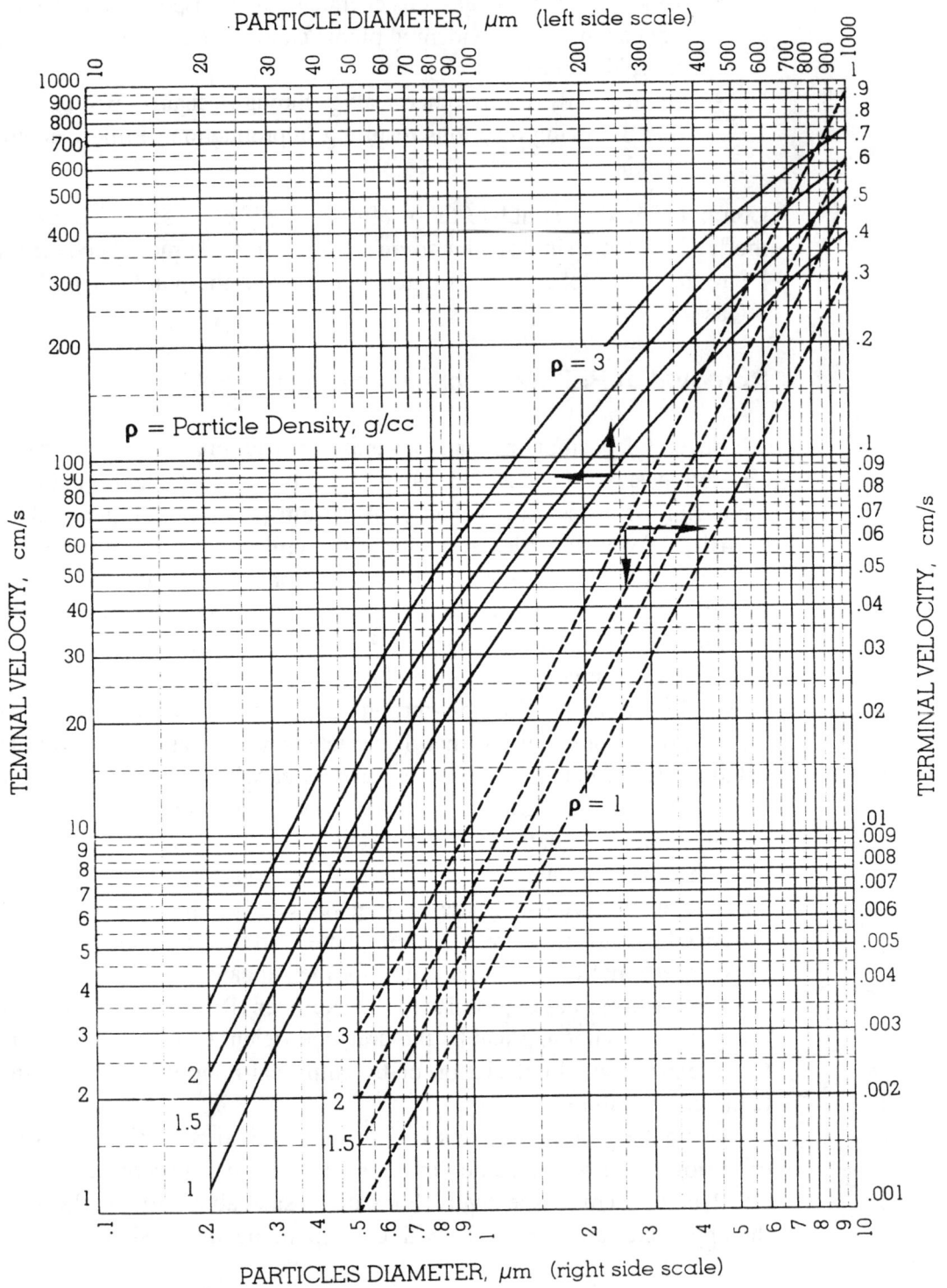

Figure 10-2 Terminal Velocity of Spherical Particles in Air at 25 °C

Example 10-3

For particulate matter of 10 μm average size, and a density of 2 g/cm³, determine the rate of deposition from the air at 25°C and 1 atm. Neglect the Cunningham Correction.

Solution:

For a 10-μm particle, Stoke's Law is applicable.
For air at 25°C, 1 atm, $\mu = 1.86 \times 10^{-4}$ poises or g/cm-sec.
$d = 10 \times 10^{-4}$ cm, $g = 981$ cm/sec²

From Equation 7-5 without ρ_f:

$$V_t = \frac{(2)(981)(10 \times 10^{-4})^2}{18\,(1.86 \times 10^{-4})}$$

$$= 0.58 \text{ cm/sec.}$$

Alternate method:
From Figure 10-2 (using the bottom and the right axes) for $d = 10$ μm,

$V_t = 0.6$ cm/sec (including the Cunningham Correction).

Dispersion of the Pollutants in the Atmosphere

Pollutants emitting from smokestacks or automobile exhausts contain gases and fine particles, the majority in the range of 0.1–10 μm. Particles larger than this settle down close to the emission source by the process discussed in the previous section. However, particles smaller than 10 μm tend to follow the motion of the gas in which they are borne. The gaseous and fine particle effluent first rises due to buoyancy and then is advected by wind and dispersed by Fickian diffusion. This is fully analogous to the groundwater plume problem discussed in Chapter 9, although mixing in air takes place due to turbulence rather than mechanical dispersion. Dispersion is very widespread in air. The assessment of the dispersion coefficients is rather involved; the coefficients are expressed in terms of the statistical parameter of standard deviations by correlating them to the diffusion coefficients. When pollutants are transported over a long distance, the equilibrium chemical reactions are also incorporated in what is known as the advection-dispersion-reaction equation similar to the application for water flow.

If the plume contains a substantial amount of particulates, they tend to settle by the gravitational effect over a long time period or distance. For average particles having a terminal velocity V_t, the free fall distance in time t is $V_t t$. To predict the particulates' pollution stream, the vertical location of the center line of the plume can be shifted downward by the amount $V_t t$.

Atmospheric Chemical Reactions

Chemical reactions contribute to many atmospheric pollution problems. Two highlights of chemical reactions in the atmosphere are that most of them are directed by the sun, i.e. they are photochemical reactions that involve a high level of energy, and that most of the chemical transformations are oxidations, since they occur in a medium containing 22% oxygen.

Forming and breaking of ozone is associated with a phenomenon of photo-dissociation, which means breaking of molecular oxygen (O-O) bond, carbon (C-C) bond, or hydrocarbon (C-H) bond. To break a molecular bond, an energy up to 500 kiloJoules (kJ) per mole may be required. The ultraviolet rays of the sun have an energy of about 600 kJ/mole. Hence, most photochemical reactions occur near the ultraviolet region. Smog and ozone depletion are related to this process, as discussed subsequently. The molecules of some chemicals excited (energized) by the sun, but not to the extent of photo-dissociation, go through molecular reactions. For example, oxides of sulfur and nitrogen are further oxidized in the atmosphere to form acids, as discussed in the context of acid rain.

Problems of Atmospheric Pollution

A common notion in the past was that the atmosphere is immense enough to dilute and disperse any substance put into it. To achieve a more effective dilution, the use of increasingly tall chimneys was a technological solution. However, many episodes of problems caused by air pollutants, including 4,000 deaths in London in 1952, exposed the vulnerability of air resources to pollution. Many air pollution problems have been recognized, ranging from a small area affected by a single industry, to a city-wide problem contributed by multiple contaminants, to the global-scale contamination by universal pollutants. Specifically, the problems pertain to:

- Acid deposition;
- Smog formation;
- Atmospheric warming or the greenhouse effect; and
- Stratospheric ozone depletion, or the hole in the sky.

Acid Rain

Pure water has a pH of 7.0. The carbon dioxide naturally present in air gets dissolved in rain to form carbonic acid (H_2CO_3)—the weak acid responsible for bubbles in soda. This pure rainwater in equilibrium with CO_2, without any other substance, has a pH of 5.6. Some other chemicals such as sulfates and nitrates are also naturally present in air and tend to further lower the pH of rainwater. Certain other alkaline substances such as ammonia, soil dust, and fly ash (from coal combustion) raise the pH level. The pH of rain reflects a resulting effect of several different substances present in the atmosphere. In remote regions unaffected by human activities, a pH of 5.0 and lower has been observed. However, the term acid rain refers to the distribution of major acidifying pollutants such as SO_2 and NO_x by human activities, their chemical transformation in atmosphere, and their deposition on earth's surface is in dry or wet form. Sulfur dioxide and nitrogen oxides, a mixture of nitric oxide (NO) and nitrogen dioxide (NO_2) reduces the pH to 4.0 or less.

In the United States, the areas most prone to acid rain are the eastern states. More than 95% of sulfur emitted in the eastern U.S. comes from anthropogenic sources. Burning of coal and oil by coal-fired power plants, industrial and commercial combustion, transportation, and metal smelters contribute to sulfur emission. Similarly, more than 90% of the NO_x emitted in the eastern U.S. comes from human sources. Fossil fuel combustion by power plants, industrial and residential boilers, and automobile sources contribute to NO_x emission. In addition, high-temperature combustion in air combines air and nitrogen to produce NO_x.

Transport, Transformation, and Deposition of Acid Rain

The fate of acidifying gases after emission from a stationary source is shown in Figure 10-3. A plume from an exhaust source such as a chimney containing sulfur dioxide or nitrogen oxides or both, spreads out into a cone shape in the direction of the wind. The plume movement is controlled by advection (wind velocity) and diffusion both horizontally and vertically. As a result, the pollutant concentration decreases downwind. After a distance, the plume loses its coherence and disrupts as a result of wind shear.

About 10–20 km downwind from the source, a substantial portion of sulfur dioxide (SO_2) and nitrogen oxide (NO_x) particles are deposited on the ground. The rest may travel hundreds or even thousands of kilometers. During this travel time, SO_2 and NO_x are converted to sulfuric acid (H_2SO_4) and nitric acid (HNO_3), either in gas-phase reactions in dry atmosphere or in liquid-phase reactions within clouds, the latter being more reactive. Eventually these acids are brought to earth's surface by snow or rainfall.

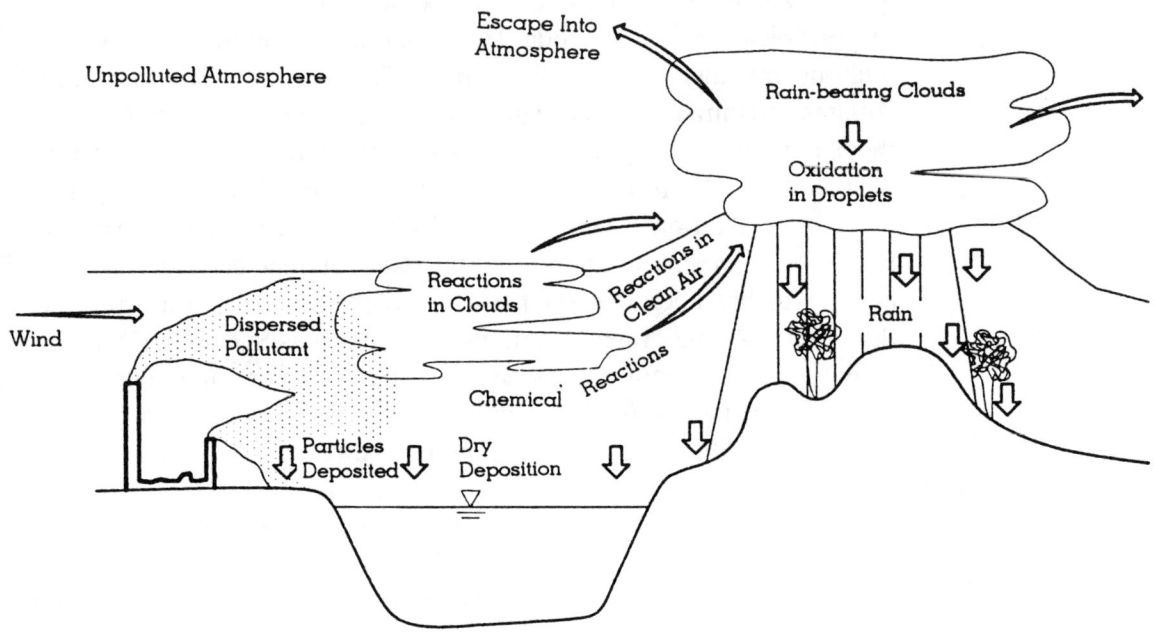

Figure 10-3 Fate of Acidifying Gases after Emission
(As modified from Mason, B.J., 1992)

There are many different pathways by which SO_2 and NO_x are converted to acid form in the atmosphere. The simple reactions are

$$2SO_2 + O_2 + 2H_2O \rightarrow 2H_2SO_4 \tag{10-2}$$

and

$$NO + NO_2 + O_2 + H_2O \rightarrow 2HNO_3 \tag{10-3}$$

The conversion of NO_x is much faster than conversion of SO_2. Some scientists argue that human activities are not a major cause of acid deposition. According to them, dozens of natural chemicals that are found in water, along with transportation of water through soil formations and sea sprays, all contribute to acidity. However, according to a study by the National Research Council in 1986, based on the mass balance of the emitted pollutants and acid deposition, there is no doubt that emissions of sulfur dioxide (and nitrogen oxides) produce acid rain and that the acid rain harms the environment.

Acid Rain Damage

The damage caused by acid deposition is widespread. It is a worldwide problem that tends to destabilize an ecosystem. The foremost among damages is a significant loss of forests. The effects of acid rain work on the tree-tops where needles or leaves turn yellow and drop prematurely from branches. It also effects the bottom of the tree where acidity leaches out nutrients from soils. Even when trees show no visible signs of damage, their growth and productivity decline.

Lakes, streams, and watersheds, and aquatic life within them are other victims of acid rain. These water bodies may tolerate acid rain to an extent because of their buffering capacity. Beyond that, the acidification decreases the population of phytoplankton and zooplankton (base-trophic-level-animals), thereby, seriously impairing the survival of high-trophic-level animals such as fish. Acid also kills microorganisms responsible for decomposing decaying plants and animals.

Some fine monuments and statues that have withstood the weathering of centuries have been disfigured by the corrosive behavior of acid rain.

Even human health is at risk because trace metals are leached into the water supply by acidified groundwater.

Acid Rain Control Measures

Requirements to control acid rain are directed toward reduction in emission of suspected precursors—SO_2 and NO_x. Among the two approaches, one relates to use of a fuel that has a lower sulfur content and that results in lower NO_x emissions; the other approach includes many technologies that are available to reduce the production of SO_2 and/or NO_x from burned gases, such as furnace modifications, scrubbing units, fluidized-bed combustion, sorbent injection in the furnace, and injection of ammonia into the flue gas. The first approach in terms of fuel switching appears to be an economical choice.

Smog Formation

Coined by a combination of *smoke* and *fog*, smog represents a cloudy formation resulting from the photochemical reaction of sunlight on the oxides of nitrogen and hydrocarbons emitted from automobiles. Originally, the term smog referred to the clouds of sulfur dioxide, sulfuric acid droplets, and heavy suspended particles being discharged by industries. That situation is now termed *gray smog*, which is quite rare. The photochemical reaction produces a high level of many oxidants; ozone is primary among them and is a major constituent of smog. Smog is formed in the lower portion of the troposphere near the ground level. It causes irritation of eyes and throat, impairs lung function, damages plants and crops and makes rubberlike products crack. Los Angeles is commonly associated with smog, although all modern cities are also affected by it.

Thermal Inversion: The Mechanism of Smog Formation

As the altitude increases, the atmosphere becomes thinner and the pressure drops. Due to this characteristic, a parcel of exhausted gas moving upward will undergo an expansion that will reduce its temperature. This natural phenomenon of reduction or lapse in the temperature is known as the *adiabatic lapse rate* (ALR), which is defined as the rate at which the temperature of the gas parcel drops due to elevation change, under the assumption that the process is adiabatic, i.e., no exchange of heat occurs between the gas parcel and the surroundings. The ALR for a dry atmosphere is 9.8 °C per 1000 m or 5.4 °F per 1000 ft. The wet ALR lies in the range of 3.6–5.5° C per 1000 m.

As noted earlier in this chapter, in general, the actual temperature of the atmosphere also drops with increased altitude in the tropospheric layer. Temperature profiles of the exhausted gas or polluted air parcel and the surrounding atmosphere are shown in Figure 10-4. The polluted air parcel at the point of release has a higher temperature than the surrounding atmosphere (exhaust gas is warmer). Two conditions can exist. In Figure 10-4a, the actual temperature of the atmosphere decreases more rapidly than the adiabatic lapse rate of the polluted parcel. At any height, the polluted air parcel will have a higher temperature than the actual surrounding temperature. Being warmer, the polluted air parcel will be lighter than the surroundings and, hence, the parcel will keep rising in the atmosphere. This is an *unstable atmosphere*.

In Figure 10-4b, the actual temperature of the atmosphere decreases at a lesser rate than the adiabatic lapse rate of the polluted air. At height X, both profiles have equal temperature. Above X, the polluted air parcel is cooler than the surroundings and, hence, heavier. This is a *stable atmosphere*. The polluted air parcel cannot rise above X. If the parcel is pushed upward by a turbulent atmosphere, it will return back to position X.

A third scenario is when the actual atmospheric temperature increases with altitude as shown in Figure 10-5, which is a *strongly stable atmosphere*. In this case, the polluted air parcel in the form of a smog will hang at height X, close to the ground. Since the atmospheric temperature has a reverse trend, this case is known as a *thermal inversion*.

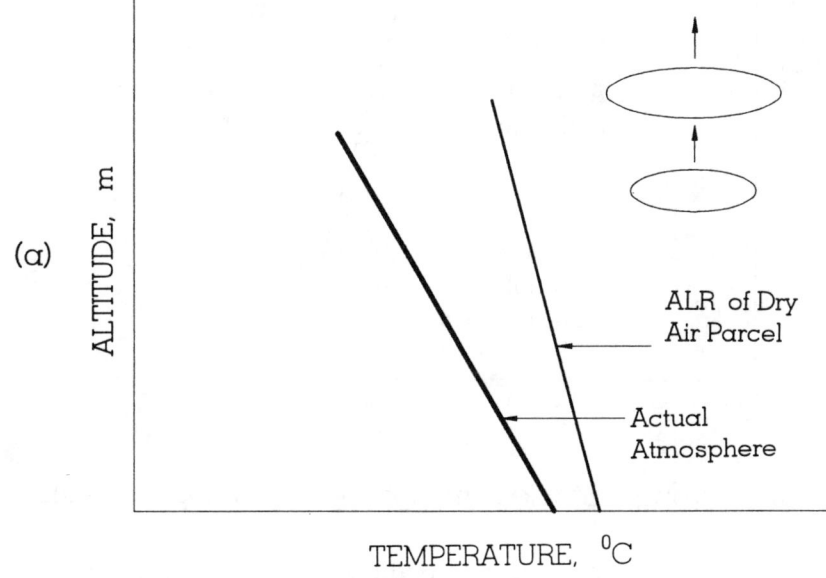

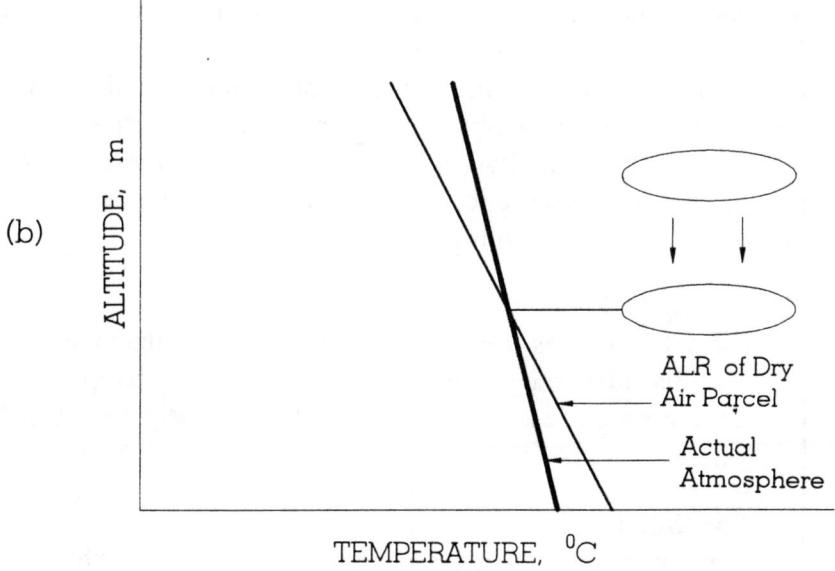

Figure 10-4 Atmospheric Stability

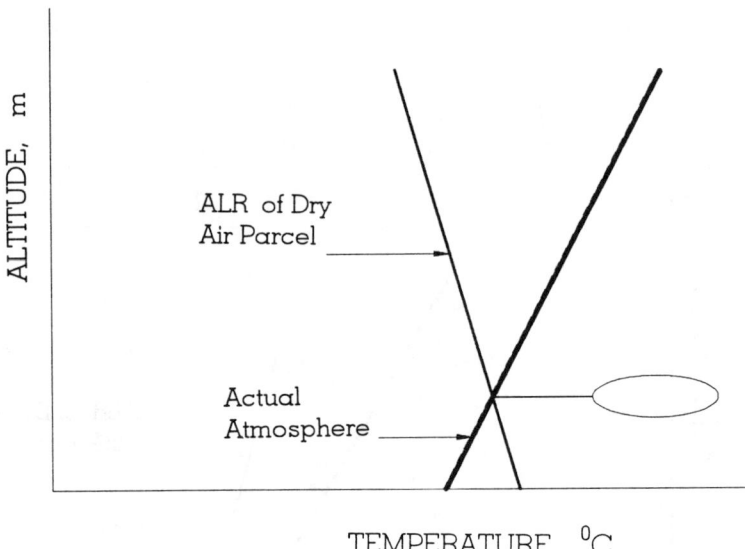

Figure 10-5 Strongly Stable Atmosphere and Smog Formation

Several meteorological conditions can bring thermal inversion. The anticyclone in which the air is compressed, and therefore warmed, produces a region of warm air over the cooler surface layer. The presence of a semipermanent anticyclone in Los Angeles is an example. The movements of hot and warm air-mass fronts, in both cases of which the warm air is forced up over the cold air, produce temperature inversions. A temporary inversion is common at night. On a clear night, the earth's surface radiates energy and cools rapidly while the upper region remains warm. These temporary inversions are responsible for morning smoke and exhaust that disappear as the sun warms the ground.

Example 10-4
The exhaust gases from a car are 10°C warmer than the surroundings. The adiabatic lapse rate is 0.65 C per 100 m. The temperature gradient of the atmosphere is +6°C per 100 m. Find the height at which smog will form.

Solution:
Suppose the atmospheric temperature is 0°C; hence exhaust will have a temperature of 10°C. Suppose the smog height is h m.

1. Temperature of air parcel = $10 - 0.65(h)/100$
2. Temperature of surrounding atmosphere. = $0 + 6(h)/100$
   Equating 1 and 2;
   $$10 - 0.65(h)/100 = 0 + 6(h)/100$$
   or $h = 150$ m

Alternatively, plot the temperature profile of atmosphere and the air parcel temperature variation as in Figure 10-6. Directly read the altitude where two lines cross each other.

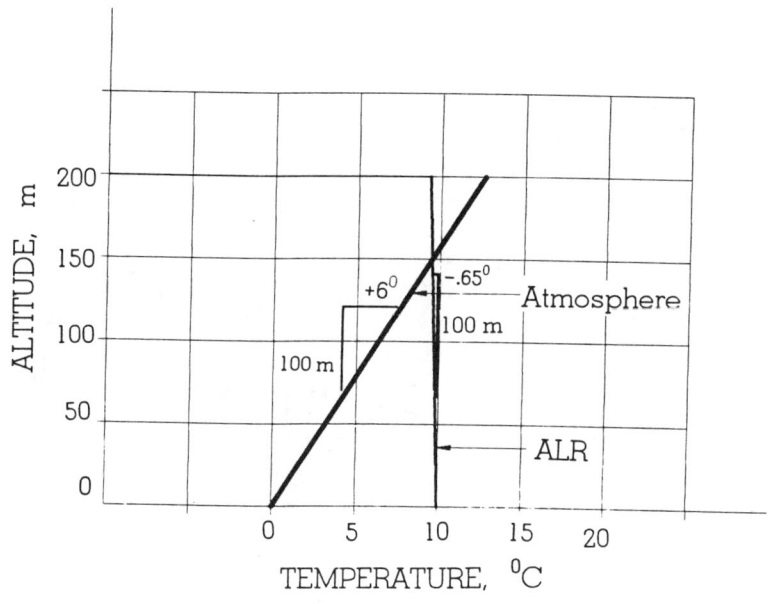

Figure 10-6 Temperature Profile of Example 10-4

Greenhouse Effect

Solar energy enters the earth's atmosphere in short wavelengths of 0.1–1 μm in the form of ultraviolet and visible light rays. It leaves the earth in wavelengths of greater than 1 μm as infrared (heat) radiation. The atmospheric absorption of solar radiation is shown in Figure 10-7. The stratospheric ozone absorbs essentially all of the ultraviolet rays of wavelengths less than 0.3 μm. This is significant as discussed in the subsequent section. Among the natural gases present in the troposphere, carbon dioxide absorbs radiation from 13 to 18 μm and water vapor of wavelengths 1–8 μm and then greater than 18 μm. This means that the atmospheric gases do not prevent the visible light from reaching the earth. However, the infrared (heat) energy radiating back from the earth is partially absorbed by CO_2 and water vapor except for the wavelengths between 7 and 12 μm. The heat energy absorbed by the atmospheric gases is reradiated back both toward the earth and to the outer atmosphere. This keeps the earth warm. These gases are called the *greenhouse gases*.

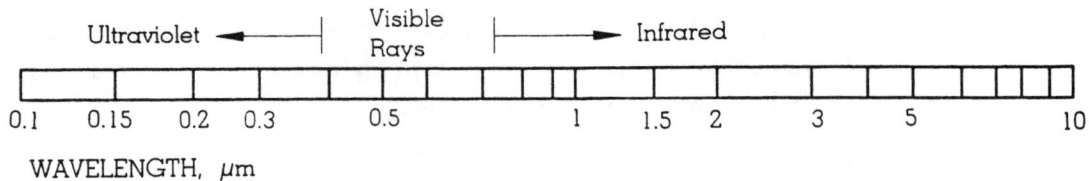

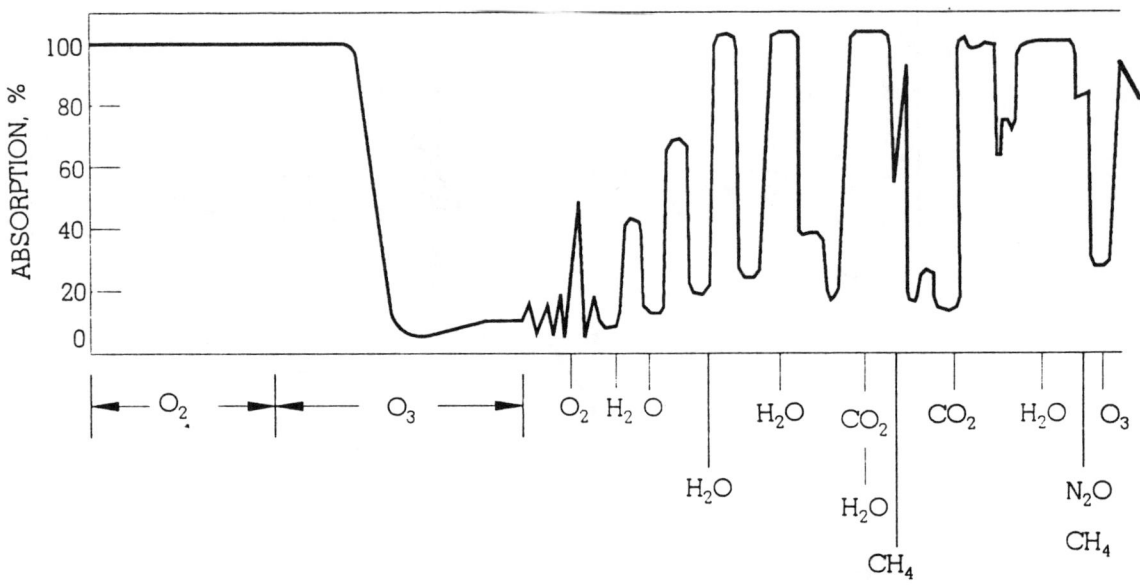

Figure 10-7 Atmospheric Absorption of Solar Radiation

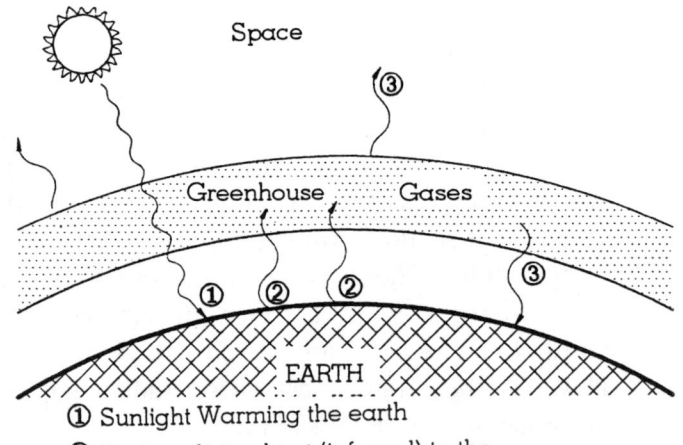

① Sunlight Warming the earth
② Earth radiates heat (infrared) to the atmosphere and some escapes into space
③ Heat absorbed by greenhouse gases reradiates back to the earth as well as out into space

Figure 10-8 A Simplified Diagram of the Greenhouse Effect

The Atmospheric Environment

The name is derived from the similar effect experienced in a greenhouse wherein the glass readily transmits the short-wavelength energy into the greenhouse but it traps the long-wavelength energy radiated by the greenhouse interior. The effect is shown in Figure 10-8.

It has been computed that the temperature of the earth is 33 °C higher than it would have been without greenhouse gases naturally present. This makes the earth liveable. The greenhouse effect is a natural phenomenon, but it is the enhanced effect by anthropogenic sources that is a cause of environmental concern.

Enhanced Greenhouse Effect or Radiative Forcing

The use of fossil fuels (coal, oil, and gas) and destruction of forests are adding more carbon dioxide into the atmosphere. In addition, some other gases that can absorb infrared radiation are also being added by human activities. Methane and ozone can absorb 7 μm and 10 μm wavelength radiation respectively. These wavelengths were not covered by the natural atmospheric gases. Table 10-5 summarizes characteristics of the greenhouse gases.

The *Global Warming Potential* (GWP), is an index that defines the time-integrated warming effect of an instantaneous release of 1 kg of a greenhouse gas in the atmosphere relative to that of carbon dioxide. The effect changes with time, hence GWP is different for a time horizon of 20, 100, and 200 years; the table values for 100 years represent the long-term effects.

The contribution to global warming, indicated by percentages in the table, is given by the following relation:

(contribution to global warming) = (GWP) × (amount of gas emitted) (10-4)

Carbon dioxide is a relatively less effective greenhouse gas but its contribution to warming is largest because of its large emission quantity. The gas having a higher GWP and a long lifetime should be given a high priority for more effective control. CFC stands out distinctly from these considerations for a high priority of control.

Example 10-5

If the global carbon dioxide emission is 26,000 giga kg in 1990, how much is the methane gas emission?

Solution:
From Equation 10-4:
CO_2 contribution to global warming = $1 \times 26000 \times 10^9$ (relative units)

This represents 61% of the total from Table 10-5.

Hence, 100% global warming = $\dfrac{26000 \times 10^9}{61} \times 100$

$= 42{,}623 \times 10^9$ (relative units)

Methane contributes 15% of the total
Hence, methane contribution to global warming =
$$(42{,}623 \times 10^9) \times \dfrac{15}{100} = 6393.5 \times 10^9$$

From Equation 10-4
For methane:
$6393.5 \times 10^9 = (21)$(amount of methane emission)
Hence, amount of methane emission = 304×10^9 kg.

Table 10-5 Characteristics of Greenhouse Gases

Gas	Atmospheric concentration (1990) (ppm)	Atmospheric lifetime (years)	Yearly increase in concentration (%)	GWP	Contribution to global warming (%)	Sources
Water vapor						Hydrologic cycle
Carbon dioxide	353	50–200	0.5	1	61	Fossil fuel, deforestation
Methane	1.72	10	0.9	21	15	Wetlands, rice, cattle
CFCs	760×10^{-6}	65–400	4	4500–7300	12	Solvents, refrigerants, foams, aerosols
Nitrous oxide	0.31	150	0.25	290	4	Fuels, fertilizers
Ozone	0.05	0.1–0.3	1		8*	Photo-oxidation of hydrocarbon and nitrogen oxides

*Includes other gases

Impacts of Greenhouse Warming

It is relatively easy to ascertain the increased radiative forcing caused by increases in greenhouse gases. However, as the earth begins to warm, many positive feedback mechanisms act to amplify the warming effect and many negative feedback mechanisms tend to reduce the temperature, thus complicating the assessment. Mathematical models have become a tool of choice to study the short- and long-terms impacts. The entire atmospheric-oceanic-terrestrial system is simulated in elaborate models. The changes in water vapor in atmosphere, cloud systems, soil moisture, air-sea interaction, biological factors, and heat retention capacity of oceans are among the feedback parameters that are difficult to model. The simplest is an energy balance model in which the only equation considered is the conservation of energy; the entire earth's surface is treated as if it has a single temperature. On the other hand, the Global Circulation Models (GCM) are three-dimensional representations that synthesize various physical and dynamic processes to the atmosphere-earth-ocean system. Extensive models may require treatment of more than 100,000 equations; still the description of many processes may be simplistic. To summarize, a factor of uncertainty exists when it comes to making accurate predictions.

Based on a detailed GCM study, the following observations were made by the Intergovernmental Panel on Climate Change (IPCC) Scientific Assessment of the World Meteorological Organization (WMO)/United Nations Environmental Program (UNEP), 1991:

- Global mean surface air temperature has increased by 0.3 to 0.6 °C over the past 100 years.
- If emissions of greenhouse gases continue according to the present pattern, the average rate of increase of the global mean temperature during next century is estimated to be 0.3 °C per decade (with uncertainty, in the range of 0.2 °C to 0.5 °C). This will result in a likely increase of about 1 °C by 2025 above the 1995 value. These changes are greater than those that have occurred naturally over the last 10,000 years.
- If emissions are subjected to strict control, including a shift to renewable energy, reversal of deforestation, and phase-out of CFCs, the average rate of increase in the global mean temperature over the next century would be about 0.1 °C per decade.
- The warming is predicted to be 50–100% greater than the global mean in high northern latitudes (North America) in winter and substantially lower than the mean in regions of sea ice.
- Precipitation is predicted to increase 0–15% in winter and decrease by 5–10% in summer in North America.

- Global warming will increase the frequency of extreme weather events, i.e., number of days with temperature exceeding 100°F will increase. The changes in the variability of weather and in the frequency of extremes will have more impact than the changes that occur in the mean climate.

Specifically, in terms of the U.S., the following changes are predicted as a result of the global warming phenomenon.

- The green belt in America will shift northward without the people, farmland, and rich soil, due to reduced summer precipitation and soil moisture in that region.
- Annual flows in several western and southwestern U.S. rivers will decrease by 40–75%, severally affecting the water supply and water quality standards.
- The reduced streamflows will affect freshwater ecosystems in the west and southwest. In the north, warming and drying will threaten marshes and peat bogs, and the species that depend on them. On the coast, salt marshes and coastal wetlands will be inundated and the coastal aquifers will be contaminated by salinity intrusion.
- Vegetation and trees species will change, effecting parks, refuges, and wilderness. Preserved areas such as Yellowstone Park will be occupied by a different kind of ecosystem.
- Every ecosystem will be impacted because, in the long term, climate and carbon dioxide are among the factors that control the ecosystem structure and composition. Based on the model results, it is difficult to make reliable estimates of regional changes in ecosystems.
- The calculations indicate a rise in sea level by 0.2 ft per decade in the next century. This will cause submersion of significant land around the coastline. Most severe effects will result from extreme events such as storm surges.

In terms of the latest observations, the average world temperature for 1995 (the hottest year on record according to the National Aeronautics and Space Administration [NASA]) was 59.8°F. This was about 0.8°F above the 1950–1980 average.

Carbon dioxide plays an important role in nature's life cycle. In spring, when the landscape turns green, plants absorb large amounts of carbon dioxide from the atmosphere which is released again when plants die off in winter. In this up-and-down pattern, the scientists at the Scripps Institution of Oceanography in La Jolla, California, have found evidence that spring is arriving about a week earlier, or in other words, the growing season is advanced by 7 days than it was 20 years ago in the Northern Hemisphere.

Ozone Depletion

Ozone plays both negative and positive roles. The section on Smog Formation in this chapter describes how ozone is formed in the troposphere by interaction of hydrocarbon, nitrogen oxides, and the sunlight, leading to smog-related problems. However, in the beginning of this chapter, it was indicated that the stratosphere contains ozone at altitudes of 10-30 km that is responsible for absorbing harmful ultraviolet rays of the sun before they hit the earth. This ozone provides us with a life protecting shield. Thus, when British scientists reported in 1985 that a hole had been occuring each spring (late August) in the ozone layer over Antarctica continuing until end of November, it caused widespread concern. Expeditions by the National Ozone Expedition Team (NOZE) comprising 150 scientists representing 19 organizations and four nations in 1987 and by NASA in 1989, and many ground-based observations, confirmed the earlier findings of damage to the ozone layer. Various theories about the causes, whether they be based on manmade chemicals and/or natural factors are being put forth.

Some of the findings based on various reports are as follows:

- Over Antarctica, average ozone concentration dropped by 50% in spring (August–October) in a hole twice as large as the U.S. In some areas, ozone has vanished completely. The blame for the Antarctic ozone hole was firmly placed on CFCs.
- Ozone losses were documented around the globe, not just over the poles. Between 30 and 60° North, where most of the world's people live, ozone had decreased by 1.7 to 3.0% between 1969 and 1986, and the winter changes were higher than this value.
- While the problem was worse over Antarctica during the spring, ozone probably decreased by 5% or more at all latitudes south of 60°South throughout the year.
- The hole alone covers about 10% of the southern hemisphere. The projections of the amount and locations of future ozone depletion are highly uncertain.

Basic Theory of Depletion

The mechanisms and specific chemical pathways of ozone depletion are quite complex and not fully understood. The basic principles related to stratospheric ozone formation and its destruction are presented here.

Above 50 mile high-energy solar radiation breaks molecular oxygen into monatomic oxygen:

$$O_2 \xrightarrow{\text{high energy}} O + O \qquad (10\text{-}5)$$

As a result, oxygen exists in monoatomic form in that region. At lower levels, in the stratosphere, the monoatomic oxygen undergoes reaction with molecules of oxygen to form ozone:

$$O + O_2 + [O_2 \text{ or } N_2] \rightarrow O_3 + [O_2 \text{ or } N_2] \tag{10-6}$$

The extra molecule in [ ] accepts the energy released in the reaction. Ozone itself is broken by high-energy radiation in the stratosphere:

$$O_3 \xrightarrow{\text{high energy}} O_2 + O \tag{10-7}$$

Reactions 10-6 and 10-7 shuttle back and forth, resulting in a balanced ozone layer that absorbs ultraviolet (UV) radiation. When any chlorine, bromine, hydroxide, or nitrogen is present in the stratosphere, it grabs oxygen released in reaction 10-7 before reaction 10-6 can take place. Thus, instead of the reaction going back to 10-6, the following reaction occurs:

$$(Cl \text{ or } Br \text{ or } OH \text{ or } NO) + O + [O_2] \rightarrow (ClO \text{ or } BrO \text{ or } HO_2 \text{ or } NO_2) + [O_2] \tag{10-8}$$

This deprives the reformation of O_3. The product (ClO, etc.) formed in reaction 10-8 photolyzes (breaks) again and again to destroy thousands of ozone molecules. The destruction stops only when ClO and NO_2 react together to form an inert molecule $ClONO_2$. However, over Antarctica, this inert molecule is reactivated again by the polar vortex, giving rise to renewed depletion in the polar spring.

Ozone-Depleting Chemicals

Chlorofluorocarbons (CFCs) constitute a family of manmade chemicals that are used in hundreds of modern-day products including aerosols, foam insulation, solvents, air conditioning systems, refrigerants, styrofoam, and flexible foams. Halons are another group of manmade chemicals that contain bromine. They are used for firefighting purposes. Both CFCs and halons do not break down in the troposphere. Within six to eight years they reach the stratosphere, where they can survive up to 100 years. They break down in the stratosphere, releasing chlorine or bromine, which is capable of destroying thousands of ozone molecules, as described.

Carbon tetrachloride and methyl chloroforms, used as solvents, are other chlorine-releasing chemicals that contribute to ozone depletion. A profile of ozone-depleting chemicals is presented in Table 10-6. Ozone Depleting

Potential (ODP) is an index similar to the GWP of the greenhouse effect. Halons are more effective ozone destroyers than CFCs.

Table 10-6 Profile of Ozone-Depleting Chemicals

Chemical	Optimum Atmospheric Life, years	Ozone Depleting Potential (ODP)	Contribution to Depletion (%)
CFCs	139	1	80
Halons	101	10	5
Carbon tetrachloride	67	1.1	10
Methyl chloroform	8	0.1	5

Impacts of Ozone Depletion

A most common effect, which is already begun to be experienced, is the rise in the incidence of skin cancer. It is projected that each 1% drop in ozone will result in a 4 to 6% increase in cases of the common, but rarely fatal, type of skin cancer, and another 1 to 2% increases in cases of melanoma, a more deadly form of skin cancer. Other health-related problems include cataracts and depressed immune systems which lower the body's resistance to diseases.

There is evidence that yields and quality of crops are substantially lowered by enhanced ultraviolet radiation. The studies indicate that aquatic ecosystems containing phytoplankton, algae, fish, and shellfish would be most threatened. Animal studies suggest that the animals' ability to fight off bacterial infection and clean out toxic substances from their systems are seriously impaired at raised UV levels. Increased UV levels also affect synthetic materials; plastics are especially vulnerable.

As a follow-up, according to a report in *Scientific American* (August, 1996), the chemists at the National Oceanic and Atmospheric Agency say that the average concentration of ozone-depleting chemicals in the lower atmosphere is decreasing rapidly, dropping about 1% by the middle of 1995.

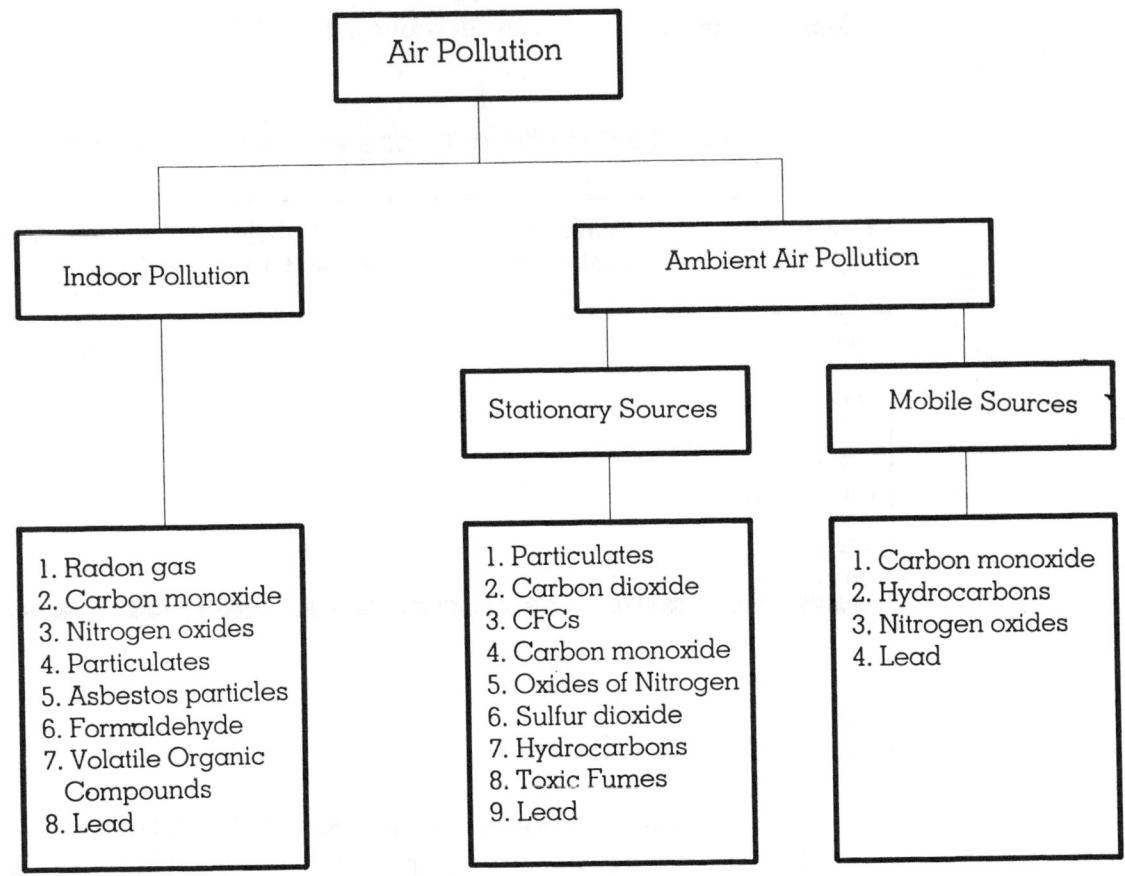

Figure 10-9 Contributory Factor Based Classification

This is a good sign that suggests that the Montreal Protocol, a treaty to ban CFCs and other halogenated compounds, is having an effect.

Air Pollution Control

Air pollution is classified in two different ways: based on the nature of pollutants, and based on the contributory factors. These classifications are illustrated in Figures 10-9 and 10-10, respectively.

Control at the source, especially for mobile pollutants by materials substitution, process modification, and equipment replacement, is the best approach to deal with air pollution. The common methods that are applied after emissions take place are briefly described below.

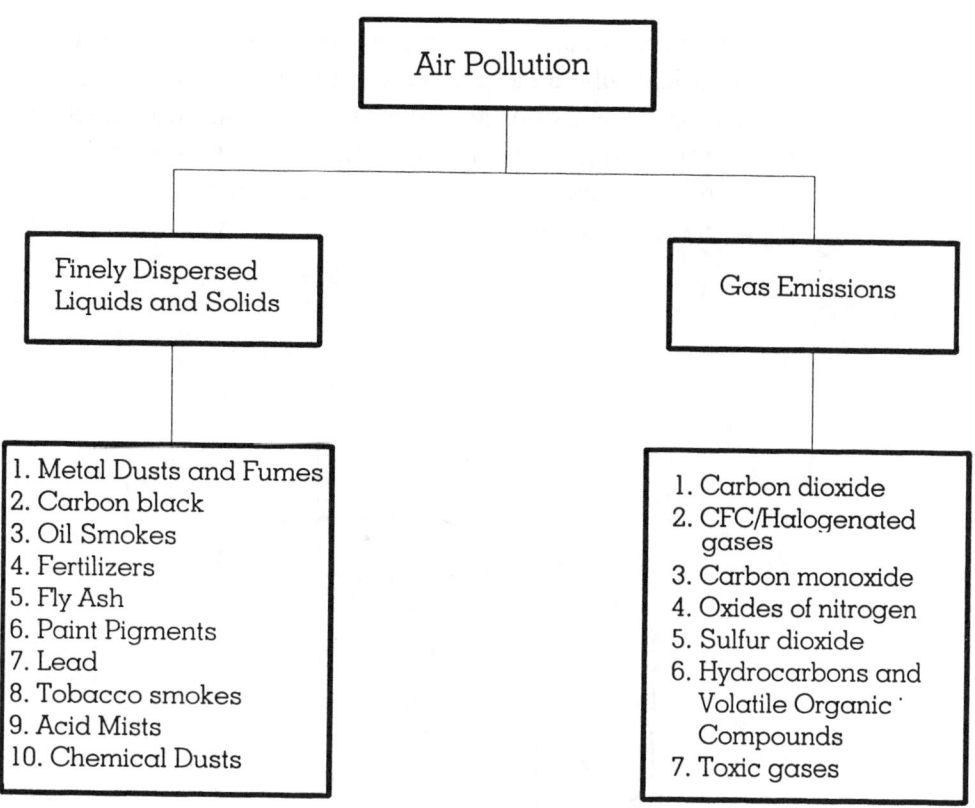

Figure 10-10 Pollutant Based Classification

Particulate Control

Devices used in industries to separate and collect particles from exhaust gases employ one of the basic principles of physics. The five classes of particulate collection equipment follow.

Gravity Settlers

These devices are suitable for particles larger than 10 μm. As the particle-laden gas enters through a narrow passage to a large chamber, its velocity is suddenly slowed to a great extent. The particles fall under the effect of gravity in accordance with Stoke's Law as gas moves horizontally. The chamber works similar to a settling tank of a water treatment process.

Cyclone Separators

These are effective for particles larger than 5 µm. The device consists of a cylindrical tube attached to a cone as shown in Figure 10-11. The gas attains a spinning motion as it enters the tube. The centrifugal force generated by the spin separates the particles which move tangentially and hit the wall of the tube, similar to what happens in a clothes dryer. The particles slide down the wall and fall to the bottom. The clean gas flows out from the top.

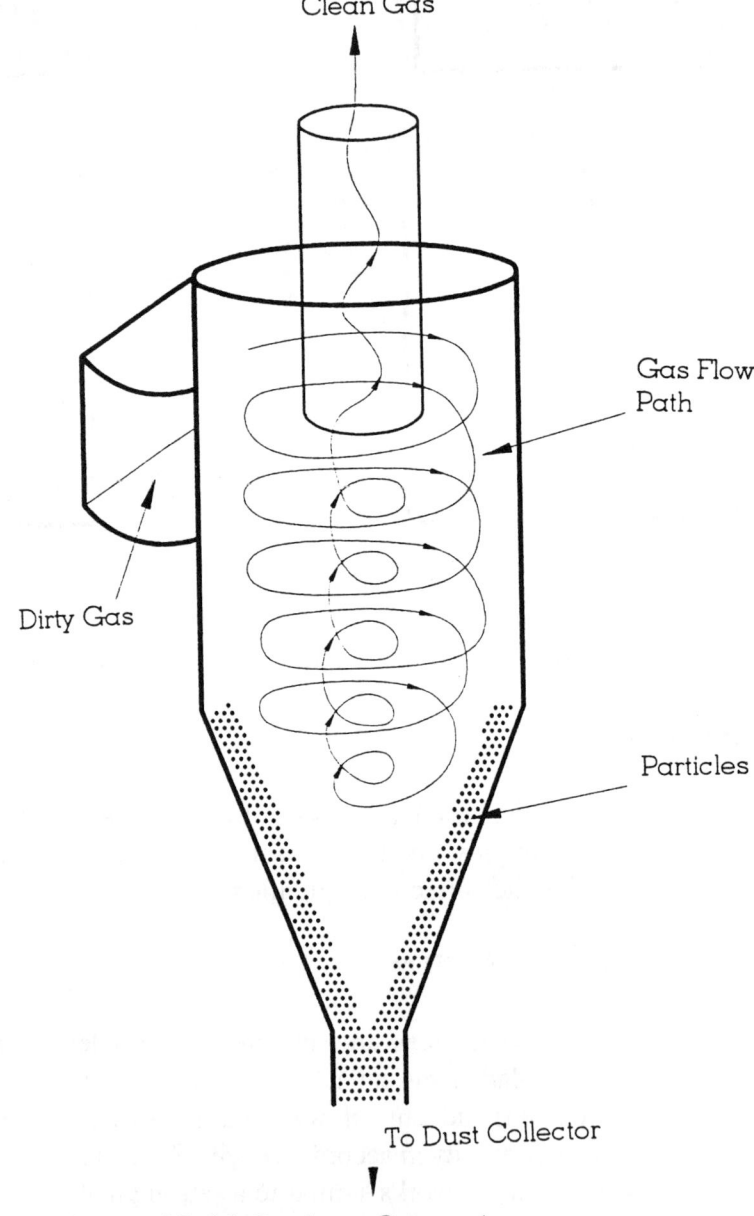

Figure 10-11 Cyclone Separator

Fabric Filters

Fabric filters are a very widely used method, effective for 0.5 μm and larger particles. A fabric filter can remove a substantial quantity of even 0.01 μm particles. The principle is simple: As the gas-carrying dust passes through the filter fabric, only clean gas goes through openings. The dust accumulates on the fabric. The filter is cleaned periodically by shaking or reversing the air flow. The device, also known as a baghouse, may contain hundreds of bags with a height of 20 ft or more and a diameter of 8–12 in.

Electrostatic Precipitators (ESPs):

ESPs can be used to handle a large volume of gas in a wide particle size range of 0.05 to 200 μm. The particles in the gas are charged by establishing a high-voltage drop between electrodes. The oppositely charge plates are arranged parallel to the direction of gas flow. The particles are attracted and collected on these plates.

Wet Scrubbers

Fine particulates ranging from 0.1 to 20 μm are effectively removed by wet scrubbing. A particulate-laden gas stream is sprayed with water. The particles are intercepted by water droplets from the gas. The water-particle slurry is collected at the bottom and has to be treated and disposed of separately.

In general, devices that handle fine particles and can achieve very high efficiency, such as fabric filters and ESPs, are very costly and less flexible in operation. Sometimes a system is composed of two or more such devices.

Collection Efficiency of Control Equipment

For each of the above devices, distinct design equations have been derived or empirically formulated to size the component dimensions. Specific relationships have also been developed for each device to compute the removal efficiency versus the particle size for a given configuration of the device. From the fractional efficiencies of different sized particles, the overall efficiency is the weighted average given by

$$\eta = \frac{\Sigma \, \eta_i m_i}{\Sigma \, m_i} \tag{10-9}$$

where η = overall collection efficiency
η_i = Collection efficiency of particles of i size (range)
m_i = mass fraction of particles of i size (range)

Example 10-6

The following data were computed for a cyclone separator. Calculate the overall collection efficiency of the particle removal.

Size range, µm	0–5	5–10	10–15	15–25	>25
Mass, mg	30	125	80	60	40
Efficiency, %	15	45	70	90	98

Solution:

Computations are arranged in Table 10-7

Table 10-7 Example 10-6 Computations

1	2	3	4	5
Size range, µm	Mass mg	m_i	η_i	$\eta_i m_i$
0-5	30	0.09	15	1.35
5-10	125	0.37	45	16.65
10-15	80	0.24	70	16.80
15-25	60	0.18	90	16.20
>25	40	0.12	98	11.76
Σ	335	1.00		62.76

column 3 = column 2 / Total of column 2
column 5 = column 3 × column 4

$$\eta = \frac{\Sigma \text{ of column 5}}{\Sigma \text{ of column 3}} = \frac{62.76}{1.00}$$

$$= 62.76\%.$$

Control of Gases and Vapors

A major portion of air pollutants are gases, such as carbon monoxide, nitrogen oxides, and sulfur dioxide, and the partially burned hydrocarbons.

The reduction of these gases to an acceptable level involves one or more chemical processes. Several methods can be used from the following approaches:

- The pollutant gas may be adsorbed on a solid surface such as activated carbon.
- The pollutant gas may be absorbed in a liquid solvent by a scrubbing operation.
- The pollutant gas may be :
  a. oxidized by incinerator,
  b. reacted chemically to another chemical that is not a pollutant, or
  c. formed in a reduced quantity by modifying the process producing the gas.

Carbon monoxide (CO) as well as hydrocarbons (HC) are products of incomplete combustion of fuel. The conditions that enhance complete combustion tend to reduce these gases. Combustion with an increased air-fuel ratio reduces CO and HC. However, excess air raises the concentration of nitric oxide (NO). Thus, the control strategies for CO (and HC) and NO are basically in conflict and have to be coordinated in a trade off for the optimal strategy.

To remove combustible air pollutants such as CO and HC, and volatile organic compounds (VOC), organic aerosols, waste gases, and odorous air pollutants, the incineration or afterburning process is used. Often, many gas or VOC sources exist within a plant and the emissions from them are gathered through a number of hoods to a common duct system leading to an incinerator or afterburner.

Sulfur dioxide, one of the abundant air pollutants emitted in the U.S., is formed whenever any fuel containing sulfur is burned; the sulfur content of natural gas and coal varies from 0.1 to 5% by weight. Very high temperatures in the flame zone of a combustion process favor formation of sulfur trioxide (SO_3). At relatively low exhaust temperatures, the main oxide of sulfur that is formed is sulfer dioxide (SO_2). However, once SO_2 is in the atmosphere, the conversion to SO_3 and sulfate (SO_4) (i.e. $MgSO_4$) and sulfuric acid (H_2SO_4) occurs by catalytic reactions.

To control SO_2 emission into the atmosphere, the two basic approaches are: to remove sulfur from fuel before it is burned, i.e., *fuel desulfurization*; and, to remove SO_2 from exhaust gases, i.e., *flue gas desulfurization* (FGD). Hundreds of methods have been proposed for FGD. Limestone ($CaCO_3$) scrubbing is a common technique in which a limestone slurry is contacted with the flue gas in a spray tower. The sulfur dioxide is absorbed in a tray scrubber resulting in formation of calcium sulfite ($CaSO_3$) and calcium sulfate ($CaSO_4$), which are precipitated.

Oxides of nitrogen NO_x have seven forms: NO, NO_2, NO_3, N_2O, N_2O_3, N_2O_4, and N_2O_5. Of these, nitric oxide (NO) and nitrogen dioxide (NO_2) are present in significant amounts.

High combustion (flame zone) temperatures favor formation of NO, not NO_2, and the rate of reactions are highly temperature-dependent being very low at exhaust temperature. Thus, during combustion, mostly NO is formed. Once the gases have moved away from the hot flame zone, the reduced temperature favors formation of NO_2, but the gases cool rapidly and the reaction rate drops by several orders of magnitude, virtually freezing the concentrations of NO and NO_2 at the levels at which they were formed during combustion. Of all the NO_x emitted from exhaust, more than 90% would be NO and the rest would be NO_2. The oxidation reaction of NO to NO_2 occurs in the atmosphere.

Oxides of nitrogen (NO_x) are formed by the following two mechanisms:

- Thermal NO_x is formed by the reaction between nitrogen and oxygen present in the air used for combustion. This is highly temperature-sensitive.
- Fuel NO_x is formed when fuel containing nitrogen on combustion forms NO_x. This is dependent on the air-fuel ratio. Small temperature changes do not affect the process, unlike the case of thermal NO_x.

The sensitivity of thermal NO_x to temperatures is highlighted in the following examples.

- At flue gas exit temperatures of 300–600°F, the NO_x concentration is less than 1 ppm and the $NO:NO_2$ ratio ranges from 10:1 to 20:1.
- At the flame-zone temperatures of 3000–3600°F, the NO_x concentration is 6000–10,000 ppm, and the $NO:NO_2$ ratio ranges from 500:1 to 1000:1.

The above properties provide a basis with which to control NO_x. The approach is to modify the combustion practice by adopting the following key strategies:

- Reduce peak temperatures of the flame zone.
- Reduce the gas residence (combustion) time in the flame zone.
- Reduce the oxygen (air) concentration in the flame zone.

Other control techniques consist of treatment of the flue (exhaust) gases by chemical methods, known as *Flue Gas Treatment Techniques* (FGT). One very advanced FGT method is the *Selective Catalytic Reduction* (SCR), which consists of using ammonia to reduce NO_x to N_2 gas in the presence of a mixture of titanium and vanadium oxides as a catalyst. The optimum temperature range for SCR is 600–800°F. Typically an 80% reduction of

NO_x is obtained by this method. However, it is believed that combustion controls alone are adequate to achieve a desirable reduction of NO_x.

PROBLEMS

1. At the earth's surface, temperature $T = 15°C$ and pressure, $p_o = 14.7$ psi. Determine the pressure at the top of the troposphere, at a height of 10 km. Neglect the temperature variation.

2. In Problem 1, if temperature decreases at a rate of $6.4°C$ per km, what will be the pressure at 10 km height? (Hint: Use the average temperature for 10 km height.)

3. Temperature and pressure at the earth's surface are $65°F$ and 13.8 psi. In a plane, the outside pressure has been recorded to be 5 psi. What is the altitude of the plane?

4. Prove that the magnitude of (MW)g/R is 0.0188 in FPS units.

5. A volume of 800 mL of carbon monoxide (CO) is added to 999 L of air at a pressure of 1.2 atm and $25°C$. Determine the concentration of CO in (1) parts per million, and (2) micrometers per cubic meter.

6. The ozone concentration at a location is 400 µg/m³ at 1 atm and $30°C$. What is the concentration in parts per million?

7. The standard value for sulfur dioxide (SO_2) is 365 µg/m³. To a volume of 1000 L of air at 1 atm and $15°C$, how much SO_2 could be added?

8. The concentration of carbon dioxide (CO_2) is 0.003 lb/ft³ at a location having a temperature of $60°F$ and pressure of 15 psi. Determine the concentration in micrograms per cubic meter and parts per million.

9. Particulate matter with an average size of 5 µm and a density of 1.5 g/cm³ is settling through the air at $25°C$ and 1 atm. Determine the settling velocity by Stoke's Law and by the graphic relation.

10. A 0.5 µm particle of density 1.5 g/cm³ settles through the air at standard room temperature and pressure ($25°C$ and 1 atmosphere). Determine the rate of settling. Cunningham factor is 1.334. Solve the problem by Stoke's Law and the graphic relation.

11. Assume a particle with unit specific gravity and the size of 5µm, 50µm, and 500 µm. Estimate the terminal velocity using the graphic approach.

12. The exhaust from an automobile takes place at a temperature $8°F$ higher than the surroundings. The adiabatic lapse rate (ALR) is $5°F$ per 1000 ft. The atmospheric temperature gradient is $+1°F$ per 100 ft. Determine whether the atmosphere is stable or unstable. What is the height of the formation of smog?

13. The smoke emitted from a 100-m tall chimney has a temperature of 30°C. The ALR is 4°C per 1000 m. The ground-level temperature is 25°C. The atmosphere has a temperature gradient of +1°C per 100 m. Determine the height of the smog formation from the ground.

14. The pollutants emit at a temperature 4°C higher than the surroundings. The ALR is 4°F per 1000 ft. The atmosphere temperature gradient for the first 150 ft is -1°F per 100 ft and then +1.5°F per 100 ft thereafter. At what height will the smog will form?

15. If the global carbon dioxide emission in 1990 is 26,000 giga kg, how much is the emission of CFC? Assume the GWP of 5500 for CFC.

16. If the global carbon dioxide emission is 26,000 giga kg, how much is the emission of nitrous oxide?

17. An ESP has the following efficiency-particle size relationship. Compute the overall collection efficiency.

Size range, μm	0–2	2–4	4–8	8–16	16–25	>25
Mass, mg	15	115	85	60	20	5
Efficiency, %	40	75	90	95	98	100

18. A cyclone separator has the following individual collection efficiencies. Calculate the overall efficiency of a system.

Size range, μm	0–2	2–4	4–7	7–10	10–15	15–25	25–40	>40
% Weight	6	20	28	17	12	8	5	4
% Efficiency	10	24	46	72	83	95	97	99

19. If the cyclone separator of Problem 18 is followed by the ESP of Problem 17, determine the combined efficiency of the system. (Hint: $(1 - \eta_o) = (1 - \eta_1)(1 - \eta_2)$; where η_1 = efficiency of the first system, and η_2 = efficiency of the second system.)

STUDY QUESTIONS

1. Has the composition of the atmosphere always been the same as given in Table 10-1? Will this be the same in the year 2100 based on present trends?

2. Suppose you make a scale model of the earth with a globe 1 foot in diameter. How far will each of the following extend from the surface?
   a. Troposphere
   b. The level below which 99% of the air is found
   c. Middle atmosphere
   d. The level above which sparse air molecules exist

3. Describe the two major types of air pollutants. List each of them.

4. What are the principal sources of air pollution?

5. Describe the effects of air pollutants on the following items, identifying specific pollutants involved with each item:
   a. Humans
   b. Animals
   c. Vegetation
   d. Nonliving matter

6. What is meant by air quality standards? What is the difference between air quality and air emission standards? What are criteria pollutants? Outline the quality standards for criteria pollutants.

7. In the U.S., where has the highest acidity level in rainfall occurred? Why? What is the mechanism that contributes to rainfall acidity?

8. What is the term used by meteorologists for the temperature change that occurs in atmosphere with altitude? Describe thermal inversion. What is its effect on air pollution?

9. What is the significance of ozone in the troposphere? What is its significance in the stratosphere? What is the effect of chorofluorocarbons entering the atmosphere? Why is it a global issue?

10. What is the greenhouse effect? What are the impacts of continued increase of greenhouse gases? What action is being taken in the U.S. against the greenhouse effect? What action is being taken at the global level?

11. What are the physical principles of different particulate control devices? What are the particle ranges for which each device is most suitable?

12. What are the main oxide forms of sulfur and nitrogen in the exhaust gases from a factory? What conversions of these take place in the atmosphere?

13. What are the approaches to control gaseous pollutants in the atmosphere. Describe a common technique of sulfur dioxide removal.

14. Define the following:
    a. National ambient air quality standards (NAAQS)
    b. New source performance standards (NSPS)
    c. Primary standards and secondary standards
    d. Troposphere, stratosphere, and mesosphere
    e. Lower, middle, and upper atmosphere
    f. Stoke's law
    g. Stable atmosphere
    h. State implementation plan (SIP)
    i. Prevention of significant deterioration (PSD)
    j. Global warming potential (GWP)
    k. Cyclone separators
    l. Wet scrubbers
    m. Flue gas desulfurization (FGD)
    n. Flue gas treatment technique (FGT)
    o. Selective catalytic reduction (SCR)

REFERENCES

Abrahamson, D. E., ed., *The Challenge of Global Warming*, Island Press, Washington, DC, 1989.

Ayers, K. W., et al., *Environmental Science and Technology Handbook,* Government Institutes, Inc., Rockville, MD, 1994.

Barry, R. G., and Chorley, R. J., *Atmosphere, Weather, and Climate*, 6th ed., Routledge, London, UK, 1992.

Boubel, R. W., et al., *Fundamentals of Air Pollution*, 3rd ed., Academy Press, San Diego, CA, 1994.

Brownwekll, F.W., *Clean Air Handbook*, 2nd ed., Government Institutes, Inc., Rockville, MD, 1993.

Butler, J. D., *Air Pollution Chemistry*, Academic Press, London, UK, 1979.

Cooper, C. D., and Alley, F. C., *Air Pollution Control: A Design Approach*, 2nd ed., Waveland Press, Prospect Heights, IL, 1994.

Corbit, R. A., ed., *Standard Handbook of Environmental Engineering*, McGraw-Hill Book Company, New York, 1990.

Ellis, J. H., "Acid Rain Control Strategies," *Environmental Science Technology,* American Chemical Society, v.22, n.11, pp. 1248-1255, 1988.

Fields, S., and Flanagan, R., "Wearing Thin," *Earth*, pp. 20-23, March, 1994.

Goldemberg, J., "How to Stop Global Warming," *Technology Review*, pp. 25-31, Nov./Dec., 1990.

Harrison, R. M., and Perry R., ed., *Handbook of Air Pollution Analysis*, Chapman and Hall, London, UK, 1986.

Hemond, H. F., and Fechner, E. J., *Chemical Fate and Transport in the Environment*, Academic Press, San Diego, CA, 1994.

Henderson, D. E., *Air Pollution and Risk Analysis*, Monograph Series of the New Liberal Arts Program, Research Foundation of State University of New York, Stonybrook, New York, 1990.

Holmes, G., Singh, B. R., and Theodore, L., *Handbook of Environmental Management*, John Wiley and Sons, New York, 1993.

Houghton, J. T., Jenkins, G. J., and Ephraums, J. J., *Climate Change: The IPCC Scientific Assessment*, Cambridge University Press, Cambridge, UK, 1991.

Knowles, P.C., ed., Dames and Moore, *Fundamentals of Environmental Science and Technology,* Government Institutes, Inc., Rockville, MD, 1992.

Layman, F., *The Greenhouse Trap*, A World Resources Institute Guide, Beacon Press, Boston, MA, 1990.

Mackenzie, J. J., *Breathing Easier: Taking Action on Climate Change, Air Pollution and Energy Insecurity*. World Resources Institute, Washington, DC, 1989.

Manzer, L. E., "The CFC - Ozone Issue: Progress on the Development of Alternative to CFCs," *Science*, v.249, pp. 31-35, July, 1990.

Mason, B.J., *Acid Rain: Its Causes and Its Effects on Inland Waters,* Clarendon Press, Oxford, UK, 1992.

Masters, G. M., *Introduction to Environmental Engineering and Science*, Prentice Hall, NJ, 1991.

Nadakavukaren, A., *Man and Environment: A Health Perspective*, 3rd ed., Waveland Press, Prospect Heights, IL, 1990.

National Research Council, *Ozone Depletion, Greenhouse Gases and Climate Changes*, National Academy Press, Washington, DC, 1989.

National Research Council, The Panel on Atmospheric Chemistry to the Committee on Atmospheric Sciences, *Atmospheric Chemistry, Problems and Scope*, National Academy of Sciences, Washington, DC, 1975.

Postel, S., *Air Pollution, Acid Rain and the Future of Forests*, Paper 58, Worldwatch Institute, Washington, DC, 1984.

Seinfeld, J. H., *Atmospheric Chemistry and Physics of Air Pollution*, John Wiley and Sons, New York, 1986.

Shea, C. P., "Protecting the Ozone Layer," *State of the World, 1989*, Worldwatch Institute, Washington, DC, 1989.

Spiro, T. G., and Stigliani, W. M., *Environmental Issues in Chemical Perspective*, State University of New York, Albany, NY, 1980.

Turco, R.P., *Earth Under Siege,* Oxford University Press, New York, 1997.

U. S. Environmental Protection Agency (EPA) Journal, Reports on *Global Warming and Ozone Hole*, March/April, 1990.

U. S. Environmental Protection Agency, *Control Techniques for VOC Emissions from Stationary Sources*, Government Institutes, Inc., Rockville, MD, 1994.

Veziroglu, T.N., ed., *Environmental Problems and Solutions*, Hemisphere Publishing Corp., New York, 1990.

Wark, K., and Warner, C. F., *Air Pollution: Its Origin and Control*, 2nd ed., Harper and Row, New York, 1981.

White, R. M., "The Great Climate Debate," *Scientific American*, v. 263, n. 1, pp. 36-43, July, 1990.

Chapter 11

THE TERRESTRIAL ENVIRONMENT

The Terrestrial System and Solid Waste

The biosphere interconnects atmospheric, aquatic, and terrestrial systems on the planet. The three systems together make life possible on the earth. A natural bond exists among the three systems through the cycling of materials and flow of energy. One environment cannot act in isolation without the others; atmospheric or aquatic imbalances affect the terrestrial system and vice-versa.

In the unpolluted environment, materials continuously cycle in a steady state, contributing to the stability of the biosphere. However, humans, as an important member of the terrestrial system, have acquired the capacity to cause significant disturbances among these cycles. As part of the cycles, these changes go through the aquatic environment, the atmospheric environment, and are then passed on to the terrestrial environment, ultimately affecting humans themselves. A model of modern society presented in Figure 3-7 is based on the concept of generating and disposing of large amounts of wastes. The wastes generated terrestrially, excluding any wastes produced in air or water waste streams but including wastes recovered as sludge after treatment of air and water waste streams, are collectively called *solid wastes*.

Whereas the Clean Air Act (1990) and the Clean Water Act (1987) control the air and water resources respectively, the Resources Conservation and Recovery Act (RCRA) (1976), with the latest amendments of the Hazardous and Solid Waste Amendments (HSWA) (1984), cover all aspects of solid wastes. The act defines solid waste as follows.

The term solid waste means any garbage or refuse, sludge from a waste treatment plant, water supply treatment plant, or air pollution control facility, and other discarded material including solid, liquid, semisolid, or contained gaseous material resulting from industrial, commercial, mining, and agriculture operations and from community activities. It does not include solid or discarded material in domestic sewage or solid and discarded dissolved material in irrigation return flows or industrial discharges, which are point sources subjected to permits. Nuclear materials and byproducts are also not covered; these materials are separately controlled by the Atomic Energy Act.

The act further defines hazardous waste as follows: Hazardous waste is a solid waste or combination of solid wastes which because of its quantity, concentration, or physical, chemical, or infectious characteristics may:

- Cause or significantly contribute to increase in mortality or an increase in serious irreversible or incapacitating reversible illness; or
- Pose a substantial present or potential hazard to human health or the environment when improperly treated, stored, transported, or disposed of or otherwise managed.

According to the above RCRA definition, hazardous waste is a subset of solid waste. However, in practice, the solid waste category covers all nonhazardous wastes and all hazardous wastes are identified as a separate entity.

Whether a solid waste belongs to the hazardous waste category is ascertained as follows.

A solid waste is deemed a hazardous waste if it contains one of the chemicals that appears on one of the three hazardous waste lists prepared by U.S. EPA. A distinct number is assigned to each listed waste. The lists cover volatile organics, semivolatile organics, heavy metals, pesticides, PCBs, cyanide, and asbestos. Understanding the categorization differences and similarities of chemicals on these lists is crucial to compliant management according to hazardous waste regulations.

If a waste is not on the U.S. EPA lists, it is still considered hazardous if it exhibits one of the four characteristics: ignitability, corrosivity, reactivity, or toxicity. These characteristics are discussed subsequently.

As the name implies, a waste is considered an unwanted and useless material that is normally disposed of on the land. The first part of this chapter describes the sources, classification, types, and properties of wastes, the interaction between wastes and soils, the management or control of wastes, and the designs of land-based disposal systems. The second part covers all aspects of hazardous wastes, including the nature and characteristics of hazardous wastes, the technology of the underground storage tanks, and the remediation of soil contamination.

Sources and Classification of Solid Wastes

Waste sources correspond to the users of virgin materials. They can be categorized into residential, commercial, institutional, construction-based, municipal street and other cleaning services, treatment plant sludge, industrial process wastes, and agriculture. The term *municipal solid wastes*

(MSW) is commonly used to designate wastes from the first six sources, i.e., excluding industrial process and agricultural wastes.

The term *refuse* is used synonymously with municipal solid waste. Two common types of municipal solid wastes are garbage, which is the animal and vegetable waste resulting from handling, cooking, and serving food, and rubbish, which consists of combustible and noncombustible nonfood wastes. Trash is the combustible component of rubbish.

A typical composition of the residential MSW in the U.S. (1990) is shown in Table 11-1.

Table 11-1 Average Composition of the Residential Municipal Solid Waste

Item	Percent (by weight)
Paper	34
Cardboard	6
Food waste	10
Yard waste	18
Glass	8
Metal	9
Plastic	7
Wood	2
Textile	2
Leather/rubber	1
Dirt/ash, etc.	3

Paper is by far the biggest contributor to waste. Conversely, plastics contribute only a small quantity that can be squeezed into a very small volume. Distribution by component, shown in Table 11-1, is essentially the same for waste as generated and for waste as collected excluding the material that is recycled or otherwise disposed of at the source. In the future, the use of plastics may increase and some reduction could occur in the amount of paper consumed. Food waste could hold steady or increase slightly as consumers switch to eating more and more unprocessed vegetables.

The residential portion makes 50–75% of the total MSW generated.

Unit Waste Factors

To determine the optimum size of disposal facilities and recovery units, it is necessary to know the gross quantity of wastes generated. A convenient means to determine this is to use the unit waste factor which indicates the daily amount of waste generated by each unit of waste producer, based on actual field measurements. For residential wastes, the factor is expressed in terms of pounds (kg) per person per day. Another appropriate unit, particularly for the design of the collection systems, is the weight in pounds (kg) per household (or stop) per week (PPHW), averaged from annual data; a typical value is 55 lbs per household per week. Unit waste factors are given in Table 11-2 for various sources.

Table 11-2 Unit Waste Factors

Source	Factor	Units
Residential—single	3–5	lb/person/day
Multiple-housing units	2–3	lb/resident/day
Commercial	6	lb/employee/day
Industrial	10–11	lb/employee/day
Agricultural	13	lbs/person/day
Office/picnic/resort area	1–2	lb/person/day
Street sweeping	0.3	lb/person/day
Restaurant	0.7	lb/meal served/day
Hospital	8–9	lb/bed/day
School/colleges	0.6	lb/student/day
Ski area	2–3	lb/visitor/day
Treatment plant sludge	0.4	lb/person/day

Waste generation is not uniform throughout the year. When seasonal variations or minimum and maximum quantities are required, an appropriate multiplier should be applied to the average waste generation tabulated. For residential waste, the monthly multiplier ranges from 0.88 for January to 1.13 for May. The multiplier is 1.27 for maximum and 0.73 for minimum residential waste generation.

Physical Properties of Waste Components

In addition to weight quantities, waste volumes are important in order to determine the size and useful life of a landfill or any other disposal facility. The weight densities (specific weights) are used for weight-to-volume conversion. The densities of different waste components and those at various stages of waste handling are given in Table 11-3. Also given in the table are the moisture contents as a percent of the dry weight of the component.

Table 11-3 Physical Properties of Wastes

Item	Waste Components Density as Discarded (lb/yd^3)	Moisture Content (%)
Paper	150	6
Cardboard	85	5
Food waste	500	200
Yard waste	170	150
Glass	500	1
Glass, crushed	1466	1
Metal, nonferrous	260	2
Metal, ferrous	500	2
Plastic	100	2
Wood	300	25
Textile	240-280	10
Leather	270	10
Rubber	220	2
Dirt, ash, etc.	800	9
Bulk Waste at Different Stages		
In container	75-150	25
In compactor truck	500-700	25
In landfill	950-1500	30

Determining Waste Generation

During the design phase of a waste management facility, whether it is a recycling, resource recovery, landfill, incinerating, or composting facility, it is essential to know the total quantity and the type of waste generated in terms of weights and volumes. A simple approach for determining the waste generation in a region is to multiply the projected population of the area by the typical unit waste factor of Table 11-2. To do this, the seasonal or minimum and maximum multipliers are applied as necessary.

$$\text{Quantity of waste generated} = \text{projected population} \times \text{unit waste factor} \quad (11\text{-}1)$$

In addition to the overall quantity of waste, it is essential to determine the types of waste generated in order to select and design appropriate processing units. The waste type is determined by multiplying the total quantity of Equation 11-1 by the fraction of that type of waste from the waste composition Table 11-1.

$$\begin{bmatrix} \text{type } X \text{ waste} \\ \text{generated} \end{bmatrix} = \begin{bmatrix} \text{total quantity} \\ \text{of waste} \end{bmatrix} - \begin{bmatrix} \text{fraction of } X \\ \text{in Table 11-1} \end{bmatrix} \quad (11\text{-}2)$$

The dry unit weight of the type X waste can be computed by:

$$\text{dry weight of type } X \text{ waste} = \frac{\text{type } X \text{ waste}}{\left(1 + \dfrac{\omega}{100}\right)} \quad (11\text{-}3)$$

where ω = percent of moisture content from Table 11-3.

The waste volumes corresponding to total quantity and different types of waste generated are determined by using appropriate density data from Table 11-3.

$$\text{volume of } X \text{ waste} = \frac{\text{quantity of } X \text{ waste}}{\text{density of } X \text{ waste}} \quad (11\text{-}4)$$

where X denotes any waste component (i.e., paper, or the overall composite waste, at any stage of compaction.)

Example 11-1

The projected population of a town is 25,000. For weekly waste collection, determine the amount of waste to be picked up, the volume of waste to be picked up, the amount of paper waste produced, and the dry weight of the paper.

Solution:

Assume a waste factor of 5 lb/person-day based on Table 11-2

From Equation 11-1:
  Total daily waste = (25000 persons) × (5 lb/person-day)
    = 125,000 lb/day
  Weekly waste = 125,000 × 7 = 0.875 × 10^6 lb

Assume a density of 100 lb/yd^3 based on Table 11-3.
From Equation 11-4:
$$\text{Volume of weekly waste} = \frac{0.875 \times 10^6}{100} = 8750 \text{ yd}^3$$

Fraction of Paper 34% from Table 11-1
From Equation 11-2:

$$\text{Amount of paper waste} = (0.875 \times 10^6) \times \frac{34}{100} = 298 \times 10^3 \text{ lb.}$$

From Equation 11-3: For paper, $\omega = 6\%$

$$\text{Dry paper weight} = \frac{298 \times 10^3}{\left(1 + \frac{6}{100}\right)} = 281 \times 10^3 \text{ lb.}$$

Fate of Pollutants in Soil Matrix

Chemical compounds introduced into soils through a waste stream go through a series of physical, chemical, and biological changes. They may be evaporated in a gaseous form, washed out by rain or irrigation, dissolved in infiltrating water, and retained within the soil mass either as insoluble products or as adsorbates on soil particles.

Migration and fate of dissolved chemicals in a soil system consists of:

- Physical movement by advection and dispersion;
- Chemical reactions such as volatization, sorption, ion exchange, ionization, hydrolysis, oxidation-reduction, and complexation; and
- Biologic actions like bioaccumulation and transformation.

By all of these processes, the chemicals can leach through the soil into the groundwater where similar processes may take place on a different scale. The fate of contaminants in the groundwater is discussed in Chapter 9.

Chemicals within the soil mass are available for plant uptake and transformed by soil organisms into simple inorganic products and water. Some materials have a short life in soils while some synthetic organics remain stored for a considerable time. Certain heavy metals are firmly bound to soils. Only a fraction of these chemicals might be bioavailable for reactions. The metabolism by soil microorganisms bring out the following biotransformations to the chemicals:

- Mineralization or complete conversion of an organic compound to inorganic products.
- Detoxication or transformation of a toxic substance to a harmless form.
- Activation or conversion of a primarily nontoxic compound into a toxic one or enhancement of its toxic potential.
- Incomplete transformation or products of partial reactions. Many times, this is a cause of environmental concern since these partial products may be more toxic or persistent than the original substance.

The application of sewage sludge to soils can lead to heavy metal concentrations far beyond the established tolerance levels of microorganisms, and thus induce a considerable reduction in a particular microorganism population and its biodegrading ability. Similarly, pesticides, particularly fungicides, can destroy a large segment of a microbial population, and thereby stop many reactions that these microorganisms catalyze. Because of prolonged persistence of these synthetic organics, the soil microflora might be suppressed for a long time before they can be reestablished.

Municipal Solid Waste (MSW) Management

RCRA defines solid waste management as the systematic administration of all activities related to solid wastes, from handling at the source to their final disposal. Resource conservation and recovery, which were small parts of the earlier management programs, are being increasingly incorporated by waste management agencies into their plans.

Resource conservation measures reduce the amounts of solid waste being produced, reduce the overall consumption, and utilize the recovered resources. Resource recovery obtains materials and energy from waste and recycles the materials.

The comprehensive solid waste management program provides for handling and separation of the waste at the source, collection and transfer, processing operations, resource conservation and recovery, and treatment and disposal.

Pursuant to subtitle D of the RCRA, along with HSWA (1984), the regulation of nonhazardous wastes is the responsibility of the states. Federal involvement is limited to establishing the minimum criteria for the best practicable controls and to setting monitoring requirements for solid waste disposal facilities.

The states have delegated the responsibility to the municipality, which means a city, town, borough, county, district, or other public body, to administer the management program.

Handling and Separating MSW at the Source

Many communities have adopted recycling programs whereby newspapers, aluminum cans, glass, and plastic are separated from the waste stream by the producers. The solid wastes remaining after recyclable materials have been separated are placed in one or more containers/bags; the separated waste is placed in special containers. The separated materials are either placed at the curbside for collection or brought to a centralized recycling center. Papers, cans, plastics, metals, cardboard, waste oils, tires, and batteries are collected in offices, commercial buildings, and workshops. The commercial facilities use large containers mounted on rollers that can be mechanically lifted and emptied into trucks.

The material separated for recycling from residential and commercial sources in the U.S. in 1992 is estimated to be about 15% of the total waste.

Collection and Transfer Operations

Collection and transfer are the most visible and costly part of the management plan, costing 75–80% of the total solid waste budget. The collection vehicles are rear loaders or side loaders, or front loaders for commercial collection, and have a receiving hopper, a compactor, and a storage compartment. The rear loaders have several operational advantages but involve higher maintenance due to relatively complex mechanisms. Rear loaders are available in 14 body sizes ranging from 9 to 32 cubic yards capacity. The compaction density on rear loaders ranges from 500 to 1100 lb/yd^3, and the average on-route efficiency is 80–85% of the manufacturer's rated density.

Side loaders have a simpler mechanism. They are suitable for one-sided collection. Side loaders are available in 28 body sizes from 6 to 40 cubic yards. Their compaction density of 300 – 800 lb/yd^3 is lower than the rear loaders. The average on-route density ranges from 85-95% of the manufacturer's rating.

In some cases, it is not economical to directly haul the waste to the disposal site in the route collection vehicle. To increase productivity, a terminal is often established where the route vehicles empty their loads and quickly return to route collection while a larger vehicle transports the loads of several vehicles to the disposal site. The terminal, called a *transfer station*, is a two- floor structure; the route vehicle unloads from the upper floor into a tractor trailer or tipping floor on the lower level.

The schematics of the route collection process are shown in Figure 11-1.

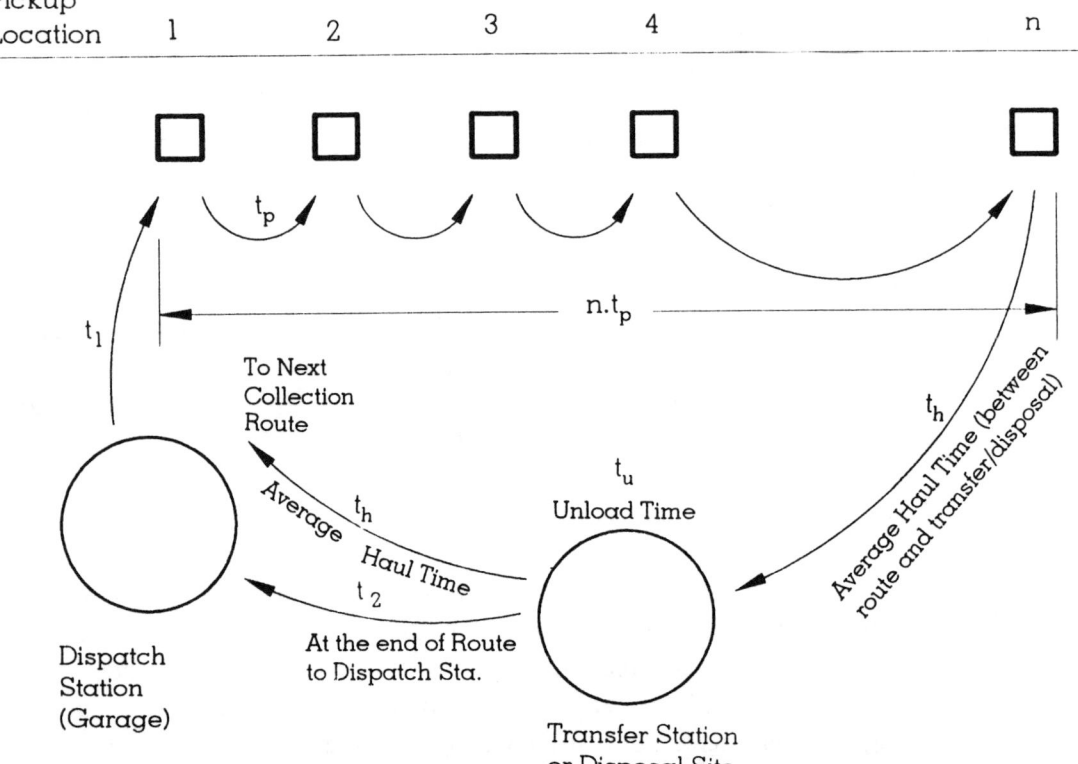

Figure 11-1 Route Collection Sequence

The vehicle capacity is matched with the route; the number of stops or pickup locations should be adequate to fill the vehicle. Once a vehicle is filled, it goes to a disposal site or transfer station to be emptied. From there, it continues to the next route within the district, and so on. At the end of the day, it returns to the dispatch station or garage. In one day, a vehicle might make a number of trips to the disposal site/transfer station. The total working time may be accounted for by

$$H = N_d (nt_p + t_u) + (2N_d - 1) t_h + t_1 + t_2 + B \qquad (11\text{-}5)$$

where H = length of a working day, in hr
N_d = Number of trips per day to disposal/transfer site
n = number of pickup locations per route (load) or per trip
t_p = mean time for collection at each stop and to move to the next stop, in hr
t_h = Mean haul time from route to disposal/transfer site, in hr
t_u = Time to empty the vehicle at disposal/transfer site, in hr
t_1 = Time to travel from dispatch station (garage) to route, in hr
t_2 = Time to travel from disposal/transfer site to dispatch station, in hr
B = Off-route time in a day, for break, maintenance, personal needs, in hr.

The number of pickup locations per route (per load of vehicle) can be given by.

$$n = \frac{aC_v C_d}{W_h} \qquad (11\text{-}6)$$

where C_v = capacity of vehicle, yd^3
C_d = compacted density in vehicle, lb/yd^3
a = compaction efficiency of vehicle
W_h = weight of waste per household or per stop, lb.

Equations 11-5 and 11-6 are used to compliment one another. The time parameter t_p in Equation 11-5 is dependent upon the size of crew, number of containers per household, and the location of pickup points on the route. Some empirical relations exist to compute, t_p. The value of n is determined from Equation 11-5 by fixing practical values to all other parameters of the equation. Using this value of n in Equation 11-6, the required size of the vehicle, C_v is determined. Alternatively, for a predecided vehicle capacity, n is computed from Equation 11-6. Substituting this n in Equation 11-5, the number of trips (in a whole number) that can be made to the site per day are computed. From the population of total number of households in a town, the number of vehicles required is ascertained.

Example 11-2

For the town in Example 11-1, design the collection system for the following data:

Average household size = 4 persons
Time from garage to route, t_1 = 13 min
Time from disposal site to garage, t_2 = 16 min
Crew takes two breaks of 15 min each and lunch break of 30 min
Average time to haul from route to disposal station, t_h = 14 min
Number of trips made to disposal site/day, N_d = 3
Time per collection stop, including reaching to next stop, t_p = 1.5 min
Time to empty vehicle at disposal, t_u = 6 min
Vehicle compaction efficiency = 80%
Vehicle compacted density = 600 lb/yd^3.

Solution:
Assume an 8-hr work day
B = 2(15) + 30 = 60 min
Total number of households = $\dfrac{25{,}000 \text{ persons}}{4 \text{ persons/house}}$ = 6,250

Weight per household = (5 lb/person/day)(4 persons/house)
= 20 lb/day
Weekly weight/household = 7 x 20 = 140 lb.

From Equation 11-5:

$$8 = 3\left\{n\left(\frac{1.5}{60}\right) + \left(\frac{6}{60}\right)\right\} + (6-1)\left(\frac{14}{60}\right) + \left(\frac{13}{60}\right) + \left(\frac{16}{60}\right) + \left(\frac{60}{60}\right)$$

hence, n = 67 pickups/trip

From Equation 11-6:
$$67 = \frac{0.8 C_v (600)}{140}$$

hence, C_v = 19.5 yd^3, use 20 yd^3 loader
Pickups per day = $(N_d)n$ = 3(67) = 201
For a 5-day work week,
Number of household served = 5(201) = 1005

Number of vehicles required = $\dfrac{\text{total households}}{\text{household served/week}}$

= $\dfrac{6250}{1005}$ = 6.2 or 7 vehicles.

Processing Operations

Waste separation and processing occur at a *material recovery facility* (MRF). The MRF objectives are to further separate and process the source-separated wastes, recover materials from common wastes, and improve the quality of recovered wastes. The unit operations used at MRFs include:

- Shredding, which is breaking waste components into smaller sizes;
- Screening and/or air classification, which results in separation of lighter combustibles or organic components from heavier noncombustibles or inorganic matter;
- Magnetic separation to remove ferrous materials; and
- Densification to reduce volume by baling, crushing, and pelletting.

Resource Conservation and Recovery

The important elements of resource conservation and recovery (RC&R) are:

- Reduction of waste generation;
- Reuse of waste;
- Recovery and recycling of waste;
- Recovery of energy; and
- Reduction of volume.

The first two strategies are key distinguishing factors between a throwaway and a sustainable society. They require concerted efforts on the part of the society to use items as many times as possible before discarding them and to cut down on materials consumption to the extent possible by making changes in use styles. These efforts necessitate a fundamental change of attitude and a shift in lifestyles.

U.S. EPA emphasizes materials recovery and recycling. The at-source separation of materials such as paper, glass, plastic, aluminum, tin, and (commercially) tires, batteries, and waste oils, and the mechanical separation and processing at MRF, are the activities directed to recover useful byproducts from MSW for recycling. In theory, most needs for paper and glass could be met by recovery from MSW. However, U.S. EPA has placed the practical limit of potential recovery from MSW, expressed as a percent of the nation's consumption, to about 14% for paper, 19% for tin, 8% for aluminum, and 7% for iron.

MSW possesses considerable energy potential; the energy content of the solid waste is one half to one-third of that of coal on a unit mass basis. It is estimated that if the entire MSW in the U.S. were to be converted to energy, it would supply 3–5% of the total U.S. need, enough for all residential lighting requirements. However, U.S. EPA has placed the potential energy recovery limit at 0.5%.

Utilizing MSW as fuel accomplishes two purposes: to produce energy, and to achieve significant volume reduction of MSW resulting in inert material. Energy is derived essentially by combustion of MSW, although anaerobic digestion, composting, and landfill gas recovery are other processes of a biological nature whereby energy is derived. MSW can be burned as received (unprocessed) as a fuel; the solid waste containing 75% moisture and a significant amount noncombustible material could be ignited. Typical heat value of as-received MSW is 4500–5000 Btu/lb. MSW that has been processed to separate out the recyclables, not only provides recovery for material, but improves the fuel quality as well because the noncombustible portion containing glass, metals, and ceramics of low heat values have been removed. The remaining portion, excluding the noncombustible portion of the solid waste, is described as *refuse-derived fuel* (RDF). RDF has a heat value of 6000–8000 Btu/lb.

The major technologies available to recover energy from MSW, in as-received state or RDF, are:

- *Waterball incineration*: Combustion of MSW in a furnace to produce steam for steam/turbine electric generation.
- *Modular incineration*: Combustion in two-stage starved-air furnaces with heat recovery boilers or heat exchangers.
- *Refuse-derived fuels*: MSW is concentrated into a combustible component and used along with other fossil fuels.
- *Pyrolysis*: MSW is processed in an oxygen-deficient condition to derive gaseous, liquid, and solid fuels.

Treatment and Disposal

The principal method of treating (stabilizing) and finally disposing of residual solid waste is by the municipal solid waste landfill (MSWLF) mode. Other methods of interest are incineration and composting.

- Municipal solid waste landfill (MSWLF) is a land disposal site designated to minimize environmental hazards resulting from the spreading of solid wastes. It can receive only RCRA subtitle D wastes. The subtitle C hazardous wastes are prohibited without treatment, as discussed subsequently. Since water seeping through the fill can contaminate groundwater and nearby surface waterbodies, proper design of landfill bottom (containment system) and top when the landfill is closed (capping) are important

considerations in this method. Important aspects of landfill designs are discussed separately.
- Incineration or thermal destruction is the process of burning the waste in furnaces at high temperatures (900°C or higher). The bulk of the solid waste is reduced to a smaller quantity of ashes that are disposed of on a landfill. However, a large amount of gaseous pollutants and particulate matter can be generated in the combustion process. Accordingly, adequate air pollution control devices need to be installed with an incinerator system, significantly increasing the cost of the incinerator.
- Composting is appropriate for the organic component of solid waste (foods, yard waste, and wastewater sludges). As such, the application of this method is limited to residential, commercial, and materials recovery facilities (MRFs). Composting is a natural biodegradation process in which microorganisms present in the pile of organic waste decompose matter aerobically into humus-like material. Heat produced during the biodegradation process results in the destruction of pathogenic bacteria. Compost is useful as a soil conditioner and improves the moisture-holding capacity of the soil; composted matter is not a valuable fertilizer.

MSW Landfill Design Considerations

Landfill Configurations

A modern landfill design includes the following steps:

- *Site selection*: Site location is a very difficult decision, since it involves many social, economic, political, engineering, and environmental variables. Opposition by local citizens might eliminate a viable site. There are certain requirements with respect to offset distances, depth to water table, and ledge that must be observed.
- *Hydrology and soils investigations of the selected site*: The investigations will include use of topographic maps, soil borings, water well logs, observation wells, groundwater contours, and soil testing.
- *Containment system*: In order to prevent the leachate from flowing out of the fill, a leak-proof bed has to be designed on which the solid waste is deposited, compacted, and then covered with a thin layer of soil at the end of each day. Thus, a cell consisting of a waste layer of 1 to 1.5 ft thick and a soil cover of 6 in. is created each day within the fill. An intermediate cover 12 in. thick is provided at the end of a lift, to begin another lift, at a monthly interval, for example.

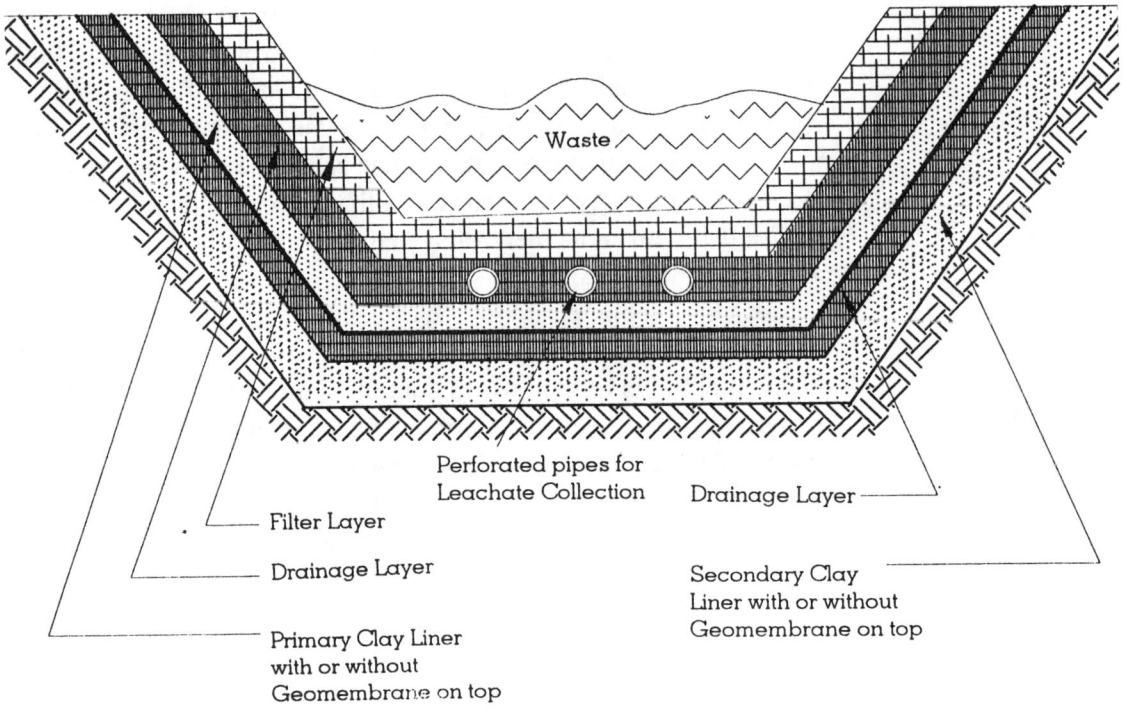

Figure 11-2 Cross-Section of a Double-Liner Containment System

A double-liner system for containment is shown in Figure 11-2. The waste is placed on the primary liner system. The secondary system provides the second leak-proof layer underneath the primary layer. Each system consists of a filter soil or geotextile, a drainage layer or geonet, and a clay layer or composite layer of the synthetic clay liner and geomembrane. In the primary system, perforated pipes are laid within the drainage layer on a grade of 2% or more to collect leachate in a sump at the end of the landfill.

- *Landfill capping*: After a site is fully filled to its design capacity, which provides for a depth of at least 20 ft and a minimum life of 10 years, a water-tight final cover 2–3 ft thick is placed over the site and seeded.

The cap consists of a compacted clay layer over the solid waste (with or without a geomembrane liner), a drainage layer, and finally a layer of topsoil. Vertical pipes penetrate through the cap into the fill to collect methane gas generated as a result of degradation of the waste. Groundwater monitoring is an integral part of the fill design. The wells are located selectively upstream from the fill boundary for baseline quality assessment, and along the downstream boundary of the fill to monitor leakage from the fill. The profile of groundwater regime is prepared by a combination of observations from up-gradient and down-gradient wells.

Landfill Size

The annual volume required for a landfill can be given by

$$V = \frac{PEC(365)}{D_c} \tag{11-7}$$

where V = annual volume of landfill, yd³
P = population served by landfill
C = unit waste factor, lb/person/day
D_c = density of compacted fill, lbs/yd³

$$E = \frac{(\text{soil cover} + \text{waste})}{\text{waste}}$$

For an average unit waste factor C of 5 lb/person/day; the compacted density of 1100 lb/yd³; and the daily soil cover of 6 in., monthly cover of 12 in., and the final cover of 2 ft over a daily waste layer of 12 in., i.e., E value of 1.52, Equation 11-7 reduces to the following:

$$V = 2.52P \quad \text{(See footnote 1)}[1] \tag{11-8}$$

For a known depth of the fill, decided from the field conditions, the size of the landfill can be ascertained from Equation 11-7.

Example 11-3
For the town in Example 11-1, determine the area of the daily cell and the area and volume of landfill in 20 years, for a depth of 40 ft.

Solution:
From Equation 11-7:
$V = 2.52(25,000) = 63,000$ yd³

---

[1] If the volume is to be measured in m³, the $V = 1.93 P$.

In a year, the landfill operates for 52 weeks, 5 days/week,

hence, daily volume = $\dfrac{63,000}{(52)(5)}$ = 242.3 yd³ or 6542 ft³

Daily cell has 12 in. waste and 6 in. soil, (i.e., depth of 18 in. or 1.5 ft)

Area of cell = $\dfrac{6542}{1.5}$ = 4361 ft² or 66 ft x 66 ft

Volume in 20 years = 63,000 × 20 = 1,260,000 yd³

Area of landfill = $\dfrac{1,260,000}{40}$ × 27 = 850,500 ft² or 19.5 acres.

Characteristics of Hazardous Wastes

Ignitability

An ignitable hazardous waste bears U.S. EPA number of D001. Any one of the following four conditions exhibit ignitable behavior:

1. A liquid (other than aqueous solution containing less than 24% alcohol by volume) having a flash point of less than 140°F.
2. A nonliquid that can catch fire by friction, absorption of moisture, or spontaneous chemical changes, and burn vigorously on catching fire.
3. A compressed flammable gas as defined by the Department of Transportation (DOT) regulations (49 C.F.R.173.300).
4. A waste defined as an oxidizer by Department of Transportation (DOT) regulations (49 C.F.R.173.151).

Corrosivity

A corrosive waste has U.S. EPA number D002. A waste is deemed to be corrosive if it exhibits one of the following conditions:

1. It is an aqueous solution that has pH of 2 or less, or 12.5 or more; or
2. It is a liquid that corrodes steel at a rate of more than 0.25 in. per year.

Reactivity

A reactive waste has U.S. EPA number D003. These wastes are extremely unstable and have a tendency to react violently or explode in a number of situations as listed by U.S. EPA.

Toxicity

These wastes are given U.S. EPA numbers D004-D043, depending upon the toxic substance present. This characteristic is designed to identify a waste that can leach a hazardous concentration of a specific substance into groundwater. This characteristic is determined following a testing procedure that extracts the toxic constituents from a waste in a manner that, U.S. EPA believes, simulates the leaching action that occurs in landfills. The test method is called the *Toxicity Characteristic Leaching Procedure* (TCLP). The TCLP tests for 25 organic chemicals, eight inorganics, and six pesticides. A waste is deemed to exhibit toxicity if, using the prescribed method, the extract from a sample of the waste contains contaminants at levels of regulatory concern.

Regulation of Hazardous Waste

The hazardous waste contained within MSW is less than 1% of the total MSW quantity. Of this 1%, 75–85% is from residential sources. The bulk of the hazardous waste produced from industrial sources can be grouped under five generic categories:

- Metals and metal finishing;
- Paints and solvents;
- Organics, includes chemicals, pharmaceuticals, rubber, plastics, and textiles;
- Petroleum; and
- Inorganics.

On an average, more than 20% of the industrial waste stream constitutes hazardous waste. Relatively, more than 70% of this is contributed by chemical and petroleum industries.

Prior to enactment of the RCRA (1976) U.S. EPA had reported that only 10% of the hazardous waste was being disposed of in an adequate manner. Accordingly, Congress enacted the RCRA, (1976) and its extension, the Hazardous and Solid Waste Amendments (HSWA, 1984) to establish proper regulation of hazardous wastes. As the RCRA/HSWA applied to only active facilities that generate hazardous wastes and did not cover the abandoned and closed disposal sites or spills, the Comprehensive Environmental Response, Compensation and Liability Act (CERCLA), commonly known as the Superfund, was enacted in 1980 to address the latter problem. The Superfund Amendments and Reauthorization Act (SARA, 1986) extended the provisions of CERCLA.

Household hazardous wastes (HHW) are excluded from RCRA. In addition, RCRA divided hazardous waste generators in the following three categories. Those households producing less than 100 kg (220 lb) per month[2] of hazardous waste are recognized as very small generators. In liquid form this amounts to about one half of a 55-gal drum per month. Those producing between 100 kg and 1000 kg per month, called small-quantity generators (SQG). Large generators are households producing 1000 kg or more per month.

The first category is conditionally exempted from RCRA; the condition being that this waste is not classified as an acutely hazardous waste. If it is, the generator is subjected to all rules of RCRA. For the second category, less strict conditions apply for temporary accumulation at the site, but in most respects, SQG are regulated in the same way as the large generators.

Thus, household and, generally, commercial (very small generators) hazardous wastes are not subjected to RCRA regulations. However, the solid waste management authorities, i.e., municipalities, generally impose conditions for special handling of hazardous waste while collecting regular garbage.

All other hazardous wastes require strict management as per RCRA/HSWA rules.

The Cradle-to-Grave Concept

RCRA/HSWA regulated hazardous wastes have to successively meet the generator, transportation, and treatment, storage, and disposal (TSD) requirements. A generator of a hazardous waste is responsible for that waste forever; the generator is held accountable even for the functions of the TSD that may be remote from the generator with respect to the generated waste. RCRA has devised a system that monitors hazardous material at all times and ensures its permanent accountability. The system requires that the generator of the waste file a notification with U.S. EPA for each waste site.[3] After the notice is filed, the generator should initiate the management of hazardous waste since it cannot be stored at the site for more than 90 days without a permit. A decision has to be made by the generator as whether to obtain a permit from U.S. EPA to treat and dispose of the waste at the site or to transport it off the site. In the latter case, the generator must complete a manifest form (an itemized list) of the waste. This is the cradle stage. The manifest has to be tracked throughout each step of waste handling until the

---

[2]There is an additional restriction in terms of the amount of acutely hazardous waste for all three cases.

[3]A transporter or an operator of TSD also has to file a notification with U.S. EPA.

ultimate disposal point (the grave); the U.S. EPA is kept informed at each step of the process. While dealing with the waste, the generators, transporters, and owners/operators of the TSD facilities have to follow the standards dictated by the RCRA/HSWA. A summary of the salient features of the standards follows:

Generator phase requirements: These include obtaining a U.S. EPA ID number, pre-transportation handling, minimizing waste to reduce volume and/or toxicity, preparing a manifest, keeping records, and reporting.

Transportation phase requirements: These include obtaining a U.S. EPA ID number, signing and keeping a manifest with waste all the times, complying with DOT provisions on labeling, marking, and placarding, using proper containers, and responding to spills.

Hazardous Waste Management

In accordance with HSWA requirements, a generator of hazardous waste is required to certify that a waste minimization program has been instituted. U.S. EPA has established the following ranking of priorities to manage hazardous waste:

- Waste reduction;
- Separation and concentration;
- Exchange and recovery;
- Treatment of waste; and
- Ultimate disposal.

The first three items relate to waste minimization.

Waste Minimization Methods

Waste minimization methods can be grouped into the following actions:

Change of raw materials purchasing and control policies. Reduce the number of different brands of products, select appropriate container sizes, keep inventory, and rotate materials to use old products before their expiration dates.

Improved housekeeping. Cut down on leakages, maintain machines, and clean machine parts.

Changed production methods. Use improved methods. Instead of a production-related approach, use a reduced-waste-oriented approach.

Use of less toxic materials. Many times, a nontoxic material can be substituted for a toxic ingredient, although the cost may be higher. The reduced cost of disposal might justify the change.

Reduced wastewater or increased concentration. Reducing of the overall quantity of water containing waste can cut down on treatment and disposal costs and can assist in materials recovery.

Separation of wastes. Individual wastes should be separated; some might need unusual treatments and others might not be a 'regulated" waste.

Exchange of wastes. Waste from an industry might be useful material for another industry or might be suitable for a different process in the same industry.

Reuse and recovery: Use materials more than once when feasible. Apply the recovery process to reclaim materials that can be recycled or sold as byproducts.

Hazardous Waste Treatment Technology

RCRA/HSWA prohibits the ultimate disposal of a hazardous waste in a landfill without its treatment first. The treatment technologies include the following.

Segregated treatment of hazardous waste streams. According to RCRA mixture rule, any quantity of a hazardous waste, when mixed with a nonhazardous waste, renders the entire mixture hazardous. In light of this provision, it might make sense to treat wastes from production units separately, especially when the hazardous stream makes up a small fraction of the total waste.

Toxicity reduction. If the constituent that imparts toxicity is captured or destroyed, a waste can be converted to a nonhazardous state. This can be achieved by either 1) biological degradation by special microorganisms, 2) chemical detoxification by adsorption, absorption, precipitation, complexation, chemical reduction, chemical oxidation, and/or electrolysis, and 3) by physical destruction or immobilization by microwave destruction, photolysis, and/or solidification in an inert material matrix.

Volume reduction. Volume can be reduced by either modifying the chemical processes so that some solid residues are not formed at the time of hazardous waste precipitation, or by thickening, dewatering, and drying the waste sludge.

Changing the waste character. When a waste is burned in an incinerator or a boiler, the combusting organics are rendered into a small volume of inert ash; this method reduces the volume, toxicity, and mobility of the waste.

However, exhaust gases containing incompletely burned organics require extensive treatment.

Disposal Technology

The disposal techniques for the treated hazardous waste can be classified into the following categories: the secured landfill, landfarming or ex-situ bioremediation, deep-well injection, and incineration.

Landfill disposal accounts for more than two-thirds of the total hazardous waste disposal. Landfill operation and incineration have been described earlier in this chapter. Landfarming is discussed later within this chapter. Deep-well injection consists of pumping the waste into geologically secured formations at great depth; 2,000 m is typical. The formations must be located quite far below the potable aquifers and should be isolated by thick impervious strata.

Problems of the Terrestrial Environment

Some common problems of the terrestrial environment originate from land use practices. These include deforestation, wetland destruction, and soil deterioration. The contamination of soil is intimately related to groundwater pollution since the passage of water is through the porous soil medium. Leachates from open dumps and landfills, and leakages from underground tanks have contributed to a very widespread contamination of the groundwater supply. RCRA discontinued open dumping practices, assured the proper design of secured landfills, and banned the land disposal of hazardous wastes. To manage the problem of leaking underground tanks, Subtitle I was created and attached to the RCRA through the HSWA of 1984, although underground tanks bear no direct relationship to the waste management issues addressed by the RCRA. The design features of a secured landfill are discussed earlier in this chapter. The underground storage tank (UST) technology, as regulated by RCRA, is presented subsequently along with remediation technology of contaminated soils.

Deforestation

Destruction of forests has taken place at an alarming pace. Forests in developing countries have declined by nearly one-half during this century. The United States has lost 90% of its ancient forests. While the forest area of many developed countries has stabilized or improved recently, the destruction continues in the developing countries. Each year 30 million acres of tropical forests are being cleared in those countries.

Deforestation is a complex problem. The factors responsible for forest destruction include: spread of agriculture, raising of livestock, use of fuelwood, demand for construction lumber, and urban development. In developing countries, the rapid growth of population, coupled with demand by developed countries for tropical timber, have accentuated these factors. The downward spiral of forest destruction has lead to distressed conditions of people in some countries and to a globally climatic problem.

Importance of Forests

Forests play pivotal roles in the natural ecological system. They regulate carbon, oxygen, and nitrogen cycles, and control rainfall and temperatures. Thus, they act as a buffer to unusual global climate changes, preventing droughts, floods, and greenhouse warming. They provide stability to the ground, preventing erosion of topsoil, sedimentation, and flash flooding of rivers.

Forests are habitats for perhaps half of the all life on the planet; they are vital for biodiversity. They hold genetic information accumulated during 4 billion years of evolution.

In nature's economy, forests make important contributions, too. Forest products are an essential source of fodder for livestock. They provide fruits and nuts, gums, honey, oils, resins, tannins, and fibers. Many life-saving medicines are derived from plants, roots, and trees. There is a growing recognition of forest viability as a small-scale enterprise. It is unfortunate that in spite of so many sustained benefits in ecological and environmental terms, forests continue to decline in developing countries in order to fulfill primary needs of food, fuel, and housing. Developed countries continue to demand tropical timber supplied by developing countries. Governmental policies of developing countries, driven to earn more foreign exchange and by pressure from lumber industries, have accelerated the depletion.

Preservation of Forests

Because agriculture and energy practices primarily lead to forest destruction, the solutions to halt deforestation have to come from outside the forestry sector; actions within the forestry sector will not be sufficient. Short-term measures will not solve the problem. The International Task Force convened by the World Resources Institute, the World Bank, and the U.N. Development Program in 1985 proposed the action program in the following areas.

Fuelwood and Agroforestry

The action plan consists of fuelwood conservation, improved fuelwood use, fuelwood substitution, protection and management of existing woodlands, expanded tree planting, and mobilization of public support.

Upland Watershed Land Use

The program focuses on establishing tree and grass cover to stabilize upland areas; to provide adequate supplies of fuelwood, fodder, and building poles; to control livestock grazing; and to develop sustainable farming systems.

Forest Management for Industrial Uses

The program centers on three major types of activities: protection and management of natural forests; more intensive use of existing resources, particularly lesser-known hardwood species; and conversion of existing logged areas into fast-growing plantations.

Conservation of Tropical Forest Ecosystems

The action plan comprises of the following measures:

- Reducing pressures on tropical forests by intensifying agriculture on nonforest lands;
- Establishing plantations on lands that are already cleared rather than cutting undisturbed forests;
- Developing and implementing national conservation strategies as part of the national plan, especially in developing countries;
- Eliminating or minimizing further destruction or conversion of national parks or other areas under severe threat of encroachment;
- Developing conservation data centers. Links should be developed between data centers, government agencies, international agencies, and nongovernmental organizations; and
- Encouraging all nations to accede to the major international conventions. They should endorse and implement the World Conservation Strategy, Biosphere Reserve Action Plan, and Bali Action Plan.

Research, Training, and Extension Programs

Major constraints to an effective forestry program are weak research programs, shortage of trained forestry personnel, and lack of forestry extension and support services. The program concentrates on:

- Supporting high-priority research topics;
- Strengthening national research, training, and education institutions;
- Involving local people in extension and outreach programs;
- Developing extension material such as pamphlets, audiovisual presentations, and mass media; and
- Integrating of agriculture and forestry in research, training, and extension.

Wetlands Encroachment

Wetlands are transitional lands between terrestrial upland and water bodies. Several definitions exist for wetlands. According to U.S. Army Corps of Engineers and U.S. EPA, wetlands are those areas that are inundated or saturated by surface or groundwater at a frequency and duration sufficient to support, and that under normal circumstances do support, a prevalence of vegetation typically adapted for life in saturated soil conditions. Wetlands generally include swamps, marshes, bogs, and similar areas.

The nearly saturated condition of soil is a distinguishing characteristic of wetlands. There are two major categories of wetlands: freshwater wetlands that include swamps, marshes, bogs, and riverines; and coastal wetlands that are affected by tides such as tidal marshes and mangroves. Wetlands can occur anywhere; small spring or seep wetlands can occur high on a hillside. Wetlands are delineated by the presence of hydrophytic vegetation, hydric soil, and/or wetland hydrology.

Wetlands, once viewed as worthless, are now recognized to be vital to environmental health. The functions and values of wetlands are outlined below.

Wildlife Habitats

Unique habitats are situated at the interface of land and water. Tidal wetlands are nursery grounds for shrimp, crabs, and other valuable species of fish and shellfish. Tidal wetlands are the most productive ecosystem on the planet and are an essential component of the estuarine food web. Freshwater wetlands are vital habitats for aquatic, terrestrial, and amphibious species including fish, reptiles, waterfowl, birds, mammals, and many rare and endangered species. They encompass a vast variety of vegetation and woody plant communities.

Water Quality

Wetlands are capable of cleansing water through the uptake of pollutants. They trap nutrients and sediments to maintain water quality of open water bodies. Some wetlands even filter out heavy metals, coliform bacteria, pesticides, and toxic chemicals.

Groundwater Recharge

Wetlands have a watertable close to the surface or at very shallow depths. Normally, the moisture supply of wetlands is hydraulically connected to groundwater in the vicinity. Accordingly, wetlands are often credited with the ability to recharge underground aquifers.

Flood Attenuation

Floodplain and depression wetlands, the areas where flood water is stored when streams and rivers overflow their banks, have flood attenuation effects by holding water temporarily at the time of peak flow. The presence of vegetation and forests enhances this function by to slowing floodwater velocities and transpiring moisture through plants.

Along the coast, wetlands buffer the shore against storm damage and erosion.

Wetland Destruction

The U.S. Fish and Wildlife Service estimates that 117 million acres of the original 221 million acres of wetlands in the U.S. have been lost, mostly by conversion to agriculture. Figures show an average loss of 458,000 acres of wetlands per year. California is severely affected; the original 5 million acres of wetlands have fallen to 500,000 acres in 1990. Before the values and functions of wetlands were fully realized, some state and federal policies encouraged draining of wetlands for farmland and urban development. The Chesapeake Bay, one of the richest estuaries, lost about 12% of its coastal wetlands in just 20 years.

Wetland Protection Program

The primary authority to protect wetlands is drawn from Section 404 of the Clean Water Act of 1972 and amendments of 1977. The program, under Section 404, prohibits discharge of any dredged or fill material from a point source into waters of the U.S. (including wetlands) without a permit. The U.S. Army Corps of Engineers has been given responsibility of administering the Section 404 Program on a day-to-day basis and can issue or deny the permit. U.S. EPA has an authority to veto a Corps permit on the grounds of "unacceptable adverse impact." The Fish and Wildlife Coordination Act requires the Corps to consult the U.S. Fish and Wildlife Service, the National Marine Fisheries Services, and state fish and wildlife departments whenever an applicant seeks a permit.

In addition to the federal program, the states have their own programs pursuant to state laws to protect coastal and freshwater wetlands. The states, under Section 401 of the Clean Water Act, can deny or place conditions on the federal Section 404 permit.

Underground Storage Tank Technology

The HSWA added subtitle I to the RCRA, thus establishing a federal program that directs the U.S. EPA to set regulations for controlling new and existing underground storage tanks (USTs). U.S. EPA issued the proposed rules in April 1987 and, adopted the final rules in September 1988. To these rules, the U.S. EPA added a financial responsibility clause in October 1988. Since 1988, the U.S. EPA has issued several amendments to the rules: the authority has been delegated to states to implement state UST programs that should be equal to or stricter than the RCRA UST program. RCRA excluded the following USTs from the regulations under Subtitle I:

- residential or farm tanks for motor oil of 1,100 gal or less capacity;
- heating oil storage tanks;
- septic tanks; and
- tanks used to store hazardous wastes regulated in Subtitle C of the RCRA.

Perhaps one half the existing tanks are exempted from UST regulations by Subtitle I of the RCRA. Still the number of tanks covered under the regulations is very high. In 1992, the U.S. EPA estimated more than 1.6 million UST systems located over 700,000 facilities nationwide. About 50% of these are petroleum storage tanks owned by gas stations. Another 45% are petroleum storage tanks owned by industries and agencies for their use. Only 5% of tanks are used for storing hazardous wastes.

More than 20% of the tanks are more than 30 years old, installed during the oil boom of 1950 through 1960. These tanks constructed of bare steel, are suspected to have extensive leaks. In addition, thousands of gas station tanks that were shut down during the oil crisis of the 1970s are not properly closed. A study suggests that cleaning up the leaking tanks could cost tens of billions of dollars and might take many decades to complete. Leaking underground storage tanks (LUST) have been recognized to be a serious problem.

The cause of failure of more than 90% of USTs is corrosion. The corrosion of a metal in contact with the ground is a natural phenomena. According to theory, an electromotive force results between dissimilar metals in close proximity to one another in the soil. The flow of electric current due to this force through the soil between the metal surfaces produces a corrosion effect at one metal surface anode, while the other metal surface cathode is not affected. The *cathode protection technique* counteracts this by inducing an electromotive force externally, which is equal and opposite in direction to the natural current flow in the soil. This technique has been successfully used to protect buried metal structures.

Until the 1960s, the tanks were made of 3/16 in. thick to 1/4 in. thick carbon steel plate coated with black asphaltum rust protector; no corrosion prevention was provided. Noncorrosive fiberglass reinforced plastic (FRP) tanks were introduced in the 1960s. In the 1970s, tanks were developed with epoxy-coated steel, sacrificial anode attachment, and high-density polyethylene (HDPE). In the 1980s, design progressed to the double-walled tanks in steel, FRP, or HDPE and combinations thereof..

Frequently, tank system leakage occurred within the piping network. Accordingly, the piping design developed apace with improved tank technology. In recent years, pipe manufacturers developed a jacketed double-walled flexible piping system to match the tank design.

U.S. EPA regulations emphasize leak detection. Leak-detection technology, which prior to mid-1970 was confined to testing of the tank tightness by filling a standpipe and noticing the drop in level, has improved to incorporate:

- Intake gaging equipment using electric, magnetic, and optic probes;
- equipment based on measuring changes in the buoyant force acting on a submerged object;
- Manometric measurements;
- Bubble-gage pressure sensing;
- Laser-beam technology; and
- Vacuum and vapor sensors.

Leak-detection technology will continue to become more effective and more economical.

RCRA UST Program

Objectives of the RCRA UST program are as follows:

- To identify all existing tanks and require that they either be brought up to the standards or be properly closed;
- To install on existing tanks a leak-detection system to determine whether they have leaked. If so, the owners/operators should take corrective action;
- To ensure that the authority is notified when a new tank is being installed. All new tanks should meet strict design and operating standards.

For closure of a petroleum UST or any corrective action under the first two items above, the rules fix the responsibility on the owner/operator of the tank. To ensure that the corrective action will be taken as and when needed, the owner/operator must obtain a guarantee that he or she or some one on his/her behalf (e.g. an insurance company) will pay up to $1 million (more if more than 12 tanks are affected).

When the Superfund Amendments and Reauthorization Act (SARA) was adopted in 1986, a new provision was added to the RCRA UST program setting up a *Leaking UST Trust Fund*. The financial responsibility amount by the owner/operator is a type of deductible. U.S. EPA is authorized to use funds from the trust for the cost of the corrective action above and beyond the amount provided by the guarantee.

The requirements for new tanks, upgrading, and closure of USTs are summarized below.

New UST System

RCRA subtitle I, after May 8, 1985, allows new installation of UST only if the following three conditions are met:

- UST system should prevent releases due to corrosion or structural failure for the life of the system;
- The system should be cathodically protected against corrosion or made of material impervious to corrosion or otherwise designed to prevent releases; and
- The system must be compatible with the substance to be stored.

U.S. EPA set extensive standards for new USTs in the 1988 regulations. Three areas emphasized in the regulations related to corrosion protection, leak control, and spill/overflow prevention. The requirements for these are indicated in Table 11-4.

Table 11-4 Requirements for UST Systems

Requirement	Standard to be Met	Timetable to Implementation
New Tanks (Installed after December, 1988)		
Corrosion Protection	Three options: - Coated and cathodically protected steel; - Noncorrosive material (FRP/HDPE); or - Steel clad with fiberglass	At the time of UST installation
Leak Control	1. Use of a double-walled UST or a secondary barrier of lining or vault. 2. Two options for detection of a leak: - Monthly monitoring; or - Monthly inventory control with tank tightness testing every 5 years (until December 1998, then first option only)	At the time of UST installation
Spill/Overflow Prevention	1. Fill box; and 2. Any one of the following: - Automatic shutoff mechanism - Ball check valve - Flapper gate in pipe - Overfill alarm	
Upgrading Existing Tanks		
Corrosion Protection	1. Same as for new tanks 2. Add cathodic protection system 3. Interior lining	To be completed by December 1998 (up to 10 years of the regulation date
Leak Control	1. Same as for new tanks	Approved form of leak detection must be installed by December 1993
Spill/Overflow Prevention	1. Same as for new tanks	To be completed by December 1993

Monthly leak detection falls into the following categories: 1) inventory control plus tank tightness testing (inventory consisting of daily record of volume measurements for input, withdrawal, and stored volume), 2) manual or automatic gaging of tank liquid levels, 3) interstitial monitoring between the primary (inner wall) and secondary (outer wall or barrier) space, and 4) external monitoring either by:

- Vapor monitoring in the soil outside the tank installation; or
- Groundwater monitoring outside the proximity of the tank for the presence of product.

Upgrading the Existing UST System

Upgrading a UST consists of the following:

- Installation of an approved form of leak detection system to all existing tanks by December 1993 and monitoring for leakage.
- Installation of the corrosion-protection fittings to tanks and piping systems by December 1998.
- Providing spill and overflow control measures by December 1998.

The requirements for upgrading are indicated in Table 11-4.

Once a leakage or release from UST system has been detected or spill/overflow of more than 25 gal has been noticed, the owner/operator must comply with various corrective measures. The initial action includes reporting the release to the implementing agency and taking immediate action to prevent further release or mitigation of any associated substance. The permanent corrective measures include the owner/operator submitting a corrective plan to the implementing agency for remediation of contaminated soil and groundwater, and commencing clean-up upon the approval of the plan. The financial responsibility clause, as described earlier, applies to owners/operators for corrective measures.

Closure of UST Systems

Tanks that are not proposed to be upgraded or not intended to be used any longer have to be closed after the state implementing agency is informed and approval is obtained. The state assigns an inspector and/or directs the owner/operator as to what tests and reports are required for evidence of no releases at the site.

For a temporary closure of up to 30 days or a specified time duration, the tank is emptied and cleaned. Flammable substances such as gasoline are neutralized by filling with CO_2 ice at a rate of 15 lb of CO_2 ice/1000-gal tank capacity. Electric service is disconnected. The tank access openings are sealed with light concrete filler. Putting the tank back in service is easy: the concrete cap is removed and electrical service is restored.

For closing beyond the specified temporary period, the permanent closure will require:

- Approval of the state agency;
- Site inspection and/or investigations for evidence of release of the stored product in the tank;
- Pulling the tank out of the ground; and
- Filling the hole with a clean fill material.

Under certain circumstances, the tank may be allowed to be closed in place, such as when it is under a structure or when its removal poses a danger to nearby structures/facilities. In such cases, the tank is filled at the site with inert material such as concrete slurry. All other conditions of the closure apply.

If any contamination occurred during the closure or existed before, the responsibility of cleaning the site rests with the owner/operator, according to the requirements discussed.

Aboveground Storage Tank Technology

Aboveground storage tanks (ASTs) are not subject to direct federal regulations as are USTs. For this reason and because of other incentives such as the lower cost of installation, lower insurance rates, and easy leakage detection, ASTs are preferred over USTs. However, almost all local fire and zoning departments follow the National Fire Protection Code according to which gasoline and other volatile flammable liquid storage tanks should be installed underground unless they are located on large sites that can allow several hundred feet of space between the tank and the adjacent structures. As a result, ASTs are used for high-volume storage, as in refineries, oil fields, and chemical plants, ranging from 20,000 to multi-million gallon capacity.

Common ASTs, to a maximum of 50,000 gal capacity, follow the Underwriters Laboratories, UL142 standards. The Steel Tank Institute has developed specifications for fabrication of an open-top, secondary containment steel structure consisting of a solid steel floor and steel vertical sidewall diking, within which a standard UL142 primary storage tank is contained. Many other innovations have been introduced, including outer concrete containment with a base slab system, and lining the inside of the tank to protect against internal corrosion.

A release from a large AST system is effectively resolved at the state level. The federal laws that exercise control over an AST system are:

- The Clean Water Act of 1987, requiring a facility located near rivers, streams, creeks, or any drainage network to prepare a spill prevention control and counter-measure plan (SPCC); and
- The Oil Pollution Act of 1990, giving the U.S. EPA authority to oversee the spill counter-measures adopted by AST owners/operators.

Table 11-5 Tests for Contaminated Soils and Groundwaters

Test	Contaminant	Method Applied
1. Total petroleum hydrocarbon (TPH)	Petroleum contamination without regard to type	Screening
2. Volatile organic carbon (VOC)	Mixed waste including organic compounds	Screening
3. Total extractable petroleum hydrocarbon (TEPH)	Diesel and related hydrocarbons	Gas chromatography (GC)
4. Total recoverable hydrocarbon (TRH or TRPH)	Lubricating and fuel oil	Screening for soil, GC for soil/water
5. Total volatile petroleum hydrocarbon (TVPII)	All gasoline	Screening for soil, GC for soil/water
6. Benzene, toluene, ethylbenzene, and xylene (BTEX)	Aromatics part of gasoline	GC/mass spectroscopy (MS)
7. Volatile organic analysis (VOA or VOC)	Volatile organic compounds	GC/MS
8. PAH	PAH component of hydrocarbon	GC/flame ionization detector (FID)
9. Pesticide/PCB	Pesticides/PCBs	GC/electron capture detector (ECD)
10. Toxic characteristic leachate procedures (TCLP)	Hazardous waste	U.S. EPA TCLP procedure

Table 11-6 Soil Treatment Technologies

Method	Applicable to Soil Type	Applicable to Contaminant Type	Advantages	Disadvantages
ABATEMENT				
Vitrification	Any type without excessive moisture	Any type including radioactive	Good for any type of soil and any type of contaminant	Suitable for small waste sites
Solidification/ Stabilization	Any type	Heavy oils, sludges, PCB, metals, radioactive materials	Good for any type of soil and a wide category of waste except volatile compounds	For small-to-moderate sites Future land use limited
Asphalt Incorporation	Any type except high moisture and organic contents	Heavy petroleum; not gasoline	Asphalt solidifies into a solid that immobilizes contaminants even though asphalt may not be suitable for use	Relatively expensive. Method remains largely empirical for deciding the mix

(Table 11-6 continued)

Method	Applicable to Soil Type	Applicable to Contaminant Type	Advantages	Disadvantages
		EX SITU		
Excavation and disposal	Any type	Any type	A very common method Relatively quick and proven technology	Suitable for small sites only When contaminant is confined on-site only, not too deep
Landfarming	Any type	Light (gasoline) and heavy hydrocarbons	Low technology, non-labor intensive	Requires large area Results in up to 40% volatization in atmosphere
Low temperature thermal stripping (LTTS)	Gravel, sand, and silt (not clay, saturated and high organic soils)	Volatile hydrocarbons and organics	Treatment time is short Soils can be used as backfill	Excessive cost of excavation and processing
High-temperature thermal stripping (HTTS)	Any type except high moisture and high organic content	All hydrocarbons and organics; not suitable for inorganics and metals	Suitable for soil, clay, and wastes–petroleum sludge and waste oil that are difficult to treat otherwise	Excessive cost of excavation and processing
		IN SITU		
Volatization or soil venting	Sand, gravel (loose, porous)	Volatile hydrocarbons (gasoline and VOC)	A proven technology Can be effectively combined for with other methods, i.e. with groundwater as air sparging	Limited application for specific soils, contaminants and site conditions
Leaching (washing) and chemical extraction	Sand, gravel (loose, porous)	Light hydrocarbons (gasoline) diesel, light fuel, VOC, metals, pesticides	Flushing solution can be changed to a variety of chemicals to suit contaminants Can be effectively combined with other methods	High cost Long treatment time Not a proven method
Bioremediation	Gravel, sand, coarse silt (permeability not below 1×10^{-4} cm/sec)	All hydrocarbons, and organic compounds; not metal contaminants	Site disturbance is minimal, the process is natural Low cost of treatment Only practical method for certain cases such as deep and extensive contamination	Extensive monitoring requirement Long clean-up time Not a standard technology—need to justify the method Viable option for gasoline—less clear cut option for heavy hydrocarbons/ persistent organics

Cleaning of Contaminated Soils

The contamination of soils and waters by aqueous pollutants and gasoline products is an interconnected phenomenon. In cleaning operations, the threefold problem relates to the recovery of free product or pure compound floating over the water table, the removal of the product held within groundwater, and the removal of the product from unsaturated soils. The first two items have been discussed in Chapter 9 in the context of groundwater treatment. The approach to soils remediation is similar to that of groundwater technology because both are a part of the same problem. The complimentary methods for both soils and water are very attractive options.

Appropriate Tests for Soil Contaminants

The majority of soil and water contamination is due to petroleum products. Petroleum is a mixture of hydrocarbons containing hundreds of compounds. Through the distillation process, these compounds are separated into groups:

- Lightest hydrocarbons containing 1 to 4 carbons (C_1 to C_4) are gases at room temperature used in heating gases;
- Next-lighter hydrocarbons in the C_4 to C_{12} range are used in gasoline formation;
- Middle distillates, from C_{12} to C_{20}, comprise kerosene, diesel, light and heavy fuel oils, and lubricating oils. These have high polyaromic hydrocarbon (PAH) concentration; and
- Heavy hydrocarbons, C_{20} to C_{30}, are residual fuel substances containing PAH.

The gasoline component can be further divided into:

light aliphates: *n*-pentane, iso-pentane, *n*-hexane, *n*-heptane, and cyclohexane; and
aromatics: benzene, toluene, ethylbenzene, and xylenes (BTEX).

The nature of compounds and their grouping play important roles in contaminants migration. The light aliphatic component of gasoline is highly volatile and exists as vapor. Aromatic gasoline components (BTEX) migrate through all phases—vapor, adsorbed (to soil particles), and aqueous (water soluble) form; it is an important contaminant. PAHs are strongly adsorbed to soil particles and, as such, exist in immobile phase.

Testing consists of screening methods to detect broad classes of contaminants and analytical methods for a specific category of contaminants or compounds. The tests to be performed are decided on the basis of contaminants to be investigated as shown in Table 11-5.

The clean-up standards vary very widely in terms of the indicator tests and contaminant limits to be met. For example, 100 ppm for TPH and 10 ppm for BTEX might be the limits for soils.

Soil Treatment Technologies

Similar to and common with groundwater treatment, the methods fall into three categories: *abatement technologies* designed to immobilize the contaminants and permanently contain them, *ex situ* or *off-site technologies* that require the contaminated soil to be excavated and then treated on site or off site, and *in situ* or *on-site technologies* that are carried out without excavation or a minimal excavation. These technologies are briefly described below and their features are summarized in Table 11-6.

Abatement technologies include:

- *Vitrification*: Electrodes are placed in an array around the contaminated area and the soil is melted to glass-like form at 1500–2000 °C.
- *Solidification/stabilization*: Contaminants are immobilized in concrete or fly ash.
- *Asphalt Incorporation*: Contaminated soils are mixed in a small amount (up to 5%) with asphalt. Subsequently, asphalt is used in road construction or crushed and applied as a road base.

Ex situ technologies include:

- *Excavation and disposal*: Contaminants are physically removed, and after treatment, are disposed in a landfill or incinerated.
- *Landfarming*: The contaminated soil is excavated and spread in 12 inch to 18 inch deep rows. Periodically, the soil is turned with a rototiller. Bioremediation occurs as a result of natural bacterial activities.
- *Low temperature thermal stripping* (LTTS): Contaminated soil is fed into a chamber or rotating kiln and heated to 200–260 °C. The VOCs are vaporized.
- *High temperature thermal stripping* (HTTS): Contaminated soil is fed into a chamber and incinerated at 800–1100 °C. Temperatures are kept lower than those to cause vitrification.

In situ technologies include:

- *Volatization or soil venting*: Hydrocarbon vapors in the pore spaces are entrained in a stream of air injected through a system of wells and the contaminated air is extracted through another system of collection wells.
- *Leaching and chemical extraction*: The arrangement of injection and extraction wells is similar to the volatization method. By injecting a stream of fresh water, steam, surfactant (detergent), or other chemical additive, the soil is gradually flushed of contaminants. Extracted contaminated water/solution is treated externally.
- *Bioremediation*: A process of aerobic microbial oxidation using naturally occurring soil bacteria. The design provides for a system of injection wells, recovery, or extraction wells and downstream barriers that control the mass movement into and out of the contaminated area. A bioremediation system is shown in Figure 11-3. The injection wells pump in water with oxygen, nutrients, substrate, and additional microbial seed. After treatment, water with oxidation products is collected by recovery wells, treated at the surface and recirculated. Many types of injection systems exist.

Bioremediation is a very recently developed and fast-growing technology, patented in 1974 and applied to sites in late 1980s. Many new techniques and combinations with other methods are emerging in the field. These developing technologies may help restore superfund sites in the future.

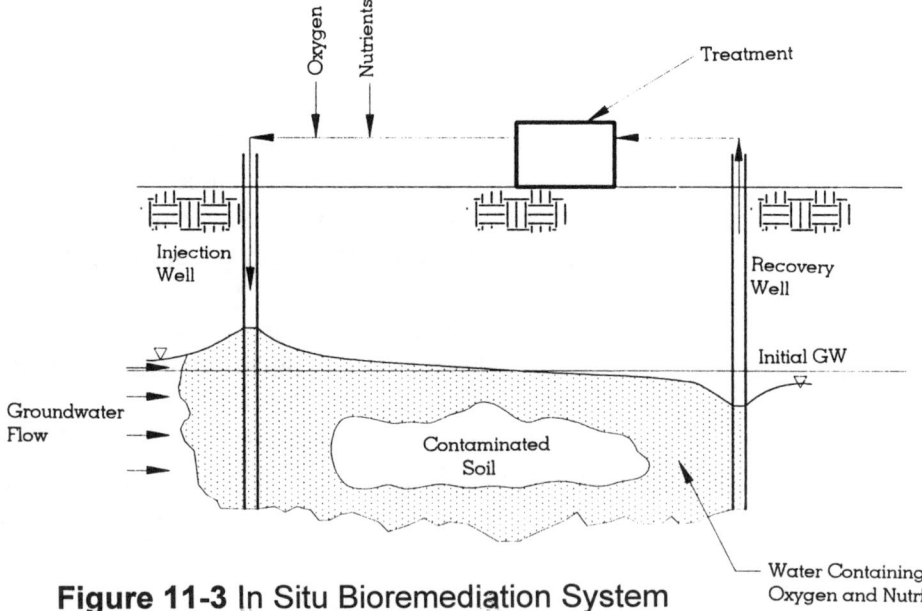

Figure 11-3 In Situ Bioremediation System

PROBLEMS

1. A town has a population of 8,000. The unit waste factor for the town is 5 lb/person/day. For weekly pick up from the town, determine the amount of waste produced, the volume of waste produced, and the weight of waste when fully dried. Use the average values given in Table 11-3 for waste in container.

2. In Problem 1, the waste is compacted in a landfill to a depth of 10 ft and it occupies an area of 1.2 acres in a year. What is the density of the fill?

3. In Problem 1, if the composition of the waste is as indicated in Table 11-1, determine the weight of each component produced and the volume of each component of the waste.

4. In Problem 3, determine the dry weight of each component.

5. The student population at a high school is 1200. For a 5-day school week, if paper waste is 50% of the total waste, determine:

    a) the amount of waste generated per week,
    b) the amount of paper waste produced.

    If the standard sizes for containers are 1.5, 3.0, 4.5, and 6 m³, select the storage container for the school. The waste density is 150 lb/yd³. Use the standard unit waste factor and moisture content.

6. A commercial establishment operating 5 days per week has 20 employees. The following waste data was recorded. Determine the rate of daily contribution (unit waste factor), and the density of waste

Annual dry waste quantity	29,700 lb
Moisture content	5%
Annual volume	45 yd³.

7. In a particular week, Mrs. X fills three containers of waste. She makes the following measurements. Determine the average density of her household waste. Volume of each container = 3 ft³, weight of each container = 8 lb, weight of containers filled with waste:

Can no.	Gross weight, in lb
1	16
2	23.5
3	19.2

8. The following time durations were observed from a collection-crew route analysis (crew works 8 hr/day). Determine the number of households visited for a pick up each day.

Average number of cans per household	3
Time from garage to route	15 min
Off-route time of crew	Two breaks of 15 min each
Time from transfer station to garage	10 min
Average haul time	20 min
Emptying time	8 min
Number of trips to transfer site	3
Collection plus drive time per stop	t_p (in hr) = 0.7 + 0.2 (no. cans/household)

9. For a unit waste production of 5 lb/person/day, an average household size of 4 persons/house, and a compaction density of 700 lb/yd^3, determine the size of the rear-loader vehicle to be bought, with a compaction efficiency of 85%.

10. Determine the number of vehicles required by the town of Problem 1. Use the data from Problem 8 and household of 4 persons/house. If the town cannot buy more than one vehicle, how many houses have to be served per trip without any change in the number of trips per day?

11. For the data of Problem 8 remaining unchanged, how long will the working day be, in hours, for the collection crew to meet the requirement of Problem 10 with one vehicle?

12. A town has bought 9 cubic-yard rear-loader vehicle that meets its needs. It has a rated density of 780 lb/yd^3 and an efficiency of 80%. If the average household waste generation is 55 lb/house/week, determine the number of houses served per trip of the vehicle.

13. In Problem 12, using the time duration data of Problem 8, determine how many trips to the transfer site the vehicle will make per day.

14. In Problem 12, determine the town population and the daily volume of uncompacted waste produced in the town. The unit waste factor is 3 lb/person/day, average household size is three persons, and the uncompacted density is 100 lb/yd^3.

15. For the town of Problem 1, design the daily landfill cell of 12 in. thickness. Determine the total volume and area of the landfill for 20 years having a depth of 30 ft.

16. A village has a population of 1700. Assuming a daily cell of 0.15 m of soil cover and 0.3 m of solid waste, determine the daily spread area, and the landfill space required for 20 years to maintain a depth of 10 m.

17. The total area of a landfill is restricted to 10 acres. The daily cell area should not exceed 50 ft × 50 ft. What population can fill this support? What will the height be in 25 years?

STUDY QUESTIONS

1. What are the operations involved in a comprehensive solid waste management plan? Which one is the most costly part of the plan? Describe it in brief.

2. Why is resource conservation and recovery (RC&R) important? Discuss its essential elements.

3. What are the disposal alternatives of solid waste? Describe the salient features of a modern landfill with a sketch.

4. What are the essential characteristics of a hazardous waste? How can a hazardous waste be identified from a solid waste?

5. What is the cradle-to-grave concept? Describe how to manage a hazardous waste?

6. Why are forests important resources? How should they be managed and conserved?

7. Why are wetlands important? How should they be protected?

8. What are the objectives of an underground storage tank program? Outline the requirements for upgrading of existing tanks and for installating of new tanks.

9. Describe the bioremediation technique for contaminated soil treatment. Distinguish it from other in situ and ex situ techniques.

10. Define the following:
    a. Municipal solid waste (MSW)
    b. Unit waste factor
    c. Refuse-derived fuel (RDF)
    d. Incineration
    e. Composting
    f. Leachate
    g. Toxicity Characteristic Leaching Procedure (TCLP)
    h. Manifest form
    i. Deep injection well
    j. Section 404 permit
    k. Venting
    l. Landfarming
    m. Low and high temperature thermal stripping

REFERENCES

Allen, J. L., ed., *Annual Editions Environment 92/93*, The Dushkin Publishing Group, Guilford, CT, 1992.

Ayers, K. W., et al., *Environmental Science and Technology Handbook*, Government Institutes, Inc., Rockville, MD, 1994.

Braddock, T., *Wetlands: An Introduction to Ecology, the Law, and Permitting*, Government Institutes, Rockville, MD, 1995.

Cahn, R., ed., *An Environmental Agenda for the Future*, Island Press, Washington, DC, 1985.

Caruana, C. M., "Hazardous Waste Management - A Top Priority with EPA.," *Chemical Engineering Progress*, pp. 50-53, July, 1986.

Cheremisinoff, P. N., *A Guide to Underground Storage Tanks: Evaluation, Site Assessment and Remediation*, Prentice Hall, NJ, 1992.

Cole, G. M., *Assessment and Remediation of Petroleum Contaminated Sites*, Lewis Publishers, Boca Raton, FL, 1994.

Cookson, J. T., Jr., *Bioremediation Engineering: Design and Application*, McGraw-Hill Book Company, New York, 1995.

Corbitt, R. A., ed., *Standard Handbook of Environmental Engineering*, McGraw-Hill Book Company, New York, 1990.

Daniel, J. E., ed., *1993 Earth Journal*, Buzzworm Books, Boulder, CO, 1992.

Daugherty, J., *Industrial Environmental Management: A Practical Approach*, Government Institutes, Inc., Rockville, MD, 1996.

Dawson, G. W., and Mercer, B. W., *Hazardous Waste Management*, John Wiley and Sons, New York, 1986.

Dennison, M. S., *Pollution Prevention Strategies and Technologies,* Government Institutes, Inc., Rockville, MD, 1996.

Durning, A. T., *Saving the Forests: What Will It Take?*, Paper 117, Worldwatch Institute, December, 1993.

Francis, C. W., and Aurbach, S. I., ed., *Environment and Solid Wastes: Characterization, Treatment and Disposal*, Butterworth Publishers, Boston, MA, 1983.

Franzle, O., *Contaminants in Terrestrial Environments*, Springer-Verlag, Berlin, 1993.

Hall, R. M., Jr., et al., *RCRA Hazardous Waste Handbook,* 11th ed., Government Institutes, Inc., Rockville, MD, 1995.

Haun, J. W., *Guide to the Management of Hazardous Waste*, Fulcrum Publishing, CO, 1991.

Holmes, G., Singh, B. R., and Theodore, L., *Handbook of Environmental Management and Technology*, Wiley-Interscience, New York, 1993.

Hopper, D. R., "Cleaning Up Contaminated Waste Sites," *Chemical Engineering*, pp. 93-110, August, 1989.

Institution for Solid Wastes of American Public Works Association, *Municipal Refuse Disposal*, Public Administration Service, Chicago, IL, 1970.

International Task Force of World Resources Institute, World Bank and U.N. Development Program, *Tropical Forests: A Call for Action*, Parts I, II, III, World Resources Institute, Washington, DC.1985.

Kaufman, D. G., and Franz, C. M., *Biosphere 2000: Protecting Our Global Environment*, Harper Collins College Publishers, New York, 1993.

Knowles, P.C., ed., Dames and Moore, *Fundamentals of Environmental Science and Technology*, Government Institutes, Inc., Rockville, MD, 1992.

Kreith, F., ed., *Handbook of Solid Waste Management*, McGraw-Hill Book Company, New York, 1994.

Martins, K., "Responding Properly to Hazardous Waste Spills," *Chemical Engineering*, pp. 87-91, January, 1988.

Miller, G. T., *Resource, Conservation and Management*, Wadsworth Publishing Company, Belmont, CA, 1990.

Nathanson, J. A., *Basic Environmental Technology*, John Wiley and Sons, New York, 1986.

Ney, R. E., Jr., *Fate and Transport of Organic Chemicals in the Environment: A Practical Guide*, 2nd ed., Government Institutes, Inc., Rockville, MD, 1995.

Oliver, T. and Kostecki, P., "State Summary of Soil and Groundwater Cleanup Standards," *Soil and Groundwater Cleanup*, Independence, MO, Nov., 1995.

Rizzo, J. A., and Young, A. D., *Aboveground Storage Tank Management: A Practical Guide*, Government Institutes, Inc., Rockville, MD, 1990.

Rizzo, J. A., *Underground Storage Tank Management: A Practical Guide,* 4th ed., Government Institutes, Inc., Rockville, MD, 1991.

Sullivan, T. F. P., ed., *Environmental Law Handbook*, 14th ed., Government Institutes, Rockville, MD, 1997.

Tchobanoglous, G., Theisen, H., and Vigil, S. A., *Integrated Solid Waste Management*, McGraw-Hill Book Company, New York, 1993.

U.S. Army, Corps of Engineers Environmental Laboratory, *Wetland Delineation Manual*, Government Institutes, Inc., Rockville, MD, 1987.

U.S. Environmental Protection Agency,*The Federal Wetlands Protection Program in New England.* Boston: U.S. EPA Region I, Boston, MA, 1991.

World Resources Institute, *The 1992 Information Please Environmental Almanac*, Houghton Mifflin Company, Boston, MA, 1991.

World Resources Institute: *The 1993 Information Please Environmental Amanac*, Houghton Mifflin Company, Boston, MA, 1993.

Young, A. D., "Underground and Aboveground Storage Tank Technology," Chapter 8 in *Environmental Science and Technology Handbook*, Ayers, K. W., ed., Government Institutes, Inc, Rockville, MD, 1994.

APPENDIX A

TABLE A-1 LENGTH EQUIVALENTS

Unit	mm	m	in.	ft	yd	mi
Millimeter	1	10^{-3}	0.0394	0.00328	0.00109	6.214×10^{-7}
Meter	10^3	1	39.37	3.281	1.0936	6.214×10^{-4}
Inch	25.4	0.0254	1	0.0833	0.02778	1.578×10^{-5}
Foot	304.8	0.3048	12	1	0.333	1.894×10^{-4}
Yard	914.4	0.9144	36	3	1	5.682×10^{-4}
Mile	1.609×10^6	1.609×10^3	6.336×10^4	5280	1760	1

TABLE A-2 AREA EQUIVALENTS

Unit	in.^{2}	ft^2	m^2	acre	mi^2
Square inch	1	6.944×10^{-3}	6.452×10^{-4}	1.59×10^{-7}	2.491×10^{-10}
Square foot	144	1	0.0929	2.30×10^{-5}	3.587×10^{-8}
Square meter	1550	10.764	1	2.50×10^{-4}	3.861×10^{-7}
Acre	6.270×10^6	43,560	4047	1	1.56×10^{-3}
Square mile	4.014×10^9	2.788×10^7	2.59×10^6	640	1

TABLE A-3 VOLUME EQUIVALENTS

Unit	in.^{3}	gal	ft^3	m^3	acre-ft	cfs-day
Cubic inch	1	0.00433	5.79×10^{-4}	1.64×10^{-5}	1.33×10^{-8}	6.70×10^{-9}
Gallon	231	1	0.134	0.00379	3.07×10^{-6}	1.55×10^{-6}
Cubic foot	1728	7.48	1	0.0283	2.30×10^{-5}	1.16×10^{-5}
Cubic meter	61,000	264	35.3	1	8.11×10^{-4}	4.09×10^{-4}
Acre-foot	7.53×10^7	3.26×10^5	43,560	1233	1	0.504
Cubic foot per second-day	1.49×10^8	6.46×10^5	86,400	2447	1.98	1

TABLE A-4 VELOCITY EQUIVALENTS

Unit	ft/sec	mi/hr	Equivalent m/sec	km/hr	kn
Feet per second	1	0.6818	0.3048	1.097	0.5925
Miles per hour	1.467	1	0.4470	1.609	0.8690
Meters per second	3.281	2.237	1	3.600	1.944
Kilometers per hour	0.9113	0.6214	0.2778	1	0.5400
Knots	1.688	1.151	0.5144	1.852	1

TABLE A-5 DISCHARGE EQUIVALENTS

Unit	gal/day	ft³/day	Equivalent gal/min	acre-ft/day	cfs	m³/s
U.S. gallons per day	1	0.134	6.94×10^{-4}	3.07×10^{-6}	1.55×10^{-6}	4.38×10^{-8}
Cubic feet per day	7.48	1	5.19×10^{-3}	2.30×10^{-5}	1.16×10^{-5}	3.28×10^{-7}
U.S. gallons per minute	1440	193	1	4.42×10^{-3}	2.23×10^{-3}	6.31×10^{-5}
Acre-feet per day	3.26×10^5	43,560	226	1	0.504	0.0143
Cubic feet per second	6.46×10^5	86,400	449	1.98	1	0.0283
Cubic meters per second	2.28×10^7	3.05×10^6	15,800	70.0	35.3	1

TABLE A-6 PRESSURE EQUIVALENTS

Unit	ft H₂O	in. Hg	mm Hg	Equivalent mbar	kPa	psi	kg/m²
Foot of water (32 °F)	1	0.883	22.42	29.89	2.989	0.4335	304.8
Inch of mercury (32 °F)	1.133	1	25.40	33.86	3.386	0.4912	345.3
Millimeter of mercury (0 °C)	0.0446	0.03937	1	1.333	0.1333	0.01934	13.60
Millibar	0.0335	0.02953	0.7501	1	0.1000	0.01450	10.20
Kilopascal (N/m² × 10³)	0.335	0.2953	7.501	10.00	1	0.1450	102.0
Pounds per square inch	2.307	2.036	51.71	68.95	6.895	1	703.1
Kilograms per square meter	.000328	0.002896	0.07356	0.09807	0.009807	0.001422	1

Appendix A

TABLE A-7 ENERGY EQUIVALENTS

Unit	Equivalent					
	Btu	cal	J	kW-hr	ft-lb	hp-hr
British thermal unit (60 °F)	1	252.0	1055	0.0002930	777.9	0.0003929
Calorie (15 °C)	0.003969	1	4.186	1.163×10^{-6}	3.087	1.559×10^{-6}
Joule	0.0009482	0.2389	1	2.778×10^{-7}	0.7376	3.725×10^{-7}
Kilowatt-hour	3413	860,100	3.600×10^6	1	2.655×10^6	1.341
Foot-pound	0.001286	0.3239	1.356	3.766×10^{-7}	1	5.051×10^{-7}
Horsepower-hour	2545	641,300	2.685×10^6	0.7457	1.980×10^6	1

TABLE A-8 POWER EQUIVALENTS

Unit	Equivalent				
	W or J/sec	kW	ft-lb/sec	hp	Btu/hr
Watts (or Joules per second)	1	0.001	0.737	0.00134	3.412
Kilowatts	1,000	1	737.6	1.314	3,412
Foot-pounds per second	1.356	0.001356	1	0.001818	4.63
Horsepower	745.5	0.7455	550	1	2,545
British thermal units per hour	0.293	2.93×10^{-4}	0.216	3.93×10^{-4}	1

TABLE A-9 DYNAMIC VISCOSITY EQUIVALENTS

Unit	Equivalent			
	N·sec/m²	g/cm-sec (poise)	lb-sec/ft²	kg/m-hr
Newtons-seconds per meter square	1	10.0	0.0209	3600
Grams per centimeter-second	0.1	1	2.089×10^{-3}	360
Pounds-seconds per foot square	47.88	478.80	1	1.724×10^5
Kilograms per meter-hour	2.778×10^{-4}	2.778×10^{-3}	5.80×10^{-6}	1

1 poise = 100 centipoise (cp)

APPENDIX B

TABLE B-1   SOME OTHER USEFUL CONVERSION FACTORS

Multiply:	By:	To obtain:
Mass (kg)	9.81	Weight in Newtons
Pound	4.448	Newton (N)
	0.4536	Kilogram
Liter	1000	Cubic centimeter
Pounds per ft^2	47.88	N/m^2 *or* pascal
Horsepower	745.7	Watt
	550	Foot-lb/sec
Standard atmosphere	101.325	Kilopascal (kPa)
U.S. or short ton	2000	Pound
Metric ton or tonne	1000	Kilogram
Short ton	0.907	Metric ton
	0.892	Long ton
Nautical mile	1852	Meter
U.S. mile	1609	Meter
Square mile	2.59	Square kilometer
Square kilometer	100	Hectare (ha)
°F	5/9 (°F − 32)	°C
Log to base e (i.e., log$_e$, where $e = 2.718$	0.434	Log to base 10 (i.e., log$_{10}$)

APPENDIX C

TABLE C-1 PHYSICAL PROPERTIES OF WATER IN METRIC UNITS[a]

Temp. (°C)	Specific gravity	Density (g/cm³)	Surface tension (N/m)	Heat of vaporization (cal/g)	Viscosity Dynamic (poise)[b]	Viscosity Kinematic (stokes)[c]	Bulk modulus of elasticity (N/m²)	Vapor pressure mm Hg	Vapor pressure Millibar	Vapor pressure g/cm²
0	0.99987	0.99984	75.6×10^{-3}	597.3	1.790×10^{-2}	1.790×10^{-2}	2.02×10^9	4.58	6.11	6.23
5	0.99999	0.99996	74.9	594.5	1.520	1.520	2.06	6.54	8.72	8.89
10	0.99973	0.99970	74.2	591.7	1.310	1.310	2.10	9.20	12.27	12.51
15	0.99913	0.99910	73.5	588.9	1.140	1.140	2.14	12.78	17.04	17.38
20	0.99824	0.99821	72.8	586.0	1.000	1.000	2.18	17.53	23.37	23.83
25	0.99708	0.99705	72.0	583.2	0.890	0.893	2.22	23.76	31.67	32.30
30	0.99568	0.99565	71.2	580.4	0.798	0.801	2.25	31.83	42.43	43.27
35	0.99407	0.99404	70.4	577.6	0.719	0.723	2.27	42.18	56.24	57.34
40	0.99225	0.99222	69.6	574.7	0.653	0.658	2.28	55.34	73.78	75.23
50	0.98807	0.98804	67.9	569.0	0.547	0.554	2.29	92.56	123.40	125.83
60	0.98323	0.98320	66.2	563.2	0.466	0.474	2.28	149.46	199.26	203.19
70	0.97780	0.97777	64.4	557.4	0.404	0.413	2.25	233.79	311.69	317.84
80	0.97182	0.97179	62.6	551.4	0.355	0.365	2.20	355.28	473.67	483.01
90	0.96534	0.96531	60.8	545.3	0.315	0.326	2.14	525.89	701.13	714.95
100	0.95839	0.95836	58.9	539.1	0.282	0.294	2.07	760.00	1013.25	1033.23

[a] SI units:
Density: kg/m³ = g/cm³ × 10³
Specific weight: N/m³ = density in kg/m³ × 9.81
Dynamic viscosity: N·s/m² = poise × 10⁻¹
Kinematic viscosity: m²/s = stokes × 10⁻⁴
Vapor pressure: N/m² = millibar × 10² or g/cm² × 98.1

[b] poise = (g/cm·s)

[c] stokes = (cm²/s)

TABLE C-2 PHYSICAL PROPERTIES OF WATER IN ENGLISH UNITS

Temp. (°F)	Specific gravity	Specific Weight (lb/ft³)	Surface tension (lb/ft)	Heat of vaporization (Btu/lb)	Viscosity Dynamic (lb-sec/ft²)	Viscosity Kinematic (ft²/sec)	Bulk modules of elasticity (psi)	Vapor pressure in. Hg	Vapor pressure Millibar	Vapor pressure lb/in.²
32	0.99986	62.418	0.518×10^{-2}	1075.5	3.746×10^{-5}	1.931×10^{-5}	293×10^3	0.180	6.11	0.089
40	0.99998	62.426	0.514	1071.0	3.229	1.664	294	0.248	8.39	0.122
50	0.99971	62.409	0.509	1065.3	2.735	1.410	305	0.362	12.27	0.178
60	0.99902	62.366	0.504	1059.7	2.359	1.217	311	0.522	17.66	0.256
70	0.99798	62.301	0.500	1054.0	2.050	1.058	320	0.739	25.03	0.363
80	0.99662	62.216	0.492	1048.4	1.799	0.930	322	1.032	34.96	0.507
90	0.99497	62.113	0.486	1042.7	1.595	0.826	323	1.422	18.15	0.698
100	0.99306	61.994	0.480	1037.1	1.424	0.739	327	1.933	65.47	0.950
120	0.98856	61.713	0.473	1025.6	1.168	0.609	333	3.448	116.75	1.693
140	0.98321	61.379	0.454	1014.0	0.981	0.514	330	5.884	119.26	2.890
160	0.97714	61.000	0.441	1002.2	0.838	0.442	326	9.656	326.98	4.742
180	0.97041	60.580	0.426	990.2	0.726	0.386	318	15.295	517.95	7.512
200	0.96306	60.121	0.412	977.9	0.637	0.341	308	23.468	794.72	11.526
212	0.95837	59.828	0.404	970.3	0.593	0.319	300	29.921	1013.25	14.696

*Maximum specific weight is 62.427 lb/ft³ at 39.2°F

APPENDIX D

TABLE D-1 PHYSICAL PROPERTIES OF AIR AT STANDARD ATMOSPHERIC PRESSURE

Temperature °F	Density, slugs/ft^3	Specific weight lb/ft^3	Dynamic viscosity, lb-sec/ft^2	Kinematic viscosity, ft^2/sec
0	0.00268	0.0862	3.28×10^{-7}	1.26×10^{-4}
20	0.00257	0.0827	3.50	1.36
40	0.00247	0.0794	3.62	1.46
60	0.00237	0.0763	3.74	1.58
80	0.00228	0.0735	3.85	1.69
100	0.00220	0.0709	3.96	1.80
120	0.00215	0.0684	4.07	1.89
150	0.00204	0.0651	4.23	2.07
200	0.00187	0.0601	4.49	2.40
300	0.00162	0.0522	4.96	3.06
400	0.00143	0.0460	5.41	3.77

Temperature °C	Density, kg/m^3	Specific weight n/m^3	Dynamic viscosity N·s/m^2	Kinematic viscosity, m^2/s
-20	1.39	13.6	1.56×10^{-5}	1.13×10^{-4}
-10	1.34	13.1	1.62	1.21
0	1.29	12.6	1.68	1.30
10	1.25	12.2	1.73	1.39
20	1.20	11.8	1.80	1.49
40	1.12	11.0	1.91	1.70
60	1.06	10.4	2.03	1.92
80	0.99	9.71	2.15	2.17
100	0.94	9.24	2.28	2.45
150	0.83	8.18	2.38	2.85
200	0.75	7.32	2.59	3.45
300	0.62	6.04	2.93	4.75

APPENDIX E

TABLE E-1 SATURATION VALUES OF DISSOLVED OXYGEN IN WATER EXPOSED TO WATER-SATURATED AIR CONTAINING 20.90% OXYGEN UNDER A PRESSURE OF 760 mm Hg

Temperature (°C)	Chloride Concentration in Water (mg/L)			Difference per 100 mg Chloride	Vapor pressure (mm)
	0	5,000	10,000		
	Dissolved Oxygen (mg/L)				
0	14.6	13.8	13.0	0.017	5
1	14.2	13.4	12.6	0.016	5
2	13.8	13.1	12.3	0.015	5
3	13.5	12.7	12.0	0.015	6
4	13.1	12.4	11.7	0.014	6
5	12.8	12.1	11.4	0.014	7
6	12.5	11.8	11.1	0.014	7
7	12.2	11.5	10.9	0.013	8
8	11.9	11.2	10.6	0.013	8
9	11.6	11.0	10.4	0.012	9
10	11.3	10.7	10.1	0.012	9
11	11.1	10.5	9.9	0.011	10
12	10.8	10.3	9.7	0.011	11
13	10.6	10.1	9.5	0.011	11
14	10.4	9.9	9.3	0.010	12
15	10.2	9.7	9.1	0.010	13
16	10.0	9.5	9.0	0.010	14
17	9.7	9.3	8.8	0.010	15
18	9.5	9.1	8.6	0.009	16
19	9.4	8.9	8.5	0.009	17
20	9.2	8.7	8.3	0.009	18
21	9.0	8.6	8.1	0.009	19
22	8.8	8.4	8.0	0.008	20
23	8.7	8.3	7.9	0.008	21
24	8.5	8.1	7.7	0.008	22
25	8.4	8.0	7.6	0.008	24
26	8.2	7.9	7.4	0.008	25
27	8.1	7.7	7.3	0.008	27
28	7.9	7.5	7.1	0.008	28
29	7.8	7.4	7.0	0.008	30
30	7.6	7.3	6.9	0.008	32

APPENDIX F

TABLE F-1 QUANTITY OF AIR POLLUTANT EMISSIONS
(in million metric tons—except lead, measured in thousand metric tons)

Year	PM-10	Sulfur dioxide	Nitrogen oxides	Volatile organic compounds	Carbon monoxide	Lead	Carbon dioxide as carbon
1940	14.04	18.10	6.86	15.53	82.43	--------	--------
1950	14.71	20.31	9.44	18.92	89.62	--------	696.1
1960	12.60	20.18	13.23	22.06	94.15	--------	799.5
1970	11.00	28.42	18.92	26.98	107.68	199.10	1165.5
1975	6.60	25.51	20.23	22.81	92.63	143.83	1179.0
1980	6.40	23.78	21.47	25.72	117.03	68.00	1261.8
1981	6.00	22.51	21.32	24.04	111.58	53.42	1213.3
1982	4.90	21.21	20.57	22.56	105.37	52.31	1152.4
1983	5.50	20.62	19.97	23.05	105.20	44.66	1157.9
1984	5.80	21.47	20.53	23.71	102.49	38.30	1194.2
1985	5.62	21.22	20.34	22.69	97.88	18.26	1217.6
1986	5.29	20.39	20.21	23.00	95.16	6.62	1237.4
1987	5.50	20.52	20.69	22.43	90.09	6.21	1283.4
1988	5.86	20.95	21.44	22.70	89.87	5.86	1357.1
1989	5.65	21.04	21.30	21.69	84.73	5.53	1369.9
1990	5.53	20.17	21.03	21.48	94.13	5.11	1346.7
1991	5.29	20.10	20.84	21.23	90.65	4.55	1346.0
1992	5.40	19.59	20.85	20.62	87.39	4.30	--------
1993	-----	19.86	21.22	------	88.12	4.43	--------

TABLE F-2 AMBIENT AIR POLLUTANT CONCENTRATIONS

Year	Sulfur dioxide, ppm	Carbon monoxide, ppm	Ozone, ppm	Nitrogen dioxide, ppm	PM-10 particulates $\mu g/m^3$	Lead, mg/m^3
1984	0.0098	7.69	0.1252	0.0220	-----	0.423
1985	0.0093	6.97	0.1237	0.0220	-----	0.291
1986	0.0091	7.11	0.1197	0.0220	-----	0.180
1987	0.0089	6.69	0.1261	0.0220	-----	0.156
1988	0.0091	6.38	0.1364	0.0223	33.2	0.103
1989	0.0088	6.34	0.1169	0.0219	33.0	0.080
1990	0.0081	5.87	0.1143	0.0207	29.8	0.079
1991	0.0079	5.55	0.1155	0.0206	29.8	0.058
1992	0.0074	5.18	0.1071	0.0199	27.2	0.050
1993	0.0073	4.88	0.1097	0.0194	26.4	0.045

1. Sulfur dioxide, nitrogen dioxide, and PM-10 data are arithmetic means.
2. Carbon monoxide and ozone data are second maximum readings over 8 hr and 24 hr periods.
3. Lead data are quarterly readings.

APPENDIX G

TABLE G-1 POPULATION, GDP AND ENERGY CONSUMPTION IN THE US

Year	Energy Consumption, Btu × 10^{15}					Consumption/ production ratio	Population in millions	Gross Domestic Product (GDP) ($ billion)
	Total consumption	Percent of consumption						
		Coal	Petroleum	Natural gas	Other			
1960	43.8	22.5	45.5	28.3	3.8	1.06	180.67	513.3
1965	52.7	22.0	44.1	29.9	4.0	1.07	194.30	702.7
1970	66.4	18.5	44.4	32.8	4.3	1.07	205.05	1010.7
1975	70.5	17.9	46.4	28.3	7.4	1.18	215.97	1585.9
1980	76.0	20.3	45.0	26.9	7.8	1.17	227.73	2708.0
1985	74.0	23.6	41.8	24.1	10.5	1.14	238.47	4038.7
1990	81.3	23.5	41.3	23.8	11.5	1.20	249.90	5546.1
1991	81.1	23.1	40.5	24.2	12.2	1.20	252.67	5724.8
1992	82.4	22.9	40.7	24.4	11.9	1.23	255.46	6020.2
1993	84.0	23.4	40.2	24.8	11.6	1.28	258.25	6343.3

APPENDIX H

TABLE H-1 QUANTITY OF MUNICIPAL SOLID WASTE (MSW) GENERATED

Item and Material	1960	1965	1970	1975	1980	1985	1986	1987	1988	1989	1990
Waste generated (million tons)	87.8	103.4	121.9	128.0	151.5	164.4	170.7	178.1	184.2	191.4	195.7
Per person per day (pounds)	2.66	3.00	3.27	3.26	3.65	3.77	3.88	4.01	4.12	4.20	4.30
					Percent distribution of generation						
Paper and paperboard	34.1	36.8	36.3	33.6	36.1	37.4	38.4	39.1	38.9	37.6	37.5
Glass	7.6	8.4	10.4	10.5	9.9	8.0	7.6	6.9	6.8	6.7	6.7
Metals	12.0	10.7	11.6	11.2	9.6	8.6	8.5	8.3	8.3	8.2	8.3
Plastics	0.5	1.4	2.5	3.5	5.2	7.1	7.2	7.5	7.8	8.0	8.3
Rubber and leather	2.3	2.5	2.6	3.0	2.8	2.3	2.5	2.5	2.5	2.4	2.4
Textiles	1.9	1.8	1.6	1.7	1.7	1.7	1.6	2.1	2.1	2.9	2.9
Wood	3.4	3.4	3.3	3.4	4.4	5.0	5.3	5.5	6.1	6.1	6.3
Food wastes	13.9	12.3	10.5	10.5	8.7	8.0	7.7	7.4	7.2	6.9	6.7
Yard wastes	22.8	20.9	19.0	19.7	18.2	18.2	17.7	17.4	17.2	18.1	17.9
Other wastes	1.6	1.8	2.2	2.9	3.4	3.6	3.4	3.3	3.1	3.1	3.1

ANSWERS TO CHAPTER PROBLEMS

CHAPTER 1

1. a) 379 kg/day   b) $625/day
2. 57.33 kg
3. 0.208 mg/hr
4. 1) 68.22 kg/day   2) 113.7 kg/day
5. 28.57 mg/L
6. $1540/day for 1.5, $625/day for 1.1, % increase 146%
7.
Year	
1980	$117.1 \text{ kg} \times 10^3$
1985	97.4
1990	83.7

8.
Period	
1980-81	−3.12 % growth of CO
1984-85	−4.28
1989-90	−0.97

9.
	$\left\{\dfrac{e}{o}\right\}$	$\{c\}$
1985-2000	-3.25	2.72
2000-2025	-3.04	0.88
2025-2050	-1.76	0.41
2050-2075	-1.54	0.65
2075-2100	-1.43	0.28

10. 176,666 persons
11. 186,093 persons
12. 167,477 persons
13. 45.05 thousands
14. 1975) 29,941 persons   1998) 39,458 persons
15. 10.65 years
16. a) 245 years   b) 755 years
17. 46.95×10^{18} Joules
18. 1) 212 years   2) 27.72 years
19. 366.8 ppm
20. 3.3 years

429

CHAPTER 2

1.
Element	AMU	Gram
Carbon	12	2×10^{-23}
Chlorine	35	5.8×10^{-23}
Gold	197	3.3×10^{-22}
Mercury	200	3.3×10^{-22}
Uranium	238	4.0×10^{-22}

2. 1) 1000 kg/m³ 2) 4000 N
3. 59.59 lbs
4. 1) 9.81 N 2) 9.83 N
5. 200×10^{18} N
6. 0
7. 20 N
8. 19.62 N
9. 1) 1 m/s² 2) 19.62 N 3) -19.62 N
10. The force will be 9 times smaller.
11. 5.33 lbs
12. 3.09×10^3 gal/min
13. 0.13 m/s
14. 10.45 ft/s, 31.82 ft/s
15. 2750 ft³/s, 2.82 ft/s
16. 6.36 mg/L
17. 237.5 mg/L
18. 220 cfs
19. 7600 m³/hr
20. a) -31.07 acre-ft b) 24.23 acre-ft
21. -0.5 ft
22. 166.67 g/min
23. 1.95×10^3 kg/day
24. 0.73 mg/L
25. 0.56 mg/L
26. 451×10^6 ft³
27. 0.023 mg/L

CHAPTER 3

1. 19.86 m/s, 3.27 m/s
2. 25.38 ft/s, 30 ft-lbs
3. 176.58 m, same
4. 12.48×10^6 ft-lbs, 16.91×10^6 Joules, 4.7 kWh

CHAPTER 3 continued

5. 192,472 Joules, 9623.6 Watts
6. 7848 Nm, 3139 Nm, 10.85 m/s
7. 0
8. 315,000 Nm
9. 13.35×10^{33} cal
10. 6%
11. 22.9×10^{22} Joules
12. 15.3 days
13. 4756 years
14. 377×10^{24} W
15. 123 °C
16. 188×10^{15} Joules

CHAPTER 4

1. 25.5%
2. 0.308 moles
3. $CH_4 + 2O_2 \longrightarrow CO_2 + 2H_2O$
4. 275 g
5. 225 g
6. 400 g
7. $2C_4H_{10} + 13O_2 \longrightarrow 8CO_2 + 10H_2O$
   $2NaCl + H_2SO_4 \longrightarrow Na_2SO_4 + 2HCl$
   $Na_2CO_3 + Ca(OH)_2 \longrightarrow 2NaOH + CaCO_3$
   $Ca_3(PO_4)_2 + 3SiO_2 + 5C \longrightarrow 3CaSiO_3 + 5CO + 2P$
8. 533.3 kg
9. a) $4C_7H_5N_3O_6 + 21O_2 \longrightarrow 28CO_2 + 6N_2 + 10H_2O$
   b) 10.5 moles of Oxygen for 2 moles of TNT
   c) 54.5 g
10. a) $Al_2(SO_4)_3 \cdot 14H_2O + 3Ca(HCO_3)_2 \longrightarrow 2Al(OH)_3 + 3CaSO_4 + 14H_2O + 6CO_2$
    b) 30 moles
    c) 0.52 kg
11. 1) 625×10^3 mg/L   2) 625×10^3 mg/L   3) 0.59 mol/L
12. 1) 160 ppm   2) 160 mg/L
13. a) 0.625   b) 1.563   c) 0.51
14. 320 mg/L
15. 0.528 g
16. 0.168×10^{-6}
17. 2.46×10^{-5}
18. 33.625×10^{-5}
19. 3.36%

CHAPTER 4 continued

20. 0.283×10^{-10}
21. 1.51×10^{-2} mol/L
22. 0.17×10^{-4} mol/L
23. 2.85
24. 0.0032
25. 10.51
26. 4.44
27. 4.48 L
28. 4004 L
29. 300 L
30. 20.71 mL
31. 799.4 ppm
32. 930.9×10^3 µg/m^3
33. 0.166 ppm
34. 1) 5000 ppm   2) 29.92 mg/L
35. 1.56 atm
36. 2.78 L
37. 23.3 L
38. 46.52 L
39. 2-methylpentane
40. 2,3-dimethylpentane
41. 1.72×10^9 kilocalorie
42. 128.6 Btu

CHAPTER 5

1. 395.80×10^6 kcal
2. 1489 kg
3. 1) 3.74 kcal/g   2) 370.8×10^6 kcal
4. 80×10^{12} kg
5. 639.5 kg/acre
6. 3302 kcal, 495 kcal
7. 0.53 kg
8. 5.33 lbs
9. 1) $4C_{11}H_{14}O_4N + 47O_2 \longrightarrow 44CO_2 + 22H_2O + 4NH_3$
   2) 3.36 lbs
10.

Level	Efficiency
1	20%
2	11%
3	20%
4	29%

Answers to Chapter Problems

CHAPTER 5 continued

12.   1) $P_{i-1} = I_i$ (general relation)
       2) $R_i = 0.85\, I_i$
       3) $I_i = 6.67\, P_i$

13.

i	P	I_i	R_i cal/m²/day
4	3	14	11
3	14	70	56
2	70	700	630
1	700	375×10³	374,300

(where $I_i = P_{i-1}$ and $R_i = I_i - P_i$)

15. 2718.6×10^3 J/mole
16.   8075 kJ, 1615 kJ
18.   1) 0.203/hr   2) 0.283/hr   3) 100 mg/L
19.   1830 mg/L, 2067 mg/L
20.   7918 mg/L
21.   1) 7000   2) 330
22.   16.2 mg/L
23.   34 mg/L
24.   45.12 mg/L, 219.6 mg/L
25.   0.227/day
26.   1.58 mL

CHAPTER 6

1.   1) 2.5 ppm/yr   2) 3.27 ppm/yr
2.   52.9%
3.   1) 23×10^{-6} ppm/yr   2) 3197 ppm
4.   1) 1.5 days   2) 3.74 MGD
5.   2.05 ft³/yr
7.   3.7 in.
8.   1) 96.8% 2) 3.2% 3) 64.8%   4) 35.2 5) 4.8%
10.   0.117 mg/L/yr
11.   1) 0.585 °F   2) 0.4%
12.   1040 lbs/yr buildup

CHAPTER 7

1.   152.4 mg/cm²-s
2. 3.9×10^{-6} mg/s
3. 2.36×10^{-6} mg/s
4.   2.52 mg/cm²-s

CHAPTER 7 continued

5. 1.47×10^{-7} mg/cm²-s
6. 2×10^{-5} mg/cm²-s
7. 0.5 mg/cm²-s
8. 9.95×10^{-6} mg/cm²-s
9. 0.086 cm/s
10. grit at 3.46 cm/s, organics at 276.6 cm/s
11. 0.84 cm/s
12. 0.044 cm/s
13. 60.56 cm/s
14. a) 2400 gal/ft²-day   b) 6.33 cm/s
15. 0.0059 cm
16. 46.53 million gallons/day
17. 2.56 ft
18. 0.22 ft
19. 6110 m³
20. 27 gpm/ft²
21. 12.84 m/min
22. 1.08 million gallons
23. 65 ft (L) x 35 ft (W) x 30 ft (H)
24. k(to base 10) = 0.0049, k(to base e) = 0.0113
25. 2 days, 2 million gallons
27. First case more effective
28. 421.7 mg/L
29. 0.5 per day
30. 170,000 gallons

CHAPTER 8

1. 12.6 mg/L
2. min) 2.1 mg/L max) 31.5 mg/L
3. 40.28 mg/L
4. a) 44.05 µg/L

    b)
Outlet	Distribution(µg/L)
1	6.29
2	9.44
3	12.59
4	15.73

5. 1.15 µg/L
6. 0.111 mg/L
7. 0.658 mg/kg/day

CHAPTER 8 continued

8. 5.7 mg/kg/day
9. 24.6 ppm, 3 g/year
10. 2×10^{-5} ppm
11. 0.07/yr
12. 1.04×10^{-3} ppm
13. 0.12 ppm
14. 74 mg/L
15. 4800 mg/L
16. 28.2 mL
17. 120 mg/L
18. 1375 mg/L, 780 mg/L
19. 1141 mg/L, 447 mg/L

CHAPTER 9

1. 6.19 mg/L
2. 11.44 mg/L
3.

t, days	DO, mg/L
1	6.11
1.5	5.99
2	6.02
3	6.29
5	7.06
8	7.87

4. 3.95 mgL, 3 days
5. 388,800 ft
6. No clear limiting factor
7. 0.096 ft/day, 1.92 ft^3/day/ft width
8. 133.3 ft^3/day
9. Water: 1000 ft^3/day, oil: 400 ft^3/day
10. 691.2 ft^3/day
11. 16 m
12. 1 m
13. 65 ft x 32 ft x 11 ft
14. 2 of 36 ft diam., 11 ft height
15. 3.58 hr, 750 gal/day/ft^2
16. 102 ft x 51 ft x 15 ft
17. 0.09/day
18. 130 ft diam., 6 ft depth
19. 172 ft diam., 5 ft height

CHAPTER 9 continued
20. 764 gal/day/ft^2, 3.52 hr, 35.7 lbs/day/ft^2
21. 47 ft diam., 7 ft height
22. 88.6%
23. 1) 35%   2) 77%   3) 85%
24. 92.6 %
25. 146,768 L, 12%

CHAPTER 10

1. 4.5 psi
2. 3.87 psi
3. 28,400 ft
5. 800.2 ppm, 1.1 x 10^6 μg/m^3
6. 0.207 ppm
7. 0.135 mL
8. 48.13 x 10^6 μg/m^3, 25,397 ppm
9. 0.11 cm/s, 0.12 cm/s
10. 0.0015 cm/s
11. 1) 0.08 cm/s   2) 7.3 cm/s   3) 205 cm/s
12. 1) Stable   2) 533.3 ft
13. 385.7 m
14. 576 ft
15. 0.93 x 10^9 kg
16. 5.9 x 10^9 kg
17. 83.56%
18. 56.9%
19. 92.9%

CHAPTER 11

1. 1) 40,000 lb/day   2) 2490 yd^3 3) 224,000 lb
2. 754 lb/yd^3
5. 1) 3600 lb   2) 1698 lb   3) 4.5 m^3
6. 1) 6 lb/employee/day   2) 693 lb/yd^3
7. 104.2 lbs/yd^3
8. 228 houses
9. 18 yd^3
10. 1) 2 vehicles   2) 134 houses/trip
11. 12 hr
12. 102

CHAPTER 11 continued
13. 2 trips
14. 1) 3060 persons   2) 31.8 yd^3/day
15. 1) 37.4 ft x 37.4 ft   2) 403,200 yd^3, 8.3 acres
16. 1) 5.3 m x 5.3 m   2) 65620 m^3, 6562 m^2
17. 1) 386,905 persons   2) 44.8 ft

GLOSSARY

Abiotic. Nonliving components of the environment or ecosystem.

Absolute Temperature. Measurement of temperature in absolute term on a thermodynamic scale. There are two units related as follows:

Kelvin, $K = C + 273.15$
Rankine, $R = F + 459.76$

Absorption. A process in which a substance or energy is dissolved into another substance or body; for example water is absorbed by clay or infrared radiation by carbon dioxide.

Abyssal Zone. In the open water of the ocean, the cold, dark, water depths near the ocean bottom. Other zones are: The Bathyal Zone is located above the abyssal zone. The Euphotic Zone is the top layer penetrated by sunlight in which primary production by photosynthesis takes place. The shallow region that borders the land is Neritic Zone where a vast population of floating and anchored plants takes place.

Acid. A substance which dissolves in water forming hydrogen ions. Such solutions have a pH of less than 7, have a sour taste, neutralize bases, and conduct electricity.

Acid Rain. Rainfall with a greater acidity than normal rain. Normal rain has a pH of about 5.6. Acid rain results from air contaminants of sulfur dioxide and nitrogen oxides that dissolve in water to form acids.

Activated Carbon. A material obtained by burning a carbon material and then increasing its absorptive capacity by steam treatment which produces an internal porous structure. It is used to remove dissolved organic matter.

Activated Sludge. The suspended solids, mostly microorganisms, present in the reactor (aeration tank) of the sewage treatment plant. This speeds breakdown of organic matter in sewage.

Activated Sludge Process. The process of using the activated sludge to breakdown the organic matter in raw sewage, known as secondary waste treatment. The activated sludge is subsequently separated from the treated wastewater by a sedimentation tank.

Adaptation. Ability of a living organism to cope with its environment. Adaptation causes change in structure or habit of the organism to produce better adjustment with environment.

Adiabatic Lapse Rate. The decline in temperature of an air parcel as it rises in the atmosphere due to pressure drop and gas expansion only without any heat exchange with the surrounding air through convection or mixing.

Adsorption. A phenomenon whereby molecules of liquids, condensed gases, or solids are attracted to and attached to the surface of another substance, usually a solid. There are two basic processes in waste treatment: activated carbon adsorption, and resin adsorption.

Advanced Waste Treatment. A treatment to wastewater employed after the biological treatment to remove toxic and heavy metals or to produce a high quality effluent. The term Tertiary Treatment is commonly used.

Advection. Transport of a pollutant due to velocity of (along with) a moving fluid; similar to Convection.

Aeration. Addition of air (oxygen) during wastewater treatment so that the dissolved oxygen content of water does not fall below a level insufficient for rapid degradation.

Aeration Period. The theoretical time in hours that the liquid in an aeration tank is subjected to aeration; it is the volume of the tank divided by the rate of flow of wastewater.

Aerobic Digestion (Treatment). A process in which microorganisms decompose complex organic compounds in the presence of oxygen and use the energy generated for reproduction and growth. Two basic aerobic biological treatment are: Activated Sludge and Trickling Filter.

Aerosols. Fine liquid and/or solid particles suspended in the air, e.g., dust, fog, and smoke.

Air. The mixture of gases that fill the lower portion of the atmosphere. The dry air consists of (by volume) 78% nitrogen, 21% oxygen, 1% argon, and trace amount of carbon dioxide, helium, hydrogen, methane, etc. On the average, water vapor to 0.7% is also present.

Air Parcel. A theoretical volume of air considered in analysis of air pollution dispersion such as smog formation.

Air Pollutants. Gases, liquids, and/or solids that in certain amount may be hazardous to humans, animals or plants. These include oxides of carbon, sulfur, nitrogen, inorganic and organic acids, particulates, and gases.

Air Stripping. Transferring (removing) dissolved contaminants from groundwater to air or vapor (steam stripping) by providing as much contact as possible between the liquid (groundwater) and the air or vapor.

Algal Bloom. Rapid growth of algae caused by addition of nutrients like nitrogen and phosphorus; algae are a group of one-celled free floating green plants.

Aliphatic Hydrocarbon. Organic compound with a straight chain structure. One of the two major groups of hydrocarbons; the other being Aromatic. Classes of aliphatics are: Alkanes—carbon atoms connected by single bonds, Alkenes—carbon atoms connected by double bonds, and Alkynes—carbon atoms connected by triple bonds.

Alkaline. Water Solution having a pH greater than 7; neutralizes acid to form a salt and water.

Alpha Particle. A positively charged particle emitted by certain radioactive materials, e.g., decay of radon-222. It is the least penetrating of alpha, beta, and gamma types.

Ambient Air. Outdoor air subjected to meteorological and climatic changes (not indoor or workplace air).

Ammonification. A process carried out by bacteria in which organic compounds are degraded with the release of ammonia into the atmosphere. For example, conversion of nitrates and nitrites to ammonium compounds by soil bacteria.

Anaerobic Digestion. An airtight tank in which anaerobic (without oxygen) microorganisms decompose organic material and produce methane gas. Commonly used to reduce the volume of sludge in Biological Wastewater Treatment.

Anion. A negatively charged ion of an electrolyte which migrates towards the anode.

Anode. The positive electrode that attracts anions. In corrosion, process anode has a tendency to go into solution.

Aquifer. An underground formation capable of storing and yielding significant groundwater supply; usually it consists of sand, gravel, and permeable rock. Two types: Artesian or Confined aquifer that carries water under pressure due to overlying rock formations and Watertable, or Unconfined aquifer, where water rises to a watertable under atmospheric pressure.

Arithmetic Growth. An increase by a constant amount per time period; for example, a population growth of 1,000 persons per year in an area.

Aromatic Hydrocarbons. Organic compounds with closed ring structure. Aromatics are usually more difficult to decompose than straight chain hydrocarbons, and pose greater danger to humans and environment.

Atmosphere. The envelope of gases surrounding the earth. It extends up to 600 mi over the entire earth surface; beyond it is an infinite expanse of space. The weight of atmosphere is 5.7×10^{15} tons, about one-millionth of the weight of the earth.

Atom. The smallest unit of an element that contains the characteristics of that element. It consists of a nucleus having protons and neutrons, and surrounding shell of electrons.

Atomic Mass Unit (AMU). One AMU is defined as one twelfth the mass of a carbon atom consisting of six protons and six neutrons. A unit is equal to 1.6604×10^{-24} g.

Atomic Number. The number of protons in the nucleus of an atom. Each element has an unique atomic number.

Atomic Weight. Approximately equal to the sum of number of protons and neutrons in an atom.

Autotroph. Organisms such as green plants that derive their food from inorganic substances.

Avogadro's Hypothesis. Equal volumes of different gases at the same pressure and temperature contain equal numbers of molecules.

Avogadro's Number. Number of molecules in one mole of a substance is always equal to 6.02×10^{23}.

Backwashing. Cleaning of rapid sand filters by upward flow of water in water and wastewater treatment.

Bacteria. Single-celled microscopic living organisms. Together with fungi, they serve as decomposers in aquatic and terrestrial ecosystems. Pathogenic bacteria cause diseases.

Baghouse. An air pollution control device similar to a home vacuum cleaner bag of much larger size. It removes particulate matter when dirty air is passed through it. Types are: Pulse jet, Reverse air, Shaker.

Base Flow. The part of a streamflow contributed by groundwater.

Basin. A river basin is the land area from which water drains into that river; *see* Catchment area, Watershed.

Batch Process. A process in which a fixed quantity of materials (e.g., wastewater) is treated at a time, in contrast to continuous process.

Bathyal Zone. *See* Abyssal Zone.

Benthic. Relating to aquatic organisms at the bottom of a water body.

Benthos. The organisms living on the bottom of an ocean, river, lake, or water body.
Others are: Plankton- the microscopic plants and animals that live suspended in water, and Nekton- the animals in aquatic system that are free-swimming, independent of current or waves.

Bernoulli's Equation. The sum of pressure (energy) head, static (datum) head, and kinetic head at any one point in a closed system is equal to the sum of similar heads at any other points. (Based on the principle of Energy Conservation).

Best Available Control Technology (BACT). Limitations based on the maximum degree of reduction achievable through application of available methods, systems, and techniques.

Beta Particle. Electrons emitted from decay of some radioactive elements. Their ionization power is less than Alpha particles but their penetration power is greater.

Bioaccumulation. The progressive increase in the amount of a chemical in an organism resulting from the uptake or absorption of the substance at a rate more than its breakdown or excretion rate.

Bioassay. To assess the adverse effects of any substance or chemical by using living organisms such as bacteria, plants, or animals. Organisms are exposed to samples in air or water environment and changes are observed.

Bioavailability. The rate and magnitude by which any chemical substance or toxicant becomes available to organisms. A non-bioavailable material does not have any effects among organisms.

Biochemical Oxygen Demand (BOD). The amount of oxygen required by aerobic bacteria to decompose the organic matter in wastewater within a specified time at a given temperature. This oxygen is derived from the dissolved oxygen in water. The greater the amount of (organic) waste material, larger will be the requirement for dissolved oxygen. Thus, the BOD is an indicator of the strength of the organic waste. It is customary to measure the oxygen demand for 5 days and it is expressed as BOD_5 in mg/L.

Bioconcentration. Increase in concentration of a chemical in an organism due to uptake exceeding excretion rate; *see* Bioaccumulation. Bioconcentration Factor measures the level of accumulation as compared to the level in the medium (such as water) in which the organism resides.

Biodegradation. The decomposition or breakdown of organic materials in the environment by living organisms. Biodegradation follows the exponential decay.

Biogeochemical Cycles. All essential materials (chemical substances) are cycled on the planet through atmosphere, hydrosphere, lithosphere, and living organisms wherein the reservoirs of these materials exist to and from which the mass flow (flux) takes place. As the materials move in the cycle, they change chemical forms.

Biological Wastewater Treatment. A common method of treatment of wastewater by using bacteria to degrade organic materials. This comprises of a reactor unit and the final clarifier (sedimentation tank). The specific type of reactor unit includes aeration tank, tricking filter, biological contactor, biodisc.

Biomagnification. The higher a trophic level in a food chain, the more biomass is consumed to balance energy needs. The concentration of heavy metals (like mercury) or organic contaminants (like chlorinated hydrocarbon) which do not readily decompose, increases successively up the food chain. Compared to bioaccumulation which is the increase in concentration of a chemical by direct uptake from the environment, biomagnification is increase through the food chain.

Biomass. Dry mass of living organisms in a specified area.

Biomes. Major ecosystem types on the earth (terrestrial systems) defined by climate and dominant vegetation.

Biosphere. The portion of the lithosphere, hydrosphere, and atmosphere that support life; also called Ecosphere.

Biota. All living things found in an area.

Biotransformation. The metabolic conversion of an absorbed chemical by an organism, usually to a less toxic form.

Blackbody. A theoretical body with a surface that reflects no light.

Black Box. A part of a system that performs a known function or role, but for which the details of how it operates are either unknown or omitted.

Blank. A quality control sample that is used as a base compared to which a contaminant is analyzed in the laboratory.

Bloom. In an aquatic system, the rapid growth of algae.

Boom. A device used on a water surface to contain or deflect an oil spill, consists of an above-water freeboard and an underwater skirt attached to a float.

Box Model. Representation of stocks or storage by boxes and arrows from box to box for the flow directions and amounts. The concentration is assumed to be uniformly distributed in the box.

Boyle's Law. The volume of a perfect gas is inversely proportional to the pressure exerted on the gas, at a given temperature.

Breakpoint Chlorination. The level when a sufficient amount of chlorine is added to water or wastewater to react with all oxidizable compounds, such as ammonia. Beyond this the level of residual chlorine increases to eliminate microorganisms that may enter after the treatment.

Calorie. The amount of heat required to raise the temperature of one gram of water by one degree Celsius. When used for food energy, it is expressed with a capital "C" and equals 1,000 calories (or kilocalories).

Carnivore. A living organism using animal substances as a source of food.

Catchment Basin. A region which collects all the rainwater that falls on it, directing it into a river or stream. The boundary of a catchment basin is defined by the ridge beyond which water flows away from the basin.

Cathode. The negative electrode that attracts cations; opposite of anode.

Cathodic Protection. A technique to protect steel tanks from corrosion by connecting a metal like zinc or magnesium to tank to serve as an anode.

Cation. A positively charged ion.

Charles Law. Volume of a gas is directly proportional to the absolute temperature at a constant pressure.

Chemical Energy. The potential energy contained in the bonds of chemical compounds. For example, energy in the form of carbohydrates in food.

Chemical Oxygen Demand (COD). A strong oxidizing agent (typically, potassium dichromate) is used to oxidize all organic matter in a sample of water without distinguishing between biodegradable and nonbiodegradable organic matter.

Chemical Precipitation. By adding chemicals to a solution forming insoluble material of the contaminant in solution.

Chlorine Demand. The amount of chlorine required to react with all dissolved and particulate materials and inorganic ammonia in water.

Chlorine Residual. Chlorine in excess of demand which is available to eliminate microorganisms that enter the water after the treatment.

Chlorofluorocarbons (CFC). A class of hydrocarbon derivatives in which chlorine and fluorine are substituted for some or all of the hydrogens, e.g., CCl_2F_2. They are commonly called Freons. They are widely used in consumer products. They are implicated in ozone depletion and the greenhouse effect.

Chromatography. A process used to separate similar compounds by allowing a solution of different compounds to move through a medium that adsorbs the compounds in such a way that they are separated in zones.

Chronic Exposure. In toxicology, the doses that extend for a long period, from six months to a lifetime.

Clarifier. A tank in which wastewater is held to allow the settling of particulate matter.

Clay Liner. A layer of clay (of permeability of less than 1×10^{-7} cm/s) placed at the bottom and sides of a containment system to prevent migration of liquids from the disposal site.

Clean Air Act (CAA). The basic federal air pollution control statute. It was first passed in 1963 and has been amended periodically. It provides for the national ambient air quality standards, the state implementation plan process, the prevention of significant deterioration program, and national emission standards for hazardous pollutants. The 1990 amendments included provisions pollutants of global concerns and stricter auto emission standards.

Clean Water Act (CWA). The basic federal water pollution control statute. The Water Quality Act of 1965 set the water quality standards. Further amendments made in 1966, 1972, 1977, and 1987 expanded the scope and set the effluent standards through the National Pollution Discharge Elimination System (NPDES).

Closed System. Scientific (physics and chemistry) studies are made on a closed system that does not exchange matter or energy with its surroundings. In ecology, a closed system exchange energy but not matter with the surroundings.

Coagulation. Binding together of individual particles in air or water to form floc and thus increase their rate of settlement.

Coastal Water. The water in the coastal zone.

Coliform Bacteria. Gram-negative, rod-shaped bacteria primarily found in the digestive tract of humans and animals but also found in soils and water. Its presence in water indicates contact with feces and potentially dangerous bacterial contamination.

Colloidal Matter. Microscopic suspended particles that do not settle in standing liquid and require special treatment such as coagulation or dialysis.

Colorimetry. Methods of chemical analysis in which concentration of impurities in water is determined by comparing the sample color with a set of known color standards.

Combustion. A process of burning resulting from the rapid oxidation of substances. Three main components of combustion are: fuel (hydrocarbon), oxidizer (oxygen), and dilutent (nitrogen, water vapor). In addition, waste is the fourth component in incineration.

Complete Mix Batch Reactor. A closed reactor system in which mixing is provided to ensure no concentration gradient.

Complete Mix Flow-through Reactor. A system which allows the fluids to flow in and out but inside the reactor the fluid is fully mixed.

Composite Sample. *See* Grab Sample.

Composting. A biological process in which solid wastes containing organic matter are converted to usable stable material by the action of microorganisms already present in the waste. Three methods comprise: Mechanical, Ventilated Cell, and Windrow.

Compound. A substance made up of two or more elements combined chemically.

Comprehensive Environmental Response, Compensation, and Liability Act (CERCLA). The statute, also called the Superfund, establishes federal authority for emergency response and cleanup of hazardous substances that have been spilled, improperly disposed, or released into the environment. It was enacted in 1980 and significantly amended in 1984 and in 1986 through the Superfund Amendments and Reauthorization Act (SARA). The primary responsibility for cleanup rests with the generator or disposer of the hazardous substance with the back up federal response using a trust fund.

Concentration. The amount of a substance in a given amount of medium (air, water, soil, food). The substance as well as the medium can be expressed as mass or volume. Thus, it is mass to mass ratio, mass to volume ratio, or volume to volume ratio.

Conditioning. Any treatment that makes dewatering easier—adding minerals or chemicals (chemical conditioning) or mechanical flocculation.

Cone of Depression. The shape of the watertable after a drop in the level in a well due to pumping groundwater at a faster rate than the recharge of the well.

Confined Aquifer. *See* Aquifer.

Confirmed Test. The second stage in the examination of coliform bacteria. The first stage is the Presumptive Test and the third stage is the Completed Test.

Conservation. Avoiding waste of and using of natural resources according to the principles to assure optimum usage.

Consumers. Living organisms that do not produce their own food but eat other organisms to gain energy. In the food chain, the place for a consumer is defined with what it eats. Herbivores, the primary consumers, eat plants, the secondary consumers eat animals that feed on plants.

Contaminant. Any unwanted physical, chemical, biological, or radiological substance or matter that has an adverse effect on air, water, or soil.

Continental Shelf. The shallow shoreline area that borders the continents and separates from the deeper oceanic regions, usually water less than 200 m deep. At the outer edge of the shelf, the sea floor drops sharply.

Continuity Equation. Based on the principle of conservation of mass, the relation balances the volumetric flow rate into a system to the volumetric rate out of the system; the volumetric rate, Q, is given by the product of the fluid velocity, V, and the cross sectional area, A.

Convection. The transfer of a substance or heat by a moving fluid (due to velocity), such as air or water; similar to advection.

Conventional Pollutants. Under the Clean Water Act, the pollutants listed are: biochemical oxygen demanding, fecal coliform, suspended solids, oil and grease, and pH. The commonly understood pollutants are: organic wastes, sediments, acids, bacteria and viruses, nutrients, oil and grease, and heat.

Corrosivity. A waste is corrosive if pH is equal to or less than 2.0 or equal to or greater than 12.5. A characteristic used to classify a hazardous waste.

Cyclone Collector or Separator. A cylindrical or cone-shaped device used to remove particulate matter from dirty air by centrifugal force. The particle-containing air enters near the top of the cyclone and spins downward, throwing the particles outward to the cyclone wall. The particles fall into a hopper. The clean air exists from the top.

Dalton's Law (of Partial Pressure). A mixture of nonreacting gases or vapors exerts a total pressure which is equal to the sum of the pressures that would be exerted individually by each gas or vapor alone.

Darcy's Law. Stated in 1856, the groundwater flow velocity, V is equal to the product of the hydraulic conductivity, K, of the aquifer, and the slope of the watertable (the hydraulic gradient), H/L.

Death Phase. The terminal stage of growth (decline) of bacteria in laboratory culture after the supply of available nutrients exhaust. The number of viable cells decreases.

Decomposers. A community of organisms, including bacteria and fungi, that breakdown the organic matter including dead animals into simpler inorganic materials.

Deep-well Injection. Pumping in liquid waste into deep pervious rock formation or fractured rocks far beneath the aquifers used for drinking and irrigation water.

Deforestation. The removal of trees and other vegetation on a large scale to expand agriculture or grazing lands or to use the timber. Deforestation gives rise to soil erosion and contributes to carbon dioxide increase.

Degradation. Decrease in the quality of the environment. Also the chemical, physical, and biological breakdown of a complex material into simpler components.

Denitrification. The process of removal of nitrate ions from soil or water by anaerobic bacteria. Nitrate is reduced to nitrite, then to ammonia, and finally to nitrogen gas which is lost to the atmosphere.

Density. The mass of a substance per unit of its volume. In ecosystem, the number of individuals of a particular species per unit area of land.

Desalination or Desalinization. The removal of salts from sea or brackish water. Two main techniques are: Distillation and Osmosis.

Desertification. A process of conversion of a land that is covered by vegetation to a desert, often through human actions.

Design Capacity. The number of tons of solid waste that a burning (thermal processing) facility is designed to handle during 24 hours of continuous operation.

Design Flow. The rate of wastewater that a treatment facility is designed to process efficiently, commonly expressed in millions of gallons per day.

Desorption. Opposite of adsorption; the removal of a substance that has been absorbed or adsorbed by another material.

Desulfurization. The removal of sulfur from flue gases produced on combustion or from fossil fuels to reduce the sulfur content.

Detention Basin. A relatively small storage that is temporarily filled during heavy rains to slow down the runoff.

Detention Time (Period). The duration for which liquid is retained in the container unit. Time for which wastewater stays in a settling tank or stormwater remains in a detention basin. Detention time equals volume of container divided by the rate of liquid inflow.

Detritus Food Chain. The food chain that is based on the consumption of dead plant biomass by the primary consumers. Opposite to grazing food chain that is based on the consumption of live plant biomass by the primary consumers.

Dewatering. Removal of water from sludge. Main techniques are centrifugation, press, and air drying.

DDT (Dichloro-diphenyl-trichloromethane). The chlorinated hydrocarbon insecticide (highly toxic to insects). It can biomagnify and is persistent. It has a half-life of 15 years. It can collect in fatty tissues of certain animals. EPA banned DDT in 1972.

Diffusion. The spreading or scattering of a gas or liquid mass, or of heat and light from a more concentrated to a less concentrated region.

Digester. In wastewater treatment, a closed tank in which biological decomposition (digestion) of the organic matter of the sludge takes place, anaerobically or aerobically, resulting in partial gasification, liquefaction, and mineralization of organic matter.

Dilution Factor. The extent to which the concentration of a solution or suspension is lowered by adding a diluent, usually water. The ratio of the diluent plus solution divided by the solution.

Dioxin. *See* TCDD.

Discharge. The volumetric rate of flow of a fluid, commonly expressed as cubic ft per second.

Disinfection. Killing of pathogens or organisms capable of causing diseases. Chlorination is a common method.

Dispersion. The scattering process similar to diffusion. In air pollution, dispersion indicates the combined action of advection and diffusion.

Dissolved Oxygen (DO). The amount of molecular oxygen dissolved in water. There is a saturation level for DO depending on the temperature and solids content of water. A single most indicator of water suitability for aquatic organisms.

Dissolved Solids (Total). The total amount of dissolved matter, inorganic and organic, contained in water. Potable water supplies with more than 500 mg/L are not recommended by the U.S. Public Health Service.

Diversity. A measure of the number of different kind of species in an area. Diversity in quantitative terms expressed by the Diversity Index.

Doubling Time. Time required for a population to double in size.

Drag Force (Coefficient). The resisting force that a particle experiences as it moves through a fluid medium.

Drainage Basin. *See* Catchment Basin.

Drawdown. The lowering of the water level in a well and in the adjacent watertable due to pumping of the well.

Ecological Pyramid. It is the diagram that shows the transfer of useful energy from one trophic level to the next or it shows the biomass of living organisms at different trophic levels in an ecosystem. The figure resembles a pyramid.

Ecology. The science of interrelations between living organisms and their abiotic surroundings.

Ecosphere. *See* Biosphere.

Ecosystem. A functional ecological unit. A unit wherein living things (biotic) and their nonliving (abiotic) surroundings interact with each to sustain the unit. For example, a fish tank.

Eddy Diffusion. Mixing of a contaminant in a medium by eddies that are produced due to turbulence of flow. It is analogous to molecular diffusion but the molecular diffusion is the property of the fluid, while turbulent diffusion is the property of the flow.

Effluent. The wastewater, treated or untreated, that flows out of a treatment plant, sewer, or any other outlet and flows into a receiving water.

Electrode. A conducting terminal to pass current, one is anode, the other is cathode.

Electrolysis. The passage of electric current through an electrolyte, causing migration of positive ions to negative electrode and negative ions to positive electrode.

Electrolyte. Any compound that dissociates into ions when dissolved in water. The resulting solution conducts electricity.

Electromagnetic Wave (Rays). All radiation energy, including from the sun, travels as electromagnetic waves. These waves consist of two vibrating electric and magnetic fields traveling together.

Electron. Negatively charged subatomic particle within an atom.

Electrostatic Precipitator (ESP). A device to collect particulate matter from dirty air. An electric charge is applied to particles which are, then collected on the surface of opposite charged plates.

Element. The simplest unit of a substance. There are 103 elements on the earth like oxygen, carbon, nitrogen, iron.

Emission. The release of pollutants in atmosphere.

Encapsulation. To enclose a hazardous waste into another material such as concrete, in such a way as to isolate the waste from external effects of air, water, and soil.

Endangered Species. Species of plants and animals that have a small population which is in danger of disappearing due to man-made or natural changes. The species are protected by Endangered Species Act of 1973.

Endothermic (Endoergic) Reaction. A reaction or process that absorbs heat from the surroundings as the reaction proceeds.

Energy. The capacity of a body to do work. There are two laws of energy: Energy cannot be destroyed or created, however the form of energy can be changed, and when one form of energy converts to another, the entropy increases. There are two main forms of energy: Potential energy and kinetic energy.

Energy Head. The basic units for energy are ft-pounds or Newton-meter. In an energy balance equation, the weight unit is dropped and the energy terms are expressed in ft or meter, known as energy head.

Energy Flow. The flow of energy is an essential requirement for an ecosystem. The path comprises the input of solar energy, energy captured by photosynthesis, the utilization of energy by various trophic levels, with the loss of heat at every level.

Energy Pyramid. *See* Ecological Pyramid.

Enthalpy. The heat content of a substance. All chemical reactions either produce or require heat. The heat evolved or absorbed in a process at constant pressure is called the change of enthalpy for the process.

Entropy. A measure of disorder within a system; the higher the entropy, the more disordered the system. As energy decreases when one form of energy is converted to another, its entropy increases. Thus, entropy is an indicator of the energy unavailable to do work.

Environment. An aggregate of all conditions that influence the form and survival of individuals and communities.

Environmental Assessment. An analysis that determines whether any action would significantly affect the environment and thus require preparation of an Environmental Impact Statement (EIS).

Environmental Audit. An investigation of a company's compliance with environmental regulations.

Environmental Impact Statement (EIS). A document required under the National Environmental Policy Act (NEPA) for major projects or legislative proposals significantly affecting the environment. It describes the benefits and adverse effects of the undertaking and lists alternative actions.

Environmental Protection Agency (EPA). The U.S. Congress created the EPA on December 2, 1970 by bringing together parts of various government agencies involved with control of pollution. There are 11 main environmental laws passed under the jurisdiction of the EPA.

Epilimnion. The upper layer of a lake in which water is warmer. The bottom layer with colder water is Hypolimnion. Between these two layers is the Thermocline.

Equilibrium. A steady-state system in which flow in equals flow out. For box models, there is no gain or loss of matter or energy in each box. In a chemical process, the forward and reverse reactions proceed at the same rate defined by the equilibrium constant.

Equivalent Weight. For an element, the atomic weight divided by the valence. For a compound, the molecular weight divided by the number of hydrogen or hydroxyl ions in the compound.

Erosion. The wearing away of the exposed land surface by wind or water. Excess runoff and land-clearing practices contribute to this.

Euphotic Zone. *See* Abyssal Zone.

Eutrophic. A river, lake, or other body of water having excessive plant nutrients, and, hence, profilic growth of plants. Other types of water body are: Mesotrophic with a moderate nutrient contents, and Oligotrophic with low content of nutrients, hence, clear water.

Eutrophication The addition of plant nutrients- nitrates and phosphates into lakes or other body of water. This leads to profilic growth of aquatic plants shortening the life of the water body.

Evaporation. Process by which a substance changes from liquid to gas or vapor. The opposite is condensation. Used in the phase separation process. Evaporation of water from land and water body is an essential component of the hydrologic cycle.

Evapotranspiration. The total amount of water that evaporates from the soil and by transpiration from the plants growing in the soil in an area.

Exothermic (Exoergic) Reaction. The reaction during which heat is liberated to the surroundings.

Exponential Growth or Decay. The growth or decline in population at a rate proportional to the total population at any instant. Similar to the growth of money in a bank. This is also applicable to the growth or decay of a pollutant, chemical or biological reactions, and radioactivity.

Extinction. Elimination of all individuals of a particular species.

Fabric Filter. Filter bags made of Teflon, nylon, glass fiber, or cotton to remove particulate matter; *see* Baghouse.

Facility. In RCRA, any place—land, structures, other appurtenances, and improvements on land—that is used to treat, store, or dispose of a hazardous waste. In CERCLA, any building, structure, installation, equipment, pipeline, well, pit, pond, lagoon, impoundment, ditch, landfill, container, or any site or area wherein a hazardous waste is deposited, stored, disposed of, placed, or otherwise come to be located.

Facultative Bacteria. Bacteria that can exist and grow under both aerobic and anaerobic conditions.

Fall Turnover. After summer, in fall, the epilimnion (surface water) of a lake cools and becomes denser. It sinks, forcing the hypolimnion (bottom water) to the surface, thus mixing the lake.

Fecal Coliform Bacteria. One of the two types of bacteria found in the colon of warm-blooded mammals, such as human; the other is Fecal Streptococci. The presence of Fecal Coliform is taken to indicate that the material is contaminated with human waste.

Fick's Law. The rate of diffusion equals the concentration gradient times the cross sectional area times a diffusion coefficient.

Filter. A porous medium through which a liquid or gas is passed to remove suspended particles.

Final Clarifier (Sedimentation). A settling tank placed after the biological reactor as part of the secondary biological treatment. The tank removes suspended solids in treated wastewater.

First Order Kinetics. The rate reaction is directly proportional to the concentration of the reactant; *see* Exponential Growth.

Fission. Prying open an atom nucleus into two or more nuclei by bombardment with neutrons, alpha particles, gamma rays, deuterons, or protons. This releases a large amount of energy.

Fixation. The fixation in the nitrogen cycle is conversion of inert nitrogen from air to a form that becomes more available to plants. In waste treatment, fixing is to increase the stability of waste material by formation of stable solid derivatives.

Flocculation. To provide movement to water or wastewater in a tank resulting in coagulation of particles.

Flood. A streamflow that greatly exceeds the average flow. The low land adjoining the stream, known as flood plain, submerges as a result of flooding.

Flow Rate. Rate of flow of a fluid moving through a system expressed as volumetric units per unit time, *see* Discharge.

Flow, Laminar and Turbulent. Laminar flow is smooth with fluid particles moving in distinct paths without mixing. It occurs when the Reynolds Number is less than a specified value; less than 2100 in a pipe, and 500 in a channel. In turbulent flow, the fluid moves in a haphazard manner and eddies are present.

Flow, Steady and Unsteady. A flow is steady if its properties such as velocity, temperature, and pressure at any place do not change with time; conversely it is unsteady if its properties change with time.

Flow, Uniform or Nonuniform. The flow is uniform if the slope of the water surface is parallel to the bottom, i.e., the cross section of the flow section and the velocity are the same throughout the length. Conversely, it is nonuniform.

Flue Gas. A mixture of combustion gases emerging from a chimney.

Flue Gas Desulfurization (FGD). *See* Desulfurization.

Fluidized Bed Combustion (FBC). Burning of a combustible material on a bed of mixed inert particles and fuel, like coal. Air or oxygen is blown through the mixed bed to hold them is suspension to increase the burning rate. The bed acts in a fluid-like state.

Flux. The flow rate of mass per unit of cross sectional area.

Food Chain. A sequence of organisms in an ecosystem in order of which the energy is transferred; each link in the chain known as trophic level, feeds on and obtains energy from the preceding it and, in turn, is eaten by the one following it. Two major categories are: Grazing food chain, and Detritus food chain.

Food to Microorganism (F/M) Ratio. The ratio of mass of organic material fed per day into an aeration tank of a wastewater treatment facility to the mass of microorganisms in the aeration tank. It is a design parameter in the wastewater system.

Food Web. Most animals eat from more than one trophic level (not linearly as depicted by a food chain). This interrelationship among the organisms as to how they derive food from different trophic levels at the same time in an ecosystem is a food web. It is a complex feeding system within an environment.

Force. The action of one body on another that causes acceleration of the second body. It equals mass of the body times its acceleration, according to Newton's Second Law.

Fossil Fuel. Coal, oil, and natural gas which is derived from the remains of plants and animals buried underground in ancient time.

Front. A zone of transition in air masses. If a cold air mass is displacing a warm one, the transition is called a cold front. Fronts are associated with low pressure areas.

Fumigation. A rapid increase in air pollution at ground level caused by turbulence of the atmosphere created by the rising morning sun.

Fungi. A group of organisms that lack chlorophyll, are nonmobile, filamentous, and multicellular. Molds, mildews, yeasts, mushrooms, and puffballs. Stabilize sewage and breakdown solid waste in composting.

Fusion. Combining nuclei of small atoms to form a larger atom. This requires very high temperatures and pressures and results in tremendous release of energy. The underlying process of hydrogen bomb.

Gamma Radiation (Ray). Electromagnetic radiation produced by some radioactive substances. It is extremely short wavelength (0.001 to 0.1 micrometer) of high energy emitted from the nucleus. It is most penetrating and more energetic than x-rays. It can be stopped by dense material like lead. Other radiations are: Alpha and Beta.

Garbage. Waste material that is biodegradable, typically from domestic and commercial sources.

Gas Chromatography (GC). Used to quantitatively or qualitatively examine for organic pollutants, like pesticides. The sample is vaporized and carried along in an inert gas like nitrogen through a column. The various components of the sample move through the column at different speed. A device is used to detect the level of each component as it exits the column.

Gas Chromatography/Mass Spectrometry (GC/MS). A sensitive analytical technique in which the effluent from a Gas Chromatograph is piped to a Mass Spectrometer for additional analysis.

Gas Laws. *See* Boyle's Law, Charles' Law, and Ideal Gas Law.

Gaussian Plume Model. A basic model for the dispersion of a plume of pollutant from a continuous point source. The distribution of the concentration of the pollutant perpendicular to the plume axis is assumed normal (Gaussian) distribution. The plume dimensions are described by dispersion parameters.

Geometric Growth. *See* Exponential Growth.

Geosphere. The solid, nonliving portion of the earth; this excludes the atmosphere, hydrosphere, and biosphere.

Geothermal. Pertaining to the heat of the interior of the earth. In some places this heat has produced hot water or steam that can be harnessed for energy needs.

Global Warming. *See* Greenhouse Effect.

Grab Sample. A sample taken at a random location at a random time or a single air or water sample. The sample does not represent the long term conditions at a site in contrast to a composite sample which comprises a series of small samples taken over a period of time and combined as one sample.

Gram Molecular Weight (GMW). The mass in grams of a substance equal to its molecular weight. For example, GMW for carbon is 12 grams. The amount of material equal to its gram molecular weight is one gram-Mole of the substance. For carbon 1 gram-Mole equals 12 grams.

Gravimetric. Based on measurements by weight or mass of materials or samples.

Gravitational Constant (Force). The constant of proportionality in Newton's Law of gravitational attraction, according to which the gravitational force between any two bodies is proportional to the product of masses of two bodies and inversely proportional to the square of the distance between the two bodies.

Gravity Separation. The treatment that uses gravitational principle to separate substances. For example, settling tank.

Gray Water. All wastewater from a household other than from toilets.

Grazing Food Chain. *Refer to* Detritus Food Chain.

Greenhouse Effect. The heating effect of the atmosphere upon the earth. The sun rays pass through atmosphere and are absorbed by the earth. The earth radiates this energy as heat (infrared) waves. The carbon dioxide in the air, along with some other gases, absorbs most of these heat waves and reradiates them towards the earth and to space. The atmosphere, thus, acts like a greenhouse, hence, the name. Higher concentration of carbon dioxide will reradiate more heat waves, increasing the earth temperature. This will have many environmental consequences.

Grit Chamber. Sand and fine gravel are settled to the bottom of a chamber installed at the beginning of a sewage treatment plant.

Gross Domestic Product. The value of all goods and services produced within the nation.

Groundwater. The water below the land surface in a zone of saturation where all pores are filled with water.

Groundwater Contamination. Degradation of quality of groundwater due to 1) leakage from septic tanks, landfills, underground storage tanks, and pipes and sewers, 2) direct migration through cracks, 3) rising watertable, 4) pumping in vicinity of contamination, 5) recharge by contaminated water, and 6) seawater intrusion.

Habitat. A place where a population of an organism or a group of organisms live, such as a tropical forest.

Half-Life. The time taken for one-half of the atoms of a radioactive element to disintegrate (give up radioactivity) or for one-half of a chemical material to be degraded.

Halocarbon. *See* CFC.

Halogen. Any one of the reactive nonmetals: bromine, fluorine, chorine, iodine, and astatine.

Halons. Carbon compounds containing bromine, used as fire-extinguishing gases. Similar to CFC, they have long life and cause ozone depletion.

Hard Water. Water that contains high concentrations of metal ions, such as calcium, magnesium, and iron. Hard water causes deposition of scale in boilers, interference in industrial processes, soap not to lather, and an objectionable taste.

Hazardous and Solid Waste Act (HSWA), 1984. An amendment of RCRA of 1976. It established policies with respect to waste minimization, land disposal restrictions for hazardous waste, deep-well injections, domestic sewage sludge.

Hazardous Material (Substance). Any material that poses a threat to human health and/or to the environment. These materials are listed by the major environmental statues and are controlled by them as follows: 1) materials are regulated as hazardous under Section 311(b)(2)(A) and Section 307(a) of the Clean Water Act (CWA), 2) under Section 112 of the Clean Air Act (CAA), 3) under Section 3001 of the Resource Conservation and Recovery Act (RCRA), 4) under Section 102 of the Comprehensive Environmental Response, Compensation and Liability Act (CERCLA), 5) under Section 7 of the Toxic Substance Control Act (TSCA), and 6) under Title 40, Section 302.4 of the Superfund Amendments and Reauthorization Act (SARA).

Hazardous Waste. In RCRA Subtitle C, hazardous waste is defined as any solid waste or combination of solid waste which may: (A) cause, or significantly contribute to an increase in mortality, or an increase in serious irreversible or incapacitating reversible illness; or (B) pose a substantial present or potential hazard to human health or the environment when improperly treated, stored, transported, disposed of, or otherwise managed. A waste is identified as hazardous, 1) if it is listed in Title 40, Part 261 of RCRA, or 2) it exhibits any one of the four characteristics—ignitability, corrosivity, reactivity, or toxicity.

Head. A term for energy at any point; *see* Energy Head.

Heat Exchanger. Any mechanical device used to transfer heat energy from one medium to another. Radiator with accessories is a heat exchanger for cars.

Heat Islands. In cities, the substantial use of light energy, heat emissions by large concentration of population, radiations by brick, concrete, and asphalt create a dome-like envelope of elevated temperature over the city. This prevents the rising hot air to cool at its normal adiabatic lapse rate, and may lead to a local trapping of air pollutants under the dome.

Heat of Combustion. The energy liberated when a compound experiences complete combustion with oxygen with both the reactants starting and the products ending at the same conditions, usually 1 atm and 25 °C or 60 °F.

Heavy Metals. Metal elements with high atomic weights, e.g., mercury, lead, arsenic, chromium, silver, cadmium. They bioaccumulate in the food chain and create toxicity.

Henry's Law. For any atmospheric gas in contact with water, the ratio of the atmospheric concentration of the gas to the gas concentration in water is a constant, given by Henry's Law Constant. Sometimes this ratio is expressed between the partial pressure exerted by the gas in the atmosphere over the water to the concentration of the gas in the water, which is another form of Henry's Constant.

Herbivore. An organism that eats only plants; a primary consumer.

Heterogeneous Reaction. A reaction is heterogeneous if it takes place in the presence of at least two phases of substances, e.g., liquid and gas. The reaction in a single phase is homogeneous.

Heterotroph. Consumers and decomposers that get their food from autotrophs.

Homeostasis. A mechanism by which an ecosystem maintains itself against external stresses. For example, when population of a species increases beyond a limit, the rate of growth tends to slow down; however, when the population declines significantly, the birth rate picks up.

Hydraulic Conductivity. An expression for ease at which water flows through porous medium. In Darcy's Law, it is the coefficient to compute the rate of flow through soil.

Hydraulic Gradient (Grade Line). Slope of watertable or the change in water level per unit of distance along the direction of flow.

Hydraulic Head. The height to which water will rise if a pipe is inserted in hydraulic system; the water pressure in length unit.

Hydraulic Loading. The volume of water applied to a treatment unit per time period.

Hydraulic Radius. The flow area divided by the wetted perimeter of pipe, channel, or any other flow system.

Hydrocarbon. A vast family of compounds containing carbon and hydrogen in various combinations. They are a major component of a fuel.

Hydrologic Cycle. The process of precipitation, evaporation, and runoff through which water circulates through the earth.

Hydrolysis. A chemical reaction of water with a substance in which hydrogen (H) and hydroxyl (OH) are added to the other substance forming two or more new compounds. A mechanism that pollutants in contact with water undergoes.

Hydrosphere. The part of the planet containing water, ocean, lakes, rivers.

Hydrostatic Pressure. Pressure exerted by standing water due to depth alone.

Hypolimnion. The bottom layer of water in a lake.

Ideal Gas (Law). A gas that would follow the ideal gas law. Real gases follow the law at normal temperatures and pressures. However, the law is applied in all air pollution calculations. According to the Ideal Gas Law, the product of the volume and pressure divided by absolute temperature is constant for any gas; the constant is known as the Universal gas constant.

Igneous Rock. Rock formed by cooling of magma that has erupted from deeper part of the earth, such as granite.

Ignitability. Capable of burning or catching on fire.

Imhoff Tank. A two-chamber sewage treatment unit which allows the sedimentation to take place in upper chamber and microbial digestion in its lower chamber.

Immiscible. Two liquids in contact with each other, neither of which acts as solvent for the other to any appreciable extent.

Impoundment. Any land area that can hold water, such as detention ponds and lagoons.

Incineration. An engineered process in which combustible solids, liquids, and gaseous wastes are burned at high temperatures (900–2900 °F) and changed to noncombustible products.

Incipient Lethal Level (LC_{50}). The level of a toxic substance beyond which 50 percent of a test population of organisms will die.

Incubation. To grow bacteria in a controlled environment that is conducive for the optimum growth.

Indicator Organisms. The microorganisms whose presence indicates contamination by human wastes.

Indicator Species. Those species in a habitat that are very sensitive to environmental changes. Their decline serves as an early warning of the endangerment.

Indoor Air Pollution. Contamination of indoor air by physical, chemical, and biological pollutants.

Inert Gas. A gas that does not react with other substances under ordinary conditions.

Infiltration. The ingress of a fluid (water or air) into a substance through pores or small openings, such as seeping of water underground.

Influent. Water, wastewater entering any treatment or processing plant.

Infrared Radiation. Heat waves or long wavelengths radiation that are larger than the visible light (0.75 micrometer) and shorter than the radio waves (300 micrometer).

Injection Well. A well used to dispose of waste below groundwater level in a liquid form.

Inland Waters. All waters on the earth, including estuaries and bays, that are not a part of ocean or sea.

Inorganic Chemical. A matter of mineral origin, that does not contain carbon atoms.

Internal Energy. The total (kinetic and potential) energy contained within atoms or molecules that form a body or a system. It is effected as the work is done on or by the system.

In Vitro/In Vivo. In vitro is a test performed in glass or test-tube. In vivo is a test performed inside living organism.

Ion. A charged atom that occurs when one atom or a group of atoms loses or gains one or more electrons during chemical reactions. When electron(s) are lost, the atom is a positively charged ion (cation) and when electron(s) are gained, it is a negatively charged ion (anion).

Ion Exchange. Water that contains chemicals in solution is treated by ion exchange. A solid material, known as resin or ion exchanger, is introduced into an aqueous solution. The ions are exchanged between the chemical in solution and the exchanger. The exchanger is regenerated (restored to original form) by treating with a different solution.

Ionization. The process by which an atom or a molecule acquires an electric charge to become an ion in a chemical reaction or by radiation.

Ionosphere. An atmospheric layer above the Mesosphere, above a distance of about 80 km from the earth. The solar radiation is very intense, causing the ionization of the sparse gas molecules present.

Isobar/Isohaline/Isohytes/Isotherm. A line drawn on a map that connects points having the same value of 1) barometric pressure for isobar, 2) salinity for isohaline, 3) rainfall for isohyte, and 4) temperature for isotherm.

Isomer. An organic compound that has the same molecular formula as another compound but a different molecular structure (atoms are arranged differently). The two compounds will have different physical and chemical properties.

Isotope. Two atoms of the same element that have the same number protons (and hence, electrons) but a different number of neutrons. Both atoms will have the same atomic number but different atomic weights.

Isotropic. Displaying equal property in all directions at a point. For example, the same hydraulic conductivity in all directions.

Kelvin. *See* Absolute Temperature.

Kinematic Viscosity. Dynamic viscosity divided by density of a fluid.

Kinetic Energy. Energy possessed by a substance due to its motion.

Laboratory Blank. *See* Blank.

Lagoon. A shallow pond used to stabilize and decompose organic materials in wastewater. Three main types are Conventional, Aerobic, and Anaerobic.

Lag Phase. Following the inoculation of laboratory culture of bacteria, there is a period for the organisms to adjust; during this time there is no increase in number of bacteria.

Lag Time. The time interval between a step change in input concentration and the corresponding first change observed in response. In hydrology, it is the time interval from the center of a rainfall pattern to the crest of the corresponding hydrograph.

Laminar Flow. *See* Flow, Laminar/Turbulent.

Land Disposal. Disposal of waste in or on land. This includes landfill, surface impoundments, waste piles, injection wells, land treatment facilities, salt domes and salt beds, underground mines, concrete vaults, and bunkers.

Land Treatment/ Land Application/ Land Farming/ Land Spreading/ Sludge Farming. The biological treatment of sludge in which the large microbial population in soil is used to degrade the organic matter.

Landfill (Sanitary). An engineered land disposal site where solid waste materials from municipal and industrial sources are buried. Fill is designed to avoid the leakage into surface and groundwaters.

Lapse Rate. *See* Adiabatic Lapse Rate.

Law of the Minimum (Liebig's). The population of a species is controlled by a chemical or physical factor that is in shortest supply relative to the level of its requirement by the species. Thus the factor limiting the growth of a population is called the Limiting Factor.

Lethal Concentration- LC_{50}. *See* Incipient LC_{50}.

Leachate. The rainwater passing through a landfill dissolves and carries in suspension materials stored in the fill; it can dissolve newspaper ink. Leachate is often toxic and can travel out of a fill that is not properly engineered.

Leaching. The process by which the percolating fluids dissolve the solid constituents in soil and carry them away with the fluid.

Life Cycle. All phases or stages through which an organism passes during its lifetime from the fertilized egg to death.

Limiting Factor. *See* Law of the Minimum.

Limnetic Zone. In an open water body, the region beyond the littoral zone to a depth at which there is sufficient penetration of the sunlight for photosynthesis.

Limnology. The scientific study of the physical, chemical, and biological conditions of lakes, ponds, and rivers.

Linear Growth. *See* Arithmetic Growth.

Linear Regression. A statistical method to fit a line through a number of given points to express an equation. The drawn line is such that the sum of the square of the vertical distances from the points to the line is minimum. This is also known as the Line of the Least Square Fit.

Liner. The material used to make a container impervious; for example the furnace lining ensures the chamber is impervious to escaping gases as does the landfill lining for leachate proofing.

Listed Waste. Chemical substances listed under Title 40, Part 261 of the Resource Conservation and Recovery Act (RCRA) regulations. In RCRA, hazardous wastes have been placed on one of three lists: nonspecific source wastes, specific source wastes, and commercial chemical products.

Lithosphere. The solid portion of earth below the surface. It is commonly extends to a depth of about 80 km from the surface.

Littoral Zone. In open water body (lake or sea), the area of shallow waters close to the shore where rooted plants exist.

Loading. An amount of materials or thermal energy that is introduced to a treatment unit or discharged into a water body. It includes river loading, waste loading, organic loading, hydraulic loading, particulate loading, dust loading.

Log Phase. *See* Exponential Growth Phase.

Macronutrients. Chemicals required by an organism in large quantity. For plants, these are nitrogen, potassium, phosphorus, calcium, and sulfur.

Magma. Molten rock found in the mantle, beneath the crust of the earth. On solidification it becomes igneous rock.

Mangrove. A tidal swamp forest developed along the coast in tropical climates that supports plant species that can survive and grow in saline conditions.

Manning's Formula. In an open channel, the volumetric flow rate is directly proportional to the cross section area of water, 2/3 power of the hydraulic radius, and channel bottom slope, and indirectly proportional to the Manning's roughness coefficient.

Mass and Weight. Mass is a measure of the content in a matter. Weight is the gravity force acting on the mass, i.e., weight equals mass times gravitational constant.

Mass Conservation. The basic principle of matter that matter or mass cannot be created or destroyed, it can simply change from one form to another.

Mass Flux. *See* Flux.

Mass Number. The number of protons plus the number of neutrons in the nucleus of an atom. This is an indicator of the atomic weight.

Mass Spectrometer (MS). An instrument that sorts out atomic or molecular ions based on their masses and electrical charges. By comparison of the mass spectrum (pattern of peaks on the chart) of the known compounds with the spectrum obtained from the sample, the identity of the materials is determined.

Mass Transfer. There are two modes for mass transfer: Convection and Diffusion; *see* corresponding terms.

Materials Balance. Based upon the principle of conservation of mass, the approach is used to estimate the pollutant quantity in the environment. By knowing input mass to a system or process, the output mass or the mass added or withdrawn from the system can be determined by a balanced equation.

Mathematical Model. Formulating a set of equations that depict a real process. The calibration or verification of the model is an important aspect that requires that the set of equations (model) duplicates the known conditions with the historic data.

Maximum Contaminant Level (MCL). The maximum permissible concentration of a drinking water contaminant set by the EPA under the Safe Drinking Water Act of 1974, and subsequent amendments. There are two categories: Primary MCL- Levels, with an adequate margin of safety, to protect the public health, and Secondary MCL- Levels, with an adequate margin of safety, to protect the public welfare and water systems.

Maximum Contaminant Level Goal (MCLG). The nonenforceable levels for several listed contaminants that will cause no known or anticipated adverse human health effects, with a margin of safety. These are desirable levels to achieve.

Maximum Permissible Concentration (MPC). The amount of toxic or radioactive materials in water, air, or food that will result in Maximum Permissible Dose.

Maximum Permissible Dose (MPD). The level of exposure to toxic or radioactive substances with no adverse health effects.

Membrane Filtration. Filtration through membranes or thin films under pressure (50-100 psi). The membranes are available in a variety of pore openings. The filters with small openings can retain bacteria and can be used to remove bacteria from a liquid sample for examination.

Membrane Processes. Used to remove fine metallic solutes from a liquid sample, such as to reverse osmosis and ultrafiltration.

Mesosphere. The atmospheric layer above the Stratosphere; it begins at about 50 km and extends to about 80 km.

Mesotrophic. *See* Eutrophic.

Metabolism. The sum of all chemical reactions occurring within a cell or an entire organism. These are: Catabolism—energy releasing breakdown of molecules, and Anabolism—synthesis of new molecules. The products of metabolism are Metabolites.

Metamorphic Rock. Rock formed by the exposure of sedimentary or igneous rock to high temperatures, high pressures, and chemical processes deep beneath the earth surface.

Micronutrients. Chemical nutrients required by an organism in a very small quantity. For plants, these are magnesium, copper, iron, zinc, manganese, vanadium, molybdenum, cobalt, boron, chlorine, and silicon.

Microorganism. Very small sized organisms which are not visible to or rarely visible to unaided eye. They pass through U.S. sieve size #30 but retained on #100. These include plants like bacteria, fungi, and algae and animals like protozoa, rotifers, crustaceans, and nematodes.

Mixed Liquor. Liquid in the aeration tank comprising of water, organic matter, and the activated sludge.

Mixed Liquor Suspended Solids (MLSS). A measure of the solid material in the aeration tank, mainly organic compounds and activated sludge.

Mixed Liquor Volatile Suspended Solids (MLVSS). The portion of MLSS that will vaporize at 600 °C. This is a measure of the biomass present in the tank.

Mixture. Two or more different elements or compounds mixed together without a chemical reaction between them; they can be separated by physical or mechanical means.

Molarity. The number of moles of a substance dissolved to make one liter of solution.

Mole. The amount of a substance (in any unit) which is numerically equal to its molecular weight. A pound-mole of carbon is 12 pounds and a gram-mole of carbon is 12 grams. One gram-mole of any substance always contains 6.02×10^{23} molecules.

Molecular Weight. The sum of the atomic weight of each atom in the molecule.

Molecule. A group of atoms that are held together by chemical bonds. They could be the atoms of the same element (O_2) or atoms of different elements, thus forming a compound (H_2O). For a compound, the smallest unit is a molecule.

Momentum. The product of the mass and the velocity of a moving body. Similar to the mass and energy, the momentum is also conserved in a system.

Monitoring. The process of measuring (by sampling, analysis or direct observations) the amount of pollutants or radioactive contaminants present in the environment (air, water, soil), on a continuous, periodical, or random basis.

Monitoring Well. A well drilled in close proximity of a waste storage or disposal site to collect groundwater samples for physical, chemical, and biological analysis, to detect for the leakage at the site.

Montreal Protocol. The 1987 international agreement to phase out the production of CFCs and Halons by the participating countries, to protect the ozone layer.

Most Probable Number (MPN). The MPN method, based on the testing of multiple tubes of geometrically increasing diluted sample, gives a statistical estimate of the number of coliform bacteria present in 100 mL of water or sewage sample. The presence of coliform, is an indicator of pollution by human feces.

Multimedia Filter. A filtration device designed to remove suspended solids consisting of three or more layers with coarsest material (anthracite) at the top, medium size (sand) in the middle, and finest (garnet) at the bottom. The rate of filtration is faster than the rapid sand filter and effluent is cleaner than the dual-media filter.

Municipal Solid Waste. The solid wastes, including trash and garbage, that originate from houses, commercial sources, and construction sites. Nonhazardous sludge from municipal sewage treatment plant can be included in this category.

National Air Monitoring System (NAMS). A national network of air monitoring stations designed by the EPA and individual state agencies to monitor the ambient air quality.

National Ambient Air Quality Standards (NAAQS). In Clean Air Act, the limits set for the ambient air concentrations of particulate matter, sulfur dioxide, nitrogen oxides, ozone, carbon monoxide, and lead to protect human health (primary standards) and public welfare (secondary standards).

National (Oil and Hazardous Substances Pollution) Contingency Plan (NCP). Pursuant to requirements under the CERCLA and the Clean Water Act, the EPA developed the policies and procedures to be followed for responding to spills and releases of hazardous substances and oils into the environment, both for sudden accidental release and intentional gradual leaks. The cleanups of the Superfund sites in done in accordance with the NCP.

National Emission Standards for Hazardous Air Pollutants (NESHAP). As part of the CAA, the technology-based limits set by the EPA for air emissions of certain pollutants that may pose a significant risk of deaths or serious illness on long term exposure. The limits are set for asbestos, benzene, beryllium, arsenic, mercury, radionuclides, and vinyl chloride; over 25 others are under consideration.

National Environmental Policy Act (NEPA). The 1969 act constituted the Council on Environmental Quality (CEQ) to administer the Act. The Act requires that all federal action plans should incorporate the environmental Impact Statement (EIS) in their proposals that examines the environmental consequences of the proposed actions along with the alternative plans.

National Pollutant Discharge Elimination System (NPDES). A provision within the Clean Water Act that all treatment plants discharging into any waters of the U.S. should obtain a permit issued by the EPA or a state agency authorized the EPA. The permit lists the amount of discharge permitted and the level of cleanup required prior to discharging.

National Primary and Secondary Drinking Water Regulations (NPDWR). *See* MCL Primary and Secondary Standards.

National Priority List (NPL). A list of most serious hazardous waste sites identified for cleanup under the Superfund. The list is updated annually by the EPA based on a hazard ranking system.

Natural Resources. Resources belonging to natural environment such as air, water, land, forests, fish, wildlife, and minerals.

Natural Selection. The process by which the organisms adapt to their environment by strengthening the desirable genetic qualities and eliminating the undesirable ones. Those fitted to the environment survive and those do not disappear. It is survival of the fittest.

Nekton. *See* Benthos.

Neritic Zone. *See* Abyssal Zone.

Neutralization. The process of adding an alkaline material to acid or to add an acidic to an alkaline material to bring pH to neutral level of 7 or to any other desired pH level. The primary products are water and salt.

Neutron. The subatomic particle in the nucleus of an atom that has no charge but has a mass equal to a proton.

Niche. The habitat that a species is better adapted to occupy.

Nitrification. Nitrification of wastewater, generally by bacteria, creates ammonia. Ammonia is oxidized to nitrite and then to nitrate by bacteria in water or soil.

Nitrogen Fixation. The conversion of the atmospheric nitrogen to reduced form (ammonia and amino acids) that can be used as a nitrogen source by organisms since all require a source of nitrogen as nutrient. This is carried out by a selected group of bacteria.

Nitrogen Oxides. These comprise: $NO, NO_2, NO_3, N_2O, N_2O_3, N_2O_4, N_2O_5$. The first two, nitric oxide and nitrogen dioxide are the primary air pollutants.

Nitrogenous BOD. When ammonia and nitrite contaminants are present in wastewater, they are oxidized by chemoautotrophic bacteria. The amount of oxygen required is the nitrogenous BOD which can sometimes complicate the interpretation of the (carboneous) BOD of the sewage.

Nitrogenous Waste. Animal or plant residues that contain large amounts of nitrogen, includes chemical waste containing nitrogen.

Nonconventional Pollutants. In the Clean Water Act, water pollutants not covered under conventional pollutants, toxic pollutants or thermal discharges. These include chloride, iron, ammonia, color, and total phenols.

Nonionizing Radiation. Electromagnetic waves that are not highly energetic to produce charged ions when strike an object but can heat an object and do biological harm, such as visible light, microwaves, radio-waves.

Nonpoint Sources. The pollution sources that are diffused and do not have a specific outlet; examples are street runoff, septic systems, agriculture discharge.

Nonrenewable Resources. The resources such as coal or mineral ores that, once consumed, cannot be replaced on human time scale (for hundreds of years).

No Observed Adverse Effect Level (NOAEL). The highest dose in an experiment that did not produce any observable adverse effect.

Normal Solution. One normal (1N) solution contains one gram equivalent weight of a substance in one Liter of solution. If it contains four-tenth of the gram equivalent weight per Liter, it is 0.4N solution.

Nuclear Energy. Energy derived from the fission of heavy element, such as uranium or by the fusion of light elements such as hydrogen.

Nuclear Reactor. An assembly wherein a fission chain reaction is produced in a controlled manner. The reactors have a core containing nuclear fuel, usually a mixture of uranium-238 and uranium-235 and/or plutonium-239 in the form of rods. The heat generated by fission is used to make steam, which, in turn, generates electricity. The spent fuel is highly radioactive.

Nucleus. The central core in an atom that contains protons and neutrons.

Nutrients. Materials used by a living organism to sustain its existence, promote growth, and provide energy. Mainly, these pertain to compounds of nitrogen and phosphorus. Although, water and oxygen are essential for life, they are not commonly called nutrients.

Nutrient Cycles. Based on the conservation of mass, the nutrients cycle through the biosphere. The main nutrient cycles are carbon, nitrogen, phosphorus, and sulfur.

Ocean Dumping. Under the Ocean Dumping Ban Act of 1988, which amended the Marine Protection, Research and Sanctuaries Act of 1972, no new permits are issued for dumping of sewage sludge and industrial waste; the existing permits ceased by December 31, 1991.

Ocean Thermal Energy Conversion (OTEC). The temperature gradient between warm surface water and cooler deep water in ocean can drive an evaporation-condensation cycle of a fluid to rotate a turbine-generator. This is an alternative renewable energy technique.

Oil Spill. An intentional or accidental discharge of oil which runs into a water body. The control measures are: mechanical containment (booms), absorption, oil skimming, ignition, oil eating bacteria, and detergents.

Oligotrophic. *Refer to* Eutrophic.

Omnivore. An organism that eats both plants (herbivore) and animals (carnivore), such as human.

Open Dumping. City dumps or landfills operated without environmental safeguards recommended by current law. The main difference between an open dump and a sanitary landfill is that the latter is engineered to minimize the pollution of the surroundings. Daily soil cover and final cover in sanitary landfill ameliorate unethical conditions.

Organic Matter. A substance that contains carbon atoms. All living matters are organic. First it was thought that all organic matter can be derived from plants and animals only, and the inorganic matter were obtained from mineral sources. Now it is known that organic matter can be derived from nonliving sources too. Organic matter could be volatile, semivolatile, and nonvolatile.

Organic Loading. The amount of biodegradable material applied to a wastewater treatment unit per time period (day).

Organism. Any living being- human, plant, animal. Ecologically divided in three classes: Producers, Consumers, and Decomposers.

Osmosis. If a semipermeable membrane (either natural or artificial) is separating two solutions of different concentrations, the diffusion of the solvent (usually water) takes place from the dilute to the more concentrated solution; the membrane allows the passage of water but prevents the substance dissolved in the water. This diffusion will continue till both solutions are equal in concentration. If the pressure is applied to the more concentrated side, the flow water will be from more concentrated to dilute solution. This is Reverse Osmosis.

Overflow. Discharge over a hydraulic device such as weir overflow.

Overland Flow. In Hydrologic Cycle, the portion of water running over the land (surface runoff) after the rainfall. In Wastewater Technique, the application of water over a sloped surface. As the water flows over a vegetated land, the nutrients are removed and the water is collected at the bottom of the slope for reuse or discharge into receiving waters.

Overturn. In stratified lakes, the mixing or turnover of water from top to bottom. This causes mixing of pollutants. This occurs in fall and/or spring.

Oxidation. It a chemical sense, it is the loss of electrons by an atoms or molecule. Oxidation is always accompanied by the reduction as some other atom or molecule gain that electron. Commonly, it refers to a process wherein a substance is converted to another form by combination with oxygen. Any material undergoing combustion in the presence of air is subject to oxidation. Organic matter decomposed by bacteria are subject to oxidation.

Oxidation Pond. A man-made lake or body of water in which organic waste is placed in to be decomposed by bacteria. The air is bubbled through the pond.

Oxygenation. To bubble in oxygen for getting it dissolved into water or wastewater.

Glossary

Oxygen Demand. The need for oxygen for biological or chemical processes in water. This supply is obtained from the oxygen that becomes dissolved in water. This quantity, which depends upon the water temperature and solids content, is limited. The biota living in water, all demand a portion of this limited resource.

Oxygen Demanding Waste. Any organic matter that stimulates metabolism (serves as food) of bacteria, such as domestic waste. It places a corresponding demand on dissolved oxygen for respiration process.

Oxygen Sag Curve. A graph of dissolved oxygen against time or distance, plotted for a stream from the point of addition of degradable organic material (waste). As bacteria decompose the waste, they consume dissolve oxygen resulting in dropping of the dissolved oxygen concentration. When the waste is exhausted, the oxygen level in the stream is replenished from atmospheric oxygen. The graph has a sag, hence the name.

Ozonation. To use ozone gas (O_3) as a disinfectant to kill pathogenic bacteria.

Ozone Hole. A large area (of the size of the United States) over Antarctica discovered to have a seasonal drop in stratospheric ozone concentration by 50% or more. This drop is linked to the release of CFCs on the earth.

Ozone Layer. A region in atmosphere, about 10 to 30 miles above the earth's surface, containing ozone. This ozone absorbs the ultraviolet radiation from the sun. Thus, it provides a measure of protection to plants and animal life since ultraviolet radiation is mutagenic.

Ozone Depletion. The destruction of ozone molecules in the ozone layer by chemical reactions with compounds released by humans, such as CFCs and Halons.

Packed Tower. An air pollution control device in which contaminated air is forced upward through a tower packed with materials having a large surface area. A liquid is sprayed downward through the tower. The air contaminants are absorbed into the liquid trickling through the packing material. A similar process is used in opposite manner for the removal of organic contaminants by capturing in an air stream, known as Air Stripping.

Partial Pressure. The pressure exerted by an individual gas in a gaseous mixture; *see* Dalton's Law.

Particle Size Distribution. The proportions of grains of different sizes (size ranges) in a sample (soil or airborne particles), expressed in percent by weight.

Particulate. Fine solid particles or liquid droplets (not gases) such as smoke, mist, fumes, smog, aerosols, found in dirty air stream.

Particulate Matter-10 (PM10). Airborne particles of diameter equal to or less than 10 micrometer. Particles of this size can penetrate lungs or absorbed by blood.

Pascal (Pa). The SI unit of pressure designating Newton per square meter.

Pathogen. The disease-causing microorganisms.

Pelagic. Marine biota that belong to open sea, free from dependence on bottom or shore.

Percolation. The downward movement of water through subsurface soil layers.

Periodic Table. An arrangement of the elements by atomic numbers to show the relationship between them. Horizontal rows are periods and vertical columns are groups.

Permanent Hardness. The hardness that cannot be removed by heating the water; this hardness is due to noncarbonate ions in water.

Permeability. *See* Hydraulic Conductivity.

Peroxyacetyl Nitrate (PAN). A component of the photochemical smog, created by reaction involving sunlight, hydrocarbons, and oxides of nitrogen. It causes eye irritation and vegetation injuries.

Persistent Chemicals. A substance that resists biodegradation and chemical oxidation when released into the environment. They accumulate in air, water, or land. For example, DDT and PCBs.

pH Scale. A measure of acidity or alkalinity of a medium. The pH value is the negative of logarithm of the hydrogen ion concentration. The scale ranges from 0 to 14; 7 being the neutral. A value less than 7 is acidic and more than 7 is alkaline or basic. One pH unit represents 10 fold difference in acidity or alkalinity.

Photic Zone. The upper portion of a lake or sea into which light penetrates to support the growth of phytoplankton.

Photochemical Oxidants. The components of photochemical smog. These include, ozone, nitrogen dioxide, PAN, and oxygenated hydrocarbons. The highest concentration is of ozone. EPA has set the standards for photochemical oxidants.

Photochemical Smog. A light haze formation consisting of photochemical oxidants caused by the effect of the sunlight on automobile exhaust gases. It causes breathing problem, coughing, chest soreness, and eye irritation. It damages plants and cracks rubber. Thermal inversion leads to formation of smog close to the surface.

Photoionization Detector (PID). An instrument to detect the amount of a specific organic matter present in a gas stream without laboratory analysis. The gas stream is exposed to the ultraviolet energy. The organic compound, if present, will be ionized by the ultraviolet energy. This will produce a current proportional to the amount of material, which will be directly read by the instrument.

Photosynthesis. The essential process whereby chlorophyll-containing (green) plants take carbon dioxide and water from the air and using sunlight converts them to sugar and oxygen.

Photovoltaic Cell. A cell consisting of a thin wafer of silicon and a small amount of metal, emits electrons when the sunlight strikes it, thus producing a feeble electric current.

Phytoplankton. Microscopic plants and animals of the Plankton community that live suspended in water and are incapable of independent movement. Phytoplankton are microscopic aquatic plants like algae, and Zooplankton are microscopic aquatic animals—like protozoa.

Piezometric Head (Surface). The level to which water will rise if a well is drilled at a particular point. This equals watertable in unconfined well and higher than underground water surface in an artesian well.

Plant Nutrients. The ingredients of fertilizer—phosphate, nitrate, and ammonia.

Plug Flow. A system or reactor in which the flow of fluid is not mixed in the axial direction and the material is discharged out in the same order as it is entering the system; opposite of a complete-mix system.

Plume. The mass of emitted contaminants spreading by diffusion in the environment—air, water, soil. The form and direction of a plume is detectable by the concentration of the contaminants.

Point Source. An identifiable discharge point for the pollutant, e.g., a pipe, ditch, chimney.

Pollutant. Any chemical substance or physical agent (heat, sound, electromagnetic radiation) introduced to the environment in an amount that threatens human health, wildlife, plants, or the orderly functioning and/or human enjoyment of any aspect of the environment.

Polychlorinated Biphenyls (PCB). A variety of products (210 compounds) of the chlorination of the hydrocarbon. They were extensively used as insulating and cooling agents in transformers, as plasticisers in paints, varnishes, inks, and sealants. They are shown to induce cancer in mammals. As little as 0.5 g harms humans. They are highly persistent chemicals. The use of PCB was banned by the Toxic Substance Control Act of 1976.

Polycyclic. Organic compounds composed of multiple units of six-carbon benzene ring.

Polycyclic Aromatic Hydrocarbons (PAH). A group of aromatic ring compounds consisting of many benzene rings. They are found in tar and petroleum and emitted by combustion. The exposure to these can cause cancer. Their concentration in water should be less than 0.2 microgram/Liter.

Polymer. The organic compounds made from linking together smaller organic molecules. They have a heavy molecular weight. They are both natural (cellulose, proteins, starches, nucleic acids) and synthetic (nylon, plastic, rubber). Certain polymers act as coagulants to enhance settlement of particles.

Polyvinyl Chloride (PVC). A strong synthetic polymer plastic used in pipes, toys, and electrical applications. It releases hydrochloric acid when burnt.

Population Equivalent (PE). An expression of the relative strength of an organic waste (usually from industries) in terms of the population that would produce the equivalent waste. Usually, 1 PE = 0.17 pound of BOD.

Porosity. The ratio of volume of pores to the volume of grains in a soil sample. If the pore spaces are poorly connected, even a highly porous soil may have low permeability.

Potable Water. Water suitable for drinking and cooking.

Potential Energy. The energy possessed by a substance because of its position (storage behind a dam) or its composition (food).

Power. Rate of producing or using energy, energy/time. The units are Watt and Horsepower; 1 HP = 746 Watts.

Precipitation. The moisture, in any form, that falls from the atmosphere to the ground, mainly rainfall and snow. In chemical treatment, it is the process whereby some or all of a substance in solution is transformed into a solid form and then removed from the solution.

Precursor. In air pollution, a contaminant that reacts with sunlight and/or with other chemicals to produce a new form of contaminants, as in photochemical smog.

Pressure. The force exerted by an enclosed fluid on unit area of the containing vessel. Typically expressed in pounds per square inch (psi) and Newton per square meter, or Pascal. One standard atmosphere exerts a pressure of 14.6959 psi.

Pressure Head. Pressure expressed in terms of the height of a column of any fluid that will exert a given pressure at the base. Pressure head equals the pressure divided by the unit weight of the fluid. It represents the energy head.

Pressure, Absolute. This equals atmospheric pressure plus gage pressure.

Pressure, Gage. Pressure measured by a gage (instrument) with reference to (over and above) the atmospheric pressure.

Pretreatment. Reducing or eliminating certain pollutants in a waste stream before discharging them into a publicly owned treatment works (POTW).

Prevention of Significant Deterioration. A Clean Air Act regulatory program under which air quality in an area can be allowed to deteriorate only by a certain fixed amount for particular pollutants, even if the ambient quality standard for the pollutant is met.

Primary Air Pollutant. A pollutant in the ambient air that is hazardous in the same form in which it is released into the atmosphere; for example, carbon monoxide. The other air pollutant is Secondary Air Pollutant, made up of substances formed in the atmosphere by reactions of primary air pollutants; for example, ozone.

Primary Consumer. In a food chain, animals at the trophic level that feed on plants only. Animals that eat other animals are secondary consumers.

Primary Settling Tank (Clarifier). In wastewater treatment, the first holding tank where the raw sewage is detained to settle 30-35% of the organic matter, before it is passed on to the next stage.

Primary Standards. *Refer to* MCL Primary Standards.

Primary Waste Treatment. A series of mechanical (physical) treatment processes which remove most of the floating and suspended solids, but have a limited effect on dissolved and colloidal material. This includes: Screens, Grit Chamber, and Primary Settling Tank.

Principal Organic Hazardous Constituents (POHCs). The chlorinated organic compounds, as specified by the EPA, that are found in chemical waste stream. POHCs have potential to form hazardous organic materials on incineration, known as the products of incomplete combustion (PIC). POHCs should be monitored during the trial burn of a hazardous waste incinerator.

Priority Pollutants. The Clean Water Act, 1977, Section 307, lists 129 pollutants in 65 classes of chemical materials as toxic pollutants, which require technology-based effluent standards for their control.

Process Water. Any water that comes in contact with any material or product being manufactured. The water is often released as wastewater after use.

Producer. The organisms that produce their own food, primarily green plants. They assimilate carbon dioxide and other inorganic nutrients into organic matter with solar energy, which serve as food to consumers.

Products of Incomplete Combustion (PIC). A complete combustion of chlorinated hydrocarbon should be carbon dioxide, water, and hydrogen chloride. However, during the incineration of chlorinated organic compounds, certain potentially hazardous organic materials are formed recognized as PICs. These could include chloroform, vinyl chloride, benzene, tetrachloroethane, and 1,1-dichloroethylene.

Profundal Zone. Deep region of a lake that lies beneath the limnetic zone. The penetration of sunlight is very limited, and, hence, green plant life is absent.

Proton. A subatomic particle carrying a positive charge. Together with neutrons, they form the nucleus of an atom. The number of protons in an atom is referred to as the atomic number.

Pyramid of Biomass/Pyramid of Energy. *See* Ecological Pyramid.

Pyrolysis. The material (coal, wood, organic matter) is destroyed by heat (temperatures 500- 1000 C) in the absence of oxygen. In incineration this is done with the help of the oxygen. This is also called the Destructive Distillation.

Qualitative Analysis. The examination of a sample to determine what chemical compounds or elements are present irrespective of the amount of those compounds or elements.

Quality Assurance(QA). A total program to assure reliability of measurement as per requirements. QA process requires 1) an understanding of what needs to be measured, 2) how it is to be measured in accordance with specifications, 3) how to systematically accomplish what needs to be done, 4) how to adequately evaluate what was done, and 5) how to report evaluated data that are technically sound and legally defensible.

Quality Control (QC). A system of monitoring, checks, audits, and corrective actions to ensure that environmental sampling, analysis, performance, reporting are of the highest achievable quality.

Quantitative Analysis. The examination of a substance or sample to determine the exact amount of certain specified compounds or elements present in the substance/sample.

Radiation. The emission of energy in electromagnetic waves or by particles. The rays that originate in the sun, radiate energy to the earth. Light and heat are emitted through radiation. *Refer to* Alpha, Beta, and Gamma Radiation. Two major divisions are: Ionizing radiation and Nonionizing radiation.

Radioactive Material. Atoms with an unstable nucleus that spontaneously emit ionizing radiation; such emission is the Radioactivity.

Radioactive Waste. Any waste that contains radioactive material in excess of that listed by the EPA (10 CRF Part 20, Appendix B, Table II, Column 2). *See* Transuranic Radioactive Waste.

Radioisotope. Radioactive isotope (form) of any naturally occurring or artificially created element that emits alpha, beta, or gamma radiation.

Radionuclides. Radioactive isotopes of various elements are collectively referred to as radionuclides.

Radius of Influence. The horizontal distance from a well to a point where watertable is not effected (lowered) by pumping of the well.

Rainforests. An ecosystem consisting of dense growth of tree species associated with high rainfall and humidity. Three major divisions are: Tropical, Subtropical, and Temperate.

Rankine. *Refer to* Absolute Temperature.

Rapid Sand Filter. A filter for water treatment. Water, after sedimentation, is applied to a 2-ft thick sand bed. The cleaned water is removed from an underdrainage system. The filter is cleaned periodically by backwashing.

Rate Constant (Coefficient). The constant of proportionality in an equation of chemical reaction, microbial kinetics, or linear or exponential growth rate. In a first order reaction, the rate of reaction equals the rate constant times the remaining concentration of the reactant.

Rational Method. A simple empirical method to compute the peak runoff caused by rainfall of a given frequency over a small area.

Raw Sewage/Raw Water. Wastewater or water before it has been treated for any application.

Reactor. A container or unit in which chemical or biological reactions take place.

Reaeration. The natural process by which streams deficient in oxygen acquire oxygen from atmosphere to attain saturated dissolved oxygen level.

Reagent. A laboratory chemical added for promoting a specific reaction.

Recarbonation. The bubbling in of the carbon dioxide to lower the pH. In water treatment, it is done after the lime-soda process, and in wastewater treatment after ammonia stripping.

Recharge. Refilling of an aquifer by natural rain percolating through soil or through artificial recharge basin infiltrating water into the ground.

Recirculation. The return of activated sludge from secondary sedimentation to the aeration tank or influent line.

Reclamation. The process of restoring a disturbed land area to the natural state or deriving the usable materials from waste, such as solids reclamation, reclamation of water.

Recoverable. The material that still has useful properties after serving its original purpose.

Recycle. Returning the wasted material to the production line again to manufacture goods. Recycling goes beyond the reuse of a product. For example, for aluminum can, it is remelting of cans to make a product again.

Reducing Agents. Any material that loses electrons in a reaction, thus, by itself it is oxidized but reduces the other material. For example, the carbon bearing materials that combines with oxygen.

Reduction. A chemical reaction during which electrons are added to an atom or molecule. Since electrons are derived from another molecule in the reaction, that molecule is subject to oxidation. For an organic compound, this is accompanied with addition of hydrogen atoms, called the hydrogenation of organic compounds.

Refuse. The complete range of wasted materials, including garbage, rubbish, and trash.

Refuse Derived Fuel (RDF). The combustible or organic portion of municipal solid waste that is separated out and used as a fuel source.

Regeneration. Recovery of spent chemical reagents; for example activated carbon grains and ion exchange resins are regenerated by the action of steam and sodium solution respectively.

Regression. A statistical analysis that establishes a relationship between two sets of data.

Relative Humidity. The ratio of the amount of water vapor present in a given volume of air to the maximum amount of vapor that can be held by the same volume of air at a specified temperature and pressure.

Remedial Action. Development of action plan and specifications for a site cleanup.

Renewable Resources. A source of energy or matter that is replaced by the nature, such as firewood, the water behind a dam.

Residence Time. The time that a substance spends in a container, such as a reactor, a lake, the atmosphere, the human body. It equals the mass of the substance in container divided by the rate of mass (flux) of the material into or out of the container.

Residual. The amount of material or energy that is leftover after a natural or technological process has been completed, e.g., the sludge remaining after the wastewater treatment.

Resource. Any source of matter and energy that is needed for living or to improve the quality of life.

Resource Conservation. The reduction of overall resource consumption and utilization of recovered resources.

Resource Conservation and Recovery Act (RCRA). The federal statute of 1976, amended in 1980, and 1984, which itself was an amendment of the Solid Waste Disposal Act of 1965. It provides for the management of solid (nonhazardous) and hazardous wastes. EPA has set minimum standards for all waste disposal facilities. For hazardous wastes, it has set standards for generation, transportation, storage and treatment of wastes. The underground storage program and medical waste tracking have been added.

Resource Recovery. The processing of residential, commercial, or industrial wastes physically, biologically, and chemically to recover paper, glass, metal, combustible material, and energy from incineration.

Resource Management. To apply practices to safeguard the future of renewable resources and to uphold the principle of sustained yield.

Respiration. The consumption of oxygen and release of carbon dioxide by animals and humans that happens when organic material (food) is broken down or digested. This is the reverse of photosynthesis. Respiration that utilizes free molecular oxygen is usually aerobic, although certain bacteria are capable of anaerobic respiration wherein inorganic compounds containing oxygen atom act as a source of oxygen.

Retention Basin. A natural storage, like a lake or pond, used to slow stormwater runoff.

Retention Time. *See* Detention Time and Residence Time.

Glossary 477

Reverse Osmosis. *Refer to* Osmosis.

Reversible Reaction. A chemical reaction that can proceed in either direction depending upon the conditions.

Reynolds Number (Re). A dimensionless number that indicates flow condition being laminar or turbulent; *refer to* Flow, Laminar/Turbulent. The number is expressed by the product of the density of fluid times the velocity of fluid times flow diameter divided by the fluid dynamic viscosity.

Riparian Right. The owner of land bordering on a stream or a lake have the right (the privilege without owning it) to the access and use of the shore and water. The right is automatic.

River Basin. The land area surrounding a river, the rain water from which drains into the river.

Rotating Biological Contactor (RBC). A type of secondary wastewater treatment unit in which a horizontal shaft is mounted by a series of discs. The growth of attached microorganism takes place on the discs. The one-half of each disc is immersed in a tank containing the wastewater, and the other half surface is exposed to the air which promotes oxygenation. These discs slowly revolve and contact the wastewater, stimulating the biodegradation.

Rubbish. Non-kitchen solid waste from residences. commercial establishments, and institutions.

Runoff. The part of rainfall, snow melt, or irrigation water that runs off the land into streams.

Safe Drinking Water Act (SDWA). A federal statute enacted in 1974, it sets standards for maximum allowable levels of certain chemicals, microbial and radioactivity pollutants in public drinking water systems- primary MCL and secondary MCL. Also it regulates underground injection systems including deep-well injection.

Safe Yield (Sustained Yield). Continuous long-term use of a renewable resource at a rate that permits the resource generation by the same amount for undiminished use of the resource in the future.

Salinity. The amount of salts, usually sodium chloride , in water. The salinity of sea water is 35 parts per thousand.

Saltwater Intrusion. The movement of sea water into a fresh water body, occurring in either surface or underground; a result of over pumping of aquifers near the shoreline.

Sampling. A sample measurement of constituants from the amount of water, air, or waste withdrawn. Major types are: Grab sample—a sample taken at random location at a random time; Composite—a series of samples taken separately over a given time period and combined as one sample; Integrated—a sample collected in a single vessel over a period of time; Replicate—a sample that is divided into two or more parts at a step in the measurement process with each part carried through the remaining steps in the measurement process; Split—a sample divided into two portions, one of which sent to a different lab and both subjected to the same measurement process.

Sanitary Landfill. The ground disposal of municipal solid waste (MSW); *see* Landfill.

Saturated Zone. The water below the watertable underground in which all pores or voids of the soil are fully filled with water.

Savanna. A type of terrestrial ecosystem where grasses, sedges, and small shrubs dominate.

Scfm (Standard Cubic feet per minute). The volumetric flow rate of gases referred under the standard conditions of 60 °F and 1 atm. Since density and other properties of gases depend on temperature and pressure, the standard conditions is a way to compare gases as different conditions.

Screening. A physical process to remove coarse floating and suspended solids from water and wastewater. In public health, any method useful to judge a toxic potential of a substance, to indicate the desirability of further testing.

Scrubber. A common device designed to remove pollutant particulate matter or gases from an exhaust stream produced by combustion in industrial processes. The major types are: cyclone scrubber, fume scrubber, ionizing scrubber, mechanical scrubber, orifice type scrubber, packed tower scrubber, plate scrubber, spray chamber, venturi scrubber, wet filter.

Second Law of Thermodynamics (Energy). Any system tends to become more disorganized. When energy converts from one form to another, the randomness (entropy) tends to increase. At every conversion, a portion of (high quality) energy is lost as heat (low quality) energy.

Secondary Air Pollution. *Refer to* Primary Air Pollutant.

Secondary Clarifier (Settling). *See* Final Clarifier.

Secondary Wastewater Treatment. A phase of wastewater treatment following the removal of floating and suspended particles in the primary treatment. In the second phase bacteria are used to decompose dissolved organic waste, hence the process is known as the Biological Process. The process consists of a biological unit and a final clarifier. The biological units can be: aeration or activated sludge process, trickling filter, or rotating biological contactor (RBC).

Section 404 Program (Permit). Under Section 404 of the Clean Water Act, for any activity of dredging or discharging of fill material in a wetland, a permit has to be obtained. The regulations for Section 404 are written by the EPA according to which the permit is granted by the State and the U.S. Army Corps of Engineers.

Secure Landfill. A modern designed landfill containment system comprising of a liner, a drainage layer, leachate collection pipes, and a system of monitoring wells.

Sedimentary. The rock formed through the consolidation of fine grained material such as limestone. It has stratification characteristics.

Sedimentation. A physical process where solid particles are made to settle down by gravitational force (their own weight). The very fine particles are flocculated to settle.

Self Purification. The cleansing property possessed by streams and lakes. Biodegradable material added to a body of water will be utilized by the microorganisms residing in water. If an excessive amount is not added, the water will undergo self-cleaning.

Septic Tank. An underground covered concrete tank to which the sanitary waste from a household is discharged. The solids settle to the bottom and decomposed by anaerobic bacteria. The liquid effluent flows to a leaching fields to seep into the soil. The sludge, known as septage, is pumped periodically.

Settleable Solids. The suspended solids that will settle out of a fluid medium over a given period of time. In the Imhoff Cone Test, it is the volume of materials that settles in one hour.

Settling Velocity. The terminal rate of falling of a particle through a fluid when the gravitational force is balanced by the fluid drag (frictional) force; *see* Stoke's Law.

Sewage. The waste produced by households and commercial establishments that is discharged out through the building sewer; the new common name is wastewater.

Sewage Treatment Plant. A facility designed to treat sewage. Most facility employ primary and secondary treatment. Chlorine is often added to the plant effluent before discharging to a receiving water.

Sewer. A pipe or a conduit to carry polluted water from the source to a treatment or disposal site. Different types are: Storm Sewer that carries runoff from a heavy rainfall; Sanitary Sewer that carries wastewater from houses, commercial sources, and public facilities; and Combined Sewer that transport both the storm runoff and sanitary wastewater.

Slag. During purification of an ore to derive metal, the nonmetallic glass-like material is formed in a blast furnace or smelter due to impurities reacting with a flux. It floats as a liquid on the top of the process which is removed, cooled, and solidified.

Sludge. The portion of the waste that contains the solid material separated during a treatment process. Usually, it has a consistency of a soft mud. It has to be treated before disposal. The main treatment of sludge consists of:

1. **Sludge Thickening:** Typically, a sewage sludge contains 2% solids (solids to water ratio) which increases to 6% during thickening process. This reduces the total sludge volume three times. Thickening is done by gravity thickener, flotation thickening, or centrifugation.
2. **Sludge Digestion (Stabilization):** The solids in raw sludge are mostly organic matter which are broken down and stabilize by bacteria to innocuous residue. Digested sludge has little biodegradable organic matter and contains about 50% inorganic solids. Sludge digestion is done by anaerobic digestion, aerobic digestion, heat treatment, and lime stabilization.
3. **Sludge Conditioning:** The objective is to produce a sludge that can readily give up water associated with sludge particles. This is done either chemically or by heat treatment.

4. **Sludge Dewatering:** Removing water to a feasible extent; typically a dewatered sludge has 20 to 35% solids content. This is done by centrifuge, pressure filter, vacuum filter, horizontal belt filter.
5. **Sludge Drying:** An operation is to dry the digested sludge. It is done in open area or in dryers. The drying results in sludge cake.
6. **Sludge Disposal:** The final disposal of the sludge cake. The alternatives are composting, thermal (incineration and pyrolysis), and land application.

Sludge Age. The ratio of the weight of volatile solids in the digester (anaerobic or aerobic) to the weight of volatile solids added per day. It is the Mean Cell (Microbial) Residence Time in the digester.

Sludge Volume Index (SVI). A laboratory test value that is used to indicate the rate at which sludge is to be returned from the discharge line of an aeration tank to the input (upstream) end. This is based on the measurement of activated sludge solids settled in a 1-Liter cylinder and the MLSS in the aeration tank.

Small Quantity Generator (SQG). As defined by the RCRA, a generator (1) who generates less than 100 kg of hazardous waste in a calendar month or accumulates less than 100 kg at any time, or (2) who generates less than 1 kg of acutely hazardous waste per month or accumulates less than 1 kg at any time. Such facility is subject to less rigorous regulations.

Smelting. An operation that melts ore to separate metals. Exhausts from smelters cause pollution.

Smog. *See* Photochemical Smog.

Solar Cell. *See* Photovoltaic Cell.

Solar Energy. Almost all energy sources are related to the sun, directly or indirectly. But the term commonly refers to conversion of direct sunlight into some usable form of energy which is most desirable, being a perpetual source. Four categories of solar energy uses are: 1) Passive system to directly heat houses without using any mechanical/electrical devices; 2) Active system that heats the water and stores it for pumping to heat a home or for other purpose; 3) Focussing the sunlight on a heating element, on a larger scale than item 2, to make steam and using a steam turbine to produce electricity; and 4) Converting the sunlight directly to electricity, using solar cells.

Solids. Various solids are: 1) Total Solids (TS)—The weight of solid residue left after drying a water sample at 105 °C; 2) Total Suspended Solids (TSS)—The solids retained on filter by passing a sample through a standard glass fiber filter and, the, drying the solids at 105 °C; 3) Total Dissolved Solids (TDS)—The difference between the total and suspended solids; 4) Volatile Solids (VS)—Material of Total Solids that is lost when heated to 550 °C; and 5) Settleable Solids—The volume of materials that settles in Imhoff Cone in one hour.

Solid Waste. All waste material, solid, fluid and gaseous, excluding discharges into surface waters via wastewater treatment plants, industrial point sources, and return irrigation flows, and air pollutants directly discharged into atmosphere. Municipal Solid Wastes and Hazardous Wastes fall into solid waste category.

Solubility Product Constant (Ks). When a solid compound partly dissolves in water, the product of the molar concentrations of the ions in the solution is the solubility product constant, which is a constant for that compound.

Solute. The substance that is dissolved in a solution.

Solution. A mixture of solute and solvent.

Solvent. The dissolving medium; water is an universal solvent.

Sorbent. A material that can absorb or adsorb solids, liquids, or gaseous substances.

Sparging. Removing the volatile material from a sample by passing bubbles of inert gas.

Species. A group of organisms whose members have a close mutual resemblance and a common origin.

Species Diversity. A measure of different species that inhabit a given location and the number of individuals of each type present. It is understood that a large number of species in a community are desirable for well-being of the living things in that community.

Specific Gravity. The ratio of density or unit weight of any substance to the density or unit weight of water at specific temperature and pressure.

Specific Heat. The ratio of the amount of heat (calories) required to raise the temperature of one gram of a substance by one degree C to the amount of heat required to raise the temperature of one gram of water by one degree C, which is one calorie.

Specific Yield. The volume of water that can be obtained from a unit volume of an aquifer.

Spectrometry. An instrument that measures the wavelength of electromagnetic radiation emitted by elements when heated to a high temperature. Since each element emits a unique spectrum, the technique can be used to detect substances.

Spectrophotometry. To analyze a water sample by means of spectrum emitted by substances in water, under exposure to light.

Spent Fuel. Fuel rods, from a nuclear reactor, that can no longer efficiently sustain a continuous fission reaction. The rods are usually removed after three years and are highly radioactive.

Standard Atmospheric Conditions. The state of a system or characteristics of a substance under the reference conditions established for temperature and pressure. In physics and chemistry, the conditions comprise of 1 atm and 32 °F (0 °C). In air pollution, the standard atmosphere (used by the EPA) is taken at 1 atm and 25 °C (room temperature).

State Implementation Plan (SIP). A state plan, as approved by the EPA, for the establishment, regulation, and enforcement of air pollution standards.

State or Local Air Monitoring Stations (SLAMSs). Similar to NAMS, these nonfederal stations make up the ambient air quality monitoring network which is required to be provided for to implement the SIP. Any combination of SLAMS and NAMS may occupy the same facility.

Stationary Source. An emission source which is a plant or an area that remains fixed geographically. The major pollution is due to combustion of fuels in furnaces and incinerators. The opposite is a Mobile Source like automobile.

Standard Method. A method of known and demonstrated accuracy issued by a competent organization.

Standard Methods. Used in a short form for the reference document "Standard Methods for the Examination of Water and Wastewater," prepared jointly by the American Public Health Association, American Water Works Association, and Water Pollution Control Federation; it serves as a primary reference for water related analytical methods.

Steady-State. In a system for flow of material or energy, the balance condition in which inflow (of matter and/or energy) into the system equals the flow out.

Stoichiometric. Calculation of exact proportion or the mass of various elements or compounds involved in a chemical reaction as reactants and products.

Stoke's Law. The law that describes the terminal velocity at which an object falls through a fluid (air or water). The velocity is directly proportional to the square of the diameter of the object, the difference of the densities of the falling object and the fluid, and inversely proportional to the viscosity of the fluid.

Stratosphere. The atmosphere surrounding the earth is stratified in zones with distinct temperature variations. The second layer above the troposphere to about 30 mi is the stratosphere. Here temperature increases with altitude. The protective ozone layer is located in the stratosphere.

Streamline. The path that a particle or dye will follow in a nonturbulent flow. The lines perpendicular to the streamlines are equipotential lines; together they form a flow network.

Stripping. Methods of removing unwanted dissolved gases from water to a flowing gas (air) or vapor stream. It involves increasing the surface area of water and flowing the air (vapor) through it. The continuing operation of the passing air strips the dissolved gases from water and carries them with it to the atmosphere or a condenser. Also called Air Stripping.

Substrate. The reactant portion of any biochemical material, such as organics, that are transformed into a product, or any substance used as a nutrient (food) by microorganisms.

Superfund. *See* the Comprehensive Environmental Response, Compensation, and Liability Act (CERCLA) of 1980.

Superfund Amendments and Reauthorization Act (SARA). 1986 amendments to the CERCLA.

Synthetic Organic Chemicals (SOCs). Manmade organic chemicals; some are volatile, some remains dissolved in water. Many are persistent and harmful to health.

Temporary Hardness. Hardness that can be removed by heating of water. Heating drives off carbon dioxide, so that carbonate ions combine with dissolved calcium or magnesium ions to form precipitate.

Temperature. *See* Absolute Temperature.

Terminal Velocity. *See* Settling Velocity.

Terrestrial Radiation. The solar radiation is absorbed by atmosphere, ground, and water. From the surface of the earth, the energy is reradiated to atmosphere as heat (long wave infrared radiation).

Territorial Waters. The line marking the seaward limit of inland waters; it extends from the low water level to a 3 mi distance seaward.

Tertiary Treatment. The wastewater treatment beyond primary and secondary treatments to further improve the effluent quality or to remove inorganic matters. Methods comprise of chemical, electrochemical, biological, and other complex procedures.

Tetrachlorodibenzofuran (TCDF). A polyhalogenated hydrocarbon found as a contaminant in PCB which is highly toxic.

Tetrachlorodibenzo-para-dioxin (TCDD). An aromatic halogenated hydrocarbon that is one of the most toxic compounds known; it effects liver and kidney. Also known as Dioxin.

Thermal Conductivity. Heat conducting property of a substance; it is the quantity of heat which flows per unit time across unit area of a surface of unit thickness when the temperature difference on the face is one degree.

Theoretical or Total Oxygen Demand (TOD). The amount of oxygen required for the complete decomposition of a substance.

Thermal Inversion. Normally the temperature decreases with altitude in troposphere. However, due to meteorological conditions, sometimes the air temperature increases with altitude. This condition results in a layer of warmer air above the cooler pollutants layer which inhibits dispersion of pollutants giving rise to smog.

Thermal Pollution. The addition of a large quantity of heat to the environment (water, air, land) that effects the ecosystem. The hot water discharged from a plant forms a Thermal Plume. The warmer environment causes changes in aquatic organisms.

Thermal Power Plant. A plant that generates electricity by rotating turbine by using steam. Coal is the usual energy source to convert water to steam.

Thermal Process (Treatment). The processing of waste materials by application of high amount of heat. Incineration of hazardous waste and combustion of organic materials are examples.

Thermocline. *See* Epilimnion.

Thermodynamics and Thermodynamic Laws. Thermodynamics is the study of energy transformation involving heat energy. The first law is the concept of energy conservation; *see* Energy, and *also see* Second Law of Thermodynamics.

Threshold. The minimum amount of a substance when its effect becomes measurable or observable.

Threshold Limit Value (TLV). The threshold dose or concentration of a substance to which workers can be exposed on a daily basis for the life time without adverse effects.

Threshold Planning Quantity (TPQ). A quantity designated for each chemical on the list of extremely hazardous substances by the Superfund Amendments and Reauthorization Act (SARA). The TPQ amount at a facility places a requirement on that facility to notify the state emergency response commission and such facility is subject to emergency planning under SARA.

Tidal Energy. The conversion of the mechanical energy of rising and falling of tides into electrical energy by turning turbines.

Titration. A laboratory method to determine concentration of a constituent in a solution by neutralizing it with a solution of known strength to a completion state as signaled by a noticeable end point.

Tolerance Limit. The numerical value or range that identifies the acceptable level for a limiting factor in an ecosystem.

Tolerance Limit Median (TLM). The concentration that kills 50% of the test organisms within specified time, usually 96 hours; *see* LC50.

Toluene. A solvent used in paints and varnishes, also present in Gasoline. It is similar to benzene but less toxic.

Total Dynamic Head. The total energy imparted by a pump comprising the total static head of the elevation difference between the suction and discharge line plus the energy losses through the system.

Total Kjeldahl Nitrogen. The sum of nitrogen present in a sample in ammonia and organic nitrogen form; it does not include nitrite and nitrate nitrogen.

Total Carbon (TC). This includes both organic and inorganic forms of carbon compounds that are soluble as well as insoluble. The analysis involves converting all forms to carbon dioxide and, then, measuring the concentration of carbon dioxide. The measurement is rapid as compared to BOD. The parameter is an indicator of the strength of wastewater when inorganic carbons are not significant.

Total Dissolved Solids. *See* Solids.

Total Organic Carbon (TOC). A measure of the organic carbons both dissolved and suspended in a sample. The analysis is similar to total carbon on acidified sample to remove inorganic carbons. It is rapid and indicator of wastewater strength. There are empirical relations with BOD.

Total Solids (TS). *See* Solids.

Total Suspended Solids (TSS). *See* Solids.

Total Toxic Organics (TTO). The summation of all defined toxic organic materials greater than 0.01 mg/L.

Toxic. Materials that can cause acute (short term) or chronic (long term) damage to living tissues on contact or absorption.

Toxic Chemical Group. The toxic chemicals that can be grouped as: 1) organic solvents, 2) toxic metals, 3) pesticides, and 4) herbicides.

Toxic Substance Control Act (TSCA). The 1976 federal law that bans manufacture, sale, and use of any new or existing chemical, including PCB that poses an unreasonable risk to human health and the environment. It requires a premanufacture notice (PMN) from chemical producers for a new chemical to regulate its production.

Toxicant and Effect. Toxicants are toxic substances. They are classified according to their effect on the respiratory system: 1) Asphyxiants that deprive body of oxygen, 2) Irritants that irritate air passages, 3) Necrosis that results in cell death and adema, 4) Fibrosis that produces fibrotic tissues, 5) Allergens that induce allergy, and 6) Carcinogens that cause cancer.

Toxicity. The degree of danger posed by a toxic substance. The toxicity rating are: 1) Extremely toxic, 2) Highly toxic, 3) Moderately toxic, 4) Slightly toxic, and 5) Practically nontoxic.

Toxicity Characteristic Leaching Procedure (TCLP). A test designed by the EPA to determine the potential for leachate formation by a waste (organic or inorganic) under simulated conditions similar to a site. During testing, if the extract obtained from the sample contains specified materials more than their allowable values, then the waste is defined as toxic waste.

Trace Elements. Elements essential for animal and plant life but required only in small quantity, such as manganese, iron, zinc, copper, molybdenum, and cobalt.

Trace Metals. Metals found in small quantities in air, water, soil, or food due to their insolubility.

Transfer Station. In solid waste management, a facility where smaller truckloads are compacted and reloaded to a larger truck for transportation to a landfill. The transfer station can also act as a Resource Recovery Facility (RRF) where recyclable materials could be separated out.

Transmissivity (T). A lumped parameter to indicate the flow characteristic of an aquifer as a whole; it equals hydraulic conductivity times the aquifer saturated thickness.

Transport. In air modeling, the movement of pollutants in the model domain. For hazardous wastes, the movement of a substance by any mode as controlled by RCRA, including any stoppage in transit.

Transuranic Radioactive Waste. Waste containing more than 100 nano-curies of alpha emitting transuranic isotopes per gram of waste, with half-life greater than 20 years.

Treatment, Storage, and Disposal (TSD). As regulated by RCRA, any facility where hazardous waste is treated, stored, and/or disposed off.

Treatment Technology. A technology that treats a waste for the purpose of resource recovery, volume reduction, detoxification, or/and disposal. The treatment processes fall under the categories of thermal treatment, chemical treatment, physical treatment, and biological treatment.

Trickling Filter. One of the approaches of the biological treatment of wastewater in which a large tank is filled with stone-bed filter on which microbial growth takes place. The wastewater from primary tank is sprayed over the stone bed and allowed to trickle through. The organic waste is biodegraded by aerobic process as it trickles downward through the filter.

Trinitrotoluene (TNT). A highly explosive compound formed by toluene reacting with nitric acid in the presence of sulfuric acid.

Trophic Level. The levels indicating the order in which energy is transferred through a food chain. All organisms belonging to the same trophic level derive energy (food) from the previous trophic level.

Tropical Rain Forest. Characterized by heavy rains (50–200 in./year), and temperature between 70 and 95 °F, these are the richest terrestrial ecosystem in terms of the amount of biomass and biomass diversity. They occupy equatorial regions.

Troposphere. The lowest portion of the atmosphere from the earth surface to about 6–11 mi above. About 75% of the total air supply is contained within the troposphere. The temperature decreases with altitude at adiabatic lapse rate (ALR) in this layer.

Turbidity. A condition caused by the suspended matter in water that results in reduction of visibility through the water. It is measured in arbitrary turbidity units based on light diffraction.

Ultrafiltration. A treatment similar to reverse osmosis except that it treats solution with larger solute particles so that the solvents can more easily filter through a membrane.

Ultrasonic. With application of acoustic waves with frequency above the sound waves audible by human ear (above 20 kHz).

Ultraviolet (UV) Radiation. The portion of the electromagnetic spectrum that extends from the violet band of visible light (wavelength of about 0.4 micrometer) to the x-rays (wavelength of about 0.001 micrometer). Radiation in this range does not ionize matter and, hence, is termed nonionizing. The spectrum between wavelengths 0.001 and 0.16 micrometer does not transmit through air and, hence, is of a little significance to the environment. The remaining spectrum is divided in three classes: 1)

UV-C Range from wavelength 0.16 to 0.29 micrometer, referred to as far-ultraviolet; this range is most effective in killing microorganisms, 2) UV-B Range from 0.29 to 0.32 micrometer; this range is most responsible for sunburn and cancer cases, 3) UV-A Range from wavelength 0.32 to 0.4 micrometer, referred to as near-ultraviolet; this range responsible for skin tanning. UV radiation between 0.13 to 0.2 micrometer generates ozone in the stratosphere.

Unconfined Aquifer. *See* Aquifer.

Underground Injection. A method of solid waste disposal wherein the waste mixed with water is discharged into a bored, drilled, driven, or dug well.

Underground Injection Control (UIC). A program required in each state under the Safe Drinking Water Act that requires that an applicant must demonstrate that the well has no reasonable chances of adversely affecting underground drinking water before a permit for underground injection could be issued.

Underground Storage Tank (UST). Under the RARA, all tanks with at least 10% of their volume beneath the ground level should meet certain performance standards, should have spill controls, and regularly monitored for leaks. When not in service, they should be closed as per prescribed procedure.

Uniform and Nonuniform Flow. *See* Flow, Uniform and Nonuniform.

Universal Gas Constant. *See* Ideal Gas Law.

Unregulated Hazardous Substance. A hazardous material for which no standards, requirements, criteria, or limitations are in effect under the Clean Air Act, the Clean Water Act, the Safe Drinking Water Act, or the Toxic Substances Control Act.

Uranium. In 1942, it was discovered that heavy atom of uranium can be split (fissioned) to derive tremendous energy. Uranium is found in natural deposits combined with other elements. Uranium-238, is the most abundant isotope that naturally occurs, but it is not fissionable. This has to be first enriched to Uranium-235 isotope which is fissionable, to be used as a fuel in nuclear reactor. Also *refer to* Spent Fuel.

Urban Runoff. The stormwater runoff from city streets and buildings that carry pollutants to a sewer system or a receiving water.

Vacuum Filtration. A process in sludge dewatering.

Vadose Water. The water that exists in unsaturated soil between the ground surface and the watertable

Valence. The number of electrons an atom gives up or receives when a chemical bond is formed in a reaction or when ions are formed from elements during ionization.

Vinyl Chloride. A cancer causing colorless gas formed by reaction of acetylene and hydrogen chloride or by cracking of ethylene dichloride.

Viruses. An intracellular parasite microorganism smaller than bacteria. They are capable of producing infection and causing diseases, and are insensitive to antibiotics. They do not multiply by division as bacteria do.

Viscosity (Dynamic). A measure of the resistance of a fluid to flow against a surface. It is a ratio of the shear stress that develops between adjacent strata of flowing fluid to the velocity gradient between the strata. It is expressed as mass per length per time. Also *refer to* Kinematic Viscosity.

Visible Range. The part of the electromagnetic spectrum in the wavelengths range of 0.4 to 0.71 micrometer, that can be seen by the human eye.

Vitrification. A process of melting a solid substance with very high temperatures and then solidifying into noncrystalline (glass like) material. This is used as a treatment for radioactive or hazardous wastes to contain them into a solid mass.

Volatile Organic Compounds (VOC). A group of organic compounds with high vapor pressures. They constitute a major category of air contaminants. Thousands of individual compounds exist within the group including those exhausted by automobiles and organic solvents lost to evaporation. Some participate in photochemical smog.

Washout. The gaseous and particulate air pollutants captured by precipitation.

Wasteload Allocation. The maximum amount of a pollutant material that a discharger of a waste is allowed to release into a particular waterway; this allocation is made so that the total pollutant from all dischargers does not exceed the expected limit.

Waste Minimization. The prevention or restriction of waste generation at its source by redesigned products or the pattern of consumption; this includes recycling activity undertaken by a generator.

Waste Minimization Policy. According to the Hazardous and Solid Waste Act, it is the national policy to reduce the generation of hazardous waste rather than controlling waste after its generation. EPA has developed regulations to meet this policy.

Wastewater. Commonly used as a synonym for sewage; it is the spent water from homes and commercial and institutional sources that mostly contains organic materials.

Wastewater Treatment System, Complete. All the treatment works necessary to meet the requirements of CWA, Title III. This comprises of 1) the transportation of wastewater from individual homes to a treatment plant, 2) the treatment of wastewater to remove pollutants, and 3) the disposal of the treated wastewater effluent and sludge.

Water Balance. A balance sheet for all waters that are entering and leaving a unit process or operation in any form via any mode, i.e., raw material, intermediate product, finished product, by-product, process leaks.

Water Budget. An accounting for quantity of water that constitute the inflow into, the outflow from, and the change of storage within an hydrologic unit, such as hydrologic cycle.

Water, Consumptive Use of. The use of water resulting in a large proportion of loss to the atmosphere by evapotranspiration, such as drinking and irrigation. The other is Nonconsumptive Use of Water in which very little portion is lost to the atmosphere by evapotranspiration and practically the entire amount returns to stream or ground, such as hydropower.

Watercourse. A natural or artificial waterway in which water flows continuously or intermittently with some degree of regularity.

Water Quality Criteria. The concentration limits for pollutants in water that are set for individual pollutants and for different water uses, such as for drinking, irrigation, industrial uses, or recreation. These limits render the water suitable for its intended use.

Water Quality Limited Segments. A portion of a stream where it is known that water quality can not meet applicable quality standards even after applying the technology-based effluent limitations.

Water Quality Standards. The ambient standards for water bodies adopted by a state and approved by the EPA. These are based on the use of the water body and the water quality criteria which must be met to protect the designated use or uses.

Water Rights. The right to divert or store water for a beneficial use such as drinking water, irrigation, hydropower.

Watershed. *See* Catchment Basin.

Watertable. The wetting surface underground below which soil is saturated with water; it is top of the saturation zone or groundwater.

Water Withdrawal. The amount of water removed from a surface or groundwater source for a use.

Wavelength. The waves, including electromagnetic, propagate as ripples forming crests and troughs. The distance between two successive crests is the wavelength. The number of times a wave will travel from one crest to the next in one second is its frequency.

Weight. The resultant force of attraction by the earth on any mass of a body; it equals the mass times the gravitational constant, g ($g = 32.2$ ft/s^2 or 9.81 m/s^2).

Weir. A low dam, barrier, or notch placed in a channel or tank, over which the water flows; it controls the flow of water and measures the amount of water flowing over it.

Wet Scrubbing. *See* Scrubber.

Wetlands. A land area adjoining a water body that is saturated or the watertable lies close to the surface. It is characterized by a prevalence of vegetation and organisms. Examples include swamps, bogs, marshes, fens, and estuaries.

Wind Profile. The variation of wind speed and direction with height. The variation follows a logarithmic equation or the power law. An empirical relation in which the wind speed varies as a power of the height (altitude) is applied to estimate the wind speed at higher elevations from wind measurements at ground level.

Work. It equals the force applied on a body times the distance that body moved or the pressure applied to a system times volume change within the system.

X-Rays. The electromagnetic radiation of a wavelength shorter than ultraviolet and longer than gamma rays. It is ionizing since it produces ions in a material it strikes and has penetration power.

Zeolite. Natural and synthetic silicates used for water softening in ion-exchange process.

Zero Discharge. No discharge of an effluent into a receiving water. This is the goal of the Clean Water Act.

Zone of Aeration. *See* Vadose Zone.

Zone of Saturation. *See* Saturated Zone.

Zooplankton. *See* Plankton.

INDEX

A

Abiotic component, 1, 129, 136
Above ground storage tank (AST), 396
Acid rain, 275, 335
Acids (acidity), 100, 263
Activated sludge process, 230, 304
    aeration tank, 306
Adaptability, 182
Adiabate lapse rate (ALR), 75, 338
Adsorption, 226, 298, 300
Advanced treatment. *See* Wastewater treatment
Advection. *See* Convection
Aeration period. *See* Detention time
Aeration tank. *See* Activated sludge process
Air composition, 321
Air stripping, 297
Alkanes, 111, 113, 114
Alkalinity, 263. *See also* Bases
Alkenes, 111, 113, 114
Alkynes, 111, 113, 114
Anaerobic digestion, 232
Analysis;
    gravimetric, 258
    volumetric, 258
    colorimetric, 259
Analytical procedures. *See* Analysis
Angstrom (°A), 36
Aquifer, 285
Aromatic, 115
Atmosphere, 1, 321
    fate of chemicals, 330
    pressure variation, 324
    stable, 338
    strongly stable, 338
    unstable, 338
    upper, 324
Atom, 31
    structure, 34

Atomic;
    dimension, 35
    mass unit (AMU), 35, 96
    number, 35
    weight, 35
Attached growth, 229, 306
Autotrophic, 135
Avogadro's hypothesis, 96, 105

B

Backwash, 216, 303
    hydraulics, 219
Bacteria. *See also* Microorganisms
    aerobes. 149
    anaerobes, 149
    autotrophs, 149
    facultate, 149
    heterotrophs, 149
Bases, 100
Best available technology (BAT), 22
Biochemical oxygen demand, 156, 229, 266
Bioconcentration, 255
Biodiversity, 181, 185, 190
Biofuels energy, 82
Biogeochemical cycles, 137, 167
Biological diversity. *See* Biodiversity
Biological processes, 230, 304
    activated sludge. *See* Activated sludge
    aerobic, 229
    anaerobic, 229
    attached growth, 229
    sludge. *See* Sludge
    suspended growth, 229
    trickling filter. *See* Trickling filter
Biomass, 134, 146
    growth rate, 151
    pyramid, 146
Biome, 196

Bioremediation, 399, 402
Biosphere, 1
Biotic component, 1, 129
Biotic potential, 179
BOD loading, 307
BOD. *See* Biochemical oxygen demand
Boyle's Law. *See* Gas

C

Carbon cycle, 171
Carnivores. *See* Consumers, secondary
Catalyst, 220
Cathode protection technique, 393
CERCLA, 383
Charles Law. *See* Gas
Chemical extraction, 399, 402
Chemical reactions, 95
Chlorofluorocarbon (CFC), 117, 329, 348
Clean Air Act, 327
Clean Water Act (CWA), 246, 276, 392
Closed boundary. *See* Controlled volume
Coal, 118. *See also* Fuel, fossil
Cogulation, 210
Coliform bacteria, 153, 266
Colorimetric. *See* Analysis
Community, 129
Compound, 31
    ionic, 100
    molecular, 100
    organic, 109, 116
Concentration, 98, 107, 276
    biomass, 151
Consumers, 134, 143. *See also* Food chain, Food Web
    primary, 134
    secondary, 134
    tertiary, 134
Containment system. *See* Landfill
Contaminants. *See* Pollutants
    soluble, 291
Contaminated groundwater, 398
Contaminated soils, 398
Contamination, 21
    groundwater, 285, 288, 398
Continental shelf. *See* Sea Zonation
Continuity equation, 42
Controlled-volume, 5, 43
Convection, 206

Corrosivity, 382
(The) Cradle to Grave Concept, 384
Critical height, 293
Cunningham correction factor, 331
Cyclone separator, 352

D

Dalton's Law. *See* Gas
Darcy's Law, 287, 293
Decomposers, 135, 143
Deforestation. 188, 387
Deserts, 198
Desulfurization;
    fuel, 355
    fuel gas, 355
Detention time, 305
    hydraulic, 221
Detritus feeders, 135
Dewatering, 312
Diffuse layer, 224
Diffusion, 205
    molecular, 209
Dilution, 154, 157, 276
    factor, 248
Disinfection process, 228
Dispersion, 300
    in atmosphere, 333
    coefficient, 209
Disposal, 312, 378, 387
Dissolved oxygen (DO), 264, 278
    saturation concentration, 277
Double-Layer Theory, 224
Double-Linear System, 380
Drag coefficient, 213, 331

E

Ecological magnification. *See* Bioconcentration
Ecological system. *See* Ecosystem
Ecology, 1
Ecosystem, 1, 129, 167
    characteristics, 182
    climax, 184
    dynamic balance, 181
    energy subsidized, 143
    forest, 389
    fresh water. *See* Fresh water ecosystem
    functioning, 167
    marine. *See* Marine ecosystem

Index

solar-powered, 143
stresses, 179
terrestrial. *See* Terrestrial ecosystem
Electrical instrument, 262
Electrolyte. *See* Ions
Electromagnetic waves;
    frequency, 70
    spectrum, 71
    speed, 70
    wavelength, 70
Electron, 35
Electrostatic precipitator (ESP), 353
Element, 31
Endogenous phase. *See* Microbial growth
    decay coefficient, 151
Endothermic. *See* Process
Energy, 3, 59
    alternative strategy, 79
    biofuels, 82
    chemistry, 117
    conservation, 63, 80
    entropy, 64, 67, 147
    free, 147
    geothermal, 82
    hydrogen, 83, 118
    hydropower, 81
    kinetic, 61
    nuclear, 86, 119
    ocean thermal, 82
    potential, 61
    pyramid, 146
    renewable, 81
    Second Law, 64
    solar, 81
    tidal, 82
    units, 60
    windpower, 81
Enthalpy, 147
Entropy. *See* Energy
Environment, 1
    pollution, 21
Environmental degradation, 9
Environmental resistance, 179
Equilibrium constant, 101
Ethanol, 118

Eutrophication, 282
    cultural, 283
Evaporation, 171, 243

Exponential phase. *See* Microbial growth
Exothermic. *See* Process

F

Fabric filters, 353
Fate of pollutants. *See* Pollutants
Fickian. *See* Diffusion
    Fickian transport, 206
Fick's Law, 206, 292
Filtration. *See* Processes
    hydraulics, 218
Final clarifier, 304, 309
Final sedimentation tank. *See* Final clarifier
Fire;
    crown, 139
    surface, 139
Fission, 37. *See* Radioactivity
Floating booms, 300
Flocculation, 210, 302
Flow;
    laminar, 218
    streamline, 218
Flue gas treatment techniques. *See* Nitrogen oxides
Fluidization principle, 219
Food chain, 134
Food to microorganism (F/M) ratio, 152, 231, 306
Food web, 134
Forests 197, 388
Free energy. *See* Energy
Freshwater ecosystem, 191
Fuel;
    fossil, 69
    recovery, 119
Fusion, 37. *See* Radioactivity

G

Gas, 105
    Charles Law, 105
    constant, 105, 324
    Dalton's Law, 108
    Henry's Law, 108, 208
    ideal gas, 106, 324
    solubility. *See* Solubility
    transfer process, 226
Gas chromatography, 262
Geothermal energy, 82
Global warming potential, 343

Grasslands, 198
Gravimetric. *See* Analysis
Gravity filtration, 216
Gravity settler, 351
Greenhouse effect, 341, 343
Greenhouse gases, 344
Groundwater, 140, 285
    contaminated. *See* Contaminated groundwater
    control, 296
    treatment techniques, 297

H

Hazardous Solid Waste Amendments Act (HSWA), 365, 383
Hazardous waste, 382
    management, 385
    treatment technology, 386
Henry's Law. *See* Gas
Herbivores. *See* Consumers, Primary
Hydraulic conductivity, 285, 287, 293
Hydraulic detention time. *See* Detention time
Hydraulic gradient, 287, 293
Hydraulic loading, 308
Hydrocarbon compounds. *See* Organic compounds
    Also see Alkane, Alkene, Alkyne, Aromatic
Hydrogen gas. *See* Energy
Hydrologic cycle. *See* Water cycle
Hydropower. *See* Energy
Hydrosphere, 1

I

Ideal gas. *See* Gas
Ignitability, 382
Incineration, 378
Ion, 37, 100
Ion exchange process, 225
Ionization constant, 102
Ionosphere, 324
ISDS, 303
Isotopes, 35
IUPAC names. *See* Organic names

K

Kozeny equation. *See* Filtration, hydraulics

L

Lag phase. *See* Microbial growth
Lake stages, 282
Lake zones, 191
Landfill, 378
    capping, 380
Landfarming, 398, 401
Law of Mass Action. *See* Le Chatelier Principle
Law of the Minimum, 136
Law of Tolerance, 136
Leachate, 380
Leak detection technology, 393
Leaking underwater storage tank (LUST), 393
Le Chatelier Principle, 101
Less developed countries, 8
Limiting factors, 137
    law, 136
Limits of tolerance, 179
Lithosphere, 1
Logistic Curve Method. *See* Population

M

Marine ecosystem, 193
Mass. *Also see* Matter
    conservation, 38
    Newton's Law, 38
Mass flux, 208
Mass transfer coefficient, 208
Materials balance, 43
Matter, 3, 31
    classification, 32
    heirarchical organization, 33
    properties, 38
    radioactive, 36
    theory, 34
Maximum containment level goal (MCLG), 249
Maximum containment level (MCL), 249
    secondary, 251
Meinzer's classification, 286
Membrane separation process, 227
Mesosphere, 323
Metabolism, 135
Methane, 330
Methanol, 118
Microbial growth, 150
    endogenous phase, 150
    exponential phase, 150
    lag phase, 150

stationary phase, 150
Microorganisms, 148. *Also see* Decomposers
   kinetics, 149
Mixed liquor, 17
   suspended solid (MLSS), 230, 306
   volatile suspended solid (MLVSS), 230
Mixing, 302
Mixture, 31
Mobile sources, 328
Model, 5
   basic environmental quality, 6
   blackbody, 72
      components, 8, 12
   extension, 9
   mathematical, 6
Molarity, 99
Mole, 96
Molecule, 31
More developed countries (MDC), 8, 25
Most probable number, 153
Municipal solid waste, 27
   collection. *See* Solid waste collection
   composition, 367
   management, 372
   processing, 377
   seperation, 372

N

National ambient air quality standards (NAAQS), 327
National pollutants discharge elimination system (NPDES), 247
Natural gas, 118
Nessler tubes. *See* Colorimetric
Neutron, 35
New source performance standards (NSPS), 327
Newton;
   action-reaction, 40
      basic laws, 38
   motion, 39
   universal gravity, 40
Nitrogen
   activation, 174
   ammonification, 174
   dentrification, 174, 229
   fixation (fixing), 174
   nitrification, 174, 229
Nitrogen cycle, 173
Nitrogen oxides;
   flue gas treatment techniques (FGT), 356
   selective catalytic reduction (SCR), 356
Nonaqueous phase liquids (NPL), 292
Nonpoint sources. *See* Pollution
No Observable Adverse Effect Level (NOAEL), 249
Nonwetting fluids, 292
Normality, 258
Nutrient cycles. *See* Biogeochemical Cycle

O

Ocean pollution, 298
Ocean thermal energy, 82
Oil spills, 299
Omnivores, 134
One-way society. *See* Throwaway society
Optical instrument, 261
Organic, 109
   compounds, 109
   formulas, 112
   matter, 277
   names, 112
Overturn, Lake;
   fall, 284
   spring, 284
Oxygen cycle, 176
Oxygen sag curve, 279
Ozone, 178, 330
Ozone depleting potential, 349
Ozone depletion, 347
   chemicals, 348

P

Partial pressure of gases. *See* Dalton's Law
Particulate matter (PM-10), 327, 351
Parts per million, 98, 107, 326
Periodic table, 96
Permeability (coefficient). *See* Hydraulic conductivity
Petroleum, 117
pH, 103, 263
Photochemical smog, 5, 176
Photoelectric device, 259
Photolysis, 179
Photosynthesis, 69, 131
Photovoltaic (PV) cells, 85, 119
Piezometric head, 285
Point sources. *See* Pollution

Pollutants, 169, 326
    concentration, 276
    fate in atmosphere, 330
    fate in groundwater, 290
    fate in soil matrix, 371
    fate in water, 254
    quality standards, 327
    units of measurement, 326
Pollution 21, 244
    acceptable level, 22
    aquatic, 276
    atmospheric water. *See* Acid rain
    control, 22
    lake, 282
    nonpoint, 22, 253
    ocean. *See* Ocean pollution
    point, 22, 253, 276
    socially optimum level, 21
    stream. *See* Stream pollution
    water resources, 252
    zero, 21
Population, 13, 129
    arithmetic growth, 14
    declining rate of growth, 15
    exponential growth, 15
    logistic curve, 17
Power, 59
Precipitation (chemical), 225, 298
Precipitation (rain), 171, 243
Prevention of significant deterioration (PSD), 327
Primary clarifier. *See* Sedimentation tank
Primary standards, 327
Primary treatment. *See* Wastewater treatment
Process;
    endothermic, 95, 138
    exothermic, 95, 138
Processes;
    activated sludge. *See* Activated
        sludge process
    adsorption. *See* Adsorption process
    biological. *See* Biological process
    chemical, 220, 223
    disinfection process. *See* Disinfection process
    filtration, 216, 302
    gas transfer. *See* Gas transfer process
    ion-exchange. *See* Ion exchange process
    membrane separation. *See* Membrane
        separation process
    sedimentation, 210, 302
    settling, 209
    titration. *See* Titration process
    transport, 205
    trickling filter. *See* Trickling filter process
    unit. *See* Unit process
Producers, 130, 143
Productivity;
    gross primary, 134
    net primary, 134
Proton, 35
Pyramid of energy. *See* Energy
Pyrolysis, 378

Q

Quality of atmosphere, 325
Quality of processed water, 249
Quality of water, 244
    characteristics, 245, 257
    gross measures, 245
    specific measures, 245
Quality standards. *See* Pollutants

R

Radiation, 36
    alpha, 36
    amount, 70
    beta, 36
    gamma, 36
    nuclear, 36
Radioactivity, 36
    induced, 37
Reaction;
    heterogeneous, 220
    homogenous, 220
    kinetic, 221
Reactivity, 382
Reactor, 220, 222
Reference dose, 249
Refuse, 367
Refuse-derived fuels, 378
Renewal time. *See* Residence time
Requirements for UST, 395
Residence time, 169
    mean, 221
Residual saturation;
    nonwetting, 292
    wetting, 292
Resource, 8

Index

base, 17
classification, 18
consumption, 17
flow, 17
nonrenewable, 17
renewable, 17
stock, 17
Resource Conservation & Recovery Act (RCRA), 365,
373, 383, 392
UST program, 394
Resource Conservation & Recovery, 377
Respiration, 134
Return sludge flow, 307
Reverse osmosis. *See* Membrane separation
Reynolds number, 213, 331

S

Safe Drinking Water Act (SDWA), 249
Sand filter, 217
Saturation concentration *See* Dissolved oxygen
Scavenger. *See* Detritus feeders
Screening, 302
Scrubbers, 353
limestone, 353
Secondary standards, 327
Secondary treatment. *See* Wastewater treatment
Sedimentation. *See* Process, sedimentation
tank, 303, 305
Sediments, 277
Sea zonation, 193
Selective catalytic reduction. *See* Nitrogen oxides
Semiarid regions, 198
Settlement. *See* Process, sedimentation
Sludge, 229, 312
processes, 312
treatment. *See* Wastewater treatment
Small quantity generators (SQG), 384
Smog, 337. *See also* Photochemical smog
Soil;
horizons, 140
profile, 140
treatment technologies, 398, 401
water, 141
Soil venting, 399, 402
Solar;
active, 83
energy, 81, 83, 85

passive, 83
spectrum. *See* Electromagnetic waves
thermal-electric engine, 84
Solid waste collection, 373
Solid wastes, 365
classification, 366
physical properties, 369
Solidification, 398, 401
Solids loading, 309
Solids;
total, 265
volatile, 265
Solubility, 99
product constant, 103
gases, 108
Solute. *See* Solution
Solution, 31
liquid, 98
solute, 98
solvent, 98
Solvent. *See* Solution
Species, keystone, 181
Specific yield. *See* Storage coefficient
Speciation, 184
Spectrophotometer, 261
Stabilization, 229, 312, 378, 398, 401
Standing crop. *See* Biomass
State implementation plans (SIP), 327
Stationary phase. *See* Microbial growth
Steady-state, 43
Stefan-Boltzmann Law, 72
Stern layer, 224
Stoichiometry, 96
Stoke's Law. *See* Terminal velocity
Storage coefficient, 285
Stratification;
atmosphere, 322
lake, 283
Stratosphere, 323
Stream pollution, 275
Streamline. *See* Flow
Streeter-Phelps Model. *See* Oxygen sag curve
Substrate, 150, 229
Succession, 184
Superfund Amendments and Reauthorization Act (SARA), 383, 394
Superfund. *See* CERCLA
Surface overflow rate (SOR), 215, 305
Suspended growth, 229, 306

Sustainable earth, 77
Sustained yield, 19
Synthetic Organic Chemicals (SOC), 246
System;
    conserved, 44
    nonconserved, 49
    steady-state, 168, 181
    stored, 46
Systematic names. *See* IUPAC

T

Terminal velocity, 212, 331
Terrestrial;
    ecosystem, 195
    environment, 365
Thermal inversion, 75, 338
Thermal stripping;
    low temperature, 399, 401
    high temperature, 399, 401
Thermodynamics, 61
    First Law, 63
    Second Law, 64
Thermosphere, 324
Thickening, 312
Throwaway society (model), 77
Tidal energy, 82
Titration process, 258
Total maximum daily load (TMDL), 248
Toxicity characteristic leaching procedure (TCLP), 383
Toxicity, 383
Toxins, 277
Treatment train, 205
Treatment, storage, disposal, 384
Trickling filter, 232, 304, 307
Trophic levels, 134, 146. *See also* Food chain
Troposphere, 323

U

Ultraviolet rays, 341, 349
Underground storage tank (UST), 387
    closure, 396
    existing system, 396
    new system, 396
    technology, 392
Unit processes, 301
    biological, 301
    chemical, 301
    physical (operation), 301
Unit waste factor, 368
Unsteady state, 43

V

Vitrification, 398, 401
Volatile organic chemicals (VOC), 246, 329
Volatization. *See* Soil venting
Volumetric. *See* Analysis

W

Waste generation, 370
Waste minimization, 385
Wastewater treatment, 301, 303
    advanced, 311
    primary, 303
    secondary, 304, 306
    sludge, 311
Water cycle, 167, 169
Water table, 285
Water treatment, 301
Weight, 3, 40
Wein's Law, 72
Wetlands, 188, 192, 390
Windpower, 81
Work, 59
    macroscopic, 59
    microscopic, 59

Y

Yield Coefficient, 151

Z

Zeta potential, 225

Government Institutes Catalog
To order a complete catalog of our books, electronic products, training materials and course, CALL (301) 921-2355.

	Edition	Price	PC #
ENVIRONMENTAL			
CFR Chemical Lists on CD ROM	1996	$110	4080
Chemical Guide to the Internet	1996	$69	519
Clean Air Handbook, 2nd Edition	1993	$89	343
Clean Water Handbook, 3rd Edition	1996	$89	512
EH&S Auditing Made Easy	1997	$79	581
▪ EMMI-Envl Monitoring Methods Index for Windows-Network	1997	$537	4082
▪ EMMI-Envl Monitoring Methods Index for Windows-Single User	1997	$179	4082
Environmental Audits, 7th Edition	1996	$79	525
Environmental Engineering and Science: An Introduction	1997	$79	548
Environmental Guide to the Internet, 3rd Edition	1997	$59	578
Environmental Law Handbook, 14th Edition	1997	$79	560
Environmental Regulatory Glossary, 6th Edition	1993	$79	353
Environmental Statutes	1997	$69	562
▪ Environmental Statutes Book/Disk Package	1997	$207	562
▪ Environmental Statutes on Disk for Windows-Network	1997	$415	4060
▪ Environmental Statutes on Disk for Windows-Single User	1997	$135	4060
Environmentalism at the Crossroads	1996	$39	570
ESAs Made Easy	1996	$59	536
▪ GI Environmental Database-Network	1995	$447	4073
▪ GI Environmental Database-Single User	1995	$149	4073
Industrial Environmental Management: A Practical Handbook	1996	$79	515
▪ IRIS Database-Network	1996	$237	4078
▪ IRIS Database-Single User	1996	$495	4078
ISO 14000: Understanding Environmental Standards	1996	$69	510
ISO 14001: An Executive Report	1996	$59	551
Lead Regulation Handbook	1996	$79	518
Principles of EH&S Management	1995	$69	478
Property Rights: Understanding Government Takings	1996	$79	554
RCRA Hazardous Waste Handbook, 11th Edition	1995	$115	503
TSCA Handbook, 3rd Edition	1997	$95	566
Wetlands: An Introduction to Ecology, the Law and Permitting	1995	$65	467
Wetland Delineation Manual	1987	$59	367
Wetland Mitigation: Mitigation Banking and Other Strategies	1996	$75	534
State Wildlife Laws Handbook	1993	$94	357
SAFETY AND HEALTH			
Construction Safety Handbook	1996	$79	547
Cumulative Trauma Disorders	1997	$59	553
Forklift Safety	1997	$65	559
Fundamentals of Occupational Safety & Health	1996	$49	539
Making Sense of OSHA Compliance	1997	$59	535
Managing Change for Safety and Health Professionals	1997	$59	563
▪ OSHA CFRs Made Easy-Network	1997	$387	4084
▪ OSHA CFRs Made Easy-Single User	1997	$129	4084
OSHA Technical Manual, Electronic Edition	1997	$99	4086
Safety & Health in Agriculture, Forestry and Fisheries	1997	$125	552
Safety & Health on the Internet	1996	$39	523
Safety Made Easy	1995	$49	463

We also carry all 50 Titles of the Code of Federal Regulations.
Call for price and availability.

Government Institutes

4 Research Place, Suite 200 • Rockville, MD 20850-3226
Tel. (301) 921-2355 • Fax (301) 921-0373
E-mail: giinfo@govinst.com • Internet: http://www.govinst.com

ⓖ ORDER FORM

Qty.	Product Code	Title	Price

Subtotal_____
MD Residents add 5% Sales Tax_____
Shipping and Handling_____

Within US: Add $6/item for 1-4 items. Add $3/item for 5+ items/
Outside US: Add $15 /item for Airmail. Add $10/item for Surface

Payment Enclosed_____

Method of Payment

❏ Check (*payable to Government Institutes in US dollars*) $ _____

❏ Purchase Order (please refer to Source Code: BK)

❏ Credit Card: Exp.___/___   ❏ MC   ❏ VISA   ❏ AMEX

   Credit Card No. _____

   Signature. _____

Name: _____
Company: _____
Address: _____
City: _____ State/Province: _____
Zip/Postal Code: _____ Country: _____
Telephone: (   )_____ Fax: (   )_____
E-mail Address: _____

Source Code: BK

Government Institutes

4 Research Place, Suite 200 • Rockville, MD 20850-3226
Tel. (301) 921-2355 • Fax (301) 921-0373
E-mail: giinfo@govinst.com • Internet: http://www.govinst.com

WITHDRAWN
O'Shaughnessy-Frey Library
University of St. Thomas
St. Paul, MN 55105
Libraries

GE 105 .G86 1997
Gupta, Ram S.
Environmental engineering
 and science

Encyclopedia of Applied Spectroscopy

Edited by
David L. Andrews

Related Titles

Hollas, J. M.

Modern Spectroscopy

2003
ISBN: 978-0-470-84416-8

Gauglitz, G., Vo-Dinh, T. (eds.)

Handbook of Spectroscopy

2003
ISBN: 978-3-527-29782-5

Bohr, H. G. (ed.)

Handbook of Molecular Biophysics Methods and Applications

2009
ISBN: 978-3-527-40702-6

Encyclopedia of Applied Spectroscopy

Edited by
David L. Andrews

WILEY-VCH Verlag GmbH & Co. KGaA

The Editor

Prof. David L. Andrews
School of Chemical Sciences
University of East Anglia
Norwich, United Kingdom

Cover
Cover image reproduced
with kind permission
of Cindy Mitchell.

All books published by **Wiley-VCH** are carefully produced. Nevertheless, authors, editors, and publisher do not warrant the information contained in these books, including this book, to be free of errors. Readers are advised to keep in mind that statements, data, illustrations, procedural details or other items may inadvertently be inaccurate.

Library of Congress Card No.: applied for

British Library Cataloguing-in-Publication Data
A catalogue record for this book is available from the British Library.

Bibliographic information published by the Deutsche Nationalbibliothek
The Deutsche Nationalbibliothek lists this publication in the Deutsche Nationalbibliografie; detailed bibliographic data are available on the Internet at ⟨http://dnb.d-nb.de⟩.

© 2009 WILEY-VCH Verlag GmbH & Co. KGaA, Weinheim

All rights reserved (including those of translation into other languages). No part of this book may be reproduced in any form – by photoprinting, microfilm, or any other means – nor transmitted or translated into a machine language without written permission from the publishers. Registered names, trademarks, etc. used in this book, even when not specifically marked as such, are not to be considered unprotected by law.

Composition Laserwords Private Ltd., Chennai, India
Printing betz-druck GmbH, Darmstadt
Bookbinding Litges & Dopf GmbH, Heppenheim

Printed in the Federal Republic of Germany
Printed on acid-free paper

ISBN: 978-3-527-40773-6

Contents

Preface *IX*

List of Contributors *XI*

1 Electromagnetic Radiation 1
   Wendell T. Hill, III

2 Gamma-Ray Spectroscopy 27
   Andrew M. Sandorfi, Christopher J. (Kim) Lister and Craig Woody

3 Mössbauer Spectroscopy 51
   De-Ping Yang

4 X-Ray Spectroscopy 87
   Thomas H. Markert[†] and Eckhart Förster

5 Positron Spectroscopy 115
   Paul G. Coleman

6 Muon Spectroscopy 153
   Upali A. Jayasooriya and Roger Grinter

7 Neutron Spectroscopy 183
   Anton Heidemann

8 Electron Scattering by Atoms, Ions, and Molecules 209
   Klaus Bartschat, Philip G. Burke and Albert Crowe

9 Photoemission and Photoelectron Spectroscopy 283
   Franz J. Himpsel and Ingolf Lindau

10 Ultraviolet and Visible Light Spectrometers 319
   John L. Hardwick

Encyclopedia of Applied Spectroscopy. Edited by David L. Andrews.
Copyright © 2009 WILEY-VCH Verlag GmbH & Co. KGaA, Weinheim
ISBN: 978-3-527-40773-6

11	Ultraviolet and Visible Absorption Spectroscopy 353 Robert H. Lipson	
12	Colorimetry 381 Robert T. Marcus	
13	Atomic Spectrometry and Elemental Analysis 421 Bernhard Welz and Daniel L.G. Borges	
14	Molecular Fluorescence 477 Bernard Valeur	
15	Resonance Energy Transfer 533 David L. Andrews and David S. Bradshaw	
16	Optical Spectroscopy of Biological Materials 555 Valery V. Tuchin	
17	Single-Molecule Fluorescence: Biophysics 627 Michael Prummer and Christian Hübner	
18	Laser Sources 655 David L. Andrews and Robert H. Lipson	
19	Linear Laser Spectroscopies 695 Stephen H. Ashworth	
20	Photophysical and Photochemical Dynamics 717 Ottó Horváth and Kenneth L. Stevenson	
21	Condensed Phase Ultrafast Dynamics 747 Stephen R. Meech and Ismael A. Heisler	
22	Ultrafast Spectroscopy 769 Oliver Kühn and Stefan Lochbrunner	
23	Surface Second Harmonic and Sum-Frequency Generation 817 Mitsumasa Iwamoto, Takaaki Manaka and Eunju Lim	
24	Raman Scattering for Speciation and Analysis 833 Karen Esmonde-White, Mekhala Raghavan and Michael Morris	
25	Spectrometers for Infrared Light 865 Yukihiro Ozaki and Shigeaki Morita	

26	Infrared Molecular Vibrational Spectroscopy 887 *Andrew B. Horn*	
27	Nuclear Magnetic Resonance Spectrometry 933 *Vladimir I. Bakhmutov*	
28	Biomolecular Structures by Solution Nuclear Magnetic Resonance 963 *Tharin M. A. Blumenschein*	
29	Mass Spectrometry 989 *Jürgen H. Gross*	
30	Chromatography 1055 *James M. Miller*	
31	Chemometrics and Multivariate Analysis 1103 *Richard G. Brereton*	
32	Fourier and Other Mathematical Transforms 1149 *Ronald N. Bracewell*[†]	

Glossary   *1177*

Index   *1191*

Preface

Modern spectroscopy is a broad and extraordinarily diverse field of activity, whose origins are readily traced to an innate inquisitiveness about our surroundings, and their wonderfully varied colours. In his poem 'The Dunce', de la Mare wrote: " 'Why is the grass so cool, fresh and green? The sky so deep, and blue?' Get to your Chemistry, you dullard, you!" Taken out of context – and I believe the irony would not have been lost on that gentle author – these lines could now be taken as a leitmotif for the subject that spectroscopy has become. But the originally intended sense of these lines also reflects a deeper truth – that it is somehow in our very nature to question the underlying reasons for colour and appearance.

From humble origins in the early science of colorimetry, a catalog of spectroscopic techniques has evolved. Now ranging from readily available and inexpensive methods such as infrared spectrometry, through to exotic and highly expensive facility-based techniques such as muon spectroscopy, the full range of this scale is duly represented within these covers. It is not claimed to be completely comprehensive; for example, space has not been found for either the uses of hyper-Raman scattering in studying changes of phase in crystalline materials, nor for electron paramagnetic resonance in the elucidation of metalloprotein structures, important as they are.

As the range of spectroscopies has developed, so too have the categories of application diversified. Validation, speciation, structure determination, trace detection, quality assurance, mechanism elucidation and diagnosis; each of these represents a type of operation in which spectroscopic methods have a well-proven and widely deployed value. Again, although the aim here has been to capture the realm of modern spectroscopy with an emphasis on applied techniques, the range of specific applications must only be regarded as indicative of spectroscopic capabilities. A focus on any one area of application – such as food, plastics, or remote sensing – could just as easily fill another volume just as weighty as this.

Many technical developments have had a significant impact on routine spectroscopic practice, not least the enormous advances in microelectronics, the fabrication of optical elements, and the development of tailor-made software packages. The latter have of course not only driven the move to PC control and operation of spectral acquisition, they have also changed the way we handle data. Some of the most important elements of analytical instrumentation have been developed from initially more obscure techniques – pulse sequence NMR and

Encyclopedia of Applied Spectroscopy. Edited by David L. Andrews.
Copyright © 2009 WILEY-VCH Verlag GmbH & Co. KGaA, Weinheim
ISBN: 978-3-527-40773-6

multiphoton absorption spectroscopy are good examples, also illustrating the fact that improvements in resolution by no means account for all spectroscopic advances. The field has progressed not only through advances in well-established methods – it has also spawned entirely new fields of spectroscopy.

This compilation would not have been possible without the keen involvement of an outstanding set of expert authors. It is a pleasure that several of my colleagues are named amongst them, but I am equally delighted that the work has attracted contributors from the four corners of the world. The success of this whole work can only compound my debt of gratitude to them. Finally, I must also add my sincere thanks to the editorial staff at Wiley-VCH for their meticulous care and dedication, and especially to Dr Christoph v. Friedeburg, who has ably and enthusiastically assisted at every stage in the development of this Encyclopedia.

Norwich, March 2009 *David L. Andrews*

List of Contributors

David L. Andrews
University of East Anglia
School of Chemistry
Norwich, NR4 7TJ
UK

Dr Stephen H. Ashworth
University of East Anglia
School of Chemistry
Norwich, NR4 7TJ
UK

Vladimir I. Bakhmutov
Texas A&M University
Department of Chemistry
P.O. Box 30012
College Station, TX 77842-3012
USA

Klaus Bartschat
Drake University
Department of Physics
and Astronomy
Des Moines, IA 50311
USA

Tharin M. A. Blumenschein
University of East Anglia
School of Chemistry
Norwich, NR4 7TJ
UK

Daniel L. G. Borges
Universidade Federal de Santa Catarina
Departamento de Química
88040-900 Florianópolis – SC
Brazil

Ronald N. Bracewell
Stanford University
Electrical Engineering Department
161 Packard Building
350 Serra Mall
Stanford, CA 94305-9505
USA

David S. Bradshaw
University of East Anglia
School of Chemistry
Norwich, NR4 7TJ
UK

Richard G. Brereton
University of Bristol
Centre for Chemometrics
School of Chemistry
Cantock's Close
Bristol BS8 1TS
UK

Philip G. Burke
The Queen's University of Belfast
University Road
Belfast BT7 1NN
Northern Ireland

Encyclopedia of Applied Spectroscopy. Edited by David L. Andrews.
Copyright © 2009 WILEY-VCH Verlag GmbH & Co. KGaA, Weinheim
ISBN: 978-3-527-40773-6

List of Contributors

Paul G. Coleman
University of Bath
Department of Physics
Bath BA2 7AY
UK

Albert Crowe
Newcastle University
School of Natural Sciences
Faculty of Science, Agriculture
and Engineering
Newcastle upon Tyne NE1 7RU
UK

Karen Esmonde-White
University of Michigan
Department of Chemistry
930 N. University Ave
Ann Arbor, MI 48109
USA

Eckhart Förster
Friedrich Schiller University
X-Ray Optics Group
Institute of Optics and
Quantum Electronics
07743 Jena
Germany

Roger Grinter
University of East Anglia
School of Chemistry
Norwich, NR4 7TJ
UK

Jürgen H. Gross
University of Heidelberg
Institute of Organic Chemistry
Im Neuenheimer Feld 270
69120 Heidelberg
Germany

John L. Hardwick
University of Oregon
Department of Chemistry
Eugene, OR 97403
USA

Anton Heidemann
250 Chemin de Pré la Côte
38410 St. Martin d'Uriage
France

Ismael A. Heisler
University of East Anglia
School of Chemistry
Norwich, NR4 7TJ
UK

Wendell T. Hill, III
University of Maryland
Joint Quantum Institute
Department of Physics
and Institute for Physical
Science and Technology
IPST Bldg
College Park, MD 20742
USA

Franz J. Himpsel
University of Wisconsin Madison
Department of Physics
5108 Chamberlin Hall
1150 University Ave.
Madison, WI 53706-1390
USA

Andrew B. Horn
University of Manchester
School of Chemistry
Oxford Road
Manchester M13 9PL
UK

Ottó Horváth
University of Pannonia
Institute of Chemistry
Department of General
and Inorganic Chemistry
P.O. Box 158
8201 Veszprém
Hungary

Christian Hübner
Universitaet zu Lübeck
Institut für Physik
Ratzeburger Allee 160
23538 Lübeck
Germany

Mitsumasa Iwamoto
Tokyo Institute of Technology
Department of Physical Electronics
2-12-1 O-okayama
Meguro-ku
Tokyo 152-8552
Japan

Upali A. Jayasooriya
University of East Anglia
School of Chemistry
Norwich, NR4 7TJ
UK

Oliver Kühn
Universität Rostock
Institut für Physik
Universitätsplatz 3
18055 Rostock
Germany

Eunju Lim
Tokyo Institute of Technology
Department of Physical Electronics
2-12-1 O-okayama
Meguro-ku
Tokyo 152-8552
Japan

Ingolf Lindau
Lund University
Department of Physics-SLF
Professorsgatan 1
S-22 100 Lund
Sweden

Robert H. Lipson
University of Western Ontario
Department of Chemistry
London, ON N6A 5B7
Canada

Christopher J. (Kim) Lister
Argonne National Laboratory
Physics Division
9700 S.Cass Ave
Argonne, IL 60439
USA

Stefan Lochbrunner
Universität Rostock
Institut für Physik
Universitätsplatz 3
18055 Rostock
Germany

Takaaki Manaka
Tokyo Institute of Technology
Department of Physical Electronics
2-12-1 O-okayama
Meguro-ku
Tokyo 152-8552
Japan

Robert T. Marcus
Sun Chemical Corporation
1701 Westinghouse Boulevard
Charlotte, NC 28273
USA

Thomas H. Markert[†]
Center for Space Research
Cambridge, MA
USA

Stephen R. Meech
University of East Anglia
School of Chemistry
Norwich, NR4 7TJ
UK

James M. Miller
Emeritus Professor of Chemistry
Drew University
36 Madison Ave
Madison, NJ 07940
USA

Shigeaki Morita
Nagoya University
EcoTopia Science Institute
Furo-cho, Chikusa-ku
Nagoya 464-8603
Japan

Michael Morris
University of Michigan
Department of Chemistry
930 N. University Ave
Ann Arbor, MI 48109
USA

Yukihiro Ozaki
Kwansei-Gakuin University
School of Science and Technology
2-1 Gakuen
Sanda 669-1337
Japan

Michael Prummer
F. Hoffmann-La Roche Ltd.
Pharmaceutical Divisions
Discovery Technologies
4070 Basel
Switzerland

Mekhala Raghavan
University of Michigan
Department of Chemistry
930 N. University Ave
Ann Arbor, MI 48109
USA

Andrew M. Sandorfi
Thomas Jefferson National
Accelerator Facility
Physics Division
12000 Jefferson Ave
Newport News, VA 23606
USA

Kenneth L. Stevenson
Purdue University Fort Wayne
Department of Chemistry
E. Coliseum Blud
Fort Wayne, IN 46805-1499
USA

Valery V. Tuchin
Saratov State University
Institute of Optics and Biophotonics
83 Astrakhanskaya str.
Saratov 410012
and
Institute of Precise Mechanics
and Control of RAS
24 Rabochaya str.
Saratov 410028
Russia

Bernard Valeur
CNAM
292 rue Saint-Martin
75141 Paris cedex 03
France

Bernhard Welz
Universidade Federal de Santa Catarina
Departamento de Química
88040-900 Florianópolis – SC
Brazil

Craig Woody
Brookhaven National Laboratory
Physics Department
Pennsylvania St. Bldg. 510
Upton, NY 11973-5000
USA

De-Ping Yang
College of the Holy Cross
Physics Department
1 College St.
Worcester, MA 01610
USA

1
Electromagnetic Radiation

Wendell T. Hill, III

1.1	**Introduction**	3
1.2	**The Spectrum of Light**	7
1.3	**Basics of Electromagnetic Waves**	8
1.3.1	Maxwell's Equations	9
1.3.2	Wave Equation	10
1.3.2.1	Plane Waves	10
1.3.2.2	Scalar Harmonic Waves	11
1.3.2.3	Waves with Curved Phase Fronts	12
1.4	**Energy, Intensity, Power, and Brightness**	12
1.5	**Polarization**	13
1.5.1	Polarization Bookkeeping	15
1.5.2	Jones Matrices	15
1.5.3	Mueller Matrices	15
1.6	**Longitudinal Field Component**	16
1.7	**Diffraction**	17
1.8	**Interference**	18
1.8.1	Superposition: Single Frequency	18
1.8.1.1	Interferometry	20
1.8.2	Superposition: Multiple Frequencies	20
1.8.3	Short Pulses	22
1.9	**Photons and Particles**	24
	References	24
	Further Reading	25

Encyclopedia of Applied Spectroscopy. Edited by David L. Andrews.
Copyright © 2009 WILEY-VCH Verlag GmbH & Co. KGaA, Weinheim
ISBN: 978-3-527-40773-6

> In the beginning ... darkness was upon the face of the deep.
> And God said, Let there be light: and there was light.
>
> Genesis 1:1-3

1.1
Introduction

Light has been a trusted probe of a variety of aspects of the universe since the beginning of scientific inquiry. Today, regardless of whether searching for gravitational waves, exploring the fundamental properties of quantum mechanics, or designing metamaterial, light continues to play a critical role in revealing nature and engineering tools to enhance life. In this chapter, we review a few key elements of classical and quantum light. A comprehensive review of light is well beyond the scope of this chapter. Thus, we have chosen to focus on the properties that are most often encountered in the laboratory while providing some context and history.

Light, an electromagnetic (EM) field, is an intimate coupling between time-dependent electric and magnetic fields. Classically, the EM field is described quantitatively through Maxwell's equations (Section 1.3.1), where it can be viewed as a wave[1] – a disturbance – satisfying the wave equation,

$$\nabla^2 \Psi = \varepsilon\mu \frac{\partial^2 \Psi}{\partial t^2} \qquad (1.1)$$

In Eq. (1.1), ε and μ are the permittivity and permeability of the medium through which the light is traversing ($\sqrt{1/\varepsilon\mu} = v$ is the speed of light or phase velocity in the medium), and ∇^2 is the Laplacian.[2] The form of the solution depends on the coordinate system – rectangular, spherical, and so on – in which ∇^2 is expressed, as we discuss in Section 1.3.2. However, in general, the solution for the so-called running wave in one dimension is

$$\Psi(\mathbf{x}, t) = f(\mathbf{x} \pm vt) \qquad (1.2)$$

---

1) Christiaan Huygens, a contemporary of Isaac Newton, viewed light as a wave prior to the mathematical formulation as we now know it. Newton, on the other hand, was convinced that light was a stream of corpuscles.

2) The Laplacian is shorthand for

$$\nabla \cdot \nabla$$

which can be written as

$$\frac{\partial^2}{\partial x^2} + \frac{\partial^2}{\partial y^2} + \frac{\partial^2}{\partial z^2}$$

in rectangular coordinates.

Encyclopedia of Applied Spectroscopy. Edited by David L. Andrews.
Copyright © 2009 WILEY-VCH Verlag GmbH & Co. KGaA, Weinheim
ISBN: 978-3-527-40773-6

where **x** describes the distance that the disturbance moves as t increases. The "−"("+") sign indicates motion in the positive (negative) **x** direction.

In vacuum, $\varepsilon \to \varepsilon_0$, $\mu \to \mu_0$, and $v \to c = \sqrt{1/\varepsilon_0\mu_0}$, the vacuum light speed. Special relativity tells us that c sets a "speed boundary" across which information cannot flow. Specifically, those of us living in a "sub-c" universe are prohibited from achieving speeds equal to or larger than c as well as those living in a "super-c" universe from speeds lower than c. Light waves in vacuum are very special and differ from other waves that we encounter in everyday life. There is no rest frame for light and light travels at the same speed in all frames.

As is true of all waves, light is characterized by a wavelength (λ) and a frequency (ν). In vacuum, these quantities are linked by c,

$$\nu \equiv \frac{c}{\lambda} \qquad (1.3)$$

In media different from vacuum $v = c/n$ where,

$$n = \sqrt{\left(\frac{\varepsilon}{\varepsilon_0}\right)\left(\frac{\mu}{\mu_0}\right)} \qquad (1.4)$$

is the index of refraction and the quantities in parentheses are the electric and magnetic dielectric constants respectively; μ differs slightly for μ_0 for most cases of interest. The more general relationship between λ and ν is

$$\nu = \frac{v}{\lambda} = \frac{c}{\lambda_0} \qquad (1.5)$$

where λ_0 is defined as the vacuum wavelength. By definition, ν is medium independent and maintains its vacuum value so

$$\lambda = \frac{\lambda_0}{n} \qquad (1.6)$$

We note that it is possible for the speed of massive particles to exceed the speed of light in media. Shock waves that result are similar to a sonic boom for sound waves. In the case of light, it is called *Cherenkov radiation*.[3] The emitted light is confined to a cone, the half angle of which is defined by

$$\alpha = \cos^{-1}\left(\frac{c}{n v_p}\right) \qquad (1.7)$$

where v_p is the particle speed.

The wavelength spectrum of light is vast, ranging from radio waves to γ-rays, with characteristic wavelengths as large as astrophysical objects to as small as nuclei, respectively (see Section 1.2). A narrow light source, such as a line-narrowed continuous wave (CW) laser, is often said to emit a "single frequency;" such light is termed *monochromatic light*. Monochromanicity, however, is a relative statement. Monochromatic as compared to what? All known sources of light emit within some bandwidth – a spread in wavelength ($\Delta\lambda$) or frequency ($\Delta\nu$) – be it as broad as the solar spectrum or as narrow as the resonance line of an atom.[4] While λ and ν are inversely proportional, it is helpful to

---

[3] The 1958 Nobel Prize in Physics went to Pavel A. Cherenkov for his discovery in 1934 that bears his name, which he shared with Ilya M. Frank and Igor Y. Tamm for their explanation (Cherenkov, Frank and Tamm, 1958).

[4] Microwave sources and state-of-the-art ultrastable lasers can have widths of a fraction of a Hertz. Even still, $\Delta\nu \neq 0$!

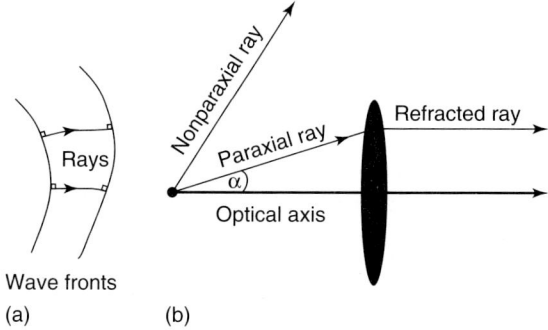

Fig. 1.1 Definition of rays and wave fronts (a) and paraxial rays (b) where $\alpha \ll 1$ rad.

recognize that

$$\left|\frac{\Delta\lambda}{\lambda}\right| = \left|\frac{\Delta\nu}{\nu}\right| \quad (1.8)$$

Classically, the treatment of light falls into two categories: geometrical and physical optics. In geometrical optics, the wave properties (e.g., diffraction) are ignored. Conceptually, we let $\lambda \to 0$ and instead discuss rays. While we do not review the usage of rays in this chapter, we do point out that ray tracing is employed extensively for designing optical systems. Rays are related to waves in that they are perpendicular lines joining the wave fronts (see Figure 1.1). The wave fronts turn out to be the surfaces of constant phase and so rays point in the direction of energy flow. Thus, rays from a point source are radial lines perpendicular to spherical surfaces. Typically, when dealing with rays, we focus on a subset of all the rays called *paraxial rays*. These rays are nearly parallel or form a small angle about a preferred direction. In the example shown in Figure 1.1, the ray traversing the center of the lens is the preferred direction and is called the *optical axis*. Paraxial rays deviate from the optical axis by such a small amount that $\sin\alpha \simeq \tan\alpha \simeq \alpha$. Rays are useful for describing refraction, the bending or redirection of light at the interface between two media with different indices of refraction, and reflection. Refraction, responsible for focusing of light by lenses and the angular spread of the $\Delta\lambda$ components after passing through prisms, is a result of momentum conservation and is succinctly stated through Fermat's principle: *light traverses a path from A to B that is an extremum of the optical path length (OPL)*.[5] That is,

$$\delta(OPL) \equiv \delta\left(\int_A^B n(s)ds\right) = 0 \quad (1.9)$$

where s is the geometric path. Fermat's Principle leads to two important properties of light. First, the law of reflection,

$$\theta_i = \theta_r \quad (1.10)$$

---

5) The principle is often stated as the shortest path, that is, $\delta(OPL)$ would be a minimum. However, the calculus of variation only uses the fact that *OPL* is stationary; the second derivative is not considered. Thus, while usually the case, the path taken is not necessarily the minimum optical path.

where θ_i (θ_r) is the incident (reflected) angle. The second is Snell's Law,

$$n_1 \sin\theta_1 = n_2 \sin\theta_2 \quad (1.11)$$

relating the incident (θ_1) and refracted (θ_2) angles for a ray refracted (i.e., bent) as it passes through an interface between two media with different indices of refraction.

It is interesting to note that Fermat's principle (ca 1657) is closely related to Maupertu's principle in mechanics (ca 1744) for self-contained systems obeying conservation laws,

$$\delta\left(\int_A^B p\,ds\right) = 0 \quad (1.12)$$

where p is the momentum. Equation (1.12) is the principle of least action[6] when formulated more generally as

$$\delta\left(\int_A^B \mathcal{L}\,dt\right) = 0 \quad (1.13)$$

with $\mathcal{L}$ being the Lagrangian.[7] These equations show that particles and rays of light assume rectilinear motions in free space or when there are no forces (fields[8]) and the index of refraction is constant. In general, the trajectories of particles and light are stationary. The index of refraction plays the role of a field causing rays to deviate from linearity when not constant just like forces (potentials) cause particle trajectories to bend.

The geometrical approximation is good when the variation of the physical features of the media are large in comparison to λ. When they become comparable to λ, the wave properties of the light must be considered. The realm of physical optics allows descriptions of elements such as apertures and grating. It further provides a framework to discuss fundamental concepts such as diffraction (the angular spread of a beam of light and the bending of light around obstacles Section 1.7), interference (the superposition of two or more waves, leading to constructive and destructive sums depending on the relative phase of the waves Section 1.8), and coherence (issues associated with how stable the phase is in time and across wave fronts).

The smallest unit of light is called the *photon*, light quanta after the German *Lichtquanten* meaning portions of light.[9] While centuries before the age of quantum physics, Isaac Newton championed the idea of light as a stream of corpuscles, photons are quantum entities whose behavior under certain conditions are well known. However, the answer to the question *"What is a photon precisely"* continues to be illusive. Over the years, the *definitions* tend to fall into one of three distinct categories:

- a fundamental particle;
- an elementary excitation of the EM field; or
- something registered by a photodetector.

---

[6] Like with Fermat's principle, least action is a bit of a misnomer; stationary action would be more appropriate as again only the first derivative is considered.

[7] We point out that Hamilton, Lagrange, Euler and others played a role in the development of the principle as well.

[8] Even in vacuum, the trajectory of light is deflected by a gravitational field. See, for example, Refs. Misner, Thorne and Wheeler (1973) and Hartle (2003) for a discussion of light in a gravitational field.

[9] Gilbert N. Lewis is given credit for coining this name (Lewis, 1926).

In this chapter, we do not argue for or against one view over another.

Massless photons, like massive particles, carry both energy, $h\nu$, and momentum, h/λ, where h is Planck's constant. However, the photon wavefunction must be constructed with care. There have been suggestions that the photon can be understood as simply a classical field plus vacuum fluctuations[10] – a semiclassical approach if you will. There are cases, however, where such an approach gives the wrong answer (as determined by experimental observation). Thus, we have two regimes: classical light and quantum light. By definition, quantum light is any behavior of light that cannot be explained by classical fields, that is, solutions to the wave equation. An example would be squeezed light (Henry and Glotzer, 1988).

The photon is considered to be a fundamental particle. It has an intrinsic spin, which is an integer of unit magnitude. Thus, it obeys Bose statistics, but it has only two states of helicity (aligned or antialigned with its direction of propagation) because being massless, it has no vacuum rest frame. In addition to spin, light has other nonclassical features, typically revealing themselves through intensity noise, correlations, and counting statistics. Finally, both classical fields as well as photons can carry orbital angular momentum and support vortices and solitons (Desyatnikov, Kivshar and Torner, 2005; Kivsha and Agrawol, 2003; Pismen, 1999).

We conclude this introduction by pointing out that if we substitute h/λ for p into Eq. (1.12), we get a different formulation of Fermat's principle,

$$\delta \left(h \int_A^B \frac{ds}{\lambda} \right) = 0 \quad (1.14)$$

The close analogy between Eqs (1.12) and (1.14) suggests an intimate connection between matter and light, from which one can postulate a wave equation for matter similar to that for light. As we discuss in Section 1.9, if we identify the wavelength of the particle as h/p (the de Broglie wavelength) and the index of refraction with $(U - V)/U$, where U and V are the total and potential energies, respectively, the time-independent Schrödiger equation emerges in a form that is not very different from the wave equation for light. Thus, photons, like matter, exhibit both wave and particle behavior.

With this overview as a backdrop, the remainder of this chapter is devoted to the details of selected characteristics of light. We start with the description of EM spectrum in Section 1.2 followed by a review of the wave equation and its solutions in Section 1.3. In Section 1.4, we consider radiometric issues and address the vector nature of light in Sections 1.5 and 1.6. We cover diffraction and interference in Sections 1.7 and 1.8 and conclude the chapter by further discussing the photon matter analogy in Section 1.9.

1.2
The Spectrum of Light

The EM spectrum is traditionally divided into the seven regions shown in Figure 1.2. It should be understood that the boundaries between these regions as well as those between subregions are

---

10) Vacuum fluctuations refer to the photons that are created spontaneously from the vacuum.

1 Electromagnetic Radiation

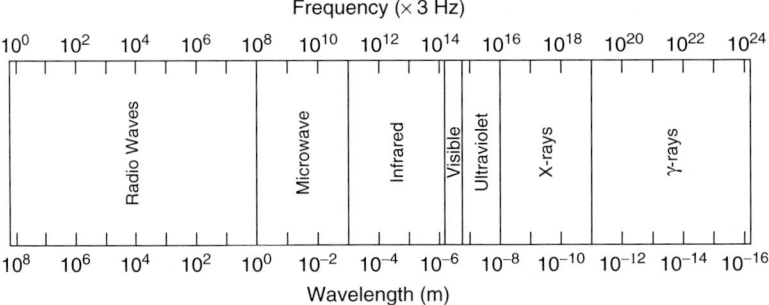

Fig. 1.2 The electromagnetic radiation spectrum.

not hard and fast, nor are the number of subregions unique. The most familiar region of the spectrum, the visible region, consists of wavelengths that range from about 0.40 µm at the blue end to 0.78 µm at the red end. Table 1.1 shows the corresponding colors for the wavelengths between. The visible subregions are a good example of the nonuniqueness of subbands of regions; for example, some references include cyan between green and blue while others insert indigo between blue and violet. Breaking the spectrum into six rather than seven or eight bands is of little consequence typically, because most objects emit a range of colors (i.e., $\Delta\lambda$ is relatively broad) or multiple colors (e.g., $\lambda_1 + \lambda_2 + \cdots + \lambda_n$ again spanning a large $\Delta\lambda$), making the identification of a pure color a rare event. Of course, when $\Delta\lambda$ is small as it often is for some lasers, our eyes in fact do perceive a pure color. For example, consider the red Helium–Neon laser at 632.8 nm or the green doubled Nd : YAG laser at 532 nm.

Subbands also exist for the other regions of the EM spectrum. Tables 1.2–1.4 give some of the more familiar subbands for the other regions. More about the spectrum of light can be found in Ref. (HyperPhysics, 2006).

1.3
Basics of Electromagnetic Waves

As mentioned in the introduction, physical optics is concerned with the

Tab. 1.1 The approximate wavelength, frequency, and energy ranges for six primary visible color bands. Energies increase from left to right.

Color	Wavelengths (nm)	Frequencies ($\times 10^{14}$ Hz)	Energies (eV)
Red	780–625	3.8–4.8	1.6–2.0
Orange	625–590	4.8–5.1	2.0–2.1
Yellow	590–565	5.1–5.3	2.1–2.2
Green	565–500	5.3–6.0	2.2–2.4
Blue	500–435	6.0–6.9	2.4–2.8
Violet	435–380	6.9–7.9	2.8–3.6

1.3 Basics of Electromagnetic Waves

Tab. 1.2 The approximate wavelengths, frequencies, and energies of key radio bands.

Designation	Wavelengths (m)	Frequencies (MHz)	Energy (meV)
AM radio	560–190	0.540–1.600	$(2.33–6.62) \times 10^{-3}$
TV	5.5–0.33	54–890	0.223–3.68
FM radio	3.40–2.78	88.1–108.1	0.364–0.447
Cell phone	0.43, 0.35, 0.18, 0.16, and 0.14	700, 850, 1700, 1900, and 2100	2.89, 3.52, 7.03, 7.85, and 8.68
Satellite radio	0.129–0.128	2320–2345	9.59 and 9.70
WiFi[a]	0.124–0.121	2412–2480	9.97–10.1
Radar	300–0.001	1–300 000	0.00414–1241

[a] The IEEE 802.11b/g/n standards communicate at 2.4 GHz. The IEEE 802.11a standard communicates at 5 GHz, but is essentially obselete.

Tab. 1.3 The approximate wavelength, frequency, and energy ranges of the three primary infrared (IR) bands.

Color	Wavelengths (μm)	Frequencies ($\times 10^{14}$ Hz)	Energies (eV)
Far IR	1000–10	3×10^{-3}–0.3	1.24×10^{-3}–0.124
IR-C	10–3.0	0.3–1.0	0.124–0.414
IR-B	3.0–1.4	1.0–2.14	0.414–0.886
IR-A	1.4–0.7	2.14–4.28	0.886–1.77

wave properties of light, which we examine in this section. We first review how Maxwell's equations lead to the wave equation for the components of the E and B fields. We examine EM wave solutions and their properties, primarily focusing on laser light in homogeneous media in the absence of charge and current sources.

1.3.1 Maxwell's Equations

As we alluded to in the introduction, a wave theory of light predates Maxwell's treatment. In the late 1800s, it was believed that the *aether*, an omnipresent elastic medium, supported light propagation much like the atmosphere or material supports sound waves. Although the theory based on the *aether* had consistency issues and required ad hoc assumptions to sustain it, its belief was so pervasive that even the null result of the experiments by Michelson and Morley could not easily dethrone it.[11] It was not until Maxwell showed that his

---

11) The Michelson-Morley experiment was designed to measure a shift in interference fringes (see Section 1.8.1 for a discussion of interference) caused by a change in the speed of light moving with and perpendicular to the *aether*. The experiment is discussed in Refs. Michelson (1881) and Michelson and Morley (1887a,b).

Tab. 1.4 The approximate wavelength, frequency, and energy ranges of the three primary ultraviolet (UV) bands.

Designation	Wavelengths (nm)	Frequencies ($\times 10^{14}$ Hz)	Energies (eV)
UVA	400–320	7.45–9.37	3.10–3.88
UVB	320–280	9.37–10.7	3.88–4.43
UVC	280–200	10.7–15.0	4.43–6.2
Vacuum UV	200–50	15.0–60.0	6.2–24.8
Extreme UV	80–2.5	$37.5–1.2 \times 10^3$	15.5–500
Soft X-ray	4.5–0.15	$(0.7–2) \times 10^4$	$275–8.3 \times 10^3$

mathematical formulation of Gauss's, Ampere's and Faraday's laws (Maxwell's equations) lead naturally to an EM wave equation that could account for observed phenomena without resorting to arbitrary assumptions that the *aether* idea was abandoned.

Maxwell's equations in a medium with sources (represented by ρ, the charge density) and currents (represented by $\mathbf{j}$, the current density) can be written as

$$\nabla \cdot \mathbf{D} = \rho \quad (1.15)$$

$$\nabla \cdot \mathbf{H} = 0 \quad (1.16)$$

$$\nabla \times \mathbf{E} = \mathbf{j} - \frac{\partial \mathbf{B}}{\partial t} \quad (1.17)$$

$$\nabla \times \mathbf{H} = \frac{\partial \mathbf{D}}{\partial t} \quad (1.18)$$

where

$$\mathbf{D} = \varepsilon \mathbf{E} \quad (1.19)$$

$$\mathbf{B} = \mu \mathbf{H} \quad (1.20)$$

These equations are simplified in the absence of sources, $\rho \to j \to 0$, and are the only cases we consider in this chapter.

1.3.2 Wave Equation

It is very straightforward to show that $\mathbf{E}$ and $\mathbf{H}$ of the EM field satisfy Maxwell's equations and the wave equation simultaneously in a homogeneous medium where $\nabla \varepsilon = \nabla \mu = 0$, by taking the curl of Eqs (1.17) and (1.18). Using the vector identity, $\nabla \times \nabla \times \mathbf{V} = \nabla(\nabla \cdot \mathbf{V}) - \nabla^2 \mathbf{V}$, and the fact that $\nabla \cdot \nabla \times \mathbf{V} = 0$ leads to

$$\nabla^2 \mathbf{E} = \varepsilon \mu \frac{\partial^2 \mathbf{E}}{\partial t^2} \quad (1.21)$$

$$\nabla^2 \mathbf{H} = \varepsilon \mu \frac{\partial^2 \mathbf{H}}{\partial t^2} \quad (1.22)$$

which, by inspection, are the same as Eq. (1.1) because $\varepsilon \mu = n^2 \varepsilon_0 \mu_0 = 1/v^2$. Of course, the general solution is of the form of Eq. (1.2), however, the specifics depend on the geometry and constraints of the problem.

1.3.2.1 Plane Waves

The simplest solution for EM waves is that of a plane wave, where each component of $\mathbf{E}$ and $\mathbf{H}$ is a function of $\xi = \hat{u} \cdot \mathbf{r} - vt$. Recall that $\hat{u} \cdot \mathbf{r} =$ constant defines a plane with $\hat{u}$ being a dimensionless unit vector perpendicular to that plane. It is straightforward then to show that

$$\frac{\partial \mathbf{E}}{\partial t} = -v \frac{\partial \mathbf{E}}{\partial \xi} \quad (1.23)$$

$$\nabla \times \mathbf{E} = \hat{u} \times \frac{\partial \mathbf{E}}{\partial \xi} \quad (1.24)$$

and similarly for **H**. Because **E** and **H** must satisfy Eqs (1.15)–(1.18) (again assuming a homogeneous medium), we can further write

$$\hat{u} \times \frac{\partial \mathbf{E}}{\partial \xi} = \sqrt{\frac{\mu}{\varepsilon}} \frac{\partial \mathbf{H}}{\partial \xi} \quad (1.25)$$

$$\hat{u} \times \frac{\partial \mathbf{H}}{\partial \xi} = -\sqrt{\frac{\varepsilon}{\mu}} \frac{\partial \mathbf{E}}{\partial \xi} \quad (1.26)$$

Integrating Eqs (1.25) and (1.26) and setting the constant to zero (no contribution from the background) leads to

$$\mathbf{E} = -\sqrt{\frac{\mu}{\varepsilon}} \, \hat{u} \times \mathbf{H} \quad (1.27)$$

$$\mathbf{H} = \sqrt{\frac{\varepsilon}{\mu}} \, \hat{u} \times \mathbf{E} \quad (1.28)$$

This implies that **E**, **H**, and $\hat{u}$ form a right-handed orthogonal triad and that light is a transverse field, that is, **E** and **H** oscillate in a plane normal to the propagation direction, in homogeneous media (including vacuum) without sources.

1.3.2.2 Scalar Harmonic Waves

The most common building block for the EM wave is a wave that is harmonic in both time and space. These exhibit sinusoidal variation. Typically, a scalar wave can be expressed as either a real quantity[12]

$$\Psi(\mathbf{r}, t) = A(\mathbf{r}, t) \cos(\mathbf{k} \cdot \mathbf{r} \pm \omega t + \varphi) \quad (1.29)$$

or a complex quantity,

$$\Psi(\mathbf{r}, t) = A(\mathbf{r}, t) e^{i(\mathbf{k} \cdot \mathbf{r} \pm \omega t + \varphi)} \quad (1.30)$$

where A is an amplitude that is a slowly varying function of position and time (compared with the rapid variation of the sinusoidal arguments), **k** is the wavevector ($|\mathbf{k}| = 2\pi/\lambda$) and ω ($= 2\pi\nu$) is the angular frequency. Now it should be clear that $\hat{u}$. The harmonic time dependence of the wave allows the wave equation to be written as

$$\nabla^2 \psi + k^2 \psi = 0 \quad (1.31)$$

Because **E** and **B** are vectors, the EM wave is actually a vector wave. Generally, each component of the field satisfies the wave equation (Eq. 1.31) and has solutions like those in Eq. (1.29) and (1.30).

The argument of the harmonic wave consists of two phase terms, $\bar{\xi}_\pm \equiv \mathbf{k} \cdot \mathbf{r} \pm \omega t$, the dimensionless version of ξ, and φ. A constant $\bar{\xi}_-$ ($\bar{\xi}_+$) defines a profile or phase of the wave that moves toward more positive (negative) r as time evolves at a speed

$$v = \frac{\omega}{|\mathbf{k}|} \quad (1.32)$$

known as *the phase velocity*. The second phase term is often referred to as the *relative phase of the wave*. It can be a fixed constant or time dependent. When it is a constant or has a well defined time dependence, it gives rise

---

12) Note, we have multiplied ξ by $|\mathbf{k}|$ ($|\mathbf{k}|v = \omega$ because $|\mathbf{k}| = n\omega/c$) to make the argument dimensionless.

to coherence. When it varies randomly with time, the light is said to be incoherent. Furthermore, when $\varphi(t) = \text{const} \cdot t^n$, the frequency changes with time. A linear dependence simply shifts the frequency while higher powers chirp the frequency – as the wave passes the frequency either increases or decreases, depending on the sign of the constant. More complicated functions are possible as well.

1.3.2.3 Waves with Curved Phase Fronts

Although plane waves are highly convenient to use, they are appropriate only when dealing with light that is effectively far from its source.[13] Many situations do not fall into this category. It is beyond the scope of this chapter to discuss nonplanar waves extensively, but we will give two examples. For a more extensive discussion, the reader is directed to the text by Cowan, 1968. First, when the fronts are not planes, the solutions in Eq. (1.30) must be modified to correspond to the Laplacian being expressed in a different coordinate system. For example, a spherical wave takes the form

$$\Psi(r, t) = \frac{A}{r} e^{i(kr \pm \omega t + \varphi)} \quad (1.33)$$

where kr = constant. The phase fronts are clearly spheres. A bit more complicated example would be a cylindrical wave, which follows

$$\Psi(\rho, z, \theta, t)$$
$$= A J_m(k\rho) e^{\pm i k_z z} e^{\pm i m\theta} e^{-i(\omega t + \varphi)} \quad (1.34)$$

where $J_m(k\rho)$ is the mth order Bessel function of the first kind (which are

regular at the origin)[14] with m being a positive integer, and

$$k^2 = \left(\frac{\omega}{c}\right)^2 + k_z^2 \quad (1.35)$$

The surfaces of constant phase are just cylinders in this case.

1.4
Energy, Intensity, Power, and Brightness

Because the EM field is composed of E and B fields, its *Energy Density* is given by

$$u = \frac{1}{2}\left(\varepsilon|\mathbf{E}|^2 + \mu|\mathbf{H}|^2\right) \quad (1.36)$$

As a wave, this energy flows as described by the Poynting Vector,

$$\mathbf{S} = \mathbf{E} \times \mathbf{H} \quad (1.37)$$

The *Intensity* of the light is defined as the time average of **S**,

$$I = |\langle \mathbf{S} \rangle| \equiv \frac{1}{2}|\mathbf{E} \times \mathbf{H}| = \frac{n}{2\mu c}|\mathbf{E}|^2 \quad (1.38)$$

which has dimensions of watt per square centimeter.[15] In Eq. (1.38) we used the fact that $\omega/|\mathbf{k}| = c/n$. In vacuum, using the fact that $\varepsilon_0 \mu_0 = 1/c^2$, we can write

$$I = \frac{1}{2\mu_0 c}|\mathbf{E}|^2 = \frac{1}{2}\varepsilon_0 c|\mathbf{E}|^2 \quad (1.39)$$

$$\simeq \frac{|\mathbf{E}|^2}{240\pi} \quad (1.39a)$$

---

13) By far field, we mean that the phase fronts are planes. This can be achieved near the source with lenses.

14) The boundary conditions of the problem might dictate a different Bessel solution. For example, if the origin were excluded, Bessel functions of the second kind, which are singular at the origin, would have to be considered as well.

15) Technically, the SI unit is watt per square meter but in the United States, it is typically expressed as watt per square centimeter.

In this form, I has dimensions of watt per square centimeter (watt per square meter) when the dimensions of $\mathbf{E}$ are volt per centimeter (volt per meter). The *Power*, P, delivered is the integrated intensity over the exposed area,

$$I = \frac{P}{A} \qquad (1.40)$$

where A is the area.[16] A related quantity, the *Brightness*, which is sometimes referred to as the *Radiance*, takes into account the solid angle, $\Delta\Omega$, through which the intensity is delivered and is given by

$$B = \frac{I}{\Delta\Omega} \qquad (1.41)$$

which has dimensions watts per steradian per square centimeter. It is interesting to note that an unfocused laser delivering 1 mW of power at 780 nm is considerably brighter than a 100 W light bulb, 1.7×10^7 W/sr-cm^2 for a typical laser beam[17] with $w_0 = 1$ mm and $\Delta\Omega = 2 \times 10^{-7}$ sr compared with 0.6 W/sr-cm^2 for a light bulb at a distance of 1 m radiating into 4π. Thus, a laser is considered very bright, which can do real damage to an unprotected eye. Finally, laser light can be further characterized by its *spectral brightness*, the brightness per unit optical bandwidth,

$$SB = \frac{B}{\Delta\nu} \qquad (1.42)$$

with units as watts per steradian per square centimeter hertz. The *brightness* and *spectral brightness* are often confused with each other as well as with the *Luminance*, a photometric quantity referring to a perceived brightness related more to how the eye responds.

1.5
Polarization

As mentioned earlier, EM waves are actually vector waves, because $\mathbf{E}$ and $\mathbf{B}$ point in specific directions. *Polarization* captures this feature, and is defined in terms of the direction of $\mathbf{E}$.[18] The most general case is elliptical polarization, which has two limiting cases, linear and circular polarization. These names are so chosen because they describe the geometric shapes $\mathbf{E}$ that sweeps out while looking at the light along (parallel or antiparallel to) $\mathbf{k}$. We have already discussed that $\mathbf{E}$, $\mathbf{B}$, and $\mathbf{k}$ form a right-handed Cartesian triad so polarization also specifies the direction of $\mathbf{B}$. We will take $\hat{k} \equiv \hat{z}$ and focus on light that is perfectly polarized in the discussion that follows.

In general $\mathbf{E}$ will have two orthogonal components,

$$\mathbf{E}_1 = \hat{x} E_{01} e^{i(kz - \omega t + \varphi_1)}$$
$$= \hat{x} E_{01} e^{i(\bar{\xi} + \varphi_1)} \qquad (1.43)$$

$$\mathbf{E}_2 = \hat{y} E_{02} e^{i(kz - \omega t + \varphi_2)}$$
$$= \hat{y} E_{02} e^{i(\bar{\xi} + \varphi_2)} \qquad (1.44)$$

We will first consider the case where E_{01}, E_{02}, φ_1, and φ_2 are all real and

---

16) The area of a laser beam is given by πw_0^2, where w_0 is the beam radius.
17) For a diffraction limited laser beam, $\Delta\Omega = \pi \theta_d^2$ where $\theta_d = \lambda/\pi w_0$.
18) Another reason for considering only $\mathbf{E}$ is the magnitude of $\mathbf{B}$ relative to $\mathbf{E}$ is down by a factor of c. Thus at low intensities, $< 10^{14}$ W/cm^2, $\mathbf{E}$ dominates the physics.

Tab. 1.5 Various electromagnetic field quantities.

Quantity	Name	SI Unit
$c = 2.99792458 \times 10^8$	Light vacuum speed[a]	m/s
$\mu_0 = 4\pi \times 10^{-7}$	Vacuum permeability[a]	T-m/A (kg-m/A^2-s^2)
$\varepsilon_0 = 8.854187817\ldots \times 10^{-7}$	Vacuum permittivity	F/m (A^2-sec^4/kg-m^3)
E	Electric field[b]	V/m (kg-m/A-s^3)
D	Electric displacement	C/m^2
B	Magnetic induction[c]	T (kg/A-s^2)
H	Magnetic field	A/m
ρ	Charge density[b]	C/m^3
j	Current density[b]	A/m^2
P	Power	W (kg/m^2-s^3)
$\mathbf{S} \equiv \mathbf{E} \times \mathbf{H}$	Poynting vector	W/m^2 (kg/m^4-s^3)
$I \equiv \langle |\mathbf{S}| \rangle$	Intensity[b]	W/cm^2

[a] All defined to be exact.
[b] In the US, the explicit length measures for these quantities are given in centimeters, for example, volt per centimeter, watt per square centimeter, and so on.
[c] Sometimes called the *magnetic-flux density*.

time independent. Taking the real part of these fields,

$$\mathbf{E}_1 = \hat{x} E_{01} \cos(\bar{\xi} + \varphi_1) \quad (1.45)$$

$$\mathbf{E}_2 = \hat{y} E_{02} \cos(\bar{\xi} + \varphi_2) \quad (1.46)$$

leads to an equation of a conic section

$$\left(\frac{|\mathbf{E}_1|}{E_{01}}\right)^2 + \left(\frac{|\mathbf{E}_2|}{E_{02}}\right)^2$$
$$- 2\left(\frac{|\mathbf{E}_1|}{E_{01}}\right)\left(\frac{|\mathbf{E}_2|}{E_{02}}\right)\cos\varphi = \sin^2\varphi \quad (1.47)$$

and $\varphi = \varphi_2 - \varphi_1$. Equation (1.47) describes an ellipse when

$$\frac{\sin^2\varphi}{E_{01}^2 E_{02}^2} \geq 0 \quad (1.48)$$

Because the numerator and denominator are positive definite, Eq. (1.48) is always true.

Special Linear Case # 1: $\varphi = 0$ or $\varphi = \pi$
Equation (1.47) reduces to

$$\frac{|\mathbf{E}_1|}{E_{01}} = \frac{|\mathbf{E}_2|}{E_{02}} \quad (1.49)$$

which describes a straight line. Because the two component oscillate in phase, this case leads to linear polarization. The case for $\varphi = 0$ and $\varphi = \pi$ are orthogonal to each other.

Special Circular Case # 2: $\varphi = \pm\pi/2$ and $E_{01} = E_{02}$
The equation reduces to

$$|\mathbf{E}_1|^2 + |\mathbf{E}_2|^2 = E_{01}^2 \quad (1.50)$$

the equation of a circle of radius E_{01}. The two components are out of phase by half a wavelength (or period) but the magnitude of the resultant, E_{01}, is constant but sweeps out a circle leading to circular polarization. The sense of rotation depends on the sign of φ, with the minus (plus) sign producing light with positive (negative) helicity, where positive helicity obeys the right-hand rule, so if you look in the direction of propagation, the E-field rotates clockwise.[19]

Special Elliptical Case # 3: $\varphi = \pm\pi/2$ *and* $E_{01} \neq E_{02}$
Equation (1.47) reduces to

$$\left(\frac{|\mathbf{E}_1|}{E_{01}}\right)^2 + \left(\frac{|\mathbf{E}_2|}{E_{02}}\right)^2 = 1 \qquad (1.51)$$

which is an ellipse with the major axis aligned with the horizontal (vertical) axis when $E_{01} > E_{02}$ ($E_{01} < E_{02}$). The sense of rotation is the same as in special case # 2.

General Elliptical Case $E_{01} \neq E_{02}$
In the general elliptical polarization case, one has a rotated ellipse where the angle, α, of the major axis away from the $\hat{E}_1$ direction is given by

$$\tan 2\alpha = \frac{2 E_{01} E_{02}}{E_{01}^2 - E_{02}^2} \cos\varphi \qquad (1.52)$$

Note, when $E_{01} = E_{02}$, Eq. (1.52) cannot be used and one must go back to Eq. (1.47) to determine α.

1.5.1
Polarization Bookkeeping

There are several approaches to keeping track of the polarization of light, which is particularly important when light interacts with media that can either decrease the intensity or delay the transit time of one polarization or helicity relative to the other. Here, we mention two matrices, namely, the Jones and Mueller matrices. A more in depth discussion can be found in Ref. Goldstein, 2003. The Jones approach involves a set of 2×2 matrices with complex elements that transform two-element vectors that describe the complex amplitude and phase of the light. Every Jones vector corresponds to a physically realizable polarization configuration. The Mueller approach uses a set of 4×4 matrices with real elements, which have values of either 0 or ± 1, to transform the Stokes vectors (see Section 1.5.3). However, some matrices do not represent real configurations.

1.5.2
Jones Matrices

In the Jones calculus Clark Jones, 1941, an initial complex field **E** is transformed to the final complex field **E**′ via matrix multiplication

$$\begin{pmatrix} E'_x e^{i\phi'_x} \\ E'_y e^{i\phi'_y} \end{pmatrix} = \begin{pmatrix} j_{11} & j_{12} \\ j_{21} & j_{22} \end{pmatrix} \cdot \begin{pmatrix} E_x e^{i\phi_x} \\ E_y e^{i\phi_y} \end{pmatrix}$$
(1.53)

where the components of the matrix **J** for common elements are given in Table 1.6.

1.5.3
Mueller Matrices

The general Muller calculus is also a matrix operation,

$$\mathbf{S}' = \mathbf{M} \cdot \mathbf{S} \qquad (1.54)$$

where **M** is a 4×4 matrix and **S** is the four-element Stokes vector, credited to Sir George Gabriel Stokes for their invention. Given the electric fields in Eqs (1.45) and (1.46), the four

---

19) It should be noted that some references define circular polarization in terms of right-hand and left-hand circular polarization. This definition traditionally corresponds to looking antiparallel to **k** so $\varphi = -\pi/2$ would lead to left-hand circular polarization.

Tab. 1.6 Jones matrices for common optical elements.

Optical element	Jones matrix
Linear polarizer $\|\hat{x}$	$\begin{pmatrix} 1 & 0 \\ 0 & 0 \end{pmatrix}$
Linear polarizer $\|\hat{y}$	$\begin{pmatrix} 0 & 0 \\ 0 & 1 \end{pmatrix}$
Linear polarizer at $\pm 45°$	$\frac{1}{2}\begin{pmatrix} 1 & \pm 1 \\ \pm 1 & 1 \end{pmatrix}$
$\frac{1}{4}$-Wave plate, Fast axis $\|\hat{x}$ (+) $\|\hat{y}$ (−)	$e^{i\pi/4}\begin{pmatrix} 1 & 0 \\ 0 & \pm i \end{pmatrix}$
Circular polarizer, $\pm$ Helicity	$e^{i\pi/4}\begin{pmatrix} 1 & \mp i \\ \pm i & 1 \end{pmatrix}$

Tab. 1.7 Mueller matrices for common optical elements.

Optical element	Mueller matrix
Linear polarizer $\hat{x}$ (+) $\hat{y}$ (−)	$\frac{1}{2}\begin{pmatrix} 1 & \pm 1 & 0 & 0 \\ \pm 1 & 1 & 0 & 0 \\ 0 & 0 & 0 & 0 \\ 0 & 0 & 0 & 0 \end{pmatrix}$
Linear polarizer at $\pm 45°$	$\frac{1}{2}\begin{pmatrix} 1 & 0 & \pm 1 & 0 \\ 0 & 0 & 0 & 0 \\ \pm 1 & 0 & 1 & 0 \\ 0 & 0 & 0 & 0 \end{pmatrix}$
$\frac{1}{4}$-Wave plate, Fast axis $\hat{x}$ (+) $\hat{y}$ (−)	$\begin{pmatrix} 1 & 0 & 0 & 0 \\ 0 & 1 & 0 & 0 \\ 0 & 0 & 0 & \pm 1 \\ 0 & 0 & \pm 1 & 0 \end{pmatrix}$
Circular polarizer, $\pm$ Helicity	$\frac{1}{2}\begin{pmatrix} 1 & 0 & 0 & \pm 1 \\ 0 & 0 & 0 & 0 \\ 0 & 0 & 0 & 0 \\ \pm 1 & 0 & 0 & 1 \end{pmatrix}$

components of **S** are defined as

$$S_0 = |E_{01}|^2 + |E_{02}|^2 \quad (1.55)$$

$$S_1 = |E_{01}|^2 - |E_{02}|^2 \quad (1.56)$$

$$S_2 = |2E_{01}E_{02}\cos\varphi| \quad (1.57)$$

$$S_3 = |2E_{01}E_{02}\sin\varphi| \quad (1.58)$$

The Mueller matrices are given in Table 1.7.

1.6
Longitudinal Field Component

Another manifestation of the vector nature of light is that it has a longitudinal component, even though it is customary to ignore it. It must be noted that a pure plane wave exists over all space and has no transverse variation. The finite extent of the field turns out to be acceptable in many cases but one runs into problems with very intense light, particularly when focused. Lax, Louisell, and McKnight (1975) showed that a purely transverse field is not an exact solution to Maxwell's equations, but rather it is the zeroth-order solution to the paraxial approximation to Maxwell's equations. The exact solutions require a longitudinal component given by

$$E_z(x,y,z) = \frac{1}{ik}\nabla_\perp \cdot \mathbf{E}_\perp \quad (1.59)$$

where $k = |\mathbf{k}| = \sqrt{k_x^2 + k_y^2 + k_z^2}$ and the $\perp$ symbol indicate the transverse components of the ∇ operator and **E** field. Using Fourier analysis of the fields, Scully and Zubairy (1991) showed a field obeying the paraxial approximation,

$$k_x, k_y \ll k \quad (1.60)$$

$$k_z \simeq k\left(1 - \frac{1}{2}\frac{k_x^2 + k_y^2}{k^2}\right) \quad (1.61)$$

implies Eq. (1.59).

1.7
Diffraction

When light passes a sharp edge, it does not produce a sharp shadow. Also, when it passes through a circular hole, it does not produce a disk of the same size. Under the right conditions, it produces not only a larger spot but also rings. Furthermore, the transverse size of a laser beam expands as it propagates. These observations are elegantly described by diffraction theory. Diffraction falls into two classes – Fraunhofer and Fresnel. Fraunhofer diffractions describes what happens when the phase fronts are near plane waves, where the curvature of the field can be ignored. Fresnel diffraction takes curvature into account.

Huygens, in the late seventeenth century, suggested a description for wave propagation as a collection of individual spherical sources called *secondary sources*, the sum of which would make up the wavefront. It is a straightforward exercise to convince oneself that Huygens's principle can be used to construct a plane as well as other simple geometries. When applied to a hole, Huygens's approach leads to an emerging spherical wave, because part of the plane wave is blocked. This would appear to account for the observed spread. However, there is a difficulty. If the secondary waves are spherical, then there should also be part of the wave going backward. Huygens had to ignore this part of the wave. It turns that when considered more mathematically, this problem is corrected by what is call the *obliquity factor*.

The mathematical statement of the principle for a wave propagating in free space is the Fresnel-Kirchhoff integral formula,[20]

$$\psi_P = -\frac{ik}{4\pi}\psi_0 \iint \frac{e^{ik(r+\bar{r})}}{r\bar{r}}$$
$$\times [\cos\theta(\mathbf{n},\mathbf{r}) - \cos\theta(\mathbf{n},\bar{\mathbf{r}})]\,dA$$

(1.62)

where the integral is over the area of the aperture. The distances, r and $\bar{r}$, between the aperture and observation point and aperture and source, respectively, are defined in Figure 1.3 as is $\mathbf{n}$, the normal to the surface, pointing toward the source. The angles between the vectors and the normal are represented by $\theta(\mathbf{n},\mathbf{r})$ and $\theta(\mathbf{n},\bar{\mathbf{r}})$.

Let's consider an example of an aperture. In the Fraunhofer limit, s and p are effectively a long way from the aperture. In this case, we can take the surface of the aperture to be a spherical cap such that F is constant. Thus, $\bar{r}$ and $\mathbf{n}$ are antiparallel always and $\cos\theta(\mathbf{n},\bar{\mathbf{r}}) = -1$.[21] Equation (1.62) then reduces to

$$\psi_P = -\frac{ik}{4\pi}A\psi_0 \int \frac{e^{ik(r+\bar{r})}}{r\bar{r}}$$
$$\times [\cos\theta(\mathbf{n},\mathbf{r}) + 1]\,dA \quad (1.63)$$

---

20) The Fresnel-Kirchhoff integral formula of Eq. (1.62) can be derived from Green's theorem (see, for example, Fowles, 1968) for two functions that are continuous, integrable and satisfy the wave equation,

$$\iint (V\mathbf{\nabla}_\perp U - U\mathbf{\nabla}_\perp V)\,dA$$
$$= \iiint (V\nabla^2 U - U\nabla^2 V)\,dV$$

where the first integral is over any closed surface and the second is over the volume enclosed.

21) When $\bar{r}$ and r are much larger than the aperture size, a spherical surface is not much different from a flat surface.

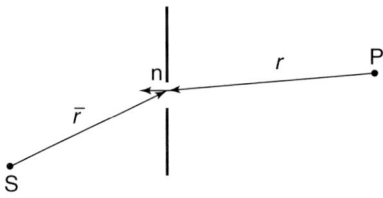

Fig. 1.3 Geometry for the Fresnel-Kirchhoff integral formula.

where $\cos\theta(\mathbf{n}, \mathbf{r}) + 1$ is the obliquity factor mentioned above, which is zero for the wave that is going backwards, towards the source. Table 1.8 gives a few key results derived from Eq. (1.63).

From Table 1.8, we draw an important conclusion. The smaller the aperture, the larger the diffraction. In general, the diffraction angle is given by

$$\text{Slit} \longrightarrow \theta \sim \lambda/b \tag{1.64}$$

$$\text{Circle} \longrightarrow \theta \sim 1.22\lambda/2r \tag{1.65}$$

1.8
Interference

Interference is concerned with the superposition of waves. In general, the sum of wave solutions is also a solution to the wave equation. Here, we consider three examples: summing two waves of the same frequency, the general case of summing multiple waves of different frequencies and generation of short pulses of light by summing many frequencies with a precise phase relationship.

1.8.1
Superposition: Single Frequency

This is straightforward to see with plane waves. Consider first two plane waves of the same frequency but with different real amplitudes and a relative phase between them,

$$A_T(\mathbf{r}, t, \varphi_o) = a_1 e^{i(\mathbf{k}\cdot\mathbf{r}-\omega t+\varphi_1)} + a_2 e^{i(\mathbf{k}\cdot\mathbf{r}-\omega t+\varphi_2)} \tag{1.66}$$

The sum produces a new sinusoidal wave,

$$A_T(\mathbf{r}, t, \varphi) = A_0 e^{i(\mathbf{k}\cdot\mathbf{r}-\omega t+\varphi_0)} \tag{1.67}$$

where

$$A_0 e^{i\varphi_0} = a_1 e^{i\varphi_1} + a_2 e^{i\varphi_0} \tag{1.68}$$

We can determine A_0 and φ_0 in terms of $a_{1,2}$ and $\varphi_{1,2}$ by expanding the exponentials such that

$$A_0(\cos\varphi_0 + i\sin\varphi_0)$$
$$= a_1(\cos\varphi_1 + i\sin\varphi_1)$$
$$+ a_2(\cos\varphi_2 + i\sin\varphi_3) \tag{1.69}$$

Tab. 1.8 Diffraction patterns from key apertures.*

Aperture	Intensity
Slit, width b	$I_0 (\sin\beta/\beta)^2$
Rectangular slit, area $a \times b$	$I_0 (\sin\alpha/\alpha)^2 (\sin\beta/\beta)^2$
Circular aperture, radius r	$I_0 \left(2J_1(\rho)/\rho\right)^2$
N slits, h spacing	$N^2 I_0 (\sin\beta/\beta)^2 (\sin N\gamma/N\sin\gamma)^2$

*$\alpha = (\pi a/\lambda)\sin\phi$
$\beta = (\pi b/\lambda)\sin\theta$
$\gamma = (\pi h/\lambda)\sin\theta$
$\rho = (2\pi r/\lambda)\sin\theta$
ϕ diffraction angle, y direction.
θ diffraction angle, x direction.

Equating the cosine (sine) terms on the left with those on the right and then dividing the sine terms by the cosine terms leads to

$$\tan \varphi_0 = \frac{a_1 \sin \varphi_1 + a_2 \sin \varphi_2}{a_1 \cos \varphi_1 + a_2 \cos \varphi_2} \quad (1.70)$$

At the same time, taking the modulus squared of Eq. (1.68) produces

$$|A_0|^2 = |a_1|^2 + |a_2|^2 + \left(a_1 a_2^* e^{i\Delta\varphi_0} + c.c.\right)$$
$$= a_1^2 + a_2^2 + 2a_1 a_2 \cos \Delta\varphi_0 \quad (1.71)$$

where $\Delta\varphi_0 = \varphi_1 - \varphi_2$. Given $a_{1,2}$ and $\varphi_{1,2}$, A_0 (the intensity) and φ_0 can be found from Eqs (1.70) and (1.71). Equation (1.71) is known as the *coherent sum of the two waves*. That is, one adds the amplitudes before squaring to get the total intensity. The intensity is proportional to the square of the amplitude so it is also possible to write Eq. (1.71) as

$$I_T = I_1 + I_2 + 2\sqrt{I_1 I_2} \cos \Delta\varphi_0 \quad (1.72)$$

The third term in (Eqs 1.71 and 1.72) is sometimes called the *interference term* and plays an important role in describing the intensity of the resultant wave. Consider the case where $a_1 = a_2$ so $I_1 = I_2 = I$. When $\Delta\varphi_0 = 2m\pi$ ($m = 0, 1, 2, \ldots$), the two waves are said to be in phase, in which case

$$I_T = (a_1 + a_2)^2 = 4I \quad (1.73)$$

When $\Delta\varphi_0 = (2m+1)\pi/2$, we have the opposite extreme,

$$I_T = (a_1 - a_2)^2 = 0 \quad (1.74)$$

When $a_1 \neq a_2$, the two extremes give resultants with maximum and minimum I_T respectively.

In the more general case of many waves, all with the same frequency, we have

$$A_T(\mathbf{r}, t, \varphi_0) = A_0 e^{i(\mathbf{k}\cdot\mathbf{r} - \omega t + \varphi_0)}$$
$$= \sum_{j=1}^{N} a_j e^{i(\mathbf{k}\cdot\mathbf{r} - \omega t + \varphi_j)} \quad (1.75)$$

where

$$I_T = |A_0|^2 = \sum_{j=1}^{N} |a_j|^2 + \frac{1}{2} \sum_{j \neq k}^{N}$$
$$\left(a_j a_k^* e^{i(\varphi_j - \varphi_k)} + c.c.\right) \quad (1.76)$$

and

$$\tan \varphi_0 = \frac{\sum_{j=1}^{N} a_j \sin \varphi_j}{\sum_{j=1}^{N} a_j \cos \varphi_j} \quad (1.77)$$

Again, the resultant is a sinusoidal wave with an intensity given by a coherent sum. In the case where all the amplitudes are the same so that each wave has an intensity I,

$$I_T = N^2 I \quad (1.78)$$

In the case where $\Delta_{j,k} = \varphi_j - \varphi_k$ is not well defined but varies randomly with time, it is straightforward to show that the second sum in Eq. (1.76) vanishes by writing the exponentials in terms of sines and cosines and using the fact that the time average of $\sin \Delta_{j,k} \to 0$ as does that of $\cos \Delta_{j,k}$. Thus, the interference terms vanish. In the case where all amplitudes are the same, the resultant wave corresponds to an incoherent sum of the contributors,

$$I_T = NI \quad (1.79)$$

For an incoherent sum, one squares first and then adds the intensities.

1.8.1.1 Interferometry

An entire field of study with industrial applications is built upon an equation similar to Eq. (1.72). The most general situation is where a beam of light is divided into two with each traveling different paths and brought back together. Because the two beams came from the same source, and if the path length difference is not too large, so that the two beams are still in phase, the resultant intensity will be the same as Eq. (1.72) except that $\Delta\varphi_0 \to \delta$ in the argument of the interference term where

$$\delta = k\Delta l \tag{1.80}$$

with Δl being the path length difference between the two arms. In this case, constructive interference occurs when $\Delta l = n\lambda$, whereas destructive interference occurs when $\Delta l = (2n+1)\lambda/2$, where n is a positive integer. Two-beam interferometry exploits interference patterns to measure inhomogeneities and defects in material.

1.8.2
Superposition: Multiple Frequencies

Superposition involving waves of different frequencies leads to some very interesting possibilities such as ultrashort busts of light. The general principle of summing waves with different frequencies can be understood in the special case where the amplitude and phase are the same for each wave:

$$A_T(\mathbf{r}, t) = A_0 \exp[i(\mathbf{k}_1 \cdot \mathbf{r} - \omega_1 t)]$$
$$+ A_0 \exp[i(\mathbf{k}_2 \cdot \mathbf{r} - \omega_2 t)]$$

$$= 2A_0 \exp\left[\frac{i}{2}(\Delta\mathbf{k} \cdot \mathbf{r} - \Delta\omega t)\right]$$
$$\times \exp\left[\frac{i}{2}(\mathbf{k}_m \cdot \mathbf{r} - \omega_m t)\right] \tag{1.81}$$

where

$$\Delta\mathbf{k} = \frac{1}{2}(\mathbf{k}_1 - \mathbf{k}_2) \tag{1.82}$$

$$\mathbf{k}_m = \frac{1}{2}(\mathbf{k}_1 + \mathbf{k}_2) \tag{1.83}$$

$$\Delta\omega = \frac{1}{2}(\omega_1 - \omega_2) \tag{1.84}$$

$$\omega_m = \frac{1}{2}(\omega_1 + \omega_2) \tag{1.85}$$

Equation (1.81) represents a wave oscillating at the mean of the two frequencies, ω_m, and modulated by a temporal and spatial envelope given by $2A_0 \exp[\frac{i}{2}(\Delta\mathbf{k} \cdot \mathbf{r} - \Delta\omega t)]$. Figure 1.4 shows examples of adding two waves with different frequencies. Unlike the case of equal frequencies, in this case, the two sinusoidal waves produce a wave that is periodic but not sinusoidal. Such waves are called *anharmonic*. For the sum of two waves, we have two different speeds. As with a single frequency, we again have a phase velocity – the ratio between the average frequency and wavenumber, $v_{ph} = \omega_m/|\mathbf{k}|$. But, we have a new speed that goes by the name of the group velocity, the speed with which the envelope moves, $v_g = \Delta\omega/\Delta|\mathbf{k}|$.

When a wave is composed of many frequencies, $\omega \to \omega(k)$. Typically, the frequencies are grouped around a central frequency, $\omega(k_0)$, allowing $\omega(k)$ to be expanded into a Taylor series,

$$\omega(k) = \omega(k_0) + (k - k_0)\left.\frac{d\omega}{dk}\right|_{k_0} + \cdots \tag{1.86}$$

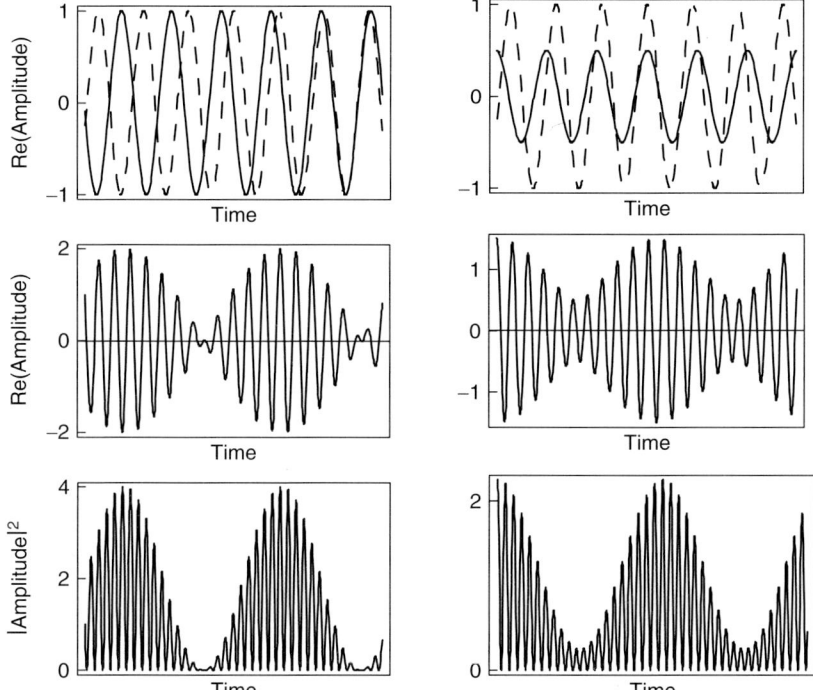

Fig. 1.4 Superposition of two waves, $A_1 \exp[i(\mathbf{k}_1 \cdot \mathbf{r} - \omega_1 t) + \varphi_1] + A_2 \exp[i(\mathbf{k}_2 \cdot \mathbf{r} - \omega_2 t) + \varphi_2]$ where the top row corresponds to the individual amplitudes (A_1 solid curves), the middle row to the resultant sum of amplitudes and the bottom row to the square of the modulus of the resultant amplitudes: (left column) $A_1 = A_2$, $\Delta\varphi = \pi/2$ and $\omega_1/\omega_2 = 0.9$; (right column) $A_1 = A_2/2$, $\Delta\varphi = 0$ and $\omega_1/\omega_2 = 0.9$, where $\Delta\varphi = \varphi_1 - \varphi_2$.

In this case, the addition of the various frequency components is more easily handled via Fourier analysis (see, for example Arfken and Weber (2000) and Boas (2006) for a review), where the amplitude $A(t)$ ($A(z)$) in the time (coordinate space) domain is linked to $\tilde{A}(\omega)$ ($\tilde{A}(k)$) in the frequency (spatial frequency) domain through

$$A(t) = \frac{1}{\sqrt{2\pi}} \int_{-\infty}^{\infty} \tilde{A}(\omega) e^{-i\omega t} d\omega \quad (1.87)$$

$$\tilde{A}(\omega) = \frac{1}{\sqrt{2\pi}} \int_{-\infty}^{\infty} A(t) e^{i\omega t} dt \quad (1.88)$$

for time-frequency and

$$A(z) = \frac{1}{\sqrt{2\pi}} \int_{-\infty}^{\infty} \tilde{A}(\mathbf{k}) e^{-ikz} dk \quad (1.89)$$

$$\tilde{A}(k) = \frac{1}{\sqrt{2\pi}} \int_{-\infty}^{\infty} A(z) e^{ikz} dz \quad (1.90)$$

for coordinate space - frequency space. Thus, in Fourier components, the electric field for a scalar wave propagating in the z-direction can be written as

$$E(z,t) = \frac{1}{\sqrt{2\pi}} \int_{-\infty}^{\infty} \tilde{E}(k) e^{-i[kz + \omega(k)t]} dk \quad (1.91)$$

In many cases, $d\omega/dk$ is the appropriate and more general expression for the group velocity. This can be seen by substituting the first two terms of Eq. (1.86) into Eq. (1.91),

$$E(z,t) = \frac{1}{\sqrt{2\pi}} e^{i[k_0(d\omega/dk)|_{k_0} - \omega(k_0)]t}$$
$$\times \int_{-\infty}^{\infty} \tilde{E}(k) e^{-i[z+(d\omega/dk)|_{k_0} t]k} dk \quad (1.92)$$

However, Eq. (1.90) implies

$$\tilde{E}(k) = \frac{1}{\sqrt{2\pi}} \int_{-\infty}^{\infty} E(z, t=0) e^{ikz} dz \quad (1.93)$$

which allows Eq. (1.92) to be written as

$$E(z,t) = \frac{e^{i[k_0(d\omega/dk)|_{k_0} - \omega(k_0)]t}}{2\pi}$$
$$\times \int_{-\infty}^{\infty} E(z',0) dz'$$
$$\times \int_{-\infty}^{\infty} e^{i(z'-z-(d\omega/dk)|_{k_0} t)k} dk \quad (1.94)$$

where we do the k integration first. The last integral is just $\delta(z' - z - \frac{d\omega}{dk}|_{k_0} t)$, from which we get

$$E(z,t) = \frac{1}{2\pi} E(z + d\omega/dk|_{k_0} t, 0)$$
$$\times e^{-i[\omega(k_0) - k_0(d\omega/dk)|_{k_0}]t} \quad (1.95)$$

By inspection, it is clear that the envelope in Eq. (1.95) moves with speed $d\omega/dk|_{k_0}$ and the carrier oscillates with frequency $\omega(k_0) - k_0 \frac{d\omega}{dk}|_{k_0}$ under the envelope. Thus, we define the group velocity as

$$v_g = \frac{d\omega}{d|\mathbf{k}|} \quad (1.96)$$

In vacuum $v_{ph} = v_g$. However, if the medium through which the wave propagates is dispersive, $n \to n(\lambda)$ so that $dn/d|\mathbf{k}| \neq 0$, the two velocities can be very different. Thus, it is often convenient to write v_g in a form that includes the dispersion explicitly,

$$v_g = \frac{c}{n} \left(1 - \frac{|\mathbf{k}|}{n} \frac{dn}{d|\mathbf{k}|}\right) \quad (1.97)$$

The group velocity is typically the speed with which information is transmitted. It is important to remember that the group velocity is actually only the first term in a series and in cases where $dn/d|\mathbf{k}|$ changes very rapidly or is anomalous (i.e., negative), higher order terms must be kept to determine the speed with which information travels correctly.

1.8.3
Short Pulses

Figure 1.4 shows the basic idea for generating pulses of light of short duration. Specifically, in this case, two frequencies with well-defined relative phase (i.e., fixed in time) are summed in the frequency domain to provide a new wave with beats in the time domain. As additional frequencies are added, the temporal width of the beat envelope narrows. To gain a better understanding of the relationship between the length of the pulse train and its *bandwidth* or number of frequencies required to sum in order to produce it, we will turn the problem around and start with an idealized pulse train in the time domain. Figure 1.5, for example, shows two finite length, idealized, pulse trains, one with three cycles and the other with six cycles.

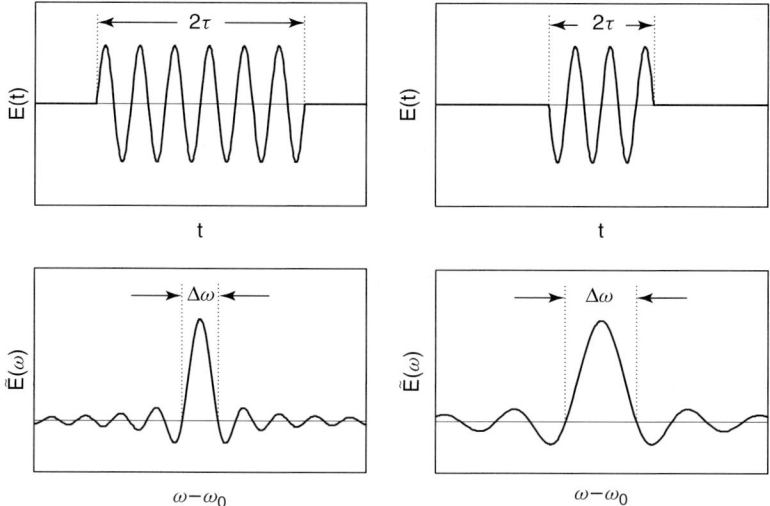

Fig. 1.5 Idealized short pulses formed by finite unit amplitude N-cycle pulse trains (top) with $N = 6$ (left) and $N = 3$ (right). Their respective Fourier transforms appear below with peak amplitudes of $\sqrt{\pi/2}N/\omega_0$ and the first zeros occurring at $\omega = \omega_0(1 \pm 1/N)$.

Mathematically, these obey

$$E(t) = \begin{cases} E_0 \sin \omega_0 t & \text{for } -\tau \leq t \leq \tau \\ 0 & \text{at other times.} \end{cases}$$

(1.98)

The length of this pulse is 2τ, where $\tau = N\pi/\omega_0$ with N being the number of cycles in the train. Using a Fourier analysis similar to that described above, the frequency spectrum is given by

$$\tilde{E}(\omega) = \frac{E_0}{\sqrt{2\pi}} \left[\frac{\sin \tau(\omega - \omega_0)}{\omega - \omega_0} - \frac{\sin \tau(\omega + \omega_0)}{\omega + \omega_0} \right]$$

(1.99)

At optical or near IR frequencies, because the second term is much smaller than the first, we can apply the Fourier transform to just the first term, which is also plotted in Figure 1.5. Clearly, the number of frequencies involved in the shorter pulse is larger than the number needed for the longer pulse. This inverse relationship between the length of the pulse in the time domain and the spread in the frequency domain is conveniently captured in the time-bandwidth product, $\tau \Delta \nu$. From Figure 1.5 and Eqs (1.98) and (1.99), it is clear that $\tilde{E}(\omega) = 0$ when $N\pi(\omega - \omega_0)/\omega_0 = \pm \pi$. Thus, $\Delta \omega = \omega_+ - \omega_- = 2\omega_0/N = 2\pi/\tau$, where $\omega_\pm = \omega_0(1 \pm 1/N)$, which leads to

$$\tau \Delta \nu = 1$$

(1.100)

It is interesting to note that if we multiply Eq. (1.100) by h, this time-bandwidth product satisfies the Heisenberg uncertainty principle,

$$\Delta t \Delta E \geq \frac{\hbar}{2}$$

(1.101)

where $\Delta E = h\Delta \nu$, $\hbar = h/2\pi$ and we substituted Δt for τ. The minimum is reached when the so-called minimum uncertainty wavepacket is prepared.[22]

Ultrashort pulses are achieved by "locking" the frequency components that extends over a wide frequency range. The minimum width achievable by this technique corresponds to one complete cycle of light. At 800 nm, near the peak of the Ti : Sapphire laser, this is ~ 2.7 femtoseconds. For a more complete discussion on mode locking and the generation of ultrashort pulses, the reader is directed to the classic text by Siegman Siegman, 1986

1.9
Photons and Particles

We conclude by discussing the analogy between light and particle waves a bit further. Equations (1.13), (1.14) and (1.31) can be used to motive the time-independent Schrödinger wave equation,

$$-\frac{\hbar^2}{2m}\nabla^2\psi + V\psi = \mathcal{E}\psi \qquad (1.102)$$

$$\nabla^2\psi + \frac{2m}{\hbar^2}(\mathcal{E}-V)\psi = 0 \qquad (1.103)$$

where $\hbar = h/2\pi$. Because λ for the particle is h/p, in free space, we postulate that k ($= 2\pi n/\lambda$) in Eq. (1.31) must be proportional to p. Thus, in the absence of a potential (when $n=1$)

$$k_{n=1}^2 = \frac{p^2}{h^2/4\pi^2} \qquad (1.104)$$

But,

$$p^2 = 2m\mathcal{E} \qquad (1.105)$$

---

[22] The minimum spread criterion applies to conjugate variables such as time frequency and position momentum.

so

$$k_{n=1}^2 = \frac{2m}{\hbar^2}\mathcal{E}, \text{ which lead to Eq. 1.102.}$$

$$(1.106)$$

To account for the potential we let $n^2 = (\mathcal{E} - V)/\mathcal{E} \neq 1$ so that $k^2 = n^2 k_{n=1}^2$ is just the coefficient of the second term in Eq 1.103.

Space does not permit a more in depth discussion of the quantum nature of light. The interested reader is directed to a recent review Smith and Raymer, 2007 and references therein.

References

Arfken, G. and Weber, H. (2000) *Mathematical Methods for Physicists*, 5th edn, Academic Press, New York. ISBN 0-12-059825-6.
Boas, M.L. (2006) *Mathematical Methods in the Physical Sciences*, 3rd edn, Wiley, New York. ISBN 978-0-471-19826-0.
Cherenkov, P.A., Frank, I.M. and Tamm, I.Y. (1958) Nobel Prize. http://nobelprize.org/nobel_prizes/physics/laureates/1958/.
Clark Jones, R. (1941) New calculus for the treatment of optical systems. I. Description and discussion of the calculus. *J. Opt. Soc. Am.*, **31**, 488.
Cowan, E.W. (1968) *Basic Electromagnetism*, Academic Press, New York. ISBN 0-12-193950-2.
Desyatnikov, A.S., Kivshar, Y.S. and Torner, L. (2005) *Progress in Optics*, Vol. **47**, North-Holland, Publishing Company Amsterdam, 291–391.
Fowles, G.F. (1968) *Introduction to Modern Optics*, Holt, Rinehart and Winston, Inc, New York. ISBN 0-03-065365-7.
Glauber, R.J. (2005) Nobel Prize. http://nobelprize.org/nobel_prizes/physics/laureates/2005/glauber-lecture%.html.
Goldstein, D. (2003) *Polarized Light*, 2nd edn, Marcel Dekker, Inc., New York. ISBN 0-8247-4053-X, Revised and Expanded.
Hartle, J.B. (2003) *Gravity: An Introduction to Einstein's General Relativity*, Addison Wesley, San Francisco. ISBN 0805386629.

Henry, R.W. and Glotzer, S.C. (1988) A squeezed-state primer. *Am. J. Phys.*, **56**, 318.

Kivsha, Y.S. and Agrawol, G.P. (2003) *Optical Solitons*, Academic Press, New York.

Lax, M., Louisell, W.H. and McKnight, W.B. (1975) From Maxwell to paraxial wave optics. *Phys. Rev. A*, **11**, 1365.

Lewis, G.N. (1926) *Nature*, **118**, 874.

Michelson, A.A. (1881) The relative motion of the earth and the luminiferous aether. *Am. J. Sci.*, **22**, 120.

Michelson, A.A. and Morley, E.W. (1887) On the relative motion of the earth and the luminiferous ether. *Am. J. Sci.*, **34**, 333.

Michelson, A.A. and Morley, E.W. (1887) On the relative motion of the earth and the luminiferous ether. *Philos. Mag.*, series 5, **524**, 449.

Misner, C.W., Thorne, K.S. and Wheeler, J.A. (1973) *Gravitation*, W. H. Freeman and Company, San Francisco. ISBN 0-7167-0344-0.

Nave, C.R. (2006) *HyperPhysics: Electricity and Magnetism; EM Waves; Electromagnetic Spectrum*, Georgia State University, http://hyperphysics.phy-astr.gsu.edu/Hbase/ligcon.html#c1.

Pismen, L.M. (1999) *Vortices in Nonlinear Fields*, Clarendon Press, Oxford. ISBN 0-935702-11-3.

Scully, M.O. and Zubairy, M.S. (1991) Simple laser accelerator: optics and particle dynamics. *Phys. Rev. A*, **44**, 2656.

Siegman, A.E. (1986) *Lasers*, University Science Books, Sausalito. ISBN 0-935702-11-3.

Smith, B.J. and Raymer, M.G. (2007) Photon wave functions, wave-packet quantization of light, and coherence theory. *New J. Phys.*, **9**, 414.

Further Reading

Bockasten, K. (1974) *Phys. Rev. A*, **9**, 1087.

Born, M. and Wolf, E. (1999) *Principles of Optics: Electromagnetic Theory of Propagation, Interference and Diffraction of Light*, 7th edn, Cambridge University Press, New York. ISBN 0 521 64222 1.

Cowan, R.D. (1981) *The Theory of Atomic Structure and Spectra*, University of California Press, Los Angeles. ISBN 0-520-03821-5.

Ditchburn, R.W. (1961) *Light*, Dover Publications, New York. ISBN.

Durnin, J. (1987) Exact solutions for diffracting beams. I. The scalar theory. *J. Opt. Soc. Am. A*, **4**, 651.

Edmonds, A.R. (1974) *Angular Momentum in Quantum Mechanics*, Princeton University Press, Princeton. ISBN 0-691-07912-9, Third Printing with Corrections.

Einstein, A. (1905) *Ann. Phys.*, **17**, 132.

Einstein, A. (1921) Nobel Prize. http://nobelprize.org/nobel_prizes/physics/laureates/1921/.

Faisal, F.H. (1987) *Theory of Multiphoton Processes*, Plenum Press, New York. ISBN 0-306-42317-0.

Gallagher, T.F. (1994) *Rydberg Atoms, Atomic, Molecular and Chemical Physics*, Cambridge University Press, New York. ISBN 0-521-38531-8 hardback Engin QC454.A8S27 1994.

Herzberg, G. (1950) *Molecular Spectra and Molecular Structure I: Spectra of Diatomic Molecules*, Van Nostrand Reinhold Company, New York.

Johnson, C.S. Jr. and Pedersen, L.G. (1986) *Problems and Solutions in Quantum Chemistry and Physics*. Dover Publications, New York.

Kim, M.S. (2008) Recent developments in photon-level operations on travelling light fields. *J. Phys. B At. Mol. Opt. Phys.*, **41**, 8, 133001.

Planck, M.K.E.L. (1918) Nobel Prize. http://nobelprize.org/nobel_prizes/physics/laureates/1918/planck-bio.html.

Messiah, A. (2000) *Quantum Mechanics*, Dover Publications, New York. ISBN 0-48-640924-4.

Planck, M. (1900) *Verh. dt. phys. Ges.*, **2**, 202.

Planck, M. (1900) *Verh. dt. phys. Ges.*, **2**, 237.

Rayeigh, L. (1900) *Phil. Mag.*, **49**, 539.

Sakurai, J.J. (1973) *Advanced Quantum Mechanics*, Addison-Wesley Publishing Company, Inc, Menlo Park.

Salpeter, E.E. and Bethe, H.A. (1977) *Quantum Mechanics of One- and Two-Electron Atoms*, Plenum Publishing Corporation, New York. ISBN 0-306-20022-8, First paperback printing.

Scully, M.O. and Suhail Zubairy, M. (1999) *Quantum Optics*, Cambridge University Press, New York. ISBN 0-521-43595-1.

Sobelman, I.I. (1979) *Atomic Spectra and Radiative Transitions*, Springer Series in Chemical Physics. Spring-Verlag, New York. ISBN 0-387-09082-7.

ter Haar, D. (1967) *The Old Quantum Theory*, Pergamon, New York. ISBN.

Turunen, J., Vasara, A. and Friberg, A.T. (1989) Realization of general nondiffracting beams with computer-generated holograms. *J. Opt. Soc. Am. A*, **6**, 1748.

Walker Jon Mathews, R.L. (1970) *Mathematical Methods of Physics*, 2nd edn, Benjamin Cummings, San Francisco. ISBN 0-8053-7002-1.

2
Gamma-Ray Spectroscopy

Andrew M. Sandorfi, Christopher J. (Kim) Lister and Craig Woody

2.1	**Introduction** 29	
2.2	**Energy Deposition of γ-Rays in Materials** 29	
2.3	**Solid-State Detectors for Low and Intermediate Energy Radiation**	30
2.3.1	Solid-State Semiconductors for γ-Ray Detection 31	
2.3.2	Requirements of Large Solid-State Detector Arrays 32	
2.3.3	Modern Detector Arrays 34	
2.4	**Shower Development** 36	
2.4.1	Longitudinal Development 37	
2.4.2	Transverse Development 37	
2.5	**Calorimetry Techniques** 39	
2.5.1	Calorimeter Types 39	
2.5.2	Energy Measurement 41	
2.6	**Factors Affecting γ-Ray Calorimeter Response** 42	
2.6.1	Energy Resolution 42	
2.6.2	Position Resolution 45	
2.6.3	Time Resolution 45	
2.7	**Readout Devices** 46	
2.7.1	Optical Detectors 46	
2.7.2	Readout Electronics 48	
	Acknowledgments 48	
	Glossary 48	
	References 49	
	Further Reading 49	

Encyclopedia of Applied Spectroscopy. Edited by David L. Andrews.
Copyright © 2009 WILEY-VCH Verlag GmbH & Co. KGaA, Weinheim
ISBN: 978-3-527-40773-6

2.1
Introduction

γ-Rays are the highest energy photons in the electromagnetic spectrum. Their energies range from tens of kilo electronvolts (10^3 eV, or KeV) to multi-giga electronvolts (10^9 eV, or GeV) and beyond. Section 2.2 provides an overview of the processes by which γ-rays interact with matter. Below a few mega electronvolts (10^6 eV, or MeV), their detection relies on photoelectric absorption, whereas at higher energies their interaction results in the creation of electron–positron pairs, which in turn interact with matter by initiating a cascade shower. The detection techniques in these two regimes are distinctly different. The solid-state devices used at low-to-medium energies are discussed in Section 2.3, together with their characteristics and the compromises adopted in modern large segmented detector arrays. In Section 2.4, we begin the discussion of the high-energy regime with the physical processes occurring within a cascade shower and the principle shower characteristics. Measurements on the shower allow determination of the γ-ray properties of total energy, arrival position, and arrival time. The various types of calorimeters used for such measurements, along with their characteristics, are discussed in Sections 2.5–2.7.

2.2
Energy Deposition of γ-Rays in Materials

γ-Rays propagate freely in vacuum at the speed of light, leaving no detectable phenomena. It is only through their interaction with matter and, specifically, the generation of charged particles that they can be detected and studied. Figure 2.1 illustrates the contributions made by the three main interaction processes in the experimentally important scintillation crystal sodium iodide. Shown here are the components of the attenuation coefficient $a(E_\gamma)$, the parameter describing the exponential attenuation of the number N of incident γ-rays as they travel through a crystal to a depth x, as a function of the γ-ray energy E_γ:

$$N = N_0 e^{-xa(E_\gamma)} \qquad (2.1)$$

The dominant process at low energy (≤ 0.2 MeV) is photoelectric absorption of the γ-ray. This results in the emission of a previously bound atomic electron with all the energy of the γ-ray minus the binding energy of the struck electron.

Encyclopedia of Applied Spectroscopy. Edited by David L. Andrews.
Copyright © 2009 WILEY-VCH Verlag GmbH & Co. KGaA, Weinheim
ISBN: 978-3-527-40773-6

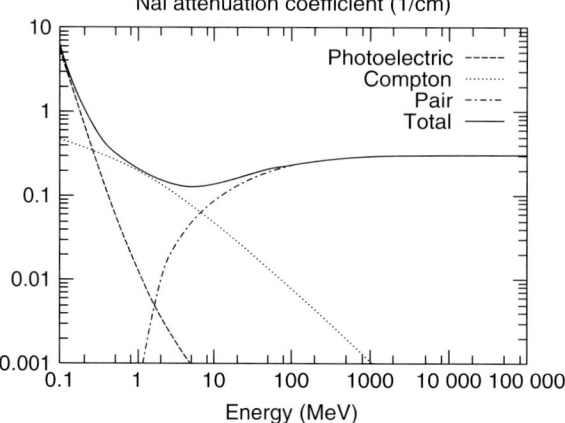

Fig. 2.1 The linear attenuation coefficient of sodium iodide and its decomposition into the contributions from the three interaction processes as a function of γ-ray energy. The results come from the EGS4 Monte Carlo computer program for electron–γ-cascade shower modeling (Nelson, Hirayama and Rogers, 1985).

As the incident energy is raised, the next phenomenon that is important is Compton scattering. In this case, an atomic electron is again struck, but the γ-ray is not absorbed. Instead, the γ-ray scatters off the electron and imparts a fraction of its energy to the struck particle, depending on the angle of scattering. This process is principally important in the transition region of energy before the onset of the third interaction phenomenon.

Pair production is the dominant process by which γ-rays interact with matter in the energy regime above the threshold energy for creating electron–positron pairs (≥ 1.02 MeV). The γ-ray is transformed into a pair of charged particles, an electron and its antiparticle, a positron. The process requires a virtual photon from the electromagnetic field around a nucleus or electron in order to conserve both energy and momentum. However, essentially all the γ-ray energy is split between the electron and positron. The pair emerges nearly symmetrically arranged along the original γ-ray direction with a small average angle between the electron and γ of $\theta_e \sim m_e c^2 / E_\gamma$, where $m_e c^2$ is the rest-mass energy of the electron, 0.511 MeV.

2.3
Solid-State Detectors for Low and Intermediate Energy Radiation

Before around 1970, γ-rays in the kilo electronvolts to few mega electronvolts range were commonly detected in scintillating materials that converted the energy deposited into visible light, one of the most common being sodium iodide – see Sections 2.4 and 2.5. A revolution came with the advent of solid-state devices, particularly Germanium counters. These detectors have evolved from tiny crystals with an active volume of a few grams to detector systems containing hundreds of kilograms of active material. Compared to scintillators, these detectors have remarkable resolution, approaching 1 part in 1000, and are the standard choice for this energy regime.

2.3.1
Solid-State Semiconductors for γ-Ray Detection

The key advantage of using semiconductor materials for detecting electromagnetic radiation lies in the efficient conversion of the photon energy into a large number of charge carriers. The initial photon interaction produces electrons through photoelectric absorption, Compton scattering, and, for energies above 1 MeV, pair production (Section 2.2). The primary electrons thermalize in the semiconductor material, exciting many other electrons from valence states into the conduction band, thus releasing them from lattice sites and allowing them to be swept to the edge of the detector and collected as a charge pulse. In practice, there are technical issues with this process, which must be addressed. First, the material needs to be very pure in order to keep the number of ambient charge carriers low, and the leakage current small. Second, an electric field needs to be established throughout the detector in order to efficiently collect all the charges in the conduction band, before the electrons fall back to valence states through recombination. Third, a rectifying junction needs to be established in order to avoid the growth of large leakage currents as the electric field is applied. Reverse biasing this diode junction can allow strong fields to be maintained, more than kilo electronvolts per centimeter, which leads to fast and efficient charge collection. All of these issues have been solved and large semiconductor counters are now commonplace for detecting photons with energies below a few mega electronvolts, where a reasonably high fraction of the γ-ray interactions can be contained within the detector material. (A comprehensive review of the operating principles of such devices has been made by Knoll (1999)).

There are trade-offs to be made in selecting the detector material. In principle, a material with very high density and very small band gap between valence and conduction bands would be ideal: the first to maximize the photon interaction cross-section, and the second to maximize the number of charge carriers generated. The high-Z aspect of the material is especially important as the γ-ray energy increases, since the probability of interaction changes significantly with both photon energy and the detector material (see Section 2.4). In practice, despite many years of effort, silicon and germanium semiconductors still dominate this regime of γ-spectroscopy, though new materials with higher atomic numbers and bigger band gaps, such as CdZnTe, HgI_2, and others, are making significant impact in photon imaging, and other fields. The band gap issue is application specific: a small band gap results in more charge carriers and, hence, better resolution, but necessitates that the detector be operated at low temperatures, for germanium usually at about 100 K. A bigger band gap semiconductor, such as CdZnTe, allows near room temperature operation, but with loss of resolution. Table 2.1 summarizes the characteristics of common solid-state detector materials.

Large-volume (>500 cc or >2.5 kg) germanium detectors, although expensive and require cooling, are the counters

Tab. 2.1 Characteristics of materials commonly used in semiconductor γ-ray detectors.

Material	Average Z	Density (g cc^{-1})	Band gap (eV)	Operating temperature
Si	14	2.33	1.10	<0 °C
Ge	32	5.32	0.67	< −170 °C pr (100 K)
CdZnTe	38	5.50	2.00	0 °C
HgI$_2$	62	6.40	2.15	0 °C

of choice for research in γ-ray spectroscopy and usually for isotope identification. They can have good efficiency, excellent energy resolution, and reasonable timing characteristics. For uniform performance, detectors are cut from a "boule," a large single crystal of germanium, grown from a molten bath of the material. The first detectors were small and required lithium drifting to keep the leakage currents low, but with improved crystal purification and growing techniques leading to larger volume and purity of material, large intrinsic detectors are now common (up to 10 cm in diameter and 14 cm in length). These intrinsic materials still retain some impurities. These can be of *p-type*, which can have the largest physical size and highest energy resolution, or *n-type*, which are more radiation-hard and can be thermally annealed to remove damage from neutron fluxes encountered in the vicinity of nuclear reactions.

There are limitations on the optimum size of a detector. As the crystals get bigger, more voltage is often needed to deplete the material and allow high internal fields. The physical drift distance for charge also grows, which leads to longer charge-collection times and more chance of charge loss in trapping centers. As a result, there has been a move toward mounting multiple crystals in a single cryostat, thus increasing the overall volume of active germanium without pushing the boundaries of crystal growth or operation. The combination of many germanium counters into large arrays has increased the efficiency and sensitivity of experiments by many orders of magnitude.

2.3.2
Requirements of Large Solid-State Detector Arrays

The low and intermediate regimes of γ-rays span the range of energies normally emitted from nuclear bound states. Such emissions arise when the nucleons rearrange themselves to find their most bound configuration. The rearrangements can arise from individual protons or neutrons changing quantum orbits, analogous to atomic X-ray emission, or due to more collective bulk motion of nuclei. The resulting radiation has energies from tens of kilo electronvolts to a few mega electronvolts. The study of discrete γ-emissions between nuclear states as the nucleons "fall" from a highly excited configuration to their lowest and most stable arrangement, emitting a burst of γ-rays, has been the most effective tool for investigating nuclear structure. The study of such bursts, which can contain more than 40 γ-rays, requires a range of detector characteristics, sometimes conflicting.

The development of arrays of photon detectors has been progressing for more than 30 years and is moving into a new generation, the era of γ-ray tracking. Many excellent review articles have been written about this progress, especially for arrays of large germanium counters (Beausang and Simpson, 1996; Lee, Deleplanque and Vetter, 2003; Gelletly and Eberth, 2006; Ebeth and Simpson, 2008).

To develop a spectrometer for investigating modest energy γ-rays, several issues must be considered and optimized. These include the efficiency for detecting photons, the energy resolution, angle resolution, count rate, the number of γ-rays emitted at one time (multiplicity), the time intervals between γ-emissions, and the overall spectral quality (see below). The choice of the detection material and the geometric configuration are largely governed by these issues. All these factors must also be weighed against cost, which can run from thousands to tens of millions of dollars.

Efficiency: γ-Rays are absorbed in material exponentially; thus, no detector is ever thick enough to stop all photons. At low-to-medium energies a compromise thickness corresponding to 80% absorption of the highest energy photons of interest is often used. The radiation from reactions is emitted in all directions; thus, a perfect 4π shell of detectors is a goal. In some special applications, for example, in decay heat measurements of reactor fuel rods, the total calorimetric sum of γ-energy is more important than individual photons. Maximizing the absolute efficiency, by making a hermetic detector, is very important in such measurements.

Energy resolution: Occasionally, just the presence of a γ-ray is important, but most often, high-energy resolution (accurately determining the exact energy of the photon) is critical. At best, with small solid-state detector arrays, energy resolutions of 1 part in 1000 can be approached across this energy regime.

Angle resolution: The angle of emission of a γ-ray from a nuclear reaction contains information about the quantum numbers of the parent and daughter states of the decay. For moving nuclei, the Doppler shift of the radiation can be used to infer the lifetime of states. Practically, there is a trade-off between angle resolution and the number of detectors in the system, and how far from the source they are placed.

Count rate: The operating mechanism of the detectors determines their count-rate limitation. While for fast scintillators, individual counters can detect photons at megahertz rates, large semiconductors are limited to only tens of kilohertz. This has to balance against the number of detectors in the system and their distance from the source to provide a viable operating regime.

Multiplicity: Some nuclear reactions produce in excess of 40 γ-rays in the cooling process. Retaining a good chance of catching individual photons, without any one crystal registering multiple hits, demands high segmentation.

Time intervals: The cooling cascade of radiation is usually fast and is often complete in less than 500 picoseconds (5×10^{-10} seconds). However, occasionally, nuclear states cannot easily decay, and the "hesitation" can be directly measured electronically, especially if the detectors have good time characteristics. Good timing is usually directly associated with high count-rate capability, but usually is anticorrelated with good energy resolution.

Spectral quality: In a perfect detector, a monoenergetic photon would always lead to a monoenergetic signal output. In practice, this never happens, either due to the primary process (e.g., loss of a Compton scattered photons from the counter) or due to detector imperfection (e.g., loss of produced light or charge). The spectrum quality is usually defined as the ratio of "full energy" or "photopeak" events to the total number of interactions. As such, it is essentially the fraction of γ-rays for which a measurement of their total energy is obtained. For low-energy radiation (~0.2 MeV) this can be high, >80%, but it falls with energy as secondary photons leak out of the detector.

2.3.3
Modern Detector Arrays

To evaluate the competing criteria, a variety of *figures of merits* have been discussed, which allow arrays to be optimized for specific applications (Beausang and Simpson, 1996; Lee Deleplanque and Vetter, 2003; Gelletly and Eberth, 2006; Ebeth and Simpson, 2008). The most challenging conditions generally involve fusion–evaporation reactions, which produce nuclei at high angular momentum from which large number of γ-rays are emitted in the de-excitation process. In such regimes, an effective *resolving power* is frequently used, which reflects the ability to

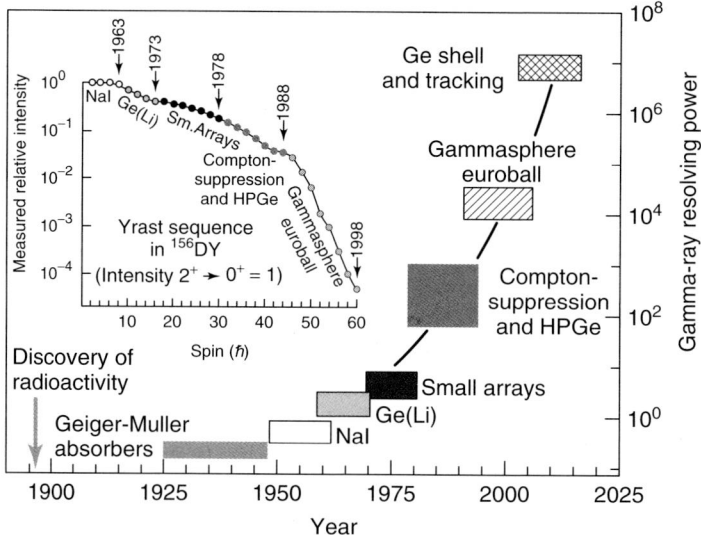

Fig. 2.2 The evolution of γ-ray detection technology from the discovery of radioactivity to the present. The *resolving power* plotted here reflects the ability to observe transitions between rare and exotic nuclear states (LRP, 2002). The insert shows a typical relative intensity for the production of the lowest energy states of a given angular momentum. Excited nuclear states decay by sequentially emitting γ-rays along this *Yrast* sequence. There is a strong inverse relationship between *resolving power* and the experimental limit for the observation of highly excited nuclear states with large angular momentum.

Tab. 2.2 Properties of crystal scintillators and Cherenkov materials. The right-most column gives the light output relative to NaI(Tl).

	Density (g cm^{-3})	Refractive index	Radiation length (cm)	Molierè radius (cm)	Emission maximum (nm)	Decay time (ns)	Light output
NaI(Tl)	3.67	1.85	2.59	4.5	410	250	1.00
BGO	7.13	2.20	1.12	2.4	480	300	0.15
LSO	7.40	1.82	1.14	2.1	420	40	0.83
GSO	6.71	1.85	1.38	2.2	430		
Fast						56	0.30
Slow						600	0.03
BaF$_2$	4.89	1.56	2.05	3.4			
Fast					220	0.7	0.05
Slow					310	620	0.20
CsI(Tl)	4.53	1.80	1.85	3.8	565	1000	1.40
CsI	4.53	1.80	1.85	3.8			
Fast					305	10,35	0.10
Slow					480	1000	0.02
LaBr$_3$(Ce)	5.29	1.90	1.88	2.85	356	20	1.30
LaCl$_3$(Ce)	3.86	1.90	2.81	3.71	335		
Fast						24	0.42
Slow						570	0.13
CeF$_3$	6.16	1.68	1.70	2.6	340	10,30	0.10
PbWO$_4$	8.28	2.16	0.90	2.2	500	2,15	0.01
PbF$_2$	7.77	1.82	0.93	2.2	Cher.	Cher.	3.0×10^{-3}
Pb Glass (SF5)	4.08	1.67	2.54	3.7	Cher.	Cher.	1.0×10^{-3}
H$_2$O	1.00	1.33	36.1	6.8	Cher.	Cher.	0.8×10^{-3}

observe the individual γ-ray transitions as the nuclei de-excite. In Figure 2.2, increases in resolving power have accompanied each generation of detector system. Detector technology is fast approaching a factor of a million improvement since γ-rays were first observed.

The current state-of-the-art detector is the *Gammasphere* array, which was built by a collaboration of US National Laboratories and Universities and has been operated since 1996 as an international facility for nuclear physics research (Gammasphere, 2008). This is a hybrid system, consisting of 110 separate modules, each with a germanium detector mounted in a bismuth germinate shield (BGO; Table 2.2) to suppress events in which a Compton scattered photon leaks out, thereby improving the *spectral quality*. The detector can be operated in many modes: as a calorimeter, as a multiplicity filter, or as 110 individual counters. In the high-resolution germanium operation, it has ~10% absolute efficiency for 1.33 MeV γ-rays.

New technologies to progress beyond the Gammasphere generation are currently under study as a project called *GRETA*, the *gamma ray energy tracking array* (Beausang, 2003). This will involve building a hermetic pure germanium shell. Each detector will have highly segmented electrodes that allow location of the charge deposition. This information allows "tracking," the reconstruction of events. Tracking has many advantages, including recovering interactions where photons scatter out of the shell. The first phase of this array, a 1/4-prototype GRETINA (Lee *et al.*, 2004), is expected to be operational in 2012. For applications involving the observation of multiple γ-rays, GRETA will surpass Gammasphere by a factor of 1000.

Whenever knowledge of the precise energy has been important, solid-state detectors have revolutionized the scientific study of moderate energy γ-rays. Nonetheless, in some investigations, such as in β-decay, the total integrated flux of γ-rays is of paramount importance, as it can reveal the "strength function" distribution of the decay. This has both spectroscopic interests, for example in investigating the distribution of Gammow–Teller decay strength, and practical interest, as it is a factor in determining "decay heat" in a reactor. For these applications, hermiticity is the key factor, and large volume scintillators still remain the best solution. For low-to-medium energy γ-rays, these are comparatively simple and inexpensive to operate. As the energy grows above the $e^{\pm}$ pair threshold, and a cascade shower begins to develop, hermiticity becomes a complex issue and the detection systems become elaborate. We begin this discussion in the following section.

2.4
Shower Development

There are two main processes by which energetic electrons (and positrons) are brought to rest:

1. collisions with atomic electrons, resulting in either ionization energy loss for soft collisions or division of the energy between the incident particle and the atomic electron for harder collisions and
2. emission of a γ-ray in the electromagnetic field of a nucleus or electron, termed *bremsstrahlung*.

(For positrons, there is the additional process of annihilation, but this mainly occurs at rest and produces two γ-rays each with half the pair creation threshold energy.) Collisional energy losses dominate at low energy, but bremsstrahlung dominates at high energy. The crossover point is at a material-dependent critical energy (Bethe and Askin, 1953)

$$E_c \simeq 1600 m_e \frac{c^2}{Z_{\text{eff}}} \qquad (2.2)$$

where $m_e c^2$ is the rest-mass energy of the electron, 0.511 MeV, and Z_{eff} is an appropriately weighted average of the nuclear charge of the material constituents. It is interesting to note that the dominant high-energy electron interaction, bremsstrahlung, is related by crossing symmetry to the dominant high-energy photon interaction, pair production. Thus, it is not surprising that the length of material in which a high-energy electron loses $1/e$ of its energy to radiation, the radiation length X_0 Tsai (1974, 1977),

$$\frac{1}{X_0} \simeq 4Z(Z+1)n\alpha r_e^2 \ln(183 Z^{-1/3})$$
$$(2.3)$$

2.4 Shower Development

is approximately the same as the mean distance for a high-energy γ to travel before producing an $e^{\pm}$ pair, λ_{pair}, where

$$\frac{1}{\lambda_{\text{pair}}} \simeq \left(\frac{7}{9}\right) 4Z(Z+1)n\alpha r_e^2$$
$$\times \ln(183 Z^{-1/3}) \qquad (2.4)$$

where Z is the charge of the nucleus, $n = \rho N_a/A$ is the atomic number density, ρ is the mass density of the absorber, N_a is Avogadro's number, A is the atomic weight of the nucleus, α is the fine-structure constant $1/137$, and r_e is the classical radius of the electron, 2.8×10^{-13} cm. Similarly the emission angle between the bremsstrahlung γ and the initial electron is characterized by the same angle, $\theta_\gamma \sim m_e c^2/E_e$, as the pair emission.

2.4.1 Longitudinal Development

The effect of combining these two processes, bremsstrahlung and pair production, is the creation of an electron–γ-cascade shower, as shown schematically in Figure 2.3. An incident γ-ray produces an electron–positron pair, which then loses energy by radiating lower-energy γs, which, in turn, generate more pairs, and so on, until the electrons and positrons are below E_c and no longer lose significant energy by radiation. The strong forward peaking of the pair and bremsstrahlung emission means that the shower tends to progress in a straight line along the initial direction. A vastly oversimplified approximation to the showering process would be that the average energy per particle is halved in each radiation length and the number of particles doubles until the average energy is reduced to the critical energy, at which point the shower terminates. Then the shower linear extent is $\sim X_0 \ln(E_\gamma/E_c)/\ln 2$, and the maximum number of particles is just $\sim E_\gamma/E_c$, with approximately equal numbers of γ-rays, electrons, and positrons at each step. Finally, the integral track length of the charged particles in this simplified model is $\sim (2/3 \ln 2)(E_\gamma/E_c)X_0$. More detailed calculations support these three qualitative features: the shower depth is logarithmic in incident energy, and the number of shower particles as well as the integral track length are both proportional to the incident energy.

2.4.2 Transverse Development

After the first few radiation lengths, the transverse extent of the shower is determined by the multiple scattering of the electrons and positrons. According to the Molière theory of multiple scattering

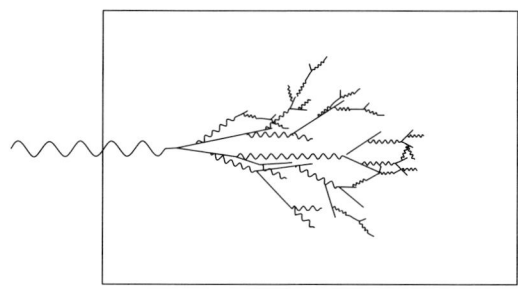

Fig. 2.3 Schematic representation of how an electron–γ-cascade shower develops. A high-energy γ-ray enters from the left and produces a pair, which in turn emits bremsstrahlung to make more γ-rays, and so on.

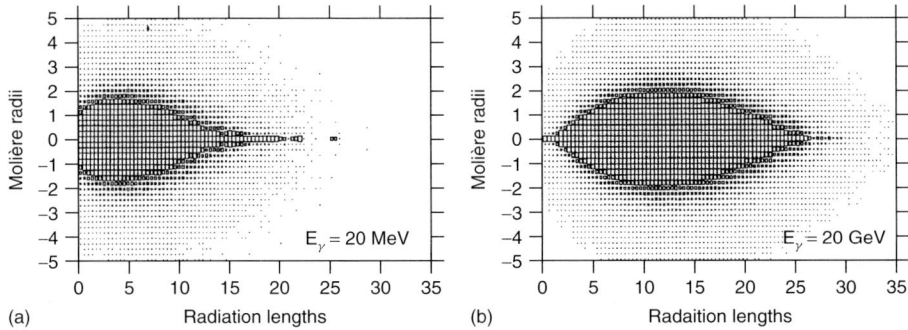

Fig. 2.4 Average relative energy deposition from an electron–γ-cascade shower. Size of the box represents the amount of energy deposited in 1-cm cubes of NaI in that position, averaged over many showers. Maximum box size occurs at a level such that the saturated boxes contain 90% of the γ-ray energy. A 20-MeV γ-ray enters from the left in (a) and a 20-GeV one in (b). The radiation length of NaI is 2.6 cm and the Molière radius is 4.5 cm (see Table 2.2).

(Scott, 1963), the lateral displacement of an electron with energy equal to the critical energy traveling a distance of one radiation length defines the so-called Molière radius,

$$R_M = \sqrt{\frac{4\pi}{\alpha}} m_e c^2 \frac{X_0}{E_c} \quad (2.5)$$

which characterizes the shower distribution perpendicular to the shower axis. Experimentally, it is found that a cylinder of radius $1 R_M$ contains $\sim$90% of the shower energy deposition, $2 R_M \sim 97\%$, and $3 R_M \sim 99\%$ (Kleinknecht, 1982).

Figures 2.4 and 2.5 present the results of detailed simulations of such

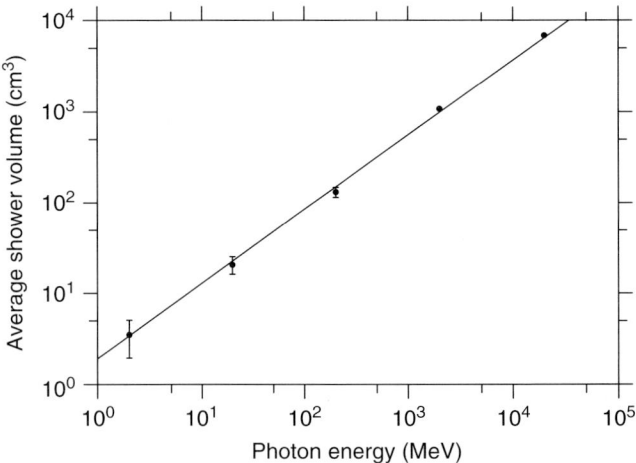

Fig. 2.5 Average number of 1-cm cubes of NaI that receive any energy deposition during an electron–γ-cascade shower as a function of shower energy. The error bars represent the variation in the number of cubes from shower to shower.

electron–γ-cascade showers in NaI with the GEANT Monte Carlo program (Brun and Carminati, 1993). Figure 2.4 shows contours of the energy deposition along and transverse to the initial direction for γ-rays of 20 MeV and 20 GeV. This is the result of averaging over many such showers. As can be seen, the factor-of-1000 increase in energy between the two showers produces a deeper penetration, by about 10 [$\simeq \log_2(1000)$] radiation lengths ($X_0 = 2.6$ cm in NaI, Table 2.2). Both showers exhibit similar radial extent, approximately equal to the Moliere radius ($R_M = 4.5$ cm in NaI, Table 2.2). Any given shower will produce a much more sparse and random energy deposition. Figure 2.5 attempts to demonstrate this by plotting the average number of 1-cm NaI cubes in which any shower energy deposition has occurred as a function of incident γ-ray energy. For the 20-MeV case, this is only 20 cubes versus the 3400 cm^3 required in Figure 2.4 to contain 90% of the energy. For the 20-GeV case, the comparison is 6820 cubes versus 7000 cm^3.

2.5 Calorimetry Techniques

2.5.1 Calorimeter Types

γ-Ray detection takes place through the fundamental interactions of photons with matter described above. At high energies, the dominant process of pair production leads to electron–positron pairs, which undergo ionization and radiative energy loss in the detector material. The calorimetry technique makes use of the fact that this energy loss is proportional to the total energy deposited in the detector in order to measure the initial photon energy. There are, in general, two classes of calorimeters that are used to measure this energy. The first consists of a *totally active* detector in which all the initial energy is deposited in a material that produces an observable signal. The second is a *sampling* detector in which only a fraction of the initial energy is deposited in an active material, and the rest is deposited in a passive material and is not observed.

Totally active detectors usually consist of a high-density, high-Z material, such as an inorganic crystal or glass, which provides a high conversion efficiency for an incident photon. Table 2.2 lists some of the commonly used inorganic scintillators. The inorganic scintillators have the highest light output and provide the best energy resolution at lower energies. The last (right-most) column of Table 2.2 gives the light output relative to NaI, which yields 38 000 optical photons per mega electronvolts of energy deposited (see Table 2.3). The incident photon undergoes an initial interaction on a high-Z atom according to the processes described above, and the resulting electron or positron produces ionization, which leads to the formation of electron–ion pairs. The electrons and ions then transfer their energy to luminescent centers in the material, which can be either intrinsic, such as in a BaF_2 crystal, or a dopant or fluor that has been added as an activator, such as thallium (Tl) in the case of NaI or CsI. The excited luminescent centers then radiate their energy in the form of scintillation light, typically in either the visible or the ultraviolet (UV) wavelength range. In this process, each step takes place with a certain efficiency that contributes to the overall

Tab. 2.3 Efficiency factors for crystal scintillators.

Scintillator	L (γs per MeV)	β	S	Q	η
NaI(Tl)	38 000	0.88	0.59	1.00	0.52
CsI(Tl)	65 000	0.97	0.99	1.00	0.96
CsI (pure)	2000	0.97	0.29	0.10–0.15	0.03
BGO	8200	0.69	1.00	0.13	0.09
BaF$_2$					
Fast	1800	<0.82	>0.9	1.00	0.08
Slow	9950	<0.72	>0.33	1.00	0.24
CeF$_3$	3200	0.61	0.13	1.00	0.08

scintillation efficiency of the material. If we denote the conversion efficiency as β, the efficiency for energy transfer to the luminescent center as S, and the radiative decay efficiency as Q, then the overall scintillation efficiency η is given by $\eta = \beta S Q$. Table 2.3 gives the respective efficiency factors for some commonly used scintillating crystals, taken from Lempicki, Wojtowicz and Berman (1993).

The Cherenkov materials, some of which are also listed in Table 2.2, produce light only by Cherenkov radiation and generally have a much lower light output than inorganic scintillators. Cherenkov radiation is the light emitted by charged particles when their velocity exceeds the phase velocity of light in the medium. It is a very small fraction of the total energy-loss process. These materials are therefore used most often at higher energies where the total amount of energy deposited is much higher. (The columns labeled "Emission maximum" and "Decay time" in Table 2.2 are not relevant for the Cherenkov materials. The emission is nonresonant and prompt, and thus entries have been made with "Cher." to identify these materials.)

A schematic representation of a sampling detector is shown in Figure 2.6. Typically, the passive medium consists of some high-density material, such as lead, which provides the high photon conversion efficiency, whereas a lower density material is used as the active medium. Because only a fraction of the total energy is measured, a sampling detector generally has poorer energy resolution than a totally active detector but can be built for a much lower cost on account of the smaller quantity of active detector material required. It is therefore widely used for large-area, high-energy calorimeters.

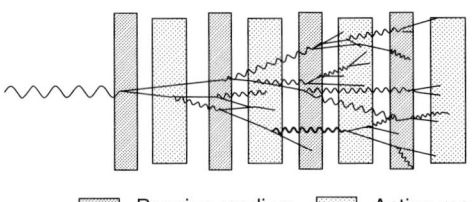

Fig. 2.6 Schematic representation of a sampling calorimeter.

2.5.2
Energy Measurement

There are several techniques that are used to measure the energy deposited in the active medium in both a totally active and a sampling detector. These are generally based on either the direct detection of the ionization charge or the detection of either scintillation or Cherenkov light produced in the active medium. Direct ionization can be detected in either gases, liquids, or solids, and the use of these materials depends largely on the energy range of interest. Because of their low density, gases are rarely used in γ-ray detection, even in sampling calorimeters. Liquids, in particular noble liquids, are often used in sampling detectors because of their relatively high-Z and the ability to collect the ionization charge easily using readout electrodes inside the detector. A high voltage is applied across the gap formed by these electrodes, which causes the charge to drift across the gap where it is collected and measured using charge-sensitive detectors. Table 2.4 lists some of the properties of noble liquids that have been used for γ-ray detection. Because of its lower cost, liquid argon has been the preferred choice. All of these liquids operate at cryogenic temperatures and are typically cooled by using liquid nitrogen. Although the implementation of such noble liquid detectors is complex, nonetheless several have been used in large-area, high-energy physics applications (Dø Collaboration, 1994; SLD Collaboration, 1984).

Plastic scintillator, again because of its low cost, is also widely used in large-area calorimeters. Table 2.5 gives a list of some of the commonly used plastic scintillators along with some of their properties. The light output is given

Tab. 2.4 Properties of noble liquids.

	Liquid Ar	Liquid Kr	Liquid Xe
Density (g cm^{-3})	1.39	2.45	3.06
Radiation length (cm)	14.3	4.76	2.77
Molière radius (cm)	7.3	4.7	4.1
dE/dx (MeV cm^{-1})	2.11	3.45	3.89
Drift velocity (cm μs^{-1})	0.5	0.5	0.3
W-value (eV per ion pair)	23.6	20.5	15.6
Dielectric constant	1.51	1.66	1.95
Triple point (K)	84	116	161

Tab. 2.5 Properties of Saint-Gobain (Bicron) plastic scintillators (Hurlbut, 1985).

Scintillator	Light output (% anthracene)	Decay time (ns)	Wavelength of max emission (nm)	Attenuation length (cm)	Application
BC 400	65	2.4	423	250	General purpose
BC 404	68	1.8	408	120	Fast counting
BC 408	64	2.1	425	380	Large area
BC 412	60	3.3	434	400	Large area
BC 418	67	1.4	391	100	Ultrafast timing
BC 428	50	12.0	490	330	Green emitting

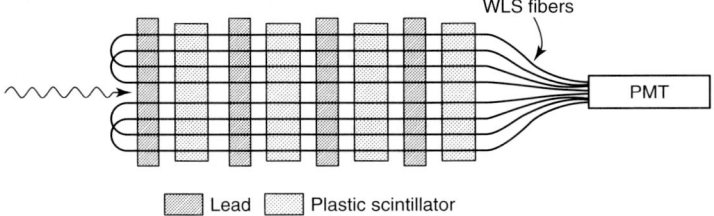

Fig. 2.7 A sampling calorimeter with wavelength-shifting (WLS) fiber-optic readout.

relative to anthracene, an early material now rarely used except as the standard for reporting light output. One aspect of sampling detectors that use scintillating plastic is the need to collect the light produced onto a readout device. This is generally accomplished through the use of light guides or by allowing the scintillation light to excite another fluorescent material known as a *wavelength shifter* that carries the light to the photodetector. The wavelength-shifting (WLS) technique has been extensively used because of its ability to provide a wide variety of readout techniques. One such method is illustrated in Figure 2.7 (PHENIX collaboration, 1993). In this example, the light from scintillation plates, which are interleaved between plates of lead, is used to excite WLS fibers that pass longitudinally through the detector. The fibers are then bundled together at the back and read out with photomultiplier tubes (PMT). This provides a convenient way of placing the readout devices at the back of the detector where they will not interfere with any of the incident particles.

Finally, sampling detectors using silicon semiconductor (Table 2.1) as the active medium have also been built and provide good energy resolution because of the large amount of charge produced. However, because of the significant cost of high-purity silicon, these detectors have only been used in rather limited, small-area applications.

2.6
Factors Affecting γ-Ray Calorimeter Response

The overall performance of a γ-ray detector can be characterized by three main categories of resolution that are important experimentally. These are energy resolution, position resolution, and time resolution. There are a number of factors that determine these properties in any calorimeter design.

2.6.1
Energy Resolution

The energy resolution of a calorimeter can be parameterized by the following expression:

$$\left(\frac{\Delta E}{E}\right)^2 = \left(\frac{a}{\sqrt{E}}\right)^2 + \left(\frac{b}{E}\right)^2 + c^2 \tag{2.6}$$

The first term is a *statistical* component and contains contributions to the energy resolution due to statistical fluctuations. These can be fluctuations in the amount of primary ionization produced (e.g., the total number of ionizing tracks produced in the electromagnetic shower) or fluctuations in the number of observed particles (either photoelectrons or ion pairs). Usually, for a high-energy shower, the number of ionizing tracks produced is large; hence, the fluctuations are rather small ($\leq 0.5\%/\sqrt{E}$), and the fluctuations in the number of observed particles dominate (Amaldi, 1981). For a sampling detector, this includes a contribution from the so-called sampling fluctuations, which are due to the fact that only a fraction of the energy is being detected, and this fraction can vary from one shower to the next. For high-Z inactive radiators of critical energy E_c, radiation length X_0, and total length L_x, the sampling fluctuation contribution to the energy resolution (full width at half maximum of the resulting signal) for γ-rays of energy E can be approximated as (Amaldi, 1981)

$$\frac{\Delta E}{E} = 7.5\% \sqrt{\frac{L_x}{X_0} \frac{E_c(\text{MeV})}{E(\text{GeV})} \frac{1}{G(Z, \cos\theta)}}$$

(2.7)

Here, $G(Z, \cos\theta)$ is a complicated function of geometry, the Z of the radiator material, and the angle θ that characterizes the mean angular spread of particles around the shower axis, but it is typically a number between 1.0 and 1.3. In a sampling calorimeter, these fluctuations generally make the largest contribution to the statistical term in the resolution.

The term of Eq. 24.6 with coefficient b contains contributions that are fixed in magnitude relative to the observed signal. This *noise term* contains contributions from various sources of noise. These include any electronic noise in the readout devices or their associated electronics, as well as any constant background noise. Since this term goes as $1/E$, it is most important at lower energies and is often negligible at higher energies.

The term with coefficient c reflects complex functions of the detector, which are rather slowly varying with energy and are often approximated by a constant. The two most important functions are the degree of shower containment and the uniformity of response throughout the active volume. As the fraction of the shower energy escaping a detector volume increases with increasing γ-ray energy, the fluctuations in this signal loss grow until a roughly constant plateau in resolution is reached (Dowell et al., 1987). This contribution can be made small with sufficiently large detector volumes, although the associated costs increase sharply with energy.

Even for large totally active detectors, such as NaI(Tl), inhomogeneities in the crystal can be a fundamental limitation to the ultimate energy resolution one can achieve. The inhomogeneities limit the performance through the variation in light generation and collection for the same amount of energy deposited in different regions of the active volume. This is a significant limitation at moderate energies – for example, the 20-MeV shower shown in Figure 2.4 – where the variations in the shower distributions are large. The effect is much smaller at high energies where each shower fills a large detector volume (see Figure 2.5). Variations are governed by crystal impurities and defects, by deviations in the local concentrations of dopant, by the reflective properties of crystal surfaces, by solid-angle effects and the placement of the light collection devices, and by nonuniformities in efficiency across the surfaces of the light collectors (Dowell et al., 1990). Controlling, or even predicting, any one of these components is quite difficult. However, their combined effect can be studied with muons and unfolded with the techniques of computer-aided tomography (CAT or μCAT here) (Dowell et al., 1990).

High-energy muons produce only narrow, well-defined ionization tracks that can completely penetrate even large detectors. Being weakly interacting, they do not undergo nuclear reactions and can be used to deposit a known amount of energy in every element of detector volume along their path. The response to many muons traversing well-defined paths can be combined with Fourier reconstruction algorithms to produce three-dimensional images of the nonuniformities in light generation and collection. As an example, a section of such a μCAT map – a two-dimensional slice through a 24-cm-diameter × 36-cm-long NaI(Tl) crystal viewed at one end by seven 7-cm photomultiplier tubes – is shown in Figure 2.8. The gray-scale image shows the shift in the net collected signal for a nominal muon energy deposition of 5 MeV cm^{-1} of track length in each cell. (It is interesting to note that a visual inspection of the interior of this crystal failed to reveal any trace of the complex patterns evident in this μCAT map.) These nonuniformity images can be combined with Monte Carlo simulations of energy deposition in electromagnetic showers to predict accurately the resolution for γ-rays of any energy (Dowell et al., 1990). On the basis of the μCAT scan of Figure 2.8, the predicted resolution for that detector is shown as the solid line in Figure 2.9, together with measured results. (For comparison, the dashed curve gives the expected behavior of a perfectly uniform detector.) Although the nonuniformities are complex and highly varying, the volume sampled by the γ-ray showers is large (see Figure 2.4) and the net effect is to produce a nearly constant resolution at high energies.

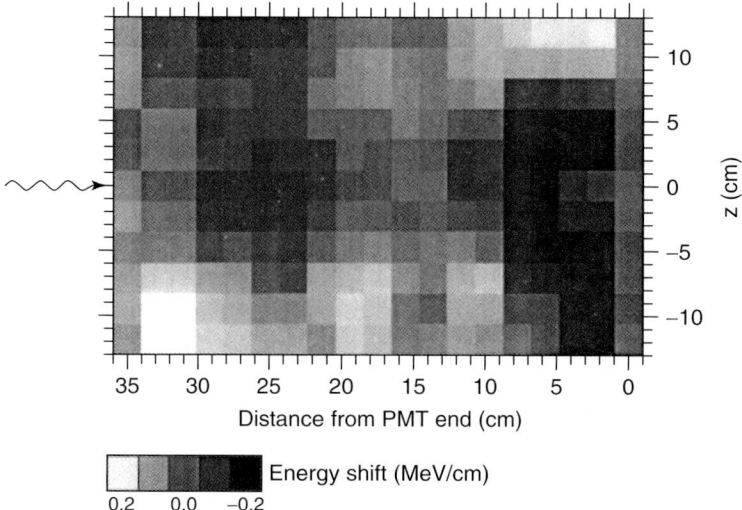

Fig. 2.8 A μCAT scan of a NaI detector, in which cosmic-ray muons deposit the same amount of energy in each cell.

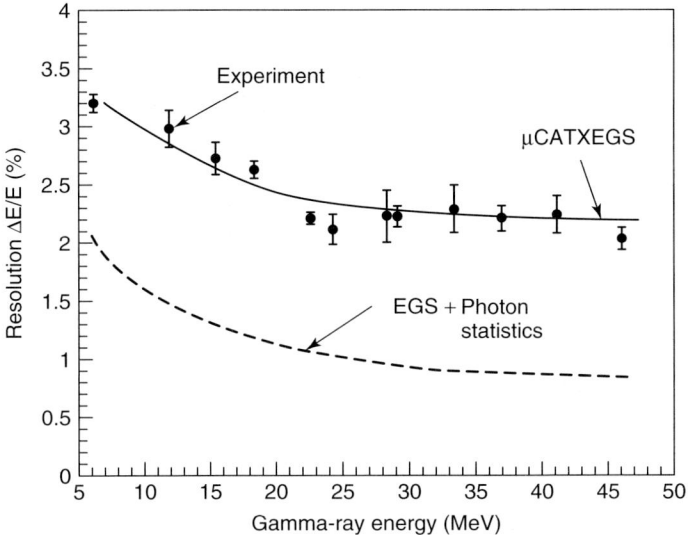

Fig. 2.9 Measured resolution of a NaI detector at several energies, the μCAT predicted resolution (solid line), and the resolution of a perfectly uniform detector (dashed curve).

2.6.2
Position Resolution

The spatial resolution of the calorimeter is determined primarily by the density of the detector materials as well as its spatial segmentation. As mentioned earlier, the containment of a shower in the transverse direction is determined by the Molière radius. Therefore, the natural unit for transverse segmentation of a calorimeter is in cells of approximately one Molière radius square. However, the position resolution of the calorimeter will be significantly better than this, since the shower position can be determined from a fit using information from many cells that contain energy from the shower. This can be done as a simple energy-weighted average given by

$$x_{\text{measured}} = \frac{\Sigma \omega_i x_i}{\Sigma \omega_i} \quad (2.8)$$

where ω_i is the energy measured in the ith calorimeter cell and x is the position of that cell for the coordinate being measured. However, more sophisticated techniques are often used where the weighting parameters depend logarithmically on the energy in order to better account for large transverse energy fluctuations (Awes et al., 1992).

2.6.3
Time Resolution

There are two main aspects of the time resolution of a γ-ray calorimeter. The first is the variation in the measured arrival time of the signal produced by an incident particle, and the second is the ability of the calorimeter to resolve two particles that are close together in time. The first aspect, called the *leading edge* time resolution, is most important in time-of-flight applications, where the calorimeter is used to measure the transit time of particles arriving from some external source. The principle factors contributing to this resolution are

1. the variation in time in the generation of the detectable signal, which can be either optical photons from a scintillator, Cherenkov photons, or ionization charge;
2. the time variation in the propagation of that signal to the readout device;
3. the variation in time due to changes in the amplitude or shape of the electronic pulse from the readout system, including contributions from electronic noise.

In most cases, the time variation in the production of the primary signal is small compared with the other effects, since the time scale involved for producing either optical photons or ion pairs is rather short (≤ 1 nanosecond). The second effect, which is mainly due to light or charge collection, can be significant, especially in a large calorimeter. It is affected by the propagation time in the detection medium (see Table 2.2 for indices of refraction in optical materials and Table 2.4 for drift velocities in noble liquids), as well as the size of the detector. The third effect includes the time variation in the detector response and readout system due to differences in energy deposition. Since the observed energy is subject to statistical fluctuations, this effect can be dominant, particularly at low energies. In general, the larger the prompt signal from the detector and the faster the response time of the readout system, the better will be the time resolution. For optical detectors producing large signals and using fast readout devices, it is possible to achieve time resolutions in the range from a few tenths down to a few hundredths of a nanosecond. For charge-collection devices, depending on the size of the signal, it is possible to obtain time resolutions on the order of a few nanoseconds.

The second aspect of a calorimeter's time resolution is its so-called pulse pair resolution or resolving time. This is the ability of the calorimeter to resolve two particles closely spaced in time. The main factor determining this resolution is the width in time of the electronic readout pulse. This depends on the response time of the active material (e.g., scintillation decay times, as in Tables 2.2 and 2.5), the propagation time in the detector (e.g., drift velocities in noble liquids, as in Table 2.4), and the response time of the readout system. For a fast device, this can be on the order of a few nanoseconds. For slower detectors, it is possible to resolve two particles close in time by analyzing the pulse shape of the measured signal and comparing it with the known response for a single particle. This technique can provide improved time separation, but may give poorer energy resolution for particles whose arrival times are within the duration of the output pulse.

2.7
Readout Devices

2.7.1
Optical Detectors

γ-Ray detectors that produce secondary scintillation photons need a light-sensitive readout devise, and a variety of options are available. The choice of the device to be used depends strongly on the type of detector (either totally absorbing or sampling) and on the type of detector material used. For scintillating materials, the most commonly used readout device is the photomultiplier tube. This device, shown schematically in Figure 2.10, consists of a photocathode that converts either UV or visible photons into electrons (or, more appropriately, photoelectrons), which are then amplified by a series of electron multipliers called *dynodes*. The photoelectrons are produced with an efficiency, known as the *quantum efficiency*, that is typically around 15–25%. The photoelectrons are then accelerated in an applied electric field and impinge on the dynode, producing several outgoing electrons for each incident electron. This process is repeated many times in subsequent stages of the photomultiplier tube until the desired gain has been achieved and the resulting charge is collected on the anode electrode. Gains of 10^6–10^8 can be readily obtained with modern photomultiplier tubes, providing the possibility of detecting even single photoelectrons and, hence, single incident photons. Photomultiplier tubes are therefore one of the most sensitive types of readout devices used for detecting scintillation light. In addition, they can be designed to have a fast response time (typically a few nanoseconds) and can therefore be used in fast timing applications. Recently, photomultiplier tubes having segmented photocathodes with separate readouts have been developed. These provide an efficient, compact way of processing signals from segmented detector arrays. The disadvantages of photomultiplier tubes are that they require rather high voltages to operate (typically 1–2 kV) and that they do not work well inside a magnetic field.

Another type of optical detector is a light-sensitive photodiode. These can be either a silicon photodiode, such as those used in photocell detectors, or a vacuum photodiode, which has a construction similar to that of a photomultiplier tube having only a photocathode and anode electrode. An inherent property of these devices is that they have no intrinsic gain and therefore rely on the incident radiation to produce a sufficiently large signal to be measured directly. However, some devices, such as avalanche photodiodes, can achieve modest gains (~10 to 100) by an amplification process within the silicon detector itself, which greatly improves their sensitivity to low light levels. The active areas of avalanche photodiodes have increased significantly in the last decade (e.g., 25 cm^2 position-sensitive units are currently available), and this has increased their use in many applications. Other types of vacuum devices, such as phototriodes or phototetrodes, are essentially low-gain photomultiplier tubes that can have large photocathode areas and can be used in

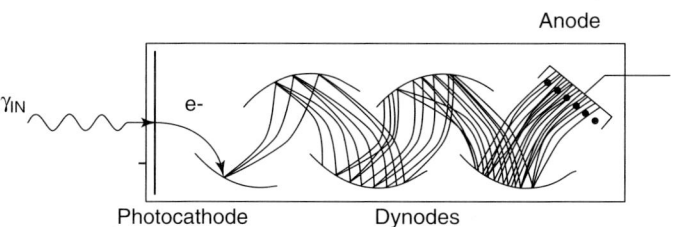

Fig. 2.10 Schematic representation of a photomultiplier tube.

other applications where modest gain is required.

2.7.2
Readout Electronics

Readout devices that have either no gain or low intrinsic gain usually require the use of a low-noise, charge-sensitive amplifier to detect the signal. The same is true for the detection of the direct ionization charge produced in noble liquids, such as those used in sampling calorimeters. In these cases, the primary signal can be on the order of only a few thousand electrons, and it is generally required to amplify this signal to the level of 10^7 electrons or more in order for it to be used for digitization and readout. This can be achieved with modern electronics through the use of low-noise amplifiers and signal-shaping techniques (Radeka, 1988). These amplifiers are used to integrate the primary charge produced in the detector and deliver a signal that has the proper pulse shape and timing characteristics for the readout system. It is also generally required to carry out this signal processing with a minimum amount of noise introduced by the electronics in order to maximize the signal-to-noise ratio. With a careful design, it is possible to achieve noise levels on the order of a few hundred electrons with signal-shaping times on the order of a few microseconds, or a few thousand electrons with shaping times on the order of a few nanoseconds. However, the choice of the appropriate signal-shaping time and noise considerations depend strongly on the specific experimental application.

Acknowledgments

We gratefully acknowledge additional contributors to the 1st edition of this chapter: Dr M. M. Lowry (Brookhaven National Laboratory and Thomas Jefferson National Accelerator facility), Dr S. Hoblit (Brookhaven National Laboratory and the University of Virginia), and Dr M. Lucas (Brookhaven National Laboratory and Ohio University). We are indebted to Dr M. Riley of Florida State University for providing a version of Figure 2.2 suitable for inclusion here.

Glossary

Bremsstrahlung: γ-Ray emitted by a high-energy electron interacting with the electric field of a nucleus or atomic electron.

Calorimeter: Device that measures the total energy it absorbs.

Cascade Shower: The sequential energy-loss process in which γ-rays convert to electron–positron pairs, which then bremsstrahlung, emitting new γ-rays that in turn pair produce, and so on until all of the energy in the initial γ-ray is dissipated.

Cherenkov Radiation: Photons produced by a charged particle when its velocity in the medium exceeds that of the photons in the medium.

Compton Scattering: Elastic scattering of a γ-ray off of an atomic electron.

Hermiticity: The degree to which a detector array is sensitive to all γ-rays, regardless of their directions or angles of emission from the source.

Pair Production: Transformation of a γ-ray interacting with the electric field of a nucleus or atomic electron into an electron–positron pair.

Photoelectric Absorption: Absorption of a γ-ray by an atomic electron.

Photomultiplier Tube: Device that generates a current pulse when struck by an optical photon, via a cascade multiplication of the secondary electron.

Spectral Quality: The fraction of γ-rays for which a measurement of their total energy is obtained.

References

Amaldi, U. (1981) *Phys. Scr.*, **23**, 409–424.
Awes, T., Obenshain, F.E., Plasil, F., Saini, S., Sorensen, S.P., and Young, G.R. (1992) *Nucl. Instrum. Methods Phys. Res., Sect. A*, **311**, 130–138.
Beausang, C.W. (2003) *Nucl. Instrum. Methods Phys. Res., Sect. A*, **B 204**, 666.
Beausang, C.W. and Simpson, J. (1996) *J. Phys.*, **G 22**, 527–228.
Bethe, H.A. and Ashkin, J. (1953) in *Experimental Nuclear Physics* (ed. E. Segrè), Vol. 1, John Wiley & Sons Inc., New York, p. 166.
Brun, R. and Carminati, F. (1993) *CERN Program Library Long Writeup W5013*.
DØ Collaboration, Abachi, S. et al. (1994) *Nucl. Instrum. Methods Phys. Res., Sect. A*, **338**, 185–253.
Dowell, D.H., Sandorfi, A.M., Baron, A.Q.R., Fineman, B.J., Kistner, O.C., Matone, G., Thorn, C.E., and Sealock, R.M. (1990) *Nucl. Instrum. Methods Phys. Res., Sect. A*, **286**, 183–201.
Dowell, D.H., Ziegler, W.P., Sandorfi, A.M., Ziegler, B., and Schoch, B. (1987) *Nucl. Instrum. Methods Phys. Res., Sect. A*, **254**, 570–577.
Ebeth, J. and Simpson, J. (2008) *Prog. Part. Nucl. Phys.*, **60**, 283–337.
Gammasphere (2008) <http://nucalf.physics.fsu.edu/~riley/gamma/>.
Gelletly, W. and Eberth, J. (2006) *Lecture Notes in Physics*, Vol. 700, Springer, pp. 79–117.
Hurlbut, C.R. (1985) *Paper Presented at the American Nuclear Society Winter Meeting, excerpted from the Bicron Catalog*, Bicron, Inc., Newbury, OH.
Kehoe, W.L. (ed.), (PHENIX Collaboration, Adare, A., et al.) (1993) in *BNL Report No. 48922* Brookhaven National Laboratory, Upton, NY.
Kleinknecht, K. (1982) *Phys. Rep.*, **84**, 85–161.
Lee, I.Y., Clark, R.M., Cromaz, M., Deleplanque, M.A., Descovich, M., Diamond, R.M., Fallon, P., Macchiavelli, A.O., Stephens, F.S., and Ward, D. (2004) *Nucl. Phys.*, **A 746**, 255.
Lee, I.Y., Deleplanque, M.A., and Vetter, K. (2003) *Rep. Prog. Phys.*, **66**, 1095–1144.
Lempicki, A., Wojtowicz, A.J., and Berman, E. (1993) *Nucl. Instrum. Methods Phys. Res., Sect. A*, **333**, 304–311.
LRP (2002) *Opportunities in Nuclear Science, A Long Range Plan for the next decade*, <http://www.sc.doe.gov/np/nsac/docs/LRP_5547_FINAL.pdf>.
Nelson, W.R., Hirayama, H., and Rogers, D.W.O. (1985) *SLAC Report No. 265*, SLAC, Stanford, CA.
Radeka, V. (1988) *Annu. Rev. Nucl. Part. Sci.*, **38**, 217–277.
Scott, W.T. (1963) *Rev. Mod. Phys.*, **35**, 231–313.
SLD Collaboration, Abe, Kenji et al. (1984) *SLAC Report No. 273*, SLAC, Stanford, CA; Haller, G.M. Fox, J.D., Smith, S.R., IEEE Trans. NS 36, 675 (1989).
Tsai, Y.S. (1974) *Rev. Mod. Phys.*, **46**, 815–851.
Tsai, Y.S. (1977) *Rev. Mod. Phys.*, **49**, 421.

Further Reading

Evans, R.D. (1955) *The Atomic Nucleus*, McGraw-Hill, New York.
Fabjan, C.W. and Ludlam, T. (1982) *Annu. Rev. Nucl. Part. Sci.*, **32**, 335–389.
Gratta, G., Newman, H., and Zhu, R.Y. (1994) *Annu. Rev. Nucl. Part. Sci.*, **44**, 453–500.
Knoll, G.F. (1999) *Radiation Detection and Measurement*, 3rd edn, John Wiley & Sons Inc., New York.
Majewski, S. and Zorn, C. (1992) in *Instrumentation in High Energy Physics* (ed. F. Sauli), World Scientific, Singapore, p. 157.

3
Mössbauer Spectroscopy

De-Ping Yang

3.1	**Introduction** 53	
3.2	**The Mössbauer Effect** 53	
3.3	**The Recoil-Free Fraction** 56	
3.4	**Sources of Mössbauer Radiation** 58	
3.5	**The Mössbauer Spectrometer** 60	
3.6	**The Shape and Intensity of a Spectral Line** 62	
3.7	**Hyperfine Interactions** 64	
3.7.1	Isomer Shift 64	
3.7.2	Magnetic Dipole Interaction 65	
3.7.3	Electric Quadrupole Interaction 68	
3.7.4	Combined Electric Quadrupole and Magnetic Dipole Interactions 71	
3.7.5	Saturation Effect in the Presence of Hyperfine Splittings 72	
3.8	**Second-Order Doppler Effect** 73	
3.9	**Applications of Mössbauer spectroscopy** 73	
3.10	**Mössbauer Spectroscopy Using Synchrotron Radiation** 76	
3.10.1	Monochromatization of Synchrotron Radiation 77	
3.10.2	Time-Domain Mössbauer Spectroscopy 78	
3.10.3	Applications of Time-Domain Mössbauer Spectroscopy 81	
3.11	**Synchrotron Methods versus Conventional Methods** 82	
	References 83	
	Further Reading 84	

Encyclopedia of Applied Spectroscopy. Edited by David L. Andrews.
Copyright © 2009 WILEY-VCH Verlag GmbH & Co. KGaA, Weinheim
ISBN: 978-3-527-40773-6

3.1 Introduction

Mössbauer spectroscopy is a unique technique based on the nuclear resonance via recoil-free emission, absorption, and scattering of γ-rays. Because of its extremely high energy resolution and its particular timescale, a Mössbauer spectrum can provide valuable information on structural, chemical, magnetic, and dynamic properties of a variety of solid materials. Mössbauer spectroscopy finds applications in various fields of scientific research, such as physics, metallurgy, materials science, chemistry, mineralogy, biology, medicine, environmental science, archeology, and art. In this chapter, the theoretical principles, experimental techniques, and various applications of Mössbauer spectroscopy are briefly described. For rigorous theoretical treatment and complete experimental details, the reader is referred to specialized books and review articles listed in the Further Reading section at the end of this chapter. There are more than 50 000 journal articles reporting research using Mössbauer spectroscopy in the literature. They have been systematically collected by the Mössbauer Effect Data Center at the University of North Carolina at Asheville and compiled in the Center's publications *The Mössbauer Effect Data Index* (1966–77) and *The Mössbauer Effect Reference and Data Journal* (since 1978).

3.2 The Mössbauer Effect

The principle of Mössbauer spectroscopy is based on the recoil-free nuclear γ-resonance, known as the *Mössbauer effect*, named after its discoverer Rudolf L. Mössbauer. γ-Rays are photons emitted from certain nuclei when they make a transition from an excited state to a lower state. These γ-rays may be absorbed by other nuclei of the same isotope in their lower state, making a transition to the same excited state, much like any other resonance absorption phenomenon. This lower state is usually the stable ground state, whereas the excited state has a typical lifetime of $\tau \approx 100$ nanoseconds. According to the uncertainty principle $(\Delta E)(\Delta t) \sim \hbar$, where the Planck constant $\hbar \approx 10^{-15}$ eV s, this lifetime corresponds to a natural linewidth of $\Gamma \approx 10^{-8}$ eV. Since the γ-photon energies are typically in the order of $E_\gamma \approx 10^4$ eV, this linewidth ($\Gamma/E_\gamma \approx 10^{-12}$) is very narrow and characterizes an extremely "monochromatic" nature of

Encyclopedia of Applied Spectroscopy. Edited by David L. Andrews.
Copyright © 2009 WILEY-VCH Verlag GmbH & Co. KGaA, Weinheim
ISBN: 978-3-527-40773-6

γ-rays emitted from a particular nuclear isotope. We are fortunate to have access to nuclear γ-rays possessing such a unique quality, but this is a double-edged sword. On the one hand, an extraordinarily high energy resolution can be achieved if γ-rays are used in resonance spectroscopy. On the other hand, a precise resonance condition within 10^{-8} eV must be strictly satisfied between the emitter and the absorber.

During the first half of the twentieth century, considerable amount of research was devoted to finding ways of matching the energies of emitted γ-photons with the energy requirement for nuclear resonant absorption. The main obstacle was recoil of the nucleus, during emission as well as absorption of γ-rays. When a free nucleus emits a photon of energy E_γ (Figure 3.1), the photon carries with it a momentum of

$$p = \frac{E_\gamma}{c} \tag{3.1}$$

where c is the speed of light. According to momentum conservation, the nucleus recoils in the opposite direction with the same moment p, which corresponds to a recoil kinetic energy of

$$E_R = \frac{p^2}{2M} \tag{3.2}$$

where M is its mass. Because of this recoil energy, the actual energy of the emitted photon is reduced by E_R, and Eq. (3.2) can be rewritten as

$$E_R = \frac{(E_\gamma/c)^2}{2M} = \frac{E_\gamma^2}{2Mc^2} \tag{3.3}$$

The absorption of a γ-photon by a free nucleus involves exactly the opposite recoil process – the nucleus will gain a momentum of p and a recoil kinetic energy of E_R. Therefore, a photon of merely an energy of E_γ does not suffice to cause a nuclear transition, but an excess energy of E_R is needed to compensate the recoil energy imparting the motion to the nucleus.

Since the rest mass $M_e c^2$ of an electron is about 0.5 MeV and a typical nucleus is about 10^5 times more massive than the electron, the nuclear rest mass Mc^2 is of the order of 10^{11} eV. Using $E_\gamma \approx 10^4$ eV, we estimate the recoil energy

$$E_R \approx \frac{(10^4)^2}{10^{11}} = 10^{-3} \text{ eV} \tag{3.4}$$

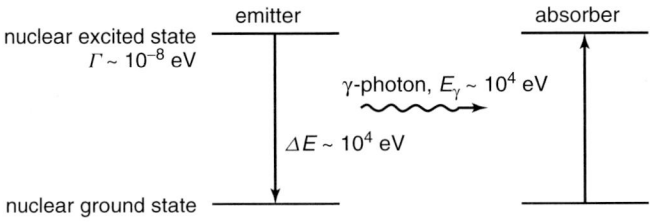

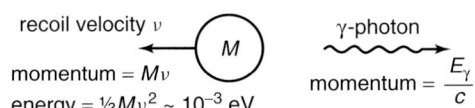

Fig. 3.1 Emission of a γ-photon and recoil of the nucleus.

This is about 100 000 times larger than the natural linewidth Γ of the photon energy; therefore, recoil of a free nucleus shifts the photon energy far beyond its "bandwidth" and completely destroys the resonance condition.

Various methods, mostly based on the Doppler effect, were used to compensate the recoil energy by raising the emitted photon energy so that the emission and absorption lines overlapped, at least partially. For example, it was well known that the Doppler broadening associated with thermal motion of the nuclei could add energy to photons emitted from those fast-moving nuclei that had the same direction as the γ-rays. Alternatively, macroscopic mechanical motion ($\sim 10^5$ cm s^{-1}) of the emitting nuclei (i.e., a source) toward the absorber could shift the photon energy to meet the resonance condition. In certain situations, a preceding nuclear reaction may result in a fast-moving nucleus so that the Doppler shift in the emitted photon could become large enough to cause resonance absorption. These techniques did allow the observation of nuclear resonance, but the recoil energy was merely compensated, often at the expense of the energy resolution and the loss of γ-ray intensity.

Mössbauer (1958) discovered a distinctively new phenomenon that, when the emitting nuclei are bound in a solid, a fraction of the emitted γ-rays have exactly the energy E_γ, corresponding to a recoil-free process. This process seemingly violates the conservation of momentum when we consider one individual emission. However, we should realize that the recoil momentum and energy can no longer be transferred to the emitting nucleus alone, but to a large number of tightly bound atoms within the entire solid, in the form of their simultaneous vibrations (i.e., creating phonons). While using γ-rays from ^{191}Ir and comparing their resonant scatterings by Ir and Pt, Mössbauer experimentally observed higher scattering intensities by Ir at lower temperatures, which contradicted the theory based on thermal broadening of the γ-rays. He then successfully observed higher resonant absorption of the 129 keV radiation from Ir at a temperature of 88 K and correctly attributed the phenomenon to the zero-phonon process (Mössbauer, 1958, 2000). Because the typical phonon frequency is in the order of 10^{13} Hz (corresponding to $\hbar\omega \sim 10^{-2}$ eV), a single recoil ($\sim 10^{-3}$ eV) does not have enough energy to create one phonon (Figure 3.2).

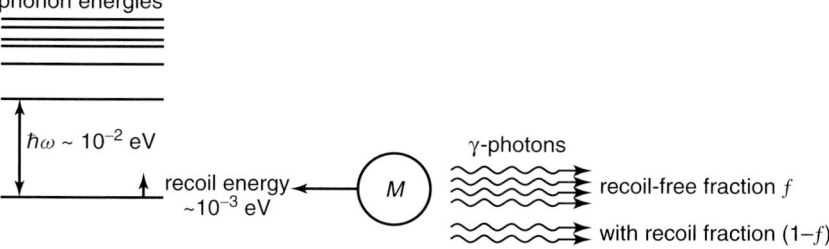

Fig. 3.2 Recoil energy compared with the phonon energy in a typical solid. Because the recoil energy due to the emission of one photon is not large enough to create one phonon, a fraction of photons are emitted in the recoil-free fashion.

Therefore, there is a sizeable probability of recoil-free emission. If all emissions (both recoil-free and with recoil) are considered collectively, momentum and energy conservation laws are completely obeyed.

When these γ-rays, emitted in the recoil-free fashion, are allowed to be absorbed, also in a similar recoil-free way, resonance experiments can be carried out without the need for energy compensation. Now we have essentially two categories of γ-photons; some are emitted, absorbed, or scattered at exactly the energy corresponding to a nuclear transition, and others carry higher or lower energies corresponding to the creation or annihilation of certain phonons. The first group (recoil-free) is wonderfully suitable for resonance experiments with extremely high energy resolution, whereas the second group is suitable for studying lattice dynamics because these γ-rays (with recoil) carry information about phonon distribution in the emitter, absorber, or scatterer. In recognition of his experimental discovery and theoretical interpretation of this recoil-free phenomenon, Mössbauer was awarded the Nobel Prize in physics in 1961 and this phenomenon has since been known as the *Mössbauer effect*.

3.3
The Recoil-Free Fraction

The recoil-free γ-photons constitute a fraction of the total γ-radiation emitted from a particular source. This parameter is known as the *recoil-free fraction f*. The value of f obviously depends on how the recoil energy E_R compares with the phonon energies $\hbar\omega$. Only when $E_R \ll \hbar\omega$ will f be reasonably large. According to Lipkin's sum rule (Lipkin, 1960), when a large number of absorption events are considered, the average energy transferred to the lattice must be exactly equal to E_R. Let a total of m γ-photons be emitted, among which n are recoil-free and, among the remaining photons, each excites a single phonon (neglecting double phonons). Then

$$mE_R = (m-n)\hbar\omega \quad (3.5)$$

On the basis of the Einstein model of lattice vibrations of a single frequency ω, we arrive at an approximate expression for the recoil-free fraction f:

$$f = \frac{n}{m} = 1 - \frac{E_R}{\hbar\omega} \quad (3.6)$$

A more precise expression for the recoilless fraction is

$$f = e^{-k^2 \langle u^2 \rangle} \quad (3.7)$$

where $\langle u^2 \rangle$ is the mean-square displacement of a nucleus along the direction of the wave vector $\boldsymbol{k}$ of the emitted γ-ray. The wave vector $\boldsymbol{k}$ is related to the photon energy E_γ by $\hbar k = E_\gamma/c$. This expression in Eq. (3.7) points out that to have a relative large f, both the γ-photon energy E_γ and the atomic vibrations $\langle u^2 \rangle$ should be small. The first condition of small $\boldsymbol{k}$ can be satisfied by selecting isotopes with low-energy γ-transitions. At present, Mössbauer effect has been observed from more than 100 nuclear isotopes (e.g., ^{57}Fe, ^{119}Sn, ^{191}Ir, etc.), many of which have transition energies less than 50 keV. The second condition of small $\langle u^2 \rangle$ requires a rigidly bound solid (characterized by a high Einstein or Debye temperature), perhaps cooled to a low temperature (small vibration amplitudes). This implies that in a liquid or a gas, Mössbauer effect is extremely difficult to observe because of the large $\langle u^2 \rangle$ values. For any γ-ray energy higher than 100 keV, the source and the absorber are usually kept at low temperatures to reduce their $\langle u^2 \rangle$ values.

The mean-square displacement $\langle u^2 \rangle$ is intimately related to lattice dynamics, that is, how the collective atomic vibrations take place at a given temperature. We adopt a particular model for lattice dynamics, for example, the Debye model with a phonon density of states (DOS) $g(\omega) = 3\omega^2/\omega_D^3$, where ω_D is related to the parameter known as the *Debye temperature* θ_D through the Planck and Boltzmann constants, $\hbar \omega_D = k_B \theta_D$. Applying the Bose–Einstein distribution of phonons among the available phonon states at temperature T, we arrive at a more specific expression for the recoil-free fraction,

$$f = \exp\left[-k^2 \frac{3\hbar^2}{Mk_B\theta_D}\left(\frac{1}{4} + \frac{T^2}{\theta_D^2}\int_0^{\theta_D/T} \frac{x\,dx}{(e^x - 1)}\right)\right] \quad (3.8)$$

where $x = (\hbar\omega)/(k_B T)$ is a dimensionless variable for the definite integral, which covers the range from 0 to θ_D/T. In the high-temperature limit ($\theta_D/T \ll 1$), x only takes small values, $e^x \approx 1 + x$, and the above expression simplifies to

$$f \approx \exp\left[-k^2 \frac{3\hbar^2}{Mk_B\theta_D}\left(\frac{1}{4} + \frac{T}{\theta_D}\right)\right]$$

$$\approx \exp\left[-k^2 \frac{3\hbar^2 T}{Mk_B\theta_D^2}\right] \quad (3.9)$$

In the low-temperature limit ($\theta_D/T \gg 1$), the integral has a value of $\pi^2/6$, and the above express simplifies to

$$f \approx \exp\left[-k^2 \frac{3\hbar^2}{Mk_B\theta_D}\left(\frac{1}{4} + \frac{\pi^2 T^2}{6\theta_D^2}\right)\right]$$

$$= \exp\left[-k^2 \frac{\hbar^2}{2Mk_B\theta_D}\left(\frac{3}{2} + \frac{\pi^2 T^2}{\theta_D^2}\right)\right] \quad (3.10)$$

At absolute zero ($T = 0$ K), it is simply

$$f = \exp\left[-k^2 \frac{3\hbar^2}{4Mk_B\theta_D}\right] \quad (3.11)$$

The Debye model assumes isotropic, harmonic vibrations only. When the lattice has lower symmetry, $\langle u^2 \rangle$ depends on the atomic vibration direction with respect to the lattice symmetry axes, resulting in an anisotropic behavior of the recoil-free fraction f. This affects the quantitative measurement of Mössbauer effect, which was first observed by Goldanskii in polycrystalline samples (Goldanskii et al., 1962) and was first explained theoretically by Karyagin (Karyagin, 1963). This phenomenon is known as the *Goldanskii–Karyagin* (G–K) effect, which can be used to study anisotropy in lattice dynamics. External pressure applied to the solid may also cause a change in the mean-square displacement $\langle u^2 \rangle$ and produce additional anisotropy, which will be reflected in a change in the recoil-free fraction. At high temperatures, if the Mössbauer effect still persists, anharmonicity in the atomic vibration may render the Debye model invalid, and the theoretical treatment of f may involve higher order terms and become a complex function of temperature. Even at low temperatures, anharmonic behavior has been detected in several solids.

It is important for resonant absorption experiments that a relatively large

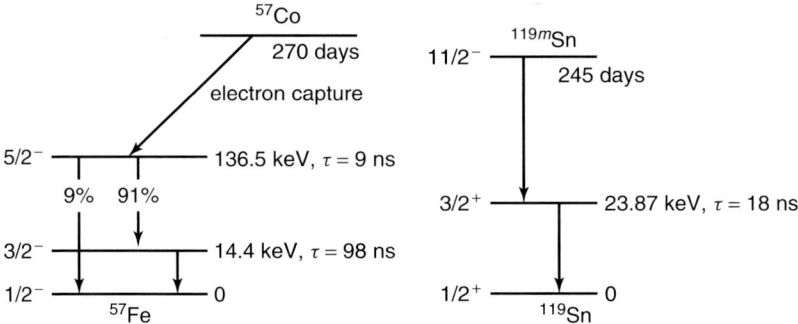

Fig. 3.3 Nuclear decay schemes of ^{57}Co and ^{119m}Sn.

fraction is available and its variation is precisely known (e.g., whether f remains constant when comparing results from a series of samples, or how f changes within the temperature range of the experiments). It is even more important for investigating lattice dynamics using Mössbauer effect because f carries all the information about phonon behavior. Special attention has always been paid to the precise measurement of the recoil-free fraction f. The precision has reached 1% when radioactive sources are used, and it can be better than 0.4% when synchrotron radiation is employed. Very often, we are more interested in how f changes with sample composition, temperature, or pressure than in the absolute value of f.

3.4
Sources of Mössbauer Radiation

Mössbauer effect has been observed in about 100 isotopes. ^{40}K is the lightest isotope, followed by ^{57}Fe. The Mössbauer isotopes are not distributed evenly in the periodic table – about 75% of them are concentrated in elements of atomic numbers between 50 and 80. There are only about 20 Mössbauer isotopes that are in practical use. ^{57}Fe and ^{119}Sn, whose decay schemes are given in Figure 3.3, are the most popular. ^{57}Fe is by far the most important Mössbauer isotope; more than 70% of the research work involves ^{57}Fe. Other frequently used isotopes are ^{151}Eu, ^{197}Au, ^{129}I, ^{121}Sb, and ^{125}Te. Increased attention has been paid to ^{237}Np, ^{155}Gd, ^{161}Dy, and especially to ^{67}Zn and ^{181}Ta, which are used for taking high-resolution spectra.

For ^{57}Fe Mössbauer spectroscopy, the source usually contains the radioactive ^{57}Co, which captures a K-electron and becomes ^{57}Fe at the 136-keV metastable state. It then makes transitions to the ground state, accompanied by γ-ray emissions, among which the 14.4-keV Mössbauer radiation is the most useful. The half-life of the ^{57}Co decay is $T_{1/2} = 270$ days.

For Mössbauer spectroscopy, the radioactive isotope is embedded in a solid host material (matrix) to form a Mössbauer source. Mössbauer sources are specially prepared via nuclear reactions, and they are commercially available. Sources can also be made by "in-beam" implantation, allowing the use of isotopes with short life times, such as ^{57}Mn ($T_{1/2} = 1.45$ minutes), which decays directly to the 14.4-keV excited state ^{57}Fe (Yoshida et al., 2002).

The quality of a Mössbauer source depends on the properties of the isotope and the selection of the host material. The Mössbauer isotope should satisfy the following criteria:

1. The γ-photon energy E_γ should be in the range 5–150 keV, preferably less than 50 keV. This is because both the recoil-free fraction f and the resonance cross section σ_0 decrease as E_γ increases, especially f, which decreases more rapidly.
2. The desired lifetime τ of the excited state is 1–100 nanoseconds. A long τ corresponds to a narrow natural width Γ_n, and a slight mechanical vibration may destroy the resonance condition. A short τ means that Γ_n may be too large to yield a high-resolution spectrum.
3. The internal conversion coefficient α should be small (<10), to ensure a relatively large probability of available γ-ray emission.
4. It is preferred that the parent nucleus has a long half-life $T_{1/2}$ and allows for easy production of a high activity source.
5. The Mössbauer isotope should not have a high nuclear spin, which would produce complicated spectra and make analysis more difficult.
6. The Mössbauer isotope should have a reasonably large natural abundance. Among the Mössbauer isotopes, ^{57}Fe fulfills the above requirements most satisfactorily, except for its low natural abundance of 2.14%. ^{119}Sn also possesses many good qualities.

A good Mössbauer source requires an appropriate host, a suitable fabrication process, and so on. The following are the main aspects that should be taken into consideration:

1. The radiation from the source should be monochromatic with an energy width as close to the natural width as possible, which requires that the host matrix be a nonmagnetic material of a cubic lattice and of high purity.
2. The host material should have a high Debye temperature to maximize the recoil-free fraction f; metals (and occasionally ionic crystals) of high melting temperatures are usually chosen for the matrix material.
3. The number of stable Mössbauer nuclei in the host material should be minimized to prevent resonant self-absorption.
4. The host material should be made very thin, to reduce the photoelectric effect and Compton scattering caused by the Mössbauer γ-rays.
5. The host should be chemically stable and properly sealed for protection against oxidation, leakage of radioactive material, and so on. For ^{57}Fe, rhodium (Rh) and palladium (Pd) are good hosts, giving an excellent f-value of about 0.78 at room temperature.

In the last two decades, researchers have also been attracted to synchrotron radiation (SR) as a new Mössbauer radiation source. SR provides polarized pulsed radiation of high intensity, high collimation, and narrow beams. The high intensity of the SR source has been utilized for scattering experiments where conventional radiation sources cannot provide adequate results. The pulsed nature of SR radiation is most suitable for measuring time spectra where the entire nuclear ensemble is excited simultaneously and its coherent decay can be observed at different time intervals. One more advantage associated with SR is that γ-rays are naturally polarized, either linearly or circularly, which allow selective absorption and therefore provide additional information. At present, SR Mössbauer radiation can only be produced in large facilities, such as the European Synchrotron Radiation Facility (ESRF) in Grenoble (France), the Advanced Photon Source (APS) in Argonne (USA), and the Super Photon ring (SPring-8) in Hyogo (Japan).

One of the drawbacks of SR radiation is its relatively wide energy distribution. Therefore, it must be collimated and monochromatized before being used in any absorption or scattering experiments. Going through a double-crystal premonochromator Si(111), followed by a high-resolution monochromator based on particular Bragg reflections, the bandwidth can be reduced to the order of milli electronvolts, or even micro electronvolts.

3.5
The Mössbauer Spectrometer

The most common technique for obtaining a Mössbauer spectrum is using a velocity-scanning spectrometer in the transmission geometry, a block diagram of which is shown in Figure 3.4. It consists of a radiation source, an absorber, a detector with its electronic recording system, a wave-form generator, a drive circuit, and a transducer. A beam of γ-rays from a source containing the Mössbauer isotope is directed toward a thin absorber containing the stable nuclei of the same isotope. The transmitted γ-rays are detected by a suitable detector and recorded in a computer system. To obtain a resonance spectrum (i.e., the amount of transmitted γ-rays versus photon energy), the energy of the incoming γ-rays must be modulated

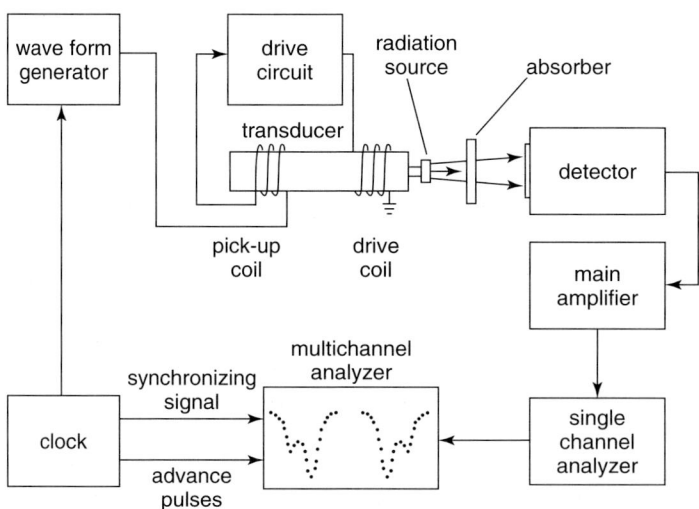

Fig. 3.4 Block diagram of a Mössbauer spectrometer in the transmission geometry.

using Doppler effect. In most cases, the source undergoes a mechanical motion with a velocity v toward or away from the absorber, whereas the absorber is at rest, so that it is easier to change its temperature or to apply an external magnetic field to the absorber. The energy of the radiation is therefore Doppler shifted to

$$E = E_0 \left(1 + \frac{v}{c}\right)\left(1 - \frac{v^2}{c^2}\right)^{-1/2} \quad (3.12)$$

where E_0 is the photon energy when the source is at rest. The source velocities that are required to cover the energy range of a typical Mössbauer spectrum are usually less than 1 m s^{-1}, thus $v/c \ll 1$ and a very good approximation of the above equation is

$$E = E_0 \left(1 + \frac{v}{c}\right) \text{ or } \Delta E = E_0 \frac{v}{c} \quad (3.13)$$

Because of the energy modulation by source velocity, the energy scale in a Mössbauer spectrum is customarily presented in terms of the source velocity. For example, 1 mm s^{-1} corresponds to an energy shift of 4.80×10^{-8} eV for a ^{57}Fe Mössbauer spectrum ($E_0 =$ 14.4 keV).

The source is driven by a velocity transducer, generally operating in the velocity scan mode. The drive coil converts an applied current to the velocity of a shaft. A pickup coil provides a signal proportional to the actual velocity; a negative feedback circuit ensures accuracy and stability. The radiation source is attached to one end of the shaft, and a prism (or a mirror) for measuring velocity is mounted on the other end. The source thus scans periodically through the velocity range of interest, the simplest way of scanning velocity being constant acceleration.

The most widely employed detectors are gas proportional counters and NaI(Tl) scintillation counters, followed by semiconductor detectors. A detector should have a high efficiency and energy resolution, as well as quick response time. Typically, a gas proportional counter has a cylindrical metal tube (cathode) filled with a mixture of gases and a metal wire on the axis (anode). Operating under a high voltage of 1500–3000 V, the gas proportional counter has a high signal-to-noise ratio and a high count rate. The NaI(Tl) scintillation counters are made of sodium iodine crystals doped with thallium with detection efficiency as high as 100% and are easy to use. The semiconductor detector is essentially a p–n junction, with a reverse-biased high voltage, creating a region that is sensitive to γ-rays. They offer the best resolution over the entire range of energies of interest in Mössbauer experiments, and their efficiency is comparable to proportional or scintillation counters.

In Mössbauer spectroscopy, the absorber is usually the sample to be investigated. In the transmission geometry, the thickness of the absorber should be carefully chosen as it significantly affects the quality of the spectrum. When the sample is too thin, it would contain too few Mössbauer nuclei, resulting in weak spectral intensity, high background, and large statistical uncertainty. When the sample is too thick, significant atomic absorption would occur, causing a decrease in resonant absorption and distortion in the spectral shape. Obviously, there should be an optimal thickness between these two extremes. The effective thickness of the absorber is defined by

$$t_a = n_a f_a \sigma_0 d \quad (3.14)$$

where n_a is the number of Mössbauer nuclei in the absorber per milligram, f_a is the recoil-free fraction of the absorber, σ_0 is the maximum resonance cross section (in centimeter squared), and d is the thickness of the absorber (in milligrams per centimeter squared). For most of the absorption experiments, the effective thickness should be $t_a \approx 1$.

Sample preparation for the absorber is quite straightforward. Metallic or alloy samples are usually wrought or roll-milled into foils of the appropriate thickness. If only small pieces of sample foils are available, they may be arranged to cover the entire sample area with as few gaps and overlaps as possible. Most of the materials are in the powder form, which can be pressed to the desired thickness between two pieces of thin plastic sheets. For low-temperature measurements, the powder sample is usually mixed with a solid chemical of light atoms such as a sugar, and pressed into a "free-standing" sample. For high-temperature measurements, the powder sample is usually placed between two pieces of boron nitride sheets. Liquid samples are usually sealed in the sample holder and refrigerated until frozen. ^{57}Fe has a relatively low natural abundance; if the ^{57}Fe content in the sample is not sufficient to give a satisfactory spectrum, ^{57}Fe enrichment in the sample may be necessary.

The Mössbauer spectrometer may also include an external magnetic field (static or high frequency) to be applied to the absorber or the source. Sample temperature may be varied by using an oven or by using a cryogenic system. Windows for γ-rays to enter and exit the sample chamber should be made from a low-atomic-number element, such as beryllium.

Emission Mössbauer spectra may also be recorded from a transmission spectrometer, where the source is the material under investigation and the absorber is a well-characterized material.

In addition to the transmission geometry, Mössbauer effect can be utilized in the scattering geometry, which does not require a thin sample but probes into the surface of bulk materials. Experiments in the scattering geometry can detect not only the elastically scattered γ-rays but also the accompanying conversion electrons (conversion electron Mössbauer spectroscopy, CEMS), the X rays following the conversion events (conversion X-ray Mössbauer spectroscopy), or the inelastically scattered γ-rays (Mössbauer diffraction).

3.6
The Shape and Intensity of a Spectral Line

Because the nuclear excited state has a certain natural width Γ_s, the emitted γ-rays from a Mössbauer source are not completely monochromatic, but follow the Lorentzian distribution around E_γ (the Breit–Wigner formula),

$$\mathcal{L}(E)dE = \frac{\Gamma_s}{2\pi} \frac{1}{(E - E_\gamma)^2 + \Gamma_s^2/4} dE$$

(3.15)

After a resonant absorption, the nuclei in the absorber are in the excited state (with a natural width of Γ_a). They decay to the ground state through internal conversion or γ-ray emission in all directions. The cross section of resonant absorption of γ-rays (as a function of photon energy E)

is also described by the Breit–Wigner formula:

$$\sigma_a(E) = \sigma_0 \frac{\Gamma_a^2/4}{(E-E_\gamma)^2 + \Gamma_a^2/4} \quad (3.16)$$

where the maximum resonance cross section is

$$\sigma_0 = \frac{\lambda^2}{2\pi} \frac{1+2I_e}{1+2I_g} \frac{1}{1+\alpha} \quad (3.17)$$

λ is the wavelength of the γ-ray, I_e and I_g are the nuclear spins of the excited and ground states, respectively, and α is the internal conversion coefficient.

A Mössbauer absorption spectrum is a record of transmitted γ-ray counts through the absorber as a function of γ-ray energy. Since the experiment involves both emission and absorption, we expect the line shape to be a convolution of the emission spectrum and the absorption cross section. Assuming a single-line emission and single-line absorption, we have

$$\sigma_a^{\exp}(E) \propto \int_{-\infty}^{+\infty} \mathcal{L}(E-x)\sigma(x)dx$$

$$= \frac{\sigma_0 \Gamma_a}{\Gamma_a + \Gamma_s} \frac{(\Gamma_s+\Gamma_a)^2/4}{(E-E_\gamma)^2 + (\Gamma_s+\Gamma_a)^2/4}$$
$$(3.18)$$

which also has a Lorentzian shape, with a full width at half maximum equal to $\Gamma_s + \Gamma_a$ (the sum of the natural widths of the Mössbauer nuclei in the source and the absorber).

Experimentally, we usually measure resonance absorption by detecting the transmitted or scattered γ-rays. The vertical axis of a Mössbauer spectrum records the counts of detected γ-photons. In the transmission geometry, the original total intensity I_0 emitted from the source includes both recoil-free and recoiled γ-rays. After going through the absorber, the intensity is reduced mainly through two absorption processes: Mössbauer resonant absorption and nonresonant atomic absorption. When the source velocity does not meet the resonance condition, only the second process takes place and we observe an ideally flat baseline of the spectrum. When the source velocity is within the linewidth of absorption, resonant absorption significantly reduces the transmitted γ-rays and the spectrum shows an intensity minimum at the center of the resonance. The amount of maximum absorption may range from a few percent to 40% of the baseline count. The inverted peak can be characterized by parameters such as position of the resonance, full width at half-maximum absorption, and depth and area of the resonance line. When spectra are obtained in the scattering geometry, similar parameters are used, except that peaks representing the detected γ-rays, X rays, or conversion electrons are now above the baseline.

The horizontal axis of a Mössbauer spectrum is customarily labeled by the source velocity in millimeters per second. Velocity calibration is usually done by certain standard samples. For example, metallic α-iron is commonly used to determine the zero of source velocity and the scale of energy shift for ^{57}Fe spectra. The energy scale may also be independently calibrated by using a Michelson interferometer installed at the end of the oscillating shaft of the velocity transducer.

Since $\Gamma_s \approx \Gamma_a \sim 10^{-8}$ eV and E_γ is typically 10^4 eV, the energy resolution of a Mössbauer line can be as good as 10^{-12}. Because of this extremely high energy resolution, Mössbauer spectroscopy is ideally suited for studying hyperfine interactions.

3.7 Hyperfine Interactions

Because of the interactions between the nucleus and the surrounding electrons, nuclear energy levels are slightly altered and certain energy degeneracies are lifted. Some of the Mössbauer spectral lines will shift and others will split because of hyperfine interactions. These interactions are mainly electromagnetic but must be treated using quantum mechanics. The energies involved may be as small as 10^{-8} eV. Hyperfine interactions with nuclei at their ground states were initially observed in optical spectroscopy, followed by more accurate measurements from resonance experiments such as nuclear magnetic resonance (NMR) and electron paramagnetic resonance (EPR). Mössbauer spectroscopy offers an extra dimension of advantage. In addition to its high energy resolution, Mössbauer effect allows the observation of interactions between electrons and the nuclei in excited states, not just those in the ground state.

There are three main types of hyperfine interactions in Mössbauer effect:

1. electric monopole interaction (E0), which causes isomer shift δ, a shift of the entire resonance spectrum;
2. magnetic dipole interaction (M1), which causes magnetic hyperfine splittings of the spectral lines (Zeeman splittings);
3. electric quadrupole interaction (E2), which causes quadrupole splittings.

The electric dipole interaction (E1) is forbidden by parity requirement, and all interactions of higher orders (M2, E3, etc.) are much weaker to be detected.

3.7.1 Isomer Shift

We assume that a nucleus of atomic number Z is a uniformly charged sphere (monopole) when it is either in the ground state (with radius R_g) or in the excited state (with radius R_e). These radii have finite values and are different ($R_g \neq R_e$), although each is nearly zero compared with the size of an electron orbit. Let $\psi(0)$ represent the wave function of an s-electron within the volume of the nucleus ($r < R_g$, $r < R_e$, or practically, $r \approx 0$). The Coulomb energy between the nuclear charge and the s-electron's presence within the nuclear volume can be calculated to be

$$E_g = S(Z)Ze \frac{2\pi}{5} R_g^2 e |\psi(0)|^2 \quad \text{and}$$

$$E_e = S(Z)Ze \frac{2\pi}{5} R_e^2 e |\psi(0)|^2 \quad (3.19)$$

for the nuclear ground state and excited state, respectively. In these expressions, $S(Z)$ is the relativity factor due to relativistic effects in heavy elements (Shirley, 1964). Since the ground state is now raised by E_g and the excited state is raised by E_e, γ-transition energy is modified accordingly by

$$\Delta E = \frac{2\pi}{5} S(Z) Ze^2 \left(R_e^2 - R_g^2 \right) |\psi(0)|^2$$

(3.20)

The same type of interaction occurs in the absorber (ΔE_a) as well as in the source (ΔE_s). Since the source and the absorber are usually not chemically identical, $\psi_a(0)$ differs from $\psi_s(0)$. As a consequence, the γ-photon energy change ΔE_a in the absorber is not identical to ΔE_s in the source. These energy changes are reflected in the Mössbauer spectrum as a shift of each line by an amount equal to

$$\delta = \frac{2\pi}{5} S(Z) Z e^2 \left(R_e^2 - R_g^2 \right)$$
$$\times \left(|\psi_a(0)|^2 - |\psi_s(0)|^2 \right) \quad (3.21)$$

This is due to two concomitant differences: one difference between the nuclear radius R_g in the ground state and its counterpart R_e in the excited state (a nuclear isomer), and the other difference between the s-electron density $\psi_a(0)$ in the absorber and its counterpart $\psi_s(0)$ in the source. This spectral shift δ is known as the *isomer shift* because of the volume difference in the nuclear isomers. It is also known as *chemical shift*, referring to its origin from the difference in chemical properties of the source and the absorber. The first difference is a fixed parameter once a nuclear isotope is chosen, whereas the second difference reflects the chemical contrast between the source and the absorber. In Mössbauer spectroscopy, we exploit the latter difference by using a series of absorbers of different chemical environments, for example, different oxidation states and coordination numbers.

Isomer shift is a relative quantity. If the nucleus increases in size when excited ($R_e > R_g$), the isomer shift increases linearly with the absorber's electron density $e|\psi_a(0)|^2$. For ^{57}Fe, the nucleus volume actually decreases when excited ($R_e < R_g$), and the isomer shift behavior is the opposite. The actual value of isomer shift depends on the choice of not only the source but also the calibration standard. Therefore, when reporting experimental results, it is important to state the type of source (e.g., ^{57}Co in rhodium matrix) and the reference point for isomer shift (e.g., with respect to the spectral center of

α-Fe at room temperature). Typical ranges of isomer shifts for iron compounds in various oxidation states and spin states are available in the literature, for example, in Gibb's *Principles of Mössbauer Spectroscopy* (Gibb, 1976).

The isomer shift has been extensively investigated in many areas of chemistry and in materials science. Using isomer shift for studying the electronic structure in solids has been considered an extremely useful experimental method. The isomer shift δ can provide important information on the character of a chemical bond, as well as on oxidation state, spin state, electronegativity of a ligand, coordination number, and so on. A large amount of experimental isomer shift data has been accumulated to date; however, the interpretation of these results is not an easy task. The main difficulty is the lack of a unified model for the chemical bonds that can satisfactorily explain the isomer shift data. For practical applications, isomer shift measurements from a series of samples of similar properties are usually required. Through a comparative study, one can extract information on the electronic structure.

It should be pointed out that Mössbauer effect is so far the only method that can measure isomer shift, because the nuclear volume of the excited state is involved. The information on the electronic structure provided by isomer shift is not available from NMR, EPR, or any other methodologies.

3.7.2
Magnetic Dipole Interaction

The interaction between the nuclear magnetic dipole moment μ and the magnetic field B produced at the site of nucleus by the surrounding electrons

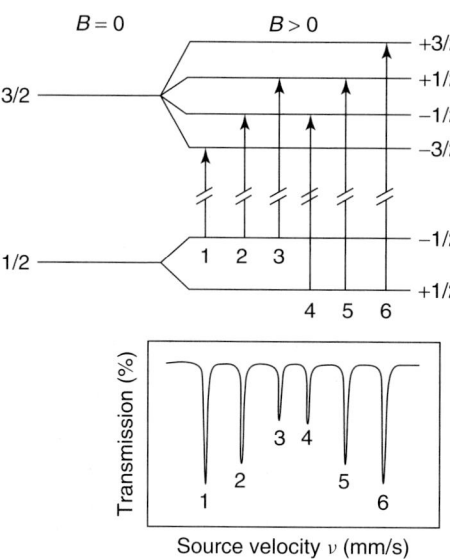

Fig. 3.5 Magnetic splittings of the ^{57}Fe nuclear energy levels and a sextet in the Mössbauer spectrum due to magnetic splittings.

or ions is called *the magnetic hyperfine interaction*. For a nucleus of spin $I \geq 1/2$, this interaction lifts the degeneracy of the energy level and splits it into $(2I + 1)$ sublevels, similar to the Zeeman effect in atomic spectroscopy. This type of splitting in the nuclear ground state had been observed in NMR and EPR. Using Mössbauer effect, nuclear Zeeman effect was first observed by Hanna et al. (1960).

The Hamiltonian of the interaction between a nuclear magnetic dipole moment $\boldsymbol{\mu}$ and a magnetic field $\boldsymbol{B}$ is

$$\mathcal{H}_M = -\hat{\boldsymbol{\mu}} \cdot \boldsymbol{B} = -g\mu_N \hat{I}_z B \quad (3.22)$$

where g is the dimensionless nuclear g-factor ($g = \mu/I\mu_N$), μ_N is the nuclear Bohr magneton (5.05×10^{-27} J T^{-1}), and the z-axis is along the direction of the magnetic field. The corresponding sublevel energies are

$$E_M = -gmB\mu_N \quad (3.23)$$

where $m = I, I - 1, \ldots, -I$.

The first excited state of ^{57}Fe has a spin of $I_e = 3/2$ and splits into four sublevels equally separated by $g_e B \mu_N$. The ground state has a spin of $I_g = 1/2$ and splits into two sublevels separated by $g_g B \mu_N$, as shown in Figure 3.5. The g_g-factor of the ground state, in general, is different from g_e of the excited state (for ^{57}Fe, $g_e = -0.1031$ and $g_g = 0.1808$). Therefore, the separation between the sublevels in the ground state is different from that in the excited state. Since the nuclear transition in ^{57}Fe is of the magnetic dipole type (M1), it can take place provided the selection rule ($\Delta m = \pm 1$ or 0) is obeyed. Thus there are only six allowed transitions, resulting in a sextet in the absorption spectrum as shown in Figure 3.5. The transitions with $\Delta m = \pm 2$ are forbidden.

The position of each line in the characteristic sextet can be easily calculated on the basis of values of $g_e B \mu_N$ and $g_g B \mu_N$. The separation between lines 3 and 4 equals $(g_g - |g_e|)B\mu_N$, while all other separations between adjacent lines are each equal to $g_g B \mu_N$.

The intensity of each absorption line is proportional to the transition probability, which can be calculated using the Clebsch–Gordan coefficients. Assuming a thin absorber with an isotropic recoil-free fraction f, we arrive at the following symmetric intensity ratio of the six absorption lines:

$$3 : \frac{4\sin^2\theta}{(1+\cos^2\theta)} : 1 : 1 : \frac{4\sin^2\theta}{(1+\cos^2\theta)} : 3 \quad (3.24)$$

where θ is the angle between the magnetic field B and the γ-ray direction. When the magnetic field is aligned with the γ-ray direction ($\theta = 0°$), the relative intensities of the sextet are in the ratio of 3 : 0 : 1 : 1 : 0 : 3. When the magnetic field is perpendicular to the γ-ray direction ($\theta = 90°$), the ratio is 3 : 4 : 1 : 1 : 4 : 3. If the magnetic field vectors at the nuclei are randomly oriented, then an integration over the entire solid angle gives an intensity ratio of 3 : 2 : 1 : 1 : 2 : 3.

A convenient way to describe the magnetic hyperfine interaction relies on the use of the effective magnetic field B_{eff}, which has two major contributions: a local magnetic field B_{loc} at the Mössbauer nucleus by the lattice and a hyperfine magnetic field B_{hf} by the Mössbauer atom's own electrons,

$$B_{\text{eff}} = B_{\text{loc}} + B_{\text{hf}} \quad (3.25)$$

The local field by the lattice may be due to the material's magnetic ordering, or may be applied externally, or both. It may have the following contributions:

$$B_{\text{loc}} = B_{\text{ext}} - DM + \frac{4\pi}{3} M \quad (3.26)$$

where B_{ext} is an external field, M is magnetization, DM represents the demagnetization field, and $4\pi M/3$ represents the Lorentz field. In general, the local field is much smaller than the hyperfine field. The hyperfine field B_{hf} has three contributions:

$$B_{\text{hf}} = B_s + B_L + B_D \quad (3.27)$$

The first term B_s is called the *Fermi contact field* produced by spin polarization of s-electrons within the volume of the nucleus, and may be expressed as

$$B_s = -\frac{4\mu_0}{3} \mu_B \sum_n [|\psi_{ns\uparrow}(0)|^2 - |\psi_{ns\downarrow}(0)|^2] \quad (3.28)$$

where μ_B is the Bohr magneton (9.27 × 10^{-4} J T^{-1}), $|\psi_{ns\uparrow}(0)|^2$ and $|\psi_{ns\downarrow}(0)|^2$ represent the ns spin-up and spin-down electron densities within the nucleus volume, respectively. The spin polarization of the s-electrons is caused by unpaired d-electrons because of an exchange interaction that slightly reduces the repulsion between electrons of the same spin. In iron compounds, B_s may range from 20 to 60 T and is the largest among the three terms in B_{hf}, owing to a large number of unpaired d-electrons.

The second term B_L is called the *orbital field* owing to the orbital motions of the unpaired electrons around the nucleus. This motion constitutes a circular current, which in turn produces a magnetic field at the nucleus:

$$B_L = -\frac{\mu_0}{2\pi} \mu_B \langle r^{-3}\rangle\langle L\rangle \quad \text{or}$$

$$B_L = -\frac{\mu_0}{2\pi} \mu_B \langle r^{-3}\rangle(g-2)\langle S\rangle \quad (3.29)$$

where $\langle L\rangle$ and $\langle S\rangle$ are the expectation values of the orbital and spin angular momenta of the unpaired electrons.

For Fe^{3+}, $g \approx 2$, and consequently $B_L \approx 0$. But for high-spin Fe^{2+}, $B_L \approx 20$ T (opposite to B_s) and, as expected, B_L values in rare-earth compounds are relatively large.

The third term B_D is the dipole field at the nucleus, produced by the total spin magnetic moment of the valence electrons. It can be written as

$$B_D = \frac{\mu_0}{8\pi}\mu_B \langle r^{-3}\rangle \langle 3\cos^2\theta - 1\rangle \langle S\rangle \tag{3.30}$$

B_D is obviously zero for a charge distribution with a cubic symmetry. For Fe group ions, B_D is small even in noncubic systems, ranging only from 0 to 8 T. However, in rare-earth compounds where the orbital momentum is not quenched, B_D can be quite large.

Since the separation between the lines of a sextet is proportional to the effective field B_{eff}, in which the magnetic hyperfine field B_{hf} usually dominates, we can easily deduce B_{hf} from the spectrum. It is often very strong ($B_{\text{hf}} \approx -33$ T in α-Fe) and it is local (within the vicinity of the nucleus). To measure the sign of the magnetic hyperfine field, we may apply an external magnetic field of 2 to 5 T to the sample and detect whether B_{eff} increases or decreases. When it increases, B_{hf} is positive; otherwise it is negative.

Mössbauer effect has found extensive applications in magnetism and in research of magnetic materials. Without requiring an applied external magnetic field, Mössbauer spectroscopy can be used to study the temperature dependence of spontaneous magnetization, the magnitude and orientation of hyperfine fields, and the magnetic structure of new materials. It can also be used to measure the ordering temperatures (T_C, T_N) and spin reorientation temperature, to determine phase transitions and phase compositions, to study magnetic lattice anisotropy and relaxation phenomena, and so on. Magnetism arises mainly from the atomic magnetic moments. Transition metals ($3d$, $4d$, $5d$), the lanthanides ($4f$), and the actinides ($5f$) all have unfilled valence electrons and have atomic magnetic moments. Fortunately, many isotopes of these elements are Mössbauer nuclei, for example, ^{57}Fe, ^{61}Ni, ^{99}Ru, ^{149}Sm, ^{151}Eu, ^{155}Gd, ^{159}Tb, ^{161}Dy, ^{165}Ho, ^{166}Er, ^{169}Tm, ^{170}Yb, ^{193}Ir, and ^{237}Np. Obviously, ^{57}Fe is most utilized because iron is the most important element in magnetism. The field of magnetism would not be as successful without Mössbauer spectroscopy.

3.7.3
Electric Quadrupole Interaction

If a nucleus (in either the ground state or an excited state) has a spin $I \geq 1$, it possesses an electric quadrupole moment eQ. For a nucleus surrounded by an asymmetric charge distribution, there is an electric field gradient (EFG), described by a tensor ∇E, at the site of the nucleus. In general, the EFG tensor contains nine second-order derivatives of electric potential ($V_{xy} = \partial^2 V/\partial x \partial y$, etc.). By choosing the principal z-axis along the highest gradient direction, we can diagonalize the EFG tensor so that $|V_{zz}| \geq |V_{xx}| \geq |V_{yy}|$. Since it is also traceless, $V_{xx} + V_{yy} + V_{zz} = 0$, only two independent parameters are needed to describe EFG. Customarily, V_{zz} is represented by eq, and the quantity $(V_{xx} - V_{yy})/V_{zz}$ is represented by η, known as the *asymmetry parameter* ($0 \leq \eta \leq 1$).

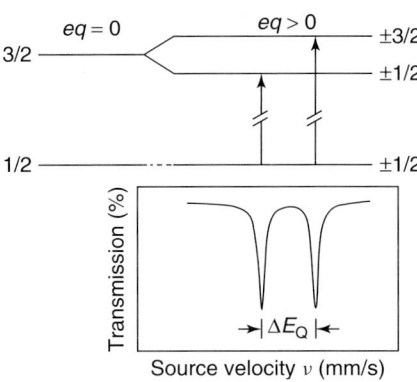

Fig. 3.6 Quadrupole splitting of the ^{57}Fe excited state and a doublet in the Mössbauer spectrum.

The Hamiltonian of this electric quadrupole interaction is

$$\mathcal{H}_Q = e\hat{Q}\nabla E = \frac{e^2qQ}{4I(2I-1)}$$
$$\times \left[3\hat{I}_z^2 - \hat{I}^2 + \frac{1}{2}\eta(\hat{I}_+^2 + \hat{I}_-^2) \right] \quad (3.31)$$

where $\hat{I}_+ = \hat{I}_x + i\hat{I}_y$ and $\hat{I}_- = \hat{I}_x - i\hat{I}_y$ are the raising and lowering operators, respectively. The eigenvalues of this Hamiltonian are

$$E_Q(m) = \frac{e^2qQ}{4I(2I-1)}[3m^2 - I(I+1)]$$
$$\times \left(1 + \frac{\eta^2}{3}\right)^{1/2} \quad (3.32)$$

where $m = I, I-1, \ldots, -I$.

The ^{57}Fe excited state (14.4 keV) has a nuclear spin of $I = 3/2$, and this energy level splits into two sublevels ($m = \pm 3/2$ and $m = \pm 1/2$) due to quadrupole interaction. Because only m^2 appears in Eq. (3.32), each sublevel is still doubly degenerate. The energy eigenvalues of the two sublevels are

$$E_0 + E_Q\left(\pm\frac{3}{2}\right)$$
$$= E_0 + \frac{e^2qQ}{4}\left(1 + \frac{\eta^2}{3}\right)^{1/2} \quad (3.33)$$

$$E_0 + E_Q\left(\pm\frac{1}{2}\right)$$
$$= E_0 - \frac{e^2qQ}{4}\left(1 + \frac{\eta^2}{3}\right)^{1/2} \quad (3.34)$$

The ground state of ^{57}Fe has $I = 1/2$, so $Q = 0$, and the energy level does not split, as shown in Figure 3.6. Therefore, in the absence of Zeeman splitting, nuclear transition may take place from the ground state to either of the two excited states, resulting in a doublet in the Mössbauer spectrum, as shown in Figure 3.6. The separation between the two resonance lines, known as *quadrupole splitting*, is

$$\Delta E_Q = \frac{e^2qQ}{2}\left(1 + \frac{\eta^2}{3}\right)^{1/2} \quad (3.35)$$

Assuming an isotropic recoil-free fraction f, the relative intensities of the spectral lines of the quadrupole doublet in a single crystal can be calculated as

$$I_{\pm 3/2} = \frac{C}{\sqrt{1 + (\eta^2/3)}}\left(4\sqrt{1 + (\eta^2/3)}\right.$$
$$\left. + 3\cos^2\theta - 1 + \eta \sin^2\theta \cos 2\phi\right)$$
$$(3.36)$$

$$I_{\pm 1/2} = \frac{C}{\sqrt{1+(\eta^2/3)}} \left(4\sqrt{1+(\eta^2/3)} \right.$$
$$\left. -3\cos^2\theta + 1 - \eta \sin^2\theta \cos 2\phi \right)$$
(3.37)

where C is a constant, and θ and ϕ are the polar and azimuthal angles of the incident γ-ray direction in the EFG principal-axis system. In the case of an axially symmetric EFG ($\eta = 0$), the intensity ratio of the two absorption lines is

$$\frac{I_{\pm 3/2}}{I_{\pm 1/2}} = \frac{1+\cos^2\theta}{\frac{5}{3}-\cos^2\theta} \quad (3.38)$$

For a polycrytalline sample or a powder sample, the crystal axes are randomly oriented. Integrating over the entire solid angle for each intensity, we obtain

$$\frac{\langle I_{\pm 3/2}\rangle}{\langle I_{\pm 1/2}\rangle} = \frac{\frac{1}{4\pi}\int I_{\pm 3/2}\sin\theta\, d\theta\, d\phi}{\frac{1}{4\pi}\int I_{\pm 1/2}\sin\theta\, d\theta\, d\phi} = 1$$
(3.39)

If the recoil-free fraction is not isotropic but $f(\theta)$ is angular dependent, the above integrals must each be weighted by $f(\theta)$ and the ratio will not be 1 (a manifestation of the Goldanskii–Karyagin effect),

$$\frac{\langle I_{\pm 3/2}\rangle}{\langle I_{\pm 1/2}\rangle} = \frac{\frac{1}{4\pi}\int I_{\pm 3/2}f(\theta)\sin\theta\, d\theta\, d\phi}{\frac{1}{4\pi}\int I_{\pm 1/2}f(\theta)\sin\theta\, d\theta\, d\phi}$$
$$\neq 1 \quad (3.40)$$

The quadrupole splitting $\Delta E_Q = 1/2 e^2 q Q(1+\eta^2/3)^{1/2}$ as measured from a Mössbauer spectrum only provides the product of eQ and eq, not their individual values or their signs. The nuclear quadrupole moment eQ is a fixed parameter for a particular isotope and may be independently measured (especially that of the ground state). The size and sign of the z-component of EFG eq ($= V_{zz}$) is determined by partially filled valence orbitals (valence) as well as by the charges on the neighboring ions and ligands surrounding the Mössbauer atom (lattice). Theoretical calculations of total V_{zz} at the site of the nucleus gave the following result:

$$V_{zz} = (1-R)(V_{zz})_{\text{val}} + (1-\gamma)(V_{zz})_{\text{lat}}$$
(3.41)

where $(V_{zz})_{\text{val}}$ and $(V_{zz})_{\text{lat}}$ are contributions from valence and lattice, respectively, each with an empirical parameter R (Sternheimer shielding factor) and a γ (Sternheimer antishielding factor) (Sternheimer, 1966). Therefore, ΔE_Q provides information on the symmetry of local chemical and electronic environment (oxidation state and spin state). Both $(V_{zz})_{\text{val}}$ and $(V_{zz})_{\text{lat}}$ are inversely proportional to r^3. Since the valence electrons are closer to the nucleus than the lattice charges, the former is much larger than the latter. For high-spin $^{57}\text{Fe}^{3+}$, the valence shell is $3d^5$ (exactly half filled), and $(V_{zz})_{\text{val}} = 0$ because of symmetry. For high-spin $^{57}\text{Fe}^{2+}$, the sixth electron in $3d^6$ is the main source of EFG. The quadrupole splitting for Fe^{2+} ($\Delta E_Q \approx 3.0$ mm s^{-1}) has been found to be much larger than that for Fe^{3+} ($\Delta E_Q \approx 0.5$ mm s^{-1}). The ΔE_Q values in low-spin $^{57}\text{Fe}^{2+}$ (<0.8 mm s^{-1}) are smaller than the ΔE_Q values in low-spin $^{57}\text{Fe}^{3+}$ (0.7–1.7 mm s^{-1}), again reflecting the one-electron difference in their d-orbitals.

$(V_{zz})_{\text{val}}$ may be strongly temperature dependent when a crystal field with lower-than-cubic symmetry splits the orbital energy levels. These suborbitals will be populated according to Boltzmann distribution, and hence the temperature dependence, which is reflected in the Mössbauer spectra taken at different temperatures (Chen et al., 1992).

3.7.4
Combined Electric Quadrupole and Magnetic Dipole Interactions

When the above three hyperfine interactions are simultaneously present, the combined Hamiltonian is

$$\mathcal{H} = \mathcal{H}_{IS} + \mathcal{H}_M + \mathcal{H}_Q \quad (3.42)$$

where the isomer shift Hamiltonian $\mathcal{H}_{IS}$ is independent of the nuclear spin I, but both the magnetic dipole $\mathcal{H}_M$ and electric quadrupole $\mathcal{H}_Q$ Hamiltonians involve the same operator $\hat{I}$. When $\mathcal{H}_M$ and $\mathcal{H}_Q$ are considered separately, the quantization axis (the z-axis) for $\hat{I}$ in $\mathcal{H}_M$ is most conveniently chosen as the direction of the magnetic field $\boldsymbol{B}$; for $\hat{I}$ in $\mathcal{H}_Q$ it is along one principal axis of the EFG tensor. When both $\boldsymbol{B}$ and EFG are present, it is likely that the magnetic field is not aligned with the EFG principal axis. The combined Hamiltonian therefore takes a more complicated form:

$$\mathcal{H} = \mathcal{H}_{IS} + \frac{e^2qQ}{4I(2I-1)}$$
$$\times \left[3\hat{I}_z^2 - \hat{I}^2 + \frac{1}{2}\eta(\hat{I}_+^2 + \hat{I}_-^2) \right]$$
$$- g\mu_N B \left\{ \left[\frac{1}{2}(\hat{I}_+ + \hat{I}_-)\cos\phi \right.\right.$$
$$\left.+ \frac{1}{2}(\hat{I}_+ - \hat{I}_-)\sin\phi \right]$$
$$\left. \times \sin\theta + \hat{I}_z \cos\theta \right\} \quad (3.43)$$

where the z-axis is along the EFG principal axis, and θ and ϕ are the Euler angles describing the magnetic field direction within the EFG principal axes system. It is possible to obtain analytical solutions of eigenvalues and eigenvectors of this Hamiltonian, which involves the mixing of all original states and consequently allows the previously "forbidden" transitions (Häggström, 1974). To obtain complete analytical solutions is a tedious procedure and the solutions are very complicated for practical purposes.

In several special cases, the analysis is greatly simplified. When the EFG is axially symmetric ($\eta = 0$) and its z-axis happens to be aligned with the magnetic field $\boldsymbol{B}$ ($\theta = 0$), $\mathcal{H}_M$ and $\mathcal{H}_Q$ commute. The Zeeman states of $\mathcal{H}_M$ are also eigenvectors of $\mathcal{H}_Q$, and the new energy levels are simply combinations of the original E_M and E_Q:

$$E(m) = E_{IS} + E_M + E_Q$$
$$= E_{IS} - gmB\mu_N$$
$$+ (-1)^{|m|+1/2}\frac{e^2qQ}{4} \quad (3.44)$$

The Zeeman energy levels are no longer equally spaced; in the case of $I = 3/2$, the inner two energy levels are shifted in one direction and the outer two in the opposite direction.

It is often the case where either the magnetic hyperfine interaction or electric quadrupole interaction is dominant while the other occurs as a perturbation. For example, electric quadrupole interaction in a magnetically ordered solid is much weaker than the magnetic hyperfine interaction. On the other hand, an externally applied magnetic field to a paramagnetic material may result in a magnetic hyperfine interaction much weaker than the existing electric quadrupole interaction.

When the magnetic hyperfine interaction dominates, the energy levels are

$$E(m) = E_{IS} - gmB\mu_N$$
$$+ (-1)^{|m|+1/2} \frac{e^2qQ}{4}$$
$$\times \frac{3\cos^2\theta - 1 + \eta\sin^2\theta\cos 2\phi}{2}$$

(3.45)

which is similar to the previous case, except that the additional shift now depends on the relative angles (θ and ϕ) between the EFG and the magnetic field B, as well as on the EFG's asymmetry parameter (η). When the electric quadrupole interaction dominates, the two states corresponding to $m = \pm 1/2$ will mix and give the following two different energy levels:

$$E(\pm\tfrac{1}{2}) = E_{IS} \pm \frac{1}{2}gB\mu_N$$
$$\times \sqrt{4 - 3\cos^2\theta} - \frac{e^2qQ}{4}$$

(3.46)

and the two states corresponding to $m = \pm 3/2$ have the following energy levels:

$$E(\pm\tfrac{3}{2}) = E_{IS} \pm \frac{3}{2}gB\mu_N\cos\theta + \frac{e^2qQ}{4}$$

(3.47)

Compared with the perfectly aligned and symmetric EFG cases, spectra from these more general cases are more difficult to analyze, but they allow us to extract more information, such as the sign of the EFG component V_{zz} and the relative direction of the hyperfine magnetic field.

3.7.5
Saturation Effect in the Presence of Hyperfine Splittings

It has been observed that in a sextet spectrum due to a magnetic hyperfine field the spectral intensities (absorption areas A) do not strictly follow the theoretical predictions of $3:2:1:1:2:3$ for powder samples. This is because the area A is not a linear function of the effective thickness t_a of the absorber, but has a smaller slope for a larger t_a, manifesting a saturation effect as shown in Figure 3.7. Among the six absorption lines, each pair comes from participation of only one portion of the total population of Mössbauer nuclei: lines 1 and 6 are due to one-half of the total population, lines 2 and 5 are due to one-third of them, and lines 3 and 4 are due to only a sixth. Therefore, each pair corresponds to a different partial effective thickness. Because of the nonlinear saturation

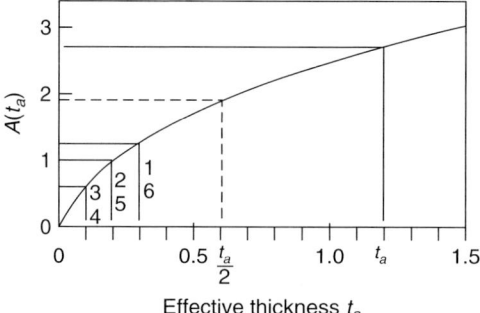

Effective thickness t_a

Fig. 3.7 Absorption area A as a function of effective thickness t_a. Because of the saturation effect, $2A\left(\dfrac{t_a}{2}\right) > A(t_a)$, and $12A\left(\dfrac{t_a}{12}\right) > 6A\left(\dfrac{t_a}{6}\right) > 4A\left(\dfrac{t_a}{4}\right)$.

effect, the absorption areas are not strictly in the ratio of 3 : 2 : 1, but $x : y : 1$, with $x < 3$ and $y < 2$.

Another important aspect is that the splitting of a single line will result in an increase of the total absorption area. Suppose a single line (effective thickness t_a) has an area of $A(t_a)$. When it splits into two sublines (each of partial effective thickness $\frac{t_a}{2}$), the total area is $2A(\frac{t_a}{2})$, which is clearly larger than $A(t_a)$, as illustrated in Figure 3.7.

When the absorption intensities are measured and used to infer any quantitative conclusions about the structural or chemical composition of the absorber, the saturation effect must be carefully taken into account.

3.8
Second-Order Doppler Effect

The second-order Doppler effect is a relativistic effect that results in the shift of the entire spectrum. Expanding Eq. (3.12) and retaining up to the second-order term, we obtain

$$\Delta E = E_0 \left(\frac{v}{c} - \frac{v^2}{2c^2} \right) \quad (3.48)$$

When atomic vibrations in the absorber are considered, their average velocity strongly depends on temperature. The corresponding Doppler shift is then

$$\langle \Delta E \rangle = -E_0 \frac{\langle v^2 \rangle}{2c^2} \quad (3.49)$$

This second-order Doppler effect modifies the energy of the γ-rays. In a Mössbauer spectrum, second-order Doppler shift δ_{SOD} is manifested as a relative quantity between the source and the absorber:

$$\delta_{\text{SOD}} = \frac{\langle \Delta E \rangle}{E_0} = \frac{\langle v^2 \rangle_s - \langle v^2 \rangle_a}{2c^2} \quad (3.50)$$

where $\langle v^2 \rangle_s$ and $\langle v^2 \rangle_a$ are the mean-square velocities of the source and absorber, respectively.

Both the isomer shift and second-order Doppler shift have the effect of displacing the entire Mössbauer spectrum. They can be separated by utilizing the fact that the second-order Doppler shift depends strongly on temperature while the isomer shift is a measure of the s-electron density at the nucleus and approximately temperature-independent. Because the specific behavior of $\langle v^2 \rangle$ is determined by lattice vibration, δ_{SOD} is also useful in studying lattice dynamics.

3.9
Applications of Mössbauer spectroscopy

Mössbauer spectroscopy is mainly used to study the properties of materials through hyperfine interactions. Isomer shift, quadrupole splitting, magnetic hyperfine field, second-order Doppler effect, and the recoil-free fraction are all important parameters in Mössbauer spectroscopy. They are determined by the positions and the areas of spectral lines, as well as the spectral shapes and linewidths. These parameters also depend on external conditions such as temperature, pressure, and magnetic field. It is clear that Mössbauer spectra can provide an abundance of information.

Using hyperfine interactions, Mössbauer effect serves as a bridge linking the nucleus with its environmental details, including the crystal structures of materials, the perfectness of a crystal, lattice periodicity, magnetic ordering, the amorphous state, oxidation state in a compound, coordination number, effect of ligands, and the vibrations and diffusions of atoms in the solid. Because of its extremely diverse applications, Mössbauer spectroscopy has developed into a truly unique interdisciplinary field of science.

The isomer shift (chemical shift) δ provides information about the chemical environment, because of its sensitivity to changes in electron density at the site of the nucleus. Any changes in oxidation state, spin state, valence orbitals, type of chemical bond, ligand electronegativity, coordination number, substitutional or interstitial impurities, and so on, will be reflected in the isomer shift. Because it is often difficult to separate these causes, an investigation of a series of similar samples is usually required, and a comparison of their isomer shift data then provides unequivocal conclusions. In ^{57}Fe Mössbauer spectroscopy, isomer shift is very sensitive to the oxidation and spin states of iron. The δ-value (relative to α-Fe) is typically in the range from $+0.8$ to $+1.5$ mm s^{-1} for high-spin Fe^{2+} ($S = 2$), but in a distinctively different range of -0.2 to $+0.5$ mm s^{-1} for low-spin Fe^{+2} ($S = 0$). For Fe^{3+}, the isomer shift values for low spin ($S = 1/2$) and high spin ($S = 5/2$) are, respectively, $-0.2 - +0.4$ mm s^{-1} and $+0.2 - +0.6$ mm $^{-1}$s (Gibb, 1976). They overlap in the region from $+0.2 - +0.4$ mm s^{-1}; if a compound's isomer shift falls in this region, an independent determination of the spin state is required. Also, the two low-spin ions Fe^{2+} and Fe^{3+} have almost identical isomer shift values, but fortunately their quadrupole splittings are quite different, allowing unambiguous determinations of these two oxidation states.

The quadrupole splitting provides information about the lack of symmetry of the local electronic environment in which the nucleus resides. The asymmetry arises from nonspherical molecular orbitals or noncubic ligand coordination from the surrounding lattice. When combined with isomer shift data, quadrupole splitting values reflect the space group of local symmetry, the types of orbitals, relative orientations of local chemical bonds, and so on. Quadrupole splitting values may also be used to distinguish long-range and short-range orders in solids of less perfect lattices, such as amorphous materials and metallic glasses. Quadrupole splittings are also observed in magnetic materials, especially when they are in the paramagnetic phase, in the form of spin glasses, or in the superparamagnetic configuration. The behavior of quadrupole splitting as a function of temperature can be used to determine the respective phase transition temperatures.

The ubiquitous sextet ^{57}Fe Mössbauer spectrum, because of magnetic hyperfine interaction, contains a wealth of information about the magnetic properties of the material. Combined with result from bulk magnetic measurements, a spectrum with a sextet or several superimposed sextets indicates whether the material is ferromagnetic, antiferromagnetic, or ferrimagnetic. Fitting the experimental spectrum according to energy values of the Hamiltonian, we obtain parameters such as hyperfine magnetic field and isomer shift, which in turn allows us to calculate the magnetic moment of the atom. Experiments over a wide range of temperatures are beneficial for detecting magnetic phase transitions and for establishing a relationship between hyperfine magnetic field B_{hf} (at the nuclear site) and the exchange field B_E (causing spontaneous magnetic ordering). For most magnetic ordered materials, B_{hf} and B_E are proportional to each other, and they have identical temperature dependence. Therefore, information from Mössbauer spectroscopy can be directly correlated to macroscopic magnetic measurements (e.g., magnetization) as well as other microscopic measurements (e.g., NMR and muon resonance). For amorphous magnetic materials, the Mössbauer sextet is usually poorly resolved and broadened; it may be necessary to use a continuous distribution of hyperfine magnetic fields, instead of a few discrete values, to fit the spectrum.

Mössbauer spectroscopy has a particular timescale, characterized by the lifetime of the excited state from which the γ-rays are emitted. Usually of the order of 0.1 microseconds, this timescale τ is fast for many atomic processes, but it is slow relative to certain fluctuating mechanisms. Therefore, it is convenient to use Mössbauer spectroscopy to study materials that involve weak interactions between atoms, molecules, or clusters where relaxation times τ_0 may be comparable to τ. An example is superparamagnetism in which nano-sized magnetic particles undergo fast thermal fluctuations, the magnetic hyperfine interaction is averaged out, and only quadrupole interaction is observed (Yang et al., 2003). At lower temperatures when fluctuation slows down, hyperfine magnetic field begins to be observed. Near the so-called blocking temperature, a Mössbauer spectrum may reveal both coexisting phases, each with broadened spectral lines.

An external magnetic field may be applied to the absorber. A constant DC magnetic field applied to a magnetic material will allow us to determine the direction of the hyperfine field with respect to the atomic magnetic moment and, in the case of ferrimagnetic material, the direction and size of each component. When a constant external magnetic field is applied to a non-magnetic material, one may observe Zeeman splittings as a perturbation to the quadrupole interaction, which will elucidate the direction of EFG relative to the crystallographic axes when the sample is a single crystal. An external high-frequency magnetic field may be applied to a soft magnetic material to induce a fast fluctuation of the hyperfine magnetic field, and therefore eliminate the Zeeman splittings in the Mössbauer spectrum, revealing only the quadrupole splittings and simplifying the analysis (Kopcewicz, 1989).

Other applications of Mössbauer spectroscopy include the following: measuring nuclear properties, testing general theory of relativity (Pound and Rebka, 1959; Kholmetskii, 2000), elucidating mechanism of catalysis (Jiang et al., 2002), identifying minerals and ores (Stevens, 1998), studying corrosion in civil engineering (Cook, 2004), classifying chemical compositions of extraterrestrial samples and meteorites (Morris et al., 2004), determining protein functions in biology (Bauminger and Harrison, 2003), analyzing iron compounds in atmospheric aerosols (Kopcewicz and Kopcewicz, 2000), exploring possibility of cancer therapy (Mills et al., 1988), characterizing painting pigments in fine arts (Kuno et al., 2004; Keisch, 1973), and archaeometry (Lazzarini et al., 1980).

Mössbauer spectroscopy has enjoyed continued success because of the following extraordinary characteristics and advantages:

1. It has decisively the highest energy resolution of all spectroscopic methods.
2. Traditional Mössbauer spectroscopy requires only a relatively simple apparatus, in contrast to other nuclear physics research systems that are usually huge in size and exorbitantly expensive.
3. It is a nondestructive method for studying a solid and obtains microscopic statistical information on the atomic scale rather than a macroscopic average.

Mössbauer spectroscopy complements other nuclear methods such as neutron scattering, perturbed angular correlation, and nuclear magnetic resonance.

Like other experimental methods, Mössbauer spectroscopy has its own limitations. Although the total number of Mössbauer isotopes now exceeds 100, some of them have very short half-lives, and others require liquid-nitrogen or even liquid-helium temperatures for the observation of Mössbauer effect. By and large, for most isotopes, either the corresponding sources are very expensive or their Mössbauer effect is difficult to observe. ^{57}Fe remains to be the best Mössbauer isotope. Iron is found in a large array of materials. Many minerals have certain amount of Fe, magnetic materials are largely based on Fe, and there are numerous iron alloys and iron compounds. Because of this economic implication of iron and iron products, research using ^{57}Fe Mössbauer spectroscopy has always assumed the leading role.

Mössbauer spectroscopy has a history of 50 years since the first experiments of resonant absorption of nuclear γ-rays. New aspects of Mössbauer effect are still being investigated; especially, as SR became available, new theories and new techniques are being developed to make use of this artificial form of radiation to the fullest.

3.10
Mössbauer Spectroscopy Using Synchrotron Radiation

Electromagnetic radiation from a synchrotron accelerator may be used to observe Mössbauer effect. SR provides polarized pulsed radiation of high intensity, high collimation, and narrow beams. The only drawback is that SR is far from monochromatic. However, its energy can be adjusted and it can cover an energy range for a majority of Mössbauer transitions. The high intensity of SR sources has been utilized for scattering experiments where conventional radiation sources cannot provide adequate results. More importantly, the pulsed nature of SR radiation is most suitable for measuring time spectra – using a short SR pulse ($<10^{-10}$ seconds) to excite a nuclear ensemble, forming an "exciton" and allowing the observation of its coherent decay at different time intervals. The method of measuring time spectra is called *time-domain Mössbauer spectroscopy*, whereas the transmission method is referred to as *energy-domain Mössbauer spectroscopy*. In the last two decades, significant progress has been made in synchrotron Mössbauer spectroscopy, especially in the time-domain method, providing a direct and efficient approach to the study of Mössbauer effect and hyperfine fields. There emerged several new research areas that are not accessible with the conventional radiation sources. One example is the direct measurement of phonon DOS using SR sources. It was known that the phonon DOS could be measured, in principle, by Mössbauer effect, but because of various difficulties it was not achieved until SR became available in the 1990s. In 1999 and 2000, the journal *Hyperfine Interactions* devoted volumes 123, 124, and 125 to a comprehensive collection of 40 review articles written by the pioneers in the theory and experiments of nuclear resonant scattering of SR (Gerdau and de Waard, 1999, 2000).

The most distinct advantage of SR is its high brilliance. The brilliance of a third-generation SR source is about 9–10 orders of magnitude higher than that from a rotating target X-ray generator, and about 12 orders of magnitude higher than a ^{57}Co source of 3.7×10^8 Bq (10 mCi). SR can be tuned to the Mössbauer transition energies not only for the most common isotope ^{57}Fe but also for others such as ^{83}Kr, ^{119}Sn, ^{151}Eu, and ^{161}Dy. However, SR sources require a large and costly facility, and will not be available in ordinary Mössbauer laboratories. Therefore, the SR sources are unlikely to replace the conventional radiation sources.

3.10.1
Monochromatization of Synchrotron Radiation

The bandwidth of an SR beam is at least 10 orders of magnitude wider than the nuclear resonant width ($\sim 10^{-8}$ eV). After going through a double-crystal premonochromator Si(111), SR bandwidth can be reduced to the order of electronvolts. Further collimation and monochromatization are necessary. There are mainly two types of monochromators, one based on the scattering by electrons and the other on resonant scattering by nuclei. The first type can offer a bandwidth within a few milli electronvolts, and the second type can provide the desired bandwidth of a few micro electronvolts for nuclear resonant experiments.

Using monochromators based on electron scattering and specifically designed detectors, many lattice dynamics experiments as well as nuclear forward scattering experiments have been performed with acceptable signal-to-noise ratios, although a milli electronvolt bandwidth is still about 10^6 times wider than the resonance linewidth of ^{57}Fe. A typical scatterer contains a large number of electrons in addition to the resonant nuclei, and those electrons will nonresonantly scatter all the SR in the meV band. Therefore, only 10^{-6} of the detected photon count is due to nuclear resonant scattering. The electron scattering process is prompt in time, whereas the nuclear resonant scattering is a time-delayed process since the typical lifetime of the nuclear isomeric state is long ($\tau \sim 100$ nanoseconds) compared to the incident SR pulse (~ 0.1 nanoseconds). Using this time difference, photons from nonresonant electron scattering can be discriminated by a detector with nanosecond rise and fall times. For certain crystalline scatterers, we may choose specific scattering angles so that the electronic Bragg reflections are canceled or forbidden, for example, using the $(2n+1\ 2n+1\ 2n+1)$ plane in α-Fe$_2$O$_3$.

Through the use of nuclear Bragg scattering, monochromators such as single crystal ^{57}FeBO$_3$, multilayers $25\times$ [^{57}Fe

(22 Å)/Sc(11 Å)/^{56}Fe(22 Å)/Sc(11 Å)], or the grazing incident antireflection (GIAR) film can further reduce the bandwidth to 10^{-6}–10^{-8} eV, approaching the natural width of the nuclear energy level. Because hyperfine interactions are involved in the above nuclear Bragg scattering, the resultant radiation may contain a few separate spectral lines. This multicomponent radiation may be directly used in experiment; but if a truly single-line monochromatic source is desired, an additional absorber may be added to filter out the unwanted spectral components.

The following description represents a specific example of monochromatization and polarization of SR (Smirnov, 2000). SR is first reflected off a double-crystal Si(1 1 1) monochromator, reducing the bandwidth to 2.8 eV at the nuclear resonance energy. A channel-cut Si(8 4 0) polarizer reduces the π-polarized component from 1% to less than 10^{-4} % and the bandwidth to milli electronvolts. It is then reflected off the (3 3 3) plane of a single-crystal ^{57}FeBO$_3$, placed in an oven (above its Neèl temperature to eliminate magnetic hyperfine interaction) with an external magnetic field in the (3 3 3) plane. The outcome is an extremely narrow, linearly polarized, completely recoil-free, and highly intense Mössbauer source for ^{57}Fe, equivalent to a ^{57}Co source of an enormous activity of $\sim 3.7 \times 10^{13}$ Bq (1000 Ci).

3.10.2
Time-Domain Mössbauer Spectroscopy

In time-domain Mössbauer spectroscopy (TDMS), coherent decays in resonant absorption, nuclear Bragg scattering, or forward scattering are observed at different times ($\sim$10–100 nanoseconds) after the nuclear system has been excited. Mössbauer parameters and hyperfine interactions can be studied by TDMS through the analysis of several new phenomena such as dynamical beats (DBs), quantum beats (QBs), and speed-up effect of initial decay.

In time-domain experiments, a strong pulse of SR simultaneously excites all Mössbauer nuclei in a sample and they decay in a coherent fashion. Each nucleus is no longer isolated, but interacts with others. It is possible for a nuclear excitation to propagate elastically throughout the entire ensemble of nuclei (a γ-photon emitted by one excited nucleus may be reabsorbed or scattered by other nuclei that are identical to the emitting nucleus). Without this interaction, individually excited nucleus would decay exponentially with its natural lifetime. However, an interacting ensemble of nuclei behaves differently. Such a collective nuclear excitation by an SR pulse is known as a *nuclear exciton*. Elastic decay of this exciton deviates from exponential behavior and contains intensity modulations characterized by beats.

In experiments using traditional Mössbauer radiation source, we may detect how the intensity of radiation, emitted from a source, evolves with time as it goes through an absorber. Typically, a delay coincidence circuit is used; a block diagram is shown in Figure 3.8 where an oscillating absorber allows absorption at a certain resonance velocities. In the case of traditional ^{57}Fe spectroscopy using a ^{57}Co source, it is convenient that 91% of the decay process follows a cascade through two excited states (136.5 keV $\rightarrow$ 14.4 keV $\rightarrow$ ground state). The time zero is set by the detection of a 122.1-keV photon (γ_1), which signifies that the nucleus

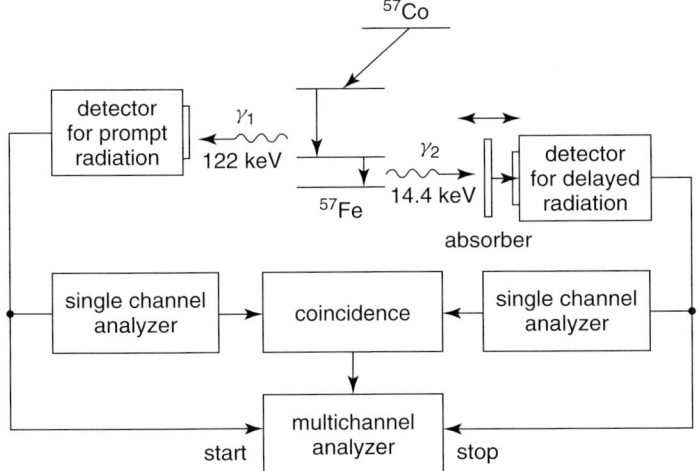

Fig. 3.8 Block diagram of the prompt and delay coincidence detection circuits.

is now in the 14.4-keV state (natural lifetime $\tau = 98$ nanoseconds). This signal starts the time measurement t until a 14.4-keV photon (γ_2) is detected. A time-domain spectrum is then constructed by plotting the number of events $N(t)$ against the time delay t. With no absorber, this curve is perfectly exponential $N(t) \propto e^{-t/\tau}$ and measures the lifetime τ. With the oscillating absorber in place, the detected γ_2 photons include those that have energies just above resonance and those just below resonance. Because these two groups of photons have very distinct dispersion characteristics (traveling through the absorber with very different speeds), $N(t)$ at the detector is an "interference pattern" instead of a purely exponential decay. It is characterized by aperiodic modulations, known as *dynamical beats* (DBs), and by the speed-up of initial decay.

Theoretically, this can be modeled as the propagation of a truncated harmonic wave through a dispersive medium, and the result is

$$N(t) \propto e^{-t/\tau} \left[J_0(\sqrt{t_a t/\tau}) \right]^2 \quad (3.51)$$

where the exponential behavior is now modulated by a Bessel function of the zeroth order J_0, oscillating aperiodically with the apparent periods increasing with time (Figure 3.9). In addition, the modulation depends on t_a, the effective thickness of the absorber. For a very thin absorber ($t_a \ll 1$) or very short times ($t \ll \tau$), we may expand the Bessel function and retain the first term. This portion of the time-domain spectrum becomes

$$N(t) \propto e^{-t/\tau} e^{-t_a t/2\tau} \quad (3.52)$$

where the second factor is also exponential, making the apparent decay rate to be $\tau/(1 + t_a/2)$, shorter than the natural lifetime τ. This phenomenon is known as *speed up* of initial decay.

DBs and speed up of initial decay can also be observed in experiments using SR, where a short pulse (~0.1

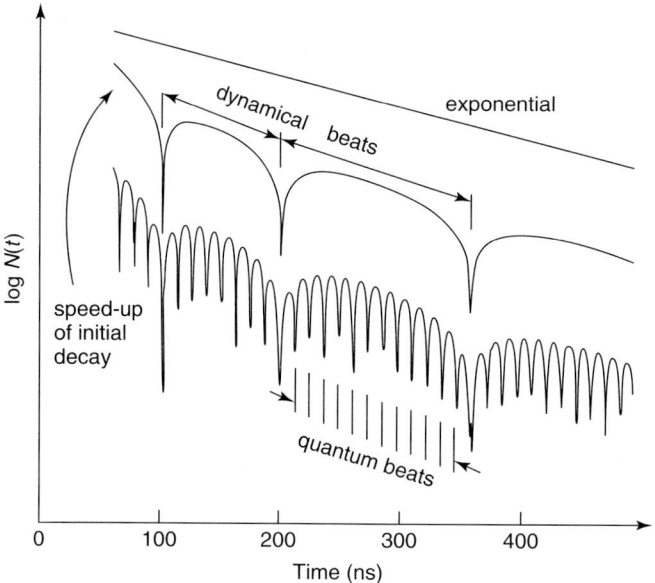

Fig. 3.9 Main features in time-domain Mössbauer spectra, showing speed up of initial decay, dynamical beats, and quantum beats.

nanosecond) of radiation of energies within a particular bandwidth is applied. The intensity of nuclear Bragg scattering or forward scattering is then observed as a function of time. (For readers familiar with pulsed NMR, this is similar to one single RF pulse followed by the observation of "free induction decay," or FID.) The theoretical treatment now requires the inclusion of all the energies within the bandwidth of the SR radiation. The resultant time-domain spectrum (DB) can be expressed in terms of a Bessel function of the first order J_1,

$$N(t) \propto e^{-t/\tau} \left[t_a \frac{J_1(\sqrt{t_a t/\tau})}{\sqrt{t_a t/\tau}} \right]^2 \quad (3.53)$$

For $t_a \ll 1$ and $t \ll \tau$, we have again the speed up of initial decay:

$$N(t) \propto t_a^2 e^{-t/\tau} e^{-t_a t/4\tau} \quad (3.54)$$

with an apparent rate of $\tau/(1 + t_a/4)$.

When the scatterer contains hyperfine interactions and therefore allows multiple resonant scatterings, another interference phenomenon takes place: periodic modulations (known as *quantum beats* (QBs)) are now superimposed onto the DBs, as shown in Figure 3.9. For example, a scatterer with two resonance energies E_1 and E_2 would give

$$N(t) \propto e^{-t/\tau} \left[t_a \frac{J_1(\sqrt{t_a t/\tau})}{\sqrt{t_a t/\tau}} \right]^2$$
$$\times \cos^2\left(\frac{E_2 - E_1}{2\hbar} t\right) \quad (3.55)$$

In this case, the QBs are periodic and they contain information about the hyperfine splitting.

Comparing time-domain Mössbauer spectroscopy with energy-domain Mössbauer spectroscopy, we find similarities and differences. Both are phenomena of nuclear resonance. The energy-domain

Mössbauer spectra are mainly based on resonance absorption. The transmitted counts of photons are measured as functions of their energies (an energy spectrum), which represents an incoherent sum of the spectral components of the transmitted radiation. In other words, the transmitted spectrum reflects the incoherent process of nuclear resonance absorption by individual nuclei. By contrast, time-domain Mössbauer spectroscopy belongs to the scattering method. A scattering spectrum measured as a function of time is a coherent sum of the spectral components of the scattered radiation from nuclei collectively excited by an SR pulse. This leads to those important interference effects in TDMS.

A time-domain spectrum may be Fourier-transformed to an energy-domain spectrum. A purely exponential time-domain spectrum gives a familiar Lorentzian energy spectrum. However, time-domain spectra with QB and DB will give energy-domain spectra with interesting shapes, such as a double-hump structure.

3.10.3
Applications of Time-Domain Mössbauer Spectroscopy

Time-domain Mössbauer spectroscopy has found many applications, among which measurements of the recoil-free fraction f and the phonon DOS are perhaps the most interesting and the easiest to understand. (Strictly speaking, one does not measure the recoil-free fraction f directly in scattering experiments, but use the Lamb–Mössbauer factor f_{LM}. In an isotropic scatterer, f_{LM} and f are identical.)

In the time domain, we measure the recoil-free fraction f through determining t_a because they are linearly related according to Eq. (3.14), $t_a = n_a f \sigma_0 d$. Using nuclear forward scattering, t_a may be extracted from the speed up of initial decay or from the minimum positions of DBs. An excellent set of experiments has been carried out using, as the scatterer, a polycrystalline α-Fe foil of thickness of 11 μm and a 95% ^{57}Fe enrichment (Bergmann et al., 1994). A magnetic field of 0.6 T was applied in the plane of the foil so that only the $\Delta m = 0$ transitions were allowed. As the scatterer temperature was varied between 9.7 and 1048 K, time-domain Mössbauer spectra were obtained using SR of bandwidth 10 meV. A large time window was open to observe as many periods of DB as possible so that t_a could be deduced more accurately. The time spectra contained both QBs and DBs, from which temperature dependence of the magnetic hyperfine field B_{hf} were determined from the QB periods and that of the Lamb–Mössbauer factor f_{LM} from the DB "periods." The results were in excellent agreement with traditional energy-domain Mössbauer spectroscopy and bulk magnetic measurements.

Using SR of energy resolution 6 meV tuned precisely at the Mössbauer transition energy, inelastic nuclear resonant scattering has been applied to measure the phonon DOS directly (Sturhahn et al., 1995). A very short pulse (<0.1 nanosecond) of SR excites essentially all electrons and nuclei in the scatterer. Nonresonantly scattered radiation (by nuclei other than the Mössbauer isotope and all electrons) appears promptly in time (within a few nanoseconds), whereas photons that are resonantly absorbed by a Mössbauer nuclei need a delayed time (comparable to the natural life time of the excited state) to be reemitted. The prompt nonresonantly scattered radiation was eliminated by only counting in a time window of 30 to 600 nanoseconds after the arrival of the SR flash. Photons with precisely the Mössbauer transition energy will naturally cause nuclear resonance and hence elastic scattering. Photons with less energy can also excite nuclear resonance by annihilation of a phonon (lattice vibration), and photons with more energy can excite nuclear resonance by creation of a phonon. The probability of annihilation or creation of phonons is proportional to the local DOS. By detecting the delayed K-fluorescence radiation due to conversion electrons caused by the deexcitation of the Mössbauer nuclei, the phonon DOS can be directly measured. This phonon spectrum can also be normalized and integrated to give the recoil-free fraction f or the Lamb–Mössbauer factor f_{LM}, which provides an alternative method independent of isotope abundance, sample thickness, resonance cross section, or hyperfine interactions.

3.11
Synchrotron Methods versus Conventional Methods

Synchrotron Mössbauer spectroscopy has attracted a significant amount of attention from researchers and become a well-established methodology in the last decade. SR provides extremely strong and narrow photon beams, which facilitates spectral measurement under special experimental conditions, such as high temperature, high pressure, high magnetic field, and working with small samples (~ 1 mm^2) or nanostructured thin films. In biological samples, the concentration of Mössbauer isotope is usually too low for conventional Mössbauer spectroscopy, but SR should be strong enough to produce detectable signals. Usually, results for the recoil-free fraction f from synchrotron Mössbauer spectroscopic have higher accuracy than results from the conventional methods, partly due to the absence of the saturation effect in synchrotron Mössbauer methods. Scattering experiments require only relatively short measurement time, ranging from a few minutes to a few hours. The conventional Mössbauer spectroscopy may require up to hundreds of hours. Hyperfine parameters measured from synchrotron experiments have accuracies comparable with those from conventional experiments. In QB experiments, sufficient data accumulation is required in the chosen time window to yield satisfactory measurements of the periods.

Perhaps the most important contribution from synchrotron Mössbauer spectroscopy is its ability to measure phonon DOS directly. So far, this has not been achieved by conventional Mössbauer spectroscopy. Phonon DOS may be deduced from inelastic neutron scattering by extracting force constants from the fitted dispersion curves. However, on using inelastic nuclear resonant scattering with SR where the Mössbauer nuclei serve as analyzers, phonon DOS can be measured without phonon dispersion relations; hence this is a model-independent and more accurate method. Furthermore, inelastic nuclear resonant scattering allows us to measure *partial* density of states (PDOS) related to a specific isotope or a particular impurity.

However, synchrotron Mössbauer spectroscopy suffers from several shortcomings. Because time-domain experiments are based on the interference phenomenon, the corresponding spectra are very complex, whereas conventional Mössbauer spectra provide certain direct visual information. If two or more hyperfine fields are involved, time spectra may be severely modulated. The second major drawback of synchrotron Mössbauer spectroscopy is obviously the high expenses in constructing and maintaining such a large centralized synchrotron facility. Synchrotron Mössbauer spectroscopy is an important supplement to the conventional Mössbauer spectroscopy. The former will never completely replace the latter, but will help solve problems that cannot be studied or solved satisfactorily by using conventional methods.

References

Bauminger, E.R. and Harrison, P.M. (2003) Ferritin, the path of iron into the core, as seen by Mössbauer spectroscopy. *Hyperfine Interact.*, **151/152**, 3–19.

Bergmann, U., Shastri, S.D., Siddons, D.P., Batterman, B.W., and Hastings, J.B. (1994) Temperature dependence of nuclear forward scattering of synchrotron radiation in α-^{57}Fe. *Phys. Rev. B*, **50**, 5957–5961.

Chen, Y.L., Xu, B.F., Chen, J.G., and Ge, Y.Y. (1992) Fe^{2+}–Fe^{3+} ordered distribution in chromite spinels. *Phys. Chem. Miner.*, **19**, 255–259.

Cook, D.C. (2004) Application of Mössbauer spectroscopy to the study of corrosion. *Hyperfine Interact.*, **153**, 61–82.

Gerdau, E. and de Waard, H. (1999) *Hyperfine Interact.*, **123/124**, reviews and references therein, 3–879.

Gerdau, E. and de Waard, H. (2000) *Hyperfine Interact.*, **125**, reviews and references therein, 3–221.

Gibb, T.C. (1976) *Principles of Mössbauer Spectroscopy*, Chapman & Hall, London.

Goldanskii, V.I., Gorodinskii, G.M., Karyagin, S.V., Korytko, L.A., Krizhanskii, L.M., Makarov, E.F., Suzdalev, I.P., and Khrapov, V.V. (1962) The Mössbauer effect in tin compounds. *Proc. Acad. Sci. USSR (Phys. Chem. Sect.)*, **147**, 766–768.

Häggström, L. (1974) *Determination of Hyperfine Parameters from 1/2 $\to$ 3/2 Transitions in Mössbauer Spectroscopy*, Uppsala University Institute of Physics Report UUIP-851, Uppsala, Sweden.

Hanna, S.S., Heberle, J., Littlejohn, C., Perlow, G.J., Preston, R.S., and Vincent, D.H. (1960) Polarized spectra and hyperfine structure in Fe^{57}. *Phys. Rev. Lett.*, **4**, 177–180.

Jiang, K.Y., Yang, X.L., Yuan, Y.T., Mao, L.S., and Yang, D.P. (2002) A Mössbauer effect study on the structural components in potassium-promoted iron oxide catalysts for dehydrogenation of ethylbenzene. *Hyperfine Interact.*, **139**, 97–105.

Karyagin, S.V. (1963) A possible cause for the doublet component asymmetry in the Mössbauer absorption spectrum of some powdered tin compounds. *Proc. Acad. Sci. USSR (Phys. Chem. Sect.)*, **148**, 110–112.

Keisch, B. (1973) Mössbauer effect studies in the fine arts. *Archaeometry*, **15**, 79–104.

Kholmetskii, A.L. (2000) Proposal for a test of the relativity principle by using Mössbauer synchrotron radiation. *Hyperfine Interact.*, **126**, 411–416.

Kopcewicz, M. (1989) in *Mössbauer Spectroscopy Applied to Inorganic Chemistry* (eds G.J. Long and F. Grandjean), Vol. III, Plenum, New York, pp. 243–287.

Kopcewicz, B. and Kopcewicz, M. (2000) Ecological aspects of Mössbauer study of iron-containing atmospheric aerosols. *Hyperfine Interact.*, **126**, 131–135.

Kuno, A., Matsuo, M., Pascual Soto, A., and Tsukamoto, K. (2004) Mössbauer spectroscopic study of a mural painting from Morgadal Grande, Mexico. *Hyperfine Interact.*, **156/157**, 431–437.

Lazzarini, L., Calogero, S., Burriesci, N., and Petrera, M. (1980) Chemical, mineralogical and Mössbauer studies of Venetian and

Paduan Renaissance sgraffito ceramics. *Archaeometry*, **22**, 57–68.

Lipkin, H.J. (1960) Some simple features of the Mössbauer effect. *Ann. Phys.*, **9**, 332–339.

Mills, R.L., Walter, C.W., Venkataraman, L., Pang, K., and Farrell, J.J. (1988) A novel cancer therapy using a Mössbauer-isotope compound. *Nature*, **336**, 787–789.

Morris, R.V. et al. (2004) Mineralogy at Gusev Crater from the Mössbauer spectrometer on the Spirit Rover. *Science*, **305**, 833–836.

Mössbauer, R.L. (1958) Kernresonanzfluoreszenz von Gammastrahlung in Ir^{191}. *Z. Phys.*, **151**, 124–143.

Mössbauer, R.L. (2000) The discovery of the Mössbauer effect. *Hyperfine Interact.*, **126**, 1–12.

Pound, R.V. and Rebka, G.A. (1959) Gravitational red-shift in nuclear resonance. *Phys. Rev. Lett.*, **3**, 439–441.

Shirley, D.A. (1964) Application and interpretation of isomer shifts. *Revs. Mod. Phys.*, **39**, 339–351.

Smirnov, G.V. (2000) Synchrotron Mössbauer source of ^{57}Fe radiation. *Hyperfine Interact.*, **125**, 91–112.

Sternheimer, R.M. (1966) Shielding and antishielding effects for various ions and atomic systems. *Phys. Rev.*, **146**, 140–160.

Stevens, J.G. (1998) Documentation and evaluation of Mössbauer data for minerals. *Hyperfine Interact.*, **117**, 71–81.

Sturhahn, W., Toellner, T.S., Alp, E.E., Zhang, X., Ando, M., Yoda, Y., Kikuta, S., Seto, M., Kimball, C.W., and Dabrowski, B. (1995) Phonon density of states measured by inelastic nuclear resonant scattering. *Phys. Rev. Lett.*, **74**, 3832–3835.

Yang, D.P., Lavoie, L.K., Zhang, Y.D., Zhang, Z.T., and Ge, S.H. (2003) Mössbauer spectroscopic and x-ray diffraction studies of structural and magnetic properties of heat-treated $(Ni_{0.5}Zn_{0.5})Fe_2O_4$ nanoparticles. *J. Appl. Phys.*, **93**, 7492–7494.

Yoshida, Y., Kobayashi, Y., Yoshida, A., Diao, X., Ogawa, S., Hayakawa, K., Yukihira, K., Shimura, F., and Ambe, F. (2002) In-beam Mössbauer spectroscopy after GeV-ion implantation at an on-line projectile-fragments separator. *Hyperfine Interact.*, **141/142**, 157–162.

Further Reading

Bancroft, G.M. (1973) *Mössbauer Spectroscopy: An Introduction for Inorganic Chemists and Geochemists*, John Wiley & Sons, Ltd, New York.

Chen, Y.L. and Yang, D.P. (2007) *Mössbauer Effect in Lattice Dynamics*, Wiley-VCH Verlag GmbH, Weinheim.

Dickson, D.P.E. and Berry, F.J. (eds) (1986) *Mössbauer Spectroscopy*, Cambridge University Press, Cambridge.

Frauenfelder, H. (ed.) (1962) *The Mössbauer Effect*, W.A. Benjamin Publishers, New York.

Gonser, U. (ed.) (1975) *Mössbauer Spectroscopy*, Springer-Verlag, New York.

Gonser, U. (ed.) (1981) *Mössbauer Spectroscopy II: The Exotic Side of the Method*, Springer-Verlag, Berlin.

Gruverman, I.J. (ed.) (1965–1975) *Mössbauer Effect Methodology*, Plenum Press, New York, Vols. 1–10.

Herber, R.H. (ed.) (1984) *Chemical Mössbauer Spectroscopy*, Plenum Press, New York.

Long, G.J. (ed.) (1984) *Mössbauer Spectroscopy Applied to Inorganic Chemistry*, Vol. 1, Plenum Press, New York.

Long, G.J. (ed.) (1987) *Mössbauer Spectroscopy Applied to Inorganic Chemistry*, Vol. 2, Plenum Press, New York.

Long, G.J. and Grandjean, F. (eds) (1989) *Mössbauer Spectroscopy Applied to Inorganic Chemistry*, Vol. 3, Plenum Press, New York.

Long, G.J. and Grandjean, F. (eds) (1993) *Mössbauer Spectroscopy Applied to Magnetism and Materials Science*, Vol. 1, Plenum Press, New York.

Long, G.J. and Grandjean, F. (eds) (1996) *Mössbauer Spectroscopy Applied to Magnetism and Materials Science*, Vol. 2, Plenum Press, New York.

Kopcewicz, M. (1994) Mössbauer effect, in *Encyclopedia of Applied Physics* (ed G.L. Trigg), Vol. 11, VCH Publishers, New York, pp. 1–22.

Kolk, B. (1984) Studies of dynamical properties of solids with the Mössbauer effect, in *Dynamical Properties of Solids* (ed G.K. Horton and A.A. Maradudin), Vol. 5, North-Holland Physics Publishing, Amsterdam, pp. 1–328.

Further Reading

Maddock, A. (1997) *Mössbauer Spectroscopy: Principles and Applications*, Horwood Publishing, Chichester.

Stevens, J.G. (1981) Mössbauer spectroscopy, in *Handbook of Spectroscopy* (ed. J.W. Robinson), Vol. III, CRC Press, Boca Raton, pp. 403–528.

Stevens, J.G. and Shenoy, G.K. (eds) (1981) *Mössbauer Spectroscopy and its Chemical Applications*, American Chemical Society, Washington, DC.

4
X-Ray Spectroscopy

Thomas H. Markert[†] and Eckhart Förster

4.1	**Introduction**	**89**
4.2	**General Concepts**	**90**
4.3	**Dispersive Spectrometers**	**91**
4.3.1	Bragg Crystal Devices	91
4.3.1.1	Bragg's Law	91
4.3.1.2	Flat Crystal Bragg Spectrometer	93
4.3.1.3	Bent-Crystal Spectrometer	95
4.3.1.4	Bragg Spectroscopy with Synthetic Crystals (Multilayers)	98
4.3.2	Diffraction Gratings	100
4.4	**Nondispersive (Energy-Dispersive) Spectrometers**	**103**
4.4.1	Semiconductor Devices	103
4.4.1.1	Lithium-Drifted Silicon [Si(Li)] Detectors	105
4.4.1.2	Germanium Detectors	105
4.4.1.3	Silicon Drift Detectors (SDDs)	106
4.4.1.4	Charge-Coupled Devices (CCDs)	106
4.4.1.5	Room-Temperature Devices	106
4.4.2	Cryogenic Devices	107
4.5	**Applications**	**109**
	Acknowledgments	110
	Glossary	110
	References	112
	Further Reading	113

Encyclopedia of Applied Spectroscopy. Edited by David L. Andrews.
Copyright © 2009 WILEY-VCH Verlag GmbH & Co. KGaA, Weinheim
ISBN: 978-3-527-40773-6

4.1
Introduction

X-ray spectrometers are devices designed to study X-ray beams, with the goals of determining the wavelengths (or equivalently the energies) and the fluxes of the incident X-rays. Although spectroscopy of a sort was done from within a few years after Röntgen's discovery of X-radiation (1895), an enormous breakthrough was achieved when crystal diffraction revealed the first detailed X-ray spectrum (Bragg and Bragg, 1913). The Bragg technique has been refined over the years but is still used extensively and, for many purposes, gives the most useful results (see Section 4.3.1). Since that time X-ray spectroscopy has been applied, with great success, in many branches of pure and applied science (Agarwal, 1991). X-ray spectrometers have improved significantly, and new classes of such instrumentation are evolving rapidly (Tsuji, Injuk and van Grieken, 2004; Janssens, Adams and Rindby, 2000). In this chapter, the tools of X-ray spectroscopy are briefly described and the most common spectrometer devices are discussed. The strengths and limitations of the various spectrometers are indicated, and brief examples of applications and the kinds of devices that might be appropriate for real-world experiments are provided.

To some degree, all X-ray detectors are spectrometers, since all have some ability to determine wavelengths and measure fluxes. However, here we concentrate on those detectors (and spectroscopic devices) that have moderate spectral resolving powers (roughly, the resolving power $E/\Delta E$ must be better than about 20 over at least some of the spectral range). This criterion excludes, in general, proportional counters, microchannel plate devices, and scintillation detectors. Useful discussions of these classes of detectors (including their spectroscopic capabilities) can be found in the books by Fraser (1989), Knoll (1989), and Michette and Buckley (1993).

The survey here is further restricted to the energy (wavelength) range of 100 eV (124 Å) to 100 keV (0.124 Å). Lower and higher energy spectrometers are discussed in articles and books on ultraviolet and γ-ray spectroscopy. Photons with energies above the creation threshold for electron–positron pairs ($E \geq 1.02$ MeV) interact with matter by initiating a cascade shower. Measurements on these hard photons allow their total energy, arrival position, and arrival time to

Encyclopedia of Applied Spectroscopy. Edited by David L. Andrews.
Copyright © 2009 WILEY-VCH Verlag GmbH & Co. KGaA, Weinheim
ISBN: 978-3-527-40773-6

be determined (Sauli, 1992; Ramana Murthy and Wolfendale, 1993). Although this chapter is mainly concerned with the energy range up to 30 keV, we briefly consider, at this point, the range beyond this up to 511 keV, that is, half the electron–positron creation threshold. The resolving power of high-purity germanium (HPGe) detectors (see Section 4.1.2) is typically several hundreds, which is excellent for resolving lines of the highest energy atomic transitions and the lower energy nuclear transitions. The disadvantage of germanium detectors, which progressively worsens as energies increase, is that the percentage of photons that convert via the photoelectric effect becomes smaller with higher energies (about 75% at 100 keV to about 10% at 511 keV). Consequently, the full energy efficiency becomes small.

Sophisticated X-ray spectrometers have been recently designed to study X-ray emission of new types of hot plasma flashes in order to learn about space- and time-dependent ion distributions as well as electron density and temperatures (see Section 3.1.3). Commercially available X-ray spectrometers register X-ray fluorescence of a wide range of samples being irradiated either by an X-ray source or directly by an electron beam in order to obtain their elemental composition. Well-developed X-ray spectrometers with powerful, water-cooled X-ray tubes of many Kilowatts power now compete with small spectrometers equipped by air-cooled X-ray minitubes of a few up to 50-W power and adapted focusing optics (e.g., polycapillaries).

Finally, X-ray spectrometers have been constructed to study X-ray absorption of species in experiments with bright synchrotron and hot plasma sources. The X-ray absorption fine structure provides information about the geometrical distribution of electrons in the vicinity of the absorbing atom. Some recent practical applications are outlined in Section 4.5.

4.2
General Concepts

The primary figure of merit of most spectrometers is the resolving power:

$$R = \frac{\lambda}{\Delta\lambda} = \frac{E}{\Delta E} \quad (4.1)$$

where $\Delta\lambda$ is the full width at half maximum (FWHM) (in wavelength units) of the spectrometer response to a monochromatic X-ray of wavelength λ. The energy E is related to λ by $E = hc/\lambda$, where h is Planck's constant and c is the speed of light. The relationship $E\lambda \sim 1.24$ (E in kilo electronvolt and λ in nanometer units) is often useful in X-ray spectroscopy. Note that X-ray spectroscopists also rely on the older units (angstroms and electronvolt or kilo electronvolt) for wavelength and energy. In this chapter, we typically use these as well, since the interested reader will find them in most of the standard references. To convert to SI units, note that 1 Å = 0.1 nm and 1 eV = 1.60218×10^{-19} J.

Besides the (i) resolving power, other relevant performance parameters of X-ray spectrometers, which may, in fact, be decisive in the selection of an instrument for a particular experiment, are as follows: (ii) sensitivity (or instrumental efficiency); (iii) linearity of (or at least information on) spectral response; (iv) maximum counting rate of the detector; (v) breadth of spectral coverage; (vi) field of view; (vii) physical size; (viii) simplicity of operation; and

(ix) cost. The strengths of various spectrometer types are illustrated as they are discussed later.

X-ray spectrometers can be divided into two general categories. The first is the dispersive spectrometers that use X-ray diffraction to spread out the incident spectrum (a prism is an optical analogy). A detector can then read off the dispersed X-rays and associate a wavelength to the position of the detected photons. Dispersive spectrometers are sometimes called *wavelength-dispersive spectrometers* (*WDSs*).

Nondispersive spectrometers, on the other hand, convert the energy of an incident X-ray to a number of "particles" (electron–ion pairs, electron–hole pairs, phonons, superconductor quasiparticles, etc.), and the energy resolution of the spectrometer is determined by the statistical fluctuations of the number of particles (this statement holds only for ideal detectors; no actual device is free of systematic uncertainties). All designs of nondispersive spectrometers suffer from the problem of "dead time," the time required to process an X-ray absorption event. Note that some authors refer to nondispersive spectrometers as *energy-dispersive spectrometers* (*EDSs*), generalizing the word "dispersive" beyond simple spatial separation.

Broadly speaking, dispersive spectrometers work by taking advantage of the wave nature of the X-rays (interference produces the diffraction) and nondispersive (energy-dispersive) spectrometers work by taking advantage of the particle nature of the X-rays (photons ionize or otherwise interact with the detectors). Of course, this comment is overly simplistic, since dispersive spectrometers always use photon detectors to read out the dispersed spectrum. In fact, it is often useful to combine dispersive and nondispersive spectrometers, such as diffraction gratings and solid-state detectors (Sections 4.3.2 and 4.4.1), so as to combine the strengths of the different devices.

In Section 3, the two types of dispersive X-ray spectrometers, Bragg crystal diffractors and diffraction gratings, are discussed. In Section 4.4, the nondispersive detectors (those with reasonably high resolving powers) are considered.

4.3 Dispersive Spectrometers

4.3.1 Bragg Crystal Devices

4.3.1.1 Bragg's Law

Figure 4.1 illustrates the elementary derivation of the fundamental equation of crystal diffraction, that is, Bragg's law,

$$n\lambda = 2d \sin\theta \left(1 - \frac{\delta}{\sin^2\theta}\right) \quad (4.2)$$

where λ is the wavelength of the incident X-ray, d is the spacing between the planes of the diffracting atoms, often called *net planes*, θ is the angle of incidence (and reflection) of the X-ray beam onto the crystal surface, and n is a positive integer, namely, the order of diffraction. Bragg's law states that only X-rays with wavelengths that satisfy Eq. (4.2) will be efficiently reflected from the crystal. Figure 4.1 shows how Bragg's law is derived: X-rays of wavelength λ are incident on a crystal surface with a glancing angle θ_{in}. Although each scattering site scatters in all directions, only those rays that satisfy $\theta_{in} = \theta_{out}$ will constructively interfere and result in a significant reflected flux (this is,

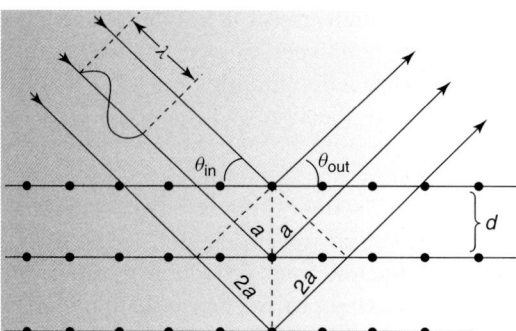

Fig. 4.1 Derivation of Bragg's law.

of course, consistent with the general law of reflection in geometrical optics). Furthermore, a second constraint on the reflected ray arises from the requirement that X-rays diffracted from each crystal plane must also interfere constructively (since the X-rays have relatively high energies, they can typically penetrate through hundreds to thousands of crystal planes). The X-rays penetrate the crystal surfaces and are scattered by the various crystal molecular planes (the planes are separated by d). The differences in the path lengths between the X-ray waves reflected off the various planes are $2a$, $4a$, $6a$, ... Clearly, there will be constructive interference between the various waves only if the path-length differences are integral multiples of the wavelength λ, that is, if $n\lambda = 2a$, where n is a positive integer. Simple trigonometry shows that $a = d \sin \theta$, thereby giving Eq. (4.2).

The crystal lattice parameter of silicon was linked to the SI meter base unit by combined optical and X-ray interferometry experiments with an experimental uncertainty error of about 10^{-8} (reviewed in Becker (2001)). For practical purposes, the lattice parameters (or net plane distances d) of any crystal can be derived at lower accuracy from a high-resolution X-ray diffractometer experiment. Here, the Bragg angles θ of the silicon gauge crystal are compared with those of the crystals being measured (Klöpfel et al., 1997). In such thorough evaluations of crystal parameters, several correction factors have to be considered, for example, refraction effects on the boundary between air and crystal. This leads to a correction of Bragg's law:

$$n\lambda = 2d \sin\theta \left(\frac{1-\delta}{\sin^2 \theta} \right) \quad (4.3)$$

where $1 - \delta$ is the real part of the index of refraction of the crystal at the wavelength of interest. Other effects must also be considered for a rigorous treatment of the Bragg diffraction – such as

1. extinction (the primary beam cannot reach deeply into the crystal since it is reflected by intervening crystal layers);
2. absorption (again, the primary beam is absorbed by intervening crystal layers); and
3. multiple reflection (crystal layers reflect both "up" (toward the surface of the crystal) and "down" (away from the surface)).

Thorough discussions of these topics are beyond the scope of this chapter but may be found in some detail in the books by Compton and Allison (1967), Pinsker (1978), James (1982), and Authier (2001).

4.3.1.2 Flat Crystal Bragg Spectrometer

Figure 4.2 is a schematic drawing of a generic X-ray spectrometer based on the Bragg principle. Here, X-rays are collimated so that they strike the diffracting crystal at the Bragg angle for X-rays of wavelength λ. Only those X-rays that satisfy Eq. (4.2) are reflected and registered in the detector. Since (presumably) the spacing d and incident angle θ are known, the wavelength of the diffracted X-ray can be determined to be within a factor of $1/n$ (n is the diffraction order). Removing order ambiguity is an issue for all dispersive spectrometers. Techniques for determining the actual X-ray wavelength (i.e., removing the $1/n$ uncertainty) are discussed later.

To be useful as a spectrometer, it is necessary for a device to cover a range of wavelengths. This is achieved for the generic device in Figure 4.2 by mounting the crystal and the detector on coaxial rotary stages, so that the crystal can present a range of incident angles to the X-ray beam. Any rotation of the crystal by an angle $\Delta\theta$ is accompanied by a rotation of the detector by $2\Delta\theta$, so that the diffracted X-rays fall on the detector. To study a single X-ray emission line, for example, the crystal must be rotated over a range of angles that cover the response of the crystal to a monochromatic line. It should be apparent, from this discussion, that a Bragg device can only measure X-rays with wavelengths in the range $0 \leq \lambda \leq 2d$ (i.e., when $0° \leq \theta \leq 90°$); in practice, however, both the short- and long-wavelength limits are restricted by geometrical constraints. To measure longer wavelengths, one requires crystals (or multilayer diffractors discussed below) with larger $2d$ spacings.

From Bragg's law, the angular dispersion ($d\lambda/d\theta$) can be easily derived: after division by the reflection order n, the wavelength λ is a product of a constant factor ($2d/n$) and $\sin\theta$. If this factor is expressed by $\lambda/\sin\theta$ in the derivative, we obtain

$$\frac{d\lambda}{d\theta} = \frac{\lambda}{\tan\theta} \qquad (4.4)$$

Thus, the angular dispersion is inversely proportional to $\tan\theta$, which means that the resolving power R is directly proportional to $\tan\theta$. If we have a set of

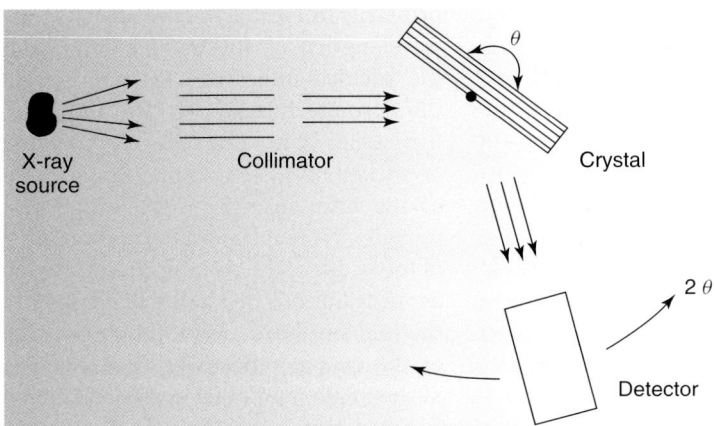

Fig. 4.2 Schematic drawing of a generic crystal spectrometer.

crystals with different net plane distances d, we should choose a plane with $2d$ slightly larger than λ to obtain the highest resolving power R.

The FWHM of the response function of an ideal crystal spectrometer is dominated by the natural resolving power of the crystal (the angular dispersion of the collimated X-rays also contributes, however, and in some cases will dominate). The response function of the crystal alone is called the *rocking curve*, because the complete response of the crystal to a monochromatic X-ray is measured by rotating (rocking) the crystal over a small angular range. The FWHM is the rocking curve width. Very early on, Compton and Allison (Becker, 2001) had derived a general theory of X-ray spectrometers. In the case of Figure 4.2, the relation between the X-ray emissivity J, the rocking curve C, and the reflected intensity P holds with some minor simplifications:

$$P(\lambda) = \int d\lambda' J(\lambda - \lambda') \int_{\alpha_{min}}^{\alpha_{max}} d\alpha$$
$$\times \int_{\phi_{min}}^{\phi_{max}} d\phi \cdot C\left(\alpha - \frac{\phi^2}{2}\tan\theta_0 - \frac{\lambda' - \lambda_0}{\lambda_0}\tan\theta_0\right) \quad (4.5)$$

where the divergence angles α and ϕ describe the angular source emission in and out of the reflection plane (Figure 4.2). To minimize the broadening of X-ray emissivity J by convolution with the rocking curve C, the FWHM of C should be much smaller than the divergence $(\alpha_{max} - \alpha_{min})$ in order not to lose X-ray photons in the Bragg reflection. The meridional divergence (multiplied by the angular dispersion of Eq. 4.4) should, in addition, be much smaller than the FWHM of X-ray emissivity J. This is normally realized in WDSs, if the resolving power is high and the collimator is well adapted (cf. Figure 4.2). For small sagittal divergences $(\phi_{max} - \phi_{min})$, this broadening influence in Eq. (4.5) can be neglected since it depends on the second order of ϕ.

If the absorption in the spectrometer crystal is very small (which holds for hard X-rays), the rocking curve C has the form of a cylinder head (Darwin curve) with a flat top of almost unity reflection (i.e., in the range of interference total reflection). For the general case of structurally perfect crystals, the rocking curves can be computed using the dynamical theory of X-ray interferences (see the book by Authier (2001) and references therein). A narrow FWHM of the rocking curves leads not only to a good spectral resolution but also eventually to a loss of X-ray photons on the detector, whereas a wide FWHM provides many photons at the cost of spectral resolution.

Finally, the sensitivity of the Bragg devices must be known if one wishes to determine the incident flux as a function of wavelength (this is not always the case; sometimes, experimenters may be primarily interested in the locations (i.e., wavelengths) of spectral features or in the relative intensities (as opposed to the absolute fluxes)). To determine the flux in an X-ray line, the crystal must be rocked over the nominal Bragg angle of the line, covering a $\Delta\theta$ range that includes the response of the crystal (most of the width of the rocking curve), as well as contributions due to the dispersion of the incident beam. With the knowledge of the crystal reflectivity as a function of the angle (and other parameters such as the detector quantum efficiency), the incident flux in the X-ray line can be computed. The parameters of interest for computing the flux are not always easy to determine, however, particularly at all energies.

As noted above, Bragg's law permits a residual uncertainty in the X-ray wavelength because of the reflectivity of various diffraction orders of n. (This uncertainty is also present for grating spectrometers.) This ambiguity may be minimized or eliminated in a number of ways. The simplest way of eliminating it is to use a detector that has a sufficient inherent spectral resolution that the orders may be distinguished by the detector. Even low-resolution devices such as proportional counters can separate orders sufficiently in a wide energy range. Other techniques are to use filters and/or X-ray optics (which do not reflect efficiently at higher energies) to minimize the contribution from higher orders.

Collection and diffraction efficiency of a flat crystal spectrometer can be improved if parallel beam X-ray optics is incorporated. A diverging beam originating from the X-ray source is parallelized by polycapillary and grazing-incidence optics. The now parallel beam impinges on the spectrometer crystal (Love, 2002). This results in a much larger solid angle of X-rays that can be collected. The wavelength-dispersive unit can, for example, be incorporated in a scanning electron microscope.

4.3.1.3 Bent-Crystal Spectrometer

Besides the industrial applications of X-ray spectrometers where a high X-ray flux is often available, there are tasks in which the detector flux is rather limited. This is the case for X-ray diagnostics of high-temperature plasmas and for electron-beam X-ray microanalysis (see Section 4.3.1.4). There is, therefore, a need for additional X-ray optic elements to collect and focus X-rays to the detector. Depending on the required resolving power and luminosity as well as the geometry of spectrometer, different X-ray optics, grazing-incidence mirrors, capillaries, bent gratings, bent crystals, and so on must be selected.

Bent-crystal spectrometers employ, in the simplest form, a cylindrically bent crystal (instead of a flat one) as the dispersing element and a detector that is positioned on the circumference of the Rowland circle (see Figure 4.3(a) and (b)). The curvature of the reflecting net planes is twice that of the radius of the Rowland circle in this (Johann-type) spectrometer. Equation (4.5) is still applicable when the rocking curve is corrected for crystal bending (Uschmann, Malgrange and Förster, 1997) and modified relations are used between the divergence angles α, ϕ, and the wavelength λ'. Since rocking curves are usually wider and bent crystals accept a diverging beam of X-rays, photon numbers on the detector can be increased by 1 or 2 orders of magnitude in comparison to the flat crystal spectrometer. Figure 4.3(c) shows the dominant resonance lines of hydrogen-like and helium-like aluminum ions together with intercombination and satellite lines on a relatively low continuum emission. The electron density and temperature can be obtained from plasma physics models from ratios of selected spectral lines (Attwood, 1999).

The highly excited aluminum ions were produced by focusing high-power Nd : glass laser pulses (200 J, 1 nanosecond) on solid aluminum target (Renner et al., 1994).

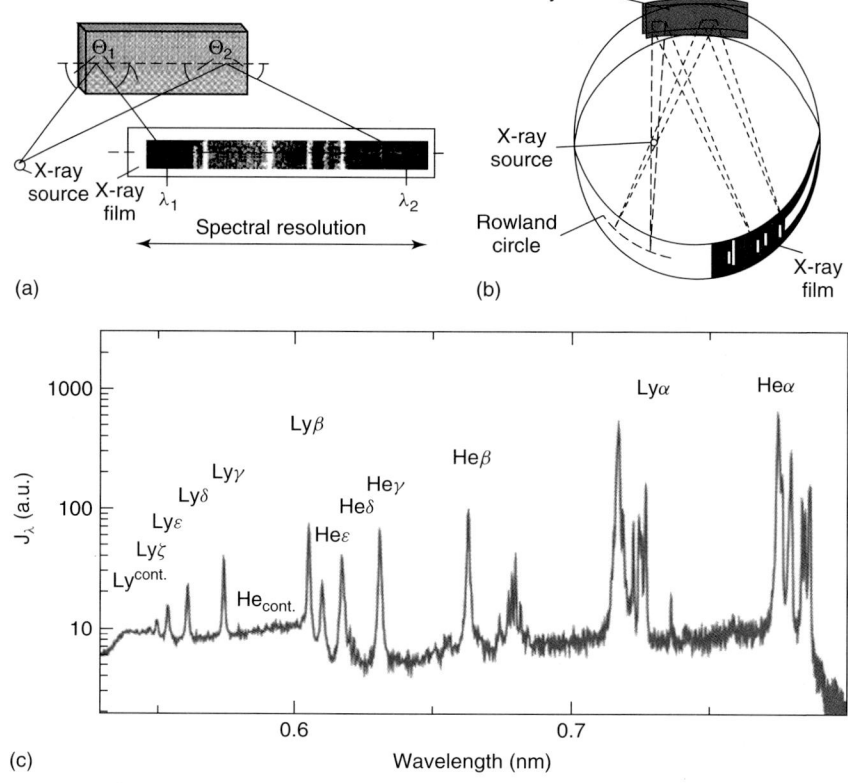

Fig. 4.3 Schematic drawings of (a) a flat crystal spectrometer, (b) a Johann-type bent-crystal spectrometer, and (c) X-ray spectrum of a laser-produced Al plasma.

If the cylindrical crystal is rotated by 90° around its central surface normal, then X-rays are collected in the ϕ angle (instead of the α angle) direction. Figure 4.4(a) shows the von Hamos-type spectrometer in which the distances from the crystal to the source and the detector are equal. By bending a thin crystalline wafer in two dimensions, spherical, toroidal, or ellipsoidal crystals can be produced with curvature radii down to 10 cm. As shown in Figure 4.4(b) and (c), luminous X-ray spectrometry and quasimonochromatic X-ray imaging can be performed. Mißalla et al. (1999) have discussed reflection and focusing properties of these schemes when short X-ray pulses in the kilo electronvolts range are considered.

A cylindrically bent analyzer crystal can be combined with a flat position-sensitive detector in the von Hamos geometry (see Section 4.4.1.4). This makes parallel data collection possible for light elements in WDSs (Love, 2002). Toroidal crystals combine focusing in both sagittal and meridional directions if the ratio of the respective radii is $\sin^2 \theta$. They are characterized by high resolving power and efficiency but require much careful preparation and testing.

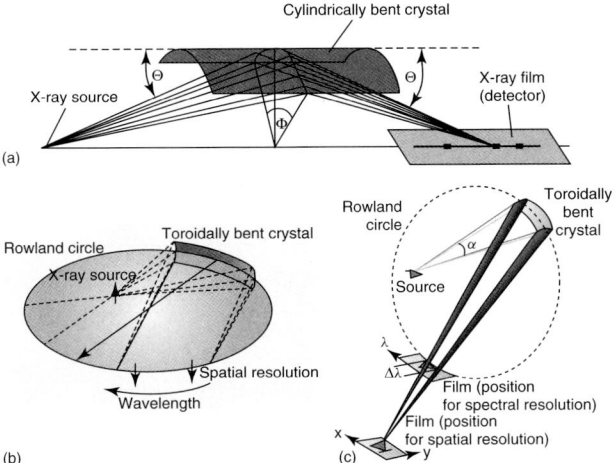

Fig. 4.4 Schematic drawings of bent-crystal spectrometers: (a) von Hamos-type, (b) toroidally bent crystal spectrometer, and (c) quasimonochromatic X-ray imaging setting.

An even higher spectrometer efficiency can be achieved by applying ellipsoidal highly oriented pyrolithic graphite (HOPG) crystals if a wider reflection curve (a 10th of a degree) is tolerated.

All the spectrometer types discussed so far are based on the Bragg equation and its corresponding angular dispersion $(d\lambda/d\theta = \lambda/\tan\theta)$. There is, however, a special variant of a Johann spectrometer in which a point source and film are both located on the Rowland circle (Figure 4.5(b)).

Assuming that the X-ray wavelength λ_0 is reflected by the cylindrical crystal in the Rowland circle plane, X-rays with nonzero sagittal angles ϕ impinge on the crystal under slightly smaller Bragg angles. Geometrical considerations lead to the following relation between

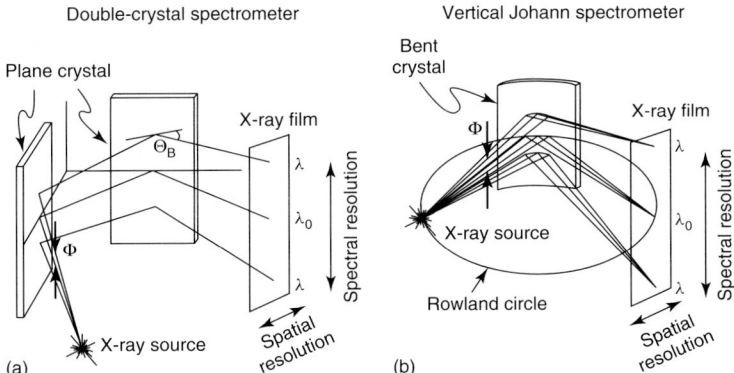

Fig. 4.5 Schematic drawings of high-resolution X-ray spectrometer with vertical (sagittal) dispersion: (a) double flat crystal and (b) Johann-type cylindrically bent-crystal setting.

Tab. 4.1 Key parameters for different X-ray crystal spectrometers.[a]

Type of spectrometer	Luminosity (mrad^2)	Spectral resolution	Range (Å)
Flat single crystal	1.74×10^{-2}	4000	1.14
Johann classical scheme	1.88×10^{-1}	12 800	1.34
Double-crystal vertical dispersion	4.36×10^{-3}	13 000	0.18
Johann vertical dispersion	2.72	10 400	0.18

[a] Quartz crystals of dimension 20×30 mm^2 either flat or with curvature radius of 76.8 mm were used. Source-to-detector distance was 130 mm; source of size 1 mm^2 emits a certain number of photons per square milliradian per square millimeter into a solid angle 1 mrad^2 (Renner et al., 1994).

wavelength λ and the divergence angle ϕ:

$$\lambda = \lambda_0 \cos \phi \qquad (4.6)$$

where the maximum wavelength λ_0 can be simply changed by moving the X-ray source and the film along the Rowland circle. This leads to a high luminosity and a high spectral resolving power (Renner et al., 1994), which is needed for sophisticated X-ray plasma diagnostics, but is limited to a small spectral coverage. There is also a simple extension of the flat crystal spectrometer to a double-crystal spectrometer in which wavelengths are determined by Eq. (4.6). It suffers, however, from a very low luminosity and is therefore applicable only for strong X-ray sources. Table 4.1 summarizes some key parameters of the X-ray spectrometers of Figures 4.3(a) and (b) and 4.5(a) and (b), respectively.

4.3.1.4 Bragg Spectroscopy with Synthetic Crystals (Multilayers)

Most Bragg diffractors are either natural crystals or crystals grown in the laboratory according to certain chemical and thermal specifications. Crystals of this kind have $2d$ spacings in the range 0.2–2.7 nm and can be used to cover energies between ~500 eV and 35 keV (Attwood, 1999). For lower energies, synthetic multilayer diffractors sometimes called *layered synthetic microstructures* (*LSMs*) or simply multilayers are frequently used, replacing in most cases, the Langmuir–Blodgett soap films used for many years. Multilayer diffractors are made by depositing layers (using the sputtering or evaporation techniques of semiconductor technology) of alternating high-Z and low-Z materials (e.g., carbon and tungsten) with precise thicknesses so as to perform the same functions as one-dimensional crystals. In the multilayer case, the effective spacing d is the sum of the thicknesses of the high-Z and low-Z materials. Up to several hundreds, layers can be laid down in this way (more layers are of little value since the relatively low-energy X-rays cannot penetrate deeper). Multilayers can now be deposited on curved substrates (up to 30-cm lateral size), even with gradients in depth and lateral thickness in order to optimize imaging performance. A comparison of focusing optics for femtosecond X-ray diffraction with Cu Kα line radiation shows differences in the key parameters, focus size, flux, and photons per second (Bargheer et al., 2005). The authors considered a toroidally bent Ge Crystal, multilayer optics, ellipsoidal monocapillary, and polycapillary X-ray optics.

A clear advantage of using multilayer films is that the spacing d can be selected to be any desired value (although d spacings smaller than about 25 Å are difficult to fabricate). Generally, the resolving power of multilayer diffractors is limited to about 50, which is superior to many other kinds of spectrometers but will not be adequate for some applications. As shown in Figure 4.6, effective methods are available to change the form of the multilayer reflection curve. By using higher reflection orders n, the FWHM of the reflection curve can be reduced to obtain higher resolution. In contrast, the FWHM can be broadened by using depth gradients in layer thicknesses to obtain higher spectrometer sensitivity on the expense of peak reflectivity.

In recent years, multilayers have been used as the reflecting surfaces of many different optical devices, providing enhanced reflection at certain soft X-ray wavelengths (i.e., those wavelengths in which the Bragg condition is met). For example, traditional X-ray telescopes utilize grazing incidence ($\theta <$ a few degrees) on polished metallic surfaces (Fraser, 1989). Such telescopes are difficult to fabricate and typically have a relatively small collecting area. More traditional telescope geometries (i.e., those with nearly normal incidence) can be used, however, if the reflecting surfaces are coated with multilayers. The resulting telescope forms sharp images over a narrow wavelength band. Figure 4.7 shows an image of the sun obtained with an X-ray telescope coated with layers of cobalt and carbon with a period of 31.8 Å. At nearly normal incidence, the mirrors reflect efficiently at a wavelength of 63.5 Å. The bandpass of 1.4 Å encompasses the wavelengths of two bright X-ray emission lines from ions of highly ionized Mg and Fe, which are sensitive diagnostics of the plasma temperature in the range $(1-3) \times 10^6$ K, in which these ions are abundant. The figure shows the high-temperature material in the solar corona, in contrast to the much cooler photosphere that appears relatively dark

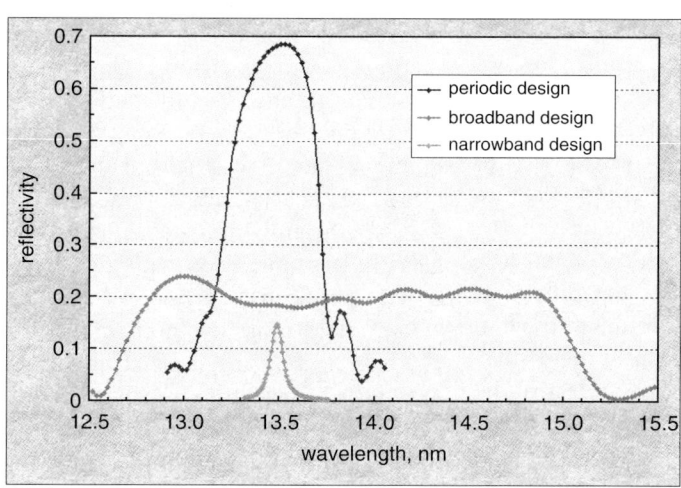

Fig. 4.6 Change in Mo/Si multilayer reflection curves to obtain higher resolution and higher sensitivity, respectively. (courtesy of Torsten Feigl and Sergij Yulin.)

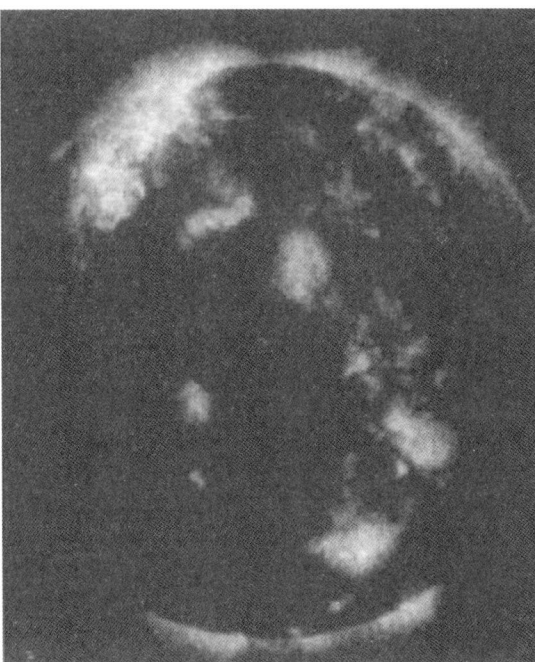

Fig. 4.7 Image of the sun taken on 11 July 1991 at $\lambda = 63.5$ Å with a bandwidth of 1.4 Å. The hot solar corona is seen clearly, while the cooler surface of the sun is relatively dark at this wavelength. (courtesy of Leon Golub.)

at this wavelength (Golub, Zirin and Wang, 1994).

4.3.2 Diffraction Gratings

Diffraction gratings are similar to crystals in that a periodic structure sets up an interference pattern that disperses the X-rays according to wavelength. They differ from crystals in that the structure is periodic only on the surface of the diffractor instead of both on the surface and within the body of the diffractor, as for crystals. As a result, diffraction gratings reflect (or transmit) incident X-rays at all wavelengths regardless of the incidence angle.

Gratings disperse light according to the grating equation:

$$n\lambda = p(\cos\beta - \cos\alpha) \qquad (4.7)$$

where the incident and diffracted angles (α and β) are defined in Figure 4.8 for the two types of gratings: reflection and transmission gratings. As was the case for crystal spectrometers, n is the integer order of diffraction (except that for gratings, n can be positive, negative, or zero). For gratings, p is the period of the grating lines or grooves. X-ray reflection gratings operate at small incidence angles ($\alpha < 2°-3°$) because surfaces reflect X-rays efficiently only at grazing angles (unless enhanced with multilayer coats; see Section 3.1.4 and (Spiller, 1994)). Transmission gratings generally operate at nearly normal incidence ($\alpha \sim 90°$) where the transmitted efficiency is greatest, although they can be used with moderate efficiency over a large range of angles.

Ordinary ruled reflection gratings were among the first X-ray spectroscopic

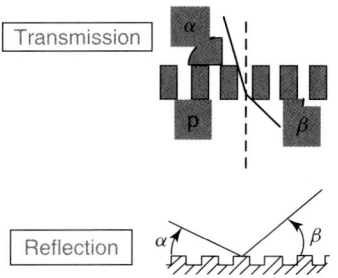

Fig. 4.8 Schematic illustration of the two kinds of X-ray gratings, the transmission and the reflection gratings. Although not indicated in the diagram, both kinds of gratings can be blazed to enhance the efficiency in some of the orders.

instruments, although they have been limited, until recent years to the longer wavelengths ($\lambda > \sim 20$ Å) since it is difficult to rule gratings with sufficiently small periods. In recent years, maturing technology (essentially, replacement of the mechanical ruling method by lithography or holography) has led to reflection gratings that can be used at wavelengths as short as ~ 5 Å. As is the case for reflection gratings used at optical wavelengths, X-ray gratings can be blazed (i.e., can have asymmetric reflecting surfaces) so as to enhance reflections into particular orders.

X-ray transmission gratings are a relatively recent development. Originally conceived for applications in X-ray astronomy (Gursky and Zehnpfennig, 1966), they have also been applied as spectrometers in synchrotrons and in X-ray lasers, and as interferometers in atomic diffraction experiments. Advantages of transmission gratings are their light weight and that they are typically used at normal incidence so that the spectrometer can be more compact than for a reflection grating. Effective transmission gratings must be thick enough so that the grating bars are at least partially opaque to the X-rays and must have small enough periods so that the dispersion angle is large enough to be useful. Figure 4.9 shows a sketch of a transmission grating developed for X-ray astronomy.

Transmission gratings are made with the techniques of semiconductor technology (microfabrication and nanofabrication). Periodic patterns are generated from holographic interference techniques or from mechanical ruling for larger period gratings. The periodic pattern is transferred to thicker metallic grating bars (gold is most often used). The first-generation transmission gratings were used primarily at lower energies ($E < 3$ keV) because the grating–bar thicknesses were too small (gratings became transparent at higher energies) and/or the bar periods were too large (dispersion angles were small). Recent developments have extended the effective energy range of transmission gratings to ~ 10 keV (Schattenburg et al., 1991).

Transmission gratings are usually fairly small (a few square centimeters, Figure 4.9), but individual gratings can be combined into a larger assembly. For X-ray astronomy applications, for example, grating areas of over 1000 cm² have been used at the objective of X-ray telescopes to provide a large collecting area for spectroscopy of cosmic sources. For laboratory applications, free-standing gold bars with up to 5000 lines mm^{-1} have been produced where the gold bars were fixed on a periodic gold support

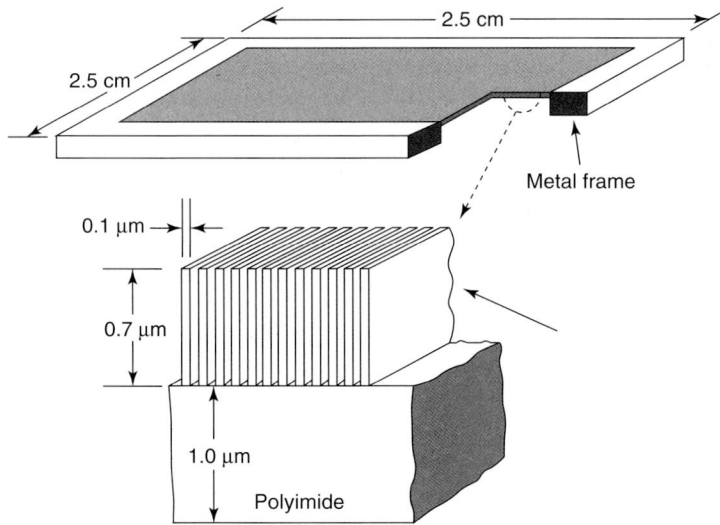

Fig. 4.9 Schematic of a high-energy grating facet being built for the US observatory Advanced X-ray Astrophysics Facility (AXAF). The gold grating bars (period = 2000 Å) are supported by a thin (1-μm) plastic film.

grid with 17 μm × 17 μm active areas. In combination with a toroidal grazing-incidence mirror, these transmission gratings were used in X-ray spectroscopy of both laser-produced plasmas and scattered radiation from a soft X-ray free electron laser (Jasny et al., 1994; Höll et al., 2007).

Like Bragg spectrometers, grating spectrometers can, in theory, achieve extremely large resolving powers. The theoretical maximum resolution for diffraction order n is given as

$$R = \frac{E}{\Delta E} = nN \qquad (4.8)$$

where N is the number of grating grooves or lines illuminated by the X-ray beam. For a grating with 1000 lines mm^{-1} (routine for transmission gratings), a 5-mm-wide X-ray beam can achieve a resolving power of 5000. It is, however, difficult to fabricate perfect periodic gratings, so that the variation in the grating period usually limits the resolving power to $p/\Delta p$, where Δp is the variation in the grating period over the illuminated surface of the grating.

Note that both classes of dispersive spectrometers (crystals and gratings) can, with relatively minor modifications, be used as monochromators as well as spectrometers, that is, they can select a narrow energy band out of a broad spectrum and transfer this band for further processing and analysis. Both grating and crystal monochromators are employed (generally, the crystals at the shorter wavelengths, where the gratings are less effective) in this way.

As was noted in the discussion of crystal spectrometers, there are potential benefits to curving the diffractors. In such a design, the incident X-rays can diverge from a point (parallel or nearly parallel beams have been tacitly assumed for the flat diffractors discussed above), and the curved gratings perform both energy dispersion and focusing. In the Rowland configuration (which is the most commonly used for curved diffractors), a point source at the entrance slit will focus to a (slightly curved) line at the exit slit, each wavelength focusing to a different position as shown in Figure 4.4(b) and (c). More complex curvatures (spheres, ellipsoids, and toroids) can improve the focusing properties at some wavelengths, although it can be more difficult to fabricate the curved diffractor elements (Michette and Buckley, 1993).

4.4 Nondispersive (Energy-Dispersive) Spectrometers

4.4.1 Semiconductor Devices

Semiconductor-based detectors, sometimes simply called *solid-state detectors*, have been used as X-ray spectrometers for more than 30 years. A basic description of how such devices work can be found in Tsuji, Injuk and van Gieken (2004); Fraser (1989); Knoll (1989). The fundamental principle is that an X-ray will liberate electron–hole pairs as it interacts within the semiconductor. The number of such pairs is given on the average by $N = E/w$, where E is the X-ray energy and w the ionization energy. For silicon and germanium, the ionization energies w are approximately 3.62 and 2.96 eV, respectively, and are slightly temperature dependent. An X-ray photon with energy 1 keV, therefore, will liberate on average several hundred electron–hole pairs. This free charge can then be gathered, as in a gas ionization detector, by applying an electric field to the ionized region. The amount of charge gathered is proportional to the energy of the incident X-ray.

Figure 4.10 shows a schematic diagram of a generic semiconductor detector. Two thin surfaces are heavily doped so as to be n-type (where electrons are the majority charge carriers) and p-type (where holes are the charge carriers). A central region, considerably thicker than the n- and p-types, is relatively free of charge carriers and is called the *depletion region*. The key feature of the depletion region is that it has a nonzero electric field, so that electrons and holes that are produced by photon interactions are easily separated. For X-ray detectors, the depletion region can consist of either an extremely pure semiconductor (typically germanium or silicon) or a volume of a semiconductor into which an element of opposite polarity (n- or p-type) has been diffused to compensate for the excess of electrons or holes (lithium drifted into p-type silicon or germanium is the most common). This compensated region (often called the *intrinsic region* since the layer has physical properties similar to pure, intrinsic semiconductors) is where the X-ray interaction takes place. To be effective as an X-ray detector over a broad energy range, the depletion region (whether naturally free of charge carriers or compensated) should be adequately thick (so that high-energy X-rays will be stopped), and any dead or absorbing layers (the n- or p-type regions or other nondetecting layers) should be thin.

When an X-ray enters the depletion layer (assuming that the layer is thick enough), it interacts with the semiconductor to produce electron–hole pairs. A voltage is applied to the surfaces to attract the electrons and holes, which migrate rapidly to the electrodes.

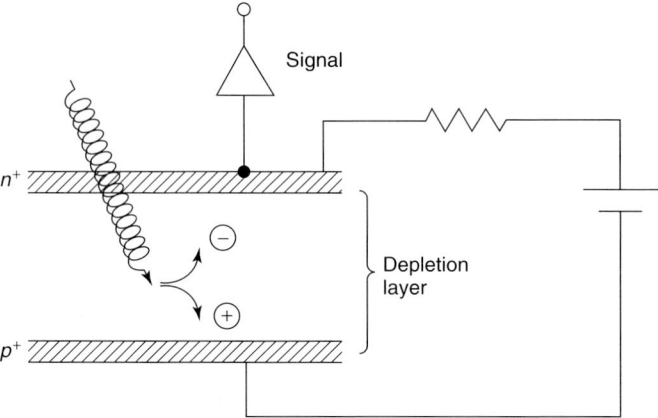

Fig. 4.10 Drawing of a (very) generic semiconductor spectrometer detector.

The heavily doped n- and p-type regions serve as barriers to any leakage current (the current that would flow in an undoped semiconductor as a result of the finite resistivity of the material). The leakage current must be suppressed for any prospect of detecting the small signal arising from an X-ray interaction. The signal goes to a preamplifier as shown in Figure 4.10 and is then processed electronically.

The collected charge following an X-ray photoelectric interaction is $N = E/w$ electrons (and holes). The deviation in the collected charge is related to X-ray energy according to

$$\Delta E = 2.354\, w \left(\frac{FE}{w} + K^2 + D^2 \right)^{0.5} \quad (4.9)$$

where ΔE is the FWHM response to a monochromatic X-ray line (see Eq. (4.1)), w is the ionization energy, E is the energy of the incident X-ray, K arises from the system electronic noise, and D is the dark current (the flow of electrons and holes that is present, owing to various impurities and imperfections in the device, even when there are no X-rays incident on the detector). Both K and D are measured in equivalent free electrons. F is the so-called Fano factor (Fano, 1947), which relates the statistical fluctuations in the number of electrons liberated to the expected number (E/w). (Fano factors for semiconductors have values of about 0.1, although there is an uncertainty in F of at least 50% due to the difficulties in precisely measuring this parameter.) Most semiconductor devices are operated at low temperatures (liquid nitrogen is typically used in the laboratory with $T = 77$ K), which helps to limit the dark current. The resolving power $E/\Delta E$ for a lithium-drifted silicon [Si(Li)] detector with system noise $\sqrt{K^2 + D^2}$ of 10 electrons is about 20 for 2 keV and grows to about 60 for 10 keV. At higher energies, the resolution is limited by the statistical fluctuations in the electron number; at low energies, the system noise dominates. Figure 4.11 shows a spectrum obtained with an Si(Li) device (see below), indicating the usefulness of a semiconductor device. The spectrum is the result of X-ray fluorescence of a sample consisting of thin films of gold,

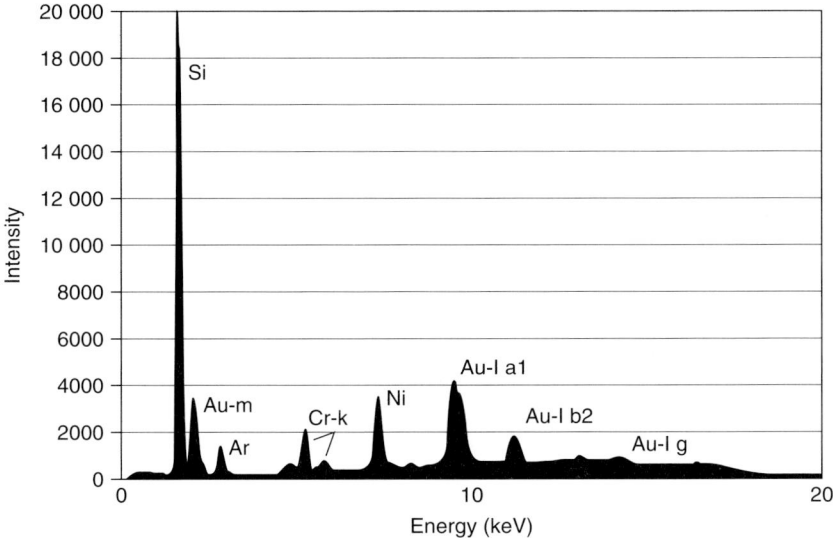

Fig. 4.11 Spectrum obtained with an Si(Li) detector.

nickel, and chromium deposited on a silicon substrate. The intensities of the various fluorescent lines can be used to determine the thicknesses of the films. Note that the resolving power of the semiconductor device is adequate for the quantitative analysis required.

There are a number of semiconductor devices that are used for X-ray spectroscopy. Here, four general types are listed, giving some of the strengths and weaknesses of the various classes. All of these devices are available commercially. See Knoll (1989) and Fraser (1989) for good introductory discussions on these devices and more detailed references.

4.4.1.1 Lithium-Drifted Silicon [Si(Li)] Detectors

Si(Li) detectors have depletion layers of thickness of a few millimeters to ~1 cm, manufactured by introducing ("drifting") a lithium dopant into slightly p-type silicon. The net result of this drifting process (which can take several weeks) is to produce a compensated depletion region that is essentially free of charge carriers. Si(Li) detectors can have significant efficiencies for energies up to about 30 keV. At lower energies, the dead layers limit the performance to energies greater than about 0.1 keV. For many commercial detectors, furthermore, a vacuum-tight window of 1-μm polymer with a cover of 30-nm aluminum is now used.

4.4.1.2 Germanium Detectors

Lithium-drifted germanium detectors [Ge(Li)] were used for many years, particularly at higher energies. In more recent years, extremely pure germanium crystals have been grown (HPGe) that have relatively few free charges and hence can be used directly as a depletion layer without the introduction of a compensating dopant. HPGe detectors complement Si(Li) detectors by having efficiencies at significantly higher energies (up to tens of mega electronvolts; see Section 4.5), but are usually not preferred at lower energies ($E < 20$ keV) since the Si(Li) detectors have better resolution and less prominent escape peaks.

4.4.1.3 Silicon Drift Detectors (SDDs)

A drift chamber detector is a large semiconductor wafer of, for example, high-resistivity n-type silicon; it is fully depleted from a small n^+ ohmic contact, which is positively biased with respect to the p^+ contacts covering both wafer surfaces. By implanting a parallel p^+ strip pattern and applying an electrical field on both wafer sides, electrons are forced to move to the central n^+ readout anode (Strüder et al., 1998). The silicon drift detector (SDD) incorporates an on-chip amplifier and has a very low capacitive loading; energy resolution is 127 eV for 6 kcal at $-20°$. A particular advantage is that commercially available SDDs (Röntec) can be employed for high X-ray counting rates in the range of 400 000 counts per second. The idea for this new detector scheme originated from Gatti and Rehak (1984); the SDDs are fabricated for a wide field of applications, for example, for several synchrotron radiation experiments (Strüder et al., 1998).

4.4.1.4 Charge-Coupled Devices (CCDs)

Charge-coupled devices (CCDs) are silicon-based detectors used extensively in television technology but are also effective (with some modifications) in the X-ray regime. While having spectral resolving powers similar (or even superior) to the best Si(Li) and HPGe devices, they are also excellent imaging detectors, with pixel sizes typically <20 μm and array dimensions 1024×1024 pixels or greater. The most common CCD design is a two-dimensional array of Metal-Oxide-Semiconductor (MOS) capacitors on the surface of a semiconductor. (An alternative, first proposed by Gatti and Rehak (1984), is to use an array of pn junctions on the surface of n-type silicon.) The charge from an X-ray interaction in the depletion layer of the semiconductor is stored, momentarily, in the nearest capacitor. By varying the surface voltages, the charge can then be passed onto adjacent capacitors until it reaches the edge of the detector, at which point it can be read out (in practice, the charge packet is read into a perpendicular register once it reaches the edge of the array, and is further transferred in this perpendicular direction; in this way, only a single amplifier circuit is required to read out an entire CCD chip). This charge transfer, or coupling, gives the device its name.

The depletion regions of CCDs are significantly shallower than that of the other semiconductor devices discussed here (of order of tens of microns), primarily because of the lower voltages required for the charge transfer. Consequently, currently available X-ray-sensitive CCDs are used most generally at energies $E < 10$ keV since higher energy X-rays simply pass through the device. The capacitor structure on the front surface, furthermore, acts as a dead layer that can absorb photons with $E < 0.5$ keV.

4.4.1.5 Room-Temperature Devices

The solid-state devices described above operate most efficiently when cooled to temperatures <77 K (liquid-nitrogen temperatures). A class of detectors composed of semiconductor compounds (HgI_2, CdS, CdTe, and GaAs) is actually more effective at room temperatures than when cooled significantly. These materials have larger band gaps (than Si or Ge), so that thermal fluctuations result in a much lower leakage current. Furthermore, because all of these compounds have a much higher effective atomic number, a viable detector can be much thinner (have a thinner depletion layer) and still detect higher energy photons. Unfortunately, the room-temperature devices are difficult to work with, have poor mobility (particularly for holes), and have a generally inferior resolving power. In spite of some clear benefits, the various room-temperature detectors are not as commonly used as the silicon- and germanium-based devices.

4.4.2 Cryogenic Devices

A relatively recent advance in X-ray spectroscopy is the development of what are often called *cryogenic detectors*. These devices convert incident X-rays into thermal-acoustic quanta (phonons) or superconductor quasiparticles (broken Cooper pairs). The energy w required to create such a particle is quite small ($10^{-5}-10^{-3}$ eV) so that extremely high resolving powers are possible in principle (the number of particles is large and thus the relative uncertainty in their number is small). To achieve the optimal resolution of such devices (theoretically ΔE can be as small as ~ 1 eV), various noise sources must be suppressed by operating at extremely low temperatures ($T < 100$ mK typically) and, hence, the generic name for such detectors – *cryogenic devices*. Although cryogenic detectors are not yet available commercially, they hold significant promise for future applications.

Two basic classes of cryogenic detectors have been developed in recent years: calorimeters and quasiparticle detectors. The calorimeters measure the temperature in a low-temperature substrate due to the absorption of a single X-ray photon (hence, they are phonon detectors). The quasiparticle detectors most often discussed for X-ray spectroscopy are superconducting tunnel junctions (STJs) for which the mean particle energies are 0.001 eV. The quasiparticles are detected by their quantum-mechanical current tunneling through a thin oxide layer. Although the STJ devices are promising, they have not achieved, to date, the resolving power of the calorimeters and thus are not discussed further (see (Labov and Young, 1993) for some recent articles on this topic).

The microcalorimeter has undergone extensive development over the past decade. Resolutions as high as $\Delta E \sim 2$ eV at 1.5 keV (Newbury et al., 2002) have been achieved. Thus, the microcalorimeter has a resolving power that is 10–50 times better than that of the best semiconductor devices. A schematic diagram of a generic calorimeter is shown in Figure 4.12. Basically, a calorimeter absorbs an incoming X-ray and converts the X-ray energy to heat (phonons), which raises the temperature of a cryogenically cooled substrate. The increase in temperature is measured by a thermometer connected to the absorber. After a period of time, the equilibrium temperature is restored through the contact of the substrate with a cold bath. The increase in temperature is proportional to the energy of the X-ray.

The energy resolution of an ideal X-ray calorimeter is given as (Moseley, Mather

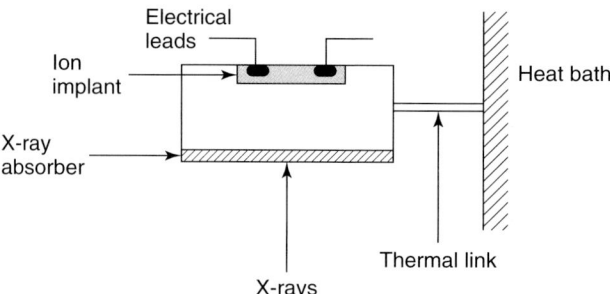

Fig. 4.12 Schematic diagram indicating the operation of an X-ray calorimeter. (Holt, 1987.)

and McCammon, 1984)

$$\Delta E = 2.35 \xi \sqrt{kT^2 C} \quad (4.10)$$

where T is the equilibrium temperature, C is the heat capacity of the device, k is Boltzmann's constant, and ξ is a nearly constant parameter with magnitude ~ 1 to 3. The parameter ξ is derived from the temperature variations of the heat capacity and thermal conductivity of the device, and the properties of the thermometer (Kelley et al., 1988). For small detectors ($0.5 \times 0.5 \times 0.1$ mm^3 are typical dimensions), cooled to ~ 0.1 K, Eq. (4.10) yields $\Delta E \sim 1$ eV, but, in practice, additional noise terms in the thermometer limit the current actual performance of calorimeters to ~ 7 eV. Note that the size of the detector is an important issue since ΔE scales as $C^{1/2}$. Also note that heat capacity often scales as T^3, so that the operating temperature is also of great importance; the best calorimeters built to date operate at about 0.1 K.

The obvious advantages of an X-ray calorimeter are as follows:

1. its grasp of a wide energy range (since it is nondispersive) and
2. its extremely high resolving power.

The resolution ΔE is essentially a constant; thus, the resolving power ($E/\Delta E$) varies as E varies. Figure 4.13 shows the resolution E_{res} as a function of the X-ray energy E_ν for several types of X-ray spectrometers. If digital processing of microcalorimeter signals is performed, energy resolutions comparable to those of the Bragg crystal devices in some spectral ranges (Newbury et al., 2002) can be obtained.

The practical disadvantages of the X-ray calorimeter are as follows:

1. its small size (although mosaic detectors have been designed) and
2. the need to have a complex cryogenic system for the detector to function optimally.

The latter is a particular problem for space applications in which weight and risk are always a matter of concern. For typical designs, a large Dewar flask of liquid helium and an additional refrigeration stage are required.

X-ray microcalorimeters may not be the optimal choice of spectrometer in an environment in which an extremely high counting rate is expected, because of the relatively long time required to process each event (several milliseconds).

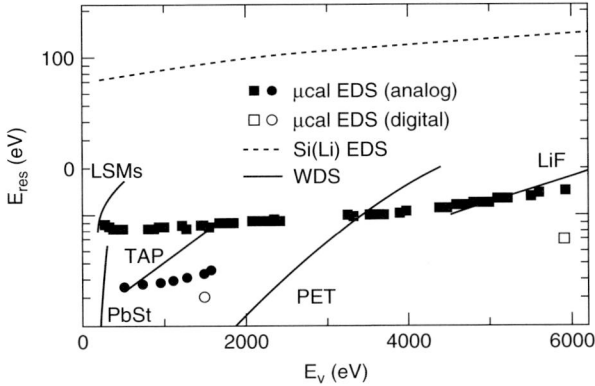

Fig. 4.13 Energy resolution (FWHM) of different types of X-ray spectrometers: microcalorimeter (energy-dispersive spectrometer (μcal EDS)) with analog and digital data processing, lithium-drifted silicon detector (Si(Li) EDS), and wavelength-dispersive spectrometer (WDS). Dispersive elements for WDS are LSMs, thallium acid phthalate (TAP), pentaerythrol (PET), lithium fluoride (LiF), and PbSt (lead stearate). (Newbury et al., 2002.)

4.5 Applications

X-ray spectrometers are used in virtually every area of pure and applied science. Extensive discussion of specific applications can be found in the references following this chapter. In a general sense, however, X-ray spectrometers are employed

- to measure the properties (such as the temperatures, ionization states, and elemental compositions) of hot ($>10^6$ K) plasmas, both cosmic and laboratory (Silver and Kahn, 1993), where in extreme conditions time evolution and spatial distribution of laboratory plasma parameters have been obtained (Golovkin et al., 2002);
- to determine compositions of various materials (qualitative and quantitative analysis via the monitoring of characteristic lines from X-ray fluorescence or direct bombardment with electrons or other charged particles; see, e.g., (Jenkins, 1988)); and
- to measure and interpret the absorption spectra of various solids at high resolution in order to determine the distribution of electrons. The study of X-ray absorption fine structure (XAFS) has blossomed in recent years with the advent of bright, monochromatic X-ray beams available from synchrotrons (Koningsberger and Prins, 1988). Ultrafast time-resolved X-ray absorption spectroscopy is a nascent technique that permits study of photoinduced processes on many systems down to the subfemtosecond time domain (Bressler and Chergui, 2004).

Depending on the application, some of the various spectrometers discussed earlier may be more appropriate than others. For example, semiconductor devices, with their moderate resolving powers and relatively broadband energy response, are often used for X-ray fluorescence analysis (see Figure 4.11), since the fine structure of the spectral features is not generally an issue. When extremely high resolution is required, at least at the lower energies, crystal spectrometers are still the instruments of choice (the spectrum in Figure 4.3, for example, would be extremely difficult to unravel with a semiconductor spectrometer). When low flux (or short exposure times) is an issue and/or a significant energy band is desired, the narrow bandpass of the crystal spectrometers argues in favor of a nondispersive device (or a grating spectrometer in which the entire dispersed spectrum can be read off simultaneously with an imaging detector). The desire for a broad bandpass and high resolution may lead the user to a calorimeter, in spite of its size, weight, and complexity. For some applications, the expected counting rate may be so high as to rule out some of the energy-dispersive detectors. For example, in many synchrotron applications, the detector of choice may be a photodiode (which measures current and not individual photons), and a dispersive device will be used as the spectrometer. Besides these complex and high-resolution experiments at synchrotron radiation centers, portable equipment for X-ray fluorescence analysis has been developed. The technological progress is based upon on miniature and air-cooled X-ray tubes, often polycapillary optics, thermoelectrically cooled X-ray detectors, multichannel analyzers, and dedicated software (Tsuji, Injuk and van Grieken, 2004). All components have small weight and are incorporated in a pistol-like housing. This portable energy-dispersive X-ray fluorescence equipment has been employed in many applications, such as archaeometry, analysis of lead in paint, environmental analysis, analysis of industrial alloys, soil analysis even on Mars, and so on.

Another factor that affects the choice of spectrometer is the nature of the optical system of which it is a component. For example, the small size of a calorimeter detector requires a tightly focused beam. Flat dispersive spectrometers (as in Figure 4.4) must receive all the radiation at the same incident angle; otherwise, the dispersed image will be blurred (thus effectively reducing the resolving power). As noted in Section 3.1.3, for some applications, crystals and gratings are curved so that the incident radiation will strike the crystal at nearly the same angle.

Acknowledgments

Thomas H. Markert is grateful to Richard Aucoin, Daniel Dewey, Keith Gendreau, George Clark, and Una Hwang for helpful discussions.

Eckhart Förster wishes to thank Ingo Uschmann, Oldrich Renner, Konrad Goetz, and Michael Wendt for their critical remarks and is also grateful to Jana Brusberg and Richard Hutcheon for their help in the preparation of the chapter.

Glossary

Blazed Gratings: Gratings for which the bar, groove, or line shape is structured so as to enhance reflection into a particular order.

Cryogenic Devices: A relatively new class of X-ray spectrometers that operates at very low temperatures ($T < 2$ K typically). Such devices are capable of extremely high resolving powers ($\Delta E \sim$ 7 eV has been demonstrated at 5.9 keV for at least one such device).

Diffraction: Modification of the intensity and/or phase of an electromagnetic wave by the presence of an object (such as a slit, hole, edge, etc.) in the path of the wave.

Dispersive Spectrometer: An X-ray spectrometer that employs wavelength dispersion. X-rays are dispersed (i.e., spatially separated) by the effects of diffraction. Dispersive spectrometers are sometimes called *wavelength-dispersive spectrometers*. The two classes of dispersive spectrometers are Bragg devices and diffraction gratings.

Escape Peak: For monochromatic incident X-ray photons with an energy E_x smaller than that of the absorption edge of detector atoms, the pulse height output is proportional to E_x. However, when the energy E_x of the monochromatic incident X-ray photons exceeds the absorption edge of the detector atoms, the output may contain two pulse height distributions. The additional or escape peak has a mean pulse height proportional to the difference between the energies of the incident photons E_x and the escaping (from active detector volume) photons $E_{K\alpha}$.

Monochromators: Devices used for producing radiation at a single wavelength (or, more realistically, in a narrow range of wavelengths). The elements of dispersive spectrometers (crystals and gratings) can also be used as X-ray monochromators.

Multilayers: Synthetic crystals formed by sputtering or evaporating alternating layers of high-Z and low-Z materials onto a substrate. Multilayers can be used as spectrometers or monochromators and can be a part of an X-ray optics system to enhance the performance of the system over a narrow band and is also called *layered synthetic microstructures* or *LSMs*.

Nondispersive Spectrometer: An X-ray spectrometer that does not employ wavelength dispersion. Such devices generally operate by converting the photon energy into some other sort of particle or quasiparticle (e.g., electron–ion pairs or phonons). Nondispersive spectrometers are often called *energy-dispersive spectrometers*.

Resolving Power: The resolving power of a spectrometer (at an energy $E = hc/\lambda$) is usually defined as the ratio of the energy (or wavelength) of interest to the width of the response function of the spectrometer to a monochromatic X-ray line, that is, $R(E) = E/\Delta E = \lambda/\Delta\lambda$. The width of the line is usually chosen to be the FWHM of the spectral response function. Note that some authors invert this definition of resolving power so that it equals $\Delta E/E = \Delta\lambda/\lambda$ and often quote it as a percentage.

Rocking Curve: In crystal spectrometers, the rocking curve is a measure of the spectral resolving power of the crystal, independent of any geometric effects introduced by the geometry of the spectrometer. Specifically, it is the FWHM in degrees of the response of the crystal (as it is rotated or rocked) to a monochromatic X-ray beam.

Rowland Circle Configuration: This configuration is common in focusing X-ray spectrometers. The diverging X-ray source, the diffraction grating, and the detector all lie on a circle (the Rowland circle, after Henry Rowland) that has a diameter equal to the radius of curvature of the curved diffractor. Such a configuration gives one-dimensional focusing and disperses the spectrum along the circle. A variation, using a curved crystal rather than a grating, allows for imaging of small fields along the circle.

References

Agarwal, B.K. (1991) *X-ray Spectroscopy*, Springer, Berlin.

Attwood, D. (1999) *Soft X-rays and Extreme Ultraviolet Radiation: Principles and Applications*, Cambridge University Press, Cambridge.

Authier, A. (2001) *Dynamical Theory of X-ray Diffraction*, Oxford University Press, Oxford.

Bargheer, M., Zhavoronkov, N., Bruch, R., Legall, H., Stiel, H., Woerner, M., and Elsaesser, T. (2005) *Appl. Phys. B*, **80**, 715–719.

Becker, P. (2001) *Rep. Prog. Phys.*, **64**, 1945.

Bragg, W.H. and Bragg, W.L. (1913) *Proc. Phys. Soc., London A*, **88**, 428–438.

Bressler, C. and Chergui, M. (2004) *Chem. Rev.*, **104**, 1781–1812.

Compton, A.H. and Allison, S.K. (1967) *X-rays in Theory and Experiment*, 2nd edn, Van Nostrand, Princeton.

Fano, U. (1947) *Phys. Rev.*, **72**, 26.

Fraser, G.W. (1989) *X-ray Detectors in Astronomy*, Cambridge University Press, Cambridge, UK.

Gatti, E. and Rehak, P. (1984) *Nucl. Instrum. Methods, A*, **225**, 608–614.

Golovkin, I., Mancini, R., Louis, S., Ochi, Y., Fujita, K., Nishimura, H., Shirga, H., Miyanaga, N., Azechi, H., Butzbach, R., Uschmann, I., Förster, E., Delettrez, J., Koch, J., Lee, R.W., and Klein, L. (2002) *Phys. Rev. Lett.*, **88**, 0450021–0450024.

Golub, L., Zirin, H., and Wang, H. (1994) *Solar Phys.*, **153**, 179–198.

Gursky, H. and Zehnpfennig, T. (1966) *Appl. Opt.*, **5**, 875–876.

Höll, A. et al. (2007) *High Energy Density Phys.*, **3**, 120–130.

Holt, S.S. (1987) *Astrophys. Lett. Commun.*, **26**, 35.

James, R.W. (1982) *The Optical Principles of the Diffraction of X-rays*, Ox Bow Press, Woodbridge, CT.

Janssens, K., Adams, F., and Rindby, A. (2000) *Microscopic X-ray Fluorescence Analysis*, John Wiley & Sons, Ltd, Chichester.

Jasny, J. et al. (1994) *Rev. Sci. Instrum.*, **65**, 1631.

Jenkins, R. (1988) *X-ray Fluorescence Spectrometry*, John Wiley & Sons, Ltd, New York.

Kelley, R.L., Holt, S.S., Madejski, G.M., Moseley, S.H., Schoelkopf, R.J., Szymkowiak, A.E., McCammon, D., Edwards, B., Juda, M., Skinner, M., and Zhang, J. (1988) in *X-ray Instrumentation in Astronomy II, SPIE Proceedings No. 982* (ed. L. Golub), SPIE, Bellingham, WA, p. 219.

Klöpfel, D., Hölzer, G., Förster, E., and Beiersdörfer, P. (1997) *Rev. Sci. Instrum.*, **68**, 3669.

Knoll, G.F. (1989) *Radiation Detection and Measurement*, 2nd edn, John Wiley & Sons, Ltd, New York.

Koningsberger, D.C. and Prins, R. (eds) (1988) *X-ray Absorption: Principles, Applications, Techniques of EXAFS, SEXAFS, and XANES*, John Wiley & Sons, Ltd, New York.

Labov, S.E. and Young, B.A. (1993) *J. Low Temp. Phys.*, **93**, 185–858.

Love, G. (2002) *Microchim. Acta*, **138**, 115.

Michette, A.G. and Buckley, C.J. (eds) (1993) *X-ray Science and Technology*, Institute of Physics Publishing, Bristol, UK.

Mißalla, T., Uschmann, I., Förster, E., Jenke, G., and von der Linde, D. (1999) *Rev. Sci. Instrum.*, **70**, 1288–1299.

Moseley, S.H., Mather, J.C., and McCammon, D. (1984) *J. Appl. Phys.*, **56**, 1257.

Newbury, D., Wollman, D., Nam, S.W., Hilton, G., Irwin, K., Small, J., and Martinis, J. (2002) *Microchim. Acta*, **138**, 265.

Pinsker, Z.G. (1978) *Dynamical Scattering of X-rays in Crystals*, Springer, Berlin.

Ramana Murthy, P.V. and Wolfendale, A.W. (1993) *Gamma–Ray Astronomy*, Cambridge University Press, Cambridge, MA.

Renner, O., Kopecky, M., Krousky, E., Förster, E., Mißalla, T., and Wark, J.S. (1994) *Laser Particle Beams*, **12**(3), 539.

Sauli, F. (ed) (1992) *Instrumentation in High Energy Physics*, World Scientific, Singapore.

Schattenburg, M.L., Canizares, C.R., Dewey, D., Flanagan, K.A., Levine, A.M., Lum, K.S., Manikkalingam, R., and Markert, T.H. (1991) *Opt. Eng.*, **30**, 1590–1600.

Silver, E. and Kahn, S. (eds) (1993) *UV and X-ray Spectroscopy of Laboratory and Astrophysical Plasmas*, Cambridge University Press, Cambridge, UK.

Spiller, E. (1994) *Soft X-ray Optics*, SPIE Optical Engineering Press, Bellingham, WA.

Strüder, L., Fiorini, C., Gatti, E., Hartmann, R., Holl, P., Krause, N., Lechner, P., Longoni, A., Lutz, G., Kemmer, J., Meidinger, M., Popp, M., Soltau, H., Weber, U., and von Zanthier, C. (1998) *J. Synchrotron Radiat.*, **5**, 268.

Tsuji, K., Injuk, J., and van Grieken, R. (2004) *X-ray Spectrometry: Recent Technological Advances*, John Wiley & Sons, Ltd, New York.

Uschmann, I., Malgrange, C., and Förster, E. (1997) *J. Appl. Crystallogr.*, **30**, 4554.

Further Reading

Bertin, E.P. (1975) *Principles and Practice of X-ray Spectrometric Analysis*, 2nd edn, Plenum, New York.

Dyson, N.A. (1990) *X-rays in Atomic and Nuclear Physics*, 2nd edn, Cambridge University Press, Cambridge, UK.

Hows, M.J. and Morgan, D.V. (eds) (1979) *Charge-Coupled Devices and Systems*, John Wiley & Sons, Ltd, Chichester.

Russ, J.C. (1984) *Fundamentals of Energy Dispersive X-ray Analysis*, Butterworths, London.

VanGrieken, R.E. and Markowicz, A.A. (eds) (1993) *Handbook of X-ray Spectroscopy: Methods and Techniques*, Marcel Dekker, New York.

5
Positron Spectroscopy

Paul G. Coleman

5.1	**Introduction**	**117**
5.2	**Fundamentals**	**118**
5.2.1	Annihilation of Free Positrons and Positronium	118
5.2.2	The Fate of Positrons in Condensed Matter	119
5.2.2.1	Positron Backscattering	119
5.2.2.2	Positron Implantation	120
5.2.2.3	Positron Diffusion	120
5.2.2.4	Positron Trapping	121
5.2.2.5	Positron-Surface Interactions	121
5.2.2.6	Positron Annihilation: Observables	123
5.2.2.7	Positronium Annihilation	124
5.3	**Experimental Methods**	**125**
5.3.1	Sources of Positrons	125
5.3.2	Lifetime Spectroscopy	126
5.3.2.1	Experimental Lifetime Systems	126
5.3.3	Angular Correlation of Annihilation Radiation	127
5.3.4	Doppler Broadening of Annihilation Radiation	129
5.3.4.1	Two-Detector DBAR	131
5.3.4.2	Age-Momentum Correlation (AMOC)	131
5.3.5	Positron Beams	132
5.3.5.1	Laboratory-Based Beams	132
5.3.5.2	High-Intensity Positron Beams	134
5.3.5.3	Beam Bunching	134
5.3.5.4	Positron Microbeams	134
5.3.5.5	Polarized Positron Beams	135
5.3.5.6	MeV Positron Beams	135
5.3.5.7	Trap-Based Beams	136
5.3.5.8	Positron-Beam-Based Spectroscopies	136
5.4	**Examples of Research Using Positron Spectroscopies**	**139**
5.4.1	Vacancy-Type Defects	139
5.4.2	Structural Changes	140

Encyclopedia of Applied Spectroscopy. Edited by David L. Andrews.
Copyright © 2009 WILEY-VCH Verlag GmbH & Co. KGaA, Weinheim
ISBN: 978-3-527-40773-6

5.4.3 Nanoparticles 141
5.4.4 Interfaces 142
5.4.5 Fermi Surfaces 142
5.4.6 Nanoporous Materials and Open Volumes in Polymers 143
5.4.7 Surfaces 144
5.4.8 Positron Microscopy 146
5.4.9 Positron and Positronium Chemistry 146
Acknowledgments 148
Glossary 148
References 148
Further Reading 150

5.1
Introduction

The history of positron annihilation spectroscopy is, in many ways, an archetypal story of scientific development, from fundamental physics to applications of industrial importance. Predicted by Dirac (1930) and discovered by Anderson (1932), and found to be emitted from artificially produced isotopes in the same decade (Joliot-Curie and Joliot-Curie, 1934), the use of positrons as a subatomic probe of matter was delayed for over a decade by world events.

In 1942, the first fundamental measurement on γ-rays emitted as a result of the annihilation of positrons in solids was performed; Behringer and Montgomery (1942) found that the angle between the two γ-rays which almost always follow annihilation events did not deviate from 180° by more than 15 minutes of arc. DeBenedetti, Cowan and Konneker (1949) were the first to demonstrate that this angle deviated from 180° by an amount linked to the electron momentum and soon afterward published their vision of a new spectroscopy of electron momentum distributions – *angular correlation of annihilation radiation* (now termed *ACAR*) (DeBenedetti *et al.*, 1950). DeBenedetti was joined in his pioneering work by Page and Heinberg (1956), Stewart (1957), and Berko and Plaskett (1958).

Shearer and Deutsch (1949) pioneered the measurement of the lifetimes of positrons in gases, and it was this technique that led to the experimental discovery (Deutsch, 1951) of the positron–electron bound state, positronium (theoretically considered and named *electrum* by Mohorovičić (1934) and renamed *positronium* by Ruark (1945)). Extension of positron lifetime spectroscopy to liquids and solids required faster timing techniques, and these were developed by Bell and Graham (1953) and exploited soon thereafter by Landes, Berko and Zuchelli (1956) and others in the study of structural changes on the atomic scale.

In the 1960s, with the development of more widely available equipment, the level of activity and sophistication increased in both lifetime and angular correlation spectroscopies – for example, yielding new information on positron slowing down and annihilation in gases (Osmon, 1965) and in metals (Weisberg and Berko, 1967). In this decade, experimental problems such as the effect of the radioactive positron source on results, the analysis of data containing

Encyclopedia of Applied Spectroscopy. Edited by David L. Andrews.
Copyright © 2009 WILEY-VCH Verlag GmbH & Co. KGaA, Weinheim
ISBN: 978-3-527-40773-6

multiple lifetime components, and struggles with electronic stability, were still being worked on. In the late 1960s, the role of positron trapping in atomic-scale open-volume point defects in solids was first realized (and explained why different laboratories often produced conflicting results for notionally the same sample). Rather than confusing the issue and burying the technique, this realization gave positron spectroscopy a new lease of life; positrons are so efficiently trapped by these defects that they soon became a sensitive probe of defect structures and open volumes in a wide variety of solids (Petersen, Thrane and Cotterill, 1974).

Further advances in instrumentation, coupled with the traditional resourcefulness and inventiveness which has characterized the positron research community over the years, led to the application of Doppler broadening spectroscopy in the 1970s – a high count-rate, lower-resolution measure of electron momentum densities – which quickly grew as its efficacy and scope for application were realized (Maier et al., 1979). In the same decade, ACAR was enhanced by the development of two-dimensional measurements, first by using an array of discrete photon detectors (Mader et al., 1976) and then by using position-sensitive gamma cameras (West, Mayers and Walters, 1981) or multiwire proportional chambers (Jeavons et al., 1978).

One of the most significant developments in positron techniques also happened in the early 1970s, with the first usable laboratory-based beams of controllable low energy (electronvolts to kilo electronvolts) (Canter et al., 1972). After a few years of application only in atomic physics (i.e., in positron scattering), these beams were applied to solids and, after a number of groundbreaking fundamental studies of positron-surface interactions (Mills, Platzman and Brown, 1978), have been used widely since the 1980s to study surface and near-surface characteristics, thin films, and interfaces (e.g., Triftshäuser and Kögel, 1982). Beam systems – both laboratory and facility based – are now available, delivering positrons with energies controllable between 0.02 eV and mega electronvolts at intensities up to almost 10^9 per second.

We focus on positron spectroscopy of condensed matter; much important fundamental research has been and continues to be performed in atomic physics, but this activity lies outside the scope of this chapter. For an excellent overview of this work, the reader is directed to the book by Charlton and Humberston (2001).

5.2
Fundamentals

5.2.1
Annihilation of Free Positrons and Positronium

The annihilation of a positron by its antiparticle, the electron, both stationary, can theoretically result in the emission of any number n of γ-rays if n is greater than 1. The γ-rays take away the rest energy $2mc^2$, where m is the mass of each particle and c is the speed of light. Zero and single γ-ray emission are forbidden for an isolated positron–electron pair at rest because both energy and momentum have to be conserved; both can theoretically occur in the presence of a third body, however. While the emission of two or more γ-photons satisfies both energy and momentum conservation, the probability of n photon emission decreases sharply as n increases (and as the number of vertices on the corresponding Feynman diagrams (Charlton and Humberston, 2001)). For example, Ore and Powell (1949) calculated that the probability of three-γ emission is about 370 times less likely than that for two-γ emission, and the contributions to experimental studies of decays with n greater than 3 are practically negligible. Consequently, almost all annihilation events involving free (or quasi-free) positrons and electrons result in the emission of two γ-rays of energy mc^2 (511 keV) in opposite directions (to conserve momentum).

A positron and electron can exist in the bound hydrogen-like system, called *positronium* (*Ps*). Because the reduced mass is half that of the electron in hydrogen, energy levels are half those in hydrogen (e.g., the ground-state binding energy is 6.8 eV) and interparticle separations are doubled. Ps annihilation events are governed by a selection rule resulting from charge parity (CP) invariance. The parity of the γ-photons is $(-1)^n$, and for ground-state Ps it is $(-1)^S$, where S is the total spin angular momentum of the Ps atom. If the positron and electron in the Ps have opposing spins (the singlet state, *para*-Ps or *p*-Ps) then the total $S = 0$ and consequently n has to be an even number. Conversely, if the positron and electron spins are parallel and $S = 1$ (the triplet state, *ortho*-Ps or *o*-Ps) then n has to be odd. Following the arguments in the previous paragraph, this means that *p*-Ps principally decays into two 511-keV antiparallel γ-rays, whereas *o*-Ps decays into three γ-rays whose total energy is $2\,mc^2$ or 1022 keV. The distribution of γ-energies from *o*-Ps decay was calculated by Ore and Powell (1949) and measured by Chang, Tang and Xi (1985), indicating that the majority of *o*-Ps decays involve the emission of two γ-rays having energies that are a large fraction of mc^2, traveling in roughly opposite directions, with a third low-energy γ-photon emitted at some angle between them. Ore and Powell also computed that the mean lifetime of *o*-Ps in vacuo is about 140 nanoseconds, over 1100 times longer than the 125 picoseconds mean life of *p*-Ps (Wheeler, 1946).

5.2.2
The Fate of Positrons in Condensed Matter

5.2.2.1 Positron Backscattering

A significant fraction of positrons incident upon a target material, whether from a radioactive source or in a monoenergetic beam, are turned around in a few collisions and leave the target with reduced energies. Measurements have shown that there is significant quasi-elastic backscattering and that the angular distribution of backscattered positrons is peaked at angles around the surface normal. Backscatter coefficients η_+ range from a few percent to almost 50%, and depend principally on the incident positron energy and the atomic number of the target material (increasing with both) (MacKenzie et al., 1973; Coleman et al., 1992). The consequences of backscattering have to be considered in experiments involving the detection of annihilation γ-radiation, where the detection of γ-radiation from the decay of backscattered positrons at sites other than the intended target can corrupt data.

5.2.2.2 Positron Implantation

Energetic positrons entering a sample material lose energy via electronic collisions, reaching electronvolt energies in a fraction of a picosecond and then thermalizing in a further time period from one to a few picoseconds, depending on the material. This second phase is dominated by phonon interactions, and, here, the shortest thermalization times are in metals and small band-gap semiconductors; in insulators and wide band-gap semiconductors several picoseconds may be taken for thermalization, and in extreme cases positrons may not reach thermal equilibrium before annihilation. This last scenario is also possible for monoenergetic positrons implanted at very low energies (i.e., below about 1 keV), which may retain $\sim$0.1–1 eV when encountering the sample surface and thus are able to pass through or interact with the surface in ways different from those positrons that have fully thermalized.

Thus, we define positron implantation (or penetration) depth as that at which an incident positron reaches thermal equilibrium. Positrons emitted from the nuclei of a radioisotope decaying via positive β-decay penetrate to depths z below the surface of condensed matter targets whose distribution is traditionally given by the implantation profile $P(z) = \alpha \exp(-\alpha z)$, with $\alpha(\text{cm}) \approx 16 \rho E_m^{-1.4}$ (ρ = density in grams per cubic centimeter, E_m = maximum β-positron energy in kilo electronvolts). This means that β-positrons can penetrate to depths of $\sim$1 mm, and are thus used to study the bulk properties of materials. A recent experimental study (Foster et al., 2007) has shown that the exponential model provides a reasonable description of β-positron implantation except at small depths, where it underestimates $P(z)$ by $\sim$10%, and very close to the surface, where $P(z)$ should tend to zero rather than its maximum value.

If the incident positrons impinge on a target normally with an energy E, however, their implantation profile has been generally taken to be well described by a Gaussian derivative:

$$P(z, E) = 2 \left(\frac{z}{z_0^2} \right) \exp \left(\frac{-z^2}{z_0^2} \right) \quad (5.1)$$

where the parameter z_0 (in nanometers) is usually taken to be related to E and to the material density ρ (in grams per cubic centimeter) by the simple expression $(40/\rho) E^{1.6}$. z_0 is 11% larger than the mean implantation depth. While the form for $P(z, E)$ has been shown to be adequate in most cases, the prefactor and the energy exponent have been found to be material dependent (Baker et al., 1991).

5.2.2.3 Positron Diffusion

The diffusive motion undergone by thermalized positrons in condensed matter may be principally considered to involve isotropic elastic scattering from acoustic phonons, and may be described by the diffusion equation because, although the positron is a quantum entity, its wavelength and mean free path (both approximately nanometers) are small compared with typical total distances traveled before annihilation. The positron diffusivity D_+ is proportional to its mobility η_+, with the constant of proportionality being (kT/e). D_+ varies with temperature as $T^{1/2}$ for scattering off acoustic phonons, as $T^{-3/2}$ for optical phonons, and $T^{3/2}$ for neutral impurities; however, the latter two dependencies are only relevant for certain materials and at very high or very low temperatures. At room temperature D_+ has values of a few square centimeters per second for most metals and semiconductors. The diffusion length of positrons $L_+ = (D_+\tau)^{1/2}$, where τ is the mean lifetime of positrons in a material (having values in the range ~100–200 picoseconds); typical values of L_+ are ~100 nm in metals to 270 nm in silicon.

5.2.2.4 Positron Trapping

Once considered a problem in positron measurements, the propensity for positrons to be trapped efficiently by open-volume point defects – vacancies, vacancy clusters, and voids – and at shallow trapping sites such as at a negatively charged impurity – is now the mainstay of applied positron annihilation spectroscopy. Although the classical picture of positrons being attracted to a region of lower than average positive charge is commonly used to explain positron trapping, the diffusing positron is a quantum particle and trapping should be considered to be quantum localization. Positrons provide a unique method for nondestructive characterization of such defects at concentrations as low as 10^{-7} per atom, saturating at ~10^{-4} per atom.

There are two main models describing positron trapping. For small point defects such as vacancies, the trapping is *transition-limited* – that is, dependent upon the probability of trapping once a defect is encountered; this is described by the specific trapping coefficient ν, which has typical values in the range of 10^{14}–10^{15} per second. The total trapping rate κ is then νC, where C is the defect concentration per atom and ν for semiconductors has a temperature dependence, which can be used to identify the defect type. The second trapping model is *diffusion-limited* – that is, determined by the probability of finding the trapping sites which, once found, have a trapping probability of unity. The total trapping rate is then $\kappa = 4\pi R D_+ N$, where R (centimeters) and N (cubic centimeters) are the radius and number density of the defects, respectively, and D_+ is the positron diffusion constant in square centimeters per second.

The charge state of open-volume point defects in semiconductors and insulators can play an important role in positron trapping. The specific trapping rate ν is very small for positively charged defects, whereas neutral sites are effective positron traps and negative charge can increase ν by a significant factor. While open-volume defects are deep traps for positrons, shallow trapping can occur around negatively charged vacancy or impurity defects – this is usually seen at low temperatures and identified by the temperature dependence of the trapping probability (Saarinen et al., 1997).

Thermally induced detrapping from defect sites can occur, depending on the Boltzmann factor $\exp(-E_b/kT)$, where E_b is the binding energy of the positron in the trap; therefore, for example, for deep traps – that is, those for which E_b is a few electronvolts – detrapping can be considered negligible.

5.2.2.5 Positron-Surface Interactions

If a positron encounters the surface of a material sample after implantation it can suffer a variety of fates, all of which have found application in various experimental studies. We shall, for the moment, assume that the positron has been thermalized before reaching the surface, and not consider those backscattered positrons that penetrate the surface without any significant interaction with it.

There are three main channels open to the positron on encountering a surface: (i) trapping in the surface potential well, (ii) binding with a surface electron and leaving as Ps, and (iii) being reemitted from the surface by a negative positron work function φ_+. For those materials for which φ_+ is positive, channel (iii) is not open.

Surface Trapping The potential encountered by a positron may be considered for our purposes to be a combination of the attractive image potential, which falls off as the reciprocal of distance from the surface, and the repulsive dipole potential, which falls off significantly more quickly above the surface (i.e., within a few angstroms). The combination of these two potentials thus creates a deep well above the surface in which positrons are efficiently trapped. On an atomically clean perfect surface, the trapped positron can be used as a probe of electronic surface states. If the temperature of the material is increased then the trapped positron can be desorbed, but only via the formation of thermal-energy Ps, which supplies its 6.8-eV binding energy; this, however, is an excellent source of low-energy Ps atoms for spectroscopic measurements.

Positronium Formation For metals and semiconductors, the surface is the only site where the electron density is low enough to permit Ps formation. The formation potential for Ps at surfaces, essentially the energy possessed by the emitted Ps atom, is the balance between the sum of the electron and positron work functions – that is, the energy required to remove both particles independently from the solid – and the 6.8 eV gained by forming the bound state. In the early days of positron-surface experimentation, Ps formation was used as a signature that the positron had diffused back to the surface; however, the probability of forming Ps is critically dependent on surface conditions, which can have a positive or negative influence on measurements. Measurement of the velocity spectra of emitted Ps was also used as a probe of electronic surface states. Of great fundamental importance is the fact that small fractions of the Ps emitted can be in an excited state or as the negative ion Ps^-.

Positron Reemission The sum of the correlation and dipole potentials outlined in Section 5.2.2.5, Surface Trapping (the correlation potential is the asymptotic level of the image potential inside the material) can be positive or negative – that is, positrons are either attracted or repelled by the surface, depending on their relative magnitudes. The difference between the resultant total potential within a solid and the reference vacuum level far from the surface is the positron work function φ_+. For a number of solids the dipole potential is greater than the correlation potential and φ_+ is negative – that is, thermalized positrons are emitted from the surface with an energy of $\sim\varphi_+$. Negative φ_+ values for metals range from close to zero to ~ 3 eV and, like their electron equivalent, are dependent on surface composition and structure. For a perfect surface, the emission is normal to the surface, smeared only by thermal effects; for a "real" surface, the emitted positrons may undergo inelastic collisions above the surface that lower their average energy and increase their angular spread; if enough energy is lost in such a collision, then the positron may fall into the surface well and be trapped. Thus, there is a correlation between the magnitude of φ_+ and the fraction of positrons reemitted from a surface.

A small fraction of positrons from a radioactive source incident on a solid with negative φ_+ possess energies low enough (i.e., below ~10 keV) to be thermalized, return to the surface, and subsequently be reemitted with work function energies. This is the basis of the generation of monoenergetic positron beams, which have been widely used to study surface and near-surface phenomena for the past 25 years.

Surface Branching Ratios The relative probabilities of these three main channels open to the positron are described by the branching ratios ε. For a solid with a positron work function of ~-1 eV, the three branching ratios are in the approximate ratio of $1:1:1$. For surfaces with a positive φ_+, approximately half the positrons form Ps and half fall into the surface trap at room temperature.

Nonthermalized Positrons If the positrons incident on a surface have energies below about 1 keV, there is a significant probability that they will return to the surface without being fully thermalized – that is, with residual energies of several hundred milli electronvolts. These epithermal positrons interact with the surface with rather different branching ratios; they can more readily overcome surface trapping and leave as positrons with energies above φ_+, and can pick up an electron and form Ps, which again leaves with higher energy than that formed by the process outlined in Section 5.2.2.5, Positronium Formation.

On encountering "real", dirty surfaces – for example, oxide-covered – the branching ratios can be hugely different from those for a clean perfect surface and, in addition, there can be significant, sometimes overwhelming, trapping at surface defects.

We next focus on the annihilation of thermal-energy positrons with electrons, and of Ps-like states, in condensed matter. Other aspects of positron and Ps interactions in solids and liquids are discussed in later sections.

5.2.2.6 Positron Annihilation: Observables

The cross section σ for annihilation is inversely proportional to positron speed v, but the annihilation rate is proportional to σv. Consequently, although annihilation during positron slowing down is possible, has been observed, and may play a role in some positron experiments (Weber *et al.*, 1999), the overwhelming fraction of annihilation events occur after thermalization, when positrons spend most of their time in diffusive motion.

Although positrons rapidly (i.e., in approximately picoseconds) thermalize in condensed matter, and so can be thought of as quasi-stationary, the electrons that annihilate them are decidedly not. Conduction electrons at the top of the Fermi sea in metals, for example, have several electronvolt energies and bound electrons can have kilo electronvolt energies. However, most positrons will perish at the hands of the lower energy conduction or valence electrons, both because of the time of interaction and because of their propensity to spend more time away from the positive ion cores.

If we initially treat the positron and electrons as independent particles (i.e., we ignore mutual interactions), then the probability $\rho(p)$ that a pair of γ-rays with total momentum p will result from the annihilation of a positron and an electron is given by

$$\rho(\boldsymbol{p}) = \left(\frac{r_0^2 c}{8\pi^2}\right) \sum_{i,j} n_i^+ n_j^- \left| \iiint \exp(-i\boldsymbol{pr}) \psi_i^+(r) \psi_j^-(r) d^3 r \right|^2 \quad (5.2)$$

where n_i^+ and n_j^- are the occupancies of the positron and electron states, the former being close to a delta function, and $\psi_i^+(r)$ and $\psi_j^-(r)$ are the positron and electron wave functions, respectively. If we assume that the positron momentum is relatively negligible, then p represents the electron momentum at the moment of annihilation.

Direct complete evaluation of the so-called momentum density $\rho(p)$ would be possible if one could perform a three-dimensional measurement of the momenta of the annihilation photons. In principle, one could achieve this by measuring the x and y components via the angles between the photons (see Section 5.3) and the z component via the Doppler shifts in the measured photon energies. In practice, we integrate over the z component and record the two-dimensional contour "map", that is, $\rho(p_x, p_y) = \int \rho(\boldsymbol{p}) dp_z$. This is the basis of the experimental technique known by the acronym 2D-ACAR (two-dimensional angular correlation of annihilation radiation).

If instead we integrate over both x and y components of p, then we arrive at the momentum density in the z direction, which gives rise to the Doppler broadening of annihilation radiation (DBAR), an extremely useful spectroscopy for following changes or differences in electronic structure: $\rho(p_z) = \iint \rho(\boldsymbol{p}) dp_x dp_y$.

Finally, the total annihilation rate is given by the integral of $\rho(p)$ over all three components of $\boldsymbol{p}$: $\lambda = \iiint \rho(\boldsymbol{p}) dp_x dp_y dp_z$. Reference to Eq. (5.2) shows that λ is proportional to electron density in the vicinity of the annihilated positron. The mean positron lifetime τ is the reciprocal of the annihilation rate λ, and measurement of the various possible lifetime components for positrons in condensed matter forms the basis of positron annihilation lifetime spectroscopy (PALS). The independent particle model leading to Eq. (5.2) is inadequate for the calculation of λ; $\rho(p)$ has to be multiplied by a positron–electron enhancement factor ε to take account of the interactions between positron and electrons and between the electrons in the material being studied (Kahana, 1963).

5.2.2.7 Positronium Annihilation

Although the quasi-stable bound-state Ps cannot form in metals or semiconductors, as a result of electron screening (except above the surface – as discussed later), it can exist in insulating solids and liquids.

A small variety of models have been proposed to describe Ps formation in such solids and liquids, and their applicability depends to a large extent on the material under study. The Ore Model (Ore, 1949) considers Ps formation to occur principally in the range of energies from $(E_i - B)$ to E_{ex}, where E_{ex} and E_i are the threshold energies for atomic excitation and ionization, respectively, and $B = 6.8$ eV is the ground-state binding energy of Ps. Above E_{ex} other interactions are assumed to outcompete Ps formation. A second prominent model is the Spur Model, due to Mogensen (1974), in which the positron binds to an electron released in a spur during the slowing-down process, under conditions of small relative momentum. An extension of this model is to consider the end of the positron track to be a "blob", rather than a spur (Stepanov and Byakov, 2002). A third, particularly considered with respect to Ps in polymers, involves the formation of Ps in open volumes or holes, the electron being picked up from the surface (Brandt, Berko and Walker, 1960); if the positron is not completely thermalized, then the Ps atom may undergo thermalizing collisions with hole walls.

There have been many groundbreaking fundamental measurements on Ps over the past 50 years, from precise measurements of vacuum decay rates to measurements on excited states, observations of the negative ion and hydride, and of Ps–Ps interactions en route to a Bose–Einstein Ps condensate (Cassidy and Mills, 2007). In Section 5.4, however, we briefly summarize only some of those applications in which Ps decay has been used to characterize condensed matter on the atomic scale, mostly involving the measurement of Ps lifetimes and intensities.

5.3
Experimental Methods

5.3.1
Sources of Positrons

The two methods for obtaining positrons for use in spectroscopic measurements are (i) radioactive sources and (ii) pair production. The most common radioisotope used in the laboratory is sodium-22 (half-life 2.7 years), with cobalt-58 sometimes used in experiments requiring higher intensities over shorter periods (half-life 71 days). These and other positron-emitting sources are created in reactor cores by nuclear bombardment; a number of steps have been taken toward the in-house creation of strong positron sources using tabletop proton and deuteron accelerators to produce short-lived but high-activity radionuclides (Hirose, Nakajyo and Washio, 1997). In-house or facility-based sources are primarily used in the production of intense positron beams, as are positrons created by pair production in a linear accelerator (LINAC). In the latter case, bremsstrahlung radiation from pulsed energetic LINAC electrons created electron–positron pairs in a target of high atomic number; the energetic positrons are then moderated to form an intense, often pulsed, positron beam.

Positrons emitted via the positive β-decay of radioactive sources are longitudinally polarized (i.e., spin-polarized in the direction of their emission) as a consequence of parity nonconservation; the weak interaction that mediates β-decay leads to a nonvanishing helicity. This property has been exploited in a number of experiments to study the bulk magnetic properties of materials, principally using ACAR (Section 5.3.3). The principle underlying its application is that positrons are ~1000 times more likely to annihilate an electron in the opposite spin state. Therefore, if the spin states of some of the electrons in a sample are changed (e.g., by measuring magnetic and nonmagnetic samples, or by changing the external magnetic field direction),

5.3.2
Lifetime Spectroscopy

The first positron spectroscopy to be developed, and still widely used currently, involves the measurement of the mean lifetimes of positrons in condensed matter. As introduced in Section 5.2.2.5, the mean lifetime of positrons in condensed matter is related directly to the electron density in the vicinity of the positrons at their moment of annihilation. For example, if positrons are annihilated while trapped in open-volume defects, the average electron density is reduced and the mean lifetime is increased, and there is, in general, a characteristic lifetime value for every different annihilation state.

In the most general case, positrons can decay from a range of states s each with annihilation rates $\lambda(s)$ (λ being the reciprocal of the mean lifetime τ), with the probability of being in state s at annihilation being $P(s)$. The resultant lifetime spectrum is

$$I(t) = \int P(s) \exp(-\lambda(s)t)\,ds \quad (5.3)$$

which, if the annihilation proceed via a small, discrete states labeled i, reduces to

$$I(t) = \sum_i I_i \exp(-\lambda_i t) \quad (5.4)$$

This spectrum can be fitted with standard programs to extract the decay rates (and thus lifetimes) and corresponding intensities, deconvoluting the time resolution of the measuring system and allowing for system-related components such as that associated with positron decay in the source. In practice, the number of discrete components that can be reliably obtained from experimental spectra is crucially dependent on the range of values of λ_i and the number of states involved; a maximum i value of 3 is typical. A common strategy in many experiments is to use the average positron lifetime $\bar{\tau}$ (or decay rate $\bar{\lambda}$), defined as

$$\bar{\lambda} = \bar{\tau}^{-1} = \int \lambda(s)P(s)\,ds \quad \text{or}$$
$$= \sum_i I_i \lambda_i \quad (5.5)$$

particularly when it is the change in λ with some external parameter that is important.

5.3.2.1 Experimental Lifetime Systems

Conventional positron lifetime spectrometers have at their heart a radioactive source, commonly sodium-22 deposited between two thin low-Z films, sandwiched between two pieces of the sample to be studied. If the sample is to be cooled or heated during the measurements, it may be mounted in a small evacuated chamber. Sodium-22 provides a "prompt" 1.28-MeV γ-ray essentially coincident with the emission of the positron, as a result of the de-excitation neon-22; this is detected by a scintillator with a fast response (e.g., barium fluoride or doped plastic) coupled to a photomultiplier. The 511-keV annihilation "death" γ-ray is recorded by a second similar detector assembly, and the fast pulses (rise time ~1 nanosecond) created are fed into a timing system, which records up to about 10^6 time intervals between birth and death γ-rays. Traditionally, the timing systems have included single channel analyzers to select pulses of appropriate

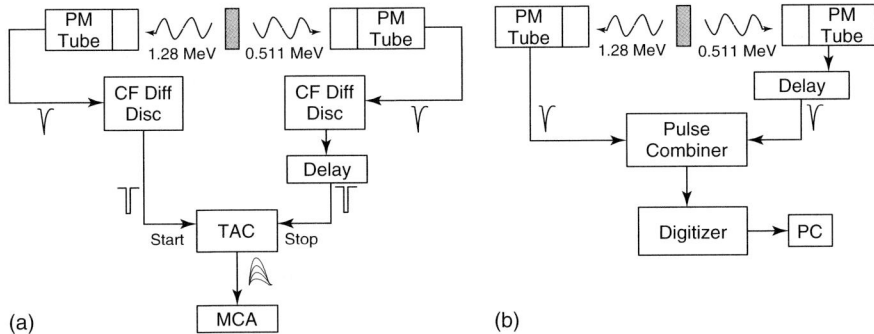

Fig. 5.1 (a) Standard fast–fast analog timing system: PM TUBE, photomultiplier tube/scintillator assembly; CF DIFF DISC, constant fraction differential discriminator; DELAY, nanosecond delay box or cable; TAC, time-to-amplitude converter; MCA, multichannel analyzer. (b) Digital equivalent of an analog timing system.

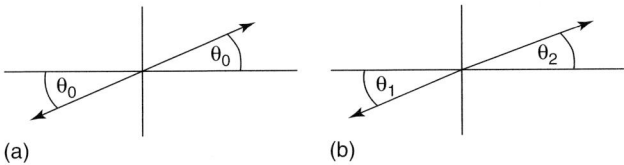

Fig. 5.2 A 2-γ annihilation in (a) center-of-mass frame and (b) laboratory frame.

size, a time-to-amplitude converter (usually with a suitable delay line before the "stop" input) and a multichannel analyzer to collect the entire range of pulse heights in parallel (i.e., the entire range of positron lifetimes). In recent years, this analog system has been replaced by a digital equivalent in which every pulse representing a time difference is digitized and stored, allowing postcollection timing optimization (Figure 5.1).

5.3.3
Angular Correlation of Annihilation Radiation

Although, as discussed in Section 5.2.1 for stationary particles, in two-γ annihilation in the center of mass frame the two photons leave the site of the annihilation with identical energies (mc^2, 0.511 MeV) in exactly opposite directions (to conserve momentum), in the laboratory frame, the momentum of the annihilating pair moving with a velocity v means that the angle between γ-rays – as shown in Figure 5.2 – is no longer 180°.

Relativistic transformation from one frame to the other yields expressions for $\tan \theta_1$ and $\tan \theta_2$. Then $\delta\theta \approx \tan(\theta_1 - \theta_2)$ which, ignoring terms in $(v/c)^2$, reduces to $(2v/c) \sin \theta_0$. Now if we set $\theta_0 \approx \theta_1 \approx \theta_2 = \theta$, then $\delta\theta \approx 2mv \sin\theta/mc = p_t/mc$, where p_t is the component of momentum of the annihilating pair in a direction transverse to the γ-emission. Measurement of $\delta\theta$ thus directly yields information on p_t. $\delta\theta$ typically ranges up to 20 mrad; in order to achieve the necessary resolution, the γ-detection regions have to be small and spaced many metres from the sample being

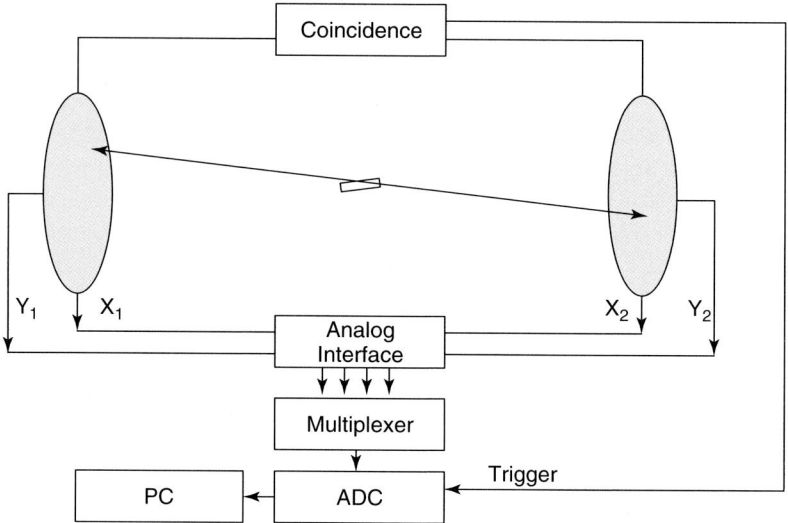

Fig. 5.3 Schematic diagram of 2D-ACAR apparatus. Positrons are guided from source to target by a strong (~1 T) magnetic field. (Courtesy Bristol positron group.)

studied. In one-dimensional studies (1D-ACAR) long thin slits – usually between lead blocks – define the γ-directions; in two-dimensional (2D-ACAR) experiments (Figure 5.3) annihilation photons are detected by position-sensitive detectors such as Anger cameras or multiwire ionization chambers. To effectively eliminate background events, positrons are guided from the approximately giga becquerel source to the samples by a strong (~1 T) magnetic field over a few centimeters from a radioisotope source, or by a guiding magnetic field in a low-energy positron beam system. About 2×10^8 coincident annihilation events are typically recorded. Samples are commonly mounted in a small vacuum chamber on a cold finger to enable measurement at temperatures down to ~10^1 K, and can be fixed at different crystalline orientations. Anger cameras have spatial resolution of 2–3 mm, so that if they are mounted at ~15 m on either side of the sample they provide angular resolution of ~0.2 mrad. A somewhat better resolution is possible with multiwire proportional counters. These high-resolution capabilities have found their place in the measurement of occupied electron states, and thus the Fermi surfaces of a number of simple and complex solids (see Section 5.4.5 for examples). The Lock–Crisp–West (LCW) procedure (Lock, Crisp and West, 1973) folds the momentum density back into the first Brillouin zone, which then gives a distribution that directly indicates the occupancy of states across the projected zone. (The Fermi surface is defined by whether a particular state is occupied or not for each band, and these occupancy breaks are revealed directly by 2D-ACAR.) The full three-dimensional Fermi surface can be reconstructed tomographically from an appropriate number of two-dimensional projections along different crystallographic orientations.

5.3.4
Doppler Broadening of Annihilation Radiation

The component of the momentum p_p of an annihilating positron–electron pair in the direction of emission of the annihilation γ-photon gives rise to a Doppler shift in the photon energy of $\pm c p_p/2$. For example, if p_p arises almost wholly from an electron moving close to the Fermi level with a kinetic energy of 5 eV, $cp/2 = 1130$ eV. This is of the same order as the energy resolution of a high-purity germanium detector. Because the electrons can have a component of momentum toward or away from the detector at the moment of annihilation, Doppler broadening around mc^2 (511 keV) is measured. Changes in the annihilation linewidth from a fraction of 1% to a few percent are typically measured and, although the technique provides a measure of the average electron momentum, which is of considerably lower resolution than ACAR, its high signal rates and relative simplicity has found wide application in the study of defects and phase changes in materials. This is referred to as *Doppler broadening spectroscopy (DBS)* or *DBAR*.

A typical system for DBAR is sketched in Figure 5.4. The Ge crystal is cooled by liquid nitrogen or by an electrical cooler to effectively eliminate thermal noise. A preamplifier is routinely mounted inside the cooled detector head. The optional biased amplifier allows expansion of the photopeak into a larger number of channels on the multichannel analyzer (MCA). Since small drifts can broaden the spectrum, stabilization is necessary, and this is achieved by temperature control and by using a digital stabilizer set to a nearby reference line – for example, the ^{7}Be 478-keV γ-line. Digital systems have been developed, which house all the elements shown in Figure 5.4.

Samples for DBAR are typically mounted in an evacuated chamber to avoid sample contamination on heating or cooling, and data contamination by annihilation signals in air; such mounting also enables temperature control. For traditional bulk measurements using a radioactive source, two pieces of the sample material sandwich a source held in low-Z films, and correction has to be

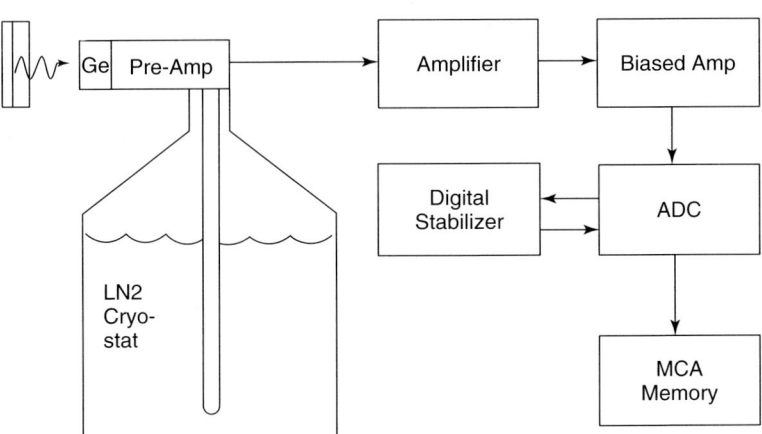

Fig. 5.4 Basic setup for Doppler broadening spectroscopy.

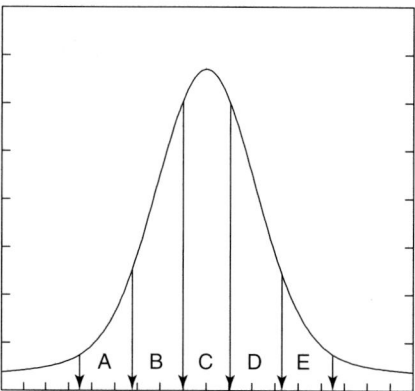

Fig. 5.5 Regions of interest defined for DBAR: sharpness parameter $S = C/T$, wing parameter $W = (A + E)/T$, where $T = A + B + C + D + E$.

made for annihilations in the source itself; if the sample is at the target end of a positron beam system, then no source correction is necessary.

Resultant "photopeak" spectra corresponding to the annihilation γ-line are quasi-Gaussian in shape; entirely free (i.e., conduction) electrons would lead to a parabolic peak, and core electrons to a number of increasingly small "side" peaks; the sum of these contributions has a Gaussian, or multi-Gaussian, appearance. The central part of the annihilation line thus preferentially contains contributions from the lower momentum conduction and/or valence electrons, while the wings contain contributions from annihilation with core electrons. Analysis routines have been developed which take account of the detailed shape of the peak, with a view to maximizing the amount of information obtained in Doppler broadening experiment, although it is more common to use the raw data to track *changes* in lineshape parameters under changing experimental conditions.

As discussed later, by decreasing background significantly, detailed analysis of the peak shape can exploit the technique's sensitivity to the elemental environment. However, we first consider the standard method for describing the Doppler-broadened linewidth – that is, the use of simple lineshape parameters. The most common parameters used are called S and W – the sharpness and wing parameters, respectively – which are defined in Figure 5.5. Three regions of the peak – here A (central) and C and E (the two wings) – are chosen, symmetrically about its centroid. The limits of these regions are chosen such that the central and total wing fractions of the peak area – which are called the S and W parameters – are ~ 0.5 and ~ 0.1–0.2, respectively. Background is first subtracted if deemed necessary. Following our argument above, S and W should principally reflect changes in the momentum density of lower and higher momentum electrons, respectively. For example, positron annihilation in open-volume defects typically leads to an increase in S and a decrease in W.

Neither S nor W has an absolute value, as it is their dependence on external parameters that is of common interest. However, it is common to express the parameters normalized to bulk values – that is, S/S_{bulk} and W/W_{bulk}, where the denominators are the parameters associated with the defect-free bulk material being studied, measured with the same apparatus. The choice of which parameter to use (S and W are not the only choices) and the limits of regions defining the parameters are system-dependent.

If S_f and W_f are the S and W parameters characteristic of free positrons, then we can define a new parameter $R = |(S - S_f)/(W - W_f)|$, which depends only on the nature of the defect, and not on its concentration. If R is found to be constant, then this points to the existence of only one kind of defect (Mantl and Trifthäuser, 1978).

5.3.4.1 Two-Detector DBAR

This technique was proposed many years ago by Lynn *et al.* (1977) but has been reintroduced and developed in the past decade. If a sample is viewed by a second γ-photon detector, and pulses from the two detectors (in coincidence) are fed into the inputs of a two-dimensional multichannel analyzer array, a photopeak results that not only has vastly reduced background (i.e., by two or more orders of magnitude) but the resolution is also decreased by a factor of $\sqrt{2}$. Example data for such a system are shown in Figure 5.6. This method enables one to study annihilations with core electrons and, in the case of recent studies with positron beams, identify the chemical environment in which the positron decays. The measurement typically involves the accumulation of single annihilation line spectra of high statistical precision whose shape is analyzed carefully and compared with calculations.

5.3.4.2 Age-Momentum Correlation (AMOC)

Simultaneous measurement of positron lifetime and the momentum of the annihilating pair (i.e., PALS + DBAR) can give information on thermalization and transitions between positron states (and hence on chemical reactions of positrons or Ps). The most recent version uses a mega electronvolt positron beam (Siegle *et al.*, 1997).

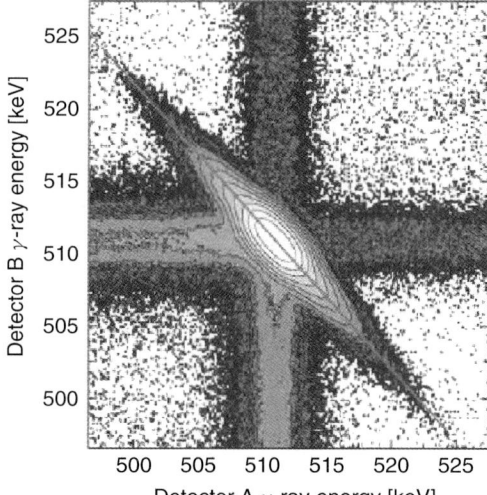

Fig. 5.6 A 2D raw data from coincidence DBAR measurement. The peak, indicated by the diagonal from top left to bottom right, is $\sqrt{2}$ narrower than a single-detector spectrum, and has essentially no background. (Courtesy Halle positron group.)

5.3.5
Positron Beams

Positron beams offer two major advantages over traditional positron systems: (i) the control of incident positron energy, and thus average implantation depths from the surface to depths of a few micrometers, and (ii) the separation of the thermalization of positrons implanted into a material from their eventual annihilation in another. This has led to new applications in surface, near-surface, interface, and thin-film studies.

The reemission of work–function–energy positrons from a solid surface was outlined in Section 5.2.2.5, Positron Reemission. In the case of simple metals, the fraction of implanted positrons reaching an exit surface is optimized if the moderator metal contains very few nonequilibrium defects. This means annealing, preferably *in situ*, to as high a temperature as possible (say $\sim 0.8\ T_m$) in as low an ambient pressure as possible (e.g., $\sim 10^{-6}$ Pa or less). Because of its high positron work function ($\varphi_+ \sim 2.7$ eV), however, tungsten operates well as a positron moderator after greatly varying preparation procedures. The same is not true, for example, with nickel or copper – whose $\varphi_+ \leq 1$ eV – for which careful surface cleaning and maintenance is required.

Many moderator geometries have been used; the simplest and most common is the mesh, which has quasi-transmission geometry. Moderator efficiencies are commonly in the $10^{-4} - 10^{-3}$ range. Choice of moderator material and geometry is governed by the application of the positron beam – for example, for many applications, the priority is simply to maximize moderator efficiency, whereas for others a well-collimated parallel beam requires a planar low-ϕ_+ surface cooled to minimize thermal smearing.

The highest moderation efficiencies (above 10^{-3}) recorded are for solid rare gases, in which positrons do not fully thermalize and have a long effective diffusion length, which can be condensed directly onto the radioactive source capsule.

It has long been recognized that moderation efficiency could be greatly enhanced by drifting a larger fraction of thermalized positrons to the exit surface by an internal electric field, but there has been little practical progress to date on the realization of such field-assisted moderators.

5.3.5.1 Laboratory-Based Beams
A standard slow positron beam system (Figure 5.7) comprises a flight tube pumped to high or ultrahigh vacuum; positrons ($\sim 10^5$ s^{-1} from a ~ 4 GBq primary source) are transported in an axial magnetic field, sometimes with focusing elements. Unmoderated (β) positrons may be filtered from the beam by a curved section or an ExB velocity filter and the positrons are accelerated to the final desired energy either by raising the source of the system to a positive potential or by holding the sample at a negative potential. Beam positioning may be fine-tuned by using a pair of trim coils. Annihilation radiation from the sample target passes through a thin foil window to the Ge detector for DBAR measurements; detection of radiation from the source is minimized by shielding and by making the system a few metres in length. These considerations are relatively unimportant if the

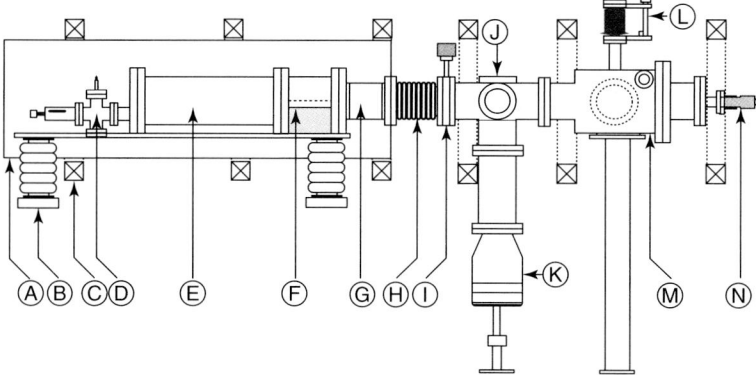

Fig. 5.7 Magnetic-transport positron beam system. A, grounded shield; B, standoff insulators; C, coils for magnetic field; D, source/moderator; E, *ExB* plates; F, lead shielding; G, accelerator; H, bellows; I, aperture; J, guiding coils; K, turbopump; L, sample manipulator; M, sample chamber; N, CEMA/CCD camera.

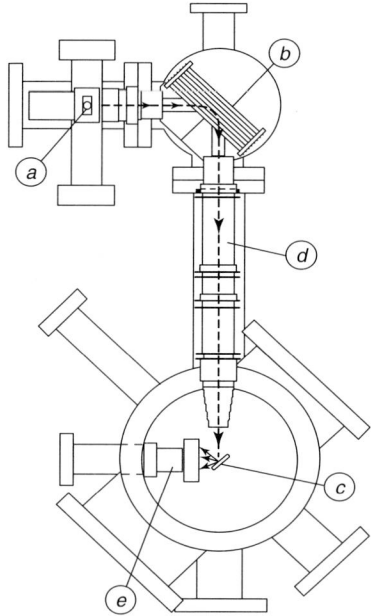

Fig. 5.8 Example of an electrostatic positron beam system. (a) Source, (b) electrostatic reflector, (c) sample, (d) electrostatic lenses, and (e) microchannel plate detector.

system is to be used for particle spectroscopies – see Figure 5.8 – especially if electrostatic transport and focusing is in use (Roach, Bakshi and Canter, 1995).

Brightness enhancement is achieved in positron beams by repeated focusing and remoderation (see Figure 5.9). Although remoderation losses may be 70%, the

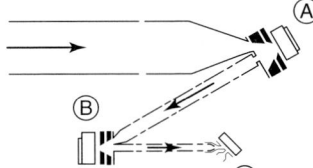

Fig. 5.9 Reflection-geometry brightness enhancement stage. A and B, remoderator surfaces; C, target.

beam area can decrease by a factor of ~50 at each stage; hence beam intensity per unit area can increase after n stages by a factor 15^n.

5.3.5.2 High-Intensity Positron Beams

Standard laboratory beam intensities up to ~10^6 controllable-energy positrons per second are currently available, and a great range of experiments have been performed with such beams. However, to extend the capabilities and realize the full potential of positrons as a spectroscopic tool, more intense beams are essential for many studies in which the probability of an outcome is very small, for example, the combination of controlled implantation with 2D-ACAR, many surface spectroscopies where rapid measurements are required, and the study of many-positron systems.

Facility-based systems – both reactor and LINAC based – are still a main focus for intense positron beam generation, but in recent years there has been a movement toward the use of small-scale laboratory-based accelerator systems to produce short-lived but intense positron sources. In the case of LINACs (Section 5.3.1), the beam is pulsed (Section 5.3.5.3); this property can be exploited in applications where timing is an advantage. In laboratories in reactor complexes, an intense positron beam is guided far from the core and possibly subjected to remoderation, focusing, and bunching for timing applications. The production of positrons in reactor cores is not solely via the creation of intense radioactive sources, but can also proceed via pair production by energetic photons; steps have to be taken to overcome radiation damage problems in the reactor core. Several intense beam systems are in operation or are currently being developed across the world.

5.3.5.3 Beam Bunching

LINAC beam pulses are of widths from several nanoseconds to microseconds; if subnanosecond timing resolution is required, the positrons must be put through a buncher. Several buncher designs have been put into practice; the first, by Mills (1980), used magnetic mirrors. Later bunchers used radiofrequency chopping and bunching techniques – for example, the system at Munich (Bauer et al., 1987), which can now achieve a timing resolution below 200 picoseconds.

5.3.5.4 Positron Microbeams

Beams of approximately micrometer dimensions have been created in the laboratory (Figure 5.10) for (i) optimum areal brightness and (ii) position-sensitive annihilation spectroscopy. Optimum areal brightness is important in positron reemission microscopy systems in which the microscope optics requires a small but very bright spot – the contrast mechanism in positron reemission microscopy (PRM) is based on the position dependence of the probability of positron

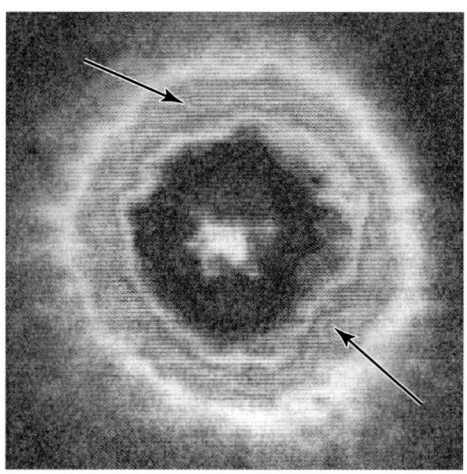

Fig. 5.10 A 10-μm square intensity cross-sectional image of a 5-keV positron microbeam at Brandeis (Canter et al., 1994). The central white peak area contains approximately 465 counts per pixel, and the ring marked with arrows contains about 180 counts per pixel. Reproduced by permission, AIP.

reemission from a surface. Position-sensitive annihilation spectroscopy is required for the building of micrometer-resolution annihilation maps of near-surface structural defects by beam rastering, as has been achieved in Munich (David et al., 2001).

5.3.5.5 Polarized Positron Beams

It was discovered in 1979 that the depolarization resulting from moderation of the β-positrons by a moderator, the basis of the creation of controllable-energy positron beams, is negligible (Zitzewitz et al., 1979). Highly polarized positron beams can, in principle, probe, nondestructively, and with a degree of depth sensitivity, spin-polarized electrons in thin films or in modified near-surface regions (to depths of a few micrometers) of materials. Standard laboratory-based beams have to be adapted in two main ways: (i) the high-Z backing behind the radioactive source has to be replaced by a low-Z material to minimize backscattering, and (ii) the source has to be positioned further away from the moderator so as to select primary positrons emitted into a small forward cone (both (i) and (ii) result in loss of beam intensity). The spin polarization of a positron beam is given by

$$P = \left(\frac{v}{2c}\right)(1 + \cos\alpha) \qquad (5.6)$$

where v is the emission velocity, c the speed of light, and α the half-angle of the cone of acceptance of β-positrons at the moderator. (v/c is called the *helicity*.) For positrons emitted in a cone of half-angle 30° from the commonly used source ^{22}Na, $P \sim 70\%$.

There has been little exploitation of this property of positron beams, apart from an early experiment that demonstrated their capacity to probe surface magnetism (Gidley, Köymen and Capehart, 1982). The principle of the measurements is based on the comparison of DBAR or annihilation line profile data for differently polarized samples.

5.3.5.6 MeV Positron Beams

High-energy (approximately mega electronvolts) monoenergetic positron beams have been built and used in recent years. Megavolt accelerators used in this work have been of Pelletron or Van de Graaff type. Source-free lifetime measurements in a wide range of materials, beam-based age-momentum correlation (AMOC), and novel annihilation spectroscopies are possible with such systems – for example, positron channeling and in-flight annihilation studies, and bulk annihilation measurements, which are free of problems associated with the presence of a positron source.

5.3.5.7 Trap-Based Beams

Surko and co-workers have developed a source of very low energy positrons with extremely low energy (approximately milli electronvolts) energy spread (Gilbert et al., 1997), which has found wide and important applications, particularly, but not solely, in atomic and molecular physics. The principle of their system is that positrons are extracted from a plasma of thermalized positrons stored in a Penning-type trap (Figure 5.11). The resulting beam can be DC or pulsed. Similar trap-based positron beams have been successfully employed in the antihydrogen work at CERN and other laboratories prosecuting new fundamental research.

5.3.5.8 Positron-Beam-Based Spectroscopies

The traditional spectroscopies (PALS, ACAR, and DBAR) have all been performed with positron beams, giving the extra information on depth dependence and the ability to probe depths up to approximately micrometers from a sample surface. The acronyms for such spectroscopies may be constructed by inserting VE (for variable energy) before those for the bulk spectroscopies. In all these cases, codes such as VEPFIT (Van Veen et al., 1990) and POSTRAP (Aers et al., 1995) are commonly used to separate and identify contributions to the measured parameters from annihilation events over a range of depths from different states such as vacancy-type defects, bulk material, interface regions, and the sample surface. The diffusion equation is solved – for example, with VEPFIT – and the positron parameter (like the lineshape parameter S) and associated effective positron diffusion length, characteristic of each chosen layer below the surface, are evaluated to achieve best fit to the experimental data. In some cases – for example, where there is one type of trapping center of concentration C_D per atom – the fitted parameter and diffusion length are linked, and self-consistency can be

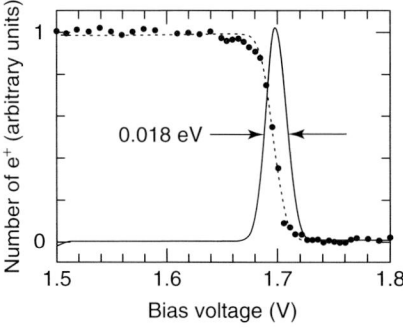

Fig. 5.11 Energy distribution of a pulse of positrons extracted from a thermalized room temperature positron plasma stored in a Penning trap (Gilbert et al., 1997). Reproduced by permission, AIP.

checked. As an example, let S be the fitted lineshape parameter for a particular layer beneath a surface (i.e., between two given or fitted depths). Then $C_D = (\lambda/\nu)(S-1)/(S_D - S)$, where λ and ν were defined in Section 5.2.2.4. Additionally C_D can also be written in terms of the effective diffusion length L as $(\lambda/\nu)[(L_+/L)^2 - 1]$, where L_+ is the positron diffusion length in the undefected "perfect" material. Equating these two expressions one arrives at $L^2 = L_+^2[(S - S_D)/(1 - S_D)]$.

The depth resolution of position beam techniques degrades as the mean positron implantation depth increases; the width of $P(z, E)$ (Section 5.2.2.2) is comparable to the mean depth. This means that, although resolutions of ~ 50 nm are possible for near-surface characterization, one relies on fitting codes or known sample parameters (like film or layer thicknesses) to model depth profiles at deeper depths. The only method for maintaining depth resolutions similar to that near the surface is to employ an etch-and-measure technique, removing sequentially thin layers of material and probing the progressively revealed new surface regions with low-energy positrons. The problem with very near surface layers or thin films is the propensity for positrons to diffuse out of them, either to the surface or to nearby material, and careful fitting and/or data interpretation is required; this problem is less severe for thin layers with a very short positron diffusion length (e.g., oxides). The effects of internal electric fields must also be considered. Finally, the influence of nonthermalized positron interactions at low incident positron energies E has to be recognized, and for this reason data for E below ~ 1 keV is often ignored.

An alternative to the use of fitting codes is the identification of different annihilation sites (e.g., layers) using the graphical method of plotting S versus W parameters for each incident energy E. An example is shown in Section 5.4.4. If there are two possible annihilation states with their characteristic S and W values – such as the sample surface (subscript S) and bulk (subscript B) – then an $S-W$ plot will be a straight line joining the two points (S_S, W_S) and (S_B, W_B) on the graph. Consider a third state, corresponding, say, to a defected layer with characteristic $S = S_D$ and $W = W_D$ (note that these are average *layer* values, not those specific to a particular defect); if there are energies E for which effectively all the positrons are annihilated in this layer, then the $S-W$ plot will have two straight lines joining (S_S, W_S), (S_B, W_B), and (S_D, W_D). Importantly, however, even if only a *fraction* of the positrons decay in the layer at any energy E, extrapolation of lines on the resulting $S-W$ plot will still identify the point (S_D, W_D) – just as a fitting code such as VEPFIT does. $S-W$ plots are thus a powerful visual tool for identifying different annihilation sites.

Positron Surface Spectroscopies Notwithstanding the fact that full realization of positron beams as a surface probe via positron diffraction, reemission, or annihilation will not come until intense positron beams are widely available, much progress has been made over the past 25 years. Positron microscopy is not mentioned in this section as it was discussed earlier in Section 5.3.5.4.

Low-energy positron diffraction (LEPD) was pioneered by Canter and co-workers at Brandeis University in 1980 (Rosenberg, Weiss and Canter, 1980) and developed by the same group over the next two decades to a level at which they could demonstrate that LEPD can be used to achieve, qualitatively and quantitatively, better agreement between experimental and theoretical $I-V$ profiles (diffracted beam intensity vs. energy), leading to significantly more reliable determinations of the surface structure than is possible using the traditional electron equivalent, low-energy electron diffraction (LEED). This improved performance is for the following reasons: (i) the phase shifts for positron scattering are less sensitive to atomic number than those for electrons, so that LEPD is more sensitive to structural parameters in multicomponent systems; (ii) the inelastic mean free path, which plays an important role in the determination of probe depth, is smaller for positrons than for electrons, and so LEPD has a greater surface sensitivity than LEED; (iii) uncertainties in the positron–electron correlation term used in LEPD calculations are less important than the equivalent electron–electron uncertainties in LEED; (iv) positrons are decelerated as they approach ion cores and so relativistic effects such as spin-orbit coupling are reduced for positron scattering from surfaces containing high-Z atoms, and LEPD $I-V$ profiles are only weakly spin-dependent when compared to LEED; and (v) as a result of the absence of spin-exchange repulsion, positrons interact more weakly with interstitial valence electrons and so the muffin-tin model works better for LEPD than for LEED for covalently bonded semiconductors.

Reflection high-energy positron diffraction (RHEPD) has been demonstrated by Kawasuso and Okada (1998). A 20-keV highly parallel positron beam is incident upon the surface being studied at glancing ($<5°$) angles. The most important difference between RHEPD and its electron equivalent, RHEED, is that the high-energy positrons can undergo total reflection from the surface because of the positive crystal potential, and are thus very sensitive to the presence of adsorbate atoms, topological irregularities, and lattice vibrations in the topmost surface layer.

Positronium reflection from LiF was observed by Weber *et al*. 1988; to date, it is the only report of experimental work in this area.

Positron-annihilation-induced Auger electron spectroscopy (PAES) was developed by Weiss *et al*. (1988) in the late 1980s. Auger electron emission in PAES results from annihilation of a surface core electron by a positron implanted with very low (~ 10 eV) energy, rather than from impact ionization as in its electron equivalent, electron-induced Auger electron spectroscopy (EAES). The two major advantages of PAES over EAES are the elimination of secondary electron background and the extremely high surface sensitivity. Ohdaira *et al*. (1997) have developed a time-of-flight PAES system at ETL, Japan.

Positronium emission spectroscopies can provide information on electronic surface states; for example, assuming that Ps formation is a sudden process, a measurement of the Ps velocity distribution should yield information on the electronic density of states, although there remains a discrepancy between theory and experiment at low Ps energies.

2D-ACAR was first used in the mid-1980s to attempt to directly observe the positronic surface state; the observed symmetrical momentum distribution may have instead been a signature of localized surface trapping (Lynn et al., 1985). 2D-ACAR was, however, successfully used to study Ps momentum distributions, and showed sensitivity to electronic structure (Chen et al., 1987).

Reemitted positron spectroscopy measures the energy spectra of positrons reemitted from a film-covered surface as a function of overlayer thickness (Gidley, 1989; Ociepa et al., 1990), which has been shown to be an excellent probe of any processes that affect the sum of the positron and electron bulk chemical potential, including alloying of the overlayer film.

Positronium formation spectroscopy, that is, monitoring variations in Ps yield (both in magnitude and dependence on incident positron energy) or in the Ps contribution to 2D-ACAR spectra, is an extremely sensitive tool for monitoring oxide growth on metals and semiconductors.

Future developments in positron surface studies may include *positron-induced ion desorption, surface barrier potential measurements* via very low energy (approximately electronvolts) positron reflection, glancing-angle RHEPD, and the dependence of the Ps formation probability on incident positron energy (coupled with LEED, for the surface atomic configuration, and angle-resolved ultraviolet photoemission spectroscopy, for information on electronic states). Similar measurements can give information on the atomic positions of the adatoms, even if these are hydrogen (in contrast to the conventional angle-resolved photoemission). In *inverse Ps formation spectroscopy*, a beam of Ps atoms bombards a surface, the electron is given up to an unfilled state, and the positron takes away information on that state; this spectroscopy should be more sensitive than existing probes.

The advance of positron surface science, however, awaits more widespread availability of intense positron beams. Examples of developments that may flow from such availability include *positron holography*, proposed as intrinsically more suitable than electron holography because of the positron's weak scattering and large damping in solids; *polarized PAES*, in which a highly polarized incident beam of positrons creates polarized core holes to enable novel studies of magnetic surfaces; *reemitted positron energy loss spectroscopy; Ps diffraction*; and *inverse Ps formation*.

5.4 Examples of Research Using Positron Spectroscopies

The following examples are designed to provide an overview rather than a detailed summary of every important result obtained using positron annihilation spectroscopy (PAS) and VEPAS techniques. A relatively small number of examples will thus be selected to demonstrate the applicability of the various techniques.

5.4.1 Vacancy-Type Defects

The propensity for diffusing positrons to trap in open-volume point defects has been the basis of one of the major applications of positron spectroscopy.

The measurement of formation enthalpies H_V^F for thermally generated equilibrium defects has long been a staple of PAS in bulk solids. Both DBAR and

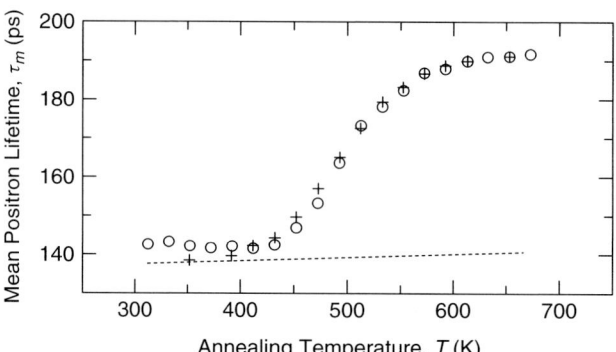

Fig. 5.12 Temperature dependence of the mean positron lifetime in Cu_3Sn. Error bars are within the points. Circles, ramping up; crosses, ramping down in temperature (Shishido et al., 2007). Reproduced by permission, Wiley-VCH.

PALS have been used for such studies. Figure 5.12 shows a recent measurement of mean positron lifetime τ_m in the alloy Cu_3Sn; the rise in τ_m above 433 K indicates the formation of vacancy traps. The positron trapping rate κ is deduced from the measured positron parameters (either lineshape or mean lifetime – called P here) normalized to the bulk defect-free value P_b (so that $P_b = 1$) as

$$\kappa = \frac{(P-1)}{(P_d - P)} \quad (5.7)$$

where P_d is the parameter associated with the defect (in Figure 5.12, the high-temperature asymptotic value). The slope of the Arrhenius plot of κ versus $1/T$ then yields the vacancy formation enthalpy H_V^F. In this example, the value of H_V^F was similar to that measured for Cu, suggesting that the vacancies are being formed on Cu sites.

Nonequilibrium vacancy defects, such as those induced by strain or implantation, have also been widely studied by PAS. Figure 5.13 shows raw VEDBAR data with a typical response to defects (principally divacancies) produced by 2 MeV Si^+ ion implantation into Si for a range of ion doses from 10^{11}–10^{15} cm^{-2} (Coleman, Burrows and Knights, 2002). Observations from this and other measurements for mega electronvolt ions of widely different masses led to the universal expression for divacancy concentrations at approximately half ion range

$$C_D = (2.79 \times 10^{10}) \phi_A^{0.63} \text{ cm}^{-3} \quad (5.8)$$

where ϕ_A is the ion dose corrected by a multiplying factor equal to the vacancy/ion/Å value provided by the widely used simulation code SRIM (www.srim.org).

PAS can give information about dislocations – both in bulk materials and in thin films (e.g., relaxed SiGe) – primarily because it is thought that vacancies exist in kinks along the dislocation line and diffusing positrons are thus trapped by them.

5.4.2
Structural Changes

PAS has been, and continues to be, exploited in the study of structural

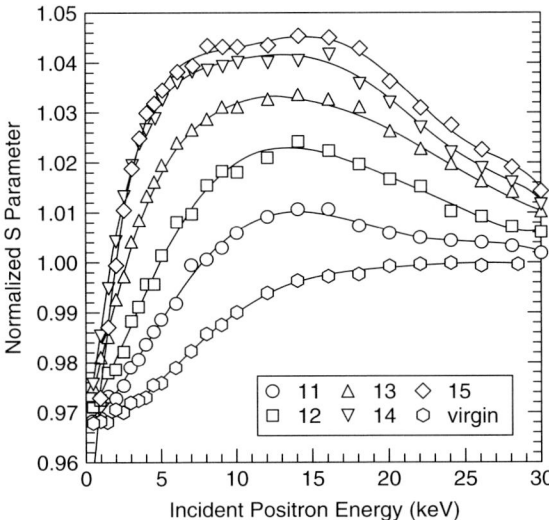

Fig. 5.13 Normalized $S(E)$ for FZ Si unimplanted and implanted with 2 MeV Si^+ ions at doses of 10^n cm^{-2} ($n = 11, 12, 13, 14,$ and 15) (Coleman, Burrows and Knights, 2002).

changes associated with phase transitions, precipitation, deformation, and so on, induced thermally (including aging at room temperature) or mechanically. For example, monitoring the S parameter as the steel Fe–Mn–Si–Cr–Ni is deformed (Mostafa et al., 2007) shows no response below a few percent strain, as the macroscopic deformation associated with reorientation of martensite plates does not create microdefects that trap positrons. Above ~5% strain S increases as vacancy sites are created, some perhaps along dislocations, until at ~16% strain and above the response starts to saturate as all positrons become trapped; this behavior is also seen in CuZnAl shape memory alloys. The nonlinearity of the corresponding S–W plot (see Section 5.3.5.8) suggests that there is a variety of defect types created, and/or that the alloy undergoes a γ–ε phase transition during deformation (as seen in microscopic images).

5.4.3
Nanoparticles

An important recent advance in PAS has resulted from the observation that positrons can act as "magic bullets" in the study of embedded nanoparticles, either residing in the open volume around the particles or being preferentially attracted to them via their greater positron affinity. In both cases, the positrons are much more sensitive to the nanoparticles than would be expected from geometrical arguments alone. An example of this, which, in addition, illustrates the use of coincidence DBAR to gain chemical information on the atomic environment of the annihilated positrons, is shown in Figure 5.14 (Nagai et al., 2000). The spectra shown are ratios of the outer

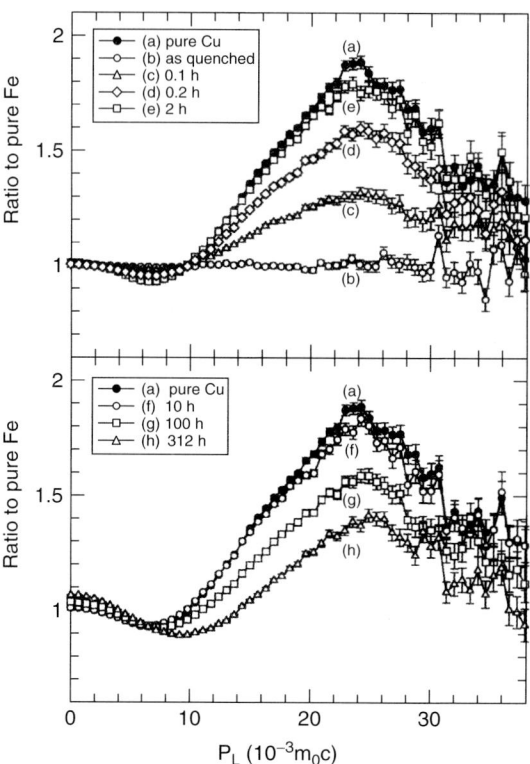

Fig. 5.14 CDBAR ratios of $Fe_{0.99}Cu_{0.01}$ to pure Fe: (a) pure Cu, (b) alloy as quenched, the after aging at 550 °C for (c) 0.1 hour, (d) 0.2 hour, (e) 2 hours, (f) 10 hours, (g) 100 hours, and (h) 312 hours (Nagai et al., 2000). Reproduced by permission, APS: http://link.aps.org/abstract/PRB/v61/p6574.

wing parts of the measured annihilation line to that of Fe. Positrons are shown to be annihilated preferentially by Cu electrons as Cu nanoparticles form in a dilute (1%) alloy of Cu in Fe, by virtue of the 1 eV-deep well created by the relative positron affinities of Cu and Fe.

5.4.4
Interfaces

Generally, positrons are significantly more sensitive to interface states than geometric models would suggest, as they are commonly trapped there during diffusion. An example of VEPAS response to interfaces is shown in Figure 5.15, which also demonstrates the usefulness of the graphical parameter–parameter method; here, two W parameters, rather than the more usual S and W, have been plotted (Coleman et al., 2007).

5.4.5
Fermi Surfaces

Knowledge of the details of Fermi surfaces can aid understanding of material properties, and 2D-ACAR represents a major tool for probing such surfaces. An example is a study of a rare-earth nickel borocarbide, which exhibits competing or coexisting antiferromagnetism and superconductivity. Dugdale et al. (1999) found evidence supporting this from the nesting behavior found in a particular sheet of the Fermi surface for $LuNi_2B_2C$ (Figure 5.16).

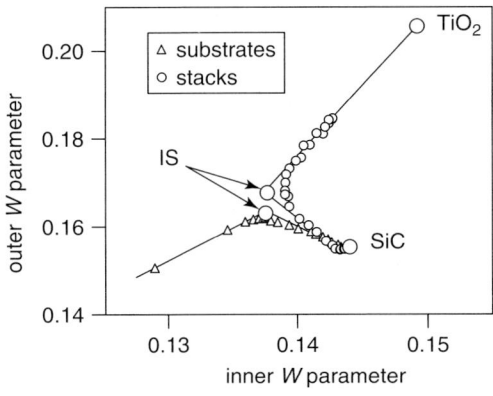

Fig. 5.15 $W_o - W_i$ map for TiO2 stacks on SiC stacks and for SiC substrates. The "pure state" points for SiC and TiO_2 are shown as circles, as are the two closely related interface state (IS) (Coleman et al., 2007).

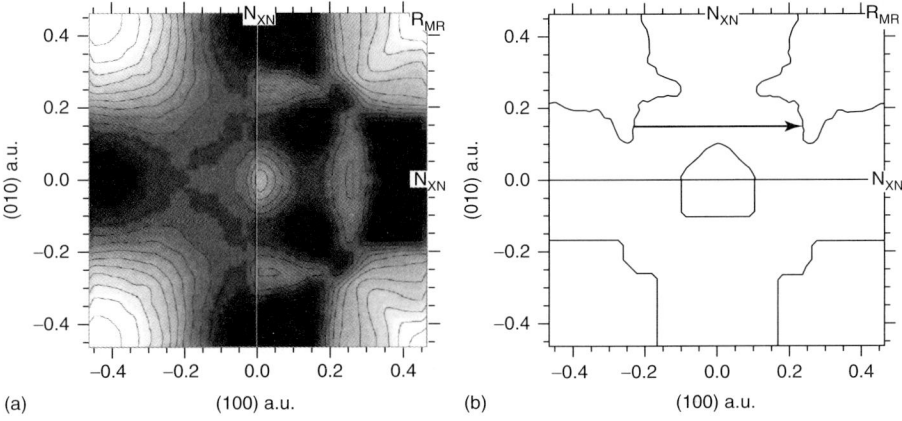

Fig. 5.16 (a): Experimental (left) and calculated electron density of $LuNi_2B_2C$ projected along the [001] direction. Black signifies holes, and white represents electrons. (b): Fermi surface topology of $LuNi_2B_2C$ – experimental (top) and theory for third band in the (001) plane through the Γ point. The arrow indicates the nesting feature. From Dugdale et al. (1999). Reproduced by permission, APS: http://link.aps.org/abstract/PRL/v83/p4824.

5.4.6
Nanoporous Materials and Open Volumes in Polymers

The application of Ps and positron lifetime measurement have long been used as a probes of open volume in polymers, often used in conjunction with other experimental techniques such as differential scanning calorimetry and ionic conductivity measurements. An example of a recent application was the study of polyethylene glycol dimethacrylate (Figure 5.17), where PALS was performed for temperatures in the range 100–370 K (Bamford et al., 2001). These measurements allow the evaluation of the glass transition temperature, coefficients of expansion of the hole volume, fractional free volume, and hole number density.

There has been much activity in recent years in the area of pore size measurement and distribution in thin

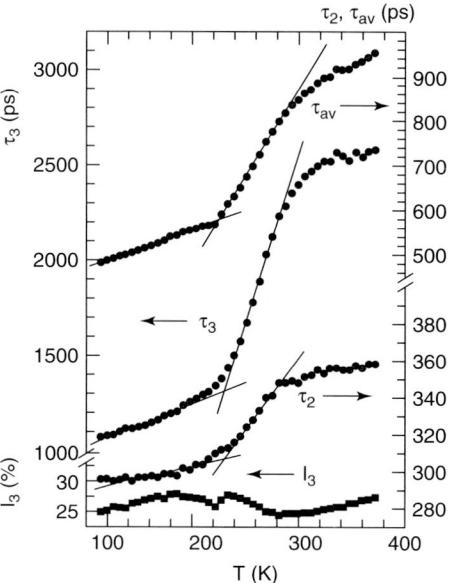

Fig. 5.17 The o-Ps lifetime τ_3, its intensity I_3, the e^+ lifetime τ_2, and the average positron lifetime τ_{av} in poly[(EG)$_{23}$DMA]; from Bamford et al. (2001). Reproduced by permission, AIP.

low-k dielectric layers grown for nanoelectronics applications. The basis of these measurements is that Ps can form in nanometer-sized pores, and its lifetime is linked to pore size. It is also possible to measure lineshape parameters and the ratio of three-γ (o-Ps) to two-γ decay events, both sensitive to the decay of Ps, to obtain some measure of the size, number density, and interconnectivity of the pores. These methods can be applied to films of thickness of approximately micrometers, and can provide some depth sensitivity, if linked to a positron beam system. Because Ps lifetimes can be approximately nanoseconds, timed beam systems can be constructed with relatively long timing resolutions, such as that of Gidley et al. (2000), Figure 5.18, who use the detection of secondary electrons to tag incident positrons. Interconnectivity can result in Ps escape as a long-lived naturally decaying entity in the vacuum space above the sample.

5.4.7
Surfaces

Although there have been several new initiatives in recent years in positron-surface studies, and activity in the ultrathin film and polymer coating field, which could be considered as surface science, we focus here on the two spectroscopies with the longest history and the most impressive results to date – that is, PAES and LEPD.

A recent example of the power of PAES, illustrating the advantages outlined in Section 5.3.5.8, Positron Surface Spectroscopies, involved the study of Se passivation layers on the Si(001) surface by time-of-flight PAES (Zhu et al., 2005). The Se monolayer was found to be stable after days of air exposure, the physisorbed oxygen on the passivated surface being desorbed below 400 °C. The Se passivation layer desorbs from the Si(001) surface above 800 °C in ultra-high vacuum (Figure 5.19).

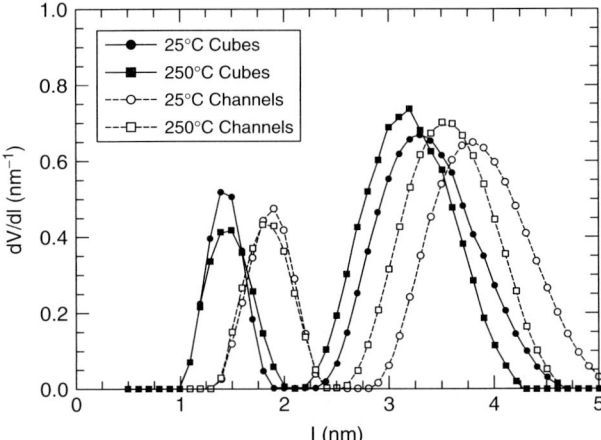

Fig. 5.18 Void size distributions derived from Ps lifetime measurements in low-dielectric thin films of methylsilsequioxane. The solid and broken lines refer to cube- or channel-shaped pores (Gidley *et al.*, 2000). Reproduced by permission, AIP.

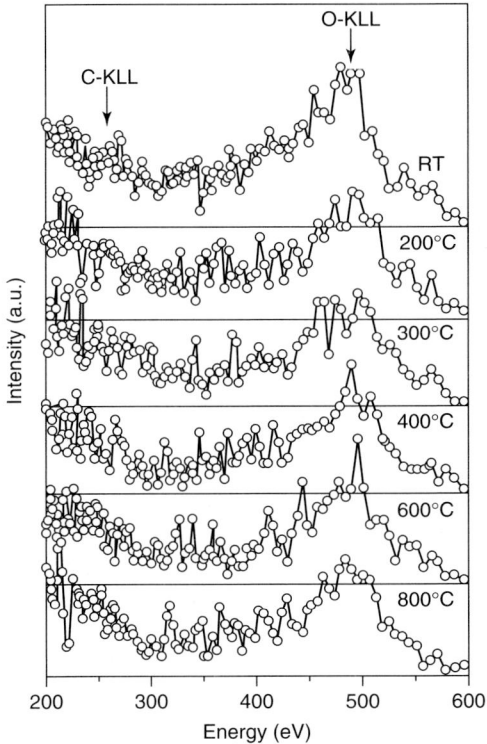

Fig. 5.19 PAES spectra for an Se-passivated Si(001) after isochronal annealing at increasing temperatures (Zhu *et al.*, 2005). Reproduced by permission, AIP.

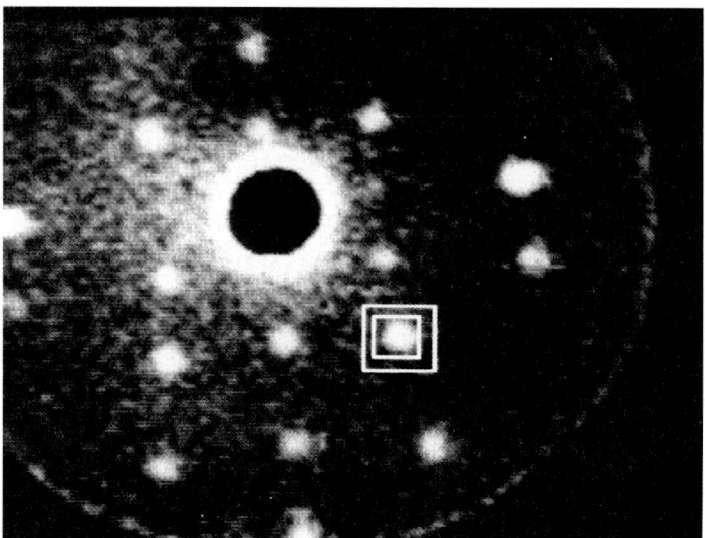

Fig. 5.20 Digital LEPD spot pattern from GaAs(110). The boxes indicate the regions used to evaluate signal intensity (Chen *et al.*, 1993). Reproduced by permission, APS: http://link.aps.org/abstract/PRB/v48/p2400.

In the clearest demonstration of the advantages of LEPD over LEED for the evaluation of the surface structure of some materials – specifically compound semiconductors – Chen *et al.* (1993) showed that the multiple-scattering theory fits LEPD data better than LEED and hence yields structural parameters with smaller uncertainties, and suggested that the differences between LEPD and LEED results were real (see Figure 5.20).

5.4.8
Positron Microscopy

The Brandeis positron reemission microscope used brightness enhancement to obtain a bright microbeam of $\sim$5-μm diameter full width half maximum and $\sim 10^6$ positrons per second. The microscope was able to image defect structures on the surface of a thin Ni(100) film with a spatial resolution of 300 ± 10 nm (Brandes, Canter and Mills, 1988) (Figure 5.21). Improved resolution should be possible with the application of intense positron beams. More recently, there has been work aimed at further exploitation of positron reemission microscopy in Japan, and in Munich an approximately micrometer-pulsed beam (David *et al.*, 2001) was rastered across a surface exhibiting a fatigue crack on the micron scale, creating the two-dimensional positron lifetime image shown in Figure 5.22, which showed the presence of vacancy defects around the crack invisible to standard microscopic methods (Egger *et al.*, 2002).

5.4.9
Positron and Positronium Chemistry

Some current areas of activity in positron and Ps chemistry were mentioned earlier

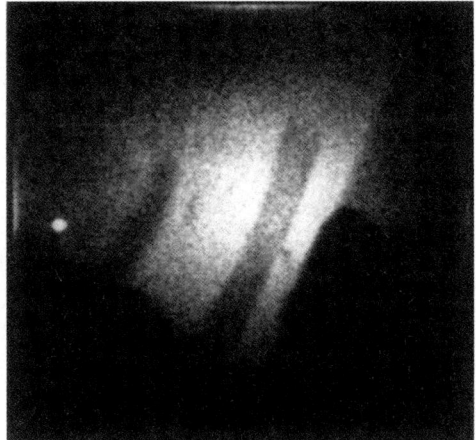

Fig. 5.21 Positron reemission microscope image of Ni foil with contrast due to positron trapping at defects. Magnification 1150. Data collection time 14 hours, white areas have ~40 counts/pixel. From Brandes, Canter and Mills, 1988. Reproduced by permission, APS: http://link.aps.org/abstract/PRL/v61/p492.

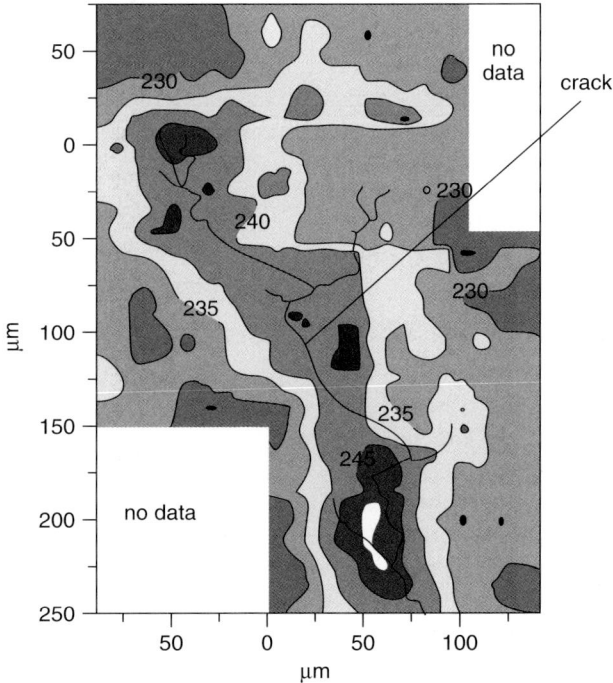

Fig. 5.22 Positron lifetime image of a fatigue crack in copper taken by the Munich scanning positron microscope; lifetimes range from 200 to 250 picoseconds. Incident positron energy = 5 keV (David et al., 2001). Reproduced by permission, Elsevier.

in Section 5.4.5. The formation of Ps in liquids, its enhancement and inhibition, electron and positron scavenging reactions, the formation of positron and Ps bound states, Ps trapping in "bubbles" in liquids, Ps oxidation and spin conversion, have been studied experimentally using PALS, DBS, and ACAR spectroscopies as well as AMOC (Section 5.3.4.2). Rate constants for Ps reactions have been deduced; because Ps is a light particle its diffusion time has to be taken into account. Further details on the chemical reactions of positrons and Ps can be found in the book edited by Jean *et al.* (see the section Further Reading).

Acknowledgments

The author is grateful for the support of the Engineering and Physical Sciences Research Council, UK, over many years, and currently under grant EP/F029829/1. He also acknowledges the guidance he obtained from the earlier contribution to this Encyclopedia of Csaba Szeles and Kelvin Lynn.

Glossary

Angular correlation: The angular distribution of two annihilation γ-photons about 180°, directly related to the momentum distribution of the annihilating pair (and hence, in condensed matter, essentially the electron momentum density).

Annihilation: The decay of a positron–electron pair, with the emission of energy in the form of γ-radiation; for free particles, two photons are most commonly emitted.

Doppler broadening: The broadening of the annihilation γ-line due to the nonzero momentum of an annihilating positron–electron pair.

Lifetime: The mean life of a positron in a material.

Positron: The antiparticle of the electron.

Positronium (Ps): The quasi-stable positron–electron bound state, existing as *ortho*-Ps (spin 1) or *para*-Ps (spin 0), decaying naturally with the emission principally of three or two photons, respectively.

Specific trapping coefficient: The probability of positron trapping by a particular type of defect per second per defect site. The product of this coefficient and the defect concentration (per atom) gives the trapping rate.

Trapping: The localization of positrons (and sometimes Ps) in defects sites; vacancy-type defects are deep traps and negatively charged impurities are shallow traps.

References

Aers, G.C., Marshall, P.A., Leung, T.C., and Goldberg, R.D. (1995) *Appl. Surf. Sci.*, **85**, 196–209.

Anderson, C.D. (1932) *Science*, **76**, 238–239.

Baker, J.A., Chilton, N.B., Jensen, K.O., Walker, A.B., and Coleman, P.G. (1991) *Appl. Phys. Lett.*, **59**, 2962–2964.

Bamford, D., Dlubek, G., Reiche, A., Alam, M.A., Meyer, W., Galvosas, P., and Rittig, F. (2001) *J. Chem. Phys.*, **115**, 7260–7270.

Bauer, W., Maier, K., Major, J., Schaefer, H.-E., Seeger, A., Carstanjen, H.-D., Decker, W., Diehl, J., and Stoll, H. (1987) *Appl. Phys. A*, **43**, 261–267.

Behringer, R. and Montgomery, C.G. (1942) *Phys. Rev.*, **61**, 222–224.

Bell, R.E. and Graham, R.L. (1953) *Phys. Rev.*, **90**, 644–654.
Berko, S. and Plaskett, J.S. (1958) *Phys. Rev.*, **112**, 1877–1887.
Brandes, G.R., Canter, K.F., and Mills, A.P. (1988) *Phys. Rev. Lett.*, **61**, 492–495.
Brandt, W., Berko, S., and Walker, W.W. (1960) *Phys. Rev.*, **112**, 1289–1295.
Canter, K.F., Coleman, P.G., Griffith, T.C., and Heyland, G.R. (1972) *J. Phys. B*, **5**, L167–L169.
Canter, K.F., Dharmavaram, V., Smirnov, A.G., Wesley, S.A., Wong, K.A., Xie, R., Brandes, G.R., and Mills, A.P. Jr (1994) *Slow Positron Techniques for Solids and Surfaces* (eds E. Ottewitte and A.H. Weiss), Conference Series 303, AIP, New York, p. 385.
Cassidy, D.B. and Mills, A.P. Jr (2007) *Nature*, **449**, 195–197.
Chang, T.-B., Tang, X.-W., and Li, Y.-Q. (1985) *Phys. Lett.*, **B157**, 357–360.
Charlton, M and Humberston, J.W. (2001) *Positron Physics*, Cambridge University Press, Cambridge.
Chen, D.M., Berko, S., Canter, K.F., Lynn, K.G., Mills, A.P. Jr, Roellig, L.O., and Sferlazzo, P. (1987) *Phys. Rev. Lett.*, **58**, 921–924.
Chen, X.M., Canter, K.F., Duke, C.B., Paton, A., Lessor, D.L., and Ford, W.K. (1993) *Phys. Rev. B*, **48**, 2400–2411.
Coleman, P.G., Albrecht, L., Walker, A.B., and Jensen, K.O. (1992) *J. Phys. Condens. Matter*, **4**, 10311–10322.
Coleman, P.G., Burrows, C.P., and Knights, A.P. (2002) *Appl. Phys. Lett.*, **80**, 947–949.
Coleman, P.G., Burrows, C.P., Mahapatra, R., and Wright, N.G. (2007) *J. Appl. Phys.*, **102**, 014106–0141-4.
David, A., Kögel, G., Sperr, P., and Triftshäuser, W. (2001) *Phys. Rev. Lett.*, **87**, 067402–067411.
DeBenedetti, S., Cowan, C.E., and Konneker, W.R. (1949) *Phys. Rev.*, **76**, 440.
DeBenedetti, S., Cowan, C.E., Konneker, W.R., and Primakoff, H. (1950) *Phys. Rev.*, **77**, 205–212.
Deutsch, M. (1951) *Phys. Rev.*, **82**, 455–456.
Dirac, P.A.M. (1930) *Proc. Cambridge Philos. Soc.*, **26**, 361–375.
Dugdale, S.B., Alam, M.A., Wilkinson, I., Hughes, R.J., Fisher, I.R., Canfield, P.C., Jarlborg, T., and Santi, G. (1999) *Phys. Rev. Lett.*, **83**, 4824–4827.
Egger, W., Kögel, G., Sperr, P., Triftshäuser, W., Rödling, S., Bär, J., and Gudladt, H.-J. (2002) *Appl. Surf. Sci.*, **194**, 214–217.
Foster, P.J., Mascher, P., Knights, A.P., and Coleman, P.G. (2007) *J. Appl. Phys.*, **101**, 043702–043703.
Gidley, D.W., Köymen, A.R., and Capehart, T.W. (1982) *Phys. Rev. Lett.*, **49**, 1779–1782.
Gidley, D.W. (1989) *Phys. Rev. Lett.*, **62**, 811–814.
Gidley, D.W., Frieze, W.E., Dull, T.L., Sun, J., Yee, A.F., Nguyen, C.V., and Yoon, D.Y. (2000) *Appl. Phys. Lett.*, **76**, 1282–1284.
Gilbert, S.J., Kurz, C., Greaves, R.G., and Surko, C.M. (1997) *Appl. Phys. Lett.*, **70**, 1944–1946.
Jeavons, A.P., Townsend, D.W., Ford, N.L., Kull, K., Manuel A.A., Fischer O., and Peter, M. (1978) *IEEE Trans. Nucl. Sci.*, **25**, 164–169.
Joliot-Curie, I. and Joliot-Curie, F. (1934) *Nature*, **133**, 201–202.
Kahana, S. (1963) *Phys. Rev.*, **129**, 1622–1628.
Kawasuso, A. and Okada, S. (1998) *Phys. Rev. Lett.*, **81**, 2695–2698.
Landes, H.S., Berko, S., and Zuchelli, A.J. (1956) *Phys. Rev.*, **103**, 828–829.
Lock, D.G., Crisp, V.H.C., and West, R.N. (1973) *J. Physics F*, **3**, 561–570.
Lynn, K.G., Mills, A.P. Jr, West, R.N., Berko, S., Canter, K.F., and Roellig, L.O. (1985) *Phys. Rev. Lett.*, **54**, 1702–1705.
Lynn, K.G., MacDonald, J.R., Boie, R.A., Feldman, L.C., Gabbe, J.D., Robbins, M.F., Bonderup, E., and Golovchenko, J. (1977) *Phys. Rev. Lett.*, **38**, 241–244.
Mader, J., Berko, S., Krakauer, H., and Bansil, A. (1976) *Phys. Rev. Lett.*, **37**, 1232–1236.
Maier, K., Peo, M., Saile, B., Scaefer, H.E., and Seeger, A. (1979) *Phil. Mag. A*, **40**, 701–728.
Mantl, S. and Triftshäuser, W. (1978) *Phys. Rev. B*, **17**, 1645–1652.
Mills, A.P. Jr (1980) *Appl. Phys.*, **22**, 273–276.
Mills, A.P., Platzman, P.M., and Brown, B.L. Jr (1978) *Phys. Rev. Lett.*, **41**, 1076–1079.
Mogensen, O.E. (1974) *J. Chem. Phys.*, **60**, 998–1004.

Mohorovičić, S. (1934) *Astron. Nachr.*, **14**, 93–108.
Mostafa, K.M., Caenegem, N.V., De Baerdemaeker, J., Segers, D., and Houbaert, Y. (2007) *Phys. Stat. Sol.*, **4**, 3554–3558.
Nagai, Y., Hasegawa, H., Tang, Z., Hempel, A., Yubuta, K., Shimamura, T., Kawazoe, Y, Kawai, A., and Kano, F. (2000) *Phys. Rev. B.*, **61**, 6574–6578.
Ociepa, J.G., Schultz, P.J., Griffiths, K., and Norton, P.R. (1990) *Surf. Sci.*, **225**, 281–291.
Ohdaira, T., Suzuki, R., Mikado, T., Ohgaki, H., Chiwaki, M., and Yamazaki, T. (1997) *Appl. Surf. Sci.*, **116**, 177–180.
Ore, A. (1949) *Univ. Bergen Aarb. Naturvit. Rekke*, **9**.
Ore, A. and Powell, J.L. (1949) *Phys. Rev.*, **75**, 1696–1699.
Osmon, P.E. (1965) *Phys. Rev.*, **138**, B216–B218.
Page, L.A. and Heinberg, M. (1956) *Phys. Rev.*, **102**, 1545–1553.
Petersen, K., Thrane, N., and Cotterill, R.M.J. (1974) *Phil. Mag.*, **29**, 9–23.
Roach, T., Bakshi, A., and Canter, K.F. (1995) *Meas. Sci. Technol.*, **6**, 496–501.
Rosenberg, I.J., Weiss, A.H., and Canter, K.F. (1980) *Phys. Rev. Lett.*, **44**, 1139–1142.
Ruark, A.E. (1945) *Phys. Rev.*, **68**, 278.
Saarinen, K., Laine, T., Kuisma, S., Nissila, J., Hautojarvi, P., Dobrzynski, L., Baranowski, J., Pakula, K., Stepniewski, R., Wojdak, M., Wysmolek, A., Suski, T., Leszczynski, M., Grzegory, I., and Porowski, S. (1997) *Phys. Rev. Lett.*, **79**, 3030–3033.
Stewart, A.T. (1957) *Can. J. Phys.*, **35**, 168–183.
Shearer, J.W. and Deutsch, M. (1949) Minutes of the semi-centennial meeting at Cambridge, *Phys. Rev.*, **76**, 462.
Shishido, I., Yasueda, H., Mizuno, M., Araki, H., and Shirai, Y. (2007) *Phys. Stat. Sol. C*, **4**, 3563–3566.
Siegle, A., Stoll, H., Castellaz, P., Major, J., Schneider, H., and Seeger, A. (1997) *Appl. Surf. Sci.*, **116**, 140–144.
Stepanov, S.V. and Byakov, V.M. (2002) *J. Chem. Phys.*, **116**, 6178–6195.
Triftshäuser, W. and Kögel, G. (1982) *Phys. Rev. Lett.*, **48**, 1741–1744.
Van Veen, A., Schut, H., de Vries, J., Hakvoort, R.A., and IJpma, M.R. (1990) *AIP Conference Proceedings*, AIP, New York, Vol. 218, 171–196.
Weber, M.H., Tang, S., Berko, S., Brown, B.L., Canter, K.F., Lynn, K.G., Mills, A.P., Roellig, L.O., and Viescas, A.J. (1988) *Phys. Rev. Lett.*, **61**, 2542–2545.
Weber, M.H., Hunt, A.W., Golovchenko, J.A., and Lynn, K.G. (1999) *Phys. Rev. Lett.*, **83**, 4658–4661.
Weisberg, H. and Berko, S. (1967) *Phys. Rev.*, **154**, 249–257.
Weiss, A.H., Mayer, R., Jibaly, M., Lei, C., Mehl, D., and Lynn, K.G. (1988) *Phys. Rev. Lett.*, **61**, 2245–2248.
West, R.N., Mayers, J., and Walters, P.A. (1981) *J. Phys. E*, **14**, 478–488.
Wheeler, J.A. (1946) *Ann. New York Acad. Sci.*, **48**, 219.
Zhu, J.G., Nadesalingam, M.P., Weiss, A.H., and Tao, M. (2005) *J. Appl. Phys.*, **97**, 103510.
Zitzewitz, P.W., Van House, J.C., Rich, A., and Gidley, D.W. (1979) *Phys. Rev. Lett.*, **43**, 1281–1284.

Further Reading

Coleman, P.G. (ed) (2000) *Positron Beams and their Applications*, World Scientific, Singapore.
Dupasquier, A. and Mills, A.P. Jr (eds) (1994) *Positron Spectroscopy of Solids, Proceedings of the International School of Physics "Enrico Fermi", Course LXXV*, North Holland, Amsterdam.
Gidley, D.W., Peng, H.-G., and Vallery, R.S. (2006) *Ann. Rev. Mater. Res.* **36**, 49–79.
Hirose, M., Nakajyo, T., and Washio, M. (1997) *Appl. Surf. Sci.*, **116**, 63–67.
Jean, Y.C., Mallon, P.E., and Schrader, D.M. (eds) (2003) *Positron and Positronium Chemistry*, World Scientific, Singapore.
Knights, A.P., Mascher, P., and Simpson, P.J. (2007) Proceedings of the 14th International Conference on Positron Annihilation. *Phys. Stat. Sol.*, **4**, 3413–4040.
Krause-Rehberg, R. and Leipner, H.S. (1999) *Positron Annihilation in Semiconductors: Defect Studies*, Springer, Berlin.
Mackenzie, I.K., Schulte, C.W., Jackman, T., and Campbell, J.L. (1973) *Phys. Rev. A*, **7**, 135–145.

Mills, A.P. Jr, Crane, W.S., and Canter, K.F. (eds) (1986) *Positron Studies of Solids, Surfaces and Atoms*, World Scientific, Singapore.

Pethrick, R.A. (1997) *Prog. Polymer Sci.* **22**, 1–47.

Surko, C.M. and Greaves, R.G. (2004) *Phys. Plasmas* **11**, 2333.nhi

Weiss, A (1995) in *The Handbook of Surface Imaging and Visualisation* (ed. A.T. Hubbard), CRC Press.

6
Muon Spectroscopy

Upali A. Jayasooriya and Roger Grinter

6.1	**Introduction** 155	
6.2	**Discovery of the Muon** 155	
6.2.1	Muon Beams 157	
6.2.1.1	Decay Muon Beams 158	
6.2.1.2	Surface Muon Beams 158	
6.2.1.3	Low-Energy Muon Beams 158	
6.3	**Muons Implanted in Matter** 159	
6.3.1	Muon Range and Range Straggling within a Sample 159	
6.4	**Theoretical Aspects of Muon Spectroscopy** 161	
6.4.1	Muonium 163	
6.4.2	Free Radials Produced by Muon Implantation 166	
6.4.3	Selection Rules 169	
6.5	**Variants of μSR Spectroscopy** 169	
6.5.1	Transverse-Field Muon Spectroscopy (TF-μSR) 170	
6.5.2	Longitudinal-Field Muon Spectroscopy (LF-μSR) 172	
6.6	**Longitudinal-Field Muon Spin Relaxation** 173	
6.7	**Avoided Level Crossing or ALC-μSR** 177	
6.8	**Conclusion** 180	
	References 180	

Encyclopedia of Applied Spectroscopy. Edited by David L. Andrews.
Copyright © 2009 WILEY-VCH Verlag GmbH & Co. KGaA, Weinheim
ISBN: 978-3-527-40773-6

6.1
Introduction

Muon spectroscopy covers a large area of scientific endeavor. There have been 11 international conferences, held once every 3 years, with the latest being held at Tsukuba in Japan in July 2008. The proceedings were always published in peer reviewed international journals. There are a few books and journal special editions also dedicated to this subject to which we refer at appropriate places in the following text. The objective of the following chapter is to provide a simple introduction to muon spectroscopy, covering some of the basic techniques used to produce muon beams and to study the properties of the implanted muons from which one infers the properties of the samples under study.

More familiar spin-spectroscopic techniques such as NMR and EPR involve the measurement of photons needed to bring about transitions between the relevant energy levels in the sample under scrutiny, the spectral responses being dependent on the populations of the levels involved. In marked contrast, muon spin relaxation (μSR) involves the direct measurement of the populations of the different energy levels. This is done by detecting the positron that is emitted by each muon when it decays radioactively, and the direction of positron emission gives the orientation of the muon spin at the time of decay. In this way it is possible, for example, to measure the numbers of muon spins parallel and antiparallel to an applied magnetic field and hence to obtain a direct measure of the populations.

6.2
Discovery of the Muon

The fundamental particles of matter consist of quarks, leptons, and the particles that mediate the interactions between them. The muon is a member of the lepton family of fundamental particles and is the dominant constituent of cosmic rays arriving at sea level, at the rate of about one muon per square centimeter of the earth's surface per minute (Caso et al., 1998). The first reported observation of a novel particle, later to be identified as a muon by Neddermeyer and Anderson (1937), was by Kunze (1933) in experiments with Wilson cloud chambers exposed to cosmic rays. This new particle was initially suspected to be that responsible for the strong nuclear

Encyclopedia of Applied Spectroscopy. Edited by David L. Andrews.
Copyright © 2009 WILEY-VCH Verlag GmbH & Co. KGaA, Weinheim
ISBN: 978-3-527-40773-6

force as postulated by Yukawa, (1935) to explain β-decay. However, Conversi et al. (1947) showed that this new particle had a very weak nuclear interaction with matter, unlike the predictions for the Yukawa particle. Theoretical work by Tanikawa et al. (1946) presented a two-meson hypothesis to explain these experimental data, whereas the Yukawa theory predicted that a strongly interacting meson (pion) decayed to a weakly interacting mesotron (muon). The experimental verification of the Tanikawa–Sakata–Inoue theory was due to Lattes et al. (1947) who reported the observation of several two-meson decay events as predicted by theory (see Figure 6.1).

However, the cosmic-ray-generated muon fluxes of $\sim$180/m^{-2} s^{-1} at sea level are insufficient for muon spectroscopy and higher intensities of muons, $>10^6$ per second, are therefore produced in central facility laboratories, the so-called meson factories, that operate powerful particle accelerators. The process of muon generation involves the bombardment of a light element target such as graphite with a high-energy proton beam, produced using a synchrotron or a cyclotron. According to the Yukawa model, the nucleons are held together by the exchange of a pion. The extremely short range of the nuclear force, of the order of 10^{-15} m, predicts the mass of the exchanged particle to be of the order of 200 electron masses. In order to produce pions external to the nucleus, it must be bombarded with other nucleons of sufficient kinetic energy such that the available centre of mass energy is greater than the pion mass of 140 MeV. The following are some nucleon–nucleon reactions that produce pions.

$$\begin{aligned} p+p &\longrightarrow p+n+\pi^+ \\ p+n &\longrightarrow p+n+\pi^0 \\ &\longrightarrow p+p+\pi^0 \\ &\longrightarrow p+p+\pi^- \\ &\longrightarrow n+n+\pi^+ \end{aligned} \quad (6.1)$$

The pions thus produced have a mean lifetime of 26.04 nanoseconds, decaying to produce a muon and a muon-neutrino.

$$\begin{aligned} \pi^+ &\longrightarrow \mu^+ + \nu_\mu \quad \text{and} \\ \pi^- &\longrightarrow \mu^- + \bar{\nu}_\mu \end{aligned} \quad (6.2)$$

The negative muon is a particle and the positive muon an antiparticle, illustrated here by overlining the antineutrinos. Finally, after a half-life of 2.197 microseconds, the muons decay into a positron

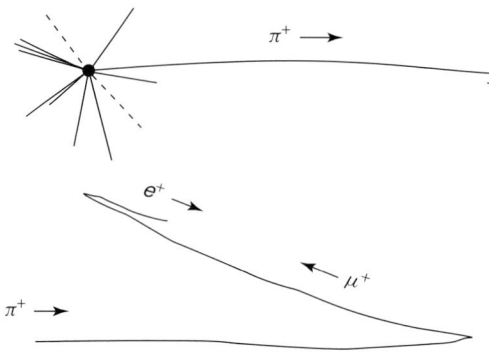

Fig. 6.1 An emulsion exposure showing a two meson event. (a) A cosmic-ray-induced nuclear interaction showing the track of an emitted positively charged pion (π^+) traveling from left to right. (b) The decay of this pion into a muon (μ^+) which moves from right to left. Eventually, the muon itself decays into a positron (e^+). Note that only the charged particles are detected.

and two neutrinos.

$$\mu^+ \longrightarrow e^+ + \nu_e + \bar{\nu}_\mu \quad \text{and}$$
$$\mu^- \longrightarrow e^- + \bar{\nu}_e + \nu_\mu \quad (6.3)$$

The possibility of using muons as a spectroscopic probe for the investigation of materials was first predicted by Garwin et al. in a publication in 1957 on their investigations into the parity violation in the nuclear weak force (Garwin, 1957), where they state that, "It seems possible that polarized positive and negative muons will become a powerful tool for exploring magnetic fields in nuclei, atoms and interatomic regions."

The distribution of the decay positron emission from the muon decay is given by the probability function

$$W(\theta) = 1 + a \cos \theta \quad (6.4)$$

where θ is the angle between the muon spin and the direction of positron emission and the asymmetry factor a increases monotonically with the positron energy up to a value of $a = 1$ for the maximum energy of 52.83 MeV. The change in the angular probability function $W(\theta)$ with positron energy is shown in Figure 6.2. The low-energy positrons normally do not get to the target, being absorbed by the intervening materials and further 'degraders' are often placed between the sample and the detectors to confine measurements to the highest energy positrons, thus increasing the measured asymmetry A. Therefore A is an empirical parameter that should be determined for each experiment. It is this characteristic asymmetry that allows one to monitor the evolution of the muon spin when implanted within a sample, thus giving accurate access to the local microscopic magnetic fields, the fundamental basis of muon spectroscopy.

6.2.1
Muon Beams

There are two main ways of generating muons from the pions that are emitted by the bombardment of the light-element target with high-energy protons. Using a continuous beam of protons from a cyclotron to produce a pseudo-continuous beam of muons as is done in the muon facilities such as those at the Paul-Scherrer Institute in Villigen, Switzerland, and TRI-University

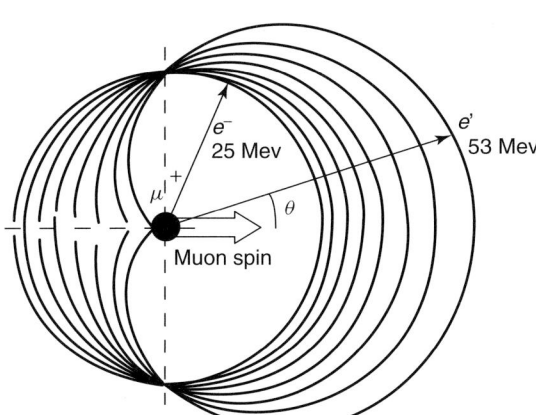

Fig. 6.2 The angular distribution of positrons from positively charged muons for various positron energies.

Meson Factory in Vancouver, Canada. In laboratories such as the ISIS-RAL (Rutherford Appleton Laboratory) facility in the United Kingdom and the most recent J-PARC (Japan Proton Accelerator Research Complex) in Japan, pulsed beams of muons are produced using bunches of high-energy protons from a proton synchrotron, resulting in bunches of muons. The muon beams produced can be further divided into three main categories dependent on the momentum of the muons.

6.2.1.1 Decay Muon Beams

Here, pions with sufficient energy to escape the production target before decay are used. These are directed into a high longitudinal magnetic field, usually provided by a superconducting solenoid a few meters long, and the muon production will then be via the pion decay in flight. Both μ^+ and μ^- over a wide selection of momenta are possible with this method. However, these beams produce muon polarizations of 70–80% due to the finite kinematical acceptance of the channel. Because of their high energy, the muons that are produced in this technique have high penetration depths and are therefore useful in studying samples such as those within high-pressure cells.

6.2.1.2 Surface Muon Beams

This method focuses on low-energy pions that come to rest within the muon target prior to decay to give muons. Because the resulting muons from the decay of pions that are at rest are of low energy (4.1 MeV), it is only those that are produced close to the surface of the target that have sufficient range to escape it. Accordingly, these are called the *surface* muons, and are 100% polarized in any chosen direction of emission from the target (Piper et al., 1976). These muon beams are of very high intensity, because the pion-stopping density near the surface of the target is very high. These muons, being of lower energy, have high stopping rates and short penetration depths of <1 mm. However, only positive muons are produced by this method. Because the negative pions, π^-, when stopped within the graphite target are immediately captured by the carbon nuclei, it is not possible to obtain negative muon, μ^-, beams from this technique.

6.2.1.3 Low-Energy Muon Beams

In order to study the technologically important thin films, surface coatings and so on, there is much interest in producing muons having sufficiently low energies to implant within these shallow depths. There are two successful methods in operation at present. The first uses moderation with a van der Waals bound solid (the most efficient μ^+ moderator up to now) of a few hundred nanometers to achieve the low energies. Surface beams from a secondary beam impinge on such a moderator on its upstream side. Very slow muons that escape from the downstream side of the moderator are collected and used as the low-energy tertiary beam (Morenzoni et al., 1997). The second method, which is confined to pulsed beams, uses re-emitted thermal muonium. In materials such as hot noble metals or low-density silica powder, a fraction (W, Pt, Ir or Re about 0.01–0.05; SiO_2 about 0.10) of the implanted surface muons, after reaching thermal energies, diffuses to the surface and is re-emitted as thermal muonium. The thermal muonium thus produced is ionized by multiphoton excitation using synchronized laser pulses to give thermal μ^+, which is used as the source of a low-energy beam (Mills et al., 1986; Beer et al., 1986; Nagamine, 1987; Chu et al., 1988; Matsushita and Nagamine, 1996).

6.3 Muons Implanted in Matter

We mainly discuss the positive muon since it is the species that is most commonly used for spectroscopic purposes. What happens to the muon from the moment of implantation into the sample? The implanted muon rapidly loses energy first by ionization of atoms and scattering of electrons followed by a series of electron capture and loss reactions, within about a nanosecond of implantation. At the end of the radiolysis track, the muon/muonium still has sufficient energy to propagate a distance of about 1 μm through a solid sample before any data acquisition can commence. Therefore, the region of the sample that is measured is further downstream from any region of radiation damage (Chappert et al., 1984). At the end of its radiolysis track it mainly exists in one of the following forms.

1. On its own as the positive ion, the muon may mimic a proton. When in solid samples, for example, it may occupy an interstitial site and precess in the local magnetic field providing one of the most sensitive microscopic magnetometers known.
2. The proton equivalent thus formed may react with a molecule as in a protonation reaction to give a positively charged *muonated* species.
3. It may hold on to the last electron picked up during the radiolysis and form *muonium*, an analog of a hydrogen atom.
4. The muonium thus formed may then add to an unsaturated center in a molecule of the sample to form a *muoniated* radical species which is charge neutral.
5. Muonium may abstract other atoms from molecules to give diamagnetic species containing the muon.

In essence, μSR spectroscopy consists of techniques to investigate all of these species, and they are discussed later.

6.3.1 Muon Range and Range Straggling within a Sample

The statistical nature of the muon energy loss processes and the momentum spread of the incident muons cause the implantation range, R, of muons in a sample to be associated with a range width or straggling, ΔR, given by the following expressions:

$$R = ap^{3.5} \tag{6.5}$$

$$\Delta R = a \left[0.008 + 12.25 \left(\frac{\Delta p}{p} \right)^2 \right]^{1/2} p^{3.5} \tag{6.6}$$

where p is the muon momentum in MeV/c. For surface muons of $p = 26.5$ MeV/c, $R\rho$ is typically 110 mg cm^{-2} (where ρ is the sample density) and $\Delta R = 20\%$ of R, which gives a range between 0.1 and 1 mm for a typical sample. Therefore, the "surface" muons stop in the bulk of the matter and not at the surface. Some typical values of the range and straggling for different muon energies are shown in Table 6.1 (Lee et al., 1998).

Tab. 6.1 Some typical values of the range and straggling for different muon energies.

Kinetic Energy E (keV)	Range R (nm)	Straggling ΔR (nm)
0.010	0.5	0.3
0.100	2.1	1.3
1.0	13.1	5.4
10.0	75.0	18.0
30.0	244.0	36.0
(surface muons) 4 000	710 000	100 000

Tab. 6.2 A comparison of the properties of the electron, muon, and proton.

	Charge/unit of electronic charge	Spin	Mass	Magnetic moment	Magnetogyric ratio[a], $\gamma/2\pi$ (kHz G^{-1})	Lifetime (microseconds)
e	-1	$\frac{1}{2}$	$m_e = 0.51$ MeV	$-657\,\mu_p$	2802.5	–
μ	± 1	$\frac{1}{2}$	$207\,m_e = 105.7$ MeV $= 0.113\,m_p$	$+3.18\,\mu_p$	13.55	2.19
p	$+1$	$\frac{1}{2}$	$1836\,m_e = 938$ MeV	$+\mu_p$	4.258	–

[a]Larmor frequency: ν (MHz T^{-1}), $= 10 \times \gamma/2\pi$ (kHz G^{-1})

Tab. 6.3 Muonium as an isotope of hydrogen.

Isotope	Mass (m_e)	Reduced mass (m_e)	Bohr radius (nm)	Ionization energy (eV)
Tritium (^{3}H)	5498	0.9998	0.05290	13.603
Deuterium (^{2}H)	3675	0.9997	0.05293	13.602
Hydrogen or protium (^{1}H)	1847	0.9995	0.05292	13.599
Muonium (Mu)	208	0.9952	0.05315	13.541

From a chemical point of view, it is useful to compare the muon with the electron and proton, which shows that the positive muon may be considered as a light proton, with approximately one-ninth the mass of a proton, Table 6.2. This analogy is further reinforced by a comparison of the properties of muonium (μ^+e^-) with that of hydrogen and other isotopes, as shown in Table 6.3.

The close similarity of the reduced masses, Bohr radii and the ionization energies, clearly justifies muonium's claim to be a light isotope of hydrogen (or protium). The light mass of muonium manifests itself, for example, in strong quantum tunneling effects and some of the largest kinetic isotope effects observed to date (Roduner and Münger, 1984).

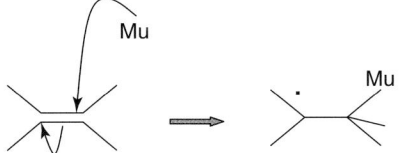

Fig. 6.3 Schematic of muonium addition to an ethene molecule.

Muonium that is formed at the end of the radiation track in materials containing unsaturated molecules may add to these unsaturated centers to give radical species as illustrated in Figure 6.3, by an example of addition to a carbon=carbon double bond.

In this simple form, μSR is essentially the study of the interaction between the resulting unpaired electron and the muon nucleus, similar to electron-nuclear double resonance, ENDOR. The definition of the acronym μSR was given in the very first issue of the μSR Newsletter in 1974 as "μSR stands for Muon Spin Relaxation, Rotation, Resonance, Research or what have you. The intention of the mnemonic acronym is to draw attention to the analogy with NMR and ESR, the range of whose applications is well known. Any study of the interactions of the muon spin by virtue of the asymmetric decay is considered μSR, but this definition is not intended to exclude any peripherally related phenomena, especially if relevant to the use of the muon's magnetic moment as a delicate probe of matter."

6.4 Theoretical Aspects of Muon Spectroscopy

Of the many types of experiment in muon spectroscopy, a large proportion require, for their interpretation, the energy levels that result from the interactions of unpaired electrons with muons and atomic nuclei, usually protons. This interaction, "coupling," is a consequence of the magnetic properties of the foregoing particles and the measurements are normally made in the presence of an applied magnetic field. Thus, the magnetic properties of these three particles are central to the interpretation of the experiments and those useful in the present context are listed in Table 6.2.

All three particles have a spin quantum number, I, of $\frac{1}{2}$ and a z-component of spin, m_I, of $\pm\frac{1}{2}$, and the relationship between the magnitude of the magnetic moment, μ, and spin is

$$\mu = \frac{g[I(I+1)]^{1/2}qh}{4\pi m} = \sqrt{\left(\frac{3}{4}\right)}g\beta \quad (6.7)$$

and for the z-component of μ,

$$\mu_z = \frac{gm_I qh}{4\pi m} = \pm\tfrac{1}{2}g\beta \quad (6.8)$$

where h is Planck's constant, m the mass of the particle, $q = |e|$, and g is the g-factor. In the cases of the electron and proton, the quantities $|e|h/4\pi m_e$ (β_e) and $|e|h/4\pi m_p$ (β_N) are known as the *Bohr magneton* and the *nuclear magneton*, respectively. A positive value for the vector μ indicates that the magnetic moment and angular momentum vectors are parallel, a negative value that they are antiparallel. The magnitudes and signs of the magnetic moments of the electron and muon can be determined theoretically, but for nuclei this is not the case and they must be found by experiment. The signs of the moments determine the orientations of higher and lower energy in a magnetic field. We may take the g-factors to be 2 for the muon and electron and 5.59 for the proton. Equation (6.8) is frequently written in the form

$$\mu_z = \frac{\pm\tfrac{1}{2}\gamma h}{2\pi} \qquad (6.9)$$

in which $g\beta$ is replaced by $\gamma h/2\pi$, where γ is the magnetogyric ratio, the factor which converts the spin angular momentum of the particle in units of $h/2\pi$ into magnetic moment.

The energies of the two possible orientations ($m_I = \pm\tfrac{1}{2}$) of each particle in a magnetic field, B, are determined by the scalar product of the two vectors **B** and **μ**. For a field along the z-axis:

$$E = \boldsymbol{\mu} \cdot \mathbf{B} = \mu_z B_z = \pm\tfrac{1}{2}g\beta B_z$$
$$= \frac{\pm\tfrac{1}{2}\gamma B_z h}{2\pi} \equiv \pm\tfrac{1}{2}\gamma B_z \hbar \qquad (6.10)$$

In Figure 6.4, the splitting of the two energy levels, which are degenerate in zero field, is illustrated with an indication of the corresponding m_I values of the three particles. The frequency, v, of the radiation required to induce transitions between these at a field of 1 T is the frequency with which the magnetic moment precesses in that field which is known as the *Larmor frequency* (Table 6.2). It is almost universal practice in muon spectroscopy to express energies in terms of the corresponding frequencies measured in hertz, (Hz), rather than in true energy units and we frequently adopt that practice here.

Thus, the energy of a particular spin state can be calculated from knowledge of the fundamental constants of the particle and h. In quantum-mechanical form, we express this by writing down the quantum-mechanical operator for the problem, known as the *Hamiltonian* in the case of energy, and solving the resulting Schrödinger equation

$$\hat{H}|\Psi\rangle = E|\Psi\rangle \qquad (6.11)$$

where $|\Psi\rangle$ is the wave function of the spin system whose energy we seek.

For a muon, μ, in a magnetic field, B_z, expressing energy in hertz rather than in energy units, this equation takes the

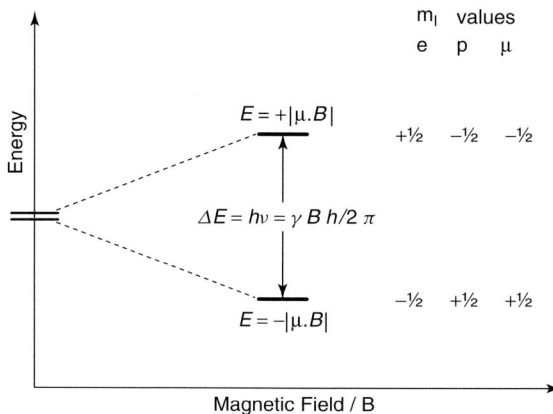

Fig. 6.4 The energies of an electron, a proton, or a muon in a magnetic field, B.

following form, which is known as the *Zeeman* contribution to the energy:

$$\hat{H}|\Psi_\mu\rangle = -B_z \nu_\mu \hat{I}_z |\Psi_\mu\rangle \quad (6.12)$$

where ν_μ is the Larmor frequency of the muon, $\hat{I}_z$ is the operator for the z-component of the spin angular momentum and $|\Psi\rangle$ is the spin function, which is discussed in more detail below. (For electrons we usually write $\hat{S}$ rather than $\hat{I}$, but this is purely a difference of notation).

The interpretation of this equation is the following: $|\Psi_\mu\rangle$ is the wave function representing the muon spin. In general, both the total spin and its z-component would be given, that is, $|I, I_z\rangle = |1/2, \pm 1/2\rangle$, but we frequently give only the value of m_I and write simply $|+1/2\rangle$ or $|-1/2\rangle$. The result of operating with $\hat{I}_z$ on these functions is

$$\hat{I}_z |-1/2\rangle = m_I |-1/2\rangle$$
$$= -1/2 |-1/2\rangle \quad \text{and}$$
$$\hat{I}_z |+1/2\rangle = m_I |+1/2\rangle$$
$$= +1/2 |+1/2\rangle \quad (6.13)$$

Thus, if $|\Psi_\mu\rangle = |+1/2\rangle$ we have

$$\hat{H}|\Psi_\mu\rangle = \hat{H}|+1/2\rangle = -B_z \nu_\mu \hat{I}_z |+1/2\rangle$$
$$= -1/2 B_z \nu_\mu |+1/2\rangle \quad (6.14)$$

The difference in energy of the two states in the presence of a magnetic field is the Zeeman energy

$$\Delta E = -B_z \nu_\mu \{\langle -1/2|\hat{I}_z|-1/2\rangle$$
$$- \langle +1/2|\hat{I}_z|+1/2\rangle\}$$
$$= -B_z \nu_\mu \{-1/2 - 1/2\}$$
$$= B_z \nu_\mu = \gamma_\mu \hbar B_z \quad (6.15)$$

Excellent accounts of the behavior of spin $=1/2$ particles in magnetic fields can be found in the texts by Carrington and McLachlan (1979) and by Harris (1983).

6.4.1
Muonium

The positively charged muon can form a stable entity with the negatively charged electron and the resulting "atom" may be regarded as a light isotope of hydrogen with a mass of one-ninth of that atom, Table 6.3. The electron occupies the orbital of lowest energy, which corresponds to the 1-s atomic orbital of the hydrogen atom. Muonium provides a useful species with which to illustrate the basic theory of the magnetic energy levels of such a system. Since it is of no particular relevance to muon spectroscopy, we ignore the potential energy that results from the Coulombic attraction of the two particles and the kinetic energy of their motion and concentrate upon the very much smaller magnetic terms. In the absence of a magnetic field, the major contribution to this energy arises from the isotropic Fermi contact interaction between the two particles. This is not the dipole–dipole interaction of the magnetic moments of the two particles. The interaction depends upon the electron density at the muon to which only s-orbitals can contribute. It has no classical counterpart and may be represented by the Hamiltonian operator

$$\hat{H} = A\hat{S} \cdot \hat{I} \quad (6.16)$$

where $\hat{S}$ and $\hat{I}$ are the electron and muon spin operators, respectively, and A is the coupling constant, measured in hertz. If the electron occupies an orbital other than an s-orbital, for example, p or d, then there is a classical dipole–dipole interaction between the magnetic moments of the two particles. However, in a freely tumbling system, this interaction averages to zero and its presence is only observable when this freedom is lost. In contrast to the Fermi-contact interaction, the dipolar interaction is anisotropic.

In order to evaluate the scalar product, $\hat{S} \cdot \hat{I}$, we express each operator in terms of its Cartesian components and write

$$\hat{H} = A\hat{\mathbf{S}} \cdot \hat{\mathbf{I}} = A(\hat{S}_x \hat{I}_x + \hat{S}_y \hat{I}_y + \hat{S}_z \hat{I}_z)$$
$$= A(\tfrac{1}{2}[\hat{S}_+ \hat{I}_- + \hat{S}_- \hat{I}_+] + \hat{S}_z \hat{I}_z) \quad (6.17)$$

where $\hat{I}_+ = (\hat{I}_x + i\hat{I}_y)$ and $\hat{I}_- = (\hat{I}_x - i\hat{I}_y)$ with similar expressions for $\hat{S}_\pm$. We rewrite the operator in this way because the results of the operation of $\hat{I}_+$ and $\hat{I}_-$ on spin functions $|+\tfrac{1}{2}\rangle$ and $|-\tfrac{1}{2}\rangle$ are particularly simple.

$$\hat{I}_+|+\tfrac{1}{2}\rangle = 0, \quad \hat{I}_-|+\tfrac{1}{2}\rangle = |-\tfrac{1}{2}\rangle,$$
$$\hat{I}_+|-\tfrac{1}{2}\rangle = |+\tfrac{1}{2}\rangle, \quad \hat{I}_-|-\tfrac{1}{2}\rangle = 0$$
$$(6.18)$$

We now construct the basis spin functions of the problem by taking all four possible combinations of the orientations of the particle spins, $|\alpha_e \alpha_\mu\rangle$, $|\beta_e \alpha_\mu\rangle$, $|\alpha_e \beta_\mu\rangle$ and $|\beta_e \beta_\mu\rangle$. We do not need to include the total spin quantum numbers. To further simplify the notation, we have adopted the widespread convention of writing α and β for $+\tfrac{1}{2}$ and $-\tfrac{1}{2}$, respectively. As an example of the operation of $\hat{H}$ upon a spin function we have

$$\hat{H}|\beta_e \alpha_\mu\rangle = A(\tfrac{1}{2}[\hat{S}_- \hat{I}_+ + \hat{S}_+ \hat{I}_-] $$
$$+ \hat{S}_z \hat{I}_z)|\beta_e \alpha_\mu\rangle = A(\tfrac{1}{2}[\hat{S}_- \hat{I}_+|\beta_e \alpha_\mu\rangle$$
$$+ \hat{S}_+ \hat{I}_-|\beta_e \alpha_\mu\rangle] + \hat{S}_z \hat{I}_z|\beta_e \alpha_\mu\rangle)$$
$$= A(\tfrac{1}{2}[0 + |\alpha_e \beta_\mu\rangle]$$
$$+ (-\tfrac{1}{2} \times \tfrac{1}{2})|\beta_e \alpha_\mu\rangle)$$
$$= \tfrac{1}{2}A|\alpha_e \beta_\mu\rangle - \tfrac{1}{4}A|\beta_e \alpha_\mu\rangle) \quad (6.19)$$

The spin functions α and β are normalized, that is, $\langle \alpha_\mu | \alpha_\mu \rangle = \langle \beta_\mu | \beta_\mu \rangle = 1$, and orthogonal, $\langle \alpha_\mu | \beta_\mu \rangle = \langle \beta_\mu | \alpha_\mu \rangle = 0$. Therefore, the energy matrix of all 16 possible combinations $\langle \Psi_a | \hat{H} | \Psi_b \rangle$, in Hz, is

	$|\alpha_e \alpha_\mu\rangle$	$|\alpha_e \beta_\mu\rangle$	$|\beta_e \alpha_\mu\rangle$	$|\beta_e \beta_\mu\rangle$
$\langle \alpha_e \alpha_\mu |$	$+\tfrac{1}{4}A$	0	0	0
$\langle \alpha_e \beta_\mu |$	0	$-\tfrac{1}{4}A$	$+\tfrac{1}{2}A$	0
$\langle \beta_e \alpha_\mu |$	0	$+\tfrac{1}{2}A$	$-\tfrac{1}{4}A$	0
$\langle \beta_e \beta_\mu |$	0	0	0	$+\tfrac{1}{4}A$

We see that the two spin functions $|\alpha_e \alpha_\mu\rangle$ and $|\beta_e \beta_\mu\rangle$ are eigenfunctions of the operator and give rise to two energy levels of $+\tfrac{1}{4}A$ Hz, while $|\beta_e \alpha_\mu\rangle$ and $|\alpha_e \beta_\mu\rangle$ interact with each other producing a 2×2 matrix, which is readily diagonalized to give energy levels of $+\tfrac{1}{4}A$ and $-\tfrac{3}{4}A$. The energies, wave functions, and $M = m_e + m_\mu$ values are given in Table 6.4.

Thus, in the absence of a magnetic field, muonium has two magnetic energy levels, a triply degenerate state at $+\tfrac{1}{4}A$ with a total spin of 1 and z-component values of $+1$, 0, and -1, and a singly degenerate state at $-\tfrac{3}{4}A$ with total spin and z-component of 0. In muonium, the energy difference between these two states corresponds to a frequency of 4463.3 MHz, which is, of course, the value of A. The transition between

Tab. 6.4 The energy levels of muonium.

Energy	Wave function	M
$+\tfrac{1}{4}A$	$|\alpha_e \alpha_\mu\rangle$	$+1$
$+\tfrac{1}{4}A$	$(1/\sqrt{2})\{|\alpha_e \beta_\mu\rangle + |\beta_e \alpha_\mu\rangle\}$	0
$+\tfrac{1}{4}A$	$|\beta_e \beta_\mu\rangle$	-1
$-\tfrac{3}{4}A$	$(1/\sqrt{2})\{|\alpha_e \beta_\mu\rangle - |\beta_e \alpha_\mu\rangle\}$	0

$\hat{H}$	$\|\alpha_e\alpha_\mu\rangle$	$\|\alpha_e\beta_\mu\rangle$	$\|\beta_e\alpha_\mu\rangle$	$\|\beta_e\beta_\mu\rangle$
$\langle\alpha_e\alpha_\mu\|$	$+\tfrac{1}{4}A - \tfrac{1}{2}B_z(\nu_\mu - \nu_e)$	0	0	0
$\langle\alpha_e\beta_\mu\|$	0	$-\tfrac{1}{4}A + \tfrac{1}{2}B_z(\nu_\mu + \nu_e)$	$+\tfrac{1}{2}A$	0
$\langle\beta_e\alpha_\mu\|$	0	$+\tfrac{1}{2}A$	$-\tfrac{1}{4}A - \tfrac{1}{2}B_z(\nu_\mu + \nu_e)$	0
$\langle\beta_e\beta_\mu\|$	0	0	0	$+\tfrac{1}{4}A + \tfrac{1}{2}B_z(\nu_\mu - \nu_e)$

the equivalent states in the hydrogen atom gives rise to the emission at 1420.4 MHz, which has been used by radio astronomers to map the galaxy.

If we now imagine our muonium atom to be in a magnetic field having a z-component B_z, we add the appropriate Zeeman terms to the Hamiltonian that becomes

$$\hat{H} = A\hat{\mathbf{S}} \cdot \hat{\mathbf{I}} + B_z\nu_e\hat{S}_z - B_z\nu_\mu\hat{I}_z \quad (6.20)$$

resulting in the new energy matrix shown above.

We note that the field only contributes to the diagonal elements of the energy matrix so that the functions $|\alpha_e\alpha_\mu\rangle$ and $|\beta_e\beta_\mu\rangle$ remain eigenfunctions of $\hat{H}$, but with different energies. The 2 × 2 matrix of the mixed functions $|\beta_e\alpha_\mu\rangle$ and $|\alpha_e\beta_\mu\rangle$ is now rather more difficult to diagonalize but the result is given in Table 6.5:

where, for the 2 × 2 matrix,
$\Delta = \tfrac{1}{2}$ (the difference between the diagonal elements) $= -\tfrac{1}{2}B_z(\nu_\mu + \nu_e)$,
$\Omega = +|\{\Delta^2 +$ the square of the off-diagonal element$\}^{1/2}| = +|\{\Delta^2 + \tfrac{1}{4}A^2\}|^{1/2}$,
$\cos\vartheta = |[(\Omega + \Delta)/2\Omega]^{1/2}|$ and $\sin\vartheta = |[(\Omega - \Delta)/2\Omega]^{1/2}|$.

The graph of the four energy levels against magnetic field is the *Breit–Rabi* diagram from which, Figure 6.5, we see that as the magnetic field is increased the triple degeneracy is lost and the levels diverge. At low fields where the coupling, A, has the greatest effect there is a pronounced curvature in two of the graphs, but at higher fields where $A \ll B_z(\nu_\mu + \nu_e)$ they become straight

Tab. 6.5 Energy levels of muonium in a magnetic field.

	Energy	Wave function	M_I
Triplet			
E1	$+\tfrac{1}{4}A - \tfrac{1}{2}B_z(\nu_\mu - \nu_e)$	$\|\alpha_e\alpha_\mu\rangle$	+1
E2	$-\tfrac{1}{4}A + \tfrac{1}{2}([B_z(\nu_\mu + \nu_e)]^2 + A^2)^{1/2}$	$\cos\vartheta\|\alpha_e\beta_\mu\rangle + \sin\vartheta\|\beta_e\alpha_\mu\rangle$	0
E3	$+\tfrac{1}{4}A + \tfrac{1}{2}B_z(\nu_\mu - \nu_e)$	$\|\beta_e\beta_\mu\rangle$	−1
Singlet			
E4	$-\tfrac{1}{4}A - \tfrac{1}{2}([B_z(\nu_\mu + \nu_e)]^2 + A^2)^{1/2}$	$-\sin\vartheta\|\alpha_e\beta_\mu\rangle + \cos\vartheta\|\beta_e\alpha_\mu\rangle$	0

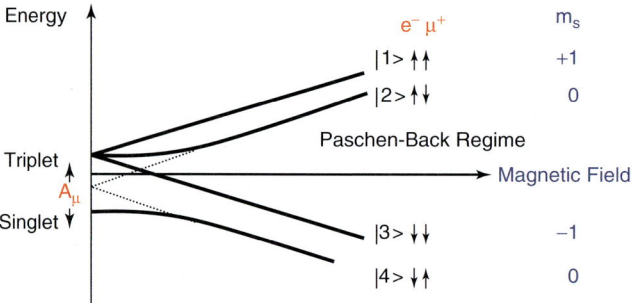

Fig. 6.5 Energy levels of a muon–electron system, Breit–Rabi diagram.

lines. We say that the coupling has been "quenched" by the magnetic field and E2 is almost entirely $|\alpha_e\beta_\mu\rangle$ while E4 is effectively pure $|\beta_e\alpha_\mu\rangle$, that is, ϑ is very close to 0. This is known as the *Paschen-Back* region after the similar effect in atomic spectroscopy where spin–orbit coupling is quenched by magnetic fields.

If muonium is irradiated by radiation of suitable frequencies, then transitions between the above states may be induced. For a muon transition, there must be a change of the z-component of the muon spin, that is, $\alpha_\mu \leftrightarrow \beta_\mu$, but no change in the electron spin. Therefore, the muon transitions $1 \leftrightarrow 2$ and $2 \leftrightarrow 3$, $1 \leftrightarrow 4$ and $3 \leftrightarrow 4$ are, in principle, possible. The probabilities of these transitions are proportional to $(\cos\vartheta)^2$, $(\sin\vartheta)^2$, $(\sin\vartheta)^2$, and $(\cos\vartheta)^2$, respectively. The possible electron transitions, $\alpha_e \leftrightarrow \beta_e$, are the same but their intensities are proportional to $(\sin\vartheta)^2$, $(\cos\vartheta)^2$, $(\cos\vartheta)^2$, and $(\sin\vartheta)^2$, respectively. However, even at modest field strengths, the energy gap between the states $|\beta_e\alpha_\mu\rangle$ and $|\alpha_e\beta_\mu\rangle$ is so much larger than the off-diagonal matrix element that mixes them that $\sin\vartheta \approx 0$ and $\cos\vartheta \approx 1$, so that the transitions with intensity proportional to $(\sin\vartheta)^2$ are extremely weak and we may regard $2 \leftrightarrow 3$ and $1 \leftrightarrow 4$ as electron transitions and $1 \leftrightarrow 2$ and $3 \leftrightarrow 4$ as muon transitions. Transitions in which the spins of both particles change, that is, $2 \leftrightarrow 4$ and $1 \leftrightarrow 3$, are forbidden in simple spectroscopy, but they make important contributions to relaxation processes in anisotropic systems.

6.4.2
Free Radicals Produced by Muon Implantation

From the point of view of μSR, a much more interesting case is the one in which an unpaired electron interacts magnetically with an implanted muon and the nuclei of one or more hydrogen atoms. The magnetic Hamiltonian must now contain terms representing the Zeeman effect upon the muon, the electron and all the protons involved. There is no Fermi-contact interaction between the muon and the protons since, once the muon has implanted, all particles, except the electron, are confined to fixed locations in the molecule. The Hamiltonian has the form

$$\hat{H} = A_{e\mu}\hat{S}_e \cdot \hat{I}_\mu + \sum_i A_{ei}\hat{S}_e \cdot \hat{I}_i$$
$$+ B_z v_e \hat{S}_z - B_z v_\mu \hat{I}_{\mu z}$$
$$- \sum_i B_z v_{ip} \hat{I}_{iz} \qquad (6.21)$$

where $\hat{S}_e$, $\hat{I}_\mu$ and $\hat{I}_i$ are the operators for the total spin angular momentum of the electron, muon, and ith proton, respectively; $\hat{S}_z$, $\hat{I}_{\mu z}$ and $\hat{I}_{iz}$ are the corresponding operators for the z-components of the angular momentum, $A_{e\mu}$ is the muon–electron coupling constant, A_{ei} the coupling between the ith proton and the electron and the last three terms give the Zeeman energies of the various particles. The problem, though simple in principle, increases rapidly in size as the number of protons increases; the number of basis states doubling with each proton added. Though the energy matrix always separates into blocks, each of which involves only those states that have the same value of M_I, the total of all the m_I values of the particles involved, these submatrices also increase rapidly in size and we need to modify our approach to the problem. We recognize that the Zeeman term for a proton is about $1/3$ of that for a muon and about $1/70$ of that for an electron (Table 6.2) and we know from experiment that the coupling between electron and muon is usually much larger than that between electron and proton. For the many problems where these criteria apply, we adopt a perturbation approach and write the Hamiltonian, Eq. (6.21), in the form

$$\hat{H} = \hat{H}_0 + \hat{H}^1 \quad (6.22)$$

where

$$\hat{H}_0 = A_{e\mu}\hat{\mathbf{S}}_e \cdot \hat{\mathbf{I}}_\mu + B_z v_e \hat{S}_z - B_z v_\mu \hat{I}_{\mu z} \quad (6.23)$$

is the Hamiltonian, which we have used above for muonium, and the perturbation is

$$\hat{H}^1 = \sum_i A_{ei}\hat{\mathbf{S}}_e \cdot \hat{\mathbf{I}}_i - \sum_i B_z v_{ip}\hat{I}_{iz} \quad (6.24)$$

We now take the eigenfunctions of the Hamiltonian $\hat{H}_0$, which we have found above, modify them to include the proton spin functions and determine their energies in the presence of the Hamiltonian $\hat{H}^1$. The sum of the energies due to $\hat{H}_0$ and $\hat{H}^1$ then gives us our result.

As an example, we take the eigenfunction $\phi 2$ from Table 6.5. The results for $\hat{H}_0$ are

$$\phi 2 = \cos\vartheta|\alpha_e\beta_\mu\rangle + \sin\vartheta|\beta_e\alpha_\mu\rangle,$$
$$E2 = -\tfrac{1}{4}A + \tfrac{1}{2}([B_z(v_\mu + v_e)]^2 + A^2)^{1/2}$$
$$(6.25)$$

If the radical also contains n protons, then there are 2^n possible proton spin-function combinations, which must be combined, one at a time, with $\phi 2$. A typical one of these augmented functions may be written as

$$\phi 2' = \cos\vartheta|\alpha_e\beta_\mu X_p\rangle + \sin\vartheta|\beta_e\alpha_\mu X_p\rangle \quad (6.26)$$

where X_p represents a sequence of n proton m_I values. To determine the effect of the perturbation, $\hat{H}^1$, we evaluate

$$\langle\phi 2'|\hat{H}^1|\phi 2'\rangle = \langle\cos\vartheta\langle\alpha_e\beta_\mu X_p|$$
$$+ \sin\vartheta\langle\beta_e\alpha_\mu X_p||\hat{H}^1|\cos\vartheta|\alpha_e\beta_\mu X_p\rangle$$
$$+ \sin\vartheta|\beta_e\alpha_\mu X_p\rangle\rangle$$
$$= \cos^2\vartheta\langle\alpha_e\beta_\mu X_p|\hat{H}^1|\alpha_e\beta_\mu X_p\rangle$$
$$+ \sin^2\vartheta\langle\beta_e\alpha_\mu X_p|\hat{H}^1|\beta_e\alpha_\mu X_p\rangle$$
$$+ \cos\vartheta\sin\vartheta\langle\alpha_e\beta_\mu X_p|\hat{H}^1|\beta_e\alpha_\mu X_p\rangle$$
$$+ \sin\vartheta\cos\vartheta\langle\beta_e\alpha_\mu X_p|\hat{H}^1|\alpha_e\beta_\mu X_p\rangle$$
$$(6.27)$$

Taking the first of the above terms, we have

$$\cos^2 \vartheta \langle \alpha_e \beta_\mu X_p | \hat{H}^1 | \alpha_e \beta_\mu X_p \rangle$$
$$= \cos^2 \vartheta \langle \alpha_e \beta_\mu X_p | \sum_i A_{ei} \hat{S}_e \cdot \hat{I}_i$$
$$- \sum_i B_z \nu_{ip} \hat{I}_{iz} | \alpha_e \beta_\mu X_p \rangle \quad (6.28)$$

and since the Hamiltonian, $\hat{H}^1$, contains no operators involving the muon we factor the muon spin function out giving

$$\cos^2 \vartheta \langle \beta_\mu | \beta_\mu \rangle \langle \alpha_e X_p | \sum_i A_{ei} \hat{S}_e \cdot \hat{I}_i$$
$$- \sum_i B_z \nu_{ip} \hat{I}_{iz} | \alpha_e X_p \rangle = \cos^2 \vartheta \langle \beta_\mu | \beta_\mu \rangle$$
$$[\langle \alpha_e X_p | \sum_i A_{ei} \hat{S}_e \cdot \hat{I}_i | \alpha_e X_p \rangle$$
$$- \langle X_p | \sum_i B_z \nu_{ip} \hat{I}_{iz} | X_p \rangle]$$
$$= \cos^2 \langle \alpha_e X_p | \sum_i A_{ei} \hat{S}_e \cdot \hat{I}_i | \alpha_e X_p \rangle$$
$$- \cos^2 \vartheta \langle X_p | \sum_i B_z \nu_{ip} \hat{I}_{iz} | X_p \rangle \quad (6.29)$$

since the spin functions α and β are normalized. The second term gives a very similar result:

$$\sin^2 \vartheta \langle \alpha_\mu | \alpha_\mu \rangle \langle \beta_e X_p | \sum_i A_{ei} \hat{S}_e \cdot \hat{I}_i$$
$$- \sum_i B_z \nu_{ip} \hat{I}_{iz} | \beta_e X_p \rangle$$
$$= \sin^2 \vartheta \langle \beta_e X_p | \sum_i A_{ei} \hat{S}_e \cdot \hat{I}_i | \beta_e X_p \rangle$$
$$- \sin^2 \vartheta \langle X_p | \sum_i B_z \nu_{ip} \hat{I}_{iz} | X_p \rangle \quad (6.30)$$

The third and fourth terms give zero because, on factoring out the muon functions, we find $\langle \alpha_\mu | \beta_\mu \rangle$ or $\langle \beta_\mu | \alpha_\mu \rangle$, both of which are zero because of the orthogonality of the α and β spin functions. Summing the two contributions, noting the fact that there is a change from α_e to β_e on going from the first to the second, we have

$$\langle \phi 2' | \hat{H}^1 | \phi 2' \rangle = (\cos^2 \vartheta - \sin^2 \vartheta)$$
$$\times \left(\langle \alpha_e X_p | \sum_i A_{ei} \hat{S}_e \cdot \hat{I}_i | \alpha_e X_p \rangle \right.$$
$$\left. - \langle X_p | \sum_i B_z \nu_{ip} \hat{I}_{iz} | X_p \rangle \right) \quad (6.31)$$

The first term above gives the contribution to the energy of all the electron–proton couplings and the second is the sum of the proton Zeeman terms. Thus the new energy, $E2'$, in the presence of the protons is the energy determined for muonium plus the calculated perturbation, that is,

$$E2' = -\tfrac{1}{4}A + \tfrac{1}{2}([B_z(\nu_\mu + \nu_e)]^2$$
$$+ A^2)^{1/2} + \langle \phi 2 | \hat{H}^1 | \phi 2 \rangle \quad (6.32)$$

and the other energies, corrected for the presence of the protons, are

$$E1' = +\tfrac{1}{4}A - \tfrac{1}{2}B_z(\nu_\mu - \nu_e)$$
$$+ \langle \phi 1' | \hat{H}^1 | \phi 1' \rangle \quad (6.33)$$
$$E3' = +\tfrac{1}{4}A + \tfrac{1}{2}B_z(\nu_\mu - \nu_e)$$
$$+ \langle \phi 3' | \hat{H}^1 | \phi 3' \rangle \quad (6.34)$$
$$E4' = -\tfrac{1}{4}A - \tfrac{1}{2}([B_z(\nu_\mu + \nu_e)]^2$$
$$+ A^2)^{1/2} + \langle \phi 4 | \hat{H}^1 | \phi 4 \rangle \quad (6.35)$$

Expressions for the corresponding wave functions are rather complicated and tedious to derive. The general method for obtaining perturbed wave functions is described in all texts on quantum mechanics.

6.4.3
Selection Rules

Although the stimulation of transitions by electromagnetic radiation plays a much smaller role in μSR than in other resonance spectroscopies, a few words on the subject is justified. The interaction between the muon magnetic moment and the fluctuating magnetic field of incident radiation is of exactly the same form as that between the muon and the applied field, which has been described above. Therefore, the selection rules are determined by the orientation of the radiation field relative to that of the muon. For a muon spin function $|\psi\rangle$, the components are given by Eq. (6.36), that is,

$$\mu_x = \gamma \hat{I}_x |\psi\rangle = \tfrac{1}{2}(\hat{I}_+ + \hat{I}_-)|\psi\rangle \quad (6.36a)$$

$$\mu_y = \gamma \hat{I}_y |\psi\rangle = -\tfrac{1}{2}i(\hat{I}_+ - \hat{I}_-)|\psi\rangle \quad (6.36b)$$

$$\mu_z = \gamma \hat{I}_z |\psi\rangle \quad (6.36c)$$

The results of these operations on the muon spin functions $|+\tfrac{1}{2}\rangle$ and $|-\tfrac{1}{2}\rangle$ have been given above (Eq. 6.18) and we see that, because $\langle\alpha|\alpha\rangle = \langle\beta|\beta\rangle = 1$ whereas $\langle\alpha|\beta\rangle = 0$, $\hat{I}_+$ and $\hat{I}_-$ link spin states in which the muon spin functions differ such that $\Delta m_\mu = \pm 1$, while $\hat{I}_z$ links those in which the muon spin function is unchanged, that is, $\Delta m_\mu = 0$. The same is true for the proton and the electron. Therefore, if all the spins are quantized along the z-axis, which also defines the direction of the applied field, radiation with its magnetic field orientated along x or y can stimulate transitions between states, which differ in their muon spin component so that the z-component of the total spin, $M = m_e + m_\mu + \Sigma_p m_p$, can change by $\Delta M = \pm 1$. In the case of radiation with its magnetic field along z, $\Delta M = 0$. In the changes of spin, which comprise the relaxation process, changes of the spin state having $\Delta M = \pm 1$ and 0 are frequently observed. The former are termed *muon spin-flip transitions* and must be accompanied by some exchange of energy with radiation or the molecular environment. The latter can occur when a muon spin change is accompanied by a simultaneous proton spin change and no exchange of energy is required. They are termed *flip-flop transitions* and are only observed where the immediate environment of the particles is anisotropic as, for example, when there is dipolar coupling.

A final observation of relevance to the calculation of muon energy levels is worth making. The energies associated with muon spectroscopy are extremely small relative to most of the other energy terms found in atoms and molecules. Consequently, energies, which in other branches of spectroscopy are frequently neglected, may be sufficiently large to be important in muon work. For example, the spin-orbit coupling in the valence p-orbitals of atoms in the first and second rows of the periodic table may well be important in determining the energy levels observed in the μSR of molecules containing such elements.

6.5
Variants of μSR Spectroscopy

There are essentially two techniques of obtaining information from muons implanted into matter: transverse-field muon spectroscopy (TF-μSR) and longitudinal-field muon spectroscopy (LF-μSR). Zero-field muon spectroscopy (ZF-μSR), avoided level crossing muon spectroscopy (ALC-μSR) or level-crossing muon spectroscopy (LC-μSR), and radio frequency muon spectroscopy (RF-μSR)

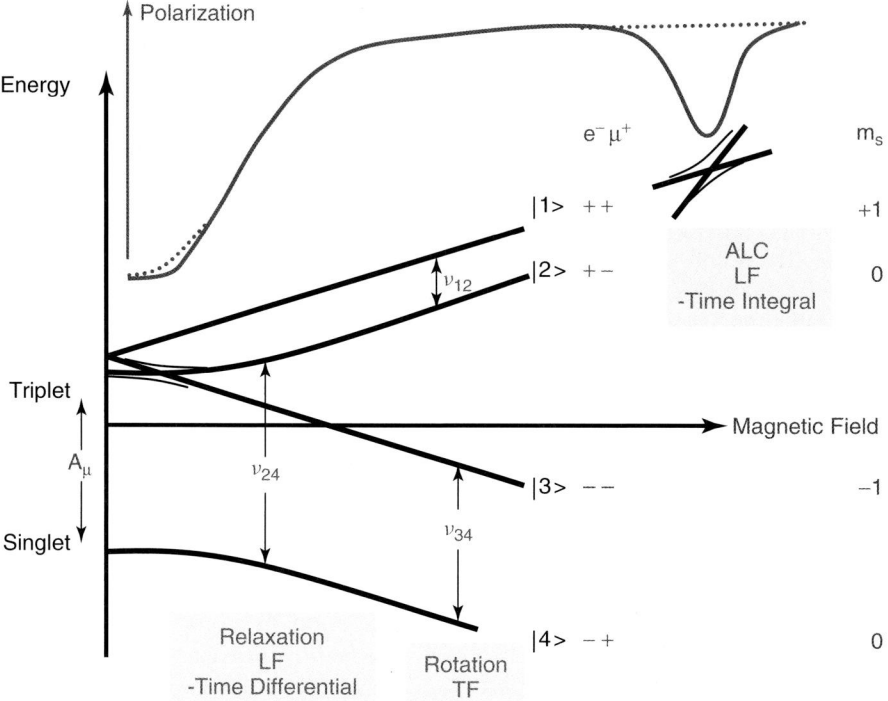

Fig. 6.6 Energy levels of a muon–electron system: Breit–Rabi diagram with a schematic representation of the transitions observed with different experimental arrangements.

are variants of the latter technique. Figure 6.6 shows the energy levels of a muon–electron system with a schematic representation of the transitions observed with different experimental arrangements. The choice of technique to be used for a particular experiment is dependent on factors such as the state of the sample and the information required. The following are descriptions of each of these techniques.

6.5.1
Transverse-Field Muon Spectroscopy (TF-μSR)

An external magnetic field that is perpendicular to the muon spin direction at the moment of implantation is used in TF-μSR, which is a time-differential technique. The implanted muon precesses around this applied field and the precession signals, when evaluated, provide information about the hyperfine fields experienced by the implanted muon. Up to four sets of detectors surrounding the sample, placed around a plane perpendicular to the applied magnetic field, are used to measure the positrons that are emitted from the radioactively decaying muons. A schematic diagram of such a spectroscopic arrangement is shown in Figure 6.7.

The experiment using a pseudocontinuous muon source consists of measuring the time interval between detecting the incoming muons and the corresponding decay positrons using a fast clock. A time

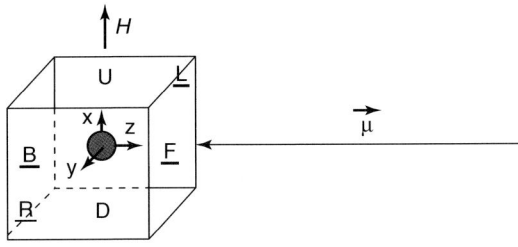

Fig. 6.7 A schematic diagram of a transverse-field muon spectrometer (TF-μSR). The muon beam arrives along the z direction of the coordinate system at the sample, which is depicted as a sphere. The muon spin direction is antiparallel to its momentum. The external magnetic field H is transverse to the muon spin direction at the point of implantation and is shown here as the x-direction. A cube around the sample shows the positions of the detectors with the faces defined with respect to the incoming muon spin direction. F – forward; B – backward; U – up; D – down; R – right and L – left. In a transverse-field muon experiment, the scintillation counters of the decay positrons are placed around the sample on the F̱, Ḇ, Ṟ, and Ḻ faces.

resolution in the range of nanoseconds is thus used to measure the lifetime of muons implanted in the sample and is stored as a histogram. Events involving more than one muon stopping in the sample are rejected using fast electronics in the counting circuit. For an unpolarized beam of muons, this shows the radioactive decay of the muons as a simple exponential with the characteristic lifetime of τ_μ. In the case of a polarized beam of muons, the anisotropy of the decay positron emission results in a modulation of the positron counts in each detector. This additional modulation is analogous to the free induction decay, (FID), observed in the case of FT-NMR or FT-EPR of a spin $= 1/2$ particle after a 90° RF-pulse (Figure 6.8). Information about the muon transitions is given by the frequency, initial amplitude, and the phase of this FID signal.

The histogram of data obtained from a TF-μSR experiment may be expressed as

$$H(t) = N_0 \left[BG + \exp\left(\frac{-t}{\tau_\mu}\right) \{1 + F(t)\} \right]$$

(6.37)

where N_0 is a measure of the total number of counts, BG is the background contribution and $F(t)$ is the contribution from the FID. In general, when there is more than one frequency due either to the presence of a number of species or several sites within the same species, the FID function may be expressed as a sum of contributions from each of these frequencies.

$$F(t) = \sum_i A_i \exp(-\lambda_i t) \cos(\omega_i t + \phi_i)$$

(6.38)

where ω_i is the angular frequency ($\omega_i = 2\pi \nu_i$) of the ith component, A_i its amplitude (asymmetry), λ_i a damping constant, and ϕ_i its initial phase.

The asymmetries A_i depend on the beam polarization P, the asymmetry coefficient a (this accounts for the relative efficiencies of the detectors), the solid

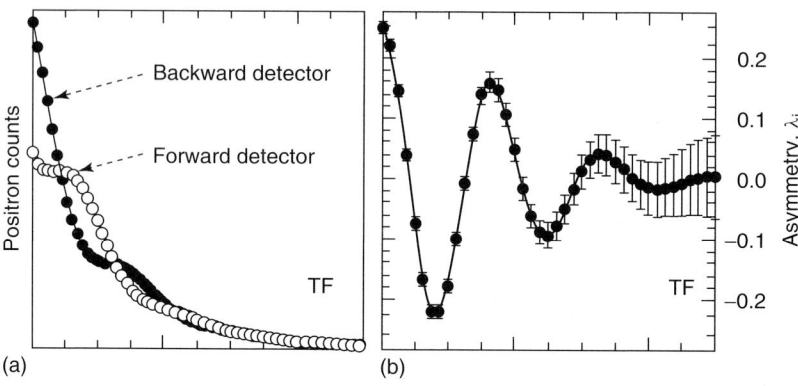

Fig. 6.8 Simulation of raw data from two of the detectors is shown in (a). The FID left over after the removal of the exponential radioactive decay is shown in (b).

angle subtended by the scintillation counters, and on any reaction and relaxation rates. The asymmetries are converted to absolute *fractional muon polarizations* P_i by calibration against a standard. Depending on the environments of the muons in the sample, it is possible to divide this further into fractions due to diamagnetic states, P_D, muonium, P_M, and radical states, P_R. The sum of these fractions is often observed to be less than the initial polarization of the muon beam. This is due to a loss of some of the polarization at shorter times compared to the time of the experiment and is denoted as the *missing or lost fraction*, P_L.

The damping of the FID signal is exponential with a damping constant λ_i, which is mainly dependent on two factors. Any chemical reactions of importance on the timescale of the experiment (λ_{ch}) and any other additional relaxation processes (λ_0) that are present. The latter may be due to a multitude of effects such as the inhomogeneities of the magnetic field, electron spin flip processes, and other physical or even chemical relaxation processes. In the presence of a solute of concentration [S]

the damping constant may be written as

$$\lambda_i = \lambda_0 + \lambda_{ch} = \lambda_0 + k[S] \quad (6.39)$$

Since all muon experiments involve either a single muon in the sample or a pulse of about a couple of thousand muons in the sample at a time, these experiments occur under ideal conditions of "infinite dilution" and hence follow pseudo-first-order kinetics.

The initial phase ϕ_i depends on experimental factors as well as reaction or relaxation in a precursor of the observed state.

6.5.2
Longitudinal-Field Muon Spectroscopy (LF-μSR)

An external magnetic field that is parallel to the muon spin direction at the moment of implantation is used in LF-μSR, which is carried out as a time-differential technique in LF-μSR studies and as a time-integral technique when used as in ALC-μSR or LC-μSR. The experimental arrangement consists of applying a magnetic field parallel to the muon spin direction and measuring the

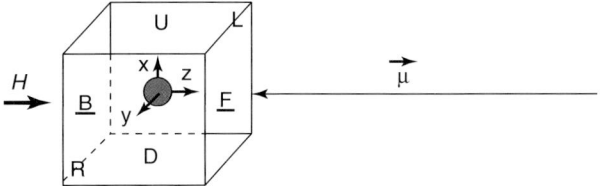

Fig. 6.9 The muon beam is along the z direction of the coordinate system at the sample, which is depicted as a sphere. The muon spin direction is antiparallel to its momentum. The external magnetic field H is parallel to the muon spin direction at the point of implantation and is shown here as the z-direction. A cube around the sample shows the positions of the detectors as in Figure 6.7. In a longitudinal-field muon experiment, the scintillation counters of the decay positrons are placed on F̲ and B̲ faces.

asymmetry of the decay at the time of implantation into the sample by placing two positron scintillation counters in the forward (F̲) and backward (B̲) directions to the muon spin. The asymmetry is measured either as a function of time for LF-μSR measurements or as a time-averaged asymmetry for the ALC-μSR or LC-μSR. A schematic diagram of an LF-μSR spectrometer is shown in Figure 6.9.

6.6 Longitudinal-Field Muon Spin Relaxation

The simplest case of a single muon in a magnetic field is used here to illustrate some of the salient features of LF-μSR. The energy levels of spin $= \frac{1}{2}$ particles in a magnetic field are shown in Figure 6.4.

It is useful for comparisons with NMR spectroscopy to note that the value γ_μ (or ν_μ) is 3.18 times that for a proton.

Muon beams are produced with very high spin polarization, Figure 6.2, with the so-called surface muon beams showing almost 100% spin polarization. We denote the fractional populations of the Zeeman energy levels shown in Figure 6.4 by n_- and n_+. At the time of implantation of the muons into the sample from a 100% spin-polarized beam of muons, these populations are either $n_- = 1$ and $n_+ = 0$ or $n_- = 0$ and $n_+ = 1$, depending on the relative directions of the longitudinal magnetic field and the muon spin direction. Therefore, the net polarization of the muons in the sample at the moment of implantation may be written as

$$P = \frac{n_- - n_+}{n_- + n_+} = \pm 1 \quad (6.40)$$

It is interesting to compare this with the situation that one encounters with EPR and NMR (except with dynamic nuclear polarization) spectroscopies where it is the difference in population between these levels at thermal equilibrium that is of relevance. Using Eq. (6.15), the population differences in those cases are determined by the Boltzmann distribution, that is,

$$\frac{n_+}{n_-} = \exp\left(\frac{-\Delta E}{k_B T}\right) = \exp\left(\frac{-B h \gamma_\mu}{2\pi k_B T}\right) \quad (6.41)$$

Thus the net polarization at thermal equilibrium, $P(\infty)$, is

$$P(\infty) = \tanh\left(\frac{Bh\gamma_\mu}{4\pi k_B T}\right) \qquad (6.42)$$

However, the Zeeman energies are very much smaller than the thermal energy of the sample, that is, $Bh\gamma_\mu \ll k_B T$, so that we require only the first term in the expansion of the tanh function and

$$P(\infty) \approx \frac{h\gamma_\mu B}{4\pi k_B T} \qquad (6.43)$$

Therefore, from the moment of implantation, with almost 100% spin polarization, the muons will try to attain the thermal equilibrium polarization. The latter at room temperature and at a magnetic field of 1 T is $\approx 0.0005\%$. However, this requires a mechanism that would assist the muon spins to flip in a longitudinal magnetic field. In a purely longitudinal field of sufficient magnitude to be away from any resonance conditions, and in the absence of any molecular processes, one would not expect any observable spin-flip, that is, relaxation. Of the several types of processes that may influence the transitions between muon spin-down and spin-up eigenstates, the effect of molecular dynamics as a mechanism is discussed here.

Assuming the upward and downward transition probabilities to be $W_\uparrow$ and $W_\downarrow$, respectively, following Cox (1998) one may write the simple rate equations for the change of the two populations as

$$\frac{dn_+}{dt} = -W_\downarrow n_+ + W_\uparrow n_- \qquad (6.44)$$

$$\frac{dn_-}{dt} = +W_\downarrow n_+ - W_\uparrow n_- \qquad (6.45)$$

Combining these two equations we have

$$\frac{d}{dt}\delta P = -(W_\uparrow + W_\downarrow)\delta P \qquad (6.46)$$

where $\delta P = P(t) - P(\infty)$ is the difference between the polarizations at time t and after reaching thermal equilibrium. Solution of this differential equation gives

$$\delta P(t) \propto e^{-(W_\uparrow + W_\downarrow)t} = e^{-\lambda t} \qquad (6.47)$$

Hence, a simple exponential decay is expected with a relaxation rate λ, equal to the sum of the upward and downward transition probabilities. In analogy with the notation used in conventional magnetic resonance, a relaxation time can also be defined as

$$T_1 = \lambda^{-1} \qquad (6.48)$$

Under experimental conditions, one often encounters situations where the initial polarization is less than 100%. This may be represented by the following equation

$$P(t) = p_1 G_{//}(t) + p_2 \qquad (6.49)$$

or in terms of the asymmetries or amplitudes of the μSR signals,

$$a(t) = a_1 G_{//}(t) + a_2 \qquad (6.50)$$

where $G_{//}(t)$ is called the *relaxation function*, defined so that $G_{//}(0) = 1$, the subscript $//$ specifying that this is longitudinal relaxation.

Now consider a molecular dynamic process that would influence the relaxation of a muon spin. This dynamic process should essentially satisfy two main conditions.

First of all it should cause a fluctuation of the muon's magnetic environment. In addition, these fluctuations must have a component transverse to the main or static average longitudinal magnetic field because any fluctuation parallel to the muon beam polarization is not effective in bringing about a transition between the two stationary states of the muon as shown in Figure 6.4.

The second condition is the need for a quantum of energy equal to the Zeeman splitting, so as to bring about a transition between the spin-up and spin-down states. The implication of this is that the power spectrum of the molecular fluctuations must contain a significant density of states at the Larmor frequency, corresponding to the Zeeman splitting.

$$\nu_\mu = \frac{\Delta E}{h} = \frac{\gamma_\mu B}{2\pi} \quad \text{or} \quad \omega_\mu = \gamma_\mu B \tag{6.51}$$

The transition probability W between the two eigenstates, assuming $W_\uparrow = W_\downarrow$ for simplicity, so that $\lambda = 2W$, is found using the time-dependent perturbation theory in the form of the Fermi's golden rule to be

$$W \approx (\gamma_\mu B_l)^2 J(\omega_\mu) \tag{6.52}$$

where B_l is the root mean square amplitude of transverse fluctuations of the local field and $J(\omega_\mu)$ is the spectral density function or power spectrum of the fluctuations. Spin-lattice relaxation samples this spectrum at the angular Larmor frequency, $\omega = \omega_\mu$. The usual form of the spectral density function is a simple Lorentzian, characterized by a single correlation time τ_c.

$$J(\omega) = \frac{\tau_c}{\pi(1 + \omega^2 \tau_c^2)} \tag{6.53}$$

This gives the relaxation rate as

$$\lambda = T_1^{-1} \propto (\gamma_\mu B_l)^2 \frac{\tau_c}{1 + \omega_\mu^2 \tau_c^2} \tag{6.54}$$

This is commonly known as the *BPP* expression because it was first formulated by Bloembergen, Purcell and Pound (1948).

The theory of the case of a muon–electron (muonium like) system was considered in detail earlier in this chapter, for an isotropic, static (or time averaged) hyperfine interaction between the muon and the electron. Cox (1998) has discussed the relaxation mechanisms in this type of system in detail with experimental results and simulations to clearly demonstrate the various aspects of such a system. The most likely candidate that would account for the relaxation due to molecular dynamics is shown to be the fluctuation of the hyperfine interaction or the so-called Fermi contact term. The hyperfine constant is, in fact, a thermal average over the different vibrational modes of the molecule. Therefore, the molecular vibrations or librations that provide a relaxation mechanism will modulate the hyperfine constant. Hyperfine interaction is in general anisotropic, except in the gas or liquid state when the fast tumbling of the molecules averages out the anisotropy leaving only the isotropic part. One can think of two separate mechanisms of relaxation depending on the modulation of either the isotropic part or the anisotropic part.

Modulation of the *isotropic* hyperfine interaction has the same spin operator $(\hat{\mathbf{I}} \cdot \hat{\mathbf{S}})$ as the time-averaged hyperfine interaction, thus making this mechanism ineffective in zero field. In an applied field, it will only induce the flip-flop transition $2 \leftrightarrow 4$ (Figure 6.6). The contribution of this mechanism to the observed relaxation rate begins by increasing quadratically with field, reaching a plateau region until eventually quenched by the $J(\omega_{2\leftrightarrow 4})$ spectral density term. The modulation of the *anisotropic* hyperfine interaction was shown to induce the other transitions $1 \leftrightarrow 2$, $1 \leftrightarrow 4$, $2 \leftrightarrow 3$, and $3 \leftrightarrow 4$ (Figure 6.6). It is useful to note that the μSR response is, in principle, a superposition of several exponential terms. However, in practice, these appear in combination such that, whenever the eigenvalues differ greatly in magnitude, only one of these will have a significant weight or contribution. Alternatively, when the different eigenvalues have comparable weight, they have very similar magnitudes. Therefore, the muon relaxation function due to molecular dynamics is, in practice, indistinguishable from a single exponential from which one can extract an effective relaxation rate, $\lambda = T_1^{-1}$.

An investigation of the dependence of the muon spin relaxation rate with temperature may be used to obtain information on molecular dynamics. Reorientation dynamics may depolarize the muons by causing anisotropic or dipolar terms in the electron–muon hyperfine interaction to fluctuate. The relaxation rate is observed to peak at temperatures when the reorientation rate of the molecule matches the frequency of the dominant transition between the coupled muon–electron spin states. The correlation time, τ, for the dynamic process at each temperature may be obtained using the equations according to Cox and Sivia (1994, 1997).

$$\lambda = 2(2\pi\delta A)^2 \frac{x^2}{1+x^2} \frac{\tau}{1+\omega^2\tau^2} \quad (6.55)$$

where $(2\pi\delta A)^2 x^2/(1+x^2)$ is the matrix element responsible for the transition, x is the reduced field in units of hyperfine field, and $\tau/(1+\omega^2\tau^2)$ is the spectral density function. The transition frequency $\omega (= 2\pi A\sqrt{1+x^2})$ used to extract the correlation time τ is that between states $|2\rangle = \cos\vartheta|\alpha_e\beta_\mu\rangle + \sin\vartheta|\beta_e\alpha_\mu\rangle$ and $|4\rangle = \cos\vartheta|\beta_e\alpha_\mu\rangle - \sin\vartheta|\alpha_e\beta_\mu\rangle$ in the Breit–Rabi diagram, Figure 6.5. This is the only transition induced in the spin relaxation mechanism involving modulation of the isotropic hyperfine interaction, thus making it possible to describe the muon spin relaxation rate in terms of a single transition. The modulation of the anisotropic part of the hyperfine interaction would make other transitions possible. However, unique activation parameters can be determined provided that one transition is dominant.

An approximate value of the hyperfine coupling constant, which is dependent on the structure of the radical species, can be deduced using repolarization curves. These are plots of the initial amplitude (asymmetry) of the muon relaxation signal against the applied magnetic field. They show the decoupling of the hyperfine interaction by the applied field, and a fit to the data reveals the hyperfine interaction. For the simplest case of an isotropic hyperfine coupling, the hyperfine field B_0 is given by the formula

$$B_0 = \frac{A_\mu}{(\gamma_e + \gamma_m)} \quad (6.56)$$

where A_μ is the muon–electron hyperfine interaction and, γ_e and γ_m are the electron and muon magnetogyric ratios, respectively.

However, this type of estimate can be distorted by factors such as the anisotropy of the interaction and motional effects, and are discussed in detail in the literature (Pratt, 1997). More accurate values of the hyperfine interaction are obtained using either TF-μSR experiments that have been already described and/or ALC-μSR. The following is a brief description of the latter technique.

6.7 Avoided Level Crossing or ALC-μSR

If one extends the Breit–Rabi diagram to much higher magnetic fields, Figure 6.6, the levels $|1\rangle$ and $|2\rangle$ approach each other and eventually their energies cross. However, if the interaction between the electron and the muon is anisotropic, or if the states are both coupled to the same, energetically remote state, these levels do not cross but mix and repel each other as they get closer. Both mechanisms may be operative at the same time. This mixing of levels gives rise to a depolarization of the muon spin, the detection of which provides a way of sensing these crossings. It was Abragam (1984) who first predicted the use of such mixing for the measurement of hyperfine splittings of paramagnetic ions in longitudinal field μSR. The field values at which these crossings occur are dependent upon the strength of the hyperfine interactions involved. The amplitude and the shape of the resulting polarization dips provide additional and sometimes complimentary information to that of TF-μSR data. Heming et al. (1986) developed the necessary theory to interpret the sharp dips in polarization observed at these level crossings in the case of coupling to a remote state.

Studies of muons implanted in ^{13}C-enriched samples of the fullerenes, C_{60} and C_{70}, provide interesting examples of the use of the ALC technique. ^{13}C has a nuclear spin of $1/2$ and when muons are implanted in enriched (^{13}C $>$ 99%) C_{60} there is only one possible Mu adduct because of the high (dodecahedral) symmetry of the molecule. The spin states may be described in terms of the four possible combinations of the m_s values of a muon and a ^{13}C nucleus; compare muonium above. Furthermore, at high fields in the Paschen–Back region, these four spin states constitute accurate descriptions of the possible spin states of the adduct. However, in longitudinal fields of the order of 1.2 T, where the levels approach each other in energy, the anisotropic coupling induces a mixing of the states $|\alpha_e \alpha_\mu\rangle$ and $|\alpha_e \beta_\mu\rangle$, which would otherwise cross, and the resulting relaxation ($\Delta M = \pm 1$) is seen as a sharp dip in the polarization. From an analysis of the position and shape of the polarization dip Roduner et al. (1995) were able to investigate the reorientational dynamics of solid C_{60}. They characterized a spherical diffusion or isotropic jump-reorientation motion with an Arrhenius activation energy of 176 meV and a frequency factor of 2.95×10^{-12} seconds. They contrast this with similar studies of C_{70}.

The symmetry of C_{70} is considerably lower than that of C_{60} and five different muonium adducts can be envisaged. ALC-μSR has been used by Macrae et al. (1995) to study the rotational dynamics of these adducts. They find that the behavior is well described by a pseudostatic model with a temperature-dependent order parameter. Below 170 K the system is frozen on a 30 nanoseconds timescale. As the sample is warmed, complex motion appears with rotation and jumping between sites. Above 390 K, the reorientational motion is essentially isotropic and the avoided level-crossing resonance disappears.

When C_{60} is dissolved in decalin, the anisotropy that induces mixing rather than crossing is no longer present, but the states can mix by another mechanism, which has been analyzed by Heming et al. (1986). We have discussed earlier the energy levels of a muon implanted in a free radical containing one or more protons with which the muon interacts. This theme is pursued a little further in order to provide more detail on this origin of the ALC phenomenon. We consider the simplest case of the interaction between an electron, a muon, and a single proton. The Hamiltonian is

$$\hat{H}_0 = A_{e\mu}\hat{\mathbf{S}}_e \cdot \hat{\mathbf{I}}_\mu + A_{ep}\hat{\mathbf{S}}_e \cdot \hat{\mathbf{I}}_p + B_z \nu_e \hat{S}_z - B_z \nu_\mu \hat{I}_{\mu z} - B_z \nu_p \hat{I}_{pz} \quad (6.57)$$

One half of the 8×8 energy matrix (shown below) is sufficient for our purposes, the other half is, of course, extremely similar.

We see at a glance that the state $|\alpha_e \alpha_\mu \alpha_p\rangle$ does not mix with any others and, on careful inspection, we also note that, because of the much larger value of ν_e, for larger values of the field the state $|\beta_e \alpha_\mu \alpha_p\rangle$ is energetically distant from those with which it interacts. Therefore, since the coupling constants A_{ep} and $A_{e\mu}$ are small, the mixing of the three coupled states is also small and the values of the diagonal elements of the matrix provide a first approximation to the energy levels of the system. Furthermore, as the field is increased, the energies of the states $|\alpha_e \alpha_\mu \beta_p\rangle$ and $|\alpha_e \beta_\mu \alpha_p\rangle$ converge, become equal and then diverge, that is, they cross. The value of the field at which the crossing is predicted is readily calculated by equating the two diagonal matrix elements. (We assume that the field is always along the z-axis and drop the subscript z to simplify the notation.)

$$+ \tfrac{1}{4}(A_{e\mu} - A_{ep}) + \tfrac{1}{2}B(\nu_e - \nu_\mu + \nu_p)$$
$$= -\tfrac{1}{4}(A_{e\mu} - A_{ep}) + \tfrac{1}{2}B(\nu_e + \nu_\mu - \nu_p) \quad (6.58)$$

or

$$B = \frac{A_{e\mu} - A_{ep}}{2(\nu_\mu - \nu_p)} \quad (6.59)$$

However, this crossing never takes place because, as the energies of the "crossing" states get closer, their coupling to $|\beta_e \alpha_\mu \alpha_p\rangle$ has an increasing effect upon them and instead of crossing they mix and then diverge again. Heming et al. (1986) obtained an equation for the field at which the levels are closest together and the energy gap between them at that point from which the frequency of the transition between them, the ALC transition frequency, can be calculated.

$\hat{H}_0$	$\|\alpha_e\alpha_\mu\alpha_p\rangle$	$\|\alpha_e\alpha_\mu\beta_p\rangle$	$\|\alpha_e\beta_\mu\alpha_p\rangle$	$\|\beta_e\alpha_\mu\alpha_p\rangle$
$\langle\alpha_e\alpha_\mu\alpha_p\|$	$+\tfrac{1}{4}(A_{e\mu} + A_{ep})$ $+\tfrac{1}{2}B_z(\nu_e - \nu_\mu - \nu_p)$	0	0	0
$\langle\alpha_e\alpha_\mu\beta_p\|$	0	$+\tfrac{1}{4}(A_{e\mu} - A_{ep})$ $+\tfrac{1}{2}B_z(\nu_e - \nu_\mu + \nu_p)$	0	$+\tfrac{1}{2}A_{ep}$
$\langle\alpha_e\beta_\mu\alpha_p\|$	0	0	$-\tfrac{1}{4}(A_{e\mu} - A_{ep})$ $+\tfrac{1}{2}B_z(\nu_e + \nu_\mu - \nu_p)$	$+\tfrac{1}{2}A_{e\mu}$
$\langle\beta_e\alpha_\mu\alpha_p\|$	0	$+\tfrac{1}{2}A_{ep}$	$+\tfrac{1}{2}A_{e\mu}$	$-\tfrac{1}{4}(A_{e\mu} + A_{ep})$ $+\tfrac{1}{2}B_z(-\nu_e - \nu_\mu - \nu_p)$

We focus attention on the 3 × 3 matrix obtained by omitting the noninteracting state, $|\alpha_e \alpha_\mu \alpha_p\rangle$, from the matrix above. By means of a mathematical transformation Heming et al. reduced this matrix to a 2 × 2 form, which they were able to diagonalize, to obtain expressions for the fields at which the $\Delta M = 0$ and $\Delta M = \pm 1$ resonances take place. The mixing of the $|\alpha_a \alpha_\mu \beta_p\rangle$ and $|\alpha_e \beta_\mu \alpha_p\rangle$ states gives rise to the depolarization of the muons and a sharp dip in the polarization. Heming et al. were able to deduce expressions for the position and shape of the polarization dips that agreed well with experiment and to show that their equations applied to any number of protons or to nuclei other than protons. The equations quoted below are slightly modified versions of those from the original papers, which were used by Roduner et al. (1998) in a later publication.

$$B_r(\Delta_0) = \frac{A_{e\mu} - A_{ep}}{2(\gamma_\mu - \gamma_p)} - \frac{A_{e\mu} + A_{ep}}{2\gamma_e} \quad (6.60)$$

and

$$B_r(\Delta_1) = \frac{|A_{e\mu}|}{2\gamma_\mu} - \frac{A_{e\mu}}{2\gamma_e} \quad (6.61)$$

Percival et al. (1995) used the first of the above equations to determine the ^{13}C hyperfine coupling constants ($A_{ep} \rightarrow A_{ec}$) and hence the spin density distribution for a solution of 99% enriched ^{13}C$_{60}$ in decalin.

The ALC technique has also been exploited to great effect by Roduner and his collaborators in a study of the reorientational dynamics of the muonium-cyclohexadienyl adduct (Roduner et al., 1998). If muons are implanted into benzene molecules adsorbed on a high-silica zeolite, the muon forms muonium, which then adds to the benzene to form a cyclohexadienyl radical in which one carbon is bound to both the muon and its original proton. The unpaired electron is delocalized over the molecule. Two ALC dips are seen, one at 1.9 T, a Δ_1 transition due to a pure muon spin flip, and absent in the isotropic case, and a second just below 2.1 T, a Δ_0 transition due to a muon–proton flip-flop transition, $|\alpha_e \alpha_\mu \beta_p\rangle \leftrightarrow |\alpha_e \beta_\mu \alpha_p\rangle$. The detailed information that can be obtained using this technique is illustrated by the analysis (Heming et al., 1986) of the potential rotational motion. If the radical is assumed to perform fast uniaxial rotation about an axis perpendicular to the molecular plane, an axial component of the hyperfine anisotropy of -6.8 MHz is indicated by simulations of the spectra. However, if it is assumed to perform fast uniaxial rotation about the long axis of the radical, the simulations suggest an axial component of $+5.8$ MHz. By detailed analysis of the shapes of the polarization dips, it was found that on a critical timescale of about 50 nanoseconds, the radical performs fast uniaxial rotations about an axis perpendicular to the molecular plane. There is also a two-site jump motion between orientations differing by an angle of 110° and 1.0 kJ mol^{-1} in energy. Above 50 K, this causes increasing averaging of the orientations which is marked, but not isotropic, between 200 and 450 K. Below 50 K, the uniaxial rotation slows down below the timescale of the experiment and below 20 K motion ceases. The authors interpret these observations as being possible only if the radicals occupy sites at the channel intersections. It is also suggested that the spectra observed above ca. 450 K indicate the presence of thermally accessible sites with much lower mobility, possibly located inside the channels.

Roduner and his co-workers have also used these techniques to obtain information about the reorientational behavior of cyclohexadienyl and similar radicals adsorbed on silica (Reid *et al.*, 1990; Schwager *et al.*, 1995; Roduner *et al.*, 1995).

6.8
Conclusion

This chapter was intended to provide a brief introduction to muon spectroscopy. Only a few examples of applications were chosen in the discussion. For the vast majority of the applications to subjects such as the study of magnetism, semiconductors, high-temperature superconductors, chemical kinetics and isotope effects and so on, the reader is referred to the proceedings of the International Conferences of Muon Spectroscopy (1996, 2000, 2003, 2006), journal special editions (Rhodes, 2000), and other documents in the literature (Chappert *et al.*, 1984; Lee *et al.*, 1998; Walker, 1983). Applications to date have been heavily weighted toward Physics, but the potential for applications to Chemistry and Biology are very large but in their infancy as yet. Of particular interest would be the time window accessible to muon spectroscopy, which partly overlaps and lies between NMR and Quasielastic Neutron Scattering (QENS) and offers new opportunities in the study of dynamic processes.

References

Abragam, A. (1984) *C. R. Acad Sc. Paris, Series II*, **299**(3), 559.

Beer, G.A., Marshall, G.M., Mason, G.R., Olin, A., Gelbart, Z., Kendall, K.R., Bowen, T., Halverston, G.P., Pifer, A.E., Fry, C.A., Warren, J.B., and Kunselman, A.R. (1986) *Phys. Rev. Lett.*, **57**, 671.

Bloembergen, N., Purcell, E.M., and Pound, R.V. (1948) *Phys. Rev.*, **73**, 679.

Carrington, A. and McLachlan, A.D. (1979) *Introduction to Magnetic Resonance*, Chapman and Hall, London.

Caso, C., et al. (1998) The 1998 review of particle physics. *Eur. Phys. J.*, **C3**, 1.

Chappert, J. (1984) in *Muons and Pions in Materials Research*, (eds J. Chappert and R.I. Grynszpan), Elsevier Amsterdam.

Chu, S., Mills, A.P., Yodh, A.G., Nagamine, K., Miyake, Y., and Kuga, T. (1988) *Phys. Rev. Lett.*, **60**, 101.

Conversi, M., Pancini, E., and Piccioni, O. (1947) *Phys. Rev.*, **71**, 209.

Cox, S.F.J. (1998) *Solid State Nucl. Magn. Reson.*, **11**, 103.

Cox, S.F.J. and Sivia, D.S. (1994) *Hyperfine Interact*, **87**, 871.

Cox, S.F.J. and Sivia, D.S. (1997) *Appl. Magn. Reson.*, **12**, 213.

Garwin, R.L., et al. (1957) *Phys. Rev.*, **105**, 1415.

Harris, R.K. (1983) *Nuclear Magnetic Resonance Spectroscopy*, Pitman, London.

Heming, M., Roduner, E., Patterson, B.D., Odermatt, W., Schneider, J., Baumeler, H., Keller, H., and Savić, I.M. (1986) *Chem. Phys. Lett.*, **128**, 100.

Kunze, P.Z. (1933) *Physics*, **83**, 1.

Lattes, C.M.G., Muirhead, H., Occhialini, G.P.S., and Powell, C.F. (1947) *Nature*, **159**, 694.

Lee, S.L., Kilcoyne S.H., and Cywinski, R. (eds) (1998) *Muon Science, Muons in Physics, Chemistry and Materials, Proceedings of the Fifty First Scottish Universities Summer School in Physics St Andrews*, Copublished by: Scottish Universities Summer School in Physics and Institute of Physics Publishing, Bristol and Philadelphia.

Macrae, R.M., Prassides, K., Thomas, I.M., Roduner, E., Niedermayer, C., Binninger, U., Bernhard, C., Hofer, A., and Reid, I.D. (1995) *Chem. Phys.*, **192**, 231.

Matsushita, A. and Nagamine, K. (1996) *Chem. Phys. Lett.*, **253**, 407.

Mills, A.P. Jr, Imazato, J., Saitoh, S., Uedono, A., Kawashima, Y., and Nagamine, K. (1986) *Phys. Rev. Lett.*, **56**, 1463.

Morenzoni, E., Prokscha, Th., Hofer, A., Matthias, B., Meyberg, M., Wutzke, Th., Birke, M., Hückler, G.I., Litterst, J., and Niedermayer, C. (1997) *Appl. J. Phys.*, **81**, 3340.

Nagamine, K. (1987) *At. Phys.*, **10**, 225.

Neddermeyer, S.H. and Anderson, C.D. (1937) *Phys. Rev.*, **51**, 884.

Percival, P.W., Addison-Jones, B., Brodovitch, J.-C., Ji, F., Horoyski, P.J., Thewalt, M.L.W., and Anthony, T.R. (1995) *Chem. Phys. Lett.*, **245**, 90.

Piper, A.E., Bowen, T., and Kendall, K.R. (1976) *Nucl. Inst. Meth.*, **135**, 39.

Pratt, F.N. (1997) *Philos. Mag. Lett.*, **75**, 371.

Reid, I.D., Azuma, T., and Roduner, E. (1990) *Hyperfine Interact*, **65**, 879.

Rhodes, C.R. (2000) *Magn Reso Chem*, **38**.

Roduner, E. and Münger, K. (1984) *Hyperfine Interact*, **17–19**, 793.

Roduner, E., Prassides, K., Macrae, R.M., Thomas, I.M., Niedermayer, C., Binninger, U., Bernhard, C., Hofer, A., and Reid, I.D. (1995) *Chem. Phys.*, **192**, 231.

Roduner, E., Schwager, M., Tregenna-Piggott, P., Dilger, H., Shelley, M., and Reid, I.D. (1995) *Ber. Bunsen-Ges. Phys. Chem.*, **99**, 1338.

Roduner, E., Stolmar, M., Dilger, H., and Reid, I. (1998) *J. Phys. Chem.*, **102**, 7591.

Schwager, M., Dilger, H., Roduner, E., and Reid, I.D. (1995) *Ber. Bunsen-Ges. Phys. Chem.*, **99**, 142.

Tanikawa, K., Sakata, S., and Inoue, T. (1946) *Prog. Theor. Phys.*, **1**, 143.

(a) Walker, D.C. (1983) *Muon and Muonium Chemistry*, Cambridge University Press; (b) Roduner, E. (1988) *The Positive Muon as a Probe in Free Radical Chemistry, Lecture Notes in Chemistry*, (eds G. Berthier, M.J.S. Dewar, H. Fischer, K. Fukui, G.G. Hall, J. Hinze, H.H. Jaffe, J. Jortnar, W. Kutzelnigg, K. Ruedenberg, and J. Tomasi). Vol. 49, Publisher, Springer-Verlag; (c) Jayasooriya, U.A. (2004) μSR studies of molecular dynamics in organometallic chemistry, in *Fluxional Organimetallic and Coordination Compounds*, (eds M. Gielen, R. Willem, and B. Wrackmeyer), John Wiley & Sons, Ltd.

Yukawa, H. (1935) *Proc. Phys. Math. Soc. Japan*, **17**, 48.

(a) (1996) Proceedings of the 7th International Conference on Muon Spin Rotation/Relaxation/Resonance (μSR'96), Hyperfine Interactions vols. 104–106; (b) (2000) Proceedings of the 8th International Conference on Muon Spin Rotation, Relaxation and Resonance (μSR'99), Physica B condensed matter, vols. 289–290; (c) (2003) Proceedings of the 9th International Conference on Muon Spin Rotation, Relaxation and Resonance (μSR'), Physica B condensed matter, vol. 326; (d) (2006) Proceedings of the 10th International Conference on Muon Spin Rotation, Relaxation and Resonance (μSR'99), Physica B condensed matter, vols. 374–375.

7
Neutron Spectroscopy

Anton Heidemann

7.1	Introduction	185
7.2	Theoretical Foundations of Neutron Scattering	187
7.2.1	Typical Response Functions	190
7.2.2	Magnetic Scattering	190
7.3	Techniques in Neutron Scattering	191
7.3.1	Neutron Sources	191
7.3.1.1	Nuclear Fission Reactors	191
7.3.1.2	Spallation Sources	191
7.3.2	Moderators	192
7.3.3	Neutron Optics	192
7.3.4	Neutron Scattering Instruments	194
7.3.4.1	Instruments without Energy Analysis	194
7.3.4.2	Instruments with Energy Analysis	194
7.4	Neutron Spectroscopy: Techniques and Applications	194
7.4.1	Time-of-Flight (TOF) Spectroscopy	194
7.4.1.1	Direct Geometry Spectrometers	195
7.4.1.2	Indirect (Inverted) Geometry Spectrometers	197
7.4.2	Three-Axis Spectroscopy (TAS)	198
7.4.2.1	Principle	198
7.4.2.2	Layout	199
7.4.2.3	The TAS Resolution Function	200
7.4.2.4	Applications	200
7.4.2.5	Comparison between TAS and TOF Spectroscopy of Coherent Excitations in Single Crystals	200
7.4.3	Back-Scattering (BS) Spectroscopy	201
7.4.3.1	Principle	201
7.4.3.2	Existing Instruments	202
7.4.3.3	Applications	202
7.4.4	Neutron Spin Echo (NSE) Spectroscopy	203
7.4.4.1	Principle	203
7.4.4.2	Implementation	204

Encyclopedia of Applied Spectroscopy. Edited by David L. Andrews.
Copyright © 2009 WILEY-VCH Verlag GmbH & Co. KGaA, Weinheim
ISBN: 978-3-527-40773-6

7.4.4.3	NSE Variants	205
7.4.4.4	NSE versus Standard Inelastic Spectrometers	206
7.5	**Conclusion**	**206**
	Acknowledgments	**207**
	Glossary	**207**
	References	**207**
	Web sites	**208**

7.1
Introduction

The neutron is composed of one up and two down quarks with charges of 2/3 and −1/3, respectively. The total charge of the neutron is zero. Its internal structure leads to an electrical charge distribution. This results in a nonzero magnetic moment and an electric polarizability. Hot, thermal, cold, and very cold neutrons have wavelengths ranging from 0.1 to 100 Å and energies from a few electronvolts to microelectronvolts or from terahertz to megahertz. They are therefore an ideal probe for the study of the atomic order and the atomic motions in condensed matter. Neutrons behave like waves. Therefore, a neutron beam can be monochromatized by Bragg reflection from crystals. Neutrons also behave like particles flying with velocities in the range from kilometers per second to meters per second. Their energy can therefore be determined by time-of-flight (TOF) measurements. The zero net charge of the neutron means that it interacts very weakly with matter and penetrates deeply into the sample and the sample container. Neutrons interact directly with the nuclei of the atoms. This short-range interaction can be considered point like characterized by a single parameter, with the scattering length a of the order of 10^{-12} cm. Figure 7.1 shows the variation of the scattering length for the different elements. There is no systematic variation across the periodic table as there is for X-rays and electrons.

The magnetic moment of the neutron interacts with the unpaired electron spins of magnetic atoms with strength comparable to that of the nuclear interaction. Therefore, the neutron is also a powerful probe of magnetic properties of solids. The effective magnetic scattering length a_m for an atom with a magnetic moment of μ Bohr magnetons is given by

$$a_m = 0.27 \times 10^{-12} \, \mu \, (\text{cm}) \qquad (7.1)$$

Neutrons are spin 1/2 particles. A neutron beam can therefore be polarized. A polarized neutron beam entering a magnetic field perpendicular to its magnetic moment will undergo Larmor precession. This property is used in the neutron spin echo (NSE) technique.

The neutron beams used in scattering experiments can be produced either by a nuclear reactor or by a neutron spallation source. Both types of sources can be either continuous or pulsed. However, the majority of research reactors are

Encyclopedia of Applied Spectroscopy. Edited by David L. Andrews.
Copyright © 2009 WILEY-VCH Verlag GmbH & Co. KGaA, Weinheim
ISBN: 978-3-527-40773-6

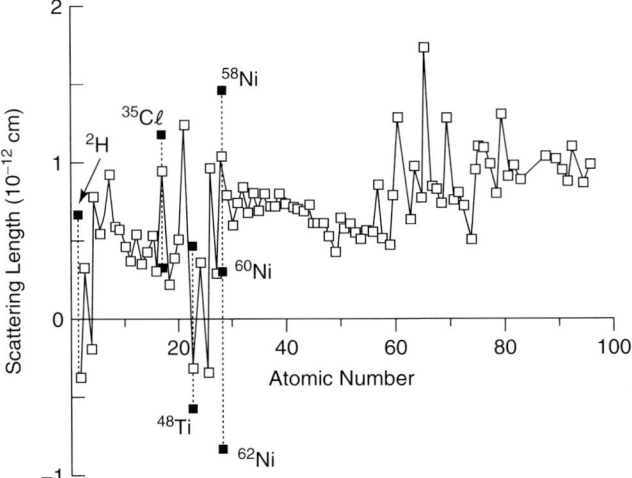

Fig. 7.1 Variation of the coherent scattering length a with atomic number (Price and Sköld, 1986).

continuous sources, and most of the existing spallation sources are pulsed. The latter are well suited to TOF techniques. In a neutron scattering experiment, one measures the wave vector transfer $\mathbf{Q}$ or the momentum transfer $\hbar\mathbf{Q}$ and the transfer of energy $\hbar\omega$ of the neutron to the sample

$$\mathbf{Q} = \mathbf{k}_i - \mathbf{k}_f \tag{7.2}$$

$$\hbar\omega = E_i - E_f = \left(\frac{\hbar^2}{2m}\right)\left(k_i^2 - k_f^2\right) \tag{7.3}$$

Equation (7.2) describes the momentum conservation and Eq. (7.3) the energy conservation. $k = 2\pi/\lambda$ is the magnitude of the wave vector and λ is the wavelength of the neutron. The subscripts i and f refer to the initial and final neutron beam.

A few additional remarks concerning the scattering triangle may be useful (see Figure 7.2). The scattering event is characterized by 6 variables: 3 from each vector $\mathbf{k}_i$ and $\mathbf{k}_f$. However, we are only interested in $\mathbf{Q}$ and $\hbar\omega$. Therefore, we have 2 redundant variables. They should be chosen to optimize the experiment. For noncrystalline samples, only $|\mathbf{Q}| = Q$ and $\hbar\omega$ are meaningful. Here, we have four redundant variables.

The following relations are valid for the energy E and the velocity v of neutrons:

$$E[\text{meV}] = 2.072 \; k^2[\text{Å}^{-2}] \tag{7.4}$$

$$v[\text{m/s}] = \frac{3956}{\lambda[\text{Å}]} \tag{7.5}$$

Energy conversion:

$$1 \text{ meV} = 0.2418 \text{ THz} = 11.6045 \text{ K}$$
$$= 8.006554 \text{ cm}^{-1} \tag{7.6}$$

We distinguish between two types of experiments:

1. Neutron scattering without energy analysis: this is the domain of diffraction, reflectometry, small-angle and diffuse scattering.
2. Neutron scattering with energy analysis. This is the domain of *neutron spectroscopy*.

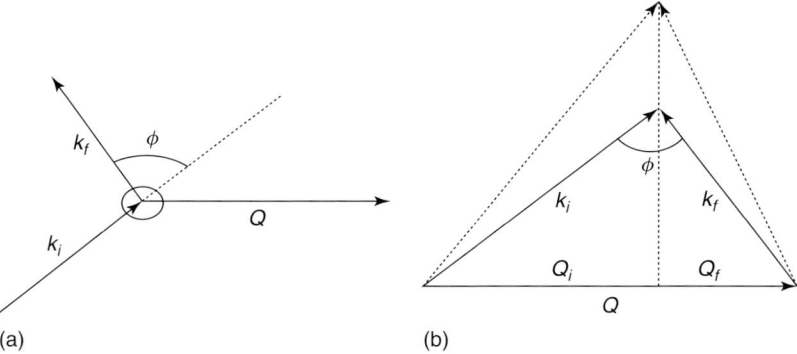

Fig. 7.2 Scattering in (a) real and (b) reciprocal space. The dotted line in (b) shows an alternate configuration leading to the same momentum and energy transfers (**Q**, ω) (Hippert et al., 2005).

In this monograph, we only briefly mention neutron scattering without energy analysis. The main part of the chapter is devoted to neutron spectroscopy covering both *inelastic neutron scattering* (INS) and *quasi-elastic neutron scattering* (QENS).

Figure 7.3 shows the areas in the (Q, ω) and (r, t) planes, which are covered by neutron spectroscopy. It also shows the areas investigated by other scattering techniques such as electromagnetic and electron scattering. It is obvious that neutron spectroscopy is complementary to these other techniques.

A very comprehensive review of neutron spectroscopy has been published recently (Hippert et al., 2005). The interested reader is referred to this book for further reading.

7.2 Theoretical Foundations of Neutron Scattering

The quantity that essentially results from neutron scattering experiments is the double differential cross section $d^2\sigma/d\Omega dE_f$. It describes the probability with which an incident neutron of energy E_i is scattered by an angle ϕ into a solid angle volume element $d\Omega$ and an energy element between E_f and $E_f + dE_f$. $d^2\sigma/d\Omega dE_f$ can be derived in first-order perturbation theory (Born approximation). van Hove (1954) has shown that $d^2\sigma/d\Omega dE_f$ is connected with correlation functions, which describe the motion of the scattering particles in space and time. Therefore, they allow the calculation of $d^2\sigma/d\Omega dE_f$ in terms of simple physical models.

In the following, we briefly describe the main ingredients of this theory. Details can be found in (Lovesey, 1984; Scherm, 1972).

The Hamiltonian of the total system, consisting of the sample and the neutron, is given by Eq. (7.7)

$$H = H_0 + \frac{p_n^2}{2m} + V \qquad (7.7)$$

Here, H_0 is the Hamiltonian of the unperturbed system, V is the interaction potential between the neutron and the scattering particles, and $p_n^2/2m$ is the kinetic energy of the neutron, $\mathbf{p}_n = \hbar \mathbf{k}$.

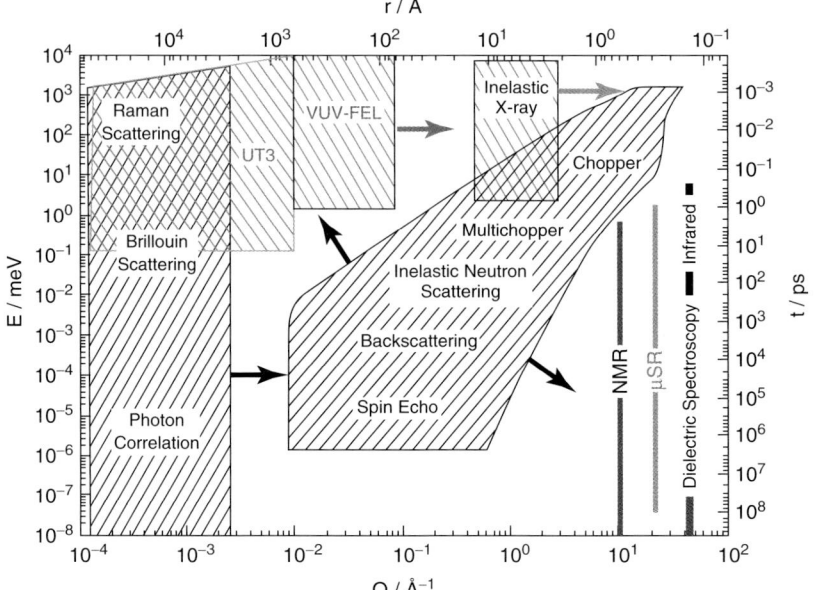

Fig. 7.3 Distance–time (r, t) and $(\mathbf{Q}, \omega)$ scales probed by neutron spectroscopy. (Pynn, unpublished.)

The incident and scattered neutrons are characterized by plane waves with spin σ: $|\mathbf{k}\sigma>$. In the following, we consider, for simplicity, only the case of unpolarized neutrons.

We denote the eigenfunctions and eigenvalues of the unperturbed system by $|n>$ and E_n with

$$H_0|n> = E_n|n> \quad (7.8)$$

Using Fermi's golden rule, the transition probability per unit time of the system from an initial state $|\mathbf{k}_i n_0>$ to a final state $|\mathbf{k}_f n_1>$ is given by

$$w(\mathbf{k}_i, n_0 \longrightarrow \mathbf{k}_f, n_1) = \frac{2\pi}{\hbar} |<\mathbf{k}_f n_1| \\ \times V|\mathbf{k}_i n_0>|^2 \delta\left(E_{n_1} - E_{n_0} - \hbar\omega\right) \quad (7.9)$$

We have to sum up over all transitions, where the wave vector changes from $\mathbf{k}_i$ to $\mathbf{k}_f$:

$$w(\mathbf{k}_i \longrightarrow \mathbf{k}_f) = \sum_{n_0} \sum_{n_1} p(n_0) \\ \times w(\mathbf{k}_i, n_0 \longrightarrow \mathbf{k}_f, n_1) \quad (7.10)$$

$p(n_0)$ is the statistical weight of the state $|n_0>$:

$$p(n_0) = \frac{\exp\left(\dfrac{-E_{n_0}}{k_B T}\right)}{\sum_{n_0} \exp\left(\dfrac{-E_{n_0}}{k_B T}\right)} \quad (7.11)$$

With normalized plane waves

$$\mathbf{k}> = L^{-3/2} \exp(i\mathbf{k} \cdot \mathbf{r}) \quad (7.12)$$

and the Fermi pseudopotential for V (Fermi, 1936)

$$V(\mathbf{r} - \mathbf{r}_i) = \left(\frac{2\pi \hbar^2}{m}\right) a_i \delta(\mathbf{r} - \mathbf{r}_i) \quad (7.13)$$

we obtain finally

$$\frac{d^2\sigma}{d\Omega\, dE_f} = \left(\frac{k_f}{k_i}\right) \cdot N^{-1} \sum_{n_0}\sum_{n_1} p(n_0)$$

$$\times \sum_i\sum_j a_i a_j <n_0|\exp(-i\mathbf{Q}\cdot\mathbf{r}_i)|$$

$$\times n_1><n_1|\exp(i\mathbf{Q}\cdot\mathbf{r}_j)|$$

$$\times n_0> \delta(E_{n_1} - E_{n_0} - \hbar\omega) \quad (7.14)$$

where a_i is the scattering length of the ith nucleus. N is the number of scattering centers.

Introducing the coherent and incoherent scattering functions (the dynamic structure factors) $S_{\mathrm{coh}}(\mathbf{Q},\omega)$ and $S_{\mathrm{inc}}(\mathbf{Q},\omega)$, we get

$$\frac{d^2\sigma}{d\Omega\, dE_f}$$

$$= \hbar^{-1}\left(\frac{k_f}{k_i}\right)\cdot[<a>^2 S_{\mathrm{coh}}(\mathbf{Q},\omega)$$

$$+ (<a^2> - <a>^2)S_{\mathrm{inc}}(\mathbf{Q},\omega)] \quad (7.15)$$

$S_{\mathrm{coh}}(\mathbf{Q},\omega)$

$$= N^{-1}\sum_{n_0}\sum_{n_1} p(n_0)\sum_i\sum_j <n_0$$

$$\times |\exp(-i\mathbf{Q}\cdot\mathbf{r}_i)|n_1><n_1|$$

$$\times \exp(i\mathbf{Q}\cdot\mathbf{r}_j)|$$

$$\times n_0> \delta(E_{n_1} - E_{n_0} - \hbar\omega) \quad (7.16)$$

$S_{\mathrm{inc}}(\mathbf{Q},\omega)$

$$= N^{-1}\sum_{n_0}\sum_{n_1} p(n_0)\sum_{j=1}^{N} |<n_1|$$

$$\times \exp(i\mathbf{Q}\cdot\mathbf{r}_j)|n_0>|^2$$

$$\times \delta(E_{n_1} - E_{n_0} - \hbar\omega) \quad (7.17)$$

We assume that the sample contains a mixture of isotopes with different scattering lengths a_j that are randomly distributed over the positions r_j. By this, we introduce the following averages:
$<a> = a_{\mathrm{coh}}$ is the coherent scattering length.
$<a^2> - <a>^2 = a^2_{\mathrm{inc}}$ is the square of the incoherent scattering length. Its origin can be either isotopic or spin incoherence: for nuclei with a spin I not equal to zero, the interaction depends on the orientation between the spin of the neutron and the nucleus, with scattering lengths a_+ and a_-. In the case where there is no correlation between the nuclear spins and their sites, we get

$$a_{\mathrm{coh}} = \frac{(I+1)a_+ + Ia_-}{(2I+1)} \quad (7.18)$$

$$a^2_{\mathrm{inc}} = \frac{I(I+1)(a_+ - a_-)^2}{(2I+1)} \quad (7.19)$$

The scattering functions depend only on the properties of the unperturbed sample and not on the properties of the scattered particle. They can be calculated if the eigenfunctions of the sample are known.

van Hove (1954) established the following relations:

$S_{\mathrm{coh}}(\mathbf{Q},\omega)$

$$= (2\pi)^{-1}\int \exp[i(\mathbf{Q}\cdot\mathbf{r} - \omega t)]$$

$$\times G(\mathbf{r},t)d\mathbf{r}dt \quad (7.20)$$

$S_{\mathrm{inc}}(\mathbf{Q},\omega)$

$$= (2\pi)^{-1}\int \exp[i(\mathbf{Q}\cdot\mathbf{r} - \omega t)]$$

$$\times G_s(\mathbf{r},t)d\mathbf{r}dt \quad (7.21)$$

G and G_s are correlation functions in space and time. S_{coh} and S_{inc} are their four-dimensional Fourier transforms in $(\mathbf{Q},\omega)$ space. The so-called intermediate scattering functions $I(\mathbf{Q},t)$ and $I_s(\mathbf{Q},t)$

are defined as

$$I(\mathbf{Q}, t) = \int \exp(i\mathbf{Q} \cdot \mathbf{r}) G(\mathbf{r}, t) d\mathbf{r} \quad (7.22)$$

$$I_s(\mathbf{Q}, t) = \int \exp(i\mathbf{Q} \cdot \mathbf{r}) G_s(\mathbf{r}, t) d\mathbf{r} \quad (7.23)$$

These functions are measured in NSE spectroscopy (see Section 7.4.4).

So far, all the calculations were done in the framework of quantum mechanics. However, using the correlation function approach of van Hove, things can be simplified in the classical limit. Examples are the calculation of S_{inc} for simple cases like translational diffusion in liquids and solids, reorientations of molecular groups in crystals at not too low temperatures, phonon spectra, and so on. The quantum mechanical approach is needed for the case of tunneling of molecular groups in condensed matter at low temperature.

Collective excitations (phonons, magnons) with dispersion, where ω depends on $\mathbf{q} = \mathbf{Q} - \boldsymbol{\tau}$ (see Section 7.4.2) and the intensity depends on $\mathbf{Q}$, are determined via the measurement of S_{coh}. Information about local single-particle motions (diffusion, reorientation of molecules) is obtained from the measurement of S_{inc}. Here, only the intensity depends on $\mathbf{Q}$ and the frequency is independent of $\mathbf{q}$.

7.2.1
Typical Response Functions

An excitation at a given frequency ω_q with an infinite lifetime can be represented by

$$S(\mathbf{Q}, \omega) = \frac{A(\mathbf{Q})}{\left(1 - \exp\left(\frac{-\hbar\omega}{k_B T}\right)\right)}$$
$$\times [\delta(\omega - \omega_q) - \delta(\omega + \omega_q)] \quad (7.24)$$

where $A(Q)$ is a dimensionless structure factor.

If the excitation has a finite lifetime, the delta functions can be replaced by Lorentzians. Overdamped excitations can be described by quasi-elastic Lorentzians:

$$S(\mathbf{Q}, \omega) = \frac{A(\mathbf{Q})}{\left(1 - \exp\left(\frac{-\hbar\omega}{k_B T}\right)\right)}$$
$$\times \frac{1}{\pi} \frac{\Gamma_Q}{\omega^2 + \Gamma^2_Q} \quad (7.25)$$

where Γ_Q is the half width at half maximum (HWHM) of the Lorentzian.

7.2.2
Magnetic Scattering

Neutrons are scattered by the unpaired electrons of atoms. The double differential magnetic cross section can be written as (Lovesey, 1984)

$$\frac{d^2\sigma}{d\Omega dE_f} = \hbar^{-1} \left(\frac{k_f}{k_i}\right) \cdot a_m^2 S_{mag}(\mathbf{Q}, \omega) \quad (7.26)$$

The dynamic structure factor is given by

$$S_{mag}(\mathbf{Q}, \omega)$$
$$= \frac{1}{2\pi} \int dt \, \exp(-i\omega t)$$
$$\times < \mathbf{M}^*_\perp(-\mathbf{Q}, 0) \cdot \mathbf{M}_\perp(\mathbf{Q}, t) > \quad (7.27)$$

The magnetic interaction operator $\mathbf{M}(\mathbf{Q}, t)$ is the Fourier transform of the total magnetization density $\mathbf{M}(\mathbf{r}, t)$. Therefore, neutrons observe the sum of spin and orbit contributions. Only magnetic moments perpendicular to $\mathbf{Q}$ are measured. This is expressed by

$$\mathbf{M}_\perp = \mathbf{Q} \times (\mathbf{M} \times \mathbf{Q}) \quad (7.28)$$

rather than by **M** occurring in Eq. (7.27). Magnons are described by $S_{\mathrm{mag}}(\mathbf{Q}, \omega)$

7.3 Techniques in Neutron Scattering

7.3.1 Neutron Sources

Neutron scattering suffers from low data rates, particularly in comparison with X-rays or infrared spectroscopy.

Much effort has therefore been concentrated on increasing the fluxes of neutron sources. Presently, there exist two types of high-flux neutron sources.

7.3.1.1 Nuclear Fission Reactors

The principle is based on the following nuclear reaction:

$$n_{\mathrm{thermal}} + U^{235} \longrightarrow 2 \text{ fission}$$
$$\text{fragments} + 2.5\ n_{\mathrm{fast}} + 180 \text{ MeV} \quad (7.29)$$

Here, the HFR of the ILL in France (ILL) is the world's best steady state reactor specifically designed for neutron scattering experiments. Its thermal flux near the reactor core is 1.5×10^{15} neutrons s^{-1} cm^{-2}. The reactor became critical in 1972. Since then no research reactor with a higher flux has become operational. There have been attempts to develop nuclear research reactors producing a significantly higher flux (ANS at Oak Ridge National Laboratory (ORNL), USA in the 1980s), but they were not built. The reasons are manyfold: very expensive to build and to operate, problems with radiation damage and heat removal (very short lifetimes of reactor components). The reactor FRMII similar to the ILL reactor with a flux about 50% of the latter has become critical in Munich (FRMII) recently.

7.3.1.2 Spallation Sources

Spallation is a common nuclear reaction occurring for high-energy protons bombarding heavy nuclei. The principle is based on the following nuclear reaction:

$$p + \text{heavy nucleus} \longrightarrow \text{spallation}$$
$$\text{fragments} + n + 24 \text{ MeV} \quad (7.30)$$

An important limiting factor in determining the maximum neutron output of a particular type of source is the rate of removal of the heat deposited in the target by the nuclear reaction. The energy released per useful neutron is about 3 times smaller in spallation compared to fission. If, in addition, the spallation source is pulsed, the heat generated in the target will again be reduced. Therefore, a modern pulsed high-flux spallation source like the future project of the European Spallation Source (ESS) can, in principle, achieve a peak flux 40 times higher than the average steady flux of the ILL reactor.

Until recently, the spallation source ISIS at the Rutherford laboratory in United Kingdom can be considered as the world leading pulsed neutron source. It has a cold *peak* flux similar to the continuous flux of the ILL reactor.

As TOF is used on the ISIS spectrometers, this means that the data rate on instruments on both sources will be similar. ISIS however gains in the high-energy epi-thermal range. With the event of the spallation source SNS becoming operational at ORNL, the situation will change: in the final state this new source should produce a significantly higher flux than ISIS. However, one has to wait until all the technical problems are solved to be able to achieve the nominal design flux of the SNS.

For the sake of completeness, we mention the Swiss Neutron Spallation Source

(SINQ), which is a *continuous* spallation source at the Paul Scherrer Institute (PSI).

7.3.2
Moderators

Both types of neutron sources produce high-energy neutrons in the megaelectronvolt range. To become useful in neutron scattering, these neutrons have to be slowed down by so-called moderators. As neutrons have a mass very similar to the mass of the proton, they can be slowed down very efficiently by interaction with the latter in hydrogen or deuterium containing materials like H$_2$O or D$_2$O. Therefore, light or heavy water at room temperature are used as moderators to produce "thermal" neutrons with energies around 20 meV. Liquid H$_2$ or D$_2$ at 20 K is used to produce cold neutrons with an energy around 2 meV. Hot neutrons with 150 meV are moderated with graphite at very high temperatures of about 2000 K.

7.3.3
Neutron Optics

Neutron scattering is an intensity-limited technique: even on a high-flux reactor like the HFR at the ILL the incident flux of a high-resolution back-scattering (BS) spectrometer like the IN10 is down to 10^4 s^{-1} cm^{-2}. All possible tricks in neutron optics must therefore be played to optimize the flux of the instruments. There exist quite a number of neutron optical beam definition devices (Anderson *et al.*, 1999). In the following, we describe the more important ones.

Neutron guides: they permit to transport neutron beams over long distances without significant loss of the order of 1% m^{-1}. This makes it possible to gain space around a nuclear reactor or to perform high-resolution TOF experiments using very long (100 m or more) distances. The implementation of a group of thermal and cold neutron guides at the early stage of construction of the ILL facility is part of the ILL's success story. The principle of neutron guides is a mirror reflection of neutrons. The refractive index n for neutrons in matter is close to but smaller than one:

$$n = 1 - \frac{\lambda^2 N_v a_{\text{coh}}}{2\pi} \qquad (7.31)$$

N_v is the number density of nuclei in the sample. The glancing angle for cold neutrons is of the order of 1°. It increases with increasing wavelength.

Good neutrons guides are made of materials with high N_v and a_{coh} values. Normal neutron guides use mirrors made of channels of glass plates with or without Nickel coating. In order to increase the transmitted flux, currently super-mirror coatings are used. Super mirrors are made out of a sequence of alternating coatings of varying thickness. In this way, the glancing angle can be increased substantially. Flux gain factors of up to one order of magnitude become reality.

Soller collimators: These are mechanical devices that permit a divergent neutron beam to be collimated. They are composed of a number of equidistant neutron-absorbing thin blades separated by spacers.

Choppers: These enable a pulsed neutron beam to be produced. This is either achieved by a rotating mechanical device like a disc with windows or for polarized neutrons with an electromagnetic device.

A Fermi chopper is basically made out of a curved Soller collimator that rotates around an axis perpendicular to the neutron beam.

Crystal monochromators: Using Bragg reflection of a collimated white neutron beam on a crystal, we can obtain a monochromatic beam with wavelength λ:

$$\lambda = 2d_{hkl} \sin \theta_B \qquad (7.32)$$

d_{hkl} and θ_B are the d spacing of the lattice plane *(hkl)* and the Bragg angle.

Two types of crystals are used: mosaic crystals and perfect crystals. The former (Cu, Be, pyrolythic graphite) are the main components of three-axis spectrometers, the latter (Si, CaF$_2$) are used on BS instruments.

Focusing devices: These are typically curved systems like crystal arrays on a cylindrical or spherical surface. They allow spatial and monochromatic focusing of a neutron beam. Again, large area (20 m^2) crystal arrays are needed in some cases (see Section 7.4.3).

Polycrystalline filters: They are used to remove unwanted radiation from the beam while maintaining as high a transmission as possible for neutrons of the required energy.

Velocity selectors: These are fast rotating mechanical devices with a helical neutron path. They allow the preparation of a monochromatic beam of cold and thermal neutrons with a typical band width of 15% in wavelength.

Polarizers: A neutron beam can be polarized by reflection from a super-mirror made of alternating magnetic and nonmagnetic coatings or by Bragg reflection from a magnetic crystal like a Heusler alloy.

Spin analyzers: The same technique is used as for the polarizers. However, spin analyzers covering a large solid angle are realized by devices filled with polarized He^3 gas.

Spin orientation devices: π and $\pi/2$ Flippers: A flipper action is based on a nonadiabatic transition of a neutron spin through a magnetic field. By using a flat coil (a few millimeters thick) perpendicular to the neutron beam with the appropriate field inside, the polarized neutron can by flipped by π or $\pi/2$.

Precession coils: In Section 7.4.4, about NSE, we discuss that the Fourier time is proportional to the field integral.

Large diameter long solenoids with strong field integrals and homogeneous windings are needed.

Fresnel coils: These are special devices used in NSE to correct for field integral inhomogeneities. Current loops with a spiral structure where the radius changes as $r \sim \phi$ are the appropriate devices.

Doppler drives: This mechanical or electromechanical device makes it possible to modulate the velocity of Bragg reflected neutrons via the longitudinal neutron Doppler effect. They are used on BS spectrometers. Velocity variations up to a few meters per second are possible at present.

Neutron detectors: The principle of neutron detection is based on neutron absorption by nuclei. Gas detectors use He^3 or B^{10}F$_3$ gas in an electric field. Scintillator detectors use Li^6 in some solid Li containing compound. Here, the photon output triggered by neutron absorption is detected by charge-coupled devices (CCDs) or photomultipliers. The challenge is to develop multidetectors that cover very large areas up to tenths of square meters.

7.3.4
Neutron Scattering Instruments

In the early days of neutron scattering, instruments were invented by the pioneers like Shull, Wollan and Koehler (1951) for diffractometers, Brockhouse (1961) for three-axis spectrometers, Maier-Leibnitz (1966) for neutron optics and Mezei (1972) for the NSE technique. Today, powerful software tools are available to optimize neutron scattering instruments consisting of the components mentioned above. We distinguish between two types of instruments.

7.3.4.1 Instruments without Energy Analysis

With these instruments, one measures the angular distribution of neutrons scattered by the sample. The incident neutron energy is defined by a crystal monochromator, a velocity selector, or by TOF. Three different types of instruments can be distinguished.

Diffractometers measure the scattered intensity over a wide range of scattering angles either simultaneously or in specific scans. The main domain of application is crystallography, the study of atomic and magnetic structures of condensed matter including liquids (Bacon, 1962; Kisi and Howard, 2008).

Small-angle neutron scattering (SANS) instruments (Williams, May and Guinier, 1994) measure the scattering of neutrons at small angles. This makes it possible to study objects and structures in the range of about 10–1000 Å at low resolution. Typical examples are microstructures in alloys and global structures of very large molecules (polymers, biological cells). The fact that the two hydrogen isotopes H and D have scattering lengths of opposite sign allows one to use isotope labeling and contrast matching methods to make visible or invisible parts of the objects under study.

In reflectometers (Daillant et al., 1999) a cold neutron beam is reflected from the sample as from a mirror. The reflected intensity (specular or nonspecular) is detected at small angles. A reflectometer is therefore similar to a SANS instrument.

Neutron reflectometry is the technique for the study of planar structures with a wide range of materials from magnetic multilayers to biological systems at the solid–liquid interface.

7.3.4.2 Instruments with Energy Analysis

Instruments with energy analysis cover the domain that we call *neutron spectroscopy*. We distinguish two types of experiments:

INS with $\hbar\omega \neq 0$: All kinds of periodic motions (collective and (or) single particle) give rise to INS.

QENS: Here, one measures the broadening of the "elastic" line around $\hbar\omega \sim 0$. All kinds of diffusive motions give rise to QENS (Springer and Lechner, 2005).

7.4
Neutron Spectroscopy: Techniques and Applications

7.4.1
Time-of-Flight (TOF) Spectroscopy

In the TOF technique, a burst of polychromatic neutrons is produced and the times taken by the neutrons to travel from the source of the burst to the detector are measured (Windsor, 1981). TOF spectrometers can be divided into two classes: direct and indirect geometry instruments. The accessible regions in $(\mathbf{Q}, \omega)$ space are different for both geometries. They can be calculated using the equations governed by the conservation laws of momentum and energy transfer (Eqs. (7.2) and (7.3)). They are shown in Figure 7.4.

7.4.1.1 Direct Geometry Spectrometers

In direct geometry spectrometers (Figure 7.4(a)), the incident pulsed beam is monochromatized by a device like a crystal or a chopper system. The final energy E_f is determined by TOF. The scattered neutrons are detected by a detector bank covering (if possible) the whole range of scattering angles with a large vertical aperture. QENS and neutron energy gain spectroscopy are the main applications.

Examples are the TOF spectrometers IN4 and BRISP for thermal neutrons and IN5, IN6 for cold neutrons at the ILL. IN5 is a high-precision instrument used to study low-energy transfer processes as a function of momentum transfer Q.

As shown in Figure 7.5, IN5 has an array of four disk choppers spinning at up to 20 000 revolutions per minute to pulse and to monochromatize the incident beam. Wavelengths from 1.8 to 20 Å are available yielding elastic energy resolutions between 1 meV and 1 μeV. IN5 is a very flexible instrument.

Applications include

- diffusion in disordered systems such as liquids, molecular crystals, amorphous solids (glasses), polymers, hydrogen-metal systems, super-ionic conductors;
- dynamics of soft matter including gels, polymers, proteins, and biological membranes;
- rotational tunneling in molecular crystals;
- crystal field splitting;
- spin dynamics in High-T_c superconductors;
- critical scattering phenomena in dense gases and solids;
- spin dynamics of molecular spin clusters.

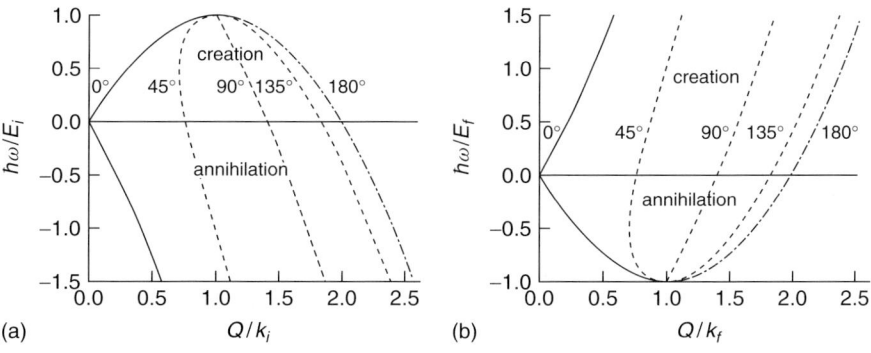

Fig. 7.4 Scattering parabolas for different scattering angles. (a) $k_i = $ constant (b) $k_f = $ constant (Hippert et al., 2005).

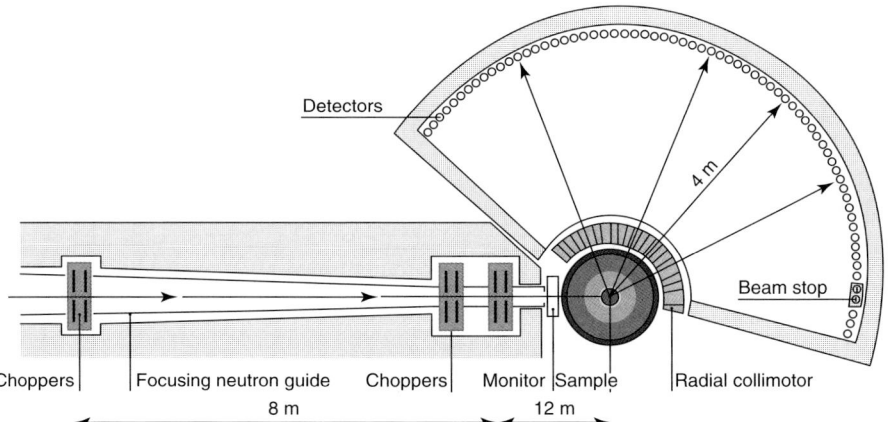

Fig. 7.5 Layout of the direct geometry TOF spectrometer IN5 at the ILL.

A beautiful example of a result obtained on IN5 from magnetic excitations in the molecular spin cluster in Mn_{12}-acetate is shown in Figure 7.6. Mn_{12}-acetate contains ferrimagnetic spin clusters with a total spin $S = 10$. The magnetic ground-state splitting is determined by magnetic INS (Mirebeau *et al.*, 1999). Figure 7.6 clearly shows the effect of temperature on the population of the magnetic sublevels.

The TOF spectrometer IN6 at the ILL employs a vertically curved triplet pyrolythic graphite crystal together with a Fermi chopper. Time focusing is used here to improve the energy resolution. Incident wavelengths between 4 and 6 Å are available with elastic resolutions from 170 to 50 μeV. The strength of IN6 is the very high sensitivity and high data rate. BRISP is a thermal neutron *Brillouin* scattering spectrometer optimized to operate at *small* scattering angles. Applications include coherent magnetic excitations and density fluctuations of disordered systems at low Q.

IN4 is a TOF spectrometer for thermal neutrons used to study mainly single-particle excitations in condensed matter. Applications include the study of energy level spacing of magnetic ions, crystal field splittings, valence fluctuations, heavy fermions, vibrational states in amorphous solids and polycrystals, molecular excitations, and matrix isolated systems.

Similar spectrometers exist at the spallation source ISIS at the Rutherford laboratory such as HET (High-Energy Transfer), MARI, and MAPS. Investigations on HET have broadened from studies of high-energy magnetic excitations and the dynamics of hydrogen-metal systems into the fields of quantum magnetism, non-Fermi liquids, and Brillouin scattering.

MAPS has changed the way the neutron community thinks about inelastic neutron scattering. Its huge array of position sensitive detectors has created a survey technique that is able to map vast areas of the Brillouin zone, making it possible to see the unexpected. It is able to reveal broad features that could be dismissed as background on a three-axis spectrometer.

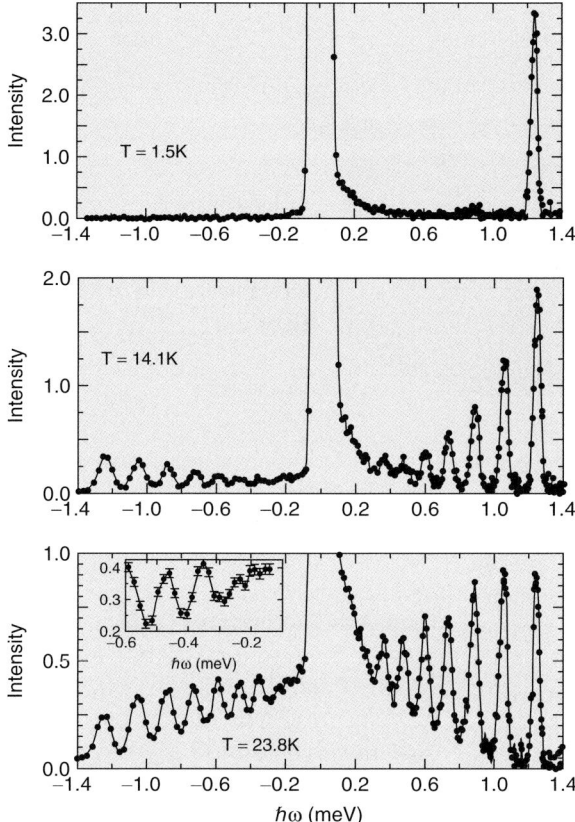

Fig. 7.6 Energy spectra from the Mn$_{12}$-acetate molecular spin cluster obtained on IN5 (Mirebeau et al., 1999).

The position-sensitive detectors give near-continuous coverage over a large solid angle detector array in the forward direction. The pixel size in reciprocal space is significantly smaller than the resolution volume defined by the other instrumental contributions. In contrast, conventional detectors on HET and MARI integrate along one direction in reciprocal space, which overwhelms the intrinsic resolution in that direction. With MAPS, there is complete freedom to construct scans along any direction in reciprocal space and project data onto any plane in reciprocal space.

MAPS is optimized to measure collective high-energy magnetic excitations in single crystals with varying energy resolution depending on choice of the monochromating chopper.

7.4.1.2 Indirect (Inverted) Geometry Spectrometers

In indirect (inverted) geometry spectrometers, an incident white pulsed beam hits the sample (Figure 7.4b). The final energy is defined by a crystal or a filter analyzer. The initial energy is determined by TOF. Again one tries to cover a large solid angle either by using

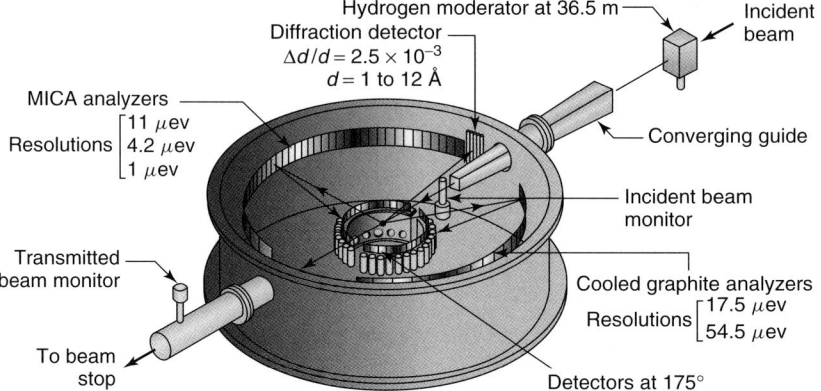

Fig. 7.7 The IRIS spectrometer at ISIS.

many analyzer crystals simultaneously or by a crystal filter. QENS and neutron scattering with energy loss are the main applications. This kind of geometry is rarely used on a continuous neutron source.

Examples on the pulsed source ISIS are the spectrometers TOSCA, PRISMA, and IRIS (Figure 7.7). PRISMA can perform scans along a defined direction in **Q** space with its multiarm analyzer system.

IRIS is a high-resolution TOF BS spectrometer: the pulsed white beam travels through a 30-m long neutron guide before hitting the sample. The scattered beam is analyzed in energy and momentum transfer Q by a large solid angle, spherically curved crystal analyzer in near-BS geometry. The long flight path allows the incident energy to be defined with the same precision as the analyzer defines the final energy, that is, $\Delta E/E = 1\%$ or 15 μeV at $\hbar\omega = 0$. Neutron energy loss spectroscopy down to 7 meV is possible. A similar instrument is in the commissioning phase at the SNS. It will have an elastic energy resolution of about 2 μeV with a final energy of 2 meV.

Neutron BS is described in more detail in Section 7.4.3.

7.4.2
Three-Axis Spectroscopy (TAS)

7.4.2.1 Principle

In order to obtain information on the excitation spectrum at a given point **Q** in reciprocal space, a more sophisticated procedure has to be adopted than just scanning the energy of the scattered neutrons. At each point of the scan the scattering triangle is modified, so that $\mathbf{k}_i$ and $\mathbf{k}_f$ will close at the same wave vector transfer **Q** but the length of one of the vectors is varied to provide for the required energy transfer. We define

$$\mathbf{Q} = \boldsymbol{\tau}_{hkl} + \mathbf{q} \qquad (7.33)$$

where $\boldsymbol{\tau}_{hkl}$ is the reciprocal lattice vector and **q** the reduced wave vector within the Brillouin zone. Therefore, in addition to the $\mathbf{k}_i$ and $\mathbf{k}_f$ modification, all the angles of the scattering triangle will also change.

There are two ways to achieve this. In the TOF technique, all energy transfers are measured simultaneously in all scattering directions, and later, offline,

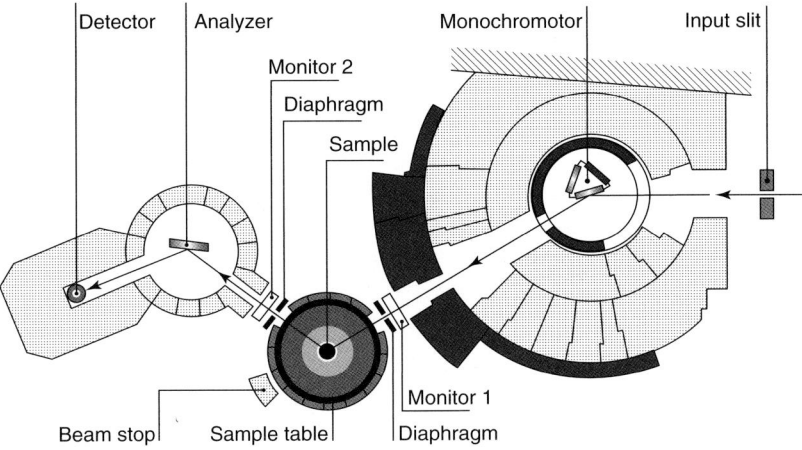

Fig. 7.8 Layout of the TAS spectrometer IN8 at the ILL.

one selects from the huge amount of data only those points that contain data in a specific **Q** direction. This procedure became feasible only with the rapid progress in computing technology. In the three-axis spectroscopy (TAS) technique, one takes data directly and only in a specific **Q** direction. There is a long ongoing discussion on which is the "better" technique. Currently, it is clear that both techniques work with their intrinsic advantages and drawbacks. At the end of this chapter, we try to compare the two.

7.4.2.2 Layout

A typical setup of a TAS spectrometer is schematically displayed in Figure 7.8.

The incident and scattered neutron wave vectors, k_i and k_f, are selected by Bragg reflection from the monochromator and analyzer crystals, respectively. Large crystals of Cu, Zn, or pyrolytic graphite (PG) with a mosaic width of 20–30 minutes have been used in combination with Soller collimators defining the beam divergence of the incident and diffracted beam. On modern instruments, the mosaic crystals are segmented into plates of a few centimeters in size and mounted on mechanical devices (benders), which permit to control their individual orientation in order to act as curved mirrors and to focus the neutron beam horizontally and/or vertically on the sample and detector. More recently, elastically bent perfect Si or Ge crystals, offering improved focusing properties, thanks to the absence of random mosaic block misorientations, have been employed for work requiring higher resolution. The monochromatic incident flux is monitored by a low-efficiency detector called *monitor 1*. Another similar monitor 2 is placed in the scattered beam to detect the possible presence of strong Bragg peaks, which might give rise to spurious signals in the inelastic spectra. The detector is usually a single He^3 proportional gas tube. The monochromator and analyzer crystals diffract, together with the nominal wavelength λ, also its harmonics, $\lambda/2$ and $\lambda/3$, which may be at the origin of spurious effects. They can be eliminated by appropriate filters. Typical examples of state-of-the-art spectrometers are IN1,

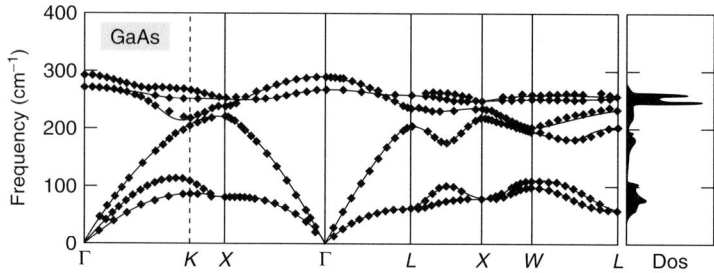

Fig. 7.9 Phonon dispersion curves for GaAs at 12 K. The calculated density of states is shown on the right.

IN8, IN14, and IN20 at the ILL. IN1 is a TAS spectrometer for hot neutrons, IN8 and IN20 for thermal neutrons, and IN14 for cold neutrons. All of them can also use polarized neutrons with or without polarization analysis.

7.4.2.3 The TAS Resolution Function

The resolution volume, that is the region of the $(\mathbf{Q}, \omega)$ space, which corresponds to the momentum end energy transfers of neutrons arriving in the detector in a given spectrometer configuration, can be approximated by a four-dimensional ellipsoid. This resolution volume can be calculated analytically in the Gauss approximation or, more recently, numerically by powerful Monte-Carlo methods with programs like RESTRAX (Saroun and Kulda, 1997) or McStas (Lefmann et al., 2000).

7.4.2.4 Applications

The main application of TAS spectroscopy is the determination of collective excitations like phonons and magnons by coherent scattering on single crystalline samples. The goal is to measure the dispersion relations $\hbar\omega(\mathbf{q})$ in single crystals, where $\hbar\omega$ is the energy and $\mathbf{q}$ the wave vector of the excitation. Dispersion-less excitations, where $\hbar\omega$ does not depend on $\mathbf{q}$, originating from single-particle excitations are better determined by TOF spectroscopy.

Figure 7.9 shows a comparison between measured (Strauch and Dorner, 1990) and calculated (Giannozzi et al., 1991) phonon dispersion curves in GaAs. The dispersion curves consist of three acoustic and three optical branches.

7.4.2.5 Comparison between TAS and TOF Spectroscopy of Coherent Excitations in Single Crystals

- **Advantage of TAS:** The energy of the excitation is determined at a selected point in reciprocal space and the scan is carried out in a single well-defined direction chosen by the experimentalist.

- **Disadvantages of TAS:** Spurious peaks may be produced by higher harmonic Bragg peaks from the monochromator and analyzer crystals.
  Unexpected excitations may be missed.

- **Advantage of TOF:** An overview over the excitations of a more or less substantial fraction of reciprocal space is obtained in one run. So unexpected features may be detected.

- **Disadvantage of TOF:** The dispersion relation in a specific **q**-direction has to be extracted by the computer out of a huge amount of data.

Globally, one could say that the two techniques are complementary to each other. Further information about TAS spectroscopy can be found in (Shirane, Shapiro and Tranquada, 2002).

7.4.3
Back-Scattering (BS) Spectroscopy

7.4.3.1 Principle

Like three-axis instruments, BS spectrometers use single crystals as monochromators and analyzers, but with *fixed* Bragg angles equal to or close to 90° (Birr, Heidemann and Alefeld, 1971). In this way, a very high energy resolution of the order of microelectronvolts is achieved. This can be shown by differentiating the Bragg equation:

$$\frac{\Delta E}{E} = 2\frac{\Delta \lambda}{\lambda} = 2\left(\frac{\Delta d}{d} + \mathrm{ctg}\theta \; \Delta\theta\right) \tag{7.34}$$

For a Bragg angle of 90°, the geometry term becomes zero in first order. $\Delta d/d$ describes a lattice parameter variation. In perfect crystals, it is determined by primary extinction and has typical values around 10^{-5}:

$$\frac{\Delta d}{d} \equiv \frac{\Delta \tau}{\tau} = \frac{16\pi \; F_\tau N_0}{\tau^2} \tag{7.35}$$

Here, F_τ is the structure factor of the reciprocal lattice vector τ, N_0 is the number density of the unit cells.

For a perfect silicon crystal with (111) orientation, one calculates an energy width $\Delta E = 2E\Delta\tau/\tau = 0.077$ µeV. On a real BS spectrometer, the total energy resolution is calculated by the convolution of the monochromator and the analyzer contribution including the geometry term in second order.

For $\theta_B = 90°$ the latter is given by

$$\left(\frac{\Delta k}{k}\right)_{\mathrm{geom}} = \frac{(\Delta\theta)^2}{8} \tag{7.36}$$

This yields a typical value for the energy resolution $\Delta E = 0.2$ µeV (full width at half maximum (FWHM)) at an incident energy $E_i = 2$ meV for the case of a BS spectrometer using silicon crystals in (111) orientation in good agreement with experimental data.

As the Bragg angle is fixed, the energy scan can only be performed by varying the length of $\mathbf{k}_i$ either by moving the monochromator periodically parallel to the lattice vector τ using the longitudinal Doppler effect or by scanning the lattice parameter d of the monochromator by temperature variation. In the former case, Doppler velocities v_{doppler} of a few meters per second yield energy transfers up to 50 µeV for 2-meV neutrons:

$$\delta E_{\mathrm{doppler}} = 2E_i \frac{v_{\mathrm{doppler}}}{v_i} \tag{7.37}$$

Equation (7.37) is valid in first-order $v_{\mathrm{doppler}} \ll v_i$.

In the latter case, temperature variations up to 300 °K on a CaF$_2$ crystal with (422) orientation yield energy transfers up to 200 µeV for thermal neutrons.

Because of the very high energy resolution, the incident flux of a BS spectrometer is low. To overcome this problem, one uses a large spherically curved multicrystal analyzer covering the whole scattering angular range with a huge solid angle. Analyzer surface areas up to 20 m^2 are not unusual. The analyzed neutrons are detected by a multidetector.

The Q resolution of a BS spectrometer is dominated by the incident and scattered beam divergence. The former depends on the geometry of the primary spectrometer, the latter on the geometry of the crystal analyzer/multidetector. The compromise between intensity and Q resolution leads to values of the latter between 5 and 10%. The problem of separating the back-scattered from the incident beam in exact BS can be solved by the use of a deflector crystal and (or) a chopper and TOF discrimination. Higher order contamination can be eliminated by the latter technique at the same time.

7.4.3.2 Existing Instruments

Typical BS instruments are IN10 and IN16 (see Figure 7.10) for cold neutrons and IN13 for thermal neutrons at the ILL.

Similar BS spectrometers exist at NIST and the FRMII reactor. Both use a so-called phase-space transformer (PST), which allows an intensity gain of a factor 2 to 4 (Schelten and Alefeld, 1984). This device transforms a collimated white beam into a divergent monochromatic beam by Bragg reflection from a mosaic crystal moving *parallel* to the scattering plane at high speed. Again the Doppler effect is used. The price to pay is the increased background produced by the proximity of the strong white beam.

The TOF BS spectrometers IRIS at ISIS and BASIS at SNS have been mentioned in section 7.4.1. Here, BS is only used in the secondary spectrometer and TOF in the primary spectrometer; this is the way to go on a pulsed source.

7.4.3.3 Applications

The main application of BS spectroscopy is high-energy resolution incoherent neutron scattering of low-energy single-particle excitations in the

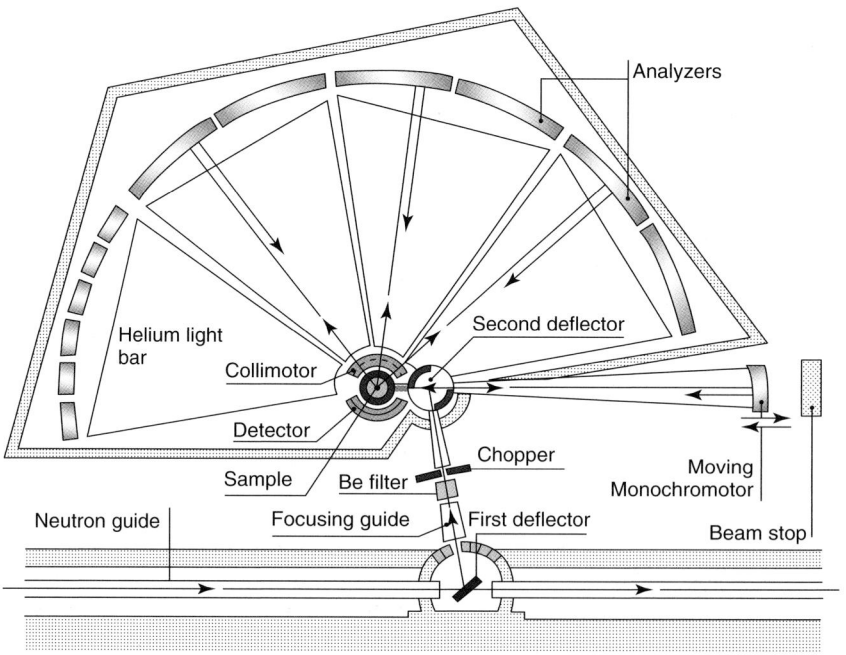

Fig. 7.10 Layout of the BS spectrometer IN16 at the ILL.

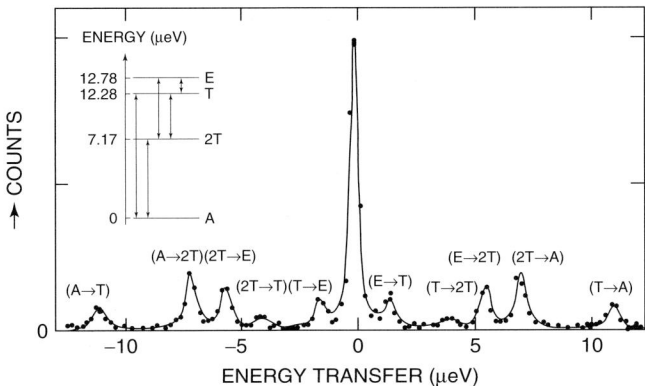

Fig. 7.11 Tunneling spectrum of NH_4^+ ions in NH_4ClO_4 at 4 K (Prager and Alefeld, 1976; Press, 1981).

microelectronvolt or gigahertz range with moderate momentum transfer resolution: tunneling spectroscopy (Prager and Alefeld, 1976; Prager and Heidemann, 1997; Press, 1981), hyperfine interactions (Heidemann, 1972; Word, Heidemann and Richter, 1977; Chatterji and Frick, 2004), slow motions in condensed matter: viscous liquids, glass dynamics (Mezei, Pappas and Gutberlet, 2003), liquid and plastic crystals, soft matter, and self diffusion. For further information, see (Hippert et al., 2005).

Figure 7.11 shows the tunneling spectrum of NH_4^+ ions in NH_4ClO_4 at 4 K measured on IN10. The NH_4^+ ions behave like quantum rotators at low temperatures. Analyzing the energies and intensities of the observed transitions yields the overlap matrix elements (Prager and Alefeld, 1976; Press, 1981).

7.4.4
Neutron Spin Echo (NSE) Spectroscopy

7.4.4.1 Principle

Neutron scattering suffers from the strong correlation between energy resolution and intensity, that is, high-energy resolution is only obtained at the expense of low intensity. In BS spectroscopy, this problem is bypassed by the use of *spatial focusing* of the scattered (and incident) neutron beam resulting in only moderately good resolution in wave vector transfer $\mathbf{Q}$.

The NSE technique (Mezei, 1972) allows the energy resolution to be decoupled from the intensity in cases whenever Q resolution is not so important or is determined by the angular resolution rather than the wavelength distribution. This is achieved by a very special "time-focusing" technique using the precession of neutron spins in a magnetic field.

A polarized neutron beam entering a region with a magnetic field B perpendicular to the magnetic moment of the neutrons will undergo precession with the Larmor frequency ω_L:

$$\omega_L = \gamma B \text{ where } \gamma = \frac{2.916 \text{ kHz}}{\text{Oe}} \quad (7.38)$$

The total precession angle after traversing a distance l with velocity v will be $\phi_1 = \omega_L \, l_1/v_1$. If v has a finite distribution, the beam will appear to be depolarized after a short distance. Assume now that this beam hits a sample.

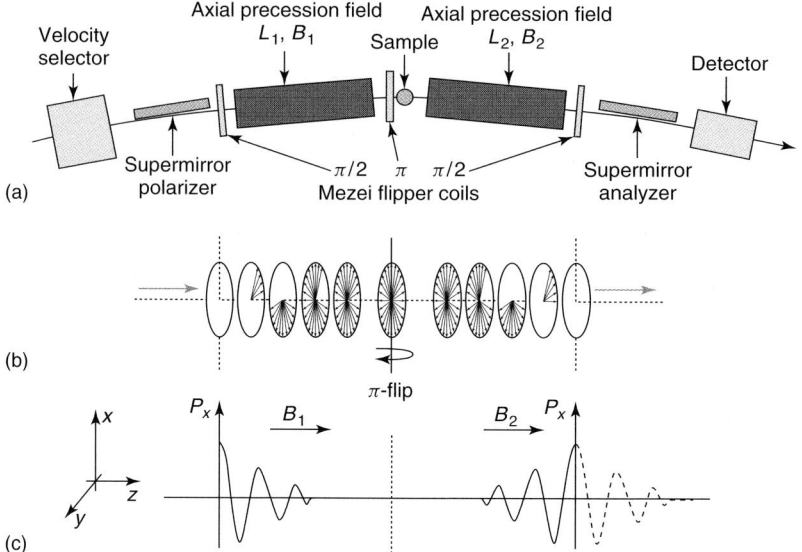

Fig. 7.12 Layout of an NSE spectrometer: (a) principal components, (b) dephasing and rephasing of the neutron spins, (c) x-component of the net neutron polarization (Hippert et al., 2005).

The scattered beam would now travel through another region with an opposite magnetic field $-B$. Then the total precession angle will be

$$\phi_{tot} = \omega_L \left(\frac{l_1}{v_1} - \frac{l_2}{v_2} \right) \quad (7.39)$$

in the case of elastic scattering and if $l_1 = l_2$, ϕ_{tot} is independent of v and we recover the incident beam polarization.

In the case of a small energy change $\hbar\omega \ll mv^2$ we obtain in first-order approximation,

$$\phi_{tot} = \left(\frac{\hbar \gamma Bl}{mv^3} \right) \omega \quad (7.40)$$

Analyzing the polarization P_{final} after the second flight path yields

$$P_{final}(\mathbf{Q}, t_F) = <\cos\Phi> = \int \cos(\omega\, t_F)$$
$$\times S(\mathbf{Q}, \omega) d\omega \equiv S(\mathbf{Q}, t_F) \quad (7.41)$$

where $t_F = \hbar\, \gamma Bl/mv^3$ is the so-called Fourier time. NSE therefore directly measures the intermediate scattering function. It is important to note that the Fourier time t_F is proportional to $1/v^3$ or to λ^3. So with very cold long-wavelength neutrons the time domain can be extended to the near microseconds range or, in terms of energy transfer, nanoelectronvolt values can be measured.

7.4.4.2 Implementation

A velocity selector with a coarse wavelength resolution of the order of 15% selects a poorly monochromatic beam, which is polarized (Figure 7.12). After passing a $\pi/2$ flipper, the beam starts precessing in a solenoid magnetic field, traverses a π flipper and hits the sample. The scattered beam precesses in the second solenoid field in the opposite direction, traverses a second $\pi/2$ flipper, enters an analyzer, and is finally detected with a counter.

This layout is used by the NSE spectrometer IN11 at the ILL. A typical field of applications of NSE spectroscopy is coherent quasi-elastic small-angle scattering. An example is the slow motion polymer dynamics in the megahertz to gigahertz range.

Figure 7.13 shows experimental data of $S(Q, t_f)$ from IN15 for polyethylene together with theoretical predictions. Solid lines represent the model of reptation by de Gennes (1981). Note that the Fourier time extends to 200 ns.

7.4.4.3 NSE Variants

A number of variants of this technique exist. The monochromatization by a velocity selector can be replaced by a TOF analysis of a white neutron beam as implemented on IN15 at the ILL. The NSE tool can also be used as an option on a TAS spectrometer with polarization analysis (TASSE) (Mezei, 1980). Here the application is the measurement of the line width of elementary excitations without and with dispersion. For the latter case, special tricks have to be played in order to turn the $(\mathbf{q}, \omega)$ ellipsoid by tilting the angles of the precession fields. The IN20 spectrometer at the ILL allows use of this technique. Another variant is the so-called resonance or zero field neutron spin echo (ZFNSE). It was invented by Gähler and Golub (1987).

Principle The neutron precession in the solenoids is replaced by the passage of the neutrons through a strong high-frequency electromagnetic field. This technique is a promising option for high-resolution inelastic scattering on TAS spectrometers. A dedicated instrument of this kind, the TRISP spectrometer, became operational recently at the FR-MII reactor.

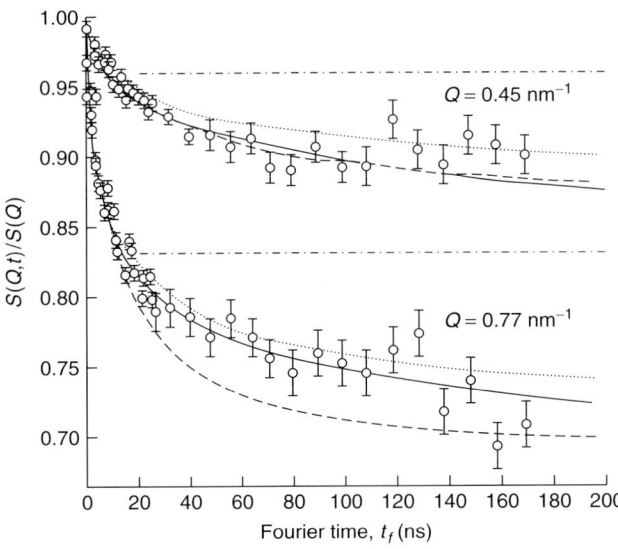

Fig. 7.13 Plot of $S(Q, t_f)$ data for polyethylene measured on IN15 (Hippert et al., 2005).

7.4.4.4 NSE versus Standard Inelastic Spectrometers

On all instruments working in $(\mathbf{Q}, \omega)$ space, the instrumental resolution function has to be deconvoluted from the measured spectrum to obtain the pure sample response $S(\mathbf{Q}, \omega)$. In $(\mathbf{Q}, t)$ space, the deconvolution becomes a simple division. This is an obvious advantage. Samples containing nuclei that scatter spin incoherently such as H, V, Co are not good candidates for NSE because they decrease the scattered polarization by a factor 1/3. Incoherent scattering, in general, is better investigated on spectrometers working in $(\mathbf{Q}, \omega)$ space. A sample with a strong elastic peak and weak inelastic excitations close to the elastic line is also not a good candidate for NSE: the resulting $S(t)$ curve would be a cosine oscillation with a small amplitude sitting on a high constant background. Good samples for NSE are those giving rise to a strong quasi-elastic signal at small momentum transfer, as from deuterated polymer solutions or from critical scattering near a phase transition. For further information, see (Mezei, Pappas and Gutberlat, 2003).

7.5 Conclusion

We have seen that neutron spectroscopy covers a large range of applications. This is again summarized in Figure 7.14, which shows the applications covered by neutron spectroscopy in the (Q, ω) plane. The region inside the trapezoidal box is

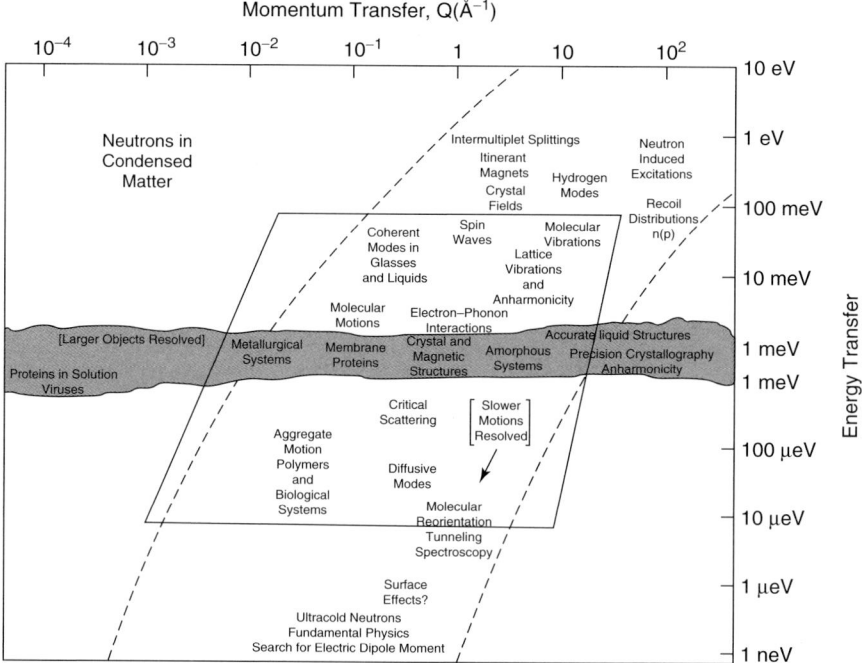

Fig. 7.14 Applications of neutron scattering in condensed matter research (Lander and Emery, 1985).

the range probed by TAS spectroscopy. The shaded region corresponds to an expanded cut in the energy scale (Lander and Emery, 1985).

Acknowledgments

The author is indebted to D. Gray for checking the English.

Glossary

Doppler Drive: Device to change the neutron velocity using Doppler effect.

Fermi Chopper: Mechanical device to pulse and monochromatize a neutron beam.

Nuclear Spallation: Process in which a heavy nucleus fragments into lighter nuclei and neutrons, when hit by highly energetic particles like protons from an accelerator.

Neutron Back-Scattering: Method to achieve very high energy resolution in neutron scattering using Bragg reflection with a Bragg angle of 90°.

Neutron Guide: Device to transport a neutron beam over long distances.

Neutron Three-Axis Spectrometer: Instrument consisting of three axes: monochromator, sample, and analyzer.

Neutron Spin Echo: Method to achieve very high energy resolution in neutron scattering using the Larmor precession of neutrons.

Phase-Space Transformer: Device to transform a white collimated neutron beam into a monochromatic divergent beam using the Doppler effect.

Super-Mirror: Mirror for neutrons with an increased glancing angle using multilayer coating.

Scattering Length: Parameter defining the strength of the neutron sample interaction.

Time-of-Flight Method: Determination of the neutron velocity by measuring its time of flight over a given distance.

Velocity Selector: Mechanical device to monochromatize a neutron beam.

References

Anderson, I.S., Brown, P.J., Carpenter, J.M., Lander, G., Pynn, R., Rowe, J.M., Schärpf, O., Sears, V.F. and Willis, B.T.M. (1999) *International Tables for Crystallography C*, 2nd edn, Kluwer Academic Publishers, p. 426.

Bacon, G.E. (1962), *Neutron Diffraction*, Clarendon Press, Oxford.

Birr, M., Heidemann, A. and Alefeld, B. (1971) *Nucl. Instrum. Methods*, **95**, 435.

Brockhouse, B.N. (1961), *Inelastic Neutron Scattering in Solids and Liquids*, IAEA, p. 113.

Chatterji, T. and Frick, B. (2004) *Solid State Commun.*, **131**, 453.

Daillant, J. and Gibaud, A. (eds) (1999) *X-Ray and Neutron Reflectometry: Principles and Applications*, Springer Verlag.

Fermi, E. (1936) *Ric. Sci. Vicotruiz.*, **7**, part 2, 13.

Gähler, R. and Golub, R. (1987) *Z. Phys.*, **B65**, 269.

de Gennes, P.G. (1981) *J. Physique*, **42**, 73.

Giannozzi, P., Gironcoli, S., Pavone, P. and Baroni, S. (1991) *Phys. Rev.*, **B43**, 7231.

Heidemann, A. (1972) *Z. Phys.*, **238**, 208.

Hippert, F., Geissler, E., Hodeau, J.L., Lelievre-Berna, E. and Regnard, J.R. (eds) (2005) *Neutron and X-Ray Spectroscopy*, Springer Verlag.

van Hove, L. (1954) *Phys. Rev.*, **95**, 249.

Kisi, E.H. and Howard, C.J. (2008) *Applications of Neutron Powder Diffraction*, Oxford Series on Neutron Scattering in

Condensed Matter, vol. 15, Oxford University Press.

Lander, G.H. and Emery, V.J. (1985) *Nucl. Instrum. Methods*, **B 12**, 525.

Lefmann, K., Nielson, K., Tennant, A. and Lake, B. (2000) *Physica B*, **276–278**, 152–153.

Lovesey, S.W. (1984) *Theory of Neutron Scattering from Condensed Matter*, Clarendon Press, Oxford.

Mezei, F. (1972) *Z. Phys.*, **255**, 146.

Mezei, F. (ed.) (1980) *Neutron Spin Echo*, Lecture Notes in Physics, vol. 128, Springer Verlag.

Mezei, F., Pappas, C. and Gutberlet, T. (eds) (2003) *Neutron Spin Echo Spectroscopy: Basics, Trends and Applications*, Lecture Notes in Physics, vol. 601, Springer Verlag.

Mirebeau, I., Hennion, M., Casalta, H., Andres, H., Güdel, H.U., Irodova, A.V. and Caneschi, A. (1999) *Phys. Rev. Lett.*, **83**, 628.

Prager, M. and Alefeld, B. (1976) *J. Chem. Phys.*, **65**, 4927.

Prager, M. and Heidemann, A. (1997) *Chem. Rev.*, **97**, 2933.

Press, W. (1981) *Single-Particle Rotations in Molecular Crystals*, Springer Tracts in Modern Physics, vol. 92.

Price, D.L. and Sköld, K. (1986) *Neutron Scattering*, Methods of Experimental Physics, vol. 23, Academic Press, London, Part **A**, p. 1.

Saroun, J. and Kulda, J. (1997) *Physica B*, **234–236**, 1102.

Schelten, J., Alefeld, B. (1984) Jülich Report.

Scherm, R. (1972) *Ann. Phys.*, **7**, 349.

Shirane, G., Shapiro, S.M. and Tranquada, J.M. (2002) *Neutron Scattering with a Triple-Axis Spectrometer, Basic Techniques*, Cambridge University Press.

Shull, C.G., Wollan, E.O. and Koehler, W.C. (1951) *Phys. Rev.*, **84**, 904.

Springer, T. and Lechner, R.E. (2005) Diffusion studies of solids by quasi-elastic neutron scattering, in *Diffusion in Condensed Matter-Methods, Materials, Models*, (eds P. Heitjans and J. Kärger), Springer Verlag.

Strauch, D. and Dorner, B. (1990) *J. Phys.: Condens. Matter*, **2**, 1457.

Williams, C., May, R.P. and Guinier, A. (1994) Small angle scattering of X-rays and neutrons, in *Characterisation of Materials*, Materials Science and Technology, vol. 2B (ed. E. Lifshin), VCH Verlagsgesellschaft, Weinheim, p. 611.

Windsor, C.G. (1981) *Pulsed Neutron Scattering*, Taylor & Francis, London.

Maier-Leibnitz, H. (1966) *Nucleonik*, **8**, 61.

Word, R., Heidemann, A. and Richter, D. (1977) *Z. Phys.*, **B28**, 23–30.

Web sites

BS review: http://wwwold.ill.fr/YellowBook/IN16/BS-review/index.htm.

ESS: http://neutron.neutron-eu.net/n_ess.

FRMII: http://www.frm2.tu-muenchen.de.

ILL: http://www.ill.eu/.

Cicognani, G., *The Yellow Book: Guide to Neutron Scattering Facilities, ILL*. Copies can be obtained from: ILL, SCO, BP156, F-38042 Grenoble Cedex 9 (France).

ISIS: http://www.isis.rl.ac.uk/.

NIST: http://rrdjazz.nist.com.

SINQ: http://sinq.web.psi.ch/.

SNS: http://neutrons.ornl.gov/.

8
Electron Scattering by Atoms, Ions, and Molecules

Klaus Bartschat, Philip G. Burke and Albert Crowe

8.1	**Introduction** 211	
8.2	**Review of Experimental Methods** 212	
8.2.1	The Electron Spectrometer 212	
8.2.2	Elastic Scattering 213	
8.2.2.1	Differential Cross Sections 213	
8.2.3	Excitation of Discrete States 215	
8.2.3.1	Differential Cross Sections 215	
8.2.3.2	Integral Cross Sections 215	
8.2.3.3	Angular and Polarization Correlations 216	
8.2.4	Ionization 220	
8.2.4.1	Total Cross Sections 220	
8.2.4.2	Double Differential Cross Sections 221	
8.2.4.3	Higher Order, including, Fully Differential Cross Sections 222	
8.2.4.4	(e,2e) Studies of Direct Single Ionization 223	
8.2.4.5	(e,3e) Studies of Double Ionization 226	
8.2.4.6	Excitation–Ionization 227	
8.2.4.7	Excitation–Autoionization 229	
8.2.4.8	Inner-Shell Ionization 230	
8.2.5	Electron–Molecule/Biomolecule Interactions 230	
8.2.6	Electron–Ion Interactions 231	
8.2.7	Spin-Polarized Electron Studies 233	
8.2.7.1	Exchange Interaction 234	
8.2.7.2	Spin–Orbit Interaction 235	
8.2.7.3	A Typical Apparatus and Some Illustrative Results 236	
8.2.8	Positron Scattering 238	
8.3	**Review of Theoretical Methods** 238	
8.3.1	The Close-Coupling Expansion 238	
8.3.1.1	The Wave Equation, Scattering Amplitude, and Cross Section 238	
8.3.1.2	Extension of the Close-Coupling Method to Intermediate and High Energies 242	
8.3.1.3	Inclusion of Relativistic Effects 243	

Encyclopedia of Applied Spectroscopy. Edited by David L. Andrews.
Copyright © 2009 WILEY-VCH Verlag GmbH & Co. KGaA, Weinheim
ISBN: 978-3-527-40773-6

8.3.1.4	Illustrative Results	244
8.3.1.5	Resonances and Quantum Defect Theory	247
8.3.1.6	Dielectronic Recombination	250
8.3.2	Born Series Methods	251
8.3.2.1	Elastic Scattering and Excitation of Atoms and Ions at Intermediate and High Energies	251
8.3.2.2	The Born Series	251
8.3.2.3	The Distorted-Wave (DW) Method	252
8.3.3	Initial-Value Methods	254
8.3.3.1	Time-Dependent Close-Coupling (TDCC)	254
8.3.3.2	Exterior Complex Scaling (ECS)	255
8.3.4	Ionization of Atoms and Ions	256
8.3.4.1	The Ionization Amplitude and Cross Section	256
8.3.4.2	Threshold Behavior of the Ionization Cross Section	258
8.3.4.3	Ionization at Intermediate and High Energies	259
8.3.4.4	Excitation–Autoionization	261
8.3.5	Scattering by Molecules	262
8.3.5.1	Laboratory Frame Representation	262
8.3.5.2	Molecular Frame Representation	263
8.3.5.3	Inclusion of the Nuclear Motion	265
8.3.5.4	Illustrative Results for Molecules	270
8.4	**Conclusions and Future Work**	**274**
	References	**276**
	Further Reading	**281**

8.1
Introduction

The experimental study of electron–atom and electron–molecule collisions dates back to the beginning of the 20th century Lenard (1903). Later in 1914, Nobel prize winners Franck and Hertz (1914) demonstrated the energy loss of electrons scattered from mercury vapor. Theoretically, electron scattering by atoms and molecules has attracted considerable attention since the earliest days of quantum mechanics, because such processes provide a means of investigating the dynamics of many-particle systems at a fundamental level. In addition, a detailed understanding of these processes is required in many other fields, particularly in astrophysics, plasma physics, laser physics, electrical discharges, ionospheric processes in planetary atmospheres, isotope separation, and radiation chemistry and physics.

In recent years, a number of important advances have been made both on the experimental and on the theoretical side. On the experimental side, these include absolute measurements of cross sections, experiments using coincidence techniques, spin-polarized projectile beams and targets, high-energy-resolution electron beams, and bright beams of positrons. On the theoretical side, many new developments have been stimulated by the increasing availability of powerful computers that allow for detailed calculations to be carried out for complex atoms, ions, and molecules.

This review is an update to the article by Burke and Crowe (1993). As previously, we first describe recent experimental work in Section 8.2, which is followed by the theory of electron scattering by atoms, ions, and molecules in Section 8.3. We further compare theoretical predictions with recent experimental data for a few representative cases. Within the page limit of this article, we can mention only a small fraction of the ongoing work in this field. Interested readers are referred to journals such as The Physical Review A or Journal of Physics B, which contain extensive sections on electron collisions, as well as to the Proceedings and Book of Abstracts of the biannual "International Conference on Photonic, Electronic, and Atomic Collisions" (ICPEAC). Finally, in Section 8.4, we draw some conclusions and mention the areas where future work is needed.

Encyclopedia of Applied Spectroscopy. Edited by David L. Andrews.
Copyright © 2009 WILEY-VCH Verlag GmbH & Co. KGaA, Weinheim
ISBN: 978-3-527-40773-6

8.2
Review of Experimental Methods

8.2.1
The Electron Spectrometer

The basic workhorse of a large number of electron scattering studies is the electron spectrometer. In principle, free electrons can be formed into a beam and their energy selected using various combinations of electrostatic and magnetic fields. In practice, the use of electrostatic fields is more common, these being more easily controlled and shielded than their magnetic counterparts. This is particularly important where it is essential to preserve the direction of low-energy electrons following the collision process.

We show in Figure 8.1 a recent example of a spectrometer that combines the characteristics of a conventional electrostatic device with a recent innovation Allan (2004). Basically, the electron gun consists of a source of electrons produced by thermionic emission from a heated filament. The electrons are collimated and focused by an electrostatic lens system onto the input aperture of a double hemispherical energy selector. Those electrons within a narrow band of energies satisfying the criteria for transmission through the selector are then focused on the gas beam produced by a nozzle arrangement. Scattered electrons from the interaction region traveling in the direction of the scattered electron analyzer are similarly focused onto the input aperture of its double hemispherical analyzer, the transmitted electrons being focused onto a single-channel electron multiplier detector.

One drawback of conventional electron spectrometers is that the angular range of the electron analyzer is limited by the physical presence of other components of the spectrometer. This was overcome by Read and Channing (1996) who applied a localized static magnetic field to the interaction region of a conventional spectrometer. The incident electron beam and the scattered electrons are respectively steered to and from the interaction region through angles set by the field (hence the common name "magnetic angle changer" ("MAC")). This steering means that electrons normally scattered into inaccessible scattering angles are rotated into the accessible angular range of the electron analyzer while the magnetic field design is such that it leaves the angular distribution of the electrons undistorted. The spectrometer shown in Figure 8.1 has a MAC fitted, thereby enabling the full angular range $0° - 180°$ to be accessed.

Another important development in the study of small and/or multidifferential cross sections is the increasing use of multidetection methods, usually in conjunction with toroidal analyzers. These enable simultaneous detection of scattering over a wide range of angles. Recent examples are the spectrometers of Catoire et al. (2007) and of Lower et al. (2007).

Perhaps the most novel and exciting recent experimental development in electron–atom/molecule scattering is the reaction microscope described by Ullrich et al. (2003). In contrast to conventional electron spectrometers, it uses recoil ion and electron momentum spectroscopy to measure the vector momenta of outgoing charged particles over large (close to 4π) solid angles.

These recent experimental innovations have been applied in studies of

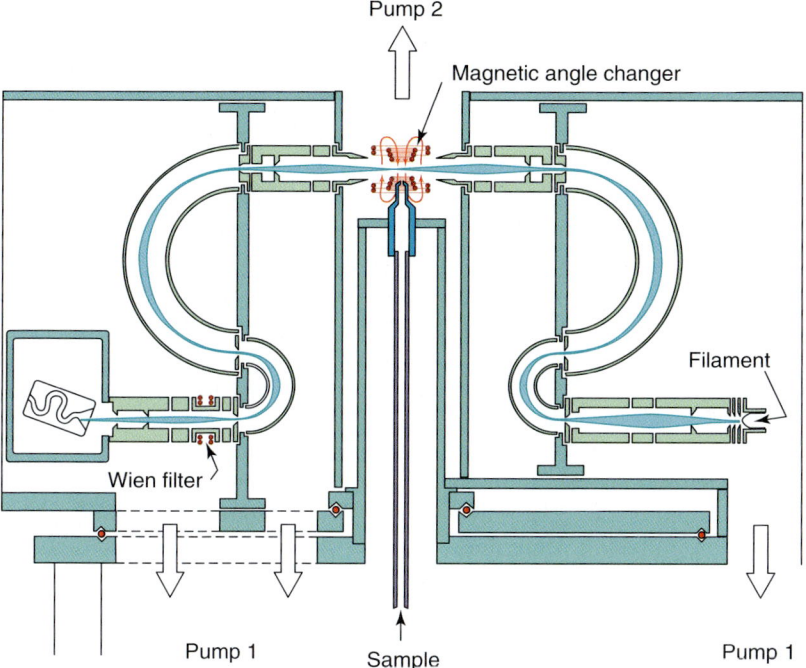

Fig. 8.1 The electron spectrometer of Allan (2004).

ionization and are discussed further in Section 8.2.4.3.

8.2.2
Elastic Scattering

Elastic scattering is characterized by the absence of energy transfer from the incident electron to the internal energy of the atom.

8.2.2.1 Differential Cross Sections

Differential elastic scattering studies are normally carried out using an electron spectrometer broadly similar to that shown in Figure 8.1. The electron analyzer is tuned to transmit only the electrons having an energy equal to the incident electron energy. Relative angular differential cross sections are determined by observation of the scattered electron signal, as the electron analyzer is rotated about the reaction region. The elastic scattering differential cross section $d\sigma(E,\theta)/d\Omega$ can be written as

$$\frac{d\sigma}{d\Omega}(E,\theta) = \frac{dN_S(E,\theta)}{N_i\,n\,l\,d\Omega} \qquad (8.1)$$

where $dN_S(E,\theta)$ is the number of electrons with energy E per unit time scattered into a solid angle $d\Omega$ at a scattering angle θ with respect to the incident electron beam direction, and N_i is the number of electrons per unit time incident on the target gas of number density n in a scattering cell of length l.

Various techniques have been used to place the relative cross sections on an absolute cross-section scale. The most direct method of making the cross section absolute is to measure all of the quantities on the right-hand side of Eq. (8.1). However, this is rarely done owing to the extreme difficulties in measuring both these quantities and the various efficiencies involved.

In recent years, the relative flow technique has been the most widely used method for placing measured relative cross sections on an absolute scale. The unknown cross sections are determined by comparing the measured electron scattering intensities against those of another species with known cross sections under identical experimental conditions. Details of the technique are given by Brunger and Buckman (2002) while the most recent developments are described by Khakoo et al. (2007).

Helium elastic scattering cross sections are the most commonly used standards. They have been accurately measured and put on an absolute cross-section scale using a phase shift analysis; see, for example, Andrick (1973). The measured values are accurately reproduced by the convergent close-coupling (CCC) theory of Fursa and Bray (1997) and the R-matrix with pseudostates (RMPS) method of Bartschat et al. (1996a). For energies up to 19 eV, that is, below the first excitation threshold, accurate variational calculations for the s-wave and p-wave phaseshifts were already performed 30 years ago by Nesbet (1979).

In Figure 8.2, we show the elastic differential scattering cross section for xenon at an electron energy of 30 eV. One of the two experiments reported by Cho et al. (2006) uses a MAC to cover the full range of scattering angles and the relative flow method has been used to put the data on an absolute cross-section scale. Their theories show

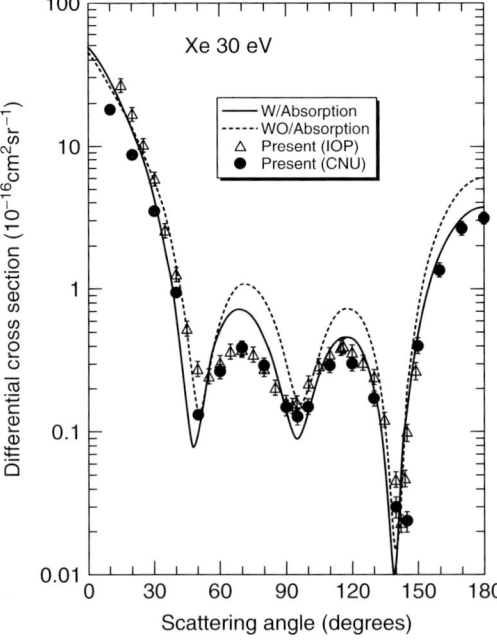

Fig. 8.2 The absolute differential elastic scattering cross section for xenon at an incident electron energy of 30 eV. (From Cho et al. (2006).)

the importance of including absorption effects in the theoretical description.

8.2.3
Excitation of Discrete States

The excitation process for an atomic target can be written as

$$e^-(E_i) + A_i \longrightarrow e^-(E_j) + A_j \qquad (8.2)$$

where, in atomic units, $E_i = \frac{1}{2}k_i^2$ is the incident electron energy while $E_j = \frac{1}{2}k_j^2$ is the scattered electron energy; k_i and k_j are the respective wave numbers. (In this and later equations, we use atomic units, $\hbar = m = e = 1$, where m is the mass of the electron and $-e$ is its charge.)

8.2.3.1 Differential Cross Sections

The scattered electron studies are conducted as for elastic scattering, except that the electron analyzer is tuned to accept electrons with energy E_j. Angular differential cross sections can then be measured for a particular excited state. An example of an angular differential cross section is shown in Figure 8.6 for the 2^1P state of helium.

For spectroscopic studies, the measurement of energy-loss spectra, where the energy of the detected electrons is varied for a fixed incident electron energy or alternatively the detected electron energy is fixed for a varying incident electron energy, is a useful technique. Electron spectroscopy is particularly valuable for the study of optically forbidden transitions.

8.2.3.2 Integral Cross Sections

When the excited atom is metastable, information on the excitation process can be obtained in a relatively straightforward way by making use of the fact that it can eject an electron from a surface whose work function is less than the internal energy of the atom. Time-of-flight (TOF) techniques are normally used to isolate the neutral-atom signal from that due to photons. The method has the advantage that the energy resolution of the measurements is determined by the incident electron beam resolution only. Hence, it has been used as a convenient method for the study of resonance features contributing to the excitation process, see Section 8.3.1.4, Figure 8.21. It has the disadvantage that it is difficult to isolate individual metastable states that are excited. The TOF technique can also be used to determine the energy distributions of metastable fragment atoms from molecular dissociation.

The traditional method of studying short-lived excited states is to monitor the photon intensity arising from the optical decay of the excited state as a function of the incident electron energy, giving the so-called "excitation function" of the state. Here individual states, particularly those radiating in the visible region of the spectrum, can often be readily isolated using high-resolution optical spectrometers. However, even for simple atoms, there is great difficulty relating the measured excitation function to the cross section for excitation of the state because of, often unknown, cascade contributions from higher excited states, inherent and instrumental polarization effects, or trapping of the emitted radiation prior to reaching the detector, in addition to the usual difficulties of making cross sections absolute. Again, many studies have concentrated on resonance features in the excitation functions; see, for example, the recent work of Stepanovic et al. (2006) in Figure 8.20 of Section 8.3.1.4. For a review of optical excitation functions for electron–atom and electron–ion excitation, see Heddle and Gallagher (1989).

Another way of measuring total cross sections takes advantage of the rapidly developing trap technology in atomic physics. In this technique, a laser is used to trap the atoms of interest, and absolute cross sections can be obtained directly from the trap loss when the laser is switched off. Examples of such measurements are the work of Schappe et al. (1995) and MacAskill et al. (2002). Lukomski et al. (2006) showed that the technique can, in principle, also be used to obtain total cross sections from laser-excited initial states that would be hard to measure otherwise. Another trap measurement involving metastable He atoms was carried out by Uhlmann et al. (2005).

Information on the excited state is also accessible through the polarization of the emitted photons. Many studies of the polarization fraction

$$P = \frac{I_\parallel - I_\perp}{I_\parallel + I_\perp} \quad (8.3)$$

where $I_\parallel$ and $I_\perp$ are the radiation intensities observed following transmission through a polarizer with optic axis parallel and perpendicular to the incident electron beam, have been carried out.

8.2.3.3 Angular and Polarization Correlations

The information from polarization fraction data has limited value because it relates to the incoherent mixture of the magnetic sublevels of the excited state due to integration over all possible kinematical processes. On the other hand, if identically prepared states are produced, a coherent mixture of the magnetic sublevels of the excited state is observed. For example, an excited $^1P^o$ state can be described as

$$|\psi\,^1P\rangle = a_{+1}|11\rangle + a_0|10\rangle + a_{-1}|1-1\rangle \quad (8.4)$$

where the a_M are the excitation amplitudes for the sublevels $|LM\rangle$ with magnetic quantum number M. Theoretical analyses of coherent excitation can be found in Blum (1996), Andersen, Gallagher and Hertel (1988), and Andersen and Bartschat (2001).

Experimentally, the conditions for coherent excitation are produced by observation of the scattered electron responsible for excitation of the particular states at a well-defined scattering angle, in coincidence with the emitted photon. This ensures that only identically prepared excited states are observed. Information on the excited state is then determined by measuring the polarization state of the emitted radiation. This can be done directly, at least for simple systems such as $^1S \rightarrow\,^1P^o$ excitation in helium, by measuring the linear and circular polarization of the radiation emitted perpendicular to the scattering plane. Alternatively, the information can be deduced from the angular correlation between the scattered electron and the emitted photon.

Eminyan et al. (1974) carried out the first electron–photon angular correlation experiment for the 2^1P state of helium. In this experiment, those electrons that lost 21.2 eV (the excitation energy of the 2^1P state of helium) were detected by the electron analyzer at a fixed scattering angle. Angular correlations were then determined by observing the coincidence signal as the photon detector was rotated about the collision center.

The first polarization correlation measurement was reported for the 3^1P state of helium by Standage and Kleinpoppen (1976). Here, the linear polarizations P_1 and P_2 and the circular polarization P_3 are measured in coincidence with the scattered electrons for radiation emitted perpendicular to the scattering plane (z-direction). In the "natural coordinate system" with the z-axis defined perpendicular to the scattering plane and the x-axis chosen along the incident beam direction, these "Stokes parameters" are defined by

$$I_z P_1 = I_z(0°) - I_z(90°) \qquad (8.5)$$

$$I_z P_2 = I_z(45°) - I_z(135°) \qquad (8.6)$$

$$I_z P_3 = I_z(\text{RHC}) - I_z(\text{LHC}) \qquad (8.7)$$

where $I_z(\theta)$ is the signal transmitted by a linear polarizer with optic axis at an angle θ to the electron beam, I is the total signal in that direction, and RHC and LHC refer to right-hand and left-hand circularly polarized radiation. A more complete description of excitation of states with higher angular momentum requires a further linear polarization measurement in the scattering plane (usually the y-direction),

$$I_y P_1 = I_y(0°) - I_y(90°) \qquad (8.8)$$

This is also true for heavier targets in which relativistic effects such as the spin–orbit interaction can no longer be ignored.

Note that many other definitions of the Stokes parameters have been used in the literature, as well as a different coordinate system. The "collision frame," often used for convenience in numerical calculations, is defined with the z-axis along the incident beam and the y-axis perpendicular to the scattering plane. For a translation between the various conventions, see Andersen, Gallagher and Hertel (1988).

We show in Figure 8.3 a schematic diagram of the polarization analysis system used in Newcastle. By rotating the in-plane photon detector, the apparatus can also measure angular correlations.

We show the most recent experimental data and theory for the reduced (after fine-structure depolarization has been taken into account) Stokes parameters for 1s − 2p excitation in e − H scattering at 54.4 eV in Figure 8.4. Note that the angular correlation data of Yalim, Cvejanovic and Crowe (1997) and the polarization correlation data of Williams and Mikosza (2006) lie consistently within one standard deviation of theory.

The measured angular correlations have been analyzed in different ways. The analysis of Andersen, Gallagher and Hertel (1988) in terms of the shape and dynamics of the excited state charge cloud gives a particularly straightforward physical picture. They define the following parameters, sometimes referred to as *electron impact coherence parameters*: the alignment angle γ, the length l and width w in the scattering plane, and the height h of the charge cloud along the axis perpendicular to the scattering plane. Furthermore, $L_\perp$ is the expectation value of the angular momentum transfer perpendicular to the scattering plane while the degree of linear polarization, P_l, of the emitted light is defined in terms of l and w. As shown in Figure 8.5, these

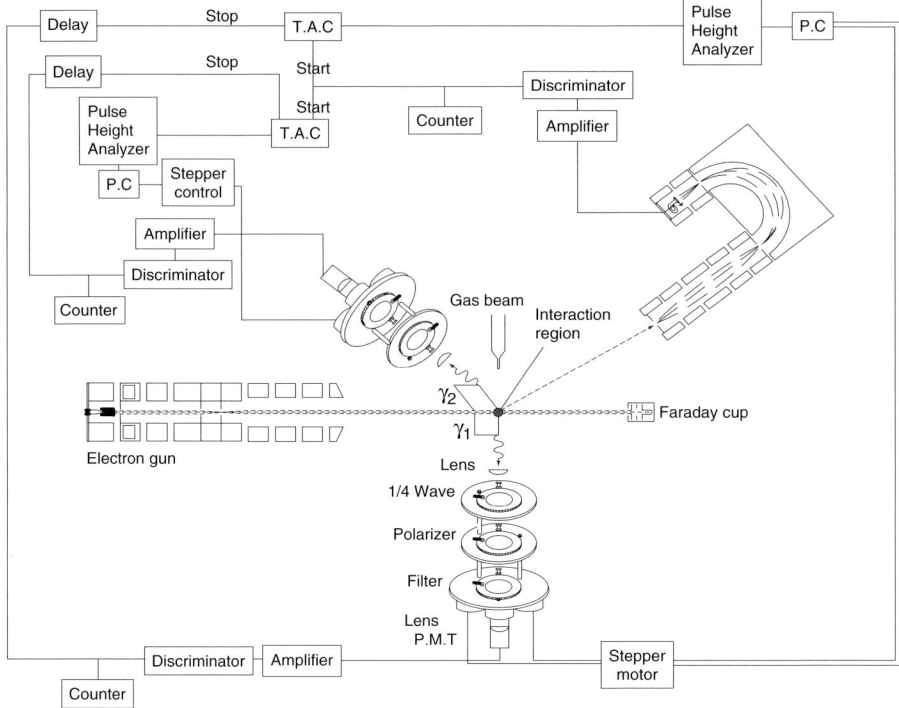

Fig. 8.3 The Newcastle scattered electron-polarized photon correlation spectrometer.

parameters are readily related to the measured Stokes parameters and angular correlations. They have the added advantage that they can be used to describe a wide range of excited states in different atoms.

The variations of $L_\perp$, P_l, and γ with scattering angle are shown in Figure 8.6 for the 2^1P state in helium at an incident electron energy of 50 eV. Once again, there is excellent agreement between experiment and the most sophisticated theories currently available for this problem.

An alternative method giving the same complete information on excitation is superelastic scattering from laser-excited states. The technique involves exciting the state of interest with a laser. Electrons are then scattered from this excited state and those scattered electrons that have gained energy corresponding to the energy of the excited state (superelastically scattered) are observed as a function of the polarization of the laser photons. Compared with the correlation methods, the technique is relatively fast because it does not involve coincidence methods. On the negative side, it can be applied only to the states that are accessible by a single laser photon. The latest developments and results from excitation of $4P_1$ state of calcium can be found in Hussey et al. (2008) and for the $(6p)^2P_{3/2}$ state of cesium in Slaughter et al. (2007). Interestingly, the latter study shows that the measured Stokes parameters are insensitive to inclusion of core and relativistic effects for a heavy atom.

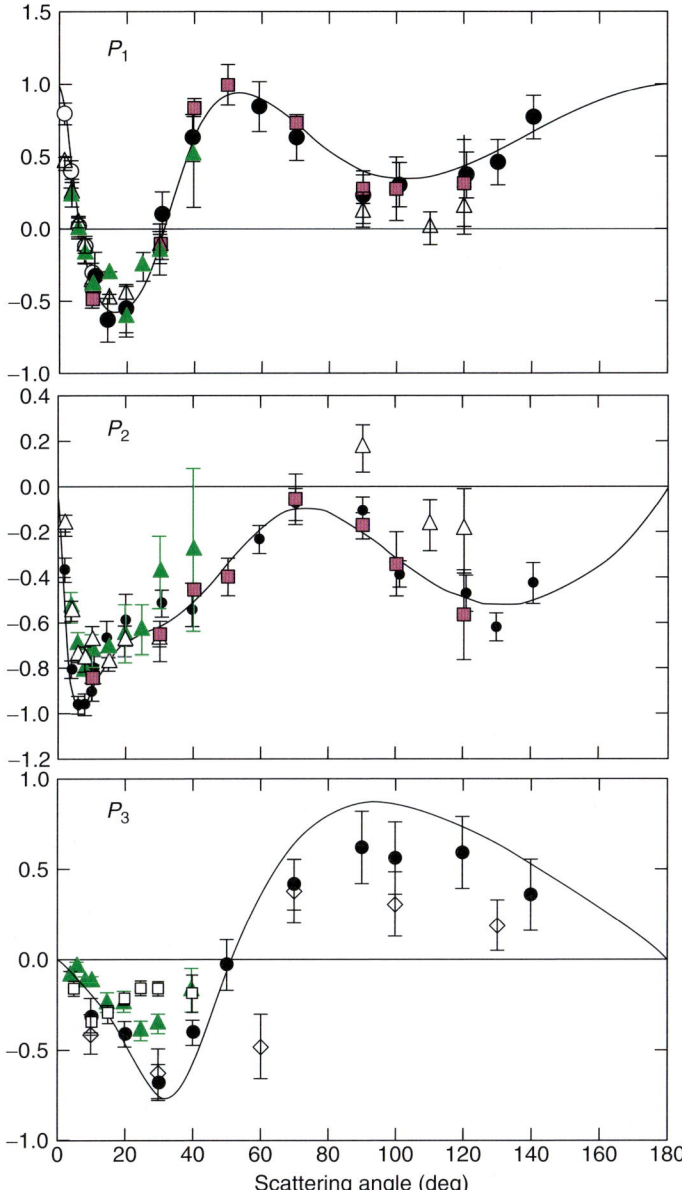

Fig. 8.4 The reduced Stokes parameters, P_1, P_2 and P_3 for the H(2p) state at an incident electron energy of 54 eV. Closed squares, Yalim, Cvejanovic and Crowe (1997); closed and open circles, Williams and Mikosza (2006); closed triangles, Gradziel and O'Neill (2004); open triangles, O'Neill *et al.* (1998); open squares, Nic Chormaic, Chwirot and Slevin (1993); open diamonds, Williams (1986); full line, the PECS theory of Bartlett *et al.* (2005). Adapted from Williams and Mikosza (2006).

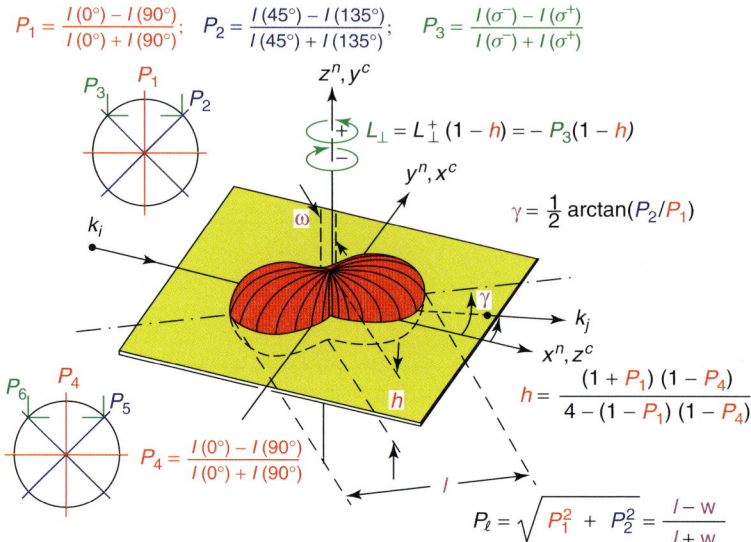

Fig. 8.5 The charge cloud parameters of Andersen et al. (see text).

More detailed reviews of angular and polarization correlation and superelastic scattering studies were given by Andersen, Gallagher and Hertel (1988) Andersen et al. (1997), and Andersen and Bartschat (2001).

8.2.4 Ionization

The simplest electron-impact ionization process arises from the ejection of a single atomic electron from an outer shell:

$$e^-(E_i) + A_i \rightarrow A_j^+ + e^-(E_A) + e^-(E_B) \tag{8.9}$$

where E_i, E_A, and E_B are the incident, scattered, and ejected electron energies, respectively. They are related through energy conservation by

$$E_i = E_A + E_B + \varepsilon \tag{8.10}$$

where ε is the binding energy of the ejected atomic electron.

8.2.4.1 Total Cross Sections

These measurements are conventionally carried out by observation of the positive ions produced in the ionization process:

$$i^+ = i_0 N \sigma_i x \tag{8.11}$$

where i_0 is the incident electron beam current; N, the number density of the target gas; x, the effective target thickness; i^+, the measured positive-ion current; and σ_i, the total ionization cross section. The work of Rapp and Englander-Golden (1965) for a range of stable target atoms and molecules is widely used as a standard. For unstable targets, crossed modulated-beam techniques have been developed involving both thermal energy and fast atomic beams. Components of the total ionization cross section, for

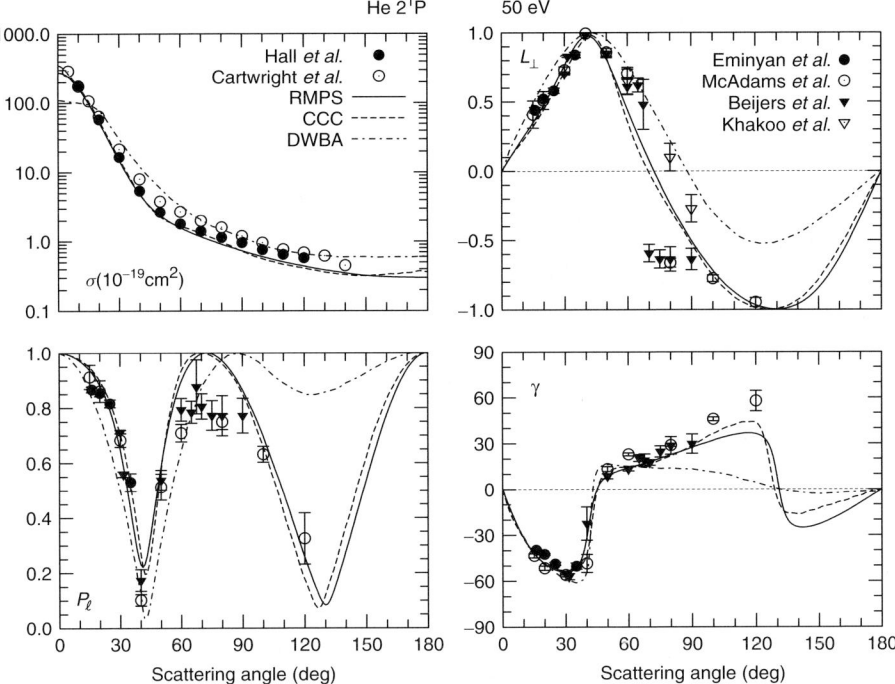

Fig. 8.6 Differential cross section (DCS) and electron impact coherence parameters $L_\perp$, γ and P_ℓ for electron impact excitation of the 2^1P^o state in helium from the ground state 1^1S at an incident electron energy of 50 eV. The theoretical curves are from: RMPS, Bartschat et al. (1996a); CCC, Fursa and Bray (1995); DWBA, Beijers et al. (1987). The sources for the experimental data are: Hall et al. (1973); Cartwright et al. (1992); Eminyan et al. (1974); Beijers et al. (1987); McAdams et al. (1980); Khakoo et al. (1986).

example, multiple ionization and dissociative ionization, can be identified by incorporating mass spectroscopic analysis in the apparatus. Accurate total and partial ionization cross sections for a wide range of atoms and molecules were recently reported by a group from Rice University; see Lindsay and Mangan (2003) and references therein.

8.2.4.2 Double Differential Cross Sections

The detection of a single outgoing electron as a function of its energy and angle is less valuable than in the case of excitation, because the available excess energy of the incoming electron is shared between the scattered and ejected electrons, thus leading to a continuum of energies for both electrons. Note that single differential cross sections as a function of one variable are difficult to measure.

Two distinct double differential cross sections (DDCS) $d^2\sigma/d\Omega dE$ for ionization can be measured. The angular distributions of outgoing electrons of fixed energy may be observed for a particular incident electron energy. For incident electron energies well above the ionization threshold, the general behavior of these cross sections is readily characterized. When the observed electron energy is high, the measured cross section is typical of a "scattered" electron, strongly peaked in the forward direction and falling by orders of magnitude as the scattering angle is increased. As the observed electron energy is decreased, the forward peak becomes less pronounced and the cross section is dominated by a peak typically around 60° scattering angle, due to the increasing dominance of "ejected" electrons.

Alternatively, the energy distribution of the outgoing electron can be monitored at particular scattering angles. Extensive DDCS data were recently reported at low electron energies for atomic hydrogen and helium by Childers *et al.* (2004) and Schow *et al.* (2005).

8.2.4.3 Higher Order, including, Fully Differential Cross Sections

In recent years, partly due to the development of more efficient coincidence electron spectrometers, a wide range of multidifferential cross sections are being measured for most direct and indirect ionization processes. These include single ionization to the ground and specific excited ion states, double excitation–autoionization, double ionization, and inner-shell ionization. Each of these is discussed in detail in the following sections.

Experimentally, these measurements are performed using electron spectrometers with some combination of electron analyzers to observe the scattered and/or ejected electrons and, if appropriate, a photon detector to observe the decay of the excited ion state. Coincidence techniques allow for identification of the ionization process of interest.

Two very different techniques have been used to dramatically increase the efficiency of spectrometers that are used to detect electrons. Catoire *et al.* (2007) describe a spectrometer using three toroidal electron analyzers in conjunction with time- and position-sensitive detectors to observe the three outgoing electrons. This setup enables the *simultaneous* observation of both the ejected electrons from double ionization over most of the scattering plane, defined by the incident electron momentum k_i and the scattered electron momentum k_A, and for a reasonable range of scattering angles. Figure 8.7 shows a schematic illustration of their triple coincidence spectrometer.

A typical reaction microscope used by Dürr *et al.* (2008) to study single ionization processes is shown in Fig. 8.8. It operates on entirely different principles from conventional electron spectrometers. Briefly, a pulsed beam of electrons crosses a supersonic atom beam. The ejected electrons and the recoiling ions are extracted in opposite directions by a weak uniform electric field that is parallel to the direction of the incident electron beam. A uniform magnetic field is also applied in this direction to confine electrons that are emitted perpendicular to the electric field. After passing through field-free drift regions, the slow ejected electrons are then detected in two time- and position-sensitive multihit detectors, allowing for the vector momenta of the two particles to be calculated. Unlike most conventional coincidence electron spectrometers, including the toroidal based systems, which only enable measurements in a single plane at any one time, this technique allows for data to be collected over a large part of the full solid angle simultaneously.

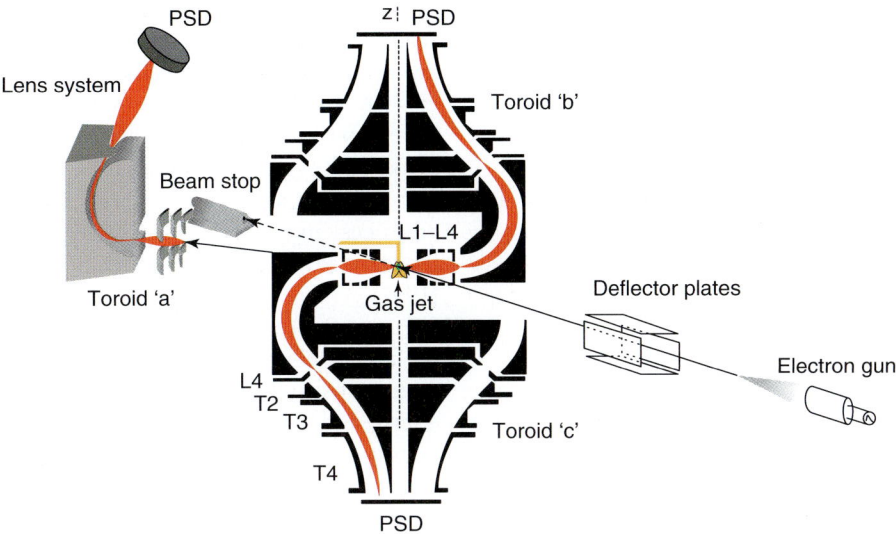

Fig. 8.7 The (e,2e)/(e,3e) spectrometer of Catoire et al. (2007).

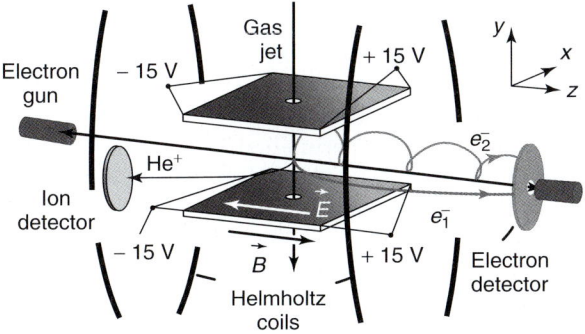

Fig. 8.8 The reaction microscope of Dürr et al. (2008).

8.2.4.4 (e,2e) Studies of Direct Single Ionization

The most detailed information on ionizing collisions involving ejection of a *single* electron and leaving the ion in its ground state comes from the measurement of triple differential cross sections (TDCS), $d^3\sigma_i/d\Omega_A d\Omega_B dE$. They are also referred to as (e,2e) cross sections.

The momentum transfer vector

$$K = k_i - k_A \quad (8.12)$$

is an important parameter in discussing this work. Two major and distinct groups of study have been developed depending on the magnitude of the momentum transfer vector K.

The magnitude of K is maximized in experiments in which the two outgoing electrons from a high-energy collision are observed with equal energies at equal angles $\theta = \theta_A = \theta_B$ relative to the incident electron beam. In such "high-energy symmetric kinematics," the interaction approximates a binary collision between two free electrons. Conservation of momentum requires

$$k_i + p = k_A + k_B \tag{8.13}$$

where p is the momentum of the atomic electron prior to the collision. The magnitude of the struck electron momentum is then given by

$$p = [(2k_B \cos\theta - k_i)^2 + 4k_B \sin^2\theta \\ \times \sin^2\varphi_A/2]^{1/2} \tag{8.14}$$

Here, in principle, variation of any of the parameters k_B, θ, or φ_A allows for the electron momentum p to be scanned. It can be shown that the (e,2e) cross section is proportional to the spherically averaged electron momentum distribution of the atomic electron. Hence a measurement of the TDCS as a function of $\pm p$ reveals the atomic electron momentum distribution, which is related to the spatial wave function through a Fourier transform. Measurements of this type were extensively reviewed, for example, by Weigold and McCarthy (1999).

Measurements of the TDCS for a range of small momentum transfers K were pioneered by Ehrhardt et al. (1969). Using low-energy incident electrons (< 600 eV), they measured angular correlations between the outgoing scattered and ejected electrons under conditions where a slow ejected electron is observed in coincidence with a fast electron scattered through small angles. Most ionizing collisions fall into this category of "low-energy asymmetric kinematics."

The early experiments by Ehrhardt's group, mostly in helium, quickly established the general trends of relative low-energy asymmetric (e,2e) cross sections measured in the scattering plane. Typical results (see Figure 8.9 for an example) are dominated by a peak in a direction close to the momentum transfer direction, referred to as the *binary encounter* peak, and a peak in the opposite direction, referred to as the *recoil peak*. Earlier advances in (e,2e) and related studies were reviewed by Lahmam-Bennani (1991).

Figure 8.10 shows the current state of the art for electron-impact ionization of atomic hydrogen close to the threshold energy and for equal sharing of the excess energy by the two outgoing electrons. The most sophisticated currently available theories for this two-electron process, convergent close-coupling (CCC), time-dependent close-coupling (TDCC), and exterior complex scaling (ECS), all of which are discussed in the theory section, yield predictions in excellent agreement with each other. These predictions also agree very well with most of the available experimental data. In fact, for the discrepancies which remain, it is not obvious that the problems lie with theory for this benchmark process.

Stevenson and Lohmann (2008) reported the first (e,2e) measurements where one of the outgoing electrons is observed over the full angular range, $0° - 360°$, using a MAC device. Their new data for ionization of the 3p electron in argon reveal structures in the previously unobserved angular ranges that are generally not well reproduced by the state-of-the-art calculations of Prideaux, Madison and Bartschat (2005) and Bartschat and Vorov (2005).

Recent reaction microscope measurements in three dimensions also revealed previously unobserved features in the helium TDCS. Typical data of Dürr et al. (2006) are shown in Figure 8.11. In the

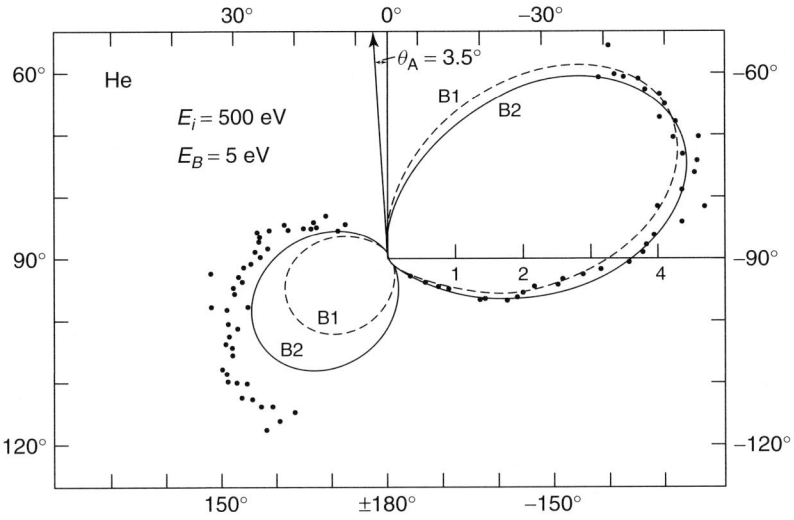

Fig. 8.9 The triple-differential cross section (in atomic units) for helium as a function of the ejected electron angle θ_B for an incident electron energy of 500 eV, an ejected electron energy of 5 eV, and a fixed detection angle of 3.5° for the fast electron. The dots are experimental data, while the curves B1 and B2 represent first-order and second-order plane-wave Born approximations. (Adapted from Ehrhardt et al. (1986).)

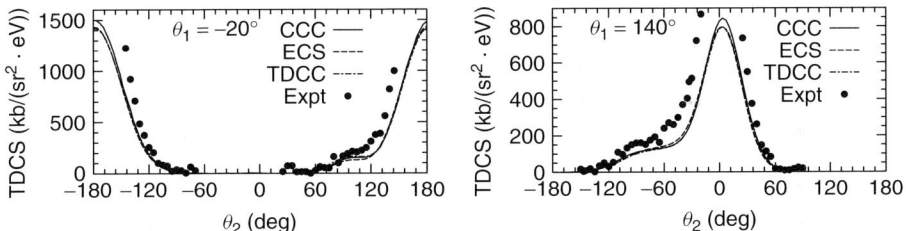

Fig. 8.10 TDCS for electron-impact ionization of H at an incident electron energy of 17.6 eV. Both outgoing electrons have an energy of 2 eV. The detection of one electron is fixed at −20° or 140°, respectively. The absolute experimental data of Röder, Baertschy and Bray (2003) are compared with predictions from CCC, time-dependent close-coupling, and exterior complex scaling models. (Adapted from Colgan and Pindzola (2006).)

scattering plane, the data are consistent with previous experiments and, after taking account of the postcollision interactions between the charged particles, with the dominance of first-order processes. However, in the plane perpendicular to the momentum transfer vector, there is a sizable cross section with maxima at angles of 70° and 290°. These are not reproduced by first-order theories Dürr et al. (2008), even at an incident energy as high as 1 keV.

There is an increasing interest in the use of the (e,2e) technique to study the dynamics of molecular ionization, including molecules of biological interest,

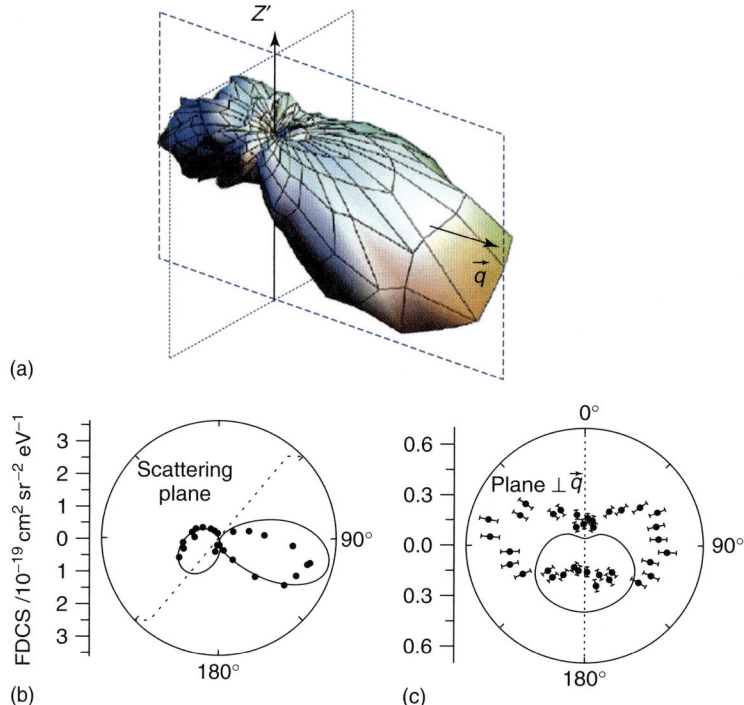

Fig. 8.11 The three-dimensional (e,2e) cross sections of Dürr et al. (2006) for helium at an incident electron energy of 102 eV, a scattering angle of 20° and slow outgoing electron energy of 10 eV. (a) 3-D cross section. (b) Cut in the scattering plane [dashed line in (a)]. (c) Cut perpendicular to the scattering plane [dotted line in (a)]. The dotted lines in (b) and (c) show the incident electron direction. The solid curves are the three Coulomb wave function (3C) theory. (From Dürr et al. (2006).)

Milne-Brownlie et al. (2004). This is despite the experimental complexities associated with the additional degrees of freedom arising from the nuclear motion. Even overlapping electronic states may remain unresolved in these experiments, and so far experiments have only been carried out for random orientations of the internuclear axis.

8.2.4.5 (e,3e) Studies of Double Ionization

Double electron ejection has been studied using the (e,3e) method, which involves detection of all three outgoing electrons in coincidence. The process can be written as

$$e^-(E_i) + A_i \longrightarrow A_j^{++} + e^-(E_A) \\ + e^-(E_B) + e^-(E_C) \quad (8.15)$$

These experiments yield fivefold differential cross sections, $d^5\sigma_i/dE_A dE_B d\Omega_A d\Omega_B d\Omega_C$.

Although the (e,3e) technique was pioneered by Lahmam-Bennani, Dupré and Duguet (1989) two decades ago, it can still be considered to be in its infancy due to the small cross sections

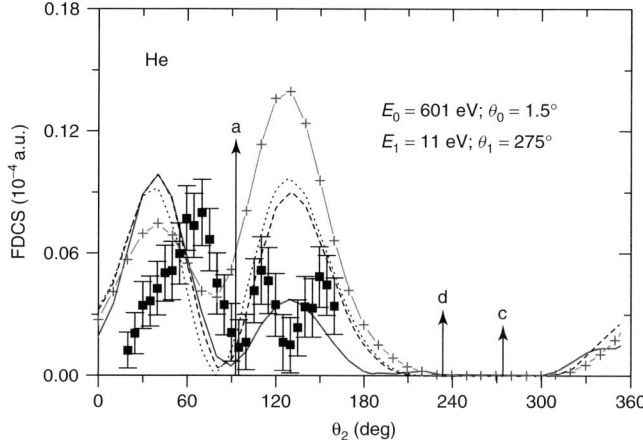

Fig. 8.12 The fivefold coplanar differential cross section in atomic units for the double ionization of helium at an incident electron energy of 601 eV. The scattered electron with an energy of 500 eV and scattered through 1.5° is observed along with an 11 eV ejected electron at 275° and the other ejected electron as a function of angle. The experimental data are from Lahmam-Bennani et al. (2003). Theory: first Born, dotted line; second Born, solid line with crosses; approximate 6C models, solid and dashed lines. According to general selection rules (Berakdar, Lahmam-Bennani and Dal Cappello (2003)), there should be minima at the angles denoted by "a," "c," and "d." (From Elazzouzi et al. (2005).)

for double ionization, the wide variation of kinematical situations arising from the presence of the second ejected electron, and the experimental complexities involved. The experimental efficiencies have been greatly enhanced by the approaches of both Catoire et al. (2007) and Dürr et al. (2007). Figure 8.12 shows typical (e,3e) data for the kinematical conditions shown. While the minima predicted by fundamental selection rules (Berakdar, Lahmam-Bennani and Dal Cappello (2003)) are reproduced by experiment, the overall agreement with the calculations shown is not good.

8.2.4.6 Excitation–Ionization

The study of ionization, where the ion is formed in an excited state, has been the subject of considerable attention recently. At least some of this work was stimulated by the desire to carry out the complete ionization experiment, analogous to the complete excitation experiments discussed in Section 8.2.3.3. Such a complete experiment would require a triple coincidence experiment, (e, 2eγ), in which the polarization of the decay photon is determined for specific kinematics by its observation in coincidence with the two outgoing electrons. Such an experiment has not been performed to date, although some progress toward its realization has been made.

Schematically, the process is written as

$$e^-(E_i) + A_i \longrightarrow A_j^{+*}$$
$$+ e^-(E_A) + e^-(E_B) \qquad (8.16)$$

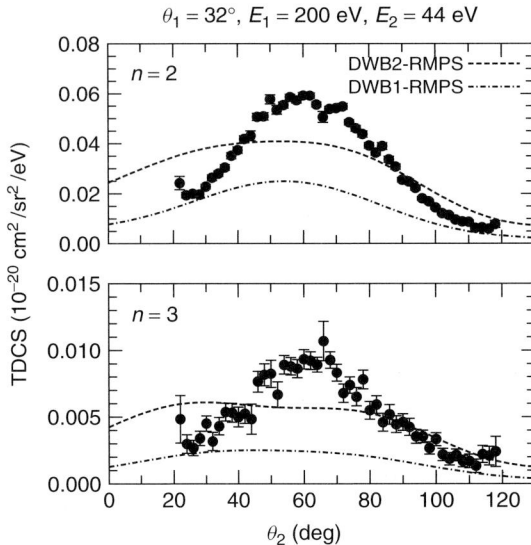

Fig. 8.13 Experimental TDCS for excitation-ionization to the $n = 2$ and $n = 3$ states of He$^+$ compared with hybrid R-matrix calculations. In each case, the outgoing scattered electron has an energy of 200 eV and the scattering angle is 32°. The ejected electron energy is 44 eV. The experimental data are from Bellm, Lower and Bartschat (2006), after combining their measured ratios with theoretical predictions for ionization without excitation. (From Bartschat et al. (2007).)

The excited ion state, A_j^{+*}, then decays by optical emission to a lower lying ion state,

$$A_j^{+*} \longrightarrow A_k^+ + h\nu \quad (8.17)$$

Three related types of studies have been reported. Stefani, Avaldi and Camilloni (1990) first reported an (e,2e) study of the He$^+(n = 2)$ states. This produces a TDCS with the disadvantages that the 2s and 2p ion states are not resolved and that the detailed dynamical information gained from the polarization of the photon is absent. Hayes and Williams (1996) first applied the single electron–photon, (e,eγ), technique to the same system. This type of study enables the 2p state to be isolated in the study and angular correlation parameters to be determined, integrated over the slow ejected electron emission angles (Schwienhorst et al., 1996). A single-electron-polarized-photon study of the $4p^2P_{3/2}$ excited state of Ca$^+$ was reported by Stevenson and Crowe (2004).

Sakhelashvili et al. (2005) carried out an (e,2eγ) experiment in helium without resolving the different He$^+(np)^2P$ states and without regard to the direction or polarization of the emitted photons. If any isotropy in the emitted radiation is neglected as well, the data correspond to the TDCS for the $(np)^2P$ states, with approximately 90% of the signal believed to be coming from the He$^+(2p)$ state.

Figure 8.13 shows TDCS for the excitation-ionization to the $n = 2$ and $n = 3$ states of He$^+$ compared with hybrid R-matrix calculations. The absolute "experimental data" presented by Bartschat et al. (2007) were generated by combining the measured ratios of these TDCS to the $n = 1$ ion state by Bellm, Lower and Bartschat (2006) with CCC results for $n = 1$. The data show broad peaks in the forward direction. Although the kinematics are not identical, the TDCS of Dogan and Crowe (2000) for these states suggest that more interesting features in the TDCS exist outside the angular range of the data shown here. The need for higher order effects to be taken into account in theoretically describing these two-active atomic electron processes is demonstrated by comparing theoretical predictions obtained through first-order (DWB1-RMPS) or second-order (DWB2-RMPS) treatments of the projectile-target interaction, combined with a convergent R-matrix with pseudostates (RMPS) model for the inital bound state and the ejected-electron–residual-ion interaction.

8.2.4.7 Excitation–Autoionization

A major contribution to the ionization cross section for complex atoms and ions often arises from the process known as excitation–autoionization. In this process the target is first excited to an autoionizing or resonant state lying above the ionization threshold by the incident electron,

$$e^- + A_i \rightarrow e^- + A_j^* \quad (8.18)$$

and the resultant resonant state A_j^* subsequently autoionizes,

$$A_j^* \rightarrow A_k^+ + e^- \quad (8.19)$$

Hence, the total effect of these two processes is to ionize the original atom or ion. A recent example involving inner-shell excitation of Na was reported by Borovik, Zatsarinny and Bartschat (2008).

In addition, the incident electron may sometimes be captured into a resonant state that decays by the emission of two (or occasionally more) electrons:

$$e^- + A_i \rightarrow A_l^{-*} \rightarrow A_k^+ + 2e^- \quad (8.20)$$

This process, known as resonant excitation double autoionization (REDA), interferes with the direct process given by Eqs. (8.16) and (8.17) and often dominates the cross section close to threshold. Excitation–autoionization was first observed in the ionization of Ba^+ by Peart and Dolder (1968) and has since been shown to be a dominant ionizing mechanism for many ions of laboratory and astrophysical interest.

Excitation–autoionization can be studied experimentally by observation of either the ejected or the scattered electron spectra. Observation of the ejected electron has the advantage that the observed spectra are independent of the inherent inhomogeneity in the incident electron energy E_i.

Autoionization spectra are characterized by asymmetric line profiles due to interference between the direct ionization and the excitation–autoionization process. The majority of single-electron spectra have been analyzed to yield the energies and widths of the resonance states.

Excitation–autoionization processes involving the $2l2l'$ states of helium have also been studied using the (e,2e) technique (see McDonald and Crowe (1993) and references therein). In these experiments, the angular variation of the Shore resonance parameters (Shore (1967)) of each of the autoionizing states was determined. In the most recent study, deHarak, Bartschat and Martin (2008) measured out-of-plane TDCS for each of these singlet states. Dogan and Crowe (2002) used the electron–photon coincidence method to observe the influence of the $3lnl'$ autoionizing states of helium on simultaneous excitation-ionization of the He^+2p state.

8.2.4.8 Inner-Shell Ionization

We are interested in processes of the type

$$e^- + A_i \longrightarrow A^+(J, M) + 2e^- \quad (8.21)$$

where the ion $A^+(J, M)$ has an inner-shell vacancy and can decay by either X-ray

$$A^+(J, M) \longrightarrow A^+(J'', M'') + h\nu \quad (8.22)$$

or Auger-electron emission

$$A^+(J, M) \longrightarrow A^{++}(J', M') + e_{\text{Auger}} \quad (8.23)$$

Except for very heavy atoms, the dominant decay mechanism is by Auger-electron emission, and hence we will concentrate on that mode of decay.

Most studies of inner-shell ionization–Auger-electron decay have involved observation of the energy-loss spectrum of a single outgoing electron. These measurements yield a spectrum of Auger-electron lines superimposed on the continuum of scattered and ejected electrons from all ionization processes. A second type of experimental study has involved measurement of the angular distribution of the Auger electrons. When an inner-shell vacancy is created with $j > \frac{1}{2}$ by a beam of particles, the initial ion A^+ will be aligned with respect to the incident beam direction, the alignment manifesting itself in an anisotropic angular distribution of the decay product. More detailed information on the alignment of the initial ion can be obtained from angular correlation measurements between the Auger electrons and a fast scattered electron, where only specific momentum transfer processes are studied. Such studies were first reported by Sewell and Crowe (1982). Recently, both the (e,2e) and (e,3e) techniques were used by Naja et al. (2007) to study inner-shell ionization, taking advantage of the increased efficiency of the electron detectors discussed above.

8.2.5 Electron–Molecule/Biomolecule Interactions

The experimental techniques used to study electron scattering from molecules are the same as for atoms and do not merit special attention. For example, the spectrometers of Allan (2004) and Catoire et al. (2007) discussed earlier have been used to study different aspects of electron–molecule scattering. In this case, the nuclear motion opens up a wealth of additional possibilities, in particular in low-energy collisions that often reveal resonance effects. Such resonance and threshold phenomena in low-energy electron collisions with molecules and clusters were reviewed by Hotop et al. (2003).

Experiments to determine electronic excitation cross sections, with typical energy resolutions in the range of 30 – 50 meV, can be very problematic due to the nuclear degrees of freedom. In this case, spectral deconvolution techniques are often necessary to obtain well-defined results, especially for polyatomic molecules. Brunger et al. (2008) give an excellent recent example of the use of deconvolution methods to obtain differential excitation cross sections for each of the six lowest lying electronic states of water.

One recent development that justifies comment is the study of biomolecules. This interest is largely generated by the important role played by low-energy electrons in radiobiology. Such low-energy (≤ 30 eV) electrons are produced by the

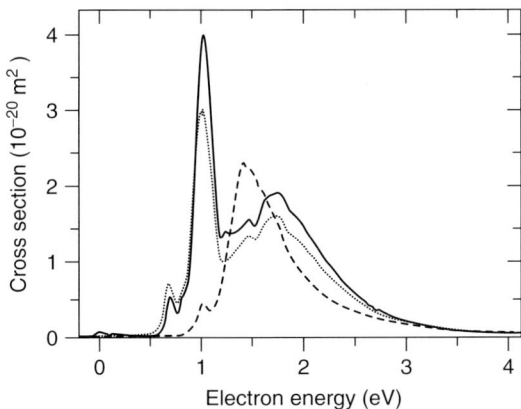

Fig. 8.14 The cross sections for formation of the most abundant negative ions formed by electron attachment to each of uracil (U), thymine (U), and cytosine (C). (U-H)$^-$ from U (dotted line); (T-H)$^-$ from T (full line and × 0.33); (C-H)$^-$ from C (dashed line). (From Denifl et al. (2004).)

interaction of fast ionizing radiation with cellular constituents. The vast majority of these electrons are responsible for the production of reactive negative ions and radical fragments through the formation and decay of resonances.

Advances in understanding the role of these low-energy electrons have relied to a considerable extent on the interplay between scattering from gaseous- and solid-state targets. For example, while Boudaiffa et al. (2000) observed single- and double-strand breaks in supercoiled DNA deposited on a cold surface, they argued that the resonances observed were similar to those seen in gas-phase electron attachment experiments. More generally the gas-phase experiments are important in distinguishing between environmental and intrinsic molecular effects.

The first gas-phase electron scattering studies on uracil and the DNA bases were performed by Aflatooni, Gallup and Burrow (1998) using their electron transmission method developed for electron–atom scattering. Another important series of systematic studies of the stable negative ions produced from uracil and the DNA bases, detected using a high-resolution quadrupole mass spectrometer, have provided considerable insight into the mechanisms of damage to DNA by low-energy electrons; see, for example, Ptasińska et al. (2005). Electron attachment to these biomolecules leads to dissociation into various fragments without any measurable amount of the stable parent ions. Figure 8.14 shows the cross sections for the formation of the most dominant negative ions produced from the RNA base uracil and the DNA bases, thymine and cytosine reported by Denifl et al. (2004). The reader is referred to the original publication for details and to the review of Sanche (2005) for a broad discussion of the effects arising from the interaction of low-energy electrons with biomolecules.

8.2.6 Electron–Ion Interactions

Experimental studies of electron-ion interactions have traditionally been sparse compared with their theoretical counterparts. Recently, this has been transformed by the adoption of heavy-ion storage rings for electron–ion studies. The technique can be considered

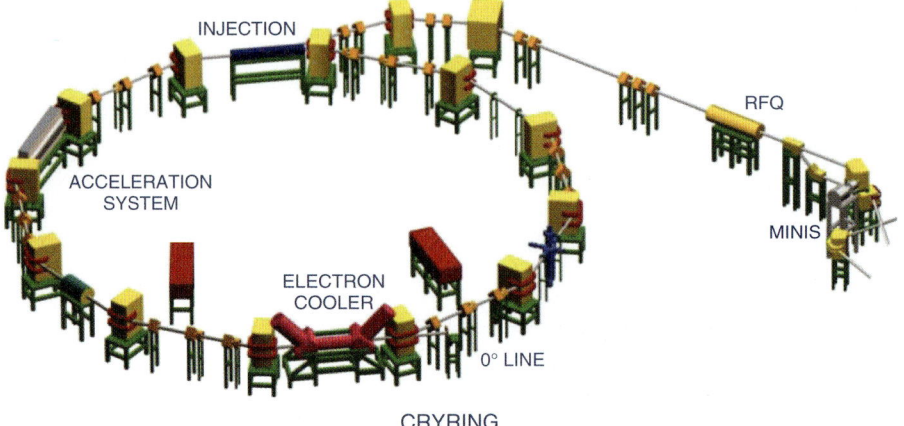

Fig. 8.15 The Stockholm ion storage ring (51.6 m circumference). (From Larsson and Thomas (2001).)

as an extension of the merged beams method, in which electron and ion beams are brought together and interact over a common path length, (Phaneuf et al., 1999). One of a number of very significant advantages of the storage rings is that the ions pass through the interaction region more than 10^5 times per second, whereas the merged beams machines are single-pass devices. One of the storage rings, which has been used in atomic/molecular physics, the CRYRING device, is shown in Figure 8.15 (Larsson, 2001). Briefly, ions are extracted from the ion source MINIS, mass selected, and then injected into the storage ring and further accelerated to the required beam energy, typically $\sim$ MeV. In the straight "electron cooler" section, the stored ion beam is merged with an electron beam with the same velocity and cooled, typically for a few seconds. Electron cooling is based on the fact that the Coulomb interaction between the electrons and ions leads to heat transfer from the warm ion beam to the cold and continuously renewed electron beam. Amongst the outcomes of cooling are an ion beam with an extremely narrow momentum spread and ions in the lowest vibrational state of a molecular ion. The electron beam in the cooler also has excellent properties for electron–ion collision studies. The electron–ion interaction energy is given by the relative velocities of the two beams. In principle, this can be made zero, but in practice the lowest energy is a few milli electron-volts due to the velocity distributions and finite angular resolutions of the beams. Overall energy resolutions are typically $\sim 1 - 2$ meV. The more recent addition of a supersonic expansion ion source has also led to studies involving rotationally cold molecular ions (McCall et al., 2003). A range of detection devices can be used to detect the collision products depending on the process studied.

Storage rings have been used to study a wide range of low-energy (0 – 100 eV) electron–ion collisions with unprecedented resolution and often of fundamental importance in fusion and astrophysical plasmas. The cross

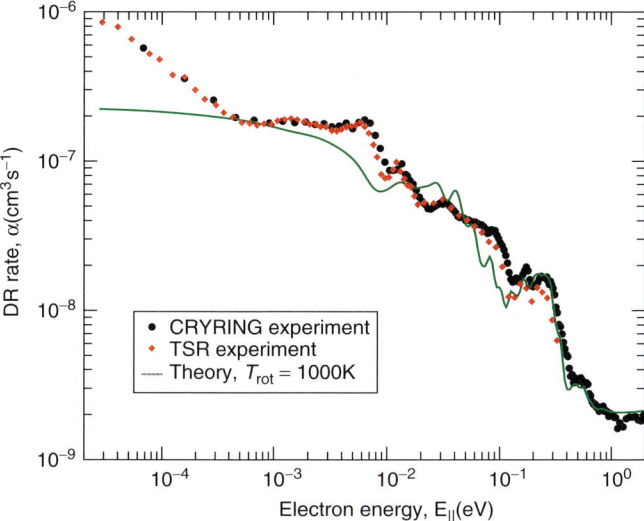

Fig. 8.16 Dissociative recombination rate of H_3^+ as a function of the incident electron energy. The theoretical predictions of dos Santos, Kokoouline and Greene (2007) are compared with experimental data from the Stockholm (CRYRING) and Heidelberg (TSR) groups.

section measurements of recombination processes are particularly important. Figure 8.16 shows an example of dissociative recombination. For many years, theory had been unable to explain the experimentally observed large rates for this process in H_3^+, which is of fundamental importance in diffuse interstellar clouds. Kokoouline and Greene (2003a) finally identified Jahn–Teller coupling as the critical mechanism that must be included in a theoretical treatment. As seen in Figure 8.16, the latest calculations of dos Santos, Kokoouline and Greene (2007) finally obtained good agreement with recent storage ring data.

8.2.7
Spin-Polarized Electron Studies

Previous discussions have involved experiments in which the electron beam was produced by thermionic emission with the spins of the electrons having arbitrary directions. Hence, for example, in discussing excitation of the 2^2P state of atomic hydrogen in Section 8.2.3.3, the analysis involves an average over all spin orientations.

It is possible to produce electron beams with a preferential orientation of the electron spin (Kessler (1985)). In general, the use of spin-polarized electron beams in electron–atom scattering provides a further definition of the initial and/or final state of the system, leading toward the "perfect" or "complete" scattering experiment. In such an experiment, all parameters that can be determined according to the limits set by quantum mechanics, are measured for a given process. This information is contained in the magnitudes and (relative) phases of the complex-valued scattering amplitudes. The concept of complete experiments was introduced by Bederson (1969). More information can be found in Andersen and Bartschat (1996); Andersen and Bartschat (2001).

Spin-polarized electron and atom beams provide a means of studying spin-dependent interactions, the details of which are often masked in experiments involving unpolarized electrons by the stronger Coulomb interaction. Two specific spin-dependent interactions have been the subject of extensive study – the exchange interaction and the spin–orbit interaction. Although both these effects are important for scattering from a wide range of atomic targets, it is useful to consider the extreme situations where only one is important.

8.2.7.1 Exchange Interaction

This is particularly important for spin-$\frac{1}{2}$, low-Z targets, including atomic hydrogen and the alkali atoms. Assuming that the total spin of the collision system (singlet and triplet) is conserved, the cross sections for scattering from such targets depend on the relative orientations of the electron and target spins, that is, $\sigma(\uparrow\downarrow) \neq \sigma(\uparrow\uparrow)$, where the arrows indicate the electron and atom spins. As an example, we give in Table 8.1 the possible scattering modes for the elastic scattering of electrons with spin parallel or antiparallel to an atom with a given spin direction ($A\uparrow$ or $A\downarrow$, together with the corresponding scattering amplitudes and cross sections. These expressions can readily be generalized to inelastic scattering (Kessler, 1985).

Scattering of spin-polarized electrons from spin-polarized atoms with spin analysis of the products yields the cross sections $|f|^2$, $|g|^2$, and $|f-g|^2$. This constitutes an extremely difficult experiment. Some information can already be obtained if either the initial electron or atom beam is spin polarized, and if either the electron or the atom polarization is measured after the collision process. However, a complete experiment requires the measurement of the relative phase of f and g as well.

Most experimental studies do not set out to obtain complete information on the scattering amplitudes. Several experiments have been carried out by scattering polarized electrons from polarized atoms and measuring the asymmetry

$$A = \frac{\sigma(\uparrow\downarrow) - \sigma(\uparrow\uparrow)}{\sigma(\uparrow\downarrow) + \sigma(\uparrow\uparrow)} \quad (8.24)$$

where $\sigma(\uparrow\downarrow)$ is the cross section for antiparallel electron and atom spins and $\sigma(\uparrow\uparrow)$ is the cross section for parallel electron and atom spins. Of particular importance for the further confirmation of the CCC theory for hydrogen-like systems (Bray (1994)) were several experiments performed by the

Tab. 8.1 Amplitudes and cross sections for scattering of spin-polarized electrons and spin-$\frac{1}{2}$ targets.

Process	Amplitude	Cross section
$e\downarrow + A\uparrow \rightarrow e\downarrow + A\uparrow$	f	$\|f\|^2$
$e\downarrow + A\uparrow \rightarrow e\uparrow + A\downarrow$	$-g$	$\|g\|^2$
$e\uparrow + A\uparrow \rightarrow e\uparrow + A\uparrow$	$f-g$	$\|f-g\|^2$

NIST group for elastic and inelastic electron scattering from Na atoms (see McClelland, Kelley and Celotta (1989), Lorentz et al. (1993), and references therein).

Exchange effects are also very important in spin-forbidden transitions, such as $^1S^e \to {}^3P^o$ in He, the heavy noble gases, and even mercury. If such a transition is achieved by an incident polarized electron beam, the change of the beam polarization is a measure of the relative cross section. Furthermore, an unpolarized electron beam can become polarized if an individual fine-structure level of the target is resolved. This so-called "fine-structure effect," originally suggested by Hanne (1976), has been seen in both excitation and ionization Hanne (1983), Jones, Madison and Hanne (1994), and Andersen and Bartschat (2001).

8.2.7.2 Spin–Orbit Interaction

This is particularly important for high-Z targets (e.g., Cs or Hg). In this case, relativistic effects cause a spin–orbit interaction between the incident electron and the target atom characterized by the scalar product $l \cdot s$, where l is the orbital and s is the spin angular momentum.

In the case of elastic scattering from closed-shell targets (like Hg), the scattering process can be described by two complex amplitudes, f, due to the Coulomb interaction, and g, due to the spin–orbit interaction (not to be confused with the direct and exchange amplitudes above). Complete determination of f and g, apart from an overall phase, requires three independent measurements. One of these is the so-called "polarization function" S_P that determines the spin polarization of an initially unpolarized electron beam after the scattering. Owing to parity conservation, the polarization vector must be perpendicular to the scattering plane. Another possibility is to measure the spin-asymmetry function S_A, which determines the left-right asymmetry in the DCS if the incident electron beam is already spin polarized. Owing to time-reversal invariance, $S_P = S_A \equiv S$ for elastic scattering (Bartschat, 1989). The other parameters, called T and U for elastic scattering from closed-shell targets, determine the change of an initial electron polarization due to the scattering process. Such experiments are highly complicated but have been performed successfully in the Münster group; see, for example, Berger and Kessler (1986). For inelastic collisions, "generalized" STU-parameters need to be defined (Bartschat, 1989), which have also been measured in several cases; see Andersen et al. (1997) for a summary of available data.

Using a spin-polarized incident electron beam also vastly increases the possibilities for the polarization properties of the emitted radiation. Not surprisingly, "generalized" Stokes parameters (Andersen and Bartschat (2001)) have to be defined that depend on the electron polarization as well as the position and the setting of the light analyzers. In addition to performing very difficult angle-resolved electron–photon coincidence experiments, it has become evident that some interesting information can already be obtained by only observing the polarization of the emitted light. For properly chosen transitions, observing the circular polarization of the emitted light can be used to optically measure the electron polarization Eminyan and Lampel (1980), Gay (1983), Goeke, Kessler and Hanne (1987), while observation of the linear polarization P_2 can give an indication about the importance of explicitly spin-dependent effects on the collision process Bartschat et al. (1981), Bartschat and Blum (1982). Recent data for the "integrated Stokes parameter" in spin-polarized electron-impact excitation of Zn were reported by Pravica et al. (2007).

8.2.7.3 A Typical Apparatus and Some Illustrative Results

Figure 8.17 shows the apparatus of Sohn and Hanne (1992) used to determine generalized Stokes parameters for electron-impact excitation of mercury. More importantly, from the point of view of this review, it illustrates the essential features of many polarized-electron studies. Today's sophisticated experiments have only become feasible with the development of highly efficient GaAs polarized-electron sources. The polarized electrons are produced from the GaAsP cathode by photoemission using circularly polarized light, the direction of polarization coinciding with the axis of the incident laser beam. The electron beam is then guided through an electron-optical system, which also includes the possibility of rotating the polarization vector via a Wien filter. After scattering from a beam of target atoms, the scattered electrons and the emitted photons (with their polarization analyzed as well) are observed in coincidence. Finally, the polarization of the incident beam is determined by measuring the left-right asymmetry of the accelerated outgoing electrons (without a target beam) scattered from a gold foil, making use of the spin–orbit effect in a so-called "Mott detector."

Similar apparatus were built by Berger and Kessler (1986) to measure the (generalized) STU-parameters for electron collisions with mercury and heavy noble gases, and by the Bielefeld group (Baum et al. (2002); Baum et al. (2004))

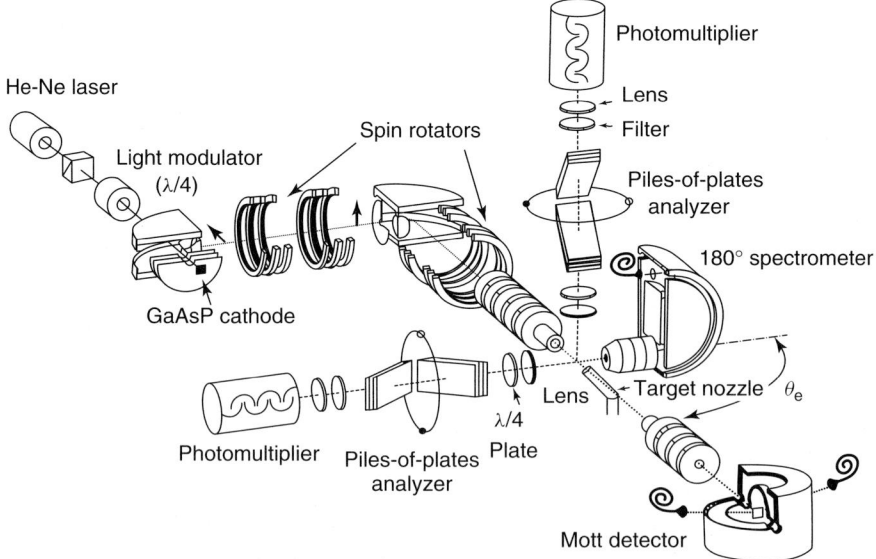

Fig. 8.17 The polarized-electron apparatus of Sohn and Hanne (1992).

to measure several spin asymmetries for elastic and inelastic e–Cs collisions. For such heavy targets, both exchange effects and the spin–orbit interaction are often important, and it is by no means trivial to disentangle the various effects. Some sophisticated recipes for performing complete, or almost complete, experiments were discussed by Andersen and Bartschat (1996;2001). The general idea is to look for parameters that would vanish if a particular interaction is not important for the outcome of the collision process. If such an observable is found to be nonzero experimentally, it is generally a good candidate to test this particular aspect of a theoretical model.

Figure 8.18 shows two such examples, namely, the integrated Stokes parameter P_2 for e–Hg excitation and the spin-asymmetry function S_A for elastic e–Cs scattering. Both observables would be zero in any nonrelativistic calculation. In both cases, we note excellent agreement with theoretical predictions that include relativistic effects. However, numerous examples remain where the agreement between theory and experiment is much less satisfactory, in particular for more complex targets such as the heavy noble gases.

Spin-polarized electron beams have also been used in the study of ionization processes, either in ionization asymmetries using a spin-polarized atomic beam as well (see, for example, Baum et al. (1993;1992)), or in studies where a p-electron is ejected from a noble gas such as Xe and the fine-structure of the residual ion is resolved (see Bellm et al. (2008) and references therein). Finally, relatively few polarized-electron experiments have been performed with molecular targets to date. In many cases, spin-effects are largely masked by the averaging over molecular orientations. However, Green et al. (2004) recently found an example of circularly polarized molecular fluorescence induced

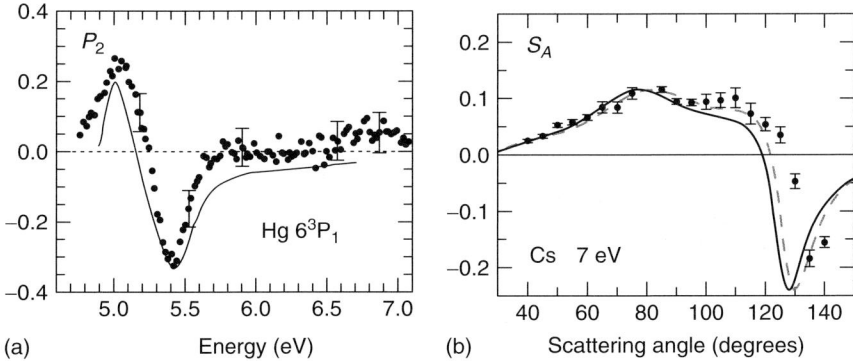

Fig. 8.18 (a) Angle-integrated Stokes parameter P_2 for polarized-electron-impact excitation of the $(6s6p)^3P_1$ state of Hg as a function of the incident electron energy. The experimental data of Wolcke et al. (1983) are compared with theoretical predictions from a Breit–Pauli R-matrix approach developed by Bartschat et al. (1984).
(b) Spin-asymmetry function S_A for elastic e–Cs scattering at an electron energy of 7 eV. The experimental data of Baum et al. (2002) are compared with fully relativistic CCC (solid line, Fursa and Bray (2008)) and fully relativistic Dirac B-spline R-matrix results (dashed line, Zatsarinny and Bartschat (2008)).

by spin exchange. Another pioneering experiment was performed by Mayer and Kessler (1995) who verified electron optic dichroism.

8.2.8
Positron Scattering

The scattering of positrons from atoms and molecules is an interesting area in itself, but is also important in testing some aspects of the theory of electron collisions as a result of the various comparisons and contrasts between the two situations. For example, the static potential associated with the Coulomb interaction is attractive for electrons and repulsive for positrons, while the polarization potential is attractive for both particles. Exchange effects occur only for electron collisions. On the other hand, for positrons, the rearrangement channel of positronium formation can become very important during the interaction.

A detailed discussion of positron-atom/molecule scattering lies outside the scope of this review. The reader is referred to the review by Surko, Gribakin and Buckman (2005) for details of the numerous recent experimental and theoretical developments in this field.

8.3
Review of Theoretical Methods

The ongoing rapid development of computational resources has resulted in a variety of new methods to treat electron/positron collisions with atoms, molecules, and ions. Some of these approaches are, at least in principle, universally applicable over the entire energy range of interest. Consequently, we will no longer use the traditional classification in terms of "low-energy" and "high-energy" methods. Instead, our first separation will be made according to whether an approach is based on the close-coupling expansion or on the Born series. While both expansions, if carried to infinity, would lead to exact results (within limits such as a nonrelativistic or a relativistic formulation), these expansions are cut off or approximated in practical applications and, via the details of the procedure, result in different classifications of a particular method.

Another important aspect concerns the traditional distinction between excitation, including elastic scattering and ionization. As is shown below, theoretical methods originally developed for excitation processes have been very successful in treating ionizing collisions as well. This is sometimes achieved by reinterpreting the results obtained for excitation of a positive-energy pseudostate as an ionization process. However, in that case, special measures are required to extract the information that is actually measured in the corresponding experiments.

8.3.1
The Close-Coupling Expansion

8.3.1.1 The Wave Equation, Scattering Amplitude, and Cross Section

We start by considering elastic scattering and excitation of atoms and atomic ions by electron impact in a nonrelativistic framework and consider the process

$$e^- + A_i \longrightarrow e^- + A_j \quad (8.25)$$

where A_i and A_j are the initial and final N-electron bound states of the target.

The Schrödinger equation for the $(N+1)$-electron collision system (target electrons plus projectile) for a target with nuclear charge Z is given by

$$H_{N+1}\Psi = E\Psi \quad (8.26)$$

where E is the total energy of the system. In atomic units, the $(N+1)$-electron Hamiltonian H_{N+1} is

$$H_{N+1} = \sum_{i=1}^{N+1}\left(-\tfrac{1}{2}\nabla_i^2 - \frac{Z}{r_i}\right) + \sum_{i>j=1}^{N+1}\frac{1}{r_{ij}} \quad (8.27)$$

Here $r_{ij} \equiv |\mathbf{r}_i - \mathbf{r}_j|$, with $\mathbf{r}_i$ and $\mathbf{r}_j$ denoting the vector coordinates of the electrons i and j, respectively. The origin of the coordinate system is taken to be the target nucleus, which is assumed to have infinite mass.

Next, we introduce the target eigenfunctions Φ_i and corresponding eigenenergies w_i by

$$\langle \Phi_i | H_N | \Phi_j \rangle = w_i \delta_{ij} \quad (8.28)$$

where H_N is the target Hamiltonian defined by Eq. (8.27) with $N+1$ replaced by N.

We look for a solution of Eq. (8.26) corresponding to the process defined by Eq. (8.25), where an electron is incident upon the target in some state Φ_i and is scattered leaving the target in some other state Φ_j. The asymptotic form of the wave function in the case of a neutral target is

$$\Psi_i \xrightarrow[r\to\infty]{} \Phi_i \chi_{m_i} e^{ik_i z}$$

$$+ \sum_j \Phi_j \chi_{m_j} \times f_{ji}(\theta,\varphi)\frac{e^{ik_j r}}{r} \quad (8.29)$$

Here χ_{m_i} and χ_{m_j} are the spin eigenfunctions of the incident and scattered electrons and f_{ji} is the scattering amplitude. In numerical calculations, the direction of spin quantization is usually taken to be the incident beam direction (the so-called "collision system"). The wave numbers k_i and k_j of the incident and scattered electrons are related to the total energy E of the collision system by

$$E = w_i + \tfrac{1}{2}k_i^2 = w_j + \tfrac{1}{2}k_j^2 \quad (8.30)$$

The outgoing-wave term on the right of Eq. (8.29) contains contributions from all target states that are energetically allowed, that is, for which $k_j^2 \geq 0$. The remaining states, for which $k_j^2 < 0$, can only occur virtually in the collision. Although they do not contribute to the asymptotic form, they can play an important role during the process giving rise, for example, to intermediate resonance states. Finally, when the target is an ion, the exponents in Eq. (8.29) must be modified by including logarithmic phase factors due to the distortion by the Coulomb field. These factors introduce no essential complications, and hence we shall not include them explicitly below.

The angle-differential cross section for a transition from a state $|\mathbf{k}_i, \Phi_i, \chi_{m_i}\rangle$ to a state $|\mathbf{k}_j, \Phi_j, \chi_{m_j}\rangle$ can be calculated by considering the incident and scattered fluxes in Eq. (8.29). One finds

$$\frac{d\sigma_{ji}}{d\Omega} = \frac{k_j}{k_i}|f_{ji}(\theta,\varphi)|^2 \quad (8.31)$$

The fundamental problem is to solve the Schrödinger equation (8.26) to obtain the scattering amplitude and the cross section corresponding to the process given by Eq. (8.25). We start by expanding the total wave function in the form

$$\Psi_i^\Gamma(X_{N+1}) =$$
$$\mathcal{A} \sum_j \overline{\Phi}_j^\Gamma(X_N; \widehat{r}_{N+1}\sigma_{N+1}) F_{ji}^\Gamma(r_{N+1})$$
$$+ \sum_j \chi_j^\Gamma(X_{N+1}) a_{ji}^\Gamma \quad (8.32)$$

Here $X_{N+1} = x_1, x_2, \ldots, x_{N+1}$ represents the space and spin coordinates of all $N+1$ electrons, $x_i \equiv r_i\sigma_i$ denotes the space and spin coordinates of the ith electron, and $\mathcal{A}$ is the operator that antisymmetrizes the first summation with respect to any pair of electrons in accordance with the Pauli exclusion principle. The channel functions $\overline{\Phi}_j^\Gamma$ are obtained by coupling the orbital and spin angular momenta of the target with those of the scattered electron to form eigenstates of $\Gamma = LSM_LM_S$ and parity π that are conserved in the collision. The role of the square-integrable, fully antisymmetrized correlation functions χ_j^Γ is further discussed below.

Equation (8.32) is the so-called "close-coupling expansion." To be complete, it should include all discrete bound states as well as the continuum states of the target. In practice, of course, exact completeness is impossible to achieve in a numerical treatment, and hence the expansion has to be cut off. In the simplest case, only one target state might be retained, allowing for elastic scattering from this state to be described in the "static exchange approximation." Traditionally, the expansion was cut after a few discrete target states, thereby enabling the calculation of transitions between all the states retained in the expansion. However, it was soon realized that such a severe cut in the expansion typically limits the applicability of the method to "low" energies, generally below the threshold of the highest target state kept in the expansion. Consequently, significant work has been devoted in recent years to account in some approximate way for the highly-lying discrete and even the continuum states of the target. This resulted in the convergent close-coupling (CCC) method, originally developed by Bray and Stelbovics (1992), and the R-matrix with Pseudostates (RMPS) method proposed by Bartschat et al. (1996b) based on the same idea. Both methods are further discussed below.

By substituting Eq. (8.32) into the Schrödinger equation (8.26), projecting onto the channel functions $\overline{\Phi}_j^\Gamma$ and onto the square-integrable functions χ_j^Γ, and eliminating the coefficients a_{ji}^Γ, one obtains coupled integro-differential equations for the reduced radial functions F_{ji}^Γ representing the motion of the scattered electron. These are of the form

$$\left(\frac{d^2}{dr^2} - \frac{l_j(l_j+1)}{r^2} + \frac{2Z}{r} + k_j^2\right) F_{ji}^\Gamma(r)$$
$$= 2 \sum_l \left(V_{jl}^\Gamma + W_{jl}^\Gamma + X_{jl}^\Gamma\right) F_{li}^\Gamma(r) \quad (8.33)$$

where l_j is the orbital angular momentum of the scattered electron and V_{jl}^Γ, W_{jl}^Γ, and X_{jl}^Γ are partial-wave decompositions of the local direct, nonlocal exchange, and nonlocal correlation potentials, respectively.

The direct potential can be written as

$$V_{ij}^\Gamma(r_{N+1}) = \left\langle \overline{\Phi}_i^\Gamma(X_N; \widehat{r}_{N+1}\sigma_{N+1}) \right|$$
$$\sum_{i=1}^N \frac{1}{r_{i\,N+1}} \left| \overline{\Phi}_j^\Gamma(X_N; \widehat{r}_{N+1}\sigma_{N+1}) \right\rangle$$
$$(8.34)$$

where the integral is taken over all $N+1$ electron space and spin coordinates, except for the radial coordinate of electron "$N+1$." This potential has the asymptotic form

$$V_{ij}^\Gamma(r) = \frac{N}{r}\delta_{ij} + \sum_{\lambda=1}^{\lambda_{max}} a_{ij}^\lambda r^{-\lambda-1}, \quad r > a \quad (8.35)$$

where a is the range beyond which the bound orbitals in $\overline{\Phi}_i^\Gamma$ and $\overline{\Phi}_j^\Gamma$ are negligible. The long-range potentials with $\lambda \geq 1$ play an important role in low-energy electron–atom scattering.

The exchange and correlation potentials, unlike the direct potential, are both nonlocal, with the exchange potential vanishing exponentially for large r. Explicit expressions for these potentials are complicated and, except for hydrogen-like targets, are not written down explicitly in practice. Instead, they are calculated by general computer programs.

To determine the cross section, we look for solutions of Eqs. (8.33) satisfying K-matrix asymptotic boundary conditions:

$$F_{ji}^\Gamma(r) \xrightarrow[r\to\infty]{} k_j^{-1/2}(\sin\theta_j \delta_{ji} + \cos\theta_j K_{ji}^\Gamma)$$

for open channels;

$$F_{ji}^\Gamma(r) \xrightarrow[r\to\infty]{} 0,$$

for closed channels (8.36)

Here

$$\theta_j = k_j r - \tfrac{1}{2} l_j \pi + \frac{z}{k_j}\ln 2k_j r + \sigma_j \quad (8.37)$$

with $z = Z - N$ as the residual charge on the ion and $\sigma_j = \arg\Gamma(l_j + 1 - iz/k_j)$ denoting the Coulomb phase shift. The (scattering) S-matrix is related to the (transition) T-matrix and the (reactance) K-matrix with elements defined by Eqs. (8.36) by the matrix equations

$$S^\Gamma = T^\Gamma - 1 = \frac{1+iK^\Gamma}{1-iK^\Gamma} \quad (8.38)$$

The hermiticity and time-reversal invariance of the Hamiltonian ensure that K^Γ is real and symmetric while S^Γ is unitary and symmetric.

The scattering amplitude defined by Eq. (8.29) and hence the differential cross section can be readily related to the S-matrix. The resultant expression is given by Burke (1985). Computer programs to calculate scattering amplitudes, the DCS, and many other observables were published by Bartschat and Scott (1983), Bartschat (1983); Bartschat (1996), and Grum-Grzhimailo (2003). The total cross section is obtained by averaging the differential cross section over initial spin states, summing over final spin states, and integrating over all scattered electron angles. One obtains

$$\sigma_{tot}(i \to j) = \frac{\pi}{k_i^2} \sum_{\substack{LS\pi \\ l_i l_j}} \frac{(2L+1)(2S+1)}{2(2L_i+1)(2S_i+1)}$$

$$\times |S_{ji}^\Gamma - \delta_{ji}|^2 \quad (8.39)$$

for a transition from an initial target state $\alpha_i L_i S_i$ to a final target state $\alpha_j L_j S_j$, where α_i and α_j represent any additional quantum numbers that are required to define the initial and final target states. It is also useful to define the collision strength by

$$\Omega(i,j) = k_i^2(2L_i+1)(2S_i+1)\sigma_{tot}$$

$$\times (i \to j) \quad (8.40)$$

It is dimensionless and symmetric with respect to interchange of the initial and final states denoted by i and j.

There are many different ways to solve the close-coupling equations in practice. These include packages based on the R-matrix method, on the reduction of Eqs. (8.33) to linear algebraic equations, on the noniterative integral equations method, and on the Kohn variational method. These computational methods and early associated program packages were reviewed by Burke and Seaton (1971) and by Burke and Eissner (1983). The R-matrix method, in particular, has been applied in numerous calculations. An overview of the method and key papers in the development can be found in Burke and Berrington (1993) and in Burke, Noble and Burke (2006). The most important recent developments of the R-matrix method include the RMPS formulation mentioned above Bartschat et al. (1996a), Gorczyca and Badnell (1997), and an entirely independent formulation developed by Zatsarinny and coworkers. The program package BSR (Zatsarinny (2006)) is based on the use of basis splines ("B-splines") as a universal and effectively complete basis to expand the wave function of the projectile inside the R-matrix box. In addition, great flexibility is achieved by allowing for the use of term-dependent, nonorthogonal sets of orbitals to describe the target structure. As is shown by the examples below, the method is particularly advantageous in the treatment of electron collisions with complex atoms, such as noble gases and target states involving several open shells. In these cases, it is well worth to pay the price of a more complicated setup of the Hamiltonian matrix, in return for a highly accurate target description with a relatively small number of individually optimized configurations.

8.3.1.2 Extension of the Close-Coupling Method to Intermediate and High Energies

As mentioned above, a severe limitation to the applicability of the close-coupling method was the traditional cut-off of the expansion after a few low-lying target states. Early attempts to address this problem augmented the expansion (8.32) by adding a few well-chosen pseudostates to represent the remaining infinity of omitted eigenstates (Burke and Schey (1962); Damburg and Karule (1967); Burke and Webb (1970); Callaway and Wooten (1973)). These pseudostates are not target eigenstates but instead are chosen to optimize the representation of the short-range correlation effects or long-range polarization effects. In this way, part of the continuum is included in expansion (8.32), and the pseudostates allow approximately for loss of flux into all open channels including the ionizing channels.

Although a number of calculations had been carried out using this approach, particularly for e—H scattering, early calculations suffered from the appearance of unphysical pseudostate thresholds and pseudoresonances in the intermediate energy region of interest. Burke, Noble and Scott (1987) proposed an intermediate-energy R-matrix (IERM) theory, in which two electrons are allowed to occupy continuum orbitals in the wave function expansion. This method gave accurate results for electron and positron scattering by hydrogen atoms, although special care had to be taken in the averaging of unphysical structures in the T-matrix elements at intermediate energies.

Bray and Stelbovics (1992) finally achieved a breakthrough by developing the CCC method. The basic idea is to include a large number of systematically generated square-integrable pseudostates in the close-coupling expansion and check for convergence of the final results. Such L^2 methods were already suggested and reviewed by Reinhardt (1979) and Broad (1983), but they met with various difficulties at the time. The CCC method was formulated in momentum space, based on the Lippman-Schwinger equation for the T-matrix. To date, the method has been applied with great success to electron and positron collisions with hydrogen-like Bray and Stelbovics (1995) and helium-like Fursa and Bray (1995) systems, including electron-impact ionization as well as single and double ionization by photon impact Bray et al. (2002). If necessary, inner shells are represented by sophisticated core potentials. The most recent extension of the method is a fully relativistic formulation for scattering from quasi-one-electron systems such as Cs Fursa and Bray (2008).

An equivalent method based on the R-matrix approach in coordinate space is the "R-matrix with Pseudostates" (RMPS) method proposed by Bartschat et al. (1996b) and further developed by Gorczyca and Badnell (1997). Once again, many square-integrable states, obtained by diagonalizing the target hamiltonian in a given basis (e.g., Sturmian or Laguerre type), are included in the close-coupling expansion. When used in connection with the standard expansion of the projectile wave function in terms of a basis set, the method is generally restricted to low and intermediate energies. However, the computer program is general and hence accurate results can readily be obtained for complex targets as well.

The above developments have, to a large extent, replaced the use of optical-potential methods based on the close-coupling expansion. However, semiempirical potentials are sometimes employed to represent the effect of inner closed shells, core polarization, and even exchange. Some examples are discussed below.

8.3.1.3 Inclusion of Relativistic Effects

As the nuclear charge Z of the target increases, relativistic effects become important even for low-energy electron scattering. There are two ways in which relativistic effects play a role. First, there is a direct effect corresponding to the relativistic distortion of the wave function describing the scattered electron by the strong nuclear Coulomb potential. Second, there is an indirect effect caused by the change in the charge distribution of the target due to the use of relativistic target wave functions.

For atoms and ions with small and intermediate Z values, the scattering calculations can first be performed in LS-coupling; the K-matrices are subsequently recoupled to yield transitions between fine-structure levels. For atoms and ions with high Z values, relativistic terms must be retained in the Hamiltonian. One method of achieving this is to use the Breit–Pauli Hamiltonian

$$H_{N+1}^{BP} = H_{N+1}^{NR} + H_{N+1}^{REL} \qquad (8.41)$$

where H_{N+1}^{NR} is the nonrelativistic Hamiltonian given by Eq. (8.27) and H_{N+1}^{REL} contains the one-body spin–orbit, Darwin, and mass correction terms, as well as possibly two-body relativistic terms. The total wave function is expanded in a form analogous to Eq. (8.32), except that the target states must also be generated using the Breit–Pauli Hamiltonian, and the wave function is an eigenstate of the total electronic angular momentum J, its z-component M_J, and the parity π rather than of LSM_LM_S and π.

Alternatively, for high-Z atoms and ions, the Dirac-Coulomb Hamiltonian

$$H_{N+1}^D = \sum_{i=1}^{N+1}\left(c\boldsymbol{\alpha}\cdot\boldsymbol{p}_i + \beta c^2 - \frac{Z}{r_i}\right)$$
$$+ \sum_{i>j=1}^{N+1}\frac{1}{r_{ij}} \qquad (8.42)$$

may be used, where $\boldsymbol{\alpha}$ and β are the usual Dirac matrices and c is the speed of light.

Figure 8.18 showed two examples of close-coupling results obtained with the Breit–Pauli R-matrix method for e − Hg scattering by Bartschat et al. (1984), who included the $(6s^2)^1S^0$, $(6s6p)^3P^o_{0,1,2}$, and $(6s6p)^1P^o_1$ target states in the expansion of the total wave function, and with the fully relativistic CCC (Fursa and Bray (2008)) and the fully relativistic B-spline R-matrix method (Zatsarinny and Bartschat (2008)) for elastic e − Cs collisions. The good agreement between experiment and theory indicates that the three main effects of strong channel coupling, electron exchange, and the spin–orbit interaction are correctly represented by the theories in these cases. However, these are still comparatively simple targets resembling quasi-one-electron and quasi-two-electron systems. Significantly, more challenges are expected for heavy noble-gas targets such as Kr and Xe, or for cases in which inner-shell excitation plays an important role.

8.3.1.4 Illustrative Results

Figure 8.19 shows the angle-integrated cross section for elastic scattering and electron-impact excitation from the $(1s)^2S$ state of atomic hydrogen. The excellent agreement between predictions from the CCC, RMPS, and IERM methods, as well as with the experimental data of Williams (1988), indicates that these sophisticated methods can indeed solve this problem to a very high degree of accuracy.

Stepanovic et al. (2006) recently measured the angle-integrated cross sections for electron-impact excitation of the 3^3S and 3^1S states of helium shown in figure 8.20. For a direct comparison with the experimental data, the 69-state B-spline R-matrix (BSR-69) results, originally generated on a very fine energy mesh with a stepsize of 10^{-5} Ry, were convoluted with a Gaussian function of width 37 meV (FWHM), corresponding to the energy resolution in the experiment. The dashed line in the top graph for the 3^3S state represents the theoretical predictions for direct excitation, while the solid lines represent the emission cross sections, that is, including the cascade contributions from the higher-lying $n = 3, 4, 5$ states. The experimental data were normalized to the BSR-69 results with cascade contribution at 23.20 eV for both the 3^3S and 3^1S states, because the cross sections exhibit only a smooth energy dependence around this energy and hence the energy resolution does not play a role. For a detailed analysis of the resonances (see Stepanovic et al. (2006)), the right part of the figure exhibits the theoretical excitation cross sections without convolution with the experimental energy resolution.

Figure 8.21 exhibits results for the production of metastable neon atoms.

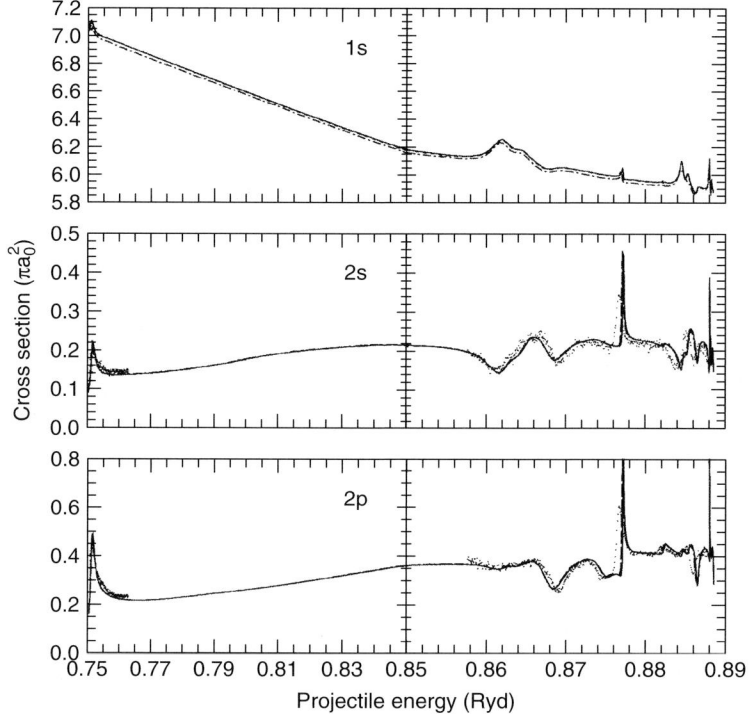

Fig. 8.19 Elastic and excitation cross sections for e−H scattering at incident electron energies between the $n = 2$ and $n = 3$ thresholds. The RMPS (solid line), IERM (dashed line), and CCC (dash-dotted line) curves are barely distinguishable. The small dots represent the experimental data of Williams (1988). (From Bartschat et al. (1996c).)

After renormalizing the published experimental data of Buckman et al. (1983) (within the given overall experimental uncertainty) and including the cascade contributions, the semirelativistic BSR method of Zatsarinny and Bartschat (2004) obtains essentially perfect agreement with experiment regarding the energy dependence in the near-threshold region. These results represent a considerable improvement over previous theoretical predictions. The predicted metastable cross sections were then used to calibrate and interpret new experimental data obtained with an energy resolution of ≈4 meV by Bömmels et al. (2005). For example, a very sharp Feshbach resonance was theoretically predicted and experimentally confirmed at 18.527 eV.

The enormous importance of FeII, the singly ionized species of the iron element, for astronomical observations is well known. A meaningful interpretation of the spectral lines critically depends on an accurate knowledge of the electron-induced collision strengths, as well as the so-called "effective collision strengths," which are obtained by integrating the energy-dependent results over a Maxwellian distribution for a range of electron temperatures. Because

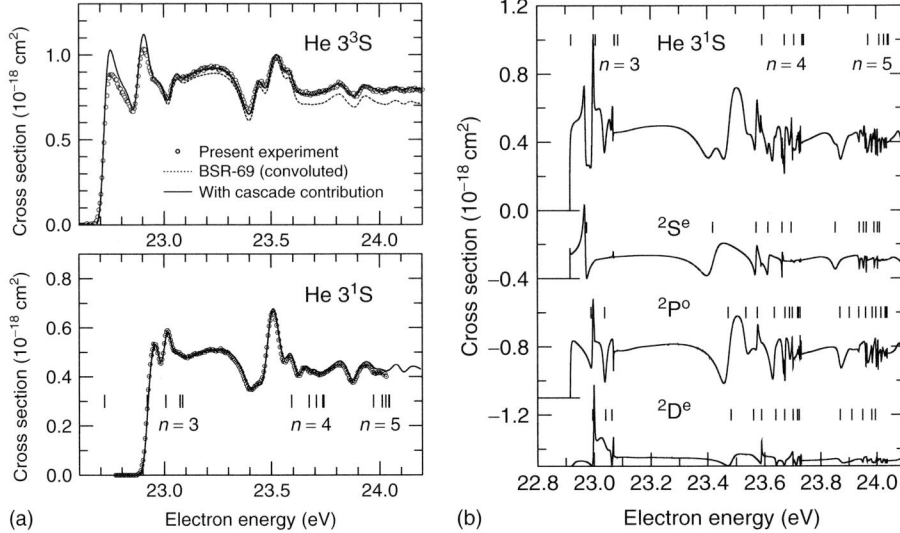

Fig. 8.20 (a) Angle-integrated cross sections for electron-impact excitation of the 3^3S and 3^1S states in helium. The theoretical results without (dashed line) and with (solid line) cascade contributions were convoluted with a Gaussian of 37 meV (FWHM). The vertical bars in the lower panel represent the thresholds for the He target states. (b) Angle-integrated cross section for electron-impact excitation of the 3^1S state. The thick solid line at the top exhibits the predictions from a 69-state BSR model summed over all partial waves, while the remaining thin solid lines (offset to increase the visibility) show the most important individual partial-wave contributions. The vertical bars represent the thresholds for the He target states at the top and prominent He^- resonances found in a detailed partial-wave analysis. (From Stepanovic et al. (2006).)

of the importance of these data and the complete lack of experiments, major theoretical and computational efforts have been devoted to this system for more than two decades.

Figure 8.22 exhibits results for the collision strength for the $(3d^6 4s)a^6D - (3d^7)a^4F$ transition. Predictions of Ramsbottom et al. (2004) and Zatsarinny and Bartschat (2005) for contributions to the collisions strength from selected partial-wave symmetries are compared and exhibit overall satisfactory agreement. The difference in the intensity of some narrow resonances may be caused by the fact that finer energy steps of 10^{-5} Ry were used in the BSR calculations than the 10^{-4} Ry energy mesh chosen by Ramsbottom et al. (2004). A notable exception is a broad structure in the $^5D^e$ partial wave around 0.15 Ry. While there are differences at such detailed level, good overall agreement among the calculations was obtained after summing up the individual partial-wave contributions and integrating over a Maxwellian distribution. Because there are no experimental data available for these cases, such independent checks are very important in assessing the reliability of the data before providing them to the astrophysical community.

Having established reliable results in a nonrelativistic framework, an important next step involves fine-structure

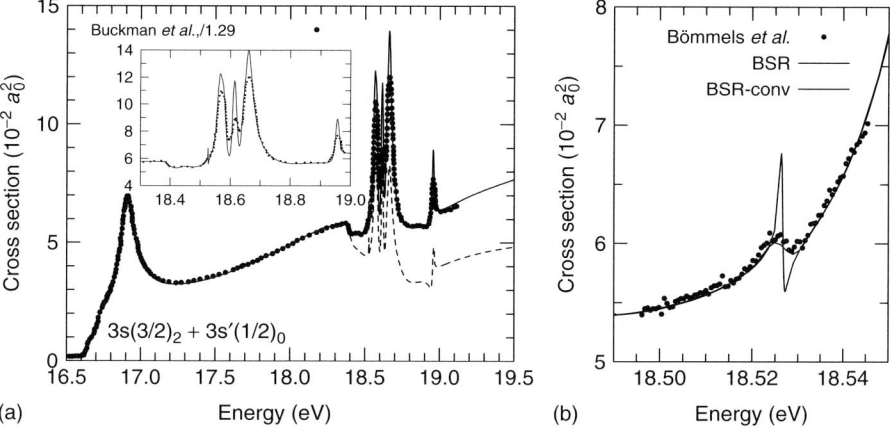

Fig. 8.21 (a) Angle-integrated cross section for production of neon atoms in the metastable $3s[3/2]_2$ and $3s'[1/2]_0$ states (from Zatsarinny and Bartschat (2004).) The experimental data of Buckman *et al.* (1983) (thick dots) were re-normalized by a factor of 1.29 to provide a good visual fit to the BSR theory at energies just above the excitation threshold. The solid line includes the cascade contributions from all the states included in the model, while the thin dotted line (starting around 18.4 eV) represents the results without cascades. (b) A very narrow resonance predicted by Zatsarinny and Bartschat (2004) and later found in the experiment of Bömmels *et al* (2005).

transitions. Ramsbottom *et al.* (2007) recently carried out a 262-state Breit–Pauli *R*-matrix calculation with about 1800 coupled channels to study fine-structure forbidden transitions in e–FeII collisions. This knowledge is required in many astrophysical applications.

FeII is just one example of a target with an open d-shell, which represents a major challenge to atomic collision theory. On the other hand, collisions systems involving such targets are very important in practical applications. An example is e – Mo, which may be a suitable candidate for a mercury-free lighting source (Petrov *et al.* (2004); Bartschat *et al.* (2004a)). Another case of interest involves electron collisions with tungsten and its ions, which are important impurities in fusion devices such as ITER.

Finally, as mentioned above, an important development has been the extension of the close-coupling method to intermediate and even high energies, and the inclusion of relativistic effects, either by recoupling of nonrelativistic results, perturbatively through the Breit–Pauli hamiltonian, or in a fully relativistic treatment. Figure 8.23 shows the total cross section for electron collisions with Cs atoms. Overall, there is good agreement between results from a nonrelativistic CCC model, a semirelativistic Breit–Pauli *R*-matrix approach, and a fully relativistic Dirac-based *B*-spline *R*-matrix treatment.

8.3.1.5 Resonances and Quantum Defect Theory

We have already seen in previous sections that resonances play an important role in low-energy scattering of electrons by atoms and ions. The general theory of resonances has been discussed by Fano (1961), Fano (1983),

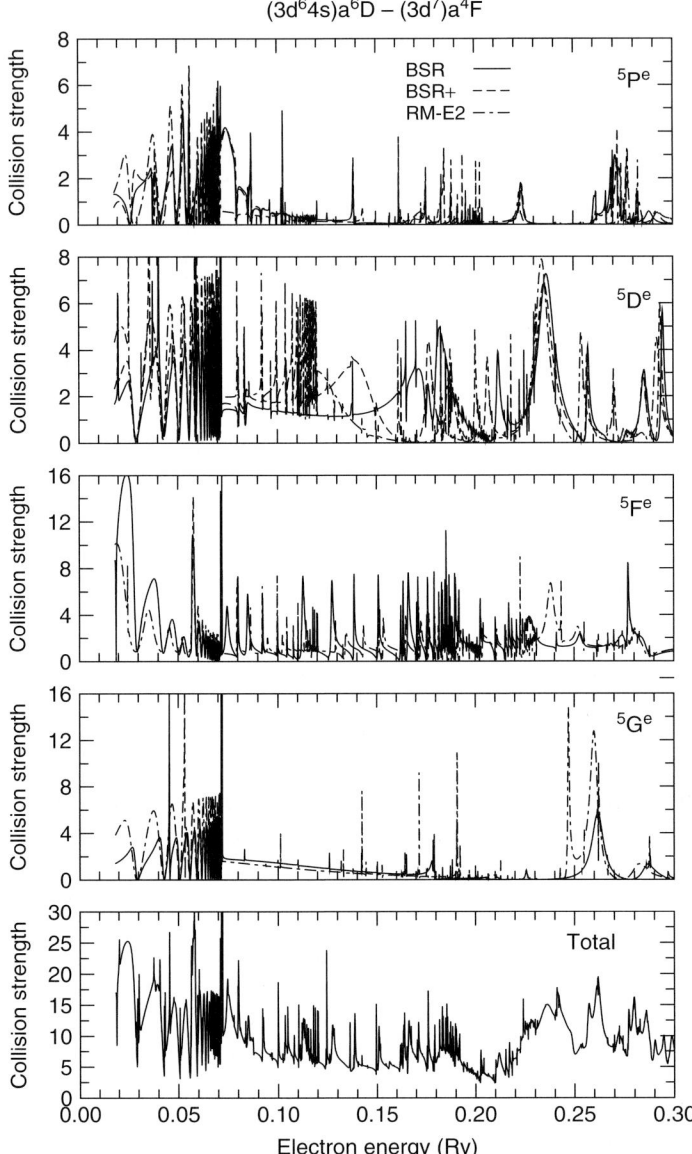

Fig. 8.22 Important partial-wave contributions, indicated in the top four panels, and total (bottom) collision strength for the electron-induced transition $(3d^6 4s)^6D - (3d^7)^4F$ in Fe^+. (From Zatsarinny and Bartschat (2005).)

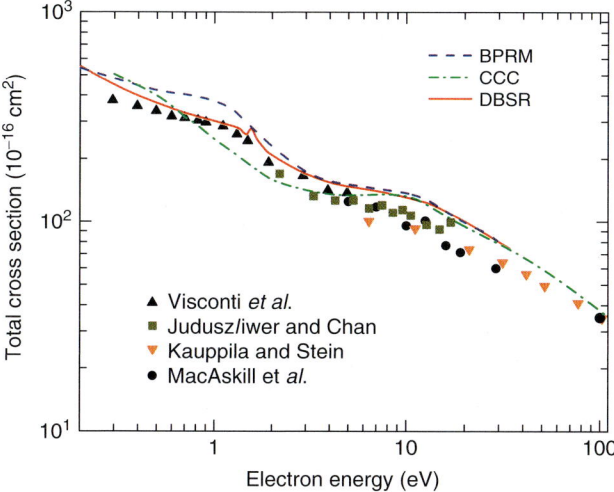

Fig. 8.23 Total electron scattering cross section from the ground $(6s)^2S$ ground state of Cs. The 12-state DBSR results of Zatsarinny and Bartschat (2008) are compared with various sets of experimental data (Visconti, Slevin and Rubin (1970), Jaduszliwer and Chan (1992), Kauppila and Stein (2001), MacAskill et al. (2002)), as well as predictions from a 40-state semirelativistic Breit–Pauli R-matrix (BPRM) calculation by Bartschat and Fang (2000) and a nonrelativistic CCC model presented by MacAskill et al. (2002). (From Zatsarinny and Bartschat (2008).)

Burke (1968), Schulz (1973a), and Schulz (1973b). Here, we review their principal features.

It is useful to distinguish two types of resonances, namely, open-channel or "shape" resonances, and closed-channel or "Feshbach" resonances. Shape resonances occur when the effective potential between the electron and the target has a characteristic shape with an inner attractive well and an outer repulsive barrier, the latter usually caused by the centrifugal repulsion. Examples of such resonances include a dominant p-wave shape resonance in low-energy elastic e − Mg collisions Bartschat et al. (2004b) and a broad d-wave shape resonance in e − CH_4 scattering Gianturco and Scialla (1987).

On the other hand, Feshbach resonances occur for incident electron energies just below excitation thresholds. The scattered electron is captured temporarily into a bound state with the target in an excited state. We have seen several examples of such resonances in the previous sections.

Calculations based on expansion (8.32) are a very convenient and appropriate way of describing Feshbach resonances, provided that the relevant open and closed channels are included in the expansion. For example, bound states caused by the long-range attractive Coulomb potential in the closed channels for electron–ion scattering give rise to Rydberg series of resonances when the coupling to the open channels is included.

One of the motivations for the development of multichannel quantum defect theory (MQDT) (Seaton (1958); Seaton (1983)) is to treat entire series of resonances rather than an individual resonance. Seaton showed that the K-matrix defined by Eq. (8.36) can be written as

$$K^\Gamma = K^\Gamma_{oo} - K^\Gamma_{oc}(K^\Gamma_{cc} + \tan \pi \nu_c)^{-1} K^\Gamma_{co} \quad (8.43)$$

where the elements of the submatrices K^Γ_{oo}, K^Γ_{oc}, K^Γ_{co}, and K^Γ_{cc} are obtained by analytically continuing the K-matrix from above the threshold to which the resonances converge to below this threshold, where "o" refers to the channels that remain open below this threshold, and "c" refers to the channels that become closed. In these closed channels ($k_i^2 < 0$), the components ν_i of the vector $\boldsymbol{\nu}_c$ are defined by

$$k_i^2 = -(Z-N)^2/\nu_i^2 \quad (8.44)$$

The factor $\tan \pi \nu_c$ causes the last term in Eq. (8.43) to have an infinite series of poles that give rise to the Feshbach resonances. When the K-matrix is calculated by solving Eqs. (8.36) at a few energies above threshold, it can be extrapolated to yield an infinite series of resonances below threshold. This approach was extended by Gailitis (1963) to predict the average behavior of electron–ion cross sections in resonance regions.

8.3.1.6 Dielectronic Recombination

It was pointed out by Burgess (1964); Burgess (1965) that Rydberg resonance series play a fundamental role in dielectronic recombination. This process is illustrated, for example, in the case of $e - C^{3+}$ scattering by the equation

$$e^- + C^{3+}(1s^2 2s) \to C^{2+*}(1s^2 2p\, nl)$$
$$\to C^{2+}(1s^2 2s\, nl) + h\nu \quad (8.45)$$

The electron is captured into a member of a Rydberg series of resonances converging to the $1s^2 2p$ excited state of the C^{3+} ion. These resonances can stabilize by radiative decay, thereby leaving the C^{2+} ion in an excited bound state. This process can often yield dielectronic recombination coefficients α^{DR} that are two orders of magnitude larger than radiative recombination coefficients, basically because of the very large number of resonance levels that contribute.

In the isolated-resonance approximation, α_i^{DR} is given by

$$\alpha_i^{DR} = \left(\frac{2\pi}{kT}\right)^{3/2} \sum_n \frac{g_n}{2g_i} \frac{\Gamma_n^a \Gamma_n^r}{\Gamma_n^a + \Gamma_n^r} e^{-E_n/kT} \quad (8.46)$$

where i refers to the initial state of the target ion and n to the intermediate resonance state. The coefficient g is the statistical weight of the state, Γ_n^a is the autoionization width, Γ_n^r is the radiative width of the resonance state, and E_n is the energy of the resonance state relative to the ground state of the target ion. The velocity distribution of the electrons is assumed to be Maxwellian corresponding to a temperature T. The summation over n in Eq. (8.46) converges for large n because Γ_n^a behaves as n^{-3} along a Rydberg resonance series while Γ_n^r becomes independent of n. In practice, very high n values often contribute.

Dielectronic recombination measurements have been made on a number of ions, while many calculations have been carried out illustrating the importance of this recombination process in hot gaseous plasmas. More information, including an extensive list of references, can be found in Pindzola, Griffin and Badnell (2006).

8.3.2
Born Series Methods

8.3.2.1 Elastic Scattering and Excitation of Atoms and Ions at Intermediate and High Energies

We now consider electron–atom and electron–ion scattering at electron impact energies greater than the ionization threshold of the target. In this case, an infinite number of channels are open, so that they cannot all be included in the expansion of the total wave function. Several approaches have been developed to treat scattering at these energies. Before the recent extensions of traditionally low-energy methods based on the close-coupling expansion, variants of the Born approximation were developed to address such problems. Being computationally much less demanding than CCC, RMPS, and the "initial-value methods" to be discussed below, the "distorted-wave"(DW) method in particular continues to be of practical importance for applications to complex targets, especially for ionization processes. The method has also proven to be an accurate way of calculating partial-wave cross sections for high angular momenta, and thus to obtain converged results with the number of partial waves. Finally, by studying the effect of different distortion potentials, including varying orders of the perturbation, and moving (parts of) the perturbation to the exact treatment by including it into the calculation of the scattering wave function, it is possible to study the importance of individual physical effects on the outcome of a collision process. Such effects may include the polarization of the atomic charge cloud due to the incident projectile, exchange, absorption (i.e., loss of flux into inelastic channels), or the long-range Coulomb interaction between three (or more) charged particles in the final state.

8.3.2.2 The Born Series

In the "high-energy" domain, which extends from several times the ionization threshold upward, methods based on the Born series are appropriate. Ignoring exchange for the moment, the Born series for the direct scattering amplitude can be written as (Walters (1984); Joachain (1987))

$$f = \sum_{n=1}^{\infty} f_{Bn} \qquad (8.47)$$

The nth Born term f_{Bn} contains the interaction V_{pt} between the scattered electron and the target atom n times and the direct Green's operator, $G_0^+ = (E - K_p - H_N + i\varepsilon)^{-1}$, $n - 1$ times. Here H_N is again the N-electron target hamiltonian while K_p is the kinetic energy operator for the projectile.

It is important to retain consistently all terms in the Born series with similar energy and momentum-transfer ($\Delta = |\mathbf{k}_i - \mathbf{k}_j|$) dependences. For elastic scattering, the scattering amplitude converges to the first Born approximation for all Δ, but at lower energies, it is necessary to include $\text{Re}\{f_{B3}\}$ as well as the second Born terms to obtain the cross section correct to k^{-2}. In the forward direction, convergence to the first Born approximation is slow because of contributions from Im{f_{B2}}, which allow for loss of flux into the open channels.

For inelastic scattering, the first Born approximation does not give the correct high-energy limit for large Δ. Instead, this comes from Im{f_{B2}}. Physically we can understand this by noting that inelastic scattering into large scattering angles involves a collision with the nucleus, followed or preceded by an inelastic collision with the bound electron to give excitation. In the forward direction, Re{f_{B3}} must again be retained to yield the high-energy limit correct to k^{-2}.

Because the third Born term is difficult to calculate directly, Byron and Joachain (see Joachain (1987)) suggested that, to third order, the scattering amplitude should be calculated using the eikonal Born series (EBS)

$$f_{EBS} = f_{B1} + f_{B2} + f_{G3} + g_{Och} \qquad (8.48)$$

where f_{G3} is the third-order term in the expansion of the Glauber amplitude (Glauber (1959)) in a power series of the projectile–target interaction V_{pt}, while g_{Och} is the Ochkur electron exchange amplitude (Ochkur (1964)). The EBS method has been very successful when perturbation theory converges rapidly, that is, at high energies, at small and intermediate angles, and for light atoms.

8.3.2.3 The Distorted-Wave (DW) Method

Distorted-Wave (DW) methods are characterized by the fact that the interaction is separated into two parts, one of which is treated exactly while the other is included using first-order perturbation theory. A semirelativistic first-order DW approximation (DWBA) was introduced by Madison and Shelton (1973) and later modified by Bartschat and Madison (1987) to treat electron collisions with heavy noble gases and also with mercury. Other variants include the first-order many-body theory (FOMBT) of the Los Alamos group Machado, Leal and Csanak (1982), the semirelativistic version developed and described by Dasgupta, Blaha and Giuliani (2000), and the relativistic distorted-wave (RDW) method of the Toronto group Zuo, McEachran and Stauffer (1991). The basic idea behind all these methods is to calculate distorted waves for the projectile electron by solving the differential equation

$$\left[\frac{d^2}{dr^2} - \frac{l(l+1)}{r^2} - 2\{U(r) + V_r(r) - E\} \right] \chi_{E,l}(r) = 0 \qquad (8.49)$$

Here $U(r)$ is the static Coulomb potential, often modified by including terms to account for electron exchange and the charge distortion (polarization) of the target, while V_r contains relativistic effects. [Instead of (8.49), the fully relativistic RDW solves the corresponding Dirac equation.] The distorted waves are then used to calculate transition matrix elements, treating to first order the part of the projectile–target interaction not explicitly included in the distortion potentials.

A detailed description of the DWBA method, including a computer program, was provided by Madison and Bartschat (1996). Here we only mention the principal advantages and disadvantages. The DWBA approach has proven particularly useful for electron scattering by highly ionized ions where it is assumed that only the two channels corresponding to

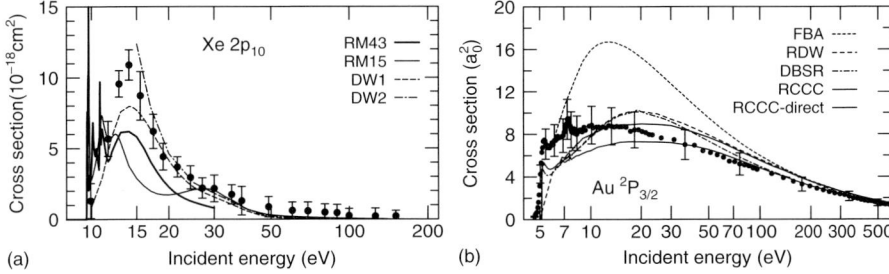

Fig. 8.24 Angle-integrated cross section for electron-impact excitation of the $(5p^56p)$ "$2p_{10}$" state in Xe (a) and the $(6p)^2P_{3/2}$ state in Au (b). The figures show the comparison of various DW and Breit–Pauli R-matrix (RM) predictions for Xe and results from a plane-wave first-order Born approximation (FBA), as well as the relativistic distorted-wave RDW, relativistic Dirac B-spline R-matrix (DBSR), and relativistic convergent close-coupling (RCCC) for Au with experimental data from the Wisconsin (Xe) and Flinders (Au) groups. The difference between the RCCC and RCCC-direct results is due to cascade effects. (Adapted from Bartschat, Dasgupta and Madison (2004) for Xe and from Maslov et al. (2008) for Au.)

the initial and final states in Eq. (8.32) need to be considered. The dominant diagonal potentials are treated exactly and the off-diagonal potentials coupling these channels are accounted for to first order. The method becomes more accurate at intermediate energies as $z = Z - N$ increases. As mentioned previously, the DWBA has also proven to be an accurate way of calculating partial-wave cross sections for high angular momenta. The angular momentum barrier in this case depresses the importance of the coupling potential.

The DWBA approach can also be a useful way of predicting the total and differential cross sections of inelastic electron–atom at intermediate energies well removed from threshold. Furthermore, the method is fast and relatively easy to implement. In principle, it allows for significant flexibility in the target description for the initial and final states because the orbitals can be specifically optimized for the states of interest in a given transition.

The major problem with the DW approach is the fact that channel coupling is neglected. Consequently, the method cannot account for near-threshold Feshbach-type resonances, it is generally much less reliable for optically forbidden than for optically allowed transitions, and further problems can occur if the calculation is performed without explicit unitarization of the scattering matrix; see Dasgupta et al. (2001).

Figure 8.24 exhibits a comparison of results for angle-integrated cross sections obtained in a variety of Born-based and close-coupling-based approximations. In both cases, the first-order DW models do quite well compared to the close-coupling approximations, while the first-order plane-wave approach generally overestimates the cross section near the maximum. However, in many cases of practical interest, the (plane-wave) FBA can still be used to get reasonably accurate results for excitation of optically allowed transitions by performing a rescaling transformation of the results as suggested by Kim (2001).

8.3.3
Initial-Value Methods

In this section, we briefly discuss the basic idea behind two highly successful methods developed using a somewhat different philosophy. These are the TDCC and the ECS approaches. The goal of both methods is the construction of a highly accurate numerical solution of the Schrödinger equation by propagating a known initial state. The matching to sometimes highly complicated boundary conditions, especially for the ionization processes to be discussed below, is avoided through a different way of extracting the relevant information from the wave function after large propagation times or sufficiently far away from the interaction region.

8.3.3.1 Time-Dependent Close-Coupling (TDCC)

With the recent development of massively parallel supercomputers, the use of more direct numerical approaches has become very popular. One example is the TDCC method recently reviewed by Pindzola *et al.* (2007). For a target with one or two electrons outside a closed shell, they derive the time-dependent close-coupling equations for the radial parts of the wave function describing the projectile and the active target electron or electrons. For each partial-wave symmetry, these equations are propagated in time over a two- or three-dimensional lattice.

The initial wave function is written as the properly symmetrized product of a bound state, represented on the lattice, and a Gaussian wavepacket for the projectile. The width of the package is chosen in such a way that any overlap between the projectile and the target wave function before the collision can be neglected. The electron wave function is then propagated for a sufficiently long time to ensure that the final state after the collision has been reached. However, note that the propagation time has to be small enough to ensure that boundary effects, due to collisions with the wall of the lattice, do not disturb the result. Cross sections for elastic scattering, excitation and ionization are obtained from projections of the final-state wave function on individual asymptotic channel functions.

The principal advantage of the method lies in the fact that the radial part of the wave function, which contains all the dynamics of the collision, is not described by a finite basis set as in the time-independent close-coupling approaches discussed above. On the other hand, the computational resources needed for such calculations are extensive. Hence, the method is so far effectively restricted to (quasi-)one- and (quasi-)two-electron targets. Also, often only the lowest few partial waves are treated in this way, while contributions from higher partial waves are handled by simpler methods, such as the first-order DWBA approximation mentioned above.

Figure 8.25 shows the development of a wavepacket propagated in the TDCC approach. This particular example is for the $^1S^e$ partial-wave symmetry in $e - Be^+$ collisions at an incident projectile energy of 50 eV. Initially, we have a properly symmetrized product of a Gauss packet for the projectile and the active 2s electron in Be^+, which then develops into a complex object containing contributions from elastic scattering, excitation, and ionization. The latter can be identified by the wavepacket having nonzero magnitude in the region where the radial

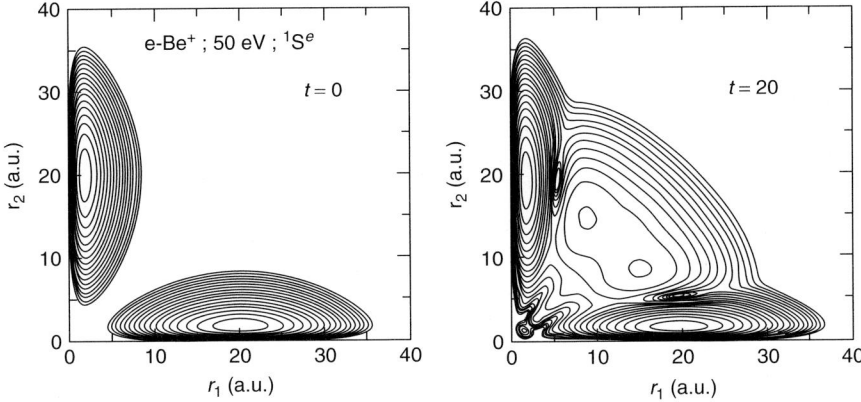

Fig. 8.25 Magnitude of the wave function for e − Be^+ collisions in the $^1S^e$ partial wave as a function of the radial coordinates of the projectile and the active electron for an incident projectile energy of 50 eV. After developing for 20 a.u. of time, the initial wavepacket on the left contains contributions from elastic scattering, excitation, and ionization. (Adapted from Pindzola et al. (1997)).

coordinates of both electrons (r_1 and r_2) are large.

8.3.3.2 Exterior Complex Scaling (ECS)

The exterior complex scaling method is another attempt to obtain the full wave function more directly than in the standard close-coupling or Born series approaches. The basic idea is to separate the wave function into two parts, namely, an incident part Φ_{in} and a scattered part Ψ_{sc}. The Schrödinger equation is rewritten as

$$(E - H)\Psi_{sc} = (H - E)\Phi_{in} \quad (8.50)$$

In other words, the incident wave function is the driving term in the equation for the scattering wave function.

The critical point in what follows is the fact that Ψ_{sc} *only* contains outgoing waves. Hence, by making the variable transformation

$$z(r) \to \begin{cases} r & \text{if } < R_0 \\ R_0 + (r - R_0)\,e^{i\theta} & \text{if } r \geq R_0 \end{cases} \quad (8.51)$$

for all radial coordinates beyond some value R_0, any outgoing oscillating wave is transformed to an exponentially decreasing function. In this way, the complicated three-body Coulomb boundary conditions can be avoided in the solution of the driven Schrödinger equation. The physical information can be extracted inside the region defined by R_0 by using a variety of formulas for the scattering amplitude and the cross section. Of course, R_0 has to be chosen sufficiently large to ensure converged and stable results.

The ECS method was applied with great success to the electron-impact ionization of atomic hydrogen by Rescigno et al. (1999). As an example, Figure 8.26 shows the scattering wave function on a 2D radial grid for a particular set of angular momentum quantum numbers. A review of other applications was given by McCurdy, Baertschy and Rescigno (2004). The numerical efficiency of the method was greatly enhanced by a propagation algorithm (the PECS approach) described by Bartlett

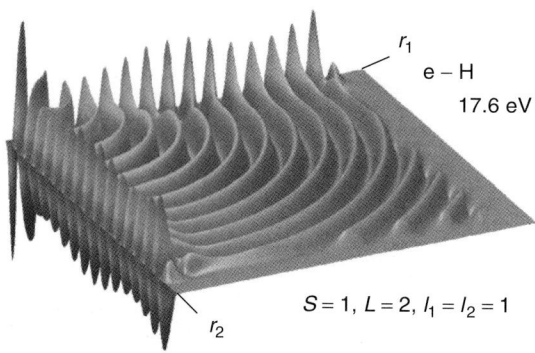

Fig. 8.26 Real part of the wave function for e − H collisions for an incident projectile energy of 17.6 eV in the triplet spin channel for individual angular momenta $l_1 = l_2 = 1$ and a coupled angular momentum $L = 2$. Note how the oscillatory parts of the scattering wave function vanish at large values of the radial coordinates r_1 and r_2. (Adapted from McCurdy, Baertschy and Rescigno (2004)).

(2006). The ECS method has recently been combined with a variety of numerical schemes, including B-splines and finite-element discrete-variable representations (FEDVR). Calculations for He, in the so-called "S-wave model" with all orbital angular momenta being zero, were reported by Bartlett et al. (2007), thus suggesting that the full problem with three active electrons will likely be treated in the near future.

8.3.4
Ionization of Atoms and Ions

In this section, we discuss the process whereby electrons are ejected from the target giving rise to ionization. We are mainly concerned with single ionization or (e,2e) reactions of the type discussed in Section 8.2.4, that is,

$$e_i^- + A_i \longrightarrow A_j^+ + e_A^- + e_B^- \quad (8.52)$$

where A_i is the initial bound state of the target atom (or ion) and A_j^+ is the final state of the ion.

8.3.4.1 The Ionization Amplitude and Cross Section

The scattering amplitude for the direct ionization process is given by the integral expression Rudge and Seaton (1965)

$$f_i(\mathbf{k}_A, \mathbf{k}_B) = -(2\pi)^{1/2} \times \exp[i\chi(\mathbf{k}_A, \mathbf{k}_B)]$$
$$\times \langle\langle H_{N+1} - E)\Phi | \Psi_i^{(+)}\rangle$$
(8.53)

Here $\Psi_i^{(+)}$ is an exact solution of Eq. (8.26) satisfying plane-wave plus outgoing-wave boundary conditions (8.29) and Φ is a solution of the equation

$$(H_{N+1} - V - E)\Phi = 0 \quad (8.54)$$

satisfying ingoing-wave boundary conditions. Also, $\mathbf{k}_A$ and $\mathbf{k}_B$ are the momenta of the two outgoing electrons defined in terms of E_A and E_B in Eq. (8.9) by $\frac{1}{2}k_A^2 = E_A$ and $\frac{1}{2}k_B^2 = E_B$.

In practice, V is chosen so that Φ has a simple form. Thus, for electron scattering by H-like ions with nuclear charge Z, V is often chosen as

$$V = -\frac{Z - Z_A}{r_1} - \frac{Z - Z_B}{r_2} + \frac{1}{r_{12}} \quad (8.55)$$

An approximate final-state wave function is then defined by

$$\Phi(\mathbf{r}_1, \mathbf{r}_2) = \psi_C^{(-)}(Z_A, \mathbf{k}_A, \mathbf{r}_1)$$
$$\times \psi_C^{(-)}(Z_B, \mathbf{k}_B, \mathbf{r}_2) \quad (8.56)$$

where $\psi_C^{(-)}$ is a one-electron Coulomb wave function satisfying ingoing-wave boundary conditions while Z_A and Z_B are the effective charges "seen" by the outgoing electrons. In order that the scattering amplitude not contain a divergent phase factor, Z_A and Z_B must satisfy the condition

$$\frac{Z_A}{k_A} + \frac{Z_B}{k_B} = \frac{Z}{k_A} + \frac{Z}{k_B} - \frac{1}{|\mathbf{k}_A - \mathbf{k}_B|} \quad (8.57)$$

in which case the phase $\chi(\mathbf{k}_A, \mathbf{k}_B)$ is given by

$$\chi(\mathbf{k}_A, \mathbf{k}_B) = \frac{Z_A}{k_A} \ln \frac{k_A^2}{k_A^2 + k_B^2} + \frac{Z_B}{k_B} \ln \frac{k_B^2}{k_A^2 + k_B^2} \quad (8.58)$$

The exchange ionization amplitude can be obtained from the theorem

$$g_i(\mathbf{k}_A, \mathbf{k}_B) = f_i(\mathbf{k}_B, \mathbf{k}_A) \quad (8.59)$$

first derived by Peterkop (1961). This result follows from the fact that the two amplitudes $f_i(\mathbf{k}_B, \mathbf{k}_A)$ and $g_i(\mathbf{k}_A, \mathbf{k}_B)$ describe the same physical process, in which the electron at $\mathbf{r}_N$ has momentum $\mathbf{k}_B$ and the electron at $\mathbf{r}_{N+1}$ has momentum $\mathbf{k}_A$. In practice, the Peterkop theorem will not be valid if approximate wave functions are used to calculate the ionization amplitudes. Therefore, a relative phase $\tau_i(\mathbf{k}_A, \mathbf{k}_B)$ between the two amplitudes is sometimes introduced by

$$g_i(\mathbf{k}_A, \mathbf{k}_B) = \exp[i\tau_i(\mathbf{k}_A, \mathbf{k}_B)] f_i(\mathbf{k}_B, \mathbf{k}_A) \quad (8.60)$$

Note that different choices for $\tau_i(\mathbf{k}_A, \mathbf{k}_B)$ can lead to significant variations in the predicted cross sections.

An important development in ionization theory was the adaptation of the continuum distorted-wave eikonal-initial-state (CDW-EIS) approach, which was put on a firm mathematical footing for heavy-particle impact by Crothers and McCann (1983), to electron impact. In particular, Brauner, Briggs and Klar (1989) used the so-called 3C wave function, which has the correct asymptotic behavior originally proposed by Redmond (1973), when all three particles are infinitely far apart from each other. In fact, Jones and Madison (2000) showed that the wavefunction is accurate when at least one particle is far away from the other two. The 3C function for a system of three free Coulomb particles (specifically two electrons and a proton) for the final state of an (e,2e) process on atomic hydrogen is given by

$$\Phi^-(\mathbf{r}_1, \mathbf{r}_2) =$$
$$(2\pi)^{-3} \exp i \times (\mathbf{k}_1 \cdot \mathbf{r}_1 + \mathbf{k}_2 \cdot \mathbf{r}_2)$$
$$\times C(\alpha_1, \mathbf{k}_1, \mathbf{r}_1) \times C(\alpha_2, \mathbf{k}_2, \mathbf{r}_2)$$
$$\times C(\alpha_{12}, \mathbf{k}_{12}, \mathbf{r}_{12}) \quad (8.61)$$

Here $\mathbf{k}_{12} \equiv \mu(\mathbf{k}_1 - \mathbf{k}_2)$ is the relative wave vector of the two particles while $\mu = 1/2$ is the reduced mass of the two electrons. Furthermore, the Sommerfeld parameters are given by $\alpha_1 = -1/k_1$, $\alpha_2 = -1/k_2$, and $\alpha_{12} = \mu/k_{12}$, respectively. Distortion effects due to the Coulomb potential are contained in the function

$$C(\alpha, \mathbf{k}, \mathbf{r}) \equiv \Gamma(1 - i\alpha) \exp(-\pi\alpha/2)_1$$
$$\times F_1(i\alpha, 1; -ikr - i\mathbf{k} \cdot \mathbf{r}) \quad (8.62)$$

where $_1F_1$ and Γ are the hypergeometric and gamma functions, respectively.

Since 1989, several further modifications to the 3C function have been proposed. These include a suggestion by Alt and Mukhamedzhanov (1993) to use locally defined rather than global wave vectors, Berakdar's "dynamically screened" DS3C function (Berakdar (1996)), the use of the full CDW-EIS function to account for initial-state effects by Jones and Madison (1998), and the development of a DW version by Prideaux, Madison and Bartschat (2005). While improvements have been achieved in some cases, especially regarding the account of the postcollision interaction, the major drawback of the approach has been the fact that 3C-type wave functions are generally not sufficiently accurate at short distances from the target center. Here "short" is typically only a few atomic units (Foster et al. (2006); Bartschat, Weflen and Guan (2007)), but that is where the actual ionization process occurs.

Ionization cross sections are obtained by taking the ratio of the number of ionization events per unit time and per unit target atom to the incident electron flux. For random electron spin orientations in targets with one active electron, for example, the triple-differential cross section (TDCS) for ionization from an initial state $|i\rangle$ is given by

$$\frac{d^3\sigma_i}{d\Omega_A d\Omega_B dE} = \frac{k_A k_B}{k_i}$$
$$\times \left[\frac{1}{4}|f_i + g_i|^2 + \frac{3}{4}|f_i - g_i|^2\right] \quad (8.63)$$

As discussed in Sections 8.2.4.2 and 8.2.4.3, various double and single differential cross sections can be formed by integrating the TDCS with respect to $d\Omega_A$, $d\Omega_B$, and/or dE. By integrating Eq. (8.63) over all outgoing electron scattering angles and energies, we obtain the total ionization cross section

$$\sigma_i = \frac{1}{k_i} \int_0^{E/2} dE \, k_A k_B \int d\Omega_A \int d\Omega_B$$
$$\times \left[\frac{1}{4}|f_i + g_i|^2 + \frac{3}{4}|f_i - g_i|^2\right] \quad (8.64)$$

Because the two electrons are indistinguishable, the upper limit of integration over the energy variable is $E/2$ to avoid double counting. For atomic targets, the plane-wave Born program of Bartlett and Stelbovics (2003) allows for a fast way of obtaining reasonably accurate total ionization cross sections that are needed in many applications. The "Binary Encounter Bethe" model of Kim and Rudd (1994), Kim et al. (2008) has also been highly successful in the production of ionization cross sections for complex atoms and molecules, including radicals, that are important for the modeling of many processes in plasmas.

8.3.4.2 Threshold Behavior of the Ionization Cross Section

In a fundamental paper, Wigner (1948) pointed out that the derivation of threshold laws does not require a detailed knowledge of the collision dynamics in the "reaction zone," where all the particles are close together. Instead, they are determined by the asymptotic form of the wave function as the particles move apart. Wigner used this result to obtain threshold laws for two particles escaping from each other.

Wannier (1953) extended Wigner's theory to single ionization of atoms or ions by electrons. He considered the division of configuration space into three regions. In analogy with Wigner's analysis, it is not necessary to know the detailed behavior of the electrons in the reaction zone ($R < R_0$), but only to assume that the distribution in phase space of the two escaping electrons is uniform (i.e., quasiergodic) as they enter the Coulomb zone ($R_0 < R < R_1$). Wannier then assumed that for large enough R_0 the Coulomb potential varied sufficiently slowly for classical mechanics to be valid in the Coulomb zone even when the total energy $E = E_A + E_B$ tends to zero. Finally, at very large $R > R_1$, the combined kinetic energies of the electrons is larger than the potential energy, and hence the electrons move freely. As $E \to 0$, the free zone recedes and the Coulomb zone extends to infinity. The threshold behavior of the ionization cross section is therefore determined by the motion of the two electrons in the Coulomb zone.

Wannier showed that the threshold ionization cross section for hydrogenlike ions is dominated by the region in the neighborhood of a saddle point in the potential energy surface at $\alpha = \pi/4$ and $\theta_{12} = \pi$, where $\tan \alpha = r_1/r_2$ and θ_{12} is the angle between the two electron vectors $\mathbf{r}_1$ and $\mathbf{r}_2$. By expanding the potential around this saddle point, he showed that for total angular momentum $L = 0$, the threshold law for the total ionization cross section is given by

$$\sigma_{\text{ion}} \sim E^m \tag{8.65}$$

where

$$m = -\frac{1}{4} + \frac{1}{4}\left(\frac{100Z - 9}{4Z - 1}\right)^{1/2} \tag{8.66}$$

When $Z = 1$, then $m = 1.127$; as $Z \to \infty$, $m \to 1$. A further important result, obtained by Vinkalns and Gailitis (1967), is that the width of the angular distribution in θ_{12} is

$$\Delta \theta_{12} \sim E^{1/4} \tag{8.67}$$

with a maximum at $\theta_{12} = \pi$. Hence the angular distribution becomes more concentrated as $E \to 0$.

Greene and Rau (1982, 1983) extended the Wannier theory to other $LS\pi$ values. They showed that in all states, except for $^3S^e$ and $^1P^e$, the wave function can remain finite at the saddle point, and hence the Wannier threshold law will pertain. In the $^3S^e$ and $^1P^e$ states, on the other hand, the symmetry of the wave function forces a node to occur at the saddle point, thereby giving rise to an exponent of $3m$ rather than m in Eq. (8.65).

Figure 8.27 shows the cross section and the spin asymmetry for electron impact ionization of atomic hydrogen in the near-threshold regime. The PECS results of Bartlett (2006) are compared to a variety of other theoretical and experimental results. As seen from the figure, the PECS theory agrees particularly well with the analytical threshold law and the experimental data. This is mostly due to the fact that the wave function can be integrated out in a stable way to very large distances.

8.3.4.3 Ionization at Intermediate and High Energies

The integral expression (8.53) for the direct ionization amplitude provides a starting point for the calculation of the cross section when some approximation is made for $\Psi_i^{(+)}$. The simplest approximation is to represent $\Psi_i^{(+)}$ by the incident plane-wave term. This gives a variety of first Born approximations, depending on the choice of phase in Eq. (8.60). However, this approximation can only be expected to be accurate at high energies.

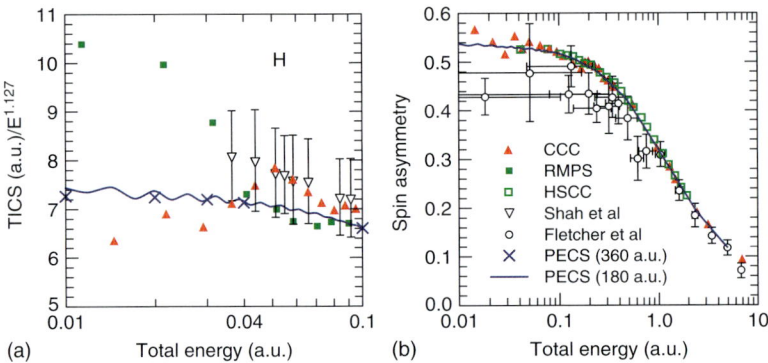

Fig. 8.27 Cross section (a) and spin asymmetry (b) for electron impact ionization of atomic hydrogen in the near-threshold regime. The PECS results of Bartlett (2006) are compared to a variety of other theoretical and experimental results. (Adapted from Bartlett (2006)).

To obtain cross sections that are reliable at lower energies, more sophisticated approximations must be made for $\Psi_i^{(+)}$. For example, $\Psi_i^{(+)}$ in Eq. (8.53) has been represented by coupled eigenstate plus pseudostate expansions of the form of Eq. (8.32). This is the basic idea behind the highly successful application of the CCC method to ionization processes, and it could be used in the RMPS and IERM methods as well. Alternatively, the ionization cross section can be obtained directly from the asymptotic form of $\Psi_i^{(+)}$. This is done, for example, in the TDCC and ECS approaches mentioned above.

Another approximation is a hybrid method introduced by Jakubowicz and Moores (1981) and later implemented into an R-matrix framework by Bartschat and Burke (1987). Starting from Eq. (8.53), the N-electron target in the initial state and the (N − 1)-electron residual ion plus ejected electron in the final state are both expanded as in Eq. (8.32), whereas the ionizing electron is represented by plane or distorted waves in both the initial and final states. This approximation is appropriate for situations in which the ejected electron moves much slower than the ionizing electron, and it becomes increasingly accurate at all energies for ionization of ions as Z increases. After further development, it has become possible to account, at least approximately, for second-order effects in the Projectile–target interaction (Reid, Bartschat, and Raeker (1998)) and to represent the initial bound-state and the ejected-electron–residual-ion interaction by a convergent R-matrix with pseudostates expansion. As seen in Figure 8.13, the method has been fairly successful in describing simultaneous ionization plus excitation of He and other quasi-two-electron systems. Bartschat and Vorov (2005) also applied it with some success to direct ionization of the outermost p and s electrons in heavy noble gases. However, the most recent experimental data of Stevenson and Lohmann (2008) suggest that further improvements are necessary to obtain good agreement with experiment over the full angular range.

At high energies, methods based on the plane-wave Born series become valid. The TDCS for e − He in Figure 8.9

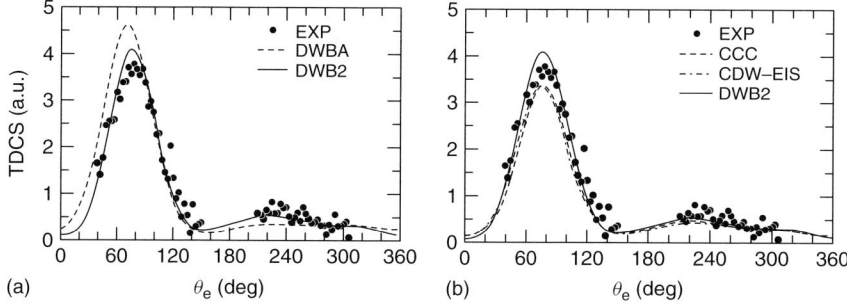

Fig. 8.28 The triple-differential cross section (in atomic units) for atomic hydrogen as a function of the ejected electron angle θ_e for an incident electron energy of 250 eV, an ejected electron energy of 5 eV, and a fixed detection angle of 8° for the fast electron. The dots are experimental data of Ehrhardt *et al.* (1986), while the various theoretical curves represent first-order and second-order B1 and B2 DW Born approximations, CCC and continuum distorted-wave plus eikonal-initial-state (CDW-EIS) model. (Adapted from Chen *et al.* (2004)).

shows that the second Born approximation correctly predicts the change in height and shift of the binary and recoil peaks from that given by the first Born approximation.

Replacing the plane waves by distorted waves allows for the Born-type methods to reach lower energies. This is shown in Figure 8.28 for 250 eV electron-impact ionization of atomic hydrogen.

8.3.4.4 Excitation–Autoionization

Excitation–autoionization discussed in Section 8.2.4.7 can be described using the hybrid method described above. In this case, it is necessary that the relevant resonant states are included in the expansion describing the scattering of the ejected electron from the $(N-1)$-electron residual ion. These resonances occur in Rydberg series converging to the excited states of the residual ion. However, this theory is appropriate only if the scattered electron is sufficiently fast that it can be adequately described by a distorted wave. In particular, this theory will fail if the incident electron forms a resonant state with the target ion as in Eq. (8.20).

When the incident electron interacts strongly with the target, it is often possible to obtain an accurate estimate of this cross section by assuming that the two processes defined by Eqs. (8.18) and (8.19) occur independently and can thus be calculated separately. This is equivalent to assuming that the lifetimes of the autoionizing states are large compared with their excitation time, and hence that these states can be treated as if they were bound states lying in the continuum. The total wave function describing the excitation process is then expanded as in Eq. (8.32), where the first summation includes the initial state of the target as well as all important autoionizing states. The ionization cross section is then obtained by solving the coupled integro-differential equations (8.33) either directly or by using a DW approximation and then summing the excitation cross section to all autoionizing states that decay into the continuum. This theory also describes the REDA process, because the intermediate states in Eq. (8.20) appear as resonances in the scattering process defined by Eq. (8.18). This approach has been widely used to obtain excitation–autoionization cross sections for complex ions (Müller, 1990).

8.3.5
Scattering by Molecules

The processes that occur in electron scattering by molecules are more varied than those that occur in electron scattering by atoms and ions because of the possibility of exciting degrees of freedom associated with the motion of the nuclei as well as electronic degrees of freedom. In addition, the multicentered and nonspherical nature of the electron–molecule interaction considerably complicates the solution of the scattering problem by reducing its symmetry.

There are two other distinctive features of electron–molecule scattering. First, there is the crucial role that resonances play in vibrational excitation and dissociative attachment (Hotop et al., 2003). For example, it is mainly near resonances that the scattered electron spends sufficient time in the neighborhood of the molecule to be able to excite the nuclear motion with significant probability. Second, the long-range interaction between an electron and a molecule is more complicated than that between an electron and an atom. Thus, for a linear molecule, the potential has the asymptotic form

$$V(r, \widehat{R}) \xrightarrow[r \to \infty]{} -\frac{\mu}{r^2} P_1(\cos\theta)$$
$$-\frac{Q}{r^3} P_2 \times (\cos\theta) - \frac{\alpha_0}{2r^4}$$
$$-\frac{\alpha_2}{2r^4} P_2(\cos\theta) \quad (8.68)$$

where $\widehat{R}$ is a unit vector along the internuclear axis, r is the coordinate of the scattered electron with respect to the center of gravity of the molecule, and $\cos\theta = \widehat{r} \cdot \widehat{R}$. For polar molecules, the dipole moment μ is nonzero and the corresponding term in Eq. (8.68) dominates the angular distribution in the forward direction and gives rise to bound and virtual states that often lead to threshold peaks in the vibrational excitation cross sections. Furthermore, the long-range nature of the quadrupole (Q) and polarization (α_0 and α_2) terms in the potential lead to enhanced rotational excitation cross sections.

8.3.5.1 Laboratory Frame Representation

We first consider the derivation of the basic equations describing electron–molecule scattering in the laboratory frame of reference. Our discussion closely parallels that given in Section 8.3.1.1 for electron–atom scattering. The Schrödinger equation describing the electron–molecule system is

$$(H_m + K + V)\Psi = E\Psi \quad (8.69)$$

where H_m is the Hamiltonian for the target molecule, K is the kinetic energy of the scattered electron, and V is the electron–molecule interaction potential

$$V(R, r_m, r) = \sum_j \frac{1}{|r - r_j|} - \sum_j \frac{Z_j}{|r - R_j|}$$
$$(8.70)$$

Here R stands symbolically for the positions R_j of all the nuclei, r_m for the set of coordinates r_j of the electrons in the target molecule, and r for the coordinate of the scattered electron. The total energy E refers to the frame of reference where the center of mass of the whole system is at rest.

We again introduce the target eigenstates and possible pseudostates Φ_i by the equation

$$\langle \Phi_i | H_m | \Phi_j \rangle = w_i \delta_{ij} \tag{8.71}$$

and then expand the total wave function Ψ, in analogy with Eq. (8.32) for atoms and ions, in the form

$$\Psi_i = \mathcal{A} \sum_j \Phi_j(\mathbf{R}, \mathbf{r}_m) \mathcal{F}_{ji}(\mathbf{r})$$

$$+ \sum_j \chi_j(\mathbf{R}, \mathbf{r}_m, \mathbf{r}) \alpha_{ji} \tag{8.72}$$

Here, we have suppressed the spin variables for notational simplicity and have not carried out a partial-wave decomposition of the wave function $\mathcal{F}_{ji}$ representing the scattered electron. The subscripts i and j now represent the ro-vibrational and electronic states of the target molecule.

Coupled equations for the functions $\mathcal{F}_{ji}$ can be obtained by substituting expansion (8.72) into Eq. (8.69) and projecting onto the target states Φ_i and the square-integrable function χ_j. After eliminating the coefficients α_{ji}, we obtain the following coupled integro-differential equations that are satisfied by the functions $\mathcal{F}_{ji}$:

$$(\nabla^2 + k_j^2) \mathcal{F}_{ji}(\mathbf{r})$$
$$= 2 \sum_l (V_{jl} + W_{jl} + X_{jl}) \mathcal{F}_{li}(\mathbf{r}) \tag{8.73}$$

Here

$$k_j^2 = 2(E - w_j) \tag{8.74}$$

and V_{jl}, W_{jl}, and X_{jl} are again the direct, nonlocal exchange, and nonlocal correlation potentials. By expanding the $\mathcal{F}_{ji}$ in partial waves we obtain a set of coupled radial integro-differential equations, analogous to Eqs. (8.33) for atoms and ions, which can, in principle, be solved numerically.

The scattering amplitude and cross section can be obtained from the asymptotic form of the total wave function, which is written in a form analogous to Eq. (8.29). The differential cross section for a transition from an initial state $|\mathbf{k}_i, \Phi_i, \chi_m\rangle$ to a final state $|\mathbf{k}_j, \Phi_j, \chi_m\rangle$ is then given by Eq. (8.31), except that the scattering amplitude $f_{ji}(\theta, \phi)$ now represents transitions between ro-vibrational states in addition to electronic states.

8.3.5.2 Molecular Frame Representation

The theory described in the previous section is completely general and has been the basis of a number of early calculations for light diatomic molecules. However, a major difficulty arises because of the very large number of ro-vibrational and electronic channels that need to be included in expansion (8.72) for all but the simplest low-energy calculations.

This difficulty can, to a large extent, be overcome by adopting an approximation similar to that used in molecular bound-state calculations, where a Born-Oppenheimer separation of the electronic and nuclear motion is made. The electronic motion is first determined with the nuclei held fixed. This is referred to as the *fixed-nuclei approximation*. The molecular rotational and vibrational motion is then included in a second step in the calculation. The entire procedure is called the *adiabatic-nuclei approximation*. It owes its validity to the large ratio of the nuclear mass to the electronic mass

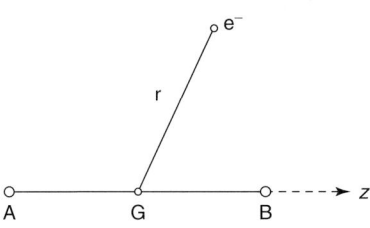

Fig. 8.29 Molecular frame for a diatomic molecule.

and can be adopted in electron–molecule scattering processes when the collision time is much shorter than the period of molecular rotation and/or molecular vibration. The approximation is thus expected to be valid when the scattered electron energy is not close to threshold or when the energy does not coincide with that of a narrow resonance. In these cases, further developments are needed to obtain reliable cross sections.

The fixed-nuclei approximation is not new, having been used in the 1930s to describe the scattering of electrons by homonuclear diatomic molecules by Stier (1932) and Fisk (1936). However, since the early 1970s, it has gained general acceptance as the basis of *ab initio* computational methods that are yielding the most accurate cross sections for complex molecules.

To formulate the collision in this representation, we adopt a frame of reference that is rigidly attached to the molecule, as illustrated in Figure 8.29 for a diatomic molecule. In this figure, A and B are the two nuclei, G is the center of gravity that is taken as the origin of coordinates, and the z-axis is chosen along the internuclear axis.

The fixed-nuclei approximation starts from the Schrödinger equation

$$(H_{el} + K + V)\psi = E\psi \quad (8.75)$$

where H_{el} is the electronic part of the target Hamiltonian defined by assuming that the target nuclei have fixed coordinates denoted by $\boldsymbol{R}$. It follows that H_{el} is related to H_m in Eq. (8.69) by

$$H_m = H_{el} + K_R \quad (8.76)$$

where K_R is the kinetic energy operator for the rotational and vibrational motion of the nuclei. The remaining quantities K and V are the same as in Eq. (8.69).

The solution of Eq. (8.75) proceeds in an analogous way to the solution of Eq. (8.26) for electron–atom or electron–ion scattering. We adopt an expansion similar to Eq. (8.32), where we now expand the function representing the motion of the scattered electron in terms of symmetry-adapted angular functions that transform as an appropriate irreducible representation (IRR) of the molecular point group (Burke, Chandra and Gianturco, 1972). Substituting this expansion into Eq. (8.75) and projecting onto the corresponding channel functions and the square-integrable functions yields the coupled integro-differential equations

$$\left(\frac{d^2}{dr^2} - \frac{l_j(l_j+1)}{r^2} + k_j^2\right) F_{ji}^{\Delta}(r)$$
$$= 2\sum_l (V_{jl}^{\Delta} + W_{jl}^{\Delta} + X_{jl}^{\Delta}) F_{li}^{\Delta}(r) \quad (8.77)$$

Now the channel indices i, j, and l represent the component of the IRR as well as the electronic state of the target, Δ refers to the conserved quantum numbers that include the IRR and the total spin, l_j is the angular momentum of the scattered electron, and

$$k_j^2 = 2(E - \varepsilon_j) \qquad (8.78)$$

Finally, V_{jl}^Δ, W_{jl}^Δ, and X_{jl}^Δ are the direct, nonlocal exchange, and nonlocal correlation potentials that depend parametrically on the internuclear coordinates $\mathbf{R}$.

The final step is to solve Eqs. (8.77), for each set of coordinates $\mathbf{R}$, subject to K-matrix boundary conditions analogous to Eqs. (8.36). The scattering matrix in the molecular frame of reference, and hence the cross section, can then be obtained in a straightforward way.

A number of general computer-program packages have been developed to solve Eqs. (8.77). As for electron–atom or electron–ion scattering, these include packages based on the reduction of these equations to linear algebraic equations, on the Schwinger variational method, on the R-matrix method, and on the complex Kohn method. All of these approaches use standard quantum-chemistry structure packages that have been appropriately modified for scattering. More details about the methods can be found in the book edited by Huo and Gianturco (1995).

For scattering calculations involving only the ground electronic state, a number of approaches have been developed that replace the nonlocal exchange and correlation potentials by local potentials and hence make the resultant equation easier to solve. See, for example, Gianturco and Jain (1986), Gianturco and Lucchese (2004), or Tonzani and Greene (2005).

8.3.5.3 Inclusion of the Nuclear Motion

We now discuss how observables involving the nuclear motion, such as rotational and vibrational excitation cross sections and dissociative attachment cross sections, can be obtained from the solutions of the fixed-nuclei equations.

One of the most widely used approaches is the adiabatic-nuclei approximation proposed by Drozdov (1955, 1956) and by Chase (1956) in studies of neutron scattering by nuclei. As an illustration of this approach, we write the relevant scattering amplitude for a diatomic molecule in a $^1\Sigma$ state as

$$A_{ivjm,i'v'j'm'}(\widehat{\mathbf{k}}, \widehat{\mathbf{r}}) =$$
$$\langle X_{iv}(R) Y_{jm}(\widehat{\mathbf{R}}) \times |A_{ii'}''(\widehat{\mathbf{k}} \cdot \widehat{\mathbf{r}}; R)|$$
$$\times X_{i'v'}(R) Y_{j'm'}(\widehat{\mathbf{R}}) \rangle \qquad (8.79)$$

where $A_{ii'}''(\widehat{\mathbf{k}} \cdot \widehat{\mathbf{r}}; R)$ is the fixed-nuclei scattering amplitude obtained by solving Eqs. (8.77) and $X_{iv}(R)$ and $Y_{jm}(\widehat{\mathbf{R}})$ are the molecular vibrational and rotational eigenfunctions. This approximation is valid provided that the collision time is short compared with the vibration and/or rotation time. It has been widely used for nonresonant scattering away from threshold.

The adiabatic-nuclei approximation breaks down close to threshold or in the neighborhood of narrow resonances as discussed by Morrison (1988). A rather straightforward way of including nonadiabatic effects that arise in vibrational excitation is to retain the vibrational terms in the Hamiltonian but still to treat the rotational motion adiabatically. Hence, instead of Eq. (8.75), we solve the equation

$$(H_{\text{el}} + K_{\text{vib}} + K + V)\widetilde{\psi} = E\widetilde{\psi} \qquad (8.80)$$

Here K_{vib} is the kinetic energy operator for the nuclear vibrational motion, while the other quantities have the same meaning as in Eq. (8.75). Adopting a frame of reference in which the molecule has fixed spatial orientation, and separating out the angular variables of the scattered electron, we can derive a set of integro-differential equations coupling the target vibrational as well as electronic eigenstates. This approach has been used with success by several authors, but it is computationally demanding because the number of coupled channels can become very large.

Earlier, we noted that vibrational excitation and dissociative attachment are particularly important in resonance regions. As a result, a number of approaches have been developed for describing these processes based on electron–molecule resonance theories (Herzenberg (1984), Domcke (1991)). The basic idea is to introduce a series of fixed-nuclei resonance states ψ_n^r for a range of values of R either by imposing Siegert (1939) resonance boundary conditions or by introducing Feshbach projection operators. The amplitude for a transition from an initial electronic-vibrational state iv to a final state $i'v'$ is then given by

$$T_{iv,i'v'} = \sum_n \langle \chi_{iv}(R)\zeta_{ni}(R)|G_n^r(R,R')| $$
$$\times \zeta_{ni'}(R')\chi_{i'v'}(R')\rangle \quad (8.81)$$

where $\chi_{iv}(R)$ are the vibrational eigenfunctions, $\zeta_{ni}(R)$ are the "entry amplitudes" from the resonance states ψ_n^r into the initial or final electronic states, and $G_n^r(R,R')$ are the Green's functions that describe the propagation in the intermediate resonance states ψ_n^r.

We now consider the frame-transformation theory of Chang and Fano (1972) which has been influential in electron–atom scattering as well as laying the foundations of work on nuclear motion effects in electron–molecule scattering. Following earlier work by Fano (1970), who considered the extension of the quantum defect theory of Seaton (1958, 1983) to molecular photoabsorption, Chang and Fano pointed out that the electron–molecule interaction exhibits qualitatively different physical features when the distance r of the scattered electron from the center of gravity of the molecule lies in different regions.

We illustrate the main features of frame-transformation theory in the upper part of Figure 8.30. We see in this figure that configuration space is partitioned into two main regions labeled A and B. If vibrational motion is not considered, then in region A, where $r \leq a_2$, the molecular frame of reference can be used to describe the scattering process and the electron–target atom complex can be described as in molecular bound-state calculations. However, at larger distances in region B, where $r > a_2$, the coupling of the scattered electron to the molecular axis no longer dominates and the laboratory frame of reference is appropriate. In this case, the rotational quantum number j of the molecule and the orbital and spin quantum numbers (ℓ, s) of the scattered electron label the scattering channels, where the total angular momentum $\boldsymbol{J} = \boldsymbol{j} + \boldsymbol{\ell} + \boldsymbol{s}$ is conserved in the collision. When the vibrational motion of the molecule is being considered, region A must be partitioned into two subregions Aa and Ab. In subregion Aa, where $r \leq a_1$, the scattered electron spends only a short time compared with the vibration time and

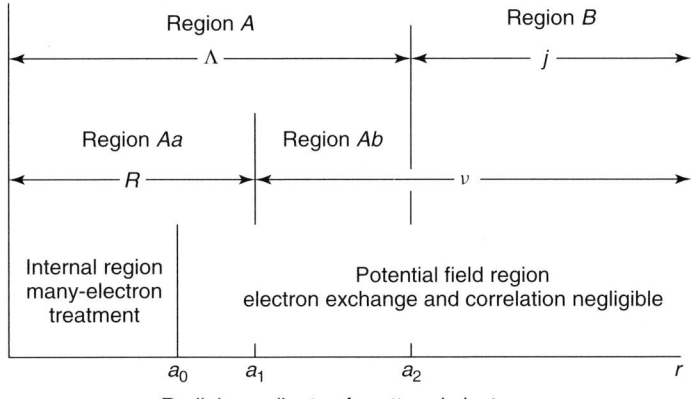

Fig. 8.30 The upper part of the figure shows the partitioning of configuration space in frame-transformation theory of electron–molecule scattering with the quantum numbers relevant to different ranges of the electron–molecule distance r. The lower part of the figure shows the partitioning of configuration space adopted in electron–molecule scattering calculations.

hence the Born-Oppenheimer separation of the nuclear and electronic motion can be adopted and the collision wave function can be determined as a function of the internuclear coordinate R. On the other hand, in regions Ab and B, where $r > a_1$, the vibrational motion must be included nonadiabatically and the vibrational channels included explicitly in the expansion of the electron–molecule scattering wave function.

In the lower part of Figure 8.30 we compare the frame-transformation partitioning of configuration space with the partitioning of configuration space adopted in electron–molecule scattering calculations using one of the general computer-program packages discussed in Section 8.3.5.2. In this case, the subdivision is determined by the need to include electron–electron exchange and correlation effects in an internal region $r \leq a_0$ where the scattered electron penetrates the electronic cloud corresponding to the target states of interest. A multicenter configuration-interaction expansion of the total electron–molecule wave function in the molecular frame of reference can be adopted in this region. However, when $r > a_0$, the scattered electron lies outside the target electronic cloud and moves in the long-range local multipole potential of the residual molecule, including molecular polarization terms, which can be conveniently represented by a single-center expansion. If the rotational and vibrational motions of the molecule are being considered, a transformation from the molecular frame of reference to the laboratory frame of reference is applied to the single-center wave function at a_0, a_1, or a_2 as appropriate. In addition, we observe that, although a_0 is shown to be less than a_1 in Figure 8.30, this is not necessarily always the case. In particular, in applications where electron collisions with excited molecular states are considered, it may be necessary for a_0 to be greater than a_1 to

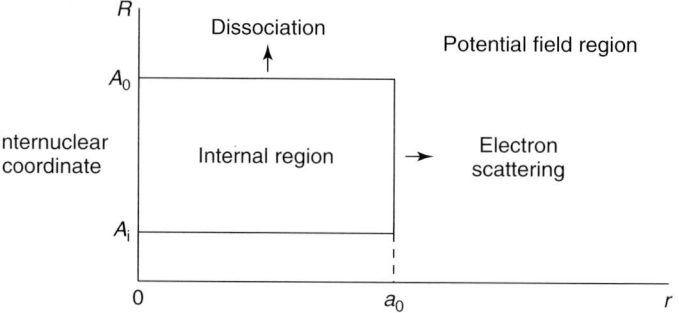

Fig. 8.31 Partitioning of configuration space in electron–molecule scattering calculations allowing for dissociative processes.

include electron exchange and correlation effects arising from the long-range tail of the wave functions corresponding to excited states. The solutions in the internal region of the potential field region can then be determined and matched on the boundary $r = a_0$, using one of the general electron–molecule computer packages discussed in Section 8.3.5.2.

The theory must be further extended to enable dissociative processes as well as rotational, vibrational, and electronic processes to be calculated. In the case of diatomic molecules, we consider the following processes:

scattering calculations allowing for dissociative processes, which generalizes the lower part of Figure 8.30, is shown in Figure 8.31. The internal region is taken to be a rectangle defined by $0 \leq r \leq a_0$ and $A_i \leq R \leq A_0$, where a_0 is defined as in Figure 8.30, A_i is chosen to exclude the nuclear Coulomb repulsion singularity at $R = 0$ where the wave function describing the nuclear motion is negligible, and A_0 is chosen so that the target vibrational states of interest have negligible amplitude for $R > A_0$. For $r > a_0$ the collision complex separates into an electron plus a residual molecule, which may be

$$e^- + AB_{iv} \longrightarrow \begin{cases} AB_{i'v'} + e^- & \text{vibronic excitation} \\ A_{i'} + B_{i''}^- & \text{dissociative attachment/recombination} \\ A_{i'} + B_{i''} + e^- & \text{dissociation} \end{cases} \quad (8.82)$$

where i, i' and i'' label the electronic states, v and v' label the vibrational states and where, for notational convenience, we have omitted the rotational quantum numbers. The partitioning of configuration space adopted in electron–molecule

rotationally, vibrationally, and electronically excited. For $R > A_0$ the molecule separates into an atom plus negative ion or into two atoms corresponding to dissociative attachment or dissociation. The solutions in the internal region and the two external regions are then matched on the boundaries between them. In the internal region, the calculation can be carried out in the molecular frame using one of the general computer program packages discussed in Section 8.3.5.2, while the calculations for the external regions can be carried out in the laboratory frame. The corresponding procedure using R-matrix theory was introduced by Schneider, Le Dourneuf and Burke (1979); Schneider, Le Dourneuf and Vo Ky Lan (1979) and has been reviewed by Burke and Tennyson (2005).

The extension by Fano (1970) of MQDT to molecular photoabsorption has led to an increasing interest in recent years in its application to the study of electron–molecule scattering and molecular photoionization and dissociation processes. As we saw in Figures. 8.30 and 8.31, configuration space can be partitioned into different regions depending on the distance of the electron from the molecule or the distance apart of the dissociating atoms or ions. In each of these regions the collision process can be analyzed in a frame of reference that reflects the physical characteristics of the region, for example the relative importance of electron–electron exchange and the relative motion of the electron and nuclear components. This enables the calculation and analysis of the process to be carried out in each region using the most appropriate frame of reference and, using MQDT, allows for the S-matrix describing the scattering process to be expressed in terms of analytic functions of the energy, thereby enabling it to be continued through thresholds as discussed in Section 8.3.1.5.

Following the application by Fano (1970) of MQDT to the high-resolution photoabsorption spectrum of H_2 near threshold, observed by Herzberg (1969), Jungen and Atabec (1977) developed and applied the method to rovibronic interactions in the photoabsorption spectrum of H_2 and D_2. This work was later extended by Jungen and Dill (1980) to treat rotational and vibrational preionization channels of H_2, obtaining good agreement with photoionization data of Dehmer and Chupka (1976). Further theoretical developments were made by Giusti (1980), Giusti-Suzor and Jungen (1984), Jungen (1984), and Stephens and Greene (1995). A review of molecular applications of MQDT was written by Greene and Jungen (1985). An *ab initio* study of vibrational inelastic processes in H_2 scattering below 5 eV was carried out by Robicheaux (1991), using the frame-transformation method proposed by Greene and Jungen (1985). Also, a noniterative eigenchannel R-matrix approach combined with MQDT was developed by Gao, Jungen and Greene (1993). It was applied to predissociation of H_2 in the $3p\pi D\ ^1\Pi_u^+$ state, and a unified MQDT treatment of both molecular ionization and dissociation was developed by Jungen and Ross (1997).

Finally, we briefly discuss recent theoretical dissociative recombination studies of the triatomic ion H_3^+ by Kokoouline and Greene (2003a, 2003b) and by dos Santos, Kokoouline and Greene (2007). Kokoouline and Greene (2003b, 2005) also considered dissociative recombination of the triatomic ions D_3^+ H_2D^+ and D_2H^+. As mentioned in Section 8.2.6, where complementary experimental studies from the Stockholm (CRYRING) and the Heidelberg (TSR) groups were discussed, dissociative recombination of the H_3^+ ion is a fundamental process in diffuse interstellar clouds. Also, as the simplest triatomic ion, detailed theoretical studies can be seen as a prototype for the study of electron scattering by more complex polyatomic atoms and ions.

In contrast to dissociative attachment/recombination in diatomic molecules and ions given by Eq. (8.82), there is an additional three-body dissociative pathway for H_3^+. Thus we have the following possibilities

$$e^- + H_3^+ \longrightarrow \begin{matrix} H_3^+ + e^- \\ H_2 + H \\ H + H + H \end{matrix} \qquad (8.83)$$

Here we have omitted the electronic, vibrational, and rotational labels, as well as processes in which H or H_2 are ionized. We see from Eq. (8.83) that dissociative recombination of H_3^+ is inherently a four-body problem. To represent the Jahn–Teller coupling between the electronic and vibrational motion, it is in principle necessary to take into account at least two degrees of freedom of vibrational motion. Finally, the third vibrational coordinate should be introduced to describe the dissociative channel.

The theoretical approach developed by Kokoouline and Greene (2003a, 2003b) combined MQDT to represent the closed channels, the adiabatic hyperspherical coordinate approach to represent the motion of the nuclei, which is discussed in detail by Kokoouline and Masnou-Seeuws (2006), and included outgoing-wave Siegert (1939) pseudostates to represent the vibrational continuum. The Siegert pseudostates, which are analogous to the pseudostates discussed in Section 8.3.1.1, are introduced to let dissociative flux escape if it reaches the hyper-radial boundary. In the later work by dos Santos, Kokoouline and Greene (2007), accurate vibrational wave functions were used and a larger number of possible rotational states of the H_3^+ ground state were included. This resulted in better agreement with the experimental storage-ring data shown in Figure 8.16. It was also shown that the Jahn–Teller coupling between the electronic and vibrational motion plays an important role in this process. In conclusion, this work has shown that recent state-of-the-art *ab initio* calculations on dissociative recombination in simple polyatomic molecules are now capable of accurately describing this complex process.

8.3.5.4 Illustrative Results for Molecules

In this section we present a few examples of recent work that illustrate what theory is capable of in this field. We will show a variety of angle-integrated and angle-differential cross sections, for elastic scattering, excitation, and finally ionization.

To begin with, Figure 8.32 shows the angle-differential cross section for vibrationally elastic electron scattering from water molecules at an energy of 4 eV. There is very good agreement between the predictions of Faure, Gorfinkiel and Tennyson (2004) and the experimental data of Cho, Lee and Park (2003). The theoretical results were obtained by the *R*-matrix method within the fixed-nuclei approximation. To compare directly with experiment, they represent the sum over rotational excitations.

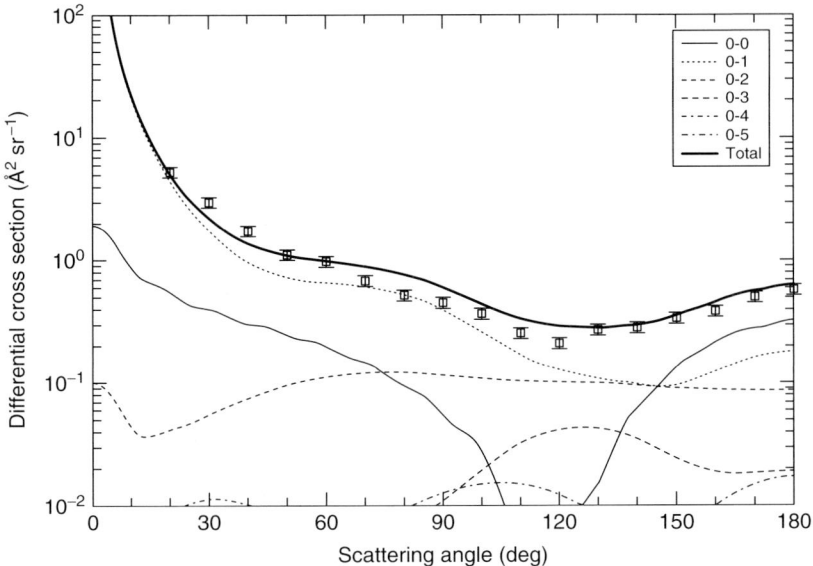

Fig. 8.32 Angle-differential cross section for vibrationally elastic electron scattering from water at an incident electron energy of 4 eV. The thick solid line is the rotationally summed result of Faure, Gorfinkiel and Tennyson (2004) while the squares represent the experimental data of Cho, Lee and Park (2003). Also shown are partial cross sections for individual rotational transitions. (From Faure, Gorfinkiel and Tennyson (2004).)

Next, Figure 8.33 exhibits results for elastic scattering and electron-induced vibrational excitation of NO. For this problem, Trevisan et al. (2005) employed a nonlocal treatment of the nuclear dynamics, which significantly improved agreement with experiment compared to results obtained in earlier treatments with a local complex potential. Nevertheless, the agreement between theory and experiment is still far from perfect, indicating the remaining difficulties even for diatomic molecules. An extensive review on the subject was published by Brunger and Buckman (2002).

The situation becomes even more complicated for polyatomic molecules. Most calculations are restricted to elastic scattering. Examples include elastic electron scattering from C_2H_4 by Winstead, McKoy and Bettega (2005) and very large molecules such as uracil treated by Gianturco and Lucchese (2004), by Tonzani and Greene (2005), and by Winstead and McKoy (2006). Examples of recent experimental work on molecules of biological interest include the papers by Allan (2007) on tetrahydrofuran and by Khakoo et al. (2008) on methanol amd ethanol.

Figure 8.34 exhibits the challenges associated with electron-impact excitation of electronic states in di-atomic and even more so poly–atomicmolecules. Both experiment and theory are very difficult. For the CO transition, there is qualitative agreement between theory and experiment, whereas the predictions and the measurements are orders of magnitude apart for the $\tilde{B}^1A_1$

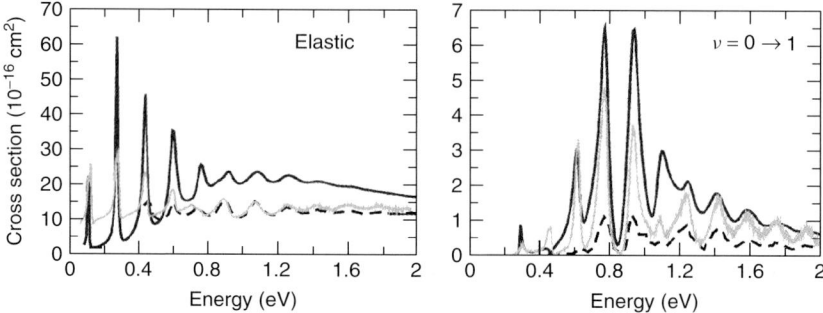

Fig. 8.33 Angle-integrated cross section for elastic scattering and excitation of the $v = 0 \rightarrow 1$ transition in NO. The theoretical predictions of Trevisan et al. (2005) (dark solid line) are compared with two sets of experimental data by Jelisavcic, Panajotovic and Buckman (2003) and Allan (2005). (Adapted from Trevisan et al. (2005).)

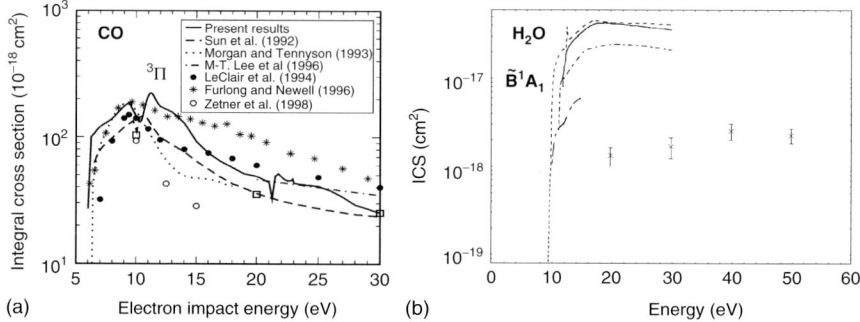

Fig. 8.34 Angle-integrated cross section for electron-impact excitation of the $^3\Pi$ state of CO (a) and the $\tilde{B}^1A_1$ state of water (b). (Adapted from da Costa et al. (2007) (CO) and Thorn et al. (2007) (H_2O).)

state in water. While there are obvious problems in reconciling the experimental and theoretical magnitudes, Gil et al. (1994) achieved impressive qualitative agreement with the (relative) electron energy loss spectra measured at a few fixed angles by Trajmar, Williams and Kuppermann (1973) for water. Further progress was made recently in the fully *ab initio* R-matrix calculation on methanol reported by Bouchiha et al. (2007). Good agreement with experiment was achieved regarding the total (elastic plus excitation) cross section. Unfortunately, no experimental data are available to test the individual predictions for the excitation processes, including some Feshbach resonances.

For optically allowed transitions, the semiempirical "Binary Encounter f-scaling" method of Kim (2001) often allows for a quick and accurate estimate of the excitation cross section. This is shown for excitation of CO in Figure 8.35.

Figure 8.36 shows the results of Gorfinkiel and Tennyson (2005) for ionization of H_2 by electron impact. These *ab initio* results were obtained using the molecular R-matrix with pseudostates

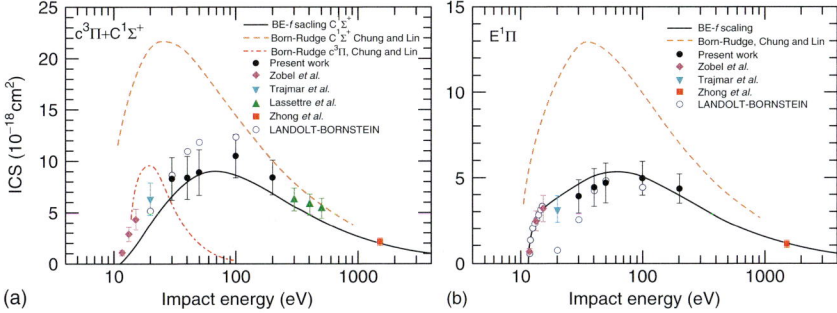

Fig. 8.35 Angle-integrated cross section for electron-impact excitation of the $C^1\Sigma + c^3\Pi$ (a) and the $E^1\Pi$ (b) states of CO. (Adapted from Kawahara et al. (2008).)

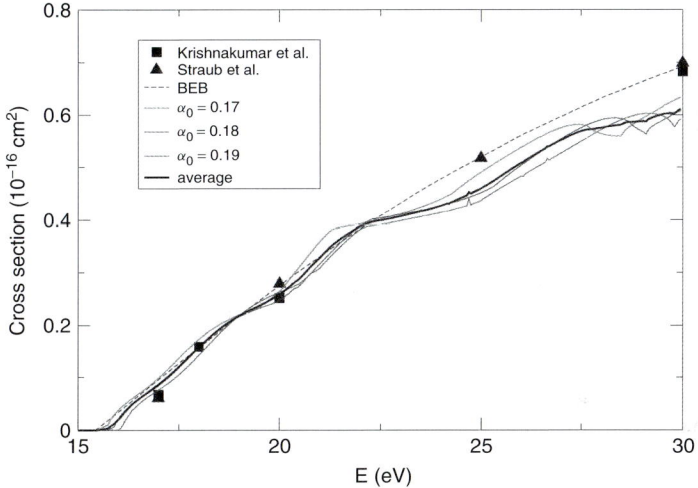

Fig. 8.36 Angle-integrated cross section for electron-impact ionization of H_2. The various values of α_0 refer to the basis used. See Gorfinkiel and Tennyson (2005) for details.

(MRMPS) method. Clearly, the results for the ionization problem are very encouraging. On the other hand, the very same model predicted a resonance structure in the excitation cross section that was not seen in most other calculations, nor in the few available experimental data.

We finish this section with the triple-differential cross section of electron-impact ionization of N_2. For this, and certainly for large molecules, such calculations are extremely difficult due to the complex molecular structure and the need to average over the molecular orientations that are usually not resolved in experiments with molecules in the gas phase. To address this problem, Gao, Madison and Peacher (2005a) suggested the "orientation averaged molecular orbital" (OAMA) approach. The major advantage of the method is the fact that it

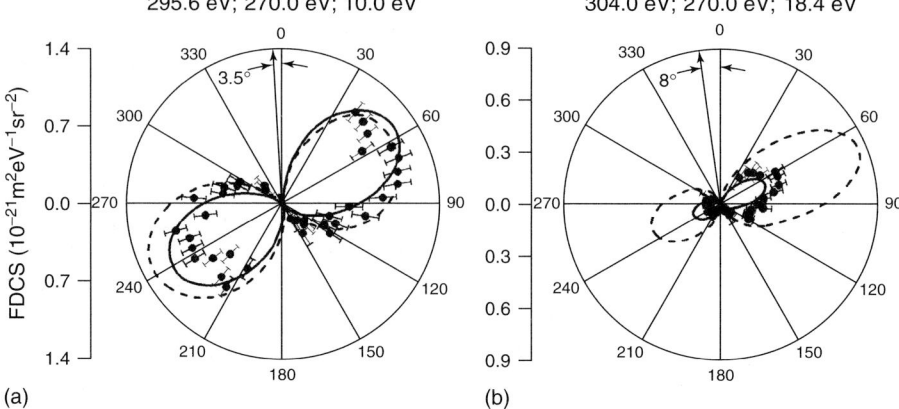

Fig. 8.37 Fully differential cross section for electron-impact ionization of the $3\sigma_g$ state in N_2 for coplanar asymmetric scattering. The incident electron energy is 295.6 eV (a) and 304.0 eV (b). The faster of the two outgoing electrons with energy 270 eV is detected at a scattering angle $\theta_a = 3.5°$ (a) or $\theta_a = 8.0°$ (b), while the detection angle of the slower electron (10.0 eV or 18.4 eV) is varied. The absolute experimental data are from Avaldi *et al.* (1992). The solid and dashed lines represent the M3DW and DWBA results, respectively, see Section 8.3.5.4, normalized to the experiment at the single point at 295.6 eV where $\theta_a = 3.5°$ and $\theta_b = 68°$. (From Gao, Madison and Peacher (2005b).)

is adaptable to the DW theory developed for (e,2e) processes with atoms. However, note that it is restricted to molecular orbitals of appropriate symmetry.

Figure 8.37 shows results for electron-impact ionization of the $3\sigma_g$ state in N_2 for the coplanar asymmetric geometry. For this system, the Rolla group extended the orbital average method further and also included the proper asymptotic three-body Coulomb boundary condition. Comparison between their results obtained in the standard DWBA approach and their molecular 3DW (M3DW) method shows that both calculations reproduce in a satisfactory way the shape of the angular distributions. However, the experimental data of Avaldi *et al.* (1992) are cross-normalized, that is, only a single normalization factor is necessary (and allowed) to put the experimental data on an absolute scale. The figure shows that the M3DW is far superior to the DWBA in reproducing the proper cross normalization. Further applications, for example, to water Kaiser *et al.* (2007) showed some encouraging qualitative agreement between theory and experiment, although discrepancies remained that call for a more sophisticated theoretical treatment.

8.4
Conclusions and Future Work

We have presented an overview of electron scattering by atoms, ions and molecules both from an experimental and from a theoretical standpoint. These areas have seen major advances during the past decade, and our knowledge of these processes is now much more secure than only ten years ago. In particular, advances in experimental methods involving multihit coincidence techniques, high-resolution electron beams, the MAC, spin-polarized projectile and target beams, and the reaction microscope are now providing critical benchmark data against which theory can be tested.

However, much work needs to be done before a complete picture emerges. In the case of electron–atom scattering, there is still an incomplete understanding of collisions involving highly excited states. While a full quantal description of low-energy ionization has been achieved for the atomic hydrogen target and also for helium without excitation, this is not the case for more complex targets or for more correlated processes such as ionization with simultaneous excitation, double ionization, and out-of-plane ionization. Another interesting example is the need to develop a complete quantal theory of excitation–autoionization near threshold. More work is also required to fully understand collisions with heavy atoms and ions, where relativistic effects are important. Examples include fine-structure transitions in collisions of astrophysical importance, such as the $e - Fe^+$ system, and generally atomic and ionic targets with open d-shells. Important examples include e–Mo or e–W collisions for mercury-free alternative lighting concepts or understanding and diagnosing the effects of impurities in fusion devices.

Turning to electron–molecule scattering, we have, in addition to the problems mentioned above for atoms and ions, the need to account fully for the nuclear motion, particularly in the role that it plays during electronic excitation. Experimentally, the resolution of well-defined states will remain a challenge while exciting new information can be expected from studies with "fixed in space" nuclear axes. Also, while new computer-program packages and rapidly increasing computational power have allowed for some *ab initio* calculations on polyatomic molecules, going beyond elastic scattering from biological molecules remains a significant challenge for the future. Some *ab initio* calculations, in the fixed-nuclei approximation, have recently become possible. The inclusion of vibrational excitation in such calculations is the next grand challenge in this field.

It is also worth pointing out that many of the techniques originally developed for steady-state electron collisions with atoms, ions, molecules, and even surfaces are now being used to treat such collisions on ultrashort time scales, involving strong-field laser-atom interactions that lead to excitation, single and multiple ionization. For recent examples, see Guan et al., (2007); Guan, Bartschat and Schneider (2008), Feist et al., (2008), and references therein.

The need for electron–atom, electron–ion, and electron–molecule collision cross sections in many applications provides an important stimulus for experimental and theoretical studies of electron collisions. Important questions involving highly ionized ions in hot plasmas that are difficult to resolve otherwise have already been theoretically explained. However, there are still plenty of unresolved topics, particularly in low-temperature discharges involving neutral atomic and molecular species, where more research is needed before final answers can be given. These questions from researchers in adjacent fields will continue to provide the motivation for work in this field for many years to come.

References

Aflatooni, K., Gallup, G.A., and Burrow, P.D. (1998) *J. Phys. Chem. A*, **102**, 6205.

Allan, M. (2004) *Phys. Scripta*, **T110**, 161.

Allan, M. (2005) *J. Phys. B*, **38**, 603.

Allan, M. (2007) *J. Phys. B*, **40**, 3531.

Alt, E.O. and Mukhamedzhanov, A.M. (1993) *Phys. Rev. A*, **47**, 2004.

Andersen, N. and Bartschat, K. (1996) *Adv. At. Mol. Opt. Phys.*, **36**, 1.

Andersen, N. and Bartschat, K. (2001) *Polarization, Alignment and Orientation in Atomic Collisions*, Springer, New York.

Andersen, N., Bartschat, K., Broad, J.T., and Hertel, I.V. (1997) *Phys. Rep.*, **279**, 251.

Andersen, N., Gallagher, J.W., and Hertel, I.V. (1988) *Phys. Rep.*, **165**, 1.

Andrick, D. (1973) *Adv. Atom. Mol. Phys.*, **9**, 207.

Avaldi, L., Camilloni, R., Fainelli, E., and Stefani, G. (1992) *J. Phys. B*, **25**, 3551.

Bartlett, P.L. (2006) *J. Phys. B*, **39**, R379.

Bartlett, P.L. and Stelbovics, A.T. (2003) *Comp. Phys. Commun.*, **154**, 159.

Bartlett, P.L., Stelbovics, A.T., Lee, G.M., and Bray, I. (2005) *J. Phys. B*, **38**, L95.

Bartlett, P.L., Stelbovics, A.T., Rescigno, T.N., and McCurdy, C.W. (2007) *J. Phys.: Conf. Ser.*, **88**, 012011.

Bartschat, K. (1983) *Comp. Phys. Commun.*, **30**, 383.

Bartschat, K. (1989) *Phys. Rep.*, **180**, 1.

Bartschat, K. (1996) in *Computational Atomic Physics* (ed. K. Bartschat), chaps. 10 and 11, Springer, Heidelberg and New York.

Bartschat, K. and Blum, K. (1982) *Z. Phys. A*, **304**, 85.

Bartschat, K., Blum, K., Hanne, G.F., and Kessler, J. (1981) *J. Phys. B*, **14**, 3761.

(a) Bartschat, K., Bray, I., Fursa, D.V., Stelbovics, A.T. (2007) *Phys. Rev. A*, **76**, 024703; (b) corrigendum (2008) *Phys. Rev. A*, **77**, 029903.

Bartschat, K., Hudson, E.T., Scott, M.P., Burke, P.G., and Burke, V.M. (1996a) *J. Phys. B*, **29**, 2875.

Bartschat, K., Hudson, E.T., Scott, M.P., Burke, P.G., and Burke, V.M. (1996b) *J. Phys. B*, **29**, 115.

Bartschat, K., Bray, I., Burke, P.G., and Scott, M.P. (1996c) *J. Phys. B*, **29**, 5493.

Bartschat, K. and Burke, P.G. (1987) *J. Phys. B*, **20**, 3191.

Bartschat, K., Dasgupta, A., Petrov, G.M., Giuliani, J.L., and Pecharek, R.E. (2004a) *New J. Phys.*, **6**, 145.

Bartschat, K., Zatsarinny, O., Bray, I., Fursa, D.V., and Stelbovics, A.T. (2004b) *J. Phys. B*, **37**, 2617.

Bartschat, K., Dasgupta, A., and Madison, D.H. (2004) *Phys. Rev. A*, **69**, 062706.

Bartschat, K. and Fang, Y. (2000) *Phys. Rev. A*, **62**, 052719.

Bartschat, K. and Madison, D.H. (1987) *J. Phys. B*, **20**, 5839.

Bartschat, K. and Scott, N.S. (1983) *Comp. Phys. Commun.*, **30**, 369.

Bartschat, K., Scott, N.S., Blum, K., and Burke, P.G. (1984) *J. Phys. B*, **17**, 269.

Bartschat, K. and Vorov, V. (2005) *Phys. Rev. A*, **72**, 022728.

Bartschat, K., Weflen, D., and Guan, X. (2007) *J. Phys. B*, **40**, 3231.

Baum, G., Blask, W., Freienstein, P., Frost, L., Hesse, S., Raith, W., Rappolt, P., and Streun, M. (1992) *Phys. Rev. Lett.*, **69**, 3037.

Baum, G., Förster, S., Pavlovic, N., Roth, B., Bartschat, K., and Bray, I. (2004) *Phys. Rev. A*, **70**, 012707.

Baum, G., Granitza, B., Grau, L., Leuer, B., Raith, W., Rott, K., Tondera, M., and Witthuhn, B. (1993) *J. Phys. B*, **26**, 331.

Baum, G., Pavlovic, N., Roth, B., Bartschat, K., Fang, Y., and Bray, I. (2002) *Phys. Rev. A*, **66**, 022705.

Bederson, B. (1969) *Comm. Atom. Mol. Phys.*, **1**, 41 and 65.

Beijers, J.P.M., Madison, D.H., van Eck, J., and Heideman, H.G.M. (1987) *J. Phys. B*, **20**, 167.

Bellm, S., Lower, J., and Bartschat, K. (2006) *Phys. Rev. Lett.*, **96**, 223201.

Bellm, S., Lower, J., Stegen, Z., Madison, D.H., and Saha, H.P. (2008) *Phys. Rev. A*, **77**, 032722.

Berakdar, J. (1996) *Phys. Rev. A*, **53**, 2314.

Berakdar, J., Lahmam-Bennani, A., and Dal Cappello, C. (2003) *Phys. Rep.*, **374**, 91.

Berger, O. and Kessler, J. (1986) *J. Phys. B*, **19**, 3539.

Blum, K. (1996) *Density Matrix Theory and Applications*, 2nd edn, Plenum, New York.

Bömmels, J., Franz, K., Hoffmann, T.H., Gopalan, A., Zatsarinny, O., Bartschat, K., Ruf, M.-W., and Hotop, H. (2005) *Phys. Rev. A*, **71**, 012704.

Borovik, A., Zatsarinny, O., and Bartschat, K. (2008) *J. Phys. B*, **41**, 035206.

Bouchiha, D., Gorfinkiel, J.D., Caron, L.G., and Sanche, L. (2007) *J. Phys. B*, **40**, 1259.

Boudaiffa, B., Cloutier, P., Hunting, D., Huels, M.A., and Sanche, L. (2000) *Science*, **287**, 1658.

Brauner, M., Briggs, J.S., and Klar, H. (1989) *J. Phys. B*, **22**, 2265.

Bray, I. (1994) *Phys. Rev. A*, **49**, 1066.

Bray, I., Fursa, D.V., Kheifets, A.S., and Stelbovics, A.T. (2002) *J. Phys. B*, **35**, R117.

Bray, I. and Stelbovics, A.T. (1992) *Phys. Rev. Lett.*, **69**, 53.

Bray, I. and Stelbovics, A.T. (1995) *Adv. At. Mol. Phys.*, **35**, 209.

Broad, J.T. (1983) in *Electron-Atom and Electron-Molecule Collisions* (ed. J. Hinze), Plenum, New York, p. 91.

Brunger, M.J. and Buckman, S.J. (2002) *Phys. Rep.*, **357**, 215.

Brunger, M.J., Thorn, P.A., Campbell, L., Diakomichalis, N., Kato, H., Kawahara, H., Hoshino, M., Tanaka, H., and Kim, Y.-K. (2008) *Int. J. Mass Spectrom.*, **271**, 80.

Buckman, S.J., Hammond, P., Read, F.H., and King, G.C. (1983) *J. Phys. B*, **16**, 4219.

Burgess, A. (1964) *Astrophys. J.*, **139**, 776.

Burgess, A. (1965) *Ann. d'Astrophys.*, **28**, 774.

Burke, P.G. (1968) *Adv. Atom. Mol. Phys.*, **4**, 173.

Burke, P.G. (1985) in *Fundamental Processes in Atomic Collision Physics* (eds H. Kleinpoppen, J.S. Briggs, and H.O. Lutz), Plenum, New York, p. 51.

Burke, P.G. and Berrington, K.A. (1993) *Atomic and Molecular Processes: An R-Matrix Approach*, Institute of Physics Publishing, Bristol.

Burke, P.G., Chandra, N., and Gianturco, F.A. (1972) *J. Phys. B*, **5**, 2212.

Burke, P.G. and Crowe, A. (1993) in *Encyclopedia of Applied Physics*, Vol. 5, VCH Publishers, Weinheim and New York, p. 499.

Burke, P.G. and Eissner, W.B. (1983) in *Atoms in Astrophysics* (eds P.G. Burke, W.B. Eissner, D.G. Hummer, and I.C. Percival), Plenum, New York, p. 1.

Burke, P.G., Noble, C.J., and Burke, V.M. (2006) *Adv. At. Mol. Opt. Phys.*, **54**, 237.

Burke, P.G., Noble, C.J., and Scott, M.P. (1987) *Proc. Roy. Soc. A*, **410**, 289.

Burke, P.G. and Schey, H.M. (1962) *Phys. Rev.*, **126**, 147.

Burke, P.G. and Seaton, M.J. (1971) *Meth. Comp. Phys.*, **10**, 1.

Burke, P.G. and Tennyson, J. (2005) *Mol. Phys.*, **103**, 2537.

Burke, P.G. and Webb, T.G. (1970) *J. Phys. B*, **3**, L131.

Callaway, J. and Wooten, J.W. (1973) *Phys. Lett. A*, **45**, 85.

Cartwright, D.C., Csanak, G., Trajmar, S., and Register, D.F. (1992) *Phys. Rev. A*, **45**, 1602.

Catoire, F., Staicu-Casagrande, E.M., Lahmam-Bennani, A., Duguet, A., Naja, A., and Ren, X.G. (2007) *Rev. Sci. Instrum.*, **78**, 013108.

Chang, E.S. and Fano, U. (1972) *Phys. Rev. A*, **6**, 173.

Chase, D.M. (1956) *Phys. Rev.*, **104**, 838.

Chen, Z., Madison, D.H., Whelan, C.T., and Walters, H.R.J. (2004) *J. Phys. B*, **37**, 981.

Childers, J.G., James, K.E., Bray, I., Baertschy, M., and Khakoo, M.A. Jr. (2004) *Phys. Rev. A*, **69**, 022709.

Cho, H., McEachran, R.P., Buckman, S.J., Filipović, D.M., Pejčev, V., Marinković, B.P., Tanaka, H., Stauffer, A.D., and Jung, E.C. (2006) *J. Phys. B*, **39**, 3781.

Cho, H., Lee, H.S., and Park, Y.S. (2003) *Radiat. Phys. Chem.*, **68**, 115.

Colgan, J. and Pindzola, M.S. (2006) *Phys. Rev. A*, **74**, 012713.

Crothers, D.S.F. and McCann, J.F. (1983) *J. Phys. B*, **16**, 3229.

Damburg, R. and Karule, E. (1967) *Proc. Phys. Soc.*, **90**, 637.

da Costa, R.F., Bettega, M.H.F., Ferreira, L.G., and Lima, M.A.P. (2007) *J. Phys.: Conf. Ser.*, **88**, 012028.

Dasgupta, A., Bartschat, K., Vaid, D., Grum-Grzhimailo, A.N., Madison, D.H., Blaha, M., and Giuliani, J.L. (2001) *Phys. Rev. A*, **64**, 052710.

Dasgupta, A., Blaha, M., and Giuliani, J.L. (2000) *Phys. Rev. A*, **61**, 012703.

deHarak, B.A., Bartschat, K., and Martin, N.L.S. (2008) *Phys. Rev. Lett.*, **100**, 063201.

Dehmer, P.M. and Chupka, W.A. (1976) *J. Chem. Phys.*, **65**, 2243.

Denifl, S., Ptasińska, S., Hanel, G., Gstir, B., Scheier, P., Probst, M., Farizon, B., Farizon, M., Matejcik, S., Illenberger, E., and Märk, T.D. (2004) *Phys. Scr.*, **T110**, 252.

Dogan, M. and Crowe, A. (2000) *J. Phys. B*, **33**, L461.

Dogan, M. and Crowe, A. (2002) *J. Phys. B*, **35**, 2773.

Domcke, W. (1991) *Phys. Rep.*, **208**, 97.

dos Santos, S.F., Kokoouline, V., and Greene, C.H. (2007) *J. Chem. Phys.*, **127**, 124309.

Drozdov, S.I. (1955) *Sov. Phys. JETP*, **1**, 591.

Drozdov, S.I. (1956) *Sov. Phys. JETP*, **3**, 759.

Dürr, M., Dimopoulou, C., Najjari, B., Dorn, A., Bartschat, K., Bray, I., Fursa, D.V., Chen, Z., Madison, D.H., and Ullrich, J. (2008) *Phys. Rev. A*, **77**, 032717.

Dürr, M., Dimopoulou, C., Najjari, B., Dorn, A., and Ullrich, J. (2006) *Phys. Rev. Lett.*, **96**, 243202.

Dürr, M., Dorn, A., Ullrich, J., Cao, S.P., Czasch, A., Kheifets, A.S., Götz, J.R., and Briggs, J.S. (2007) *Phys. Rev. Lett.*, **98**, 193201.

Ehrhardt, H., Jung, K., Knoth, G., and Schlemmer, P. (1986) *Z. Phys. D*, **1**, 3.

Ehrhardt, H., Schulz, M., Tekaat, T., and Willmann, K. (1969) *Phys. Rev. Lett.*, **22**, 89.

Elazzouzi, S., Dal Cappello, C., Lahmam-Bennani, A., and Catoire, F. (2005) *J. Phys. B*, **38**, 1391.

Eminyan, M. and Lampel, G. (1980) *Phys. Rev. Lett.*, **45**, 1171.

Eminyan, M., MacAdam, K.B., Slevin, J., and Kleinpoppen, H. (1974) *J. Phys. B*, **7**, 1519.

Fano, U. (1961) *Phys. Rev.*, **124**, 1866.

Fano, U. (1970) *Phys. Rev. A*, **2**, 353.

Fano, U. (1983) *Rep. Prog. Phys.*, **46**, 97.

Faure, A., Gorfinkiel, J.D., and Tennyson, J. (2004) *J. Phys. B*, **37**, 801.

Feist, J., Nagele, S., Pazourek, R., Persson, E., Schneider, B.I., Collins, L.A., and Burgdörfer, J. (2008) *Phys. Rev. A*, **77**, 043420.

Fisk, J.B. (1936) *Phys. Rev.*, **49**, 167.

Foster, M., Peacher, J.L., Schulz, M., Madison, D.H., Chen, Z., and Walters, H.R.J. (2006) *Phys. Rev. Lett.*, **97**, 093202.

Franck, J. and Hertz, G. (1914) *Verh. Dtsch. Phys. Ges.*, **16**, 457.

Fursa, D.V. and Bray, I. (1995) *Phys. Rev. A*, **52**, 1279.

Fursa, D.V. and Bray, I. (1997) *J. Phys. B*, **30**, 757.

Fursa, D.V. and Bray, I. (2008) *Phys. Rev. Lett.*, **100**, 113201.

Gailitis, M. (1963) *Sov. Phys. JETP*, **17**, 1328.

Gao, H., Jungen, Ch., and Greene, C.H. (1993) *Phys. Rev. A*, **47**, 4877.

Gao, J., Madison, D.H., and Peacher, J.L. (2005a) *Phys. Rev. A*, **72,**, 020701(R).

Gao, J., Madison, D.H., and Peacher, J.L. (2005b) *J. Chem. Phys.*, **123**, 204314.

Gay, T.J. (1983) *J. Phys. B.*, **16**, L553.

Gianturco, F.A. and Jain, A. (1986) *Phys. Rep.*, **143**, 347.

Gianturco, F.A. and Lucchese, R.R. (2004) *J. Chem. Phys.*, **120**, 7446.

Gianturco, F.A. and Scialla, S. (1987) *J. Phys. B*, **20**, 3171.

Gil, T.J., Rescigno, T.N., McCurdy, C.W., and Lengsfield, B.H. III (1994) *Phys. Rev. A*, **49**, 2642.

Giusti, A. (1980) *J. Phys. B*, **13**, 3867.

Giusti-Suzor, A. and Jungen, Ch. (1984) *J. Chem. Phys.*, **80**, 986.

Glauber, R.J. (1959) in *Lectures in Theoretical Physics* (ed. W.E. Britten), Interscience, New York, p. 315.

Goeke, J., Kessler, J., and Hanne, G.F. (1987) *Phys. Rev. Lett.*, **59**, 1413.

Gorczyca, T.W. and Badnell, N.R. (1997) *J. Phys. B*, **30**, 3897.

Gorfinkiel, J.D. and Tennyson, J. (2005) *J. Phys. B*, **38**, 1607.

Gradziel, M.L. and O'Neill, R.W. (2004) *J. Phys. B*, **37**, 1893.

Green, A.S., Gallup, G.A., Rosenberry, M.A., and Gay, T.J. (2004) *Phys. Rev. Lett.*, **92**, 093201.

Greene, C.H. and Jungen, Ch. (1985) *Adv. At. Mol. Phys.*, **21**, 51.

Greene, C.H. and Rau, A.R.P. (1982) *Phys. Rev. Lett.*, **48**, 533.

Greene, C.H. and Rau, A.R.P. (1983) *J. Phys. B*, **16**, 99.

Grum-Grzhimailo, A.N. (2003) *Comp. Phys. Commun.*, **152**, 101.

Guan, X., Zatsarinny, O., Bartschat, K., Schneider, B.I., Feist, J., and Noble, C.J. (2007) *Phys. Rev. A*, **76**, 053411.

Guan, X., Bartschat, K., and Schneider, B.I. (2008) *Phys. Rev. A*, **77**, 043421.

Hall, R.I., Joyez, G., Mazeau, Y., Reinhard, J., and Schermann, C.S. (1973) *J. Physique*, **34**, 827.

Hanne, G.F. (1976) *J. Phys. B*, **9**, 805.

Hanne, G.F. (1983) *Phys. Rep.*, **95**, 95.

Hayes, P.A. and Williams, J.F. (1996) *Phys. Rev. Lett.*, **77**, 3098.

Heddle, D.W.O. and Gallagher, J.W. (1989) *Rev. Mod. Phys.*, **61**, 221.

Herzberg, G. (1969) *Phys. Rev. Lett.*, **23**, 1081.

Herzenberg, A. (1984) in *Electron Molecule Collisions* (eds I. Shimamura and K. Takayanagi), Plenum, New York, p. 351.

Hotop, H., Ruf, M.-W., Allan, M., and Fabrikant, I.I. (2003) *Adv. At. Mol. Opt. Phys.*, **49**, 85.

Huo, W.M. and Gianturco, F.A. (eds) (1995) *Computational Methods for Electron-Molecule Collisions*, Plenum, New York.

Hussey, M., Murray, A., MacGillivray, W., and King, G. (2008) *J. Phys. B*, **41**, 055202.

Jaduszliwer, B. and Chan, Y.C. (1992) *Phys. Rev. A*, **45**, 197.

Jakubowicz, H. and Moores, D.L. (1981) *J. Phys. B*, **14**, 3733.

Jelisavcic, M., Panajotovic, R., and Buckman, S.J. (2003) *Phys. Rev. Lett.*, **90**, 203201.

Joachain, C.J. (1987) in *Collision Theory for Atoms and Molecules* (ed. F.A. Gianturco), Plenum, New York, p. 59.

Jones, S., Madison, D.H., and Hanne, G.F. (1994) *Phys. Rev. Lett.*, **72**, 2554.

Jones, S. and Madison, D.H. (1998) *Phys. Rev. Lett.*, **81**, 2886.

Jones, S. and Madison, D.H. (2000) *Phys. Rev. A*, **62**, 042701.

Jungen, Ch. (1984) *Phys. Rev. Lett.*, **53**, 2394.

Jungen, Ch. and Atabec, O. (1977) *J. Chem. Phys.*, **66**, 5584.

Jungen, Ch. and Dill, D. (1980) *J. Chem. Phys.*, **73**, 3338.

Jungen, Ch. and Ross, S.C. (1997) *Phys. Rev. A*, **55**, R2503.

Kaiser, C., Spieker, D., Gao, J., Hussey, M., Murray, A., and Madison, D.H. (2007) *J. Phys. B*, **40**, 2563.

Kauppila, W.E. and Stein, T.S. (2001) unpublished; cited in MacAskill *et al.* (2002).

Kawahara, H., Kato, H., Hoshino, M., Tanaka, H., and Brunger, M.J. (2008) *Phys. Rev. A*, **77**, 012713.

Kessler, J. (1985) *Polarized Electrons*, 2nd edn., Springer, Berlin.

Khakoo, M.A., Becker, K., Forand, J.L., and McConkey, J.W. (1986) *J. Phys. B*, **19**, L209.

Khakoo, M.A., Keane, K., Campbell, C., Guzman, N., and Hazlett, K. (2007) *J. Phys. B*, **40**, 3601.

Khakoo, M.A., Blumer, J., Keane, K., Campbell, C., Silva, H., Lopes, M.C.A., Winstead, C., McKoy, V., da Costa, R.F., Ferreira, L.G., Lima, M.A.P., and Bettega, M.H.F. (2008) *Phys. Rev. A*, **77**, 042705.

Kim, Y.K. (2001) *Phys. Rev. A*, **64**, 032713.

Kim, Y.K. and Rudd, M.E. (1994) *Phys. Rev. A*, **50**, 3954.

Kim, Y.K., Irikura, K.K., Rudd, M.E., Ali, M.A., Stone, P.M., Coursey, J.S., Dragoset, R.A., Kishore, A.R., Olsen, K.J., Sansonetti, A.M., Wiersma, G.G., Zucker, D.S., and Zucker, M.A. (2008) http://physics.nist.gov/Phys RefData/Ionization/Xsection.html.

Kokoouline, V. and Greene, C.H. (2003a) *Phys. Rev. Lett.*, **90**, 133201.

Kokoouline, V. and Greene, C.H. (2003b) *Phys. Rev. A*, **68**, 012703.

Kokoouline, V. and Greene, C.H. (2005) *Phys. Rev. A*, **72**, 022712.

Kokoouline, V. and Masnou-Seeuws, F. (2006) *Phys. Rev. A*, **73**, 012702.

Lahmam-Bennani, A. (1991) *J. Phys. B*, **24**, 2401.

Lahmam-Bennani, A., Duguet, A., Dal Cappello, C., Nebdi, H., and Piraux, B. (2003) *Phys. Rev. A*, **67**, 010701.

Lahmam-Bennani, A., Dupré, C., and Duguet, A. (1989) *Phys. Rev. Lett.*, **63**, 1582.

Larsson, M. (2001) *Gas Phase Ion Chem.*, **4**, 179.

Larsson, M. and Thomas, R. (2001) *Phys. Chem. Chem. Phys.*, **3**, 4471.

Lenard, P. (1903) *Ann. Phys.*, **12**, 714.

Lindsay, B.G. and Mangan, M.A. (2003) in *Landolt-Börnstein I-17C* (ed. Y. Itikawa), Springer, Berlin.

Lorentz, S.R., Scholten, R.E., McClelland, J.J., Kelley, M.H., and Celotta, R.J. (1993) *Phys. Rev. A*, **47**, 3000.

Lower, J., Panatović, R., Bellm, S., and Weigold, E. (2007) *Rev. Sci. Instrum.*, **78**, 111301.

Lukomski, M., Sutton, S., Kedzierski, W., Reddish, T.J., Bartschat, K., Bartlett, P.L., Bray, I., Stelbovics, A.T., and McConkey, J.W. (2006) *Phys. Rev. A*, **74**, 032708.

MacAskill, J.A., Kedzierski, W., McConkey, J.W., Domyslawska, J., and Bray, I. (2002) *J. Elect. Spect. Rel. Phen.*, **123**,(2002), 173.

Machado, L., Leal, E.P., and Csanak, G. (1982) *J. Phys. B*, **15**, 1773.

Madison, D.H. and Bartschat, K. (1996) in *Computational Atomic Physics* (ed. K. Bartschat), chap. 4, Springer, Heidelberg and New York.

Madison, D.H. and Shelton, W.N. (1973) *Phys. Rev. A*, **7**, 499.

Maslov, M., Brunger, M.J., Teubner, P.J.O., Zatsarinny, O., Bartschat, K., Fursa, D.V., Bray, I., and McEachran, R.P. (2008) *Phys. Rev. A*, **77**, 062711.

Mayer, S. and Kessler, J. (1995) *Phys. Rev. Lett.*, **74**, 4803.

McAdams, R., Hollywood, M.T., Crowe, A., and Williams, J.F. (1980) *J. Phys. B*, **13**, 3691.

McCall, B.J., Huneycutt, A.J., Saykally, R.J., Geballe, T.R., Djuric, N., Dunn, G.H., Semaniak, J., Novotny, O., Al-Khalili, A., Ehlerding, A., Hellberg, F., Kalhori, S., Neau, A., Thomas, R., Österdahl, F., and Larsson, M. (2003) *Nature*, **422**, 500.

McClelland, J.J., Kelley, M.H., and Celotta, R.J. (1989) *Phys. Rev. A*, **40**, 2321.

McCurdy, C.W., Baertschy, M., and Rescigno, T.N. (2004) *J. Phys. B*, **37**, R137.

McDonald, D.G. and Crowe, A. (1993) *J. Phys. B*, **26**, 2887.

Milne-Brownlie, D.S., Cavanagh, S.J., Lohmann, B., Champion, C., Hervieux, P.A., and Hanssen, J. (2004) *Phys. Rev. A*, **69**, 032701.

Morrison, M.A. (1988) *Adv. At. Mol. Phys.*, **24**, 51.

Müller, A. (1990) in *The Physics of Electronic and Atomic Collisions* (eds A. Dalgarno, R.S. Freund, P.M. Koch, M.S. Lubell, and T.B. Lucatorto), (AIP Conf. Proc. 205), p. 418.

Naja, A., Staicu-Casagrande, E.M., Ren, X.G., Catoire, F., Lahham-Bennani, A., Dal Cappello, C., and Whelan, C.T. (2007) *J. Phys. B*, **40**, 2871.

Nesbet, R.K. (1979) *Phys. Rev. A*, **20**, 58.

Nic Chormaic, S., Chwirot, S., and Slevin, J. (1993) *J. Phys. B*, **26**, 139.

Ochkur, V.I. (1964) *Sov. Phys. JETP*, **18**, 503.

O'Neill, R.W., van der Burgt, P.J.M., Dziczek, D., Bowe, P., Chwirot, S., and Slevin, J.A. (1998) *Phys. Rev. Lett.*, **80**, 1630.

Peart, B. and Dolder, K.T. (1968) *J. Phys. B*, **1**, 872.

Peterkop, R.K. (1961) *Proc. Phys. Soc.*, **77**, 1220.

Petrov, G.M., Giuliani, J.L., Dasgupta, A., Bartschat, K., and Pecharek, R.E. (2004) *J. Appl. Phys.*, **95**, 5284.

Phaneuf, R.A., Dunn, G.H., Havener, C.C., and Müller, A. (1999) *Rep. Prog. Phys.*, **62**, 1143.

Pindzola, M.S., Griffin, D.C., and Badnell, N.R. (2006) in *Springer Handbook of Atomic, Molecular, and Optical Physics* (ed. G. W. F. Drake), chap. 55, Springer, Heidelberg.

Pindzola, M.S., Robicheaux, F., Badnell, N.R., and Gorczyca, T.W. (1997) *Phys. Rev. A*, **56**, 1994.

Pindzola, M.S., Robicheaux, F., Loch, S.D., Berengut, J.C., Topcu, T., Colgan, J., Foster, M., Griffin, D.C., Ballance, C.P., Schultz, D.R., Tinami, T., Badnell, N.R., Witthoeft, M.C., Plante, D.R., Mitnik, D.M., Ludlow, J.A., and Kleiman, U. (2007) *J. Phys. B*, **40**, R39.

Pravica, L., Cvejanovic, D., Williams, J.F., and Napier, S.A. (2007) *J. Phys.: Conf. Ser.*, **88**, 012067.

Prideaux, A., Madison, D.H., and Bartschat, K. (2005) *Phys. Rev. A*, **72**, 032702.

Ptasińska, S., Denifl, S., Grill, V., Märk, T.D., Scheier, P., Gohlke, S., Huels, M.A., and Illenberger, E. (2005) *Angew. Chem. Int. Ed.*, **44**, 1647.

Ramsbottom, C.A., Noble, C.J., Burke, V.M., Scott, M.P., and Burke, P.G. (2004) *J. Phys. B*, **37**, 3609.

Ramsbottom, C.A., Hudson, C.E., Norrington, P.H., and Scott, M.P. (2007) *Astron. Astrophys.*, **475**, 765.

Rapp, D. and Englander-Golden, P. (1965) *J. Chem. Phys.*, **43**, 1464.

(a) Redmond, P.J. (1973) unpublished; (b) quoted in Rosenberg, L. (1973) *Phys. Rev. D*, **8**, 1833.

Read, F.H. and Channing, J.M. (1996) *Rev. Sci. Instrum.*, **67**, 2372.

(a) Reid, R.H.G., Bartschat, K., and Raeker, A. (1998) *J. Phys. B*, **31**, 563; (b) corrigendum (2000) *J. Phys. B*, **33**, 5261.

Reinhardt, W.P. (1979) *Comp. Phys. Comm.*, **17**, 1.

Rescigno, T.N., Baertschy, M., Isaacs, W.A., and McCurdy, C.W. (1999) *Science*, **286**, 2474.

Robicheaux, F. (1991) *Phys. Rev. A*, **43**, 5946.

Röder, J., Baertschy, M., and Bray, I. (2003) *Phys. Rev. A*, **67**, 010702.

Rudge, M.R.H. and Seaton, M.J. (1965) *Proc. Roy. Soc. A*, **283**, 262.

Sakhelashvili, G., Dorn, A., Hörr, C., Ullrich, J., Kheifets, A.S., Lower, J., and Bartschat, K. (2005) *Phys. Rev. Lett.*, **95**, 033201.

Sanche, L. (2005) *Eur. Phys. J. D*, **35**, 367.

Schappe, R.S., Feng, P., Anderson, L.W., Lin, C.C., and Walker, T. (1995) *Europhys. Lett.*, **29**, 439.

Schneider, B.I., Le Dourneuf, M., and Burke, P.G. (1979) *J. Phys. B*, **12**, L365.

Schneider, B.I., Le Dourneuf, M., and Lan, Vo Ky (1979) *Phys. Rev. Lett.*, **43**, 1926.

Schow, E., Hazlett, K., Childers, J.G., Medina, C., Vitug, G., Bray, I., Fursa, D.V., and Khakoo, M.A. (2005) *Phys. Rev. A*, **72**, 062717.

Schulz, G.J. (1973a) *Rev. Mod. Phys.*, **45**, 378.

Schulz, G.J. (1973b) *Rev. Mod. Phys.*, **45**, 423.

Schwienhorst, R., Raeker, A., Bartschat, K., and Blum, K. (1996) *J. Phys. B*, **29**, 2305.

Seaton, M.J. (1958) *Mon. Not. R. Astron. Soc.*, **118**, 504.

Seaton, M.J. (1983) *Rep. Prog. Phys.*, **46**, 167.

Sewell, E.C. and Crowe, A. (1982) *J. Phys. B*, **15**, L357.

Shore, B.W. (1967) *Rev. Mod. Phys.*, **39**, 439.

Siegert, A.J.F. (1939) *Phys. Rev.*, **56**, 750.

Slaughter, D.A., Karaganov, V., Brunger, M.J., Teubner, P.J.O., Bray, I., and Bartschat, K. (2007) *Phys. Rev. A*, **75**, 062717.

Sohn, M. and Hanne, G.F. (1992) *J. Phys. B*, **25**, 4627.

Standage, M.C. and Kleinpoppen, H. (1976) *Phys. Rev. Lett.*, **36**, 577.

Stefani, G., Avaldi, L., and Camilloni, R. (1990) *J. Phys. B*, **23**, L227.

Stepanovic, M., Minic, M., Cvejanovic, D., Jureta, J., Kurepa, J., Cvejanovic, S., Zatsarinny, O., and Bartschat, K. (2006) *J. Phys. B*, **39**, 1547.

Stephens, J.A. and Greene, C.H. (1995) *J. Chem. Phys.*, **103**, 5470.

Stevenson, M. and Crowe, A. (2004) *J. Phys. B*, **37**, 2493.

Stevenson, M.A. and Lohmann, B. (2008) *Phys. Rev. A*, **77**, 032708.

Stier, H.C. (1932) *Z. Phys.*, **76**, 439.

Surko, C.M., Gribakin, G.F., and Buckman, S.J. (2005) *J. Phys. B*, **38**, R57.

Thorn, P.A., Brunger, M.J., Kato, H., Hoshino, M., and Tanaka, H. (2007) *J. Phys. B*, **40**, 697.

Tonzani, S. and Greene, C.H. (2005) *J. Chem. Phys.*, **124**, 054312.

Trajmar, S., Williams, W., and Kuppermann, A. (1973) *J. Chem. Phys.*, **58**, 2521.

Trevisan, C.S., Houfek, K., Zhang, Z., Orel, A.E., McCurdy, C.W., and Rescigno, T.N. (2005) *Phys. Rev. A*, **71**, 052714.

Uhlmann, L.J., Dall, R.G., Truscott, A.G., Hoogerland, M.D., Baldwin, K.G.H., and Buckman, S.J. (2005) *Phys. Rev. Lett.*, **94**, 173201.

Ullrich, J., Moshammer, R., Dorn, A., Dörner, R., Schmidt, L.Ph.H., and Schmidt-Böcking, H. (2003) *Rep. Prog. Phys.*, **66**, 1463.

Vinkalns, I. and Gailitis, M. (1967) in *Abstracts of the Vth ICPEAC* (eds I.P. Flaks and E.S. Solovyov), Nauka, Leningrad, p. 648.

Visconti, P.J., Slevin, J.A., and Rubin, K. (1970) *Phys. Rev. A*, **3**, 1310.

Walters, H.R.J. (1984) *Phys. Rep.*, **116**, 1.

Wannier, G.H. (1953) *Phys. Rev.*, **90**, 817.

Weigold, E. and McCarthy, I.E. (1999) *Electron Momentum Spectroscopy*, Kluwer, Dordrecht/Plenum, New York.

Wigner, E.P. (1948) *Phys. Rev.*, **73**, 1002.

Williams, J.F. (1986) *Aust. J. Phys.*, **39**, 621.

Williams, J.F. (1988) *J. Phys. B*, **21**, 2107.

Williams, J.F. and Mikosza, A.G. (2006) *J. Phys. B*, **39**, 4113.

Winstead, C. and McKoy, V. (2006) *J. Chem. Phys.*, **125**, 174304.

Winstead, C., McKoy, V., and Bettega, M.H.F. (2005) *Phys. Rev. A*, **72**, 042721.

Wolcke, A., Bartschat, K., Blum, K., Borgmann, H., Hanne, G.F., and Kessler, J. (1983) *J. Phys. B*, **16**, 639.

Yalim, H.A., Cvejanovic, D., and Crowe, A. (1997) *Phys. Rev. Lett.*, **79**, 2951.

Zatsarinny, O. (2006) *Comp. Phys. Commun.*, **174**, 273.

Zatsarinny, O. and Bartschat, K. (2004) *J. Phys. B*, **37**, 2173.

Zatsarinny, O. and Bartschat, K. (2005) *Phys. Rev. A*, **72**, 020702(R).

Zatsarinny, O. and Bartschat, K. (2008) *Phys. Rev. A*, **77**, 062701.

Zuo, T., McEachran, R.P., and Stauffer, A.D. (1991) *J. Phys. B*, **24**, 2853.

Further Reading

Balashov, V.V., Grum-Grzhimailo, A.N., and Kabachnik, N.M. (2000) *Polarization and Correlation Phenomena in Atomic Collisions*, Springer, New York.

Bartschat, K. (ed.) (1996) *Computational Atomic Physics*, Springer, Heidelberg and New York.

Becker, U., Moshammer, R., Mokler, P., and Ullrich, J. (eds) (2007) *XXV International Conference on Photonic, Electronic and Atomic Collisions, J. Phys. Conf. Ser.*, **88**, (entire issue).

Bransden, B.H. (1983) *Atomic Collision Theory*, 2nd edn., Benjamin, New York.

Burke, P.G. and Joachain, C.J. (1995) *Theory of Electron-Atom Collisions. Part One: Potential Scattering*, Springer, Heidelberg and New York.

Burke, P.G. and Moiseiwitsch, B.L. (eds) (1976) *Atomic Processes and Applications*, North-Holland, Amsterdam.

Burke, P.G. and West, J.B. (eds) (1988) *Electron-Molecule Scattering and Photoionization*, Plenum, New York.

Fainstein, P.D., Lima, M.A.P., Miraglia, J.E., Montenegro, E.C., and Rivarola, R.D. (eds) (2006) *Photonic, Electronic and Atomic Collisions*, World Scientific, Singapore.

Drake, G.W.F. (ed.) (2006) *Springer Handbook of Atomic, Molecular, and Optical Physics*, Springer, Heidelberg.

Ehrhardt, H., Fischer, M., Jung, K., Byron, F.W., Joachain, C.J., and Piraux, B. (1982) *Phys. Rev. Lett.*, **48**, 1807.

Gianturco, F.A. (ed.) (1987) *Collision Theory for Atoms and Molecules*, Plenum, New York.

Grant, I.P. (2007) *Relativistic Quantum Theory of Atoms and Molecules*, Springer, Heidelberg.

Herzenberg, A. (ed.) (1990) *Aspects of Electron-Molecule Scattering and Photoionization*, American Institute of Physics, New York.

Massey, H.S.W. and Bates, D.R. (eds) (1982) *Applied Atomic Collision Physics*, Vol. 1: Atmospheric Physics and Chemistry, Academic, New York.

Massey, H.S.W. and Burhop, E.H.S. (1969) *Electronic and Ionic Impact Phenomena*, 2nd edn., Vols. 1 and 2, Clarendon, Oxford.

McCarthy, I.E. and Weigold, E. (1995) *Electron-Atom Collisions*, University Press, Cambridge.

McGuire, J.H. (1997) *Electron Correlation Dynamics in Atomic Collisions*, University Press, Cambridge.

Mott, N.F. and Massey, H.S.W. (1965) *The Theory of Atomic Collisions*, 3rd edn., Clarendon, Oxford.

Mšrk, T.D. and Dunn, G.H. (eds.) (1985) *Electron-Impact Ionization*, Springer, Berlin.

McDaniel, E.W. (1969) *Atomic Collisions: Electron and Photon Projectiles*, John Wiley & Sons, New York.

Nesbet, R.K. (1980) *Variational Methods in Electron-Atom Scattering Theory*, Plenum, New York.

Peterkop, R.K. (1977) *Theory of Ionization of Atoms by Electron Impact*, Colorado Associated University Press, Boulder.

Shimamura, I. and Takayanagi K. (eds) (1984) *Electron-Molecule Collisions*, Plenum, New York.

Ullrich, J. and Shevelko, V.P. (eds) (2003) *Many-Particle Quantum Dynamics in Atomic and Molecular Fragmentation*, Springer, Heidelberg.

Watanabe, T., Shimamura, I., Shimizu, M., and Itikawa, Y. (eds) (1990) *Molecular Processes in Space*, Plenum, New York.

9
Photoemission and Photoelectron Spectroscopy

Franz J. Himpsel and Ingolf Lindau

9.1	**Introduction** 285	
9.1.1	History 285	
9.1.2	Phenomena 286	
9.1.3	Theory 288	
9.2	**Instrumentation** 293	
9.2.1	Light Sources 293	
9.2.2	Detectors 295	
9.3	**Photoelectric Yield and Absorption Spectroscopy** 298	
9.3.1	Mechanism 298	
9.3.2	Unoccupied Atomic and Molecular Orbitals 298	
9.3.3	Solids 299	
9.3.4	Magnetic Dichroism 300	
9.3.5	Microspectroscopy 300	
9.4	**Core-Level Spectroscopy** 301	
9.4.1	Origin of Core-Level Shifts 301	
9.4.2	Atoms, Shake-Up, and Shake-Off Satellites 302	
9.4.3	Adsorbates and Surfaces 303	
9.4.4	Solids 305	
9.5	**Valence Electron Spectroscopy** 305	
9.5.1	Occupied Atomic and Molecular Orbitals 307	
9.5.2	Surfaces and Quantum Wells 308	
9.5.3	Band Mapping 309	
9.5.4	Time-Resolved, Pump-Probe Measurements 311	
9.5.5	Inverse Photoemission 311	
9.6	**Structural Methods** 312	
9.6.1	EXAFS and Photoelectron Diffraction 312	
9.6.2	Photoelectron Diffraction 313	
9.7	**Summary and Outlook** 314	
9.8	**Databases** 315	
	Glossary 315	
	References 317	

Encyclopedia of Applied Spectroscopy. Edited by David L. Andrews.
Copyright © 2009 WILEY-VCH Verlag GmbH & Co. KGaA, Weinheim
ISBN: 978-3-527-40773-6

9.1
Introduction

9.1.1
History

Photoemission, that is, the emission of electrons by photons, was discovered as an unwanted side effect in the study of electromagnetic waves by Hertz (1887). At that time, neither the electron nor the photon was known as *particles*. Einstein (1905) explained the photoelectric effect as a quantum phenomenon and derived the correlation

$$E_{max} = h\nu - \Phi \tag{9.1}$$

between the maximum kinetic energy E_{max} of the photoelectrons, the photon energy $h\nu$, and the work function Φ of the emitting solid (compare Figure 9.1). Many of the following decades were spent in finding clear experimental evidence for or against the Einstein formula (Millikan, 1916). The time was not ripe for photoelectron spectroscopy as we know it today, because photoemission turned out to be a rather surface-sensitive effect. The mean free path of photoelectrons in a solid limits the probing depth to a few atomic layers. The prerequisites for the preparation of reproducible, clean, and well-ordered surfaces were not in place until the sixties, when the necessary ultrahigh vacuum could be produced and measured on a routine basis. At a residual gas pressure of 10^{-6} torr (1.3×10^{-4} Pa), a surface atom is being hit by a residual gas molecule about once in every second, thus requiring pressures in the 10^{-10} torr range for controlled photoemission studies. When surface characterization techniques, such as low-energy electron diffraction (LEED) and Auger spectroscopy became generally available, photoelectron spectroscopy took a big leap forward. Photoelectron spectroscopy is now an indispensable technique for the determination of the electronic structure of solids, surfaces, and gas-phase species. Tunable and polarized synchrotron radiation became the light source of choice in the 1970s, and gave rise to the construction of a series of storage rings dedicated to the production of light from the ultraviolet to the X-ray regime. In the 1980s, the measurement of a complete set of quantum numbers became feasible for electrons in solids, which involves measuring the momentum and spin polarization of photoelectrons as a function of the photon energy and photon polarization. In the 1990s, measurements with an energy resolution

Encyclopedia of Applied Spectroscopy. Edited by David L. Andrews.
Copyright © 2009 WILEY-VCH Verlag GmbH & Co. KGaA, Weinheim
ISBN: 978-3-527-40773-6

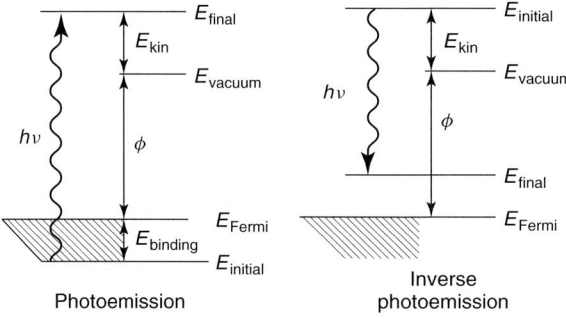

Fig. 9.1 Energy diagram for photoemission (photon in, electron out) and inverse photoemission (electron in, photon out). The energy of the lower state is determined by subtracting the photon energy $h\nu$ from the energy of the upper state. The latter corresponds to the kinetic energy of the emitted or the incoming electron.

better than the thermal energy became practical, which provided an opportunity to examine the electronic states that are relevant to magnetism, superconductivity, and other transport properties. Most recently, temporal resolution was introduced with pump–probe experiments for observing dynamics down to tens of femtoseconds, and spatial resolution down to tens of nanometers was achieved in photoelectron microscopes.

9.1.2
Phenomena

The basic energy diagram of the photoemission process is given in Figure 9.1. An electron is lifted from an occupied initial state into an unoccupied final state. Energy conservation allows it to obtain the desired energy $E_{initial}$ of the initial state from the measured energy E_{final} of the photoelectron in the final state by simply subtracting the photon energy $h\nu$. As reference energies, one can use either the vacuum level E_{vacuum}, corresponding to a photoelectron with zero velocity, or the Fermi level E_{Fermi}, corresponding to the highest occupied state. From there one obtains energy differences, such as the kinetic energy E_{kin}, the work function Φ, and the binding energy $E_{binding}$. The Einstein relation for the maximum kinetic energy $E_{max} = h\nu - \Phi$ follows directly for photoemission from the highest occupied state at E_{Fermi}.

Figure 9.1 also shows the energy diagram for the time-reversed process, that is, an incident electron dropping down into a lower unoccupied state by emission of a photon. This process is equivalent to the quantum description of bremsstrahlung, but in the low-energy regime of typically 10-eV kinetic energy, it is known today as *inverse photoemission*. It characterizes unoccupied states in a fashion complementary to the information provided by photoemission about occupied states. The counterpart to the Einstein relation is the Duane–Hunt equation

$$h\nu_{max} = E_{kin} + \Phi \qquad (9.2)$$

for the highest photon energy of the bremsstrahlung emitted by electrons with kinetic energy E_{kin}. Inverse photoemission as a surface technique got a late start compared to photoemission because its cross section is down by 4–5 orders of magnitude, requiring extremely sensitive photon detectors.

There are other electron and photon emission phenomena related to photoemission. As shown in Figure 9.2, a core hole created in the photoemission process can decay by an Auger process (a), which is most likely for shallow core levels, or by fluorescence (b), which is prevalent for deep core levels (Keski-Rahkonen and Krause, 1974). In addition, the photoelectron can lose energy by creating electron–hole pairs or plasmons during its escape. These processes actually limit the escape depth. The end result is a spectrum of emitted photoelectrons that contains several secondary features in addition to the directly emitted photoelectrons (Figure 9.3). For valence spectroscopy (often labeled as ultraviolet photoelectron spectroscopy (UPS)), the main disturbance is a tail of secondary electrons at low kinetic energies, caused by energy losses to electron hole pairs (Figure 9.3a). For core-level spectroscopy (often labeled as X-ray photoelectron spectroscopy (XPS), or electron spectroscopy for chemical analysis (ESCA)), there are additional Auger peaks and plasmon loss satellites complicating the spectra (Figure 9.3b). With tunable synchrotron radiation, it is straightforward to separate out the Auger peaks, because they do not change their kinetic energy with increasing photon energy, in contrast to elastic core lines and their plasmon loss satellites.

An important consideration in photoemission experiments is the probing depth, which is governed by the mean free path of the photoelectrons (Figure 9.4). Electrons penetrate solids farther at high kinetic energies, but their mean free path increases again at very low energies, because they do not have enough energy to excite plasmons (in metals) or electron–hole pairs (in insulators). In the 30- to 70-eV range, the mean free path reaches a minimum of only a few atomic distances, making it possible to separate surface

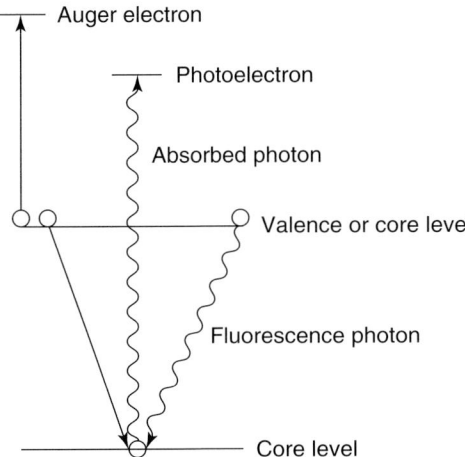

Fig. 9.2 Energy diagram for core-level photoemission, including the decay of the core hole by an Auger process (a) or by fluorescence (b).

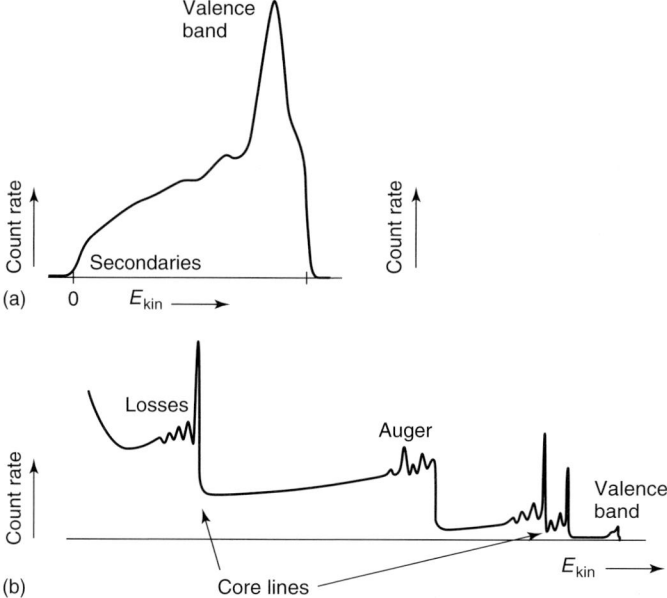

Fig. 9.3 (a) Typical valence photoelectron spectrum, showing photoelectrons directly emitted from the valence band and secondary electrons emitted after electron–hole pair production. The two energy cutoffs are determined by the vacuum level (zero kinetic energy E_{kin}) and the Fermi level (i.e., excitation of the highest occupied state). (b) Typical core-level photoelectron spectrum, showing core-level peaks, their plasmon loss satellites, and Auger electron features (Feuerbacher, Fitton and Willis, 1978).

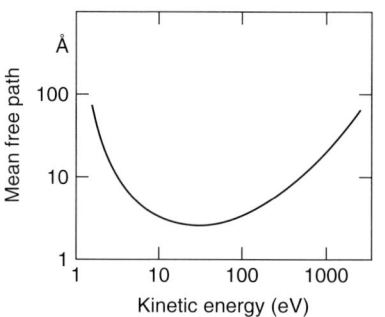

Fig. 9.4 Schematic dependence of the mean free path of photoelectron on their kinetic energy. The minimum at 30–70 eV can be used to probe surface or interface layers.

atoms from the bulk. The tunability of synchrotron radiation provides the ability of producing photoelectrons with the desired kinetic energy, which has spawned the field of surface core-level spectroscopy.

9.1.3
Theory

Photoemission is unique in providing the complete set of quantum numbers that characterize electrons in solids.

This made photoemission the preferred technique for investigating electrons in a solid and led to Philip Anderson's famous characterization of photoemission as "smoking gun" that might give away the secrets of electrons in high-temperature superconductors (Anderson and Schrieffer, 1991). For electrons in a periodic crystal lattice, these quantum numbers are energy E, momentum $\mathbf{p} = \hbar\mathbf{k}$, spin, and angular symmetry (represented by the point group in a solid). Polycrystalline solids can be characterized by the same quantum numbers, because each crystallite typically contains thousands to millions of unit cells. However, when the size of a crystallite reaches dimensions in the single digit nanometer regime (tens of atom diameters), the situation changes: instead of a continuum of energies, one obtains quantized states with discrete energies. This alters the fundamental electronic properties of a material, such as the band gap of a semiconductor, and leads to the capability of tailoring the electronic properties of materials in nanotechnology (Himpsel et al., 1998).

The quantum numbers are obtained by using conservation laws during the photoemission process (see the general reviews Himpsel (1983), Kevan (1992) and Hüfner (2007)). We have already depicted the energy conservation law in Figure 9.1. A similar conservation law holds for the momentum parallel to the surface $\hbar k_\|$, where $k_\|$ is the corresponding wave vector. The wave vector of the photon is usually negligible compared to that of the electron for the energy range in question, because its wavelength is very much longer. The wavelength λ and wave vector $k = 2\pi/\lambda$ of the photon are given by the photon energy $h\nu$ via

$$\left(\frac{\lambda}{\text{Å}}\right) = \frac{12399}{(h\nu/\text{eV})} \quad (9.3)$$

and the parallel wave vector of the electron by its kinetic energy E_{kin} and polar escape angle ϑ via

$$\left(\frac{k_\|}{\text{Å}^{-1}}\right) = 0.51 \times \left(\frac{E_{\text{kin}}}{\text{eV}}\right)^{1/2} \sin\vartheta \quad (9.4)$$

(1 Å = 0.1 nm). The momentum component perpendicular to the surface is not conserved during the passage of the photoelectron across the surface energy barrier, but it can be varied by tuning the photon energy. Angular symmetry can be inferred from polarization selection rules for the point group at a particular point in momentum space. Spin is conserved.

The complete set of quantum numbers is usually displayed in the form of energy-versus-momentum band dispersion, labeled with the appropriate symmetry and spin symbols (see Figure 9.17 in Section 9.5.3). Another way to represent the quantum numbers is in the form of constant energy surfaces with two momentum components as variables (see Figure 9.18 in Section 9.5.3). The Fermi surface is particularly important, because electrons near the Fermi level determine most electronic properties of a solid, such as conductivity, superconductivity, magnetism, and electronic phase transitions. The relevant electron energies are within a thermal energy range around the Fermi level, which is given by the width $3.5\ k_B T$ of the Fermi–Dirac distribution (k_B = Boltzmann constant, T = absolute temperature, $k_B T \approx 25$ meV at room temperature).

The theoretical description of photoemission is provided by an expression similar to Fermi's Golden Rule for a differential cross section (for details see Schattke et al. (2003)). It contains the absolute square of a matrix element M_{if} for the optical transition between the initial state $|\psi_i\rangle$ and the final state $\langle\psi_f|$ of the photoelectron, an average over all possible initial states, and a sum over all final states that are detected. The matrix element takes the following form:

$$M_{if} \propto \langle\psi_f|A(r)\cdot p|\psi_i\rangle \quad (9.5)$$

$A(r)$ is the vector potential of the photon, which becomes a constant in the widely used dipole approximation.

The initial state is $|\psi_i\rangle$ is a Bloch function in a solid or at a surface. For core levels and gas-phase photoemission, the initial state simply becomes an atomic/molecular wave function. In many-body theory, the initial state is not an electron but a hole. This seems to be at odds with the fact that the initial state was occupied by an electron before the photoemission process. However, to measure the binding energy of this electron, one needs to eject it by photoemission. That leaves a hole in the solid. In the following, we keep using the notion of electrons for both initial and final states with the understanding that initial state electrons should really be considered as holes.

The final state $\langle\psi_f|$ is often called a *time-reversed LEED state* (LEED is low energy electron diffraction), because an energy- and angle-resolving detector accepts only photoelectrons with well-defined momentum. Those are similar to the incident electrons in a diffraction experiment, except for the fact that the $\langle\psi_f|$ is a complex conjugate with an inverted propagation direction. The diffracted waves in LEED correspond to incoming electron waves in the final state of photoemission, which is somewhat counterintuitive but mathematically correct. Codes for such "single-step" photoemission calculations are now widely available (see Section 2.10 in Schattke et al. (2003)). The earlier "three-step" model separates the Golden Rule expression into three simpler but approximate steps, that is, optical excitation, scattering on the way out, and transmission/refraction at the surface.

For a quick interpretation of photoemission spectra, one has to mainly consider the various selection rules inherent in the optical matrix element M_{if}, such as the conservation laws for energy and momentum parallel to the surface (modulo a reciprocal surface lattice vector, see Himpsel (1983) and Kevan (1992)). The momentum perpendicular to the surface is not conserved, because the periodicity of a solid is destroyed in this direction. In the three-step model, one can keep perpendicular momentum conservation for the first step, which takes place completely inside the solid. However, transitions from bulk states to evanescent final states are not covered by this approximation. Another useful selection rule holds for parity of the wave functions with respect to a mirror plane. The product of the parities of the initial, the operator $A\cdot p$, and the final state has to be even.

Although selection rules provide clear yes–no decisions, there are also more subtle effects due to the wave functions of the electronic states involved in the matrix element. The atomic symmetry character determines the energy dependence of the cross section (Yeh and Lindau, 1985), allowing a selection of specific orbitals by varying the photon energy. For example, the s, p-states in transition and noble metals dominate the spectra near the photoelectric threshold, while the d-states turn on at 10 eV above threshold. It takes photon energies of 30 eV above threshold to make the f-states in rare earths visible. Resonance effects at a threshold for a core-level excitation can also enhance particular orbitals. Conversely, the cross section for states with a radial node exhibits so-called Cooper minima, where they become almost invisible.

The wave functions of the electronic states in solids and at surfaces are usually approximated by the ground state wave functions obtained from a variety of schemes, for example, tight binding, local density approximation (LDA), or Hartree–Fock. Strictly speaking, one should use the excited-state wave functions and energies that represent the hole created in the photoemission process, or the extra electron added to the solid in the case of inverse photoemission. Such excited-state quasiparticle calculations have now become feasible, and provide the most accurate band dispersions to date. Particularly in the case of semiconductors, the traditional ground state methods are unable to determine the fundamental band gap from first principles, with Hartree–Fock overestimating it and LDA underestimating it, typically by a factor of 2.

Photoemission is able to provide even more information than the complete set of quantum numbers. In many-body theory, an electron is described by the Green's function G(A, B), which describes the propagation of the electron from point A to point B – obviously a very fundamental description of an electron. Photoemission essentially measures the imaginary part of the Green's function in momentum space. From the Greens function, one can extract the self-energy Σ of an electron, whose imaginary part quantifies the scattering of electrons in a solid. The real part of Σ describes an energy shift due to many-body interactions, which is not contained in single-electron band calculations. The real and imaginary parts of Σ are related to each other by an integral transform, the Kramers–Kronig relation.

Experimentally, the information about the Green's function is contained in the lineshape of the transitions observed in angle-resolved photoemission. Measuring the intrinsic lineshape requires high energy and momentum resolution together with high-quality samples. This has become possible in recent years (see Figures 9.5 and 9.6 and Hüfner (2007)). The energy width ΔE of a spectral peak determines the scattering time τ and the momentum width Δp determines the scattering length l:

$$\tau = \frac{\hbar}{\Delta E} \quad l = \frac{\hbar}{\Delta p} \tag{9.6}$$

For example, one can alloy 20% Fe into the Ni crystal to form permalloy, a key material for magnetoelectronics. Taking spectra similar to those in Figure 9.7, one finds that the momentum width greatly increases for the spin-down electrons but not for the spin-up electrons (not shown; see Altmann et al. (2001)). This leads to a scattering length l of 0.6 nm for spin-down electrons, which agrees with the result of magnetoresistance measurements.

The real part of the self-energy manifests itself in a transition from "dressed" electrons close to the Fermi level to "bare" electrons farther away from it. For example, an electron can be "dressed" by a phonon cloud due to electron–phonon interaction (Plummer et al., 2003). If the

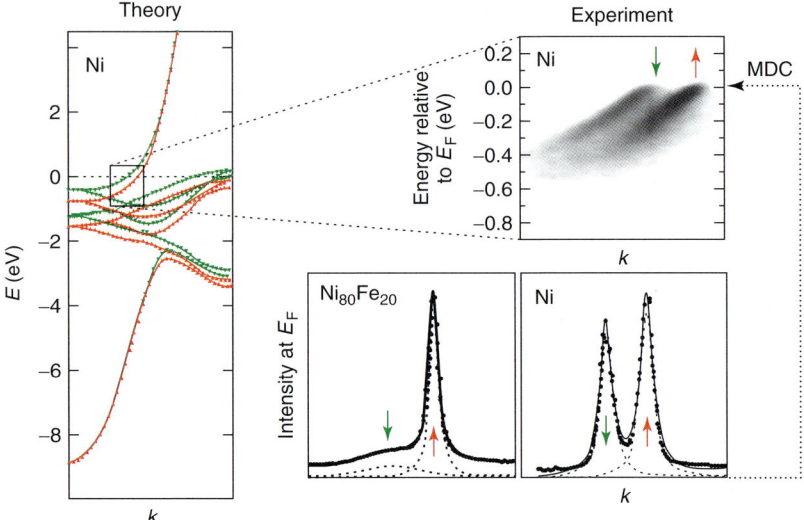

Fig. 9.5 Energy-versus-momentum band dispersions of ferromagnetic nickel, showing two sets of bands for the two spin directions (calculations on the left, data on the right). Modern electron spectrometers with energy and angle multidetection can zoom into the bands close to the Fermi level and image the band dispersion (upper right, with high photoemission intensity shown in dark shade). A momentum distribution curve (MDC) reveals the momentum broadening due to the finite scattering length (lower right). After Altmann et al. (2001).

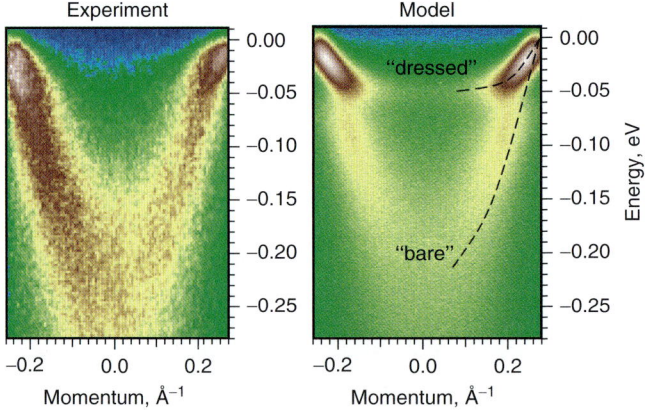

Fig. 9.6 Experimental and calculated plots of the photoemission intensity versus E and k for a high-temperature superconductor. There is a kink at about -0.04 eV, which indicates the transition from "dressed" to "bare" electrons. From Inosov et al. (2007).

electron moves too fast, the phonon cloud cannot follow and leaves a bare electron. This transition from dressed to bare electrons happens about one phonon energy away from the Fermi level. It manifests itself as a kink in the

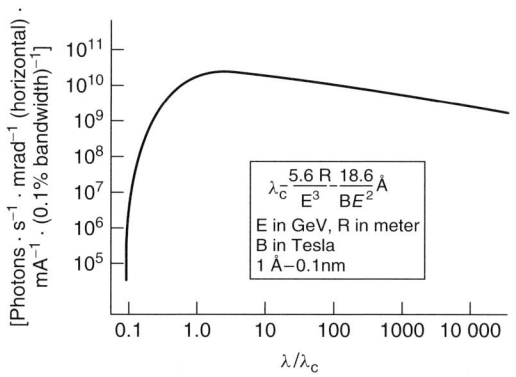

Fig. 9.7 The universal curve for the synchrotron radiation flux, plotted with λ/λ_c on the horizontal axis. λ_c is the critical wavelength, given by the electron energy E, the radius R of the electron orbit, or the magnetic field of the bending magnet.

$E(k)$ band dispersion where band curvature changes. The curvature is related to the inverse effective mass, and the mass changes from heavy for dressed electrons to light for bare electrons. An example is shown in Figure 9.6. This effect is particularly interesting for superconductors. Electrons are paired by exchanging phonons in conventional superconductors. The big question in high-temperature superconductors is the pairing mechanism, that is, the nature of the particle exchanged between pairs. The search for this particle has led to an intense study of kinks in the band-structure of high-temperature superconductors. The example in Figure 9.6 shows how to model such a kink self-consistently in terms of the real and imaginary parts of the self-energy (Inosov et al., 2007).

9.2 Instrumentation

9.2.1 Light Sources

To excite photoelectrons, a minimum photon energy of 5–10 eV is required. This is one of the reasons why conventional lasers have been of limited applicability in photoelectron spectroscopy. A detailed discussion of the possible sources for photoelectron excitation is given in Siegbahn and Karlsson (1982). In the ultraviolet radiation region, the most common source is based on a capillary glow discharge. The most convenient line is the so-called He I, which provides monochromatic radiation with a narrow line-width at 21.2 eV. This radiation originates from the 2p to 1s transition in neutral He atoms. Emission can also be produced from the He ions, the primary emission line being He II at an energy of 40.8 eV. Similarly, emission lines of Ne I at 16.8 eV and Ne II at 26.9 eV have frequently been used in photoelectron spectroscopy. Recent designs produce very efficient and highly monochromatic radiation by electron cyclotron resonance (ECR) in low-pressure He gas.

When a target is bombarded with an electron beam of sufficient energy and intensity, X-rays will be emitted via core-level fluorescence or bremsstrahlung. In principle, a vast amount of X-ray emitting materials should be available, but there are very few practical X-ray sources with high intensity, narrow energy width, and without satellite structures. The dominant radiation sources are Mg K$_\alpha$ at 1253.6 eV and Al K$_\alpha$ at 1486.7 eV. Because both magnesium and aluminum are easy to handle and their inherent line-width is as narrow as 1 eV, most commercial X-ray anodes are based on Al and/or Mg as the target material. It is also possible to produce monochromatized Al K$_\alpha$ and Mg K$_\alpha$ radiation by using, for instance, a quartz crystal monochromator. An excitation line-width of a few tenths of an electronvolt can be achieved, but with considerable loss of intensity. These technical approaches of achieving higher resolution have found their way into commercial instruments.

During the last two decades, synchrotron radiation has emerged both as a powerful and convenient excitation source in photoelectron spectroscopy (Winick and Doniach 1980; Koch, 1983; Winick et al., 1989). The term *synchrotron radiation* is usually associated with the radiation emitted by relativistic electrons accelerated in a circular orbit. The basic properties of synchrotron radiation were first presented in an elegant paper by Schwinger (1949), although they had been treated much earlier by Schott (1912). With E being the electron energy and R the radius of the circular orbit, the total synchrotron radiation energy emitted per revolution by one electron is given by

$$\Delta E(R, E) = \frac{4\pi}{3} \times \frac{e^2}{R} \times \left(\frac{E}{m_0 c^2}\right)^4 \quad (9.7)$$

The spectral radiation distribution is characterized by a critical energy ε_c, given by

$$\varepsilon_c = \frac{3hc}{2R} \times \left(\frac{E}{m_0 c^2}\right)^3 \quad (9.8)$$

The physical significance of this equation is that half of the total power is radiated above the critical energy and half below. The photon intensity rapidly decreases for photon energies above the critical energy ε_c. A universal curve for the synchrotron radiation photon flux is shown in Figure 9.7. Synchrotron radiation has a number of desirable properties. It provides (i) a continuous spectral distribution from the infrared region into the X-ray region, (ii) high intensity, (iii) high degree of collimation, (iv) high degree of polarization (completely linearly polarized in the plane of the orbital and elliptically out of the plane), and (v) a pulsed time structure given by the orbital frequency of the circulating electron beam. An important development occurred with the realization that magnetic structures inserted in straight sections of the storage ring, so-called undulators and wigglers, could drastically improve the radiation characteristics of conventional bending magnets. Presently, severalz facilities are in operation and under construction that are optimized for these insertion devices, so-called third-generation sources. Figure 9.8 (Winick, 1994) shows the brightness of X-ray sources as a function of time since the X-ray tube was first introduced in 1895. Synchrotron radiation provides a few orders of magnitude higher brightness than conventional sources. Intense efforts are presently under way for the next step of source developments, that is, vacuum ultraviolet (VUV) and X-ray free-electron lasers (XFEL), also based on undulators but using linear accelerators instead of storage rings and using seed lasers. In addition to the properties listed above, such sources will provide femtosecond time resolution, full transverse coherence, and about ten orders of

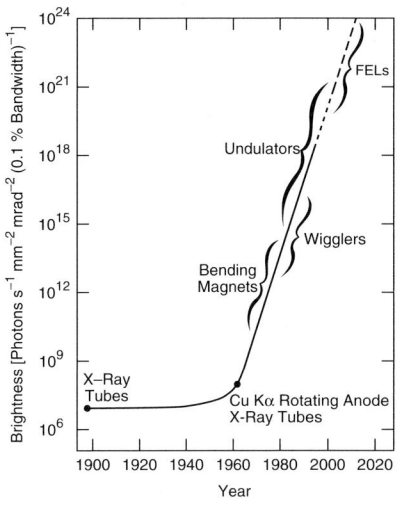

Fig. 9.8 Spectral brightness of X-ray sources as a function of time. From Winick (1994).

magnitude increase in peak brightness. Coupled with the rapid developments of conventional lasers toward even shorter wavelengths, the next decade will undoubtedly see spectacular improvements in light sources for photoemission (Chattopadhyay et al., 2001).

An important advancement in laser-based sources for photoelectron spectroscopy has recently been achieved by developing a UV laser that can reach a photon energy of 6.994 eV while producing 10^{15} photons per second (Kiss et al., 2005). It operates in a quasi-continuous-wave (quasi-CW) mode at a repetition frequency of 80 MHz. The high photon energy of 6.994 eV is achieved via the second harmonic of a frequency-tripled Nd : YVO$_3$ laser by using an optically contacted prism-coupled KBBF (KBe$_2$BO$_3$F$_2$) crystal (Togashi et al., 2003). The setup is shown schematically in Figure 9.9(a). The electrons emitted from the sample have low kinetic energies, typically a few electron volts, and their energy and angle distributions are determined with a high-precision hemispherical electron analyzer. The laser source with unprecedented performance characteristics in combination with a high-resolution spectrometer has made it possible to measure spectra with a total resolution down to 360 µeV for electrons plus photons (Figure 9.9b). This points to the importance of developing novel sources in parallel with improvements in detector technology (as discussed in Section 9.2.2). The impact of these new laser sources is also related to the basic photoemission phenomena discussed in Section 9.1.2. The mean free path of photoelectrons increases with the decrease in kinetic energy, with the implication that the laser source discussed here makes it possible to study bulk phenomena. This is of particular importance for materials with large unit cells. The combination of ultrahigh energy resolution and bulk sensitivity has opened up new avenues to study complex phenomena, such as high-temperature superconductivity.

9.2.2
Detectors

The most common type of photoelectron spectrometers has been the hemispherical analyzer, which is of the electrostatic type. A key feature of efficient spectrometers is parallel detection. Energy

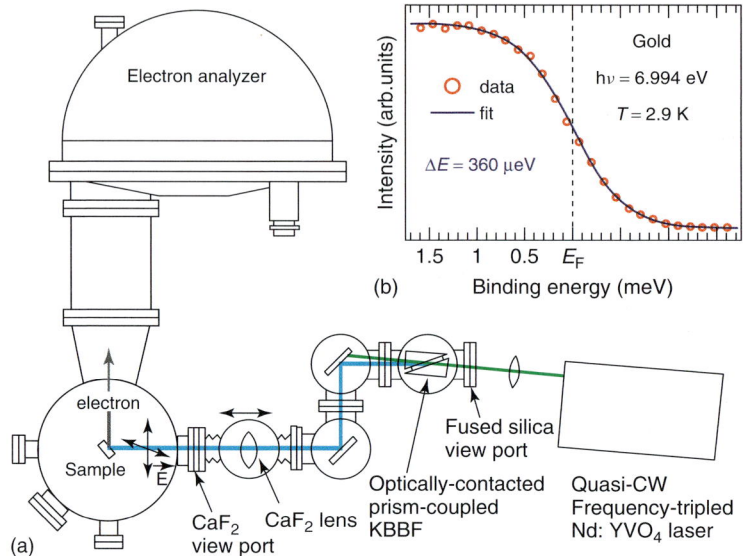

Fig. 9.9 (a) Schematic of a laser-based setup for high-resolution photoemission measurements. (b) Fermi edge of gold at low temperature, demonstrating 360 μeV overall resolution. From Kiss et al. (2005).

multidetection in hemispherical analyzers has been widely used for core-level spectroscopy (Section 9.4). There exist custom designs with more than one variable being detected simultaneously, such as energy and one angle in toroidal analyzers and two angles in display spectrometers. Simultaneous energy and angle multidetection has become available commercially with hemispherical spectrometers (Martensson et al., 1994), which has led to a renaissance of angle-resolved photoemission from valence states. The states very close to the Fermi level E_{Fermi} can now be mapped with an energy resolution comparable to the thermal energy and a momentum resolution comparable to the inverse mean free path. This is illustrated in Figure 9.5 for ferromagnetic nickel, in which the energy-versus-momentum band dispersion is plotted (Altmann et al., 2001). The $E(k)$ relation can be viewed directly on a television monitor, and the data in Figure 9.5 (upper right) are simply the time average of such a screen image. Currently, a third dimension is added to the multidetection method by incorporating time-of-flight energy detection into a spectrometer that displays both in-plane momentum components. Combined with short laser pulses, one can achieve very high energy resolution.

ZEKE (zero kinetic energy spectroscopy) was invented about 25 years ago (Schlag, 1998) and has developed rapidly since then (Cockett, 2005). Its main impact has been on high-resolution studies of molecular ions where it has yielded information on transition states in chemical reactions. ZEKE can provide energy resolutions 2–3 orders of magnitude higher than conventional photoelectron spectroscopy. The ZEKE technique is based on laser ionization of the molecule followed by a delayed

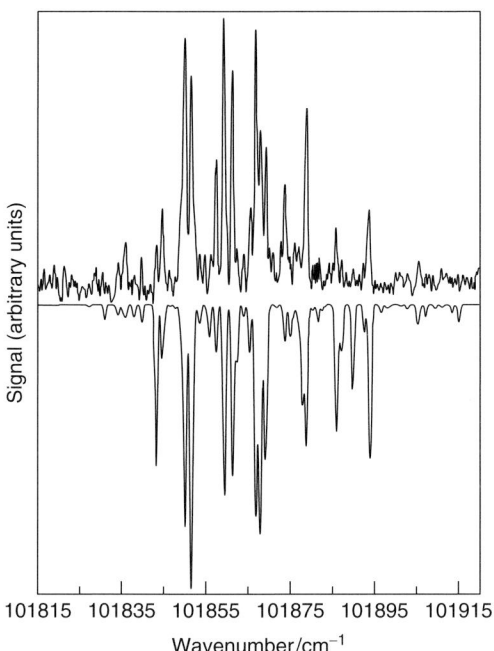

Fig. 9.10 Top: Rotationally resolved ZEKE photoelectron spectrum of a single vibrational structure in deuterated methane (CD$_2$H$_2$). Bottom: A simulated spectrum (inverted), broadened by a Gaussian line shape (0.7 cm^{-1} = 87 µeV full width half maximum). From Signorell and Merkt (2000).

electric pulse. In this way, electrons with zero kinetic energy can be separated out from the rest of the emitted electrons. Thus in a time-of-flight arrangement, the zero kinetic energy electrons can be collected separately. The physics behind ZEKE is intimately connected to Rydberg states. Rydberg states have hydrogenic characteristics with a very large principal quantum number n. The electrons in Rydberg states reside in very large orbitals and thus are much less sensitive to decay processes. This is the key for achieving high resolution. The ZEKE method is based on field ionization from the large-n Rydberg states. Even though conventional photoelectron spectroscopy with HeI radiation has been used routinely to study the vibrational fine structure of molecular energy levels, ZEKE makes it possible to resolve the entire set of vibrational levels, not only for the electronic ground state but also for the first excited state (Signorell and Merkt, 2000). It has also been demonstrated that fully resolved rotational spectra provide unprecedented information on the structure of molecular ions, for example, in a number of isotopomers of the methane ion (Keski-Rahkonen and Krause, 1974) (see Figure 9.10). Most of them exhibited such a small Jahn–Teller effect that structural changes can be ascribed to tunneling phenomena. For a normal Jahn–Teller effect, by contrast, the structure will be frozen to a specific minimum.

For detecting the spin of photoelectrons, one typically measures the right/left asymmetry in the electron scattering cross section at heavy nuclei (Mott scattering), after accelerating photoelectrons to 100 keV. A less bulky detector uses scattering at the electron cloud surrounding a heavy nucleus, typically at 150-eV electron energy. It lacks the absolute calibration of a Mott detector, however. In general, spin detection costs 3–4 orders of magnitude in detection efficiency.

9.3
Photoelectric Yield and Absorption Spectroscopy

9.3.1
Mechanism

This section starts with the simplest measurement in photoemission, which consists of determining the total photocurrent versus photon energy. Subsequent sections will deal with energy- and angle-resolved differential cross sections.

Before sophisticated measurements became routine, the determination of the quantum yield near the photoelectric threshold dominated the field. Such measurements were used to confirm the Einstein formula (Millikan, 1916), and to obtain the work function Φ of various materials by finding the photon energy for the onset of photoemission. Accurate determinations involved various extrapolation schemes, predicting certain power laws for the photoelectric yield near threshold (Fowler, 1931; Cardona and Ley, 1978). The photoelectric yield increases roughly like the square of the excess energy, because photoemission is dominated near threshold by the narrowing of an "escape cone" in momentum space that contains the electronic states with sufficient momentum perpendicular to the surface to overcome the surface potential step. Quantum yield measurements had technological consequences in the development of photocathodes for photomultipliers and night vision goggles.

The bulk of today's yield measurements is performed in a regime far above the photoelectric threshold, that is, at core-level absorption edges. In this case, the total electron yield (TEY) is closely related to the absorption coefficient, as long as the escape depth of the photoelectrons is short compared to the absorption length of the photons (Gudat and Kunz, 1972). Essentially, the number of photoelectrons is proportional to the number of core holes generated, which in turn is proportional to the absorption coefficient in the linear absorption regime near the surface. In the following, we will discuss applications of near edge X-ray absorption fine structure (NEXAFS) measurements. They involve transitions from a core level to the lowest unoccupied states. These are typically atomic/molecular orbitals that characterize the local chemical bonding and the magnetism at a specific atom. In semiconductors one observes the lowest conduction bands. Electron yield measurements over a wider photon energy range (hundreds of electronvolts) are usually labeled as EXAFS (extended X-ray absorption fine structure). The related technique of photoelectron diffraction is discussed in Section 9.6.1.

9.3.2
Unoccupied Atomic and Molecular Orbitals

Many molecules or polymers have characteristic unoccupied orbitals that dominate the absorption fine structure within the first tens of electron volts above a core-level threshold, most notably the lowest unoccupied orbital (LUMO), which serves as the electron acceptor in chemical reactions. (Its occupied counterpart, the highest occupied molecular

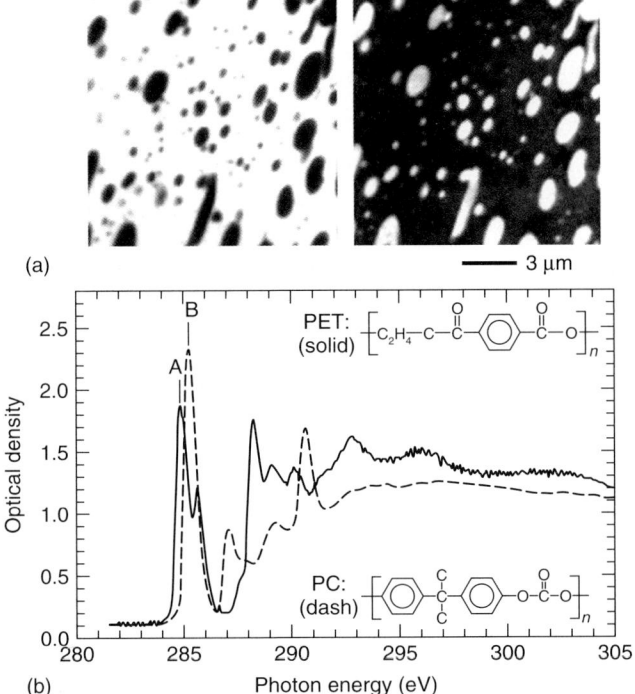

Fig. 9.11 Identification of different polymers in a blend by their π^* orbitals. The carbon 1s core-level absorption spectra (a) are used to identify the C1s to π^* transition energies of two polymers. By tuning into the respective π^* resonances, the two components can be distinguished in the micrograph (b). From Ade et al. (1992).

orbital (HOMO), can be measured by valence electron spectroscopy, which is discussed in Section 9.5.1.) The pattern of antibonding π^* and σ^* orbitals often serves as a fingerprint of a molecule. A large database of gas-phase spectra (Hitchcock and Mancini, 1994) helps in identifying surface species in NEXAFS (Stöhr, 1992). The polarization dependence of the absorption from different orbitals provides the orientation of adsorbed molecules. Similar applications exist with polymers. Figure 9.11 shows that two different polymers in a blend can easily be distinguished by their π^* orbitals (Ade et al., 1992).

9.3.3 Solids

Absorption spectroscopy in solids provides information about the unoccupied states, like in the gas phase. In contrast to molecular systems, in which the Coulomb interaction between the excited electron and the core hole significantly distorts the energy scale, there is enough screening in many solid state systems to use the absorption spectrum directly as a measure of the density of unoccupied states. By using different core levels, it is possible to project the density of states onto a particular atomic species and to select states with

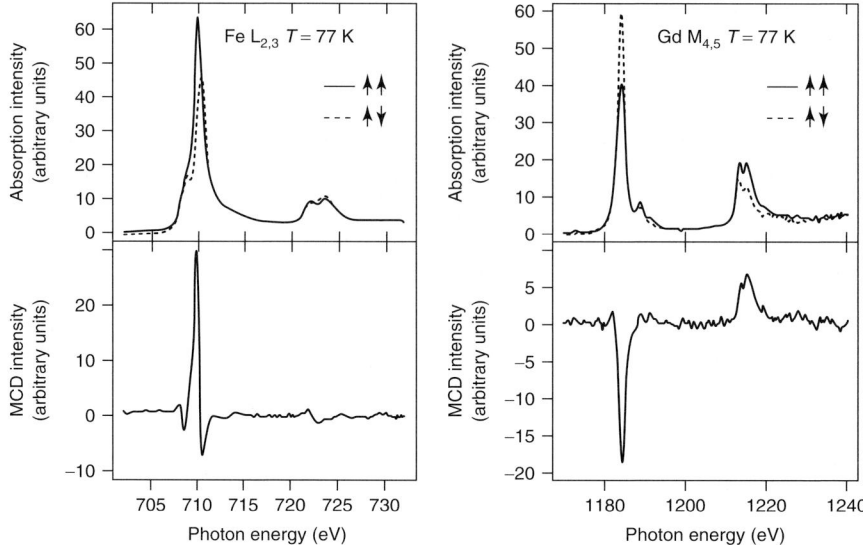

Fig. 9.12 Use of magnetic circular dichroism for determining the magnetization of the elemental species Fe (a) and Gd (b) in a garnet. Two absorption spectra for opposite magnetization directions are shown on top, and their difference at the bottom. From Rudolf et al. (1992).

a specific angular momentum l, using the $l \rightarrow (l \pm 1)$ selection rule. A more thorough analysis has to include the electron–hole interaction, which lowers the transition energy and leads to a characteristic multiplet structure, particularly for the localized $2p \rightarrow 3d$ and $3d, 4d \rightarrow 4f$ transitions.

9.3.4
Magnetic Dichroism

Using circularly polarized light with magnetic samples, it is possible to determine magnetic properties of specific atoms in magnetic alloys, at surfaces, and in molecules. Atomic sum rules produce an estimate of both the spin and the orbital contribution to the local magnetic moment. The two pieces of independent information come from the dichroism of the two spin-orbit partner lines of a core level, that is, the change in absorption induced by flipping the helicity of the light relative to the magnetization of the sample. In addition, one can play all the other selection rules, such as projecting the magnetic moment onto an atomic species or separating magnetism of d-electrons from that of f- and s-, p-electrons. Figure 9.12 gives an example, in which the magnetization of different constituents in a gadolinium iron garnet is determined (Rudolf et al., 1992). This ability of separating the magnetism of different elements is very helpful for tailoring magnetic materials in magnetic nanostructures (Himpsel et al., 1998).

9.3.5
Microspectroscopy

The recent development of third-generation synchrotron light sources with high brilliance makes it possible to perform absorption measurements with a spatial resolution down to tens of nanometers. The two pictures of a polymer blend in Figure 9.8 were taken at two different photon energies, where the two components absorb differently because of their different π^* orbital energies (Ade et al., 1992). In this case, the absorption coefficient was measured directly in a scanning transmission X-ray microscope (STXM). Such experiments are frequently performed with an imaging photoemission electron microscope (PEEM). Microspectroscopy promises to become a versatile method that identifies not only the elemental species in a sample by the choice of the core-level absorption edge but also their chemical state by selecting transitions into individual unoccupied valence states. There are many applications in magnetism, biology, geology, tribology, and environmental science.

9.4
Core-Level Spectroscopy

9.4.1
Origin of Core-Level Shifts

The photoelectric effect shown in Figures 9.1 and 9.2 applies not only to valence electrons but also to the deeper core levels, provided the incoming photons have sufficient energy to overcome the core-level binding energy. Analogous to Eq. (9.1), it is possible to obtain the binding energy of a core electron from the photon energy and the kinetic energy. In a simplified picture, the electrons will thus appear at discrete energies characteristic for each element. XPS is thus element-specific. The binding energy is also sensitive to the chemical environment of an atom, which led to the acronym ESCA.

Quantitative calculations of core-level shifts need to take into account both the initial state electronic structure (the chemical shift) and the relaxation in the final state (which is caused by the removal of an electron) (Fadley, 1978). In the most simplified picture, the magnitude of the chemical shift is correlated with the amount of charge transfer in the initial state. For instance, the binding energy of a core level increases with the oxidation state of an element, because the increasing positive charge on the atom binds a core electron more strongly. This is demonstrated for the oxidation states of silicon in Figure 9.13 (Himpsel, 1990). All oxidation states of silicon are resolvable, with the three peaks between Si and SiO_2 originating from intermediate oxidation states, Si^{1+}, Si^{2+}, and Si^{3+}. An electropositive ligand gives up electrons to silicon, causing a chemical shift in the opposite direction, as shown on the right side of Figure 9.13 for calcium. It is interesting to note that the chemical shift in SiO_2 increases slightly with increasing oxide thickness (not shown), a manifestation of decreased screening in the final state. The intensity of the different oxidation state peaks in Figure 9.13 can be used to estimate the amount of suboxides in the interfacial $Si-SiO_2$ region (Himpsel et al., 1988). This information has been quite valuable as a diagnostic tool for improving the gate oxide in silicon-base field-effect transistors and for developing high-k dielectrics to replace SiO_2 in the most recent generation of silicon microchips.

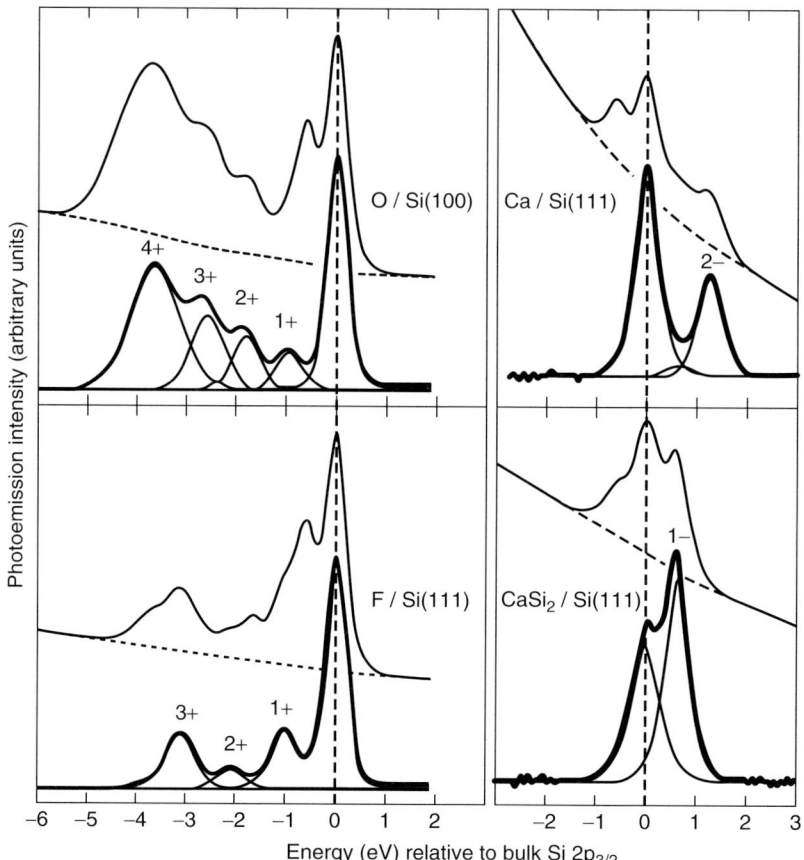

Fig. 9.13 Discrete chemical shifts, observed for different oxidation and reduction states of silicon. Electronegative ligands (O, F) give rise to chemical shifts toward higher binding energy and electropositive ligands (Ca) induce shifts to lower binding energy. Raw data are given on top of each panel. At the bottom, the secondary electron background has been subtracted and the Si $2p_{1/2}$ component removed. After Himpsel et al. (1988).

9.4.2
Atoms, Shake-Up, and Shake-Off Satellites

The core-level spectra from a closed-shell noble gas atom should in principle be very simple with discrete peaks corresponding to the different atomic subshells. However, final-state effects give rise to multielectron excitations, and atomic core-level spectra are characterized by a complex satellite structure, first discussed in Carlson (1967). The most likely multielectron process is a two-electron transition. However, higher order transitions and configuration interactions also contribute to the satellite spectra. It is common to distinguish between shake-up states and shake-off states, in which the second electron is excited to a higher bound state or an unbound continuum, respectively. The

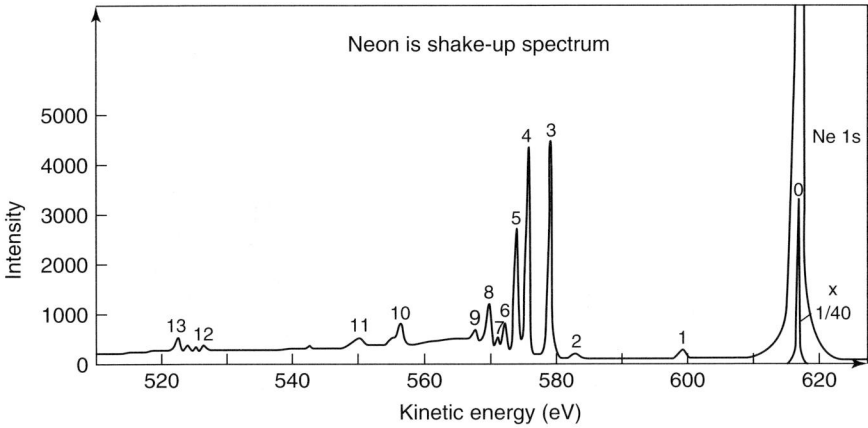

Fig. 9.14 The Ne 1s core-level photoelectron spectrum with accompanying satellites, labeled 1 through 13. From Gelius (1974).

Ne 1s core-level spectrum and the magnified satellite structure of the one-electron 1s line are shown in Figure 9.14. A detailed discussion and compilation of atomic core-level spectra and satellites can be found in Siegbahn and Karlsson (1982) and Sonntag and Zimmermann (1992).

9.4.3
Adsorbates and Surfaces

The chemical core-level shift of molecules adsorbed at surfaces provides information about the nature of chemisorption, dissociative or nondissociative, as well as structural information. Figure 9.15 illustrates the results for carbon monoxide on Pt(111) (Björneholm et al., 1993) by showing spectra for all relevant core levels (O 1s, C 1s, and Pt 4f). The clean Pt 4f spectrum contains two components, one for the topmost atomic layer, and the other for the underlying bulk Pt (Johansson and Mårtensson, 1980). CO forms a number of ordered phases on Pt (111), depending on the coverage. All of these have distinct arrangements of the CO molecule at the surface (on top, bridge site, and so on), and each of these structures leads to a characteristic chemical shift of the O 1s and C 1s levels. Chemically shifted peaks can thus provide information about both the bonding and the structure of adsorbed systems (Andersen et al., 1991). They are used extensively by companies involved in semiconductor materials processing, developing catalysts, tribology, and surface coating. Tunable synchrotron radiation makes it possible to optimize the surface sensitivity of core-level spectroscopy by detecting photoelectrons near the escape depth minimum (Figure 9.4).

Third-generation synchrotron radiation sources have opened up new avenues in core-level studies thanks to higher brightness and improved energy resolution. Time-resolved core-level spectroscopy is now feasible, as demonstrated for CO overlayers on Mo(110) in Figure 9.16 (Jaworowski et al., 2001). As the exposure to CO is increased, the C 1s level first shifts toward higher

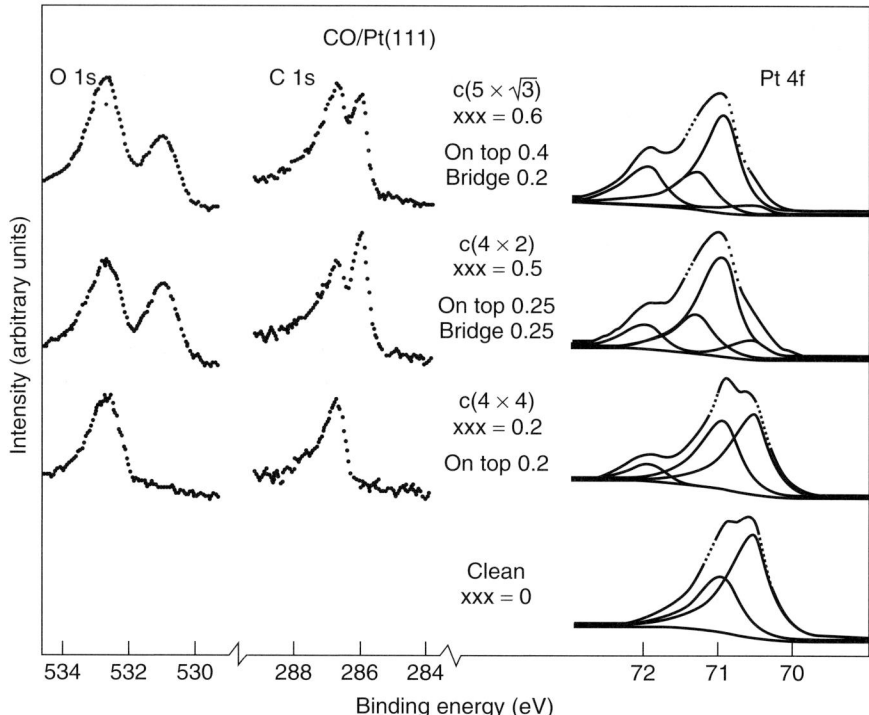

Fig. 9.15 The O 1s, C 1s, and Pt 4f core-level photoelectron spectra for various structures of CO chemisorbed on Pt(111). Adsorbate and substrate core-level shifts provide bonding site information. From Björneholm et al., (1993).

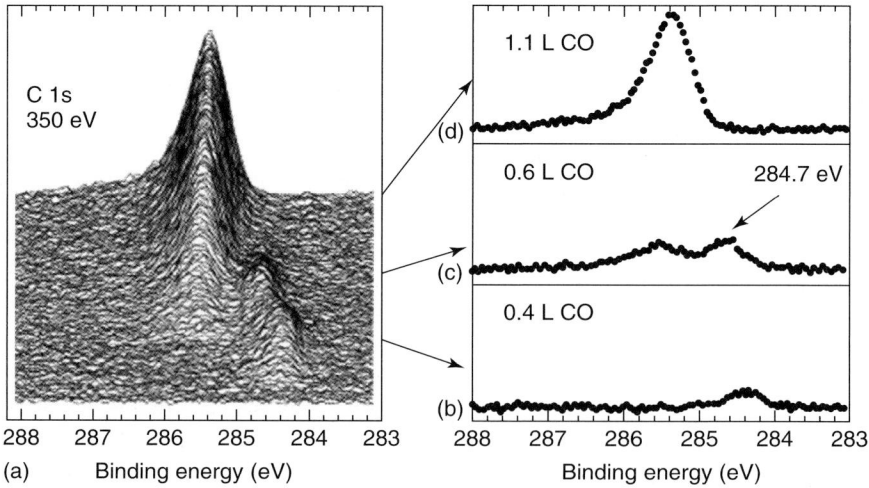

Fig. 9.16 Time-resolved core-level spectra from the C 1s region as a Mo(110) surface is exposed to CO. The recording time for each spectrum is 16 s. From Jaworowski et al. (2001).

binding energy and then develops a second component. A further increase of the CO coverage reduces the first component until it eventually disappears at saturation coverage. This evolution can be understood by a change in the tilt angle of the CO molecules at the surface. Such studies are valuable for understanding the precursor state before the molecular dissociates at higher temperature.

9.4.4
Solids

In Section 9.4.1, it was pointed out how core-level spectroscopy can be used for both qualitative (elemental and chemical specific) and quantitative analysis. Here, two other aspects of studies of solids are illustrated. The first example relates to the growth of thin films and their interfaces, a topic of particular importance for semiconductor structures (for an overview, see Weaver (1988)). The growth of a II-VI compound on a III-V semiconductor, that is, CdTe on InSb(100) is characterized by quantitative core-level spectroscopy using synchrotron radiation in Figure 9.17 (Mackey et al., 1986). Two different growth conditions are shown, that is, room temperature (a) and 500 K (b). The core-level spectra show the evolution of the intensity of the 4d levels of In and Sb (substrate), and Cd and Te (overlayers) with increasing CdTe deposition. At room temperature, the overlayer appears to be stoichiometric (no relative change in the Cd and Te core-level intensities), and the In–Sb 4d core levels are attenuated monotonically. The interface is obviously quite abrupt because the substrate core levels are covered up by 7–15 Å of CdTe. The situation is dramatically different for deposition at 500 K. The In 4 d signal does not decrease as much as Sb 4d, the Te 4d is broadened, and the Cd signal is barely detectable. In this case, chemical reactions take place at the interface and an interfacial layer with indium telluride forms with segregated Sb. It can be concluded that the interface is neither abrupt nor stoichiometric on an atomic scale.

A second example deals with core-level spectra, which are completely dominated by atomic final-state effects and are very insensitive to the chemical environment. The most prominent representatives can be found among rare-earth compounds. The 4f core-level spectra for a few rare-earth antimonides are shown in Figure 9.18 (Campagna, Wertheim and Bucher, 1976). All the three spectra can be described very well with final-state multiplets without any consideration of the effects from differences in the chemical environment. From the multiplet structure, one can directly read off the valence of the rare-earth atoms.

9.5
Valence Electron Spectroscopy

Valence states are more complex than core levels, because their wave functions extend out to neighbor atoms and interact with them. Consequently, the set of quantum numbers characterizing valence states is richer, containing the three momentum components as additional quantum numbers (in solids) and forming bonding–antibonding orbital combinations (in molecules). The extra quantum numbers are determined by using additional variables, such as the angular pattern of the photoelectron

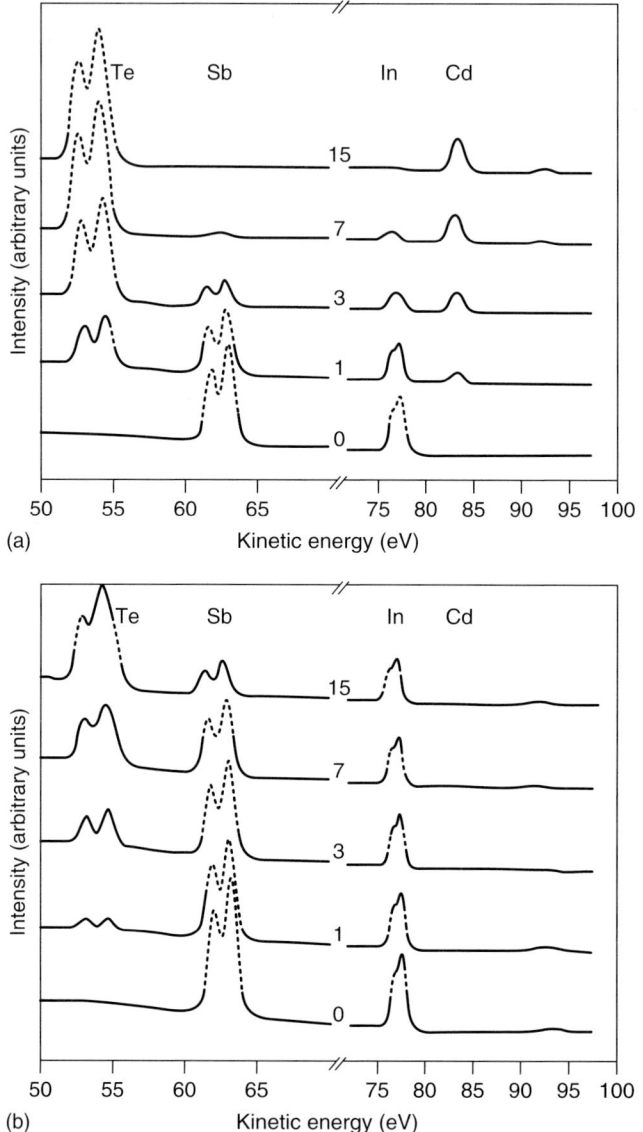

Fig. 9.17 Growth of CdTe on InSb(100), studied via the intensities of the 4d core levels (film thickness given in angstroms). At room temperature (a) the overlayer is stochiometric with a sharp interface. At 500 K (b) an interfacial layer of InTe is formed, together with segregated Sb. From Mackey et al. (1986).

and the photon polarization. The optimum energy range for mapping valence states has an upper limit of about 50 eV, because the cross section of valence states decays rapidly at higher photon energies.

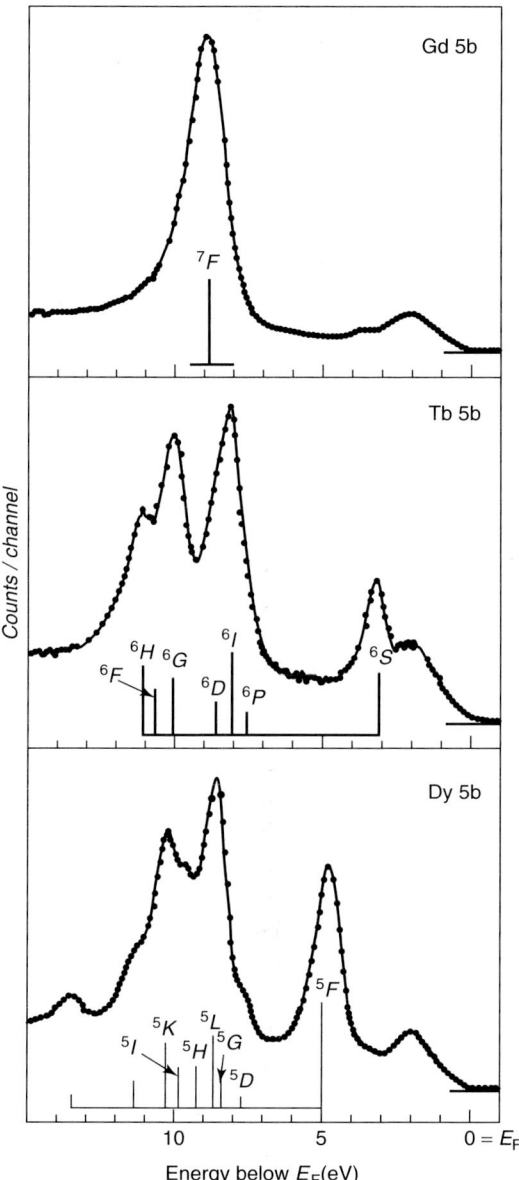

Fig. 9.18 Core-level spectra of the 4f levels for rare-earth antimonides. The agreement with calculated $4f^{n-1}$ multiplets shows that the multiplet structure reflects the number of 4f electrons in the final state. From Campagna, Wertheim and Bucher (1976).

9.5.1
Occupied Atomic and Molecular Orbitals

The valence orbitals of free molecules can be resolved in great detail by UPS (Turner et al., 1970), including vibrational and rotational fine structure. During adsorption at a surface, for example, in catalysis-oriented studies, the molecular orbitals can be used for fingerprinting the adsorbate (Figure 9.19 (Demuth and Eastman, 1974)). Orbitals

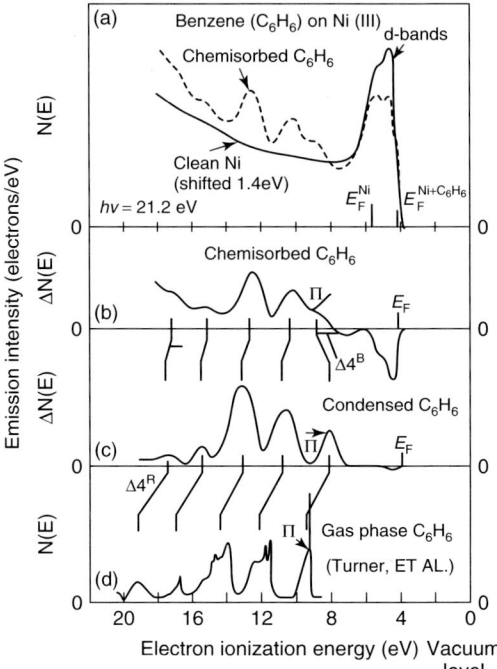

Fig. 9.19 Fingerprinting of adsorbed molecules by photoelectron spectroscopy. Benzene in the gas phase exhibits a characteristic pattern of valence orbitals (Turner et al., 1970), which is shifted rigidly in energy after condensing it at a surface. The stronger chemisorption bond on a nickel surface causes the uppermost bonding orbital to fall out of place. From Demuth and Eastman (1974).

not involved in the surface-to-adsorbate bond experience only a rigid, upward energy shift, which is due to dielectric screening at the surface. Therefore, the pattern of orbital energies at a surface is often sufficient to identify the adsorbate, to decide whether or not a molecule has decomposed, and to pinpoint the orbitals that are involved in the surface bond by their additional energy shift.

Angular effects are averaged out in randomly oriented molecules except for only a single parameter β, which determines the angular distribution $d\sigma/d\Omega$ of photoelectrons in terms of the polar emission angle θ relative to the polarization of the light:

$$\frac{d\sigma}{d\Omega} = \left(\frac{\sigma}{4\pi}\right) \times (1 + \beta P_2(\cos\vartheta)) \quad (9.9)$$

$P_2(x) = (3x^2 - 1)/2$ is the $l = 2$ Legendre polynomial. The asymmetry parameter β can take values between -1 and 2. At the magic angle $\vartheta = 54°$ the emission intensity is independent of β, making this a useful geometry for quantitative measurements of the cross section σ of a given orbital.

Molecules adsorbed at surfaces can be oriented by a crystalline substrate. In such a case, it becomes possible to determine their orientation by changing the polarization of the light and using dipole selection rules. For example, a p_z orbital is excited with the electric field vector along the z-direction, and a $p_{x,y}$ orbital with the component along the x, y- direction.

9.5.2
Surfaces and Quantum Wells

The surface disrupts the potential for electrons in a solid enough to give rise to separate surface states. They can have localized, broken-bond character, such as in semiconductors, or delocalized, evanescent wave character, such as s, p-states in metals. The d-states at metal surfaces lie somewhere in between. Surface states can be ideally probed with photoelectron spectroscopy when working at kinetic energies of 20–70 eV for the photoelectrons, where their escape depth goes through a minimum that is only a few atomic layers deep. Another way to become more sensitive to surface states is by placing the momentum and the kinetic energy of the photoelectrons in a band gap of bulk states. Both methods require tunable synchrotron radiation.

There are several criteria that help in distinguishing surface states from bulk states. The "crud" test is based on the assumption that a surface state is quenched by an adsorbate, which is true in most, but not all, cases. A cleaner method utilizes the two-dimensional nature of surface states, which implies that the energy of a surface state does not vary with the momentum perpendicular to the surface. This is tested by varying the photon energy at fixed parallel momentum. The most thorough test for a surface state consists of mapping all the bulk bands and ensuring that the state falls into a symmetry gap of the band structure. There are many surface resonances, however, that do not pass this last test but still exhibit strong localization in the outermost layer.

Quantum well states are two-dimensional, like surface states, but they are confined on both sides. Typically they are observed in highly perfect thin films where they are confined on the outer side by vacuum and on the inner side by a step in the inner potential. These films may be viewed as electron interferometers similar to a Fabry–Perot interferometer in optics. The standing electron waves inside the inerferometer correspond to quantum well states. Magnetic quantum well states play a role in magnetoelectronics, for example in the effect of giant magnetoresistance (GMR) that is used in reading heads for magnetically-stored data (compare Himpsel et al. (1998) for electronic states in magnetic nanostructures).

Quantum well states can be confined laterally in quantum wires and quantum dots, leading to one- and zero-dimensional electronic states. Currently, it is only possible to measure an ensemble of wires and dots where size inhomogeneities lead to an inhomogeneous broadening of the electronic states. Instruments for angle-resolved photoemission with a spatial resolution in submicrometer regime are under development that eventually will make it possible to study the sharp levels of a single nano object.

9.5.3
Band Mapping

The electronic structure of a crystalline solid is characterized mainly by the quantum numbers E and $\mathbf{p} = \hbar \mathbf{k}$, plus spin and point group symmetry (see Section 9.1.3). This information is traditionally compounded into $E(\mathbf{k})$ band dispersion plots with the appropriate symmetry labels for each band (Figures 9.5 and 9.20). The experimental techniques that can probe the complete set of quantum numbers are photoemission and inverse photoemission, with photoemission probing occupied states and inverse photoemission unoccupied states (see Figure 9.1). There are just

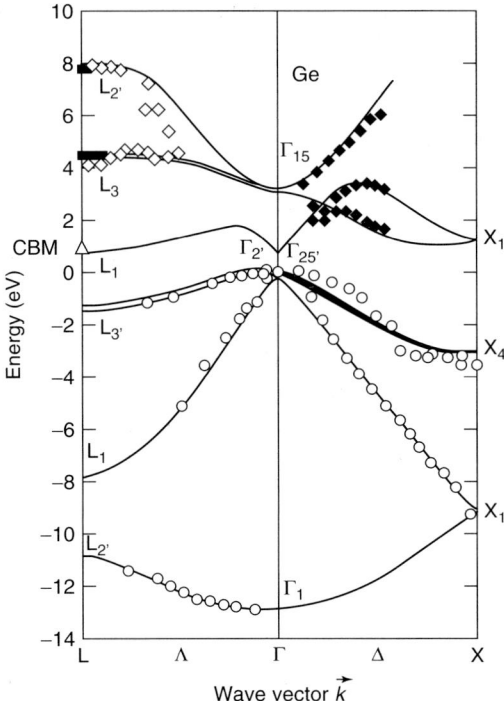

Fig. 9.20 Energy-versus-momentum band dispersions for germanium obtained from angle-resolved photoemission, inverse photoemission, and quasiparticle band calculations. Such band dispersions provide the complete information about electrons in solids. The labels denote the point group symmetry. From Wachs et al. (1985), Ortega and Himpsel (1994), and Hybertsen and Louie (1986).

enough variables in these experiments to cover all the unknowns. Roughly speaking, E is determined from the kinetic energy of the electrons, the two momentum components parallel to the surface, $k_{\parallel}$, from the polar and azimuthal angles of the electrons, the perpendicular momentum from the photon energy, spin from the spin of the electrons, and the point group symmetry from the photon polarization. A more thorough analysis, particularly concerning the perpendicular momentum, can be found in various reviews (Himpsel, 1983; Kevan, 1992). Figure 9.20 shows the results from band mapping for germanium, both experimentally and theoretically. Angle-resolved photoemission and inverse photoemission results are compared with state-of-the-art quasiparticle calculations (Wachs et al., 1985; Ortega and Himpsel, 1994; Hybertsen and Louie, 1986). Traditional local density calculations would not give the correct band gap for germanium, and in fact predict it to be metallic. A substantial body of experimental and theoretical band dispersions has been accumulated, and is being compiled as the basic set of electronic structure data for solids (Goldmann and Koch, 1989).

Multidetection allows it to determine the photoemission intensity versus E and $\mathbf{k}$ in quasi-continuous fashion and thereby make a connection to the imaginary part of the Greens function that describes electronic states in the presence of many-electron interactions (see Section 9.1.3). The photoemission intensity $I(E, \mathbf{k})$ can be displayed versus E, k_x (as shown in Figures 9.5 and 9.20) or versus k_x, k_y (as shown in Figure 9.21;

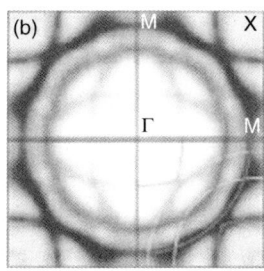

Fig. 9.21 Fermi surface of the two-dimensional material strontium ruthenate. From Damascelli et al. (2000).

(Damascelli et al., 2000)). The former represents the $E(k_x)$ band dispersion, the latter the Fermi surface.

Current developments are directed toward high-energy resolution, to map the electronic states at the Fermi level that determine transport phenomena and phase transitions, such as superconductivity (Damascelli et al., 2000; Inosov et al., 2007) and magnetism (Altmann et al., 2001).

9.5.4 Time-Resolved, Pump-Probe Measurements

The lifetime of electronic excitations and their decay mechanisms are at the heart of many technologies, such as optoelectronics and solar cells where long lifetimes are desirable. While optical experiments determine joint density of states, time-resolved photoemission is able to resolve the momentum of the states. The lifetime of electrons and holes strongly increases when approaching the Fermi momentum. It can also become quite long near local band minima in momentum space.

The experiments are typically performed with pulsed lasers in a pump–probe mode, which leads to the name two-photon photoemission (for an overview see Fauster (2003). Typically (but not exclusively), low-energy photons pump electrons from the valence band into long-lived intermediate states and higher energy ultraviolet photons ionize them from there (Höfer et al., 1997; Fauster, 2003). These photons originate from the same laser, either directly or after frequency multiplication, and thus can be synchronized within tens of femtoseconds.

An example is provided in Figure 9.22, in which electrons are pumped into image states and then ionized by the probe pulse (Höfer et al., 1997). Image states are electrons bound to a metal surface by their own image potential, which generates a hydrogen-like series of states converging to the vacuum level, but with 16 times smaller binding energies than in hydrogen. Coherent pumping of several image states with an ultrashort pulse causes quantum oscillations between different image states. These show the time dependence of the photoemission intensity as in Figure 9.19.

9.5.5 Inverse Photoemission

Inverse photoemission or bremsstrahlung isochromat spectroscopy (BIS) may be viewed as the time-reversed photoemission process (see Figure 9.1), with an incident electron and an emitted photon. It complements photoemission by probing unoccupied states, and in

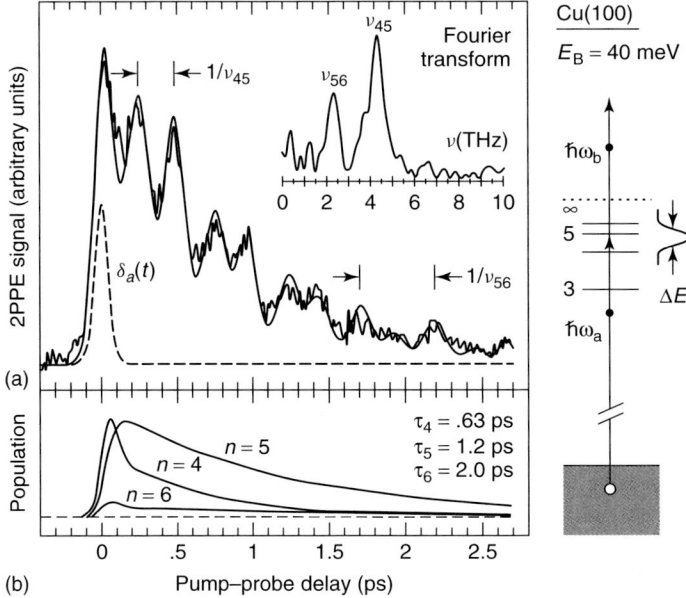

Fig. 9.22 Quantum beats in time-resolved photoemission. A femtosecond laser excites several quantum well states coherently, leading to an oscillatory population change between different states. From Höfer et al. (1997).

particular, the unoccupied states between the Fermi level and the vacuum level that are inaccessible to photoemission, neither as initial nor as final state. This gap contains many interesting states, for example, minority spin states in ferromagnets, broken-bond orbitals at semiconductor surfaces, and antibonding orbitals of adsorbed molecules. To visualize the territory covered by photoemission and inverse photoemission, one may examine the analogy between occupied and unoccupied states, bonding versus antibonding states, or donor versus acceptor states. Inverse photoemission poses experimental challenges, because its cross section is 4–5 orders of magnitude lower than that for photoemission, due to a phase space factor in the cross section. Typical elastic photoelectron yields are in the range of 10^{-3} electrons per photon, while typical inverse photoemission yields are 10^{-8} photons per electron. Nevertheless, this technique has become a routine complement to photoemission work, as exemplified by the conduction band data in Figure 9.17. Results on semiconductors are reviewed by Himpsel (1990), results on metals and adsorbates by Dose (1985) and Smith (1988).

9.6
Structural Methods

9.6.1
EXAFS and Photoelectron Diffraction

As mentioned in Section 9.3.1, EXAFS provides structural information. The ejected photoelectron from an absorbing atom can be described as an outgoing spherical wave. The outgoing photoelectron wave will be backscattered by neighboring atoms, which thereby will create incoming waves, one from each atom. The interference between the outgoing and incoming waves gives rise to modulation of the absorption coefficient for photon energies above the threshold. The periodicity of this modulation contains information about the distance between the absorbing atom and its neighbors, and the amplitude of the modulation is a measure of the number of neighbors, that is, the coordination number. The EXAFS technique is used extensively for structural determinations of solids, liquids, and biomolecules (Prins and Koningsberger, 1985; Teo and Jay, 1981; Teo, 1986). A surface-sensitive version of EXAFS, with the acronym SEXAFS, surface extended X-ray absorption fine structure can be applied to determine the surface structure of monolayers of adsorbates on surfaces (Prins and Koningsberger, 1985). Experimentally, two parameters are measured: adsorbate-substrate bond lengths and absorption sites.

9.6.2
Photoelectron Diffraction

Although an EXAFS experiment integrates over all outgoing electron energies and emission angles, there is additional information to be gained from the energy and the angular distribution of the diffracted photoelectrons. The outgoing waves of photoelectrons emitted from a specific core-level scatter at nearby atoms and produce intricate diffraction patterns. This information is used in photoelectron diffraction (PhD) (Fadley, 1993; Fadley *et al.*, 1994; Woodruff, 2001).

Experimentally, the intensity of a core level is measured either as function of the direction, or as the energy of the photoelectron, providing what is termed as *scanned-angle or scanned-energy data*. Considerable theoretical effort has been made into establishing accurate models relating photoelectron diffraction data to the surface structure (Fadley *et al.*, 1994). Compared to X-ray diffraction, the analysis is complicated by multiple scattering, which is inherent in the strong interaction of electrons with matter. That, on the other hand, provides high surface sensitivity. For photoelectrons with kinetic energies above 500 eV, it turns out that the scattering amplitude is highly peaked in the forward direction (forward focusing), which provides a direct determination of the bond directions for absorbed molecules and epitaxial relationships in thin films.

In addition to measure the momentum parallel to the surface via the angular distribution, it is also possible to vary the magnitude of the momentum by varying the photon energy, like in EXAFS. This requires the access to synchrotron radiation so that the photoelectron energy can be tuned continuously. This method is illustrated by a structure determination for the acetate species, CH_3COO- on the Cu (110) surface (Woodruff, 2001) (see Figure 9.23). Because the two carbon atoms are in different chemical environments, they can be distinguished by their chemical shift. The intensity modulations of the two chemically shifted peaks were measured in a well-defined emission direction as a function of the photoelectron kinetic energy. Such data can be analyzed by a multiple scattering model using trial structures (Woodruff, 2001), analogous to the determinaton of surface structures in LEED.

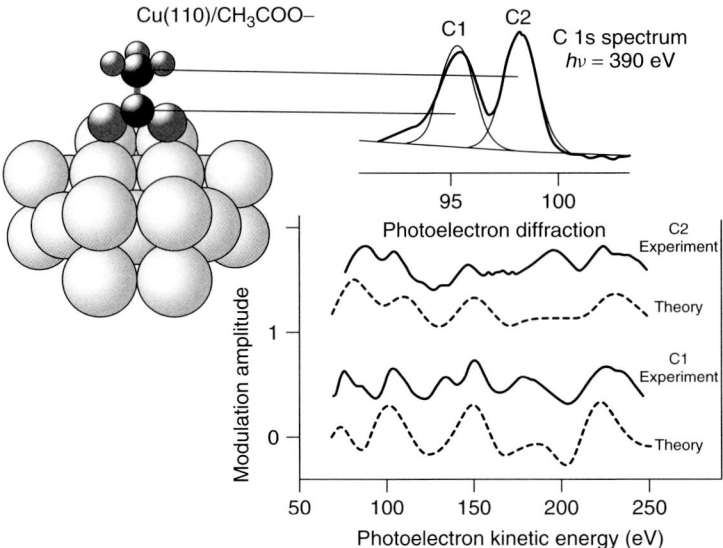

Fig. 9.23 Adsorption geometry of acetate (CH_3COO-) on the Cu(110) surface determined by photoelectron diffraction (PhD). From Woodruff (2001).

9.7 Summary and Outlook

Photoelectron spectroscopy has made significant contributions to the basic understanding of the electronic structure of atoms, molecules, solids, and surfaces. The technique has also been adopted into commercial instrumentation for routine qualitative and quantitative chemical analysis and materials characterization. Presently, thousands of research papers are published each year involving photoelectron spectroscopy in basic research or applications.

The development of the field of photoelectron spectroscopy, both for basic research and applications, has been closely tied to advancements in instrumentation, that is, better light sources and detectors. This trend will certainly continue in the foreseeable future. The third-generation synchrotron radiation sources are presently being implemented and will provide orders of magnitude improvements in brightness. It will be possible to study atoms, molecules, and solids with unprecedented spectral resolution. This will have a major impact for a better theoretical understanding of the complex photoelectron spectra of atoms and molecules. For solids and surfaces, major advances can be expected for studies of the electronic states around superconducting gaps in high-T_c materials, of charge density waves, surface transitions, and surface magnetism. Major breakthroughs should be anticipated in photoelectron microscopy in which all the assets of photoelectron spectroscopy will be applied with a spatial resolution down to a few tens of nanometers. Therefore, it is reasonable to predict that the tremendous growth of photoelectron spectroscopy during the last 40 years will continue well into the future.

9.8 Databases

Core-Level Binding Energies: Cardona and Ley (1978).

Atomic Core-Level Spectra: Siegbahn and Karlsson (1982).

Absorption and Core-Level Spectra of Metal Atoms: Sonntag and Zimmermann (1992).

Molecular Core-Level Absorption: Hitchcock and Mancini (1994). Avialable online at: http://unicorn.chemistry.mcmaster.ca/corex/cedb-title.html

Molecular Core-Level Energies: Bakke, Chen and Jolly (1980).

Molecular Valence Spectra: Turner *et al.* (1970).

Solid State Core and Valence Spectra: Goldmann and Koch (1989).

Cross sections: Yeh and Lindau (1985) and Keski-Rahkonen and Krause (1974).

Various databases are published regularly in Surface Science Spectra, published by the American Institute of Physics for the American Vacuum Society.

Glossary

Al K_α: X-ray line at 1.49-keV photon energy, emitted from the 1s level of aluminum. It is frequently used for core-level photoelectron spectroscopy (see Section 9.2.1).

ARPES: Angle-resolved photoelectron spectroscopy.

Auger Spectroscopy: Use of the Auger electrons emitted during the decay of core holes for elemental analysis at surfaces (see Figure 9.2).

BIS: Bremsstrahlung isochromat spectroscopy. A form of inverse photoemission with constant photon energy and variable electron energy, originally used in the X-ray regime (see Section 9.5.5).

ESCA: Electron spectroscopy for chemical analysis. Mapping of the core-level energies and their chemical shifts, typically using the K_α lines of Al or Mg at 1.49 keV and 1.25 keV, respectively (Siegbahn *et al.*, 1967). Synonym: XPS.

EXAFS: Extended X-ray absorption fine structure, determination of nearest-neighbor distances by interference effects in the core-level absorption spectrum (see Section 9.6.1).

GMR: Giant magnetoresistance, a strong variation of the electrical resistance in a magnetic field induced in sandwiches of magnetic and nonmagnetic metal layers.

GW: Calculation method for quasiparticles (see below), based on solving the Dyson equations with the full electron and photon propagators G and W (therefore the GW acronym), and a free particle approximation for the vertex. It goes beyond the LDA in determining the electronic structure for the excited state, which is measured in photoemission and inverse photoemission (see Section 9.1.3).

He I: Resonance line of helium gas at 21.2-eV photon energy, frequently used for photoelectron spectroscopy of valence states (see Section 9.2.1).

HOMO: Highest occupied molecular orbital in molecules. Together with the LUMO, the HOMO is responsible for most chemical reactions of a molecule, in particular charge transfer reactions, such as in a battery or in a solar cell.

LDA: Local density approximation, representing one of the most common methods for determining electronic states in solids and at surfaces (see Section 9.1.3).

LEED: Low-energy electron diffraction, often used in conjunction with photoelectron spectroscopy to characterize the ordering at surfaces.

LUMO: Lowest unoccupied molecular orbital in molecules. Together with the HOMO, the LUMO is responsible for most chemical reactions of a molecule.

Many-Body Theory: Many-body theory stands for a theory dealing with an infinite number of interacting particles. This is mathematically simpler than dealing with 10^{20} electrons in a macroscopic solid.

MCD: Magnetic circular dichroism, change of absorption between left- and right-handed, circularly polarized light, used for the determination of the magnetization at specific atomic sites (see Section 9.3.4).

NEXAFS: Near Edge X-ray absorption fine structure, mapping of unoccupied states via transitions from core levels (see Section 9.3). Synonyms: XANES, XAS, Partial Yield Spectroscopy.

PhD: Photoelectron diffraction, determination of the local atomic structure of surfaces (see Section 9.6.2).

Propagator: The propagator (or Green's function) describes the propagation of a particle from point A to point B in many-body theory. Mathematically, it is the expectation value of a particle creation operator at point A by an annihilation operator at point B.

Quasiparticle: In many-body theory, an electron loses its simple particle character by dragging other particles alon, such as phonons or other electrons in a solid. This whole particle cloud moves together and becomes a quasiparticle.

SEXAFS: Surface extended X-ray absorption fine structure, determination of nearest-neighbor distances at surfaces by interference effects in the core-level absorption spectrum (see Section 9.6.1).

Synchrotron Radiation: Versatile source of ultraviolet and X-ray light, emitted by high-energy electrons in a storage ring deflected by a magnetic field (see Section 9.2.1).

TEY: Total electron yield. One of the detection modes in XAS/NEXAFS/XANES spectroscopy where all electrons originating from the decay of the core hole are collected (see Section 9.3).

UPS: Ultraviolet photoelectron spectroscopy, mapping of the valence band structure of solids and surfaces with photon energies in the 5 to 50-eV range (see Figure 9.3).

XANES: X-ray absorption near edge structure, mapping of unoccupied states via transitions from core levels (see Section 9.3). Synonyms: NEXAFS, Partial Yield Spectroscopy.

XAS: X-ray absorption spectroscopy. Related terms: NEXAFS, XANES, Partial yield spectroscopy (see Section 9.3).

XPS: X-ray photoelectron spectroscopy. Mapping of the core-level energies and their chemical shifts, typically using the K_α lines of Al or Mg at 1.49 keV and 1.25 keV, respectively. See also ESCA.

XSW: X-ray standing wave, determination of the local atomic structure of surfaces (see Section 9.6.1).

ZEKE: Zero kinetic energy spectroscopy, a high-resolution spectroscopy for studies of molecular ions (see Section 9.2.2).

References

Ade, H., Zhang, X., Cameron, S., Costello, C., Kirz, J., and Williams, S. (1992) *Science*, **258**, 972.

Altmann, K.N., Gilman, N., Hayoz, J., Willis, R.F., and Himpsel, F.J. (2001) *Phys. Rev. Lett.*, **87**, 137201.

Andersen, J.N., Qvarford, M., Nyholm, R., Sorensen, S.L., and Wigren, C. (1991) *Phys. Rev. Lett.*, **67**, 2822.

Anderson, P.W. and Schrieffer, R. (1991) *Phys. Today*, **44**(June), 54.

Bakke, A.A., Chen, H.-W., and Jolly, W. (1980) *J. Electron. Spectrosc.*, **20**, 333.

Björneholm, O., Nilsson, A., Tillborg, H., Bennich, P., Sandell, A., Hernnäs, B., Puglia, C., and Mårtensson, N. (1993) in *MAX-Lab Activity Report 1992* (eds J.N. Andersen, L.S.O. Johansson, and R. Nyholm), National Laboratory, Lund, pp. 58–59.

Campagna, M., Wertheim, G.K., and Bucher, E. (1976) *Struct. Bond.*, **30**, 99.

Cardona, M. and Ley, L. (1978) *Photoemission in Solids I, Topics in Applied Physics*, Vol. 26. Springer, Berlin.

Carlson, T.A. (1967) *Phys. Rev.*, **156**, 142.

Chattopadhyay, S., Cornacchia, M., Lindau, I., and Pellegrini C. (eds) (2001) *Physics of and Science with the X-Ray Free-Electron Laser*, Vol. 581, American Institute of Physics, New York, Conference Proceedings.

Cockett, M.C.R. (2005) *Chem. Soc. Rev.*, **34**, 933.

Damascelli, A., Lu, D.H., Shen, K.M., Armitage, N.P., Ronning, F., Feng, D.L., Kim, C., Shen, Z.-X., Kimura, T., Tokura, Y., Mao, Z.Q., and Maeno, Y. (2000) *Phys. Rev. Lett.*, **85**, 5194.

Demuth, J.E. and Eastman, D.E. (1974) *Phys. Rev. Lett.*, **32**, 1123.

Dose, V. (1985) *Surf. Sci. Rep.*, **5**, 337.

Einstein, A. (1905) *Annn. Phys.*, **17**, 132.

Fadley, C.S. (1978) in *Electron Spectroscopy: Theory, Techniques and Applications*, Vol. 2 (eds C.R. Brundle and A.D. Baker), Academic Press, New York, pp. 1–156.

Fadley, C.S., Thevuthasan, S., Kaduwela, A.P., Westpal, C., Kim, Y.J., Ynzunza, R., Len, P., Tober, E., Zhang, F., Wang, Z., Ruebush, S., Budge, A., and Van Hove, M.A. (1994) *J. Electron. Spectrosc. Relat. Phenom.*, **68**, 19.

Fadley, C.S. (1993) in *Synchrotron Radiation Research: Advances in Surface Science* (ed. R.Z. Bachrach), Plenum, New York.

Fauster, Th. (2003), in *Solid State Photoemission and Related Methods*, Chapter 8 (eds W. Schattke and M.A. Van Hove), Wiley-VCH Verlag Gmbh, Weinheim.

Feuerbacher, B., Fitton, B., and Willis, R.F. (1978) *Photoemission and the Electronic Properties of Surfaces*, John Wiley & Sons, Inc., New York.

Fowler, R.F. (1931) *Phys. Rev.*, **38**, 45.

Gelius, U. (1974) *J. Electron. Spectrosc. Relat. Phenom.*, **5**, 985.

Goldmann, A. and Koch, E.-E. (1989) *Numerical Data and Functional Relationships in Science and Technology, New Series, Group III, Electronic Structure of Solids: Photoemission Spectra and Related Data*, Vols. 23a,b,c, Springer, Berlin.

Gudat, W. and Kunz, C. (1972) *Phys. Rev. Lett.*, **29**, 169.

Hertz, H. (1887) *Ann. Phys.*, **32**, 983.

Himpsel, F.J. (1983) *Adv. Phys.*, **32**, 1.

Himpsel, F.J. (1990) *Surf. Sci. Rep.*, **12**, 1.

(a) Himpsel, F.J., McFeely, F.R., Taleb-Ibrahimi, A., Yarmoff, J.A., and Hollinger, G. (1988) *Phys. Rev. B*, **38**, 6084; (b) Himpsel, F.J., Meyerson, B.S., McFeely, F.R., Morar, J.F., Taleb-Ibrahimi, A., and Yarmoff, J.A. (1990) in *Photoemission and Absorption Spectroscopy of Solids and Interfaces with Synchrotron Radiation* (eds M. Campagna and R. Rosei), North Holland, Amsterdam, pp. 203–236.

Himpsel, F.J., Ortega, J.E., Mankey, G.J., and Willis, R.F. (1998) **47**, 511.

Hitchcock, A.P. and Mancini, D.C. (1994) *J. Electron. Spectrosc.*, **67**, 1.

Höfer, U., Shumay, I.L., Reuss, Ch., Thomann, U., Wallauer, W., and Fauster, Th. (1997) *Science*, **277**, 1480.

Hybertsen, M.S. and Louie, S.G. (1986) *Phys. Rev. B*, **34**, 5390.

Hüfner, S. (ed.) (2007) *Very High Resolution Photoelectron Spectroscopy, Lecture Notes in Physics, Vol.715*, Springer, Heidelberg.

Inosov, D.S., Borisenko, S.V., Eremin, I., Kordyuk, A.A., Zabolotnyy, V.B., Geck, J., Koitzsch, A., Fink, J., Knupfer, M., Büchner, B., Berger, H., and Follath, R. (2007) *Phys. Rev. B*, **75**, 172505.

Jaworowski, A.J., Smedh, M., Borg, M., Sandell, A., Beutler, A., Sorensen, S.L., Lundgen, E., and Anderson, J.N. (2001) *Surf. Sci.*, **492**, 185.

Johansson, B. and Mårtensson, N. (1980) *Phys. Rev. B*, **21**, 4427.

Keski-Rahkonen, O. and Krause, M.O. (1974) *At. Data Nucl. Data Tables*, **14**, 139.

Kevan, S.D. (ed.) (1992) *Angle-Resolved Photoemission*, Elsevier/North Holland, Amsterdam.

Kiss, T., Kanetaka, F., Yokoya, T., Shimojima, T., Kanai, K., Shin, S., Onuki, Y., Togashi, T., Zhang, C., Chen, C.-T., and Watanabe, S. (2005) *Phys. Rev. Lett.*, **94**, 057001.

Koch, E.E. (ed.) (1983) *Handbook on Synchrotron Radiation*, North Holland, Amsterdam.

Mackey, K.J., Allen, P.M.G., Herrenden-Harker, W.G., Williams, R.H., Whitehouse, C.R., and Williams, G.M. (1986) *Appl. Phys. Lett.*, **49**, 354.

Martensson, N., Baltzer, P., Bruhwiler, P.A., Forsell, J.O., Nilsson, A., Stenborg, A., and Wannberg, B. (1994) *J. Electron. Spectrosc. Relat. Phenom.*, **70**, 117.

Millikan, R.A. (1916) *Phys. Rev.*, **7**, 18 and 355.

Ortega, J.E. and Himpsel, F.J. (1994) *Phys. Rev. B*, **47**, 2130.

Plummer, E.W., Shib, J., Tanga, S.-J., Rotenberg, E., and Kevan, S.D. (2003) *Prog. Surf. Sci.*, **74**, 251.

Prins, R. and Koningsberger, D. (eds) (1985) *X-Ray Absorption: Principles, Applications, Techniques of EXAFS, SEXAFS and XANES*, John Wiley & Sons, Inc., New York.

Rudolf, P., Sette, F., Tjeng, L.H., Meigs, G., and Chen, C.T. (1992) *J. Magn. Magn. Mater.*, **109**, 109.

Schattke, W., Van Hove, M.A., Garcia de Abajo, F.J., Diez Muino, R., and Manella, N. (2003) in *Solid State Photoemission and Related Methods*, Chapter 2 (eds W. Schattke and M.A. Van Hove, Wiley-VCH Verlag GmbH.Weinheim.

Schlag, E.W. (1998) *ZEKE Spectroscopy*, University Press, Cambridge.

Schott, G.A. (1912) *Electromagnetic Radiation*, Cambridge University Press, Cambridge.

Schwinger, J. (1949) *Phys. Rev.*, **75**, 798, 1912.

Siegbahn, H. and Karlsson, L. (1982) in *Encyclopedia of Physics*, Vol. XXXI (ed. S. Flügge), Springer, New York, pp. 215–468.

Siegbahn, K., Nordling, C., Fahlman, A., Nordberg, R., Hamrin, K., Hedman, J., Johansson, G., Bergmark, T., Karlsson, S.-E., Lindgren, I., and Lindberg, B. (1967) *ESCA: Atomic, Molecular, and Solid State Structure Studied by Means of Electron Spectroscopy*, Almquist and Wiksells, Stockholm.

Signorell, R. and Merkt, F. (2000) *Faraday Discuss.*, **115**, 205.

Smith, N.V. (1988) *Rep. Prog. Phys.*, **51**, 1227.

Stöhr, J. (1992) *NEXAFS Spectroscopy*, Springer Series in Surface Sciences, Vol. 25, Springer, Heidelberg.

Sonntag, B. and Zimmermann, P. (1992) *Rep. Prog. Phys.*, **55**, 911.

Teo, B.K. (1986) *EXAFS: Basic Principles and Data Analysis*, Springer, New York.

Teo, B.K. and Jay, D.C. (eds) (1981) *EXAFS Spectroscopy: Techniques and Applications*, Plenum, New York.

Togashi, T., Kanai, T., Sekikawa, T., Watanabe, S., Chen, C.-T., Zhang, C., Xu, Z., and Wang, J. (2003) *Opt. Lett.*, **28**, 254.

Turner, D.W., Baker, C., Baker, A.D., and Brundle, C.R. (1970) *Molecular Photoelectron Spectroscopy*, Wiley-Interscience, London.

Yeh, J.J. and Lindau, I. (1985) *At. Data Nucl. Data Tables*, **32**, 1.

Wachs, A.L., Miller, T., Hsieh, T.C., Shapire, A.P., and Chiang, T.C. (1985) *Phys. Rev. B*, **32**, 2326.

Weaver, J.H. (1988) in *Analytical Techniques for Thin Films, Treatise on Materials Science and Technology*, Vol. 27 (eds K.N. Tu and R. Rosenberg), Academic Press, New York, pp. 15–63.

Winick, H. (ed.) (1994) *Synchrotron Radiation Sources – A Primer*, World Scientific, Singapore.

Winick, H. and Doniach, S. (Eds) (1980) *Synchrotron Radiation Research*, Plenum Press, New York.

Winick, H., Xian, D., Ye, M.-H., and Huang, T. (eds) (1989) *Applications of Synchrotron Radiation*, Gordon & Breach, New York.

Woodruff, D.P. (2001) *Surf. Sci.*, **485**, 49.

10
Ultraviolet and Visible Light Spectrometers

John L. Hardwick

10.1	**Introduction**	**321**
10.2	**Design Considerations**	**322**
10.2.1	Luminosity or Étendue	322
10.2.2	Resolution	323
10.2.3	Dispersion	323
10.2.4	Efficiency	323
10.3	**Elements of a Generic Spectrometer**	**323**
10.3.1	Entrance Aperture	323
10.3.2	Collimating Optics	324
10.3.3	Wavelength Selecting Element	324
10.3.4	Focusing Optics	324
10.3.5	Exit Aperture	325
10.3.6	Detector	325
10.4	**Prism Spectrometers**	**325**
10.4.1	Dispersion	325
10.4.2	Resolution	326
10.4.3	Luminosity	326
10.4.4	Specialized Designs	326
10.4.5	Aberrations	328
10.4.5.1	Spherical Aberration	328
10.4.5.2	Coma	329
10.4.5.3	Astigmatism	329
10.4.5.4	Curvature of Field	330
10.4.5.5	Chromatic Aberration	330
10.5	**Grating Spectrometers**	**330**
10.5.1	Dispersion	330
10.5.2	Resolution	331
10.5.3	Luminosity	331
10.5.4	Design and Manufacture of Gratings	332
10.5.4.1	Transmission Grating	332
10.5.4.2	Reflection Grating	332

Encyclopedia of Applied Spectroscopy. Edited by David L. Andrews.
Copyright © 2009 WILEY-VCH Verlag GmbH & Co. KGaA, Weinheim
ISBN: 978-3-527-40773-6

10.5.4.3	Echelles	333
10.5.4.4	Holographic Gratings	333
10.5.5	Plane Grating Spectrometers	334
10.5.5.1	Aberrations	334
10.5.5.2	Types of Plane Grating Spectrometers	336
10.5.6	Concave Grating Spectrometers	338
10.5.6.1	Optical Aberrations of Concave Mirrors	338
10.5.6.2	Rowland Circle Mounts	339
10.5.6.3	Non-Rowland Circle Mounts	341
10.6	**Interferometric Spectrometers**	**342**
10.6.1	Michelson	342
10.6.2	Fabry–Perot	343
10.7	**Hybrid and Compound Instruments and Imaging Filters**	**345**
10.7.1	Order-Sorter for a Grating Spectrometer	345
10.7.2	Harrison Echelle	345
10.7.3	Fabry–Perot/Grating Instrument	346
10.7.4	Double and Triple Monochromators	346
10.7.5	Imaging Filters	347
10.7.6	The Sisam Spectrometer	347
10.8	**Detectors**	**347**
10.8.1	Ideal Detector	347
10.8.2	Types of Detectors	348
	Glossary	**351**
	References	**351**
	Further Reading	**352**

10.1
Introduction

The visible part of the electromagnetic spectrum extends from about 400 to 750 nm. At longer wavelengths, the retina loses sensitivity, whereas at shorter wavelengths, the lens of the eye absorbs most of the incident radiation. The ultraviolet spectrum extends from 400 to 180 nm. At wavelengths shorter than 180 nm atmospheric oxygen strongly absorbs light. Wavelengths between 180 and about 1 nm are classified as vacuum ultraviolet; wavelengths shorter than this are classified as X-rays.

A spectrometer is an instrument for discriminating among different wavelengths of light, producing a record of intensity as a function of wavelength or frequency from a particular source. Spectrometers are commonly used as tools in chemical analysis, atomic and molecular physics, astronomical measurements, and shaping and filtering optical radiation in both the time and the frequency domains.

Dispersion of light by a prism was reported by Johannes Marcus Marci in 1647. In 1666, Isaac Newton used a prism to disperse a beam of sunlight onto the wall of a darkened room; by using a lens to focus the light from the prism, he was able to disperse the visible spectrum over a length of about 25 cm. The first observations of spectral lines occurred 140 years later, when W. H. Wollaston and, independently, Joseph Fraunhofer restricted the entrance aperture of Newton's basic device with an entrance slit, allowing images of that slit to be formed as single separate colors. This innovation allowed the first observation of the "Fraunhofer lines" of the solar spectrum. Fraunhofer cataloged several hundred of these lines for use as in defining colors of the visible spectrum so that the refractive index of glasses at those colors could be more carefully defined. Fraunhofer also ruled the first diffraction grating and using it, he measured for the first time the wavelength of the yellow light from the sodium D lines. In addition, he was the first to observe stellar and planetary spectra using a prism placed in front of an astronomical telescope.

Forty years later, in 1859, Gustav Kirchhoff and Robert Bunsen developed the first practical spectroscope and showed that it could be used for chemical analysis. They demonstrated that atoms have characteristic absorption and emission spectra and recognized that the absorption and emission lines were identical in

Encyclopedia of Applied Spectroscopy. Edited by David L. Andrews.
Copyright © 2009 WILEY-VCH Verlag GmbH & Co. KGaA, Weinheim
ISBN: 978-3-527-40773-6

wavelength for a given atom; the appearance of a spectral line in absorption or emission depended only on the temperature of the sample. These developments mark the origin of optical spectroscopy as a branch of both analytical chemistry and chemical physics.

In 1862, G. G. Stokes used a quartz prism as a dispersing element and a fluorescent screen as a detector to make the first observations of ultraviolet atomic emission spectra. Although Stokes was unable to make wavelength measurements, he was able to observe spectra out to 186 nm. Extension of spectroscopic measurements into the vacuum ultraviolet was made by Victor Schumann in 1893. Using a fluorite prism and an evacuated spectrograph, Schumann was able to extend ultraviolet measurements to 120 nm, where absorption by the fluorite prism limited its useful range.

The production of diffraction gratings was greatly advanced in the late nineteenth century by the work of Henry Rowland on the development of an accurate ruling engine. Rowland also invented the concave grating – a grating ruled on a spherical mirror blank, which greatly simplified the construction of spectrographs. Using these gratings, Rowland was able to make accurate measurements of wavelengths of the ultraviolet spectrum down to 210 nm. In 1906, Theodore Lyman used such a concave grating to extend measurements in the vacuum ultraviolet down to 50 nm.

Theoretical understanding of atomic and molecular spectra had to await the rise of quantum mechanics in the early twentieth century. Beginning with the work of Niels Bohr, it became clear that spectra in the visible and ultraviolet regions depended on the internal electronic structure of atoms and molecules, and that the absorption and emission of radiation in this region corresponded to transitions of electrons between allowed energy levels.

10.2
Design Considerations

There are two figures of merit for any spectrometer, and generally they are mutually exclusive. They are *resolution*, the ability to distinguish between wavelengths of light that are nearly the same, and *luminosity*, the ability to accept light from a source and transmit all of it to the detector. Because improvement of one of these properties is generally at the expense of the other, a real spectrometer is always a compromise between the two.

10.2.1
Luminosity or Étendue

The light gathering power of a spectrometer is a function of both the area of the entrance aperture and the solid angle with which the instrument can accept incident light. The product of these two quantities is referred to as the *étendue of the instrument*. Terms that are often used synonymously are "throughput" and "luminosity." The étendue is defined as $L = a\Omega$, here a is the area of the entrance aperture and Ω is the solid angle subtended by the aperture stop of the instrument at the entrance aperture. For an optical system without losses (e.g., from scattering or absorption), L remains a constant through the system.

10.2.2 Resolution

The resolution of an instrument is the ability to discriminate between two adjacent wavelengths (or frequencies) of light. Resolution is most easily understood in the case of a monochromator, which transmits a certain range of wavelengths and rejects all others. The minimum attainable bandwidth of the instrument is its resolution, $\Delta\lambda$. A closely related quantity is the resolving power, R, defined as

$$R = \frac{\lambda}{\Delta\lambda} \tag{10.1}$$

In a dispersive instrument, the maximum possible resolution is most conveniently calculated as the product of the reciprocal angular dispersion $d\lambda/d\theta$ and the diffraction-limited angular resolution $\Delta\theta$ from the aperture stop imposed by the dispersing element; the angle θ is the deviation of the dispersed light. For a nondispersing spectrometer, the resolution in wavenumber units is the reciprocal of the difference in optical retardation of rays.

The resolution desired depends on the experiment to be performed. Ideally, the resolution of the spectrometer will be matched to the line width of the sample to be examined. For solids or liquids at moderate temperatures, absorption lines have a width of several nanometers or more, and the resolution requirements on the spectrometer are modest. For gases, the width of an absorption or emission line is usually limited by the Doppler width, which is typically 2×10^{-6} times the wavelength. A high-resolution spectrometer is therefore defined as one whose resolving power is Doppler limited, that is, $R \geq 5 \cdot 10^5$.

10.2.3 Dispersion

The angular dispersion of a grating or prism is the change of the displacement angle with wavelength, $d\theta/d\lambda$. The linear dispersion of a spectrometer is the change of the distance along the focal plane with wavelength $dl/d\lambda$; this will determine the physical separation of two spectral lines at the slit or detector. A more frequently used quantity is the reciprocal linear dispersion, $d\lambda/dl$, also called the "plate factor."

10.2.4 Efficiency

The product of the resolving power and the étendue (the "resolution-luminosity product") is sometimes referred to as the *efficiency of the spectrometer*. It is a property of the instrument in question and, sometimes, the wavelength of operation. For a grating spectrometer operating at a given wavelength, for example, the efficiency is fixed, but it is possible within limits to trade resolution for throughput by opening the entrance and exit slits.

10.3 Elements of a Generic Spectrometer

There are several elements which are common to most spectrometers, and these are illustrated in Figure 10.1.

10.3.1 Entrance Aperture

The *entrance aperture* is used to define the source of light to be analyzed. In a prism or grating spectrometer, it is ordinarily a

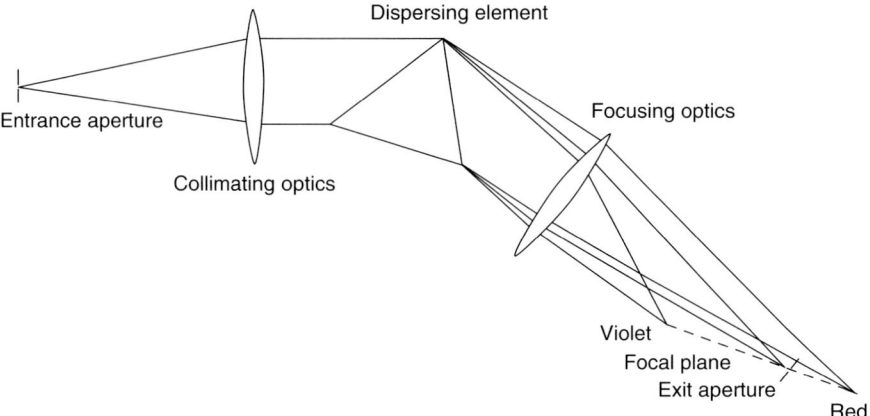

Fig. 10.1 A minimum deviation prism spectrometer, illustrating the elements common to most spectrometers.

slit, although a more complicated shape is sometimes used to enhance throughput. In an interferometric spectrometer, the entrance aperture is usually circular.

10.3.2
Collimating Optics

Collimating optics take the rays of light from the entrance aperture, which are typically diverging, and make the rays collinear. This is generally desirable so that all of the rays strike the dispersing element at the same angle; if they do not, often either the resolution is degraded, or the focus of the instrument changes with wavelength, or both. The collimating optics may be either refractive or reflective (i.e., either lenses or concave mirrors), although mirrors are generally used in modern instruments because of their lack of chromatic aberration.

10.3.3
Wavelength Selecting Element

The kind of wavelength selector in a spectrometer is usually classified as dispersive or nondispersive. Dispersive elements separate the wavelengths along some spatial dimension, while nondispersive elements pass some wavelengths and reject others without altering the optical path of the transmitted light. The most commonly used dispersing elements are the prism and the diffraction grating, although some interferometers, such as the Fabry–Perot, are also used in a dispersing mode. The most versatile nondispersing spectrometer is the Michelson interferometer, which is the basis of the modern Fourier transform spectrometer. Other examples of nondispersive elements are filters, which are used in simple colorimeters or in imaging spectrometers.

10.3.4
Focusing Optics

The focusing or condensing optics (in spectrographs often called the *camera* optics) take the collimated light passed by the wavelength selecting element and focus it onto an exit aperture (or several apertures) or a detector.

Together with the collimating optics, the focusing optics form an image of the entrance aperture at the exit aperture. The usual goal of spectrometer design is to make the collimating and focusing optics sufficiently aberration-free that the performance of the spectrometer will be limited only by the wavelength selecting element. As with the collimating optics, either lenses or mirrors may be used, with the preference being given to mirrors for most instruments.

10.3.5
Exit Aperture

The exit aperture must be matched to the image of the entrance aperture in such a way that it provides maximum discrimination of the wavelengths of interest while providing maximum allowable throughput of light. In a conventional grating or prism monochromator, for example, the exit aperture is a single exit slit in the shape of the image of the entrance slit.

10.3.6
Detector

The detector transforms the light into either a permanent image or an electronic signal. In addition to the requirement that it must be sensitive in the wavelength region of interest, the detector is chosen so as to minimize unwanted noise and provide a reproducible (ideally linear) response. Spectrometers are often distinguished from one another on the basis of the method of detection: an instrument which allows a spectrum to be observed visually is a *spectroscope*, one which makes a photographic record is called a *spectrograph*, and the term *spectrometer* refers to an instrument which records the spectrum photoelectrically. If photometric accuracy is of paramount importance, the device is called a *spectrophotometer*. In this chapter, the term *spectrometer* is used as an inclusive term covering all these types of instrument.

10.4
Prism Spectrometers

The earliest spectrometers were prism spectrometers, and the most commonly used grating spectrometer designs have their origins in prism spectrometers. Prism spectrometers, although they have generally poor optical performance in comparison with most modern designs, have the advantages of being inexpensive, rugged, easy to align, and technically undemanding.

10.4.1
Dispersion

Light incident on a prism with an angle α (measured normal to the face of the prism, as shown in Figure 10.2) will be refracted according to Snell's law, $\sin \alpha / \sin \alpha' = n_{\text{prism}} / n_{\text{air}}$, where n_{prism} and n_{air} are the refractive indices of air and the prism material at the wavelength of the incident light. On exiting the prism, the light will be refracted again by an angle β, for a total deviation θ of the beam of light. The deviation θ of the beam depends on the refractive index n of the prism, while the dispersion of the beam, $d\theta/d\lambda$, depends on the *change* of refractive index of the prism material with wavelength, $dn/d\lambda$. The angular dispersion of a prism is defined as the change of the angle of deviation θ with

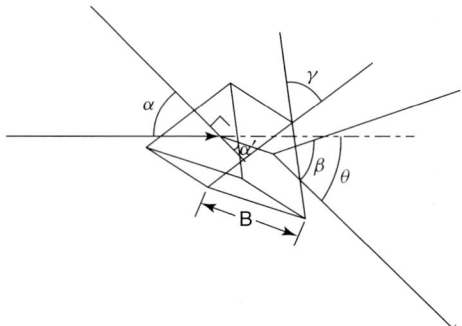

Fig. 10.2 Coordinates used in describing the dispersion of a prism.

wavelength and is given by the formula

$$\frac{d\theta}{d\lambda} = \frac{d\theta}{dn}\frac{dn}{d\lambda} \quad (10.2)$$

The first factor may be evaluated by differentiating Snell's law, while the second factor is a property of the material of the prism. If the angle of incidence α is nearly equal to the angle of departure and the refractive index of air is taken as 1, $d\theta/dn$ is given in terms of the apex angle γ of the prism as

$$\frac{d\theta}{dn} = \frac{2\sin(\gamma/2)}{\sqrt{1 - n^2 \sin^2(\gamma/2)}} \quad (10.3)$$

The linear dispersion of the spectrometer $dl/d\lambda = fd\theta/d\lambda$ is the angular dispersion times the focal length of the condensing optics. Note, however, that the focal length of the condensing optics can often change considerably over the spectrum as a result of the chromatic aberration of the condensing lens.

10.4.2
Resolution

The angular resolution of a prism spectrometer, in the absence of optical aberrations of the lenses, will be determined only by the prism itself. It may be written in terms of the reciprocal angular dispersion as

$$\Delta\lambda = \frac{d\lambda}{d\theta}\Delta\theta \quad (10.4)$$

where $\Delta\theta$ is the diffraction limit imposed by the wavelength of light and the aperture of the prism itself. The resolving power for a prism at minimum deviation (Figure 10.2) is given in terms of the base B as (Sawyer, 1963)

$$R = B \left| \frac{dn}{d\lambda} \right| \quad (10.5)$$

10.4.3
Luminosity

The effective area of the aperture stop of a prism instrument is the product of the area of the prism face and the cosine of the angle of incidence. The solid angle of acceptance of the instrument will be this area divided by the square of the focal length of the collimating lens. The product of this acceptance angle and the area of the entrance slit will constitute the luminosity.

10.4.4
Specialized Designs

The design of a prism spectrometer depends on whether a large flat focal

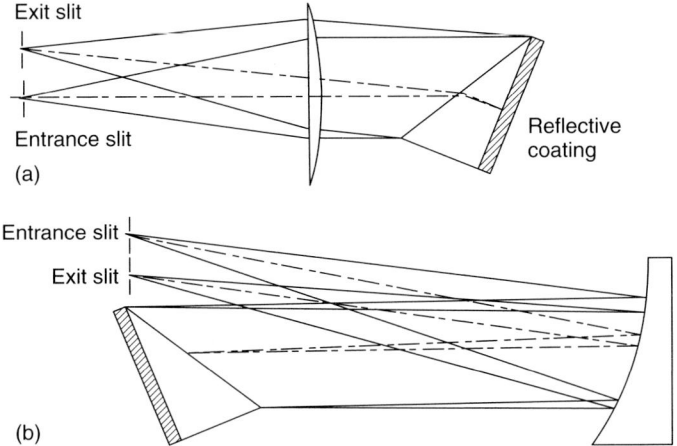

Fig. 10.3 Littrow mount prism spectrometers, (a) using lenses for imaging and (b) using an off-axis parabolic mirror for imaging. The entrance and exit slits would ordinarily be placed as close together as possible.

plane is most desirable (in the case of a spectrograph) or whether it is deemed necessary to scan the wavelength easily by rotating the prism (as in the case of a spectroscope or a spectrometer).

To minimize reflection losses, the total deviation of the light as it passes through the prism is kept to a minimum. This consideration is in conflict with the requirement that the base of the prism must be large to maximize both dispersion and resolution, and usually a 60° angle for the apex of the prism is chosen as a compromise between these two requirements. A typical minimum deviation spectrograph is illustrated in Figure 10.1. Light is collimated and focused by lenses in this design. If the lenses are of the same refractive index as the prism, then the focal surface is, to a first approximation, a plane that passes through the entrance slit (James and Sternberg, 1969).

A more compact design is the Littrow configuration, as shown in Figure 10.3(a). Here the prism is replaced by a prism having half the original apex angle whose rear surface is coated with a reflective coating. Light passes forward and then backward through the prism, and a single lens serves as both the collimating and focusing optic. The entrance and exit slits may be offset horizontally or, more frequently, vertically. This arrangement lends itself to scanning by simply rotating the prism, or an image of a spectrum may be formed as in the minimum deviation instrument. A variation which employs reflective optics is also used (Figure 10.3b). The mirror used to image the slit is chosen to be an off-axis paraboloid to minimize the astigmatism and spherical aberrations associated with a spherical mirror used off-axis. There are so many variations on the Littrow design that any spectrometer which disperses light back along the original path is often referred to as a *Littrow mount*.

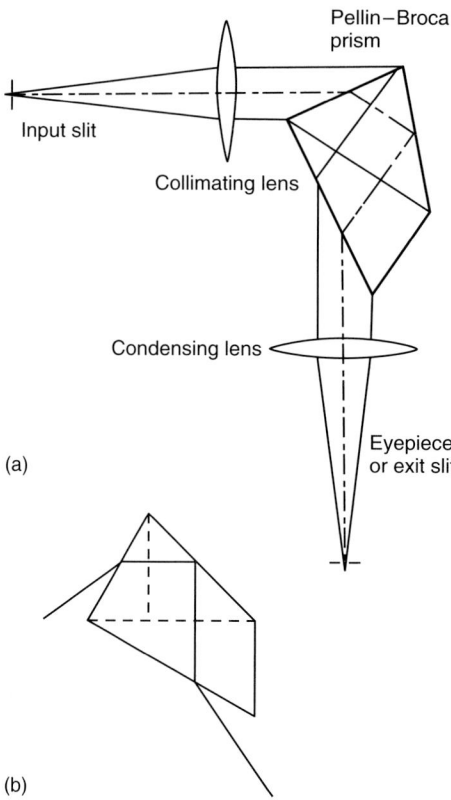

Fig. 10.4 (a) A Pellin–Broca spectrometer. Light is refracted at a constant 90° angle as the prism is rotated. (b) A Pellin–Broca prism, illustrating its conception as a composite of three separate prisms.

One of many *constant deviation* mounts of a prism is the Pellin–Broca spectrometer, illustrated in Figure 10.4(a). This design has the advantage that the prism may be rotated in such a way that the minimum deviation beam is always refracted at right angles to the incident beam. As indicated in Figure 10.4(b), the Pellin–Broca prism may be viewed as a pair of 30–60–90° prisms joined by a 45° total internal reflection prism.

10.4.5
Aberrations

The aberrations of the imaging optics that limit the resolution of a prism spectrometer include spherical aberration, coma, astigmatism, curvature of field, and, in some designs, chromatic aberration.

10.4.5.1 Spherical Aberration

In a spherical lens, the rays passing through the center of the lens do not focus in exactly the same point as the rays, which pass through the outer portion of the lens, as illustrated in Figure 10.5. The rays from the various parts of the lens converge to a minimum spot size, the *circle of least confusion*, which determines the resolution limit of a spherically figured lens. To a good approximation, a pair of plano-convex lenses used as collimating and condensing optics will

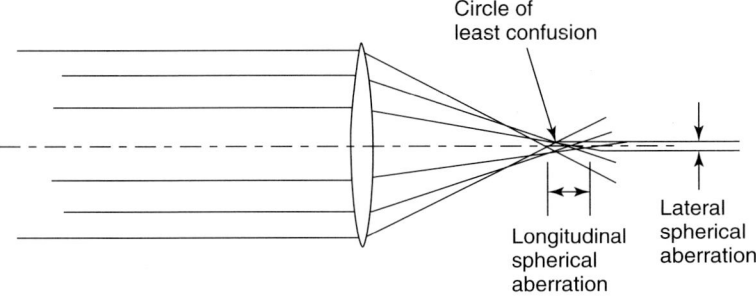

Fig. 10.5 Spherical aberration in a lens.

provide the best form for reducing spherical aberration, with the plane side of the lens facing the image in each case.

10.4.5.2 Coma

Coma is an aberration, which affects only off-axis rays. As illustrated by Figure 10.6, coma is a result of rays passing through different annular rings of the lens being imaged, not into points, but into small circles of different size and located at different centers. The change in size with the changing focal point results in a *comatic flare* of the image resembling the shape of a comet tail, which is the origin of the term. Coma increases rapidly with the off-axis angle, severely restricting the size of the focal plane for wide-aperture spectrographs.

10.4.5.3 Astigmatism

Similar to coma, astigmatism is an off-axis aberration. If the rays striking the lens are off-axis in the vertical direction, the rays in the vertical plane will form a focus at a different point than those in the horizontal plane. The former will be focused into a line tangent to a circle about the principal optic axis (the tangential focus) and the latter into a line pointed toward the optic axis (the saggital focus). Astigmatism, which spreads a point source into a line parallel to the slit, will not necessarily

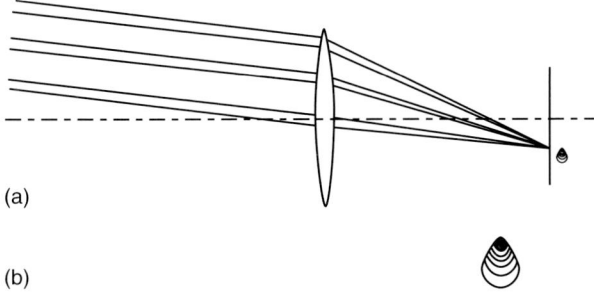

Fig. 10.6 Coma produced by a simple lens. (a) Off-axis marginal rays are imaged into different positions than the central ray. (b) Annular sections of the lens are imaged into circles whose size and position depend on the distance of the annuli from the center of the lens.

degrade the resolution, although it will reduce the brightness of the image; astigmatism, which spreads a point source perpendicular to the direction of the slit, will degrade both the throughput and the resolution.

10.4.5.4 Curvature of Field

The focal surface of an optical system is not usually planar, but instead is a curved surface. Whether this curvature is a problem depends in part on whether the curvature is fixed or changes according to the setting of the prism, and in part on whether the instrument is intended for use with an exit slit, an array detector, or a photographic plate. If the curvature changes with the prism angle, as it may with a minimum deviation mount, the prism cannot be rotated without defocusing the instrument, and the instrument is suitable only as a fixed prism spectrograph. If, on the other hand, the curvature and focus remain constant with the prism angle, the prism can be rotated to scan the spectrometer.

10.4.5.5 Chromatic Aberration

The high dispersion of a glass which makes it suitable for use as a prism means that lenses made from the same material will have different focal lengths at different wavelengths. If the spectrometer is to be used with photographic detection, as in the minimum deviation spectrograph of Figure 10.1, this is not a problem, because the plateholder can be adjusted once for optimum focus and fixed in that position. For a constant deviation mount such as the Pellin–Broca mount of Figure 10.4, both the collimating and the focusing lenses must be corrected for chromatic aberration if the spectrometer is to remain in focus as the prism is rotated.

10.5 Grating Spectrometers

Most modern ultraviolet and visible spectrometers are based on the diffraction grating, which offers improved resolution, luminosity, and dispersion compared with prism instruments of comparable size and expense. These may be classified as plane grating spectrometers, of which the Czerny–Turner spectrometer is an example, and concave grating spectrometers, which include the Eagle and Seya–Namioka mounts. The concave grating design is usually preferred whenever the number of optical surfaces must be minimized, as in the vacuum ultraviolet region, because the concave grating serves as both the dispersing element and the imaging optics. Plane grating instruments must rely on concave mirrors to image the spectrum; these mirrors lead to absorption losses at each optical surface, but the additional design freedom afforded by the imaging optics allow far greater control over aberration corrections than is possible for concave gratings. Accordingly, plane grating instruments are used whenever absorption losses at the mirrors are negligible and a high quality stigmatic image is desirable, as in the visible and near infrared.

10.5.1 Dispersion

Diffraction of light by a grating is determined by the grating equation:

$$m\lambda = a(\sin\alpha + \sin\beta) \quad (10.6)$$

where m, the order of diffraction, must be an integer; λ is the diffracted wavelength; α is the angle of incidence of the light with respect to the normal; and β is the angle of diffraction with respect to the normal. The reciprocal angular dispersion, $d\lambda/d\beta$, is given as $a\cos\beta/m$; the tuning of wavelength as the grating is turned is twice this number. The linear dispersion at the focus of the spectrometer is the angular dispersion times the focal length of the condensing optics.

10.5.2 Resolution

As with a prism, the resolution of a diffraction grating is the product of the dispersion and the diffraction-limited angular resolution imposed by the aperture of the grating. Unlike prism instruments, however, the dispersion is independent of the material of the grating, and so the resolution of a grating may be stated in several rather simple ways. Taking the ruled width of the grating as W and the angular resolution of light leaving the grating as $\Delta\beta = \lambda/(W\cos\beta)$, the resolution of the grating is found to be

$$\Delta\lambda = \frac{d\lambda}{d\beta}\Delta\beta$$

$$= \frac{a}{m}\cos\beta \cdot \frac{\lambda}{W\cos\beta} \qquad (10.7)$$

$$= \frac{a}{m} \cdot \frac{\lambda}{W}$$

The resolving power, in turn, is given by

$$R = \frac{\lambda}{\Delta\lambda} \qquad (10.8)$$

$$= m\frac{W}{a}$$

which is just the order of diffraction times the number of grooves in the grating. Clearly, for a grating spectrometer to achieve its maximum resolving power, the entire width of the grating must be illuminated.

If $\alpha \approx \beta$, the resolution can be rewritten in terms of the wave number $\tilde{\nu}$ as

$$\Delta\tilde{\nu} = \frac{\partial\tilde{\nu}}{\partial\lambda}\Delta\lambda \qquad (10.9)$$

$$= \frac{1}{W \cdot 2(\sin\alpha)}$$

which is just the reciprocal of the optical path difference of rays on opposite sides of the grating.

10.5.3 Luminosity

The solid angle of acceptance of the grating, Ω, is given to a good approximation by the equation

$$\Omega = \frac{A\cos(\alpha)}{f^2} \qquad (10.10)$$

where A is the area of the grating, α is the angle of incidence, and f is the focal length of the collimating optics. The luminosity of a grating spectrometer is thus equal to the product of Ω and the area a of the entrance slit, or

$$L = \frac{Aa\cos(\alpha)}{f^2} \qquad (10.11)$$

The "speed" of a spectrometer is often quoted in terms of its focal ratio, $F = f/d$, where d is the maximum diameter of the aperture stop dictated by the size of the grating. Thus, a spectrometer with a focal length of 1 m and a grating whose diagonal measures 12.5 cm will have a speed of $f/8$. This number obviously does not accurately measure the luminosity of the instrument as it is ordinarily used; it does, however, specify the minimum aperture of any external optics that are guaranteed to illuminate the entire surface of the grating. The luminosity of a spectrometer varies as the inverse square of the focal ratio.

10.5.4
Design and Manufacture of Gratings

Until recently, all gratings were ruled mechanically by inscribing parallel grooves on an optically polished master blank. The tolerances involved in the manufacture of such gratings generally require interferometric control of the ruling engine to compensate for mechanical errors in the lead screw; typically, modern ruling engines use a frequency stabilized laser for this purpose. Eliminating periodic errors in the rulings is extremely important, because periodic errors will result in diffraction at angles other than the design angle. The grating "ghosts" produced in this way can cause lines to appear at apparent wavelengths far removed from the true wavelength, creating problems ranging from nuisance to disaster depending on the experiment being performed.

More recently, gratings have been produced by holographically recording the groove pattern on a photosensitive surface. The main advantage of a holographic grating is the absence of the inevitable periodic errors that are found in mechanically ruled gratings. The principal advantage claimed for ruled gratings is a greater control over the shape of the groove and, consequently, a more sharply defined blaze angle and higher efficiency. These issues have been addressed in some detail theoretically (Loewen et al., 1977). Gratings for use in the visible and ultraviolet are available with groove spacings from 300 grooves/mm to 4800 grooves/mm.

10.5.4.1 Transmission Grating

Conceptually, the simplest grating is the transmission grating. Transmission gratings may be fabricated either by imprinting a transparent substrate with closely spaced lines or by ruling a transparent substrate with shaped grooves. As schematically shown in Figure 10.7, the transmission grating is simply an extension of multiple slit interference to a very large number of slits. If two waves emerging from adjacent slits have a total path length which differs by an integral number of waves, constructive interference will occur and the intensity will be nonzero. For an angle of incidence α with respect to the grating normal and an angle of diffraction β, this condition requires $m\lambda = a(\sin\alpha + \sin\beta)$. This equation is always satisfied for $\alpha = -\beta$; therefore, transmission straight through the grating is allowed with $m = 0$ for all wavelengths. In addition, a wavelength may be diffracted at angles corresponding to nonzero integral values of m.

10.5.4.2 Reflection Grating

A more commonly used type of ruled grating is the *reflection grating*, which is typically ruled on a polished glass optical flat coated with a thin film of aluminum. Once again, the condition for constructive interference is that $m\lambda = a(\sin\alpha + \sin\beta)$. The reflection grating is preferred over the transmission grating for much the same reasons that mirrors

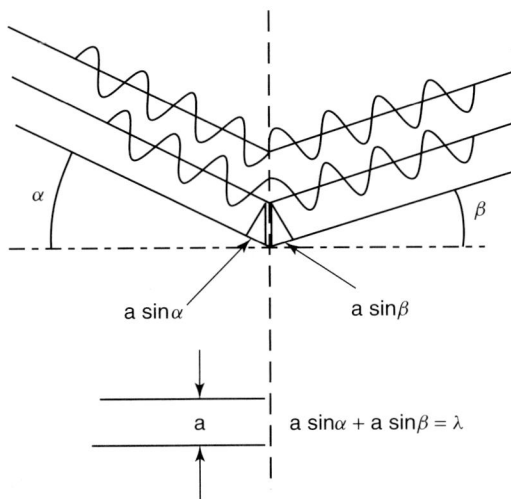

Fig. 10.7 Multiple-slit interference as observed in a segment of a transmission grating. If the waves are out of phase by an integral number of wavelengths, constructive interference will occur and intensity will be maximum.

are preferred over lenses as imaging optics: low absorption and scattering losses, high efficiency in the ultraviolet, and insensitivity to the homogeneity of the substrate.

By controlling the shape and angle of the diamond ruling tool, it is possible to control the shape of the grooves in such a way that light striking the grating normal to the face of the groves is diffracted with very high efficiency. When a grating is viewed at this angle, the grating appears to blaze with color, and this angle of highest efficiency is referred to as the *blaze angle* of the grating.

Ruled gratings are not ordinarily sold as originals, but are instead used as masters to produce replica gratings. This process allows a single ruling to be duplicated inexpensively, greatly reducing the cost of diffraction gratings in commercial spectrometers.

10.5.4.3 Echelles

If the grooves of a grating are ruled so that the sides are flat, the grating will have a sharply defined angle of maximum efficiency where the angle of diffraction coincides with the specular reflection from the groove facets. Such a grating is called an *echelle* or an *echelette*. It is ruled very coarsely so that the shape of the grooves can be controlled accurately, and therefore it is used in a high order of diffraction (typically $m = 100$). The free spectral range of such a grating is relatively small, so additional dispersing optics must be used to separate overlapping orders.

10.5.4.4 Holographic Gratings

Gratings may also be produced optically, without the use of a ruling engine. In this process, a pair of light beams is used to produce a pattern of interference fringes at a surface that has been coated with photoresist. Although the earliest such gratings predate the invention of the laser, the process did not become practical before the availability of high powered ion lasers (Hutley, 1982). Such gratings are generally referred to as *holographic gratings*.

Two intersecting coherent beams of light will produce interference fringes within their volume of intersection with a spacing D given by

$$D = \frac{\lambda}{2\sin(\theta/2)} \quad (10.12)$$

where θ is the angle of intersection. If a grating blank is placed at an angle ϕ to the bisector of θ, fringes will be produced with a spacing a of

$$a = \frac{\lambda}{2\sin\theta \sin\phi} \quad (10.13)$$

The blank, previously coated with a thin layer of photoresist, is subsequently developed to produce a corrugated surface with the spacing of the interference fringes. This surface is generally coated with a reflective metal layer and used as a reflection grating.

10.5.5
Plane Grating Spectrometers

10.5.5.1 Aberrations
Plane grating spectrometers almost invariably use spherical concave mirrors or, less frequently, paraboloids to image the entrance slit into a spectrum. Similar to lenses, the mirrors that form images in a typical grating spectrometer are subject to spherical aberration, coma, astigmatism, and curvature of plane; and, as is the case with prism instruments, the practical resolution of a grating spectrometer is frequently limited by optical aberrations. The only major aberration, which is completely removed using reflective optics, is chromatic aberration. As these aberrations usually define the limits of performance of an instrument, much of the design of grating spectrometers involves arranging the optics in such a way that aberrations induced by one optical element are corrected in part by others.

Spherical Aberration The rays reflecting from the center of a spherical mirror do not focus in exactly the same point as the rays, which are reflected from the outer portion of the mirror. As with a lens, the rays from the various parts of the lens converge to a minimum spot size, the *circle of least confusion*, which determines the resolution limit. For a spherical mirror, the diameter of this circle is $f/64F^3$, where F is the focal ratio and f is the focal length of the mirror. Thus, a 1-m focal length instrument with a 125 mm × 125 mm diffraction grating will have a spherical aberration amounting to 86 µm at the focal plane if uncorrected. If the same grating were placed in a 2-m focal length instrument, the uncorrected spherical aberration would be only 11 µm. For a parabolic reflector, the spherical aberration is zero: all rays parallel to the principal optical axis focus to a diffraction-limited point (about 4 µm for the above example of a 1-m focal length instrument with a 125 mm × 125 mm grating).

Coma Coma in a mirror, as in a lens, affects only off-axis rays. As with a lens, annular rings of a spherical mirror images off-axis rays into circles whose radius increases with the radius of the annular ring. The comatic flare is often more pronounced in reflective imaging systems than in systems which use lenses, because geometrical constraints require the entrance and exit slits to be placed off the principal optical axes of the imaging mirrors. For a spherical mirror, collimated light is imaged into a pattern whose height and width are given as (Harwit and Sloane, 1979)

$$H = \frac{3}{16} \frac{\theta f}{F^2} \left(1 - \frac{d}{2f}\right) \tag{10.14}$$

$$W = \frac{1}{8} \frac{\theta f}{F^2} \left(1 - \frac{d}{2f}\right)$$

where f is the focal length, F is the focal ratio, d is the distance from the mirror to the aperture stop (i.e., the grating), and θ is the angle the incident light makes with respect to the principal optical axis of the mirror. For the case of a 1-m focal length instrument with a 125-mm diffraction grating, the uncorrected coma at the ends of a 20-mm slit would increase the size of the image by about 10 µm. If the instrument is required to form an extended image of the spectrum, the coma would increase with the off-axis angle: at the edge of a 250-mm focal plane, the coma would increase to 100 µm.

The increase of comatic flare with the aperture of the imaging optics is the most serious practical problem preventing the design of a high resolution Ebert or Czerny–Turner grating spectrometer with a low focal ratio, because a wide collection aperture inevitably involves a large off-axis angle which limits the resolution due to unacceptably large coma.

Astigmatism Similar to a simple lens, a spherical mirror forms an astigmatic image if illuminated off-axis, as illustrated in Figure 10.8. The tangential and saggital foci, f_T and f_S, for a spherical mirror illuminated with collimated light at an

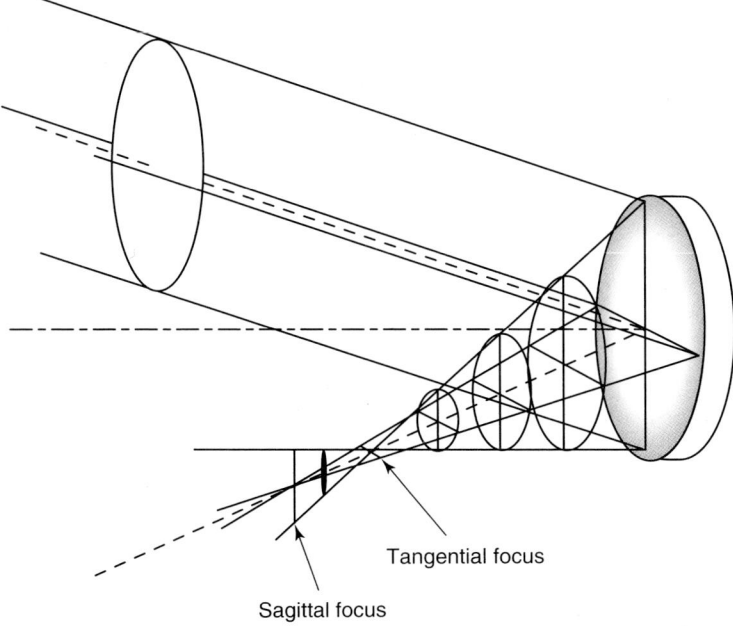

Fig. 10.8 Astigmatism of a spherical mirror. Off-axis rays of collimated light converge in different positions to form tangential and saggital foci.

angle θ to the principal axis are given as (Jenkins and White, 1976)

$$f_T = f \cos\theta \tag{10.15}$$

$$f_S = \frac{f}{\cos\theta}$$

While a paraboloidal mirror eliminates spherical aberration, it significantly exaggerates astigmatism, limiting its useful angular aperture.

Curvature of Field The focal surface of a spherical mirror is not in general a plane; this fact becomes important if an extended spectrum is to be imaged onto a photographic plate or an array detector. The exact curvature of the surface is determined by the curvature of the condensing optics, the distance of the condensing mirror from the grating, and the angle of the light with respect to the principal optical axis. It has been known since the work of Fastie (1952b) that, if the grating in an Ebert or Czerny–Turner spectrometer is placed at a distance of $2f/\sqrt{3}$ from the center of curvature of the condensing mirror, the focal surface is very nearly a plane. Reader (1969) has identified two other nearby positions which offer further corrections to the field curvature. The "corrected flat field" offers a somewhat flatter field than the Fastie position and insures that the focal plane is nearly perpendicular to the direction of propagation of light. Reader's "superflat field" provides the most nearly flat focal surface but inclines the plane slightly with respect to the direction of propagation. For the 1-m Czerny–Turner spectrometer, the offset of the grating from the Fastie flat-field position to the corrected flat-field position is 10 mm; the distance from the Fastie position to the superflat field position is 26 mm (Scheeline et al., 1991).

10.5.5.2 Types of Plane Grating Spectrometers

Ebert-Fastie A particularly simple grating spectrometer was designed by Hermann Ebert in 1898. Dismissed and ignored by the spectroscopic community as flawed and impractical, Ebert's design was reinvented and improved in 1952 by William Fastie, who demonstrated its simplicity and optical superiority over any other instrument in use at that time (Fastie, 1952b, 1952a, 1991). The mount uses a single spherical mirror as both collimator and condenser, as illustrated in Figure 10.9(a). The grating is invariably mounted with its grooves aligned vertically, so Figure 10.9(a) is a view from above the spectrograph.

There are two implementations of the Ebert–Fastie mount: in the original design, the entrance and exit slit were placed side by side on either side of the grating (the "horizontal" Ebert mount, Figure 10.9b), while in some later versions, the slits were placed above and below the grating (the "vertical" Ebert mount, Figure 10.9c). The former has the advantage that the curvature or tilt of the image remains unchanged as the grating is rotated; the latter has the advantage of reducing stray light and thus simplifying the internal layout of the optics. In either case, both the entrance and exit slits must lie in the same plane to keep the instrument in focus over all angles of the grating; this point has been misunderstood in some commercial implementations of the Ebert mount.

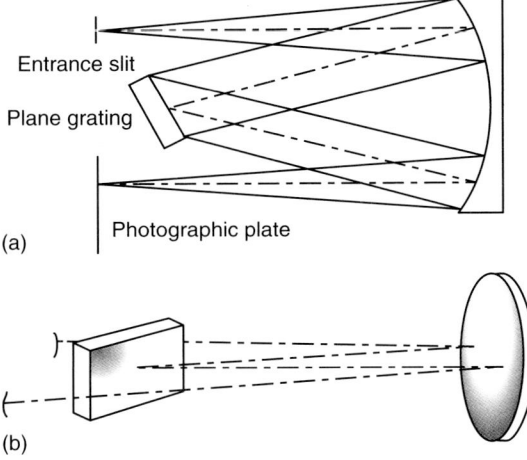

Fig. 10.9 The Ebert–Fastie grating mount, illustrating (a) a top view of an Ebert spectrograph, (b) a perspective view of a "horizontal" Ebert spectrometer, and (c) the "vertical" Ebert spectrometer.

Both designs have the effect of partially compensating for both spherical aberration and coma. In addition, Fastie has shown that the use of curved entrance and exit slits can reduce or eliminate astigmatism. The optimum shape for the slits of the horizontal Ebert spectrometer involves a pair of slits with a common center of curvature located on the line passing through the center of the grating and the center of curvature of the mirror.

Czerny–Turner If the single mirror of the Ebert–Fastie spectrometer is replaced by two independent mirrors, one for collimation and one for focusing, the Czerny–Turner spectrometer is obtained. The most commonly used design for this spectrometer is one in which two spherical mirrors are placed side by side as illustrated in Figure 10.10. This mount has all of the optical virtues of the Ebert mount (Czerny and Turner, 1930), but the extra degree of freedom afforded by the additional mirror allows superior correction of the image aberrations.

Shafer et al. (1964) and Reader (1969) have demonstrated that the performance of the Czerny–Turner spectrometer can be optimized by using mirrors of different focal lengths as collimator and focusing mirror. These authors have shown how the residual coma, astigmatism, and spherical aberration are affected by the choice of focal lengths. Reader has also provided equations for the focal surface of the focusing mirror in terms of the focal length and the distance of the mirror from the grating. Because of its high performance and flexibility, the Czerny–Turner mount in its many varieties forms the basis of most commercial spectrometers.

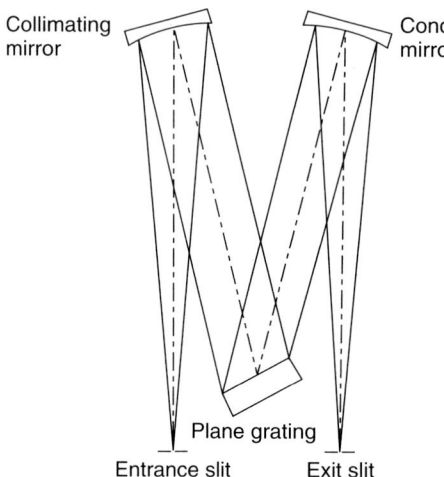

Fig. 10.10 The basic Czerny–Turner grating spectrometer.

Newtonian The most straightforward way to reduce the off-axis aberrations of coma and astigmatism is to bring the optical path as close to the principal optical axis as possible. This is the principle behind the Newtonian telescope, for instance, and it is possible to design a spectrometer based on this principle. A small plane diagonal mirror is used to bring the rays from the center of the entrance slit onto the principal optical axis of the collimating mirror, and a second plane diagonal mirror diverts the image of the spectrum from the condensing mirror to the focal plane. The problem of reducing optical aberrations then becomes the same as reducing optical aberrations in an astronomical telescope. Variations on this approach use the Schmidt or Cassegrain telescopes as collimator or condenser.

10.5.6
Concave Grating Spectrometers

If a grating is produced on a concave spherical mirror blank, the concave grating thus formed serves as both the dispersing element and as the image-forming optics. With rare exceptions, it is the only element in the optical train of a spectrometer in which it is used. There is one major advantage and one major disadvantage that follow from this choice: the advantage is that absorption losses at optical surfaces are considerably reduced, whereas the disadvantage is that the opportunity to correct optical aberrations is severely limited.

Concave grating spectrometers are of two general types: those which form images on the Rowland circle and those which do not. The former mounts include the Paschen–Runge mount, the Eagle mount, and some grazing incidence spectrometers. The latter include the Seya–Namioka and related mounts, the Wadsworth mount, and several modern designs which employ specially designed holographic gratings which compensate for certain aberrations.

10.5.6.1 Optical Aberrations of Concave Mirrors

The concave grating is ruled on a spherical mirror, and so inherits the spherical mirror's optical aberrations. Chief among these is astigmatism, because the grating is always used off the principal optical axis. In the vertical plane (i.e., the plane in which the light is dispersed), meridional rays will be focused according to the equation

$$\frac{1}{r_1} + \frac{1}{r_2} = \frac{2}{R} \tag{10.16}$$

where R is the radius of curvature of the spherical surface and r_1 and r_2 are the distances from the grating to the entrance and exit slits. The condition for horizontal focus, on the other hand, is

$$\frac{\cos \alpha}{R} - \frac{\cos^2 \alpha}{r_1} + \frac{\cos \beta}{R} - \frac{\cos^2 \beta}{r_2} = 0 \tag{10.17}$$

where, as usual, α and β are the angles of incidence and diffraction at the grating.

For wide-aperture concave gratings, spherical aberration can also become a problem. In such a grating, rays diffracted from the extreme ends of the grooves do not come to a focus in exactly the same line as meridional rays. The effect of spherical aberration increases as the fourth power of the vertical acceptance angle, requiring a compromise between luminosity, which increases with the square of the aperture, and resolution, which decreases as the fourth power of the aperture.

10.5.6.2 Rowland Circle Mounts

If both the grating and the entrance slit are placed on a circle whose diameter is the radius of curvature of the concave grating, as indicated in Figure 10.11, then images of the slit will also be formed on this circle at diffraction angles which satisfy the grating equation, $m\lambda = a(\sin \alpha + \sin \beta)$, just as a plane grating. The horizontal focal equation for the concave grating is solved by setting

$$\frac{\cos \alpha}{R} - \frac{\cos^2 \alpha}{r_1} = 0 \tag{10.18}$$

and

$$\frac{\cos \beta}{R} - \frac{\cos^2 \beta}{r_2} = 0 \tag{10.19}$$

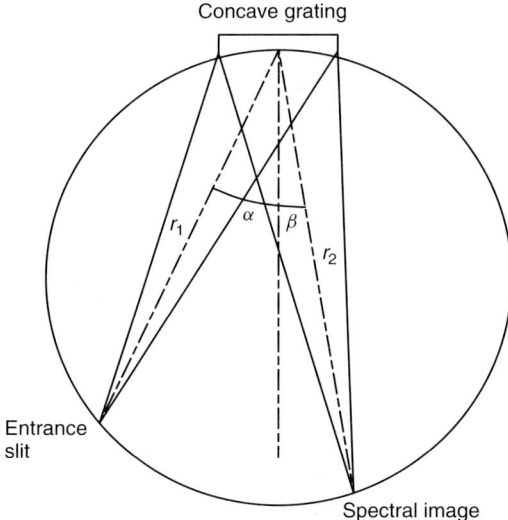

Fig. 10.11 The Rowland circle. Conjugate images of a concave grating are formed on a circle whose diameter equals the radius of curvature of the grating.

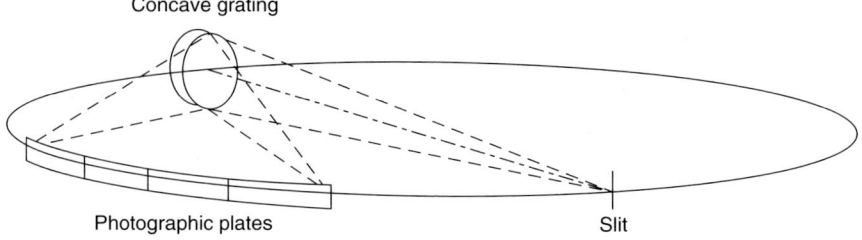

Fig. 10.12 The Paschen–Runge mount of a concave grating, illustrated for use as a spectrograph.

Paschen–Runge The most straightforward design using the Rowland circle is the Paschen–Runge spectrograph, illustrated in Figure 10.12. Here an entrance slit is placed at one point on the Rowland circle and photographic plates are located at points all around the circle wherever the spectrum is to be obtained. Using an instrument of this type, it is possible to cover the entire visible and ultraviolet spectrum at high resolution in a single exposure. The instrument occupies a very large area, and in addition is very difficult to isolate from building vibrations. As a result, it is used mostly in specialized circumstances requiring a high information bandwidth; an example would be to record the spectrum of a unique or precious sample such as a rare isotope.

Eagle Mount Among the many more compact mounts based on the Rowland circle, the Eagle mount is probably the most widely used (Eagle, 1910). In this mount, the entrance slit, grating and exit slit or plateholder are placed along a single optical axis, as shown in Figure 10.13. Variations of this mount place the plateholder in the same horizontal plane with the grating and entrance slit. In the Eagle mount, the plateholder is a segment of the Rowland circle, and it

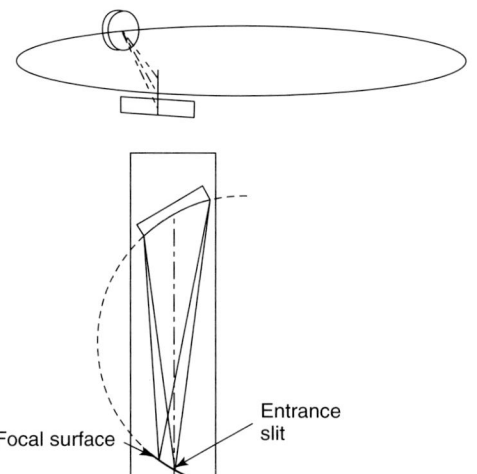

Fig. 10.13 The Eagle mount of a concave grating. The Rowland circle is indicated in both the perspective and the top view.

must be readjusted to fall on the focal surface of the grating each time the grating angle is changed. Moreover, the distance between the grating and the entrance slit must also be changed as the grating is rotated to keep the instrument in focus. Accordingly, the Eagle mount is not well suited for use as a scanning spectrometer. However, it functions very well as a high resolution spectrograph, and the compact design makes it suitable for a vacuum instrument.

Grazing Incidence Even the single surface of the concave grating produces unacceptable absorption losses at wavelengths shorter than 50 nm when used at normal incidence. At these wavelengths, use is made of the fact that short wavelength light striking a surface tangentially is reflected much more efficiently than light striking the same surface at normal incidence. This observation led to the *grazing incidence* mount of the concave grating, which is yet another mount making use of the Rowland circle. In this mount, the angle of incidence ranges from 60° to slightly less than 90°, and the angle of diffraction is only slightly less and on the opposite side of the grating normal.

At such high angles of incidence, the aberrations of the grating are quite pronounced. It has been shown that, in grazing incidence, the optimum width of a concave grating is much smaller than that for normal incidence; increasing the width above the optimum value degrades resolution without significantly increasing the central intensity of emission lines. Similarly, the useful length of a grating groove is limited by astigmatism.

10.5.6.3 Non-Rowland Circle Mounts

Seya–Namioka Other mounts of the concave grating are not based on the Rowland circle, but instead employ fixed entrance and exit slits while the grating is rotated about a point (Miyake, 1959). Such a mount is substantially simpler to use as a scanning monochromator than one based on the Rowland circle, and considerable effort has been made to find the best design to maintain focus of such an instrument as the grating is rotated.

Seya (1951) designed such a mount based on the requirement that the grating pivot about the center of its face. This design was refined by Namioka (1959), who used the condition that there be a minimum change of focus as the grating is rotated. The difference between the angle of incidence and the angle of diffraction is found to be 70°15′, and the distance of the slits from the surface of the grating is 0.8156 times the radius of curvature of the grating.

The Seya–Namioka solution of the focal equation for the concave grating involves taking

$$\frac{\cos\alpha}{R} - \frac{\cos^2\alpha}{r_1} + \frac{\cos\beta}{R} - \frac{\cos^2\beta}{r_2}$$
$$= F(r_1, r_2, \alpha - \beta, \alpha + \beta) \quad (10.20)$$

and setting F and its first three derivatives with respect to $(\alpha + \beta)$ equal to zero at some particular angle of incidence. While this condition does not guarantee a good focus over a wide range, the simplicity of the design makes it highly useful as a scanning vacuum monochromator.

Wadsworth A different kind of non-Rowland circle mount was devised by

Wadsworth to produce a stigmatic image using a concave grating. The grating is illuminated with collimated light from a concave mirror just as in a plane grating spectrometer, and the concave grating produces a stigmatic image at the focus. The image is brighter than the astigmatic image produced by the various Rowland circle mounts, provided the additional loss due to absorption at the first mirror surface can be tolerated. The focal surface is not flat, however, and both the focus and the curvature of the focal surface change as the grating is rotated, making the mount inconvenient as either a scanning spectrometer or a spectrograph. As a result, the Wadsworth mount is not widely used.

Compensated Holographic Yet another class of mounts has been designed around holographically recorded concave gratings which are specifically designed to compensate for one or more of the aberrations inherent in concave gratings. These aberration-corrected holographic gratings are not recorded with collimated light sources, and so the projection of the grooves onto a chord will not be equally spaced as is the case with a ruled grating or an ordinary concave grating. The greatest flexibility in correcting aberrations is achieved by allowing the focal surface of the grating to depart from that of the ruled concave grating.

10.6
Interferometric Spectrometers

10.6.1
Michelson

The Michelson interferometer, which has in the past four decades seen a rapid rise in popularity for use as a Fourier transform infrared spectrometer, has seen relatively little use as in the visible and ultraviolet regions. The reason for this is threefold: First, the requirement that the mirrors be kept in alignment to within a fraction of a wavelength during the full scan is a much more stringent condition in the ultraviolet than in the infrared. Second, the signal/noise ratio is not usually detector limited in the visible or ultraviolet, so that the multiplex advantage of a Fourier transform spectrometer is less important than in the infrared. Finally, the requirement for resolution of 0.001 cm^{-1} or better, which is important in the infrared and can most easily be achieved with a Fourier transform instrument, is less important in the visible or ultraviolet because of the much greater Doppler width.

Despite these obstacles, the Fourier transform spectrometer has seen some limited use in the shorter wavelength regions. There are two strategies for maintaining alignment during a scan. One of these involves passive stabilization of the fixed and moving mirrors using a self-aligning optical element. This may be either a corner cube or a cats-eye mirror, each of which is designed to reflect light back along the original path or parallel to it. The other strategy makes use of active stabilization: a monochromatic source such as a single frequency laser is passed through the interferometer, and the interference fringes are monitored to provide a feedback signal which corrects the tilt of one of the mirrors.

In addition to its use as a Fourier transform spectrometer, the Michelson interferometer has achieved considerable popularity in the measurement of the

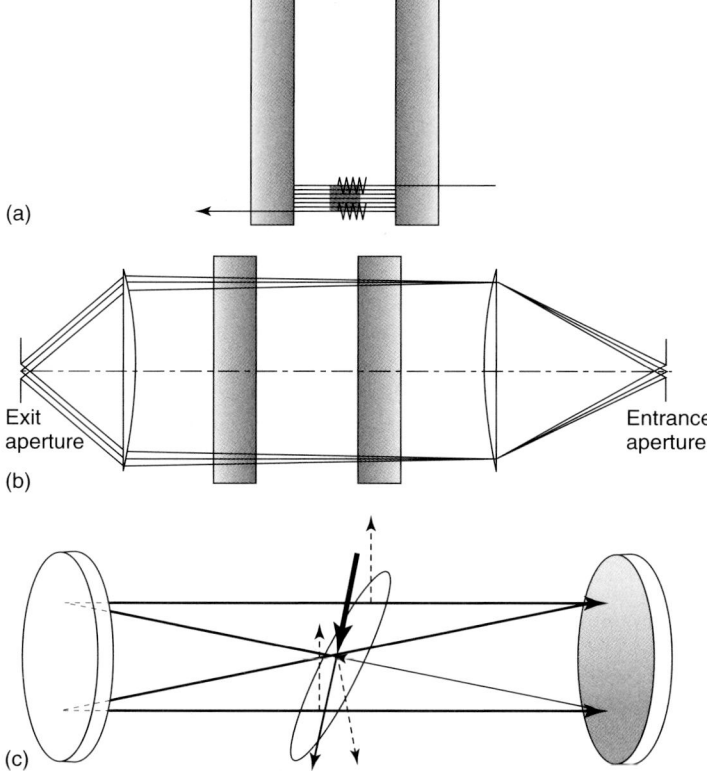

Fig. 10.14 The Fabry–Perot étalon. (a) The path of light indicating the resonance condition. (b) Illumination of the interferometer for use as a spectrometer. (c) The internally coupled confocal Fabry–Perot.

wavelength of single frequency lasers. In this application, the Fourier transform is unnecessary, as the wavenumber of an unknown laser can be obtained simply by counting the fringes during a sweep of the interferometer and comparing the number with the fringes from a laser of known frequency (Kowalski et al., 1978; Fox et al., 1999).

10.6.2
Fabry–Perot

The highest resolution spectrometer commonly used in the visible and ultraviolet is the Fabry–Perot interferometer or étalon (Fabry and Perot, 1899). Invented in 1899, the basic Fabry–Perot interferometer consists of a pair of partially reflective flat plates held parallel at a fixed distance d from each other (Figure 10.14a). This arrangement forms a resonant cavity which passes through light if the wavelength satisfies the condition $m\lambda = 2d\cos\theta$, where θ is the angle between the incident ray and the normal to the plate surface and m is an integer. At wavelengths which do not satisfy this condition, light is reflected. For a Fabry–Perot interferometer with perfectly flat plates and no absorption losses, the transmitted intensity is described in terms of the incident intensity I_0 and the reflectance r by the Airy formula:

$$I(\lambda) = \frac{I_0(1-r)^2}{1 - 2r\cos\left(\frac{4\pi d}{\lambda}\cos\theta\right) + r^2} \quad (10.21)$$

The Fabry–Perot interferometer can be used as the dispersing element of a spectrometer by using suitable collimating and condensing optics, as indicated in Figure 10.14(b). The entrance aperture is circular, and spectrometer produces at the output focal plane a set of concentric rings.

The resolving power R is given as

$$R = \frac{m\pi\sqrt{r}}{1-r} \quad (10.22)$$

while the free spectral range is found to be simply

$$\Delta\lambda = \frac{\lambda}{m} \quad (10.23)$$

The ratio of the free spectral range to the resolution, called the *finesse* of the Fabry–Perot interferometer, is given as

$$F_r = \frac{\pi\sqrt{r}}{1-r} \quad (10.24)$$

where the symbol F_r is used to emphasize that this is the part of the finesse due to the reflectance. The finesse is the most straightforward figure of merit for the Fabry–Perot interferometer, because it is the width of adjacent peaks of the instrument function divided by their separation. If absorption losses are considered, the peak intensity passed by the interferometer drops to

$$I_{max} = I_0(1 - (A/(1-r))^2 \quad (10.25)$$

where A is the absorption coefficient of the reflectors.

If the interferometer plates are not perfectly flat, there will be both loss of intensity and a degradation of the resolution compared with the formulas above, and the instrument function will be a convolution of the Airy formula with these additional losses (Jacquinot, 1960). A contribution to the effective finesse can be determined from the surface irregularities if they are known: if the irregularities are characterized by a width Δd, then the finesse is limited to $F_d = \lambda/(2\Delta d)$. For example, if the surface imperfections are of the order $\Delta d = \lambda/200$, the flatness finesse F_d will be limited to 100. Increasing the reflectance to improve the reflectance finesse above this value will only introduce additional absorption and scattering losses.

In a similar way, a nonzero size for the entrance aperture of the collimating optics will result in a loss of resolution which can be cast into the form of a lower finesse. The solid angle of acceptance Ω will restrict the resolving power to a maximum value of $R_\Omega = 2\pi/\Omega$, so the aperture-limited finesse will be

$$F_\Omega = \frac{R_\Omega}{m} = \frac{2\pi}{\Omega m} \quad (10.26)$$

A variation on the plane-mirror Fabry–Perot is one which uses concave spherical mirrors placed so that their focal points coincide. This *confocal* Fabry–Perot has as its condition for constructive interference that $m\lambda = 4d$; consequently, it has a free spectral range (and resolving power) which is twice that of a plane Fabry–Perot of the same spacing. In addition, the confocal Fabry–Perot has a luminosity which exceeds that of the plane interferometer, since the interference condition is independent of the angle of incidence.

A Fabry–Perot interferometer is ordinarily used in an order $m = 10^5-10^6$, so that a modest finesse of 10–20 can easily produce a resolving power of a few million. However, the free spectral range is correspondingly limited so that this instrument is ordinarily used for uncluttered and isolated spectra, such as the nuclear hyperfine splitting of single atomic emission lines.

The combination of high resolution and luminosity make the Fabry–Perot interferometer the instrument of choice for monitoring and stabilizing high-resolution lasers such as dye lasers or diode lasers. One particularly useful configuration is an "internally coupled" confocal interferometer, Figure 10.14(c) (Reich et al., 1986). The "internal coupling" refers to the means of coupling light into the cavity via a nearly transparent beam splitter placed near the common foci of the two spherical mirrors that make up the cavity. This method of coupling the light in and out of the cavity eliminates the need for specialized partially reflective dielectric coatings on the mirrors, greatly increasing the useful spectral range of the device: a single instrument can typically be operated without modification from the mid-infrared into the visible spectrum.

10.7
Hybrid and Compound Instruments and Imaging Filters

There are a large number of innovative designs for spectrometers which do not fit neatly into any of the categories above. Some of these are compound instruments with more than one dispersing element; others are hybrids which make use of diffraction gratings in what are essentially interferometers.

10.7.1
Order-Sorter for a Grating Spectrometer

The most common type of compound instrument involves a low-resolution spectrometer which serves as a predisperser or order-sorter for a high-resolution spectrometer with a limited free spectral range. As an example, consider a grating spectrometer employing a 300 groove/mm grating blazed at 65°. A 250-mm-wide grating of this type will produce a resolving power of 750 000 in the tenth order when used to observe a wavelength of 600 nm. It will also, however, diffract light at 545 and 666 nm in the ninth and eleventh orders, and these orders must be removed to record an absorption spectrum. This is usually accomplished by illuminating the slit of the grating spectrometer with a small prism monochromator centered at the wavelength of interest. The prism instrument can, at the modest resolution required to separate the adjacent orders, be designed to have high throughput so that the luminosity of the overall instrument is only slightly reduced.

10.7.2
Harrison Echelle

If the grating of the Czerny–Turner spectrometer is chosen to be an echelle, the spectrometer can be very efficient and deliver very high resolution. The disadvantage of such an arrangement is that the echelle must operate in a high order (typically $m = 50-100$ in the visible), and wavelengths from many overlapping orders are diffracted at the same angle. To separate the orders,

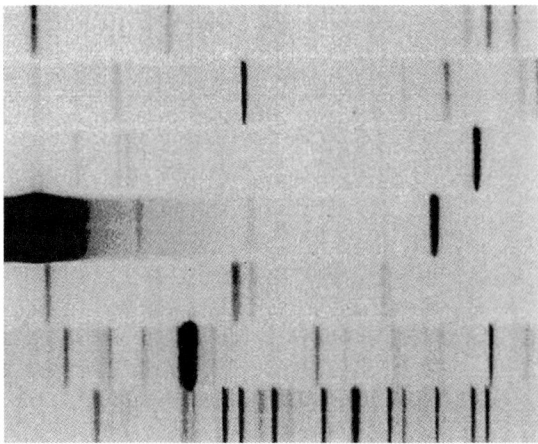

Fig. 10.15 A portion of an echelle spectrogram. Wavelength is dispersed horizontally within an order, while adjacent orders are separated vertically.

some other dispersing element must be used. One ingenious solution to this problem is to use a prism placed immediately in front of the grating as an order-sorting cross disperser. The prism-grating pair can be matched in such a way that the spectrum appears as short strips of adjacent orders as illustrated in Figure 10.15. A second solution which produces much the same result is to replace the condensing mirror of the echelle spectrometer with a concave grating whose grooves are perpendicular to those of the echelle.

10.7.3
Fabry–Perot/Grating Instrument

A second such tandem instrument is the Fabry–Perot interferometer in series with a grating spectrometer. In this case, it is the grating instrument which acts as a low-resolution order-sorter for the much higher resolution étalon. If the free spectral range of the étalon is matched to the resolution of the grating spectrometer, the maximum theoretical resolution is increased by a factor of the finesse of the Fabry–Perot. The instrument can be used either as a scanning spectrometer, by scanning the grating and the interferometer synchronously, or it can be used as an imaging spectrograph, in which case the image is a two-dimensional pattern of Fabry–Perot rings superimposed on the spectrometer slit.

10.7.4
Double and Triple Monochromators

In many instances, a tandem spectrometer combination is not intended to increase the resolution but rather to improve the signal/noise ratio. Double or triple monochromators are used to reduce scattered light from a prominent emission line and so improve the ability to observe a nearby faint feature. The exit slit of the first monochromator serves as the entrance slit of the second monochromator, which is scanned synchronously with the first. If the dispersion is additive, the entrance slit of the first and last monochromator determine the spectral resolution to a first approximation, while the central slit determines the rejection of scattered light. A typical use of such a spectrometer would be to observe low-frequency Raman or Brillouin scattering in the presence of a nearby exciting line. If the dispersion is subtractive, the double monochromator is essentially a sharp cutoff bandpass filter whose width is determined by the width of the central slit. The subtractive dispersion arrangement has the advantage in time-resolved experiments that all rays are transmitted through the instrument in the same time interval, because any optical retardation of a marginal ray through one monochromator is cancelled by the other.

10.7.5
Imaging Filters

Imaging spectrometers have long been of interest to astronomers, but they are also useful for any application requiring position-sensitive wavelength discrimination. One promising technology for imaging spectrometers is the acousto-optic tunable filter. These devices make use of Bragg diffraction by acoustic waves in an optoelastic medium. These acoustic waves are produced by a piezoelectric element in contact with the optoelastic element; and by tuning the acoustic excitation frequency, it is possible to tune the transmitted wavelength. Filters with a resolving power of a few hundred have been demonstrated that are tunable through much of the visible and into the near infrared (Georgiev et al., 2002).

10.7.6
The Sisam Spectrometer

A hybrid instrument originally described by Connes (1959) attempted to capture the advantages of Michelson interferometer without the necessity of implementing the Fourier transform. The instrument consists of a Michelson interferometer in which the two mirrors have been replaced by diffraction gratings. The most significant technical problem with this arrangement is that the gratings are required to rotate in tandem; otherwise, the resolution of the instrument is degraded. This problem was effectively solved by Till *et al.* (1975), who constructed an optical system in which the two diffraction gratings were mounted back-to-back on a single rotating platform. Although the availability of fast and inexpensive computers has eliminated the main advantage of the Sisam spectrometer, it remains a useful design for highly specialized tasks; chief among these is the high resolution detection of a weak emission signal in the presence of a much stronger signal.

10.8
Detectors

10.8.1
Ideal Detector

The ideal detector would have unit quantum efficiency (that is, one photoelectric or photochemical event per photon), no dark current or readout noise, zero response time, and a flat response over a wide spectral range. In addition to these features, a detector which can increase the *information bandwidth* either by multiplexing a signal or by allowing parallel detection of several signals will improve the performance of a spectrometer enormously. Naturally, no single detector type embodies all of these features, and different compromises among the various desirable properties of detectors must be made depending on the nature of the experiment and the spectral region to be covered.

10.8.2
Types of Detectors

Photographic Plate The first detector used for making a permanent record of a spectrum was the photographic plate. Photographic plates can be obtained which cover the spectral region in the near infrared to about 1200 nm or through the vacuum ultraviolet. Photographic plates are especially suitable for high resolution measurements because of the ease of superimposing a wavelength calibration on a spectrum using an emission source which produces known wavelengths. In addition, the information bandwidth of a photographic plate is extremely high: if a single emission line is 10 μm wide (a typical width), a 250-mm-wide photographic plate has an advantage of 25 000 over a spectrometer using a single photoelectric detector with comparable quantum efficiency. If the spectrometer is of the Harrison echelle type, the advantage becomes even greater and can easily reach several million. This advantage is mitigated somewhat by a relatively low quantum efficiency (typically 0.01 or lower) and is severely compromised by the highly nonlinear response of photographic emulsions. The main difficulty with photographic detection is, however, the inconvenience of waiting several hours between performing an experiment and measuring the recorded spectrum. In addition, there is no way of adapting photographic recording to exploit modern time-dependent techniques. Despite these inconveniences, there are certain spectrograph designs such as the Paschen–Runge mount, for which the photographic plate has no cost-effective replacement.

Photomultipliers The photomultiplier tube (PMT) is the detector of choice for most spectroscopic applications. This detector consists of a vacuum tube containing a photoemissive surface (a *photocathode*) which ejects electrons when exposed to light. These electrons are accelerated by an electric field toward elements of a *dynode chain*, each of which will emit several electrons when struck by an energetic electron. The photocathode determines the spectral sensitivity and quantum efficiency of the photomultiplier, while the dynode chain serves to amplify the signal of a single photoelectron to a measurable signal. Photomultipliers can be obtained commercially with useful sensitivity from 200 to 900 nm. Quantum efficiency is typically 0.1–0.3 at peak sensitivity, and the internal amplification of the photomultiplier boosts the signal by a factor of about 10^6.

The photomultiplier produces a burst of current in response to each photon, and this feature may be used to count photons in favorable circumstances. The rise time of a pulse of electrons is a few nanoseconds, while the transit time is approximately 10 times that value; accordingly, the counting rate of photomultiplier pulses may range upward of several million per second before overlapping pulses begin to become a problem. The photon noise is simply the square root of the number of photons, so that the signal/noise ratio is about 10^3 for a 1-second sampling time under these circumstances. The dark count in selected photomultipliers may be as low as 10 counts per second, leading to photon noise limited behavior at count rates as low as a few hundred per second.

The impact of the availability of these detectors on spectrometer design cannot be overstated. In the infrared, even the best detectors have responses which are limited by the noise of the detector itself. In consequence, infrared spectrometers have always been designed to provide maximum efficiency of detecting a small signal in the presence of a large amount of noise. This has routinely been achieved by multiplexing the signal, leading to the popularity of Fourier transform or Hadamard transform spectrometers in the infrared. In the visible and ultraviolet regions, such techniques do not provide any great benefit because the detectors are of such high quality.

An additional advantage of electronic detectors such as photomultipliers is that they can be used for *phase sensitive detection*. If the source of light incident on a spectrometer is modulated (either in amplitude or in wavelength), the signal emerging from the spectrometer will also be modulated. A modern lock-in amplifier can then be used to amplify only that part of the signal from the detector that has both the same frequency as the source modulation and a constant phase with respect to that modulation. This scheme of detection can greatly reduce the bandwidth of the noise spectrum, improving the signal/noise ratio by several orders of magnitude.

Photodiode Several types of solid state detector have become available recently. Most of these depend on the ability of a photon to promote an electron in a semiconductor from one energy level to another where it is more mobile. The simplest of these is the photodiode, which is the solid state equivalent of an unamplified vacuum tube photocell. Each incoming photon with an energy in excess of the semiconductor band gap promotes one electron from the valence band to the conduction band, and this may be detected in the form of a change in potential or a change in current. The difficulty with this device is that the current carried by a single electron is well below the noise level of an ammeter, so the device is insensitive to a low light level. A typical silicon photodiode produces $1 \text{ A W}^{-1}/$ of incident radiant power compared with 10^5 A W^{-1} for a good photomultiplier. At a radiant power of 10^{-12} W, the signal/noise of a photomultiplier is about 10^4 times higher than a photodiode.

The two advantages the photodiode have over the PMT are speed and infrared sensitivity. The photodiode can be configured so as to have a very large frequency bandwidth (cutting off at frequencies above 1 GHz). Silicon photodiodes have peak quantum efficiency in the red of 0.8 or better, retaining quantum efficiency as high as 0.1 out to 1000 nm. Indium gallium arsenide (InGaAs) photodiodes are sensitive to wavelengths as long as 1700 nm.

Avalanche Photodiode One improvement on the photodiode that makes it suitable for low light level detection is the avalanche photodiode (APD). This device has a built-in amplification similar in principle to the photomultiplier: the diode is run at a high reverse bias so that a photoelectron is accelerated by the applied field, ejecting a cascade of secondary electrons. This internal amplification boosts the current produced by a factor of about 100, improving the overall signal/noise ratio by a factor of 50. The amplification is high enough to allow APDs to be adapted for use in photon counting, and photon counting units with matched amplifier/discriminators are now available. With a quantum efficiency of 0.8 and a dark count as low as 25 counts/s, the APD is becoming competitive with the photomultiplier, especially at wavelengths longer than 800 nm.

Diode Array A second technique for improving the performance of the photodiode is to place a large number of photodiodes (128 up to 2048) in a linear array at the focal plane of a spectrometer. This tactic improves the information bandwidth of the instrument as a whole by a factor up to the number of parallel detectors. In spectrometers employing a diode array, sensitivity is further improved by the use of an image intensifier to amplify the image of the spectrum before it is imaged onto the diode array. The exit aperture is fixed at the size of a single photodiode, which is ordinarily 25 μm wide and up to 1 mm high.

The availability of inexpensive solid state array detectors such as the diode array and charged coupled device (CCD) has prompted a dramatic miniaturization of laboratory spectrometers. Instruments based on the Czerny–Turner mount are now commercially available which easily fit in the palm of the hand and weigh less than 300 g, including the detector and readout electronics. These offer coverage of the visible and UV spectrum at a resolving power of roughly 10^3.

CCD A different type of array detector is the CCD, which is commonly employed in video cameras. The CCD is similar to the photodiode in that the photon is used to promote an electron, producing an electron–hole pair. Instead of using the electron to produce a current, however, the CCD captures the electron on a small capacitor (a "gate"), allowing many electrons to accumulate before the charge packet is converted to an output voltage and digitized. Each pixel consists of such a gate, and a single CCD chip may contain several million pixels arranged as a rectangular array. The advantage of a CCD is that, by accumulating a charge over an exposure of several seconds or minutes, the integrated signal may be made to surpass the readout noise, even though the light level is low. For best performance, therefore, the dark current must be kept to a minimum, so that the dark current does not saturate the gate before the pixel is read; this is most often accomplished by cooling the CCD chip. Like the diode array, a CCD fixes the spectrometer exit aperture at the size of a single pixel. This size ranges from 25 μm × 25 μm down to 9 μm × 9 μm for CCDs, which are presently available commercially.

The simplest use of a CCD is to place it in the focal plane of the instrument in place of a photographic plate (Florek et al., 1996). The CCD is becoming a sufficiently popular detector, however, that customized spectrometers are now being designed to take advantage of the properties of this detector (Scheeline et al., 1991; Den Hartog and Holly, 1997; Furenlid and Cardona, 1988).

Conventional front-illuminated CCDs are silicon-based devices, and so are limited in their blue and ultraviolet sensitivity due to the absorption of short wavelength light by silicon before the photon can penetrate to the gate. This problem has been partly overcome by the use of thinned back-illuminated CCDs, with the result that CCDs are now available with useful sensitivity throughout the ultraviolet. As with photodiodes, CCDs based on indium gallium arsenide are now becoming available; these InGaAs detectors are sensitive from about 900 nm well into the near infrared.

Glossary

Aberration: The Gaussian approximation for a ray passing through a lens or reflected from a spherical mirror is that the angle of deviation θ is small enough for $\sin \theta$ to be replaced by θ. Any departure from this approximation is an *optical aberration*. For practical purposes, any departure from geometric optics that causes a point source to be imaged into a nonpoint image is an aberration.

Dark Current: Current from a photodetector, which is present even in the absence of illumination.

Information Bandwidth: The speed with which information channels may be recorded. In a spectrometer, the information bandwidth may be increased either by increasing the efficiency of recording a single wavelength or by recording several wavelengths simultaneously.

Monochromator: An instrument used to isolate a single wavelength of light.

Polychromator: An instrument capable of isolating several monochromatic rays simultaneously from an incident beam.

Spectrograph: An instrument which produces a graphical record (usually a photograph) of an optical spectrum.

Spectrometer: Any instrument that measures the intensity of light as a function of wavelength.

Spectrophotometer: A spectrometer specialized for accurately recording the intensity of light.

References

Connes, P. (1959) *Rev. d'Opt.*, **38**, 158.
Czerny, M. and Turner, A.F. (1930) *Z. Phys.*, **61**, 792.
Den Hartog, D.J. and Holly, D.J. (1997) *Rev. Sci. Instrum.*, **68**, 1036.
Eagle, A. (1910) *Astrophys. J.*, **31**, 120.
Fabry, C. and Perot, A. (1899) *Annal. Chim. Phys*, **16**, 115.
Fastie, W.G. (1952a) *J. Opt. Soc. Am.*, **42**, 641.
Fastie, W.G. (1952b) *J. Opt. Soc. Am.*, **42**, 647.
Fastie, W.G. (1991) *Phys. Today*, **44**, 37.
Florek, S., Beckerross, H., and Florek, T. (1996) *Fresenius Z. Anal. Chem.*, **355**, 269.
Fox, P.J., Scholten, R.E., Walkiewicz, M.R., and Drullinger, R.E. (1999) *Am. J. Phys.*, **67**, 624.
Furenlid, I. and Cardona, O.U.S.A. (1988) *Publ. Astronom. Soc. Pac.*, **100**, 1001.
Georgiev, G., Glenar, D.A., and Hillman, J.J. (2002) *Appl. Opt.*, **41**, 209.
Harwit, M. and Sloane, N.J.A. (1979) *Hadamard Transform Optics*, Academic Press, New York.
Hutley, M.C. (1982) *Diffraction Gratings*, Vol. 6, Academic Press, London, p. 330.
Jacquinot, P. (1960) *Rep. Prog. Phys.*, **23**, 268.
James, J.F. and Sternberg, R.F. (1969) *The Design of Optical Spectrometers*, Chapman and Hall, Ltd., London.
Jenkins, F.A. and White, H.E. (1976) *Fundamentals of Optics*, 4th edn, McGraw-Hill, New York, p. 111.
Kowalski, F.V., Teets, R.E., Demtröder, W., and Schawlow, A.L. (1978) *J. Opt. Soc. Am.*, **68**, 1611.
Loewen, E.G., Nevière, M., and Maystre, D. (1977) *Appl. Opt.*, **16**, 2711.
Miyake, K.P. (1959) *Sci. Light*, **8**, 39.
Namioka, T. (1959) *J. Opt. Soc. Am.*, **49**, 951.
Reader, J. (1969) *J. Opt. Soc. Am.*, **59**, 1189.
Reich, M., Schieder, R., Clar, H.J., and Winnewisser, G. (1986) *Appl. Opt.*, **25**, 130.
Sawyer, R.A. (1963) *Experimental Spectroscopy*, 3rd edn, Dover Publications, New York, p. 358.

Seya, M. (1951) *Sci. Light*, **2**, 8.
Scheeline, A., Bye, C.A., Miller, D.L., Rynders, S.W., and Owen, R.C. (1991) *Appl. Spectrosc.*, **45**, 334.
Shafer, A.B., Megill, L.R., and Droppelman, L. (1964) *J. Opt. Soc. Am.*, **54**, 879.
Till, S.M., Jones, W.J., and Shotton, K.C. (1975) *Proc. R. Soc. London, A*, **346**, 395.

Further Reading

Hutley, M.C. (1982) *Diffraction Gratings*, Academic Press, London.
James, J.F. (2007) *Spectrograph Design Fundamentals*, Cambridge University Press, Cambridge.
James, J.F. and Sternberg, R.F. (1969) *The Design of Optical Spectrometers*, Chapman and Hall, Ltd., London.
Jenkins, F.A. and White, H.E. (1976) *Fundamentals of Optics*, 4th edn, McGraw-Hill, New York.
Kitchin, C.R. (1995) *Optical Astronomical Spectroscopy*, Institute of Physics Publishing, Bristol and Philadelphia.
Loewen, E.G., Nevière, M., and Maystre, D. (1977), **16**, 2711–2721.

11
Ultraviolet and Visible Absorption Spectroscopy

Robert H. Lipson

11.1	Introduction	355
11.2	Spectral Background	355
11.2.1	State Labels	355
11.2.1.1	Atoms	355
11.2.1.2	Molecules	356
11.2.2	Electronic Selection Rules	357
11.2.3	Born–Oppenheimer Approximation and the Franck–Condon Principle	358
11.2.4	Rotational Fine Structure	361
11.3	Absorption Spectra	362
11.3.1	Beer–Lambert Law	362
11.3.2	Chemometrics	363
11.3.3	Derivative Spectroscopy	363
11.3.4	Circular Dichroism Spectroscopy	363
11.4	The Origin of UV/Visible Transitions in Organic Molecules	364
11.4.1	Chromophores	365
11.5	The Origin of UV/Visible Transitions in Inorganic Molecules	368
11.6	The Origin of UV/Visible Transitions in Solids	370
11.7	Experimental Techniques for UV/Visible Absorption Spectroscopy	371
11.7.1	Light Sources	371
11.7.1.1	Thermal Sources	371
11.7.1.2	Luminescent Sources	371
11.7.1.3	Light-Emitting Diodes	372
11.7.1.4	Synchrotrons	372
11.7.2	Wavelength Dispersive Instruments	372
11.7.2.1	Fastie–Ebert Configuration	373
11.7.2.2	Czerny–Turner Configuration	374
11.7.2.3	Single-Beam Versus Double-Beam Spectrometers	374
11.7.3	Detectors	374
11.7.3.1	Photodiodes	374
11.7.3.2	Photomultipliers (PMTs)	375

Encyclopedia of Applied Spectroscopy. Edited by David L. Andrews.
Copyright © 2009 WILEY-VCH Verlag GmbH & Co. KGaA, Weinheim
ISBN: 978-3-527-40773-6

11.7.3.3	Microchannel Plates (MCP)	375
11.7.3.4	Charge Coupled Devices (CCD)	376
11.8	**Laser-Based Techniques**	**376**
11.8.1	Fluorescence Excitation Spectroscopy	376
11.8.2	Excitation Spectroscopy with Mass Detection	377
11.8.3	Cavity Ring-Down Spectroscopy	378
	References	**378**
	Further Reading	**379**

11.1
Introduction

An overview of spectroscopy fundamentals for atoms and molecules is provided as they pertain to the ultraviolet (UV) and visible spectral regions. Topics include state labels, electronic selection rules, the Born–Oppenheimer approximation, and the Franck–Condon principle. The Beer–Lambert law, chemometrics, derivative spectroscopy, and circular dichroism (CD) spectroscopy are explained as are the origins of UV/visible transitions in organic and inorganic compounds as well as solids. Relevant experimental methods and instrumentation such as light sources, monochromators, and detectors, which can be used to record UV/visible absorption spectra, are described. A brief discussion on laser-based techniques, including fluorescence excitation spectroscopy, excitation spectroscopy with mass detection, and cavity ring-down spectroscopy, is also given.

11.2
Spectral Background

Absorption occurs when electromagnetic radiation interacts with matter resulting in a loss of beam intensity after passing through the sample. Light absorption can take place in the gas phase, liquid phase, solid phase, or in a plasma, and can involve a single species or a complex mixture.

The UV spectral region lies between ~200 and 400 nm while the longer wavelength visible region lies between ~400 and 700 nm. The conversions to frequency (ν), cm^{-1} wave number ($\tilde{\nu}$), and energy (E) units are presented in Table 11.1.

UV/visible photons are usually sufficiently energetic to promote electrons from the atomic or molecular ground states to higher lying valence or Rydberg states. In the case of molecules, the energy can also excite excited-state vibrational and rotational motions.

11.2.1
State Labels

11.2.1.1 Atoms
The orbital angular momentum of an electron in an atom can only adopt integer values : $\ell = 0, 1, 2$, and so on. The total orbital angular momentum of a many-electron system **L** can be calculated from the vector sum $\mathbf{L} = \sum_i \vec{\ell}_i$. Although **L** can only have integer

Encyclopedia of Applied Spectroscopy. Edited by David L. Andrews.
Copyright © 2009 WILEY-VCH Verlag GmbH & Co. KGaA, Weinheim
ISBN: 978-3-527-40773-6

Tab. 11.1 Wavelength, λ, frequency, ν, wave number, $\tilde{\nu}$, and energy, E, ranges for the ultraviolet and visible spectral regions.

Range	λ (nm)	ν (Hz)	$\tilde{\nu}$ (cm^{-1})	E (J)
UV	200–400	1.5×10^{15}–7.5×10^{14}	50 000–25,000	9.94×10^{-19}–4.97×10^{-19}
Visible	400–780	7.5×10^{14}–3.8×10^{14}	25 000–12820.5	4.97×10^{-19}–$2.52 \times 10^{1-19}$

values = 0, 1, 2, and so on, it should be appreciated that a single multielectron configuration can often lead to many values of **L**. The terms are called $S, P, D, F\ldots$ when $L = 0, 1, 2, 3\ldots$, respectively (Bernath, 1995).

Each electron has spin $s = 1/2$ and the total electron spin of the system can be calculated from the vector sum $\mathbf{S} = \sum_i \mathbf{s}_i$, where $\mathbf{s}_i$ is the electron spin of electron, i. A singlet state arises when the total electron spin is zero. This terminology comes from the definition of the spin multiplicity $= 2S + 1$. If the total spin of a state is $S = 1 (2S + 1 = 3)$, then the term is called a *triplet state*. One-electron systems or atoms with unpaired electrons can have terms with half-integer total electron spin; that is, $S = 1/2, 3/2, 5/2$, and so on, leading to doublet states, quartet state, sextet states, and so on. These arguments hold for molecules as well.

The total angular momentum, **J**, can be calculated in two ways. For light atoms with relatively few electrons (typically, atoms with atomic numbers $Z < 40$), **J** can be deduced from the vector sum of **L** and **S** (Russell–Saunders coupling scheme). Again one term with specific values of **L** and **S** can give rise to many J-levels. Russell–Saunders atomic terms are labeled by $^{2S+1}L_J$.

For heavier atoms, the individual electron orbital and spin angular momenta, $\vec{\ell}_i$ and $\mathbf{s}_i$ first vector-couple to form total angular momenta $\mathbf{j}_i$. The total angular momentum for the system is found from the vector sum $\mathbf{J} = \sum_i \mathbf{j}_i$ (j–j coupling scheme). The total angular momentum **J** can have integer or half-integer values depending on whether **S** is integer or half integer, and their numerical value serves as the label for the term.

The parity, I, of a one-electron state can be determined from $I = (-1)^\ell$ where ℓ is the orbital angular momentum of the electron involved. For a many-electron system, the parity is determined not by the total angular momentum L, but by a sum of the individual electron orbital angular momenta in the configuration; that is, by $I = (-1)^{\sum_i \ell_i}$. A state or term is considered even or odd depending on whether I is $+1$ or -1, respectively. The term symbols for an electron state would then be denoted by g (gerade; even) or u (ungerade, odd).

11.2.1.2 Molecules

The electronic states of diatomic or linear molecules are usually labeled by the component of orbital angular momentum along the bond axis Λ. When $\Lambda = 0, 1, 2, 3, \ldots$, the states are called $\Sigma, \Pi, \Delta, \Phi, \cdots$, respectively. Because linear systems have cylindrical symmetry, each state with $\Lambda > 0$ is twofold degenerate, corresponding physically to clockwise and counterclockwise electron orbital motion about the molecular axis. This degeneracy can be broken,

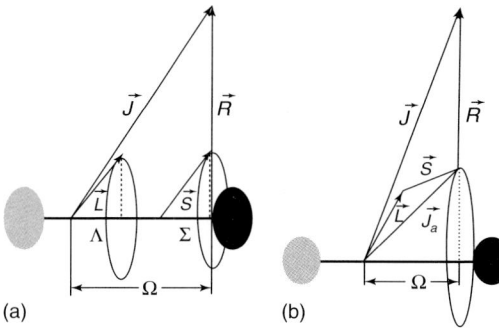

Fig. 11.1 (a) Hund's coupling case (a); (b) Hund's coupling case (c) for a diatomic molecule. **R, L, S,** and **J** refer to the rotational, orbital, spin, and total angular momenta, respectively. J_a is the vector sum of **L** and **S** in Hund's case (c). The precession of **J** about the internuclear axis in part a) is not indicated. In part (b) the precession of **L** and **S** about J_a or **J** about the internuclear axis is not shown. Λ, Σ, and Ω are the projections of **L, S** and **J** along the bond axis.

however, by molecular rotation in a process called Λ-doubling (Herzberg, 1950). The symmetry of Σ-states with respect to reflection in a plane containing the molecular axis σ_v must be considered. Σ-states that are symmetric with respect to σ_v are labeled Σ^+, while those that are antisymmetric with respect this symmetry operation are called Σ^-. These superscripts are not included for states with $\Lambda > 0$ since they already include +and − components. Molecular terms are denoted by $^{2S+1}\Lambda_\Omega$ where $\Omega = |\Lambda + \Sigma|$ is the component of total angular momentum along the molecular axis. Here Σ does not mean $\Lambda = 0$, but is instead the component of electron spin S along the bond axis, which ranges from $+S$ to $-S$ in integer steps.

The labels discussed above are appropriate for Hund's coupling case (a) (Figure 11.1a). There are five common coupling cases (a)–(e) (Herzberg, 1950). Hund's case (a) resembles the Russell–Saunders scheme presented above for light atoms. Hund's case (c) (Figure 11.1b) is the molecular analog of j−j coupling when spin–orbit coupling is important. Here Λ and Σ are no longer well defined, and the states are labeled only by $\Omega = 0, 1/2, 1, 3/2$, and so on, depending on the total electron spin. Again, the + and − labels must be added to the 0 states, and g and u subscripts included if needed.

The electronic states of nonlinear molecules are labeled by the Mulliken label for their irreducible representation, Γ, within their particular point group (Cotton, 1990), and again the g and u superscripts are used if appropriate.

11.2.2
Electronic Selection Rules

UV/visible transitions are induced when an incident electromagnetic field interacts with the different electric or magnetic multipole moments, $\tilde{M}$, of the electron distribution of the molecule. Quantum-mechanical selection rules dictate transitions which will be observed. If ψ_e' and ψ_e'' are the upper and lower electronic states, respectively, a particular transition between these states will be allowed if the transition matrix element $\int \psi_e'^* \hat{M} \psi_e'' d\tau_e$ is nonzero (Herzberg, 1966). Here $\hat{M}$ is the operator corresponding to the particular charge distribution in the molecule, and the integration is over the electronic coordinates, τ_e. The intensity of the transition is proportional to the square of the transition matrix element although this can be very difficult to calculate. As a rule of thumb, the strongest transitions (sometimes referred to as *E1 transitions*) involve the electric dipole moment: $\hat{M} = -\sum_i e_i r_i$. E2 and E3 transitions involve the electric quadrupole and electric octupole moments, respectively, but are considerably weaker than E1 transitions. M1, M2, and M3 transitions involve the magnetic dipole, magnetic quadrupole, and the magnetic octupole moments, respectively, and are also relatively weak. Typically, M1 and E2 transitions are comparable in intensity. Both require high sensitivity to be detected, and usually will only be observed in frequency regions where E1 transitions are forbidden. The rest of this chapter will focus on single-photon electric dipole transitions.

There are additional symmetries that govern selection rules. Single-photon electric dipole transitions are allowed when there is a change in parity. The selection rule governing parity, the so-called Laporte selection rule, can be written for one-electron systems as $\Delta\ell = \pm 1$. For many-electron systems, the selection rule becomes $\Delta L = 0, \pm 1$.

Electron spin does not usually change during an electronic transition that leads to a $\Delta S = 0$ selection rule. The spin selection rule can be relaxed, however, when the spin–orbit interaction becomes important. Then the selection rules for L and S becomes less rigorous, and those for the total angular momentum, J must be considered. It can be shown that the selection rules are $\Delta J = 0, \pm 1$; $J' = 0 \leftrightarrow J'' = 0$ forbidden. The selection rule forbidding single-photon transitions between states with $J = 0$ can be understood as a consequence of conserving angular momentum in an absorption process involving photons that have their own intrinsic nonzero angular momentum.

The selection rules for linear molecules in Hund's case (a) are $\Delta\Lambda = 0$, ± 1, $\Delta S = 0$, g $\leftrightarrow$ u, $\left\{ \begin{matrix} + \\ - \end{matrix} \right\} \leftrightarrow \left\{ \begin{matrix} + \\ - \end{matrix} \right\}$. In Hund's case (c) the rules are $\Delta\Omega = 0$, ± 1, g $\leftrightarrow$ u, $\left\{ \begin{matrix} + \\ - \end{matrix} \right\} \leftrightarrow \left\{ \begin{matrix} + \\ - \end{matrix} \right\}$.

In general, a molecular transition will be allowed if the direct product of the irreducible representations of the ground-state wave function, excited-state wave function, and the multipole moment contains the totally symmetric representation of the particular point group, $\Gamma(A_1)$, that is, $\Gamma(\psi'_e) \otimes \Gamma(\hat{M}) \otimes \Gamma(\psi''_e) \subset \Gamma(A_1)$. Equivalently, a transition will be allowed if the direct product $\Gamma(\psi'_e) \otimes \Gamma(\psi''_e) \subset \Gamma(\hat{M})$.

11.2.3
Born–Oppenheimer Approximation and the Franck–Condon Principle

Vibrations play a significant role in determining the intensity of transitions between molecular electronic states (Hollas, 1996). An appropriate starting point is the Born–Oppenheimer approximation, which allows the electronic and nuclear motions to be considered separately because of the large mass differences between the nuclei of a molecule and an electron. Within this approximation, the total wave function of a state, Ψ, can be expressed as $\Psi(r, R) = \psi(r, R)\chi(R)$ where χ is a nuclear wave function that depends on nuclear coordinates $\{R\}$ and ψ is an electronic wave function that depends on the electronic coordinates $\{r\}$ and parametrically on the position of the nuclei. The electronic wave function is therefore found by solving the Schrodinger wave equation at fixed nuclear positions.

A potential energy surface can then be constructed by finding the electronic energy at different nuclear coordinates.

The Franck–Condon principle states that electronic transitions will occur on a much faster timescale than that it takes the nuclei to adjust to their new equilibrium positions. A single-photon electric dipole transition matrix element is defined as $\int \Psi' \hat{\mu} \Psi'' dr dR$, where Ψ' and Ψ'' are the wave functions of the excited and ground state, respectively, and $\hat{\mu}$ is the electric dipole operator. As a consequence of the Born–Oppenheimer approximation, the matrix element can be written as a product of an electronic component $R_e = \int \psi' \hat{\mu} \psi'' dr$ and a vibrational overlap integral $S_{v',v''} = \int \chi'_{v'} \chi_{v''} dR$. The intensity of an electronic transition will only be appreciable if the electronic transition probability $|R_e|^2$ is nonzero. This requires that the transition obey the appropriate selection rules for the orbital and spin quantum numbers. The square of $S_{v',v''}$, called the *Franck–Condon factor*, however, modulates the intensity distribution of the vibronic bands observed for a specific electronic transition.

For the sake of simplicity, the Franck–Condon principle is illustrated here for diatomic molecules because they have only one vibrational motion (bond stretching), and therefore their potential energy surfaces reduce to potential energy curves. Quantum mechanically, except for the lowest vibrational level ($v = 0$), which has its maximum wave function amplitude at the equilibrium bond length of the molecule in that electronic state, the largest amplitudes of the vibrational wave functions for $v > 0$ are found at the classical turning points. These correspond to bond lengths where the total energy is potential energy only; that is, the turning points are bond lengths where the nuclear velocities are zero and classically the bond is either at its maximum extension (outer turning point) or its maximum compression (inner turning point). As shown in Figure 11.2(a) the ($v' = 0, v'' = 0$) vibronic transition will have the largest Franck–Condon factor (vibrational wave function overlap) when the ground-state equilibrium bond length r''_e, is $\approx$ the excited equilibrium bond length r'_e. The Franck–Condon principle predicts that this transition will be the most intense feature in the absorption spectrum. However when $r'_e > r''_e$ (Figure 11.2b) or $r'_e < r''_e$ the intensity maximum will no longer be the (0,0) transition. In both cases, the spectrum will exhibit a (v',0) progression with a Gaussian intensity distribution maximizing at a value of v', which depends on Δr_e.

In some molecular systems, the Franck–Condon factors are largest for vibronic transition where $v' \gg 0$. One interesting example is the transitions between the ground and ion-pair states of the diatomic halogens. There are 20 ion-pair states of a diatomic halogen XY (X and Y = Cl, Br, I) that dissociates to X^+ ($^3P, ^1D, ^1S$) + Y^- (^1S). The strongest transitions from their ground states terminate at the first tier $D0_u^+$ or $E0^+$ state depending on whether the halogen is homonuclear or heteronuclear, respectively. A molecular orbital analysis shows that these excited ion-pair states have antibonding character. However since the long-range binding is coulombic in nature, the resultant excited states have deeper potential energy curves than the ground states, longer bond lengths, and smaller vibrational frequencies. As a result, the most intense Franck–Condon transitions from

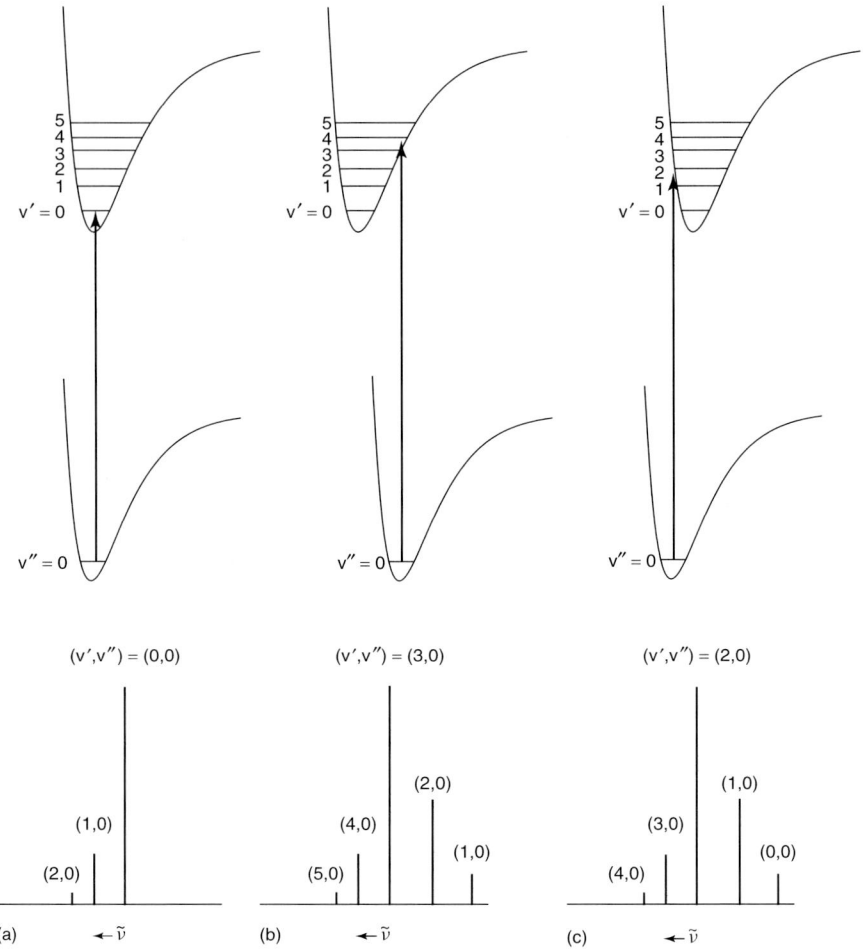

Fig. 11.2 Potential energy curves and resultant stick spectra for the case where (a) $r'_e \approx r''_e$; (b) $r'_e > r''_e$, and (c) $r'_e < r''_e$.

$v'' = 0$ of the IBr ground state, for example, probe the inner wing of the excited E0$^+$ state potential energy curve in the vacuum UV spectral region exciting high vibrational levels between $v' = 260$ and 383 (Lipson and Hoy, 1989).

In general, a vibronic transition will be allowed if $\int \psi'^*_e \chi_{v'} \hat{M} \psi''_e \chi_{v''} d\tau_r d\tau_R$ is nonzero, where $\hat{M}$ is the electric dipole operator. This condition holds if $\Gamma(\psi'_e) \otimes \Gamma(\chi_{v'}) \otimes \Gamma(\hat{M}) \otimes \Gamma(\psi''_e) \otimes \Gamma(\chi_{v''}) \subset \Gamma(A_1)$. Within the Born–Oppenheimer approximation, the vibronic transition matrix element will be a product of $3N - 6$ (nonlinear molecules) or $3N - 5$ (linear molecules) vibrational overlap integrals. At room temperature, the ground-state population resides primarily in $v'' = 0$ of the normal modes. These states transform as the total symmetric representation of the molecular point group. If the excited-state vibration is totally symmetric the selection rule on v is $\Delta v_i = 0, \pm 1, \pm 2, \ldots$. If the vibration is nontotally symmetric the Franck–Condon factor vanishes when $\Delta v_i = \pm 1, \pm 3, \pm 5, \ldots$ because $\Gamma(\chi_{v'}) \otimes \Gamma(\chi_{v''})$ does not contain the totally symmetric representation. The selection rule therefore becomes $\Delta v_i = \pm 2, \pm 4, \pm 6, \ldots$.

A series of vibronic transitions originating from a common lower vibrational state and probing a range of excited vibrational states is called a *progression*. For example, the progression originating from the lowest ground-state vibrational level would be $(v' = 0, 1, 2, \ldots, v'' = 0)$, and the intensities of the observed bands are determined by their Franck–Condon factors. Vibronic transitions where $\Delta v = v' - v'' = $ constant (a positive or negative integer) form *sequences*. For example, a $\Delta v = 0$ sequence would be made up of the $(0,0), (1,1), (2,2), \ldots$ transitions. Since the population in $v'' > 0$ is expected to be small for most normal modes, hot-band progressions (originating from $v'' > 0$) and sequence transitions involving $v'' > 0$ are not expected to be particularly intense.

11.2.4
Rotational Fine Structure

Usually, rotational fine structure can only be resolved for molecules in the gas phase, where, typically, the linewidths found in UV/visible spectra are Doppler-broadened. Rotationally resolved spectra are much more difficult to obtain for liquid- or solid-phase samples where other line broadening mechanisms are operative.

A full discussion regarding the rotational fine structure associated with electronic transitions of molecules is beyond the scope of this chapter. Instead, a short discussion pertaining to diatomic molecules is presented here. The rotational selection rules for diatomic molecules (Hund's case (a), when $\Delta \Lambda = 0, \pm 1$) are $\Delta J = J' - J'' = 0, \pm 1$ where J' and J'' are the upper and lower state rotational quantum numbers. This leads to three branches labeled P ($\Delta J = -1$), Q ($\Delta J = -1$), and R ($\Delta J = +1$). However, only P and R branches are found for electronic transitions where $\Delta \Lambda = 0$, $\Lambda' = \Lambda'' = 0$. This can be understood by conservation of angular momentum arguments. On the other hand, if $\Delta \Lambda = 0$, $\Lambda' = \Lambda'' \neq 0$, the three branches are allowed but the Q-branch will be weak, while the P and R branches will be strong.

The wave number ($\tilde{v}$) expressions for these branches ignoring centrifugal distortion effects, are given by

$$\tilde{v}_P = \tilde{v}_o - (B_{v'} + B_{v''})J''$$
$$+ (B_{v'} - B_{v''})J''^2$$
$$\tilde{v}_Q = \tilde{v}_o + (B_{v'} - B_{v''})J''(J'' + 1)$$
$$\tilde{v}_R = \tilde{v}_o + 2B_{v'} + (3B_{v'} - B_{v''})J''$$
$$+ (B_{v'} - B_{v''})J''^2 \quad (11.1)$$

where $B_{v'}$ and $B_{v''}$ are the upper and lower state rotational constants in states v' and v'', respectively, and $\tilde{v}_o$ is the "pure" vibronic band origin. A B_v-value is inversely proportional to bond length, r_v, in a particular vibronic state, and hence a detailed rotational analysis can provide quantitative details about the upper and lower state vibronic geometries. These values can be used to find the equilibrium structure of the molecule.

The equations in Eq. (11.1) are quadratic in J. If $B_{v'} > B_{v''}$, that is, $r_{v'} < r_{v''}$, the rotational line spacings will decrease in the P-branch and increase in the R-branch. At some value of

J'' a band head will form, which is characteristic edge structure (in low resolution) composed of many rotational lines. The observed vibronic bands will be degraded to the higher energies (blue-shaded). Conversely, if $B_{v'} < B_{v''}$, that is, $r_{v'} > r_{v''}$, the spacings will decrease in the R-branch and increase in the P-branch. The resultant vibronic bands will then be red-shade.

11.3 Absorption Spectra

11.3.1 Beer–Lambert Law

In general, a UV/visible absorption spectrum is found by recording a plot of transmitted light versus wavelength. Experimentally, the amount of absorption or absorbance, A, is determined by measuring the transmitted light intensity, I, relative to the incident light intensity I_o using

$$A = -\log_{10}\left(\frac{I}{I_o}\right) \qquad (11.2)$$

The absorbance, which is unitless and is sometimes called the *optical density*, is proportional to the number of molecules in the path of the beam, and can therefore in the solution phase be calculated using the Beer–Lambert law:

$$A = \varepsilon(\lambda)[C]\ell \qquad (11.3)$$

where $\varepsilon(\lambda)$, which is the absorption coefficient of the sample at a specific wavelength, λ, $[C]$ is the sample concentration, and ℓ is the path length. ε is also called the *molar absorptivity* or *molar extinction coefficient* and has units of L mol^{-1} cm^{-1} when $[C]$ and ℓ are expressed in molarity and centimeters, respectively. An analogous expression can be written for gas-phase samples using number densities (in molecules per centimeter cubed) instead of concentrations. Then, the absorption coefficient is identified as an absorption cross section, σ (centimeter squared). The molar extinction coefficient is related to σ by

$$\varepsilon = \sigma(N_0 - N_1) \qquad (11.4)$$

where N_0 and N_1 are the ground-state and excited-state number densities, respectively.

Typically, ε (in liters per mole per centimeter) can vary between 0 and 10^6 where values $\geq 10^4$ are related to high-intensity absorptions while values $<10^3$ are considered low-intensity absorptions. Forbidden transitions will have molar absorptivities ranging from 0 to $\sim 10^3$.

If more than one species exists in solution, the overall absorbance will be equal to the sum of the absorbances of the individual species. For an N-component mixture, A can be written as

$$A(\lambda) = \ell \sum_{i=1}^{N} \varepsilon(\lambda)[C_i] \qquad (11.5)$$

The spectrum of a solution containing a pair of absorbing species can exhibit one or more isosbestic points, which correspond to wavelengths where the molar absorptivities of the species are identical. Even when a 1 : 1 reaction occurs the absorbance of a solution of the two reactants and product at an isosbestic point will remain invariant to the degree of reaction; that is, independent of their particular concentrations.

The Beer–Lambert law assumes that A will be a constant if the product of $[C]$ and ℓ is constant. However, this may not always be the case. For example, there could be molecular association at higher concentrations, or the species could undergo photochemistry upon irradiation. The Beer–Lambert law may also fail for acids, bases, and salts that exhibit concentration-dependent ionization, for samples that are strongly fluorescent, or if a thermal equilibrium exists between the ground state and a low-lying excited state. As a result, the linearity associated with the Beer–Lambert law should always be confirmed experimentally before an analysis.

11.3.2
Chemometrics

As noted above the Beer–Lambert law is additive. This allows the composition of a mixture of compounds to be determined if their individual absorption spectra are known. This technique lends itself very well to analysis by computer software. The field where mathematical or statistical methods are applied to data analysis is called *chemometrics*. The basic approach can be illustrated by considering a mixture of three compounds, a, b, and c with concentrations $[C_a]$, $[C_b]$, $[C_c]$ respectively (Rouessac and Rouessac, 2007). The absorbance of this mixture is measured at three wavelengths λ_1, λ_2, and λ_3 resulting in three measurements A_1, A_2, and A_3, respectively. By using a path length of $\ell = 1$ cm the absorbances at the three wavelengths can be written in matrix form as

$$\tilde{A} = \tilde{\varepsilon}\tilde{C} \qquad (11.6)$$

where

$$\tilde{A} = \begin{pmatrix} A_1 \\ A_2 \\ A_3 \end{pmatrix}, \quad \tilde{\varepsilon} = \begin{pmatrix} \varepsilon_a^1 & \varepsilon_b^1 & \varepsilon_c^1 \\ \varepsilon_a^2 & \varepsilon_b^2 & \varepsilon_c^2 \\ \varepsilon_a^3 & \varepsilon_b^3 & \varepsilon_c^3 \end{pmatrix}$$

and $\tilde{C} = \begin{pmatrix} [C_a] \\ [C_b] \\ [C_c] \end{pmatrix}$

If the nine absorbances in the $\tilde{\varepsilon}$ matrix are known, $\tilde{C}$ can then be found by computing $\tilde{A}\tilde{\varepsilon}^{-1}$. The precision of the composition determination can be improved for mixtures that have similar spectra by using many tens of data points, even though, mathematically, the system of equations in Eq. (11.6) would be overdetermined.

11.3.3
Derivative Spectroscopy

It can be helpful, instead of plotting A versus λ, to mathematically plot the absorbance derivatives $dA/d\lambda$ as a function of wavelength. This can be done for higher order derivatives as well. Such plots serve to amplify weak slope variations of the zeroth-order absorption spectrum. First-derivative plots will pass through zero at the location of each peak maximum, while second-derivative plots produce zero values at wavelengths corresponding to both the minima and maxima of the original absorption spectrum. This technique is useful when the zeroth-order spectrum has a uniformly increasing baseline due to a strong absorption at shorter wavelengths or because of light scattering. It is also helpful when there are weak absorbances that are overshadowed by stronger bands in the original spectrum.

11.3.4
Circular Dichroism Spectroscopy

CD spectroscopy measures the differential absorption of left-hand and right-hand circularly polarized light as a

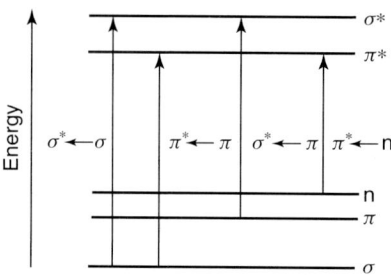

Fig. 11.3 An energy level diagram for a generic organic molecule indicating the possible types of electronic transitions, and their relative energy ordering.

function of exciting wavelength (Berova, Di bari and Pescitelli, 2007). The technique is sensitive to absolute configurations and therefore is able to discriminate different enantiomers (stereoisomers whose mirror images are not superimposable). Formally, a chiral molecule is one that does not possess an improper axis of rotation. CD is defined as

$$CD = A^L - A^R \qquad (11.7)$$

where A^L and A^R are the absorptions of left and right-hand circularly polarized light, respectively. The difference in the molar absorption coefficients, $\Delta\varepsilon$, at a specific wavelength can be expressed in terms of CD by

$$\Delta\varepsilon = \varepsilon^L - \varepsilon^R = \frac{CD}{[C]\ell} \qquad (11.8)$$

CD values for enantiomers are always equal and opposite, and therefore, when scanning through an absorption band $\Delta\varepsilon$ will change rapidly in one direction, pass through zero at the peak maximum, and then change rapidly in the opposite direction. This is known as the *Cotton effect* after its discoverer. Positive and negative Cotton effects refer to the cases where $\Delta\varepsilon$ is >0 and <0, respectively.

Although electric dipole moments, μ_{ij}, are larger than magnetic dipole moments, $\mathbf{m}_{ij}$, the rotational strength of a CD signal, R_{ij}, which is proportional to $\int \Delta\varepsilon d\nu$, depends on the product $\mu_{ij} \cdot \mathbf{m}_{ij}$. R_{ij} is a signed quantity, and as such can be used to interpret CD spectra and to make configurational assignments. The assignments depend strongly on the chromophores (Section 11.4.1) and their arrangements in the molecule. In general, related complexes that have the same CD sign for a given absorption have the same absolute configuration.

11.4
The Origin of UV/Visible Transitions in Organic Molecules

A generic energy level diagram of an organic molecule is presented in Figure 11.3.

The filled molecular orbitals of an organic molecule are designated as σ-bonding, π-bonding, or nonbonding, n, depending on the nature of the atomic orbitals involved. The most probable electronic transitions will take place from the highest occupied molecular orbital (HOMO) to the lowest occupied molecular orbital (LUMO), which are typically antibonding in nature (σ^* or π^*). In general, the energies E of the possible transitions are ordered: $E(\pi^* \leftarrow n) < E(\pi^* \leftarrow \pi) < E(\sigma^* \leftarrow n) < E(\pi^* \leftarrow \sigma) < E(\sigma^* \leftarrow \sigma)$. Of these transitions, $\pi^* \leftarrow \pi$ and $\pi^* \leftarrow n$ dominate in the UV/visible region.

11.4.1
Chromophores

Although molecular orbitals involved in absorption extend over the entire molecule, it is often possible to identify a specific atom or group of atoms in the molecule as a chromophore; that is, as a functional group that acts as the dominant light-absorbing unit. A list of chromophore properties for the most common functional groups is presented in Table 11.2 (Pavia, Lampman and Kriz, 1996)

Most $\pi^* \leftarrow n$ transitions involving heteroatoms are weak because the spatial orthogonality of the n-bonding orbitals relative to the π^*-system make the transition's symmetry forbidden. Similarly, there is no overlap between σ and π orbitals, and therefore they can be considered separately. The $\pi^* \leftarrow \pi$ transitions listed in Table 11.2 are not particularly intense. However, their transition wavelengths can shift dramatically to longer wavelengths and grow in intensity (hyperchromism) with increasing degree of conjugation owing to a reduction in the HOMO–LUMO energy gap in much the same way that the energy level spacing of a one-dimensional particle in a box decreases with increasing box size. This trend can be examined in Table 11.3 for various polyenes although similar trends have been observed for many other unsaturated systems. The absorption intensity growth with conjugation reflects an increasing transition dipole that lies approximately along the direction of the π-system. The deviations evident in Table 11.3 arise because the polyene backbones are not completely delocalized but, instead, consist of alternating single

Tab. 11.2 Absorption wavelength maxima, λ_{max}, and molar absorption coefficient maxima, ε_{max}, for various transitions of different chromophores.

chromophore	λ_{max} (nm)	ε_{max} (L mol^{-1} cm^{-1})	Transition type
C–C	<180	1000	$\sigma^* \leftarrow \sigma$
C–H	<180	100	$\sigma^* \leftarrow \sigma$
R–OH	180	316	$\sigma^* \leftarrow n$
R–O–R	180	3162	$\sigma^* \leftarrow n$
R–NH$_2$	190	3162	$\sigma^* \leftarrow n$
R–SH	210	1000	$\sigma^* \leftarrow n$
R$_2$C=CR$_2$	175	1000	$\pi^* \leftarrow \pi$
R–C≡C–R	170	1000	$\pi^* \leftarrow \pi$
R–C≡N	160	<10	$\pi^* \leftarrow n$
R–N=N–R	340	<10	$\pi^* \leftarrow n$
R–NO$_2$	271	<10	$\pi^* \leftarrow n$
R–CHO	190	100	$\pi^* \leftarrow \pi$
	290	10	$\pi^* \leftarrow n$
R$_2$CO	180	1000	$\pi^* \leftarrow \pi$
	280	32	$\pi^* \leftarrow n$
RCOOH	205	32	$\pi^* \leftarrow n$
RCOOR'	205	32	$\pi^* \leftarrow n$
RCONH$_2$	210	32	$\pi^* \leftarrow n$

and double bonds. Furthermore, the absorption λ_{max} and $\varepsilon(\lambda_{max})$ are sensitive to cis–trans isomerization (Klessinger and Michl, 1995).

Substituents that increase the intensity and often the λ_{max} of absorption peaks when attached to a chromophore are called *auxochromes*. Woodward (1941, 1942) and Fieser, Fieser and Rajagopalan (1948) established a set of empirical rules for dienes that allow the absorption maximum of a compound to be determined for different additional chromophores present. Similar rules have also been deduced for conjugated carbonyl compounds, and mono- and di-substituted benzenes. These rules can be useful for structure determination.

A proper solvent for solution-phase UV/visible spectroscopy should be non-absorbing at wavelengths absorbed by the solute of interest. A list of common solvents and their short wavelength transmission cutoffs is presented in Table 11.4 (Gauglitz and Vo-Dinh, 2003). While water and ethanol are good solvents for many substances, hexane and other hydrocarbons are appropriate for less polar compounds. The latter solvents also interact more weakly with the solute, allowing a finer structure to be resolved.

Solvent polarity can play a role in determining the position of peak maxima in an absorption spectrum. This effect, called *solvatochromism*, arises because the HOMO and LUMO orbitals involved in a UV/visible absorption have different polarities and therefore interact differently with the solvent. Bathochromic (red) shifts result when the LUMO is stabilized in the solvent relative to the ground state. This effect is common for $\pi^* \leftarrow \pi$ transitions. Conversely, hypsochromic (blue) shifts arise when the HOMO is preferentially stabilized by the solvent. This type of shift is commonly observed for $\pi^* \leftarrow n$ transitions.

Peptide bonds have intense $\pi^* \leftarrow \pi$ transitions between 190 and 210 nm while aromatic amino acids absorb strongly between 205 and 280 nm (Table 11.5, Gauglitz and Vo-Dinh, 2003). The λ_{max} values for the UV absorptions of tryptophan, tyrosine and phenylalanine come between 255 and 280 nm, and are often used to determine

Tab. 11.4 Ultraviolet cutoff wavelength for various commonly used solvents.

Solvent	Cutoff wavelength (nm)
Hexane, C_6H_{14}	200
Ethanol, C_2H_5OH	210
Water, H_2O	210
Methanol, CH_3OH	210
Acetonitrile, CH_3CN	215
Cyclohexane, C_6H_{12}	215
Chloroform, $CHCl_3$	250
Carbon tetrachloride, CCl_4	280
Benzene, C_6H_6	280
Pyridine, C_5H_5N	310

Tab. 11.3 Absorption wavelength maxima, λ_{max}, and molar absorption coefficient maxima, ε_{max}, for various polyenes.

$H(CH=CH)_n H$	λ_{max} (nm)	ε (L mol^{-1} cm^{-1})
1	162	10 000
2	217	21 000
3	258	35 000
4	296	52 000
5	335	118 000
8	415	210 000
11	470	185 000
15	547	150 000

Tab. 11.5 Absorption wavelength maxima, λ_{max}, for various amino acid residues.

Amino acid	Structure	λ_{max} (nm)
Histidine		210
Tryptophan		220, 280
Tyrosine		195, 222, 275
L-Cysteine		235
Phenylalanine		190, 205, 255

the total protein concentration in solution. Furthermore these wavelengths tend not to degrade the sample. The phenolic chromophore on tyrosine is sensitive to the pH of the solution in that its absorption maximum shifts to the red with increasing solution acidity. Similarly, increases in the nonpolar environment will also red-shift the absorption maxima of tyrosine and tryptophan. The UV/visible spectra of these residues can therefore be used to probe the conformational differences of different protein states.

11.5
The Origin of UV/Visible Transitions in Inorganic Molecules

Transition metal complexes are known to be important in biological systems, as catalysts, as components of solar cells, as light emitters, nonlinear media, and so on. One of their most striking features is the wide variety of colors they can exhibit both as solids and in solution. The energies of the five d-orbitals for a given principal quantum number $n \geq 3$ in a free atom or ion are essentially degenerate. Upon ligation, however, the degeneracy of the d-orbital set is lifted in a manner determined by the nature of the ligands and the symmetry of the complex. Although ligand-field theory (through a group theoretical analysis) is the appropriate formalism needed to understand the nature of the molecular orbitals that arise because of coordination, the simpler crystal-field theory also provides insight into the visible absorptions of the complexes (Figgis, 1966)

Consider a ML_6 octahedral complex where M and L are the metal and ligand, respectively. In crystal-field theory, the ligands are approximated as point-negative charges although similar results can be deduced even if the ligands are neutral but dipolar. If the metal ion is placed at the origin of a coordinate system, the d_{z^2} and $d_{x^2-y^2}$ orbitals form a twofold degenerate set (with e_g symmetry) that have lobes that point along the axes directly toward the ligands, which are raised in energy by Coulombic repulsion. The remaining orbitals (d_{xy}, d_{yz}, d_{zx}) form a threefold degenerate set with t_{2g} symmetry that lies lower in energy than $\{e_g\}$ because they are orientated between the axes and therefore, do not point directly at the ligands. Ligand-field theory also predicts this splitting pattern. However, ligand-field theory shows explicitly that the e_g orbitals unlike the t_{2g} set are not pure metal d-orbitals but have ligand character as well.

The d-orbital splitting patterns for complexes of lower symmetries can be deduced by removing specific ligands, and noting if this yields a net stabilization of some of the d-orbitals relative to the others. As shown in Figure 11.4, for example, forming a square planar ML_4 complex by removing the 2 ligands orientated along the z-axis lowers the energy of the d_{z^2} orbital in the e_g set relative to $d_{x^2-y^2}$, as well as the energy of the d_{yz} and d_{xz} orbitals relative

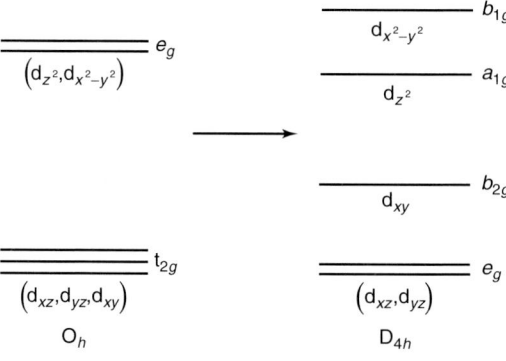

Fig. 11.4 Crystal-field splitting of the d-orbitals in an octahedral (O_h) and square planar (D_{4h}) geometry. The d-orbitals are labeled by their irreducible representations in the appropriate point group.

to d_{xy} in the t_{2g} set. Thus, the energy ordering of the d-orbitals in such a case might be $E(d_{xy}, d_{yz}) < E(d_{z^2}) < E(d_{xy}) < E(d_{x^2-y^2})$. Alternatively, the energy of d_{z^2} may even fall below that of the d_{xy}, d_{yz} pair depending on the metal ion and the nature of the ligands.

The splitting between the e_g and t_{2g} set in an octahedral complex is denoted by Δ_o and its magnitude is a strong function of the ligands making up the complex. This splitting is not trivial. For example, Δ_o for the $Ti(H_2O)_6^{3+}$ ion is 20 000 cm^{-1}. Ligands that yield a large Δ_o value are called *strong-field ligands* while those that result in small splittings are called *weak-field ligands*. The ligands can be empirically ordered in this regard for a given metal and geometry in a so-called spectrochemical series, which can be rationalized by considering ligand electronegativities, polarizabilities, permanent dipole moments, π-bonding characteristics, and so on. A partial list from strong field to weak field is given by

$$CO \approx CN^- > NO_2^- > NCS^- > H_2O$$
$$> OH^- > F^- > SCN^- > Cl^-$$
$$> Br^- > I^-$$

The magnitude of Δ_o also depends on the metal ion and its charge when the geometry and the ligands are held constant, although the rationale for the ordering is not as obvious. A partial spectrochemical series ranging from strong-field to weak-field ions is give by

$$Pt^{4+} > Ir^{3+} > Rh^{3+} > Co^{3+} > Cr^{3+}$$
$$> Fe^{3+} > Fe^{2+} > Co^{2+} > Ni^{2+}$$
$$> Mn^{2+}$$

The d-orbitals fill according to Hund's rules, which states that the terms of highest spin multiplicity lies lowest in energy. Electrons will spin-pair in the t_{2g} set before filling the e_g orbitals if the magnitude of Δ_o is large (strong-field ligands). Complexes of this sort are termed *low-spin systems*. Conversely, electrons will fill both the t_{2g} and e_g orbitals before spin-pairing if the ligands are weak field. These complexes are designated as high-spin systems.

UV visible transitions in a ML_6 complex involve promoting a d-electron from t_{2g} to e_g. Formally, however, d–d transitions are electric-dipole Laporte forbidden. However, this rule can be relaxed through vibronic coupling because the total wave function of a term includes not only an electronic component but also a vibrational normal mode. If that normal mode has the correct symmetry (in this case u-symmetry) the transition can become allowed. This is known as the *Herzberg–Teller effect*. Physically, the vibrational motion breaks the inversion symmetry of the molecule. Even so, octahedral complexes tend to have small molar absorptivities. On the other hand, the Laporte selection rule does not apply to noncentrosymmetric metal complexes geometries such as tetrahedral or trigonal bipyramidal, and so the d–d transitions tend to be stronger although the absorption coefficients are only in the range $\varepsilon \approx 100-200$.

The strongest metal complex absorptions usually lie in the UV and can be assigned to metal–ligand charge-transfer (g ↔ u) transitions. If an electron is transferred from an orbital lying principally on the ligand to an orbital lying principally on the metal, or vice versa, the transitions are termed a *ligand-to-metal* or *metal-to-ligand* charge transfer, respectively. These can occur for d^0 and d^{10} systems as well as open d-shell complexes. Indeed, all polyatomic molecules exhibit such transitions, although they often come at much shorter wavelengths than the UV/visible. A charge-transfer transition, however, can occur at longer wavelengths if the metal is easily oxidized and the ligand easily reduced, or vice versa. These absorptions tend to be very intense and can often mask the crystal-field transitions. Notable in this regard are metal iodide complexes, because I^- is readily oxidized. Typical ε values for octahedral hexahalo anions range from ~600 to 60 000 (Lever, 1968). Charge-transfer transitions can be used in the analysis of ions such as Fe and Mg in biologically important systems such as hemoglobin and chlorophyll, respectively.

11.6
The Origin of UV/Visible Transitions in Solids

While transition metal crystal-field splittings are also responsible for many of the striking colors observed in mineral gemstones, the broad subject of UV/visible transitions in solids is limited here to two additional examples – semiconductors and color centers.

UV/visible radiation can excite many semiconductors from their valence band to their conduction band. These can be classified as direct or indirect (phonon-assisted) transitions. A partial list of direct band-gap transitions for some binary semiconductors, and their band gaps and absorption edges at room temperature are given in Table 11.6 (Berger, 1997).

Tab. 11.6 Electronic band gap and wavelength absorption edge for several common semiconductor solids.

Semiconductor	Band gap (eV)	λ_{edge} (nm)
ZnO	3.2	387
ZnS	2.8	443
ZnTe	2.4	517
CdS	2.4	517
HgS	2.1	590
CuCl	3.4	365
CuBr	2.9	428
CuI	2.9	428
AgI	2.9	428
BN	5.8	214
AlN	6.2	200
GaN	3.4	365
InN	2.0	605

Band gaps increase in energy (decrease in wavelength) at lower temperatures. Indirect band-gap materials tend to have shorter wavelength absorption edges. Furthermore, if defect levels are located within the band gap they can absorb light at longer wavelengths than the band edge.

Color centers are formed by first producing vacancies in a crystal lattice, usually by irradiating the material with X-rays. Electrons or holes are then trapped in the lattice defects. The number of color centers produced is a function of the irradiation time. Depending on the crystal, color centers can persist for a period of time ranging from days to years. The alkali halide crystals have been the subject of considerable scrutiny in this regard (Schulman and Compton, 1962), although defects can also be generated in minerals including diamond. In an F-center, one or more free electrons located in an anionic vacancy become trapped within the electrostatic field of the enclosure. They can therefore undergo energy level transitions by the absorption of UV/visible wavelengths, thus imparting color to the crystal. Defect electronic transitions in alkali halides have been the basis of several commercial infrared lasers (Silfvast, 1996).

11.7
Experimental Techniques for UV/Visible Absorption Spectroscopy

In addition to the sample, the basic UV/visible absorption spectrometer consists of three parts: a light source, a wavelength dispersive instrument such as a monochromater, and a detector. Each is considered separately.

11.7.1
Light Sources

The most common UV/visible light sources can be grouped into two categories: thermal sources and luminescent sources. The merits of light-emitting diodes (LEDs) and synchrotrons are also discussed (Schmidt, 2005).

11.7.1.1 Thermal Sources

Thermal sources convert electrical power into heat and emit light according to Planck's Law. The simple incandescent light bulb is the most common example. In these devices a tungsten filament is placed in an evacuated glass bulb and is resistively heated by an electric current. The resultant Planck Law temperature is typically ~ 2500 K. Over time, small amounts of the tungsten will vaporize from the filament, coating and blackening the bulb. This effect can be reduced by adding an inert gas to the bulb such as Ar, Kr, or Xe.

Halogens lamps essentially operate the same way as incandescent lamps except here a halogen gas (usually a bromide) is added to the bulb. This additive will react with the evaporated tungsten vapor to form a tungsten bromide compound that falls back onto the filament where it is converted to elemental tungsten. The sizes of these sources, which operate at T ~ 3450 K, are usually small, and therefore, make good point-light sources.

11.7.1.2 Luminescent Sources

The mechanism behind luminescent sources involves electronic excitation of atoms that emit light upon relaxation. Gas-discharge lamps are the most common devices in this group and, depending on the pressure of the device, they can operate as continuum sources (broad wavelength outputs) or as atomic line sources (specific wavelengths).

Hg lamps consist of a bulb containing an inert gas and a trace amount of mercury. When a high voltage is applied and the gas breaks down, electrons are accelerated to kinetic energies that cause the rare gas to be ionized. Ion–electron recombination excites the rare gas, which subsequently transfers its energy to the Hg atoms, which, in turn, emit light. At low pressures ($\sim 10^2$ Pa), most the light comes out at the Hg resonance line at 253.7 nm. At medium pressure ($\sim 10^5$ Pa) the output has many spectral lines, and the device usually requires higher currents to operate. At high pressures ($> 3 \times 10^6$ Pa) the spectral output consists of broad overlapping bands that extend from the UV to the near infrared. However, the flux in the red and blue–green regions of the visible spectrum is relatively small.

Xe arc lamps operate at a Planck's Law temperature of ~6000 K. They put out less radiant energy than an Hg lamp in the UV region, but more in the visible region. Their chief strength is that the spectral output is relatively smooth in the visible although there are some spectral lines in the blue and near infrared regions.

Among the best continuum UV light sources are the hydrogen lamps where emission results from the recombination reaction: $H_2 \leftarrow 2H$, at a surface. Heat conduction can be minimized in the device and the radiant energy increased by using deuterium instead of hydrogen. The emission from these sources maximizes at 290 nm. Often, the one strong atomic H or D Balmer series emission line at 486.12 or 485.99 nm, respectively, can be used for wavelength calibration.

In most spectrophotometers, two lamps are needed to cover the entire UV/visible spectral range. Typically, a D_2 lamp is used for the UV and a halogen lamp is used for the visible ranges. There is an obvious complexity to the instrumentation that will arise when one lamp is switched to the other.

11.7.1.3 Light-Emitting Diodes

LEDs are solid-state devices where the recombination of electrons and holes generates light. Depending on the semiconductor material, they can emit light between ~350 and 1550 nm. While their efficiencies and radiant energies are usually low, they have long lifetimes and are very inexpensive. White LEDs have now been developed, which operate on one of two mechanisms. The most dominant approach is to combine a 450–470-nm gallium nitride (GaN) LED with an yttrium aluminum garnet (YAG) phosphor. The blue output of the GaN LED causes the phosphor to glow white out to ~700 nm. The second method involves combining the outputs of red, green, and blue LEDs in the proper proportion to obtain white light.

11.7.1.4 Synchrotrons

Synchrotron emission from an electron storage ring provides wavelength coverage from the infrared to γ-rays. Each pulse of synchrotron light is highly polarized and short ($<10^{-10}$ seconds), and the repetition rate of the pulse train can reach 5×10^8 s^{-1}. Despite this wavelength versatility, synchrotrons are "big science" and as such, would probably not be the first choice for work in the UV/visible where cheaper tabletop sources abound. They are, however, indispensable for high-resolution spectral studies and crystallography in the X-ray region.

11.7.2
Wavelength Dispersive Instruments

Dispersive instruments spatially separate a light beam on the basis of wavelength. The simplest optic for performing this feat is a prism, usually made of quartz or glass, whose dispersive power is a function of its wavelength-dependent index of refraction $n(\lambda)$. A geometrical analysis shows that the wavelength-dependent beam deflection, δ through a prism is given by $d\delta/d\lambda = -B/H \, dn/d\lambda$, where $dn/d\lambda$ is the change in index of refraction with wavelength, B the dimension of the base of the prism, and H is the height of the incident beam.

Higher resolution dispersion can be obtained using diffraction gratings. Diffraction gratings are ruled or holographically produced substrates containing several hundred to $>10^3$ lines per millimeter. A grating gives rise to wavelength-dependent interference patterns by Bragg diffraction when illuminated. The directions of constructive interference, or intensity maxima, arise when the path differences between scattered rays are an integer multiple of the wavelength. These integers, m, are called *diffraction orders*.

Gratings can operate in transmission (slits) or in reflection (grooves), and their resolving power R is defined as

$$R = \frac{\lambda}{d\lambda} \qquad (11.9)$$

R is a measure of how well it can separate two close-lying wavelengths, and is related to the number of grooves, N through

$$R = mN \qquad (11.10)$$

Although the linear dispersion (spatial separation in a plane), $d\lambda/dx$, will increase with order m, at a fixed grating orientation light, for example, at 800 nm in first order, will overlap 400 nm light in second order, 266.7 nm light in third order, and so on. The different orders can be separated using filters, or, alternatively, by modifying (blazing) the shape of the grooves so that the scattered energy is concentrated primarily into one order.

Diffraction gratings are the main dispersion elements in monochromators. These instruments are used to measure the intensity of different wavelengths sequentially as a grating is rotated. The two common geometries used for UV/visible spectroscopy are the Fastie–Ebert configuration and a Czerny–Turner design.

11.7.2.1 Fastie–Ebert Configuration

The optical arrangement of a Fastie–Ebert spectrograph is shown in Figure 11.5. The main components are a spherical mirror and one plane diffraction grating. The entrance and exit slits of the instrument are both placed at the focal plane of the spherical mirror. Light entering the first slit and reflected off the mirror is dispersed by the grating. The

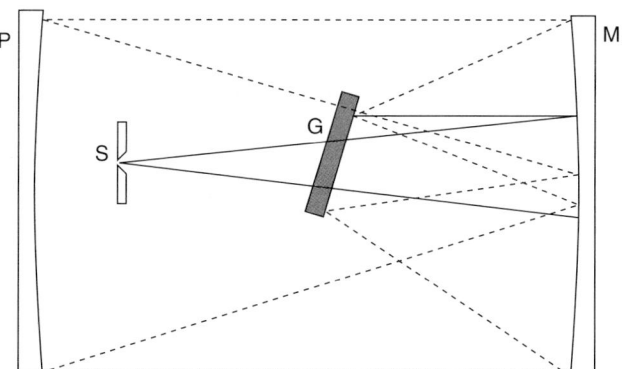

Fig. 11.5 A schematic of a Fastie–Ebert spectrograph. P, photographic plate; S, slit; G, grating; and M, concave mirror. The solid and dashed lines indicate the light path above and in line with the grating.

light is then rereflected and collimated by the spherical mirror to form an image at the exit slit. A UV/visible spectrum can be obtained by detecting the output light intensity photographically, or with a detector at the exit slit as a function of grating rotation.

11.7.2.2 Czerny–Turner Configuration

Czerny–Turner monochromators operate in much the same manner as the Fastie–Ebert design, in that the entrance and exit slits are fixed, and the monochromator is tuned by rotating a plane diffraction grating. The main difference is that the single spherical mirror in the former design is now replaced by two concave mirrors. This provides more flexibility in the geometry of the mirrors and can eliminate certain optical aberrations such as coma. This is also the preferred design when designing larger instruments where a single spherical mirror with good optical properties may be difficult to manufacture.

11.7.2.3 Single-Beam Versus Double-Beam Spectrometers

Single-beam spectrometers disperse one beam of light passed from a source; the beam is then passed through the sample and onto the detector. The measurement can be then repeated for a reference solvent. Double-beam spectrometers, on the other hand, use a split dispersed light beam that is passed through the sample and reference solvent at the same time (Skoog, 1985).

Single-beam instruments are technically simpler and often produce spectra with a better signal-to-noise ratio because the full radiant light energy of the lamp is used. However, the quality of the spectrum is also determined by the constancy of experimental conditions such as the output power of the lamp and electrical noise.

Conversely, dual-beam instrument operation compensates for lamp and electrical fluctuations, but the spectra often have poorer signal-to-noise ratios because only half the light beam is used for either the sample or the reference solvent. The instruments are also technically more sophisticated.

In general, single-beam instruments are adequate for absorbance measurements done at a fixed wavelength, while dual-beam instruments may be more appropriate for recording absorption spectra over a broad wavelength range.

11.7.3 Detectors

Many UV/visible detectors have been developed. A short survey of the most common devices is presented here.

11.7.3.1 Photodiodes

UV/visible light can be measured using an Si-based device known as a *reverse bias photodiode*. Here a voltage depletes the conductance at the junction of the p–n junction to nearly zero. When light falls on the depleted layer, holes and electrons are formed, which provide a current proportional to the light intensity. The equivalent quantum efficiency of a photodiode is considerably higher than that of a photomultiplier (PMT) tube (Section 5.3.2), but has a lower sensitivity because there is no signal amplification. The diode sensitivity can be increased using a p–i–n or PIN structure. Here an intrinsic (undoped) layer, i, is placed between the p- and n-doped regions, which, when illuminated, also creates hole and electrons that contribute to the photocurrent. The quantum efficiency of this arrangement is larger because the depletion layer is thicker. The output of a diode provides an indication of the impinging light energy over a time interval.

11.7.3.2 Photomultipliers (PMTs)

Many UV/visible detectors rely on the photoelectric effect to convert optical signals to an electrical output. Here, a material (termed a *photocathode*) with an appropriate work function will eject electrons upon illumination. Typical photocathodes are made of mixed alkalis (Table 11.7, Lytle, 1974). Under an applied voltage the elected electrons are measured as a photocurrent. The quantum efficiency of the photocathode is defined as the probability that a photon will eject an electron, and is usually less than unity.

In a PMT tube, the electrons ejected from the photocathode strike a chain of additional photoemissive electrodes, or dynodes, each at a successively higher positive potential. The first dynode will emit several electrons for each electron striking it. Those electrons cause the second dynode to emit several electrons, rapidly leading to a cascade multiplication of electrons at the end of the chain. The total gain in many devices can exceed 10^6. PMT-tube sensitivity is usually limited by their dark-current emission, but this can be almost eliminated by cooling the tube to low temperatures.

11.7.3.3 Microchannel Plates (MCP)

A microchannel is a small glass tube ~1-mm long, with typical inner diameters between 8 and 20 µm, whose inner surface is coated with a semiconductor material (Dereniak and Crowe, 1984). Like a dynode stage of a PMT, the inner surface will emit secondary electrons when struck by a primary electron. The gain in a channel under an applied voltage of $\approx 10^3$ V is on the order of 10^4. Microchannel plates (MCPs) are arrays of these tubes sandwiched between metal electrodes, and there can be between 10^4–10^7 microchannels per plate. The primary electrons come from a photocathode placed in front of the array. The gain can be increased to $\approx 10^7$ by placing two MCPs in series. Each channel can produce an intensified picture element

Tab. 11.7 Absorption maximum, λ_{max}, and quantum efficiency for several typical photocathode materials.

Nominal Composition of Photocathode material (response[a])	λ_{max} (nm)	Quantum efficiency (%)
Cs_8Sb (S-11)	440	16
Cs_8Sb (S-4)	400	12.4
K–Cs–Sb (117)	400	25
K–Cs–Sb (115)	400	25
(Cs)Na_2KSb (S-20)	420	20
(Cs)Na_2KSb (112)	420	18

[a] trade name; all "S" numbers are (Radio Corporation of America) RCA designations.

or pixel in an imaging system or the total current can be collected and plotted as a function of wavelength to yield a spectrum.

11.7.3.4 Charge Coupled Devices (CCD)

Although charge coupled devices (CCDs) were first invented as a computer memory circuit, they have become prevalent as light detectors in imaging applications. CCD cameras are either one-dimensional or two-dimensional pixel arrays of metal oxide semiconducting capacitors (based on Si) (Eastman Kodak, 2001). The number of electrons and holes produced within a pixel by an incident light beam is linearly proportional to the light level and exposure time at that location. There is also a nonlinear dependence on wavelength because of a changing penetration depth into the photocapacitor. When a voltage is applied to the pixel, a potential well is formed, which "collects," or integrates, the charge. The charge formed in each pixel is then transferred serially within the silicon substrate using a series of voltage gates, converted to a voltage signal, amplified, and then processed (by other equipment) to give an image. These devices can be employed in an absorption spectrometer by removing the exit slit and imaging the output off the diffraction grating onto a pixel array.

Overall, PMTs have excellent sensitivity, linearity, and dynamic range for measuring instantaneous light powers, and are reasonably priced. MCPs and CCDs are also excellent detectors with the added advantage that they can be used for spatial imaging. While these latter devices have recently become more affordable, they are usually more expensive than most PMTs.

11.8
Laser-Based Techniques

Although lasers have had a profound impact on the field of spectroscopy, they are surprisingly poorly suited for direct absorption measurements in the UV/visible range. The reason for this is that a direct absorption measurement entails measuring a small decrease in the output of an intense but usually fluctuating monochromatic source due to sample absorption. Nevertheless, several powerful laser-based methods have been developed, which circumvent these difficulties and allow UV/visible absorption spectra to be recorded particularly for gas-phase species.

11.8.1
Fluorescence Excitation Spectroscopy

Laser-induced fluorescence involves detecting optical emission from a sample that has been excited by the absorption of laser radiation. As shown in Figure 11.6, fluorescence excitation spectra can be recorded by measuring the total undispersed laser-induced emission as a function of the tunable exciting light wavelength using an optical detector such as a PMT. It can be shown that if the light collection efficiency and fluorescence quantum yield are independent of wavelength, the resultant spectrum is equivalent to an absorption spectrum (Lipson and Shi, 2002). This technique is considerably more sensitive than direct absorption because, in principle, optical detection is measured against a zero-light background. Nevertheless, care must be taken to eliminate scattered light, which can be orders of magnitude more intense than the fluorescence. Spatial pinhole filters before the PMT can

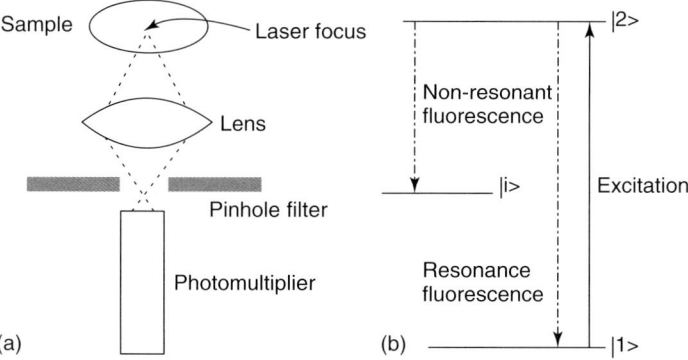

Fig. 11.6 (a) Schematic diagram of a typical experimental arrangement required to record laser-induced fluorescence excitation spectra. (b) A three-level diagram showing a single-photon transition between states $|1>$ and $|2>$, and possible resonance fluorescence and nonresonant fluorescence to an intermediate level $|i>$.

help in this regard. Sub-Doppler resolution gas-phase spectra can be obtained by sample delivery in a molecular beam or a supersonic jet expansion (Smalley, Wharton and Levy, 1977), and detecting fluorescence in a direction perpendicular to the gas flow.

11.8.2
Excitation Spectroscopy with Mass Detection

Not every molecule will fluoresce when excited. In such cases, one can use one or more laser beams to ionize the molecule of interest for detection in a time-of-flight mass spectrometer (Ashfold, 1998). The process of resonant two-photon ionization (R2PI) is described by

$$M \xrightarrow{h\nu} M^* \xrightarrow{h\nu'} M^+ + e^- \quad (11.11)$$

where the first photon at frequency ν resonantly excites a molecule M while a second photon at frequency ν' ionizes the excited molecule. The ionization reaction is a one-colour process if the photons involved come from the same laser and have the same frequency. If photons are used from different lasers, and ν and ν' are different, the process is called a *two-colour ionization*. Mass selected, vibronically resolved excited-state spectra can be obtained by varying the laser excitation wavelength while monitoring the molecular mass peak. R2PI is the simplest variant of $(m + n')$ resonance enhanced multiphoton ionization (REMPI) where m and n corresponds to the number of photons required to excite the molecule and ionize the excited state, respectively, and the prime indicates that the ionizing photons come from a laser different from the excitation source.

If the molecule predissociates or undergoes some photochemical reaction, it is possible to monitor the mass of a daughter fragment ion produced to obtain an absorption spectrum of the parent. This general approach of detecting a consequence of absorption to obtain the spectrum of a parent molecule is called *action spectroscopy*.

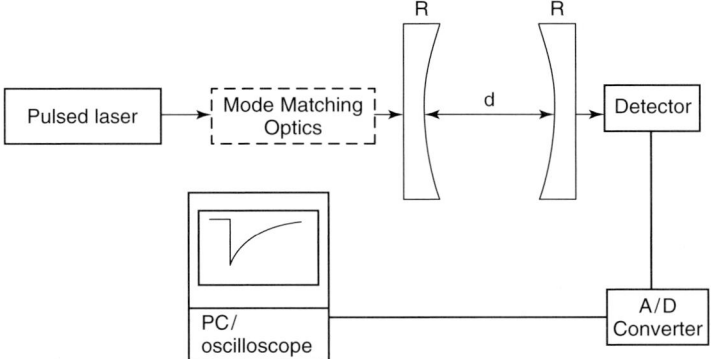

Fig. 11.7 Schematic of a typical cavity ring-down spectroscopy setup. R is the reflectivity of the mirrors making up the cavity of length d. The mode-matching optics may be optional.

11.8.3
Cavity Ring-Down Spectroscopy

Absorption spectra in cavity ring-down spectroscopy are obtained by measuring the rate of decay of light intensity inside a high-Q resonator (Figure 11.7) (Paldus and Kachanov, 2005). The intensity of light injected into the cavity will decay exponentially with a rate constant, $R(\lambda)$, that depends on the optical losses in the cavity.

The optical loss in an empty cavity, $L_{cav}(\lambda)$, is the sum of the round-trip scattering losses, $L_{scat}(\lambda)$ and the round-trip transmission losses through the high-reflectivity mirrors, $L_{trans}(\lambda)$. When a sample is added to the cavity, there is an additional loss due to absorption, $A(\lambda)$, which is governed by Beer's Law (Eq. 11.2). These factors are related to $R(\lambda)$ through

$$R(\lambda) = \frac{L_{cav}}{t_{rt}} + c\varepsilon(\lambda)[C] \qquad (11.12)$$

where $t_{rt} = \ell_{rt}/c$ is the photon round-trip transit time in the resonator, and ℓ_{rt} and c are the cavity round-trip length and the speed of light, respectively. $R(\lambda)^{-1}$ is called the *cavity ring-down time*. Measuring $R(\lambda)$ from the exponential intensity decay provides an accurate measurement of $\varepsilon(\lambda)$. An absorption spectrum is acquired by measuring $\varepsilon(\lambda)$ as a function of exciting laser wavelength, λ.

If high-reflectivity mirrors are used (>99.99%) that have low scattering losses (<0.001%), the effective total path length, which is determined by the average photon lifetime in the cavity, can be enhanced by factors of $>10^4$. This makes cavity ring-down spectroscopy very sensitive, even relative to setups that employ a multipass configuration. Cavity ring-down measurements are independent of the initial light intensity inside the cavity and the path length. Both pulsed and continuous wave sources can be used, and the technique is chemically specific because of the narrow linewidths available for many laser sources.

References

Ashfold, M.N.R. (1998) Multiphoton ionization (MPI) and Resonance enhanced multiphoton (REMPI) spectroscopy, in

Nonlinear Spectroscopy for Molecular Structure Determination (eds R.W. Field, E. Hirota, J.P. Maier, and S. Tsuchiya), Blackwell Science, Malden, pp. 127–145.

Berger, L.I. (1997) *Semiconductor Materials*, CRC Press, Boca Raton.

Bernath, P.F. (1995) *Spectra of Atoms and Molecules*, Oxford University Press, New York.

Berova, N., Di bari, L., and Pescitelli, G. (2007) *Chem. Soc. Rev.*, **36**, 914–931.

Cotton, F.A. (1990) *Chemical Applications of Group Theory*, 3rd edn, John Wiley & Sons, Inc., New York.

Dereniak, E.L. and Crowe, D.G. (1984) *Optical Radiation Detectors*, John Wiley & Sons, Inc., New York.

Eastman Kodak (2001) *Primer on CCDs*, http://www.kodak.com/ezpres/business/ccd/global/plugins/acrobat/en/support docs/chargeCoupleDevice.pdf.

Fieser, L.F., Fieser, M., and Rajagopalan, S. (1948) *J. Org. Chem.*, **13**, 800–806.

Figgis, B.N. (1966) *Introduction to Ligand Fields*, John Wiley & Sons, Inc., New York.

Gauglitz, G. and Vo-Dinh, T. (2003) *Hanbook of Spectroscopy*, Vol. 1, Wiley-VCH Verlag GmbH & Company kGaA, Weinheim.

Herzberg, G. (1950) *Molecular Spectra and Molecular Structure I. Spectra of Diatomic Molecules*, Van Nostrand Reinhold Company, New York.

Herzberg, G. (1966) *Molecular Spectra and Molecular Structure III. Electronic Spectra and Electronic Structure of Polyatomic Molecules*, Van Nostrand Reinhold Company, New York.

Hollas, J.M. (1996) *Modern Spectroscopy*, 3rd edn, John Wiley & Sons, Ltd, Chichester.

Klessinger, M. and Michl, J. (1995) *Excited States and Photochemistry of Organic Molecules*, VCH Publisher, New York.

Lever, A.B.P. (1968) *Inorganic Electronic Spectroscopy*, Elsevier Publishing Company, Amsterdam.

Lipson, R.H. and Hoy, A.R. (1989) *Mol. Phys.*, **68**, 1311–1319.

Lipson, R.H. and Shi, Y.J. (2002) Ultraviolet and vacuum ultraviolet laser spectroscopy using fluorescence and time-of-flight mass detection, in *Ultraviolet Spectroscopy and UV Lasers, Practical Spectroscopy Series, Vol. 30* (eds P. Misra and M.A. Dubinskii), Marcel Dekker, Inc., New York.

Lytle, F.E. (1974) *Anal. Chem.*, **46**, 545A–557A.

Paldus, B.A. and Kachanov, A.A. (2005) *Can. J. Phys.*, **83**, 975–999.

Pavia, D.L., Lampman, G.M., Kriz, G.S. (1996) *Introduction to Spectroscopy: A guide for Organic Chemistry*, Saunders Golden Sunburst Series, 2nd edn, Harcourt Brace College Publishers, Fort Worth.

Rouessac, F. and Rouessac, A. (2007) *Chemical Analysis*, 2rd edn, John Wiley & Sons, Ltd, Chichester.

Schmidt, W. (2005) *Optical Spectroscopy in Chemistry and Life Sciences*, Wiley-VCH Verlag GmbH & Company KGaA, Weinheim.

Schulman, J.H. and Compton, W.D. (1962) *Color Centers in Solids*, Pergamon Press, New York.

Silfvast, W.T. (1996) *Laser Fundamentals*, Cambridge University Press, Cambridge.

Skoog, D.A. (1985) *Principles of Instrumental Analysis*, 3rd edn, Saunders College Publishing, Philadelphia.

Smalley, R.E., Wharton, L., and Levy, D.H. (1977) *Acc. Chem. Res.*, **10**, 139–145.

Woodward, R.B. (1941) *J. Am. Chem. Soc.*, **63**, 1123–1126.

Woodward, R.B. (1942) *J. Am. Chem. Soc.*, **64**, 72–77.

Further Reading

General Spectroscopy Textbooks

Bernath, P.F. (2005) *Spectra of Atoms and Molecules*, 2nd edn, Oxford University Press, New York.

Herzberg, G. (1950) *Molecular Spectra and Molecular Structure*, Vols. I-III, VonNostrand Reinhold Company, New York (classic texts on molecular spectroscopy).

Hollas, J.M. (2004) *Modern Spectroscopy*, 4rd edn, John Wiley & Sons, Ltd, Chichester.

Karplus, M. and Porter, R.N. (1970) *Atoms and Molecules: An Introduction for Students of Physical Chemistry*, The Benjamin/Cummings Publishing Company, Menlo Park.

Group Theory Texts with Spectroscopy Applications

Bishop, D.M. (1993) *Group Theory and Chemistry*, Dover Publications, Inc., New York.

Carter, R.L. (1998) *Molecular Symmetry and Group Theory*, John Wiley & Sons, Inc., New York.

Cotton, F.A. (1990) *Chemical Applications of Group Theory*, 3rd edn, John Wiley & Sons, Inc., New York.

Jaffé, H.H. and Orchin, M. (2002) *Symmetry in Chemistry*, Dover Publications, Inc., Mineola, New York.

Organic Spectroscopy

Lambert, J.B., Shurvell, H.F., Lightner, D., and Cooks, R.G. (1987) *Introduction to Organic Spectroscopy*, Macmillan Publishing Company, New York.

12
Colorimetry

Robert T. Marcus

12.1	**Introduction** 383	
12.2	**Color Perception** 383	
12.2.1	Color Constancy and Metamerism 384	
12.3	**The CIE System of Colorimetry** 385	
12.3.1	Light and Light Sources 385	
12.3.1.1	Visible Light 385	
12.3.1.2	Standard Light Sources and Illuminants 386	
12.3.1.3	Color-Matching Booths 388	
12.3.2	Characterizing Objects 388	
12.3.3	The Standard Observers 390	
12.3.3.1	The Eye and Color Vision 390	
12.3.3.2	Flare and Metamerism 391	
12.3.3.3	Characterizing Observers 391	
12.3.3.4	Calculating Tristimulus Values 393	
12.3.3.5	Chromaticity Coordinates and Chromaticity Diagrams 394	
12.4	**Color Spaces** 395	
12.4.1	The Munsell Color Space 395	
12.4.2	The 1976 CIE $L^*a^*b^*$ (CIELAB) Color Space 396	
12.4.3	The 1976 CIE $L^*u^*v^*$ (CIELUV) Color Space 397	
12.5	**Color Differences and Tolerances** 398	
12.5.1	Single-Number Color Scales 400	
12.5.1.1	Whiteness Scales 400	
12.5.1.1	Yellowness Indices 401	
12.5.1.2	Scales for Liquids 401	
12.5.2	The CIELAB and CIELUV Color-Difference Equations 401	
12.5.2.1	The CIELAB Color-Difference Equation 401	
12.5.2.2	The CIELUV Color-Difference Equation 402	
12.5.3	The CMC($l:c$) Color-Tolerance Equation 403	
12.5.4	The CIEDE 2000 Color-Difference Equation 403	
12.5.5	The DIN 99 Color-Tolerance Equation 405	
12.5.6	Setting Instrumental Color Tolerances 406	

Encyclopedia of Applied Spectroscopy. Edited by David L. Andrews.
Copyright © 2009 WILEY-VCH Verlag GmbH & Co. KGaA, Weinheim
ISBN: 978-3-527-40773-6

12 Colorimetry

12.6 **Instrumental Color Measurement** 406
12.6.1 Measuring Reflectance 406
12.6.1.1 Specular and Diffuse Reflectance 406
12.6.1.2 Illuminating and Viewing Geometries 407
12.6.1.3 Spectrophotometers, Colorimeters, Spectrocolorimeters, and LED-Based Instruments 410
12.6.1.4 Standardization 411
12.6.1.5 Repeatability, Reproducibility, and Accuracy 411
12.6.1.6 Measuring Fluorescent Materials 412
12.6.1.7 Measuring Materials Containing Metallic Flakes and Interference Pigments 412
12.6.2 Measuring Transmittance 415
12.6.2.1 Regular and Diffuse Transmittance 415
12.6.2.2 Illuminating and Viewing Geometries 415
12.6.2.3 Standardization 415
12.6.2.4 Measuring Transmittance 416
12.6.3 Radiometric Measurements 416
12.7 **Summary** 416
References 417

12.1
Introduction

The Earth, our home, is a blue planet. Color abounds from the blue sky, the striking sunsets, and the lush green of the rain forests to the bright reds and oranges of the fall foliage. Since prehistoric times man has used color to brighten life. Ancient civilizations created brightly dyed fabrics and decorated pottery. Michelangelo, Renoir, and Picasso are among the master artists who used color to enhance their art. Imagine how dull the world would be without color. Until recently, products were available in a limited variety of colors. You could buy a Model T automobile in any color you wanted as long as it was black. However, times have changed and today's consumers demand products in a wide variety of colors. Modern materials provide us with products in a multitude of colors that could only have been dreamt of in the past. Materials colored with metallic and interference pigments change colors as the illuminating and viewing conditions change. Materials colored with thermochromic pigments change color with temperature. The use of fluorescent pigments results in materials with eye-popping colors.

Scientist and technicians use colorimetry, the measurement of color, to ensure that consumers get the colors they expect consistently over time. In colorimetry, we try to simulate the human experience to quantify and control color.

In this chapter we consider color perception, the mathematics of simulating the perception of color, the color difference between two objects, and the instruments used to measure color. Several special measurement cases are also discussed.

12.2
Color Perception

In conversation we speak about a yellow school bus or a green traffic light as if the object's color was an intrinsic property of the object. Color, however, is a perception. Light falls upon an object. The object reflects and/or transmits that light depending upon the pigments used to color the object. The reflected and/or transmitted light enters the observer's eyes and the brain translates the resulting nerve impulses into the perception of color. Color is but one aspect of an object's appearance. Gloss and texture are two other important aspects of appearance. Color perception

Encyclopedia of Applied Spectroscopy. Edited by David L. Andrews.
Copyright © 2009 WILEY-VCH Verlag GmbH & Co. KGaA, Weinheim
ISBN: 978-3-527-40773-6

is an extremely complex phenomenon. For example, the background on which an object is viewed can have a major effect on the perceived color of that object. The ambient light to which the eye becomes adapted also influences the color of objects. When we speak of color in the remainder of this chapter, we are speaking of the perception of a single, independent color by an individual with normal color vision (99.5% of women and 92% of men).

Color is three dimensional; hence, three terms are needed to describe a specific color. Typically, colors are described with reference to their lightness, chroma, and hue. Hue would be the general name for a color, for example, blue, green, yellow, red, and purple. Colorists think of hue as a circular dimension around a central axis of neutral (white to black) colors. Lightness is the dimension used to distinguish among white, gray, and black. Pure white would approach the highest possible lightness, whereas jet black would approach the lowest possible lightness. Grays range between white and black. Lightness can distinguish between colors with distinctive hues, for example, the difference between a light blue and a dark blue. Chroma is a more difficult concept. Chroma distinguishes a color's colorfulness or difference from a neutral gray. This would be the difference between a brick red and a red cherry. If the brick and the cherry have the same hue (the brick is neither yellower nor bluer than the cherry) and the same lightness, then the brick would be described as being a low-chroma red and the cherry would be described as being a high-chroma red. The terms dull (low chroma) and saturated (high chroma) are sometimes applied to chroma. Colors that differ only in lightness (white to gray to black) are called *achromatic colors*. Those colors with perceptible hue and chroma are called *chromatic colors*.

12.2.1
Color Constancy and Metamerism

The color of an object may change as the lighting conditions change. This change in color is called *flare*. For example, a red ball may look bluer outside in daylight than in a living room lighted with incandescent lamps. We would say that the ball flared red when illuminated by incandescent light. The magnitude of the change is dependent upon the pigments used to color the object. Colorists judge the effect of different illuminating conditions using industrial light booths (cabinets). Most light booths provide a simulated daylight lamp, an incandescent lamp, and a fluorescent lamp. Some light booths also have ultraviolet lamps for determining whether an object is fluorescent. The eye will try to adjust to different lamps so that they appear to be "white" (chromatic adaptation) when looked at directly.

In most industrial situations, it is preferred to keep an object's color the same as the lamp is changed. That effect is called *color constancy*. Changes in color resulting from changes in the light source can cause some very undesirable results. For example, two yellow objects are colored using different pigments but have the same color when illuminated by a simulated daylight lamp. When an incandescent lamp is used to illuminate the two objects, one flares green and the other flares red. The colors of the two objects no longer match. This effect is called *metamerism*, and we describe the two objects as being metameric to one another. Metamerism is a major problem for industrial color matchers, particularly when trying to match a target color obtained in one material (such as a dyed bath towel) to that made in a different material (such as a plastic toothbrush holder). Can you imagine buying a suit whose jacket only matches the pants when illuminated by incandescent light?

12.3
The CIE System of Colorimetry

The Commission Internationale de l'Éclairage (International Commission of Illumination), which is usually referred to by its French initials – CIE, is an organization of countries that is concerned with light and its measurement. The scientists and engineers whose countries were members of the CIE did the pioneering work in color measurements. The CIE, by the consent of all its member countries, specifies the basic standards for use in colorimetry. The first major recommendations regarding colorimetric standards were made by the CIE in 1931 and formed the basis of modern colorimetry. Since then many additional recommendations have been made by the CIE and adopted in the field of color measurement (CIE, 2004)

12.3.1
Light and Light Sources

12.3.1.1 Visible Light
The Earth is constantly bombarded with electromagnetic radiations. This includes cosmic rays, gamma rays, visible light, and radio lights. Visible light makes up only a small portion of the electromagnetic spectrum. Visible light is the electromagnetic energy that stimulates the cells in the human eye. In colorimetry, we characterize visible light by its wavelength in nanometers (nm). The CIE considers visible light to range from 360 to 830 nm. Not everybody can perceive light throughout that entire wavelength range. Most humans can perceive light between 400 and 700 nm and many color measuring instruments can only measure light within that more restrictive wavelength range.

Each of the wavelengths (or a narrow band of wavelengths) of light is perceived as a color. The wavelengths between 360 and 500 nm will appear to be shades of violet and blue; the wavelengths between 500 and 580 nm will appear to be shades of green; and the wavelengths between 580 and 830 nm will appear to move from shades of yellow to shades of orange to shades of red.

A light source emitting a broad band of wavelengths may appear to have color even though the eye adapts and tries to make it appear white. The sun appears red, yellow, or white depending upon the time of day and the scattering of the light by particles in the atmosphere. An incandescent bulb will be described as yellow, and fluorescent lamps will usually look white. If you take a walk around dusk, your eyes will adapt to the bluish color of daylight. If you pass a house with a window to an interior room lit by an incandescent lamp, you will immediately notice how yellow the incandescent lamp is compared to outside daylight.

If you heat a block of pure carbon, it will change color from black to yellow, to white, and then blue as the temperature increases. The carbon block can be heated until its color matches the color of the light source. The temperature of the carbon block (in kelvin, K) when its color matches the light source would be the color temperature of the light source. An incandescent light bulb that appears to be yellow has a color temperature of about 2850 K, whereas north sky daylight has a color temperature of about 7500 K. If a match between the heated carbon and the light source cannot be made, the temperature of the closest color to the light source would be used and is called the *correlated color temperature*. In conversation, and often in writing, many people do not distinguish between the color temperature and the correlated color temperature and use the terms interchangeably.

A light source can be characterized by its spectral power distribution (SPD), which is the relative power emitted by the light source at each wavelength of interest. Some light sources emit radiation outside of the region of visible light. Wavelengths of interest for those light sources may extend below the 360 nm visible range into the ultraviolet region and above the 830 nm visible range into the infrared region.

12.3.1.2 Standard Light Sources and Illuminants

For hundreds of years, north sky daylight has been the favorite light source for color matchers. Unfortunately, the use of natural daylight limits the time available for color matching. In our modern world, teams of color matchers may have to work round the clock to keep up with the demands for their products. Natural daylight is not constant and changes during the course of the day. In the morning, natural daylight will have a color temperature of about 3000 K and at noontime about 5000 K. Average daylight has a color temperature of 6500 K and north sky daylight has a color temperature of 7500 K. Average daylight and north sky daylight would appear bluish if it were not for the chromatic adaptation of the eye, which makes daylight appear to be white. The bluish cast of daylight can sometimes be seen in photographs.

Recognizing the problem with natural daylight, the CIE defined three reproducible light sources for use in colorimetry. CIE source A was a tungsten filament light bulb with a color temperature of 2854 K. CIE sources B and C were created by filtering the light from source A. Source B approximated noon sunlight and source C approximated average daylight. Source B is no longer recommended for colorimetry.

A distinction must be made between the terms *light source* and *illuminant*. A light source is a real entity that can be turned on and off to illuminate objects. An illuminant is an SPD representing a light source. CIE standard illuminants A, B, and C were created by measuring the SPDs of standard sources A, B, and C. Illuminants are used in colorimetric calculations.

Although CIE source C is an approximation to average daylight, there are some significant differences. Since source C is based on an incandescent lamp, it does not have much power in the blue and ultraviolet regions of the spectrum, whereas natural daylight does. When measuring fluorescent materials, it is important to simulate the amount of ultraviolet radiation in natural daylight.

In 1968, the CIE adopted the D series of standard illuminants duplicating the SPDs of various phases of natural daylight (CIE, 2004; ASTM 308). The name of the various D illuminants is based upon their correlated color temperature, such as D50 (5000 K) and D65 (6500 K).

D65 is commonly used in the textile, paint, and plastics industries, whereas D50 is typically used in the graphic arts. Very few real light sources simulate any of the D illuminants satisfactorily; hence, the CIE recommended procedures for assigning the quality of daylight simulators (CIE, 2004).

Fluorescent lamps are widely used in stores and offices. As a result, the CIE defined a series of fluorescent (F) illuminants to represent commonly used fluorescent lamps (CIE, 2004). F illuminants are grouped according to the types of phosphors used in the manufacture of lamps. F1–F6 represent the traditional fluorescent lamps, which use two semibroadband emissions of antimony and manganese activations in a calcium halo-phosphate phosphor. F2 represents a typical cool white fluorescent lamp and is the most used of the F illuminants. F7–F9 represent the "broad-band" group of fluorescent lamps, which are made using multiple phosphors in order to have better color-rendering properties than the standard group. Color rendering is the ability of the illuminant to keep a group of colors the same as the reference illuminant, typically D65. F10–F12 represent the three narrow-band group of fluorescent lamps, which were developed to save energy and still had excellent color-rendering properties. Each of these illuminants has a narrow-band emission in the red, green, and blue wavelength regions. Rare-earth phosphors are used to make lamps representing these illuminants. Illuminants D65, A, and F2 are illustrated in Figure 12.1 and illuminants F2 and F11 are illustrated in Figure 12.2.

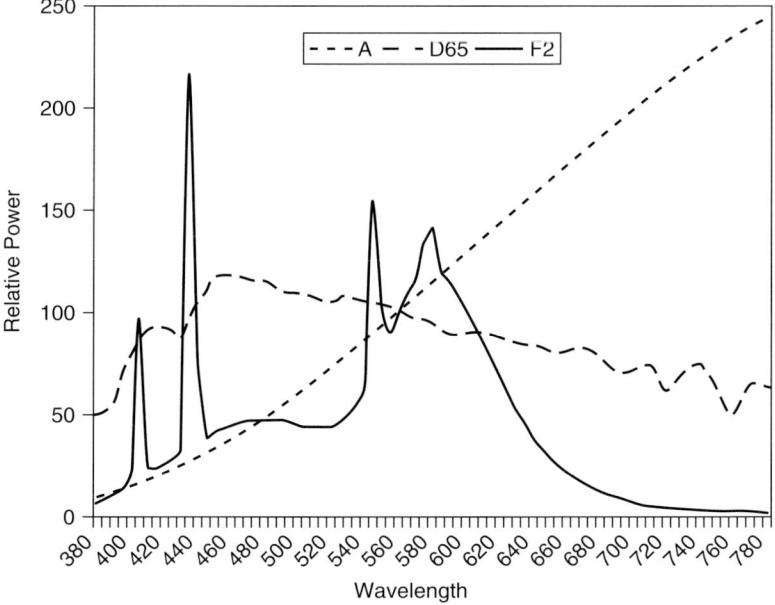

Fig. 12.1 The relative spectral power distribution of CIE standard illuminants A, D65, and F2. The curves are for shape comparison and normalized so that they have a value of 100.00 at 560 nm.

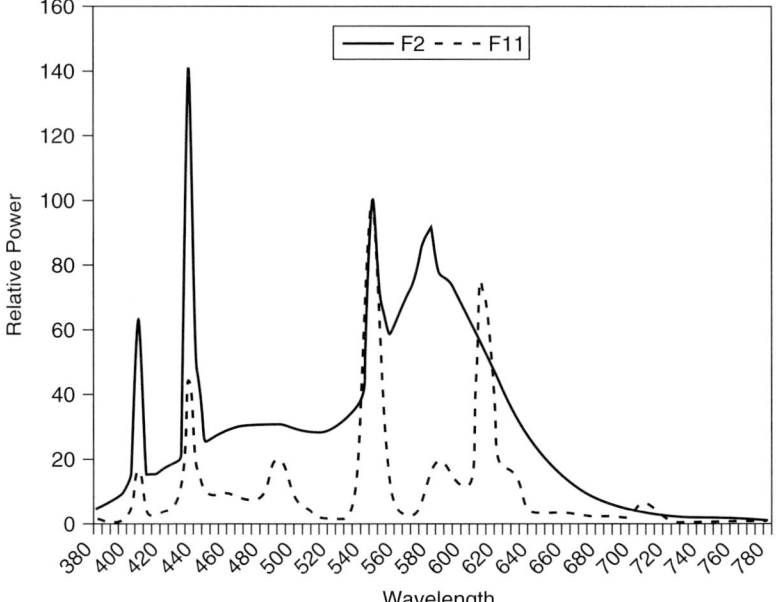

Fig. 12.2 The relative spectral power distribution of CIE standard illuminants F2 and F11. The curves are for shape comparison and normalized so that they have a value of 100 at 540 nm.

12.3.1.3 Color-Matching Booths

Commercial color matching takes place during the day and at night. Daylight is not available at night and other light sources are required to determine metamerism. It is essential that the visual evaluation of color be done under standardized illumination, such as that provided by a color-matching light booth. The light sources in the booth attempt to duplicate the CIE standard illuminants. Some manufacturers do better than others and it is extremely difficult to come close to achieving the SPD of real daylight with an artificial light source. However, a carefully manufactured and maintained light booth permits a colorist to make a visual evaluation of color with the confidence that the illumination stays constant over time.

12.3.2 Characterizing Objects

When light illuminates an object, the object will absorb, reflect, emit, and/or transmit that light. The events that take place depend upon the object and the pigments used to color it. If all of the light is reflected from an object, it will appear white. If all of the light is transmitted through an object, it will appear clear. If all of the light is absorbed by an object, it will appear black. Most objects will selectively absorb some wavelengths of light and will reflect or transmit the other wavelengths. The color of the object depends upon which wavelengths are absorbed and which are reflected or transmitted. Fluorescent objects absorb some wavelengths of light and then (in addition to reflection or transmission) emit light at longer wavelengths. Phosphorescent objects will emit light after the initial light no longer illuminates them and will "glow in the dark".

In colorimetry, we represent objects by their spectral reflectance or transmittance curves, the amount of light reflected or transmitted at each wavelength of the visible spectrum. The reflectance of an object is the ratio of reflected radiant flux to that of the incident flux. In common usage, the term *reflectance* is often used as an abbreviation for the term *reflectance factor*. The reflectance factor is the ratio of the intensity of reflected radiant energy to that reflected from the perfect reflecting diffuser under the same geometric and spectral conditions of measurement. The perfect reflecting diffuser is defined as an ideal reflecting surface that neither absorbs nor transmits light, but reflects diffusely, with the radiance of the reflecting surface being the same for all reflecting angles, regardless of the angular distribution of the incident light (ASTM E284). The perfect reflecting diffuser is the theoretical standard used for the instrumental measurement of reflectance. For the remainder of this chapter, we generally refer to the reflectance of an object since the calculations do not distinguish between the two. The transmittance of an object is the ratio of transmitted flux to that of the incident flux under specified geometric and spectral conditions (ASTM E284).

Figure 12.3 shows the spectral reflectance curves of black ink and of a white ceramic tile. Figure 12.4 shows the spectral reflectance of a blue ink, a green ink, and a red ink. Black (and gray) objects absorb all of the wavelengths of light. A perfect white would reflect all of

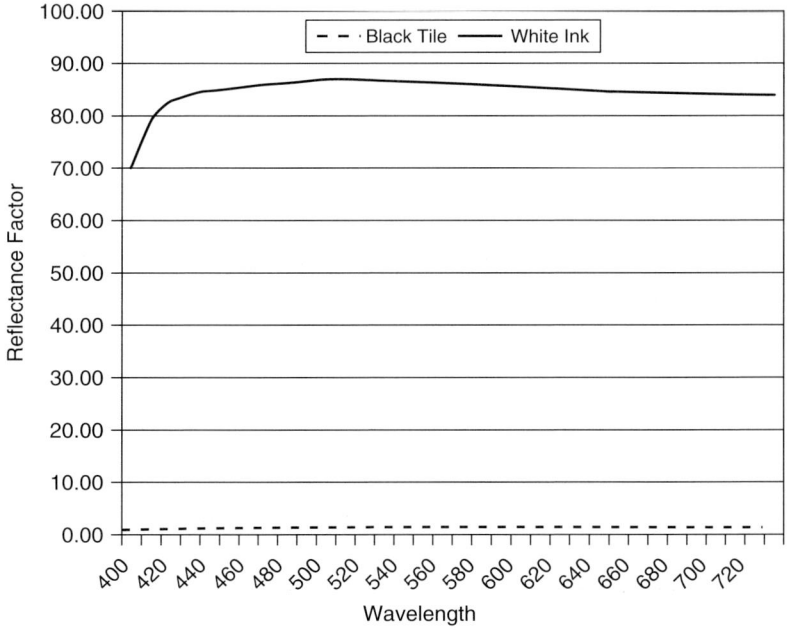

Fig. 12.3 The spectral reflectance factor curves of black ink and of a white tile.

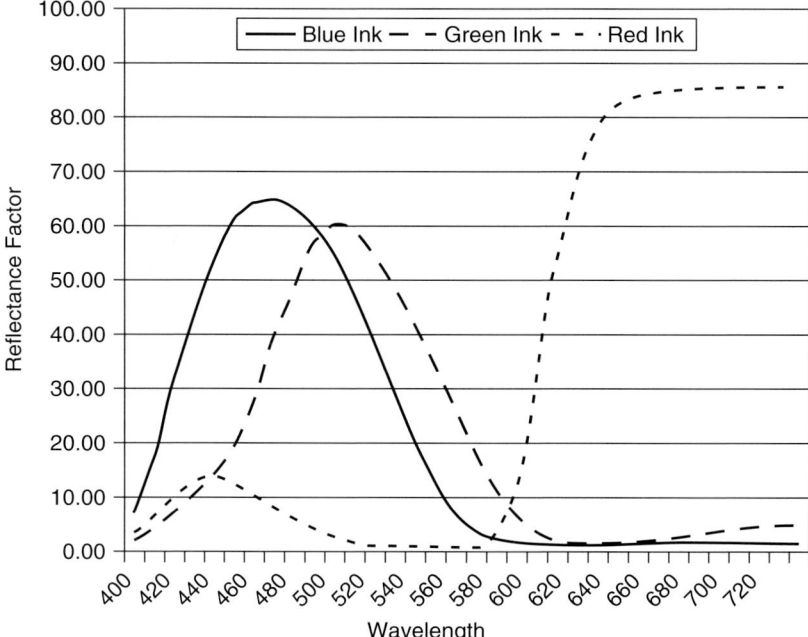

Fig. 12.4 The spectral reflectance factor curves of a blue ink, a green ink, and a red ink.

the wavelengths of light, but white objects in the real world are never perfect and will have some absorption. Many white objects have significant absorption below about 420 nm because they are colored with rutile titanium dioxide, a white pigment that absorbs strongly in that region of the spectrum. Blue objects absorb the green and red wavelengths of light, green objects absorb the blue and red wavelengths of light, and red objects absorb the blue and green wavelengths of light.

A spectrophotometer is used to measure the reflectance or transmittance factor of an object as a function of wavelength. The measured reflectance or transmittance factor of an object depends on the instrument used to make the measurements and the conditions under which the measurements are made. Several standard measuring geometries were developed for colorimetry by the CIE. These geometries are discussed in greater detail later in this chapter.

12.3.3
The Standard Observers

12.3.3.1 The Eye and Color Vision

The human eye is much like a camera. The lens images the scene on the light-sensitive retina. Two types of receptors, rods and cones, in the retina absorb a portion of the incident light and generate a signal, which is processed and interpreted by the brain. Rods detect small amounts of light and are responsible for night vision. They are not sensitive to color and are inactive at higher light levels. Cones have a much lower sensitivity to light but become active at daytime light levels. There are three types of cones called S, M, and L. S cones are sensitive to the short wavelengths of light, M cones are sensitive to the medium wavelengths of light, and L cones are sensitive to the long wavelengths of light. Although the spectral sensitivities of the cones slightly overlap, they are sensitive to wavelengths that roughly correspond to blue, green, and red. The cones reduce the entire spectrum of light into three signals. The distribution of rods and cones varies throughout the retina, with the cones distributed most densely at the center of the eye in an area called the *fovea*. Light triggers impulses from the cone cells that are either subtracted or added to create the black-white, red-green, or yellow-blue signals that are sent to the brain. These opponent signals travel to the brain via the optic nerve where they are interpreted as a colored image.

Most modern theories of color vision combine impulses from the rods and the three cone receptors of the eye into opponent color signals in the brain. So far, a theory of color vision that accounts for all the effects related to color perception has not been proposed.

12.3.3.2 Flare and Metamerism

Flare occurs when an object illuminated by one light source produces a different set of opponent signals than when that object is illuminated by a different light source.

The color of two objects matches when each object produces the same opponent signals. Since the cones in the eye reduce the spectrum into three signals, the two objects do not have to have the same spectral reflectance distribution to produce the same opponent signals. Metamerism occurs when the two objects have different spectral reflectance distributions but produce the same opponent signals when illuminated by one light source but produce different opponent signals when illuminated by a different light source.

12.3.3.3 Characterizing Observers

Since all color vision results from the opponent signals of the cones, shining combinations of red, green, and blue lights on the cones can produce all the colors. This was the principle used to characterize the two CIE Standard Observers.

If a single wavelength (or very narrow band of wavelengths) of light from the visible spectrum is projected upon one half of a screen, an observer should be able to match the perceived color by varying the intensities of a red, a green, and a blue light projected on the other half of the screen as shown in Figure 12.5. The amounts of each

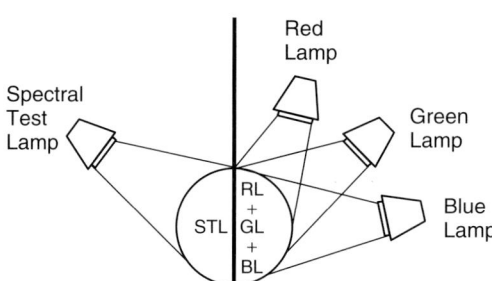

Fig. 12.5 Matching spectral colors with combinations of red, green, and blue lights.

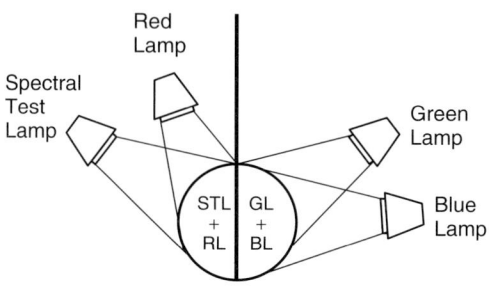

Fig. 12.6 Some spectral colors could not be matched using the combinations of red, green, and blue lights so that one of the lights had to be moved to make a match.

the three lights needed to match the spectral color are called *color-matching functions*. Observers are characterized by determining their color-matching functions for each wavelength of the visible spectrum.

Ideally, the red, green, and blue lights used in these experiments should match the responses of the three cones; unfortunately, they do not. As a result, some of the spectral colors could not be matched using combinations of real lights. When this occurred, one of the three matching lights was moved so that it illuminated the same side of the screen as the spectral color and could then be matched with the remaining two lights as shown in Figure 12.6. Moving a matching lamp so that it illuminates the same side of the observer's field of view as the test lamp results in a negative color-matching function. To avoid using negative numbers in the colorimetric calculations, the color-vision scientists created three "imaginary" lights, X, Y, and Z, to replace the real red, green, and blue lights used in the experiment. The three sets of color-matching functions (one for each light) were transformed mathematically in three sets of color-matching functions, $\bar{x}(\lambda)$, $\bar{y}(\lambda)$, and $\bar{z}(\lambda)$, that are always positive numbers. The symbol λ is used to indicate that the color-matching functions are wavelength dependent.

In the experiments that resulted in the 1931 CIE 2° Standard Observer, a group of observers viewed a circular field that covered 2° of their visual field, which is about the size of a dime viewed at 18 inches. Most of the cones are contained within this field of view. A 2° field of view is similar to that used to observe signal lights, flares, and other small objects.

Often a color is viewed using a much larger field of view. For example, industrial color matchers may often use samples that are at least 4 × 6 inches. Rods and the distribution of cones affect the perception of these colors. In 1964, the CIE recommended the CIE 1964 Supplemental Standard Observer. The experiments that resulted in the 1964 CIE 10° Standard Observer were based on observers viewing a circular field that covered 10° of their visual field, which is about the size of a 3-in. circle at 18 in. The three sets of color-matching functions for the CIE 10° Standard Observer are designated $\bar{x}_{10}(\lambda)$, $\bar{y}_{10}(\lambda)$, and $\bar{z}_{10}(\lambda)$ and are shown in Figure 12.7.

Since the two CIE Standard Observers were based on averaging the color-matching functions of a number of observers, it is unlikely that they will be the same as any single observer. The differences from the CIE Standard Observer usually do not present a

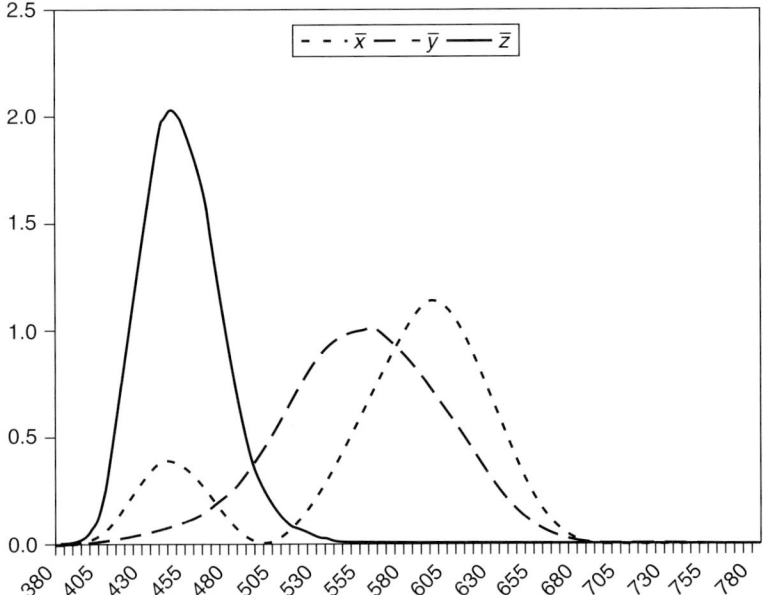

Fig. 12.7 The CIE color-matching functions for the 10° standard observer.

problem. However, they can affect the evaluation of highly metameric samples.

12.3.3.4 Calculating Tristimulus Values

Tristimulus values are the amounts of the X, Y, and Z lights needed to match a color. Each tristimulus value is calculated by finding the integral of the product of the spectral power distribution of the illuminant, $S(\lambda)$, the reflectance (or transmittance) of the object, $R(\lambda)$, and the appropriate color-matching function of the observer over the visible wavelength region as shown in Figure 12.8. The integral is approximated by taking the sum of those products at 1-nm intervals from 360 to 830 nm and then normalizing the sum as shown in Eqs (12.1)–(12.4). For transmitting objects substitute the transmittance factor, $T(\lambda)$, for the reflectance factor $R(\lambda)$. The factor k normalizes Y so that it is equal to 100.00 for the perfect reflecting diffuser or perfect transmitter, that is, $R(\lambda)$ or $T(\lambda)$ is equal to 1.00 for all wavelengths of the visible spectrum:

$$X = k \sum S(\lambda)R(\lambda)\bar{x}(\lambda) \qquad (12.1)$$

$$Y = k \sum S(\lambda)R(\lambda)\bar{y}(\lambda) \qquad (12.2)$$

$$Z = k \sum S(\lambda)R(\lambda)\bar{z}(\lambda) \qquad (12.3)$$

$$k = \frac{100}{\sum S(\lambda)\bar{y}(\lambda)} \qquad (12.4)$$

Reflectance and transmittance factors are often displayed as percentages ranging from 0.00 to 100.00. When calculating tristimulus values, the reflectance and transmittance factors should be expressed as a decimal ranging from 0.00 to 1.00. The above-mentioned summations are also valid for reflectance or transmittance measurements made at 5 nm intervals.

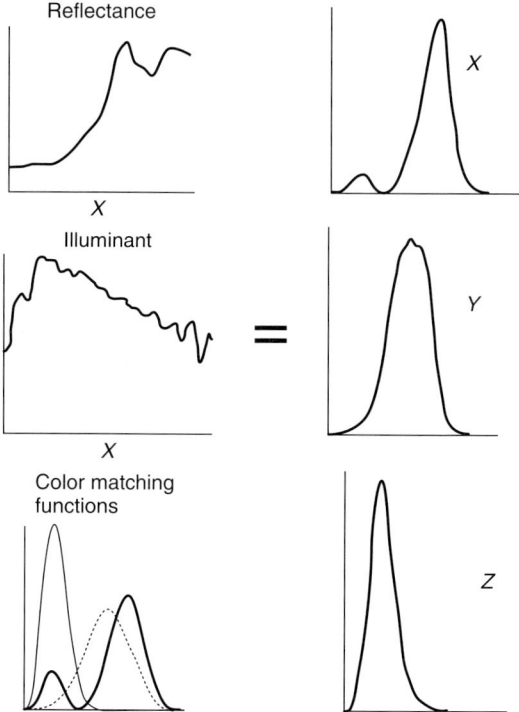

Fig. 12.8 Calculating the tristimulus values of a color.

Most commercial spectrophotometers designed for color measurement measure reflectance or transmittance factor at 10 nm intervals and some instruments measure at 20 nm intervals. Tristimulus values cannot be correctly calculated by only using the table values at 10 or 20 nm intervals. To facilitate the calculation of tristimulus factors from measurements made at 10 and 20 nm intervals, tables of normalized weighting factors, $W_x(\lambda)$, $W_y(\lambda)$, and $W_z(\lambda)$, have been published by the ASTM (ASTM 308). Weighting factors are derived from the combination of the illuminant and observer data along with other adjustments based on the wavelength interval and range selected. The tristimulus values are computed by summing the product of the weighting factor and the reflectance at the same wavelength as indicated in Eqs (12.5)–(12.7):

$$X = \sum W_X(\lambda) R(\lambda) \qquad (12.5)$$

$$Y = \sum W_Y(\lambda) R(\lambda) \qquad (12.6)$$

$$Z = \sum W_Z(\lambda) R(\lambda) \qquad (12.7)$$

12.3.3.5 Chromaticity Coordinates and Chromaticity Diagrams

Most people cannot visualize a color based upon its tristimulus values. Thus, it is very useful to be able to make a graph to help visualize colors and their relationship with other colors, such as a standard and a batch. Graphing tristimulus values would require a three-dimensional plot that is very difficult to make. Separating a color into two components, its lightness and its chromaticity, allows for

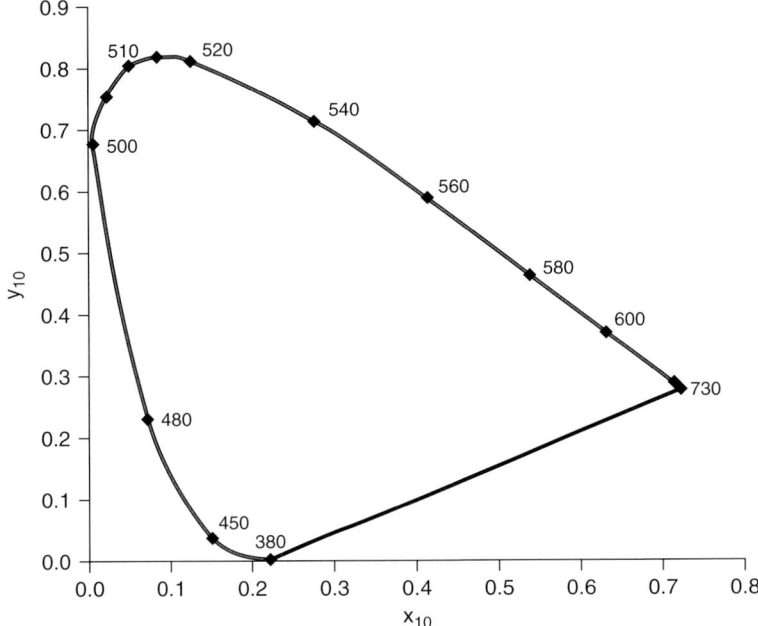

Fig. 12.9 The CIE 1964 chromaticity diagram.

a two-dimensional plot. A color's chromaticity is made up of its hue and chroma. The Y tristimulus value defines the color's lightness. The color's chromaticity coordinates, x and y, are calculated from the tristimulus values as shown in Eqs (12.8) and (12.9):

$$x = \frac{X}{X+Y+Z} \qquad (12.8)$$

$$y = \frac{Y}{X+Y+Z} \qquad (12.9)$$

A chromaticity diagram based on the 1964 Standard Observer is shown in Figure 12.9. The horseshoe shape contains the chromaticities of the wavelengths of the visible spectrum. The nonspectral purples are created by mixing blue light with a wavelength of 380 nm and red light with a wavelength of 780 nm. These colors fall on the straight line connecting the ends of the horseshoe.

Chromaticity diagrams are very useful for people working with colored lights and phosphors. Mixtures of two lights or phosphors will fall on the line connecting the chromaticities of each at a position predicted by the relative luminance of each light in the mixture. Plotting the chromaticities of accepted batches relative to the standard color helps define industrial color tolerances (ASTM 3134).

12.4 Color Spaces

12.4.1 The Munsell Color Space

Albert Munsell, an artist and educator, developed a visually uniform color space in the early 1900s. Munsell's system had three coordinates – hue, value, and chroma.

Value is an artist's term for lightness. A perfect black would have a value of 0, and a perfect white would have a value of 10. Each step (0–1, 1–2, 3–4,..., 9–10) appears to be equally different in lightness.

Munsell defined a hue circle consisting of 10 hues (red, R; yellow-red, YR; yellow, Y; green-yellow, GY; green, G; blue-green, BY; blue, B; purple-blue, PB; purple, P; and red-purple, RP). Each pair of adjacent hues in the hue circle will appear to have the same color difference as any other pair of adjacent hues in the circle when the samples have the same lightness and same chroma.

Chroma would designate a color's distance from the central neutral axis. When Munsell created his system, he used his brightest vermilion pigment to anchor the chroma at 10. Colors can have chromas greater than 10. Munsell created an atlas to exemplify his system. In his first atlas, he divided the chroma scale in visually equal steps of 2.0 chroma units.

On a perfect, visually uniform chromaticity diagram, the Munsell hue circle would plot as a perfect circle.

12.4.2
The 1976 CIE $L^*a^*b^*$ (CIELAB) Color Space

Figure 12.10 illustrates a Munsell hue circle at a value of 5 and a chroma of 8 plotted on the 1931 CIE chromaticity diagram. The hue circle looks more like a pear than a circle. The members of the CIE wanted a more uniform color space than either the 1931 space or the 1964 color space. Work in the late 1960s and early 1970s resulted in the CIE 1976 $L^*a^*b^*$ (CIELAB) color space (CIE, 2004). CIELAB has gained wide acceptance in industry.

CIELAB is an opponent-type color space. Values for the lightness axis L^*,

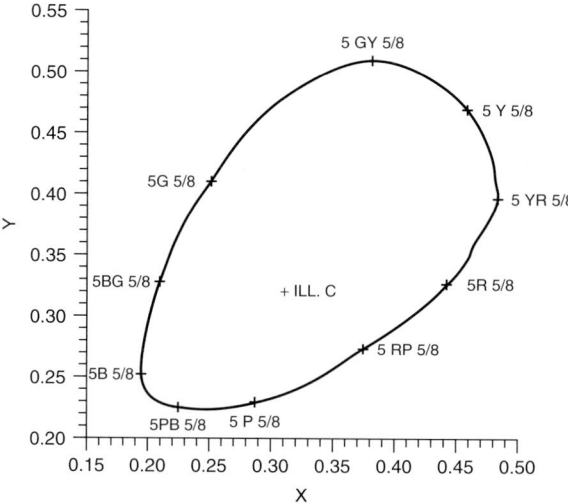

Fig. 12.10 The Munsell hue circle using the 1931 CIE chromaticity coordinates.

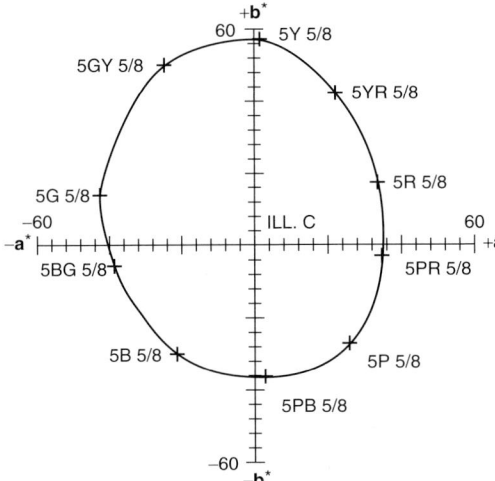

Fig. 12.11 The Munsell hue circle plotted in CIELAB color space.

the redness (positive values)-greenness (negative) axis a^*, and the yellowness (positive)-blueness (negative) axis b^* are calculated from Eqs (12.10)–(12.12):

$$L^* = 116 f\left(\frac{Y}{Y_n}\right)^{1/3} - 16 \quad (12.10)$$

$$a^* = 500\left[f\left(\frac{X}{X_n}\right)^{1/3} - f\left(\frac{Y}{Y_n}\right)^{1/3}\right] \quad (12.11)$$

$$b^* = 200\left[f\left(\frac{Y}{Y_n}\right)^{1/3} - f\left(\frac{Z}{Z_n}\right)^{1/3}\right] \quad (12.12)$$

When X/X_n, Y/Y_n, $Z/Z_n \geq 0.008856$, then $f(X/X_n) = (X/X_n)^{1/3}$, $f(Y/Y_n) = (Y/Y_n)^{1/3}$ and $f(Z/Z_n) = (Z/Z_n)^{1/3}$, where X_n, Y_n, and Z_n, are the tristimulus values of the illuminant or white point.

When X/X_n, Y/Y_n, $Z/Z_n < 0.008856$, then $f(X/X_n) = 7.787(X/X_n) + 16/116$, $f(Y/Y_n) = 7.787(Y/Y_n) + 16/116$, and $f(Z/Z_n) = 7.787(Z/Z_n) + 16/116$.

For workers who prefer working with lightness, chroma, and hue coordinates, the CIE defined chroma and hue angle as shown in Eqs (12.13) and (12.14). Note that the hue angle is measured in degrees.

$$C^* = \left(a^{*2} + b^{*2}\right)^{1/2} \quad (12.13)$$

$$h = \tan^{-1}\left(\frac{b^*}{a^*}\right) \quad (12.14)$$

These correlate well with visual judgments of lightness, chroma, and hue, respectively.

Figure 12.11 illustrates the Munsell hue circle plotted in CIELAB color space. CIELAB is much closer to a visually uniform color space than the 1931 or 1964 color spaces.

A chromaticity diagram was not defined for the CIELAB color space.

12.4.3
The 1976 CIE $L^*u^*v^*$ (CIELUV) Color Space

Since some industries, such as television and video displays, find chromaticity diagrams useful, the CIE also recommended the CIELUV color space that does have a chromaticity diagram. Values for the CIELUV coordinates are calculated from Eqs (12.15)–(12.22):

$$L^* = 116 \left(\frac{Y}{Y_n}\right)^{1/3} - 16 \text{ when}$$

$$\frac{Y}{Y_n} > 0.008856 \quad (12.15)$$

$$L^* = 903.3 \left(\frac{Y}{Y_n}\right)^{1/3} \text{ when}$$

$$\frac{Y}{Y_n} \leq 0.008856 \quad (12.16)$$

$$u^* = 13L^*(u' - u'_n) \quad (12.17)$$

$$v^* = 13L^*(v' - v'_n) \quad (12.18)$$

where the chromaticity coordinates of the color (u', v') are defined as

$$u' = \frac{4X}{X + 15Y + 3Z} \quad (12.19)$$

$$v' = \frac{9X}{X + 15Y + 3Z} \quad (12.20)$$

and the chromaticity coordinates of the illuminant or white point are defined as

$$u'_n = \frac{4X_n}{X_n + 15Y_n + 3Z_n} \quad (12.21)$$

$$v'_n = \frac{9X_n}{X_n + 15Y_n + 3Z_n} \quad (12.22)$$

The CIE also defined chroma, C^*_{uv}, and hue angle, h_{uv}, coordinates for CIELUV that are calculated using Eqs (12.23) and (12.24):

$$C^*_{uv} = \left(u^{*2} + v^{*2}\right)^{1/2} \quad (12.23)$$

$$h_{uv} = \arctan\left(\frac{v^*}{u^*}\right) \quad (12.24)$$

In addition, the CIE defined a correlate to perceived saturation, S^*_{uv}, that is calculated using either Eq. (12.25) or its equivalent Eq. (12.26):

$$S^*_{uv} = \frac{C^*_{uv}}{L^*} \quad (12.25)$$

$$S^*_{uv} = 13\left[(u' - u'_n)^2 + (v' - v'_n)^2\right]^{1/2} \quad (12.26)$$

The perceived saturation, S^*_{uv}, remains constant for a series of colors of constant chromaticity as the lightness increases or decreases (Hunt, 1991).

12.5
Color Differences and Tolerances

Most of the time, industrial color matching and production involve the evaluation and description of color differences. A colorist may be evaluating two lots of different pigments or dyes, comparing a target color with a laboratory sample or comparing a product standard with a production batch. Qualitatively describing the color differences is essential and being able to quantify the color difference is extremely useful. People are most sensitive to differences in hue, more sensitive to differences in chroma, and least sensitive to differences in lightness. This is the result of how colors change in nature. For example, a shadow falling across a fence results in the fence being darker in the shadow area. Although this is an actual color difference based on the difference in lightness, a person attributes the difference to the shadow and not to the color of the fence itself.

Distinguishing between the perceptibility of two colors and the acceptability of a color match are extremely important. A color difference between two samples is perceptible if you can see the color difference between them. Although you may be able to perceive the color difference between a product standard and a production batch, the difference in color may still be acceptable to the user or consumer. A color tolerance tells the colorist to what extent a color difference is acceptable between the standard and the batch. The size of the color tolerance depends on the product and how it is being used. For example, the color difference between two bolts of dyed fabric must be very small if you intend to make a suit using both bolts. However, the tolerance may be much larger if the bolts would not be used together and will be viewed as separate products.

Colorists and designers have used many terms for describing color differences. They describe the differences between two colors as being stronger, duller, slightly blue, and too gray. Designers have asked colorists to make a color "happier." Unfortunately, most of those terms do not help the colorist decide how to adjust the batch since one person's "slightly" may be another person's "strongly." And how do you make a color "happier"? For chromatic colors, the author recommends using the terms lightness, chroma, and hue to describe color differences. The batch is either equal in lightness to the standard or darker or lighter. The batch may have the same, a higher, or a lower chroma than the standard. Hue differences are usually described in terms of redness, yellowness, greenness, or blueness. A red or green sample may be yellower or bluer than the standard, and a yellow or blue sample may be redder or greener than the standard. For example, a red sample may be moderately yellower, slightly darker, and lower in chroma than the standard. For neutral- and low-chroma colors, it is often useful to describe color differences in terms of lightness, redness-greenness, and yellowness-blueness. The visual color difference between a low-chroma color standard and a batch can often be larger and still acceptable if both are in the same red-green and yellow-blue quadrants of color space.

Although many people rely upon an instrumental evaluation of color difference, the color of the final product is often evaluated visually. Regardless of whether or not an instrumental color difference is within a specified tolerance, there will be problems if the color does not look right. Therefore, before instrumental measurements are made, color and gloss differences between a standard and a sample should be evaluated visually. Visual evaluations should be made in light booths using standardized viewing conditions (ASTM 4449; ASTM 1729; ASTM 3928; ASTM 2616). Small-to-moderate color differences can be quantitatively evaluated visually by using a gray scale (ASTM 2616). Visually prepared color-tolerance sets consist of samples having acceptable color differences at the extremes of lightness–darkness, high- and low-chroma and the redness-greenness or yellowness-blueness differences in hue. A visual color-tolerance set can be measured to provide a set of instrumental color tolerances.

To obtain the most accurate instrumental evaluation of color difference, the standard and the batch should have the same texture and gloss. Some instruments will make measurements that minimize the affect of texture and gloss differences while others will accentuate the differences. The human eye–brain evaluation of color difference has a perceptibility tolerance as does instrumental measurements. An instrumental

12.5.1
Single-Number Color Scales

measurement, and the resulting color-difference calculation, may produce differences that are below those perceptible to a person. Care must be taken not to set an instrumental tolerance below which a person can perceive.

Some research has taken place on the optimum methods for calculating color differences. This has led to a number of published color-difference equations. Some of the equations correlate better with human perception than others. The goal of a color-difference equation is generally to determine a single number that will determine whether a batch is acceptable to a standard. Even though most industrial colorists tend to use a single-number color-difference tolerance, better results are usually obtained when tolerances are set for each component of the color difference.

12.5.1
Single-Number Color Scales

In some cases, the color of a sample varies along a single direction in color space, which can be adequately described by a single scale value. This occurs in the evaluation of whiteness and yellowness and in the evaluation of some liquids.

12.5.1.1 Whiteness Scales

Whiteness scales indicate the deviation from an ideal white, which is typically assigned a whiteness value of 100. Equation (12.27) represents a commonly used whiteness scale equation (ASTM 313):

$$WI = Y + (WI, x)(x_n - x) + (WI, y)(y_n - y) \quad (12.27)$$

where Y, x, and y are the luminance factor and the chromaticity coordinates of the batch; x_n and y_n are the chromaticity coordinates for the CIE standard illuminant and source used; and WI, x and WI, y are numerical coefficients that are specified in Table 12.1. The whiteness index is valid only for specimens with $40 < WI < (5Y - 280)$.

A tint index often accompanies a whiteness index to indicate the direction in which the white varies. Equation (12.28) represents the tint index for the whiteness scale in Eq. (12.27):

$$T = (T, x)(x_n - x) + (T, y)(y_n - y) \quad (12.28)$$

where x and y are the chromaticity coordinates of the batch; x_n and y_n are the chromaticity coordinates for the CIE standard illuminant and source used;

Tab. 12.1 Coefficients for the equation for CIE whiteness index and tint.

Value	CIE Standard Illuminant and Observer					
	C, 1931	D$_{50}$, 1931	D$_{65}$, 1931	C, 1964	D$_{50}$, 1964	D$_{64}$, 1964
x_n	0.3101	0.3457	0.3127	0.3104	0.3477	0.3138
y_n	0.3161	0.3585	0.3290	0.3191	0.3595	0.3310
WI, x	800	800	800	800	800	800
WI, y	1700	1700	1700	1700	1700	1700
T, x	1000	1000	1000	900	900	900
T, y	650	650	650	650	650	650

Tab. 12.2 Coefficients of the equation for yellowness index.

Value	CIE Standard Illuminant and observer			
	C, 1931	D_{65}, 1931	C, 1964	D_{64}, 1964
C_X	1.2769	1.2985	1.2871	1.3013
C_Z	1.0592	1.1335	1.0781	1.1498

and T, x and T, y are numerical coefficients that are specified in Table 12.1. The tint index is valid only for specimens with $-3 < T < +3$.

12.5.1.1 Yellowness Indices

Yellowness indices are often used to describe the yellowing of clear plastics and white paints. They show the departure of a batch from neutral toward yellow. Equation (12.29) represents a commonly used yellowness index equation (ASTM 313):

$$YI = \frac{100\,(C_X X - C_Z Z)}{Y} \quad (12.29)$$

where X, Y, and Z are the tristimulus values of the batch and C_X and C_Z are numerical coefficients that are specified in Table 12.2.

12.5.1.2 Scales for Liquids

When only a limited range of color is involved, such as in the testing of the color of lacquer used in the paint industry, a simple comparison of the color of the batch with standard colored solutions or glasses ranging from colorless to highly colored can be made. A standardized series of these solutions or glasses is used to provide a specialized color scale. The color is often a measure of concentration of ingredient. One difficulty in using these special color scales is that the color of the specimen may not match with that of the standard; this can make rating on a single-number scale difficult. Nevertheless, their simplicity, low cost, and adaptability to special situations have resulted in wide use of single-number scales for certain applications (ASTM 156; ASTM 1209; ASTM 1500; ASTM 6045; ASTM 1544; ASTM 6166; ASTM 1686).

12.5.2 The CIELAB and CIELUV Color-Difference Equations

In addition to recommending the CIELAB and CIELUV color spaces, in 1976 the CIE also recommended color-difference equations for those spaces (CIE, 2004).

12.5.2.1 The CIELAB Color-Difference Equation

The popularity of this color space increased through the 1980s and became the most commonly used color space in industry. To calculate CIELAB color difference, the L^*, a^*, and b^* values of the batch or sample color are subtracted from those of the standard or target color (ASTM 2244):

$$\Delta L^* = L^*_{\text{batch}} - L^*_{\text{standard}} \quad (12.30)$$

$$\Delta a^* = a^*_{\text{batch}} - a^*_{\text{standard}} \quad (12.31)$$

$$\Delta b^* = b^*_{\text{batch}} - b^*_{\text{standard}} \quad (12.32)$$

A negative value of ΔL^* indicates that the batch is darker than the standard, whereas a positive value indicates that the batch is lighter than the standard. A negative value of Δa^* indicates that the batch is greener than the standard, whereas a positive value of Δa^* indicates that the batch is redder than the standard. A negative value of Δb^* indicates that the batch is bluer than the standard, whereas a positive value indicates that the batch is yellower than the standard.

The total color difference between the batch and standard is given by Eq. (12.33):

$$\Delta E_{ab}^* = \left[(\Delta L^*)^2 + (\Delta a^*)^2 + (\Delta b^*)^2\right]^{1/2} \quad (12.33)$$

The chroma difference between the batch and standard is given by

$$\Delta C_{ab}^* = C_{\text{batch}}^* - C_{\text{standard}}^* \quad (12.34)$$

The hue difference is not the difference in the hue angle. It is that part of the color difference that is left after accounting for the lightness and chroma differences. The hue difference between the batch and standard, ΔH^*_{ab}, is calculated using Eq. (12.35):

$$\Delta H_{ab}^* = \left[(\Delta E_{ab}^*)^2 + (\Delta L^*)^2 + (\Delta C_{ab}^*)^2\right]^{1/2} \quad (12.35)$$

The sign of ΔH^*_{ab} is taken as positive if the hue angle of the batch is greater than that of the standard and is taken as negative if the hue angle is less than that of the standard.

The notations DL^*, Da^*, Db^*, DC^*, DH^*, and DE^* are often substituted for ΔL^*, Δa^*, Δb^*, ΔC_{ab}^*, ΔH_{ab}^*, and ΔE_{ab}^*.

12.5.2.2 The CIELUV Color-Difference Equation

Calculating the CIELUV color differences is very similar to calculating the CIELAB color differences (ASTM 2244):

$$\Delta L^* = L^*_{\text{batch}} - L^*_{\text{standard}} \quad (12.36)$$

$$\Delta u^* = u^*_{\text{batch}} - u^*_{\text{standard}} \quad (12.37)$$

$$\Delta v^* = v^*_{\text{batch}} - v^*_{\text{standard}} \quad (12.38)$$

A negative value of ΔL^* indicates that the batch is darker than the standard, whereas a positive value indicates that the batch is lighter than the standard. Δu^* and Δv^* represent differences in the chromaticity coordinates but are not associated with color names as are Δa^* and Δb^*.

The total color difference between the batch and standard is given by Eq. (12.39):

$$\Delta E_{ab}^* = \left[(\Delta L^*)^2 + (\Delta u^*)^2 + (\Delta v^*)^2\right]^{1/2} \quad (12.39)$$

The chroma difference between the batch and standard is given by

$$\Delta C_{uv}^* = C_{uv,\text{batch}}^* - C_{uv,\text{standard}}^* \quad (12.40)$$

The hue difference is not the difference in the hue angle. It is that part of the color difference that is left after accounting for the lightness and chroma differences. The hue difference between the batch and the standard, ΔH^*_{uv}, is calculated using Eq. (12.41):

$$\Delta H_{uv}^* = \left[(\Delta E_{uv}^*)^2 + (\Delta L^*)^2 + (\Delta C_{uv}^*)^2\right]^{1/2} \quad (12.41)$$

The sign of ΔH^*_{uv} is taken as positive if the hue angle of the sample is greater than that of the standard and is taken as negative if the hue angle is less than that of the standard.

The notations DL^*, Da^*, Db^*, DC^*, DH^*, and DE^* are often substituted for ΔL^*, Δa^*, Δb^*, ΔC^*_{uv}, ΔH^*_{uv}, and ΔE^*_{uv}.

12.5.3
The CMC(*l* : *c*) Color-Tolerance Equation

A number of new color-difference and color-tolerance equations have been developed since the adoption of the CIELAB equations. One of these, known as the *CMC(l : c) equation* (CIE, 2004; ASTM 2244; McDonald, 1980; AATCC, 173–1992), is a modification of CIELAB that has improved uniformity of visual perception of its color differences. The CMC(*l* : *c*) equation has found widespread use in the industry and is now more popular than CIELAB. The CMC(*l* : *c*) equation is more appropriately called a *color-tolerance equation* rather than a color-difference equation. A color-difference equation calculates the same numeric value regardless of which specimen is used – the batch or the standard. A color-tolerance equation will calculate different values depending upon the position of the standard in color space. The CMC(*l* : *c*) equation modifies the ΔL^*, ΔC^*, and ΔH^* components of the CIELAB color difference before calculating the total color difference:

$$\Delta E_{\text{CMC}}(l:c) = cf *$$
$$\sqrt{\left(\frac{\Delta L^*}{l * S_L}\right)^2 + \left(\frac{\Delta C^*}{c * S_C}\right)^2 + \left(\frac{\Delta H^*}{S_H}\right)^2}$$
(12.42)

$$S_L = \frac{0.040975 * L^*}{1 + (0.01765 * L^*)}$$
for $L^* \geq 16$ (12.43)

$$S_L = 0.511 \quad \text{for } L^* < 16 \quad (12.44)$$

$$S_C = \frac{0.0638 * C^*}{1 + (0.0131 * C^*)} + 0.638$$
(12.45)

$$S_H = S_C * ((F * T) + 1 - F) \quad (12.46)$$

in which

$$F = \left\{\frac{(C)^4}{(C)^4 + 1900}\right\}^{1/2} \quad (12.47)$$

$$T = 0.56 + \text{abs}\,|0.2 * \cos(h + 168°)|$$
if $164° < h < 345°$ (12.48)

or

$$T = 0.36 + \text{abs}\,|0.4 * \cos(h + 35°)|$$
for other values of h (12.49)

where "abs" indicates the absolute, that is, positive value of the term inside the brackets.

The parameters *l* and *c* weight the relative contributions of lightness to chromaticity in the color difference and compensate for systematic bias or parametric effects such as texture and sample separation. The most common values for *l* : *c* are 2 : 1. The parameter *cf* is a commercial factor (AATCC 173–1992) used to adjust the total volume of the tolerance region so that accept/reject decisions can be made on the basis of a unit value of the tolerance.

12.5.4
The CIEDE 2000 Color-Difference Equation

Research was continued to find a better color-difference equation than CMC (*l* : *c*). The CIE developed an equation that correlates better with human perception than the CMC(*l* : *c*) equation. This equation was recommended by the CIE in 2001 and is called *CIEDE2000* (CIE, 2001). Equations (12.50)–(12.66) are required to calculate CIEDE2000 (ASTM 2244):

$$L' = L^* \tag{12.50}$$

$$a^* = (1+G)*a^* \tag{12.51}$$

$$b' = b^* \tag{12.52}$$

$$C' = \sqrt{(a')^2 + (b')^2} \tag{12.53}$$

$$h' = \arctan\left(\frac{b'}{a'}\right) \tag{12.54}$$

$$G = 0.5 * \left(1 - \sqrt{\frac{(\overline{C^*})^7}{(\overline{C^*})^7 + 25^7}}\right) \tag{12.55}$$

where $\overline{C^*}$ is the arithmetic mean of the CIELAB C* values for the pair of specimens (standard and batch).

$$\Delta L' = L'_B - L'_S \tag{12.56}$$

$$\Delta C' = C'_B - C'_S \tag{12.57}$$

$$\Delta H = \frac{(a'_S * b'_B) - (a'_B * b'_S)}{\sqrt{0.5 * [(C'_S * C'_B) + (a'_S * a'_B) + (b'_S * b'_B)]}} \tag{12.58}$$

$$\Delta E_{00}(K_L, K_C, K_H)$$
$$= \sqrt{\left(\frac{\Delta L^*}{K_L * S_L}\right)^2 + \left(\frac{\Delta C^*}{K_C * S_C}\right)^2 + \left(\frac{\Delta H^*}{K_H * S_H}\right)^2 + R_T * \left(\frac{\Delta C' * \Delta H'}{K_C * S_C * K_L * S_H}\right)} \tag{12.59}$$

The factors K_L, K_C, and K_H are correction terms for variation in perceived color-difference due to the viewing conditions. To obtain color-differences similar to CMC(2 : 1), set $K_L = 2$ and $K_C = K_H = 1$:

$$S_L = 1 + \frac{0.015 * (\overline{L'} - 50)^2}{\sqrt{20 + (\overline{L'} - 50)^2}} \tag{12.60}$$

where $\overline{L'}$ is the arithmetic mean of the CIELAB L' values for the pair of specimens (standard and batch).

$$S_C = 1 + \left(0.045 * \overline{C'}\right) \tag{12.61}$$

where $\overline{C'}$ is the arithmetic mean of the CIELAB C' values for the pair of specimens (standard and batch).

$$S_H = 1 + \left(0.015 * \overline{C'} * T\right) \tag{12.62}$$

$$T = 1 - \left[0.17 * \cos\left(\overline{h'} - 30°\right)\right]$$
$$+ \left[0.24 * \cos\left(2\overline{h'}\right)\right]$$
$$+ \left[0.32 * \cos\left(3\overline{h'} + 6°\right)\right]$$
$$- \left[0.20 * \cos\left(4\overline{h'} - 63°\right)\right] \tag{12.63}$$

where $\overline{h'}$ is the arithmetic mean of the CIELAB L' values for the pair of specimens (standard and batch). All angles are in degrees. Care should be taken while calculating the mean hue angle if the color-difference pair has samples in different quadrants. For example, a color-difference pair has hue angles of 30° and 300°. The mean hue angle for this example is 345°. To determine the mean correctly, calculate the absolute difference of the hue angles. If the absolute difference is larger than

180°, then add 360° to the smaller hue angle and divide that sum by 2:

$$R_T = -R_C * \sin(2 * \Delta\theta) \qquad (12.64)$$

$$R_C = 2 * \sqrt{\frac{\left(\overline{C}'\right)^7}{\left(\overline{C}'\right)^7 + 25^7}} \qquad (12.65)$$

$$\Delta\theta = 30 * \exp\left\{-\left[\frac{\overline{h}' - 275°}{25}\right]^2\right\} \qquad (12.66)$$

Although the CIEDE2000 equation correlates better with the existing data sets, it is not as popular as the CMC($l : c$) equation.

12.5.5
The DIN 99 Color-Tolerance Equation

The concept behind the CMC($l : c$) and CIEDE2000 equations involved changing the color-difference components of the CIELAB color-difference equation to make the final equation agree more closely with human perception. A committee of the Deutsches Institut für Normung e.V. (DIN) in Berlin, Germany, took a different approach for developing an improved color-difference equation. The idea behind DIN99 is to retain a Euclidean color-difference formula by correcting the nonuniformity of CIELAB color space (DIN 6176). Equations (12.67)–(12.78) define the new color space and the steps for the calculation of a DIN99 color difference:

Red-green axis:

$$e = a^* \cos(16°) + b^* \sin(16°) \qquad (12.67)$$

Yellow-blue axis:

$$f = 0.7\left(-a^* \sin(16°) + b^* \cos(16°)\right) \qquad (12.68)$$

Chroma:

$$G = \sqrt{e^2 + f^2} \qquad (12.69)$$

Hue angle:

$$h_{ef} = \arctan\left(\frac{f}{e}\right) \qquad (12.70)$$

which are used to develop a new coordinate system defined as

$$L_{99} = 105.51\frac{\ln(1 + 0.0158\, L^*)}{k_e} \qquad (12.71)$$

$$a_{99} = C_{99}\cos(h_{99}) \qquad (12.72)$$

$$b_{99} = C_{99}\sin(h_{99}) \qquad (12.73)$$

where

$$C_{99} = \frac{\ln(1 + 0.045\, G)}{0.045 k_{CH} k_e} \qquad (12.74)$$

$$h_{99} = h_{ef}\frac{180}{\pi} \qquad (12.75)$$

and k_{CH} and k_e are adjustable parameters. The color difference is given by the Euclidean distance:

$$\Delta E_{99} = \sqrt{(\Delta L_{99})^2 + (\Delta a_{99})^2 + (\Delta b_{99})^2} \qquad (12.76)$$

or

$$\Delta E_{99} = \sqrt{(\Delta L_{99})^2 + (\Delta C_{99})^2 + (\Delta H_{99})^2} \qquad (12.77)$$

where ΔH_{99} is defined as

$$\Delta H_{99} = \sqrt{(\Delta a_{99})^2 + (\Delta b_{99})^2 - (C_{99})^2} \qquad (12.78)$$

which is similar to the corresponding CIELAB definition.

It remains to be seen which approach to the color-difference problem will gain acceptance. The CIE continues to have groups working on the problem of a more uniform color space as well as improved color-difference equations.

12.5.6
Setting Instrumental Color Tolerances

As with all measurements, instrumental color measurement has variation. The sources of the variation include not only the instrument itself but also variation in the sample itself. Some samples are more homogeneous than others and some instruments have less variation than others. In addition, all manufacturing processes cause variation in color. A colorist must consider the variation of the whole manufacturing and measurement system when setting instrumental color tolerances. Unfortunately, there is a tendency among a number of colorists to expect the total color difference to be zero. Making multiple measurements on real samples demonstrate that this is not possible. Setting a tolerance below the perceptibility limit is a waste of time because the customers will not see that color difference. The closer the tolerance is to the perceptibility limit, the more expensive it is to produce the product. A compromise must be made in determining the limits of an acceptable, instrumental color-difference tolerance.

The first step in determining an instrumental tolerance is to know the variation in the manufacturing and measurement processes. Measuring multiple samples of multiple batches can provide this information. Unless the seller wants to select, or blend, batches of product, the tolerances cannot be set to limits less than those found in this step. Preliminary tolerances are usually set based on past experience where it is not practical, or too expensive, to determine actual production limits initially for a new color. Final tolerances are determined after the seller gets some experience with making the color. Measuring a number of batches that the customer has judged relative to the standard helps to more accurately determine the acceptability limits (ASTM 3134).

Another method of setting instrumental tolerances is from measurements of a visually prepared color-tolerance set. Designers and marketing personnel usually determine these color-difference limits of these sets before the color is actually manufactured. It is possible that the color-tolerance set might be tighter than the manufacturing variations. The colorist preparing the color-tolerance set must caution the buyer not to set the tolerances so tight that the product cannot be produced within the tolerances.

12.6
Instrumental Color Measurement

12.6.1
Measuring Reflectance

12.6.1.1 Specular and Diffuse Reflectance

If we shine a beam of light onto a perfect mirror from the left at 45° from its normal, the light will be reflected to the right at 45° from the normal. The light does not penetrate the surface of the mirror. The mirrorlike reflection from the surface of a sample is called *specular reflection*.

If we shine a beam of light onto a very good diffuser, such as a compressed pellet of barium sulfate, the light will enter the material and get scattered (redirected) many times by the barium sulfate particles. Eventually the light will find its way out of the interior of the pellet and the barium sulfate tablet will appear to reflect light in all directions as shown in Figure 12.12.

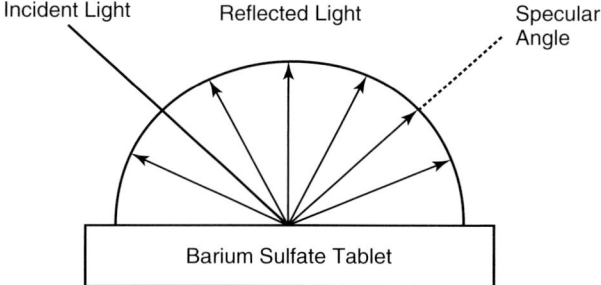

Fig. 12.12 Diffuse light reflected from a barium sulfate tablet.

Most materials have some specular reflection and some diffuse reflection as shown in Figure 12.13. Materials with very high gloss will have very narrow cones at the specular angle. As the gloss decreases, the specular cone widens. About 4% of the light reflected from glossy paint samples and glossy plastic samples will be specular reflection. The remaining 96% will be diffusely reflected.

12.6.1.2 Illuminating and Viewing Geometries

Color-measuring instruments consist of an illuminator, a specimen holder, and a receiver. The illuminator contains the light source and associated optics necessary to illuminate the specimen. An illuminator may also contain a diffraction grating or filters, diffusers, and various electronics. The receiver contains the components and optics necessary to gather and analyze the light reflected from the specimen. A receiver may also contain a diffraction grating or filters, diffusers, and various electronics. These components may be arranged in a variety of ways depending upon what the optical designer wants to measure. The CIE has recommended four illuminating and viewing geometries for color measurement (CIE, 2004; CIE, 1995). Two of them are bidirectional and two of them are diffuse. A number of factors should be considered when deciding upon which illuminating and viewing geometry is best for a given type of color

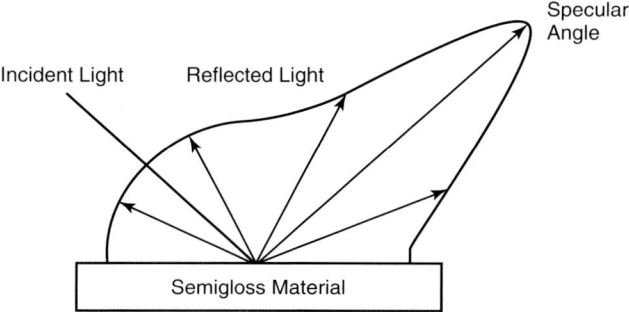

Fig. 12.13 Light reflected from a semigloss material with both a diffuse component and a specular component.

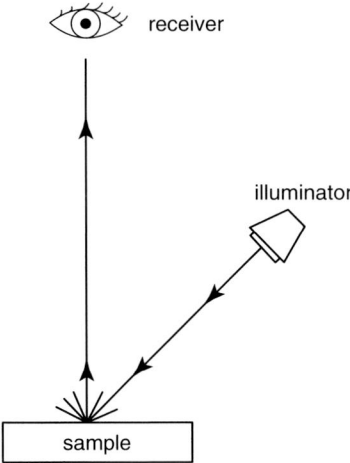

Fig. 12.14 The 45:0 illuminating and viewing geometry.

measurement (ASTM 179; Rich, 1988; Mabon, 1992).

Bidirectional geometries illuminate the specimen with a narrow beam of light and gather the reflected light with a receiver having a narrow field of view. One of the CIE bidirectional recommendations is to illuminate the specimen at 45° from its normal and view the reflected light along the specimen's normal. This illuminating and viewing geometry is called 45:0 and is shown in Figure 12.14. The CIE also recommends the reverse geometry (0:45) in which the specimen is illuminated along its normal and viewed at 45° from its normal. Both geometries produce equivalent results for nonfluorescent samples that do not contain metallic flakes or interference pigments (ASTM 179; Billmeyer and Marcus, 1969). The 45:0 and 0:45 geometries do not measure the specular reflection from a specimen and simulate the viewing conditions under which many specimens are evaluated visually. These measurements are sensitive to the texture of the specimen and any polarization affects that may occur when light enters the specimen. Multiple measurements should be made to average the differences due to the sample's texture. Using annular or circumferential illumination (or viewing) also minimizes texture and polarization affects. Circumferential illumination provides illumination (or the receiver will gather the reflected light) in many beams distributed at uniform intervals around the 45° cone about the sample's normal, c : 45 (or 45 : c). Annular illumination provides illumination (or the receiver will gather the reflected light) continuously and uniformly around the 45° cone about the specimen's normal, a : 45 (45 : a). Because of the surface texture influence, measurements made with bidirectional geometries are sometimes described as measuring appearance rather than color. Instruments with bidirectional geometries are very popular in the graphics arts community.

Figure 12.15 illustrates the most popular of the CIE diffuse geometries called *diffuse/normal* or d : 0. The illuminator shines light on the wall of an integrating sphere. An integrating sphere is a hollow metal sphere coated with an efficient white diffusing material. The most

12.6 Instrumental Color Measurement

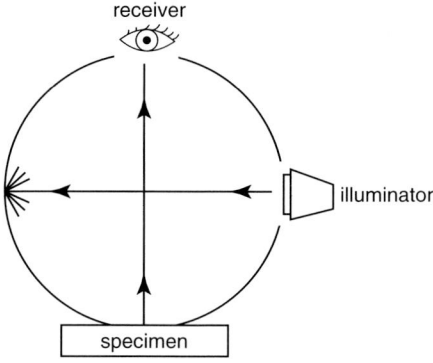

Fig. 12.15 The diffuse-normal (d : 0) illuminating and viewing geometry.

common integrating sphere coatings are barium sulfate (BaSO$_4$) and polytetrafluoroethylene (PTFE). Hollowing a block of PTFE has been used to create some spheres. Ports are cut into the wall of the integrating sphere, which are needed for the illuminator and the receiver. The light from the illuminator is scattered in all directions, which uniformly illuminates the integrating sphere that then acts as a diffuse hemispherical illuminator to illuminate the sample. The receiver views the specimen along its normal. Baffles are used to prevent light from the integrating sphere wall from entering the receiver directly. In the other CIE diffuse geometry (normal/diffuse or 0 : d), the specimen is illuminated directly along its normal and the diffuse light from the specimen uniformly illuminates the integrating sphere. The receiver then views the wall of the integrating sphere. Both geometries produce equivalent results.

Neither of the two recommended diffuse geometries measures the specular reflection from the sample. By making a slight modification to the diffuse geometries, the specular reflection from the sample can also be measured. The modification, shown in Figure 12.16, involves moving the receiver or illuminator slightly off the sample's normal (6–8°). When this modification is used, a third port (the specular port) is cut in the

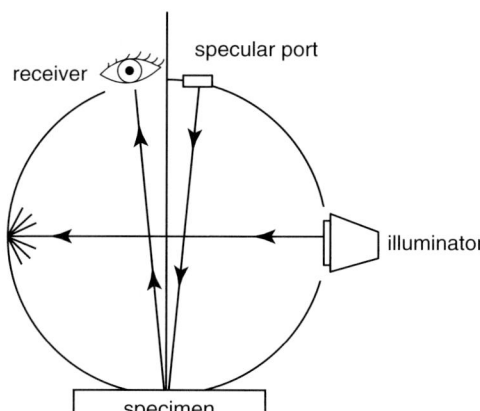

Fig. 12.16 The diffuse near-normal (d : 0 : i) illuminating and viewing geometry.

integrating sphere wall. To include the sample's specular reflection, a white cap is used to cover the specular port. The white cap should have the same reflectance as the wall of the integrating sphere. The two geometries including the specimen's specular reflection are called, respectively, *total/near-normal* ($d:0:i$) and *near-normal/total* ($0:d:i$). To exclude the specimen's specular reflection, a light trap is used for the specular port. The two geometries excluding the specimen's specular reflection are called *diffuse/near-normal* ($d:0:e$) and *near-normal/diffuse* ($0:d:e$). The gloss and texture of the specimen will have little effect on the color measurements when the specimen's specular reflection is included. Thus, many colorists prefer this geometry for computerized color matching. Measuring highly glossy specimens using one of the diffuse geometries with the specular reflection excluded will correlate with a bidirectional 45:0 instrument. This correlation will also occur for perfectly matte specimens for which there are not a noticeable specular reflection peak. Since the shape of the specular reflection varies by material and specimen and is poorly defined, designing the perfect universal light trap, without affecting the measurement of the diffuse reflectance, is impossible (Budde, 1980). The amount of the specular reflection trapped is also dependent upon the design of the instrument and of the integrating sphere.

12.6.1.3 Spectrophotometers, Colorimeters, Spectrocolorimeters, and LED-Based Instruments

Spectral reflectance values are required to calculate tristimulus values and color coordinates. Spectrophotometers measure the reflection (or transmission) characteristics of a specimen at different wavelengths of the visible spectrum. Spectrophotometers have been manufactured in all of the recommended illuminating and viewing geometries. Although some spectrophotometers can measure the reflectance continuously at all wavelengths of the spectrum, others (abridged spectrophotometers) can only measure selected wavelengths. Many commercial spectrophotometers measure the reflectance in 10 nm increments.

Reflectance is the ratio of the amount of light reflected by a specimen relative to the amount of light illuminating the specimen. Most spectrophotometers actually measure reflectance factor rather than reflectance. Reflectance factor is the ratio of the amount of light reflected by a specimen relative to the amount of light that would be reflected from a perfect reflecting diffuser under the same geometric and spectral conditions of measurement. Many colorists use the term reflectance when they mean reflectance factor. This distinction becomes very important when measuring fluorescent materials, materials with metallic flakes, and materials with interference pigments. Although the reflectance from a specimen can never be greater than 1.0 (100%), the reflectance factor may exceed 1.0 by a significant amount for some materials.

Colorimeters measure broad areas of the light reflected from a specimen using broad-band filters or some other spectral defining device. The amount of light reflected is converted directly into tristimulus values or color coordinates such as L^*, a^*, and b^*. Because colorimeters usually are customized for only one illuminant/observer combination, they cannot measure metamerism.

Colorimeters are generally not as accurate as spectrophotometers because of the difficulties in matching the light source and filters to the combination of CIE illuminant spectral power distributions and color-matching functions. Many of the current colorimeters are based on video camera technology. These instruments are coupled with advanced image technology to make color measurements on small areas of a complex scene. This technology is particularly useful for materials whose appearance is naturally or intentionally nonuniform such as woven and patterned fabrics. Video-based instruments may also be used for online applications such as controlling color on web printing application.

Spectrocolorimeters are spectrophotometers that output only tristimulus values or color coordinates.

Several instruments that illuminate a specimen with light emitting diodes (LEDs) have been developed. Spectral reflectance factors, tristimulus values, and color coordinates are calculated from the reflected light.

12.6.1.4 Standardization

The first step in making accurate color measurements is to standardize the instrument. This means the zero point of the instrument and the high end of the photometric scale must be set. A light trap is often used to set the zero point of the instrument. Most instrument manufacturers provide a white ceramic tile for setting the high end of the photometric scale. The values of the white are traceable to a national standardizing laboratory such as the US National Institute for Standards and Technology (NIST). A number of other materials have been used to set the high end of the photometric scale. They include optical grade barium sulfate (BsSO4), optical grade PTFE, and opal glasses (ASTM 259; Carter, Billmeyer and Rich, 2000). Instruments should be standardized at regular intervals. In most industrial settings, they are standardized once per 8-hour shift.

12.6.1.5 Repeatability, Reproducibility, and Accuracy

Top-of-the-line color measuring instruments claim a short-term repeatability of 0.01 CIELAB units. This is based upon multiple measurements of a white standard. The interinstrument agreement (reproducibility) is claimed to be 0.15 CIELAB units (maximum), with an average of 0.08 CIELAB units. This claim is based upon reading a set of 12 colored tiles. It should be noted that the very low repeatability and reproducibility numbers could only be achieved with stable standards at a tightly controlled temperature. Accuracy is determined by comparing instrumentally measured results with those of a known standard. There are no known standards available for color measurement with "absolute" values. Fortunately most industrial color measurement is concerned with the comparative measurement of a standard (or target) to a batch (or trial). As long as the instruments are using the same illuminating and viewing conditions and have comparable repeatability and reproducibility, these measurements are adequate. Long-term instrument stability should be monitored as part of a good quality assurance program. A set of stable ceramic standards was designed for this purpose (Clark, 1971), but any set of stable standards can be used. Some instrument manufacturers and private laboratories have programs to help the colorist monitor his/her instruments.

The repeatability and reproducibility claimed by instrument manufacturers are tested under tightly controlled conditions with highly uniform samples. It is extremely unlikely that a colorist would be able to duplicate these claims in the field. More variation in a color measurement comes from the sample rather than from the instrument. Careful sampling and sample preparation can help reduce the variability. The colorists should determine the repeatability and reproducibility using their own materials under their typical measurement conditions. Knowing how the instrument performs under field conditions helps determine practical color tolerances. The variability in measuring real specimens in the field can be reduced by making multiple measurements (ASTM 1345).

12.6.1.6 Measuring Fluorescent Materials

Fluorescent materials do more than just reflect light. They absorb radiation at some wavelengths and emit light at wavelengths longer than those that were absorbed. This is in addition to the light reflected in the same region of the spectrum. For example, many white papers will absorb ultraviolet radiation (below 400 nm) and emit blue light. The higher energy in the blue region of the spectrum makes the paper appear whiter than paper that does not fluoresce.

If an instrument with monochromatic light illuminates a fluorescent specimen, and a receiver sensitive to the entire visible spectrum is used to measure the reflectance, the receiver will not be able to distinguish the amount of light reflected at that wavelength from the light emitted at the longer wavelength so that the reflectance will appear abnormally high. Monochromatic light illuminating the sample at the wavelength where light is emitted will result in the measurement of the true reflectance at that wavelength. Spectrofluorimeters that both illuminate the sample with monochromatic light and view the sample at a single (narrow band) of wavelengths for the complete measurement of fluorescent samples were developed.

Most industrial fluorescent materials are measured with instruments that illuminate the sample with a broadband polychromatic illuminator simulating CIE Illuminant D65. Some of these instruments have special filters to control the amount of ultraviolet illumination from the source to better simulate D65 as the lamp in the illuminator ages. These instruments measure the combined reflectance and emittance, which is the total spectral reflectance factor. Because of the emittance, the total spectral reflectance factor can be greater than 1.0.

To properly measure fluorescent samples, only the 45 : 0 or 0 : 45 illuminating/viewing geometry should be used (ASTM 991). Emitted fluorescent light in an integrating sphere lowers the efficiency of the integrating sphere, resulting in lower total spectral reflectance factors than when using the other geometries (McKinnon, 1987).

Gundlach and Terstiege (1994) recommended using fluorescent calibration standards to standardize single polychromator instruments to ensure the proper measurement of fluorescent specimens.

12.6.1.7 Measuring Materials Containing Metallic Flakes and Interference Pigments

Metallic flakes and interference pigments are called *effect pigments*. When a

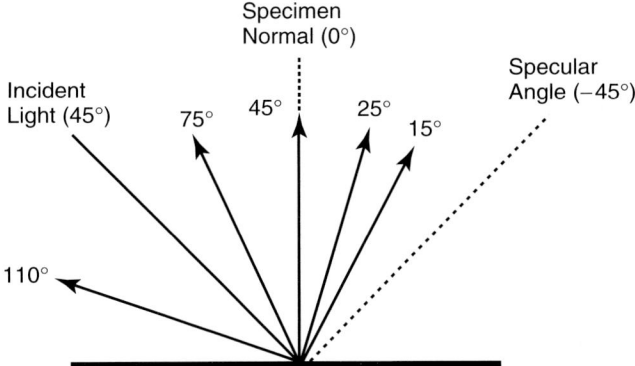

Fig. 12.17 The illuminating and viewing geometries used for the measurement of metal flake pigmented materials. The ASTM recommends using the aspecular angles of 15, 45, and 110°.

material is colored with these pigments, one or more of the attributes of color change as the illumination or viewing angle is changed. This is called *gonioapparance*.

Materials colored with only metallic flakes appear lighter when viewed near the specular angle than when viewed at angles farther away from the specular angle. This change in lightness is called *flop* in the coatings industry. The larger the metallic flakes, the greater the flop. Metallic flakes also add a sparkling appearance to the material. Multiangle instruments (sometimes called *goniospectrophotometers*) were designed to measure materials containing metallic flakes. These instruments are modifications of the common 45 : 0 bidirectional instruments. The ASTM specifies the angles of illumination and viewing by the illumination anormal angle, the viewing (detection) anormal angle, and the viewing aspecular angle enclosed in parenthesis (ASTM 2175). An anormal angle is an angle measured from the specimen normal in the illuminator plane unless otherwise specified.

Positive anormal angles are usually diagrammed to the left of the specimen normal on the same side as the illuminator. An aspecular angle is a viewing angle measured from the specular direction in the illuminator plane unless otherwise specified (ASTM E284). Positive aspecular angles are in the direction toward the illuminator axis. Research instruments let the colorist specify the angles of illumination and view. Portable instruments illuminate the sample at 45° from the normal. Portable instruments view the specimen at three to five aspecular angles (see Figure 12.17). Most five-angle instruments would view the specimen at aspecular angles of 15, 25, 45, 75, and 110°. A three-angle instrument would use either 15° (preferred) or 25, 45° and either 75 or 110° (preferred) (ASTM 2194). The aspecular angle of 45° corresponds to an illuminating/viewing geometry of 45 : 0 or 45° : 0° (as 45) using the ASTM notation.

Measuring materials colored with interference pigments is a more difficult problem. The color of an interference

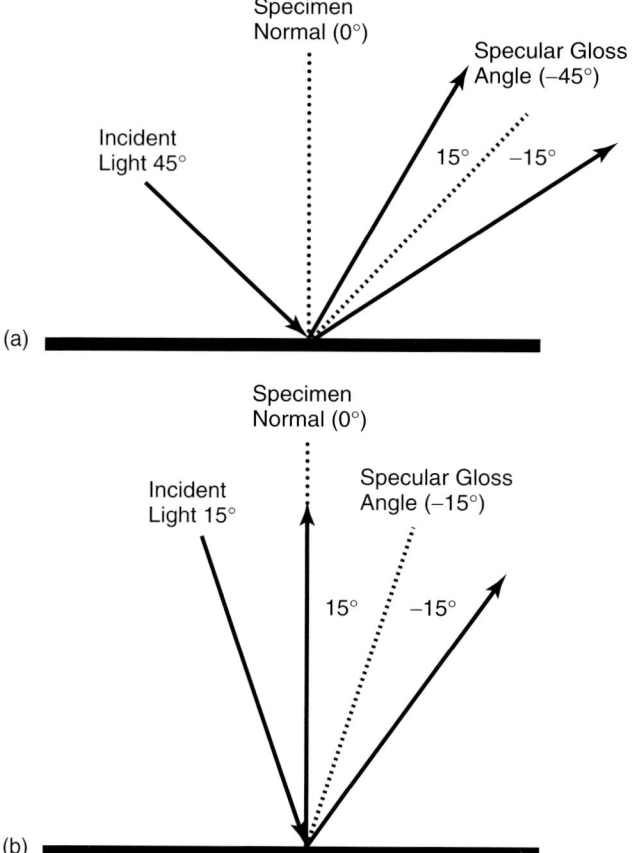

Fig. 12.18 The ASTM recommended viewing geometries used for the multiangle color measurement of interference pigments. (a) The specimen is illuminated by incident light at an anormal angle of 45°. (b) The specimen is illuminated by incident light at an anormal angle of 15°. The specimen normals and specular gloss angles are shown as anormal angles. The two viewing angles for each illumination angle are shown as aspecular angles.

pigment is seen only near the angle of the specular reflection. As the angle of the illuminator changes, the angle of specular reflection changes. The hue and chroma of the interference pigment change as the specular angle changes. Thus, an observer can see extreme changes in hue as the specular angle changes. To measure this effect, you need more than one angle of illumination in addition to multiple angles of view (see Figure 12.18a and b). The ASTM specifies that the minimum requirement for an instrument measuring interference pigments is to have two illuminators, one at an anormal angle of 45° and one at an anormal angle of 15°. Two aspecular viewing angles (−15 and +15°) are required for each illumination angle for a total of four angles (ASTM 2539). Some commercial

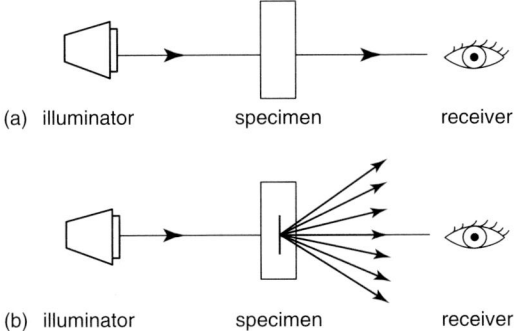

Fig. 12.19 The 0 : 0 illuminating and viewing geometry for measuring transmittance. This is only accurate for samples with no scattering (a). Hazy samples (b) cannot be measured using this illuminating and viewing geometry.

multiangle instruments have a third illuminator at an anormal angle of 65°.

Significant visual effects can be obtained by coloring a material using both metallic flakes and interference pigments. These specimens can be measured with multiangle instruments with multiple angles of illumination and multiple angles of view.

12.6.2
Measuring Transmittance

12.6.2.1 Regular and Diffuse Transmittance

Regular transmittance occurs in materials that can only absorb light. They may appear crystal clear or colored but totally transparent. Hazy materials not only absorb light but also scatter it as it passes through. They will appear to be almost, but not totally, transparent and may be clear or colored. Translucent materials reflect, transmit, and scatter light as it passes through. They will appear to be milky and may be white or colored.

12.6.2.2 Illuminating and Viewing Geometries

Although the CIE has recommended four geometries for transmittance, only three of them are commonly used for color measurement.

Spectrophotometers used for analytical chemistry measure transmittance using 0 : 0 illuminating/viewing geometry as shown in Figure 12.19(a) and (b). The illuminator illuminates the specimen along its normal and the receiver views the sample on the other side along it normal. This geometry measures transmittance accurately only for specimens that have regular transmission.

The diffuse-normal ($d : 0$) shown in Figure 12.15 and its reverse geometry normal-diffuse ($0 : d$) can be used to measure materials that transmit regularly or diffusely. To measure regular transmission, the specimen is placed away from the instrument's integrating sphere. To measure diffuse transmission, the specimen is placed in contact with the instrument's integrating sphere. When integrating sphere instruments designed for measuring reflectance are used, a white material with reflectance as close to that of the integrating sphere's wall is placed in the sample port.

12.6.2.3 Standardization

The instrument's zero point is usually set by blocking the light beam with an opaque material.

There are several ways of setting the high end of the photometric scale. The easiest technique is to make a measurement with nothing in the sample holder and set the transmittance factor to 1.0 at each wavelength. Transmittance measurements will then be relative to air. A totally colorless blank can be measured and the transmittance factor is set to 1.0 at each wavelength. It is important that this blank specimen be made of the same material and has the same thickness as the specimen to be measured. Transmittance measurements will then be relative to the clear specimen.

Liquid samples must be put into a sample holder before they are measured. An empty holder can be measured and the transmittance factor is set to 1.0 at each wavelength. Although the measurements would still be relative to air, the affect on transmission of the sample holder will be eliminated. Lastly, the liquid sample holder can be filled with water or a solvent of similar optical properties as that of the specimen. The transmittance measurements will then be relative to water or the solvent.

12.6.2.4 Measuring Transmittance

Specimens being measured should be flat with parallel sides. Curved specimens, such as lenses, are very difficult to measure because the curvature changes the path of the light. Liquid sample holders should be made of optical glass to minimize the effect of the sample holder on the measurement. Transmission changes with thickness and concentration. If two specimens are being compared, they must be prepared in the same way if the comparison is to be valid. This is particularly important if a numeric standard is used. A haze index can be calculated from multiple measurements of transmittance (ASTM 1003).

12.6.3
Radiometric Measurements

Spectral radiometers and radiometric colorimeters were designed to measure lamps, light sources, and visual display units. These instruments are closely related to spectrophotometers and colorimeters, but they need not supply an illuminator because the object being measured is the source of light.

Lamps and light sources can be measured either directly or indirectly by measuring the reflectance of a stable white reflecting surface being illuminated by the lamp or light source (ASTM 1341). Measuring televisions and visual display units usually requires that the emitted light be directly imaged on the optics of the instrument (ASTM 1336; ASTM 1455).

12.7
Summary

Color is an integral part of commerce and entertainment. Color perception depends upon the light source illuminating an object, the pigments and dyes used in manufacturing the object, and the eye of the observer whose impulses are translated by the brain as a color. Color is three dimensional and thus three terms (lightness, chroma, and hue) are needed to describe a color. Color is only one aspect of appearance. Members of the CIE developed a system to simulate the color perception. They standardized illuminants (mathematical light sources) and color-matching functions (simulating the eye–brain combination). Spectrophotometers are used to characterize the object completely. Tristimulus values and color coordinates can be calculated from the spectrophotometric measurements, the standardized illuminants, and the color-matching functions. Color differences between a target or standard color and a trial or batch color can be calculated from the measurements. For some situations, a single-number color scale is sufficient. For other situations, a three-component color difference is required. There are a number of equations available for calculating color differences.

An object may absorb light, scatter and reflect light, and/or transmit light. Reflectance factors are usually measured for opaque and translucent objects. Transmittance is usually measured for hazy and transparent objects. Spectrophotometers measure reflectance and/or transmittance at given wavelengths of the visible spectrum and calculations are done to determine tristimulus values and color coordinates. Colorimeters determine tristimulus values or color coordinates directly during the measurement using broad-band spectral devices. Colorimeters cannot usually detect metamerism. Fluorescent samples not only reflect light but also emit light. Special techniques are required to measure those samples. Measuring materials containing only traditional colorants and metallic flakes require instruments with multiple viewing angles. Measuring materials containing interference pigments require instruments with multiple illumination angles and multiple viewing angles. Measuring lamps and other light emitters require spectral radiometers and radiometric colorimeters.

References

CIE (2004) *Colorimetry*, 3rd edn, Publication No. 15:2004, Central Bureau of the Commission International de l'Éclairage (CIE), Vienna. Available from U.S. National Committee of the CIE (International Commission on Illumination), c/o Thomas M. Lemons, TLA-Lighting Consultants, Inc., 7 Pond St., Salem, MA 01970, www.cie-usnc.org.

ASTM ASTM E 308 (2006) *Practice for computing the colors of objects by using the CIE System, Annual Book of ASTM Standards*, ASTM International, West Conshohocken. www.astm.org.

CIE (2004) *Method for Assessing the Quality of Daylight Simulators for Colorimetry*, Publication No. 51.2–1981, Central Bureau of the Commission International de l'Éclairage (CIE), Vienna. Available from U.S. National Committee of the CIE (International Commission on Illumination), c/o Thomas M. Lemons, TLA-Lighting Consultants, Inc., 7 Pond St., Salem, MA 01970, www.cie-usnc.org.

ASTM ASTM E 284 (2008) *Standard terminology of appearance, Annual Book of ASTM Standards*, ASTM International, West Conshohocken. www.astm.org.

ASTM ASTM D 3134 (1997) *Standard practice for establishing color and gloss tolerances, Annual Book of ASTM Standards*, ASTM International, West Conshohocken. www.astm.org.

Hunt, R.W.G. (1991) *Measuring Colour*, 2nd edn, Ellis Horwood, Chichester, West Sussex, England.

ASTM ASTM D 4449 (2008) *Standard test method for visual evaluation of gloss differences between surfaces of similar appearance, Annual Book of ASTM Standards*, ASTM International, West Conshohocken. www.astm.org.

ASTM ASTM D 1729 (1996) *Standard practice for visual appraisal of colors and color differences of diffusely-illuminated opaque materials, Annual Book of ASTM Standards*, ASTM International, West Conshohocken. www.astm.org.

ASTM ASTM D 3928 (2000) *Standard test method for evaluation of gloss or sheen uniformity, Annual Book of ASTM Standards*,

ASTM International, West Conshohocken. www.astm.org.

ASTM ASTM D 2616 (1996) *Standard test method for evaluation of visual color difference with a gray scale*, Annual Book of ASTM Standards, ASTM International, West Conshohocken. www.astm.org.

ASTM ASTM E 313 (2005) *Standard practice for calculating yellowness and whiteness indices from instrumentally measured color coordinates*, Annual Book of ASTM Standards, ASTM International, West Conshohocken. www.astm.org.

ASTM ASTM D 156 (2007) *Standard test method for Saybolt color of petroleum products (Saybolt Chromometer Method)*, Annual Book of ASTM Standards, ASTM International, West Conshohocken. www.astm.org.

ASTM ASTM D 1209 (2005) *Standard test method for color of clear liquids (Platinum-Cobalt Scale)*, Annual Book of ASTM Standards, ASTM International, West Conshohocken. www.astm.org.

ASTM ASTM D 1500 (2007) *Standard test method for ASTM color of petroleum products (ASTM Color Scale)*, Annual Book of ASTM Standards, ASTM International, West Conshohocken. www.astm.org.

ASTM ASTM D 6045 (2004) *Standard test method for color of petroleum products by the automatic tristimulus method*, Annual Book of ASTM Standards, ASTM International, West Conshohocken. www.astm.org.

ASTM ASTM D 1544 (2004) *Standard test method for color of transparent liquids (Gardner Color Scale) [Historical standard – no longer active]*, Annual Book of ASTM Standards, ASTM International, West Conshohocken. www.astm.org.

ASTM ASTM D 6166 (2008) *Standard test method for color of naval stores and related products (Instrumental Determination of Gardner Color)*, Annual Book of ASTM Standards, ASTM International, West Conshohocken. www.astm.org.

ASTM ASTM D 1686 (1996) *Standard test method for color of solid aromatic hydrocarbons and related materials in the molten state (Platinum Cobalt Scale)*, Annual Book of ASTM Standards, ASTM International, West Conshohocken. www.astm.org.

ASTM ASTM D 2244 (2007) *Standard practice for calculation of color tolerances and color differences from instrumentally measured color coordinates*, Annual Book of ASTM Standards, ASTM International, West Conshohocken. www.astm.org.

McDonald, R. (1980) Industrial pass/fail colour matching. *J. Soc. Dyers Colour.*, **96**, Part I, 372–376; Part II, 418–433; Part III, 486–495.

AATCC (1992) AATCC Test Method 173–1992, "CMC: calculation of small color differences for acceptability", *AATCC Technical Manual*, AATCC, Research Triangle Park. www.aatcc.org.

CIE (2001) *Improvement to Industrial Colour-difference Evaluation*, Publication No. 142–2001, Central Bureau of the Commission International de l'Éclairage (CIE), Vienna. Available from U.S. National Committee of the CIE (International Commission on Illumination), c/o Thomas M. Lemons, TLA-Lighting Consultants, Inc., 7 Pond St., Salem, MA 01970, www.cie-usnc.org.

DIN DIN 6176. Farbmetrische Bestimmung von Farbabständen bei Körperfarben nach der DIN99-Formel (Colorimetric evaluation of colour differences of surface colours according to DIN99 formula – available in English), DIN Deutsches Institut für Normung e. V., Burggrafenstraße 6, 10787 Berlin, Germany. www.din.de.

CIE (1995) *Geometric Tolerances for Color Measurement*, Publication No. 176–2006, Central Bureau of the Commission International de l'Éclairage (CIE), Vienna. Available from U.S. National Committee of the CIE (International Commission on Illumination), c/o Thomas M. Lemons, TLA-Lighting Consultants, Inc., 7 Pond St., Salem, MA 01970, www.cie-usnc.org.

ASTM ASTM E 179 (1996) *Standard guide for selection of geometric conditions for measurement of reflection and transmission properties of materials*, Annual Book of ASTM Standards, ASTM International, West Conshohocken. www.astm.org.

Rich, D.C. (1988) The effect of measuring geometry on computer color matching. *Color Res. Appl.*, **13**, 113–118.

Mabon, T.J. (1992) Color measurement of plastics: which geometry is best. Presented at the Regional Technical Conference of the Society of Plastics Engineers, Inc., in *Color Tolerances: Measuring up to Today's*

Standards, Cherry Hill, New Jersey. September 14–16.

Billmeyer, F.W. and Marcus Jr., R.T. (1969) Effect of illuminating and viewing geometry on the color coordinates of samples with various surface textures. *Appl. Opt.*, **8**, 1763–1768.

Budde, W. (1980) The gloss trap in diffuse reflectance measurements. *Color Res. Appl.*, **5**, 73–75.

ASTM ASTM E 259 (2006) *Standard practice for preparation of pressed powder white reflectance factor transfer standards for hemispherical and bi-directional geometries, Annual Book of ASTM Standards*, ASTM International, West Conshohocken. www.astm.org.

Carter, E.C., Billmeyer Jr., F.W., and Rich, D.C. (2000) Guide to Material Standards and their Use in Color Measurement, *ISCC Technical Report 89-1*. Available from the Inter-Society Color Council, Reston. www.iscc.org.

Clark, F.J.J. (1971) Ceramic colour standards. *Die. Farbe.*, **20**, 299–306.

ASTM ASTM E 1345 (1998) *Standard practice for reducing the effect of variability of color measurement by use of multiple measurements, Annual Book of ASTM Standards*, ASTM International, West Conshohocken. www.astm.org.

ASTM ASTM E 991 (2006) *Standard practice for color measurement of fluorescent specimens using the one-monochromator method, Annual Book of ASTM Standards*, ASTM International, West Conshohocken. www.astm.org.

McKinnon, R.A. (1987) Methods of measuring the colour of opaque fluorescent materials. *Rev. Prog. Coloration*, **17**, 56–60.

Gundlach, D. and Terstiege, H. (1994) Problems in measurement of fluorescent materials. *Color Res. Appl.*, **19**, 427–436.

ASTM ASTM E 2175 (2001) *Practice for specifying the geometry of multiangle spectrophotometers, Annual Book of ASTM Standards*, ASTM International, West Conshohocken. www.astm.org.

ASTM ASTM E 2194 (2003) *Practice for multiangle color measurement of metal flake pigmented materials, Annual Book of ASTM Standards*, ASTM International, West Conshohocken. www.astm.org.

ASTM ASTM E 2539 (2008) *Standard practice for multiangle color measurement of interference pigments, Annual Book of ASTM Standards*, ASTM International, West Conshohocken. www.astm.org.

ASTM ASTM D 1003 (2007) *Standard test method for haze and luminous transmittance of transparent plastics, Annual Book of ASTM Standards*, ASTM International, West Conshohocken. www.astm.org.

ASTM ASTM E 1341 (2006) *Practice for obtaining spectroradiometric data from radiant sources for colorimetry, Annual Book of ASTM Standards*, ASTM International, West Conshohocken. www.astm.org.

ASTM ASTM E 1336 (1996) *Test method for obtaining colorimetric data from a visual display unit by spectroradiometry, Annual Book of ASTM Standards*, ASTM International, West Conshohocken. www.astm.org.

ASTM ASTM E 1455 (2003) *Practice for obtaining colorimetric data from a visual display unit using tristimulus colorimeters, Annual Book of ASTM Standards*, ASTM International, West Conshohocken. www.astm.org.

13
Atomic Spectrometry and Elemental Analysis

Bernhard Welz and Daniel L.G. Borges

13.1	**Introduction**	**423**
13.2	**General Considerations**	**424**
13.2.1	Atomic Spectra	424
13.2.2	Interferences	425
13.3	**Optical Atomic Spectroscopy**	**426**
13.3.1	Basic Principles	426
13.3.1.1	Physics of Atomization and Excitation	426
13.3.1.2	Atomic Spectra and Spectral Lines	427
13.3.1.3	Measurement of Radiation, Calibration, and Evaluation	428
13.3.2	Optical Emission Spectrometry	429
13.3.2.1	Excitation Sources	430
13.3.2.2	Spectrometers	436
13.3.2.3	Detectors	438
13.3.3	Atomic Absorption Spectrometry	439
13.3.3.1	Atomizers	440
13.3.3.2	Spectrometers for Line Source AAS	446
13.3.3.3	Spectrometers for High-Resolution Continuum Source AAS	452
13.3.4	Atomic Fluorescence Spectrometry	460
13.3.4.1	General Principles	460
13.3.4.2	Instrumentation	461
13.4	**X-Ray Spectroscopy**	**463**
13.4.1	Theoretical Principles	464
13.4.1.1	Emission of X-Rays	464
13.4.1.2	Absorption of X-Rays	466
13.4.1.3	X-Ray Fluorescence (XRF)	467
13.4.2	Instrumentation	467
13.4.2.1	X-Ray Fluorescence	467
13.4.2.2	Total-Reflection (TR) XRF	469
13.4.2.3	Particle-Induced X-Ray Emission (PIXE)	471
13.4.2.4	Portable Instrumentation	471
13.5	**Outlook**	**471**

Encyclopedia of Applied Spectroscopy. Edited by David L. Andrews.
Copyright © 2009 WILEY-VCH Verlag GmbH & Co. KGaA, Weinheim
ISBN: 978-3-527-40773-6

Acknowledgments 472
Acronyms 472
References 473

13.1
Introduction

The early history of atomic spectroscopy is closely connected with the observation of the sunlight. Wollaston (1802) reported seeing dark lines in the sun's spectrum, but he could not give a satisfactory explanation for them. Fifteen years later, Fraunhofer designed the first grating ruling engines and succeeded in producing transmission gratings with up to 600 grooves per centimeter; he noted the same lines and proceeded to map the dark lines. As spectra produced by gratings, unlike prism spectra, are nearly linear with wavelength, Fraunhofer (1817) succeeded in calculating the wavelength of the dark lines with remarkable accuracy.

During the first half of the nineteenth century, a good deal of experimentation took place with colored flames generated by injecting salts into alcohol burners or oil lamps as excitation sources. Herschel (1823) and Talbot (1826) succeeded to identify alkali metals using this technique. A breakthrough occurred in 1855, when Bunsen designed an efficient burner for coal gas, in which fuel and air were mixed before the mixture was ignited. The new flame was much steadier and gave a higher temperature, making possible a much more efficient study of the emitted spectra.

Soon after, a close collaboration began between Kirchhoff, professor of physics, and Bunsen, professor of chemistry, both at the University of Heidelberg, Germany. Kirchhoff compared the Fraunhofer lines in the sun's spectrum to those obtained in the laboratory, and showed that they arose from the same elements. He explained the reversed appearance of the lines as due to a process of absorption as the emission rays passed through the cooler outer layer of the sun's atmosphere, which caused them to show up dark against the bright background. Kirchhoff (1860) also concluded that absorption occurs only at the same wavelengths as emission, a phenomenon called *resonance*.

Among the most important stages in the development of emission spectroscopy are the first measurements in the ultraviolet (UV) made by Mascart (1864) with a photographic plate, which have later been extended to the low-vacuum UV. Balmer (1885), with his equation for the Balmer series of hydrogen lines, started the search for an explanation for the origin of atomic spectra, which was continued by several

Encyclopedia of Applied Spectroscopy. Edited by David L. Andrews.
Copyright © 2009 WILEY-VCH Verlag GmbH & Co. KGaA, Weinheim
ISBN: 978-3-527-40773-6

other researchers. It was the work of Bohr (1913), with his concept of the "astronomical" atom that placed the origin of spectra on a firm theoretical foundation. Later, Sommerfeld (1916) proposed elliptical orbits and the Schrödinger wave equation finally permitted calculation of the quantum states necessary to describe the energy positions of electrons in atoms. The development of X-ray spectroscopy is mainly based on the work of Barkla, Moseley, and Siegbahn, who received the Nobel price for physics in 1924.

13.2
General Considerations

13.2.1 Atomic Spectra

Atomic spectra are due to radiative transitions of electrons between the ground state and/or excited states of an atom. Three forms of radiative transition are possible: (i) *spontaneous emission* for the transition from a higher excited state to a lower state after the transfer of kinetic or electrical energy to the atom; (ii) *absorption of radiation* with a corresponding transition of a lower state to a higher excited state; and (iii) *induced emission* for the transition from a higher excited state to a lower state after the absorption of external radiation of corresponding frequency (the reversed process to absorption of radiation); the latter is generally referred to as *atomic fluorescence*. Atomic spectra are line spectra and are specific to the absorbing or emitting atoms (elements), that is, the spectra contain information on the atomic structure. An electron in an atom can only absorb and emit energy at well-defined wavelengths, which correspond to the energy difference between the two levels of the transition. Absorption spectra are generally simpler than emission spectra, as absorption usually starts from the ground state of the neutral atom according to selection rules, whereas all kinds of excited states can be reached in thermal and electrical excitation, resulting in a whole series of de-excitation pathways, and a correspondingly greater number of emission lines, as shown in Figure 13.1 for the spectrum of sodium.

Transitions of electrons of the outer shell are found in the spectral range between 100 nm and 1 mm, and are referred to as *optical spectroscopy*; transitions of electrons from inner (K, L, and M) shells are found in the spectral range between about 10 pm and 10 nm, and are referred to as *X-ray spectroscopy*.

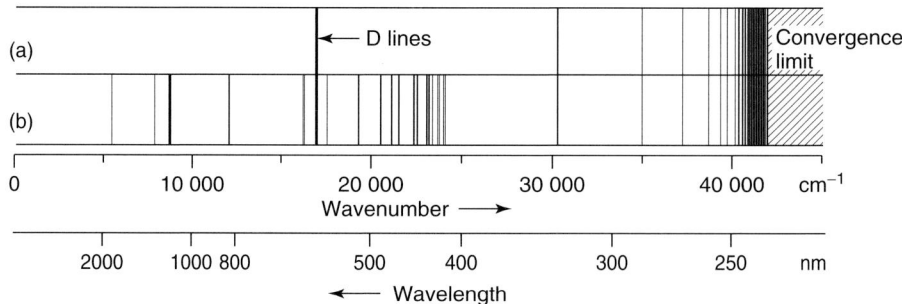

Fig. 13.1 The spectrum of the sodium atom; (a) in absorption and (b) in emission. (From Welz and Sperling (1999), Wiley-VCH, p. 67; reproduced with permission.)

The term *spectroscopy* – at least in atomic spectroscopy – refers to the visual evaluation of spectra using a dispersive system; however, it is also used as the general term for all spectroscopic measurement and observation procedures. The term *spectrometry* refers to the photoelectric measurement and evaluation of spectral radiation units using a dispersive spectral apparatus.

13.2.2
Interferences

The presence of concomitants (matrix) in the test sample can cause interferences in the determination of the analyte. The interference can influence both the attainable precision and the trueness. However, interference only leads to a measurement error if it is not eliminated, or if it is not taken into consideration in the evaluation process, by suitable measures. Such measures can be instrumental, such as background correction, or the choice of a suitable calibration technique, such as the analyte addition technique, or the use of chemical additives.

Interferences in spectrochemical analysis can be classified according to a number of aspects. The various categories are not mutually exclusive but rather complement each other in characterizing the interference. Frequently, the distinction is made between *specific* (for the analyte) and *nonspecific* interferences or between *additive* (the measured signal is increased or reduced by the same value independent of the analyte concentration or mass) and *multiplicative* interferences (the slope of the calibration curve is affected by concomitants). In the following, we shall largely adhere to International Union of Pure and Applied Chemistry (IUPAC) recommendations (IUPAC, 1978), according to which interferences are divided generally into *spectral* and *nonspectral interferences*.

Spectral interferences are due to the incomplete isolation of the radiation emitted or absorbed by the analyte from other radiation or radiation absorption detected and processed by the electrical measuring system. Spectral interferences may be due to (i) direct or partial overlapping of the analytical line with the emission or absorption line of another element; (ii) emission of radiation by concomitants that is not overlapping with the analyte line, but is not separated by the monochromator/polychromator; (iii) emission or absorption of the radiation of the analytical line by gaseous molecules; and (iv) radiation scattering caused by particles in the observation volume. Spectral interferences are always additive interferences; details are discussed with the specific techniques, where the interferences appear to a significantly different degree.

Nonspectral interferences are those that cause an influence on the number of analyte atoms (absolute or per time unit) in the observation volume, and thus on the measurement value. Nonspectral interferences are most conveniently classified according to the place or time of their occurrence. The most common types of nonspectral interferences are (i) *transport interferences*, that is, interferences that occur during the transport of a (liquid) sample from the sample container to the atomizer/excitation source due to the physical properties of the solution; (ii) *solute-volatilization interferences*, which are due to incomplete transfer of the sample into gaseous species, an interference that is typically limited to atomizers/excitation sources of relatively low temperature; (iii) *vapor-phase interferences*, which include all the equilibrium processes that may occur in the vapor phase, such as *dissociation, ionization, and excitation interferences*. If an interference cannot be classified because its cause is unknown or it is of complex nature, we speak of a *matrix effect*. Other interferences that are typical for specific atomizers and excitation sources are discussed in connection with these atomizers.

13.3
Optical Atomic Spectroscopy

13.3.1
Basic Principles

13.3.1.1 Physics of Atomization and Excitation

In the atomizer, the sample to be analyzed is dissociated into atoms; the *degree of dissociation* depends on the temperature of, and on chemical reactions within the atomizer. The *ground state* is the lowest energy state of an atom. By absorption of additional thermal, electrical, or optical energy, a neutral atom can be transferred into an excited state or it can be ionized. The *excited state* is unstable; the atom returns to a lower energy state after a lifetime of typically 10^{-9}–10^{-8} seconds. The transitions are limited by selection rules. *Ionization* refers to the complete separation of one or more electrons from an atom, so that the remaining atom becomes electrically charged. The degree of ionization is the ratio between the number of ions and that of the originally present number of atoms of the element under consideration.

If thermal equilibrium is established, the fraction of atoms N_j in an excited stage j can be calculated by statistical mechanics. The fraction of excited atoms at a certain energy level in comparison to the number of atoms in the ground state N_0 is given by the *Boltzmann distribution*

$$\frac{N_j}{N_0} = \frac{P_j}{P_0} e^{-E_j/kT} \qquad (13.1)$$

where P_j is the statistical weight of the selected excited state j, P_0 the statistical weight of the ground state, E_j is the energy of excitation, k is the Boltzmann constant, and T the absolute temperature. In Eq. (13.1), the exponent is inversely proportional to the absolute temperature, which means that the increase in the relative number of excited atoms N_j/N_0 is exponential with increasing temperature. Walsh (1955) calculated the ratio N_j/N_0 for a number of elements at various temperatures, as is shown in Table 13.1. For most of the elements and temperatures, the number of atoms in the ground state N_0 is virtually identical to the total number of atoms N.

Tab. 13.1 Temperature and wavelength dependence of the ratio N_j/N_0 (Walsh, 1955).

Element	Excitation energy eV	Wavelength nm	2000 K	N_j/N_0 3000 K	4000 K
Zn	5.80	213.9	7.29×10^{-15}	5.58×10^{-10}	1.48×10^{-7}
Ca	2.93	422.7	1.21×10^{-7}	3.69×10^{-5}	6.03×10^{-4}
Na	2.11	589.0	0.86×10^{-4}	5.88×10^{-4}	4.44×10^{-3}
Cs	1.46	852.1	4.44×10^{-4}	7.24×10^{-3}	2.98×10^{-2}

13.3.1.2 Atomic Spectra and Spectral Lines

Dispersion of the radiation emitted or absorbed by the atomizer/excitation source results in a spectrum that is characteristic of the emitting or absorbing species. Atomic spectra are line spectra; molecular spectra are band spectra that extend over a wider spectral range. A *spectral line* is defined as the monochromatic radiation that is emitted or absorbed at the transition of an atom or ion between different energy states E_1 and E_2. The wavelength λ of the transition is characteristic for the two energy states according to

$$\lambda = \frac{hc}{E_2 - E_1} \qquad (13.2)$$

where h is Planck's constant and c is the speed of light. Spectral lines, the radiation of which can be absorbed by atoms in the ground state or that are generated by a transition into the ground state, are called *resonance lines*.

Atomic lines are not strictly monochromatic, but characterized by a distribution of the radiation intensity over wavelength that can be described by a profile function, as is schematically shown in Figure 13.2. The width of an atomic line is usually given as the full width at half maximum (FWHM), or briefly *half width*. The part of the line that is within the FWHM is called the *line core*; the parts that are outside are called the *line wings*. The *natural line width* is the smallest possible line width, which is caused by the residence time of the atom in the involved energy states and the resulting uncertainty of the measurement. The natural width of atomic lines is typically of the order of 0.01–0.1 pm, which is, however, not of practical significance due to the various broadening effects that occur under normal analytical conditions. The most important effects that finally determine the *physical line width* are Doppler and pressure broadening. The *Doppler broadening* is caused by the random movement of the emitting or absorbing atoms; it is proportional to the square root of the temperature and results in a symmetric broadening of the line. *Collision broadening* is caused by collision of the emitting or absorbing atoms with other neutral or charged particles; it depends on the pressure and temperature of the environment and results in an asymmetric broadening. Depending on the mass of the atom and the collision partner this effect may shift the line

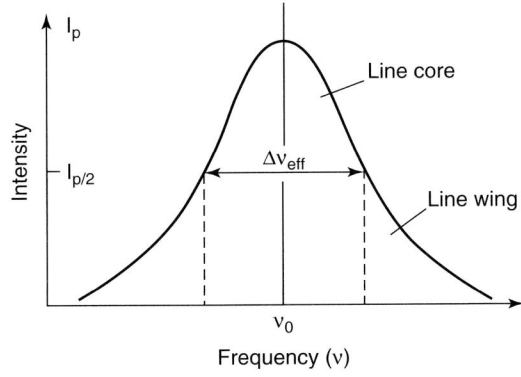

Fig. 13.2 Typical line profile and half width (FWHM, $\Delta\nu_{\text{eff}}$) of a spectral line. (From Welz and Sperling (1999), Wiley-VCH, p. 74; reproduced with permission.)

maximum to lower or higher wavelength, an effect called *Lorentz shift*.

The physical width of an atomic line depends on a variety of factors (besides individual differences in the hyperfine structure). Lines in the vacuum UV are approximately one order of magnitude narrower than lines in the beginning infrared. Lines emitted by low-pressure lamps, such as hollow cathode lamps (refer to Section 13.3.3.2, Radiation Sources for LS AAS), essentially only exhibit Doppler broadening, no Lorentz shifts, and typically have half widths between 0.3 and 1.5 pm. Lines emitted by a high-temperature plasma (refer to Section 13.3.2.1, Plasma Sources) are typically one order of magnitude broader, and the absorption lines in flames are in between with half widths of 1.5–5 pm.

13.3.1.3 Measurement of Radiation, Calibration, and Evaluation

The intensity of the emitted or absorbed radiation is measured using a *photoelectric detector* that converts the flux of photons into a flux of electrons using the outer or inner photo effect. The most common detectors are the photocathode detector, commonly denominated as photomultiplier tube (PMT) that uses the outer photo effect and the semiconductor detectors that use the inner photo effect. The latter ones come in a variety of array detectors that are increasingly used in modern equipment.

The measurement of signals in optical emission spectrometry (OES) and atomic fluorescence spectrometry (AFS) is typically done in emission or fluorescence intensity with the dimension counts per second (s^{-1}). In atomic absorption spectrometry (AAS), in contrast, measurement is done in absorbance, A, according to the Beer–Lambert law, which, for AAS, is usually written in the form:

$$A \equiv \log \frac{\Phi_0(\lambda)}{\Phi_{tr}(\lambda)} = 0.43 N\ell\kappa(\lambda) \quad (13.3)$$

where $\Phi_0(\lambda)$ is the incident radiant power, $\Phi_{tr}(\lambda)$ is the transmitted radiant power, N is the total number of free atoms, ℓ is the length of the absorbing layer, and $\kappa(\lambda)$ is the spectral atomic absorption coefficient. In addition to the absorbance A, which is typically used for steady-state signals, for example in flame atomic absorption spectrometry (FAAS) the time-integrated absorbance, A_{int} (which corresponds to the area under the absorption pulse), is used for the evaluation of transient signals, as they are typical for graphite furnace atomic absorption spectrometry (GF AAS). Whereas the absorbance is a dimensionless measure, the integrated absorbance has the dimension seconds.

None of the techniques used for atomic spectroscopy is an absolute technique, which means that, for quantitative determinations, all of them require calibration using appropriate *calibration standards*. These could be, in an ideal case, aqueous solutions with a known content of the analyte element, or, in the case of severe matrix effects, have to be reference materials with analyte content and matrix composition close to that of the test sample to be analyzed. A *calibration function* that describes the mathematical dependence of the measured value on the concentration or mass of the analyte has to be established for quantitative determination. The slope of the calibration function is termed *sensitivity*.

Table 13.2 shows the different measurement units, their dimension and the resulting units for the sensitivity of different atomic spectrometric techniques.

Tab. 13.2 Examples for the dimension of sensitivity in different atomic spectrometric techniques.

Technique	Measured quantity	Dimension	Concentration or mass[a]	Sensitivity
OES	Intensity	I	s^{-1}	$mg \cdot L^{-1}$	$L \cdot mg^{-1} \cdot s^{-1}$
FAAS	Absorbance	A	–	$mg \cdot L^{-1}$	$L \cdot mg^{-1}$
GF AAS	Integrated absorbance	A_{int}	s	ng	$s \cdot ng^{-1}$

[a] Other concentration or mass units might be used as well.

This comparison clearly demonstrates that "sensitivity" is not an appropriate parameter to compare the analytical performance of different systems due to the entirely different definitions. In addition, sensitivity is only one aspect, but noise is at least as important. Hence, the comparison of different systems should be on the basis of the signal-to-noise (S/N) ratio exclusively. Appropriate measures of the S/N ratio are the limit of detection (LOD), and the limit of quantification (LOQ), which are usually defined as 3 times and 10 times, respectively, the standard deviation of a blank, divided by the slope of the calibration function. Obviously, LOD and LOQ should be compared for real samples, not only for matrix-free calibration solutions. Another important parameter for OES is the signal-to-background ratio, and the figure for comparison is usually the background equivalent concentration (BEC), that is, the analyte concentration that gives the same signal as the background.

The measurement of the emission intensity, particularly in OES, is subject to relatively great day-to-day variations due to minor changes in the temperature of the excitation source and in detector sensitivity. The measurement of the absorbance in AAS, in contrast, is relatively independent of such fluctuations, firstly, as the number of atoms in the ground state is essentially independent on minor temperature changes (refer to Section 13.3.1.1) and the intensity of the radiation from the primary source has no influence on the sensitivity either, as the absorbance only depends on the ratio $\Phi_0(\lambda)/\Phi_{tr}(\lambda)$. Measurement quantities, such as the characteristic concentration, c_0, the concentration of an analyte that gives an absorbance reading of $A = 0.0044$ in FAAS, and the characteristic mass, m_0, the mass of an analyte that gives an integrated absorbance reading of $A_{int} = 0.0044$ seconds in GF AAS may therefore be used to test the proper function of the equipment and of analytical procedures on a day-to-day basis.

13.3.2
Optical Emission Spectrometry

OES, often referred to as atomic emission spectrometry (AES) – an acronym the use of which has been discouraged by IUPAC because it is reserved for Auger electron spectroscopy (AES) – is the oldest of the atomic spectrometric techniques. A great deal of research has been carried out in this area in the late nineteenth and the early twentieth century, including the design of ever more advanced mono- and polychromators. Modern quantitative methods of spectrochemical analysis using OES, however,

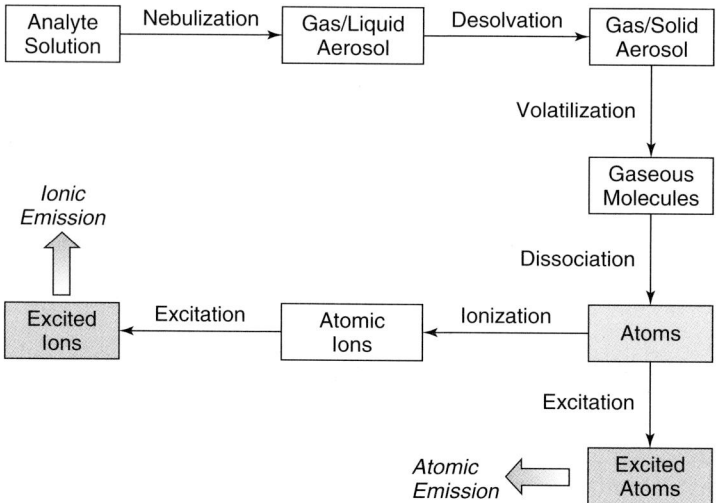

Fig. 13.3 The process of atomization and excitation for a liquid/dissolved sample.

did not begin until about 1925, when the first commercial optical equipment began to be manufactured in Europe and the United States. A photographic plate was used as detector in these instruments, a practice that continued for about half a century.

An optical emission spectrometer consists basically of an atomization and excitation source with a sample introduction system, a wavelength-dispersing system, such as a monochromator or polychromator, and a detector, amplifier, and readout system. The kind of samples that can be analyzed and the elements that can be determined depend mostly on the excitation source.

13.3.2.1 Excitation Sources

The role of the excitation source in OES is to transfer the sample into the gaseous state, to dissociate molecular species, and to provide sufficient energy to excite the analyte atoms, promoting the emission of photons that will later be detected.

A liquid sample, when introduced into the excitation source, is submitted to several steps that are shown schematically in Figure 13.3. Initially, the sample aerosol, which is formed by a nebulizer, is desolvated, that is, the solvent is evaporated. The desolvated particles then are vaporized, that is, converted to gaseous species, and finally dissociated in the atomization process. The atoms formed in this process are in equilibrium with their respective excited (neutral) and ionic species, which may also undergo excitation (Boss and Freeden, 1999). In the cases of direct solid sample and gas analysis, some of the above stages might be omitted or altered.

Electrical Arcs and Sparks Many analytical applications of OES produced their spectra by arc or spark excitation techniques. The historical development in this area is most difficult to document since almost from the start, after observations of the sun's spectrum, the attempt was to use high-energy sources. The first commercial instruments for OES that have been introduced in the first half of the twentieth century also used electrical arcs and sparks for the direct vaporization and excitation of solid samples. This kind of equipment is the working horse in most metallurgical laboratories even today. An *arc* is an electrical discharge that takes place between two or more conducting electrodes. The sample usually stands as one of the electrodes, either as a powder, a solid mixture, or a solution residue.

One of the most common arc types used in OES is the direct current (*dc*) *arc*. The dc arc consists of a continuous discharge of 1–30 A between a pair of metal or graphite electrodes, one of which contains the sample. The dc arc provides good detection power and generates relatively high amounts of neutral (excited) atoms, although it is also characterized by relatively poor precision and by an effect denominated *selective volatilization*, which occurs because the electrodes are only gradually heated by the arc. The consequence is that the most volatile materials vaporize first, which may result in an erroneously high result for volatile analytes. A *spectrochemical buffer* is frequently used to reduce or control the effects of selective volatilization; substances that are typically employed include graphite powder or alkali and alkaline–earth compounds (Ingle and Crouch, 1988; Mika and Török, 1974; Zhou, Zhou and Hou, 2005).

The alternating current (*ac*) *arcs* may be either operated at high (2–4 kV) or low (100–400 V) voltages. Unlike the dc arc, the ac arc is extinguished at the end of each half-cycle, which allows the sampling of different portions of the electrode at each cycle. Compared to dc arcs, ac arcs lead to better precision and more efficient sample volatilization, whereas the sensitivity is usually lower and the analysis of refractory materials may become more difficult (Ingle and Crouch, 1988).

Sparks are intermittent electrical discharges of a few microseconds that occur under application of high electrical potentials. Spark sampling usually consists of two stages: firstly, a low-energy discharge is produced to ionize the noble gas (usually Ar) in the sampling chamber; later, after the conducting plasma has been generated, a high-energy discharge (spark) is induced, causing vaporization of the sample in the location where the spark impacts, where temperatures as high as 50 000 K can be reached (Ingle and Crouch, 1988; Mika and Török, 1974). Spark sources are characterized by relatively high precision, as they are able to vaporize several distinct portions of the sample electrode in successive discharges. Spark-source OES instruments are, however, usually less sensitive than arc and discharge instruments, mostly because the high-peak currents and high-power density in sparks lead to a high degree of single and even multiple ionization of analyte species. In addition, the mass of sample that is vaporized in each cycle is relatively low. Despite those drawbacks, spark sources have been successfully applied to the analysis of samples, such as environmental, geological, and biological materials, but the main focus has been, and is, the analysis of metal alloys (Zhou, Zhou and Hou, 2005).

One of the limitations of arc and spark OES is the need of reference materials with a matrix composition and analyte content as close as possible to the sample for calibration due to the significant matrix effects encountered with these techniques. This requirement might increase the cost of analysis significantly.

Glow Discharges Glow discharges (GDs) have been used in OES since the beginning of the twentieth century. Usually, the sample to be analyzed was filled into a hollow cathode, which was placed inside a quartz tube through which an inert gas (usually Ar) was pumped at low pressure. A voltage potential was then applied between the anode (usually made of graphite) and the hollow cathode, causing electrons to migrate from the cathode to the anode. In their path they collide with Ar atoms, generating positively charged metastable argon species, which are accelerated to the sample holder, held at negative potential. Collision with the sample surface will result in the ejection of atoms and electrons from the sample, a process called *sputtering*. Sputtered atoms collide with metastable Ar atoms or high-energy electrons, and are excited. With relaxation, the photons emitted from the excited species produce a small plasma (GD), which is used for analytical purposes (Pisonero *et al.*, 2006; Angeli *et al.*, 2003; Jakubowski *et al.*, 2007). In the second half of the twentieth century glow discharge optical emission spectrometry (GD OES) became quite popular after Grimm (1968) had proposed several modifications of the setup. He used a flat cathode (the sample) and a ring anode, as shown in Figure 13.4, which significantly facilitated the preparation and change of samples. GD OES is ideally suited for surface analysis and depth profiling, as it produces a crater with a flat bottom, which is in contrast to other ablation techniques. Some typical applications of GD OES include the analysis of thin films (Pisonero *et al.*, 2006; Angeli *et al.*, 2003) and metal alloys (Galindo and Albella, 2008).

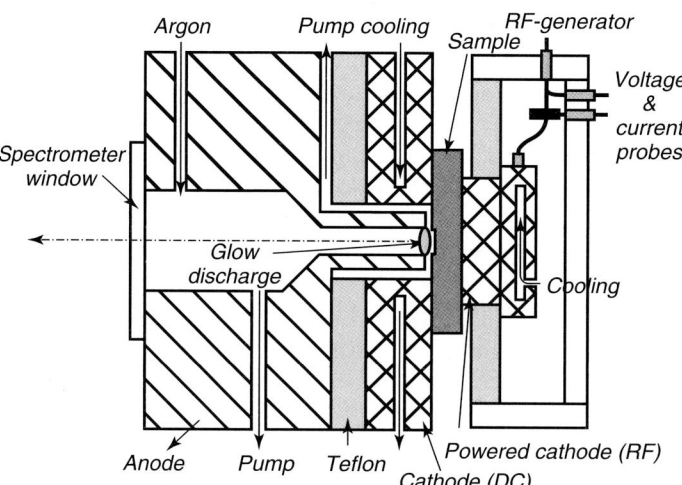

Fig. 13.4 Schematic design of a Grimm discharge lamp for GD OES. (From Hodoroba *et al.* (2001), p. 31; reproduced with permission from Elsevier.)

Flames The simplest excitation source is a *flame*, a mixture of at least two gases, a fuel and an oxidant, and the first burner designed for atomic spectroscopy was the burner of Bunsen. The work that gave principal impetus to flame excitation was that of Lundegårdh (1936). In his method, the sample solution was sprayed from a nebulizer into a condensing chamber and then into an air–acetylene flame. Since spectra produced by flames exhibit significantly less lines compared to those produced by arc and spark emission, simple devices for spectral isolation could be used. Barnes *et al.* (1945) reported a simple filter photometer for alkali metal determinations; instruments of this kind are still in use nowadays, mainly in clinical laboratories, and the technique is often referred to as *flame photometry*.

Flame OES is limited to the analysis of liquid or dissolved/digested samples, and a nebulizer is typically required for sample introduction into the flame. The nebulizer-burner unit might be similar to the design used for FAAS (refer to Section 13.3.3.1, Flame Atomizers) or it might be of the *total consumption* design that makes possible to use oxygen as the oxidant and hence to achieve higher flame temperatures. Among the advantages of flames as excitation sources are their stability and the relatively simple and inexpensive operation. However, the temperatures of flames are, in most cases, not sufficient to promote the excitation of a significant fraction of analyte atoms, as can also be deduced from the data in Table 13.1 (Ingle and Crouch, 1988). The use of flames for OES is restricted to a few isolated applications nowadays.

Plasma Sources One of the most widely used excitation sources in OES nowadays is the *inductively coupled plasma* (ICP), which has been investigated independently by Greenfield, Jones and Berry (1964) in the UK and Wendt and Fassel (1965) in the USA. The ICP is a partially ionized gas (usually argon) that is formed from an electrical discharge inside a quartz torch and maintained by an induction coil to which a radiofrequency (RF) power is applied. The schematic design of an ICP torch, which consists of three concentric cylinders, is shown in Figure 13.5. The main gas flow (usually at flow rates above $15\,l\,min^{-1}$), which is the main flow responsible for plasma formation, is in the outer cylinder. A secondary gas flow (usually at flow rates of $1-1.5\,l\,min^{-1}$) runs between the inner and the middle cylinder and is responsible for cooling the torch and maintaining a distance between the plasma and the torch walls. The *nebulizer gas* flows inside the inner cylinder, being responsible for the transport of the sample aerosol to the plasma itself. To initiate the plasma, a discharge takes place, generating Ar ions and electrons. An RF power is then applied to the metallic load coil located in the front part of the plasma torch, originating a magnetic field that maintains the electrons on the torch, promoting their collision with other neutral Ar atoms. The collisions generate more electrons and ionized gas species in a chain reaction, and the otherwise neutral gas becomes a plasma, constituted by neutral atoms, ions, and electrons. Owing to the high temperatures that can be attained in this system (up to about 10 000 K, depending on the plasma region), basically all elements can be excited and even ionized to a great extent in an ICP. The same ICP source can be used with a

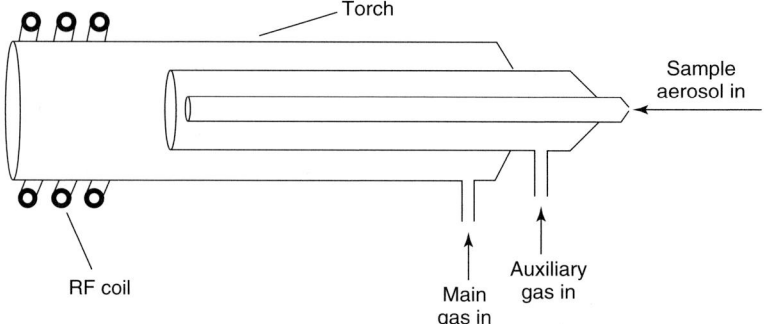

Fig. 13.5 Schematic design of an ICP torch; for details see text.

mass spectrometer, in a technique called *inductively coupled plasma mass spectrometry, (ICP-MS)* (Montaser and Golightly, 1992; Becker, 2007).

In inductively coupled plasma optical emission spectrometry (ICP OES), liquid samples are used almost exclusively. They are introduced into the plasma using a pneumatic nebulizer and a spray chamber in order to obtain an aerosol with very small droplet size. A wide variety of different nebulizers and spray chambers have been proposed for ICP OES (Nölte, 2003), but they are not discussed in detail here. The most commonly used one is the Meinhard nebulizer, a concentric nebulizer of very simple design, in combination with a Scott spray chamber, as shown in Figure 13.6. Another frequently used nebulizer is the *cross-flow* nebulizer, where a capillary with a high-pressure gas flow is positioned at a 90° angle with the capillary tip through which the

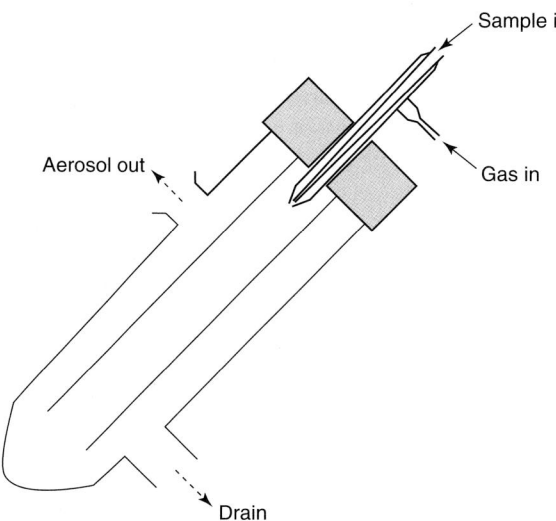

Fig. 13.6 Schematic design of a Meinhard concentric nebulizer in a Scott spray chamber for ICP OES.

sample flows. Both nebulizers have an uptake rate of about 2 ml min^{-1} and are typically operated in combination with a peristaltic pump for sample introduction in order to reduce the effect of viscosity of the solution on the flow rate. As a consequence of the requirement for very low volume droplets in order to maintain plasma stability, the sample introduction efficiency is around 2%, so that most of the aspirated sample is discarded through the drain (Boss and Freeden, 1999).

There are basically two different configurations possible for ICP OES, the *radially viewed plasma* (or side-on), where the ICP is positioned at a 90° angle with the detector, and the *axially viewed plasma* (or end-on), where the ICP and the detector are assembled in the same optical axis. The latter configuration provides higher sensitivity due to the longer path length that is exposed to the detector, although the risk of interference with this configuration is considerably higher (Boss and Freeden, 1999). The major problem in ICP OES, similar to arc and spark OES, is the extremely great number of emission lines that is produced at the high plasma temperatures, which obviously increases the risk of spectral interferences proportionally. Modern simultaneous ICP OES instruments offer computer programs to correct for spectral interferences, measuring the interferent emission at a suitable line, calculating and subtracting its contribution to the emission at the analytical line using a least-squares algorithm.

An alternative for sample introduction in ICP OES is the use of a graphite furnace specially designed to carry the sample aerosol from the graphite tube to the plasma source. This technique is known as *electrothermal vaporization* (ETV), and works under the same basic principles as the graphite furnace technique used in AAS (refer to Section 13.3.3.1, Electrothermal Atomizers). Among its advantages are the higher transport efficiency and the greater tolerance to matrix components, allowing even the analysis of solid samples directly or prepared as slurries. Most ICP OES instruments, however, have difficulties to handle transient signals, which are generated by ETV. As a consequence, the use of ETV devices became more popular in ICP-MS.

Another technique that is occasionally used in ICP OES is hydride generation (refer to Section 13.3.3.1, Atomizers for Chemical Vapor Generation) in order to separate the analyte from the matrix and introduce it into the plasma in the form of a gaseous compound. In this case, the plasma conditions have to be adapted in order to tolerate the hydrogen that is developed during the chemical reaction.

Other plasma sources have also been investigated as excitation sources in OES. The *direct current plasma* (DCP) was the first plasma source described and commercialized. In this system, the plasma is generated when a direct current is established between two fixed electrodes, and the sample aerosol is carried to the plasma region by an argon flow. The *microwave-induced plasma* (MIP) is also an important plasma source. In this case, the plasma is generated by application of microwave power, generated by a magnetron, to a cavity with an inert gas, usually Ar or He (Rosenkranz and Bettmer, 2000). Microplasmas with dimensions of 1 mm or less, usually microwave-induced, generated by high frequency or dc discharges, are a promising field of research in OES, and have been experiencing increasing popularity in the past few years (Foest, Schmidt and Becker, 2006; Broekaert and Siemens, 2004).

Lasers The most recent excitation sources in OES are lasers used in *laser-induced breakdown spectroscopy* (LIBS). In a very simplified definition, Sallé, Mauchien and Maurice (2007) described LIBS as a technique based on the focusing of high-power laser pulses onto a sample surface leading to the generation of a plasma composed of excited species, which emit radiation. Nd : YAG lasers are typically used at 1064 nm, but the results obtained with UV lasers are not significantly different (Fantoni et al., 2008). The technique uses short-pulsed lasers (in the nanosecond range) to induce the dielectric breakdown of a gas located at the laser focal point. The microplasma that is generated as a consequence can reach temperatures as high as 20 000 K, which is sufficient to break chemical bonds present in the sample and to produce excited atoms and ions. The results obtained using LIBS depend on several factors, which include the surrounding gas, sample matrix, and laser parameters, such as wavelength, power density, and pulsing frequency (Rusak et al., 1998; Kennedy, Hammer and Rockwell, 1997). The applications of LIBS for qualitative and quantitative chemical analysis range from depth profile studies (Das et al., 2008) to the analysis of paints (Kim et al., 2007) and meteorites (De Giacomo et al., 2007). Fantoni et al. (2008) published a review article about semiquantitative and quantitative analysis using LIBS. Portable LIBS spectrometers have also been described and applied successfully to a number of elements (Harmon et al., 2005; Mohamed, 2008; Goujon et al., 2008).

13.3.2.2 Spectrometers

A spectrometer is an apparatus for the conduction and dispersion of the emitted radiation and a system for the measurement of spectral radiation quantities. We distinguish between *sequential spectrometers*, where the radiation of different wavelengths is measured one after the other, and *simultaneous spectrometers*, where the radiation of different wavelengths is measured at the same time, using more than one detector or a single space-resolving detector. The sequential spectrometer uses a *monochromator*; the simultaneous spectrometer uses a *polychromator*. Lenses and mirrors are used for the conduction and collimation of the radiation; prisms and gratings are used for their dispersion. Monochromators have one entrance and one exit slit; polychromators have one entrance slit and no exit slit in case a space-resolving detector is used, or multiple exit slits in case where more than one detector is used. Because of the great number of lines that are emitted in OES, particularly from high-temperature excitation sources, the quality of the spectrometer and the relative freedom from spectral interferences is directly related to its resolving power.

The most popular mounting used for monochromators in OES is the Czerny–Turner mounting, which is shown in Figure 13.7; the Czerny–Turner mounting offers the best correction for coma and astigmatism, and hence the lowest stray light level. Monochromator-based instruments are usually compact, show greater spectral flexibility and allow background correction to be performed more accurately compared to polychromators.

In the case of *polychromators*, we have to distinguish between three different

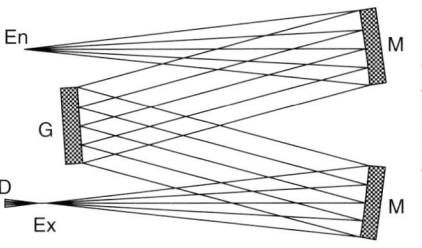

Fig. 13.7 Schematic design of a monochromator in Czerny–Turner mounting with flat grating. (From Welz and Sperling (1999), p. 62; reproduced with permission from Wiley-VCH.)

designs, the classical spectrograph, the conventional prism or grating polychromator, and the echelle grating polychromator. The *spectrograph* used a prism or a grating as the dispersive element, usually in Littrow, Ebert, or Czerny–Turner mounting, and it often had a focal length of several meters in order to obtain high resolution. It had no exit slit and a photographic plate or film as the detector, which recorded the entire spectral range of interest. Spectrographs have played a decisive role in research and routine analysis up to the 1970s, but are no longer in use because of their complexity and the time-consuming evaluation of the photographic plates.

In *conventional polychromators* with a prism or a grating and multiple exit slits, each exit slit is assigned to a specific wavelength (element) and equipped with its own detector, hence making simultaneous multielement determination possible. One of the most popular configurations used for polychromators in OES is the Paschen–Runge mount assembled in a *Rowland circle*, as shown in Figure 13.8. In this system, radiation from the excitation source is conducted toward a concave diffraction grating, and the resolved spectrum is directed to several exit slits located on the opposite side of the circle. An individual detector is associated to each exit slit. Conventional polychromator-based instruments usually have higher sample throughput, although there are restrictions with regard to the number of elements that can be determined, which have to be specified when the equipment is purchased. There are also limitations with respect to background correction, as the lines to be used for correction have to be defined in advance as well, which increases the risk of spectral interferences (Broekaert, 2005; Boss and Freeden, 1999). Simultaneous spectrometers with Rowland circle mounting and a series of space-resolving detectors instead of PMT obviously do not show the above limitations.

Many modern ICP OES instruments use an *echelle grating polychromator* for wavelength dispersion. Echelle gratings are used because of their much higher resolving power compared to conventional holographic gratings. According to the general grating formula

$$R = n\,O \tag{13.4}$$

where R is the resolution, n the number of grating grooves that are illuminated, and O is the grating order. The resolution of conventional holographic gratings, which are generally used in the first order, only depends on the ruling density and the grating size. Echelle gratings, in contrast, exhibit maximum intensity for grating orders around 100. This means that, although their ruling density is typically an order of magnitude lower than that of holographic gratings, their

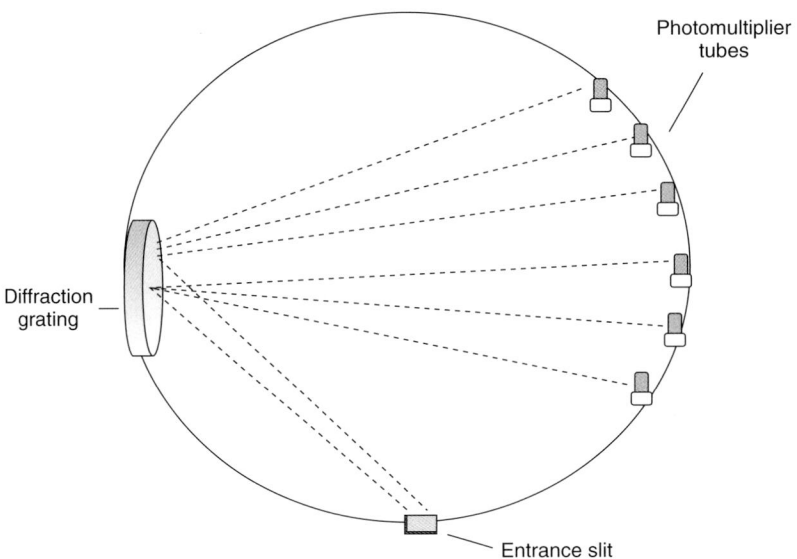

Fig. 13.8 Schematic design of a polychromator in Paschen–Runge mounting, assembled in a Rowland circle with concave grating.

resolution is still an order of magnitude higher. In this case, a second dispersive system, either a diffraction grating or a prism is used and positioned perpendicularly to the echelle grating. The echelle grating separates the polychromatic radiation by wavelength and overlapping spectral orders that are later separated or *cross dispersed* by the second optical component, generating a two-dimensional pattern. Detection of the bi-dimensional spectrum is usually achieved by a solid-state detector, such as a charge-coupled device (CCD) or charge-injection device (CID). A schematic representation of an echelle spectrometer is shown in Figure 13.9.

13.3.2.3 Detectors

The intensity of the radiation emitted by excited atoms or ions and separated by the wavelength-dispersing device is measured using a detector. For about half a century, photographic plates and films were used for that purpose in OES; they were increasingly replaced by PMTs in the second half of the twentieth century. In a PMT, an anode, a radiation-sensitive electrode (known as the *photocathode*) and a series of emission cathodes (known as *dynodes*), typically 10–12, are assembled and held at an increasingly positive potential. Radiation that passes the exit slit reaches the photocathode, which will emit electrons upon the impact of photons. The emitted electrons will be accelerated to the first dynode, which will release a greater number of secondary electrons that will be accelerated to the next dynode, where the process is repeated, resulting in a cascade effect. The multiplication of electrons along a PMT depends on the amplitude of the high voltage applied. At the end of the process, an electrical current is generated related to the number of photons that impacted the photocathode.

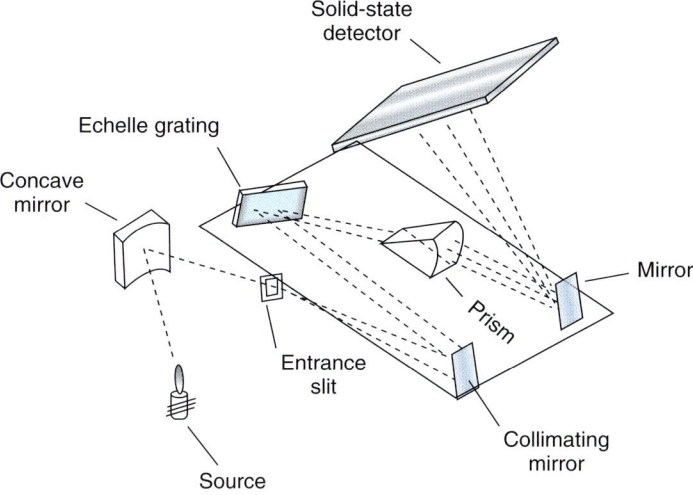

Fig. 13.9 Schematic design of an echelle spectrometer.

Modern echelle grating polychromators are equipped with solid-state detectors, such as CCDs or CIDs, which are based on the light-sensitive properties of solid silicon. In a CCD, the impact of photons generates electrons that are collected and accumulated by positively biased capacitors. The charge packets are transferred in discrete time increments by the controlled movement of potential wells to preamplifier and to the readout. In a linear CCD, the charge is moved in a stepwise fashion from element to element and is detected at the end of the line. A two-dimensional array CCD consists of an assembly of interconnected linear CCDs. The on-chip summing of charges in adjacent pixels along rows or columns is called *binning* (Laqua *et al.*, 1995). In the CID, photons striking the silicon wafer generate positive charges, and these charges are stored below negatively biased capacitor plates. The amount of stored charge is proportional to the intensity of the radiation and to the integration time. When a pixel is selected for readout, the row capacitor associated to it is maintained at a ground potential (zero volts), resulting in the injection of charge into the substrate and in an output signal (Ingle and Crouch, 1988).

13.3.3
Atomic Absorption Spectrometry

After the effectiveness of AAS for atomic spectrometric investigations had been demonstrated by Kirchhoff and Bunsen, the technique was used by other researchers as a qualitative method. Wood (1905) performed the definitive experiments that elucidated atomic absorption by resonance in gases. The absorption of the radiation from a mercury discharge lamp by mercury vapor, which appears as shadow on a screen, became a common experiment in physics lectures in colleges. Woodson (1939) applied the procedure to a quantitative method for the determination of mercury – the first application of atomic absorption to quantitative measurement. Besides these isolated examples, however, OES was preferred as an analytical technique during the first half of the twentieth century. This was to a great extent due to the fact that continuum radiation sources, as they have been used in the nineteenth century, were not really suited for AAS measurements. Particularly with a photographic plate as a detector, it was much easier to detect and measure a small emission, that is, a small increase in the opacity of the photographic layer than a small reduction of a strong emission over a narrow spectral range of a few picometers.

A turning point came in 1955 when Walsh reintroduced AAS to chemical analysis. He recognized the importance of a high-intensity, narrow linewidth source for high sensitivity and suggested the use of hollow cathode lamps (HCL). This way the high resolution that is required for atomic absorption measurements was provided by the line radiation source; in this case, a monochromator of medium resolution could be used, which only had to separate the analytical line from other lines emitted by the source. Walsh also proposed to modulate the HCL radiation and to use an amplifier tuned to the same modulation frequency (Russell, Shelton and Walsh, 1957), so that measurements could be made without interference from flame emission. This concept made AAS highly specific and selective for the element to be determined, but at the same time it limited the technique to the determination of one element at a time. The use of continuum radiation sources for AAS measurement could only be realized toward the end of the twentieth century due to the significant progress in the technology of components and in the design of spectrometers (Harnly, 1999; Heitmann et al., 1996), as is discussed in Section 13.3.3.3.

An atomic absorption spectrometer consists of an atomizer with a sample introduction system and a spectrometer. The latter consists of a radiation source, a monochromator, and a detector, and it usually also contains a device for background correction. As there are fundamental differences between spectrometers for line source and continuum source AAS, these two spectrometer types are treated in separate sections.

13.3.3.1 Atomizers

The role of an atomizer in AAS, similar to the excitation source in OES (Section 13.3.2.1), is to transfer the sample into the gaseous state and to dissociate molecular species into ground-state atoms. Ionization of the analyte atoms has to be avoided, as ionic species have a different absorption spectrum and are therefore not available for atomic absorption measurement.

Flame Atomizers Walsh (1955) made use of a premixed, laminar-flow *air–acetylene flame*, similar to that of a Lundegårdh burner to produce atoms in the ground state. This flame, which has a temperature between 2200 and 2300 °C, and can be operated under reducing (fuel-rich) as well as oxidizing (fuel-lean) conditions, allowed the sensitive determination of some 35 elements. It still is the most popular flame used in AAS; however, it is not suitable for the determination of refractory elements and those with high affinity to oxygen. The introduction of the hot (∼2700 °C) and reducing *nitrous oxide–acetylene flame*

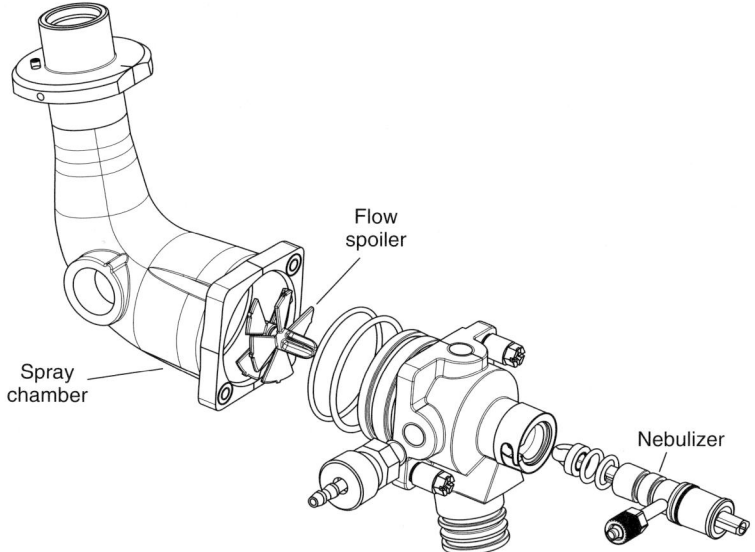

Fig. 13.10 Schematic design of a premix burner for FAAS with concrntric nebulizer, mixing chamber and impact devices. (With kind permission of Analytik Jena AG, Jena, Germany.)

(Willis, 1965) brought the solution for this limitation and added another 30 elements to the list of those that could be determined by AAS. Other gas combinations, such as the *air–hydrogen flame*, found only limited interest for special applications. Flames with oxygen as the oxidant cannot be used in a premix burner due to their high burning velocity; total consumption burners that have reached some popularity in OES have been found unsuitable for AAS (Welz and Sperling, 1999). Next to the temperature it is mostly the chemical environment, that is, the components of the flame gases, that plays a decisive role in the atomization process.

Samples are usually aspirated by a concentric *pneumatic nebulizer*, using the oxidant as the nebulizer gas; the uptake rate of these nebulizers is typically 5–10 ml min^{-1}. The generated aerosol is introduced into a mixing chamber, where it is mixed with the fuel gas and additional oxidant (Figure 13.10). The mixing chamber usually also contains impact devices, such as impact beads and flow spoilers, for postnebulization and separation of big droplets. The ''conditioned'' aerosol that finally reaches the flame typically has a maximum droplet size of <10 μm; only about 5% of the aspirated sample finally reaches the flame. Nevertheless, the amount of sample that is introduced into a flame in AAS is about an order of magnitude greater than that in ICP OES (refer to Section 13.3.2.1, Plasma Sources).

In the flame, the aerosol is submitted to several successive processes, including *desolvation* (evaporation of the solvent), *solute volatilization* (transfer of the dry particles into the gas phase), and *dissociation* of the molecules into atoms (refer to Section 13.3.2.1). The completeness of these three processes is decisive for the absence of nonspectral interferences; however, as the process of dissociation is an equilibrium process, it is sufficient that the same degree of dissociation is reached for the analyte in the sample and in the calibration solution to obtain freedom from interference. Because of its simplicity and freedom from spectral interferences, flame AAS was, to an extent, replacing flame emission spectrometry in the market, except for the flame photometers that were used mostly in clinical laboratories.

The major sources of interferences in flames are transport interferences due to differences in viscosity, surface tension, and so on, between the solution, which result in different aspiration rates and nebulization properties, as well as dissociation and ionization interferences. The former ones can be controlled using the analyte addition technique for calibration, whereas the latter ones can be removed by adding appropriate buffers or releasing reagents (Welz and Sperling, 1999).

Electrothermal Atomizers The term *electrothermal atomization (ETA)* refers to an atomizer that is made of a material of relatively high specific resistance, which is heated by application of a high current. Several ET atomizers have been described, differing mainly in the design and in the material from which the atomizer was made. Nowadays, essentially all of the commercially available ET atomizers are made of graphite, although other devices, such as tungsten coils, have also been used successfully (Berndt and Schaldach, 1988; Silva et al., 1996). For simplicity reasons, graphite tube atomizers are discussed exclusively in this section. The GF atomizer has originally been proposed by L'vov (1961) and later been simplified by Massmann (1968); the first commercial atomizer of this type was introduced in 1970. The main advantages of this atomizer are that the entire sample portion that is introduced into the graphite tube is atomized typically within 1–2 seconds and the analyte atoms remain in the absorption volume much longer, resulting in an increase in sensitivity of 2–3 orders of magnitude, compared to flame atomizers (Welz and Sperling, 1999). Another important feature of this atomizer is that not only liquid and dissolved samples can be analyzed but also gaseous and solid samples directly (Welz et al. 2007b), resulting in further increase in sensitivity and greatly simplified sample preparation.

A GF atomizer consists of a cylindrical graphite tube, typically 20–25 mm long with a diameter of about 5 mm. A small orifice is usually present, through which the (liquid) sample is introduced. The tube is positioned between two cooled graphite electrodes, and acts as a resistance, reaching temperatures as high as 2700 °C. An inert gas (argon) constantly flows around the tube, protecting it from contact with the outside air and, consequently, from burning at high temperatures. An additional internal gas flow is used to eliminate the solvent and matrix vapors during thermal pretreatment of the sample. A *temperature program* is usually applied, consisting of four essential stages: *drying*, where a sufficiently high temperature is applied to eliminate the solvent; *pyrolysis*, where the intention is to eliminate the matrix components without vaporizing the analyte; *atomization*, where the temperature is increased rapidly in order to promote the formation of the maximum concentration of analyte atoms in the absorption volume. The internal gas flow is normally interrupted in this stage to increase the residence time of analyte atoms in the absorption volume. The last stage is a *cleaning* stage, where all matrix and analyte residues are eliminated prior to the introduction of the next sample.

While the temperature, heating ramp, and hold time of all stages of the temperature program require optimization, the two most critical parameters are pyrolysis and atomization. Firstly, a *pyrolysis curve* is established plotting the integrated absorbance signals obtained at a set atomization temperature against the pyrolysis temperature, which is increased in increments of typically 100 °C. This curve usually exhibits a plateau up to a maximum temperature above which the signal begins to drop, indicating analyte losses in the pyrolysis stage as shown in Figure 13.11(a). Secondly, using the maximum pyrolysis temperature at which no analyte losses occur, the atomization temperature is increased gradually, and at least for the more volatile elements a maximum is obtained, as shown in Figure 13.11(b).

During the first decade after its introduction, GF AAS was considered a technique that is subject to numerous difficult-to-control interferences, which was mostly due to the fact that it was largely unknown how to use the technique properly. Slavin, Manning and Carnrick (1981), based on research of L'vov (1978), introduced the stabilized temperature platform furnace (STPF) concept, which was the key to essentially interference-free GF AAS analysis. The most important components of the concept were (i) atomization of the sample from a graphite *platform* inserted into the graphite (instead of from the tube wall) in order to delay atomization until the tube has reached its final temperature; (ii) integration over the *peak area* of the transient signal (instead of using peak height) in order to compensate for matrix influences on the kinetics of atom release; (iii) use of *chemical modifiers* in order to avoid low-temperature analyte losses due to matrix components. The last topic has been treated in a large number of publications (Tsalev, Slaveykova and Mandjukov, 1990); a mixture of Pd and Mg nitrates, co-injected with the sample, was proposed as a *universal chemical modifier* (Schlemmer and Welz, 1986), and its use has been demonstrated for over 20 elements (Welz, Schlemmer and Mudakavi, 1992). In the meantime, *permanent chemical modifiers* are investigated increasingly, where noble metals, such as Pd, Ir, Ru, or Rh, and carbide-forming elements, such as W, Zr, Th, and Ta, or mixtures of both, are used as a kind of surface coating of the graphite platform to stabilize the analyte for a large number of consecutive determinations.

One aspect that was not included in the original STPF concept, but which turned out later to be another very decisive parameter, was the spatial isothermality of the graphite atomizer. In the original concept of Massmann, which has been adopted by all GF manufacturers in the following years, the graphite tube was contacted at its ends and heated longitudinally. As the contacts were cooled, this resulted in a significant temperature gradient along the graphite tube, which was the reason for condensation of matrix and analyte residues on the cool ends, causing severe background absorption and memory effects. *Transversally heated atomizers* solved this problem and should

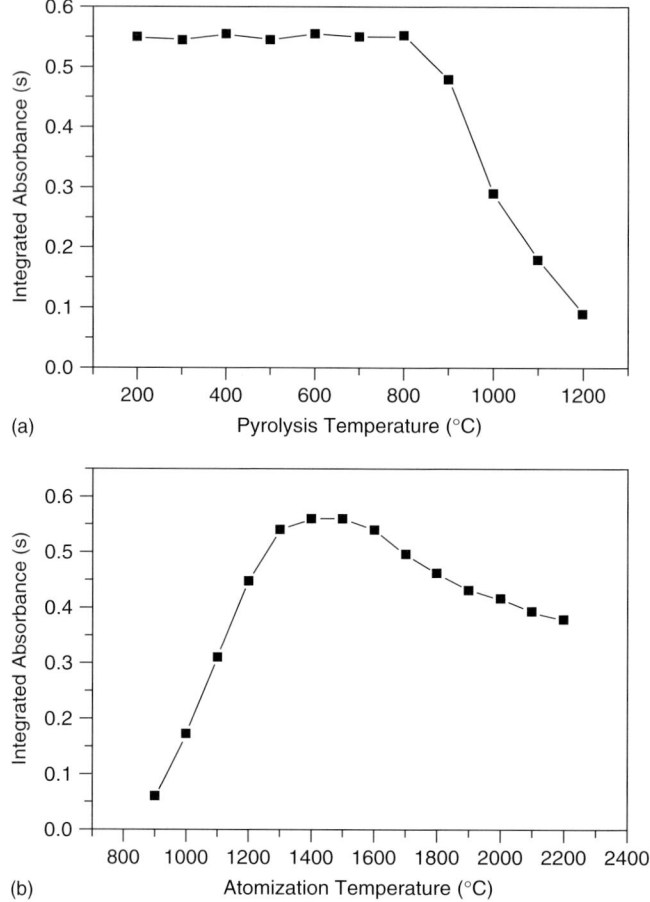

Fig. 13.11 Typical pyrolysis and atomization curves for GF AAS; (a) pyrolysis curve and (b) atomization curve; for details see text.

be preferred over longitudinally heated ones. The schematic design of such an atomizer tube is shown in Figure 13.12.

Atomizers for Chemical Vapor Generation
Chemical vapor generation (CVG) is actually a general term for several sample preparation and introduction techniques, where the analyte is converted into a gaseous compound using a chemical reaction. This gaseous compound may then be introduced into different atomizers for atomization and measurement of the absorbance. The two most important techniques are the cold vapor (CV) technique for the determination of mercury and the hydride generation (HG) technique for the determination of elements that form volatile hydrides, mainly As, Bi, Sb, Se, Sn, and Te. In the former technique, mercury (II) is reduced to the element using stannous chloride according to

$$Hg^{++} + Sn^{++} \rightarrow Hg^0 + Sn^{4+} \quad (13.5)$$

Fig. 13.12 Transversally heated graphite tube for GF AAS. (With kind permission of Analytik Jena AG, Jena, Germany.)

Other reducing agents, such as sodium tetrahydroborate (THB), may be used as well. As mercury is reduced to the element in this reaction, and as mercury has a significant vapor pressure already at room temperature, cold vapor atomic absorption spectrometry (CV AAS) does not need an atomizer. The mercury vapor can simply be transferred to an absorption cell using a flow of gas, where the absorbance can be measured, a procedure that has been described by Hatch and Ott (1968). Alternately, it can be collected by amalgamation on a precious metal, such as gold gauze, and liberated by heating in order to increase the sensitivity. Another way of preconcentrating mercury is in a graphite tube treated with a permanent modifier and atomization using GF AAS. This preconcentration also eliminates interferences due to different release kinetics of the vapor from the solution.

The formation of volatile hydrides of arsenic and a few other elements upon reaction with zinc in acid solution is known as the *Marsh and Gutzeit tests* since the nineteenth century. Holak (1969) proposed this technique for the determination of As by AAS. A major breakthrough came with the introduction of sodium THB as the reducing agent (Braman, Justen and Forebank, 1972), which increased the number of elements that could be determined to eight. The mechanism of hydride formation was, for a long time, believed to be due to "nascent hydrogen," a mechanism that has now been ruled out (D'Ulivo *et al.*, 2004a; D'Ulivo, Onor and Pitzalis, 2004b) In fact, it is due to direct interaction between the analyte and THB, resulting in hydroborane intermediates that will later decompose to generate the volatile species. Another recent advance in this technique is that the number of elements that form volatile compounds with THB has increased to more than 15 (Guo *et al.*, 2004; Sturgeon, Guo and Mester, 2005), although not all of these reactions might be of practical analytical importance.

Initially, the generated hydrides were introduced into a flame for atomization, which, however, did not result in sufficiently high sensitivity. For about two decades, the heated quartz tube atomizer (QTA), an electrically or flame-heated quartz tube has been the preferred atomizer for hydride generation atomic absorption spectrometry (HG AAS), mostly because of its high sensitivity. It took some time until the mechanism of atomization in the QTA was understood, as thermal dissociation could be excluded as the mechanism, because atoms of As, Se, and so on, could not exist at temperatures $<1000\,°C$ thermodynamically. Later, it has been shown that atomization occurs in a cloud of hydrogen radicals that is formed in the entrance part of the T-tube in a reaction between hydrogen that is generated in the reaction of the acid sample solution with THB and traces of oxygen (Dědina, 1988).

The QTA has been increasingly replaced by the GF atomizer over the past decade, mostly because preconcentration in a graphite tube, treated with a permanent modifier, increases the sensitivity and eliminates kinetic interferences due to the release of the hydride from the solution. The major advantage of HG AAS is the separation of the analyte from the matrix, which largely eliminates interferences in the atomization stage. The major problems are interferences in the liquid phase that impede the formation or liberation of the gaseous compound from the solution. These interferences are mostly caused by transition elements of Group VIII and Group I, which are reduced by THB, for example, to the finely dispersed metal that adsorbs and decomposes the hydrides (Welz and Melcher, 1984(a, b, c); Welz and Schubert-Jacobs, 1986). Another problem of CVG is that the analyte has to be present in ionic form and in a well-defined oxidation state, as, for example, only tetravalent selenium and tellurium form a hydride, and in the case of As and Sb the trivalent form reacts faster and gives better sensitivity. This means that CVG generally requires complete sample digestion and a prereduction step in order to give reliable results.

13.3.3.2 Spectrometers for Line Source AAS

In line source atomic absorption spectroscopy (LS AAS), the high resolution required to measure atomic absorption is provided by the radiation source that emits radiation only within the narrow spectral interval that can be absorbed by the analyte atoms. As there is practically only the spectrum of the analyte element emitted by the source, the only task of the monochromator is to isolate the analytical line from other lines emitted by the source.

Radiation Sources for LS AAS The line source used most frequently in AAS is the *hollow cathode lamp* (HCL), a lamp that has been described as an emission source as early as 1916 (Paschen, 1916). However, in order to become applicable for routine use in AAS, this lamp had to be modified significantly (Russell and Walsh, 1959, Jones and Walsh, 1960), and was optimized further in the following years (Manning and Vollmer, 1967). The schematic design of an HCL is shown in Figure 13.13. An HCL is a spectral lamp with a hollow, cylindrical cathode made of, or containing the element of interest; the anode is mostly made of tungsten or nickel. The glass cylinder of the lamp is filled with an inert gas, usually neon or argon, at a pressure of about 1 kPa; for elements that have their main analytical line in the UV range, the front window is mostly made of quartz. HCL often also contain some insulation in order to concentrate the discharge to the inside of the cathode. When a voltage of 100–200 V is applied across the electrodes, a GD takes place in the reduced-pressure gas atmosphere, as described for the GD excitation source (refer to Section 13.3.2.1, Glow Discharges). As a result, the emission spectrum of the element of interest is obtained, and the radiation beam can be directed toward the absorption volume

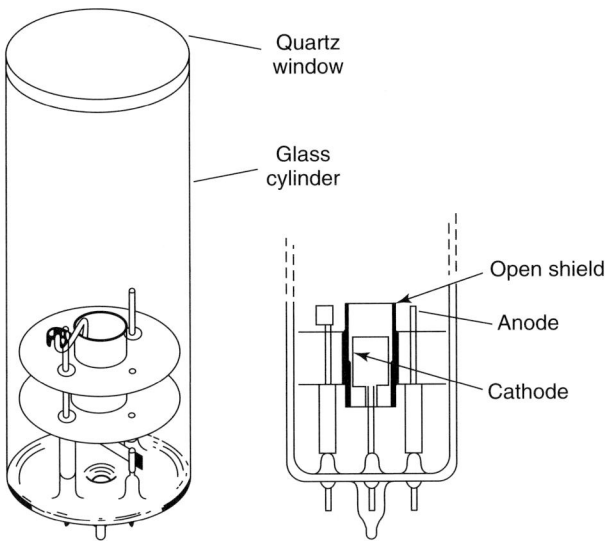

Fig. 13.13 Schematic design of a hollow cathode lamp for AAS. (From Welz and Sperling (1999), p. 107; reproduced with permission from Wiley-VCH.)

in the atomizer, where the absorption process takes place.

For some of the volatile elements, such as Hg, and those that have their main lines in the vacuum-UV range, particularly As and Se, the quality of HCL and their emission intensity are not always satisfactory. In these cases, *electrodeless discharge lamps* (EDLs) might be used with advantage. An EDL consists of a sealed quartz bulb filled with the element of interest (or a compound of it, such as an iodide) and an inert gas. The bulb is placed inside a metallic coil, to which an RF power is applied, generating an ICP discharge. This promotes volatilization of the element inside the quartz bulb; further dissociation and excitation will be induced by the plasma, resulting in the emission of the wavelengths characteristic for the element. The emission intensity of an EDL is typically 20–50 times higher than that of an HCL, resulting in better precision and lower LOD. The disadvantages of this lamp type are the need of an additional power supply and the long warmup times until they reach stable emission.

Other radiation sources, such as diode lasers (Zybin et al., 2005; Hergenröder and Niemax, 1988), have also been investigated with some success, but their use remains restricted to research instruments.

Monochromators for LS AAS The role of the monochromator in LS AAS is to isolate the analytical line from all other lines emitted by the radiation source. Medium-resolution grating monochromators in Czerny–Turner (refer to Section 13.3.2.2) or the simpler Littrow mounting are used most frequently. A particularly important aspect is the *spectral bandwidth*, $\Delta\lambda$, which corresponds to the radiation interval that effectively reaches the detector. The spectral bandwidth is determined by the width of the entrance slit, the *geometric slit width*, s, which controls the amount of radiation that enters the monochromator, and by a characteristic of the dispersing device, the *reciprocal linear dispersion*, $d\lambda/dx$, (dimension: nanometers per millimeter) according to

$$\Delta\lambda = s\frac{d\lambda}{dx} \quad (13.6)$$

The reciprocal linear dispersion depends on the ruling density and the ruled area of the grating; the smaller the numerical value of the reciprocal linear dispersion, the wider can be the geometric slit width in order to obtain a given spectral bandwidth. A wide geometric slit width means that a great amount of radiation can enter the monochromator, resulting in a good S/N ratio. The spectral bandwidth is determined by the spectral environment of the analytical line; it should be chosen as wide as possible, and as narrow as necessary to exclude other lines emitted by the radiation source. Typical values to fulfill this requirement are between 0.2 and 2 nm.

Detectors for LS AAS The role of detectors in AAS is to proportionally convert optical radiation (photons) into an electrical signal; PMT are mostly used for that purpose, although solid-state detectors provide better S/N ratios (for details refer to Section 13.3.2.3).

Spectral Interferences and Background Correction in LS AAS Spectral interferences due to direct or partial *line overlap* are relatively scarce in AAS due to the much smaller number of absorption lines compared to the number of emission lines, particularly in high-temperature excitation sources. However, there are a few well-known examples, such as the overlap of the 213.859-nm secondary iron line with the main analytical line for zinc at 213.857 nm. There is essentially no way to correct for this kind of interference in LS AAS, which means that the determination of Zn in the presence of high iron concentration is always affected by a systematic error.

Other spectral interferences encountered in AAS are *radiation scattering* at particles in the absorption volume, an interference that is mostly limited to GF AAS, and *molecular absorption* due to nondissociated matrix components. The latter has to be classified into dissociation continua, that is, molecules are dissociated by the incident radiation, and electron excitation spectra. In the latter case, we again have to distinguish between electron excitation spectra of diatomic molecules, which exhibit a pronounced rotational fine structure, and those of tri-atomic molecules, the fine structure of which cannot be resolved, so that they appear as broadband spectra.

There is no way to measure only atomic absorption in the presence of any kind of background absorption. In LS AAS, it is necessary, first to measure total absorption and then background absorption, and to calculate atomic absorption by difference in a fast sequential process.

The first system that was proposed to correct for background absorption in AAS used a *deuterium lamp* as a second radiation source in addition to the line source (Koirtyohann and Pickett, 1965). The radiation of the two sources was passed alternately through the flame, typically at a frequency of 50–60 Hz. This system assumed that background absorption is spectrally continuous, at least within the section of the spectrum that is passing the exit slit (0.2–2 nm), whereas atomic absorption occurs only within a very narrow spectral interval (a few picometers). Under this premise, essentially only the background will attenuate the continuum radiation emitted from the deuterium lamp, as atomic absorption represents a fraction of <1% of the spectral interval. The line source, in contrast, will measure total absorption, that is, atomic and background absorption. A "subtraction" of the signal obtained from the D_2 lamp from the signal of the line source will provide the net atomic absorption signal; Figure 13.14 shows a schematic representation of this system.

Background correction with a deuterium lamp is relatively simple and does not add significantly to the cost of the instrument. However, it increases the noise level, as two radiation sources are used, hence deteriorating the S/N ratio and LOD. Although rapidly changing background signals, as they typically occur in GF AAS, could cause artifacts, the system works reasonably well in the case of continuous background absorption. However, the system fails to correct for any *structured background*, such as the electron transition spectra of diatomic molecules that exhibit rotational fine structure. In this case, the deuterium lamp makes an average of the background absorption over the spectral bandwidth, which is in most cases different from the actual background at the analytical wavelength, resulting in over- or under-correction. Deuterium background correction can actually even be the source of errors when the absorption line of a concomitant element is within the spectral bandwidth without overlapping with the analytical line. If this concomitant element is present at high concentration, it can absorb a significant portion of the continuous radiation without absorbing radiation of the line source, resulting in over-correction, which appears as a negative absorbance signal. The emission intensity of deuterium lamps decreases significantly above 350 nm, so that it

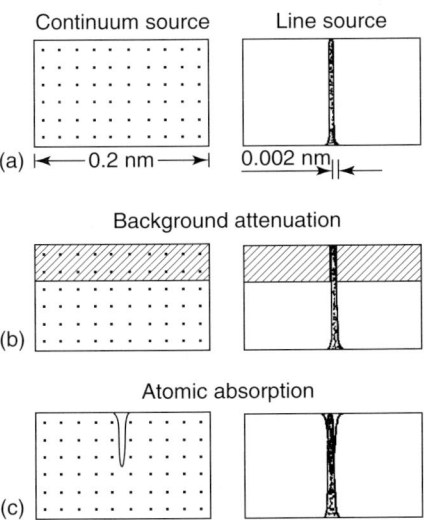

Fig. 13.14 Mode of function of deuterium lamp background correction; (a) the radiant intensity, represented schematically by dots, for the deuterium lamp is distributed over the entire width of the spectral band isolated by the slit (e.g., 0.2–2 nm), while for the line source it is limited to a few picometers; (b) broad-band background absorption attenuates the radiation emitted by both sources to equal degrees; (c) atomic absorption, which is limited to a few picometers, in the first approximation attenuates only the radiation from the line source. (From Welz and Sperling (1999), p. 122; reproduced with permission from Wiley-VCH.)

cannot be used for correction at higher wavelengths.

A more efficient, but also more complex approach for background correction makes use of the *Zeeman effect*, that is, the splitting of electronic levels of atoms (and consequently of spectral lines) under the influence of a magnetic field (De Loos-Vollebregt and de Galan, 1985). The spectral lines split into three components or groups of components: a π component of the same frequency as the original line, and two σ components (σ^+ and σ^-), with slightly higher and lower frequency, as shown in Figure 13.15. In addition to the splitting the components are also polarized, the π components are linearly polarized in a direction parallel to the magnetic field, while the σ components are circularly polarized in a direction perpendicular to the magnetic field. For *Zeeman-effect background correction* (ZBC), an alternating magnetic field is usually applied at the atomizer, and all measurements are made using the line profile emitted by the spectral lamp. The total absorbance (atomic and background) is measured with the magnetic field off, and the background signal is measured with the magnetic field on. In this case, the σ components of the analyte atoms are shifted out of the emission profile of the lamp, and the π components are usually eliminated using a polarizer. Alternately, the magnetic field can be applied parallel to the radiation beam; in this case, the π components become "invisible," so that no polarizer is necessary.

ZBC is almost exclusively used for GF AAS and can be successfully employed for most analytical applications, as it

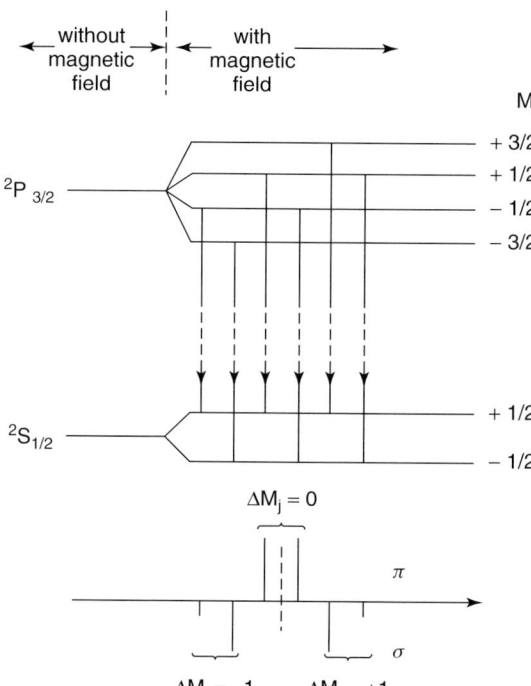

Fig. 13.15 Splitting of the energy levels in a magnetic field with the resulting splitting pattern for the electron transition $^2S_{1/2} \leftrightarrow {}^2P_{3/2}$. (From Welz and Sperling (1999), p. 97; reproduced with permission from Wiley-VCH.)

efficiently corrects for most types of background absorption, including structured background. The only notable exceptions are when the concomitant species, an atom or a diatomic molecule, is also affected by the magnetic field; in this case, the background without and with magnetic field is not the same, which inevitably results in background correction errors. In the case of a near-line coincidence with an absorption line of a concomitant element, the latter obviously also shows Zeeman splitting; in this case, the σ components of the concomitant element might overlap with the emission line even more severely than without magnetic field, causing an even more pronounced spectral interference. The same is true for the spectra of some diatomic molecules, such as the phosphorus suboxide PO (Becker-Ross, Florek and Heitmann, 2000), which also undergo changes in the magnetic field, causing background correction errors.

The sensitivity of AAS instruments with ZBC is lower when compared to other instruments, as the separation of the $\sigma^\pm$ components is usually not complete, which means that some atomic absorption will always be measured together with the background and "subtracted" from the total absorption, resulting in lower integrated absorbance signals. For higher analyte concentrations, this may result in the so-called *rollover effect*, which can also be noticed on transient signals that start to split and exhibit two maxima at high concentrations (Figure 13.16). More recently, instruments using a "three-field" mode have been introduced, where in addition to the "zero" field (magnet off) and the high field strength, an adjustable intermediate field can be used in order to deliberately reduce sensitivity and

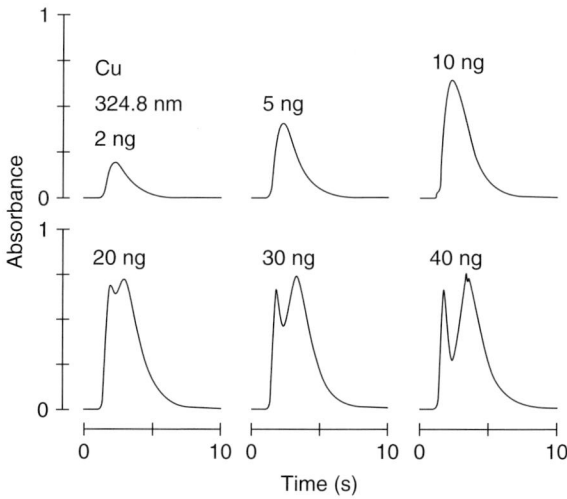

Fig. 13.16 Effect of rollover of the calibration curve on the signal shape for GF AAS; determination of increasing masses of copper in an alternating magnetic at the atomizer. (From Welz and Sperling (1999), p. 131; reproduced with permission from Wiley-VCH.)

increase linearity, making possible the determination of higher concentrations of the analyte without the need for dilution (Gleisner, Eichardt and Welz, 2003).

13.3.3.3 Spectrometers for High-Resolution Continuum Source AAS

Although LS AAS is an established, reliable, and robust analytical technique for a large number of applications, it has been questioned already in the late 1980s that if AAS could survive in the face of strong competition mainly from ICP OES and ICP-MS without a renovation of the basic concept (Hieftje, 1989). The approach that has been proposed was the use of a continuum radiation source, high-resolution spectral sorting devices and completely new detection systems. Although continuum radiation sources have been investigated as an alternative to line sources since the early 1960s (Welz et al., 2005), it was only some 30 years later, when finally components became available, that made it possible to design equipment for high-resolution continuum source atomic absorption spectroscopy (HR-CS AAS) with a performance that was superior to that of LS AAS (Heitmann et al., 1996, Harnly, 1999). The new approach includes a xenon short-arc lamp, a high-resolution double monochromator and a linear CCD array detector, as is shown schematically in Figure 13.17.

Radiation Source for HR-CS AAS The radiation source, a specially-designed high-pressure Xe short-arc lamp with a nominal power of 300 W is operating in a hot-spot mode, as shown in Figure 13.18(a). The hot spot with a diameter of about 0.2 mm reaches a temperature of about 10 000 K, which is the precondition for providing sufficient radiation intensity over the entire spectral range from 190 to 900 nm. Figure 13.18(b) shows that the emission

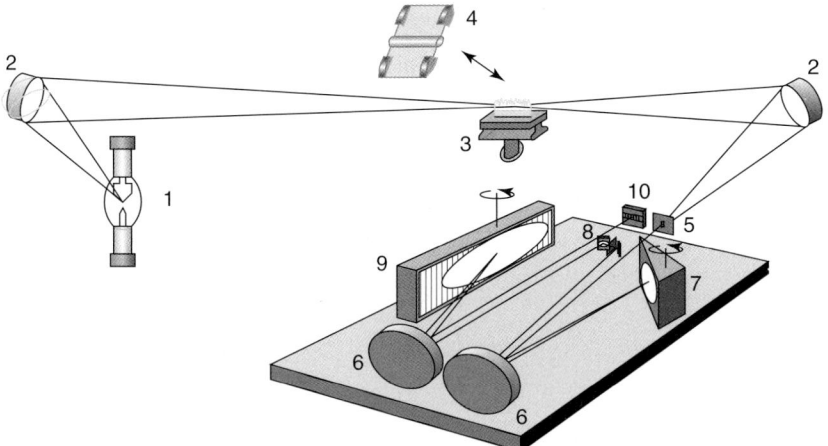

Fig. 13.17 Schematic design of an instrument for HR-CS AAS; 1 – xenon short-arc lamp; 2 – elliptical mirrors; 3 – flame atomizer; 4 – graphite tube atomizer; 5 – entrance slit; 6 – parabolic mirrors; 7 – prism; 8 – intermediate slit; 9 – echelle grating; 10 – CCD detector. (From Welz et al. (2003), p. 223; reproduced with permission from the Brazilian Chemical Society.)

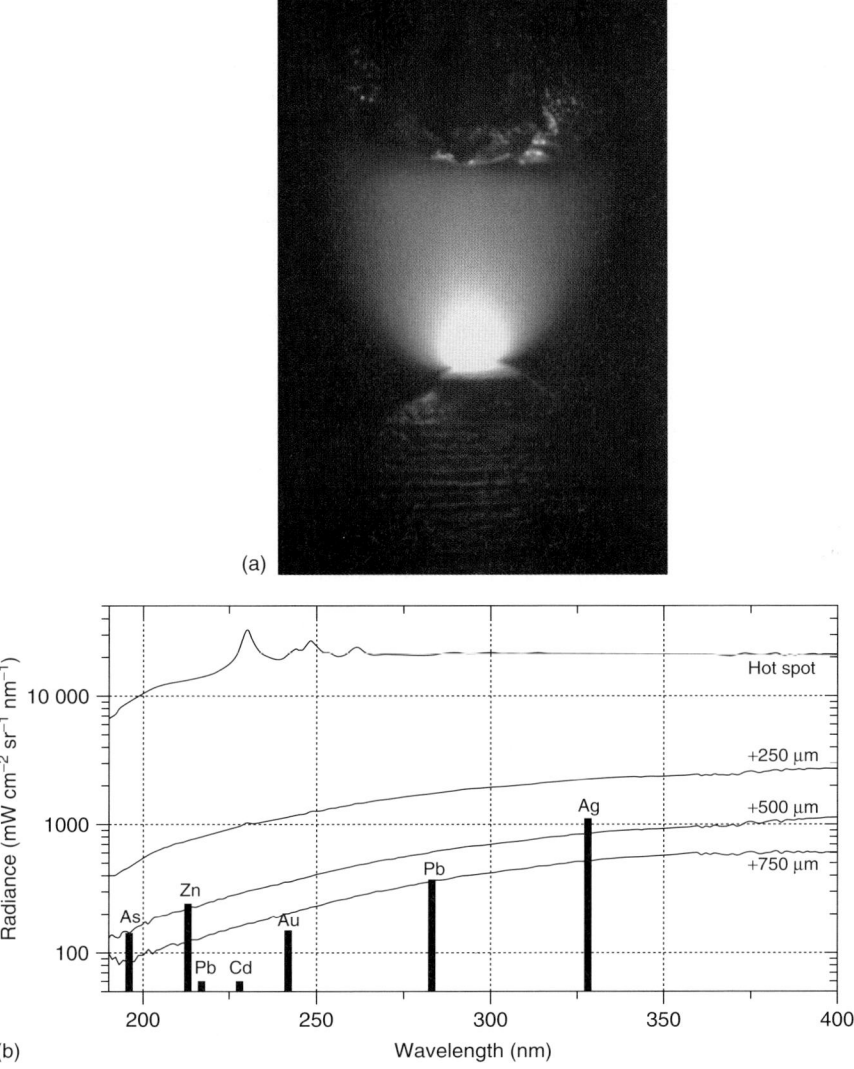

Fig. 13.18 High-pressure Xe short-arc lamp as radiation source for HR-CS AAS; (a) hot spot with a diameter of about 0.2 mm and a temperature of ~10 000 K; (b) wavelength-dependent spectral radiance in the hot spot in comparison to some selected emission lines from HCL. (From Welz et al. (2005), pp. 32–33; reproduced with permission from Wiley-VCH.)

intensity of this lamp in the UV range is typically 2 orders of magnitude higher than that of selected HCL and conventional continuum radiation sources. This high radiation intensity has a direct influence on the S/N ratio, and hence on the LOD, as in AAS the noise is inversely proportional to the square root of the radiation intensity. As the position of the hot spot is not stable over time, the radiation emitted by the Xe short-arc lamp is directed toward the atomizer by a set of mirrors, and it is actively stabilized by computer software to assure that the radiation beam is in a stable (central) position in the absorption volume.

Monochromator for HR-CS AAS The high-resolution double monochromator consists of a predispersing prism monochromator and a high-resolving echelle grating monochromator, both in Littrow mounting. The radiation enters the monochromator, is reflected by a parabolic mirror onto the Littrow prism, which has a mirror on its other side, so that the radiation passes the prism twice. The prism is rotated in a way that only that part of the spectrum that contains the analytical line will be reflected to pass the intermediate slit and enter the second monochromator. A second parabolic mirror directs the radiation toward the echelle grating, which provides the high resolution of the small spectrum interval selected by the intermediate slit, which is then reflected onto the detector without passing an exit slit. This double monochromator provides a resolution of $\lambda/\Delta\lambda \approx 140\,000$, corresponding to a pixel resolution better than 2 pm at 200 nm. This resolution is some 2 orders of magnitude higher than that of a conventional LS AAS monochromator, and also better than that of typical ICP OES mono- and polychromators.

Detector and Measurement Principle in HR-CS AAS After the radiation has been highly resolved by the echelle grating, it is directed to the detector, which is typically a UV-sensitive linear CCD array with 512 or 588 pixels, 200 of which are used for analytical purposes. All pixels are illuminated for typically 1–10 milliseconds and read out simultaneously, and as each pixel is equipped with an individual amplifier, the system actually operates with 200 individual detectors. Each pixel detects a small portion of the highly resolved spectrum around the analytical line, as shown in Figure 13.19 for the primary absorption line of cadmium. The center pixel measures the absorbance at the line core and the neighbor pixels measure the absorbance at the line wings. This means that in HR-CS AAS the measured absorbance becomes a function of the pixel width, whereas in LS AAS it is a function of the width (profile) of the line emitted by the radiation source. In contrast to LS AAS, however, there is the possibility in HR-CS AAS to add the absorbancies measured at several pixels and arrive at the *wavelength selected absorbance* (WSA), $A_{\Sigma n}$, for steady-state signals, and the *peak volume selected absorbance* (PVSA), $A_{\Sigma n,\text{int}}$, for transient signals, where n is in both cases the number of pixels used for measurement (Heitmann *et al.* 2007). The number of pixels that are finally used for measurement has to be optimized, as each pixel also has a contribution to the noise. For narrow absorption lines, the optimum S/N ratio is usually obtained with three pixels, whereas for wider lines the sum of up to seven pixels might be used.

The availability of 200 independent and simultaneously operating detectors adds a third dimension, the spectral resolution, to the traditional absorbance-over-time measurement obtained by LS AAS, which means that three signal evaluation modes are available: first, the traditional absorbance over time can be visualized for each individual pixel of the detector; secondly, the absorbance as a function of the wavelength can also

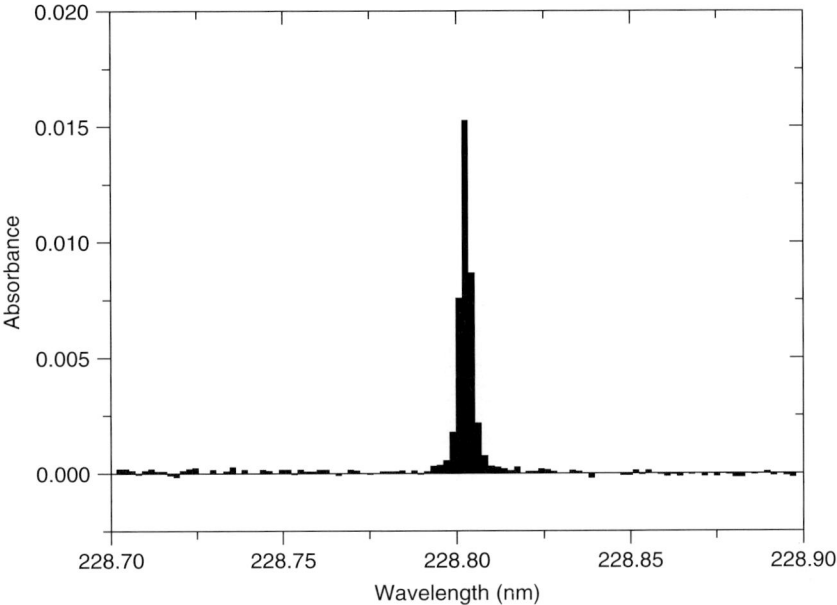

Fig. 13.19 Absorption spectrum in the vicinity of the Cd resonance line at 228.802 nm recorded simultaneously with 200 pixels of a CCD array detector. (From Welz et al. (2003), p. 223; reproduced with permission from the Brazilian Chemical Society.)

be visualized, showing all the spectral events in the vicinity of the analytical line; and thirdly, a time-resolved three-dimensional spectrum can be obtained, showing the entire highly resolved spectrum around the analytical line, as shown in Figure 13.20. With such information, any spectral event occurring in addition to analyte absorption, such as absorption by concomitant elements or structured molecular absorption can be detected, and appropriate measures can be taken in order to avoid or correct for spectral interferences (Welz et al., 2005).

Background and Baseline Correction In HR-CS AAS, no additional instrumental devices, such as secondary radiation sources or magnetic fields, are necessary for background correction; any kind of background correction, and also the correction for baseline noise due to lamp flicker noise, and so on, is carried out with detector pixels that are not used to measure analyte absorption. This means that background measurement, background correction, and baseline correction are strictly simultaneous with the measurement of atomic absorption, which is in contrast to LS AAS, where these measurements and corrections are sequential.

Initially, we have to distinguish between continuous (within the spectral section that reaches the detector) and discontinuous absorption (refer to Section 13.3.3.2, Spectral Interferences and Background Correction in LS AAS). Any spectrally continuous increase or decrease of the radiant flux over time is measured identically at all pixels of the detector, and can be eliminated

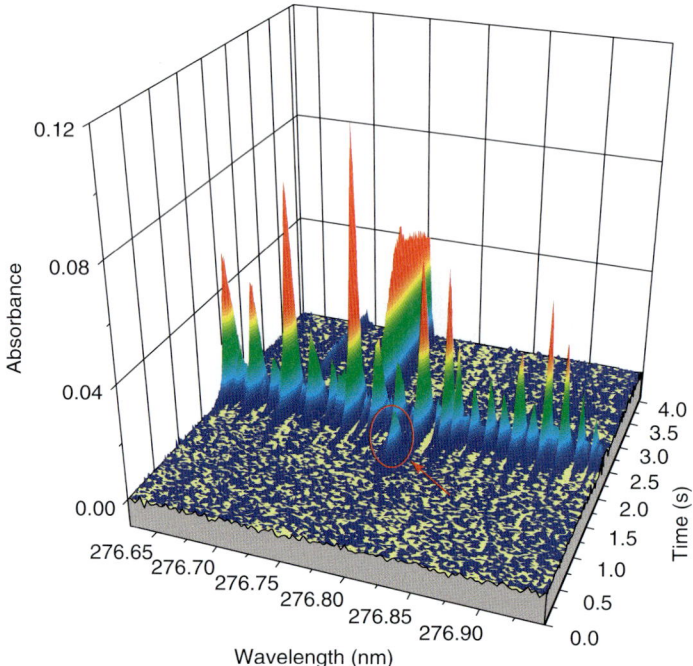

Fig. 13.20 Time- and wavelength-resolved absorption spectrum for a marine sediment recorded in the vicinity of the Tl absorption line at 276.787 nm using HR-CS GF AAS; the Tl absorption, which is marked by a circle, is temporally and spectrally resolved from the fine-structured molecular absorption and from an iron absorption line that appears only late in the atomization cycle.

using correction pixels. This includes any fluctuation in the lamp intensity (arcs are notoriously instable), but it also includes any continuous background, such as radiation scattering and molecular absorption without fine structure, as the entire radiation is passing through the atomizer. Any events affecting all pixels at the same time will be readily identified and corrected automatically, resulting in a smooth baseline and an absorption signal with an extremely low noise level, as shown in Figure 13.21 for the determination of Pb using high-resolution continuum source graphite furnace atomic absorption spectrometry (HR-CS GF AAS).

After correction for any continuous change in the radiant flux remains the discontinuous absorption of radiation. Analyte absorption is obviously discontinuous, as it occurs within a few picometers, and 3–5 pixels are typically sufficient for its measurement. If more than one absorption line of the same or of different elements is within the wavelength interval that reaches the detector, this absorption is measured as well and appears in the wavelength-resolved image. No measure is necessary if there is no spectral overlap between the analyte line and a concomitant line, as at high resolution only the signals in the center of the analytical line are

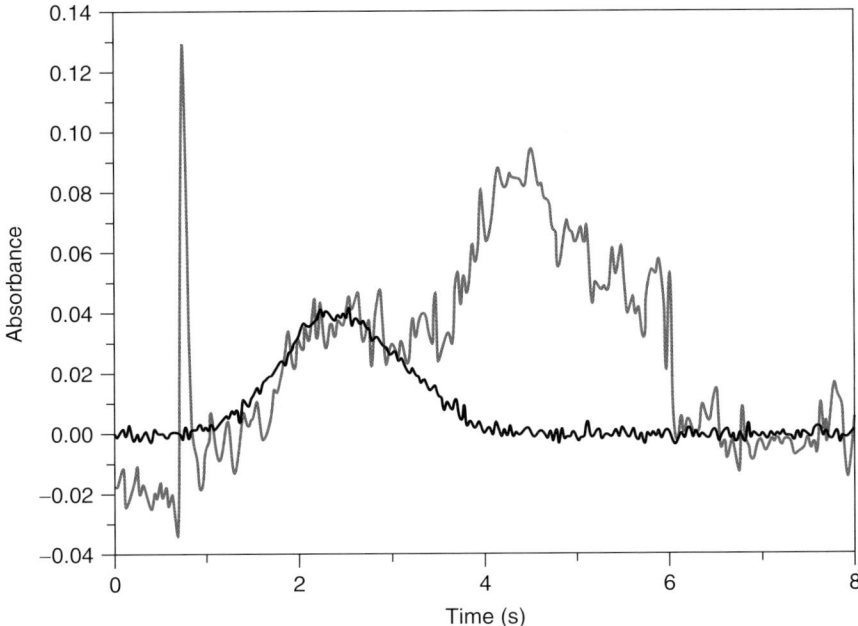

Fig. 13.21 Absorbance-over-time signal recorded for Bovine Muscle certified reference material at the Pb resonance line at 217.001 nm using one pixel only; *gray line*: without background correction; *black line*: after automatic correction for all spectrally continuous fluctuations of the intensity. (From Borges et al. (2006).)

considered. In the case of GF AAS there is an additional possibility, even in the case of direct line coincidence, to separate the absorbance in time, in case the analytes have significantly different volatility. Other sources of discontinuous absorption are electron excitation spectra of diatomic molecules, such as NO, PO, CS, and SiO, which exhibit pronounced rotation fine structure. These spectra can be easily recognized due to visualization of the spectral region in the vicinity of the analytical line. In this case, the same rules apply as in the case of a concomitant absorption line, i.e., no measure is necessary if there is no spectral or temporal overlap between the analyte line and the molecular absorption "lines."

In the case that analyte and concomitant atomic or molecular absorption overlap spectrally and cannot be separated in time, the software allows the user to correct for the interference using a least-squares algorithm. Figure 13.22 shows the well-known coincidence of the primary resonance line for Zn at 213.857 nm with a secondary Fe line at 213.859 nm, which cannot be separated spectrally, as the line width is greater than their distance. This interference cannot be corrected at all in classical LS AAS. Fortunately, Fe has a second absorption line at 213.970 nm within the spectral interval that reaches the detector, which can be used for correction. As the absorbance ratio between the two lines is constant and known, it is sufficient to measure the absorbance at the 213.970-nm line, calculate the absorbance at the 213.859-nm Fe line, and subtract this

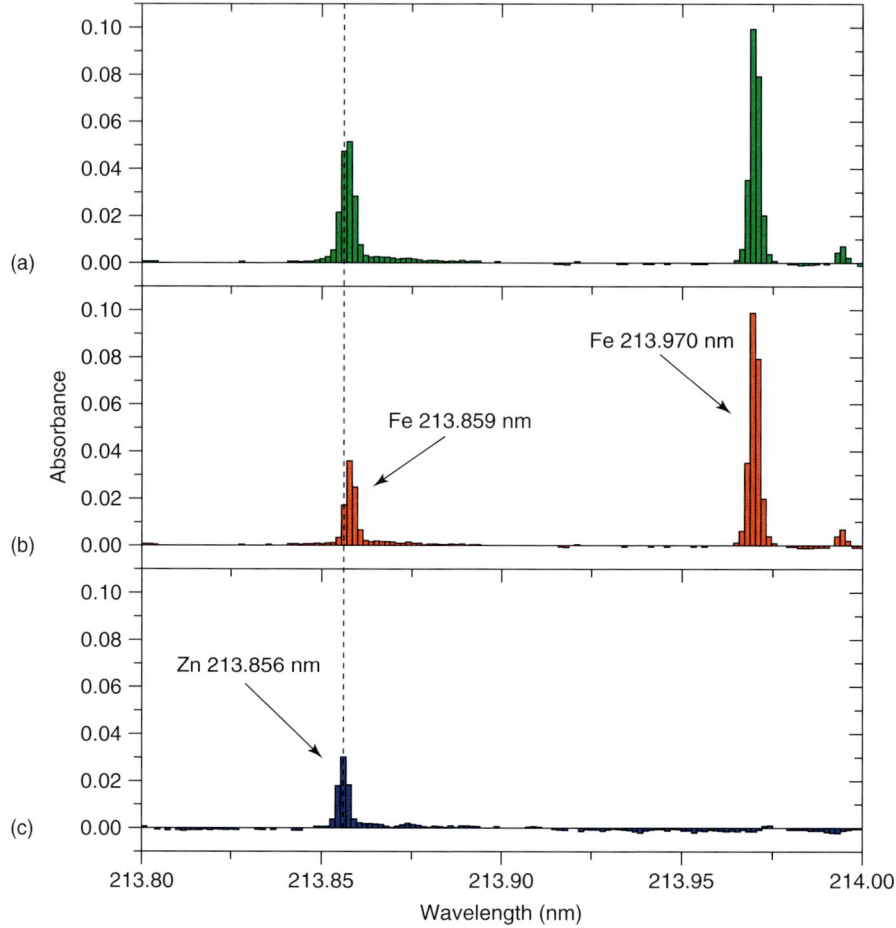

Fig. 13.22 Least-squares correction of the spectral interference due to direct-line overlap between the Zn resonance line at 213.856 nm and the secondary Fe absorption line at 213.859 nm: determination of 0.3 mg L^{-1} Zn in the presence of 1000 mg L^{-1} Fe; (a) integrated absorbance spectrum for Zn in iron matrix without correction; (b) spectrum for pure iron used for correction; (c) resulting spectrum after correction. (From Heitmann, U., ISAS – Institute for Analytical Sciences, Department of Interface Spectroscopy, Berlin, Germany, reproduced with permission.)

value from the total absorbance to get the net absorbance due to Zn.

The same algorithm can be used to correct for molecular spectra with rotational fine structure. In this case, a synthetic solution is prepared with a compound that generates the desired reference spectrum in the "atomization" cycle (such as $NH_4H_2PO_4$ to generate PO). This reference spectrum is stored in the software and "subtracted" from the sample spectrum (Welz et al., 2005). This process can be repeated in case there is more than one diatomic molecule or concomitant atom causing spectral interference.

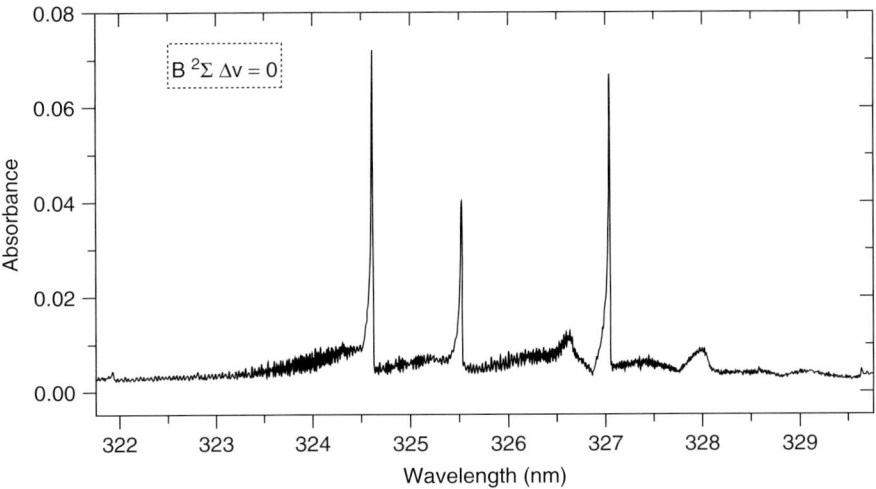

Fig. 13.23 Detail of the molecular spectrum of the phosphorus monoxide PO in the range between 322 and 329 nm, corresponding to the electronic transition X $^2\Pi$ → B $^2\Sigma$. (From Welz et al. (2005), p. 195; reproduced with permission from Wiley-VCH.)

Advantages and Applications of HR-CS AAS There are several advantages of HR-CS AAS over conventional LS AAS, which obviously include the use of a single radiation source for all elements instead of a separate source for each analyte. The high emission intensity of the Xe short-arc lamp has no influence on the measured absorbance, which, according to Eq. (13.4) is proportional to the ratio of the incident and the transmitted radiant power; however, the emission intensity has an influence on the S/N ratio, and hence on the LOD, as the noise is inversely proportional to the square root of the radiant power. The visibility of the spectral environment around the analytical line at high resolution is providing a new dimension of information that has previously been unknown, and which can readily be used for method development and to avoid interferences. The truly simultaneous correction for all wavelength-independent fluctuations of the radiant power, including continuous background absorption, have been discussed in Section 13.3.3.3, Background and Baseline Correction, together with the correction capabilities for discontinuous background. Obviously, HR-CS AAS can be used for any application traditionally assigned to LS AAS, but it has also been shown to improve and facilitate the analysis of complex samples, especially using direct solid sampling (Welz et al., 2007b). As the use of a continuous radiation source makes any wavelength between 190 and 900 nm available for analytical purposes, HR-CS AAS is not limited to atomic absorption, but can also be used for the determination of nonmetals, such as phosphorus, sulfur, and the halogens using the absorption spectra of diatomic molecules, such as PO, which is shown in part in Figure 13.23, CS, AlF, and so on (Huang et al., 2006(a, b, c); Huang et al., 2008; Heitmann et al., 2006). The analytical applications of HR-CS AAS have been recently reviewed by Welz et al. (2007a).

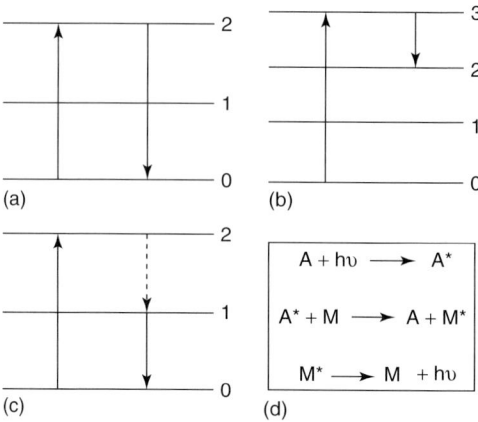

Fig. 13.24 Schematic presentation of important processes for AFS; (a) resonance fluorescence; (b) direct-line fluorescence; (c) stepwise-line fluorescence; (d) sensitized fluorescence. (Adapted from Greenfield (1994), p. 566, reproduced with permission of The Royal Society of Chemistry.)

13.3.4
Atomic Fluorescence Spectrometry

Atomic fluorescence was studied as early as 1902 by Wood (1902), and Nichols and Howes (1924) considered fluorescence in flames, but neither of these publications dealt with analytical applications. Winefordner and Vickers (1964) were the first to investigate the possibilities of using AFS as a practical analytical technique. Veillon et al. (1966) demonstrated the usefulness of continuum radiation sources for AFS, whereas West (1967) investigated EDL. Further information about the research activities during that period might be obtained from a review article by Winefordner and Mansfield (1967). The major advantage of AFS was the significantly better S/N ratio, and hence better LOD compared to FAAS. Nevertheless, AFS has never found general application as an analytical technique. Among the reasons for that might be that the best S/N ratios were obtained in low-temperature flames that were more subject to matrix effects. The main reason, however, was that the research in AFS was coinciding in time with the development of GF AAS and ICP OES, and none of the instrument manufacturing companies was prepared to invest in yet another technique. Nowadays, AFS has found a niche in combination with chemical vapor generation for the determination of mercury and some hydride-forming elements, such as arsenic and selenium.

13.3.4.1 General Principles
Atomic fluorescence is a radiational deactivation process, which occurs after the excitation of free atoms by the absorption of radiation of characteristic wavelength from an excitation source. Atomic fluorescence can occur by a number of different processes, but only four of them are considered to be of importance and are schematically represented in Figure 13.24 (Browner, 1974; Robinson, 1990; Greenfield, 1995).

Resonance fluorescence (Figure 13.24a) occurs when atoms absorb and reemit radiation of the same wavelength. Lines corresponding to transitions between excited electronic states and the ground state of an atom constitute most analytically useful examples of resonance fluorescence, since the transition probabilities for resonance transitions are

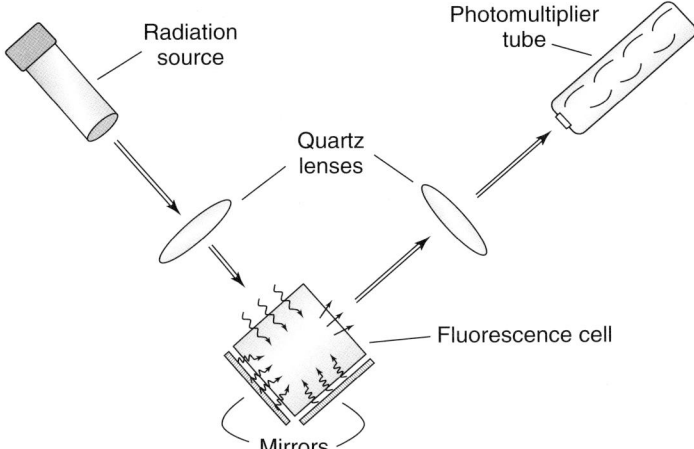

Fig. 13.25 Schematic design of an atomic fluorescence spectrometer.

usually much greater than those for other transitions.

Nonresonance fluorescence occurs when the exciting line and the observed fluorescent line are of different wavelengths. *Stokes direct-line fluorescence* results when the exciting and the fluorescent lines originate from optical transitions with a common upper level (Figure 13.24b). An atom is excited (usually from the ground state) to a higher excited state, and then undergoes a direct transition to a metastable level above the ground state, emitting a fluorescent line at longer wavelength, compared to that of the exciting radiation. When the upper level of the exciting and the emitted lines are different, *Stokes stepwise-line fluorescence* is said to occur (Figure 13.24c). Atoms excited by radiation lose part of their energy, usually by collisional deactivation, before emitting fluorescence radiation of longer wavelength. *Anti-Stokes fluorescence* (not shown here) occurs when the wavelength of the fluorescent line is shorter than that of the exciting line. The deficit in photon energy is usually supplied thermally, and the term *thermally-assisted* fluorescence can be used.

In addition to the above-discussed types of atomic fluorescence, another, although less usual type exists, called *sensitized fluorescence*. In this process, an atom or molecule excited by absorption of electromagnetic radiation (donor) transfers its excitation energy to the sample atom (acceptor) by collision. The acceptor then undergoes radiative deactivation, resulting in atomic fluorescence. The process is schematically represented in Figure 13.24(d).

13.3.4.2 Instrumentation

The basic setup of an AFS instrument consists in a radiation source, an atomizer, a spectral sorting device, a detector, and devices for signal processing and readout, as shown in Figure 13.25. The detection of the fluorescence radiation should always be misplaced in relation to the incident radiation, that is, for instance, be positioned at a 90° angle, in order to prevent that photons emitted by the radiation source reach the detector.

Atomizers for AFS The physical and chemical environments within an atomizer have a profound effect upon the magnitude of the fluorescent flux emitted from the atomizer in addition to its influence on atomization efficiency. In practice, the excited atom is almost inevitably subject to radiationless deactivation due to collisions with other species present in the atomizer. These so-called *quenching collisions* are characterized by a quantum yield, Y, which is the number of quanta emitted as fluorescence to the number of quanta absorbed and capable of producing fluorescence. Quantum yields were calculated for several analytes, gas mixtures, and temperatures (Sychra, Svoboda and Rubeška, 1975); for air–acetylene and air–hydrogen flames Y was found to be typically below 0.1; values close to $Y = 1$ could only be obtained in hot argon environment, which would make ET atomizers an ideal source for AFS.

A variety of atomizers have been used for AFS, including flames (Sychra, Svoboda and Rubeška, 1975), graphite furnaces (Massmann, 1968), and ICP (Greenfield, 1994); however, there are no instruments commercially available with such configurations. The only instrumentation that is available nowadays for AFS are specialized apparatus for the determination of mercury (Cai, 2000) using the cold-vapor technique, which does not require any atomizer, and for the hydride-forming elements (refer to Section 13.3.3.1, Atomizers for Chemical Vapor Generation). In the latter case special atomizers are used, such as small hydrogen diffusion flames, where the atomization occurs by a chemical process rather than a thermal one. The interferences that might be observed during generation and separation of the gaseous analyte species from the liquid phase are identical to those observed for AAS (refer to Section 13.3.3.1, Atomizers for Chemical Vapor Generation).

Radiation Sources The intensity of the fluorescence signal is directly proportional to the intensity of the incident radiation, that is, the excitation source has a major influence on the sensitivity that can be obtained by AFS. This dependence can be mathematically expressed by means of the simplified Eq. (13.8), where Φ_0 and Φ_F represent the incident and the fluorescence (emitted) radiant power, κ_0 corresponds to the atomic absorption coefficient, ℓ is the length of the absorption cell, and φ is the quantum efficiency of the fluorescence process (Robinson, 1990).

$$\Phi_F = \Phi_0(1 - e^{-\kappa_0 \ell})\varphi \qquad (13.7)$$

Significant research in AFS was therefore done for the development of high-intensity radiation sources, particularly electrodeless discharge lamps (Sychra, Svoboda and Rubeška, 1975). Modern instruments usually employ a line radiation source, such as a hollow cathode lamp or an electrodeless discharge lamp; these lamps are similar to those used for AAS (refer to Section 13.3.3.2, Radiation Sources for LS AAS). Other sources, such as specially designed mercury vapor lamps, might be used as well.

When lasers are used as an excitation source in AFS, the technique is called *laser-excited atomic fluorescence spectrometry* (*LEAFS*). Among the advantages that are associated with the use of lasers are the higher radiant power compared to conventional line sources, resulting in high sensitivity, and the extended linear dynamic range, which may cover up to 7 orders of magnitude, and the possibility to perform sequential multielement analysis, using oscillator-based lasers (Stchur et al., 2001; Hou et al., 1998). Disadvantages include the high cost and complexity of laser systems.

Spectral Sorting Devices and Detectors One of the advantages of AFS is the selective character of the fluorescence radiation to be monitored, that is, usually only the analyte element can absorb the radiation emitted by the primary source if a line source is used. This means that the monochromator might even be a simple nondispersive system, that is, a filter. The monochromator used in AFS has the sole function of isolating the analytical fluorescence wavelength from other wavelengths that might be emitted by concomitants or by the analyte. Lasers require more sophisticated optical devices, such as beam expanders and diffraction gratings (Hou et al., 1998).

The detectors used in AFS are similar to those used in AAS and OES. The PMT is by far the most frequently used detection device, although solid-state detectors, such as the CCD, have also been employed (for details refer to Section 13.3.2.3).

Interferences Nonspectral interferences are obviously independent of the detection technique and specific for the sample introduction and atomization technique. Reference can therefore be made to similar atomizers used in AAS. Spectral interferences due to direct-line or near-line overlap is extremely unlikely with line radiation sources, as usually only the analyte can absorb the radiation and emit fluorescence. The most serious problem, which cannot be eliminated by improving spectral resolution or by modulating the radiation source, is radiation scattering at nonvolatilized aerosol particles, particularly in flame atomizers. This problem is most serious for resonance fluorescence and might be avoided using a nonresonance fluorescence line; this, however, is usually associated with a significant loss of sensitivity and therefore not generally applicable. Another, although less frequently encountered problem is broadband fluorescence emitted by molecules in the same spectral interval where atomic fluorescence is monitored. This interference also cannot be eliminated by modulating the radiation source, as the fluorescence is caused by the radiation source. Both interferences are much more serious if a continuum radiation source is used instead of a line source. The problem of thermal emission from the atomizer can be solved by modulating the primary radiation source and using a selective amplifier.

13.4
X-Ray Spectroscopy

X-ray spectroscopy (XRS), like optical spectroscopy, is based on the measurements of emission, absorption, fluorescence, scattering, and diffraction of electromagnetic radiation. X-rays are defined as radiation of short wavelength, typically between 10 pm and 10 nm (0.1–100 Å), which is generated by *deceleration* of high-energy electrons or by electron transitions in the *inner orbitals* of atoms (Jenkins and de Vries, 1970; Jenkins, 1988). Upon ejection of an inner electron from an atom, the remaining electrons in the outer orbitals will decay to lower-energy states to compensate for the electron loss and to fill in the vacancy in the inner orbital. These transitions are characterized by the emission of the excess energy in the form of X-ray photons with characteristic energies corresponding to the energy of the electron transition; therefore, each element has its own characteristic X-ray spectrum (Carr-Brion and Payne, 1970). These transitions are named after the orbital from where the original electron has been ejected. For instance, ejection of an electron from the innermost orbital generates "K" X-rays; ejection from the next orbital generates "L" X-rays, and so on, which is in accordance to the atom model of Bohr. Each emission is also associated to a determined transition probability, which is indicated by Greek letters (α, β, γ). For instance, a K-series spectrum, corresponding to ionization from a K-shell, consists of a strong K_α and a weaker K_β line due to a less-frequent electron transition (Müller, 1972).

13.4.1
Theoretical Principles

13.4.1.1 Emission of X-Rays

For analytical purposes, X-rays may be produced in different ways: (i) by bombardment of a metal target with a high-energy electron beam; (ii) by exposing a substance to a primary X-ray, generating a secondary X-ray fluorescence beam; (iii) by using a radioactive source that emits X-rays upon its decay; and (iv) by using synchrotron radiation. X-ray sources, like sources for optical radiation, often generate both, continuous and discontinuous (line) spectra. The continuous radiation is known as *white radiation* or *bremsstrahlung*, as it is the result of the deceleration of high-energy electrons.

The most usual X-ray sources are *X-ray tubes*, which are operated at controlled (high) voltages up to 100 kV applied between a cathode and an anode. In an X-ray tube, a hot filament, acting as a cathode, emits electrons that are accelerated in a high-voltage field to impact against a metal anode (target). The impact of the electrons with the anode generates heat and electromagnetic radiation in the form of X-rays, as a result of the deceleration of the electrons and the conversion of kinetic energy. Under certain conditions, only a continuous spectrum is obtained, as shown in Figure 13.26(a), which under other conditions may be superimposed by a line spectrum, as shown in Figure 13.26(b). The continuous X-ray spectra are characterized by an exactly defined short-wavelength limit, λ_0, which only depends on the acceleration voltage, but is independent of the target material. The maximum photon energy corresponds to an abrupt deceleration of the electron to the kinetic energy 0 in a single impact; the continuum at longer wavelengths corresponds to a stepwise deceleration in multiple collisions. The superimposed line spectra are due to electron transitions within the atom, and are characteristic for the element, as is discussed in the context of X-ray fluorescence (refer to Section 13.4.1.3). The line spectra are simple and consist of a few lines only.

Radioisotopes are also common excitation sources; however, they are some 4 orders of magnitude weaker compared to X-ray tubes. In this case, a γ-ray-emitting radioisotope, such as ^{241}Am, ^{109}Cd, ^{153}Gd, ^{155}Eu, or ^{145}Sm is used, and γ-rays induce the emission of

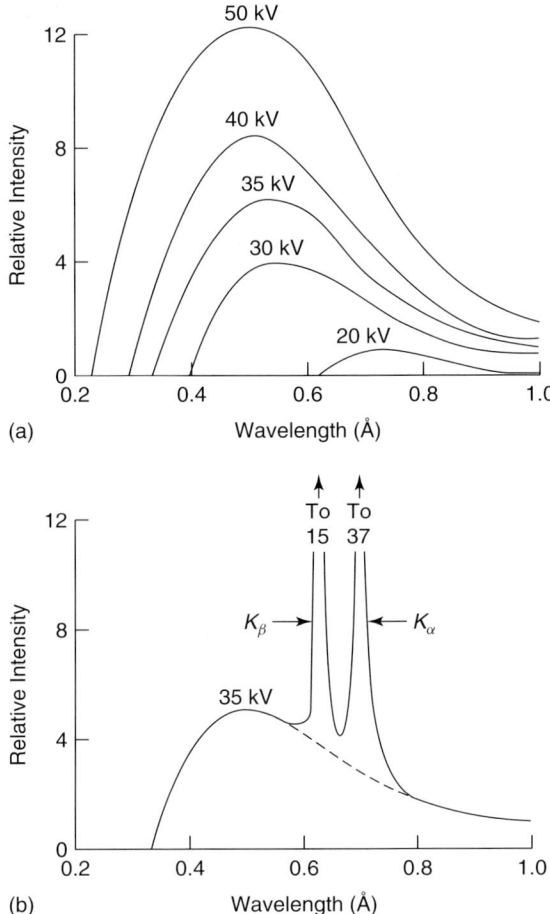

Fig. 13.26 Typical spectra emitted by X-ray tubes under different conditions; (a) continuous spectra emitted by an X-ray tube with a tungsten target using different acceleration voltages; (b) continuous spectrum with superimposed line spectrum emitted by an X-ray tube with a molybdenum target. (From Skoog et al. (1998), Brooks & Cole, P. 273; reproduced with permission from Cenage learning Inc.)

X-rays either by direct impact on the sample surface or by impact on a secondary target, which will then emit X-rays that are used to excite sample components (known as the γ-X source). Several advantages are ascribed to radioisotope excitation sources, such as high stability and low cost (Jenkins, 1988).

The acceleration of high-energy electrons in a high vacuum tube under the influence of a strong magnetic field results in the emission of the so-called *synchrotron radiation*, which can also be used as an excitation source in XRS. Synchrotron radiation can be considered as magnetic *bremsstrahlung*, and it can be used as a source either directly or after modifications, which include selection by an absorber and the use of diffraction crystals and mirrors. Owing to the high intensity of synchrotron radiation, the analytical sensitivity of X-ray fluorescence (XRF) is significantly increased, as well as the S/N ratio (Saisho, 1989).

13.4.1.2 Absorption of X-Rays

When an X-ray penetrates a thin layer of a solid substrate, its intensity is reduced due to absorption and scattering. Except for the lightest elements the latter effect can be neglected, at least in the wavelength range of strong absorption. Similar to the emission spectrum, the absorption spectrum of an element is simple and consists of a few lines only. The absorption spectrum of lead and silver, which are shown in Figure 13.27, consist of a few peaks only with the first one at 0.14 Å. A peculiarity of X-ray absorption spectra is their discontinuous increase with so-called absorption edges. One of the inner electrons of the atomic shell is expelled upon the absorption of an X-ray photon, resulting in an excited ion. In this process, the entire energy $h\nu$ of the radiation is distributed to the kinetic energy of the electron and the potential energy of the excited ion. The probability of absorption is highest when the photon energy corresponds exactly to the energy that is necessary to expel the electron from the atom, that is, if the kinetic energy of the expelled electron is close to zero. For higher photon energy the probability decreases gradually, and for lower photon energy abruptly, as the photon does not have enough energy to expel an ion from the K-shell any more, and the next peaks are reached only when the photon energy reaches a level that corresponds to the energy of the L electrons.

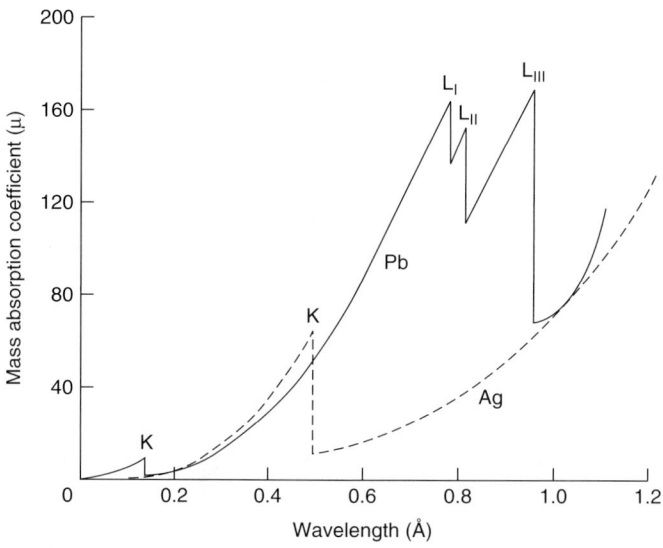

Fig. 13.27 X-ray absorption spectrum of silver and lead. (From Skoog et al. (1998), Brooks & Cole, P. 277; reproduced with permission from Cenage learning Inc.)

13.4.1.3 X-Ray Fluorescence (XRF)

The absorption of X-rays produces electronically excited ions that return into their ground state by electron transitions from higher energy levels. These transitions are associated by the emission (fluorescence) of characteristic wavelengths, which are identical to those that are generated by electron bombardment. However, the wavelengths observed in XRF are always longer than those of the corresponding absorption edge, as the complete liberation of the electron, that is, ionization of the atom, is necessary for absorption, whereas fluorescence only requires transition of an electron from a higher to a lower energy level. The intensity of the emitted X-ray photon will depend on the effectiveness of the de-excitation process within a given atom, and the *fluorescence yield* can be defined as the fraction of all electron transitions that are associated with the emission of an X-ray photon (Müller, 1972). The fluorescence yield increases dramatically with increasing atomic number, and is close to unity for elements with atomic numbers above 70.

13.4.2 Instrumentation

13.4.2.1 X-Ray Fluorescence

The most commonly used technique is XRF, which has a wide range of application in the analysis of solid materials and is characterized as a nondestructive technique. Commercially available XRF instruments can be classified into wavelength-dispersive and energy-dispersive instruments. *Wavelength-dispersive* (WD) instruments, which are schematically represented in Figure 13.28(a), generally use an X-ray tube as the source in order to compensate the energy losses that are encountered when an X-ray is first collimated onto a crystal and then dispersed into its wavelength components. Analyzing crystals made of a wide variety of materials have been used, including LiF, Si, pentaerythritol (PE), topaz, CaF_2, and many

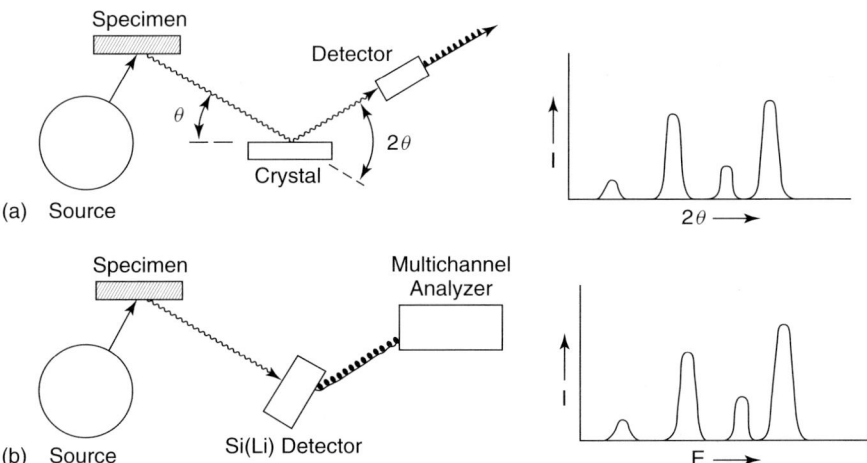

Fig. 13.28 Schematic design of typical XRF instruments; (a) wavelength-dispersive X-ray fluorescence equipment; (b) energy-dispersive X-ray fluorescence equipment. (From Jenkins (1988), p. 54; reproduced with permission from Wiley-Interscience.)

others, depending mainly on the atomic number of the element to be determined and on the specific application. We distinguish between sequential single channel and simultaneous multichannel instruments.

Single-channel instruments can be operated manually or automatically. The former ones can be used for the *quantitative determination* of a few elements. In this case, crystal and detector are adjusted at an appropriate angle (θ and 2θ) and the photons counted for a long enough time to produce a sufficiently high signal. Automatic instruments are better suited for *qualitative analysis*, where the entire spectrum has to be registered. For this purpose, analyzing crystal and detector are rotated synchronously maintaining their relative angular position (Lachance and Claisse, 1994).

Wavelength-dispersive *multichannel instruments* are big and expensive instruments that allow the simultaneous detection and determination of up to 24 elements. The individual channels, each of which consists of a crystal and a detector, are grouped radially around an X-ray source and a sample holder at an appropriate angle for a given analyte line. Each detector has its own amplifier, analyzer, and counter or integrator. The simultaneous determination of all elements may take between a few seconds and a few minutes. Instruments of this type are made for a specific application and offer little flexibility for other purposes.

In *energy dispersive (ED) instruments*, illustrated in Figure 13.28(b), the fluorescence radiation beam coming from the sample is immediately directed toward a semiconductor detector, which is coupled to several electronic devices that are capable of selecting the signal according to small differences in the energy, depending on their wavelength. In this case, a spectrum of voltage pulses is obtained, which is directly proportional to the spectrum of X-ray photon energies entering the detector, as shown in Figure 13.29 (Jenkins 1988; Lachance and Claisse, 1994).

The obvious advantages of ED systems are their simpler design and the absence of movable parts, which also simplifies their operation. The absence of collimators and diffracting crystals, and the short distance between sample and detector increase the signal intensity by about 2 orders of magnitude, which allows the use of much weaker (and less expensive) radiation sources (refer to Section 13.4.2.1). Although this then results in lower sensitivity compared to wavelength-dispersive instruments, their much lower price and their ability to measure all the elements within the range at the same time makes them an attractive alternative.

There are three main types of X-ray detectors to convert X-ray photon energy into voltage pulses. A gas flow *proportional counter* is a tube filled with a mixture of an inert gas (typically Ar) and a quenching gas (such as CH_4) with a wire along its radial axis. The tube wall is grounded and a high potential (usually around 1.8 kV) is applied to the wire. As the X-ray photons enter the tube, the inert gas is ionized and the electrons generated in the ionization process are accelerated toward the wire (anode), colliding with other neutral gas species and generating more ionized gas species in a cascade effect. A voltage pulse is therefore produced, which is proportional to the energy of the incident X-ray photon. The proportional counter is ideally suited for measurement of longer wavelengths (above 1.5 Å).

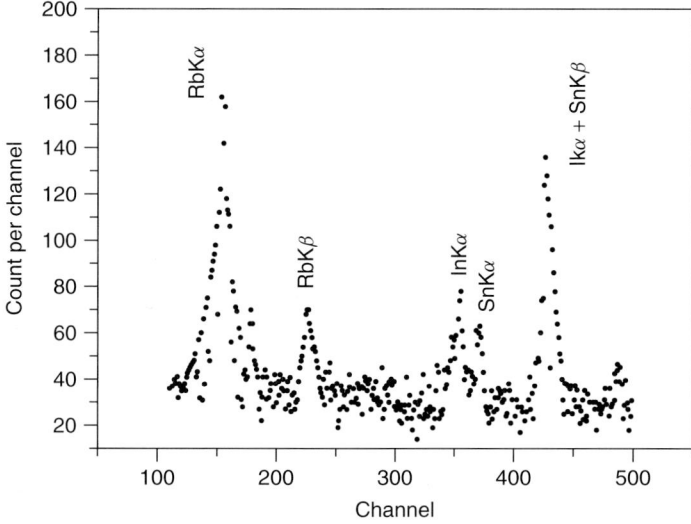

Fig. 13.29 Spectrum of an iron sample recorded with an ED XRF equipment. (From Ekinci and Ìngeç (2008), p. 1120, reproduced with permission from Elsevier.)

A *scintillation counter* can be used for shorter wavelengths. In this detector, the incident X-ray photons impact against a phosphor substance, such as a NaI crystal doped with Tl, resulting in the emission of lower energy photons, which are then accelerated to a PMT, generating electrons in a cascade effect, similar to the PMT used in optical spectroscopy. The generated current is proportional to the energy of the incident X-ray photon. A variation of this detector is the *gas scintillation counter*, where a gas is used instead of a solid phosphor.

Semiconductors are also frequently employed in detectors, particularly in energy-dispersive instruments. The most usual is the *Si(Li) detector*, where a small Si cylinder, compensated by Li in order to increase the electrical resistivity, is used. In this case, a p–i–n-type diode is assembled and, upon impact, X-ray photons generate a number of electron-hole pairs and, consequently, a determined electric charge. Application of a high voltage to the Si disk results in transfer of the accumulated charge to a charge-sensitive preamplifier, producing a charge pulse, the voltage amplitude of which is proportional to the energy of the incident X-ray photon. Electronic noise is reduced by cooling the detector, usually with liquid nitrogen.

The major limitation of all XRF techniques is the need of solid reference materials – usually certified reference materials – with an analyte content and a matrix composition as close as possible to the composition of the sample for calibration in order to compensate for matrix effects.

13.4.2.2 Total-Reflection (TR) XRF

Total reflection XRF is a variation of energy dispersive X-ray fluorescence (ED XRF), which is based on the fact that at an incident angle below the critical angle, the primary radiation beam is completely reflected. The reflection angle is exactly the same as the incidence angle, and the intensity is also roughly the same, except for a small fraction of the radiation that may be refracted by the sample (Wobrauschek, 2007; Yoneda and Horiuchi, 1971). As at a low incidence angle the X-ray photons practically do not penetrate the sample substrate or support, the scattering of radiation, which is a major source of background in XRF, is remarkably reduced, which ultimately results in an improvement in the S/N ratio and, consequently, in the LOD. In addition, the sample itself is irradiated twice by the X-ray photons, as both the primary incident and the reflected beam strike the sample surface. The result is a highly sensitive XRF technique with LOD in the picogram per gram range for selected applications (Knoth and Schwenke, 1978; Wobrauschek, 2007).

The instrumental setup for total-reflection X-ray fluorescence (TR XRF) is not much different from that for ED XRF, except for the fact that the primary radiation beam has to be adjusted to enter the sample at a glancing angle of a few arc seconds. This can be achieved by placing reflection units and controlled apertures in the beam path between the source and the sample chamber. In addition, in TR XRF, the detector is positioned directly above the sample, as represented schematically in Figure 13.30. As in most ED XRF instruments, X-ray tubes, and Si(Li) detectors are commonly used.

Prior to TR XRF analysis, samples must be placed on a reflective, flat polished surface. Measurements can be performed in air, in a He atmosphere or in a vacuum chamber, which has the advantage of avoiding the scattering of primary X-ray photons from atmospheric gas molecules and is mandatory

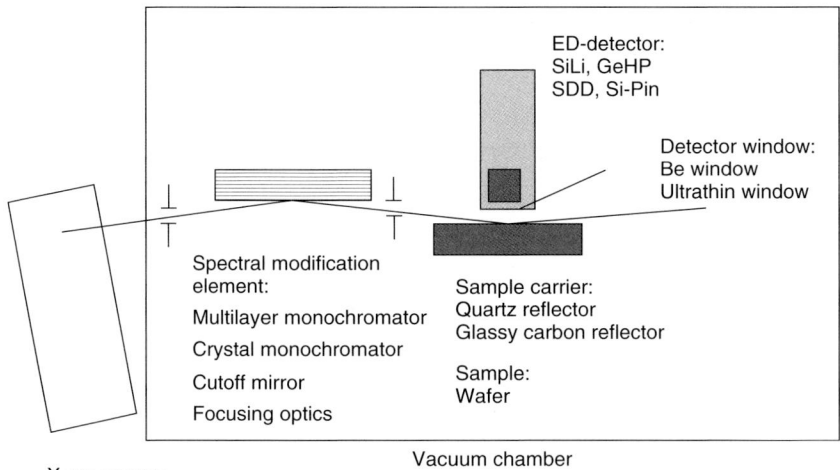

Fig. 13.30 Instrumental setup for TR XRF. (From Wobrauschek (2007), p. 293; reproduced with permission from Wiley.)

for the detection of low-atomic-number elements. Samples are more suitably analyzed as aqueous solutions, and solid samples should typically be digested prior to analysis (Klockenkämper and von Bohlen, 1999). The application of TR XRF for trace element determination covers a wide range of sample types; a complete review has been published by Wobrauschek (2007).

13.4.2.3 Particle-Induced X-Ray Emission (PIXE)

Particle (or proton) induced X-ray emission (PIXE) is a technique based on the emission of X-ray photons by sample constituents excited by collision with charged particles, usually protons. The sequence of events is identical to the one described for XRF, that is, emission of X-ray photons following the filling of electron vacancies by outer-shell electrons (Johansson and Campbell, 1988). Instrumentation for PIXE is rather similar to that for ED XRF, except for the excitation source. In PIXE, a proton beam with energies varying typically between 1 and 10 MeV and beam currents of a few microamperes is generated by means of a small particle accelerator. The proton beam is collimated, usually by a magnetic field, and the proton energy is stabilized and uniformed by passing through a slit, before entering the sample chamber, which is under vacuum. An Si(Li) detector is typically employed (Johansson and Campbell, 1988; Khan and Crumpton, 1981).

Several advantages are associated with the use of PIXE, such as the ability to perform fast multielement analysis, the high sensitivity for elements with atomic numbers above 11, the suitability for small samples (micro- or μ-PIXE) and the lower *bremsstrahlung*, due to the smaller deceleration experienced by protons upon collision with electrons (Khan and Crumpton, 1981). The application of PIXE ranges from the analysis of pigments in historic paintings (Neelmeijer *et al.*, 2000) to biological samples (Bertrand, Weber and Schoefs, 2003).

13.4.2.4 Portable Instrumentation

Portable (field) XRF instruments are commercially available and have been gaining importance for a range of on-site applications. The development of portable XRF instruments has become possible mostly due to the development of small X-ray tubes and to the use of radioisotopes as X-ray sources. Portable instruments have been mostly developed for "conventional" XRF, but instruments for PIXE (Pappalardo *et al.*, 2007) and TR XRF (Mages *et al.*, 2003), among others, have been described as well. Applications range from archaeometrical studies (Zarkadas and Karydas, 2004) to coating thickness evaluation (Carapelle *et al.*, 2007).

13.5 Outlook

Most of the instrumental techniques of atomic spectrometry can nowadays be considered established techniques, so that no dramatic innovations should be expected. Miniaturization is an interesting trend that not only reduces the size of the equipment but also often reduces sample consumption and measurement time, and improves LOD and LOQ. Microplasmas with dimensions of 1 mm or less, usually microwave-induced, generated by high-frequency or dc discharges, are a promising field of research in OES, and have been experiencing increasing popularity in the past few years (Foest, Schmidt and Becker, 2006; Broekaert and Siemens, 2004). Yu, Du and Wang (2007) proposed a miniature AFS system in a lab-on-valve for mercury determination, and Wormhoudt et al. (2005) described a determination of carbon in steel by LIBS using a microchip laser and miniature spectrometer. It is not easy to predict, however, if such systems will find entrance into routine laboratories, or if they will remain restricted to research labs.

The most recent development in atomic spectrometry, HR-CS AAS, has been received extremely well in the market, and will have a significant impact not only on classical LS AAS but also on alternate techniques, such as ICP OES and ICP-MS. The most obvious next step will be the development of simultaneous HR-CS AAS, which will make it even more competitive. Although this development depends on further progress in detector technology, it will, for sure, take less than a decade until this step becomes technically feasible.

Acknowledgments

The authors are grateful to Conselho Nacional de Desenvolvimento Científico e Tecnológico (CNPq) for scholarships and financial support.

Acronyms

AAS	atomic absorption spectrometry
ac	alternating current
AES	auger electron spectroscopy
AFS	atomic fluorescence spectrometry
BEC	background-equivalent concentration
CCD	charge-coupled device
CID	charge-injection device
CV	cold vapor (technique)
CV AAS	cold vapor atomic absorption spectrometry
CVG	chemical vapor generation
dc	direct current
DCP	direct current plasma
ED	energy dispersive
EDL	electrodeless discharge lamp
ED XRF	energy dispersive X-ray fluorescence
ETA	electrothermal atomization
ETV	electrothermal vaporization
FAAS	flame atomic absorption spectrometry
FWHM	full width at half maximum (half width)
GD	glow discharge
GF	graphite furnace
GF AAS	graphite furnace atomic absorption spectrometry
HCL	hollow cathode lamp
HG	hydride generation (technique)
HG AAS	hydride generation atomic absorption spectrometry
HR-CS AAS	high-resolution continuum source atomic absorption spectrometry
HR-CS GF AAS	high-resolution continuum source graphite furnace atomic absorption spectrometry
ICP	inductively coupled plasma
ICP-MS	inductively coupled plasma mass spectrometry
ICP OES	inductively coupled plasma optical emission spectrometry

IUPAC	International Union of Pure and Applied Chemistry	
LEAFS	laser-excited atomic fluorescence spectrometry	
LIBS	laser-induced breakdown spectroscopy	
LOD	limit of detection	
LOQ	limit of quantification	
LS AAS	line source atomic absorption spectrometry	
MIP	microwave-induced plasma	
OES	optical emission spectrometry	
PIXE	particle-induces X-ray emission	
PMT	photomultiplier tube	
PVSA	peak volume selected absorbance	
QTA	quartz tube atomizer	
RF	radio frequency	
S/N	signal-to-noise (ratio)	
STPF	stabilized temperature platform furnace	
THB	tetrahydroborate	
TR XRF	total-reflection X-ray fluorescence	
UV	ultraviolet	
WD	wavelength dispersive	
WD XRF	wavelength dispersive X-ray fluorescence	
WSA	wavelength selected absorbance	
XRF	X-ray fluorescence	
XRS	X-ray spectroscopy	
ZBC	Zeeman-effect background correction	

References

Angeli, J., Bengtson, A., Bogaerts, A., Hoffmann, V., Hodoroaba, V.D., and Steers, E. (2003) *J. Anal. At. Spectrom.*, **18**, 670.

Balmer, J.J. (1885) *Ann. Phys.*, **25**, 80.

Barnes, R.B., Richardson, D., Berry, J.W., and Hood, R.L. (1945) *Ind. Eng. Chem. Anal. Ed.*, **17**, 605.

Becker, J.S. (2007) *Inorganic Mass Spectrometry*, John Wiley & Sons, Ltd, New York.

Becker-Ross, H., Florek, S., and Heitmann, U. (2000) *J. Anal. At. Spectrom.*, **15**, 137–141.

Berndt, H. and Schaldach, G. (1988) *J. Anal. At. Spectrom.*, **3**, 709–712.

Bertrand, M., Weber, G., and Schoefs, B. (2003) *Trends Anal. Chem.*, **22**, 254.

Bohr, N. (1913) *Philos. Mag.*, **26**, 476.

Borges, D.L.G., Silva, A.F., Curtius, A.J., Welz, B., and Heitmann, U. (2006) *J. Anal. At. Spectrom.*, **21**, 763–769.

Boss, C.B. and Fredeen, K.J. (1999) *Concepts, Instrumentation and Techniques in Inductively Coupled Plasma Optical Emission Spectrometry*, Perkin-Elmer Co., Norwalk, USA.

Braman, R.S., Justen, L.L., and Forebank, C.C. (1972) *Anal. Chem.*, **44**, 2195–2199.

Broekaert, J.A.C. (2005) *Analytical Atomic Spectrometry with Flames and Plasmas*, Wiley-VCH Verlag GmbH, Weinheim, New York.

Broekaert, J.A.C. and Siemens, V. (2004) *Anal. Bioanal. Chem.*, **380**, 185–189.

Browner, R.F. (1974) *Analyst*, **99**, 617.

Cai, Y. (2000) *TrAC, Trends Anal. Chem.*, **19**, 62.

Carapelle, A., Fleury-Frenette, K., Collette, J.P., Garnir, H.P., and Harlet, P. (2007) *Rev. Sci. Instrum.*, **78**, 123109.

Carr-Brion, K.G. and Payne, K.W. (1970) *Analyst*, **95**, 977.

Das, D.K., McDonald, J.P., Yalisove, S.M., and Pollock, T.M. (2008) *Spectrochim. Acta, Part B*, **63**, 27.

Dedina, J. (1988) *Prog. Anal. Spectrosc.*, **11**, 251–360.

De Giacomo, A., Dell'Aglio, M., De Pascale, O., Longo, S., and Capitelli, M. (2007) *Spectrochim. Acta, Part B*, **62**, 1606.

De Loos-Vollebregt, M.T.C. and de Galan, L. (1985) *Prog. Anal. Spectrosc.*, **8**, 47–81.

D'Ulivo, A., Baiocchi, C., Pitzalis, E., Onor, M., and Zamboni, R. (2004a) *Spectrochim. Acta, Part B*, **59**, 471–486.

D'Ulivo, A., Onor, M., and Pitzalis, E. (2004b) *Anal. Chem.*, **76**, 6342–6352.

Ekinci, N. and İngeç, M. (2008) *Appl. Radiat. Isot.*, **66**, 1117–1122.

Fantoni, R., Caneve, L., Colao, F., Fornarini, L., Lazic, V., and Spizzichino, V. (2008) *Spectrochim. Acta, Part B*, **63**, 1097–1108.

Foest, R., Schmidt, M., and Becker, K. (2006) *Int. J. Mass Spectrom.*, **248**, 87–102.

Fraunhofer, J. (1817) *Ann. Phys.*, **56**, 264.

Galindo, R.E. and Albella, J.M. (2008) *Spectrochim. Acta, Part B*, **63**, 422.

Gleisner, H., Eichardt, K., and Welz, B. (2003) *Spectrochim. Acta, Part B*, **58**, 1663–1678.

Goujon, J., Giakoumaki, A., Piñon, V., Musset, O., Anglos, D., Georgiou, E., and Boquillon, J.P. (2008) *Spectrochim. Acta, Part B*, **63**, 1091–1096.

Greenfield, S. (1994) *J. Anal. At. Spectrom.*, **9**, 565.

Greenfield, S. (1995) *TrAC, Trends Anal. Chem.*, **14**, 435.

Greenfield, S., Jones, I.L.W., and Berry, C.T. (1964) *Analyst*, **89**, 713.

Grimm, W. (1968) *Spectrochim. Acta, Part B*, **23**, 443.

Guo, X., Sturgeon, R.E., Mester, Z., and Gardener, G.J. (2004) *Anal. Chem.*, **76**, 2401–2405.

Harmon, R.S., De Lucia, F.C., Miziolek, A.W., McNesby, K.L., Walters, R.A., and French, P.D. (2005) *Geochem.: Explor. Environ., Anal.*, **5**, 21–28.

Harnly, J.M. (1999) *J. Anal. At. Spectrom.*, **14**, 137.

Hatch, W.R. and Ott, W.L. (1968) *Anal. Chem.*, **40**, 2085–2087.

Heitmann, U., Becker-Ross, H., Florek, S., Huang, M.D., and Okruss, M. (2006) *J. Anal. At. Spectrom.*, **21**, 1314–1320.

Heitmann, U., Schütz, M., Becker-Ross, H., and Florek, S. (1996) *Spectrochim. Acta, Part B*, **51**, 1095–1105.

Heitmann, U., Welz, B., Borges, D.L.G., and Lepri, F.G. (2007) *Spectrochim. Acta, Part B*, **62**, 1222–1230.

Hergenröder, R. and Niemax, K. (1988) *Spectrochim. Acta, Part B*, **43**, 1443–1449.

Herschel, J.F.W. (1823) *Trans. R. Soc. Edinb.*, **9**, 445.

Hieftje, G.M. (1989) *J. Anal. At. Spectrom.*, **4**, 117–122.

Hodoroba, V.D., Unger, W.E.S., Jenett, H., Hoffmann, V., Hagenhoff, B., Kayser, S., and Wetzig, K. (2001) *Appl. Surf. Sci.*, **179**, 30–37.

Holak, W. (1969) *Anal. Chem.*, **41**, 1712–1713.

Hou, X., Stchur, P., Yang, K.X., and Michel, R.G. (1998) *TrAC, Trends Anal. Chem.*, **17**, 532.

Huang, M.D., Becker-Ross, H., Florek, S., Heitmann, U., and Okruss, M. (2006a) *Spectrochim. Acta, Part B*, **61**, 181–188.

Huang, M.D., Becker-Ross, H., Florek, S., Heitmann, U., and Okruss, M. (2006b) *Spectrochim. Acta, Part B*, **61**, 572–578.

Huang, M.D., Becker-Ross, H., Florek, S., Heitmann, U., and Okruss, M. (2006c) *Spectrochim. Acta, Part B*, **61**, 959–964.

Huang, M.D., Becker-Ross, H., Florek, S., Heitmann, U., and Okruss, M. (2008) *Spectrochim. Acta, Part B*, **63**, 566–570.

Ingle Jr, J.D. and Crouch, S.R. (1988) *Spectrochemical Analysis*, Prentice Hall, New Jersey, USA.

IUPAC. (1978) *Spectrochim. Acta, Part B*, **33**, 247–269.

Jakubowski, N., Dorka, R., Steers, E., and Tempez, A. (2007) *J. Anal. At. Spectrom.*, **22**, 722.

Jenkins, R. (1988) *X-Ray Fluorescence Spectrometry*, Wiey-Interscience, New York.

Jenkins, R. and de Vries, J.L. (1970) *Practical X-Ray Spectrometry*, MacMillan, London.

Johansson, S.A.E. and Campbell, J.L. (1988) *P.I.X.E. – A Novel Technique for Elemental Analysis*, John Wiley & Sons, Ltd, Chichester.

Jones, W.G. and Walsh, A. (1960) *Spectrochim. Acta*, **16**, 249–254.

Kennedy, P.K., Hammer, D.K., and Rockwell, B.H. (1997) *Prog. Quant. Electron.*, **21**, 155.

Khan, M.R. and Crumpton, D. (1981) *CRC Crit. Rev. Anal. Chem*, **11**, 103.

Kim, T., Nguyen, B.T., Minassian, V., and Lin, C.T. (2007) *J. Coat. Technol. Res.*, **4**, 241.

Kirchhoff, G.R. (1860) *Philos. Mag.*, **20**, 1.

Klockenkämper, R. and von Bohlen, A. (1999) *J. Anal. At. Spectrom.*, **14**, 571.

Knoth, J. and Schwenke, H. (1978) *Fresen. Z. Anal. Chem.*, **291**, 200.

Koirtyohann, S.R. and Pickett, E.E. (1965) *Anal. Chem.*, **37**, 601–603.

Lachance, G.R. and Claisse, F. (1994) *Quantitative X-Ray Fluorescence Analysis – Theory and Application*, John Wiley & Sons, Ltd, Chichester.

Laqua, K., Schrader, B., Hoffmann, G.G., Moore, D.S., and Vo-Dinh, T. (1995) *Pure Appl. Chem.*, **67**, 1745–1760.

Lundegårdh, H. (1936) *Lantbruks-Hogskol. Ann.*, **3**, 49.

L'vov, B.V. (1961) *Spectrochim. Acta*, **17**, 761.

L'vov, B. (1978) *Spectrochim. Acta, Part B*, **33**, 153–193.

Mages, M., Woelfl, S., Óvári, M., and Jun, W.V.T. (2003) *Spectrochim. Acta, Part B*, **58**, 2129–2138.

Manning, D.C. and Vollmer, J. (1967) *At. Absorp. Newsl.*, **6**, 38–41.

Mascart, E. (1864) *Compt. Rend.*, **58**, 1111.

Massmann, H. (1968) *Spectrochim. Acta, Part B*, **23**, 215–226.

Mika, J. and Török, T. (1974) *Analytical Emission Spectroscopy Fundamentals*, Butterworths, London.

Mohamed, W.T.Y. (2008) *Opt. Laser Technol.*, **40**, 30–38.

Montaser, A. and Golightly, D.W. (1992) *Inductively Coupled Plasmas in Analytical Atomic Spectrometry*, Wiley-VCH Verlag GmbH, New York.

Müller, R.O. (1972) *Spectrochemical Analysis by X-Ray*, Plenum Press, New York.

Neelmeijer, C., Brissaud, I., Calligaro, T., Demortier, G., Hautojärvi, A., Mäder, M., Martinot, L., Schreiner, M., Tuurnala, T., and Weber, G. (2000) *X-Ray Spectrom.*, **29**, 101.

Nichols, E.L. and Howes, H.L. (1924) *Phys. Rev.*, **23**, 472.

Nölte, J. (2003) *ICP Emission Spectrometry – A Practical Guide*, Wiley-VCH Verlag GmbH, Wienheim.

Pappalardo, L., de Sanoit, J., Marchetta, C., Pappalardo, G., Romano, F.P., and Rizzo, F. (2007) *X-Ray Spectrom.*, **36**, 310–315.

Paschen, A. (1916) *Ann. Phys.*, **50**, 901–940.

Pisonero, J., Fernández, B., Pereiro, R., Bordel, N., and Sanz-Medel, A. (2006) *TrAC, Trends Anal. Chem.*, **25**, 11.

Robinson, J.W. (1990) *Atomic Spectroscopy*, Marcel Dekker, New York.

Rosenkranz, B. and Bettmer, J. (2000) *TrAC, Trends Anal. Chem.*, **19**, 138.

Rusak, D.A., Castle, B.C., Smith, B.W., and Winefordner, J.D. (1998) *TrAC, Trends Anal. Chem.*, **17**, 453.

Russell, B.J., Shelton, J.P., and Walsh, A. (1957) *Spectrochim. Acta*, **8**, 317.

Russell, B.J. and Walsh, A. (1959) *Spectrochim. Acta*, **15**, 883–885.

Saisho, H. (1989) *Trends Anal. Chem.*, **8**, 209.

Sallé, B., Mauchien, P., and Maurice, S. (2007) *Spectrochim. Acta, Part B*, **62**, 739.

Schlemmer, G. and Welz, B. (1986) *Spectrochim. Acta, Part B*, **41**, 1157–1165.

Silva, M.M., Krug, F.J., Oliveira, P.V., Nóbrega, J.A., Reis, B.F., and Penteado, D.A.G. (1996) *Spectrochim. Acta, Part B*, **51**, 1925–1934.

Skoog, D.A., Holler, F.J., and Niemann, T.A. (1998) *Principles of Instrumental Analysis*, Harcourt College Publishers, Philadelphia, USA.

Slavin, W., Manning, D.C., and Carnrick, G.R. (1981) *At. Spectrosc.*, **2**, 137–145.

Sommerfeld, A. (1916) *Ann. Phys.*, **51**, 1.

Stchur, R., Yang, K.X., Hou, X., Sun, T., and Michel, R.G. (2001) *Spectrochim. Acta, Part B*, **56**, 1565.

Sturgeon, R.E., Guo, X., and Mester, Z. (2005) *Anal. Bioanal. Chem.*, **382**, 881–883.

Sychra, V., Svoboda, V., and Rubeška, I. (1975) *Atomic Fluorescence Spectroscopy*, Van Nostrand Reinhold, London.

Talbot, W.H.F. (1826) *Brewster's J. Sci.*, **5**, 77.

Tsalev, D.L., Slaveykova, V.I., and Mandjukov, P.B. (1990) *Spectrochim. Acta Rev.*, **13**, 225–274.

Veillon, C., Mansfield, J.M., Parsons, M.L., and Winefordner, J.D. (1966) *Anal. Chem.*, **38**, 204.

Walsh, A. (1955) *Spectrochim. Acta*, **7**, 108–117.

Welz, B., Becker-Ross, H., Florek, S., and Heitmann, U. (2005) *High-Resolution Continuum Source AAS – The Better Way to Do Atomic Absorption Spectrometry*, Wiley-VCH Verlag GmbH, Weinheim.

Welz, B., Becker-Ross, H., Florek, S., Heitmann, U., and Vale, M.G.R. (2003) *J. Brazil. Chem. Soc.*, **14**, 220–229.

Welz, B., Borges, D.L.G., Lepri, F.G., Vale, M.G.R., and Heitmann, U. (2007a) *Spectrochim. Acta, Part B*, **62**, 873–883.

Welz, B. and Melcher, M. (1984a) *Analyst*, **109**, 569–572.

Welz, B. and Melcher, M. (1984b) *Analyst*, **109**, 573–575.

Welz, B. and Melcher, M. (1984c) *Analyst*, **109**, 577–579.

Welz, B., Schlemmer, G., and Mudakavi, J.R. (1992) *J. Anal. At. Spectrom.*, **7**, 1257–1271.

Welz, B. and Schubert-Jacobs, M. (1986) *J. Anal. At. Spectrom.*, **1**, 23–27.

Welz, B. and Sperling, M. (1999) *Atomic Absorption Spectrometry*, Wiley-VCH Verlag GmbH, Weiheim.

Welz, B., Vale, M.G.R., Borges, D.L.G., and Heitmann, U. (2007b) *Anal. Bioanal. Chem.*, **389**, 2085–2095.

Wendt, R.H. and Fassel, V.A. (1965) *Anal. Chem.*, **37**, 920.

West, T.S. (1967) *Endeavour*, **26**, 44.

Willis, J.B. (1965) *Nature*, **207**, 715.

Winefordner, J.D. and Mansfield, J.M. (1967) *Appl. Spectrosc. Rev.*, **1**, 1.

Winefordner, J.D. and Vickers, T.J. (1964) *Anal. Chem.*, **36**, 161.

Wobrauschek, P. (2007) *X-Ray Spectrom.*, **36**, 289.

Wollaston, W.H. (1802) *Philos. Trans. R. Soc. London*, **92**, 365.

Wood, R.W. (1902) *Philos. Mag.*, **3**, 128.

Wood, R.W. (1905) *Philos. Mag.*, **10**, 513.

Woodson, T.T. (1939) *Rev. Sci. Instrum.*, **10**, 308.

Wormhoudt, J., Iannarilli Jr, F.J., Jones, S., Annen, K.D., and Freedman, A. (2005) *Appl. Spectrosc.*, **59**, 1098–1102.

Yoneda, Y. and Horiuchi, T. (1971) *Rev. Sci. Instrum.*, **42**, 1069.

Yu, Y.-L., Du, Z., and Wang, J.-H. (2007) *J. Anal. At. Spectrom.*, **22**, 650–656.

Zarkadas, C. and Karydas, A.G. (2004) *Spectrochim. Acta, Part B*, **59**, 1611–1618.

Zhou, Z., Zhou, K., and Hou, X. (2005) *Appl. Spectrosc. Rev.*, **40**, 165.

Zybin, A., Koch, J., Wizemann, H.D., Franzke, J., Niemax, K. (2005) *Spectrochim. Acta, Part B*, **60**, 1–11.

14
Molecular Fluorescence

Bernard Valeur

14.1	**Introduction** 479	
14.2	**Characteristics of Fluorescence Emission** 481	
14.2.1	Radiative and Nonradiative Transitions between Electronic States 481	
14.2.1.1	Internal Conversion 482	
14.2.1.2	Fluorescence 482	
14.2.1.3	Intersystem Crossing and Subsequent Processes 483	
14.2.2	Lifetimes and Quantum Yields 484	
14.2.2.1	Excited-State Lifetimes 484	
14.2.2.2	Quantum Yields 485	
14.2.3	Emission and Excitation Spectra 486	
14.2.3.1	Emission Spectra 486	
14.2.3.2	Excitation Spectra 487	
14.2.3.3	Stokes Shift 488	
14.2.4	Fluorescence Anisotropy 488	
14.3	**Effects of Intermolecular Photophysical Processes on Fluorescence** 489	
14.3.1	Main Photoinduced Intermolecular Processes 489	
14.3.2	Fluorescence Quenching 490	
14.3.2.1	Dynamic Quenching 490	
14.3.2.2	Static Quenching 492	
14.3.3	Photoinduced Electron Transfer 494	
14.3.4	Formation of Excimers and Exciplexes 495	
14.3.4.1	Excimers 495	
14.3.4.2	Exciplexes 498	
14.3.5	Photoinduced Proton Transfer 498	
14.3.5.1	Equations for Excited-State Deprotonation 499	
14.3.5.2	Determination of the Excited-State pK^* 500	
14.3.5.3	pH Dependence of Absorption and Emission Spectra 501	
14.3.6	Excitation Energy Transfer 501	
14.4	**Applications of Molecular Fluorescence** 502	
14.4.1	Historical Introduction 502	

Encyclopedia of Applied Spectroscopy. Edited by David L. Andrews.
Copyright © 2009 WILEY-VCH Verlag GmbH & Co. KGaA, Weinheim
ISBN: 978-3-527-40773-6

14.4.2	Estimation of Local Physical Parameters	503
14.4.2.1	Viscosity 503	
14.4.2.2	Temperature 508	
14.4.2.3	Pressure 511	
14.4.3	Chemical Sensing 513	
14.4.3.1	Fluorescent Molecular Sensors 513	
14.4.3.2	Microfabricated Analysis Systems 519	
14.4.4	Tracers in Biology 521	
14.4.4.1	Fluorescent Proteins 522	
14.4.4.2	Semiconductor Nanocrystals 523	
14.4.5	Clinical Diagnostics 523	
14.4.5.1	Critical Care Analysis 523	
14.4.5.2	Angiography 524	
14.4.5.3	Bladder Tumor Detection 524	
14.4.6	Miscellaneous 526	
14.4.6.1	Fluorescence LIDAR 526	
14.4.6.2	Food Science 527	
14.4.6.3	Forensics 527	
14.4.6.4	Counterfeit Detection 528	
	References 529	
	Further Reading 531	

14.1
Introduction

Molecular fluorescence is extensively used in physical, chemical, material, biological, and medical sciences as a tool of detection/analysis, visualization, investigation of local properties, diagnostics, and so on. In fact, fluorescent compounds can be used not only for mere visualization but also as probes, indicators, sensors, and tracers for providing information on local physical or chemical parameters: pressure, temperature, viscosity, polarity, pH, concentrations of ionic or neutral species, and so on (Valeur, 2002; Lakowicz, 2006; Schulman, 1985–1993; Lakowicz, 1991–2007; Valeur and Brochon, 2001; Kraayenhof and Visser, 2003; Hof, Hutterer and Fidler, 2005; Berberan-Santos, 2007; Jameson, Croney and Moens, 2003; Valeur, Berberan-Santos and Martin, 2007).

Once a molecule is excited by absorption of a photon, it can return to the ground state via many pathways (Figure 14.1). The radiative pathway, that is, emission of photons with retention of spin multiplicity, is called *fluorescence*. Another radiative pathway is *phosphorescence*, but this process first requires an intersystem crossing from the initially singlet excited state to a triplet state. This is the major distinction to be made between fluorescence and phosphorescence, and not only the fact that the decay of fluorescence intensity following light pulse excitation is generally much faster than that of phosphorescence. The nonradiative pathway of de-excitation called *internal conversion* is the direct return to the ground state without emission of photon. In the excited state, the molecule may undergo intramolecular processes: charge transfer, conformational change, and so on.

Interactions in the excited state with other molecules may lead to de-excitation: electron transfer, proton transfer, energy transfer, excimer or exciplex formation, and photochemical reaction. These de-excitation pathways may compete with fluorescence emission if they take place at a time scale comparable with the average time (called *lifetime*) during which the molecules stay in the excited state. This average time represents the *experimental time window* for observation of dynamic processes. The characteristics of fluorescence (spectrum, quantum yield, and lifetime), which are affected by any excited-state process

Encyclopedia of Applied Spectroscopy. Edited by David L. Andrews.
Copyright © 2009 WILEY-VCH Verlag GmbH & Co. KGaA, Weinheim
ISBN: 978-3-527-40773-6

14 Molecular Fluorescence

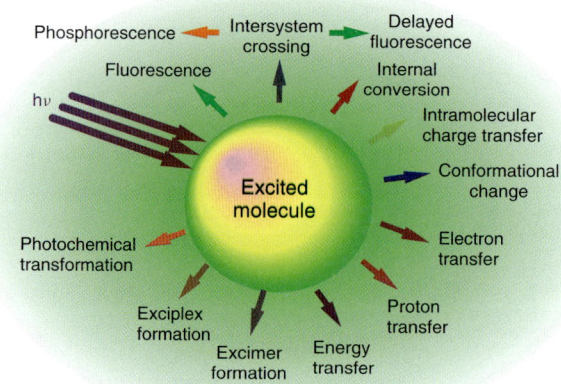

Fig. 14.1 Excited-state processes following excitation of a molecule. (Reproduced with permission from Valeur (2002). Copyright Wiley-VCH Verlag GmbH & Co. KGaA.)

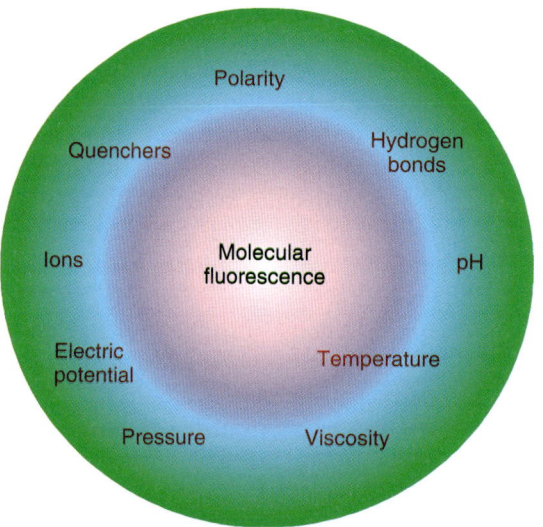

Fig. 14.2 Physical and chemical parameters affecting the emission of fluorescence. (Reproduced with permission from Valeur (2002). Copyright Wiley-VCH Verlag GmbH & Co. KGaA.)

involving interactions of the excited molecule with its close environment, can then provide information on such a microenvironment. Figure 14.2 shows the physical and chemical parameters that characterize a microenvironment and can thus affect the fluorescence characteristics of a molecule.

In this chapter, in the second section the characteristics of fluorescence emission are described. In the third section, the effects of photoinduced intermolecular processes on fluorescence are discussed. The fourth section presents a selection of applications of molecular fluorescence in various fields.

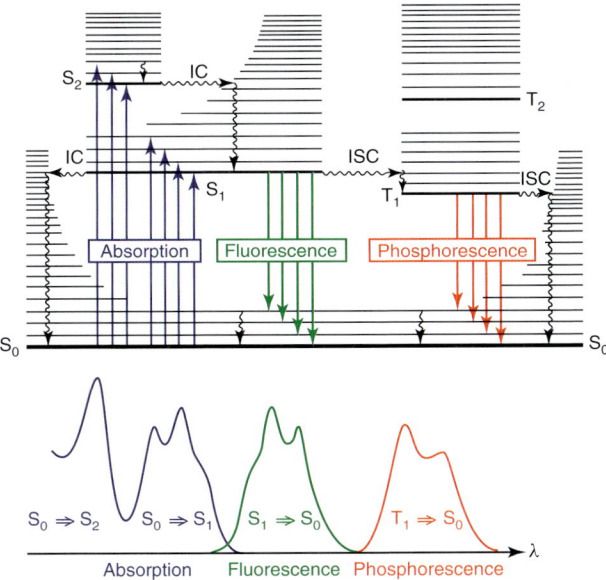

Fig. 14.3 Perrin–Jablonski diagram and relative position of the spectra. IC, internal conversion; ISC, intersystem crossing. (Reproduced with permission from Valeur (2002). Copyright Wiley-VCH Verlag GmbH & Co. KGaA.)

14.2 Characteristics of Fluorescence Emission

In this section, the characteristics of fluorescence emission are described for molecules in dilute solution in the absence of any interaction with other molecules (Valeur, 2002) chap. 3. The photophysical processes resulting from such interactions are discussed in Section 14.3.

14.2.1 Radiative and Nonradiative Transitions between Electronic States

The processes that are involved in the excitation and de-excitation of a fluorescent molecule are conveniently visualized by means of the Perrin–Jablonski diagram (Figure 14.3): photon absorption, internal conversion, fluorescence, intersystem crossing, phosphorescence, and so on. Vibrational levels are associated with all electronic states. The singlet states are denoted S_0 (fundamental electronic state), $S_1, S_2, \ldots$ and the triplet states, $T_1, T_2, \ldots$. It is important to note that absorption is very fast ($\approx 10^{-15}$ seconds) with respect to all other processes (so that there is no concomitant displacement of the nuclei according to the Franck–Condon principle).

In a solution at room temperature, the majority of molecules are in the 0 (lowest) vibrational energy level of S_0. Therefore, the vertical arrows corresponding to absorption start from this level. Absorption of a photon can bring a molecule to one of the vibrational levels of $S_1, S_2, \ldots$. The subsequent possible de-excitation processes are described.

Tab. 14.1 Characteristic times involved in the excitation and de-excitation of a molecule. (From Valeur (2002).)

Process or lifetime	Characteristic time
Absorption	10^{-15} s
Vibrational relaxation	$10^{-12} - 10^{-10}$ s
Lifetime of the singlet excited state S_0	$10^{-10} - 10^{-7}$ s $\Rightarrow$ fluorescence
Intersystem crossing	$10^{-10} - 10^{-8}$ s
Internal conversion	$10^{-11} - 10^{-9}$ s
Lifetime of the triplet excited state T_1	$>10^{-6}$ s $\Rightarrow$ phosphorescence

The characteristic times of these processes are given in Table 14.1.

14.2.1.1 Internal Conversion

Internal conversion is a nonradiative transition between two electronic states of the same spin multiplicity.

When a molecule is excited to an energy level higher than the lowest vibrational level of the first electronic state, vibrational relaxation – and internal conversion $S_n \rightarrow S_1$ if the singlet excited state is higher than S_1 – leads the excited molecule toward the lowest vibrational level of the S_1 singlet state. The excess of vibrational energy is transferred to the solvent during collisions of the excited molecule with the surrounding solvent molecules.

Internal conversion from S_1 to S_0 is less efficient than that from S_2 to S_1 owing to the much larger energy gap between S_1 and S_0. Therefore, internal conversion from S_1 to S_0 can compete with emission of photons (fluorescence) and intersystem crossing to the triplet state.

14.2.1.2 Fluorescence

Emission of photons from an excited molecule without change in multiplicity is called *fluorescence*. Besides a few exceptions, fluorescence emission occurs from S_1 and therefore its characteristics (except polarization) do not depend on the excitation wavelength.

The fluorescence spectrum is located at higher wavelengths (lower energy) than the absorption spectrum because of the energy loss in the excited state by vibrational relaxation (Figure 14.3). The gap (expressed in wavenumbers) between the maximum of the first absorption band and the maximum of fluorescence is called *Stokes shift*.

In general, the differences between the vibrational levels are similar in the ground and excited states, so that the fluorescence spectrum often resembles the first absorption band ("mirror image" rule).

Emission of a photon is a process that is as fast as absorption of a photon ($\approx 10^{-15}$ second), but excited molecules stay in the S_1 state for a certain time (a few tens of picoseconds to a few hundreds of nanoseconds, according to the type of molecule and the medium) before emitting a photon or undergoing other de-excitation processes (internal conversion and intersystem crossing). Thus, after excitation of a population of molecules by a very short pulse of light, the fluorescence intensity decreases exponentially with a characteristic time, called *excited-state lifetime*, reflecting the average time during which the molecules stay in the S_1 excited state (10^{-10}–10^{-7} seconds) before returning to the ground state (see Section 14.2.2.1). The fluorescence intensity decay is formally comparable with a radioactive decay that is also exponential, with a characteristic time called *radioactive period*, characterizing the average lifetime of a radioelement before disintegration.

The emission of fluorescence photons is a spontaneous process. Another emission process, called *stimulated emission*, can be observed under certain conditions, for instance, in dye lasers.

14.2.1.3 Intersystem Crossing and Subsequent Processes

A third possible de-excitation process from S_1 is intersystem crossing toward the T_1 triplet state, followed by other processes (Figure 14.3).

Intersystem Crossing *Intersystem crossing* is a nonradiative transition between two isoenergetic vibrational levels belonging to electronic states of different multiplicities.

Once an excited molecule has relaxed to the 0 vibrational level of the S_1 state, it can move to the isoenergetic vibrational level of the T_n triplet state; then, vibrational relaxation brings it into the lowest vibrational level of T_1. Intersystem crossing may compete with the other pathways of de-excitation from S_1 (fluorescence and internal conversion $S_1 \to S_0$).

Crossing between states of different multiplicity is in principle forbidden, but spin–orbit coupling can be large enough to make it possible. The presence of heavy atoms (such as I, Br, Pb, etc.) favors intersystem crossing because it increases spin–orbit coupling.

Phosphorescence versus Nonradiative De-excitation The emission of photons from the triplet state T_1 is called *phosphorescence*. Although the transition $T_1 \to S_0$ is forbidden, phosphorescence can be observed because of spin–orbit coupling. The lifetime of the triplet state T_1 is longer ($>10^{-6}$ second) than that of the singlet state S_1.

In solution at room temperature, the numerous collisions of the molecules in the triplet state state T_1 with solvent molecules favor intersystem crossing and vibrational relaxation to S_0. Nonradiative de-excitation from the triplet state T_1 is thus predominant. On the contrary, at low temperature and/or in a rigid medium, phosphorescence can be observed. Under these conditions, the lifetime of the triplet state may be long enough to observe phosphorescence on a time scale up to seconds, minutes, or even more.

The phosphorescence spectrum is located at wavelengths higher than the fluorescence spectrum (Figure 14.3) because the energy of the lowest vibrational level of the triplet state T_1 is lower than that of the singlet state S_1.

Delayed Fluorescence

- **Thermally activated delayed fluorescence**: When the energy difference between S_1 and T_1 is small and when the lifetime of T_1 is long enough, reverse intersystem crossing $T_1 \to S_1$ can occur. Thus, the emission spectrum is identical to that of normal fluorescence, but the decay

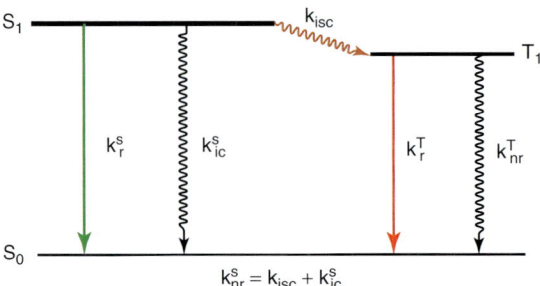

Scheme 14.1 De-excitation processes with relevant rate constants.

time constant is much longer because the molecules stay in the triplet state before emitting from S_1. The efficiency of delayed fluorescence increases with increasing temperature, because the process is thermally activated. It is generally called *delayed fluorescence of E-type* because it has been observed for the first time with eosin. E-type-delayed fluorescence is usually a weak phenomenon, but it is exceptionally intense in the case of fullerene C_{70} because the quantum yield for triplet formation is close to 1, the S_1-T_1 energy gap is quite small, and the intrinsic triplet lifetime is in the order of several milliseconds (Berberan-Santos and Garcia, 1996).

- **Triplet–triplet annihilation:** In concentrated solutions, a collision between two molecules in state T_1 can provide enough energy to allow one of them to return to state S_1. Such a triplet–triplet annihilation thus leads to a delayed fluorescence emission. It is called *delayed fluorescence of P type* because it has been observed for the first time with pyrene. The decay time constant of this type of delayed fluorescence is half the lifetime of the triplet state in dilute solution, and the intensity has a characteristic quadratic dependence with excitation light intensity.

14.2.2
Lifetimes and Quantum Yields

14.2.2.1 Excited-State Lifetimes

The various processes of de-excitation are represented in Scheme 14.1 with the relevant rate constants. The de-excitation processes that result from intermolecular interactions are not considered in this section. They are described in Section 14.3.

The rate constants are denoted as follows:

- k_r^S is the rate constant for radiative deactivation $S_1 \to S_0$ with emission of fluorescence.
- k_{ic}^S is the rate constant for internal conversion $S_1 \to S_0$.
- k_{isc} is the rate constant for intersystem crossing.
- k_r^T is the rate constant for radiative deactivation $T_1 \to S_0$ with emission of phosphorescence.
- k_{nr}^T is the rate constant for nonradiative deactivation (intersystem crossing) $T_1 \to S_0$.

It is convenient to introduce the overall rate constant k_{nr}^S for nonradiative deactivation from S_1: $k_{nr}^S = k_{ic}^S + k_{isc}$ (14.1)

Let us consider a dilute solution of a fluorescent molecule M whose concentration is [M] (in mole per liter). A pulse of light supposed to be infinitely short (i.e., a δ-function (Dirac)) at time 0 brings a certain number of molecules M to the S_1 excited state by absorption of photons. Then, these excited molecules return to S_0 either radiatively or nonradiatively or undergo intersystem crossing. As in classical chemical kinetics, the rate of disappearance of excited molecules is expressed by the following differential equation:

$$-\frac{d[^1M^*]}{dt} = (k_r^S + k_{nr}^S)[^1M^*] \quad (14.2)$$

The time evolution of the concentration in excited molecules $[^1M^*]$ is obtained by integration of this equation:

$$[^1M^*] = [^1M^*]_0 \exp\left(-\frac{t}{\tau_S}\right) \quad (14.3)$$

where $[^1M^*]_0$ is the concentration in excited molecules at time 0 resulting from pulse light excitation, and τ_S is the lifetime of excited state S_1, given as

$$\tau_S = \frac{1}{k_r^S + k_{nr}^S} \quad (14.4)$$

The fluorescence intensity is defined as the amount of photons (in mol, or its equivalent, in einstein; 1 einstein = 1 mol of photons) emitted per unit time (second) and per unit volume of solution (liter) according to

$$M^* \xrightarrow{k_r^S} M + photon \quad (14.5)$$

The fluorescence intensity i_F at time t after excitation by a very short pulse of light at time 0 is proportional, at any time, to the instantaneous concentration of molecules still excited $[^1M^*]$; the proportionality factor is the rate constant for radiative de-excitation k_r^S:

$$i_F(t) = k_r^S[^1M^*]$$
$$= k_r^S[^1M^*]_0 \exp\left(-\frac{t}{\tau_S}\right) \quad (14.6)$$

$i_F(t)$, the δ-pulse response of the system, decreases according to a single exponential.

In practice, the measured signal I_F representing the fluorescence intensity is proportional to i_F, the proportionality factor depending on instrumental conditions. The numerical value of I_F is thus generally obtained in an arbitrary scale.

If the only way of de-excitation from S_1 to S_0 was fluorescence emission, the lifetime would be $1/k_r^S$: it is called *radiative lifetime* (in preference to *natural lifetime*) and denoted τ_r.

The lifetime of a homogeneous population of fluorophores is in principle independent of the excitation wavelength because internal conversion and vibrational relaxation are always very fast in solution and emission arises from the lowest vibrational level of state S_1.

If a fraction of excited molecules reach the triplet state from which they return to the ground state either radiatively or nonradiatively, the concentration of molecules in the triplet state decays exponentially with a time constant τ_T representing the lifetime of the triplet state:

$$\tau_T = \frac{1}{k_r^T + k_{nr}^T} \quad (14.7)$$

14.2.2.2 Quantum Yields

The *fluorescence quantum yield* Φ_F is the fraction of excited molecules that return to the ground state S_0 with emission of fluorescence photons.

$$\Phi_F = \frac{k_r^S}{k_r^S + k_{nr}^S} = k_r^S \tau_S \quad (14.8)$$

In other words, the fluorescence quantum yield is the ratio of the number of emitted photons (over the whole duration of the decay) and the number of absorbed photons. According to Eq. (14.6), the ratio of the δ-pulse response $i_F(t)$ and the number of absorbed photons is given by

$$\frac{i_F(t)}{[^1M^*]_0} = k_r^S \exp\left(-\frac{t}{\tau_S}\right) \quad (14.9)$$

The fluorescence quantum yield is then obtained by integration of this relation over the whole duration of the decay (mathematically from 0 to infinity):

$$\frac{1}{[^1M^*]_0} \int_0^\infty i_F(t) = k_r^S \tau_S = \Phi_F \quad (14.10)$$

The quantum yields of intersystem crossing (Φ_{isc}) and phosphorescence (Φ_P) are given by

$$\Phi_{isc} = \frac{k_{isc}}{k_r^S + k_{nr}^S} = k_{isc}^S \tau_S \quad (14.11)$$

$$\Phi_P = \frac{k_r^T}{k_r^T + k_{nr}^T} \Phi_{isc} \quad (14.12)$$

Using the radiative lifetime, as previously defined, the fluorescence quantum yield can also be written as

$$\Phi_F = \frac{\tau_S}{\tau_r} \quad (14.13)$$

It is interesting to note that when the fluorescence quantum yield and the excited-state lifetime of a fluorophore are measured under the same conditions, the nonradiative and radiative rate constants can be easily calculated by means of the following relations:

$$k_r^S = \frac{\Phi_F}{\tau_S}, \quad k_r^S = \frac{1}{\tau_S}(1 - \Phi_F) \quad (14.14)$$

14.2.3
Emission and Excitation Spectra

14.2.3.1 Emission Spectra

The variations in fluorescence intensity as a function of the observation wavelength λ_E, for a fixed excitation wavelength λ_F, represents the *emission spectrum*. It is convenient to express the steady-state fluorescence intensity (i.e., under continuous illumination) per absorbed photon as a function of the wavelength of the emitted photons, denoted $F_\lambda(\lambda_F)$ (in m^{-1} or nm^{-1}), normalized so that the integral over the whole spectrum is equal to the fluorescence quantum yield Φ_F:

$$\int_0^\infty F_\lambda(\lambda_F) d\lambda_F = \Phi_F \quad (14.15)$$

$F_\lambda(\lambda_F)$ represents the *fluorescence spectrum* or *emission spectrum*: it reflects the distribution of the probability of the various transitions from the lowest vibrational level of S_1 to the various vibrational levels of S_0.

Assuming that the number of absorbed photons follows the Beer–Lambert law, the intensity of the transmitted light is given by

$$I_T(\lambda_E) = I_0(\lambda_E) \exp[-2.3 \, \varepsilon(\lambda_E) \, l \, c] \quad (14.16)$$

where $I_0(\lambda_E)$ is the intensity of the incident light, $\varepsilon(\lambda_E)$ denotes the molar absorption coefficient of the fluorophore at wavelength λ_E (in liter per mole per centimeter), l is the optical path in the sample (in centimeter), and c is the concentration (in mol per liter).

The quantity $\varepsilon(\lambda_E)lc$ represents the absorbance $A(\lambda_E)$ at wavelength λ_E.

The difference between the intensity of the incident light $I_0(\lambda_E)$ and the intensity of the transmitted light $I_T(\lambda_E)$ represents the absorbed intensity $I_A(\lambda_E)$.

The fluorescence intensity can be written as

$$I_F(\lambda_E, \lambda_F) = k\, F_\lambda(\lambda_F)\, I_A(\lambda_E) \quad (14.17)$$

where k is a proportionality factor that depends on several parameters and in particular on the optical configuration for observation (i.e., the solid angle through which the instrument collects fluorescence that is, in fact, emitted in all directions) and on the bandwidth of the monochromators (i.e., the widths of the entrance and exit slits).

Finally, the general expression for the fluorescence intensity is the following:

$$I_F(\lambda_E, \lambda_F) = k\, F_\lambda(\lambda_F)\, I_0(\lambda_E)$$
$$\times \{1-\exp[-2.3\varepsilon(\lambda_E)lc]\}$$
$$(14.18)$$

In highly diluted solutions, the exponential term can be approximated to $1 - 2.3\varepsilon lc$, so that the fluorescence intensity becomes proportional to the concentration:

$$I_F(\lambda_E, \lambda_F) \cong k\, F_\lambda(\lambda_F)\, I_0(\lambda_E)$$
$$\times [2.3\varepsilon(\lambda_E)lc]$$
$$= 2.3\, k\, F_\lambda(\lambda_F)\, I_0(\lambda_E)\, A(\lambda_E)$$
$$(14.19)$$

It should be emphasized that such a proportionality is valid only for low absorbances. Deviation from a linear variation increases with increasing absorbance.

In practice, the variations in I_F measured as a function of wavelength λ_F, for a fixed excitation wavelength λ_E, reflect the variations in $F_\lambda(\lambda_F)$ and thus provide the fluorescence spectrum. Because the proportionality factor k is generally unknown, the numerical value of the measured intensity I_F has no meaning, and generally speaking, I_F is expressed in arbitrary units. It should be noted that k depends on the wavelength because the transmission of the monochromator and the sensitivity of the detector are wavelength dependent. Therefore, correction of spectra is necessary for quantitative measurements.

Spectrofluorometry is a very sensitive technique – up to 1000 times more sensitive than spectrophotometry. This is because the fluorescence intensity is measured above a low background level, whereas in the measurement of low absorbances, two large signals that are slightly different are compared. Thanks to outstanding progress in instrumentation, it is now possible in some cases to even detect a single fluorescent molecule.

14.2.3.2 Excitation Spectra

The variations in fluorescence intensity as a function of the excitation wavelength λ_E, for a fixed observation wavelength λ_F, represent the *excitation spectrum*. In Eq. (14.19), when the observation wavelength λ_F is fixed, the variations in fluorescence intensity reflect the evolution of the product $I_0(\lambda_E)\, A(\lambda_E)$. Provided that the wavelength dependence of the incident light $I_0(\lambda_E)$ is compensated, the sole term to be taken into consideration is $A(\lambda_E)$, which represents the absorption spectrum. The corrected excitation spectrum is thus identical in shape with the absorption spectrum provided that there is a single species in the ground state.

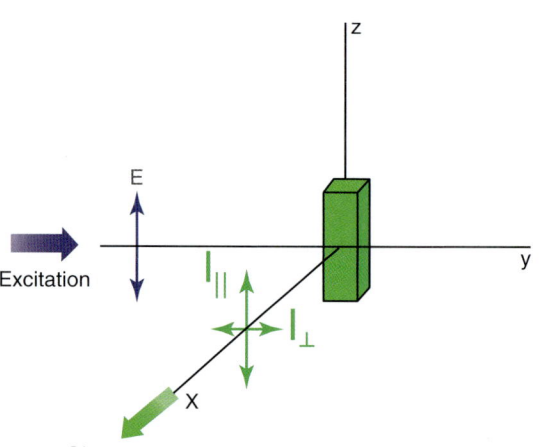

Scheme 14.2 Usual optical configuration for fluorescence anisotropy measurements.

In contrast, when several species are present, or when a sole species exists in different forms in the ground state (aggregates, complexes, tautomeric forms, etc.), the excitation and absorption spectra are no longer superimposable. Comparison between absorption and emission spectra often provides useful information.

14.2.3.3 Stokes Shift

The *Stokes shift* is the difference, expressed in wavenumbers, between the maximum of the first absorption band and the maximum of fluorescence spectrum: $\Delta \bar{\nu} = \bar{\nu}_a - \bar{\nu}_f$.

This parameter depends on the structure of the molecule and on its microenvironment. For instance, when the dipole moment of a fluorescent molecule is higher in the excited state than in the ground state, the Stokes shift increases with solvent polarity. This dependence is a means to estimate the polarity of an unknown environment.

From the practical point of view, the detection of a fluorescent species is easier when the Stokes shift is larger.

14.2.4
Fluorescence Anisotropy

Fluorescence anisotropy, also called *emission anisotropy*, is used to characterize the polarization of fluorescence resulting from photoselection when an assembly of molecules is illuminated by a linearly polarized light (Valeur, 2002 Chapter 5). It is defined as

$$r = \frac{I_{||} - I_{\perp}}{I_{||} + 2I_{\perp}} \quad (14.20)$$

where $I_{||}$ and $I_{\perp}$ are the intensities measured with a linear polarizer for emission parallel and perpendicular, respectively, to the electric vector of linearly polarized incident light (which is often vertical) (Scheme 14.2). The quantity $I_{||} + 2I_{\perp}$ is proportional to the total fluorescence intensity I.

Fluorescence polarization may also be characterized by the polarization ratio, also called the *degree of polarization*, defined as

$$p = \frac{I_{||} - I_{\perp}}{I_{||} + I_{\perp}} \quad (14.21)$$

The relation between r and p is

$$r = \frac{2p}{(3-p)} \quad (14.22)$$

For parallel absorbing and emitting transition moments, the (theoretical) values are $(r, p) = (1/2, 2/5)$; when the transition moments are perpendicular, the values are $(r, p) = (-1/3, -1/5)$.

In many cases, it is preferable to use emission anisotropy because it is additive; the overall contribution of n components r_i, each contributing to the total fluorescence intensity with a fraction $f_i = I_i/I$, is given by

$$r = \sum_{i=1}^{n} f_i r_i \quad \text{with} \quad \sum_{i=1}^{n} f_i = 1 \quad (14.23)$$

On continuous illumination, the measured emission anisotropy is called *steady-state emission anisotropy* and is given by

$$\bar{r} = \frac{\int_0^\infty r(t) I(t) dt}{\int_0^\infty I(t) dt} \quad (14.24)$$

where $r(t)$ is the instantaneous anisotropy and $I(t)$ is the fluorescence intensity at time t following a δ-pulse excitation.

The term *fundamental emission anisotropy* describes a situation in which no depolarizing events occur subsequent to the initial formation of the emitting state, such as those caused by rotational diffusion or energy transfer. It also assumes that there is no overlap between differently polarized transitions. The (theoretical) value of the fundamental emission anisotropy depends on the angle α between the absorption and emission transition moments according to the relation $r_0 = <3\cos^2\alpha>/5$ where $<>$ denotes an average over the orientations of the photoselected molecules.

r_0 can take on values ranging from $-1/5$ for $\alpha = 90°$ (perpendicular transition moments) to $2/5$ for $\alpha = 0°$ (parallel transition moments). In spite of the severe assumptions, the expression is frequently used to determine relative transition moment angles.

In time-resolved fluorescence with δ-pulse excitation, the theoretical value at time zero is identified with the fundamental emission anisotropy.

14.3
Effects of Intermolecular Photophysical Processes on Fluorescence

14.3.1
Main Photoinduced Intermolecular Processes

The interaction between an excited molecule M* and another molecule Q may occur of course as long as M is in its excited state (Valeur, 2002, Chapter 4). The key parameter is thus the lifetime of the excited state of M*, which is the experimental time window through which processes of similar duration, and competing with the intrinsic de-excitation, can be observed. It is subsequently denoted τ_0 and is given by Eq. (14.4) rewritten as

$$\tau_0 = \frac{1}{k_r + k_{nr}} \quad (14.25)$$

Scheme 14.3 illustrates the competition between the intermolecular interaction characterized by the rate constant k_q for the bimolecular process and the intrinsic de-excitation of M* whose rate constant is $k_M = 1/\tau_0$.

The main intermolecular photophysical processes are presented in Table 14.2.

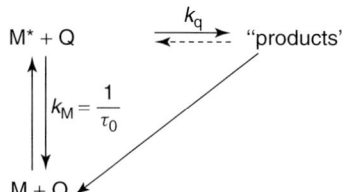

Scheme 14.3 Competition in the excited state between intrinsic de-excitation and interaction with another molecule Q.

Tab. 14.2 Main photophysical processes. (From Valeur (2002).)

Photophysical process	M* + Q → products	Donor	Acceptor
Collision with a heavy atom (e.g., I$^-$, Br in CBr$_4$) or a paramagnetic species (e.g., O$_2$, NO)	M* + Q → M + Q + heat		
Electron transfer	D* + A → D$^{\cdot+}$ + A$^{\cdot-}$	D*	A
	A* + D → A$^{\cdot-}$ + D$^{\cdot+}$	D	A*
Excimer formation	M* + M → (MM)*		
Exciplex formation	D* + A → (DA)*	D*	A
	A* + D → (DA)*	D	A*
Proton transfer	AH* + B → A^{-*} + BH$^+$	AH*	B
	B* + AH → BH^{+*} + A$^-$	AH	B*
Energy transfer			
Heterotransfer	D* + A → D + A*	D*	A
Homotransfer	M* + M → M + M*	M*	M

It should be noted that most of them involve a fast transfer process from a donor to an acceptor: electron transfer, proton transfer, and energy transfer.

Because of the additional way of de-excitation due to the interaction with Q, the fluorescence quantum yield of M is decreased. The loss of fluorescence intensity is called *fluorescence quenching* irrespective of the nature of the competing intermolecular process and even if this process leads to a fluorescent species (the term quenching applies only to the initially excited molecule). Another consequence is the faster decrease in the fluorescence intensity after excitation by a light pulse.

If Q is in large excess, the excited-state process occurs under nondiffusive conditions. However, if Q is not in large excess and if mutual approach of M* and Q is possible during the excited-state lifetime, the bimolecular excited-state process is then *diffusion controlled*, and the rate constant is time dependent.

14.3.2
Fluorescence Quenching

14.3.2.1 Dynamic Quenching

Stern–Volmer Kinetics In the first approach, let us assume, that the experimental quenching rate constant k_q is time independent. According to the

14.3 Effects of Intermolecular Photophysical Processes on Fluorescence

simplified Scheme 14.3, the time evolution of the concentration of M* following a δ-pulse excitation obeys the following differential equation:

$$\frac{d[M^*]}{dt} = -(k_M + k_q[Q])\,[M^*]$$

$$= -\left(\frac{1}{\tau_0} + k_q[Q]\right)[M^*]$$

(14.26)

With the initial condition $[M^*] = [M^*]_0$ at $t = 0$, integration of this differential equation leads to

$$[M^*] = [M^*]_0 \exp\left\{-\left(\frac{1}{\tau_0} + k_q[Q]\right)t\right\}$$

(14.27)

The fluorescence intensity is obtained by multiplying the concentration of M* by the radiative rate constant:

$$i(t) = k_r[M^*]$$
$$= k_r[M^*]_0 \exp\left\{-\left(\frac{1}{\tau_0} + k_q[Q]\right)t\right\}$$
$$= i(0) \exp\left\{-\left(\frac{1}{\tau_0} + k_q[Q]\right)t\right\}$$

(14.28)

where k_r is the radiative rate constant of M*. The fluorescence decay is thus a single exponential whose time constant is given by

$$\tau = \frac{1}{\frac{1}{\tau_0} + k_q[Q]} = \frac{\tau_0}{1 + k_q\tau_0[Q]}$$

(14.29)

Hence,

$$\frac{\tau_0}{\tau} = 1 + k_q\tau_0[Q]$$

(14.30)

The fluorescence quantum yields in the absence and in the presence of quencher are, respectively,

$$\Phi_0 = k_r\tau_0$$

(14.31)

and

$$\Phi = \frac{k_r}{k_r + k_{nr} + k_q[Q]}$$
$$= \frac{k_r}{1/\tau_0 + k_q[Q]}$$

(14.32)

The ratio leads to the Stern–Volmer relation:

$$\frac{\Phi_0}{\Phi} = \frac{I_0}{I} = 1 + k_q\tau_0[Q] = 1 + K_{SV}[Q]$$

(14.33)

where I_0 and I are the steady-state fluorescence intensities (for a couple of wavelengths λ_E and λ_F) in the absence and in the presence of quencher, respectively. $K_{SV} = k_q\tau_0$ is the Stern–Volmer constant. Generally, the ratio I_0/I is plotted versus the quencher concentration (Stern–Volmer plot). If the variation is found to be linear, the slope yields the Stern–Volmer constant. Then, k_q can be calculated if the excited-state lifetime in the absence of quencher is known.

Time-resolved experiments in the absence and in the presence of quencher allow us to check whether the fluorescence decay is, in fact, a single exponential and provide directly the value of k_q.

Two cases are to be distinguished:

- If the bimolecular process is not diffusion-limited: $k_q = pk_1$, where p is the probability of reaction for the encounter pair (often called efficiency), and k_1 is the diffusional rate constant.
- If the bimolecular process is diffusion-limited: k_q is identical to the diffusional rate constant k_1, which can be written in the following simplified form (proposed for the first

time by Smoluchowski):

$$k_1 = 4\pi N R_c D$$
$$(\text{in l mol}^{-1} \text{ s}^{-1}) \quad (14.34)$$

where R_c is the distance of closest approach (in centimeter), D is the mutual diffusion coefficient (in square centimeter per second), N is equal to $N_A/1000$, N_A being Avogadro's constant. The distance of closest approach is generally taken as the sum of the radii of the two molecules (R_M for the fluorophore and R_Q for the quencher). The mutual diffusion coefficient D is the sum of the translational diffusion coefficients of the two species, D_M and D_Q, which can be expressed by the Stokes–Einstein relation:

$$D = D_M + D_Q = \frac{kT}{f \pi \eta}\left(\frac{1}{R_M} + \frac{1}{R_Q}\right) \quad (14.35)$$

where k is Boltzmann's constant, η is the viscosity of the medium, and f is a coefficient that is equal to 6 for "stick" boundary conditions and 4 for "slip" boundary conditions.

If R_M and R_Q are of the same order, the diffusional rate constant is approximately equal to $8RT/3\eta$.

In liquids, k_1 is about $10^9 - 10^{10}$ l mol^{-1} s^{-1}. The diffusion coefficient of molecules in the usual solvents at room temperature is generally of the order of 10^{-5} cm^2 s^{-1}.

Quenching of fluorescence (and phosphorescence) by dioxygen deserves attention because it is always present in solution. The effect on quantum yields and lifetimes strongly depends on the nature of the compound and the medium. Oxygen quenching is a collisional process and is therefore diffusion controlled. Consequently, compounds of long lifetime, such as naphthalene and pyrene, are particularly sensitive to the presence of oxygen. Moreover, oxygen quenching is less efficient in media of high viscosity. Oxygen quenching can be reduced by bubbling with nitrogen or argon; the most efficient method is, however, to use the freeze-pump-thaw technique (especially for phosphorescence studies).

Transient Effects In reality, the diffusional rate constant is time dependent and should be written as $k_1(t)$. Several models have been developed to express the time-dependent rate constant. Using Smoluchowski's theory, the fluorescence decay can be written in the following form:

$$i(t) = i(0) \exp\left(-at - 2b\sqrt{t}\right) \quad (14.36)$$

where

$$a = \frac{1}{\tau_0} + 4\pi N R_c D[Q] \quad (14.37)$$

$$b = 4\sqrt{\pi D} N R_c^2 [Q] \quad (14.38)$$

It should be noted that, in media of low viscosity, the transient term is significant only at short times (less than 100 picoseconds at viscosities similar to that of water) and can be neglected, whereas in viscous media, this term cannot be ignored (the fluorescence decay is then no longer a single exponential).

14.3.2.2 Static Quenching

Sphere of Effective Quenching In rigid matrices or in media so viscous that M* and Q cannot change their positions in space relative to one another during the excited-state lifetime of M*, Perrin has proposed a model in which quenching of a fluorophore is assumed to be complete if a quencher molecule Q is located inside a sphere (called *sphere of effective quenching, active sphere*, or *quenching sphere*) of volume V_q surrounding the fluorophore M. If a quencher is outside the active sphere, it has no effect at all on M. Therefore, the fluorescence intensity of the solution is decreased by the addition of Q, but the fluorescence decay after pulse excitation is unaffected.

According to this model, the ratio I_0/I is given as

$$\frac{I_0}{I} = \exp(V_q N[Q]) \qquad (14.39)$$

At low quencher concentrations, $\exp(V_q N[Q]) \cong 1 + V_q N[Q]$: the plot of I_0/I is thus almost linear (as in the case of the Stern–Volmer plot), but at high concentrations, it shows an upward curvature.

A plot of $\ln(I_0/I)$ versus $[Q]$ yields V_q. The values of $V_q N$ are often found to be in the range of 1–3 l mol^{-1}. This corresponds to a quenching sphere radius of about 10 Å, which is slightly larger than the van der Waals contact distance between M and Q.

Perrin's model has been used in particular for the interpretation of nonradiative energy transfer in rigid media.

Formation of a Ground-State Nonfluorescent Complex In some cases, M can form a nonfluorescent 1 : 1 complex according to the equilibrium

$$M + Q \rightleftharpoons MQ \qquad (14.40)$$

The stability constant of the complex is

$$K_S = \frac{[MQ]}{[M][Q]} \qquad (14.41)$$

The excited-state lifetime of the uncomplexed fluorophores M is unaffected. The fluorescence intensity of the solution decreases with the addition of Q, but the fluorescence decay after pulse excitation is unaffected, in contrast to dynamic quenching. Quinones, hydroquinones, purines, and pyrimidines are well-known examples of molecules responsible for static quenching.

Using the mass conservation law (where $[M]_0$ is the total concentration of M)

$$[M]_0 = [M] + [MQ] \qquad (14.42)$$

the fraction of uncomplexed fluorophores is

$$\frac{[M]}{[M]_0} = \frac{1}{1 + K_S[Q]} \qquad (14.43)$$

In dilute solutions, the fluorescence intensities being proportional to the concentrations, we obtain

$$\frac{I_0}{I} = 1 + K_S[Q] \qquad (14.44)$$

A linear relation is thus obtained, as in the case of Stern–Volmer plot (Eq. (14.33)), but there is no change in excited-state lifetime for static quenching, whereas in the case of dynamic quenching, the ratio I_0/I is proportional to the ratio τ_0/τ of the lifetimes.

Simultaneous Dynamic and Static Quenching When both static and dynamic quenching occur simultaneously, a deviation of the plot of I_0/I versus $[Q]$ from linearity is observed.

In the case of static quenching resulting from the formation of a nonfluorescent complex, the ratio I_0/I obtained for dynamic quenching must be multiplied by the fraction of fluorescent molecules (i.e., uncomplexed):

$$\frac{I}{I_0} = \left[\frac{I}{I_0}\right]_{dyn} \times \frac{[M]}{[M]_0} \quad (14.45)$$

Using Eqs (14.33) and (14.43), the ratio I_0/I is given by

$$\frac{I_0}{I} = (1 + K_{SV}[Q])(1 + K_S[Q])$$
$$= 1 + (K_{SV} + K_S)[Q]$$
$$+ K_{SV} K_S [Q]^2 \quad (14.46)$$

An upward curvature is thus observed. K_{SV} and K_S can be determined by curve fitting using Eq. (14.46), or alternatively from the plot of $(I_0/I - 1)/[Q]$ versus $[Q]$, which should be linear.

Alternatively, using the model of sphere of effective quenching, we obtain the following relation:

$$\frac{I}{I_0} = (1 + K_{SV}[Q]) \exp(V_q N_A [Q]) \quad (14.47)$$

It should be noted that an upward curvature can also be due to transient effects, which may superimpose the effect of static quenching.

For unambiguous assignment of the dynamic and static quenching constants, it is recommended to carry out time-resolved experiments.

14.3.3
Photoinduced Electron Transfer

Photoinduced electron transfer (PET) is involved in many organic photochemical reactions and is often responsible for fluorescence quenching. It plays a major role in photosynthesis and in artificial systems for the conversion of solar energy based on photoinduced charge separation.

It is well known that the oxidative and reductive properties of molecules can be enhanced in the excited state. Oxydative and reductive electron transfer processes occur according to the following reactions:

$$D^* + A \rightarrow D^{\cdot +} + A^{\cdot -} \quad (14.48)$$
$$A^* + D \rightarrow A^{\cdot -} + D^{\cdot +} \quad (14.49)$$

Examples of donor and acceptor molecules are given in Table 14.3.

In the gas phase, the variations in standard free enthalpy ΔG^0 for the above reactions can be expressed by using the redox potentials E^0 and the excitation energy ΔE_{00}, that is, the difference in

Tab. 14.3 Examples of electron donors and acceptors. (From Valeur (2002).)

Electron donors	Electron acceptors
X—⟨phenyl⟩	⟨phenyl⟩—Y
X = H−, (CH₃)₂N−, CH₃O−, HS−	Y = −CN, −C(=O)−R, −N(=O)O
Naphthalene	
Anthracene	9,10-dicyanoanthracene
Phenanthrene	
Pyrene	
Perylene	

energy between the lowest vibrational levels of the excited state and the ground state:

$$\Delta G^0 = E^0_{D^+/D^*} - E^0_{A/A^-}$$
$$= E^0_{D^+/D} - E^0_{A/A^-} - \Delta E_0(D) \quad (14.50)$$
$$\Delta G^0 = E^0_{D^+/D} - E^0_{A^*/A^-}$$
$$= E^0_{D^+/D} - E^0_{A/A^-} - \Delta E_0(A) \quad (14.51)$$

These equations are called *Rehm–Weller equations*. If the redox potentials are expressed in volts, then ΔG^0 is given in volts. Conversion into joule per mole requires multiplication by the Faraday constant ($F = 96\,500$ C mol^{-1}).

In solution, two terms should be added for taking into account the solvation effect (enthalpic term ΔH_{solv}) and the Coulombic energy of the formed ion pair:

$$\Delta G^0 = E^0_{D^+/D} - E^0_{A/A^-} - \Delta E_0(D)$$
$$- \Delta H_{solv} - \frac{e^2}{4\pi\varepsilon r} \quad (14.52)$$
$$\Delta G^0 = E^0_{D^+/D} - E^0_{A/A^-} - \Delta E_0(A)$$
$$- \Delta H_{solv} - \frac{e^2}{4\pi\varepsilon r} \quad (14.53)$$

where e is the electron charge, ε is the dielectric constant of the solvent, and r is the distance between the two ions.

The redox potentials can be determined by electrochemical measurements or by theoretical calculations using the energy levels of the lowest unoccupied molecular orbital (LUMO) and the highest occupied molecular orbital (HOMO).

14.3.4
Formation of Excimers and Exciplexes

Excimers are formed by collision of an excited molecule and an identical unexcited molecule:

$$M^* + M \rightleftarrows (MM)^* \text{ (or } E^*) \quad (14.54)$$

Excimers are dimers in the excited state (the term excimer results form the contraction of "*exc*ited d*imer*"). The notation $(MM)^*$ expresses that the excitation energy is delocalized over the two moieties.

Exciplexes are formed by collision of an excited molecule (electron donor or acceptor) with an unlike unexcited molecule (electron acceptor or donor):

$$D^* + A \rightleftarrows (DA)^* \quad (14.55)$$
$$A^* + D \rightleftarrows (DA)^* \quad (14.56)$$

Exciplexes are excited-state complexes (the term exciplex comes from "*exc*ited com*plex*").

Excimer and exciplex formations are diffusion-controlled processes. The photophysical effects are thus detected at relatively high concentrations of the species so that a sufficient number of collisions can occur during the excited-state lifetime. Temperature and viscosity are important parameters.

14.3.4.1 Excimers

Many aromatic hydrocarbons such as naphthalene or pyrene can form excimers in which the two aromatic rings are facing at a distance of approximately 3–4 Å. The fluorescence band corresponding to excimer is located at wavelengths higher than that of the monomer and does not show vibronic bands because the lowest state is dissociative and can thus be considered as a continuum (Figure 14.4).

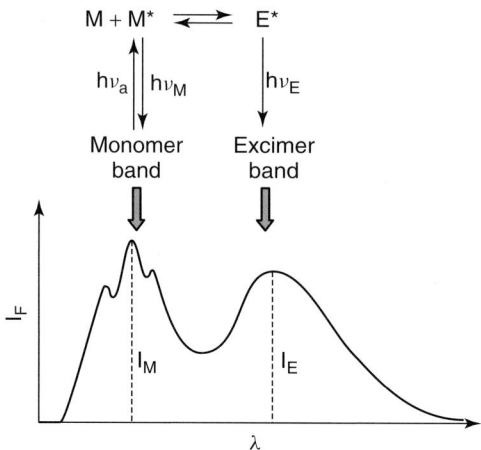

Fig. 14.4 Excimer formation and typical fluorescence spectrum showing the monomer and excimer bands. (Reproduced with permission from Valeur (2002). Copyright Wiley-VCH Verlag GmbH & Co. KGaA.)

Let k_M and k_E be the reciprocals of the excited-state lifetimes of the monomer and the excimer, respectively, and k_1 and k_{-1} are rate constants for the excimer formation and dissociation processes, respectively (Scheme 14.4).

The time evolution of the fluorescence intensity of the monomer M and the excimer E following a δ-pulse excitation can be obtained from the differential equations expressing the evolution of the species according to Scheme 14.4:

$$\frac{d[M^*]}{dt} = -k_M [M^*] - k_1[M][M^*]$$
$$+ k^*_{-1}[E^*] \quad (14.57)$$

$$\frac{d[E^*]}{dt} = k_1 [M][M^*] - (k_D + k_{-1})[E^*]$$
$$(14.58)$$

Formation of excimer E^* is a diffusion-controlled process. Under the approximation that k_1 is time independent, the δ-pulse responses, under the following initial conditions (at $t = 0$), $[M^*] = [M^*]_0$ and $[E^*]_0 = 0$, are

$$i_M(t) = k_r [M^*]$$
$$= \frac{k_r [M^*]_0}{\beta_1 - \beta_2} \big[(X - \beta_2)e^{-\beta_1 t}$$
$$+ (\beta_1 - X)e^{-\beta_2 t}\big] \quad (14.59)$$

$$i_E(t) = k_r k_1 [E^*]$$
$$= \frac{k_r k_1 [M][M^*]_0}{\beta_1 - \beta_2} \big[e^{-\beta_2 t} - e^{-\beta_1 t}\big]$$
$$(14.60)$$

where k_r and k_r' are the radiative rate constants of M^* and E^*, respectively, and

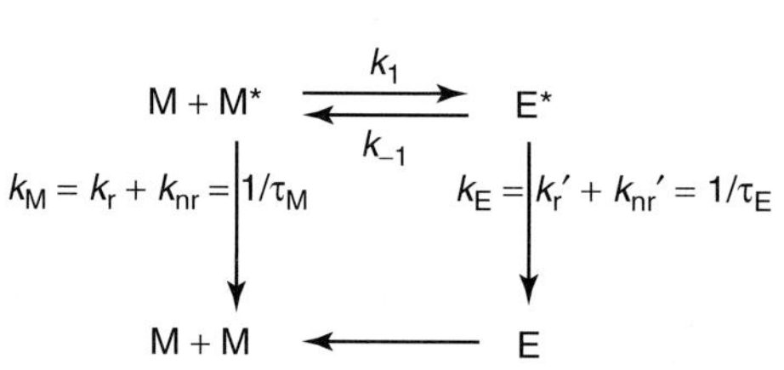

Scheme 14.4 Excimer formation and relevant rate constants.

β_1 and β_2 are given by

$$\beta_{2,1} = \frac{1}{2}\{X + Y \pm [(Y-X)^2 + 4k_1 k_{-1}[M]]^{1/2}\} \quad (14.61)$$

where

$$X = k_M + k_1[M] = \frac{1}{\tau_M} + k_1[M] \quad \text{and}$$

$$Y = k_E + k_{-1} = 1/\tau_E + k_{-1} \quad (14.62)$$

The decay of monomer emission is thus a sum of two exponentials, whereas the time evolution of the excimer emission is a difference of two exponentials in which the preexponential factors are of opposite signs. The time constants are the same in the expressions of $i_M(t)$ and $i_E(t)$ (β_1 and β_2 are the eigenvalues of the system). The negative term in $i_E(t)$ represents the rise in intensity corresponding to excimer formation; the fluorescence intensity starts indeed from zero since excimer does not absorb light and can only be formed from the monomer.

If the dissociation of the excimer cannot occur during the lifetime of the excited state ($k_{-1} = 0$), we have

$$i_M(t) = k_r [M^*]_0 \, e^{-Xt} \quad (14.63)$$

$$i_E(t) = \frac{k_r k_1 [M][M^*]_0}{X - k_E} \left[e^{-k_E t} - e^{-Xt} \right] \quad (14.64)$$

In this case, the fluorescence decay of the monomer is a single exponential with a decay time $(1/X)$ equal to the risetime of the excimer fluorescence.

The steady-state fluorescence intensities are obtained by integration of Eqs (14.59) and (14.60). The ratio of the fluorescence intensities of the excimer and monomer bands I_E/I_M (Figure 14.4) is often used for characterizing the efficiency of excimer formation. This ratio is given as

$$\frac{I_E}{I_M} = \frac{k'_r}{k_r} \frac{k_1[M]}{k_{-1} + 1/\tau_E} \quad (14.65)$$

Under conditions where the dissociation rate of the excimer is slow with respect to de-excitation, this equation is reduced to

$$\frac{I_E}{I_M} = \frac{k'_r}{k_r} \tau_E k_1[M] \quad (14.66)$$

These equations show that the ratio I_E/I_M is proportional to the rate constant k_1 for excimer formation.

When the two monomers are linked by a short flexible chain, intramolecular excimers can be formed. This process is still diffusion controlled, but is no longer translational; it requires a close approach between the two molecules via internal rotations during the excited-state lifetime. Because intramolecular excimer formation is independent of the total concentration, Eqs (14.59)–(14.66) are still valid provided that $k_1[M]$ is replaced by k_1.

Assuming that the Stokes–Einstein relation (Eq. (14.35)) is valid, k_1 is proportional to the ratio T/η, η being the viscosity of the medium. A well-known example of application is the estimation of the local fluidity of a medium (see Section 14.4.2.1, Method Based on Intramolecular Excimer Formation).

Another example of application is the determination of micellar aggregation numbers based on self-quenching of pyrene by excimer formation within micelles (Kalyanasundaran, 1987).

14.3.4.2 Exciplexes

The kinetic scheme for exciplex formation is slightly similar to that of excimer formation. Formation of an exciplex with anthracene and N,N-diethylaniline is a classical example: it results from the transfer of an electron from an amine molecule to an excited anthracene molecule. The quenching of anthracene fluorescence in a nonpolar solvent such as hexane is accompanied by the appearance of a broad structureless emission band of the exciplex at higher wavelengths than anthracene. When the solvent polarity increases, the exciplex band is red shifted. The intensity of this band decreases as a result of the competition between de-excitation and dissociation of the exciplex.

Exciplexes can be considered in some cases as intermediate species in electron transfer from a donor to an acceptor (see Section 14.2.3).

14.3.5 Photoinduced Proton Transfer

Only reactions in aqueous solutions are considered here. Thus, an acid or a base undergoing excited-state deprotonation or protonation, respectively, will always be in close contact with water molecules acting as either a proton acceptor or a proton donor. Therefore, these excited-state reactions will not be diffusion controlled.

Excitation may trigger a photoinduced proton transfer when acids and bases are stronger in the excited state than in the ground state. For instance, the acidic character of a proton-donor group (e.g., OH substituent of an aromatic ring) can be enhanced upon excitation so that the pK^* of this group in the excited state is much lower than the pK in the ground state (Table 14.4). In the same manner, the pK^* of a proton-acceptor group (e.g.,

Tab. 14.4 Examples of pK and pK^* values. (Adapted from Valeur (2002).)

Reaction		Compound	pK	pK^*
Excited-state deprotonation	$ArOH \xrightarrow{-H^+} ArO^-$	Phenol	10.6	3.6
		2-Naphthol	9.3	2.8
		2-Naphthol-6-sulfonate	9.12	1.66
		2-Naphthol-6, 8-disulfonate	9.3	<1
		8-Hydroxypyrene-1,3,6-trisulfonate (pyranine)	7.2	1.3
	$ArNH_2 \xrightarrow{-H^+} ArNH^-$	2-Naphthylamine	7.1	12.2
Excited-state protonation	$ArCO_2^- \xrightarrow{+H^+} Ar\,CO_2H$	Anthracene-9-carboxylate	3.7	6.9
	$ArN \xrightarrow{+H^+} ArNH^+$	acridine	5.5	1.6
		6-methoxyquinoline	5.2	11.8

heterocyclic nitrogen atom) in the excited state can be much higher than in the ground state (pK) (Table 14.4).

14.3.5.1 Equations for Excited-State Deprotonation

The general equations are given only in the most extensively studied case where the excited-state process is proton ejection ($pK^* \ll pK$); the proton donor is thus an acid, AH*, and the proton acceptor is a water molecule. Assuming that there is no geminate proton recombination, the processes are presented in Scheme 14.5, where τ_0 and τ_0' are the excited-state lifetimes of the acidic (AH*) and basic (A^{-*}) forms, respectively, and k_1 and k_{-1} are the rate constants for deprotonation and reprotonation, respectively. k_1 is a pseudo-first-order rate constant, whereas k_{-1} is a second-order rate constant. The excited-state equilibrium constant is $K^* = k_1/k_{-1}$.

The probability that the back reaction can take place during the excited-state lifetime of A^{-*} is of utmost importance. The rate of this reaction is pH dependent because the process is diffusion controlled ($k_{-1} \approx 5 \times 10^{10}$ l mol^{-1} s^{-1}). If pH ≤ 2, $k_{-1}[H_3O^+] > \approx 5 \times 10^8$ s^{-1}: the back reaction must then be taken into account. The reciprocal of this value is indeed of the order of the excited-state lifetime of most organic bases.

According to Scheme 14.5, the differential equations expressing the time evolution of the concentration of the species are

$$\frac{d[AH^*]}{dt} = -\left(k_1 + \frac{1}{\tau_0}\right)[AH^*]$$
$$+ k_{-1}[A^{-*}][H_3O^+] \quad (14.67)$$

$$\frac{d[A^{-*}]}{dt} = k_1 [AH^*]$$
$$- \left(k_{-1}[H_3O^+] + \frac{1}{\tau_0'}\right)[A^{-*}] \quad (14.68)$$

When AH is selectively excited by a δ-pulse of light, the responses of the fluorescence intensities, under the following initial conditions (at $t = 0$), $[AH^*] = [AH^*]_0$ and $[A^{-*}]_0 = 0$, are

$$i_{AH^*}(t) = k_r [AH^*] = \frac{k_r [AH^*]_0}{\beta_1 - \beta_2}$$
$$\times \left[(X - \beta_2)e^{-\beta_1 t}\right.$$
$$\left. + (\beta_1 - X)e^{-\beta_2 t}\right] \quad (14.69)$$

$$i_{A^{-*}} = k_r' [A^{-*}] = \frac{k_r' k_1 [AH^*]_0}{\beta_1 - \beta_2}$$
$$\times \left[e^{-\beta_2 t} - e^{-\beta_1 t}\right] \quad (14.70)$$

where k_r and k_r' are the radiative rate constants of AH* and A^{-*}, respectively, and β_1 and β_2 are given by

$$\beta_{2,1} = \frac{1}{2}\left\{X + Y \pm \left[(Y - X)^2 \right.\right.$$
$$\left.\left. + 4k_1 k_{-1}[H_3O^+]\right]^{1/2}\right\} \quad (14.71)$$

where $X = k_1 + 1/\tau_0$ and $Y = k_{-1}[H_3O^+] + 1/\tau_0'$.

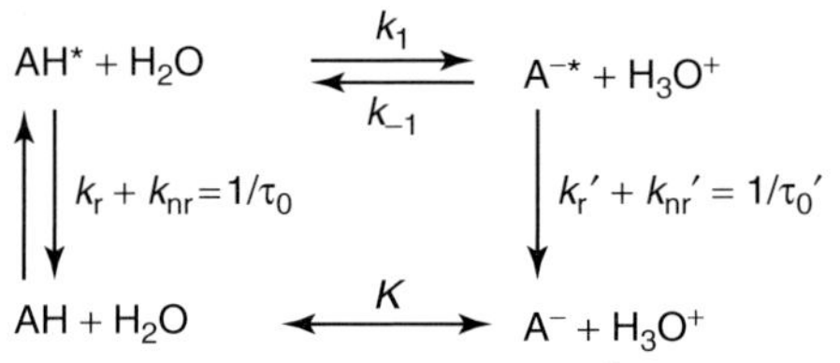

$$\begin{array}{ccc}
AH^* + H_2O & \underset{k_{-1}}{\overset{k_1}{\rightleftharpoons}} & A^{-*} + H_3O^+ \\
\Big\updownarrow k_r + k_{nr} = 1/\tau_0 & & \Big\downarrow k_r' + k_{nr}' = 1/\tau_0' \\
AH + H_2O & \underset{}{\overset{K}{\longleftarrow}} & A^- + H_3O^+
\end{array}$$

Scheme 14.5 Excited-state deprotonation of an acid AH.

Under continuous illumination, the steady-state intensities can be easily calculated by integration of Eqs (14.69) and (14.70):

$$I_{AH^*} = C\Phi_0 \frac{1 + k_{-1}\tau_0'[H_3O^+]}{1 + k_1\tau_0 + k_{-1}\tau_0'[H_3O^+]} \quad (14.72)$$

$$I_{A^{-*}} = C\Phi_0' \frac{k_1\tau_0}{1 + k_1\tau_0 + k_{-1}\tau_0'[H_3O^+]} \quad (14.73)$$

where Φ_0 and Φ_0' are the fluorescence quantum yields of AH and A^-, respectively, in the absence of excited-state reaction ($\Phi_0 = k_r\tau_0$; $\Phi_0' = k_r'\tau_0'$). C is a multiplication factor that is introduced to take into account the experimental conditions (total concentration, choice of excitation and emission wavelengths, bandpasses for absorption and emission intensity of the incident light, and sensitivity of the instrument).

Under conditions where the back reaction cannot take place during the excited-state lifetime ($k_{-1} = 0$), β_1 and β_2 become equal to X and Y, respectively, and Eqs (14.69) and (14.70) are reduced to

$$i_{AH^*}(t) = [AH^*]_0 \, k_r \, e^{-Xt} \quad (14.74)$$

$$i_{A^{-*}}(t) = \frac{k_r' \, k_1 [AH^*]_0}{X - Y} \left[e^{-Yt} - e^{-Xt} \right] \quad (14.75)$$

where $X = k_1 + 1/\tau_0$ and $Y = 1/\tau_0'$. Therefore, $i_{AH^*}(t)$ becomes a single exponential and $i_{A^{-*}}(t)$ is still a difference of two exponentials with a rise time $(1/X)$ (equal to the decay time of AH^*) and a decay time $(1/Y)$ (equal to the lifetime of the basic form).

The steady-state intensities reduce to

$$I_{AH^*} = C\Phi_0 \frac{1}{1 + k_1\tau_0} \quad (14.76)$$

$$I_{A^{-*}} = C\Phi_0' \frac{k_1\tau_0}{1 + k_1\tau_0} \quad (14.77)$$

14.3.5.2 Determination of the Excited-State pK^*

Estimation by Means of the Förster Cycle pK^* can be theoretically predicted by means of the Förster cycle associated with spectroscopic measurements. Details on the theory can be found in Valeur (2002) p. 103. Only the final equation is given here at 298 K:

$$pK^* - pK = 2.1 \times 10^{-3}(\bar{\nu}_{A^-} - \bar{\nu}_{AH}) \quad (14.78)$$

where $\bar{\nu}_{AH}$ and $\bar{\nu}_{A^-}$ are the wavenumbers (in cm^{-1}) corresponding to the 0–0 transitions of AH and A^-. The determination of $\bar{\nu}_{AH}$ and $\bar{\nu}_{A^-}$ is the major cause of inaccuracy in the estimation of pK^*. They are usually estimated by means of the average between the wavenumbers corresponding to the maxima of absorption and emission:

$$\nu_0 = \frac{\bar{\nu}_{abs}^{max} + \bar{\nu}_{em}^{max}}{2} \quad (14.79)$$

However, the best approximation is to use the intersection point of the mutually normalized absorption and emission spectra (as it was the practice in Förster's laboratory). In the case of large Stokes shift, it may be difficult to determine an intersection point and it is then preferable to take the average of the wavenumbers corresponding to half-heights of the absorption and emission bands, which is a better approximation than the average of the wavenumbers corresponding to the maxima.

The simplicity of the Förster cycle method explains why it has been extensively used. One of the important features of this cycle is the possibility to use it even in the case where the equilibrium is not established within the excited-state lifetime.

Steady-State Measurements The value of the excited-state pK^* can be determined by fluorometric titration but only when the equilibrium is established in the excited state. Rewriting Eqs (14.72) and (14.73) as $I_{AH^*} = C\Phi$ and $I_{A^{-*}} = C\Phi'$, it is easy to obtain the following relation:

$$\frac{\Phi/\Phi_0}{\Phi'/\Phi'_0} = \frac{1}{k_1\tau_0} + \frac{k_{-1}\tau'_0}{k_1\tau_0}[H_3O^+] \quad (14.80)$$

Consequently, if the excited-state lifetimes τ_0 and τ_0' are known, the plot of $(\Phi/\Phi_0)/(\Phi'/\Phi_0')$ versus $[H_3O^+]$ yields the rate constants k_1 and k_{-1}. The ratio k_1/k_{-1} yields K^*.

Time-Resolved Experiments The most reliable method for the determination of k_1 and k_{-1} is based on time-resolved experiments. They provide the values of the decay times from which the rate constants k_1 and k_{-1} are determined by means of Eqs (14.69) and (14.70).

14.3.5.3 pH Dependence of Absorption and Emission Spectra

The absorption and fluorescence spectra are pH dependent when the acido-basic properties are not the same in the ground and excited states. The discussion is restricted again to the most important case of photoinduced proton ejection ($pK^* \ll pK$). Let us recall that for pH values less than approximately 2, the back reaction occurs. Moreover, distinction should be made according to the value of pK^*.

1. If pK^* is greater than approximately 2, a plateau is observed for the relative fluorescence quantum yield of the acidic form and the basic form between pK^* and pK (Figure 14.5) because of the absence of diffusional recombination (see Eqs (14.75) and (14.76)). A typical example is 2-naphthol ($pK = 9.3$, $pK^* = 2.8$).

2. If pK^* is smaller than approximately 2, the acid is very strong in the excited-state and, in general, k_1 is much larger than the reciprocal of the excited-state lifetime so that the fluorescence of the acidic form is not observed at $pH > pK^* + 2$ but only for lower pH values (Figure 14.5). This is the case of pyranine ($pK = 7.7$, $pK^* = 1.3$).

14.3.6
Excitation Energy Transfer

Distinction between radiative and nonradiative transfer is of major importance (Valeur, 2002 Chapter 4, p. 110, and Chapter. 9); Andrews and Demidov, 1999. In radiative transfer a photon emitted by a molecule D is absorbed by a molecule A (or D). This process is observed when the average distance between D and A (or D) is larger than the wavelength. It does not require any interaction between the partners, but it depends on the spectral overlap and on the concentration. In contrast, nonradiative transfer occurs without emission of photons at distances lower than the wavelength and results from short- or long-range interactions between molecules. For instance, nonradiative transfer by dipole–dipole interaction is possible at distances up to 8–10 nm. Consequently, such a transfer provides a tool for determining distances of a few nanometers between chromophores. It should be

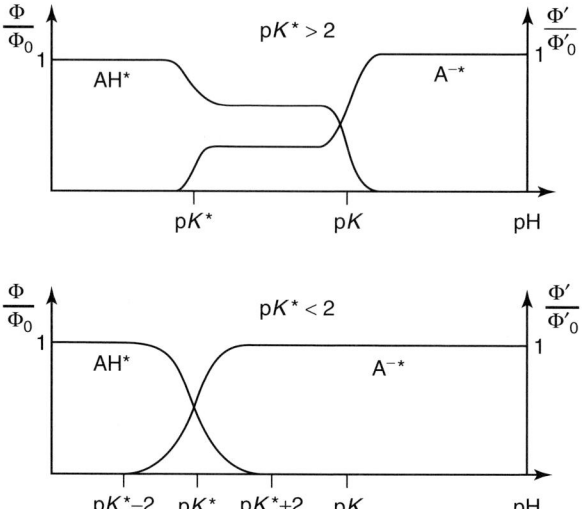

Fig. 14.5 Variations in relative quantum yields of the basic and acidic forms as a function of pH. (Reproduced with permission from Valeur (2002). Copyright Wiley-VCH Verlag GmbH & Co. KGaA.)

noted that the transfer of excitation energy plays a major role in photosynthetic systems.

Nonradiative transfer of excitation energy requires some interaction between a donor molecule and an acceptor molecule, and it can occur if the emission spectrum of the donor overlaps the absorption spectrum of the acceptor so that several vibronic transitions in the donor have practically the same energy as the corresponding transitions in the acceptor. Such transitions are coupled, that is, are in *resonance* and, hence, the term *resonance energy transfer* (*RET*). Fluorescence resonance energy transfer (FRET) is often employed, but this expression is not correct because it is not the fluorescence that is transferred but the electronic energy of the donor. Therefore, it is recommended to use either EET (*excitation energy transfer* or *electronic energy transfer*) or RET.

Different interaction mechanisms are involved in RET. The interactions may be Coulombic and/or due to intermolecular orbital overlap. The Coulombic interactions consist of long-range dipole–dipole interactions (Förster's mechanism) and short-range multipolar interactions. The interactions due to intermolecular orbital overlap, which include electron exchange (Dexter's mechanism) and charge resonance interactions, are only short range.

14.4
Applications of Molecular Fluorescence

14.4.1
Historical Introduction

Fluorescence has long been used as an analytical tool for the determination of the concentrations of neutral or ionic species (Valeur, 2007). In his famous book *History of Luminescence*, Harvey (1957) reports that Victor Pierre, who was a professor in Prague and later in Vienna, published a paper in 1862 in which he described the investigation of solutions of single fluorescent compounds and mixtures. He noted that bands of fluorescent spectra were characteristic of a particular substance. He also noted the effect of solvent and acidity or alkalinity. G. G. Stokes – who invented the term fluorescence – had certainly the same idea in mind. In fact, one of the topics of his famous Burnett lectures was " Fluorescence – Its use as a means of discrimination." However, no specific example was given in the text.

The term *fluorescence analysis* was employed for the first time in the paper by Göppelsröder (1868): the drastic enhancement of fluorescence intensity accompanying the complexation of morin (a hydroxyflavone derivative) with aluminum provides a straightforward way to detect this metal.

The use of fluorescent tracers in hydrology is also an old application of fluorescence. Uranin (the disodium salt of fluorescein) was used for the first time in 1877 as a tracer for monitoring the flow of Danube. On all maps, it is shown that the Danube springs in the Black Forest and, after many hundreds of kilometers, flows into the Black Sea. However, because of big sink (swallow holes) in the bed of Danube, it was demonstrated, thanks to the fluorescent tracer, that only a small part of the water from Danube spring arrives at the Black Sea. Most of it flows into the North Sea. Fluorescence tracing is now currently used in hydrology, especially to simulate pollutions.

Numerous applications of fluorescence were developed at the beginning of the twentieth century and reported in several books Radley and Grant (1933); Dake and De Ment (1941); Pringsheim and Vogel (1943). For instance, in the second part of the book of Radley and Grant (1933), the list of applications of fluorescence analysis is impressive: agriculture, bacteriology, botany, ceramics, slags and cements, drugs, foods and food products, fuels, inorganic chemistry, leather and tanning, legal work, medical and biological sciences, minerals and gems, museum work, organic chemistry, paints and varnishes, paper and cellulose and its derivatives, the rubber industry, textiles, and waters and sewage. Pringsheim and Vogel (1943) also gave various examples of application including fluorescent carpets and ceilings in theaters or fluorescent ballets.

The applications of fluorescence that are currently used are now presented.

14.4.2
Estimation of Local Physical Parameters

Physical parameters, such as viscosity, temperature, and pressure, can be locally estimated by means of fluorescent probes, that is, fluorescent compounds whose emissive properties are affected by one of these parameters.

14.4.2.1 Viscosity

Basic Concepts It is important to first recall that viscosity is a macroscopic parameter that loses its physical meaning at a molecular scale (Valeur, 2002 Chapter 8; Valeur, 1993). The term *microviscosity* must be used but with caution. The term *fluidity* is appropriate to characterize, in a very general way, the effects of viscous drag and cohesion of the probed microenvironment (polymers, micelles, gels, lipid bilayers of vesicles or biological membranes, etc.).

In all fluorescence techniques permitting evaluation of the fluidity of a microenvironment by means of a fluorescent probe, the underlying physical quantity is a diffusion coefficient expressing the viscous drag of the surrounding molecules (Figure 14.6).

Let us first recall that the diffusion coefficients for translational and rotational diffusions are, respectively,

$$D_t = \frac{kT}{6\pi \eta r} \quad (14.81)$$

$$D_r = \frac{kT}{8\pi \eta r^3} = \frac{kT}{6\eta V} \quad (14.82)$$

where k is the Boltzman constant, T is the absolute temperature, η is the viscosity, r is the hydrodynamic radius of the sphere, and V is its hydrodynamic volume.

The use of these equations to estimate the viscosity η of an unknown environment from the experimental value of the diffusion constant is questionable. In fact, these equations are valid only for a rigid sphere being large compared to the molecular dimensions, moving in a homogeneous newtonian fluid and obeying the Stokes hydrodynamic law. Many other relations have been proposed, but it should be pointed out that *there is no satisfactory relationship between diffusion and bulk viscosity for probes*. The main reason is that the size of probes is comparable to that of the surrounding molecules forming the microenvironment to be probed. Another difficulty arises in the case of organized assemblies such as micellar systems, biological membranes, and so on, because the microenvironment is not isotropic.

In other words, *viscosity is a macroscopic parameter and any attempt to get absolute values of the viscosity of a medium from measurements using a fluorescent probe is hopeless*.

It is always observed that the diffusion of molecular probes is more rapid than predicted by the theory. Therefore, the "slip" boundary condition is often introduced, and sometimes a mixture of "stick" and "slip" boundary conditions is assumed. Relation (14.82) can then be

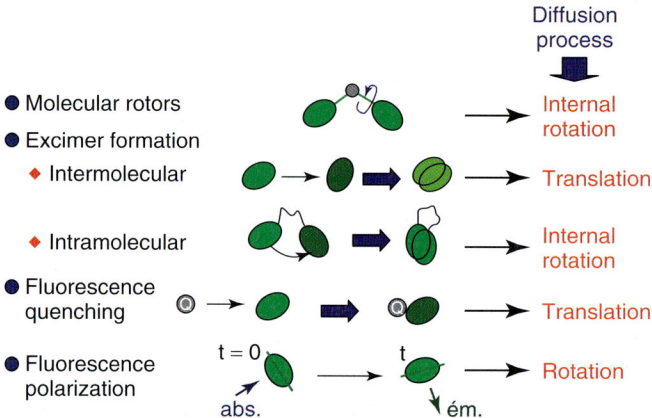

Fig. 14.6 Fluorescence techniques for the evaluation of the fluidity of a microenvironment.

rewritten as

$$D_r = \frac{kT}{6\eta V s g} \quad (14.83)$$

where s is the coupling factor ($s = 1$ for "stick" and $s < 1$ for "slip" boundary conditions) and g is the shape factor. In the case of charged molecules, an additional friction force should be introduced as a result of the induced polarization of the surrounding solvent molecules.

A faster diffusion than predicted can be explained by the fact that a solute molecule moves according to two diffusional processes: a viscous process with displacement of solvent molecules (*Stokes diffusion*) and a process associated with migration into holes of the solvent (*free volume diffusion*). The importance of free volume should be emphasized at a molecular level. An empirical relation between viscosity and free volume was proposed by Doolittle:

$$\eta = \eta_0 \exp\left(\frac{V_0}{V_f}\right) \quad (14.84)$$

where η_0 is a constant and V_0 and V_f are the Van der Waals volume and the free volume of the solvent, respectively.

The choice of the method for probing fluidity depends on the system to be investigated. The use of intermolecular quenching and intermolecular excimer formation is not recommended for probing fluidity of microheterogeneous media because of possible perturbation of the translational diffusion process. The methods of molecular rotors and intramolecular excimer formation are convenient and rapid, but the time-resolved fluorescence polarization technique provides much more detailed information, including an order parameter if the medium is anisotropic.

Use of Molecular Rotors A *molecular rotor* – as a fluorescent probe – is a molecule that undergoes internal rotation(s) resulting in viscosity-dependent changes of its emissive properties. They belong to the family of diphenylmethane dyes (e.g., auramine O), triphenylmethanes dyes (e.g., crystal violet), and twisted intramolecular charge transfer (TICT) compounds (e.g., p-dimethylaminobenzonitrile (DMABN)). Examples of molecular rotors of practical interest are given in Figure 14.7.

In solvents of medium and high viscosity, an empirical relation based on free volume considerations has been proposed to account for the viscosity dependence of the fluorescence quantum yield:

$$\Phi_F \cong b \left(\frac{\eta}{T}\right)^x \quad (14.85)$$

where x is a constant for a particular probe.

In polymers above the glass transition temperature, the fluorescence quantum yields were, in fact, found to be a power function of the bulk viscosity with values of the exponent x lower than 1 (e.g., for p-N,N-dimethylaminobenzylidenemalononitrile, $x = 0.69$ in glycerol and 0.43 in dimethylphtalate).

Molecular rotors are a powerful tool to study changes in free volume of polymers as a function of molecular weight, stereoregularity, crosslinking, polymer chain relaxation, and flexibility. Monitoring of polymerization reactions is possible in this way (Loufty, 1986).

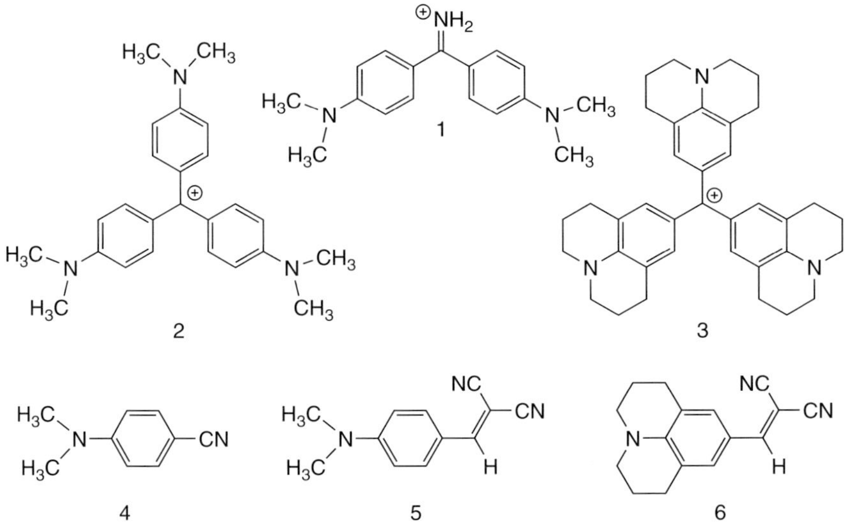

Fig. 14.7 Examples of molecular rotors. 1, auramine O; 2, crystal violet; 4, p-N,N-dimethylaminobenzonitrile (DMABN); 5, p-N,N-dimethylaminobenzylidenemalononitrile; 6, julolidinebenzylidenemalononitrile.

Method Based on Intramolecular Excimer Formation Bifluorophoric molecules consisting of two identical fluorophores linked by a short flexible chain may form an excimer. Examples of such bifluorophores currently used in investigations of fluidity are given in Figure 14.8.

The efficiency of intramolecular excimer formation does not depend on the concentration of fluorophores so that Eq. (14.66) should be rewritten by replacing $k_1[M]$ by k_1:

$$\frac{I_E}{I_M} = \frac{k'_r}{k_r} \tau_E k_1 \quad (14.86)$$

It should be noted that when effects of temperature on fluidity are investigated, difficulties may arise from the possible temperature dependence of the excimer lifetime. It is then recommended to perform time-resolved fluorescence experiments.

As in the case of molecular rotors, free volume effects are important, but the free volume fraction measured by intramolecular excimers is smaller than for molecular rotors. The volume swept out during the conformational change required for excimer formation is in fact larger, and consequently these probes do not respond in frozen media or polymers below the glass transition temperature.

The following empirical relation is currently used for the rate constant for excimer formation:

$$k_1 = \alpha \, \eta^{-x} \quad (14.87)$$

It is important to note that in addition to viscosity, the response of a probe depends on several parameters: chemical nature of the solvent, length of the chain, temperature, and oxygen quenching (Viriot et al., 1983).

Intramolecular excimers have been used for probing bulk polymers, micelles, vesicles, and biological membranes (Viriot et al., 1983; Bokobza and Monnerie, 1986; Bokobza, 1990;

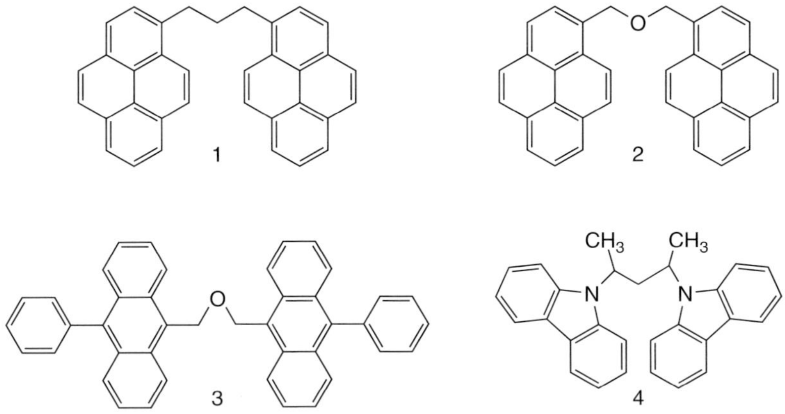

Fig. 14.8 Examples of bifluorophores forming excimers for the study of fluidity. 1, α,ω-di-(1-pyrenyl)propane; 2, α,ω-di-(1-pyrenyl)methylether; 3, 10,10'-diphenyl-*bis*-9-anthrylmethyloxide (DIPHANT); 4, meso-2,4-di(*N*-carbazolyl)pentane.

Georgescauld et al., 1980; Vauhkonen et al., 1990; Zachariasse and Kozankiewicz, and Kühnle, 1983).

Fluorescence Polarization Method The preferred orientation of fluorescent probes resulting from photoselection at time zero is gradually affected as a function of time by the rotational Brownian motions if the medium is fluid enough to permit rotation of the probes during the excited-state lifetime. This causes a depolarization of fluorescence, from which information can be obtained on the molecular motions that depend on the size and the shape of molecules and on the fluidity of their microenvironment.

A molecule undergoes isotropic rotation if it is spherical, but this is also the case of a rodlike molecule whose directions of its absorption and emission transition moments coincide with the long molecular axis (e.g., diphenylhexatriene): the rotations can be considered as isotropic because any rotation about this long axis has no effect on the emission anisotropy.

For isotropic motions in an isotropic medium, the values of the instantaneous and steady-state emission anisotropies are given by

$$r(t) = r_0 \exp(-6D_r t) \qquad (14.88)$$

$$\frac{1}{\bar{r}} = \frac{1}{r_0}(1 + 6D_r \tau) \qquad (14.89)$$

which allows one to determine the diffusion constant D_r. In principle, a value of the viscosity η could be calculated from the Stokes–Einstein relation (Eq. (14.82) or (14.83)) provided that the hydrodynamic volume V of the probe is known. It should be again emphasized that any "microviscosity" value calculated in this way would be questionable and thus useless. Nevertheless, the changes in D_r upon an external perturbation (e.g., temperature, pressure, additive, etc.) reflect well the changes in fluidity of a medium.

In ordered systems such as lipid bilayers and liquid crystals, the rotational motions of a probe are hindered and the emission anisotropy does not decay to zero but to a steady value r_∞. Equations (14.88) and (14.89) are thus no longer valid. For isotropic rotations (rodlike probe), the emission anisotropy can often be written in the following approximated form:

$$r(t) = (r_0 - r_\infty) \exp\left(-\frac{t}{\tau_c}\right) + r_\infty \quad (14.90)$$

where τ_c is the effective relaxation time of $r(t)$, that is, the time with which the initially photoselected distribution of orientations approaches the stationary distribution.

Time-resolved emission anisotropy experiments yield τ_c, r_0, and r_∞. The data can be processed according to the *wobble-in-cone model* (Kinosita, Kawato and Ikegami, 1977; Lipari and Szabo, 1980) in which the rotations of a rodlike probe (with the direction of its absorption and emission transition moments coinciding with the long molecular axis) are restricted within a cone. The rotational motions are described by the rotational diffusion coefficient D_w around an axis perpendicular to the long molecular axis and an order parameter (half angle of the cone θ_c) reflecting the degree of orientational constraint due to the surrounding paraffinic chains. θ_c can be determined from the ratio r_∞/r_0, and D_W from τ_c and θ_c.

The rotations of the probe are thus describe by (i) the wobbling diffusion constant D_w reflecting the fluidity of the medium, and more precisely the chain mobility and (ii) the order parameter θ_c, half angle of the cone, expressing the degree of orientational constraint.

More general theories have been developed (Van der Meer, Kooyman and Levine, 1982; Van der Meer et al., 1984; Zannoni, Arcioni and Cavatorta, 1983; Fisz, 1985; Szabo, 1984; Pottel et al., 1986).

14.4.2.2 Temperature

Fluorescent molecular thermometers are based on the temperature dependence of the photophysical properties of fluorescent compounds (fluorescence spectrum, fluorescence quantum yield, and excited-state lifetime) (Chandrasekharan and Kelly, 2004; Uchiyama, de Silva and Iwau, 2006). Temperature is indeed a key parameter of several photophysical processes: intersystem crossing (Section 14.2.1.3), delayed fluorescence (Section 14.2.1.3), and excimer or exciplex formation (Section 14.3.4).

The advantages of fluorescent molecular thermometers over contact temperature sensors are the following: absence of sensitivity to electromagnetic radiation, fast response, and spatial resolution from the nanoscale to the macroscale (temperature-sensitive paints).

The compounds whose fluorescence quantum yield is more or less strongly affected by temperature changes are either organic (e.g., perylene, acridine yellow, perylenedicarboximide, rhodamine B, pyronine B, pyronine Y) or organometallic (e.g., ruthenium or europium complexes : Ru(bpy), Ru(phen)$_3$, Europium thenoyltrifluoroacetonate (EuTTA)).

For instance, the temperature dependence of the fluorescence quantum yield of a solution of rhodamine B has been exploited for spatial and temporal mapping of temperature within the microchannel networks of microfluidic devices. Changes in fluorescence intensity were used to determine intrachannel temperature variations with a precision of $\pm 3\,°C$, a spatial resolution of 1 µm, and a temporal resolution of 33 milliseconds (Ross, Gaitan and Locascio, 2001). With the same aim, fluorescence lifetime imaging (FLIM) has been used to quantitatively image temperature distributions in three spatial dimensions within microchannels environments (Benninger et al., 2006). The advantage of lifetime measurements is that, unlike fluorescence intensity, excited-state lifetime is independent of fluorophore concentration, fluctuations of the light source, and photobleaching effects.

Ratiometric measurements, that is the ratio of the fluorescence intensities at two observation wavelengths, offer the same advantage. This requires a change in the shape of the fluorescence spectrum as a function of temperature. This is, for instance, the case of exciplex formation with perylene and N-allyl-N-methylaniline (NA), leading to the appearance of a broad fluorescence band centered at 550 nm (Chandrashekaran and Kelly, 2001):

$$\text{Perylene (blue fluorescence)} + \text{NA} \rightleftharpoons \text{Perylene-NA (green fluorescence)}$$

A film consisting of perylene and NA encapsulated and anchored, respectively, in a soft elastic polystyrene matrix is a temperature sensor whose working range is 25–85 °C. Figure 14.9 shows the effect of temperature changes on the emission spectra.

An interesting application of fluorescent molecular thermometers concerns real-time polymer processing monitoring. Measurements by thermocouples and infrared imaging are not accurate in this case. Benzoxazolyl stilbene and

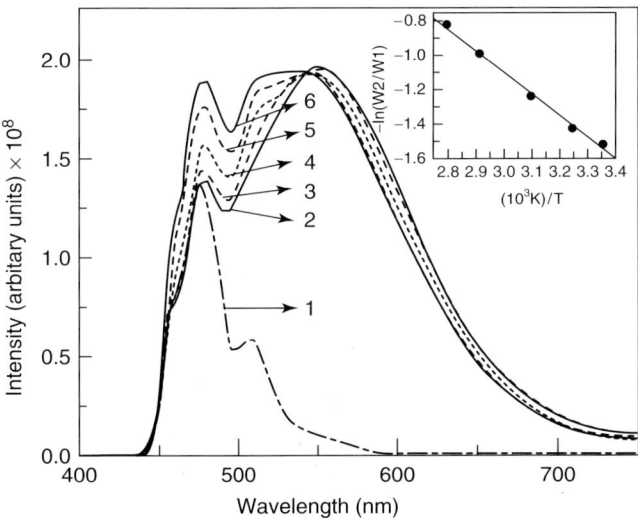

Fig. 14.9 Effect of temperature on the emission spectra of a polystyrene film composed of 0.017 wt % of perylene and 18.96 wt % of N-allyl-N-methylaniline : (2) 298 K; (3) 304 K; (4) 323 K; (5) 343 K; (6) 358 K; (1) film of perylene alone at 298 K. Excitation wavelength: 386 nm. (Reproduced with permission from Chandrashekaran and Kelly (2001). Copyright 2001 American Chemical Society.)

perylene were chosen because of their compatibility with polymer matrices and their high thermal stability. Morever, the relative intensities of the vibronic bands are temperature dependent, which allow ratiometric measurements. Plots of the intensity ratio versus temperature are linear with a slope of 0.09–0.2% per degree. Temperatures as high as 300 °C can be measured (Bur, Wangel and Roth, 2002).

Optical thermometers based on the thermally activated delayed fluorescence have been proposed. Acridine yellow dissolved in a rigid saccharide glass exhibits a temperature-dependent ratio of fluorescence to phosphorescence intensity, but the effect is weak (Fister, Rank and Harris, 1995). A much stronger effect has been observed with fullerene C_{70}, which shows very intense delayed fluorescence (see Section 14.2.1.3). For C_{70} dispersed in a polymer matrix (poly(*ter*-butyl methacrylate)), the working range extends from −80 to 140 °C (Baleizao *et al.*, 2007). Figure 14.10 shows the increase in intensity of red fluorescence. The working range is even wider when monitoring the lifetime of the delayed fluorescence.

Tab. 14.5 Temperature ranges of temperature-sensitive paints.

Compound	Temperature range (°C)
Perylene	0–100
Perylenedicarboximide	50–100
Rhodamine B	0–80
Pyronine B	50–100
Pyronine Y	0–100
Ruthenium complex Ru(bpy)	0–90
Europium complex EuTTA[a]	−20 to 80

[a] Europium theonyltrifluoroacetonate.

Surface temperature mapping can be obtained by using paints containing temperature-sensitive fluorescent compounds (Liu et and Sullivan, 2005). Table 14.5 gives examples of such compounds together with the temperature ranges. In aerodynamics, temperature-sensitive paints are employed in conjunction with pressure-sensitive paints (PSPs) (see the following section) (Liu *et al.*, 1997). Both paints are also used in advanced turbomachinery applications

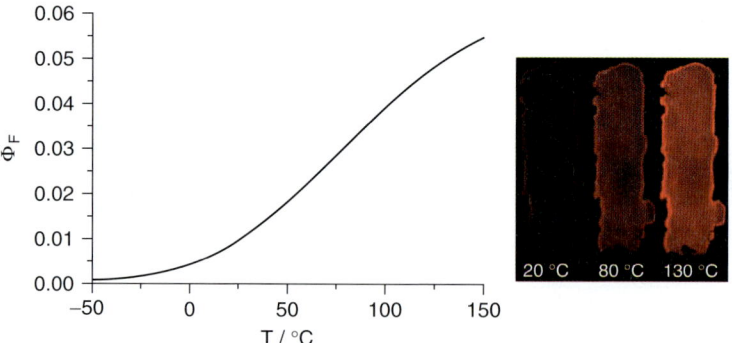

Fig. 14.10 Fluorescence quantum yield versus temperature for C_{70} dispersed in a polymer matrix (poly(*ter*-butyl methacrylate)) and color picture at three temperatures. (Reproduced with permission from Baleizao *et al.* (2007). Copyright Wiley-VCH Verlag GmbH & Co. KGaA.)

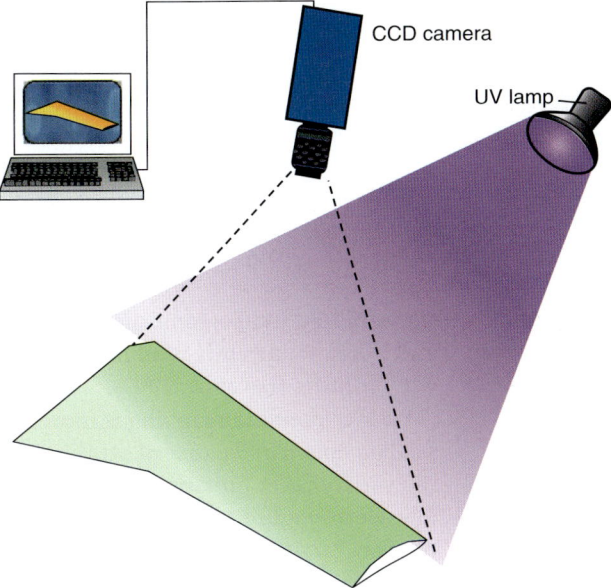

Fig. 14.11 Experimental set-up for temperature and pressure surface mapping using temperature- and pressure-sensitive paints.

(Navarra *et al.*, 2001). Figure 14.11 shows the experimental set-up.

At a microscopic scale, *fluorescence microthermography* is a powerful technique for generating high-resolution thermal maps of integrated circuits and for hot-spot localization in electrical failure analysis (Kolodner and Tyson, 1982; Herzum *et al.*, 1998). The europium complex EuTTA is well suited for these measurements. The lateral resolution is 0.5 μm and the thermal resolution is in the millikelvin range. This technique is thus superior to other routine thermography methods (liquid crystal thermography, infrared thermography, or laser thermoreflectance).

14.4.2.3 Pressure

An interesting application of fluorescence quenching is the use of fluorescent paints sensitive to air pressure, with the aim at studying the aerodynamism of aircraft or car models in wind tunnels. Such PSPs sprayed on a model provide a surface pressure mapping, that is, the instantaneous two-dimensional pressure distribution on the surface. They are more convenient and cheaper than the numerous pressure sensors that would be required.

PSPs are oxygen sensitive: they contain fluorophores whose fluorescence is quenched by oxygen (Figure 14.12). Therefore, the larger the air pressure, the larger the partial pressure of oxygen, and the higher the efficiency of quenching of the fluorophores that are embedded in a transparent polymer. Fluorophores should have a long excited-state lifetime: pyrene derivatives and ruthenium complexes are thus suitable for this application.

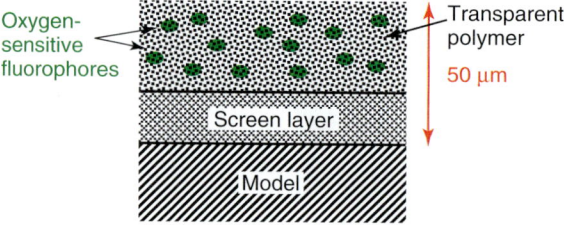

Fig. 14.12 Pressure-sensitive paint on a model.

The Stern–Volmer equation that holds for collisional quenching of fluorescence (Eq. (14.33)) must be rewritten in the following alternative form:

$$\frac{I_0}{I} = A + B\frac{P}{P_0} = \frac{\tau_0}{\tau} \tag{14.91}$$

where I, τ, and P are the fluorescence intensity, the excited-state lifetime, and the pressure, respectively. The subscripts 0 refer to the atmospheric pressure. In some cases, excited-states lifetimes are measured instead of fluorescence intensities.

Since the 1980s the PSP technique has been mainly used in aeronautics for transonic and supersonic flow studies, and it is now currently used in industrial wind tunnels (Liu et al., 1997).

Surface pressure mapping is more difficult to obtain in the case of cars because the variations in pressure are smaller at low speed. Figure 14.13 shows an investigation carried out in ONERA (Engler et al., 2002). The PSP is made of pyrene with gadolinium oxysulfide as reference dye (insensitive to oxygen) that permits correction for the nonhomogeneous distribution of illumination. The pressure map accurately reveals the variations in pressure at the roof/rear window junction.

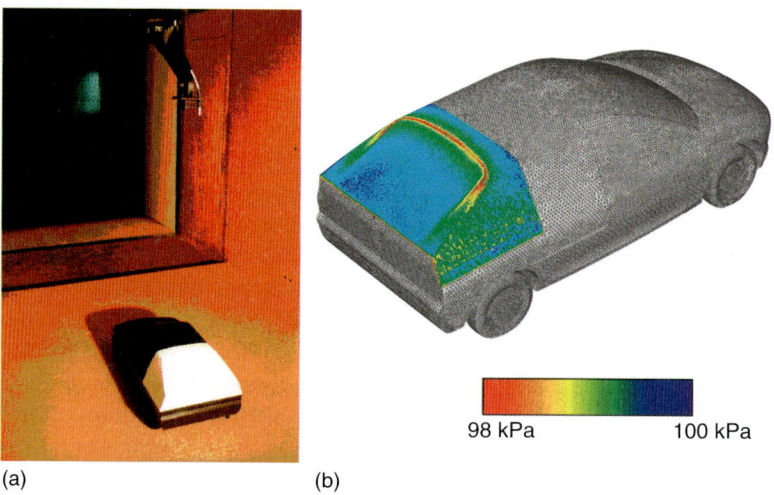

Fig. 14.13 Model of a Peugeot 206 car in a wind tunnel (a) and pressure map (b). (Courtesy of M.C. Merienne, ONERA, France.)

14.4.3
Chemical Sensing

Fluorescence is extensively used for chemical sensing in many fields: analytical chemistry, biology, biotechnology, clinical diagnostics, toxicology, environment, forensics, and chemical oceanography (Lakowicz, 1994; de Silva and Tecilla, 2005; Narayanaswamy and Wolfbeis, 2004). The success of fluorescence can be explained by the distinct advantages that this technique offers in terms of sensitivity, selectivity, response time, local observation under microscope, and remote detection by means of optical fibers.

In principle, a *fluorescent sensor* is the complete optical sensing device containing the light source, the analyte-responsive (supra)molecular moiety properly immobilized (e.g., in plastified polymers, sol–gel matrices, etc.), the optical system (involving an optical fiber or not), and the light detector (photomultiplier or photodiode) connected to appropriate electronics for displaying the signal. When considering only the fluorescent analyte-responsive (supra)molecular moiety, the term *fluorescent molecular sensor* should be used.

14.4.3.1 Fluorescent Molecular Sensors

The design of a fluorescent molecular sensor that is selective of a given analyte is actually a work of molecular engineering that involves many disciplinary fields: photophysics, photochemistry, analytical chemistry, physical chemistry, coordination chemistry, and supramolecular chemistry (Valeur, 2002 Chapter 10; Desvergne and Czarnik, 1997).

The fluorophores that are implied in fluorescent molecular sensors are numerous: naphthalene, anthracene, pyrene, aminonaphthalimide, diaminonaphthylsulfonyl, coumarins, fluorescein, eosin, rhodamines, benzidine, alizarin, seminaphthofluorescein, oligo- and polyphenylenes, porphyrins, ruthenium complexes, and so on.

Much progress has been made in sensitivity and selectivity toward a given analyte. The main targets are the following:

- cations: H_3O^+ (pH), alkali, alkaline-earth, transition, and post-transition;
- anions: fluoride, chloride, carboxylates, phosphates, (adenosine triphosphate) ATP, nitrates, and so on; and
- molecules: hydrocarbons, aminoacids, sugars, urea, ammonia, amines, alcohols, O_2, CO_2, H_2O_2, and so on.

Three classes of fluorescent molecular sensors are to be distinguished (Figure 14.14):

- Class 1: fluorophores that undergo quenching upon collision with an analyte (e.g., O_2, Cl^-).
- Class 2: fluorophores that can reversibly bind an analyte. If the analyte is a proton, the term *fluorescent pH indicator* is often employed. If the analyte is an ion, the term *fluorescent chelating agent* is appropriate. Fluorescence can be either quenched or enhanced upon binding. In the case of enhancement, the compound is said to be *fluorogenic* (e.g., 8-hydroxyquinoline (oxine)).
- Class 3: fluorophores linked, via a spacer or not, to a receptor that ensures appropriate affinity and selectivity toward a given analyte. The changes in fluorescence characteristics of the fluorophore upon interaction with the bound analyte result from the perturbation by the latter of photoinduced processes such as electron transfer, charge transfer, energy transfer, and

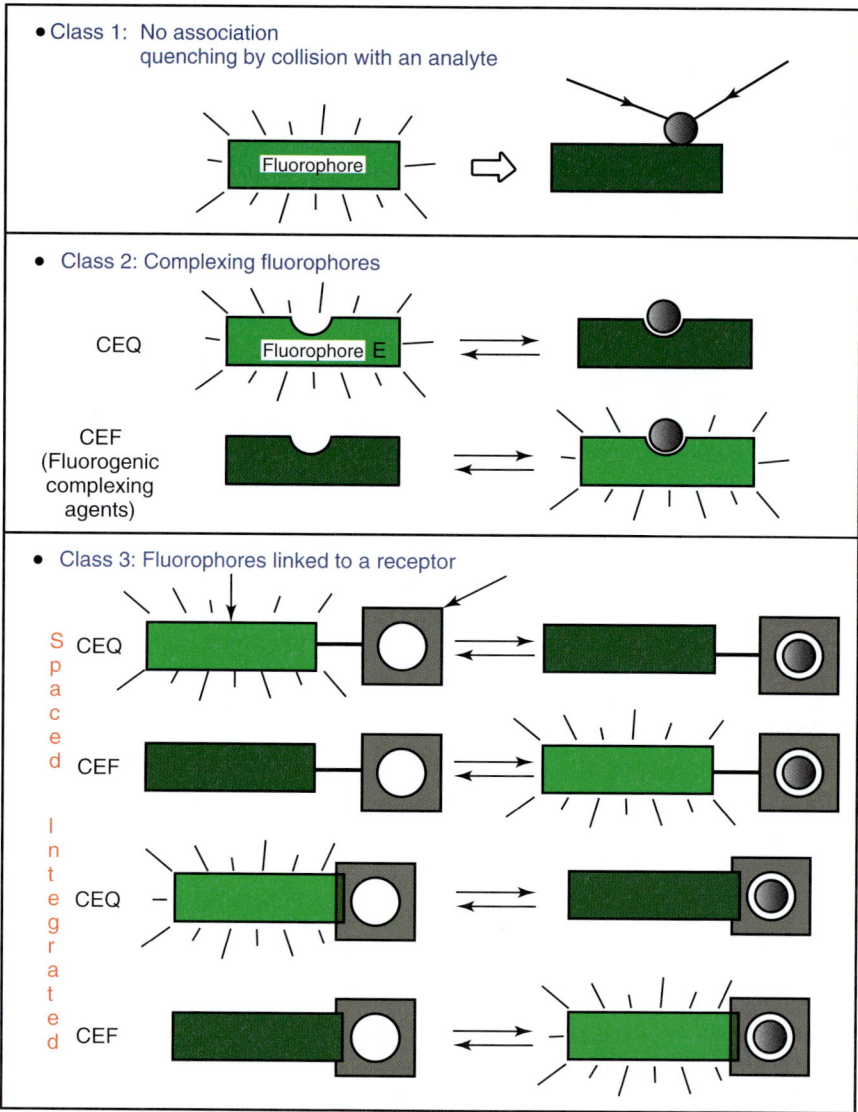

Fig. 14.14 Classes of fluorescent molecular sensors.

excimer or exciplex formation or disappearance. The design of such sensors is relevant to the fields of supramolecular chemistry and photophysics. In the case of ion recognition, the receptor is called *ionophore*, and the whole molecular sensor is called *fluoroionophore*.

Fluorescent pH Indicators Fluorescent pH indicators offer a much better sensitivity than the classical dyes, such as phenolphthalein, thymol blue, and so on, based on color change. They are thus widely used in analytical chemistry, bioanalytical chemistry, cellular biology (for measuring intracellular pH), and medicine (for monitoring pH and pCO_2 in blood; pCO_2 is determined via the bicarbonate couple). In microscopy, fluorescent pH indicators provide spatial information on pH. Moreover, remote sensing of pH is possible by means of fiber-optic chemical sensors.

A fluorescent indicator AH can undergo deprotonation according to the following equilibrium:

$$AH + H_2O \rightleftharpoons A^- + H_3O^+ \quad (14.92)$$

The true value of the acidity constant (that depends only on temperature) must be written with activities:

$$K_A = \frac{a_{A^-} a_{H_3O^+}}{a_{AH}} \quad (14.93)$$

Using the definition of pH

$$pH = -\log a_{H_3O^+} \quad (14.94)$$

Equation (14.93) can be rewritten in the following form:

$$pH = pK_A + \log \frac{a_{A^-}}{a_{AH}} \quad (14.95)$$

$$pH = pK_A + \log \frac{f_{A^-}}{f_{AH}} + \log \frac{[A^-]}{[AH]} \quad (14.96)$$

where f_{A^-} and f_{AB} are the activity coefficients relative to molarities.

In very dilute aqueous solutions, the activity coefficients are close to 1 (reference state: solute at infinite dilution), and Eq. (14.96) reduces to the well-known Henderson–Hasselbach equation:

$$pH = pK_A + \log \frac{[A^-]}{[AH]} \quad (14.97)$$

Otherwise, pK_A should be replaced by an apparent pK:

$$pK_{app} = pK_A + \log \frac{f_{A^-}}{f_{AH}} \quad (14.98)$$

Hence,

$$pH = pK_{app} + \log \frac{[A^-]}{[AH]} \quad (14.99)$$

pK_{app} depends on ionic strength, specific interactions between the indicator and the medium, and structural changes of the medium. For instance, pK_{app} differs significantly from pK_A at the interface of micelles or lipid bilayers.

For fluorometric titrations, the following relation can be used (Valeur, 2002 p. 338):

$$pH = pK_{app} + \log \frac{I - I_A}{I_B - I} \quad (14.100)$$

where I is the fluorescence intensity at a given wavelength, and I_A and I_B are the fluorescence intensities measured at the same wavelength when the indicator is only in the acidic form or only in the basic form, respectively.

It is recommended, whenever possible, to carry out ratiometric measurements, that is, to determine the ratio of the fluorescence intensities measured at two excitation wavelengths or two emission wavelengths. The ratiometric measurements are indeed preferable because the ratio of the fluorescence intensities at two wavelengths is independent of the total concentration of the dye, photobleaching, fluctuations of the source intensity, and sensitivity of the instrument. Examples of fluorescent

Tab. 14.6 Examples of fluorescent pH indicators allowing ratiometric measurements. (From Valeur (2002).)

Fluorophore	pK$_A$	Type of measurement
4-MU	7.8	excitation ratio 365/335 nm
Pyranine	7.2	excitation ratio 450/400 nm
fluorescein	2.2, 4.4, 6.4	excitation ratio 490/435 nm
BCECF	7.0	excitation ratio 505/439 nm
SNAFL dyes	7.0–7.8	excitation ratio 490/540 nm or emission ratio 540/630 nm
SNARF dyes	7.0–7.8	emission ratio 580/630 nm
CNF	7.5	excitation ratio 600/510 nm or emission ratio 550/670 nm

4-MU, 4-methylumbelliferone; BCECF, 2′,7′-bis(carboxyethyl)-5(or 6)-carboxyfluorescein; SNAFL, semi-naphthofluoresceins; SNARF, semi-naphthorhodafluors; CNF, carboxynaphthofluorescein.

pH indicators allowing ratiometric measurements are given in Table 14.6.

For ratiometric measurements, the following equation may be used to fit the calibration curve (Valeur, 2002 p. 338):

$$pH = pK_{app} + \log \frac{R - R_A}{R_B - R} + \log \frac{I_A(\lambda_2)}{I_B(\lambda_2)}$$

(14.101)

where R is the ratio $I(\lambda_1)/I(\lambda_2)$ of the fluorescence intensities at two excitation wavelengths (or two emission wavelengths) λ_1 and λ_2. R_A and R_B are the values of R when only the acidic form or the basic form is present, respectively. $I_A(\lambda_2)/I_B(\lambda_2)$ is the ratio of the fluorescence intensity of the acidic form alone to the intensity of the basic form alone at the wavelength λ_2 chosen for the denominator of R. It should be emphasized that R, R_A, and R_B are very sensitive to the wavelengths chosen and also to instrumental settings.

According to the preceding relations, the usual working range is about 2 pH units around pK_A. However, the working range is broader when the indicator possesses more that one pK_A (e.g., fluorescein).

Fluorescent Molecular Sensors of Cations
Detecting cations is of great interest in many fields: chemistry, biology, clinical biochemistry, and environment (de Silva et al., 1997; Valeur, 1994; Valeur and Leray, 2000). Sodium, potassium, magnesium, and calcium are involved in biological processes such as transmission of nerve impulse, muscle contraction, regulation of cell activity, and so on. Zinc is an essential component of many enzymes (e.g., in carbonic anhydrase and zinc finger proteins); it plays a major role in enzyme regulation, gene expression, neurotransmission, and so on. In medicine, monitoring of metal ions (e.g., Na$^+$, K$^+$, Mg^{2+}, Ca^{2+}, Li$^+$) in blood and urines is of major importance for diagnosis. In medicine, it is necessary to control the lithium level in the serum for patients under treatment for maniac depression and potassium in the case of high blood pressure. In environment, early detection of toxic metal ions such

as mercury, lead, and cadmium is of interest. In chemical oceanography, it has been demonstrated that some nutrients required for the survival of microorganisms in sea water contain zinc, iron, and manganese as enzyme cofactors.

In the design of a fluorescent molecular sensor of cations, called *fluoroionophore*, much attention is to be paid to the characteristics of the ionophore moiety and to the expected changes in fluorescence characteristics of the fluorophore moiety upon binding.

The ionophore can be a chelator, an open-chain structure (podand), a macrocycle (coronand, e.g., crown ether), a macrobicycle (cryptand), a calixarene derivative, and so on. The characteristics of the ionophore, that is, the ligand topology and the number and nature of the complexing heteroatoms or groups, should match the characteristics of the cation, that is, ionic diameter, charge density, coordination number, and intrinsic nature (e.g., hardness of metal cations, nature and structure of organic cations, etc.) according to the general principles of supramolecular chemistry.

The ionophore may be linked to the fluorophore via a spacer, but in many cases some atoms or groups participating in the complexation belong to the fluorophore. Therefore, the selectivity of binding often results from the whole structure involving both signaling and recognition moieties.

The changes in emissive properties of the fluorophore upon cation binding are due to the cation control of photoinduced electron transfer (PET), photoinduced charge transfer (PCT), excimer formation or disappearance, and excitation energy transfer. The discussion is restricted to the most popular PET and PCT sensors.

In a PET sensor (Bissell et al., 1992), excitation of the fluorophore induces the promotion of an electron of the HOMO to the LUMO, which enables PET from the HOMO of the donor (cation-free receptor) to that of the fluorophore, causing fluorescence quenching of the latter (see Section 14.3.3). Upon cation binding, the redox potential of the donor is raised so that the relevant HOMO becomes lower in energy than that of the fluorophore. Therefore, PET is not possible any more and fluorescence quenching is suppressed. Examples of PET sensors are given in Figure 14.15.

In a PCT sensor (Valeur and Leray, 2001), the fluorophore contains an electron-donating group (often an amino group) conjugated to an electron-withdrawing group and thus undergoes intramolecular charge transfer from the donor to the acceptor upon excitation by light. The consequent change in dipole moment results in a Stokes shift that depends on the microenvironment of the fluorophore. Therefore,

- when a cation is in close interaction with the donor moiety (such as an amino group), it reduces the efficiency of intramolecular charge transfer and the fluorescence spectrum is blue-shifted.
- when a cation is in close interaction with the acceptor moiety, it enhances the efficiency of intramolecular charge transfer and the fluorescence spectrum is red shifted.

In addition to these spectral shifts, changes in quantum yields and lifetimes are often observed. All the photophysical effects are obviously dependent on the charge and the size of the cation, and selectivity of these effects is expected.

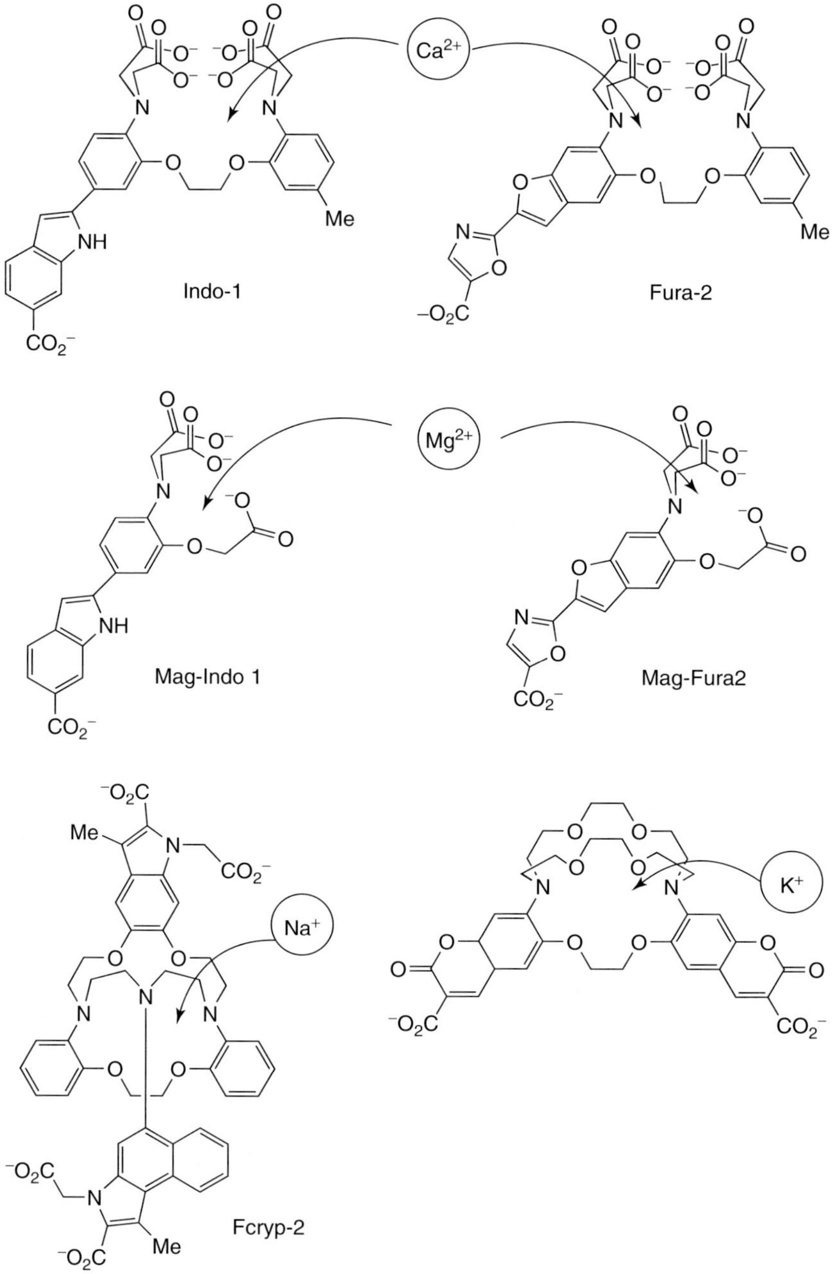

518 | *14 Molecular Fluorescence*

Fig. 14.15 Examples of PET sensors.

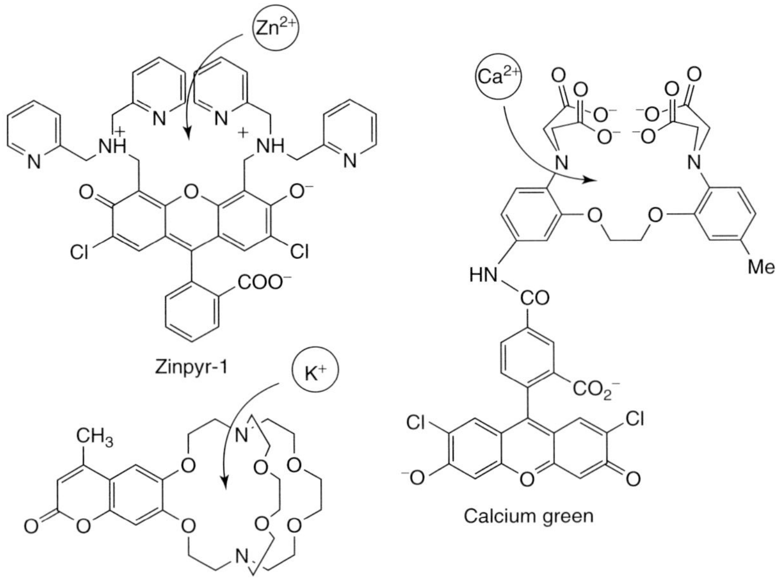

Fig. 14.16 Examples of PCT sensors.

Examples of PCT sensors are given in Figure 14.16.

Fluorescent Molecular Sensors of Anions
Anions play key roles in chemical and biological processes. Many anions act as nucleophiles, bases, redox agents, and phase transfer catalysts. Most of enzymes bind anions as either substrates or cofactors. The chloride ion is of special interest because it is crucial in several phases of human biology and in disease regulation. Moreover, it is of great interest to detect anionic pollutants such as nitrates and phosphates in ground water.

Anion molecular sensors are either based on collisional quenching (in general, they exhibit a poor selectivity) or based on recognition by an anion receptor linked to a fluorophore (fluoroionophore), which is described later.

Many fluorescent molecular sensors for halide ions (except F^-) are based on collisional quenching of a dye. In particular, the determination of chloride anions in living cells is achieved according to this principle. Examples of halide ion sensors are given in Figure 14.17. The drawback of these molecular sensors is their lack of selectivity. Moreover, the absence of spectral change precludes ratiometric measurements. However, dual-wavelength Cl^- sensors have been designed.

Anion sensors containing an anion receptor have a much better selectivity. Sensors of phosphate and pyrophosphate groups have attracted much attention because of their biological relevance. Figure 14.18 shows some examples.

14.4.3.2 Microfabricated Analysis Systems

Miniaturization of sensing devices has been the object of considerable effort. Microfabricated systems indeed offer many advantages that are of particular

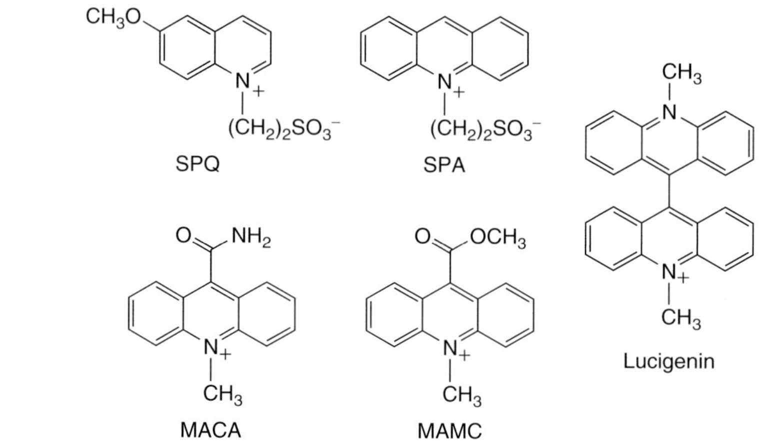

SPQ
SPA
Lucigenin
MACA
MAMC

Fig. 14.17 Examples of halide sensors.

Weak fluorescence
High fluorescence
PPi
Weak fluorescence
High fluorescence

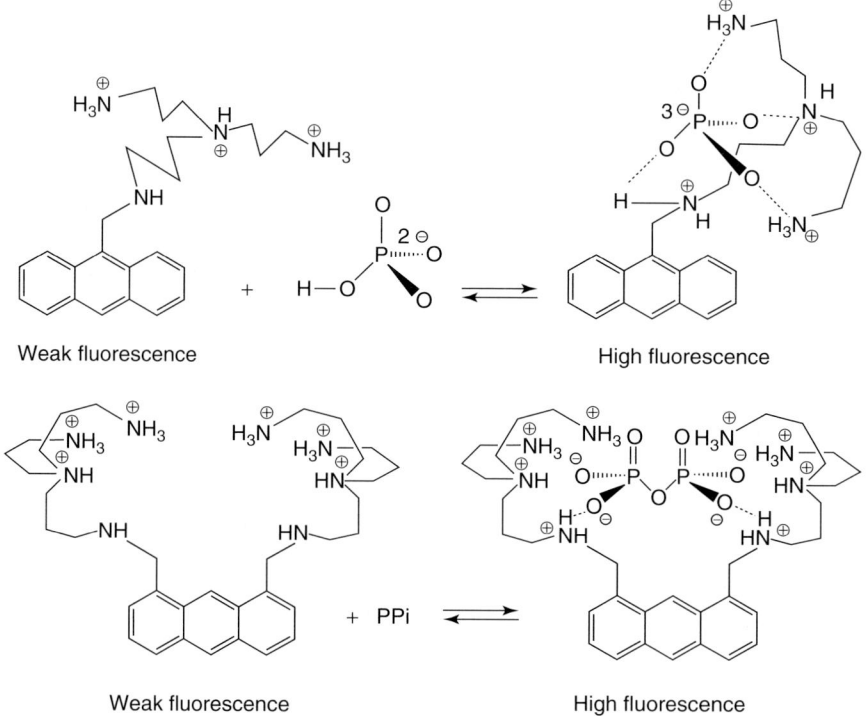

Fig. 14.18 Examples of sensors of phosphate groups.

interest in the case of biological samples: high speed, online monitoring, and low consumption of samples and reagents. The development of lab-on-chips with, in particular, the DNA gene chips is impressive. Microfabricated analysis systems, called *micro-total analysis systems* (μ-*TASs*), are based on microfluidic technology using microchipchannels, microreactors, valves, and pumps (Reyes et al., 2002; Auroux et al., 2002). Among the detection methods employed in these systems, fluorescence deserves attention. The applications are mainly relevant to life sciences: clinical diagnostics, immunoassays, DNA separation and analysis, sequencing, and so on.

A nice example is the very sophisticated microfluidic device developed by Richard Mathies and coworkers (University of California at Berkeley) for the detection of aminoacids in future missions on Mars (Skelley et al., 2005) (Figure 14.19). It is based on capillary electrophoresis and the use of fluorescamine as a fluorescence marker for aminoacids. The first step of the analysis is sublimation of amines and aminoacids onto an aluminum disk spin-coated with fluorescamine. Successful tests were carried out in Atacama desert in Chile, which can be considered as the best Mars analog site because it is one of the most arid regions in the world. Sensitivity ranges from micromolar to 0.1 nM, corresponding to part-per-trillion sensitivity.

14.4.4
Tracers in Biology

Molecular fluorescence is widely used in biology for cellular imaging using (confocal) fluorescence microscopy. pH imaging and ion imaging (calcium, magnesium, zinc, chloride, etc.) are possible by using appropriate fluorescent pH indicators or fluorescent molecular sensors for ions (see Section 14.4.2). With the aim at tracing cellular components during biological processes, fluorescent species are used to visualize these components.

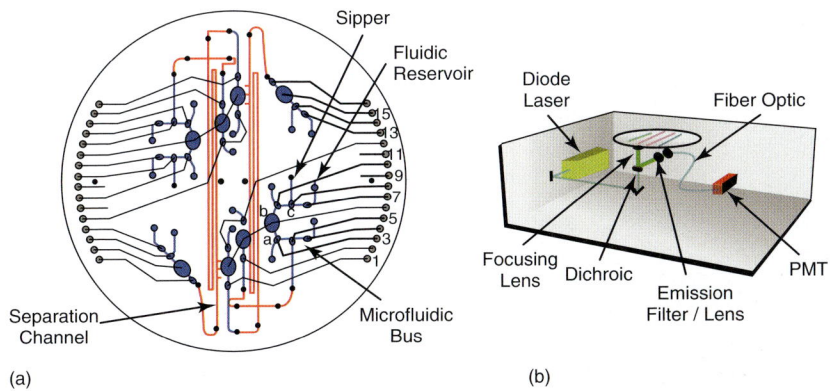

Fig. 14.19 Microdevice for amino acid biomarker detection and analysis on Mars. (a) Top view of the microdevice showing registration of the capillary electrophoresis channel (red), pneumatic manifold (black), and fluidic bus wafers (blue). (b) Schematic of the instrument showing confocal excitation and detection optics. (Reproduced with permission from Skelley et al. (2005), Copyright 2005 National Academy of Sciences, USA.)

In contrast to fluorescent probes, they should be insensitive to their microenvironment as much as possible. The four classes of fluorescent tracers in biology are organic fluorophores, fluorescent antibodies, fluorescent proteins, and semiconductor nanocrystals. The first two classes are well known and are not described here, whereas the use of fluorescent proteins and semiconductor nanocrystals is more recent and deserves particular attention.

14.4.4.1 Fluorescent Proteins

Fluorescent proteins become more and more popular as tracers in biology (Zimmer, 2002; Lippincott-Schwartz and Patterson, 2003). The first of them is green fluorescent protein (GFP) from *Aequora Vitoria,* a bioluminescent jelly fish. The green fluorescence is due to a partially cyclic tripeptide (serine, tyrosine, and glycine) in an α-helix. A fusion between a cloned gene and GFP can be created using standard techniques, which leads to a chimera that can be expressed in a cell. Thus, GFP acts as a fluorescent label of a protein of interest whose localization or activity is not affected. It is thus possible to visualize dynamic cellular events and to monitor protein localization.

Spectral variants of GFP emitting in the blue-cyan and the yellowish green have been generated, but the wavelength of the emission maximum is not greater than approximately 530 nm. However, corals (Anthozoa) can provide fluorescent proteins emitting in the red, such as DsRed.

The simultaneous use of several colors is thus possible. A beautiful example of multicolor imaging is shown in Figure 14.20.

At a larger scale, fluorescent proteins also offer an outstanding tool for imaging muscle fibers in vivo, so that one can see the effect of new therapies for muscle diseases.

At an even larger scale, a whole animal can be rendered fluorescent if all cells express the gene of GFP. This has been applied to rabbits, mice, rats, pigs, and silkworms. The interest is

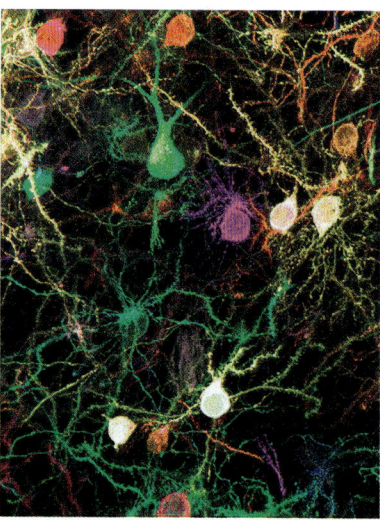

Fig. 14.20 Multicolor imaging of mossy fibers in the cerebellum of a transgenic mouse. This image taken by confocal microscopy shows mossy fibers in one folium of a young adult transgenic mouse. Spectral variants of GFP are expressed in a combinatorial way following genetic recombination mediated by cre-recombinase. The various colors are produced by the particular subset of fluorescent proteins expressed in each axon and their individual concentrations. In this animal, at least 12 distinct spectral fingerprints in different cells can be discerned. (Courtesy of J. Lichtman, Harvard University.)

that GFP makes it possible to follow the fate of cells in embryos, in normal organs, and in tumors, without requiring destruction of the cells. A transgenic fluorescent rabbit generated in the French *Institut National de la Recherche Agronomique* (INRA) appears green under UV illumination, even the eyes show green fluorescence. A bioartist, Eduardo Kac, from the Chicago University wished to breed these rabbits to be used as exotic pets, but this proposition was rejected by the French researchers.

In 2003, a company located in Texas put in the market a genetically modified aquarium fish having the capacity to be red fluorescent under UV illumination. This announcement was followed by protests coming from both activists and the scientific community. Indeed, this event raises several unanswered questions. Are these fish a threat for environment? Are they suffering from their transgene and from UV illumination? Is it ethical to use genetic engineering to generate this particular kind of pets? (Knight, 2003)

14.4.4.2 Semiconductor Nanocrystals

The fluorescence of semiconductor nanocrystals is not strictly relevant to molecular fluorescence, but their growing interest deserves attention here. The emissive properties of these nanocrystals, also called *quantum dots*, are due to quantum confinement. The emission wavelength ranges from the UV to the infrared depending on the nature of the semiconductor and the size of the nanocrystal. For quantum dots made of cadmium selenide with a protective coating of zinc sulfide, the emission wavelength ranges from the blue (diameter of 2 nm) to the red (diameter of 7 nm).

Quantum dots offer distinct advantages for tracing owing to their specific characteristics:

- broad absorption spectrum allowing a common excitation wavelength for nanocrystals having different emission wavelengths;
- narrow emission spectrum (gaussian) : 12–15 nm;
- fluorescence quantum yield greater than 50%;
- excellent photostability;
- coatable for water solubility and biocompatibility.

It should be emphasized that the photostability of quantum dots is much better than that of organic dyes, and they are much brighter in tissue. It is thus possible to monitor the trajectories of nanocrystals for 20 minutes, instead of 5 seconds with an antibody coupled to an organic fluorophore such as Cy3 (Dahan *et al.*, 2003).

Quantum dots have also been successfully used for in vivo cancer targeting and imaging (Gao *et al.*, 2004). Applications to encoding are also of great interest. For this purpose, nanocrystals are embedded in various proportions into polymer microspheres with millions of possible combinations. These microspheres can then be attached to DNA sequences, peptides, and so on, and are thus useful for high throughput screening (genomics, proteomics, and drug screening).

14.4.5
Clinical Diagnostics

14.4.5.1 Critical Care Analysis

The Osmetech/Roche OSPI critical care analyzer can measure up to eight critical care analytes in whole blood (Tusa and He, 2005): pH, CO_2/Hb, O_2, Na^+, K^+, Ca^{2+}, or Cl^-. Six appropriate fluorescent molecular sensors are embedded in separate disks on a disposable cartridge. It takes not more than 2 minutes to get the result from 120 µl of whole blood. The list of the analytes, the normal range, the pathological range, and the relevant molecular sensors are reported in Table 14.7. Pyranine is used as a fluorescent pH indicator. CO_2 is determined in the following way: a membrane blocks the passage of H_3O^+ and other ions into the sensor but allows CO_2 through, where it forms carbonic acid, increasing the local concentration of H_3O^+ that is determined by pyranine. Cation recognition is achieved by fluoroionophores consisting of the naphthalimide fluorophore linked to an azacrown for Na^+, a cryptand for K^+, and a chelating moiety for Ca^{2+} (Figure 14.21). The dissociation constants are consistent with the pathological ranges.

14.4.5.2 Angiography

Angiography is a good example of application of fluorescence for in vivo diagnostics. Fluorescent tracers are used to visualize the blood vessels of the retina in order to detect anomalies: lesions, aneurisms, and occlusions resulting from specific eye diseases.

A fluorescent dye (fluorescein or indocyanine green) is injected into a vein in the arm of the patient. The dye is conveyed by the circulatory system and reaches the retina vessels.

Figure 14.22 shows an angiogram observed after injection of fluorescein: the existence of bright points leads to the diagnostic of an early diabetic retinopathy showing microaneurisms.

Indocyanine green provides information on deeper tissues, as compared to fluorescein. Because it absorbs near 800 nm and emits in the near infrared, it can reveal deeper vessels called *choroidal vessels*. In the case of an aged-related macular degeneration, leaky vessels, that is, choroidal neovascularization, can be visualized, while the angiogram with fluorescein also shows fluid leakage for the same patient.

14.4.5.3 Bladder Tumor Detection

The measurement of relative fluorescence intensities of tryptophan and NADH by endoscopy permits early detection of bladder tumors in situ. Light pulses from an excimer laser (308 nm) are conveyed by a catheter containing several optical fibers. Fluorescence is

Tab. 14.7 Fluorescent sensors for critical care analytes. (Adapted from Tusa and He (2005).)

	Normal range	Pathological range	Fluorescent sensor
pH	7.35–7.45	6.7–7.7	pyranine
P(CO_2) (torr)	30–45	10–200	pyranine
P(O_2) (torr)	70–100	20–600	Oxygen-sensitive dye
Na^+ (mM)	135–145	100–180	Naphthalimide dye – azacrown[a]
K^+ (mM)	3–5	1–10	Naphthalimide dye – cryptand[a]
Ca^{2+} (mM)	1.0–1.4	0.3–2.5	Naphthalimide dye – chelator[a]

[a] See Figure 14.21.

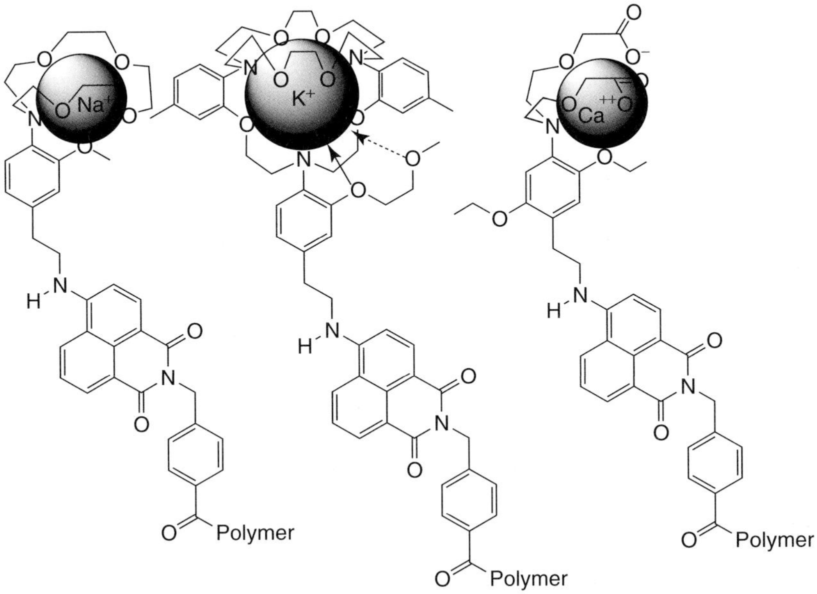

Fig. 14.21 Fluoroionophores for the selective detection of Na$^+$, K$^+$, and Ca^{2+}. (Reproduced from Tusa and He (2005) by permission of the Royal Society of Chemistry.)

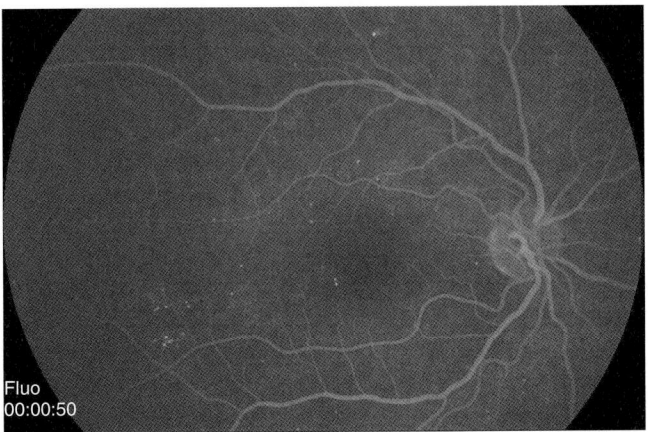

Fig. 14.22 Angiogram showing microaneurisms (bright points). (Courtesy of Dr J.C. Hache, Lille hospital, France.)

collected by other fibers and analyzed by a monochromator and a photomultiplier (Figure 14.23) (Avrillier *et al.*, 1997). The fluorescence spectra are different according to the state of the tissue. For tumors and carcinoma in situ (CIS), the ratio of the fluorescence intensities at 360 and 440 nm is statistically greater than 2. This test is reliable and can be run during routine visit.

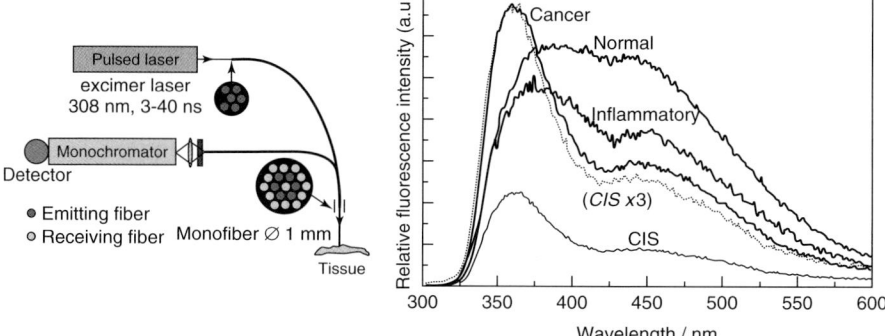

Fig. 14.23 Experimental set-up for the detection of bladder tumors. Fluorescence spectra of normal, inflammatory and cancerous tissues upon excitation at 308 nm (CIS, carcinoma in situ). (Adapted from Avrillier *et al.* (1997).)

14.4.6
Miscellaneous

14.4.6.1 Fluorescence LIDAR

LIDAR is the acronym of light detection and ranging (which is a transposition of RADAR for radiowave detection and ranging). LIDAR is a useful tool for atmospheric monitoring and, in particular, for the detection of SO_2 and NO_2 in atmosphere. This technique that is based on back-scattering and absorption of species can be extended to fluorescence detection by using appropriate emission filters. The applications concern the status of vegetation, the estimation of plankton in sea water, and imaging of the façades of historical monuments.

A Nd:YAG laser is often used for these applications: it provides excitation pulses of about 10 nanoseconds duration at 532 and 355 nm, thanks to frequency doubling and tripling, respectively.

Vegetation It is of great interest to monitor the chemical activities and the status of trees and forests. Early detection and mapping of damaged vegetation as a consequence of pollution can also be made. Remote detection of chlorophyll by a LIDAR is possible. Upon excitation at 532 nm, the chlorophyll concentration can be estimated as a function of time by taking the ratio of the intensity of the images at 740 and 685 nm (Saito *et al.*, 2000). The test is validated by the comparison of the average intensity ratio and the amount of chlorophyll determined by chromatography. The average intensity ratio follows the chlorophyll concentration that increases in summer and decreases in autumn.

Sea Water The detection of fluorescence from chlorophyll in phytoplankton and from phycoerythrin-containing plankton (e.g., cyanobacteria) (both excited at 532 nm) is possible from aircrafts or ships, thanks to a LIDAR. The biomass can thus be estimated. In addition, the fluorescence of dissolved organic substances can be detected upon excitation at 355 nm. NASA has developed an airborne oceanographic LIDAR and a shipboard laser fluorometer to achieve such evaluations.

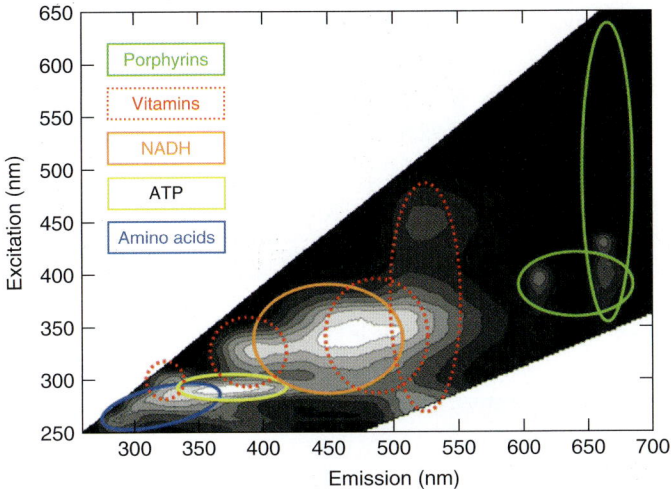

Fig. 14.24 Two-dimensional excitation–emission patterns of typical food-relevant fluorophores. (Reproduced with permission from Christensen et al. (2006). Copyright 2006 American Chemical Society.)

Façades of Historical Monuments Svanberg and coworkers have developed a system for imaging the laser-induced fluorescence from the façades of historical monuments (Weibring et al., 2001). The aim is the visualization of areas of biodeterogen (lichens, green algae) and the identification of different stone types with natural surface aging or crust and pollution deposited layers. Façade status assesment is thus obtained in view of restauration planning.

The LIDAR system is placed in a truck at a distance of about 60 m from the façade. A 355-nm pulsed laser beam is swept over the façade row by row. The spectrally resolved fluorescence signals are recorded on each point.

14.4.6.2 Food Science

Food systems (meat, fish, dairy products, oils, cereals, beer, fruits, vegetables, etc.) contain several fluorescent substances such as proteins, vitamins, secondary metabolites, pigments, toxins, and flavoring compounds. The intrinsic fluorescence from intact food can thus provide valuable information on the quality of food products, their authenticity, and the effect of processing and aging. The emitted fluorescence is quite complex because of the presence of several intrinsic fluorophores. Moreover, the fluorescence signals are affected by phenomena such as Rayleigh and Raman scattering, inner filter effects, self-quenching, and so on. The complex signals can be decomposed from a series of two-dimensional excitation–emission patterns into the pure constituent signals using multiway chemometrics (Christensen et al., 2006). Figure 14.24 shows the spectral properties of selected food-relevant fluorophores.

14.4.6.3 Forensics

On a crime scene, the visualization of fingerprints by means of powders is well known. The fluorescent powders are the most efficient.

 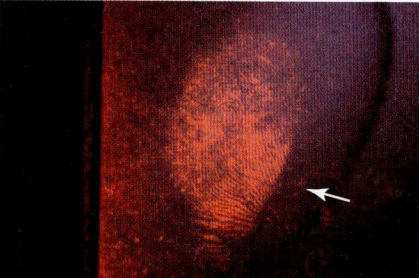

Fig. 14.25 Fingerprint on a piece of paper revealed by fluorescence after treatment with diaza-9-fluorenone. (Courtesy of the Identité judiciaire de la Préfecture de police de Paris.)

Fig. 14.26 A bill of 50 Euros shows specific color under UV illumination (photograph by the author.)

Fingerprints on papers are more difficult to detect, especially when they are old. In this case, the procedure is to soak the paper in a solution of diaza-9-fluorenone (that reacts with aminoacids present in the sweat deposited by the fingers), and to dry it. Observation under UV illumination can reveal the fluorescent fingerprints provided that the blue fluorescence of paper due to whitening agents is rejected by means of an appropriate filter (Figure 14.25).

14.4.6.4 Counterfeit Detection

In many situations, the luminescent pattern serves as a security check for documents, especially for bills. Various luminescent compounds either organic or inorganic are employed. For example, the Euros bills show different colors (e.g., stars, Europe map, etc.) under UV illumination (Figure 14.26). Small fluorescent fibers can also be seen.

The use of semiconductor nanocrystals (Section 14.4.4.4) to counterfeit detection is to be expected in the near future (bank notes, ID cards, badges, etc.).

Fluorescence is also a means to detect counterfeited art works: paintings, ceramics, glassware, sculptures, ivories, old marbles, and so on. In fact, fluorescence depends on ageing. Moreover, traces of restoration, overpainting, falsified signatures, and so on can also be easily detected.

In conclusion, fluorescence is a very powerful tool in both fundamental and applied research. The technological and industrial applications are numerous in variegated fields. The success of fluorescence as a tool of investigation of the structure and dynamics of matter or living systems arises from the high sensitivity of fluorometric techniques, the specificity of fluorescence characteristics due to the microenvironment of the emitting molecule, and the ability of the latter to provide spatial and temporal information. After a long history, no doubt that fluorescence still has a bright future.

References

Andrews, D.L. and Demidov, A.A. (eds) (1999) *Resonance Energy Transfer*, John Wiley & Sons, Ltd, New York.

Auroux, P.A., Iossifidis, D., Reyes, D., and Manz, A. (2002) *Anal. Chem.*, **74**, 2637.

Avrillier, S., Tinet, E., Ettori, D., and Anidjar, M. (1997) *Phys. Scr.*, **72**, 87.

Baleizao, C., Nagl, S., Borisov, S.M., Schäferling, M., Wolfbeis, O., and Berberan-Santos, M.N. (2007) *Chem. Eur. J.*, **13**, 3643.

Benninger, R.K.P., Koç, Y., Hofmann, O., Requejo-Isidro, J., Neil, M.A.A., French, P.M.W., and deMello, A.J. (2006) *Anal. Chem.*, **783**, 2272.

Berberan-Santos, M.N. (ed.) (2007) *Fluorescence of Super Molecules, Polymers and Nanosystems*, Springer Series on Fluorescence, Vol. 4, Springer, Berlin.

Berberan-Santos, M.N. and Garcia, J.M.M. (1996) *J. Am. Chem. Soc.*, **118**, 9391.

Bissell, R.A., de Silva, A.P., Gunaratne, H.Q.N., Lynch, P.L.M., Maguire, G.E.M., and Sandanayake, K.R.A.S. (1992) *Chem. Soc. Rev.*, **21**, 187.

Bokobza, L. (1990) *Prog. Polym. Sci.*, **15**, 337.

Bokobza, L., Monnerie, L. (1986) in *Photophysical and Photochemical Tools in Polymer Science* (ed. M.A. Winnik), D. Reidel Publishing Company, pp. 449–466.

Bur, A.J., Wangel, M.G., and Roth, S. (2002) *Appl. Spectrosc.*, **56**, 174.

Chandrashekaran, N. and Kelly, L.A. (2001) *J. Am. Chem Soc.*, **123**, 9898.

Chandrashekaran, N. and Kelly, L.A. (2004) in *Reviews in Fluorescence* (eds C.D. Geddes and J.R. Lakowicz), Kluwer Academic, New York, p. 21.

Christensen, J., Norgaard, L., Bro, R., and Engelsen, S. (2006) *Chem. Rev.*, **106**, 1979.

Dahan, M., Lévi, S., Luccardini, C., Rostaing, P., Riveau, B., and Triller, A. (2003) *Science*, **302**, 442.

Dake, H.C. and De Ment, J. (1941) *Fluorescent Light and its Applications*, Chemical Publishing Co., Brooklyn.

Desvergne, J.P. and Czarnik, A.W. (eds) (1997) *Chemosensors of Ion and Molecule Recognition*, Kluwer, Dordrecht.

Engler, R.H., Mérienne, M.C., Klein, C., and Le Sant, Y. (2002) *Aerosp. Sci. Techn.*, **6**, 313.

Fister, J.C., Rank, D., and Harris, J.M. (1995) *Anal. Chem.*, **67**, 4269.

Fisz, J.J. (1985) *Chem. Phys.*, **99**, 177; ibid. 1989, **132**, 303; ibid. 1989, **132**, 315.

Gao, X., Cui, Y., Levenson, R.M., Chung, L.W.K., and Nie, S. (2004) *Nat. Biotechnol.*, **22**, 969.

Georgescauld, D., Desmasez, J.P., Lapouyade, R., Babeau, A., Richard, H., and Winnik, M. (1980) *Photochem. Photobiol.*, **31**, 539–545.

Göppelsröder, F. (1868) *J. Prakt. Chem.*, **104**, 10.

Harvey, E.N. (1957) *A History of Luminescence*, The American Philosophical Society, Philadelphia.

Herzum, C., Boit, C., Kölzer, J., Otto, J., and Weiland, R. (1998) *Microelectron. J.*, **29**, 163.

Hof, M., Hutterer, R., and Fidler, V. (eds) (2005) *Fluorescence Spectroscopy in Biology*, Springer Series on Fluorescence Vol. 3, Springer, Berlin.

Jameson, D.M., Croney, J.C., and Moens, P.D.J. (2003) *Methods Enzymol.*, **360**, 1.

Kalyanasundaran, K. (1987) *Photochemistry in Microheterogeneous Systems*, chap. 2, Academic Press, Orlando.

Kinosita, K., Kawato, S., and Ikegami, A. (1977) *Biophys. J.*, **20**, 289.

Knight, J. (2003) *Nature*, **426**, 372.

Kolodner, P. and Tyson, J. (1982) *Appl. Phys. Lett.*, **409**, 782; 1983, **421**, 117.

Kraayenhof, R. and Visser, A.J.W.G. (eds) (2003) *Fluorescence Spectroscopy, Imaging and Probes*, Springer Series on Fluorescence, Vol. 2, Springer, Berlin.

Lakowicz, J.R. (ed.) (1991–2007) *Topics in Fluorescence Spectroscopy*, Vol. 1–11, Plenum Press, New-York.

Lakowicz, J.R. (ed.) (1994) *Probe Design and Chemical Sensing, Topics in Fluorescence Spectroscopy*, Vol. 4, Plenum, New-York.

Lakowicz, J.R. (2006) *Principles of Fluorescence Spectroscopy*, 3rd edn, Springer, New York.

Lipari, G. and Szabo, A. (1980) *Biophys. J.*, **30**, 489.

Lippincott-Schwartz, J. and Patterson, G.H. (2003) *Science*, **300**, 87.

Liu, T., Campbell, B.T., Burns, S.P., and Sullivan, J.P. (1997) *Appl. Mech. Rev.*, **50**, 227.

Liu et, T. and Sullivan, J.P. (2005) *Pressure and Temperature Sensitive Paints*, Springer, Berlin.

Loufty, R.O. (1986) *Pure Appl. Chem.*, **58**, 1239.

Narayanaswamy, R. and Wolfbeis, O.S. (eds) (2004) *Optical Sensors, Industrial, Environmental and Diagnostic Applications*, Springer Verlag, Berlin.

Navarra, K.R., Rabe, D.C., Fonov, S.D., Goss, L.P., and Hah, C. (2001) *J. Turbomachinery.*, **123**, 823.

Pottel, H., Herreman, W., Van der Meer, B.W., and Ameloot, M. (1986) *Chem. Phys.*, **102**, 37.

Pringsheim, P. and Vogel, M. (1943) *Luminescence of Liquids and Solids and its Practical Applications*, Interscience, New York.

Radley, J.A. and Grant, J. (1933) *Fluorescence Analysis in Ultraviolet Light*, Van Nostrand Co., New York.

Reyes, D., Iossifidis, D., Auroux, P.A., and Manz, A. (2002) *Anal. Chem.*, **74**, 2623.

Ross, D., Gaitan, M., and Locascio, M.E. (2001) *Anal. Chem.*, **73**, 4117.

Saito, Y., Saito, R., KawaHara, T.D., Nomura, A., and Takeda, S. (2000) *For. Ecol. Manag.*, **128**, 129.

Schulman, S.G. (ed.) (1985–1993) *Molecular Luminescence Spectroscopy*, Parts 1–3, Wiley Interscience, New York.

de Silva, A.P., Gunaratne, H.Q.N., Gunnlaugsson, T., Huxley, A.J.M., McCoy, C.P., Rademacher, J.T., and Rice, T.E. (1997) *Chem. Rev.*, **97**, 1515.

de Silva, A.P. and Tecilla, P. Guest editors, (2005) Fluorescent sensors, special issue of *J. Mater. Chem.*, **15**, 2617.

Skelley, A.M., Cherer, J.R., Aubrey, A.D., Grover, W.H., Ivester, R.H.C., Ehrenfreund, P., Grunthaner, F.J., Bada, J.L., and Mathies, R.A. (2005) *Proc. Natl. Acad. Sci. U.S.A.*, **102**, 1041.

Szabo, A. (1984) *J. Chem. Phys.*, **81**, 150.

Tusa, J.K. and He, H. (2005) *J. Mater. Chem.*, **15**, 2640.

Uchiyama, S., de Silva, A.P., and Iwau, K. (2006) *J. Chem. Educ.*, **83**, 720.

Valeur, B. (1993) in *Molecular Luminescence Spectroscopy. Methods and Applications*, Part 3, (ed. S.G. Schulman), Wiley-Interscience, pp. 25–84.

Valeur, B. (1994) in *Probe Design and Chemical Sensing, Topics in Fluorescence Spectroscopy*, Vol. 4, (ed. J.R. Lakowicz), Plenum, New-York, pp. 21–48.

Valeur, B. (2002) *Molecular Fluorescence. Principles and Applications*, Wiley-VCH Verlag GmbH, Weinheim.

Valeur, B. (2007) in *Fluorescence of Super Molecules, Polymers and Nanosystems*, Springer Series on Fluorescence, Vol. 4, (ed. M.N. Berberan-Santos), Springer, pp. 21–43.

Valeur, B., Berberan-Santos, M.N., and Martin, M.M. (2007) in *Analytical Methods in Supramolecular Chemistry* (ed. C.A. Schalley), Wiley-VCH Verlag GmbH, pp. 220–264.

Valeur, B. and Brochon, J.-C. (eds) (2001) *New Trends in Fluorescence Spectroscopy. Applications to Chemical and Life Sciences*, Springer Series on Fluorescence, Vol. 1, Springer, Berlin.

Valeur, B. and Leray, I. (2000) *Coord. Chem. Rev.*, **205**, 3.

Valeur, B. and Leray, I. (2001) in *New trends in Fluorescence Spectroscopy. Application to Chemical and Life Sciences* (eds B. Valeur and J.C. Brochon), Springer, Berlin, pp. 187–207.

Van der Meer, W., Kooyman, R.P.H., and Levine, Y.K. (1982) *Chem. Phys.*, **66**, 39.

Van der Meer, W., Pottel, H., Herreman, W., Ameloot, M., Hendrickx, H., and Schröder, H. (1984) *Biophys. J.*, **46**, 515.

Vauhkonen, M., Sassaroli, M., Somerharju, P., and Eisinger, J. (1990) *Biophys. J.*, **57**, 291.

Viriot, M.L., Bouchy, M., Donner, M., and André, J.C. (1983) *Photobiochem. Photobiophys.*, **5**, 293.

Weibring, P., Johansson, T., Edner, H., Svanberg, S., Sundner, S., Raimondi, V., Cecchi, G., and Pantani, L. (2001) *Appl. Opt.*, **40**, 6111.

Zachariasse, K.A., Kozankiewicz, B., and Kühnle, W. (1983) in *Photochemistry and Photobiology*, Vol. II, (ed. A.H. Zewail), Harwood, London, pp. 941–960.

Zannoni, C., Arcioni, A., and Cavatorta, P. (1983) *Chem. Phys. Lipids*, **32**, 179.

Zimmer, M. (2002) *Chem. Rev.*, **102**, 759.

Further Reading

Baeyens, W.R.G., deKeukeleire, D. and Korkidis, K. (eds) (1991) *Luminescence Techniques in Chemical and Biochemical Analysis*, Marcel Dekker, New York,.

Becker, R.S. (1969) *Theory and Interpretation of Fluorescence and Phosphorescence*, Wiley Interscience, New York.

Berlman, I.B. (1965, 1971) *Handbook of Fluorescence Spectra of Aromatic Molecules*, Academic Press, New York.

Birks, J.B. (ed.) (1975) *Organic Molecular Photophysics*, Vol. 1 and 2, John Wiley & Sons, Londres.

Cundall, R.B. and Dale, R.E. (eds) (1983) *Time-Resolved Fluorescence Spectroscopy in Biochemistry and Biology*, Plenum Press, New York.

Czarnik, A.W. (ed.) (1992) *Fluorescence Chemosensors for Ion and Molecule Recognition*, American Chemical Society, Washington.

Demas, J.N. (1983) *Excited State Lifetime Measurement*, Academic Press, New York.

Guilbault, G. (ed.) (1973, 1990) *Practical Fluorescence*, 1st edn, 2nd edn, Marcel Dekker, New York.

Jameson, D.M. and Reinhart, G.D. (eds) (1989) *Fluorescent Biomolecules*, Plenum Press, New York.

Mielenz, K.D. (ed.) (1982) *Measurement of Photoluminescence*, Academic Press, Washington, DC.

Miller, J.N. (ed.) (1981) *Standards in Fluorescence Spectrometry, Ultraviolet Spectrometry Group*, Chappman and Hall, Londres.

O'Connor, D.V. and Phillips, D. (1984) *Time-Correlated Single Photon Counting*, Academic Press, Londres.

Parker, C.A. (1968) *Photoluminescence of Solutions*, Elsevier, Amsterdam.

Rettig, W., Strehmel, B., Schrader, S., and Seifert, H. (eds) (1999) *Applied Fluorescence in Chemistry, Biology and Medicine*, Springer, Berlin.

Slavik, J. (ed.) (1996) *Fluorescence Microscopy and Fluorescent Probes*, Plenum Press, New York.

Wolfbeis, O.S. (ed.) (1993) *Fluorescence Spectroscopy. New Methods and Applications*, Springer-Verlag, Berlin.

15
Resonance Energy Transfer

David L. Andrews and David S. Bradshaw

15.1	**Introduction**	**535**
15.2	**History of RET**	**535**
15.2.1	The First Experiments	535
15.2.2	Early Developments of Theory	536
15.2.3	Förster Theory	536
15.3	**The Photophysics of RET**	**537**
15.3.1	Primary Excitation Processes	537
15.3.2	Coupling of Electronic Transitions	538
15.3.3	Dissipation and Line-Broadening	538
15.3.4	The Förster Equation	539
15.3.5	Orientation Dependence	540
15.3.6	Förster Radius	541
15.3.7	Polarization Features	542
15.3.8	Diffusion Effects	543
15.3.9	Long-Range Transfer	544
15.3.10	Dexter Transfer	544
15.4	**Applications of RET to Molecular Biology**	**545**
15.4.1	Spectroscopic Ruler	545
15.4.2	Conformational Change	546
15.4.3	Intensity-Based Imaging	547
15.4.4	Lifetime-Based Imaging	548
15.4.5	Other Applications	550
	Acknowledgments	**550**
	References	**550**
	Further Reading	**553**

Encyclopedia of Applied Spectroscopy. Edited by David L. Andrews.
Copyright © 2009 WILEY-VCH Verlag GmbH & Co. KGaA, Weinheim
ISBN: 978-3-527-40773-6

15.1
Introduction

Resonance energy transfer[1] (RET) is a spectroscopic process whose relevance in all major areas of science is reflected both by the wide prevalence of the effect and through numerous technical applications. It is an optical near-field mechanism that effects the transportation of electronic excitation between physically distinct atomic or molecular components, based on transition dipole–dipole coupling. In this chapter, a comprehensive survey of the process is presented, beginning with an outline of the history and highlighting the early contributions of Perrin and Förster. This is followed by a review of the photophysics and then a discussion of some prominent applications of RET. Particular emphasis is given to techniques used in molecular biology, ranging from the "spectroscopic ruler" measurements of functional group separation to fluorescence lifetime microscopy. Finally, applications to synthetic polymers and chemical sensors are examined.

---

1) RET is also known as Förster – or fluorescence – resonance energy transfer (FRET), or electronic energy transfer (EET).

15.2
History of RET

15.2.1
The First Experiments

RET is a process by means of which the energy of an excited atom or molecule (usually called the *donor*, but known historically as the *sensitizer*) is transferred nonradiatively to an acceptor molecule ("activator"), through intermolecular dipole–dipole coupling. The origins of its discovery can be traced back to 1922, when the phenomenon of RET ("sensitized fluorescence") was first experimentally observed by Franck (1922); Cario (1922); Cario and Franck (1922) in the gas phase. This spectroscopic experiment involved illuminating a mixture of mercury and thallium vapors at a wavelength absorbed solely by the mercury; the resulting fluorescence spectra was proven to include frequencies that could only be emitted from thallium. Such energy transfer in vapors was at first assumed to be uniquely associated with interatomic collisions, but a discovery that transfer could occur at larger separations than the collision radii showed that this was not necessarily the case. Soon, RET was also being observed in solutions (Gaviola

---

Encyclopedia of Applied Spectroscopy. Edited by David L. Andrews.
Copyright © 2009 WILEY-VCH Verlag GmbH & Co. KGaA, Weinheim
ISBN: 978-3-527-40773-6

and Pringsheim, 1924), and over the following years in many other physical systems.

15.2.2
Early Developments of Theory

The first theoretical explanation of the phenomenon was proposed by the Nobel laureate Perrin (1927). He recognized that energy could be transferred from an excited molecule to its neighbors amongst closely spaced molecules through dipole interactions; he named this process as "transfert d'activation," and his paper on the subject became the earliest attempt to describe nonradiative (near-field) energy transfer. However, despite its initial success, Perrin's model incorrectly predicted that nonradiative energy transfer should be possible between dye molecules up to an intermolecular distance of 1000 Å, deriving from an inaccurate assumption that the molecules would act as Hertzian oscillators with exactly defined resonance frequencies. Five years later (Perrin, 1932), Perrin's son Francis developed a corresponding quantum mechanical theory of RET, based on the results of Kallman and London (1928). In this work, he recognized a "spreading of absorption and emission frequency" due to the interactions of the dye with the solvent, thus reducing the probability of energy transfer. As a result, efficient transfer was calculated to occur up to 150–250 Å, still approximately a factor of 3 greater than the experimentally observed. A detailed and readable survey of these early contributions of J. and F. Perrin can be found in a review by Berberan-Santos (2001).

15.2.3
Förster Theory

Extending the ideas of J. and F. Perrin, Förster developed the first essentially correct theoretical treatment of RET (Förster, 1946, 1948, 1959). Förster determined that energy transfer, through dipolar coupling between molecules, mostly depends on two important quantities: spectrum overlap and intermolecular distance. Following the observation that "the absorption and fluorescence spectra of similar molecules are far from completely overlapping," he found a means to quantify the spectral overlap integral. The dipole–dipole interaction was known to have an inverse proportionality on the cube of the molecular separation. Because the rate of energy transfer is proportional to the square of this coupling, it thus depends on the sixth power of the separation, that is, the famous R^{-6} distance-dependence law. Moreover, the acceptor distance at which this rate equates to that of spontaneous emission by the donor, now termed the *Förster radius* R_0, has been calculated to be between 10 and 100 Å, in agreement with experimental observations.

Much later, the distance dependence predicted by Förster was fully verified by fluorescence studies of donor–acceptor pairs at known separations (Latt, Cheung and Blout, 1965; Stryer and Haugland, 1967), leading to the suggested employment of RET as a "spectroscopic ruler" by Stryer and Haugland (1967); hence, a technique to measure the proximity relationships and conformational change in macromolecules was realized (see Section 15.3). With the advent of the laser, in the 1960s, the modern understanding of RET led to a raft of modern applications. An excellent in-depth review on the history of RET is given by Clegg (2006).

15.3
The Photophysics of RET

Resonance energy transfer is a mechanism that is now known to operate across a diverse and extensive range of physical systems, encompassing not only gases and dye solutions but also protein complexes, doped crystals, polymers, and so on. Nonetheless, at a fundamental level, it is possible to identify numerous common features in the underlying photophysics.

15.3.1
Primary Excitation Processes

To study the subject in detail, let us commence with the photoexcitation process that creates the conditions for RET to occur. When resonant ultraviolet or visible radiation impinges on any non-homogeneous dielectric material, the primary result of photon absorption is the population of electronic excited states in individual atomic, molecular, or other nanoscale centers – henceforth, to be grouped together under the generic term "chromophore." Typically, such absorption is immediately followed by a rapid but partial degradation of the acquired energy, the associated losses (largely due to vibrational dissipation) ultimately to be manifest in the form of heat. This effect owes its origin to the principle that the release of electronic energy by fluorescence generally occurs from the lowest vibrational level of the excited state. However, if any nearby chromophore has a suitably disposed electronic state, of a similar or slightly lower energy, that neighbor may acquire the major part of the electronic excitation

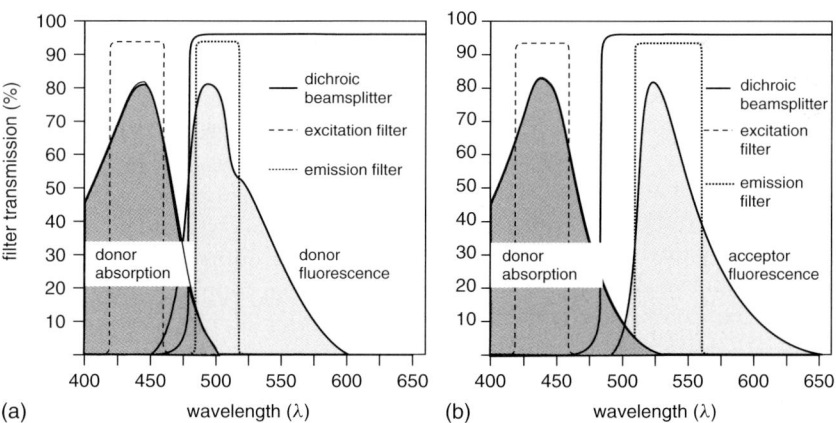

Fig. 15.1 Typical spectral discrimination between the fluorescence from donor and acceptor species (here notionally based on a cyan fluorescent protein donor and a yellow fluorescent protein acceptor): (a) the transmission characteristics of a short-wavelength filter ensure initial excitation of only the donor; a dichroic beam-splitter and another narrow emission filter ensure that only the (Stokes-shifted) fluorescence from the donor reaches a detector; (b) in the same system, a longer-wavelength emission filter ensures capture of only the acceptor fluorescence, following RET.

through RET – a process that takes place well before any further thermal degradation of the excited state energy occurs. The mechanism is most commonly studied through spectrometric differentiation of fluorescence emerging from the initially excited donor and from the acceptor species, as illustrated in Figure 15.1. As is shown in the following, the propensity for energy to be transferred between any two chromophores is severely restricted by distance, and if no suitable acceptor is within reach, the donor will generally shed its energy by fluorescence or local dissipation.

15.3.2
Coupling of Electronic Transitions

In systems where RET occurs, the donors and acceptors are usually fluorophores, that is, chromophores that have the capacity to exhibit fluorescence decay. Moreover, in RET, the transitions of donor decay and acceptor excitation are generally electric dipole-allowed – although other possibilities occasionally arise. Accordingly, the theory of energy transfer, for donor–acceptor displacements beyond the region of significant wavefunction overlap, is traditionally conceived in terms of an electrodynamical coupling between transition dipoles.

Consider the pairwise transfer of excitation between two chromophores D and A. In the context of this elementary mechanism, D is designated as the donor and A as the acceptor. Specifically, let it be assumed that prior excitation of the donor generates an electronically excited species D^*. Forward progress of the energy is then accompanied by donor decay to the ground electronic state. Acquiring the energy, D undergoes a transition from its ground to its excited state. The complete RET process is generally a singlet–singlet coupling mechanism[2], due to the constraints of spin conservation (although this does not apply to Dexter exchange, see below), and its entirety may be expressed by the following chemical equation:

$$^1D^* + {}^1A \xrightarrow{\text{RET}} {}^1D + {}^1A^* \qquad (15.1)$$

The excited acceptor, A^*, subsequently decays either in a further transfer event or by another means such as fluorescence. Because the D^* and A^* excited states are real, with measurable lifetimes, the core process of energy transfer itself is fundamentally separable from the initial electronic excitation of D and the eventual decay of A; the latter processes do not, therefore, enter into the theory of the pair transfer.

15.3.3
Dissipation and Line-Broadening

To delve more deeply into the nature of the process, it needs to be recognized that Eq. (15.1) tells only part of the story, dealing as it does with only electronic excitations. In general, other dissipative processes are also engaged. In a solid, the linewidth of optical transitions manifests the influence of local electronic environments which, in the case of strong coupling, may lead to the production of phonon side-bands. Similar effects in solutions or disordered solids represent inhomogeneous interactions with a solvent or host, while the broad bands exhibited by chromophores in complex molecular systems signify extensively overlapped vibrational levels,

---

2) Triplet-triplet energy transfer is also allowed by the Förster mechanism.

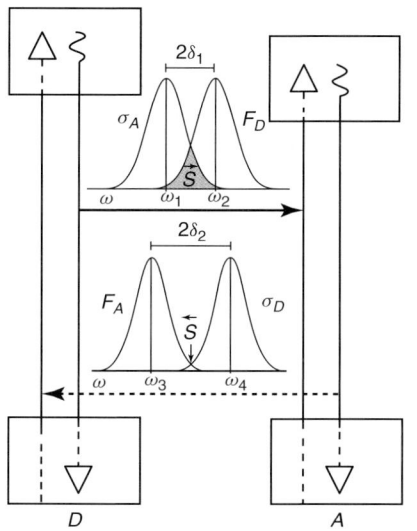

Fig. 15.2 Energetics and spectral overlap features (top) for energy transfer from D to A (and below, potentially backward transfer from A to D). For each chromophore, F denotes the fluorescence spectrum and σ the absorption. Wavy downward lines denote vibrational dissipation.

including those associated with skeletal modes of the superstructure. In each case, energy level broadening can allow pair transfer to occur at any point within the region of overlap between the donor emission and the acceptor absorption bands, as illustrated in Figure 15.2.

15.3.4
The Förster Equation

The Förster theory delivers an expression for the rate of pairwise energy transfer, w_F, valid for any donor–acceptor separation, R, that is substantially smaller than the wavelengths of visible radiation. For systems where the common host material for the donor and acceptor has refractive index n, at the optical frequency corresponding to the mean transferred energy, the Förster result is as follows (Demidov and Andrews, 2001):

$$w_F = \frac{9\kappa^2 c^4}{8\pi \tau_{D^*} n^4 R^6} \int F_D(\omega) \sigma_A(\omega) \frac{d\omega}{\omega^4} \quad (15.2)$$

In this expression, $F_D(\omega)$ denotes the normalized fluorescence spectrum of the donor, $\sigma_A(\omega)$ represents the linear absorption cross section of the acceptor, and ω is an optical frequency in radians per unit time; the specific form of the integral within which they appear is known as the *spectral overlap* – one of the key determinants of energy transfer efficiency (Andrews and Rodríguez, 2007). Also in Eq. (15.2), c is the speed of light and τ_{D^*} is the associated radiative decay lifetime in the absence of transfer. The latter is related to the measured fluorescence lifetime τ_{fl} through the fluorescence quantum yield $\eta = \tau_{fl}/\tau_{D^*}$, where τ_{fl}^{-1} is explicitly expressible as

$$\frac{1}{\tau_{fl}} = \frac{1}{\tau_{D^*}} + \frac{1}{\tau_{D^*}} \left(\frac{R_0}{R}\right)^6 \quad (15.3)$$

The last term on the right-hand side of Eq. (15.3) is expressed with reference to the Förster radius R_0 – the distance at which the rates of donor deactivation by RET and by spontaneous fluorescence become equal. As is evident from Figure 15.2, the propensity for forward transfer is usually significantly greater than that for backward transfer, due to a sizeable difference in the spectral overlaps for the two processes.

15.3.5
Orientation Dependence

The κ factor in Eq. (15.2) depends on the orientations of the donor and acceptor, both with respect to each other and with respect to their mutual displacement unit vector $\hat{R}$, as follows:

$$\kappa = (\hat{\boldsymbol{\mu}}_D \cdot \hat{\boldsymbol{\mu}}_A) - 3(\hat{R} \cdot \hat{\boldsymbol{\mu}}_D)(\hat{R} \cdot \hat{\boldsymbol{\mu}}_A)$$

(15.4)

For each chromophore, $\hat{\boldsymbol{\mu}}$ designates a unit vector in the direction of the appropriate transition dipole moment. The possible values of κ^2, as featured in Eq. (15.2), lie in the range (0, 4). It is evident that in the case of fixed chromophore positions and orientations, the result delivered by Eq. (15.4) is a function of three independent angles, as shown and defined in Figure 15.3:

$$\kappa = \cos\theta_T - 3\cos\theta_D \cos\theta_A \quad (15.5)$$

Unfavorable orientations can thus reduce the rate of energy transfer to zero; other configurations, including many of those found in photobiological systems, optimize the transfer rate. The angular disposition of chromophores is therefore a very important facet of energy transfer. It is important to note that transfer is not *necessarily* precluded when the transition moments of the donor and acceptor lie in perpendicular directions – provided that neither is also disposed orthogonally to $R(=R\hat{R})$.

In any, at least partially, fluid or disordered system, the relative orientation of all donor–acceptor pairs may not be identical, and it is then the distributional average of κ^2 that determines the overall measured response. In the isotropic case (completely uncorrelated orientations), the κ^2 factor averages to $2/3$; departures from this value provide the quantitative signature of a degree of orientational correlation. In molecules of sufficiently high symmetry, it can also happen that either the donor or the acceptor transition moment is not unambiguously identifiable with a particular direction in the corresponding chromophore reference frame. Specifically, the electronic transition may then relate to a transition involving a degenerate state – as can occur with square planar complexes, for example (Galli et al., 1993). Alternatively, the same

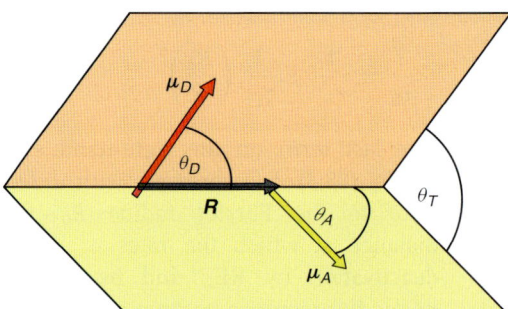

Fig. 15.3 Relative orientations and positions of the donor and acceptor and their transition moments: Here, angles θ_D and θ_A subtended by donor and acceptor transition moments ($\boldsymbol{\mu}_D$ and $\boldsymbol{\mu}_A$, respectively) against the inter-chromophore displacement vector, **R**; the symbol θ_T is the angle between the transition moments.

observational features might indicate rapid but orientationally confined motions. The considerable complication that each of these effects brings into the trigonometric analysis of RET has been extensively researched and reported by van der Meer (1999).

15.3.6
Förster Radius

In applications of RET to spectroscopy, as noted above, it is usually significant that the electronically excited donor can in principle release its energy by spontaneous decay, the ensuing fluorescence also being amenable to detection by any suitably placed photodetector. Because the alternative possibility (that of energy being transferred to another chromophore within the system) has such a sharp decline in efficiency as the distance to the acceptor increases, it is common to invoke the critical distance R_0. The Förster rate equation is itself often cast in an alternative form, exactly equivalent to Eq. (15.2), explicitly exhibiting this critical distance (Lakowicz, 1999):

$$w_F = \frac{3\kappa^2}{2} \frac{1}{\tau_{D^*}} \left(\frac{\overline{R}_0}{R}\right)^6 \quad (15.6)$$

Here $\overline{R}_0$ is defined as the Förster radius for which the orientation factor κ^2 would assume its isotropic average value, $2/3$ (Valeur, 2002). For complex systems, the angular dependence is quite commonly disregarded and the following simpler expression employed:

$$w_F = \frac{1}{\tau_{D^*}} \left(\frac{R_0}{R}\right)^6 \quad (15.7)$$

leading to a transfer efficiency Φ_T expressed as

$$\Phi_T = \frac{1}{1 + (R/R_0)^6} = 1 - \frac{\tau_{fl}}{\tau_{D^*}}$$

$$= 1 - \frac{I_{fl}}{I_{D^*}} \quad (15.8)$$

where, I_{fl} and I_{D^*} are the intensities of the donor fluorescence with the acceptor present and excluded, respectively, and τ_{fl} specifically denotes the fluorescence lifetime of the donor

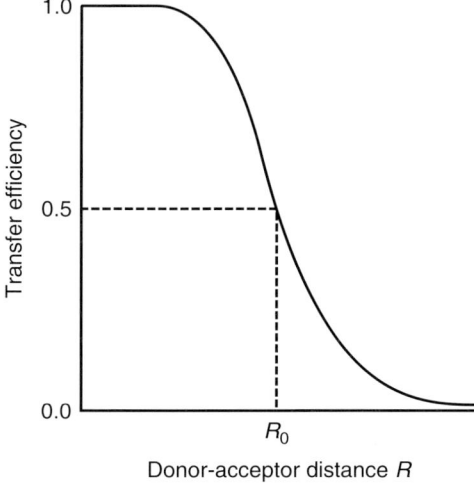

Fig. 15.4 Distance dependence of the transfer efficiency between a pair of chromophores, calculated according to Eq. (15.8).

measured within its RET environment. As graphically depicted in Figure 15.4, a donor–acceptor displacement equal to R_0 corresponds to a transfer efficiency of 50%.

The final equality on the right-hand side of Eq. (15.8), which holds provided decay processes follow single-exponential decay kinetics, provides a formula cast in terms of easily measurable quantities. This is particularly useful because it allows energy transfer efficiencies to be calculated simply on the basis of intensity measurements (for example using a fluorimeter), obviating the separate time-resolved measurements that are otherwise generally necessary for the evaluation of the characteristic decay lifetimes τ_A and τ_{D^*}. When a given electronically excited chromophore is within a distance R_0 of a suitable acceptor, RET will generally be the dominant decay mechanism; conversely, for distances beyond R_0, spontaneous decay (usually fluorescence) will be the primary means of donor deactivation.

15.3.7
Polarization Features

When linearly polarized laser light is used to excite any specific species within a complex disordered solid or liquid system, the probability for excitation of any particular molecule is proportional to $\cos^2 \theta$, where θ is the angle between the appropriate excitation transition moment and the electric polarization vector of the input radiation. Consequently, the population of excited molecules has a markedly anisotropic distribution, a phenomenon associated with the term *photoselection*. If radiative decay were to ensue instantaneously, that is, from precisely the initially populated excited level, then the fluorescence would carry the full imprint of that anisotropy and itself exhibit a degree of polarization – the highest value possible. Accounting for the necessary three-dimensional rotational average (Andrews and Thirunamachandran, 1977), it is readily shown that the fluorescence intensity components polarized parallel to and perpendicular to the polarization of the excitation beam, $I_\parallel$ and $I_\perp$, respectively, would then lie in the ratio 3 : 1. Commonly observed departures from this result thus signify the extent to which the orientation of the emission dipole differs from that of the prior, initial excitation – which may be due to intervening decay, molecular motion, or intermolecular energy transfer.

The two most widely used quantitative expressions of polarization retention are the *fluorescence anisotropy*, r, or the *degree of polarization*, P. Both convey the same information; they are defined and related as follows:

$$r = \frac{I_\parallel - I_\perp}{I_\parallel + 2I_\perp}, \quad P = \frac{I_\parallel - I_\perp}{I_\parallel + I_\perp} \Rightarrow$$

$$r = \frac{2P}{3 - P} \qquad (15.9)$$

The denominator of the expression for r designates the net fluorescence intensity. In a specific situation where the donor and acceptor have transition dipole moments oriented in parallel, then $r = 0.4$ and $P = 0.5$.

A key molecular factor for determining any loss in polarization is the angle θ between the directions of the absorption and emission transition dipole moments. In terms of this parameter and its influence on the measured fluorescence anisotropy, the case where internal decay intervenes between excitation and fluorescence decay within a single molecule is not different from that of a donor–acceptor pair where the absorption and emission processes are spatially separated – provided the donor and acceptor in the latter case have a fixed mutual orientation (the orientation of the pair being random). The following result, derived by (Levshin, 1925) and Perrin (1929) can be applied in both situations:

$$P = \frac{3\cos^2\theta - 1}{3 + \cos^2\theta} \quad (15.10)$$

In the case of a donor–acceptor pair, θ has to be interpreted as the angle θ_T shown in Figure 15.3. Equation (15.10) thus allows direct calculation of this microscopic parameter, through measurement of the macroscopic quantity P. Moreover, when P proves to exhibit a time-dependent decay, a study of the kinetics provides information on the extent of rotational motion intervening between the absorption and the emission events.

Very different behavior is observed for RET systems in which the donor and acceptor are orientationally uncorrelated, that is, where both are independently, randomly oriented. In such cases, there is a very rapid loss of polarization "memory," and it transpires that the associated degree of anisotropy is precisely 1/25, that is, $r = 0.04$ (Agranovich and Galanin, 1982); two or more energy transfer jumps will therefore usually, to all intents and purposes, totally destroy any polarization in any ensuing fluorescence. However, it should be noted that there is a surprising recovery in the anisotropy at distances approaching the transfer wavelength. The effect is sufficiently strong to warrant attention in dilute solution studies.

15.3.8
Diffusion Effects

Till now, only RET between a donor–acceptor pair has been considered. The discussion is now extended to an ensemble of donors D and acceptors A, all units of which are distributed randomly within an n-dimensional volume. For systems in which translational diffusion is extremely slow compared to the rate of energy transfer, the time-dependence of the donor intensity decay at time t, $I_{D^*}(t)$, as obtained by Förster (1959), is given by the following expression:

$$I_{D^*}(t) = I_{D^*}(0)$$
$$\times \exp\left[-\frac{t}{\tau_{D^*}} - 2\gamma\left(\frac{t}{\tau_{D^*}}\right)^{n/6}\right]$$
$$(15.11)$$

The most commonly applied form of this expression is when n equals 3, that is, RET in three dimensions. In Eq. (15.11), the parameter γ is explicitly written as

$$\gamma = \frac{2}{3}\pi^{3/2} C_A R_0^3 \quad (15.12)$$

in which C_A is the concentration of acceptors (number per unit volume) and $(4/3)\pi R_0^3 C_A$ represents the average number of acceptor chromophores in a sphere of radius R_0; the orientational factor is again set as $2/3$.

The case where diffusion is comparable to the transfer rate is very complicated, and calculations by Butler and Pilling (1979) have shown that large errors arise on using Förster theory for systems with diffusion coefficients in excess of 10^{-5} cm^2 s^{-1}. To address such systems, a successful approximation was developed by Gösele et al. (1975) This approach involves the insertion of a multiplier G within the second term in the exponential of Eq.(15.11). With $n = 3$, the parameter G is given by the following equation:

$$G = \left(\frac{1 + 5.47x + 4.00x^2}{1 + 3.34x}\right)^{3/4} \quad (15.13)$$

in which $x = D(R_0^6/\tau_{D^*})^{-1/3}t^{2/3}$, where D is the mutual diffusion coefficient. In contrast to the Förster theory, the above method provides an excellent approximation – as was fully verified by Butler and Pilling (1979).

15.3.9
Long-Range Transfer

Förster theory is found to be increasingly inaccurate for RET as donor–acceptor distances extend over and beyond 100 Å. It was originally assumed that a "radiative" process accounted for energy transfer over such donor–acceptor separations, and some recent literature on the subject still perpetuate this initial overstatement. Certain sources wrongly treat Förster "radiationless" energy transfer as exact, distinct, and separable from "radiative" energy transfer – the latter signifying successive but independent processes of fluorescence emission by a donor and capture of the ensuing photon by an acceptor.

Although that certainly is the observed character of RET over very long distances – as for example between donor and acceptor components in a dilute solution – it is now known that both "radiative" and Förster transfer are simply the long- and short-range limits of one powerful, all-pervasive mechanism. The latter, determined from quantum electrodynamical calculations, is the outcome of the unified theory of RET (Andrews, 1989). This theory not only embraces Förster and "radiative" energy transfer, but also addresses the intermediate range in which neither of these mechanisms are fully valid. An expression for the total pairwise energy transfer rate, ranging from molecular dimensions up to interstellar distances, is written as

$$w = w_F + w_I + w_{rad} \quad (15.14)$$

where w_F represents the Förster rate of Eq. (15.2), w_{rad} is the rate of "radiative" energy transfer – explicitly given by the following equation:

$$w_{rad} = \frac{9\kappa'^2}{8\pi \tau_D R^2} \int F_D(\omega) \sigma_A(\omega)\, d\omega \quad (15.15)$$

and w_I is the intermediate term that is expressed as

$$w_I = \frac{9c^2}{8\pi \tau_D n^2 R^4}(\kappa^2 - 2\kappa\kappa')$$

$$\times \int F_D(\omega)\sigma_A(\omega)\frac{d\omega}{\omega^2} \quad (15.16)$$

In both Eqs (15.15) and (15.16), the symbol κ' denotes an orientation factor identical to Eq. (15.4) but with the "3" omitted from the second term. In summary, the unified theory of RET contains not only the R^{-6} term of Förster theory and the R^{-2} term denoting the inverse-square law of "radiative" transfer, but also a previously unidentified R^{-4} intermediate term.

15.3.10
Dexter Transfer

Before concluding this section, it is worth observing that other forms of donor–acceptor coupling are also possible, although considerably less relevant to the systems of interest in the following focused account on applications. For example, the transfer of energy between atomic or molecular components with significantly overlapped wavefunctions is usually described in terms of Dexter theory (Dexter, 1953) – where the coupling involves electron exchange and carries an exponential decay with distance, directly reflecting the radial form of overlapping wavefunctions and electron distributions. Unlike Förster transfer, singlet–triplet energy exchange ($^3D^* +\,^1A \rightarrow\,^1D +\,^3A^*$) may also be allowed by the Dexter mechanism. This is because Dexter transfer does not involve transition dipole moments and, thus, is unaffected by the dipole-forbidden character of the transitions $T_1 \rightarrow S_0$ and $S_0 \rightarrow T_1$ within chromophores D and A, respectively. Compared to materials in which the donor and acceptor orbitals do not spatially overlap, such systems are of less use for either device or analytical applications. This is largely because the coupled chromophores lose their electronic and optical integrity, the Dexter mechanism being operational only at very short distances (<10 Å).

15.4
Applications of RET to Molecular Biology

The field in which measurements of RET have undoubtedly had the greatest impact is molecular biology. The importance of RET to this subject, especially in application to biological macromolecules, was first realized following the construction of spectroscopic equipment for routine fluorescence measurements (Steinberg, 1971; Stryer, 1978; dos Remedios, Miki and Barden, 1987; Wu and Brand, 1994). Toward the turn of the twenty-first century, RET underwent a period of significant redevelopment as a spectroscopic technique (Selvin, 2000). This resurgence arose mainly due to the advent of new experimentation methods, for example single-pair RET (Ha et al., 1996), and further advances in instrumentation. The key advantage of RET techniques over others is that fluorescence measurements are highly sensitive, being made against a zero background; moreover, the UV/visible signals are relatively easy to detect, they are specific and the required instrumentation is noninvasive.

15.4.1
Spectroscopic Ruler

A major use of RET, based on its strong distance dependence, exploits its capacity to supply accurate spatial information about molecular structures. This derives from the quantitative assessment of the interchromophore separations, based on comparisons between the corresponding RET efficiencies (Hohng, Joo and Ha, 2004; Schuler, 2005; Koushik et al., 2006; Zhang and Allen, 2007). Such a technique is popularly known as a "spectroscopic ruler." The elucidations of molecular structure by such means usually lack information on the relative *orientations* of the groups involved, and as an expedient the calculations usually ignore the kappa parameter (Eq. (15.4)). The apparent crudeness of this approach becomes more defensible on realizing that, even if it were to introduce a factor of two inaccuracy, the deduced group spacing would still be in error by only 12% (because $2^{1/6} = 1.12$). Refinements to the theory to accommodate the effect of fluctuations in position or orientation of the participant groups introduce considerable complexity, although progress is being made in several areas (dos Remedios and Moens, 1995; Tanaka, 1998; Yu, 2007; Isaksson et al., 2007; Jang, 2007).

15.4.2
Conformational Change

Through identification of motions in macromolecules, that is, the variation in proximity of one chromophore with respect to another, a number of valuable RET applications arise; including the detection of conformational changes and folding in proteins (Schuler, 2005; Talaga et al., 2000; Heyduk, 2002; Schuler and Eaton, 2008) and the inspection of intracellular protein-protein (Sekar and Periasamy, 2003; Zal and Gascoigne, 2004; Parsons, Vojnovic and Ameer-Beg, 2004; You et al., 2006) and protein-DNA (Hillisch, Lorenz and Diekmann, 2001; Cremazy et al., 2005) interactions (see, for example, Figures 15.5 and 15.6). These and other such processes can be registered by selectively exciting one chromophore using laser light, and monitoring either the decrease in fluorescence from that chromophore or the rise in the generally longer-wavelength fluorescence from the other chromophore as it adopts the role of acceptor. The judicious use of optical dichroic filters can make this RET technique perfectly straightforward (see Figure 15.1). In cases where the two material components of interest do not display suitably overlapped absorption and fluorescence features in an optically accessible wavelength range, molecular tagging with site-specific "extrinsic" (i.e., artificially attached) chromophores can solve the problem. Located at a molecular site of interest, and being selected on the basis of a significant spectral overlap with the counterpart component, such tags can act either in the capacity of donor or acceptor. Lanthanide ions, with their characteristically prominent and linelike absorption features, prove to be particularly valuable in this connection (Selvin, 2002). The semiconductor nanocrystals known as *quantum dots* are also useful in this regard. These crystalline nanoparticles offer several unique traits, including size- and composition-tunable emission from visible to infrared wavelengths, the possibility of a single light source simultaneously exciting different-sized dots, large absorption coefficients across a wide spectral range, and very high level of

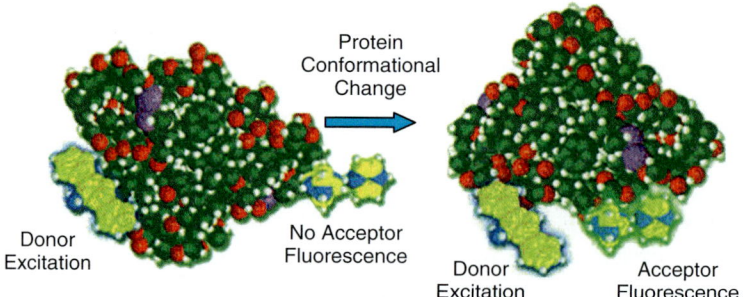

Fig. 15.5 Graphical depiction of RET detection of protein conformational change. (Adapted from Olympus Corporation website.)

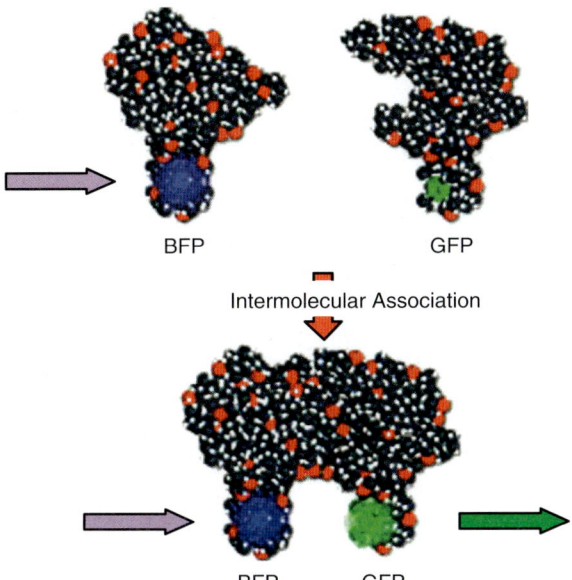

Fig. 15.6 Graphical depiction of RET detection of *in vivo* protein–protein interactions. Purple arrow denotes input laser of wavelength 380 nm and green arrow indicates protein emission at 510 nm. BFP and GFP are acronyms for blue and green fluorescent proteins, respectively. (Adapted from Olympus Corporation website.)

photostability (Chan *et al.*, 2002; Clapp, Medintz and Mattoussi, 2006).

15.4.3
Intensity-Based Imaging

Over the past two decades, there has been burgeoning interest in microscopy based on RET (Clegg, 1996; Day, Periasamy and Schaufele, 2001; Wouters, Verveer and Bastiaens, 2001; Hoppe, Christensen and Swanson, 2002; Jares-Erijman and Jovin, 2003), typical instrumentation for which is illustrated in Figure 15.7. There are three specific types of RET method routinely used in the production of biological images. The principles of sensitized-emission RET have already been described (Figure 15.1). For microscopy purposes, this method is fairly inaccurate; no RET donor–acceptor pair is ideal, that is, there will almost always be some overlap between the donor and the acceptor absorbance bands and also the donor and the acceptor emission spectra. Therefore, filters that completely separate these kinds of spectrum are difficult to design. Various calculational algorithms (Gordon *et al.*, 1998; Xia and Liu, 2001; van Rheenen, Langeslag and Jalink, 2004) have been proposed to compensate for this problem, although the methods are complex and no single procedure has received universal acceptance.

A widely used alternative, experimental approach (Karpova *et al.*, 2003; van

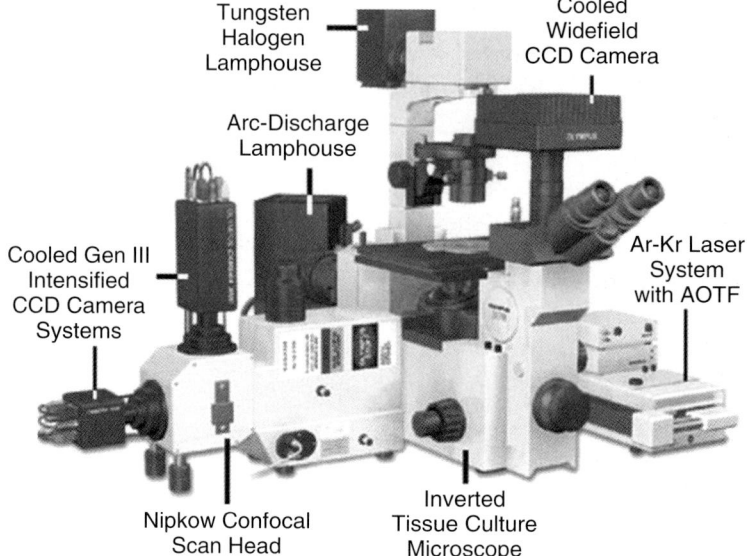

Fig. 15.7 Typical commercial setup of a microscope based on RET. (Adapted from Olympus Corporation website.)

Munster et al., 2005) involves deliberately photobleaching the acceptor, the result of which is complete exclusion of RET. In this method, the donor emission is analyzed before and after the acceptor is bleached by the input of an intense laser beam (at a suitable wavelength). The difference between the donor intensities, with and without the laser input, enables the determination of the transfer efficiency by employing Eq. (15.8). Here, account is taken of spectral bleed-through between the two absorbance bands, and equally between the two emission bands. Signal contamination is still not entirely eliminated, due to a small amount of back-transfer through donor excitation by acceptor emission. Often the main disadvantage in prolonged illumination of the acceptor is the possibility of damage to the sample. Therefore, in practice, photobleaching is seldom appropriate for *in vivo* studies.

15.4.4
Lifetime-Based Imaging

Fluorescence need not be characterized from excitation and emission spectra alone; highly significant information can also be secured from lifetime measurements. Thus, when suitable time-resolved instrumentation is available, the determination of decay kinetics (usually on the nanosecond timescale) enables analysis through resonance energy transfer-based fluorescence lifetime imaging microscopy (FRET-FLIM) (Bastiaens and Squire, 1999; Duncan et al., 2004; Wallrabe and Periasamy, 2005; Peter et al., 2005). In this method, spectral bleed-through is no longer an issue because measurements are made only for the determination of donor lifetimes; back-transfer is usually extremely low, within the noise level. The presence of the acceptor within the local

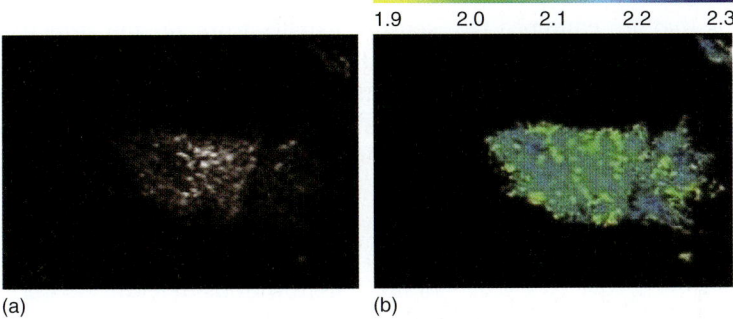

(a) (b)

Fig. 15.8 MDA-MB-231 cancer cell images recorded with argon laser two-photon excitation, RET microscope based on; (a) intensity and (b) fluorescence lifetime (nanosecond). In the latter image, areas of locally reduced lifetime signify clustered intracellular vesicles. (Adapted from Peter et al. (2005).)

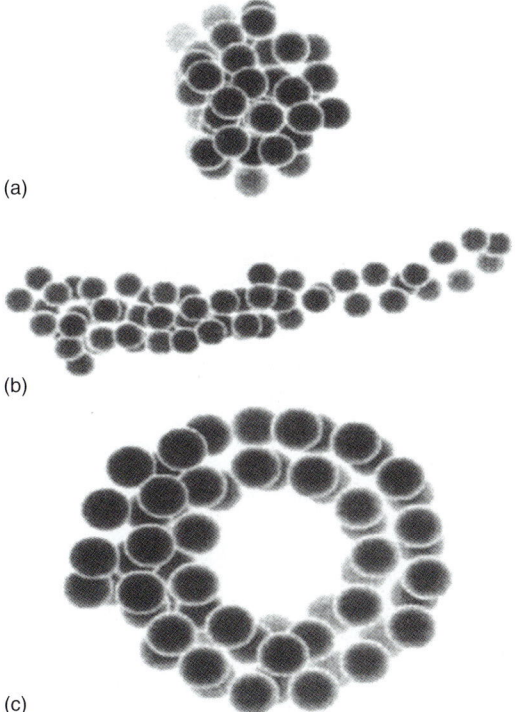

Fig. 15.9 Snapshots of various morphological constructions of a homopolymer chain. The structures (a)–(c) are spherical, rod, and toroidal, respectively. (Adapted from Srinivas and Bagchi (2002).)

environment of the donor influences the fluorescence lifetime of the donor. By measuring the donor lifetime in the presence and the absence of the acceptor, one can accurately calculate the transfer efficiency by use of Eq.(15.8). Drawbacks to FRET-FLIM are the technical challenges the technique presents and the expense of the equipment. Nonetheless, in optical systems that are equipped to provide both intensity and lifetime measurements, a comparison of the two types of image affords a particularly rich source of information, as illustrated by the cancer cell images of Figure 15.8.

15.4.5
Other Applications

Beyond the realm of molecular biology, RET has value in a number of more specifically chemical applications. Two prominent examples are to be found in the fields of synthetic macromolecules and chemical sensors. In polymer science, building on the pioneering principles of Morawetz (1988), RET is now used to determine morphological information on polymer interfaces. Such studies have, for instance, enabled the quantitative characterization of interfacial thickness in polymers of various structures (Farinha and Martinho, 2008). Moreover, RET has been utilized in the study of polymer conformational dynamics. One especially interesting application is the effective differentiation between various collapsed and/or ordered homopolymer chain conformations (spherical, rod and toroidal; as depicted in Figure 15.9) through the associated distribution of transfer effiencies (Srinivas and Bagchi, 2002; Saini, Singh and Bagchi, 2006).

The fabrication of RET-based, analyte-specific sensors has enabled the detection of a variety of species, including dimers of functionalized calixarenes in organic solutions (Castellano et al., 2000), copper(II) in aqueous solution (Cano-Raya, Fernández-Ramos and Capitán-Vallvey, 2006), hydrogen peroxide (Albers, Okreglak and Chang, 2006), phosgene (Zhang and Rudkevich, 2007) and many others. These chemical sensors usually work on the principle of a donor–acceptor system designed such that the presence of the analyte causes the acceptor chromophore to move within closer proximity to a donor, enabling the implementation of an RET process that is not observed in the analyte's absence. Therefore, on irradiation of the system with the relevant chemical present, a strong emission from the acceptor signals the presence of the analyte.

Acknowledgments

We gratefully acknowledge helpful comments from Professor Steve Meech. Research on the theory of RET, at the University of East Anglia, is currently supported by the Leverhulme Trust.

References

Agranovich, V.M. and Galanin, M.D. (1982) *Electronic Excitation Energy Transfer in Condensed Matter*, Elsevier/North-Holland, Amsterdam, The Netherlands.

Albers, A.E., Okreglak, V.S., and Chang, C.J. (2006) A FRET-based approach to ratiometric fluorescence detection of hydrogen peroxide. *J. Am. Chem. Soc.*, **128**, 9640–9641.

Andrews, D.L. (1989) A unified theory of radiative and radiationless molecular-energy transfer. *Chem. Phys.*, **135**, 195–201.

Andrews, D.L. and Rodríguez, J. (2007) Resonance energy transfer: spectral overlap, efficiency and direction. *J. Chem. Phys.*, **127**, 084509.

Andrews, D.L. and Thirunamachandran, T. (1977) On three-dimensional rotational averages. *J. Chem. Phys.*, **67**, 5026–5033.

Bastiaens, P.I.H. and Squire, A. (1999) Fluorescence lifetime imaging microscopy:

spatial resolution of biochemical processes in the cell. *Trends Cell Biol.*, **9**, 48–52.

Berberan-Santos, M.N. (2001) Pioneering contributions of Jean and Francis Perrin to molecular luminescence, in *New Trends in Fluorescence Spectroscopy. Applications to Chemical and Life Sciences* (eds B. Valeur and J.-C. Brochon), Chapter 2, Springer, Berlin.

Butler, P.R. and Pilling, M.J. (1979) The breakdown of Förster kinetics in low viscosity liquids. An approximate analytical form for the time-dependent rate constant. *Chem. Phys.*, **41**, 239–243.

Clegg, R.M. (1996) Fluorescence resonance energy transfer, in *Fluorescence Imaging Spectroscopy and Microscopy* (eds X.F. Wang and B. Herman), Chapter 7, John Wiley & Sons, Ltd, New York.

Clegg, R.M. (2006) The history of FRET: From conception through the labors of birth, in *Reviews in Fluorescence* (eds C.D. Geddes and J.R. Lakowicz), Vol. 3, Chapter 1, Springer, New York.

Cano-Raya, C., Fernández-Ramos, M.D., and Capitán-Vallvey, L.F. (2006) Fluorescence resonance energy transfer disposable sensor for copper(II). *Anal. Chim. Acta*, **555**, 299–307.

Cario, G. (1922) Über Entstehung wahrer Lichtabsorption un scheinbare Koppelung von Quantensprüngen. *Z. Phys.*, **10**, 185–199.

Cario, G. and Franck, J. (1922) Über Zerlegugen von Wasserstoffmolekülen durch angeregte Quecksilberatome. *Z. Phys.*, **11**, 161–166.

Castellano, R.K., Craig, S.L., Nuckolls, C., and Rebek, J. (2000) Detection and mechanistic studies of multicomponent assembly by fluorescence resonance energy transfer. *J. Am. Chem. Soc.*, **122**, 7876–7882.

Chan, W.C.W., Maxwell, D.J., Gao, X., Bailey, R.E., Han, M., and Nie, S. (2002) Luminescent quantum dots for multiplexed biological detection and imaging. *Curr. Opin. Biotechnol.*, **13**, 40–46.

Clapp, A.R., Medintz, I.L., and Mattoussi, H. (2006) Förster resonance energy transfer investigations using quantum-dot fluorophores. *Chem. Phys. Chem.*, **7**, 47–57.

Cremazy, F.G.E., Manders, E.M.M., Bastiaens, P.I.H., Kramer, G., Hager, G.L., van Munster, E.B., Verschure, P.J.,

Gadella, T.W.J., and van Driel, R. (2005) Imaging in situ protein–DNA interactions in the cell nucleus using FRET–FLIM. *Exp. Cell Res.*, **309**, 390–396.

Day, R.N., Periasamy, A., and Schaufele, F. (2001) Fluorescence resonance energy transfer microscopy of localized protein interactions in the living cell nucleus. *Methods*, **25**, 4–18.

Demidov, A.A. and Andrews, D.L. (2001) in *Encyclopedia of Chemical Physics and Physical Chemistry* (eds J.H. Moore and N.D. Spencer), Vol. 3, Institute of Physics, Bristol, pp. 2701–2715.

Dexter, D.L. (1953) A theory of sensitized luminescence in solids. *J. Chem. Phys.*, **21**, 836–850.

Duncan, R.R., Bergmann, A., Cousin, M.A., Apps, D.K., and Shipston, M.J. (2004) Multi-dimensional time-correlated single photon counting (TCSPC) fluorescence lifetime imaging microscopy (FLIM) to detect FRET in cells. *J. Microsc.*, **215**, 1–12.

Farinha, J.P.S. and Martinho, J.M.G. (2008) Resonance energy transfer in polymer interfaces. *Springer Ser. Fluoresc.*, **4**, 215–255.

Förster, T. (1946) Energiewanderung und fluoreszenz. *Naturwissenschaften*, **33**, 166–175.

Förster, T. (1948) Zwischenmolekulare Energiewanderung und Fluoreszenz. *Ann. Phys.*, **2**, 55–75.

Förster, T. (1959) 10th Spiers Memorial Lecture. Transfer mechanisms of electronic excitation. *Discuss. Faraday Soc.*, **27**, 7–17.

Franck, J. (1922) Einige aus der Theorie von Klein und Rosseland zu ziehende Folgerungen über Fluorescence, photochemische Prozesse und die Electronenemission glühender Körper. *Z. Phys.*, **9**, 259–266.

Galli, C., Wynne, K., Lecours, S.M., Therien, M.J., and Hochstrasser, R.M. (1993) Direct measurement of electronic dephasing using anisotropy. *Chem. Phys. Lett.*, **206**, 493–499.

Gaviola, E. and Pringsheim, P. (1924) Über den einfluß der konzentration auf die polarisation der fluoreszenz von farbstofflösungen. *Z. Phys.*, **24**, 24–36.

Gordon, G.W., Berry, G., Liang, X.H., Levine, B., and Herman, B. (1998) Quantitative fluorescence resonance energy transfer measurements using fluorescence microscopy. *Biophys. J.*, **74**, 2702–2713.

Gösele, U., Hauser, M., Klein, U.K.A., and Frey, R. (1975) Diffusion and long-range energy transfer. *Chem. Phys. Lett.*, **34**, 519–522.

Ha, T., Enderle, T., Ogletree, D.F., Chemla, D.S., Selvin, P.R., and Weiss, S. (1996) Probing the interaction between two single molecules: fluorescence resonance energy transfer between a single donor and a single acceptor. *Proc. Natl. Acad. Sci. USA*, **93**, 6264–6268.

Heyduk, T. (2002) Measuring protein conformational changes by FRET/LRET. *Curr. Opin. Biotechnol.*, **13**, 292–296.

Hillisch, A., Lorenz, M., and Diekmann, S. (2001) Recent advances in FRET: distance determination in protein–DNA complexes. *Curr. Opin. Struct. Biol.*, **11**, 201–207.

Hohng, S., Joo, C., and Ha, T. (2004) Single-molecule three-color FRET. *Biophys. J.*, **87**, 1328–1337.

Hoppe, A., Christensen, K., and Swanson, J.A. (2002) Fluorescence resonance energy transfer-based stoichiometry in living cells. *Biophy. J.*, **83**, 3652–3664.

Isaksson, M., Norlin, N., Westlund, P.-O., and Johansson, L.B.-Å. (2007) On the quantitative molecular analysis of electronic energy transfer within donor-acceptor pairs. *Phys. Chem. Chem. Phys.*, **9**, 1941–1951.

Jang, S. (2007) Generalization of the Förster resonance energy transfer theory for quantum mechanical modulation of the donor-acceptor coupling. *J. Chem. Phys.*, **127**, 174710.

Jares-Erijman, E.A. and Jovin, T.M. (2003) FRET imaging. *Nat. Biotechnol.*, **21**, 1387–1395.

Kallman, H. and London, F. (1928) Über quantenmechanische Energieübertragung zwischen atomaren Systemen. *Z. Phys. Chem. B*, **2**, 207–243.

Karpova, T.S., Baumann, C.T., He, L., Wu, X., Grammer, A., Lipsky, P., Hager, G.L., and McNally, J.G. (2003) Fluorescence resonance energy transfer from cyan to yellow fluorescent protein detected by acceptor photobleaching using confocal microscopy and a single laser. *J. Microsc.*, **209**, 56–70.

Koushik, S.V., Chen, H., Thaler, C., Puhl, H.L. III, and Vogel, S.S. (2006) Cerulean, venus and venus$_{Y67C}$ FRET reference standards. *Biophys. J.*, **91**, L99–L101.

Lakowicz, J.R. (1999) *Principles of Fluorescence Spectroscopy*, 2nd edn, Chapter 10, Kluwer Academic, New York, USA.

Latt, S.A., Cheung, H.T., and Blout, E.R. (1965) Energy transfer. A system with relatively fixed donor-acceptor separation. *J. Am. Chem. Soc.*, **87**, 995–1003.

Levshin, W.L. (1925) Polarisierte Fluoreszenz und Phosphoreszenz der Farbstofflösungen. IV. *Z. Phys.*, **32**, 307–326.

van der Meer, B.W. (1999) in *Resonance Energy Transfer* (eds D.L., Andrews and A.A., Demidov), John Wiley & Sons, Ltd, New York, pp. 151–172.

Morawetz, H. (1988) Studies of synthetic polymers by nonradiative energy transfer. *Science*, **240**, 172–176.

van Munster, E.B., Kremers, G.J., Adjobo-Hermans, M.J.W., and Gadella, T.W.J. (2005) Fluorescence resonance energy transfer (FRET) measurement by gradual acceptor photobleaching. *J. Microsc.*, **218**, 253–262.

Parsons, M., Vojnovic, B., and Ameer-Beg, S. (2004) Imaging protein–protein interactions in cell motility using fluorescence resonance energy transfer (FRET). *Biochem. Soc. Trans.*, **32**, 431–433.

Perrin, J. (1927) Fluorescence et induction moléculaire par résonance. *C. R. Acad. Sci.*, **184**, 1097–1100.

Perrin, F. (1929) La fluorescence des solutions. *Ann. Phys.*, **12**, 169–275.

Perrin, F. (1932) Théorie quantique des transferts d'activation entre molecules de même espéce. Cas des solutions fluorescents. *Ann. Phys.*, **17**, 283–314.

Peter, M., Ameer-Beg, S.M., Hughes, M.K.Y., Keppler, M.D., Prag, S., Marsh, M., Vojnovic, B., and Ng, T. (2005) Multiphoton-FLIM quantification of the EGFP-mRFP1 FRET pair for localization of membrane receptor-kinase interactions. *Biophys. J.*, **88**, 1224–1237.

dos Remedios, C.G., Miki, M., and Barden, J.A. (1987) Fluorescence resonance energy transfer measurements of distances in actin and myosin. A critical evaluation. *J. Muscle Res. Cell Motil.*, **8**, 97–117.

dos Remedios, C.G. and Moens, P.D.J. (1995) Fluorescence resonance energy transfer spectroscopy is a reliable 'ruler' for measuring structural changes in proteins – dispelling the problem of the unknown orientation factor. *J. Struct. Biol.*, **115**, 175–185.

van Rheenen, J., Langeslag, M., and Jalink, K. (2004) Correcting confocal acquisition to optimize imaging of fluorescence resonance energy transfer by sensitized emission. *Biophys. J.*, **86**, 2517–2529.

Saini, S., Singh, H., and Bagchi, B. (2006) Fluorescence resonance energy transfer (FRET) in chemistry and biology: Non-Förster distance dependence of the FRET rate. *J. Chem. Sci.*, **118**, 23–35.

Schuler, B. (2005) Single-molecule fluorescence spectroscopy of protein folding. *ChemPhysChem*, **6**, 1206–1220.

Schuler, B. and Eaton, W.A. (2008) Protein folding studied by single-molecule FRET. *Curr. Opin. Struct. Biol.*, **18**, 16–26.

Sekar, R.B. and Periasamy, A. (2003) Fluorescence resonance energy transfer (FRET) microscopy imaging of live cell protein localizations. *J. Cell Biol.*, **160**, 629–633.

Selvin, P.R. (2000) The renaissance of fluorescence resonance energy transfer. *Nat. Struct. Bio.*, **7**, 730–734.

Selvin, P.R. (2002) Principles and biophysical applications of lanthanide-based probes. *Annu. Rev. Biophys. Biomol. Struc.*, **31**, 275–302.

Srinivas, G. and Bagchi, B. (2002) Detection of collapsed and ordered polymer structures by fluorescence resonance energy transfer in stiff homopolymers: Bimodality in the reaction efficiency distribution. *J. Chem. Phys*, **116**, 837–844.

Steinberg, I.Z. (1971) Long-range nonradiative transfer of electronic excitation energy in proteins and polypeptides. *Annu. Rev. Biochem.*, **40**, 83–114.

Stryer, L. (1978) Fluorescence energy transfer as a spectroscopic ruler. *Ann. Rev. Biochem.*, **47**, 819–846.

Stryer, L. and Haugland, R.P. (1967) Energy transfer: a spectroscopic ruler. *Proc. Natl. Acad. Sci.*, **58**, 719–726.

Talaga, D.S., Lau, W.L., Roder, H., Tang, J., Jia, Y.W., DeGrado, W.F., and Hochstrasser, R.M. (2000) Dynamics and folding of single two-stranded coiled-coil peptides studied by fluorescent energy transfer confocal microscopy. *Proc. Natl. Acad. Sci. USA*, **97**, 13021–13026.

Tanaka, F. (1998) Theory of time-resolved fluorescence under the interaction of energy transfer in a bichromophoric system: effect of internal rotations of energy donor and acceptor. *J. Chem. Phys.*, **109**, 1084–1092.

Valeur, B. (2002) *Molecular Fluorescence: Principles and Applications*, Chapter 9, Wiley-VCH Verlag GmbH, Weinheim, Germany.

Wallrabe, H. and Periasamy, A. (2005) Imaging protein molecules using FRET and FLIM microscopy. *Curr. Opin. Biotechnol.*, **16**, 19–27.

Wouters, F.S., Verveer, P.J., and Bastiaens, P.I.H. (2001) Imaging biochemistry inside cells. *Trends Cell Biol.*, **11**, 203–211.

Wu, P. and Brand, L. (1994) Resonance energy transfer: methods and applications. *Anal. Biochem.*, **218**, 1–13.

Xia, Z. and Liu, Y. (2001) Reliable and global measurement of fluorescence resonance energy transfer using fluorescence microscopes. *Biophys. J*, **81**, 2395–2402.

You, X., Nguyen, A.W., Jabaiah, A., Sheff, M.A., Thorn, K.S., and Daugherty, P.S. (2006) Intracellular protein interaction mapping with FRET hybrids. *Proc. Natl. Acad. Sci. USA*, **103**, 18458–18463.

Yu, Z.G. (2007) Fluorescent resonant energy transfer: correlated fluctuations of donor and acceptor. *J. Chem. Phys.*, **127**, 221101.

Zal, T. and Gascoigne, N.R.J. (2004) Using live FRET imaging to reveal early protein–protein interactions during T cell activation. *Curr. Opin. Immunol.*, **16**, 418–427.

Zhang, J. and Allen, M.D. (2007) FRET-based biosensors for protein kinases: Illuminating the kinome. *Mol. Biosyst.*, **3**, 759–765.

Zhang, H. and Rudkevich, D.M. (2007) A FRET approach to phosgene detection. *Chem. Commun.*, 1238–1239.

Further Reading

Andrews, D.L. (2008) Mechanistic principles and applications of resonance energy transfer. *Canad. J. Chem.*, **86**, 855–870.

Andrews, D.L. and Demidov, A.A. (eds) (1999) *Resonance Energy Transfer*, John Wiley & Sons, Ltd, New York.

van der Meer, B.W., Coker, G., and Chen, S.-Y. (1994) *Resonance Energy Transfer Theory and Data*, Wiley-VCH Verlag GmbH, New York.

Periasamy, A. (2001) *Methods in Cellular Imaging*, Oxford University Press, New York.

Periasamy, A. and Day, R.N. (2005) *Molecular Imaging: FRET Microscopy and Spectroscopy*, Oxford University Press, New York.

Review of FRET microscopy on Olympus Corporation's web site: http://www.olympusfluoview.com/applications/fretintro.html (accessed July 2008).

Sapsford, K.E., Berti, L., and Medintz, I.L. (2006) Materials for fluorescence resonance energy transfer analysis: beyond traditional donor–acceptor combinations. *Angew. Chem. Int. Ed.*, **45**, 4562–4588.

Scholes, G.D. (2003) Long-range resonance energy transfer in molecular systems. *Annu. Rev. Phys. Chem.*, **54**, 57–87.

16
Optical Spectroscopy of Biological Materials

Valery V. Tuchin

16.1	**Introduction** 557	
16.2	**Tissue Optics** 557	
16.2.1	Tissue Structure 557	
16.2.1.1	Tissue Optical Models 557	
16.2.1.2	Sizing of Tissue and Cell Compartments 558	
16.2.2	Light Interaction with Tissues 559	
16.2.2.1	Light Absorption and Scattering 559	
16.2.2.2	Theoretical Description for CW 561	
16.2.2.3	Index of Refraction of Tissue Compartments 565	
16.2.2.4	Short Pulse Interaction with Biomaterials 566	
16.3	**Measurements of Optical Properties of Tissues and Blood** 568	
16.3.1	Introduction 568	
16.3.2	Integrating Sphere Technique 570	
16.3.3	Multiflux Approach 570	
16.3.4	The Inverse Adding–Doubling Method 571	
16.3.5	Inverse Monte Carlo Method 571	
16.3.6	Spatially Resolved Measurements 573	
16.3.7	Optical Coherence Tomography 574	
16.3.8	Measurement of the Scattering Phase Function 574	
16.3.9	Human Tissue Optical Properties Analysis 575	
16.3.10	Optical Properties of Blood 613	
16.3.11	Refractive Index Measurements 613	
16.3.12	Optical Projection Tomography 616	
	Glossary 617	
	References 622	
	Further Reading 625	

Encyclopedia of Applied Spectroscopy. Edited by David L. Andrews.
Copyright © 2009 WILEY-VCH Verlag GmbH & Co. KGaA, Weinheim
ISBN: 978-3-527-40773-6

16.1
Introduction

Optics of biological materials determines diagnostic and therapeutic facilities of conventional and laser light. Principle noninvasiveness of low-intensity visible and infrared light, its substantial permeation into biological tissues, and dependence of its propagating properties on tissue morphology and functioning allow for providing structural and functional imaging of many body organs without any risk and limitations. The knowledge of optical properties of tissues and blood is also important for prediction of light–tissue interaction with various follow-up therapeutic effects. Light dosimetry is the key problem in any kind of phototreatment, including cancer treatment using photodynamic or photothermal technologies, as well as laser precise surgery.

In this chapter, optics of biological tissues and blood is briefly discussed. Methods and algorithms for solving the inverse problem of finding tissue and blood optical parameters, such as absorption and scattering coefficients, anisotropy factor, and refractive index are presented. Measuring techniques, such as integrating sphere (IS), spatially resolved, time-resolved, and angular-resolved, optical coherence tomography (OCT), as well as inverse methods, such as inverse adding–doubling (IAD) and inverse Monte Carlo (IMC) are reviewed. Data on optical properties of human tissue and blood measured *in vitro, ex vivo*, and *in vivo* are presented.

16.2
Tissue Optics

16.2.1
Tissue Structure

16.2.1.1 Tissue Optical Models
Biological tissues are optically inhomogeneous and absorbing media. Their average refractive index is higher than that of air, thus partial reflection of the radiation at the tissue/air interface (Fresnel reflection) takes place, while the remaining part of radiation penetrates the tissue. Multiple scattering and absorption are responsible for light beams broadening and eventual decay as they travel through a tissue. The bulk scattering causes a large portion of radiation to be scattered in the backward direction. Light propagation within a tissue depends on the scattering and absorption properties of its compartments:

Encyclopedia of Applied Spectroscopy. Edited by David L. Andrews.
Copyright © 2009 WILEY-VCH Verlag GmbH & Co. KGaA, Weinheim
ISBN: 978-3-527-40773-6

cells, cell organelles, and various fiber structures (Tuchin, 2002; Vo-Dinh, 2003; Tuchin, 2007). The size, shape, and density of these structures, their refractive indices play important roles in the propagation of light in tissues.

The great diversity and structural complexity of tissues require the development of adequate optical models accounting for the scattering and absorption properties of tissues. Two major approaches are currently used: (i) tissue modeled as a medium with a continuous random spatial distribution of optical parameters and (ii) tissue considered as a discrete ensemble of scatterers. The choice of the approach is defined by both the structural specificity of the tissue under study and the kind of light-scattering characteristics that are to be described.

Many tissues can be represented as a random continuum of the inhomogeneities of the refractive index with a varying spatial scale. This approach is applicable for tissues with no pronounced boundaries between elements that feature significant heterogeneity. The process of scattering by these structures may be described under certain conditions using the model of a phase screen (Goodman, 1985; Rytov, Kravtsov and Tatarskii, 1989; Tuchin, 2007).

In accordance with the second approach, biological media are often modeled as ensembles of particles, since many cells and microorganisms are close in shape to spheres, ellipsoids, or rods. A system of noninteracting homogeneous spherical particles is the simplest tissue model. Mie theory rigorously describes the diffraction of light by a spherical particle (Bohren and Huffman, 1983; Born and Wolf, 1999). The further development of this model involves accounting for more complex structure of the spherical particles, namely, the multilayered spheres and the spheres with radial nonhomogeneity, anisotropy, and optical activity (Mishchenko, Hovenier and Travis, 2000; Mishchenko, Travis and Lacis, 2002). Since a connective tissue consists of fiber structures, a system of long cylinders is the most appropriate model for it. Muscular tissue, skin dermis, cerebral membrane (*dura mater*), eye cornea, and sclera belong to this type of tissue formed essentially by collagen fibrils. The solution of the problem of light diffraction in a single homogeneous or multilayered cylinder is also well described (Bohren and Huffman, 1983). Attempts to describe light scattering by a system of interacting particles as a more realistic tissue model (quasi-ordered spherical particles or cylinders) have also been made (see Tuchin, 2007).

16.2.1.2 Sizing of Tissue and Cell Compartments

The sizes of cells and tissue structure elements vary in size from a few tenths of nanometers to hundreds of micrometers (Dyson, 1974; Hogan, Alvardo and Weddel, 1971; Silver, 1987; Kessel, 1998; Tuchin, 2002; Vo-Dinh, 2003; Tuchin, Wang and Zimnyakov, 2006; Tuchin, 2007). Blood cells (erythrocytes, leukocytes, and platelets) exhibit the following parameters. A normal erythrocyte in plasma has the shape of a concave–concave disc with a diameter varying from 7.1 to 9.2 µm, a thickness of 0.9–1.2 µm in the center, 1.7–2.4 µm on the periphery, and a volume of $\sim$90 µm^3. Leukocytes are formed like spheres with a diameter of 8–22 µm. Platelets in the blood stream are biconvex disklike particles with diameters ranging from 2 to 4 µm. Normally, blood has about 10 times as many erythrocytes as platelets and about 30 times as many platelets as leukocytes.

Most other mammalian cells have diameters in the range of 5–75 µm. In the epidermal layer, the cells are large (with an average cross-sectional area of about 80 µm^2) and quite uniform in size. Fat cells, each containing a single lipid droplet that nearly fills the entire cell and therefore results in eccentric placement of the cytoplasm and nucleus, have a wide range of diameters from a few microns to 50–75 µm. Fat cells may reach a diameter of 100–200 µm in pathological cases.

There are a wide variety of structures within cells that determine tissue light scattering. Cell nuclei are on the order of 5–10 µm in diameter; mitochondria, lysosomes, and peroxisomes have dimensions of 1–2 µm; ribosomes are on the order of 20 nm in diameter; and structures within various organelles can have dimensions up to a few hundred nanometers. Usually, the scatterers in cells are not spherical. The models of prolate ellipsoids with a ratio of the ellipsoid axes between 2 and 10 are more typical.

The hollow organs of the body are lined with a thin, highly cellular surface layer of epithelial tissue, which is supported by underlying, relatively acellular connective tissue. In healthy tissues, the epithelium often consists of a single, well-organized layer of cells with en-face diameter of 10–20 µm and height of 25 µm. In dysplastic epithelium, cells proliferate and their nuclei enlarge and appear darker (hyperchromatic) when stained. Enlarged nuclei are primary indicators of cancer, dysplasia, and cell regeneration in most human tissues.

In fibrous tissues or tissues containing fiber layers (cornea, sclera, *dura mater*, muscle, myocardium, tendon, cartilage, vessel wall, retinal nerve fiber layer, etc.) and composed mostly of microfibrils and/or microtubules, typical diameters of the cylindrical structural elements are 10–400 nm. Their length is in a range from 10–25 µm to a few millimeters.

For some tissues, the size distribution of the scattering particles may be essentially monodispersive and for others it may be quite broad. Examples for the two are transparent eye cornea stroma, which has a sharply monodispersive distribution, and turbid eye sclera, which has a rather broad distribution of collagen fiber diameters.

16.2.2
Light Interaction with Tissues

16.2.2.1 Light Absorption and Scattering

Absorbed light is converted to heat or radiated in the form of fluorescence, it is also consumed in photobiochemical reactions. The absorption spectrum depends on the type of predominant absorption centers and water content of tissues. Absolute values of absorption coefficients for typical tissues lie in the range of 10^{-2}–10^4 cm^{-1} (Anderson and Parrish, 1982; Chance, 1989; Duck, 1990; Frank and Kessler, 1992; Henderson and Dougherty, 1992; Welch and van Gemert, 1992; Niemz, 1996; Tuchin, 2002; Berlien and Mueller, 2003; Vo-Dinh, 2003; Tuchin, 2007). In the ultraviolet (UV) and infrared (IR) ($\lambda \geq$ 2000 nm) spectral regions, light is readily absorbed, which accounts for the small contribution of scattering and inability of radiation to penetrate deep into tissues (only through one or two cell layers). Short-wave visible light penetrates typical tissues as deep as 0.5–2.5 mm, whereupon it undergoes an *e*-fold decrease of intensity. In this case, both scattering and absorption occur, with 15–40% of the incident radiation being reflected. In the wavelength range of 600–1600 nm, scattering prevails over absorption, and light penetrates to a depth of 8–10 mm. Simultaneously, the intensity of the reflected radiation increases to 35–70% of the total incident light (due to backscattering).

Absorption spectrum of biological material is expressed usually in terms of the wavelength dependence of absorption coefficient. Such spectra for some tissues and tissue components: water (75%), aorta, skin, epidermis, melanosome, and whole blood are presented in Figure 16.1. The often-used lasers and their wavelengths are presented on the crossings of water-absorption curve, the corresponding laser wavelengths showing what absorption coefficient is expected on absorption in water. Since water is the major component of any soft tissue, presented absorption coefficients will be only slightly corrected for real tissues. As shown in Figure 16.1, for bloodless skin in the visible range, absorption is equal to water (75%), since dermis as the main skin component that is well supplied by water defines absorption coefficient of whole skin in this range.

Light interaction with multilayer and multicomponent tissues, such as skin

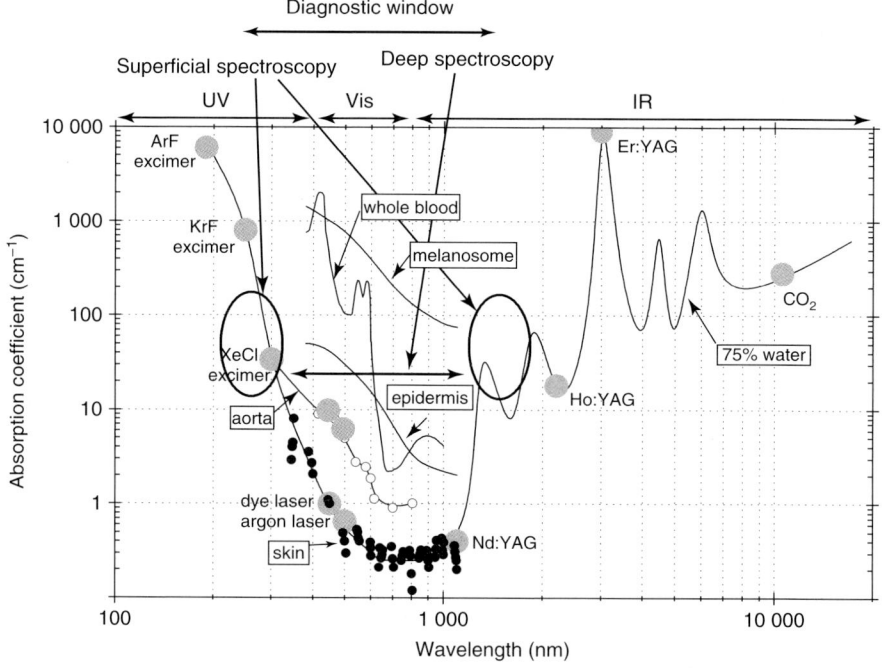

Fig. 16.1 Absorption spectra of some tissues and tissue components: water (75%), aorta, skin, epidermis, melanosome, and whole blood; lasers and their wavelengths as well as diagnostic window and wavelength ranges suitable for superficial and deep spectroscopy are also shown (Tuchin, 2007).

of female breast, is a very complicated process. In skin, the horny layer (stratum corneum (SC)) reflects about 5–7% of the incident light. A collimated light beam is transformed to a diffuse one by microscopic inhomogeneities at the air/horny layer interface. A major part of reflected light results from backscattering in different skin layers (SC, epidermis, dermis, blood, and percutaneous fat). The absorption of diffuse light by skin pigments is a measure of bilirubin content, hemoglobin concentration and its saturation with oxygen, and the concentration of pharmaceutical products in blood and tissues; these characteristics are widely used in the diagnosis of various diseases (Tuchin, 2002; Vo-Dinh, 2003; Tuchin, 2007). Certain diagnostic modalities take advantage of ready transdermal penetration of visible and near-infrared (NIR) light inside the body in the wavelength region corresponding to the diagnostic/therapeutic window (600–1600 nm).

16.2.2.2 Theoretical Description for CW

A collimated (laser) beam is attenuated in a thin tissue layer of thickness d in accordance with the exponential law, Bouguer-Beer-Lambert law (Tuchin, 2002; Vo-Dinh, 2003; Tuchin, 2007):

$$I(d) = (1 - R_F) I_0 \exp(-\mu_t d) \quad (16.1)$$

where $I(d)$ is the intensity of transmitted light measured using a distant photodetector with a small aperture (online or collimated transmittance) in W cm^{-2}; R_F is the coefficient of Fresnel reflection; at the normal beam incidence, $R_F = [(n-1)/(n+1)]^2$; n is the relative mean refractive index of tissue and surrounding media; I_0 is the incident light intensity in W cm^{-2};

$$\mu_t = \mu_a + \mu_s \quad (16.2)$$

is the extinction coefficient (interaction or total attenuation coefficient) in cm^{-1}, where μ_a is the absorption coefficient in cm^{-1}, and μ_s is the scattering coefficient in cm^{-1}. Strictly speaking, Eq. (16.1) is valid only for a highly absorbing biological media, when $\mu_a \gg \mu_s$.

The extinction coefficient is connected with the extinction cross section σ_{ext} as

$$\mu_t = \rho_s \sigma_{\text{ext}} \quad (16.3)$$

where ρ_s is the density of particles (tissue and cell compounds). For a system of particles with absorption,

$$\sigma_{\text{ext}} = \sigma_{\text{sca}} + \sigma_{\text{abs}} \quad (16.4)$$

and

$$\mu_s = \rho_s \sigma_{\text{sca}} \quad \mu_a = \rho_s \sigma_{\text{abs}} \quad (16.5)$$

The average scattering cross section per particle can be presented in the form suitable for experimental evaluations (Bohren and Huffman, 1983):

$$\sigma_{\text{sca}} = \frac{\lambda^2}{2\pi} \frac{1}{I_0} \int_0^\pi I(\theta) \sin\theta \, d\theta \quad (16.6)$$

where I_0 is the intensity of the incident light, $I(\theta)$ is the angular distribution of the scattered light by a particle, and θ is the scattering angle.

The mean free path (MFP) length between two interactions is denoted by

$$l_{\text{ph}} = \mu_t^{-1} \quad (16.7)$$

To analyze light propagation under multiple scattering conditions, it is assumed that absorbing and scattering centers are uniformly distributed across the tissue.

Visible and NIR radiation is normally subject to anisotropic scattering characterized by a clearly apparent direction of photons undergoing single scattering, which may be due to the presence of large cellular organelles (mitochondria, lysosomes, and inner membranes (Golgi apparatus)) (Tuchin, 2002, 2007).

When the scattering medium is illuminated by unpolarized light and/or only the intensity of multiply scattered light needs to be computed, a sufficiently strict mathematical description of continuous wave (CW) light propagation in a medium is possible in the framework of the scalar stationary radiation transfer theory (RTT) (Chandrasekhar, 1960; van de Hulst, 1980, 1981; Ishimaru, 1997; Mishchenko, Travis and Lacis, 2000, 2002, 2006; Müller et al., 1993; Niemz, 1996; Sobolev, 1974; Thomas and Stamnes, 1999; Tuchin, 2002, 2007; Tuchin, Wang and Zimnyakov, 2006; Vo-Dinh, 2003; Wang and Wu, 2007; Welch and van Gemert, 1992; Yanovitskij, 1997; Zege, Ivanov and Katsev, 1991). The theory is valid for an ensemble of scatterers located far from one another and has been successfully used to work out many practical problems of tissue optics. The main stationary equation of RTT for monochromatic light has the form

$$\frac{\partial I(\bar{r}, \bar{s})}{\partial s} = -\mu_t I(\bar{r}, \bar{s}) + \frac{\mu_s}{4\pi}$$
$$\times \int_{4\pi} I(\bar{r}, \bar{s}') p(\bar{s}, \bar{s}') d\Omega' \quad (16.8)$$

where $I(\bar{r}, \bar{s})$ is the radiance (or specific intensity) – average power flux density at a point $\bar{r}$ in the given direction $\bar{s}$ (W cm^{-2}sr); $p(\bar{s}, \bar{s}')$ is the scattering phase function, 1/sr; and $d\Omega'$ is the unit solid angle about the direction $\bar{s}'$, sr. It is assumed that there are no radiation sources inside the medium.

For practical purposes, integrals of the function $I(\bar{r}, \bar{s})$ over certain phase space regions $(\bar{r}, \bar{s})$ are of greater value than the function itself. Specifically, optical probes of tissues frequently measure the outgoing light distribution function at the medium surface, which is characterized by the radiant flux density or irradiance (W cm^{-2}):

$$F(\bar{r}) = \int_{(\bar{s}\bar{N})>0} I(\bar{r}, \bar{s})(\bar{s}\bar{N}) d\Omega \quad (16.9)$$

where $\bar{N}$ is the outside normal vector to the domain boundary surface.

In problems of optical dosimetry in tissues, the measured quantity is actually the total radiant energy fluence rate $U(\bar{r})$. It is the sum of the radiance over all angles at a point $\bar{r}$ and is measured by watts per square centimeter:

$$U(\bar{r}) = \int_{4\pi} I(\bar{r}, \bar{s}) d\Omega \quad (16.10)$$

The phase function $p(\bar{s}, \bar{s}')$ describes the scattering properties of the medium and is, in fact, the probability density function for scattering in the direction $\bar{s}'$ of a photon traveling in the direction $\bar{s}$; in other words, it characterizes an elementary scattering act. If scattering is symmetric relative to the direction of the incident wave, then the phase function depends only on the scattering angle θ (angle between directions $\bar{s}$ and $\bar{s}'$), that is,

$$p(\bar{s}, \bar{s}') = p(\theta) \quad (16.11)$$

The assumption of random distribution of scatterers in a medium (i.e., the absence of spatial correlation in the tissue structure) leads to normalization

$$\int_0^\pi p(\theta) 2\pi \sin\theta \, d\theta = 1 \quad (16.12)$$

In practice, the phase function is usually well approximated with the aid of the postulated Henyey–Greenstein function:

$$p(\theta) = \frac{1}{4\pi} \cdot \frac{1 - g^2}{(1 + g^2 - 2g \cos\theta)^{3/2}} \quad (16.13)$$

where g is the scattering anisotropy parameter (mean cosine of the scattering angle θ).

$$g \equiv \langle \cos\theta \rangle$$
$$= \int_0^\pi p(\theta) \cos\theta \cdot 2\pi \sin\theta \, d\theta \quad (16.14)$$

The value of g varies in the range from -1 to 1: $g = 0$ corresponds to isotropic (Rayleigh) scattering, $g = 1$ to total forward scattering (Mie scattering at large particles), and -1 to total backward scattering.

The integro-differential Eq. (16.8) is too complicated; therefore, it is frequently simplified by representing the solution in the form of spherical harmonics. Such simplification leads to a system of $(N+1)^2$ connected differential partial derivative equations known as the P_N approximation. This system is reducible to a single differential equation of order $(N+1)$. For example, four connected differential equations reducible to a single diffusion-type equation are necessary for $N = 1$. It has the following form for an isotropic medium:

$$(\nabla^2 - \mu_{\text{eff}}^2) U(\bar{r}) = -Q(\bar{r}) \quad (16.15)$$

where

$$\mu_{\text{eff}} = [3\mu_a(\mu_s' + \mu_a)]^{1/2} \quad (16.16)$$

is the effective attenuation coefficient or inverse diffusion length, $\mu_{\text{eff}} = 1/l_d$ in cm^{-1};

$$Q(\bar{r}) = (cD)^{-1} q(\bar{r}) \quad (16.17)$$

$q(\bar{r})$ is the source function (i.e., the number of photons injected into the unit volume),

$$D = \frac{1}{3(\mu_s' + \mu_a)} \quad (16.18)$$

is the photon diffusion coefficient in cm$^2/c$;

$$\mu_s' = (1-g)\mu_s \quad (16.19)$$

is the reduced (transport) scattering coefficient in cm^{-1}, and c is the velocity of light in the medium. The transport mean free path (TMFP) of a photon (centimeters) is defined as

$$l_t = \frac{1}{\mu_t'} = (\mu_a + \mu_s')^{-1} \quad (16.20)$$

where $\mu_t' = \mu_a + \mu_s'$ is the transport coefficient. The TMFP is the distance over which the photon loses its initial direction. It is worthwhile to note that the TMFP in a medium with anisotropic single scattering significantly exceeds the MFP in a medium with isotropic single scattering $l_t \gg l_{\text{ph}}$.

The diffusion approximation is inapplicable for light beam input near the object's surface where single or low-step scattering prevails. When a narrow light beam normally incident upon a semi-infinite turbid medium with anisotropic scattering, it can be considered as converted into an isotropic point source at the depth of one TMFP below the surface.

The refractive index variation in tissues is quantified by the ratio $m \equiv n_s/n_0$, where n_s and n_0 are the index of refraction of scattering particles and background material, respectively. This ratio determines light-scattering efficiency. For example, in a simple monodisperse tissue model, such as dielectric spheres of equal diameter $2a$, the reduced scattering coefficient (Graaff et al., 1992):

$$\mu'_s = 3.28\pi a^2 \rho_s \frac{2\pi a}{\lambda}^{0.37} (m-1)^{2.09} \quad (16.21)$$

where λ is the light wavelength in the scattering medium. This equation is valid for noninteracting Mie scatterers, $g > 0.9$; $5 < 2\pi a/\lambda < 50$; $1 < m < 1.1$. It gives dependencies of reduced scattering coefficient on refractive index mismatch (m), particle size (a), density (ρ_s), and wavelength (λ).

Basing on diffusion approximation attenuation of a light (laser) beam of intensity I_0 at depths $z \geq l_t$ in a thick tissue may be described as

$$I(z) \approx I_0 b_s \exp(-\mu_{\text{eff}} z) \quad (16.22)$$

where b_s accounts for additional irradiation of upper layers of a tissue due to backscattering (photon recycling effect). Respectively, the depth of light penetration into a tissue is

$$l_e = l_d[\ln b_s + 1] \quad (16.23)$$

Typically, for tissues $b_s = 1-5$ for beam diameter of 1–20 mm (Star, Wilson and Patterson, 1992). Thus, when wide laser beams are used for irradiation of highly scattering tissues with low absorption, light energy is accumulated in tissue due to high multiplicity of chaotic long-path photon migrations. The light-power density within the superficial tissue layers may substantially (up to fivefold) exceed the incident power density.

Measurement of diffusely reflected light is often used to evaluate bulk tissue optical properties. To solve diffusion equation, the boundary conditions are derived by considering Fresnel's laws of reflection and balancing the fluence rate and photon flow crossing the interface. For the source term modeled as a point scattering source and extrapolated boundary approach satisfying the boundary condition, the spatially resolved steady-state reflectance per incident photon, $R(r_{\text{sd}})$, is expressed as (Farrell and Patterson, 2001):

$$R(r_{\text{sd}}) = \frac{F_U}{4\pi} \left\{ l_t \left(\mu_{\text{eff}} + \frac{1}{r_1} \right) \right.$$
$$\times \frac{\exp(-\mu_{\text{eff}} r_1)}{r_1^2} + (l_t + 2z_b)$$
$$\times \left(\mu_{\text{eff}} + \frac{1}{r_2} \right) \frac{\exp(-\mu_{\text{eff}} r_2)}{r_2^2} \right\}$$
$$+ \frac{F_F}{4\pi D} \left\{ \frac{\exp(-\mu_{\text{eff}} r_1)}{r_1} \right.$$
$$\left. - \frac{\exp(-\mu_{\text{eff}} r_2)}{r_2} \right\} \quad (16.24)$$

where r_{sd} is the distance between light source and detector at the tissue surface (source–detector separation), cm; $r_1 = \sqrt{l_t^2 + r_{\text{sd}}^2}$; $r_2 = \sqrt{(l_t + 2z_b)^2 + r_{\text{sd}}^2}$; $z_b = 2AD$ is the distance to the extrapolated boundary, $A = (1 + R_{\text{eff}})/(1 - R_{\text{eff}})$, R_{eff} is the effective reflection coefficient, which can be found by integrating the Fresnel reflection coefficient over all incident angles; and D is the diffusion coefficient. The parameters F_U and F_F represent the fractions of the fluence rate and the flux, which exit the tissue across the interface. These values are obtained

16.2.2.3 Index of Refraction of Tissue Compartments

The mean refractive index $\bar{n}$ of a tissue is defined by the refractive indices of its scattering particle material n_s and ground (surrounding) matter n_0. The refractive indices of tissue structure elements and the tissue itself, can be derived using the law of Gladstone and Dale, which states that the resulting value represents an average of the refractive indices of the components related to their volume fractions as (Tuchin, 2002, 2007):

$$\bar{n} = \sum_{i=1}^{N} n_i f_i \qquad \sum_i f_i = 1 \qquad (16.25)$$

where n_i and f_i are the refractive index and volume fraction of the individual components, respectively, and N is the number of components.

Soft tissue is composed of closely packed groups of cells entrapped in a network of fibers through which water percolates. At a microscopic scale, the tissue components have no pronounced boundaries. They appear to merge into a continuous structure with spatial variations in the refractive index. To model such a complicated structure as a collection of particles, it is necessary to resort to a statistical approach.

It has been shown that the tissue components that contribute most to the local refractive-index variations are the connective tissue fibers (bundles of elastin and collagen), cytoplasmic organelles (mitochondria, lysosomes, and peroxisomes), cell nuclei, and melanin granules. The average background index is defined as the weighted average of refractive indices of the cytoplasm and the interstitial fluid, n_{cp} and n_{is}, as

$$\bar{n}_0 = f_{cp} n_{cp} + (1 - f_{cp}) n_{is} \qquad (16.26)$$

where f_{cp} is the volume fraction of the fluid in the tissue contained inside the cells. Accounting for typical experimental data as $n_{cp} = 1.367$ and $n_{is} = 1.355$, and that approximately 60% of the total fluid in soft tissue is contained in the intracellular compartment, it follows from Eq. (16.26) that $\bar{n}_0 = 1.362$. The refractive index of scattering particle material can be determined as the sum of the background index and index variation components (Tuchin, 2007):

$$\bar{n}_s = \bar{n}_0 + f_f(n_f - n_{is}) + f_{nc}(n_{nc} - n_{cp})$$
$$+ f_{or}(n_{or} - n_{cp}) \qquad (16.27)$$

where the subscripts f, is, nc, cp, and or refer to the fibers, interstitial fluid, nuclei, cytoplasm, and organelles, which are the major contributors to index variations. The terms in parentheses in this expression are the differences between the refractive indices of the three types of tissue component and their respective backgrounds; the multiplying factors are the volume fractions of the elements in the solid portion of the tissue. The refractive index of the connective-tissue fibers is about 1.47, which corresponds to about 55% hydration of collagen, its main component. Taking into account that $n_{nc} \cong n_{or} = 1.40$, the mean index variation can be expressed in terms of the fibrous-tissue fraction c_f only:

$$\bar{n}_s = \bar{n}_0 + f_f(n_f - n_{is}) + (1 - f_f)$$
$$\times (n_{nc} - n_{cp}) \qquad (16.28)$$

Collagen and elastin fibers compose approximately 70% of the fat-free dry weight of the dermis, 45% of the heart, and 2–3% of the nonmuscular internal organs (Schmitt and Kumar, 1998). Therefore, depending on tissue type, f_f may be as small as about 0.02 or as large as 0.7. For $n_f - n_{is} = 1.470 - 1.355 = 0.115$ and $n_{nc} - n_{cp} = n_{or} - n_{cp} = 1.400 - 1.367 = 0.033$, the mean index variations that correspond to these two extremes are $\bar{n}_s = 1.397$ and 1.452. Indeed, for some hard tissues or melanin granules in skin or eye iris, refractive index of scatterers could be as high as 1.5–1.7.

Measuring refractive indices in tissues and their constituent components is an important focus of interest in tissue optics because index of refraction determines light reflection and refraction at the interfaces between air and tissue, detecting fiber and tissue, and tissue layers; it also strongly influences light propagation and distribution within tissues, defines speed of light in tissue, and governs how the photons migrate (Tuchin, 2007). Although these studies have a rather long history (Duck, 1990), the mean values of refractive indices for many tissues are missing in the literature. According to Duck, most of them have refractive indices for visible light in the 1.335–1.620 range (e.g., 1.55 in the SC, 1.620 in the enamel, and 1.386 at the lens surface). It is worthwhile noting that *in vitro* and *in vivo* measures may differ significantly. For example, the refractive index in rat mesenteric tissue *in vitro* was found to be 1.52 compared with only 1.38 *in vivo*. This difference can be accounted for by the decreased refractivity of ground matter, n_0, due to impaired hydration.

The optical properties of tissues, including refractive indices, are known to depend on water content. Tissue is modeled as a mixture of water and a bioorganic compound of a tissue. For instance, the refractive index of human skin can be approximated by a 70/30 mixture of water and protein (Troy and Thennadil, 2001). Assuming that protein has a constant refractive index value of 1.5 over the entire wavelength range, the following expression for estimation of skin index of refraction was suggested:

$$n_{skin}(\lambda) = 0.7(1.58 - 8.45 \times 10^{-4}\lambda + 1.10 \times 10^{-6}\lambda^2 - 7.19 \times 10^{-10}\lambda^3 + 2.32 \times 10^{-13}\lambda^4 - 2.98 \times 10^{-17}\lambda^5) + 0.3 \cdot 1.5 \quad (16.29)$$

where wavelength λ is in nanometers.

16.2.2.4 Short Pulse Interaction with Biomaterials

When probing the plane-parallel layer of a scattering medium (biomaterial) with an ultrashort laser pulse, the transmitted pulse consists of a ballistic component, a group of photons having zigzag trajectories, and a highly intensive diffuse component (Müller et al., 1993; Tuchin, 2007). Both unscattered photons and photons undergoing forward-directed single-step scattering contribute to the intensity of the ballistic component. This component is subject to exponential attenuation with increasing sample thickness (see Eq. (16.1)), thus, has a limited utility for practical diagnostic purposes in medicine. The group of snake photons with zigzag trajectories includes photons that have experienced only a few collisions each. They propagate along trajectories that deviate only slightly from the direction of the incident beam and form the first-arriving to detector part of the diffuse component.

These photons carry information about the optical properties of the biomaterial and parameters of any foreign object (i.e., tumor), which they may happen to come across during their travel.

Typically for many scattering biomaterials, the diffuse component is very broad and intense since it contains the bulk of incident photons after they have participated in many scattering acts and therefore migrate in different directions and have different path lengths. Moreover, the diffuse component carries information about the optical properties of the scattering medium, and its deformation may reflect the presence of local inhomogeneities in the medium. The resolution obtained by this method at a high light-gathering power is much lower than in the method measuring straight-passing photons. Two probing schemes are conceivable, one recording transmitted photons and the other taking advantage of their backscattering. If in the diffusion approximation (valid at $\mu_a \ll \mu_s'$), the tissue is homogeneous and semi-infinite, the size of both the source and the detector is small compared with the distance r_{sd} between them at the tissue surface, and the pulse may be regarded as single, then the light distribution is described by the time-dependent diffusion equation (Müller et al., 1993; Tuchin, 2002, 2007):

$$\left(\nabla^2 - c\mu_a D^{-1} - D^{-1}\frac{\partial}{\partial t}\right) \cdot U(\bar{r}, t) = -Q(\bar{r}, t) \quad (16.30)$$

which is, in fact, the generalization of the CW Eq. (16.15). Solving Eq. (16.30) yields the following relation for the number of backscattered photons at the surface for unit time and from unit area $R(r_{sd}, t)$ (Jacques, 1989; Patterson, Chance and Wilson, 1989)

$$R(r_{sd}, t) = \frac{z_0}{(4\pi D)^{3/2}} t^{-5/2}$$

$$\times \exp\left(-\frac{r_{sd}^2 + z_0^2}{2Dt}\right) \exp(-\mu_a ct) \quad (16.31)$$

and correspondingly for transmittance

$$T(r_{sd}, d, t) = (4\pi D)^{-3/2} t^{-5/2} \exp\frac{-r_{sd}^2}{4Dt}$$

$$\times \left\{(d - z_0)\exp\left[-\frac{(d - z_0)^2}{4Dt}\right]\right.$$

$$- (d + z_0)\exp\left[-\frac{(d + z_0)^2}{4Dt}\right]$$

$$+ (3d - z_0)\exp\left[-\frac{(3d - z_0)^2}{4Dt}\right]$$

$$\left. - (3d + z_0)\exp\left[-\frac{(3d + z_0)^2}{4Dt}\right]\right\}$$

$$\times \exp(-\mu_a ct) \quad (16.32)$$

where $z_0 = (\mu_s')^{-1}$ and d is the tissue thickness.

In practice, μ_a and μ_s' are estimated by fitting Eq. (16.31) or Eq. (16.32) with the shape of a pulse measured by the time-resolved photon counting technique. Experimentally measured optical parameters of many tissues and model media obtained by the pulse method can be found in Cheong, Prahl and Welch (1990); Welch and Gemert (1992); Müller et al. (1993); Roggan et al. (1995); Tuchin (2002); Vo-Dinh (2003); Tuchin (2007). An important advantage of the pulse method is its applicability to *in vivo* studies owing to the possibility of the separate evaluation of μ_a and μ_s' using a single measurement in the backscattering or transillumination regimes.

Many studies are also devoted to image transfer in tissues and the evaluation of the resolving power of optical tomographic schemes making use of the first-transmitted photons of ultrashort pulses (Müller et al., 1993; Tuchin, 2002; Vo-Dinh, 2003; Tuchin, 2007). A contrast image of an object in a scattering medium can be provided by electronical or optical time-gating of the earliest-arriving, minimally scattered light (ballistic and snake photons), which contains geometric information. The first group of schemes uses the electronic time-gating procedure. The time-correlated single-photon counting technique explores a high-repetition rate picosecond laser (e.g., a cavity-dumped mode-locked dye laser). At the detection of the firstly arrived photons, the time delay is measured with a time-to-amplitude converter and a histogram of the arrival times is built up using a large number of low-energy pulses. The time resolution of such a technique is limited to about 50 picoseconds. For more energetic pulses from lasers with a lower repetition rate, the usage of the streak cameras allows for a time resolution down to 1 picosecond. If a synchroscan streak camera is employed, even a high-repetition rate source with low-energy pulses can be used.

The second group of techniques uses the optical nonlinear effects to select photons. For a scheme with an optical Kerr gate, an energetic laser is used. Part of the pulse is transmitted into the tissue and part is used to open the shutter by utilizing the optical Kerr effect (i.e., the cell with CS_2). Since the gate width is determined by the length of the laser pulse only, subpicosecond gate times can be achieved. The Raman amplifier gating technique also uses energetic laser pulses. A Stokes wave generated by stimulated Raman scattering in a gas cell is used to probe the tissue. The low-intensity transmitted light is amplified in a Raman amplifier, which in turn is pumped by an ultrashort laser pulse. This pulse has the proper time delay to strobe on the desired early temporal part of the light under investigation. The third group of schemes uses a time-correlated frequency doubling technique and is frequently used in optical autocorrelators for monitoring laser pulse characteristics. It can be used directly for optical gating of signal photons.

16.3
Measurements of Optical Properties of Tissues and Blood

16.3.1
Introduction

Methods for determining the optical parameters of tissues can be divided into two large groups, direct and indirect methods (Tuchin, 2007). Direct methods include those based on some fundamental concepts and rules such as the Bouguer–Beer–Lambert law (see Eq. (16.1)), the single-scattering phase function (see Eqs. (16.11) and (16.13)) for thin samples, or the effective light penetration depth for slabs. The parameters measured are the collimated light transmission T_c and angular dependence of the scattered light intensity $I(\theta)$ in W cm^{-2}sr, for thin samples or the fluence rate inside a slab. The normalized scattering angular dependence is equal to the scattering phase function $I(\theta)/I(0) \equiv p(\theta)$, 1/sr. These methods are advantageous in that they use very simple analytic expressions for data processing. Their disadvantages are related to the necessity to strictly fulfill experimental conditions dictated by the selected model (single scattering in thin samples, exclusion of the effects of light polarization and refraction at cuvette edges, and so on; in the case of slabs with multiple scattering, the recording detector must be placed far from both the light source and the medium boundaries.

Indirect methods obtain the solution of the inverse scattering problem using a theoretical model of light propagation in a medium. They are in turn divided into iterative and noniterative models. The noniterative models use equations in which the optical properties are defined through parameters directly related to the quantities being evaluated. They are based on the two-flux or multiflux models. In indirect iterative methods, the optical properties are implicitly defined through measured parameters. Quantities determining the optical properties of a scattering medium are enumerated until the estimated and measured values for reflectance and transmittance coincide with the desired accuracy. These methods are cumbersome, but the optical models currently in use may be even more complicated than those underlying noniterative methods (examples include the diffusion theory (DT), IAD, and IMC methods) (see Tuchin, 2007).

The optical parameters of tissue samples (μ_a, μ_s, and g) are measured by different methods. *In vitro* evaluation is most often achieved by the double-integrating sphere (DIS) method combined with collimated transmittance measurements. This approach implies either sequential or simultaneous determination of three parameters: collimated transmittance $T_c = I(d)/I(0)$ (see Eq. (16.1)), total transmittance $T_t = T_c + T_d$ (T_d being diffuse transmittance), and diffuse reflectance R_d. The optical parameters of the tissue are deduced from these measurements using different theoretical expressions or numerical methods (two-flux and multiflux models, the IMC or IAD methods) relating μ_a, μ_s, and g to the parameters being investigated.

Any three measurements from the following five are sufficient for the evaluation of all three optical parameters (Cheong, Prahl and Welch, 1990): (i) total (or diffuse) transmittance for collimated or diffuse radiation; (ii) total (or diffuse) reflectance for collimated or diffuse radiation; (iii) absorption by a sample placed inside an IS; (iv) collimated transmittance (of unscattered light); (v) angular distribution of radiation scattered by the sample. Iterative methods normally take into account discrepancies between refractive indices at sample boundaries as well as the multilayer nature of the sample. The following factors are responsible for the errors in the estimated values of optical coefficients and need to be borne in mind in a comparative analysis of optical parameters obtained in different experiments: (i) physiological conditions of tissues (the degree of hydration, homogeneity, species-specific variability, frozen/thawed or fixed/unfixed state, *in vitro/in vivo* measurements, smooth/rough surface); (ii) geometry of irradiation; (iii) matching/mismatching interface refractive indices; (iv) orientation of detecting optical fibers inside the sample relative to the source fibers; (v) numerical aperture of the recording fibers; (vi) angular resolution of photodetectors; (vii) separation of radiation experiencing forward scattering from unscattered radiation; and

(viii) theory used to solve the inverse problem.

16.3.2
Integrating Sphere Technique

One of the indirect methods to determine optical properties of tissues *in vitro* is the IS technique. Diffuse reflectance R_d, total transmittance T_t, and collimated transmittance T_c are measured. In general, absorption coefficient μ_a, scattering coefficient μ_s, and anisotropy factor g can be obtained from these data using an inverse method based on the radiative transfer theory (see Eq. (16.8)). When the scattering phase function $p(\theta)$ is available from goniophotometry, g can be readily calculated. In this case, for the determination of μ_a and μ_s, it is sufficient to measure R_d and T_t only. Sometimes in experiments with tissue and blood samples, a DIS configuration is preferable, since in this case both reflectance and transmittance can be measured simultaneously and less degradation of the sample is expected during measurements.

The IS technique was used by a number of investigators to determine the absorption coefficient, the scattering coefficient, the anisotropy factor, and/or the reduced scattering coefficient of tissues and blood (see Tuchin, 2007). Barium-sulfate or Spectralon® ISs are used in the experiments. As monochromatic light sources, a laser, an Xe-lamp, and/or a Hg-lamp combined with monochromator are used, while a photomultiplier or an Si-photodiode is employed as a detector. Sometimes, a white light source is used as irradiator and charge-coupled device (CCD) fiber-optic spectrometer as a detector.

16.3.3
Multiflux Approach

To separate the light beam attenuation due to absorption from the loss due to scattering, the one-dimensional, two-flux Kubelka-Munk model (KMM) can be used as the simplest approach to solve the problem. This approach has been widely used to determine the absorption and scattering coefficients of biological tissues, provided the scattering is significantly dominant over the absorption (see Tuchin, 2007). The KMM assumes that light incident on a slab of tissue because of interaction with the scattering media can be modeled by two fluxes, counterpropagating in the tissue slab. The optical flux, which propagates in the same direction as the incident flux, is decreased by absorption and scattering processes and is also increased by backscattering of the counterpropagating flux in the same direction. Changes in counterpropagating flux are determined in an analogous manner. The fraction of each flux lost by absorption per unit path length is denoted as K, while the fraction lost due to scattering is called S. The main assumptions of the KMM: K and S parameters are assumed to be uniform throughout the tissue slab; all light fluxes are diffuse; and the amount of light lost from the edges of the sample during reflectance measurements is negligible. Basic KMM does not account for reflections at boundaries at which index of refraction mismatches exist. Following the KMM and diffusion approximation of the RTT, the KMM parameters were expressed in terms of light transport theory: the absorption and scattering coefficients and scattering anisotropy factor (Cheong, Prahl and

Welch, 1990). Often, such simple methods as the KMM, are used as the first step of the inverse algorithm for estimation of the optical properties of tissues.

16.3.4
The Inverse Adding–Doubling Method

The IAD method provides a tool for the rapid and accurate solution of inverse scattering problem. It is based on the general method for the solution of the transport equation (see Eq. (16.8)) for plane-parallel layers suggested by van de Hulst (1980) and introduced to tissue optics by Prahl, van Gemert and Welch (1993). An important advantage of the IAD method when applied to tissue optics is the possibility of rapidly obtaining iterative solutions with the aid of up-to-date microcomputers; moreover, it is flexible enough to take into account anisotropy of scattering and the internal reflection from the sample boundaries. The method includes the following steps: (i) the choice of optical parameters to be measured; (ii) counting reflections and transmissions; (iii) comparison of calculated and measured reflectance and transmittance; and (iv) repetition of the procedure until the estimated and measured values coincide with the desired accuracy.

The method allows any intended accuracy to be achieved for all the parameters being measured, provided the necessary computer time is available. An error of 3% or less is considered acceptable (Prahl, van Gemert and Welch, 1993). Also, the method may be used to directly correct experimental findings obtained with the aid of ISs. The term *doubling* in the method IAD means that the reflection and transmission estimates for a layer at certain ingoing and outgoing light angles may be used to calculate both the transmittance and reflectance for a layer twice as thick by means of superimposing one upon the other and summing the contributions of each layer to the total reflectance and transmittance. Reflection and transmission in a layer having an arbitrary thickness are calculated in consecutive order, first for the thin layer with the same optical characteristics (single scattering), then by consecutive doubling of the thickness, for any selected layer. The term *adding* indicates that the doubling procedure may be extended to heterogeneous layers for modeling multilayer tissues or taking into account internal reflections related to abrupt change in refractive index.

The IAD is a numerical method for solving the one-dimensional transport equation in slab geometry. It can be used for media with an arbitrary phase function and arbitrary angular distribution of the spatially uniform incident radiation. The IAD method has been successfully applied to determine optical parameters of blood; human and animal dermis; ocular tissues such as retina, choroids, sclera, conjunctiva, and ciliary body; aorta; and other soft tissues in the wide range of the wavelengths. The IAD provides accurate results in cases when the side losses are not significant, but it is less flexible than the IMC technique.

16.3.5
Inverse Monte Carlo Method

Both the real geometry of the experiment and the tissue structure may be complicated. Therefore, the MC method should be used if reliable estimates are to be obtained. A number of algorithms that use the IMC method are available now in the literature (see Tuchin, 2007).

Many researchers use the MC simulation program provided by Jacques.

The MC technique is employed as a method to solve the forward problem in the inverse algorithm for the determination of the optical properties of tissues. The MC method is based on the formalism of the RTT (see Eq. (16.8)), where the absorption coefficient is defined as a probability of a photon to be absorbed per unit length, and the scattering coefficient is defined as the probability of a photon to be scattered per unit length. Using these probabilities, a random sampling of photon trajectories is generated.

The basic algorithm for the generation of photon trajectories can be shortly described as follows (Tuchin, 2007). A photon described by three spatial coordinates and two angles (x, y, z, θ, ϕ) is assigned its weight $W = W_0$ and placed in its initial position, depending on the source characteristics. When an incident photon enters a scattering layer, it is allowed to travel a free path length, l. The l value depends on the particle concentration, ρ, and extinction cross section σ_{ext}. The free path length l is a random quantity that takes any positive values with the probability density $p(l)$:

$$p(l) = \rho\sigma_{ext}e^{-\rho\sigma_{ext}l} \quad (16.33)$$

The particular realization of the free path length l is dictated by the value of a random number ξ that is uniformly distributed over the interval [0, 1]:

$$\int_0^l p(l)dl = \xi \quad (16.34)$$

Substituting Eq. (16.33) into Eq. (16.34) yields the value l of the certain realization in the form

$$l = -\frac{1}{\rho\sigma_{ext}}\ln\xi \quad (16.35)$$

If the distance l is larger than the thickness of the scattering system, then this photon is detected as transmitted without any scattering. If, having passed the distance l, the photon remains within the scattering volume, then the possible events of photon-particle interaction (scattering or absorption) are randomly selected.

The direction of the photon's next movement is determined by the scattering phase function substituted as the probability density distribution. Several approximations for the scattering phase function of tissue and blood have been used in MC simulations. These include the two empirical phase functions widely used to approximate the scattering phase function of tissue and blood, Henyey–Greenstein phase function (HGPF) (see Eq. (16.13)) and the Gegenbauer kernel phase function (GKPF) (Barber and Hill, 1990), and theoretical Mie phase function (Bohren and Huffman, 1983).

When the photon reaches the boundary, part of its weight is transmitted according to the Fresnel equations. The amount transmitted through the boundary is added to the reflectance or transmittance. Since the refraction angle is determined by the Snell's law, the angular distribution of the outgoing light can be calculated. The photon with the remaining part of the weight is specularly reflected and continues its random walk.

When the photon's weight becomes lower than a predetermined minimal value, the photon can be terminated using "Russian roulette" procedure.

This procedure saves time, since it does not make sense to continue the random walk of the photon, which will not essentially contribute to the measured signal. On the other hand, it ensures that the energy balance is maintained throughout the simulation process.

The MC method has several advantages over the other methods because it may take into account mismatched medium-glass and glass–air interfaces, losses of light at the edges of the sample, any phase function of the medium, and the finite size and arbitrary angular distribution of the incident beam. If the collimated transmittance is measured, then the contribution of scattered light into the measured collimated signal can be accounted for. The only disadvantage of this method is the long time needed to ensure good statistical convergence, since it is a statistical approach. The standard deviation of a quantity (diffuse reflectance, transmittance, etc.) approximated by MC technique decreases proportionally to $1/\sqrt{N}$, where N is the total number of launched photons. It is worthy to note that stable operation of the algorithm was maintained by generation from 10^5 to 5×10^5 photons per iteration. Two to five iterations were usually necessary to estimate the optical parameters with approximately 2% accuracy.

16.3.6
Spatially Resolved Measurements

For many tissues, *in vivo* measurements are possible only in the geometry of the backscattering. The spatially resolved reflectance (SRR) $R(r_{sd})$ is defined as the power of the backscattered light per unit of area detected by a receiver at the surface of the tissue at a distance r_{sd} from the source. $R(r_{sd})$ depends on the optical properties of the sample, that is, the absorption coefficient μ_a, the scattering coefficient μ_s, and the phase function $p(\theta)$, the refractive index, and the numerical aperture (NA) of the receiving system (Kumar and Schmitt, 1997; Farrell and Patterson, 2001). The corresponding relation for the backscattering intensity as a function of a source and detector positions and optical parameters can be written on the basis of a diffusion approximation (see Eq. (16.24)).

For a semi-infinite medium and source and detector probes (e.g., optical fibers) separated by a distance r_{sd} and normally oriented to the sample surface, the reflecting flux is given by Kumar and Schmitt (1997)

$$R = \frac{z_0 A}{2\pi} \left[\frac{\mu_{\text{eff}}}{r_{sd}^2 + z_0^2} + \frac{1}{(r_{sd}^2 + z_0^2)^{3/2}} \right]$$
$$\times \exp\left[-\mu_{\text{eff}}(r_{sd}^2 + z_0^2)^{1/2}\right] \quad (16.36)$$

where $z_0 = K/\mu_s'$ is the extrapolation length, K is a dimensionless constant with a magnitude that depends on the anisotropy parameter of the scatterers and the reflection coefficient at the surface, A is the area of detector, μ_{eff} is defined by Eq. (16.16).

The measurement of the intensity of a back-reflected light from a tissue for different source–detector separations, r_{sd}, is the basis of the spatially resolved technique, which allows one to evaluate the absorption and the scattering coefficients using, for example, the analytical expressions (16.24) or (16.36), valid for highly scattering thick tissues.

When optical parameters of skin or mucosa are under investigation, the small source–detector separations should be used, where the diffusion approximation is not valid due to proximity to the tissue boundary. In that case, more sophisticated approximations of the RTT solution should be employed; in particular, a numerical solution of the inverse problem by the MC method is prospective. Such an SRR measurement can be implemented using multifiber probes with a number of fixed source–detector separations or using a CCD with a special optical system, allowing for the depth profiling of tissue optical properties if enough fibers or pixels are employed.

16.3.7
Optical Coherence Tomography

OCT is a newly developed modality that allows one to evaluate the scattering and absorption properties of tissue *in vivo* within the limits of an OCT penetration depth of 1–3 mm (see Tuchin, 2007). The use of OCT to measure the single-scattering coefficient of tissues μ_s has been described by Schmitt, Knüttel and Bonnar (1993). In its simplest form, this method assumes that backscattered light from a tissue decreases in the intensity according to

$$I_b \cong I_0 \exp[-2(\mu_a + \mu_s)z] \quad (16.37)$$

where $(\mu_a + \mu_s)$ is the total attenuation coefficient and $2z$ is the round-trip distance of light backscattered at a depth z. For most tissues in the NIR, $\mu_a \ll \mu_s$; thus, μ_s can be estimated roughly as

$$\mu_s \cong \frac{1}{2}z\left\{\ln\left[\frac{I_b(z)}{I_0}\right]\right\} \quad (16.38)$$

or as the gradient of a graph $\ln[I_b(z)/I_0]$ versus z. Experimental data for tissue OCT images show that the logarithmically scaled average for multiple in-depth scans' backscattered intensity, $\ln[I_b(z)]$, decays exponentially; thus, by performing a linear regression on this curve, the scattering coefficient can be determined. More comprehensive algorithms accounting for multiple scattering effects and properties of a small-angle scattering phase function are available in the literature (Tuchin, 2007; Levitz et al., 2004).

16.3.8
Measurement of the Scattering Phase Function

Direct measurement of the scattering phase function $p(\theta)$ is important for the choice of an adequate model for the tissue being examined. The scattering phase function is usually determined from goniophotometric measurements (GPMs) in relatively thin tissue samples (Jacques, Alte and Prahl, 1987). Measured scattering intensity angular dependence is approximated either by the HGPF (see Eq. (16.13)) or by a set of HGPFs, with each function characterizing the type of scatterers and specific contribution to the angular dependence (Marchesini et al., 1989). In the limiting case of a two-component model of a medium containing large and small (compared with the wavelength) scatterers, the angular dependence is represented in the form of anisotropic and isotropic components. Other approximating functions are equally useful, for example, those obtained from the Rayleigh–Gans approximation, ensuing from the Mie theory, or a two-parameter GKPF (HGPF is a special simpler case of this phase function) (see Tuchin, 2002, 2007). Some of these types of approximations were used to find the dependence of the scattering anisotropy factor g for dermis and epidermis on the wavelengths in the range of 300–1300 nm, which proved to coincide fairly well with the empirical formula (van Gemert et al., 1989),

$$g_e \sim g_d \sim 0.62 + \lambda \times 0.29 \times 10^{-3} \quad (16.39)$$

on the assumption of a 10% contribution of isotropic scattering (at least in the spectral range of 300–630 nm). The wavelength λ is given in nanometers.

16.3.9
Human Tissue Optical Properties Analysis

The above-discussed methods and techniques were successfully applied for estimation of optical properties of a wide number of tissues. Measurements done *in vitro, ex vivo*, and *in vivo* for human tissues are summarized in Table 16.1. It is clearly seen that attention was primarily focused on female breast and head/brain optical properties investigations because of great importance and perspectives of optical mammography and optical monitoring and treatment of mental diseases. Skin and underlying tissues are also well studied.

The most detailed *in vitro* investigations of normal and coagulated brain tissues (gray matter, white matter, cerebellum, pons, and thalamus), as well as of native tumor tissues (astrocytoma WHO grade II and meningioma), using single-IS spectral measurements in the spectral range from 360 to 1100 nm and IMC algorithm for data processing are described by Yaroslavsky et al. (2002). All brain tissues under study shared qualitatively similar dependencies of the optical properties on the wavelength. The scattering coefficient decreased and the anisotropy factor increased with the wavelength, which can be explained by the lowering of the contribution of Rayleigh scattering and growing of the contribution of Mie scattering with the wavelength. The wavelength-dependent absorption coefficient behavior of all brain tissues resembled a mixture of oxy and deoxyhemoglobin absorption spectra. This means that in spite of careful preparation of the samples, it was not possible to remove all blood residuals from the tissue sections.

At the same time, the differences in the spectral characteristics of the brain tissues have been observed. For example, the total attenuation coefficients ($\mu_t = \mu_a + \mu_s$) of white matter are substantially higher than those of gray matter. The two brain stem tissues (pons and thalamus) also have different optical properties. The tumors are generally macroscopically less homogeneous than any normal tissues; thus, their scattering coefficients and anisotropy factors are slightly higher than those of normal gray matter. The same tendency of scattering coefficients growing is typical for breast tumors (carcinomas) (Peters et al., 1990).

After coagulation, the values of absorption and scattering coefficients increased for all tissues. The extent of this increase, however, is different for each tissue type, and is characterized by factors from 2 to 5. It was shown (Yaroslavsky et al., 2002) that a significant increase of both interaction coefficients is a result of substantial structure changes, caused mostly by tissue shrinkage and condensation, as well as collagen swelling and homogenization of the tissue. Tissue shrinkage caused by losing water at coagulation makes tissue more dense, which leads to increase of both scattering and absorption coefficients in the spectral range

Tab. 16.1 Optical properties of human tissues measured *in vitro*, *ex vivo*, and *in vivo* (root mean square (rms) values are given in parentheses).

Tissue	λ, nm	μ_a, cm^{-1}	μ_s, cm^{-1}	μ_s', cm^{-1}	g	Remarks
			In vitro measurements			
Aorta:						
Normal	308	33	–	77	–	*Postmortem* (6 hr), excised, in 4 °C saline, slab, water bath (85 °C), integrating sphere (IS) technique, IAD Cheong, Prahl and Welch, (1990)
Normal coagulated	308	44	–	270	–	
Fibrous plaque	308	24	–	81	–	
Fibrous plaque coagulated	308	34	–	272	–	
Normal	1064	0.53(0.09)	239(45)	23.9	0.9	*Postmortem*, slab, 70 °C water bath, 10 min, IS, IMC, goniophotometric measurements (GPM) Cheong, Prahl and Welch (1990)
Coagulated	1064	0.46(0.18)	293(73)	29.3	0.9	
Fibro-fatty	355	17.7	–	64.9	–	*Postmortem*, resected, slab (24 hr), photoacoustic (PA) Cheong, Prahl and Welch (1990)
	532	3.6	–	24.8	–	
	1064	0.09	–	7.7	–	
Normal	633*	0.52	316	41	0.87	*Postmortem*, slab, IS, GPM, *diffusion theory (DT), **IMC Cheong, Prahl and Welch (1990)
	1064**	0.5	239	23.9	0.9	
	1064*	0.7	–	22.4	–	
	1320**	2.2	233	23.3	0.9	
	1320*	4.3	–	17.8	–	

16.3 Measurements of Optical Properties of Tissues and Blood

Tissue	λ (nm)				Reference/Notes
Normal	470	5.3(0.9)	—	—	Thin sections (250 μm, intima and media), kept in saline, IS, DT Cilesiz and Welch (1993); corrected data (see Tuchin (1994), p. 379)
	476	5.1(0.9)	—	—	
	488	4.5(0.9)	—	—	
	514.5	3.7(0.9)	—	—	
	580	2.8(0.9)	—	—	
	600	2.6(0.9)	—	—	
	633	2.6(0.9)	—	—	
	1064	2.7(0.5)	—	—	
Adventitia	476	18.1	267	42.6(6.0)	0.74 Frozen sections, IS, DT Keijzer et al. (1989)
	580	11.3	217	41.9(5.9)	0.77
	600	6.1	211	39.9(5.6)	0.78
	633	5.8	195	36.9(5.4)	0.81
	1064	2.0	484	—	0.97 Double IS (DIS), DT Roggan et al. (1995)
Intima	476	14.8	237	31.1(4.9)	0.81 Frozen sections, IS, DT Keijzer et al. (1989)
	580	8.9	183	29.6(4.7)	0.81
	600	4.0	178	27.4(4.4)	0.81
	633	3.6	171	15.5(2.8)	0.85
	1064	2.3	165	—	0.97 DIS, DT Roggan et al. (1995)
Media	476	7.3	410	45.0	0.89 Frozen sections, IS, DT Keijzer et al. (1989)
	580	4.8	331	34.8	0.90
	600	2.5	323	33.8	0.89
	633	2.3	310	25.7	0.90
	1064	1.0	634	—	0.96 DIS, DT Roggan et al. (1995)

(continued overleaf)

Tab. 16.1 (Continued)

Tissue	λ, nm	μ_a, cm^{-1}	μ_s, cm^{-1}	$\mu_s{}'$, cm^{-1}	g	Remarks
Bladder:						
Integral	633	1.40	88.0	3.52	0.96	Excised, kept in saline Cheong, Prahl and Welch (1990)
Integral	633	1.40	29.3	2.64	0.91	DIS, DT Müller and Roggan (1995); Roggan et al. (1995)
Mucous	1064	0.7	7.5	–	0.85	
Wall	1064	0.9	54.3	–	0.85	
Integral	1064	0.4	116	–	0.90	
Blood:						
HbO$_2$ (Hct = 0.41)	665	1.30	1246	6.11	0.995	Whole blood; absorbance, radial reflectance, and/or GPM; Mie theory, transport theory, or IMC Cheong, Prahl and Welch (1990); Roggan et al. (1995)
HbO$_2$ (Hct = 0.41)	685	2.65	1413	14.13	0.990	
HbO$_2$ (Hct = 0.41)	960	2.84	505	3.84	0.992	
HbO$_2$ (Hct = 0.4)	810	4.5	–	6.6	–	
HbO$_2$ (Hct = 0.4)	1064	3.0	–	3.4	–	
Hb (Hct = 0.41)	960	16.8	668	5.08	0.992	
Hb (Hct = 0.4)	810	4.5	–	3.9	–	
Hb (Hct = 0.4)	1064	0.3	–	6.6	–	
Hct = 0.47 (partially oxygenated)	450	381	2940	8.3	0.9972	Whole blood; IAD and Beer's law; data by Jacques Cheong, Prahl and Welch (1990)
	488	133	3190	4.0	0.9987	
	514	116	3320	4.1	0.9988	
	577	301	3140	7.3	0.9977	
	630	14.3	3660	8.9	0.9976	
	760	15.5	2820	7.9	0.9972	

(Hct = 0.45–0.46, oxygenation >98%)	633	15.5	644.7	—	0.982	DIS, Henyey–Greenstein phase function (HGPF), IMC, whole blood Yaroslavsky et al. (1999)
	710	4(0.8)	737(75)	—	0.986(0.006)	
	810	6.5(0.5)	690(80)	—	0.989(0.002)	
	910	8.9(0.4)	649(25)	—	0.992(0.002)	
	1010	8.3(0.4)	645(25)	—	0.992(0.001)	
	1110	4.2(0.3)	630(20)	—	0.993(0.001)	
	1210	5.5(0.5)	654(20)	—	0.995(0.001)	
(Hct = 0.421, oxygenation >99%)	260	375.5(9.0)	631.5(57.6)	136.4(28.0)	0.784(0.030)	IS, fresh erythrocytes from a healthy blood donor diluted in PBS, pH 7.4, hemoglobin concentration 129 g l^{-1}, the temperature was kept constant at 20 °C, turbulence-free cuvette with a laminar flow and a sample thickness of 116 μm, constant wall share rate of 600 s^{-1}; in the wavelength region around 415 nm a cuvette of 40 μm in thickness was used; Reynolds–McCormick phase function ($\alpha = 1.7$), IMC, data were presented by Friebel et al. (2006).
	350	368.1	559.5	82.5	0.852	
	415	782.5(62.9)	390.3(61.2)	129.5(17.0)	0.668(0.008)	
	520	120.4(6.9)	766.2(42.4)	24.9(7.6)	0.967(0.009)	
	540	232.3	655.6	35.8	0.945	
	555	178.9	709.3	33.0	0.953	
	575	231.6	658.0	31.7	0.952	
	585	160.2(10.3)	751.7(46.1)	33.5(9.7)	0.955(0.007)	
	630	2.51(0.09)	894.6(28.6)	22.3(3.3)	0.975(0.004)	
	670	1.22	892.3	21.5	0.976	
	780	2.85	821.5	20.5	0.975	
	800	3.27(0.12)	809.9(66.4)	20.2(5.4)	0.975(0.003)	
	830	4.90	798.7	20.1	0.975	
	870	5.10	784.4	20.1	0.974	
	950	6.15(0.35)	712.0(69.8)	20.8(2.7)	0.971(0.002)	
	1050	4.91(0.12)	661.3(12.8)	19.91(0.67)	0.9699(0.0006)	
	1100	3.74	639.5	18.85	0.970	

(continued overleaf)

Tab. 16.1 (Continued)

Tissue	λ, nm	μ_a, cm^{-1}	μ_s, cm^{-1}	μ_s', cm^{-1}	g	Remarks
Brain:						
Glioma (male, 65 yr, 4-hr *postmortem*)	415	16.6	—	6	—	Sterenborg et al. (1989); data from graphs
	488	12.5	—	3	—	
	630	3.0	—	3	—	
	800–1100	≈1.0	—	>1–2	—	
Gray matter (male, 71 yr, 24-hr *postmortem*)	514	19.5	—	85	—	
	585	14.5	—	63	—	
	630	4.3	—	52	—	
	800–1100	≈1.0	—	45–20	—	
Melanoma (male 71 yr, 24-hr *postmortem*)	585	2	—	158	—	
	630	20.0	—	75	—	
	800	8.0	—	40	—	
	900	4.0	—	30	—	
	1100	2.0	—	25	—	
White matter (female, 32 yr, 24-hr *postmortem*)	415	2.1	—	24	—	
	488	1.0	—	60	—	
	630	0.2	—	32	—	
	800–1100	0.2–0.3	—	40–20	—	
White matter (female, 63 yr, 30-hr *postmortem*)	488	2.7	—	25	—	
	630	0.9	—	22	—	
	800–1100	1.0–1.5	—	20–10	—	

Tissue	λ (nm)	μ_a	μ_s	g	Method/Reference
Gray matter	633	2.7(2)	354(37)	20.6(2)	0.94(0.004)	Freshly resected slabs Cheong, Prahl and Welch (1990)
	1064	5.0(5)	134(14)	11.8(9)	0.90(0.007)	
White matter	633	2.2(2)	532(41)	91(5)	0.82(0.01)	
	1064	3.2(4)	469(34)	60.3(2.5)	0.87(0.007)	
Gray matter (n = 7)	360	3.33(2.19)	141.3(42.6)	—	0.818(0.093)	DIS, IMC Yaroslavsky et al. (1996); Schwarzmaier et al. (1997)
	640	0.17(0.26)	90.1(32.5)	—	0.89(0.04)	
	1060	0.56(0.7)	56.8(18.0)	—	0.90(0.05)	
Gray matter coagulated (n = 7)	360	9.39(1.70)	426(122)	—	0.868(0.031)	DIS, IMC; 2 hr, 80 °C Yaroslavsky et al. (1996); Schwarzmaier et al. (1997)
	740	0.45(0.27)	—	—	—	
	1100	1.0(0.45)	179.8(32.6)	—	0.954(0.001)	
White matter (n = 7)	360	2.53(0.55)	402.0(91.8)	—	0.702(0.093)	DIS, IMC (Yaroslavsky et al., 1996; Schwarzmaier et al., 1997)
	640	0.8(0.2)	408.2(88.5)	—	0.84(0.05)	
	860	0.97(0.4)	353.1(68.1)	—	0.871(0.028)	
	1060	1.08(0.51)	299.5(70.1)	—	0.889(0.010)	
White matter coagulated (n = 7)	360	8.3(3.65)	604.2(131.5)	—	0.800(0.089)	DIS, IMC; 2 hr, 80 °C Yaroslavsky et al. (1996); Schwarzmaier et al. (1997)
	860	1.7(1.3)	417.0(272.5)	—	0.922(0.025)	
	1060	2.15(1.34)	363.3(226.8)	—	0.930(0.015)	

(continued overleaf)

Tab. 16.1 (Continued)

Tissue	λ, nm	μ_a, cm^{-1}	μ_s, cm^{-1}	μ_s', cm^{-1}	g	Remarks
Gray matter (n = 7)	450	0.7	117	14.04	0.88	IS, IMC, quasi-Newton inverse algorithm, HGPF; hemoglobin-free cryosections (<48-hr *postmortem*): gray matter – 100–200 µm; white matter – 80–150 µm; coagulation: saline bath 80 °C coagulation: saline bath 80 °C, 2 hr Yaroslavsky *et al.* (2002)
	510	0.4	106	12.72	0.88	
	630	0.2	90	9.9	0.89	
	670	0.2	84	8.4	0.90	
	1064	0.5	57	5.7	0.90	
White matter (n = 7)	450	1.4	420	92.4	0.78	
	510	1.0	426	80.94	0.81	
	630	0.8	409	65.44	0.84	
	670	0.7	401	60.15	0.85	
	850	1.0	342	41	0.88	
	1064	1.0	296	32.56	0.89	
White matter coagulated (n = 7)	850	0.9	300	36.0	0.88	
	1064	0.1	270	29.7	0.89	
Astrocytoma (grade II WHO, n = 4)	400	18.8(11.3)	198.4(55.6)	–	0.93(0.03)	IS, IMC, quasi-Newton inverse algorithm, HGPF; hemoglobin-free cryosections of normal tissues (<48-hr *postmortem*): cerebellum, gray matter, pons, and thalamus – 100–200 µm; white matter – 80–150 µm; and tumors excised from patients of ≈300 µm in thickness; coagulation: saline bath 80 °C, 2 hr Yaroslavsky *et al.* (2002)
	490	2.5(0.9)	158.5(53.7)	–	0.96(0.02)	
	600	1.2(0.7)	132.4(49.0)	–	0.96(0.02)	
	700	0.5(0.3)	113.2(41.8)	–	0.96(0.02)	
	800	0.7(0.2)	96.7(41.8)	–	0.96(0.01)	
	900	0.3(0.2)	86.4(34.6)	–	0.96(0.01)	
	1000	0.5(0.3)	79.0(34.2)	–	0.96(0.01)	
	1100	0.6(0.2)	73.8(29.6)	–	0.96(0.01)	
Cerebellum (n = 7)	400	4.7(0.8)	276.7(19.1)	–	0.80(0.03)	
	500	1.4(0.2)	277.5(32.6)	–	0.85(0.02)	
	600	0.8(0.2)	272.1(12.3)	–	0.87(0.02)	
	700	0.6(0.1)	266.8(12.1)	–	0.89(0.01)	
	800	0.6(0.1)	250.3(17.2)	–	0.90(0.01)	
	900	0.7(0.1)	229.6(15.8)	–	0.90(0.01)	
	1000	0.8(0.1)	215.4(14.7)	–	0.90(0.01)	
	1100	0.7(0.1)	202.1(13.9)	–	0.90(0.01)	

Cerebellum coagulated ($n = 7$)	400	19.3(7.7)	560.0(25.5)	—	0.61(0.01)
	500	5.1(1.7)	512.2(47.8)	—	0.77(0.02)
	600	2.9(1.4)	458.2(65.6)	—	0.78(0.01)
	700	1.7(0.4)	489.9(70.1)	—	0.85(0.01)
	800	1.1(0.2)	458.2(54.0)	—	0.87(0.02)
	900	1.1(0.3)	458.2(65.6)	—	0.89(0.02)
	1000	1.0(0.4)	419.1(49.4)	—	0.90(0.03)
	1100	1.1(0.5)	428.5(40.0)	—	0.91(0.03)
Gray matter ($n = 7$)	400	2.6(0.6)	128.5(18.4)	—	0.87(0.02)
	500	0.5(0.2)	109.9(13.0)	—	0.88(0.01)
	600	0.3(0.1)	94.1(13.5)	—	0.89(0.02)
	700	0.2(0.1)	84.1(12.0)	—	0.90(0.02)
	800	0.2(0.1)	77.0(11.0)	—	0.90(0.02)
	900	0.3(0.2)	67.3(9.6)	—	0.90(0.02)
	1000	0.6(0.3)	61.6(5.7)	—	0.90(0.02)
	1100	0.5(0.3)	55.1(6.5)	—	0.90(0.02)
Gray matter coagulated ($n = 7$)	400	7.5(0.4)	258.6(18.8)	—	0.78(0.04)
	500	1.8(0.2)	326.5(7.7)	—	0.85(0.03)
	600	0.7(0.1)	319.0(15.2)	—	0.87(0.03)
	700	0.7(0.1)	319.0(7.5)	—	0.88(0.03)
	800	0.8(0.1)	252.7(18.3)	—	0.87(0.02)
	900	0.9(0.1)	214.6(10.3)	—	0.87(0.02)
	1000	1.4(0.2)	191.0(18.7)	—	0.88(0.03)
	1100	1.5(0.2)	186.6(13.5)	—	0.88(0.03)

(continued overleaf)

Tab. 16.1 (Continued)

Tissue	λ, nm	μ_a, cm^{-1}	μ_s, cm^{-1}	μ_s', cm^{-1}	g	Remarks
Meningioma (n = 6)	410	4.1(0.5)	197.4(19.8)	–	0.88(0.02)	
	490	1.3(0.2)	188.2(18.8)	–	0.93(0.01)	
	590	0.7(0.2)	171.1(12.7)	–	0.95(0.01)	
	690	0.3(0.1)	155.5(15.6)	–	0.95(0.01)	
	790	0.2(0.1)	141.3(14.2)	–	0.96(0.01)	
	910	0.2(0.1)	116.8(8.6)	–	0.95(0.01)	
	990	0.4(0.2)	163.5(15.3)	–	0.96(0.01)	
	1100	0.6(0.2)	133.7(19.2)	–	0.97(0.01)	
Pons (n = 7)	400	3.1(0.7)	163.5(15.3)	–	0.89(0.02)	
	500	0.9(0.3)	133.7(19.2)	–	0.91(0.01)	
	600	0.6(0.2)	109.4(18.5)	–	0.91(0.01)	
	700	0.5(0.2)	93.5(20.9)	–	0.91(0.01)	
	800	0.6(0.3)	83.6(21.0)	–	0.91(0.01)	
	900	0.7(0.3)	74.8(18.7)	–	0.92(0.01)	
	1000	1.0(0.4)	69.9(17.5)	–	0.91(0.01)	
	1100	0.9(0.4)	64.0(17.8)	–	0.92(0.01)	
Pons coagulated (n = 7)	410	17.2(1.6)	685.7(63.7)	–	0.85(0.02)	
	510	8.5(0.8)	627.5(73.6)	–	0.89(0.01)	
	610	7.7(0.5)	510.5(70.5)	–	0.89(0.01)	
	710	6.9(0.6)	402.5(67.7)	–	0.89(0.01)	
	810	6.5(0.6)	329.7(55.4)	–	0.89(0.01)	
	910	5.9(1.0)	276.0(46.4)	–	0.88(0.01)	
	1010	5.7(1.0)	241.6(34.4)	–	0.88(0.01)	
	1100	6.5(0.9)	221.1(31.5)	–	0.88(0.01)	

Thalamus (n = 7)	410	3.2(1.0)	146.7(49.4)	—	0.86(0.03)
	510	0.9(0.3)	188.7(31.9)	—	0.87(0.03)
	610	0.6(0.2)	176.3(34.5)	—	0.88(0.02)
	710	0.5(0.3)	169.0(28.7)	—	0.89(0.03)
	810	0.7(0.3)	158.5(35.3)	—	0.89(0.02)
	910	0.7(0.3)	155.4(22.3)	—	0.90(0.02)
	1010	0.8(0.3)	139.3(34.9)	—	0.90(0.02)
	1100	0.8(0.3)	146.0(36.6)	—	0.91(0.02)
Thalamus coagulated (n = 7)	400	15.0(3.3)	391.1(56.1)	—	0.83(0.04)
	500	4.2(0.9)	399.9(67.7)	—	0.90(0.01)
	600	1.6(0.6)	365.7(43.2)	—	0.92(0.01)
	700	1.4(0.3)	327.0(30.6)	—	0.92(0.01)
	800	1.1(0.3)	286.0(33.8)	—	0.93(0.01)
	900	1.1(0.3)	267.4(31.6)	—	0.93(0.01)
	1000	1.4(0.4)	233.8(39.7)	—	0.93(0.01)
	1100	1.5(0.4)	223.6(32.1)	—	0.94(0.01)
White matter (n = 7)	400	3.1(0.2)	413.5(21.4)	—	0.75(0.03)
	500	0.9(0.1)	413.5(43.9)	—	0.80(0.02)
	600	0.8(0.1)	413.5(21.4)	—	0.83(0.02)
	700	0.8(0.1)	393.1(30.9)	—	0.85(0.02)
	800	0.9(0.1)	364.5(28.6)	—	0.87(0.01)
	900	1.0(0.1)	329.5(35.0)	—	0.88(0.01)
	1000	1.2(0.2)	305.4(15.9)	—	0.88(0.01)
	1100	1.0(0.2)	283.2(22.2)	—	0.88(0.01)

(continued overleaf)

Tab. 16.1 (Continued)

Tissue	λ, nm	μ_a, cm^{-1}	μ_s, cm^{-1}	μ_s', cm^{-1}	g	Remarks
White matter coagulated ($n = 7$)	410	8.7(1.7)	568.7(111.9)	–	0.83(0.03)	IS, IAD, fixed the anisotropy factor $g = 0.85$ and the refractive index $n = 1.40$ were assumed for every wavelength and for every sample; brain tissue samples were acquired during open craniotomy for tumor resection or temporal lobectomy, hemoglobin-free cryosections with thickness from 0.22 to 1.25 mm were studied, measurements were done at 25 °C, pH 7.4; data were presented by Gebhart, Lin and Mahadevan-Jansen (2006).
	510	2.9(0.6)	513.2(116.9)	–	0.87(0.02)	
	610	1.7(0.4)	500.2(129.9)	–	0.90(0.02)	
	710	1.4(0.5)	475.2(108.3)	–	0.91(0.01)	
	810	1.5(0.5)	440.0(114.3)	–	0.92(0.01)	
	910	1.7(0.6)	407.4(92.8)	–	0.93(0.01)	
	1010	1.9(0.6)	367.7(95.5)	–	0.93(0.01)	
	1100	2.4(0.5)	358.4(81.6)	–	0.93(0.01)	
Gray matter ($n = 25$)	400	9.778	–	25.878	–	
	428	16.722	–	26.709	–	
	488	2.272	–	15.957	–	
	550	2.955	–	13.315	–	
	632	0.925	–	10.370	–	
	670	0.809	–	9.480	–	
	750	0.599	–	8.481	–	
	830	0.485	–	7.707	–	
	900	0.503	–	7.055	–	
	1000	0.585	–	6.059	–	
	1064	0.502	–	5.333	–	
	1150	0.815	–	5.070	–	
	1300	0.894	–	4.560	–	

White matter (n = 19)	400	9.134	88.611	—
	428	15.417	80.905	—
	488	1.869	70.112	—
	550	2.584	62.383	—
	632	0.801	53.179	—
	670	0.711	50.067	—
	750	0.649	45.061	—
	830	0.626	40.634	—
	900	0.684	37.607	—
	1000	0.883	32.603	—
	1064	0.752	30.161	—
	1150	1.135	27.951	—
	1300	1.274	24.250	—
Glioma (n = 39)	400	12.393	39.009	—
	428	16.124	37.076	—
	488	2.592	28.933	—
	550	2.768	25.300	—
	632	0.846	21.068	—
	670	0.741	19.608	—
	750	0.679	17.343	—
	830	0.662	15.481	—
	900	0.707	14.138	—
	950	0.768	13.646	—
	1000	0.938	11.588	—
	1064	0.822	10.344	—
	1150	1.231	9.654	—
	1300	1.412	8.523	—

(continued overleaf)

Tab. 16.1 (Continued)

Tissue	λ, nm	μ_a, cm^{-1}	μ_s, cm^{-1}	μ_s', cm^{-1}	g	Remarks
Breast (female):						
Fatty normal ($n = 23$)	749	0.18(0.16)	8.48(3.43)	–	–	Excised, kept in saline, 37 °C Troy, Page and Sevick-Muraca (1996)
	789	0.08(0.10)	7.67(2.57)	–	–	
	836	0.11(0.10)	7.27(2.40)	–	–	
Fibrous normal ($n = 35$)	749	0.13(0.19)	9.75(2.27)	–	–	
	789	0.06(0.12)	8.94(2.45)	–	–	
	836	0.05(0.08)	8.10(2.21)	–	–	
Infiltrating carcinoma ($n = 48$)	749	0.15(0.14)	10.91(5.59)	–	–	
	789	0.04(0.08)	10.12(5.05)	–	–	
	836	0.10(0.19)	9.10(4.54)	–	–	
Mucinous carcinoma ($n = 3$)	749	0.26(0.20)	–	6.15(2.44)	–	
	789	0.016(0.072)	–	5.09(2.42)	–	
	836	0.023(0.108)	–	4.78(3.67)	–	
Ductal carcinoma in situ ($n = 5$)	749	0.076(0.068)	–	13.10(2.85)	–	
	789	0.023(0.034)	–	12.21(2.45)	–	
	836	0.039(0.068)	–	10.46(2.65)	–	
Glandular tissue ($n = 3$)	540	3.58(1.56)	–	24.4(5.8)	–	Homogenized tissue Peters et al. (1990)
	700	0.47(0.11)	–	14.2(3.0)	–	
	900	0.62(0.05)	–	9.9(2.0)	–	

Tissue	λ (nm)					Reference / Notes
Fatty tissue (n = 7)	540	2.27(0.57)	—	10.3(1.9)	—	Das, Liu and Alfano, (1997)
	700	0.70(0.08)	—	8.6(1.3)	—	Spatially resolved reflectance (SRR); DT; source–detector separation, r_{sd} > 1.2 mm; fibers with core diameter 400 μm; tissue slices of thickness ~10 mm; GPM, tissue slices of 20 μm, HGPF; *double HGPF Ghosh et al. (2001)
	900	0.75(0.08)	—	7.9(1.1)	—	
Fibrocystic (n = 8)	540	1.64(0.66)	—	21.7(3.3)	—	
	700	0.22(0.09)	—	13.4(1.9)	—	
	900	0.27(0.11)	—	9.5(1.7)	—	
Fibroadenoma (n = 6)	540	4.38(3.14)	—	11.1(3.0)	—	
	700	0.52(0.47)	—	7.2(1.7)	—	
	900	0.72(0.53)	—	5.3(1.4)	—	
Carcinoma (n = 9)	540	3.07(0.99)	—	19.0(5.1)	—	
	700	0.45(0.12)	—	11.8(3.1)	—	
	900	0.50(0.15)	—	8.9(2.6)	—	
Fatty tissue	625	0.06(0.02)	—	14.3(2.1)	—	
Benign tumor	625	0.33(0.06)	—	3.8(0.3)	—	
Invasive ductal carcinoma (n = 10; 9 in the age group 55–65 yr and 1–35 yr)	450	2.55(0.30)	—	31.5(2.5)	—	
	500	2.22(0.22)	—	29.5(2.2)	—	
	550	2.13(0.23)	—	28.4(2.0)	—	
	600	1.90(0.19)	—	26.8(1.8)	—	
	630	1.64(0.17)	—	26.2(1.5)	—	
	650	1.48(0.15)	—	25.7(1.3)	—	
	633	—	—	—	0.96(0.01)	
	633	—	—	—	0.86(0.02)*	

(continued overleaf)

Tab. 16.1 (Continued)

Tissue	λ, nm	μ_a, cm^{-1}	μ_s, cm^{-1}	μ_s', cm^{-1}	g	Remarks
Adjacent healthy tissue ($n = 10$; 9 in the age group 55–65 yr and 1–35 yr)	450	1.45(0.22)	–	21.7(2.1)	–	SRR; DT; $r_{sd} > 1.2$ mm; fibers with core diameter 400 μm; tissue slices of thickness ~10 mm;
	500	1.24(0.21)	–	20.1(1.8)	–	
	550	1.16(0.26)	–	18.2(1.6)	–	
	600	1.00(0.12)	–	16.4(1.5)	–	
	630	0.82(0.07)	–	15.7(1.3)	–	
	650	0.74(0.08)	–	15.3(1.2)	–	
	633	–	–	–	0.88(0.01)	GPM, tissue slices of 20 μm; HGPF; *double HGPF
	633	–	–	–	0.76(0.01)*	Ghosh et al. (2001)
Colon:						
Muscle	1064	3.3	238	–	0.93	Roggan et al. (1995)
Submucous	1064	2.3	117	–	0.91	
Mucous	1064	2.7	39	–	0.91	
Integral	1064	0.4	261	–	0.94	
Esophagus	633	0.4	–	12	–	2.5-mm slab Cheong, Prahl and Welch (1990)
Esophagus (mucous)	1064	1.1	83	–	0.86	Roggan et al. (1995)
Fat:						
Abdominal	1064	3.0	37	–	0.91	Roggan et al. (1995)
Subcutaneous	1064	2.6	29	–	0.91	

16.3 Measurements of Optical Properties of Tissues and Blood

Gallstones:				
Porcinement	351	102(16)	—	Dehydrated, embedded in plastic, and sliced in 1-mm slab, pulsed photothermal radiometry technique Cheong, Prahl and Welch (1990)
	488	179(28)	—	
	580	125(29)	—	
	630	85(11)	—	
	1060	121(12)	—	
Cholesterol	351	88(7)	—	—
	488	62(15)	—	—
	580	36(7)	—	—
	630	44(10)	—	—
	1060	60(9)	—	—
Head (adult):				
Dura mater ($n = 8$), postmortem, <24 h	400	3.08(0.15)	22.35(0.89)	IS, IAD; excised tissue slabs; in the spectral ranges 480–550 and 600–700 nm: $\mu_s' = 4.54 \times 10^4 \lambda^{-1.23}$, [$\lambda$] in nanometers Genina et al. (2005)
	450	1.51(0.08)	22.89(0.92)	
	500	1.09(0.05)	21.60(0.86)	
	550	1.10(0.05)	18.48(0.74)	
	600	0.80(0.04)	17.11(0.68)	
	650	0.70(0.04)	15.51(0.62)	
	700	0.74(0.04)	13.99(0.56)	

(continued overleaf)

Tab. 16.1 (Continued)

Tissue	λ, nm	μ_a, cm^{-1}	μ_s, cm^{-1}	μ_s', cm^{-1}	g	Remarks
Scalp ($n = 3$)	805	0.52(0.04)	—	14.09(1.74)	—	Adult scalp *postmortem* (<12 hr), excised, slab, IS, IAD; data averaged for three tissue samples with thicknesses of 6 ± 0.5 mm, 3.5 ± 0.15 mm, and 3.5 ± 0.12 mm Bashkatov et al. (2006)
	1000	0.33(0.03)	—	16.83(2.77)	—	
	1100	0.19(0.04)	—	17.10(2.69)	—	
	1200	0.65(0.04)	—	16.70(2.89)	—	
	1300	0.50(0.07)	—	14.70(2.59)	—	
	1400	1.98(0.31)	—	14.28(3.69)	—	
	1430	2.19(0.29)	—	13.15(3.07)	—	
	1600	1.43(0.22)	—	14.16(3.41)	—	
	1700	1.87(0.28)	—	14.71(3.51)	—	
	1800	1.73(0.22)	—	13.36(2.91)	—	
	1900	2.57(0.28)	—	12.15(3.05)	—	
	2000	2.09(0.29)	—	12.00(2.91)	—	
Skull bone ($n = 8$)	801	0.11(0.02)	—	19.48(1.52)	—	Adult head *postmortem* (24 hr), excised, slab from the occipital part, IS, IAD; data averaged for 8 tissue samples $\mu_s' = 1.53 \times 10^3 \lambda^{-0.65}$, [$\lambda$] in nanometers (spectral range from 1130 to 1910 nm is excluded) Bashkatov et al. (2006)
	980	0.23(0.03)	—	17.38(1.01)	—	
	1100	0.16(0.03)	—	15.92(0.76)	—	
	1200	0.67(0.07)	—	16.77(0.85)	—	
	1300	0.54(0.05)	—	14.78(0.80)	—	
	1400	2.43(0.24)	—	17.22(1.73)	—	
	1465	3.33(0.31)	—	16.84(1.88)	—	
	1600	2.47(0.40)	—	15.84(3.05)	—	
	1700	2.77(0.46)	—	16.12(3.72)	—	
	1800	2.97(0.62)	—	15.42(3.98)	—	
	1930	4.97(1.52)	—	10.92(2.17)	—	
	2000	4.47(1.18)	—	11.48(2.01)	—	

Tissue	λ (nm)				Reference
Heart:					
Endocard	1060	0.07	136	—	Excised, kept in saline, Cheong, Prahl and Welch (1990)
Epicard	1060	0.35	167	0.97	Roggan et al. (1995)
Myocard	1060	0.3	177.5	0.98	
Epicard	1060	0.21	127.1	0.96	
Aneurysm	1060	0.4	137	0.93	
Trabecula	1064	1.4	424	0.98	
				0.97	
Kidney:					
Pars convalescent	1064	2.4	72	0.86	Roggan et al. (1995)
Medulla renal	1064	2.1	77	0.87	
Liver	515	18.9(1.7)	285(20)	—	Frozen sections
	635	2.3(1.0)	313(136)	0.68	Marchesini et al. (1989)
	1064	0.7	356	0.95	
Lung	515	25.5(3.0)	356(39)	—	
	635	8.1(2.8)	324(46)	0.75	
	1064	2.8	39	0.91	
Muscle	515	11.2(1.8)	530(44)	—	
	1064	2.0	215	0.96	
Meniscus	360	13	—	108	Frozen, thawed, slab
	400	4.6	—	67	Cheong, Prahl and Welch, (1990)
	488	1	—	30	
	514	0.73	—	26	
	630	0.36	—	11	
	800	0.52	—	5.1	
	1064	0.34	—	2.6	

(continued overleaf)

Tab. 16.1 (Continued)

Tissue	λ, nm	μ_a, cm^{-1}	μ_s, cm^{-1}	μ_s', cm^{-1}	g	Remarks
Prostate:						
Normal	850	0.6(0.2)	100(20)	–	0.94(0.02)	Shock frozen sections of 60–500 μm, 0.5–3-hr postmortem; coagulation in water bath (75 °C, 10 min) Roggan et al. (1995)
	980	0.4(0.2)	90(20)	–	0.95(0.02)	
	1064	0.3(0.2)	80(20)	–	0.95(0.02)	
Coagulated	850	7.0(0.2)	230(30)	–	0.94(0.02)	
	980	5.0(0.2)	190(30)	–	0.95(0.02)	
	1064	4.0(0.2)	180(30)	–	0.95(0.02)	
Normal	1064	1.5(0.2)	47(13)	0.64	0.862	Freshly excised, slab, water bath (70 °C, 10 min) Cheong, Prahl and Welch (1990)
Coagulated	1064	0.8(0.2)	80(12)	1.12	0.861	
Sclera ($n = 5$)	404	5.00(0.50)	–	81.40(8.14)	–	IS, IAD; excised tissue slabs, <24-hr postmortem, stored in saline at 4 °C; measurements at room temperature; $\mu_s' = 8.95 \times 10^4 \lambda^{-1.16}$, [$\lambda$] in nanometers Tuchin (2007)
	499	2.96(0.30)	–	65.17(6.52)	–	
	599	1.95(0.19)	–	53.16(5.32)	–	
	699	1.67(0.17)	–	44.15(4.42)	–	
	749	1.65(0.17)	–	40.10(4.01)	–	
	799	1.58(0.16)	–	37.64(3.76)	–	
Skin:						
	250	1150	2600	260	0.9	Data from graphs; μ_s' is calculated van Gemert et al. (1989)
	308	600	2400	240	0.9	
	337	330	2300	230	0.9	
	351	300	2200	220	0.9	
	400	230	2000	200	0.9	

Tissue	λ (nm)	μ_a	μ_s	g	Remarks
Epidermis	250	1000	2000	0.69	Data from graphs; μ_s' and g are calculated using van Gemert et al. (1989)
	308	300	1400	0.71	
	337	120	1200	0.72	
	351	100	1100	0.72	
	415	66	800	0.74	
	488	50	600	0.76	
	514	44	600	0.77	
	585	36	470	0.79	
	633	35	450	0.80	
	800	40	420	0.85	
Dermis	250	35	833	0.69	Data from graphs; values are transformed in accordance with data for $\lambda = 633$ nm (bloodless tissue, hydration ~85%), μ_s' and g are calculated Jacques, Alter and Prahl (1987); van Gemert et al. (1989)
	308	12	583	0.71	
	337	8.2	500	0.72	
	351	7	458	0.72	
	415	4.7	320	0.74	
	488	3.5	250	0.76	
	514	3	250	0.77	
	585	3	196	0.79	
	633	2.7	187.5	0.80	
	800	2.3	175	0.85	
Dermis	749	0.24(0.19)	23.1(0.75)	–	Frozen sections, DIS Troy, Page and Sevick-Muraca (1996)
	789	0.75(0.06)	22.8(1.29)	–	
	836	0.98(0.15)	15.9(2.16)	–	

(continued overleaf)

Tab. 16.1 (Continued)

Tissue	λ, nm	μ_a, cm^{-1}	μ_s, cm^{-1}	$\mu_s{'}$, cm^{-1}	g	Remarks
Caucasian male skin ($n = 3$)	500	5.1	—	50	—	IS, IAD Chan et al. (1996)
	810	0.26	—	15.8	—	
Caucasian female skin ($n = 3$)	500	5.2	—	23.9	—	
	810	0.97	—	8.2	—	
Hispanic male ($n = 3$) skin	500	3.8	—	24.2	—	
	810	0.87	—	7.5	—	
Caucasian skin ($n = 21$)	400	3.76(0.35)	—	71.79(9.42)	—	IS, IAD; tissue slabs, 1–6 mm; postmortem; <24 hr after death; stored at 20 °C in saline; measurements at room temperature; in the spectral range 400–2000 nm: $\mu_s{'} = 1.1 \times 10^{12} \lambda^{-4} + 73.7 \lambda^{-0.22}$, [$\lambda$] in nanometers Bashkatov et al. (2005a)
	600	0.69(0.13)	—	21.78(2.98)	—	
	800	0.43(0.11)	—	14.02(1.89)	—	
	1000	0.27(0.03)	—	16.83(2.77)	—	
	1200	0.54(0.04)	—	16.71(2.89)	—	
	1400	1.64(0.31)	—	14.28(3.69)	—	
	1600	1.19(0.22)	—	14.16(3.41)	—	
	1800	1.44(0.22)	—	13.36(2.91)	—	
	2000	1.74(0.29)	—	12.01(2.91)	—	
Epidermis	370	1.35(0.16)	—	11.56(1.25)	0.8	IS, IMC, slabs Salomatina et al. (2006)
	514	0.63(0.07)	—	6.67(0.66)	0.8	
	633	0.26(0.07)	—	4.76(0.45)	0.8	
	830	0.14(0.06)	—	3.56(0.35)	0.8	
	1064	0.02(0.02)	—	2.97(0.32)	0.8	
	1220	0.07(0.04)	—	2.63(0.31)	0.8	
	1270	0.06(0.04)	—	2.62(0.31)	0.8	
	1470	2.96(0.42)	—	3.08(0.45)	0.8	
	1570	1.01(0.20)	—	2.39(0.34)	0.8	

16.3 Measurements of Optical Properties of Tissues and Blood

Tissue	λ (nm)			
Dermis	370	0.98(0.14)	—	8.76(1.36)	0.8
	514	0.31(0.04)	—	4.32(0.41)	0.8
	633	0.15(0.02)	—	2.99(0.27)	0.8
	830	0.11(0.02)	—	2.15(0.23)	0.8
	1064	0.05(0.02)	—	1.80(0.21)	0.8
	1220	0.13(0.02)	—	1.65(0.20)	0.8
	1270	0.10(0.02)	—	1.63(0.20)	0.8
	1370	0.48(0.04)	—	1.66(0.19)	0.8
	1470	2.19(0.20)	—	2.13(0.21)	0.8
	1570	0.85(0.07)	—	1.65(0.19)	0.8
Subcutaneous fat	370	1.18(0.21)	—	5.27(0.69)	0.8
	514	0.47(0.07)	—	3.37(0.43)	0.8
	633	0.14(0.03)	—	2.54(0.30)	0.8
	830	0.10(0.02)	—	1.96(0.20)	0.8
	1064	0.07(0.02)	—	1.69(0.15)	0.8
	1220	0.15(0.03)	—	1.61(0.15)	0.8
	1270	0.10(0.03)	—	1.59(0.14)	0.8
	1370	0.27(0.04)	—	1.60(0.15)	0.8
	1470	1.08(0.18)	—	1.81(0.19)	0.8
	1570	0.43(0.07)	—	1.60(0.16)	0.8

(continued overleaf)

Tab. 16.1 (Continued)

Tissue	λ, nm	μ_a, cm^{-1}	μ_s, cm^{-1}	μ_s', cm^{-1}	g	Remarks
Infiltrative basal cell carcinoma	370	0.68(0.08)	–	6.52(0.92)	0.8	
	514	0.26(0.06)	–	4.04(0.30)	0.8	
	633	0.15(0.05)	–	2.81(0.28)	0.8	
	830	0.09(0.04)	–	1.92(0.25)	0.8	
	1064	0.08(0.04)	–	1.26(0.09)	0.8	
	1220	0.17(0.09)	–	1.09(0.10)	0.8	
	1270	0.18(0.12)	–	1.05(0.11)	0.8	
	1370	0.69(0.27)	–	1.09(0.10)	0.8	
	1470	2.75(0.54)	–	1.66(0.32)	0.8	
	1570	1.12(0.31)	–	1.11(0.16)	0.8	
Nodular basal cell carcinoma	370	0.87(0.29)	–	4.62(0.61)	0.8	
	514	0.28(0.11)	–	3.27(0.18)	0.8	
	633	0.12(0.06)	–	2.27(0.12)	0.8	
	830	0.02(0.01)	–	1.49(0.07)	0.8	
	1064	0.00(0.00)	–	1.16(0.06)	0.8	
	1220	0.02(0.01)	–	1.01(0.04)	0.8	
	1270	0.01(0.01)	–	1.00(0.04)	0.8	
	1370	0.32(0.03)	–	1.03(0.06)	0.8	
	1470	1.86(0.16)	–	1.59(0.15)	0.8	
	1570	0.67(0.04)	–	1.06(0.08)	0.8	

Tissue	λ (nm)				Reference
Squamous cell carcinoma	370	0.94(0.20)	—	4.36(0.61)	0.8	
	514	0.32(0.04)	—	2.80(0.39)	0.8	
	633	0.13(0.02)	—	1.88(0.25)	0.8	
	830	0.05(0.02)	—	1.22(0.15)	0.8	
	1064	0.04(0.02)	—	0.88(0.12)	0.8	
	1220	0.11(0.03)	—	0.81(0.11)	0.8	
	1270	0.11(0.03)	—	0.78(0.11)	0.8	
	1370	0.43(0.05)	—	0.85(0.11)	0.8	
	1470	2.35(0.21)	—	1.44(0.23)	0.8	
	1570	0.92(0.12)	—	0.92(0.13)	0.8	
Spleen	1064	6.0	137	—	0.90	Roggan et al. (1995)
Stomach:						Roggan et al. (1995)
Muscle	1064	3.3	29.5	—	0.87	
Mucous	1064	2.8	732	—	0.91	
Integral	1064	0.8	128	—	0.91	
Tooth:						
Dentin	543	4	180	—	—	IS, GPM* Zijp and ten Bosch (1991, 1997); Fried et al. (1995)
Enamel	633	4	130	—	—	
	633	6.0*	1200*	672*	0.44*	
	543	<1	45	—	—	
	633	<1	25	—	—	

(continued overleaf)

Tab. 16.1 (Continued)

Tissue	λ, nm	μ_a, cm^{-1}	μ_s, cm^{-1}	μ_s', cm^{-1}	g	Remarks
Dentin	543	3–4	280(84)	–	0.93(0.02)	GPM, double HGPF, fractions of isotropic scatterers are 0–2% for dentin and 60–35% for enamel; polished plane-parallel sections of 30–2000 μm Fried, et al. (1995)
	633	3–4	280(84)	–	0.93(0.02)	
	1053	3–4	260(78)	–	0.93(0.02)	
Enamel	543	<1	105(30)	–	0.96(0.02)	
	633	<1	60(18)	–	0.96(0.02)	
	1053	<1	15(5)	–	0.96(0.02)	
Enamel	200	≈10	≈450	–	–	Compiled data of a few papers, from graphs Fried (2003)
	300	≈5	≈270	–	–	
	400	≈1	≈150	–	–	
	500	<1	≈73	–	–	
	600	<1	≈64	–	–	
	700	<1	≈50	–	–	
	800	<1	≈33	–	–	
	1000	<1	≈16	–	–	
Dentin	2940	2200	–	–	–	Time-resolved radiometry, data from graphs Fried (2003)
	2790	1500	–	–	–	
	9600	6500	–	–	–	
	10600	800	–	–	–	
Enamel	2940	800	–	–	–	
	2790	400	–	–	–	
	9600	8000	–	–	–	
	10600	800	–	–	–	

Tissue	λ (nm)	μa (cm⁻¹)	μs (cm⁻¹)	μs' (cm⁻¹)	g	Remarks
Dentin	2790	988(111)	—	—	—	Transmission measurements Fried (2003)
	10300	1198(104)	—	—	—	
	10600	813(63)	—	—	—	
Enamel	2940	768(27)	—	—	—	
	2790	451(29)	—	—	—	
	10300	1168(49)	—	—	—	
	10600	819(62)	—	—	—	
Uterus	635	0.35(0.1)	394(91)	122	0.69	Frozen sections Marchesini et al. (1989)
Vein (femoral)	1064	3.2	487	—	0.97	Roggan et al. (1995)

Ex vivo measurements

Tissue	λ (nm)	μa	μs	μs'	g	Remarks
Aorta:						
Normal (n = 4)	1300	—	150–360	—	0.9–1	Optical coherence tomography (OCT); 4 hr of autopsy; $g_{\mathrm{eff}} = \cos\theta_{\mathrm{rms}}, \theta_{\mathrm{rms}} -$ rms scattering angle, $g_{\mathrm{eff}} \geq g$ Levitz et al. (2004)
Lipid rich (n = 4)	1300	—	0–200	—	0.6–1	
Fibrous (n = 3)	1300	—	50–400	—	0.6–1	
Fibrocalcific (n = 3)	1300	—	0–200	—	0.8–1	
Fat:						
Abdominal (n = 2)	360	3.12(0.78)	—	32.59(3.26)	—	IS, IAD; tissue slabs, <12 hr after surgery, stored at 4 °C in saline; measurements at room temperature; in the spectral range 600–1600 nm: $\mu'_s = 1.23 \times 10^3 \lambda^{-0.59}$, [λ] in nanometers Bashkatov et al. (2005b)
	500	2.37(0.59)	—	28.23(2.82)	—	
	600	1.90(0.47)	—	28.58(2.86)	—	
	800	1.87(0.47)	—	25.74(2.57)	—	
	1000	1.77(0.44)	—	20.14(2.01)	—	
	1200	1.79(0.45)	—	17.55(1.76)	—	
	1400	1.75(0.44)	—	17.15(1.71)	—	
	1600	1.47(0.37)	—	16.41(1.64)	—	
	1800	1.92(0.48)	—	17.40(1.74)	—	
	1900	2.48(0.62)	—	20.38(2.04)	—	
	2200	1.65(0.41)	—	18.95(1.89)	—	

(continued overleaf)

Tab. 16.1 (Continued)

Tissue	λ, nm	μ_a, cm^{-1}	μ_s, cm^{-1}	μ_s', cm^{-1}	g	Remarks
Subcutaneous (n = 6)	400	2.26(0.24)	–	13.39(2.78)	–	IS, IAD; tissue slabs, 1–3 mm; <6 hr after surgery; stored at 20 °C in saline; measurements at room temperature; in the spectral range 600–1500 nm: $\mu_s' = 1.05 \times 10^3 \lambda^{-0.68}$, [λ] in nanometers Bashkatov et al. (2005a)
	600	1.18(0.02)	–	13.39(4.65)	–	
	800	1.07(0.11)	–	11.62(4.63)	–	
	1000	1.06(0.06)	–	9.39(3.32)	–	
	1200	1.06(0.07)	–	7.91(3.17)	–	
	1400	1.08(0.03)	–	7.51(3.31)	–	
	1600	0.89(0.04)	–	7.16(3.21)	–	
	1800	1.21(0.01)	–	7.50(3.48)	–	
	1900	1.62(0.06)	–	8.72(4.15)	–	
	2000	1.43(0.09)	–	8.24(4.03)	–	
Mucous of maxillary sinus at antritis	400	4.89(0.92)	–	36.01(6.41)	–	IS, IAD; tissue slabs, 1–2 mm; <6 hr after surgery; stored at 20 °C in saline; measurements at room temperature; in the spectral range 600–1300 nm: $\mu_s' = 4.4 \times 10^5 \lambda^{-1.62}$, [λ] in nanometers Bashkatov et al. (2005a)
	600	0.45(0.23)	–	13.81(2.43)	–	
	800	0.13(0.16)	–	9.79(1.68)	–	
	1000	0.27(0.21)	–	6.14(0.74)	–	
	1200	0.57(0.31)	–	4.43(0.43)	–	
	1400	4.84(1.79)	–	5.07(0.71)	–	
	1600	2.83(1.01)	–	3.13(0.55)	–	
	1800	3.04(1.15)	–	3.04(0.57)	–	
	1900	9.23(2.69)	–	7.01(3.57)	–	
	2000	9.31(2.28)	–	6.26(3.56)	–	

Sample	λ (nm)	μ_s'	μ_s	g	Notes
Skin:					
Caucasian dermis ($n = 12$)	633	0.33(0.09)	—	27.3(5.4)	0.9	A single integrating sphere "comparison" method, IMC; samples from abdominal and breast tissue obtained from plastic surgery, $g = 0.9$ is supposed value in calculations Laufer et al. (1998); Simpson et al. (1998)
	700	0.19(0.06)	—	23.2(4.1)	0.9	
	900	0.13(0.07)	—	16.3(2.5)	0.9	
Negroid dermis ($n = 5$)	633	2.41(1.53)	—	32.1(20.4)	0.9	
	700	1.49(0.88)	—	26.8(14.1)	0.9	
	900	0.45(0.18)	—	18.1(0.4)	0.9	
Subdermis (primarily globular fat cells) ($n = 12$)	633	0.13(0.05)	—	12.6(3.4)	0.9	
	700	0.09(0.03)	—	12.1(3.2)	0.9	
	900	0.12(0.04)	—	10.8(2.7)	0.9	
Muscle ($n = 1$)	633	1.21	—	8.9	0.9	
	700	0.46	—	8.3	0.9	
	900	0.32	—	5.9	0.9	
Sample/subject −01/01; female (F), age = 51 yr; back of knee, left leg; moderate inflammation in dermis; SC = 40–70 μm; E = 40–150 μm; D = 300 μm	1460	17.88(1.12)	—	10.74(0.49)	—	DIS, IAD; slabs containing stratum corneum (SC), epidermis (E), and dermis (D), taken from 14 subjects; measured within 24 hr of excision; heated to 37 °C; three measurements on each side of the sample; 2.5-cm-diameter sample ports on the setup; total data for 52 wavelengths in the range 1000–2200 nm are available Troy and Thennadil (2001)
	1600	5.35(0.24)	—	8.06(0.29)	—	
	2200	7.46(0.56)	—	7.17(0.26)	—	
07/04; male (M), age = 64 yr; thigh, right leg; mild chronic dermatitis; SC = 20–30 μm; E = 50–90 μm; D = 300 μm	1000	0.69(0.01)	—	10.45(0.61)	—	
	1460	16.64(0.95)	—	10.75(0.81)	—	
	1600	4.96(0.27)	—	7.72(0.40)	—	
	2200	13.04(2.36)	—	9.42(1.57)	—	

(continued overleaf)

Tab. 16.1 (Continued)

Tissue	λ, nm	μ_a, cm^{-1}	μ_s, cm^{-1}	μ_s', cm^{-1}	g	Remarks
15/10; M, age = 70 yr; scalp; mild chronic dermatitis w/solar elastosis; SC = 4–15 μm; E = 8–10 μm; D = 200 μm	1000	1.04(0.02)	—	12.26(0.44)	—	
	1460	15.95(0.99)	—	10.75(1.20)	—	
	1600	5.09(0.23)	—	8.83(0.92)	—	
	2200	12.65(0.52)	—	8.83(1.94)	—	
20/13; F, age = 53 yr; scalp/facial tissue; mild solar damage; SC = 4 μm; E = 10 μm; D = 200 μm	1000	1.53(0.02)	—	12.89(0.77)	—	
	1460	16.82(1.13)	—	12.01(0.81)	—	
	1600	5.57(0.19)	—	9.47(0.60)	—	
	2200	13.46(0.58)	—	10.41(0.71)	—	
21/14; F, age = 52 yr; abdomen; mild chronic inflammation; SC = 4–5 μm; E = 10 μm; = 200 μm	1000	0.88(0.03)	—	14.96(1.28)	—	
	1460	18.21(2.51)	—	14.20(0.71)	—	
	1600	5.74(0.68)	—	10.58(0.44)	—	
	2200	11.33(0.76)	—	10.40(0.47)	—	
Uterus:						
Postmenopausal	630	0.515(0.054)	—	9.1(1.7)	—	Frequency-domain (FD); intact uteri were obtained by hysterectomy; during measurement period (3–4 hr) wet gauze was applied Madsen et al. (1994)
Premenopausal	630	0.193(0.013)	—	7.3(0.9)	—	
	630	0.314(0.030)	—	8.9(1.5)	—	
	630	0.213(0.024)	—	6.0(0.8)	—	
Fibroid	630	0.197(0.030)	—	7.3(1.5)	—	
	630	0.0824(0.0075)	—	7.2(0.9)	—	

16.3 Measurements of Optical Properties of Tissues and Blood

In vivo measurements

Tissue	λ				Ref.
Adenocarcinoma (multiple subcutaneous large-cell, male 62 yrs):					FD, r_{sd} = 2.2 cm Fishkin et al. (1997); Tromberg et al. (1997)
Abdominal, normal tissue	674	0.0589(0.0036)	—	8.94(0.19)	—
	811	0.0645(0.0032)	—	8.82(0.18)	—
	849	0.0690(0.0025)	—	8.77(0.14)	—
	956	0.1110(0.015)	—	7.00(0.62)	—
Abdominal, tumor	674	0.169(0.02)	—	8.48(0.73)	—
	811	0.190(0.015)	—	8.30(0.49)	—
	849	0.276(0.03)	—	9.93(0.87)	—
	956	—	—	—	—
Back, normal tissue	674	0.0883(0.006)	—	10.7(0.4)	—
	811	0.0892(0.005)	—	9.99(0.27)	—
	849	0.0915(0.0030)	—	9.65(0.15)	—
	956	0.127(0.03)	—	6.3(0.9)	—
Back, tumor	674	0.174(0.02)	—	10.4(0.9)	—
	811	0.177(0.013)	—	9.23(0.5)	—
	849	0.190(0.01)	—	9.20(0.33)	—
	956	0.186(0.16)	—	4.7(2.7)	—
Brain:					
Normal cortex, temporal and frontal lobe	674	>0.2	—	10(1)	0.92	SRR; measurements during brain surgery Bevilacqua et al. (1997)
	849	>0.2	—	9.2(1)	0.92
	956	>0.2	—	8.5(1)	0.92
Normal optic nerve	674	0.60(0.25)	—	18(1)	0.92
	849	0.75(0.25)	—	17(1)	0.92
	956	0.65(0.25)	—	16(1)	0.92
Astrocytoma of optic nerve	674	1.6(1)	—	14(1)	0.92
	849	1.1(1)	—	8.5(1)	0.92
	950	1.8(1)	—	8.5(1)	0.92

(continued overleaf)

Tab. 16.1 (Continued)

Tissue	λ, nm	μ_a, cm^{-1}	μ_s, cm^{-1}	μ_s', cm^{-1}	g	Remarks
Normal cortex, frontal lobe	674	<0.2	–	10(0.5)	–	SRR; measurements during brain surgery Bevilacqua et al. (1999); Mobley and Vo-Dinh (2003)
	811	<0.1	–	9.1(0.5)	–	
	849	<0.1	–	9.2(0.5)	–	
	956	0.15(0.1)	–	8.9(0.5)	–	
Normal cortex, frontal lobe	674	0.2(0.1)	–	10(0.5)	–	
	811	0.2(0.1)	–	8.2(0.5)	–	
	849	<0.1	–	8.2(0.5)	–	
	956	0.25(0.1)	–	8.2(0.5)	–	
Normal optic nerve	674	0.6(0.3)	–	17.5(2)	–	
	849	0.8(0.3)	–	16(2)	–	
	956	0.7(0.3)	–	15.2(2)	–	
Astrocytoma of optic nerve	674	1.4(0.3)	–	12.5(1)	–	
	811	1.2(0.3)	–	9.5(1)	–	
	849	0.9(0.3)	–	7.6(1)	–	
	956	1.5(0.3)	–	7.3(1)	–	
Normal white matter	674	2.5(0.5)	–	13.5(1)	–	
	849	0.95(0.2)	–	8.5(1)	–	
	956	0.9(0.2)	–	7.8(1)	–	
White matter with scar	674	<0.2	–	6.5(0.5)	–	
	849	<0.2	–	8(0.5)	–	

Tissue	λ (nm)	μ_a (cm⁻¹)	μ_s (cm⁻¹)	μ_s' (cm⁻¹)	Remarks, Reference
Medulloblastoma	674	2.6(0.5)	—	14(1)	
	849	1(0.2)	—	10.7(1)	
	956	0.75(0.2)	—	4(1)	
Breast (female):					
Normal (30 Japanese women, averaged for all ages)	753	0.046(0.014)	—	8.9(1.3)	Time domain (TD), μ_a (cm⁻¹) ≈0.087–8.31 × 10⁻⁴x, μ_s' (cm⁻¹) ≈13–0.08x, where x = age (20–80 yr) Suzuki et al. (1996)
Normal (6 women, 26–43 yr)	800	0.017–0.045	—	7.2–13.5	TD, μ_s' (cm⁻¹) ≈16.7–7.9 × 10⁻³λ, λ = 500–1060 nm Heusmann Kolzer and Mitic (1996)
Normal (6 women, tissue thickness, 33–49 mm at light compression)	580	0.70(0.12)	—	—	Measurements of transmission, g ≈ 0.92–0.95, μ_s' = 12–13 cm⁻¹ Key et al. (1991)
	780	0.23(0.02)	—	—	
	850	0.27(0.03)	—	—	
Breast cancer (5 patients)	630	0.305(0.16)	—	9.41(7.35)	SRR; relapsed cancer, HPD (72 hr) Driver, Lowdell and Ash (1991)

(continued overleaf)

Tab. 16.1 (Continued)

Tissue	λ, nm	μ_a, cm^{-1}	μ_s, cm^{-1}	μ_s', cm^{-1}	g	Remarks
Normal (56 yr)	674	0.04	–	8.5	–	FD, $r_{sd} = 2.2$ cm Fishkin et al. (1997); Tromberg et al. (1997)
	811	0.035	–	7.4	–	
	849	0.035	–	7.0	–	
	956	0.085	–	6.5	–	
Fibroadenoma with ductal hyperplasia (56 yr)	674	0.055	–	9	–	
	811	0.06	–	8	–	
	849	0.055	–	7.6	–	
	956	0.12	–	7.5	–	
Normal (27 yr)	674	0.035	–	11.1	–	
	811	0.03	–	9.6	–	
	849	0.038	–	9.6	–	
	956	0.09	–	9.7	–	
Fluid-filled cyst (27 yr)	674	0.07	–	7.9	–	
	811	0.07	–	7.0	–	
	849	0.08	–	7.0	–	
	956	0.16	–	7.0	–	
Papillary cancer (55 yr)	690	0.084(0.014)	–	15.0(0.3)	–	FD, DT Fantini et al. (1998)
	825	0.085(0.017)	–	12.7(0.3)	–	
Normal (67 yr)	674	0.057	–	9.5	–	FD, diffusion approximation; tumor 1.8 × 0.9 cm; segmented reconstruction Holboke et al. (2000)
	782	0.050	–	9.4	–	
	803	0.047	–	9.0	–	
	849	0.054	–	8.9	–	
Ductal carcinoma in situ (67 yr)	674	0.17	–	4.1	–	
	782	0.18	–	3.6	–	
	803	0.15	–	4.2	–	
	849	0.21	–	3.3	–	
Calf (11 subjects, 14 measurements)	800	0.17(0.05)	–	9.4(0.7)	–	μ_s'(cm^{-1}) $\approx 16 - 8.9 \times 10^{-3}\lambda$, $\lambda = 760$–900 nm Matcher, Cope and Delpy (1997)

16.3 Measurements of Optical Properties of Tissues and Blood

Tissue	λ (nm)					Reference / Notes
Cervical stromal tissue	849	0.34	61.1	–	0.9	Hayakawa et al. (2001)
Cervical tissue:					μ_{bs}	$\mu_{bs} = \mu_s p_b$, p_b – probability of backscattering
Epithelium ($n = 36$)	1300	–	10–140	–	0.1–11	OCT, two-layered model, genetic inverse algorithm Turchin et al. (2003)
Stroma ($n = 36$)	1300	–	30–290	–	1.5–12	
Dysplasia II–III	1300	–	40–65	–	1.4–3.6	OCT, single-layered model, genetic inverse algorithm Dolin et al. (2004)
Leukoplakia	1300	–	16–32	–	1.3–2.0	
Epithelium	1300	–	80(25)	–	0.28(0.08)	OCT, genetic inverse algorithm two-layered model single-layered model Dolin et al. (2004)
Stroma	1300	–	210(30)	–	3.0(2)	
Cancer	1300	–	300(20)	–	1.2(0.6)	
Gastrointestinal tract:					$\gamma = (1 - g_2)/(1 - g_1)$	
Mucosa in the antrum	500	2.5(0.8)	–	16.8(3.4)	1.98(0.20)	Endoscopic SRR, r_{sd} from 0.3 to 1.35 mm; IMC, two-moments HGPF; 35 patients (21 females and 14 males, age from 23 to 87), for each patient, four sites were usually selected – two in the antrum and two in the fundus, average data for normal tissue Thueler et al. (1993)
	550	3.6(1.3)	–	13.8(3.4)	1.93(0.15)	
	600	1.0(0.6)	–	12.8(2.1)	1.90(0.12)	
	650	0.5(0.5)	–	11.7(1.8)	1.87(0.12)	
	700	0.4(0.4)	–	10.7(1.6)	1.86(0.12)	
	750	0.45(0.45)	–	9.7(1.6)	1.85(0.12)	
	800	0.5(0.4)	–	9.0(1.6)	1.85(0.12)	
	850	0.7(0.5)	–	8.8(1.6)	1.85(0.12)	
	900	0.8(0.5)	–	8.3(1.5)	1.86(0.12)	

(continued overleaf)

Tab. 16.1 (Continued)

Tissue	λ, nm	μ_a, cm^{-1}	μ_s, cm^{-1}	μ_s', cm^{-1}	g	Remarks
Mucosa in the fundus	500	3.3(0.8)	–	21.7(3.1)	2.12(0.26)	
	550	4.4(1.5)	–	18.6(3.1)	2.04(0.24)	
	600	1.8(0.5)	–	16.5(2.1)	2.00(0.19)	
	650	0.8(0.4)	–	15.0(2.1)	1.97(0.17)	
	700	0.7(0.4)	–	13.8(1.7)	1.98(0.15)	
	750	0.7(0.5)	–	12.4(1.6)	1.98(0.15)	
	800	0.7(0.3)	–	11.7(1.6)	1.95(0.12)	
	850	0.7(0.3)	–	11.0(1.4)	1.92(0.12)	
	900	0.8(0.4)	–	10.3(1.4)	1.94(0.12)	
Head (7 subjects, 10 measurements)	800	0.16(0.01)	–	9.4(0.7)	–	μ_s'(cm^{-1}) ≈ 14.5 − 6.5 × 10^{-3}λ, λ = 760–900 nm Matcher, cope and Delpy (1997)
Forearm (5 subjects, 14 measurements)	800	0.23(0.04)	–	6.8(0.8)	–	TD, μ_s'(cm^{-1}) ≈ 11 − 5.1 × 10^{-3}λ, λ = 760–900 nm Troy, Page and Sevick-Muraca (1996)
Forearm	715	0.18	–	3.7	–	FD; r_{sd} – a few centimeters Gratton et al. (1997)
	825	0.24	–	3.0	–	
Forehead	715	0.16	–	7.3	–	
	825	0.16	–	6.9	–	
Skin:						
Dermis	660	0.07–0.2	–	9 – 14.5	–	SRR, IMC Graaff et al. (1993)
Skin	633	0.62	–	32	–	Doornbos et al. (1999)
	700	0.38	–	28.7	–	
Skin (0–1 mm)	633	0.67	–	16.2	–	Kienle, Lilge and Patterson (1994)
Skin (1–2 mm)	633	0.026	–	12.0	–	
Skin (>2 mm)	633	0.96	–	5.3	–	
Epidermis and dermis	750	0.375	–	15	–	SRR; DT Farrell, Patterson and Essenpreis (1998)
Subcutaneous fat	750	0.03	–	10	–	

Tissue	λ (nm)	μ_a (cm⁻¹)	μ_s' (cm⁻¹)	Remarks, Reference
Arm	633	0.17(0.01)	—	SRR, 9 detecting 600-μm fibers; mean separation 1.7 mm; Mie phase function Doornbos et al. (1999)
	660	0.128(0.005)	—	
	700	0.090(0.002)	—	
Foot sole	633	0.072(0.002)	—	
	660	0.053(0.003)	—	
	700	0.037(0.001)	—	
Forehead	633	0.090(0.009)	—	
	660	0.052(0.003)	—	
	700	0.0240(0.002)	—	
Abdominal skin:				
chosen direction perpendicular direction (along collagen fibers)	810	0.014	—	SRR; CCD detector, $r_{sd} \leq 10$ mm; diffusion approximation Nickell et al. (2000)
	810	0.07	—	
Forearm (light skin, $n = 7$):				
skin temperature −22 °C	590	2.372(0.282)	9.191(0.931)	SRR; MC-generated grid Khalil et al. (2003)
	750	0.966(0.110)	7.340(0.901)	
	950	0.981(0.073)	6.067(0.847)	
skin temperature −38 °C	590	2.869(0.289)	9.613(0.894)	
	750	1.157(0.106)	7.649(0.971)	
	950	1.135(0.123)	6.234(0.928)	
Dermis of a lower arm	1300	—	47	OCT (Schmitt, Knüttel and Bonnar (1993)
Stratum corneum of finger	1300	—	12	

(continued overleaf)

Tab. 16.1 (Continued)

Tissue	λ, nm	μ_a, cm^{-1}	μ_s, cm^{-1}	μ_s', cm^{-1}	g	Remarks
Volar side of lower arm:						OCT Knüttel and Boehlau-Godau (2000)
Epidermis	1300	–	15–20	–	–	
upper dermis	1300	–	80–100	–	–	
Palm of hand:						
stratum corneum	1300	–	10–15	–	–	
epidermis:						
grandular layer	1300	–	60–70	–	–	
basal layer	1300	–	40–50	–	–	
upper dermis	1300	–	50–80	–	–	
Volar side of lower arm (epidermis and dermis)						OCT; depth up to 350 μm; skin treated with a detergent solution (2% of anionic tensides in water) Knüttel, Bonev and Knaak (2004))
Normal	1300	–	140	–	–	
Treated	1300	–	80	–	–	
Skull	674	<0.2	–	9(1)	–	SRR; measurements during brain surgery Bevilacqua et al. (1999)
	849	<0.1	–	9(1)	–	
	956	0.15(0.1)	–	8.5(1)	–	

where water absorption is weak (up to 1100–1300 nm). The refractive index microscopic redistribution of a tissue due to cellular and fiber proteins' denaturation and homogenization at thermal action also may have a strong inclusion in alteration of scattering and absorption properties.

16.3.10
Optical Properties of Blood

Fresh human blood placed in calibrated thin cuvette (thickness from 0.01 to 0.5 mm, slab geometry) is usually used for the determination of blood optical parameters. Before the optical measurements, standard clinical tests are necessary to determine the concentration of red and white blood cells, concentration of platelets, hematocrit, mean corpuscular volume and hemoglobin, and the other parameters of interest. If blood sample oxygenation level is of interest, it may be controlled using a conventional blood gas analyzer (Yaroslavsky et al., 1999). In most cases, the experiments are performed with either completely oxygenated or completely deoxygenated blood (Hammer et al., 1998; Roggan et al., 1999; Yaroslavsky et al., 1999; Tuchin, 2002). To obtain complete oxygen saturation, the sample is exposed to air or O_2. To completely deoxygenate blood, sodium dithionite ($Na_2S_2O_4$) is added. To be sure that neither the volume nor the surface area of the blood particles changes during the experiments, the pH of the samples should be maintained in the range of physiological values, at approximately 7.4.

In reality, blood is flowing through the blood vessels and, therefore, it is preferable to study the optical properties of flowing blood. The red blood cells (RBCs) in flow are subject to deformation and orientation. At lower shear rates, reversible aggregation occurs; while under the higher shear rates, erythrocytes are deformed into ellipsoids. The experiments with flowing undiluted and diluted blood are reported (Nilsson et al., 1997; Roggan et al., 1999; Steenbergen, Kolkman and de Mul, 1999).

16.3.11
Refractive Index Measurements

Refractivity measurements in a number of strongly scattering tissues and blood were performed using various techniques (see Tuchin, 2007). Experimental values of mean refractive index for some tissues measured for selected wavelengths are summarized by Tuchin (2007). One of the techniques is a fiber-optic refractometer based on a simple concept that the cone of light issuing from an optical fiber is dependent on the indices of the cladding material (tissue) (Bolin et al., 1989). Using this simple and sensitive technique, it was found that at 633 nm, fatty tissue has the largest refractive index (1.455), followed by kidney (1.418), muscular tissue (1.410), then blood and spleen (1.400). The lowest refractive indices were found in lungs and liver (1.380 and 1.368, respectively). Also, it turned out that tissue homogenization does not significantly affect the refractive indices (the change does not exceed a measurement error equal to 0.006), whereas coagulated tissues have higher refractive indices than native ones (e.g., for egg white, $\bar{n}$ changing from 1.321 to 1.388). Moreover, there is a tendency for refractive indices to decrease with increasing light wavelength from 390 to 700 nm (e.g., for bovine muscle in the limits 1.42–1.39), which is characteristic of the majority of related abiological materials.

The principle of total internal reflection at laser beam irradiation is also used for tissue and blood refraction measurements (Liu and Xie, 1996; Cheng et al., 2002). A thin tissue sample is sandwiched between two right-angled prisms that are made of ZF5 glass with a high refractive index $n_0 = 1.70827$ and angle $\alpha = 29°55'41.4''$. Measurements for fresh animal tissues and human blood at four laser wavelengths 488, 632.8, 1079.5, and 1341.4 nm and room temperature were presented in a form of Cauchy dispersion equation as (Cheng et al., 2002)

$$\bar{n} = A + B\lambda^{-2} + C\lambda^{-4} \qquad (16.40)$$

with λ in nanometers, values of the Cauchy coefficients are given in the following table (Table 16.2):

An expression for human blood plasma received by Cheng et al. (2002) was extrapolated to shorter wavelengths from 400 to 1000 nm (Tuchin et al., 2004):

$$n_{bp}(\lambda) = 1.3254 + 8.4052 \times 10^3 \lambda^{-2}$$
$$- 3.9572 \times 10^8 \lambda^{-4}$$
$$- 2.3617 \times 10^{13} \lambda^{-6} \qquad (16.41)$$

For modeling of the behavior of refractive index of tissues, blood, and their components, one may use a remarkable property of proteins: that equal concentrations of aqueous solutions of different proteins all have approximately the same refractive index, n_{pw} (Barer, Ross and Tkaczyk, 1953). Moreover, the refractive index varies almost linearly with concentration, C_p

$$n_{pw}(\lambda) = n_w(\lambda) + \beta_p(\lambda) \cdot C_p \qquad (16.42)$$

Tab. 16.2 Values of the Cauchy coefficients for some tissues.

Tissue sample	A	B × 10^{-3}	C × 10^{-9}
Porcine muscle$^\|$	1.3694	0.073223	1.8317
Porcine muscle$^\perp$	1.3657	1.5123	1.5291
Porcine adipose	1.4753	4.3902	0.92385
Porcine small intestine	1.3563	4.3905	0.92379
Ovine muscle$^\|$	1.3716	5.8677	0.43999
Ovine muscle$^\perp$	1.3682	8.7456	−0.16532
Human whole blood	1.3587	1.4744	1.7103
Human blood plasma	1.3194	14.578	−1.7383

$^\|$ Porcine (or ovine) muscle samples with the tissue fibers oriented in parallel to the interface.
$^\perp$ Same sample with the tissue fibers oriented perpendicular to the interface.

where n_w is refractive index of water and β_p is the specific refractive increment; C_p is measured in grams per 100 ml (grams per deciliter). For example, the refractive index of human erythrocyte cytoplasm, defined by the cell-bounded hemoglobin solution, can be found from this equation at $\beta_p = 0.001942$ valid for a wavelength of 589 nm; that is, for normal hemoglobin concentration in cytoplasm of 300–360 g l^{-1}, the RBC refractive index $n_{RBC} = 1.393$–1.406 (Roggan et al., 1999). Values of specific refractive increment β_p for some other proteins measured by Abbe refractometer at a wavelength 589 nm are given in the following table (Table 16.3) (Barer, Ross and Tkaczyk, 1953):

Other materials of specific biological interest are the carbohydrates, lipids, and nucleic acid compounds. The first two usually have low values of β; in

Tab. 16.3 Values of specific refractive increment β_p for some proteins measured by Abbe refractometer at a wavelength 589 nm.

Protein	β_p, dl/g
Lipoprotein	0.00170–0.00171
Total serum (human)	0.00179
Total albumin	0.00181
Egg albumin	0.001813
Euglobulin	0.00183

the region of 0.0014–0.0015, and nucleic acids have higher values, 0.0016–0.0020 (Barer, Ross and Tkaczyk, 1953).

Some other techniques of refractive index measurements of biological liquids are also available. One of them allows for refractive index measurements at a few separate wavelengths and uses an equilateral small angle (of 10°) hollow prism made from thin quartz slides (Sardar and Levy, 1998).

Since the refractive index of tissue and blood components defines their scattering properties, measured scattering parameters may have an advantage to evaluate the refractive index of biomaterial components and their mean values (Tuchin, 2007). Determination of the reduced scattering coefficient of a tissue sample using IS or spatially resolved techniques and corresponding algorithms for extraction of the scattering coefficient, such as IAD or IMC, the knowledge of the refractive indices of the scatterers and the ground material at one of the wavelengths, as well as experimental or theoretical estimations for mean radius of the scatterers, allows one to solve the inverse problem and reconstruct the spectral dependence of the refractive index of the scatterers for a given spectral dependence of the refractive index of the ground material.

Similar measurements and theoretical estimations done for a tissue sample before and after its prolonged bathing in saline or other biocompatible liquid with known optical characteristics allow one to evaluate the spectral dependencies both of refractive index of the scatterers and the ground material.

A short pulse time delay technique was also successfully applied for refractive index estimation of normal breast tissue and malignant breast tissue (Das, Liu and Alfano, 1997). Using the known thickness of the sample and the measured shift Δt of the transmitted pulse peak relative to the delay time measured through a layer of air of the same thickness, the mean phase refractive index $\bar{n}$ of a tissue sample can be calculated. Very short pulses should be used in such measurements; thus, a group of different wavelengths propagates in a media and the material dispersion $(d\bar{n}/d\lambda)$ should be accounted for by introducing the group refractive index

$$\bar{n}_g = \bar{n} - \lambda \frac{d\bar{n}}{d\lambda} \qquad (16.43)$$

The time delay in the pulse arrival for a tissue sample of thickness d is

$$\Delta t = \frac{d}{c_0}(\bar{n}_{g1} - n_{g2}) \qquad (16.44)$$

where c_0 is the light velocity in a free space; $\bar{n}_{g1}$ is the effective (mean) group refractive index of a tissue, and n_{g2} is the group refractive index of the homogeneous reference medium (air). The effective group refractive index of a tissue is

$$\bar{n}_{g1} = f_s n_{gs} + (1 - f_s) n_{g0} \qquad (16.45)$$

where f_s is the volume fraction of the scatterers composing a tissue, n_{gs} is the group refractive index of the scatterers, and n_{g0} is the group refractive index of the ground material of a tissue. The values of the phase refractive index of the above-mentioned two samples were calculated to be $\bar{n} = 1.403$ for normal and 1.431 for malignant tissue.

OCT provides simple and straightforward measurements of the index of refraction both *in vitro* and *in vivo* (Tearney et al., 1995; Wang et al., 1995; Knüttel and Boehlau-Godau, 2000; Knüttel, Bonev and Knaak, 2004; Ohmi et al., 2000; Tuchin, Xu and Wang, 2002; Tuchin, 2007). The in-depth scale of OCT images is determined by the optical path length Δz_{opt} between two points along the depth direction. Because a broadband light source is used, the optical path length is proportional to the group refractive index n_g and geometrical path length Δz as

$$\Delta z_{opt} = n_g \Delta z \quad (16.46)$$

Usually, $n_g \cong n$. This simple relation is valid for a homogeneous medium and can be used in *in vitro* studies when geometrical thickness of a tissue sample Δz is known.

Sometimes both refractive index and thickness of a tissue sample should be measured simultaneously. In that case, a two-step procedure can be applied. First, a stationary mirror is placed in the sample arm of an interferometer to get the geometric position of the mirror supposing that the group refractive index of air is 1 (z_1). Then a tissue sample with unknown index n_g, and thickness d should be placed before the mirror in the sample arm. Two peaks from the anterior (z_2) and the posterior (z_3) surfaces of the sample will appear with the distance between them equal to a sample optical thickness, and the position of the mirror (z_4) will be shifted by $(n_g - 1)d$ due to the sample whose group refractive index is greater than that of air. Thus, the calculation of the geometrical thickness and the group refractive index proceeds as follows:

$$d = (z_3 - z_2) - (z_4 - z_1)$$
$$n_g = \frac{z_3 - z_2}{d} \quad (16.47)$$

For *in vivo* measurements of the index of refraction, a focus-tracking method that uses OCT to track the focal-length shift that results from translating the focus of an objective along the optical axis within a tissue was introduced (Tearney et al., 1995) and further developed (Knüttel and Boehlau-Godau, 2000; Knüttel, Bonev and Knaak, 2004; Zvyagin et al., 2003).

16.3.12
Optical Projection Tomography

Optical projection tomography (OPT), based on ballistic and/or quasi-ballistic photon detection, allows for imaging inhomogeneities and quantifying optical properties of biological materials with thicknesses from tens of microns to centimeters (Darrell et al., 2008; Oldham et al., 2008). Major applications are in developmental embryology and gene expression studies, as well as in three-dimensional imaging and quantification within organs of small animals and xenograft tumors. Typically, light from a uniform backlight transverses through the sample to form projection images captured by a CCD camera or fluorescence from fluorophores distributed in biological material and excited by a light orthogonal to the imaging axis is detected at rotation of the sample around its axis. There have been substantial developments in modeling, reconstructing algorithms, reflection, refraction and scattering reduction, and in enhancements of the optical systems used in OPT. As it was noted by Darell et al. (2008), several applications of OPT would benefit from quantitative reconstructions, that is, reconstructions in which, for example, the intensities of two areas of gene expression could be reliably compared, thus, they described methods of accounting for the quantitative effects of the isotropic emission and blurring in a fluorescence OPT imaging system.

Glossary

Absorption Coefficient: In a nonscattering medium absorption coefficient μ_a is defined as the reciprocal of the distance d over which light of intensity $I(d=0) = I_0$ is attenuated (due to absorption) to $I(d) = I_0/e \approx 0.37 I_0$; the units are typically cm^{-1}; behind this definition is a fundamental process of photon absorption that is characterized by a photon absorption cross section.

Anisotropic Scattering: A scattering process characterized by a clearly apparent direction of photons that may be due to the presence of large scatterers.

Aorta: The great artery that leaves the left ventricle; it conducts the whole of the arterial blood supply to all parts of the body other than the lungs; in humans, it carries blood at the rate of 4 dm^3 per minute.

Astrocytoma: A rather well-differentiated glioma (a *tumor* of the brain), which consists of cells that look like astrocytes (a starlike cell of the macroglia of nerve tissue).

Bilirubin: Yellow bile pigment formed in the breakdown of heme (an iron-containing substance; the basic unit of the *hemoglobin* molecule; mammals have four heme units in their hemoglobin).

Cancer: A general term applied to a carcinoma or a sarcoma; the typical symptoms are a *tumor* or swelling, a discharge, pain, an upset in the function of an organ, general weakness, and loss of weight.

Cartilage: A strong, resilient, skeletal tissue; its simplest and most common form consists of a matrix of a polysaccharide-containing protein in which are embedded cartilage cells (chondroblasts); the matrix is without structure and without blood vessels; this type is known as *hyaline cartilage*, is translucent and clear, and occurs in the cartilaginous rings of the trachea and bronchi; elastic cartilage (yellow fibrocartilage) contains yellow fibers in the matrix; it occurs in the external ear and in the epiglottis; white fibrocartilage contains white fibers in the matrix; it occurs in the disks of cartilage between the vertebrae; all types of cartilage contain chondroblasts, which deposit the matrix and become enclosed in the matrix as chondrocytes.

CCD: Charge-coupled device – a solid-state electronic device that serves as an imaging chip and is used in video cameras and fast spectrometers.

Cell Organelle: A part of a cell that is a structural and functional unit, for example, a flagellum is a locomotive organelle, a *mitochondrion* is a respiratory organelle; organelles in a cell correspond to organs in an organism.

Cerebellum: A region of the brain that plays an important role in the integration of sensory perception and motor output; many neural pathways link the cerebellum with the motor cortex, which sends information to the muscles causing them to move and the spinocerebellar tract that provides feedback on the position of the body in space; the cerebellum integrates these pathways, using the constant feedback on body position to fine-tune motor movements.

Cerebral Membrane (*dura mater*): Or pachymeninx, is the tough and inflexible outermost of the three layers of the *meninges* surrounding the brain and spinal cord; the *dura mater* itself has two layers: a superficial layer, which is actually the skull's inner *periosteum*, and a deep layer, the *dura mater proper*.

Collagen: A tough, inelastic, fibrous protein; on boiling it forms gelatin; on adding acetic acid it swells up and dissolves; collagen is formed and maintained in tissues by fibroblasts; it forms white fibers in *connective tissue*; the tropocollagen or "collagen molecule" subunit is a rod about 300-nm long and 1.5 nm in diameter, made up of three polypeptide strands, subunits spontaneously self-assemble, with regularly staggered ends, into even larger arrays in the extracellular spaces of tissues; there is some covalent crosslinking within the triple helices, and a variable amount of covalent crosslinking between tropocollagen helices, to form the different types of collagen found in different mature tissues – similar to the situation found with the α-keratins in hair; a distinctive feature of collagen is the regular arrangement of amino acids in each of the three chains of these collagen subunits; in bone, entire collagen triple helices lie in a parallel, staggered array; 40-nm gaps between the ends of the tropocollagen subunits probably serve as nucleation sites for the deposition of long, hard, fine crystals of the mineral component, which is (approximately) hydroxyapatite with some phosphate; it is in this way that certain kinds of cartilage turn into bone; collagen gives bone its elasticity and contributes to fracture resistance.

Connective Tissue: Various body tissues that bind together and support organs and other tissues, for example, connective tissue surrounds muscles and nerves, connects bones and muscles and underlies the *skin; cartilage* and bone are also connective tissues; typical connective tissue consists of cells scattered in an amorphous mucopolysaccharide matrix in which there are varying amounts of connective tissue fibers (mainly *collagen*, but also *elastin* and *reticulin*)

CW: Continuous wave

Cytoplasm: All the protoplasm of a cell exclusive of the nucleus; it is not just a simple, slightly viscous, fluid; various structures, called *organelles*, each concerned with different functions of the cell, are situated in it; the plasma membrane is part of the cytoplasm.

Cytoplasmic Organelles: See *cell organelle*.

Dermis: The inner layer of the *skin*; it is composed of *connective tissue*, blood and lymph vessels, muscles and nerves; *collagen* fibers are abundant in the dermis and run parallel to the surface of the skin; they give the skin elasticity; sweat glands and hair follicles are scattered throughout the dermis; the dermis is much thicker than the epidermis and is developed from mesoderm.

Diffusion Approximation (Diffusion Theory): The approximated diffusion-type solution of the *RTT*, which is accurate for describing of photon migration in infinite, homogeneous, highly scattering media.

Dysplasia: An abnormality in the appearance of cells indicative of an early step toward transformation into a neoplasia. It is therefore a preneoplastic or precancerous change; this abnormal growth is restricted to the originating system or location, for example, a displasia in the epithelial layer will not invade into the deeper tissue, or a displasia solely in a red blood cell line (refractory anemia) will stay within the bone marrow and cardiovascular systems; the best known form of displasia is the precursor lesions to cervical *cancer*, called cervical intraepithelial neoplasia (CIN); this lesion is usually caused by an infection with the human papilloma virus (HPV).

Elastin: An elastic fibrous protein resistant to boiling and to acetic acid; it forms highly elastic yellow fibers in *connective tissue*; elastin is formed and maintained in tissues by fibroblasts.

Epidermis: The outer layer of the *skin* is a stratified *epithelium* that varies relatively little in thickness over most of the body (between 75 and 150 µm), except on the palms and soles, where its thickness may be 0.4–0.6 mm; the epidermis is conventionally subdivided into (i) stratum basale, a basal cell layer of keratinocytes, which is the germinative layer of the epidermis; (ii) the stratum spinosum, which consists of several layers of polyhedral cells lying above the germinal layer; (iii) the stratum granulosum, which is a layer of flattened cells containing distinctive cytoplasmic inclusions and keratohyalin granules; and (iv) the overlying *stratum corneum*, consisting of lamellae of anucleate thin, flat squames that are terminally differentiated keratinocytes.

Epithelium: A sheet of epithelial tissue; epithelium is derived from ectoderm and endoderm.

Eye Cornea: The transparent covering at the front of the eyeball; it is the modified continuation of the *sclera*; it refracts light and is the most important element in the refractive system of the eye.

Eye Sclera: A dense, white matter, fibrous membrane which, with the *cornea*, forms the external covering of the eyeball; scleral regions: limbal, equatorial, and posterior pole region.

Extinction Coefficient (Attenuation Coefficient): The reciprocal of the distance over which light of intensity I is attenuated to $I/e \approx 0.37I$; attenuation or interaction coefficient, $\mu_t = \mu_a + \mu_s$, where μ_a is the *absorption coefficient* and μ_s is the *scattering coefficient*; the units are typically cm^{-1}.

Ex vivo: Taken from the living organism; pertaining to experiments on animal or human organs that are excised from the living body and kept in conditions very close to the natural ones.

Fat (Adipose Tissue): A modification of areolar *tissue* in which globules of oil are deposited in some of the cells (fat cells); the cells tend to be grouped together and in mammals occur in the tissues under the skin and around the abdominal organs (kidneys, liver, etc.).

Fibrous Tissue: A tissue mainly consisting of conjunctive *collagen* (or *elastin*) fibers, often packed in lamellar bundles.

GKPF: Gegenbauer kernel phase function.

Golgi Apparatus: A netlike mass of material in the *cytoplasm* of animal cells, believed to function in cellular secretion.

Gray Matter: A nervous tissue found in the central nervous system; it contains numerous cell bodies (cytons), dendrites, synapses, terminal processes of axons, blood vessels, and neuroglia; it is internal to *white matter* in the spinal cord and some other parts of the brain; it is external to white matter in the cerebral hemisphere and in the *cerebellum*; coordination in the central nervous system is effected in gray matter; brain nuclei and nerve centers are composed of gray matter.

Hemoglobin: A red iron-containing respiratory pigment found in the blood; it conveys oxygen to the tissues and occurs in reduced form (deoxyhemoglobin) in venous blood and in combination with oxygen (oxyhemoglobin) in arterial blood; it consists of *heme* combined with globin, a blood protein; it is chemically related to chlorophyll, cytochrome, hemocyanin, and myoglobin.

HGPF: Henyey–Greenstein phase function.

Horny Layer: The same as stratum corneum, see *epidermis*.

Interstitial Fluid: A solution that bathes and surrounds the cells of multicellular animals; it is the main component of the extracellular fluid, which also includes plasma and transcellular fluid; on average, a subject has about 11 l of interstitial fluid providing the cells of the body with nutrients and a means of waste removal.

IAD: Inverse adding-doubling.

IMC: Inverse Monte Carlo.

In vitro: In medicine, pertaining to experiments on dead tissue.

In vivo: In medicine, pertaining to experiments on living animals and humans.

Isotropic Scattering: An equality of scattering properties along all axes.

Lysosomes: Membrane-bound particles, smaller than a *mitochondrion*, occurring in large numbers in the *cytoplasm* of cells; they contain hydrolytic enzymes that are released when the cell is damaged; these enzymes assist in the digestion and removal of dead cells, the digestion of food and other substances, and the destruction of redundant *organelles*.

Melanin: A dark-brown or black pigment; melanin in *melanosomes* of normal *skin* is an extremely dense, virtually insoluble polymer of high molecular weight and is always attached to a structural protein; mammalian melanin pigments have one of two chemical compositions: eumelanin, a brown polymer, and pheomelanin, a yellow–reddish alkali-soluble pigment.

Melanosome (Melanin Granular): The cytoplasmic **organelles** on which melanin pigments are synthesized and deposited; normal human skin color is primarily related to the size, type, color, and distribution of melanosomes; melanosomes are the product of specialized exocrine glands: melanocytes.

Meningioma: The most common benign *tumor* of the brain (95% of benign tumors); however, they can also be malignant; they arise from the arachnoidal cap cells of the meninges and represent about 15% of all primary brain tumors; they are more common in females than in males (2:1) and has a peak incidence in the sixth and seventh decades.

Mesentry: Sheets of thin **connective** *tissue* by which the stomach and intestines are suspended from the dorsal wall of the abdominal cavity; the mesenteries carry blood, lymph vessels, and nerves to the organs of the alimentary canal.

Mitochondrion (*pl.* Mitochondria): A threadlike, or rodlike, granular *organelle* in the *cytoplasm* of cells, about 0.5 μm in width, and up to 10 μm in length for threadlike mitochondria; mitochondria are bounded by a double membrane; the inner membrane is folded inward at a number of places to form cristae; mitochondria contain phosphates and numerous enzymes that vary in different tissues; their function is cellular respiration and the release of chemical energy in the form of adenosine triphosphate (ATP) for use in most of the cell's biological functions; the cells of all organisms, except bacteria and blue–green algae, contain mitochondria in varying numbers mitochondria are especially numerous in cells involved in significant metabolic activity, such as **liver** cells; mitochondria are self-replicating.

MFP: Mean free path of a photon.

Myocardium: The muscular substance of the heart.

OCT: Optical coherence tomography.

Percutaneous: Pertains to any medical procedure where access to inner organs or other tissue is done through the *skin*, for instance, via needle puncture of the skin, rather than by using an "open" approach where inner organs or tissue are exposed; phototherapy is another example of percutaneous treatment.

Peroxisome: A specialized *organelle* containing the oxidizing enzymes that degrade peroxides.

Photon Diffusion Coefficient: The proportionality coefficient between mean-square displacement of a photon within time interval τ: $\langle \Delta r^2 \rangle \sim D\tau$.

Pons: A structure located on the brain stem; in humans it is above the medulla, below the midbrain, and anterior to the cerebellum.

Reduced (Transport) Scattering Coefficient: A lumped property incorporating the *scattering coefficient* μ_s and the *scattering anisotropy parameter* g: $\mu_s' = \mu_s(1-g)$ (cm^{-1}); μ_s' describes the diffusion of photons in a random walk of step size of $1/\mu_s'$ (cm), where each step involves isotropic scattering; this is equivalent to description of photon movement using many small steps $1/\mu_s$ that each involve only a partial (anisotropic) deflection angle if there are many scattering events before an absorption event, that is, $\mu_a \ll \mu_s'$ (diffusion regime, see *diffusion approximation*); μ_s' is useful in the diffusion regime, which is commonly encountered when treating how visible and near-infrared light propagates through tissues; for many *tissues*, the reduced scattering coefficient obeys a power law, $\mu_s' = q\lambda^{-h}$(cm^{-1}, λ in μm).

Ribosomes: In the *cytoplasm* of a cell any several minute, angular, or spherical particles composed of protein and RNA.

RTT: Radiation transfer theory.

Scattering Anisotropy Parameter: A measure of the amount of forward direction retained after a single scattering event; if a photon is scattered by a particle so that its trajectory is deflected by a deflection angle θ, then the component of the new trajectory that is aligned in the forward direction is presented as $\cos\theta$; there is an average deflection angle and the mean value of $\langle\cos\theta\rangle$ is defined as the anisotropy, $g \equiv \langle\cos\theta\rangle$; the value of g varies in the range from -1 to 1: $g = 0$ corresponds to isotropic (Rayleigh) scattering, $g = 1$ to total forward scattering (Mie scattering at large particles), and -1 to total backward scattering.

Scattering Coefficient: A particle with a particular geometrical size redirects incident photons into new directions and so prevents the forward on-axis transmission of photons, this process constitutes scattering; the scattering coefficient μ_s (cm^{-1}) describes a medium containing many scattering particles at a concentration described as a volume density ρ (cm^3); the scattering coefficient is essentially the cross-sectional area σ_{sca} (cm^{-1}) per unit volume of medium: $\mu_s = \rho \sigma_{sca}$; a power law for dependence of the scattering coefficient on the wavelength is typical for many tissues: $\mu_s \propto \lambda^{-h}$, for different tissue structures parameter h is ranging from 1 to 2.

Scattering Phase Function: The function that describes the scattering properties of the medium and is, in fact, the probability density function for scattering in the direction $\vec{s}'$ of a photon traveling in the direction $\vec{s}$; it characterizes an elementary scattering act: if scattering is symmetric relative to the direction of the incident wave, then the phase function depends only on the scattering angle θ (angle between directions $\vec{s}$ and $\vec{s}'$).

Skin: The external protective covering on a body, joined by *connective tissue* to the muscles; it consists of an inner *dermis* and an outer *epidermis*.

Stratum Corneum: see *epidermis stroma* The supporting framework, usually of *connective tissue*, of an organ, as distinguished from the parenchyma, for example, the main layer of the *eye sclera* and *cornea*.

Tendon: A cord of white *fibrous tissue*; it usually attaches muscle to bones.

Thalamus: A pair and symmetric part of the brain; it constitutes the main part of the diencephalons; in the caudal (tail) to oral (mouth) sequence of neuromeres, the diencephalons is located between the mesencephalon (cerebral peduncule, belonging to the brain stem) and the telencephalon.

TMFP: Transport mean free path of a photon.

Tumor: An abnormal or diseased swelling in any part of the body, especially a more or less circumscribed overgrowth of new *tissue* that is autonomous, differs more or less in structure from the part in which it grows, and serves no useful purpose; neoplasm.

White Matter: The nervous *tissue* found in the central nervous system; it consists of tracts of medullated nerve fibers in the brain and spinal cord; it also contains blood vessels and neuroglia; it is mainly external to *gray matter*, but is internal to gray matter in the cerebral hemispheres and in the *cerebellum*; the medullated fibers give the tissue its shiny white appearance.

WHO: World Health Organization.

Whole Blood: Blood containing all its natural components: blood cells and plasma.

References

Anderson, R.R. and Parrish, J.A. (1982) in *The Science of Photomedicine*, (eds J.D. Regan and J.A. Parrish), Plenum Press, New York, pp. 147–194.

Barber, P.W. and Hill, S.C. (1990) *Light Scattering by Particles: Computational Methods*, World Scientific, Singapore.

Barer, R., Ross, K.F.A. and Tkaczyk, S. (1953) *Nature* **171**(4356), 720.

Bashkatov, A.N., Genina, E.A., Kochubey, V.I., and Tuchin, V.V. (2005a) *J. Phys. D: Appl. Phys.*, **38**, 2543.

Bashkatov, A.N., Genina, E.A., Kochubey, V.I., and Tuchin, V.V. (2005b) *Opt. Spectrosc.*, **99**, 836.

Bashkatov, A.N., Genina, E.A., Kochubey, V.I., Lakodina, N.A., and Tuchin, V.V. (2006) *Proc. SPIE*, **6163**, 616310.

Berlien, H.-P. and Mueller, G.J. (eds) (2003) *Applied Laser Medicine*, Springer-Verlag, Berlin.

Bevilacqua, F., Piguet, D., Marquet, P. et al. (1997) *Proc. SPIE*, **3194**, 262.

Bevilacqua, F., Piguet, D., Marquet, P. et al. (1999) *Appl. Opt.*, **38**, 4939.

Bohren, C.F. and Huffman, D.R. (1983) *Absorption and Scattering of Light by Small Particles*, John Wiley & Sons, Ltd, New York.

Bolin, F.P., Preuss, L.E., Taylor, R.C., and Ference, R.J. (1989) *Appl. Opt.*, **28**, 2297.

Born, M. and Wolf, E. (1999) *Principles of Optics*, 7th edn., Cambridge University, Cambridge.

Chan, E.K., Sorg, B., Protsenko, D. et al. (1996) *IEEE J. Select. Tops Quant. Electr.*, **2**, 943.

Chance, B. (ed.) (1989) *Photon Migration in Tissue*, Plenum Press, New York.

Chandrasekhar, C. (1960) *Radiative Transfer*, Dover, Toronto.

Cheng, S., Shen, H.Y., Zhang, G. et al. (2002) *Proc. SPIE*, **4916**, 172.

Cheong, W.-F., Prahl, S.A., and Welch, A.J. (1990) *IEEE J. Quantum Electr.*, **26**, 2166. updated by W.-F. Cheong; further additions by L. Wang and S. L. Jacques, August 6, 1993).

Cilesiz, I.F. and Welch, A.J. (1993) *Appl. Opt.*, **32**, 477.

Darrell, A., Meyer, H., Marias, K. et al. (2008) *Phys. Med. Biol.*, **53**, 3863.

Das, B.B., Liu, F., and Alfano, R.R. (1997) *Rep. Prog. Phys.*, **60**, 227.

Dolin, L.S., Feldchtein, F.I., Gelikonov, G.V. et al. (2004) in *Coherent-Domain Optical Methods: Biomedical Diagnostics, Environmental and Material Science*, Vol. 2, (ed. V.V. Tuchin), Kluwer Academic Publishers, Boston, p. 211.

Doornbos, R.M.P., Lang, R., Aalders, M.C. et al. (1999) *Phys. Med. Biol.*, **44**, 967.

Driver, I., Lowdell, C.P., and Ash, D.V. (1991) *Phys. Med. Biol.*, **36**, 805.

Duck, F.A. (1990) *Physical Properties of Tissue: a Comprehensive Reference Book*, Academic Press, London.

Dyson, R.D. (1974) *Cell Biology: a Molecular Approach*, Allyn and Bacon, Boston.

Genina, E.A., Bashkatov, A.N., Kochubey, V.I., and Tuchin, V.V. (2005) *Opt. Spectrosc.*, **98**, 470.

Fantini, S., Walker, S.A., Franceschini, M.A. et al. (1998) *Appl. Opt.*, **37**, 1982.

Farrell, T.J., Patterson, M.S., and Essenpreis, M. (1998) *Appl. Opt.*, **37**, 1958.

Farrell, T.J. and Patterson, M.S. (2001) *J. Biomed. Opt.*, **6**, 468.

Fishkin, J.B., Coquoz, O., Anderson, E.R. et al. (1997) *Appl. Opt.*, **36**, 10.

Frank K. and Kessler, M. (eds) (1992) *Quantitative Spectroscopy in Tissue*, pmi Verlag, Frankfurt am Main.

Fried, D. (2003) in *Biomedical Photonics Handbook*, (ed. T. Vo-Dinh), CRC Press, Boca Raton, pp. 50–51.

Fried, D., Featherstone, J.D.B., Glena, R.E., and Seka, W. (1995) *Appl. Opt.*, **34**, 1278.

Friebel, M., Roggan, A., Müller, G., and Meinke, M. (2006) *J. Biomed. Opt.*, **11**, 034021.

Gebhart, S.C., Lin, W.C., and Mahadevan-Jansen, A. (2006) *Phys. Med. Biol.*, **51**, 2011.

van Gemert, M.J.C., Jacques, S.L., Sterenborg, H.J.C.M., and Star, W.M. (1989) *IEEE Tranc. Biomed. Eng.*, **36**, 1146.

Ghosh, N., Mohanty, S.K., Majumder, S.K., and Gupta, P.K. (2001) *Appl. Opt.*, **40**, 176.

Goodman, J.W. (1985) *Statistical Optics*, Wiley-Interscience Publication, New York.

Graaff, R., Aarnoudse, J.G., Zijp, J.R. et al. (1992) *Appl. Opt.*, **31**, 1370.

Graaff, R., Dassel, A.C.M., Koelink, M.H. et al. (1993) *Appl. Opt.*, **32**, 435.

Gratton, E., Fantini, S., Franceschini, M.A. et al. (1997) *Phil. Trans. R. Soc. Lond. B*, **352**, 727.

Hammer, M., Schweitzer, D., Michel, B. et al. (1998) *Appl. Opt.*, **37**, 7410.

Hayakawa, C.K., Spanier, J., Bevilacqua, F. et al. (2001) *Opt. Lett.*, **26**, 1335.

Heusmann, H., Kolzer, J., and Mitic, G. (1996) *J.Biomed. Opt.*, **1**, 425.

Henderson, B.W. and Dougherty, T.J. (eds.) (1992) *Photodynamic Therapy: Basic Principles and Clinical Applications*, Marcel-Dekker, New York.

Hogan, M.J., Alvarado, J.A., and Weddel, J. (1971) *Histology of the Human Eye*, W.B. Sanders Co, Philadelphia.

Holboke, M.J., Tromberg, B.J., Li, X. et al. (2000) *J. Biomed. Opt.*, **5**, 237.

van de Hulst, H.C. (1980) *Multiple Light Scattering. Tables, Formulas and Applications*, Academic Press, New York.

van de Hulst, H.C. (1981) *Light Scattering by Small Particles*, Dover, New York.

Ishimaru, A. (1997) *Wave Propagation and Scattering in Random Media*, IEEE Press, New York.

Jacques, S.L. (1989) *IEEE Trans. Biomed. Eng.*, **36**, 1155.

Jacques, S.L. (2003) http://omlc.ogi.edu/software/index.html.

Jacques, S.L., Alter, C.A., and Prahl, S.A. (1987) *Lasers Life Sci.*, **1**, 309.

Keijzer, M., Richards-Kortum, R.R., Jacques, S.L., and Feld, M.S. (1989), *Appl. Opt.* **28**(20), 4286.

Kessel, R.G. (1998) *Basic Medical Histology: The Biology of Cells, Tissues, and Organs*, Oxford University Press, New York.

Key, H., Davies, E.R., Jackson, P.C., and Wells, P.N.T. (1991) *Phys. Med. Biol.*, **36**, 579.

Khalil, O., Yeh, S.-J., Lowery, M.G., Wu, X. et al. (2003) *J. Biomed. Opt.*, **8**, 191.

Kienle, A., Lilge, L., and Patterson, M.S. (1994) *Proc SPIE*, **2326**, 212.

Knüttel, A. and Boehlau-Godau, M. (2000) *J. Biomed. Opt.*, **5**, 83.

Knüttel, A., Bonev, S., and Knaak, W. (2004) *J. Biomed. Opt.*, **9**, 265.

Kumar, G. and Schmitt, J.M. (1997) *Appl. Opt.*, **36**(10), 2286.

Liu, H. and Xie, S. (1996) *Appl. Opt.*, **35**, 1793.

Laufer, J., Simpson, C.R., Kohl, M. et al. (1998) *Phys. Med. Biol.*, **43**, 2479.

Levitz, D., Thrane, L., Frosz, M.H. et al. (2004) *Opt. Exp.*, **12**, 249.

Madsen, S.J., Wyst, P., Svaasand, L.O., et al. (1994) *Phys. Med. Biol.*, **39**(8), 1191.

Marchesini, R., Bertoni, A., Andreola, S. et al. (1989) *Appl. Opt.*, **28**, 2318.

Matcher, S.J., Cope, M., and Delpy, D.T. (1997), *Appl. Opt.* **36**(1), 386.

Mishchenko, M.I., Hovenier, J.W., and Travis, L.D. (eds.) (2000) *Light Scattering by Nonspherical Particles*, Academic Press, San Diego.

Mishchenko, M.I., Travis, L.D., and Lacis, A.A. (2002) *Scattering, Absorption, and Emission of Light by Small Particles*, Cambridge University Press, Cambridge.

Mishchenko, M.I., Travis, L.D. and Lacis, A.A. (2006) *Multiple Sacattering of Light by Particles: Radiative Transfer and Coherent Backscattering*, Cambridge University Press, New York.

Mobley, J. and Vo-Dinh, T. (2003) in *Biomedical Photonics Handbook* (ed. T. Vo-Dinh), CRC Press, Boca Raton, pp. 2–1.

Müller, G., Chance, B., Alfano, R., et al. (eds) (1993) *Medical Optical Tomography: Functional Imaging and Monitoring*, **IS11**, SPIE Press, Bellingham, WA.

Müller, G. and Roggan, A. (eds) (1995) *Laser–Induced Interstitial Thermotherapy*, **PM25**, SPIE Press, Bellingham, WA.

Nickell, S., Hermann, M., Essenpreis, M. et al. (2000) *Phys. Med. Biol.*, **45**, 2873.

Niemz, H. (1996) *Laser-Tissue Interactions. Fundamentals and Applications*, Springer Verlag, Berlin.

Nilsson, A.M.K., Lucassen, G.W., Verkruysse, W. et al. (1997) *Photochem. Photobiol.*, **65**, 366.

Ohmi, M., Ohnishi, Y., Yoden, K., and Haruna, M. (2000) *IEEE Trans. Biomed. Eng.*, **47**, 1266.

Oldham, M., Sakhalkar, H., Oliver, T. et al. (2008) *J. Biomed. Opt.*, **13**, 021113.

Patterson, M.S., Chance, B., and Wilson, B.C. (1989) *Appl. Opt.*, **28**, 2331.

Peters, V.G., Wyman, D.R., Patterson, M.S., and Frank, G.L. (1990) *Phys. Med. Biol.*, **35**, 1317.

Prahl, S.A. (2007) http://omlc.ogi.edu/software/iad/index.html.

Prahl, S.A., van Gemert, M.J.C., and Welch, A.J. (1993) *Appl. Opt.*, **32**, 559.

Roggan, A., Dörschel, K., Minet, O., et al. (1995) in *Laser–Induced Interstitial Thermotherapy*, **PM25** (eds G. Müller and A. Roggan), SPIE Press, Bellingham, WA, pp. 10–44.

Roggan, A., Friebel, M., Dörschel, K. et al. (1999) *J. Biomed. Opt.*, **4**, 36.

Rytov, S.M., Kravtsov, Yu.A., and Tatarskii, V.I. (1989) *Wave Propagation through Random Media*, Principles of Statistical Radiophysics 4, Springer-Verlag, Berlin.

Salomatina, E., Jiang, B., Novak, J., and Yaroslavsky, A.N. (2006) *J. Biomed. Opt.*, **11**, 064026.

Sardar, D.K. and Levy, L.B. (1998) *Lasers Med. Sci.*, **13**, 106.
Schmitt, J.M., Knüttel, A., and Bonnar, R.F. (1993) *Appl. Opt.*, **32**, 6032.
Schmitt, J.M. and Kumar, G. (1998) *Appl. Opt.* **37**(13), 2788.
Schwarzmaier, H.-J., Yaroslavsky, A.N., Yaroslavsky, I.V. et al. (1997) *Proc. SPIE*, **2970**, 492.
Silver, F.H. (1987) *Biological Materials: Structure, Mechanical Properties, and Modeling of Soft Tissues*, New York, New York University Press.
Simpson, C.R., Kohl, M., Essenpreis, M., and Cope, M. (1998) *Phys. Med. Biol.*, **43**, 2465.
Sobolev, V.V. (1974) *Light Scattering in Planetary Atmospheres*, Pergamon Press, Oxford.
Star, W.M., Wilson, B.C., and Patterson, M.C. (1992) in *Photodynamic Therapy, Basic Principles and Clinical Applications*, (eds B.W. Henderson and T.J. Dougherty), Marcel-Dekker, New York, p. 335.
Steenbergen, W., Kolkman, R., and de Mul, F. (1999) *J. Opt. Soc. Am. A*, **16**, 2967.
Sterenborg, H.J.C.M., van Gemert, M.J.C., Kamphorst, W. et al. (1989) *Lasers Med. Sci.*, **4**, 221.
Suzuki, K., Yamashita, Y., Ohta, K. et al. (1996) *J. Biomed. Opt.*, **1**, 330.
Tearney, G.J., Brezinski, M.E., Southern, J.F. et al. (1995) *Opt. Lett.*, **20**, 2258.
Thomas, G.E. and Stamnes, K. (1999) *Radiative Transfer in the Atmosphere and Ocean*, Cambridge University Press, New York.
Thueler, F., Charvet, I., Bevilacqua, F. et al. (2003) *J. Biomed. Opt.*, **8**, 495.
Tromberg, B.J., Coquoz, O., Fishkin, J.B. et al. (1997) *Phil. Trans. R. Soc. Lond. B.*, **352**, 661.
Troy, T.L., Page, D.L., and Sevick-Muraca, E.M. (1996) *J. Biomed. Opt.*, **1**, 342.
Troy, T.L. and Thennadil, S.N. (2001) *J. Biomed. Opt.*, **6**, 167.
Tuchin, V.V. (ed.) (1994) *Selected Papers on Tissue Optics: Applications in Medical Diagnostics and Therapy*, **MS102**, SPIE Press, Bellingham, WA.
Tuchin, V.V. (ed.) (2002) *Handbook of Optical Biomedical Diagnostics*, **PM107**, SPIE Press, Bellingham, WA.
Tuchin, V.V. (2007) *Tissue Optics: Light Scattering Methods and Instruments for Medical Diagnosis*, **PM 166**, 2nd edn, SPIE Press, Bellingham, WA.
Tuchin, V.V., Wang, L.V., and Zimnyakov, D.A. (2006) *Optical Polarization in Biomedical Applications*, Springer-Verlag, New York.
Tuchin, V.V., Xu, X., and Wang, R.K. (2002) *Appl. Opt. – OT*, **41**(1), 258.
Tuchin, V.V., Zhestkov, D.M., Bashkatov, A.N., and Genina, E.A. (2004) *Opt. Exp.*, **12**, 2966.
Turchin, I.V., Sergeeva, E.A., Dolin, L.S., and Kamensky, V.A. (2003) *Laser Phys.*, **13**, 1524.
Vo-Dinh T. (ed.) (2003), *Biomedical Photonics Handbook*, CRC Press, Boca Raton.
Wang, X.J., Milner, T.E., Dhond, R.P. et al. (1995) *Opt. Lett.*, **20**, 524.
Wang, L.V. and Wu, H.-I. (2007) *Biomedical Optics: Principles and Imaging*, Wiley-Interscience, Hoboken, NJ.
Welch, A.J. and van.Gemert, M.C.J. (eds.) (1992) *Tissue Optics*, Academic Press, New York.
Yanovitskij, E.J. (1997) *Light Scattering in Inhomogeneous Atmospheres*, Springer-Verlag, Berlin.
Yaroslavsky, A.N., Schulze, P.C., Yaroslavsky, I.V. et al. (2002) *Phys. Med. Biol.*, **47**, 2059.
Yaroslavsky, A.N., Yaroslavsky, I.V., Goldbach, T., and Schwarzmaier, H.-J. (1996) *Appl. Opt.*, **35**, 6797.
Yaroslavsky, A.N., Yaroslavsky, I.V., Goldbach, T., and Schwarzmaier, H.-J. (1999) *J. Biomed. Opt.*, **4**, 47.
Zege, E.P., Ivanov, A.P., and Katsev, I.L. (1991) *Image Transfer through a Scattering Medium*, Springer-Verlag, New York.
Zijp, J.R. and ten Bosch, J.J. (1991) *Archs Oral Biol.*, **36**, 283.
Zijp, J.R. and ten Bosch, J.J. (1997) *Appl. Opt.*, **36**, 1671.
Zvyagin, A.V., Silva, K.K.M.B.D., Alexandrov, S.A. et al. (2003) *Opt. Exp.*, **11**, 3503.

Further Reading

Ahluwalia, G. (ed.) (2008) *Light Based Systems for Cosmetic Application*, William Andrew, Inc., Norwich, NY.

Baron, E. (ed.) (2009) *Light-Based Therapies for Skin of Color*, Springer, NY.

Drexler, W. and Fujimoto, J.G. (eds.) (2008) *Optical Coherence Tomography: Technology and Applications*, Springer, Berlin.

Drezek, R., Guillaud, M., Collier, T. et al. (2003) *J. Biomed. Opt.*, **8**, 7.

Splinter, R. and Hooper, B.A. (2007) *An Introduction to Biomedical Optics*, Taylor & Francis, New York, London.

Tuchin, V.V. (ed.) (2009) *Handbook of Optical Sensing of Glucose in Biological Fluids and Tissues*, CRC Press, Taylor & Francis Group, London.

17
Single-Molecule Fluorescence: Biophysics

Michael Prummer and Christian Hübner

17.1	Introduction 629	
17.2	**Photophysical Basics** 630	
17.2.1	Energy Levels and Transitions 630	
17.2.2	Energy Transfer 632	
17.2.3	The Dye Molecule as a Dipole 634	
17.3	**Techniques** 635	
17.3.1	Signal-to-Noise Considerations 635	
17.3.2	Wide-Field Fluorescence Microscopy Techniques 636	
17.3.2.1	Epifluorescence Microscopy 637	
17.3.2.2	TIRF Microscopy 637	
17.3.3	Confocal Optical Microscopy 638	
17.4	**Applications** 640	
17.4.1	Protein Folding 640	
17.4.1.1	The Unfolded State 640	
17.4.1.2	Single Molecule Folding/Unfolding Transitions 642	
17.4.2	Enzymes 643	
17.4.2.1	Static and Dynamic Heterogeneity 643	
17.4.2.2	Conformational Changes and Catalytic Activity 646	
17.5	**Conclusions** 650	
	References 651	

Encyclopedia of Applied Spectroscopy. Edited by David L. Andrews.
Copyright © 2009 WILEY-VCH Verlag GmbH & Co. KGaA, Weinheim
ISBN: 978-3-527-40773-6

17.1 Introduction

The first detection of the fluorescence light of a single molecule dates back to 1976 (Hirschfeld, 1976). At that time, the molecule under investigation contained about 100 fluorescent chromophores. With the advent of extremely sensitive photon detectors and effective filters, the detection of the fluorescence of just a single fluorophore became possible in 1990 (Orrit and Bernard, 1990), however, in a crystalline host at liquid helium temperature. In the same year, single fluorophores were observed in solution at room temperature (Shera et al., 1990), paving the way to the application of single-molecule fluorescence on biological systems in their native surrounding. The next milestones for the perception of single-molecule fluorescence as a useful biophysical technique were the demonstration of fluorescent resonance energy transfer on an individual basis (Ha et al., 1996) and the direct observation of the diffusion of a dye-labeled protein in a lipid bilayer (Schmidt et al., 1995). These two key experiments sparked a tremendous development of the technique over the following years, and already in 2001 the coming of age of single-molecule spectroscopy was stated (Kelley, Michalet and Weiss, 2001). Today, the number of groups performing single-molecule fluorescence detection as well as the number of publications in the field is rapidly growing. The pioneers of single-molecule fluorescence have set out their children, and the grand children are starting their own groups. This chapter therefore cannot be comprehensive, but rather a selection of prototype experiments demonstrating the enormous potential of single-molecule fluorescence for molecular protein biophysics, with an emphasis on protein conformation in the context of protein folding and catalysis. Owing to space limitations, the fields of single-molecule observations of molecular motors and of membrane diffusion are omitted.

Thereby, the aspects of the respective applications will be balanced by the methodological developments. The aim of this approach is to provide the reader with a framework of knowledge that allows for deciding whether a single-molecule approach to a certain problem might be advantageous, and which method is to be preferred.

The basic photophysical principles underlying fluorescence detection, particularly at the high excitation rates necessary

Encyclopedia of Applied Spectroscopy. Edited by David L. Andrews.
Copyright © 2009 WILEY-VCH Verlag GmbH & Co. KGaA, Weinheim
ISBN: 978-3-527-40773-6

to observe the faint light of a single dye molecule, are outlined in the second section. The third section briefly introduce the experimental layouts used, emphasizing their advantages and drawbacks. Finally, selected applications are discussed in the fourth section.

17.2 Photophysical Basics

17.2.1 Energy Levels and Transitions

The fluorescence of single dye molecules is based on the same energy level scheme as ordinary molecular bulk fluorescence, consisting of, in the simplest case, an electronic ground state, an excited state, and a triplet state. In each electronic level, several vibronic states exist. The high repetition rates needed to obtain a measurable signal, however, demand a second look at the energy levels and transitions involved. Besides the levels found in every textbook, higher excited states are shown in the Jablonski diagram in Figure 17.1. Those higher levels exist both in the singlet as well as in the triplet manifold. At high laser intensities, in particular in pulsed excitation, further excitations from the states S_1 and T_1 are possible.

First taking only transitions between the electronic ground state S_0, the first excited singlet state S_1, and the first excited triplet state T_1 into account and neglecting stimulated emission, the following set of rate equations can be derived:

$$\dot{n}_1 = -I\sigma n_1 + k_{rad} n_2 \\ + k_{ic} n_2 + k_{risc} n_3$$

$$\dot{n}_2 = I\sigma n_1 - k_{rad} n_2 - k_{ic} n_2 - k_{isc} n_2$$

$$\dot{n}_3 = k_{isc} n_2 - k_{risc} n_3 \qquad (17.1)$$

where n_1, n_2, and n_3 are the occupation probabilities of S_0, S_1, and T_1, respectively, k_{rad} the radiative rate, k_{ic} the rate of internal conversion, k_{isc} the intersystem crossing rate, k_{risc} the reverse intersystem crossing rate, I the excitation intensity, and σ the absorption cross section of the fluorophore.

The quantum yield of fluorescence Φ_f is the ratio of radiative transitions from

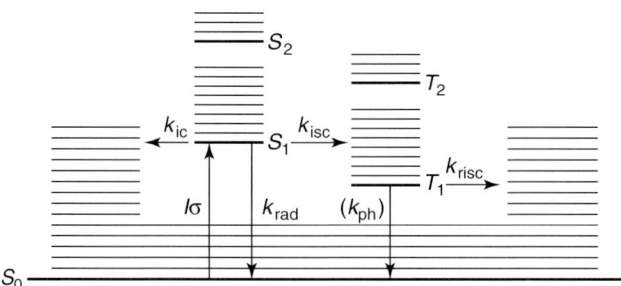

Fig. 17.1 Jablonski diagram showing the electronic ground state S_0, the first electronically and second excited singlet states S_1 and S_2, the first and the second electronically excited triplet states T_1 and T_2, and the most important transitions between the states as detailed in the text.

S_0 to all other transitions from that state:

$$\Phi_f = \frac{k_{rad}}{k_{ic} + k_{isc} + k_{rad}} = k_{rad}\tau_f \quad (17.2)$$

where τ_f is the fluorescence lifetime, which is the experimentally accessible time constant of the decay of fluorescence after pulsed excitation. Therefore, the quantum yield of fluorescence can be determined by measuring the fluorescence lifetime and deriving k_{rad} from the absorption and emission spectra of the dye making use of the relation between the Einstein coefficients for absorption and spontaneous emission (Strickler and Berg, 1962):

$$k_{rad} = 2.88 \times 10^{-9} n^2 \frac{\int F(\bar{\nu}) d\bar{\nu}}{\int F(\bar{\nu}) d\bar{\nu}/\bar{\nu}^3}$$

$$\times \int \frac{\varepsilon(\bar{\nu})}{\bar{\nu}} d\bar{\nu} \quad (17.3)$$

where $F(\bar{\nu})$ is the fluorescence and $\varepsilon(\bar{\nu})$ the absorption spectrum of the dye, with $\bar{\nu}$ the wave number in cm^{-1}. Fluorophores for single-molecule spectroscopy should have quantum yields of fluorescence close to unity to ensure a strong enough fluorescence signal to be detected. Therefore, the intersystem crossing rate as well as the rate of internal conversion should be small.

Under stationary conditions (all time derivatives are zero), the emission rate R as a function of the excitation rate from the set of rate equations (17.1) is given by

$$R = n_2 k_{rad}$$

$$= \frac{k_{rad}}{1 + \frac{k_{isc}}{k_{risc}} + \frac{k_{rad} + k_{ic} + k_{ies}}{I\sigma}} \quad (17.4)$$

Owing to increasing population of the triplet state T_1 at high excitation rates, the maximum achievable emission rate with infinitely strong excitation is R_∞:

$$R_\infty = \frac{k_{rad}}{1 + \frac{k_{isc}}{k_{risc}}} \quad (17.5)$$

The emission rate R can be rewritten as

$$R = R_\infty \frac{1}{\left(1 + \frac{I_{sat}}{I}\right)} \quad (17.6)$$

with the saturation intensity I_{sat}, which is the excitation intensity where half of the maximum emission rate is reached, given by

$$I_{sat} = \frac{k_{risc}}{\sigma} \frac{k_{rad} + k_{ic} + k_{isc}}{k_{risc} + k_{isc}} \quad (17.7)$$

The saturation of the average emission rate due to triplet excursions is known as the *triplet bottleneck*.

The solution to the set of rate equations 17.1 with the initial condition of $n_1 = 1$, $n_2 = n_3 = 0$ provides the time evolution of the occupation probability of the first electronically excited singlet state $n_2(t)$, which is related to the fluorescence intensity autocorrelation function or second-order autocorrelation function $g^{(2)}(t)$ defined as

$$g^{(2)}(t) = \frac{\langle n(\tau)n(t+\tau)\rangle_\tau}{\langle n(\tau)\rangle_\tau^2} = \frac{n_2(t)}{\langle n_2 \rangle}$$

$$(17.8)$$

and reads

$$g^{(2)}(t) = 1 - \frac{k_{isc}^{eff} + k_{risc}}{k_{risc}} e^{-(k_{rad}+2I\sigma)t}$$

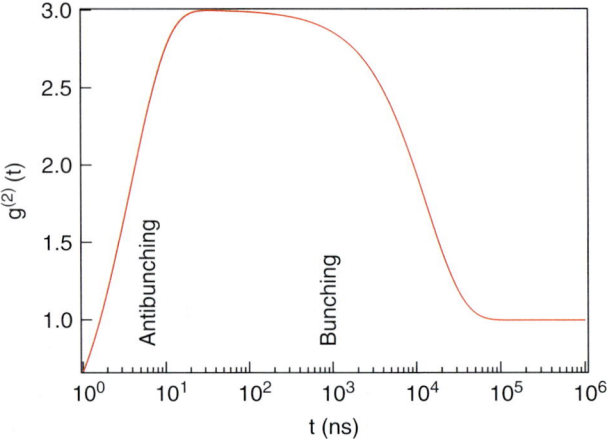

Fig. 17.2 Intensity autocorrelation function of the fluorescence of a single dye molecule under room-temperature excitation.

$$+ \frac{k_{\text{isc}}^{\text{eff}}}{k_{\text{risc}}} e^{-(k_{\text{isc}}^{\text{eff}}+k_{\text{risc}})t} \quad (17.9)$$

where the effective intersystem crossing rate $k_{\text{isc}}^{\text{eff}}$ is given as

$$k_{\text{isc}}^{\text{eff}} = \langle n_2 \rangle k_{\text{isc}} \quad (17.10)$$

The theoretical autocorrelation function of a single dye molecule according to Eq. 17.10 is shown in Figure 17.2 on a logarithmic timescale. This function drops to zero at the temporal origin, which is intuitively understood taking into consideration that a single quantum system can never emit two photons simultaneously. This behavior is referenced to as photon antibunching. Thereafter, the autocorrelation function takes a value above 1, indicating that the occupation probability of S_1 at this time is higher than on average. This is because the system will always be excited first to S_1, but on average the occupancy $\langle n_2 \rangle$ is limited by the triplet bottleneck. Because of triplet excursions, a single molecule emits its photons in bunches (cycling in the singlet manifold under photon emission), separated by dark periods (when the molecule is shelved in the triplet state). Therefore, the second term in Eq. 17.10 is referred to as the *photon bunching*. If dynamical processes are to be studied by single-molecule fluorescence, care has to be taken that photophysical effects do not lead to misinterpretations.

17.2.2
Energy Transfer

Energy transfer between two or more fluorescent entities is a major subject in single-molecule fluorescence. It plays an important role in light harvesting complexes and, on the other hand, is the basis of one of the most popular single-molecule techniques, fluorescence resonance energy transfer (FRET), which is referred to as a *molecular ruler* (Stryer and Haugland, 1967) owing to its capability to measure distances between two fluorophores by spectral means.

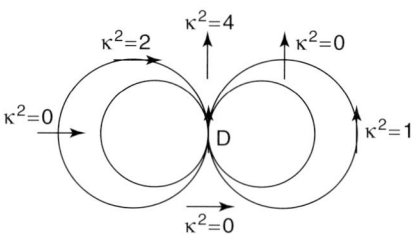

Fig. 17.3 Orientation factors κ^2 for different cases of relative donor–acceptor placement.

In FRET, one fluorophore, called *donor*, is excited, and the excitation energy is radiationless transferred to a nearby fluorpohore, called *acceptor*. The two chromophores may be chemically identical (homotransfer) or distinct (heterotransfer). The transfer mechanism is based on the dipole–dipole interaction between the two fluorophores in the weak coupling regime, that is, the interaction energy is small compared to the transition energies. The transfer efficiency theoretically derived by Förster (Förster, 1946) depends on the relative location and orientation of the transition dipoles of the fluorophores, their distance R, the refractive index of the medium n in the vicinity of both, and the overlap integral J. The latter takes account for the resonance condition and is the integral of the product of the normalized emission spectrum of the donor $F_D(\lambda)$ and the absorption spectrum (molar extinction coefficient in cm^{-1}M^{-1}) $\varepsilon_A(\lambda)$ of the acceptor:

$$J = \int_0^\infty F_D(\lambda)\varepsilon_A(\lambda)\lambda^4\, d\lambda \quad (17.11)$$

The transfer rate, that is, the rate with which the excited donor transfers its energy to the acceptor, is given as

$$k_t = 8.8 \times 10^{-23} \text{mol} \frac{1}{\tau_D} \frac{1}{R^6} \frac{\kappa^2 \Phi_D}{n^4} J$$

$$= \frac{1}{\tau_D}\left(\frac{R_0}{R}\right)^6 \quad (17.12)$$

with the distance R taken in Ångstrom and R_0 the Förster radius. Here, the relative position and orientation of the dyes are taken into account via the orientation factor κ^2. Values of κ^2 for some cases are shown in Figure 17.3. The orientation dependence of the energy transfer rate may jeopardize distance measurements. Therefore, a free rotation of the fluorophores on a timescale faster than the fluorescence lifetime is aimed at, leading to an average orientation factor $\langle\kappa^2\rangle = 2/3$. Linkers with several carbon atoms are used to assure unhindered rotation of the dye. However, interactions between the dye and the protein may lead to blocking of the free rotation. Fluorescence anisotropy measurements provide information on the rotational flexibility of the dyes and should therefore always be conducted as a control. Fortunately, even a partial blocking of rotational freedom does not significantly alter the average orientation factor. If, for example, one of the dyes cannot rotate at all while the other one is free to rotate, the average orientation factor takes the value $\langle\kappa^2\rangle = 0.63$ (Berberan-Santos and Prieto, 1988).

In the experiment, the transfer rate is not directly accessible but rather the transfer efficiency, which is the fraction of transferred excitation quanta, and which is related to the distance R as

$$E_t = \frac{1}{1 + \left(\dfrac{R}{R_0}\right)^6} \qquad (17.13)$$

Although in ensemble experiments the transfer efficiency is measured via comparison of donor fluorescence in the presence and absence of the acceptor (e.g., by photobleaching the acceptor), in single-molecule experiments the fluorescence signals of the donor F_A and the acceptor F_D are utilized:

$$E_t = \frac{F_A}{\gamma F_D + F_A} \qquad (17.14)$$

The correction factor γ takes into account the quantum efficiencies of fluorescence of the donor and the acceptor and their respective detection efficiencies. If there is direct excitation of the donor (i.e., not FRET mediated), the acceptor signal needs to be corrected by subtraction of that signal. A very robust method to determine the transfer efficiency is based on the fluorescence lifetime of the donor τ_{D0} and τ_{DA} in the absence and presence of the acceptor:

$$E_t = 1 - \frac{\tau_{DA}}{\tau_{D0}} \qquad (17.15)$$

The fluorescence lifetime of the donor without the acceptor is determined in a control sample.

In single-molecule FRET experiments, the photophysics of the dyes has to be treated carefully. Although triplet excursions of the donor simply render the whole donor–acceptor pair dark, the situation is more complex if the acceptor is in the triplet state. If only one excited singlet and triplet states are considered, one would expect that no energy transfer occurs and that all excitation quanta lead to an emission of the donor in this case, resulting in an anticorrelated fluctuation of the donor and acceptor fluorescence. The existence of higher excited states for the acceptor, however, in the singlet as well as in the triplet system, gives rise to other pathways for the excitation energy. In particular, FRET-type energy transfer can result in an excitation of the acceptor to a higher triplet state T_n. Depending on the efficiency of this transfer process, the donor fluorescence may increase, stay constant, decrease, or even completely vanish when the acceptor is triplet shelved (Maus et al., 2001). The latter process is known as *singlet–triplet annihilation (STA)*.

17.2.3
The Dye Molecule as a Dipole

Classically, a dye molecule can be treated as a dipole antenna. Therefore, its absorption and emission has the same anisotropy. Only the component of the electric field parallel to the absorption dipole axis can excite the fluorophore, and the probability of photon emission has the same shape as the emission intensity of a dipole antenna. This holds in an isotropic medium as well as in the vicinity of an interface between media with different refractive indices. If a fluorophore is, for example, located close to an interface to a medium with a higher refractive index (imagine a fluorophore in an aqueous surrounding close to a glass coverslip), most of its fluorescence is emitted toward the interface (Figure 17.4). Additionally, the

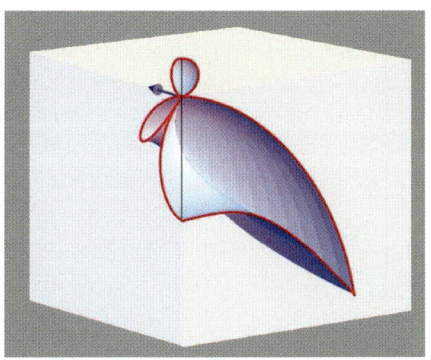

Fig. 17.4 Fluorescence emission characteristics of a dipole close to an interface to a medium with higher refractive index.

light emitted by a fluorophore is linearly polarized in the dipole direction. The anisotropy as well as the polarization of the emission can be exploited for the determination of the orientation of the dipole in space or even for the relative orientation of the dipoles in a donor–acceptor pair (Hübner et al., 2004).

17.3 Techniques

17.3.1 Signal-to-Noise Considerations

The key prerequisite for successful single-molecule fluorescence experiments is a high signal-to-noise (SNR) ratio. The fluorescence signal in single molecule detection (SMD) is typically the photon count within a given time interval. Owing to the stochastic nature of photon emission, this photon number is Poisson distributed. The noise of the signal therefore is the shot-noise of the total number of detected photons. The total number of detected photons is composed of the signal itself and background contributions. The latter can be further divided into background, induced by the excitation laser, and the background signal coming from the detectors in the absence of any source of light, the dark signal.

The laser-induced background originates from elastic Rayleigh and inelastic Raman scattering, as well as from fluorescence in the optical system and the solvent where the molecule under observation is embedded. Its contribution is linearly dependent on the laser intensity. The dark signal, on the other hand, is a constant. The total SNR ratio can be given as

$$SNR = \frac{\eta_{det} \cdot R \cdot T_{int}}{\sqrt{\eta_{det} \cdot R \cdot T_{int} + C_b \cdot I_L \cdot T_{int} + N_d \cdot T_{int}}} \quad (17.16)$$

$$= \frac{\eta_{det} \cdot R}{\sqrt{\eta_{det} \cdot R + C_b \cdot I_L + N_d}} \cdot \sqrt{T_{int}}$$

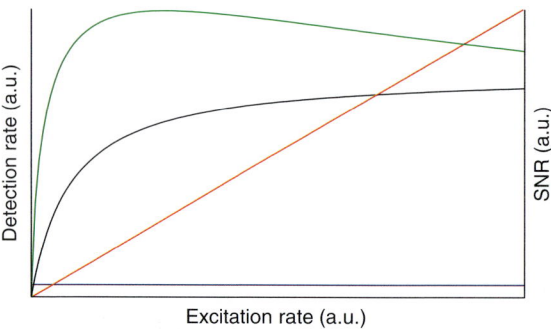

Fig. 17.5 Signal-to-noise ratio as a function of exitation laser intensity according to Eqs 17.16 and 17.6 (green). The fluorescence intensities of the single molecule (black), the dark signal (blue), and the laser-induced background (red) are also plotted.

where R is the emission rate of the fluorophore, T_{int} is the integration time, η_{det} is the detection efficiency, C_b is the constant of proportionality between laser intensity I_L and laser-induced background photon rate, and N_d is the dark count rate of the detector.

Saturation of the fluorescence signal leads to an optimum excitation rate for a given integration time where the SNR reaches its maximum, as shown in Figure 17.5.

17.3.2
Wide-Field Fluorescence Microscopy Techniques

In wide-field fluorescence microscopy, the whole field of view of an optical microscope is illuminated to excite fluorescence, and the whole field of view is also imaged onto a camera. The great advantage of wide-field microscopy is that, on the one hand, many molecules can be observed in parallel and that, on the other hand, the trace of a moving molecule can be followed. However, the time resolution is limited by the cameras in use.

For single-molecule fluorescence imaging, a very sensitive camera is necessary. If slow processes are to be followed, back-thinned cooled charge-coupled device (CCD) cameras are the best choice. Those CCDs convert incoming photons into electrons with an efficiency close to unity. However, the readout process of the charges introduces noise of typically a few electrons. Therefore, single-photon detection is not possible, and longer exposure times are necessary.

If single photons are to be detected with a camera and a short integration time is required, intensified cameras or electron multiplying CCDs (EMCCDs) can be used. Intensified CCDs have an image intensifier coupled to the CCD camera. The image intensifier is a multichannel plate photomultiplier (MCP-PM) with a phosphor behind the anode. Thus, an impinging photon is first converted into a photoelectron, which is multiplied in the PM, and the electron avalanche is back converted into light. The light from the phosphor is then detected by the CCD. Typical quantum efficiencies of the photocathodes of

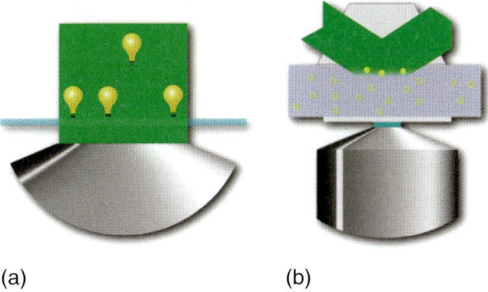

Fig. 17.6 Sketch of epifluorescence (a) and TIRF (b) illumination modes in wide-field microscopy.

image intensifiers are in the 10–20% range.

A great breakthrough in camera technology was the introduction of an electron multiplying or gain register in the CCD. In this register, the single charges are multiplied in an avalanche process principally similar to a PM. Thus, the single electron generated by a single photon is multiplied to a number that is higher than the readout noise and can therefore be detected. EMCCDs have the same high quantum efficiency of photon detection as CCDs close to unity.

The time resolution of a CCD camera is typically limited by the readout process. However, if only a short sequence of images is to be taken, the charges can be first shifted on the CCD chip and subsequently read out. Depending on the size of the image, exposure times in the submillisecond range are possible this way.

With respect to excitation of fluorescence, two different illumination modes can be applied, namely, conventional epifluorescence and total internal reflection (TIR) excitation as detailed later.

17.3.2.1 Epifluorescence Microscopy

The simplest way to observe the fluorescence of single fluorophores is by using a conventional epifluorescence microscope. Here, the whole field of view is illuminated by parallel laser light (Figure 17.6 (a)). Therefore, the laser is focused into the back focal plane of the microscope objective. This technique, however, requires a sample that is thinner than the focal depth of the microscope; otherwise, out-of-focus molecules would give rise to a high background signal. Such a thin sample is, for example, molecules tethered on the surface of a coverslip or embedded in a thin polymer film. A cell membrane also represents a thin sample, so that epifluorescence microscopy is the method of choice to study, for example, membrane diffusion (Schütz et al., 2000).

17.3.2.2 TIRF Microscopy

If the requirement of a thin sample for epifluorescence microscopy is not fulfilled, illumination of out-of-focus molecules can be circumvented by TIR excitation. The excitation laser is totally reflected typically at a glass–water interface and only fluorophores in the evanescent field at the interface are excited. Depending on the angle of incidence and the wavelength of the laser, the evanescent field penetrates some 10 nm into the aqueous medium. The small penetration depth is advantageous with respect to lowering the background, but, on the other hand, prevents observation of features further away from the interface.

Two different realizations for TIRF excitation are used: objective-type TIRF and prism-type TIRF.

In objective-type TIRF, illumination is carried out similar to epifluorescence by focusing the laser in the back focal plane of an oil immersion microscope objective. In contrast to epifluorescence, this focus is offset from the optical axis so that the emerging parallel light is tilted by an angle larger than the critical angle for total reflection at the glass–water interface of 61°. The highest achievable tilt with an NA = 1.4 oil immersion objective is 67°. Dedicated TIRF objectives have NA = 1.45 corresponding to a maximum tilt angle of 73°. The advantage of objective-type TIRF is that the fluorescence is collected trough the glass cover slip, in a direction where most of the fluorescence is emitted (cf. Figure 17.4). Thus, a high detection efficiency is assured.

In prism-type TIRF (Figure 17.6(b)), the laser is directed into a prism and the excitation occurs on the prism surface with TIR. In contrast to objective TIRF, the fluorescence is collected through the aqueous medium. Thus, the detection efficiency is considerably lower. The advantage of prism-type TIRF is the higher flexibility in the choice of the angle of incidence of the laser light.

17.3.3
Confocal Optical Microscopy

The main principle behind confocal optical microscopy invented by Marvin Minsky in 1957 (Minsky, 1957) is that a point light source and a point detector are both being imaged in the same position in the sample. The point spread function (PSF) of the microscope, that is, the image that is obtained from a point in the sample, therefore is the product of the two PSFs of illumination and detection. Typically, the microscope objective is used for both focussing the laser and detecting the fluorescence in an epi-illumination scheme, and excitation and fluorescence light are separated in a dichroic beam splitter as can be seen in Figure 17.7. The product of the excitation and detection PSFs defines the confocal volume of about 0.1 fl. The shape of the confocal volume depends on the illumination conditions, but can be approximated quite well by a three-dimensional Gaussian intensity distribution. The advantage of the smaller PSF is not primarily the slightly higher resolution but rather a considerable reduced background. Laser-induced background signal is in a first approximation proportional to the detection volume. Compared to wide-field microscopy, this volume is reduced by a factor of about $2^{3/2}$.

To obtain an image, the confocal volume is raster scanned across the sample. In contrast to wide-field microscopy, where all points of the image are obtained in parallel in a single exposure, in confocal microscopy the image is built in a serial manner, point by point. This results in a drastically prolonged time to record a single image as compared to wide-field microscopy.

To build the image, either the sample or the confocal volume is moved. For sample scanning, piezo-driven scan stages are used, whereas beam or laser scanning is realized by galvanometer mirror scanners. For the latter, a telecentric lens system generates an image of the

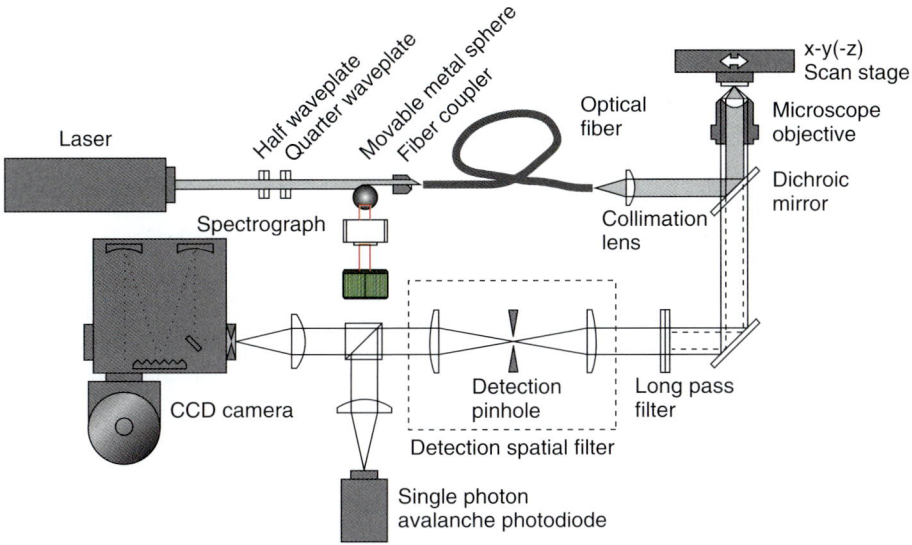

Fig. 17.7 Schematic drawing of a sample scanning confocal optical microscope.

scanning mirrors in the back focal plane of the microscope objective. Thus, a tilt of the scanning mirror results in a shift to the confocal volume in the sample.

Sample scanning has the clear advantage that the light always propagates on the optical axis through the microscope objective. Therefore, aberrations due to tilt of the beam with respect to the optical axis such as coma or astigmatism are avoided. The allowed size of the image is only limited by the scan range. In contrast, in beam-scanning microscopes the size of the image is limited by the microscope objective. On the other hand, beam scanning also has several advantages: the achievable scan speed is considerably higher, the field of view can be changed by using microscope objectives with different magnifications, and the sample can be mounted in a cryostat.

The pointlike light source required for confocal imaging is realized either by the use of an excitation pinhole or by feeding the laser through a single-mode optical fiber. A fiber typically guarantees a perfect Gaussian mode, whereas a pinhole is advantageous with respect to polarization control.

The other element defining the confocal volume is the detection pinhole. For single-molecule fluorescence experiments, the size of the confocal pinhole is chosen with respect to minimal loss of fluorescence signal. A diameter corresponding to the diameter of the airy disc turned out to be a good choice.

Key elements of the confocal microscope are the dichroic or multichroic beam splitter and the emission filters. Although in the past holographic notch filters were primarily used to block the back-scattered laser light, the development of even better dielectric filters has rendered them a cost-effective alternative. In particular, sputtered dielectric filters convince with steep edges, high optical density at the laser wavelength, high transmission in the pass band, and their ruggedness. Care should be taken that the filters are subpixel registered, that is, that they do not introduce beam deviations.

The confocal optical microscope is used in two different regimes for single-molecule experiments: in solution experiments, the confocal volume is placed stationary in the sample solution, and freely diffusing molecules lead to fluorescence bursts when they cross it. Thus, distributions of fluorescence properties of the molecules are obtained. If there is dynamics present on a timescale faster than the transit time, this dynamics can be studied. Dynamics on longer timescales, however, is not accessible with this method.

In immobilized experiments, the Brownian motion is prohibited by appropriate immobilization techniques such as embedding in a gel or tethering to a surface. The confocal volume is placed on a particular molecule localized earlier by an image scan, and the (transient) fluorescence is recorded. In this manner, dynamical processes on all timescales can be studied.

17.4
Applications

17.4.1
Protein Folding

The mechanism by which the unstructured polypeptide chain of a protein finds its unique structure remains one of the great mysteries in protein biophysics. It is believed that there is not one single pathway but rather a bunch of ways from the ensemble of unstructured chains to the final three-dimensional structure. This heterogeneity is an appealing target for single-molecule studies. Its ability to resolve subensembles allows for the study of the unfolded state and of folding kinetics and chain dynamics under equilibrium conditions, where either of the folded or unfolded state is weakly populated.

17.4.1.1 The Unfolded State

The best-studied subject in protein folding with single molecules is the unfolded state. The interest for the unfolded state arises from the idea that knowledge about its conformation and dynamics under nativelike conditions provides information about the initial steps on the folding pathway. In the first experimental single-molecule FRET study, the well-investigated protein chymotrypsin inhibitor 2 was specifically labeled with tetramethylrhodamine as the donor and Cy5 as the acceptor (Figure 17.8(a)) (Deniz et al., 2000). The confocal volume was placed in the sample solution, and brownian diffusion gave rise to a burst of fluorescence each time a molecule was passing by. Each burst was analyzed with respect to the transfer efficiency and added to a histogram. In this manner, a distribution of FRET efficiencies is obtained. The histograms shown in Figure 17.8(b) consist of two distinct, rather broad peaks, corresponding to the folded state (high transfer efficiency) and the unfolded state (low transfer efficiency). From the areas under the peaks, the concentrations of the folded and unfolded state can be directly obtained. Strikingly, the center position of the unfolded peak shifts to higher transfer efficiencies with decreasing denaturant concentration, as can be seen in Figure 17.8(c). This shift is indicative of a compaction of the unfolded polypeptide chain under nativelike conditions.

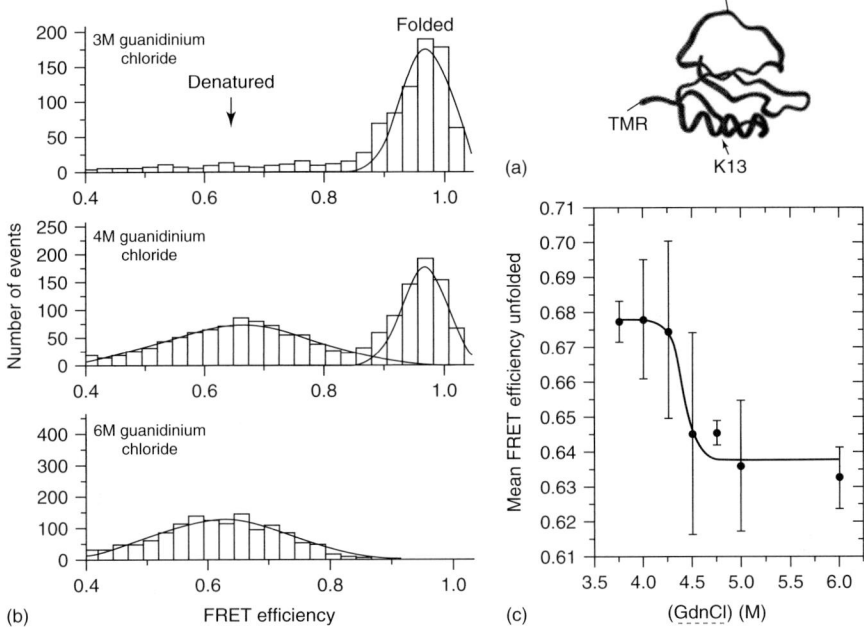

Fig. 17.8 (a) A structural sketch of chymotrypsin inhibitor 2 showing the labeling sites with tetramethylrhodamine as the donor and Cy5 as the acceptor. (b) Energy transfer histograms for different GdnCl concentrations. (c) Mean FRET efficiency as a function of denaturant concentration. (Adapted from Deniz et al. (2000).)

Using the same methodology, Schuler et al. studied the folding transition of the cold shock protein from *Thermotoga maritima* (Schuler, Lipman and Eaton, 2002). CspTm was labeled with a donor and an acceptor dye (Figure 17.9(a) and (b)), and single-molecule bursts were analyzed for their FRET efficiency and compiled in a histogram shown in Figure 17.9(c). Again, a compaction of the unfolded chain is observed (Figure 17.9(d)). Furthermore, from the finding that the width of the transfer efficiency distribution is shot-noise limited, it can be concluded that the reconfiguration time is considerably shorter than the transit time of the molecule through the focus. Thus, an upper limit for the reconfiguration time for the polypeptide chain is obtained, which allows for estimating a lower bound for the energy barrier height between the unfolded and folded state of the protein.

Recently, Hofmann et al. have studied the influence of different denaturants, namely, guanidinium chloride (GdnCl) and urea, on the collapse of the unfolded chain of the model protein barstar (Hofmann et al., 2008). Using the same FRET pair as Schuler et al., the subpopulation-resolved transfer efficiencies were measured as a function of denaturant concentration. It was found that for low denaturant concentrations, in the salt GdnCl the chain is more compact than in urea. One possible explanation for this finding is that the charges in the polypeptide chain are shielded in GdnCl, thus reducing electrostatic repulsion. This

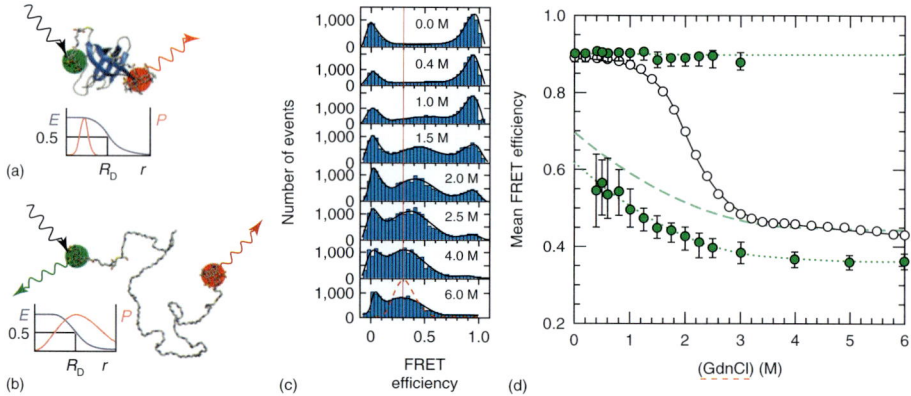

Fig. 17.9 The doubly labeled protein CsPTm in the folded (a) and unfolded (b) state additionally showing the label-distance distribution and the distance-dependent FRET efficiency. The donor dye AlexaFluor 488 is shown in green and the acceptor AlexaFluor 594 in red. (c) The distributions of apparent FRET efficiencies of single CsPTm molecules diffusing through the confocal volume for increasing GdnCl concentrations. The three peaks can be associated to (from left to right) molecules missing: an acceptor dye, unfolded, and folded molecules, respectively. (d) Mean transfer efficiencies derived from the histograms for the folded and unfolded population (the open circles). (Adapted from Schuler, Lipman and Eaton (2002).)

interpretation was further evidenced by unfolding data in urea and the addition of salt, showing a clear compaction of the chain upon addition of salt. This experiment shows that electrostatic interactions are important determinants of the conformation of the unfolded peptide chain.

The dynamics of the unfolded chain of donor–acceptor-labeled CspTm was directly monitored by Nettels et al., with a sophisticated technique based on a Hanburry-Brown and Twiss detection scheme (Nettels et al., 2007). With this method, the intensity autocorrelation function can be measured down to the nanosecond timescale by circumventing the dead time of the detectors using a pair of them. Chain dynamics associated with fluctuations of the donor–acceptor distance correspondingly results in intensity fluctuations of the donor and acceptor. Thus, the intensity autocorrelation shows a decay on the timescale of the fluctuations. Figure 17.10 shows the donor and acceptor interphoton-time histograms for the unfolded population featuring photon antibunching on the nanosecond timescale typical for a single emitting quantum system. Strikingly, an additional decay of the correlation function is found. The time constant of this exponential decay is 44 nanoseconds, being the reconfiguration time of the chain, which is related to the attempt frequency to cross the energy barrier separating the unfolded and folded state. Thus, single-molecule photon statistics is a valuable tool for the study of chain dynamics associated with the folding reaction.

17.4.1.2 Single Molecule Folding/Unfolding Transitions

A direct observation of folding/unfolding transitions succeeded for the protein adenylate kinase from *Escherichia coli*, immobilized by encapsulation in

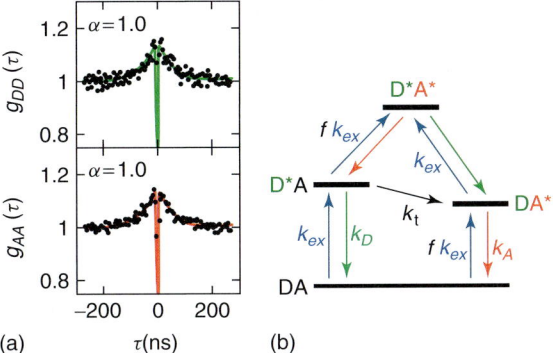

Fig. 17.10 (a) Interphoton-time histogram of donor (top) and acceptor (bottom) fluorescence showing antibunching at short times and a correlation on the 100-nanoseconds timescale. (b) Photophysical model with the ground states A and D and the respective excited states A* and B*. Fluctuations of k_t lead to the decay in the correlation functions shown in (a). (Adapted from Nettels et al., (2007).)

surface-tethered vesicles (Rhoades, Gussakovsky and Haran, 2003). The protein is labeled with a donor–acceptor pair, and their fluorescence is transiently recorded as a function of time. Therefore, the confocal volume is centered on a vesicle with an encapsulated protein. Surprisingly, rather slow transitions between the two states are occasionally observed (Figure 17.11(a)). These slow transitions might indicate the search for the folding pathway in the rugged energy landscape between the folded and unfolded state. Further evidence for the existence of local minima in this landscape is provided by the finding of small steps in the transfer efficiency indicating partial folding/unfolding events (Figure 17.11(b)). Furthermore, there is a broad distribution of step sizes for both, folding and unfolding transitions (Figure 17.11(c)). The stochastic nature of the walk on the energy landscape renders it impossible to observe those in ensemble experiments, clearly pinpointing the advantage of the single-molecule approach.

17.4.2 Enzymes

Single-molecule investigations of enzymatic activity are very appealing for a number of reasons. The most obvious one is common to all single-molecule experiments: the study of heterogeneity. Single-molecule fluorescence, on the other hand, also allows measurements of enzyme kinetics under equilibrium conditions. Furthermore, under nonequilibrium conditions, the system may remain stationary, that is, the concentration of educts and products stays almost constant, due to the low enzyme concentration in single-molecule studies. Eventually, the investigation of conformational changes associated with enzymatic activity may elucidate the mechanisms of catalysis.

17.4.2.1 Static and Dynamic Heterogeneity

The prototype experiment for the study of static and dynamic heterogeneity in enzyme activity was performed by Xie and

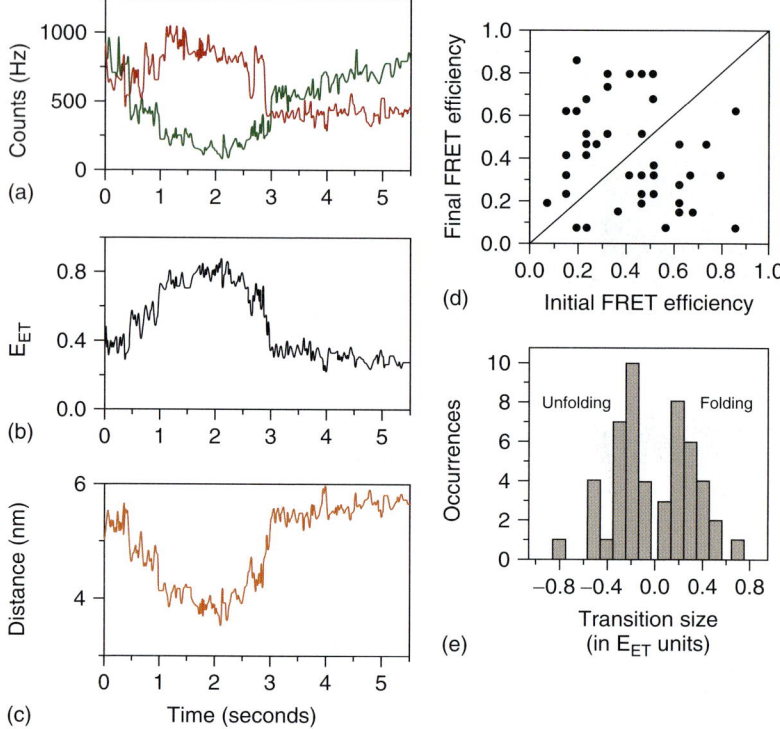

Fig. 17.11 (a) Transient donor and acceptor intensities for a single adenylate kinase molecule inside a vesicle. The anticorrelated behavior is indicative of a change in FRET efficiency shown in (b), which can, in turn, be related to a donor–acceptor distance change shown in (c). (d) Distribution of initial/final FRET efficiencies for folding (below the diagonal line) and unfolding (above the diagonal line) transitions. (e) Distribution of FRET efficiency step sizes of folding/unfolding. (Adapted from Rhoades, Gussakovsky and Haran (2003).)

coworkers with the enzyme cholesterol oxidase (COx) (Lu, Xun and Xie, 1998). FAD, the cofactor of this enzyme, is fluorescent in the oxidized and nonfluorescent in the reduced state. During enzymatic turnover, FAD switches between the oxidized and reduced states.

The COx molecules are embedded in an agarose gel (Figure 17.12(a)), thus freezing their translational mobility while preserving rotational freedom. The latter is very important to avoid fluctuations in the fluorescence signal due to slow reorientation. The confocal volume was centered at one COx molecule, and FAD fluorescence recorded as a function of time.

Binary jumps in the transient fluorescence intensity as shown in Figure 17.12(b) are indicative of cycling between the fluorescent and nonfluorescent state of FAD. Thus, the catalytic activity is directly accessible, where one turnover is just one cycle of high and low fluorescence intensity. The steady-state kinetics can be obtained from the distributions of the on and off states of the fluorescence corresponding to the two modifications of FAD within the reaction cycle. Thereby, both half-cycles of

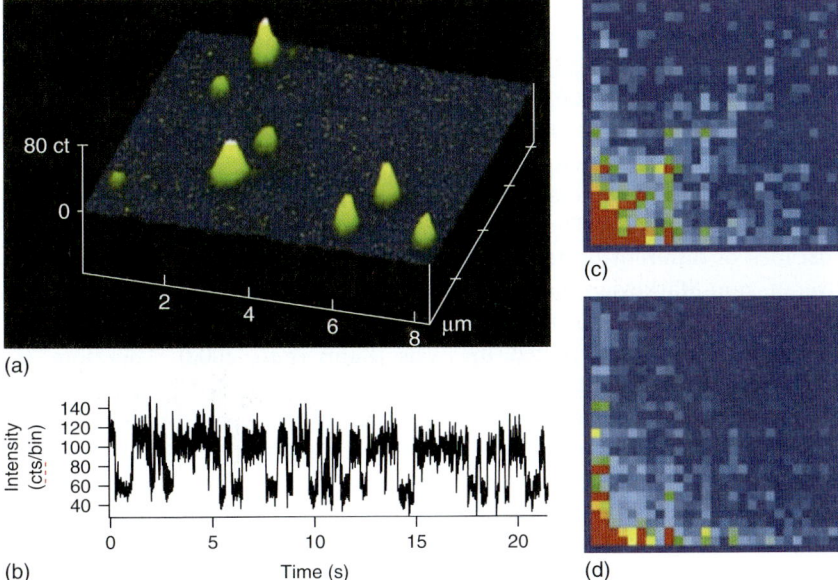

Fig. 17.12 (a) Fluorescence intensity image showing single-Cox molecules. (b) Transient fluorescence intensity of a single-Cox molecule showing binary on/off behavior. (c) and (d) Two-dimensional conditional probability distribution for a pair of two adjacent on-times (a) and for a pair of on-times separated by 10 turnovers. (Adapted from Lu, Xun and Xie (1998).)

the reaction can be analyzed within one single-molecule experiment.

Strikingly, there is a rather broad distribution of the on-times corresponding to the FAD reduction half-cycle for 33 COx molecules, indicative of static disorder, that is, different enzymatic activities of the different enzyme molecules. Now the question arises if there is also dynamic disorder, which is a fluctuation in enzymatic activity over time. To address this question, pairs of on-times separated by a defined number of turnovers are considered and their conditional probabilities calculated. If an on-time of a certain duration is with a high probability followed by an on-time of the same length after this certain number of turnovers, the conditional probability is high. If there is no correlation between the length of the pairs of on-times seperated by the number of turnovers, the conditional probability is low. In a two-dimensional conditional probability distribution histogram, a diagonal feature is evidence for a correlation of the on-time durations of on-states. Such a feature is visible for adjacent on-times (Figure 17.12(c)), but absent for on-times separated by 10 turnover cycles (Figure 17.12(d)). A covariance analysis even allows for the determination of the characteristic timescale for the fluctuations in catalytic activity. Fluctuations of the spectral mean of FAD emission in the absence of cholesterol on the same timescale are interpreted in terms of conformational changes, with the simplest model of interchange between two conformational substates.

Edman and Rigler have used a similar approach for the study of catalytic activity of the enzyme horseradish peroxidase (Edman and Rigler, 2000). Using dihydrorhodamine 6G as a substrate that is nonfluorescent, the catalytic turnover can be followed by the generation of the product rhodamine 6G. The memory landscapes of different proteins are indicative of non-Markovian behavior. Furthermore, oscillations in these memory landscapes can be interpreted by a model where the molecule passes through a number of inactive (conformational) substates and one active substate.

Recently, static heterogeneity was also reported for single-enzyme molecules using a fluorescence-based activity assay with much larger statistics (Rissin, Gorris and Walt, 2008). However, this experiment was based on the determination of product concentration and is therefore not capable of accessing the turnovers of a single-enzyme molecule.

17.4.2.2 Conformational Changes and Catalytic Activity

The prime example for the study of conformational dynamics of single-enzyme molecules is the rotation of F_0F_1 ATPase (Kagawa and Racker, 1966). This membrane protein synthesizes ATP from ADP and phosphate driven by a proton gradient across the membrane. The stepwise rotary motion was first directly observed by videomicroscopy of a fluorescent actin filament attached to the F_1 part (Noji et al., 1997). Although ensemble methods are able to show that rotation occurs, only single-molecule experiments allow for resolution of a sequence of steps. After this key experiment, the rotation was investigated by different single-molecule fluorescence methods Rondelez et al. (2005); Yasuda et al. (2003); Yasuda et al. (2001); Hisabori, Kondoh and Yoshida (1999); Iino et al. (2005); Itoh et al. (2004); Masaike et al. (2000); Nishizaka et al. (2004); Adachi et al. (2007); Kaim et al. (2002); Börsch et al. (2002); Diez et al. (2004); Zimmermann et al. (2006); Steigmiller et al. (2004). Kaim et al. show by polarized detection the stepwise rotation of complete F_0F_1 ATPase from *Propionigenium modestum* both in synthesis and hydrolysis (Kaim et al., 2002). Therefore, the c subunit was labeled by a fluorescent dye with two linkers in order to assure rigid joining (Figure 17.13(a)). From the transient fluorescence detected in two channels corresponding to orthogonal polarizations the polarization P can be calculated according to

$$P = \frac{I_\parallel - I_\perp}{I_\parallel + I_\perp} \quad (17.17)$$

The time-dependent polarization in the presence of ATP and Na$^+$ in Figure 17.13(b) shows stepwise changes between three levels, indicating rotational motion in three steps. A data analysis by means of the autocorrelation function of the transient polarization revealed that a model with three pairs of 90°/30° steps as proposed by Yasuda et al. (Yasuda et al., 2001) was most consistent with the data.

Single-molecule FRET experiments were also able to show stepwise rotation of F_0F_1 ATPase. Incorporating doubly labeled ATPase (Figure 17.13) in liposomes freely diffusing in solution, Diez et al. recorded FRET efficiency transients during passage of the liposomes through the focal volume (Diez et al., 2004). Those transients show stepwise changes both in hydrolysis and in synthesis between three levels (Figure 17.13(d)

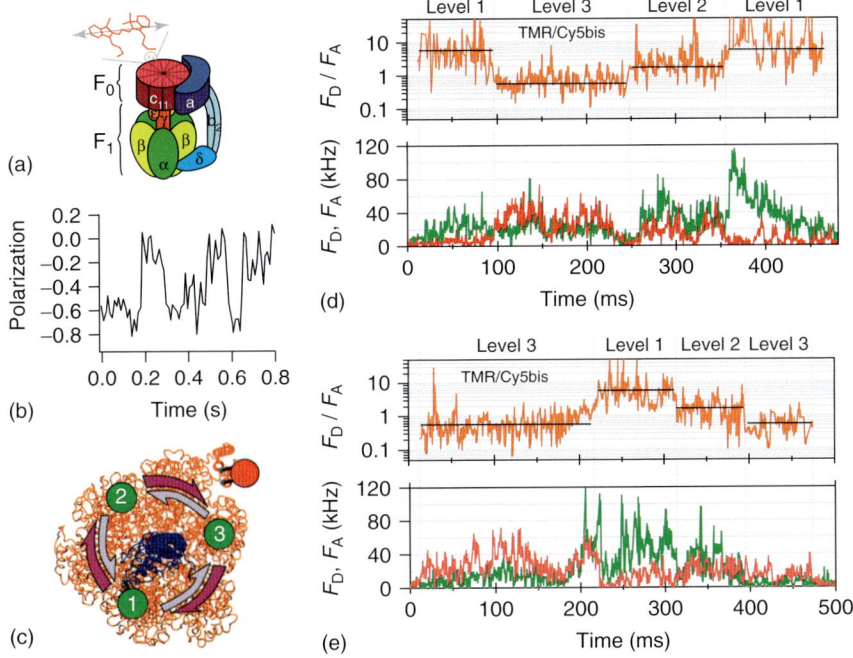

Fig. 17.13 F_0F_1 ATPase labeled at the c subunit (a) shows steps in the polarization transient (b). Donor–acceptor-labeled ATPase (c) exhibits jumps between three levels in the ratio between the fluorescence of the donor and acceptor both in hydrolysis (d) as well as in synthesis (e) with reversed order. (Adapted from Kaim et al. (2002) and Diez et al. (2004).)

and (e)). Strikingly, the order of the levels is reversed in hydrolysis/synthesis, thus elegantly demonstrating the reverse of rotation. Recently, Yasuda et al. were able to identify the transiently populated ATP waiting conformation by single-molecule FRET experiments on immobilized ATPase molecules (Yasuda et al., 2003).

The first single-molecule FRET demonstration of conformational dynamics associated with enzymatic activity succeeded with the enzyme *staphylococcal nuclease* (SNase) in the group of Shimon Weiss (Ha et al., 1999). SNase labeled with a donor–acceptor pair showed fluctuations of the transfer efficiency that can be quantitatively analyzed in terms of the autocorrelation function of the transfer efficiency shown in Figure 17.14. The decay in the E_t autocorrelation function can be fit by an exponential with time constants ranging from some tens to thousands of milliseconds, with an average of 41 milliseconds. In the presence of pTp, a known inhibitor of SNase activity, this mean is shifted to 133 milliseconds. Therefore, it can be concluded that the inhibitor affects the conformational flexibility that is essential for SNase function. By labeling of the SNase with donor and the DNA substrate with the acceptor, SNase activity can be directly monitored via the association time, which is the time period of high FRET efficiency. A clear difference in association times is observed if the DNA is labeled either at the 5′ or 3′ end. This difference is

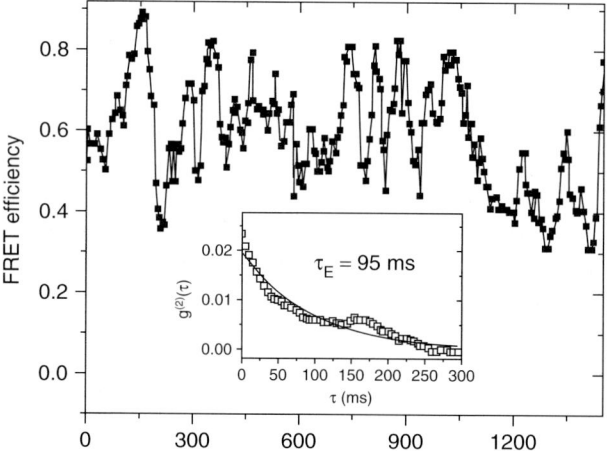

Fig. 17.14 FRET efficiency fluctuations of a single SNase molecule immobilized on a coverslip. The inset shows the autocorrelation function of the the FRET efficiency together with an exponential fit. (Adapted from Ha et al. (1999).)

in agreement with SNase processivity in the 3′–5′ direction.

Quenching of the dye AlexaFluor 488 was exploited for the investigation of conformational transitions of the enzyme dihydrofolate reductase (DHFR) (Zhang et al., 2004). DHFR is a two-substrate enzyme catalyzing the reduction in dihydrofolate (DHF) to tetrahydrofolate (THF) by NADPH. To this end, AlexaFluor 488 was specifically attached to the structural loop that closes over the substrates after binding. Single- DHFR molecules were tethered on biotynilated cover slips via a biotin–streptavidin–biotin sandwich. In a TIRF illumination wide-field set-up, the transient intensities of many DHFR molecules could be recorded in parallel by an intensified CCD camera. Upon addition of DHF, conformational dynamics manifests itself in intensity fluctuations of the Alexa dye. In the simplest model, DHFR is switching between two conformational states with an opening and closing rate. Thus, using a threshold criterion, open and closed states can be assigned, delivering the dwell times in the respective states. The histograms of the dwell times provide the rates of opening and closing. From the concentration dependence of the closing rate in conjunction with transient ensemble kinetics measurements a model can be derived where rapid binding of DHF is followed by a conformational change.

Henzler-Wildman et al. have investigated the conformational heterogeneity and dynamics of Adk from *aquifex aeolicus* by X-ray crystallography, molecular dynamics simulations, NMR spectroscopy, and smFRET (Henzler-Wildman et al., 2007). Adk is a two-substrate enzyme catalyzing phosphotransfer between AMP and ATP in the presence of Mg^{2+}. From X-ray crystallography in the presence of the substrates it is known that a conformational change occurs when the substrates are bound. Adk was specifically labeled with AlexaFluor 488 as the donor and AlexaFluor

633 as the acceptor at the two substrate-binding lids. First, freely diffusing Adk was investigated, and burstwise histograms of the transfer efficiencies were derived. Surprisingly, even in the absence of any substrate, Adk shows at least two subpopulations, where one corresponds to the conformation where both lids are opened and the other one corresponds to the closed conformation, which can be concluded by comparing the distances from the FRET experiments with crystallographic data. Upon addition of the inhibitor diadenosin pentaphosphate (AP5A), the equilibrium between the two conformations shifts toward the closed conformation.

To study the dynamics of conformational exchange, labeled Adk was immobilized on a functionalized cover slip via a His$_6$ tag. In spite of the considerable noise level, which results from the short integration time dictated by the required time resolution, transitions between the open and closed states can be observed. The overlap of the transfer efficiency histograms of two conformations renders a single threshold criterion for the discrimination of the two states unusable. Instead, a threshold zone was defined, where a change-of-state required a complete crossing of the zone. After assingment of the open and closed states, respectively, the dwell times in these states are compiled in histograms. From the histograms shown in Figure 17.15(a) and (b) the rate constants for opening and closing are derived. The values found for both cases, $k_{open} = 6500 \pm 500\,\mathrm{s}^{-1}$ and $k_{close} = 2000 \pm 200\,\mathrm{s}^{-1}$ for the substrate-free and $k_{open} = 40 \pm 20\,\mathrm{s}^{-1}$ and $k_{close} = 350 \pm 200\,\mathrm{s}^{-1}$ for the inhibitor-loaded enzyme, turn out to be very similar to the rates of chemical exchange found by NMR.

The assignment of open/closed states based solely on the apparent FRET efficiency is error-prone due to photophysical processes, which may also lead to fluctuations in the apparent FRET efficiency. It needs to be checked explicitly whether anticorrelated intensity changes of the donor and acceptor fluorescence signals occur. Therefore, a correlation analysis was performed. Clear evidence for anticorrelated behavior of donor–acceptor fluorescence intensity is provided by an anticorrelation in the donor–acceptor cross-correlation function as can be seen in Figure 17.15(c). At shorter times, a correlated fluctuation of donor and acceptor fluorescence is found, which is attributed to triplet dynamics of the donor and/or triplet dynamics of the acceptor and efficient STA. A global fit of the three correlation functions, namely, the donor and acceptor autocorrelation and their respective cross-correlation functions, provides the overall rate of distance dynamics and triplet kinetics. The overall rate of distance dynamics of $k_{tot} = 7000 \pm 2000\,\mathrm{s}^{-1}$ is in good agreement with the rate from NMR and from the states analysis. During catalysis, opening and closing rates of $k_{open} = 40 \pm 10\,\mathrm{s}^{-1}$ and $k_{close} = 1000 \pm 500\,\mathrm{s}^{-1}$ are found matching the NMR data on chemical exchange. Taking into account that NMR can only provide information about the occurrence and timescale of dynamics, but no spatial information, these experiments nicely demonstrate that single-molecule FRET experiments perfectly complement other biophysical methods.

Single-molecule FRET experiments have also been performed on E. coli AdK, where opening and closing rates of $k_{open} = 120 \pm 40\,\mathrm{s}^{-1}$ and $k_{close} = 220 \pm$

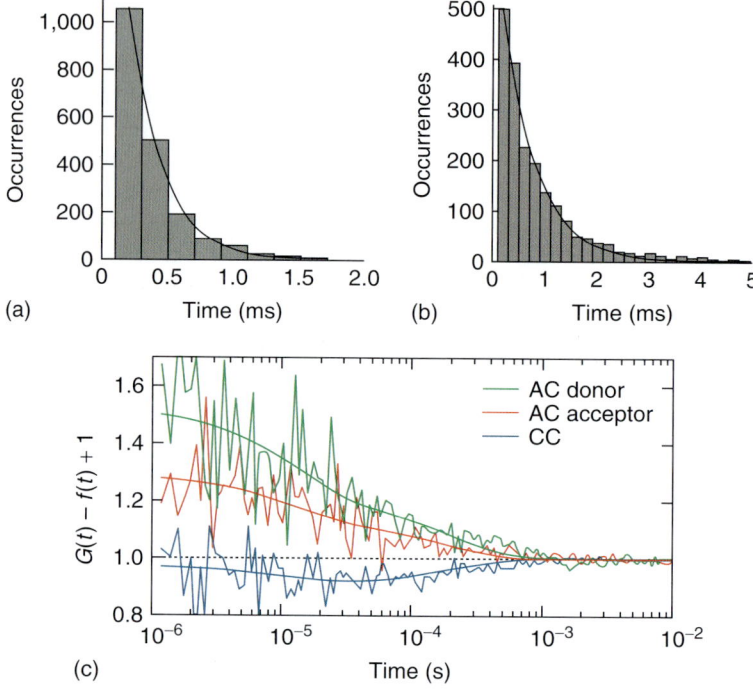

Fig. 17.15 Histograms of (a) low FRET efficiency and (b) high FRET efficiency states in the transient fluorescence of doubly labeled adenylate kinase from aquifex aeolicus attributed to an open and closed conformation. (c) Donor (green) and acceptor (red) autocorrelation and donor–acceptor cross-correlation function (blue) of the same transients. (Adapted from Henzler-Wildman et al. (2007).)

$70\,s^{-1}$ were found for substrate-free Adk (Hanson et al., 2007).

17.5 Conclusions

In recent years, single-molecule fluorescence has proven to be a valuable tool to obtain information on the conformational dynamics and heterogeneity of proteins that is not accessible by ensemble methods. However, the method is still far from being routinely applicable, mostly for problems with photophysics and photochemistry of the dye molecules. There have been several attempts made to improve the photostability of the dyes, but what works out for a particular dye may be even counterproductive for another. More effort has to be made on both synthesizing new dyes with better stability (Weil et al., 2005) and improving the photostability of existing dyes (Rasnik, McKinney and Ha, 2006).

On the other hand, the noisy signal of a single molecule remains a challenge for data analysis, which might be confronted by methods such as Hidden–Markov modeling (McKinney, Joo and Ha, 2006).

References

Adachi, K., Oiwa, K., Nishizaka, T., Furuike, S., Noji, H., Itoh, H., Yoshida, M. and Kinosita, K. Jr (2007) Coupling of rotation and catalysis in F-1-ATPase revealed by single-molecule imaging and manipulation. *Cell*, **130**, (2), 309–321. DOI: 10.1016/j.cell.2007.05.020.

Berberan-Santos, M. and Prieto, M. (1988) Monte carlo simulation of orientational effects on direct energy transfer. *J. Chem. Phys.*, **88**, (10), 6341–6349.

Börsch, M., Diez, M., Zimmermann, B., Reuter, R. and Gräber, P. (2002) Stepwise rotation of the gamma-subunit of EF0F1-ATP synthase observed by intramolecular single-molecule fluorescence resonance energy transfer. *FEBS Lett.*, **527**, (1–3), 147–152.

Deniz, A., Laurence, T., Beligere, G., Dahan, M., Martin, A., Chemla, D., Dawson, P., Schultz, P. and Weiss, S. (2000) Single-molecule protein folding: Diffusion fluorescence resonance energy transfer studies of the denaturation of chymotrypsin inhibitor 2. *PNAS*, **97**, (10), 5179–5184.

Diez, M., Zimmermann, B., Börsch, M., König, M., Schweinberger, E., Steigmiller, S., Reuter, R., Felekyan, S., Kudryavtsev, V., Seidel, C.A.M. *etal* (2004) Proton-powered subunit rotation in single membrane-bound F0F1-ATP synthase. *Nat. Struct. Mol. Biol.*, **11**, (2), 135–141.

Edman, L. and Rigler, R. (2000) Memory landscapes of single-enzyme molecules. *Proc. Natl. Acad. Sci. U.S.A.*, **97**, (15), 8266–8271.

Förster, T. (1946) Zwischenmolekulare Energiewanderung und Fluoreszenz. *Ann. Phys. (Berlin)*, **6**, 166–175.

Ha, T., Enderle, T., Ogletree, D., Chemla, D., Selvin, P. and Weiss, S. (1996) Probing the interaction between two single molecules: Fluorescence resonance energy transfer between a sinlge donor and a single acceptor. *Proc. Natl. Acad. Sci.*, **93**, 6264–6268.

Ha, T.J., Ting, A.Y., Liang, J., Caldwell, W.B., Deniz, A.A., Chemla, D.S., Schultz, P.G. and Weiss, S. (1999) Single-molecule fluorescence spectroscopy of enzyme conformational dynamics and cleavage mechanism. *Proc. Natl. Acad. Sci. U.S.A.*, **96**, (3), 893–898.

Hanson, J.A., Duderstadt, K., Watkins, L.P., Bhattacharyya, S., Brokaw, J., Chu, J.W. and Yang, H. (2007) Illuminating the mechanistic roles of enzyme conformational dynamics. *Proc. Natl. Acad. Sci.*, **104**, (46), 18–055–18–060.

Henzler-Wildman, K.A., Thai, V., Lei, M., Ott, M., Wolf-Watz, M., Fenn, T., Pozharski, E., Wilson, M.A., Petsko, G.A., Karplus, M. *etal* (2007) Intrinsic motions along an enzymatic reaction trajectory. *Nature*, **450**, (7171), 838–844. DOI: 10.1038/nature06410.

Hirschfeld, T. (1976) Optical microscopic observation of single small molecules. *Appl. Opt*, **15**, (12), 2965–2966.

Hisabori, T., Kondoh, A. and Yoshida, M. (1999) The gamma subunit in chloroplast F-1-ATPase can rotate in a unidirectional and counter-clockwise manner. *FEBS Lett.*, **463**, (1–2), 35–38.

Hofmann, H., Golbik, R.P., Ott, M., Hübner, C.G. and Ulbrich-Hofmann, R. (2008) Coulomb forces control the density of the collapsed unfolded state of barstar. *J. Mol. Biol.*, **376**, (2), 597–605. DOI: 10.1016/j.jmb.2007.11.083.

Hübner, C.G., Ksenofontov, V., Nolde, F., Müllen, K. and Basché, T. (2004) Three-dimensional orientational colocalization of individual donor-acceptor pairs. *J. Chem. Phys.*, **120**, (23), 10867–10870.

Iino, R., Rondelez, Y., Yoshida, M. and Noji, H. (2005) Chemomechanical coupling in single-molecule F-type ATP synthase. *J. Bioenerg. Biomembr.*, **37**, (6), 451–454. DOI: 10.1007/s10863-005-9489-5.

Itoh, H., Takahashi, A., Adachi, K., Noji, H., Yasuda, R., Yoshida, M. and Kinosita, K. (2004) Mechanically driven ATP synthesis by F-1-ATPase. *Nature*, **427**, (6973), 465–468. DOI: 10.1038/nature02212.

Kagawa, Y. and Racker, E. (1966) Partial resolution of enzymes catalyzing oxidative phosphorylation. 9. reconstruction of oligomycin-sensitive adenosine triphosphatase. *J. Biol. Chem.*, **241**, (10), 2467–2474.

Kaim, G., Prummer, M., Sick, B., Zumofen, G., Renn, A., Wild, U.P. and Dimroth, P. (2002) Coupled rotation within single F0F1 enzyme complexes during ATP synthesis or hydrolysis. *FEBS Lett.*, **525**, (1–3), 156–163.

Kelley, A.M., Michalet, X. and Weiss, S. (2001) Chemical physics - single-molecule spectroscopy comes of age. *Science*, **292**, (5522), 1671–1672.

Lu, H.P., Xun, L.Y. and Xie, X.S. (1998) Single-molecule enzymatic dynamics. *Science*, **282**, (5395), 1877–1882.

Masaike, T., Mitome, N., Noji, H., Muneyuki, E., Yasuda, R., Kinosita, K. and Yoshida, M. (2000) Rotation of F-1-ATPase and the hinge residues of the beta subunit. *J. Exp. Biol.*, **203**, (1), 1–8.

Maus, M., De, R., Lor, M., Weil, T., Mitra, S., Wiesler, U.M., Herrmann, A., Hofkens, J., Vosch, T. and Müllen, K. etal (2001) Intramolecular energy hopping and energy trapping in polyphenylene dendrimers with multiple peryleneimide donor chromophores and a terryleneimide acceptor trap chromophore. *J. Am. Chem. Soc.*, **123**, (31), 7668–7676.

McKinney, S.A., Joo, C. and Ha, T. (2006) Analysis of single-molecule FRET trajectories using hidden Markov modeling. *Biophys. J.*, **91**, (5), 1941–1951. DOI: 10.1529/biophysj.106.082487.

Minsky M. (1957) Microscopy apparatus. US patent 3013467.

Nettels, D., Gopich, IV, Hoffmann, A. and Schuler, B. (2007) Ultrafast dynamics of protein collapse from single-molecule photon statistics. *Proc. Natl. Acad. Sci.* 2007; **104**, (8): 2655–2660.

Nishizaka, T., Oiwa, K., Noji, H., Kimura, S., Muneyuki, E., Yoshida, M. and Kinosita, K. (2004) Chemomechanical coupling in F-1-ATPase revealed by simultaneous observation of nucleotide kinetics and rotation. *Nat. Struct. Mol. Biol.*, **11**, (2), 142–148. DOI: 10.1038/nsmb721.

Noji, H., Yasuda, R., Yoshida, M. and Kinosita, J.K. (1997) Direct observation of the rotation of F/sub 1/-ATPase. *Nature*, **386**, (6622), 299–302.

Orrit, M. and Bernard, J. (1990) Single pentacene molecules detected by fluorescence excitation in a p-terphenyl crystal. *Phys. Rev. Lett.*, **65**, (21), 2716–2719.

Rasnik, I., McKinney, S. and Ha, T. (2006) Nonblinking and long-lasting single-molecule fluorescence imaging. *Nat. Methods*, **3**, (11), 891–893.

Rhoades, E., Gussakovsky, E. and Haran, G. (2003) Watching proteins fold one molecule at a time. *Proc. Natl. Acad. Sci. U.S.A.*, **100**, 3197–3202.

Rissin, D.M., Gorris, H.H. and Walt, D.R. (2008) Distinct and long-lived activity states of single enzyme molecules. *J. Am. Chem. Soc.*, **130**, (15), 5349–5353.

Rondelez, Y., Tresset, G., Nakashima, T., Kato-Yamada, Y., Fujita, H., Takeuchi, S. and Noji, H. (2005) Highly coupled ATP synthesis by F-1-ATPase single molecules. *Nature*, **433**, (7027), 773–777. DOI: 10.1038/nature03277.

Schmidt, T., Schütz, G.J., Baumgartner, W., Gruber, H.J. and Schindler, H. (1995) Characterization of photophysics and mobility of single molecules in a fluid lipid membrane. *J. Phys. Chem.*, **99**, (49), 17662–17668.

Schuler, B., Lipman, E.A. and Eaton, W.A. (2002) Probing the free-energy surface for protein folding with single-molecule fluorescence spectroscopy. *Nature*, **419**, 743–747.

Schütz, G.J., Kada, G., Pastushenko, V.P. and Schindler, H. (2000) Properties of lipid microdomains in a muscle cell membrane visualized by single molecule microscopy. *EMBO J.*, **19**, (5), 892–901.

Shera, E.B., Seitzinger, N.K., Davis, L.M., Keller, R.A. and Soper, S.A. (1990) Detection of single fluorescent molecules. *Chem. Phys. Lett.*, **174**, 553–557.

Steigmiller, S., Zimmermann, B., Diez, M., Börsch, M. and Gröber, P. (2004) Binding of single nucleotides to H+-ATP synthases observed by fluorescence resonance energy transfer. *Bioelectrochemistry*, **63**, (1–2), 79–85. DOI: 10.1016/j.bioelechem.2003.08.008.

Strickler, S.J. and Berg, R.A. (1962) Relationship between absorption intensity and fluorescence lifetime of molecules. *J. Chem. Phys.*, **37**, (4), 814–822.

Stryer, L. and Haugland, R. (1967) Energy transfer: a spectroscopic ruler. *PNAS*, **58**, 719–726.

Weil, T., Abdalla, M.A., Jatzke, C., Hengstler, J. and Müllen, K. (2005) Water-soluble rylene dyes as high-performance colorants for the staining of cells. *Biomacromolecules*, **6**, (1), 68–79.

Yasuda, R., Masaike, T., Adachi, K., Noji, H., Itoh, H. and Kinosita, K. (2003) The ATP-waiting conformation of rotating F-1-ATPase revealed by single-pair fluorescence resonance energy transfer. *Proc. Natl. Acad. Sci. U.S.A.*, **100**, (16), 9314–9318.

Yasuda, R., Noji, H., Yoshida, M., Kinosita, K. and Itoh, H. (2001) Resolution of distinct rotational substeps by submillisecond kinetic analysis of F-1-ATPase. *Nature*, **410**, (6831), 898–904.

Zhang, Z.Q., Rajagopalan, P.T.R., Selzer, T., Benkovic, S.J. and Hammes, G.G. (2004) Single-molecule and transient kinetics investigation of the interaction of dihydrofolate reductase with NADPH and dihydrofolate. *Proc. Natl. Acad. Sci. U.S.A.*, **101**, (9), 2764–2769.

Zimmermann, B., Diez, M., Börsch, M. and Gröber, P. (2006) Subunit movements in membrane-integrated EF0F1 during ATP synthesis detected by single-molecule spectroscopy. *Biochim. Biophys. Acta Bioenerg.*, **1757**, (5–6), 311–319. DOI: 10.1016/j.bbabio.2006.03.020.

18
Laser Sources

David L. Andrews and Robert H. Lipson

18.1	**Introduction** 657	
18.2	**Optically Pumped Solid-State Lasers** 657	
18.2.1	Ruby Laser 659	
18.2.2	Neodymium Lasers 660	
18.2.3	Widely Tunable Lasers 662	
18.2.4	Color-Center Lasers 662	
18.2.5	Fiber Lasers 662	
18.3	**Diode Lasers** 664	
18.3.1	Vertical Cavity Surface Emitter Lasers 666	
18.3.2	Visible Laser Diodes 666	
18.3.3	Quantum Dot Lasers 667	
18.3.4	Quantum Cascade Lasers 667	
18.3.5	Quantum Microcavity Lasers 668	
18.4	**Atomic and Ionic Gas Lasers** 669	
18.4.1	Helium–Neon Laser 669	
18.4.2	Argon and Krypton Lasers 671	
18.5	**Molecular Gas Lasers** 672	
18.5.1	Carbon Dioxide Laser 672	
18.5.2	Nitrogen Laser 675	
18.5.3	Iodine Laser 675	
18.5.4	Excimer Lasers 676	
18.5.5	Dye Lasers 677	
18.6	**Pulsing Techniques** 680	
18.6.1	Cavity Dumping 680	
18.6.2	Q-Switching 681	
18.6.3	Mode-Locking 681	
18.6.4	Ultrafast Pulse Compression 683	
18.6.5	Continuum Generation 684	
18.7	**Short Wavelength Coherent Radiation** 684	
18.7.1	Frequency Conversion 684	
18.7.2	Frequency Doubling 685	

Encyclopedia of Applied Spectroscopy. Edited by David L. Andrews.
Copyright © 2009 WILEY-VCH Verlag GmbH & Co. KGaA, Weinheim
ISBN: 978-3-527-40773-6

18.7.3 Optical Parametric Oscillators 687
18.7.4 Frequency Tripling and Four-Wave Mixing 687
18.7.5 Higher Order Harmonic Generation 689
18.7.6 Free-Electron Laser 691
18.7.7 Tabletop X-ray Lasers 692
References 692

18.1
Introduction

The uses of lasers in spectroscopy are extensive. The key characteristics of laser light – particularly the unparalleled combination of an essentially monochromatic output at high intensity levels – are extremely well suited to a wide variety of spectroscopic techniques. Often, high laser powers are exploited in the determination of weak or transient spectral features, or to produce optically nonlinear processes; equally, the narrow bandwidth of many laser sources is utilized for high-resolution studies. A wide variety of optical materials is employed in the lasers themselves, and the detailed nature of these materials determines the principal output characteristics such as wavelength and achievable power. Since the construction of the first laser based on ruby, widely ranging materials have been adopted as laser media. The range is still being extended to provide output at new wavelengths and, as can be seen from Table 18.1, the output available from commercial lasers now covers most of the electromagnetic spectrum.

By a variety of means, laser light can also be tailored to specific applications by frequency-conversion and pulse-generation techniques; the facility to modify these characteristics is again important for many spectroscopic investigations. This chapter reviews the basic mechanisms and principles of operation of the most important laser sources and beam management techniques, focusing on systems that are especially well suited to spectroscopic applications. Space does not permit a thorough description of the entire range of laser systems now available; the more limited scope of the following account is a discussion of representative laser instrumentation.

18.2
Optically Pumped Solid-State Lasers

The active medium in most solid-state lasers, capable of delivering 1 W or more power in the UV to near-IR range, is generally a transparent crystal or glass doped with a small amount of transition metal; transitions in the metal ions are responsible for the laser action. The two most common dopants are chromium (in the ruby and alexandrite lasers) and neodymium (Nd : YAG and Nd : glass lasers), with

Encyclopedia of Applied Spectroscopy. Edited by David L. Andrews.
Copyright © 2009 WILEY-VCH Verlag GmbH & Co. KGaA, Weinheim
ISBN: 978-3-527-40773-6

Tab. 18.1 Wavelengths available from commercial lasers.[a]

λ/nm	Laser	λ/nm	Laser	λ/nm	Laser
157	Fluorine	428	Nitrogen	568	Krypton
174	Ruby × 4	437	Argon	578	Copper
193	Argon fluoride	442	Helium–cadmium	596	Xenon
213	Nd:YAG × 5	455	Argon	628	Gold
222	Krypton chloride	458	Krypton	633	Helium–neon
231	Ruby × 3	458	Argon	647	Krypton
244	Argon × 2	462	Krypton	657	Krypton
248	Krypton fluoride	463	Krypton	677	Krypton
257	Argon × 2	466	Argon	680	GaAlInphosphide
263	Nd:YLF × 4	468	Krypton	687	Krypton
266	Nd:YAG × 4	473	Argon	694	Ruby
275	Argon	476	Krypton	723	Lead
305	Argon	477	Argon	753	Krypton
308	Xenon chloride	477	Krypton	799	Krypton
312	Gold	483	Krypton	852	Calcium
325	Helium–cadmium	485	Krypton	866	Calcium
334	Argon	488	Argon	904	Gallium arsenide
337	Nitrogen	496	Xenon	1053	Nd:YLF
347	Ruby × 2	497	Argon	1060	Nd:glass
351	Krypton	502	Argon	1064	Nd:YAG
351	Nd:YLF × 3	511	Copper	1092	Argon
351	Xenon fluoride	515	Argon	1130	Barium
351	Argon	521	Krypton	1152	Helium–neon
353	Xenon fluoride	527	Nd:YLF × 2	1290	Manganese
355	Nd:YAG × 3	529	Argon	1300	In Ga As phosphide
356	Krypton	531	Krypton	1313	Nd:YLF
364	Argon	532	Nd:YAG × 2	1315	Iodine
407	Krypton	534	Manganese	1319	Nd:YAG
413	Krypton	540	Xenon	1500	In Ga As phosphide
415	Krypton	544	Helium–neon	1523	Helium–neon

[a] Emission lines of commonly available discrete-wavelength lasers over the range 100–1600 nm, harmonics indicated by × 2, × 3 etc. The box indicates wavelengths encompassed by the tuning range of a titanium:sapphire source.

the dopant concentration typically 1% or less. Such lasers are optically pumped by a broadband flash-lamp source, and can be pulsed to produce the high intensities.

18.2.1
Ruby Laser

The ruby laser has an important place in history as it was the first type ever constructed (in 1960). In construction, the modern ruby laser comprises a rod of commercial ruby (0.05% Cr_2O_3 in an Al_2O_3 lattice) of between 3 and 25 mm diameter and up to 20 cm length. The chromium ions are excited by the emission from a flash-lamp coiled around it, or placed alongside it within an elliptical reflector as shown in Figure 18.1.

From the energy level scheme shown in Figure 18.2, it is apparent that the ruby laser may be regarded as a *pseudo three-level system*, because in the course of excitation, decay, and emission, only three Cr^{3+} levels are involved; $^4A_2$, $^4T_1$ or $^4T_2$, and ^2E. These energy levels are quite different from those of a free chromium atom, due to crystal field splitting. The initial flash-lamp excitation takes the Cr^{3+} ions up from the ground state E_1 ($^4A_2$) to one of the two E_3 (^4F) levels. Both of these levels have subnanosecond lifetimes, and rapidly decay into the E_2 level (^2E). The decay takes place by nonradiative processes that channel energy into lattice vibrations, heating the crystal. Because the E_2 level has a much longer decay time of around 4 ms, a population inversion is created between the E_2 and E_1 states, leading to laser emission at a wavelength of 694.3 nm as the majority of Cr^{3+} ions cascade down to the ground state. The lasing process thus generates a single, intense pulse of light, typically between 0.3 and 3 ms duration; the delay between successive pulses usually lasts several seconds, and can be as long as a full minute. A very different kind of pulsing can be created by the technique known as *Q-switching*, discussed in Section 18.6.2

A problem with the ruby laser, common to all lasers of this type, is the damage caused by the repeated cycle of heating and cooling associated with the generation of each pulse, which

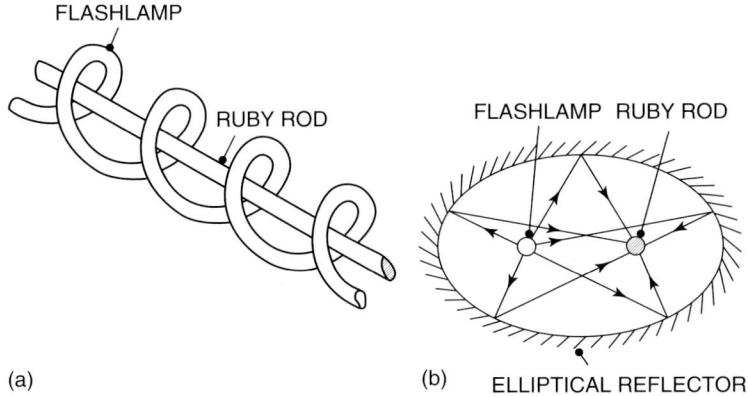

(a) (b)

Fig. 18.1 Arrangement of flash-lamp tube and ruby rod in a ruby laser.
(a) Helical flash-lamp option; (b) confocal flash-lamp and laser rod within an elliptical reflector.

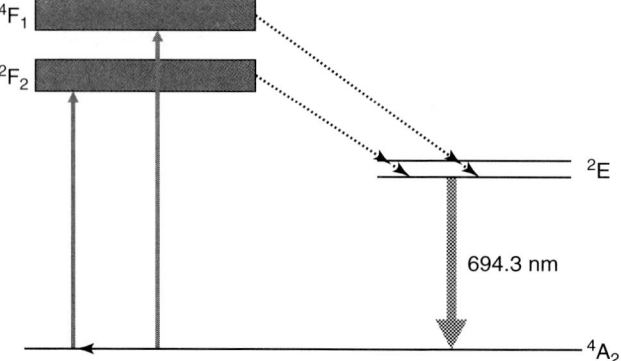

Fig. 18.2 Energetics of the three-level ruby laser. The vertical arrows upward on the left hand side refer to pumping transitions. The dashed lines refer to nonradiative relaxation from the pump bands into the upper lasing level. The broad vertical arrow downward corresponds to the laser line at 649.3 nm.

ultimately necessitates the replacement of the ruby rod. To improve performance, the rod usually needs to be cooled by circulation of water in a jacket around it. Despite its drawbacks, with pulse energies as high as 200 J, the ruby laser represents a powerful source of monochromatic light in the optical region. The emission bandwidth is typically about 0.5 nm, but this can be reduced by a factor of up to 10^4 by suitable optics. The beam diameter of a low-power ruby laser can be as little as 1 mm, with 0.25 mrad divergence; the most powerful lasers may have beams up to 25 mm in diameter and a divergence of several mrad.

18.2.2
Neodymium Lasers

Neodymium lasers are of two main types. In one, the host lattice for the neodymium ions is a crystal – usually yttrium aluminum garnet crystal $(Y_3Al_5O_{12})$ – and in the other, the host is an amorphous glass. These are referred to as Nd : YAG and Nd : glass, respectively. Although transitions in the neodymium ions are responsible for laser action in both cases, the emission characteristics differ because of the influence of the host lattice on the neodymium energy levels. Also, glass does not have the excellent thermal conductivity properties of YAG crystal, so that it is more suitable for pulsed than for continuous-wave (CW) operation. Like the ruby laser, both types of neodymium laser are normally pumped by a flashlamp arranged confocally alongside a rod of the laser material in an elliptical reflector. The energy levels of neodymium ions involved in laser action are split by interaction with the crystal field, as illustrated in Figure 18.3. As a result, transitions between components of the $^4F_{3,2}$ and $^4I_{11/2}$ states, forbidden in the free state, are permitted and can give rise to laser emission. The $^4F_{3,2}$ levels are initially populated following nonradiative decay from higher energy levels excited by the flash-lamp, and because the terminal $^4I_{11/2}$ laser level lies above

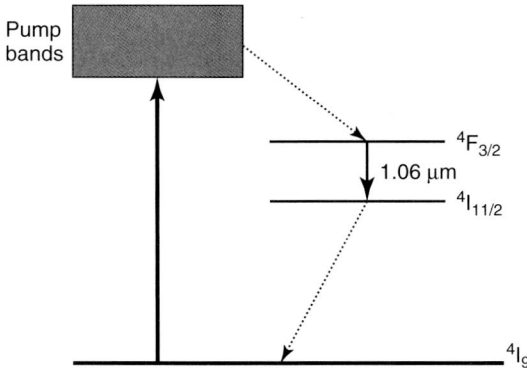

Fig. 18.3 Simplified energy levels of a four-level neodymium laser lasing at 1.06 µm.

the $^4I_{9/2}$ ground state, we thus have a pseudo-four-level system.

The principal emission wavelength for both types of neodymium laser is around 1.064 µm, in the near-IR (or 1.053 µm in the less common Nd:YLF lasers, where lithium replaces aluminum in the host lattice). Some commercial lasers can also operate on a different transition, producing 1.319 µm output. The YAG and glass host materials impose very different characteristics on the emission, however. Quite apart from the differences in thermal conductivity, which determine the question of continuous or pulsed operation, one of the main differences appears in the line width. Because glass has an amorphous structure, the crystal field splitting also varies from site to site. Hence, the line width is much broader than in the Nd:YAG laser, where the lattice is much more regular and the field splitting is more uniform. However, the concentration of neodymium dopant in glass may be as large as 6%, compared with 1.5% in a YAG host, so that a much higher energy output can be obtained. For both these reasons, the Nd:glass laser is ideally suited for the production of extremely high intensity ultrashort pulses by the technique of mode-locking (see Section 18.6.3). In fact, amongst all commercially available lasers, it is the neodymium lasers that deliver the highest available laser beam intensities.

The output power of a typical CW Nd:YAG laser is several watts, and can exceed 200 W. When operating in a pulsed mode, the energy per pulse varies according to the method of pulsing and the pulse repetition rate, but can be anywhere between a small fraction of a joule and 100 J for a single pulse. The majority of Nd:YAG research applications in spectroscopy make use not of the 1.064 µm radiation as such, but rather the high-intensity visible light that can be produced by frequency-conversion methods. Particularly important in this respect are the wavelengths of 532, 355, and 266 nm obtained by harmonic generation (see Section 18.7.1).

Diode-pumped solid-state lasers represent a very different laser technology. These miniature-scale devices, which can often fit comfortably into the hand, incorporate a semiconductor diode laser (Section 18.3) to directly pump a small Nd:YAG crystal. Cavity modifications enable line-widths of less than 5 kHz (2×10^{-7} cm^{-1}) to be attained in certain models. While CW output powers are generally low (milliwatt scale), the

1-W target is exceeded in some commercially available systems. Also important is the common inclusion of frequency-conversion crystals in these devices, so providing milliwatt emission in the visible region, especially a 532-nm line in the green. Such integrated solid-state lasers have the advantages of smaller size, higher efficiency, better stability, and lower noise levels than most gas discharge lasers.

18.2.3
Widely Tunable Lasers

Recently, other types of solid-state laser have been developed which, although again based on optical transitions in transition metal ions embedded in a host ionic crystal, produce *tunable* radiation. For a number of spectroscopic applications, such lasers represent an attractive alternative to tunable lasers based on organic dyes. Their chief advantages are an appreciably more rugged and compact construction and operation without use of toxic chemicals.

The recently developed titanium : sapphire laser, commonly pumped by a primary laser beam such as the all-lines output of an ion gas laser, is the best known example in this category, and is today one of the most widely used commercially available solid-state lasers. Crucially, the coupling of electronic levels with lattice vibrations here leads to a broad, continuous band of *vibronic* energy levels. On the basis of the Al_2O_3 crystal doped with approximately 0.1% Ti_2O_3, lasing occurs on vibronic $^2E \rightarrow {}^2T_2$ transitions in the Ti^{3+} ions, offering high-power tunable CW or pulsed emission over the exceptionally large wavelength range 650–1100 nm in the near-infrared, peaking near 800 nm. At that wavelength, over 5 W CW outputs can be achieved. The titanium : sapphire laser is finding an increasingly wide range of applications, particularly in spectroscopy. Other commercially available vibronic lasers offering pulsed output in the infrared are Co^{2+}-doped MgF_2 (the first tunable solid-state laser) tunable over the range 1.75–2.50 µm, and chromium-doped Mg_2SiO_4 (forsterite), in which the laser-active species is Cr^{4+} and the corresponding emission range 1.15–1.35 µm.

18.2.4
Color-Center Lasers

Another type of solid-state laser in which the active medium is a regular ionic crystal does not involve transition metal ions. This is the so-called *F-center* (or *color center*) laser that operates on optical transitions at defect sites in alkali halide crystals, as for example in KCl doped with thallium; the color centers are typically excited by a pump Nd : YAG or argon/krypton ion laser. Such lasers produce radiation that is tunable over a small range of wavelengths in the overall region 0.8–3.4 µm; different crystals are required for operation over different parts of this range. One disadvantage of this type of laser is that the crystals need to be held at cryogenic temperatures. Color center lasers have found limited applications, although they have proven use in the high-resolution spectroscopy of small molecules.

18.2.5
Fiber Lasers

As the name suggests, the gain medium of these devices are silica or fluoride glass fibers doped with active rare-earth ions

(King, 2004). Some of dopants used include Nd, Tm, or Yb. The fiber, which has an external cladding with an index of refraction which is lower than that of the core, confines the pump and laser light to the core. Mirrors that are attached at the ends of the fiber or Fresnel reflections at each end complete the cavity. The devices are usually end-pumped through one of the mirrors (longitudinal pumping), although side pumping is possible. The long interaction lengths and wave guiding nature of the cladded fibers translates into devices with high gain, low thresholds, high efficiencies, and compactness. Unlike other solid-state lasers, the beam quality in these devises is determined here mainly by the refractive index profile of the cladding materials. Gain broadening in the fiber allows for limited tunability. Coherent emission can be obtained between ≈ 0.3 and $4.0\,\mu m$ using different ions and different fibers.

A particularly important dopant is erbium, Er^{3+}, which under laser-diode pumping at 0.81, 0.98, or $1.48\,\mu m$ can produce coherent radiation between 1.52 and $1.62\,\mu m$. This output range overlaps $\lambda = 1.55\,\mu m$, which corresponds to the wavelength of lowest transmission losses in fiber-optic systems. This makes Er^{3+}-doped fiber amplifiers a key component in optical communications systems. A schematic energy level diagram of Er^{3+} is shown in Figure 18.4. Overall, the behavior of the $^4I_{15/2}$ lower laser state is similar to that of a dye laser molecule (Section 18.5.5). The higher energy M_J components are sparsely populated at typical operating temperatures, which results in a population inversion with respect to the $^4I_{13/2}$ upper lasing state. The inversion is maintained by rapid lower-level nonradiative relaxation.

One parasitic effect that can reduce the infrared gain in a fiber laser is the process of upconversion where light is emitted at higher photon energies than that of the pump photon generating the excitation. However, at the same time, this has led to another class of solid-state fiber lasers, called *upconversion lasers*, that operate at different wavelengths in the visible, depending on the rare-earth ion dopant(s) involved (Auzel, 2004).

Several mechanisms can lead to upconversion. The first is sequential absorption of pump laser photons through metastable (relatively long lived) ion excited states. In this case, low dopant concentrations but high pump powers

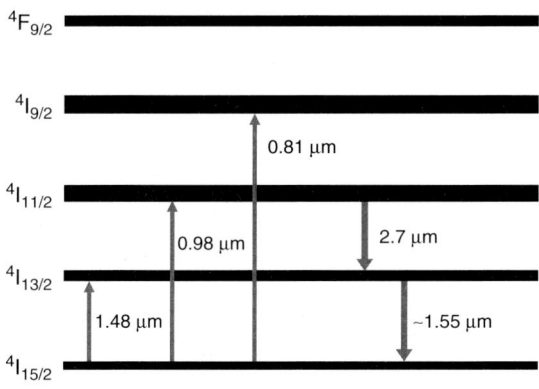

Fig. 18.4 Energy level diagram of Er^{3+} showing pumping transitions as vertical upward arrows and lasing transitions as vertical downward arrows. The thickness of the lines indicates unresolved fine structure.

are required. A second mechanism involves sequential energy transfers between different rare-earth ions. This could involve a simple radiative transfer where the emission of a donor ion is absorbed by an acceptor ion. Such an interaction has an R^{-2} interatomic distance dependence. Alternatively, the process could involve a nonradiative mechanism such as a Forster resonant energy transfer (FRET) or a Dexter transfer. Again, the emission of the donor ion must overlap the absorption profile of the acceptor species. However, the Forster mechanism is mediated by a dipole–dipole interaction between the donor and acceptor, and has a steep R^{-6} distance dependence. Dexter energy transfers are controlled by even higher order multipole interactions or very short-range exchange forces, and therefore are expected to more important in systems with forbidden transitions. In both cases, the donor and acceptor ion concentrations need to be relatively high and the energy transfer can be resonant or phonon assisted. A third mechanism involving cooperative upconversion is possible when the donor and acceptor ions are the same, and the dopant concentration has led to clustering. Here, a resonant dipole-dipole interaction between the two excited ions having an R^{-3} distance dependence cause one ion to be promoted to an even higher excited state, whereas the other is de-excited.

The choice of fiber is important for upconversion lasers. First, they must be transparent at the wavelengths used to pump the ions. Perhaps more importantly, they should have relatively large phonon energies to minimize phonon-assisted nonradiative relaxation of the metastable ions states involved in the upconversion process. Some of the more successful devices have been constructed using fibers made from the five component fluoride glass ZBLAN: (53 mol % ZrF_4; 10 mol % BaF_2; 4.5 mol % LaF_3, 3.5 mol % AlF_3, and 20 mol % NaF) (Jordan, Jha and Ryan, 1993). On the other hand, Si fibers can be problematic because they tend to absorb at wavelengths longer than around 2 μm.

18.3
Diode Lasers

In the solid-state lasers considered above, the energy levels are associated with low concentrations of dopant. Because, under these conditions, the dopant atoms are essentially isolated from each other, their energy levels remain discrete and the emission exhibits the same kind of line spectrum that we associate with isolated atoms or molecules. In the case of semiconductor diode lasers, however, it is energy bands, rather than discrete energy levels that are involved in the laser action.

Solid-state electronic devices generally make use of junctions between p-type and n-type semiconductors. The most familiar semiconductor materials are Class IV elements such as silicon and germanium; however, binary compounds between Class III and Class V elements, such as gallium arsenide (GaAs), exhibit similar behavior. In the latter cases, the lattice consists of two interpenetrating face-centered cubic lattices, the p- and n-type crystal reflecting a stoichiometry that departs from precisely 1:1. In p-type material, some arsenic atoms are replaced by gallium; in n-type crystal, the converse applies. On the basis of a junction between these materials, diode lasers operate in a broadly similar way to light-emitting diodes. When an electrical potential is applied across the very thin diode junction formed between p- and n-type semiconductor layers, the electrons crossing the semiconductor boundary recombine with holes and drop down from the conduction band to the valence band. In doing so, they emit radiation in the process as illustrated in Figure 18.5. The emission is most often in the infrared, and the optical properties of the crystal at such wavelengths make it possible for the crystal end faces to form the confines of the resonant cavity. One advantage of this kind of edge emitter is its extremely small size, usually about half a millimeter. However, the geometry of the gain region does result in very poor beam quality and collimation; divergences of 10° are by no means unusual. However, this does not preclude edge emitters from being used to pump fiber or other solid-state lasers to obtain better beam outputs, albeit, at the expense of overall efficiency.

There are two main types of semiconductor lasers: those operating at fixed wavelengths and others that are tunable. The three most common fixed-wavelength types are gallium arsenide, gallium aluminum arsenide, and indium gallium arsenide phosphide. Gallium arsenide lasers emit at a wavelength of around 0.904 µm; the wide range of flexibility in the stoichiometry of the other types makes it possible to produce a range of lasers operating at various fixed wavelengths in the region 0.8–1.3 µm. "Lead salt" diode lasers, which are derived from nonstoichiometric binary compounds of lead, cadmium, and tin with tellurium, selenium, and sulfur, emit in the range 2.8–30 µm (3500–330 cm^{-1}) depending on the exact composition. Although these lasers require a very low operating temperature, typically in the range 15–90 K, the operating wavelength is very temperature-dependent, and the wavelength can be tuned by varying the temperature. The tuning range for a lead

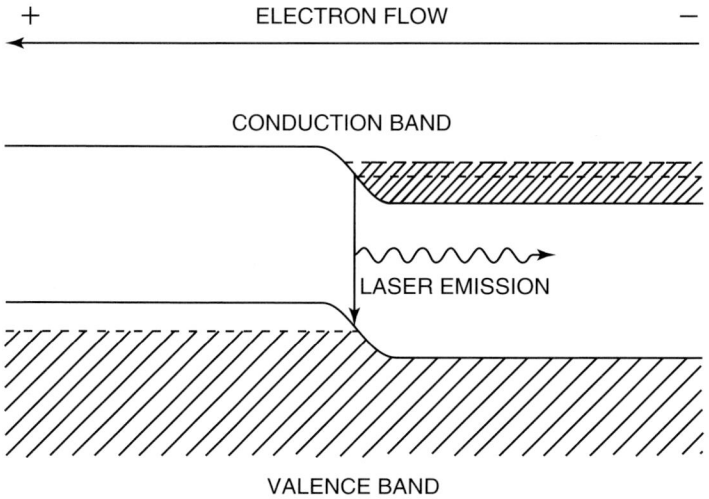

Fig. 18.5 Energetics of a diode laser; energy levels of p-type material on the left and n-type on the right.

salt diode laser of a particular composition is typically about 100 cm^{-1}.

Modes in a diode laser are typically separated by 1–2 cm^{-1}, and each individual mode generally has a very narrow line width, 10^{-3} cm^{-1} or less. The output power of CW diode lasers is generally on the milliwatt scale, though some array devices are capable of emitting as much as 10 W. Diode lasers are very well suited to high-resolution IR spectroscopy, because the line width is sufficiently small to enable the rotational structure of vibrational transitions to be resolved for many small molecules. This method has proved particularly valuable in characterizing short-lived intermediates in chemical reactions. More diverse applications can be expected as the range of diode laser emission wavelengths encroaches further into the visible region – devices emitting at the blue end of the spectrum are now available, in some cases through frequency-doubling methods (Section 7.1).

18.3.1
Vertical Cavity Surface Emitter Lasers

Vertical cavity surface emitter lasers (VCSEL) are diode lasers where the axis of coherent beam propagation is perpendicular to the gain region at the p–n junction. In this geometry, the width of the active area exceeds the height, so that beam divergence due to diffraction is much reduced (Hecht, 2005). However, because the active volume is small, high-reflectivity mirrors must be attached to the ends of the cavity as part of the fabrication process. Alternatively, distributed-Bragg-reflection gratings can be used to deflect light oscillating in the horizontal plane of an edge emitter out through the surface. Grating-coupled surface emitters have also been designed where a second-order grating in the active layer diffracts the light through the surface. The coherent emission from this configuration is bright with good spatial quality.

18.3.2
Visible Laser Diodes

Doped III–V nitride-based semiconductors have a wide direct band gap for blue emission (Leinen et al., 2000). For example, the band gap of AlGaInN can be varied between 2 and 6.2 eV at room temperature depending on its composition. Commercial edge and surface emitting lasers based on InGaN emit light by the recombination of injected electrons and holes in an active layer (Nakamura, 1998). The In allows a strong conduction band to valence band emission at room temperature which otherwise would be difficult to obtain using GaN alone. These devices are expected to have a large impact on the development of full color laser displays, which require at least three primary colors: red, green, and blue to produce any other visible wavelength. Blue-emitting diode lasers also provide an approximately three time enhancement in the capacity of digital versatile disks (DVDs) over those read using red diode lasers. Presently, "Blu-ray" discs and "HD–DVDs" are competing formats that both capitalize on the use of blue-emitting diode lasers for reading and writing, and for providing digital movies with a higher resolution than conventional DVDs.

18.3.3
Quantum Dot Lasers

Quantum dots are nanometer-sized semiconducting assemblies whose energy levels are fully discrete and atomic-like (Asada, Miyamoto and Suematsu, 1986). The energy level spacings depend on the radius of the dot. Whereas bulk semiconductors are three dimensional, quantum films are two dimensional, and quantum wires are one dimensional, quantum dots are often classified as zero dimensional. Considerable effort has been expended on fabricating arrays of uncapped (no organic surface terminators) InAs quantum dot lasers grown on GaAs that can lase at 1.3 µm (Mowbray, 2005; Fiore *et al.*, 2006). InAs self assembles on a GaAs substrate into a quantum dot array upon epitaxial growth because the two semiconductors have significantly different lattice constants. Although the individual dots have narrow spectral lines, size dispersion in an assembly of dots results in an inhomogeneous broadening of several tens of nanometers. The resultant quantum dot arrays have low gain for stimulated emission due to their lower density of states and large spectral widths. However, this problem can be overcome by stacking several layers of quantum dots within one assembly. On a positive note, these devices have minimal sensitivity to operating temperatures between 20 and 85 °C because their energy level spacings are larger than the Boltzmann energy, kT. They require ultra low threshold current densities for lasing, and the inhomogeneous broadening provides some degree of tunability.

18.3.4
Quantum Cascade Lasers

Quantum cascade laser emission is a result of electronic transitions between the discrete energy levels of a quantum well within the conduction band of a semiconductor. A quantum well is an ultrathin (typically nanometer scale) layer of one semiconductor sandwiched between two layers of a

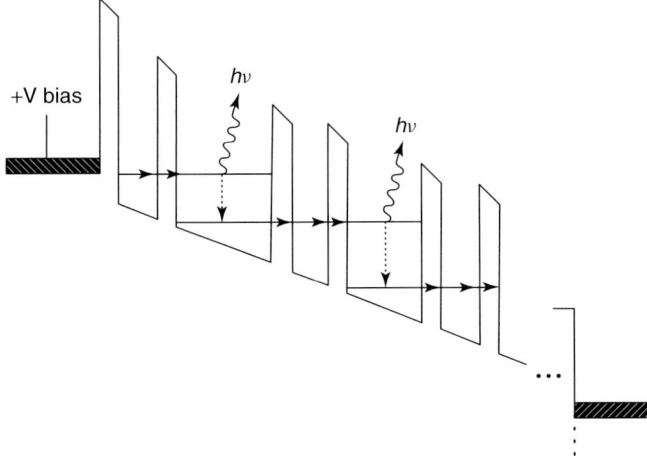

Fig. 18.6 Well structure of a quantum cascade laser.

second semiconductor with a wider band gap. This two-dimensional arrangement effectively confines injected electrons within the well. Because the discrete energy level spacing with the quantum well can be varied by changing its thickness, lasers based on this geometry are tuned between 4 and 24 µm in the mid- to far-infrared.

Although these devices require sophisticated molecular beam epitaxy to be fabricated, they are unique in that unlike other semiconductor devices the laser wavelength of the continuous output is not dependent on the chemical properties or the composition of the medium (Faist et al., 1994). As shown in Figure 18.6, a typical device consists of a series of quantum well. Under an applied voltage, an electron injected into the first quantum well will emit a photon when it drops from higher to lower energy. Because the well is within the conduction band, the electron can tunnel into an identical adjacent well where it produces a second photon. This process is repeated for each well. The effect of multiple quantum wells is that one electron can produce many photons as it cascades down an energy "staircase" created by the applied voltage. This avalanche effect results in output lasers powers which are ~1000 X that of a diode laser.

18.3.5
Quantum Microcavity Lasers

Microcavities are resonators that are characterized by high quality factors, Q, and small volumes, V. Like all resonators, microcavities support discrete optical mode frequencies for photons propagating along its axis. The lowest frequency mode supported by the resonator is called the *fundamental resonator mode*. Because the resonator mode frequency spectrum is size-dependent, microcavities, by virtue of their small dimensions, have mode frequency distributions that are considerably sparser than those for larger cavities.

There are three common geometries associated with microcavities (Vahala, 2003). The first is a Fabry–Perot resonator with two coplanar mirrors separated by one of more multiples of $\lambda/2$ where λ is the emission wavelength. Here, the Q-factor, (which is related to the photon lifetime in the cavity) is determined primarily by the reflectivity of the mirrors and is typically of the order of 2×10^3. The typical volume of a Fabry–Perot microcavity of refractive index n, operating at wavelength λ is $V = 5(\lambda/n)^3$.

The second configuration consists of spherical or a disk–shaped dielectric structures that confine optical surface waves by continuous total internal reflection. In general, the resultant atomic-like optical modes can be classified by three indices: n, ℓ, and m (Righini et al., 2005). The radial mode numbers, n, corresponds to the number of maxima in the radial component, while the angular momentum index, ℓ, is related to the number of wavelengths contained in the equatorial length. The combination $\ell - |m| + 1$ gives the number of maxima in the polar component. The smallest volume modes, called *whispering gallery modes*, are low order radial modes with $\ell \approx |m|$ that orbit in the equatorial ring. Typical Q-factors and volumes for these so-called whispering gallery resonators are 12×10^3 and $6(\lambda/n)^3$, respectively. The Q-factor is limited by losses due to surface roughness.

The third type of microcavity is based on photonic crystals. Photonic crystals are materials with an index of refraction periodicity on the order of λ. The photonic band gap which opens if the index contrast of the periodicity is sufficiently large corresponds to a frequency range where light with wavelengths $\sim 2\lambda$ will not propagate (Yablonovitch, 1987). This is the optical analog of the electronic band gap in semiconductors. However, microcavities can be formed by introducing a defect into the periodic structure (for example by drilling an additional hole or omitting a hole). These cavities can be very small ($V \sim 1.2(\lambda/n)^3$), and their Q-factors can very large ($\sim 13 \times 10^3$) provided there is an adequate spatial extent of the index periodicity. However, photonic crystals can be challenging to fabricate for operation in the visible region of the spectrum.

When an optically active medium is placed inside a microcavity, its spontaneous emission can be dramatically changed because the spontaneous emission rate depends on the optical mode density at the emission wavelength. For example, media in cavities having a fundamental mode frequency that is resonant with its emission frequency will exhibit an enhanced emission rate. This is the so-called Purcell effect. Conversely, a cavity can be designed to suppress spontaneous emission. Similarly, media can experience an enhanced or reduced absorption depending on whether or not the medium's absorption frequency is resonant with a mode of the cavity.

Microcavity physics has led to the development of several laser sources. Lasing has been observed in dye-filled water droplets, dye-doped silica and polystyrene spheres, semiconductor microdisks, vertical cavity micropillars containing a quantum dot, and doped photonic crystals. One unique property of lasers based on microcavities is their low thresholds relative to conventional devices. Indeed, it has been argued that an ideal microcavity laser, where spontaneous emission has been suppressed, is actually thresholdless (Rice and Carmichael, 1994).

18.4
Atomic and Ionic Gas Lasers

The class of lasers in which the active medium is a gas covers a wide variety of devices. Generally, the gas is monatomic, or else it is composed of very simple molecules. In both cases, because laser emission results from optical transitions in free atoms or molecules, usually at low pressures, the emission line width is small. The gas is often contained in a sealed tube, with the initial excitation provided by an electrical discharge, so that in many cases, the innermost part of the laser bears a superficial resemblance to a conventional fluorescent light. The tube can be constructed from various materials, and need not necessarily be transparent; silica is commonly employed – and also beryllium oxide, which has some advantages in its high thermal conductivity. It is quite a common feature to have the laser tube contain a mixture of two gases, one of which is involved in the pump step and the other in the laser emission. Such gas lasers are extremely reliable, and they are the most widely used type for routine purposes.

18.4.1
Helium–Neon Laser

The helium–neon laser was the first CW laser ever constructed, and it was also the first laser to be made available to a commercial market, in 1962. The active medium is a mixture of the two gases contained in a glass tube at low pressure; the partial pressure of helium is approximately 1 mbar, and that of neon 0.1 mbar. The initial excitation, provided by an electrical discharge, primarily serves to excite helium atoms by electron impact. The excited helium atoms subsequently undergo a process of collisional energy transfer to neon atoms; certain levels of helium and neon are very close in energy, so that this transfer takes place with a high degree of efficiency. Because the levels of neon that are so populated lie above the lowest excited states, a population inversion is created relative to these levels, enabling laser emission to occur as shown in Figure 18.7.

Three distinct wavelengths can be produced in the laser emission stage; one visible wavelength, typically of milliwatt power, appears in the red at 632.8 nm, and there are two IR wavelengths of somewhat lower power at 1.152 and 3.391 μm. Following emission, the lasing cycle is completed as the neon undergoes a two-step radiationless decay back down to its ground state. This involves transition to a metastable $2p^5 3s^1$ level, followed by collisional deactivation at the inner surface of the tube. The last step has to be rapid for the laser to operate efficiently; for this reason, the surface/volume ratio of the laser tube has to be kept as large as possible, which generally means keeping the tube diameter small. In practice, tubes are commonly only a few millimeters in diameter. Other very weak transitions are utilized in a 1-mW helium–neon laser emitting at various wavelengths including 543.5 nm in the green; its principal virtue is that it is substantially cheaper than any other green laser.

Helium–neon lasers operate continuously, and despite their low output power, they have the twin virtues of being

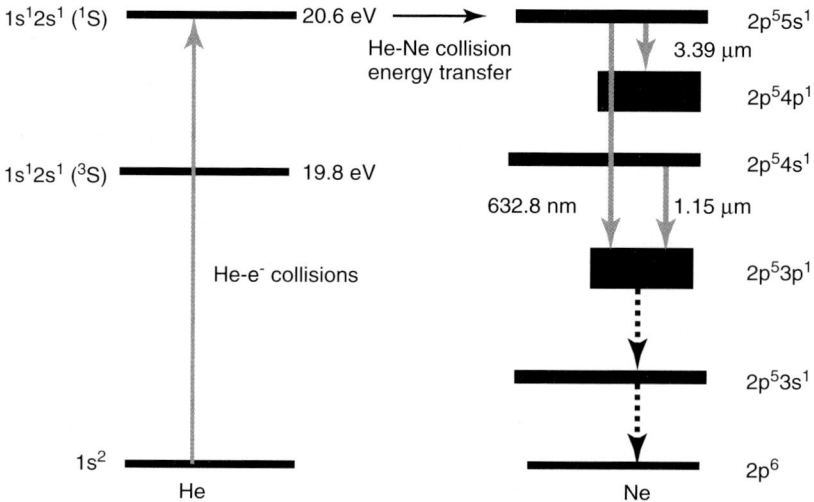

Fig. 18.7 Simplified energy levels involved in He–Ne lasing. The broadness of the levels is to indicate unresolved fine structure.

both small and relatively inexpensive. Consequently, they have more widely ranging applications than any other kind of laser. Their principal applications are those that hinge on the typically narrow laser beam width, where power is not very important. Another laser of similar type is the helium–cadmium laser, in which transitions of cadmium atoms result in milliwatt emission at 442 nm in the blue and 325 nm in the ultraviolet.

18.4.2
Argon and Krypton Lasers

The argon laser is the most common example of a family of ion lasers in which the active medium is a single-component inert gas. The gas, at a pressure of approximately 0.5 mbar, is contained in a plasma tube of 2–3 mm bore, and is excited by a continuous electrical discharge. Argon atoms are ionized and further excited by electron impact; the pumping process populates several ionic states. The population inversion between these states and others of lower energy results in emission at a series of discrete wavelengths over the range 350–530 nm (see Table 18.1), the two strongest lines appearing at 488.0 and 514.5 nm. These two wavelengths are emitted as the result of transitions from the singly ionized states with electron configuration $3s^2 3p^4 4p^1$ down to the $3s^2 3p^4 4s^1$ state. Further radiative decay to the multiplet associated with the ionic ground configuration $3s^2 3p^5$ then occurs, and the cycle is completed either by electron capture or by further impact excitation. Doubly ionized Ar^{2+} ions contribute to the near-ultraviolet laser emission.

Because several wavelengths are produced by this laser, dispersive optics are generally placed between the end mirrors to select one particular wavelength for amplification; the output can thus be varied. By selecting a single longitudinal mode, an output line width as small as 0.0001 cm^{-1} is obtainable. The pumping of ionic levels required for laser action requires a large and continuous input of energy, and the relatively low efficiency of the device means that a large amount of thermal energy has to be dissipated. Cooling is commonly achieved by the circulation of water in a jacket around the tube. The output power of a CW argon laser usually ranges from milliwatt emission up to about 25 W. Such lasers have found widespread research applications in spectroscopy, where they are often employed to pump dye lasers (see Section 18.5.5).

The other common member of the family of ion lasers is the krypton laser. In most respects, it is very similar to the argon laser, emitting wavelengths over the range 350–800 nm, although because of its lower efficiency the output takes place at somewhat lower power levels, up to around 5 W. The strongest emission is at a wavelength of 647.1 nm. In fact, the strong similarity in physical requirements and performance between the argon and krypton lasers makes it possible to construct a laser containing a mixture of the two gases, which provides a very good range of wavelengths across the whole of the visible spectrum – see Table 18.1. Such lasers emit different wavelengths useful for biomedical applications, the blue-green argon lines being especially useful in many biological applications.

Laser operation does not by itself automatically produce polarized light, although the output is commonly at least partially polarized as a result of

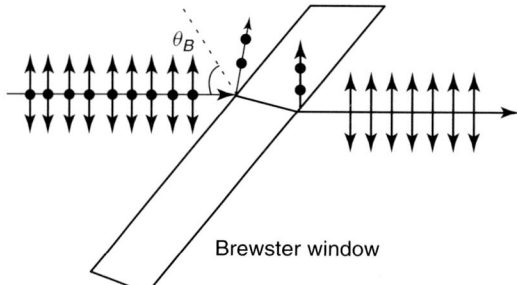

Fig. 18.8 Brewster angle laser window as used in a gas laser. Light polarized in the plane of the figure (double headed arrows) has a low reflection loss at the Brewster angle (θ_B) while light polarized perpendicular to the plane of the figure (indicated by dots) is highly reflected off both surfaces of the window. Only the polarization in the plane of the figure will be strongly amplified in the laser medium.

interactions with the numerous optical components. In these lasers (and others), it is desirable to obtain highly polarized emission, and it is usual to incorporate optics into the laser cavity to accomplish this. The standard method is to use a Brewster angle window. This is a parallel-sided piece of glass set at the Brewster angle, defined as $\theta_B = \tan^{-1} n$, where n is the refractive index of the glass for the appropriate wavelength. The Brewster angle is the angle at which reflected light is completely plane polarized. When such elements are used for the end windows of a gas laser tube, for example, one selected polarization experiences gain and is amplified within the cavity. Polarized emission thus ensues, as shown in Figure 18.8.

An alternative is to make use of an optically anisotropic or *birefringent* crystal such as calcite, in which orthogonal polarizations propagate in slightly different directions and can thus be separated. Calcite is a good example of such a material. Crystals of this type have a number of other important uses in controlling laser polarization; mostly, these stem from the fact that in such media, different polarizations propagate at slightly different velocities. By passing *polarized* light in a suitable direction through such a crystal, the birefringence effects a change in the polarization state, the extent of which depends on the distance the beam travels. With a slice of crystal of the correct thickness, plane polarized light thus emerges circularly polarized, or *vice versa*, and with a crystal of twice the thickness, plane polarized light emerges with a rotated plane of polarization, or circularly polarized light reverses its handedness. The two optical elements based on these principles are known as a *quarter-wave plate* and a *half-wave plate*, respectively. These optical elements can be used intracavity or external to the resonator.

18.5
Molecular Gas Lasers

18.5.1
Carbon Dioxide Laser

In the CO_2 laser, the energy levels involved in laser action are rotation–vibration levels, and emission therefore occurs at much longer wavelengths, well into the IR. The lasing medium consists of a mixture of CO_2, N_2, and He gas approximately in the ratio of partial pressures $1:4:5$; the helium being added to improve the lasing efficiency. The sequence of excitation is illustrated in Figure 18.9. The first step is the excitation, by electron impact, of

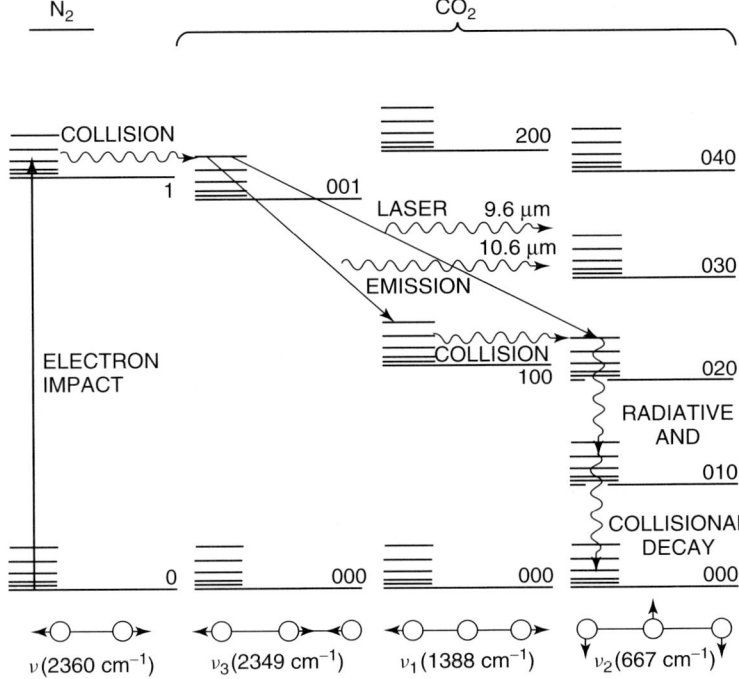

Fig. 18.9 Energetics of the carbon dioxide laser, rotational structure of the vibrational levels shown schematically. (Adapted from Andrews (1997).)

various metastable rotational sublevels in the first vibrationally excited level of nitrogen. The rotational sublevels of one vibrationally excited state (001) of carbon dioxide (having one quantum of energy in the ν_3 vibration, the antisymmetric stretch) has almost the same energy as the vibrationally excited nitrogen. Consequently, collision between the two molecules results in very efficient transfer of energy to the CO_2. Laser emission then occurs by two routes, involving radiative decay to sublevels of the (100) and (020) states. The two laser transitions result in emission in bands centered on wavelengths at around 10.6 and 9.6 μm, respectively. Because various rotational sublevels can be involved in the emissive transitions, use of suitable optical elements enables the laser to be operated at various discrete frequencies within the 10.6 and 9.6 μm bands as listed in Table 18.2. Because the CO_2 energy level positions vary for different isotopic species, other wavelengths may be obtained by using isotopically substituted gas.

A small carbon dioxide laser, with a discharge tube about half a meter in length, may have an efficiency rating as high as 30%, and produce a continuous output of 20 W; even a battery-powered hand-held model can produce 10 W CW output. Much higher powers are available from longer tubes; although efficiency drops, outputs in the kilowatt range are obtainable from the largest devices. Apart from extending the cavity

Tab. 18.2 Emission of the carbon dioxide laser. The lines (wave number units $\tilde{\nu}$) result from rotation–vibration transitions, the rotational quantum number J changing by one unit. Transitions $J \to J-1$ are denoted as P(J); transitions $J \to J+1$ are denoted as R(J).

Line	$\tilde{\nu}$ (cm^{-1})	Line	$\tilde{\nu}$ (cm^{-1})	Line	$\tilde{\nu}$ (cm^{-1})
10.6 µm band (001 → 100 transitions)					
P(56)	907.78	P(22)	942.38	R(18)	974.62
P(54)	910.02	P(20)	944.19	R(20)	975.93
P(52)	912.23	P(18)	945.98	R(22)	977.21
P(50)	914.42	P(16)	947.74	R(24)	978.47
P(48)	916.58	P(14)	949.48	R(26)	979.71
P(46)	918.72	P(12)	951.19	R(28)	980.91
P(44)	920.83	P(10)	952.88	R(30)	982.10
P(42)	922.92	P(8)	954.55	R(32)	983.25
P(40)	924.97	P(6)	956.19	R(34)	984.38
P(38)	927.01	P(4)	957.80	R(36)	985.49
P(36)	929.02	R(4)	964.77	R(38)	986.57
P(34)	931.00	R(6)	966.25	R(40)	987.62
P(32)	932.96	R(8)	967.71	R(42)	988.65
P(30)	934.90	R(10)	969.14	R(44)	989.65
P(28)	936.80	R(12)	970.55	R(46)	990.62
P(26)	938.69	R(14)	971.93	R(48)	991.57
P(24)	940.55	R(16)	973.29	R(50)	992.49
9.6 µm band (001 → 020 transitions)					
P(50)	1016.72	P(20)	1046.85	R(16)	1075.99
P(48)	1018.90	P(18)	1048.66	R(18)	1077.30
P(46)	1021.06	P(16)	1050.44	R(20)	1078.59
P(44)	1023.19	P(14)	1052.20	R(22)	1079.85
P(42)	1025.30	P(12)	1053.92	R(24)	1081.09
P(40)	1027.38	P(10)	1055.63	R(26)	1082.30
P(38)	1029.44	P(8)	1057.30	R(28)	1083.48
P(36)	1031.48	P(6)	1058.95	R(30)	1084.64
P(34)	1033.49	P(4)	1060.57	R(32)	1085.77
P(32)	1035.47	R(4)	1067.54	R(34)	1086.87
P(30)	1037.43	R(6)	1069.01	R(36)	1087.95
P(28)	1039.37	R(8)	1070.46	R(38)	1089.00
P(26)	1041.28	R(10)	1071.88	R(40)	1090.03
P(24)	1043.16	R(12)	1073.28	R(42)	1091.03
P(22)	1045.02	R(14)	1074.65	R(44)	1092.01

length, another means used to increase the output power is to increase the pressure of carbon dioxide. Carbon dioxide lasers can be made to operate at or above atmospheric pressure. To produce sufficiently strong fields without using dangerously high voltages, it is necessary to apply a potential across, rather than along the tube; such a laser is referred to as a *transverse excitation atmospheric* (*TEA*) *laser*. In devices where the gas pressure is raised to around 15 atmospheres, pressure broadening results in the formation of a quasi-continuum of emission frequencies, enabling the laser output to be continuously tuned over a range of approximately 910–1100 cm^{-1}.

18.5.2
Nitrogen Laser

The nitrogen laser operates on electronic, rather than vibrational transitions. A high-voltage electrical discharge populates the $C^3\Pi_u$ triplet electronic excited state, and the laser transition is to the lower energy metastable $B^3\Pi_g$ state. The upper laser level has a very short lifetime of only 40 nanoseconds, and consequently population inversion cannot be sustained; essentially all the excited nitrogen molecules undergo radiative decay together over a very short period of time, effectively emptying the cavity of all its energy. This *super-radiant* emission is sufficiently powerful that a highly intense pulse is produced without the need for repeatedly passing the light back and forth between end mirrors. The laser automatically operates in a pulsed mode, producing pulses of about 10 nanoseconds or less duration at a wavelength of 337.1 nm. Bandwidth is approximately 0.1 nm, and pulse repetition frequency is 1–200 Hz. Pulses tend to have a relatively unstable temporal profile as a result of the low residence time of photons in the laser cavity. Because the laser can produce peak intensities in the 10^{10} Wm^{-2} range, the nitrogen laser is one of the most powerful commercial sources of ultraviolet radiation, and it is frequently used in photochemical studies. It has also been a popular choice as a pump for dye lasers, although here it is now often displaced by the more powerful wavelengths from excimer lasers or the second or third harmonics of the Nd : YAG laser.

18.5.3
Iodine Laser

Another laser operating on broadly similar principles is the iodine laser, more correctly termed as the *atomic iodine photodissociation laser*. Unusually, polyatomic chemistry is involved in its operation, but the crucial laser transition takes place in free atomic iodine, so that it is arguable whether or not it is truly designated as a molecular laser. The laser is activated by the photolysis of an iodohydrocarbon or iodofluorocarbon gas such as 1-iodoheptafluoropropane, C_3F_7I, by ultraviolet light from a flashlamp. The gas is introduced into a silica laser tube at a pressure of 30–300 mbar; the ensuing mechanisms are illustrated in Figure 18.10. Laser emission involves transition between an excited metastable $^2P_{1/2}$ state and the ground $^2P_{3/2}$ state of atomic iodine; this produces narrow line width output at a wavelength of 1.315 μm (7605 cm^{-1}), consisting of closely spaced hyperfine components encompassing a range of less than 1 cm^{-1}. In the absence of any other pulsing mechanism, the laser typically produces pulses of microsecond duration, and pulse energies of several joules; however the output is often modified by Q-switching or mode-locking (see Section 8) to produce pulse trains of nanosecond or subnanosecond duration.

An important advantage of the iodine laser is the fact that the active medium is comparatively cheap and available in large quantities. Moreover, its emission wavelength lies in a region offering excellent atmospheric transmission. One variation is a device known as the *chemical oxygen iodine laser*, in which collisional energy transfer occurs from electronically excited ($^1\Delta$) oxygen

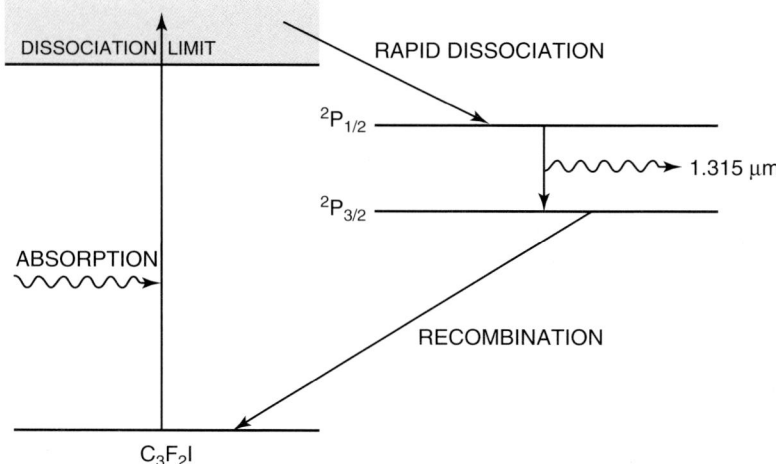

Fig. 18.10 Energetics of the atomic iodine photodissociation laser (not showing fine structure or side reactions).

molecules to atomic iodine, here formed by dissociation of molecular iodine, populates the $^2P_{1/2}$ state and so leads to the same 1.315 μm laser emission. By employing supersonically flowing gases, multikilowatt CW outputs can be achieved.

18.5.4
Excimer Lasers

In excimer lasers, the active medium is an *exciplex* or an excited complex. The crucial feature of an exciplex is that only when it is electronically excited, does it exist in a bound state with a well-defined potential energy minimum; the ground electronic state generally has no potential energy minimum, or else only a very shallow one. Most examples are found in the inert gas halides such as KrF. In principle, it is only homonuclear diatomics, such as Xe_2, that should be termed *excimers*, but in connection with lasers the designation is widely applied to heteronuclear systems. The exciplex is generally formed by chemical reaction between inert gas and halide ions produced by an electrical discharge. In the case of krypton fluoride, KrF, for example, the exciplex has a lifetime of about 2.5 nanoseconds. The decay transition can be stimulated to produce laser emission with a high efficiency, typically around 20%.

Excimer lasers are super-radiant, producing radiation with pulse durations of 10–20 nanoseconds and pulse repetition frequencies generally in the 1–500 Hz range. Pulse energies can be up to 1 J, with peak powers in the milliwatt region and average power between 20 and 100 W. The emission wavelengths of commercially available systems are: F_2 157 nm; ArF 193 nm; KrCl 222 nm; KrF 248 nm; XeCl 308 nm; XeF 351 and 353 nm. These short ultraviolet wavelengths lie in a region of the electromagnetic spectrum absorbed by a wide range of materials; photoabsorption often results in the rupture of chemical bonds and some degree of sample vaporization.

18.5.5
Dye Lasers

Dye lasers are essentially different from every other common category of laser. Major differences arise due to the unusual nature of the active medium, a solution of a chemically complex organic dye. A wide range of more than 200 dyes can be used for this purpose; the only general requirements are an absorption band in the visible and/or the ultraviolet, and a broad fluorescence spectrum. The kind of compounds that best satisfy these criteria consist of comparatively large polyatomic molecules with extensive electron delocalization. One of the most widely used systems is the dye known as Rhodamine 6G ($C_{28}H_{31}N_2O_3Cl$), whose broad absorption and fluorescence are shown in Figure 18.11.

Generally, the absorption of visible light first results in a transition from the ground singlet state S_0 to the energy continuum belonging to the first excited singlet state S_1. This is immediately followed by rapid radiationless decay to the lowest energy level within the S_1 continuum, as illustrated in Figure 18.12; in the case of Rhodamine 6G, this process is complete within 20 picoseconds of the initial excitation. Fluorescent emission then results in a downward transition to levels within the S_0 continuum, followed by further radiationless decay. It is the fluorescent emission process that has made the basis of laser action, provided a population inversion is set up between the upper and lower levels involved in the transition; this is essentially a four-level laser. Clearly, because the emitted photons have less energy, the fluorescence must always occur at longer wavelengths than the initial excitation.

One significant loss mechanism is quenching of the excited states through interaction with other molecules; triplet quenchers such as dimethylsulfoxide (DMSO) are often added to increase the output power by repopulating singlet

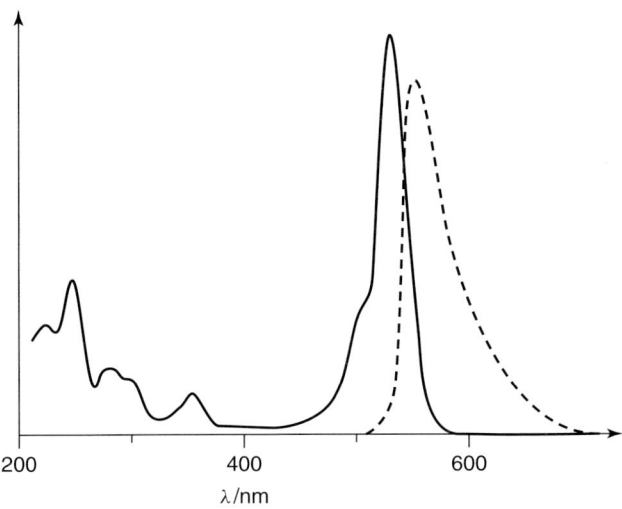

Fig. 18.11 Solution spectra of Rhodamine 6G in ethanol. The solid curve shows absorption and the dotted curve shows the fluorescence at longer wavelength.

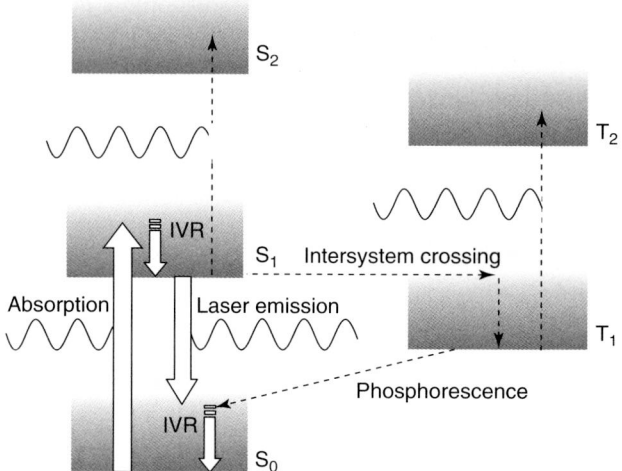

Fig. 18.12 Jablonski diagram for a laser dye. Solid vertical arrows show transitions involved in laser action; dotted arrows illustrate some competing transitions. IVR: intramolecular vibrational redistribution.

states. Another problem is the heat produced by the radiationless decay transitions, a necessary part of the laser cycle. This can rapidly degrade a dye, and therefore it is common practice to continuously circulate the dye solution, to facilitate cooling. A common setup is shown in Figure 18.13. The pump radiation from a flash-lamp, or primary laser source with emission in the visible or near-ultraviolet range, is focussed at a point traversed by a jet of dye solution, which typically has a concentration in the range 10^{-2}–10^{-4} mol l^{-1}. The solvent is usually based on ethylene glycol, which provides the viscosity needed to maintain an optically flat jet stream.

Fluorescent emission from the dye jet is stimulated by the formation of an optical cavity, with two parallel end mirrors placed on either side of the jet. However, because the fluorescence occurs over a range of wavelengths, monochromatic laser emission can only be obtained by the use of additional dispersive optics within the cavity; rotation of this element can change the wavelength amplified within the dye laser, so that *tunable emission* is obtained. A dye laser based on a solution of Rhodamine 6G in methanol, for example, is continuously tunable over the range 570–660 nm. The full range of commercially available dye lasers provides complete coverage of the range 200 nm–1 µm: typical tuning curves for some of the more important dyes are shown in Figure 18.14.

The dye laser is often used as a means of providing continuously tunable radiation Commonly, an inert gas or nitrogen laser is used as the pump, and the output is tunable over a range of wavelengths (determined by the choice of dye), all longer than the pump. The facility for tuning provides a very useful method of obtaining laser emission

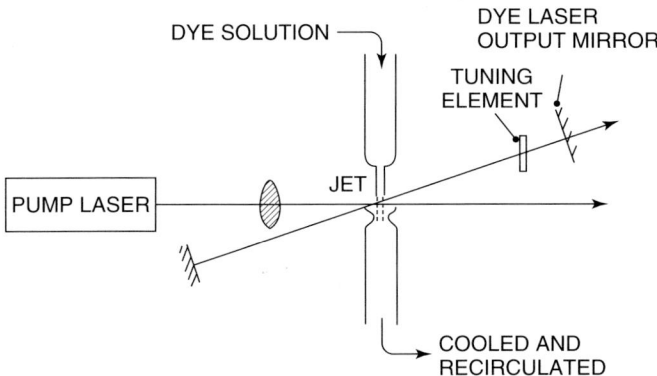

Fig. 18.13 Schematic diagram of a laser-pumped dye laser. (Adapted from Andrews (1997).)

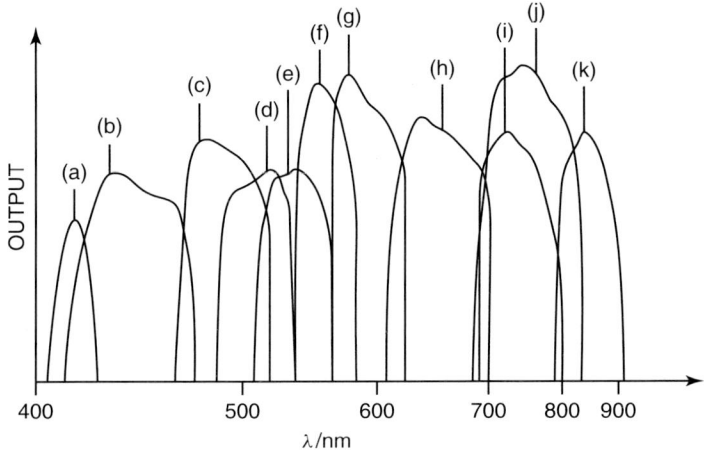

Fig. 18.14 Tuning curves for a dye laser pumped by lines from a krypton/argon laser. The dye and pump wavelength for each curve are given in Table 18.3. (Reproduced by kind permission of Coherent Radiation Ltd.)

at a chosen wavelength. Conversion efficiencies are generally somewhat low; 5–10% is common, although with certain dyes figures of 20% or more are possible. The highest powers are obtained by pumping with harmonics of a Nd laser. CW dye lasers produce emission with line width in the range 10–20 GHz (around 0.5 cm^{-1}), although with suitable optics this can be reduced to about 1 GHz. The combination of narrow line width, good frequency stability and tunability is particularly attractive for spectroscopic applications. One disadvantage of the dye laser is that it tends to have rather poorer amplitude stability than a gas laser, so that indirect spectroscopic methods such as fluorescence or the photoacoustic effect are often the most appropriate.

Tab. 18.3 Dyes and pump wavelengths for the tuning curves of Figure 18.14.

Curve	Dye	Pump laser; wavelength λ (nm)
(a)	Stilbene 1	Argon; 333.6, 351.1, 363.8
(b)	Stilbene 3	Argon; 333.6, 351.1, 363.8
(c)	Coumarin 102	Krypton; 406.7, 413.1, 415.4
(d)	Coumarin 30	Krypton; 406.7, 413.1, 415.4
(e)	Coumarin 6[a]	Argon; 488.0
(f)	Rhodamine 110	Argon; 514.5
(g)	Rhodamine 6G	Argon; 514.5
(h)	DCM[b]	Argon; 488.0
(i)	Pyridine 2	Argon; 488.0, 496.5, 501.7, 514.5
(j)	Rhodamine 700	Krypton; 647.1, 657.0, 676.5
(k)	Styryl 9M	Argon; 514.5

[a] With cyclo-octatetraene and 9-methylanthracene.
[b] 4-dicyanomethylene 2-methyl 6-(p-dimethylaminostyrol) 4H-pyran.

One important variation on the concept is the *ring dye laser*, in which the laser radiation travels around a cyclic route between a series of mirrors, rather than simply back and forth between two. When both clockwise and anticlockwise traveling waves are present, their frequencies are normally identical. However, any rotation of the laser itself results in a small difference between these two frequencies, and detection of this difference can be used as the basis for very accurate measurement of the rotation; this is the principle behind the ring laser gyroscope. Alternatively, an optical diode can be placed within the cavity to select one particular direction of propagation (clockwise or anticlockwise). In this case, the ring laser has an emission line width typically at least a factor of ten smaller than a conventional dye, but in state-of-the-art actively stabilized cases the line width may be as small as 10^{-15} cm^{-1}.

18.6
Pulsing Techniques

There are several reasons why it is common practice to make use of a pulsing device, either to convert a CW laser to pulsed output or to shorten the duration of the pulses emitted by a naturally pulsed laser. One objective is to obtain high peak intensities, to optimize multiphoton or other nonlinear optical processes. Another motive is to produce pulses of very short duration to make measurements of processes that occur on the same timescale. There are three widely used pulsing methods, reviewed below.

18.6.1
Cavity Dumping

Cavity dumping refers to a method of rapidly emptying a laser cavity of the energy stored within it. The simplest means of achieving this with a CW laser would be to have both of the end mirrors fully reflective, but with a third coupling mirror able to be switched into the beam to reflect light out of the cavity. This configuration would deliver a single pulse of light, terminating laser action until the coupling mirror was again switched out of the beam. In practice, an acousto-optic modulator is usually employed. This double-pass device is driven by a radio-frequency electric field, and generates an acoustic wave that produces off-axis diffraction of the laser beam. The cavity-dumped output of the laser in this case consists of an essentially sinusoidal temporal profile; the frequency of oscillation is twice the acoustic frequency of the cavity dumper, typically of the order of megahertz. This method of pulsing is often used in conjunction with mode locking (Section 18.6.3).

18.6.2
Q-Switching

The term *Q-switching* refers to a method that essentially involves first reducing, and then swiftly increasing the quality, or Q-factor of the laser, formally defined as $Q = 2\pi \times$ energy stored in cavity/energy loss per optical cycle (where the optical cycle length is the inverse of the resonant frequency). Q-switching thus represents the effect of suddenly reducing the rate of energy loss within the cavity. In practice, the pumping rate has to exceed the rate of spontaneous decay to build the required population inversion, and the time taken for Q-switching has to be sufficiently short to produce a single pulse. There are several commonly used methods. One of the most common is an electro-optical method based on a Pockels cell, which rapidly switches the cavity transmissivity of a selected polarization, effecting a shutter action when voltage is applied. A Pockels cell consists of a crystalline material such as potassium dihydrogen phosphate (KDP) that exhibits the Pockels effect, essentially a proportional change in refractive index on application of an electric field. A Pockels cell can be used to rotate the plane of polarization of light passing through it, or to reverse the handedness of circularly polarized light. However, its particular advantage is that, because the effect takes place only while the electric field is applied, modulation of the field at a suitable frequency results in the corresponding modulation of the beam polarization. The Q of the resonator can be *actively* switched from low to high when the Pockel cell is used in conjunction with a polarizer by applying a voltage to the Pockel cell. Pockel cells are also a particularly useful for increasing detection sensitivity in many techniques based on optical activity, such as polarimetry and circular differential Raman scattering.

Another common method for Q-switching involves the use of a *saturable absorber* dye such as cryptocyanine. With a cell containing a saturable absorber solution placed inside the laser cavity, the intensity of emission increases to a point where the dye is bleached only after a large population inversion has been achieved in the active medium; stimulated emission therefore results in the emission of a pulse of light. This last method of Q-switching is known as *passive*, because the switching time is not determined by external constraints. Q-switching generally produces pulses with a duration of the order of 10^{-9}–10^{-8} s.

18.6.3
Mode-Locking

The third, widely used method of producing laser pulses is known as *mode locking*. This produces pulses of much shorter duration than those produced by either cavity dumping or Q-switching, the output typically being measured on the picosecond or femtosecond scale and referred to as *ultrashort*. The technique has very important applications in chemistry in the study of ultrafast reactions, for example.

The mechanism of mode locking is as follows. Laser emission generally occurs over a small range of line frequencies within the emission line width of the active medium, separated $\Delta v = c/2L$, (the free spectral range, determined by the cavity length L), which is typically around 10^8 Hz. In the case of solid-state or dye lasers, where the fluorescence is broad, the number of longitudinal modes N sustained within the laser cavity may thus be as many as $10^3 - 10^4$ if frequency-selective elements are absent. Mode locking establishes a phase relationship that leads to constructive interference between all the modes at just one point, with destructive interference everywhere else. Consequently, a single pulse of travels back and forth between the end mirrors, giving a shot of output each time it encounters the output mirror. The pulse duration (the pulse-width at half power) is given by $\Delta t = 4\pi L/(2N+1)c$, and the interval between successive pulses is the cavity round-trip time. Hence, mode-locked pulses are typically separated by an interval of around 10 nanoseconds, and have a duration of 1 – 10 picoseconds; the greater the number of modes, the shorter the pulse length, in accordance with Fourier principles. The temporal

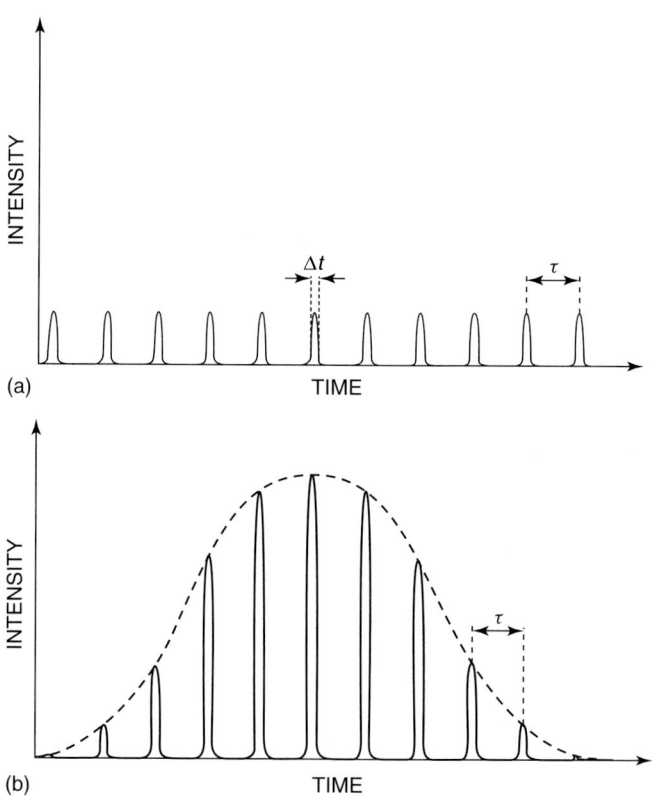

Fig. 18.15 Temporal profile of mode-locked emission from: (a) CW laser; (b) pulsed laser. In the second case, several pulses than illustrated would normally be found within the overall pulse envelope.

profile of the emission from mode-locked CW and pulsed lasers is as shown in Figure 18.15, and in each case consists of a pulse train. The instantaneous intensity of a mode-locked laser pulse can be readily computed on the assumption that the overall energy emitted within the duration of one laser round-trip is the same before and after mode-locking. Thus, with a pulse-length reduction from 10 nanoseconds to 1 picosecond, the pulse power can be increased by a factor of approximately 10^4, increasing the power of GW pulses up to the region of 10^{13} W.

As with Q-switching, there are several means of accomplishing mode-locking, both active and passive methods. The active methods generally involve modulating the gain of the laser cavity using an electro-optic or acousto-optic switch driven at the frequency $c/2L$. Passive mode-locking is generally accomplished with a saturable absorber cell – although, in the exceptional case of the titanium : sapphire laser, automatic mode-locking can be observed without the inclusion of any such element.

Another mode-locking technique principally employed with tunable dye lasers is *synchronous pumping*. Here, a mode-locked primary laser pumps a dye laser whose optical cavity length is adjusted to be equal to that of the pump, so that the round-trip time of the dye laser matches the interval between pump pulses. In this way, frequency-converted pulses propagating within the dye laser cavity arrive at the dye jet at the same rate as the pump pulses. The period over which amplification is effective in the dye laser is thus very short, and once again picosecond pulses emerge. In practice, the dye laser cavity length needs to be kept constant to within a few microns, so that special mounting and temperature control are necessary. Finally, there are various means by which pulse lengths can be reduced, as for example in *colliding-pulse* ring dye lasers.

18.6.4
Ultrafast Pulse Compression

Mode-locking commercially available lasers in conjunction with pulse compression techniques can now provide pulses of a few picoseconds duration in the infrared, and ultrashort pulses measured on the femtosecond (10^{-15}s) timescale in the UV/visible range. A common approach in generating ultrashort pulses is to first pass a longer pulse (on the order of 1–100 picoseconds) through an optical fiber to impose a chirp on the pulse carrier, and then compress the output using a dispersive delay line often made from two gratings (Figure 18.16).

A chirped pulse is one in which the wavelength of the carrier changes continuously throughout the pulse. A chirp

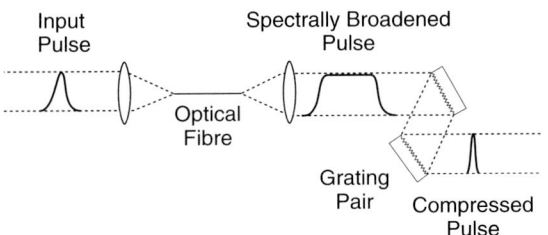

Fig. 18.16 Experimental arrangement used for pulse compression.

can occur in an optical fiber due to the nonlinear process of self-phase modulation, where the index of refraction of the fiber is instantaneously modified in the presence of an intense light pulse through the nonlinear optical Kerr effect (Monerie, 1989). This variation in the index of refraction produces a shift in the instantaneous phase of the pulse. The effect is that as the pulse passes through the fiber, its leading edge experiences a shift to lower frequencies, while its trailing edge experiences a shift to higher frequencies. In a normally dispersive medium, the front of the pulse will move faster than the back of the pulse, resulting in both a spectral and temporal broadening. The effect of the grating pair after the fiber is to produce a phase shift that is opposite to the chirp (Treacy, 1969). This compresses the output of the fiber. Geometrically, the phase shift is a result of time delays that arise because the optical path length due to Bragg diffraction off the gratings increases with increasing wavelength.

Early work using a fiber/grating pair combination yielded 30-femtosecond pulses at 619 nm, corresponding to 14 optical cycles (Shank et al., 1982). More sophisticated arrangements involving prisms and diffraction gratings have led to pulses as short as 6 femtoseconds at 620 nm (Fork et al., 1987). Recently, chirped mirrors have been developed, which have the same function as a grating pair (Steinmeyer, 2006). These are dielectric mirrors multilayered with low and high index coatings whose thicknesses increase monotonically from the surface to the substrate. The reflectivity of such a structure takes place at a depth within the mirror stack that depends on wavelength. This imposes a negative group delay on the pulse that compensates for any material dispersion in the laser system leading to ultrashort pulse generation.

18.6.5
Continuum Generation

It is a notable feature of ultrashort (picosecond) laser pulses that, on passage through certain media, a broad continuum of light is generated, typically covering a range of about $10\,000$ cm^{-1} of the laser frequency. The facility to produce such a broad continuum of light with such short duration has a number of distinctive uses, as for example in the flash photolysis study of ultrafast reactions. The light produced by this method, which results from a mechanism known as *self-phase modulation*, is referred to as an *ultrafast supercontinuum laser source* or as *picosecond continuum*. A typical spectral profile is shown in Figure 18.17. Such a source is an invaluable component of the instrumentation for ultrafast pump-probe spectroscopy.

18.7
Short Wavelength Coherent Radiation

18.7.1
Frequency Conversion

In tailoring the wavelength of laser output for different applications, a variety of methods can be utilized. One method involves the use of dye lasers, as discussed above. Another is Raman shifting, in which the stimulated Raman effect is employed for conversion to either shorter or longer wavelength. The latter process is

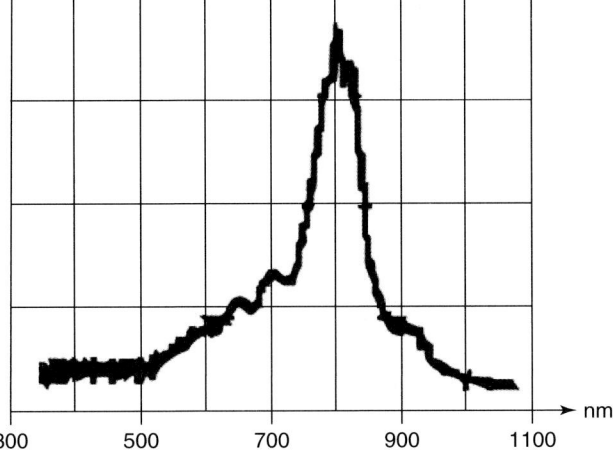

Fig. 18.17 Typical spectrum of the supercontinuum generated by a titanium : sapphire laser, drawn on a logarithmic vertical scale. (Redrawn from an original, courtesy of Coherent Laser Group.)

usually accomplished by the passage of the laser light through a suitable crystal or a stainless steel cell containing gas at a pressure of several atmospheres. Conversion efficiency for the principal Stokes shift to longer wavelength can be as high as 35%. The nature of the crystal or gas determines the frequency increment; $Ba(NO_3)_2$ gives a shift of 1047 cm^{-1}, while the most commonly used gases H_2, D_2, and CH_4 produce shifts of 4155, 2987, and 2917 cm^{-1} respectively. However, by far the most widely adopted method of frequency conversion involves frequency doubling.

18.7.2
Frequency Doubling

Frequency doubling is the best known example of a *nonlinear optical* process. The latter term refers to a wide range of frequency-conversion effects that are nonlinearly dependent on laser intensity, such that their conversion efficiencies are generally improved when laser power is increased. The energetics of frequency doubling (second harmonic generation) are illustrated in Figure 18.18.

Two photons of laser light with frequency ν are assumed to be absorbed by a substance in its ground state, and a single photon of frequency ν' is emitted through a return transition to the ground state. However, this process is parametric. The high intensity of electric field of the incident laser forces the electrons to oscillate in the ground state at the harmonic frequency. This induced nonlinear polarization is a source of *coherent* emission at the doubled frequency. As frequency doubling is a concerted process; there is no intermediate excited state with a measurable lifetime; hence, the energy-time uncertainty principle allows the process to take place regardless of whether there are energy levels of the substance at energies $h\nu$ or $2h\nu$ above the ground state. Indeed, it is usually better if there are not, because the presence of such levels can lead to competing absorption processes. Symmetry

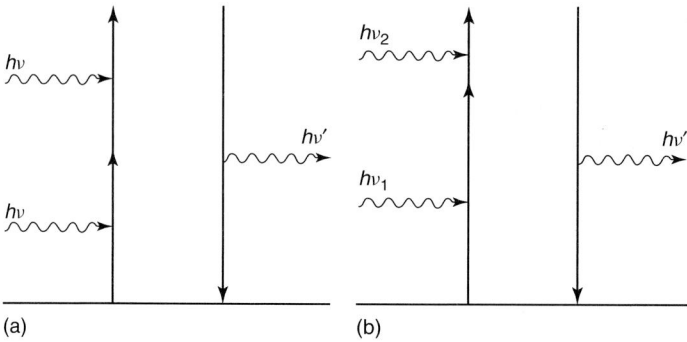

Fig. 18.18 Energetics of (a) frequency doubling, where the emitted photon has frequency $v' = 2v$, and (b) frequency mixing, where the emitted photon has frequency $v' = v_1 + v_2$.

considerations reveal that the frequency-doubling process can take place only in media that lack a center of symmetry – generally a noncentrosymmetric crystal is employed. Although liquids can very weakly generate second harmonics where their bulk symmetry is broken, a feature that can incidentally prove useful for the characterization of liquid surfaces, they are ineffective for efficient frequency conversion.

One singular advantage of harmonic generation is that it is a *coherent* process, so that with laser input the emission also has typically laserlike characteristics even though no population inversion was created in the material. To sustain coherence, however, it is necessary for the conversion process to conserve photon momentum. For harmonic emission in the forward direction, momentum is conserved only if the harmonic photons have wave vector $\mathbf{k}' = 2\mathbf{k}$, where $\mathbf{k}$ is the wave vector of the laser photons; this leads to a requirement for the refractive indices of the pump beam and the harmonic to be equal. Fulfillment of this wave-vector matching condition can only be accomplished in birefringent solids where refractive index is a function of the polarization and direction of propagation. Here, it is often possible to obtain index matching by careful orientation of the crystal. Refractive indices are often very heat sensitive, and fine control of the temperature is normally required; the crystal is accordingly mounted inside a temperature-controlled heating cell. Either the crystal orientation or its temperature can be varied for any particular incident wavelength, to achieve maximum efficiency. Conversion efficiencies may be as high as 20–30%, and under optimum conditions, it is still higher. Frequency doubling is particularly useful for generating powerful visible (532 nm) radiation using an infrared Nd laser pump. Two of the most widely used crystals are KDP (KH_2PO_4) and lithium niobate (Li_3NbO). The better choice for tunable UV generation is β-barium borate, with a transparency range of 190–3500 nm.

The generally high conversion efficiency of frequency doubling makes it possible to use a series of crystals to produce $4\times, 8\times, 16\times, \ldots$, the pump frequency, and so to obtain coherent radiation at very short wavelengths. Similarly, third harmonic generation (which can take place in either centrosymmetric or noncentrosymmetric media) can be used to generate 355 nm radiation from Nd : YAG sources. More commonly, however, suitable crystals can be used to generate odd harmonics (3×, 5×, 7×, etc.) by two-colour mixing (see below). Indeed 355 nm is usually generated in Nd: YAG systems by mixing 532 nm +1.06 micron light. For some spectroscopic applications, it is also significant, that, a doubling crystal coupled with a dye laser provides tunable emission over a range of wavelengths, below the operating wavelength of the pump.

18.7.3
Optical Parametric Oscillators

In frequency mixing, a variation of the frequency-doubling process permits a coupling between two beams of laser radiation with different frequencies to produce output at the sum frequency, as illustrated in Figure 18.18(b). Because wavelength is inversely proportional to frequency, the emitted wavelength is given by $1/\lambda' = 1/\lambda_1 + 1/\lambda_2$, hence shorter than either of the input wavelengths. A closely related application of nonlinear optics concerns the *optical parametric oscillator*, essentially a difference-frequency technique. Often the third harmonic at 355 nm from a Nd : YAG laser pumps the nonlinear crystal, placed along with suitable frequency-selective elements in a secondary cavity (as in a dye laser). The result is that two new frequencies (the signal and idler frequency) are formed whose sum equals that of the pump. Wide tunability and high conversion efficiencies can be obtained in the IR by this method because phase matching for a specific wavelength can be achieved by rotating the nonlinear crystal to specific angles with respect to the optic axis. The same principles can be applied in the visible and UV by the adoption of a different pump, such as one of the other harmonics of a Nd : YAG laser. For example, pumping beta-barium borate crystal, with fourth-harmonic 266 nm radiation, can produce tunable output covering the range 300 nm to 2.5 μm.

18.7.4
Frequency Tripling and Four-Wave Mixing

Higher order frequency mixing can be employed as means of generating shorter wavelength radiation. Two methods, which exploit third-order nonlinearities in centrosymmetric media (media where second order effects are absent), are nonresonant third harmonic generation, THG, and two-photon resonance enhanced four-wave mixing (Lipson, Shi and Lacey, 2002). The energy level schemes for these processes are shown in Figure 18.19. Radiation in the vacuum ultraviolet (VUV, 100 nm $\leq \lambda \leq$ 200 nm) or the extreme ultraviolet (XUV, $\lambda <$ 100 nm) can be obtained by using visible and UV fundamentals. Although technically the radiation produced by THG or four-wave mixing is not the result of a population inversion, the radiation has all the characteristics of a laser; namely, it is coherent, directional, monochromatic, and, even more significant for molecular spectroscopy, is widely tunable. There are no other "real" VUV or XUV laser sources that are competitive in this regard.

Nonresonant THG arises when intense lasers pulses at a frequency ν induce a nonlinear polarization in an atomic medium that acts as a source of radiation at 3ν. Solids and liquids

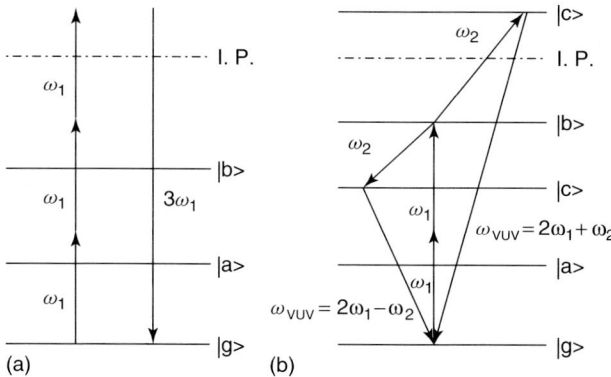

Fig. 18.19 Energetics of (a) frequency tripling, where the emitted photon has frequency $\omega' = 3\omega_1$, and (b) two-photon resonance-enhanced four-wave mixing, where the emitted photon has frequency $\omega' = 2\omega_1 \pm \omega_2$.

are usually never used as they become strongly absorbing below 200 nm. In almost all cases, pulsed lasers are used to generate the third harmonic radiation because its intensity is proportional to the cube of the input fundamental laser power, P_ω^3. The medium must also be negatively dispersive; that is, the wave vector mismatch between the light at 3ν and the light at ν must be <0. This limits the tunability of the generated light to the red of allowed Rydberg ← ground state transitions in the medium. Still as shown in Figure 18.20, significant portions of the VUV and XUV can be covered using rare gases and Hg as nonlinear media. Typical conversion efficiencies range from 10^{-6} to 10^{-3}%, which while low, still translate into intense coherent VUV and XUV outputs given that commercially available dye lasers for example can readily provide fundamental beams with high peak powers (1–10 MW; ~10-nanosecond pulse durations). One particularly simple and popular source is VUV light at ~118 nm radiation generated by tripling the third harmonic of a Nd:YAG laser at 355 nm. Such a source can have wide applications in the soft-ionization of organic molecules, that is, ionization without fragmentation.

Two-photon resonantly enhanced four-wave mixing is a two-color third-order nonlinear process where one laser at frequency ν_1 is tuned to be in resonance with a two-photon allowed transition in a nonlinear medium. A second laser operating at ν_2 can be tuned to generate coherent radiation at the sum and difference frequencies $2\nu_1 \pm \nu_2$. The expression four-wave mixing indicates that three input waves generate a fourth output signal. The nonlinear output wavelengths lie in the VUV and XUV when using visible and UV fundamentals.

The coherent output radiation is widely tunable because there is no inherent restriction on the frequency ν_2. Phase matching for sum-frequency mixing requires the medium to be negatively dispersive. However, the medium can be either positively or negatively dispersive for four-wave difference mixing. Most atomic media that have been used for four-wave mixing

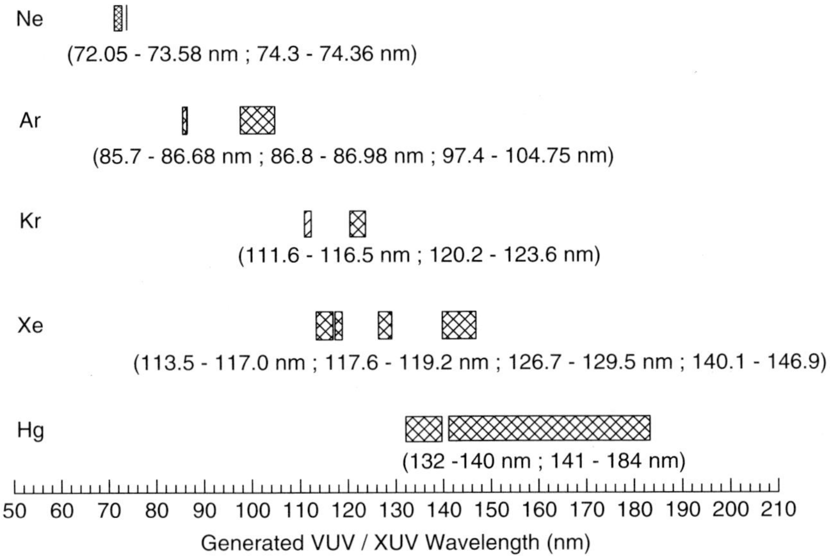
Ne
(72.05 - 73.58 nm ; 74.3 - 74.36 nm)
Ar
(85.7 - 86.68 nm ; 86.8 - 86.98 nm ; 97.4 - 104.75 nm)
Kr
(111.6 - 116.5 nm ; 120.2 - 123.6 nm)
Xe
(113.5 - 117.0 nm ; 117.6 - 119.2 nm ; 126.7 - 129.5 nm ; 140.1 - 146.9)
Hg
(132 -140 nm ; 141 - 184 nm)
50 60 70 80 90 100 110 120 130 140 150 160 170 180 190 200 210
Generated VUV / XUV Wavelength (nm)

Fig. 18.20 VUV and XUV spectral tuning regions for nonresonant THG in Ne, Ar, Kr, Xe, and Hg.

are intrinsically negatively dispersive because the generated light lies higher in energy than the energy of the main resonance line of the medium, whereas the fundamental energies lie below.

Pulsed fundamental laser beams are almost always used since both sum- and difference-mixing scale as $P_1{}^2P_2$. The advantage of two-photon resonance enhanced four-wave mixing over nonresonant THG however, is that the third-order nonlinear susceptibility of the process can be enhanced many orders of magnitude by tuning one of the fundamental laser frequencies to be in resonance with a two-photon allowed transition in the nonlinear medium. In this way, intense VUV and XUV radiation can be generated with only moderate incident laser powers.

The tuning ranges available using Kr gas, Xe gas, Hg vapor and Mg vapor are shown in Figure 18.21. Metal vapors need to be generated at high temperatures, usually in heat pipes, while rare gas are easily handled in a gas cell. Rare gases have the additional advantage that they can also be expanded as supersonic expansions in a vacuum system. These gas jets are spatially localized and therefore can be supported in a windowless arrangement, provided the vacuum system has adequate differential gas pumping capacity. The elimination of windows is ideal for XUV generation as there are no broadly transmitting materials in these spectral regions.

18.7.5
Higher Order Harmonic Generation

The high intensities ($10^{13}-10^{16}$ W cm^{-2}) associated with femtosecond sources such as Ti : sapphire lasers can be used to extend the generation of short wavelength light into the soft X-ray region by

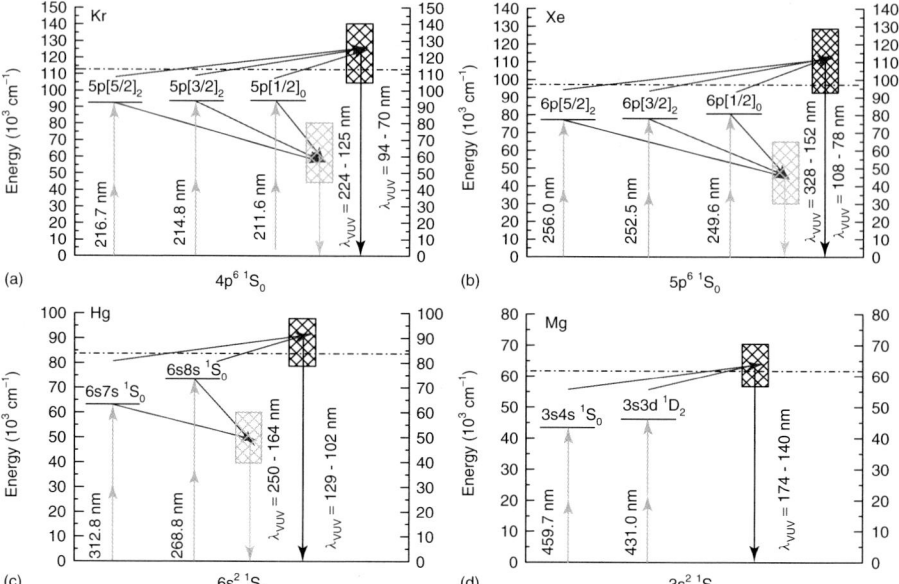

Fig. 18.21 VUV and XUV spectral tuning regions for two-photon resonance-enhanced four-wave sum and difference mixing in (a) Kr, (b) Xe, (c) Hg, and (d) Mg.

high-order harmonic generation. When the intensity of the input laser pulses is sufficiently high, electrons from a gas-phase atomic medium are ionized and begin to move away from the resultant ions at relativistic velocities. However, because the distance they travel is small within one or more laser periods, they still respond to the electric field by undergoing ponderomotive free space oscillation. Those electrons that recombine with the ions within a fraction of one optical cycle give up their energy as high order multiples of the input laser frequency. However, a plateau exists on the frequency of harmonic light that can be generated, which is determined by the ionization potential of the atom being ionized and the maximum kinetic energy of the recombining electron (Corkum, 1993). Remarkably, the pulse duration of the light generated by this strong field interaction is on the attosecond (10^{-18} s) timescale that provides new opportunities to probe complex electron dynamics in molecules and materials (Kapteyn et al., 2007). The output efficiency scales with the gas density, and phase matching is approximately independent of the order of the process. Wavelengths as short as 2.7 nm have been generated in He using 26-femtosecond 800-nm pulses from a Ti : sapphire laser (Chang et al., 1997). Coherent harmonic generation up to order 221 and shorter continuum range of wavelengths up to order 297 were resolved. The shortest wavelength lies within the "water window" region of X-ray transmission, making these sources useful for the study of biological samples.

18.7.6
Free-Electron Laser

The free-electron laser has a highly accelerated electron beam as its active medium. One of the most common experimental arrangements, illustrated in Figure 18.22, involves passing a beam of very high energy (10^7–10^8 MeV) relativistic electrons from the accelerator of a synchrotron ring between the poles of a series of regularly spaced magnets of alternating polarity called an *undulator*. The electrons are repeatedly accelerated and decelerated in a direction perpendicular to their direction of travel, resulting in an oscillatory path, as shown. The effect of this process is to produce emission of coherent *bremsstrahlung* radiation along the axis of the laser at a wavelength that depends on the periodicity of the undulator and the energy of the electrons. The radiation is then trapped between parallel mirrors and stimulates further emission.

High-energy electrons travel at an appreciable fraction of the speed of light, and can be characterized by a parameter f, denoting the ratio of their relativistic total energy to their rest-mass energy. If the magnet separation is d, then the laser emission wavelength is given by the simple formula $\lambda = d/2f^2$. Because the energy of the electrons emerging from the accelerator can be continuously varied, the result is again a laser with tuning capability. There is, in principle, no limit to the range of tuning right across the infrared, visible, and ultraviolet regions of the electromagnetic spectrum. Moreover, this kind of laser has been shown to produce high average powers and to be capable of a reasonable efficiency. For example, a setup in the Naval Research Laboratory in the USA. can produce 75 MW pulses of 4 mm radiation with an efficiency of about 6%. The efficiency in the visible and ultraviolet is generally lower. The downside of these devices is that they are expensive and require access to a large synchrotron facility.

Plans have been made to develop free-electron lasers that can operate in the soft and hard X-ray regions. For example, a second generation free-electron laser has been proposed for the BESSY beamline in Germany where a seed laser pulse copropagates with the electron bunch through the undulator to allow other laser sources to be synchronized for pump-probe experiments (Follath, 2007). The seed laser pulses at 51–1.24 nm are generated by higher harmonic generation of a tunable laser operating between 230 and 460 nm. Nonlinear interactions generate higher harmonics of the seed laser, and therefore additional insertion

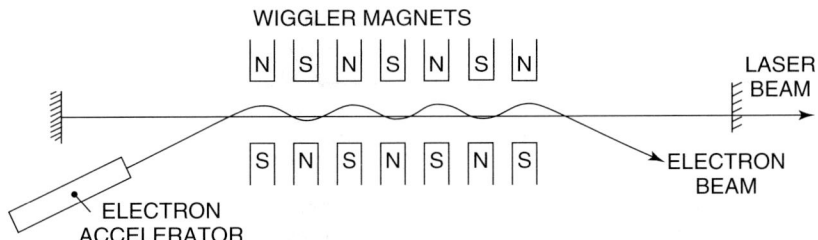

Fig. 18.22 Configuration of a free-electron laser.

devices are required to extract our one particular wavelength. Proof of principle of this approach has been realized by researchers at Brookhaven National Laboratory who demonstrated a high-gain harmonic-generation free-electron laser operating in the ultraviolet at 266 nm (Yu et al., 2003).

18.7.7
Tabletop X-ray Lasers

Despite the obstacles inherent in generating a population inversion that can lead to gain in the X-ray region, there have been many notable successes (Rocca, 1999). While the field is too broad to cover in this article, it is worth noting that Rocca and coworkers have developed fixed-wavelength soft X-ray lasers that use fast pulsed discharge excitation of capillary channels to create plasmas that exhibit gain by collisional recombination. For example, this scheme was used to generate coherent radiation on the 46.9 nm 3p-3s ($J = 0-1$) line of Ne-like Ar (Rocca et al., 1994). These lasers are attractive because of their small size, earning them the moniker "tabletop" devices.

References

Andrews, D.L. (1997) Lasers in Chemistry, 3rd edn. Springer-Verlag, Berlin.
Asasda, M., Miyamoto, Y., and Suematsu, Y. (1986) IEEE J. Quantum Electron., **QE-22**, 1915–1921.
Auzel, F. (2004) Chem. Rev., **104**, 139–173.
Chang, Z., Rundquist, A., Wang, H., Murnane, M.M., and Kapteyn, H.C. (1997) Phys. Rev. Lett., **79**, 2967–2970.
Corkum, P.B. (1993) Phys. Rev. Lett., **71**, 1994–1997.
Faist, J., Capasso, F., Sivco, D.L., Sirtori, C., Hutchinson, A.L., and Cho, A.Y. (1994) Science, **264**, 553–556.
Fiore, A., Markus, A., Rossetti, M., and Li, L.H. (2006) Laser World Focus, **42**, 124–127.
Follath, R. (2007) Proc. SPIE, **6586**, 658604.
Fork, R.L., Brito Cruz, C.H., Becker, P.C., and Shank, C.V. (1987) Opt. Lett., **12**, 483–485.
Hecht, J. (2005) Laser Focus World, **41**, 143147.
Jordan, W.G., Jha, A., and Ryan, J. (1993) J. Mat. Sci. Lett., **11**, 771–773.
Kapteyn, H., Cohen, O., Christov, I., and Murnane, M. (2007) Science, **317**, 775–778.
King, T.A. (2004) Fibre lasers, in Comprehensive Series in Photochemistry and Photobiology, **4** (eds G. Palumbo and R. Pratesi), Royal Society of Chemistry, Cambridge.
Leinen, H., Gläβner, D., Metcalf, H., Wynands, R. Haubrich, D., and Meschede, D. (2000) Appl. Phys. B, **70**, 567–571.
Lipson, R.H., Shi, Y.J., and Lacey, D. (2002) in An Introduction to Laser Spectroscopy, 2nd edn (eds D.L. Andrews and A.A. Demidov), Chapter 9, Kluwer Academic Plenum Publishers, New York, pp. 257–309.
Monerie, M. (1989) Phys. Scripta., **T29**, 218–222.
Mowbray, D. (2005) Laser Focus World, **41**, 157–159.
Nakamura, S. (1998) Science, **281**, 956–961.
Rice, P.R. and Carmichael, H.J. (1994) Phys. Rev. A, **50**, 4318–4329.
Righini, G.C., Brenci, M., Chiasera, A., Feron, P., Ferrari, M., Nunzi, J.-M., Conti, G., and Pelli, S. (2005) Proc. SPIE, **6029**, 603903.
Rocca, J.J. (1999) Rev. Sci. Instrum., **70**, 3799–3827.
Rocca, J.J., Shlyaptsev, V., Tomasel, F.G., Cortázar, O.D., Hartshorn, D., and Chilla, J.L.A. (1994) Phys. Rev. Lett., **73**, 2192–2195.
Shank, C.V., Fork, R.L., Yen, R., and Stolen, R.H. (1982) Appl. Phys. Lett., **40**, 761–763.
Steinmeyer, G. (2006) Appl. Opt., **45**, 1484–1490.
Treacy, E.B. (1969) IEE J. Quantum Electron., **QE-5**, 454–458.
Vahala, K.J. (2003) Nature, **424**, 839–846.

Yablonovitch, E. (1987) *Phys. Rev. Lett.*, **58**, 2059–2062.

Yu, L.H., DiMauro, L., Doyuran, A., Graves, W.S., Johnson, E.D., Heese, R., Krinsky, S., Loos, H., Murphy, J.B., Rakowsky, G., Rose, J., Shaftan, T., Sheehy, B., Skaritka, J., Wang, X.J., and Wu, Z. (2003) *Phys. Rev. Lett.*, **91**, 074801-1–074801-4.

19
Linear Laser Spectroscopies

Stephen H. Ashworth

19.1	**Introduction**	**697**
19.2	**Direct Techniques**	**698**
19.2.1	Direct Absorption	698
19.2.2	Multipass Techniques	698
19.2.3	Cavity Ring Down	699
19.2.4	Cavity-Enhanced Absorption	699
19.2.5	Intracavity Techniques	700
19.2.6	Laser Magnetic Resonance Spectroscopy	700
19.2.7	Raman Spectroscopy	701
19.2.8	Polarization Spectroscopy	703
19.3	**Proxy Techniques**	**703**
19.3.1	Laser-Induced Fluorescence	704
19.3.1.1	Imaging	705
19.3.1.2	Dispersed Fluorescence	705
19.3.1.3	Stimulated Emission Pumping	706
19.3.1.4	Intermodulated Fluorescence	706
19.3.1.5	Saturation Dip Spectroscopy	706
19.3.1.6	Quantum Beat Spectroscopy	707
19.3.1.7	Single-Molecule Spectroscopy	707
19.3.1.8	Light Detection and Ranging (LIDAR)	707
19.3.2	Photoacoustic Spectroscopy	708
19.3.3	Photoelectron Spectroscopy	708
19.3.4	Resonance-Enhanced Multiphoton Ionization	709
19.3.5	Optogalvanic Spectroscopy	710
19.3.6	Frequency Modulation Spectroscopy	711
19.3.7	Thermal Lensing Spectroscopy	712
19.4	**Generally Applicable Techniques**	**712**
19.4.1	Velocity Modulation	712
19.4.2	Double Resonance	713

Encyclopedia of Applied Spectroscopy. Edited by David L. Andrews.
Copyright © 2009 WILEY-VCH Verlag GmbH & Co. KGaA, Weinheim
ISBN: 978-3-527-40773-6

19.4.3 Stark Spectroscopy 713
References 713
Further Reading 716

19.1
Introduction

As lasers have developed, they have become a common and important tool for spectroscopy. Applications range from probing samples remotely at great distance to generating transient species. In this chapter, we confine ourselves to linear spectroscopic techniques. In general, therefore, the signal response is proportional to the laser power. The only exceptions to this are when saturation spectroscopies and other two-photon techniques are considered. Here, the overall signal response is no longer directly proportional to the laser power but the absorption of each individual photon is still in the linear regime.

It is the unique properties of laser radiation that have made it such a powerful spectroscopic tool. Continuous wave (CW) lasers can be constructed such that their linewidth is very narrow (below 1 MHz), thus enabling high-resolution measurements. The high spectral brightness of lasers increases the probability that two or more photons may interact with the same absorber, which also makes multiple photon spectroscopies possible. In addition, the very well defined temporal envelopes produced by pulsed laser systems allow time-resolved measurements to be carried out on a range of timescales. Finally, the highly directional nature of laser radiation makes optical manipulation very simple and allows some measurements to be made remotely.

No distinction has been made here between samples. The techniques outlined can be applied to a variety of samples: not exclusively atomic or molecular. Many of the techniques, however, lend themselves to application in the gas phase where the full resolution of laser sources may be applied. Liquid, solid, or solution phases have, however, not been excluded.

This chapter is divided into two main sections. The first deals with applications that detect laser intensity directly. To monitor a small change on what is often very high laser intensity reliably, a detector has to be capable of a very high dynamic range. For this reason, many laser interactions are detected indirectly by means of a side effect. These techniques are here referred to as *proxy techniques* and are described in the second section.

A short concluding section deals with examples of named techniques that straddle these rather artificial boundaries or may be used with a variety of

Encyclopedia of Applied Spectroscopy. Edited by David L. Andrews.
Copyright © 2009 WILEY-VCH Verlag GmbH & Co. KGaA, Weinheim
ISBN: 978-3-527-40773-6

approaches. This is by no means meant to be an exhaustive list and neither is it designed to describe these methods in great depth. The reader can use the citations and review articles to obtain further details.

19.2
Direct Techniques

19.2.1
Direct Absorption

The very high spectral brightness of laser sources means that direct absorption measurements are generally limited by the dynamic range of the detector. In many cases, the signal in which we are interested is a small change in a large detected intensity, hence limiting the overall sensitivity. Methods have been developed to overcome this disadvantage, which either increase the proportion of the sample response or detect the response as a function of time.

One method to increase the proportion of sample response is simply to increase the length over which the sample and laser radiation interact. This is of special importance in dilute samples of low-pressure gases. Pressure broadening studies or measurements of small concentrations of atmospheric absorbers might require such an approach. The length of the sample cell can only be extended if the sample itself is stable and the physical dimensions of the laboratory allow this. Frequency measurements of pure rotational transitions in vibrationally excited H_2O molecules (Matsushima *et al.*, 2006) have used the latter approach with a single pass in a 1.8-m sample cell.

19.2.2
Multipass Techniques

In many cases, it is not possible simply to increase the length of the sample. This may be a result of the physical limitations of a laboratory or the spatial extent of the sample. Few laboratories can accommodate sample cells much longer than 2 m and transient species are often confined to a small region of space. In order to overcome such limitations, or to maximize path length when the sample can only be produced in a limited region, the laser radiation may be passed through the sample many times. In the simplest of these, a pair of plane mirrors may be arranged such that the light is redirected through the sample as many times as possible before reaching the detector. The low divergence of a laser source is a great benefit in these experiments (Qian and Howard, 1997).

More sophisticated approaches use systems of curved mirror cells such as the White (White, 1942) and Herriott (Herriott, Kogelnik and Kompfner, 1964; Herriott, 1963) cells. These optical arrangements are designed to minimize losses in order to produce a much longer effective path length than with plane mirrors. The beam path, however, imposes some geometrical constraints with the result that a given solution is not universally appropriate: a laser beam in a traditional White cell arrangement, for example, tends to probe a single plane. A White or Herriott cell would be a good choice if the sample were sufficiently stable to fill the cell. On the other hand, a multipass arrangement with fewer reflections from plane mirrors may well prove to be a more sensitive solution for a transient sample, which is confined to a small region of space.

19.2.3
Cavity Ring Down

Cavity ring-down spectroscopy (CRDS) might be considered close to the ultimate multipass technique, which is reviewed in Berden, Peeters and Meijer (2000). The sample is contained in a high-finesse optical cavity that is designed to minimize losses from diffraction and scattering. Laser light is coupled into the cavity and then quickly switched off. CW lasers are generally coupled to the cavity using an active control to scan or lock the CRDS cavity and often require active switching of the light. On the other hand, the broad bandwidth of a pulsed laser generally overlaps with at least one of the closely spaced cavity modes without the need for active cavity length adjustment. As a result, what the arrangement lacks in resolution it more than makes up for in simplicity.

Once the cavity excitation stops, the stored intensity leaking through one of the mirrors may be monitored as a function of time. A compromise has often to be reached between the wavelength range over which the mirrors are highly reflective and the reflectivity of the mirrors. In general, the more highly reflective the mirrors are, the smaller the wavelength range over which they are highly reflective.

In a CRDS cavity, the number of round trips, n, required for the light to reach $1/e$ of the original intensity is given by $n = -1/[2\ln(R)]$, where R is the mirror reflectivity. The intensity of light confined in a cavity bounded by mirrors of reflectivity $R = 0.9999$ will fall to $1/e$ of the original intensity after 5000 round trips. Thus, a cavity with such mirrors separated by 1 m and filled with sample produces an effective path length of around 10 km.

Active switching of light out of (or into) the cavity is obviously not necessary if pulsed lasers are used; for CW lasers, the switching is usually effected by deflecting the beam with an acousto-optic modulator. The intensity leaking from the cavity decays exponentially with a time constant that is determined by cavity losses and may be as long as tens of microseconds. The difference in time constant between an empty cavity and one containing sample gives a direct measure of the absorption coefficient of the sample. One elegant feature of this technique is that the signal is, at least in principle, immune to power fluctuations in the laser. Part of the absorption spectrum of the iodine monofluoride radical recorded using CRDS is shown in Figure 19.1.

19.2.4
Cavity-Enhanced Absorption

The technique of CRDS has been developed further to avoid switching the laser out of the cavity. This has become known as *cavity-enhanced absorption spectroscopy (CEAS)* or *integrated cavity output spectroscopy (ICOS)*. The laser may be scanned rapidly and sometimes repeatedly over a portion of the sample spectrum and the absorption recorded is the integrated output of the cavity. This has the advantage of simplicity compared to cavity ring-down as the decay of intensity is not recorded as a function of time. It does, however, lose the immunity to power fluctuations of the light source. Off-axis cavity designs have recently become popular as the alignment becomes very robust, which is attractive when designing field instruments (Moyer *et al.*, 2008).

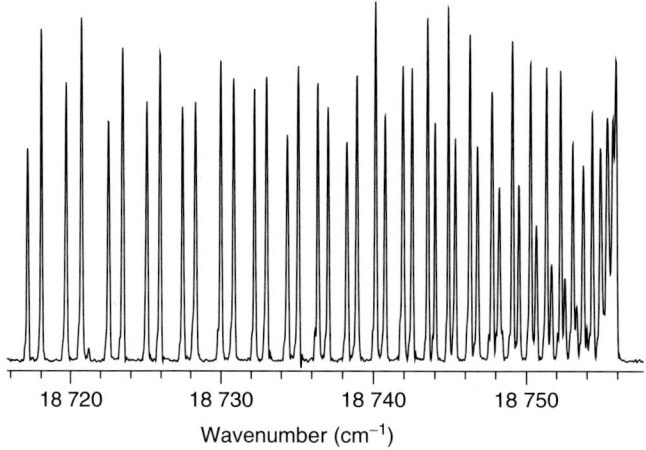

Fig. 19.1 A portion of the absorption spectrum of the iodine monofluoride radical produced by the flash photolysis of CF_3I recorded using cavity ring-down spectroscopy. (Ashworth S.H., Joseph D.M. and Plane J.M.C. *unpublished*.)

The integration of the cavity output and the need for stable lasers means that CEAS requires CW lasers or at least quasi-CW lasers. In common with CRDS, the technique may be applied over a wide wavelength range. It has been applied to make a portable spectrometer to measure ratios of carbon isotopes of CO_2 in the 2 µm region (Wehr *et al.*, 2008) and also to detect atmospherically important species in the ultraviolet region (Mejean, Kassi and Romanini, 2008).

19.2.5
Intracavity Techniques

In some cases, the sample may be usefully placed within a CW laser cavity. The technique is often known as *intracavity laser absorption spectroscopy* (*ICLAS*). These samples may be thought of as being exposed to an infinite number of round trips of the laser light. A small loss due to absorbance has a large effect on the gain of the laser (Peterson *et al.*, 1971). Such a description is, however, deceptively simple. Although the signal is obtained simply by monitoring the laser output, sensitive measurements are only possible with very stable laser sources. The laser itself needs to have been modified. The cavity quality factor, Q, has probably been reduced and the alignment changed by the introduction of a sample cell. Second, the signal itself is a result of a further change in Q as the measurement is taken. Consequently, active stabilization of the laser can become challenging. Nevertheless, the technique is capable of extremely sensitive measurements (Ding *et al.*, 2005).

19.2.6
Laser Magnetic Resonance Spectroscopy

Laser magnetic resonance (*LMR*) is a technique used in spectral regions where laser radiation is not easily tunable. This is the case in the mid- and far infrared. The mid-infrared carbon monoxide laser and a number of far infrared gas lasers operate over a range of discrete frequencies. The gases are lasing on one of many molecular transitions so the output may be one of a range of discrete frequencies. Therefore, these discretely tunable lasers are known as *line tunable lasers*.

The samples under investigation rarely have absorptions that coincide with individual laser transitions, so in the absence of tunable laser light the molecular transitions themselves are tuned. The sample must, therefore, be a radical species in order that the magnetic field may interact with the sample. This selectivity may be used to advantage to detect radical species in a complex mixture. The sample is arranged so that it may be exposed to a magnetic field either between the poles of an electromagnet or within a solenoid. The spectrum is recorded as a function of field. A small, sinusoidal magnetic field is superimposed on the scanned magnetic field and phase-sensitive detection is generally used.

In the far-infrared, gases such as methanol, hydrazine, and CF_2H_2 are used as lasing media. They are generally pumped by CO_2 lasers and give a range of discrete lines with wavelengths in the 50–500-μm range. The gases are lasing on rotational transitions and isotopic variants give access to a range of alternative laser frequencies. The frequencies of known laser transitions are independently measured and periodically tabulated (Inguscio *et al.*, 1986; Zink *et al.*, 2008). New laser transitions are occasionally discovered and have to be precisely measured before they may be used for high-resolution measurements. The development of frequency combs has meant that complicated frequency chains are no longer necessary for accurate frequency determinations (Beverini *et al.*, 2005).

Most LMR spectrometers are designed with the sample within the laser cavity and thus LMR is usually an intracavity technique. The signal is obtained by monitoring the laser output as a function of magnetic field. Figure 19.2 shows part of an LMR spectrum recorded during a study of the HSS radical in the far infrared.

The spectrometer can, however, be designed so that the laser is remote from the sample. The intracavity advantage is lost in this case, so it may be necessary to construct a multipass cell around the magnet (Hinz *et al.*, 1985). On the other hand, this arrangement allows the Faraday (*vide infra*) or Voigt effects to be exploited, which increases the detection sensitivity (Gillett, Cooksy and Brown, 2006).

19.2.7
Raman Spectroscopy

The narrow linewidth and intensity of laser sources has enabled *Raman spectroscopy* to become a mainstream technique. Raman spectroscopy enables vibrational spectra to be recorded using visible radiation albeit with rather different selection rules than apply to infrared absorption. Energy may both be removed from (Stokes) or added to (anti-Stokes) the incident radiation field from ground state and excited vibrational modes, respectively. These inelastic scattering processes may conveniently be pictured as absorption to, and emission from, a virtual state. A single-frequency CW laser is generally used and scattered light is collected and dispersed. The shifts in frequency are generally small as

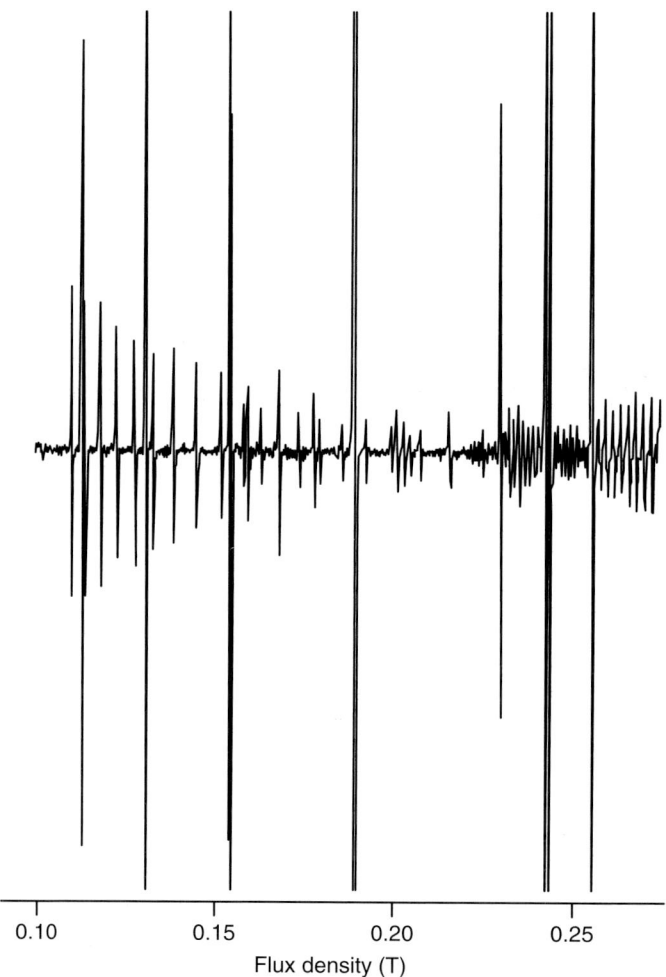

Fig. 19.2 Part of a far-infrared LMR spectrum recorded during a study of the HSS radical using the 203 μm laser line of $^{13}CH_3OH$. The first derivative lineshape is a result of the phase-sensitive detection used. (Ashworth S.H., Brown J.M. and Evenson K.M. *unpublished*.)

they usually correspond to vibrational spacings in the sample.

Rayleigh scattering is far more intense than the Raman shifted components of the scattered light and care must be taken to reduce the intensity of this light reaching the detector. Double and even triple monochromators may be employed for this. Advances in coating technologies, however, mean that filters can now be produced with extremely sharp cutoffs between high and low optical density, which makes experimental arrangements much simpler.

Naturally, the region where the Stokes lines are sought in a Raman spectrum is exactly the region where one might expect intense fluorescence. This is a problem, which often occurs because fluorescence is generally orders of magnitude more intense than the Raman shifted signals. This can be avoided by depositing the sample on a suitable substrate (e.g., graphite), which efficiently quenches fluorescence, or by using a sufficiently long excitation wavelength that the sample is not electronically excited. The latter approach cannot, of course, be applied to more sophisticated variants of the technique such as resonance Raman spectroscopy (Mak and Kincaid, 2008).

A suitable surface, such as rough silver, may increase the signal by orders of magnitude. This constitutes *surface-enhanced resonance Raman spectroscopy* (*SERRS*) (Smith, Faulds and Graham, 2006). Coherent anti-Stokes Raman spectroscopy (CARS) is strictly a nonlinear technique but should at least be mentioned here (Zheltikov, 2000). The CARS signal has the advantage of being highly directional and is therefore relatively straightforward to manipulate optically, to achieve high rejection of unwanted light.

19.2.8
Polarization Spectroscopy

Polarization spectroscopy may be implemented in a number of ways. In the simplest form, the sample is placed in a strong magnetic field, directed along the laser axis, between crossed polarizers. The Faraday effect in an isotropic sample, such as a gas, causes absorptions to rotate the plane of polarization of the laser light. Signals are thus detected against a very low background making this potentially a very sensitive technique.

In a second application, counterpropagating beams of unequal intensity from a narrow bandwidth laser may be used. The higher intensity pump beam is circularly polarized and the weaker probe beam is linearly polarized. The sample is placed between crossed polarizers through which the probe beam is directed. Absorption of the pump beam produces in a nonequilibrium population of magnetic sublevels. This disequilibrium is referred to as an *orientation* and in turn leads to a rotation in the polarization of the probe beam. The detector, behind the crossed polarizer, therefore, only detects signal when both pump and probe interact with the same population. This only occurs when the interaction is with the fraction of the population, which has no component of velocity along the laser axis, and is thus an inherently sub-Doppler technique (Demtröder, 2008).

An extension of this is known as polarization labelling spectroscopy. In this variant, the two beams are no longer derived from the same laser (and as a result the technique is not sub-Doppler). Like other double-resonance techniques (see below), it is usually applied to spectroscopic problems to use secure assignments from one set of bands to gain information about the assignment of a second set (Adohi-Krou *et al.*, 2008). The known transition is pumped with the first, fixed, laser and only transitions in the second set of bands with common electronic states are observed on scanning the second laser.

19.3
Proxy Techniques

The techniques considered up to now have required that a change in laser power be detected in some manner after interaction with a sample. An alternative and sometimes more sensitive approach is to detect a process that results from the absorption of a photon. Care must be exercised with what are here referred to as *proxy techniques*, as they will tend to yield different information from, and have a more limited range of applicability than, direct absorption.

19.3.1
Laser-Induced Fluorescence

The technique of *laser-induced fluorescence* (*LIF*) relies on the fate of the excited state formed on absorption. The word fluorescence itself implies that the upper state is an excited electronic state. As a result, the majority of LIF applications are in the visible and ultraviolet regions of the spectrum.

If the sample has a nonzero quantum yield of fluorescence, the total fluorescence intensity measured as a function of wavelength produces an excitation spectrum, which is a proxy for the absorption spectrum. The spectrum thus formed is a measure of the production of fluorescing, excited state molecules. The fluorescence signal may be shown to be proportional to the number of molecules present in the sample and, in the absence of other wavelength-dependent effects, the signal faithfully reproduces the absorption spectrum.

The absorption spectrum is not faithfully reproduced if other pathways to dispose of the excitation are available. In the gas phase, at low pressure, examples of potential pathways are dissociation and predissociation; the energy of the photon is sufficient to cause bond rupture. Often the fragments are not electronically excited and thus unable to emit, thereby reducing or even completely quenching the overall fluorescence intensity. At higher pressures in the gas phase or solution phase, collisional quenching may become sufficiently important that the efficiency of LIF is reduced.

A great advantage of LIF is that it is essentially a zero background technique. Fluorescence emission is collected off-axis from the excitation source (usually at right angles), so a highly sensitive detector may be used. Unfortunately, a lot of effort and careful experimental design are required to collect more than only a small proportion of the emission. In most cases, the detection sensitivity is sufficient so that it is unnecessary to make this additional effort. The most sophisticated arrangements implement collection optics that not only increase the efficiency of collection but may also provide spatial filtering, which helps discriminate against other sources of light. One of the most efficient systems reported uses spherical mirrors and collects around 30% of the emitted light (Korter *et al.*, 2001).

Scattered light from the excitation source is often the hardest to reject but LIF may also be applied in environments where other sources of light are visible to the detector. LIF is readily applied to furnaces, plasmas, and flames. Such sources may generate a continuum of frequencies where others may generate a structured emission or a combination of the two. A variety of techniques are used to reject unwanted signal at the detector. Some, such as time gating, lock-in detection, and phase-sensitive detection, use a well-defined time dependence of signal – either a modulation or a pulse – to discriminate against both noise and unwanted light. These techniques are, however, of no use if the detector itself is swamped by unwanted illumination.

The sensitivity of LIF, the fact that the signal is proportional to the number of emitters, and above all the selectivity of the technique means that it is easily applied to analytical and kinetic measurements. When observing changes in concentration using LIF, the laser source is not scanned and the signal will often be normalized to take account of variations in the laser power. Once again background light, including laser scatter, may be a problem. In such cases, bandpass filters are effective means to reduce the unwanted light intensity reaching the detector. A monochromator is a very convenient adjustable filter, both in terms of wavelength and bandpass.

When the band of wavelengths detected includes the laser wavelength, this is known as *resonant fluorescence detection*. Nonresonant fluorescence detection is possible if there is a well-defined transition to another energy level. The band of wavelengths required to detect fluorescence emission now no longer includes the laser wavelength. The great advantage of nonresonant fluorescence detection is, therefore, the high rejection of scatter from the excitation source.

19.3.1.1 Imaging

One of the benefits of a laser source is that the radiation need not be in the form of a beam on interaction with the sample. If the laser light is transformed into a sheet, fluorescence may be excited in a plane in the sample. The fluorescence may then be imaged onto a two-dimensional detector allowing dynamic processes, such as combustion, to be probed (Luong *et al.*, 2008). If the sheet of light is arranged so that it may be scanned in a direction perpendicular to the plane of light, a three-dimensional image may be built up.

19.3.1.2 Dispersed Fluorescence

A natural extension to the technique of nonresonant LIF detection mentioned above is to measure the fluorescence intensity as a function of emission wavelength. This is known as *dispersed* or *resolved fluorescence* and gives information on the ground state of the sample, shown schematically in Figure 19.3. Given that the collection efficiency of fluorescence is generally rather low, dispersed fluorescence usually requires quite careful alignment and sensitive equipment.

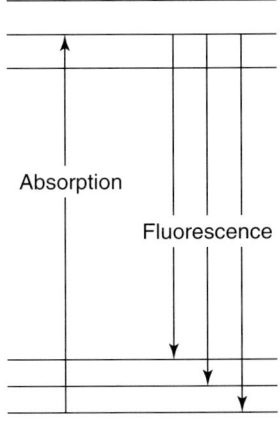

Fig. 19.3 The principle of laser-induced fluorescence. The excitation spectrum is recorded by scanning the absorbed photon and recording the total fluorescence. The wavelength dependence of the emission recorded with a fixed absorption wavelength forms the dispersed (or resolved) fluorescence spectrum.

The resolution is generally determined by the recording instrument. Any systems that are capable of resolving the fluorescence may be used to record dispersed fluorescence. Compact charge-coupled device (CCD) spectrometers offer low resolution rugged devices, whereas the highest resolution is more likely to be afforded by Fourier transform spectrometers (Hodges et al., 2007), but these are much less rugged. The instrument of choice is often dictated by a compromise between the required resolution, fluorescence intensity, and the robustness that is required of the equipment.

19.3.1.3 Stimulated Emission Pumping

An alternative to dispersed fluorescence is the technique of *stimulated emission pumping* (*SEP*). A first photon fixed at one wavelength, as in dispersed fluorescence, is used to excite fluorescence. When a second photon incident on the sample is resonant with a downward transition, emission is stimulated or pumped. This leads to a decrease in the excited population and hence a decrease in the fluorescence intensity. The signal level is monitored as a function of the wavelength of a second photon. The signal is usually displayed as the ratio between fluorescence intensity before and after the pump laser as shown by the shaded areas in Figure 19.4. A good example is given in Mukarakate et al. (2008) where the normally downward going lines have also been reversed for clarity. The additional complexity of the technique in that it requires two lasers is compensated by the resolution, which is generally limited only by the pump laser.

19.3.1.4 Intermodulated Fluorescence

Intermodulated fluorescence is one of the techniques that may be used to reach sub-Doppler resolution. It involves single photon absorption from counterpropagating laser beams. The beams are chopped at different frequencies ω_1 and ω_2. When the beams are probing the same population, a signal is produced at the sum frequency $\omega_1 + \omega_2$. The only population that is probed by both beams simultaneously is that which has no velocity component along the laser beams. Thus the signal is inherently sub-Doppler and is often used to investigate hyperfine structure (Manke et al., 2008).

19.3.1.5 Saturation Dip Spectroscopy

Saturation dip spectroscopy has many names and incarnations. In simple terms, the sample is exposed to counterpropagating beams of sufficient intensity that at line center the signal intensity drops because the lower state population has been depleted. This

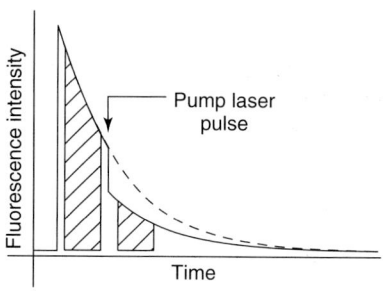

Fig. 19.4 The shaded areas show the areas to be compared in stimulated emission pumping. When the second (pump) laser interacts with the initially excited population, the fluorescence intensity is reduced.

sort of spectroscopy is variously referred to as *Lamb dip, saturation dip,* and *hole burning spectroscopy* in different fields and applications. It is treated in depth in Demtröder's classic book (Demtröder, 2008).

Saturation spectroscopy may be relatively straightforward if the sample is within the cavity of a standing wave CW laser as it is naturally exposed to collinear, counterpropagating radiation.

19.3.1.6 Quantum Beat Spectroscopy

The observation of quantum beats may be thought of as an analog of the observations in a Young's double slit experiment. Closely spaced molecular energy levels may be excited simultaneously by a single, usually pulsed, laser. The laser bandwidth obviously has to be sufficiently broad to excite two or more levels and form a coherent superposition (Ishii, Takeuchi and Tahara, 2008). Modulation at a frequency corresponding to the difference between the excited energy levels is superimposed on the exponential decay of fluorescence. The principle is illustrated in Figure 19.5. A Fourier transform of the beat frequency reveals the energy difference of the levels involved and produces a Doppler-free spectrum even when the energy difference is smaller than the Doppler width of the fluorescence.

19.3.1.7 Single-Molecule Spectroscopy

In many cases, spectroscopic signals are averaged over a large number of molecules, each of which is in a slightly different environment. This averaging effect is known as *heterogeneous broadening* as the signal is made up of contributions from many different individual absorptions. In environments where absorbers are constrained to a small region of space, it is possible to overcome this by recording signal from a single molecule or a single emitter, such as a quantum dot. Obviously, the fluorescence signal will tend to be weak in comparison with ensemble spectroscopy so a single molecule has to be pumped many times to its excited state in a single measurement. The field of view is constrained and the concentration of emitters kept low so that on average only one emitter may be detected at a time. Single-molecule spectroscopy may be used to investigate guest–host interactions and the homogeneity of the local environment. Wustholz *et al.* (2008) and Wirth and Legg (2007) have reviewed this extensive area.

19.3.1.8 Light Detection and Ranging (LIDAR)

Backscattered radiation from a laser directed through the atmosphere may be

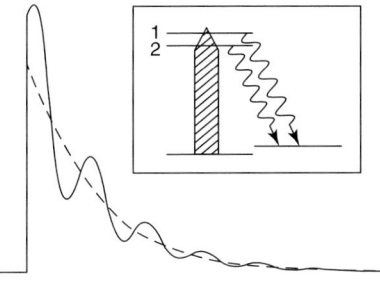

Fig. 19.5 Quantum beats (solid line) superimposed on a simple exponential decay (broken line). The inset shows the associated energy level scheme. The laser excites states 1 and 2 simultaneously and the interference results in quantum beats.

collected and, under favorable circumstances, used to measure the quantity and range of airborne substances. Rayleigh scatter is the strongest signal and is the result of inelastic scattering from all molecules in the atmosphere. This scatter is often carefully screened from detectors in order to allow detection of atomic and molecular signals. On the other hand, the Rayleigh scatter itself may be used to measure important quantities (Schoch, Baumgarten and Fiedler, 2008). The intensity of the scatter is proportional to gas density, the width of the scattered peak is related to gas temperature, and, in addition, one component of bulk velocity results in a shift of the peak wavelength of the scatter.

Resonant atomic fluorescence is perhaps the next strongest source. This allows the concentration of species such as iron and sodium to be measured in the mesosphere, at a range of around 85 km, from ground-based measurements (Gardner et al., 2005). Even a highly collimated laser irradiates a substantial area at this range and very large and efficient collection optics are required to detect the very weak signals involved. Indeed, the signals are often so weak that astronomical telescopes are employed to collect the signal. Closer to home LIDAR is routinely mounted on mobile platforms and may be used to map atmospheric constituents, for example, plumes from volcanoes (Neri et al., 2008) or chimneys.

19.3.2
Photoacoustic Spectroscopy

In an appropriate sample, the absorption of laser radiation results in a pressure wave. This can be detected using the simple expedient of a microphone. Care must be taken, however, that windows and other optical components in the sample cell do not also contribute to the signal. In order to avoid these effects when using a pulsed laser, the cell must be carefully designed (Rabasovic et al., 2008). A CW laser obviously has to be modulated in order to produce a pressure pulse. This may be turned to advantage by ensuring that the sample chamber is resonant at the modulation frequency of the laser. Further isolation may be achieved by using buffer volumes as acoustic filters (Li et al., 2008). This method has been combined with fluorescence detection to enable the quenching mechanisms in solution to be investigated (Kamath, Kartha and Mahato, 2008). *Photoacoustic spectroscopy* appears to be particularly suited as a detection method of absorption in the midlinfrared region (5–12 µm) (Patel, 2008).

19.3.3
Photoelectron Spectroscopy

Photon absorption may transfer enough energy to the sample to liberate an electron. The excess kinetic energy of the electron so liberated may be measured and thus the energy of the resultant ion state inferred. The resolution is limited by the electron energy analyzer and the radiation source but is usually high enough to resolve vibrational structure. The difference between the excitation energy and the kinetic energy of the electron gives information on the electronic structure of the resulting ion as shown in Figure 19.6. *Anion photoelectron spectroscopy*, on the other hand, allows the electronic state of the resulting neutral to be determined.

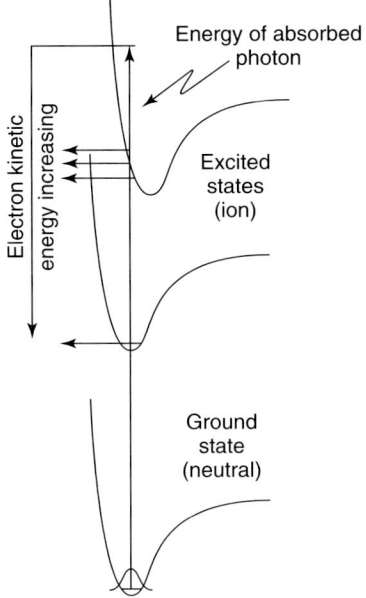

Fig. 19.6 In photoelectron spectroscopy, the ground state of the absorber undergoes a vertical transition to the ionization continuum. The electron carries away the difference in kinetic energy between the absorbed photon and the resultant ion state as indicated by arrows. The pattern of peaks gives information on the geometry change on ionization.

Traditionally, the hard ultraviolet light from a helium discharge has been used to generate photoelectron spectra. The advent of lasers and the availability of synchrotrons have widened the wavelength range over which photoelectrons may be generated and also reduced the linewidth of the exciting light.

The high resolution and tunability of laser sources may be employed to measure the yield of photoelectrons generated just at the ionization threshold. These electrons have a single value of kinetic energy, which is essentially zero, and makes the electron analyzer redundant. This is known as *zero electron kinetic energy* or *ZEKE* spectroscopy (Cockett, 2005). In ZEKE spectroscopy, the electrons produced with nonzero kinetic energy are allowed to disperse before an electric field is applied to the sample region to extract those that remain. The principle of ZEKE was first demonstrated by Chewter and coworkers (Chewter, Müller-Dethlefs and Schlag, 1987). The advantage of this approach is that the resolution is no longer limited by the electron analyzer; instead, it is limited by the laser resolution and the electric fields used to extract the electrons.

19.3.4
Resonance-Enhanced Multiphoton Ionization

Detection of ions is generally much more efficient than the detection of photons. In the simplest implementation of *resonance-enhanced multiphoton ionization* (*REMPI*), the sample is arranged so that it is exposed to a laser whose energy is capable of ionizing the

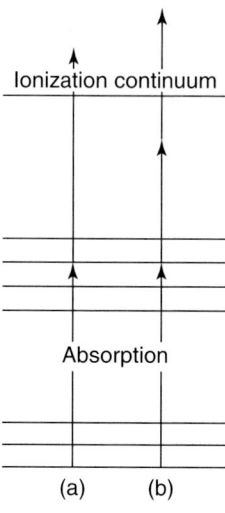

Fig. 19.7 The principle of REMPI. When absorption results in population in an upper state further photons are able to ionize the sample so absorption is detected as an increase in ion current. The second photon may be (a) the same wavelength as the first or (b) a different wavelength.

sample from the upper state of a transition of interest but not directly from its lower state. Another photon, which may result from the same source or a different source, excites the transition of interest results in an increased ion current. Both situations are illustrated in Figure 19.7. The spectrum is recorded as ion current as a function of laser wavelength.

Not only is detection of ions more efficient than the detection of photons, but also time of flight techniques may be used to assist in the identification of ions and fragments. In fact, ionic fragments may be detected by imaging using a two-dimensional detector at the end of a flight tube (Ashfold *et al.*, 2006). These techniques allow the kinetic energy release of the fragments to be reconstructed so that both energetic and dynamic information may be obtained.

19.3.5
Optogalvanic Spectroscopy

A plasma, such as that found in a hollow cathode lamp, is a complex system of excited states and energetic electrons. With a given electrical power input, the system remains in equilibrium. If light whose energy corresponds to an absorption in one of the plasma species, be it an ionic or neutral species is incident on the plasma, the equilibrium is perturbed. This perturbation leads to a change in the power demand from the power supply. The effect was first described in the mid-1920s (Penning, 1928) but not used with lasers until the 1970s Green *et al.* (1976a,b).

Plasmas are generally run at moderately high voltages and a very simple method to obtain the optogalvanic signal is to AC couple the voltage. Thus, the small voltage fluctuations as a result of irradiation are readily detectable. Given that the change in voltage on absorption is often on the order of millivolts, the stability of the power supply is important. Examples of one- and two-photon absorption along with the waveform observed in a commercial hollow cathode lamp are given in Omidyan *et al.* (2008).

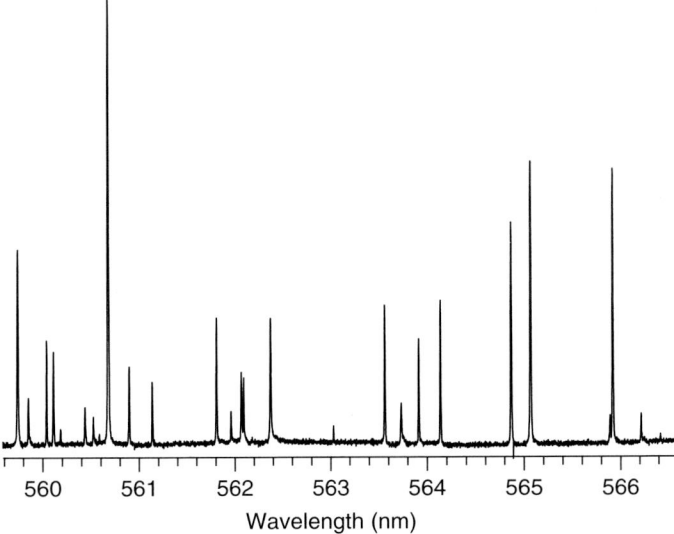

Fig. 19.8 Part of the spectrum of argon atoms recorded using the optogalvanic effect in a commercial hollow cathode lamp. (Greenacre C. and Ashworth S.H. *unpublished*.)

The technique is very straightforward to implement with commercial hollow cathode lamps and is routinely used as a calibration method for pulsed tunable lasers. Part of the spectrum of atomic argon in a commercial hollow cathode lamp is shown in Figure 19.8. Optogalvanic detection can also be applied to plasmas excited by radiofrequency discharges and has been used to detect attomole quantities of $^{14}CO_2$ (Murnick, Dogru and Ilkmen, 2008).

19.3.6
Frequency Modulation Spectroscopy

A laser beam may have sidebands superimposed on it by the use of an electro- or acousto-optic modulator or, in some cases, by direct modulation of the laser itself. The dominant sidebands are 180° out of phase and shifted in frequency from the carrier $\pm \omega$ Hz, where ω is the frequency of the local oscillator. Both sidebands and carrier are directed through the sample and incident on the detector. Off resonance, there is no absorption so the sidebands cancel exactly. This is also the case with the carrier at the absorption line center as both sidebands are absorbed to the same extent – given that the absorption line is symmetric. With the carrier frequency at any other position in the absorption one sideband is preferentially absorbed. The asymmetric absorption of the sidebands results in an AC signal at the detector as shown schematically in Figure 19.9.

One of the first demonstrations of frequency modulation (FM) spectroscopy was on sodium vapor and molecular iodine (Bjorklund, 1980). It is now routinely applied to high-resolution and high-sensitivity measurements in both

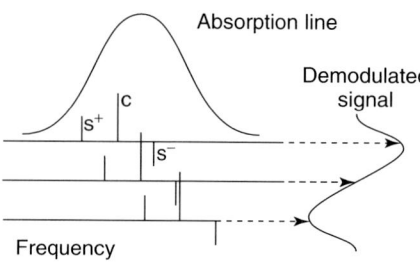

Fig. 19.9 An illustration of FM spectroscopy. The carrier frequency (c) and the sidebands (s+ and s−) which are 180° out of phase are scanned across an absorption line. The resultant demodulated signal is a first derivative lineshape.

spectroscopy and kinetics (Friedrichs, 2008). Combined with cavity enhancement it is known as *noise-immune cavity-enhanced optical heterodyne molecular spectroscopy* or *NICE-OHMS* (Foltynowicz et al., 2008).

19.3.7
Thermal Lensing Spectroscopy

The *thermal lensing* technique relies on the change in density of a sample on interaction with laser radiation. For this reason, it is usually applied to dilute samples in the solid or solution phase. Laser energy is absorbed by the chromophore under investigation, which subsequently undergoes energy transfer. Energy released to the immediate environment results in local heating and thus a change in density. The thermally induced lens causes the laser beam downstream of the target to change its divergence. The irradiance is measured as a function of time by measuring the total energy passing through a small aperture. Thermal lensing may be detected in the excitation beam or in a second probe beam, which is not absorbed by the sample. It has been shown that the influence of solvent (Proskurnin et al., 2008) and the experimental matrix (Georges, 2008) may be important in the interpretation of results. Recent reviews detail the application to detection in flow injection analysis and separation (Franko, 2008) and food analysis and environmental research (Franko, 2001).

19.4
Generally Applicable Techniques

The remaining techniques may find application in situations where either direct or proxy detection schemes are used.

19.4.1
Velocity Modulation

The study of molecular ions either by direct or proxy techniques is often complicated by the presence of neutral species. Such interferences are especially prevalent in the spectroscopy of plasmas and discharges where concentrations of ions may be orders of magnitude lower than that of neutral species. This difficulty was first overcome by Gudeman et al. (1983) by the expedient of modulating the electric field responsible for the discharge. The Doppler shift imparted by the sinusoidal variation of the drift velocity acts as a selective frequency modulation. By detecting at the same frequency, by either lock-in or phase-sensitive detection, only the signals resulting from ions are detected.

19.4.2
Double Resonance

SEP *wide infra* is an example of a double-resonance technique. Signal is only observed when two resonant conditions are fulfilled. Such applications may have many different names and involve alternative radiation sources. SEP is one example of *optical–optical double resonance* (*OODR*) where the population of a quantum state is detected by one photon and modulated by a second (Ye, Pang and Cheung, 2007). Population transfer may, however, be achieved by the absorption of a microwave photon that produces either enhancement or reduction of a laser signal: *microwave-optical double resonance* (*MODR*) (Barnett, Ramsay and Zhu, 2005). Similarly, ion dip spectroscopy measures the influence of a second, usually laser, photon on the ion current usually produced by photoionization by a laser (Sakota, Kageura and Sekiya, 2008).

19.4.3
Stark Spectroscopy

Stark spectroscopy is the electric field analog to LMR spectroscopy but tends to be less limited in its range of application. The Stark splitting patterns of transitions may be used to assist in assignments. In combination with tunable lasers, the shift of a single transition may be measured as a function of electric field, which allows the dipole moment in both ground and excited states to be determined (Chen and Steimle, 2008).

References

Adohi-Krou, A., Jastrzebski, W., Kowalczyk, P., Stolyarov, A.V., and Ross, A.J. (2008) Investigation of the $D\,^1\Pi$ state of NaK by polarisation labelling spectroscopy. *J. Mol. Spectrosc.*, **250**(1), 27.

Ashfold, M.N.R., Nahler, N.H., Orr-Ewing, A.J., Vieuxmaire, O.P.J., Toomes, R.L., Kitsopoulos, T.N., Garcia, I.A., Chestakov, D.A., Wu, S.M., and Parker, D.H. (2006) Imaging the dynamics of gas phase reactions. *Phys. Chem. Chem. Phys.*, **8**(1), 26.

Barnett, M., Ramsay, D.A., and Zhu, Q. (2005) Studies of collisional selection rules in thioformaldehyde (H_2CS) by microwave-optical double resonance. *J. Chem. Phys.*, **123**(15), 154310.

Berden, G., Peeters, R., and Meijer, G. (2000) Cavity ring-down spectroscopy: experimental schemes and applications. *Int. Rev. Phys. Chem.*, **19**(4), 565.

Beverini, N., Carelli, G., De Michele, A., Maccioni, E., Moretti, A., Nyushkov, B., and Sorrentino, F. (2005) High-accuracy frequency measurements in the far infrared. *Laser Phys.*, **15**(7), 1014.

Bjorklund, G.C. (1980) Frequency-modulation spectroscopy – new method for measuring weak absorptions and dispersions. *Opt. Lett.*, **5**(1), 15.

Chen, J.H. and Steimle, T.C. (2008) A molecular beam optical Stark study of nickel monohydride, NiH. *Chem. Phys. Lett.*, **457**(1–3), 23.

Chewter, L.A., Müller-Dethlefs, K., and Schlag, E.W. (1987) Determination of the ionization energy of the benzene-argon complex by zero kinetic energy photoelectron spectroscopy. *Chem. Phys. Lett.*, **135**(3), 219.

Cockett, M.C.R. (2005) Photoelectron spectroscopy without photoelectrons: twenty years of ZEKE spectroscopy. *Chem. Soc. Rev.*, **34**(11), 935.

Demtröder, W. (2008) *Laser Spectroscopy: Experimental Techniques*, 4th revised ed., Springer-Verlag GmbH, Berlin and Heidelberg.

Ding, Y., Campargue, A., Bertseva, E., Tashkun, S., and Perevalov, V.I. (2005) Highly sensitive absorption spectroscopy of carbon dioxide by ICLAS-VeCSEL between 8800 and 9530 cm^{-1}. *J. Mol. Spectrosc.*, **231**(2), 117.

Foltynowicz, A., Schmidt, F.M., Ma, W., and Axner, O. (2008) Noise-immune cavity-enhanced optical heterodyne molecular spectroscopy: current status and future potential. *Appl. Phys. B-Lasers Optics*, **92**(3), 313.

Franko, M. (2001) Recent applications of thermal lens spectrometry in food analysis and environmental research. *Talanta*, **54**(1), 1.

Franko, M. (2008) Thermal lens spectrometric detection in flow injection analysis and separation techniques. *Appl. Spectrosc. Rev.*, **43**(4), 358.

Friedrichs, G. (2008) Sensitive absorption methods for quantitative gas phase kinetic measurements. Part 1: Frequency modulation spectroscopy. *Z. Phys. Chem.-Int. J. Res. Phys. Chem. Chem. Phys.*, **222**, 1.

Gardner, C.S., Plane, J.M.C., Pan, W.L., Vondrak, T., Murray, B.J., and Chu, X.Z. (2005) Seasonal variations of the Na and Fe layers at the South Pole and their implications for the chemistry and general circulation of the polar mesosphere. *J. Geophys. Res.-Atmosph.*, **110**(D10), D10302.1–D10302.13.

Georges, J. (2008) Matrix effects in thermal lens spectrometry: Influence of salts, surfactants, polymers and solvent mixtures. *Spectrochim. Acta Part A-Mol. Biomol. Spectrosc.*, **69**(4), 1063.

Gillett, D.A., Cooksy, A.L., and Brown, J.M. (2006) Infrared laser magnetic resonance spectroscopy of the ν_3 fundamental and associated hot bands of the NCO free radical. *J. Mol. Spectrosc.*, **239**(2), 190.

Green, R.B., Keller, R.A., Luther, G.G., Schenck, P.K., and Travis, J.C. (1976a) Galvanic detection of optical absorptions in a gas-discharge. *Appl. Phys. Lett.*, **29**(11), 727.

Green, R.B., Keller, R.A., Schenck, P.K., Travis, J.C., and Luther, G.G. (1976b) Opto-galvanic detection of species in flames. *J. Am. Chem. Soc.*, **98**(26), 8517.

Gudeman, C.S., Begemann, M.H., Pfaff, J., and Saykally, R.J. (1983) Velocity-modulated infrared-laser spectroscopy of molecular-ions - the ν_1 band of HCO$^+$. *Phys. Rev. Lett.*, **50**(10), 727.

Herriott, D.R. (1963) Spherical-mirror oscillating interferometer. *Appl. Opt.*, **2**(8), 865.

Herriott, D., Kogelnik, H., and Kompfner, R. (1964) Off-axis paths in spherical mirror interferometers. *Appl. Opt.*, **3**(4), 523.

Hinz, A., Zeitz, D., Bohle, W., and Urban, W. (1985) A Faraday laser magnetic resonance spectrometer for spectroscopy of molecular radical ions. *Appl. Phys. B: Lasers Optics*, **36**(1), 1.

Hodges, P.J., Ross, A.J., Crozet, P., Salami, H., and Brown, J.M. (2007) On the spin-orbit splitting of CuCl$_2$ in its $^2\Pi_g$ ground state. *J. Chem. Phys.*, **127**(2), 024309.

Inguscio, M., Moruzzi, G., Evenson, K.M., and Jennings, D.A. (1986) A review of frequency measurements of optically pumped lasers from 0.1 to 8 THz. *J. Appl. Phys.*, **60**(12), R161.

Ishii, K., Takeuchi, S., and Tahara, T. (2008) Pronounced non Condon effect as the origin of the quantum beat observed in the time-resolved absorption signal from excited-state cis-stilbene. *J. Phys. Chem. A*, **112**(11), 2219.

Kamath, S.D., Kartha, V.B., and Mahato, K.K. (2008) Dynamics of l-tryptophan in aqueous solution by simultaneous laser induced fluorescence (LIF) and photoacoustic spectroscopy (PAS). *Spectrochim. Acta Part A: Mol. Biomol. Spectrosc.*, **70**(1), 187.

Korter, T.M., Borst, D.R., Butler, C.J., and Pratt, D.W. (2001) Stark effects in gas-phase electronic spectra. Dipole moment of aniline in its excited S_1 state. *J. Am. Chem. Soc.*, **123**(1), 96.

Li, J.S., Liu, K., Zhang, W.J., Chen, W.D., and Gao, X.M. (2008) Pressure-induced line broadening for the (30012) ← (00001) band of CO^2 measured with tunable diode laser photoacoustic spectroscopy. *J. Quant. Spectrosc. Radiat. Transfer*, **109**(9), 1575.

Luong, M., Zhang, R., Schulz, C., and Sick, V. (2008) Toluene laser-induced fluorescence for in-cylinder temperature imaging in internal combustion engines. *Appl. Phys. B-Lasers Optics*, **91**(3–4), 669.

Mak, P.J. and Kincaid, J.R. (2008) Resonance Raman spectroscopic studies of hydroperoxo derivatives of cobalt-substituted myoglobin. *J. Inorg. Biochem.*, **102**(10), 1952.

Manke, K.J., Vervoort, T.R., Kuwata, K.T., and Varberg, T.D. (2008) Electronic spectrum of TaO and its hyperfine structure. *J. Chem. Phys.*, **128**(10), 104302.

Matsushima, F., Tomatsu, N., Nagai, T., Moriwaki, Y., and Takagi, K. (2006) Frequency measurement of pure rotational transitions in the $v_2 = 1$ state of H_2O. *J. Mol. Spectrosc.*, **235**(2), 190.

Mejean, G., Kassi, S., and Romanini, D. (2008) Measurement of reactive atmospheric species by ultraviolet cavity-enhanced spectroscopy with a mode-locked femtosecond laser. *Opt. Lett.*, **33**(11), 1231.

Moyer, E.J., Sayres, D.S., Engel, G.S., Clair, J.M.S., Keutsch, F.N., Allen, N.T., Kroll, J.H., and Anderson, J.G. (2008) Design considerations in high-sensitivity off-axis integrated cavity output spectroscopy. *Appl. Phys. B-Lasers Optics*, **92**(3), 467.

Mukarakate, C., Tao, C., Jordan, C.D., Polik, W.F., and Reid, S.A. (2008) Stimulated emission pumping spectroscopy of the $X\,^1A'$ state of CHF. *J. Phys. Chem. A*, **112**, 466.

Murnick, D.E., Dogru, O., and Ilkmen, E. (2008) Intracavity optogalvanic spectroscopy. An analytical technique for C^{14} analysis with subattomole sensitivity. *Anal. Chem.*, **80**(13), 4820.

Neri, M., Mazzarini, F., Tarquini, S., Bisson, M., Isola, I., Behncke, B., and Pareschi, M.T. (2008) The changing face of Mount Etna's summit area documented with Lidar technology. *Geophys. Res. Lett.*, **35**(9), L09305.

Omidyan, R., Fathi, F., Farrokhpour, H., and Tabrizchi, M. (2008) One- and two-photon laser optogalvanic spectroscopy of neon in the 570-626 nm region. *Opt. Commun.*, **281**(22), 5555.

Patel, C.K.N. (2008) Laser photoacoustic spectroscopy helps fight terrorism: high sensitivity detection of chemical Warfare Agent and explosives. *Eur. Phys. J.-Special Top.*, **153**, 1.

Penning, F.M. (1928) Demonstratie van een nieuw photoelectrisch effect. *Physica*, **8**, 137.

Peterson, N.C., Kurylo, M.J., Braun, W., Bass, A.M., and Keller, R.A. (1971) Enhancement of absorption spectra by dye-laser quenching. *J. Opt. Soc. Am.*, **61**(6), 746.

Proskurnin, M., Bendrysheva, S., Kuznetsova, V., Zhirkov, A., and Zuev, B. (2008) Effect of a solvent on the parameters of the analytical signal, detection limit, and analytical range of the determination in analytical thermal lens spectrometry. *J. Anal. Chem.*, **63**(8), 741.

Qian, H.B. and Howard, B.J. (1997) High resolution infrared spectroscopy and structure of $CO-N_2O$. *J. Mol. Spectrosc.*, **184**(1), 156.

Rabasovic, M.D., Nikolic, J.D., Markushev, D.D., and Jovanovic-Kurepa, J. (2008) Pulsed photoacoustic gas cell design for low pressure studies. *Opt. Mater.*, **30**(7), 1197.

Sakota, K., Kageura, Y., and Sekiya, H. (2008) Cooperativity of hydrogen-bonded networks in 7-azaindole$(CH_3OH)_n$(n=2,3) clusters evidenced by IR-UV ion-dip spectroscopy and natural bond orbital analysis. *J. Chem. Phys.*, **129**(5), 054303.

Schoch, A., Baumgarten, G., and Fiedler, J. (2008) Polar middle atmosphere temperature climatology from Rayleigh lidar measurements at ALOMAR (69° N). *Ann. Geophys.*, **26**(7), 1681.

Smith, W.E., Faulds, K., and Graham, D. (2006) Quantitative surface-enhanced resonance Raman spectroscopy for analysis. *Surf.-Enhanc. Raman Scatter.: Phys. Appl.*, **103**, 381.

Wehr, R., Kassi, S., Romanini, D., and Gianfrani, L. (2008) Optical feedback cavity-enhanced absorption spectroscopy for in situ measurements of the ratio $^{13}C : ^{12}C$ in CO_2. *Appl. Phys. B-Lasers Optics*, **92**(3), 459.

White, J.U. (1942) Long optical paths of large aperture. *J. Opt. Soc. Am.*, **32**(5), 285.

Wirth, M.J. and Legg, M.A. (2007) Single-molecule probing of adsorption and diffusion on silica surfaces. *Annu. Rev. Phys. Chem.*, **58**, 489.

Wustholz, K.L., Sluss, D.R.B., Kahr, B., and Reid, P.J. (2008) Applications of single-molecule microscopy to problems in dyed composite materials. *Int. Rev. Phys. Chem.*, **27**(2), 167.

Ye, J.J., Pang, H.F., and Cheung, A.S.C. (2007) Optical-optical double resonance spectroscopy of YBr and YCl. *Chem. Phys. Lett.*, **442**(4–6), 251.

Zheltikov, A.M. (2000) Coherent anti-Stokes Raman scattering: from proof-of-the-principle experiments to femtosecond CARS and higher order wave-mixing generalizations. *J. Raman Spectrosc.*, **31**(8–9), 653.

Zink, L.R., Willcutt, A., Murphy, M., and Jackson, M. (2008) Frequencies of cw FIR laser lines for use in laser magnetic resonance spectroscopy. *Appl. Phys. B-Lasers Optics*, **92**(1), 5.

Andrews, D.L. and Demidov, A.A. (eds) (2002) *An Introduction to Laser Spectroscopy*, 2nd edn, Kluwer Academic/Plenum, New York.

Ellis, A.M., Feher, M., and Wright, T.G. (2005) *Electronic and Photoelectron Spectroscopy: Fundamentals and Case Studies*, Cambridge University Press, Cambridge.

Hollas, J.M. (1998) *High Resolution Spectroscopy*, 2nd, John Wiley & Sons, Ltd, Chichester.

Stephenson, S.K. and Saykally, R.J. (2005) Velocity modulation spectroscopy of ions. *Chem. Rev.*, **105**(9), 3220.

Further Reading

Andrews, D.L. (1997) *Lasers in Chemistry*, 3rd edn, Springer-Verlag, Berlin.

20
Photophysical and Photochemical Dynamics

Ottó Horváth and Kenneth L. Stevenson

20.1	**Introduction** 719	
20.2	**Fundamental Principles of Photophysics and Photochemistry** 719	
20.2.1	Light and Energy 719	
20.2.2	Light-Absorption Principles 720	
20.2.3	Quantum Yields 721	
20.2.4	Quantum-Mechanical Principles of Absorption 721	
20.2.5	Energy Partitioning and Dissipation Pathways 725	
20.2.5.1	Photophysical Pathways: Luminescence and Nonradiative Decay 725	
20.2.5.2	Photochemical Pathways 727	
20.3	**Types of Photochemical and Photophysical Experiments** 729	
20.3.1	Continuous (cw) Photolysis 729	
20.3.2	Laser Flash Photolysis 730	
20.3.3	Laser Spectroscopy 732	
20.4	**Femtochemistry** 733	
20.5	**Reactions of Excited Species** 734	
20.5.1	Electronic Energy Transfer 734	
20.5.2	Photochemistry of Organic Compounds 735	
20.5.2.1	Electronic Transitions and Spectra of Organic Molecules 735	
20.5.2.2	Correlation Rules and Symmetry Conservations 736	
20.5.2.3	Intramolecular Processes 736	
20.5.2.4	Intermolecular Processes 737	
20.5.3	Photochemistry of Inorganic and Coordination Compounds 738	
20.5.3.1	Ligand-Field (LF) Transitions 738	
20.5.3.2	Ligand-to-Metal (LMCT) and Metal-to-Ligand (MLCT) Charge-Transfer Transitions 739	
20.5.3.3	Charge-Transfer-to-Solvent (CTTS) Transitions 740	
20.5.3.4	Other Charge-Transfer Transitions 740	
20.6	**Conclusion** 741	
	Acknowledgments 741	

Encyclopedia of Applied Spectroscopy. Edited by David L. Andrews.
Copyright © 2009 WILEY-VCH Verlag GmbH & Co. KGaA, Weinheim
ISBN: 978-3-527-40773-6

Glossary 741
References 743
Further Reading 743

20.1
Introduction

The interaction of light with matter is of fundamental importance. Indeed, all of the energy we derive from fossil fuels was deposited over vast periods of time by photochemical processes, which converted energy from the sun into chemical energy. The study of chemical reactions induced by light (RIL) is usually termed *photochemistry*, whereas laser photochemistry (LP) refers to the study of reactions promoted with laser light. Photochemistry is concerned specifically with the ways that light can bring about changes in materials and has become very important in the investigation of chemical kinetics or chemical dynamics since photoinitiation is a much more selective and, in the case of the use of laser pulses, a much faster process than thermal initiation of reactions, providing greater control over the reactions. Moreover, lasers have proved invaluable not only to initiate reactions but also to analyze their products. Reaction intermediates, normally prepared thermally via precursors or collisions, can be formed directly by a selective photoprocess, permitting a vast array of reactive intermediates of known composition and energy to be available for new syntheses or experimental studies.

20.2
Fundamental Principles of Photophysics and Photochemistry

Changes in chemical structure caused by light are the subject of *photochemistry*, whereas light-induced changes in the ways electrons are distributed within a chemical entity while the atomic nuclei remain intact are the realm of *photophysics*. The initial event in any photochemical or photophysical process is the absorption of a photon by an atom, molecule, or complex ion, resulting in an electronically excited state. What happens after this event determines whether or not a photochemical reaction will occur, or whether some other photophysical process occurs such as fluorescence, phosphorescence, or the simple conversion of the light energy to heat. The following discussion describes the fundamental principles governing these processes.

20.2.1
Light and Energy

To understand the process of photolysis, that is, how light causes changes in the structure and reactivity of chemical systems, it is necessary to describe some of the properties of light. On the one hand, it is convenient to think of light as an electromagnetic wave, summarized by the expression

$$\lambda = \frac{c}{\nu} \qquad (20.1)$$

relating the wavelength, λ, to the frequency, ν, through the speed of propagation, c. On the other hand, light consists of quanta, or photons, each of which has energy given by

$$E = h\nu \qquad (20.2)$$

where h, known as Planck's constant, can be interpreted as the proportionality constant between the energy and frequency of the wave associated with the photon.

Laser light has certain advantages over conventional, thermal light sources for use in photochemistry, such as high-power flux density, low angular spread, narrow frequency (or wavelength) distribution, and extremely short pulsewidths in pulsed lasers.

When chemical systems absorb light, individual molecules or atoms absorb only those photons with the right amount of energy to effect an excitation of the molecule from a lower to a higher energy level, if such a transition is allowed. Because light is an oscillating electric field, a transition is allowed for a normal absorption process only if the excited state has a different dipole moment than the ground state. Energy is distributed in molecules in the various levels of molecular motions, that is, translation, rotation, vibration, electronic, and so on, and the kind of electromagnetic radiation absorbed will depend on the kind of motion that is affected (see Chapter 11 by Lipson).

20.2.2
Light-Absorption Principles

The first law of photochemistry from Grotthus (1817) and Draper (1843) states that only the light that is absorbed by a molecule can produce photochemical change within the molecule. The amount of light absorbed in a homogeneous medium is defined by the Beer–Lambert law:

$$I = I_0 10^{-\varepsilon cl} \qquad (20.3)$$

where I_0 and I are the light intensities at the front of the absorber and at distance l within the absorbing medium, respectively, and ε is the extinction coefficient in units consistent with the pathlength l (usually in centimeter) and the concentration c (usually in mole per liter). If one is to do quantitative continuous photolysis studies, it is important that an accurate value of the light absorbed by a photoactive sample can be measured since this determines the rate of the reaction. From Eq. (20.3) one can show that the absorbed light intensity, or radiant flux J, is given as

$$J = I_0(1 - 10^{-\varepsilon cl}) \qquad (20.4)$$

Photochemists generally measure the incident light intensity, I_0, falling on a sample, and then use this equation to calculate J. The accurate measurement of light intensities, or actinometry, is very important in determining the photokinetics of the system. One convenient method for doing this is the photolysis of a chemical actinometer, which has been calibrated for the wavelength of the photoreaction under study. Almost any photoreaction system that has a well-defined quantum yield at the irradiating wavelength can be used as an actinometer, but for continuous photolysis work in the wavelength range 250–550 nm, the potassium trioxalatoferrate(III) system, developed by Hatchard and Parker (1956), is the most widely used and reliable actinometer available. For longer wavelengths to 750 nm, the photoaquation of aqueous Reinecke's salt, $KCr(NH_3)_2(NCS)_4$, works quite well (Wegner and Adamson, 1966). The common unit for J and I_0 in the aforementioned equations is einstein per second, where an einstein is 1 mol or 6.02×10^{23} photons. Occasionally it is expressed per unit volume of reactant, such as einstein per second per liter, analogous to the method for expressing rates of reactions.

20.2.3
Quantum Yields

The quantum yield or quantum efficiency of a photochemical reaction is normally defined as the fraction of photoactivated molecules that undergo some change. If there are a number of parallel pathways resulting in different products, then each pathway will have its own quantum yield, ϕ. Sometimes the primary step in a photoreaction is the initiator of subsequent thermal reactions resulting in quantum yields, which may exceed unity. The second law of photochemistry states that the absorption of light by a molecule is a one-quantum process, so that the sum of the primary process quantum yields, ϕ, must be unity, that is, $\Sigma \phi_i = 1$, where ϕ_i is the quantum yield of each of the primary processes, which may include, but are not limited to, dissociation, isomerization, charge-separation, luminescence, or radiationless deactivation.

For a reaction,

$$A \xrightarrow{h\nu} B + C + \cdots \quad (20.5)$$

the rate can be expressed as

$$\frac{-d[A]}{dt} = \phi_A J_A \quad (20.6)$$

where $[A]$ is the concentration of photoactive species. One thus determines quantum yields by measuring the ratio of the reaction rate to the absorbed light intensity, J_A.

20.2.4
Quantum-Mechanical Principles of Absorption

If A and A^* represent the ground and excited states, respectively, the process can be written as the reaction

$$A \xrightarrow{h\nu} A^* \quad (20.7)$$

For the absorption to occur, two criteria must be met by the absorbing species:

1. the difference between the energy of the excited state, E_2, and the energy of the ground state, E_1, must be exactly equal to the energy of the photon, that is,

$$h\nu = E_1 - E_2 \quad (20.8)$$

and
2. there must be a finite probability for the transition, as defined by the electronic structure of the molecule.

The Einstein transition probability for absorption, B_{12}, is given by the following expression:

$$B_{12} = \frac{8\pi^3}{3h^2 c} g_2 (\mu_{12})^2 \quad (20.9)$$

where g_2 is the degeneracy of the excited state, and μ_{12} is the transition moment given by the integral,

$$\mu_{12} = \int \Psi_1 \left(e \sum_i r_i\right) \Psi_2 \, dq \quad (20.10)$$

In this equation, Ψ_1 and Ψ_2 are the wave functions of the ground and excited states, respectively; the sum is over all of the lengths of the dipole-moment vectors, r_i, of the electrons; e is the charge on the electron; and the integral is taken over the three-dimensional coordinates, q, of the molecule.

While the exact solution of Eq. (20.10) is impossible for anything but the hydrogen atom, approximate wave functions can be used to derive for various transitions qualitative selection rules that reveal when the transition moment, μ_{12}, is nonzero and thus results in an allowed transition, as opposed to one that is forbidden. Two selection rules are particularly important and relate directly to photochemical processes: the spin conservation rule and the Laporte rule. For systems containing atoms with relatively few electrons (i.e., light atoms), a reliable approximation allows for the separation of the spin wavefunctions, S_1 and S_2, from the total wavefunctions Ψ_1 and Ψ_2, such that the transition moment can be written as the product of two integrals:

$$\mu_{12} = \int \phi_1 \left(e \sum_i r_i\right) \phi_2 dq$$
$$\times \int S_1 S_2 \, dq \quad (20.11)$$

where ϕ_1 and ϕ_2 are the orbital wavefunctions of ground and excited states, respectively. The spin part of the transition moment vanishes when $\Delta S \neq 0$, that is, when the total spin quantum numbers, or spin multiplicities, of the excited and ground states are not the same, and the transition is said to be "spin-forbidden." The Laporte rule applies to species with a center of symmetry, such as octahedral complexes, and requires that transitions between orbitals must be either $u \to g$ or $g \to u$ (g, gerade or even and u, ungerade or uneven, that is, symmetric and antisymmetric, respectively, for inversion of the orbital about the center of symmetry), but not $g \to g$ or $u \to u$, such transitions being "parity-forbidden." The methods of group theory are often necessary in applying the Laporte rule.

In practice, not only allowed transitions but also sometimes forbidden transitions may be observed because of subtle perturbations in the system, such as spin–orbit coupling, not taken into account in the approximation procedures used to derive the selection rules. The allowed transitions, of course, are much stronger, as indicated by the bandwidth and extinction coefficient of the absorption band. The classical oscillator strength f of the band is related to the Einstein absorption probability by the following expression:

$$f = \left(\frac{m_e c^2 \omega_{12}}{\pi e^2}\right) B_{12} \quad (20.12)$$

where m_e is the mass of the electron and ω_{12} is the frequency of the transition in cm^{-1}. Experimentally, the oscillator strength can be determined from the integral of the absorption band, that is,

$$f = 4.32 \times 10^{-9} F \int_{\omega_1}^{\omega_2} \varepsilon \, d\omega \quad (20.13)$$

where the integration is over all frequencies from the onset to the disappearance of the band in question, and the value of F is related to the index of refraction, n, through $F = 9n/(n^2 + 2)^2$. As a very crude approximation, the oscillator strength can be estimated from

$$f \cong 4.32 \times 10^{-9} F\varepsilon_{max}\Delta\omega_{1/2}, \quad (20.14)$$

where $\Delta\omega_{1/2}$ is the width of the band at $\varepsilon = 1/2\omega_{max}$ (half-bandwidth).

Since the oscillator strength for a single electron in a fully allowed transition should be unity, the highest value of ε_{max} one should observe is about 5×10^4 M^{-1} cm^{-1}, for a half-bandwidth of 5000 cm^{-1}. Parity- and spin-forbidden transitions have much smaller oscillator strengths, resulting in values of ε on the order of 10^2 or less, but they are easily measured with the crudest of spectrometers. Thus, measurement of the oscillator strengths of various bands observed in a spectrum can be of help in making the correct assignments to the transitions.

The foregoing discussion has been based on the assumption that nuclear motions in ordinary molecular vibrations are so slow that they do not affect the electronic energy levels of the molecule, an approximation known as the *Born–Oppenheimer principle*. This means that the total wavefunction for the molecule can be separated into the product of an electronic part and a nuclear part:

$$\psi_{total} = \psi_{elec} \times \psi_{nucl} \quad (20.15)$$

and each part is treated independently. This permits the construction of the familiar energy diagrams of diatomic molecules, which show the electronic potential energy as a function of internuclear distance, a general example of which is shown in Figure 20.1. From a purely classical viewpoint, one would expect the molecule to reside at the exact bottom of either well, but, in fact, this is not the case because it would imply that the atoms would be stationary, that is, not vibrating.

The quantum-mechanical treatment of the simple harmonic oscillator (which approximates a diatomic molecule to the extent that the bottom parts of the potential wells are approximated by a parabola) requires that the lowest energy of the vibrating molecule must be $1/2h\nu_0$ above the minimum (a result of the uncertainty principle, which does not allow both the position and momentum to be completely defined as would be the case if the zero-point energy were 0), where ν_0 is the fundamental frequency of vibration in this ground state. Several vibrational levels for each electronic state are represented in the wells by horizontal lines that show not only the value of vibrational energy but also the turning points, that is, limits of compression and extension of the atoms during the vibrations.

Figure 20.1 shows what happens during an electronic excitation by absorption of light. Because nuclear motions are vastly slower than electronic transitions, the latter occur vertically, that is, the excited molecule has the same internuclear separation as the ground-state molecule, a rule known as the *Franck–Condon principle*. At temperatures in the vicinity of 25 °C most molecules in a thermally equilibrated system are in the ground vibrational state as a result of the Boltzmann distribution law; therefore, nearly all of the electronic transitions will arise out of this ground vibrational state, as shown in the diagram. However, because the excited electronic potential well is wider and slightly displaced to the right as a consequence of having a weaker

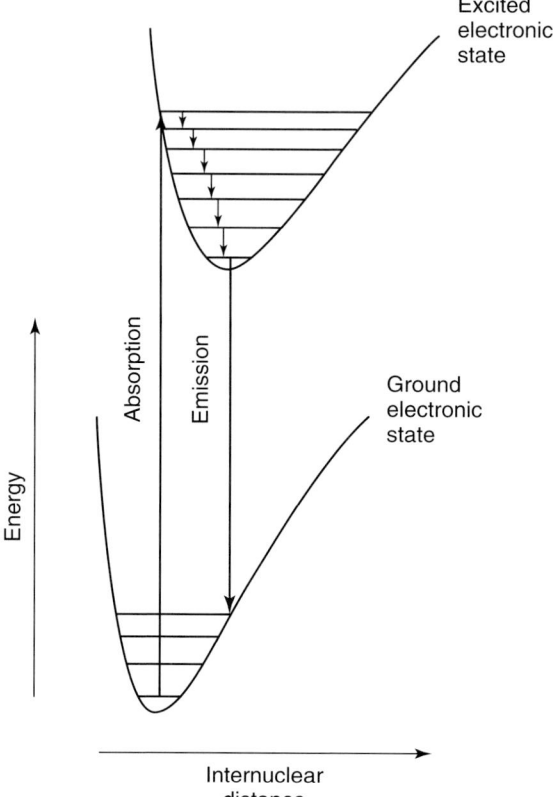

Fig. 20.1 Typical ground- and excited-state potential-energy wells for a molecule, showing vibrational relaxation and emission.

bond than in the ground electronic state, the vertical transitions end in excited vibrational levels in the upper well. Because of the necessity for optimum mixing of the initial and final vibrational wavefunctions in the transitions, those with highest probability occur between vibrational levels in which the molecule is near a turning point in its vibration (except for ground-state vibrations where this highest probability occurs at the center, or equilibrium internuclear distance). This results in a series of transitions to excited vibrational levels in the upper well, giving rise to a set of closely spaced lines in the absorption spectrum of a gas at low pressure or to a broad absorption band in a condensed medium in which the lines are broadened and overlap.

In polyatomic systems, there are numerous vibrational modes and internuclear distances, requiring a multidimensional analog of Figure 20.1 to describe the energy states. Nevertheless, one can still imagine minimum electronic potential-energy wells with superimposed vibrational levels in the hyperspace of n dimensions, where n is the number of atoms in the system. Thus, the simple principles of the energetics of diatomic molecules are routinely conceptually applied to more complex systems.

20.2.5
Energy Partitioning and Dissipation Pathways

What happens to the photoactive species after initial electronic excitation by absorption determines the eventual outcome of the photophysical or photochemical process. Since the excited system has a large number of its molecules in excited vibrational levels, it usually undergoes a very rapid thermal equilibration with its surroundings resulting in a Boltzmann population distribution, which means that most molecules drop back to the lowest vibrational level of the excited electronic state, a process called *vibrational relaxation* and shown by the short, downward arrows in Figure 20.1.

Certain bonds between atoms give rise to characteristic frequencies in the infrared (IR), corresponding to bond stretches and bends. Thermal sources of radiation are ineffective in breaking specific bonds, since energy cannot be supplied fast enough to stretch a bond to breaking, in competition with intramolecular and intermolecular energy relaxation.

In recent years, the topic of mode-specific photochemistry has revived in the context of a better understanding of the energy redistribution that may occur inside molecules.

20.2.5.1 Photophysical Pathways: Luminescence and Nonradiative Decay

After the vibrational relaxation of the excited electronic state, the system may return to the ground-state potential well by spontaneously emitting a photon, also shown in Figure 20.1. Note, however, that in this case, since the excited-state potential well is shifted slightly to the right, the most likely electronic transitions are from the lowest vibrational level in the excited electronic state to vibrationally excited states in the ground potential well. This gives rise to an emission band, said to be *Stokes-shifted*, which is at lower energies than, and almost an exact mirror image of, the absorption band. This form of luminescence is often referred to as *fluorescence*, a term which is generally applied to emissions of very short lifetimes between singlet energy levels. From quantum mechanics, the Einstein probability for spontaneous emission, A_{21}, is related to the oscillator strength of the absorption band, through

$$A_{21} = \left(\frac{8\pi^2 \omega_{21}^2}{m_e c}\right)\left(\frac{g_1}{g_2}\right) f \quad (20.16)$$

where g_1 and g_2 are the degeneracies of the ground and excited electronic states, respectively, and the other variables have been previously defined. If emission is the only mode of decay, then

$$\tau_0 = \frac{1}{A_{21}} \quad (20.17)$$

where τ_0 is the mean radiative lifetime for the emission, a quantity which is thus inversely proportional to the oscillator strength of the absorption.

Nonradiative decay from excited electronic states may occur if there is sufficient overlap between the excited vibrational levels of the lower electronic level and the lowest vibrational states of the excited electronic state. If this occurs between electronic levels of the

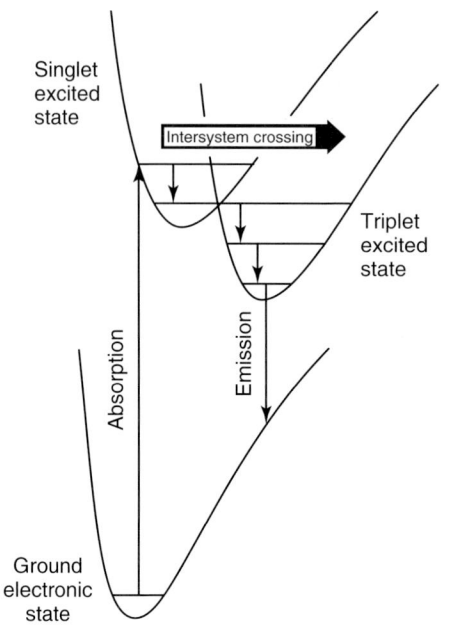

Fig. 20.2 Potential-energy diagram showing excited singlet and triplet states.

same multiplicity, it is called *internal conversion*; if electronic levels of different multiplicities are involved, it is termed *intersystem crossing*. Figure 20.2 illustrates the situation in which there are two excited electronic states of differing multiplicities, for example, singlet and triplet, which have considerable overlap in their vibrational manifolds with each other.

Intersystem crossing from the uppermost potential well to the one that crosses it should be relatively easy from the viewpoint of vibrational relaxation, but forbidden because of the spin conservation rule. However, such processes occur readily because of spin–orbit coupling. Decay of either of the two excited electronic states to the ground state is less likely from the viewpoint of vibrational relaxation because of the poor mixing of vibrational levels, but it can occur through collisional deactivation in which other molecules absorb the energy difference and behave as quenchers of the excited states without actually undergoing chemical change. Since this process is controlled by the collision rate in the system, it is easy to see why this kind of deactivation increases in liquid media at higher temperatures, causing competing luminescence processes to be decreased.

Decay from the excited state of different multiplicity back to the ground state may also occur by emission or nonradiative processes. In both cases, however, the transition has a low probability because of the different spin multiplicities of the two states and the wide separation between the two electronic levels; thus, the excited state has a rather long lifetime. Emission from such a state is often called *phosphorescence* and is characterized by lifetimes as long as seconds, minutes, or even hours after the absorption of light.

Tab. 20.1 Photophysical processes, typical rate constants, and lifetimes in liquid media.

Process	Rate constants (s^{-1})	Lifetimes
Absorption of photon	$\sim 10^{15}$	1 fs
Vibrational relaxation	10^{11}–10^{12}	1–10 ps
Fluorescence	10^8–10^9	1–10 ns
Intersystem crossing between excited states	0–10^{10}	>0.1 ns
Phosphorescence	0–10^3	>1 ms
Radiationless deactivation		
Internal conversion	0–10^9	>1 ns
Intersystem crossing	0–10^3	>1 ms

Table 20.1 summarizes the various photophysical processes that may occur in liquid media and the typical ranges in rate constants and lifetimes for each process. Not all emission processes fall neatly into the definitions for fluorescence and phosphorescence and therefore all types of emission can be generally referred to as *luminescence*.

In the condensed phase, vibrational relaxation typically occurs on the order of tens of femtoseconds to hundreds of picoseconds, as shown in laser pump–probe experiments (Elsaesser and Kaiser, 1991). However, it is interesting to note that in some simple liquids, such as liquid nitrogen or inert gases, vibrational relaxation times on the order of minutes have been found (Brueck and Osgood, 1976).

20.2.5.2 Photochemical Pathways

The extent to which photochemical products are formed will depend on how well the photochemical pathways compete with the various photophysical decay processes described above. This, in turn, will depend on the reactivity of the excited states, which are usually chemically very different from the ground states. Figure 20.3 is another way of outlining the many choices of pathways through which the excited system can dissipate the energy it gained in absorbing a photon. Nearly all types of reactions can be classified as either unimolecular or bimolecular. The excited states that are reactive may be those which were originally excited in the absorption of light, or they may be other states, labeled *reactive intermediates* in Figure 20.3, accessible by radiationless decay.

There are two special types of reactive intermediates that are formed by bimolecular processes: *excimers* that are formed by dimerization of an excited state, and *exciplexes* that are formed by a complexation reaction of an excited state with ligands. In any case, the potential-energy well of the excited reactive intermediate relative to the ground and excited states of the photoreactant can also be represented by the diagram in Figure 20.2. Photophysical processes are those which result in a return of the system to the ground state while photochemical processes result in different products.

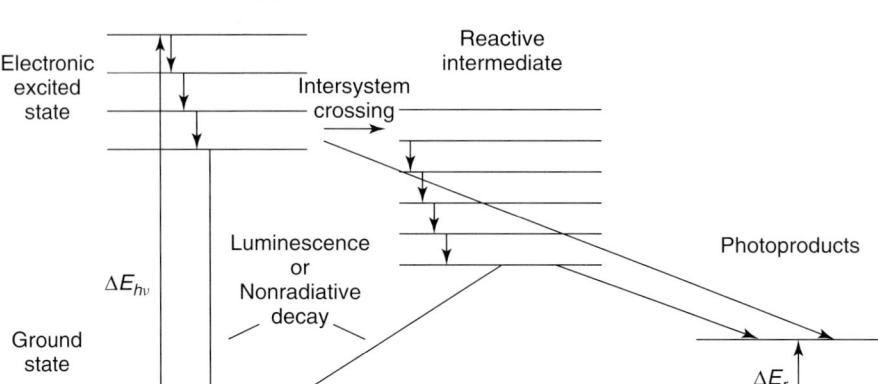

Fig. 20.3 Energy diagram (so-called Jablonski diagram) summarizing the various pathways available to an excited molecule or complex.

Many of these processes occur simultaneously and compete with each other such that one usually observes a mixture of outcomes, each with its own quantum yield. Each of the various decay pathways leading away from the excited photoreactant has its own rate, and the quantum yield for a specific pathway, ϕ_j, is the ratio of an individual rate of decay for that pathway to the sum of the rates of all of the decay pathways, such that (assuming first-order reactions);

$$\phi_j = \frac{k_j}{\sum_i k_i} \qquad (20.18)$$

where k_j is the rate constant for the pathway of interest and k_i represents the rate constant for each of the other pathways.

An often-cited quantity is the lifetime, τ, of an excited state or reactive intermediate. This is defined as the time required for a sample to decay to $1/e$ of its original concentration and can be shown to be equal to the reciprocal of the sum of the first-order rate constants for all decay pathways:

$$\tau = \left(\sum_i k_i\right)^{-1} \qquad (20.19)$$

It is often the case that excited states may react through bimolecular collisions with other species, thus causing the excited states to be "quenched." Such reactions offer additional pathways for the decay of the excited states, and they must be accounted for in any quantum-yield expression. Since these reactions are second-order processes, their rates will be proportional to the concentrations of both the excited state and the quencher; thus, the quantum-yield expression, Eq. (20.18), will contain a second-order rate constant times quencher concentration for each of these quenching pathways. For example, suppose that the only pathways for decay of an excited photoreactant are luminescence, nonradiative decay, or reaction with a quencher Q, to form a product, as in

$$A \underset{k_L, k_{NR}}{\overset{h\nu}{\rightleftarrows}} A^* \xrightarrow{K_P[Q]} \text{Products} \quad (20.20)$$

where k_L and k_{NR} are the first-order luminescence and nonradiative decay constants, and k_P is the second-order constant for the bimolecular quenching, which forms the product. The quantum yield of product formation, ϕ_P, would then be

$$\phi_P = \frac{k_P[Q]}{k_L + k_{NR} + k_P[Q]} \quad (20.21)$$

where $[Q]$ is the concentration of the quencher. One can establish that a second-order quenching process occurs by obtaining a linear plot of $1/\phi_P$ versus $1/[Q]$ since the reciprocal of Eq. (20.21) is

$$\frac{1}{\phi_P} = 1 + \frac{k_L + K_{NR}}{k_P[Q]} \quad (20.22)$$

The slope of such a plot, known as a *Stern–Volmer plot*, can be used to calculate the second-order quenching constant if the lifetime in the absence of quencher is known. An alternative method of obtaining the second-order rate constant directly is from the slope of measured pseudo-first-order decay constants versus $[Q]$.

Many photoreaction schemes are considerably more complex than that shown in Figure 20.3. For an excellent detailed analysis of photochemical kinetics, the reader can consult the book Demas (1983).

20.3 Types of Photochemical and Photophysical Experiments

There are usually two kinds of experiments performed by photochemists: continuous (wave) cw or photolysis, and flash photolysis. Each type is useful for gaining certain kinds of information.

20.3.1 Continuous (cw) Photolysis

In cw photolysis, a continuous source of light, such as a halogen or an arc lamp, is directed or focused onto a sample in a cuvette or photochemical reactor. For the determination of quantum yield, one generally utilizes a well-collimated beam of light impinging on the sample in a cuvette. The beam may first be passed through a monochromator or filter if wavelength information is desired. The rate of reaction may be determined by monitoring at various time intervals properties that are functions of concentration, such as absorbance, pH, conductance, luminescence, or volume of gas emitted or absorbed. The light-absorption rate can be determined from actinometry (see Section 2.2).

Since photochemical processes are sensitive to specific wavelengths of light, careful attention must be paid to the spectral output of the light source selected for experiments.

For quantitative cw experiments, the usual type of lamp used is a high-pressure arc lamp containing mercury, xenon, or a mixture of mercury and xenon. These types of lamps are efficient converters of energy; they have a wide spectral response from the ultraviolet (UV) to the infrared; and because of their short-arc lengths, it is easy to focus

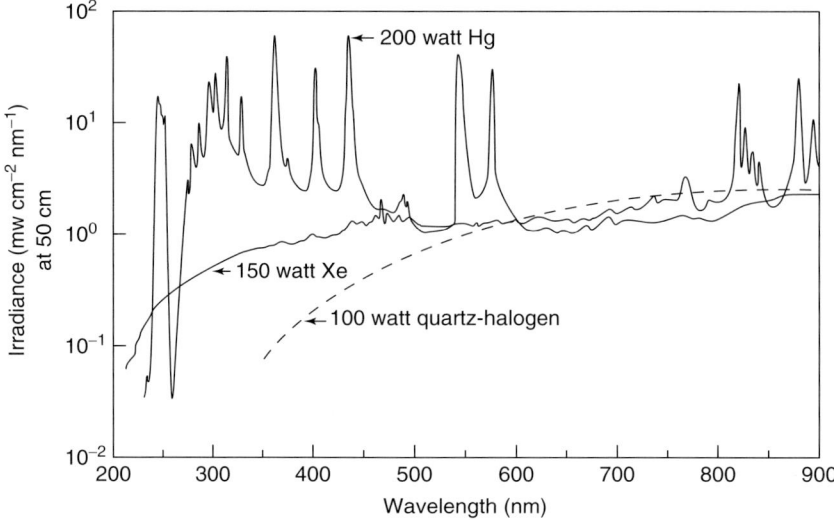

Fig. 20.4 The energy-output spectra for (solid curve) 200-W mercury, (dot-dashed curve) 150-W xenon high-pressure short-arc lamps, and (dashed curve) 100-W tungsten–halogen lamp.

their outputs through monochromators and onto a cuvette. The energy output of a high-pressure mercury arc lamp is concentrated primarily on the series of emission lines of the mercury spectrum, excluding the resonance line at 254 nm, whereas xenon arc lamps tend to give a more continuous output from the UV to the IR (Figure 20.4). The features of both types of lamp can be obtained in an arc lamp envelope filled with a mixture of xenon and mercury. Figure 20.4 also shows the visible portion of the output from a tungsten–halogen lamp, which behaves as a blackbody radiator.

20.3.2
Laser Flash Photolysis

In flash photolysis, a pulsed light source delivers a short burst of light to the sample. This generates short-lived intermediates whose properties, such as absorption or luminescence, can be monitored as a function of time and displayed on an oscilloscope screen. The kinds of information obtained for the intermediates are lifetimes and decay constants, absorption and luminescence spectra, and quantum yields of formation. Such pieces of information lead to an understanding of the intermediate steps and the mechanism of the photochemical reaction. Thus, flash photolysis can "observe" intermediates directly, even those of very fast lifetimes on the nanosecond or picosecond timescale. Both cw and flash photolysis yield important complementary information about photochemical systems.

Modern flash-photolysis experiments rely almost exclusively on pulsed lasers to produce the required high-energy and short pulsewidth flash. Lasers may operate in either the pulsed or cw mode, depending on the type. In all cases, they have extremely narrow bandwidths and virtually nonexistent beam divergences compared to noncoherent light sources. The energy output is centered only at one wavelength and depends on the material in the optical cavity. Availability of a wide selection of wavelengths and the ability to tune lasers over a limited wavelength range have emerged in the past several decades by the use of a variety of cavity materials, dye lasers, frequency modulating crystals, and optical parametric oscillators. Another type of laser, the *free electron laser*, produces coherent light at wavelengths from the far IR to the deep X-ray region, by the relativistic acceleration of a cluster of electrons in a storage ring. Such radiation is called *synchrotron radiation* (see above) and can be a very useful source of tunable laser light at almost any given pulse width, provided one has access to the rather expensive equipment required for its generation. For a complete description of the many laser configurations possible for laser flash photolysis, the reader is referred to Andrews (1992).

A simple laser flash-photolysis setup is diagramed in Figure 20.5. When the sample is excited by absorption of the laser pulse, the oscilloscope is triggered to receive the signal coming from the photomultiplier (PM) attached to the monochromator. The monochromator selects the wavelength of light of interest, and if the transient produced by the flash absorbs at this wavelength, a signal like that shown may appear as the transient decays over a selected time span. This signal can be computer-processed to yield absorbance versus time plots, and the decay constant and lifetime can be calculated. Alternatively, if the monochromator has a photodiode array detector, or if the sample is exposed to series of flashes, each one at a different wavelength, then it is possible to obtain a time-resolved spectrum, that is, an absorption spectrum of the short-lived transient at a given time after the flash. The setup also allows the user to obtain analogous luminescence decay information with the analyzing lamp turned off.

Owing to the limitations of PM rise times and oscilloscope time resolution, a rather different technique has to be used for experiments in the picosecond or femtosecond time regime. In this case, a *pump–probe* configuration is used in which time resolution is controlled by varying the pathlength of the probe pulse. In such experiments, the laser pump and probe pulses are produced in synchrony, but the probe pulse is diverted through an adjustable optical

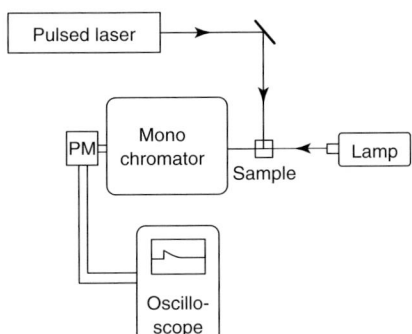

Fig. 20.5 A typical laser flash-photolysis experimental setup.

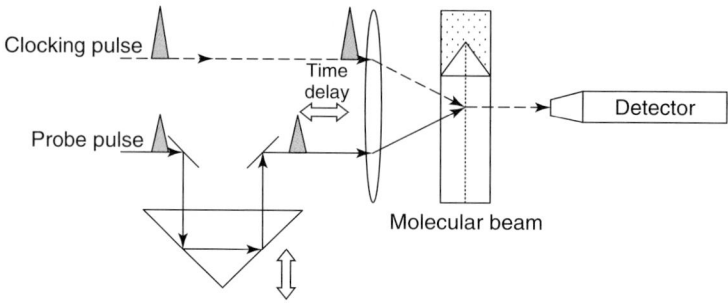

Fig. 20.6 A simplified scheme of the experimental layout for clocking in femtochemistry apparatus.

path length (Figure 20.6). Considering that light travels only about 0.3 mm in 1 picosecond and 0.3 µm in a femtosecond, accurate and precise time delays can be made by very small adjustments in the probe pathlength with respect to the pump pulse.

20.3.3
Laser Spectroscopy

The extremely high spectral purity of a laser beam, coupled with the advances in tunability of lasers, has made them indispensable tools for obtaining high-resolution spectral information of materials. Figure 20.1 shows two electronic energy levels depicted as potential-energy wells, with several vibrational levels superposed on each one. The more closely spaced molecular rotational levels that should be superposed on top of each vibrational level are not shown. In normal spectroscopy, the exciting source for the various transitions between energy levels is some kind of lamp–monochromator combination that has a fairly wide bandwidth and thus limits the degree of resolution of the energies of the transitions. On the other hand, lasers have an extremely narrow bandwidth as an excitation source, so that high-resolution spectra can be obtained. Several important laser spectroscopy techniques are briefly described. For more details on these techniques, the reader is referred to the book by Andrews (1992).

In *electronic photoabsorption spectroscopy*, a sample is excited by a visible or UV laser such that electronic excited states can be probed in the absorption mode. Determining the absorption spectrum of an excited state can yield important information about photochemical intermediates and the various reaction pathways available to a photoexcited species. Alternatively, because the bandwidth of a laser may be considerably smaller than the line width of the electronic transition, which absorbs the laser beam, replacing the conventional monochromater of a spectrometer with a tunable laser can result in highly resolved spectra of the ground state, yielding precise energies and transition probabilities of the excited electronic states. This is especially applicable to gaseous systems at low temperatures, in which the rotational and vibrational features of the spectra become minimized and the electronic transitions become "purer." One common method of significantly lowering the temperature of a gas under study is by supersonic jet-cooling, that is, ejecting the gas from a nozzle at supersonic speeds.

Laser-induced fluorescence (LIF) spectroscopy is a technique that produces a fluorescence spectrum by exciting the sample with a laser. In conventional fluorescence spectroscopy, one can obtain two types of spectra: an *excitation spectrum* in which the intensity of emission at a fixed wavelength is measured as a function of excitation wavelength and an *emission spectrum* in which intensity of emission is measured as a function of emission wavelength while exciting at a fixed wavelength. Figure 20.1 shows how the fluorescence process occurs and indicates that the excitation spectrum gives the energy changes occurring in the absorption process, whereas the emission spectrum gives the energy released when the excited species returns to the ground state. The use of a tunable laser as the excitation source significantly increases the resolution of an excitation spectrum so that vibrational and rotational fine structure can be observed in samples such as jet-cooled gases, crystals, and molecular beams. Important temporal information can also be obtained by using ultrashort pulsed lasers as the excitation source.

The absorption of laser light can sometimes occur in a multistep process, because of the very high light intensities of a laser beam. For example, a molecule can be excited to an electronic energy level that has a value of E above the ground state by the absorption of two photons of energy, $E/2$, at precisely the same time. This nonlinear effect gives rise to multiphoton absorption spectroscopy. One important advantage of this technique over conventional spectroscopy is that the electronic transitions do not obey the Laporte rule; in two-photon absorption, for example, the transitions must be $g \leftrightarrow g$ or $u \leftrightarrow u$, rather than $g \leftrightarrow u$ or $u \leftrightarrow g$. This increased access to excited states results in a greater understanding of the electronic and vibrational structure of molecules.

20.4 Femtochemistry

In femtochemistry, studies of physical, chemical, or biological changes can be made on the same timescale as that of molecular vibrations, that is, the actual nuclear motions. The femtosecond timescale is unique for the creation of coherent molecular wave packets on the atomic scale of length, a basic problem rooted in the development of quantum mechanics and duality of matter. Molecular wave functions are spatially diffuse and exhibit no motion, but superposition of a number of separate wave functions of appropriately chosen phases can produce the spatially localized and moving coherent wave packet. The powerful concept of coherence lies at the core of femtochemistry and was the key advance in observing the dynamics. Thus, the coupling of ultrafast lasers with molecular beams was essential for the initial developments in this field (Zewail, 2000a,b). The generation, amplification, and characterization of ultrashort pulses are a major part of femtochemistry experiments. Further details can be found in Zewail (1994, 2000a,b).

Such fundamental processes in chemical reactions as energy dissipation, molecular motions, and structural changes occur in the picosecond (10^{-12} seconds) to femtoseconds (10^{-15} seconds) timescale. With the advent of femtosecond lasers, it has become possible to measure events that occur on a timescale shorter than a single vibrational oscillation of a molecule. Thus, ultrafast "snapshots" can be taken of molecules, crystal lattices, unstable intermediates, or other structures in various stages of distortion, yielding valuable information about the structures of molecules and the dynamics of crystal lattices, liquids, individual molecules, and biological systems.

Many systems have recently been examined via femtosecond timescale experiments by Polanyi and Zewail (1995). For example, the dissociation of the ionic compound NaI through a covalent intermediate to the atomic products, Na + I, upon absorption of energy, occurs in less than 10 picoseconds. During this time, the activated intermediate oscillates on the average about eight times, and through an ultrafast series of probe laser pulses the amounts of activated intermediate, ionic $Na^+ I^-$ species, and product sodium atoms can be monitored, with a time resolution down to about 7 femtoseconds. The results have given direct information about the reaction time, the probability of dissociation, and the extent of ionic and covalent character in the NaI bond, and have confirmed many theoretical predictions about reaction dynamics. First, it was shown experimentally that the wave packet was highly localized in space (ca. 10 pm), thus establishing the concept of dynamics at atomic-scale resolution. Second, spreading of the wave packet was minimal up to a few picoseconds, thus establishing the concept of single-molecule trajectory. Third, vibrational (rotational) coherence was observed during the entire course of the reaction, thus establishing the concept of coherent trajectories in reactions, from reactants to products. Finally, the NaI case was the first to demonstrate the resonance behavior, in real time, of a bond converting from being covalent to being ionic along reaction coordinates (Zewail, 2000).

20.5
Reactions of Excited Species

The apparent reactivity of an excited species may be determined by the intrinsic reactivity of its electronic arrangement, by the excitation energy, and by the lifetime of the particular excited state.

The intrinsic reactivity of a species largely depends on the way in which its electrons are distributed in the available orbitals. Electronic excitation alters this arrangement and thus influences the geometry, the dipole moment, the electron-donating or -accepting ability, and the related acid–base properties.

20.5.1
Electronic Energy Transfer

An important overall mechanism of quenching in which the excitation energy is transferred to the quencher in a bimolecular process

$$M^* + Q \longrightarrow M + Q^* \qquad (20.23)$$

is also called *photosensitization*, where Q^* is some excited state of Q usually not directly accessible by optical absorption. The quencher, Q, is said to be sensitized by M, that is, although Q is by itself normally transparent at this wavelength, it is able to become excited by the excited sensitizer, M^*, whose energy should be greater than (or equal to) that of Q^*.

Energy transfer can be either radiative or nonradiative. In the previous case, the emission from M^* is reabsorbed by Q.

(Of course, the absorption spectrum of Q should overlap the emission spectrum of M^*.) Since the S_0-T_1 transition is spin-forbidden, the dominant radiative energy-transfer processes are from a singlet to a singlet or from a triplet to a singlet state.

Nonradiative energy transfers involve the mutual perturbation of the electronic structures of M^* and Q. In long-range coulombic energy transfer (Förster, 1959), the donor and acceptor molecules may be coupled through an electrostatic interaction at large (up to 10 nm) intermolecular separations. The energy-transfer processes favored by the coulombic mechanism are from a singlet to a singlet or from a singlet to a triplet state. In many sensitization reactions, triplet–triplet energy transfer plays the key role.

20.5.2 Photochemistry of Organic Compounds

If excitation affects the nature of the bonding, the shape of the molecule may also be changed. For example, while the ground state of an alkene (e.g., ethene) is planar, in the equilibrium structure of the (π, π^*) excited state the two CH_2 groups lie in perpendicular planes and only a σ bond remains between the carbon atoms.

Excited species are generally both better electron donors and better electron acceptors than the corresponding ground-state species. Therefore, excitation both decreases the ionization energy and increases the electron affinity of the molecule.

Generally, the excited singlet may be more reactive than the triplet. However, since for most organic molecules the ground state is a singlet, the triplet excited state may survive radiative and radiationless quenching so much better than the more reactive singlet. Thus, generally, the triplet is the more important excited species in terms of the overall reactivity in photoinduced organic reactions.

20.5.2.1 Electronic Transitions and Spectra of Organic Molecules

Absorption of a photon by an organic molecule causes a change in its electronic state in a variety of ways that depend on the type of the substrate and the energy of the photon. The highest energy occupied molecular orbitals (HOMOs) of an alkene are the π orbitals in the C–C bond. Absorption of a photon of wavelength around 180 nm promotes an electron from this HOMO to the lowest unoccupied molecular orbital (LUMO) π^*-orbital. The excited state is termed the (π, π^*) *excited state*, and the electronic transition leading to it is a $\pi \rightarrow \pi^*$ transition. While simple alkenes absorb at wavelengths below 200 nm, conjugation of double bonds can shift the absorption maximum into the visible range; for example, $\lambda_{max} = 476$ nm for $CH_3(CH=CH)_{10}CH_3$.

The HOMOs of molecules containing heteroatoms (such as N, O, or halogen) are usually nonbonding, and the lowest energy transitions for such molecules are $n \rightarrow \pi^*$ (in the presence of multiple bonds) or $n \rightarrow \sigma^*$ (for saturated molecules). Since $n \rightarrow \pi^*$ transitions, ketones and aldehydes have the lowest energy absorption maxima at 280 and 290 nm, respectively. Acids, anhydrides, and esters have absorption bands at shorter wavelengths (<250 nm), whereas aromatic substitution causes a considerable redshift (e.g.,

$\lambda_{max} \approx 340$ nm for benzophenone). The allowed $\sigma \rightarrow \sigma^*$ transitions of alkanes are less important, giving rise to bands in the vacuum UV.

Most organic molecules are singlets in their ground states; thus, their first excited triplet state is of lower energy than the corresponding singlet state, because of the repulsive nature of interactions between electrons of the same spin.

20.5.2.2 Correlation Rules and Symmetry Conservations

Electron-spin correlation rules are also valid for reactions in which chemical change occurs. Thus, in the permitted adiabatic reactions the total electron spin is conserved, that is, the sum of the spins of the reactants should be the same as that for the products. Conservation of symmetry is especially important to the understanding of pericyclic reactions, or reactions that pass through a cyclic transition state in a concerted manner, that is, with bond-breaking and -forming processes occurring essentially simultaneously. The well-known Woodward–Hoffmann rules (Woodward and Hoffmann, 1970), using the concepts of orbital symmetry, give a good compass for prediction or interpretation of these reactions.

20.5.2.3 Intramolecular Processes

Dissociation Dissociation of a diatomic molecule can give only two atomic fragments with unambiguous chemical identity, whereas that of a polyatomic molecule can sometimes yield many sets of products. For a detailed discussion of the methods used to elucidate primary dissociative mechanisms, the reader can consult the books by Calvert and Pitts (1966) and by Okabe (1978).

Photoisomerization The perpendicular configuration of ethene and its derivatives in the lowest energy (π, π^*) excited state is geometrically equivalent whether it is derived from a cis or trans ground-state molecule. Thus, the probability ratio for the formation of cis and trans isomers depends on the conversion rates of the excited state to the two (isomeric) ground-state forms.

In accordance with the Franck–Condon principle, the significant difference between the geometries of the ground and excited states forbid emission from the latter one. Therefore, in the case where the quantum yields are similar for both cis $\rightarrow$ trans and trans $\rightarrow$ cis, prolonged irradiation of either isomer results in the composition of the isomeric mixture at the photostationary state being predominantly determined by the individual absorbances of each isomer at the excitation wavelength. For simple alkenic systems, trans isomers generally possess higher extinction coefficients at longer wavelengths. Thus, irradiation in that range results in the predominance of the cis isomer in the photostationary state. Electrocyclic ring-closure reactions of dienes and trienes are well-known examples of valence and structural isomerizations. These processes leading to the formation of cycloalkenes require the s-cis conformation.

The reactions of polyenes in the excited singlet state often differ from those in the triplet state. Although direct excitation (to S_1) results in internal cyclization, triplet sensitization usually leads to (cyclic) dimerization. Photocyclization of aromatic compounds often results in nonaromatic products. Cyclic conjugated enones and dienones can also undergo structural photoisomerizations with ring rearrangement. These concerted stereospecific reactions proceed via a triplet (π, π^*) state. Another important type of photoisomerization is the di-π-rearrangement, which is formally a 1,2-shift with ring closure. This reaction, however, is probably a nonconcerted process, and promoted via the excited singlet.

Photochemical sigmatropic shift is also a significant type of photoisomerization. In these pericyclic reactions, a σ-bond migrates with respect to a system of π-electrons, resulting in a switching of double and single bonds. A more detailed discussion of these topics is given by Wagner and Hammond (1967).

Hydrogen Abstraction Intramolecular hydrogen atom abstractions, in contrast to photoisomerization processes, involve molecules almost exclusively in the (n, π^*) excited state. This type of reaction is especially significant in the photochemistry of most carbonyl compounds (Wagner, 1971).

20.5.2.4 Intermolecular Processes

Hydrogen abstraction can also take place from (n, π^*) excited states (especially of carbonyl compounds, see Scaiano (1973)) in intermolecular processes.

Addition Reactions Both homo- and heteroaddition reactions can be photochemically induced. The latter type of reaction can take place between unsaturated hydrocarbons and water, alcohols, and carboxylic acids or between aromatic compounds and amines. Cycloaddition of benzene and its derivatives can take place across 1,2-, 1,3-, or 1,4-positions. The 1,3-addition predominates with alkenes having only alkyl substituents on the double bonds. 1,2-addition is the main mode of reaction if the electron donor–acceptor properties of the aromatic species and the alkene are significantly different. The 1,4-cycloaddition occurs least frequently and may take place through an exciplex interaction as do the 1,2-additions as well. Cyclobutane derivatives are the main products in the photoaddition of alkenes to the unsaturated bond in α, β-unsaturated ketones. Cyclic enones undergo photocyclic addition more readily than their acyclic analogs because their reduced flexibility inhibits intersystem crossing from the excited triplet state to the ground state. The addition of the carbonyl group itself (in the excited ketones or aldehydes) to suitable alkenes can also occur, producing oxetanes (Arnold, 1968).

Oxygen is one of the most important reactants in photochemical addition reactions of unsaturated compounds. Very efficient photosensitized oxidations are observed with alkenes, dienes, dienoid heterocycles, and polycyclic aromatic compounds. The first products of these reactions are often peroxides or hydroperoxides, which may undergo subsequent oxidation steps. In most of these photosensitized oxidations, the very reactive excited singlet state of oxygen, O_2 ($^1\Delta_g$), is favored via energy transfer from the triplet state of the sensitizers (Foote, 1968). Hence, carotenoids (e.g., β-carotene) are very important photobiological reagents, because as efficient quenchers of singlet oxygen they can protect photosynthetic organisms against photooxidation sensitized by their own chlorophyll.

20.5.3
Photochemistry of Inorganic and Coordination Compounds

To illustrate the kinds of electronic transitions that occur in inorganic photochemical reactions, Figure 20.7 shows a generic molecular-orbital diagram for an octahedral transition–metal coordination complex and the various kinds of electronic transitions that may occur between the orbitals. These are categorized as follows:

1. intraligand (IL) bands, which are transitions between energy levels of the ligands (as discussed earlier about organic photochemistry) and are relatively unaffected by the metal centers;
2. ligand-field (LF) bands, which are the low-intensity transitions, usually in the visible region, caused by the splitting of the metal d orbitals (hence the other often-used term, d–d bands) by the field generated by the coordination of ligands to the metal;
3. ligand-to-metal-charge-transfer (LMCT) bands, which are transitions in which electronic charge is essentially transferred from the ligands toward the coordinating metal;
4. metal-to-ligand-charge-transfer (MLCT) bands, in which electronic charge is transferred from the coordinating metal to the ligands; and
5. charge-transfer-to-solvent (CTTS) transitions, in which electronic charge moves to the solvent. Charge-transfer bands are characterized by their tendency to produce charge separations, which may result in oxidation–reduction reactions.

Charge-transfer bands are characterized by their tendency to produce charge separations, which may result in oxidation–reduction reactions.

20.5.3.1 Ligand-Field (LF) Transitions

The transition metals are in the middle part of the periodic table where the elements have partially filled $(n-1)d$ electron subshells (where n is the period or main shell number). There are five d orbitals in a given subshell, each with its

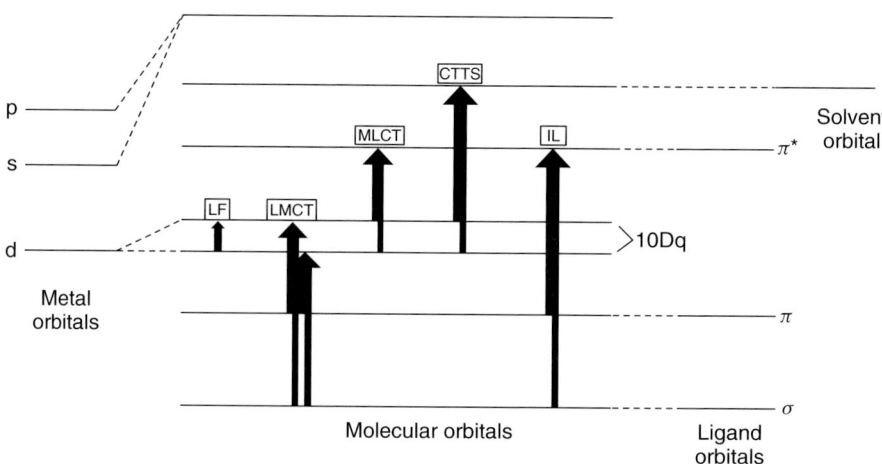

Fig. 20.7 Molecular-orbital diagram for an octahedral coordination complex and the various kinds of electronic excitation transitions that may occur between the orbitals.

own geometrical orientation in space, as shown in Figure 20.8.

In a neutral atom, the five d orbitals of Figure 20.8 are of the same energy, but in a metallic ion coordinately bound to electron-donor ligands there is a splitting of the energies, termed *crystal-field splitting*, caused by the interaction of the ligand electrons with the metal d orbitals. Which atomic orbitals undergo the strongest interaction with the ligands depends on the geometry of the complex: for an octahedral complex, the two e_g orbitals point in the directions of the incoming ligands, thus raising their energy, whereas the three t_{2g} orbitals point away from the ligands resulting in no change in energy. For a tetrahedral complex the situation is reversed, and for other geometries the splitting becomes more complicated. A molecular-orbital picture gives the same splitting patterns between metal orbitals, and these are shown for an octahedral complex in Figure 20.7 by the commonly used symbol for this difference, $10Dq$.

The split atomic orbitals of the metal ion give rise to electronic transitions resulting in LF or d–d bands. Since these transitions are Laporte-forbidden, they are rather weak such that extinction coefficients rarely exceed several hundreds, and the energies of the transitions are usually in the visible to near-IR region of the spectrum, giving a noticeable color to many transition–metal complexes.

For a transition to occur, the coordinating metal ion must have at least one d electron and there must be at least one unfilled orbital in the higher energy state to which an electron can be excited. Excitation of a metal center changes the geometrical distribution of electrons and hence may change the strength of the metal–ligand bond. Thus, LF transitions may result in ligand-exchange, isomerization, or rearrangement reactions without changing the oxidation state of the metal ion. A large number of photochemical reactions induced by LF transitions have been identified and are reviewed in the books by Balzani and Carassiti (1970), Adamson and Fleischauer (1975), Geoffroy and Wrighton (1979), and Ferraudi (1988).

20.5.3.2 Ligand-to-Metal (LMCT) and Metal-to-Ligand (MLCT) Charge-Transfer Transitions

Normally, charge-transfer bands are observed in the UV and short-wavelength visible region in complexes consisting of a donor and acceptor of electronic charge. If the donor is an electron-rich ligand, such as NH_3 or halo, and the acceptor a readily reduced metal, such as

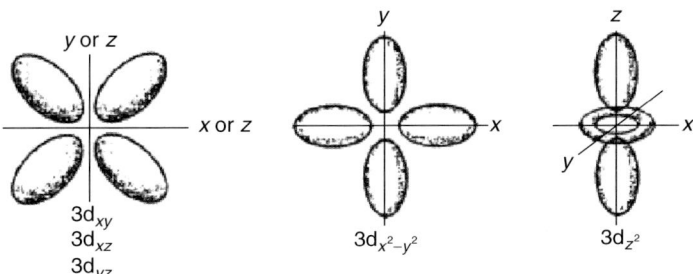

Fig. 20.8 Electron probability pictures of the three d orbitals of an atom.

Co(III) or Fe(III), the transition is LMCT; if the donor is an oxidizable metal, such as Cu(I) or Fe(II), and the acceptor an electron-poor ligand, such as 1,10-phenanthroline or bipyridine, the transition is MLCT. Such bands are Laporte- and (usually) spin-allowed yielding absorption coefficients on the order of 10^4.

One should expect that the excited species following charge-transfer absorption is at once a stronger oxidizing and reducing agent than the ground state because of the shift in electron density between metal and ligand. Thus, the photochemistry resulting from such transitions should involve redox processes of metal and ligand. This certainly seems to be the case following MLCT absorption, as exemplified by the well-known tris(bipyridyl) ruthenium (II) system (Gafney and Adamson, 1972), which has been heavily researched as a possible photocatalytic agent for water splitting (Creutz and Sutin, 1976) since the transitions occur in the visible portion of the spectrum. In addition to redox processes, MLCT transitions often result in reductive elimination.

In general, MLCT transitions usually result in processes that maintain the integrity of the metal–ligand bonds because they originate in filled or nearly filled metal d orbitals, which do not have much effect on the bonding, and they terminate in ligand-localized orbitals, which also are not usually involved in the metal–ligand bond.

By contrast, LMCT transitions often occur between ligand orbitals involved in the ligand–metal bond and a metal in a higher oxidation state with unfilled d orbitals, resulting in a coordinatively unstable, reduced metal–radical ion pair, which usually decomposes further to a complicated array of products. A useful list, prepared by Endicott (1975), tabulates the primary reduced metal and radical formed from the LMCT excitation of each of some 39 complexes of Co(III), Rh(III), Ir(III), Pt(IV), Fe(III), and Ru(III) in aqueous solution.

20.5.3.3 Charge-Transfer-to-Solvent (CTTS) Transitions

CTTS transitions are favorable when the irradiated species is easily oxidized and electron-rich, and there is some low-lying unoccupied orbital on the solvent molecule in the first or second coordination sphere of the complex. Since the first criterion for such transitions is the same as that which favors MLCT transitions, it is common for the lower oxidation states of such metals as Cu, Ru, and Fe to exhibit both MLCT and CTTS bands, sometimes with both types of transitions occurring in the same complex when an acceptor ligand is present, as has been observed, for example, in the case of the cyano complexes of copper(I) and iron(II). The most obvious indicator of a CTTS transition is the ejection by the excited species of a solvated electron into the bulk of the solvent.

20.5.3.4 Other Charge-Transfer Transitions

The number of di- and polynuclear complexes exhibiting metal–metal bonding has increased dramatically in the past few years. In such compounds, it is expected that there will be significant orbital overlap such that new metal–metal delocalized molecular orbitals will be formed. These molecular orbitals can then take part in transitions resulting in charge-transfer between the metals themselves, that is, metal–metal charge-transfer (MMCT) or between the metal–metal orbitals and ligand orbitals (perhaps labeled M_2LCT or LM_2CT, depending on the direction of the charge-transfer). For a detailed compilation of charge-transfer photochemistry of coordination compounds, see the book by Horváth and Stevenson (1993).

20.6 Conclusion

As this concise introduction demonstrates, photochemistry has great theoretical and practical significance in the natural sciences, industry, and nature. A multitude of phenomena, which have been observed and interpreted using photochemical theory and techniques, have strong connections to physics and biology, as well as to the material sciences. The development of more and more sophisticated experimental techniques in photochemistry makes possible the elucidation of the dynamics of elementary reactions. Perhaps even the energy needs of humankind will one day be met by an environmentally friendly photochemical solar energy conversion process.

Acknowledgments

This work was supported by the Hungarian Scientific Research Fund (OTKA K63494).

Glossary

Absorbance: $\log_{10}(I_0/I)$, where I_0 is the incident light intensity falling on an absorbing sample and I is the transmitted light intensity.

Actinometry: Measurement of light intensities for quantum-yield determinations.

Bond-Specific Photochemistry: Photochemistry having products determined by a specific bond or combination of bonds set into vibrational motion.

Chromophore: Light-absorbing chemical group.

Cis: A molecular structural configuration in which two like functional groups are adjacent to each other on the molecular skeleton.

Concerted Reaction: A reaction in which bond-breaking and -forming occur essentially simultaneously.

Conformation: The (possibly unstable) arrangement of atoms inside a molecule.

Conformers: Different, relatively stable, arrangements of atoms, each consisting of identical numbers and types of atoms. A dissociating or rearranging molecule may pass through different conformers. Conformers need not be capable of being physically or chemically isolated.

Conjugation: Alternating double and single bonds within a molecular chain or ring, resulting in bond stabilization via electron delocalization.

Dipole Moment: A measure of the separation between centers of positive and negative charges in polar molecules.

Excimer: A complex formed from the excited state of a molecule with an identical molecule in the ground state.

Exciplex: A complex formed from an excited molecule and a different molecule in the ground state.

Extinction Coefficient: Specific absorbance of 1-cm pathlength of solution at 1 mol l^{-1} concentration.

Flash Photolysis: Experimental study of photochemical or photophysical processes initiated by a short pulse of light.

HOMO: Highest occupied molecular orbital.

Internal Conversion: Energy loss within a given electronic excited or ground-state manifold.

Intersystem Crossing: Transition of an excited electron from a given multiplicity (such as the singlet) to a different one (such as the triplet).

Isomers: Different molecular conformations, consisting of identical atoms, that are stable, or metastable long enough to be physically or chemically separated.

Lifetime: The time required for an excited species to decay to $1/e$ of its initial concentration.

LUMO: Lowest unoccupied molecular orbital.

Mechanism: Sequence of physical and chemical steps leading to a chemical change. Sometimes used to refer to the most unique or characteristic step of an overall mechanism.

Mode-Specific Photochemistry: Photochemistry having products determined by the vibrational mode set into motion.

Monomers: Low-molecular-weight molecules or atoms that can join together to form a polymer.

Multiphoton Absorption: Direct multiphoton absorption occurs when two or more photons are absorbed at once, mediated by higher order perturbation terms of the molecular Hamiltonian interaction with the radiation field. Sequential one-photon events are sometimes called by this name, but more properly may be called *multiple-photon absorption*.

Nonradiative Energy Transfer: Loss of energy from an excited state by any process that does not produce light.

Normal Mode: Classically, molecular vibration in which (nearly) all atoms move concertedly and periodically with the same frequency, but possibly not with the same phase. Quantum mechanically, eigenfunction of the vibrational Hamiltonian.

Organometallic: A compound containing organic carbon and metal atoms.

Oscillator Strength: A measure of the area under an absorption band, or the degree to which a transition is allowed.

Pericyclic Reactions: Concerted reactions proceeding via a cyclic transition state.

Photodetachment: Ejection of an electron from a negative ion by the effect of light.

Photodissociation: Fragmentation of a molecule into two or more pieces by the effect of light.

Polarizability: The measure of charge displacement by an externally applied electric field.

Radiative Energy Transfer: Loss of energy of an excited state by emission of light.

Sensitizer: Compound that absorbs light at the irradiation wavelength, then transfers its excited-state energy to a substrate that cannot be directly excited.

Singlet: An electron configuration in which the sum of the electron-spin quantum numbers is 0, that is, all the electrons are paired with respect to their spins.

Spin–orbit Coupling: Interaction of the orbital and spin angular momenta of an electron in an atom or polyatomic species.

Trans: A molecular structural configuration in which two like functional groups are opposite to each other on the molecular skeleton.

$u \longrightarrow g$ or $g \longrightarrow u$ Transitions: Laporte-allowed transitions between wave functions that are even (gerade or g) or odd (ungerade or u) with respect to inversion about a center of symmetry.

Triplet: An electron configuration in which the sum of electron-spin quantum numbers is 1, that is, there are two unpaired electrons of parallel spin.

Vibrational Relaxation: Loss of internal vibrational energy into other modes of motion.

References

Adamson, A.W. and Fleischauer, P.D. (1975) *Concepts of Inorganic Photochemistry*, John Wiley & Sons, Ltd, New York.

Arnold, D.R. (1968) *Adv. Photochem.*, **6**, 301–423.

Andrews, D.L. (1992) *Applied Laser Spectroscopy*, Wiley-VCH Verlag GmbH, New York.

Balzani, V. and Carassiti, V. (1970) *Photochemistry of Coordination Compounds*, Academic Press, New York.

Brueck, S.R.L. and Osgood, R.M. Jr (1976) *Chem. Phys. Lett.*, **39**, 568–572.

Calvert, J.G. and Pitts, J.N. Jr (1966), *Photochemistry*, Chaps. 3–5, John Wiley & Sons, Ltd, Chichester, New York.

Creutz, C. and Sutin, N. (1976) *J. Am. Chem. Soc.*, **98**, 6384, 6385.

Demas, J.N. (1983) *Excited State Lifetime Measurements*, Academic Press, New York.

Elsaesser, T. and Kaiser, W. (1991) *Annu. Rev. Phys. Chem.*, **42**, 83–107.

Endicott, J.F. (1975) in *Concepts of Inorganic Photochemistry*, (eds A.W. Adamson and P.D. Fleischauer), Chap. 3, John Wiley & Sons, Ltd, New York.

Ferraudi, G.J. (1988) *Elements of Inorganic Photochemistry*, John Wiley & Sons, Ltd, New York.

Foote, C.S. (1968) *Acc. Chem. Res.*, **1**, 104–110.

Förster, Th. (1959) *Discuss. Faraday Soc.*, **27**, 7–17.

Gafney, H.D. and Adamson, A.W. (1972) *J. Am. Chem. Soc.*, **94**, 8238–8239.

Geoffroy, G.L. and Wrighton, M.S. (1979) *Organometallic Photochemistry*, Academic Press, New York.

Hatchard, C.G. and Parker, C.A. (1956) *Proc. R. Soc. London, Ser. A*, **235**, 518–536.

Okabe, H. (1978) *Photochemistry of Small Molecules*, Chaps. 2, 4–7, John Wiley & Sons, Ltd, Chichester, New York.

Polanyi, J.C. and Zewail, A.H. (1995) *Acc. Chem. Res.*, **28**, 119–132.

Scaiano, J.C. (1973) *J. Photochem.*, **2**, 81–118.

Wagner, P.J. (1971) *Acc. Chem. Res.*, **4**, 168–177.

Wagner, P.J. and Hammond, G.S. (1967) *Adv. Photochem.*, **5**, 21–156.

Wegner, E.E. and Adamson, A.W. (1966) *J. Am. Chem. Soc.*, **88**, 394–404.

Woodward, R.B. and Hoffmann, R. (1970) *The Conservation of Orbital Symmetry*, Wiley-VCH Verlag GmbH, Deerfield Beach, FL.

Zewail, A.H. (1994) *Femtochemistry Ultrafast Dynamics of the Chemical Bond*, Vols. I and II, World Scientific, Singapore.

Zewail, A.H. (2000a) *J. Phys. Chem.*, **104**, 5660–5694.

Zewail, A.H. (2000b) *Les Prix Nobel*, Almquist and Wiksel International, Stockholm.

Further Reading

Andrews, D.L. (1992) *Applied Laser Spectroscopy*, Wiley-VCH Verlag GmbH, New York.

Andrews, D.L. (1997) *Lasers in Chemistry*, 3rd edn, Springer-Verlag, Berlin.

Balzani, V. and Campagna, S. (2007) *Photochemistry and Photophysics of Coordination Compounds I, II (Topics in Current Chemistry)*, Springer, Berlin.

Bauerle, D. (1986) *Chemical Processing with Lasers*, Springer-Verlag, Berlin.

Boule, P. (ed.) (1999) Environmental photochemistry, in *The Handbook of*

Environmental Chemistry, (ed. O. Hutzinger), (Ed.-in-Chief), Vol. 2, Part L, Springer, Berlin.

Braun, A.M. (1991) Photochemical Technology, John Wiley & Sons, Ltd, New York.

Coyle, J.D. (1986) Introduction to Organic Photochemistry, John Wiley & Son, Ltd, Chichester, New York.

Coyle, J.D. and Hill, R.R. (eds) (1982) Light, Chemical Change and Life: a Source Book in Photochemistry, The Open University Press, Walton Hall, Milton Keynes, UK.

Demas, J.N. (1983) Excited State Lifetime Measurements, Academic Press, New York.

DeSchryver, F.C., DeFeyter, S., and Schweitzer, G. (eds) (2001) Femtochemistry, Wiley-VCH Verlag GmbH, Weinheim.

Ferraudi, G.J. (1988) Elements of Inorganic Photochemistry, John Wiley & Sons, Ltd, New York.

Fleming, G.R. (1986) Chemical Applications of Ultrafast Spectroscopy, Oxford University Press, New York.

Frimer, A.A. (ed) (1985) Singlet Oxygen, Vols. 1–4, CRC Press, Boca Raton, FL.

Gilbert, A. and Bagott, J. (1991) Essentials of Molecular Photochemistry, Blackwell Scientific, Oxford, UK.

Grigoropoulos, C.P. (1998) Experimental Methods in the Physical Sciences, Vol. 30, Academic Press, NewYork, pp. 173–223.

Griesbeck, A. and Mattay, J. (eds) (2004) Synthetic Organic Photochemistry (Molecular and Supramolecular Photochemistry), CRC Press, Boca Raton, FL.

Grossweiner, L.I. (1994) The Science of Phototherapy, CRC Press, Boca Raton, FL.

Hollenberg, J.L. (1970) J. Chem. Educ., **47**, 214.

Horspool, W.M. (ed) (1984) Synthetic Organic Photochemistry, Plenum, New York.

Horspool, W.M. and Lenci, F. (2003) CRC Handbook of Organic Photochemistry and Photobiology, Vols. 1 and 2, 2nd edn, CRC Press, Boca Raton, FL.

Horváth, O. and Stevenson, K.L. (1993) Charge Transfer Photochemistry of Coordination Compounds, Wiley-VCH Verlag GmbH, New York.

Ibbs, K.G. and Osgood, R.M. (eds) (1989) Laser Chemical Processing for Microelectronics, Cambridge University Press, Cambridge, UK.

Kimble, M. (2006) Femtochemistry VII: Fundamental Ultrafast Processes in Chemistry, Physics, and Biology, Elsevier, Amsterdam.

Klessinger, M. and Michl, J. (1995) Excited States and Photochemistry of Organic Molecules, Wiley-VCH Verlag GmbH, New York.

Kopecky, J. (1992) Organic Photochemistry: A Visual Approach, Wiley-VCH Verlag GmbH, New York.

Kuteladze, A.G. (2005) Computational Methods in Photochemistry (Molecular and Supramolecular Photochemistry), CRC Press, Boca Raton, FL.

Letokhov, V.S. (1983) Nonlinear Laser Chemistry, Springer-Verlag, Berlin.

Luxon, J.T. and Parker, D.E. (1987) Lasers in Manufacturing, IFS, Kempston, Bedford, UK.

Manz, J. and Wöstle, L. (eds) (1995) Femtosecond Chemistry, Wiley-VCH Verlag GmbH, Weinheim, Germany.

Martin, M.M. and Haynes, J.T. (eds) (2004) Femtochemistry and Femtobiology: Ultrafast Events in Molecular Science, Elsevier, Amsterdam.

Matsumoto, J. (2005) in Organic Photochemistry and Photophysics (Molecular and Supramolecular Photochemistry), (eds Ramamurthy, V. and Schanze, K.), CRC Press, Boca Raton, FL.

Monroe, B.M. and Weed, G.C. (1993) Chem. Rev., **93**, 435–448.

Montalti, M., Credi, A., Prodi, L., and Gandolfi, M.T. (2006) Handbook of Chemistry, CRC Press, Oxford, UK.

Murov, S.L., Carmichael, I., and Hug, G.L. (1993) Handbook of Photochemistry, Marcel Dekker, New York.

Olivucci, M. (2005) Computation Photochemistry, Vol. 16 (Theoretical and Computation Chemistry), Elsevier Science, Amsterdam.

Phillips, R. (1983) Sources and Applications of Ultraviolet Radiation, Academic Press, New York.

Rabek, F.F. (1982) Experimental Methods in Photochemistry and Photophysics, John Wiley & Sons, Ltd, New York.

Roffey, C.G. (1982) Photopolymerization of Surface Coatings, John Wiley & Sons, Ltd, Chichester.

Shipley, T. and Crescitelli, F. (1979) Visual Photochemistry: The Beginnings, Pergamon Press, Oxford, UK.

Singh, J. (2006) *Photochemistry and Pericyclic Reactions*, New Age International (P) Ltd, New Delhi.

Turro, N.J. (1991) *Modern Molecular Photochemistry*, Chaps. 10–14, University Science Books, Sausalito, CA.

Wayne, R.P. (1988) *Principles and Applications of Photochemistry*, Oxford University Press, Oxford, UK.

Wayne, C.E. and Wayne, R.P. (1996) *Photochemistry (Oxford Chemistry Primers, 39)*, Oxford University Press, Oxford, UK.

21
Condensed Phase Ultrafast Dynamics

Stephen R. Meech and Ismael A. Heisler

21.1	**Introduction** 749	
21.2	**Ultrafast Lasers** 749	
21.3	**Ultrashort Pulse Width Measurements** 750	
21.3.1	Autocorrelation Measurements 751	
21.3.2	Autocorrelation 752	
21.4	**Ultrafast Experiments** 753	
21.4.1	Transient Absorption 753	
21.4.2	Ultrafast Fluorescence Up-Conversion 754	
21.4.3	Polarization Spectroscopy 756	
21.4.4	Nonlinear Optics and Ultrafast Spectroscopy 759	
21.4.5	Photon Echoes 763	
21.5	**The Future** 765	
	References 766	

Encyclopedia of Applied Spectroscopy. Edited by David L. Andrews.
Copyright © 2009 WILEY-VCH Verlag GmbH & Co. KGaA, Weinheim
ISBN: 978-3-527-40773-6

21.1
Introduction

Ultrafast laser spectroscopy has evolved over the past 30 years from a somewhat esoteric area on the borders of chemistry and physics to become one of the major tools in molecular, materials, and life sciences. Much of this evolution has been driven by advances in technology, particularly, the development of the Kerr lens mode-locked titanium sapphire laser and subsequent progress in the design of optical parametric amplifiers (OPA). In this chapter, we briefly focus on these developments and outline the common methods of laser pulse characterization. We then describe some of the most widespread methods of modern ultrafast spectroscopy and indicate some of their capabilities.

21.2
Ultrafast Lasers

Short pulses are generated from a laser by modulating the gain at the round trip frequency of the laser. There are many methods of achieving such a modulation, including intracavity acousto-optic modulators and saturable absorbers (Svelto, 1998). The result is a series of pulses at a repetition frequency, which is the inverse of the round trip time and a temporal width, which is ultimately limited by the gain bandwidth of the lasing medium. The earliest studies of molecular dynamics were made with flash lamp pumped mode-locked Nd : YAG and Nd glass lasers (Kaufmann *et al.*, 1975). These sources were, by modern standards, unstable, difficult to use, and dangerous, but nevertheless readily produced picosecond pulses of very high intensities at low repetition rates. However, the first truly ultrafast experiments were made with mode-locked dye lasers. Subpicosecond pulses could be generated from dye lasers because the gain bandwidth of the laser medium is much larger than that for Nd lasers. Although very short pulses were obtained (<50 fs), the technical difficulties associated with these devices were considerable and their stability poor, making ultrafast spectroscopic measurements challenging (Kaufmann *et al.*, 1975; Fork *et al.*, 1981).

Important progress in ultrafast pulse generation was made with the discovery of solid state materials with very broad gain spectra, most notably titanium sapphire, which has a bandwidth capable of supporting pulse widths

Encyclopedia of Applied Spectroscopy. Edited by David L. Andrews.
Copyright © 2009 WILEY-VCH Verlag GmbH & Co. KGaA, Weinheim
ISBN: 978-3-527-40773-6

<10 fs. Titanium sapphire lasers were first mode-locked by traditional intracavity gain modulation methods, but the remarkable discovery that, once initiated, the mode-locking would continue without active modulation was the trigger for the development of a new generation of all solid state ultrafast lasers (Spence et al., 1991). It was quickly discovered that the mechanism for self mode-locking was a nonlinear interaction in the gain medium, which becomes more effective with shorter pulses. With careful correction for group velocity dispersion by intracavity prisms, pulses as short as 15 fs can be extracted directly from the Ti : sapphire laser.

Ti : sapphire also has suitable energy storage properties, which allow it to act as a gain medium for pulse amplification. Amplifying such short pulses to high energies creates enormous peak powers (peak power = energy per pulse per pulse width), which would ultimately destroy the amplifier cavity. To overcome this problem, the method of chirped pulse amplification was developed (Perry and Mourou, 1994). The broad spectrum of the input pulse is spread in time using gratings and prisms such that the pulse width reaches several hundred picoseconds. The broad pulse can then be safely amplified and recompressed to yield pulses of up to millijoule scale energy and tens of femtosecond's width. With careful focusing, terawatt powers are achievable.

Short high peak power amplified pulses from solid state sources also have excellent beam profiles. Such pulses are suitable for pumping OPAs, which, when seeded with broad bandwidth pulses generated through self-phase modulation, can convert approximately 800 nm output of the Ti : sapphire laser into ultrafast pulses at any wavelength between the visible and near IR (Cerullo and De Silvestri, 2003; Wilhelm et al., 1997). Such pulses may thus be used to study the electronic spectroscopy of almost any molecule. The OPA outputs two frequencies, called the *signal* and the *idler*, which may themselves be mixed in another nonlinear crystal to generate the difference frequency in the mid IR region (Towrie et al., 2003). In this way, ultrafast spectroscopy can be performed in the vibrational energy region. The availability of short stable pulses at any frequency has stimulated a range of novel and imaginative range of ultrafast experiments, some of which are mentioned below.

21.3
Ultrashort Pulse Width Measurements

Pulses with tens of optical cycles in the near infrared spectral region are now commonly encountered in research laboratories worldwide. These pulses, which are on the femtosecond time scale, are much faster than any electronic measuring equipment can possibly capture. New methods to time resolve events on the ultrashort time scale have to be developed. In reality, the knowledge related to the generation of ultrashort pulses and the methods for their characterization evolved simultaneously, because it is essential to know the distortions and constraints that limit the production of even shorter pulses. In addition, for experiments utilizing ultrashort pulses, it is necessary at least to know the time duration of such pulses to determine the time resolution of the experiment.

Presently, there are very sophisticated methods that cover a wide range of time scales and produce a complete characterization of ultrashort pulses, that is, not only information about temporal duration but also about phase and amplitude in time and frequency domains. These methods, of which frequency resolved optical gating (FROG) (Trebino *et al.*, 1997) is the best known, are not treated here. Rather we describe the most fundamental technique, which estimates the time duration of ultrashort pulses with good precision. This technique relies on the condition that the ultrashort pulse itself is used to determine its time duration; an autocorrelation process (Jean-Claude and Rudolph, 1996). The setup most widely used is based on the well-known Michelson interferometer. However, linear processes such as a simple interferogram can only produce information about the coherence of the pulse, which is related to the spectral content. To access information about the time duration, it is necessary to work with nonlinear processes, such as second harmonic generation (SHG) in a nonlinear crystal, two-photon absorption, optical Kerr gate, and so on (Boyd, 2003), which enable the measurement of the intensity autocorrelation (Rulliere, 1998).

21.3.1
Autocorrelation Measurements

A typical setup for an intensity autocorrelator is shown in Figure 21.1. A beam splitter divides an incoming pulse into two equal intensity pulses which are then focused to a common spot where a crystal with, for example, a second-order nonlinearity is positioned. The path length difference and consequently the relative delay time (τ) between the two pulses can be adjusted mechanically via a variable delay stage. If the arm length difference is made small, so that the pulses overlap temporally inside the nonlinear crystal, the process of second harmonic (or more generally sum frequency) generation occurs in the phase matched direction separated spatially from the incoming beams and their second harmonics. The intensity of the newly generated beam can be measured

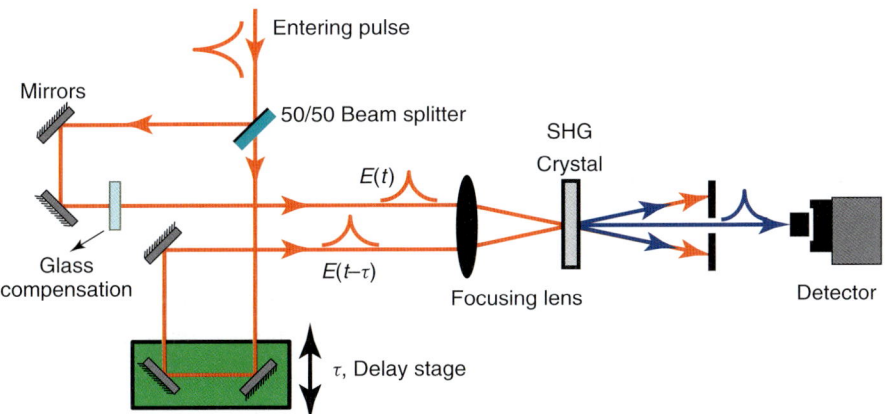

Fig. 21.1 Scheme for an intensity autocorrelator. SHG, second harmonic generation. Solid black lines are the laser fundamental, the dashed line is the second harmonic generated when the pulses overlap.

with a time integrating detector at each delay step introduced by the delay stage. If the relative time delay is increased, so that the overlap of the two pulses in the crystal is reduced, the second harmonic signal becomes weaker and ultimately disappears. The overall result is the intensity of the second harmonic signal as a function of delay time, which carries information about the incoming pulse duration.

21.3.2
Autocorrelation

The mathematical description of an autocorrelation process can be written as follows:

$$G^{(n)}(\tau) = \int_{-\infty}^{\infty} f^n(t) f^{n*}(t-\tau) dt \quad (21.1)$$

where $f(t)$ is some time-dependent function and $f^*(t)$ is its complex conjugate (Jean-Claude and Rudolph, 1996). The function $G^{(n)}(\tau)$ is called the nth-order autocorrelation. Taking $f(t)$ as the resulting electric field from the two pulses inside the crystal, $f(t) = E(t) + E(t - \tau)$, and assuming a second-order process, it is possible to write the resulting signal as

$$G^{(2)}(\tau) = \int_{-\infty}^{\infty} \left| (E(t) + E(t-\tau))^2 \right|^2 dt \quad (21.2)$$

where $G^{(2)}(\tau)$ is the second-order autocorrelation as a function of the delay time, τ. This equation can be separated into three terms: one centered at zero frequency, which is known as the *intensity autocorrelation plus a constant background*; the other two terms are centered at the optical frequency, ω, and second harmonic, 2ω, and are known as the *interferometric autocorrelation* (Jean-Claude and Rudolph, 1996). An example of an interferometric second-order autocorrelation measurement is presented in Figure 21.2(b). Most often the intensity autocorrelation, which is easier to measure than an interferometric autocorrelation, produces sufficient information to determine the time resolution of a given experiment. The intensity autocorrelation expression can be written as

$$I^{(2)}(\tau) = \int_{-\infty}^{\infty} I(t)I(t-\tau)dt \quad (21.3)$$

where $I(t) \propto |E(t)|^2$ is the time-dependent intensity of the pulse. An example

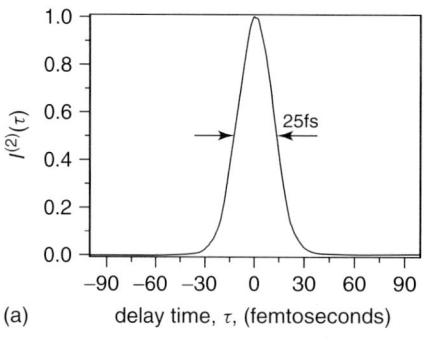
(a)
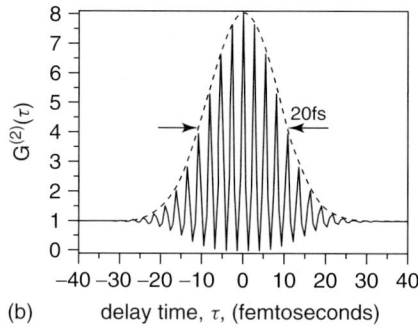
(b)

Fig. 21.2 (a) Intensity autocorrelation, without background, showing a FWHM (full width at half maximum) of 25 fs. This corresponds to a pulse with time duration of 16.3 fs, assuming a hyperbolic secant squared pulse profile. (b) Interferometric autocorrelation showing the optical field oscillations. This corresponds to a pulse of 14.8 fs.

of such a measurement can be seen in Figure 21.2(a). To recover the pulse duration, it is necessary to fit the measured data by assuming some functional form for the intensity. Usually ultrashort pulses originating from mode-locked lasers have a temporal shape described by a squared hyperbolic secant function (Jean-Claude and Rudolph, 1996). For a given spectral bandwidth, the attainable theoretical shortest temporal pulse is called *Fourier transform limited* (or simply *transform limited*). For such pulses, the spectral phase difference between the wavelengths that compose the spectrum is constant or equal to zero. This can be tested using the expression: $\Delta \nu \Delta \tau_P \geq c_B$, where $\Delta \nu$ is the spectral bandwidth of the pulse (in Hertz), $\Delta \tau_P$ is the temporal duration, and c_B is a constant that is determined by the model assumed to fit the temporal intensity autocorrelation. For Gaussian and squared hyperbolic secant functions, c_B is equal to 0.44 and 0.31, respectively. The equality in the above expression is achieved for transform limited pulses (Rulliere, 1998).

For the shortest pulses, it is found that transmission of the pulse through any dense medium leads to pulse broadening. This naturally arises from the normal dispersion of the medium, which causes the red portion of the broadband laser spectrum to travel faster than the blue, creating a spread in time (a chirp) (Jean-Claude and Rudolph, 1996). Chirped pulse amplifiers exploit this phenomenon. For high time resolution optical experiments, a correction for this effect must be made. To first order, this is achieved using a simple prism pair. Routing the chirped beam twice through the prism pair, oriented at Brewster's angle, causes the red light to travel further than the blue (negative dispersion). By carefully matching the positive and negative dispersion by controlling the prism separation, the experimenter can deliver the shortest pulses to the sample.

21.4
Ultrafast Experiments

21.4.1
Transient Absorption

This experiment is essentially an extension of the flash photolysis experiment developed by Porter and coworkers (Porter and Wilkinson, 1961) into the ultrafast time domain. It shares with that method the principle of optical excitation of the sample by one pulse and then measurement of the entire spectrum as a function of time with a second broadband pulse. In accessing the subpicosecond time domain, one encounters the same problem as was mentioned in Section 21.3, that no conventional electronic detector is capable of real-time measurements on the subpicosecond timescale. Thus the requirement for the transient absorption experiment is an ultrafast pump and an ultrafast probe pulse, and the latter should have as broad a spectrum as possible. A broad spectrum with an ultrafast temporal width may be generated by the simple expedient of focusing an intense titanium sapphire laser pulse into a transparent medium (a water cell works well). Through a highly nonlinear interaction called self-phase modulation, a fraction of this incident light is converted into a broad band source stretching from the UV to the near IR with a temporal width as short as the input pulse (Shen, 1984).

The pump and probe pulses are then combined in essentially the same geometry as was shown for the pulse width measurement (Figure 21.1), with the sample taking the place of the nonlinear crystal. The pump pulse strikes the sample at a time defined as $t = 0$, where it is absorbed, transferring population from the ground to the excited electronic state. In general, the excited state will have a distinct absorption spectrum. The spectrally broad probe pulse is incident on the same point in the sample at a time τ later, where τ is controlled by the path length difference introduced between the pump and probe paths. On transmission through the sample the probe beam spectrum is measured by a polychromator. For the most accurate measurements, a correction must be made for the dispersion which causes the red and blue components of the probe spectrum to travel at slightly different velocities, thus reducing the time resolution.

Typically, the change in transmission induced by the pump pulse is small – a small percentage at most. To obtain reliable measurements, it is desirable to record an identical reference spectrum taken when the pump pulse is blocked. From these two measurements, a difference spectrum may be created, and the reference spectrum can be used to normalize out intensity fluctuations. The logarithm of the resulting ratio yields the transient optical density difference spectrum, where negative peaks correspond to bleaching of the ground state and positive peaks show transient absorption spectra of newly created states, which may be formed either during the excitation or during subsequent chemical reactions. The transient absorption method has been used by many groups over the years to reveal photo-induced processes in molecules ranging from simple dissociation or ring opening reactions to the dynamics of the photosynthetic system. Changes in optical densities lesser than 10^{-4} can be measured.

21.4.2
Ultrafast Fluorescence Up-Conversion

Fluorescent molecules are widely used as probes of natural and synthetic materials for numerous reasons: fluorescence is very sensitive to its environment; the probe can be introduced to the medium in a site specific fashion; it may be detected with high (single molecule) sensitivity; it affords a noncontact method of probing the material (Lakowicz, 2006). Time resolved fluorescence has long been used to add an additional dimension to fluorescence spectroscopy (Maroncelli and Fleming, 1987). The classical time-correlated single photon counting method has been refined to the extent that excited state decay times of tens of picoseconds can be resolved (Phillips and O'Connor, 1983). However, for true ultrafast time resolution, the fluorescence up-conversion method must be used (Fleming, 1986).

The up-conversion method is essentially an ultrafast sampling of the temporal profile of the sample fluorescence. The experimental geometry is shown in Figure 21.3 (Shah, 1988).

Two ultrafast laser pulses are required, exactly correlated in time. Often the fundamental and second harmonic of the laser are used. The higher energy pulse electronically excites the sample, which relaxes through fluorescence decay at an energy ν_f. The fluorescence is focused into a nonlinear crystal where it is mixed with the second ultrafast laser pulse of frequency ν_p. The nonlinear

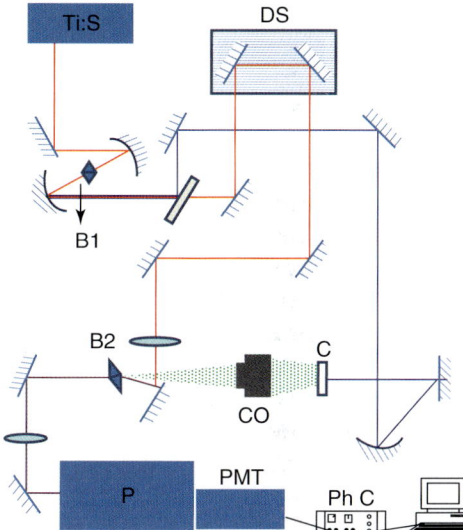

Fig. 21.3 Femtosecond fluorescence up-conversion experimental setup: Ti : S, titanium–sapphire laser; B1, 50-μm BBO crystal; B2, 100-μm BBO crystal; DS, delay stage; CO, collection optics; C, cell containing sample; P, polycromator; PMT, photomultiplier tube; Ph C, photon counter.

interaction in the crystal generates the sum frequency $\nu_{\text{sum}} = \nu_f + \nu_p$ in the phase matching direction. This signal can clearly only be generated when both fluorescence and laser pulse are present in the crystal, and its intensity is a product of their two instantaneous intensities. Thus for constant laser pulse energy, the nonlinear interaction samples the intensity of the fluorescence at the instant of the arrival of the second pulse. By scanning the arrival time of the pulse and measuring the sum frequency intensity as a function of time, the fluorescence decay profile is mapped out. As an example, Figure 21.4 presents a semilog plot of the up-converted fluorescence intensity relaxation of an aqueous solution of the dye molecule auramine O (van der Meer et al., 2000). The signal can be described by a sum of two exponential decays.

As in all ultrafast experiments, the highest time resolution is only achieved if the dispersion is minimized. For up-conversion, this requires independent dispersion compensation in the excitation and up-converting pulses, and as far as possible all reflective optics should be used (Joo and Rhee, 2005).

Fluorescence up-conversion is a particularly powerful tool as it allows detailed studies of a diverse range of phenomena. Excited state reactions (isomerization, electron and proton transfer) (Kang et al., 1990; Todd et al., 1990) are among the simplest known and their dynamics have been characterized in great detail through fluorescence. Similarly, the dynamics of solvation can be measured from the temporal evolution of the fluorescence spectrum (Maroncelli and Fleming, 1987). Such studies have provided very detailed views of condensed phase dynamics, and have been used as benchmarks to test molecular dynamics simulations. Having been well characterized in solution, similar experiments are being conducted in complex media including living cells.

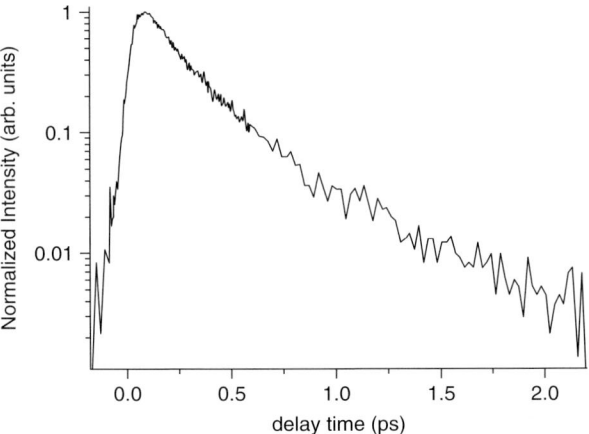

Fig. 21.4 Up-converted signal intensity showing the fluorescence relaxation of dye auramine O in water solution.

21.4.3
Polarization Spectroscopy

A number of techniques that relate spectroscopic information to the polarization properties of electromagnetic radiation are known as *polarization spectroscopy* (Kano and Levenson, 1988). In general, the initial polarization state of the light source is well defined. This source interacts with an anisotropic material resulting in a new polarization state which is analyzed. The anisotropy can be either inherent to the medium (e.g., a crystalline or liquid crystal material) or externally imposed (e.g., the Kerr effect in the presence of strong electric field). Within the realm of ultrafast spectroscopy, a long established and widely applied time resolved technique is the optical Kerr effect (OKE) (Smith and Meech, 2002). In this account, we focus on the OKE technique applied to transparent media, and its analog applied to absorbing media.

One great appeal of the OKE technique is that it measures the low-frequency Raman spectrum (0–500 cm^{-1}) of the medium with extremely high signal-to-noise ratio. This spectrum contains detailed information about the medium dynamics. The OKE signal (Figure 21.5 shows the usual experimental setup) originates from an intense linearly polarized optical pulse, which interacts with an initially isotropic sample (Lotshaw *et al.*, 1995). This pulse, called the *pump*, induces a transient anisotropy through the molecular polarizability, which is probed by a second, weak, time-delayed pulse, incident at the same point in the sample. The probe pulse is linearly polarized and has its plane of polarization at 45° with respect to the pump pulse. The pump-induced anisotropy depolarizes the probe pulse, which emerges from the sample slightly elliptically polarized. A polarizer, crossed with the plane of polarization of the probe pulse, is placed in front of the detector. The pump-induced depolarization causes a fraction of the probe pulse to reach the detector (homodyne signal). The signal is measured as a function of delay time between the pump and probe pulses and so

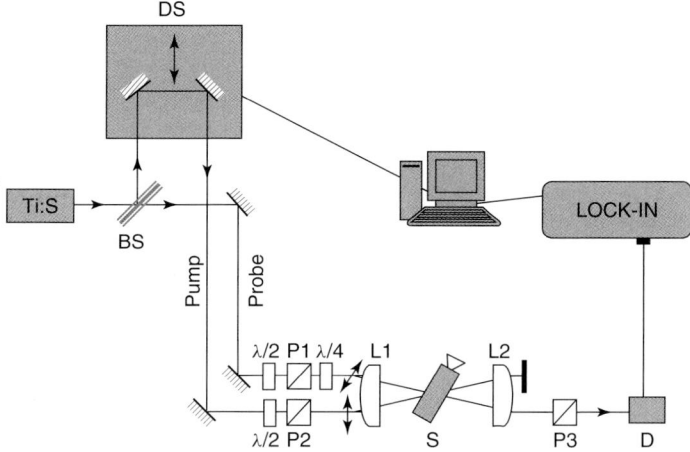

Fig. 21.5 OKE typical experimental setup. Ti : S, Ti : sapphire femtosecond laser; BS, beam splitter; DS, delay stage; P1-3, polarizers; L1-2, lenses; S, sample; and D, detector.

the relaxation of the induced anisotropy in the sample is measured. For a homodyne measurement, the detected signal is quadratically and linearly proportional to the pump and probe pulse intensities, respectively. Additionally, the measured signal is quadratically proportional to the impulse response function of the material, $R_{ijkl}(t)$ (discussed further in Section 21.4.4 below). The response function is related to the frequency-dependent third-order nonlinear susceptibility, $\chi^{(3)}_{ijkl}(\omega)$, through a Fourier transform relationship, and contains both real (dispersive) and imaginary (resonant) parts.

To improve the signal-to-noise ratio and to avoid the complexity of analyzing a quadratic signal, a heterodyned measurement is employed (Kano and Levenson, 1988). This is achieved by the addition of a local oscillator (LO) to the OKE signal. The LO is derived from the probe pulse in either of two ways: (i) by a slight rotation of the analyzer polarizer (P3), which introduces a component *in phase* with the OKE signal; (ii) by the introduction of a quarter wave plate ($\lambda/4$) in the probe beam with its fast axis initially aligned with the probe polarization; after a slight rotation of P1 (generally $\leq 1°$) a small component *in quadrature* with the homodyne OKE signal will leak through the analyzer polarizer. The signal measured by the square law detector is:

$$I_D = I_{LO} + I_S + \frac{nc}{8\pi}\left(E_S^* E_{LO} + E_S E_{LO}^*\right)$$

(21.4)

where E_i and I_i are the field and the intensity of the LO or homodyne OKE signal field, respectively. The term in brackets is the *heterodyne term*, which scales linearly with $R_{ijkl}(t)$. The LO field amplitude is set to produce a heterodyne signal much larger than the homodyne intensity, but small enough to avoid detector saturation. The LO signal is easily eliminated from measurement using lock-in detection. It is important to note that the real and imaginary contributions to the signal field, E_S, are associated, respectively, with the imaginary (dichroism) and real

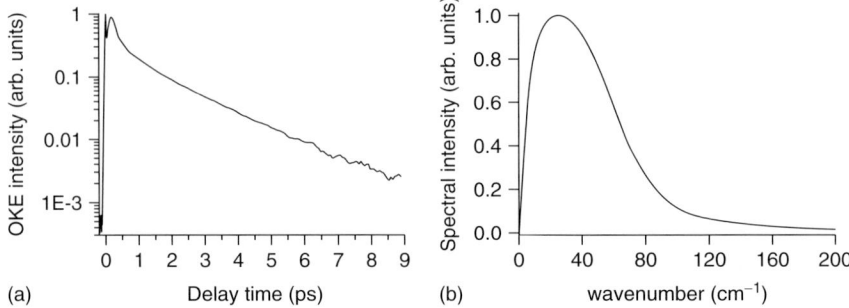

Fig. 21.6 (a) OKE signal for carbon disulfide (CS_2) on a semilog plot. (b) Spectral intensity content of CS_2 OKE signal after subtraction of the diffusive exponential component.

(birefringence) parts of the material response function, $R_{ijkl}(t)$. These may be measured separately depending on the phase introduced in the local oscillator field, E_{LO}.

When the sample is transparent, the real (birefringence) part of the response dominates. The data may be analyzed in either the time or the frequency domain. The latter are accessed through a Fourier transform deconvolution procedure resulting in the Raman spectral density (Lotshaw et al., 1995). Figure 21.6 shows a typical OKE signal for carbon disulfide (CS_2) liquid at room temperature.

CS_2 has one the largest OKE signals, due to its high polarizability anisotropy, and shows many features shared with various other molecular liquids for which OKE measurements have been performed (Castner et al., 2007; Cong et al., 1995; Fourkas, 2002; Friedman and She, 1993; Heisler et al., 2005; Hunt et al., 2007; Hunt and Meech, 2004). At time zero, there is a symmetric instantaneous response which follows the laser pulse autocorrelation. This signal arises from the electronic hyperpolarizability. The nuclear dynamics response of most interest appears with a risetime (around 100 fs) due to the inertial molecular motion that cannot follow the impulsive force exerted by the strong ultrashort laser pump field. The signal then shows a complex subpicosecond relaxation followed by a picosecond time scale, exponential response (Figure 21.6a). This component is well described by a Stokes–Einstein–Debye model of diffusive orientational relaxation (McMorrow et al., 1988). The subpicosecond decay component is usually called *nondiffusive* but scales with the diffusive relaxation (Loughnane et al., 1999). Its origin remains a matter of discussion. On an ultrafast timescale, the single-molecule contribution will take the form of a hindered molecular reorientation or librational motion, which is well fit by a molecule rattling in a cage model. This signal relaxes approximately as a Gaussian function but usually the analysis is made in the frequency domain to avoid model-dependent fitting (Figure 21.6b). Such detailed pictures of liquid state molecular dynamics provide data against which to test both theories and molecular dynamics simulations of the liquid state. In more complex fluids, additional low-frequency intramolecular modes may contribute to the Raman spectral density, and these may be the characteristic of macromolecular structure (Hunt and Meech, 2004, 2003).

When the pump and probe pulses are resonant with a strong electronic transition, then the transient dichroism (imaginary component of $R_{ijkl}(t)$) dominates the response, which may be selectively measured by using an in-phase LO. If the absorption is due to a molecular solute, then the time-dependent signal probes both the solute orientational relaxation and the ground state recovery (because both will ultimately lead to a recovery of the initial isotropic distribution). The experiment can then be used to access information on both molecular orientational motion and excited state kinetics, and has been applied to diverse samples (Cho et al., 1993; Waldeck et al., 1981).

21.4.4
Nonlinear Optics and Ultrafast Spectroscopy

It is apparent that the experiments described above report on multiple interactions of the laser pulses with the sample. Thus it is evident that a natural theoretical picture of such interactions will involve a time domain description invoking nonlinear optics (Mukamel, 1995). It is in general possible to separate optical measurements into linear and nonlinear regimes according to the electric field $E(t)$ power-law dependence of the process. In such measurements, to properly calculate the signal, one has to start with the Maxwell equations. In these, the only material quantity that appears as a source is given by the optical polarization $P(t)$ induced by some applied external electric field $E(t)$. Therefore, a complete knowledge of the optical polarization is essential for the interpretation of any time or frequency domain spectroscopic measurement. Usually a constitutive relation is used to describe the induced polarization in a medium when an electric field is applied. The polarization may be expanded as a power series in the electric field vector (Boyd, 2003):

$$\mathbf{P}(t) = \chi^{(1)}\mathbf{E}(t) + \chi^{(2)}\mathbf{E}(t)\,\mathbf{E}(t)$$
$$+ \chi^{(3)}\mathbf{E}(t)\,\mathbf{E}(t)\,\mathbf{E}(t) + \cdots$$

(21.5)

where $\chi^{(n)}$ are material susceptibility coefficients, which in general are tensors (Butcher and Cotter, 1990). As a consequence, the orientation of the induced polarization may be different from that of the applied field. Therefore, it is possible to rewrite the polarization as:

$$\mathbf{P}(t) = \mathbf{P}^{(1)}(t) + \mathbf{P}_{NL}(t) \quad (21.6)$$

where $\mathbf{P}^{(1)}(t)$ is the linear polarization, accounting for processes such as absorption, reflection, refraction, and so on, and $\mathbf{P}_{NL}(t)$ is the nonlinear polarization describing higher order processes such as second harmonic generation, optical Kerr effect, two-photon absorption, self-phase modulation, and so on. For time resolved experiments, a better approach to describing optical measurements is through response functions $R(t)$, which have their counterpart in the frequency domain description through the susceptibilities $\chi(\omega)$. The relation between these quantities is given by a Fourier transform:

$$\chi(\omega) = \int_{-\infty}^{\infty} dt\, R(t)e^{i\omega t} \quad (21.7)$$

The response function formalism can be applied to a broad range of linear and nonlinear optical measurements differing in terms of the applied temporal sequence of pulses, frequencies or propagation direction (wavevectors) (Mukamel, 1995). Thus, this formalism enables one to describe and calculate the signals expected for different ultrafast experiments and ultimately compare their information content.

The response function contains information on how a system alters in response to an external applied perturbation. It has been shown that if changes involve small departures from equilibrium, then the equilibrium fluctuations dictate the nonequilibrium response. Thus knowledge of the equilibrium dynamics is useful for predicting nonequilibrium processes.

The polarization at a given time t is related to the applied electric field at all earlier times and can be written, for first and second order, as

$$P_i^{(1)}(t) = \int_{-\infty}^{t} dt_1 \, R_{ij}^{(1)}(t, t_1) E_j(t_1) \quad (21.8)$$

$$P_i^{(2)}(t) = \int_{-\infty}^{t} dt_2 \int_{-\infty}^{t_2} dt_1 \, R_{ijk}^{(2)}(t, t_1, t_2) \\ \times E_j(t_1) E_k(t_2) \quad (21.9)$$

and for the nth-order as

$$P_i^{(n)}(t) = \int_{-\infty}^{t} dt_n \ldots \int_{-\infty}^{t_3} dt_2 \int_{-\infty}^{t_2} dt_1 \\ \times R_{ijk\ldots m}^{(n)}(t, t_1, t_2, \ldots, t_n) \\ \times E_j(t_1) E_k(t_2) \ldots E_m(t_n) \quad (21.10)$$

with the convention of summation over repeated indices. The newly defined nth-order response function, $R^{(n)}$, is a tensor of rank $n + 1$, and a real and causal function of time (Mukamel, 1995). Also, the response is a function of the relative time between perturbation and observation, that is, $R^{(n)}(t, t_0) = R^{(n)}(t - t_0)$. Ordering the fields and labeling the relative time as in Figure 21.7, it is possible to rewrite the expressions for the polarization in a more intuitive fashion as

$$P_i^{(1)}(t) = \int_0^{\infty} d\tau_1 \, R_{ij}^{(1)}(\tau_1) E_j(t - \tau_1) \quad (21.11)$$

$$P_i^{(2)}(t) = \int_0^{\infty} d\tau_2 \int_0^{\infty} d\tau_1 \, R_{ijk}^{(2)}(\tau_1, \tau_2) \\ \times E_j(t - \tau_2) E_k(t - \tau_2 - \tau_1) \quad (21.12)$$

$$P_i^{(n)}(t) = \int_0^{\infty} d\tau_n \ldots \int_0^{\infty} d\tau_2 \int_0^{\infty} d\tau_1 \\ \times \begin{pmatrix} R_{ijk\ldots m}^{(n)}(\tau_1, \tau_2, \ldots, \tau_n) \\ E_j(t - \tau_n) E_k(t - \tau_n - \tau_{n-1}) \\ \times \ldots E_m(t - \tau_n - \tau_{n-1} \\ \times \ldots - \tau_2 - \tau_1) \end{pmatrix} \quad (21.13)$$

The optical electric field couples with the material via the microscopic charge distribution. As mentioned before, for spectroscopic measurements, the signal

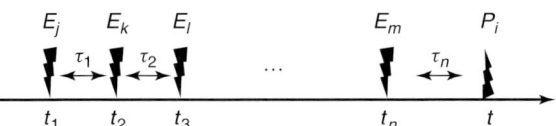

Fig. 21.7 Electric fields time ordering and resulting polarization.

is determined by the macroscopic polarization $\mathbf{P}(t)$, which in turn, is given in terms of the average over an ensemble of particles contained in a volume V, of the expectation value of the dipole operator $\boldsymbol{\mu}$:

$$\mathbf{P}(t) = \frac{1}{V}\overline{\langle\boldsymbol{\mu}(t)\rangle} = \frac{1}{V}\text{Tr}\{\boldsymbol{\mu}\rho(t)\} \quad (21.14)$$

In the above definition, the dipole operator is given by the instantaneous positions of all charged particles of the ensemble contained in volume V (nuclei and electrons):

$$\boldsymbol{\mu} = \sum_i e_i \mathbf{r}_i \quad (21.15)$$

The last term on the right hand side of Eq. (21.14) introduces the density matrix $(\rho(t))$ formalism to calculate the polarization, where Tr is the trace over the matrix representation. The density matrix evolution is determined by the Liouville equation:

$$\frac{d\rho}{dt} = -\frac{i}{\hbar}[H_I(t), \rho] \quad (21.16)$$

Here we are assuming the interaction picture description where $H_I(t)$ accounts for the light-matter time-dependent interaction, which in the electric dipole approximation can be written as

$$H_I(t) = -\boldsymbol{\mu} \cdot \mathbf{E}(t) \quad (21.17)$$

The general solution for the evolution equation can be written as:

$$\rho(t) = \rho(t_0) - \frac{i}{\hbar}\int_{t_0}^{t} dt_1 [H_I(t_1), \rho(t_1)] \quad (21.18)$$

It is not possible to solve this equation exactly. The procedure usually adopted is to expand the density matrix operator and then to solve the evolution equation iteratively.

$$\rho(t) = \rho^{(0)}(t) + \rho^{(1)}(t) + \rho^{(2)}(t) + \ldots \quad (21.19)$$

Applying this perturbative approach, solutions for the different orders are

$$\rho^{(0)}(t) = \rho(t_0) = \rho_{eq} \quad (21.20)$$

$$\rho^{(1)}(t) = -\frac{i}{\hbar}\int_{-\infty}^{t} dt_1 \left[H_I(t_1), \rho_{eq}\right] \quad (21.21)$$

$$\rho^{(2)}(t) = \left(-\frac{i}{\hbar}\right)^2 \int_{-\infty}^{t} dt_2 \int_{-\infty}^{t_2} dt_1$$
$$\times \left[H_I(t_2), \left[H_I(t_1), \rho_{eq}\right]\right] \quad (21.22)$$

and for the nth-order:

$$\rho^{(n)}(t) = \left(-\frac{i}{\hbar}\right)^n \int_{-\infty}^{t} dt_n \ldots \int_{-\infty}^{t_3} dt_2$$
$$\times \int_{-\infty}^{t_2} dt_1 [H_I(t_n), [H_I(t_{n-1}), \ldots$$
$$[H_I(t_3), [H_I(t_2), [H_I(t_1),$$
$$\rho_{eq}]]]\ldots]] \quad (21.23)$$

The system, before any interaction, is in the equilibrium state; therefore, the assumption $t_0 = -\infty$ may safely be made. Again, to present the expressions in a more physical manner, it is helpful to perform a change of variables so that the time intervals between successive interactions of the electric field with the dipole moment appear explicitly. As an example, we present the expression for the third-order polarization, after developing the commutator terms of Eq. (21.23) and inserting them into Eq. (21.14), resulting in

$$P_i^{(3)}(t) = \frac{1}{V}\left(-\frac{i}{\hbar}\right)^3 \int_0^\infty d\tau_3 \int_0^\infty d\tau_2$$

$$\times \int_0^\infty d\tau_1 \begin{pmatrix} \text{Tr}\{[[[\mu_i(\tau_1+\tau_2+\tau_3), \\ \mu_l(\tau_1+\tau_2)], \mu_k(\tau_1)], \\ \mu_j(0)]\rho_{eq}\} \\ E_j(t-\tau_3)E_k(t-\tau_3-\tau_2) \\ \times E_l(t-\tau_3-\tau_2-\tau_1) \end{pmatrix}$$

(21.24)

From this expression, the third-order response function is given by

$$R_{ijkl}^{(3)}(\tau_3, \tau_2, \tau_1) = \theta(\tau_3, \tau_2, \tau_1)\left(-\frac{i}{\hbar}\right)^3$$

$$\times \text{Tr}\{[[[\mu_i(\tau_1+\tau_2+\tau_3), \mu_l(\tau_1+\tau_2)], \mu_k(\tau_1)], \mu_j(0)]\rho_{eq}\} \quad (21.25)$$

where $\theta(\tau)$ is the Heaviside function used to impose the causality condition. This expression makes the connection between the microscopic material quantity (the dipole moment which couples to the electric field) and the response function measured through the macroscopic induced polarization. It can be further particularized to a given experiment by choosing the configuration of the applied electric fields and also the particular multipolar moment which is interacting with the field (i.e., fixed dipole moment, polarizability, hyperpolarizability, etc). By working out the commutator expressions, it is possible to rewrite the response functions as

$$R_{ijkl}^{(3)}(\tau_3, \tau_2, \tau_1) = \theta(\tau_3)\theta(\tau_2)\theta(\tau_1)$$

$$\times \left(-\frac{i}{\hbar}\right)^3 \sum_\alpha^4 R_\alpha(\tau_3, \tau_2, \tau_1)$$

$$\times -R_\alpha^*(\tau_3, \tau_2, \tau_1) \quad (21.26)$$

where

$$R_1(\tau_3, \tau_2, \tau_1) = \text{Tr}\{\mu_i(\tau_1+\tau_2+\tau_3) \\ \times \mu_l(\tau_1+\tau_2)\mu_k(\tau_1) \\ \times \mu_j(0)\rho_{eq}\} \quad (21.27)$$

$$R_2(\tau_3, \tau_2, \tau_1) = \text{Tr}\{\mu_j(0)\mu_l(\tau_1+\tau_2) \\ \times \mu_i(\tau_1+\tau_2+\tau_3) \\ \times \mu_k(\tau_1)\rho_{eq}\} \quad (21.28)$$

$$R_3(\tau_3, \tau_2, \tau_1) = \text{Tr}\{\mu_j(0)\mu_k(\tau_1) \\ \times \mu_i(\tau_1+\tau_2+\tau_3)\mu_l \\ \times (\tau_1+\tau_2)\rho_{eq}\} \quad (21.29)$$

$$R_4(\tau_3, \tau_2, \tau_1) = \text{Tr}\{\mu_k(\tau_1)\mu_l(\tau_1+\tau_2) \\ \times \mu_i(\tau_1+\tau_2+\tau_3) \\ \times \mu_j(0)\rho_{eq}\} \quad (21.30)$$

The complex terms can easily be worked out in the same fashion.

Third-order nonlinearities describe many of the coherent nonlinear experiments commonly used in research laboratories worldwide: pump-probe, transient grating, photon echoes, coherent antistokes Raman spectroscopy (CARS), degenerate four-wave mixing are some of the most important of these. These experiments are described by some or all of the eight correlation functions that contribute to $R_{ijkl}^{(3)}(\tau_3, \tau_2, \tau_1)$. This shows how complete and general the response formalism can be, and how it can be applied to essentially any ultrafast experiment (Mukamel, 1995). Roughly speaking, once the configuration of electric fields used in the experiment is established, one can refer to these expressions to determine the measured signal outcome. The next step would be to include the relaxation terms involved in a given sample system. This can be done with various

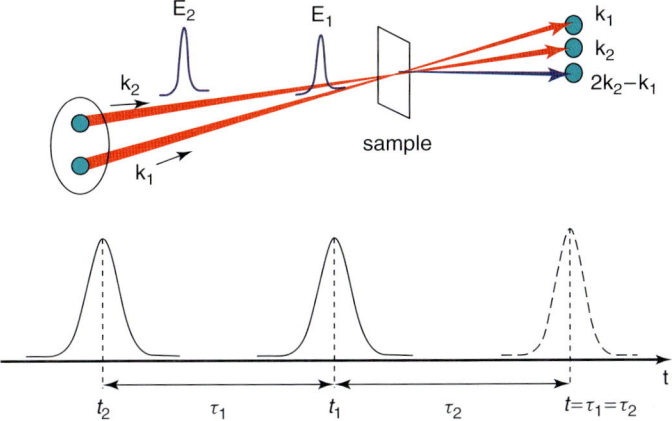

Fig. 21.8 Two-pulse photon echo time ordering and spatial beam geometry. (Adapted from Andrei Tokmakoff, 5.74 Introductory Quantum Mechanics II, Spring 2007. (MIT OpenCourseWare: Massachusetts Institute of Technology). http://ocw.mit.edu/OcwWeb/Chemistry/5-74Spring-2007/CourseHome. Accessed Oct. 1, 2008.)

degrees of complexity, and the outcome depends somewhat on the nature of the sample under study. A number of examples exist in the literature (Mukamel, 1995).

21.4.5
Photon Echoes

The photon echo technique is an interesting example of the kind of multipulse optical experiment, which is currently being developed in ultrafast spectroscopy. The original experiments were an important demonstration of coherent optical phenomena (Kurnit et al., 1964). Measurements on the picosecond timescale revealed the photon echo as being an important tool in molecular spectroscopy, particularly, in allowing the direct measurement of the homogeneous optical lineshape free of inhomogeneous broadening (de Boeij et al., 1998). In the original two-pulse photon echo, two pulses are incident on the sample at times t_1 and t_2 (Figure 21.8).

At a time $\tau_1 = t_1 - t_2$, the sample emits a signal – the photon echo – at the same frequency as the input pulses and in the phase matched direction, $\mathbf{k}_{PE} = 2\mathbf{k}_2 - \mathbf{k}_1$. The analogy with spin echoes in NMR is obvious and quite appropriate. By scanning the pulse separation and measuring the echo intensity, the homogeneous lifetime of the transition can be measured. A further level of sophistication is the three pulse photon echo (Cho et al., 1992). Figure 21.9 shows the time orderings of the pulses and the emitted signal, related to the third-order nonlinear polarization, $P^{(3)}(t')$, together with the 'box' geometry usually applied. The times τ and T denote the center to center distances between the pulse pairs E_1, E_2 and E_2, E_3, respectively. Also, t_1, t_2, and t_3 denote field–matter interaction points related to the response function

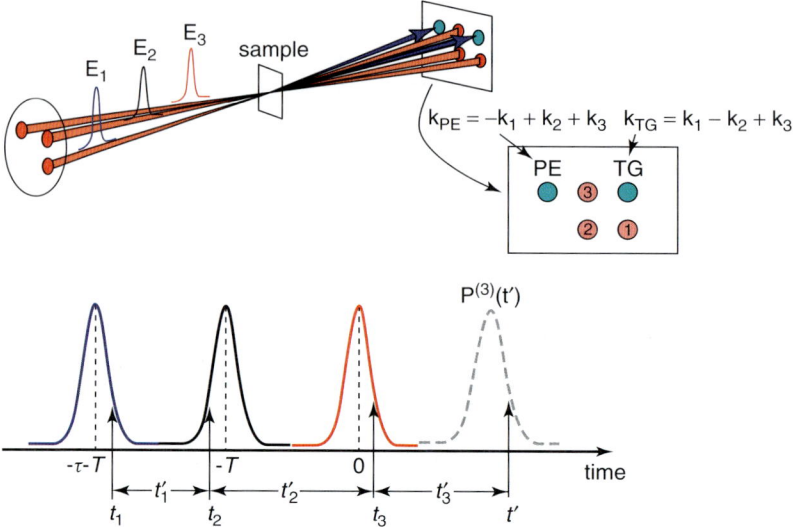

Fig. 21.9 Three pulse photon echo time ordering and spatial beam geometry. Also shown is the box geometry highlighting the phase matched direction for the photon echo signal (PE) and the transient grating signal (TG). (Adapted from Andrei Tokmakoff, 5.74 Introductory Quantum Mechanics II, Spring 2007. (MIT OpenCourseWare: Massachusetts Institute of Technology) http://ocw.mit.edu/OcwWeb/Chemistry/5-74Spring-2007/CourseHome. Accessed Oct. 1, 2008.)

formalism Eq. (21.10), while t'_1, t'_2, and t'_3 denote the relative delays between the interaction points. In this experiment, there are three pulses involved and two delays to be controlled. Changing the delays independently yields quite different dynamic information.

Varying t'_1 yields the homogeneous lifetime, while scanning t'_2 yields the population relaxation time. In analogy with NMR, the measured times are then T_2 and T_1, respectively, although the factors influencing the optical relaxation times are quite different to the nuclear spin case (and the relaxation times in molecules are many orders of magnitude faster). The phase matching direction for the emitted photon echo signal is now given by $\mathbf{k}_{PE} = -\mathbf{k}_1 + \mathbf{k}_2 + \mathbf{k}_3$, as illustrated in Figure 21.9. Also, when $t_1 - t_2 = 0$ and measuring the wave vectors phase matched beam in the direction given by $\mathbf{k}_{TG} = \mathbf{k}_1 - \mathbf{k}_2 + \mathbf{k}_3$, the experiment is analogous to transient grating scattering. The photon echo and transient grating techniques are also known as *rephasing and non-rephasing methods*, respectively. In terms of the response function contributions (Eqs (21.27)–(21.30)), it is possible to show that the terms that describe the photon echo signal are given by the sum of $R_2 + R_3$ and for the transient grating the terms that contribute are $R_1 + R_4$. Again, this illustrates the generality and completeness of the response functions formalism (Mukamel, 1995).

Unfortunately, as discussed in literature, the photon echo intensity measurements as a function of the delays

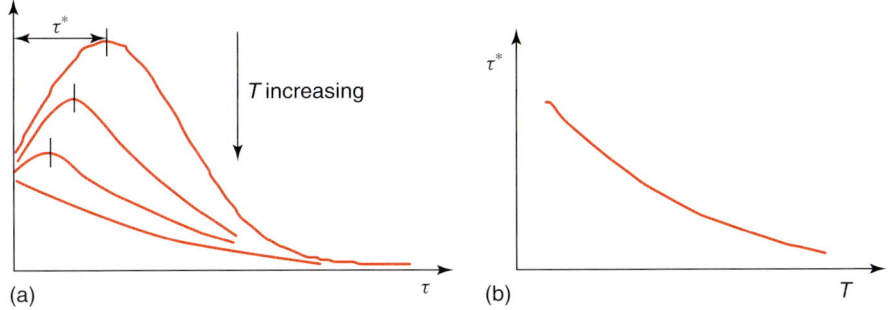

Fig. 21.10 (a) Drawing representing the evolution of the photon echo signal as the population time, T, is increased. (b) Plotting the photon echo peak against T results in the echo peak shift measurement. (Adapted from Andrei Tokmakoff, 5.74 Introductory Quantum Mechanics II, Spring 2007. (MIT OpenCourseWare: Massachusetts Institute of Technology) http://ocw.mit.edu/OcwWeb/Chemistry/5-74Spring-2007/CourseHome. Accessed Oct. 1, 2008.)

t'_1 and t'_2 provide no direct information on the system-bath interaction dynamics, unless there is a clear and strict separation of timescales as, for example, in the Bloch limit (Bloch, 1946; Nibbering et al., 1994). Usually when working with solutions, the dynamics cannot be classified simply as homogeneous or inhomogeneous. Recently for a probe parameter that could follow the dynamics of the frequency correlation function was sought. An apparently excellent candidate is the measurement of the echo peak shift (Desilvestri et al., 1984; Weiner et al., 1985). With the advent of amplified laser systems and optical parametric oscillator/amplifiers, which in conjunction are able to deliver pulses shorter than 20 fs in a broad wavelength region, the subject was revived and the echo peak shift technique has attracted new interest. In addition, very recently, it has been shown that there is a direct connection between the echo peak shift measurement and the amplitude of the frequency correlation function fluctuations due to the system-bath interaction (Fleming and Cho, 1996). In the three pulse echo peak shift (3 PEPS) experiment, the integrated photon echo intensity is measured for a fixed population delay time, T, while scanning the coherence delay time, τ. The shift of the photon echo maximum with respect to zero delay along the τ axis constitutes the observable, the so-called echo peak shift. It was demonstrated that a plot of the echo peak shift as a function of waiting time T, (Figure 21.10) tracks the frequency correlation function, which carries information about the system-bath interaction (de Boeij et al., 1996). Such experiments represent a powerful tool for unraveling optical dynamics in complex condensed phase systems.

21.5 The Future

The technology and the theory of ultrafast spectroscopy have entered into a mature phase. The future now would appear to lie in establishing the range and power of these new methods. One particularly striking example currently under intense investigation is multidimensional ultrafast spectroscopy. In preceding sections, the analogy of optical and NMR measurements was noted. Extending this idea to the multipulse experiments described above suggests the possibility of a suite of ultrafast experiments to match those associated with multipulse NMR. Numerous researchers are investigating the possibility of exploiting two-dimensional experiments, characterized my multiple pulses and time delays (of which 3PEPS is one example) to determine molecular structural evolution on an ultrafast time scale. In many respects, this is the holy grail of molecular science – a method that demonstrates where the nuclei are during the course of a chemical reaction. Important progress is being made, particularly, in terms of two-dimensional vibrational spectroscopy (Demirdoven et al., 2004; Khalil et al., 2003; Wright, 2002; Zheng et al., 2007).

References

Bloch, F. (1946) *Phys. Rev.*, **70**, 460.
de Boeij, W.P., Pshenichnikov, M.S., and Wiersma, D.A. (1996) *Chem. Phys. Lett.*, **253**, 53.
de Boeij, W.P., Pshenichnikov, M.S., and Wiersma, D.A. (1998) *Annu. Rev. Phys. Chem.*, **49**, 99.
Boyd, R.W. (ed.) (2003) *Nonlinear Optics*, 2nd edn, Academic Press, San Diego.
Butcher, P.N. and Cotter, D. (eds) (1990) *The Elements of Nonlinear Optics*, 1st edn, Cambridge University Press, New York.
Castner, E.W., Wishart, J.F., and Shirota, H. (2007) *Acc. Chem. Res.*, **40**, 1217.
Cerullo, G. and De Silvestri, S. (2003) *Rev. Sci. Instrum.*, **74**, 1.
Cho, M.H., Fleming, G.R., and Mukamel, S. (1993) *J. Chem. Phys.*, **98**, 5314.
Cho, M.H., Scherer, N.F., Fleming, G.R., and Mukamel, S. (1992) *J. Chem. Phys.*, **96**, 5618.
Cong, P., Deuel, H.P., and Simon, J.D. (1995) *Chem. Phys. Lett.*, **240**, 72.
Demirdoven, N., Cheatum, C.M., Chung, H.S., Khalil, M., Knoester, J., and Tokmakoff, A. (2004) *J. Am. Chem. Soc.*, **126**, 7981.
Desilvestri, S., Weiner, A.M., Fujimoto, J.G., and Ippen, E.P. (1984) *Chem. Phys. Lett.*, **112**, 195.
Fleming, G.R. (1986) *Chemical Applications of Ultrafast Spectroscopy*, Oxford University Press, New York.
Fleming, G.R. and Cho, M.H. (1996) *Annu. Rev. Phys. Chem.*, **47**, 109.
Fork, R.L., Greene, B.I., and Shank, C.V. (1981) *Appl. Phys. Lett.*, **38**, 671.
Fourkas, J. (ed.) (2002) *Liquid Dynamics: Experiment, Simulation and Theory*, ACS Symposium, American Chemical Society.
Friedman, J.S. and She, C.Y. (1993) *J. Chem. Phys.*, **99**, 4960.
Heisler, I.A., Correia, R.R.B., Buckup, T., and Cunha, S.L.S. (2005) *J. Chem. Phys.*, **123**, 054509.
Hunt, N.T., Jaye, A.A., and Meech, S.R. (2007) *Phys. Chem. Chem. Phys.*, **9**, 2167.
Hunt, N.T. and Meech, S.R. (2003) *Chem. Phys. Lett.*, **378**, 195.
Hunt, N.T. and Meech, S.R. (2004) *Chem. Phys. Lett.*, **400**, 368.
Jean-Claude, D. and Rudolph, W. (eds) (1996) *Ultrashort Laser Pulse Phenomena*, 4th edn, Academic Press, San Diego.
Joo, T. and Rhee, H. (2005) *Opt. Lett.*, **30**, 96.
Kang, T.J., Jarzeba, W., Barbara, P.F., and Fonseca, T. (1990) *Chem. Phys.*, **149**, 81.
Kano, S. and Levenson, M.D. (1988) *Introduction to Nonlinear Laser Spectroscopy*, Oxford University Press, New York.
Kaufmann, K.J., Dutton, P.L., Netzel, T.L., Leigh, J.S., and Rentzepis, P.M. (1975) *Science*, **188**, 1301.
Khalil, M., Demirdoven, N., and Tokmakoff, A. (2003) *J. Phys. Chem. A*, **107**, 5258.
Kurnit, N.A., Abella, I.D., and Hartmann, S.R. (1964) *Phys. Rev. Lett.*, **13**, 567.
Lakowicz, J.R. (ed.) (2006) *Principles of Fluorescence Spectroscopy*, 3rd edn, Springer-Verlag, New York.
Lotshaw, W.T., McMorrow, D., Thantu, N., Melinger, J.S., and Kitchenham, R. (1995) *J. Raman Spectrosc.*, **26**, 571.

Loughnane, B.J., Scodinu, A., Farrer, R.A., Fourkas, J.T., and Mohanty, U. (1999) *J. Chem. Phys.*, **111**, 2686.

Maroncelli, M. and Fleming, G.R. (1987) *J. Chem. Phys.*, **86**, 6221.

McMorrow, D., Lotshaw, W.T., and Kenney-Wallace, G.A. (1988) *IEEE J. Quantum Electron.*, **24**, 443.

van der Meer, M.J., Zhang, H., and Glasbeek, M. (2000) *J. Chem. Phys.*, **112**, 2878.

Mukamel, S. (ed.) (1995) *Principles of Nonlinear Optical Spectroscopy*, 1st edn, Oxford University Press, Mineola, New York.

Nibbering, E.T.J., Wiersma, D.A., and Duppen, K. (1994) *Chem. Phys.*, **183**, 167.

Perry, M.D. and Mourou, G. (1994) *Science*, **264**, 917.

Phillips, D. and O'Connor, D.V. (1983) *Time-Correlated Single Photon Counting*, Academic Press, London.

Porter, G. and Wilkinson, F. (1961) *Trans. Faraday Soc.*, **57**, 1686.

Rulliere, C. (ed.) (1998) *Femtosecond Laser Pulses*, 1st edn, Springer-Verlag, Berlin.

Shah, J. (1988) *IEEE J. Quantum Electron.*, **24**, 276.

Shen, Y.R. (ed.) (1984) *The Principles of Nonlinear Optics*, 4th edn, John Wiley & Sons, Inc., New York.

Smith, N.A. and Meech, S.R. (2002) *Int. Rev. Phys. Chem.*, **21**, 75.

Spence, D.E., Kean, P.N., and Sibbett, W. (1991) *Opt. Lett.*, **16**, 42.

Svelto, O. (1998) *Principles of Lasers*, Plenum Press, New York.

Todd, D.C., Jean, J.M., Rosenthal, S.J., Ruggiero, A.J., Yang, D., and Fleming, G.R. (1990) *J. Chem. Phys.*, **93**, 8658.

Towrie, M., Grills, D.C., Dyer, J., Weinstein, J.A., Matousek, P., Barton, R., Bailey, P.D., Subramaniam, N., Kwok, W.M., Ma, C.S., Phillips, D., Parker, A.W., and George, M.W. (2003) *Appl. Spectrosc.*, **57**, 367.

Trebino, R., DeLong, K.W., Fittinghoff, D.N., Sweetser, J.N., Krumbugel, M.A., Richman, B.A., and Kane, D.J. (1997) *Rev. Sci. Instrum.*, **68**, 3277.

Waldeck, D., Cross, A.J., McDonald, D.B., and Fleming, G.R. (1981) *J. Chem. Phys.*, **74**, 3381.

Weiner, A.M., Desilvestri, S., and Ippen, E.P. (1985) *J. Opt. Soc. Am. B*, **2**, 654.

Wilhelm, T., Piel, J., and Riedle, E. (1997) *Opt. Lett.*, **22**, 1494.

Wright, J.C. (2002) *Int. Rev. Phys. Chem.*, **21**, 185.

Zheng, J., Kwak, K., and Fayer, M.D. (2007) *Acc. Chem. Res.*, **40**, 75.

22
Ultrafast Spectroscopy

Oliver Kühn and Stefan Lochbrunner

22.1	**Introduction** 771	
22.2	**Principles and Theoretical Concepts** 772	
22.2.1	Potential Energy Surfaces 772	
22.2.2	Wave Packets 774	
22.2.3	Theoretical Nonlinear Spectroscopy 776	
22.2.3.1	Response Function Formalism 776	
22.2.3.2	Nonperturbative Methods 778	
22.3	**Experimental Methods** 778	
22.3.1	Generation of Ultrashort Pulses 778	
22.3.2	Probing with Linear Techniques 781	
22.3.3	Probing Coherences 783	
22.3.4	Pulse Shaping 784	
22.4	**Applications** 786	
22.4.1	Wave-Packet Dynamics in Small Molecules 786	
22.4.1.1	Vibrational Wave Packets 786	
22.4.1.2	Reactive Wave-Packet Dynamics 786	
22.4.1.3	Influence of the Environment 787	
22.4.2	Dynamics of Polyatomic Molecules 788	
22.4.2.1	Vibrational Energy Redistribution and Cooling 788	
22.4.2.2	Solvation 788	
22.4.2.3	Electronic Dynamics: Internal Conversion, Conical Intersections, and Photochemistry 789	
22.4.2.4	Coherent Wave-Packet Dynamics 790	
22.4.2.5	Photoinduced Electron Transfer 792	
22.4.3	Clusters and Liquids 793	
22.4.3.1	Atomic Clusters 793	
22.4.3.2	Water and Hydrogen-Bonded Networks 795	
22.4.4	Biological Systems 795	
22.4.4.1	Primary Steps of Photosynthesis 796	
22.4.4.2	Rhodopsins 796	
22.4.4.3	Peptide Dynamics 798	

Encyclopedia of Applied Spectroscopy. Edited by David L. Andrews.
Copyright © 2009 WILEY-VCH Verlag GmbH & Co. KGaA, Weinheim
ISBN: 978-3-527-40773-6

22.4.4.4	DNA	799
22.4.5	Solid State and Surfaces	801
22.4.5.1	Semiconductors	801
22.4.5.2	Polymers and Organic Materials	802
22.4.5.3	Surfaces	804
22.4.6	Laser Control	805
22.4.7	New Frontiers	807
22.4.7.1	Generation of High Harmonics and Attosecond Pulses	807
22.4.7.2	Application of High Harmonics and Attosecond Pulses	808
22.4.7.3	Carrier Envelope Phase and Frequency Comb	809
22.4.7.4	Time-Resolved X-Ray and Electron Diffraction	810
	Acknowledgments	811
	Glossary	811
	References	811
	Further Reading	815

22.1
Introduction

Ultrafast spectroscopy denotes the investigation of very fast processes with electromagnetic radiation at an extremely high time resolution. It covers the pico- and femtosecond timescale and has meanwhile reached the attosecond regime. The high impact of the field results from the fact that changes in all common forms of matter are related to the motion and rearrangement of nuclei and electrons. If we assume that an atom with a mass of 10 amu has a kinetic energy of 1 eV, then the atom needs about 2×10^{-14} seconds = 20 femtoseconds to transverse a distance of 1 Å. This shows that the femtosecond timescale is the natural scale of atomic motions and elementary processes in matter. This is also reflected by the time periods of vibrations in molecules and of phonons in condensed systems, which are in the range of 10–1000 femtoseconds.

The pump-probe technique is the most direct realization of ultrafast spectroscopy. An ultrashort light pulse, the pump pulse, excites the system under investigation and starts the process of interest (see Figure 22.2). A second ultrashort pulse, the probe pulse, arrives with a certain delay at the sample and makes a snapshot of the current stage of the process. To characterize the complete evolution of the process, the experiment has to be repeated with varying delay times. Using ultrashort light pulses instead of fast detectors and electronics results in an about 10^5 higher time resolution. To understand the observed dynamics, a description of an intrinsic nonstationary situation has to be given, usually on a quantum mechanical level, at least for a part of the system. The key elements that have to be considered are couplings between subsystems and superpositions of eigenstates and their evolution.

The development of this field started about 40 years ago when picosecond laser pulses became available. They were immediately applied to study photoinduced molecular processes, energy redistribution between vibrations, and vibrational coherences (Eisenthal, 1975; Laubereau and Kaiser, 1978; Zewail, 2000). The development of mode-locked dye lasers and in particular of the much more reliable Ti : sapphire lasers resulted in pulses with durations down to a few femtoseconds. As these systems were available, people started

Encyclopedia of Applied Spectroscopy. Edited by David L. Andrews.
Copyright © 2009 WILEY-VCH Verlag GmbH & Co. KGaA, Weinheim
ISBN: 978-3-527-40773-6

to investigate basic photochemical processes on the subpicosecond timescale like dissociation of small molecules (Rosker, Dantus and Zewail, 1988; Rose, Rosker and Zewail, 1989; Zewail, 2000), electronic relaxation (Elsaesser and Kaiser, 1991), electron (Barbara, Walker and Smith, 1992) and proton transfer (Elsaesser and Kaiser, 1986) as well as isomerization reactions (Åberg et al., 1994). These investigations promoted the understanding of molecular processes to a completely new level and were recognized by the award of the Nobel Prize in Chemistry to Ahmed Zewail in 1999 (Zewail, 2000). Owing to the development of new nonlinear conversion schemes (Riedle et al., 2000) and the refinement of detection techniques, it is now possible to investigate almost every system with a time resolution down to 20 femtoseconds or even less and research dealing with such different topics like isolated molecules, solid-state physics, or biological systems make extensive use of ultrafast spectroscopy. Accordingly, almost every technique known from steady-state spectroscopy has meanwhile been transferred into the time domain. Broadband pulses provide, in addition, the unique possibility to construct very complex pulse shapes by manipulating the amplitude and phase of the frequency components in Fourier space. Using such appropriately designed electromagnetic fields for excitation allows to influence and control processes (Brixner and Gerber, 2003). By making attosecond pulses available, the time resolution was recently again dramatically improved (Corkum and Krausz, 2007). Those pulses are generated by recolliding electrons after an ultrashort photoionization event and the shortest pulse length reported so far is only 0.13 femtoseconds. Another current development is the implementation of femtosecond X-ray (Bargheer et al., 2006) and electron diffraction techniques (Siwick et al., 2003), which can measure the change of structures and might open the way to make a direct movie of ultrafast processes with subatomic spatial resolution.

Owing to the vastness of the field, it is not possible to give a comprehensive overview. We only describe a few selective examples to give an impression of the possibilities that the technique provides as well as the concepts that are important for theoretical modeling. Accordingly, we also restrict ourselves to weak field phenomena and have to exclude the fascinating area of high-intensity physics. The remainder of the article is organized as follows. Section 22.2 introduces the basic theoretical concepts used to describe ultrafast processes and experiments. In Section 22.3, some of the essential experimental tools and methods are sketched. In Section 22.4, different applications are described with a focus on molecular processes.

22.2
Principles and Theoretical Concepts

22.2.1
Potential Energy Surfaces

The quantum dynamics of a molecular system characterized by electronic coordinates **r** and nuclear coordinates **R** is determined by the time-dependent Schrödinger equation

$$i\hbar \frac{\partial}{\partial t} \Psi(\mathbf{r}, \mathbf{R}; t) = \hat{H}(t) \Psi(\mathbf{r}, \mathbf{R}; t) \quad (22.1)$$

where the Hamilton operator consists of a molecular part and an interaction term accounting for the coupling to the radiation field, $\mathbf{E}(t)$, here taken in dipole approximation (**d**: dipole moment vector) :

$$\hat{H}(t) = \hat{H}_{\text{mol}} + \hat{V}_{\text{mol-field}}(t)$$
$$= \hat{T} + \hat{H}_{\text{el}} - \hat{\mathbf{d}} \cdot \mathbf{E}(t) \qquad (22.2)$$

Here, we introduced the nuclear kinetic energy, $\hat{T}$, and the electronic Hamiltonian, $\hat{H}_{\text{el}}$, which also contains the nuclear–nuclear repulsion. Using the Born–Oppenheimer separation of electronic and nuclear motions, the molecular wave function is written as

$$\Psi(\mathbf{r}, \mathbf{R}; t) = \sum_a \varphi_a(\mathbf{r}, \mathbf{R}) \psi_a(\mathbf{R}; t)$$
$$(22.3)$$

where we introduced the adiabatic electronic states $\varphi_a(\mathbf{r}, \mathbf{R})$ and the nuclear wave packet on a given electronic state $\psi_a(\mathbf{R}; t)$. The latter is obtained from the set of coupled equations (May and Kühn, 2004)

$$i\hbar \frac{\partial}{\partial t} \psi_a(\mathbf{R}; t)$$
$$= \left(\hat{T} + \hat{V}_a(\mathbf{R}) - \hat{\mathbf{d}}_a(\mathbf{R}) \cdot \mathbf{E}(t) \right) \psi_a(\mathbf{R}; t)$$
$$+ \sum_{b \neq a} \left(\hat{T}_{ab} - \hat{\mathbf{d}}_{ab}(\mathbf{R}) \cdot \mathbf{E}(t) \right)$$
$$\times \psi_b(\mathbf{R}; t) \qquad (22.4)$$

The first term describes the nuclear motion on the so-called potential energy hypersurface (PES) $V_a(\mathbf{R})$. This motion can be triggered, for example, by the coupling of an infrared (IR) laser field to the coordinate-dependent dipole operator of the considered electronic state. The laser field also triggers transitions between different electronic states via the transition dipole operator whose dependence on the nuclear coordinates is often neglected (Condon approximation), that is, $\hat{\mathbf{d}}_{ab}(\mathbf{R}) \approx \mathbf{d}_{ab}$. Transitions between adiabatic electronic states can also be caused by nonvanishing matrix elements of the nuclear kinetic energy. Such nonadiabatic transitions can become particularly important in situations where PESs of different electronic states come energetically close to each other. Nonadiabatic processes are at the heart of photochemistry. They cause internal conversion (IC) transforming electronic energy into vibrational energy, which comes along with a considerable shortening of the excited state lifetimes. Their description still provides a challenge to theory (Domcke, Yarkony and Köppel, 2004).

If the electronic states are well separated from each other, the transition probability follows the energy gap law, which results from the dependence of the Franck–Condon factors on the change of the vibrational quanta. On the other hand, nonadiabatic processes can be extremely efficient if a conical intersection is involved (Domcke and Stock, 1997; Domcke, Yarkony and Köppel, 2004). In this case, two electronic states are degenerate along a hyperline in the multidimensional space spanned by the nuclear coordinates. Of particular interest is the dynamics in the branching space, which is spanned by the coupling mode and the tuning mode. The nonadiabatic electronic coupling varies along the former and the energy gap between the corresponding so called diabatic states varies along the latter. This branching space is orthogonal to the hyperline of the intersection, while the shape of the adiabatic PESs in this space looks like a double cone (Figure 22.1). A conical intersection can be regarded as the extension of an avoided crossing into

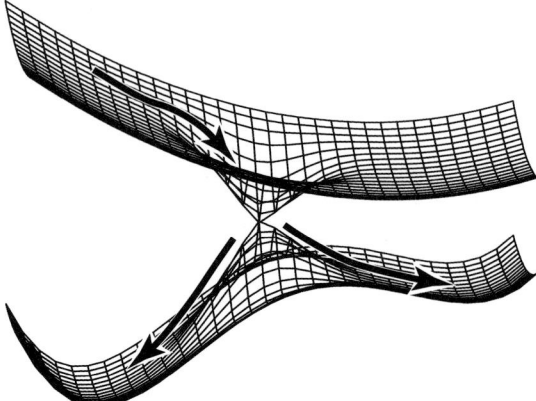

Fig. 22.1 Two PESs connected by a conical intersection. The relaxation pathway splits at the intersection and two minima corresponding to two different species are populated.

two dimensions. Owing to the double cone topology, a branching of the relaxation pathway at the intersection is quite likely. In this case, a part of the population usually reaches a minimum of the ground-state PES that differs from the original one and that is associated with a photoproduct. The rest of the population returns to the original minimum and completes the IC path. The speed of the process is given by the accessibility of the conical intersection, which depends on its energetic location and on whether the evolution of the excited-state population involves also coordinates that lead to the intersection. The quantum yield of a photoreaction depends on the dynamics at the intersection itself and how the wave packet propagates on the lower part of the double cone.

22.2.2
Wave Packets

As a consequence of the interaction with the laser field, the nuclear wave function will no longer describe a stationary molecular eigenstate on the considered PES, $\chi_{a,\xi}(\mathbf{R})$. Instead one has a coherent superposition (with coefficients $c_{a,\xi}(t)$) of such eigenstates called a *molecular wave packet*:

$$\psi_a(\mathbf{R};t) = \sum_\xi c_{a,\xi}(t)\, \chi_{a,\xi}(\mathbf{R}) \quad (22.5)$$

Wave packets have been known since the early days of quantum mechanics. However, their observation in real time had to await the development of ultrafast spectroscopies during the last decades. In ultrafast spectroscopy, the molecule interacts with a sequence of N laser fields, that is,

$$\mathbf{E}(\mathbf{r};t) = \sum_{p=1}^{N} \mathbf{e}_p E_p(t - t_p) e^{-i(\omega_p t - \mathbf{k}_p \cdot \mathbf{r})} + \text{c.c.} \quad (22.6)$$

Here, $\mathbf{e}_p$, ω_p, and $\mathbf{k}_p$ are polarization vector, carrier frequency, and wave vector, respectively, of the pth field. The pulse envelope, $E_p(t - t_p)$, is usually assumed to be of Gaussian shape centered at t_p.

In pump-probe spectroscopy, for instance, a first pump-pulse may excite a vibrational wave packet from the electronic ground to some excited

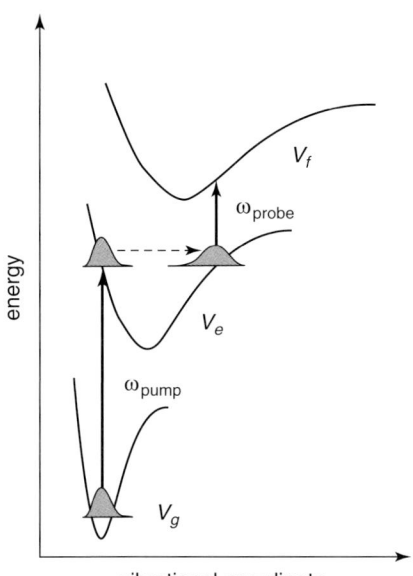

Fig. 22.2 Pump-probe spectroscopy of wave-packet dynamics along a vibrational coordinate. The pump-pulse with frequency ω_{pump} promotes the electronic ground-state (V_g) wave packet to the excited-state PES V_e, where it is probed by a time-delayed pulse with frequency ω_{probe} via a transition to a higher excited-state PES V_f.

state (see Figure 22.2). Owing to the different electronic structure, the nuclei are displaced from equilibrium and the wave packet no longer corresponds to a stationary state and propagates, that is, the bond is elongated. The ongoing dynamics can be interrogated by a time-delayed second pulse probing, for example, the transition to another electronic state from where fluorescence can be observed. The signal will reflect the oscillatory motion and dispersion of the wave packet in the intermediate state as well as concurring processes like photodissociation or nonadiabatic transitions to other electronic states.

Wave-packet dynamics as a modern research area has been initiated by Jortner and Berry's prediction of molecular quantum beats as a means for characterizing intramolecular radiationless transitions due to the breakdown of the Born–Oppenheimer approximation (Jortner and Berry, 1968). The development of numerical methods such as the split-operator approach and the utilization of fast-Fourier transforms for the calculation of the action of the kinetic energy operator onto the wave packet allowed for the efficient propagation of wave packets on general potentials in the early 1980s (Kosloff and Kosloff, 1982; Feit, Fleck and Steiger, 1982) marks a theoretical breakthrough, which enabled simulation of reactive processes at least for one-dimensional situations.

Straightforward extensions of these propagation methods face the limit of exponential scaling with the number of coupled degrees of freedom. This restriction is somewhat relaxed in the multiconfiguration time-dependent Hartree (MCTDH) method whose development has been a milestone toward the simulation of ultrafast multidimensional wave-packet dynamics (Meyer, Manthe and Cederbaum, 1990). Here, the wave packet for a *D*-dimensional system is expanded as follows:

$$\psi(\mathbf{R};t) = \sum_{j_1..j_D} A_{j_1..j_D}(t)\, \chi_{j_1}^{(1)}(R_1;t)$$

$$\times \chi_{j_2}^{(2)}(R_2;t) \ldots \chi_{j_D}^{(D)}(R_D;t) \quad (22.7)$$

The use of a time-dependent basis set facilitates a flexible representation of the moving wave packet such that a considerably smaller number of basis functions is required as compared to the case of a fixed basis. Using this approach, the quantum dynamics of some tens of coordinates can be treated numerically exact; for special systems some thousand degrees of freedom are possible. Still, the propagation of the wave packet requires the knowledge of the complete PES including possible nonadiabatic couplings. Since its determination for complex systems is not feasible, methods that are based on classical trajectories with electronic structure theory-based forces determined on the fly have been developed. Quantum effects such as nonadiabatic transitions can be built into these simulations using the surface hopping method or semiclassical approaches (Stock and Thoss, 2005).

Despite the recent success of these all-atom simulations of complex systems at finite temperature the dynamics are more rigorously described by statistical methods such as the Liouville–von Neumann equation for the density operator $\hat{\rho}$. This facilitates, for instance, a reduced description of a relevant subsystem in interaction with a heat bath (May and Kühn, 2004). As a consequence of this, interaction energy and phase relaxation occurs, captured by the diagonal and off-diagonal elements of the reduced density matrix expressed in terms of the states of the relevant system {a}. The related damped wave-packet dynamics can be described in the simplest case of a Bloch model by

$$\frac{\partial \rho_{ab}}{\partial t} = -i(\omega_{ab} - i\gamma_{ab})\rho_{ab}$$

$$- \delta_{ab} \sum_c \Gamma_{ac}\rho_{cc}$$

$$+ i\mathbf{E}(t) \cdot \sum_c (\mathbf{d}_{ac}\rho_{cb} - \rho_{ac}\mathbf{d}_{cb})$$

(22.8)

Here, ω_{ab} denotes the frequencies of transitions that are subject to dephasing with the rate γ_{ab}. The second term on the right-hand side describes the population redistribution with rates Γ_{ab} and the final term is due to the interaction with the laser field.

22.2.3
Theoretical Nonlinear Spectroscopy

22.2.3.1 Response Function Formalism

The different nonlinear spectroscopic techniques depend on the time delay between the pulses as well as on the phase-matching direction $\mathbf{k}_s$ for the detection of the macroscopic polarization field, which is defined as the expectation value of the dipole operator (n_{mol}: volume density of molecules in sample):

$$\mathbf{P}(\mathbf{r};t) = n_{\text{mol}}\langle \hat{\mathbf{d}}(\mathbf{r};t)\rangle \quad (22.9)$$

The response function formalism is based on an expansion of the polarization in powers of the electric field strengths ("tr" denotes the trace operation)

$$\mathbf{P}(\mathbf{r};t) = \sum_n \mathbf{P}^{(n)}(\mathbf{r};t)$$

$$= n_{\text{mol}} \sum_n \text{tr}\left[\mathbf{d}\hat{\rho}^{(n)}(\mathbf{r};t)\right] \quad (22.10)$$

Thus the nth order polarization is determined by a perturbation expansion of the density operator with respect to $\hat{V}_{\text{mol-field}}(t)$. This leads to

$$\mathbf{P}^{(n)}(\mathbf{r};t) = n_{\text{mol}} \int_0^\infty dt_n \ldots \int_0^\infty dt_1$$
$$\times S^{(n)}(t_n, \ldots, t_1) \mathbf{E}(\mathbf{r}; t - t_n)$$
$$\times \ldots \times \mathbf{E}(\mathbf{r}; t - t_n - \ldots - t_1) \quad (22.11)$$

where the material properties are completely described by the response functions $S^{(n)}$, which are given by nth order correlation functions of the molecular dipole operator (Mukamel, 1995). For instance, the first-order response function that contains the information about linear spectroscopy is given by

$$S^{(1)}(t_1) = \frac{i}{\hbar} \theta(t_1) \, \text{tr} \left([\hat{\mathbf{d}}(t_1), \hat{\mathbf{d}}(0)] \hat{\rho}_{\text{eq}} \right) \quad (22.12)$$

where $[\hat{A}, \hat{B}]$ denotes the commutator and $\hat{\rho}_{\text{eq}}$ is the equilibrium density operator. For the case of an electronic two-level system coupled to intra- and intermolecular degrees of freedom Eq. (2.12) becomes (ω_{eg} is the transition frequency between states g and e, d_{eg} the transition dipole matrix element, and the expectation value is taken with respect to the ground state density)

$$S^{(1)}(t_1) = \frac{i}{\hbar} \theta(t_1) |d_{\text{eg}}|^2$$
$$\times \left(\exp\left[-i\langle\omega_{\text{eg}}\rangle t_1 - g(t_1)\right] - \text{c.c.} \right) \quad (22.13)$$

with the line shape function

$$g(t) = \frac{1}{\hbar^2} \int_0^t dt' \int_0^{t'} dt'' \langle \delta\omega_{\text{eg}}(t'') \delta\omega_{\text{eg}}(0) \rangle \quad (22.14)$$

Here, it has been assumed that the fluctuation of the transition frequency due to the interaction of the two-level system with its surroundings, $\omega_{\text{eg}}(t) = \langle\omega_{\text{eg}}\rangle + \delta\omega_{\text{eg}}(t)$, obey Gaussian statistics. In this formulation, the line shape function is the central quantity. It also enters the nonlinear response functions and is the starting point for simulations. Apart from models such as the multimode Brownian oscillator (Mukamel, 1995), the correlation functions can also be obtained from classical molecular dynamics simulations.

Comparing Eqs (22.6) and (22.11), we notice that the polarization can be written as

$$\mathbf{P}^{(n)}(\mathbf{r},t) = \sum_L \mathbf{P}_L^{(n)}(t) e^{-i(\Omega_L t - \mathbf{K}_L \cdot \mathbf{r})}$$
$$(22.15)$$

Here, $L = \{l_1, l_2, \ldots, l_n\}$ comprises the numbers indicating possible combinations of fields for a given order n, $\Omega_L = \sum_p l_p \omega_p$, and $\mathbf{K}_L = \sum_p l_p \mathbf{k}_p$. In principle, a self-consistent solution of Schrödinger and Maxwell equations is required since the polarization enters as a source term into the latter. However, for optically thin samples, changes of incident fields in the medium can be neglected. Expanding the signal field into a Fourier series analogous to Eq. (22.15), one finds that the Fourier component along a certain direction $\mathbf{k}_s$ is proportional to the respective polarization term in Eq. (22.15) with wave vector selectivity introduced by the phase-matching condition, $\mathbf{k}_s \approx \mathbf{K}_L$. As an example, consider a third-order pump-probe experiment with $\mathbf{k}_1 = \mathbf{k}_2 = \mathbf{k}_{\text{pump}}$ and $\mathbf{k}_3 = \mathbf{k}_{\text{probe}}$. If the signal is detected in phase-matched probe direction, one has only terms with $L = \{\pm 1, \mp 1, 1\}$. The transient absorption signal can be calculated in slowly varying envelope approximation via

$$S(\omega_{\text{probe}}) = \frac{4\pi \omega_{\text{probe}}}{cn(\omega_{\text{probe}})}$$
$$\times \frac{\text{Im} \int dt\, E^*_{\text{probe}}(t) P^{(3)}_L(t)}{\int dt\, |E_{\text{probe}}(t)|^2}$$
(22.16)

where n is the index of refraction and we have skipped the vector character (Mukamel, 1995).

22.2.3.2 Nonperturbative Methods

In more complex situations that neither can be mapped onto the Brownian oscillator Hamiltonian nor can a diagonalization be performed, calculation of the multitime response functions becomes too tedious and a direct propagation approach is used (Domcke and Stock, 1997). This has also the advantage that it is, in principle, nonperturbative, that is, strong field effects can be considered. The idea can be illustrated for a pump-probe experiment with two incident pulses. To arbitrary order in the two fields, the polarization can be written as

$$\mathbf{P}(\mathbf{r}, t) = \sum_{n_1, n_2 = -\infty}^{+\infty} \mathbf{P}_{n_1, n_2}(t) e^{i(n_1 \mathbf{k}_{\text{pump}} + n_2 \mathbf{k}_{\text{probe}}) \cdot \mathbf{r}}$$
(22.17)

Defining the phases $\mathbf{k}_{\text{pump}} \mathbf{r} = \varphi_{\text{pump}}$ and $\mathbf{k}_{\text{probe}} \mathbf{r} = \varphi_{\text{probe}}$, the Schrödinger equation is solved using the time-dependent Hamiltonian where the wave vector dependence in the fields (22.6) is replaced by these phases. Calculating the polarization from Eq. (22.9), a certain component of the polarization can be extracted by inverting Eq. (22.17). For instance, $\mathbf{P}_{0,1}(t)$, contains the polarization component for the $\mathbf{k}_{\text{probe}}$ direction in any order with respect to the fields.

In combination with efficient propagation methods such as MCTDH, this direct approach is an important step toward first-principle quantum simulations of ultrafast spectroscopy (see, e.g., study of ultrafast electron transfer (Wang and Thoss, 2008)).

22.3
Experimental Methods

Figure 22.3 shows a typical layout of an ultrafast spectroscopic experiment. A mode locking oscillator generates ultrashort laser pulses that are subsequently amplified and transferred to other spectral regions by nonlinear optical processes. At some point of the setup, the laser beam is split into two arms to obtain a pump and a probe pulse, which are perfectly synchronized to each other. The pump pulse excites the sample while the interaction of the probe pulse is used to measure pump-induced coherences and population changes in the sample. Typically, a motorized delay line is incorporated in the path of the pump beam. By changing the pump delay, the time between the arrival of the pump and the probe pulses at the sample is systematically varied. The temporal evolution of a system after photo excitation is then reconstructed from the single point measurements taken at different delays. In the following, we shortly describe the components and techniques applied in such experiments.

22.3.1
Generation of Ultrashort Pulses

Ti : sapphire and fiber oscillators are commonly used to generate ultrashort

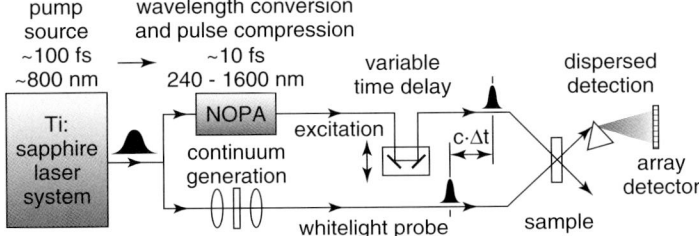

Fig. 22.3 Scheme of an ultrafast pump-probe experiment.

laser pulses at high repetition rates. Their output is directly used in applications that need only low energy pulses in a restricted wavelength range. Their layouts are shown in Figure 22.4. In these resonators, locking of longitudinal cavity modes and the formation of pulses occur since the pulses experience a higher gain or lower losses compared to less intense continuous-wave radiation. In the case of the Ti : sapphire oscillator, the intensity-dependent Kerr lens in the sapphire crystal causes for pulses a better spatial overlap with the tightly focused pump beam resulting in a higher gain (Spence, Kean and Sibbett, 1991). In the fiber oscillator, intensity-dependent birefringence of the fiber leads, in combination with appropriately aligned polarization tuning elements for pulses, to a higher transmission through the polarizer insulator (Tamura et al., 1993). To generate short pulses, the total group delay dispersion of the resonator must be set to zero. In the Ti : sapphire oscillator, a prism sequence or chirped mirrors compensate for the dispersion of the other elements. The output pulses have a center wavelength of about 800 nm, energies of several nanojoules and can be as short as a few femtoseconds. In the fiber oscillator, dispersion compensation results from the combination of sections of a standard fiber that exhibits negative dispersion with an Erbium-doped fiber, which is the gain medium and has positive dispersion. The output is in the spectral region around 1550 nm and is usually frequency doubled if it serves as seed of an amplifier system.

To achieve pulse energies in the millijoules region, the oscillator output is amplified, typically with a regenerative amplifier operating at a repetition rate of 1 kHz (see Figure 22.5) (Backus et al., 1998). Using a thin film polarizer (TFP), the seed pulse is coupled into the resonator of the amplifier. If a suitable voltage step is applied to the Pockels cell, the polarization of the pulse changes and it is captured in the resonator where it makes several round trips. A Ti : sapphire crystal pumped by a pulsed and frequency-doubled YLF or YAG laser serves as gain medium. When the inversion of the crystal is depleted, the pulse is released from the resonator by a second voltage change at the Pockels cell. The seed pulses are stretched to about hundred picoseconds by a grating stretcher before they enter the resonator. Otherwise, the high intensities of the amplified pulses would lead to nonlinear distortions of the pulse shape and to damage of the optical elements within the resonator. After the amplifier, a grating compressor removes the chirp introduced by the stretcher and the

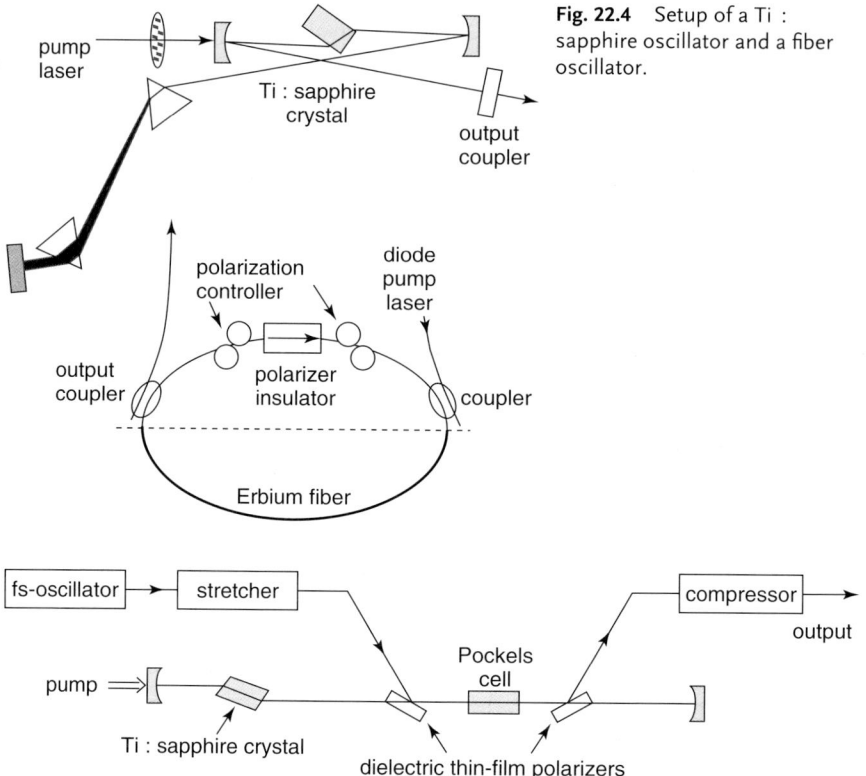

Fig. 22.4 Setup of a Ti : sapphire oscillator and a fiber oscillator.

Fig. 22.5 Scheme of a regenerative Ti : sapphire amplifier.

dispersion of the resonator resulting in pulse duration down to a few tens of femtoseconds.

Many different schemes have been successfully implemented to access other wavelength regions and to shorten the pulse duration further. Here, we restrict ourselves to an approach that is notably versatile and is being extensively applied in spectroscopic experiments. Its central element is the optical parametric amplification of a white light contiuum (Riedle et al., 2000; Cerullo and De Silvestri, 2003), see Figure 22.6. Pulses from a Ti : sapphire amplifier system with a pulse length in the order of 100 femtoseconds are typically used as input. A small fraction in the order of 2 μJ is split off and tightly focused into a sapphire substrate where self-focusing and self-phase modulation lead to the generation of a white light continuum. The remainder of the input beam is frequency doubled and serves as pump light for the parametric amplifier. The nonlinear medium of the amplifier is a $\beta - B_aB_2O_4$ (BBO) crystal cut for type I phase matching. The white light and the pump beam cross each other at a small angle of about $3°$ in the nonlinear crystal. Choosing the appropriate nonlinear geometry results in the remarkable situation where the phase-matching condition becomes, to first order, independent of the seed wavelength (Riedle et al., 2000). Then a very broad amplification bandwidth is

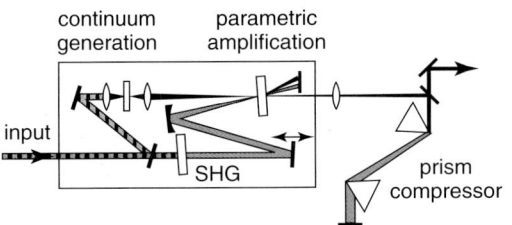

Fig. 22.6 Layout of a noncollinearly phase-matched optical parametric amplifier (NOPA) (Riedle et al., 2000).

possible even with a thick crystal and a high gain for broadband pulses and pulse energies up to several 10 µJ can be achieved.

The dispersion of the amplified pulses is controlled with a prism compressor resulting in pulse durations down to 10 femtoseconds (see Figure 22.6). Using more sophisticated compression schemes, pulse durations as short as 5 femtoseconds have been demonstrated (Cerullo and De Silvestri, 2003). Since the white light has some chirp, the output of the noncollinearly phase-matched optical parametric amplifier (NOPA) can be easily tuned over the whole visible region and even far into the near infrared (NIR) by slightly changing the path length of the pump beam and thereby shifting its temporal overlap with the white light. The NOPA pulses can be efficiently frequency doubled extending the accessible wavelength range also far into the ultraviolet (UV).

22.3.2
Probing with Linear Techniques

Meanwhile, many different techniques are available to probe the pump pulse-induced changes of a system and their evolution with time. In this paragraph, techniques that are linear in the energy of the probe pulse are considered. The most straightforward one is perhaps the transient absorption spectroscopy (see Figure 22.3). A probe pulse transverses the sample in the same region as the pump and its energy loss due to absorption within the sample is measured. Very often, a white light continuum generated in a sapphire or a CaF_2-substrate is used as probe since the white light covers the complete spectral region from the NIR to the near UV and only very low pulse energies are necessary for probing (Kovalenko et al., 1999). After the sample, the probe beam is dispersed by an imaging spectrograph and transmission changes over a broad spectral range are measured simultaneously with an array detector. An interesting feature of the technique is that absorption changes in the sample can be extracted in absolute numbers using Eq. (22.18).

$$\frac{Sig(\lambda)}{Sig_0(\lambda)} = \frac{T(\lambda)}{T_0(\lambda)} = e^{-n_{ex} \cdot (\sigma(\lambda) - \sigma_0(\lambda)) \cdot d}$$

(22.18)

$Sig(\lambda)$ is the dispersed probe signal measured with pump pulses hitting the sample and $Sig_0(\lambda)$ the signal when the pump beam is blocked. $T(\lambda)$ and $T_0(\lambda)$ stand for the corresponding sample transmissions. n_{ex} is the number density of excited particles, $\sigma(\lambda)$ is the time-dependent absorption cross section of the excited particle, $\sigma_0(\lambda)$ is its absorption cross section in the original state, and d is the sample thickness. The signal $S(\omega_{probe})$ introduced in Section 2.3.1 is directly linked to these quantities by Eq. (22.19) and the experimental results can be compared to simulations by means of Eq. (22.16).

$$S(\omega_{\text{probe}} = 2\pi c/\lambda)$$
$$= \frac{n_{\text{ex}}}{n_{\text{mol}}} \cdot (\sigma_0(\lambda) - \sigma(\lambda)) \qquad (22.19)$$

Here, n_{mol} is the number density of the particles.

In Figure 22.7, the transient spectrum after excitation of a perylene bisimide dye is compared to the steady-state absorption and fluorescence spectrum (Schlosser and Lochbrunner, 2006). The depicted change of the optical density $\Delta OD = -\log(T/T_0)$ clearly shows the three contributions to the signal: (i) the bleach of the electronic ground state due to the excitation, (ii) stimulated emission from the electronically excited state to the ground state, and (iii) the excited state absorption caused by electronic transitions from the excited state to higher lying states. The bands contributing to the transient absorption in the visible and UV reflect the electronic state of a system.

Changes in vibrational structure and population show up as variations in the band shapes. However, those can be more specifically measured using an infrared pulse as probe pulse (Elsaesser and Kaiser, 1991). Another possibility is to measure the Raman signal associated with the probe pulse (Laubereau and Kaiser, 1978) or to perform stimulated Raman measurements by applying an additional picosecond pulse for pumping the Raman transition and by recording the stimulated emission at the Raman frequencies with the femtosecond probe pulse (Kukura, McCamant and Mathies, 2007).

Usually, several contributions of the transient absorption overlap spectrally. Then it is useful to look at the time-dependent emission. In most cases, this is done by fluorescence up-conversion (Yoshihara, Tominaga and Nagasawa, 1995; Schanz et al., 2001).

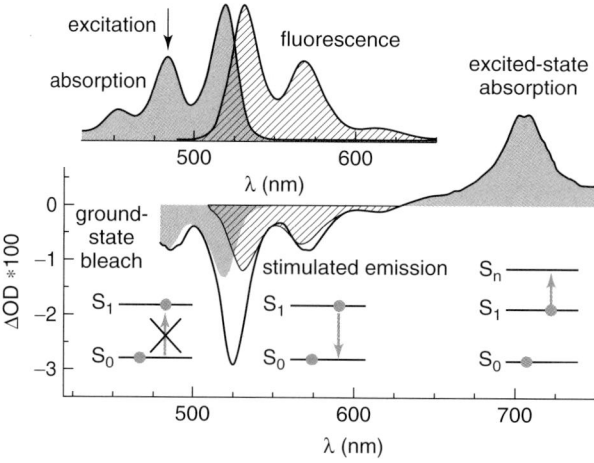

Fig. 22.7 Steady-state absorption and fluorescence spectrum of a perylene bisimide dye dissolved in chloroform and the transient absorption spectrum measured 5 picoseconds after optical excitation at 480 nm.

The spontaneous emission of the excited sample is imaged onto a nonlinear crystal. There, it is spatially overlaid with the probe pulsethat functions as a gate pulse in this case. The fraction of the emission that overlaps temporally with the gate pulse is subject to sum-frequency mixing and is transferred to shorter wavelengths. The strength and the spectrum of the sum frequency signal are measured with a spectrograph in dependence on the arrival time of the gate pulse at the crystal. The evolution of the emission spectrum is then directly reproduced by the time-dependent sum frequency signal.

For experiments in the gas phase or in a molecular beam, a very sensitive probe process is necessary due to the low sample densities. In this case, ionization techniques are used since ions and electrons can be detected with a very high efficiency (Stolow, Bragg and Neumark, 2004). One observable is the time-dependent mass spectrum, which reveals changes in the ionization probability and fragmentation pattern. Another one is the time-dependent photoelectron spectrum. It reflects the shape of the absorption bands associated with transitions from the populated electronic states to ionic states. Meanwhile powerful imaging and coincidence techniques were developed. They allow to assign each observed photoelectron to specific particles and to extract, besides the kinetic energy, the angular distribution, which is sensitive to the orbital from which the electron is emitted.

22.3.3
Probing Coherences

The probe schemes described above are not suitable to investigate electronic coherences. This is the domain of four-wave mixing (FWM) processes and interferometric techniques. In addition, nonlinear techniques of higher order have been applied. However, their discussion is beyond the scope of this article.

The most common realizations of ultrafast FWM spectroscopy are photon echo and two-dimensional spectroscopy. Three pulses are focused from somewhat different directions onto a common spot of the sample (see Figure 22.8) (Weiner, De Silvestri and Ippen, 1985). The first pulse induces a coherence between the ground and the excited state. The second pulse interacts with the coherence and creates a population in the excited-state the coherence or removes depending on the relative phase. A spatially modulated excited-state population results and constitutes a grating with the grating vector $\mathbf{k}_1 - \mathbf{k}_2$. The third pulse is diffracted by this grating and as signal a photon

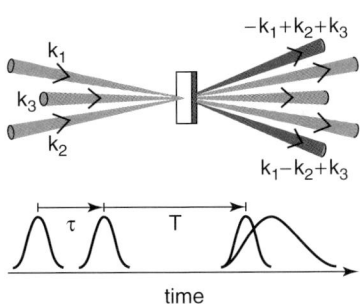

Fig. 22.8 Geometry and time diagram of a photon echo experiment.

echo with the wavevector $\mathbf{k}_1 - \mathbf{k}_2 + \mathbf{k}_3$ is generated. The time τ between the first and second pulse is often called *coherence time* since the signal strength depends on the coherence decay during this time interval. The time T between the second and the third pulse is named *population time* since during this interval the evolution of the populations is relevant. The role of the first and the second pulse can be exchanged and a second echo is emitted in the direction of the wavevector $-\mathbf{k}_1 + \mathbf{k}_2 + \mathbf{k}_3$. Its dependence on the coherence time is mirrored compared to the former echo signal. In transient grating experiments, the first two pulses are simultaneously applied, that is, the coherence time is set to zero and not varied, and only the dependence on the population time is measured. Photon echo studies focus on the coherence time and deal predominantly with dephasing phenomena.

In photon echo experiments, usually the three applied pulses are identical. However, the arrangement can also be used for coherent Raman scattering. In the anti-Stokes configuration, the center frequency of the second pulse is shifted to lower frequencies with respect to the first pulse (Schmitt et al., 1998). If the difference frequency fits to a vibrational mode, the vibrational resonance leads to a strongly increased diffracted signal at frequencies on the anti-Stokes side of the third pulse.

To extend the photon echo measurement to two-dimensional spectroscopy, a fourth laser pulse acting as local oscillator is interferometrically superimposed on the photon echo (Jonas, 2003). The resulting heterodyne signal is spectrally resolved and the coherence time τ is scanned keeping the population time T fixed. For each echo wavelength, the time-dependent signal is Fourier transformed. The result is a two-dimensional complex valued plot of the echo signal in dependence of the excitation and the signal frequencies. It shows which photon energies in the excitation contribute to the echo signal at a certain wavelength and if coherence transfer between excitation and echo generation occurs. The same approach using IR pulses is applied to investigate couplings between vibrational modes (Hamm, Helbing and Bredenbeck, 2008).

Wave-packet interference is another approach to look at the evolution of excitation-induced coherences. The sample is excited by two phase-locked collinear propagating pump pulses usually generated by means of a Michelson interferometer (Scherer et al., 1991). The first pulse generates a wave packet that should at least partially return to the Franck–Condon region when the second pulse hits the sample. Then the degree of coherence between excited and ground state and the relative phase with respect to the field of the second pulse determine whether constructive, destructive, or incoherent superposition occurs. The corresponding signature is the appearance of interference fringes in scans of the delay between the excitation pulses.

22.3.4
Pulse Shaping

Ultrashort laser pulses constitute a broad band of phase-locked frequency components. If one has access to each of these components and one can tune its amplitude and phase, then it is possible to

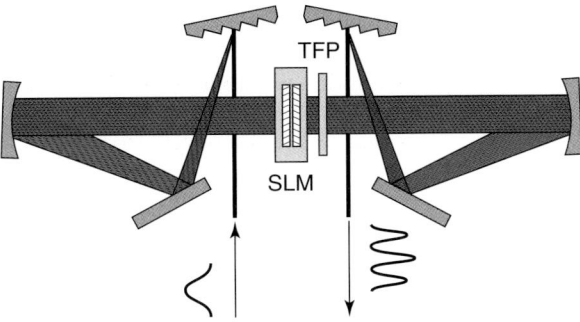

Fig. 22.9 Setup of a pulse shaper. SLM denotes a spatial light modulator containing two liquid crystal arrays and TFP a thin film polarizer.

generate completely arbitrary waveforms and pulse shapes. Only the carrier frequency would be similar and the shortest possible features are given by the original band width. Figure 22.9 shows a popular $4f$-setup, which allows manipulating with high resolution amplitude and phase of the spectral components of a pulse. (Weiner, 2000). The incoming pulse is spectrally dispersed by a grating. A cylindrical mirror or lens placed in a distance to the grating equal to its focal length f collimates the beam and then focuses each spectral component on its own spot. The focal plane represents now a Fourier plane of the pulse. A modulator, typically a liquid crystal mask, consisting of two layers is placed in this plane. The birefringence of each mask pixel is separately controlled via applied voltages and thereby the phase shift of the transmitted light and, in combination with a TFP, also its amplitude. The spectral components are recombined via a second cylindrical lens or mirror and a second grating. The pulse spectrum is now composed of independent tunable components that can be as many as there are modulator pixels.

Pulse-shaping techniques opened the way to new experimental approaches that use optimized laser pulses to drive and control photoinduced processes. Usually they are implemented by means of a feedback loop. Using a shaperpulses with different shapes are applied to excite the system under investigation. The results of the excitations are analyzed and compared to each other. The pulse shapes giving the best results are selected, in different ways modified, and again applied. The procedure is repeated until no more significant improvement is observed. Meanwhile, many light-driven processes have been controlled in this way (Brixner and Gerber, 2003). This includes nonlinear frequency conversion, high harmonic generation, molecular processes, and imaging. In case several pathways interfere with each other, coherences in the system are relevant for the result. If the phase properties of the light pulse influence the relative phases between the pathways, the resulting interference is very sensitive on the pulse and coherent control of the process can be achieved (Rabitz et al., 2000).

22.4 Applications

22.4.1 Wave-Packet Dynamics in Small Molecules

22.4.1.1 Vibrational Wave Packets

Small molecules in the gas phase provide good test cases for the concepts of nonstationary quantum mechanics. The time evolution of one-dimensional wave packets was, for example, studied in Na_2 (Baumert *et al.*, 1991). A 70-femtosecond-long laser pulse at 627 nm excites Na_2-molecules in a supersonic beam. A duplicate of the pulse is used to ionize the molecule via two-photon absorption at varying delay times (see Figure 22.10).

The Na_2^+-ion signal detected with a time of flight mass spectrometer shows a strongly oscillating time dependence, which is a clear signature of a wave-packet propagation in a bound potential. The spectrum of the excitation pulse overlaps with several vibrational levels of the $A^1\Sigma$ state and creates a coherent superposition of the vibronic states with the vibrational numbers $v' = 10-14$. The ionization probability is at the inner turning point of the $A^1\Sigma$ potential higher than at the one outer due to resonances with higher electronic states. Each recurrence of the wave packet at the inner turning point results in a transient rise of the Na_2^+ signal. The time delay between subsequent ionization maxima is 306 femtoseconds, which matches very well with the vibrational period that is given by the energetic separation of 109 cm^{-1} between the vibrational states. The long lifetime of the oscillations indicates that the wave packet broadens only slowly due to the quite moderate anharmonicity of the $A^1\Sigma$ potential at the involved vibrational levels.

22.4.1.2 Reactive Wave-Packet Dynamics

Couplings between electronic states and associated reaction channels strongly modify the wave-packet dynamics. The

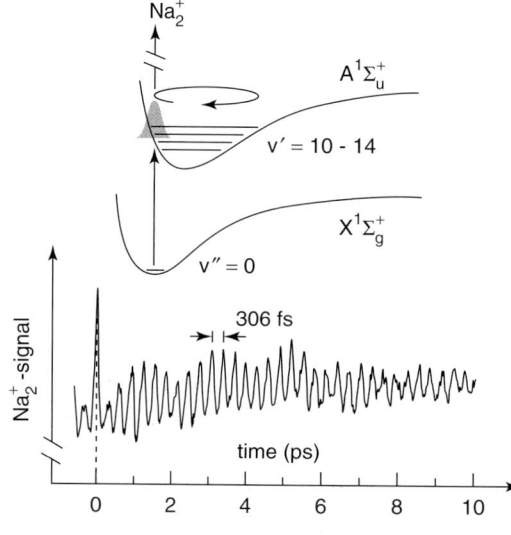

Fig. 22.10 Generation of a wave packet in Na_2 and oscillations in the ion signal due to the wave-packet propagation. (Adapted from Baumert, et al., 1991.)

observed dynamics can then be used to analyze the reaction mechanisms. A one-dimensional example is the photodissociation of NaI (Zewail, 2000; Rose, Rosker and Zewail, 1989). A pump pulse at 307 nm launches in the first electronically excited state a wave packet, which then propagates toward the outer turning point (Figure 22.11). Thereby, it passes through an avoided crossing resulting from the diabatic covalent and the diabatic ionic PES. According to the Landau–Zener mechanism, the wave packet splits into two with one staying in the upper state while the other propagates along the lower covalent PES resulting in the formation of atomic iodide and sodium. The former is reflected by the binding ionic PES, returns to the Franck–Condon region, and is again reflected there. In this way, the wave-packet splitting is repeated several times until the excited state of NaI is depopulated.

With laser-induced fluorescence, the population of atomic sodium and electronically excited NaI can be followed in time (Zewail, 2000; Rose, Rosker and Zewail, 1989). At a probe wavelength of 589 nm, the resonance fluorescence of atomic sodium is dominant and one observes steplike increases of the sodium population every time the wave packet splits at the avoided crossing. In case the probe wavelength is off resonant the fluorescence of NaI dominates. The data shows recurrences of the wave packet which decay exponentially due to the losses at the avoided crossing.

If a laser pulse prepares a wave packet on a repulsive PES, the wave packet propagates in a ballistic fashion. The motion of its center of gravity follows almost purely classical dynamics and is given by the inertia of the involved nuclei and the slope of the PES. Accordingly, the dissociation products appear with some delay after the optical excitation. In addition, the wave packet exhibits broadening as a result of its wave nature. All these features have been demonstrated, for example, in the case of the photodissociation of ICN into I and CN (Rosker, Dantus and Zewail, 1988).

22.4.1.3 Influence of the Environment
In a dense environment, the wave-packet dynamics is strongly influenced by the interaction of the excited molecule with adjacent atoms or molecules. Several effects can occur (Gühr et al., 2007). The forces exerted by the environment change the molecular PESs

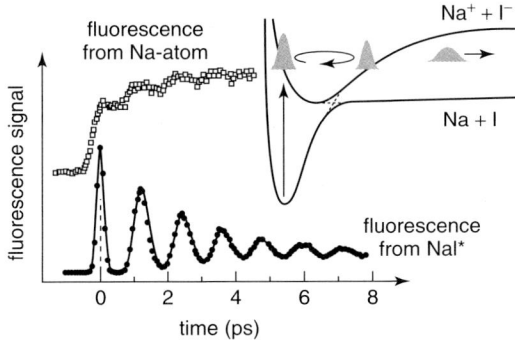

Fig. 22.11 Time-resolved observation of the photodissociation of NaI and the corresponding potential energy diagram (Zewail, 2000; Rose, Rosker and Zewail, 1989). The resonance fluorescence of the Na-atoms (squares) shows a stepwise buildup and the NaI-fluorescence a decay strongly modulated by the recurrences of the wave packet launched by the pump pulse in the electronically excited state of NaI. (Adapted from Zewail, 2000.)

and thereby the vibrational frequencies. The dynamics on originally purely repulsive PESs can even exhibit wave-packet recurrences due to reflections from adjacent particles. Moreover, energy is dissipated into the environment. Wave-packet oscillations are thereby efficiently damped and the population is collected at the potential minimum on a timescale in the order of 10 picoseconds. Fluctuations and inhomogeneities of the environment cause a dephasing of the coherent superposition generated by the optical excitation even if no energy exchange takes place. The decay of coherence-induced signal oscillations is therefore much faster than in the case of isolated molecules. On the other hand, it is possible to generate, in a matrix, coherently excited phonons with a short pulse excitation of an embedded chromophore.

22.4.2
Dynamics of Polyatomic Molecules

Larger molecules have usually several or many degrees of freedom that are involved in the dynamics, and the manifold of contributing states is very dense. This has two major consequences. First, the couplings between the intramolecular degrees of freedom have a strong impact. They cause energy redistribution between vibrational modes and often efficient radiationless electronic transitions. Secondly, the wave functions have a complex structure and observables are averages over population distributions as long as the temperature is not extremely low. Wave-packet signatures are therefore typically much less pronounced as in the case of small molecules. Quite often the dynamics is then described as a single or a series of transformations between states or configurations and the speed of the transformations is characterized by means of rates.

22.4.2.1 Vibrational Energy Redistribution and Cooling

Anharmonicities of the PES cause couplings between the vibrational modes of a molecule and thereby intramolecular vibrational energy redistribution (IVR). Energy transfer was measured for a number of molecules by IR pump and Raman probe spectroscopy (Laubereau and Kaiser, 1978; Hartl and Zinth, 2000). Transfer times in the range of 1 picosecond to almost 10 picoseconds were observed. This indicates that after several picoseconds, the energy is distributed over all vibrational modes and the molecule can be regarded as its own heat bath with an internal temperature.

Large molecules are often investigated in condensed phase. The interaction with the environment then causes an energy flow from the excited molecule to the environment and cooling of the molecule. This process was characterized by Raman measurements as well as via the shape of transient absorption spectra (Elsaesser and Kaiser, 1991; Schwarzer et al., 2002). The intramolecular temperature decays exponentially to the temperature of the environment and the corresponding cooling times vary with the solvent and solute from slightly less than 10 picoseconds up to 40 picoseconds.

22.4.2.2 Solvation

The change of electronic states is usually accompanied by a change of the charge distribution in the molecule and of the static dipole moment. The environment reacts by adjusting its polarization so that the free enthalpy is again minimized. In solution, two contributions to the solvent polarization have to be considered. One is the electronic polarizability, which reacts instantaneous on changes of the charge distribution of the solute and is characterized by the index of refraction n. The second contribution results from the orientation and arrangement of the solvent molecules in the solvation shell around the solute. The energies associated with the two contributions $E_{\text{solv, elec}}$ and $E_{\text{solv, nuc}}$ are given by Eqs (22.20) and (22.21).

$$E_{\text{solv, elec}} = \frac{\mu^2}{4\pi\,\varepsilon_0\,a^3} \cdot \frac{n^2 - 1}{2n^2 + 1} \quad (22.20)$$

$$E_{\text{solv, nuc}} = \frac{\mu^2}{4\pi\,\varepsilon_0\,a^3} \cdot \left(\frac{\varepsilon - 1}{2\varepsilon + 1} - \frac{n^2 - 1}{2n^2 + 1} \right) \quad (22.21)$$

μ is the static dipole moment of the solute, a is its Onsager radius, ε is the static dielectric constant of the solvent, and ε_0 is the dielectric constant of vacuum. To adjust the second contribution, the nuclei of the solvent have to move. This includes vibrational motions, librations, and changes in the arrangement of the solvent molecules in the solvation shell. Accordingly, this response occurs on several timescales. Its evolution can be monitored by measuring the time-dependent shift of the emission spectrum after optical excitation of a chromophore that exhibits a reasonable strong change of the electric dipole moment like, for example, coumarin 153 (Horng et al., 1995). The solvation dynamics is characterized by multiple exponential components with time constants in the range of a few hundred femtoseconds to several ten picoseconds. Smaller solvent molecules react faster and solvents, which form networks like the hydrogen-bonded networks of alcohols, exhibit slow response components associated with the rearrangement of the network. For certain solvents and a sufficient time resolution provided, the response can also exhibit oscillatory contributions caused by coherently excited motions of the solvent molecules (Pérez-Lustres et al., 2005). It was also shown that the time dependence of the solvent response can be reconstructed from the frequency-dependent dielectric constant.

22.4.2.3 Electronic Dynamics: Internal Conversion, Conical Intersections, and Photochemistry

Most molecules exhibit excited-state lifetimes that are much shorter than the radiative lifetimes. Owing to the many nuclear degrees of freedom, nonadiabatic processes are likely to occur resulting in efficient IC and in photochemical products.

A key issue to understand molecular dynamics and photochemical processes is to locate conical intersections and to identify the involved coordinates. In an interplay of experimental (Fuß et al., 1998) and theoretical work (Bernardi, Olivucci and Robb, 1996), several structural motifs were identified, which facilitate conical intersections. One important motif is, for example, a kink between a double and an adjacent single bond. It can lead to an energetic degeneracy between the electronic ground state and the first doubly excited singlet state. The molecular backbone is strongly twisted at this geometry and after the passage through the intersection, the twist motion can continue to a complete cis–trans isomerization. The rotation around the double bond is accompanied by a twist motion around the adjacent single bond. This can result in a Hula-twist isomerization where the two moieties on both sides of the double bond perform a bending motion instead of a rotation by 180° relative to each other. This allows for cis–trans isomerizations even in environments with low flexibility like proteins or organic matrices.

22.4.2.4 Coherent Wave-Packet Dynamics

Ultrashort excitation pulses generate, also in larger molecules, a coherent superposition of vibronic states. The wave packet launched by the pulse oscillates in those vibrational modes that contribute to the superposition. In time-resolved experiments with sufficient resolution, this results in oscillatory shifts of the bands contributing to the transient spectra and oscillations in kinetic traces measured at fixed probe wavelengths (see Figure 22.12). The frequencies of the oscillations are the frequencies of the contributing vibrations. Compared to small molecules, usually more

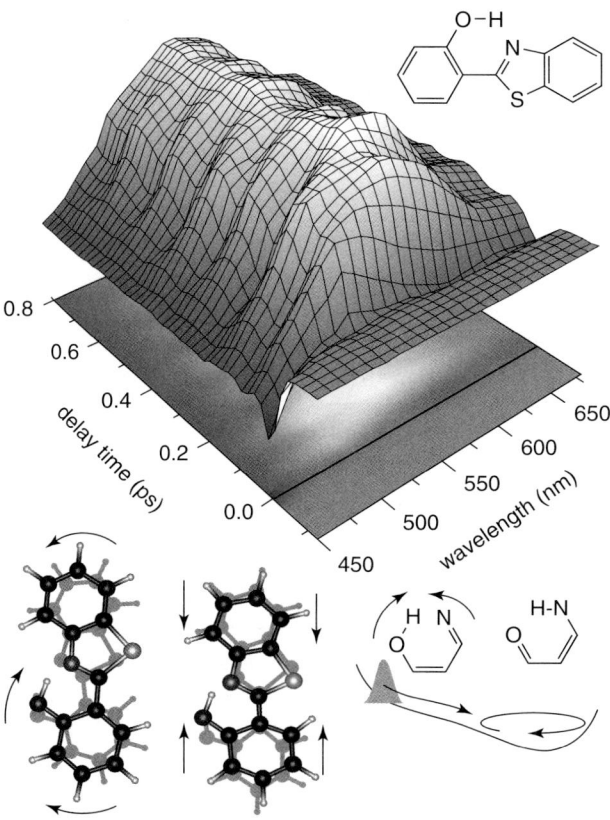

Fig. 22.12 Time-resolved transmission change after optical excitation of 2-(2'-hydroxyphenyl)benzothiazole. The excitation induces an intramolecular proton transfer, which proceeds as a ballistic wave-packet motion and excites coherently the depicted vibrational modes. (Lochbrunner, Wurzer and Riedle, 2003.)

vibrational modes and shorter damping times are observed due to the larger number of degrees of freedom.

A superposition of vibrational levels can be directly generated by the optical excitation. This is possible for Raman active modes and the superposition can then be reconstructed from the resonance Raman cross sections and the pulse spectrum (Takeuchi and Tahara, 2000). The optical excitation launches a wave packet not only in the electronically excited state but also very often in the electronic ground state via stimulated impulsive Raman scattering (SIRS). During the excitation process, population is pumped into the electronically excited state. The slope of the excited-state PES accelerates the corresponding wave packet along the Raman active modes. The ongoing interaction with the pulse can stimulate a part of the wave packet back into the electronic ground state where it oscillates in the Raman modes according to the gained momentum.

A coherent excitation of a vibrational mode can also result from the dynamics of a wave packet when the Franck–Condon region is already left and is not always caused by a direct optical excitation. For example, in azulene and in [2,2′-bipyridyl]-3,3′-diol, the coherent excitation of vibrational modes is observed that are optically inactive due to their symmetry (Wurzer et al., 1999; Stock et al., 2008). Accordingly, the analysis of oscillations reveals which coordinates contribute also beyond the Franck–Condon region to the path of the wave packet – information that can hardly be obtained by other methods.

The analysis of the coherent wave-packet motion is particularly valuable to reveal the mechanisms of photo reactions. Two examples are given here to demonstrate this approach. In the case of 2-(2′-hydroxyphenyl)benzothiazole, an excited-state intramolecular proton transfer takes place after optical excitation of the S_1-state. Femtosecond absorption experiments find that the proton transfer results in a coherent excitation of vibrational modes, which modulate the distance between the hydrogen donor and acceptor atom (Figure 22.12) (Lochbrunner, Wurzer and Riedle, 2003). This shows that prior to the transfer the donor–acceptor distance is shortened by a deformation of the molecular backbone until the hydrogen atom can be bound to the acceptor atom without the need for extra energy. Thereby the whole process can proceed along a barrierless and adiabatic pathway. The general notion is that if a process proceeds as a wave-packet motion, momentum is gained along coordinates that contribute strongly to the reaction path. After the process, the momentum results in oscillations in the associated normal modes of the product. This ringing of the product can be analyzed by time-resolved experiments and reveals which coordinates contribute to the reaction path.

In the case of the photodissociation of the metal hexacarbonyl compound $Cr(CO)_6$, oscillatory signal contributions with a low frequency of 95 cm^{-1} and with a higher frequency of about 400 cm^{-1} are observed in time-resolved, mass-selective dissociation experiments (Trushin et al., 2008). The oscillatory component with the higher frequency is observed from the beginning of the dissociation process and results from a dominant optical excitation of the totally symmetric Cr-C stretch vibration. However, the low-frequency component is only observed in measurements that monitor the return of the wave packet at the electronic ground state. The frequency points to the coherent excitation of a totally symmetric C-Cr-C bending vibration. The results indicate that the electronic relaxation proceeds through a conical intersection between the S_1 and the S_0 state, which has a trigonal–bipyramidal geometry since after such an intersection the PES must have a slope along the C-Cr-C bending coordinate, resulting in an acceleration of the wave packet along this coordinate. This example demonstrates that the analysis of coherent wave-packet oscillations can give valuable information on the location of conical intersections.

22.4.2.5 Photoinduced Electron Transfer

The fundamental process of electron transfer can take place on timescales ranging from very slow for media in a glassy state to ultrafast for photoinduced reactions (Barbara, Meyer and Ratner, 1996). The charge transfer between a donor (D) and an acceptor (A) is strongly coupled to its environments. This necessitates a reorganization of intra- or intermolecular vibrational coordinates as well as of the polarization state of the solvent. Both effects can be described in a model where the potential or free energy is plotted along a reaction coordinate resulting, for example, in a double minimum potential, where D and A configurations are separated by a barrier. Possible transfer mechanisms encompass the limits of thermal activation and nuclear tunneling. Ultrafast photoinduced electron transfer occurs on timescales where the nonequilibrium effects with respect to the surroundings are important and the concepts underlying, for example, transition state theory break down. This might relate to the solvent which typically involves two timescales, that is, femtosecond inertial and picosecond diffusional motion. Measured intramolecular electron transfer times in $Ru^{(II)}-Ru^{(III)}$ mixed valence compounds in water yielded 80 femtoseconds, which is comparable to the inertial motion, but much faster than the diffusional one (Barbara, Meyer and Ratner, 1996). Furthermore, oscillatory components in the pump-probe signal have been demonstrated. They are a consequence of a coherent vibrational wave-packet motion, which is impulsively excited upon the Franck–Condon transition $D-A \rightarrow D^+-A^-$ (see also simulation in Wang and Thoss, 2008). Other examples are intermolecular D–A complexes in solution which have been studied, for example, by fluorescence up-conversion (Rubtsov and Yoshihara, 1999). Figure 22.13 shows data giving clear evidence for wave-packet motion, which has been assigned to an out-of-plane intramolecular vibration of the acceptor.

Owing to its technological importance, ultrafast heterogeneous electron transfer has attracted considerable attention. In particular, electron injection from a photoexcited adsorbed chromophore into a semiconductor is studied as a model for the primary process occurring in dye-sensitized electrochemical solar cells (Frischkorn et al., 2007; Prezdho, Duncant and Prezhdo, 2008). In the case of perylene as dye and TiO_2 as semiconductor, injection times as short as 10 femtoseconds were observed. This fast dynamics results from the fact that the electron is transferred to a continuum of acceptor states in the conduction band of TiO_2. The insertion of bridge-anchor groups between the surface and the chromophore results in a slow down of the transfer rate, which is very sensitive to

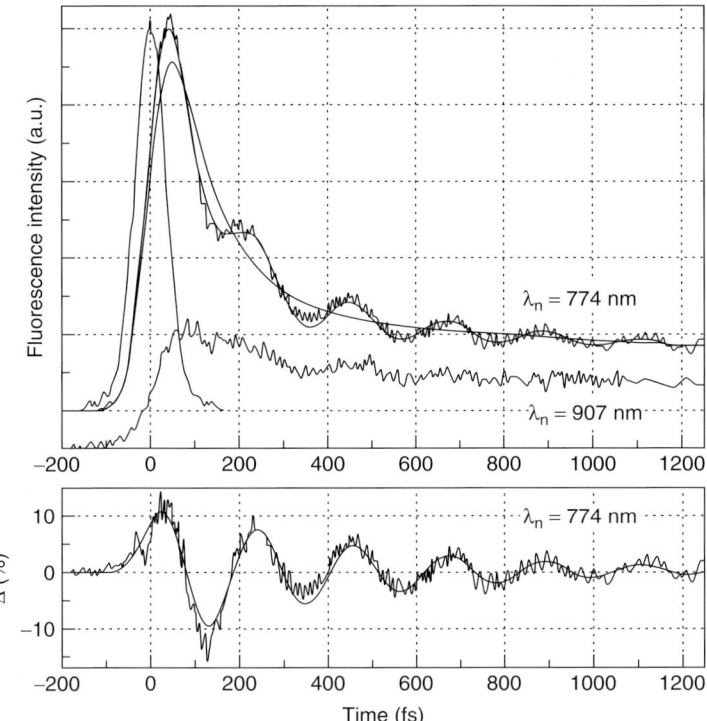

Fig. 22.13 Fluorescence decay of the hexamethylbenzene-tetracyanoethylene D–A complex in CCl$_4$ excited in the blue and red side of the time-integrated fluorescence spectrum. The oscillatory part ($\nu_{osc} = 155$ cm^{-1}) is shown separately in the lower panel. The upper panel also contains the instrument response function (85 femtoseconds FWHM) as well as result of a biexponential plus oscillatory function fit. (Reprinted from (Rubtsov and Yoshihara, 1999), with permission from the ACS.)

the group type. If the group consists of conjugated bonds, it acts like a wire and only a moderate decrease of the rate is observed. If saturated bonds interrupt the transfer pathway, they induce a tunneling barrier and suppress the transfer strongly.

22.4.3
Clusters and Liquids

22.4.3.1 Atomic Clusters

Clusters are interesting objects since they allow to study how material properties change by going from the isolated atom or molecule to the condensed phase. They are generated in supersonic expansions of atomic or molecular gases. Typically, an inert gas is used as carrier to achieve low temperatures. Depending on the parameters of the expansion, clusters of various sizes between a few atoms or molecules up to many thousands of them can be generated (Hertel and Radloff, 2006). To measure the dynamics and evolution of electronically excited clusters, mass-selective ion and photoelectron spectroscopy are used in combination with the pump-probe technique (Stolow, Bragg and Neumark, 2004).

Atomic metal cluster anions are investigated to understand the transition from a single atom to a metal. They show very fast electronic relaxation processes even if the cluster contains only a few atoms (Dermota, Zhong and Castleman, 2004). The fast dynamics results from inelastic electron scattering, which is very similar as in the solid state although somewhat slower. The dynamics at stronger excitation intensities is governed by the finite size of the cluster. Multiphoton absorption results in multiple ionization and above a certain threshold in a Coulomb explosion of the cluster driven by the positive excess charges (Dermota, Zhong and Castleman, 2004). In the cluster a plasmon can be excited, that is, a collective oscillation of electrons. Its frequency depends on the density of quasi-free electrons and shifts to lower frequencies as the cluster expands during the Coulomb explosion (Fennel et al., 2007). At some time, the plasmon frequency becomes resonant with the optical frequency of the exciting laser pulse if the pulse duration is sufficiently long. In this case, dramatically enhanced absorption occurs heating the cluster further up and resulting in extremely fast nonthermal electrons due to various acceleration processes.

In a supersonic expansion, molecules can form clusters that are typically stabilized by intermolecular hydrogen and van-der-Waals bonds. Optical excitation and photoionization of such clusters result often in complex fragmentation patterns and time dependences, and the interpretation relays on a careful analysis of the experimental findings. Ammonia clusters are a typical example (Hertel and Radloff, 2006). Nevertheless, it was possible to show for the ammonia dimer $(NH_3)_2$ that, after excitation to the $\tilde{A}$ state but prior to dissociation, a hydrogen transfer in the electronically excited state can occur leading to the NH_4-NH_2 complex. Interestingly, fast hydrogen transfer can also occur in mixed clusters containing one solute and a few solvent molecules like NH_3 or H_2O. For example, transfer in the electronically excited state from the solute to the solvent was observed on the subpicosecond timescale for indole ammonia and methanol water clusters. Such clusters give interesting insights into solvation processes since one can regard them as nanodroplets. The following two examples indicate that, in many cases, quite moderate cluster sizes result already in efficient solvation. The spectral and dynamical signatures of a solvated electron in water are almost fully developed when the cluster exceeds a size of about only 25 water molecules (Hertel and Radloff, 2006). In the case of the photodissociation of I_2^- with NIR photons, it was shown that one solvation shell of CO_2 or OCS results in complete caging and recombination of the I_2^- molecule (Sanov and Lineberger, 2004). Time-resolved experiments reveal that in the case of a full solvation shell the iodine atoms are directly reflected by the shell, whereas in the case of a partial solvation shell the weaker forces lead to a slowdown of the dynamics. If the excitation is performed with near-UV photons a higher lying dissociative state of I_2^- with the spin-orbit excited $I^- + I^*(^2P_{1/2})$ asymptote is populated. Then, quenching of the spin-orbit excitation is necessary for recombination. This is only possible if the environment strongly disturbs the electronic structure of the I_2^- anion. Therefore, the recombination yield and the dynamics depend strongly on the structure of the solvation shell and do not vary monotonically with the cluster size. However, almost complete caging is observed for about two solvation shells.

22.4.3.2 Water and Hydrogen-Bonded Networks

Disordered hydrogen-bonded networks are showing structural fluctuations on a broad range of timescales and have been targeted by ultrafast infrared spectroscopy (Nibbering et al., 2007). Owing to its relevance, for example, for aqueous solvation, water has attracted most attention, although there are some efforts devoted, for example, to alcohols such as solvated methanol-OD oligomers (Asbury et al., 2003). Due to technical limitations (high optical density of OH stretching fundamental transition in H_2O), most studies focused on isotopically diluted samples, that is, OH in HOD/D_2O or OD in HOD/H_2O. Upon excitation of the OH stretching vibration, one observes an ultrafast vibrational energy relaxation ($\sim$0.7 picoseconds), which has been assigned to be due to a Fermi resonance coupling with the HOD bending overtone vibration. Since the efficiency of this relaxation pathway depends on the proper resonance, it is not only influenced by the composition of the system but also on the fluctuation of the transition frequencies. These fluctuations are described by the correlation function of the energy gap, for example, for the OH stretch fundamental $\langle\delta\omega_{OH}(t)\delta\omega_{OH}(0)\rangle$, which determines the line shape function and therefore the nonlinear response (cf Eq. (22.14)). Three-pulse photon echo peak shift measurements on HOD/D_2O revealed a correlation function, which shows signatures of an underdamped motion (170 femtoseconds) related to the hydrogen-bond and a 1.2-picosecond decay assigned to be due to structural reorganization of the network (Fecko et al., 2003). The subpicosecond dynamics of neat liquid water itself has been studied only recently (Cowan et al., 2005). Here, one finds that correlations decay on a timescale of about 50 femtoseconds. This ultrafast loss of memory of the initially excited configuration was ascribed to the anharmonic coupling to librational motions in the hydrogen-bonded network.

The dynamics of an excess proton in water is another important issue that has been addressed by ultrafast IR spectroscopy (Woutersen and Bakker, 2006). Essentially there are two limiting structures the Eigen ($H_9O_4^+$) and the Zundel ($H_5O_2^+$) cations, which are related to proton migration. They can be distinguished by the absorption of the terminal OH stretching vibrations. Exciting the Eigen-like structure and probing for the Zundel-like one, it was found that the interconversion between these structures is faster than 100 femtoseconds.

22.4.4 Biological Systems

At first glance, stepping from molecules in solution to chromophores in biosystems seems to provide merely a different type of environment. However, biological systems are designed to fulfill a certain task, which is often made possible only by the concerted interplay of highly specific interactions. Although the biologically relevant response time to some external stimulus is in the range of microseconds or longer, the individual steps involving, for example, barrier crossings, can be substantially faster. These elementary steps and their importance for the structure–function relationship are in the focus of ultrafast spectroscopy.

22.4.4.1 Primary Steps of Photosynthesis

The conversion of sunlight into a transmembrane potential, which drives biochemical reactions in higher plants and certain bacteria builds on the ultrafast primary steps of excitation energy and electron transfer in pigment–protein complexes. Here, a variety of structures realize the same highly efficient cascade of energy funneling and charge separation. Purple bacteria such as *Rhodobacter (Rb.) sphaeroides* and *Rhodopseudomonas (Rps.) acidophila* have been the first organisms for which high-resolution X-ray structures became available. Here, light harvesting is performed by two types of ring-shaped antenna complexes. The peripheral antenna (LH2) consists of two pools of Bacteriochlorophyll a molecules, which are fixed by the protein scaffold. The closely packed B850 ring (16–18 pigments with 9 Å distance) gives rise to absorption around 850 nm, while a second absorption band around 800 nm originates from the B800 ring of 8–9 pigments being well separated (21 Å). These two pigment pools show an ultrafast dynamics, which can be described in terms of Frenkel exciton theory (Renger, May and Kühn, 2001). It has been extensively studied by various spectroscopic methods (Grondelle and Novoderezhkin, 2006; Sundström, 2008) leading to the following picture (time constants for 77 K): Excitation energy transfer among the B800 molecules takes place on timescales from 0.4 to 2 picoseconds, the transfer to the B850 pool covers the range from 0.4 to 1.1 picoseconds, while the intra-B850 transfer shows timescales of 150 femtoseconds and faster. This fast dynamics localizing the excitation energy in the red part of the spectrum is only possible due to the coupling between the electronic excitations to vibrations of the pigment and the protein, which causes energy dissipation (Renger, May and Kühn, 2001). The core antenna (LH1) receives the excitation energy from LH2 within about 5 picoseconds. Its structure is similar to that of the B850 ring, but it hosts the reaction center that traps the excitation energy within 30–40 picoseconds. The mechanism of the subsequent charge separation in the reaction center involves the initially excited special pair (P) of two strongly coupled bacteriochlorophyll molecules, a monomeric bacteriochlorophyll (B), a bacteriopheophytin (H), and a quinone (Q). Much effort had been focused on the role of B in the electron transfer chain, but by now there is clear evidence that at ambient temperatures the state P^+B^- forms within 3 picoseconds, suggesting a stepwise transfer mechanism. This intermediate decays within 0.9 picoseconds to give P^+H^- and after further 200 picoseconds P^+Q^- (Zinth and Wachtveitl, 2005) (see Figure 22.14).

Ultrafast two-dimensional spectroscopy in the visible range has been applied to the Fenna–Matthews–Olson light harvesting complex of sulfur bacteria to unravel the coupling patterns between its seven pigments, which are not arranged in some regular order (Brixner et al., 2005).

22.4.4.2 Rhodopsins

Rhodopsins are photoreceptors where the absorption of a photon leads to a

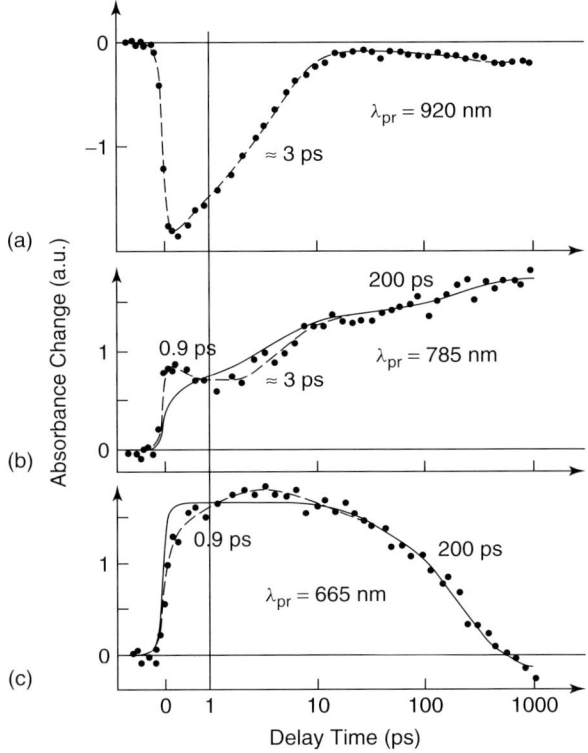

Fig. 22.14 Transient absorption of the reaction center of Rb. sphaeroides. At 920 nm, the signal is dominated by the decay of the primary donor (P*), at 785 and 665 nm the 0.9-picosecond component gives evidence for the P^+B^- to P^+H^- transfer. The 200-picosecond component corresponds to the electron transfer to the quinone. The solid and dashed lines represent fits with and without, respectively, the 0.9-picosecond component. Note that the axis is linear until 1 picosecond. (Reprinted from (Zinth and Wachtveitl, 2005), with permission from Wiley-VCH Verlag GmbH & Co. KgaA.)

conformational change, which as the result of a signaling cascade, eventually triggers some biological response. In rhodopsins, one has an 11-cis to all-trans conformational change in the retinal chromophore, which is key for the process of primary vision. In unicellular organisms, on the other hand, bacteriorhodopsin provides the means for ion transport across a membrane. The cis–trans isomerization of retinal has been one of the early targets of ultrafast spectroscopy (Sundström, 2008), see Figure 22.15. In fact, the formation of the primary photoproduct, bathorhodopsin, via an S_1–S_0 conical intersection occurs with high efficiency on a timescale of about 200 femtoseconds (Schoenlein et al., 1991). The nature of the reaction coordinate has long been disputed. Here, valuable insight about the evolution on the excited S_1 state

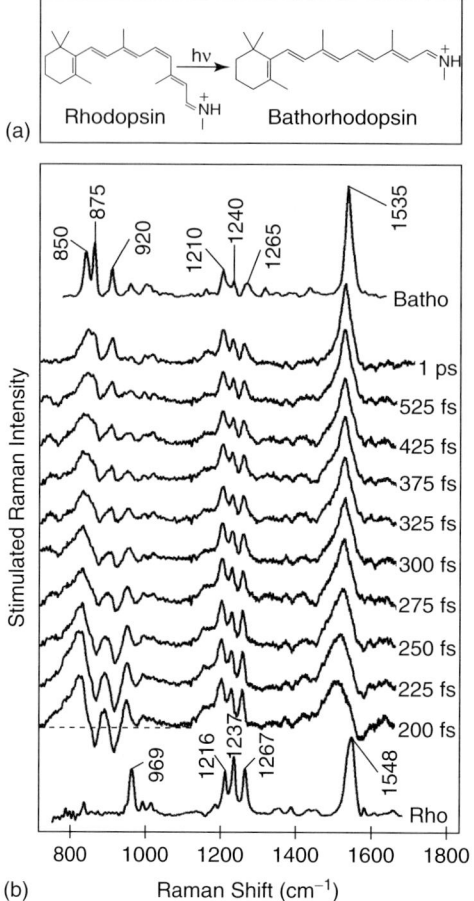

Fig. 22.15 (a) Photoinduced 11-cis to all-trans isomerization of the retinal chromophore of rhodopsin.
(b) Characterization of this process after S_1-S_0 transition by means of time-resolved femtosecond-stimulated Raman spectra; lower and uppermost curve show resonance Raman spectra of ground-state rhodopsin and bathorhodopsin, respectively. (From (Kukura et al., 2005), reprinted with permission from the AAAS.)

surface toward the conical intersection comes from on-the-fly adiabatic molecular dynamics simulations (Frutos et al., 2007). It has been found that the retinal up to about 60 femtoseconds after excitation deforms to prepare for an asynchronous crankshaft type motion, which leads to the conical intersection after about 110 femtoseconds. The coherent evolution of the vibrations after the S_1-S_0 transition has been followed by femtosecond-stimulated Raman experiments as shown in Figure 22.15 (Kukura et al., 2005). These data allow conclusions to be derived on the redistribution of the excess energy and the motions involved in forming the photoproduct. Dispersive features indicate changes in vibrational frequency along the reaction coordinate, which are particularly pronounced for the hydrogen out-of-plane motions observed around 800–900 cm^{-1}.

22.4.4.3 Peptide Dynamics

The function of peptides and proteins is largely determined by their three-dimensional structure. The latter is not static but fluctuates on rather different timescales, related to the intrinsic length scales. The fastest fluctuations in the picosecond and femtosecond range are not accessible to the well-established structure determination methods of nuclear magnetic resonance spectroscopy. However, they are likely to be of importance, for example, for energy transport or in the process of opening reaction channels leading to conformational changes. The most prominent signature in the IR spectrum of peptides is the amide I vibration (C=O stretch) of the peptide unit, which absorbs around 1620 cm^{-1}. Amide I modes on different peptide units are coupled leading to a delocalization of this vibration and therefore contain information on the secondary structure of the peptide and its fluctuations. This information, however, is congested in the linear IR spectrum. Two-dimensional (2D) IR spectroscopy can be used to unravel this information by spreading the spectrum into a second dimension. For peptides, this was first demonstrated by Hochstrasser and coworker (Hamm, Lam and Hochstrasser, 1998). They used a dynamic hole-burning approach where a narrow band laser excites the sample and a broadband laser probes the spectral changes. If the dispersed signal is plotted as a function of the frequency of the pump laser, one obtains a 2D signal, which on the diagonal contains information on the amide I eigenstates, disentangling homogeneous (antidiagonal) and inhomogeneous (diagonal) contributions to the line width. Furthermore, in the off-diagonal part of the spectrum, cross peaks appear, which are related to the coupling of the considered transitions. The localized nature of the individual amide I vibrations and the sensitivity of the coupling strengths on the backbone structure allow for a simulation of these spectra in terms of vibrational Frenkel excitons (Mukamel and Abramavicius, 2004). The above-mentioned paper triggered host of investigations of the equilibrium conformational dynamics of various peptides in solution using different 2D IR spectroscopies (Woutersen and Hamm, 2002). More recently, the focus has been extended to membrane bound peptides (Mukherjee et al., 2006).

One might ask whether this scheme can be applied to proteins containing hundreds of local amide I modes. It turns out that the 2D IR spectrum is rather sensitive to the structural motif as illustrated in Figure 22.16 (Ganim et al., 2008). This gives direct access to the observation of large conformational changes, that is, protein folding and unfolding. In order to trigger these changes, the temperature-jump method has been used (Ganim et al., 2008). An ultrafast alternative is the use of photoswitches built into the peptide structure. Photoexcitation of azobenzene, for instance, leads to a reversible cis–trans isomerization, which is capable of triggering a conformational change, for example, of small cyclic peptides (Hamm, Helbing and Bredenbeck, 2008).

22.4.4.4 DNA

The UV photochemistry and photophysics of nucleic acid bases has been studied extensively in gas and solution phase with the prospect that it contains the key for the understanding of DNA photoprotection and UV-damage (Crespo-Hernández et al., 2004). Singlet excitation of DNA bases gives rise to a rapid radiationless deactivation,

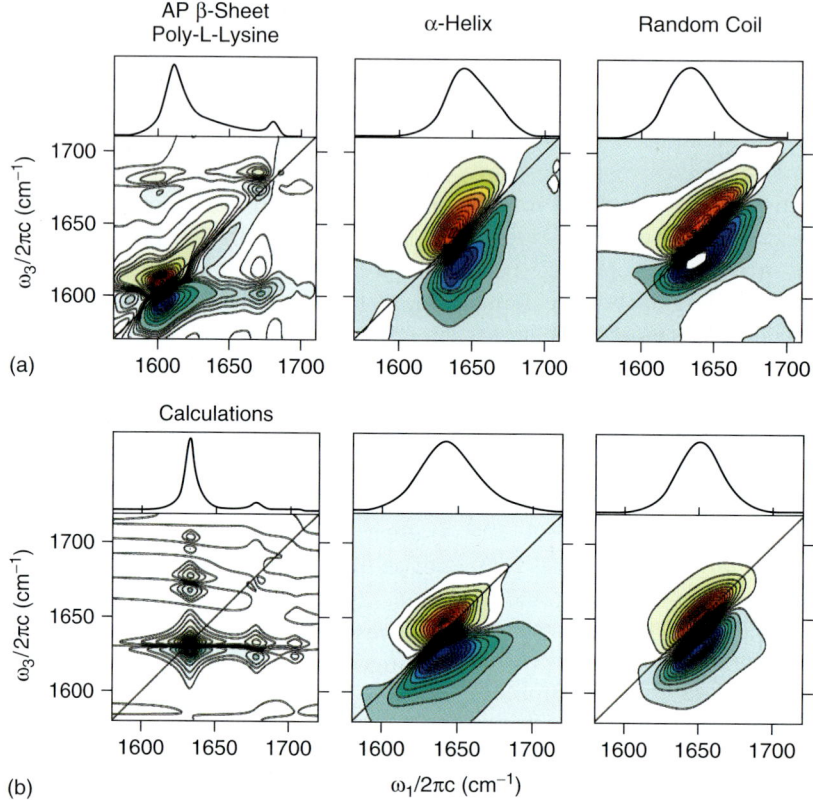

Fig. 22.16 Sensitivity of the 2D IR spectrum in the amide I region with respect to the secondary structure of poly(L-lysine) (panel (a) measured spectra, (b) model calculation). The upper/lower panels show the 2D IR spectra. The antiparallel (AP) β-sheet has two diagonal peaks as well as off-diagonal ones due to coupling pattern of amide I vibrations. These peaks merge into a single one for the α-helix and random coil, which can only be distinguished in the 2D IR spectrum. Notice that along ω_1 one has overlapping induced absorption and stimulated emission leading to an asymmetry with respect to the diagonal. (Reprinted from (Ganim et al., 2008), with permission from the ACS.)

which necessitates the use of femtosecond methods for unraveling the coupled electron-vibrational dynamics. Quantum chemical calculations predict the relevant singlet manifold of the bases to be comprised of $\pi\pi^*$ and $n\pi^*$ excitations, where the former are carrying most of the oscillator strength. Time-resolved photoelectron spectroscopy of a supersonic jet expansion of adenine has been utilized to identify a 50-femtosecond $\pi\pi^*$ to $n\pi^*$ decay followed by a 750-femtosecond internal conversion into the electronic ground state for excitation well above the $\pi\pi^*$ origin (Ullrich et al., 2004).

Upon base pairing new relaxation pathways become possible. For isolated base pairs, evidence for a deactivation along a reaction coordinate involving the hydrogen-bonded proton motion has been found. This leads to the stabilization of a charge transfer state, which converts into the S_0 ground state (Schultz et al., 2004). Upon base stacking, the situation changes and the formation of an excimer has been observed. For double strands of adenine–thymine oligomers, the excimer formation is as fast as 400 femtoseconds and the excimer state decays within 50 picoseconds (Crespo-Hernández, Cohen and Kohler, 2005).

Base pairing does have a great impact on the vibrational dynamics in the electronic ground state. The formation of hydrogen bonds leads to substantial anharmonicities of the PES. For adenine–uracil pairs in deuterochloroform solution, pump-probe IR spectroscopy revealed a decrease of the lifetime of the hydrogen-bonded NH-stretching vibration by a factor of 3 as compared to monomeric uracil (Woutersen and Cristalli, 2004). DNA oligomers are investigated in the presence of water. For the IR spectrum, this gives rise to a broad absorption around 3000–3600 cm^{-1}, masking the fundamental transitions of the hydrogen-bonded NH and NH$_2$ stretching modes. Using two-color IR pump-probe spectroscopy, the symmetric NH$_2$ stretching band of adenine–thymine oligomers could be assigned by virtue of its anharmonic coupling to the NH$_2$ bending and carbonyl stretching modes in the fingerprint range (Heyne, Krishnan and Kühn, 2008). The anharmonic couplings related to base pair stacking and secondary structure of guanine–cytosine oligomers have been studied with the help of two-dimensional IR spectroscopy in the carbonyl stretching region (amide I modes) (Krummel and Zanni, 2006).

22.4.5
Solid State and Surfaces

22.4.5.1 Semiconductors

Ultrafast spectroscopy is applied to semiconductors and semiconductor nanostructures like single and multiple quantum wells to study many-body dynamics (Chemla and Shah, 2001; Axt and Kuhn, 2004). Pump probe and, in particular, FWM experiments are powerful methods to investigate coherences, correlations, and quantum kinetics in these systems. Deviations from the uncoupled quasi-particle picture are found to be quite strong and for many aspects the theory has to be extended beyond the mean-field level. The following examples illustrate these findings.

Early experiments on GaAs found already surprisingly efficient exciton–exciton scattering at mean exciton separations of more than 100 nm (Schultheis, Kuhl and Honold, 1986). Exciton scattering by free carriers was found to be even 10 times stronger. This demonstrates that in dense matter Coulomb forces lead even for formally neutral quasi-particles like excitons to strong interaction with other quasi-particles. In this respect, it is interesting to note that the formation of quasi-particles is not instantaneous. Femtosecond experiments on GaAs with terahertz-probe pulses found that Coulomb screening of free carriers takes several tens of femtoseconds (Huber et al., 2001). This indicates that in the case of fast collisions screening is less efficient as for slow dynamics. The strong Coulomb interaction between the charge carriers leads, in general, to an ultrafast relaxation and equilibration

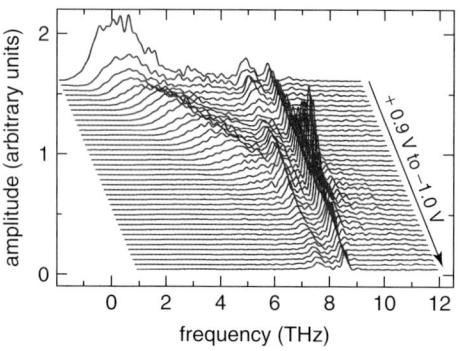

Fig. 22.17 Fourier transformations of time-resolved reflectivity measurements on a III-V semiconductor superlattice. The measurements were performed with different bias voltages as indicated at the right. (Reprinted from (Först et al., 2007), with permission from Wiley-VCH Verlag GmbH & Co. KgaA.)

(Först et al., 2007). However, for efficient recombination of electrons and holes, disorder and traps are necessary.

The analysis of quantum beats in time traces reveals information about carrier-phonon coupling. Time-resolved reflectivity experiments on III–V semiconductor superlattices showed, for example, that the coupling between optical phonons and Bloch oscillations can be efficiently tuned by an applied bias voltage (Först et al., 2007) (Figure 22.17). The bias voltage changes the Bloch frequency linearly. If it is in resonance with an optical phonon, a strong increase of the phonon signal is observed. The largest signal is found for frequencies slightly below the frequency of the uncoupled phonon at 8.8 THz. This was interpreted as an indication that the dephasing rate of the Bloch oscillations increases by resonantly enhanced electron–phonon scattering. FWM experiments on GaAs quantum wells found clear evidence for two processes contributing to the dephasing of excitons (Carter, Chen and Cundiff, 2007): migration of the excitons and the interaction with acoustic phonons. The second contribution is identified on the basis of oscillatory features in time-dependent peak shifts of photon echo signals, which have a frequency in the expected range.

Ultrafast spectroscopy has also revealed strong evidence for biexcitons giving insight in Coulomb-induced correlations between electron hole pairs (Chemla and Shah, 2001; Axt and Kuhn, 2004). In several experiments, quantum beats with a frequency corresponding to the binding energy of biexcitons have been observed. Biexcitonic signatures might also be identified by means of their Stark shift, which is a red shift, if the excitation is just below the biexciton resonance contrary to mean-field contributions, which are blue-shifted (Axt and Kuhn, 2004).

22.4.5.2 Polymers and Organic Materials

Polymers and organic materials are very interesting for photonic applications like solar cells and organic light-emitting diodes (OLEDs). Ultrafast spectroscopy has contributed substantially to understand the exciton dynamics and the photoinduced charge generation in these materials (Lanzani et al., 2005). The degree of disorder is much higher than in the case of semiconductors and the varying electronic structure of different molecular moieties has a strong impact on the dynamics (Scheblykin et al., 2007). Electron hole pairs created by an optical excitation form Frenkel excitons, which are more or less localized on molecular chromophores and migrate by hopping processes through the material. Femtosecond absorption measurements show that after optical excitation, energy transfer on two timescales takes place. Within the first picoseconds, a red shift of the emission is observed indicating a relaxation toward the lowest state in the spatial vicinity of the excited chromophore. Thereafter, energy transfer occurs via thermally activated hopping. The transfer mechanism is of Förster type in both cases (Scheblykin et al., 2007). Exciton migration becomes more efficient if the homogeneity of the site energies is improved. This was, for example, demonstrated for a polymer matrix highly doped with dye molecules, which interact only weakly with their environment but which can facilitate exciton migration by Förster energy transfer (Schlosser and Lochbrunner, 2006). Ultrashort pump pulses lead typically to exciton densities, which are high enough that bimolecular processes and exciton–exciton annihilation become important. If two excitons approach each other until they occupy two adjacent sites, they can interact and one of them is then usually annihilated (Scheblykin et al., 2007). The associated time dependence of the exciton population can be analyzed to characterize the exciton mobility.

Photogeneration of charges in polymers is intensively investigated by ultrafast spectroscopy. The application of an external electric field increases the probability for the appearance of free charges (Kersting et al., 1994). This allows to identify of the features and signatures of the generation process. The results show that primarily excitons are optically generated, which can then dissociate into free charges. The efficiency depends critically on the binding energy of the Frenkel excitons. In polymers with a significant binding energy charge pair generation occurs only if dissociation sites with an appropriate electronic structure are available, a sufficiently strong external field is applied, or the excitation brings a large amount of excess energy into the system. In the latter case, charge generation is restricted to a time window in the order of 100 femtoseconds limited by vibrational energy redistribution (Scheblykin et al., 2007). For organic solar cells, the charge generation due to the interaction with electron acceptors like fullerenes is of particular importance. Experiments applying photoconductivity as probe process showed that the charge generation depends sensitively on the electron donating polymer (Müller et al., 2005). In the case of methyl substituted poly(p-phenylene), coulombically bound polaron pairs are formed at the polymer fullerene interface. They show a high recombination probability resulting in a low photocurrent yield. In contrast, with poly(2-methoxy,5-(3,7-dimethyloctyloxy)-1,4-phenylenevinylene as donor free carriers are generated within 500 femtoseconds leading to a high yield.

J-aggregates of dye molecules show photoinduced dynamics, again determined by Frenkel excitons. Time-resolved absorption experiments found a clear signature for the transition from the one-exciton to a two-exciton state (van Burgel, Wiersma and Duppen, 1995). They also showed that efficient exciton–exciton annihilation can occur due to the mobility of the excitons. Calculations indicate that transient spectra should also exhibit features due to coherences between several exciton states (Heijs, Dijkstra and Knoester., 2007).

22.4.5.3 Surfaces

In the field of surface science, ultrafast spectroscopy plays an increasing role in order to understand the relevance of electronic surface states, the interaction with adsorbates, and the evolution of catalytic processes (Güdde, Berthold and Höfer, 2006). For most pump-probe techniques, the bulk contribution would lead to an overwhelming background signal. However, meanwhile several techniques have been developed, which are particularly sensitive to the surface. In the case of two-photon photoemission (2PPE), the kinetic energy of photoelectrons generated by the probe pulse is measured. Only electrons from the surface have a significant kinetic energy since photoelectrons from the bulk are subject to inelastic scattering before they reach the surface (Güdde, Berthold and Höfer, 2006). Other probe techniques rely on nonlinear second-order mixing processes like second harmonic generation (SHG) and sum frequency generation (SFG) (Bonn, Kleyn and Kroes, 2002). These processes are only possible if the medium exhibits no inversion symmetry. While for most materials this condition is not fulfilled in the bulk, it is an intrinsic property of surfaces and interfaces. Then the signal results only from the surface or the interface.

Image potential states are typical for metal surfaces. An electron is bound to the surface by attractive Coulomb forces resulting from the image charge in the metal. For well-ordered crystals, the electron can freely move parallel to the surface. However, the wavefunction extends into the bulk and couples strongly to the bulk states. This leads to ultrashort lifetimes in the range of 10–60 femtoseconds (Güdde, Berthold and Höfer, 2006) (Figure 22.18). If rare-gas adlayers are adsorbed on surfaces of metal crystals like copper or ruthenium, the lifetime increases strongly. The rare-gas layer can influence the lifetime in two ways. It pushes the wavefunction of the image potential state out off the bulk reducing the coupling to the conduction band (Figure 22.18). In addition, the state can be shifted in energy resulting in a larger energetic separation from the relevant bands of the bulk.

The chemistry on surfaces is characterized by a sequence of steps including adsorption of reactants, frequently dissociation, rearrangement, bonding, and desorption of products. Ultrafast spectroscopy provides new insight into the mechanisms of these processes (Bonn, Kleyn and Kroes, 2001). In addition, the laser pulse excitation of surfaces with adsorbates can also lead to new reaction pathways. Pump pulses can influence the dynamics in several ways. They can promote reactions if they are directly absorbed by the reactants. They can also be absorbed by the substrate and energy is subsequently transferred to the reactants by hot electrons or phonons. Electron-mediated processes are only possible within a short time window of about ~ 1 picosecond due to the strong electron–phonon coupling that cools the electrons efficiently whereas cooling of the phonon bath occurs on a timescale of ~ 50 picoseconds. The relevance of electron-mediated processes was demonstrated for the recombination of hydrogen atoms adsorbed on an Ru(100) surface (Frischkorn et al., 2007). After the absorption of ultrashort laser pulses, hot electrons from the substrate cause a strong excitation in coordinates associated with the motion of

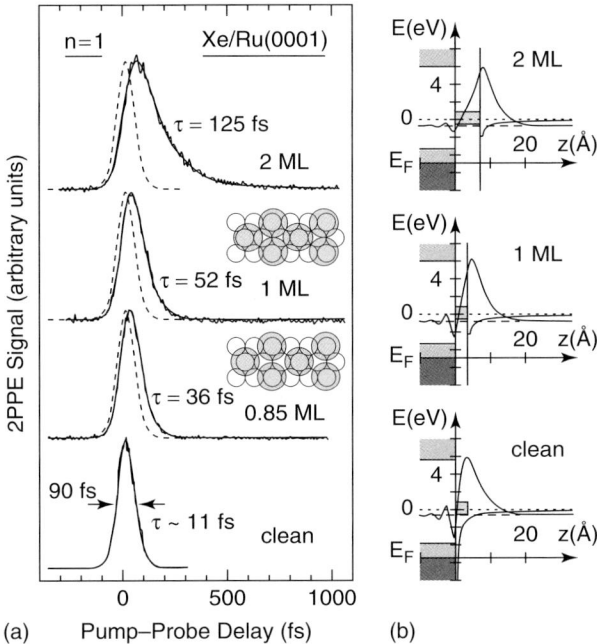

Fig. 22.18 (a) 2PPE lifetime (τ) measurements of the lowest image potential state of ruthenium(0001) for the clean surface (lowest panel and dashed curves) and with Xe coverages up to two monolayers (MLs). (b) Schematic energy diagrams and wavefunctions of the image potential state for no adlayer, one ML and two MLs of Xe. (Reprinted from (Güdde, Berthold and Höfer, 2006), with permission from the ACS.)

H-atoms parallel to the surface. This allows the formation of H–H complexes with some excess energy. Redistribution of the energy causes the excitation of vibrational motions perpendicular to the surface and finally desorption of H_2. Electron-mediated energy transfer was also found for the desorption of CO_2 from an Ru(0001) surface covered with CO and atomic oxygen (Bonn, Kleyn and Kroes, 2001).

22.4.6
Laser Control

Traditional chemistry controls reactions using a statistical approach, that is, changing equilibria via modifying external conditions. Ultrafast lasers open the possibility for controlling the wavepacket dynamics involved in elementary reaction steps. The field emerged in the 1980s triggered by the suggestion of two-photon pump-dump control of branching reactions (Tannor and Rice, 1985). Here, it is assumed that there are two exit channels on the electronic ground-state PES, $AB + C \leftarrow ABC \rightarrow A + BC$. Upon electronic excitation, the wave packet samples the available configurations on the excited-state potential and if dumped back onto the ground state, in the right moment, one can preferentially populate either of the two exit channels. A next breakthrough in terms of theoretical modeling has been the development of optimal control theory by Rabitz et al. (Peirce, Dahleh and Rabitz, 1988), where the problem of finding the proper field for populating a certain target is solved by means of functional optimization. Predicting laser fields, either by optimal control or by optimization of parameterized field forms and achieving control by tuning laser parameters in the laboratory, is known as open loop control. There are many examples for experimental realization of open loop control in atoms and diatomic molecules (Wollenhaupt, Engel and Baumert, 2005), but also in more complex systems such as the selective ground-state bondbreaking of diazomethane by chirped pulses (Windhorn et al., 2003), the population of high vibrational states of the CO group in carboxyhemoglobin (Ventalon et al., 2004), or the control of fragmentation products in the Coulomb explosion of silver clusters (Döppner et al., 2005).

From the experimental point of view it would, however, be desirable to define the target and then let the molecule decide what kind of field is required. The appropriate approach is optimized feedback closed loop control and has been suggested by Rabitz et al. (Judson and Rabitz, 1992). Here, the laser pulse is iteratively optimized using a pulse shaper by incorporating the feedback from the experiment, that is, the reaction yield. For molecular reactions, the first successful demonstration has been by the group of Gerber in 1998 (see Figure 22.19) (Assion et al., 1998).

Meanwhile, there are a number of applications (Brixner and Gerber, 2003) even to complex biological systems (Wohlleben et al., 2005). From the theoretical point of view, this approach represents a serious challenge since even for simpler systems such as polyatomics no first-principle simulation is feasible. In certain cases, however, insight can be gained from appropriately chosen reduced model systems (Daniel et al., 2003).

Using shaped laser pulses to control wave-packet dynamics is a rapidly developing field holding potential in different areas. Some examples include coherently

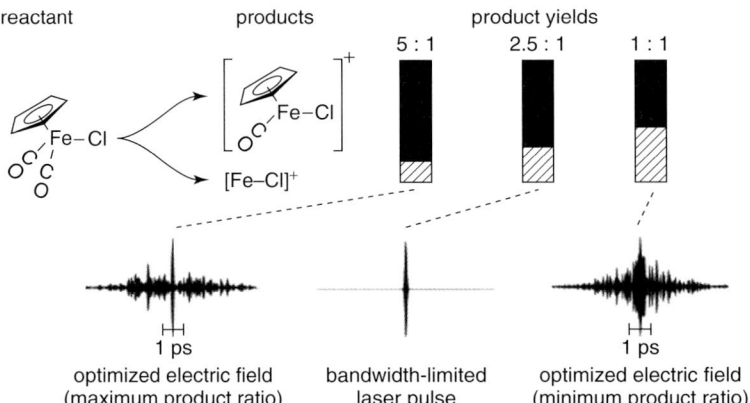

Fig. 22.19 Coherent feedback control of the $[CpFe(CO)Cl]^+/[FeCl]^+$ branching ratio. (Reprinted from (Brixner and Gerber, 2003), with permission from Wiley-VCH Verlag GmbH & Co. KgaA.)

controlled laser microscopy employing the coherent anti-Stokes Raman effect with a single pulse (Dudovich, Oron and Silberberg, 2002), the use of coherent control theory to perform logic operations with vibrational qubits in quantum information (de Vivie-Riedle and Troppmann, 2007), or even the manipulation of certain features in nonlinear optical spectra (Abramavicius and Mukamel, 2004).

22.4.7
New Frontiers

22.4.7.1 Generation of High Harmonics and Attosecond Pulses

If ultrashort high energy pulses interact with matter, coherent deep ultraviolet (XUV) and soft X-ray radiation is generated. In a typical experiment, 50–5-femtosecond short laser pulses with a center wavelength of 800 nm are focused into a gas jet of noble gas atoms or small molecules. After the jet XUV radiation, propagating collinear with the NIR-beam can be observed (Kapteyn et al., 2007). A capillary filled with gas can be used instead of a jet. A schematic spectrum of the generated radiation is shown in Figure 22.20. It consists of many odd harmonics of the fundamental (Agostini and DiMauro, 2004; Pfeifer, Spielmann and Gerber, 2006). Under suitable conditions, harmonics with order numbers up to 100 can be observed and the shortest available wavelengths are only a few nanometers. In the low-frequency region, the intensity decreases with the order until the order number reaches a plateau region in which the intensity shows only a weak dependence on the order. At high frequencies when the photon energy is $\approx 3.2 \cdot U_P + I_P$, the plateau region ends and a "cutoff" region follows in which the intensity decays rapidly with the order number. I_P is the ionization potential of the medium and U_P the ponderomotive energy associated with the pulse field strength E_0 and the angular frequency ω of the fundamental via $U_P = e_0^2 \, E_0^2 / (4 \, m_e \, \omega^2)$.

The appearance of this radiation is explained by the following three-step model (Corkum and Krausz, 2007). In the case of very intense laser pulses, the electric field strength, approaches for the valence electrons, the strength of the binding Coulomb potential. When the laser field adopts its maximal strength during an optical cycle, the tunneling probability becomes quite large and a significant fraction of the electrons is set free. Those electrons are subsequently accelerated by the laser field. Since the field reverses its direction in the next half cycle, the electrons are driven back to the atom or molecule. When they

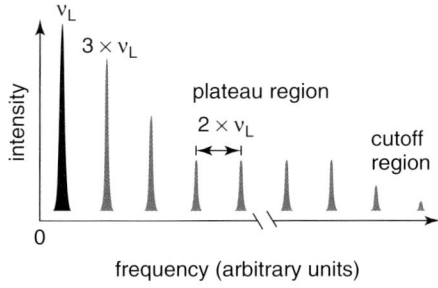

Fig. 22.20 Schematic spectrum of high harmonics generated by an ultrashort laser pulse with the center frequency ν_L. (Adapted from Agostini and DiMauro, 2004.)

return, they have a kinetic energy of about 50–1000 eV. Inelastic scattering and recombination of the electrons occur with high probability and transfer their kinetic energy into electromagnetic radiation. The recombination events themselves and the thereby emitted XUV pulses are extremely short, that is, in the order of 100 as. Since these events take place every half optical cycle, a train of attosecond pulses is generated and the spectrum consists of odd harmonics. In the maximum of the driving pulse, the recolliding electrons have the highest kinetic energy and lead to the contributions with the shortest wavelengths. In the case of a very short driving pulse, only a single attosecond pulse contributes to the highest frequencies. A single attosecond pulse can then be isolated from the train by selecting only the short wavelength contributions with an appropriate band pass filter. This approach led to isolated attosecond pulses with durations down to 130 as (Sansone et al., 2006).

22.4.7.2 Application of High Harmonics and Attosecond Pulses

High harmonic generation is a promising technique for coherent radiation in the XUV and soft X-ray region. For example, in time-resolved photoelectron spectroscopy, such probe pulses provide access to high lying core electrons and their chemical shifts. Several techniques like quasi-phase matching and pulse shaping are applied to increase the high harmonic yield (Pfeifer, Spielmann and Gerber, 2006; Kapteyn et al., 2007).

Attosecond time resolution is of particular interest for extremely fast electronic processes, like the decay of atomic core excitations. Such processes can be expected if no rearrangements or motions of nuclei are involved. Several experiments have been carried out using an attosecond pulse for excitation and the field of the associated NIR pulse for probing (Kling and Vrakking, 2008). The XUV attosecond pulse results in photoemission of valence as well as core-level electrons. The NIR field accelerates the photoelectrons and changes the photoelectron spectrum. This effect depends sensitively on the strength of the NIR field at the moment of the electron release and thereby on the phase of the NIR carrier wave at the time of the attosecond excitation. Accordingly, this "streaking" technique provides a time resolution comparable to the attosecond pulse length. Even more interesting, the ejection time of secondary photoelectrons, which result from relaxation processes like, Auger decays, can be precisely determined (Corkum and Krausz, 2007). In a similar way, tunnel ionization induced by the NIR field can be used as probe process with a time resolution much below the optical period.

Another approach uses the high harmonic spectrum to investigate the dynamics in small molecules during the generation process (Itatani et al., 2004). After the tunnel ionization, the remaining ion is usually in a nonstationary configuration and its wavefunction evolves while the photoelectron first propagates away and then returns to the core. Since the whole process is coherent, the interference between the wavefunctions of the returning electron and the core has a strong impact on the emission of the XUV light (Kling and Vrakking, 2008). The characterization of the XUV spectrum, in dependence on the molecular orientation, can therefore provide a tomographic picture of the electronic wavefunction and its evolution.

22.4.7.3 Carrier Envelope Phase and Frequency Comb

In the case of very short laser pulses and nonlinear interactions, the carrier envelope phase (CEP), that is, the relative phase between the pulse envelope and the carrier wave becomes important. For a linearly polarized Gaussian pulse, a CEP of 0 indicates that the maximum of the envelope coincides with a field maximum of the carrier wave whereas a CEP of $\pm\pi/2$ indicates a zero crossing of the field at the maximum of the envelope (see Figure 22.21). The maximal field strength is therefore slightly higher in the former case than in the latter.

In a femtosecond oscillator, the phase delay and the group delay of a roundtrip are somewhat different and subsequent pulses exhibit different CEPs. In the ideal case of constant phase and group delay per roundtrip the shift of the CEP between the pulses is always the same. The spectrum of such a pulse train consists of equally spaced frequency components with a spacing given by the repetition rate of the oscillator (see Figure 22.21). Extrapolating this frequency comb toward lower frequencies results in a lowest component, which deviates from zero by the carrier envelope offset frequency (f_{CEO}) given by the CEP shift per pulse (Hänsch, 2006). f_{CEO} can be measured with a 1 f to 2 f interferometer (see Figure 22.21). This requires a pulse spectrum with reasonably strong components separated by an octave from each other. If the width of the pulse spectrum is insufficient, it can be broadened by self-phase modulation in a microstructured fiber. The low-frequency components are frequency-doubled with a nonlinear crystal and the SHG signal is overlaid with the high-frequency components on a fast photodetector. The interference between the frequency-doubled low-frequency components and the high-frequency components results in a beat signal with the carrier envelope offset frequency f_{CEO}. This radiofrequency as well as the pulse repetition rate f_{Rep} can be accurately measured and compared to high-precision standards like a cesium atomic clock. The frequency of a comb mode with the integer mode number m is strictly linked to these two frequencies by

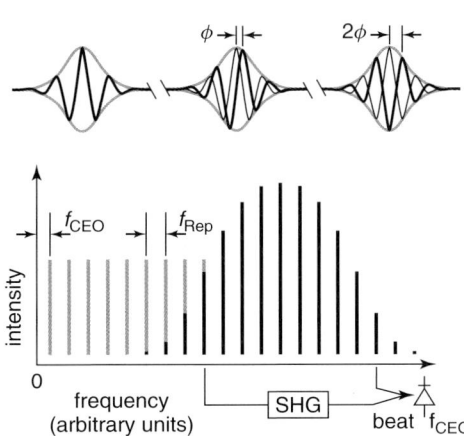

Fig. 22.21 A train of femtosecond pulses as it is generated by an oscillator. They differ only in their CEP ϕ. The lower panel shows the spectrum of the pulse train (black sticks). The gray sticks indicate an extrapolation of the observed frequencies toward zero. (Adapted from Hänsch, 2006.)

the relation $f_m = m \cdot f_{\text{Rep}} + f_{\text{CEO}}$. Accordingly, the frequency of each comb mode is known with an extremely high accuracy and can be used for high-precision spectroscopy. For this elegant approach, Theodor W. Hänsch was awarded the Nobel price in Physics in 2005 (Hänsch, 2006). The technique was, for example, applied to measure the 1S–2S transition frequency of the hydrogen atom with an accuracy of 1.4 parts in 10^{14} (Fischer et al., 2004). Such experiments are highly relevant to test fundamental physical laws. The frequency comb allows also implementing atomic clocks based on optical transitions. They have the potential for a much higher accuracy than clocks operating at radiofrequencies and measurements with the same level of accuracy can be performed in much shorter times due to the high optical frequencies.

It is also possible to stabilize and control the f_{CEO} by this technique and thereby to generate pulses with deterministic CEP and electric field (Baltuška et al., 2003). Adjusting the CEP to 0 or π results in the maximal field strength (see Figure 22.21), which is particularly useful in the case of nonlinear interactions of few cycle pulses. A prominent example is the generation of single attosecond pulses. In this case, the high-frequency components of the high harmonic spectrum are only generated during the central and strongest cycle of the electric field. Selecting these spectral components with a filter results in a single attosecond pulse (see above) (Sansone et al., 2006).

22.4.7.4 Time-Resolved X-Ray and Electron Diffraction

One central vision in ultrafast spectroscopy is to follow, in real time, the structural and geometric changes during an ultrafast process. Ultimately, this should lead to a movie with femtosecond time resolution. Since structure and function are strongly related to each other, a much deeper understanding of the relevant mechanisms is expected for most processes.

To measure time-resolved structural changes, X-ray and electron diffraction are applied. X-ray diffraction itself is an extremely well established technique and the generation of short X-ray pulses has made large progress. Since the interaction of X-ray photons with matter is comparably weak and accordingly the diffraction efficiency low, a high X-ray flux is needed to observe a diffraction pattern. In several experiments, laser-driven plasmas have been used as X-ray sources (Bargheer et al., 2006). An energetic laser pulse is focused onto a metal target like, for example, a copper tape. The high peak intensity in the order of 10^{17} W cm^{-2} results in a plasma with very energetic electrons. They penetrate into the metal and generate incoherent Bremsstrahlung and characteristic line emission, for example, K_α radiation. From this X-ray radiation, only the fraction that propagates toward the sample is used as probe beam. An optical pump pulse excites the sample and the X-ray diffraction pattern is recorded with a charge coupled device (CCD) camera in dependence on the delay time between the excitation and the X-ray pulse. Mostly, crystalline samples have been investigated so far since they provide the highest diffraction efficiencies. In this way, it was already possible to observe nonthermal melting, the dynamics of optical phonons, and the propagation of acoustic sound and shock waves. In the future, accelerator-based sources like the free electron laser will provide much higher X-ray fluxes (Pfeifer, Spielmann and Gerbr, 2006). This will make ultrafast X-ray diffraction applicable to a much wider class of samples and problems.

As an alternative to X-rays, electrons can be used as probe in ultrafast diffraction experiments (Zewail, 2006; Siwick et al., 2003). If their kinetic energy is in the order of 30 keV, their de Broglie wavelength matches internuclear distances in the solid phase and results in diffraction patterns containing rich information. Since electrons interact very strongly with matter, their diffraction efficiency is several orders of magnitude higher than that of X-rays and quite low fluxes are needed to obtain a good signal-to-noise ratio. Owing to the Coulomb repulsion between the electrons, space charge effects are usually quite severe and electron pulses spread strongly in space and time when they consist of thousand or more electrons. This makes the experiments challenging and the available time resolution is about few hundred femtoseconds (Baum et al., 2007). The electron pulses are generated by irradiating a photocathode with femtosecond laser pulses. The photoemitted electrons are accelerated by a high voltage to the appropriate kinetic energy. The sample is placed as near as possible to the exit of the acceleration section since short propagation distances minimize space charge effects. A femtosecond pump pulse initiates the dynamics in the sample and a CCD camera records the diffraction pattern behind it. In the case of dense matter the sample has to be extremely thin. Otherwise it cannot be penetrated by the electrons. If this is not possible, a grazing incidence geometry is used and the diffraction pattern is measured in reflection. Meanwhile, a number of pioneering experiments have been carried out and demonstrate that, for example, melting processes (Siwick et al., 2003), phase transitions (Baum et al., 2007), and structural changes during chemical reactions (Zewail, 2006) can be studied in detail.

Acknowledgments

This work has been financially supported by the Deutsche Forschungsgemeinschaft.

Glossary

Nonadiabatic Transition: The transition between two adiabatic electronic states triggered by the motion of the nuclei.

Pump-Probe Technique: A pump pulse initiates a process and a probe pulse measures the associated transient changes of observables.

Response Function: The function that describes the material response to the applied external fields.

Wave Packet: The superposition of quantum mechanical states that does not correspond to an eigenstate of the Hamiltonian.

References

Åberg, U., Åkesson, E., Alvarez, J.-L., Fedchenia, I., and Sundström, V. (1994) *Chem. Phys.*, **183**, 269–288.

Abramavicius, D. and Mukamel, S. (2004) *J. Chem. Phys.*, **120**, 8373–8378.

Agostini, P. and DiMauro, L.F. (2004) *Rep. Prog. Phys.*, **67**, 813–855.

Asbury, J.B., Steinel, T., Stromberg, C., Gaffney, K.J., Piletic, I.R., and Fayer, M.D. (2003) *J. Chem. Phys.*, **119**, 12981–12897.

Assion, A., Baumert, T., Bergt, M., Brixner, T., Kiefer, B., Seyfried, V., Strehle, M., and Gerber, G. (1998) *Science*, **282**, 919–922.

Axt, V.M. and Kuhn, T. (2004) *Rep. Prog. Phys.*, **67**, 433–512.

Backus, S., Durfee, C.D., Murnane, M.M., and Kapteyn, H.C. III (1998) *Rev. Sci. Instrum.*, **69**, 1207–1223.

Baltuska, A., Udem, Th., Uiberacker, M., Hentschel, M., Goulielmakis, E., Gohle, Ch., Holzwarth, R., Yakovlev, V.S., Scrinzi, A., Hänsch, T.W., and Krausz, F. (2003) *Nature*, **42**, 611–615.

Barbara, P.F., Walker, G.C., and Smith, T.P. (1992) *Science*, **256**, 975–981.

Barbara, P.F., Meyer, T.J., and Ratner, M.A. (1996) *J. Phys. Chem.*, **100**, 13148–13168.

Bargheer, M., Zhavorokov, N., Woerner, M., and Elsaesser, T. (2006) *ChemPhysChem*, **7**, 783–792.

Baum, P., Yang, D.-S., and Zewail, A.H. (2007) *Science*, **318**, 788–792.

Baumert, T., Grosser, M., Thalweiser, R., and Gerber, G. (1991) *Phys. Rev. Lett.*, **67**, 3753–3756.

Bernardi, F., Olivucci, M., and Robb, M.A. (1996) *Chem. Soc. Rev.*, **25**, 321–328.

Bonn, M., Kleyn, A.W., and Kroes, G.J. (2002) *Surf. Sci.*, **500**, 475–499.

Brixner, T. and Gerber, G. (2003) *ChemPhysChem*, **4**, 418–438.

Brixner, T., Stenger, J., Vaswani, H.M., Cho, M., Blankenship, R.E., and Fleming, G.R. (2005) *Nature*, **434**, 625–628.

van Burgel, M., Wiersma, D.A., and Duppen, K. (1995) *J. Chem. Phys.*, **102**, 20–33.

Carter, S.G., Chen, Z., and Cundiff, S.T. (2007) *Phys. Rev. B*, **76**, 121303-1–121303-4.

Cerullo, G. and De Silvestri, S. (2003) *Rev. Sci. Instrum.*, **74**, 1–18.

Chemla, D.S. and Shah, J. (2001) *Nature*, **411**, 549–557.

Corkum, P.B. and Krausz, F. (2007) *Nat. Phys.*, **3**, 381–387.

Cowan, M.L., Bruner, B.D., Huse, N., Dwyer, J.R., Chugh, B., Nibbering, E.T.J., Elasaesser, T., and Miller, R.J.D. (2005) *Nature*, **434**, 199–202.

Crespo-Hernández, C.E., Cohen, B., Hare, P.M., and Kohler, B. (2004) *Chem. Rev.*, **104**, 1977–2019.

Crespo-Hernández, C.E., Cohen, B., and Kohler, B. (2005) *Nature*, **436**, 1141–1144.

Daniel, C., Full, J., González, L., Lupulescu, C., Manz, J., Merlin, A., Vajda, S., and Wöste, L. (2003) *Science*, **299**, 536–639.

de Vivie-Riedle, R. and Troppmann, U. (2007) *Chem. Rev.*, **107**, 5082–5100.

Dermota, T.E., Zhong, Q., and Castleman, A.W. Jr. (2004) *Chem. Phys.*, **104**, 1861–1886.

Domcke, W. and Stock, G. (1997) *Adv. Chem. Phys.*, **100**, 1–169.

Domcke, W., Yarkony, D.R., and Köppel, H. (eds) (2004) *Conical Intersections*, World Scientific, New Jersey.

Döppner, T., Fennel, T., Diederich, T., Tiggesbäumker, J., and Meiwes-Broer, K.H. (2005) *Phys. Rev. Lett.*, **94**, 013401.

Dudovich, N., Oron, D., and Silberberg, Y. (2002) *Nature*, **418**, 512–514.

Eisenthal, K.B. (1975) *Acc. Chem. Res.*, **8**, 118–124.

Elsaesser, T. and Kaiser, W. (1986) *Chem. Phys. Lett.*, **128**, 231–237.

Elsaesser, T. and Kaiser, W. (1991) *Annu. Rev. Phys. Chem.*, **42**, 83–107.

Fecko, C.J., Eaves, J.D., Loparo, J.J., Tokmakoff, A., and Geissler, P.L. (2003) *Science*, **301**, 1698–1702.

Feit, M.D., Fleck, J.A., and Steiger, A. (1982) *J. Comp. Phys.*, **47**, 412–433.

Fennel, Th., Döppner, T., Passig, J., Schaal, Ch., Tiggesbäumker, J., and Meiwes-Broer, K.-H. (2007) *Phys. Rev. Lett.*, **98**, 143401-1–143401-4.

Fischer, M., Kolachevsky, N., Zimmermann, M., Holzwarth, R., Udem, Th., Hänsch, T.W., Abgrall, M., Grünert, J., Maksimovic, I., Bize, S., Marion, H., Pereira Dos Santos, F., Lemonde, P., Santarelli, G., Laurent, P., Clairon, A., Salomon, C., Haas, M., Jentschura, U.D., and Keitel, C.H. (2004) *Phys. Rev. Lett.*, **92**, 230802-1–230802-4.

Först, M., Nagel, M., Awad, M., Wächter, M., Dekorsy, T., and Kurz, H. (2007) *Phys. Stat. Sol. (b)*, **244**, 2971–2987.

Frischkorn, C., Wolf, M., Höfer, U., Güdde, J., Saalfrank, P., Nest, M., Klamroth, T., Willig, F., Ernstdorfer, R., Gundlach, L., May, V., Wang, L., Duncan, W.R., and Prezhdo, O.V. (2007) in *Analysis and Control of Ultrafast Photoinduced Reactions*, (eds O. Kühn and L. Wöste), Springer, Heidelberg, pp. 387–484.

Frutos, F.G., Andruniów, T., Santoro, F., Ferré, N., and Olivucci, M. (2007) *Proc. Natl. Acad. Sci.*, **104**, 7764–7769.

Fuß, W., Kompa, K.-L., Lochbrunner, S., Müller, A.M., Schikarski, T., Schmid, W.E., and Trushin, S.A. (1998) *Chem. Phys.*, **232**, 161–174.

Ganim, Z., Chung, H.S., Smith, A.W., Deflores, L.P., Jones, K.C., and Tokmakoff, A. (2008) *Acc. Chem. Res.*, **41**, 432–441.

Grondelle van, R. and Novoderezhkin, V.I. (2006) *Phys. Chem. Chem. Phys.*, **8**, 793–807.

Güdde, J., Berthold, W., and Höfer, U. (2006) *Chem. Phys.*, **106**, 4261–4280.

Gühr, M., Bargheer, M., Fushitani, M., Kiljunen, T., and Schwentner, N. (2007) *Phys. Chem. Chem. Phys.*, **9**, 779–801.

Hamm, P., Helbing, J., and Bredenbeck, J. (2008) *Annu. Rev. Phys. Chem.*, **59**, 291–317.

Hamm, P., Lim, M., and Hochstrasser, R.M. (1998) *J. Phys. Chem. B*, **102**, 6123–6138.

Hänsch, T.W. (2006) *ChemPhysChem*, **7**, 1170–1187.

Hartl, I. and Zinth, W. (2000) *J. Phys. Chem.*, **104**, 4218–4222.

Heijs, D.-J., Dijkstra, A.G., and Knoester, J. (2007) *Chem. Phys.*, **341**, 230–239.

Hertel, I.V. and Radloff, W. (2006) *Rep. Prog. Phys.*, **69**, 1897–2003.

Heyne, K., Krishnan, G.M., and Kühn, O. (2008) *J. Phys. Chem. B*, **112**, 7909–7915.

Horng, M.L., Gardecki, J.A., Papazyan, A., and Maroncelli, M. (1995) *J. Phys. Chem.*, **99**, 17311–17337.

Huber, R., Tauser, F., Brodschelm, A., Bichler, M., Abstreiter, G., and Leitenstorfer, A. (2001) *Nature*, **414**, 286–289.

Itatani, J., Levesque, J., Zeidler, D., Niikura, H., Pepin, H., Kieffer, J.C., Corkum, P.B., and Villeneuve, D.M. (2004) *Nature*, **432**, 867–871.

Jonas, D.M. (2003) *Annu. Rev. Phys. Chem.*, **54**, 425–463.

Jortner, J. and Berry, R.S. (1968) *J. Chem. Phys.*, **48**, 2757–2766.

Judson, R.S. and Rabitz, H. (1992) *Phys. Rev. Lett.*, **68**, 1500–1503.

Kapteyn, H., Cohen, O., Christov, I., and Murnane, M. (2007) *Science*, **317**, 775–778.

Kersting, R., Lemmer, U., Deussen, M., Bakker, H.J., Mahrt, R.F., Kurz, H., Arkhipov, V.I., Bässler, H., and Göbel, E.O. (1994) *Phys. Rev. Lett.*, **73**, 1440–1443.

Kling, M.F. and Vrakking, M.J.J. (2008) *Annu. Rev. Phys. Chem*, **59**, 463–492.

Kosloff, D. and Kosloff, R. (1982) *J. Comp. Phys.*, **52**, 35–53.

Kovalenko, S.A., Dobryakov, A.L., Ruthmann, J., and Ernsting, N.P. (1999) *Phys. Rev. A*, **59**, 2369–2384.

Krummel, A. and Zanni, M. (2006) *J. Phys. Chem. B*, **110**, 13991–14000.

Kukura, P., McCamant, D.W., and Mathies, R.A. (2007) *Annu. Rev. Phys. Chem.*, **58**, 461–488.

Kukura, P., McCamant, D.W., Yoon, S., Wandschneider, D.B., and Mathies, R.A. (2005) *Science*, **310**, 1006–1009.

Lanzani, G., Cerullo, G., Polli, D., Gambetta, A., Zavelani-Rossi, M., and Gadermaier, C. (2005) in *Physics of Organic Semiconductors* (ed. W. Brütting), Wiley-VCH, Weinheim.

Laubereau, A. and Kaiser, W. (1978) *Rev. Mod. Phys*, **50**, 607–665.

Lochbrunner, S., Wurzer, A.J., and Riedle, E. (2003) *J. Phys. Chem. A*, **107**, 10580–10590.

May, V. and Kühn, O. (2004) *Charge and Energy Transfer Dynamics in Molecular Systems*, Wiley-VCH Verlag GmbH, Weinheim.

Meyer, H.-D., Manthe, U., and Cederbaum, L. (1990) *Chem. Phys. Lett.*, **165**, 73–78.

Mukamel, S. and Abramavicius, D. (2004) *Chem. Rev.*, **104**, 2073–2098.

Mukherjee, P., Kass, I., Arkin, I., and Zanni, M.T. (2006) *Proc. Natl. Acad. Sci.*, **103**, 3528–3533.

Müller, J.G., Lupton, J.M., Feldmann, J., Lemmer, U., Scharber, M.C., Sariciftci, N.S., Brabec, C.J., and Scherf, U. (2005) *Phys. Rev. B*, **72**, 195208.

Nibbering, E.T.J., Dreyer, J., Kühn, O., Bredenbeck, J., Hamm, P., and Elsaesser, T. (2007) in *Analysis and Control of Ultrafast Photoinduced Reactions* (eds O. Kühn, L. Wöste), Springer, Heidelberg, pp. 619–687.

Peirce, A., Dahleh, M., and Rabitz, H. (1988) *Phys. Rev. A*, **37**, 4950–4494.

Perez Lustres, J.L., Kovalenko, S.A., Mosquera, M., Senyushkina, T., Flasche, W., and Ernsting, N.P. (2005) *Angew. Chem. Int. Ed.*, **44**, 2–6.

Pfeifer, T., Spielmann, C., and Gerber, G. (2006) *Rep. Prog. Phys.*, **69**, 443–505.

Prezhdo, O.V., Duncant, W.R., and Prezhdo, V.V. (2008) *Acc. Chem. Res.*, **41**, 339–348.

Rabitz, H., de Vivie-Riedle, R., Motzkus, M., and Kompa, K. (2000) *Science*, **288**, 824–828.

Renger, T., May, V., and Kühn, O. (2001) *Phys. Rep.*, **343**, 137–254.

Riedle, E., Beutter, M., Lochbrunner, S., Piel, J., Schenkl, S., Spörlein, S., and Zinth, W. (2000) *Appl. Phys. B*, **71**, 457–465.

Rose, T.S., Rosker, M.J., and Zewail, A.H. (1989) *J. Chem. Phys.*, **91**, 7415–7436.

Rosker, M.J., Dantus, M., and Zewail, A.H. (1988) *Science*, **241**, 1200–1202.

Rubtsov, I.V. and Yoshihara, K. (1999) *J. Phys. Chem. A*, **103**, 10202–10212.

Sanov, A. and Lineberger, W.C. (2004) *Phys. Chem. Chem. Phys.*, **6**, 2018–2032.

Sansone, G., Benedetti, E., Calegari, F., Vozzi, C., Avaldi, L., Flammini, R., Poletto, L., Villoresi, P., Altucci, C., Velotta, R., Stagira, S., De Silvestri, S., and Nisoli, M. (2006) *Science*, **314**, 443–446.

Schanz, R., Kovalenko, S.A., Kharlanov, V., and Ernsting, N.P. (2001) *Appl. Phys. Lett.*, **79**, 566–568.

Scheblykin, I.G., Yartsev, A., Pullerits, T., Gulbanis, V., and Sundström, V. (2007) *J. Phys. Chem. B*, **111**, 6303–6321.

Scherer, N.F., Carlson, R.J., Matro, A., Du, M., Ruggiero, A.J., Romero-Rochin, V., Cina, J.A., Fleming, G.R., and Rice, S.A. (1991) *J. Chem. Phys.*, **95**, 1487–1511.

Schlosser, M. and Lochbrunner, S. (2006) *J. Phys. Chem. B*, **110**, 6001–6009.

Schmitt, M., Knopp, G., Materny, A., and Kiefer, W. (1998) *J. Phys. Chem. A*, **102**, 4059–4065.

Schoenlein, R.W., Peteanu, L.A., Mathies, R.A., and Shank, C.V. (1991) *Science*, **254**, 412–415.

Schultheis, L., Kuhl, J., and Honold, A. (1986) *Phys. Rev. Lett.*, **57**, 1635–1638.

Schultz, T., Samoylova, E., Radloff, W., Hertel, I.V., Sobolewski, A.L., and Domcke, W. (2004) *Science*, **306**, 1765–1768.

Schwarzer, D., Hanisch, C., Kutne, P., and Troe, J. (2002) *J. Phys. Chem. A*, **106**, 8019–8028.

Siwick, B.J., Dwyer, J.R., Jordan, R.E., and Miller, R.J.D. (2003) *Science*, **302**, 1382–1385.

Spence, D.E., Kean, P.N., and Sibbett, W. (1991) *Opt. Lett.*, **16**, 42–44.

Stock, G. and Thoss, M. (2005) *Adv. Chem. Phys.*, **131**, 243–375.

Stock, K., Schriever, C., Lochbrunner, S., and Riedle, E. (2008) *Chem. Phys.*, **349**, 197–203.

Stolow, A., Bragg, A.E., and Neumark, D.M. (2004) *Chem. Rev.*, **104**, 1719–1757.

Sundström, V. (2008) *Annu. Rev. Phys. Chem.*, **59**, 53–57.

Takeuchi, S. and Tahara, T. (2000) *Chem. Phys. Lett.*, **326**, 430–438.

Tamura, K., Ippen, E.P., Haus, H.A., and Nelson, L.E. (1993) *Opt. Lett.*, **18**, 1080–1082.

Tannor, D.J. and Rice, S.A. (1985) *J. Chem. Phys.*, **83**, 5013–5018.

Trushin, S.A., Kosma, K., Fuß, W., and Schmid, W.E. (2008) *Chem. Phys.*, **347**, 309–323.

Ullrich, S., Schultz, T., Zgierski, M.Z., and Stolow, A. (2004) *J. Am. Chem. Soc.*, **126**, 2262–2263.

Ventalon, C., Fraser, J.M., Vos, M.H., Alexandrou, A., Martin, J.-L., and Joffre, M. (2004) *Proc. Natl. Acad. Sci.*, **101**, 13216–13220.

Wang, H. and Thoss, M. (2008) *Chem. Phys.*, **347**, 139–151.

Weiner, A.M. (2000) *Rev. Sci. Instrum.*, **71**, 1929–1960.

Weiner, A.M., De Silvestri, S., and Ippen, E.P. (1985) *J. Opt. Soc. Am. B*, **2**, 654–661.

Windhorn, L., Yeston, J.S., Witte, T., Fuss, W., Motzkus, M., Proch, D., Kompa, K.-L., and Moore, C.B. (2003) *J. Chem. Phys.*, **119**, 641–645.

Wohlleben, W., Buckup, T., Herek, J.L., and Motzkus, M. (2005) *ChemPhysChem*, **6**, 850–857.

Wollenhaupt, M., Engel, V., and Baumert, T. (2005) *Annu. Rev. Phys. Chem.*, **56**, 25–56.

Woutersen, S. and Bakker, H.J. (2006) *Phys. Rev. Lett.*, **96**, 138305.

Woutersen, S. and Cristalli, G. (2004) *J. Chem. Phys.*, **121**, 5381–5386.

Woutersen, S. and Hamm, P. (2002) *J. Phys.: Cond. Matt.*, **14**, R1035–R1062.

Wurzer, A.J., Wilhelm, T., Piel, J., and Riedle, E. (1999) *Chem. Phys. Lett.*, **299**, 296–302.

Yoshihara, K., Tominaga, K., and Nagasawa, Y. (1995) *Bull. Chem. Soc. Jpn.*, **68**, 696–712.

Zewail, A.H. (2000) *J. Phys. Chem. A*, **104**, 5660–5694.

Zewail, A.H. (2006) *Annu. Rev. Phys. Chem*, **57**, 65–103.

Zinth, W. and Wachtveitl, J. (2005) *Chem. Phys. Chem*, **6**, 871–880.

Further Reading

Cho, M. (2008) *Chem. Rev.*, **108**, 1331–1418.

Diels, J.-C. and Rudolph, W. (2006) *Ultrashort Laser Pulse Phenomena: Fundamentals, Techniques, and Applications on a Femtosecond Time Scale*, Academic Press, Burlington.

Fayer, M.D. (ed.) (2001) *Ultrafast Infrared and Raman Spectroscopy*, Marcel Dekker, New York.

Jortner, J. and Bixon, M. (eds) (1999) *Adv. Chem. Phys.*, **106–107**.

Kühn, O. and Wöste, L. (eds) (2007) *Analysis and Control of Ultrafast Photoinduced Reactions*, Springer, Heidelberg.

Manz, J. (1997) in *Femtochemistry and Femtobiology* (ed. V. Sundström), World Scientific, Singapore, p. 80.

Meier, T., Thomas, P., and Koch, S.W. (2007) *Coherent Semiconductor Optics*, Springer, Heidelberg.

Meyer, H.-D., Gatti, F., and Worth, G.A. (eds) (2009) *Multidimensional Quantum Dynamics: MCTDH Theory and Applications*, Wiley-VCH Verlag GmbH, Weinheim.

Mukamel, S. (1995) *Principles of Nonlinear Optical Spectroscopy*, Oxford University Press, Oxford.

23
Surface Second Harmonic and Sum-Frequency Generation

Mitsumasa Iwamoto, Takaaki Manaka and Eunju Lim

23.1	Introduction 819	
23.2	Order Parameters and Surface Layer 819	
23.3	Nonlinear Polarization and Surface Optical Second Harmonic Generation 820	
23.3.1	SHG as a Tool for Characterizing Surface Structure 821	
23.3.2	EFISHG as a Tool for Probing Excess Surface Charges 823	
23.3.3	Spectroscopic Consideration of the SHG 823	
23.3.4	Sum-Frequency Generation 825	
23.4	Measurement Details of SHG and EFISHG 826	
23.4.1	Determination of Surface Structure and Orientational Order Parameters 826	
23.4.1.1	Fundamentals 826	
23.4.1.2	Experimental Configuration of SHG Measurement 826	
23.4.1.3	Experimental Configuration of SFG Measurement 826	
23.4.2	Probing of the SHG Intensity Profile as a Convolution Beam Pattern 827	
23.4.2.1	Fundamentals 827	
23.4.2.2	Experimental Configuration of EFISHG Measurement from OFET Structure 828	
23.5	Results of SHG and EFISHG 829	
23.5.1	Determination of Surface Structure and Orientational Order Parameters 829	
23.5.2	Probing of Electric Field $E(0)$ by EFISHG Measurement 830	
	References 831	
	Further Reading 832	

Encyclopedia of Applied Spectroscopy. Edited by David L. Andrews.
Copyright © 2009 WILEY-VCH Verlag GmbH & Co. KGaA, Weinheim
ISBN: 978-3-527-40773-6

23.1 Introduction

The use of optical second harmonic generation (SHG) measurement for the exploration of dielectric polarization and related surface phenomena is described. A classification of materials using orientational order parameters shows that a surface monolayer is the most generalized system for discussing SHG enhancement from materials, and a theory developed for a monolayer system is broadly extended to other systems such as bulk system materials. On the basis of this background, SHG enhancement from an organic monolayer system is described with a schematic illustration of a typical experimental system. It is shown that SHG is generated from surface monolayers due to their noncentrosymmetric structure and is available for the determination of orientational order of surface layers, and so on. It is also shown that SHG is generated from layers with centrosymmetric structure that are subjected to a high static electric field, owing to induced effective dipoles. Results show that enhanced SHG is available for probing excess surface charges. In addition, optical sum-frequency generation (SFG) is briefly described, and we show that this can be available as spectroscopic method for detecting the motion of molecules at the molecular level.

23.2 Order Parameters and Surface Layer

The orientational order parameter is specially important to specify the dielectric property of surface layers that generate SHG and SFG. Material systems composed of rod molecules are classified using orientational order parameters S_n ($n = 1, 2, 3$) defined by the thermodynamic average of the Legendre polynomials, $P_n(\cos\theta)$ ($n = 1, 2,$ and 3), of the orientational angle θ of molecules with respect to the surface. For isotropic bulk materials, rod molecules are randomly distributed in space. Thereby $S_1 = \langle P_1(\cos\theta)\rangle = \langle\cos\theta\rangle = 0$, $S_2 = \langle P_2(\cos\theta)\rangle = \langle(3\cos^2\theta - 1)/2\rangle = 0$, and $S_3 = \langle P_3(\cos\theta)\rangle = \langle(5\cos^3\theta - 3\cos\theta)/2\rangle = 0$. On the other hand, orientational direction of molecules is restricted for ordered materials. For example, constituent molecules in nematic phase bulk liquid crystals (LCs) are totally distributed in "up" and "down" directions, without distinction between them. Therefore, $S_1 = 0$, $S_3 = 0$, but $S_2 \neq 0$. Interestingly, for a monolayer composed of rod molecules, all molecules point upward owing to symmetry breaking. Therefore, S_1, S_2, $S_3 \neq 0$. This simple classification suggests that the physical properties of materials are expressed in terms of orientational order parameters. For example, nematic LCs are characterized using $S_2 \neq 0$. Similarly, the physical properties of surface layers such as monolayers are described using nonzero order parameters S_1, S_2, and S_3. Obviously, nonzero values for the parameters, S_1 and S_3 are specific parameters, are connected to symmetry breaking. In other words, the specific physico-chemical properties of surface layers, for example, monolayers are expressed using nonzero S_1 and S_3. Hence, the generation of SH from surface layers can be discussed in terms of orientational order parameters S_1 and S_3.

23.3
Nonlinear Polarization and Surface Optical Second Harmonic Generation

In the absence of a static electric field, that is, a DC field, the SHG signal originates from a nonlinear dielectric polarization produced by the quantum interaction between the electrons in molecules and the external electric field, that is, laser light Bloembergen, 1965; Shen, 1984. Under the electric dipole approximation, SHG is allowed from the noncentrosymmetric system. On account of symmetry breaking, organic monolayers on metal, film, and water surface possess a noncentrosymmetric structure. Hence, SHG is available for characterizing orientational structure of such monolayers.

Interestingly, SHG is also allowed from a centrosymmetric system if the electric quadrupole effect is prominent. SHG due to the quadrupole effect is a bulk phenomenon and observed in thin films composed of centrosymmetric molecules such as phthalocyanine (Pc) Chollet, Kajzar and Le Moigne, 1990, fullerene (C_{60}) Wang et al., 1992, and other organic molecular films. Fortunately, the resonant conditions are different between the electric dipole mechanism and the electric quadrupole mechanism, and one can distinguish the two mechanisms by appropriately choosing a wavelength of laser used in the SHG experiment.

On the other hand, in the presence of a static electric field, the electron distribution of any centrosymmetric molecule is distorted, and an effective dipole can be induced (DC-field-induced dipole). In this situation, a different type of nonlinear polarization is produced in the centrosymmetric molecular system, and SH can be freshly generated. This is called *electic-field-induced SHG* (*EFISHG*). In organic electronic devices, there are many origins leading to the formation of local static electric field Sze, 1981; Sessler, 1987. Obviously external DC biasing used for the organic device operation is one of the possible origins. A displacement of excess charges at the metal–organic material system, injection of carriers form the metal electrode into the organic device, and so on, are other possible origins. In the following sections, our discussion is focused on SHG due to the electric dipole effect and EFISHG, because these SHG processes are very important for probing surface polarization and related phenomena.

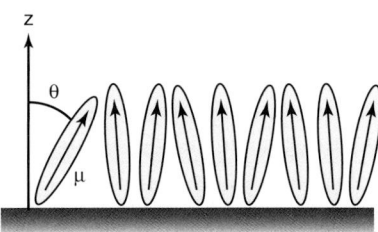

Fig. 23.1 Schematic image of the molecular orientation on a material surface.

23.3.1
SHG as a Tool for Characterizing Surface Structure

The orientational order of organic monolayers comprised of rodlike molecules is specified by using nonzero orientational order parameters, S_n ($n = 1, 2, 3$), defined by the thermodynamic average of the Lagendre polynomials, $P_n(\cos\theta)$ ($n = 1, 2, 3$), of the orientational angle θ of molecules from the normal to the surface (Figure 23.1) Iwamoto and Wu, 2000; Luckhurst and Gray, 1979. Nonlinear dielectric polarization contributing to the SHG is characterized by the second-order nonlinear electronic polarizability $\langle \overleftrightarrow{\alpha}^{(2)} \rangle$ at an angular frequency ω of the external electric field. This polarization is expressed approximately as

$$\mathbf{P}^{(2)} = \chi^{(2)} : \mathbf{EE} = N_s \langle \overleftrightarrow{\alpha}^{(2)} \rangle : \mathbf{EE} \quad (23.1)$$

The second-order susceptibility (SOS) $\chi^{(2)} \equiv [\chi_{ijk}]$ of the monolayer is related to the components of the nonlinear molecular electronic polarizability $\alpha_{i'j'k'}$ by $\chi_{ijk} = N_s \langle T_{ijk}^{i'j'k'} \rangle \alpha_{i'j'k'}$, where $T_{ijk}^{i'j'k'} = R_i^{i'} R_j^{j'} R_k^{k'}$ with the usual Euler rotation matrix $R(\phi, \theta, \psi) = [R_i^{i'}]$ describing the coordinate transformation between the molecular system (i', j', k') and the laboratory system (x, y, z) Heinz, 1991. The complete expression of the macroscopic SOS tensor $\chi^{(2)}$ of a monolayer with C_∞-symmetry on a material surface is obtained as a function of the components of the molecular nonlinear electronic polarizability $\alpha_{i'j'k'}$ and the orientational order parameters $S_n \equiv \langle P_n(\cos\theta) \rangle$. With SHG one can deduce $\chi^{(2)}$.

Nonlinear polarization $\mathbf{P}^{(2)} = \mathbf{P}^N$ is a sum of the polarization $\mathbf{P}_{ch}^N$ associated with the chirality of the monolayer and the polarizations $\mathbf{P}_{ach}^N$ associated with the nonchirality of the monolayer. The two polarizations in vector form are Iwamoto, Wu and Ou-Yang, 2000

$$\mathbf{P}_{ch}^N = \frac{1}{2} s_{14} [(\mathbf{E} \cdot \mathbf{n})(\mathbf{F} \times \mathbf{n}) + (\mathbf{F} \cdot \mathbf{n}) \\ \times (\mathbf{E} \times \mathbf{n})] + \frac{1}{2} a_{14}(\mathbf{E} \times \mathbf{F}) \\ + \frac{1}{2}(a_{36} - a_{14})\mathbf{n} \cdot (\mathbf{E} \times \mathbf{F})\mathbf{n}$$

(23.2)

$$\mathbf{P}_{ach}^N = (s_{33} - s_{15} - s_{31})(\mathbf{n} \cdot \mathbf{E})(\mathbf{n} \cdot \mathbf{F})\mathbf{n} \\ + s_{31}(\mathbf{E} \cdot \mathbf{F})\mathbf{n} + \frac{1}{2} s_{15}[(\mathbf{n} \cdot \mathbf{F})\mathbf{E} \\ + (\mathbf{n} \cdot \mathbf{E})\mathbf{F}] + \frac{1}{2} a_{15}(\mathbf{E} \times \mathbf{F}) \times \mathbf{n}$$

(23.3)

with seven independent nonzero elements given as

$$s_{14} = \frac{N_s}{2} S_2(\sigma_{14} - \sigma_{25}) = \chi_{123} + \chi_{132}$$

$$s_{15} = \frac{N_s}{5}(S_1 - S_3)(2\sigma_{33} - \sigma_{32} - \sigma_{31})$$
$$+ \frac{N_s}{10}(3S_1 + 2S_3)(\sigma_{24} + \sigma_{15})$$
$$= \chi_{131} + \chi_{113}$$

$$s_{31} = \frac{N_s}{10}(S_1 - S_3)(2\sigma_{33} - \sigma_{24} - \sigma_{15})$$
$$+ \frac{N_s}{10}(4S_1 + S_3)(\sigma_{32} + \sigma_{31})$$
$$= \chi_{311}$$

$$s_{33} = \frac{N_s}{5}(S_1 - S_3)(\sigma_{32} + \sigma_{31} + \sigma_{24}$$
$$+ \sigma_{15}) + \frac{N_s}{5}(3S_1 + 2S_3)\sigma_{33}$$
$$= \chi_{333}$$

$$a_{14} = \frac{N_s}{3}(\lambda_{14} + \lambda_{25} + \lambda_{36})$$
$$+ \frac{N_s}{6}S_2(\lambda_{14} + \lambda_{25} - 2\lambda_{36})$$
$$= \chi_{123} - \chi_{132}$$

$$a_{15} = \frac{N_s}{2}S_1(\lambda_{15} - \lambda_{24}) = \chi_{131} - \chi_{113}$$

$$a_{36} = \frac{N_s}{3}(\lambda_{14} + \lambda_{25} + \lambda_{36})$$
$$+ \frac{N_s}{3}S_2(2\lambda_{36} - \lambda_{14} - \lambda_{25})$$
$$= \chi_{312} - \chi_{321} \quad (23.4)$$

where the two 3×6 matrices, (σ_{ij}) and (λ_{ij}), are defined by the components of the nonlinear molecular polarizability $\alpha_{i',j',k'}$ in the same way as the conventional contracted notation of (s_{ij}) and (a_{ij}) defined from $\chi^{(2)}$ Shen, 1984; Nye, 1957, that is, $\sigma_{11} = \alpha_{111}$, $\sigma_{14} = \alpha_{123} + \alpha_{132}$, $\lambda_{14} = \alpha_{123} - \alpha_{132}$, and so on, and $\mathbf{E}$ and $\mathbf{F}$ are the external electric fields. $\mathbf{n}$ is unit vector normal to the water surface. In Eqs (23.2) and (23.3), two kinds of external electric fields $\mathbf{E}$ and $\mathbf{F}$ are used. In the SHG measurement, the same electric field $\mathbf{E} = \mathbf{F}$ from an optical source is used, but in the SFG measurement, and so on, two independent electric field sources $\mathbf{E} \neq \mathbf{F}$ are used. Here, we confine our discussion to the case of SHG, but we can carry similar discussion on SFG, and so on, using Eqs (23.2) and (23.3). In the following analysis, therefore $\mathbf{E} = \mathbf{F}$ is assumed.

From a nonlinear polarization point of view, the orientational order parameters S_1 and S_3 are related to the nonlinear polarization associated with the nonchirality of the surface monolayers, whereas S_2 is related to the nonlinear polarization associated with the chirality of the surface layers. In the case of monolayers with $C_{\infty v}$-symmetry, where the constituent molecules are achiral, that is, $s_{14} = a_{14} = a_{36} = 0$, the nonlinear electric polarization in SHG ($\mathbf{E} = \mathbf{F}$) reduces to

$$\mathbf{P}^N = (s_{33} - s_{31} - s_{15})(\mathbf{n} \cdot \mathbf{E})^2 \mathbf{n}$$
$$+ s_{31}(\mathbf{E} \cdot \mathbf{E})\mathbf{n} + s_{15}(\mathbf{n} \cdot \mathbf{E})\mathbf{E}$$
$$(23.5)$$

The nonlinear dielectric polarization is a function of the order parameters S_1 and S_3 together with the three elements related to the nonchirality, s_{31}, s_{33}, and s_{15}.

Similarly, in case of monolayers with C_∞-symmetry, where the constituent molecules are chiral, that is, $s_{14} \neq a_{14} \neq a_{36} \neq 0$, the nonlinear electric polarization in SHG ($\mathbf{E} = \mathbf{F}$) is given by:

$$\mathbf{P}^N = s_{14}(\mathbf{E} \cdot \mathbf{n}) \cdot (\mathbf{E} \times \mathbf{n})$$
$$+ (s_{33} - s_{31} - s_{15})(\mathbf{n} \cdot \mathbf{E})^2 \mathbf{n}$$
$$+ s_{31}(\mathbf{E} \cdot \mathbf{E})\mathbf{n} + s_{15}(\mathbf{n} \cdot \mathbf{E})\mathbf{E}$$
$$(23.6)$$

The first term on the right-hand side of Eq. (23.6) as a function of s_{14} with

S_2 (see Eq. (23.4)) represents the chiral effect. As mentioned earlier, the quantum interaction between the electrons in molecules and the external electric field makes the main contribution to SHG, and thus the nonlinear dielectric polarization given by Eqs (23.2) and (23.3) becomes the main contribution to the SHG. Equations (23.5) and (23.6) suggest that one can determine the orientational order parameters of S_1, S_2, and S_3 by means of SHG caused by the electric dipole effect.

23.3.2
EFISHG as a Tool for Probing Excess Surface Charges

The nonlinear polarization for the SHG in the medium is expressed not only as Eq. (23.1) under the electric dipole approximation but also in well-known tensor form as Shen, 1984

$$P_i = \chi^D_{ijk} E_j(\omega) E_k(\omega) \tag{23.7}$$

where χ^D_{ijk} is the second-order nonlinear susceptibility due to the electric dipole mechanism, and E_j and E_k are the electric fields of fundamental laser light. The effective polarization due to the electric quadrupole also contributes to the second-order nonlinear polarization, and it is expressed as

$$P_i = \chi^Q_{ijkl} \partial_j E_k(\omega) E_l(\omega) \tag{23.8}$$

where χ^Q_{ijkl} is the second-order nonlinear susceptibility due to the electric quadrupole mechanism.

On the one hand, second-order nonlinear polarization is activated by the external static field in case of the DC EFISHG. Since three electric fields $E_j(0)$, $E_k(\omega)$, and $E_l(\omega)$, are concerned with the generation of the nonlinear polarization, the nonlinear optical susceptibility becomes fourth-order rank tensor that does not vanish even in the centrosymmetric system. Thus, the effective polarization is expressed as

$$P_i = \chi^{DC}_{ijkl} E_j(0) E_k(\omega) E_l(\omega) \tag{23.9}$$

where $E_j(0)$ is the external static electric field induced by the local electric field, and χ^{DC}_{ijkl} is the third-order nonlinear susceptibility due to the DC-induced mechanism and $\chi^{DC}_{ijkl} E_j(0)$ corresponds to the effective dipole induced. Note that although this contribution is the third-order nonlinear optical process, double frequency of the fundamental light is generated because one of the related field is static. As mentioned earlier, there are many possible origins leading to the generation of nonzero local electric field $E_j(0) \neq 0$. Among them are external electric field, injected carriers, trapped charges, and so on. Hence, we can probe these local electric fields $E_j(0)$ by means of EFISHG.

23.3.3
Spectroscopic Consideration of the SHG

SHG originates from the quantum interaction between the electrons in molecules and the external electric field. In this section, we briefly summarize the SHGs from the viewpoint of energetics based on the quantum theory. As mentioned in the earlier sections, we should take into account the following three contributions to the SHG:

- electric dipole mechanism (χ_D);
- electric quadrupole mechanism (χ_Q); and
- induced electric dipole mechanism (χ_{DC})

and the nonlinear optical susceptibilities originated from these contributions show the different dependence on the wavelength of each related electric field, owing to the difference in the quantum interaction between the electrons in the molecules and the electric field. Thus, we can distinguish each mechanisms from a view point of spectroscopy. There are many ways to evaluate the frequency-dependent nonlinear susceptibilities. According to the quantum mechanical perturbation theory, the nonlinear susceptibilities for the SHG can be calculated as the second-order contribution to the induced dipole moment. Some of the molecular orbital (MO) calculation programs also enable us to estimate the frequency-dependent nonlinear susceptibilities. First, we discuss the resonant condition for the SHG schematically.

Schematic images for the resonance conditions are shown in Figure 23.2. In this figure, state $|g\rangle$ represents a ground state and $|n\rangle$, $|m\rangle$, and $|p\rangle$ represent the virtual excited states. Selection rules are employed to discuss the SHG, that is, the transitions between states of the same parity are forbidden for the electric dipole transition, whereas transitions between states of the same parity are allowed for the electric quadrupole transition. Considering these diagrams and the selection rules, we can estimate the resonant conditions for the SH frequency, 2ω. Clearly, for the electric dipole mechanism (Figure 23.2(a)), the lack of the centrosymmetry is required to radiate the SH wave, and resonance enhancement occurs only when the SH photon energy is coincident with the allowed excited state. On the other hand, resonance enhancement occurs only when the SH photon energy is coincident with the forbidden excited state for the electric quadrupole mechanism (Figure 23.2(b)). The case of the induced electric dipole mechanism is slightly different. This process is activated by the external field, as mentioned above. As shown in Figure 23.2(c), state $|n\rangle$ overlies $|p\rangle$ state. This indicates that the resonance enhancement occurs when the SH photon energy is coincident with both the allowed and forbidden excited states for this mechanism. Taking into account the resonant conditions discussed above, we can discuss the origin of the SHG from organic film by the spectroscopic observation. In more detail, χ_D, χ_Q and χ_{DC} are described as

$$\chi^D_{ijk} \propto \sum_{m,n} \left(\frac{\langle g|\hat{\mu}_i|n\rangle \langle n|\hat{\mu}_j|m\rangle \langle m|\hat{\mu}_k|g\rangle}{(\omega_{ng} - 2\omega)(\omega_{mg} - \omega)} \right)$$

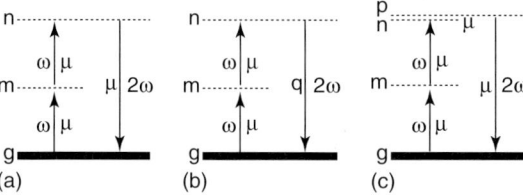

Fig. 23.2 Transition diagrams for the SH process: (a) electric dipole, (b) electric quadrupole, and (c) induced electric dipole.

$$+\frac{\langle g|\hat{\mu}_j|n\rangle\langle n|\hat{\mu}_i|m\rangle\langle m|\hat{\mu}_k|g\rangle}{(\omega_{ng}+\omega)(\omega_{mg}-\omega)}$$

$$+\frac{\langle g|\hat{\mu}_k|n\rangle\langle n|\hat{\mu}_j|m\rangle\langle m|\hat{\mu}_i|g\rangle}{(\omega_{ng}+\omega)(\omega_{mg}+2\omega)}\Bigg)$$

(23.10)

$$\chi^Q_{ijkl} \propto \sum_{m,n}\Bigg(\frac{\langle g|\hat{q}_{ij}|n\rangle\langle n|\hat{\mu}_k|m\rangle\langle m|\hat{\mu}_l|g\rangle}{(\omega_{ng}-2\omega)(\omega_{mg}-\omega)}$$

$$+\frac{\langle g|\hat{\mu}_k|n\rangle\langle n|\hat{q}_{ij}|m\rangle\langle m|\hat{\mu}_l|g\rangle}{(\omega_{ng}+\omega)(\omega_{mg}-\omega)}$$

$$+\frac{\langle g|\hat{\mu}_l|n\rangle\langle n|\hat{\mu}_k|m\rangle\langle m|\hat{q}_{ij}|g\rangle}{(\omega_{ng}+\omega)(\omega_{mg}+2\omega)}\Bigg)$$

(23.11)

and

$$\chi^{DC}_{ijkl} \propto$$

$$\sum_{m,n,o}\Bigg(\frac{\langle g|\hat{\mu}_i|p\rangle\langle p|\hat{\mu}_j|n\rangle\langle n|\hat{\mu}_k|m\rangle\langle m|\hat{\mu}_l|g\rangle}{(\omega_{pg}-2\omega)(\omega_{ng}-2\omega)(\omega_{mg}-\omega)} + \frac{\langle g|\hat{\mu}_j|p\rangle\langle p|\hat{\mu}_i|n\rangle\langle n|\hat{\mu}_k|m\rangle\langle m|\hat{\mu}_l|g\rangle}{\omega_{pg}(\omega_{ng}-2\omega)(\omega_{mg}-\omega)}$$

$$+\frac{\langle g|\hat{\mu}_j|p\rangle\langle p|\hat{\mu}_k|n\rangle\langle n|\hat{\mu}_i|m\rangle\langle m|\hat{\mu}_l|g\rangle}{\omega_{pg}(\omega_{ng}+\omega)(\omega_{mg}-\omega)} + \frac{\langle g|\hat{\mu}_j|p\rangle\langle p|\hat{\mu}_k|n\rangle\langle n|\hat{\mu}_l|m\rangle\langle m|\hat{\mu}_i|g\rangle}{\omega_{pg}(\omega_{ng}+\omega)(\omega_{mg}+2\omega)}\Bigg)$$ (23.12)

where $\hat{\mu}_i$ and $\hat{q}_{ij}$ are the electric dipole and the electric quadrupole transition moment operator and are expressed as $\hat{\mu}_i = -e\hat{r}_i$ and $\hat{q}_{ij} = -\frac{1}{6}e(3\hat{r}_i\hat{r}_j - \mathbf{r}^2\delta_{ij})$, respectively. $|g\rangle$ represents a wave function of the ground state and $|m\rangle$ and $|n\rangle$ are those of the excited state. The notation ω_{mg} means the frequency difference between $|m\rangle$ state and $|g\rangle$ state, $\omega_m - \omega_g$.

However, we do not need to go into the details of χ_D, χ_Q, and χ_{DC}, when we use the SHGs for probing the orientational order of surface molecules and for probing the local electric field.

Equations (23.10)–(23.12) suggest that it is important to know the origins of SHG enhancement in terms of the resonant frequency.

23.3.4
Sum-Frequency Generation

In this section, SFG is described. SFG also becomes a powerful tool for investigating the surface and interface phenomena of materials, especially on the molecular level. For the SFG process, three optical waves with different frequencies are interacting. There is an energy conservation between two input waves (ω_1 and ω_2) and output wave (ω_3) as $\omega_1 + \omega_2 \to \omega_3$. As mentioned, both SHG and SFG are second-order nonlinear optical processes, and in other words, SFG is a generalization of SHG (ω_1 and ω_2 are degenerate for the SHG process). Accordingly, the SFG output vanishes from the materials with inversion symmetry under the electric dipole approximation as well as the SHG process. We can realize the surface selective vibrational spectroscopy measurement by using SFG, letting the input wave with a frequency ω_1 be visible and the wave with a frequency ω_2 be infrared light.

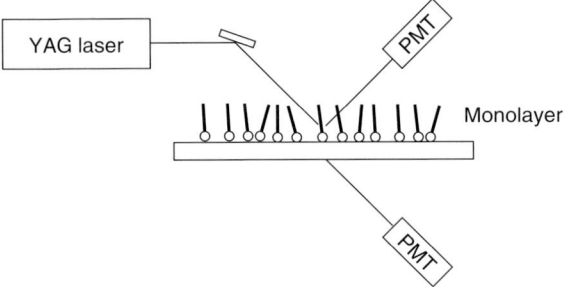

Fig. 23.3 Schematics of experimental setup for SHG measurement system is also attached. The chemical structure of 8CB.

23.4
Measurement Details of SHG and EFISHG

23.4.1
Determination of Surface Structure and Orientational Order Parameters

23.4.1.1 Fundamentals

SH light is generated from the monolayer by laser irradiation. The input light consists of the s- and the p-polarized waves and is expressed as

$$\mathbf{E}_{in} = S_\omega \mathbf{s}_{in} + P_\omega \mathbf{p}_{in} \tag{23.13}$$

where $\mathbf{s}_{in}$ and $\mathbf{p}_{in}$ are unit vectors for the s- and the p-polarized direction of the waves, and S_ω and P_ω are the amplitude of the s- and the p-polarized waves, respectively. Substituting $\mathbf{E}_{in}$ into Eqs (23.5) and (23.6), $\mathbf{P}^N$ generated in surface monolayers is obtained. The output light intensity $I_{2\omega}$ is proportional to $\mathbf{e}_{out}(2\omega) \cdot \mathbf{P}^N$, where $\mathbf{e}_{out}$ is unit vector for the output light direction of the wave. Since $\mathbf{e}_{out}(2\omega) \cdot \mathbf{P}^N$ is a function of S_1, S_2, and S_3, these orientational order parameters of surface monolayers can be determined with the enhanced SHG by the electric dipole effect Zhu, Suhr and Shen, 1987.

23.4.1.2 Experimental Configuration of SHG Measurement

Figure 23.3 shows a schematic diagram of the SHG measuring system for probing nonlinear polarization of surface layers. The SHG experiments can be carried out with the experimental setup illustrated. Briefly, the experimental setup consists of an optical measurement arrangement with a Q-switched Nd : YAG laser (wavelength 0.532 μm, pulse duration <7 nanoseconds, fundamental pulse rate ≤ 15 Hz) for the SHG measurement. The laser light is irradiated onto the surface layer with an energy of about 6 mJ at a pulse rate of 2 Hz. The size of the laser spot is about 56.0 mm².

23.4.1.3 Experimental Configuration of SFG Measurement

We can observe the SFG output of visible light, and a typical photomultiplier tube (PMT) can be used as a detector. Figure 23.4 represents the typical setup for the SFG measurement. Owing to the photon momentum conservation, the direction of SFG output is not

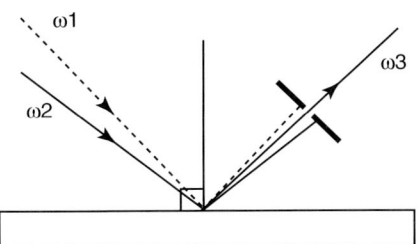

Fig. 23.4 Paths of light rays in SFG measurement.

coincident with those either of reflected fundamental waves (see Figure 23.4).

Experimentally, surface vibrational spectroscopy was first introduced by Zhu, Suhr and Shen 1987. They successful recorded a vibrational spectrum of a monolayer of molecules on a substrate by SFG. They chose second harmonics of YAG laser as a visible input and coumarin dye so that the resonance enhancement of the SFG signal occurred. They also mentioned the possibility to detect the transient surface phenomena because pulsed laser probing in SFG enables us to monitor the transient species appearing on a surface. As mentioned, SHG can also detect the orientation of molecules on a surface. Since the origin of SHG process involves electronic transitions in the molecule, the orientation of molecules directly reflects the orientation of the transition moments. In contrast, SFG can detect the orientation of the functional group of the molecule. In this sense, detailed information of the surface can be extracted from the SFG results.

23.4.2
Probing of the SHG Intensity Profile as a Convolution Beam Pattern

23.4.2.1 Fundamentals

For EFISHG, the SHG intensity is proportional to the square of local static electric field as

$$I(2\omega) \propto \left| \chi^{(3)}(-2\omega; 0, \omega, \omega) \right. $$
$$\left. \times E(0)E(\omega)E(\omega) \right|^2 \quad (23.14)$$

where $\chi^{(3)}(-2\omega; 0, \omega, \omega)$ represents a third-order nonlinear optical (NLO) susceptibility for the EFISHG process, and $E(0)$ and $E(\omega)$ represent the static electric field and electric field of light, respectively. External electric fields, trapped charges, and injected carriers are origins of the static electric field, $E(0)$. Hence, the electric field distribution in materials as well as in organic devices can be obtained at various experimental conditions by the EFISHG measurement. In more detail, SHG intensity profile can be considered as a convolution of the beam pattern of the fundamental laser and the actual electric field distribution formed in organic films as Tojima, Manaka and Iwamoto, 2001

$$I^{2\omega}(x) \propto \left| \int_{-\infty}^{\infty} E(\xi) I^{\omega}(x - \xi) d\xi \right|^2 \quad (23.15)$$

$I^{2\omega}(x)$ represents the SHG intensity at position x, and $E(\xi)$ and $I^{\omega}(\xi)$ represent the electric field distribution and the intensity profile of the fundamental laser, respectively. If the laser beam

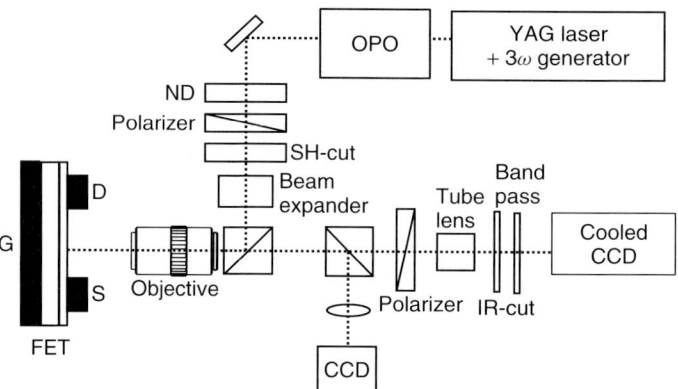

Fig. 23.5 Optical setup for the EFISHG measurement of OFET devices.

profile is approximated by δ-function as $I^\omega(x) \equiv \delta(x)$, $\sqrt{I^{2\omega}(x)}$ directly gives the local electric field $E(x)$. However, to obtain the accurate electric field distribution from the SHG, intensity profile requires a deconvolution process taking into consideration the beam profile of the fundamental laser, $I^\omega(x)$. Such deconvolution is, in general, a laborious task, because a beam profile and the SHG intensity profile are not expressed by simple mathematical functions, and the evaluated electric field from a scanning measurement sometimes loses accuracy, for example, spatial resolution. We can employ the deconvolution process for the determination of electric field distribution. On the other hand, an SHG imaging technique for visualizing the distribution of the electric field in organic materials can overcome these problems Manaka et al., 2006. The evaluation of the electric field distribution from SHG images does not require the complicated deconvolution process, and the spatial resolution is significantly improved. Moreover, omission of the scanning process results in the reduction of the measurement time.

23.4.2.2 Experimental Configuration of EFISHG Measurement from OFET Structure

A detailed configuration about the EFISHG measurement is described in Manaka et al., 2006. Figure 23.5 shows a schematic diagram of the EFISHG measuring system. The light source for the SHG measurement is an optical parametric oscillator (OPO: Continuum Surelite OPO), pumped by a third-harmonic light of Q-switched Nd–YAG laser (Continuum: SureliteII-10), a fundamental wavelength of 1120 or 1320 nm is chosen. Fundamental light from the OPO passes through a prism polarizer, long-pass filters, and a beam expander. Then it is focused on the sample with normal incidence, using a long working distance objective lens (Mitutoyo: M Plan Apo SL20 ×, NA = 0.28, WD = 20.5 mm). SH light generated from the sample is filtered by a fundamental-cut filter and an interference filter to remove fundamental and other unnecessary light. Finally, SH light is detected by a cooled charge coupled device (CCD) camera (Andor technology: DU420-BV). In this configuration, the polarization

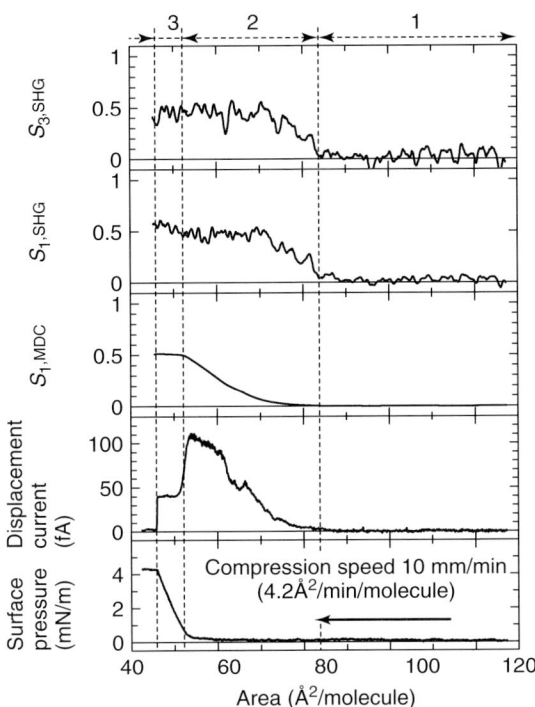

Fig. 23.6 A typical example of SHG and MDCs of 8CB monolayers during the course of monolayer compression. Order parameters, S1 and S3, of 8CB monolayer determined from the MDC-SHG measurement.

direction of the light is chosen in the direction parallel to the sample surface.

23.5
Results of SHG and EFISHG

23.5.1
Determination of Surface Structure and Orientational Order Parameters

In a manner as described in Section 23.3.1, we can carry out SHG experiment for probing orientational order of surface layers. Figure 23.6 shows an example of the determination of the order parameters S_1 and S_3 for 4-n-octyl-4′-cyanobiphenyl (8CB) monolayers with $C_{\infty v}$-symmetry on water surface. In region 1, S_1 and S_3 of SHG are very small, although fluctuations coming from the formation of domains are observed. In region 2, S_1 and S_3 increase gradually and show the maximum with decrease in the molecular area. In region 3, they are nearly constant, about 0.5. In contrast, S_1 determined by electrical Maxwell-displacement current (MDC) measurement is also plotted. S_1 is nearly zero in region 1 and increases monotonically in region 2. S_1 is about 0.5 in region 3. These results suggest that 8CB molecules lying on a water surface in large molecular area progressively stand up with a decrease in the molecular area and with an average tilting angle of about 60° in region 2. As mentioned earlier, surface polarization phenomena can be detected and the orientational orders of the surface layers can be well explored by the SHG caused by the electric dipole effect.

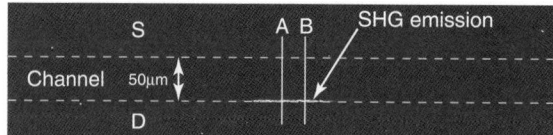

Fig. 23.7 SHG image from the channel of pentacene FET under the application of negative pulse. Channel region lies between two gold electrodes, and edges of the electrode are indicated by dashed lines. SHG emission was observed at the edge of the drain electrode.

23.5.2
Probing of Electric Field $E(0)$ by EFISHG Measurement

As mentioned in Section 23.3.2, the SHG intensity profile obtained on the basis of a scanning measurement does not display an actual spread of SHG emission from the material surface. Direct observations of the SHG image can show a more realistic distribution of the SHG emission within the limits of the system resolution. Figure 23.7 shows the SHG image from the channel of pentacene field effect transistor (FET) under the application of negative pulse.

As shown in this figure, strong SHG emission is observed at an edge of drain electrode. High electric fields between the drain and gate electrodes produce a large in-plane component of the electric field, which activates the SHG at the drain edge. It is noteworthy that the width of the SHG emission in the channel clearly decreases compared with a distribution obtained on the basis of a scanning measurement.

Figure 23.8 shows the line scan of the SHG intensity profile across the channel (represented as open squares and filled diamonds) and the in-plane component of the in-plane electric field

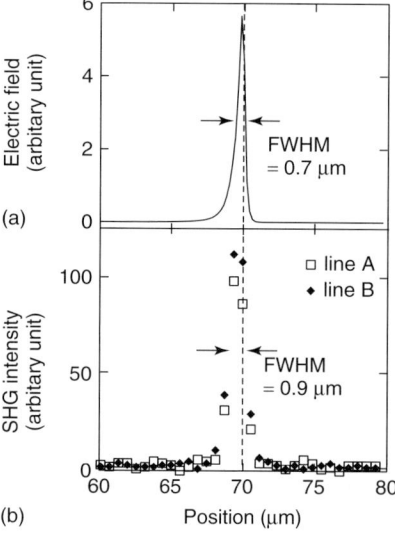

Fig. 23.8 (a) The figure represents the in-plane component of the in-plane electric field distribution in pentacene layer calculated based on a finite element method. (b) The figure shows the line scan of the SHG intensity profile across the channel. Open squares and filled diamonds, respectively, represent SHG intensity profile at line scan A and B as shown in Figure 23.7.

distribution (solid line) in the pentacene layer. Open squares and filled diamonds, respectively, represent the SHG intensity profile at line scan A and B (see Figure 23.7) as shown in Figure 23.8. The edge of the electrode is located at a position of 70 μm in these figures. It is found that the SHG intensity profile is quite sharp. The sharpness of emission is quantitatively estimated using full width at half maximum (FWHM) values of the profile. FWHM values of SHG profile and the electric field distribution are evaluated as 0.9 and 0.7 μm, respectively. Note that FWHM values of the SHG profile obtained based on the scanning measurement depended on the magnification of the objective lens and have the value of approximately 15 μm for 20 × objective.

For the electric field calculation when there are no excess charges in the device, only the electrode configuration and potential of the electrode are taken into account. In such a case, a Laplacian electric field is formed in the device. A Laplacian field is the electric field caused by the potentials of electrodes in the absence of any charges in the devices between the electrodes. Under negative bias, carrier injection from the electrode into pentacene is prohibited because the injection barrier for electrons at the pentacene/Au interface is quite high, that is, energy difference between the lowest unoccupied molecular orbital (LUMO) of pentacene and the work function of Au electrode is evaluated as 2.7 eV. Thus, the strong electric field around the edge of the electrode is maintained during bias application, and the SHG emission is concentrated around the edge of the drain electrode as shown in Figure 23.7. In other words, the sharp emission of the SHG at the edge of the electrode indicates that the SHG imaging technique successfully visualizes the electric field in the device with a spatial resolution of approximately 1 μm. As mentioned above, the electric field generated by the Laplacian electric field can be detected as the EFISHG, and the distribution of the electric field can be explored using the profile of EFISHG enhanced. Similarly, the electric field distribution caused by injected carriers and trapped charges can be evaluated by the EFISHG Zhu, Suhr and Shen, 1987. It is instructive to note that the migration of the peaks of electric field caused by injected carries is visualized by means of TRM-SHG measurement Manaka et al., 2007a. Further, OFET operation has been successfully probed by SFG, where conformational change of the molecular structure, due to dopings of the molecules by carrier injection from electrodes, is probed Manaka et al., 2007b.

References

Bloembergen, N. (1965) in *Nonlinear Optics* (ed. W.A. Benjamin.), Singapore, World Scientific, pp. 1–61.

Chollet, P.A., Kajzar, F. and Le Moigne, J. (1990), *Proceedings of SPIE*, Vol. 1273, p. 87.

Heinz, T.F. (1991) in *Nonlinear Surface Electromagnetic Phenomena* (eds H.E. Ponath and G.I. Stegemen), Elsevier Science, New York, pp. 397–398.

Iwamoto, M. and Wu, C.X. (2000), *The Physical Properties of Organic Monolayers*, World Scientific, Singapore, pp. 1–200.

Iwamoto, M., Wu, C.X. and Ou-Yang, Z.C. (2000) *Chem. Phys. Lett.*, **325**, 545–551.

Luckhurst, G.R. and Gray, G.W. (1979) *The Molecular Physics of Liquid Crystals*, Academic Press, London, pp. 51–83.

Manaka, T., Lim, E., Tamura, R. and Iwamoto, M. (2006) *Appl. Phys. Lett.*, **89**, 072113-1–072113-3.

Manaka, T., Nakao, M., Yamada, D., Lim, E. and Iwamoto, M. (2007a) *Opt. Express*, **15**, 15964–15971.

Manaka, T., Lim, E., Tamura, R. and Iwamoto, M. (2007b) *Nat. Photonics*, **1**, 581–584.

Nye, J.F. (1957) *Physical Properties of Crystals*, Oxford, pp. 68–169.

Sessler, G.M. (1987), *Electrets*, Springer-Verlag, New York, pp. 1–80.

Shen, Y.R. (1984) *The Principles of Nonlinear Optics*, John Wiley & Sons, Inc., New York, pp. 1–85.

Sze, S.M. (1981) *Physics of Semiconductor Devices*, John Wiley & Sons, Inc., New York, pp. 362–512.

Tojima, A., Manaka, T. and Iwamoto, M. (2001) *J. Chem. Phys.*, **115**, 9010–9017.

Wang, X.K., Zhang, T.G., Lin, W.P., Liu, S.Z., Wong, G.K., Kappes, M.M., Chang, R.H. and Ketterson, J.B. (1992) *Appl. Phys. Lett.*, **60**, 810–812.

Zhu, X.D., Suhr, H. and Shen, Y.R. (1987) *Phys. Rev. B*, **35**, 3047–3050.

Further Reading

Ye, H., Abu-Akeel, A., Huang, J., Katz, H.E. and Gracias, D.H. (2007) *J. Am. Chem. Soc.*, **128**, 6528–6529.

24
Raman Scattering for Speciation and Analysis

Karen Esmonde-White, Mekhala Raghavan and Michael Morris

24.1	Introduction	835
24.2	Raman Spectroscopy Basics	835
24.2.1	Theory	835
24.2.2	Enhancement	836
24.2.3	Subsurface Measurements – Spatially Offset Raman Spectroscopy (SORS)	837
24.2.4	Instrumentation	837
24.2.4.1	Laser Safety	837
24.2.4.2	Basic Instrumentation	838
24.2.4.3	Fiber Optic Probes	839
24.2.4.4	Microprobes	839
24.2.5	Data Reduction	841
24.3	**Identification and Quantification of Inorganic Materials**	**842**
24.3.1	Introduction	842
24.3.2	Minerals	842
24.3.3	Art, Archaeology, and Architecture	842
24.3.4	Forensics	844
24.3.5	Nanoscale Semiconductors	845
24.3.6	Industrial Process Control	848
24.4	**Identification and Quantification of Organic Materials**	**848**
24.4.1	Pharmaceutical Applications	848
24.4.2	Polymer Chemistry and Physics	850
24.4.3	Forensic and Homeland Security Applications	850
24.5	**Biological Applications of Raman Spectroscopy**	**851**
24.5.1	Subcellular Level	851
24.5.2	Tissue Level	852
24.5.2.1	Epithelial Tissue	852
24.5.2.2	Muscular Tissue	854
24.5.2.3	Nervous Tissue	854
24.5.2.4	Connective Tissue	855
24.5.3	Microorganisms	855

Encyclopedia of Applied Spectroscopy. Edited by David L. Andrews.
Copyright © 2009 WILEY-VCH Verlag GmbH & Co. KGaA, Weinheim
ISBN: 978-3-527-40773-6

Glossary 858
References 859
Further Reading 863

24.1
Introduction

The basic principles and instrumentation of Raman spectroscopy are reviewed. Emphasis is placed on fiber-optic probes and microprobes. Applications to a wide range of problems in several industries, as well as to forensic applications are surveyed. Applications to the arts and sciences, including the life sciences, are also reviewed. Literature references provide the reader with links to more detailed discussion of the theory of Raman spectroscopy, to the methodology used in each of the surveyed areas, and to the details of the Raman spectra of many materials.

24.2
Raman Spectroscopy Basics

24.2.1
Theory

Raman spectroscopy is a widely employed tool in physical science, life science, and engineering. Consequently, there are numerous monographs and chapters in monographs that address the fundamentals and applications for different audiences. Good introductory surveys are available (Smith and Dent, 2005; Lewis and Edwards, 2001).

Raman scattering is a form of inelastic light scattering. In general, the lasers that are used to excite Raman scatter have photon energies lesser than the energy of the first electronic excited state. The weak interaction between the exciting laser photons and the molecule results in scattering, with light scattered at a new, lower frequency (expressed in wave numbers) according to Eq. (24.1) as follows:

$$\bar{\nu}_{\text{scattered}} = \bar{\nu}_{\text{exciting}} - \bar{\nu}_{\text{vibration}} \quad (24.1)$$

Where $\bar{\nu} = 1/\lambda$. The scattered light is red-shifted and is called *Stokes-scattered*. It is assumed that all of the molecules are in the ground vibrational state. That is not quite true, especially for low frequency vibrations. If the scatter originates from a molecule in an excited vibrational state, then light is blue-shifted by the same frequency. Because this effect called *anti-Stokes Raman scattering* is weak, it is rarely used. Generally, most workers report Raman spectra as plots of intensity vs. $\bar{\nu}_{\text{vibration}}$ of scattered light.

The quantum efficiency for Raman scattering is low. Generally, it is between 10^{-7} and 10^{-9}. Although this number

Encyclopedia of Applied Spectroscopy. Edited by David L. Andrews.
Copyright © 2009 WILEY-VCH Verlag GmbH & Co. KGaA, Weinheim
ISBN: 978-3-527-40773-6

sounds low, it should be noted that a 1-W laser operating at 800 nm outputs about 4×10^{18} photons per second. Raman spectroscopic instruments use lasers that provide 0.01–0.5 W at the sample. Therefore, even with 0.01-W laser power and 10^{-9} quantum efficiency, there should be about 10^7 Raman-scattered photons generated per second. Even with very low collection efficiency, the experimental problem is often not the absolute intensity of Raman scatter but rather the presence of an intense background signal from fluorescing molecules. Because fluorescence quantum efficiencies are often at least 0.01 and can approach 1, fluorescence intensity even from impurities in a matrix can exceed Raman scatter intensity by 2 or 3 orders of magnitude.

The wavelength dependence of the intensity of Raman scatter is approximated by Eq. (24.2).

$$\overline{v}^4_{exciting} = \frac{1}{\lambda^4_{exciting}} \quad (24.2)$$

If only the intensity of the Raman signal governed the choice of exciting wavelength, spectroscopists would choose as blue a wavelength as possible. In fact, the choice of wavelength is usually governed by the need to circumvent fluorescence background. If the material to be analyzed or its matrix is fluorescent, then red or near-infrared lasers are used. This is usually the case for most industrial organic compounds and most polymers and virtually all tissue specimens and human, animal, or bacterial cells. On the other hand, many inorganics, including ceramics and semiconductors, have little or no fluorescence. In this case, green lasers are usually used. Ultraviolet lasers are used only in special cases, both to avoid photochemical damage and to preserve compatibility with glass optics and sample containers.

The selection rules for Raman scatter are different from those for infrared absorption. A vibration is Raman active if the vibration causes a change in the dipole moment of the molecule induced in an external electric field. This quantity is proportional to a parameter called *polarizability*, P. By contrast, a vibration is infrared active if there is a change in the dipole moment of the molecule when the vibration is excited. These different selection rules mean that different types of vibrations are intense in the Raman and in the infrared. In many cases, a vibration that has an intense band in one mode of observation will have a weak band in the other. For example, stretches and bends of highly polar functional groups are intense in the infrared and weak in Raman scatter. On the other hand, stretches of unsaturated moieties are intense in Raman scatter and weak in infrared absorption. Significantly, the OH stretching and bending modes of water are strong in the infrared, but weak in Raman scatter. The stretching and bending modes of Si–O moieties are also intense in infrared absorption and weak in Raman scatter. Indeed, compatibility with aqueous solutions, glass, or fused silica containers and optics has long been major reasons for use of Raman spectroscopy rather than infrared spectroscopy.

24.2.2
Enhancement

There are two general enhancement mechanisms that result in great increases in Raman intensity, resonance enhancement, and surface enhancement. The corresponding methodologies are usually called *resonance Raman spectroscopy* (*RRS*) and surface-enhanced Raman spectroscopy (SERS) (Smith and Dent, 2005; Lewis and Edwards, 2001).

Resonance enhancement occurs when the exciting laser wavelength is coincident or nearly coincident with the electronic absorption band of a molecule. For colored molecules, resonance enhancement occurs at visible wavelengths. Examples of practical interest include hemoglobin, and dyes, especially those that are red or orange and absorb 532 nm light, the frequency-doubled wavelength of the very common Nd : YAG and Nd : VO$_4$ lasers. Of course, several molecules absorb in the ultraviolet. Resonance enhancement occurs at UV wavelengths; but for practical reasons, only a small number of specialist laboratories make use of this effect.

Surface enhancement occurs when the wavelength of the exciting laser is coincident with the wavelength of the surface plasmon of a metal on which the molecule is absorbed. For specially structured surfaces of silver or gold, the surface plasmon wavelengths are in the visible or near infrared. In theory, enhancement factors can exceed 10^{10}. In practice, enhancement factors of 10^3-10^5 are observed. Although surface enhancement can be observed with easily prepared gold or silver colloids, current practice has been moving toward structured surfaces prepared by photolithography because these provide more reproducible enhancement. Increasingly, gold is preferred over silver because gold surfaces are more stable than silver surfaces. The potential user is cautioned that the technique is limited to the monolayer of molecules that are actually in contact with the surface or, but with greatly reduced enhancement, to the monolayer and one or two layers directly above it. The dynamic range of SERS is, therefore, limited and it is largely a trace or an ultratrace analytical method.

24.2.3
Subsurface Measurements – Spatially Offset Raman Spectroscopy (SORS)

An important recent development has been the use of spatially separated laser illumination and collection regions to probe beneath the surface of a turbid medium, such as a polymer, a powdered material, or tablet or a human or animal tissue (Matousek, 2007). Generally, delivery of laser and collection of Raman scatter is by optical fibers, which may be arranged in any number of different geometries. The spatial separation of illumination and collection regions allows the collection of light that has been multiply scattered as it propagates through the turbid medium. Low definition images have been acquired with SORS-type systems (Schulmerich *et al.*, 2006; Schulmerich *et al.*, 2007) and even Raman tomography has been reported (Schulmerich *et al.*, 2008).

24.2.4
Instrumentation

24.2.4.1 **Laser Safety**
Standard laser safety protocols should always be enforced. In particular, users should wear appropriate laser safety glasses. Depending on the laser power, door interlocks and warning signs may be necessary. American laboratories follow the guidelines of the ANSI Z136.1–2007 standard (Laser Institute of America, 2007). Similar standards for laser safety apply in other countries.

24.2.4.2 Basic Instrumentation

A standard Raman spectroscopy system comprises a laser source, a sample illumination and collection system, a single stage spectrograph with a holographic or dielectric prefilter to reject laser light, and a charge-coupled device (CCD) imaging detector. Many types of devices are available to illuminate the sample and to collect the scattered light. The majority of users employ either fiber-optic probes or microscope-based microprobes. A common accessory is a motorized stage to facilitate spectroscopy of multiple samples, including samples in microliter plates and similar systems.

Frequency-doubled Nd : YAG and Nd : VO$_4$ lasers, which operate at 532 nm, are preferred for nonfluorescent or weakly fluorescent samples. They are compact, stable, and have no special electrical service requirements. For fluorescent materials, diode lasers operating at 785 nm, or for many biomedical applications, those operating at 830 nm are preferred. Where the excited fluorescence is low enough, 785 nm is preferred. It is the deepest red standard wavelength diode laser for which Raman shifts of C–H stretches (i.e., below about 3100 cm^{-1}) can be observed with a CCD or other silicon photodetector. The recent development of InGaAs-based imaging detectors has made Nd : YAG 1064 nm lasers a viable choice.

Several different spectrograph configurations are used. Czerny–Turner systems, usually with toroidal mirrors for astigmatism correction, are common. On-axis designs are frequently used for their superior imaging performance. The axial transmissive spectrograph (Battey et al., 1993), which uses lenses instead of mirrors for collimation and focusing and a transmission grating instead of a reflection grating, is the most prevalent.

The detector in most Raman spectroscopy systems is a slow-scan (also called *scientific*) back-illuminated CCD (Janesick, 2001). The deep depletion version, which is fabricated from high resistivity silicon, has superior quantum efficiency above 800 nm, and is used in conjunction with 785 and 830 nm lasers. Replacement of conventional output amplifiers with a chain of avalanche breakdown stages yields an electron-multiplying charge-coupled device (EM-CCD) (Coates et al., 2003; Denvir and Conroy, 2003). The EMCCD has high gain (up to 1000 X), is capable of single photon detection and can be operated at high read-out rates (greater than 1 MHz) with very little excess noise. The electron-multiplying technology has not yet been successfully implemented on deep depletion CCD's and so has not been useful with deep red or NIR diode lasers.

As the size of solid state lasers and CCD detectors has decreased, so has the size of laboratory bench-top Raman spectroscopy systems. Modern systems generally have a footprint that is slightly higher than 0.5×0.5 m and many are smaller. Of course, the addition of a microscope or a stand for a fiber-optic probe or some other specialized accessory may increase the overall size of the instrument.

The increasing use of Raman spectroscopy for chemical process control has encouraged vendors to develop versions of their instruments for use in the factory environment. These instruments are usually packaged in enclosures for high dust, vibration, and electrical noise environments and may contain rugged versions of some components and subsystems.

Recently, several companies have introduced handheld, battery-powered instruments for off-site use in incoming materials inspection and for geological, environmental, homeland security and forensic applications. These instruments can operate for several hours without recharging. They typically are low-resolution (8–10 cm^{-1}) instruments and may have on-board databases and software for matching spectra to commonly encountered materials.

24.2.4.3 Fiber Optic Probes

An obvious advantage of the fiber optic probes is that the laser, spectrograph, and CCD can be some distance away from the sample. In the laboratory, 3–5 m is considered as the typical fiber length. Often, longer lengths are found in industrial process control to minimize the effects of noise and vibration on the instrumentation. Two general designs of fiber-optic Raman probes are in widespread use: the N-around-1 probe and the bifurcated filtered probe.

In the N-around-1 design laser, light is transmitted through a central optical fiber and Raman scatter is collected from a circular bundle of fibers that surrounds it. These are cemented into a metal, ceramic, or plastic holder and the ends are cut and polished. However, as the laser light propagates through the excitation fiber, it generates a silica Raman signal. This usually causes no problem in measurements of clear liquids. If the sample is a turbid fluid or a solid, some silica Raman scatter is reflected back into the collection fibers and becomes a background, which increases with the length of the fiber used. The background problem can be reduced if filters are placed at the ends of the optical fibers. An example is shown in Figure 24.1 (Motz et al., 2004). Of course, the filters add cost and complexity to the probe.

In the bifurcated filtered probe, a single excitation fiber and a single collection fiber are used. A dichroic mirror, notch filter, or other wavelength-selective optic are used to combine the excitation and collection light. Delivery of laser light to the sample and collection of Raman-scattered light from the sample are through a single lens. Auxiliary filters may be used to remove silica Raman scatter from the excitation light and to prevent laser light from entering the collection fiber, resulting in very low silica background. In addition to the basic, general purpose designs, available probes include models with sampling optics designed for immersion in liquids or for operation at elevated or low temperatures.

24.2.4.4 Microprobes

Raman microprobes are used because they are applicable to opaque materials and water-containing materials (such as plant or animal tissue or cells) and offer lateral spatial resolution that is similar to resolution obtainable with light microscopes, about 0.5–1 μm. Operation in this wavelength range gives the Raman microscope better spatial resolution than is available with infrared microscopes. Additionally, most employ a confocal geometry to improve axial resolution and reduce background scatter and fluorescence. Depth profiling by Raman microscopy is well established (Reinecke et al., 2001; Adar, 2001; Fleming and Kazarian, 2004). It should be cautioned that the spherical aberration induced by refractive index changes can distort

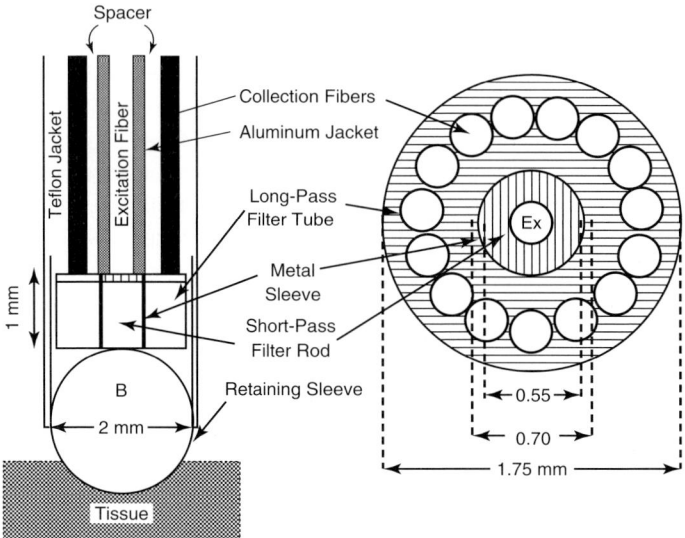

Fig. 24.1 An N-around-1 fiber-optic Raman probe. In this probe design, dielectric filters are used to block silica Raman scatter generated in the excitation (laser delivery) fiber and to block backscattered laser light from entering the collection fibers. A ball lens is used to focus the sample onto the specimen, which is skin tissue. (Reprinted by permission from reference (Motz et al., 2004).)

depth measurements (Everall, 2000a,b). Index matching and the use of immersion objectives are needed to minimize this problem (Everall et al., 2007).

Almost all commercial systems are constructed around standard epi-fluorescence microscope frames made by the major microscope makers. Coupling optics are added, but the microscope hardware is usually unchanged. For safety reasons, it is not common for a Raman microprobe to include binocular eyepieces. A video camera is used for viewing the sample and focusing the instrument. Because Raman microprobes operate in the visible or near-infrared, the spectroscopist has ready access to the wide range of specialized objectives used for light microscopy. Temperature-controlled stages and many other accessories made for the microscope frame can also be used.

A schematic Raman microprobe is shown in Figure 24.2. Coupling optics bring laser light to the microscope objective and direct Raman scatter to the spectrograph. Direct (i.e., with lenses and mirrors) coupling is common. Some vendors offer fiber optic coupling, which provides greater flexibility in component positioning and facilitates using the same spectroscopy optics with other sampling systems. However, to retain the spatial resolution of the direct-coupled instrument, it is necessary to use a single mode optical fiber to deliver the laser light to the microscope. A multimode fiber can be used to direct the Raman scatter to the spectrograph.

A basic instrument operates as a single point microprobe. Mapping is

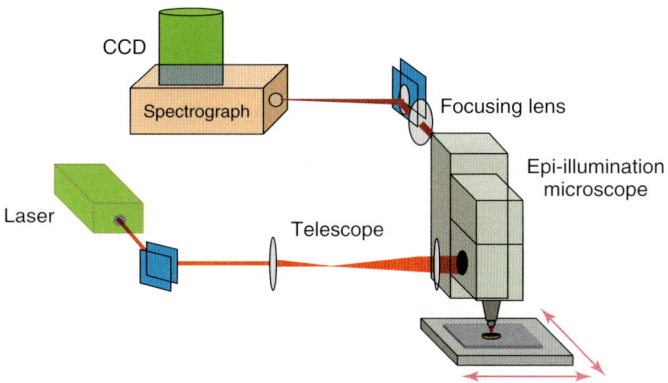

Fig. 24.2 Schematic of a Raman microprobe. Light from the laser is shaped by the telescope and focused on the sample by a microscope objective. Raman scatter is collected through the objective and focused onto the entrance slit of the spectrograph. In this simple design, a manual or motor-driven stage can be used to build up a Raman map pixel by pixel.

provided by motor-driven or piezoelectric XY stages, or by scanning mirror systems similar to those used in confocal microscopy. Motor-driven XY stages have a settling time that can approach 1 second, and this dead time can limit the speed at which point–point images can be acquired. Piezoelectric stages settle in less than 1 ms. The combination of a piezoelectric stage and an EMCCD allows very fast mapping, at least with green laser excitation. Line-focusing of the exciting laser can be used to acquire an entire line of spectra simultaneously, while increasing the total power that can be safely focused on the sample. In this case, only one-dimensional scanning is needed to generate an image, and the total image acquisition time can be much shorter than if point-point scanning is used.

24.2.5 Data Reduction

Raman spectroscopy is often used for material or contaminant identification. Visual inspection or computer-aided matching to a spectrum in a database of Raman spectra is often all that is needed. Because the intensity of Raman scatter is directly proportional to the number of molecules interrogated, quantitative analysis using band heights or areas and standards of known concentration is satisfactory. However, inaccurate subtraction of fluorescence or background can introduce errors into this seemingly simple operation. The merits and problems of background subtraction protocols remain contentious.

In industrial process control, small changes in complex materials are frequently sought. Tracking the change in the height or area of a single band or of a band height or area ratio is often not satisfactory. In such cases, a multivariate model, typically a partial least squares model, is used instead. Models contain contributions from throughout the Raman spectrum and thus strongly depend on reproducible instrument and process conditions. They can yield erroneous results if the process conditions or instrument parameters change, even if the change is small.

24.3
Identification and Quantification of Inorganic Materials

24.3.1
Introduction

Characterization of inorganic materials by Raman spectroscopy has been reported in topics ranging from surfaces of nanocrystals to characterization of bulk geological minerals. Since the late 1990s, there have been many applications of Raman to new classes of materials and novel approaches to examining existing classes of materials. More recently, fundamental studies of phonons in reduced dimensions have been investigated using Raman spectroscopy. Most reports fall under three broad topics: chemical identification, quantification, and spectroscopic response to mechanical or chemical perturbation (Lyon *et al.*, 1998; Pitt *et al.*, 2005). A comprehensive review of the theory of molecular vibrational spectroscopy and Raman spectroscopy of small inorganic molecules provides both a theoretical and experimental basis for examining a wide variety of inorganic molecules (Nakamoto, 1997). A wide variety of materials, including silicate glasses, cements, catalysts, clay minerals and ceramics (Potgieter-Vermaak, Potgieter and Grieken, 2006; McMillan, 1984; Bartlett and Cooney, 1987; Butler and Frost, 2006), have been examined using Raman spectroscopy. Mineral-based inorganic pigments, forensic physical evidence, nanoscale semiconductors and carbon nanotubes have been recently explored using Raman. In this chapter, Raman identification and quantification in these broad classes of inorganic materials are reviewed. Finally, a case study using Raman spectroscopy as a real-time industrial process control tool is presented.

24.3.2
Minerals

Examination of geological minerals is one of earliest applications of Raman spectroscopy, and it remains a vibrant field of study. A review of the major Raman bands from a wide variety of geology samples was provided by (Griffith, 1987). Various modes of Raman (resonance Raman, Raman microscopy, isotopic substitution) were reviewed and spectroscopic data was presented from two broad families: nonsilicates and silicates. A wide variety of nonsilicate minerals, such as diamond, carbonate minerals, and gypsum, and silicate minerals, such as clay minerals, micas, and olivine, were included in this chapter. Modern Raman applications of geology minerals include astrobiology remote sensing and planetary exploration (Villar and Edwards, 2006; Lipschutz *et al.*, 2007; Wang, Haskin and Cortez, 1998; Sharma *et al.*, 2002; Haskin *et al.*, 1997). Effects of high temperature and/or high pressure on Raman spectra of minerals have been extensively studied (Gillet *et al.*, 1993; Butler and Gilson, 1997; Gillet, Hemley and McMillan, 1998; Butler and Frost, 2006).

24.3.3
Art, Archaeology, and Architecture

Raman spectroscopy is used to examine mineral-based inorganic pigments in works of art, architectural buildings, and archaeology samples. Recent reviews highlighted the many applications of Raman spectroscopy in art and archaeology (Vandenabeele, Edwards and Moens, 2007; Edwards, 2004). The primary usage of Raman spectroscopy is the identification of pigments, corrosion products, or contaminants in manuscript and paintings. Ceramics, glasses, enamels, glazes, minerals, and gem artifacts are also examined with Raman spectroscopy. Spectral libraries of reference inorganic pigments are available in journals (Bell, Clark and Gibbs, 1997; Vandenabeele, Edwards and Moens, 2007) and online. Table 24.1 presents some spectral libraries available online.

Raman spectra of artifact pigment samples are typically collected in a confocal microscopy mode, because of sample heterogeneity and minute sample sizes. The choice of laser wavelength is influenced by sample fluorescence and the nature of the artifact. Many studies use multiple wavelengths to examine a single artifact. Pigments are typically examined using a near-infrared laser while metal corrosion is examined using 514.5 or 532 nm lasers. White pigments were commonly used as wall paint and as a base for the addition of colored pigments. Figure 24.3 shows the commonly used white pigments: (a) gypsum, (b) calcite, (c) barium white, and (d) bone white, adapted from UCL spectra library as described by Bell et al. (Bell, Clark and Gibbs, 1997).

Mixtures of mineral particles, resulting from artist blending pigments in the original composition or subsequent restoration efforts, complicate the interpretation of Raman spectra taken from paintings. Carbonaceous particles from soot or dirt contaminate artifacts and are a source of fluorescence. However, they are easily distinguishable in light microscopy images because they are often black particles and Raman spectra of carbon particles do not interfere with the identification of mineral-based inorganic pigments. A complicating issue is pigment degradation, either through environmental-induced or laser-induced pathways. Mixtures of lead corrosion products, such as lead oxide, lead sulfate and lead nitrate, were quantified using Raman spectroscopy (Black, Allen and Frost, 1995). Raman spectra of lead carbonates identified the presence of

Tab. 24.1 Online libraries of Raman spectra for inorganic minerals, pigments, and reference materials.

Source	URL site
University of Catania	http://www.ct.infn.it/~archeo/
Universita di Siena	http://www.dst.unisi.it/geofluids/raman/spectrum_frame.htm
University of Parma	http://www.fis.unipr.it/phevix/ramandb.html
University of Arizona (RRUFF)	http://rruff.info/
Infrared and Raman users group	http://www.irug.org/default.asp
Advanced Industrial Science and Technology (AIST)	http://riodb.ibase.aist.go.jp/rasmin/E_index.html
University College London	http://www.chem.ucl.ac.uk/resources/raman/index.html

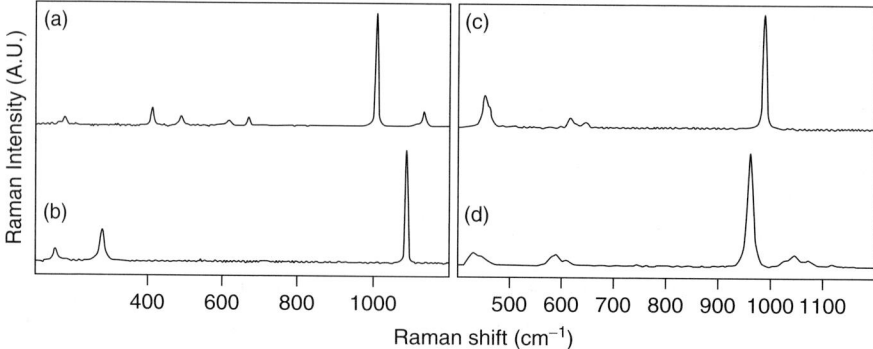

Fig. 24.3 Spectra of white-colored pigments used in paints. Figure 24.3(a) and (b) show gypsum and calcite, and Figure 24.3(c) and (d) are barium white and bone white. Spectra 3a–3d were published previously and reproduced from the University College London online spectral library (Bell, Clark and Gibbs, 1997). (http://www.chem.ucl.ac.uk/resources/raman/index.html.)

adducts and unstable corrosion products (Brooker et al., 1983; Ciomartan et al., 1996).

Raman spectra of paint layers complement historical data on the pigment. These data can be used to estimate the painting's age or direct restoration strategies (Edwards, Farwell and Brooke, 2005). In a study of architectural paint samples, cross-sections of eight paint layers were examined using Raman spectroscopy, light microscopy, and scanning electron microscopy with energy dispersive spectroscopy (SEM–EDS) (Kendix, Nielsen and Christensen, 2004). Raman spectra indicated that each paint layer had its own mixture of pigments, and the Raman data correlated with results from SEM-EDS. The dominant white pigments were calcium carbonate, white lead, titanium dioxide, and barites.

In another study, layered samples paint and stucco were recovered from a preserved Mayan temple found within the Rosalila building (Goodall et al., 2006). Paint/stucco samples from the temple and a stucco mask were examined using Raman, micro attenuated total reflectance infrared spectroscopy (ATR-IR and scanning electron microscopy (SEM) (Goodall et al., 2006). The nature of the green paint was a key finding. Raman bands associated with plagioclase and celadonite were found in green particles, but the bands were too broad and noisy to be definitive. ATR-IR of the same particles reveal a celadonite-based mixture (also called *Terra Verte*), which was confirmed by SEM and micro X-ray. Spectroscopic studies of the Rosalila samples revealed the use of the green Terra Verte pigment, and this pigment was not found in samples collected from another Mayan temple (Vandenabeele et al., 2005).

24.3.4
Forensics

Applications of Raman spectroscopy for nonbiological forensic evidence primarily fall under sample identification. Physical evidence in criminal cases is often in trace amounts, or on a large support structure such as a car panel. Many reports cite flexibility in sampling modes (e.g., microscopy or fiber optics systems), small sample requirements and the nondestructive nature of Raman spectroscopy as an advantage of the method for analysis of trace-level evidence. A 2007 applications review on forensic science highlighted the use of Raman spectroscopy to identify automobile paint, pen ink, and explosives (Brettell, Butler and Almirall, 2007). Raman spectra are primarily used to distinguish physical evidence found at a crime scene from evidence collected from a suspect. Raman spectra of automobile paints identified inorganic fillers or binders, such as magnesium silicate or talc, and inorganic pigments, such as titanium dioxide, lead chromate, or calcite (Buzzini, Massonnet and Sermier, 2006; Zieba-Palus and Borusiewicz, 2006; Gelder et al., 2005). Pigments from spray paint or house paint were also examined by Raman spectroscopy. In one case study, Raman spectra collected from red paint splattered on a suspect's shoe was undifferentiated from spectra of red paint stain on a wall. Raman spectra identified two red paints: (i) Iron oxide using 785 nm excitation and (ii) either chrome orange (lead (II) chromate) or molybdate orange (lead molybdate chromate) in stains at the crime scene. Raman spectra of paint droplets collected from the suspect's shoe also contained bands from the two paint pigments and it was concluded that the shoe was in contact with the paint used to damage the wall (Buzzini, Massonnet and Sermier, 2006). As seen in Figure 24.4, Raman spectra collected from a suspect's shoe indicated a mixture of two red paints, which were similar to the spectra collected from the crime scene.

Proof of concept for metals and anions identification in gunshot residue was shown by (Stich et al., 1998). Gunshot residues were produced from 10, 20, or 30 cm from the paper substrate. Residues left on a paper substrate were examined under a Raman microscope ($\lambda = 633$ nm) in three domains surrounding the impact point. Raman spectra showed heterogeneity in the chemical composition of residues, which was dependent on the distance of the particle from the impact site. In the region closest to the impact site, barium carbonate was the dominant component. In regions farther from the impact site, a mixture of lead-based and barium-based particles were found, and the oxidation state (carbonate or oxide) also varied.

24.3.5
Nanoscale Semiconductors

A review of Raman studies from 1995 to 1997 that characterized semiconductor surfaces was provided by (Lyon et al., 1998). Later reviews in 2003 and 2004 provide a theoretical basis of analyzing fundamental semiconductor properties such as phonon interactions, doping, strain, crystallinity, and surface composition; resonance Raman and polarized Raman were also reviewed (Jimenez, Wolf and Landesman, 2003; Nakashima, 2004). As seen in Figure 24.5, Raman bands are sensitive to local environments and differences in band shape, intensity, or width can be used to differentiate regions on a semiconductor surface.

Since 1998, two nanoscale materials have emerged that have unique electronic properties: single-wall nanotubes (SWNTs) made from carbon and quantum dots (QDs) composed of semiconductor materials. QDs are nanoparticles composed of semiconductor materials, typically with a

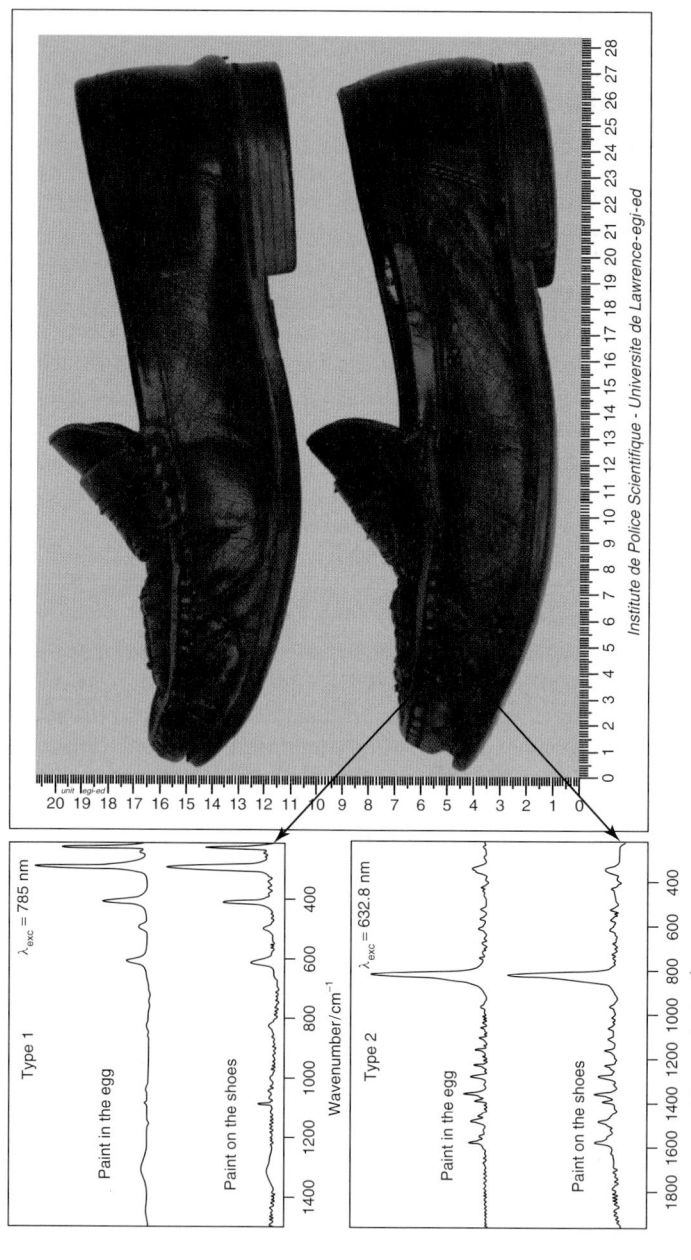

Fig. 24.4 Raman spectroscopy used in forensic evidence analysis shows correlations between evidence collected at a crime scene and evidence collected from a suspect. Paint was thrown onto the wall of a home in small "eggs," and Raman spectra of the splattered paint showed a unique mixture of two red/orange pigments. Raman spectra collected from paint evidence on a suspect shoe were undifferentiated from spectra collected from a crime scene. (Reprinted by permission from reference (Buzzini, Massonnet and Sermier, 2006).)

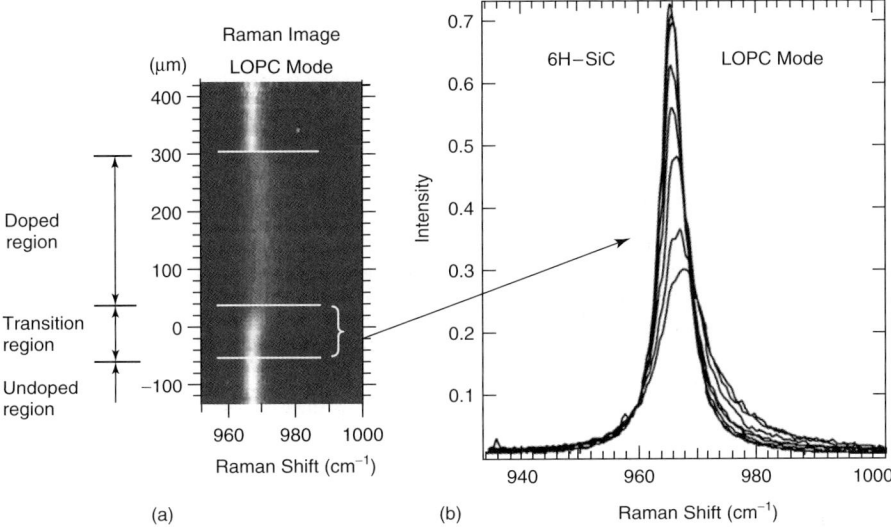

Fig. 24.5 (a) shows a Raman image of a modulation-doped 6H-SiC semiconductor surface and (b) shows Raman spectra of the longitudinal optical phonon plasmon coupled (LOPC) mode in the transition region. Raman band position and band width varied in the transition zone as a result of starting and stopping of carrier gas used in modulated doping experiments. (Figure is reproduced with permission from reference (Nakashima, 2004), copyright Institute of Physics.)

cadmium/selenium core, with potential applications as laser sources, storage materials, fluorescent dyes, and sensors (Voronin, 2006; Cottingham, 2005; Murphy, 2002). Exciting applications in diffusion, flow cytometry, and biomechanics experiments show the broad applicability of QDs in biological fluorescence or luminescence imaging (Dahan et al., 2003; Gao and NIe, 2004; Golcuk et al., 2007).

Raman spectroscopy has proven valuable in characterizing the chemical and electronic properties of QD's. Raman spectroscopy was used to understand the fundamental phonon behavior in reduced dimensions, characterize QD surfaces after chemical doping or mechanical strain, and measure the effects of temperature on longitudinal optical (LO) or transverse optical (TO) phonon frequency. Rolo and Vasilevskiy reviewed theoretical and experimental results of optical phonon modes using Raman spectroscopy (Rolo and Vasilevskiy, 2007). Resonance Raman spectroscopy and nuclear magnetic resonance (NMR) spectroscopy monitored exciton-phonon interactions; results showed improved sensitivity at the molecular level for single QDs and QDs in an ensemble (Gammon et al., 1997).

Carbon nanotubes, multiwalled or single-walled, are frequently investigated by Raman spectroscopy because graphitic structures have intense Raman bands (Barbarossa et al., 1991; Everall and Lumsdon, 1991; Lespade et al., 1984; Gibson, Ayorinde and Wen, 2007). These nanotubes are proposed as reinforcing materials in polymer composites (Coleman et al., 2006; Bokobza, 2007). Raman spectroscopy has emerged as a preferred method for characterizing these materials and has been suggested as a method for monitoring their manufacture (Degamber and Fernando, 2002). In particular, single-walled carbon nanotubes (SWCNTs) have unique electronic, mechanical, and chemical properties; their uses as a composite material or electronic device have been reviewed (Popov, 2004). Raman spectra of SWCNTshave provided valuable insight into the electronic and the phonon properties of SWCNTs. Raman spectra of SWCNTs are sensitive to tube diameter, functionalization, and mechanical strain (Dresselhaus et al., 2002, 2005; Graupner, 2007). A recent study of whole-body Raman imaging using surface-enhancing carbon nanotubes indicates exciting applications of SWCNTs in biomedical imaging (Keren et al., 2008).

24.3.6
Industrial Process Control

High specificity, spatial resolution, molecular identification, and ease of use are features that make Raman an attractive technology for monitoring inorganic chemical production (Workman et al., 2001; Workman, Koch and Veltkamp, 2003). Raman has been used to examine chlorosilane streams as a complementary technique to gas chromatography (GC) analysis (Lipp and Grosse, 1998). A fiber-optics Raman system has been used for direct monitoring of the corrosive stream, where several species were identified and quantified. Raman PAT monitored fluctuations in stream components that indicated poor performance of distillation columns.

Direct monitoring of titanium dioxide production using a fiber-optic Raman spectroscopy instrument was reported by (Clegg et al., 2001). Two forms of TiO_2, anatase and rutile, are made during one step (calciner discharge) of the synthetic route. However, the amount of anatase must be controlled so that the final product retains good optical properties. Powder X-ray diffraction (XRD) had been used as an off-line measurement of quality; however, measurements were performed infrequently. More frequent quality measurements were hypothesized to enable a more efficient calciner discharge step, and Raman spectroscopy was chosen as the analytical tool. Raman spectra of anatase and rutile in the $150-800\,cm^{-1}$ region have unique bands. Raman spectra of anatase/rutile mixtures show that the intensity ratio of $142:610\,cm^{-1}$ bands correlated to ratios measured by XRD. Effects of temperature, crystal size, and window cleanliness on anatase measurement accuracy were also discussed.

24.4
Identification and Quantification of Organic Materials

24.4.1
Pharmaceutical Applications

Raman spectroscopy is a major component of process analytical technology (PAT) (Bakeev, 2005), the pharmaceutical industry's term for process control analytical chemistry and related topics. The most important PAT application of Raman spectroscopy is crystallization monitoring and control (Févotte, 2007; Yu et al., 2004, 2007). The application is important because the polymorph determines such important properties as

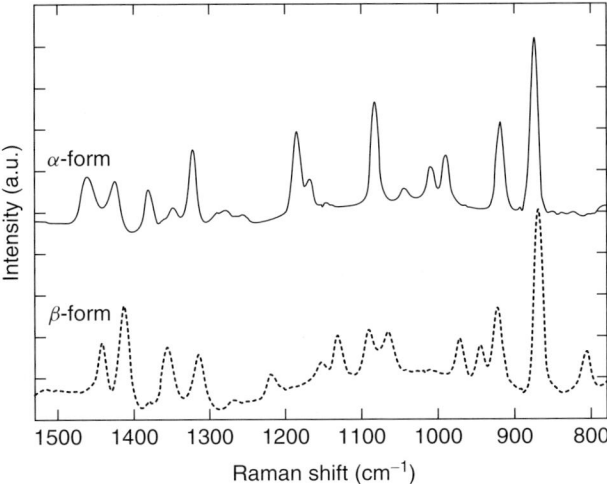

Fig. 24.6 Raman spectra of the metastable α-form and the stable β-form of L-glutamic acid. (Reprinted with permission from reference (Scholl et al., 2006).)

solubility and rate of dissolution. Establishment of the theoretical and the experimental frameworks that define conditions under which the desired polymorph is the only or at least the dominant form are the major goals of researchers (Rodrìguez-Spong et al., 2004; Llinàs and Goodman, 2008).

Crystallization monitoring depends on the differences in the Raman spectra of different polymorphs. L-Glutamic acid, which has two polymorphs, offers a dramatic example (Figure 24.6) (Scholl et al., 2006). The metastable α-form is kinetically favored and is slowly converted to the thermodynamically more stable β-polymorph. In some cases, the differences in the spectra may be more subtle than those for L-glutamic acid, but they will be easily found by PCA, if not by visual inspection (Falcon and Berglund, 2004; Okumura and Otsuka, 2005). However, some authors consider that polymorph spectra differ sufficiently that univariate calibration is almost always adequate (Rantanen et al., 2005).

Because of the ease of use and the ability to monitor changes online (O'Sullivan et al., 2003), Raman spectroscopy has increasingly displaced powder X-ray diffraction in crystallization monitoring. There are also differences in the spectra of amorphous materials and crystalline forms, enabling on-line measurement of crystallinity (Nakano, Tokumura and Umakoshi, 2002). However, there are complications caused by presence of highly scattering materials, whose number and size are changing throughout the process. Sampling a large volume through use of a nonconfocal probe with large spot size, such as the Kaiser Optical Systems PhAT probe is an effective approach to this problem (Wikström, Lewis and Taylor, 2005). Another approach to improved quantification has been to combine Raman spectroscopy with image analysis using a conventional fiber probe and a small video camera simultaneously (Caillet et al., 2007).

Other major pharmaceutical applications include monitoring of homogenization (De Beer et al., 2006), inspection of incoming materials, and inspection of tablets for composition and uniformity. Transmission Raman spectroscopy may be nearly ideal for composition monitoring in formulated products, because it is rapid and provides accurate measures of average composition of a tablet (Matousek and Parker, 2007; Eliasson et al., 2008).

24.4.2
Polymer Chemistry and Physics

Raman spectroscopy plays major roles in polymer processing and analysis. Even today, the technique remains a major tool in the characterization of entire classes of polymers (Billes and Mohammed-Ziegler, 2007). Most commercially important polymers do not themselves fluoresce under visible light, but many formulations contain additives or impurities that do. Excitation with deep red or near-infrared lasers is often needed to circumvent a fluorescence background. Many major applications are to measurement of physical properties such as crystallinity, porosity, stress, and film or fiber orientation. These are the properties that must be adjusted to obtain the desired application performance. Except in contamination studies, a known chemical composition is usually the starting point of the measurement.

It has long been known that the Raman spectra of crystalline and amorphous forms of most polymers are different (Stuart, 1996). The polymer conformation usually determines crystallinity, and markers can include intensity, position, and band width changes. Applications continue to appear (Rasha, 2006). Although most applications use single point spectroscopy only, high definition imaging has been used to map crystallinity in syndiotactic polystyrene (Zhang et al., 1998).

Because polymer bands undergo small shifts when a polymer is stretched or bent, Raman spectroscopy can be used to probe the conformational changes associated with these deformations (Colomban, 2002; Young and Eichhorn, 2007). Polarized spectroscopy is generally used, because more detailed information is available (Sourisseau and Talaga, 2006). The technique has proven useful in measurement of orientation in films (Sourisseau, 2004; Reinecke et al., 2001) as well as in fibers. The same techniques that have been developed for polymers and polymer blends can be applied to composites (Colomban, 2002; Tomba, Carella and Pastor, 2006).

24.4.3
Forensic and Homeland Security Applications

The Raman microprobe has long been used for the analysis of small particles, such as fibers, paint chips, and hairs (Adar, 1988). It is used primarily with samples for which the spatial resolution of an infrared microprobe is inadequate. Recently, there has been extensive application to forensic analysis, for example (Buzzini, Massonnet and Sermier, 2006; Massonnet et al., 2003, 2005; Thomas et al., 2003a,b). Analysis of paint chips from automobile accidents and from burglaries has been especially important. Buzzini and coworkers have presented case studies that illustrate the range of problems that are encountered in these two important areas (Buzzini, Massonnet and Sermier, 2006). Identification of illicit drugs (Hodges et al., 1989; Huong, 1986) is a closely related example.

With increasing interest in detection of explosives, Raman spectroscopy is a candidate for rapid detection (Carter et al., 2005; Cheng et al., 1995; Hayward et al., 1995; Lewis et al., 1997). If low duty cycle pulsed lasers and gated intensified CCD detectors are used, stand-off systems that can operate in ambient light are feasible (Carter et al., 2005). Recently developed fiber-optic spatial offset techniques that can retrieve spectra through translucent containers may provide a convenient means for the detection of concealed explosive materials or component materials for making explosives (Eliasson, Macleod and Matousek, 2007).

24.5
Biological Applications of Raman Spectroscopy

Raman spectroscopy has evolved as a versatile biochemical analysis tool with applications ranging from the analysis of subcellular components to the diagnosis of diseases and *in situ* evaluation of whole tissues. The positions and relative intensities of various Raman spectral bands can be used to quantitatively probe the structure and function of important biological molecules.

24.5.1
Subcellular Level

Raman band assignments, secondary structure signatures (Bandekar, 1992), and markers of side-chain environment (George, 2002) have been established for proteins. The Raman spectrum of a protein is largely dominated by bands associated with the peptide main chain, aromatic side chains, acidic residues, and sulfur-containing side chains. Amino acids that are important in primary structure are tyrosine with peaks at 830 and 850 cm^{-1}, tryptophan with peaks at 1014, 1338, 1361, and 1553 cm^{-1}, and phenylalanine with a peak at 1006 cm^{-1}. Amide I (C=O stretching) and amide III (C–H stretching and N–H in-plane bending) modes are usually used to characterize the secondary structure of the peptide backbone. The wavenumbers at which the amide I and amide III Raman bands occur for α-helical, β-sheets, and random-coiled structures are given in Table 24.2. Apart from the study of folded active proteins, Raman spectroscopy has the ability to provide a variety of probes into the structure of natively unfolded proteins (see Figure 24.7) (Maiti et al., 2004). Protein-based enzymes have been studied in real time by Raman crystallography (Carey, 2006).

The characteristic Raman bands of phospholipids arise from the vibrations of the hydrocarbon chains. The hydrocarbon chains can exist with a high degree of ordering (trans form) or in relative disorder (gauche form)

Tab. 24.2 Raman wavenumbers (per centimeter) for amide I and amide III bands for various polypeptide conformations.

Protein secondary structure	Amide I	Amide III
α-Helix	1645–1660	1265–1300
β-Sheet	1665–1680	1230–1240
Random coil	1660–1670	1240–1260

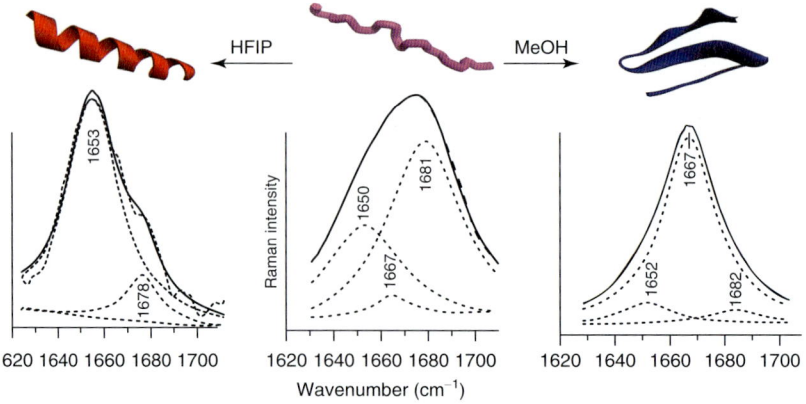

Fig. 24.7 Raman spectra of α-synuclein, a natively unfolded protein, in the presence of hexafluoro-2-propanol (HFIP) and methanol (MeOH). The secondary structure becomes largely α-helical in HFIP (1653 cm^{-1}) and predominantly β-sheet (1667 cm^{-1}) in 25% MeOH in water. (Reprinted with permission from reference (Maiti et al., 2004), copyright American Chemical Society.)

(Mendelsohn and Moore, 1998). The phospholipids membrane conformation is temperature sensitive and can be probed with Raman spectroscopy. Sample Raman spectrums of cholesterol, a lipid, and lecithin, a phospholipid, are shown in Figure 24.8 (Tantipolphan et al., 2006).

The Raman spectrum of a nucleic acid is dominated primarily by bands caused by vibrations localized either within the heterocyclic bases (500–800 cm^{-1}) or in backbone phosphate groups (800–1100 cm^{-1}). These Raman bands are employed as sensitive indicators of local structure, global conformation, intermolecular interaction, and molecular dynamics. Table 24.3 indicates the location of various Raman peaks in the spectra of different conformations of DNAs (Mahadevan-Jansen and Richards-Kortum, 1996). The Raman spectra of DNA–protein complexes yield information about the composition, secondary structure, and interactions of these molecules, including the chemical microenvironment of molecular subgroups.

24.5.2
Tissue Level

In recent years, Raman spectroscopy has seen a surge in applications to the study and diagnosis of diseases. The main aspect of Raman spectroscopy of cells and tissue is that it is capable of probing the subcellular milieu of intact cells. In this section, we describe the applications of Raman spectroscopy by tissue type. For a review on surface-enhanced Raman scattering and its biological applications, refer to these articles (Katrin et al., 2002; Cotton, Kim and Chumanov, 1991).

24.5.2.1 Epithelial Tissue
The majority of the cancers originate in the epithelial tissue, such as esophagus, colon, larynx, bladder, prostate, cervix, breast, skin, lung, and lymph nodes. As these are readily accessible to fiber-optic probes in many cases,

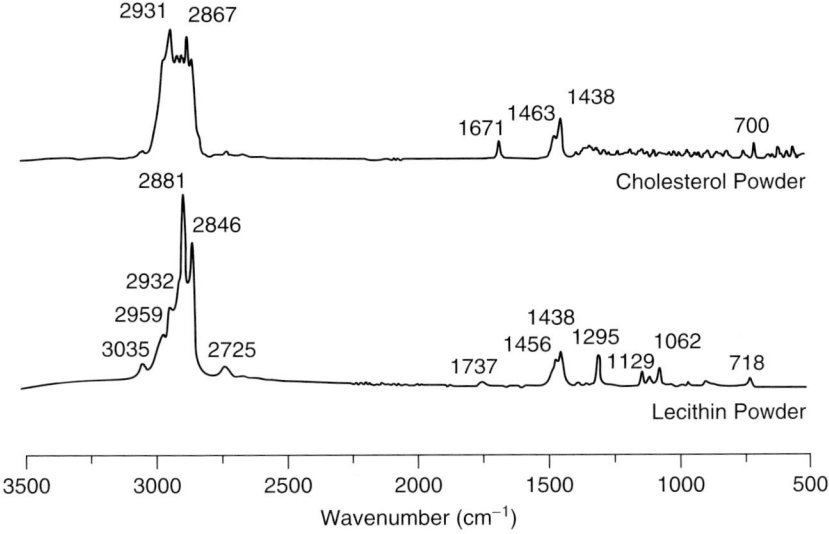

Fig. 24.8 Raman spectra of soybean lecithin and cholesterol powder. (Reprinted with permission from reference (Tantipolphan et al., 2006), copyright Elsevier.)

Tab. 24.3 Raman bands (per centimeter) for the A, B, C, and Z backbone configurations of DNA.

Nucleic acid (form)	A	C	G	T	OPO	PO$_2^-$	Deoxy-ribose
DNA (A)	727 1335	780	666 1318	642 777 1239	806–813	1099	
DNA (B)	727 1339	782 1250	682 1333	665 748 1208	825–842	1091	917 975 1448 1462
DNA (C)			670		785	1090	
DNA (Z)	624 729		640 1316		742–748	1095	

Mahadevan-Jansen and Richards-Kortum (1996).

Raman spectroscopy has the potential to identify molecular and cellular changes associated with cancer cell proliferation (Short et al., 2005). The use of multivariate spectral predictive models has enabled discrimination between benign and malignant tissue to a high degree of accuracy both *in vitro* and *in vivo*. In esophageal cancer, Raman spectroscopy has demonstrated higher levels of guanine, adenine, cytosine, uracil, O–P–O, β-sheet and disordered proteins and lower concentrations of glycogen, tryptophan, proline, and α-helix proteins in cancerous tissue as compared to benign tissue (Kendall et al., 2003). In colon cancer, Raman spectroscopy has reported higher levels of guanine, adenine, cytosine, uracil, O–P–O, lipids, disordered proteins, phenylalanine, and lower levels of tyrosine and carotenoids than in benign tissue (Boustany et al., 1999). In breast cancer, malignant tissue has shown greater concentrations of hydroxyproline and collagen and lower levels of glucose, lipid, and carotenoids than benign tissue (Redd et al., 1993). Neoplastic prostate tissue has been demonstrated to have higher concentrations of guanine, adenine, cytosine, uracil, O–P–O, beta-sheet proteins, lipids, porphyrins, and lower levels of tryptophan, tyrosine, α-helix proteins, phenylalanine, and water compared to benign tissue (Crow et al., 2003). In malignant lung tumor tissue, Raman spectroscopy has identified higher levels of nucleic acid, tryptophan and phenylalanine and lower levels of phospholipids, proline, and valine compared to normal tissue (Huang et al., 2003). Raman spectroscopic studies on skin have looked at chemical enhancing agents for the percutaneous absorption of drugs (Williams et al., 2006), the effect of UV-radiation protective agents in sunscreens for skin (Schallreuter and Wood, 2001) and the effect of aging on the water and protein structures in skin (Williams, Edwards and Barry, 1995).

24.5.2.2 Muscular Tissue

Raman spectroscopy has been applied to the study of contraction-induced conformational changes in muscle fibers and proteins of the skeletal muscles. Studies on the effect of Ca^{2+}, Mg^{2+}, and ATP on the contraction of muscle fibers showed that the contraction is characterized by a decrease in the intensity of Raman bands of protein side chains and an increase in the intensity of amide I and C–C stretch bonds (Caillé, Pigeon-Gosselin and Pézolet, 1983). A recent Raman spectroscopic study monitored changes in the relative levels of NADH and NAD in the bicep muscles of mice under oxidative stress (Sriramoju et al., 2008). This approach could be useful in the study of age- and stress-induced changes in muscles and disorders such as muscular dystrophy and ischemia. Raman spectroscopy has been used to identify and classify different layers of tissue in bladder pathology and bronchial studies. Studies on bladder walls showed that the spectra of urothelium (lipids), lamina propria (collagen), and smooth muscle layer (myosin and actin) are easily distinguishable (de Jong et al., 2002). Raman spectroscopy of bronchial tissue distinguishes between fibrocollagenous stroma tissue and smooth muscle, and could be used to diagnose histological changes in the lung tissue due to lung cancer (Koljenovic et al., 2004).

24.5.2.3 Nervous Tissue

Raman spectroscopy has been used to study the human brain and some of the associated tumors. In normal brain tissue, the spectra of white matter show a greater contribution from lipids, proteins, and cholesterol, whereas the spectra of gray matter exhibit a large contribution from water (Mahadevan-Jansen and Richards-Kortum, 1996). In general, both gray and white matter spectra exhibit peaks from proteins and lipids (Mizuno et al., 1994). Intracranial tumors such as meningeoma, glioma, and schwannoma, which originate from the meninges, glial cells, and Schwann cells, respectively, have been studied (Krafft et al., 2005). Lipid concentration and composition in the brain have been observed to change indicative of brain tumors. Calcification of tissue has also been reported in the case of brain tumors (Mizuno et al., 1994). A Raman map from a human brain tumor section is shown in Figure 24.9 (Krafft, 2004). Raman spectroscopy has been proposed as a water concentration monitoring tool for brain edema in addition to the intracranial pressure measurements using MRI and CT scans (Wolthuis et al., 2001). Because protein misfolding underlies the pathology of several neurodegenerative conditions, Raman microscopy has been used to characterize the biochemistry of transmissible spongiform encephalopathies (Carmona et al., 2004) and Alzheimer's disease (Sudworth et al., 2006).

24.5.2.4 Connective Tissue

Raman spectroscopy has the ability not only to characterize tissue but also to offer quantitative information about tissue components. Quantitative measurements on analytes in blood serum (Rohleder, Kiefer and Petrich, 2004) and whole blood (Enejder et al., 2002) have been made. Raman spectroscopy has also been used to identify microbial pathogens (Maquelin et al., 2003) in blood. Further, it can accurately quantify the relative amounts of cholesterol, calcium salts, triglycerides, and phospholipids in homogenized arterial tissue (Brennan et al., 1997) and hence, can be used to identify atherosclerotic plaque (van de Poll et al., 2002).

Raman spectroscopy has the advantage of being sensitive to both the mineral and the organic components of mineralized tissue and is applicable to wet, thick solid specimens of bone. Raman spectroscopy has been used to study the composition (Penel et al., 2005), chemical changes accompanying mechanical deformation (Morris et al., 2004), aging (Freeman et al., 2001; Ager et al., 2005), and disorders (McCreadie et al., 2006) of bone tissue and to characterize interactions between prosthetic implants and bone tissue (Bertoluzza et al., 1994; Dippel et al., 1998). Raman spectroscopy has also been used to study cartilage and its associated disorders (McCreadie et al., 2006). Dental enamel has a similar composition as that of bone mineral. Polarized Raman spectroscopic studies have been performed to study the orientation of crystallites in dental enamel (Tsuda and Arends, 1994) and in dental caries (Ko et al., 2006). Noninvasive Raman spectroscopy and imaging of biological specimens is a nascent, rapidly advancing field. Transcutaneous measurements on cadaveric and ex vivo bone specimens using spatially separated optical fibers were first reported in 2006 (Schulmerich et al., 2006). In these measurements, depths of 3–4 mm below the skin were reached. Recently, diffusive Raman tomography has been demonstrated on canine bone tissue, also showing that Raman tomography can be integrated with CT and MRI (see Figure 24.10) (Schulmerich et al., 2008).

24.5.3 Microorganisms

Raman spectroscopy has been developed as a simple and rapid identification and characterization tool for pathology. Single bacteria (Rösch et al., 2005), yeast, (Rösch et al., 2005) and cellular components of single spores (Esposito et al., 2003) have been studied. Raman spectroscopy has been able to detect trace levels of microorganisms in the presence

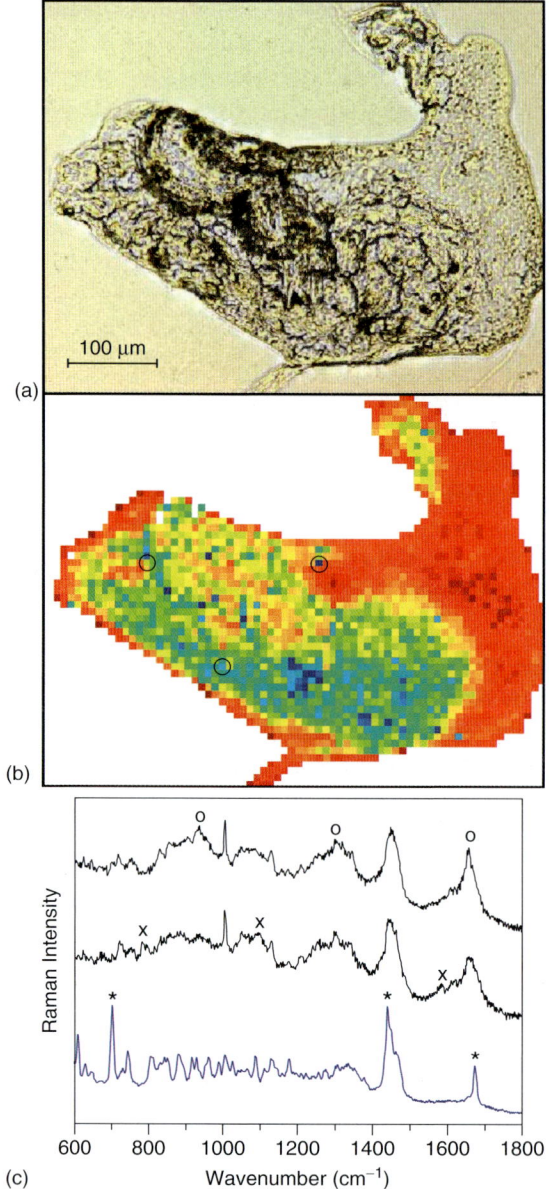

Fig. 24.9 (a) Photomicrograph of dried astrocytoma WHO grade 2 human brain tumor section. (b) Color scaled plot based on a Raman map of 74 × 57 pixels acquired with a step width of 7.5 μm and 1-minute exposure time per spectrum. (c) Selected Raman spectra from B (indicated by circles) in the wavenumber range 600–1800 cm^{-1}. Protein bands characteristic of α-helix or collagen helix (top spectrum) are marked by open dots, DNA bands (middle spectrum) by crosses, and cholesterol bands (bottom spectrum) by asterisks. (Reprinted with permission from reference (Krafft, 2004), copyright Springer-Verlag.)

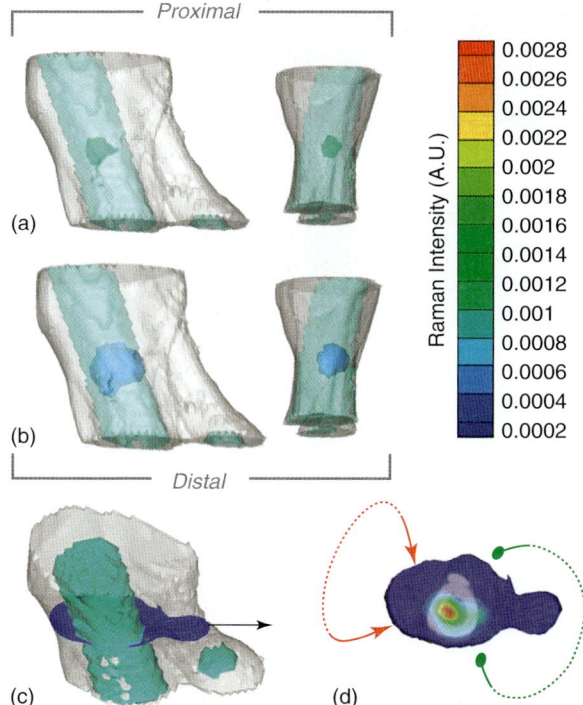

Fig. 24.10 Raman tomographic images of canine bone tissue. (a) Medial and anterior views of soft tissue mesh (white) and bone surface mesh of tibia and calcaneus (turquoise) overlaid with 50% contrast isosurface of the reconstructed three-dimensional (3-D) Raman image of bone (green). (b) Same view as (a) overlaid with 10% contrast isosurface of the reconstructed Raman image of bone (blue). (c) 3-D mesh of limb section (white), including bone (turquoise), illustrating location of the cross section (blue) containing the highest Raman scatter intensity. (d) Raman intensity at cross section in (c) in pseudocolor overlaid on the micro-CT image of the bone, showing range of illumination (red arrows) and collection (green dots) positions. (Reprinted with permission from reference (Schulmerich et al., 2008), copyright Society of Photo-Optical Instrumentation Engineers.)

of complex environmental backgrounds when used in combination with fluorescence spectroscopy and digital imaging (Kalasinsky et al., 2007). The spatial distribution of different molecular species inside a living yeast cell has been probed (see Figure 24.11) (Huang et al., 2005). In another interesting study, gram levels of anthrax spores in envelopes were detected on a mail sorting system using Raman spectroscopy (Farquharson et al., 2004). Building Raman spectral reference libraries of all microorganisms of interest has great potential in the study

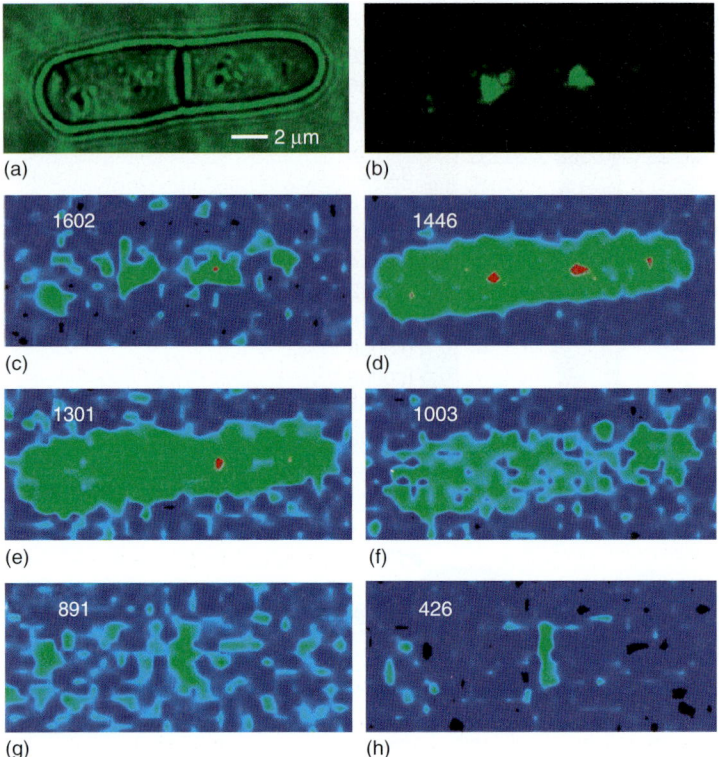

Fig. 24.11 Raman mapping of a living yeast cell – *S. pombe* cell in the G1/S phase: (a) microscopic optical image, (b) GFP image, (c) 1602 cm^{-1}, (d) 1446 cm^{-1}, (e) 1301 cm^{-1}, (f) 1003 cm^{-1}, (g) 891 cm^{-1}, and (h) 426 cm^{-1}. (Reprinted with permission from reference (Huang *et al.*, 2005), copyright American Chemical Society.)

of surface contaminants, bioaerosols, water-borne, and food-borne pathogens.

Glossary

Epi-Fluorescence: A microscopy technique in which the same objective is used for the illumination of a specimen and collection of the excited fluorescence.

Exciton: A bound state of an excited electron and a hole in the lattice of a semiconductor or insulator. Excitons can move around the crystal lattice of the material in which it is formed.

Holographic Filter: An optical filter in which a three-dimensional diffraction pattern is formed in a photosensitive material. Depending on the pattern, the holographic filter can function as a wavelength sensitive mirror (also called a *notch filter*), diffraction grating or other type of filter.

Phonon: A vibrational mode of a crystal lattice.

Surface Plasmon: A surface electromagnetic wave that propagates parallel to the surface of a metal. Surface plasmons are excited at wavelengths that depend on the metal and the three-dimensional surface structure. Silver plasmons can be excited with visible radiation. Gold and copper plasmons can be excited with deep red and near-infrared radiation.

Tomography: A three-dimensional imaging technique in which the image is constructed from a series of sections obtained at different points in a solid material.

References

Adar, F. (1988) *Microchem. J.*, **38**, 50–79.
Adar, F. (2001) Evolution and revolution of raman instrumentation – application of available technologies to spectoscopy and microscopy, in *Handbook of Raman Spectroscopy* (eds I.R. Lewis and H.G.M. Edwards), Marcel Dekker, Inc, New York.
Ager, J.W., Nalla, R.K., Breedon, K.L., and Ritchie, R.O. (2005) *J. Biomed. Opt.*, **10**(3), 034012.
Bakeev, K.A. (ed.) (2005) *Process Analytical Technology*, Blackwell, Oxford.
Bandekar, J. (1992) *Biochim. Biophys. Acta: Protein Struct. Mol. Enzymol.*, **1120**(2), 123–143.
Barbarossa, V., Galluzzi, F., Tomaciello, R., and Zanobi, A. (1991) *Chem. Phys. Lett.*, **185**, 53–55.
Bartlett, J.R. and Cooney, R.P. (1987) Raman spectroscopic studies in chemisorption and catalysis, in *Spectroscopy of Inorganic-Based Materials* (eds R.J.H. Clark and R.E. Hester), John Wiley & Sons, Inc, New York.
Battey, D.E., Slater, J.B., Wludyka, R., Owen, H., Pallister, D.M., and Morris, M.D. (1993) *Appl. Spectrosc.*, **47**, 1913–1919.
Bell, I.M., Clark, R.J.H., and Gibbs, P.J. (1997) *Spectrochim. Acta A Mol. Biomol. Spectrosc.*, **53**, 2159–2179.
Bertoluzza, A., Fagnano, C., Tinti, A., Morelli, M.A., Tosi, M.R., Maggi, G., and Marchetti, P.G. (1994) *J. Raman Spectrosc.*, **25**(1), 109–114.
Billes, F. and Mohammed-Ziegler, I. (2007) *Appl. Spectrosc. Rev.*, **42**(4), 369–441.
Black, L., Allen, G.C., and Frost, P.C. (1995) *Appl. Spectrosc.*, **49**(9), 1299–1304.

Bokobza, L. (2007) *Polymer*, **48**(17), 4907–4920.
Boustany, N.N., Crawford, J.M., Manoharan, R., Dasari, R.R., and Feld, M.S. (1999) *Lab. Invest.*, **79**(10), 1201–1214.
Brennan, J.F. III, Romer, T.J., Lees, R.S., Tercyak, A.M., Kramer, J.R. Jr., and Feld, M.S. (1997) *Circulation*, **96**(1), 99–105.
Brettell, T.A., Butler, J.M., and Almirall, J.R. (2007) *Anal. Chem.*, **79**, 4365–4384.
Brooker, M.H., Sunder, S., Taylor, P., and Lopata, V.J. (1983) *Can. J. Chem.*, **61**, 494–501.
Butler, I.S. and Frost, R.L. (2006) *Appl. Spectrosc. Rev.*, **41**, 449–471.
Butler, I.S. and Gilson, D.F.R. (1997) *J. Mol. Struct.*, **408/409**, 39–45.
Buzzini, P., Massonnet, G., and Sermier, F.M. (2006) *J. Raman Spectrosc.*, **27**, 922–931.
Caillé, J.P., Pigeon-Gosselin, M., and Pézolet, M. (1983) *Biochim. Biophys. Acta*, **758**(2), 121–127.
Caillet, A., Rivoire, A., Galvan, J.M., Puel, F., and Fevotte, G. (2007) *Cryst. Struct. Des.*, **7**(10), 2080–2087.
Carey, P.R. (2006) *Annu. Rev. Phys. Chem.*, **57**(1), 527–554.
Carmona, P., Monleón, E., Monzón, M., Badiola, J.J., and Monreal, J. (2004) *Chem. Biol.*, **11**(6), 759–764.
Carter, J.C., Angel, S.M., Lawrence-Snyder, M., Scaffidi, J., Whipple, R.E., and Reynolds, J.G. (2005) *Appl. Spectrosc.*, **59**(6), 769–775.
Cheng, C., Kirkbride, T.E., Batchelder, D.N., Lacey, R.J., and Sheldon, T.G. (1995) *J. Forensic Sci.*, **1**, 31–37.
Ciomartan, D.A., Clark, R.J.H., McDonald, L.J., and Odlyha, M. (1996) *J. Chem. Soc., Dalton Trans.*, **1996**, 3639–3645.
Clegg, I.M., Everall, N.J., King, B., Melvin, H., and Norton, C. (2001) *Appl. Spectrosc.*, **55**(9), 1138–1150.
Coates, C.G., Denvir, D.J., Conroy, E., McHale, N., Thornbury, K., and Hollywood, M. (2003) *Proc. SPIE*, **4962**, 319–328.
Coleman, J.N., Khan, U., Blau, W.J., and Gun'ko, Y.K. (2006) *Carbon*, **44**(9), 1624–1652.
Colomban, P. (2002) *Adv. Eng. Mater.*, **4**(8), 535–542.
Cottingham, K. (2005) *Anal. Chem.*, **77**, 354A–357A.

Cotton, T.M., Kim, J.-H., and Chumanov, G.D. (1991) *J. Raman Spectrosc.*, **22**(12), 729–742.

Crow, P., Stone, N., Kendall, C.A., Uff, J.S., Farmer, J.A., Barr, H., and Wright, M.P. (2003) *Br. J. Cancer*, **89**(1), 106–108.

Dahan, M., Lévi, S., Luccardini, C., Rostaing, P., Riveau, B., and Triller, A. (2003) *Science*, **302**, 442–445.

De Beer, T.R.M., Baeyens, W.R.G., Ouyang, J., Vervaet, C., and Remon, J.P. (2006) *Analyst*, **131**(10), 1137–1144.

Degamber, B. and Fernando, G.F. (2002) *MRS Bull.*, **27**(5), 370–380.

Denvir, D. and Conroy, E. (2003) *Proc. SPIE*, **4877**, 55–68.

Dippel, B., Mueller, R.T., Pingsmann, A., and Schrader, B. (1998) *Biospectroscopy*, **4**(6), 403–412.

Dresselhaus, M.S., Dresselhaus, G., Jorio, A., Souza Filho, A.G., and Saito, R. (2002) *Carbon*, **40**(12), 2043–2061.

Dresselhaus, M.S., Dresselhaus, G., Saito, R., and Jorio, A. (2005) *Phys. Rep.*, **409**(2), 47–99.

Edwards, H.G.M. (2004) *Analyst*, **129**(10), 870–879.

Edwards, H., Farwell, D., and Brooke, C. (2005) *Anal. Bioanal. Chem.*, **383**(2), 312–321.

Eliasson, C., Macleod, N.A., Jayes, L.C., Clarke, F.C., Hammond, S.V., Smith, M.R., and Matousek, P. (2008) *J. Pharm. Biomed. Anal.*, **47**(2), 221–229.

Eliasson, C., Macleod, N.A., and Matousek, P. (2007) *Anal. Chem.*, **79**(21), 8185–8189.

Enejder, A.M.K., Koo, T.-W., Oh, J., Hunter, M., Sasic, S., Feld, M.S., and Horowitz, G.L. (2002) *Opt. Lett.*, **27**(22), 2004–2006.

Esposito, A.P., Talley, C.E., Huser, T., Hollars, C.W., Schaldach, C.M., and Lane, S.M. (2003) *Appl. Spectrosc.*, **57**(7), 868–871.

Everall, N.J. (2000a) *Appl. Spectrosc.*, **54**(6), 773–782.

Everall, N.J. (2000b) *Appl. Spectrosc.*, **54**(10), 1515–1520.

Everall, N., Lapham, J., Adar, F., Whitley, A., Lee, E., and Mamedov, S. (2007) *Appl. Spectrosc.*, **61**, 251–259.

Everall, N.J. and Lumsdon, J. (1991) *J. Mater. Sci.*, **26**, 5269–5274.

Falcon, J.A. and Berglund, K.A. (2004) *Cryst. Growth Des.*, **4**(3), 457–463.

Farquharson, S., Grigely, L., Khitrov, V., Smith, W., Sperry, J.F., and Fenerty, G. (2004) *J. Raman Spectrosc.*, **35**(1), 82–86.

Févotte, G. (2007) *Chem. Eng. Res. Des.*, **85**(7), 906–920.

Fleming, O. and Kazarian, S.G. (2004) *Appl. Spectrosc.*, **58**(4), 390–394.

Freeman, J.J., Wopenka, B., Silva, M.J., and Pasteris, J.D. (2001) *Calcif. Tissue Int.*, **68**(3), 156–162.

Gammon, D., Brown, S.W., Snow, E.S., Kennedy, T.A., Katzer, D.S., and Park, D. (1997) *Science*, **277**(5322), 85–88.

Gao, X. and Nie, S. (2004) *Anal. Chem.*, **76**(8), 2406–2410.

Gelder, J.D., Vandenabeele, P., Govaert, F., and Moens, L. (2005) *J. Raman Spectrosc.*, **36**, 1059–1067.

George, J.T. Jr. (2002) *Biopolymers*, **67**(4–5), 214–225.

Gibson, R.F., Ayorinde, E.O., and Wen, Y.-F. (2007) *Compos. Sci. Technol.*, **67**(1), 1–28.

Gillet, P., Biellmann, C., Reynard, B., and McMillan, P. (1993) *Phys. Chem. Miner.*, **20**(1), 1–18.

Gillet, P., Hemley, R.J., and McMillan, P.F. (1998) Vibrational properties at high pressures and temperatures, in *Ultrahigh-Pressure Mineralogy: Physics and Chemistry of the Earth's Deep Interior* (ed. R.J. Hemley), Mineralogical Society of America, Washington, DC.

Golcuk, K., Vanasse, T.M., Morris, M.D., and Goldstein, S.A. (2007) *Proc. SPIE*, **6448**, 644815.

Goodall, R.A., Hall, J., Viel, R., Agurcia, F.R., Edwards, H.G.M., and Fredericks, P.M. (2006) *J. Raman Spectrosc.*, **37**(10), 1072–1077.

Graupner, R. (2007) *J. Raman Spectrosc.*, **38**(6), 673–683.

Griffith, W.P. (1987) Advances in the Raman and infrared spectroscopy of minerals, in *Spectroscopy of Inorganic-Based Materials* (eds R.J.H. Clark and R.E. Hester), John Wiley & Sons, Inc, New York.

Haskin, L.A., Wang, A., Rockow, K.M., Joliff, B.L., Korotev, R.L., and Viskupic, K.M. (1997) *J. Geophys. Res.*, **102**, 19293–19306.

Hayward, I.P., Kirkbride, T.E., Batchelder, D.N., and Lacey, R.J. (1995) *J. Forensic Sci.*, **40**, 883–884.

Hodges, C.M., Hendra, P.J., Willis, H.A., and Farley, T. (1989) *J. Raman Spectrosc.*, **20**, 745–749.

Huang, Y.S., Karashima, T., Yamamoto, M., and Hamaguchi, H. (2005) *Biochemistry*, **44**(30), 10009–10019.

Huang, Z., McWilliams, A., Lui, H., McLean, D.I., Lam, S., and Zeng, H. (2003) *Int. J. Cancer*, **107**(6), 1047–1052.

Huong, P.V. (1986) *J. Pharm. Biomed. Anal.*, **4**, 811–823.

Janesick, J.R. (2001) *Scientific Charge-Coupled Devices*, SPIE–The International Society for Optical Engineering, Bellingham.

Jimenez, J., Wolf, I.D., and Landesman, J.P. (2003) Micro-Raman spectroscopy of semiconductors: principles and applications, in *Microprobe Characterizations of Optoelectronic Materials* (ed. J. Jimenez), Taylor & Francis Books, Inc, New York.

de Jong, B.W.D., Bakker Schut, T.C., Wolffenbuttel, K.P., Nijman, J.M., Kok, D.J., and Puppels, G.J. (2002) *J. Urol.*, **168**(4), Suppl 1, 1771–1778.

Kalasinsky, K.S., Hadfield, T., Shea, A.A., Kalasinsky, V.F., Nelson, M.P., Neiss, J., Drauch, A.J., Vanni, G.S., and Treado, P.J. (2007) *Anal. Chem.*, **79**(7), 2658–2673.

Katrin, K., Harald, K., Irving, I., Ramachandra, R.D., and Michael, S.F. (2002) *J. Phys.: Condens. Matter*, **18**, 597–624.

Kendall, C., Stone, N., Shepherd, N., Geboes, K., Warren, B., Bennett, R., and Barr, H. (2003) *J. Pathol.*, **200**(5), 602–609.

Kendix, E., Nielsen, O.F., and Christensen, M.C. (2004) *J. Raman Spectrosc.*, **35**(8–9), 796–799.

Keren, S., Zavaleta, C., Cheng, Z., de la Zerda, A., Gheysens, O., and Gambhir, S.S. (2008) *Proc. Natl. Acad. Sci. U.S.A.*, **105**(15), 5844–5849.

Ko, A.C.T., Choo-Smith, L.-Pi., Hewko, M., Sowa, M.G., Dong, C.C.S., and Cleghorn, B. (2006) *Opt. Express*, **14**(1), 203–215.

Koljenovic, S., Schut, T.C.B., van Meerbeeck, J.P., Maat, A.P.W.M., Burgers, S.A., Zondervan, P.E., Kros, J.M., and Puppels, G.J. (2004) *J. Biomed. Opt.*, **9**(6), 1187–1197.

Krafft, C. (2004) *Anal. Bioanal. Chem.*, **378**(1), 60–62.

Krafft, C., Neudert, L., Simat, T., and Salzer, R. (2005) *Spectrochim. Acta A: Mol. Biomol. Spectrosc.*, **61**(7), 1529–1535.

Laser Institute of America (2007) *American National Standard for Safe Use of Lasers (ANSI Z136.1–2007)*, Laser Institute of America, Orlando.

Lespade, P., Marchand, A., Couzi, M., and Cruege, F. (1984) *Carbon*, **22**, 375–385.

Lewis, I.R., Daniel, J., Nelson, W., and Griffiths, P.R. (1997) *Appl. Spectrosc.*, **51**(12), 1854–1867.

Lewis, I.R. and Edwards, H.G.M. (eds) (2001) *Handbook of Raman Spectroscopy*, Marcel Dekker, Inc, New York.

Lipp, E.D. and Grosse, R.L. (1998) *Appl. Spectrosc.*, **52**(1), 42–46.

Lipschutz, M.E., Wolf, S.F., Culp, F.B., and Kent, A.J.R. (2007) *Anal. Chem.*, **79**, 4249–4274.

Llinàs, A. and Goodman, J.M. (2008) *Drug Discov. Today*, **13**(5–6), 198–210.

Lyon, L.A., Keating, C.D., Fox, A.P., Baker, B.E., He, L., Nicewarner, S.R., Mulvaney, S.P., and Natan, M.J. (1998) *Anal. Chem.*, **70**, 341R–361R.

Mahadevan-Jansen, A. and Richards-Kortum, R.R. (1996) *J. Biomed. Opt.*, **1**(1), 31–70.

Maiti, N.C., Apetri, M.M., Zagorski, M.G., Carey, P.R., and Anderson, V.E. (2004) *J. Am. Chem. Soc.*, **126**(8), 2399–2408.

Maquelin, K., Kirschner, C., Choo-Smith, L.P., Ngo-Thi, N.A., van Vreeswijk, T., Stammler, M., Endtz, H.P., Bruining, H.A., Naumann, D., and Puppels, G.J. (2003) *J. Clin. Microbiol.*, **41**(1), 324–329.

Massonnet, G., Buzzini, P., Jochem, G., Staube, M., Coyle, T., Roux, C., Thomas, J., Leijenhorst, H., van Zanten, Z., Griffin, R., Wiggins, K., and Chabli, S. (2003) *Forensic Sci. Int.*, **136**, 124.

Massonnet, G., Buzzini, P., Jochem, G., Stauber, M., Coyle, T., Roux, C., Thomas, J., Leijenhorst, H., Van Zanten, Z., Wiggins, K., Russell, C., Chabli, S., and Rosengarten, A. (2005) *J. Forensic Sci.*, **50**(5), 1028–1038.

Matousek, P. (2007) *Chem. Soc. Rev.*, **36**(8), 1292–1304.

Matousek, P. and Parker, A.W. (2007) *J. Raman Spectrosc.*, **38**(5), 563–567.

McCreadie, B.R., Morris, M.D., Chen, T.-C., Sudhaker Rao, D., Finney, W.F., Widjaja, E., and Goldstein, S.A. (2006) *Bone*, **39**(6), 1190–1195.

McMillan, P. (1984) *Am. Mineral.*, **69**, 622–644.

Mendelsohn, R. and Moore, D.J. (1998) *Chem. Phys. Lipids*, **96**(1–2), 141–157.

Mizuno, A., Kitajima, H., Kawauchi, K., Muraishi, S., and Ozaki, Y. (1994) *J. Raman Spectrosc.*, **25**(1), 25–29.

Morris, M.D., Finney, W.F., Rajachar, R.M., and Kohn, D.H. (2004) *Faraday Discuss.*, **126**, 159–168.

Motz, J.T., Hunter, M., Galindo, L.H., Gardecki, J.A., Kramer, J.A., Dasari, R.R., and Feld, M.S. (2004) *Appl. Opt.*, **43**(5), 542–554.

Murphy, C.J. (2002) *Anal. Chem.*, **74**(19), 520A–526A.

Nakamoto, K. (1997) *Infrared and Raman Spectra of Inorganic and Coordination Compounds Part A: Theory and Applications in Inorganic Chemistry*, 5th edn, John Wiley & Sons, Inc, New York.

Nakano, T., Tokumura, A., and Umakoshi, Y. (2002) *Metallurg. Mater. Transact. A*, **33A**, 521–528.

Nakashima, S. (2004) *J. Phys.: Condens. Matter*, **16**(2), S25–S37.

Okumura, T. and Otsuka, M. (2005) *Pharm. Res.*, **22**(8), 1350–1357.

O'Sullivan, B., Barrett, P., Hsiao, G., Carr, A., and Glennon, B. (2003) *Org. Process Res. Dev.*, **7**(6), 977–982.

Penel, G., Delfosse, C., Descamps, M., and Leroy, G. (2005) *Bone*, **36**(5), 893–901.

Pitt, G.D., Batchelder, D.N., Bennett, R., Bormett, R.W., Hayward, I.P., Smith, B.J.E., Williams, K.P.J., Yang, Y.Y., Baldwin, K.J., and Webster, S. (2005) *IEE Proc. Sci. Meas. Tech.*, **152**(6), 241–318.

Popov, V.N. (2004) *Mater. Sci. Eng. R: Rep.*, **43**(3), 61–102.

Potgieter-Vermaak, S.S., Potgieter, J.H., and Grieken, R.V. (2006) *Cement Concr. Res.*, **36**, 656–662.

Rantanen, J., Wikström, H., Rhea, F.E., and Taylor, L.S. (2005) *Appl. Spectrosc.*, **59**(7), 942–951.

Rasha, M.K. (2006) *J. Polym. Sci., part B: Polym. Phys.*, **44**(15), 2173–2182.

Redd, D.C.B., Feng, Z.C., Yue, K.T., and Gansler, T.S. (1993) *Appl. Spectrosc.*, **47**, 787–791.

Reinecke, H., Spells, S.J., Sacristan, J., Yarwood, J., and Majangos, C. (2001) *Appl. Spectrosc.*, **55**(12), 1660–1664.

Rodrìguez-Spong, B., Price, C.P., Jayasankar, A., Matzger, A.J., and Rodrìguez-Hornedo, N. (2004) *Adv. Drug Deliv. Rev.*, **56**(3), 241–274.

Rohleder, D., Kiefer, W., and Petrich, W. (2004) *Analyst*, **129**(10), 906–911.

Rolo, A.G. and Vasilevskiy, M.I. (2007) *J. Raman Spectrosc.*, **38**(6), 618–633.

Rosch, P., Harz, M., Schmitt, M., Peschke, K.-D., Ronneberger, O., Burkhardt, H., Motzkus, H.-W., Lankers, M., Hofer, S., Thiele, H., and Popp, J. (2005) *Appl. Environ. Microbiol.*, **71**(3), 1626–1637.

Rösch, P., Harz, M., Schmitt, M., and Popp, J. (2005) *J. Raman Spectrosc.*, **36**(5), 377–379.

Schallreuter, K.U. and Wood, J.M. (2001) *J. Photochem. Photobiol. B: Biol.*, **64**(2–3), 179–184.

Scholl, J., Bonalumi, D., Vicum, L., Mazzotti, M., and Muller, M. (2006) *Cryst. Growth Des.*, **6**(4), 881–891.

Schulmerich, M.V., Finney, W.F., Fredericks, R.A., and Morris, M.D. (2006) *Appl. Spectrosc.*, **60**(2), 109–114.

Schulmerich, M.V., Morris, M.D., Vanasse, T.M., and Goldstein, S.A. (2007) Paper read at Advanced Biomedical and Clinical Diagnostic Systems V, San Jose, CA.

Schulmerich, M.V., Srinivasan, S., Cole, J.H., Kreider, J., Dooley, K.A., Goldstein, S.A., Pogue, B.W., and Morris, M.D. (2008) *J. Biomed. Opt.*, **13**, 020506.

Sharma, S.K., Angel, S.M., Ghosh, M., Hubble, H.W., and Lucey, P.G. (2002) *Appl. Spectrosc.*, **56**(6), 699–705.

Short, K.W., Carpenter, S., Freyer, J.P., and Mourant, J.R. (2005) *Biophys. J.*, **88**(6), 4274–4288.

Smith, E. and Dent, G. (2005) *Modern Raman Spectroscopy: A Practical Approach*, John Wiley & Sons, Ltd, Chichester.

Sourisseau, C. (2004) *Chem. Rev.*, **104**(9), 3851–3892.

Sourisseau, C. and Talaga, D. (2006) *Appl. Spectrosc.*, **60**(12), 1368–1376.

Sriramoju, V., Alimova, A., Chakraverty, R., Katz, A., Gayen, S.K., Larsson, L., Savage, H.E., and Alfano, R.R. (2008) Paper read at biomedical optical spectroscopy, San Jose, CA.

Stich, S., Bard, D., Gros, L., Wenz, H.W., Yarwood, J., and Williams, K. (1998) *J. Raman Spectrosc.*, **29**, 787–790.

Stuart, B.H. (1996) *Vib. Spectrosc.*, **10**(2), 79–87.

Sudworth, C.D., Archer, J.K.J., Black, R.A., and Mann, D. (2006) Paper read at biomedical vibrational spectroscopy III: advances in research and industry, San Jose, CA.

Tantipolphan, R., Rades, T., Strachan, C.J., Gordon, K.C., and Medlicott, N.J. (2006) *J. Pharm. Biomed. Anal.*, **41**(2), 476–484.

Thomas, J., Buzzini, P., Massonnet, G., and Roux, C. (2003a) *Forensic Sci. Int.*, **136**, 118.

Thomas, J., Buzzini, P., Roux, C. and Reedy, B. (2003b) *Forensic Sci. Int.*, **136**, 125.

Tomba, J.P., Carella, J.M., and Pastor, J.M. (2006) *Appl. Spectrosc.*, **60**(2), 115–121.

Tsuda, H. and Arends, J. (1994) *J. Dent. Res.*, **73**(11), 1703–1710.

van de Poll, S.W., Sweder, W.E., Römer, T.J., Tjeerd, J., Puppels, G.J., and van der Laarse, A. (2002) *J. Cardiovasc. Risk*, **9**(5), 255–261.

Vandenabeele, P., Bodé, S., Alonso, A., and Moens, L. (2005) *Spectrochim. Acta A: Mol. Biomol. Spectrosc.*, **61**(10), 2349–2356.

Vandenabeele, P., Edwards, H.G.M., and Moens, L. (2007) *Chem. Rev.*, **107**(3), 675–687.

Villar, S.E.J. and Edwards, H.G.M. (2006) *Anal. Bioanal. Chem.*, **384**, 100–113.

Voronin, M.L. (ed.) (2006) *Frontiers in Quantum Dots Research*, Nova Science Publishers, Inc, New York.

Wang, A., Haskin, L.A., and Cortez, E. (1998) *Appl. Spectrosc.*, **52**(4), 477.

Wikström, H., Lewis, I.R., and Taylor, L.S. (2005) *Appl. Spectrosc.*, **59**(7), 934–941.

Williams, A.C., Edwards, H.G.M., and Barry, B.W. (1995) *Biochim. Biophys. Acta Protein Struct. Mol. Enzymol.*, **1246**(1), 98–105.

Williams, A.C., Edwards, H.G.M., Lawson, E.E., and Barry, B.W. (2006) *J. Raman Spectrosc.*, **37**(1–3), 361–366.

Wolthuis, R., van Aken, M., Fountas, K., Robinson, J.S. Jr., Bruining, H.A., and Puppels, G.J. (2001) *Anal. Chem.*, **73**(16), 3915–3920.

Workman, J., Creasy, K.E., Doherty, S., Bond, L., Koch, M., Ullman, A., and Veltkamp, D.J. (2001) *Anal. Chem.*, **73**(12), 2705–2718.

Workman, J., Koch, M., and Veltkamp, D.J. (2003) *Anal. Chem.*, **75**(12), 2859–2876.

Young, R.J. and Eichhorn, S.J. (2007) *Polymer*, **48**(1), 2–18.

Yu, Z.Q., Chew, J.W., Chow, P.S., and Tan, R.B.H. (2007) *Chem. Eng. Res. Des.*, **85**(7), 893–905.

Yu, L.X., Lionberger, R.A., Raw, A.S., D'Costa, R., Wu, H., and Hussain, A.S. (2004) *Adv. Drug Deliv. Rev.*, **56**(3), 349–369.

Zhang, S., Pezzuti, J.A., Morris, M.D., Appawedula, A., Hsiung, C.M., Leugers, M.A., and Bank, D. (1998) *Appl. Spectrosc.*, **52**(10), 1264–1268.

Zieba-Palus, J. and Borusiewicz, R. (2006) *J. Mol. Struct.*, **792–793**, 286–292.

Further Reading

Aroca, R. (2006) *Surface-Enhanced Vibrational Spectroscopy*, John Wiley & Sons.

Chalmers, J.M. and Griffiths, P.R. (eds) (2002) *Handbook of Vibrational Spectroscopy*, John Wiley & Sons, Inc, New York.

Edwards, H.G.M. and Chalmers, J.M. (eds) (2005) *Raman Spectroscopy in Archaeology and Art History*, RSC Analytical Spectroscopy Monographs, Royal Society of Chemistry, Cambridge.

Everall, N.J., Chalmers, J.M., and Griffiths, P.R. (eds) (2007) *Vibrational Spectroscopy of Polymers: Principles and Practice*, John Wiley & Sons.

Kneipp, K., Moskovits, M., and Kneipp, H. (eds) (2006) *Surface-Enhanced Raman Scattering: Physics and Applications*, Topics in Applied Physics, Springer.

Lasch, P. and Kneipp, J. (eds) (2008) *Biomedical Vibrational Spectroscopy*, John Wiley & Sons.

Slobodan, S. (ed.) (2007) *Pharmaceutical Applications of Raman Spectroscopy*, Wiley Series on Technologies for the Pharmaceutical Industry, John Wiley & Sons.

Smith, E. and Dent, G. (2005) *Modern Raman Spectroscopy: A Practical Approach*, John Wiley & Sons.

25
Spectrometers for Infrared Light

Yukihiro Ozaki and Shigeaki Morita

25.1	**Introduction** 867	
25.2	**General Principles of Infrared Spectrometers** 869	
25.2.1	Principles of Dispersive IR Spectrometers 870	
25.2.1.1	Diffraction Grating 870	
25.2.1.2	Monochromators 871	
25.2.2	Principles of FT-IR Spectrometers 872	
25.2.3	Advantages of FT-IR Spectrometers 877	
25.2.3.1	Fellgett Advantage (Multiplex Advantage) 877	
25.2.3.2	Jacquinot Advantage (Throughput Advantages) 877	
25.2.3.3	Connes Advantage 877	
25.2.4	Phase Correction 877	
25.3	**Components of FT-IR Spectrometers** 878	
25.3.1	Light Sources 878	
25.3.2	Interferometers 878	
25.3.3	Sample Compartments 879	
25.3.4	Detectors 880	
25.3.5	Beamsplitters and Mirrors 881	
25.3.6	Amplifiers and AD Convertors 881	
25.4	**Various Types of FT-IR Interferometers** 881	
25.4.1	Continuous-Scanning Interferometers 881	
25.4.2	Step-Scan Interferometers 882	
25.4.3	Asynchronous Time-Resolved FT-IR Spectrometer 883	
	References 885	
	Further Reading 886	

Encyclopedia of Applied Spectroscopy. Edited by David L. Andrews.
Copyright © 2009 WILEY-VCH Verlag GmbH & Co. KGaA, Weinheim
ISBN: 978-3-527-40773-6

25.1
Introduction

Infrared (IR) spectroscopy is spectroscopy in the wavelength region of 0.78 μm–1 mm; the corresponding wavenumber region is 12 800–10 cm^{-1} (Griffiths and de Haseth, 2007; Christy, Ozaki and Gregoriou, 2001). The IR region lies between visible and microwave regions and is subdivided into three: near-infrared (NIR) (0.78–2.5 μm; 12 800–4000 cm^{-1}), mid-IR (2.5–25 μm; 4000–400 cm^{-1}), and far-IR (25 μm–1 mm; 400–10 cm^{-1}) regions. In this chapter, we are concerned only with mid-IR region.

IR spectroscopy provides detailed information about vibrations of a molecule (Griffiths and de Haseth, 2007; Christy, Ozaki and Gregoriou, 2001). Since molecular vibrations reflect the physical and chemical features of a molecule, such as the arrangement of nuclei and chemical bonds within the molecule, IR spectroscopy is very useful not only for identification of the molecule but also for study of the molecular structure. Furthermore, interaction of a molecule with the surrounding environment also causes a change in molecular vibrations, and hence, IR spectroscopy contributes considerably in studying the interaction, too. IR spectroscopy has many uses from basic research to various applications.

The history of IR spectroscopy stretches back to the nineteenth century (Sheppard, 2002). In the 1880s, IR spectra of many basic compounds were measured, and the first commercial IR spectrometer was put on the market as early as 1913. In the beginning of the twentieth century, the idea of group frequencies of IR absorptions for spectral analysis was established. Thus, it can be said that IR spectroscopy is very old, but it has always remained one of the most useful spectroscopic techniques ever in molecular science and analytical science.

When irradiated with IR radiation, a molecule absorbs it under some conditions. The energy $h\nu$ of the absorbed IR radiation is equal to the difference between two energy levels E_m and E_n of vibration of the molecule. In the form of an equation,

$$h\nu = E_n - E_m \qquad (25.1)$$

In other words, absorption of IR radiation occurs principally on the basis of a transition between energy levels of molecular vibration. This is why an IR absorption spectrum is a vibrational spectrum of a molecule.

Encyclopedia of Applied Spectroscopy. Edited by David L. Andrews.
Copyright © 2009 WILEY-VCH Verlag GmbH & Co. KGaA, Weinheim
ISBN: 978-3-527-40773-6

Satisfying Eq. (25.1) is not always the only criterion for IR absorption to occur. There are transitions that are allowed by a selection rule (i.e., allowed transition) and those that are not allowed by the same rule (i.e., forbidden transition). In general, transitions with a change in the vibrational quantum number by ±1 are allowed and others are forbidden. This is known as a *selection rule* with respect to IR absorption. Another selection rule is one that is defined by the symmetry of a molecule. This selection rule is "IR radiation is absorbed when the transition dipole moment of a molecule changes as a whole in accordance with a molecular vibration."

Since most molecules are in the ground vibrational state at room temperature, a transition from the state $v'' = 0$ to the state $v'' = 1$ (first excited state) is possible. Absorption corresponding to this transition is called the *fundamental* band. Although most bands that are observed in IR absorption spectra arise from the fundamental, in some cases, we can find bands that correspond to transitions from the state $v'' = 0$ to the state $v'' = 2, 3, 4, \cdots\cdots\ldots$ (i.e., *overtone* transitions). However, since overtones are forbidden, overtone bands are very weak.

Figure 25.1 shows typical IR spectra in (a) a transmittance unit and (b) an absorbance unit. The abscissa and the ordinate of an IR absorption spectrum must now be explained. Frequency is indicated along the abscissa in units of wavenumber (cm^{-1}) (with higher wavenumbers always on the left-hand side), while *transmittance T* (%) (Figure 25.1a) or an *absorbance E* (Figure 25.1b) is expressed along the ordinate. While IR spectra include reflectance spectra and emission spectra in addition to absorption spectra, reflectance, emission intensity, and so on are expressed along the ordinate in the case of a reflectance spectrum, an emission spectrum, and so on.

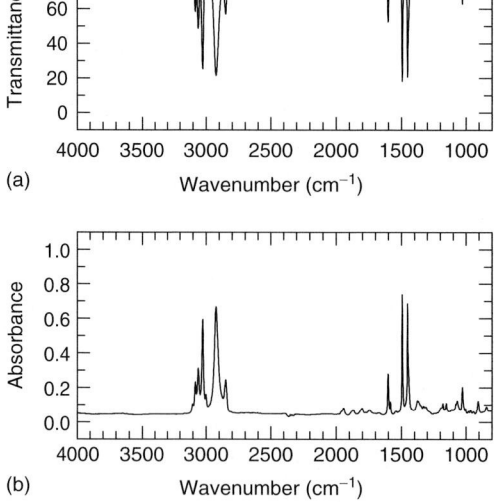

Fig. 25.1 Examples of IR spectra: (a) transmittance spectrum and (b) absorbance spectrum. (Reproduced from Christy, Ozaki and Gregoriou (2001) with permission. Copyright (2001) Elsevier.)

25.2
General Principles of Infrared Spectrometers

IR spectrometers have a long history (Sheppard, 2002). They have made remarkable progress after the Second World War. There were three revolutions in the history of IR spectrometers after the war. The first one, which occurred just after the war because of various technical developments during the war, was concerned with the development of dispersive IR spectrometers using prisms as dispersing elements. During the 1950s, diffraction gratings, detectors, and optical elements developed rapidly, and in 1960s, high-performance dispersive spectrometers with diffraction gratings became widely available commercially. This was the second revolution. The third and the biggest revolution was the advent of Fourier-transform (FT) spectrometers in the 1970s. Indeed, the advances in the field of FT-IR spectroscopy in the past 30 years have been quite remarkable.

In this chapter, we do not describe the detailed history of FT-IR spectrometers, for which many good references are available (Griffiths and de Haseth, 1986, 2007; Ferraro and Basile, 1990; Sheppard, 2002; Kauppinen and Hollberg). However, three important historical events need special mention. In 1952, Fellgett pointed out that, in principle, interferometry also provides a multiplex advantage using broadband sources in that an interferogram records the superimposed signals from all the different wavelengths presented to the detector (Sheppard, 2002). In 1958, Jacquinot emphasized the additional energy throughput advantage of interferometers over dispersive spectrometers on the basis of their circular aperture in comparison with the entrance slits of dispersive spectrometers. Another breakthrough came from the discovery of the *fast Fourier transform* (FFT) by Cooley and Tukey in 1964. The FFT is superior to the discrete FT.

Nowadays, there are two types of spectrometers used to measure IR spectra, namely, dispersive IR spectrometer and FT-IR spectrometer, although the latter is much more popular than the former. Dispersive IR spectrometer is a tool that disperses IR radiation spatially into its constituent wavelengths with a diffraction grating, a prism, or the like. Meanwhile, FT-IR spectrometer utilizes interference of radiation for the purpose of spectral separation of radiation. As each one of the two methods has its own pros and cons, no one can categorically conclude that one is superior to the other. However, FT-IR spectrometer has recently become a principal method for the following reasons:

1. It permits one to measure spectra simultaneously in the entire wavenumber region.
2. Accuracy of the wavenumbers of spectra is extremely high.
3. It secures a constant and high spectral resolution across the whole spectral region.
4. The amount of radiation reaching a detector is high as it utilizes radiation at a very high efficiency.

While FT-IR spectrometer will certainly remain a prominent IR spectroscopic method at least for the time being, dispersive IR spectroscopy could well attract a lot of attention in the future, again if multichannel detectors operable in the IR region advance and improve further. Noting that many researchers today chose FT-IR spectroscopy in almost

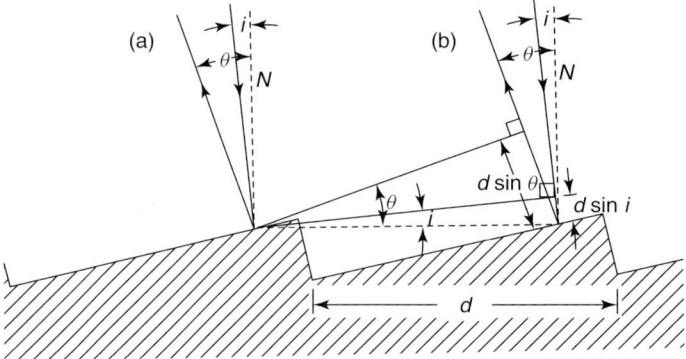

Fig. 25.2 An echelette grating, a typical reflection grating used in a dispersive IR monochromator. (Reproduced from Kauppinen et al. (1997) with permission. Copyright (1997) Wiley-VCH.)

all of their research activities that use IR spectroscopy, we discuss FT-IR spectrometer more in detail than dispersive IR spectrometer in this chapter.

25.2.1
Principles of Dispersive IR Spectrometers

A typical IR spectrometer consists of a radiation source, a wavelength selector, a sample compartment, a detection system, and an electronic and computer unit (Kauppinen and Hollberg; Griffiths and de Haseth, 1986, 2007; Christy, Ozaki and Gregoriou, 2001). In dispersive IR spectrometers, prisms and gratings are used as wavelength selectors. They measure a single wavelength at a time, that is, they scan over the wavelength region so that the wavelength through the selector is changed by rotating the prism or grating.

In the following part, we focus on grating spectrometer because prism spectrometer is rarely used in recent years.

25.2.1.1 Diffraction Grating
A grating acts like a multislit source when collimated radiation strikes it. Different wavelengths are diffracted and constructively interfaced at different angles. There are two types of diffraction gratings: transmission and reflection gratings. In grating IR monochromators, plane reflection gratings are used because transmission gratings have very poor transmission in the IR region.

Figure 25.2 shows an echelette grating, a typical reflection grating used in a dispersive IR monochromator. It has alternating reflecting and nonreflecting areas called *facets* or *lines*. The constructive interference condition for a reflection diffraction grating with the grating constant d is given by

$$k\lambda = d(\sin i + \sin \theta),$$
$$k = 0, \pm 1, \pm 2, \ldots \quad (25.2)$$

where i is the angle between the incident radiation and the normal N of the grating (incident angle), θ is the angle between the diffracted radiation and the normal N (diffracted angle), and k is the order of diffraction.

Angular dispersion, D, which represents a variation in the diffracted angle $d\theta$ against a variation in wavelength ($d\lambda$)

is given by

$$D = \frac{d\beta}{d\lambda} = \frac{k}{d\cos\beta} = \frac{\sin i + \sin \beta}{\lambda \cos \beta} \quad (25.3)$$

The resolving power, R, is yielded by

$$R = \frac{\nu}{\Delta\nu} = \frac{\lambda}{\Delta\lambda} = \frac{k\pi}{\pi/N} = kN = hD \quad (25.4)$$

where, h is an order of diffraction and N is a total number of facets of a dispersive grating. $\Delta\lambda$ and $\Delta\nu$ are defined by a line width that becomes zero first from the center of a spectral line (Rayleigh's criteria).

In an echelette grating when an incident angle and a diffracted angle satisfy the following relation

$$2\alpha = i + \theta \quad (25.5)$$

where α is an angle between the facet plane and the line plane, called *blaze angle*, the intensity of diffracted radiation becomes strong. For the radiation that illuminates and diffracts with the normal of facet ($i = \theta = \alpha$), one can obtain the following relation:

$$k\lambda = 2d\sin\alpha \quad (25.6)$$

when $k = 1$ and λ_b is called the *blaze wavelength*. This is the wavelength with $k = 1$ where the diffracted radiation becomes the strongest for the grating.

25.2.1.2 Monochromators

Figure 25.3 depicts a typical grating monochromator used in an IR spectrometer (Kauppinen and Hollberg). By rotating the grating G or by changing the angle δ, one can scan the wavelength. For a given δ, a certain wavelength is allowed to go through the slit S_2 onto the detector D.

To judge the quality of IR monochromators, optical slit width (ΔV) and mechanical slit width (w) are important. The relation between the two slit widths is given by

$$(\Delta V)^2 = (\Delta V_1)^2 + (\Delta V_0)^2 \quad (25.7)$$

$$\Delta V_1 = \frac{V^2 wd}{kf}\cos i, \quad \Delta V_0 = \frac{V}{Nk}$$

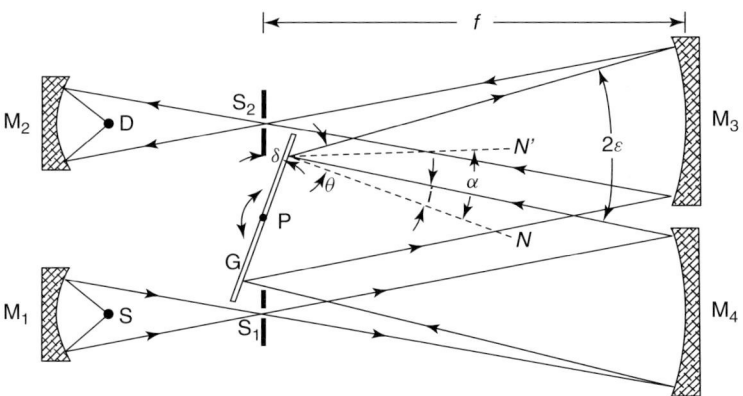

Fig. 25.3 A typical grating monochromator used in an IR spectrometer. (Reproduced from Kauppinen and Hollberg (1997) with permission. Copyright (1997) Wiley-VCH.)

where ΔV_1 and ΔV_0 are determined by a finite slit width and an infinite slit width, respectively, and f is a focal length of a collimator lens.

25.2.2
Principles of FT-IR Spectrometers

The design of many interferometers used for IR spectrometers today is based on that of the two-beam interferometers originally designed by Michelson in 1891 (Griffiths and de Haseth, 1986, 2007). Let us take a look at the principles of FT-IR spectrometers with reference to a Michelson interferometer. It comprises of a half mirror (beamsplitter) and two mutually perpendicular plane mirrors, as shown in Figure 25.4. The beamsplitter is an optical element that separates the incident ray into transmitted ray and reflected ray. Of the two plane mirrors, one is fixed (fixed mirror) and the other is movable parallel to an optical axis (movable mirror). A collimator lens shapes the rays from a light source into parallel rays, which will then impinge upon the beamsplitter. The beamsplitter separates the rays into those heading for the fixed mirror and those heading for the movable mirror (Figure 25.4). As the two rays return to the beamsplitter after being reflected, respectively, by the fixed mirror and the movable mirror, the two rays interfere with each other at the beamsplitter. The intensity of the interference wave is dependent upon how different the distance between the beamsplitter and the fixed mirror is from the distance between the beamsplitter and the movable mirror.

Let us now consider the case when monochromatic radiation is incident upon the beamsplitter. Where the fixed mirror and the movable mirror are equidistant from the beamsplitter, the optical path difference between the two rays is zero, and therefore, the two rays intensify each other. Assuming now that the movable mirror has moved by $\lambda/4$ from its position when the optical path difference is zero, the optical path difference between the two rays becomes $\lambda/2$ and the two rays weaken each other. As the movable mirror further moves by $\lambda/4$, the optical path difference becomes λ, whereby the two rays intensify each other again. The way the two rays

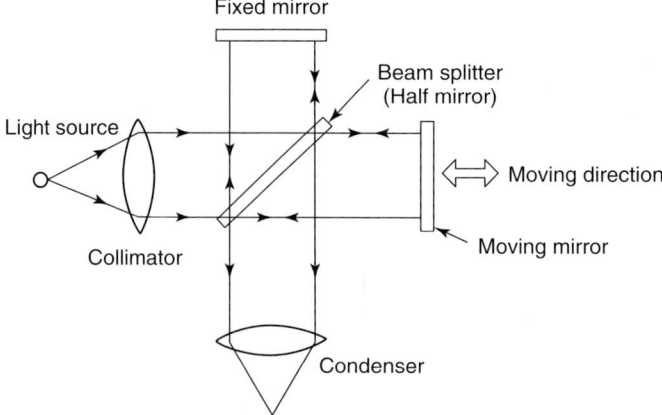

Fig. 25.4 Optical layout of a Michelson interferometer.

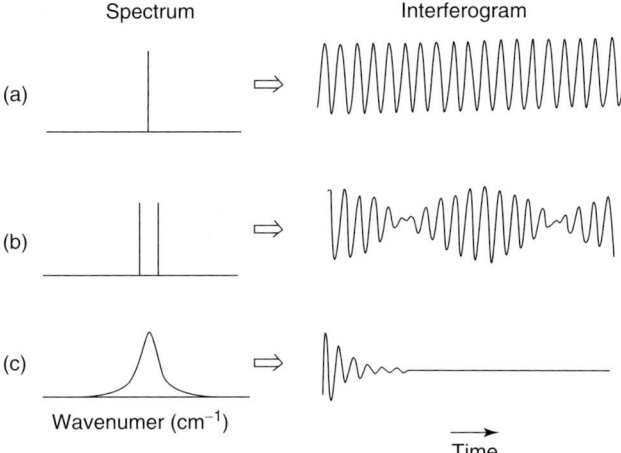

Fig. 25.5 Simple light source spectra and the corresponding interferograms from an interferometer: (a) single infinitesimally narrow line; (b) two infinitesimally narrow lines of equal intensity; (c) continuous light source.

interfere with each other thus changes as the movable mirror moves. In general, two rays intensify each other when their optical path difference is $n\lambda$ (where n is an integer) but weaken each other when their optical path difference is $(n+1/2)\lambda$. The intensity I of radiation that we observe is expressed as a function of the optical path difference x:

$$I(x) = 0.5P(\lambda)\left(1 + \cos\frac{2\pi x}{\lambda}\right) \quad (25.8)$$

where the $P(\lambda)$ indicates the intensity of the light source. Assuming that the speed of the movable mirror is v and the time is t, x can be expressed as

$$x = 2vt \quad (25.9)$$

Substituting formula (25.9) in formula (25.8) and replacing the wavelength λ with the wavenumber $\tilde{v}$, we obtain

$$I(t) = 0.5P(\tilde{v})(1 + \cos 4\pi v\tilde{v}t) \quad (25.10)$$

Formula (25.10) indicates that the intensity I of radiation consists of a direct current component $0.5P(\tilde{v})$ and an alternating current component $0.5P(\tilde{v})\cos 4\pi v\tilde{v}t$. The alternating current component indicates that radiation of the wavenumber $\tilde{v}$ has been amplitude-modulated to a frequency $2v\tilde{v}$. If translated into a drawing, the second term in formula (25.10) is as shown in the right-hand side section in Figure 25.5(a).

The component $0.5P(\tilde{v})$ in formula (25.10) is what we could obtain if our interferometer is ideal. In reality, it is $0.5P(\tilde{v})$ times $H(\tilde{v})$, which is the efficiency of the beamsplitter. Let us therefore decide that $0.5P(\tilde{v})H(\tilde{v}) \equiv B(\tilde{v})$.

Let us now consider what happens when two rays of monochromatic radiation that have the wavenumbers $\tilde{v}_1$ and $\tilde{v}_2$ impinge upon an interferometer at the same time. In this instance, the respective rays are modulated to frequencies $2v\tilde{v}_1$ and $2v\tilde{v}_2$ and a signal indicative of the sum of them is obtained as an output (Figure 25.5b).

Where continuous ray is incident, the intensity of outgoing ray can be expressed as a function of an optical path difference.

$$I(x) = \int_0^\infty B(\tilde{\nu})\,(1 + \cos 2\pi \tilde{\nu} x)\,d\tilde{\nu} \quad (25.11)$$

The second term in this formula, namely, $F(x) = \int_0^\infty B(\tilde{\nu}) \cos 2\pi \tilde{\nu} x\,d\tilde{\nu}$ is an *interferogram*. As an output corresponding to each wavenumber is expressed as a cosine function as shown in Figure 25.5(a), the sum of these outputs $F(x)$ takes a maximum value at the position of $x = 0$. When x shifts from $x = 0$, the outputs corresponding to the respective wavenumbers interfere with each other in various phases, whereby $F(x)$ rapidly becomes weak while winding. Further, an interferogram has a symmetrical profile as $F(x)$ is an even function (i.e., $F(x)$ does not change even when x is replaced with $-x$).

Introducing functions $Be(\tilde{\nu}) = B(\tilde{\nu})/2$ ($\tilde{\nu} \geq 0$) and $Be(-\tilde{\nu}) = Be(\tilde{\nu})$, we obtain

$$F(x) = \int_{-\infty}^\infty Be(\tilde{\nu}) \cos 2\pi \tilde{\nu} x\,d\tilde{\nu} \quad (25.12)$$

Hence,

$$Be(\tilde{\nu}) = \int_{-\infty}^\infty F(x) \cos 2\pi \tilde{\nu} x\,dx \quad (25.13)$$

Formulae (25.12) and (25.13) form a famous FT pair. In short, the interferogram $F(x)$ is obtained through Fourier transformation of $Be(\tilde{\nu})$ and the spectrum $Be(\tilde{\nu})$ is obtained through Fourier transformation of the interferogram $F(x)$.

While FT-IR thus requires measuring an interferogram first and Fourier transformation of the interferogram on a computer thereafter for rendering of a spectrum, an actual spectroscopic system is inclusive of a few more complex processes. First, although integration in formula (25.13) is from $-\infty$ to $+\infty$ of an optical path difference x, this is not realistic. An *apodizing function* is therefore introduced for integration within a finite range after measurement of $I(x)$ while moving the movable mirror within a finite range.

Figure 25.6 shows typical examples of apodizing functions. We introduce an apodizing function $A(x)$ whose profile is shaped as a rectangular wave,

$$A(x) = \begin{cases} 1 & (|x| \leq L) \\ 0 & (|x| > L) \end{cases} \quad (25.14)$$

and we obtain the following definite integration to replace formula (25.13):

$$Be'(\tilde{\nu}) = \int_{-\infty}^\infty A(x) F(x) \cos 2\pi \tilde{\nu} x\,dx \quad (25.15)$$

Since $A(x)$ has a finite value only within the range of $|x| \leq L$, $A(x)$ limits the range of integration. $Be'(\tilde{\nu}) = \widetilde{A(x)F(x)}$ (where the symbol $\sim$ is used to denote Fourier transformation) is a convolution of $\widetilde{A(x)}$ and $\widetilde{F(x)}$ ($= Be'(\tilde{\nu})$) according to the theorem of convolution.

$$Be'(\tilde{\nu}) = \widetilde{A(x)} * \widetilde{F(x)} = \widetilde{A(x)} * Be(\tilde{\nu}) \quad (25.16)$$

Formula (25.16) indicates that the resultant spectrum is multiplication of spectrum $Be'(\tilde{\nu})$ with an instrument function $\widetilde{A(x)}$. It is therefore very important to know the profile of the instrument function $\widetilde{A(x)}$. Calculating $\widetilde{A(x)}$, we obtain

$$\widetilde{A(x)} = \frac{\sin(2\pi L \tilde{\nu})}{\pi \tilde{\nu}} \quad (25.17)$$

This function is similar to the function shown in Figure 25.6(a). The half width

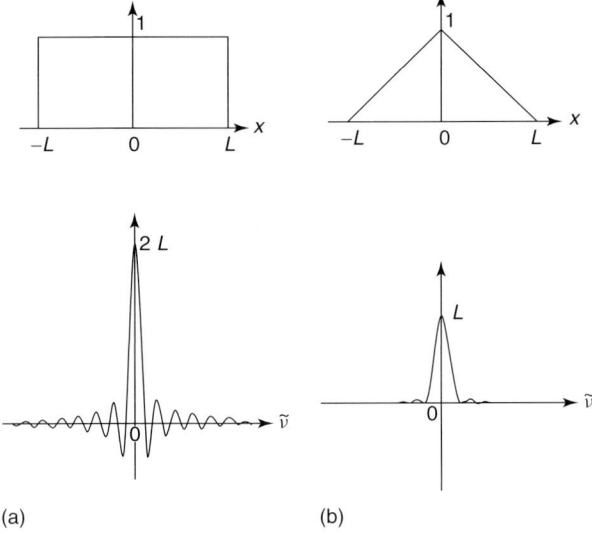

Fig. 25.6 Apodization functions and their corresponding instrument lineshape functions: (a) boxcar truncation and (b) triangular apodization.

of this function has a value $0.6/L$. As a half width determines the spectral resolution, the longer the distance L through which the movable mirror moves, the better the spectral resolution becomes. An instrument function, as that shown in Figure 25.6(a), has a major disadvantage that it includes associated ripples at which it yields positive and negative values. Using the following function $A'(x)$ (Figure 25.6b), we have

$$A'(x) = \begin{cases} 1 - |x/L| & (|x| \leq L) \\ 0 & (|x| > L) \end{cases}$$
(25.18)

as an apodizing function instead of the function shown in Figure 25.6(a); the instrument function becomes as shown in Figure 25.6(b). In the case of this instrument function, although the half width is $0.88/L$ and the spectral resolution is slightly inferior to what it is in Figure 25.6(a), problems of an associated ripple, a rebound toward the negative side, and the like are suppressed better. Several apodizing functions are implemented in FT-IR spectrometers that are available in the market. One needs to choose a suitable apodizing function in accordance with their purpose.

The second issue on an actual spectrometer is that an interferogram is not an even function but is asymmetric in reality for various reasons. A process called *phase correction* is necessary to solve this.

The flow of an FT-IR signal is described here. Figure 25.7 shows the structure of a spectroscopic unit of an FT-IR spectrometer and the behaviors of the signal within components that form the spectrometer. Radiation from a light source (denoted in [1] in Figure 25.7) impinges upon an interferometer via an incident hole. At the interferometer, the radiation is modulated into frequencies that are

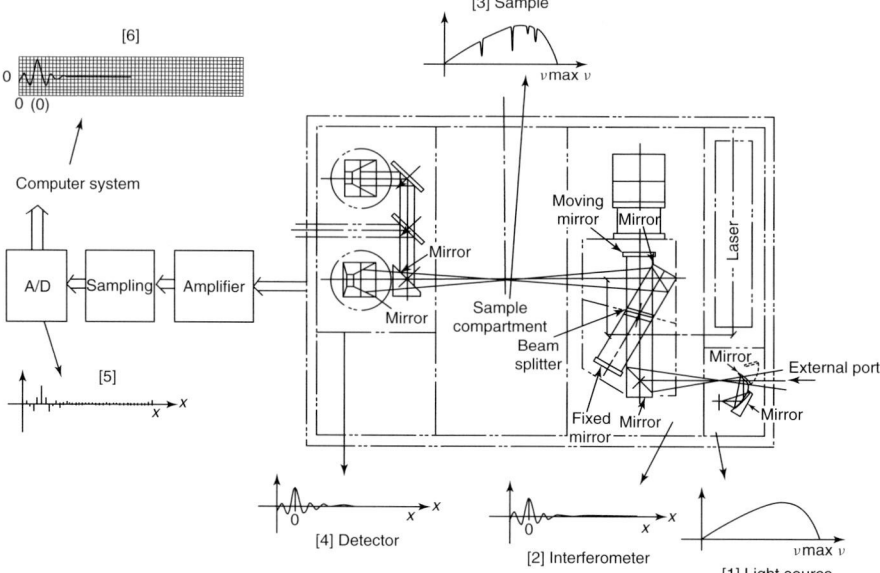

Fig. 25.7 Optical layout of a typical FT-IR spectrometer and a series of FT-IR signals.

proportional to the respective wavenumbers (denoted in [2] in Figure 25.7).

Leaving the interferometer, the radiation impinges upon a measurement sample that absorbs radiation whose wavenumbers are unique to the sample. Section [3] in Figure 25.7 illustrates absorption of radiation having the characteristic wavenumbers by the sample (Section [3] in Figure 25.7 shows the signal as a spectrum for the simplicity of illustration although an actual signal has the shape of an interferogram). A detector converts the radiation from the sample into an electric signal (the section [4] in Figure 25.7). After being amplified by an amplifier, the electric signal is converted into a digital signal by an A/D converter via a sample hold circuit (which samples the signal) (Section [5] in Figure 25.7). The AD-converted interferogram is fed a set number of times to a computer and accumulated (Section [6] in Figure 25.7). Thus, the accumulated interferogram has the shape of a Fourier-transformed spectrum as shown in Figure 25.8(a).

Fourier spectroscopy is bright because it requires simultaneous capturing of incident radiation and simultaneous measurement of the spectrum in the entire wavenumber region. As the spectrum thus measured is inclusive of the emission spectral characteristic of a light source, the spectroscopic characteristics of an interferometer and a detector and the like, without any processing, it is necessary to measure a background spectrum (Figure 25.8b) and divide the observed spectrum by the background spectrum (Figure 25.8c). The resultant spectrum is a transmission spectrum $T(\tilde{v})$. We can calculate an absorbance spectrum $E(\tilde{v})$ (Figure 25.8d) using the following formula:

$$E(\tilde{v}) = -\log_{10} T(\tilde{v}) \qquad (25.19)$$

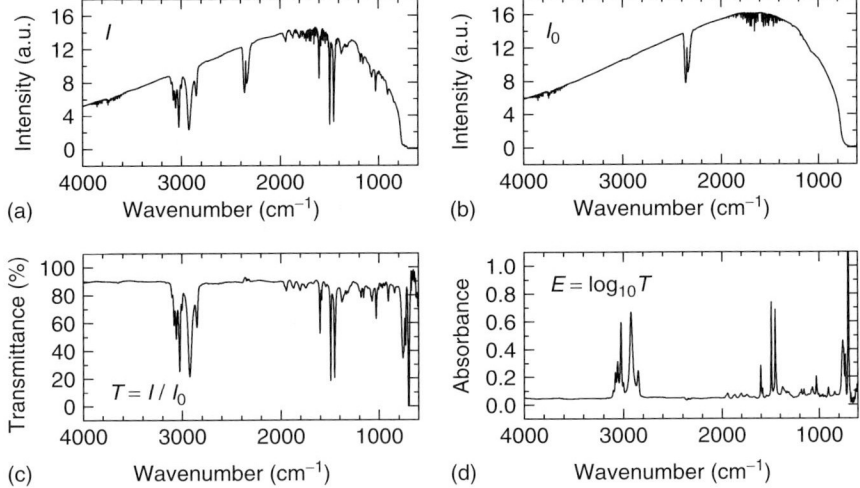

Fig. 25.8 FT-IR spectra: (a) transmission spectrum (before the division by background; (b) background spectrum; (c) transmission spectrum after the division; and (d) absorbance spectrum.

25.2.3
Advantages of FT-IR Spectrometers

It is worth summarizing the three major advantages of FT-IR spectrometers.

25.2.3.1 Fellgett Advantage (Multiplex Advantage)

The biggest advantage of FT-IR spectrometer is the simultaneous detection of the whole spectrum at once. Even though a factor of 2 in signal strength is lost because half the beam is reflected back to the source, the multichannel advantage is nevertheless 10^4 or higher. That is, theoretically, an interferometer can achieve a signal-to-noise (S/N) ratio comparable to that of a dispersive monochrometer 10^4 times faster.

25.2.3.2 Jacquinot Advantage (Throughput Advantages)

Total radiation power on a detector is much higher in a Michelson interferometer than in a grating instrument because at the optimum resolution the area of a circular aperture (the Jacquinot stop) in an FT-IR spectrometer is about 200 times larger than a slit in a grating spectrometer.

25.2.3.3 Connes Advantage

Connes advantage arises from the ability of interferometry using a monochromatic source to accurately and precisely index the retardation, resulting in a superior determination of the retardation sampling position.

25.2.4
Phase Correction

An interferogram measured in the real world, however, has an asymmetric shape because of various reasons although it is supposed to be symmetric with respect to a position of zero optical path difference, and it is therefore impossible to obtain an expected spectrum through mere Fourier cosine transformation of the interferogram (Griffiths and de Haseth, 1986, 2007; Christy, Ozaki and Gregoriou, 2001). The causes of the asymmetric shape include those attributable to a thickness difference between a half mirror and a compensator plate of a beamsplitter, those attributable to the frequency characteristics of an electric circuit system, those attributable to a deviation of the optical path of the main interferometer and that of the associated (laser) interferometer, and so on. Thus, we need what is called *phase correction*. Phase correction is the correction of an asymmetric interferogram into a symmetric interferogram. Equation (25.20) shows the modified relation for the intensity of the interferogram.

$$I(x) = \int_{-\infty}^{\infty} B(\tilde{v}) \left[1 + \cos(2\pi \tilde{v} x + \Phi_{BS}(\tilde{v}))\right] d\tilde{v} \qquad (25.20)$$

where $\Phi_{BS}(\tilde{v})$ is the wavelength-dependent phase shift introduced by the beamsplitter.

Phase correction is the mathematical procedure to remove the sine components from the interferogram. There are two popular phase correction algorithms used in single-sided interferograms: the Mertz algorithm and the *Forman* algorithm. In the former, the largest data point in the interferogram is assigned as the zero retardation point and the amplitude spectrum is calculated with respect to this point. A short double-sided interferogram is measured and its corresponding phase array is used to phase correct the entire single-sided spectrum. The *Forman* correction is essentially equivalent to the Mertz routine but it is performed in the retardation space. Besides phase effects, one must consider the effects of beam divergence, of mirror misalignment, and poor mirror drive for the performance of FT spectrometers (Griffiths and de Haseth, 2007).

25.3
Components of FT-IR Spectrometers

The basic components of dispersive spectrometers and FT spectrometers are the same with the exception of the interferometer (Kauppinen and Hollberg; Chalmers and Griffiths, 2002; Griffiths and de Haseth, 2007; Christy, Ozaki and Gregoriou, 2001). Any differences are within the elements and are the result of the different ways in which the source radiation is detected. For example, sensitive thermocouple detectors are commonplace in dispersive instruments, whereas they are not appropriate for rapid-scanning instruments.

25.3.1
Light Sources

Requirements for an IR light source are that the spectrum distribution of the light source is close to that of black body radiation, the light source offers stable emission for long periods of time, the lifetime of the light source is long, and so on. A globar lamp, Nernst heat generator, a nichrome heat generator, and the like have often been used; however, nowadays, ceramic sources have become more popular.

25.3.2
Interferometers

As shown in Figure 25.9, an interferometer of an FT-IR spectrometer is formed by a main interferometer and an associated

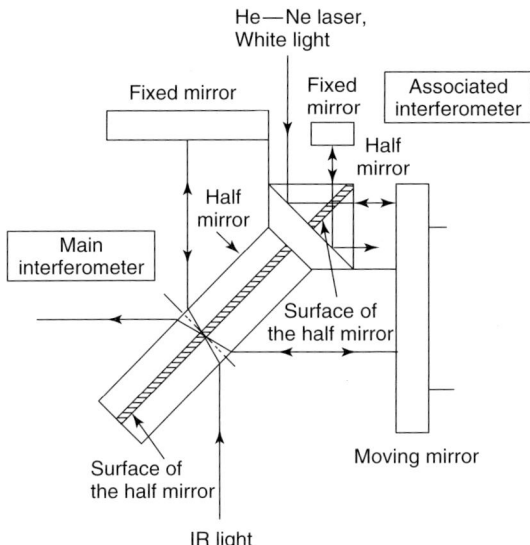

Fig. 25.9 Main and attached interferometers.

25.3 Components of FT-IR Spectrometers

interferometer. The associated interferometer is attached for monitoring of an optical path difference x of the main interferometer. Like the main interferometer, the associated interferometer is formed by a half mirror, a fixed mirror, and a movable mirror (which is shared by the main interferometer). The associated interferometer accepts both white light and He–Ne laser radiation and outputs interferometer output signals corresponding to the two types of radiations. The interferometer output signal resulting from the white light is used to denote when to start sampling the interferogram that the main interferometer yields. Meanwhile, the interferometer output signal resulting from the laser radiation is used to monitor the location of the movable mirror and instruct sampling of the interferogram, which is obtained by the main interferometer.

Spectra measured by FT-IR spectrometers have extremely high (on the order of 7 or more digits) wavenumber accuracy. This is because a Michelson interferometer directly reads the frequency of incident radiation while referring to an optical path difference determined by the emission wavelength of the He–Ne laser whose accuracy is on the order of 7 or more digits.

25.3.3
Sample Compartments

Sample compartments of commercial FT-IR spectrometers are quite different from each other depending upon models. We must pay extra attention to the shape of a ray, to where the ray converges, and to a method for measuring the ray. Cross section of the ray is usually circular and converges in the central part of the sample compartment (when the sample compartment is of the center-focus type) or on a wall surface of the sample compartment (when the sample compartment is of the side-focus type) or remains as a parallel beam without converging. We need to adequately utilize the size and the shape of a ray to enjoy the full capability of an FT-IR spectrometer. When setting a cell or accessory to a sample compartment, therefore, we must make sure that the cell or accessory will not block the ray as much as possible.

FT-IR measurement methods include those of a single-beam type and those of a double-beam type. The former is a method for measuring spectra with a measurement sample and a reference sample set alternatively on the same optical path, and is superior in terms of ordinate accuracy because of the use of the same optical path. When we use a single-beam method, we must be careful about any change in measurement environment with time. This is because any change in measurement environment with time will deteriorate the reproducibility of an ordinate. When choosing a double-beam method, we set a measurement sample and a reference sample on different optical paths at the same time and measure the samples alternately for a designated number of integration times. While the measurement environments are considered to be approximately the same, a difference with respect to the optical capabilities between the two optical paths, if any, will give rise to a problem with ordinate accuracy. The two methods thus each have their own pros and cons. This has led to the development of what is known as a *sample switching double-beam method* that enjoys the advantages of the two methods. Where the sample switching double-beam method is used, measurements are taken principally by a single-beam method while a measurement sample and a reference sample are set simultaneously on separate optical paths. After measurements are taken up to a designated number of integration times, an automatic switch mechanism switches the measurement sample and the reference sample and measurements are taken again. Resultant spectra are extremely reliable as the measurement sample and the reference sample are measured on the same optical path in the same environment. The air-tightness of the environment against external air is always critical to any measurement. IR absorption by carbon dioxide, vapor, and the like contained in the atmosphere is strong, and therefore, no matter how strong the spectral correction using a reference sample may be, the influence of the IR absorption will remain persistently. It is important to close the sample compartment airtight, fill the sample compartment with nitrogen gas, dry air, or the like and take the measurement with vapor and carbon dioxide sufficiently removed.

25.3.4
Detectors

Requirements for an FT-IR detector include, among others, that (i) the sensitivity of the detector be high; (ii) the detector remain sensitive in a wide wavenumber region; (iii) a radiation-receiving surface of the detector be large to a certain extent (so as to leverage on the "an optical system is bright" is a characteristic of FT-IR); (iv) the detector be sensitive to a wide frequency region; and (v) the detector exhibit a linear response to the intensity of incident radiation. IR detectors can be divided into two types: thermal detectors and quantum detectors. The former detectors sense the variation in temperature of a detector material due to absorption of the IR radiation. In quantum detectors, incident IR photons result in the production of free charge carriers in the responsive element. No serious temperature change in the element takes place during this process. TGS (tryglycine sulfate) detectors and MCT ($Hg_{1-x}Cd_xTe$; $x \sim 0.2$) detectors are often used in FT-IR spectrometers. The former detectors, utilizing the pyroelectric effect, are characterized by their relative inexpensiveness and sensitivity to a wide IR region. The latter detectors are semiconductor detectors that utilize the photovoltaic power caused by optical absorption by semiconductors or the photoconductivity effect, and are more sensitive than TGS detectors.

25.3.5
Beamsplitters and Mirrors

A typical beamsplitter IR instrument is of the germanium (Ge)-on-potassium bromide (KBr) substrate type. Recently, semiconductor film beamsplitters, which are formed from self-supporting semiconductors, including carbon films, have been described. Preferably, the beamsplitters are formed from silicon, germanium, or diamond films.

25.3.6
Amplifiers and AD Convertors

An interferogram converted into an electric signal by a detector is fed to a computer via a preamplifier, a main amplifier, an electric filter, a sample-hold circuit, and an analog-to-digital (A/D) convertor.

The amplifiers amplify the interferogram up to the input level of the A/D convertor without distortion. The electric filter removes unwanted signals belonging to a high-frequency region and noises that arise from sampling. These electric systems are designed so that their noises are smaller than those caused by the detector, permitting the detector to exhibit its capability adequately.

The A/D convertor is one of the factors that restrict the S/N ratio of FT-IR. Where a high-sensitivity detector such as an MCT is used for measurement, the S/N ratio of an incoming interferogram to the A/D convertor may be significantly high and exceed the dynamic range (the solution power to convert a signal) of the A/D convertor in some instances.

25.4
Various Types of FT-IR Interferometers

25.4.1
Continuous-Scanning Interferometers

Most commercially available FT-IR spectrometers use the continuous-scan mode of operation, where the movable mirror scans at constant velocity. This type of scanning works very well for routine measurements. Also, the fact that continuous-scanning interferometers can measure an interferogram in a time of 1 second or less makes them excellent tools to follow time-dependent processes. In the continuous-scan mode of interferometry, the laser fringe counter is used to sense the accuracy of the scanning velocity. If a deviation is sensed, correction signals are generated that assure proper operation (constant velocity). The consequence of this mode of operation is that each IR wavelength (λ) is modulated at its own particular Fourier frequency, given by Eq. (25.21):

$$f(\lambda) = \frac{2v}{\lambda} \quad (25.21)$$

where v is the mirror velocity.

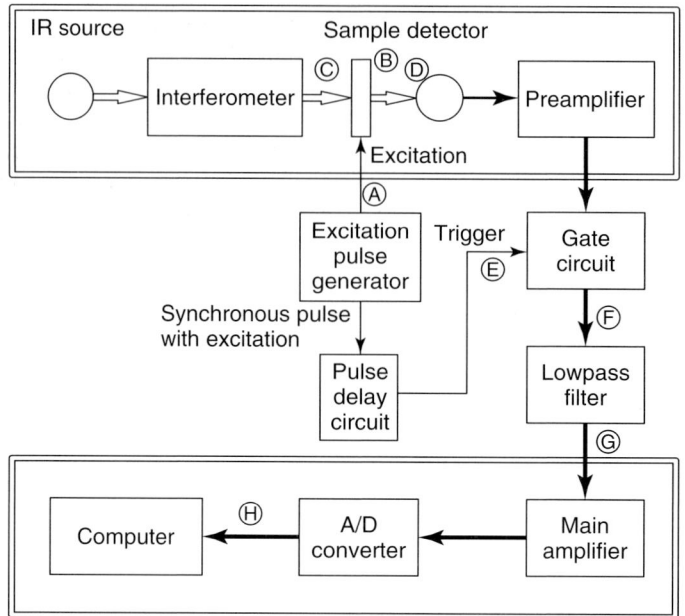

Fig. 25.10 A schematic diagram for the initial type of the asynchronous time-resolved measurement. The signal-processing assembly for time-resolved measurements is shown in the middle, while the units constituting a conventional FT-IR spectrophotometer are depicted in doubly lined boxes at the top and bottom. (Reproduced from Masutani et al. with permission. Copyright (1992) Society for Applied Spectroscopy.)

25.4.2
Step-Scan Interferometers

Step-scan interferometers are powerful for time-resolved FT-IR measurements (Griffiths and de Haseth, 2007; Christy, Ozaki and Gregoriou, 2001). By using a step-scan FT-IR spectrometer, one can monitor much faster reactions if a reaction or sample perturbation is repeatable. In the step-scan interferometer, the optical path difference is held constant over the course of the reaction or perturbation. The optical path difference is then changed, typically by one wavelength of the He–Ne laser, and the process is repeated. This procedure is reiterated at constant intervals of retardation until the optical path difference is sufficiently high that the desired spectroscopic resolution has been reached.

In step-scan FT-IR data are collected while the retardation is held constant or is oscillated about a fixed value. Therefore, to apply the technique to mid-IR and shorter wavelength measurements, a method for controlling the retardation and implementing a special sampling rate comparable to that achieved in modern continuous-scan instruments is required. In recent years, different control methods have been reported. However, all of these basically rely on the use of the He–Ne laser fringe pattern to generate the control signal and to determine the stop size. The biggest advantage of the

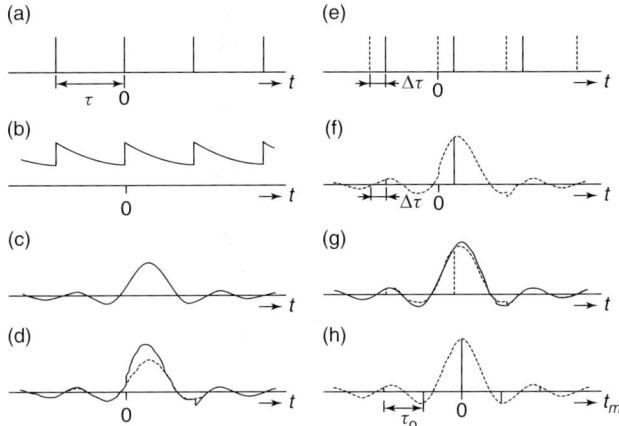

Fig. 25.11 Shape of the signals in processes (a)–(h):
(a) Excitation pulses with period τ; (b) responses of the sample to the excitations in (a); (c) ordinary interferogram without excitations; (d) interferogram modulated by the excitations in (a) (full curve) and the unmodulated interferogram (broken curve, same as in (c)); (e) trigger pulses (full line) delayed from the preceding excitation pulses (broken line) by a fixed time interval Dt; (f) discrete interferogram time resolved at time delay Dt from the excitation (full line) and the modulated analog interferogram (broken curve, the same one as in (d)); (g) time-resolved analog interferogram obtained from the low-pass filter (full curve), the modulated analog interferogram (broken curve, the same one as in (f)), and the discrete interferogram (broken line, the same one as in (f)); (h) discrete interferogram sampled by the A/D converter at its own sampling period t_0 (full time) and the corresponding analog interferogram (broken curve, the same one as in (g)). As for the definitions of t and t_m, see the figure caption for Figure 25.13 of Masutani et al. (Reproduced from Masutani et al. with permission. Copyright (1992) Society for Applied Spectroscopy.)

step-scan mode is the separation of the time of the experiment from the time of the data collection.

25.4.3
Asynchronous Time-Resolved FT-IR Spectrometer

In 1992, Masutani *et al.* (1992) proposed a time-resolving method for a conventional FT-IR spectrophotometer, which does not require synchronization between the timing for time resolving and that for the sampling of the AD converter. The system was named as an *asynchronous time-resolving system*. This method has the following advantages: (i) the signal-proceeding assembly for time-resolving measurements can be attached to any kinds of commercial FT-IR spectrometers and (ii) in principle, there is no shortest limit in time for transient phenomena to be measured.

Figure 25.10 shows a typical set up for the asynchronous time-resolved

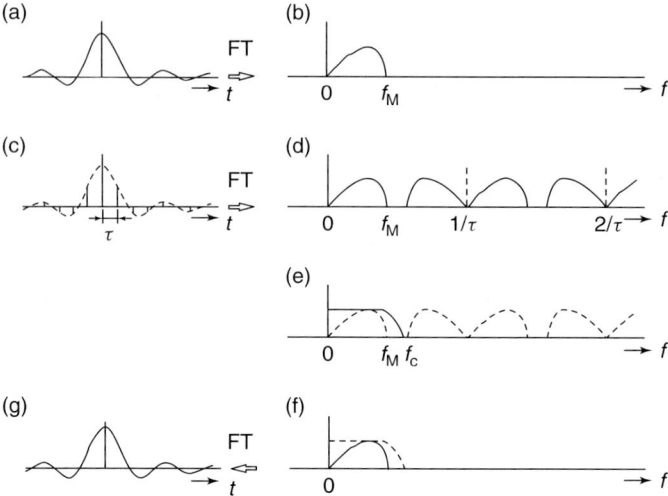

Fig. 25.12 Role of the low-pass filter. (a) An analog interferogram; (b) the spectrum corresponding to the analog interferogram in A with a maximum modulation frequency f_M; (c) a discrete interferogram (full line) with a sampling time interval t and its envelope (broken curve) corresponding to the analog interferogram in (a); (d) the spectrum obtained by inversely Fourier-transforming the discrete interferogram in (c); (e) frequency response of the low-pass filter with a cutoff frequency f_c (full curve) and the same spectrum as in (d) (broken curve); (f) output spectrum from the low-pass filter (full curve) and the same frequency response as in (e) (broken curve); (g) output analog interferogram from the low-pass filter (to be obtained by Fourier-transforming the spectrum in (f)). Horizontal arrows with FT mean the operation of Fourier or inverse Fourier transformation. Variable f in the abscissa is the modulation frequency, equal to $2us$, where u is the velocity of the moving mirror and s is the wavenumber of light. (Reproduced from Masutani et al. with permission. Copyright (1992) Society for Applied Spectroscopy.)

measurement (Masutani, 2002). Figure 25.11 depicts shapes of the signals in processes A–H in Figure 25.10 in alphabetical order: The main feature of this method is the use of a low-pass filter placed between the gate circuit and the A/D converter. The role of the low-pass filter is explained in more detail in Figure 25.3.

Figure 25.12 (a) and (b) shows an analog interferogram for radiation from the light source and a corresponding spectrum obtained by inversely Fourier-transforming the interferogram. When a discrete interferogram (c) obtained by sampling this analog interferogram by the gate circuit is inversely Fourier transformed, the spectrum obtained (d) consists of the original component and others that appear at the position determined by multiplying the sampling frequency $1/t$ by integers. Therefore, when this discrete interferogram is passed through a low-pass filter with an appropriate cutoff frequency f_c (e), the high-frequency components

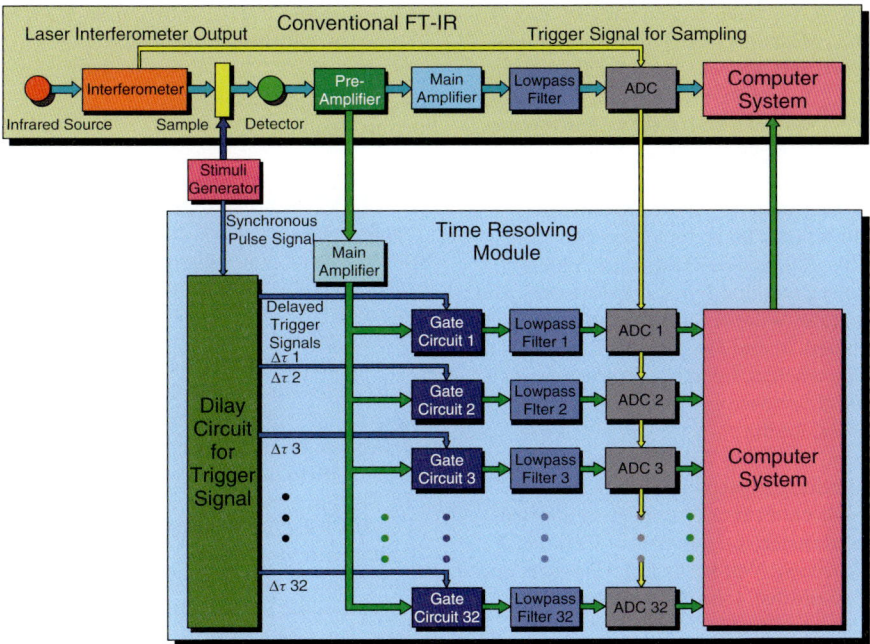

Fig. 25.13 A typical setup of the multichannel asynchronous time-resolving system.

are eliminated (f). The output spectrum (f) is identical to the spectrum (b) for the analog interferogram before sampling. Consequently, the analog interferogram (g) is obtained from the low-pass filter. Although the above Fourier transformations are not actually done by the low-pass filter, it has the function, in effect, as described above.

Figure 25.13 shows a typical setup of the multichannel asynchronous time-resolving system. The time-resolving module of the system has 32 time-resolving units, each of which consists of a gate circuit, a low-pass filter, and an A/D converter. In this system, time resolving is performed by setting each gate circuit at a different time delay. The sample is excited by repetitive stimuli at constant intervals, which do not depend upon the sampling timing of the FT-IR spectrometer. The sample excitations result in transient transmission differences, and thus the analog interferogram is modulated at very high frequency because of the repetitive excitation of the sample. The low-pass filters eliminate high-frequency components from the discrete time-resolved signals generated by the gate circuit and convert them to analog signals. The analog signals thus obtained are processed by the A/D converter at the sampling rate of the FT-IR spectrophotometer itself, as shown by the trigger signal for the sampling in Figure 25.13, which has no relation to the gate timing for time resolution.

References

Chalmers, J.M. and Griffiths, P.R. (2002) *Handbook of Vibrational Spectroscopy*, Vol. 1, John Wiley & Sons, Ltd, Chichester.

Christy, A.A., Ozaki, Y., and Gregoriou, V.G. (2001) *Modern Fourier Transform Infrared Spectroscopy*, Elsevier, Amsterdam.

Ferraro, J.R. and Basile, L.J. (Eds) (1978, 1979, 1982, 1985) *Fourier Transform Spectroscopy*, Vols.1–4, Academic Press, New York.

Griffiths, P.R. and de Haseth, J.A. (1986) *Fourier Transform Infrared Spectroscopy*, John Wiley & Sons, Inc., New York.

Griffiths, P.R. and de Haseth, J.A. (2007) *Fourier Transform Infrared Spectroscopy*, 2nd edn, John Wiley & Sons, Inc., New York.

Kauppinen, Y. and Hollberg, M.R. (1997) Spectrometers, infrared, *Encyclopedia of Applied Spectroscopy*, Wiley-VCH Verlag GmbH, Berlin.

Masutani, K. (2000) Time-resolved mid-infrared spectroscopy using an asynchronous fourier transform infrared spectrometer, in *Handbook of Vibrational Spectroscopy*, Vol. 1 (eds J.M. Chalmers and P.R. Griffiths), John Wiley & Sons, Ltd, Chichester, pp. 655–665.

Masutani, K., Sugisawa, H., Yokota, A., Furukawa, T. and Tasumi, M. (1992) Asynchronous time-resolved fourier transform infrared spectroscopy. *Appl. Spectrosc.*, **46**, 560–567.

Sheppard, N. (2002) The historical development of experimental techniques in vibrational spectroscopy, in *Handbook of Vibrational Spectroscopy*, Vol. 1 (eds J.M. Chalmers and P.R. Griffiths), John Wiley & Sons, Ltd, Chichester, pp. 1–32.

Further Reading

Bell, R.J. (1972) *Introductory Fourier Transform Spectroscopy*, Academic Press, New York.

Bourchareine, P. and Connes, P. (1963) Interferometer with compensated field for fourier transform spectroscopy. *J. Phys. (Paris)*, **24**, 134–138.

Bracewell, R. (1965) *The Fourier Transform and its Applications*, McGraw-Hill, New York.

Brigham, E.O. (1974) *The Fast Fourier Transform*, Prentice-Hall, Inc, New Jersey.

Buijs, H. (1979) in *Multiplex and/or High Throughput Spectroscopy*, SPIE Proceedings, Vol. 191 (ed. G.A. Vanasse), SPIE, Bellingham, p. 116.

Chamberlain, J. (1979) *The Principles of Interferometric Spectroscopy*, John Wiley & Sons, Ltd, Chichester.

Ferraro, J.R. and Krishmam, K. (Eds) (1990) *Practical Fourier Transform Infrared Spectroscopy, Industrial and Laboratory Chemical Analysis*, Academic Press, New York.

Mackenzie, M.W. (Ed.) (1988) *Advances in Applied Fourier Transform Infrared Spectroscopy*, John Wiley & Sons, Inc., New York.

26
Infrared Molecular Vibrational Spectroscopy

Andrew B. Horn

26.1	**Principles of Vibrational Spectroscopy**	**889**
26.1.1	Introduction 889	
26.1.2	Classical and Quantum-Mechanical Description of Infrared Vibrational Excitation in Diatomic Molecules	889
26.1.3	Group Frequencies and Symmetry in Infrared Vibrational Spectroscopy 893	
26.1.4	Gas-Phase Spectra and Rotational Fine Structure 896	
26.2	**Measurement of Infrared Vibrational Spectra**	**897**
26.2.1	Principles of Infrared Spectrometry 897	
26.2.2	Instrumentation and Spectral Processing 897	
26.2.2.1	Interferometry in Infrared Spectroscopy 897	
26.2.2.2	Processing of Infrared Spectra 901	
26.2.2.3	Infrared Detectors 902	
26.2.3	Basic Transmission Sampling Methods 902	
26.2.3.1	Gases 903	
26.2.3.2	Liquids 903	
26.2.3.3	Solids 904	
26.2.4	Reflection-Based Sampling Methods 905	
26.2.4.1	Specular Reflection 905	
26.2.4.2	Reflection–Absorption 906	
26.2.4.3	Diffuse Reflectance 912	
26.2.4.4	Attenuated Total Internal Reflection 914	
26.2.5	Infrared Microspectroscopy 921	
26.2.6	Extinction Methods from Particles 924	
26.3	**Interpretation of Infrared Spectra**	**926**
26.3.1	Empirical Interpretation from Group Frequencies 926	
	Acknowledgments 931	
	References 931	
	Further Reading 932	

Encyclopedia of Applied Spectroscopy. Edited by David L. Andrews.
Copyright © 2009 WILEY-VCH Verlag GmbH & Co. KGaA, Weinheim
ISBN: 978-3-527-40773-6

26.1
Principles of Vibrational Spectroscopy

26.1.1
Introduction

Infrared spectroscopy is routinely used in analytical laboratories for the identification of materials, molecules, and functional groups through their characteristic absorption patterns in the mid-infrared (4000–400 cm^{-1}) region of the electromagnetic spectrum. In its original conception, analytical infrared spectroscopy was based upon simple dispersive instrumentation measuring the transmitted proportion of light as a function of wavelength through a sample. Over the last 30 years, the development of interferometric methods for the measurement of frequency-domain spectra, sophisticated reflection-based sampling methods applicable to non-transparent materials, and computer-based processing methods has transformed the field into a widely applicable, quantitative method for materials analysis. In this chapter, the basic principles of infrared spectroscopy are outlined, along with a brief account of the measurement and interpretation of infrared spectra. In addition, some topical examples of the use of modern infrared spectroscopy in physics, chemistry, biology, and atmospheric science are also presented.

26.1.2
Classical and Quantum-Mechanical Description of Infrared Vibrational Excitation in Diatomic Molecules

In any form of absorption spectroscopy, the energy absorbed in a *transition* between energy levels is measured. In the mid-infrared region, these transitions correspond to changes in vibrational quantum number. Before treating the system in a quantum-mechanical manner, however, a simpler classical treatment is instructive in order to give a phenomenological perspective. The most basic classical approach relating vibrational transitions to visualizable molecular parameters treats a vibrating diatomic molecule as two masses m_1 and m_2 (in kilograms) connected by a flexible bond with a force constant k (newton per meter) using Hookes' Law, as shown in Figure 26.1.

When the masses are at equilibrium, $r = r_{eqm}$ and there are no forces acting. However, when the bond is elongated

Encyclopedia of Applied Spectroscopy. Edited by David L. Andrews.
Copyright © 2009 WILEY-VCH Verlag GmbH & Co. KGaA, Weinheim
ISBN: 978-3-527-40773-6

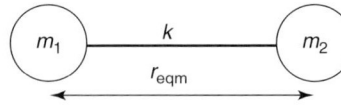

Fig. 26.1 Schematic of the basis of the simple harmonic oscillator model for a heteronuclear diatomic model. The terms are defined in the text.

(or compressed), a restoring force acts to pull (or push) the particles:

$$F(r) = -k(r - r_{eqm}) \tag{26.1}$$

Thus, the masses oscillate around the equilibrium distance. It can be shown that the oscillation frequency is given as

$$\nu = \frac{1}{2\pi}\sqrt{\frac{k}{\mu}} \quad (\text{s}^{-1}) \tag{26.2}$$

where k is the force constant as defined above and μ is the reduced mass (in kilograms):

$$\frac{1}{\mu} = \frac{1}{m_1} + \frac{1}{m_2} \quad \text{or} \quad \mu = \frac{m_1 m_2}{m_1 + m_2} \tag{26.3}$$

Replacing the masses and force constants with representative atomic values, the oscillation frequencies obtained are typically between 1.2×10^{13} and 1.2×10^{14} Hz.

The nature of the interaction between the electric field of the infrared radiation and the atoms of the molecule can best be envisaged by considering a heteronuclear diatomic molecule, where the atoms at either end of the bond have different electronegativities. This polarization results in a dipole moment, the magnitude of which oscillates with the changing bond length during vibrational motion. The electric field E associated with electromagnetic radiation also oscillates in amplitude with time like a dipole with frequency ν (s^{-1}):

$$E(t) = E_0 \cos(2\pi \nu t) \tag{26.4}$$

When the frequency of oscillation of the molecule is resonant with the frequency of the infrared radiation, absorption of energy by the molecule can occur. In practice, the requirement for a permanent dipole moment exists only for diatomic molecules and for more complicated polyatomic molecules; several atoms (connected by a number of bonds) may move at once, resulting in a changing (*dynamic*) dipole moment during the course of the vibration even if a static dipole moment does not exist. Figure 26.2 illustrates this for the antisymmetric stretching mode of CO_2. This phenomenon is often expressed as an empirical selection rule for the infrared activity of a given vibration, that is, during the course of a particular vibration the molecular dipole moment must change.

From a quantum-mechanical perspective, the problem of quantifying vibrational absorption activity and of calculating energy levels and intensities relies upon the identification of appropriate operators and wavefunctions to describe

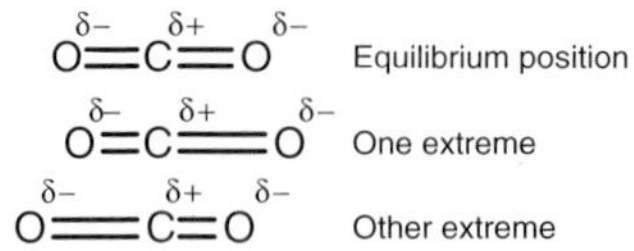

$\overset{\delta-}{O}=\overset{\delta+}{C}=\overset{\delta-}{O}$	Equilibrium position	
$\overset{\delta-}{O}=\overset{\delta+}{C}=\overset{\delta-}{O}$	One extreme	
$\overset{\delta-}{O}=\overset{\delta+}{C}=\overset{\delta-}{O}$	Other extreme	

Fig. 26.2 Schematic of the atomic displacements contributing to the antisymmetric stretching mode of CO_2.

the system. Naturally, each energy level of the quantized vibrational motion will be described by a wavefunction, which must be an eigenfunction equation of the Schrödinger equation:

$$\hat{H}\Psi = E\Psi \quad (26.5)$$

The Hamiltonian has both kinetic and potential energy components describing the vibrational motion. A simple treatment that serves to extend the classical treatment into the quantum world treats the potential energy as deriving from Hookes' law, Eq. (26.1). The potential energy operator takes the following form:

$$\hat{V} = \frac{1}{2}kx^2 \quad (26.6)$$

where $x = r - r_{eqm}$. The kinetic energy operator can be taken directly from the known result:

$$\hat{T} = -\frac{\hbar^2}{2m}\frac{\partial^2}{\partial x^2} \quad (26.7)$$

Combining these equations together and using the fact that $m = \mu$ for diatomics, we obtain

$$\hat{H}\Psi = -\frac{\hbar^2}{2\mu}\frac{\partial^2 \Psi}{\partial x^2} + \frac{1}{2}kx^2\Psi \quad (26.8)$$

This treatment is usually referred to as the *simple harmonic oscillator*, and suitable wavefunctions take the following form:

$$\Psi_v = \sqrt{\left(\frac{1}{2^v v!\sqrt{\pi}}\right)} \times H_v(y)$$

$$\times \exp\left(\frac{-y^2}{2}\right) \quad (26.9)$$

where v is the *vibrational quantum number* and $H_v(y)$ are called *Hermite polynomials*,

v	$H_v(y)$
0	1
1	$2y$
2	$4y^2 - 2$
3	$8y^3 - 14y$

where y is given as

$$y = \left(\frac{4\pi^2 \nu \mu}{h}\right)^{1/2} \times (r - r_{eqm}) \quad (26.10)$$

Transitions between these energy levels then correspond to vibrational absorption. Solving Eq. (26.8) with these eigenfunctions results in the well-known value for the energy of a given level v:

$$E = h\nu\left(v + \frac{1}{2}\right) \quad (26.11)$$

where ν is the classical frequency given in Eq. (26.2). Also note that these wavefunctions are normalized and orthogonal. Several interesting quantum-mechanical phenomena appear from this treatment. First, the probability of finding a particle at a position x is given by $|\Psi_v(x)|^2$, suggesting that in the ground state ($v = 0$), the oscillator spends most of its time at $x = 0$ (i.e., at the center), directly at odds with classical theory. Second, this also points to the existence of zero-point energy since the lowest state of the oscillator has energy $1/2h\nu$.

The quantum-mechanical treatment gives direct access to the magnitude of the spacing between the energy levels and also the selection rules, since the vibrational transition probability can be derived directly from the wavefunctions. The gross selection rule for the classical treatment given earlier still applies: the electric dipole moment must change during the course of a vibration. This is reinterpreted quantum mechanically through the values of the *transition dipole moment* μ_{fi} on going from an initial state Ψ_i to final state Ψ_f, that is, for an allowed transition, $\mu_{fi} \neq 0$. Clearly,

$$\mu_{fi} = \int_{-\infty}^{+\infty} \Psi_f^* \hat{\mu}_i.d\tau = \langle f|\hat{\mu}|i\rangle \quad (26.12)$$

where the electric dipole moment operator $\hat{\mu}$ has the form

$$\hat{\mu} = \sum_j q_j \mathbf{r}_j \quad (26.13)$$

where q_j is the charge on the *j*th particle and $\mathbf{r}_j$ is its position vector with respect to the center of mass of the molecule. Evaluation of these parameters for the wavefunctions of the simple harmonic oscillator leads to the selection rule $\Delta v = \pm 1$. In real molecules, the vibrational potential $V(x)$ is not accurately described by a simple harmonic oscillator: the real potential is anharmonic, resembling a *Morse potential*. The energy levels of the anharmonic oscillator are given by

$$E_v = h\nu \left(v + \frac{1}{2}\right) - x_e \left(v + \frac{1}{2}\right)^2 \quad (26.14)$$

where x_e is called the *anharmonicity constant*. Note that the energy levels become more closely spaced as v increases. This allows transitions across quantum numbers and in reality the selection rule is found to be $\Delta v = \pm 1, \pm 2, \pm 3$, and so on. The first of these transitions is referred to as the *fundamental*, with the higher values being the *harmonics* or *overtones*. Furthermore, the reduction in spacing between the energy levels in the anharmonic oscillator leads to the appearance of the harmonics at a slightly lower frequency than twice that of the fundamental. Nothing in the foregoing treatment forbids excitation from energy levels above the ground state, giving rise to what are know as *hot bands*. The name derives from the fact that the populations of the higher vibrational states $N(v)$ follow the Boltzmann distribution:

$$N(v) = N_0 \exp\left(-\frac{vh\nu}{kT}\right) \quad (26.15)$$

where N_0 is the ground state population and T is the temperature (kelvin). Given the typical energy spacing of vibrational levels, the population of the first vibrationally excited state is small at room temperature for all but the lowest frequency vibrations and significant populations are only achieved in hot samples. Again, the reduction in spacing between the energy levels in the anharmonic oscillator leads to the appearance of the hot band at a slightly lower frequency that the fundamental.

The intensity I of absorption upon transition from one energy level to another is proportional to the square of the transition dipole moment:

$$I \propto (\mu_{fi})^2 \quad (26.16)$$

In infrared transitions, this value depends upon both the amplitude of the vibration (since large excursions give rise to large changes in the relative positions of the atoms) and the relative polarization of the atom(s) at either end of the chemical bond. For example, as will be seen later, vibrations involving oxygen atoms tend to be intense because of the electronegativity of oxygen, and many vibrations involving hydrogen atoms tend to be intense because small, light atoms often experience large displacements. For further details, see Atkins and de Paula (2009) and Hollas (1992).

26.1.3 Group Frequencies and Symmetry in Infrared Vibrational Spectroscopy

In a simple diatomic molecule, vibration involves only the motion of two atoms. Any movement of one atom affects the atom, and since the center of mass must be conserved only one vibration can occur. This is not the case for polyatomic molecules, for which several combinations of atomic displacement are possible. These combinations are not random, since there are constraints in the retention of a fixed center of mass. For a free atom, there are three degrees of freedom, represented by translation in the x, y, and z directions. For two atoms connected together (taking the z axis as the direction between the two atom centers), a total of six degrees of freedom must therefore exist: three translations of the whole molecule (along x, y, and z axes), two rotations (around x and y axes), and one vibration (along the z axis). By extension of this argument, it can easily be seen that for a polyatomic molecule with N component atoms there must be $3N$ degrees of freedom, from which there must be three translations of the whole molecule (along x, y, and z axes) and three rotations (around the x, y, and z axes). This leaves $3N - 6$ degrees of freedom, which must therefore be accounted for by a series of (different) vibrational motions. It should also be noted, however, that if the molecule is linear, there can only be two rotations; hence, there will be $3N - 5$ vibrations in this case.

Each vibration consists of a number of atomic displacements, since it is impossible for only one bond to stretch without affecting all the other atoms and the center of mass of the molecule must remain stationary (otherwise the displacements lead to translational motion or rotation). This is illustrated for H_2O in Figure 26.3. Since there are three atoms in water, there are three vibrational combinations. The simplest motion involves a displacement of each of the H atoms simultaneously and in phase along their respective O—H bonds. To compensate for the motion of the H atoms, it is necessary for the O atom to move in the opposite direction. To retain the center of mass, the vector distance/displacement products must sum to zero; hence, the O atom will move much less than the H atoms for this to occur. This vibration is referred to as the *symmetric stretch*, for reasons that are discussed later. The second identifiable vibrational combination again involves simultaneous displacement of the H atoms along the O—H bond, but this time out of phase. To compensate for this overall motion of the H atoms to the right, the O atom must move proportionally to the left. This vibration is the *antisymmetric stretch*. The remaining vibration involves a different kind of motion: *bending*. In this case, the O atom must move proportionally upward to counteract the motion of the H atoms.

A unique combination of atomic displacements can be found for each vibrational degree of freedom of any

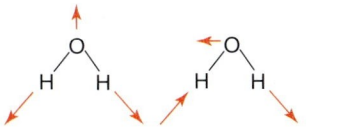
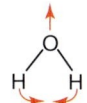

Fig. 26.3 Atomic displacements for the three vibrational modes of H_2O. Note that any allowed combination of displacements must retain the center of mass.

molecule. Each unique combination is referred to as a *normal mode*. As is shown later, some combinations may result in displacements that only differ in the directions in which they occur (e.g., in the x, y plane or the x, z plane) but which occur at the same frequency: such modes are described as *degenerate*. For example, there are three identifiable components of the antisymmetric stretching modes of methane (CH_4), polarized along the x, y, and z axes: all these have the same frequency and are therefore triply degenerate.

In general, for a given normal mode in polyatomic molecules, the individual combination of atomic displacements falls into one of two broad categories: those that are mainly localized on a small subset of atoms and those that involve more complex coupled motion of a significant proportion of the atoms. The former are often readily identifiable with various functional groups of the molecule, the so-called *functional group modes*. A particularly useful example is the stretching of the carbonyl ($>C=O$) bond. In many molecules, the vibration of this group is largely decoupled from the rest of the molecule and the mode, therefore, absorbs at roughly the same frequency in a wide range of molecule types. However, functional groups with similar absorption frequencies can couple, a clear example being the functional group modes related to the various components of a secondary amide (containing the $-CO-NH-$ functionality). In this case, the carbonyl frequency is close to that of the bending vibration of the NH, and hence strong coupling occurs to give the characteristic *amide I* and *amide II* functional group modes. The former is predominantly, but not exclusively, related to the carbonyl and the latter is related to the NH bending mode. Together, the two bands form a uniquely identifiable motif that strongly features in the infrared spectra of biological materials.

At the other end of the scale, there are normal modes that involve the simultaneous displacement of many atoms within a given molecule. Often, it is very difficult to uniquely assign these modes. However, the combined overlapping collection of these normal modes is a unique characteristic for a given molecule and is responsible for its "fingerprint." Spectra of polyatomic molecules are, therefore, comprised of a complex, unassignable fingerprint spectrum overlaid with more readily identifiable functional group modes.

For simpler polyatomic molecules, the collective displacements of the individual atoms can be described by considering the symmetry properties of the molecule. This leads to a simpler way of determining the vibrational activity of a given normal mode. The first step in this process is the determination of the symmetry properties of the molecule, following the principles of group theory, described in detail elsewhere. This basically involves identifying the collection of symmetry operations (rotations, reflections, improper rotations, centers of inversion, etc.), which can be performed upon the molecule, from which the point group can be determined. For a detailed description of the method, see Vincent (2001).

Once the point group of the molecule in question has been determined, the symmetries of the combinations of atomic displacements for each normal mode can be derived. Three Cartesian displacement vectors are placed on each atom (one for each x, y, and z

component). The effect of the symmetry operations of the point group on these collections of displacement vectors is then determined. The resulting combination is a *reducible representation*, which can then be resolved into a combination of *irreducible representations* by a simple mathematical procedure. Each irreducible representation describes a unique set of transformations of the molecule under each of the operations of the point group, each operation identifying, for example, whether or not a displacement along a given bond within the molecule remains in phase or out of phase when transferred to another (identical) bond in the same molecule. Consequently, the symmetries of each of the normal modes are determined.

In the example below, the symmetries of the normal modes of iodomethane (CH_3I) are determined. The molecule has C_{3v} point group symmetry, the operations of which are E (the identity), two C_3 (threefold) rotation axes, and three σ_v (vertical, containing the principal axis) mirror planes. Defining the principal axis as z, collinear with the C_3 axes, Cartesian displacement vectors are placed on each atom. The effect of each operation of the point group on the full set of Cartesian vectors is then determined, that is, the transformation of each of the x, y, and z vectors is calculated for each atom under the operations of the point group. The sum of these values under each class of operation in the point group then gives the reducible representation, as shown below:

C_{3v}	E	$2C_3$	$3\sigma_v$
χ (15 x, y, z vectors)	15	0	3

In this example, this reducible representation reduces to irreducible representations of $4A_1 + 1A_2 + 5E$ using conventional group theory methods. However, the inclusion of all three degrees of freedom for each atom means that this set of symmetry values accounts for the translations and rotations of the molecule as well as the vibrational normal modes. These nonvibrational degrees of freedom are eliminated by reference to the character table, which for C_{3v} reveals that the three translational degrees of freedom have $A_1 + E$ symmetry, while the three rotational degrees of freedom have $A_2 + E$ symmetry:

Removing these reveals the symmetries of the vibrational degrees of freedom (the normal modes) to be $3A_1 + 3E$. A final check of the numbers shows that this accounts for a total of nine normal modes (E symmetries are doubly degenerate, accounting for two modes simultaneously).

Once the symmetries of the normal modes have been established, the activity of the normal modes in the infrared spectrum can be determined. As described previously, not all modes will give rise to

C_{3v}	E	$2C_3$	$3\sigma_v$	Rotations and translations	Cartesian axes and their products
A_1	1	1	1	T_z	$z; x^2 + y^2; z^2$
A_2	1	1	−1	R_z	
E	2	−1	0	$(T_x, T_y); (R_x, R_y)$	$(x,y); (xz, yz); (x^2 − y^2)/\sqrt{2}$

an appropriate dynamic dipole change. For energy levels characterized by wavefunctions Ψ_i and Ψ_f, a nonzero transition dipole moment will only arise if the expectation value of the dipole moment operator for the transition is nonzero, that is,

$$\mu_{fi} = \int_{-\infty}^{+\infty} \Psi_f^* \hat{\mu} \Psi_i . d\tau \neq 0 \quad (26.17)$$

For this to be the case, the symmetry product of the wavefunctions and the operator should contain the totally symmetric irreducible representation in the point group of the molecule, that is,

$$\Gamma(\Psi_f) \times \Gamma(\mu) \times \Gamma(\Psi_i)$$
$$\supset A, A_1, A', A_g, \cdots \quad (26.18)$$

There are a range of procedures for this, which rely upon a knowledge of the symmetries of the various components. For $v = 0-1$ transitions (fundamental), the ground state wavefunction Ψ_i is totally symmetric, for example, A_1 in C_{3v}. The final state wavefunction Ψ_f has the symmetry of one of the normal modes, that is, taken from the set of symmetries determined in the example above. The dipole moment operator has the same symmetry as the x, y, and z axes in the point group of the molecule, for example, $A_1 + E$ in C_{3v}, as shown in the character table.

The problem generally simplifies further, since multiplying any symmetry species by the totally symmetric irreducible representation has no effect, for example, in C_{3v}, $A_1 \times E = E$. Consequently, it is only necessary to consider the symmetry of the normal mode and of the dipole operator, that is, to find out if $\Gamma(\Psi_f) \times \Gamma(\mu) \supset A, A_1, A', A_g$, and so on. The process of identifying whether a vibrational transition is infrared active therefore involves determining the molecular point group and identifying the symmetries of the normal modes and of the x, y, and z axes. Where the vibrational normal mode symmetry is the same as one of the axes, the vibration is infrared active. In the iodomethane example, all of the vibrational normal modes are infrared active.[1] Further information on these procedures can be found in Atkins and de Paula (2009) and Vincent (2001).

26.1.4
Gas-Phase Spectra and Rotational Fine Structure

The foregoing discussion has dealt entirely with transitions between vibrational energy levels. However, transitions between rotational levels can also occur simultaneously. As described elsewhere, rotational energy levels typically have spacing of the order of a few cm^{-1}; hence, in cases where natural linewidths are narrow (such as in low-pressure gases of simple molecules), rotational fine structure can be observed. In infrared absorption spectra, this structure is observed as a series of fine lines corresponding to a change of $+1$ in the vibrational quantum number and ± 1 in the rotational quantum number. The intensity and extent of this rotational fine structure depend upon the population

---

1) The procedure also applies to the determination of normal mode activity in vibrational Raman spectroscopy, except that the comparison is made between the symmetries of the normal modes and those of the polarizability operator α, which corresponds to the symmetries of the products of any two Cartesian axes. These are also normally shown in character tables. For molecules possessing a center of inversion, infrared and Raman activity are mutually exclusive.

of the rotational energy levels in the vibrational ground state (the spacing of the rotational levels is small; hence, the Boltzmann distribution of population in the rotational energy levels typically extends to $J = 15$ or more). A fuller description is given in Hollas (1992).

26.2 Measurement of Infrared Vibrational Spectra

26.2.1 Principles of Infrared Spectrometry

The measurement of absorption spectra requires a light source emitting the appropriate range of frequencies, a dispersing element to separate the different frequencies of light, and a detector that responds quickly over the appropriate range of frequencies. At any particular frequency, the incident light intensity I_0 and the transmitted light intensity I are related to the amount of absorption by the sample via the Beer–Lambert law:

$$-\log_{10}\left(\frac{I}{I_0}\right) = \varepsilon c \ell \qquad (26.19)$$

where ε is the molar absorption coefficient (meter square per mole), c is the concentration (mole per cubic meter), and ℓ is the path length (meter). The dimensionless logarithm term on the right-hand side (RHS) of Eq. (26.8) is the *absorbance*, A, a dimensionless quantity. Infrared spectra are usually presented as a graph of absorbance versus wavenumber.

As an alternative to absorbance, spectra can be presented as percentage *transmittance*, %T.

$$\%T = \frac{I_0 - I}{I_0} \times 100 \qquad (26.20)$$

Clearly, a transmittance of 100% represents no absorption, while a transmission of 10% represents an absorbance of 1.0. This measurement arose historically because transmittance is easier to measure experimentally in dispersive (diffraction grating) instruments. However, absorbance is linear in concentration and path length and is generally preferred for quantitative work.

26.2.2 Instrumentation and Spectral Processing

The basic requirement for benchtop infrared spectroscopy comprises a polychromatic infrared light source, a dispersion method to separate out the individual frequency components, and a suitable detector. In most instruments, infrared light is generated with a blackbody source optimized for output in the mid-infrared. These are generally electrically heated, low-current incandescent devices. Output characteristics of various sources in common use for mid- and near-infrared spectroscopy are reviewed in detail by Buijs (2002).

26.2.2.1 Interferometry in Infrared Spectroscopy

Originally, infrared spectrometers utilized rock salt prisms to disperse the frequency components, eventually moving to diffraction gratings for more flexible operation and improved longevity of the components (Workman, 1998). As for all grating-based monochromators, there are significant trade-offs between the frequency bandpass (a function of the slit widths used at the input and output of the monochromators) and the optical throughput. As a consequence, the measurement of high-resolution infrared spectra (of gases, for instance) was

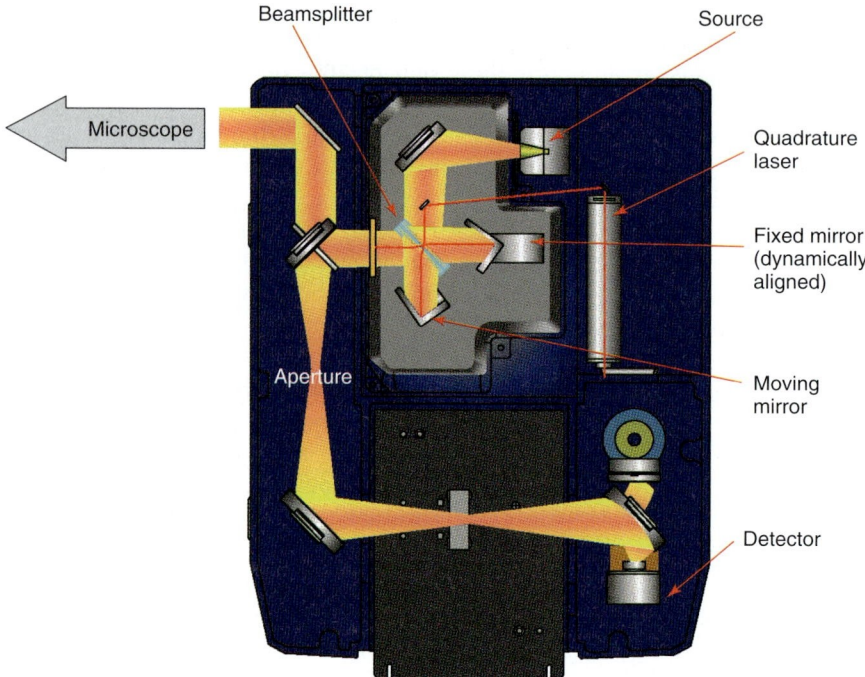

Fig. 26.4 Schematic representation of a Michelson interferometer in a commercial FTIR spectrometer. Reproduced by kind permission of JASCO UK.

a considerable experimental challenge. However, with the advent of improved computational facilities in the 1970s, methods based upon interferometry became the norm. The basic principles of interferometry as applied to infrared spectroscopy involve the separation of the source radiation into two beams using an appropriate beamsplitter. These beams then pass along separate light paths before being recombined (usually at the same beamsplitter). Any path difference experienced by the two beams leads to interference upon recombination, with different frequency components combining with different phases. The superposition of these interferences is recorded at discrete (normally equally spaced) intervals of path difference δ to give a path-difference-dependent signal referred to as an *interferogram*. The most common apparatus used to perform this separation and recombination is the Michelson interferometer, shown schematically in Figure 26.4, although other possibilities are also in widespread use. In many modern spectrometers, the fixed mirror is dynamically aligned with the moving mirror for improved stability and precision. This is normally affected using piezo-electric transducers. The output interferogram, $I(\delta)$, is measured using a detector operating over the appropriate frequency range. After digitization, the interferogram is inverted using a discrete Fourier transform to produce a frequency-domain spectrum, $I(\nu)$, and hence the common acronym *FTIR* (*Fourier-transform infrared*) spectrometry. The spectral resolution (i.e., the spacing between the data points) in the computed spectrum is inversely related to the maximum path difference δ_{max} in the interferometer. Typical general purpose benchtop FTIR spectrometers with a maximum resolution of 0.5 cm^{-1} therefore require path differences of 2 cm, which is easily attainable with relatively unsophisticated apparatus. The path difference between the two optical paths is usually measured to an appropriate precision by interference from two orthogonally polarized HeNe laser beams by quadrature. Increased resolution requires very high mechanical stability over path lengths that can be several meters. Further experimental details can be found in sources such as Griffiths and de Haseth (1986) and Jackson (2002).

The acquisition of intensity data in the temporal domain and its transformation into the spectral domain brings a number of interesting features and artifacts that need to be understood and in many cases incorporated into the post-acquisition processing of the resultant data. A typical interferogram is shown in Figure 25.5(a). The mechanical, discrete nature of the temporal domain data acquisition leads to a range of effects. First, the path differences in the two optical paths are usually created by movement (either translation or rotation) of an optical component (usually a mirror) from a path difference of zero to δ_{max} at a certain mirror velocity The detector, therefore, observes a signal that contains interference components modulated over a range, with the highest frequency component being modulated at the highest rate. The detector, therefore, needs to be capable of responding to changing signal intensity at this maximum rate. Furthermore, the sampling of this highest frequency component needs to be made at least twice its frequency (according to the Nyquist criterion). This requires suitably optimized amplification and digitization electronics. The collected data also contain a further artifact of scanning δ with a moving optical component: the data are usually "chirped" in the direction of the movement. This effect can be removed by a process called *phase correction*. In practice, interferograms are often sampled double-sided (i.e., from $-\delta_{max}$ to $+\delta_{max}$) and in both forward and backward directions, with the information required for effective phase correction being obtained from the asymmetry of the interferogram either side of the zero path difference point (Griffiths and de Haseth, 1986).

The second substantial effect of the acquisition of temporal domain data over a limited range of δ occurs upon Fourier transformation of the data to the temporal domain. Collection of data from $\delta = 0$ to δ_{max} is effectively the equivalent of multiplying the full, infinite set of temporal domain data (from $\delta = 0$ to ∞) by a boxcar function (i.e., unity from $\delta = 0$ to δ_{max} and 0 beyond δ_{max}). Consequently, the lineshape observed in the spectral domain is a convolution of the Fourier transform of the temporal data with the Fourier transform of a boxcar function. For example, a pure sine wave signal would transform to a delta function from infinite data, whereas in practice a truncated sine wave (from $\delta = 0$ to δ_{max}) transforms to a spectral feature with both a finite bandwidth and with additional side-lobes. This effect can be significant for spectral data containing sharp, narrow features (especially in gas-phase spectra). It can be significantly reduced by a process referred to as *apodization*, where the recorded temporal data are

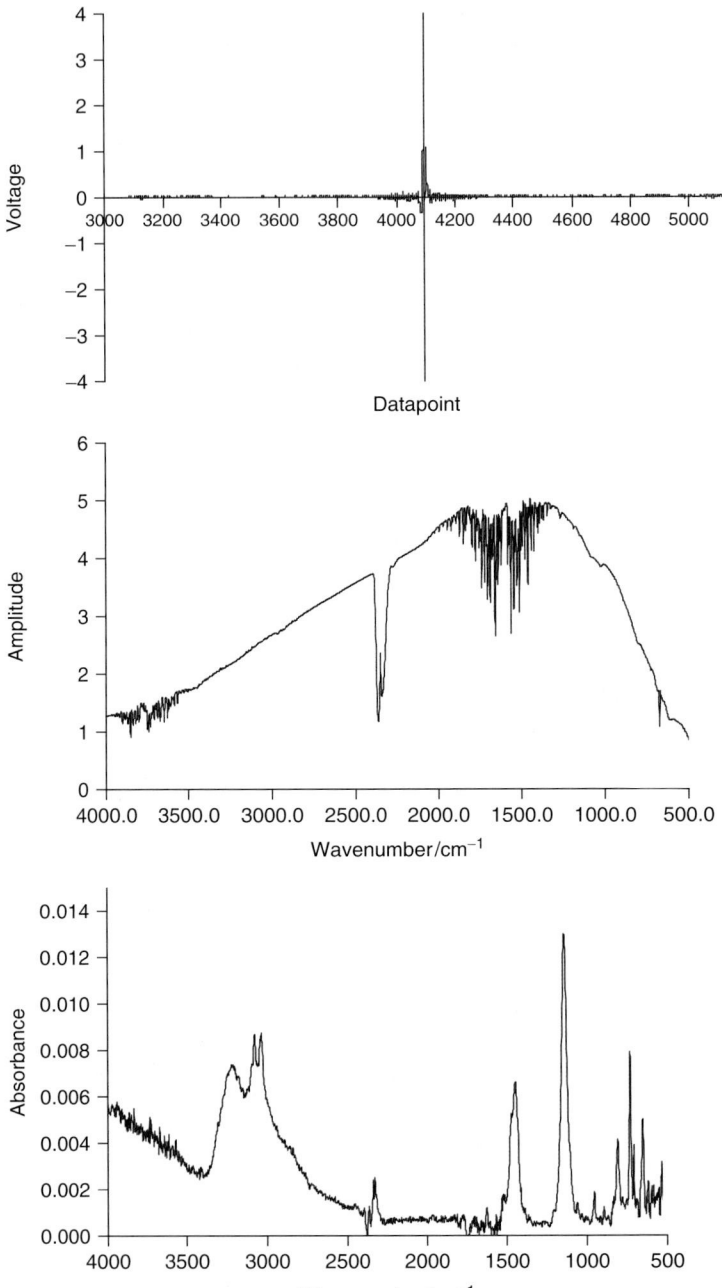

Fig. 26.5 (a) Example of a typical interferogram; (b) single-beam spectrum resulting from the Fourier transform; and (c) absorbance spectrum resulting from the ratio of two single-beam spectra, one containing background information only and one containing additional spectral information.

multiplied by a specific function with a less intrusive spectral domain lineshape. These and other FT-related artifacts are described in more detail in Griffiths and de Haseth (1986) and Kauppinen and Partanen (2001).

A final processing technique, which is frequently encountered, is zero-filling, in which the temporal interferogram data are extended by the addition of further data points (with zero intensity) beyond δ_{max}. This has the effect of placing additional spectral domain data points between the real values. The additional values are interpolations between the real values, smoothing the data, although they clearly do not add any additional information.

In practice, Fourier-transform-based methods have three main advantages over conventional, dispersive methods. First, the data points are measured temporally with a high degree of accuracy; therefore, spectral accuracy (frequency precision) is high in the computed spectrum. Furthermore, the throughput of light is not limited by the presence of slits as for monochromator-based methods. This leads to much higher working signal levels at the detector, with improved signal-to-noise (S/N) ratios. S/N ratios are also improved by the multiplex advantage, since all frequency components are effectively sampled together. In general, the S/N in FTIR spectra is noise limited and can be improved by the coaddition of interferograms before processing. Typically, the S/N improvement scales with the square root of the number of scans.

26.2.2.2 Processing of Infrared Spectra

As described above, FTIR methods therefore involve collection of data as interferograms and subsequent processing to produce a single-beam spectrum, an example of which is shown in Figure 26.5(b). During the course of this processing, a number of mathematical procedures can be performed in order to improve the visual appearance of the output, remove artifacts, and optimize the spectrum. Following the Fourier transformation of the interferogram to produce a single-beam spectrum, the simplest operation is the ratioing of two single-beam spectra to produce an absorbance spectrum. This is necessary because the single-beam operation of an interferometer-based spectrometer produces spectra that are a convolution of the source output profile, the detector response profile, and any limiting optical transmission effects from components in the light path (such as beamsplitters, mirrors, lenses, and sample-handling apparatus). Figure 26.5(b) shows this overall response profile with absorptions from CO_2 and water vapor within the optical path. The absorbance (or transmittance) spectrum of a given sample is obtained from the ratio of single-beam spectra obtained in the presence and absence of a sample. This operation is performed in software with various interventions including the reduction or elimination of water vapor and CO_2 spectral features. Figure 26.5(c) shows a typical absorbance spectrum obtained in such a manner.

A variety of more sophisticated post-processing methods are commonly available in the software accompanying commercial benchtop FTIR spectrometers for further processing of the absorbance (transmittance) spectra thus obtained. This typically ranges from basic mathematical processing such as spectral subtraction/addition and smoothing to more advanced methods such as Fourier deconvolution and spectral bandshape fitting. Databases of infrared spectra are also widely available, and algorithmic methods for comparison of spectra with these library references are used to suggest a range of possibilities and calculated goodness of fit. At a more sophisticated level, chemometric software is also increasingly available for sophisticated quantitative analysis. The specific details of these methods and packages vary widely between manufacturers. Details of these and other related methods can be found in the "Quantitative Analysis" section of (Chalmers and Griffiths, 2002, vol. 3).

26.2.2.3 Infrared Detectors

The detection of infrared radiation in interferometric applications requires apparatus that can detect photons over the relevant mid-infrared range and which is also capable of registering changes in the intensity of such radiation at the maximum modulation frequency of the output of the interferometer (Theocharous and Birch, 2002). In conventional FTIR spectrometers, this is normally achieved using either a bolometer or a photon detector. The commonest example of the former used deuterated triglycine sulfate (DTGS) as the active medium. Thermal changes in the detector upon incidence of infrared radiation are measured electrically and amplified. The response of this material is such that only relatively slow modulation frequencies (a few kilohertz) are possible, but the main advantage is that the linearity of such detectors is high. For this reason, they are the most common choice for quantitative work.

For more sensitive detection of lower levels of infrared radiation, semiconductor-based photon detectors are also in common use. These rely upon the presence of an appropriate bandgap in the detection material, with detection in one form or another of the electrons excited by interaction with the incident infrared photons. To prevent thermal excitations, it is necessary to cool the materials, usually by placing them in thermal contact with a liquid nitrogen reservoir. Although highly sensitive, such detectors have an inherently nonlinear response and care should be taken in quantitative work. The main usage of such detectors is in applications where low light levels occur, especially in sampling accessories with low light throughput. Photon detectors respond much more rapidly to changes in infrared intensity; hence, considerably higher modulation frequencies are accessible. This permits much higher scan rates within the interferometer. The most common material used in modern detectors is mercury cadmium telluride (MCT). There is a trade-off between the range of frequency response of MCT detectors and the sensitivity; with careful bandgap tuning, "wide band" MCT detectors can be made to operate across most of the mid-infrared range, although detectivity falls markedly and cost increases. For specific applications in a restricted range within the mid-infrared, other materials such as indium antimonide (InSb) are also available.

26.2.3 Basic Transmission Sampling Methods

In its simplest form, infrared spectroscopy relies upon the preparation of samples through which the infrared beam from a spectrometer can be passed with minimal loss. In some cases, this requires the use of accessories for the presentation of samples to the infrared beam (Belton and Wilson, 1990). For appropriately tractable samples, infrared spectra can be obtained by direct transmission of light. Typical values of the molar absorption coefficient ε require the use of either low concentrations or short path lengths to keep measured absorbance values in the linear region (below ca. 1.2). In practice, an appropriate amount of material is normally supported between two accurately spaced optical windows. The choice of appropriate window material largely depends on application, but some of the most common are shown in Table 26.1, along with their range of practical transmission and also some limiting physical properties.

26.2.3.1 Gases
Apparatus for the acquisition of the infrared spectra of gases normally involves path lengths of 10 cm and upward. High-resolution infrared spectra of gases are often recorded at reduced pressure to reduce the effects of collision-induced (pressure) line broadening, requiring the use of extended path lengths to attain appropriate individual signal levels. A schematic of a typical cell is shown in Figure 26.6. Multiple passes are achieved using White multipass optics, with commercial accessories available with optical path lengths up to 20 m.

26.2.3.2 Liquids
Thin films are often necessary for the acquisition of transmission infrared spectra from liquids. For pure liquids, cells with accurately spaced windows are widely available, with path lengths of the order of a few microns upward. Spectra can also be obtained from solutions, again with an appropriate choice of path length. The contrasting need to avoid solvent absorptions blacking out (reaching absorbance values greater than 1.2) while achieving suitable signals

Tab. 26.1 Properties of infrared transparent materials in common usage.

Material	Transmission range/cm^{-1}	Properties
NaCl	40,000–625	Very soft, easily fogged by water vapour; low cost
KBr	40,000–400	Very soft, easily fogged by water vapour; low cost.
ZnSe	17,000–720	Hard and brittle crystal, pale yellow in colour; sometimes known as Irtran-1.
KRS-5	20,000–250	Very soft, deforms under pressure; soluble in bases and insoluble in acids. Toxic.
CaF$_2$	77,000–1,110	Strong, acid and alkali resistant; withstands high pressure; insoluble in water. Poor range.
Si	8,300–660	Very hard but quite brittle; relatively inert; insoluble in water.
Ge	5,500–600	Very hard and brittle with a high refractive index (4); insoluble in water.
Diamond	25,000–33	Very hard, resistant to most types of chemicals. Phonon bands around 1900–2600.

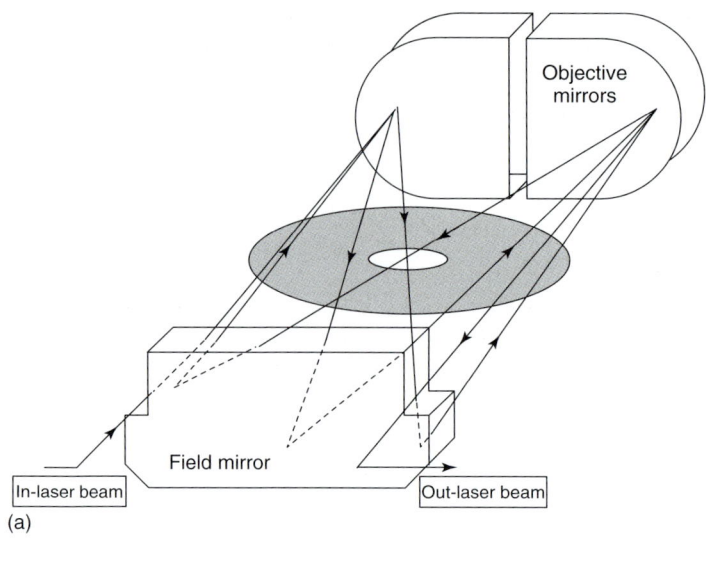

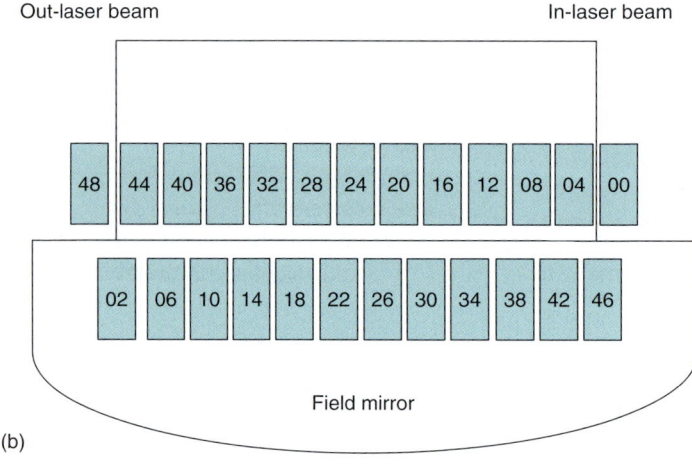

Fig. 26.6 (a) Schematic of the optical configuration of a multipass long path cell for recording infrared spectra of gases. The number of passes can be altered by movement of the objective mirrors. (b) Sequence of images on the field mirror for a 48 pass cell. Every fourth pass is imaged on the top row and the position of every fourth image translates from the entrance (00) to the exit cut-out (48). Taken from (Najera et al., 2009).

from dilute solutes requires a degree of flexibility.

26.2.3.3 Solids

Infrared spectra of solid materials are accessible in a variety of ways, usually obtained after grinding the relevant solid to a fine powder and suspending the particles thus obtained in a suitable medium. Hydrocarbon or halocarbon oils were traditionally used for this purpose, the most common being Nujol.

The resulting oily suspension is then placed between two transparent windows and mounted in a holder in the sample compartment of a spectrometer. Blank windows are normally used as a background in order to nullify the effects of scattering and reflection from imperfect windows. The main intrinsic absorption bands in Nujol arise from aliphatic CH_2 and CH_3 features and thus the material is of little use for detailed examination of organic material containing such functionality, but useful spectral windows are to be found between these resonances. Halocarbon alternatives to Nujol are frequently used to access the CH stretching regions, although these materials do have absorptions in other regions.

A more practical alternative in many organic applications are pressed discs of KBr. In this method, the sample is ground to a fine powder and mixed with dry, powdered KBr and compressed in a die with a hydraulic press. The particles of KBr flow under such pressure and a semitransparent disc is obtained. With careful control of amounts of KBr and sample, this method can be used quantitatively. A significant advantage over mulls is the absence of absorption bands due to the suspending agent, with the principal problem being the hygroscopicity of KBr, which can lead to significant water uptake unless due care with handling and storage of the materials and discs is taken. Another potential problem with this method when applied to inorganic and organic salts is the possibility of ion exchange (e.g., Br^-) between the matrix material and the sample. Care must also be exercised when studying organic solids that exhibit polymorphism, as the pressure required to form KBr discs may also be sufficient to drive a polymorphic phase transition, with a concomitant effect of the infrared spectra of the material. These and other sampling methods in common use are reviewed in detail in Coates (1998), Belton and Wilson (1990), and Hannah (2002).

26.2.4
Reflection-Based Sampling Methods

Reflection methods are finding increasing application in physical sciences as a way of examining materials with a minimum of sample preparation. Using modern sampling accessories, it is possible to acquire infrared spectra from internal, external, and diffusely reflected infrared radiation, although there are many diverse and unusual perturbations to the spectra in reflection configurations that need careful consideration.

26.2.4.1 Specular Reflection

Specular reflection refers to reflection from relatively flat materials such that the angles of incident and reflected beams are the same with respect to the surface normal. The optics of light reflection and transmission from interfaces between both absorbing and nonabsorbing media alike are well understood and accurately described by Maxwell's equations or by a Fresnel equation formalism (Heavens, 1955; McIntyre and Aspnes, 1971). Generally, the reflection behavior is different for light polarized perpendicular and parallel to the plane of incidence (defined as the plane that includes the incident electromagnetic beam and the surface normal), referred to as s- and p-polarization, respectively. In general, the optical properties of each medium are described by their complex refractive index,

n, or the dielectric constant, ε, that is,

$$\mathbf{n} = n - ik \quad (26.21)$$

where n is the refractive index and k is the absorption coefficient. When $k = 0$, the refractive index is real in all cases. Similarly, the dielectric constant has real and imaginary components:

$$\boldsymbol{\varepsilon} = \varepsilon' - i\varepsilon'' \quad (26.22)$$

The two are related.

$$\varepsilon' = (n^2 - k^2)$$
$$\varepsilon'' = 2nk \quad (26.23)$$

For the interface between two phases i and j, the Fresnel coefficients for s- and p-polarization are given as

$$r_{ij}^s = \frac{\mu_j \xi_i - \mu_i \xi_j}{\mu_j \xi_i + \mu_i \xi_j}, \quad r_{jk}^p = \frac{\varepsilon_j \xi_i - \varepsilon_i \xi_j}{\varepsilon_j \xi_i + \varepsilon_i \xi_j} \quad (26.24)$$

where μ_j is the magnetic permeability (ca. 1 in the mid-infrared) for phase j. The function ξ_j for each interface is effectively a complex angle of incidence/refraction term on the interface $(j-1)|j$.

$$\xi_j = \sqrt{\mathbf{n}_j^2 - \mathbf{n}_1^2 \sin^2 \theta_1} \quad (26.25)$$

where θ_1 is the angle of incidence of the light upon the first layer and $\mathbf{n}_j$ is the complex refractive index of the phase j (for multilayer systems, discussed later). The substrate reflectance can be determined as a function of θ using Eqs (26.4) and (26.5). Figure 26.7 shows a comparison of the reflectivity of a typical nonabsorbing dielectric, silicon ($n_{Si} = 3.44$, $k_{Si} = 0$), and a metal, platinum ($n_{Pt} = 3.0$, $k_{Pt} = 30.0$) (Horn, 1998).

There are large differences between these examples in several ways, which affect the nature of the optical configuration required to acquire spectra from the interfacial region. Most noticeably, the reflectivity of silicon is lower than for a metal (except at 90°) for both polarizations, starting from around 35% at normal incidence. The s-polarized reflectivity rises monotonically from its normal incidence value to 100% at 90° (i.e., for a given angle, less light will be reflected from dielectrics than from metals), whereas the p-polarized reflectivity of silicon falls to zero at the polarizing or Brewster angle θ_B, where

$$\theta_B = \tan^{-1}\left(\frac{n_{Si}}{n_1}\right) \quad (26.26)$$

When a substrate has frequency-dependent optical constants (i.e., absorption bands), this behavior can be used to collect reflectance spectra. The reflectivity of the substrate is modified across the absorption band by absorption and by dispersion due to the changing refractive index. This gives rise to characteristic lineshapes in reflectance spectra, an example of which is shown in Figure 26.8. The frequency-dependent n and k behavior can be separated out mathematically using a Kramers–Kronig transform, usually a built-in feature of most modern spectrometer software packages. The k-spectrum thus obtained is directly analogous to the absorption spectrum of the material (Willis and Powell, 1990).

26.2.4.2 Reflection–Absorption

In the case of metals, the reflectivity of the interface can be used to record infrared spectra of adsorbed monolayers and condensed ultrathin films. It was first demonstrated theoretically (Greenler, 1966) and then practically (Pritchard and Sims, 1970) that, under certain

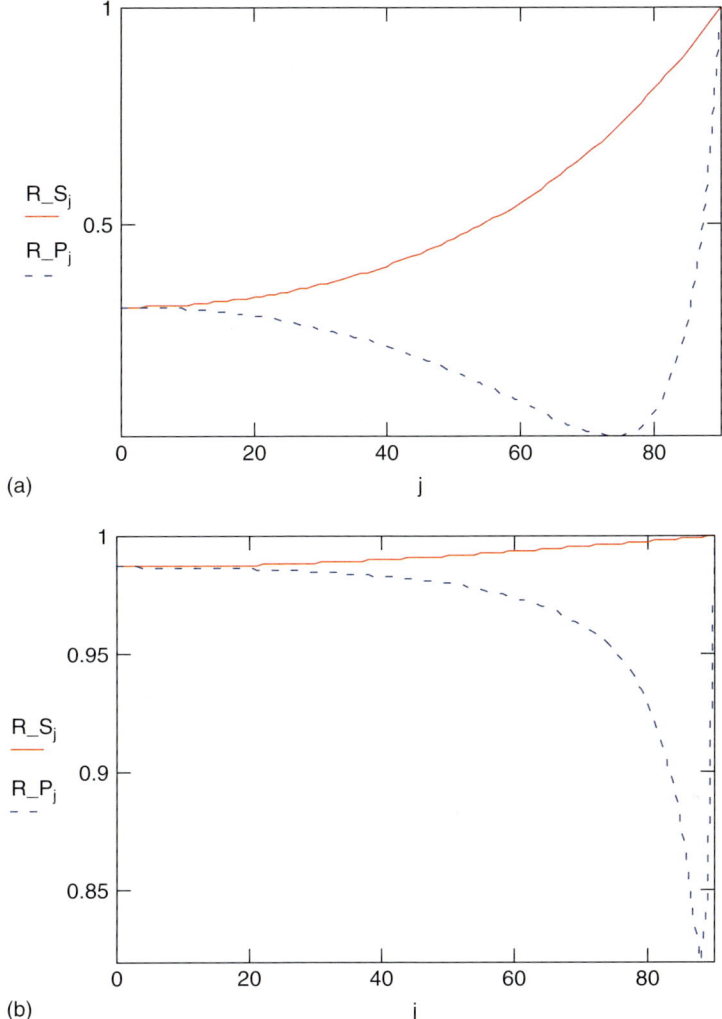

Fig. 26.7 Modeled reflectivity of the vacuum/substrate interface for (a) a typical metal with optical constants $n = 3.0$, $k = 30.0$ and (b) a typical nonabsorbing dielectric with $n = 3.44$, $k = 0$.

circumstances, infrared spectra of monolayers in metals could be obtained in a "reflection–absorption" configuration. The method, referred to variously as *reflection–absorption infrared spectroscopy* (*RAIRS*) or infrared reflection–absorption spectroscopy (IRRAS), has subsequently been used extensively to study adsorption and reaction of a substantial range of adsorbates on a variety of transition metal surfaces that are of interest in catalysis, corrosion, and electrochemistry. Practically, the experiment requires light to be incident upon the metal surface containing the adsorbate at a grazing incidence. This behavior

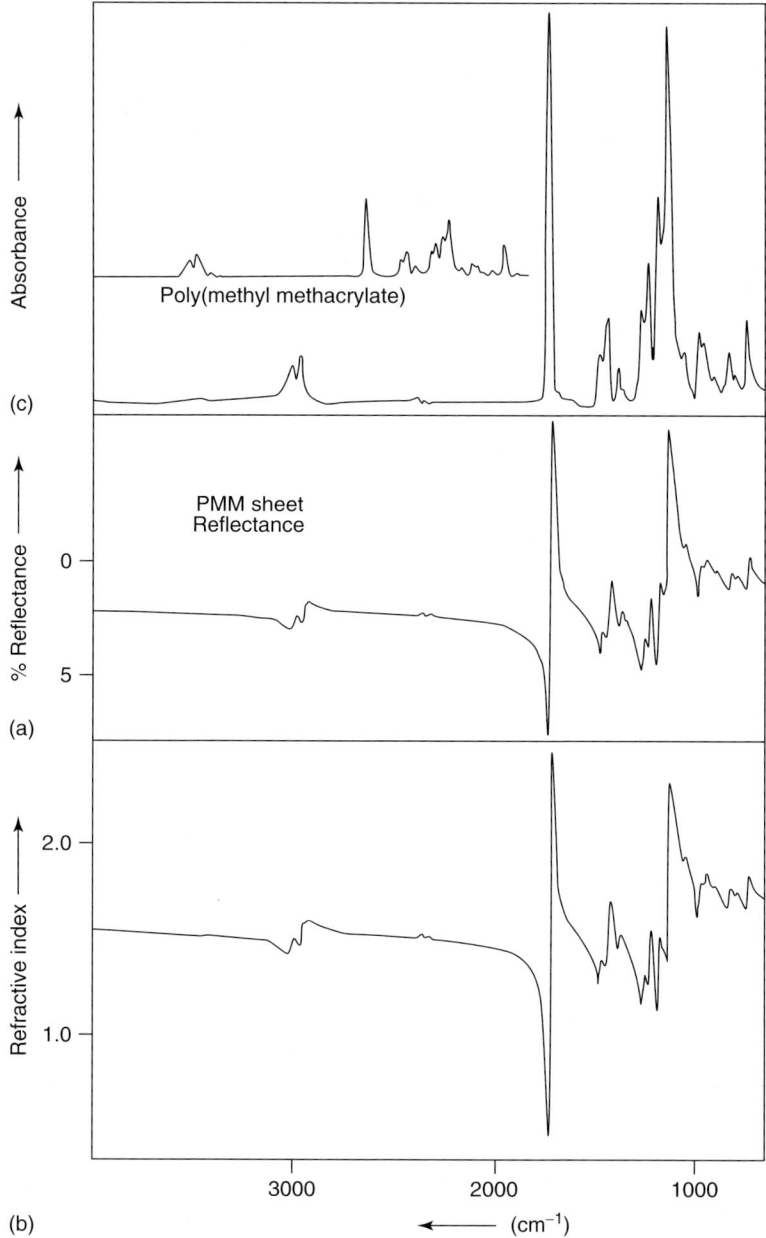

Fig. 26.8 Reflectance spectrum of poly(methyl methacrylate), showing deconvolution of the pure reflectance (a) into the *n*- (b) and *k*- (c) spectral components. Reproduced from (Willis and Powell, 1990.)

again has its origins in the optics of the vacuum/metal interface, and using an extension of the Fresnel model described above it can be shown that only vibrations with a component of their dynamic dipole, which is perpendicular to the surface, are seen. Grazing incidence is therefore necessary both to maximize the surface-normal component of the infrared beam and to interact favorably with the optical properties of the surface. Optically, coupling of the dynamic dipole of vibrational motion parallel to the surface to the infrared radiation is suppressed. A phenomenological description of this behavior was first given by Pearce and Sheppard (1976) who invoked the idea of "image charge" in the surface to show that vibrational dynamic dipoles parallel to the surface cancel out, whereas those perpendicular are reinforced, leading to the so-called metal-surface selection rule (MSSR). A detailed description of the optics and practicalities of RAIRS measurements are given elsewhere by Hollins (2000) and Bradshaw and Schweizer (1988).

Typical, RAIRS is used to monitor the adsorption of gaseous species on reactive metal surfaces. This frequently involves the use of a highly polished metal single crystal or polycrystalline metal film as a substrate, normally supported in ultrahigh vacuum to prevent adsorption of contaminants. The infrared beam from a spectrometer is coupled into the vacuum chamber through an infrared-transparent window and is incident upon the sample at a highly grazing angle. The reflected light is recollimated and focused onto a detector externally (in some cases, the source is the external component rather than the detector).

The MSSR is ideal for the identification of adsorbate geometries on metal surfaces, since only those vibrational modes with transition dipole moments perpendicular to the surface can be observed in the resulting RAIR spectrum. If accurate assignment of the normal modes to specific directions with respect to the molecular axes is possible, the presence or absence of individual modes can thus be used to deduce orientation with respect to the surface. For example, the symmetric C—H stretching mode of a methyl group attached to another moiety has its transition dipole oriented the C_3 rotation axis, whereas the antisymmetric CH stretch is polarized in a plane perpendicular to the axis. Simultaneous absence of antisymmetric mode and presence of the symmetric mode in the RAIR spectrum would therefore imply that the C_3 axis for the methyl group is perpendicular to the plane of the surface. The opposite configuration, that is, where the C_3 axis is parallel to the surface, would reverse the activity of the modes in the resulting RAIR spectrum. For whole molecules, the chemical structure of the adsorbates and their orientation with respect to the surface can often be determined by examining the positions and intensities of the absorption bands in comparison to reference spectra from isotropic bulk samples and gases. As an example, Figure 26.9 shows a RAIR spectrum from the adsorption of methyl silane on Cu(111) at 50 K. In this case, the adsorbate remains intact but lies predominantly with the Si–C axis parallel to the surface in the monolayer. Increasing coverage beyond a single monolayer results in a more disordered multilayer with a range of orientations more redolent of an amorphous thin film or liquid (Menard, Tear and Horn, 2008). Information such as this can be vital to the development of the understanding

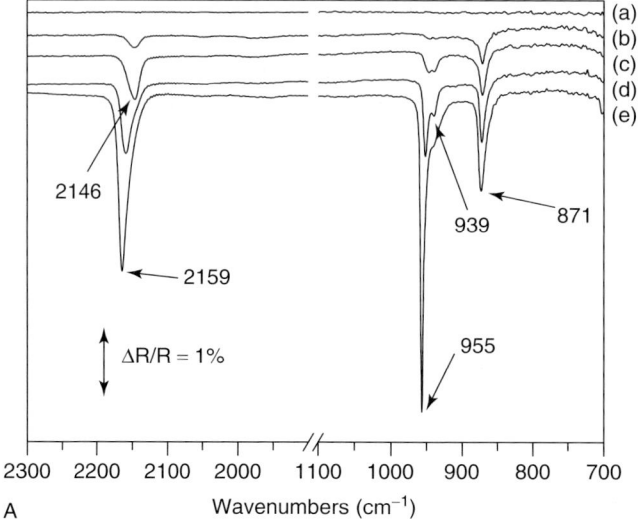

Fig. 26.9 Reflection–absorption infrared spectrum of CH_3SiH_3 (methyl silane) adsorbed on a copper Cu(111) surface at 50 K. The spectra show increasing coverage of the surface from (a) to (e), in which monolayer features are first observed, followed by multilayer features. Under the terms of the metal-surface selection rule, the bands at 870 and 2148 cm^{-1} indicate that the molecule lies with its C_3 axis flat to the surface in the monolayer; further, as the material is deposited, a disordered multilayer forms and many more bands become active. Taken from (Menard, Tear and Horn, 2008).

of catalytic processes at a microscopic level. In less specialized experiments, RAIR spectra can also be obtained from self-assembled monolayers and Langmuir–Blodgett films on metal surfaces under ambient conditions (Horn, 1998). In general, care should be taken in the interpretation of RAIR spectra from thicker films as the optical properties of the ambient phase/thin layer/substrate become less straightforward with increasing thickness. This can lead to a range of band position and shape perturbations, for instance, through the excitation of longitudinal optical resonances (Berreman, 1963). However, with a detailed knowledge of the directionality of the vibrational dynamic dipole moments, orientational information can be extracted with a very high degree of sophistication (Allara and Nuzzo, 1985).

RAIR spectra can also be obtained from nonmetallic surfaces under certain conditions. To a large extent, the MSSR does not apply except for some semimetallic substrates, with vibrational dynamic dipoles in all orientations being observed (although with different intensities). Once again, the behavior is adequately described by a simple Fresnel model. Once again using silicon as a typical example covered with a 1-nm film of isotropic adsorbate (i.e., dynamic dipoles in all directions), the following behavior is observed at the peak of an absorption band. At normal incidence, both the s- and p-polarized responses are equal (as they must be, since both are parallel to the surface at normal incidence) and negative in absorbance units, that is, the presence of an adsorption band results in an *increase* in surface reflectivity compared to the clean surface. With increasing angle, the s-polarized RAIRS response falls monotonically reaching zero at grazing incidence. The p-polarized behavior falls, crosses zero, becomes large and positive at angles of incidence just below θ_B, inverts to become large and negative above θ_B, and again reaches zero at grazing incidence. The p-polarized behavior is largely determined by the reflectivity of the underlying substrate, with the discontinuity around θ_B being a "divide by zero" error. Figure 26.10 shows this behavior schematically.

Reflection–absorption is not confined to gas/substrate interfaces. Recently, there has been much progress in the acquisition of infrared spectra from the surfaces of electrochemical cells or from substrates under high pressures of ambient gas. A number of methods are required to make this possible, largely due to strong absorptions by the electrolyte or gas. In the case of high gas pressures, this can be achieved using the s-polarized (non-interacting) response of a surface as a normalizing signal. In this case, the s-polarized component contains information about the ambient absorption with no spectral information from the surface due to the MSSR rule. The p-polarized response contains the same (isotropic) absorptions from the electrolyte and also the p-polarized surface RAIR information. Polarization modulation and normalization to the s-polarized spectra again allows the gas-phase and

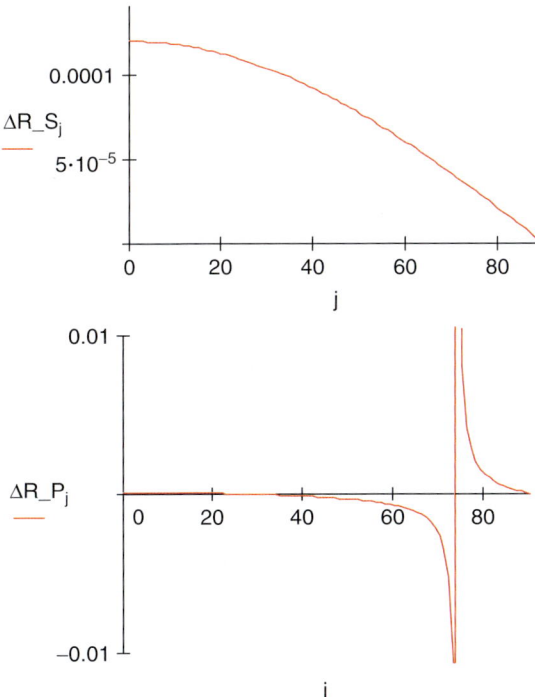

Fig. 26.10 Modeled reflectivity change due to the presence of a thin absorbing layer on a dielectric (silicon) surface as a function of angle of incidence for s- and p-polarized radiation.

surface components to be separated. For electrochemical surfaces, reduction of the ambient (electrolyte) absorption can be achieved using ultrathin electrolyte films above metal surfaces, although transport effects often dominate under such conditions. The use of thicker electrolyte layers requires the use of more advanced methods to remove the effect of electrolyte absorption. Normalized difference spectra are obtained as a function of potential, and the method is frequently referred to as *surface-normalized FT infrared spectroscopy*, (*SNIFTIRS*), and can be very sensitive to very small quantities of adsorbates on electrode surfaces (see, e.g., (Chen and Lipkowski, 1999)). Positive and negative-going bands are seen in the spectra, indicating adsorption and desorption of species from the surface. This differential reflectance behaviour can be modeled to obtain orientational information from the adsorbates under certain circumstances (Pettinger et al., 2001).

In practical terms, accessories are available to fit within the sample compartment of conventional benchtop FTIR spectrometers for recording RAIR spectra from thin films on different substrates. The simplest are fixed angle devices using 80 or 85° for thin films on metallic surfaces. More sophisticated accessories incorporate the ability to change the angle of incidence. A schematic of such a device is shown in Figure 26.11.

26.2.4.3 Diffuse Reflectance

Powdered material, whether comprised of regular crystallites or random amorphous particles, scatters light in a range of directions. This light scattering arises from a range of sources, some of which contain spectroscopic information. Where the individual crystallite surfaces are optically flat, some of the scatter arises from specular reflection (from these flat surfaces in the sample), although the range of orientations of the crystallites means that the scattered light exits in all directions. For optically rough surfaces, the scatter is again diffuse. Both of these contributions may contain some spectroscopic information from resonant interaction with material at the surface of the particles, but the effects are usually relatively weak. From a spectroscopic viewpoint, the most useful component arises from light that has passed through the particles before escaping. This component is often internally reflected several times, and again, the scattered light is emitted randomly over a wide solid angle. During its transit through the particles, resonant absorption can occur; hence, comparison of this component to diffusely scattered light from a nonabsorbing reference sample may give considerable spectroscopic information.

The theory of diffuse reflectance (DR) has been described in detail elsewhere, with various structures and nomenclatures used to identify and quantify the contributions to the total collected diffuse scatter (Milosevic and Berets, 2002). The most common treatment is the Kubelka–Munk model, for which a relatively simple solution exists for the case of an infinitely (practically, a few millimeters in a suitable metal cup) thick sample. From an experimental viewpoint, the measurement of spectroscopic information from diffusely reflected light from a sample revolves around the separation of the diffuse specular reflection

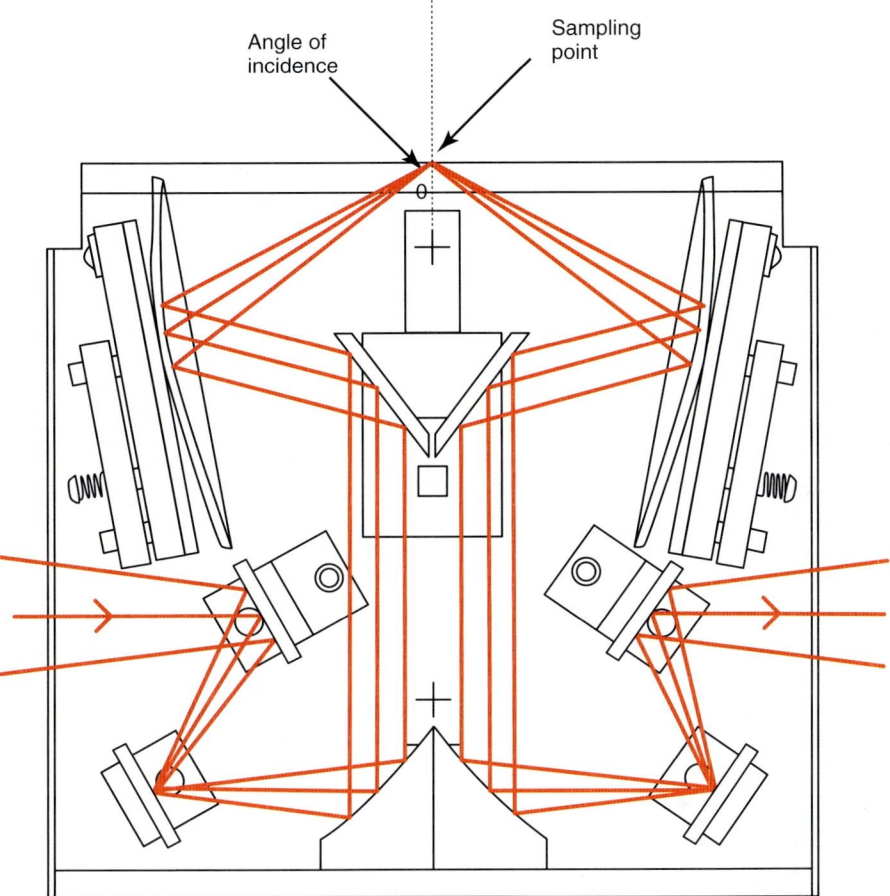

Fig. 26.11 Schematic of the optical path through a variable angle reflection accessory. Reproduced by kind permission of Pike Technologies.

from the internally reflected diffuse component. Practically, sampling utilizes a focused beam incident upon a cup containing the powdered sample. The diffusely reflected light is collected over a large solid angle using a specially shaped mirror, the purpose of which is to recollimate a maximal amount of the internally diffusely scattered radiation and return it to the light path of the spectrometer. The spectral parameter commonly obtained is R_∞, the ratio of the diffuse scatter as a function of wavenumber from a given sample and that from a powdered, nonabsorbing sample. From the Kulbelka–Munk model for an infinitely thick sample with minimal diffuse specular scatter, the absorption k and the scatter s are related to R_∞ such that

$$\frac{k}{s} = \frac{(1 - R_\infty)^2}{2 R_\infty} \qquad (26.27)$$

A number of sampling methods are in widespread use. The most common application simply requires a powdered sample to be placed in a cup at the focus point of the optics. However, in samples with very strong absorption bands, distortion occurs and it is necessary to dilute the powder. This is normally done by diluting the sample (typically to about 5%) in an inert matrix such as KBr. The sample remains loosely packed (compared to pressed KBr discs for transmission measurements) and thus there are fewer problems with pressure-induced effects on the sample. Another method for the preparation of samples for diffuse reflectance measurements utilizes a SiC disc, which is used to abrade material from otherwise intractable solid samples. The DR spectrum can be recorded directly from this disc, although the optical arrangement is far from the ideal Kubelka–Munk situation and is at best a qualitative measurement.

Quantitation in diffuse reflectance spectroscopy requires great care, as the effects of scattering from the matrix, concentration of the absorber, and dispersion in the optical constants of the sample are complex. For "ideal" powdered samples, the penetration depth of the radiation before diffuse scattering is dependent upon k, which means that diffuse reflectance measurements are particularly sensitive to weak absorptions, for which the spectroscopic path length is effectively longer.

The design of apparatus to collect diffuse scatted from powdered samples is nontrivial, and a number of designs are in use. Broadly speaking, the challenges are threefold: to illuminate the sample with as much incident radiation from the spectrometer as possible, to collect as much of the diffusely scattered light as possible, and to minimize the amount of diffuse specular reflection collected. There is a further trade-off between the solid collection angle and the area of the sample illuminated, which requires optimization in any optical configuration. The high sensitivity of Fourier-transform spectrometers is essential for diffuse reflectance measurements, as the restrictions on collection area and geometry, as well as the inclusion of various methods to reduce the specular scatter, reduce the throughput of diffuse reflectance accessories to as little as 5% of the total power. The optical configuration of a typical diffuse reflection accessory is shown in Figure 26.12. The accessory is located in the sample compartment of the spectrometer, intercepting the spectrometer beam and directing in onto specially shaped mirrors. The diffusely reflected light is collected and returned to the spectrometer light path.

In such a configuration, controlled environment chambers can be added at the sample point, allowing the investigation of reactions on particles of environmental and catalytic interest to be studied. Figure 26.13 shows a modified diffuse reflectance accessory used to observe the uptake of gas-phase NO_2 onto NaCl particles representative of material present in the atmosphere in the marine boundary layer. In this particular case, the adsorption process shows the formation of nitrate during in the reaction, which remains adsorbed onto the particle surface (Vogt and Finlayson-Pitts, 1994).

26.2.4.4 Attenuated Total Internal Reflection

The most significant development in the applicability of infrared spectroscopy to the widest variety of samples without significant preparation over recent years is the advent of attenuated total internal reflection (ATR) based methods. The ATR effect was first described by Newton in *Opticks*, and the seminal work by

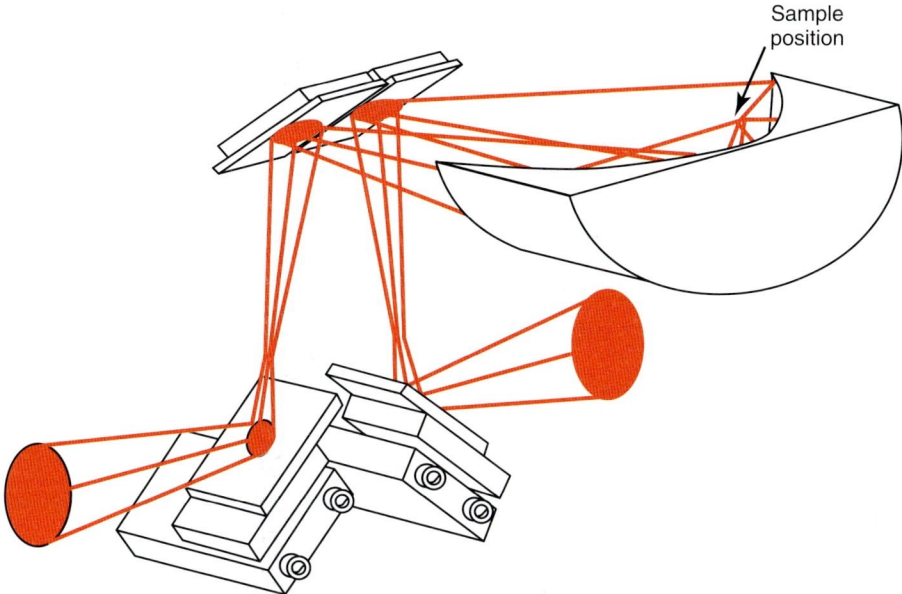

Fig. 26.12 Schematic of the light path through a typical commercial diffuse reflectance accessory. A single complex mirror directs the infrared beam onto the sample and collects the diffusely reflected light for return into the spectrometer light path. Reproduced by kind permission of Pike Technologies Inc.

Harrick (1967) in the 1960s set the groundwork for the wide-scale use of ATR methods in optical spectroscopy.

When light is incident upon the interface between two nonabsorbing media with refractive indices n_1 and n_2 from within that with the higher refractive index, total internal reflection is observed above a certain critical angle, θ_c, given as

$$\theta_c = \tan^{-1}\left(\frac{n_2}{n_1}\right) \qquad (26.28)$$

For a germanium substrate in contact with a vacuum, $n_1 = 4$, $n_2 = 1$ and hence $\theta_c = 14°$ from the surface normal. Below this angle, a proportion of the light is internally reflected and the remainder is transmitted through the interface. The variation of the internal component with angle of incidence is again adequately described by the Fresnel equations, as shown in Figure 26.14(a). Subtly different angle-dependent behavior is observed for s- and p-polarized infrared radiation, although the critical angle is the same for both.

The light that is internally reflected at the interface penetrates beyond it in the form of an evanescent wave. The amplitude of the evanescent wave falls exponentially with distance from the interface, such that the field strength at a distance z, denoted as $E(z)$, is given as

$$E(z) = E_0 \exp\left(\frac{-z}{d_p}\right) \qquad (26.29)$$

The term d_p is the "depth of penetration", a parameter that depends upon the relative refractive indices of the two

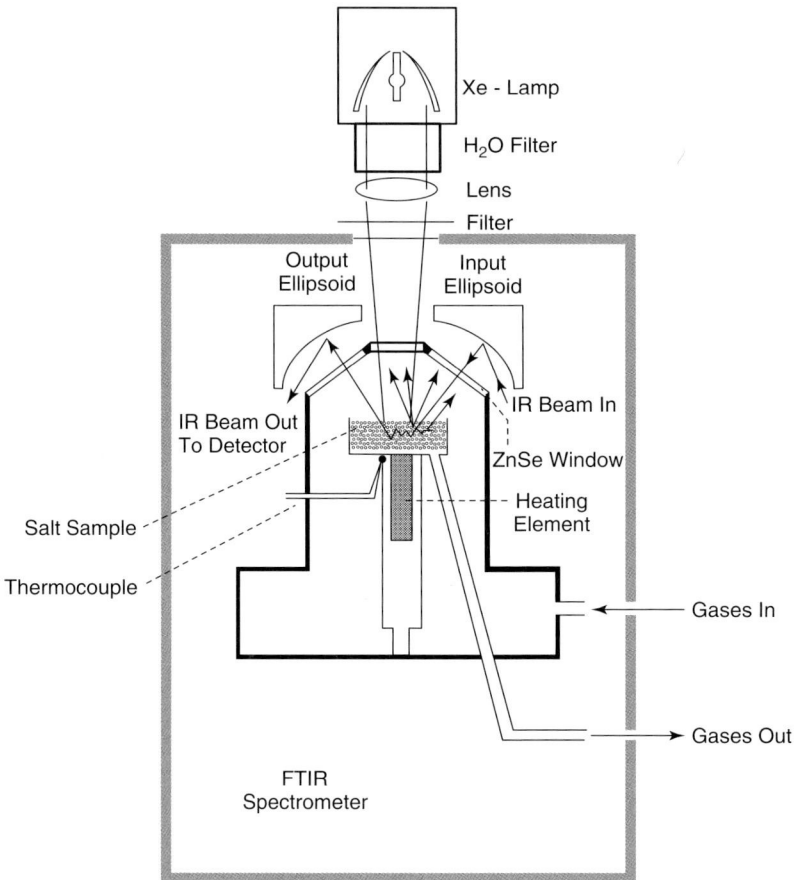

Fig. 26.13 Schematic of apparatus used to collect diffuse reflectance infrared spectra of salt particles during reaction with ambient gases. Taken from Vogt and Finlayson-Pitts (1994).

media (n_1 and n_2), the wavelength (λ), and the angle of incidence (θ), and is effectively the distance at which the evanescent wave falls to $1/e$ of its value at the interface:

$$d_p = \frac{\lambda}{2\pi n_1 \left(\sin^2 \theta - \left(\frac{n_2}{n_1} \right)^2 \right)^{1/2}}$$

(26.30)

The evanescent wave can couple resonantly with vibrational dynamic dipole moments close to the interface. In such cases, energy is lost from the internally reflected wave, the process being referred to as *attenuated total internal reflection*. The magnitude of the attenuation is related to k for the material in contact; hence, a spectrum recorded using an ATR geometry is directly related to the true absorption spectrum (Mirabella, 2002). Figure 26.14(b) shows an illustration of the magnitude of the attenuation of total internal reflection for a semi-infinitely thick film of a

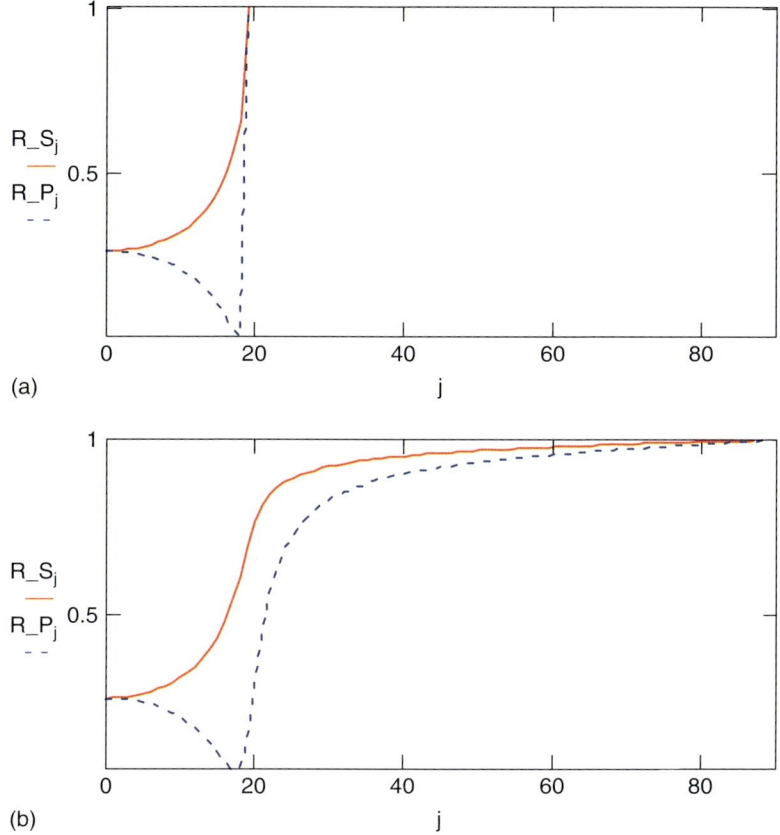

Fig. 26.14 (a) Modeled internal reflectivity of the interface between two transparent media (Ge, $n_1 = 4.0$ and vacuum, $n_2 = 1$) as a function of the angle of incidence for s- and p-polarization. (b) Modeled reflectance of the same interface with an infinitely thick layer of a weakly absorbing material replacing the vacuum ($n_2 = 1.2$, $k_2 = 0.03$).

moderately strong absorber ($n_2 = 1.2$, $k_2 = 0.03$) on a Ge internal reflection element (IRE). The magnitude of the attenuation is greatest close to the critical angle and decreases with increasing angle. There is also a difference in the attenuation of s- and p-polarized light. In practice, angles between 45 and 60° and unpolarized radiation are used, depending upon the nature of the materials to be studied and the application.

From the above equation, it can be clearly seen that there is a linear dependence of d_p upon the wavelength. At 2000 cm^{-1} for ZnSe (a common choice of material for ATR) at 60°, $d_p \approx 2$ μm for example. The concomitant effect of the wavelength dependence of d_p on the evanescent field strength versus distance above the internally reflecting interface across the spectrum is that the intensity of absorption bands across the spectrum is therefore scaled linearly with λ. Most spectrometer software packages contain routines for correction of this phenomenon. However, care is needed while using these corrections, not least because spectrometer optics normally produce a 7–10° cone of incident angles at the IRE surface.

The advent of FTIR spectrometers has made a variety of experimental configurations viable for the collection of ATR infrared spectra, since both the sensitivity is higher and the optical configuration of accessories is not constrained by slits, as was the case in the original ATR-IR apparatus used by Harrick and Fahrenfort (Harrick, 1967) in the earliest practical demonstrations of the technique. The basis of all accessories is the IRE, and arrangements using prisms, trapezoids, parallelograms, cylinders, and hemispheres are or have been available in commercial devices. A primary choice is between multiple reflection and single reflection configurations, each having their own applications and advantages (Fitzpatrick and Reffner, 2002).

Multiple reflection ATR accessories for FTIR applications tend to use horizontal trapezoidal IREs. The effective path length in these IREs is hence a multiple of d_p for each reflection, with the number of reflections being readily calculated from the dimensions of the IRE. Given the limited range of the evanescent wave above the IRE surface, good contact is essential. High surface area horizontal IREs preclude the application of the high pressures necessary to ensure good contact between harder solids and the surface and so are of limited application to powders. However, these devices are very useful for solutions and soft solid materials. Typically, the IRE is mounted in a flat plate or trough. Light from the spectrometer compartment is coupled onto the beveled end of the IRE, undergoes multiple internal reflections, exits the end of the crystal, and is returned to the spectrometer light path. For high-index materials such as Ge, antireflection coatings on the input and output faces of the IRE are essential to increase optical throughput. Figure 26.15

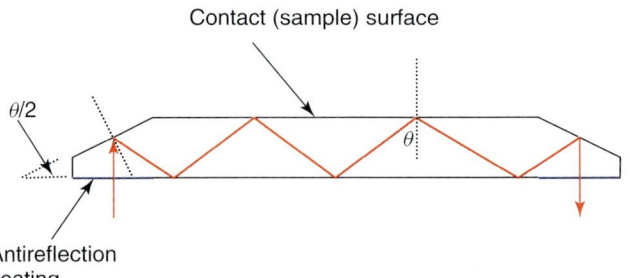

Fig. 26.15 Schematic of multiple internal reflection on a shaped prism (two reflections at the sampling interface). The angle of the beveled ends dictates the internal reflection angle θ. An antireflection coating is often used with high-index materials (e.g., Ge) to reduce reflection losses. In practice, IREs of this type are normally bonded into metal plates for location in optical carriages that translate the infrared beam to and from the sample compartment of a conventional FTIR spectrometer.

shows a horizontal ATR configuration in schematic. Flat plates and troughs are usually widely available, with choice depending upon application. In practice, these are normally removable for cleaning. Multiple internal reflection measurements of adsorption onto surfaces are also feasible, as demonstrated in the experiments of Chabal and coworkers, in which a polished and shaped silicon crystal was used as both a substrate for adsorption and an IRE (Chabal, 1986). Sensitivity well beyond that achievable for reflection–absorption or transmission measurements was achieved as a result of both the surface sensitivity and multiple reflections.

In recent years, single reflection ATR accessories have become more widely available. Figure 26.16 shows an example of the light path through a commercial sample-compartment-mounted single bounce ATR accessory. In this case, the IRE consists of a shaped ZnSe element with a thin diamond as the sample contact patch. Often such IREs also act as focusing elements, optimizing the amount of light directed onto the small diamond element (usually only a few millimeters in diameter). Light is coupled into and out of the IRE by plane mirrors, making a single internal reflection at the diamond/air interface. To improve sample contact between harder materials and the IRE, pressure can be applied with a clamp. ATR infrared spectra of liquids can also be obtained by placing small quantities of the liquid (or solution) analyte onto the diamond ATR, although it is often necessary to take steps to prevent evaporation from volatile media. ATR

Fig. 26.16 Light path through a commercial single reflection diamond ATR accessory. Reproduced by kind permission of Pike Technologies Inc.

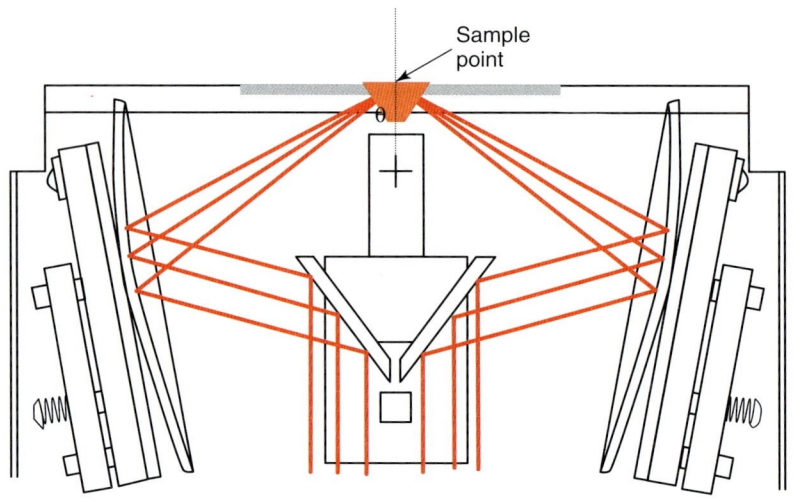

Fig. 26.17 Schematic of the light path through a commercial variable angle ATR accessory (based upon the ATR version of the Pike VeeMax II accessory). Reproduced by kind permission of Pike Technologies Inc.

accessories using composite (diamond-tipped) IREs are also widely used for in situ process monitoring, with the light from the spectrometer being directed to and from the IRE by fiber optics or light pipes. Single reflection accessories are also available with the ability to vary the internal reflection angle. These are useful for a number of purposes, including depth profiling and optical constant measurement. Figure 26.17 shows a schematic of a typical accessory, in which the infrared light from a spectrometer is coupled into and out of a prism IRE in such a way that the angle of incidence can be varied within certain limits.

ATR-based methods are now widely used as the method of choice for routine infrared analysis in a wide range of applications. However, the optics of the configuration also lend themselves to more esoteric applications. The effect of restricted spectroscopic penetration into an ambient medium using the ATR technique can be used effectively to study rates of diffusion in thin films deposited upon the surface of an IRE, as demonstrated by Pereira and Yarwood (1996). The rationale relies upon the diffusion process bringing material from an area beyond the sensitivity of the probe beam toward the IRE surface. As the material approaches the interface, it gives rise to spectral features whose intensity increases with time at a rate determined by the optical properties of the IRE/thin film and the rate of diffusion. The technique can be very sensitive when diffusion occurs upon an appropriate timescale for infrared measurements and has been applied to both polymers and condensed ice films. The rate of this diffusion can be calculated by assuming that the rate of diffusion is Fickian, in which case the absorbance versus time plot convolving spectral intensity with mass transfer will have the following form:

$$\frac{A(t) - A(0)}{A(\infty) - A(0)} = 1 - \frac{8}{\pi d_p \left(1 - \exp\left[\frac{-2L}{d_p}\right]\right)} \sum_{n=0}^{\infty} \frac{1}{2n+1}$$

$$\times \left[\frac{\exp\left[\frac{-2L}{d_p}\right]\left(\frac{2n+1}{2L}\pi\right) + (-1)^n \frac{2}{d_p}}{\left(\frac{2n+1}{2L}\pi\right)^2 + \frac{4}{d_p^2}} \right] \exp\left[-\left(\frac{2n+1}{2L}\pi\right)^2 Dt\right] \quad (26.31)$$

where $A(t)$ is the absorbance at time t, $A(0)$ is the absorbance at $t = 0$, $A(\infty)$ is the saturation absorbance, d_p is the penetration depth (as defined in Eq. (26.30)), L is the thickness of the film, and D is the Fickian diffusion coefficient. Figure 26.18 shows a plot of the intensity of one of the characteristic features of ionized HCl diffusing through an ice film along with a fitted curve for a film thickness of 1.5 μm. A diffusion coefficient of $1.5 \times 10^{-15} m^2 s^{-1}$ is obtained from this data (Horn and Sully, 1997).

26.2.5
Infrared Microspectroscopy

Infrared microspectroscopy (simply infrared microscopy in some references) relies upon the focusing of the infrared beam from an interferometer onto a microscopic sample and subsequent measurement of the infrared spectra of selected sections of the magnified image sampled from the focal plane of the collection optics. The optical considerations for infrared microspectroscopy are practically identical to those of optical microscopes. However, for infrared radiation with a wide range of constituent wavelengths, the use of reflection-based optical components is essential to minimize chromatic aberration. The smallest area from which individual nonoverlapping spectra can be obtained is of the order of the longest wavelength component of the radiation convolved with the numerical aperture of the condenser/objective elements, with about 25×25 μm^2 being a typical practical limit. Trade-offs between wavelength range, magnification, detector element size, and optical throughput must all be considered when selecting instrumentation. For single-point measurements, sample area (size and position) is selected from the magnified image in the focal plane of the objective element with an aperture. Since the amount of light reaching the detector through these various optical stages is relatively small (some 1–2 orders of magnitude less than conventional accessories), the use of high sensitivity, cooled MCT detectors is essential. Further details of the construction and design of conventional infrared microspectrocopy apparatus are given by Sommer (2002). Figure 26.19 shows a plain view of a commercial FTIR microscope, showing the complex light path needed for focusing, imaging, and magnification.

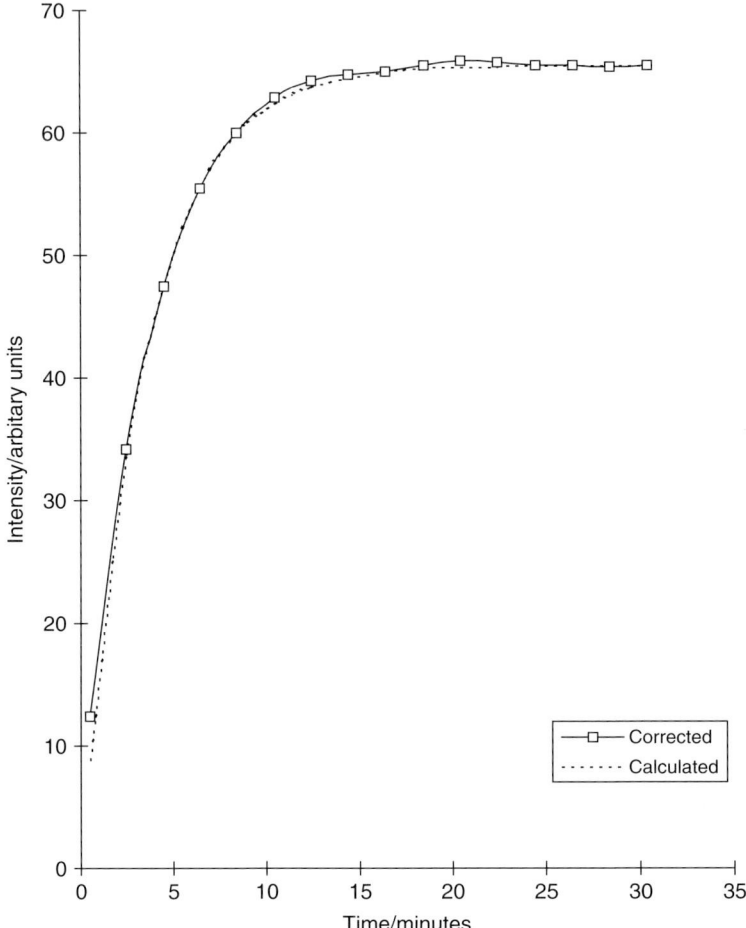

Fig. 26.18 Fickian diffusion of dissociated HCl through an ice film, measured by ATR. Taken from (Horn and Sully, 1997). The solid lines show measured integrated absorbance under the $\nu(H_3O^+)$ band, while the dotted line shows a fit using Eq. (26.31).

A variety of configurations are possible, including transmission, reflection, and ATR. Transmission measurements rely upon the production of films of sample that are sufficiently thin to keep absorbances below 1.0, which can be achieved by either microtoming samples or compression between optically transparent plates. When supported upon optically transparent substrates, corrections for focal length are needed because of refraction effects. Near-normal incidence reflection measurements are very similar in essence to transmission, differing only in the use of a reflective rather than optically transparent substrate and the simultaneous use of the objective as a condenser. For reflection at angles other than near normal, specialized objectives are needed to focus and collect

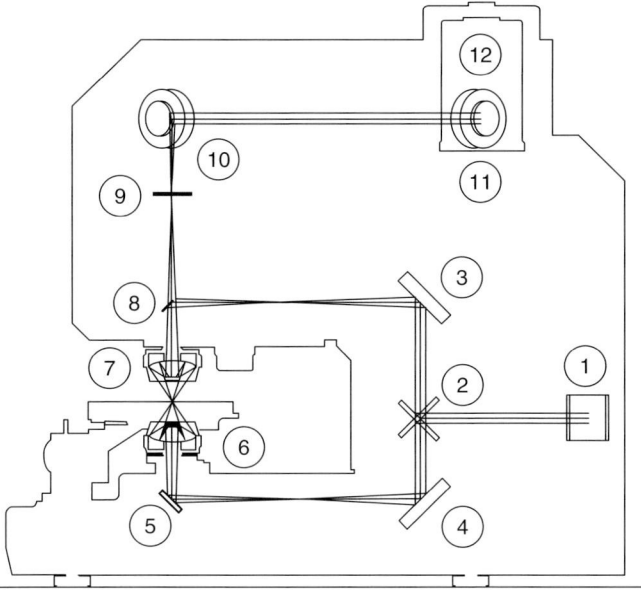

① — ⑤ : Mirror
⑥ : Cassegrain objective mirror
⑦ : Cassegrain converging mirror
⑧ : Beam splitter
⑨ : Aperture
⑩ ⑪ : Mirror
⑫ : MCT detector

Fig. 26.19 Schematic of the light path through a typical commercial FTIR microscope. Reproduced by kind permission of JASCO UK.

the infrared light. ATR objectives rely upon shaped IREs with the sampling area being determined by the size of the contact patch of the IRE. In practice, such objectives also need to incorporate an optical light path for the alignment and sample point selection. For further examples, see Humecki (1995) and references therein.

Major improvements in the sensitivity and usefulness of infrared microspectroscopy have arisen from two ways. The first is the use of synchrotron radiation sources for high-intensity infrared light. For example, an FTIR microscope recently developed at the Daresbury Laboratory in NW England utilizes an $8 \times 8\,\mu m^2$ spot from the synchrotron at high brightness. This has found application to a wide variety of microscopic samples, including prostate cancer cells with unique spectral features (Gazi *et al.*, 2007). Imaging of samples is also possible, with significant advantages over conventional sources where collection times are long due to the lower throughput.

Imaging the infrared image from the focal plane requires an array detector, and a number of such instruments are now widely available with focal plane array (FPA) detectors from 64 × 64 pixels upward. Step-scan interferometers are required for such applications in order to facilitate the recovery of interferograms from such an array. Technology is continuously improving in this field, with rapid increase in detector pixel number and density. The applications of such technology are widespread, especially in biological sciences, where the ability to extract chemical images from samples by integrating specific absorption bands provides unprecedented detail. In Figure 26.20, images of a eucalyptus leaf section from conventional microscopy, an imaged synchrotron microscope, and an FPA-based detector are shown (Heraud et al., 2006). The authors compare the spatial resolution and the signal-to-noise ratio of the FPA imaging and synchrotron mapping methods. The quality of the images is roughly equal, with the main advantage being that the FPA images were obtained in about 2 minutes, compared to 8 hours using the synchrotron source, although the latter has superior signal-to-noise ratio and potentially better spatial resolution. The major features of the internal structures of the leaf are clearly seen from the integrated amide I band.

26.2.6
Extinction Methods from Particles

When infrared light is incident upon particles with diameters similar to that of the wavelength of the light, scattering occurs. This results in an extinction spectrum that is the sum of a scattering contribution that scales with wavenumber and particle radius and an absorption contribution due to vibrational resonances, which is dependent on concentration, absorption coefficient, and path length of the absorbing species. Typically, Mie theory can be used to relate the observed infrared spectrum to aerosol size and concentration. A number of assumptions must be made for simplification: (i) that scattering is elastic (frequency of scattered light is equal to that of incident light); (ii) only single-scattering occurs (i.e., the particles do not affect each other); (iii) the complex refractive index completely describes the optical properties of the particles; and (iv) the number of particles is large with random separations such that there is no systematic phase relation between waves scattered by individual particles (Bohren and Huffman, 1983). The intensity of light, I, passing through an aerosol is given as

$$\frac{I}{I_0} = e^{(-b_{ext} z)} \qquad (26.32)$$

where I_0 is the incident intensity, z is the path length, and b_{ext} is the extinction coefficient. Typical treatments are expressed in terms of τ_{ext}, the light removed from the incident beam by the combined effect of scattering and absorption:

$$\tau_{ext} = \log_{10}\left(\frac{T_o}{T}\right) = \log_{10}\left(\frac{I_o}{I}\right)$$
$$= \frac{b_{ext} z}{2.303} \qquad (26.33)$$

For a distribution of particles with a range of diameters, it can be shown that

$$\tau_{ext} = \frac{1}{4}\frac{\pi d_p^2}{2.303} N(d_p) Q_{ext}(d_p) z \qquad (26.34)$$

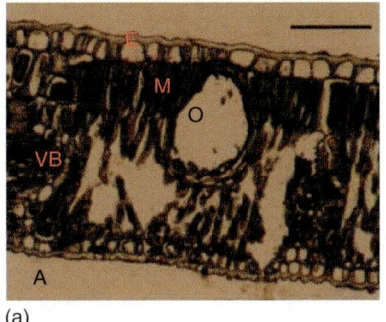
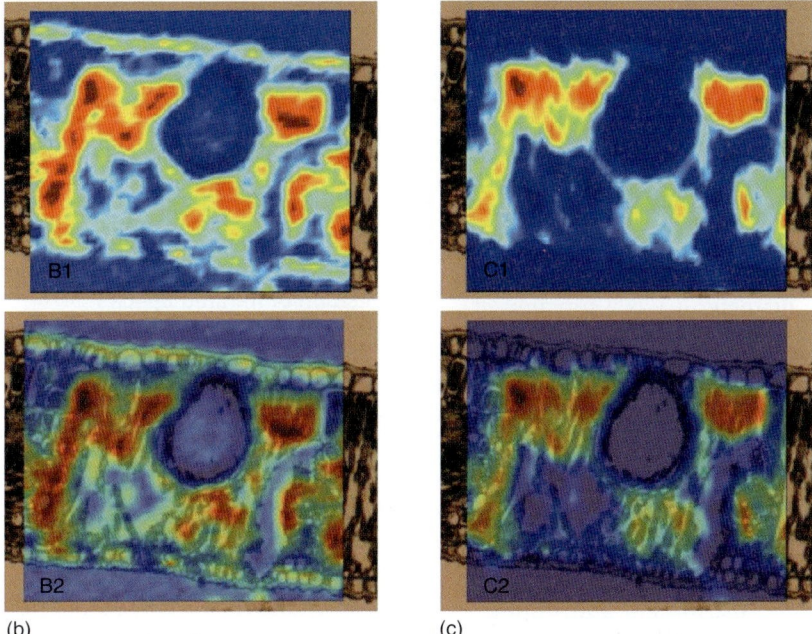

Fig. 26.20 Micrographs and infrared maps of eucalyptus leaves: (a) an optical micrograph and (b) and (c) comparisons between chemical maps of integrated absorbance from 1700 to 1580 cm^{-1} (amide I band) using a point-to point method with synchrotron radiation as a focused source (b) and with an FPA. In each case, the upper panels show the infrared maps at full opacity and the lower panels at 50% opacity. A rainbow scheme has been used to denote absorbance, with the warmest colors (red end of the spectrum). Reproduced from (Heraud et al., 2006).

where d_p is particle diameter, $N(d_p)$ is the size-dependent number concentration, and $Q_{ext}(d_p)$ is the size-dependent extinction efficiency. Light extinction by small particles within the infrared light path has scattering and absorption contributions such that

$$Q_{ext} = Q_{sca} + Q_{abs} \quad (26.35)$$

Mie theory gives a basis for the calculation of the scattering and

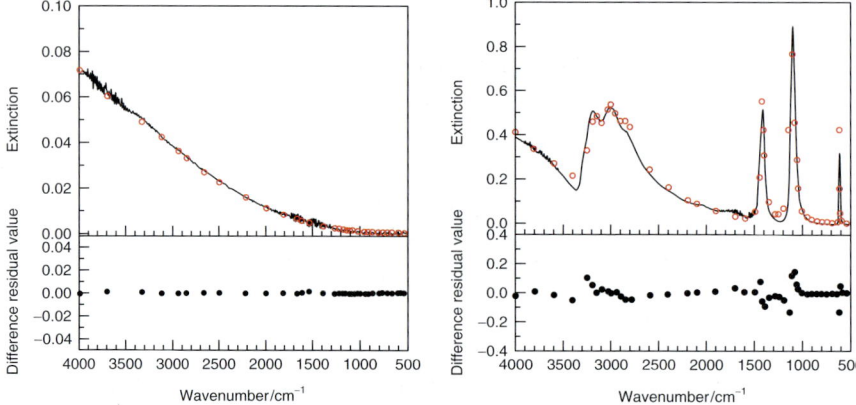

Fig. 26.21 Infrared extinction spectra from aerosols. (a) NaCl aerosols with average diameter 2.1 μm and number density 3.2×10^5 particles cm^{-3}. (b) (NH$_4$)$_2$SO$_4$ aerosols with 2.0 μm and $N = 1.1 \times 10^6$ particles cm^{-3}. In each case, the upper panels shown the FTIR extinction spectra (solid lines) and the best fit results based on Mie scattering calculations (circles). The lower panel shows the residual difference between the two spectra. Taken from Najera et al. (2009) with permission from the American Institute of Physics.

absorption of light by any spherical particle (with diameter d_p) as a function of the size distribution, number density of the aerosol particles, and the complex refractive index. Q_{ext} and Q_{sca} are usually determined iteratively. In practice, for materials with known wavelength-dependent complex refractive index, a test spectrum can be generated using these extinction coefficients based on an input size distribution and number density, with these parameters iteratively fitted to achieve a match to the real spectrum (Zasetsky et al., 2007). Figure 26.21 shows a fit for two examples using published optical constants. For pure sodium chloride aerosols, there is no absorption component and the spectrum is entirely due to scattering. For ammonium sulphate particles, both scattering and absorption are present. The lower panels show the residual from the calculated versus measured spectra (Najera et al., 2009).

26.3 Interpretation of Infrared Spectra

26.3.1 Empirical Interpretation from Group Frequencies

The interpretation and analysis of infrared spectra draws upon a number of tools, some empirical and some more rigorous. Databases of library spectra have been available for many years (from Sadtler, Aldrich, Hummel among others), and algorithms for the comparison of spectra with those contained in reference libraries for the identification of compounds are constantly evolving. However, to use infrared spectroscopy as a black box method in such a manner would be to miss a substantial part of the richness of molecular information and complexity contained within an infrared spectrum. Therefore, this section concentrates on the methodology of first principles interpretation of infrared spectra using group frequencies and correlation tables. The interpretation of infrared spectra requires both a systematic approach to the assignment of functional group modes and an understanding of the nature of the effect of the intramolecular structure on the precise position of absorption bands. In particular, it is often considered essential to make correlations between the appearances of different types of group modes.

As discussed in the preceding sections, although many individual displacements make up a vibrational mode, the combinations which dominate a given normal mode tend to be of similar frequencies. Adjacent parts of a molecule often move independently of the rest of the molecule when masses are very different or when force constants are very different. A normal mode may then be mostly made up from the displacement of the atoms of a single functional group, with a characteristic absorption frequency almost regardless of the rest of the molecules. Normal modes tend to be dominated by a specific type of motion. For each individual atomic displacement in a polyatomic molecule, two possibilities exist: displacement along the bond direction (stretch) and displacement perpendicular to the bond direction (deformation). A variety of nomenclature exists, but in general functional group modes tend to be labeled ν if they involve mainly stretching, δ if they mainly involve bond bending, ρ if they mainly involve rocking, and τ if they involve twisting. It is common to label symmetric and antisymmetric vibrations with subscripts, for example, $\nu_s(H_2O)$ for the symmetric OH stretching mode of water.

The most useful functional group modes for direct assignment tend to be those which occur in regions of the spectra where there are few other resonances. However, functional groups in spectrally congested regions can frequently be identified when their presence is correlated with other types of vibration of the same group. This might occur in several ways, the simplest being when there are either two functional group modes within a single functionality or there are two types of vibration (e.g., stretching and deformation) of the same functionality.

Aldehydes (R—CHO) provide an excellent example of the former. The stretching mode of the aldehydic C—H bond, written as ν (C—H), typically occurs between 2900 and 2700 cm^{-1} and is usually quite weak, although often clearly distinguishable because the region of the spectrum in which it occurs is normally uncluttered by other resonances. The related aldehydic carbonyl stretching mode, labeled as ν (C=O), typically occurs between 1740 and 1720 cm^{-1} and is normally very strong. These features do not vary much between different aldehydes. Consequently, the presence of both together is a clear marker for the presence of an aldehydic functionality, regardless of the composition of the rest of the molecule. Furthermore, as discussed later, the band position of an aldehydic C=O is often sufficiently different from that of other carbonyl species.

An example of the latter is as follows: the presence of two types of vibration of the same set of atoms occurs widely in organic molecules and in biological materials. The stretching modes of methyl (—CH$_3$) and methylene (—CH$_2$—) groups can be observed in the 3000–2800 cm^{-1} range, a relatively uncluttered region. The corresponding deformation modes occur in the 1500–1350 cm^{-1} range, which is usually highly congested in all but the simplest molecules. Specifically, methylene groups in long alkane chains show symmetric $\nu_s(CH_2)$ and antisymmetric $\nu_{as}(CH_2)$ stretching modes at approximately 2860 and 2930 cm^{-1} and a deformation mode $\delta(CH_2)$ at 1470 cm^{-1}. The former are often readily identifiable: their presence allows confidence in the assignment of a spectral feature around 1470 cm^{-1} to $\delta(CH_2)$. Similarly for a methyl group, the symmetric $\nu_s(CH_3)$ and antisymmetric $\nu_{as}(CH_3)$ stretches occur at approximately 2870 and 2970 cm^{-1}, accompanied by symmetric $\delta_s(CH_3)$ and antisymmetric $\delta_{as}(CH_3)$ deformations at approximately 1375 and 1460 cm^{-1}. At a slightly higher level of complexity, bands of this type also show predictable patterns of relative band intensities and characteristic band positions.

The variation of group frequencies from one molecule to another provides a further challenge to the spectroscopist. Some group modes are relatively immobile, regardless of their molecular environment (e.g., methyl and methylene stretching modes). Others, particularly those derived from functionalities that couple strongly through either inter- or intramolecular interactions, exhibit substantial ranges.

In the latter case, shifts in position (and intensity) arise because of chemical and physical interactions. Double bonds can conjugate, resulting in reduced electron density within a specific region of the molecule, thereby reducing the force constant. For example, in isolated C=C and C=O double bonds, absorptions are observed from $\nu(C=C)$ at approximately 1650 cm^{-1} and from $\nu(C=O)$ at ca. 1720 cm^{-1}. In conjugation, for example, in C=C−C=O, the $\nu(C=C)$ absorption shifts to approximately 1600 cm^{-1} and the $\nu(C=O)$ absorption shifts to approximately 1675 cm^{-1}. Intermolecular forces also affect vibrational frequencies, again through perturbation of the electron density within a chemical bond (or set of bonds). For example, H-bonding lengthens and weakens O−H bonds such that $\nu(O-H)$ values vary widely (as does the band width). Interactions such as this tend to increase the frequency of deformation modes. Solution effects may also alter frequencies. Being highly polarized, carbonyl compounds interact with solvents, H-bonding with water and alcohols, and so on, with a resultant small shift (usually to lower wavenumber) of the band position.

Carbonyl compounds provide an excellent example of the effects of intramolecular phenomena such as induction (donation or loss of electron density from substituents), angle strain, and resonance upon bond strengths and therefore the positions of absorption bands. Inductive effects due to electron-withdrawing (−I) groups attached to the C atom compete with the C$^+$−O$^-$ resonance form and increase bond strength; hence, $\nu(C=O)$ increases, for example, $\nu(>C=O)$ in aldehydes $\approx$ 1720 cm^{-1} while $\nu(-COCl) \approx$ 1800 cm^{-1}. Electron-donating (+I) groups attached to the C atom stabilize the C$^+$−O$^-$ resonance form and reduce bond strength; hence, $\nu(C=O)$ decreases. Strain effects, where the angle formed between the two substituents is forced away from the preferred 120°, increase $\nu(C=O)$. This is particularly prevalent in cyclic systems. Mesomeric effects are those where one or some other forms of the resonance are stabilized by substituents or through interaction with other species. H-bonding may preferentially stabilize one of the mesomeric forms, for example (C=O$^-$ ... H−X ↔ C$^+$−O$^-$ ...

26.3 Interpretation of Infrared Spectra

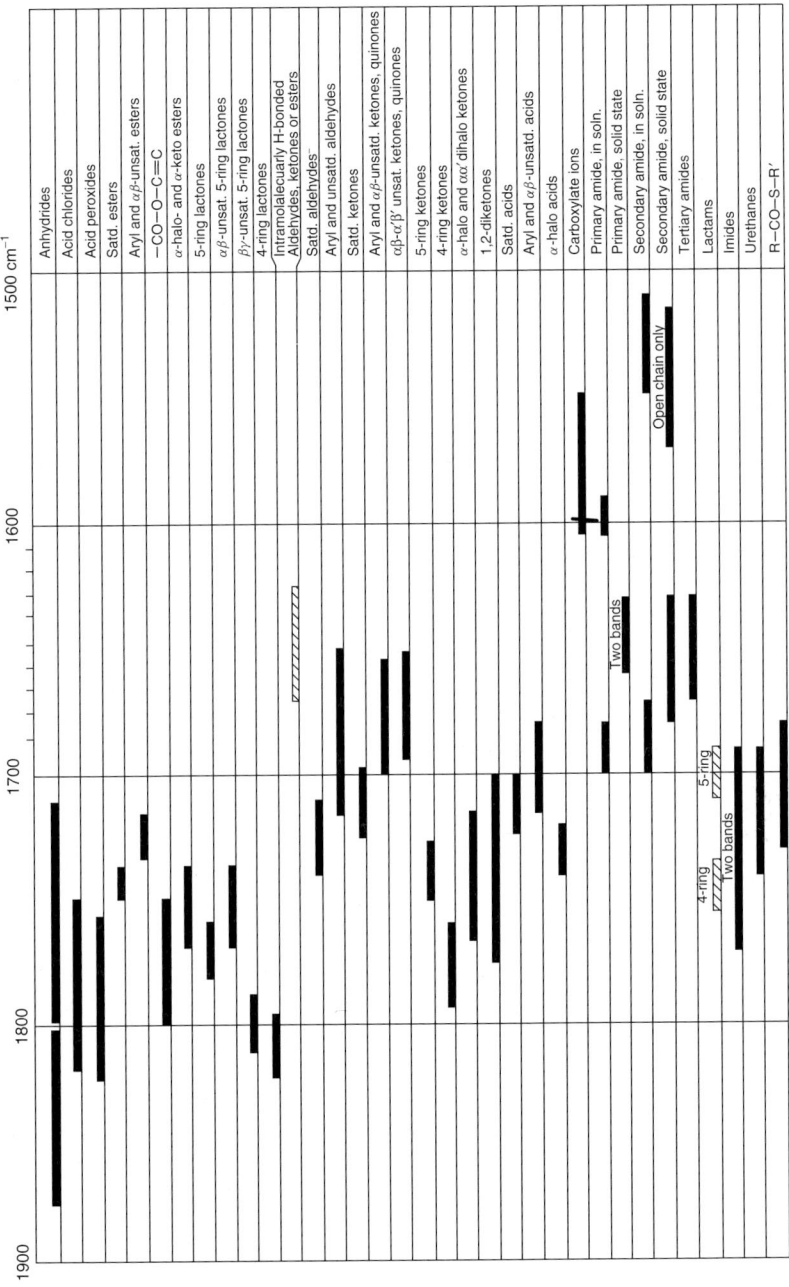

Fig. 26.22 Group frequency correlation table for the carbonyl region. Taken from Williams and Fleming (1995).

H—X), which again lowers the $\nu(C=O)$ frequency.

The above-mentioned factors are described in much greater depth in a range of textbooks (e.g., Williams and Fleming, 1995; Socrates, 2004). The information is often summarized in *group frequency correlation tables* such as that shown in Figure 26.22. A systematic approach is required for confidence in assignment. It is often easier to make correlations between the appearance of group modes if the spectrum is considered in a number of well-defined regions. Table 26.2 summarizes the regions in which the main functional group modes occur.

In larger molecules, a substantial proportion of the absorption bands is not easily assigned to vibrational modes of a specific functional group. These absorptions mainly occur below 1500 cm^{-1}, resulting in a congested spectral region. Bands that can sometimes be identified in this region include the C—C skeletal modes of organic molecules. The absorption pattern of molecules in this region is normally strongly characteristic, leading to its use as a fingerprint. However, there are functional group modes that occur in this region, some of which are readily identifiable by virtue of their intensity.

Of particular interest are the absorption bands that arise from the deformation modes of aromatic C—H moieties, especially the out-of-plane ring $\delta(CH)$ modes. These show large intensity as a result of both high amplitude of vibration of the H atoms and the large dynamic dipole that arises from the aromatic electron density. Characteristic patterns are observed for the various substitution patterns around aromatic rings. Typically, they lie between 985 and 750 cm^{-1}, with some modes being quite intense. Although they are generally rather insensitive to the nature of the substituent, the pattern of modes observed in the spectrum usually depends upon the number of adjacent hydrogen atoms on the ring (i.e., the number of substituents and where they are with respect to one

Tab. 26.2 Spectral regions in which typical functional group modes can be identified.

Spectral Region (cm^{-1})	Type	Comments/examples
4000–2300	Hydrogenic stretching modes	H lightest atom, no X=H bonds
2400–1900	Triple bonds and cumulated double bonds	Nitriles, $\nu(CN) \approx 2250$ cm^{-1}
2000–1400 cm^{-1}	Double bonds	Allenes, $\nu(-C=C=CH_2) \approx 1940$ cm^{-1} Cyano, $\nu(>C=N-) \approx 1680$ cm^{-1} Benzene rings $\approx 1600, 1580, 1500, 1470$ cm^{-1}
1600–1000 cm^{-1}	Deformation (bending) modes to hydrogen	Carbonyls, $\nu(>C=O) \approx 1900–1500$ cm^{-1} Methylene, $\delta(CH_2) \approx 1450–1480$ cm^{-1} Methyl, $\delta(CH_3) \approx 1450, 1380$ cm^{-1} amine, $\delta(NH_2) \approx 1600$ cm^{-1}
1500–400 cm^{-1}	Heavier atom stretches or light atom deformations	Aromatic out of plane $\approx 985–750$ cm^{-1} C—O single bonds $\approx 1300–1050$ cm^{-1}

Tab. 26.3 Vibrational frequencies for the δCH(op) in phase wagging mode.

Substitution	Wavenumber (cm^{-1})	Adjacent H
Mono	766–736	5 adjacent H[a]
o-di	758–744	4 adjacent H
m-di	791–773	3 adjacent H[a]
p-di	830–804	2 adjacent H

[a]Monosubstituted and m-disubstituted benzenes usually show the out-of-plane sextant ring bending mode at approximately 700 cm^{-1} as well.

another). This behavior is summarized in Table 26.3.

In summary, the interpretation of infrared spectra of single component species relies strongly upon the expertise of the analyst. However, once assigned, variations in band position, intensity, linewidth, or band shape from known reference materials in well-defined states convey a large quantity of subtle information.

Acknowledgments

The author gratefully acknowledges the cooperation of JASCO UK and PIKE Technologies for supplying the images used in a range of figures for this work. He also thank Juan Najera of the University of Manchester for prepublication copies of Figures 26.6 and 26.21.

References

Allara, D.L. and Nuzzo, R.G. (1985) *Langmuir*, **1**, 52.
Atkins, P.W. and de Paula, J. (2009) *Physical Chemistry*, 9th edn. Oxford University Press, Oxford.
Belton, P. and Wilson, R.H. (1990) Infrared sampling methods, in *Perspectives in Modern Chemical Spectroscopy*, (ed. D.L. Andrews), Springer-Verlag, Berlin.
Berreman, D.W. (1963) *Phys. Rev.*, **130**, 2193.
Bradshaw, A. and Schweizer, E.K. (1988) IRRAS studies of metal Surfaces, in *Spectroscopy of Surfaces*, (eds R.J.H. Clark and R.E. Hester), John Wiley & Sons, Ltd, Chichester.
Bohren, C.F. and Huffman, D.R. (1983) *Absorption and Scattering of Light by Small Particles*, John Wiley & Sons, Ltd, New York.
Buijs, H. (2002) in *Handbook of Vibrational Spectroscopy*, Vol. 1, (eds J. Chalmers and P.R. Griffiths), John Wiley & Sons, Ltd, Chichester.
Chabal, Y.J. (1986) *Surf. Sci.*, **168**, 594.
Chalmers, J. and Griffiths, P.R. (eds) (2002) *Handbook of Vibrational Spectroscopy*, John Wiley & Sons, Ltd, Chichester.
Chen, A. and Lipkowski, J. (1999) *J. Phys. Chem. B.*, **103**, 682.
Coates, J. (1998) in *Applied Spectroscopy: A Compact Reference For Practitioners*, (eds J.R. Workman Jr and Art.W. Springsteen), Academic Press, San Diego.
Fitzpatrick, J. and Reffner, J.A. (2002) in *Handbook of Vibrational Spectroscopy*, Vol. 2, (eds J. Chalmers and P.R. Griffiths), John Wiley & Sons, Ltd, Chichester.
Gazi, E., Dwyer, J., Lockyer, N.P., Gardner, P., Shanks, J.H., Roulson, J., Hart, C.A., Clarke, N.W., and Brown, M.D. (2007) *Anal. Bioanal. Chem.*, **387**, 1621.
Greenler, R.G. (1966) *J. Phys. Chem.*, **44**, 310.
Griffiths, P.R. and de Haseth, J.A. (1986) *Fourier Transform Infrared Spectrometry*, John Wiley & Sons, Ltd, New York.
Hannah, R.W. (2002) in *Handbook of Vibrational Spectroscopy*, Vol. 2, (eds J. Chalmers and P.R. Griffiths), John Wiley & Sons, Ltd, Chichester.
Harrick, N.J. (1967) *Internal Reflection Spectroscopy*, John Wiley Interscience, New York.
Heavens, O.S. (1955) *Optical Properties of Thin Solid Films*, Butterworths, London.
Heraud, P., Caine, S., Sanson, G., Gleadow, R., Wood, B.R., and McNaughton, D. (2006) *New Phytol.*, **173**, 216.
Hollas, J.M. (1992) *Modern Spectroscopy*, 2nd edn, John Wiley & Sons, Ltd, Chichester.
Hollins, P. (2000) Infrared reflection-absorption spectroscopy, in *Encyclopedia of Analytical Chemistry*, (ed. R.A. Meyers), John Wiley & Sons, Ltd, Chichester.

Horn, A.B. (1998) Infrared spectroscopic techniques for the study of thin interfacial films, in *Spectroscopy for Surface Science* (eds R.J.H., Clark and R.E., Hester), John Wiley & Sons, Ltd, Chichester.

Horn, A.B. and Sully, K.J. (1997) *J. Chem. Soc.-Faraday Trans.*, **93**, 2741.

Humecki, H.J. (ed.) (1995) *Practical Guide to Infrared Microspectroscopy. Practical Spectroscopy Series*, Vol. 19, Marcel Dekker, New York.

Jackson, R.S. (2002) in *Handbook of Vibrational Spectroscopy*, Vol. 1, (eds J. Chalmers and P.R. Griffiths), John Wiley & Sons, Ltd, Chichester.

Kauppinen, J. and Partanen, J. (2001) *Fourier Transforms in Spectroscopy*, Wiley-VCH Verlag GmbH, Berlin.

McIntyre, J.D.E. and Aspnes, D.E. (1971) *Surf. Sci.*, **24**, 417.

Menard, H., Tear, S.P., and Horn, A.B. (2008) *J. Phys. Cond. Matter.*, **20**, 355002.

Milosevic, M. and Berets, S.L. (2002) Accessories and sample handling for mid-infrared diffuse reflection spectroscopy, in *Handbook of Vibrational Spectroscopy*, Vol. 2, (eds J. Chalmers and P.R. Griffiths), John Wiley & Sons, Ltd, Chichester.

Mirabella, F.M. (2002) Principles, theory and practice of internal reflection spectroscopy, in *Handbook of Vibrational Spectroscopy*, Vol. 2, (eds J. Chalmers and P.R. Griffiths), John Wiley & Sons, Ltd, Chichester.

Najera, J.J., Fochesatto, J.G., Last, D.J., Percival, C.J., and Horn, A.B. (2009) *Rev. Sci. Instrum.*, **75**, 124102.

Pearce, H.A. and Sheppard, N. (1976) *Surf. Sci.*, **59**, 205.

Pereira, M. and Yarwood, J. (1996) *J. Chem. Soc. Faraday Trans.*, **92**, 2737.

Pritchard, J. and Sims, M.L. (1970) *Trans. Faraday Soc.*, **66**, 427.

Socrates, G. (2004) *Infrared and Raman Characteristic Group Frequencies: Tables and Charts*, 3rd edn, John Wiley & Sons, Ltd, Chichester.

Sommer, A.J. (2002) Mid-infrared transmission microspectroscopy, in *Handbook of Vibrational Spectroscopy*, Vol. 1, (eds J. Chalmers and P.R. Griffiths), John Wiley & Sons, Ltd, Chichester.

Theocharous, E. and Birch, R. (2002) in *Handbook of Vibrational Spectroscopy*, Vol. 1, (eds J. Chalmers and P.R. Griffiths), John Wiley & Sons, Ltd, Chichester.

Vincent, A. (2001) *Molecular Symmetry and Group Theory*, 2nd edn, John Wiley & sons, Ltd, Chichester.

Vogt, R. and Finlayson-Pitts, B.J. (1994) *J. Phys. Chem.*, **98**, 3747.

Williams, D.H. and Fleming, I. (1995) *Spectroscopic Methods in Organic Chemistry*, McGraw Hill Publishing, New York.

Willis, H.A. and Powell, D.B. (1990) Recent advances in vibrational spectroscopy, in *Perspectives in Modern Chemical Spectroscopy* (ed. D.L. Andrews), Springer-Verlag, Berlin.

Workman, J. Jr (1998) in *Applied Spectroscopy: A Compact Reference For Practitioners* (eds J.R. Workman Jr and Art.W. Springsteen), Academic Press, San Diego.

Zasetsky, A.Y., Earle, M.B., Cosic, B., Schiwon, R., Grishin, I.A., McPhail, R., Panescu, R.G., Najera, J.J., Khaliziv, A.F., Cook, A.B., and Sloan, J.J. (2007) *J. Quant. Spectrosc. Rad. Transf.*, **107**, 294.

Further Reading

Chalmers, J. and Dent, G. (1997) *Industrial Analysis with Vibrational Spectroscopy*, Royal Society of Chemistry, Cambridge.

Nakamoto, K. (1997a) *Infrared and Raman Spectra of Inorganic and Coordination Compounds. Part A: Theory and Applications in Inorganic Chemistry*, Wiley Interscience, New York.

Nakamoto, K. (1997b) *Infrared and Raman Spectra of Inorganic and Coordination Compounds. Part B: Applications in Coordination, Organometallic and Bioinorganic Chemistry*, Wiley Interscience, New York.

Sodeau, J.R. (1995) Atmospheric cryochemistry, in *Spectroscopy in Environmental Science* (eds R.J.H. Clark and R.E. Hester), John Wiley & Sons, Ltd, Chichester.

Tranter, G.E. and Holmes, J.L. (eds.) (2007) *Encyclopedia of Spectroscopy and Spectrometry*, Elsevier, Amsterdam.

Workman, J.R. Jr and Springsteen, Art.W. (eds) (1998) *Applied Spectroscopy: A Compact Reference For Practitioners*, Academic Press, San Diego.

27
Nuclear Magnetic Resonance Spectrometry

Vladimir I. Bakhmutov

27.1	Introduction	935
27.2	**NMR: General Considerations**	936
27.2.1	Nuclei in the External Magnetic Field	936
27.2.2	Detection of an NMR Signal – General Principles	938
27.2.3	NMR Equipment	939
27.2.3.1	Magnets	939
27.2.3.2	NMR Probes	940
27.3	**NMR Spectral Parameters**	940
27.3.1	Chemical Shift	940
27.3.1.1	Anisotropy of Chemical Shift	942
27.3.1.2	Quadrupolar Splitting	942
27.3.2	Influence of an Unpaired Electron on the Chemical Shift	942
27.3.3	Spin–Spin Coupling	943
27.3.4	Nuclear Relaxation	945
27.3.4.1	Relaxation Mechanisms in Liquids	945
27.3.4.2	Paramagnetic Relaxation	947
27.3.4.3	Nuclear Relaxation in Solids: Spin Diffusion	947
27.3.4.4	Relaxation Time Measurements	948
27.4	**Spectroscopic NMR Experiments in Liquids**	948
27.4.1	Routine One-Dimensional (1D) NMR spectra	948
27.4.2	Polarization Transfer in NMR (SPI, INEPT, and DEPT)	949
27.4.3	Two-Dimensional (2D) NMR Spectra	950
27.4.4	Chemical Exchanges: the NMR Time Scale	950
27.4.5	Translational Molecular Diffusion via NMR	951
27.4.6	Isomeric and Conformational Analysis in Solutions by NMR	953
27.5	**NMR Spectroscopic Experiments in the Solid State**	954
27.5.1	Wide-Line NMR	954
27.5.1.1	Dead Time in Wide-Line NMR	955
27.5.1.2	Relaxation Experiments in Solids	955
27.5.2	High-Resolution Solid-State NMR	956
27.5.2.1	High-Power Decoupling and Double Resonance	956

Encyclopedia of Applied Spectroscopy. Edited by David L. Andrews.
Copyright © 2009 WILEY-VCH Verlag GmbH & Co. KGaA, Weinheim
ISBN: 978-3-527-40773-6

27.5.3   Cross Polarization Technique   957
27.5.4   Internuclear Distances from the Solid-State NMR Data   957
27.5.5   Zero-Field Solid-State NMR Experiments   958
27.6   Sensitivity Enhancement by Dynamic Nuclear Polarization (DNP)   959
27.7   Imaging NMR   960
            References   961
            Further Reading   962

27.1
Introduction

The NMR spectrometry is an extraordinarily powerful analytical method, which originally had applications in physics and later to chemistry, biochemistry, biology, pharmacy, geophysics, materials science, and archaeology. Nowadays, this method is widely used in veterinary science and medicine, particularly in clinical research for routine human brain NMR imaging (magnetic resonance imaging (MRI)).

As a method, the NMR spectroscopy is based on resonance phenomenon independently discovered by Bloch et al. (1946) and Purcell, Torrey, and Pound (1946), which involves placing nuclear spins in an external magnetic field and their excitation by a radio-frequency irradiation. The energy, thus absorbed by the nuclei, measured in frequency units, is characteristic of the nuclei and their environments and can be used for the structural analysis of the compounds investigated.

Initially, the NMR experiments were performed by physicists to experimentally verify the basic postulates of quantum mechanics that require detailed knowledge and accurate measurements of nuclear magnetic moments. However, the development of the relaxation theory (Bloembergen, Purcell, and Pound, 1948) and also the discovery of chemical shift (Proktor and Yu, 1950) and spin–spin coupling patterns of nuclei have shown the great potential of NMR to describe solids, liquids, and gaseous matter in terms of structural and dynamic properties. Then, fast successes in the radio-frequency technology resulted in the appearance of the first commercial NMR spectrometers (1953), which strongly impacted the research in chemistry and biochemistry. In turn, more complex compounds, such as biochemical molecular systems and aggregates, required more powerful NMR spectrometers, modifications in magnet technology, and developments of new spectroscopic approaches. Therefore, the modern commercial NMR spectrometers represent multiple-pulse devices, capable of experiments in solutions and the solid state, which can operate at very high magnetic fields up to 21.14 T and detect nuclei with a small natural abundance (rare nuclei). In addition, increasing the speed of computers, performing very fast Fourier transformations (FTs), has resulted in routine use of the two- or three-dimensional NMR experiments.

Encyclopedia of Applied Spectroscopy. Edited by David L. Andrews.
Copyright © 2009 WILEY-VCH Verlag GmbH & Co. KGaA, Weinheim
ISBN: 978-3-527-40773-6

There are numerous techniques and approaches in NMR spectrometry. They are constantly modified and published in original articles, books, and the NMR Encyclopedia. Even a small portion of the NMR methods cannot be covered in this relatively short chapter, the aim of which is to present only the general principles of NMR experiments and their practical applications.

27.2
NMR: General Considerations

After publications of the pioneering works, describing NMR, its practical applications can be grouped under two categories: NMR spectrometry and NMR relaxation. In the first group, an NMR spectrum is studied, which is a frequency-dependent pattern, showing a number of nonequivalent nuclei and their environments. In contrast, the resulting data, collected by the relaxation measurements, are time-dependent and are rather related to dynamics in molecular systems. The first application is more popular among practical chemists or biochemists while the relaxation aspects attract researchers working in molecular physics, physics of solids, and biophysics. However, it should be emphasized that the frequency- and time-dependent data are the essence of the same phenomenon, and knowledge of nuclear relaxation is needed even for recording a simplest NMR spectrum because obtaining a spectrum itself can become difficult or impossible, when relaxation times are extremely long. It is probable that the first attempt to observe NMR for ^{1}H and ^{7}Li nuclei in 1936 was unsuccessful (Gorter et al., 1942) because of long relaxation times. In contrast, short relaxation times broaden NMR lines and often mask their fine structures. Sometimes, the broadenings are very strong and lines become "invisible." This is obviously important even for users applying NMR only for an express structural analysis.

27.2.1
Nuclei in the External Magnetic Field

According to Abragam (1961), a nucleus with nonzero angular momentum, P, possesses a nuclear magnetic moment, $\mu = \gamma P$, capable of interaction with the external magnetic field, where γ is the nuclear magnetogyric ratio, which depends on the nature of nuclei (Table 27.1). The P and the μ are quantized. For example, the projections of the angular nuclear momentum, P_Z, on the OZ axis can be written as

$$P_Z = \hbar \times m_I \qquad (27.1)$$

where h ($\hbar = h/2\pi$) is the Planck's constant ($h/2\pi = 1.05457266 \times 10^{-34}$ J s) and m_I is the magnetic quantum number. The latter depends on nuclear spin I and takes values from I to $-I$: $I, I-1, I-2, \ldots, -I$. In turn, I is a multiple of $\frac{1}{2}$, for example, at $I = \frac{1}{2}$, the angular momentum and the magnetic moment are

$$P_Z = \pm \left(\frac{1}{2}\right) \hbar \text{ and } \mu_Z = \pm \left(\frac{1}{2}\right) \gamma \hbar$$

$$(27.2)$$

Then, the μ_Z (or P_Z) projections, represented as vectors, can be parallel or antiparallel relative to the above Z axis. If the external magnetic field B_0 is applied along this axis, the μ_Z (or P_Z) orientations become energetically nonequivalent. This nonequivalency leads to the, so-called, Zeeman's energy splitting ΔE

Tab. 27.1 NMR properties of some nuclei.

Nucleus	Spin I	Natural abundance (%)	NMR frequency, ν_0, in megahertz at $B_0 = 2.3488$ T	γ (10^7 rad T^{-1} s^{-1})	Sensitivity relative to ^{1}H
^{1}H	1/2	99.98	100	26.752	1.000
^{2}H	1	0.016	15.35	4.107	1.45×10^{-6}
^{11}B	3/2	80.42	32.08	8.584	0.133
^{13}C	1/2	1.108	25.14	6.728	1.76×10^{-4}
^{14}N	1	99.63	7.22	1.934	1.00×10^{-3}
^{19}F	1/2	100	94.08	25.18	0.834
^{31}P	1/2	100	40.48	10.84	0.065
^{17}O	5/2	0.037	13.557	-3.628	1.08×10^{-5}
^{93}Nb	9/2	100	24.549	6.567	0.487
^{117}Sn	1/2	7.61	35.63	-9.578	3.49×10^{-3}
^{119}Sn	1/2	8.58	37.29	-10.02	4.51×10^{-3}
^{199}Hg	1/2	16.84	17.91	4.815	9.82×10^{-4}
^{205}Tl	1/2	70.5	57.63	15.589	0.140

(Figure 27.1) proportional to the μ_Z magnitude and the strength of the applied magnetic field, B_0:

$$\Delta E = 2\,\mu_Z B_0 \qquad (27.3)$$

Then, formulating ΔE, as an energy quantum, $\Delta E = h\nu$, leads to the resonance condition written as

$$h\nu_0 = 2\,\mu_Z B_0 = \gamma \hbar B_0$$
$$\nu_0 = \gamma B_0 \qquad (27.4)$$

This condition shows that the nuclear spin, placed into the external magnetic field, B_0, undergoes a single-quantum transition (i.e., the m_I number changes from $-1/2$ to $+1/2$), when an irradiation frequency reaches ν_0 (the Larmor's frequency). The transition results in an energy absorption by lattice, which is registered as an NMR signal. Table 27.1 shows the ν_0 frequencies for different nuclei.

An ensemble of nuclei will populate the Zeeman's energy levels according to factor $\exp(-\Delta E/kT)$, where T is the temperature and k is the Boltzmann's constant (1.380658×10^{-23} J K^{-1}). Thus the low-energy levels are more populated than the high levels, which give raise to macroscopic magnetization, undergoing precession around the B_0 direction

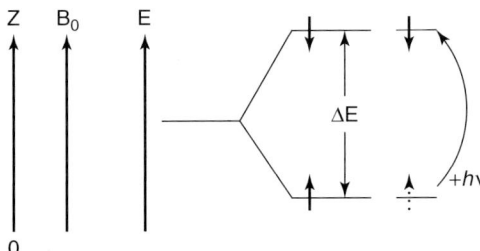

Fig. 27.1 The Zeeman energy levels of nuclear spins $I = 1/2$ in the external magnetic field B_0.

at the ν_0 frequency, responsible for observations of NMR signals. It should be emphasized that the population difference, ΔN, compared with a total number of the nuclei, N, is less: a $\Delta N/N$ ratio is of 8×10^{-6} for ^{1}H nuclei in a field of 2.35 T at 300 K. This condition defines relatively low sensitivity of NMR experiments. As seen in Table 27.1, ^{1}H and ^{19}F nuclei are mostly preferable as target nuclei in this context.

Nuclei with spin $> \frac{1}{2}$ behave similarly in the external magnetic filed. However, a number of the Zeeman's levels increases accordingly. (Such nuclei have nonspherical charge distributions responsible for appearance of nuclear quadrupole moments Q). For example, ^{2}H or ^{14}N nuclei ($I = 1$) give three energetically equidistant levels, corresponding to the μ_Z values of $+\gamma\hbar(1)$, 0 and $-\gamma\hbar(1)$. Because the double quantum transitions (the m_I value changes from +1 to −1) are not observed, the nuclei again produce a single line in an NMR spectrum.

27.2.2
Detection of an NMR Signal – General Principles

Technically, observation of an NMR signal can be realized at the ν_0 irradiation of a sample by the continuous wave (CW) method. The principle schemes of such CW spectrometers can be found in Harris (1986), Sanders and Hunter (1994), or Duel (2002). Here, a spinning sample is situated between the poles of a magnet, radio frequency is applied via an antenna coil, connected to a transmitter, and a receiver coil, surrounding the sample, registers the absorbed energy by electronic devices and a computer. The CW spectrometers imply a sweeping of the magnetic field or the radio frequency when the external magnetic field remains constant.

A simple NMR experiment, where the ν_0 irradiation is performed by a radio-frequency pulse (RFP), is shown in Figure 27.2. The experiment includes three time sections: (i) relaxation delay (RD), (ii) action of RFP, and (iii) recording the NMR data as free induction decays (FID)s, during acquisition time (AT). According to the scheme, the NMR data can be accumulated, when sections from (i) to (iii) are repeated. The resulting FIDs are time-dependent, $f(t)$, and should be converted into the NMR spectra, as the frequency-dependent data, $F(\nu)$, by FT:

$$F(\nu) = \int f(t) \, \exp(-i2\pi \nu t) \, dt$$

$$f(t) = \int F(\nu) \, \exp(+i2\pi \nu t) \, d\nu \quad (27.5)$$

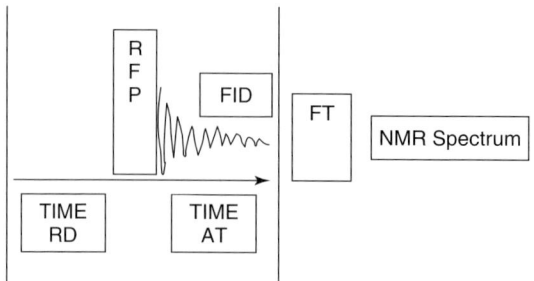

Fig. 27.2 Schematic presentation of the 1D NMR experiment where typical durations of radio-frequency pulse (RFP), relaxation delay (RD), and acquisition time (AT) are of 5–10 microseconds, 1–4 seconds, and 1–3 seconds, respectively.

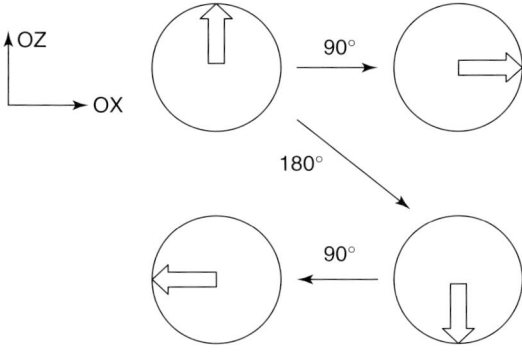

Fig. 27.3 The schematic presentation of the behavior of the macroscopic magnetization vector, **M**, in a rotating coordinate system after action of 90°- and 180°-radio-frequency pulses or two consequent 180°- and 90°-radio-frequency pulses.

Such an experiment is realized on an FT NMR spectrometer equipped with a computer capable of fast FT. It should be emphasized that durations and amplitudes of RFP should be sufficient to rotate the macroscopic magnetization, for example, by an angle α from the magnetic field B_0. According to Ernst et al. (1966), the pulse angle, α, is defined as $\alpha = \gamma B_1 t_P$, where B_1 is the power of the pulse and t_p is its duration. Then, the magnetization goes to thermal equilibrium. Figure 27.3 schematically shows the behavior of the macroscopic magnetization vector, **M**, in a rotating coordinate system due to actions of 90°- or 180°-RFPs. Here, the **M** orientation along the positive OX axis corresponds to the maximal intensity of a registered NMR signal. Thus, after a 180° pulse, the signal is not observed or its intensity becomes negative after a combination of 180 and 90° pulses.

27.2.3
NMR Equipment

As shown above, an NMR spectrometer includes the following basic units: a radio-frequency-reference generator, a transmitter, a receiver, a pulse generator, a computer, a magnet, and an NMR probe. All of the units remarkably affect the quality of the NMR spectra and the signal/noise ratio as a measure of sensitivity. However, for experimentalists, a magnet and an NMR probe are most important.

27.2.3.1 Magnets

A magnet is a source of strong, stable, and homogeneous magnetic field to obtain the high-resolution NMR spectra. When characteristics of transmitters, receivers, and other devices of NMR spectrometers do not change, then increasing the magnetic field, B_0, increases the energy, ΔE, population difference $\Delta N/N$ and hence the sensitivity of an NMR spectrometer. Iron magnets (permanent or electromagnets) are relatively unstable, sensitive to the temperature, and create moderate field strengths, corresponding to the working ^{1}H frequency of ≤100 MHz. In contrast, modern superconducting electromagnets show excellent stability, good homogeneity and their fields are capable of working at the proton frequency up to 900 MHz. For both types of magnets, stability of the magnetic field needed for the high-resolution NMR experiments can be additionally increased by field-locking. Because deuterium NMR signals coming from deuterated solvents can be well filtered from signals of ^{1}H or other nuclei, they are successfully applied for additional field stabilization.

27.2.3.2 NMR Probes

Generally, an NMR probe represents a single tank circuit needed for irradiation and detection of an NMR signal. According to Kisman and Armstrong (1974), the radio-frequency, coming from the transmitter, must be distinguished from the signal, coming from the probe. In other words, a weak signal has to be measured immediately after the action of a high-power pulse. Such a switching problem gives rise to the, so-called, deed time of the spectrometer. If the deed time is long and the FID is short, the NMR signal loses intensity. Technically, this problem is minimized in modern NMR spectrometers and their standard components reduce the deed times to few microseconds.

The design of NMR probes significantly depends on desired NMR experiments. For example, variable-temperature measurements (usually between -120 and $+120\,°C$) require the corresponding heating/cooling equipments, which provide accurate temperature settings and the absence of significant temperature gradients. Double or triple resonance and gradient-field experiments are also available when NMR probes have a special design. Finally, an NMR probe should be equipped with tools for the mechanical rotation of samples. Low spinning rates (approximately 20 Hz) are generally used for liquids to compensate the nonhomogeneity of the external magnetic field. Very high spinning rates (tens of kilohertz) are needed to suppress strong dipole–dipole interactions in solids. In the latter case, the NMR rotors are oriented at the, so-called, magic angle setting relative to the external magnetic field (see below).

27.3
NMR Spectral Parameters

Nuclear spins in molecules are not isolated and this circumstance gives rise to the phenomena of chemical shift, spin–spin coupling, and relaxation, directly related to chemical structures. In turn, these spectral parameters lead to reliable structural determinations even in the complex cases providing, thus, wide applications of the NMR spectroscopy as an analytical method.

27.3.1
Chemical Shift

Electrons, circulating around nuclei, placed into the external magnetic field, B_0, create a new magnetic field, opposing the initial field. Then, a local magnetic field, B_{loc}, modifies the resonance frequency ν_0, which leads to shielding constant σ, $B_{loc} = (1 - \sigma)B_0$, or chemical shift $\delta = -\sigma$. Chemical shift δ (Figure 27.4) is measured in parts per million and referred to *internal (or external) standards*: $\delta = \{\nu \text{ (sample)} - \nu \text{ (reference)}\} 10^6/\nu \text{ (reference)}$ (generally to TMS for ^{1}H and ^{13}C nuclei, H_3PO_4 for ^{31}P nuclei, CF_3COOH for ^{19}F nuclei, and so on).

According to Abragam, three main terms contribute to total constant σ: diamagnetic term σ^{local}(dia), paramagnetic term σ^{local}(para), and magnetic anisotropy term σ^*. The first contribution, caused by the circulation of

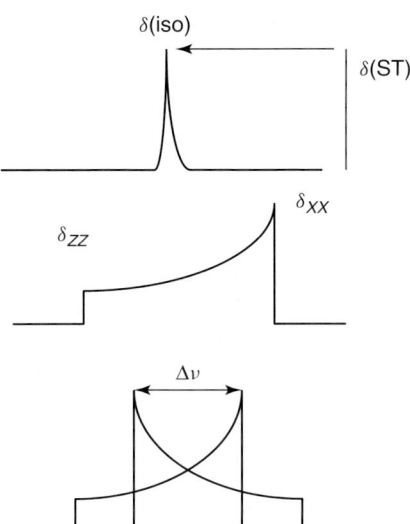

Fig. 27.4 The line shape of an NMR resonance typical of a liquid (a), a static solid with an axially symmetrical screening tensor, $\sigma_{ZZ} \neq \sigma_{XX} = \sigma_{YY}$, (b) and a powder ^{2}H sample with an axially symmetric electric field gradient at deuterium.

unperturbed spherical electrons (s) with mass m_e, is expressed in a simplest form by Lamb's equation:

$$\sigma^{\text{local}}(\text{dia}) = \left(\frac{\mu_0 e^2}{8\pi m_e}\right) \int r\rho_e \, dr \quad (27.6)$$

where μ_0 is the permeability constant, ρ_e is the charge density, and r is the electron-nucleus distance. The second term, formulated by McConnel et al. (1958), appears due to perturbed non-spherical electrons (p or d) and it can be described via the average energy of the electron excitation, ΔU, as

$$\sigma^{\text{local}}(\text{para}) = -\left(\frac{\mu_0 e^2 \hbar^2}{6\pi m^2 \Delta U}\right) \langle r^{-3} \rangle P_U$$

(27.7)

Here P_U is named as the p-electron "imbalance" expressed via charge densities and bond orders (Jameson and Gutowsky, 1964). The last (relatively small) term, σ^*, comes from electrons situated at neighboring groups. These

Tab. 27.2 Chemical shift ranges, δ (SCS), for different nuclei.

Nucleus	SCS (parts per million)[a]
^{1}H	−50 to +20
^{11}B	−130 to +95
^{13}C	−300 to +250
^{19}F	−250 to +400
^{31}P	−125 to +500
^{15}N	−500 to +850
^{195}Pt	−1370 to +12 000

[a] The positive values correspond to low-field displacements.

groups (for example, bonds C=C or C≡C) create magnetically anisotropic fields and therefore the observed effects depend on orientations of observed nuclei. Combinations of these terms lead to large variations in chemical shifts δ, which depend on the nature of nuclei (Table 27.2). A correct quantum-chemical analysis of chemical shifts is possible, but the calculations are not simple. Therefore, in practice, structural determinations or predictions of the δ values for known structures are based on numerous spectrum/structure relationships or increment equations, which can be found in Harris (1986) or Sanders and Hunter (1994). Finally, the so-called, isotopic chemical shifts (for example, for ^{1}H at ^{12}C and ^{1}H at ^{13}C) are relatively small and their role in the structural analysis is not significant.

27.3.1.1 Anisotropy of Chemical Shift

Screening constants σ are three dimensional and can be characterized by components σ_{XX}, σ_{YY}, and σ_{ZZ} that are also associated with chemical structures (Strub et al., 1983). However, generally, this coordinate system does not correspond completely to a molecular coordinate system, though the Z coordinate is often situated along chemical bonds.

Generally, the σ components differ leading to chemical shift anisotropy, $\Delta\sigma$, defined as $\Delta\sigma = \{2\sigma_{ZZ} - (\sigma_{XX} + \sigma_{YY})\}/3$. The high-amplitude molecular motions in liquids produce the Lorenz-shaped resonances, where the σ_{XX}, σ_{YY}, and σ_{ZZ} components are not observed. Here, the resonances are characterized by isotropic chemical shifts $\delta(ISO)$ (Figure 27.4) or isotropic screening constants $\sigma(ISO)$ determined as $\sigma(ISO) = (1/3)(\sigma_{XX} + \sigma_{YY} + \sigma_{ZZ})$. More complex lines are observed in rigid solids and their shapes directly depend on the σ (or δ) components that are determined from static NMR patterns (Duel, 2002). Figure 27.4 shows a solid-state NMR signal, typical of an axial-symmetrical screening-tensor where $\sigma_{XX} = \sigma_{YY} \neq \sigma_{ZZ}$. A lower symmetry ($\sigma_{XX} \neq \sigma_{YY} \neq \sigma_{ZZ}$) obviously leads to more complicated signals. This is particularly important for heavy nuclei such as ^{19}F, ^{31}P or ^{117,119}Sn.

27.3.1.2 Quadrupolar Splitting

Spins of nuclei, having quadrupole moments Q (for example ^{2}H, ^{14}N, ^{81}Br, ^{127}I, and ^{35}Cl), interact not only with the external and local magnetic fields but also with electric field gradients (EFG) at these nuclei. The energy of interactions is expressed through nuclear quadrupole coupling constant (NQCC), $NQCC = e^2q_{zz}Q/h$, where eq_{ZZ} is the principal component of the EFG tensor. In rigid solids, the quadrupolar interaction directly affects the shapes of NMR resonances. Figure 27.4 shows a typical ^{2}H NMR spectrum, recorded in a static powder sample, where the signal exhibits the, so-called, quadrupolar splitting, measured in frequency units (kilohertz) and marked as $\Delta\nu$. This shape is observed for systems with the axially symmetric EFG. In the absence of motions, the splitting provides determinations of the NQCC values via $\Delta\nu = 3/4(e^2q_{ZZ}Q/h)$. In liquids, however, the quadrupolar interaction is not observed because of the fast molecular motions and NMR signals become again Lorenz-shaped.

27.3.2 Influence of an Unpaired Electron on the Chemical Shift

The very high magnetic moment of an unpaired electron affects an NMR frequency via effective magnetic field B_{EF}:

$$B_{EF} = 2\pi A \mu_B g B_0 S \frac{(S+1)}{3\gamma_N kT}$$

$$+ \mu_0 \mu_B^2 g^2 B_0 S(S+1)$$

$$\times \frac{(3\cos^2\theta - 1)}{3kTr^3} + B_{BMS} \quad (27.8)$$

Here, according to Poople (1983), A is the isotropic electron/nucleus hyperfine coupling constant, μ_B is the Bohr magneton (9.2741×10^{-24} J T^{-1}), r is the electron/nucleus distance, θ is the angle between the electron-nucleus vector and the external magnetic field B_0, and g is the electron g factor. The first term produces the, so-called, Fermi-contact (FC) chemical shifts (McConnel et al., 1958). Owing to the direct delocalization of the unpaired electron density on the nuclei, the FC shifts in solids can reach 3000 ppm or higher as a function of the nature of the nuclei and paramagnetic centers. Generally, the chemical shift tensors are axially symmetric at large anisotropies covering ranges up to 3500–4000 ppm (Mali et al., 2005). In a combination with short relaxation times, the nuclei affected by FC interaction are difficult to observe experimentally. In liquids, in spite of the fast molecular motions, the FC shifts can be also significant (up to 100 ppm), mainly due to the very large electron/nucleus coupling.

The second field, causing smaller shifts, corresponds to dipolar electron/nucleus interaction. The interaction strongly depends on distance r and includes an angle-dependent part, ($3\cos^2\theta - 1$). In liquids, where molecular reorientations with respect to the external magnetic filed are fast, θ is defined as an angle formed by the r vector and an axis of the symmetry of the g factor (Harris, 1986). The corresponding chemical shifts are named as *pseudo-contact*, which are again associated with chemical structure. The last term, B_{BMS}, represents a demagnetization field appearing in a sample due to bulk magnetic susceptibility (BMS). This effect, generally insignificant in diamagnetic samples, plays a very important role in paramagnetic systems (Kubo, Spaniol, and Terao, 1998).

27.3.3
Spin–Spin Coupling

According to the theory, which can be found in Abragam (1961) or Harris (1986), the spin Hamiltonian, describing NMR in isotropic fluids in the limits of the quantum-mechanical formalism, takes the form:

$$h^{-1}\hat{H} = -\sum_j (2\pi)^{-1}\gamma_j B_0 (1-\sigma_j)\hat{I}_j^z$$
$$+ \sum_{j<k} J_{jk} \hat{I}_j \hat{I}_k \quad (27.9)$$

The first part, modified by screening constant σ, corresponds to isolated nuclear spins in the field, B_0 (see Eq. (27.4)). The second term is spin–spin coupling and J_{jk} is the spin–spin coupling constant. Because the J_{jk} constants are relatively small (typically from 1–2 to 200–300 Hz), spin–spin coupling is considered as perturbation of the first term.

Effects of perturbation on a number of Zeeman's energy levels can be easily shown by attributing the symbols α and β to the parallel and antiparallel mutual orientations for nuclear spins of $1/2$. Then, the spin states for a system, consisting of two nuclei, are $\beta\beta(m_I = -1)$, $\alpha\beta(m_I = 0)$, $\beta\alpha(m_I = 0)$, and $\alpha\alpha(m_I = 1)$. Thus, the first nuclear spin can "see" the second one in two states and its energy levels undergo an additional spitting by energy J measured in hertz. The J constant is a scalar magnitude, has a sign, and is independent of the external magnetic field B_0.

Two magnetically nonequivalent nuclear spins, A and X, form a simple spin system, AX, when the chemical shift difference $\Delta\delta^{AX}$ (measured in hertz) is significantly larger than the spin–spin coupling constant, $J(A-X)$: $\Delta\delta^{AX}$(hertz)/J^{AX}(hertz) > 10. Generally, line multiplicity N of the A resonance is $N = 2nI + 1$, where n is the number of the X nuclei with spin I. Relative intensities within multiplets are distributed according to the so-called Pascal's triangle. For two protons ($I = 1/2$), the AX spin system shows two doublets at 1:1 intensity ratio for each doublet. In these terms, group CH_2-CH_3 should exhibit two ^1H resonances as a quartet and triplet, respectively. Decreasing the $\Delta\delta^{AX}$(hertz)/J^{AX}(hertz) ratio transforms the AX spin system to a strongly coupled system, named as AB, to emphasize the small chemical shift difference. Similarly, simple spin systems AX_2 or AXX' are converted to strongly coupled systems ABX, AMX, AB_2 or ABC, which show complex NMR patterns. Owing to the presence of combinational transitions, their intensities and multiplicities depend on values $\Delta\delta$ and J. Such systems can be analyzed via simulations or fittings to the experimental data by convenient computer programs, using quantum-mechanical formalism. Because the J constant does not depend on B_0, the strongly coupled spin systems are simplified in higher external magnetic fields.

Spin–spin coupling is transmitted via electrons, surrounding the nuclei, and therefore the J constant is a function of chemical structures: electron distributions, atomic hybridizations, and interbond angles. Generally, the total J energy consists of three contributions coming from the spin-orbital (SO), spin-dipolar (SD), and FC interactions (Abragam, 1961) written as

$$H_{SO} = \left(\frac{\mu_0}{4\pi}\right) g_L \mu_B \gamma_N \hbar \sum 2 L_e I_N r_{eN}^{-3}$$
(27.10)

$$H_{SD} = -\left(\frac{\mu_0}{4\pi}\right) g_S \mu_B \gamma_N \hbar \sum \{r^{-3}{}_{eN} I_N S_e$$
$$- 3 r^{-5}{}_{eN}(I_N r_{eN})(S_e r_{eN})\} (27.11)$$

$$H_{FC} = \left(\frac{2}{3}\right) \mu_0 g_S \mu_B \gamma_N \hbar \sum \delta(r_{eN}) I_N S_e$$
(27.12)

where e and N correspond to electron and nucleus, respectively, L_e is the orbital angular moment operator, r_{eN} is the distance between electron and nucleus, and $\delta(r_{eN})$ is the Dirac delta function. The function takes a value of 1 at $r_{eN} = 0$ and 0 otherwise, requiring, thus, the presence of electron density directly on the nuclei. Calculations show that FC contribution usually dominates (Table 27.3) and hence J constants can be interpreted in terms of electron densities.

Spatial proximity of some nuclei, for example, ^{19}F, leads to the, so-called, through-space spin–spin coupling constants (sometimes reaching 100 Hz and more). Here, according to Contreras et al. (2000), again FC interaction dominates

Tab. 27.3 Contributions to total spin–spin coupling constants (in hertz).

Coupling constant	Molecule	FC	SO	SD
$^1J(C-C)$	C_2H_6	11.99	−0.55	0.3
	C_2H_2	69.43	4.13	2.84
$^1J(F-C)$	CH_3F	−126.30	−4.15	6.92
$^2J(F-F)$	CF_4	−1.09	6.20	16.73

but the coupling goes via nonbonding electron orbitals.

27.3.4
Nuclear Relaxation

The nuclear spins, excited by a radio-frequency irradiation, are capable of an energy exchange with their environments to reach the initial equilibrium state. There are two principally different relaxation mechanisms. The first mechanism, named as *spin-lattice relaxation*, represents the energetic exchange between nuclear spins and lattice, formulated as a continuum of nuclear and/or electron magnetic moments of any sort. The second mechanism, named as *spin–spin relaxation*, occurs when two (or more) spins undergo "flip–flop motions," which shorten, thus, the spin lifetimes while the total energy of the system does not change (for this reason, such flips are often named as the *energy-conserving spin transitions*. Finally, the relaxation of macroscopic magnetization is given by the Bloch's equations:

$$\frac{dM_Z}{dt} = \frac{-(M_Z - M_Z^0)}{T_1} \text{ and}$$

$$\frac{dM_{X,Y}}{dt} = \frac{-M_{X,Y}}{T_2} \quad (27.13)$$

and formulated as recovery of the Z-(longitudinal) component with a time constant T_1 and the Y- and X-(transverse) components with a time constant T_2.

27.3.4.1 Relaxation Mechanisms in Liquids

Nuclear relaxation will be effective if magnetic fields in the spin environments fluctuate at frequencies close to the Larmor's resonance frequency ν_0 (or $\omega_0 = 2\pi\nu_0$). In liquids, such fluctuations are directly associated with thermal molecular motions: rotational reorientations, translational motions, or their combinations. Rapid intramolecular motions also contribute to nuclear relaxation. Molecular motions are characterized by correlation times τ and activation energies ΔE that can be determined from the relaxations data if the mechanism of the process is known.

According to the theory of Bloembergen, Purcell and Pound (1948), nuclei can relax via dipole–dipole, quadrupole, spin-rotation, scalar, and chemical shift anisotropy interactions. Generally, the relaxation rates are written as $1/T_{1,2} = CJ(\omega_0, \tau)$, where C is the coupling strength and $J(\omega_0, \tau)$ is the spectral density function. Relaxation due to dipolar interaction of two identical nuclei can be expressed as

$$\frac{1}{T^{DD}_1} = \left(\frac{2}{5}\right) \gamma^4 \hbar^2 r^{-6} I(I+1)$$

$$\times \left\{ \frac{\tau}{(1+\omega_I^2\tau^2)} + \frac{4\tau}{(1+4\omega_I^2\tau^2)} \right\} \quad (27.14)$$

$$\frac{1}{T^{DD}_2} = \left(\frac{1}{5}\right) \gamma^4 \hbar^2 r^{-6} I(I+1)$$

$$\times \left\{ 3\tau + \frac{5\tau}{(1+\omega_I^2\tau^2)} + \frac{2\tau}{(1+4\omega_I^2\tau^2)} \right\} \quad (27.15)$$

where ω_I is the resonance frequency. This mechanism is particularly powerful in strong dipolar coupling, $\gamma^4\hbar^2 r^{-6}I(I+1)$, particularly at short internuclear distances r and at molecular motional frequencies close to ω_I. Similarly, relaxation of nucleus A via hetero-nuclear dipolar coupling by nucleus B is described by the following equations:

$$\frac{1}{T_1} = \left(\frac{4}{30}\right)\left(\frac{\mu_0}{4\pi}\right)^2 r(A-B)^{-6}\gamma_A^2\gamma_B^2$$

$$\times \hbar^2 I_B(I_B+1)\left\{\frac{3\tau}{(1+\omega_A^2\tau^2)}\right.$$

$$+\frac{6\tau}{(1+(\omega_A+\omega_B)^2\tau^2)}$$

$$\left.+\frac{\tau}{(1+(\omega_A-\omega_B)^2\tau^2)}\right\} \quad (27.16)$$

$$\frac{1}{T_2} = \left(\frac{4}{30}\right)\left(\frac{\mu_0}{4\pi}\right)^2 r(A-B)^{-6}\gamma_A^2\gamma_B^2$$

$$\times \hbar^2 I_B(I_B+1)\left\{4\tau\right.$$

$$+\frac{3\tau}{(1+\omega_A^2\tau^2)}+\frac{6\tau}{(1+\omega_B^2\tau^2)}$$

$$+\frac{\tau}{(1+(\omega_A-\omega_B)^2\tau^2)}$$

$$\left.+\frac{6\tau}{(1+(\omega_A+\omega_B)^2\tau^2)}\right\} \quad (27.17)$$

Because the molecular motion correlation times τ are temperature-dependent, that is, $\tau = \tau_0 \exp(E_a/RT)$, the T_1 plots of Eqs. (27.14) and (27.16) in semilogarithmic coordinates are symmetrical, V-shaped and go through minima. The slops of the curves correspond to the E_a values and the τ-independent minima provide determinations of internuclear distances. Finally, if target nuclei interact with a number of neighboring nuclei, then the corresponding contributions should be summed up.

Nuclei with spins $> \frac{1}{2}$, interacting with the EFG, eq_{ZZ}, relax by the quadrupolar mechanism:

$$\frac{1}{T_1}(Q) = \left(\frac{3}{50}\right)\pi^2(2I+3)(I^2(2I-1))^{-1}$$

$$\times \left(\frac{e^2 q_{zz} Q}{h}\right)^2 \left(1+\frac{\eta^2}{3}\right)$$

$$\left(\frac{\tau}{(1+\omega_Q^2\tau^2)}+\frac{4\tau}{(1+4\omega_Q^2\tau^2)}\right) \quad (27.18)$$

where η is the asymmetry parameter (corresponding to 0 at an axial symmetry). The $T_1(Q)$ times can change in large limits ($10^{-4} - 10^4$ ms), strongly depend on the nature of nuclei and particularly on the symmetry of their environments. For example, the ^{14}N T_1 time is as long as 10^4 ms in molecule Me$_4$^{14}NBr due to the completely symmetrical environment of ^{14}N nuclei.

The chemical-shift-anisotropy mechanism (CSA)

$$\frac{1}{T_1}(CSA) = \left(\frac{1}{15}\right)\gamma_I^2 B_0^2 (\Delta\sigma)^2$$

$$\times \left(\frac{2\tau}{(1+\omega_I^2\tau^2)}\right) \quad (27.19)$$

$$\frac{1}{T_2}(CSA) = \left(\frac{1}{90}\right)\gamma_I^2 B_0^2 (\Delta\sigma)^2$$

$$\times \left(8\tau + \frac{6\tau}{(1+\omega_I^2\tau^2)}\right) \quad (27.20)$$

is effective for nuclei with large values $\Delta\sigma$ (^{13}C, ^{15}N, ^{19}F and ^{31}P, for example). As it is seen, at $\omega^2\tau^2 \ll 1$ the relaxation rates become field-dependent and proportional to B_0^2. This is an experimental test for domination of the CSA mechanism. According to Abragam (1961), the spin-rotation mechanism is remarkable only for small molecules undergoing fast rotation in nonviscous media. This mechanism is valuable when the T_1 times reduce at increasing the temperature.

Generally, different relaxation mechanisms can operate simultaneously. In such cases, contributions that are important for studying the molecular mobility or structural features should be accurately evaluated (Bakhmutov, 2005). It should be emphasized, however, that two "independent" mechanisms can interfere to give cross-relaxation contributions.

27.3.4.2 Paramagnetic Relaxation

Fluctuating magnetic fields created by the large magnetic moment of an unpaired electron, $\mu_S = -g_e\beta eS$, can strongly reduce time of nuclear relaxation via dipolar and/or contact interactions. The dipolar mechanism, formulated by Solomon (Solomon, 1955) as controlled by reorientations of electron-nucleus vectors r, is written as (for $S = 1/2$)

$$\frac{1}{T_1^{DD}} = 0.1\gamma_I^2\gamma_S^2\hbar^2 r^{-6}$$

$$\times \left\{ \frac{\tau}{(1+(\omega_I-\omega_S)^2\tau^2)} \right.$$

$$+ \frac{\tau}{(1+\omega_I^2\tau^2)}$$

$$\left. + \frac{\tau}{(1+(\omega_I+\omega_S)^2\tau^2)} \right\} \quad (27.21)$$

$$\frac{1}{T_2^{DD}} = \left(\frac{1}{20}\right)\gamma_I^2\gamma_S^2\hbar^2 r^{-6}$$

$$\times \left\{ 4\tau + \frac{\tau}{(1+(\omega_I-\omega_S)^2\tau^2)} \right.$$

$$+ \frac{3\tau}{(1+\omega_I^2\tau^2)} + \frac{6\tau}{(1+\omega_S^2\tau^2)}$$

$$\left. + \frac{6\tau}{(1+(\omega_I+\omega_S)^2\tau^2)} \right\} \quad (27.22)$$

Here, τ is the molecular motion correlation time at rapid molecular motions or the electron relaxation time at slow reorientations. Because the nuclei, situated closer to paramagnetic centers, relax much faster (see the r^{-6} factor), these $T_{1,2}$ times can be rationalized in structural terms.

Bloembergen (1957) has shown that strong FC coupling can lead to this situation when electron-spin flips are directly accompanied by nuclear-spin flips. This mechanism, arising due to delocalization of the unpaired spin density in a nucleus, is given by the following equations:

$$\frac{1}{T_1^{CON}} = \left(\frac{2}{3}\right)S(S+1)\left(\frac{A}{\hbar}\right)^2$$

$$\times \frac{(\tau_{E2})}{(1+\omega_I^2\tau_{E2}^2)} \quad (27.23)$$

$$\frac{1}{T_2^{CON}} = \left(\frac{1}{3}\right)S(S+1)\left(\frac{A}{\hbar}\right)^2$$

$$\times \left(\tau_{E1} + \frac{(\tau_{E2})}{(1+\omega_I^2\tau_{E2}^2)}\right)$$

$$(27.24)$$

where τ_{E1} and τ_{E2} are the longitudinal and transverse electron-spin relaxation times, respectively. The FC mechanism dominates if under conditions $\omega_S\tau > 1 > \omega_I\tau$ the T_2 time is much shorter than T_1.

27.3.4.3 Nuclear Relaxation in Solids: Spin Diffusion

When solids exhibit some kind of molecular motions, then nuclei can relax by all of the above mechanisms. However, the relaxation process is usually nonexponential and described by a stretched exponential, $\exp(-(\tau/T_1)^\beta)$. Tse and Hartmann (1968) have shown that β of 0.5–0.6 is typical of powder samples. In rigid solids with very restricted motions, nuclear relaxation should be extremely slow. However, even at very small concentrations of paramagnetic impurities, the relaxation rates can increase by the, so-called, spin diffusion, suggested by

Bloembergen (1949). Owing to energy-conserving transitions of spins (mutual spin flips), belonging to dipolar-coupled nuclei, a target spin is "transferred" to a paramagnetic center, where it relaxes effectively by electron-nuclear dipolar interaction. Then, relaxation times are controlled by spin-diffusion coefficients D:

$$\frac{1}{T_1^{SD}} = \left(\frac{1}{3}\right) 8\pi N p C^{1/4} D^{3/4} \quad (27.25)$$

$$C = \left(\frac{2}{5}\right) \gamma_I^2 \gamma_S^2 \hbar^2 \, S(S+1)$$

In turn, D is defined as $D = (\mathbf{M}^{1/2}/30)a^2$, where $C = (2/5)\gamma_I^2\gamma_S^2\hbar^2 \, S(S+1)$, Np is the paramagnetic center density, $\mathbf{M}$ is the second moment of the dipolar internuclear interaction, and a is the internuclear distance. There are two distinguished cases: rapid spin diffusion and diffusion-limited relaxation. T_1^{SD} relaxation via fast spin-diffusion is always exponential, whereas diffusion-limited relaxation is nonexponential and can be treated by a stretched exponential at $0.5 < \beta < 1$. Kessemeier and Norberg (1967) have shown that the spin-diffusion coefficient D decreases in spinning solids. Thus the $T_{1,2}^{SD}$ times can depend on rates of mechanical rotation. Recently, Grutzer and Dybowski (Grutzer et al., 2001) have found that heavy nuclei, such as ^{207}Pb, can relax via a spin-phonon Raman scattering mechanism.

27.3.4.4 Relaxation Time Measurements

Modern commercial NMR spectrometers provide simple and convenient procedures for measurements of relaxation times. Relaxation times T_1 can be measured by the inversion-recovery ($180°$-τ-$90°$ at the varied delays τ), saturation-recovery or progressive saturation experiments. To improve the quality of spin inversion, the $180°$ pulse can be replaced with a composite pulse cluster $90°_\phi \, 240°_\phi \, 90°_\phi$ (see, for example, Harris, 1986). According to Sanders and Hunter (1994), the Hahn-echo ($90°$-τ-$180°$-τ-FID) or Carr–Purcell pulse sequences ($90°x'$-τ-$180°y'$-τ (first echo)-τ-$180°y'$-τ (second echo), and so on) can be applied to determine the spin–spin relaxation times T_2.

Spin-locking experiments in a rotating coordinate system create conditions providing nuclear magnetization to be aligned along the direction of the applied radio-frequency field. Because this field is smaller by several orders than the external magnetic field B_0, the magnetization decays with a specific time constant, $T_{1\rho}$, determined by significantly slower molecular motions. In solids, where molecular motions are strongly restricted, $T_{1\rho}$ values can differ strongly from T_1 and T_2 times. In contrast, the T_1, T_2, and $T_{1\rho}$ times are practically identical in solutions and nonviscous liquids, where molecular tumbling is fast.

27.4 Spectroscopic NMR Experiments in Liquids

27.4.1 Routine One-Dimensional (1D) NMR spectra

Generally, the routine 1D NMR spectra imply a direct excitation of nuclei with a single-pulse sequence (Figure 27.3). The best sensitivity in such experiments, often requiring a large number of accumulations at long delay times, can be reached when the acting angle of

RF pulses is optimized. In accordance with the Ernst rule (Ernst et al., 1966), at the total RD of 1 s, the angle should be of 8, 25, 33, 68, 86, and 90° if the T_1 time is 100, 10, 4, 1, 0.4, and 0.1 s, respectively.

Rare nuclei, for example, ^{13}C, are often spin–spin coupled by other nuclei, having a high natural abundance, for example, ^{1}H. The ^{13}C lines splitting and their observations are complicated. The NMR spectra of such nuclei are simplified with a broad-band-decoupler technique, which irradiates ^{1}H nuclei during the (RD + AT) time and suppresses, thus, the spin–spin coupling. Besides ^{1}H-decoupling, the signal intensities of carbons attached to protons (for example, groups CH, CH$_2$ or CH$_3$) increase due to the nuclear overhauser effect (NOE) (Overhauser, 1953). This effect connects with a perturbation of an equilibrium magnetization of X nuclei at irradiation of A nuclei, when nuclei A and X are coupled by dipole–dipole interaction. Commonly, when signs of γ_A and γ_X are identical, the X intensity increases according to: $I^X\{A\}/I^X{}_0 = 1 + \gamma_A/2\ \gamma_X$. To avoid the NOE, the ^{1}H- decoupler is applied during the AT time only.

Owing to the numerous and well-established relationships "spectrum/structure," even the routine 1D NMR spectra (multinuclear NMR particularly) are good tools for the determinations of chemical structures. The NOE measurements are also useful in this context. For example, Harris (1986) notes that in a system, consisting of three protons, H(A)···H(B)···H(C), coupled by dipolar interaction, irradiating the H nucleus results in the positive and negative NOE enhancement for H(B) and H(C) nuclei, respectively.

27.4.2
Polarization Transfer in NMR (SPI, INEPT, and DEPT)

When rare nuclei, for example, ^{13}C, ^{15}N, ^{29}Si, and so on are coupled by high-γ–nuclei, for example, ^{1}H, their equilibrium population difference can be perturbed by a polarization transfer technique to increase the NMR sensitivity. The physical principles of this technique can be found in the book of Friebolin (1993). Generally, an 180° pulse (selective or nonselective), which inverts the level populations, plays a main role in such experiments, named as signal enhancement by polarization transfer (SPI), insensitive nuclei enhanced by polarization transfer (INEPT) and distortionless enhancement by polarization transfer (DEPT). Among them, the DEPT sequence, shown in Figure 27.5, is most important, particularly for ^{13}C nuclei. Here an evolution time, τ, is calculated as $1/2^1J(C-H)$ and pulse $\theta = 135°$ leads to an ^{1}H-decoupled ^{13}C NMR spectrum with "negative" signals for CH$_2$ groups and "positive" signals for CH and CH$_3$ groups.

^{1}H
90°$_{X'}$ —τ— 180°$_{X'}$ —τ— θ$_{Y'}$ —τ— BBdec

^{13}C
———————— 90°$_{X'}$ —τ— 180° ———————— FID

Fig. 27.5 Presentation of the DEPT pulse sequence.

27.4.3
Two-Dimensional (2D) NMR Spectra

The pulse sequences, applied for the 2D NMR experiments, are various and are dependent on target tasks. Many of them, for example, the 2D heteronuclear J-resolved ^{13}C NMR spectra, the 2D homonuclear J-resolved ^{1}H NMR spectra, the 2D heteronuclear correlated NMR spectra, the 2D relayed NMR spectra, the 2D exchange NMR spectra, and the 2D INADEQUATE NMR spectra can be found in the book of Friebolin (1993). The more complex pulse trains are applied for NMR studies of biomolecules in solutions or the solid state (Duel, 2002). General principles of 2D NMR can be demonstrated by one of the simplest ^{1}H NMR experiments, ^{1}H–^{1}H COSY (correlated spectroscopy), shown in Figure 27.6.

After the first 90° pulse, a nuclear system is developed by proton–proton spin–spin coupling during time t_1. Thus, the first set of time-dependent data, $f(t_2)$, can be recorded after the action of the second RFP during time t_2. The $f(t_1)$ sets are collected at various t_1 values. Then, a double FT, performed for domains t_1 and t_2, leads to two frequency-axes. The resulting spectrum, shown in Figure 27.7, is a square plot, where the diagonal corresponds to the 1D ^{1}H NMR spectrum and cross peaks appear due to scalar spin–spin coupling. Similarly, one can perform a 2D NOE experiment, NOESY, (Figure 27.7) which is a good tool to establish spatial proximity of nuclei or an exchange process between them.

27.4.4
Chemical Exchanges: the NMR Time Scale

Besides nuclear relaxation and spin–spin coupling, reversible exchange processes between distinguishable spin states can also affect line shapes in NMR. These effects are particularly strong when life times (τ) of these states are in a range from 10^{-4} to 10^1 s, named as the NMR time scale.

Commonly, a chemical exchange between spins A and X perturbs shapes of their resonances, if the exchange frequency, ν_E, is comparable to chemical shift difference ($\nu_A - \nu_X$), expressed in Hertz. Then, the NMR spectra show a typical temperature evolution (Figure 27.8): the signals broaden and coalesce. A fast exchange, $\nu_E \gg (\nu_A - \nu_X)$, results in a single resonance. If spin states A and X are equally populated, this resonance appears at frequency $\nu = (\nu^A + \nu^X)/2$.

Generally, spin states can have different populations, they can belong to the same molecule (an intramolecular exchange) or different molecules (an intermolecular exchange) and finally,

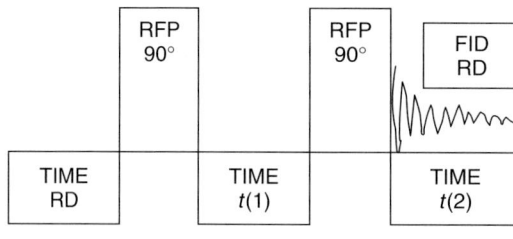

Fig. 27.6 The pulse sequence for a homonuclear proton–proton correlation (COSY) NMR experiment: the second pulse can be 90 or 45°.

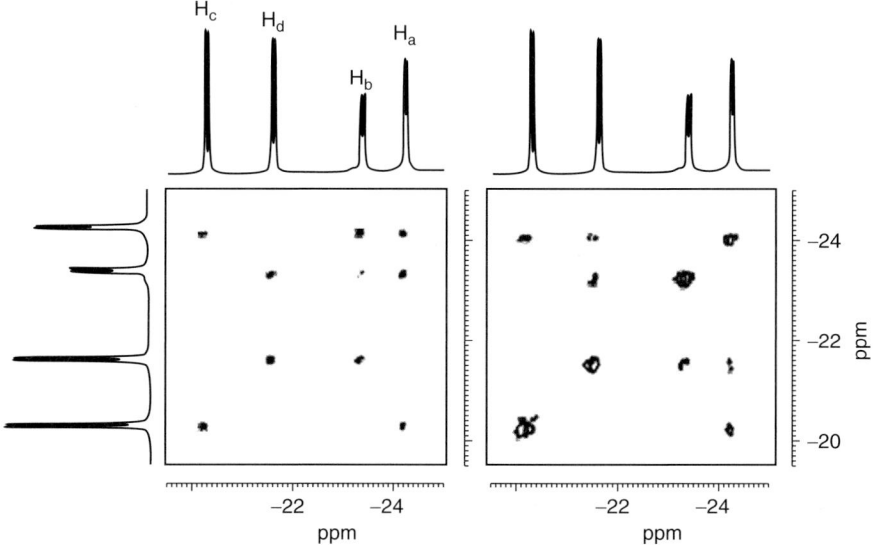

Fig. 27.7 A high-field region of the ^{1}H COSY (left) and NOESY (right) NMR spectrum of the complex [Ir$_2$(μ-H^a)(μ-Pz)$_2$(H^b)(H^c)(H^d)(NCCH$_3$)(PPr$^i_3$)$_2$] in C$_6$D$_6$: pairs of nuclei H^a/H^c, H^a/H^b and H^b/H^d are spin coupled, while pairs H^a/H^c and H^b/H^d are spatially approximated.

multicenter exchanges are also possible. Because the NMR line shapes in liquids are well established, exchange rates are accurately determined by a full line-shape-computer analysis, which can be found in NMR books cited in this chapter.

At slow exchanges, $v_E < v_A - v_X$, the NMR signals are well separated and remain narrow. However, relaxation times of the exchanging spin states can be averaged partially. Then, the FID's collected by the inversion-recovery experiments will depend on life times of these spin states which, thus, can be determined. Slow exchanges in ^{1}H NMR spectra can be easily established by saturation transfer experiments: irradiation of X nuclei leads to the saturation of the A resonance. For nuclei other than protons, 2D NOESY NMR, known as *two-dimensional exchange spectroscopy* (*EXSY*), is more preferable. Some examples of applications of the above techniques can be found in the books of Friebolin (1993), Harris (1986), and Bakhmutov (2005). It should be emphasized that the effects of chemical exchanges are particularly strong when one of the exchanging states is paramagnetic. In such cases, quantitative characterizations are reached by relaxation time measurements.

27.4.5
Translational Molecular Diffusion via NMR

The influence of magnetic-field inhomogeneity on forming the spin echo has been noted by Hahn (1950). If a static field gradient G is actually present, then a coherent transverse magnetization, created by the first 90° pulse in the Hahn-echo pulse sequence (90°-G(τ)-180°-G(2τ)-FID), undergoes an evolution in

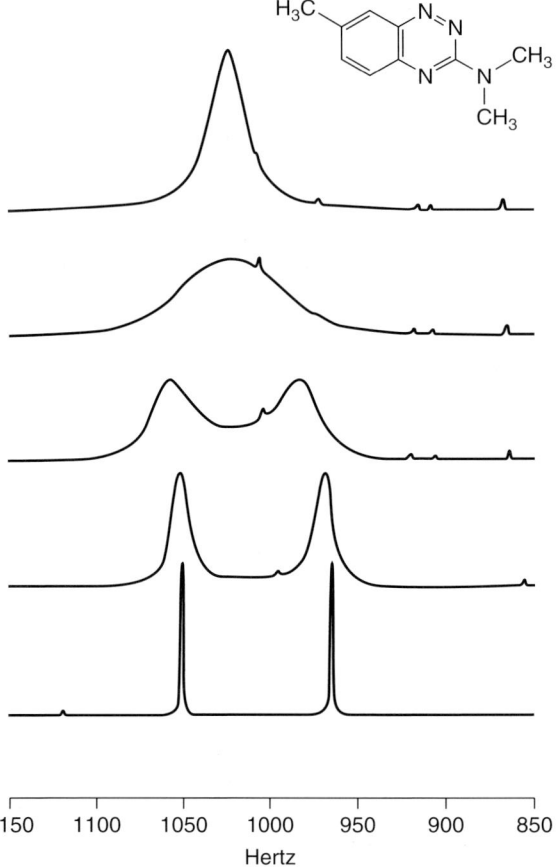

Fig. 27.8 Temperature evolution of the NCH$_3$ lines in the variable-temperature ^{1}H NMR spectra typical of a rotation around C−N bond.

this gradient. After a dephasing time, τ, the second 180° pulse inverts the sign of the net accumulated phase and the spins precess at the Larmor frequencies, reducing their accumulated phases. When the rephasing and dephasing rates are equal, then the magnetization reaches its coherent initial state after time 2τ, forming, thus, a spin echo. Then, the spin echo decays due to T_1/T_2 relaxation and additional factors are directly related to the molecular displacement on the time scale of the spin-echo experiments. At well-defined field gradients, the time correlation functions, describing the spin-echo decays, can be applied for studies of translational molecular diffusion.

Technically, these NMR experiments can be performed with the static-field gradients (SFG) or the pulsed-field gradients (PFG) suggested by McCall *et al.* (1963). As noted by Geil (1998), a combination of large static field gradients with short radio-frequency-pulse-spacings provides measurements of very small diffusion coefficients (10^{-15} m² s⁻¹) at a high spatial resolution (10 nm).

27.4.6
Isomeric and Conformational Analysis in Solutions by NMR

One of the key areas of NMR applications in solutions is the molecular conformational/configurational analysis. Because generally the analysis of cyclic molecules involves various NMR methods, such as the variable-temperature NMR spectra, the NOESY or other 2D NMR experiments, T_1 measurements, and determinations of the spin–spin coupling constants and their calculations by a convenient computer program, this aspect of the NMR spectrometry can also be a good illustration of its practical application. A methodology, typical of these studies, is demonstrated here by five-membered cycles 1 and 2, depicted in Scheme 27.1 (Guizado-Rodriguez et al., 2001).

Initial structural formulations of these compounds can be easily carried out on the basis of the ¹H, ¹³C, ¹¹B, ¹⁹F, and 2D ¹H–¹³C HETCOR NMR spectra. However, *a priori*, the cycle molecules can exist as trans-(1A, 2A) and cis-isomers (1B, 2B) and their cycles can take an envelope (E) or twist (T) conformations. In addition, usually five-membered cycles undergo fast pseudorotation with insignificant transformation energies and therefore the envelope and twist conformations can be equilibrated leading to observations of the average room-temperature (RT) NMR spectra. According to the RT multinuclear NMR spectra, recorded in C_6D_6, compound 1 gives two isomeric products, 1A and 1B, in a ratio of 9:1. Protons 2 and 2′ in the RT ¹H NMR spectrum of the major product are equivalent, whereas protons 5(4′) and 5′(4) are nonequivalent to show a complicated pattern in Figure 27.9. Simulation procedures for the AA′BB′ pattern show that the multiplet can be calculated when the geminal ¹H–¹H couplings are assumed to be negative and the J(5–4′) and J(5′–4) constants take positive but different values: 9.5 and 2.5 Hz, respectively. In terms of the modified Karplus equation (Haasnoot et al., 1980), the corresponding dihedral angles should be strongly different. The NOESY ¹H NMR spectra of 1A in toluene-d_8 recorded at room-temperature and −90°C reveal cross peaks 2(2′)-CH₃, 5′(4)-CH₃ and 5–5′, showing spatial proximity of the above protons. Protons 2 and 2′ remain equivalent even at low temperatures. In addition, calculations of the 5(4′)–5′(4) patterns, observed in toluene-d_8 at 25 and −90°C, lead to

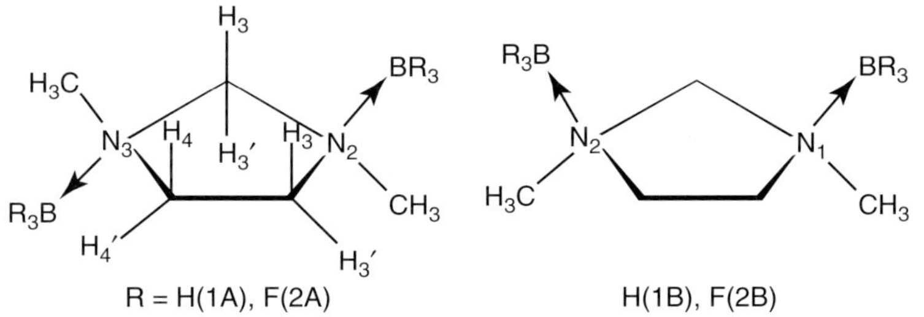

R = H(1A), F(2A)   H(1B), F(2B)

Scheme 27.1

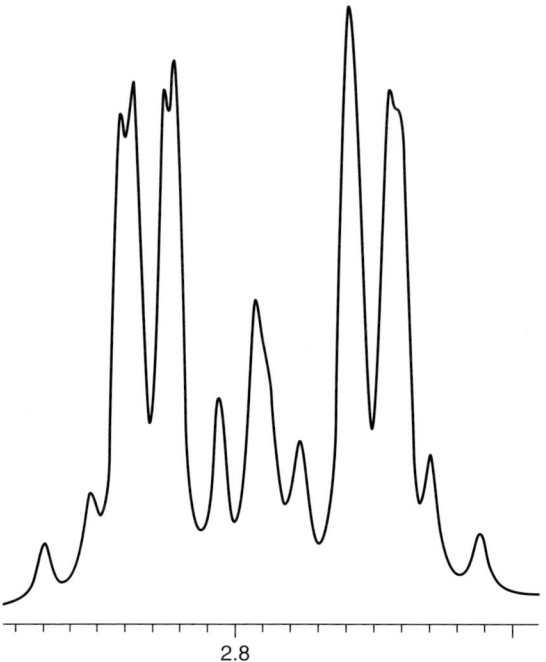

Fig. 27.9 Pattern AA'BB' observed for protons 5(4') and 5'(4) in the RT ^{1}H NMR spectrum of 1A in C_6D_6.

identical results: $^2J(5-5') = -10.5$ Hz, $^3J(5-4') = 9.5-10.0$ Hz, $^3J(5'-4) = 2.0-3.0$ Hz, $^3J(5-4) = 7.0$ Hz, $^3J(5'-4') = 7.0$ Hz, and $^2J(4-4') = -10.5$ Hz. Thus, only a trans BH_3/BH_3 arrangement is possible in 1A and its cycle takes twist conformation $T-1$ in Scheme 27.2, which is preferable even at room temperature. Compound 2 shows identical spatial properties. The same (or similar) methodology can be successfully used in the studies of other molecules.

27.5
NMR Spectroscopic Experiments in the Solid State

27.5.1
Wide-Line NMR

Resonance lines in NMR spectra of liquids or solutions are sharp due to the fast high-amplitude molecular motions. For example, line-width of a ^{1}H resonance in liquid water is as small as 0.1 Hz. In contrast, the resonance of ice is dramatically broadened up to $\sim 10^5$ Hz. According to

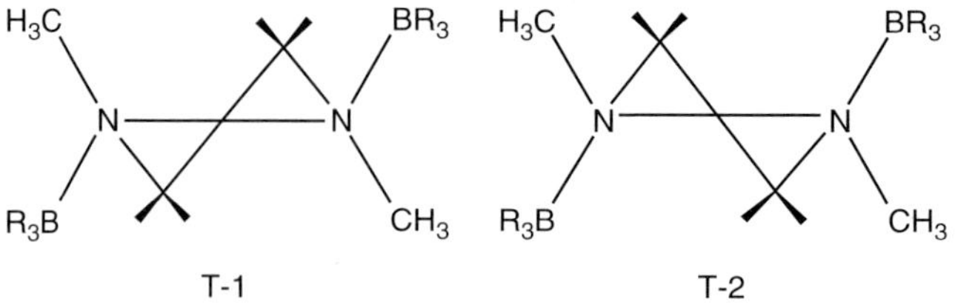

Scheme 27.2

Andrew and Szczesniak, 1995, a primary source of this broadening is magnetic dipolar proton–proton interactions, governed by a factor $(3\cos^2\theta - 1)/r^3$, where θ is the angle between the external magnetic field and internuclear vector r. It is obvious that the broadening factor is particularly large for powder samples. A single crystal sample can be oriented in the external magnetic field and its NMR spectrum is analyzed better. For example, a single crystal of $CaSO_4 \cdot 2\,H_2O$ with the relatively isolated protons shows the so-called Pake doublet providing determination of the proton–proton distance (Pake, 1948). Because dipolar interactions depend on the nature of nuclei, line widths of their resonances change drastically. For example, $^{29}Si-^{29}Si$ dipolar coupling produces line widths of 2–3 kHz only. Similarly, hetero-nuclear dipolar and CSA interactions also lead to wide lines in solid NMR. Finally, quadrupolar interactions exceed dipolar interactions by a factor e^2Q/μ^2, where eQ and μ are electric quadrupole moment and nuclear magnetic moment, respectively. All of these effects, particularly important for powder samples, give rise to the solid-state NMR spectra, where directional information "nucleus-nucleus" is lost and line widths mask the presence of chemically different nuclei.

Nevertheless, wide-line NMR can provide structural conclusions based on an approach suggested by Van Vleck (1948). In the limits of the approach, the dipolar second moment (or square width) is calculated in terms of the sum $\Sigma (3\cos^2\theta_{jk} - 1)^2/r_{jk}^6$ over all nuclei pairs in crystals or the sum Σr_{jk}^6 in polycrystalline samples. Because of the r_{jk}^6 factor, nearest dipole–dipole contacts dominate and thus structural information about molecules, groups, or ions can be obtained by comparing the experimental and calculated NMR spectra.

27.5.1.1 Dead Time in Wide-Line NMR

When a magnetization, created by RFP, decays rapidly with respect to the dead time of a spectrometer, FT converts the FID into an NMR spectrum where signal intensities are remarkably reduced and the base line shows strong distortions. This problem, particularly important for solid-state NMR, can be minimized by the applications of pulse sequences Hahn-echo ($90°_X$-τ-$180°_Y$-τ-FID) or solid echo ($90°_X$-τ-$90°_Y$-τ-FID), where τ is echo-delay time. According to Duel (2002), the FIDs can be collected immediately after $180°$ pulses. For spinning samples, the experiments require echo-delays synchronized with rotor periods.

In paramagnetic samples, resonance lines can be too broad and even unobservable in regular NMR spectra. In such cases, as noted by Mali et al. (2005), carrier frequencies in the Hahn-echo experiments can be varied and the resulting Hahn-echo NMR spectra can be summed up to show these wide lines.

27.5.1.2 Relaxation Experiments in Solids

Despite the relatively poor structural information, the wide-line NMR technique is successfully applied for the studies of molecular motions and phase transitions in solids by relaxation time measurements and/or a line-width analysis, briefly reviewed by Andrew and Szczesniak, 1995. The experiments can be performed in a very large temperature range, which is sufficient to characterize the molecular dynamics in materials of all kinds: amino acids, polypeptides, proteins, polymers, rubber, metabolites and steroids, and so on. In solid proteins, for example, it is possible to identify and characterize energetically a number of motions contributing to relaxation rates such as methyl-group reorientations, segmental motions, side-chain motions, and bound-water reorientations. It should be emphasized that solid-state NMR measurements, including T_1 and T_{1D} times and also the rotating-frame relaxation time, $T_{1\rho}$, is sensitive to motions covering the range of frequencies from 10^{-1} to 10^{11} Hz.

27.5.2
High-Resolution Solid-State NMR

At an angle θ of 54.4°, called as *magic angle* by Andrew, factor $(3\cos^2\theta - 1)/r^3$, characterizing dipole–dipole interaction, transforms to 0. It has been shown that the broadening effects, observed in static solids, can be reduced by their spinning when the given samples are situated at this angle relative to the external magnetic field. The narrowing effects will be more pronounced at spinning rates close or larger than line widths in static NMR. This principle is basic in the magic-angle-spinning (MAS) NMR spectroscopy leading to high resolution in the solid-state spectra. In practice, the collected spectra show isotropic resonances accompanied by spinning sidebands, which can be quite intense. However, positions of these bands can be changed by rotation at different rates and thus isotropic chemical shifts can be determined. In addition, there are special pulse sequences, which are capable of suppressing the sidebands. According to Duel (2002), one of such scheme is the TOSS pulse sequence consisting of four 180° pulses in a period prior to acquisition.

At intermediate spinning rates ν_R with respect to static line widths $\Delta\nu$ (for example, $\nu_R \sim \Delta\nu/4$) and domination of homonuclear dipolar coupling, the lines observed in MAS NMR spectra are still broadened due to the remaining time-dependent terms. In contrast, the MAS NMR spectra of molecular systems with dominant hetero-dipolar and CSA interactions show sharp lines even at intermediate spinning rates. According to Strub et al. (1983), distributions of sidebands in such cases can be used to determine magnetic screening tensors.

Nowadays, the modern commercial NMR spectrometers are equipped with MAS probes, where the spinning rates of 30–50 kHz are routinely achieved. Under these conditions, broadenings caused by the magnetic dipolar interactions, chemical shift anisotropy, Knight-shifts, pseudodipolar interactions, and quadrupolar interactions are canceled and the resulting spectra are determined by the isotropic chemical shifts and spin–spin coupling constants as in liquids and solutions.

27.5.2.1 High-Power Decoupling and Double Resonance

Typically, even in relatively simple molecules, NMR detections of rare nuclei, such as ^{13}C or ^{15}N, and so on, are seriously problematic because of the very strong dipolar interactions at short bond distances $^{1}H-^{13}C$ or $^{1}H-^{15}N$. Generally, these interactions are suppressed by a high-power decoupling technique producing ^{1}H irradiation when the ^{13}C or ^{15}N NMR FIDs are recorded. The decoupling can be realized by a continuous irradiation at a power of 100–1000 W or by special pulse sequences operating at proton frequency. Among them, sequences Two pulse phase modulation (TPPM), Waugh-Huber-Haeberlen (WAHUHA), and MREV-8 are most applicable (Duel, 2002).

Although, in rigid solids, the homonuclear dipolar interaction ^{1}H–^{1}H is particularly strong, lines in the ^{1}H MAS NMR spectra, recorded even at high spinning rates, are still broadened due to remaining dipolar interaction. However, such broadenings can be weakened or completely removed by the application of combined rotation and multiple pulse spectroscopy (CRAMPS), pioneered by Gerstein (1981). The CRAMPS combines sample spinning and multiple-pulse-decoupling sequences (for example, MREV-8).

27.5.3
Cross Polarization Technique

This technique helps strongly to improve a signal/noise ratio in the NMR spectra of ^{13}C, ^{15}N, and other nuclei with long relaxation times, low natural abundance, and low γ values. The task of a pulse sequence

$$^1\text{H } 90°_X \text{ - contact pulse}_{-Y} \; ^1\text{H decoupling}$$
$$^{13}\text{C - contact pulse}_{-Y} \text{ -}^{13}\text{C FID} \quad (27.26)$$

is a transfer magnetization from the ^{1}H spins with a high natural abundance to the ^{13}C (or ^{15}N) spins via the dipolar coupling between them. The amplitudes of two contact pulses, operating at proton and carbon frequencies, should be adjusted to achieve the Hartmann–Hahn conditions (Hartmann and Hahn, 1962):

$$\gamma_H B_1(^1\text{H}) = \gamma_C B_1(^{13}\text{C}) \quad (27.27)$$

Under these conditions, the energy gaps between the ^{1}H and ^{13}C rotating-frame spin states are equal and the transition requiring energy on ^{1}H spin can be exactly compensated by the transition releasing energy on ^{13}C spin. This pulse sequence in combination with the MAS technique leads to rapid accumulations of the ^{13}C NMR spectra because the recycling delay is decided by the ^{1}H (and not ^{13}C) relaxation.

27.5.4
Internuclear Distances from the Solid-State NMR Data

When molecular systems are amorphous or insoluble in common solvents, solid state NMR is the most efficient tool in their structural studies. Traditional NMR applications, based on the solid-state 1D and 2D NMR spectra, can be found in Duel (2002). Technically, collections of the NMR spectra in the solid state and solutions remarkably differ but their interpretations are very similar. It should be noted that most chemists are satisfied with obtaining such data. The other way of structural characterizations by NMR techniques is the high-precision measurement of interatomic distances, illustrated in this section.

Theoretically, information about distances between nuclei can be evaluated from the magnitudes of long-range dipole–dipole interactions. The classical Pake's powder spectrum, (Pake, 1948), expected for two nuclei in a static sample, is shown in Figure 27.10. Here, the shape of the resonance is formed by the superposition of two subspectra (dashed lines) and the distance between the two singularities corresponds to the dipolar coupling constant ω_D. Recalculations of ω_D gives finally a distance between the nuclei. A weak point in this simple approach is obviously connected with a low NMR

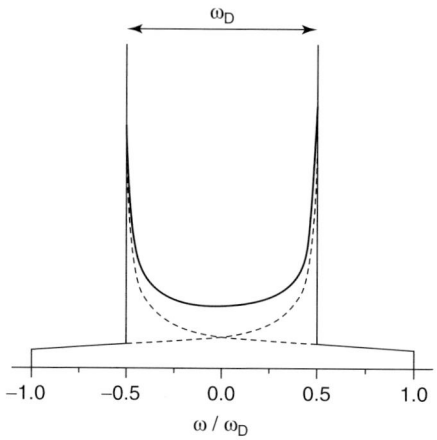

Fig. 27.10 The Pack dipolar powder spectrum. (Reproduced with permission from Lee, J-A and Khitrin, AK (2008) *Concepts Magn. Reson. Part A*, **32A**, 56.)

sensitivity particularly important for rare nuclei. Development of less direct NMR techniques, such as *SLF, PISEMA PITANSEMA, REDOR, TEDOR*, and so on, simplifies the determinations of dipolar couplings and hence internuclear distances. Some methods, their physical basis, and applications can be found, for example, in the book of Duel (2002). Lee and Khitrin (2008) have suggested a technique, specially developed for static samples, which is based on a two-dimensional single-echo pulse sequence, where adiabatic cross-polarization increases sensitivity. This technique, applied for isotopically labeled molecules, for example, glycine, provides the accuracy in measuring internuclear distances C—C and C—N, which is well compared with that in the X-ray and neutron-diffraction methods. For example, distance C_α—N is determined as 1.496 ± 0.002 by NMR and 1.476 ± 0.001 by ND.

27.5.5
Zero-Field Solid-State NMR Experiments

An initial nuclear polarization, needed for detecting NMR signals, is generally created by the strong external magnetic field. On the other hand, nuclei can be polarized by internal magnetic fields, which appear in samples due to their nature. For example, such a zero-field NMR spectrum has been observed by Reif and Purcell (1953) in solid hydrogen due to very strong proton–proton dipolar interactions. Similarly, even in the absence of an external magnetic field, the internal magnetic fields polarize the ^{57}Fe and ^{59}Co nuclei in ferromagnetic iron and cobalt, again resulting in the zero-field NMR spectra.

Other physical principles, considered by Andrew and Szczesniak (1995), are used in the zero-field NMR experiments directed to enhancements of resolution in the NMR spectra of polycrystalline samples. In such experiments, an initial nuclear polarization is created in a sample when it is placed into a strong external magnetic field. This polarized sample is transferred mechanically to an intermediate magnetic field, which, however, is still remarkably larger than local fields existing in the sample due to dipolar interactions. Then, this intermediate field is rapidly switched off and the nuclear spins undergo an evolution, during time τ_1, initiated obviously by the dipolar forces. After this time, the intermediate field is turned on and the sample is again placed into the high magnetic field, where a high-field NMR signal is detected as $I(\tau_1)$. FT of $I(\tau_1)$ gives, thus, a zero-Field NMR spectrum. In other words, this experiment leads to the high-magnetic-field NMR spectrum, recorded, at a high sensitivity, after an evolution of the nuclear spin system in the absence of the magnetic field. Because during the evolution period all directions are equivalent, numerous crystallites in polycrystalline samples will behave identically and the resulting NMR spectra will show sharp lines. Under these conditions even powder samples exhibit dipolar couplings with high resolution, usually reached only in field-oriented single crystals. Andrew and Szczesniak, 1995 note the obvious limitations: the zero-field technique can be successfully used when the spin-lattice relaxation time, T_1, in a low magnetic field is long enough to complete the experimental cycle. Similar zero-Field experiments can be performed for spinning samples. However, acting pulses and rotation rates should be synchronized in such cases.

27.6
Sensitivity Enhancement by Dynamic Nuclear Polarization (DNP)

Applications of the solution and solid-state NMR spectroscopy for many fields of chemistry, biology, and materials science often require a very high sensitivity in NMR spectra. When probing rare nuclei, for example ^{13}C, this is particularly critical when using the techniques considered in the previous sections is not enough. It should be recalled that the sensitivity depends on the natural initial nuclear polarization, which can be enhanced by the specially developed approaches based on optical pumping (Walker and Happer, 1997), para-hydrogen-induced-polarization (Bargon and Natterer, 1977), and dynamic nuclear polarization (DNP) (Wind et al., 1985). For example, Bifone et al. (1996) have found that optical pumping with laser light can enhance the nuclear polarization of gaseous xenon by 5 orders of magnitude. Among different techniques, the DNP seems to be most important and applicable to solids and solutions containing species with unpaired electrons (paramagnetic metal ions, organic groups, and so on). It is obvious that the systems investigated can be specially doped by such paramagnetic species.

Generally, an electron-nucleus system is described via the Hamiltonian:

$$H = -\omega_e\, S_Z - \omega_I\, I_Z + H_{ee}$$
$$+ H_{en} + H_{nn} \qquad (27.28)$$

where the first two terms correspond to the Zeeman's interactions of electron and nuclear spins, respectively, H_{ee} and H_{nn} describe electron–electron and nucleus–nucleus coupling, and H_{en} describes the hyperfine electron-nucleus interaction. In the context of DNP generation, the latter term plays a main role.

Irradiating the samples at (or near) the electron Larmor frequency, ω_e, can lead to simultaneous electron-nucleus spin flips due to electron-nuclear coupling H_{en}. In turn, these flips enhance the absolute value of the nuclear polarization via three mechanisms: an Overhauser effect, a solid effect, and a thermal mixing effect, described in the review of Wind et al. (1985). The enhancement factors observed in the nuclear polarization depend on the number of unpaired electrons, the ESR line widths, the nuclear relaxation rates in the absence of electrons, and the amplitude of the microwave field, applied to the sample. Generally, in solids, the enhancement factors can reach 660 and 2600 for ^{1}H and ^{13}C nuclei, respectively.

Technically, the DNP experiments require a special design of an NMR spectrometer and particularly an NMR probe, which should be equipped with a coil that can be double-tuned to ^{13}C and ^{1}H frequencies, for example, and a horn antenna to transmit the microwaves. The simplest enhancement experiments with ^{13}C nuclei imply their direct polarization. A proton polarization via DNP and its transfer to the ^{13}C spins via ^{1}H–^{13}C cross-polarization is also possible.

Addenkjaer-Larsen *et al.* (2003) have applied the similar principle for the NMR spectra in liquids to increase the signal/noise ratio by a factor of $>10^3$. This method is based on a low-temperature high-field nuclear polarization performed by a DNP polarizer where a frozen sample is irradiated with microwaves. This instrument is equipped with a superconducting magnet, a liquid-helium cooled sample space, and a special injection tool inside the DNP magnet. This injection tool provides a rapid dissolving of the solid sample in a suitable solvent and thus the resulting solution contains molecules with hyperpolarized nuclear spins. Then the hyperpolarized solution is rapidly transferred into a high-resolution NMR spectrometer where an enhanced NMR signal is registered by a convenient procedure. This powerful method can be successfully applied for different DNP-enhanced NMR experiments *in vitro* and *in vivo*.

27.7
Imaging NMR

High resolution in the NMR spectra, very important for a detailed structural analysis, can be obtained when the strong external magnetic fields are maximally homogenous. Regularly, such fields are created by magnets with bores of a few centimeters, capable of keeping the resonance condition (Eq. (27.4)) in each point through the volume of a sample investigated. In large and heterogeneous samples, requiring obviously wider bore magnets, the resonance condition cannot be satisfied throughout the whole object. Nevertheless, in such cases, the magnetic field can be homogenized only in a small volume of the object and thus a high-resolved NMR signal can be still recorded. Then, this small volume can be "spatially shifted" to collect the next signal. Repeating these experiments finally leads to the distribution of detected nuclei through the volume of the object. This principle is basic in the imaging NMR spectroscopy. The first imaging NMR experiments have been performed more than 30 years ago by the research group of Lauterbur (1973) and resulted in a special design of magnets, NMR probes and also methods for computer treatments of the collected data to construct the image.

Among the various modifications, one of the central places belongs to the variable-field-gradient technique, which focuses the magnetic filed. This technique defines the volume selectivity, important in the practical NMR imaging, and hence the spatial resolution depending on object sizes. At present, the highest spatial resolution, which can be obtained in small objects, corresponds to an element volume of 10^{-5} mm^3.

Because the time needed for the collection of NMR data from such a small element volume significantly increases, the experiments should be highly sensitive. Therefore, ^{1}H NMR plays a main role in the imaging spectroscopy particularly when the objects of studies are living systems. Owing to the discovery that relaxation times ^{1}H T_1 and ^{1}H T_2 in water molecules depend on the characteristics of their bonding, the development of the imaging technique was strongly accelerated to result in magnetic resonance tomography. As it is emphasized by Minati and Weglarz (2007), Echo-imaging, based on the ^{1}H T_2 time measurements by spin-echo sequences, leads to quality of the T_2 images well comparable to that obtained by X-ray tomography.

It is known that at a given gradient strength, a wider NMR spectrum produces a poorer image resolution. Since, for example, ^{1}H NMR spectra in static solids are generally broader than in liquids by a factor of $10^5 - 10^6$, the solid-state imaging experiments require the much stronger gradients focusing the field (Andrew and Szczesniak, 1995). Therefore, solid-state NMR imaging includes the MAS technique and the multipulse coherent averaging technique or their combination reducing the dipolar, CSA, and quadrupolar interactions. However, again, spinning rates and the field-gradient pulses should be synchronized.

References

Abragam, A (1961) *Principles of Nuclear Magnetism*, Oxford University Press, Oxford.
Addenkjaer-Larsen, J.H, Fridlund, B, Gram, A, Hansson, G, Hansson, L, Lerche, M.H, Servin, R, Thanong, M, and Goldman, K (2003) *PNAS*, **100**, 10158–10163.
Andrew, E.R and Szczesniak, E (1995) *Progr. Nucl. Magn. Reson. Spectr.*, **28**, 11–36.
Bakhmutov, V.I (2005) *Practical NMR Relaxation for Chemists*, John Wiley & Sons, Ltd, Chichester.
Bargon, J and Natterer, J (1977) *Progr. Nucl. Magn. Reson. Spectr.*, **31**, 293–315.
Bifone, A, Song, Y.O, Seydoux, R, Taylor, R.E, Goodson, B.M, Pietrass, T, Budinger, T.F, and Navon, G, Pines, A(1996) *Proc. Natl. Acad. Sci. USA*, **93**, 12932–12936.
Bloch, F, Hansen and W.W, Packard, M (1946) *Phys. Rev.*, **70**, 474–485.
Bloembergen, N, Purcell, and E.M, Pound, R.V (1948) *Phys. Rev.*, **73**, 679–712.
Bloembergen, N (1957) *J. Chem. Phys.*, **27**, 572–573.
Bloembergen, N (1949) *Physica*, **15**, 386–426.
Contreras, R.H and Peralta, J.E (2000) *Progr. Nucl. Magn. Reson. Spec.*, **37**, 321–425.
Duel, M.J (2002) *Solid-State NMR Spectroscopy: Principles and Applications*, Blackwell Sciences.
Ernst, R.R and Anderson, W.A (1966) *Rev. Sci. Instr.*, **37**, 93–102.
Friebolin, H (1993) *Basic One- and Two-Dimensional NMR Spectroscopy*, Weinheim, VCH.
Geil, B (1998) *Concepts Magn. Reson.*, **10**, 299–321.
Gerstein, B.C (1981) *Philos. Trans. R. Soc. Lond. Ser. A*, **299**, 521–546.
Gorter, C.J and Broer, L.J.F (1942) *Physica*, **9**, 591–596.
Grutzer, J.B, Stewart, K.W, Wasilishen, R.E, Lurnsden, M.D, Dybowski, C, and Beckmann, P.A (2001) *J. Am. Chem. Soc.*, **123**, 7094–7100.
Guizado-Rodriguez, M, Ariza-Castolo, A, Merino, G, Vela, A, Noth, H, Bakhmutov, and V.I, Contreras, R (2001) *J. Am. Chem. Soc.*, **123**, 9144–9152.
Haasnoot, C.A.G, Leeau, F.A.M, Altona, C (1980) *Tetrahedron*, **36**, 2783–2785.
Hahn, E.L (1950) *Phys. Rev.*, **80**, 580–594.
Harris, R.K (1986) *Nuclear Magnetic Resonance Spectroscopy: A Physicochemical View*, The Bath Press, Avon.
Hartmann, S.R and Hahn E.L (1962) *Phys. Rev.*, **128**, 2042–2053.
Jameson, C.J and Gutowsky, H.S (1964) *J. Chem. Phys.*, **40**, 1714–1724.

Kessemeier, H and Norberg, R.E (1967) *Phys. Rev.*, **155**, 321–337.

Kisman, K.E and Armstrong, R.L (1974) *Rev. Sci. Instr.*, **45**, 1159–1163.

Kubo, A, Spaniol, T.P, and Terao, T (1998) *J. Magn. Reson.*, **133**, 330–340.

Kubo, R and Tomita, H (1954) *J. Phys. Soc. Jpn.*, **9**, 888–919.

Lauterbur, P.C (1973) *Nature*, **242**, 190–193.

Lee, J.-A and Khitrin, A.K (2008) *Concepts Magn. Reson. A*, **32A**, 56–67.

Mali, G, Ristic, A and Kaucic, V (2005) *J. Phys. Chem. B.*, **109**, 10711–10716.

McCall, D.W, Douglass, D.C, and Anderson, E.W (1963) *Ber. Bunsenes. Physik. Chem.*, **67**, 336–340.

McConnel, H.M and Chesnut, D.B (1958) *J. Chem. Phys.*, **28**, 107–117.

Minati, L and Weglarz, W.P (2007) *Concepts Magn. Reson. A*, **30A**, 278–307.

Overhauser, A.W (1953) *Phys. Rev.*, **92**, 411–415.

Pake, G.E (1948) *J. Chem. Phys.*, **16**, 327–336.

Poople, C.P (1983) *Electron Spin Resonance*, John Wiley & Sons, Ltd, New York.

Proktor, W.G and Yu, F.C (1950) *Phys. Rev.*, **77**, 717–717.

Purcell, E.M, Torrey, H.C, and Pound, R.V (1946) *Phys. Rev.*, **69**, 37–38.

Reif, F and Purcell, E.M (1953) *Phys. Rev.*, **91**, 631–641.

Sanders, J.KM and Hunter, B.K (1994) *Modern NMR Spectroscopy: A Guide for Chemists*,Oxford University Press, Oxford.

Solomon, I (1955) *Phys. Rev.*, **99**, 559–565.

Strub, H, Beeler, A.J, Grant, D.M, Michel, J, Cutts, P.W, and Zilm, K.W (1983) *J. Am. Chem. Soc.*, **105**, 3333–3334.

Tse, D and Hartmann S.R (1968) *Phys. Rev. Lett.*, **21**, 511–514.

Van Vleck, J.H (1948) *Phys. Rev.*, **74**, 1168–1183.

Walker, T.G and Happer, W (1997) *Rev. Mod. Phys.*, **69**, 629–642.

Wind, R.A, Duijvestun, M.J, Van Der Lugt, C, Manenschijn, A, and Vriend, J (1985) *Prog. NMR Spectrosc.*, **17**, 33–67.

Further Reading

In addition to the very limited number of references cited, advances in magnetic resonance spectroscopy can be found in Malcolm Levitt (2001) *Spin Dynamics: Basics of Nuclear Magnetic Resonance*, John Wiley & Sons, Ltd, Chichester and James Keeler (2005) *Understanding NMR Spectroscopy*, John Wiley & Sons, Ltd, Chichester and also within the periodicals: *Journal of Magnetic Resonance, Magnetic Resonance in Chemistry, Magnetic Resonance in Medicine, Progress in NMR spectroscopy, Concepts in Magnetic Resonance, Journal of Magnetic Resonance Imaging* and *the NMR Encyclopedia*.

28
Biomolecular Structures by Solution Nuclear Magnetic Resonance

Tharin M. A. Blumenschein

28.1	Introduction 965	
28.2	**Spectra Acquisition** 966	
28.2.1	Nuclei 966	
28.2.2	Sample Preparation 967	
28.2.3	Pulse Sequences and Processing Strategies 968	
28.3	**Chemical Shift Assignment** 969	
28.3.1	Chemical Shift Assignment in Proteins 970	
28.3.2	Chemical Shift Assignment in Nucleic Acids 973	
28.3.3	Chemical Shift Assignment in Carbohydrates 973	
28.4	**Structural Restraints** 975	
28.4.1	Nuclear Overhauser Effect 975	
28.4.2	Torsion Angles 977	
28.4.3	Residual Dipolar Couplings (RDCs) 978	
28.4.4	Hydrogen Bonds 979	
28.4.5	Paramagnetic Centers 980	
28.4.6	Structural Information from Chemical Shifts 980	
28.5	**Molecular Modeling and Structure Calculation** 981	
	References 982	
	Further Reading 986	

Encyclopedia of Applied Spectroscopy. Edited by David L. Andrews.
Copyright © 2009 WILEY-VCH Verlag GmbH & Co. KGaA, Weinheim
ISBN: 978-3-527-40773-6

28.1
Introduction

Molecular structures are essential to understand the function and mechanism of molecules. The use of nuclear magnetic resonance (NMR) spectroscopy to determine the structure of biomolecules in solution has the advantage of reproducing the environment in which these molecules are naturally found.

The aim of this chapter is to provide an introduction to how NMR spectroscopy is applied to structural studies of biomolecules, following the steps described below. It is not the aim to provide an extensive review of structure determination of biological macromolecules using solution NMR, and although NMR spectroscopy can be successfully used to study the dynamics of molecules and molecular interactions (see, for example, Palmer, 2001; Kay, 2005; Latham *et al.*, 2005), these techniques are beyond the scope of this introduction.

Biological macromolecules have a number of characteristics in common, and they present a number of common challenges. The process of structure determination by NMR usually starts with the assignment of chemical shift values to the relevant nuclei in the molecule, followed by the determination of a number of structural restraints. Specific pulse sequences are available for each of those steps. The structural restraints are then used for molecular simulations, using simulated annealing techniques and molecular dynamics. Each type of structural restraint is added to the molecular energy representation as a separate pseudo-energy term, and the simulation explores the conformational space to find the minimal energy structures.

As a solution technique, NMR-determined structures reflect the heterogeneity of conformations presented by biomolecules in solution. Specific challenges in this aspect are intrinsically disordered proteins (see Dyson and Wright, 2005, for a review), which do not have a well-defined tertiary structure in their native state in the absence of their binding partner, and larger oligosaccharides and polysaccharides, that tend to be highly flexible and dynamic (Martin-Pastor and Allen Bush, 1999). In these cases, the structures determined reflect a distribution of conformations, rather than a single structure (Yi *et al.*, 2004; Lindorff-Larsen *et al.*, 2004; Osterhout *et al.*, 1989).

When compared to small organic molecules, the acquisition of NMR spectra of biological macromolecules

Encyclopedia of Applied Spectroscopy. Edited by David L. Andrews.
Copyright © 2009 WILEY-VCH Verlag GmbH & Co. KGaA, Weinheim
ISBN: 978-3-527-40773-6

presents a number of challenges. Experiments are generally performed in water, making good suppression of the solvent signal essential. In addition, the number of signals from the different ^{1}H nuclei in a biological macromolecule inevitably leads to overlapping in the one-dimensional spectrum, requiring high-field spectrometers and multidimensional spectra to resolve the individual resonances.

28.2
Spectra Acquisition

28.2.1
Nuclei

The most common atoms in biological macromolecules are carbon, hydrogen, nitrogen, and oxygen. The magnetic properties of their stable isotopes are summarized in Table 28.1.

The most abundant hydrogen isotope, ^{1}H, has the highest sensitivity, and is the most widely used nucleus for NMR. The most abundant carbon isotope, ^{12}C, has no nuclear spin, whereas the most abundant nitrogen isotope, ^{14}N, has a spin number of 1 and strong quadrupolar moment, leading to fast relaxation. Less abundant isotopes of both, however, have spin number $^1/_2$. ^{13}C is present as about 1% of the total carbon, and natural abundance spectra are routinely acquired for small molecules. For larger molecules such as proteins, the weak signal would make the acquisition of multidimensional spectra overly long, and ^{13}C labeling of proteins is a standard technique. The same is true of ^{15}N, even less abundant naturally. The only stable magnetically active oxygen isotope is ^{17}O, with a spin number of $^5/_2$, and it is not normally used for solution NMR of biomolecules. ^{31}P is naturally present in all nucleic acids and in some carbohydrates and proteins, and is often used as an additional source of information. Although other nuclei can be used to study biomolecules, structural determination is usually performed using

Tab. 28.1 Magnetic properties of the stable isotopes of biologically abundant atoms. γ is the gyromagnetic ratio.

Isotope	Spin	Natural abundance (%)	γ (10^7 rad T^{-1} s^{-1})	Sensitivity		Frequency at 11.7 T
				Relative to proton	At natural abundance	
^{1}H	$^1/_2$	99.985	26.75	1.00	1.00	500.000
^{2}H	1	0.015	4.11	9.65×10^{-2}	1.45×10^{-6}	76.753
^{12}C	0	98.89				
^{13}C	$^1/_2$	1.108	6.73	1.59×10^{-2}	1.76×10^{-4}	125.721
^{14}N	1	99.63	1.93	1.01×10^{-3}	1.01×10^{-3}	36.118
^{15}N	$^1/_2$	0.366	−2.71	1.04×10^{-3}	3.85×10^{-6}	50.664
^{16}O	0	99.75				
^{17}O	$^5/_2$	0.037	−3.63	2.91×10^{-2}	1.08×10^{-5}	67.784
^{18}O	0	0.205				
^{31}P	$^1/_2$	100	10.83	6.63×10^{-2}	6.63×10^{-2}	202.405

multidimensional ^{1}H, ^{15}N, and ^{13}C spectra. ^{15}N and ^{13}C are indirectly detected, with the magnetization being transferred back to protons for ^{1}H direct detection. These experiments require triple-resonance probes, capable of pulsing and decoupling ^{13}C and ^{15}N, sometimes simultaneously, while pulsing and detecting ^{1}H (Cavanagh et al., 1996).

28.2.2
Sample Preparation

Solution NMR of biological macromolecules is usually performed in 90% H$_2$O/10% D$_2$O, with the buffer and salt concentrations depending on the requirements for stability of the molecules being studied. Where possible, sample pH should be kept below 7.5 to slow down the exchange rate of the backbone amide protons with the solvent, which leads to loss of signal, and salt concentration should be kept low and buffers of low conductivity should be chosen, as higher buffer conductivity will decrease the signal-to-noise ratio (Kelly et al., 2002).

The large number of atoms in biological macromolecules results in heavy overlapping of resonances in the ^{1}H spectra, requiring uniform labeling with ^{15}N and/or ^{13}C for resolution of the overlapped peaks in multidimensional spectra. For larger molecules, additional uniform labeling with ^{2}H is required to decrease transverse relaxation rates (Grzesiek et al., 1993). Deuterated protein samples are usually purified in ^{1}H$_2$O-based buffers, resulting in substitution of the deuterium bound to the main-chain nitrogen with protons.

The requirement for labeled protein means that protein samples for NMR spectroscopy are normally obtained by recombinant expression in an organism that can be grown in a medium of controlled composition, to which the only source of nitrogen or carbon can be added with defined isotopic composition. Most commonly, proteins are expressed in *Escherichia coli* grown in minimal media, having glucose as the only source of carbon, and NH$_4$Cl or (NH$_4$)$_2$SO$_4$ as the only source of nitrogen. Uniformly labeled ^{13}C-glucose and ^{15}NH$_4$Cl and (^{15}NH$_4$)$_2$SO$_4$ are available commercially. For deuterated proteins, the minimal media can be prepared with D$_2$O instead of ^{1}H$_2$O, and conditioning protocols may be required for good bacterial growth and protein expression (Li, Corson and Sykes, 2002). Rich media containing ^{15}N, ^{13}C, and ^{2}H are also commercially available, and generally produce much better bacterial growth than minimal media, but are significantly more expensive. When a protein cannot be successfully expressed in bacterial cells, eukaryotic expression systems can be used, such as the yeast *Pichia pastoris* (Morgan, Kragt and Feeney, 2000) or insect cells (Brüggert et al., 2003). More recently, cell-free expression systems have been developed (Vinarov et al., 2004; Torizawa et al., 2004).

For larger proteins or specific experimental requirements, alternative labeling schemes can be required, instead of uniform labeling. Examples include different schemes for selective protonation of methyl groups in amino acid side chains, while the rest of the protein is deuterated (Gardner and Kay, 1998).

For nucleic acids, it is also necessary to obtain pure ^{13}C, ^{15}N-labeled samples in milligram amounts. Most RNA samples are produced *in vitro* by transcription of a DNA template by T7 polymerase (Milligan et al., 1987; reviewed by Fürtig et al.,

2003; Flinders and Dieckmann, 2006; Dayie, 2008). Template DNA often includes the sequence for a self-cleaving ribozyme, which will aid in purification and help in overcoming the limitations of T7 polymerase. Uniform labeling is commonly performed by using labeled NTPs as the substrate for the T7 polymerase reaction, but for large RNAs (over 20–30 nucleotides), specific labeling approaches may be required (reviewed in Latham et al., 2005; see Nelissen et al., 2008, for a recent example). Chemical synthesis is an alternative for small RNAs, and has been reviewed by Lagoja and Herdewijn (2002). DNA samples can be obtained by purifying a suitable plasmid containing multiple copies of the desired oligonucleotide from a bacterial culture grown in ^{15}N, ^{13}C-labeled medium, or from a polymerase chain reaction (PCR)-based approach (Louis et al., 1998; René et al., 2006).

Labeled carbohydrate samples for NMR studies can be even more difficult to produce, because the residue sequence in glycans often depends on species-specific synthetic pathways. NMR of small oligosaccharides can be performed on natural abundance ^{13}C, and is often used in combination with other techniques. Glycans from species that can be grown in labeled medium, such as bacteria and fungi, can be normally purified. A review of successful examples of uniform labeling in carbohydrates was done by Duus, Gotfredsen and Bock (2000), but a labeling method generally applicable to carbohydrates is not yet available. Alternative approaches for partial labeling include the enzyme-catalyzed exchange of acetyl groups with ^{13}C-enriched acetic anhydride (Yu and Prestegard, 2006).

28.2.3
Pulse Sequences and Processing Strategies

While specific pulse sequences used to determine chemical shift assignments or structural restraints are mentioned in the following sections, a few general methods can be applied to a range of pulse sequences to make them more effective or faster. Pulse sequences for biological experiments also need to contain solvent suppression pulses, because the experiments are normally performed in 90% H_2O/10% D_2O. Different water suppression techniques can be used in different pulse sequences, and depending on the characteristic of the sample. When observing protons that can easily exchange with the solvent, in nucleic acids for instance, it is important to avoid methods involving presaturation of water (more commonly used in protein studies), and utilize pulse sequences that use jump-and-return techniques (Hore, 1983; Sklenár and Bax, 1987) or the WATERGATE method (Piotto, Saudek and Sklenár, 1992).

Original multidimensional pulse sequences relied on phase cycling to eliminate unwanted signals and minimize artifacts. By changing the phase of specific pulses and the receiver, unwanted signals can be cancelled. This requires repeating the experiment with the different phase combinations, and can result in a number of repetitions much larger than needed for a good signal-to-noise ratio. In probes with pulsed-field gradient capabilities, part of the phase cycling can be replaced by the use of gradient pulses. Field gradient pulses are pulses that create a varying field along the axis of the sample, resulting in defocusing of the magnetization, which can be recovered by applying another gradient pulse, of opposite signal. If those pulses are applied at the right point in the pulse sequence, only the desired signals will be selected. Phase cycling is then used to minimize artifacts, with a smaller number of repetitions.

For larger biomolecules, one of the limiting factors is the fast transverse relaxation, which increases with the rotational correlation time. Fast transverse relaxation broadens the spectral lines, and leads to loss of signal due to relaxation during the pulse sequence. However, the four lines in an undecoupled peak quartet in a heteronuclear correlation spectrum do not relax at the same rate. Transverse relaxation optimized spectroscopy (TROSY) selects the multiplet component with slower relaxation rate, eliminating the other lines (Pervushin et al., 1997; Wider and Wüthrich, 1999). Although only part of the signal is being recovered, the spectral lines show better resolution and signal-to-noise than the comparison standard PEP-HSQC (heteronuclear single quantum coherence) spectrum used for heteronuclear correlation spectroscopy, and this effect is intensified for large molecules. The TROSY effect is field-dependent, being more pronounced at higher fields. It is normally used in conjunction with deuterated samples, which also help to increase the transverse relaxation times. TROSY elements have been incorporated in a number of multidimensional pulse sequences, for use with large molecules (Wider and Wüthrich, 1999).

While TROSY and other techniques aim to increase sensitivity and extend NMR experiments to large and otherwise troublesome proteins, there are various techniques to decrease experimental time when signal is abundant. They include SOFAST-HMQC experiments (Schanda, Kupce and Brutscher, 2005), Hadamard NMR (Kupce, Nishida and Freeman, 2003; Brutscher, 2004), single-scan NMR (Frydman, Scherf and Lupulescu, 2002), projection NMR (Szyperski et al., 1993; Brutscher et al., 1994; Kupce and Freeman, 2004), hyperdimensional NMR (Kim and Szyperski, 2003; Kupce and Freeman, 2008), nonlinear data recording and processing (Mandelshtam, 2000; Hoch and Stern, 2001; Rovnyak et al., 2004), and covariance spectroscopy (Brüschweiler and Zhang, 2004). A number of these techniques were reviewed a few years ago (Freeman and Kupce, 2003).

28.3
Chemical Shift Assignment

There is no single unified method for obtaining chemical shift assignments for polypeptides, polynucleotides, and polysaccharides. Proteins, nucleic acids, and carbohydrates have different characteristics, and the methods for chemical shift assignment differ, even though some of the pulse sequences used are the same. Most of the pulse sequences required for the assignment of biomolecules are provided in the standard installation of modern high-field NMR spectrometers. Chemical shift assignments of biological macromolecules are deposited in the Biological Magnetic Resonance Bank (BMRB) (Ulrich et al., 2008), which is also a useful source for other resonance data, such as average chemical shift values for each amino acid type. Once the chemical shift values seen in the spectra have been assigned to specific nuclei in the biomolecule, structural restraints can be defined for different atoms in the molecule.

28.3.1
Chemical Shift Assignment in Proteins

There are two stages in assigning chemical shift values to individual nuclei in proteins. First, the backbone is assigned using the sequential assignment method, created by Kurt Wütrich (Nobel prize winner in Chemistry in 2002) and coworkers (Wagner and Wüthrich, 1982; Wüthrich, 1986; reviewed in Ferentz and Wagner, 2000; Kanelis, Forman-Kay and Kay, 2002). The side chains of each residue are assigned in a second stage. Modern experiments used for chemical shift assignment are mostly ^{15}N-edited three-dimensional spectra, and are intimately connected to the labeling scheme used for the sample. For most proteins, ^{1}H, ^{13}C, and ^{15}N information is required, from a sample double-labeled with ^{13}C and ^{15}N. For large proteins, it is often not possible to perform three-dimensional NMR experiments in a fully protonated protein, due to the short transversal relaxation time (T_2), which leads to very fast signal decay and line broadening. Main-chain amide protons are usually back-exchanged into deuterated proteins, and specific suites of triple-resonance experiments are the basis for chemical shift assignment (Gardner and Kay, 1998).

Figure 28.1 shows schemes of the magnetization transfer in the most commonly used pulse sequences for backbone sequential assignment in proteins. Sequential assignment starts with a reference spectrum, a 2D {^{1}H, ^{15}N}-HSQC in which a peak corresponds to each main-chain amide, or a 3D HNCO (Kay et al., 1990; Grzesiek and Bax, 1992b), which correlates the proton and nitrogen of a main-chain amide with the carbonyl carbon from the previous amino acid residue. An ideal HSQC or HNCO spectrum will have a peak for every residue in the protein, but N-terminal residues often cannot be seen due to fast exchange with water protons, and loss of magnetization during the water suppression scheme. The faster exchange in this region is caused by the positive charge in the amino terminus of the protein. Conversely, C-terminal residues often give much stronger signals than the rest of the protein. Conformational exchange can also lead to peak broadening beyond detection, and peaks missing from the reference spectrum. Conformational exchange in the protein can also lead to the loss of peaks for some residues.

The basic experiments for backbone assignment of a medium-sized protein are HNCACB (Wittekind and Mueller, 1993) and CBCA(CO)NH (Muhandiram and Kay, 1994; Grzesiek and Bax, 1992a), HNCO and HN(CA)CO (Clubb, Thanabal and Wagner, 1992). HNCACB correlates the amide proton and nitrogen of each residue with the α- and β-carbons of the same amino acid residue more strongly, and with the α- and β-carbons of the previous residue more weakly, while CBCA(CO)NH correlates the amide proton and nitrogen of each residue with the α- and β-carbons of the previous residue only. Sequential assignment is done by matching the weaker peaks in each strip of the HNCACB spectrum with the peaks in a strip of the CBCA(CO)NH spectrum, forming a "chain" of spin systems. Residues with characteristic chemical shift values, such as serine, threonine, glycine, and alanine, help anchor this chain in relation to the amino acid sequence in the protein.

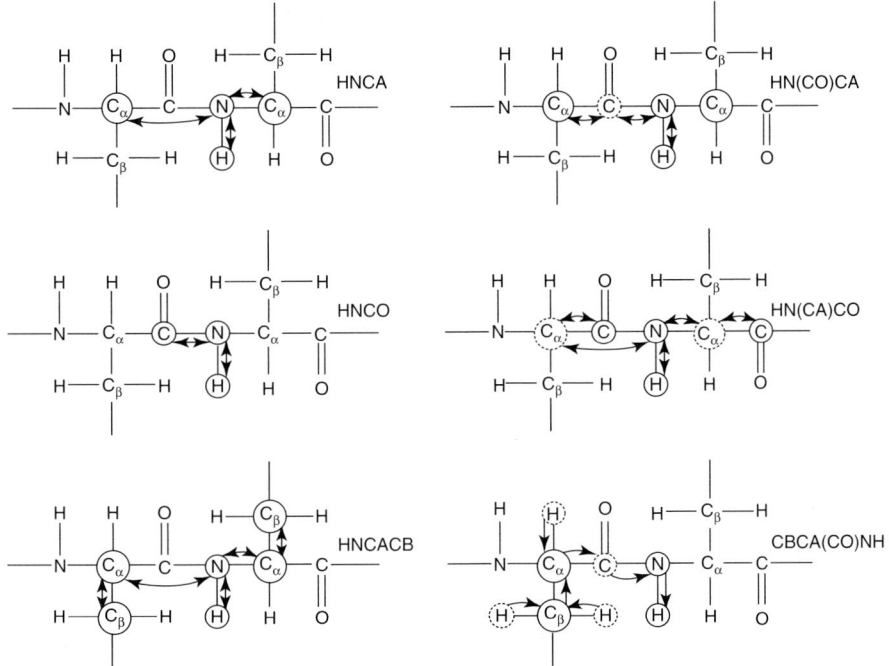

Fig. 28.1 Magnetization transfer steps in pulse sequences used for protein sequential assignment. Nuclei that are not correlated in the final spectra are represented in dashed circles, nuclei correlated in the final spectra are inside solid circles, and magnetization transfer is represented by arrows. With the exception of CBCA(CO)NH, the magnetization in all pulse sequences in the figure starts from the amide proton is transferred to the heteronuclei and back to the amide proton for data acquisition. In CBCA(CO)NH, the magnetization starts in the protons bound to C_β and C_α.

An example of sequential strips in HNCACB and CBCA(CO)NH spectra can be seen in Figure 28.2.

The CBCA(CO)NH experiment is not applicable to deuterated proteins, because it starts by transferring magnetization from the hydrogen atoms bound to the β-carbon (Figure 28.1). When the ^{1}H is replaced with deuterium, this pulse sequence is not applicable anymore. Instead, the HN(CO)CACB pulse sequence can be used. In large proteins (over 30 kDa), however, the signal that can be obtained with HNCACB and HN(CO)CACB is often too weak, and the combination of HNCA (Grzesiek and Bax, 1992b; Stonehouse et al., 1995) and HN(CO)CA (Bax and Ikura, 1991) experiments can used. Only correlating the α-carbon with the amide, these experiments have better signal-to-noise ratios than HNCACB and HN(CO)CACB under the same conditions. The disadvantage, however, is that they contain less information. Other labeling schemes and pulse sequences have been developed to extend chemical shift assignment procedures to even larger proteins, up to 82 kDa (Tugarinov, Hwang and Kay, 2004).

A number of semiautomated and automated backbone assignment programs

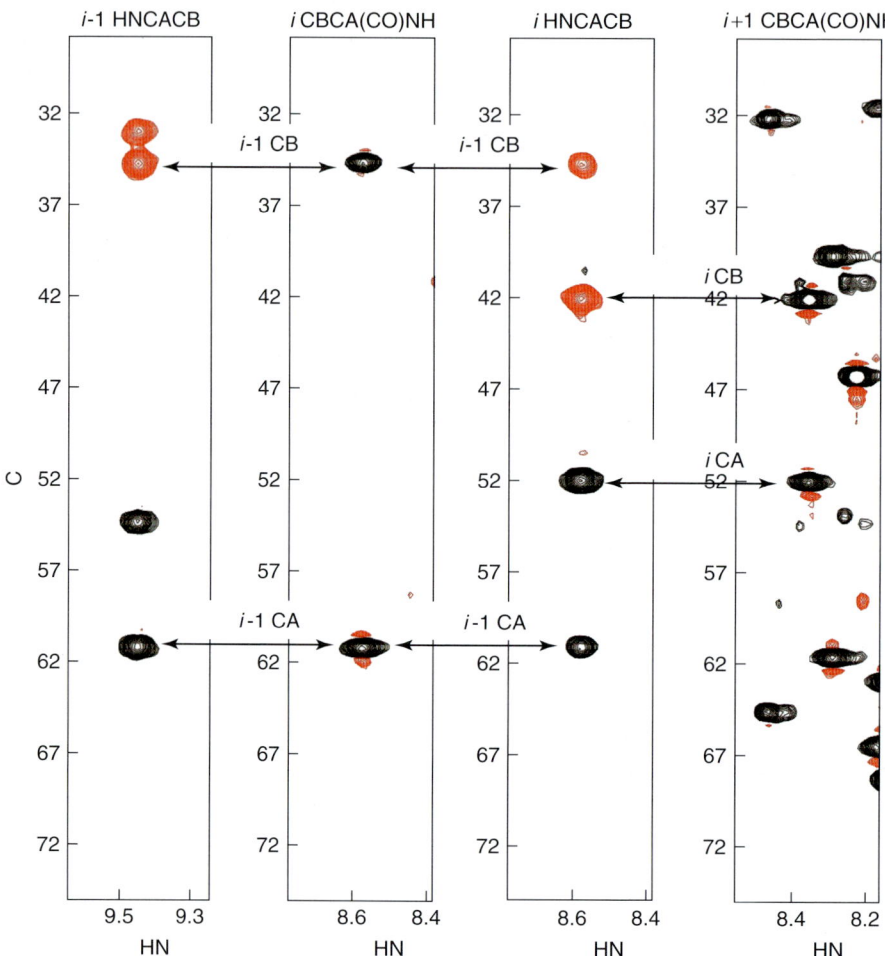

Fig. 28.2 Strips from three-dimensional HNCACB and CBCA(CO)NH spectra for sequential assignment, showing the connectivities between three sequential residues, $i-1$, i, and $i+1$. Each strip is a region of a spectrum corresponding to the amide nitrogen plane and proton resonance for the indicated residue. As seen in Figure 28.1, HNCACB spectra correlate the amide of each residue with the α- and β-carbons from the same and previous residues, while CBCA(CO)NH spectra correlate the amide of each residue with the α- and β-carbons from the previous residue. Data was kindly provided by Prof. Brian Sykes, University of Alberta, Canada.

have been developed. Semiautomated assignment programs, such as ANSIG (Helgstrand et al., 2000) and Smartnotebook (Slupsky et al., 2003), suggest possible sequential assignments to be visually verified and confirmed by the user. Automated assignment programs produce the assignment for spin systems in the protein from its amino acid sequence and a peak list. Examples are PASTA (Leutner et al., 1998), TATAPRO (Atreya, Chary and Govil, 2002), MARS (Jung and Zweckstetter, 2004), and MATCH (Volk, Hermann and Wüthrich, 2008).

Such methods still require some level of user input, and in most cases not all amino acid residues in a protein can be assigned automatically.

Once the backbone has been assigned, the side chains in the protein need to be assigned for structural determination. This is achieved through total correlation spectroscopy TOCSY-based experiments that correlate, through bonds, the nuclei already assigned to the rest of the side chain.

28.3.2
Chemical Shift Assignment in Nucleic Acids

Methods for chemical shift assignment of nucleic acids have been reviewed by Flinders and Dieckmann (2006), Tzakos et al. (2006), Fürtig et al. (2003), Cromsigt et al. (2001), Wijmenga and Van Buuren (1998), and Varani, Aboul-Ela and Allain (1996). These reviews also cover structural determination methods for nucleic acids.

Assignment of nucleic acids starts with the nucleobase spin systems, followed by correlation of each nucleobase to the corresponding sugar spin system, using HCN triple-resonance experiments in ^{13}C, ^{15}N-labeled nucleic acids. Assignment of the sugar spin system is then completed using a combination of HCCH-TOCSY and HCCH-COSY-based experiments (Flinders and Dieckmann, 2006). Sequential connections between nucleotides are made using nuclear overhauser effect spectroscopy (NOESY) experiments, or HCP and HCP-TOCSY experiments, which correlate the backbone ^{31}P nuclei to hydrogen and carbon atoms.

Base pairings can be assigned from the imino proton of guanines and uracyls, which have characteristic chemical shifts between 10 and 15 ppm. These protons exchange with water, and therefore can only be observed when involved in hydrogen bonding. A homonuclear 2D NOESY experiment correlates those resonances with neighboring protons, both from the nucleotide involved in the base pairing and from sequential nucleotides in the same strand (Tzakos et al., 2006). HNN-COSY experiments correlate the nucleotides in a base pair using scalar couplings between the nitrogen atom of the hydrogen bond acceptor and the nitrogen and hydrogen of the hydrogen bond donor (Dingley and Grzesiek, 1998; Pervushin et al., 1998). Both methods are illustrated in Figure 28.3. ^{13}C, ^{15}N edited or filtered 2D and 3D NOESY experiments are used to decrease overlap in NOESY spectra, and identify NOEs for assignment and structure determination (Fürtig et al., 2003; Flinders and Dieckmann, 2006).

In single-stranded RNA, the experiments based on base pairing are not possible, and assignment is much more difficult. The lack of base stacking also leads to decreased spectral dispersion. Isotopic labeling of RNA is essential for sequential assignment of single-stranded regions, such as bulges in the RNA structure.

28.3.3
Chemical Shift Assignment in Carbohydrates

Carbohydrates are more diverse than either proteins or nucleic acids in their basic components and connections. A large number of sugars form the possible basic units of carbohydrates and the potential for branched carbohydrate chains indicate that there is not always a primary sequence in the way that polypeptides

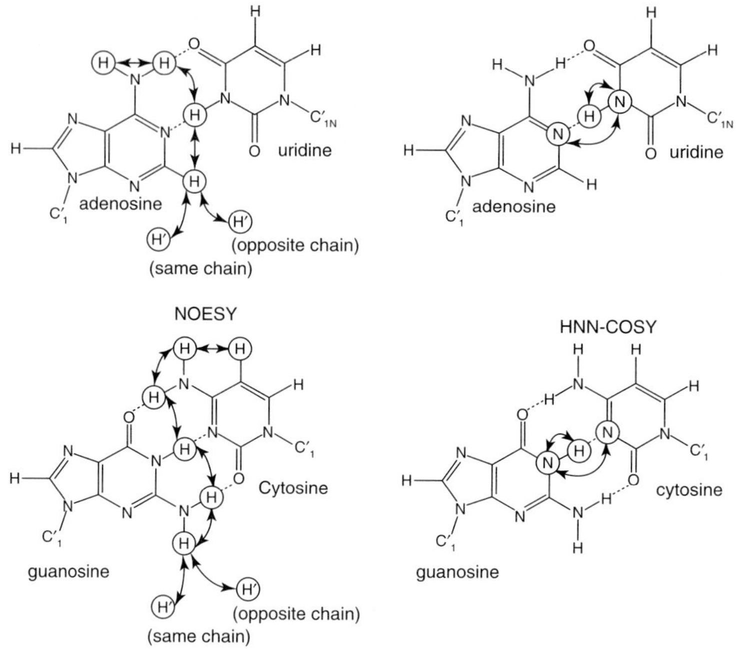

Fig. 28.3 Determination of base pairing in nucleic acids using NOESY or HNN-COSY spectra. Magnetization transfer steps are represented by arrows. In the NOESY experiment, magnetization transfer between protons occurs simultaneously in both directions of each arrow. During the HNN-COSY experiment, magnetization is transferred from the imino proton to the covalently bonded nitrogen, from there to the hydrogen-bonding nitrogen, and back to the imino proton for data acquisition.

and polynucleotides possess. This has led to different approaches to assignment being developed by different groups (Duus, Gotfredsen and Bock, 2000).

Some of the issues in chemical shift assignment of carbohydrates are related to the difficulty in obtaining labeled samples of carbohydrates. A number of experiments can be done using the natural abundance of ^{13}C, but this limitation greatly restricts the size of the molecules that can be studied. The labeling schemes described for carbohydrates in Section 28.2.1 can help to overcome this obstacle.

Unlike proteins and nucleic acids, the sequence of monomers in a carbohydrate is not clearly coded in the organism's genome. This means that the process of assignment often starts with the use of multiple techniques in addition to NMR, such as mass spectroscopy, to determine the sequence of sugar monomers in a carbohydrate.

A summary of the steps used for chemical shift assignment in carbohydrates was reviewed by Duus, Gotfredsen and Bock (2000). In short, the process for oligosaccharides starts by finding a characteristic chemical shift for each sugar monomer in the ^{1}H, and using experiments that correlate nuclei through bonds to assign the other atoms in each monomeric unit. Such experiments can include HSQC/HMQC/HMBC to correlate carbon and protons, TOCSY, DQF-COSY, and ^{13}C-edited HSQC-TOCSY. Intraresidue NOEs or ROEs (rotating-frame NOEs) can also be useful to assign individual resonances inside a monomer, and the anomeric conformation of a sugar can be determined from ^{13}C chemical shifts, and from a number of scalar coupling constants.

More recently, the H2BC (heteronuclear 2-bond correlation) experiment has been shown to provide unambiguous assignments and simplify carbohydrate spectra by correlating protons and carbons over two bonds (Petersen et al., 2006).

A strategy that can be used for carbohydrate assignment is to decrease the dimensionality of the experiment, decreasing acquisition time and focusing on a desired region. With the use of selective pulses, a 1D spectrum can be acquired corresponding to the trace of the 2D spectrum in a specific frequency, reflecting the same connectivities. Similarly, a 2D plane of a 3D spectrum can be acquired, focusing in a specific region of the spectrum (Uhrin and Barlow, 1997; Roumestand et al., 1999).

28.4
Structural Restraints

28.4.1
Nuclear Overhauser Effect

The nuclear overhauser effect (NOE) (Overhauser, 1953) arises from cross-relaxation processes that allow the transfer of magnetization from a nucleus to other nearby nuclei, through dipolar relaxation mechanisms. The build-up of an NOE between two spins i and j is given by the cross-relaxation rate, σ_{ij}, which is the difference between the rate for the flip-flip transition, W_2 (in which both spins are flipped in the same sense), and the rate for the flip-flop transition, W_0 (in which the spins are flipped in opposite senses):

$$\sigma_{ij} = (W_2 - W_0) \qquad (28.1)$$

In the homonuclear case, the resonance frequency of the two spins can be considered to be the same, ω_0, and the cross-relaxation rate will be

$$\sigma_{ij} \propto \frac{1}{r_{ij}^6}\{6J(2\omega_0) - J(0)\} \qquad (28.2)$$

where r_{ij} is the internuclear distance, $J(2\omega_0)$ is the spectral density function at twice the resonance frequency of the spins (for W_2), and $J(0)$ is the spectral density function at frequency zero (for W_0).

As it can be seen above, NOE build-up rates are proportional to the inverse of the sixth power of the distance, and therefore can be used to calculate distance restraints for structural calculations. NOESY experiments are used to measure the NOE build-up rates, and have a mixing time during which the magnetization evolves according to the NOE. However, the linearity between NOE build-up rates and distance is only complete when the mixing time is zero. In reality, the mixing time is kept small enough to allow this linearity approximation to be used. As the size of the molecule increases, and with it the rotational correlation time, dipolar relaxation mechanisms become more efficient, and the linear approximation is valid for shorter times. Spin diffusion (in which a peak in the NOESY spectrum is created via an indirect relaxation path) also increases with mixing time, as does intensity of the NOE signal before other relaxation processes cause it to diminish. Therefore, mixing times need to be optimized according to the characteristics of the molecule.

Multidimensional NOESY spectra correlate nuclei that are close in space, independently of being connected by bonds. Two-dimensional NOESY spectra correlate protons through space, and are useful for the structural determination of relatively small molecules, including small proteins. As the number of hydrogens in a molecule grows, so does the number of peaks, which start to overlap in a two-dimensional spectrum, and three-dimensional ^{15}N- or ^{13}C-edited NOESY spectra become necessary for resolving the signals. In the three-dimensional spectrum, the peaks from a two-dimensional NOESY spectrum are spread into planes according to the chemical shift of the nitrogen or carbon to which one of the protons is attached (Marion et al., 1989a; Marion et al., 1989b; Zuiderweg and Fesik, 1989).

Each peak in a NOESY spectrum represents a correlation through space between two nuclei, independently of a connection through bonds, and can be used to determine distance restraints for structure calculation. After the peaks in a NOESY spectrum are assigned to the corresponding atoms in the molecule, distance restraints are created for the nuclei correlated by each peak. This step can be time consuming, and often requires multiple rounds of interaction, because commonly many NOEs cannot be unambiguously assigned with chemical shift information only. This problem worsens with increasing protein size, or chemical shift imprecision (Mumenthaler et al., 1997). To circumvent this problem, a preliminary structure is often calculated based on the unambiguously assigned residues, and this structure is then used as additional information to resolve NOE assignment ambiguities, based on atom proximity in the preliminary structure. Alternatively, ambiguous assignments can be used if the different assignment possibilities are taken into account (Nilges, 1995).

This process can be automated in available software for the structural determination of proteins using iterative rounds of automated NOESY assignment, distance restraint determination, and structure calculation. Two examples of this type of software are CANDID (Herrmann, Güntert and Wüthrich, 2002) and ARIA (Linge et al., 2003), which differ amongst other things in their strategies for dealing with multiple assignment possibilities in the calculation process.

Owing to the relationship between distance and build-up rates, the strength of the NOE peaks decreases rapidly as the distance between nuclei increases (Wüthrich, 1986). In practice, that means that peaks in a NOESY spectrum of a biological macromolecule represent a correlation between two nuclei that are no more than 6 Å apart, and that peaks in the same spectrum can be calibrated according to their intensity to reflect the distance between the two nuclei being correlated. A common method for calibrating the distance according to the peak volume involves dividing the peaks into three classes (weak, medium, strong), each corresponding to a range of distances (Clore et al., 1985; Williamson et al., 1985). A different method involves assigning peaks corresponding to nuclei with known distances (such as between the amide proton of one residue, and the α-proton of the previous residue), and use that as a reference to calibrate the remaining distances (Gagné et al., 1997). In either method, it is better to set the upper limit of each distance too large than too small. Distance restraints that are too tight result in distorted molecular structures, while the presence of enough restraints will lead to the correct structure, even if they are too loose compared to the actual distances (Wijmenga and Van Buuren, 1998; Clore and Gronenborn, 1991).

Distance restraints based on NOE correlations are the staple of protein structure determination by solution NMR, and many proteins structures have been determined relying exclusively on them. For nonglobular macromolecules, however, they often do not offer enough information for a well-defined structure. NOE distance restraints do not extend beyond 6 Å and a good structure requires a large number of distance restraints between atoms that are distant from each other in the original residue sequence, but close in space. For elongated structures, cumulative uncertainties in the short-distance NOE restraints can lead to large uncertainties in the structure as a whole, requiring the use of other structural restraints. Examples of this situation are nucleic acids, in which pairwise interactions do not provide the same "long-range" restraints, carbohydrates, and some elongated helical proteins. In those cases, the use of other structural restraints discussed below becomes essential.

When the molecule needs to be deuterated to increase the transverse relaxation time, ^{1}H nuclei will only be available at positions that exchange with water. In this case, the low number of hydrogen nuclei limits the number of NOE restraints. To increase the number of protons available for distance restraints in proteins, while retaining the relaxation advantages of a mostly deuterated protein, different labeling schemes have been developed, that lead to deuteration of most of the protein, with hydrogens present in methyl groups on side chains (Tugarinov, Hwang and Kay, 2004; Tugarinov and Kay, 2004).

28.4.2
Torsion Angles

Using the Karplus relationship (Karplus, 1959), torsion angles can be calculated from three-bond scalar coupling between the nuclei involved. A number of pulse sequences have been developed to measure scalar couplings between relevant nuclei, and the corresponding Karplus relationships have been calibrated.

The scalar couplings $^3J_{HN\alpha}$, $^3J_{HNCO}$, $^3J_{HNC\beta}$, and $^3J_{Ci-1H\alpha i}$ correlate to backbone ϕ angles, and $^3J_{NiH\alpha i-1}$ correlates to ψ angles in proteins (discussed in Smith et al., 1996). $^3J_{HN\alpha}$ can be measured from the HNHA experiment to generate ϕ structural restraints (Barnwal et al., 2007; Kuboniwa et al., 1994; Pardi et al., 1984). Restraints for side chain χ_1 angles can be derived from $^3J_{NH\beta}$, measured from the HNHB experiment (Archer et al., 1991). Restraints for ψ angles can be derived from the intensity ratio $d_{N\alpha}/d_{\alpha N}$ in a ^{15}N NOESY-HSQC experiment (Gagné et al., 1994).

Alternatively, main-chain torsion angles can be calculated from chemical shift values by computer programs that use a database of protein structures and chemical shift values, such as TALOS (Cornilescu et al., 1999), SHIFTOR (Neal et al., 2006), and the more recent DANGLE (http://dangle.sourceforge.net/).

Numerous experiments can be used to measure scalar coupling constants in carbohydrates, as reviewed by Bush et al. (1999) and Duus, Gotfredsen and Bock (2000), and new experiments and Karplus correlations are still being determined (see Coxon, 2007; Mobli and Almond, 2007, for recent examples). Three-bond and one-bond coupling constants can be used to determine the anomeric configuration of the sugar (Duus, Gotfredsen and Bock, 2000).

In nucleic acids, the sugar pucker mode and a number of torsion angles can be determined from scalar couplings. The correlations between a large number of angles and the relevant coupling constants, and the experiments needed to determine such coupling constants, have been reviewed by Wijmenga and Van Buuren (1998). Phosphodiester backbone angles α and ζ in oligonucleotides can be calculated from cross-correlated relaxation rates of ^{1}H, ^{13}C-dipole, dipole and ^{31}P-chemical shift anisotropy (Richter et al., 2000).

28.4.3
Residual Dipolar Couplings (RDCs)

While interaction between two nuclei connected through bonds leads to line splitting because of scalar coupling, the interaction of nuclei through space leads to dipolar coupling, which arises from the interaction between the two magnetic dipoles of the nuclei. The magnitude of an observed dipolar coupling is given by the following equation:

$$D_{ij} = \frac{-\mu_0 \gamma_i \gamma_j \hbar}{4\pi^2 r^3} \left\langle \frac{3\cos^2\theta - 1}{2} \right\rangle \quad (28.3)$$

where γ is the gyromagnetic ratio of the interacting spins i and j, r is the distance between the nuclei, and θ is the angle between the applied magnetic field and the internuclear vector.

In solution NMR, the splitting due to dipolar coupling is not seen, being averaged to zero by rotational diffusion, unlike scalar coupling. (The splitting caused by scalar coupling is normally eliminated by decoupling techniques, but can be seen in undecoupled spectra.) When the molecules in the sample are partially aligned, some splitting due to dipolar coupling can be observed, named as residual dipolar coupling (RDC). By keeping the degree of alignment small, but nonzero, it is possible to have measurable effects from RDCs, and still keep the simplified spectrum obtained from motionally averaged molecules. From RDCs, orientational restraints can be derived, and are widely used in structural determination of proteins, carbohydrates, and nucleic acids. Different aspects of the use of RDCs in structural determination of biological macromolecules have been reviewed by Bax and Grishaev (2005), Blackledge (2005), Lipsitz and Tjandra (2004), and Prestegard et al. (2005). The influence of dipolar coupling on peak splitting in biological macromolecules was first shown for the main-chain amides of cyanometmyoglobin, a protein that spontaneously shows detectable partial alignment within the magnetic field from an NMR spectrometer (Tolman et al., 1995).

The effect of alignment on main-chain amide ^{15}N splitting in proteins provides a good example of RDC measurements. In a splitting-enhanced {^{1}H, ^{15}N}-HSQC spectrum (undecoupled in the ^{15}N dimension) (Tolman and Prestegard, 1996), splitting of the ^{15}N resonance in protein main-chain amides can be observed, and the coupling constant can be measured from the distance between the peaks. In solution NMR, when the molecules are tumbling isotropically, this splitting will reflect only the scalar coupling constant $^1J_{HN}$ between the proton and nitrogen. When the molecules are partially aligned by any method, the splitting will reflect the sum of the scalar coupling, $^1J_{HN}$, and the residual dipolar coupling, D_{HN}. Another method to measure the splitting involves the use of in-phase/antiphase (IPAP) spectra, in which the two components of the doublet are seen in separate spectra, therefore not increasing spectral complexity compared to the decoupled spectrum (Ottiger et al., 1998). In J-resolved spectroscopy (Aue et al., 1976), the proton resonances evolve with the chemical shift in t_2, and with J-coupling in t_1. Unlike the HSQC experiments, that take advantage of heteronuclear signal dispersion in the indirectly detected dimension, J-resolved spectroscopy requires good dispersion in the proton spectrum.

From Eq. (28.3) above, the angle between the bond vector and the magnetic field can be determined. It is important to notice that there are always two opposite orientations of the bond vector that satisfy the equation for any RDC value, and therefore RDC information should be obtained from multiple alignment media for structural calculations, or supplemented with other experimental distance information (Blackledge, 2005). RDC information can be used for structural calculation using simulated annealing programs (Tjandra and Bax, 1997). Orientational restraints are very important for the structural determination of carbohydrates, nucleic acids, and large deuterated proteins, for which it is usually not possible to acquire enough NOE distance restraints to resolve the structure. Numerous pulse sequences are available for the determination of RDCs between different nuclei in nucleic acids (reviewed by Latham et al., 2005).

Alignment of molecules with the magnetic field can be generated with the use of paramagnetic metals (Tolman et al., 1995), or by restricting the movement of the molecule with alignment media, which will result in alignment according to the shape of the molecule. Different alignment media that can be used in biological samples include liquid crystal media (Tjandra and Bax, 1997), a suspension of filamentous bacteriophage (Hansen et al., 1998), lipid bicelles (Sanders et al., 1994; Bax and Tjandra, 1997), or stretched polyacrylamide gels (Sass et al., 2000; Tycko et al., 2000).

28.4.4
Hydrogen Bonds

Experimentally detected hydrogen bonds can provide useful structural restraints. In proteins, hydrogen bonds patterns can reveal secondary structure, while in nucleic acids, hydrogen bonds are present in every base pairing. They can be measured either indirectly, from the exchange with the solvent (for example, in Krishna et al., 2004), or directly, via experiments that transfer magnetization across the hydrogen bond through the two-bond scalar couplings (Grzesiek et al., 2004). For structural determination of nucleic acids, the use of NMR experiments that transfer magnetization through H-bond scalar couplings has become widespread.

28.4.5
Paramagnetic Centers

Paramagnetic centers in biological macromolecules can occur naturally, as in metalloproteins, or be added by different methods, such as by replacing a naturally-occurring diamagnetic ion with a paramagnetic ion (Lee and Sykes, 1983), adding a nitroxide spin label (Battiste and Wagner, 2000; Gaponenko et al., 2000), or by engineering a binding site for a paramagnetic ion (for examples, see Ma and Opella, 2000; Donaldson et al., 2001; Wöhnert et al., 2003; Su et al., 2008). Different techniques have been developed to incorporate paramagnetic centers in nucleic acids, for instance using 4-thiouracil or 4-thiothymidine, which is then coupled to a spin label via the sulfur group (Ramos and Varani, 1998), or using site-specifically 2′-amino-modified nucleic acids (Edwards and Sigurdsson, 2007). The use of chemical modifications to create lanthanide-binding sites for paramagnetic structural restraints is becoming increasingly popular (Otting, 2008).

Contact shifts caused by through-bond interaction between the nucleus and the paramagnetic center can be a powerful source of structural information. The presence of an unpaired electron in an NMR sample will also affect the chemical shifts and relaxation of other nuclei close enough to the unpaired electron to present dipolar interactions with it. In practice, this can be as much as 20 Å, which is a longer distance than can be observed by NOE interactions. Having a paramagnetic center in a known position of the molecule, then, can be an effective way to determine structural constraints for nuclei located at close or medium distance from that position. The effect of a paramagnetic center on nearby nuclei can be used to calculate a number of constraints for structural determination, based on pseudocontact shifts, relaxation enhancement, contact shifts, and cross-relaxation (Bertini et al., 2002; Bertini et al., 2005). Pulse sequences may have to be modified when characterizing biomolecules containing paramagnetic centers.

Paramagnetic centers also cause the molecules to align along the external magnetic field, and this alignment can used to extract residual dipolar couplings (Tolman et al., 1995). The combination of RDC and paramagnetic restraints has been used to determine a protein structure without NOE restraints (Hus et al., 2000). As with RDC, the use of paramagnetic restraints can be very beneficial to molecules for which it is not possible to obtain sufficient NOE restraints for accurate structural determination, such as nucleic acids.

28.4.6
Structural Information from Chemical Shifts

Different conformations in biomolecules can have strong effects on the chemical shift values of nuclei in specific positions, and this information can be used to help structural determination. In proteins, regions of secondary structure can be predicted from the chemical shift index (Wishart and Nip, 1998; Wishart and Case, 2001). In brief, chemical shift values of $^1H_\alpha$, $^1H_\beta$, ^1H_N, $^{13}C_\alpha$, $^{13}C_\beta$, ^{13}CO, and ^{15}N for each residue can be compared to typical random coil values, and the difference is usually indicative of the presence or absence of helical or sheet secondary structure. For an element of secondary structure to be predicted reliably, it should be predicted by at least three amino acid residues in a row. The use of multiple nuclei also increases the reliability of the method. Methods to extract information from chemical shift values for proteins have been reviewed by Szilágyi (1995) and Case (2000). A number of uses of chemical shift values to aid in assignment and structural determination of nucleic acids have been described by Wijmenga et al. (1997), Altona et al. (2000), Cromsigt et al. (2001) and Fürtig et al. (2003).

In partially oriented samples, chemical shift anisotropy (CSA) can also yield structural information (Lipsitz and Tjandra, 2001; Lipsitz and Tjandra, 2003; Ying et al., 2006; Choy et al., 2001; Cornilescu and Bax, 2000; Wu et al., 2001; Yu and Prestegard, 2006). Chemical shifts in a molecule are usually anisotropic by nature, depending on the orientation in the magnetic field. When a molecule tumbles isotropically in solution NMR, the averaging of chemical shift values gives rise to the sharp lines seen in solution NMR. In solid state NMR, the anisotropy can be seen in the "powder pattern" if the molecules are not oriented. In partially aligned samples, the effect of CSA can be seen as an offset of the chemical shift in relation to its value in an isotropic sample.

Recently, methods for determining protein structures based on chemical shift values alone have been developed, CHESHIRE (Cavalli et al., 2007), and CS-ROSETTA (Shen et al., 2008). Both methods use molecular fragment replacement approaches, used previously on the ROSETTA *ab initio* structure prediction method (Bradley et al., 2005) and to analyze RDCs (Rohl and Baker, 2002; Delaglio et al., 2000), amongst other applications. By eliminating the need to acquire separate spectra for structural restraints, the experimental time needed to determine protein structures can be greatly reduced, and structural determination by NMR can be extended to less stable proteins. A current limitation of this approach is protein size; both methods are effective only for proteins up to 15 kDa. The newly developed Cam-Dock procedure extends this approach to protein-protein complexes (Montalvão et al., 2008).

28.5
Molecular Modeling and Structure Calculation

Structural restraints determined as described above are used to calculate the structure by molecular modeling. Simulated annealing and molecular dynamics, either in the Cartesian space of atomic positions or in the torsion angle space, are the most common methods for structure calculation from restraints derived from NMR experimental data, starting from an extended structure or from a preliminary model (Clore and Gronenborn, 1998). The different experimental restraints are incorporated into the calculations as pseudo-energy functions added to the molecular energy representations, forming a hybrid energy term $E_{\text{hybrid}} = E_{\text{phys}} + w_{\text{data}} E_{\text{data}}$, where E_{phys} is the sum of physical energy terms such as bond length and nonbonded interactions, and E_{data} assesses the consistency between the model structure and the different types of experimentally measured restraints (Brünger and Nilges, 1993).

The initial high-temperature stage of the simulated annealing calculations allows for wider exploration of the conformational space by overcoming energy barriers, followed by slow cooling while sampling the conformational space by molecular dynamics. Molecular modeling in Cartesian coordinates is based on the numerical solution of Newton's equations of motion, and the position of each atom is modeled with energy representations for bond angles and lengths (Nilges et al., 1988). Molecular dynamics in the torsion angle space assumes the distance and bond angle between the pair of atoms connected by a bond to be fixed, and allows them to rotate around the bond. The reduced number of degrees of freedom and larger time steps in the simulated annealing lead to faster calculations and convergence (Stein et al., 1997). Common programs that use molecular dynamics for structural calculations from NMR data include Xplor-NIH (Schwieters et al., 2003), CNS (Brünger, 1998; Brunger, 2007) and CYANA (Güntert, 2004), or its predecessor DYANA (Güntert et al., 1997). While computational power was a limiting factor when these methods were first developed, now these calculations are trivial and can be performed in most PCs with the appropriate software.

The iterative steps of NOESY peaks assignment and structural calculation have been automated together in a number of programs, such as ARIA (Nilges et al., 1997; Linge et al., 2003), CANDID (Herrmann et al., 2002), CYANA (Güntert, 2004), and others (Güntert, 2003). These programs require a chemical shift list as input, with the resonance values for the atoms that will be involved in NOE contacts.

More recently, methods based on molecular fragment replacement have been developed for proteins, in which model structures for each fragment are retrieved from a database according to the fragment sequence and experimental data (Bowers, Strauss and Baker, 2000). This is the same approach used for *ab initio* protein structure calculation by the program ROSETTA (Simons et al., 1997; Simons et al., 1999) and has been applied to structure calculation based on chemical shifts only (Cavalli et al., 2007; Shen et al., 2008; Montalvão et al., 2008) and to the analysis of RDC data (Rohl and Baker, 2002). These calculations are more computationally demanding than molecular dynamics-based methods.

Biomolecular structures calculated using any of the methods above need to be validated, and there are a number of programs for structural validation such as VADAR (Willard et al., 2003), PROCHECK-NMR and AQUA (Laskowski et al., 1996), and WHAT IF (Doreleijers et al., 1999; Vriend, 1990).

References

Altona, C., Faber, D.H. and Westra Hoekzema, A.J.A. (2000) *Magn. Reson. Chem.*, **38**(2), 95–107.

Archer, S.J., Ikura, M., Torchia, D.A. and Bax, A. (1991) *J. Magn. Reson.*, **95**(3), 636–641.

Atreya, H.S., Chary, K.V.R. and Govil, G. (2002) *Curr. Sci.*, **83**(11), 1372–1376.

Aue, W.P., Karhan, J. and Ernst, R.R. (1976) *J. Chem. Phys.*, **64**(10), 4226–4227.

Barnwal, R.P., Rout, A.K., Chary, K.V.R. and Atreya, H.S. (2007) *J. Biomol. NMR*, **39**(4), 259–263.

Battiste, J.L. and Wagner, G. (2000) *Biochemistry*, **39**(18), 5355–5365.

Bax, A. and Grishaev, A. (2005) *Curr. Opin. Struct. Biol.*, **15**(5), 563–570.

Bax, A. and Ikura, M. (1991) *J. Biomol. NMR*, **1**(1), 99–104.

Bax, A. and Tjandra, N. (1997) *J. Biomol. NMR*, **10**(3), 289–292.

Bertini, I., Luchinat, C. and Parigi, G. (2002) *Concepts Magn. Reson. A Bridging Educ. Res.*, **14**(4), 259–286.

Bertini, I., Luchinat, C., Parigi, G. and Pierattelli, R. (2005) *Chem. Bio. Chem.*, **6**(9), 1536–1549.

Blackledge, M. (2005) *Prog. Nucl. Magn. Reson. Spectrosc.*, **46**(1), 23–61.

Bowers, P.M., Strauss, C.E.M. and Baker, D. (2000) *J. Biomol. NMR*, **18**(4), 311–318.

Bradley, P., Misura, K.M.S. and Baker, D. (2005) *Science*, **309**(5742), 1868–1871.

Brüggert, M., Rehm, T., Shanker, S., Georgescu, J. and Holak, T.A. (2003) *J. Biomol. NMR*, **25**(4), 335–348.

Brünger, A.T. (1998) *Acta Crystallogr. D Biol. Crystallogr.*, **54**(5), 905–921.

Brünger, A.T. (2007) *Nat. Protoc.*, **2**(11), 2728–2733.

Brünger, A.T. and Nilges, M. (1993) *Q. Rev. Biophys.*, **26**(1), 49–125.

Brüschweiler, R. and Zhang, F. (2004) *J. Chem. Phys.*, **120**(11), 5253–5260.

Brutscher, B. (2004) *J. Biomol. NMR*, **29**(1), 57–64.

Brutscher, B., Simorre, J.P., Caffrey, M.S. and Marion, D. (1994) *J. Magn. Reson. B*, **105**(1), 77–82.

Bush, C.A., Martin-Pastor, M. and Imberty, A. (1999) *Annu. Rev. Biophys. Biomol. Struct.*, **28**, 269–293.

Case, D.A. (2000) *Curr. Opin. Struct. Biol.*, **10**(2), 197–203.

Cavalli, A., Salvatella, X., Dobson, C.M. and Vendruscolo, M. (2007) *Proc. Natl. Acad. Sci. U.S.A.*, **104**(23), 9615–9620.

Cavanagh, J., Fairbrother, W.J., Palmer, A.G. and Skelton, N.J. (1996) *Protein NMR Spectroscopy: Principles and Practice*, Academic Press, San Diego.

Choy, W.Y., Tollinger, M., Mueller, G.A. and Kay, L.E. (2001) *J. Biomol. NMR*, **21**(1), 31–40.

Clore, G.M. and Gronenborn, A.M. (1991) *Annu. Rev. Biophys. Biophys. Chem.*, **20**, 29–63.

Clore, G.M. and Gronenborn, A.M. (1998) *Proc. Natl. Acad. Sci. U.S.A.*, **95**(11), 5891–5898.

Clore, G.M., Gronenborn, A.M., Brünger, A.T. and Karplus, M. (1985) *J. Mol. Biol.*, **186**(2), 435–455.

Clubb, R.T., Thanabal, V. and Wagner, G. (1992) *J. Magn. Reson.*, **97**(1), 213–217.

Cornilescu, G. and Bax, A. (2000) *J. Am. Chem. Soc.*, **122**(41), 10143–10154.

Cornilescu, G., Delaglio, F. and Bax, A. (1999) *J. Biomol. NMR*, **13**(3), 289–302.

Coxon, B. (2007) *Carbohydr. Res.*, **342**(8), 1044–1054.

Cromsigt, J., Van Buuren, B., Schleucher, J. and Wijmenga, S. (2001) *Methods Enzymol.*, **338**, 371–399.

Dayie, K.T. (2008) *Int. J. Mol. Sci.*, **9**(7), 1214–1240.

Delaglio, F., Kontaxis, G. and Bax, A. (2000) *J. Am. Chem. Soc.*, **122**(9), 2142–2143.

Dingley, A.J. and Grzesiek, S. (1998) *J. Am. Chem. Soc.*, **120**(33), 8293–8297.

Donaldson, L.W., Skrynnikov, N.R., Choy, W.Y., Muhandiram, D.R., Sarkar, B., Forman-Kay, J.D. and Kay, L.E. (2001) *J. Am. Chem. Soc.*, **123**(40), 9843–9847.

Doreleijers, J.F., Vriend, G., Raves, M.L. and Kaptein, R. (1999) *Proteins Struct. Funct. Genet.*, **37**(3), 404–416.

Duus, J.Ø., Gotfredsen, C.H. and Bock, K. (2000) *Chem. Rev.*, **100**(12), 4589–4614.

Dyson, H.J. and Wright, P.E. (2005) *Nat. Rev. Mol. Cell Biol.*, **6**(3), 197–208.

Edwards, T.E. and Sigurdsson, S.T. (2007) *Nat. Protoc.*, **2**(8), 1954–1962.

Ferentz, A.E. and Wagner, G. (2000) *Q. Rev. Biophys.*, **33**(1), 29–65.

Flinders, J. and Dieckmann, T. (2006) *Prog. Nucl. Magn. Reson. Spectrosc.*, **48**(2–3), 137–159.

Freeman, R. and Kupce, E. (2003) *J. Biomol. NMR*, **27**(2), 101–113.

Frydman, L., Scherf, T. and Lupulescu, A. (2002) *Proc. Natl. Acad. Sci. U.S.A.*, **99**(25), 15858–15862.

Fürtig, B., Richter, C., Wöhnert, J. and Schwalbe, H. (2003) *Chem. Bio. Chem.*, **4**(10), 936–962.

Gagné, S.M., Li, M.X. and Sykes, B.D. (1997) *Biochemistry*, **36**(15), 4386–4392.

Gagné, S.M., Tsuda, S., Li, M.X., Chandra, M., Smillie, L.B. and Sykes, B.D. (1994) *Protein Sci.*, **3**(11), 1961–1974.

Gaponenko, V., Howarth, J.W., Columbus, L., Gasmi-Seabrook, G., Yuan, J., Hubbell, W.L. and Rosevear, P.R. (2000) *Protein Sci.*, **9**(2), 302–309.

Gardner, K.H. and Kay, L.E. (1998) *Annu. Rev. Biophys. Biomol. Struct.*, **27**, 357–406.

Grzesiek, S., Anglister, J., Ren, H. and Bax, A. (1993) *J. Am. Chem. Soc.*, **115**(10), 4369–4370.

Grzesiek, S. and Bax, A. (1992a) *J. Am. Chem. Soc.*, **114**(16), 6291–6293.

Grzesiek, S. and Bax, A. (1992b) *J. Magn. Reson.*, **96**(2), 432–440.

Grzesiek, S., Cordier, F., Jaravine, V. and Barfield, M. (2004) *Prog. Nucl. Magn. Reson. Spectrosc.*, **45**(3–4), 275–300.

Güntert, P. (2003) *Prog. Nucl. Magn. Reson. Spectrosc.*, **43**(3–4), 105–125.

Güntert, P. (2004) *Methods Mol. Biol. (Clifton)*, **278**, 353–378.

Güntert, P., Mumenthaler, C. and Wüthrich, K. (1997) *J. Mol. Biol.*, **273**(1), 283–298.

Hansen, M.R., Mueller, L. and Pardi, A. (1998) *Nat. Struct. Biol.*, **5**(12), 1065–1074.

Helgstrand, M., Kraulis, P., Allard, P. and Härd, T. (2000) *J. Biomol. NMR*, **18**(4), 329–336.

Herrmann, T., Güntert, P. and Wüthrich, K. (2002) *J. Mol. Biol.*, **319**(1), 209–227.

Hoch, J.C. and Stern, A.S. (2001) *Methods Enzymol.*, **338**, 159–178.

Hore, P.J. (1983) *J. Magn. Reson.*, **55**(2), 283–300.

Hus, J.C., Marion, D. and Blackledge, M. (2000) *J. Mol. Biol.*, **298**(5), 927–936.

Jung, Y.S. and Zweckstetter, M. (2004) *J. Biomol. NMR*, **30**(1), 11–23.

Kanelis, V., Forman-Kay, J.D. and Kay, L.E. (2002) *IUBMB Life*, **52**(6), 291–302.

Karplus, M. (1959) *J. Chem. Phys.*, **30**(1), 11–15.

Kay, L.E. (2005) *J. Magn. Reson.*, **173**(2), 193–207.

Kay, L.E., Ikura, M., Tschudin, R. and Bax, A. (1990) *J. Magn. Reson.*, **89**(3), 496–514.

Kelly, A.E., Ou, H.D., Withers, R. and Dötsch, V. (2002) *J. Am. Chem. Soc.*, **124**(40), 12013–12019.

Kim, S. and Szyperski, T. (2003) *J. Am. Chem. Soc.*, **125**(5), 1385–1393.

Krishna, M.M.G., Hoang, L., Lin, Y. and Englander, S.W. (2004) *Methods*, **34**(1), 51–64.

Kuboniwa, H., Grzesiek, S., Delaglio, F. and Bax, A. (1994) *J. Biomol. NMR*, **4**(6), 871–878.

Kupce, E. and Freeman, R. (2004) *J. Am. Chem. Soc.*, **126**(20), 6429–6440.

Kupce, E. and Freeman, R. (2008) *Prog. Nucl. Magn. Reson. Spectrosc.*, **52**(1), 22–30.

Kupce, E., Nishida, T. and Freeman, R. (2003) *Prog. Nucl. Magn. Reson. Spectrosc.*, **42**(3–4), 95–122.

Lagoja, I.M. and Herdewijn, P. (2002) *Synthesis*, **2002**(3), 301–314.

Laskowski, R.A., Rullmann, J.A.C., MacArthur, M.W., Kaptein, R. and Thornton, J.M. (1996) *J. Biomol. NMR*, **8**(4), 477–486.

Latham, M.P., Brown, D.J., McCallum, S.A. and Pardi, A. (2005) *Chem. Bio. Chem.*, **6**(9), 1492–1505.

Lee, L. and Sykes, B.D. (1983) *Biochemistry*, **22**(19), 4366–4373.

Leutner, M., Gschwind, R.M., Liermann, J., Schwarz, C., Gemmecker, G. and Kessler, H. (1998) *J. Biomol. NMR*, **11**(1), 31–43.

Li, M.X., Corson, D.C. and Sykes, B.D. (2002) *Methods Mol. Biol. (Clifton)*, **173**, 255–265.

Lindorff-Larsen, K., Kristjansdottir, S., Teilum, K., Fieber, W., Dobson, C.M., Poulsen, F.M. and Vendruscolo, M. (2004) *J. Am. Chem. Soc.*, **126**(10), 3291–3299.

Linge, J.P., Habeck, M., Rieping, W. and Nilges, M. (2003) *Bioinformatics*, **19**(2), 315–316.

Lipsitz, R.S. and Tjandra, N. (2001) *J. Am. Chem. Soc.*, **123**(44), 11065–11066.

Lipsitz, R.S. and Tjandra, N. (2003) *J. Magn. Reson.*, **164**(1), 171–176.

Lipsitz, R.S. and Tjandra, N. (2004) *Annu. Rev. Biophys. Biomol. Struct.*, **33**, 387–413.

Louis, J.M., Martin, R.G., Clore, G.M. and Gronenborn, A.M. (1998) *J. Biol. Chem.*, **273**(4), 2374–2378.

Ma, C. and Opella, S.J. (2000) *J. Magn. Reson.*, **146**(2), 381–384.

Mandelshtam, V.A. (2000) *J. Magn. Reson.*, **144**(2), 343–356.

Marion, D., Driscoll, P.C., Kay, L.E., Wingfield, P.T., Bax, A., Gronenborn, A.M. and Clore, G.M. (1989a) *Biochemistry*, **28**(15), 6150–6156.

Marion, D., Kay, L.E., Sparks, S.W., Torchia, D.A. and Bax, A. (1989b) *J. Am. Chem. Soc.*, **111**(4), 1515–1517.

Martin-Pastor, M. and Allen Bush, C. (1999) *Biochemistry*, **38**(25), 8045–8055.

Milligan, J.F., Groebe, D.R., Witherell, G.W. and Uhlenbeck, O.C. (1987) *Nucleic Acids Res.*, **15**(21), 8783–8798.

Mobli, M. and Almond, A. (2007) *Org. Biomol. Chem.*, **5**(14), 2243–2251.

Montalvão, R.W., Cavalli, A., Salvatella, X., Blundell, T.L. and Vendruscolo, M. (2008) *J. Am. Chem. Soc.*, **130**(47), 15990–15996.

Morgan, W.D., Kragt, A. and Feeney, J. (2000) *J. Biomol. NMR*, **17**(4), 337–347.

Muhandiram, D.R. and Kay, L.E. (1994) *J. Magn. Reson. B*, **103**(3), 203–216.

Mumenthaler, C., Güntert, P., Braun, W. and Wüthrich, K. (1997) *J. Biomol. NMR*, **10**(4), 351–362.

Neal, S., Berjanskii, M., Zhang, H. and Wishart, D.S. (2006) *Magn. Reson. Chem.*, **44**(S1), S158–S167.

Nelissen, F.H.T., van Gammeren, A.J., Tessari, M., Girard, F.C., Heus, H.A. and Wijmenga, S.S. (2008) *Nucleic Acids Res.*, **36**(14), e89.

Nilges, M. (1995) *J. Mol. Biol.*, **245**(5), 645–660.

Nilges, M., Clore, G.M. and Gronenborn, A.M. (1988) *FEBS Lett.*, **239**(1), 129–136.

Nilges, M., MacIas, M.J., O'Donoghue, S.I. and Oschkinat, H. (1997) *J. Mol. Biol.*, **269**(3), 408–422.

Osterhout, J.J. Jr, Baldwin, R.L., York, E.J., Stewart, J.M., Dyson, H.J. and Wright, P.E. (1989) *Biochemistry*, **28**(17), 7059–7064.

Otting, G. (2008) *J. Biomol. NMR*, **42**(1), 1–9.

Ottiger, M., Delaglio, F. and Bax, A. (1998) *J. Magn. Reson.*, **131**(2), 373–378.

Overhauser, A.W. (1953) *Phys. Rev.*, **92**(2), 411–415.

Palmer, A.G. III (2001) *Annu. Rev. Biophys. Biomol. Struct.*, **30**, 129–155.

Pardi, A., Billeter, M. and Wüthrich, K. (1984) *J. Mol. Biol.*, **180**(3), 741–751.

Pervushin, K., Ono, A., Fernández, C., Szyperski, T., Kainosho, M. and Wüthrich, K. (1998) *Proc. Natl. Acad. Sci. U.S.A.*, **95**(24), 14147–14151.

Pervushin, K., Riek, R., Wider, G. and Wüthrich, K. (1997) *Proc. Natl. Acad. Sci. U.S.A.*, **94**(23), 12366–12371.

Petersen, B.O., Vinogradov, E., Kay, W., Wurtz, P., Nyberg, N.T., Duus, J.Ø. and Sorensen, O.W. (2006) *Carbohydr. Res.*, **341**(4), 550–556.

Piotto, M., Saudek, V. and Sklenár, V. (1992) *J. Biomol. NMR*, **2**(6), 661–665.

Prestegard, J.H., Mayer, K.L., Valafar, H. and Benison, G.C. (2005) *Methods Enzymol.*, **394**, 175–209.

Ramos, A. and Varani, G. (1998) *J. Am. Chem. Soc.*, **120**(42), 10992–10993.

René, B., Masliah, G., Zargarian, L., Mauffret, O. and Fermandjian, S. (2006) *J. Biomol. NMR*, **36**(3), 137–146.

Richter, C., Reif, B., Griesinger, C. and Schwalbe, H. (2000) *J. Am. Chem. Soc.*, **122**(51), 12728–12731.

Rohl, C.A. and Baker, D. (2002) *J. Am. Chem. Soc.*, **124**(11), 2723–2729.

Roumestand, C., Delay, C., Gavin, J.A. and Canet, D. (1999) *Magn. Reson. Chem.*, **37**(7), 451–478.

Rovnyak, D., Frueh, D.P., Sastry, M., Sun, Z.Y.J., Stern, A.S., Hoch, J.C. and Wagner, G. (2004) *J. Magn. Reson.*, **170**(1), 15–21.

Sanders, C.R. II, Hare, B.J., Howard, K.P. and Prestegard, J.H. (1994) *Prog. Nucl. Magn. Reson. Spectrosc.*, **26**(PART V), 421–444.

Sass, H.J., Musco, G., Stahl, S.J., Wingfield, P.T. and Grzesiek, S. (2000) *J. Biomol. NMR*, **18**(4), 303–309.

Schanda, P., Kupce, E. and Brutscher, B. (2005) *J. Biomol. NMR*, **33**(4), 199–211.

Schwieters, C.D., Kuszewski, J.J., Tjandra, N. and Clore, G.M. (2003) *J. Magn. Reson.*, **160**(1), 65–73.

Shen, Y., Lange, O., Delaglio, F., Rossi, P., Aramini, J.M., Liu, G., Eletsky, A., Wu, Y., Singarapu, K.K., Lemak, A., Ignatchenko, A., Arrowsmith, C.H., Szyperski, T., Montelione, G.T., Baker, D. and Bax, A. (2008) *Proc. Natl. Acad. Sci. U.S.A.*, **105**(12), 4685–4690.

Simons, K.T., Bonneau, R., Ruczinski, I. and Baker, D. (1999) *Proteins Struct. Funct. Genet.*, **37**(Suppl. 3), 171–176.

Simons, K.T., Kooperberg, C., Huang, E. and Baker, D. (1997) *J. Mol. Biol.*, **268**(1), 209–225.

Sklenár, V. and Bax, A. (1987) *J. Magn. Reson.*, **74**(3), 469–479.

Slupsky, C.M., Boyko, R.F., Booth, V.K. and Sykes, B.D. (2003) *J. Biomol. NMR*, **27**(4), 313–321.

Smith, L.J., Bolin, K.A., Schwalbe, H., MacArthur, M.W., Thornton, J.M. and Dobson, C.M. (1996) *J. Mol. Biol.*, **255**(3), 494–506.

Stein, E.G., Rice, L.M. and Brünger, A.T. (1997) *J. Magn. Reson.*, **124**(1), 154–164.

Stonehouse, J., Clowes, R.T., Shaw, G.L., Keeler, J. and Laue, E.D. (1995) *J. Biomol. NMR*, **5**(3), 226–232.

Su, X.C., McAndrew, K., Huber, T. and Otting, G. (2008) *J. Am. Chem. Soc.*, **130**(5), 1681–1687.

Szilágyi, L. (1995) *Prog. Nucl. Magn. Reson. Spectrosc.*, **27**(4), 325–443.

Szyperski, T., Wider, G., Bushweller, J.H. and Wüthrich, K. (1993) *J. Am. Chem. Soc.*, **115**, 9307–9308.

Tjandra, N. and Bax, A. (1997) *Science*, **278**(5340), 1111–1114.

Tolman, J.R., Flanagan, J.M., Kennedy, M.A. and Prestegard, J.H. (1995) *Proc. Natl. Acad. Sci. U.S.A.*, **92**(20), 9279–9283.

Tolman, J.R. and Prestegard, J.H. (1996) *J. Magn. Reson. B*, **112**(3), 269–274.

Torizawa, T., Shimizu, M., Taoka, M., Miyano, H. and Kainosho, M. (2004) *J. Biomol. NMR*, **30**(3), 311–325.

Tugarinov, V., Hwang, P.M. and Kay, L.E. (2004) *Annu. Rev. Biochem.*, **73**, 107–146.

Tugarinov, V. and Kay, L.E. (2004) *J. Biomol. NMR*, **28**(2), 165–172.

Tycko, R., Blanco, F.J. and Ishii, Y. (2000) *J. Am. Chem. Soc.*, **122**(38), 9340–9341.

Tzakos, A.G., Grace, C.R.R., Lukavsky, P.J. and Riek, R. (2006) *Annu. Rev. Biophys. Biomol. Struct.*, **35**, 319–342.

Uhrin, D. and Barlow, P.N. (1997) *J. Magn. Reson.*, **126**(2), 248–255.

Ulrich, E.L., Akutsu, H., Doreleijers, J.F., Harano, Y., Ioannidis, Y.E., Lin, J., Livny, M., Mading, S., Maziuk, D., Miller, Z., Nakatani, E., Schulte, C.F., Tolmie, D.E., Kent Wenger, R., Yao, H. and Markley, J.L. (2008) *Nucleic Acids Res.*, **36**(Suppl. 1), D402–D408.

Varani, G., Aboul-Ela, F. and Allain, F.H.T. (1996) *Prog. Nucl. Magn. Reson. Spectrosc.*, **29**(1–2), 51–127.

Vinarov, D.A., Lytle, B.L., Peterson, F.C., Tyler, E.M., Volkman, B.F. and Markley, J.L. (2004) *Nat. Methods*, **1**(2), 149–153.

Volk, J., Herrmann, T. and Wüthrich, K. (2008) *J. Biomol. NMR*, **41**(3), 127–138.

Vriend, G. (1990) *J. Mol. Graph.*, **8**(1), 52–56.

Wagner, G. and Wüthrich, K. (1982) *J. Mol. Biol.*, **155**(3), 347–366.

Wider, G. and Wüthrich, K. (1999) *Curr. Opin. Struct. Biol.*, **9**(5), 594–601.

Wijmenga, S.S., Kruithof, M. and Hilbers, C.W. (1997) *J. Biomol. NMR*, **10**(4), 337–350.

Wijmenga, S.S. and Van Buuren, B.N.M. (1998) *Prog. Nucl. Magn. Reson. Spectrosc.*, **32**(4), 287–387.

Willard, L., Ranjan, A., Zhang, H., Monzavi, H., Boyko, R.F., Sykes, B.D. and Wishart, D.S. (2003) *Nucleic Acids Res.*, **31**(13), 3316–3319.

Williamson, M.P., Havel, T.F. and Wüthrich, K. (1985) *J. Mol. Biol.*, **182**(2), 295–315.

Wishart, D.S. and Case, D.A. (2001) *Methods Enzymol.*, **338**, 3–34.

Wishart, D.S. and Nip, A.M. (1998) *Biochem. Cell Biol.*, **76**(2–3), 153–163.

Wittekind, M. and Mueller, L. (1993) *J. Magn. Reson. B*, **101**(2), 201–205.

Wöhnert, J., Franz, K.J., Nitz, M., Imperiali, B. and Schwalbe, H. (2003) *J. Am. Chem. Soc.*, **125**(44), 13338–13339.

Wu, Z., Tjandra, N. and Bax, A. (2001) *J. Am. Chem. Soc.*, **123**(15), 3617–3618.

Wüthrich, K. (1986) *NMR of Proteins and Nucleic Acids*, John Wiley & Sons, Inc., New York.

Yi, X., Venot, A., Glushka, J. and Prestegard, J.H. (2004) *J. Am. Chem. Soc.*, **126**(42), 13636–13638.

Ying, J., Grishaev, A., Bryce, D.L. and Bax, A. (2006) *J. Am. Chem. Soc.*, **128**(35), 11443–11454.

Yu, F. and Prestegard, J.H. (2006) *Biophys. J.*, **91**(5), 1952–1959.

Zuiderweg, E.R.P. and Fesik, S.W. (1989) *Biochemistry*, **28**(6), 2387–2391.

Further Reading

Cavanagh, J., Fairbrother, W.J., Palmer, III A.G., Skelton, N.J. and Rance, M. (2006) *Protein NMR Spectroscopy: Principles and Practice*, 2nd edn, Academic Press, San Diego.

Evans, J.N.S. (1995) *Biomolecular NMR Spectroscopy*, Oxford University Press, New York.

Kristina Downing, A. (ed.) (2004) *Protein NMR Techniques*, Methods in Molecular Biology, Vol. 278, Humana Press, Clifton.

James, T.L. (ed.) (2005) *Nuclear Magnetic Resonance of Biological Macromolecules*, Methods in Enzymology, Vol. 394, Academic Press, San Diego.

Markwick, P.R.L., Malliavin, T. and Nilges, M. (2008) *PLoS Comput. Biol.* **4**(9).

Marius Clore, G. and Schwieters, C.D. (2002) *Curr. Opin. Struct. Biol.*, **12**(2), 146–153.

29
Mass Spectrometry

Jürgen H. Gross

29.1	**Introduction**	**991**
29.1.1	Concept of Mass Spectrometry	991
29.1.2	Goal of Mass Spectrometry	991
29.1.3	Mass Spectrometry at a Glance	992
29.1.4	Aims and Scope	993
29.2	**The Role of Isotopes in Mass Spectrometry**	**994**
29.2.1	Isotopes	994
29.2.2	Atomic Mass Scale	994
29.2.3	Isotopic Patterns	994
29.2.4	Accurate Mass	995
29.2.5	Calculation of Atomic Weights and Molecular Weights	996
29.3	**Mass Spectrometric Methods**	**996**
29.4	**Ionization Processes**	**999**
29.4.1	Formation of Positive Radical Ions	999
29.4.1.1	Electron Ionization	999
29.4.1.2	Photoionization	999
29.4.1.3	Penning Ionization	1000
29.4.1.4	Charge Transfer	1000
29.4.1.5	Field Ionization	1001
29.4.2	Formation of Positive Ions	1001
29.4.2.1	Proton Transfer	1001
29.4.2.2	Electrophilic Addition	1001
29.4.2.3	Anion Abstraction	1002
29.4.3	Formation of Negative Radical Ions	1002
29.4.3.1	Electron Capture	1002
29.4.4	Formation of Negative Ions	1002
29.4.4.1	Nucleophilic Addition	1002
29.4.4.2	Cation Abstraction	1002
29.5	**Isotopic and Elemental Analysis**	**1003**
29.5.1	Thermal Ionization	1003
29.5.2	Spark Source	1005

Encyclopedia of Applied Spectroscopy. Edited by David L. Andrews.
Copyright © 2009 WILEY-VCH Verlag GmbH & Co. KGaA, Weinheim
ISBN: 978-3-527-40773-6

29.5.3	Glow Discharge	1006
29.5.4	Inductively Coupled Plasma	1007
29.5.5	Accelerator Mass Spectrometry	1010
29.6	**Ionization of Gaseous Molecules in Vacuum**	**1011**
29.6.1	Electron Ionization	1012
29.6.2	Chemical Ionization	1014
29.6.3	Field Ionization	1017
29.7	**Desorption/Ionization**	**1019**
29.7.1	Field Desorption	1019
29.7.2	Fast Atom Bombardment and Liquid Secondary Ion Mass Spectrometry	1021
29.7.3	Laser Desorption/Ionization	1024
29.7.4	Matrix-Assisted Laser Desorption/Ionization	1026
29.8	**Ion Formation at Atmospheric Pressure**	**1028**
29.8.1	Atmospheric Pressure Ionization	1030
29.8.2	Electrospray Ionization	1032
29.8.3	Atmospheric Pressure Chemical Ionization and Atmospheric Pressure Photoionization	1036
29.8.4	Atmospheric Pressure MALDI	1039
29.8.5	Desorption Electrospray	1041
29.8.6	Direct Analysis in Real Time	1041
29.9	**Activation Methods for Tandem Mass Spectrometry**	**1042**
	Acronyms	**1046**
	References	**1047**
	Further Reading	**1052**

29.1
Introduction

29.1.1
Concept of Mass Spectrometry

Mass spectrometry (MS) includes all techniques for the determination of the masses of single atoms, molecules and their fragments or their associates. The mass spectrometric experiment comprises the sequence of (i) generation of gas phase ions from either inorganic or organic compounds by any suitable method, (ii) separation of these ions by their *mass-to-charge ratio* (m/z), and (iii) detection of the ions qualitatively and quantitatively by their respective m/z and abundance. Each of these steps can be performed in various ways. The first step, *ionization*, can be achieved thermally, by electric fields or by impact of energetic particles such as electrons, small ions or heavy cluster ions, neutral atoms, or photons. The second step, *ion separation by m/z*, is effected by static or dynamic electric or magnetic fields and combinations thereof, forcing the ions to follow a trajectory characteristic of their respective m/z value. Finally, the detection of the ions is accomplished by transforming their impact onto a detector or the image current of their motion into an electric current that is then translated into a signal, a so-called peak, in the output of the *mass spectrometer*, the *mass spectrum* (Figure 29.1). Commonly, the signals are normalized to *relative intensity* by setting the most intensive peak, the *base peak*, to 100 %. A mass spectrum can be represented as bar graph, profile data, or m/z versus intensity listing.

According to these functional requirements, a mass spectrometer can be seen as composed of three basic moieties: (i) the *ion source*, (ii) the *mass analyzer*, and (iii) the *detector*. In contemporary mass spectrometers, these moieties are operated under the common control of a computer system. The principles, types, and operational modes of mass spectrometers are outlined in Chapter 23.

29.1.2
Goal of Mass Spectrometry

The ultimate goal of MS is to identify an analyte from the molecular or atomic mass(es) of its constituents. Knowing just the mass can be sufficient for the identification of elements and for the determination of the molecular formula

Encyclopedia of Applied Spectroscopy. Edited by David L. Andrews.
Copyright © 2009 WILEY-VCH Verlag GmbH & Co. KGaA, Weinheim
ISBN: 978-3-527-40773-6

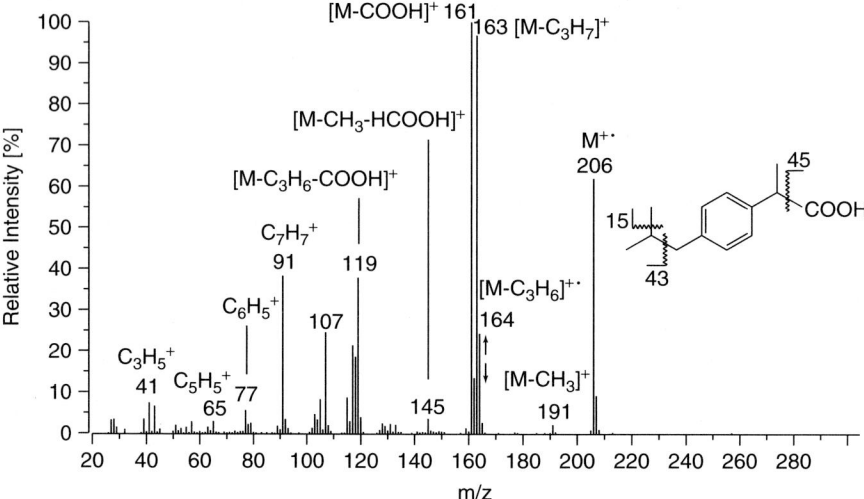

Fig. 29.1 The 70-eV electron ionization mass spectrum of the over-the-counter drug ibuprofen illustrates the principle of MS. The ionized molecule, that is, the molecular ion, fragments along several competing pathways to yield fragment ions that may themselves undergo further decomposition. The m/z values observed can – with more or less ease – be correlated to ionic compositions and eventually structures. Putting the puzzle together may reveal the structure of the analyte.

of an analyte. The relative abundance of isotopomers delivers strong supportive information to include or to rule out certain elements in such a formula and to estimate the number of atoms of a contributing element. Fragmentation of molecular species under the conditions of certain mass spectrometric experiments can deliver information on structure such as the connectivity of atoms within smaller molecules, the identification of functional groups, the (average) number and eventually the sequence of constituents of macromolecules, and in some cases even on their three-dimensional structure.

29.1.3
Mass Spectrometry at a Glance

MS has undergone tremendous transformations since J. J. Thomson's work marked its beginning in the early twentieth century (Grayson, 2002). MS developed from a mere mass measurement technique of physicists to one of the major analytical methods in chemistry and life sciences, while still preserving its importance in its fields of origin. Obviously, the fact that MS can be adapted in a very flexible manner to answer analytical questions of vast diversity allows it to become a whole branch of spectroscopy.

- MS serves for elemental identification and isotopic abundance measurement of both short-lived and stable species, thereby gaining relevance not only in physics and radiochemistry but also in geochemistry and more recently in life sciences.
- MS allows for the identification and structural characterization of small

molecules or for very large molecules as provided by either physiological processes or polymer chemistry.
- Mass spectrometric experiments can be stacked. Thus, a mass-selected ion can be studied by *tandem MS* and any product ion of this MS/MS (or MS^2) experiment may eventually be subjected to a third level (MS^3) and so forth (MS^n).
- MS, and especially tandem MS, also provides an elegant means for the study of unimolecular as well as bimolecular reactions of gas phase ions and for the determination of ion energetics.
- MS can be directly coupled to separation methods such as gas chromatography (GC) and all types of liquid chromatography (LC). *Coupling* or *hyphenation* to yield GC-MS or LC-MS instrumentation delivers high selectivity and low detection limits for the analysis of trace compounds in complicated matrices or the deconvolution of complex mixtures, making the hyphenated methods to become a major analytical tool.
- Mass spectra can be obtained from micrometer-sized areas on surfaces. This way, the lateral distribution of compounds on surfaces ranging from microelectronics to slices of tissue can be translated into images that in turn can be correlated to optical images of those samples. This is known as *mass spectral imaging*.
- Mass spectrometers can be very small. Miniaturization has led to portable instruments for environmental on-site analysis, detection systems for explosives and warfare chemicals, and last but not the least for many space missions.

In particular, the life sciences had and still have great impetus on the development of the manifold approaches to expand the mass range to higher molecular weights and increasingly fragile molecules. Environmental and pharmaceutical research present the driving force for the ever-improving limits of detection. There is ongoing research on methods of ion sampling, ion generation, and subsequent ion transfer into mass analyzers with better performance. Every application has unique requirements for the exact way in which the mass spectrometric experiment is performed, that is, on how to acquire a mass spectrum of a given analyte. MS has many facets and does not reveal itself from a single perspective. Like a view onto a globe does not reveal the complete surface of our planet, but roughly just one continent at a time, MS needs to be explored from various vantage points.

29.1.4
Aims and Scope

This chapter is intended to complement the seemingly abstract description of mass spectrometers by putting them into a real-world context of analytical applications. Nonetheless, the vast number of methods in MS and the plethora of applications of MS cannot be comprehensively covered within a single chapter. However, this chapter is intended to provide a convenient entry into the realm of MS by highlighting basics, relevant techniques, and applications. Therefore, the chapter deals with

- isotopes, isotopic abundance, and isotopic mass;
- ionization processes;

- ionization methods and the corresponding ion sources;
- the behavior of the ions formed by these ionization processes; and
- applications of tandem MS.

29.2
The Role of Isotopes in Mass Spectrometry

29.2.1
Isotopes

Elements are classified by (i) their *atomic number* Z (subscripted prefix at the element symbol, e.g., $_6$C), which is equal to the number of protons in the nucleus and also is equal to the number of electrons of a neutral atom and (ii) their *mass number* A (superscripted prefix at the element symbol, e.g., ^{12}C). The mass number is given by the sum

$$A = Z + N \qquad (29.1)$$

where N is the number of neutrons in the nucleus.

An element is termed *monoisotopic* if there exists only one nuclide of this element, that is, all of its atoms have equal A. This is the case for 20 of 83 naturally occurring stable elements, for example, F, Na, P, I, or Cs. If radioactive isotopes are also taken into account, there is no element left as monoisotopic. Elements with two naturally occurring stable nuclides are termed *di-isotopic*, for example, Li, B, Cl, Cu, Br, Ag, or Ir, and the others are termed *polyisotopic*, for example, Ru, Sn, or Xe. However, isotopes are distinguished by their mass number, the difference in mass being due to the varying number of neutrons in the nuclei of an element.

Isotopes of different elements sharing the same mass number are termed *isobars*, for example, ^{40}Ar, ^{40}K, and ^{40}Ca. The term also applies to molecules having different compositions yielding equal (nominal) mass, for example, CO, N_2, and C_2H_4.

29.2.2
Atomic Mass Scale

MS has been the key in establishing the atomic mass scale from the beginning (de Laeter, De Bièvre and Peiser, 1992). The atomic mass scale has been fixed relative to carbon. According to the Union of Pure and Applied Chemistry (IUPAC), the *unified atomic mass* [u] is defined as 1/12 of the mass of one atom of the nuclide ^{12}C in its electronic ground state. By defining the molar mass of carbon-12 exactly as 0.012 kg mol^{-1}, the mass of one atom is obtained by dividing this by the Avogadro constant. Therefore, 1 u equaling 1/12 of the atom's mass is obtained as 1 u $= 1/12 \times 0.012$ kg mol^{-1}/6.022136 $\times 10^{23} = 1.660540 \times 10^{-27}$ kg. Instead of "u" the unit *dalton* (Da) (to honor J. Dalton) is also in use.

29.2.3
Isotopic Patterns

MS separates isotopes by mass, and thus, the isotopic composition of an element is directly reflected by its mass spectrum. In molecules, the isotopes of all atoms are contributing to the isotopic pattern, some more like chlorine and some less like carbon or hydrogen. Consequently, the detection of a molecular ion or fragment translates into a spectrum representing the contribution of all isotopes (Figure 29.2). It is possible

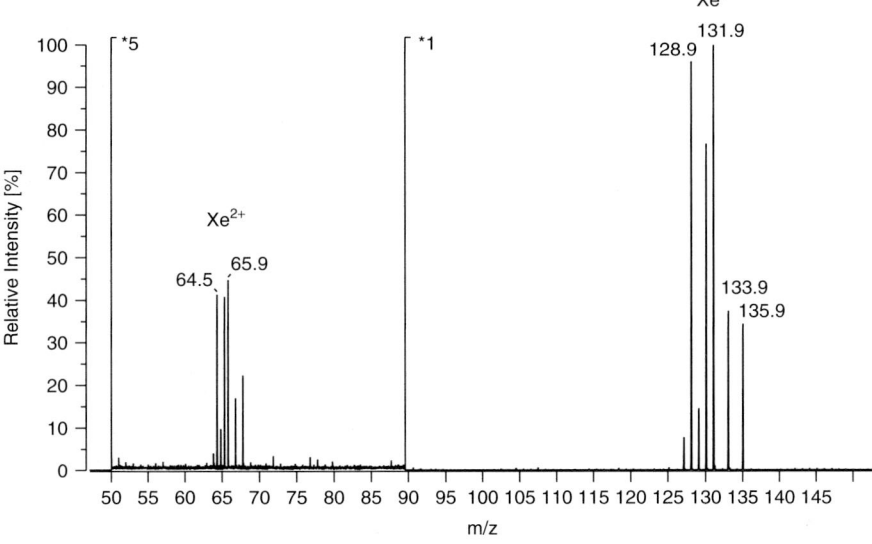

Fig. 29.2 Seventy electronvolts electron ionization mass spectrum of xenon gas. The singly as well as the doubly charged atomic ions reflect the isotopic composition of Xe from their isotopic patterns. The position of Xe^{2+} is at half m/z of $Xe^{+\bullet}$ and the distance between adjacent peaks of Xe^{2+} ion is half that of $Xe^{+\bullet}$ because $z = 2$ in m/z. Note that the intensity scale is increased by a factor of 5 for the doubly charged ion.

to calculate the isotopic pattern of a molecule based on statistical methods, because the isotopic composition of the elements is accurately known in most cases. Vice versa the experimental isotopic patterns can be fitted to theoretical distributions. In fortunate cases, in some cases, the molecular formulas of small molecules, say up to 200–300 u, can be assigned from their isotopic pattern alone.

29.2.4
Accurate Mass

Mass numbers are integers. The *accurate mass* of an isotope, however, slightly deviates from the nominal value of *A* due to the conversion of mass into energy during nucleation. The binding energy per nucleon steeply increases along the mass numbers from ^2H to its maximum around ^{56}Fe and then decreases again slightly up to ^{238}U (Figure 29.3).

Translation into isotopic mass reveals that for light elements mass is by some 10^{-3} u above the nominal value (^1H or ^{14}N), whereas it is by some 10^{-3} u (^{19}F) to 10^{-1} u (^{127}I) below for heavier elements. This *mass defect* was unveiled by Aston who had already discovered 212 of all 287 stable isotopes (Table 29.1).

To calculate the accurate mass of an ion, the electron mass ($m_e = 9.10939 \times 10^{-31}$ kg or 0.0005486 u) needs to be subtracted from the accurate mass of the corresponding neutral in the case of positive ions or added in the case of negative ions (cf. Figure 29.27).

Measuring the mass of ions with sufficient accuracy (often 0.5–10 ppm) allows to identify isotopes and molecules by

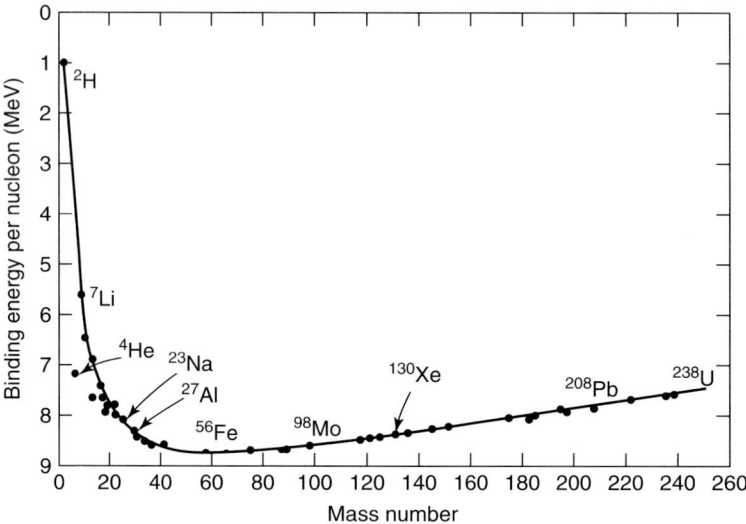

Fig. 29.3 Plot of binding energy per nucleon versus mass number. Reproduced from de Laeter, De Bièvre and Peiser (1992) by permission. © John Wiley & Sons, 1992.

their accurate mass alone (Bristow and Webb, 2003; Bristow, 2006). Assigning molecular formulas unequivocally depends on the mass of the molecule and the diversity of elements that has to be taken into account. For simple organic molecules of the general formula $C_cH_hN_nO_o$, an accuracy of 1 ppm serves this purpose up to about 400 u (Spengler, 2004). In other words, there is no second (reasonable) combination of these elements left that falls within the same mass window of ± 0.0004 u. (For an example of the use of high-resolution and accurate mass also refer to Figure 29.23.)

29.2.5
Calculation of Atomic Weights and Molecular Weights

In general, chemistry does not deal with isotopic mass but with *relative atomic mass* A_r or *atomic weight* that is obtained as the average mass of the isotopic mixture of an element:

$$A_r = \frac{\sum_i m_i \times I_i}{\sum_i I_i} \quad (29.2)$$

where m_i is the accurate mass of the isotope i and I_i is its abundance. For molecules, the *relative molecular mass* M_r or *molecular weight* (MW) is obtained analogously.

29.3
Mass Spectrometric Methods

Irrespective of the mass analyzer used, the method depends on the delivery of gaseous ions from an ion source, that is, the formation of ions is a prerequisite for mass spectral analysis of any sample. As MS approaches its 100th anniversary, there is a large variety

Tab. 29.1 Illustrative examples of isotopic mass, isotopic composition, and relative atomic mass (u) of nonradioactive elements. © IUPAC 2001.

Atomic symbol	Name	Atomic number	Mass number	Isotopic mass	Isotopic composition	Relative atomic mass
H	Hydrogen	1	1	1.007825	100	1.00795
			2	2.014101	0.0115	
He	Helium	2	3	3.016029	0.000137	4.002602
			4	4.002603	100	
C	Carbon	6	12	12.000000	100	12.0108
			13	13.003355	1.08	
N	Nitrogen	7	14	14.003074	100	14.00675
			15	15.000109	0.369	
O	Oxygen	8	16	15.994915	100	15.9994
			17	16.999132	0.038	
			18	17.999116	0.205	
F	Fluorine	9	19	18.998403	100	18.998403
Na	Sodium	11	23	22.989769	100	22.989769
P	Phosphorus	15	31	30.973762	100	30.973762
Cl	Chlorine	17	35	34.968853	100	35.4528
			37	36.965903	31.96	
Fe	Iron	26	54	53.939615	6.37	55.845
			56	55.934942	100	
			57	56.935399	2.309	
			58	57.933280	0.307	
Br	Bromine	35	79	78.918338	100	79.904
			81	80.916291	97.28	
I	Iodine	53	127	126.904468	100	126.904468
Xe	Xenon	54	124	123.905896	0.33	131.29
			126	125.904270	0.33	
			128	127.903530	7.14	
			129	128.904779	98.33	
			130	129.903508	15.17	
			131	130.905082	78.77	
			132	131.904154	100	
			134	133.905395	38.82	
			136	135.907221	32.99	
Cs	Cesium	55	133	132.905447	100	132.905447
Au	Gold	79	197	196.966552	100	196.966552
U	Uranium*	92	234	234.040946	0.0055	238.0289
			235	235.043923	0.73	
			238	238.050783	100	

of mass spectrometric methods at hand to accomplish this task for almost any sample and almost any analytical task.

Basically, it is much about selecting a suitable *ionization method*, the choice being the key for the outcome of the experiment. Although some mass spectrometers are built in a way to allow for comparatively rapid changes between several ionization methods, others are tailored to fit exactly one type of ion source.

Ionization methods in MS can be grouped in many ways, the most basic being placing them into coordinates of relative softness and state of aggregation in which ionization is effected (Figure 29.4). Under plasma conditions, for example, any inorganic or organic analyte will immediately be dissociated into its constituent atomic species. It is considerably softer to expose sample vapors to energetic electrons, whereby the molecules are ionized. However, owing the amount of energy imparted during this process, the majority of the molecular ions formed may undergo fragmentation. Such ionization methods are termed *hard ionization*. On the other hand, *soft ionization* methods do not impact high levels of energy onto the analyte. Instead, they supply just enough to achieve ionization and eventually a low level of fragmentation of the ionized molecules. The softness of a method increases when vaporization of the sample is not anymore required *before* ionization. Therefore, techniques employing *desorption/ionization*, that is, the direct transition from the condensed phase into the state of isolated ions in the gas phase, are the softest available. Accordingly, the ions detected in MS span from atomic ions on the hard ionization and low-mass side, over small molecules up to large macromolecules by far exceeding 100 000 u.

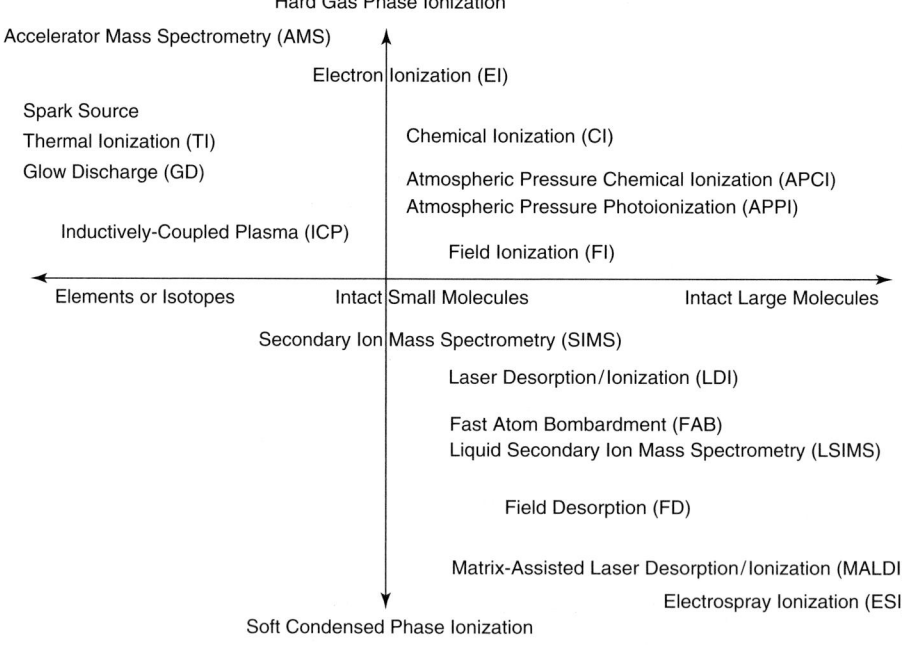

Fig. 29.4 Classification of common mass spectrometric methods. Although most of these methods, in particular those of organic and bioorganic mass spectrometry, in fact represent ionization methods that can be run on several types of mass analyzers, some require dedicated instrumentation, for example, SIMS as employed for imaging or AMS, which demands a large facility of its own.

29.4 Ionization Processes

There are many processes leading to the formation of charged species. Although some of these are unique to a certain ionization method, others may occur with several ionization methods. The interaction with electrons, for example, plays a role not only in *electron ionization* (EI) but also in *chemical ionization* (CI), *glow discharge* (GD), and less obvious in other technical implementations of ion generation. This section provides an overview of the most relevant ionization processes in MS.

29.4.1 Formation of Positive Radical Ions

29.4.1.1 Electron Ionization

EI (formerly termed *electron impact*) employs the interaction of energetic electrons with gas phase molecules. For *ionization efficiency* reasons, the *primary electrons* are routinely supplied at 70-eV kinetic energy. The energy transferred to the neutral upon collision with the electron forms a short-lived (<1 femtoseconds) electronically excited state. The excited neutral instantaneously emits one of its own electrons, thereby transforming into a positive radical ion, the *molecular ion* (radical ions are marked with • after the charge sign to indicate the unpaired electron).

$$M + e^- \rightarrow M^{+\bullet} + 2e^- \quad (29.3)$$

If the energy uptake does not exceed the *ionization energy* (IE), that is, the minimum energy required to ionize a neutral in its ground state (generally 7–15 eV), it may subsequently undergo relaxation by either radiative emission or collisions with thermal neutrals (unlikely under the typical operation conditions of an EI ion source).

Multiple ionization can occur, provided sufficient energy has been imparted during interaction with the primary electron:

$$M + e^- \rightarrow M^{2+} + 3e^- \quad (29.4)$$
$$M + e^- \rightarrow M^{3+\bullet} + 4e^- \quad (29.5)$$

EI must not necessary lead to molecular ions. If no minimum on the energy hypersurface exists, EI can cause immediate dissociation, the so-called dissociative ionization:

$$M + e^- \rightarrow [M - B]^+ + B^\bullet + 2e^- \quad (29.6)$$

Another pathway not leading to molecular ions is *ion-pair formation*:

$$M + e^- \rightarrow [M - C]^- + C^+ + e^- \quad (29.7)$$
$$M + e^- \rightarrow [M - A]^+ + A^- + e^- \quad (29.8)$$

Note that ion-pair formation simultaneously yields both cations and anions from the analyte molecule. A typical EI mass spectrum is shown in Figure 29.1. (For practical aspects of EI cf. Section 29.6.1.)

29.4.1.2 Photoionization

The absorption of UV light by a neutral can result in electronically excited states that undergo relaxation either by emission of light or by emission of an electron. Thus, the photon in *photoionization* (PI) serves the same purpose as the energetic electron in EI:

$$M + h\upsilon \rightarrow M^* \rightarrow M^{+\bullet} + e^- \quad (29.9)$$

As with the electron before, the energy absorbed must at least provide the IE of the neutral, that is, lead into a continuum state.

However, ionization with a single photon is technically difficult, because frequency-quadrupled Nd:Yag lasers deliver photons of 266 nm wavelength (4.6 eV) and even ArF excimer lasers just yield 193 nm (6.3 eV), both being well below the IE of most molecules. Fortunately, the absorption of energy has not to be a one-photon process with a highly energetic UV photon. The stepwise accumulation of energy from less energetic photons, that is, *multiphoton ionization* (MUPI) is also feasible.

Nonresonant ionization has rather low cross sections and consequently requires high power densities of the electromagnetic radiation, or results in poor ionization efficiency. Resonant absorption of photons is more effective by orders of magnitude (Wendt, 2002). The most desirable situation is when resonant absorption of the first photon leads to an intermediate state from where absorption of a second photon, preferably but not necessarily of the same wavelength, can promote the molecule into the continuum (1 + 1 process). This approach is termed *resonance-enhanced multiphoton ionization* (REMPI). Proper selection of the laser wavelengths provides compound-selective analysis at extremely low detection limits (Boesl, 2000; Zenobi, Zhan and Voumard, 1996).

It has been shown that typical nanosecond laser pulses can be inferior to picosecond laser pulses, because the latter do not allow the molecules to undergo unwanted relaxation or fragmentation before molecular ion formation, for example, mass spectra of diphenylmercury exhibit a strong molecular ion peak when subpicosecond laser ionization is employed, whereas $M^{+\bullet}$ is completely absent at nanosecond pulse width (Weickhardt, Grun and Grotemeyer, 1998).

29.4.1.3 Penning Ionization

Particles occupying long-lived electronically excited states may transfer energy to ground state neutrals. Ionization of the neutral occurs if its IE is lower than the energy received upon interaction with the excited species. This is known as *Penning ionization* (Penning, 1927):

$$M + X^* \rightarrow X + M^{+\bullet} + e^- \quad (29.10)$$

Penning ionization requires the presence of at least two different species in the gas phase and the chance for at least one collision between those particles within their dwell time in the ionization volume. The species X, for example, a noble gas, is supplied in excess to achieve a reasonable ionization efficiency for the analyte M. The energy to form the excited particles X^* may be delivered by electrons, photons, or an electric discharge. The associative variant of Penning ionization

$$M + X^* \rightarrow MX^{+\bullet} + e^- \quad (29.11)$$

is also known as *Hornbeck–Molnar* process.

29.4.1.4 Charge Transfer

Soft collisions of radical ions with neutrals present another way of ion creation. The neutrals may be ionized by *charge transfer* (CT) if their IE is lower than the *recombination energy* (RE) of the neutral reagent gas X used to deliver the $X^{+\bullet}$ ions. In CT, an electron is removed from the analyte molecule M to neutralize the charge of the *reactant ion* $X^{+\bullet}$:

$$M + X^{+\bullet} \rightarrow M^{+\bullet} + X \quad (29.12)$$

The overall process is exothermal, that is, the criterion RE > IE needs to be met. In practice, the reactant ions are generated by EI of a large excess of *reagent gas*, which is jointly admitted to the analyte gas.

29.4.1.5 Field Ionization
Strong electric fields in the order of 10–25 V nm^{-1} effectively enable the tunneling of an electron from a gaseous molecule into a solid electrode, because the electric field substantially lowers the potential barrier for the electron's escape. The molecule is thereby transformed into its molecular ion, $M^{+\bullet}$, but in contrast to EI this is achieved without significant deposition of internal energy onto the ion. This process is known as *field ionization* (FI). Thus, positive radical ion formation by FI is given as

$$M_{\text{in electric field}} \rightarrow M^{+\bullet} + e^- \quad (29.13)$$

To achieve reasonable ionization efficiency, high electric field is required, because electron loss may then proceed within about 1 femtosecond. However, a minor degree of ionization is already achieved at significantly lower field strength.

29.4.2 Formation of Positive Ions

29.4.2.1 Proton Transfer
All ionization processes discussed so far delivered positive radical ions, because they essentially caused the loss of one electron from the prospective analyte in the gas phase. If a reactant ion can serve as a Brønsted acid, $[BH]^+$, it can act as proton donor, thus yielding a protonated analyte molecule, $[M+H]^+$, by exchange of the proton:

$$M + [BH]^+ \rightarrow [M+H]^+ + B \quad (29.14)$$

The $[BH]^+$ reactant ions are supplied by multistep reactions of primary reactant ions created by EI of the respective reagent gas with neutrals of the same sort. $[M+H]^+$ ion formation is the widest known ionization process in *CI*. Proton transfer must be exothermal to occur, that is, the *proton affinity* (PA) of M must exceed the PA of X. The $[M+H]^+$ ions formed are termed *quasimolecular ions*, because they occur instead of a molecular ion, but are still intimately related to the molecule M. Different from $M^{+\bullet}$ ions, $[M+H]^+$ quasimolecular ions have intact orbitals. They are *even-electron* (or *closed-shell*) ions as opposed to their *odd-electron* (or *open-shell*) radical ion counterparts. Even-electron ions are less prone to fragmentation than radical ions.

29.4.2.2 Electrophilic Addition
There are various possibilities of what can happen when an ion encounters a molecule in the gas phase. An elastic collision accompanied by some exchange of kinetic or vibrational energy does not chemically alter the colliding particles. If the collision is soft enough, it can result in adhesion of both particles, that is, *electrophilic addition* in the case of a positive ion (electrophile) attaching to a neutral. This can also be regarded as solvation of the ion in the gas phase. The resulting *ion–neutral complex*, however, bears a charge and becomes accessible for mass analysis:

$$M + X^+ \rightarrow [M+X]^+ \quad (29.15)$$

As with $[BH]^+$ ions, such X^+ ions are typically provided by multistep reactions of reactant ions initially created by EI of the respective gas. Thus, electrophilic addition presents another CI process.

29.4.2.3 Anion Abstraction

Anion abstraction may also yield a positive analyte ion, the simplest process of anion abstraction being hydride abstraction. Other anionic substituents (A) such as halogenide ions can also be removed:

$$M + X^+ \rightarrow [M - H]^+ + XH \quad (29.16)$$

$$M + X^+ \rightarrow [M - A]^+ + XA \quad (29.17)$$

Again, the abstracting species X^+ is provided by multistep reactions of the ionized reagent gas.

29.4.3 Formation of Negative Radical Ions

29.4.3.1 Electron Capture

Thermal electrons and electrons with energies up to a few electronvolts act in much different manner from energetic electrons. Instead of highly exciting the particle by their passage, they become attached to it, creating a negative radical ion:

$$M + e^- \rightarrow M^{-\bullet} \quad (29.18)$$

Electron capture (EC) is a resonance process requiring that the *electron affinity* (EA) of the neutral is positive, that is, negative ions may occur with analytes containing highly electronegative elements (Harrison, 1992a).

The internal energy of the anion formed is governed by the heat of reaction of the process. Typically, EC yields molecular anions of low internal energy; in other words, it presents rather soft ionization. Nonetheless, the product of AC may spontaneously dissociate if it does not represent an energetic minimum on the energy hypersurface

$$M + e^- \rightarrow [M - B]^- + B^{\bullet} \quad (29.19)$$

which is termed *dissociative EC*.

Finally, the EC can also cause ion-pair formation (cf. EI):

$$M + e^- \rightarrow [M - C]^- + C^+ + e^- \quad (29.20)$$

$$M + e^- \rightarrow [M - A]^+ + A^- + e^- \quad (29.21)$$

29.4.4 Formation of Negative Ions

29.4.4.1 Nucleophilic Addition

Nucleophilic addition means the attachment of a negative ion (nucleophile) to an electron-deficient neutral. In analogy to electrophilic addition, this may be regarded as solvation of the ion (nucleophile) by the neutral (analyte) in the gas phase:

$$M + A^- \rightarrow [M + A]^- \quad (29.22)$$

The *negative-ion–neutral complex* formed can be mass-analyzed (Figure 29.5).

29.4.4.2 Cation Abstraction

Cation abstraction can lead to negative analyte ions, the simplest process of cation abstraction being proton abstraction by an anion A^-:

$$M + A^- \rightarrow [M - H]^- + AH \quad (29.23)$$

Dissociation of acidic molecules can directly result in the corresponding anions:

$$MH \rightarrow [M - H]^- + H^+ \quad (29.24)$$

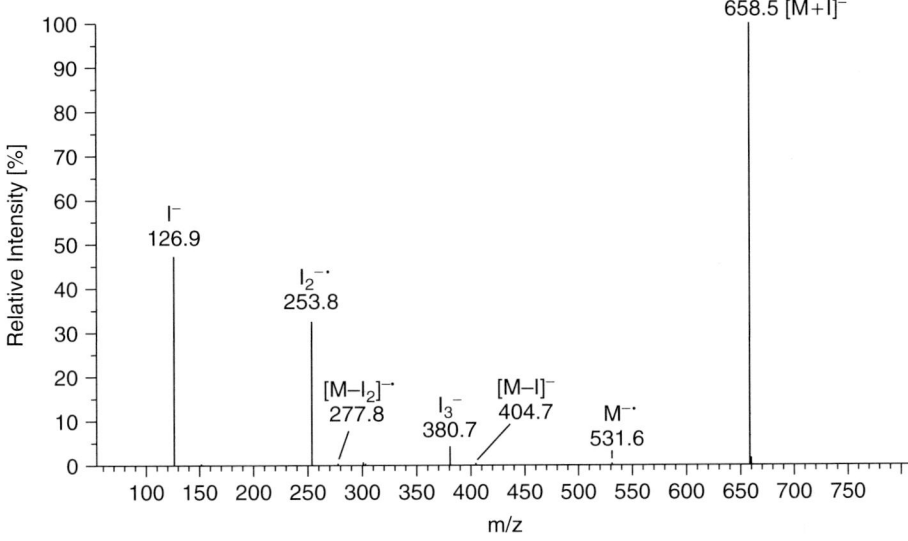

Fig. 29.5 Negative-ion chemical ionization mass spectrum of tetraiodoethene with isobutane reagent gas. The negative molecular ion at m/z 531.6 has a relative intensity of only 0.15%, whereas the iodide adduct [M+I]⁻ at m/z 658.5 leads to the base peak of the spectrum. Losses of I• and I$_2$ from M⁻• are also observed. The series of peaks at m/z 126.9, 253.8, and 380.7 correspond to traces of iodine present as impurity of tetraiodoethene. The iodine is also ionized by both EC and iodide addition. This spectrum also shows how spectra of two components of a mixture are superimposed. It is not always trivial to tell the corresponding peaks apart. Then, additional experiments such as accurate mass measurements or tandem mass spectrometry are required.

29.5
Isotopic and Elemental Analysis

29.5.1
Thermal Ionization

When a metal salt, metal oxide, or metal placed onto the surface of a rhenium or tungsten filament is heated in vacuum to 400–2300 °C, atomic and eventually molecular ions can be generated by *thermal ionization* (TI), also known as *surface ionization*. Although TI represents another ionization process, and could have been discussed in Section 29.4, it is rather peculiar, because the TI process does not occur with any other ionization method. Basically, its application is limited to *isotope ratio mass spectrometry (IR-MS)* of metals, and moreover, TI ion sources are only mounted to dedicated magnetic sector instruments, usually equipped with multicollector systems to measure isotope ratios with highest accuracy (Chapters 3.3.2 and 7.2 in Platzner et al., 1997).

The setup for TI consists of one to three ribbon filaments in close proximity (Figure 29.6). The analyte is supplied on the first filament and evaporated. The neutral atoms or molecules may hit the second filament where they are ionized by electron transfer to (positive ions) or from the bulk metal (anions). With triple filament assemblies, the third filament

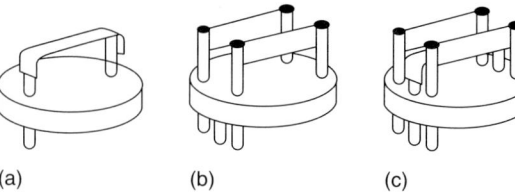

Fig. 29.6 (a) Single filament, (b) double filament, and (c) triple filament assemblies for thermal ionization of metal salts and oxides. Adapted from Platzner et al. (1997) p. 27 with permission. © John Wiley & Sons Ltd, 1997.

is normally used to supply a standard in order to sequentially measure unknown and reference under identical ion source conditions. In double and triple filament assemblies, the evaporation and ionization temperatures can be independently controlled. The former is preferably set to low values in order to obtain long-lasting signals and minor isotopic fractionation. The latter is set to high temperature for better ionization efficiency. Consequently, nanograms to micrograms of sample yield signals for hours as evaporation rates are in the order of picograms per second. To avoid frequent breaking of the vacuum, 10–20 filament assemblies are mounted on a carousel-like sample turret.

Blurring of the discrete electronic states of the neutral at high temperature is a key for TI, thereby allowing the Fermi levels of adsorbed neutral and bulk metal to merge. Then, the charge state of the weakly surface-bound particle is purely determined by Fermi statistics, because electrons move freely between the atom and the surface. The Saha–Langmuir equation describes the degree of ionization α^+ expressed as the ratio of ions and neutrals, leaving the surface under the assumption of ideal conditions:

$$\alpha^+ = \frac{N^+}{N^0} = \frac{g^+}{g^0} \exp\left[\frac{e(\Phi - IE)}{kT}\right]$$
(29.25)

$$= \frac{g^+}{g^0} \exp\left[1.16 \times 10^4 \frac{(\Phi - IE)}{T}\right]$$
(29.25a)

where g^+/g^0 is the ratio of electronic states of the ion and the neutral, Φ is the work function of the filament material, IE is the ionization energy of the atom or molecule to be ionized, and T is the temperature of the filament. As a consequence of this equation, good ion yields are obtained at high temperature with metals of low IE and filament materials with large Φ, which is fulfilled for rhenium ($\Phi_{Re} = 4.98$ eV, mp $= 3180\,°C$) and tungsten ($\Phi_W = 4.58$ eV, mp $= 3410\,°C$) (Figure 29.7).

Nonmetallic elements with high IE and metal oxides may form negative ions (Kawano and Page, 1983; Kawano, Hidaka and Page, 1983). The degree of ionization α^- is then obtained from the following expression:

$$\alpha^- = \frac{g^+}{g^0} \exp\left[1.16 \times 10^4 \frac{(EA - \Phi)}{T}\right]$$
(29.26)

where EA is the electron affinity of the adsorbed atom.

The low spatial and energetic spread of ions from TI is advantageous for the combination with single-focusing magnetic sector mass analyzers. On the other hand, TI is not suitable for mixture analysis and multielement determinations where ICP-MS is normally employed.

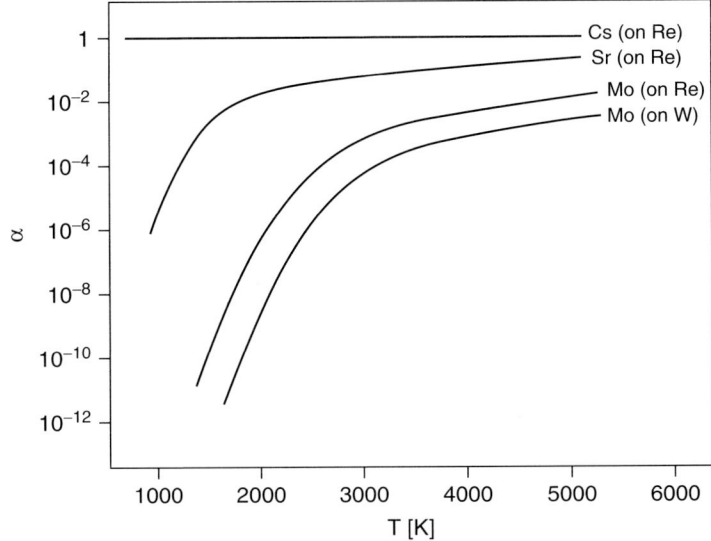

Fig. 29.7 Thermal ionization efficiency α versus filament temperature. The curves for molybdenum reflect the dependence of α on the filament material. The advantage of the Re over the W filament is in accordance with Eq. (29.25). Adapted from Becker (2008) with permission. © John Wiley & Sons Ltd, 2008.

29.5.2
Spark Source

Developed by Dempster in 1935, the spark source provided the first truly multielement and isotopic trace element tool (Koppenaal, 1990). *Spark source mass spectrometry* (SS-MS) is best suited for conducting samples, because it employs a high-voltage electric discharge between two pin-shaped electrodes (about 10 mm long and 1–2 mm in diameter) in vacuum. The electrodes, directly prepared from the solid to be analyzed, need careful cleaning, for example, by etching, before being inserted into the ion source (Chapter 5.2 in de Laeter, 2001, Chapter 2.2 in Becker, 2008) (Figure 29.8).

Nonconducting samples are ground and mixed with high-purity graphite, silver, or gold powder. The mixture is then pressed to yield the electrodes. It is obvious from this procedure that SS-MS is not suited for elemental analysis of small sample amounts; nonetheless, it plays to its strengths for simultaneous analysis of many elements including those present at trace level in alloys, ores, and similar samples.

From the cathode, electrons are emitted by field emission (opposite process to FI), which is possible due to the high local field strengths (10^8–10^9 V m^{-1}) at edges of the microscopically rough surface. The electrons are accelerated toward the anode where they cause evaporation of sample that is then ionized in the gas phase by action of the arriving electrons.

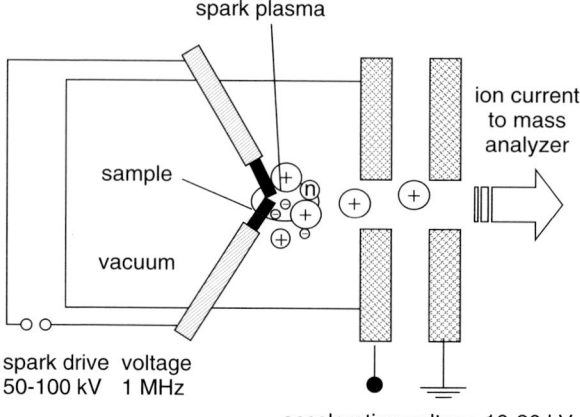

Fig. 29.8 Schematic of a RF spark source. Adapted from Becker (2008) with permission. © John Wiley & Sons Ltd, 2008.

Although operation of a spark source appears simple initially, it is quite demanding. First, it requires a high-voltage supply of several tens of kilovolts, normally at 1 MHz radiofrequency (RF) to achieve uniform ablation of both electrodes. Second, the electrodes need to be carefully adjusted and dynamically readjusted during the measurement (0.1–0.5 mm tip to tip) to maintain a stable ion current. Then, the ions emerge from the discharge with a wide kinetic energy distribution (keV range) that makes a double-focusing mass analyzer mandatory for SS-MS. Still, photoplate detection and subsequent densiometry on a series of exposures, that is, on subsequent spectra acquired for increasing periods of time, are still widespread in SS-MS (Koppenaal, 1990; Verlinden, Gijbels and Adams, 1986). However, multi-ion counting (MIC) systems are promising as they deliver improved sensitivity and precision (Jochum, 1997; Jochum et al., 2001).

29.5.3
Glow Discharge

In 1886, Goldstein was the first to report the formation of positive ions of gases from a GD in his so-called channel ray tube (p. 25 in Becker, 2008).

In a GD-MS ion source (Jakubowski et al., 2007; Nelis and Pallosi, 2006; Hoffmann et al., 2005; Stuewer, 1990; King and Harrison, 1990), a discharge current of 1–5 mA is maintained through a dilute argon atmosphere (10^1–10^3 Pa) by applying a DC voltage of 500–1000 V across the ionization volume. The cathode is represented by the solid sample that may be either a disk or a stick and an anode present as the housing and the ion source exit plate (Chapter 5.3 in de Laeter, 2001, Chapter 7.5 in Platzner et al., 1997).

The low-temperature Ar plasma serves as a source of primary ions, excited atoms, and electrons. In the cathode dark space, a region where a large potential drop exists, the positive argon ions are

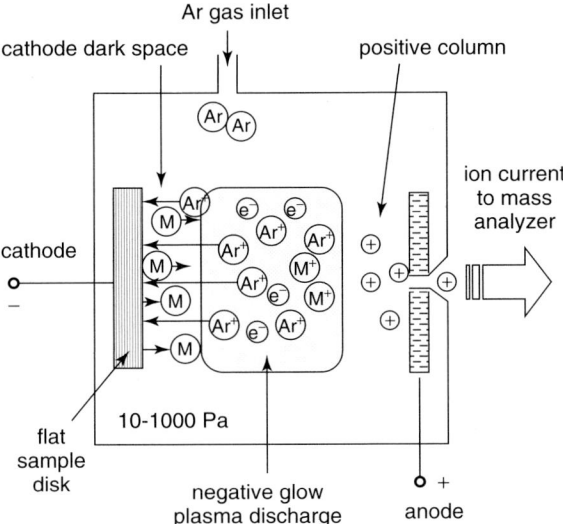

Fig. 29.9 GD ion source with a disk-shaped cathode. Adapted from Becker (2008) with permission. © John Wiley & Sons Ltd, 2008.

accelerated toward the sample cathode (Figure 29.9).

By bombarding the electrode's surface, the impinging ions provide continuous sputtering of neutrals into the gas phase. As the liberated analyte atoms move into the negative-glow region, they may be ionized by either EI or Penning ionization. The negative-glow region emits a bright glow, a phenomenon that led to the construction of the neon lamp. The electron density in the plasma can reach up to 10^{14} cm^{-3} (Penning, 1927) and their energy covers a wide distribution from thermal to energetic. Although their presence in the plasma is less by about a factor of 1000, metastable Ar* atoms are still effective for ionization because they carry 11.55 eV.

Nonconducting samples would rapidly cause the breakdown of the discharge due to electrostatic charging. RF discharges circumvent this problem (Winchester and Payling, 2004). Typically, voltages of 2 kV at RF of 13 MHz are employed.

29.5.4
Inductively Coupled Plasma

Currently, the *inductively coupled plasma* (ICP), developed by Houk *et al.* (1980), probably presents the most frequently employed method of element MS (Chapter 4 in de Laeter, 2001, Chapter 5 in Becker, 2008, Chapter 7.3 in Platzner *et al.*, 1997). Its widespread use is due to a highly versatile sampling mode that, in addition, is comparatively simple and robust because atomization and ionization are achieved in an argon plasma at atmospheric pressure. The ICP provides high ionization efficiency for elements of low IE (0.98 for 7 eV). Although these values drastically drop (0.01 for 13 eV), it is still sufficient to access elements such as P and even Cl.

The plasma is maintained by coupling electric energy (1–2 kW) from a surrounding RF (27 MHz) coil into the gas. The ion motion induced by the magnetic field causes the gas phase to heat up until it forms a continuously flowing plasma of about 8000 K at the center. The analyte is introduced axially into this plasma by a nebulizer that dissipates solutions and suspensions or by admitting gaseous samples.

The plasma torch consists of a coaxially aligned set of three quartz tubes, the outer of which is about 20 mm in diameter and is surrounded by the water-cooled RF coil. An argon stream of 12–15 l min^{-1} is used to cool the walls. The middle tube supplies a stream of Ar (1–2 l min^{-1}) as nebulizer gas to assist the formation of droplets at a size of tens of micrometers from the sample aerosol that is admitted by another gentle Ar stream from the 1–2 mm i.d. center tube (Figure 29.10).

The ions are transferred into the mass analyzer via a differentially pumped interface. A small portion of the plasma enters the first pumping stage through a

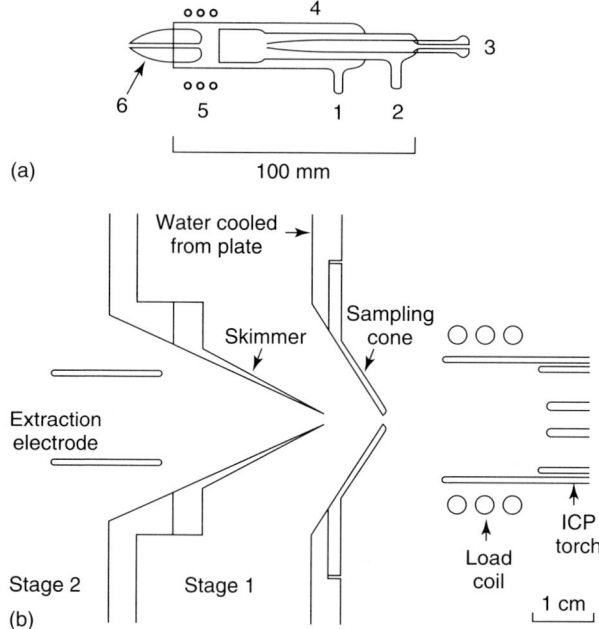

Fig. 29.10 Schematic diagram of (a) an ICP torch where 1 and 2 are inlets for argon sheath and nebulizer gas, 3 inlet for the sample, 4 the outer quartz tube of the torch, 5 the RF coil, and 6 the plasma. (b) An ion extraction interface for ICP-MS. Ions arrive from the right, a small portion is transmitted through the water-cooled sampling cone into the first pumping stage (rotary pump), the center of the free-jet expansion is then transmitted through the second skimmer cone, behind which it is guided by electric potentials through the second pumping stage (turbo pump) into the mass analyzer. Reprinted with permission from Platzner et al. (1997) p. 160. © John Wiley & Sons Ltd., 1997.

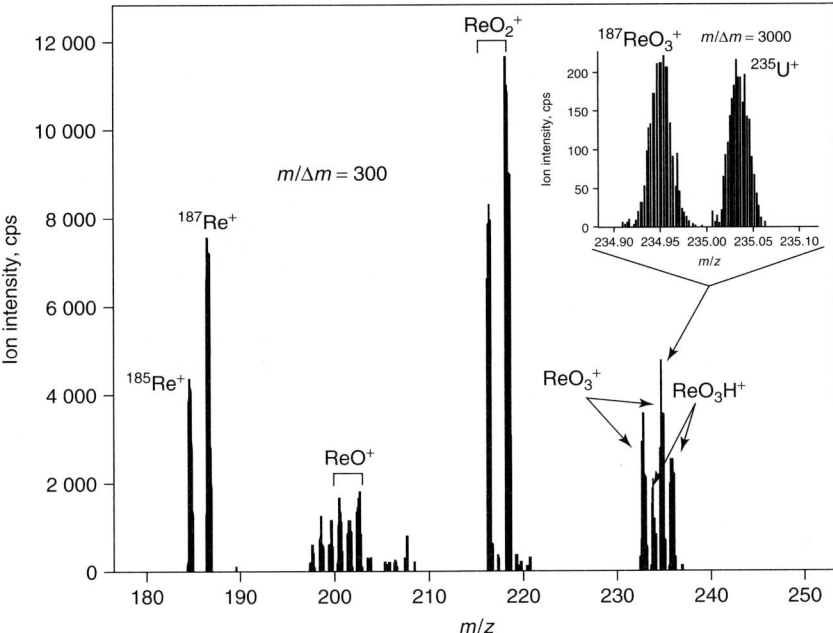

Fig. 29.11 Partial ICP mass spectrum of a radioactive waste solution in the range m/z 180–250 as obtained in low-resolution mode. The insert showing the m/z 235 range is taken from a spectrum as obtained from the same magnetic sector instrument when operated in high-resolution mode at $R = 3000$. Then, the signals due to $^{187}\text{ReO}_3^+$ and $^{235}\text{U}^+$ can easily be distinguished. Adapted from Becker and Dietze (1999) with permission. © The Royal Society of Chemistry, 1999.

hole in the water-cooled sampling cone. Application of suitable electric potentials guides the ions through the entrance of the skimmer, whereas neutrals are pumped off the supersonic expansion in this region. The first implementation of ICP-MS used a quadrupole analyzer due to its acceptance of moderate vacuum conditions (Houk et al., 1980). In between, magnetic sector (Figure 29.11) (Becker and Dietze, 1999), time of flight (TOF), and Fourier-transform ion cyclotron resonance (FT-ICR) mass analyzer are also employed.

To further complete the sampling repertoire, a gas stream may be employed to take up neutrals generated by *laser ablation* (LA). LA-ICP-MS can be used for sampling from solid surfaces of bulk material or alternatively for imaging of element distributions in the upper layers of tissues and similar samples (Figure 29.12). Thus, LA-ICP-MS is being used in biomedical research (Guenther and Hattendorf, 2005; Chery et al., 2006; Becker et al., 2007).

Another advantage of the acceptance of liquid introduction is the coupling of LC and gel electrophoresis to ICP instruments to allow for *element speciation*. This way not only the presence of an element in a sample can be detected but also the assignment to molecular species becomes feasible. This is relevant for the identification of non-C,H,N,O elements in proteins, for example (Ray et al., 2004).

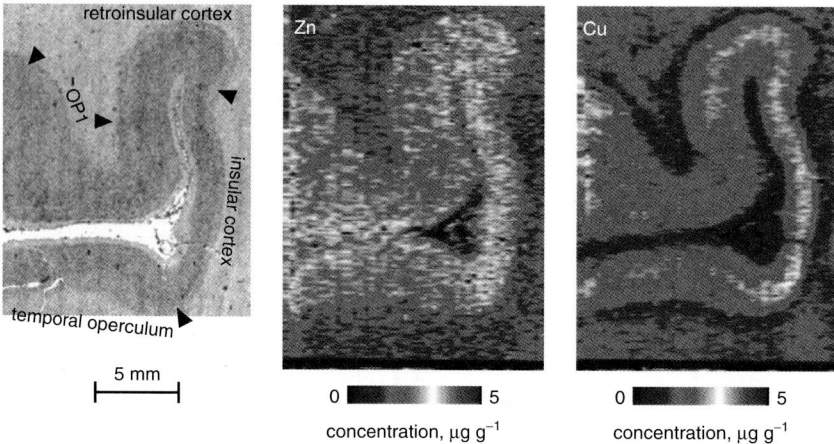

Fig. 29.12 Quantitative images of Zn and Cu in the hippocampus of human brain. A slice of 20 μm thickness was measured by LA-ICP-MS as compared to the light photograph (left). Reproduced from Becker et al. (2007) with permission. © The Royal Society of Chemistry, 2007.

29.5.5
Accelerator Mass Spectrometry

Accelerator mass spectrometry (AMS) presents an extremely sensitive technique for isotopic analysis (Tuniz et al., 1998). AMS allows to measure isotope ratios for some elements to a level of 1 in 10^{15}, which is by a factor of 10^5 lower than in other mass spectrometers. Moreover, this can be achieved for samples of about 1 mg, that is, only one million atoms of the isotope of interest are present, making it ideal for isotope ratio measurements of low-abundant isotopes, especially those having too long half-lives for counting radioactive decay from small sample amounts such as carbon-14.

The methodology of AMS differs greatly from all other approaches to the analysis of atomic mass. The main characteristic of AMS lies in the energies to which the ions are accelerated. Although other mass analyzers deal with ions having 10 eV–10 keV, in AMS they are at several megaelectronvolts. To achieve this, negative ions are created in an ion source and are mass-analyzed before entering a linear *Van de Graaff accelerator* or a *tandem accelerator*. In a tandem accelerator, negative ions are attracted by positive voltage toward the so-called high-voltage terminal at the center of the vessel. By passing the beam through a thin carbon foil (foil stripper) or a gas collision cell (gas stripper), electrons are stripped off the anion, thereby converting it to a positive ion. The atomic cations are then pushed away from the high-voltage terminal toward the exit of the accelerator. These are then mass-analyzed in a unit that basically represents a large magnetic sector instrument. The typical AMS setup requires some hundred square meters (Figure 29.13).

The higher energies enable ambiguities in the identification of atomic ions to be avoided, because molecular species are completely destroyed at the stage of the stripper. Furthermore, the selection

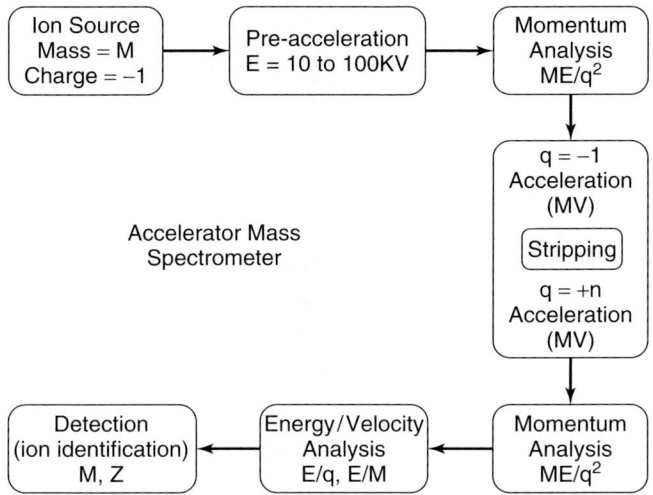

Fig. 29.13 Schematic layout of equipment for AMS. Ions experience marked changes in kinetic energy and charge state during the experiment. In total, the sample is ionized, and ions are preselected in a magnetic sector, accelerated to mega electronvolt energies, charge-inversed in the stripper, separated according to their momentum, charge, and energy, and finally counted individually (Tuniz et al., 1998).

of a suitable high charge state, characteristic of the isotope of interest, permits unequivocal identification, for example, ^{14}C, ^{26}Al, and ^{129}I are already separated from isobaric ^{14}N, ^{26}Mg, and ^{129}Xe due to the instability of the respective negative ions.

Carbon-14 dating was the application that gave birth to AMS (Nelson, Korteling and Stott, 1977; Bennett et al., 1977). Today, AMS is also employed for the determination of isotopes such as ^{10}B, ^{26}Al, ^{36}Cl, ^{41}Ca, ^{129}I, and others including stable isotopes. This versatility has led to applications of AMS in many areas of research including life sciences, geosciences, archeology, and extraterrestrial materials (Brown, Dingley and Turteltaub, 2005; Brown, Tompkins and White, 2006; Ikeda, 2005; Kutschera, 2005).

29.6
Ionization of Gaseous Molecules in Vacuum

All classical ionization methods of organic MS imply ionization in the diluted gas phase. They are applied for small molecule analysis ($M < 800$ u), provided the analyte does not contain too many heteroatoms. Molecules with more than a few oxygen, nitrogen, or sulfur atoms (highly functionalized molecules) are normally very polar – a property preventing their evaporation without prior thermal decomposition. Furthermore, molecular ions of such species often undergo facile fragmentation making them difficult to recognize in the mass spectrum. Thus, the important assignment of molecular mass is rendered impossible and the structure elucidation based

Tab. 29.2 Methods of sample evaporation.

Inlet system	Principle of operation	Cases where applicable	Limitations
Reservoir inlet (also termed *reference inlet, gas inlet*)	Injection into a heated reservoir and admission to the ion source via a needle valve. Inlet is emptied by evacuation	Gases and liquids	Risk of thermal decomposition of the sample. Memory from previous samples
Direct probe (also termed *direct insertion probe or direct inlet system*)	Evaporation from a small vial placed onto the temperature-controlled tip of a probe which is inserted through a vacuum lock	High-boiling liquids, waxes, solids	Risk of thermal decomposition of the sample
Gas chromatograph	Sample solutions are injected into GC and elute from the chromatographic column into the ionization volume	Compounds that can pass through a GC column	Chemical derivatization can be required. Time span of elution is only about 1 s

on fragment ions alone normally fails. Table 29.2 compares methods of sample evaporation.

29.6.1
Electron Ionization

EI, formerly termed *electron impact (EI)*, employs the interaction of energetic electrons with gas phase molecules. The electrons are generated by thermionic emission from a heated tungsten or rhenium filament. In EI, ions isolated in the gas phase should be formed to avoid ion–molecule reactions. Therefore, the pressure in the ionization volume of an EI ion source is maintained in the range of 10^{-5}–10^{-6} mbar so that their mean free path (about 1 m) is by far longer than the average distance traveled within the ionization zone of ion source (in the order of 1 mm). Their ion source dwell time is about 1 microsecond due to the extracting electric field accelerating them from the zone of creation to generate a continuous ion beam into the mass analyzer (Figure 29.14).

Quantitative energy transfer is required for ionization to occur, when an electron possessing energy just equal to IE collides with a neutral. Such an event is of low probability and, therefore, the *ionization efficiency* is then close to zero. However, a slight increase in electron energy brings about a steady increase in ionization efficiency. Every molecule has an ionization efficiency curve of its own (Figure 29.15). The overall efficiency of EI depends on the intrinsic properties of the ionization process as well as on the ionization cross section of the analyte. Fortunately, the curves of ionization cross section versus electron energy are very similar in shape, exhibiting a maximum at electron energies around 70 eV. In addition, the plateau of the ionization efficiency curve around

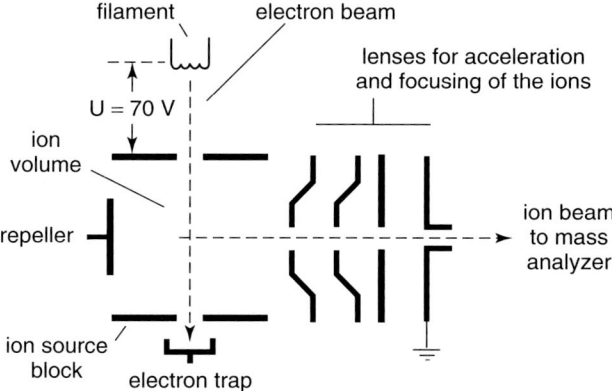

Fig. 29.14 Schematic drawing of an EI ion source. Adapted with permission from Schröder (1991). © Springer-Verlag, 1991.

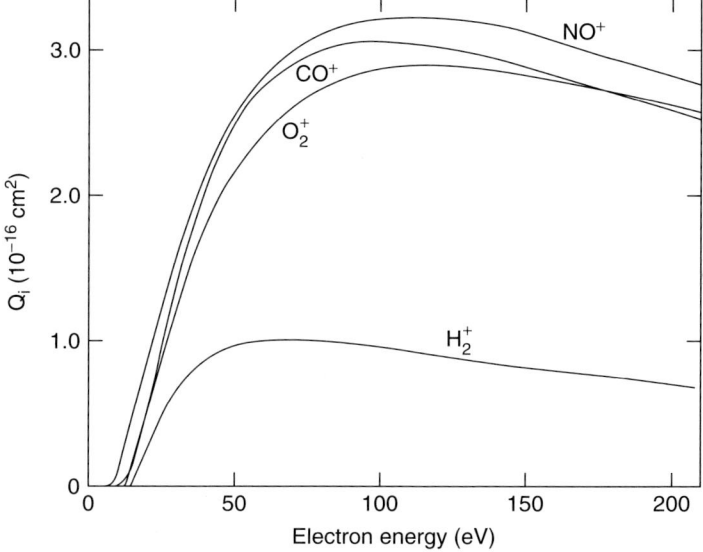

Fig. 29.15 Ionization efficiency curves for NO, CO, O$_2$, and H$_2$. Adapted with permission from Kiser (1965). © Prentice-Hall, 1965.

70 eV makes variations in electron energy negligible; in practice, EI works equally well at 60–80 eV. This assures better reproducibility of spectra and, therefore, allows comparison of spectra obtained from different mass spectrometers or from mass spectral databases. Therefore, EI spectra are almost exclusively acquired at 70 eV.

Once formed, the molecular ions "experience a rich life" during this short time span (Figure 29.16).

It takes less than 10^{-15} seconds for an electron of 70-eV kinetic energy to

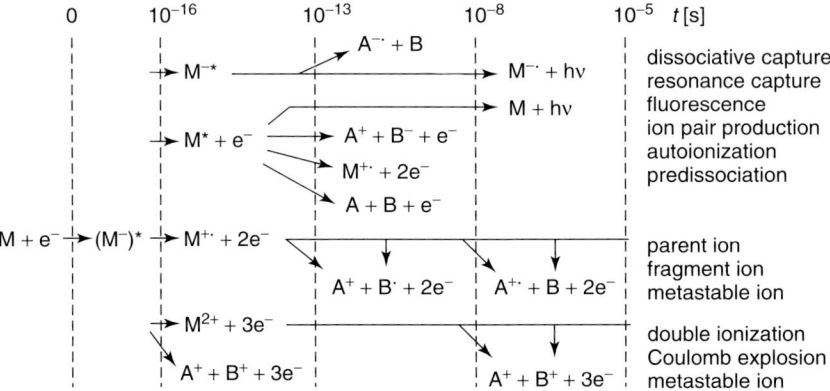

Fig. 29.16 Schematic time chart of events possible upon electron ionization. Adapted from Märk (1986) with permission. © John Wiley & Sons Inc, 1986.

pass through a molecule. Vibrations, the fastest internal motions of a molecule, are slower by at least two orders of magnitude, for example, a C–H stretching vibration takes roughly 1.1×10^{-14} seconds per cycle. Thus, the electronic motion and the nuclear motion are in fact separated, that is, the *Born–Oppenheimer approximation* is valid (Born and Oppenheimer, 1927; Seiler, 1969). Now, the *Franck–Condon principle* states that the probability for an electronic transition is highest where the electronic wave functions of both ground state and ionized state have their maxima. Combining these tenets explains the occurrence of the so-called *vertical transitions* (Franck, 1925; Condon, 1926). (Also cf. Section 29.4.1.1 for processes under EI conditions.) Although there is a wealth of chemical reactions, all of them are strictly unimolecular, because the mean free paths are long enough to prevent bimolecular interactions within the lifetime of the ions. Ions of very low internal energy reach the detector and are thus termed *stable ions* (within the below timescale); those that have life times long enough to leave the ion source but decompose during their flight through the analyzer are termed *metastable ions*, and those that fragment inside the ion source are called *unstable ions*.

The kinetics of these reactions are commonly described by the *quasi-equilibrium theory* (QET) (Wahrhaftig, 1986; Märk, 1986; Rosenstock and Krauss, 1963; Wahrhaftig, 1959).

Ions with moderate internal energy undergo decomposition reactions along well-defined pathways, leading to *fragment ions* having structures that are still closely related to the intact molecule. Interpretation of EI spectra following these routes and some well-established rules can therefore lead to the identification of a compound (McLafferty and Turecek, 1993, Chapter 6 in Gross, 2004). A sample EI spectrum of the over-the-counter drug ibuprofen elucidates the basic principle (Figure 29.1).

29.6.2
Chemical Ionization

There are various possibilities of what can happen when an ion encounters a molecule in the gas phase. An elastic collision accompanied by some exchange of kinetic or vibrational energy does not chemically alter the colliding particles. However, if the collision is soft enough it can result in adhesion of both particles or in exchange of charge via mutual transfer of either an electron or some (small) ions such as a proton. All these processes contribute to the manifold approaches to CI.

To enable bimolecular processes within the microsecond timeframe, the pressure inside the ion source has to be substantially raised to 10^{-4}–10^{-3} mbar by the introduction of a reagent gas that is then present in more than 1000-fold excess over the analyte. The analyte molecules undergo in the order of 50 collisions (as opposed to none in EI) and, thus, become thermalized and accessible to ion–molecule reactions. To avoid excessive leakage into the remaining mass spectrometer, the ionization volume of a CI ion source has only small holes for the entrance and exit of electrons, reagent gas, sample vapor, and ions produced (Figure 29.17). Sufficient pumping capacity is required at the ion source housing to maintain high vacuum conditions.

Nowadays, there are numerous variations of CI introduced in the mid-1960s (Field and Munson, 1965; Harrison, 1992b). *Positive-ion chemical ionization* (PICI or PCI) (Harrison, 1992b; Munson, 2000) employs either proton transfer (Section 29.4.2.1), electrophilic addition (Section 29.4.2.2), and anion abstraction (Section 29.4.2.3) or CT (Section 29.4.1.4) to generate analyte ions. The reagent gases methane and isobutane are preferred for $[M + H]^+$ ion formation, ammonia tends to yield $[M + NH_4]^+$ by electrophilic addition, and gases such as benzene, carbondisulfide, or noble gases serve for the production of $M^{+\bullet}$ ions by CT (Allgood, Ma and Munson, 1991; Hsu and Qian, 1993; Roussis, 1999).

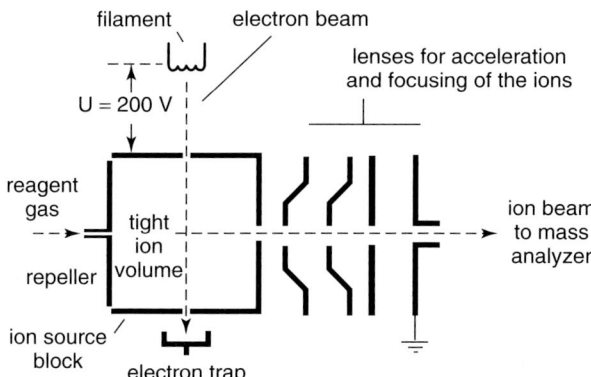

Fig. 29.17 A CI ion source can be readily obtained by replacing the ionization volume of an EI ion source with a tighter one. Many mass spectrometers have EI/CI combination ion sources switching between these methods by pushing a cylinderlike insert into the EI volume in order to close its vents (cf. Figure 29.14). Adapted with permission from Schröder (1991). © Springer-Verlag, 1991.

Negative-ion chemical ionization (NICI or NCI) (von Ardenne, Steinfelder and Tümmler, 1971; Dillard, 1973) makes use of cation abstraction (Section 29.4.4.2), nucleophilic addition (Section 29.4.4.1 and Figure 29.5), or EC (Section 29.4.3.1), the latter being no CI in the strict sense, because the reagent gas (often isobutane) solely moderates the electrons to thermal energy instead of involving any reaction (Knighton and Grimsrud, 1995). Alternatively, the electrons may be delivered from an electron monochromator (Laramée et al., 1996).

The goal of CI is to generate quasi-molecular ions or molecular ions of low internal energy in order to reduce the degree of fragmentation. The corresponding ions generally are well recognized in CI spectra of both polarities while the moderate fragmentation still assists structure elucidation of unknowns. It is common to combine the information obtained from both EI and CI spectra (Figure 29.18), or from positive-ion plus negative-ion CI spectra to improve the reliability of the mass spectral analysis.

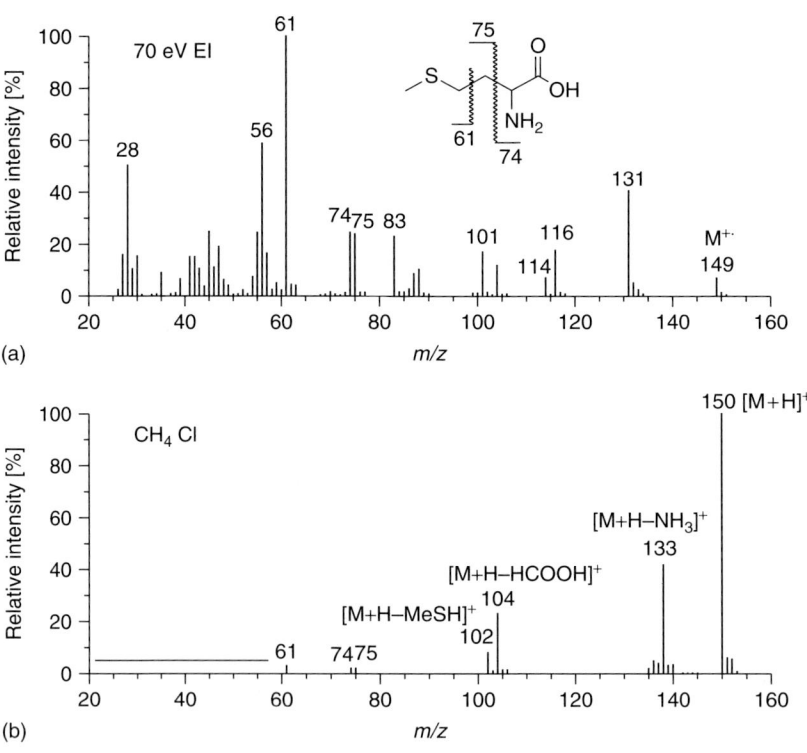

Fig. 29.18 Comparison of (a) 70-eV EI mass spectrum and (b) methane reagent gas CI mass spectrum of the amino acid methionine. The range up to m/z 60 has been excluded from the CI spectrum, because it is dominated by peaks due to the reagent gas. The reduction in fragmentation as effected by the generation of the closed-shell quasimolecular ions, $[M + H]^+$, in CI is obvious. This also results in easier identification of the sulfur isotopic pattern. Reprinted from Gross (2004) with permission. © Springer-Verlag, 2004.

29.6.3
Field Ionization

In FI, molecular ions, $M^{+\bullet}$, are softly created from gas phase neutrals by action of strong electric fields causing tunneling of an electron from a gaseous molecule into a solid electrode, the field anode (Section 29.4.1.5) (Gomer and Inghram, 1955; Beckey, 1962; Beckey, 1971; Lattimer and Schulten, 1989; Gomer, 1994). Although some micrometer-thin metal wires were employed as electrodes in pioneering work on FI, the use of activated emitters – occasionally specially optimized for FI usage – has become standard in FI-MS. The process of *emitter activation* transforms a tungsten wire of 5–13 μm in diameter into a rather fragile object resembling a hairy caterpillar species (Beckey, Krone and Röllgen, 1968; Beckey, Hilt and Schulten, 1973; Okuyama *et al.*, 1977; Rabrenovic, Ast and Kramer, 1981). These *whiskers* grown on the central wire cause an enormous increase in the electric field strength at their tips as compared to bare wires, because the radii of curvature are drastically reduced. Thus, the activated emitter provides a large number of high electric field sites for FI to occur.

The setup of an FI ion source comprises the field emitter opposed to a counter electrode set to a potential of 10–12 kV negative relative to the emitter (Figure 29.19). The emitter can be resistively heated up to about 1000 °C by passing a current of 50–90 mA through it in order to bake off any condensed sample residues between subsequent measurements. During operation, the resistive heating is normally switched off or set to low levels. The sample vapor can be introduced by any of the means listed in Table 29.2.

As long as the process of sample evaporation is not limiting, FI spectra are mainly characterized by the molecular ion peak of the analyte, eventually accompanied by few generally low-abundant fragment ions (Figure 29.20).

Owing to some degree of surface adhesion in FI, there are two accompanying processes influencing the spectra. On one side, ions formed may encounter a second step of ionization, leading to M^{2+} ions. On the other side, bimolecular processes can cause the formation of $[M + H]^+$ ions analogous to the generation of protonated

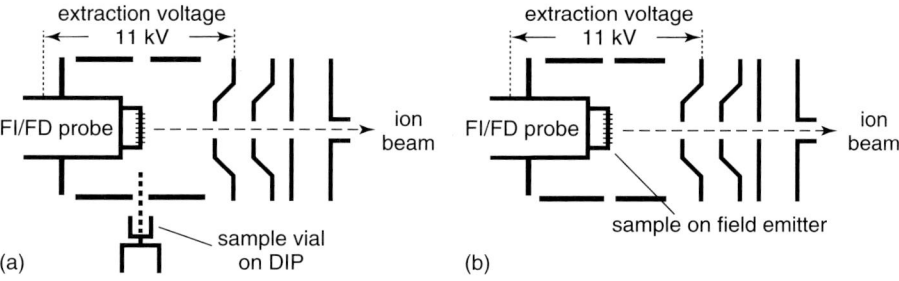

Fig. 29.19 Schematic drawings of (a) FI and (b) FD ion sources. Field ionization has lead to the development of *field desorption* (FD) where the sample is directly placed onto the emitter surface from dilute sample solution by evaporation of the solvent (Section 29.7.1). Adapted from Schröder (1991) with permission. © Springer-Verlag, 1991.

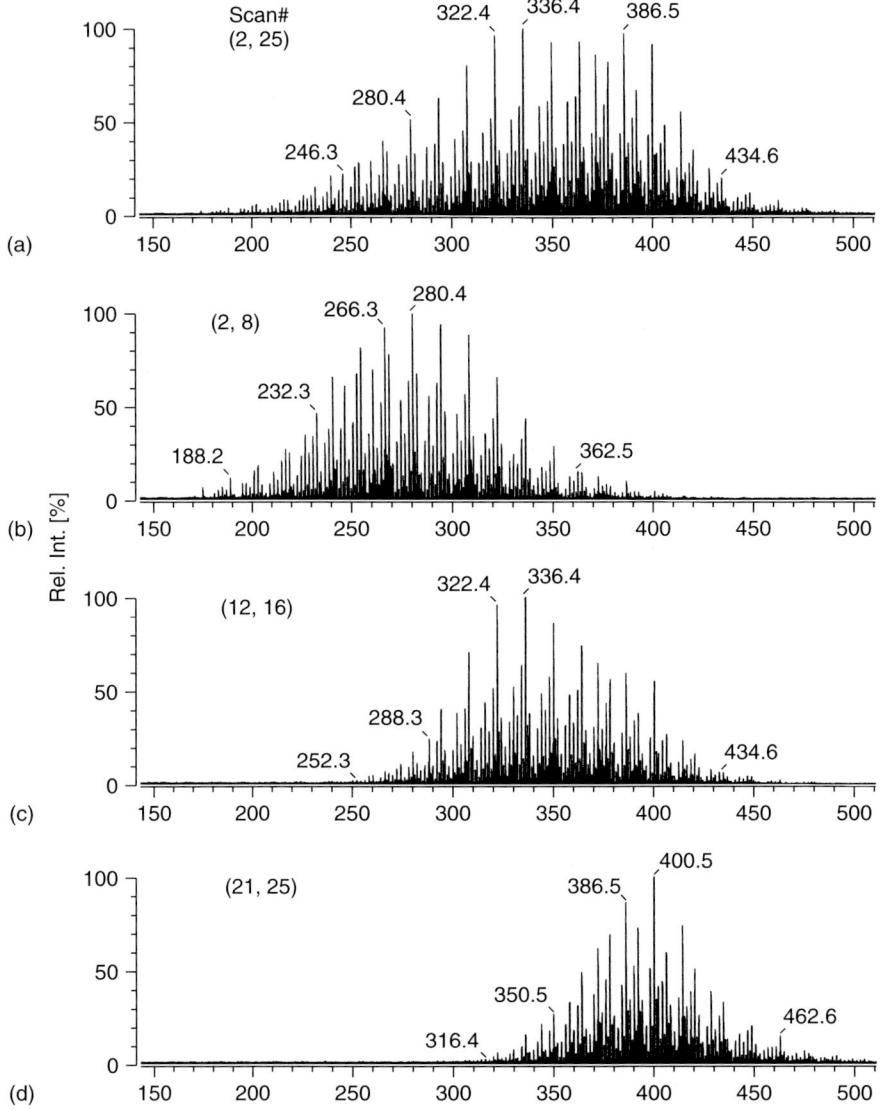

Fig. 29.20 Typical fractionation during the acquisition of the FI mass spectrum of a standard sewing machine oil when evaporated from a direct insertion probe. The majority of peaks corresponds to molecular ions. (a) A spectrum from summing up scans 2–25.
(b)–(d) Spectra are sums of a few scans during evaporation of the oil upon heating of the sample vial on the probe tip. Compounds of lower molecular weight evaporate earlier than those of higher molecular weight.

reagent gas molecules in CI (Schulten and Beckey, 1974). Although doubly charged ions are easily recognized in mass spectra by their spacing of isotopic peaks (cf. Section 29.2.3 and Figure 29.2), the partial protonation can lead to misinterpretation of isotopic patterns.

29.7
Desorption/Ionization

When the generation of gas phase ions can be accomplished by combining both evaporation and ionization into a single process, thermal decomposition before ionization can be avoided for the most part. Although CI and FI already belong to the so-called soft ionization methods, these do not possess this key property, and many of their applications have soon been replaced by *desorption/ionization* methods when they became commercially available. In contrast to EI, CI, and FI, desorption/ionization methods handle ionic analytes very well. In fact, ionic analytes are even beneficial, because they only require transfer into the gas phase, while the actual ionization step is not necessary anymore.

The numerous types of ions formed by the various desorption/ionization methods can be conveniently organized by their relevance for analytes of given polarity and charge (Table 29.3). FD (Section 29.7.1) covers the table's left column only, while LSIMS and FAB (Section 29.7.2) as well as LDI (Section 29.7.3) apply to the whole table with a compound-dependent high-mass limit of about 2000–5000 u. Finally, MALDI (Section 29.7.4) can access the whole polarity–analyte mass range, although it is not ideally suited for unpolar analytes and molecules below 500 u.

29.7.1
Field Desorption

FI has been further developed to *field desorption* (FD) where the sample is directly placed onto the emitter surface

Tab. 29.3 Ions formed by desorption/ionization.

Analytes	Positive ions	Negative ions
Nonpolar	$M^{+\bullet}$	$M^{-\bullet}$
Medium polarity	$M^{+\bullet}$ and/or $[M + H]^+$, $[M + alkali]^+$, less frequently also, *clusters* $[2M]^{+\bullet}$ and/or $[2M + H]^+$, $[2M + alkali]^+$, *adducts* $[M + Ma + H]^+$, $[M + Ma + alkali]^+$	$M^{-\bullet}$ and/or $[M - H]^-$, less frequently also, *clusters* $[2M]^{-\bullet}$ and/or $[2M - H]^-$, *adducts* $[M + Ma]^{-\bullet}$, $[M + Ma - H]^-$
Polar	$[M + H]^+$, $[M + alkali]^+$, *exchange* $[M - H_n + alkali_{n+1}]^+$, *clusters* $[nM + H]^+$, $[nM + alkali]^+$, *adducts* $[M + Ma + H]^+$, $[M + Ma + alkali]^+$, High-mass analyte $[M + nH]^{n+}$, $[M + nalkali]^{n+}$	$[M - H]^-$, *exchange* $[M - H_n + alkali_{n-1}]^-$, Clusters $[nM-H]^-$, *adducts* $[M + Ma - H]^-$, High-mass analyte $[M - nH]^{n-}$, $[M + nalkali]^{n-}$
Ionic	C^+, $[C_n + A_{n-1}]^+$, rarely $[CA]^{+\bullet}$	A^-, $[C_{n-1} + A_n]^-$, rarely $[CA]^{-\bullet}$

from dilute sample solution by a microliter syringe. After evaporation of the solvent, the emitter is inserted into the ion source. Therefore, a FD ion source is basically identical to an FI ion source (Figure 29.19). In FD, the analyte can be field ionized (the mechanism of ion creation) while still adsorbed to the surface or while just being on the leave. This leads to a significant reduction in thermal decomposition of the analyte and thereby allows for the analysis of highly polar molecules up to about m/z 2000 (Beckey, 1977; Wood, 1982; Prókai, 1990a; Prókai, 1990b). To achieve sufficient surface mobility, the emitter is resistively heated by ramping a current at 1–16 mA min^{-1} up to 40–60 mA during the measurement. High *emitter heating currents* (EHC) are employed to remove residual sample. Analytes of low polarity are even successfully analyzed up to higher molecular weights, for example, polystyrenes are accessible by FD up to m/z 10.000 (Lattimer, 2001).

The type of ions produced by FD depends on the analyte's polarity. Compounds of low polarity yield M$^{+\bullet}$ ions, medium polarity molecules are transformed into either M$^{+\bullet}$ ions or quasimolecular ions such as [M + H]$^+$ or [M + alkali]$^+$, and finally, highly polar compounds only form quasimolecular ions. Ionic analytes (cation C$^+$ and anion A$^-$) are desorbed from the surface without prior FI. C$^+$ is then detected as the most intensive peak, followed by [C$_n$A$_{n-1}$]$^+$ cluster ions. Often, nonionic analytes also form cluster ions, for example, [2M + H]$^+$, [2M + alkali]$^+$, to some degree (Figure 29.21).

Negative-ion FD is a rare exception (Daehling *et al.*, 1982; Dähling *et al.*, 1983), because emission of electrons,

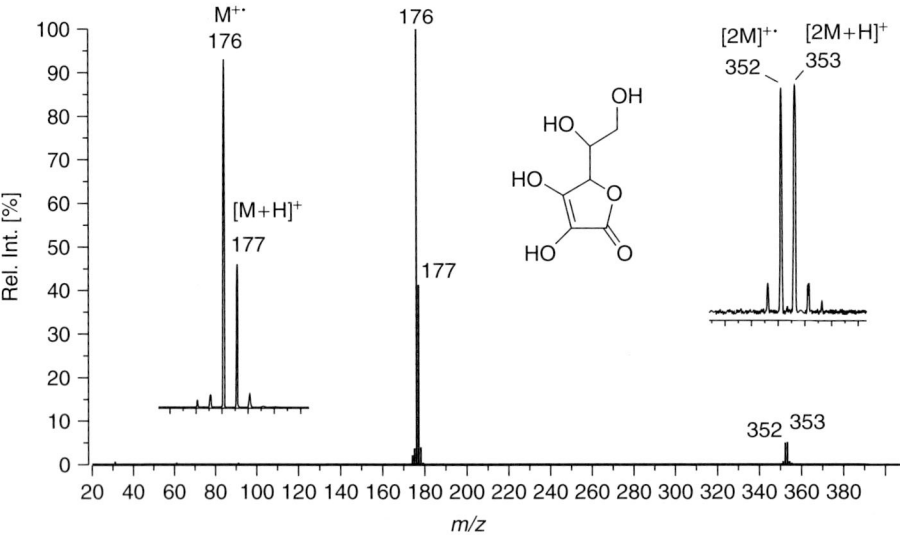

Fig. 29.21 FD mass spectrum of ascorbic acid showing M$^{+\bullet}$ and concomitant [M+H]$^+$ ions. This can be inferred from the overgrown isotopic pattern that cannot be due to C$_6$H$_8$O$_6$ alone, because then [M + 1] should only exhibit 6.6% relative to M$^{+\bullet}$ instead of the observed 41%. The cluster ion signal shows this phenomenon even more obviously.

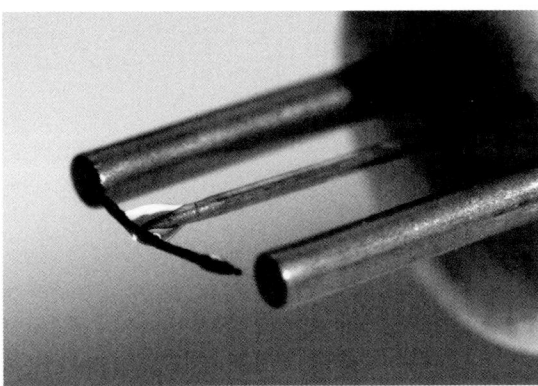

Fig. 29.22 The basic, nonetheless crucial, difference between LIFDI and standard FI or FD is the addition of the capillary. In LIFDI, the field emitter is loaded with analyte solution via this capillary positioned close to the activated area. The liquid spreads over the entire length and the solvent evaporates in vacuum. The thin central tungsten wire connecting this section to the posts is barely visible.

detrimental for the emitter, occurs from activated emitters before analyte ion formation.

The advent of *liquid injection field desorption/ionization* (LIFDI) gave new impetus for the use of FD. In LIFDI, the analyte solution (0.1–0.2 mg ml^{-1}) is transferred onto the emitter inside the ion source by a fused silica capillary (10–75 µm i.d.) (Figure 29.22). In this manner, LIFDI (i) offers inert conditions (Linden, 2004; Gross et al., 2006), (ii) speeds up the measurement process by reducing the number of tuning cycles, and (iii) even admits continuous-flow applications (Schaub et al., 2004). The combination of LIFDI with FT-ICR instruments is gaining importance in industrial analysis, for example, of nonpolar compounds in petroleum products (Figure 29.23) (Schaub et al., 2005).

29.7.2
Fast Atom Bombardment and Liquid Secondary Ion Mass Spectrometry

Energetic particles impacting onto a surface are capable of both sputtering ions as well as neutrals into the selvedge region and ionization of those gas phase neutrals (Figure 29.24). The development of the title methods started with the introduction of ion beams for the production of secondary ions from inorganic surfaces and hence the name *secondary-ion mass spectrometry* (SIMS)(Benninghoven and Kirchner, 1963). Nonconduction materials build up electric charges that prevent further useful ion production by Coulombic repulsion. Thus, primary ion beams were replaced by atom beams (*fast atom bombardment*, FAB) (Devienne, 1967; Barber, Vickerman and Wolstenholme, 1980) that can be obtained by passing rare gas ions through a collision cell where they are neutralized while their kinetic energy and direction are basically retained (Figure 29.25).

The introduction of a *liquid matrix* to dissolve the analyte made FAB a method for the soft ionization of highly polar and eventually large organic or bioorganic molecules (Morris et al., 1981; Barber et al., 1982). In addition, the matrix allowed to employ primary ion beams again, because it circumvented electrostatic charging of the target, thereby giving rise to *liquid secondary-ion mass spectrometry* (LSIMS) (Benninghoven, 1983). FAB and LSIMS work equally well for the

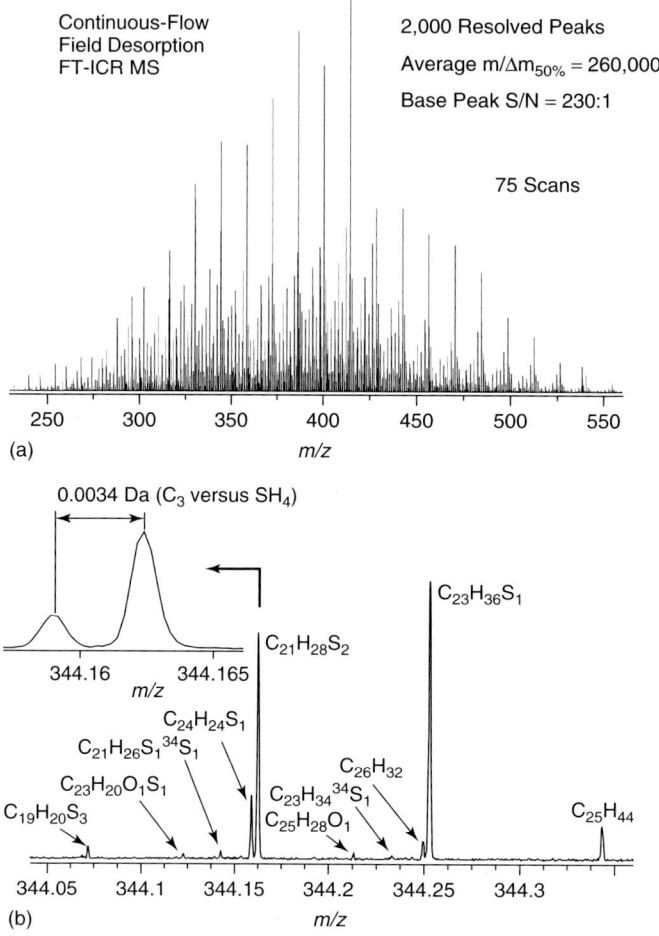

Fig. 29.23 (a) Continuous-flow FD-FT-ICR mass spectrum of a refinery process stream sample; the spectrum is the average of 75 single spectra accumulated during 1 hour. (b) Mass scale expansion at nominal m/z 344 from the broadband mass spectrum. The resolution of the 3.4×10^{-3} u distant C_3 versus SH_4 duplet can be observed at m/z 344.25 and separately in the inset at m/z 344.16. Adapted from Schaub et al. (2004) with permission. © John Wiley & Sons, 2004.

whole range of sample polarities and for negative-ion and positive-ion generation (Chapter 9 in Gross (2004) and Table 29.3).

The use of a liquid matrix, for example, glycerol, thioglycerol, 3-nitrobenzylalcohol, 2-nitrophenyloctylether, is key to the abilities of FAB and LSIMS. The matrix absorbs the primary energy, avoids intermolecular interactions of analyte molecules by dissolution, serves as a source of protons for $[M + H]^+$ ion formation by its natural acidity, and mediates the ionization in solution by other

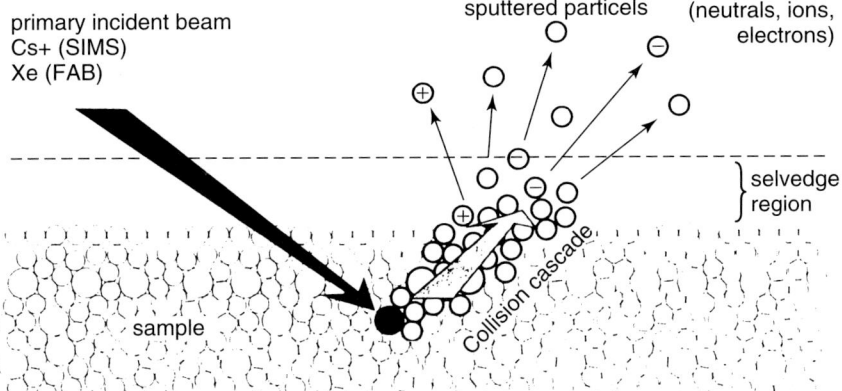

Fig. 29.24 Simplified illustration of an instantaneous collision cascade generated as a result of primary particle impact in FAB or SIMS. Adapted from Busch (1995) with permission. © John Wiley & Sons Ltd, 1995.

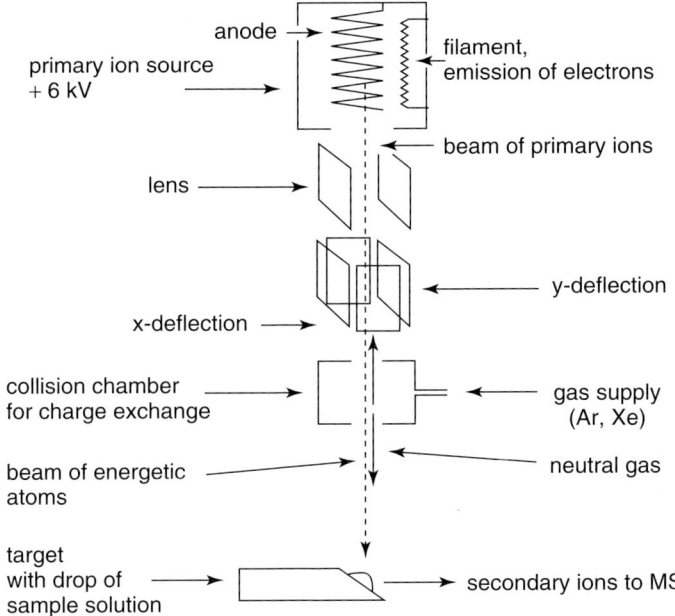

Fig. 29.25 Schematic representation of a FAB gun. The supply of neutral gas also serves as a collision chamber for charge exchange to transform the energetic primary ions generated by EI into a beam of fast neutrals hitting the sample solution on the target. In a FAB ion source, the target takes the same position as the FI/FD probe in Figure 29.19 or the reagent gas inlet in Figure 29.17, respectively; further changes as compared the previously shown ion sources are not required. Therefore, EI/CI/FAB combination ion sources are common. Ion volumes exchangeable through a vacuum lock can reduce ion source contamination.

ionic species, either present as traces such as the ubiquitous alkali ions or by deliberately added ones such as organic acids, salts, and so on (De Pauw, Agnello and Derwa, 1986). Unfortunately, evaporation of the matrix accelerates ion source contamination and limits the time span for sampling of spectra. In addition, abundant ions are produced from the matrix, delivering a strong background of peaks at every m/z with $[n\text{Ma} + \text{H}]^+$ or $[n\text{Ma} - \text{H}]^-$ matrix cluster ions typically dominating those spectra. Therefore to be identified, the analyte ions need to yield strong signals clearly distinguished from that background. Continuous-flow FAB (CF-FAB) (Caprioli, Fan and Cottrell, 1986) and frit-FAB (Ito et al., 1985) not only improve the signal-to-background ratio but also present a means for LC coupling and real-time monitoring of physiological processes by FAB (Caprioli, 1990).

Although still in use, FAB and LSIMS have widely been replaced by more recent desorption/ionization and related methods (see below). One of the advantages of FAB and LSIMS is their ability to actively generate ions from neutrals in unpolar and aprotic matrix solutions, at cryogenic temperatures even from typical organic chemistry solvents (LT-FAB) (Wang et al., 1998; Gross, 1998).

29.7.3
Laser Desorption/Ionization

Different from FAB and LSIMS, *laser desorption/ionization* (LDI) does not provide a continuous beam of ions. Owing to the pulsed nature of a laser, ions are generated in extremely short bursts, the duration of which being slightly longer than the pulse width of the laser. Typically, nitrogen lasers, that is, UV lasers at a wavelength of 337 nm delivering 3 nanosecond pulses are employed. Laser and optics are normally located on top of the ion source housing from where the light is guided onto a sample plate (target) through a quartz window (Figure 29.26). The energy provided by the laser is then absorbed by a (normally solid) layer of the analyte where it causes evaporation

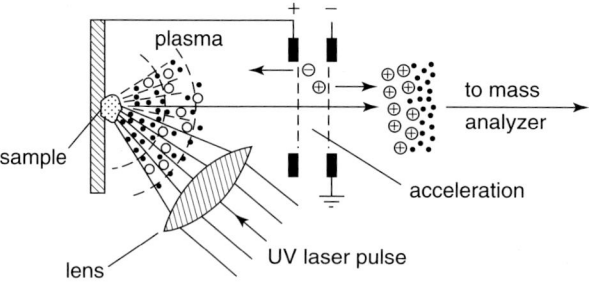

Fig. 29.26 Schematic of an ion source for laser desorption/ionization. The solid sample layer is irradiated with a laser pulse in vacuum. Ions created during this pulse and in the immediate aftermath are extracted from the laser-generated plume and accelerated into a (time-of-flight) mass analyzer. Reproduced from Mamyrin (1994) with permission. © Elsevier Science Publishers, 1994.

and ionization in analogy to the effects of impacting particles as described in Section 29.7.2 (Fenner and Daly, 1966; Vastola and Pirone, 1968; Vastola, Mumma and Pirone, 1970; Posthumus et al., 1978). In addition, nonresonant MUPI in the gas phase can occur (cf. Section 29.4.1.2).

The direct transfer of energy to the analyte, however, creates a significant portion of ions excited well above their dissociation threshold. Therefore, LDI is not as soft as FAB or especially FD. Nonetheless, the fact that LDI is an "energy-sudden" method can compensate for those levels of internal energy, because dissociation kinetics is slower by some orders of magnitude than ion formation (cf. time scale of EI, Figure 29.16).

Stable, UV-absorbing molecules such as fullerenes, polycyclic aromatic hydrocarbons (PAHs), or low-mass organic or inorganic salts are easily analyzed by LDI. In fact, LDI-MS has played an important role in the discovery of the fullerenes (Figure 29.27) (Kroto, 1997).

Although still rather soft and well applicable to analytes up to about m/z 2000, LDI has its drawbacks in (i) the necessity of an UV chromophore in the analyte molecules and (ii) the conservation of intermolecular forces within the analyte layer that together pose limits upon the application of LDI to highly polar compounds such as oligosaccharides, peptides, and proteins (Posthumus et al., 1978; McCrery, Ledford and Gross, 1982; Cotter, 1987).

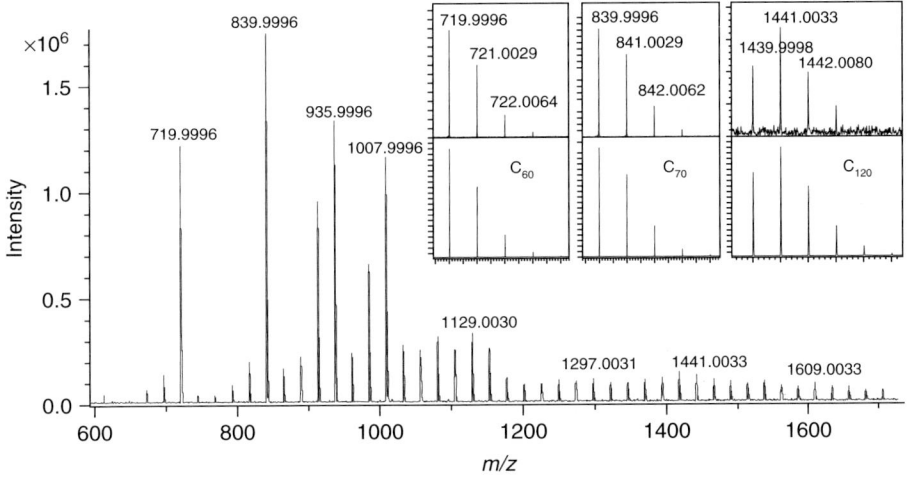

Fig. 29.27 Positive-ion LDI mass spectrum of a fullerene soot synthesized by the Krätschmer–Huffman method (Krätschmer et al., 1990). The spectrum was obtained on a Fourier-transform ion cyclotron resonance (FT-ICR) mass spectrometer; hence, the resolution is 175.000 (1.75×10^5) at m/z 840, that is, almost by a factor of 1000 higher than on early time-of-flight instruments as used for the discovery of C_{60} and larger fullerenes (Kroto, 1997) The insets show expanded views of the $M^{+\bullet}$ ions for C_{60}, C_{70}, and C_{120} (upper traces) together with the corresponding calculated isotopic patterns. Note that the accurate masses are lower than the nominal values by the mass of an electron, while the difference in mass of 1.0033 u between ^{12}C and ^{13}C can be recognized from the mass increment of the first and second isotopic peaks.

29.7.4
Matrix-Assisted Laser Desorption/Ionization

Tanaka's approach to establish a matrix in LDI was the use of glycerol with an admixture of ultrafine cobalt powder (particle size about 30 nm) serving to transform the UV-transparent liquid into a UV-absorbing medium (Tanaka et al., 1988; Tanaka, 2003) However, the restricted variability of this matrix limited its use to proteins and polar synthetic polymers, and moreover, the poisonous cobalt prevented its adoption as a routine technique.

The other approach, introduced by Hillenkamp and Karas, involved a crystalline organic matrix and started a revolution in terms of what kind of analytes became accessible by MS. Thus, *matrix-assisted laser desorption/ionization (MALDI)* with its capability to analyze intact proteins as large as antibodies (150.000 u) meant that MS had really entered the realms of biology and medicine (Figure 29.28)

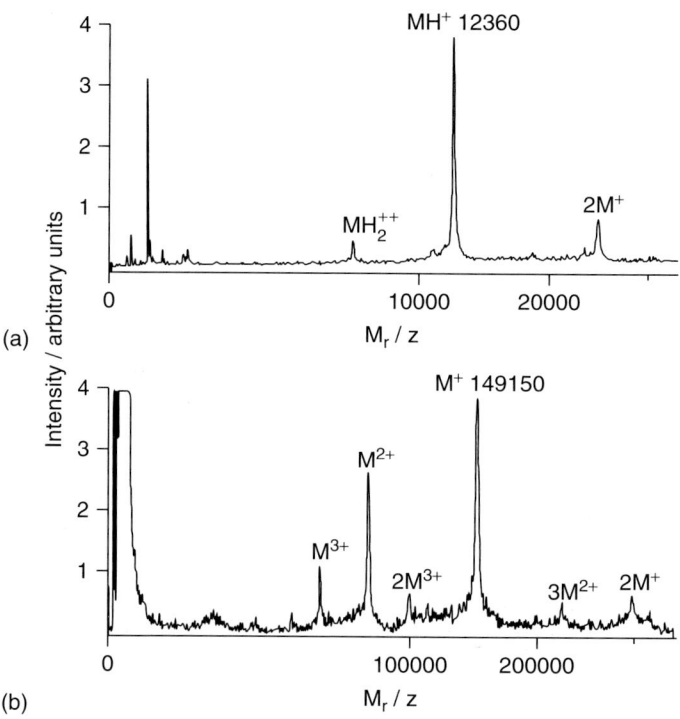

Fig. 29.28 Two UV-MALDI spectra of the early days of MALDI immediately demonstrating the potential of the method. (a) The small protein porcine cytochrome C from 2,5-dihydroxybenzoic acid matrix after accumulation of 20 laser shots and (b) a large monoclonal antibody from nicotinic acid matrix after accumulation of 30 laser shots. Although assigned as $M^{+\bullet}$ or $[M + H]^+$ ions, it may be assumed that the signals are formed by the whole palette of quasimolecular ions possible. Reproduced from Karas, Bahr and Gießmann (1991) with permission. © John Wiley & Sons Inc., 1992.

(Karas et al., 1989a; Karas et al., 1989b; Karas, Bahr and Gießmann, 1991) A majority of portions of the research done in these fields today rely on MALDI as a (high-throughput) method for fingerprinting (Welham et al., 2000), identifing, and sequencing of various classes of biopolymers (Hillenkamp and Peter-Katalinic, 2007).

The tremendous development of TOF instruments was initiated and put forward to a large extent by the demand for increasingly powerful MALDI instrumentation. Indeed, MALDI-TOF was the almost indivisible implementation of MALDI in its first decade.

The instrumental setup for MALDI does not differ from LDI; the matrix is the key to the method's success. The matrix provides effective absorption of the laser energy independent of whether the analyte possesses a chromophore or not. It separates the molecules before evaporation and supports the ionization process, either by acid/base reactions upon mixing with the analyte or during the actual desorption via ion–molecule reactions in the laser-generated plasma (e.g., cf. Sections 29.4.1.4, 29.4.2.1, 29.4.2.2) (Zenobi and Knochenmuss, 1999; Karas and Krüger, 2003).

Nitrogen UV lasers ($\lambda = 337$ nm) are most frequently employed to supply power in excess of the threshold irradiance for ion production ($>10^6$ W cm^{-2}). Nonetheless, excimer lasers ($\lambda = 293$, 248, 308, and 351 nm), frequency-tripled ($\lambda = 355$ nm), and quadrupled Nd:Yag lasers ($\lambda = 266$ nm) are also in use. In IR-MALDI, used for a minority of applications where penetration depth is important, for example, to desorb samples from gels or thin layer chromatographic (TLC) plates, Er:Yag lasers ($\lambda = 2.94$ µm) and CO_2 lasers ($\lambda = 10.6$ µm) are common (Overberg et al., 1990; Overberg, Karas and Hillenkamp, 1991).

Although many standard applications of MALDI do not present very special requirements, MALDI sample preparation can be an art in itself. For certain types of analytes such as oligonucleotides or high-mass (bio)polymers, great care needs to be excised to produce suitable crystal layers where analyte and matrix are preferentially cocrystallized from solution in molar ratios in the 1:1000–100.000 (10^5) range. The same applies when ultimate sensitivity is required. Under optimal conditions, the sample consumption is in the low femtomole range, and several attomoles have been demonstrated (Vorm, Roepstorff and Mann, 1994). Special hydrophobic targets with hydrophilic spots effecting the concentration of the preparation on a small diameter spot (200–600 µm) deliver substantially improved sensitivity (Nordhoff et al., 2003).

By selecting a matrix to the analyte's requirements, MALDI can be applied to a large variety of analytes ranging from nonpolar polymers such as (low MW) polyethylene to highly polar polymers such as poly(toluenesulfonic acid) resins. It is suitable for peptides, proteins, oligonucleotides, and DNA, for oligosaccharides (Figure 29.29), and – sometimes overlooked – for small molecule analysis, for example, for dyes, lipids, and porphyrins.

MALDI-MS presents a powerful method for the analysis of proteins and peptides. It may be used for the in-depth analysis of an individual protein to reveal its primary structure, that is, its amino acid sequence, as well as eventual (posttranslational) modifications (Figure 29.30). Proteomics, the

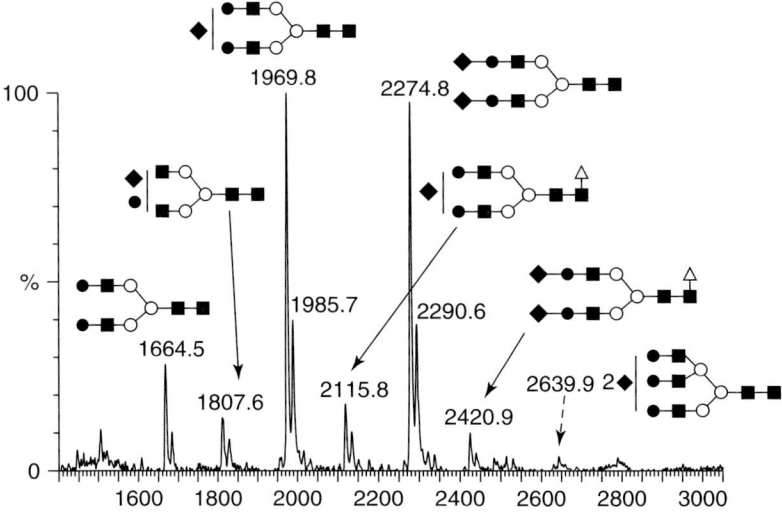

Fig. 29.29 Positive-ion linear mode MALDI-TOF spectrum from 2,4,6-trihydroxyacetophenone (THAP) matrix of native N-glycans obtained by in-gel PNGase F digestion from the two-dimensional gel of α 1-antitrypsin spots of a CDG-type-IIx patient. The sialic acid residues were methylated before MALDI analysis. Symbols: ■, GlcNAc; ●, galactose; ○, mannose; ♦, sialic acid; and △, fucose. Adapted with permission from Šagi et al. (2005) © Wiley-VCH, 2005.

study of the whole complexity of a protein mixture expressed by given type of cells, is another widespread application of MALDI, which indeed serves as a propellant for the steady improvements in sensitivity, resolution, mass accuracy, and speed of modern MALDI instrumentation. It is a common procedure to obtain survey MALDI spectra of protein tryptic digests and to subsequently measure MALDI tandem mass spectra of the tryptic peptides produced. The sequence of the whole protein may then be derived by putting the subsequences of those peptides together (Lehmann, 1996; Kinter and Sherman, 2000; Snyder, 2000).

In polymer chemistry, MALDI spectra provide identification of the monomer (repeat unit), of the average molecular weight, and of polydispersity, and – provided the low-mass end of the envelope is not beyond m/z 1000 – allow for end group analysis (Figure 29.31) (Nielen, 1999; Murgasova and Hercules, 2003; Montaudo and Lattimer, 2001; Pasch and Schrepp, 2003).

29.8
Ion Formation at Atmospheric Pressure

It is a common feature of the ionization methods discussed so far that the process of ion formation occurs under vacuum conditions, varying from the high vacuum in EI or FI to the low-pressure limit of medium vacuum in the case of CI. Although passing through a steep pressure gradient across the selvedge region, even the deliberation of ions in desorption/ionization methods proceeds from a surface located in vacuum. It is inherent to these methods that they cannot deal with a noteworthy liquid flow, a fact

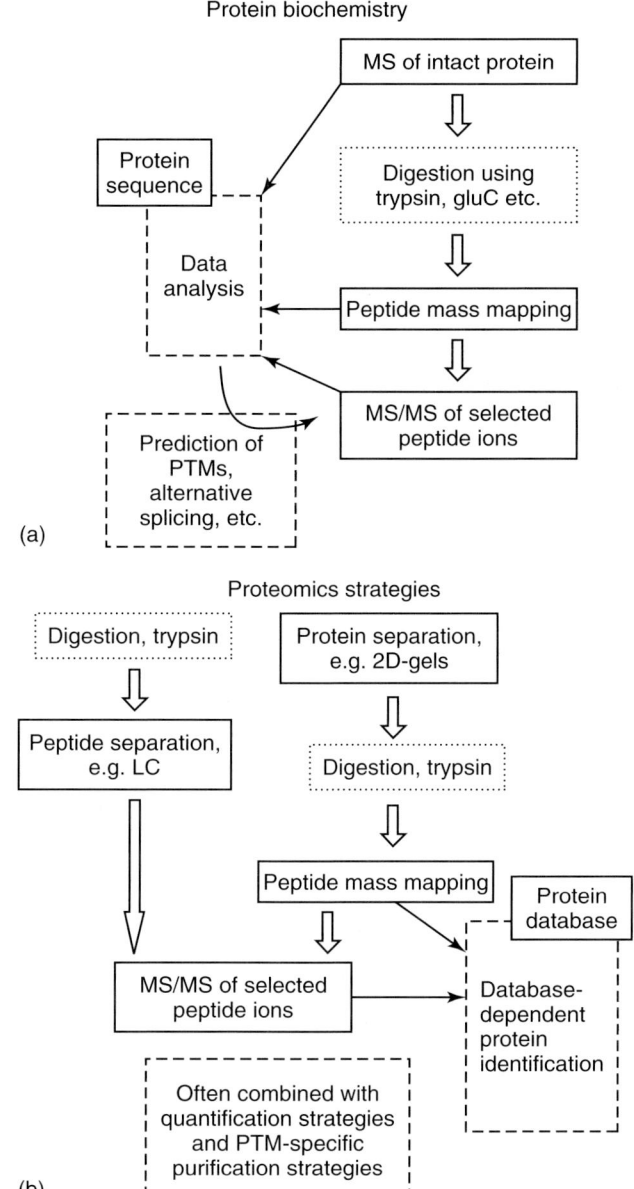

Fig. 29.30 Methodologies for protein analysis involving MALDI-MS. (a) Analysis of a purified protein to validate its identity and to detect eventual modifications. (b) Proteomics approach to detect and identify all proteins expressed by a given cell type. Here, two strategies apply, one starting with digestion of the protein before LC-MS/MS and the other employing separation on a two-dimensional gel, followed by digestion and mass analysis (p. 85 in (Hillenkamp and Peter-Katalinic, 2007)).

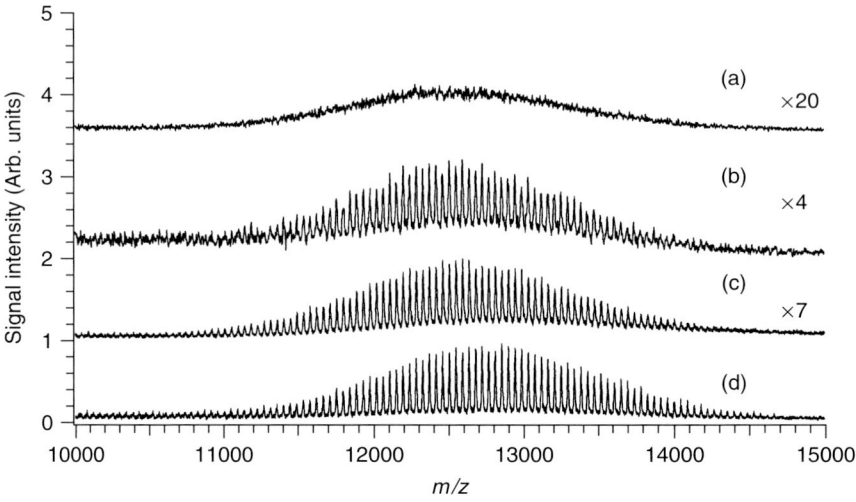

Fig. 29.31 MALDI-TOF spectra of poly(ethyleneglycol) as obtained with different matrices and in different modes of TOF operation. (a) and (d) are measured using 2-(4-hydroxyphenylazo)benzoic acid (HABA) matrix, but (a) without and (b)–(d) with time-lag focusing (TLF); (b) with 2,5-dihydroxybenzoic acid (DHB), and (c) with *trans*-3-indoleacrylic acid (IAA) matrix. These spectra (i) exemplify the enormous improvement in resolving power of TOF instruments upon introduction of TLF (delayed-extraction, DE, and pulsed ion extraction, PIE, have the same effect) and (ii) the importance of the choice of a suitable matrix. Adapted with permission from Whittal, Schriemer and Li (1997). © American Chemical Society, 1997.

preventing – or at least strongly limiting (cf. CF-FAB, Section 29.7.2) – their successful combination with LC. Nonetheless, there was a long-standing desire to develop *liquid-chromatography-mass spectrometry coupling* (LC-MS) in order to address highly polar analyte mixtures.

Atmospheric pressure ionization (API) methods take the very step of ion formation virtually to the outside of the mass spectrometer, or they expand the mass spectrometer into the laboratory atmosphere.

29.8.1
Atmospheric Pressure Ionization

The first implementation of a continuous liquid flow coupled to a mass analyzer was presented by Horning et al. as early as 1973 and was termed API (Horning et al., 1973; Horning et al., 1974). In API, the sample solution is evaporated at atmospheric pressure by a heated stream of nitrogen gas. The vapor is then passed along a ^{63}Ni foil, the electrons emitted from which start a series of ion–molecule reactions, leading to $[(H_2O)_n+H]^+$ cluster ions that finally cause $[M+H]^+$ ions to be formed. Although most of the gas is removed by a rotary pump, a portion is admitted to a quadrupole mass analyzer through a small orifice (Figure 29.32). Although API itself is not anymore in use, the method has coined the name for a whole family of ionization methods.

To improve ionization efficiency, the same group soon introduced *atmospheric*

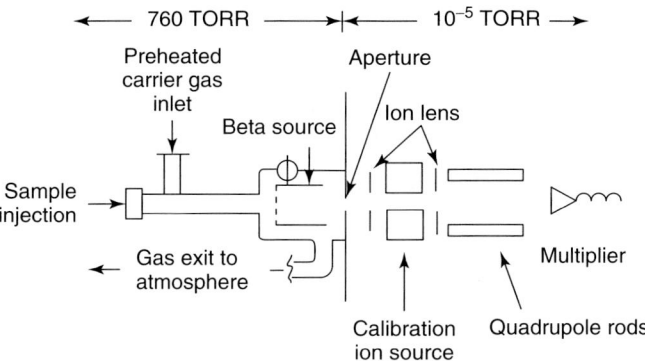

Fig. 29.32 Atmospheric pressure ionization (API) as introduced in 1973. The β-source initiates ion formation via a series of ion–molecule reactions involving water cluster ions. A small portion of the ions is transmitted into a quadrupole mass analyzer through an orifice. The quadrupole instrument shown in this first publication on API even had an EI ion source that could be used for calibration and testing of the instrument. Reproduced from Horning et al. (1973) with permission. © American Chemical Society, 1973.

pressure chemical ionization (APCI) where the formation of ions is achieved without the aid of radioactive ^{63}Ni. Here, electrons are delivered by a corona discharge from a needle tip pointing into the sample vapor or by admitting electrons produced from a heated filament as in CI (Carroll et al., 1975; Dzidic et al., 1976). The fact that the ion–molecule reactions take place at atmospheric pressure led to the term. Highly improved setups of APCI still present common LC-MS equipment (cf. Section 29.8.3) (Niessen and Voyksner, 1998; Niessen, 1998).

Another important approach to LC-MS was the *thermospray* (TSP) interface (Blakley, Carmody and Vestal, 1980; Blakley and Vestal, 1983). In TSP, the analyte solution flowing at 1–2 ml min^{-1} is rapidly vaporized by passing it through strong heater. Positive ions are formed by protonation or electrophilic addition of cations, negative ions by deprotonation or nucleophilic addition, for example, by 0.1 M ammonium acetate added to the solution before evaporation. The ions are extracted orthogonally from the vapor phase and enter the mass analyzer through an aperture in the ion exit plate. The ionization efficiency of TSP interfaces can also be increased by the addition of a source for electrons, that is, by changing its operation basically to APCI mode.

In addition to these methods, some others were rather short-lived. The *particle beam interface* (PBI) employed conventional EI or CI ion sources to effect ionization of micrometer-sized solid particles produced upon room-temperature evaporation of low-polarity solvents, followed by removal of the vapor by a jet separator (Willoughby and Browner, 1984; Brauers and von Bünau, 1990). The *moving-belt interface* transported the solid residues of solvent evaporation inside an oven into an EI or CI ion source where they were evaporated by heating the metal belt in analogy to the sample vial in standard EI or CI applications

(McFadden, Schwartz and Evans, 1976; Karger et al., 1979; McFadden, 1979).

29.8.2
Electrospray Ionization

Electrolytic solutions can be dissipated to form a mist of highly charged micrometer-sized droplets by the action of a strong electric field. This process, known as *electrospray*, is based on Zeleny's work on the behavior of surfaces under the influence of electric fields (Zeleny, 1917) and Taylor's observation that spheric droplets then take the shape of a cone (*Taylor cone*) from which a fine liquid jet is emitted (Taylor, 1964) when Coulombic repulsion exceeds the surface tension, that is, when the *Rayleigh limit* is passed (Lord Rayleigh, 1882). Experiments by Dole demonstrated the production of gaseous ions from polymer solutions by electrospray (Dole et al., 1968a; Dole et al., 1968b), and finally Fenn's group developed an interface to a quadrupole mass analyzer. Their *electrospray ionization* (ESI) device was able to deliver multiply charged ions of large biomolecules, a discovery meaning the breakthrough for ESI (Whitehouse et al., 1985; Fenn et al., 1989; Fenn, 2003). Nowadays, both ESI and MALDI are essential tools of biomedical research (Figure 29.33) (Dole, 1997; Pramanik, Ganguly and Gross, 2002).

When an electrolytic solution (typically $10^{-6}-10^{-4}$ M at $1-5\,\mu l\ min^{-1}$) reaches the end of a capillary (10–100 μm i.d.) that is connected to high voltage

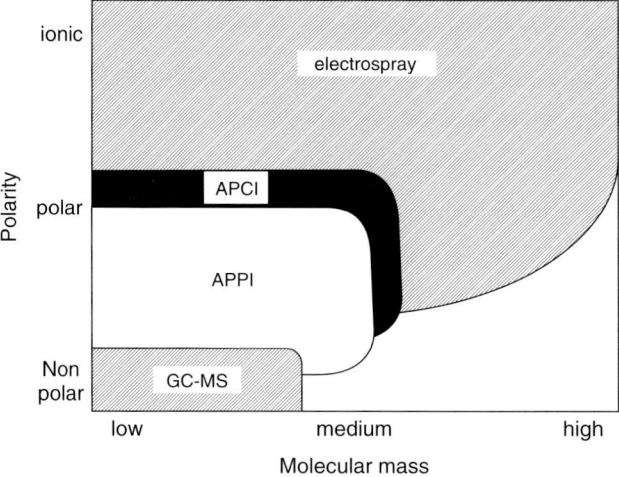

Fig. 29.33 The importance of ESI among chromatography-mass spectrometry applications can be inferred from its wide range of analyte acceptance in terms of both polarity and mass. ESI may deal with medium polar to ionic analytes ranging from as low as 10 u to as high as 10^6 u in mass. APCI and APPI are very similar to each other in coverage, both accessing only the lower left segment of the polarity-mass plane in this illustration. Adapted with permission of U. Karst from cover of *Anal. Bioanal. Chem.* **378(4)** (2004).

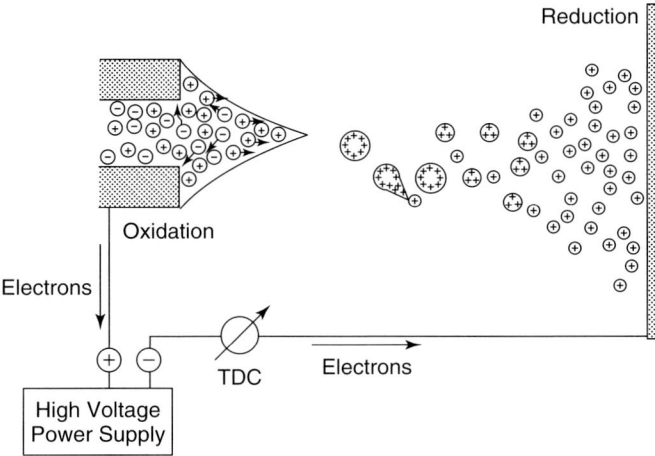

Fig. 29.34 The electrospray process is induced by charge separation at the tip of the spray capillary forcing the emerging droplet to form a cone. The increased electric field strength at cone's apex causes emission of a fine jet that breaks up to form micrometer-sized highly charged droplets that shrink upon evaporation of solvent. The overall process presents an electrolytic cell. To adapt ESI to a mass spectrometer, the counter electrode has a small hole that admits a portion of the aerosol. The sample flow may be supplied by either a liquid chromatograph or a syringe pump. Adapted from Kebarle and Tang (1993) with permission.

(3–5 kV), the electrospray process is induced. First, charge separation at the tip of the spray capillary forces the emerging droplet to form a cone. The increased electric field strength at the apex of the cone then causes emission of a fine jet that soon breaks up to yield micrometer-sized highly charged droplets shrinking upon evaporation of solvent. The overall process represents an electrolytic cell (Figure 29.34) (Kebarle and Tang, 1993). ESI can be adapted to a mass spectrometer when the counter electrode has a small hole to admit a portion of the aerosol via a differentially pumped interface. The principle of a differentially pumped interface has already been outlined (Section 29.5.4, Figure 29.10). However, opposed to the plasma conditions in ICP, the ESI sprayer is operated at room temperature, in fact, in the laboratory atmosphere. To compensate for the heat of evaporation, that is, to prevent droplets from freezing during desolvation while traveling through the interface, some external heating is required. Heating may be realized in either way, using a slow counter current of hot nitrogen gas (1–2 l min^{-1} at 150–250 °C), or transfer of the aerosol through a heated capillary (ca. 0.5 mm i.d., 10 cm length, 120–220 °C), before the first pumping stage.

The shrinking of the droplets causes ejection of much smaller droplets when the critical charge density is reached. Those off-spring droplets carry only a minor portion of the mass, but a significant portion of the charges of their precursor. After repeated *droplet jet*

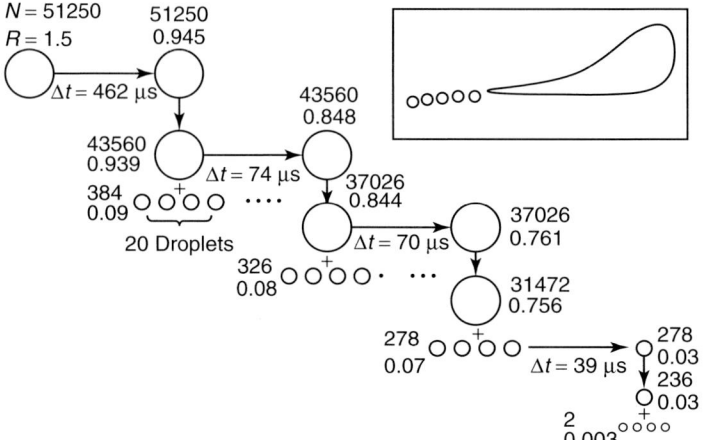

Fig. 29.35 Illustration of droplet jet fission that leads to a stepwise reduction in droplet size. The inset shows the fission process whereby a series of smaller, nonetheless highly charged, off-spring droplets is released from a precursor droplet. Adapted from Kebarle and Tang (1993) with permission.

fission, it comes to the step of production of isolated gas phase ions (Figure 29.35) (Kebarle and Tang, 1993; Cole, 2000; Kebarle, 2000). The ultimate step of ESI, the deliberation of unsolvated gas phase ions, may proceed on one of the following ways: (i) macromolecules can undergo multiple protonation or cationization in solution up to a level where Coulombic repulsion between the charges poses a limit. Those ions may then remain while all solvent molecules evaporate (*charged residue model*). (ii) For small ions, *ion evaporation* may be taken into account, that is, the shrinking droplet may reduce the charge density by emission of ions from its surface.

In recent interface designs, the sprayer is not in line of sight with the analyzer, but roughly at right angle to the entrance (orthogonal sprayer), and eventually the skimmer may be at right angle to the beam direction of the first pumping stage (z-spray). All these adjustments aim at improving the ion transmission, while reducing the load of neutrals (residual gas, solvent vapor) and unwanted ions (buffer salts) onto the system, thereby optimizing the robustness of the interface (Fenn et al., 1990; Ikonomou and Kebarle, 1994; Abian, 1999). To adapt ESI to higher flow rates (10–1000 µl min^{-1}), most modern ESI sprayers offer pneumatic assistance, that is, a nebulizer gas flow emerging from a fine gap around the spray capillary.

On the other hand, to reduce sample flow and thus the extent of use of precious samples, nano-ESI may be employed. In nano-ESI, the analyte solution is supplied in a (single use) glass capillary, drawn to form a fine tip where a small aperture (1–5 µm) allows the generation of a particularly fine spray. High voltage is connected via the metal-coated outside of the capillary. By adjusting the nano-ESI sprayer very close to the entrance of the interface, the ratio of

29.8 Ion Formation at Atmospheric Pressure

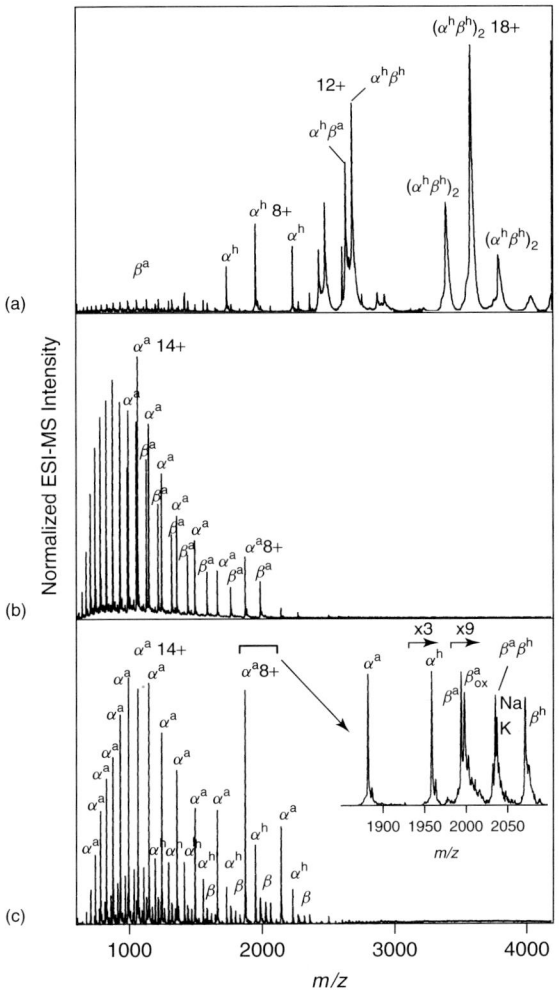

Fig. 29.36 ESI mass spectra of hemoglobin recorded under different solvent conditions. (a) "Native", aqueous solution, pH 8.0; (b) acid unfolded, aqueous solution, pH 2.0; and (c) unfolded protein at pH 10.0 in water/acetonitrile (60:40 v/v). Notation: α^h and α^a represent holo- and apo-α-globin, respectively; β-globin is labeled analogously. Inset in (c) expansion of region around m/z 2000, revealing the presence of both α and β subunits in their apo and holo forms. β^a_{ox} represents an [M + 32] species. The expanded spectrum also shows signals corresponding to $\beta^a\beta^h$ ions carrying Na$^+$ and K$^+$ adducts. Reproduced from Boys and Konermann (2007) with permission. © Elsevier Science Publishers, 2007.

admitted to rejected aerosol is significantly improved (Wilm and Mann, 1994; Wilm et al., 1996; Wilm and Mann, 1996). Thus, 1 µl of sample can deliver a 20-min lasting spray at 50 nl min^{-1} while the ion current of the mass spectrometer does not remarkably drop as compared to conventional ESI operation. Beside lower sample consumption, nano-ESI offers operation free of memory from previous samples and access to low volatility solvent systems such as 100 % water or water/dimethylsulfoxide mixtures. On the other side, the metal plated, sometimes precision-cut, capillaries are expensive and fragile and require an experienced operator.

The application of ESI to biopolymers such as proteins, oligosaccharides, or DNA delivers a distribution of charge states, either positive or negative as required by the analyte. Normally, the charge state distribution represented by all peaks belonging to one analyte species forms an even envelope. In the case of proteins for example, it has a maximum at one charge per 1000 u. This way, ESI "folds" high-mass analytes into the m/z range of most mass spectrometers (Peng and Gygi, 2001; Hanson et al., 2003). The charge state distribution of a protein is not independent of the experimental conditions; on the contrary, it is subject to remarkable changes upon pH and, thus, protein tertiary structure. However, ESI not only provides molecular mass but also allows information on the shape of macromolecules to be derived (Figure 29.36) (Boys and Konermann, 2007). Combined with the extraordinary softness of ion generation, ESI can indeed supply "wings to molecular elephants" (Fenn, 2003).

1. It is possible to calculate charge states for all peaks of the distribution from which, in turn, the mass of the neutral analyte can be deduced (Covey et al., 1988; Mann, Meng and Fenn, 1989).
2. This is due to the fact that the charge state of the ions forming the signal at the adjacent lower m/z is higher by exactly one.
3. Today, software is available to do automated *charge deconvolution*.

ESI is soft enough to yield isolated gas phase ions of rather labile molecular assemblies such as host–guest complexes, adducts, or solvated species. Therefore, ESI is a well-established analytical tool in supramolecular chemistry (Chapter 5 in Schalley, 2007). In combination with tandem MS, ESI can also deliver information on the stability of such complexes (Figure 29.37).

Advantageous on one side, the inherent softness of ESI yields mass spectra highly reliable for molecular mass and molecular formula determination, but often without fragment ions that could serve for structure elucidation. Therefore, ESI is frequently combined with tandem mass spectrometric methods to compensate for that lack (activation methods cf. Section 29.9). Among these, *collision-induced dissociation* (CID) is the most standardized and widespread technique, for example, CID is the standard tool for peptide sequencing (Figure 29.38) (McLafferty, 2001).

Finally, ESI is well applicable to (ionic) metal complexes. This way, ESI nicely complements the capabilities of FAB and FD, in particular LIFDI. Especially, syringe pump infusion delivers a signal highly constant in time, which provides the basis for advanced experiments, for example, catalyst screening by ion–molecule reactions (Figure 29.39) (Volland et al., 2001; Chen, 2003; Adlhart and Chen, 2003).

29.8.3
Atmospheric Pressure Chemical Ionization and Atmospheric Pressure Photoionization

Modern ion sources for APCI (Keski-Hynnilä et al., 2002) and *atmospheric pressure photoionization* (APPI) (Robb, Covey and Bruins, 2000; Kauppila et al., 2002; Raffaeli and Saba, 2003) are based on ESI interfaces. For APCI, the complete interface from atmospheric pressure to analyzer entrance remains unaltered, whereas the ESI sprayer is exchanged for the heated vaporizer and a corona needle is placed in the space between vaporizer exit and interface entrance. For APPI, the same vaporizer may be used, but the corona needle is replaced by a rare gas discharge UV lamp; for example, a krypton lamp delivers 10-eV photons (Figure 29.40)

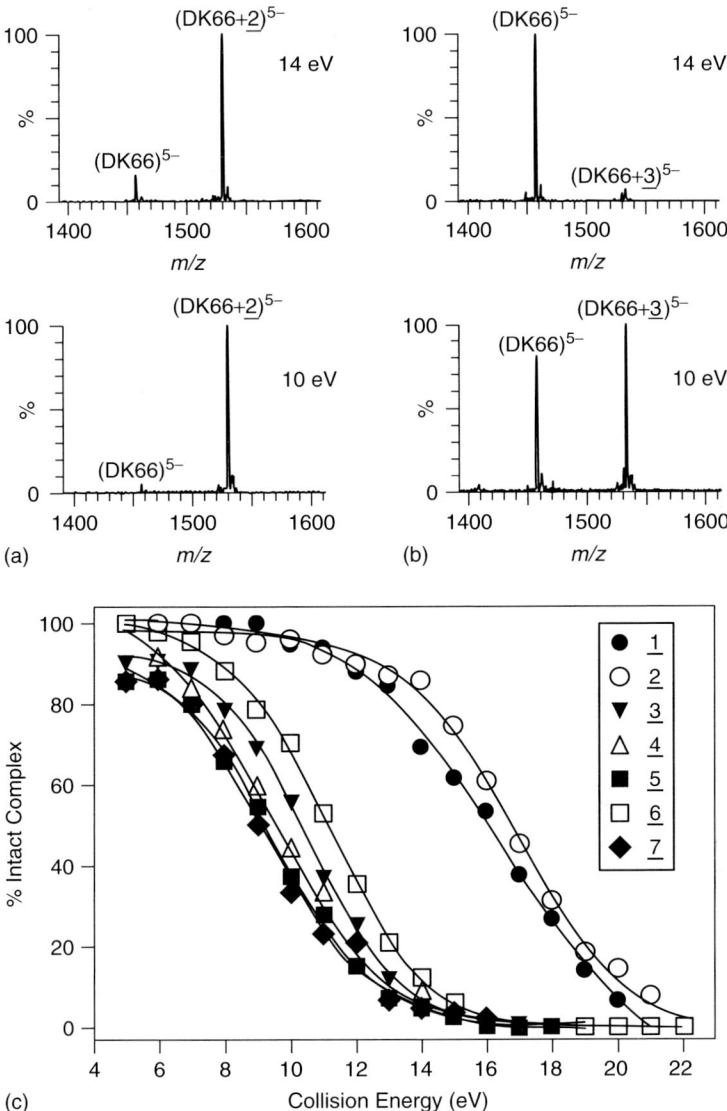

Fig. 29.37 Negative-ion ESI tandem MS experiments on [1:1]$^{5-}$ complexes with the d(CGCGAATTCGCG)$_2$ duplex (DK66). Representative tandem mass spectra obtained with different ligands: (a) benzopyridoindole drug (**2**) and (b) benzopyridoquinoxaline drug (**3**), using collision energies of 10 and 14 eV. The increased dissociation upon more energetic collisions is obvious. The argon pressure inside the RF-only hexapole collision cell was 3×10^{-5} mbar. (c) Derived breakdown graphs of different intercalator drugs with DK66 allow to derive the relative binding strengths of those drugs to the duplex; stronger binding causes the curves to shift to the right. Reprinted from Rosu *et al.* (2007) with permission. © Elsevier Science Publishers, 2007.

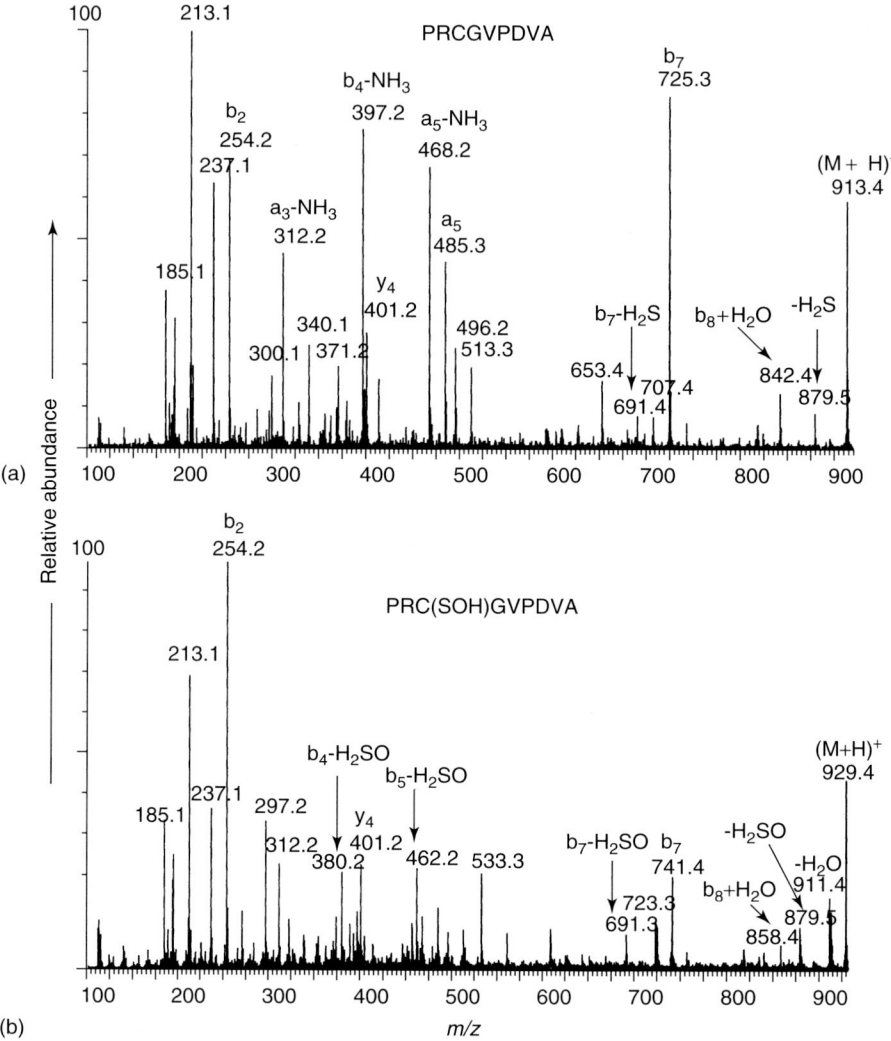

Fig. 29.38 Tandem mass spectra of the ESI-generated [M + H]$^+$ ions of (a) the unmodified cysteine switch peptide PRCGVPDVA, m/z 913.4 and (b) the cysteine sulfenic acid product PRC(SOH)GVPDVA, m/z 929.4. The b-type fragments containing the sulfenic acid residue are accompanied by peaks due to sulfenic acid loss (−50 u). Reproduced from Shetty, Spellman and Neubert (2007) with permission. © Elsevier Science Publishers, 2007.

(Keski-Hynnilä et al., 2002; Raffaeli and Saba, 2003).

The types of ions are formed depend on the solvent and the analyte. In positive-ion mode, [M + H]$^+$ ions and products of electrophilic addition are preferred with polar combinations, whereas CT will occur to yield M$^{+\bullet}$ ion for unpolar systems (cf. PICI). Although PI of the analyte also plays a role in APPI, it is less

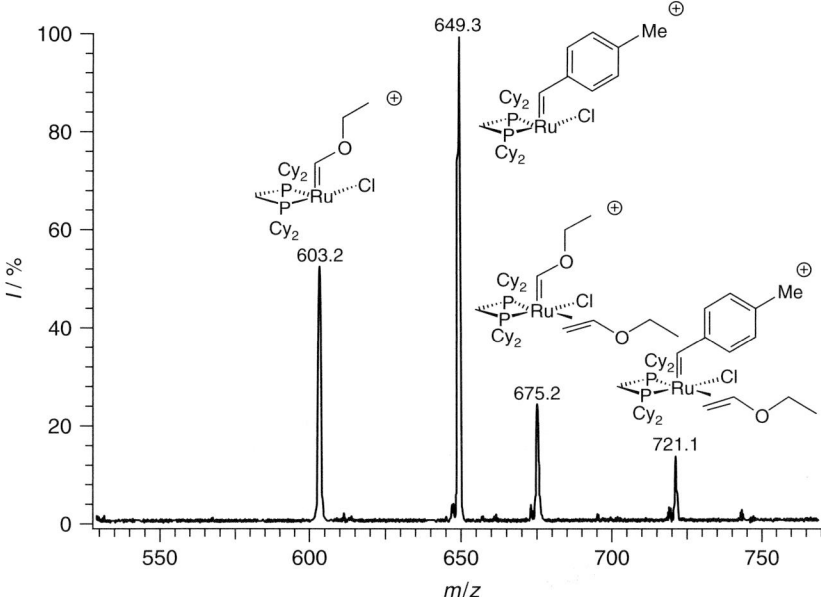

Fig. 29.39 ESI tandem MS experiment for catalyst screening via ion–molecule reactions. The mass-selected precursor ion, [(dcpm-$\kappa^2 P$)ClRu=CH(p-C$_6$H$_4$Me)]$^+$, m/z 649, is reacted with ethyl vinyl ether upon collision in the RF-only octopole collision cell of a triplequadrupole instrument at a collision energy of 1 eV. The product ions incorporating the ether molecule are observed at m/z 603, 675, and 721. Reproduced from Volland et al. (2001) with permission. © Wiley-VCH, 2001.

pronounced than one would expect. Unfortunately, there is a tendency in APPI to form radical ions and closed-shell at the same time so that isotopic patterns suffer from superimposition of both species. Analogously to positive-ion mode, operation in negative-ion mode resembles conventional NICI. However, the AP versions of PICI and NICI are slightly softer than their vacuum counterparts, because vaporization from solution reduces "thermal stress" of the analyte and the large number of low-energy collisions at atmospheric pressure serves to keep the ions thermalized.

Both APCI and APPI serve for similar types of analyses. Mostly they are employed as ionization methods in LC-MS applications for low-to-medium polarity analytes, that is, when active ionization of neutrals is required because ion formation in solution is ineffective (Abian, 1999; Tsuchiya, 1995; Reemtsma, 2003). Illustrative examples are the characterization of natural polyamines in spider venom (Figure 29.41) (Chesnov, Bigler and Hesse, 2002), metabolites of pharmaceuticals in human plasma (Yang and Henion, 2002) and PAHs (Hayen and Karst, 2003).

29.8.4
Atmospheric Pressure MALDI

Classical MALDI ion sources are operated under high vacuum conditions, which is a direct consequence of their initial development on TOF instruments.

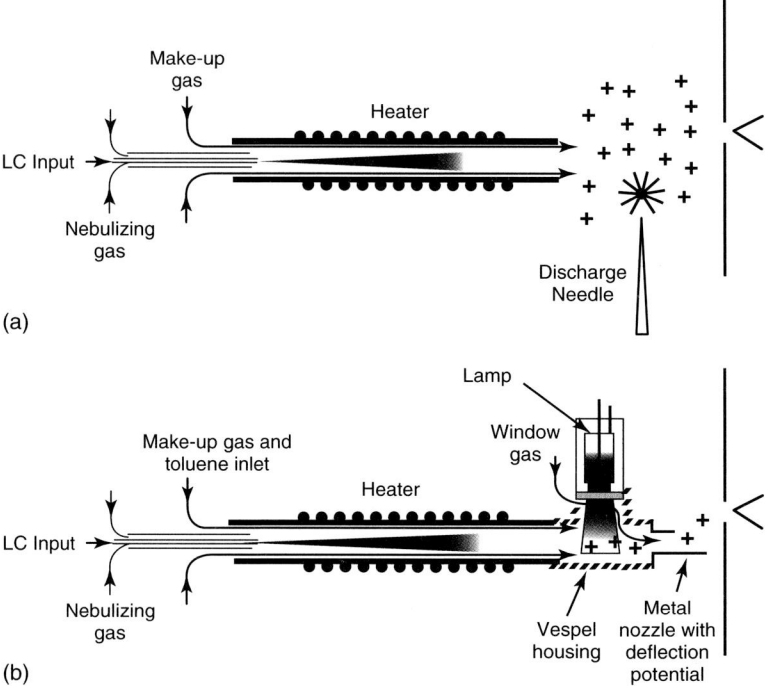

Fig. 29.40 Schematic representations of APCI and APPI interfaces. (a) Conventional heated pneumatic nebulizer interface for APCI. Ionization occurs via the corona discharge needle placed in the vaporized plume of mobile phase and volatilized analytes. (b) Schematic representation of an APPI interface, which is a modification of the probe shown in (a). It incorporates a krypton lamp affixed to the end of the heated nebulizer probe to initiate ionization by photoionization with 10.0-eV photons. Reprinted with permission from Yang and Henion (2002). © Elsevier Science Publishers, 2002.

The desire to have MALDI attached to other analyzers made it necessary to reduce the large energy spread (tens of electron volts) of the laser desorbed ions. This can be accomplished by thermalization with gas, either by having a pulsed valve admitting nitrogen or argon before each laser shot or by moving the event of desorption/ionization into a region of medium vacuum. It turned out that MALDI can even be realized at atmospheric pressure (Laiko, Baldwin and Burlingame, 2000a; Laiko, Moyer and Cotter, 2000b; O'Connor and Costello, 2001; Doroshenko et al., 2002). By positioning a MALDI target in front of the interface, the freshly generated and thermalized ions are guided directly into the instrument (Figure 29.42). Different from APCI and APPI, switching the mode of operation to AP-MALDI is slightly more demanding due to the requirement for accurate alignment of target and laser optics. In addition, sensitivity of AP-MALDI is generally lower than that of vacuum MALDI. The true advantage of atmospheric pressure MALDI (AP-MALDI) is its compatibility with AP

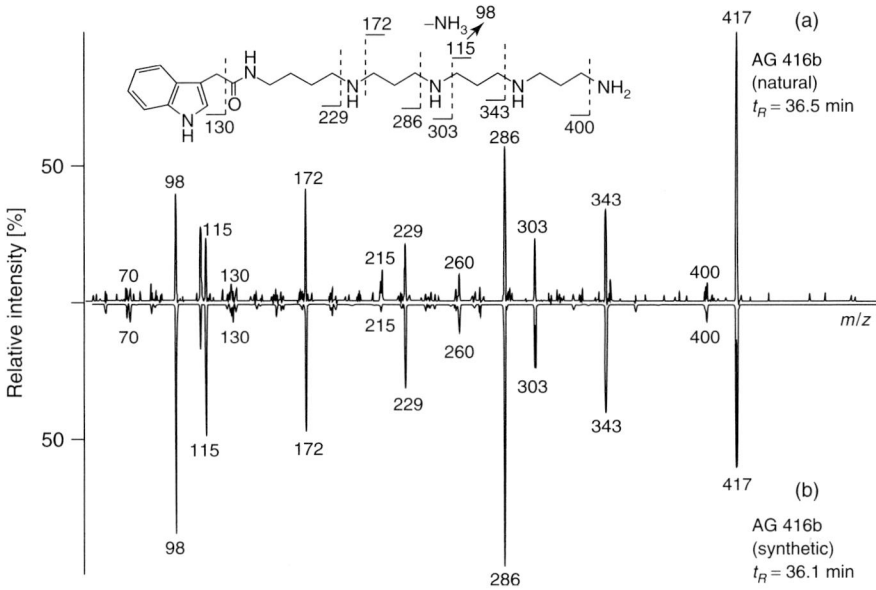

Fig. 29.41 Data from the HPLC-APCI-MS/MS analyses for (a) [M + H]⁺ ion, m/z 417, of the natural acylpolyamine of a spider and (b) the synthetic compound for reference. Adapted from Chesnov, Bigler and Hesse (2002) with permission. © IM Publications, 2002.

sources, which offers more flexibility for the use of expensive equipment.

29.8.5
Desorption Electrospray

For ESI, the analyte is normally supplied in dilute electrolytic solution. Nonetheless, Takats et al. discovered that the processes of electrospraying and sampling can be separated in a way that the electrostatically charged aerosol is blown onto a surface from which analyte molecules are sampled and ionized (Takats et al., 2004; Chen et al., 2005). This method was termed *desorption electrospray/ionization* (DESI). DESI transforms the step of sample preparation into a mere positioning of sample between the sprayer and entrance. DESI obviously plays to its strengths for the quick analysis of surface (adsorbed) compounds such as alkaloids in plants (Figure 29.43), pesticides on fruit, explosives on cloths and similar items, control of bulk compounds, and other applications where the target compound class is defined (Popov et al., 2005; Takats et al., 2005a; Takats, Wiseman and Cooks, 2005; Talaty, Takats and Cooks, 2005).

29.8.6
Direct Analysis in Real Time

Another recent addition to the repertoire for rapid, noncontact analysis at ambient pressure is presented by *direct analysis in real time* (DART) (Cody, Laramee and Durst, 2005) DART differs much from the aforementioned techniques. It is based on the reactions of electronic or vibronic excited-state species with

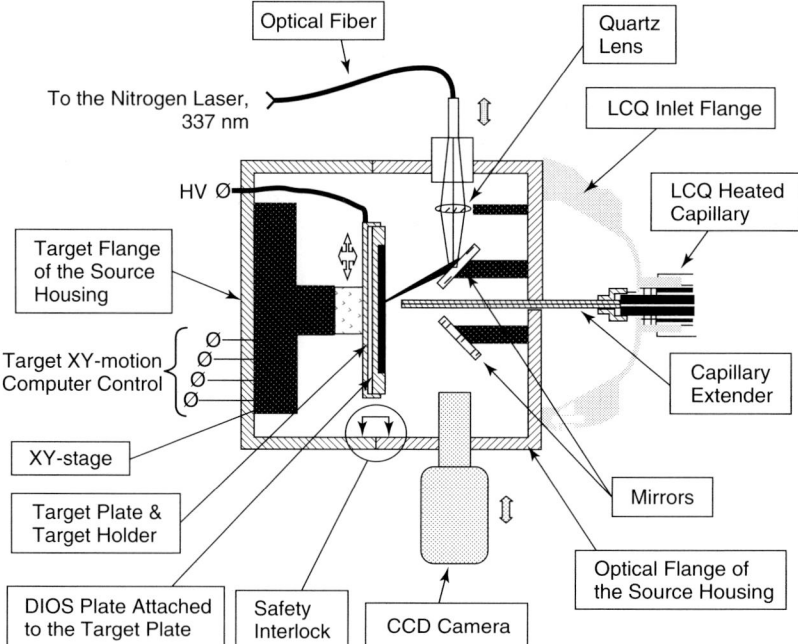

Fig. 29.42 AP-MALDI ion source. The heated capillary of an ESI ion source is extended to allow enough space for the laser optics. LCQ is a tradename of Thermo Fisher Scientific for a LC-quadrupole ion trap instrument. This specific drawing incorporates a target for *desorption/ionization on porous silicon* (DIOS) (Wei, Buriak and Siuzdak, 1999) instead of a standard MALDI target. Adapted from Laiko *et al.* (2002) with permission. © John Wiley & Sons, 2002.

reagent molecules, that is, basically on Penning ionization. A stream of nitrogen or helium gas is passed through an high-voltage electrical discharge to generate the excited molecules or atoms that then exit as a directed stream at 1 l min^{-1}. The analyte is exposed to this ionizing gas stream in proximity to the entrance of an API interface.

DART can be applied to the analysis of gases, liquids, and solids on various surfaces. As such, DART provides a very sensitive tool for the detection of explosives, the characterization of inks on paper, and the analysis of residual components from hygiene products (Figure 29.44) (Jones, Cody and McClelland, 2006; Yew, Cody and Kravitz, 2008; Haefliger and Jeckelmann, 2007; Williams *et al.*, 2006).

29.9
Activation Methods for Tandem Mass Spectrometry

The term *tandem mass spectrometry* refers to all techniques where ions mass selected by a first stage are subjected to some type of activation (or less frequently chemical reaction) before the product(s) of this process are mass-analyzed on a second stage of the instrument (McLafferty, 1983; Busch, Glish and McLuckey, 1988). On beam instruments,

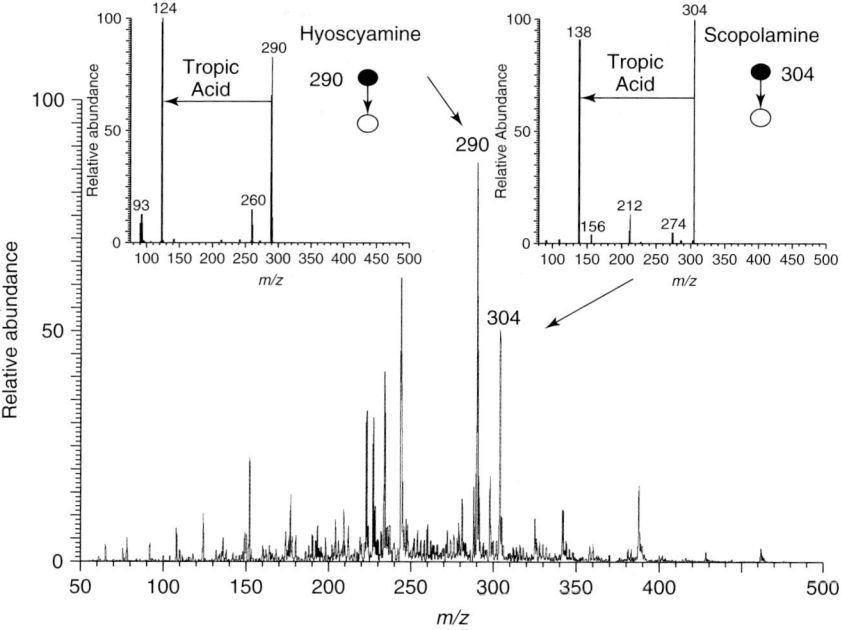

Fig. 29.43 DESI mass spectrum of deadly nightshade (*atropa belladonna*) seed using methanol : water (1 : 1) as spray solvent. Insets show tandem mass spectra of the [M + H]$^+$ ions of hyoscyamine, m/z 290, and scopolamine, m/z 304. Both alkaloid ions have the loss of tropic acid in common. Reproduced from Talaty, Takats and Cooks (2005) with permission. © The Royal Chemical Society, 2005.

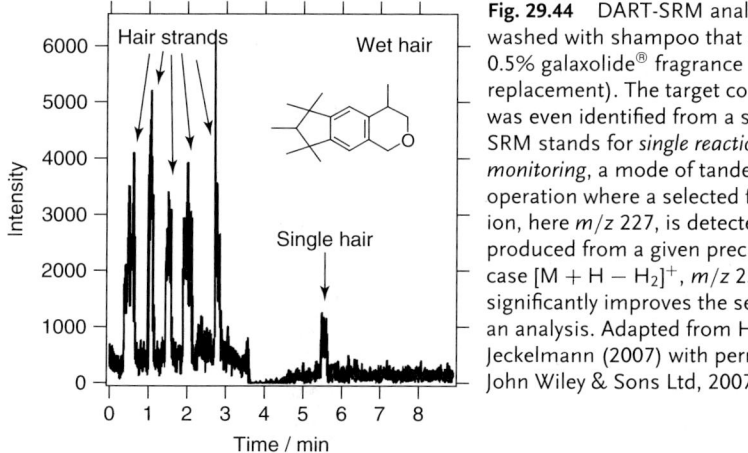

Fig. 29.44 DART-SRM analysis of hair washed with shampoo that contained 0.5% galaxolide® fragrance (musk replacement). The target compound was even identified from a single hair. SRM stands for *single reaction monitoring*, a mode of tandem MS operation where a selected fragment ion, here m/z 227, is detected only if produced from a given precursor; in this case [M + H − H$_2$]$^+$, m/z 257; SRM significantly improves the selectivity of an analysis. Adapted from Haefliger and Jeckelmann (2007) with permission. © John Wiley & Sons Ltd, 2007.

tandem MS requires a second mass analyzer behind the zone of activation; hence, this is sometimes referred to as *tandem in space*. On ion traps of any type, the sequence of selection, activation, and product analysis is usually effected inside the same unit by applying different modes of operation. Owing to its sequential character, this is then termed *tandem in time*. The activation step can be performed in many ways; the most common techniques are listed in Table 29.4. Table 29.5 lists the most important symbols in MS.

Tab. 29.4 Common activation methods for gas phase ions.

Method	Acronym	Basic principle of ion excitation	Instrumentation
Collision-induced dissociation (outdated terms: collisional activation, collisionally activated dissociation)	CID (outdated acronyms: CA, CAD)	Kinetic energy of ions is converted into vibrational excitation upon collisions; high-energy CID refers to keV ions typically once collided with He; low-energy CID refers to ions of 1–100 eV typically multiply collided with Ar (McLafferty, 1983; Busch, Glish and McLuckey, 1988). Suitable neutrals provided, collisions at very low energy can lead to chemical reactions (cf. Figure 39)	Suitable (scan) techniques for all types of mass analyzers available; independent of ion polarity
Infrared multiphoton dissociation	IRMPD	Absorption of IR photons provides vibrational excitation (Woodin, Bomse and Beauchamp, 1978; Watson, Baykut and Eyler, 1987)	On FT-ICR instruments; independent of ion polarity
Electron capture dissociation	ECD	Multiply charged positive ions undergo bond cleavage upon partial neutralization by capturing a thermal electron (Zubarev, Kelleher and McLafferty, 1998; Axelsson *et al.*, 1999)	On FT-ICR instruments; for multiply charged positive ions only
Electron transfer dissociation	ETD	Same effect as ECD, but negative radical ions are supplied (reagent ions) as the source of electrons (Coon *et al.*, 2005; Syka *et al.*, 2004)	Linear and three-dimensional quadrupole ion trap, orbitrap; for multiply charged positive ions only

Tab. 29.5 The most important symbols in mass spectrometry.

Symbol	Term	Explanation
$M^{+\bullet}$ ($M^{-\bullet}$)	Molecular ion	Positive (negative) ion formed from a molecule by removal (addition) of an electron
$[M + H]^+$, $[M + Na]^+$, $[M + NH_4]^+$, and so on	Quasimolecular ion	Ions formed instead of the molecular ion, but very closely related to it. The term also applies to $[M + Na]^+$, $[M + K]^+$ or $[M + Cl]^-$ ions, for example
MS	Mass spectrometry	Refers to the method; do not use it for a spectrum or instrument
MS/MS or MS^2	Tandem mass spectrometry	Mass-selected ions are subjected to fragmentation or reactions, the products are analyzed on a second stage
MS^3, MS^n	Tandem mass spectrometry	Higher order tandem MS experiments
m/z	Mass-to-charge ratio	Dimensionless ratio of the mass number and the number of elementary charges of an ion. The position of peaks on the abscissa of a mass spectrum is stated as "at m/z x"
u	Unified atomic mass	SI unit for atomic mass defined as 1/12 of the mass of the nuclide ^{12}C; 1 u = 1.66055×10^{-27} kg
B	Magnetic sector	Of a sector instrument
E	Electrostatic sector	Also termed ESA, electrostatic analyzer
ICR	Ion cyclotron resonance	To denote the method as well as the ICR cell of an instrument
LIT	Linear quadrupole ion trap	
Q and q	Quadrupole	Q for mass separating, q for RF-only ion guides
QIT	Quadrupole ion trap	A "three-dimensional" trap
TOF	Time of flight	To denote the method as well as the TOF analyzer of an instrument

Acronyms

AMS	Accelerator mass spectrometry
AE	Appearance energy
APCI	Atmospheric pressure chemical ionization
API	Atmospheric pressure ionization
AP-MALDI	Atmospheric pressure matrix-assisted laser desorption/ionization
APPI	Atmospheric pressure photoionization
CE	Capillary electrophoresis
CE	Charge exchange (equivalent to CT)
CF	Continuous flow
CF-FAB	Continuous-flow fast atom bombardment
CI	Chemical ionization
CID	Collision-induced dissociation
CT	Charge transfer
DART	Direct analysis in real time
DESI	Desorption electrospray/ionization
DIOS	Desorption/ionization on porous silicon
EA	Electron affinity
EC	Electron capture
ECD	Electron capture dissociation
EHC	Emitter heating current
EI	Electron ionization
ESI	Electrospray ionization
ETD	Electron transfer dissociation
FAB	Fast atom bombardment
FD	Field desorption
FI	Field ionization
FT-ICR	Fourier-transform ion cyclotron resonance
GC-MS	Gas chromatography-mass spectrometry
GD	Glow discharge
ICP	Inductively coupled plasma
IE	Ionization energy
IR-MS	Isotope ratio mass spectrometry
IR-MALDI	Infrared matrix-assisted laser desorption/ionization
IRMPD	Infrared multiphoton dissociation
LA	Laser ablation
LC-MS	Liquid-chromatography-mass spectrometry
LDI	Laser desorption/ionization
LIFDI	Liquid injection field desorption/ionization
LSIMS	Liquid secondary-ion mass spectrometry
MALDI	Matrix-assisted laser desorption/ionization
M_r:	Relative molecular mass (equal to MW)
MS	Mass spectrometry (not to be used for mass spectrometer or mass spectrum)
MS/MS	Mass spectrometry/mass spectrometry or tandem mass spectrometry
MUPI	Multiphoton ionization
MW	Molecular weight
Nano-ESI	Nanoelectrospray ionization
NCI	Negative-ion chemical ionization
NICI	Negative-ion chemical ionization
PA	Proton affinity
PCI	Positive-ion chemical ionization
PICI	Positive-ion chemical ionization
ppb	Parts per billion
ppm	Parts per million
QET	Quasi-equilibrium theory

RE	Recombination energy
REMPI	Resonance-enhanced multiphoton ionization
RF	Radio frequency
SIM	Selected ion monitoring
SIMS	Secondary-ion mass spectrometry
SRM	Selected reaction monitoring
SS-MS	Spark source mass spectrometry
TI	Thermal ionization
TOF	Time of flight
TSP	Thermospray
UV-MALDI	Ultraviolet matrix-assisted laser desorption/ionization

References

Abian, J. (1999) *J. Mass Spectrom.*, **34**, 157–168.

Adlhart, C. and Chen, P. (2003) *Helv. Chim. Acta*, **86**, 941–949.

Allgood, C., Ma, Y.C. and Munson, B. (1991) *Anal. Chem.*, **63**, 721–725.

von Ardenne, M., Steinfelder, K. and Tümmler, R. (1971) *Elektronenanlagerungs-Massenspektrographie Organischer Substanzen*, Springer-Verlag, Heidelberg, p. 403.

Axelsson, J., Palmblad, M., Håkansson, K. and Håkansson, P. (1999) *Rapid Commun. Mass Spectrom.*, **13**, 474–477.

Barber, M., Bordoli, R.S., Elliott, G.J., Sedgwick, R.D., Tyler, A.N. and Green, B.N. (1982) *J. Chem. Soc. Chem. Commun.*, 936–938.

Barber, M., Vickerman, J.C. and Wolstenholme, J. (1980) *J. Chem. Soc., Faraday Trans. 1*, **76**, 549–559.

Becker, J.S. (2008) *Inorganic Mass Spectrometry: Principles and Applications*, 1st edn, John Wiley & Sons, Ltd, Chichester, p. 514.

Becker, J.S. and Dietze, H.J. (1999) *J. Anal. At. Spectrom.*, **14**, 1493–1500.

Becker, J.S., Zoriy, M., Becker, J.S., Dobrowolska, J. and Matusch, A. (2007) *J. Anal. At. Spectrom.*, **22**, 736–744.

Beckey, H.D. (1962) *Z. Naturforsch.*, **17A**, 1103–1111.

Beckey, H.D. (1971) *Field-Ionization Mass Spectrometry*, 1st edn, Pergamon, Elmsford, p. 359.

Beckey, H.D. (1977) *Principles of Field Desorption and Field Ionization Mass Spectrometry*, 1st edn, Pergamon Press, Oxford, p. 335.

Beckey, H.D., Hilt, E. and Schulten, H.-R. (1973) *J. Phys. E: Sci. Instrum.*, **6**, 1043–1044.

Beckey, H.D., Krone, H. and Röllgen, F.W. (1968) *J. Sci. Instrum.*, **1**, 118–120.

Bennett, C.L., Beukens, R.P., Clover, M.R., Grove, H.E., Liebert, R.B., Litherland, A.E., Purser, K.H. and Sondheim, W.E. (1977) *Science*, **198**, 508–510.

Benninghoven, A. (1983) *Int. J. Mass Spectrom. Ion Phys.*, **46**, 459–462.

Benninghoven, A. and Kirchner, F. (1963) *Z. Naturforsch.*, **18A**, 1008–1010.

Blakley, C.R., Carmody, J.J. and Vestal, M.L. (1980) *J. Am. Chem. Soc.*, **102**, 5931–5933.

Blakley, C.R. and Vestal, M.L. (1983) *Anal. Chem.*, **55**, 750–754.

Boesl, U. (2000) *J. Mass Spectrom.*, **35**, 289–304.

Born, M. and Oppenheimer, J.R. (1927) *Annalen der Phys.*, **84**, 457–484.

Boys, B.L. and Konermann, L. (2007) *J. Am. Soc. Mass Spectrom.*, **18**, 8–16.

Brauers, F. and von Bünau, G. (1990) *Int. J. Mass Spectrom. Ion Process*, **99**, 249–262.

Bristow, A.W.T. (2006) *Mass Spectrom. Rev.*, **25**, 99–111.

Bristow, A.W.T. and Webb, K.S. (2003) *J. Am. Soc. Mass Spectrom.*, **14**, 1086–1098.

Brown, K., Dingley, K.H. and Turteltaub, K.W. (2005) *Methods Enzymol.*, **402**, 423–443.

Brown, K., Tompkins, E.M. and White, I.N.H. (2006) *Mass Spectrom. Rev.*, **25**, 127–145.

Busch, K.L. (1995) *J. Mass Spectrom.*, **30**, 233–240.

Busch, K.L., Glish, G.L. and McLuckey, S.A. (1988) *Mass Spectrometry/Mass Spectrometry*, 1st edn, Wiley VCH, New York.

Caprioli, R.M. (1990) *Anal. Chem.*, **62**, 477A–485A.

Caprioli, R.M., Fan, T. and Cottrell, J.S. (1986) *Anal. Chem.*, **58**, 2949–2954.

Carroll, D.I., Dzidic, I., Stillwell, R.N., Haegele, K.D. and Horning, E.C. (1975) *Anal. Chem.*, **47**, 2369–2373.

Chen, P. (2003) *Angew. Chem. Int. Ed.*, **42**, 2832–2847.

Chen, H., Talaty, N.N., Takats, Z. and Cooks, R.G. (2005) *Anal. Chem.*, **77**, 6915–6927.

Chery, C.C., Moens, L., Cornelis, R. and Vanhaecke, F. (2006) *Pure Appl. Chem.*, **78**, 91–103.

Chesnov, S., Bigler, L. and Hesse, M. (2002) *Eur. J. Mass Spectrom.*, **8**, 1–16.

Cody, R.B., Laramee, J.A. and Durst, H.D. (2005) *Anal. Chem.*, **77**, 2297–2302.

Cole, R.B. (2000) *J. Mass Spectrom.*, **35**, 763–772.

Condon, E.U. (1926) *Phys. Rev.*, **28**, 1182–1201.

Coon, J.J., Shabanowitz, J., Hunt, D.F. and Syka, J.E.P. (2005) *J. Am. Chem. Soc. Mass Spectrom.*, **16**, 880–882.

Cotter, R.J. (1987) *Anal. Chim. Acta*, **195**, 45–59.

Covey, T.R., Bonner, R.F., Shushan, B.I. and Henion, J.D. (1988) *Rapid Commun. Mass Spectrom.*, **2**, 249–256.

Daehling, P., Röllgen, F.W., Zwinselmann, J.J., Fokkens, R.H. and Nibbering, N.M.M. (1982) *Fresenius Z. Anal. Chem.*, **312**, 335–337.

Dähling, P., Ott, K.H., Röllgen, F.W., Zwinselmann, J.J., Fokkens, R.H. and Nibbering, N.M.M. (1983) *Int. J. Mass Spectrom. Ion Phys.*, **46**, 301–304.

De Pauw, E., Agnello, A. and Derwa, F. (1986) *Mass Spectrom. Rev.*, **5**, 191–212.

Devienne, F.M. (1967) *Entropie*, **18**, 61–67.

Dillard, J.G. (1973) *Chem. Rev.*, **73**, 589–644.

Dole, R.B. (ed.) (1997) *Electrospray Ionization Mass Spectrometry - Fundamentals, Instrumentation and Applications*, 1st edn John Wiley & Sons, Ltd, Chichester.

Dole, M., Hines, R.L., Mack, L.L., Mobley, R.C., Ferguson, L.D. and Alice, M.B. (1968) *Macromolecules*, **1**, 96–97.

Dole, M., Mack, L.L., Hines, R.L., Mobley, R.C., Ferguson, L.D. and Alice, M.B. (1968) *J. Chem. Phys.*, **49**, 2240–2249.

Doroshenko, V.M., Laiko, V.V., Taranenko, N.I., Berkout, V.D. and Lee, H.S. (2002) *Int. J. Mass Spectrom.*, **221**, 39–58.

Dzidic, I., Stillwell, R.N., Carroll, D.I. and Horning, E.C. (1976) *Anal. Chem.*, **48**, 1763–1768.

Fenn, J.B. (2003) *Angew. Chem. Int. Ed.*, **42**, 3871–3894.

Fenn, J.B., Mann, M., Meng, C.K. and Wong, S.F. (1990) *Mass Spectrom. Rev.*, **9**, 37–70.

Fenn, J.B., Mann, M., Meng, C.K., Wong, S.F. and Whithouse, C.M. (1989) *Science*, **246**, 64–71.

Fenner, N.C. and Daly, N.R. (1966) *Rev. Sci. Instrum.*, **37**, 1068–1070.

Field, F.H. and Munson, M.S.B. (1965) *J. Am. Chem. Soc.*, **87**, 3289–3294.

Franck, J. (1925) *Trans. Faraday Soc.*, **21**, 536–542.

Gomer, R. (1994) *Surf. Sci.*, **299/300**, 129–152.

Gomer, R. and Inghram, M.G. (1955) *J. Am. Chem. Soc.*, **77**, 500.

Grayson, M.A. (ed.) (2002) in *Measuring Mass – from Positive Rays to Proteins*, ASMS and CHF, Santa Fe and Philadelphia, p. 149.

Gross, J.H. (1998) *Rapid Commun. Mass Spectrom.*, **12**, 1833–1838.

Gross, J.H. (2004) *Mass Spectrometry – A Textbook*, Springer-Verlag Heidelberg, Heidelberg, p. 536.

Gross, J.H., Nieth, N., Linden, H.B., Blumbach, U., Richter, F.J., Tauchert, M.E., Tompers, R. and Hofmann, P. (2006) *Anal. Bioanal. Chem.*, **386**, 52–58.

Guenther, D. and Hattendorf, B. (2005) *Trends Anal. Chem.*, **24**, 255–265.

Haefliger, O.P. and Jeckelmann, N. (2007) *Rapid Commun. Mass Spectrom.*, **21**, 1361–1366.

Hanson, L., Fucini, P., Ilag, L.L., Nierhaus, K.H. and Robinson, C.V. (2003) *J. Biol. Chem.*, **278**, 1259–1267.

Harrison, A.G. (1992a) *Chemical Ionization Mass Spectrometry*, CRC Press, Boca Raton, p. 26.

Harrison, A.G. (1992b) *Chemical Ionization Mass Spectrometry*, 2nd edn, CRC Press, Boca Raton.

Hayen, H. and Karst, U. (2003) *J. Chromatogr. A*, **1000**, 549–565.

Hillenkamp, F. and Peter-Katalinic, J. (eds) (2007) *MALDI-MS. A Practical Guide to Instrumentation, Methods and Applications*, 1st edn Wiley-VCH Verlag GmbH, Weinheim.

Hoffmann, V., Kasik, M., Robinson, P.K. and Venzago, C. (2005) *Anal. Bioanal. Chem.*, **381**, 173–188.

Horning, E.C., Carroll, D.I., Dzidic, I., Haegele, K.D., Horning, M.G. and Stillwell, R.N. (1974) *J. Chromatogr. Sci.*, **12**, 725–729.

Horning, E.C., Horning, M.G., Carroll, D.I., Dzidic, I. and Stillwell, R.N. (1973) *Anal. Chem.*, **45**, 936–943.

Houk, R.S., Fassel, V.A., Flesch, G.D., Svec, H.J., Gray, A.L. and Taylor, C.E. (1980) *Anal. Chem.*, **52**, 2283–2289.

Hsu, C.S. and Qian, K. (1993) *Anal. Chem.*, **65**, 767–771.

Ikeda, T. (2005) *Radioisotopes*, **54**, 15–21.

Ikonomou, M.G. and Kebarle, P. (1994) *J. Am. Soc. Mass Spectrom.*, **5**, 791–799.

Ito, Y., Takeuchi, T., Ishii, D. and Goto, M. (1985) *J. Chromatogr.*, **346**, 161–166.

Jakubowski, N., Dorka, R., Steers, E. and Tempez, A. (2007) *J. Anal. At. Spectrom.*, **22**, 722–735.

Jochum, K.P. (1997) *Modern Analytical Geochemistry*, (ed R. Gill), Chap. 11 Addison Wesley Longman, Harlow, 188–199.

Jochum, K.P., Stoll, B., Pfänder, J.A., Seufert, M., Flanz, M., Maissenbacher, P., Hofmann, M. and Hofmann, A.W. (2001) *Fresenius J. Anal. Chem.*, **370**, 647–653.

Jones, R.W., Cody, R.B. and McClelland, J.F. (2006) *J. Forensic Sci.*, **51**, 915–918.

Karas, M., Bahr, U. and Gießmann, U. (1991) *Mass Spectrom. Rev.*, **10**, 335–357.

Karas, M., Bahr, U., Ingendoh, A. and Hillenkamp, F. (1989) *Angew. Chem.*, **101**, 805–806.

Karas, M., Ingendoh, A., Bahr, U. and Hillenkamp, F. (1989) *Biomed. Environ. Mass Spectrom.*, **18**, 841–843.

Karas, M. and Krüger, R. (2003) *Chem. Rev.*, **103**, 427–439.

Karger, B.L., Kirby, D.P., Vouros, P., Foltz, R.L. and Hidy, B. (1979) *Anal. Chem.*, **51**, 2324–2328.

Kauppila, T.J., Kuuranne, T., Meurer, E.C., Eberlin, M.N., Kotiaho, T. and Kostiainen, R. (2002) *Anal. Chem.*, **74**, 5470–5479.

Kawano, H., Hidaka, Y. and Page, F.M. (1983) *Int. J. Mass Spectrom. Ion Phys.*, **50**, 35–75.

Kawano, H. and Page, F.M. (1983) *Int. J. Mass Spectrom. Ion Phys.*, **50**, 1–33.

Kebarle, P. (2000) *J. Mass Spectrom.*, **35**, 804–817.

Kebarle, P. and Tang, L. (1993) *Anal. Chem.*, **65**, 972A–986A.

Keski-Hynnilä, H., Kurkela, M., Elovaara, E., Antonio, L., Magdalou, J., Luukkanen, L., Taskinen, J. and Koistiainen, R. (2002) *Anal. Chem.*, **74**, 3449–3457.

King, F.L. and Harrison, W.W. (1990) *Mass Spectrom. Rev.*, **9**, 285–317.

Kinter, M. and Sherman, N.E. (2000) *Protein Sequencing and Identification Using Tandem Mass Spectrometry*, John Wiley & Sons, Ltd, Chichester.

Kiser, R.W. (1965) *Introduction to Mass Spectrometry and Its Applications*, Prentice-Hall, Englewood Cliffs, p. 356.

Knighton, W.B. and Grimsrud, E.P. (1995) *Mass Spectrom. Rev.*, **14**, 327–343.

Koppenaal, D.W. (1990) *Anal. Chem.*, **62**, 303R–324R.

Krätschmer, W., Lamb, L.D., Fostiropoulos, K. and Huffman, D.R. (1990) *Nature*, **347**, 354–358.

Kroto, H. (1997) *Rev. Mod. Phys.*, **69**, 703–722.

Kutschera, W. (2005) *Int. J. Mass Spectrom.*, **242**, 145–160.

de Laeter, J.R. (2001) *Applications of Inorganic Mass Spectrometry*, 1st edn, John Wiley & Sons, Inc., New York, p. 474.

de Laeter, J.R., De Bièvre, P. and Peiser, H.S. (1992) *Mass Spectrom. Rev.*, **11**, 193–245.

Laiko, V.V., Baldwin, M.A. and Burlingame, A.L. (2000) *Anal. Chem.*, **72**, 652–657.

Laiko, V.V., Moyer, S.C. and Cotter, R.J. (2000) *Anal. Chem.*, **72**, 5239–5243.

Laiko, V.V., Taranenko, N.I., Berkout, V.D., Musselman, B.D. and Doroshenko, V.M. (2002) *Rapid Commun. Mass Spectrom.*, **16**, 1737–1742.

Laramée, J.A., Mazurkiewicz, P., Berkout, V. and Deinzer, M.L. (1996) *Mass Spectrom. Rev.*, **15**, 15–42.

Lattimer, R.P. (2001) in *Mass Spectrometry of Polymers* (eds G. Montaudo and R.P. Lattimer), CRC Press, Boca Raton, pp. 237–268.

Lattimer, R.P. and Schulten, H.-R. (1989) *Anal. Chem.*, **61**, 1201A–1215A.

Lehmann, W.D. (1996) *Massenspektrometrie in der Biochemie*, 1st edn, Spektrum Akademischer Verlag, Heidelberg.

Linden, H.B. (2004) *Eur. J. Mass Spectrom.*, **10**, 459–468.

Mamyrin, B.A. (1994) *Int. J. Mass Spectrom. Ion Process*, **131**, 1–19.

Mann, M., Meng, C.K. and Fenn, J.B. (1989) *Anal. Chem.*, **61**, 1702–1708.

Märk, T.D. (1986) in *Gaseous ion Chemistry and Mass Spectrometry* (ed. J.H. Futrell), John Wiley & Sons, Inc., New York, pp. 61–93.

McCrery, D.A., Ledford, E.B. Jr and Gross, M.L. (1982) *Anal. Chem.*, **54**, 1435–1437.

McFadden, W.H. (1979) *J. Chromatogr. Sci.*, **17**, 2–16.

McFadden, W.H., Schwartz, H.L. and Evans, S. (1976) *J. Chromatogr.*, **122**, 389–396.

McLafferty, F.W. (ed.) (1983) *Tandem Mass Spectrometry*, 1st edn, John Wiley & Sons, Inc., New York.

McLafferty, F.W. (2001) *Int. J. Mass Spectrom.*, **212**, 81–87.

McLafferty, F.W. and Turecek, F. (1993) *Interpretation of Mass Spectra*, 4th edn, University Science Books, Mill Valley.

Montaudo, G. and Lattimer, R.P. (eds) (2001) *Mass Spectrometry of Polymers*, 1st edn CRC Press, Boca Raton, p. 600.

Morris, H.R., Panico, M., Barber, M., Bordoli, R.S., Sedgwick, R.D. and Tyler, A.N. (1981) *Biochem. Biophys. Res. Commun.*, **101**, 623–631.

Munson, M.S.B. (2000) *Int. J. Mass Spectrom.*, **200**, 243–251.

Murgasova, R. and Hercules, D.M. (2003) *Int. J. Mass Spectrom.*, **226**, 151–162.

Nelis, T. and Pallosi, J. (2006) *Appl. Spectrosc. Rev.*, **41**, 227–258.

Nelson, D.E., Korteling, R.G. and Stott, W.R. (1977) *Science*, **198**, 507–508.

Nielen, M.F.W. (1999) *Mass Spectrom. Rev.*, **18**, 309–344.

Niessen, W.M.A. (1998) *J. Chromatogr. A*, **794**, 407–435.

Niessen, W.M.A. and Voyksner, R.D. (eds) (1998) *Current Practice of Liquid Chromatography-Mass Spectrometry*, Elsevier, Amsterdam, p. 438.

Nordhoff, E., Schürenberg, M., Thiele, G., Lübbert, C., Kloeppel, K.-D., Theiss, D., Lehrach, H. and Gobom, J. (2003) *Int. J. Mass Spectrom.*, **226**, 163–180.

O'Connor, P.B. and Costello, C.E. (2001) *Rapid Commun. Mass Spectrom.*, **15**, 1862–1868.

Okuyama, F., Hilt, E., Röllgen, F.W. and Beckey, H.D. (1977) *J. Vac. Sci. Technol.*, **14**, 1033–1035.

Overberg, A., Karas, M., Bahr, U., Kaufmann, R. and Hillenkamp, F. (1990) *Rapid Commun. Mass Spectrom.*, **4**, 293–296.

Overberg, A., Karas, M. and Hillenkamp, F. (1991) *Rapid Commun. Mass Spectrom.*, **5**, 128–131.

Pasch, H. and Schrepp, W. (2003) *MALDI-TOF Mass Spectrometry of Synthetic Polymers*, Springer-Verlag, Heidelberg, p. 298.

Peng, J. and Gygi, S.P. (2001) *J. Mass Spectrom.*, **36**, 1083–1096.

Penning, F.M. (1927) *Naturwissenschaften*, **15**, 818.

Platzner, I.T., Habfast, K., Walder, A.J. and Goetz, A. (1997) in *Modern Isotope Ratio Mass Spectrometry*, 1st edn (ed. I.T. Platzner), John Wiley & Sons, Ltd, Chichester, p. 514.

Popov, I.A., Chen, H., Kharybin, O.N., Nikolaev, E.N. and Cooks, R.G. (2005) *Chem. Commun.*, 1953–1955.

Posthumus, M.A., Kistemaker, P.G., Meuzelaar, H.L.C. and Ten Noever de Brauw, M.C. (1978) *Anal. Chem.*, **50**, 985–991.

Pramanik, B.N., Ganguly, A.K. and Gross, M.L. (eds) (2002) *Applied Electrospray Mass Spectrometry*, Marcel Dekker, New York, p. 434.

Prókai, L. (1990) *Pract. Spectrosc.*, **9**, 1–44.

Prókai, L. (1990) *Pract. Spectrosc.*, **9**, 169–225.

Rabrenovic, M., Ast, T. and Kramer, V. (1981) *Int. J. Mass Spectrom. Ion Phys.*, **37**, 297–307.

Raffaeli, A. and Saba, A. (2003) *Mass Spectrom. Rev.*, **22**, 318–331.

Ray, S.J., Andrade, F., Gamez, G., McClenathan, D., Rogers, D., Schilling, G., Wetzel, W. and Hieftje, G.M. (2004) *J. Chromatogr. A*, **1050**, 3–34.

Rayleigh, L. (1882) *The London, Edinburgh and Dublin Philosophical Magazine and Journal of Science*, **14**, 184–186.

Reemtsma, T. (2003) *J. Chromatogr. A*, **1000**, 477–501.

Robb, D.B., Covey, T.R. and Bruins, A.P. (2000) *Anal. Chem.*, **72**, 3653–3659.

Rosenstock, H.M. and Krauss, M. (1963) in *Mass Spectrometry of Organic Ions* (ed. F.W. McLafferty), Academic Press, London, pp. 1–64.

Rosu, F., Nguyen, C.H., De Pauw, E. and Gabelica, V. (2007) *J. Am. Soc. Mass Spectrom.*, **18**, 1052–1062.

Roussis, S. (1999) *Rapid Commun. Mass Spectrom.*, **13**, 1031–1051.

Šagi, D., Kienz, P., Denecke, J., Marquardt, T. and Peter-Katalinic, J. (2005) *Proteomics*, **5**, 2689–2701.

Schalley, C.A. (ed.) (2007) *Analytical Methods in Supramolecular Chemistry*, 1st ed., Wiley-VCH Verlag GmbH.

Schaub, T.M., Linden, H.B., Hendrickson, C.L. and Marshall, A.G. (2004) *Rapid Commun. Mass Spectrom.*, **18**, 1641–1644.

Schaub, T.M., Rodgers, R.P., Marshall, A.G., Qian, K., Green, L.A. and Olmstead, W.N. (2005) *Energy Fuels*, **19**, 1566–1573.

Schröder, E. (1991) *Massenspektrometrie – Begriffe und Definitionen*, 1st edn, Springer-Verlag, Heidelberg.

Schulten, H.-R. and Beckey, H.D. (1974) *Org. Mass Spectrom.*, **9**, 1154–1155.

Seiler, R. (1969) *Int. J. Quantum Chem.*, **3**, 25–32.

Shetty, V., Spellman, D.S. and Neubert, T.A. (2007) *J. Am. Soc. Mass Spectrom.*, **18**, 1544–1551.

Snyder, A.P. (2000) *Interpreting Protein Mass Spectra*, 1st edn, Oxford University Press, New York, p. 544.

Spengler, B. (2004) *J. Am. Soc. Mass Spectrom.*, **15**, 703–714.

Stuewer, D. (1990) *Fresenius J. Anal. Chem.*, **337**, 737–742.

Syka, J.E.P., Coon, J.J., Schroeder, M.J., Shabanowitz, J. and Hunt, D.F. (2004) *Proc. Natl. Acad. Sci. U.S.A.*, **101**, 9528–9533.

Takats, Z., Cotte-Rodriguez, I., Talaty, N., Chen, H. and Cooks, R.G. (2005) *Chem. Commun.*, 1950–1952.

Takats, Z., Wiseman, J.M. and Cooks, R.G. (2005) *J. Mass Spectrom.*, **40**, 1261–1275.

Takats, Z., Wiseman, J.M., Gologan, B. and Cooks, R.G. (2004) *Science*, **306**, 471–473.

Talaty, N., Takats, Z. and Cooks, R.G. (2005) *Analyst*, **130**, 1624–1633.

Tanaka, K. (2003) *Angew. Chem. Int. Ed.*, **42**, 3861–3870.

Tanaka, K., Waki, H., Ido, Y., Akita, S., Yoshida, Y. and Yhoshida, T. (1988) *Rapid Commun. Mass Spectrom.*, **2**, 151–153.

Taylor, G.I. (1964) *Proc. R. Soc. Lond. A*, **280**, 383–397.

Tsuchiya, M. (1995) *Adv. Mass Spectrom.*, **13**, 333–346.

Tuniz, C., Bird, J.R., Fink, D. and Herzog, G.F. (1998) *Accelerator Mass Spectrometry - Ultrasensitive Analysis for Global Science*, CRC Press, Boca Raton, p. 371.

Vastola, F.J., Mumma, R.O. and Pirone, A.J. (1970) *Org. Mass Spectrom.*, **3**, 101–104.

Vastola, F.J. and Pirone, A.J. (1968) *Adv. Mass Spectrom.*, **4**, 107–111.

Verlinden, J., Gijbels, R. and Adams, F. (1986) *J. Anal. At. Spectrom.*, **1**, 411–419.

Volland, M.A.O., Adlhart, C., Kiener, C.A., Chen, P. and Hofmann, P. (2001) *Chem. Eur. J.*, **7**, 4621–4632.

Vorm, O., Roepstorff, P. and Mann, M. (1994) *Anal. Chem.*, **66**, 3281–3287.

Wahrhaftig, A.L. (1959) in *Advances in Mass Spectrometry* (ed. J.D. Waldron), Pergamon Press, Oxford, pp. 274–286.

Wahrhaftig, A.L. (1986) in *Gaseous ion Chemistry and Mass Spectrometry* (ed. J.H. Futrell), John Wiley & Sons, Inc., New York, pp. 7–24.

Wang, C.H., Huang, M.-W., Lee, C.-Y., Chei, H.-L., Huang, J.P. and Shiea, J. (1998) *J. Am. Soc. Mass Spectrom.*, **9**, 1168–1174.

Watson, C.H., Baykut, G. and Eyler, J.R. (1987) *Anal. Chem.*, **59**, 1133–1138.

Wei, J., Buriak, J.M. and Siuzdak, G. (1999) *Nature*, **399**, 243–246.

Weickhardt, C., Grun, C. and Grotemeyer, J. (1998) *Eur. Mass Spectrom.*, **4**, 239–244.

Welham, K.J., Domin, M.A., Johnson, K., Jones, L. and Ashton, D.S. (2000) *Rapid Commun. Mass Spectrom.*, **14**, 307–310.

Wendt, K.D.A. (2002) *Eur. J. Mass Spectrom.*, **8**, 273–285.

Whitehouse, C.M., Robert, R.N., Yamashita, M. and Fenn, J.B. (1985) *Anal. Chem.*, **57**, 675–679.

Whittal, R.M., Schriemer, D.C. and Li, L. (1997) *Anal. Chem.*, **69**, 2734–2741.

Williams, J.P., Patel, V.J., Holland, R. and Scrivens, J.H. (2006) *Rapid Commun. Mass Spectrom.*, **20**, 1447–1456.

Willoughby, R.C. and Browner, R.F. (1984) *Anal. Chem.*, **56**, 2625–2631.

Wilm, M.S. and Mann, M. (1994) *Int. J. Mass Spectrom. Ion Process*, **136**, 167–180.

Wilm, M. and Mann, M. (1996) *Anal. Chem.*, **68**, 1–8.

Wilm, M., Shevshenko, A., Houthaeve, T., Breit, S., Schweigerer, L., Fotsis, T. and Mann, M. (1996) *Nature*, **379**, 466–469.

Winchester, M.R. and Payling, R. (2004) *Spectrochim. Acta B At. Spectrosc.*, **59B**, 607–666.

Wood, G.W. (1982) *Mass Spectrom. Rev.*, **1**, 63–102.

Woodin, R.L., Bomse, D.S. and Beauchamp, J.L. (1978) *J. Am. Chem. Soc.*, **100**, 3248–3250.

Yang, C. and Henion, J.D. (2002) *J. Chromatogr. A*, **970**, 155–165.

Yew, J.Y., Cody, R.B. and Kravitz, E.A. (2008) *Proc. Natl. Acad. Sci. U.S.A.*, **105**, 7135–7140.

Zeleny, J. (1917) *Phys. Rev.*, **10**, 1–7.

Zenobi, R. and Knochenmuss, R. (1999) *Mass Spectrom. Rev.*, **17**, 337–366.

Zenobi, R., Zhan, Q. and Voumard, P. (1996) *Mikrochim. Acta*, **124**, 273–281.

Zubarev, R.A., Kelleher, N.L. and McLafferty, F.W. (1998) *J. Am. Chem. Soc.*, **120**, 3265–3266.

Further Reading

Ardrey, R.E. (2003) *Liquid Chromatography-Mass Spectrometry – An Introduction*, John Wiley & Sons, Ltd, Chichester.

Assamoto, B. (ed.) (1991) *Analytical Applications of Fourier Transform Ion Cyclotron Resonance Mass Spectrometry*, Wiley-VCH Verlag GmbH, Weinheim.

Becker, J.S. (2008) *Inorganic Mass Spectrometry: Principles and Applications*, 1st edn, John Wiley & Sons, Ltd, Chichester.

Beckey, H.D. (1977) *Principles of Field Desorption and Field Ionization Mass Spectrometry*, 1st edn, Pergamon Press, Oxford.

Budde, W.L. (2001) *Analytical mass spectrometry*, 1st edn, ACS and Oxford University Press, Washington, DC and Oxford.

Busch, K.L., Glish, G.L. and McLuckey, S.A. (1988) *Mass Spectrometry/Mass Spectrometry*, 1st edn, Wiley-VCH Verlag GmbH, New York.

Caprioli, R.M. (ed.) (1990) *Continuous-Flow Fast Atom Bombardment Mass Spectrometry*, John Wiley & Sons, Ltd, Chichester, p. 189.

Castleman, A.W. Jr, Futrell, J.H., Lindinger, W., Märk, T.D., Morrison, J.D., Shirts, R.B., Smith, D.L. and Wahrhaftig, A.L. (1986) in *Gaseous ion Chemistry and Mass Spectrometry* (ed. J.H. Futrell), John Wiley & Sons, Inc., New York.

Chapman, J.R. (ed.) (2000) *Mass Spectrometry of Proteins and Peptides*, Humana Press, Totowa.

Cooks, R.G., Beynon, J.H. and Caprioli, R.M. (1973) *Metastable Ions*, 1st edn, Elsevier, Amsterdam.

Dass, C. (2007) *Fundamentals of Contemporary Mass Spectrometry*, John Wiley & Sons, Inc., Hoboken.

De Hoffmann, E. and Stroobant, V. (2007) *Mass Spectrometry – Principles and Applications*, 2nd edn, John Wiley & Sons, Ltd, Chichester.

Dole, R.B. (ed.) (1997) *Electrospray Ionization Mass Spectrometry – Fundamentals, Instrumentation and Applications*, 1st ed., John Wiley & Sons, Chichester.

Duckworth, H.E., Barber, R.C. and Venkatasubramanian, V.S. (1986) *Mass Spectroscopy*, 2nd edn, Cambridge University Press, Cambridge.

Grayson, M.A. (ed.) (2002) *Measuring Mass – from Positive Rays to Proteins*, ASMS and CHF, Santa Fe and Philadelphia.

Gross, J.H. (2004) *Mass Spectrometry – A Textbook*, Springer-Verlag Heidelberg, Heidelberg.

Harrison, A.G. (1992) *Chemical Ionization Mass Spectrometry*, 2nd edn, CRC Press, Boca Raton.

Henderson, W. and McIndoe, S.J. (2005) *Mass Spectrometry of Inorganic and Organometallic Compounds*, John Wiley & Sons, Ltd, Chichester.

Hillenkamp, F. and Peter-Katalinic, J. (eds) (2007) *MALDI-MS. A Practical Guide to Instrumentation, Methods and Applications*, 1st edn, Wiley-VCH Verlag GmbH, Weinheim.

Kaltashov, I.A. and Eyles, S.J. (2005) *Mass Spectrometry in Biophysics: Conformation and Dynamics of Biomolecules*, John Wiley & Sons, Inc., Hoboken.

Kinter, M. and Sherman, N.E. (2000) *Protein Sequencing and Identification Using Tandem Mass Spectrometry*, John Wiley & Sons, Ltd, Chichester.

de Laeter, J.R. (2001) *Applications of Inorganic Mass Spectrometry*, 1st ed, John Wiley & Sons, Inc., New York.

Lehmann, W.D. (1996) *Massenspektrometrie in der Biochemie*, Spektrum Akademischer Verlag, Heidelberg.

Levsen, K. (1978) *Fundamental Aspects of Organic Mass Spectrometry*, 1st edn, Verlag Chemie, Weinheim.

March, R.E. and Hughes, R.J. (1989) *Quadrupole Storage Mass Spectrometry*, John Wiley & Sons, Ltd, Chichester, p. 496.

March, R.E. and Todd, J.F.J. (eds) (1995) *Practical Aspects of Ion Trap Mass Spectrometry*, Vol. 1, CRC Press, Boca Raton.

March, R.E. and Todd, J.F.J. (eds) (1995) *Practical Aspects of Ion Trap Mass Spectrometry*, Vol. 2, CRC Press, Boca Raton.

March, R.E. and Todd, J.F.J. (eds) (1995) *Practical Aspects of Ion Trap Mass Spectrometry*, Vol. 3, CRC Press, Boca Raton.

March, R.E. and Todd, J.F.J. (2005) *Quadrupole Ion trap Mass Spectrometry*, John Wiley & Sons, Inc., Hoboken.

McLafferty, F.W. (ed.) (1983) *Tandem Mass Spectrometry*, 1st edn, John Wiley & Sons, Inc., New York.

McLafferty, F.W., Turecek, F. (1993) *Interpretation of Mass Spectra*, 4th edn, University Science Books, Mill Valley.

Montaudo, G. and Lattimer, R.P. (eds) (2001) *Mass Spectrometry of Polymers*, 1st edn, CRC Press, Boca Raton.

Niessen, W.M.A. and Voyksner, R.D. (eds) (1998) *Current Practice of Liquid Chromatography-Mass Spectrometry*, Elsevier, Amsterdam.

Pasch, H. and Schrepp, W. (2003) *MALDI-TOF Mass Spectrometry of Synthetic Polymers*, Springer-Verlag, Heidelberg.

Platzner, I.T., Habfast, K., Walder, A.J. and Goetz, A. (1997) *Modern Isotope Ratio Mass Spectrometry*, 1st edn (ed. I.T. Platzner), John Wiley & Sons, Ltd, Chichester, p. 514.

Pramanik, B.N., Ganguly, A.K. and Gross, M.L. (eds) (2002) *Applied Electrospray Mass Spectrometry*, Marcel Dekker, New York.

Prókai, L. (1990) *Field Desorption Mass Spectrometry*, 1st edn, Marcel Dekker, New York.

Roboz, J. (1999) *Mass Spectrometry in Cancer Research*, 1st edn, CRC Press, Boca Raton.

Rossi, D.T. and Sinz, M.W. (eds) (2002) *Mass Spectrometry in Drug Discovery*, Marcel Dekker, New York.

Schalley, C.A. (ed.) (2003) *Modern Mass Spectrometry*, Springer, New York.

Schalley, C.A. (ed.) (2007) *Analytical Methods in Supramolecular Chemistry*, 1st edn, Wiley-VCH Verlag GmbH.

Schlag, E.W. (ed.) (1994) *Time of Flight Mass Spectrometry and its Applications*, 1st edn Elsevier, Amsterdam.

Snyder, A.P. (2000) *Interpreting Protein Mass Spectra*, 1st edn, Oxford University Press, New York.

Sparkman, O.D. (2006) *Mass Spectrometry Desk Reference*, 2nd edn, Global View Publishing, Pittsburgh.

Splitter, J.S., and Turecek, F. (eds) (1994) *Applications of Mass Spectrometry to Organic Stereochemistry*, 1st edn, Verlag Chemie, Weinheim.

Taylor, H.E. (2000) *Inductively Coupled Plasma Mass Spectroscopy*, 1st edn, Academic Press, London.

Tuniz, C., Bird, J.R., Fink, D. and Herzog, G.F. (1998) *Accelerator Mass Spectrometry – Ultrasensitive Analysis for Global Science*, CRC Press, Boca Raton.

Watson, J.T. and Sparkman, O.D. (2007) *Introduction to Mass Spectrometry*, 4th edn, John Wiley & Sons, Ltd, Chichester.

Wilson, R.G., Stevie, F.A. and Magee, C.W. (1989) *Secondary Ion Mass Spectrometry: A Practical Handbook for Depth Profiling and Bulk Impurity Analysis*, John Wiley & Sons, Ltd, Chichester.

Yinon, J. (ed.) (1994) *Forensic Applications of Mass Spectrometry*, CRC Press, Boca Raton.

30
Chromatography

James M. Miller

30.1	**Introduction** 1057	
30.1.1	Historical Summary 1057	
30.2	**Basic Concepts** 1059	
30.2.1	Definition of Chromatography 1059	
30.2.2	Classification 1059	
30.2.3	Mechanism of Operation 1059	
30.2.4	Other Definitions and Symbols 1061	
30.2.5	Qualitative and Quantitative Analysis 1064	
30.3	**Gas Chromatography (GC)** 1065	
30.3.1	Instrumentation 1065	
30.3.1.1	Carrier Gas Supply 1065	
30.3.1.2	Sample Introduction 1065	
30.3.1.3	Columns 1067	
30.3.1.4	Detectors 1068	
30.3.2	Modes of Separation 1069	
30.3.2.1	Gas–Liquid Chromatography (GLC) 1069	
30.3.2.2	Gas–Solid (GSC) 1071	
30.4	**Liquid Chromatography** 1072	
30.4.1	Instrumentation 1073	
30.4.1.1	Solvent Delivery Systems 1073	
30.4.1.2	Injection Valves 1073	
30.4.1.3	Columns 1074	
30.4.1.4	Detectors 1074	
30.4.2	Modes of Separation 1076	
30.4.2.1	Liquid–Solid Chromatography (LSC) 1076	
30.4.2.2	Bonded-Phase Chromatography (BPC) 1078	
30.4.2.3	Reversed Phase Chromatography (RPC) 1078	
30.4.2.4	Ion-Pair Chromatography (IPC) 1079	
30.4.2.5	Gradient Elution 1080	
30.4.2.6	Ion-Exchange Chromatography (IEC) and Ion Chromatography (IC) 1080	

Encyclopedia of Applied Spectroscopy. Edited by David L. Andrews.
Copyright © 2009 WILEY-VCH Verlag GmbH & Co. KGaA, Weinheim
ISBN: 978-3-527-40773-6

30.4.2.7	IC Instrumentation	1081
30.4.2.8	Size-Exclusion Chromatography (SEC)	1082
30.4.3	Preparative Liquid Chromatography	1084
30.4.4	Thin-Layer Chromatography	1085
30.5	**Supercritical Fluid Chromatography (SFC)**	**1085**
30.5.1	Operating Characteristics	1086
30.5.2	Modes of Separation	1087
30.5.2.1	Isobaric SFC	1087
30.5.2.2	Pressure Programming	1087
30.6	**Spectrochemical Detection**	**1087**
30.6.1	GC/MS	1091
30.6.2	LC/MS	1091
30.6.3	Other Hyphenated Methods	1092
30.7	**Selected Other Topics**	**1093**
30.7.1	Multidimensional Separations	1093
30.7.1.1	GC/GC	1094
30.7.1.2	LC/GC	1094
30.7.1.3	LC/LC	1094
30.7.1.4	Other Combinations	1095
30.7.2	Chiral Separations	1095
30.7.3	Capillary Electrochromatography (CEC)	1096
	Acknowledgments	**1098**
	Glossary	**1098**
	References	**1100**
	Further Reading	**1101**

30.1
Introduction

Chromatography is the most widely used separation method for chemical analysis. While it is not a spectroscopic method, it is often combined with or used with spectroscopic methods such as ultraviolet/visible (UV/Vis) absorption spectroscopy or mass spectroscopy (MS). A spectrometer can be an integral part of some chromatographs; for example, a UV/Vis spectrometer is most commonly used as a detector for high-performance liquid chromatography (HPLC). Alternatively, chromatographs are coupled to, or used with, spectrometers such as MS, and the combined technique, gas chromatography–mass spectrometry (GC–MS or GC/MS), is an example of so-called hyphenated methods.

This chapter focuses on chromatography, but it includes relevant information about spectroscopic detection systems, including, in some cases, the interface between the chromatograph and the spectrometer. The outline of the remainder of the article is as follows. The first section is devoted to a description of basic concepts that apply to all forms of chromatography. This is followed by sections devoted to the two most common types of chromatography, GC and LC, followed by a shorter section on the third type, supercritical fluid chromatography (SFC). The final sections elaborate on spectroscopic detection, mainly mass spectrometry (MS), and a few special topics.

For each type of chromatography, a description of the general uses of the technique is followed by sections on instrumentation and modes of separation. In Section 30.4.1, the four essential components of a chromatograph, namely, the mobile-phase (MP) delivery system, the sample injector, the separation column (i.e., the stationary phase (SP)), and the detector, are described. In Section 30.4.2, a discussion of separation principles (i.e., the physical/chemical mechanism of resolving a mixture into its individual components) is combined with examples of a few applications that illustrate the usefulness and limitations of the techniques.

30.1.1
Historical Summary

In 1903, the Russian botanist Tswett devised a separation method that is now called *liquid-column chromatography* (*LC*). He separated structurally similar yellow and green chloroplast pigments

Encyclopedia of Applied Spectroscopy. Edited by David L. Andrews.
Copyright © 2009 WILEY-VCH Verlag GmbH & Co. KGaA, Weinheim
ISBN: 978-3-527-40773-6

by placing a leaf extract onto the top of a column of calcium carbonate particles and then washing the column with carbon disulfide (by gravity flow of the liquid). The colored pigments formed visible bands that became separated as they moved down the column. Tswett coined the term *chromatography* (from Greek, meaning "color writing"), although he also realized that the method was applicable to colorless compounds. Significant achievements in biochemistry and the chemistry of complex natural organic compounds were made possible by this method. However, the technique was time consuming and tedious, thereby restricting more widespread use in chemical separations and analysis. Faster and more efficient separations would only come years later, as a result of the theoretical and instrumental advances.

In a classic, Nobel-prize-winning paper, Martin and Synge (1941) advanced the theory of chromatography and predicted that high-speed LC would require very small column particles and high pressure. The technology for doing this was not available for many years, but, in the interim, James and Martin (1952) published the first paper on GC as had been suggested by Martin and Synge 10 years before. The potential of the technique was quickly recognized by other workers, who provided engineering advances that made gas chromatographs practical and useful, and GC became a major separation tool, providing additional advances in the theory of chromatography. It took about 10 more years until it was suggested that the principles of GC should be applied to LC. Giddings, among others, published important papers (Giddings, 1963) that gave rise to high-pressure liquid chromatography (HPLC), resulting in improved or higher performance LC, also known by the label *HPLC*. As a result of advances in instrumentation, HPLC methods have gradually approached the quality of GC methods in terms of speed and resolution. Further information about the history of chromatography can be found in the many publications by Ettre and Zlatkis (1979), Ettre (2002) and Ettre (2003).

Unfortunately, GC has some inherent limitations; it requires samples to be in the vapor phase, which limits its usefulness to compounds that can be volatilized. Current estimates are that only about 20% of known organic compounds can be satisfactorily separated by GC without some chemical modification (such as esterification of organic acids, which produces more volatile analytes). Modern LC, (HPLC), is suitable for analyzing most of the remaining 80% of known organic compounds, as well as many inorganic compounds.

Chromatography, primarily GC and HPLC, is used for rapid, efficient separation and analysis of complex mixtures of organic and inorganic compounds in the pharmaceutical, chemical, biomedical, biochemical, food, energy, cosmetics, and environmental industries, to name a few. With over 160 000 chromatographers around the world, chromatography forms one of the largest fields in the physical sciences. Clearly, it is not possible to do justice to all areas of such a large field in this short chapter. Therefore, in the sections that follow, the emphasis is on the most important theoretical, chemical, and procedural aspects of chromatography, along with an emphasis on spectroscopic detection.

30.2
Basic Concepts

30.2.1
Definition of Chromatography

Chromatography is a separation method in which the components of a sample partition between two phases: one phase is a stationary bed with a large surface area, and the other is a gas or liquid (MP), which percolates through the stationary bed. The sample is carried by the MP through the column. Samples partition (equilibrate) between the MP and the SP based on their solubilities or their relative tendencies to sorb on/in each of the respective phases. For this generalized introduction, we refer to the phases using these general terms and the symbol S for stationary and M for mobile. The components of the sample (called *solutes* or *analytes*) separate from one another based on their *relative* affinities for the two phases. This type of chromatographic process is called *elution*.

The "official" definition of the International Union of Pure and Applied Chemistry (IUPAC) is "Chromatography is a physical method of separation in which the components to be separated are distributed between two phases, one of which is stationary (SP) while the other (the MP) moves in a definite direction. Elution chromatography is a procedure in which the MP is continuously passed through or along the chromatographic bed and the sample is fed into the system as a finite slug" (Ettre, 1993).

30.2.2
Classification

Chromatography is classified according to the physical state of the MP. Thus, if the MP is a *gas*, the technique is called GC, and if it is a *liquid*, it is called LC. A subclassification can be made according to the state of the SP. If the SP is a solid, the GC technique is called *gas–solid chromatography* (*GSC*), and if it is a liquid, *gas–liquid chromatography* (*GLC*). The names used in liquid chromatography are more diverse and do not usually follow this simple pattern. A complete classification scheme is shown in Figure 30.1.

30.2.3
Mechanism of Operation

The objective in any chromatographic separation is to move individual materials through a column at different velocities so that they exit from the column at different times. In a typical experiment, a mixture of materials is injected into a flowing stream of a carrier medium (the MP) at the top of a chromatographic column, containing the SP. The MP is either a gas, a liquid, or a supercritical fluid. Common SPs are either solids or liquids. Liquids are kept stationary in the column by coating them on the inner wall of the column or by coating them on solid supports that are then packed (like solids) in the column. An alternative way to classify columns is to differentiate between packed columns and nonpacked, or open tubular (OT) columns.

A component in a sample mixture can interact with the mobile and SPs in a number of given ways. Operative partition mechanisms in chromatography are often discussed in terms of the nonpolar or polar nature of the sample, MP, and SP. A material is said to have nonpolar character if the electronic charge is uniformly distributed in the individual

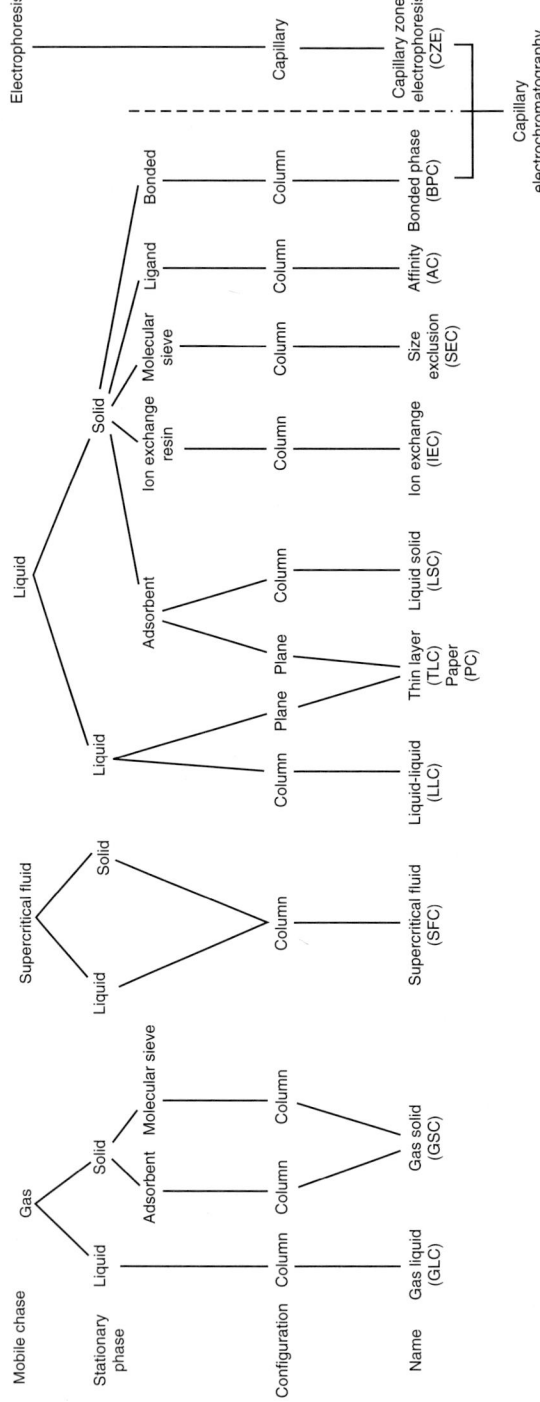

Fig. 30.1 Classification of chromatographic techniques (Miller, 2005); reprinted with permission.

molecules that compose the material. An example of a nonpolar molecule is methane (CH_4), which has a zero dipole moment because of the symmetrical arrangement of the individual dipoles constituted by the carbon–hydrogen bonds. In contrast, a material is said to have polar character if any region of the individual molecules that compose the material possesses an excess or deficiency of electronic charge relative to the rest of the molecule. Many electrically neutral molecules have polar character that can be exploited in chromatographic separations. As an illustration, consider the polar nature of the molecules benzene, butanol, 2-pentanone, nitropropane, and pyridine (Miller, 2005). The π-electron cloud of the benzene molecule can interact with the mobile and SPs through an electron donation mechanism. Butanol can interact via hydrogen bond formation with proton-accepting sites on the liquid or SP. The 2-pentanone molecule can interact through interactions of its dipoles with those of the mobile or SP. The nitro group on nitropropane is an electron-accepting group, which will be retained by SPs that possess electron-donating character. Finally, the unpaired electrons on the pyridine molecule can participate in the formation of hydrogen bonds with mobile or SPs that possess proton-donating character.

A final example demonstrates the concept of relative polarity. Consider a family of aliphatic, straight-chain alcohol molecules, each member of which is classified as polar. As the length of the carbon chain is increased, the polarity of the molecule decreases (relative polarity index: methanol > ethanol > propanol > butanol, etc.). In the examples provided in this chapter, we observe that polar sample molecules have a strong tendency to partition into polar mobile and/or SPs through one or more of the mechanisms described above. Conversely, nonpolar molecules will tend to partition into nonpolar phases. Ionic species are separated using familiar ion-exchange reactions as well as other methods that exploit their ionic nature.

30.2.4
Other Definitions and Symbols

The hypothetical chromatogram shown in Figure 30.2 displays the response of a detector as a function of time, t. The sample (composed of species 1 and 2) is applied to the top of the column as a narrow band. If a species has no interaction with the SP, it elutes from the end of the column when the total volume of MP introduced at the top of the column equals the void volume of the column. The time for elution of unretained species is t_M. The molecules of species 1 and 2 are swept along at the velocity of the MP, and they would all exit from the column at the same time were it not for the differences in their interactions with the SP. The time that each species actually spends in the SP, the adjusted retention time t_R', determines when it will exit from the column. In practice, retention is often defined in terms of the retention time t_R:

$$t_R = t_R' + t_M \quad (30.1)$$

With reference to Figure 30.2, species 1 spends less time in the SP and therefore has a shorter retention time than species 2. Thus, the retention time serves to characterize a given analyte in a given chromatographic system. The x-axis in Figure 30.2 can also represent the

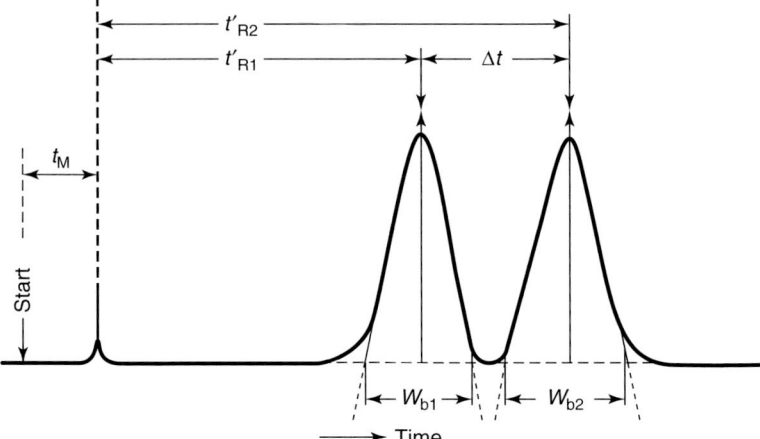

Fig. 30.2 Hypothetical chromatogram showing the separation of two analytes. The adjusted retention time is shown for each analyte as t'_R, and the retention time for a nonretained species as t_M. Chromatographic peak widths at base are labeled w_b. (Yost et al., 1980); reprinted with permission.

volume V of MP that has flowed in a given time t, because, at a constant MP flow rate, F, the following relationship holds

$$V_R = t_R \times F \qquad (30.2)$$

Consequently, volume and time are often used interchangeably on chromatograms and chromatographic data.

Another parameter elated to retention time is the retention factor, k:

$$k = \frac{V'_R}{V_M} \qquad (30.3)$$

It too can be used for qualitative analysis for a specified chromatographic system.

While the components in the sample are being separated and eluted, the once-narrow bands broaden. This is primarily the result of three effects: (i) the differences in the path that each component travels through the column; (ii) the longitudinal diffusion effects; and (iii) slow equilibration of the components between the mobile and SPs. As a result of this band spreading, the components elute as Gaussian-like peaks like the one shown in Figure 30.3. They are broadened in proportion to the square root of the retention time, so the latter peaks are the broadest. It is this square root relationship that makes it possible to separate peaks even though the peaks are themselves being broadened, because the peaks are being separated from each other on a linear basis.

As shown in Figure 30.3, the peak width is denoted as w_b when measured at the baseline. For a Gaussian peak, that width is four times the standard deviation of the curve, 4σ. To get an estimate of the efficiency of a given separation, one can compare the retention time with the peak width. The parameter that has been defined for that purpose is the plate height, N.

$$N = \left(\frac{V_R}{\sigma}\right)^2 = 16\left(\frac{V_R}{w_b}\right)^2 \qquad (30.4)$$

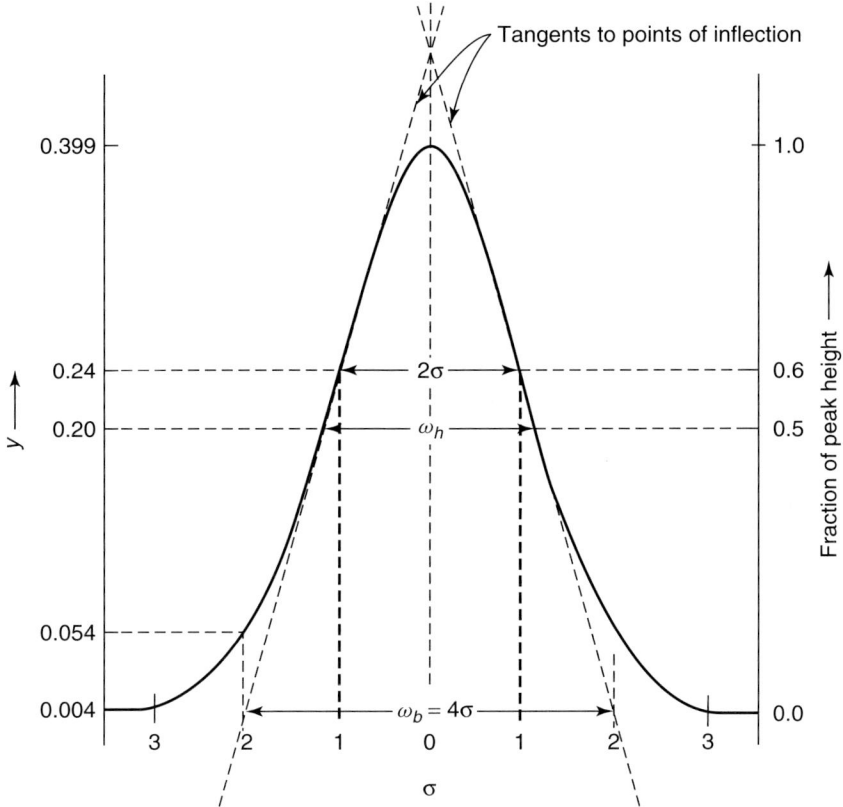

Fig. 30.3 Idealized Gaussian or normal distribution (Miller, 2005); reprinted with permission.

The name plate height came from the concept of theoretical plates that was in use in distillation processes when GC was first applied to petroleum separations. Briefly, the term *theoretical plate* referred to an imaginary unit in the chromatographic column wherein one complete equilibration of an analyte between the stationary and MPs occurred, as was visualized in distillation columns. In the entire column, there would be N of these units, but of course they do not exist in reality. For a given set of chromatographic conditions, N is approximately constant for all the peaks, and it is used in the industry to specify the quality of a column. It is a useful measure of column efficiency because it conveys the ability of the column to provide narrow peaks and good separations, and it can be easily calculated from chromatograms. This measure of column efficiency is frequently normalized to the column length L to give the plate height, H:

$$H = \frac{L}{N} \qquad (30.5)$$

Other fundamental parameters are used not only to measure the quality of a separation scheme but also in the design of a separation. The separation factor α is a measure of the ability of a particular separation scheme to separate two components, and it can be defined in terms of the quantities represented in Figure 30.2:

$$\alpha = \frac{k_2}{k_1} = \frac{V_{R2}'}{V_{R1}'} \tag{30.6}$$

If the separation factor α is 1.0, the peaks coincide and there is no separation.

The efficiency of separation depends on the difference in retention times and the sharpness of the peaks, and is commonly described in terms of the resolution (R_s) of a particular method:

$$R_s = \frac{2(t_{R2}' - t_{R1}')}{(W_{b1} + W_{b2})} \tag{30.7}$$

Narrow peaks with very different retention times are said to be separated with high resolution. A well-resolved peak arising from a pure substance is very desirable from both an analytical and a preparative viewpoint. R_s can also be expressed in terms of N, k, and α (Purnell, 1960).

$$R_s = \frac{1}{4}(N)^{\frac{1}{2}}\left[\frac{(\alpha-1)}{\alpha}\right]\left[\frac{k_2}{(1+k_2)}\right] \tag{30.8}$$

A fundamental knowledge of the physical and chemical processes occurring on the column is critical to the successful design and performance of a separation. The experimental variables that control N, k, and α (and consequently R_s) are different for GC, LC, and SFC, and are briefly discussed in the sections that follow. Chromatographers experienced in these techniques know how to adjust the experimental conditions to maximize the resolution of a given separation.

30.2.5
Qualitative and Quantitative Analysis

Since most chromatographs are equipped with detectors, it is possible to use the chromatographic data (the chromatogram) for both qualitative and quantitative analysis. The chromatographic retention parameters such as retention time or retention factor do serve to characterize a given solute and can be used for making qualitative identification when compared with standards. Correct peak assignments can be made by comparison with a literature chromatogram or by comparison with a chromatogram from injection of a known standard. However, for positive confirmations of the identities of unknown components in a mixture, spectral measurements are required, most commonly by MS. Chromatographic data alone are not sufficient for identifying a true unknown chemical. Alternatively, identification can be done by collection of effluent fractions (corresponding to individual components) followed by off-line spectroscopic analysis.

Quantitative analysis is usually based on measurements of the heights or areas of the chromatographic peaks and comparison with heights or areas resulting from known standards in accordance with normal analytical practice. Especially in GC, it is common to use the internal standard method or the method of standard addition. When standards are not available, or when peaks have not been identified, it is common practice to calculate an area percentage by dividing the peak area of the desired peak by the sum of the areas of all the peaks. Such a calculation provides only an approximate quantitative analysis because the detection sensitivity of various compounds varies and only true calibration with standards can account for these differences. Also, it is often the case that some compounds may not have been detected by the particular detector being used (a nonuniversal detector) so the true sum of the peak areas cannot be found. Modern chromatographs do have good data acquisition and processing equipment (e.g., computers with integration and calibration software) for fast and accurate quantitation.

30.3 Gas Chromatography (GC)

GC was introduced as an analytical and separation technique in the 1940s and 1950s, in large part because of the work of Martin and Synge, referred to earlier. The first commercial chromatographs became available in 1955. The first ones were made to use packed columns, so all of the early literature uses them. When OT capillary columns were invented, the proponents attempted to show that they were superior, but these columns were patent-protected and not too widely used. Also, they required some modifications of the early instruments, resulting in two types of GC instruments. Today, the instruments designed to accept capillary or OT columns are the more popular ones and they are featured in this discussion.

GC is used for analysis of mixtures containing inorganic or organic compounds that have vapor pressures greater than about 0.1 Torr (10 Pa) at instrument operating temperatures. GC can be used to analyze solids, liquids, and gases. Amounts ranging from 10^{-2} to 10^{-12} g can be introduced and analyzed.

30.3.1
Instrumentation

A schematic of a capillary GC instrument is shown in Figure 30.4. Each of the individual parts is briefly described.

30.3.1.1 Carrier Gas Supply
The most popular carrier gases used as GC MPs are helium, and hydrogen, followed by nitrogen and argon. Where available, helium is preferred over hydrogen because of the latter's explosive characteristic. Carrier gases must be very pure, and gas scrubbing techniques for removal of oxygen, hydrocarbons, and water are used to assure purity at the point of use. Flow rates are accurately controlled and, for those cases where the pressure may vary, modern instruments are equipped with automatic pressure regulation to achieve constant flow rates. Typical flow rates are 20–100 ml min^{-1} for packed columns and 0.1–10 ml min^{-1} for capillary columns.

30.3.1.2 Sample Introduction
Microliter syringes are most commonly used for the introduction of liquid samples and solutions. Automated systems using microsyringes are highly desirable and readily available. For gases, gas-tight milliliter syringes or valves are used. Injector temperatures are regulated so as to be high enough to vaporize the injected sample, but not so high as to decompose it. Injectors usually have a rubber septum over the top of the column, which is pierced by the syringe during sample introduction. Direct introduction of the sample onto the column is possible and is known as *on-column injection*.

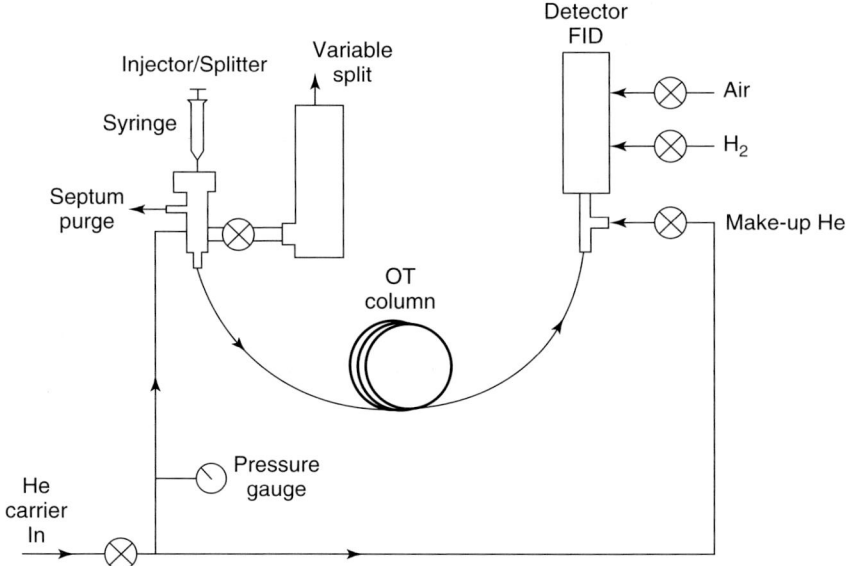

Fig. 30.4 Schematic of a typical gas chromatograph (McNair and Miller, 1998); reprinted with permission.

Because of the difficulty in injecting very small volumes of samples as required for capillary column GC, a sample inlet splitter is usually used. Small samples from split injection maintain separation efficiency and resolution. Sample splitters have a mixing chamber in which the injected sample and carrier gas are thoroughly mixed prior to being split. Typical splitting involves sending 50–99.9% of the mixture to waste through a vent tube, with the remainder being introduced onto the column. It is also possible to make a so-called splitless injection, whereby the entire mixture is directed to the column using the same apparatus, but a more involved procedure is required. A stated earlier, on-column injection is a third alternative.

The introduction of gas samples is most often accomplished with a gas sampling valve. In one valve position, a sample loop of known volume is filled with the sample gas at a known pressure. When the valve is activated, a precise amount of sample is swept out of the sample loop and onto the column by the carrier gas. This is the same type of valve that finds use in HPLC.

Other techniques can be used to introduce samples from unusual matrices. Injection of volatile components present in a relatively nonvolatile liquid can be performed by the headspace (HS) method. In one form of HS, called *purge-and-trap technique*, the liquid is sparged with an inert gas, and the volatile components are trapped by adsorption on a solid support or by condensation in a cold trap below the boiling point of the most volatile component. The trapped species are then volatilized and directed into the carrier gas stream.

Pyrolysis GC is useful for nonvolatile samples such as organic polymers. In this method, the sample is thermally pyrolyzed under controlled conditions, and the decomposition products are then analyzed by GC.

30.3.1.3 Columns

There are two types of GC columns: packed and capillary. Typical packed columns are constructed from 2–4-m lengths of stainless-steel or glass tubing with inside diameters (i.d.) of 2–4 mm. Columns are packed with 100–1000-μm-diameter porous particles. The particles are left untreated for separation by GSC and are coated with a liquid phase for GLC. Typical loading levels of liquid phases on the solid supports are about 0.5–10.0% by weight.

Typical capillary columns are 10–100-m lengths of 0.1–1.0-mm-i.d. metal or glass tubing. These columns have no packing material, and are often referred to as *OT columns*. The inner wall of the tubing has a coating that functions as the SP. GSC capillary columns have a porous layer deposited on the inner wall (porous-layer open tubular, (PLOT)). In GLC, the liquid phase may be directly applied to the inner wall (wall-coated open tubular, (WCOT)) or to a support material coating the inner wall (support-coated open tubular, (SCOT)). The thickness of a typical liquid SP film in a capillary column ranges from 0.2 to 0.5 μm. The tremendous popularity of capillary columns is due to the extremely high resolution separations that can be obtained. SPs with a variety of chemical and physical properties are available. As is seen in Section 30.3.2, the choice of an SP is dictated by the separation problem at hand.

Accurate control of column temperature is essential for reproducible GC. Column ovens that regulate temperatures within 0.2 °C between 25 and 450 °C are adequate for this purpose. Cryogenic equipment is used for temperature regulation down to about −100 °C for those applications requiring lower temperatures.

Temperature Programming (PTGC) Isothermal GC, the maintenance of a constant temperature on the column, can result in poor resolution of early-eluting peaks and significant band spreading of late-eluting peaks. Consequently, it is often desirable to use temperature programming whereby the column temperature is raised from a low initial temperature to a high final temperature during the GC run. Most GC ovens have the capability to provide one or more temperature gradients for this purpose.

A comparison of the effect of temperature programming can be seen by comparing Figure 30.5, an isothermal separation and the corresponding programmed temperature separation of the same sample. In the latter case, the early peaks are much better separated, the late peaks are eluted earlier, the peak shapes are nearly the same and are symmetrical, the total time required is decreased, and six additional compounds are eluted after the C_{15} peak. Good quantitation is obtained using temperature programming gas chromatography (PTGC) and it is also ideal for initial screening of new samples. There are a few disadvantages, the major one being the requirement that the SP have a wide temperature range of stability. In some cases, the SP may show increased bleeding during the run, since its vapor pressure increases with the temperature increase, causing the baseline to drift upward. However, in most cases, the advantages outweigh the disadvantages, and PTGC is very popular.

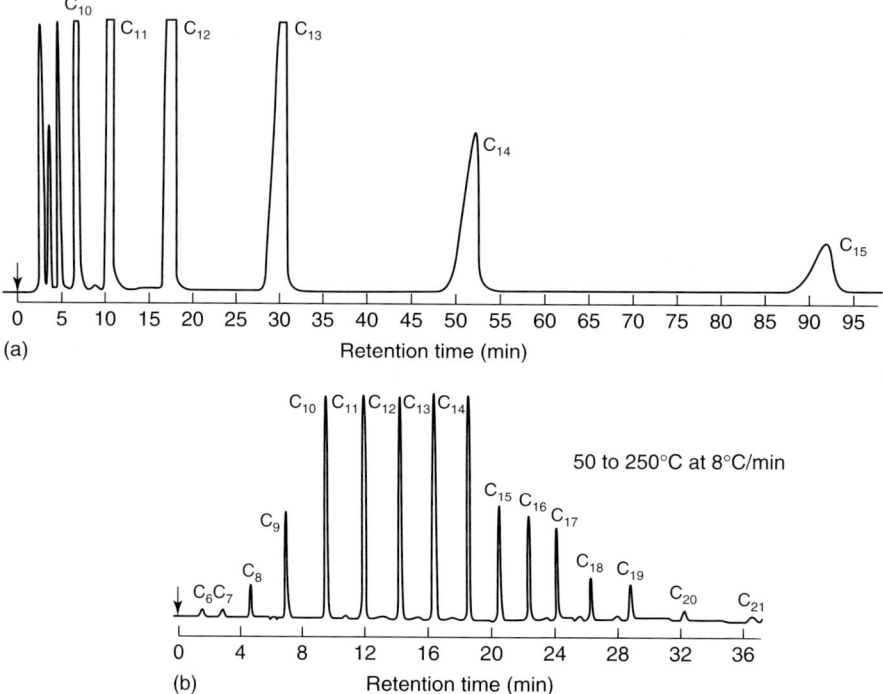

Fig. 30.5 A comparison of (a) isothermal and (b) temperature-programmed separation of n-paraffins (McNair and Miller, 1998); reprinted with permission.

30.3.1.4 Detectors

Most of the detectors used in GC were invented for this use, and many of them are based on ionization phenomena. The main exception is the thermal conductivity detector (TCD), which was already in use to measure the thermal conductivity of gases and, with minor modifications, became an ideal GC detector in the early days. For optimal use, a carrier gas of high thermal conductivity, like helium or hydrogen, is required; of course, safety precautions must be taken if hydrogen is used. It is a universal, nondestructive detector, but one with only a modest detectivity (about 10^{-8} g). The most popular one at present is the flame ionization detector (FID), which uses an oxy-hydrogen flame to burn the eluents from the GC column, thereby producing ions and electrons to form a current.

Only a few of the detectors are spectroscopic, but MS is a detector that is ideally suited to provide data for both qualitative and quantitative analysis. It is discussed later along with its use for HPLC. Infrared (IR) detection is similarly useful and it is also discussed later. The only other spectroscopic detector that finds use is the flame photometric detector (FPD). It employs a photomultiplier tube in combination with optical filters to monitor light emission generated in a flame similar to that used in the FID. Bandpass filters that transmit 394- and 526-nm radiation make this detector specific for luminescence

Tab. 30.1 Gas chromatography detectors.

Detector	Detection limit (g)	Selectivity
Thermal conductivity (TCD)	10^{-8}	Universal
Flame ionization (FID)	10^{-11}	Organics
Electron capture (ECD)	10^{-13}	Halogens, nitro-groups
Flame photometric (FPD)	$10^{-10}-10^{-11}$	Sulfur, phosphorus
Mass spectrometric (MS)	10^{-12}	Organics, others
Infrared spectrometric (FTIR)	10^{-8}	Most organics, others
Photoionization (PID)	10^{-10}	Aromatic organics, arsenic, sulfur, phosphorus
Hall electrolytic conductivity (HECD)	10^{-7}	Sulfur, nitrogen, phosphorus, halogens
Nitrogen phosphorus (NPD)	10^{-9}	Sulfur, nitrogen, phosphorus, arsenic
Discharge ionization (DID)	10^{-9}	Universal for liquids and gases
Gas density (GADE)	10^{-5}	High molecular weight
Helium ionization (HID)	10^{-9}	Universal for gases
Ultrasonic	10^{-7}	Universal for gases

from sulfur- and phosphorus-containing species, respectively. The detection limit of the FPD is about 10^{-11} g for phosphorus-containing and 10^{-10} g for sulfur-containing analytes. A few characteristics of GC detectors are given in Table 30.1.

30.3.2
Modes of Separation

30.3.2.1 Gas–Liquid Chromatography (GLC)

In GLC, the important parameters are the nature of the liquid SP and the temperature. The MP is inert and relatively unimportant in determining the mode of separation. Historically, GLC is also referred to as *partition chromatography* since the separation mechanism is based on partition equilibration of the analyte between the liquid and gas phase.

Hundreds of high boiling liquids and some solids have been used as SPs, providing a wide variety of available chemical properties (e.g., polarity, chemical functionality) and physical properties (e.g., particle size, film thickness). In fact, there are too many to enable one to select the very best one for a given analysis. However, with the current emphasis on the very efficient capillary columns, the need for high selectivity has decreased and only a few carefully chosen SPs are necessary to handle most analyses. A family of silicone polymers that differ in the functional groups they contain has become the most popular SPs. The least polar one is a polydimethylsiloxane polymer abbreviated as PDMS. Replacement of the methyl groups with more polar ones like phenyl, cyano, or fluoro results in more polar phases. Another common polar SP is a polyethyleneglycol like Carbowax 20M. Interest in using the class of compounds known as *room temperature ionic liquids* (*RTILs*) is increasing. Often, phases are chemically bonded to the inside wall of a silica capillary tube to make WCOT columns. For a more complete description of the range of available choices and the criteria used in their selection, the reader is referred to works by Grob and Barry (2004) and McNair and Miller (1998) which cover this subject in detail. In this section, some examples are provided to illustrate the capabilities of GLC.

The gas chromatograms in 30.5, which were used to illustrate PTGC, show the separation of a mixture of aliphatic hydrocarbon standards as described earlier. The hydrocarbons are typical of those found in gasoline-range petroleum products. The subscript i in the designation C_i refers to the number of carbon atoms found on the straight-chain backbone of the molecule. These chromatograms were obtained on a capillary column (coated with a nonpolar liquid phase) using column temperature programming and an FID. The nonpolar hydrocarbons have a high affinity for the nonpolar SP. A given hydrocarbon will reside in the SP until the temperature is raised high enough to generate a partial vapor pressure sufficient to increase its equilibrium concentration in the carrier gas, relative to its concentration in the SP. Therefore, the separation mechanism is primarily based on the differences in boiling points among the different hydrocarbons. The PTGC analysis begins with the column at a low temperature, below the boiling point of all the hydrocarbons. As the temperature of the column is gradually increased, hydrocarbons with successively higher boiling points are successively eluted. This type of analysis is a standard test method for determining the boiling range distribution of petroleum products, a so-called simulated distillation.

Columns are often developed and optimized for specific applications. For example, a polar SP, consisting of methyl phenyl cyanopropyl silicone bonded to a fused-silica open tubular (FSOT) column, was developed specifically for the trace analysis of tetrachlorodibenzo-p-dioxins. Analysis of these compounds is very important because of their extreme toxicity, particularly 2,3,7,8-tetra-chlorodibenzo-p-dioxin, which is classified as a supertoxic material. The chromatogram shown in Figure 30.6 is for nine isomers of tetra-chlorodibenzo-p-dioxin that differ only in the position of the chlorine atoms on the aromatic ring structure. The GC effluent is monitored with mass spectrometric detection, using selective monitoring of ions with mass-to-charge ratios of 320 and 322. These values correspond to the molecular weights of ions formed by ionization of the tetra-chlorodibenzo-p-dioxin molecule (i.e., no fragmentation), taking into account the natural relative abundance of the chlorine isotopes, ^{35}Cl and ^{37}Cl. The location of the chlorine atoms influences the relative polarities and boiling points of the individual isomers. Both properties are exploited in the separation shown in Figure 30.6.

The effect of column polarity on retention in GC is further illustrated by the example shown in Figure 30.7. Here, the injected sample, containing 25 different organic solvents, is split and chromatographed simultaneously on two capillary columns. The upper trace is the chromatogram obtained on a nonpolar column, while the lower trace is the chromatogram obtained on a polar column. Detection is achieved by the use of flame ionization detectors for effluent from each column. Notice that polar compounds such as ethanol, isopropanol, n-butanol, and 3-methyl-1-butanol elute more quickly from the nonpolar column than they do from the polar column, where their affinity

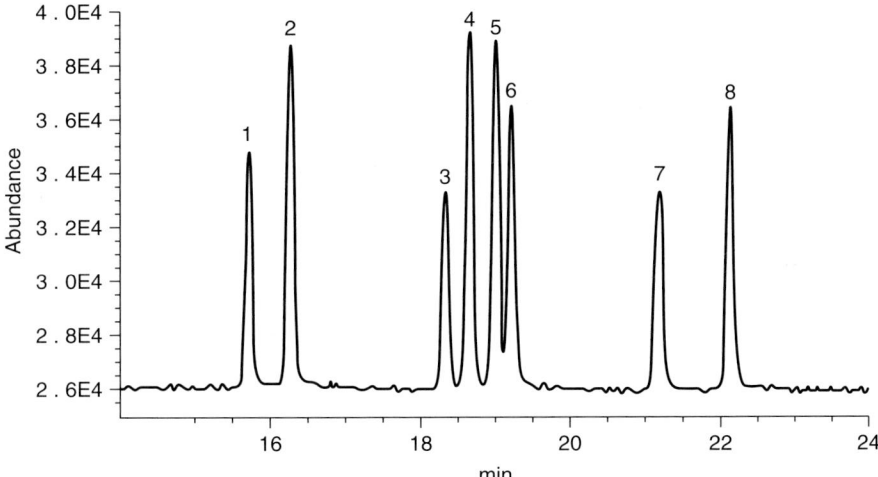

Fig. 30.6 Gas chromatogram of nanogram mixture of dioxins. Column: 35-m-long × 0.32-mm-i.d. fused-silica column, coated with 0.25-μm film of methyl phenyl cyanopropyl silicone. Column temperature: held for 0.5 min at 70 °C, then at 30 °C min^{-1} to 200 °C, then at 5 °C min^{-1} to 270 °C. Carrier gas: helium. Sample introduction: splitless injection. Detector: mass spectrometer, selected ion monitoring at m/z 320 and m/z 322. Peak assignments as noted in legend. (Quadrex Corp., 1990); reprinted with permission.

for the SP results in longer retention times. Also notice that within a series of like compounds (e.g., methyl, ethyl, isopropyl, and butyl acetate), elution is in the order of increasing boiling point on both columns, similar to the effect observed in Figure 30.5.

30.3.2.2 Gas–Solid (GSC)

GSC refers to GC performed on solid SPs. Whereas in GLC every attempt is made to minimize adsorption of solutes on active sites on the support (so that partition is based solely on liquid–liquid interactions), adsorption is the principal separation mechanism in GSC. Typical materials used as SPs include porous polymers, silica gel, diatomaceous earth (a natural form of silica), crushed firebrick, porous carbon, and zeolites (also commonly referred to as *molecular sieves*). One unique GSC separation using a molecular sieve can separate oxygen and nitrogen, as well as the other permanent gases (e.g., hydrogen, nitrogen, oxygen, argon, carbon monoxide, and carbon dioxide) and low-boiling compounds (e.g., methane, ethane, ethylene, and acetylene).

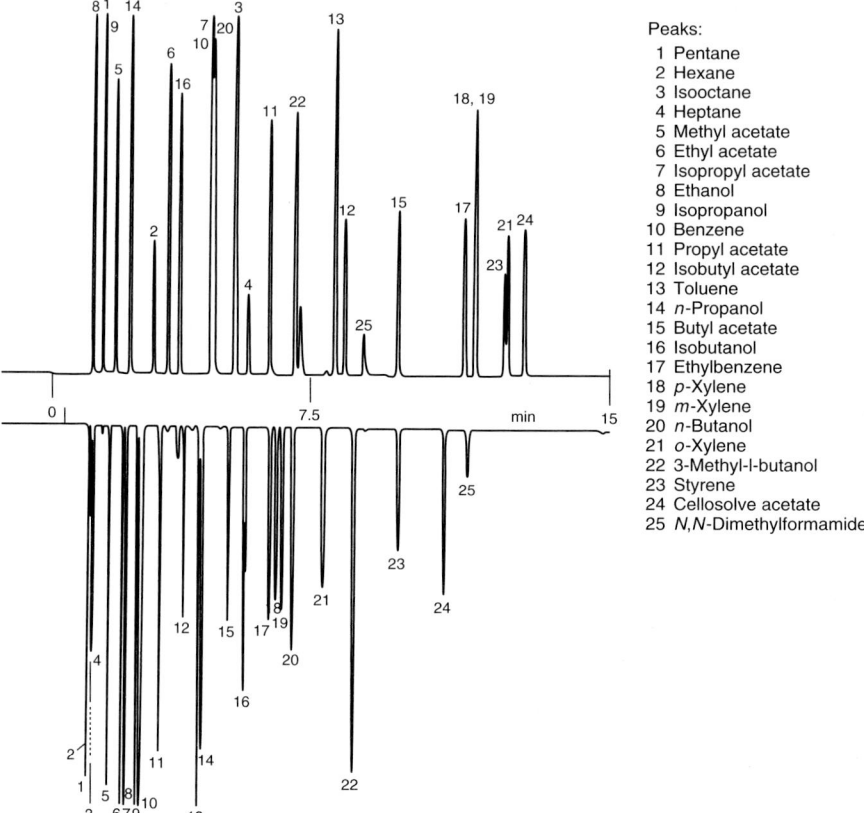

Fig. 30.7 Gas chromatograms of a mixture of organic solvents. Upper chromatogram column: 25-m-long × 0.32-mm-i.d. fused-silica column, coated with 1.0-µm phenyl vinyl methyl silicone film. Lower chromatogram column: 25-m-long × 0.32-mm-i.d. fused-silica column, coated with 0.50-µm polyethylene glycol film. Column temperature: held for 3 min at 35 °C, then at 5 °C min^{-1} to 100 °C. Carrier gas: helium, 1.5 mL min^{-1}. Sample introduction: split injection. Detectors: dual flame ionization (FID). Peak assignments as noted in legend. (HNU/Nordion Systems, 1990); reprinted with permission.

30.4
Liquid Chromatography

Modern LC, also referred to as HPLC, was introduced as an analytical separation technique in the early 1970s, and it has become the most widely used chromatographic technique. HPLC complements GC in that it provides rapid analysis of thermally labile, nonvolatile compounds that would be difficult or impossible to do by GC. Some typical examples of applications of HPLC are separation and quantitation of individual components in complex mixtures, analysis of nominally pure compounds for trace impurities, and isolation of pure compounds for synthetic and/or identification purposes. It can be used to analyze solids or liquids. A typical sample is 10 µl of a 0.1–100-mg/ml solution of the sample in a suitable solvent, although quantities as small as 10^{-6} g can be analyzed.

30.4.1 Instrumentation

A typical HPLC instrument is shown schematically in Figure 30.8. A basic, simple instrument is depicted; additional solvents, mixing valves, precolumns, and filters are often necessary, in addition. The following discussion is limited to the components shown in the figure.

30.4.1.1 Solvent Delivery Systems

Unlike GC where the MP is inert, the HPLC MP actively competes with the SP for analytes. A partitioning takes place between the two phases, so the composition of the MP is very important and may require mixtures of liquids. In one type of solvent delivery system, proportioning valves deliver precisely controlled volumes of solvent from reservoirs to the mixing chamber. A liquid chromatograph may have as many as four reservoirs, enabling solvent mixtures to be used as the MP.

The liquid MP is delivered from the mixing chamber to the top of the chromatographic column by a high-pressure pumping system with output capabilities as high as 40 MPa and flow rates in the range of 0.1–10 ml min^{-1}. Gradient elution techniques (*see* Section 30.4.2.5), where the composition of a two-, three-, or four-component MP is changed in a stepwise or continuous manner, are achieved by microprocessor control of the solvent proportioning valves. Gradient elution techniques in LC are used in much the same way that temperature programming is used in GC to enhance resolution. MPs commonly used in LC include water, dilute aqueous pH buffer solutions, acetonitrile, tetrahydrofuran, methanol, chloroform, hexane, and methylene chloride, as well as miscible mixtures composed of one or more of these solvents.

30.4.1.2 Injection Valves

Injector valves are designed to accommodate sample introduction of microliter or milliliter sizes. Sample loops are completely filled with sample while the MP

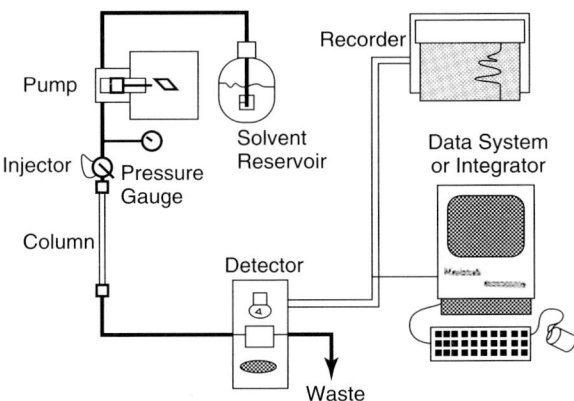

Fig. 30.8 HPLC schematic (Miller and Crowther, 2000); reprinted with permission.

is pumped through a different port on the valve. When the port to the sample loop is opened, the sample is flushed out of the loop by the MP and a precise amount of sample is delivered to the top of the chromatographic column. A typical injection volume is 1–100 µl. Automated systems based on these injection valves are commercially available.

30.4.1.3 Columns

The heart of the liquid chromatograph is the high-performance column on which the separation takes place. Typical columns are constructed from 3- to 25-cm lengths of 2–5-mm-i.d. stainless-steel tubing, although larger diameter columns (e.g., 60 cm) are used for preparative work and smaller diameter "microbore" columns (1 mm) for applications requiring advanced detectors (such as MS or Fourier-transform infrared (FTIR) spectroscopy). Columns are packed with small (3–50 µm) particles that constitute the SP. SPs with a variety of chemical and physical properties are available. As seen in Section 30.4.2, the choice of a SP/MP combination is dictated by the separation problem at hand.

Since chromatographic theory indicates that packed column performance will improve as the size of the SP particles is decreased, there has been a trend in column development toward smaller and smaller particles. There is a limit beyond which it is too difficult to pack these tiny particles and it is reached around 2- or 3-µm diameter. The smaller diameters that produce better performance also require higher pressures for fast analysis, and this situation has produced the latest version of HPLC commonly known as (*UHPLC*), *ultrahigh performance liquid chromatography*. UHPLC columns work at the limits of current technology, but they are the most efficient packed columns in use.

Another, alternative, commercial column introduced in 2001 is the monolithic column. Manufactured by a sol–gel process into a single stable rod, the monolithic columns so produced are enclosed in a plastic PEEK sheath. Compared to conventional microparticulate columns, the monolithic columns usually exhibit a lower pressure drop and can be operated with high efficiencies at higher flow rates.

30.4.1.4 Detectors

Most HPLC detectors are adapted from existing detection systems, many of which are spectroscopic. By far, the most common is based on the classic UV/Vis absorption spectrophotometer. As is the case with GC, MS is one of the most powerful and useful detectors to be used with HPLC, and it is discussed together with GC/MS. Table 30.2 summarizes some of the characteristics of the most common

Tab. 30.2 Common HPLC Detectors.

Spectroscopic
Ultraviolet absorption (UV/Vis)
Fixed wavelength UV
Variable wavelength UV
Photodiode array (PDA), also called
  diode array detector (*DAD*)
Fluorescence
Mass spectrometer (MS)

Electrical
Electrochemical detector (ECD);
  amperometric
Conductivity detector

Other
Refractive index (RI)
Evaporative light scattering detector
  (ELSD)

detectors and a brief discussion of some of them follows.

The refractive index (RI) detector functions as a nearly universal detector, that is, a detector that responds to nearly all analytes. This results from the fact that with the large number of available LC solvents, it is usually possible to design an experiment so that the analytes have refractive indices different from the MP. RI detectors are nondestructive and typical detection limits are about 10^{-6} g for a component in the injected sample.

UV/Vis absorbance detectors are among the most popular LC detectors because most organic compounds absorb 190–700-nm radiation, while many other compounds that do not absorb can be chemically reacted to form absorbing species. UV/Vis detectors are of three types: (i) fixed-wavelength detectors, which operate at only one wavelength (usually 254 or 280 nm, where a large percentage of organic compounds exhibit at least some absorbance); (ii) variable-wavelength detectors, which operate at any selected wavelength between 190 and 700 nm; and (iii) photodiode array (PDA) detectors, which can monitor the entire wavelength region between 190 and 700 nm.

PDA detectors have found increasing use in the past several years because of their utility in qualitative identification of unknowns by their characteristic UV/Vis absorbance spectra. UV/Vis detectors are nondestructive, and for a strongly absorbing compound, the lower limit of detection is about 10^{-9} g. Further details on UV/Vis absorbance can be found elsewhere in the Encyclopedia.

Fluorescence detectors are more specific than UV/Vis detectors because not all compounds that absorb light subsequently fluoresce. The use of a fluorescence detector restricts the choice of a MP to one that does not fluoresce, and minute traces of fluorescent materials commonly present in solvents can cause problems. Fluorescence detectors are nondestructive and have detection limits of about 10^{-10} g. Further details on molecular fluorescence can be found elsewhere in the Encyclopedia.

There are several electrochemical detectors (ECDs). One measures the electrical conductivity of the LC eluent by applying an alternating voltage between two electrodes and measuring the resulting current. Conductivity is related to the current–voltage ratio and represents the contribution of the individual equivalent conductivities of all ions in solution. Another is amperometric in principle and causes a chemical transformation of the analyte. The effluent from the column passes a working electrode (one of three electrodes in the detector cell) set at a specific potential with respect to the reference electrode. If a species in the effluent can be oxidized or reduced at this potential, the transfer of electrons at the electrode–effluent interface serves as the detection signal. The main advantages of the amperometric detector are sensitivity (10^{-12} g) and selectivity (by prudent choice of the potential).

In the light-scattering detector (LSD), also known as the *evaporative light-scattering detector* (*ELSD*), the effluent from the column is nebulized and the resulting fine mist is introduced into a stream of preheated gas. The mist is passed through a high-temperature drift tube where the volatile solvent is vaporized, leaving very fine particles of solute in the carrier gas stream. Laser light scattered by these particles is detected by a PDA. These detectors are meeting the need for a universal detector, which has lower detection limits than the RI detector and is compatible with gradient elution techniques, unlike RI.

Other detectors are based on specific characteristics of the analyte (radioactivity, rotation of plane of polarized light). It should also be noted that the effluent leaving a nondestructive detector (RI, UV/Vis, fluorescence, FTIR) can be collected (in effluent fractions corresponding to separated components) and analyzed off-line by other analytical techniques, such as MS, FTIR spectroscopy, or nuclear magnetic resonance (NMR) spectroscopy. This very common approach for combining powerful separation and identification tools can allow unambiguous identification of unknown components in a complex mixture.

30.4.2
Modes of Separation

As was shown in Figure 30.1, the classification of HPLC methods is not primarily according to the states of the two phases, as was the case with GC, but is more extensive. The exception is liquid–solid chromatography (LSC). Other modes, as shown in Figure 30.9, usually carry the name of the mechanism by which the separation is achieved. Also note in Figure 30.9, that another difference in the classifications, compared with GC, is the identification of those methods performed on plane surfaces, not in columns. The major one of this type is thin-layer chromatography (TLC), which is discussed separately from the column methods in Section 30.4.4.

30.4.2.1 Liquid–Solid Chromatography (LSC)

LSC is also called *adsorption chromatography* since the mechanism of the separation is mainly adsorption. It is the oldest of the various LC modes and was the one used by Tsweet. It combines a polar SP (silica or alumina) with a nonpolar or slightly polar MP (e.g., heptane, hexane, or pentane, possibly mixed with a more polar solvent like methylene chloride, chloroform, or isopropanol). Variable amounts of trace

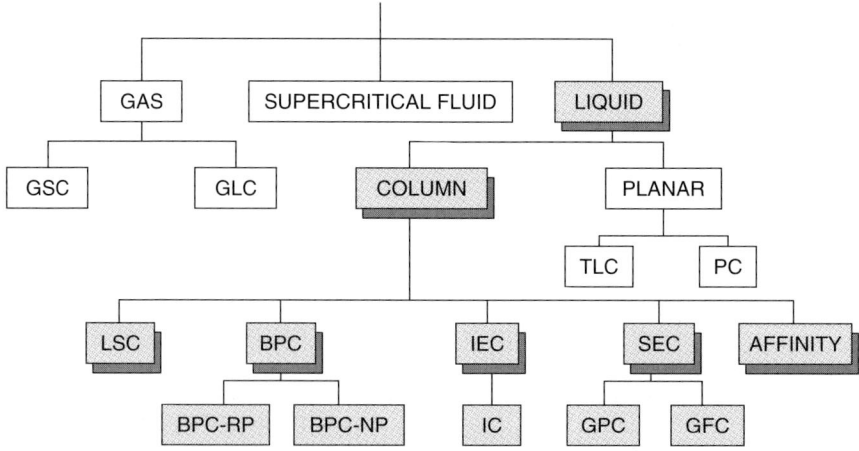

Fig. 30.9 Classification of liquid chromatographic techniques (Miller and Crowther, 2000); reprinted with permission.

water in these commonly used solvents presented a major problem with stability and reproducibility, which led to its loss of popularity. Since this system was originally the major one in use, it is also referred to as *normal phase* (NP) HPLC. Ironically, that name is still used even though LSC is currently not the normal mode of operation.

In general, polar molecules are strongly adsorbed on these SPs; therefore, they are the last to elute from the column. The opposite is true for nonpolar molecules. In Figure 30.10, a NP

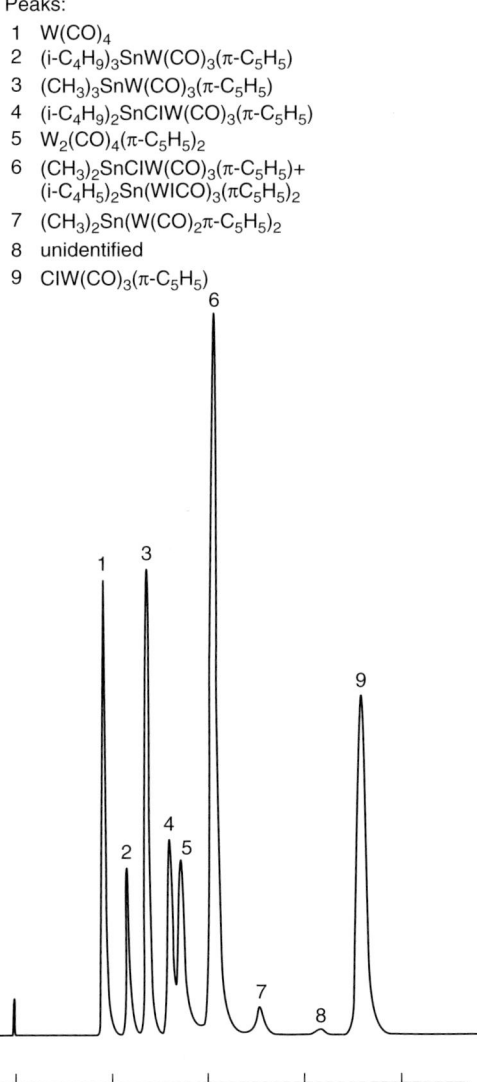

Peaks:
1 $W(CO)_4$
2 $(i\text{-}C_4H_9)_3SnW(CO)_3(\pi\text{-}C_5H_5)$
3 $(CH_3)_3SnW(CO)_3(\pi\text{-}C_5H_5)$
4 $(i\text{-}C_4H_9)_2SnClW(CO)_3(\pi\text{-}C_5H_5)$
5 $W_2(CO)_4(\pi\text{-}C_5H_5)_2$
6 $(CH_3)_2SnClW(CO)_3(\pi\text{-}C_5H_5)+$
   $(i\text{-}C_4H_5)_2Sn(WICO)_3(\pi C_5H_5)_2$
7 $(CH_3)_2Sn(W(CO)_2\pi\text{-}C_5H_5)_2$
8 unidentified
9 $ClW(CO)_3(\pi\text{-}C_5H_5)$

Fig. 30.10 Normal-phase liquid chromatogram of a mixture of tungsten carbonyl and tungsten tin carbonyl complexes. Column: 15-cm long × 4.6-mm i.d., 3-μm particles, cyano. Mobile phase: 10 : 90 mixture of tetrahydrofuran:hexane, 2 mL min^{-1}. Detector: UV absorbance at 220 nm. Peak assignments as noted in legend (Dewaele, 1990); reprinted with permission.

chromatogram of a mixture of tungsten carbonyl complexes and tungsten tin carbonyl complexes is shown (Dewaele, 1990). This chromatogram was obtained on a cyano column with detection based on UV absorbance at 220 nm. While LC is most commonly used for separation of organic materials, this example demonstrates its utility in resolving a complex synthetic mixture of organometallic compounds. While normal-phase chromatography (NPC) is particularly suited to nonpolar or moderately polar compounds that are not water soluble, it is not restricted to these cases.

30.4.2.2 Bonded-Phase Chromatography (BPC)

LSC and LLC have been replaced for the most part by bonded-phase chromatography (BPC) in which a liquid phase is chemically bonded to a solid support to become the SP. The main advantage of BPC compared with LLC is that the SP is stable and not susceptible to physical or chemical removal by action of the MP. The most common SPs are those bonded to siloxanes (see Table 30.3), although organic resins are returning to use after a decline in popularity in the 1970s and 1980s. Siloxane-based packing materials are prepared by silanization of the surface hydroxyl groups on the silica particles with alkylchlorosilane compounds. For example, an octadecyl column would be prepared by reaction of the silica particles with a solution of dimethyloctadecylchlorosilane to produce a chemically modified surface. It will be seen that the functional group bonded to the SP critically affects the types of separations that can be performed; some are for NPC, and others are for phases of opposite polarity, called *reversed phase* (*RP*).

30.4.2.3 Reversed Phase Chromatography (RPC)

The production and use of BPC in which the SP is nonpolar has become very popular. Since the polarity of such columns is the opposite or reverse of those used in LSC, their use has been designated as reversed-phase chromatography (RPC). This definition applies to nonpolar columns, whether or not they are the bonded type. The MP used with them is polar, making RPC complementary to NPC.

Typically in RPC, a hydrocarbon-bonded support, such as methyl-, octyl-, or octadecyl-siloxane (ODS), is used with an MP composed of water, methanol, acetonitrile (ACN), tetrahydrofuran, or a mixture of some or all of these solvents.

Tab. 30.3 Bonded-phase chromatography stationary phases.

Designation	Surface chemical formula
Silica (Si)	Si-OH
Amino (NH_2)	Si-O-Si(OH)$_2$-(CH$_2$)$_3$-NH$_2$
Cyano (CN)	Si-O-Si(CH$_3$)$_2$-(CH$_2$)$_n$-CN
Methyl (C_1)	Si-O-Si-(CH$_3$)$_3$
Phenyl (C_6)	Si-O-Si(CH$_3$)$_2$-C$_6$H$_5$
Octyl (C_8)	Si-O-Si(CH$_3$)$_2$-(CH$_2$)$_7$-CH$_3$
Octadecyl (C_{18})	Si-O-Si(CH$_3$)$_2$-(CH$_2$)$_{17}$-CH$_3$

In RPC, the more polar materials have greater affinities for the MP and so elute first.

Consider, for example, the RP chromatogram of a mixture of sugars shown in Figure 30.11. This chromatogram was obtained on an amino column with gradient elution using an acetonitrile/water MP and detection based on laser light scattering (Letter, 1991). The individual sugars elute in three groups, with the low molecular weight (MW = 180) fructose and glucose molecules first, the higher molecular weight (MW = 342) sucrose, maltose, and lactose molecules next, and the highest molecular weight (MW = 504) raffinose molecule last. For these sugar molecules, polarity (and order of elution) is a function not only of the molecular weight but also of the relative position of the polar hydroxyl and hydroxymethyl groups on the molecule. The latter accounts for the excellent separation even among molecules with the same molecular weight.

30.4.2.4 Ion-Pair Chromatography (IPC)

When ions are present in samples run by RPC, they typically elute very quickly (with little or no retention) because they do not interact appreciably with the nonpolar SP. As a consequence, they are not separated from one another. One way to remedy this situation is to add a counterion to the MP and form ion pairs that are uncharged. Reversed-phase ion-pair chromatography (RPIPC) is performed with conventional RP LC columns and with polystyrene-divinylbenzene resins that have no ion-exchange sites. The MP contains ions that combine with the sample ions to create neutral ion pairs.

For example, the separation of a mixture of carboxylic acids by RP could be done using a MP that has a low pH, so that the acids would be present

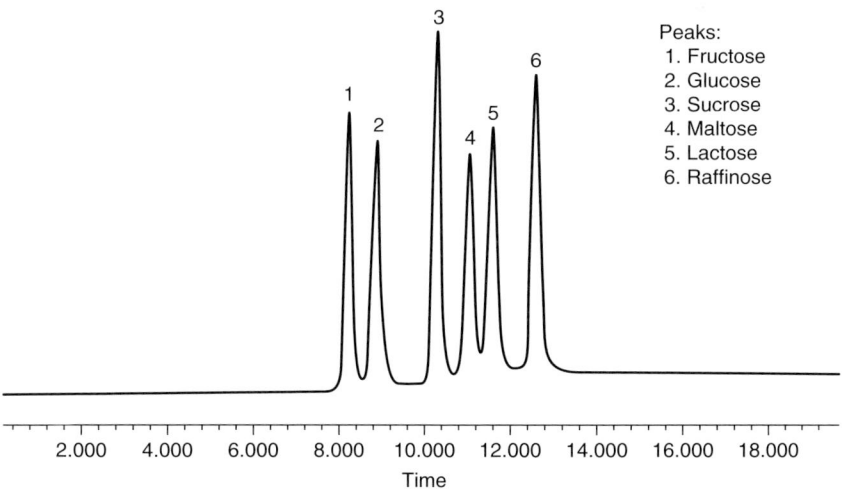

Fig. 30.11 Reversed-phase liquid chromatogram of a mixture of sugars. Column: 25-cm long × 4.6-mm i.d., 10-μm particles, amino. Mobile phase: initially 85 : 15 acetonitrile:water, then linear gradient to 60 : 40 acetonitrile:water over 10 min; 1.5 mL min^{-1}. Detector: laser light scattering. Injection size: 20 μL. Peak assignments as noted in legend (Letter, 1991); reprinted with permission.

as neutral carboxylic acids. However, if it were necessary to run the sample at a pH above that of the pK_a of the acids, the acids would be present as the carboxylate anions. To make them neutral, the addition of a cationic tetraalkylammonium ion would form neutral ion pairs with the carboxylate anions and they would likely be separated by this methodology, RPIPC.

30.4.2.5 Gradient Elution

Isocratic separations, where the MP is unchanged throughout the chromatographic process, often give poor resolution of early-eluting peaks, significant band spreading of late-eluting peaks, and unnecessarily long separation times. The solution to these problems is gradient elution, where the MP composition is continuously changed during the separation. It is used to speed the elution of strongly retained compounds or to retard the elution of early-eluting compounds. For optimum resolution and speed of analysis time, it is necessary to begin with a MP just strong enough to dissolve the most soluble species in the sample, and finish with a MP strong enough to remove the most insoluble species from the column. The resulting separation is optimized similar to the way GC separations are optimized by programmed temperature operation.

30.4.2.6 Ion-Exchange Chromatography (IEC) and Ion Chromatography (IC)

IC and ion-exchange chromatography (IEC) are used for the separation of ionic compounds, inorganic and organic. IEC is performed on SPs that possess ionic sites on the surface, using aqueous, ion-containing MPs. Columns are packed with small (5–25 μm) particles of polystyrene-divinylbenzene resin or (less commonly) silica whose surfaces are chemically modified for specific applications. For anion analysis, the SP particles have surface cationic sites such as amines (NR_3^+). For cation analysis, the SP has surface anionic sites such as sulfonates (SO_3^-).

In classical IEC, separation is based on ion exchange of sample ions (X) and MP ions (Y) with the charged groups (S) of the SP. If anions are to be separated (anion exchange), the SP contains cationic groups and the reaction in the column follows Eq. (30.9).

$$X^- + S^+Y^- \longrightarrow Y^- + S^+X^-$$

(anion exchange) (30.9)

Conversely, to separate cations (cation exchange), the stationary contains anionic groups as shown in Eq. (30.10).

$$X^+ + S^-Y^+ \longrightarrow Y^+ + S^-X^+$$

(cation exchange) (30.10)

Sample ions are in competition with MP ions for sites on the ion-exchange resin that serves as the SP. Sample ions that do not bind as strongly to the resin as MP ions will elute early in the chromatogram, while sample ions that interact more strongly will be retained and elute later in the chromatogram.

A significant advance in the practice of IEC resulted from the work of Small, Stevens and Bauman (1975) who modified IE methods so that universal detection of ions at very low concentrations became possible. The method, called *IC*, uses a conductivity detector and a second column that permits the reduction of the background conductivity of the MP to reduce the magnitude of the baseline signal. The sample must be in aqueous form for introduction into the chromatograph. A typical sample is 50 μL of a 0.1–100-mg/L solution of the sample in a suitable solvent.

MPs commonly used in IC are dilute (approximately 10^{-3}–10^{-5} molar) aqueous solutions of ionizable materials such as sodium hydroxide, hydrochloric acid, mixtures of sodium carbonate and sodium bicarbonate, and mixtures of organic acids such as oxalic acid, citric acid, and tartaric acid. In RP IC, water-soluble organic solvents such as methanol and acetonitrile can be added to the MP. Quantities as small as 10^{-6} g can be analyzed.

30.4.2.7 IC Instrumentation

Measuring the small conductivity due to an eluting ion in the presence of the high background conductivity of the MP is difficult. Suppression of the conductivity of the MP, or eluent, dramatically improves the sensitivity (Small, Stevens and Bauman, 1975). Consider the example where a sodium hydroxide MP is used to separate anions. A low concentration of sample anions exits from the column containing the SP along with a high concentration of sodium and hydroxide ions. The column effluent is then passed through a second cation-exchange column (suppressor column), with hydrogen ions attached to the ion-exchange sites on the resin. Sodium ions displace hydrogen ions from these sites, with two beneficial results. First, water is formed by the reaction

$$Na^+OH^- + S^-H^+$$
$$\longrightarrow H_2O + S^-Na^+ \quad (30.11)$$

Therefore, the effluent from the suppressor column has the very low conductivity of deionized water (as opposed to the high conductivity of a sodium hydroxide solution) and provides an excellent background for sensitive conductivity measurements on the anions. Second, the sample anions are paired with hydrogen ions instead of sodium ions as a consequence of the reaction

$$Na^+X^- + S^-H^+$$
$$\longrightarrow H^+X^- + S^-Na^+ \quad (30.12)$$

As a result, detector sensitivity to the analyte anion is maximized because the equivalent conductivity of the hydrogen ion is approximately seven times as large as that of the sodium ion. The situation for cation analysis is completely analogous. Here, an anion-exchange suppressor column loaded with hydroxide ions is used to suppress the conductivity of the MP. The development of fiber suppressors and membrane-type suppressors simplified the procedures for this type of detection and provided improvements in resolution and sensitivity compared to the use of column suppressors.

Nonsuppressed conductivity uses the same conductivity detector just described, but without suppression of the MP conductivity. This technique for ion analysis is sometimes referred to as *single-column IC*. Low background conductivity is achieved by use of low concentrations (10^{-3}–10^{-5} molar) of MPs composed of ions that have a high affinity for the SP. UV/Vis absorbance detectors can be used to detect ions that absorb radiation in the 190–700-nm range. Alternatively, light-absorbing species produced by post-column reaction of a nonabsorbing ion with a complexing reagent can be detected. An example of this is absorbance detection at 520 nm of the complexes formed between 4-(2-pyridylazo)-resorcinol (PAR) and transition-metal ions such as Fe(II), Fe(III), Cu(II), Ni(II), Zn(II), Co(II), and Pb(II). Finally, absorbance detection can be accomplished using a UV/Vis-absorbing MP (e.g., phthalate buffers) to detect nonabsorbing ions. Here, eluting ions decrease the measured absorbance and are observed as negative peaks in the chromatogram. Limits of detection for electrochemical and UV/Vis detectors are approximately the same as for LC.

In Figure 30.12, the ion chromatogram of an aqueous mixture of 34 anions, both inorganic and organic, is shown. This chromatogram was obtained with suppressed conductivity using a ternary solvent system of sodium hydroxide.

In Figure 30.13, the separation of alkali-metal cations, as well as of the ammonium ion, is achieved using a cation-exchange resin. This chromatogram was obtained on a surface-sulfonated cation-exchange resin using dilute hydrochloric acid as the MP and eluent-suppressed conductivity detection. The observed retention behavior results from the increasing affinity of the cations for the SP as the ionic radius is increased.

30.4.2.8 Size-Exclusion Chromatography (SEC)

Size-exclusion chromatography (SEC) is used to separate molecules on the basis of their physical size. SEC with aqueous MPs is sometimes referred to as *gel filtration chromatography* (*GFC*), while SEC with nonaqueous MPs is sometimes called *gel permeation chromatography* (*GPC*). SPs are porous silica microspheres (e.g., 5-μm particles) or polymer

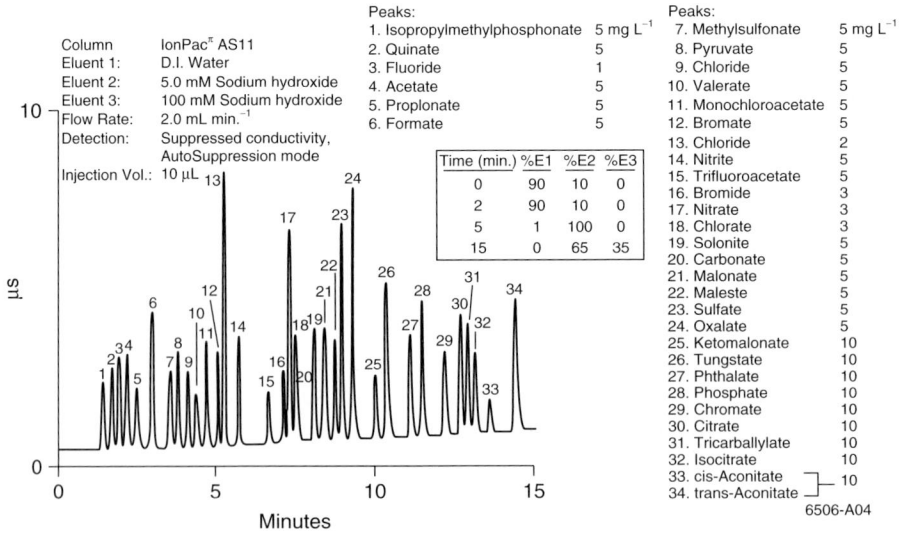

Fig. 30.12 Ion-exchange separation of anions; reprinted with permission of Dionex Corporation, Sunnyvale, CA.

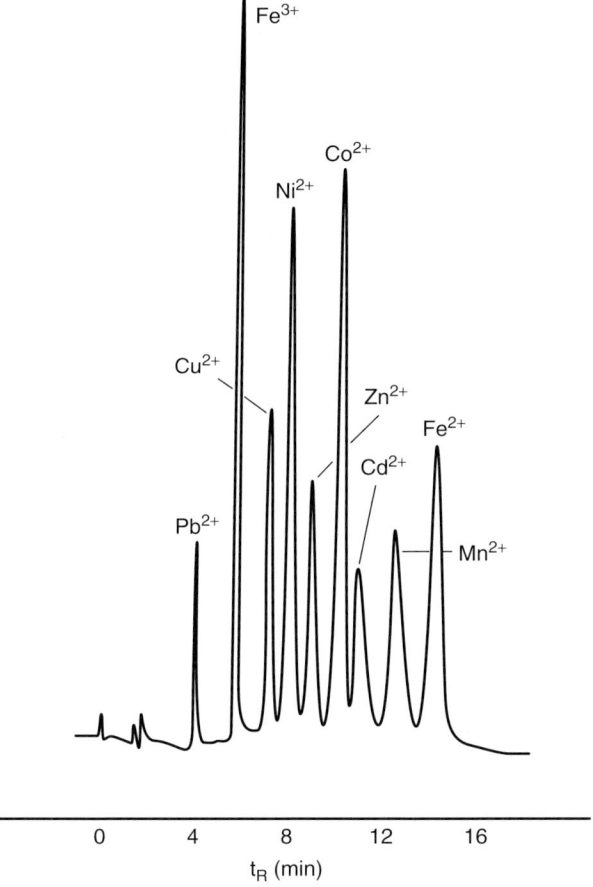

Fig. 30.13 Separation of nine transition-metal cations by ion chromatography; reprinted with permission of Dionex Corporation, Sunnyvale, CA.

beads that have porous microstructures with very well controlled pore size distributions. Typical pore sizes in these particles range from 6 to 100 nm. Molecules that are too large to permeate the pores in the particle travel around the particles and appear first in the chromatogram. Small molecules that can permeate the pores travel a long and tortuous path through the particles; therefore, they have longer retention times. MPs are chosen on the basis of their solvating power for the sample. In order to restrict the retention behavior to size-exclusion effects, it is essential that the components in the sample have a very high solubility in the MP. In the size-exclusion chromatogram shown in Figure 30.14, the components elute in the order of decreasing molecular weight (Yost, Ettre and Conlon, 1980). An important application of SEC is the analysis of molecular weight distributions of polymeric materials.

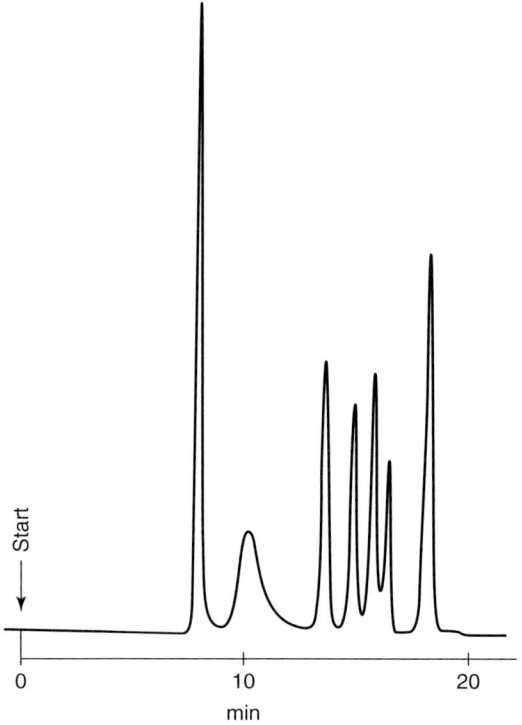

Fig. 30.14 Size-exclusion chromatogram of a mixture of organic molecules and polystyrene molecular weight standards. Two columns in series: 25-cm long × 8-mm i.d., divinylbenzene gel stationary phase. Mobile phase: tetrahydrofuran, 1 mL min^{-1}. Detector: UV absorbance at 260 nm. Peak assignment in order of elution: 1, polystyrene (molecular weight (MW) = 2, 400); 2, polystyrene (MW = 2, 100); 3, dioctyl phthalate (MW = 391); 4, dibutyl phthalate (MW = 278); 5, diethyl phthalate (MW = 222.2); 6, dimethyl phthalate (MW = 194); 7, benzene (MW = 78); (Yost et al., 1980); reprinted with permission.

30.4.3
Preparative Liquid Chromatography

The objective in preparative LC is to isolate significant quantities of chemical compounds in a high-purity form. The amount of material can be very large (grams), as might be required for synthetic applications, or quite small (milligrams) for the purpose of chemical identification and analysis by other instrumental methods. For small-scale preparative LC, a standard liquid chromatograph can be used, with the addition of high-sample-capacity columns (diameters of 0.7–2.5 cm) if necessary. Large-scale industrial preparative liquid chromatographs have large MP reservoirs, large flow-rate capabilities, large-capacity columns (some as high as 60 cm in diameter), and automated fraction collection systems. In preparative LC, the capacity of the column is frequently exceeded, so that resolution is much poorer than would be realized in a normal analytical separation. A common strategy is to load as much sample onto the column as possible without degrading the resolution to the point where it is not possible to isolate a pure fraction. This strategy minimizes the number of repeated injections and fraction collections necessary to obtain the required quantity of purified material. Prep LC continues to develop and some recent papers can be found in the *Journal of Chromatography A* (2003, Vol. 1006. Seidel-Morgenstern, A.).

Preparative separations are especially useful in chiral separations whereby production quantities of chiral drugs can be produced. For more information about chiral separations, see Section 30.7.2.

30.4.4
Thin-Layer Chromatography

Most of the techniques shown in the chromatographic classification schemes in Figures 30.1 and 30.9 are carried out in columns. The exception is the one-labeled planar, consisting of two techniques, paper chromatography (PC) and TLC. PC was the older technique and it has been largely replaced by TLC. In classical TLC, the SP is dispersed on the surface of a planar surface like a square glass plate, and the samples are applied as spots on one edge of the plate using a pipet or an eye-dropper. The plate is then dipped in a solvent reservoir and MP-flow through the SP occurs as a result of capillary forces. The components in the sample are spatially separated on the plate and are detected by charring, chemical staining, UV/Vis reflectance, radiographic, or fluorescence techniques. The physical and chemical mechanisms for separation in TLC are the same as those in modern column LC, but there are some small differences due to the fact that the MP is moving on a dry bed and the flow is not controlled. The most common SP is silica gel, which means that the separation mode is most often NP. However, RP TLC plates are also available, and most of them are available in a small particle size, giving rise to higher performance, so-called high-performance thin-layer chromatography (HPTLC).

Instrumental and automated methods for TLC have been developed. New techniques for applying the sample to the plate have also dramatically improved the resolutions that can be achieved. Instrumental methods for recording chromatograms such as densitometry, UV/Vis absorbance, and fluorescence have reduced the time required for analysis, while providing detection limits as low as 10^{-12} g (using fluorescence detection). Most dramatic are instrumental methods for forced flow of the MP, which can reduce separation times and improve separation efficiencies, because smaller SP particles and larger plate sizes can be used. The most popular of these methods is called *overpressurized layer chromatography* (*OPLC*) the instrumentation for which has been available for about 25 years. Excellent details are given in the chapter on OPLC in Sherma and Fried's, 2003 *Handbook*, a useful reference for all of TLC. Instrumental methods are more popular in Europe than in the United States, where most TLC is still performed manually. The determination of acrylamide in drinking water by HPTLC combined with MS is a typical example of instrumental TLC (Alpmann and Morlock, 2008); the authors show that their results are comparable to those obtained by HPLC/MS.

30.5
Supercritical Fluid Chromatography (SFC)

SFC is not as popular as GC or HPLC. The name implies that the MP is a supercritical fluid and has properties between a gas and a liquid. The classification system shown earlier in Figure 30.1 appropriately shows SFC placed between GC and LC. It was introduced in the early 1960s by Klesper, Corwin and Turner (1962) at a time when HPLC was being developed and some chromatographers used sufficiently high pressures to create supercritical fluids. Advocates of SFC claim that it has the best attributes of GC and HPLC, and that it has the performance characteristics of a *unified* chromatographic system. Carbon dioxide is the most commonly used MP, and therefore SFC more closely resembles normal-phase HPLC.

In an age of concern about the environment, its use is highly desirable and has caused SFC to be labeled as a green process (Anon, 2007). This characteristic is especially important for preparative work.

Numerous separations and analyses of high-molecular-weight, thermally unstable, relatively nonvolatile compounds not suitable for analysis by GC were pursued during the remainder of the decade. However, with the advent of the more convenient method of HPLC in the 1970s, interest in SFC subsided rapidly. There was renewed interest with the introduction of the first commercial SFC instrument in 1982, and since then there have been technological advances in SFC instrumentation that have further popularized the technique. Applications that would be difficult or impossible to do by either GC or HPLC have been shown to be amenable to analysis by SFC (Griffiths, 1988). Examples include the separation of lipids and phospholipids, pharmaceutical products, as well as fuel oil distillates that can be separated into fractions corresponding to particular chemical families (e.g., paraffins, olefins, naphthelenes, and aromatic compounds). SFC can be used to analyze solids and liquids in amounts ranging from 10^{-2} to 10^{-12} g.

30.5.1
Operating Characteristics

A supercritical fluid is a substance that is above its critical temperature and pressure. By far the most common supercritical fluid used as a chromatographic MP is carbon dioxide, which has a critical point of 31 °C and 7.4 MPa. At the critical point, carbon dioxide has a density of 0.448 g/mL. Since carbon dioxide is relatively nonpolar, modifiers such as methanol, ethanol, acetonitrile, propylene carbonate, and water are sometimes added to increase MP polarity and therefore the solvating power for some samples. Other less commonly used MPs in SFC include ammonia, nitrous oxide, and sulfur dioxide.

SFC requires well-regulated control of both temperature and pressure. The high pressures are usually achieved by compression of a gas (e.g., carbon dioxide) with a high-pressure syringe pump. The condensed MP is then heated to supercritical temperatures. Some properties of supercritical fluids, such as solvating power, diffusion, and viscosity, are functions of density (Smith, Wright and Yonker, 1988). Density programming (also called *pressure programming*, since MP density is a function of pressure) is therefore used to improve resolution in SFC separations in an analogous way to that in which temperature programming and gradient elution are used in GC and LC, respectively. Pressure programming is achieved by microprocessor control of the pressure in the fluid delivery system.

SFC instrumentation can be modeled after either GC or LC instrumentation and both types are available commercially; the columns used can be packed or OT. Packed-column instruments resemble HPLC instruments and OT column instruments resemble GCs. In either case, a restrictor is required at or near the exit to keep the pressure high enough to keep the fluid above its critical pressure. The simplest restrictor is a 10-μm-i.d. capillary tube joined to the column outlet by a zero-dead-volume union fitting. Instruments using packed columns have become more popular, and, like HPLC, the most popular detector is UV. SFC is flexible: preparative separations are possible as well as hyphenated SFC/MS instrumentation.

30.5.2
Modes of Separation

30.5.2.1 Isobaric SFC

The simplest form of SFC is performed at constant pressure, isobaric. As in GC and LC, the selection of a MP/SP combination is based on the requirements of the separation. Among the more important factors are sample polarity and solubility of the sample in the supercritical fluid. In Figure 30.15, the supercritical fluid chromatogram of a mixture of hydroxybenzoic acids is shown (Berger and Deye, 1990). This chromatogram was obtained on a NP (cyano) LC column with a carbon-dioxide-based MP. Methanol and citric acid are added to the supercritical carbon dioxide to increase the solubility of the polar hydroxybenzoic acids in the MP. In the absence of these additives, the elution of the more polar, strongly retained components would occur with poorer resolution and only after much longer times.

30.5.2.2 Pressure Programming

Isobaric separations often suffer from the same problems that occur in isothermal GC and isocratic LC. The solution to these problems is pressure programming, where the density of the MP (and therefore the solubility of a given solute in the MP) is continuously increased during the separation by increasing the pressure. In Figure 30.16, three supercritical fluid chromatograms of a solution of the surfactant Triton X-114 are shown (Giorgetti et al., 1989). These chromatograms were obtained on an RP (C18) LC column with a carbon-dioxide-based MP. The designator n in the figure denotes the number of monomer units in each polymeric surfactant molecule. Methanol was added to the supercritical carbon dioxide to increase the solubility of the surfactants in the MP. Comparison of Figure 30.16(a) and (c) clearly illustrates the beneficial effect that pressure programming can have on resolution. In addition, comparison of Figure 30.16(a) and (b) demonstrates the improvement in resolution that can be realized by increasing the methanol content in the MP in a linear fashion, much like gradient elution in LC.

30.6
Spectrochemical Detection

In prior sections, several spectroscopic detectors have been described, but one of them is of special importance: MS. The combination of a chromatographic instrument interfaced to MS is often referred to as a *hyphenated method* and designated as GC–MS or GC/MS, HPLC/MS, and so on. Multiple hyphenation, as in GC/MS/MS, is also common and sometimes called *hypernation*. These combinations can be considered to be multidimensional as well, and that topic is covered in Section 30.7.1.

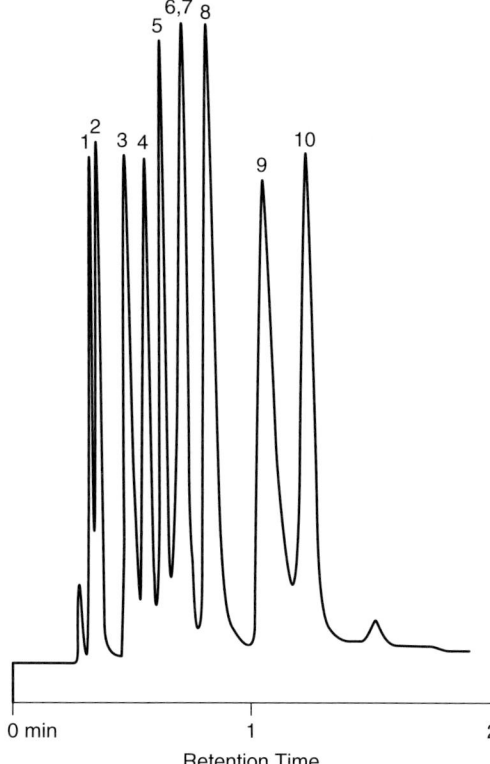

Peaks:
1. Benzoic acid
2. 2-Hydroxybenzoic acid
3. 2,6-Dihydroxybenzoic acid
4. 3-Hydroxybenzoic acid
5. 4-Hydroxybenzoic acid
6. 2,4-Dihydroxybenzoic acid
7. 2,5-Dihydroxybenzoic acid
8. 2,3,4-Trihydroxybenzoic acid
9. 2,4,6-Trihydroxybenzoic acid
10. 3,5-Dihydroxybenzoic acid

Fig. 30.15 Supercritical fluid chromatogram of a mixture of hydroxybenzoic acids. Column: 10-cm long × 2-mm i.d., normal phase, 7-µm particles, cyano. Column temperature: 60 °C. Mobile phase: carbon dioxide, containing 7.4% added methanol and 0.04% added citric acid. Pressure: 20 MPa. Flow rate 1.0 mL min^{-1}. Sample introduction: 0.2 µL on-column injection. Detector: UV/Vis absorbance at 230 and 430 nm. (Berger and Deye, 1990); reprinted with permission.

An alternative way of describing these combinations is simply to consider the mass spectrometer to be the detector attached to the chromatographic system. Thus, it is sometimes called the *MSD*, or the *mass selective detector*. Mass spectroscopists could also consider the chromatograph to be a special inlet system for MS. No significance is attached to either concept in this section, which is mainly concerned with GC/MS and LC/MS. Later, some discussion is included on the other hyphenated MS methods and other chromatographic combinations such as those with FTIR and NMR spectrometry.

With MS, it is possible to determine the molecular weight and molecular formula of an unknown compound.

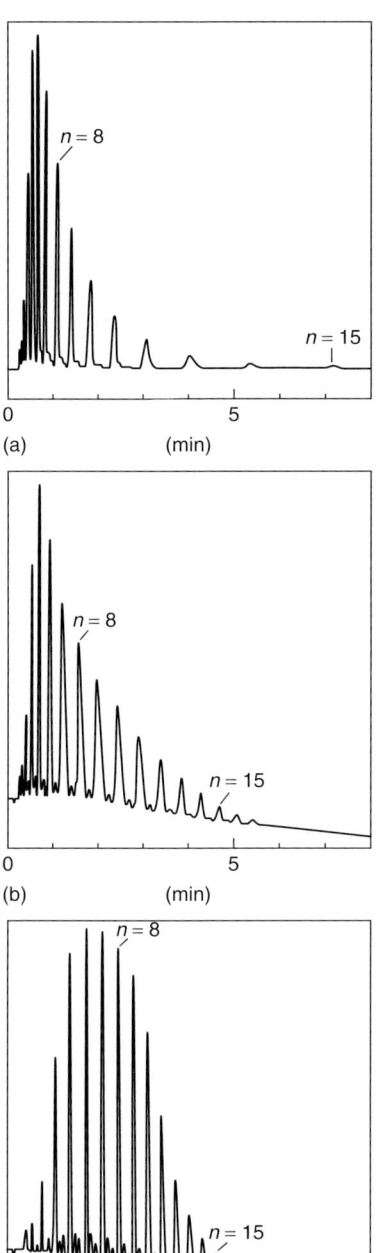

Fig. 30.16 Supercritical fluid chromatograms of the surfactant Triton X-114. Column: 10-cm long × 2-mm i.d. reversed-phase, 3-μm particles, C_{18}. Column temperature: 170 °C. Mobile phase: carbon dioxide with methanol modifier. Flow rate: 2 mL min^{-1}. Sample injection loop. Detector: UV absorbance at 278 nm. (a) Isobaric at 21.0 MPa, constant modifier flow (0.125 mL min^{-1}). (b) Isobaric at 21.0 MPa, modifier gradient (0.025–0.4 mL min^{-1} over 8 min, linear). (c) Pressure program (13.0–37.5 MPa in 8 min), constant modifier flow (0.125 mL min^{-1}) (Giorgetti et al., 1989); reprinted with permission.

In addition, characteristic fragmentation patterns (produced by sample ionization) can be used to deduce molecular structure. Spectrometers in common use with chromatographic systems include quadrupole analyzers, quadrupole ion-trap mass spectrometers, and magnetic/electrostatic dual-focusing instruments. Time-of-flight (TOF) mass spectrometers are becoming increasingly popular. Typical detection limits of about 10^{-12} g can be realized with MS.

Data from a conventional mass spectrometer can be collected in two ways, total ion chromatogram (TIC) and single ion, called *single ion monitoring* (*SIM*). In TIC (also called *scan mode*), all ions detected are displayed as one signal, so the output looks just like a chromatogram from any other chromatographic detector. Any peak in this TIC can be examined to see what ions it contains, and this information is used primarily to identify unknowns as just described. By comparison, the SIM output is set to display only those ions (of a given mass-to-charge ratio) that have been chosen for collection. So, for example, if the mass of a particular analyte or analyte fragment is chosen as the single ion, the only peaks in the chromatogram will be for peaks that contain that ion. If a characteristic fragment ion is chosen, all analytes that fragment to give that ion, and only that ion, will appear in the chromatogram. Up to six ions can be acquired using conventional MS software. SIM is more selective and sensitive than TIC. Consider the difference between the acquisition of data over a range, which covers all possible ions (say 43–500), for a TIC, compared to one that collects data for only six ions, for a SIM. The former collects data for 76 times as many ions (457/6), so much more time is spent on each ion in SIM, increasing its sensitivity. Consequently, SIM is used primarily for quantitative analyses, often at trace levels.

Conventional chromatographic columns are normally operated so that the column exits are at atmospheric pressure, although it is possible to run a GC with a vacuum outlet. By comparison, mass spectrometers must be run under high vacuum. Thus, the process of joining the two involves an interface that can take a chromatographic effluent at atmospheric pressure and reduce it to a vacuum condition (about 10^{-6} torr). As will be seen, this process is different in GC and HPLC, but has been ideally achieved in both cases. It is much easier to couple a GC effluent to an MS than an HPLC effluent, which contains a liquid MP and often a buffer both of which have to be removed or enriched or exploited (solvent-assisted ionization).

Consequently, GC/MS was the first hyphenated technique to be successfully produced (1959) and it has become a relatively simple and cost-effective instrumental method. Bench-top instruments have become smaller and mass production has allowed the costs of instruments to decrease. Both parts of the instrument as well as all data are handled by computer, and these too have improved and become less expensive.

The actual identification of an unknown analyte can be accomplished by matching its mass spectrum from the TIC with reference spectra available in the MS software. The largest MS data base, containing over 130 000 spectra, is available from NIST.

30.6.1
GC/MS

The basic incompatibility between a gas chromatograph and a mass spectrometer is the pressure difference. The chromatograph separates components at column pressures of about 10^5 Pa with carrier gas flow rates of 0.1–100 ml min^{-1}, while the ion source of the mass spectrometer operates at pressures of about 10^{-4} Pa (10^{-6} torr). OT columns up to 0.32 mm i.d. are well suited for use in a GC coupled to an MS because of their low flow rates, typically 1–2 ml min^{-1}. The type of interface, called *direct*, is used with OT (capillary) columns. The end of the OT column is extended from the GC directly into the ion source of the MS. The GC flows are low enough, and the vacuum pumping high enough, that the necessary vacuum required by the mass spectrometer can be easily maintained. In this arrangement, the outlet of the GC column is operated at vacuum. This is not common, but it has been shown that vacuum operation can, in fact, be advantageous (Cramers, Scherpenzeel and Leclercq, 1981).

Further details on mass spectrometers can be found elsewhere in the Encyclopedia as well as in books by Oehme (1999) and McMaster (2008).

Today, GC/MS has become almost as routine as GC itself, and a GC/MS system is an essential part of most GC research labs. For the analysis of volatile samples, there is no better instrument.

30.6.2
LC/MS

MS can also be interfaced with a liquid chromatograph for perhaps the ultimate in identification capabilities. The past few years have seen major changes and improvements in LC/MS. Interfacing was initially more difficult, but a major innovation was the realization that interfaces could also serve as the ionization devices. This field is now in a growth spurt and has been adapted to handle biological samples that were previously considered impossible to be run by MS. Currently, LC/MS plays a major role in biotechnology due to its capability to perform rapid, high-resolution separations and identifications of large biomolecules. While LC/MS itself is not new, books covering the new advances and applications continue to be written (Ardrey, 2003; Niessen, 1999).

The key advances in instrumentation that made LC/MS possible were the development of narrow-bore (1-mm-i.d.) LC columns that can be operated at very low MP flow rates (as low as 0.1 mL min^{-1}), and interfaces for introducing liquids into the ion source of the mass spectrometer. In HPLC/MS, the ion sources also serve as the interfaces; thermospray ionization (TSI), electrospray ionization (ESI), and atmospheric pressure ionization (API) are the most popular. The latter two can be carried out at atmospheric pressure, a method of sampling that seemed impossible only a few years ago.

In adapting a HPLC method to LC/MS operation, several changes in MP composition need to be considered. These include

- increasing the volatility of the buffer;
- changing the pH;
- adding adducts, such as Na.$^+$

The MP needs to be volatile. Reversed-phase methods are usually ideal because the major components, water and methanol or ACN, meet this requirement. However, when buffers are used in RPLC, they also need to be volatile. Ammonium acetate is one example, and trifluoroacetate (TFA) can be used to form ion pairs with basic analytes, thereby making them neutral and decreasing their interference with MS detection.

30.6.3
Other Hyphenated Methods

Other spectroscopic instruments that have been interfaced to chromatographs are FTIR and NMR. Like the mass spectrum, an IR spectrum is a highly characteristic property of an organic compound. The complementary nature of IR and MS data makes the combination of them in series the most powerful method for analysis of unknowns by GC. In the case of GC/FTIR, large samples are desired because they facilitate detection by IR, but the vapor-state analytes are not easily sampled by IR. A light pipe has been designed that can be heated to keep the analytes in the vapor state and also meet the following requirements of a good IR cell for use with GC: small volume, long path length, and high transmission. A typical light pipe is 50 cm × 1 mm in size and has a reflecting gold coating. Its description and details of its use including the data handling requirements have been discussed (Griffiths, de Haseth and Azarraga, 1983). The other interface is a type of matrix isolation in which the analytes are frozen on a rotating band, which moves into the IR beam. A comparison of these two types has been published (Schreider, Demirian and Stickler, 1986).

The liquid from a HPLC is compatible with normal IR sampling and is less of a problem. However, LC MPs may not be transparent in the IR, and water is a particularly difficult solvent to handle. For volatile organic solvents, evaporation is possible, and the remaining nonvolatile analytes can be deposited into KBr and pressed into pellets. Further discussion can be found in reference (McDonald, 1986).

Mixed tandem combinations like GC/FTIR/MS are also possible since IR is nondestructive; it is possible to combine the three instruments into one. The special requirements and some applications have been described (Wilkins, 1983). Other tandem combinations are also possible. A common one is HPLC/UV/MS.

NMR does not have the low detectability limits of the other spectroscopies used in tandem with chromatography, so it has not been used as extensively. Also, the high magnetic field of current NMR instruments requires that the chromatograph be located at a considerable distance from the NMR probe. There are only a few reports of its use with GC, and, for HPLC, special glass flow cells are needed. For proton studies, the MP should ideally have no protons, so for RP work, deuterium oxide is used. Other deuterated solvents are quite expensive. LC/NMR is finding use in the pharmaceutical industry for impurity analysis (Lindon, Nicholson and Wilson, 2000). At least one commercial instrument offers LC/MS/NMR.

Other chromatographic methods that have been coupled with MS include TLC, SFC, capillary electrochromatography CEC) and capillary zone electrophoresis (CZE). SFC is handled much like GC, and CEC much like HPLC. TLC is quite different and requires special techniques. Details can be found in Poole's book (Poole, 2003).

30.7
Selected Other Topics

30.7.1
Multidimensional Separations

When a conventional one-dimensional separation is not adequate to resolve all the components of a sample, it may be possible to devise a multidimensional process that will yield better resolution. The concept is to spread out the separation in two dimensions rather than just one. The basic theory of two-dimensional separations was presented by Giddings, first in a paper (Giddings, 1984) and then in Cortes' book (Giddings, 1990) on the subject. Although the latter covers multiple dimensions, most of the actual applications are only two-dimensional. These two dimensions need to be orthogonal (at right angles) in order to maximize the two-dimensional space created. Indeed, Giddings' definition is based on that premise (Giddings, 1984). Orthogonal methods are those that use different mechanisms or modes of action in the separation. For multidimensional column methods, each column should provide a very different type of separation. For example, in GC/GC, one column could be very polar and the other nonpolar. The combination of LC and GC would be considered two-dimensional by virtue of the large difference in their separating mechanisms. A two-dimensional LC separation might combine RP in one column with SEC in the second, thereby attaining true orthogonality. A second requirement for true orthogonality is that components separated in the first dimension must not be recombined in the second dimension. The number of peaks that can be resolved in a two-dimensional separation is the product of the peak capacities of the two operations, thus showing the advantage of two-dimensional methods. As opposed to planar methods, in which the components are distributed in two-dimensional *space*, two-dimensional column separations give components distributed in two-dimensional *time*. Column separations have the added advantage of on-line detection.

A two-dimensional separation can be run off-line or on-line, the latter being preferred for convenience, ease of automation, and improved data handling. The disadvantage to the on-line mode is that the MPs may be incompatible, especially in LC, as in a combination of RP with NP HPLC; the former MP is usually aqueous and the latter non-aqueous. This problem seldom occurs in GC, where the MP can be helium for either polar or nonpolar column separations, and GC/GC is a very common two-dimensional method.

Optimization of the separation is usually called *tuning*, which can be achieved in several ways, including

- using one column of normal length and one of very short length;
- adjusting column variables such as temperature and flow rate;
- using innovative valving or modulation between the columns;
- using cryogenic trapping; and
- using original software to make optimal use of complex data such as that obtained by MS detection.

Descriptions of the different methods and combinations of multidimensional separations and some applications can be found in the volume edited by Mondello, Lewis and Bartle (2002).

30.7.1.1 GC/GC

Much of the early GC/GC work involved heart cutting, which refers to the process of isolating a poorly separated portion (a cut) of the molecules from the first dimensional process and running it on the second dimension. Heart cutting is not a true two-dimensional process since the whole sample is not run in the second dimension. It is still a valuable procedure and one that provided the first GC examples of improved resolution by the use of two dissimilar columns.

For a true two-dimensional GC separation, a cryotrap or modulator is usually required (Ledford, TerMaat and Billesbach, 2007; Hinshaw, 2004). A fraction from the first column is isolated with the modulator and transferred to the second column. Since only a single detector is used, the separation in the second column has to occur in less time than it takes to elute a single peak from the first column. Usually it is desirable to take three or four samples from each peak from the first column, so the run time for the second column can be only a few seconds at most. Since the second separation must be very fast, the second column is short and is operated at a high temperature and fast flow rate. One of the most impressive applications has been the analysis of cigarette smoke; in one case, over 30 000 peaks were isolated using GC/GC/TOFMS (Dalluge, Beens and Brinkman, 2003). In effect, the use of an MS detector to give a GC/GC/MS system results in a separation that is indeed three dimensional.

The interface/modulator between the two columns can take several forms; the two most common ones are a thermal sweeper and a cryogenic trap. At least two commercial instruments are available, including a retrofit for an existing GC and a complete GC/GC/TOFMS instrument. Papers from the First International Symposium on Comprehensive Multidimensional GC, which include reports of modulation devices and many applications have been published (Brinkman, Th., U.A. and Vreuls, R.J.J. 2003).

30.7.1.2 LC/GC

The combination of LC and GC is usually performed in the order indicated – LC being the first dimension. The interface between the systems must be capable of transferring the relatively large amount of MP from the LC into the GC. The popular transfer techniques are (i) retention gap, also called *on-column*; (ii) concurrent eluent evaporation, employing a loop type interface; and (iii) hot injectors, most often with programmed temperature vaporization (PTV). To divert the large volume of gas generated by vaporizing the LC solvent, types 1 and 2 are usually operated with a solvent vapor exit (SVE) placed between the precolumn and the analytical column. The relatively volatile solvents often used in NP LC can be handled, but the aqueous phases used in RP, which may also contain nonvolatile salts, are more difficult, and as a consequence there has not been much development of RPLC/GC.

30.7.1.3 LC/LC

A variety of two-dimensional LC methods have been described, many of which show similarities to the two-dimensional GC methods, including heart cutting. In designing an LC/LC procedure, however, one has to deal with the fact that the MPs differ between modes. In LC, unlike in GC, the MP, as well as the SP, participates in the process of defining the retention characteristics of the system. Thus, the combination of two LC columns of differing separating characteristics is likely to require the use of two different MPs. Consequently, a truly orthogonal LC/LC system is often limited by the necessity to have compatible MPs. For example, NP followed by RP might present difficulties with immiscibility of the NP fraction injected on the RP column. Some LC combinations that do work are

- IEC and RPLC;
- SEC and RPLC;
- SEC and NPLC; and
- combinations of RPLC with electrophoretic methods.

30.7.1.4 Other Combinations

These GC and LC examples of multidimensional chromatography serve as a fitting introduction to many other combinations. The major ones are with SFC and capillary electrophoresis (CE). Since SFC is intermediate in characteristics between GC and LC (*see* Section 30.5), and some consider it to be the ideal unified technique, it is a natural to combine with GC or LC or both. Also, the CE methods (CZE and CEC) are very similar to capillary RPLC and IEC and make ideal combinations with HPLC.

30.7.2
Chiral Separations

Chromatography is even capable of separating optical isomers, a process that is becoming increasingly important, especially in the pharmaceutical industry. For many drugs, only one optical isomer is pharmacologically active, and a total analysis that does not separate and quantitate the enantiomers is unsatisfactory. Enantiomers have identical physical properties and can rarely be separated by conventional chromatographic systems. Thus, new approaches to the separation of chiral compounds have been developed.

This separation challenge has been addressed by various methods, the oldest being the formation of diastereomers. A racemic mixture of optical isomers (designated R and S) cannot be separated chromatographically unless a chiral element is introduced into the procedure. One way is to react the mixture with a chiral reagent (designated, e.g., R'), to form reaction products that will have two chiral centers: one isomer will generate an RR' reaction product, and the other the SR' combination. These two diastereomers will have different physical properties and can be separated chromatographically with nonchiral phases. This approach was the major one used in GC when the field began to develop over 30 years ago. However, it has some disadvantages, and since not all analytes are amenable to analysis by GC, other approaches to chiral separations have been developed. These can be divided into two groups: those using chiral stationary phases (CSPs), and those using chiral mobile phases (CMPs). The use of CSPs is the most popular method for achieving chiral separations because chiral groups can be bonded to the surface of a stationary support. For GC, CSPs are the only option, since it has not been found possible to operate GC with CMPs. Considerable development work on CSPs has been done for HPLC phases. Pirkle (1986) has discussed the early work on CSPs for HPLC, giving examples and describing the way they work. Much less use has been made of CMPs in chiral HPLC separations, where it is assumed that the analytes form strong associations with a chiral component in the MP and that these complexes interact differently with the achiral SP. Although the mechanisms are not all clear, intermolecular associations that seem to form the basis of chiral separations are hydrogen bonding, metal chelation, ligand exchange, and ion-pair formation.

It should also be noted that SFC is of great utility in chiral separations, a major advantage being its relatively low pressure drop over the analytical column, which allows the use of multiple columns in series to achieve higher efficiencies. These SFC columns do not have to contain identical packing materials; use of different column trains can provide unique selectivities. In addition, SFC is ideal for preparative chiral separations since the MP is volatile, easily removed, and nontoxic. A single reference for more information about chiral separations is the book by Beesley and Scott (1998).

30.7.3
Capillary Electrochromatography (CEC)

Electrophoresis is not a chromatographic method, but a technique often used by chromatographers, and one that has been combined with HPLC to produce a technique called *CEC*. The common use of capillary columns brought electrophoresis and HPLC together. The first reported use of electrophoresis in OT capillary columns was published in 1967, but it was later, in 1979–1981, that a real impetus was given to the field by the use of fused-silica columns similar to those used in capillary GC. Later, when a pseudo-SP was introduced as a secondary phase, the technique took on some characteristics of LC and became known as *electrokinetic chromatography* (*EKC*) or, if the second phase was micellular, *micellar electrokinetic chromatography* (*MEKC*). Although scientists call MEKC a chromatographic process, others do not. Subsequently, however, the use of WCOT columns, and packed capillaries with SPs of the type used in RP HPLC, launched the operational mode now called *CEC*, a bona fide chromatographic method, one which has features of both RPLC and capillary zone electrophoresis, CZE (see Figure 30.1).

The CEC apparatus most closely resembles CZE, an illustration of which is shown in Figure 30.17. A packed column similar to one that could be used in HPLC, having a small (50–250 µm) i.d. is used along with a electrophoretic voltage gradient (up to 35 kV) and a flow electroosmotic flow (EOF) of the MP (or run buffer) caused by the resulting electroosmotic pressure. Since the highest electrophoretic mobilities are obtained with acetonitrile–aqueous buffer mixtures, this mixture has become the most popular; a good buffer concentration is about 1 mM. In the most common usage, the detector (usually a UV detector similar to one that would be used in HLC) is located near the cathode end and samples are introduced into the capillary at the anode end. Ionic analytes are driven by the electrostatic field, and, in addition, ions and molecules are transported by the electroosmotic flow, which performs the same function as the pump needed to move the solvent in HPLC. The SP in the column serves as a partitioning agent just as it does in HPLC, and the combination of these forces causes a separation of the analytes.

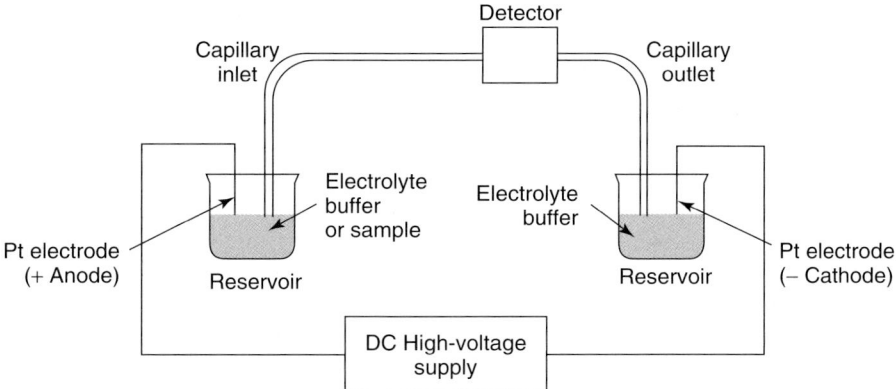

Fig. 30.17 Typical apparatus for capillary electrochromatography (CEC) and capillary electrophoresis. Reproduced with permission from Christian, G. D. *Analytical Chemistry*, 6th edition, John Wiley & Sons. Copyright 2004, John Wiley & Sons.

Tab. 30.4 Advantages and disadvantages of capillary electrochromatography (CEC). Reprinted from reference (McNair and Miller, 1998) with permission from John Wiley & Sons.

Advantages	Disadvantages
1. Greater efficiency	1. Poor detectivity
2. Wide range of applications	2. Poor quantitative precision
3. Faster	3. Poor migration time precision
4. No troublesome pumps	4. High-voltage hazard
5. Less expensive for chiral analyses	

The advantages and disadvantages of CEC listed in Table 30.4 are mainly comparisons with HPLC. The main advantage is the higher efficiency that results from the ideal flow pattern obtained in electrophoresis. The combination of HPLC and CZE provides CEC separations based on both effects for ionized materials and thus can effect separations not possible by HPLC alone. And, since the electrophoretic effects are predictable, the combined effects should be predictable if one has separate chromatographic data. Of the disadvantages, the poor detectivity and precision are sufficiently severe to discourage many chromatographers from the use of CEC, although CE is finding use for chiral separations.

The 1981 paper by Jorgenson and Lukacs (1981) is generally considered to mark the beginning of CEC, and the fundamentals have been summarized by Pyell (2001) who reported that efficiencies are about 10 times better for CEC than for HPLC. An entire volume of the *Journal of Chromatography A* is devoted to CEC (Horvath, 2000), and two books are available (Deyl and Svec, 2001; Krull et al., 2000).

Acknowledgments

Portions of this chapter have been adapted by the author from other works that have been published previously by John Wiley. Some of the text has been updated and revised from the original edition written by Michael J. Kelly.

Glossary

Bonded-Phase Chromatography: Liquid chromatography performed with a stationary phase that is chemically bonded to a particulate supporting material such as silica powder. See also normal-phase chromatography, reversed-phase chromatography.

Capacity Factor: An old term that is no longer used. See Retention Factor.

Capillary Column: In gas or supercritical fluid chromatography, the small-diameter, open tubular column that contains the stationary phase. Inner wall of the tubing is coated with a solid or liquid film that functions as the stationary phase. The major type is a wall-coated open tubular (WCOT) column.

Carrier Gas: Mobile phase in gas chromatography.

Chromatogram: The output from a chromatographic separation that shows the response of the detector as a function of time.

Chromatograph: The instrumentation used in chromatography (noun). To separate and analyze components in a mixture by chromatography (verb).

Column: In chromatography, the glass, metal, or plastic tubing that contains the stationary phase.

Detector: In chromatography, the device used for dynamic monitoring of the chemical compounds leaving the column.

Gas Chromatography (GC): Chromatography with a gaseous mobile phase.

Gas–Liquid Chromatography (GLC): Gas chromatography with a liquid stationary phase that is dispersed on or bonded to a particulate packing material or to the inner wall of the column.

Gas–Solid Chromatography (GSC): Gas chromatography with a solid stationary phase.

Gel Filtration Chromatography: Size-exclusion chromatography with an aqueous mobile phase.

Gel Permeation Chromatography: Size-exclusion chromatography with a non-aqueous mobile phase.

Gradient Elution: In liquid chromatography, a technique for improving resolution, which consists of continuously changing the composition of the mobile phase during the separation.

Height Equivalent of a Theoretical Plate: Originally, a quantity used to describe the efficiency of a particular separation scheme, but no longer used. See plate height.

High-Performance Liquid Chromatography (HPLC): Modern liquid chromatography using mobile phases pumped at high pressures, stationary phases consisting of columns containing densely packed, small-diameter particles, and dynamic detection of the components eluting from the column.

Injector: In chromatography, the device used to introduce a sample onto the top of the column.

Ion Chromatography (IC): A group of liquid chromatography techniques used for the separation and analysis of ions.

Ion-Exchange Chromatography (IEC): Ion chromatography with an ion-exchange resin as the stationary phase.

Ion-Pair Chromatography: Ion chromatography with a reversed-phase or polymeric stationary phase, and a mobile phase containing chemical reagents that form neutral ion pairs with the sample ions.

Liquid Chromatography (LC): Chromatography with a liquid mobile phase. See also high-performance liquid chromatography.

Liquid–Solid Chromatography (LSC): Liquid chromatography with a solid stationary phase.

Mobile Phase (MP): In chromatography, the flowing gas, liquid, or supercritical fluid that carries the individual components of a mixture through the stationary phase.

Nonpolar: A property of a material resulting from the distribution of electronic charge in the individual molecules. A material is said to have nonpolar character if the electronic charge on each molecule is uniformly distributed through the molecule. See also polar.

Normal-Phase Chromatography (NPC): Liquid–solid chromatography or bonded-phase chromatography with a polar stationary phase and a nonpolar mobile phase.

Number of Theoretical Plates: A measure of the efficiency of a particular separation scheme, now called *plate number*.

Open Tubular Columns (OT): See capillary column.

Packed Column: A column that contains the stationary phase as a packed powder or resin.

Plate Height (H): A measure of column efficiency equal to the column length (L) divided by the plate number (N).

Plate Number (N): A measure of column efficiency that is based on a comparison of an analyte's retention volume and its peak width.

Polar: A property of a material resulting from the distribution of electronic charge in the individual molecules. A material is said to have polar character if any region of the individual molecules that compose the material possesses an excess or deficiency of electronic charge relative to the rest of the molecule. See also nonpolar.

Resolution (R_S): In chromatography, a quantity which describes the efficiency and degree of separation achieved in the chromatographic separation of two chemical compounds.

Retardation Factor for Column Chromatography (R): A measure of chromatographic retention behavior, equal to the average speed of a given analyte through the column divided by the average mobile phase velocity. Also equal to V_M/V_R.

Retardation Factor for Thin-Layer Chromatography (R_F): A measure of chromatographic retention behavior, equal to the distance traveled on a TLC plate by a given analyte divided by the distance traveled by the solvent (MP) front.

Retention Factor (k): A measure of chromatographic retention behavior equal to V'_R/V_M. Formerly called *capacity factor, capacity ratio,* and *partition ratio,* none of which should now be used.

Retention Time (t_R): In chromatography, the length of time a chemical compound spends in the column.

Reversed-Phase Chromatography (RPC): Bonded-phase chromatography using a nonpolar stationary phase and polar mobile phase.

Separation Factor (α): In chromatography, a measure of the ability of a separation scheme to separate two components. Also referred to as the *selectivity*.

Size-Exclusion Chromatography (SEC): Liquid chromatography technique that separates chemical compounds on the basis of their physical size. See also gel filtration chromatography, gel permeation chromatography.

Stationary Phase (SP): In chromatography, the solid or liquid phase that resides in the column. The stationary phase is a powder or resin packed into the column or a film deposited on or bonded to the inner wall of the column.

Supercritical Fluid: A material that is above its critical temperature and pressure.

Supercritical Fluid Chromatography (SFC): Chromatography with a supercritical fluid as the mobile phase.

Temperature Programming (PTGC): In gas chromatography, a technique for improving resolution, which consists of continuously changing the column temperature during the separation.

Thin-Layer Chromatography (TLC): Liquid chromatography with a stationary phase dispersed on a solid planar surface. Mobile-phase flow through the stationary phase occurs as a result of capillary forces.

References

Alpmann, A. and Morlock, G. (2008) *J. Sep. Sci.*, **31**, 71–77.
Anon (2007) *Chem. Eng. News*, **85**(43), 49.
Ardrey, B. (2003) *Liquid Chromatography-Mass Spectrometry: An Introduction*, John Wiley & Sons, Hoboken, NJ.
Beesley, T.E. and Scott, R.P.W. (1998) *Chiral Chromatography*, John Wiley & Sons, Chichester, England.
Berger, T.L. and Deye, J.F. (1990) *J. Chromatogr. Sci.*, **28**, 446.
Th. Brinkman, U.A. and Vreuls, R.J.J. (2003) *J. Chromatogr. A*, **1019**, 1–285.
Cramers, C.A., Scherpenzeel, G.J., and Leclercq, P.A. (1981) *J. Chromatogr.*, **203**, 207.
Dalluge, J., Beens, J., and Brinkman, U.A.Th. (2003) *J. Chromatogr. A*, **1000**, 69–108.
Dewaele, C. (1990) *J. Chromatogr. Sci.*, **28**, 649.
Deyl, Z. and Svec, F. (eds) (2001) *Capillary Electrochemistry*, Elsevier, Amsterdam.
(a) Ettre, L.S. (1993) *Pure Appl. Chem.*, **65**, 819–872; (b) Ettre, L.S. (1993) *LC-GC No. Am.*, **11**, 502.
Ettre, L.S. (2002) *LC-GC N. Am.*, **20**, 128, 452.
Ettre, L.S. (2003) *LC-GC N. Am.*, **21**, 458.
Ettre, L.S. and Zlatkis, A. (eds) (1979) *75 Years of Chromatography – A. Historical Dialog*, Vol. 17, J. Chromatogr., Library, Elsevier, Amsterdam.
Giddings, J.C. (1963) *Anal. Chem.*, **35**, 2215.
Giddings, J.C. (1984) Two-dimensional separations: concept and promise. *Anal. Chem.*, **56**, 1258A–1270A.
Giddings, J.C. (1990) in *Multidimensional Chromatography: Techniques and Applications* (J.H. Cortes), Marcel Dekker, New York, pp. 1–27.
Giorgetti, A., Pericles, N., Widmer, H.M., Anton, K., and Datwyler, P. (1989) *J. Chromatogr. Sci.*, **27**, 318–324.
Griffiths, P.R. (1988) *Anal. Chem.*, **60**, 593A–597A.
Griffiths, P.R., de Haseth, J.A., and Azarraga, L.V. (1983) *Anal. Chem.*, **55**, 1361A.
Grob, R.L. and Barry, E.F. (2004) *Modern Practice of Gas Chromatography*, 4th edn, Wiley Interscience, Hoboken, NJ.
Hinshaw, J.V. (2004) *LC-GC No. Am.*, **22**, 32–40.

HNU/Nordion Systems (1990) *J. Chromatogr. Sci.*, **28**, 389.

Horvath, C. (ed.) (2000) *J. Chromatogr. A*, **887**, 1–513. 38 papers on CEC.

James, A.T. and Martin, A.J.P. (1952) *Biochem. J.*, **50**, 679.

Jorgenson, J.W. and Lukacs, K.D. (1981) *J. Chromatogr.*, **218**, 209–216.

Klesper, E., Corwin, A.H., and Turner, D.A. (1962) *J. Org. Chem.*, **27**, 700.

Krull, I.S., Stevenson, R.L., Mistry, K., and Swartz, M.E. (2000) *Capillary Electrochromatography and Pressurized Flow Capillary Electrochromatography*, HNB Publishing, New York.

Ledford, E.B. Jr, TerMaat, J.R., and Billesbach, C.A. (2007) *Introduction to GC x GC*, Zoex Technical Note KT 039595-1, Zoex Corp., Lincoln, NE; available at: www.zoex.com/html/technote_kt030505-1.html.

Lindon, J.C., Nicholson, J.K., and Wilson, I.D. (2000) *J. Chromatogr. B*, **748**, 233.

Martin, A.J.P. and Synge, R.L.M. (1941) *Biochem. J.*, **35**, 1358.

McDonald, R.S. (1986) *Anal. Chem.*, **58**, 1906.

McMaster, M. (2008) *GC/MS: A Practical User's Guide*, 2nd edn, Wiley-VCH Verlag GmbH, Weinheim, Germany.

McNair, H.M. and Miller, J.M. (1998) *Basic Gas Chromatography*, Wiley Interscience, Hoboken, NJ.

Miller, J.M. (2005) *Chromatography: Concepts and Contrasts*, John Wiley & Sons, Hoboken, NJ.

Miller, J.M. and Crowther, J.B. (eds) (2000) *Analytical Chemistry in a GMP Environment*, Wiley Interscience, Hoboken, NJ.

Mondello, L., Lewis, A.C., and Bartle, B.D. (eds) (2002) *Multidimensional Chromatography*, John Wiley & Sons, Chichester, England.

Niessen, W.M.A. (1999) *Liquid Chromatography-Mass Spectrometry*, 2nd edn, Marcel Dekker, New York.

Oehme, M. (1999) *Practical Introduction to GC-MS Analysis with Quadrupoles*, John Wiley & Sons, New York.

Pirkle, W.H. (1986) in *Chromatography and Separation Chemistry* (ed. S. Ahuja), ACS Symposium Series 297, American Chemical Society, Washington, DC.

Poole, C.F. (2003) *The Essence of Chromatography*, Elsevier, Amsterdam, Netherlands.

Purnell, J.H. (1960) *J. Chem. Soc.*, **1960**, 1268.

Pyell, U. (2001) Fundamentals of capillary electrochromatography, in *Advances in Chromatography* (P.R. Brown and E. Grushka), Vol. 41, Chapter 1, Marcel Dekker, New York.

Quadrex Corp. (1990) *J. Chromatogr. Sci.*, **28**, 216.

Schreider, J.F., Demirian, J.C., and Stickler, J.C. (1986) *J. Chromatogr. Sci.*, **24**, 330.

Seidel-Morgenstern, A. (2003) *J. Chromatogr. A*, **1006**, 1–293.

Sherma, J. and Fried, B. (eds) (2003) *Handbook of Thin-Layer Chromatography*, 3th edn, Marcell Dekker, New York.

Small, H., Stevens, T.S. and Bauman, W.C. (1975) *Anal. Chem.*, **47**, 1801–1809.

Smith, R.D., Wright, B.W., and Yonker, C.R. (1988) *Anal. Chem.*, **60**, 1323A–1336A.

Snyder, L.R. and Kirkland, J.J. (1979) *Introduction to Modern Liquid Chromatography*, John Wiley & Sons, New York.

Wilkins, C.L. (1983) *Science*, **222**, 291.

Yost, R.W., Ettre, L.S., and Conlon, R.D. (1980) *Practical Liquid Chromatography: An Introduction*, The Perkin–Elmer Corporation, Norwalk, CT.

Further Reading

Cunico, R.L., Gooding, K.M., and Wehr, T. (1998) *Basic HPLC and CE of Biomolecules*, Bay Bioanalytical Laboratory Inc., Richmond, CA.

Dong, M.W. (2006) *Modern HPLC for Practicing Scientists*, John Wiley & Sons, Hoboken, NJ.

Eeltink, S., Desmet, G., Vivo-Truyols, G., Rozing, G.P., Schoenmakers, P.J. and Kok, W.Th. (2006) *J. Chromatogr. A*, **1104**, 256–262.

Grob, R.L. and Barry, E.F. (2004) *Modern Practice of Gas Chromatography*, 4th edn, Wiley Interscience, Hoboken, NJ.

Hahn-Deinstrop, E. (2007) *Applied Thin-Layer Chromatography: Best Practice and Avoidance* of Mistakes, 2nd edn, Wiley-VCH Verlag GmbH, Weinheim, Germany.

Hunt, B.J. and Holding, S.R. (eds) (1989) in Size Exclusion Chromatography, Chapman and Hall, New York.

Josic, D. and Clifton, J.G. (2007) J. Chromatogr. A, **1144**, 2–13.

Kromidas, S. (ed.) (2006) HPLC Made to Measure, Wiley-VCH Verlag GmbH, Weinheim, Germany.

Letter, W.S. (1991) J. Chromatogr. Sci., **29**, 129.

McCalley, D.V. (2002) J. Chromatogr. A, **965**, 51–64.

McMaster, M. (2005) LC/MS: A Practical User's Guide, Wiley Interscience, Hoboken, NJ.

McMaster, M. (2006) HPLC: A Practical User's Guide, 2nd edn, Wiley-VCH Verlag GmbH, Weinheim, Germany.

McNair, H.M. and Miller, J.M. (1998) Basic Gas Chromatography, Wiley Interscience, Hoboken, NJ.

Meyer, V.R. (2004) Practical High-Performance Liquid Chromatography, Wiley-VCH Verlag GmbH, Weinheim, Germany.

Miller, J.M. (2005) Chromatography: Concepts and Contrasts, John Wiley & Sons, Hoboken, NJ.

Poole, C.F. (2003) The Essence of Chromatography, Elsevier, Amsterdam, Netherlands.

Rood, D. (2007) The Troubleshooting and Maintenance Guide for Gas Chromatographers, 4th edn, Wiley-VCH Verlag GmbH, Weinheim, Germany.

Schmidt-Traub, H. (ed.) (2005) Preparative Chromatography: of Fine Chemicals and Pharmaceutical Agents, Wiley-VCH Verlag GmbH, Weinheim, Germany.

Smith, J.H. and McNair, H.M. (2003) J. Chromatogr. Sci., **41**, 209–214.

Snyder, L.R. and Kirkland, J.J. (1979) Introduction to Modern Liquid Chromatography, John Wiley & Sons, New York.

Weiss, J. and Weiss, T. (eds) (translator) (2005) Handbook of Ion Chromatography, 3rd edn, Two Volumes, Wiley-VCH Verlag GmbH, Weinheim, Germany.

31
Chemometrics and Multivariate Analysis

Richard G. Brereton

31.1	**Introduction**	**1105**
31.1.1	Origins	1105
31.1.2	Development	1106
31.1.3	Software and Methodology	1107
31.1.4	Multivariate Matrices	1108
31.1.5	Notation	1108
31.2	**Multivariate Curve Resolution**	**1109**
31.2.1	Motivation	1109
31.2.2	Principal Components Analysis	1110
31.2.3	How Many Components?	1112
31.2.4	Resolution	1113
31.2.4.1	Noniterative Methods	1113
31.2.4.2	Iterative Methods	1115
31.2.4.3	Combining Information	1116
31.3	**Calibration**	**1117**
31.3.1	Univariate Calibration	1117
31.3.1.1	Classical Calibration	1118
31.3.1.2	Inverse Calibration	1119
31.3.1.3	Determining the Relationship between Variables	1119
31.3.2	Multiple Linear Regression	1120
31.3.3	Principal Components Regression	1122
31.3.4	Partial Least Squares	1123
31.3.5	Model Optimization and Validation	1125
31.4	**Exploratory Methods**	**1126**
31.4.1	Principal Components Analysis	1127
31.4.2	Cluster Analysis	1129
31.5	**Supervised Methods**	**1133**
31.5.1	Developing and Assessing Models	1133
31.5.1.1	**Modeling the Training Set**	1133
31.5.1.2	Test Sets, Cross Validation, and the Bootstrap	1134
31.5.1.3	Application	1136

Encyclopedia of Applied Spectroscopy. Edited by David L. Andrews.
Copyright © 2009 WILEY-VCH Verlag GmbH & Co. KGaA, Weinheim
ISBN: 978-3-527-40773-6

31.5.2	Classical Discriminant Analysis 1136
31.5.2.1	Univariate Classification 1136
31.5.2.2	Bivariate and Multivariate Discriminant Models 1137
31.5.3	Discriminant Partial Least Squares 1139
31.5.4	K-Nearest Neighbors 1140
31.6	**Multivariate Statistical Process Control 1141**
31.7	**Conclusions 1143**
	Glossary 1143
	References 1145
	Further Reading 1147

31.1
Introduction

31.1.1
Origins

The word *chemometrics* is said to have been coined in the early 1970s by the Swedish organic chemist Svante Wold in submitting a grant application. Over that period there were many groups working in what might in modern parlance be referred to as *chemometrics*, but which was, at the time, called by other names such as *chemical pattern recognition, machine learning,* or *laboratory optimization*. These diverse groups often published the subject in different journals: emerged as a discipline seriously in 1980s, with the organization of a School in Cosenza, Italy, in 1983 (Kowalski, 1984), the founding of two journals in 1986 (Chemometr. Intell. Lab Syst.) and 1987 (J. Chemometrics), and the publication of the first text books (Sharaf, Illman and Kowalski, 1986; Massart *et al.*, 1988; Martens and Næs, 1989; Massart and Kaufmann, 1983; Malinowski and Howery, 1980). These books have been widely used; for example, by January 2008 Martens and Næs (1989) had been cited 3145 times, Malinowski and Howery (1980) including two subsequent editions 2164 times, and Massart *et al.* (1988) 1432 times. As an example of the rapid growth of interest over the past two decades, citations per year for Martens and Næs (1989) are shown in Figure 31.1; now there are two sequels from these authors that are cited more often (Naes *et al.*, 2002; Martens and Martens, 2000). In the next two decades, there were many very significant books written on the subject (Brereton, 2003; Massart *et al.*, 1997; Vandeginste *et al.*, 1998; Otto, 1998; Beebe, Pell and Seasholtz, 1998; Kramer, 1998; Gemperline, 2006; Brereton, 2007; Massart *et al.*, 1990; Brereton *et al.*, 1992; Adams, 2004; Meloun, Militky and Forina, 1992; Meloun *et al.*, 1994; Brereton, 1992; Mark and Workman, 2003; Haswell, 1992; Esbensen, 2002; Eriksson *et al.*, 2001) often from different viewpoints.

The earliest practitioners of the subject were not primarily regarded as chemometrics experts, but worked in areas such as pharmacy, analytical chemistry, process chemistry, and organic chemistry. New subjects are often slow to develop in academia, as they often have to work within the boundaries of quite well established disciplines to fit within historic departmental structures, and a

Encyclopedia of Applied Spectroscopy. Edited by David L. Andrews.
Copyright © 2009 WILEY-VCH Verlag GmbH & Co. KGaA, Weinheim
ISBN: 978-3-527-40773-6

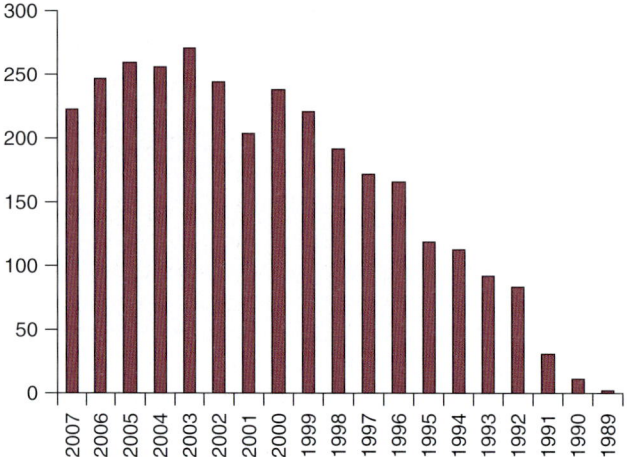

Fig. 31.1 Annual citations for Martens and Næs (1989).

research base needs to be established according to funding patterns that dictate applications.

31.1.2 Development

Historically, however, there was a major development in near infrared (NIR) spectroscopy in the 1980s as cheap, cost-effective probe technology emerged, and there were several important areas of application including process monitoring especially in petrochemical and later in pharmaceutical industries and food chemistry that catalyzed early chemometrics applications. NIR data often involve several highly overlapping peaks that cannot be easily separated or deconvoluted to build a picture of the composition of a mixture. However, NIR is ideal for looking at the overall properties of mixtures and many industrial products, such as food or petrol, that consist of a mixture of compounds. Hence, a method of spectroscopy that is cost effective and can rapidly (and often noninvasively without the need to sample) measure properties that depend on the mixture as a whole has significant advantages. NIR instruments are often stable and have low noise levels. This, however, means that statistically based approaches are required to relate the NIR spectra to the characteristics of a product or process such as acceptability and origins, and this provided a major early driving force for chemometrics. Almost all NIR instruments have a chemometrics capability and this classical area of application is well established with well-described algorithms and a high standard of software. Other spectroscopies have followed suit, especially in cases where technology has improved; for example, cost-effective probes for ultraviolet/visible (UV/vis) spectroscopy have been developed over the past decade, and with most pharmaceuticals having prominent chromophores, the application of chemometrics in reaction monitoring and product quality assessment using UV/vis and chemometrics has grown significantly.

At the other end of the scale, there is a large amount of exploratory work that can be performed using larger instruments such as acoustic spectroscopy, NMR (Nuclear Magnetic Resonance), and MS (Mass Spectrometry), which also provide rich pickings for the chemometrics expert.

31.1.3
Software and Methodology

Often there is a confusion between software capability and chemometrics methodology. In many cases, there are a large number of methods available for the analysis of spectroscopic data but only a small fraction are incorporated into user-friendly software. This is because of the economics of developing software packages, often with a quite specialist market. A package that is tailored for a less common application, such as acoustic spectroscopy, may take as long to develop and cost as much in resources as one for a widespread area such as NIR. Hence, people who are not good programmers are often restricted to prepackaged methods that are primarily targeted toward quite a restricted set of applications. Although basic methods such as principal components analysis (PCA) are widely available, most spectroscopies pose quite specific data analytical problems and require quite specific strategies before pattern recognition. Another problem is that the originators of many of the chemometric-based software companies, most of which were founded in the 1980s, were primarily focused on common methods such as PCA and partial least squares (PLS), and also developed approaches, such as cross validation, that were very efficient in usage of computer power. Moore's law states that computing power doubles approximately every 18 months: on that basis, in a period of 20 years, computing power has increased by a factor of 2^{13} or about 10 000. This means that a calculation that could be performed in a week on a desktop PC in the mid-1980s required only 1 minute 20 years later. Hence, new and much more computationally intense, methods can be conceived and implemented and can provide an assessment of a spectrum within a very short time span compared to those used two decades ago, when chemometrics methodologies started to become commercially viable. The problem here though is that software developers need to keep pace – and the human investment in programming has not changed significantly – programmers are not 10 000 times faster at producing code : with new development tools perhaps code can be developed 2 or 3 times faster than 20 years ago but not more. Therefore, a large amount of the commercial packaged software that is available is based on the philosophy of the mid-1980s, and many spectroscopists who do not have the time or expertise to develop their own code are trapped within this environment, even though if one were to start from the beginning again perhaps quite different methods would become widespread. Areas such as bioinformatics have a more recent vintage and users often do not require real-time desktop solutions. They are happy to perform data mining on a remote mainframe and most users are programmers in their own rights. Hence new methods based on machine learning can easily be incorporated. However, this problem means that there is a gap opening up between the application-based laboratory user of packaged chemometrics software who is often quite limited in the methodology that can be applied and the pioneering research groups that can implement code rapidly but do not have the resources (or investment) to develop user-friendly benchtop software.

31.1.4
Multivariate Matrices

Most of chemometrics is characterized by signals that are multivariate, that is, they are best described in two dimensions and are commonly represented by matrices as in Figure 31.2. Chemometric methods are most valuable when there are several spectra, which are normally represented as rows (often called *samples*) of the multivariate matrix. The columns of the matrix represent the sensors, normally wavelengths (or frequencies). The elements of the matrix are the raw spectroscopic measurements, usually the intensity of a spectrum at a specific wavelength.

We do not discuss experimental design or signal preprocessing (e.g., derivatives and smoothing) in this chapter but these are normally important parts of the data analytical procedure.

31.1.5
Notation

At this point, we digress to give a short introduction to matrices. Matrix notation is generally useful throughout chemometrics.

In this chapter, we employ recommended notation as follows.

- A scalar (or a single number, e.g., a single concentration) is denoted in italics, for example, x.
- A vector (or a row/column of numbers, e.g., a single spectrum) is denoted by lower case bold roman, for example, **x**.
- A matrix (or a two-dimensional array of numbers, e.g., a set of spectra) is denoted by upper case bold roman, for example, **X**. Scalars and vectors can also be regarded as matrices, one of whose dimensions is equal to 1.

The dimensions of a matrix are characterized by the number of rows (e.g., the number of spectra) and the number of columns (e.g., the number of wavelengths). Hence, a dataset that consists of 20 spectra recorded at 1000 wavelengths can be arranged as a 20×1000 matrix, with 20 rows and 1000 columns.

The inverse of a matrix is denoted by "-1" so that $\mathbf{X}^{-1}$ is the inverse of **X**. Only square matrices (where the number of rows and columns are equal) can have inverses. A few square matrices do not have inverses and this happens when the columns are correlated. The product of a matrix and its inverse is the identity matrix.

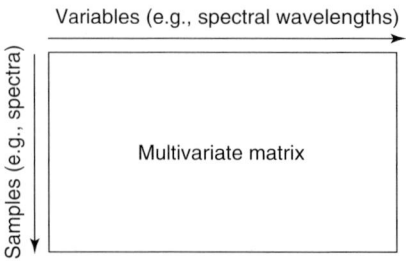

Fig. 31.2 Multivariate matrix as obtained in spectroscopy.

The transpose of a matrix is often denoted by "'" and involves swapping the rows and columns; hence, for example, the transpose of a 5×8 matrix is an 8×5 matrix.

Most people in chemometrics use the dot ("·") product of matrices (and vectors). For the dot product $\mathbf{X} \cdot \mathbf{Y}$ to be viable, the number of columns in $\mathbf{X}$ must equal the number of rows in $\mathbf{Y}$ and the dimensions of the product equal the number of rows in the first matrix and number of columns in the second one. Note that in general $\mathbf{Y} \cdot \mathbf{X}$ (if allowed) is usually different to $\mathbf{X} \cdot \mathbf{Y}$. For simplicity, sometimes the dot symbol is omitted, although it is always the dot product that is computed.

Finally, we introduce the concept a pseudoinverse of a vector or matrix, which is denoted as $\mathbf{X}^+$ so that if

$$\mathbf{Z} = \mathbf{X} \cdot \mathbf{Y} \tag{31.1}$$

$$\mathbf{Y} = \mathbf{X}^+ \cdot \mathbf{X} \cdot \mathbf{Y} = \mathbf{X}^+ \cdot \mathbf{Z} \tag{31.2}$$

It is important to note that only square matrices (where the number of columns and rows are equal) have mathematical inverses, but for matrices or vectors whose dimensions are not the same, there are two types of pseudoinverse, both of which can be denoted by $\mathbf{X}^+$, the right being

$$\mathbf{X}' \cdot (\mathbf{X} \cdot \mathbf{X}')^{-1}$$

and the left being

$$(\mathbf{X}' \cdot \mathbf{X})^{-1} \cdot \mathbf{X}'$$

The prime (') implies a transpose, in which rows and columns are interchanged; for example, the transpose of a 5×8 matrix is an 8×5 matrix. When using vectors, the pseudoinverse mentioned above simplifies still further, but this is not discussed here.

31.2
Multivariate Curve Resolution

31.2.1
Motivation

An early and important application of chemometrics to spectroscopy involves signal resolution; this usually involves the determination of the number, characteristics, and relative amounts of compounds in a series of mixture spectra. Some authors call this *factor analysis*, some *multivariate curve resolution*, and others *deconvolution*. There is a vast array of techniques available to do this; however, only a few are summarized in this chapter. A historic text is by Malinowski and Howery (Malinowski and Howery, 1980) who developed a large array of approaches. This has been updated in a more recent edition (Malinowski, 2002). Multivariate signals often occur in, for example, reactions and involve recording a series of spectra as a function of time as different compounds change in relative concentration and in equilibria studies where each spectrum (e.g., taken at different pHs) is composed of several different spectroscopically active species. The objective is to obtain rate constants or to obtain equilibrium constants, as illustrated in Figure 31.3. The methods described in this section involve the determination of the (relative) concentrations of compounds in mixtures without the addition of external standards.

The initial step is to prepare the data, for example, by smoothing, baseline correction/derivatives, alignment if necessary, and a number of other approaches

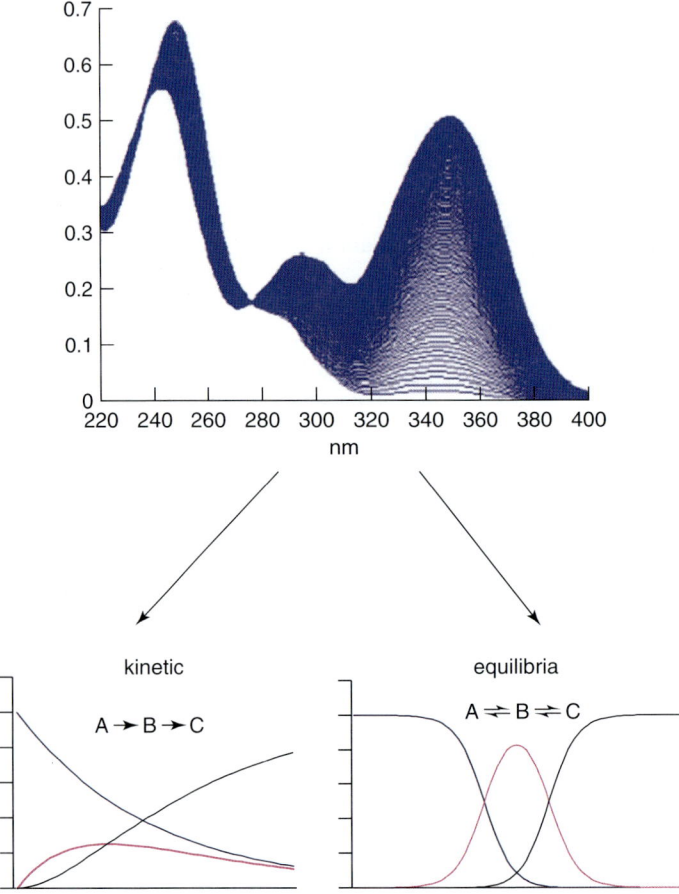

Fig. 31.3 Common motivations for resolving a series of spectra.

to reduce noise. However, these are not discussed in this chapter.

31.2.2
Principal Components Analysis

The first step is normally to perform PCA (Wold, Esbensen and Geladi, 1987; Joliffe, 1987). This common technique aims at converting an experimental data matrix from a series of spectra, which is arranged as a matrix $\mathbf{X}$ whose rows correspond to spectra (often taken at a specified time in a reaction pathway or pH in an equilibrium study) and whose columns correspond to spectroscopic wavelengths to a number of principal components (PCs) defined as follows:

$$\mathbf{X} = \mathbf{T} \cdot \mathbf{P} + \mathbf{E} \tag{31.3}$$

where

- $\mathbf{T}$ are called the *scores* and have as many rows as the original data matrix;
- $\mathbf{P}$ are the *loadings* and have as many columns as the original data matrix;

31.2 Multivariate Curve Resolution

- the number of columns in the matrix **T** is equal to the number of rows in the matrix **P**; and
- **E** is an error matrix.

PCs are abstract mathematical entities with certain specific mathematical properties, the most important being the one that scores and loadings are orthogonal, that is, the vector product of any two different scores or loadings vectors is 0. There are several different algorithms for PCA including singular value decomposition (SVD) and nonlinear iterative partial least squares (NIPALS). Using NIPALS (which is being adopted in this chapter), the loadings matrices are also normal, that is, the sum of squares of their elements is equal to 1. Each column of **T** and row of **P** is a characteristic of a PC. The size of each successive PC is often called an *eigenvalue* and reduces for subsequent components. Eigenvalues can be defined mathematically as the sum of squares of each successive scores vector or column of **T**, but this definition depends on the algorithm used to compute PCs (the NIPALS method being very common).

An important function of factor analysis is to relate PCs (sometimes called *abstract factors*) to chemical factors that can be directly interpreted as illustrated in Figure 31.4. Each chemical factor could be regarded as a chemical compound in the spectroscopic mixture, which has various characteristics including a spectrum and a concentration profile as follows:

$$\mathbf{X} = \mathbf{C} \cdot \mathbf{S} + \mathbf{E} \quad (31.4)$$

- **C** is the matrix consisting of the elution profiles of each compound and
- **S** is the matrix consisting of the spectra of each compound.

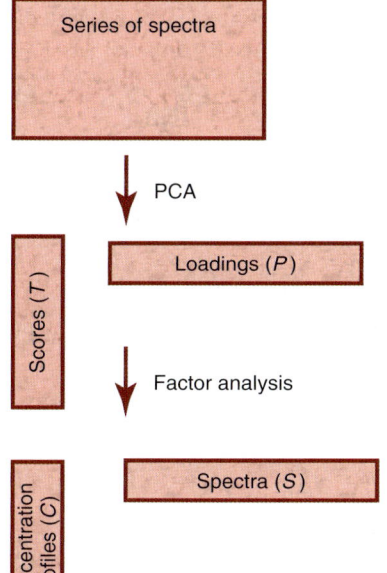

Fig. 31.4 Relationship between PCA and factor analysis.

31.2.3
How Many Components?

Normally, the first stage in factor analysis is the determination of the number of significant components in a series of mixture spectra. The number of significant PCs should be usually equal to the number of compounds in a mixture under most circumstances, although there may be problems. The first is if the concentration of two or more compounds are perfectly correlated, the number of significant PCs or factors will be less than the number of compounds in the mixture. For example, a reaction of the form $A + B \rightarrow C$ will appear to consist of two factors and not three because there are correlations between the reactants. The second is if there are additional effects such as due to variable baseline, temperature, instrumental drift, or significant noise levels, these could increase the apparent number of factors. There will always be some noise from an instrument; hence, even if there are, for example, four uncorrelated species in a mixture, a four-component model will not exactly fit the data, as there will be a small amount of error (matrix **E** above).

There are numerous methods for the determination of the number of significant components in a series of spectra. There are several groups of methods.

1. The first group of methods uses the size of each successive PC. This is measured by its eigenvalue, which is sometimes defined by the sum of squares of the scores. For each additional PC, the eigenvalue decreases. There are numerous approaches for determining the cut-off threshold, many of which have been developed and described by Malinowski (Malinowski, 2002).

2. The second group of methods is based on cross validation (Wold, 1978). These approaches involve removing one or more samples (or portions of the data) and using the rest of the samples to predict their properties. For example, if one performs PCA on 19 out of 20 spectra, one could determine the loadings of the data matrix using 19 samples. The score of the 20th sample could be predicted and this would then be used to estimate the spectrum of this sample and to see how close it is to the true spectrum. If the PCs are real PCs, they should model factors that are actually in the data, but if too many have been calculated, the later ones will be influenced primarily by noise and as such will provide poor prediction. The procedure is repeated until all the spectra are left out at least once. An optimum number of components will result in a minimum error.

3. The third set of approaches is based on the bootstrap (Efron and Tibshirani, 1993). This approach is not common in traditional spectroscopic data analysis because it is computationally intensive, but as computing power has become more cost effective and is more accessible, it is likely to be adopted more in the future. These methods involve removing several samples at a time, and then iteratively repeating this often a hundred times or more. The PC model is applied to the samples that remain each time and then tested on the samples that are left out. The number of components in the data is usually assessed by majority vote, for example, if the optimum is found to be 4 PCs in 85

out of 100 iterations, then this value is employed.
4. There are a number of techniques, such as orthogonal projection approach and SIMPLISMA (De Braekeleer and Massart, 1997; Windig *et al.*, 1992), and eigenvalue plots (Gampp *et al.*, 1985) that primarily rely on graphical inspection of data.

Then, this first step is used to determine the dimensions of the scores and loadings matrices. Hence, if the original data matrix has dimensions 30×1000 and if the number of significant PCs is denoted by A, then

- the dimensions of **T** will be $30 \times A$ and
- the dimensions of **P** will be $A \times 1000$.

The dimensions of **C** and **S** will be the same as **T** and **P**, respectively. Naturally, if the number of significant components is incorrectly estimated, then there can be problems with the reconstruction of the spectra and concentration profiles, but often this is evident visually.

31.2.4 Resolution

Once the number of components has been established, it is necessary to resolve the spectra into individual factors. One important principle behind most methods

$$\mathbf{X} \approx \mathbf{T} \cdot \mathbf{P} = \mathbf{C} \cdot \mathbf{S} \quad (31.5)$$

$$\mathbf{C} = \mathbf{T} \cdot \mathbf{R} \text{ and } \mathbf{S} = \mathbf{R}^{-1} \cdot \mathbf{P} \quad (31.6)$$

so to convert from scores to concentration profiles and from loadings to spectra, it is necessary to find a matrix **R** that is variously called a *transformation, rotation,* or *regression matrix*. This method is sometimes referred to as *principal components regression* (*PCR*). Alternatively, knowing **X** and either **C** or **S**, one can find the missing matrix, or if one has a guess of both **C** and **S**, it will be considered a good guess if one can predict **X** well.

There are a plethora of approaches that are summarized below.

31.2.4.1 Noniterative Methods

Many methods for resolution of mixtures involve finding selective variables for each compound: these could be specific wavelengths unique to each particular compound or specific points in time where one species (e.g., in a reaction mixture) is uniquely present. The first set of methods is where every component in a mixture contains some selective or unique information, which is used as a handle for factor analysis, and the second where there are some components without selective information.

Where all components have some selective information, the first step is to find selective or key variables for each component in the mixture. These could be found visually or it is possible to look at small regions (windows) in the mixture spectra and see which regions (if any) are characterized by one significant eigenvalue and thus are likely to correspond to the absorption of a single compound.

Once diagnostic variables have been found, the question is, what is the next stage? Multiple linear regression (MLR) is a common way to complete the picture. If concentration profiles are known (or can be approximated) using key variables such as m/z values, or wavenumbers, we can estimate the spectra by

$$S \approx C^+ \cdot X \qquad (31.7)$$

where superscript + stands for the pseudoinverse.

Some more mathematically oriented chemometricians prefer PCR. There are several sophisticated algorithms in the literature, but a common basis of most of the approaches is to try to find a rotation, transformation, or regression matrix between the scores of a portion of the data and the known information. The first step involves performing PCA over the entire region of the spectra, and keeping a scores matrix T that has as many columns as compounds are suspected to be in the mixture. For example, if we know the pure spectral regions for each of the compounds, then we know their concentration profiles C from these regions and thus R and S can be estimated from the scores by

$$R = T^+ \cdot C \qquad (31.8)$$

$$S \approx R^{-1} \cdot P \qquad (31.9)$$

Some PCA-based approaches are also called *evolving* (or *evolutionary*) *factor analysis*, but there is a significant terminology problem in the literature, as EFA also denotes the use of eigenvalue plots to determine how many components are in a mixture. There is a lot of history in the names chemometricians use for methods in this area, but the main thing is to keep a very cool head and understand the steps in each approach rather than become confused by the huge variety of names.

When only some components exhibit selectivity, the situation become slightly difficult. A typical example is a UV/vis spectrum of a mixture in which some of the compounds are buried within the spectra and do not have any characteristic wavelengths. Fortunately, PCA can be used here. There are several different ways of exploiting this. One approach uses the idea of a zero-intensity window. The first step is to determine regions for each compound where they are *not* present either in the spectrum or in the profile (e.g., pH profile). This information may be obtained from a number of approaches such as eigenvalue plots. In this region, we expect the intensity of the specific compound to be zero; hence, it is possible to find a vector r for each component so that

$$0 = T_0 \cdot r \qquad (31.10)$$

where 0 is a vector of zeros, if the zero-composition region is for the samples, and T_0 is the portion of the scores in this region. However, there is a problem, that is, one of the elements in the vector r must be set to an arbitrary number, usually the first coefficient is set to 1, but this does not have a serious effect on the algorithm. So long as there is always a region of the spectra or samples where at least one component in a mixture is absent, which can be identified, this procedure should always be possible to perform. If one is uncertain about the true number of compounds (or PCs), it is possible simply to change this and to see what happens with different guesses and see whether sensible results are obtained. The matrix R can be set up as mentioned above, and the estimated profiles of all the compounds are obtained over the entire region by $\hat{C} = T \cdot R$ and the spectra by $\hat{S} = R^{-1} \cdot P$.

31.2.4.2 Iterative Methods

The methods discussed in Section 31.2.4.1 are noniterative, that is, they have a single step that leads to a single solution. However over the years, chemometricians have refined these approaches to develop iterative methods that are based on many of the aforementioned principles but do not stop at the first solution and continue until the solution converges. These methods are particularly useful where only some analytes exhibit selective regions.

First, it is common to produce an initial estimate of the spectrum of each pure species in a series of mixture spectra. In equilibrium studies, there are often regions, for example, of pH, where one compound predominates or is uniquely present. One simple method is called the *needle* search. For each component in a mixture, the datapoints of maximum purity are taken as the initial guesses of the spectra of each component; these are simply spikes at the first stage, that is, one spike for each component: remember that when studying mixtures there may not be any samples that are characteristic of just one compound (e.g., a point in time of reaction mixture), but nevertheless there will be regions of maximum purity for each component in the mixture. Methods such as eigenvalue plots, OPA, and SIMPLISMA can be employed to find these regions, each pure variable being the starting guess of each component. None of these approaches are likely to provide perfect estimates of the spectra and result in initial estimates or guesses.

A feature of most of these methods is that different criteria can then be employed to assess the desirability of the model. Alternating least squares (ALS) (Tauler, Kowalski and Fleming, 1993) is a popular choice. Consider estimating, perhaps imperfectly, the above-mentioned spectra. We can now estimate, in turn, the concentration profiles by

$$\mathbf{C} = \mathbf{X} \cdot \mathbf{S}^+ \quad (31.11)$$

If a good model is obtained, the concentration profiles should obey certain properties: the most common is non-negativity, but other constraints can be imposed such as unimodality if desired, especially if there is some knowledge of the equilibrium process. Hence, $\mathbf{C}$ is altered and a new estimate of $\mathbf{S}$ can be obtained using these modified versions of concentration profiles and so on, until both $\mathbf{C}$ and $\mathbf{S}$ obey the constraints and they are stable (i.e., do not change much each iteration), to get a final solution. The method can be turned on its head in that instead of first guessing pure spectra it is also possible to first guess pure concentration profiles, and then following similar steps, usually iteratively, to obtain a correct solution.

Iterative target transform factor analysis (ITTFA) (Vandeginste, Derks and Kateman, 1985) is an alternative approach used to improve these guesses. PCA is performed first on the dataset, and a rotation or transformation matrix $\mathbf{R}$ is found to relate the scores to the first guesses of the spectra, just as mentioned above. This, in turn, allows the spectra to be predicted. However, the next stage is the difficult part. The predicted spectral matrix $\hat{\mathbf{S}}$ is examined. In the initial stages, this matrix will usually contain elements that are not physically meaningful, for example, negative regions. These negative regions are changed, normally by the simple procedure of substituting the negative numbers by zero, to provide new estimates of the spectra. This, in turn, results in fresh estimates of the elution profiles that can then be corrected for negativity and so on. This iteration continues until the estimated concentration profiles **C** show very little change. The performance of ITTFA depends on the quality of the initial guesses, but can be an effective algorithm. Nonnegativity is the most common condition, but other constraints can be added.

ALS is based on similar principles, but the criterion for convergence is slightly different and looks at the error between the reconstructed data from the product of the estimated concentration profiles and spectra (**C** · **S**) and the true data. In addition, implemented guesses of either the initial spectra or concentration profiles can be used to start the iterations, whereas for ITTFA it is normal to start with the concentration profiles. The ALS procedure is often considered slightly more general and ITTFA was developed very much with analytical applications in spectroscopy coupled to chromatography or another technique, whereas the idea of ALS was borrowed from general statistics. In practice, neither method is radically different, the main distinction being the stopping criteria.

31.2.4.3 Combining Information

Often different types of information can be combined, for example, we may know the order of a reaction and may want to combine known kinetics with multivariate information, rather than treat each type separately.

There are a wide variety of ways to combine both types of information, but most involve using two alternating criteria. Consider knowing the pure spectra of the reactants and products, as well as the starting material. We can then "guess" a value of k (the rate constant), which gives the profiles of each compound in the mixture, often represented by the matrix **C**. We can then use this to "guess" a value of the original data matrix **X** since $\mathbf{X} \approx \mathbf{C} \cdot \mathbf{S}$ or, in other words, the observed reaction spectra are the product of the concentration profiles and pure compound spectra. If our estimated rate constant is accurate, then the estimate of **X** is quite good. However, if this is not very accurate, we need to change our rate constant, change the profiles, and continue until we get a good fit.

This theoretically simple approach takes into account both the "hard" kinetic models used to obtain **C** and the "soft" spectroscopic models used to see how well the estimated spectra equal the observed spectra. Life, though, is never so simple. We may not know the spectra of every compound perfectly, and certainly not in a quantitative way, we may be uncertain of the starting concentrations, and there may be unknown impurities or some side reactions. Hence, many sophisticated elaborations of these approaches have been proposed involving changing several parameters simultaneously including the initial concentrations.

PCA also can be incorporated into these methods via PCR. If we know the scores **T** of the data matrix and guess **C** using our kinetics models, then we can find a transformation matrix $\mathbf{R} = \mathbf{T}^{+} \cdot \mathbf{C}$. Now we can see how well the scores are predicted, using matrix **R** and adjust k (and any other parameters) accordingly and continue in an iterative way. The most common techniques used are ALS and ITTFA, which been already been introduced, although several tailor-made elaborations have been developed specifically for reaction monitoring.

Kinetics knowledge is not the only information that can be obtained about a series of spectra and these models are by no means restricted to kinetics. There are many other constraints that can be incorporated. Simple ones are those where spectra and concentration profiles are positive, but more elaborate ones can be incorporated. For example, concentration profiles with time are likely to be smooth and unimodal; the profile of the starting materials will decrease with time and for all other compounds it can increase and then sometimes decrease. These can all be incorporated into algorithms, allowing the models to contain as many constraints as required, although the convergence of the algorithms can be quite tricky if a large number of factors are taken into consideration.

31.3
Calibration

Calibration plays an important role in spectroscopy. It involves using a spectral response to predict the concentrations or characteristics of individual components. In chemometrics or analytical chemistry, the calibration problem is often formulated as an *inverse* problem. For example, it is often important to be able to predict a concentration from a spectrum. In classical physical chemistry, often the problem is reversed; for example, can we predict a rate constant from a temperature and thus understand the causal relationship between temperature and observed kinetics? In analytical chemistry, the rate constant could be used as a type of thermometer to predict a temperature, rather than vice versa. In addition, there is a notation difference between traditional analytical chemistry and chemometrics, as illustrated in Figure 31.5, where the x variable, in chemometrics, normally represents spectroscopic data, and the y variable represents a predicted property, often the concentration of a compound in a mixture.

To clarify this, we use c as the property to be predicted and x as the data measured to predict this property, usually spectra; hence, the aim of calibration is to predict c from x. The spectroscopy data is called the x *block* and the concentration (or other properties) that is to be deduced from the spectra is called the c *block*.

Note that although the aim of calibration is to predict, for example, the concentrations of compounds in a spectroscopic mixture, there are usually two stages. The first one is calibration or modeling, which uses a set of samples of known composition to form a relationship between the spectral characteristics and the composition or other properties of the mixture. The second one is called *prediction* where the calibration model is used to predict the composition of unknown samples. A recent review discusses calibration in greater detail (Brereton, 2000) and the text by Martens and Næs (Martens and Næs, 1989) is a classic in the subject.

31.3.1
Univariate Calibration

The most traditional, but still very widely used, method of calibration in chemistry is univariate, involving calibrating a

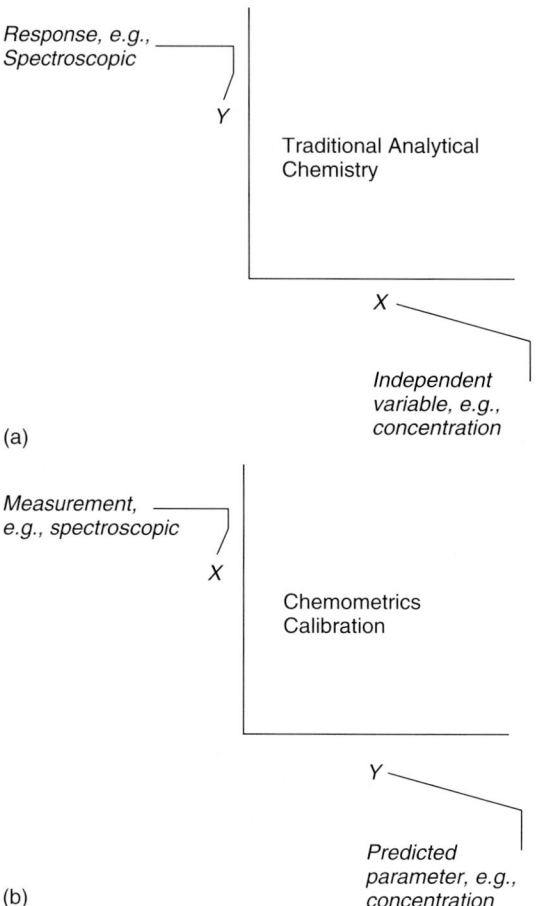

Fig. 31.5 Difference in notation in analytical chemistry (a) and chemometrics (b).

single variable (e.g., a spectroscopic intensity) to another variable (e.g., a concentration). The literature goes back a century; however, with the ready availability of matrix-based computer packages ranging from spreadsheets to more sophisticated programming environments such as Matlab, we can approach this area in new ways.

31.3.1.1 Classical Calibration

Most chemists are familiar with this approach. An equation of the form

$$x = bc \tag{31.12}$$

is obtained relating, for example, a spectroscopic intensity to a concentration. Sometimes an intercept term is introduced:

$$x = b_0 + b_1 c \tag{31.13}$$

and it is also possible to include squared and higher order terms.

There are several difficulties with classical methods. First, the aim of calibration is to predict c from x and not vice versa: the classical method involves obtaining a model x from c; it may be what a physical chemist wants but not necessarily what an analytical chemist wants to do. The second, and more serious, difficulty is that the greatest source of error is generally in the preparation of sample; hence, the measurement of a concentration (even when determining a calibration model) is likely to be less certain than the measurement of spectral intensity. Twenty years ago, many instruments produced less reproducible results than now, but the significant increase in instrumental performance has not been accompanied by an equally significant improvement in the quality of measurement cylinders, pipettes, and balances.

31.3.1.2 Inverse Calibration

An alternative approach is to assume that the errors are in the c block. This means fitting an equation of the form

$$c = ax \tag{31.14}$$

or

$$c = a_0 + a_1 x \tag{31.15}$$

In terms of both error analysis and scientific motivation, this is more rational than by employing classical calibration, but most chemists still use classical methods even though they are often less appropriate.

If, however, a calibration dataset is well behaved and does not have any significant outliers, straight lines obtained using both forms of calibration should be roughly similar, as illustrated in Figure 31.6. Sometimes there are outlying points, which may come from atypical or erroneous samples, and these can have considerable influence on the calibration lines and there are many technical ways of handling these outliers, which are often said to have high leverage: generally the two types of best fit straight line are influenced in different ways, and this discrepancy can be eliminated by removing any outliers.

31.3.1.3 Determining the Relationship between Variables

Instead of deriving quite complex equations, it is easier to obtain best fit

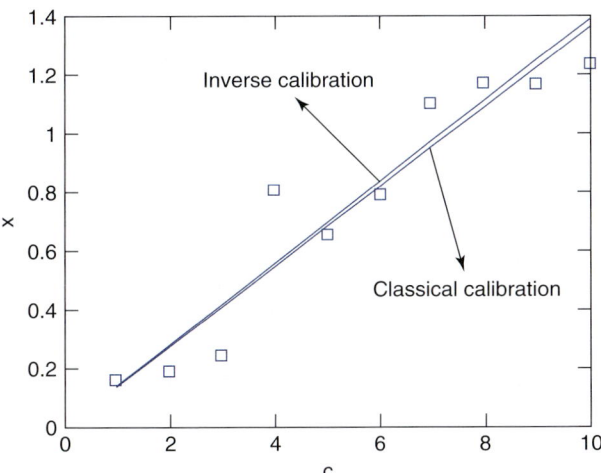

Fig. 31.6 Classical and inverse calibration lines.

calibration parameters using simple vector arithmetic. For inverse calibration, we can use the left pseudoinverse; hence,

$$a \approx (\mathbf{x}' \cdot \mathbf{x})^{-1} \cdot \mathbf{x}' \cdot \mathbf{c} = \mathbf{x}^{+} \cdot \mathbf{c} \quad (31.16)$$

where $\mathbf{x}$ is a vector of spectroscopic responses (often peak heights at a single wavelengths) and $\mathbf{c}$ is the corresponding concentrations. Note that $(\mathbf{x}' \cdot \mathbf{x})^{-1}$ is actually a scalar or a single number in this case, equal to the inverse of the sum of squares of the elements of $\mathbf{x}$.

Sometimes it is desirable to include an intercept (or baseline term) of the form

$$c = a_0 + a_1 x \quad (31.17)$$

Now the power of matrices and vectors becomes evident. Instead of using a single column vector for $\mathbf{x}$, we use a matrix whose first column consists of ones and second of the experimentally determined x values, and $\mathbf{a}$ becomes a row vector of two elements, each element being one coefficient in the equation. The equation can now be expressed as

$$\mathbf{c} = \mathbf{X} \cdot \mathbf{a} \quad (31.18)$$

so that

$$\mathbf{a} \approx \mathbf{X}^{+} \cdot \mathbf{c} \quad (31.19)$$

31.3.2
Multiple Linear Regression

MLR is one of the commonest techniques for calibration and regression both in chemistry and statistics.

In the context of the spectroscopy of mixtures, the principle is that a series of mixture spectra can be characterized by

- the spectrum of each individual component;
- the concentrations of these components; and
- noise (or experimental error).

Mathematically, this information can be expressed in a matrix format:

$$\mathbf{X} = \mathbf{C} \cdot \mathbf{S} + \mathbf{E} \quad (31.20)$$

This equation has already been discussed in Section 2.2. The $\mathbf{X}$ matrix might, for example, consist of a series of uv spectra of a reaction of the form

$$A \rightarrow B + C \quad (31.21)$$

consisting of three compounds whose concentrations we wish to monitor as a function of time. Each spectrum will be a row of the matrix $\mathbf{X}$ and each corresponding row of $\mathbf{C}$ will consist of concentrations of each component in the reaction mixture at a specific sampling time.

To study the reaction, it is usually necessary to determine $\mathbf{C}$. There are many ways of using MLR to obtain this information. The simplest is not really calibration, but involves knowing the spectra of a unit concentration of each pure component, $\mathbf{S}$. Using the pseudoinverse, we can write

$$\mathbf{C} \approx \mathbf{X} \cdot \mathbf{S}^{+} \quad (31.22)$$

However, this supposes that we can actually obtain the spectra of all components in a mixture, that we know all the components and have isolated them, and also that the conditions in the reaction mixture are similar to those of the pure components. An alternative is to take some mixtures, quench them (where necessary), and then measure the concentrations of the components using an off-line method such as chromatography, where each is separable and peak areas can easily be related to the concentration as illustrated in Figure 31.7. We can then determine and estimate $\hat{\mathbf{S}}^+$ from this training or calibration set, where the notation "^" means predicted rather than observed, as we know $\mathbf{C}$ from the off-line chromatographic peak areas. This can be used to set up a calibration model and in the future predict the concentrations of compounds from their spectra in the absence of chromatography. If we are satisfied with the calibration model, we can then predict the concentrations in the reaction mixture

$$\hat{\mathbf{C}} = \mathbf{X}_{reaction} \cdot \hat{\mathbf{S}}^+ \quad (31.23)$$

and thus obtain the profiles of each compound with time.

MLR can be extended to include nonlinear such as with squared terms, which can be useful under some circumstances, for example, if there are nonlinearities due to overloaded spectra (which may occur if concentrations in a reaction mixture, for example, are outside the region where the Beer–Lambert law is obeyed). Extra columns are simply added to the $\mathbf{X}$ matrix equal to squares of the intensities. Another common modification is weighted regression, where certain wavelengths have more significance than others.

However, MLR has the disadvantage that all significant compounds in a mixture should be known. If we have information on two out of five compounds in a series of spectra, then the two predicted spectra will also contain features of the spectra of the remaining three compounds, distributed between the two known components (dependent on the nature of the calibration set) and the concentration estimates will contain large and fairly unpredictable errors. It is sometimes possible to overcome this problem if there are selective wavelengths where only two compounds absorb, so that the $\mathbf{X}$ matrix consists of only certain selective regions of the spectrum; however, there are alternative approaches, discussed later, where good models can be obtained using information when only a proportion of the compounds in a mixture can be characterized.

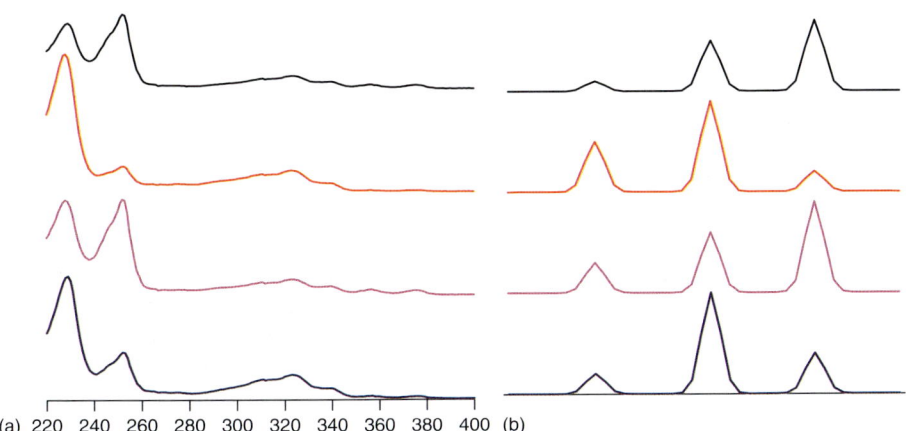

(a) 220 240 260 280 300 320 340 360 380 400 (b)

Fig. 31.7 Calibrating spectra (a) to chromatograms (b).

31.3.3
Principal Components Regression

Some consider PCR as a form of factor analysis. Although the two concepts are interchangeable according to some chemometricians, especially spectroscopists, it is important to recognize that there are widespread differences in the usage of the term *factor analysis* by different authors, groups, and software vendors.

PCR has a major role in the spectroscopy of mixtures. It is most useful where only some compounds can be identified in the mixture. For example, we may be interested in the concentration of a carcinogenic chemical in a sample of seawater, but have no direct knowledge or interest in the identities and concentrations all the other chemicals. Can we determine this by spectroscopy of mixtures? MLR works best when all the significant components are identified.

The first step, as always, is to obtain a series of spectra (the calibration or training set), for example, the UV spectra of 30 samples recorded at 200 wavelengths, represented by a 30 × 200 matrix **X**. The next step is to perform PCA as discussed in Section 31.2.2 to obtain scores and loadings matrices, **T** and **P**. A crucial decision is the determination of the number of PCs that should be retained (Section 31.2.3). Ideally, this should equal the number of compounds in a mixture. For example, if there are seven compounds we should retain the first seven PCs, so that the matrix **T** is of dimensions 30 × 7 and **P** of dimensions 7 × 200, in this context. However, in real-world situations, this information is rarely known exactly. In many cases, the number of PCs needed to model the data is very different from the number of real components in a series of mixtures. This is because of correlations between compound concentrations, spectral similarities, and experimental design that often reduce the number of components, and noise and instrumental features such as baseline problems that may increase the number of components.

The next step is variously called *regression, transformation,* or *rotation*. A common approach is to find a relationship between the scores and the true concentrations of a compound of the form

$$\mathbf{c} = \mathbf{T} \cdot \mathbf{r} \quad (31.24)$$

where **c** corresponds to the known concentrations in the mixture spectra and **r** is the column vector whose length equals the number of PCs (seven in this example). This is represented diagrammatically in Figure 31.4. Equation (31.24) can also be expressed as follows:

$$c = t_1 r_1 + t_2 r_2 + \cdots + t_7 r_7 \quad (31.25)$$

if there are seven PCs, where t_1 is the score of the first PC, and r_1 is its corresponding coefficient obtained using regression.

Knowing **c** and **T**, it is not very difficult to estimate **r** using the pseudoinverse by

$$\mathbf{r} = \mathbf{T}^+ \cdot \mathbf{c} \quad (31.26)$$

This permits the concentration of an individual compound to be estimated by knowing the scores.

If there are several compounds of interest, it is a simple procedure to expand "*c*" from a vector to a matrix, each column corresponding to a compound, so that

$$C = T \cdot R \quad (31.27)$$

As usual, we can determine R by

$$R = T^+ \cdot C \quad (31.28)$$

The extreme case is when the number of PCs is exactly equal to the number of known compounds in the mixture, in which case R becomes a square matrix.

The quality of prediction is then demonstrated on a test set (see Section 31.3.5) and finally the method is used on real data of unknowns. It is also useful to be able to preprocess the data under certain circumstances before PCA. This process often involves mean centering or standardizing the spectral matrix before PCA, rather than using the raw spectroscopic data, for example

$$x_{ij} = \frac{x_{ij} - \bar{x}_j}{s_j} \quad (31.29)$$

where x_{ij} is the spectroscopic intensity in spectrum i and wavelength j, $\bar{x}_j$ is the mean intensity over all spectra in the calibration set at wavelength j, and s_j is the corresponding standard deviation: this is valuable if it is desired to weight all wavelengths to be of equal significance, which is often useful in techniques such as NIR and UV/vis where PCR represents a significant improvement over MLR when only partial information is available about the number of components in a mixture, as one can have as many or as few components in the regression model as desired, so long as the number of PCs is determined well.

31.3.4
Partial Least Squares

PLS has a long history in chemometrics. This technique was originally proposed by the Swedish statistician, Herman Wold, whose interests were primarily in economic forecasting. In the 1960s his son Svante Wold, together with a number of Scandinavian scientists, advocated its use in chemistry. Possibly no technique in chemometrics is so mired in controversy, with some groups advocating PLS for almost everything (including classification) with numerous extensions over the last two decades, with other groups agreeing that PLS is useful, but one of very many techniques in the toolbox of the chemometrician. Probably, one confusion is between the availability of good software (there is currently some excellent public domain software for PLS) and the desirability (independently of software) for using PLS. In many areas of calibration, PCR serves almost the same purpose; however, despite this, PLS has been the subject of more extensions and articles than probably any other chemometrics method, and there is a large hard-core of believers who solve most problems (usually quite successfully) by PLS. An important early review of the method was published by Geladi and Kowalski in 1986 (Geladi and Kowalski, 1986).

One principle guiding PLS is that modeling the c block of data is as important as modeling the experimental x information. A drawback of PCR is that the PCs are calculated exclusively on the x block and do not take into account the c block. In PLS, components are obtained using both x and c data simultaneously. Statisticians like to think about modeling covariance as opposed to variance. Most conventional least squares approaches involve finding a variable that maximizes the value of x^2 or the square of the modeled experimental data or variance. PLS finds a variable that maximizes xc or the product of the modeled experimental data with the concentrations, often called the *covariance*. In physical terms, PLS assumes that there are errors in both types of information, which are of equal importance. This makes some sense: the concentrations used in calibration are subject to error (e.g., dilution and weighing) just as much as the spectra or chromatograms. MLR and PCR as commonly applied in chemistry assume that all the errors are in the spectroscopic data and that the concentrations in the calibration set are known exactly.

Often PLS is presented in the form of two equations. There are a number of ways of expressing these, a common one being

$$\mathbf{X} = \mathbf{T} \cdot \mathbf{P} + \mathbf{E} \quad (31.30)$$

$$\mathbf{c} = \mathbf{T} \cdot \mathbf{q} + \mathbf{f} \quad (31.31)$$

as illustrated in Figure 31.8. Equation (31.30) appears to be similar to that of PCA, but the scores matrix also models the concentrations, and the vector $\mathbf{q}$ has some analogy to a loadings vector. The matrix $\mathbf{T}$ is common to both the equations. $\mathbf{E}$ is an error matrix for the x block and $\mathbf{f}$ an error vector for the c block. The scores are orthogonal, but the loadings ($\mathbf{P}$) are not. There are various algorithms for PLS, and in many the loadings are not normalized. PLS scores and loadings differ numerically from PCA scores and loadings and are dependent both on experimental measurements (e.g., spectra) and the concentrations or properties of the spectral matrices. In PCA, the scores and loadings depend only on the spectra or x block. In PLS, the product $\mathbf{T} \cdot \mathbf{P}$ does not provide the best fit estimate of the x block. The size of each successive component (as assessed by the sum of squares of each successive vector in $\mathbf{T}$) does not always decline; later vectors may be larger than earlier ones, unlike in PCA. In addition to matrices $\mathbf{T}$ and $\mathbf{P}$ in PLS, there is another matrix called a *weights matrix* $\mathbf{W}$ which is often used, which has analogies to the $\mathbf{P}$ matrix and is a consequence of the algorithm. Some like to use $\mathbf{W}$ instead of or in addition to $\mathbf{P}$. In the PLS literature, it is quite important to read quite closely each paper and determine precisely which algorithm is being employed, there being a much wider range of terminology than in PCA.

The quality of the models can be determined by the size of the errors,

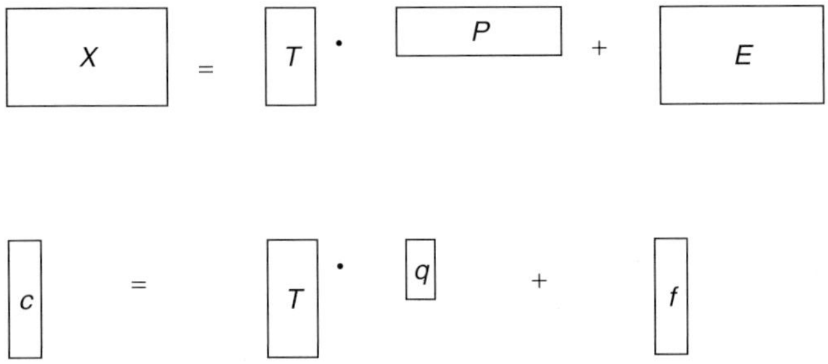

Fig. 31.8 PLS principles.

normally the sum of squares of **E** and **f** is calculated. The number of significant PLS components can be estimated based on these errors, often using cross validation or the bootstrap, although it can be done on the training set (often called *autoprediction*) as discussed in Section 31.3.5. It is important to recognize that there are a large number of criteria for determining the number of significant components that can cause major disagreements between packages and different algorithms.

There are several extensions to PLS.

The standard algorithm described above is sometimes called *PLS1*. One extension, first proposed in the 1980s, is *PLS2* (Hoskuldsson, 1988). In this case, the concentration (or *c*) block consists of more than one variable, an example being a case in which there are three components in a mixture and hence **C** becomes a matrix with three columns, and Eq. (31.31) changes to

$$\mathbf{C} = \mathbf{T} \cdot \mathbf{Q} + \mathbf{F} \qquad (31.32)$$

In analytical spectroscopy, PLS2 is not very popular and often gives worse results than PLS1. However, if there are interactions between variables that correspond to the columns of **C**, this method can have advantages. In most cases in analytical chemistry, however, mixture spectra or chromatograms are linearly additive and independent. An original advantage of PLS2 was also that it was faster in performing several PLS1 calculations, but with modern computing power this is no longer serious.

Another common extension is to include nonlinear (in most case squared) terms in the PLS equation. This is very similar in effect to extending MLR models by including extra terms. Often, however, if a problem is highly non-linear, completely different approaches such as neural networks may be more appropriate for calibration.

PLS has advantages over MLR in that only some of the components need to be known and over PCR in that the errors can be in both the x and c blocks, and is particularly good when the aim is to determine a statistical or bulk measurement, for example, the amount of wheat in protein. In Section 31.3.5, validation of the models is discussed in more detail. However, if all the components in a mixture are known and are linearly additive, more conventional approaches such as MLR may perform equally well or even better.

31.3.5
Model Optimization and Validation

One of the most important problems in the area of multivariate calibration is determining its effectiveness. The main aim of the model is to use a mathematical function of experimental data to predict a parameter such as concentration. In PCR and PLS, we have to choose the number of components to be retained in the model, which represent one of the dimensions (A) of the scores and loadings matrices. The size of these matrices influences the quality of predictions.

There are two objectives of these methods. The first is to optimize the model and thus determine the optimum number of components (in PLS and PCR). The second is to have an independent assessment about its performance. Often these two aims are mixed up and many people in the area of calibration just cite the calibration and prediction error, the latter being obtained while optimizing the model. We have discussed several approaches in the context of PCA (Section 31.2.3) and briefly introduce this in the context of calibration.

Autoprediction involves looking at the residual, often root mean square, error of calibration. This can be calculated

$$RMSEC = \frac{\sum_{i=1}^{I}(c_i - \hat{c}_i)^2}{D} \quad (31.33)$$

for the c block, where D is defined as the number of degrees of freedom, which is usually $I-A$, where A is the number of components, I is the number of samples in the calibration or training set, and $\hat{c}_i$ is the predicted value for the tth sample. If the columns are centered, $D = I - A - 1$. This usually decreases the more the variables but the number of components that result in a percentage threshold for the residual error such as 5% could be established for the number of significant PCs. There are variants on this including using the x block, in the literature.

For cross validation (Wold, 1978), similar equations can be established except that the divisor is I, the number of samples rather than the number of degrees of freedom. The root mean square of cross validation (RMSECV) should reach a minimum when the correct number of components is determined. Similar types of equations can be obtained using the bootstrap (Efron and Tibshirani, 1993), likewise searching for a minimum. In both the cases, sometimes there are problems with flat minima or noisy minima that do not have an obvious optimum, and various elaborate rules are available for choosing the number of components to be employed in the model. Both these errors (from cross validation and the bootstrap) should be larger than the autopredictive error.

An external test set is often employed to determine the model performance. This involves samples that have not been used for optimization and have no influence on the number of components in the model, which gives the root mean square error of prediction (RMSEP). Ideally the RMSEP and RMSECV should be calculated on completely different data as one is about using an independent test set to validate the model and the others to optimize the model. Usually $RMSEC < RMSECV < RMSEP$, although there can be exceptions to the rule in unusual situations.

There are several other elaborations to optimizing and validating the models, according to the author and software package, and it is usually important to understand how this is done: most packages are based on different philosophies. However, the above-mentioned methods are some of the most common.

31.4 Exploratory Methods

The approaches in Sections 31.2 and 31.3 aim primarily at determining characteristics of components primarily in a mixture, for example, the concentrations of individual compounds in a series of spectra or the spectra of these components. However, there are other reasons for employing chemometrics in spectroscopy.

Exploratory data analysis involves determining whether spectra form groups or are related. For example, it might be possible to determine the MIR spectra of extracts of the urine of patients with and without a disease and to see if these cluster differently and thus explore whether there are differences and why.

31.4.1
Principal Components Analysis

The most widespread methods are based on PCA. We have already described the theory in Section 31.2.2. In the context of calibration or resolution, the numerical values of the PCs are used to determine physical properties of components in a mixture.

In exploratory data analysis, the main aim is to determine whether spectra fall into groups and why. Most of the output is in the form of graphs. We illustrate this in an example from atomic spectroscopy in which the concentrations of 11 elements obtained using atomic spectroscopy are measured for 58 types of pottery (Bruno et al., 2000) from two origins (or groups). The aim is to determine whether the atomic spectroscopic data can be used to distinguish the pottery and if so which elements are most characteristic of each group.

The first step is normally preprocessing, which involves putting the data on a sensible scale. There are several common procedures.

1. *Using the raw data without any preprocessing*: If all the spectra are roughly on the same scale and the amount of sample in each analysis is similar, this is acceptable. However, usually some additional steps are necessary.
2. *Row scaling*: If the amount of sample to be analyzed is difficult to control, then summing the rows to a constant total by $^{nor}x_{ij} = \dfrac{x_{ij}}{\sum_{j=1}^{J} x_{ij}}$ is a common procedure.
3. *Mean centering*: Often the columns (variables) are mean-centered. There is no inherent requirement to do this, but the majority of statistical packages perform this automatically as PCA often looks at variation around a mean.
4. *Standardizing*: If all the variables are on quite different scales, then this procedure is quite important. For example, some peaks may be small, but their variation is significant relative to large peaks. This is done by

$$^{stn}x_{ij} = \frac{x_{ij} - \bar{x}_j}{\sqrt{\sum_{i=1}^{I}(x_{ij} - \bar{x}_j)^2 / I}} \quad (31.34)$$

Note that mean centering and standardizing must be performed after any row scaling. The disadvantage with standardizing is that if there are noisy regions of a spectrum where there is no real information, these will degrade the model. This can be overcome in various ways, for example, by variable selection just keeping useful regions of the spectrum or by an offset on the bottom of the equation.

For the example in this section, we only show the results of standardizing the data.

The scores plot involves plotting the scores of one PC against an other (or in three dimensions three PCs against each other), or in algebraic terms, different columns of **T** against each other. These are represented in Figure 31.9. It can be seen from inspection that there are likely to be groupings. When plotting the scores there are lots of options, for example, any set of PCs can be chosen. Usually, the most significant ones (with the largest eigenvalues) are chosen, but

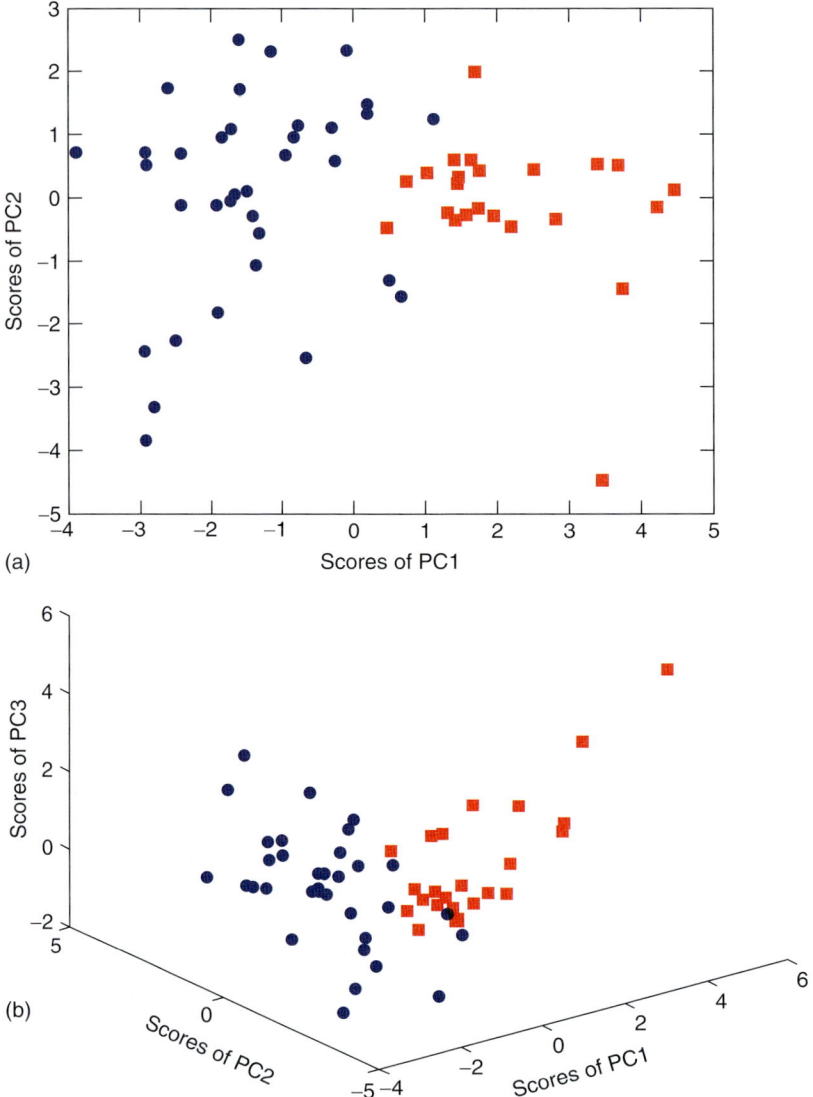

Fig. 31.9 A PC scores plots in which there are two groups in a dataset, in both two (a) and three dimensions (b).

sometimes smaller PCs are important for discriminating between samples, as the largest ones just provide overall information about the data. Often the most suitable PCs can be selected by graphical inspection of various plots.

Parallel to the scores plots are the loadings plots. These involve plotting the rows of **P** against each other, exactly analogous to the scores plots. A loadings plot is illustrated in Figure 31.10 for our example. We can see several features.

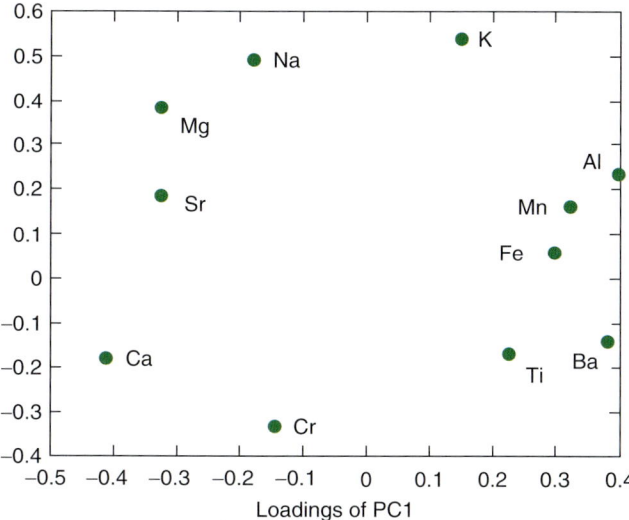

Fig. 31.10 Loadings of the elements corresponding to the example in Section 31.4.1.

The first is that some elements are clustered close together, meaning that they behave in a similar way, for example, Na and Mg. This basically means that if Na is found in high quantities so is Mg, and probably they come from the same source. The second is that one can compare the scores plot, for example, elements on the right-hand side are associated with the group represented by red squares; hence, Al and Mn are likely to be markers for this group of samples, that is, observed in high abundance.

The loadings can be of different forms, such as wavelengths in a spectrum, and would indicate the spectral features associated with specific groups of samples, for example, there may be groups of patients with and without a disease, and if these are separated in a scores plot, the corresponding loadings plot will tell whether there are spectral features diagnostic of specific groups of patients. When comparing scores and loadings plots, it is necessary to ensure that the same PCs are used in each case and that the preprocessing is the same.

PCA is not only used to see groupings but could also be used to see whether there are similarities between samples without any firm groupings, for example, whether there are outliers or whether samples taken as a function of time change with time. There are numerous graphical methods for presenting information about PCs of which the types of graphs, in this section, are the most popular.

31.4.2 Cluster Analysis

Exploratory data analysis, such as PCA, is used primarily to determine general relationships between data (Massart and Kaufmann, 1983; Everitt, Landau and Leese, 2001). Sometimes more complex questions need to be answered, such as, which samples are more similar to each other? Cluster analysis is a well-established approach that was developed primarily by biologists to determine similarities between organisms. Numerical taxonomy emerged from a desire to determine relationships between different species, for example, genera, families, and phyla. However, the spectroscopists also wish to relate samples in a similar manner. Which samples measured by spectroscopy are similar? Cluster analysis involves using a number of methods to group different samples (or objects) using chemical measurements such as spectroscopic intensities or peaks.

The first step is to determine the similarity between spectra. Often the spectra may need to be preprocessed before these distances are measured, for example, standardized, and the results depend on how the data are treated before cluster analysis, just as in the case of PCA.

A number of common numerical measures of similarity are available.

1. *Correlation coefficient between samples*: A correlation coefficient of 1 implies that the spectra have identical characteristics. This value will vary between -1 and $+1$.
2. *Cosine distance*: This measure involves determining one minus the cosine between the two spectra (treated as vectors).
3. *Euclidean distance*: The distance between spectra k and l is defined as

$$d_{kl} = \sqrt{\sum_{j=1}^{J}(x_{kj} - x_{lj})^2} \quad (31.35)$$

where there are j measurements, and x_{ij} is the jth measurement on the sample i, for example, x_{23} is the third wavelength of the second spectrum in a series. The smaller this value, the more similar the spectra; thus, this distance measure works in an opposite manner to the correlation coefficient.

4. *Manhattan distance*: This is defined slightly differently to the Euclidean distance and is given as

$$d_{kl} = \sum_{j=1}^{J} |x_{kj} - x_{lj}| \quad (31.36)$$

Once a distance measure has been chosen, a similarity (or dissimilarity) matrix can be drawn up. This consists of a symmetric matrix whose dimensions are equal to the number of samples. In a dataset of I samples, the matrix will have dimensions $I \times I$ and is illustrated in Figure 31.11.

The next step is to link the objects. The most common approach is called *agglomerative* clustering, whereby single objects are gradually connected to each other in groups.

- From the raw data, find the two most similar objects.
- Next form a "group" consisting of the two most similar objects. The original I objects are now reduced to $I - 1$ groups.
- It is difficult to decide how to represent this new grouping. There are quite a few approaches, but it is common to change the data matrix from one consisting of six rows to a new one of five rows, four corresponding to original objects and one to the new group. The numerical similarity values between this new group and the remaining objects have

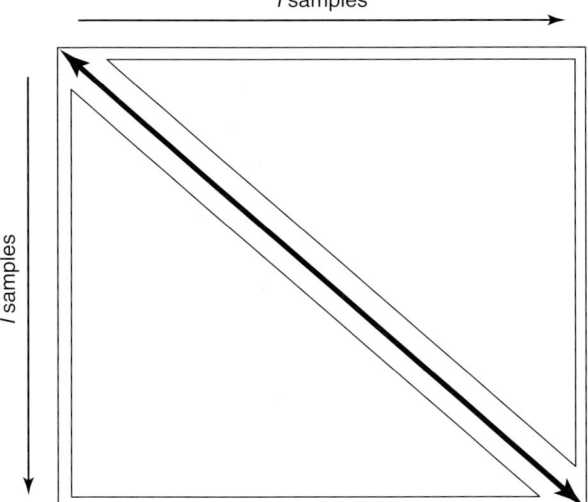

Fig. 31.11 (Dis)similarity matrix produced during cluster analysis. The diagonal values represent maximum similarity (or minimum dissimilarity) and the matrix is symmetrical.

to be recalculated. There are three principal ways of doing this.

1. *Nearest neighbor*: The similarity of the new group from all other groups is given by the *highest* similarity of either of the original objects to each other object.
2. *Farthest neighbor*: This is the opposite to nearest neighbor, and the *lowest* similarity is used.
3. *Average linkage*: The average similarity between two groups is used. There are, in fact, two different ways of doing this, differing according to the size of each group. For the raw data, we compare a group consisting of two members to one consisting of a single object; hence, the calculation is easy. However, in later steps we may want to combine a group consisting of two members with one of four members, and the two methods for calculating similarity indices differ if the number of objects in each group is also different.

The next steps consist of continuing to group the data just as above, until only one group, consisting of all the original objects, remains. It is normal to then determine at what similarity measure each object joined a larger group, and thus which objects resemble each other most. Often the result of hierarchical clustering is presented in a graphical form called a *dendrogram*: note that many biologists call this a *phylogram* and it differs from a cladogram where the size of all the branches are the same. The objects are organized in a row, according to their similarities: the vertical axis represents the similarity measure at which each successive object joins a group. Using nearest neighbor linkage and cosine distance, the dendrogram for the data of Section 31.4.1 is presented

31 Chemometrics and Multivariate Analysis

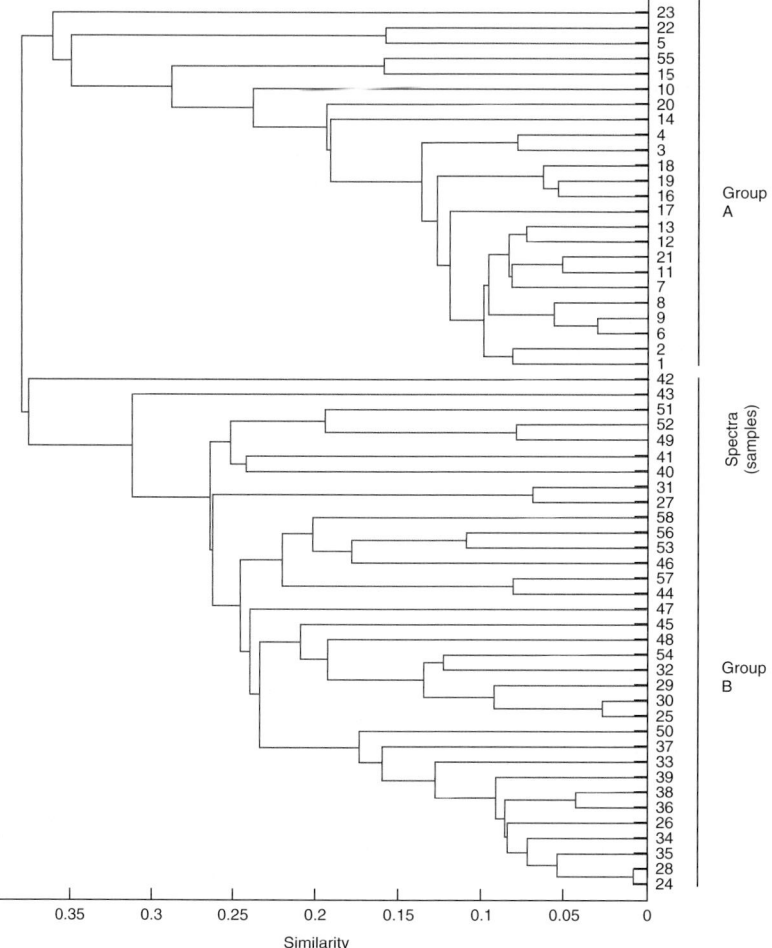

Fig. 31.12 Dendrogram produced by performing cluster analysis on the dataset of Section 31.4.1.

in Figure 31.12. Note that the first group (represented by squares in Figure 31.9) is distinct from the second group. The dendrogram also indicates the samples that are more similar to each other.

Dendrograms do have limitations. There are many clustering methods and each can produce different results. The samples are organized in a linear manner and sometimes the relationships are more complicated and better represented in two or three dimensions. The order of the samples can be variable as it is possible to swap some of the main branches around; hence, parts of the diagram are reflected. However, cluster analysis as described in this section is a well-known and established approach for simplifying and visualizing multivariate data and is often employed in chemometrics.

31.5 Supervised Methods

Classification (often called *supervised pattern recognition*) is at the heart of chemometrics. Can a mid-infrared (MIR) spectrum be used to determine whether a compound is a ketone or an ester? Can the nuclear magnetic resonance spectrum of a tissue sample be used to determine whether a patient is cancerous or not? Can we record the NIR spectrum of an orange juice and decide its origin? Is it possible to monitor a manufacturing process and decide whether the product is acceptable or not? Supervised pattern recognition is used to assign samples to a number of groups (or classes). It differs from exploratory data analysis where, although the relationship between samples is important, there are no predefined groups. There are a very large number of methods for classification and this section summarizes only a few of the most widespread used in chemometrics and spectroscopy; for more comprehensive reviews of methods, see some of the texts referenced in the introduction. However, most common applications in analytical spectroscopy are based on a relatively small number of approaches.

31.5.1 Developing and Assessing Models

Although there are numerous algorithms in the literature, chemometricians have developed a common strategy for developing classification models.

31.5.1.1 Modeling the Training Set

Normally, the first step is to produce a mathematical model between some measurements (e.g., spectra) on a series of objects and their known groups. These objects are called a *training set*. For example, a training set might consist of the NIR spectra of 150 orange juices, 50 known to be from Cyprus, 50 known to be from Spain, and 50 known to be adulterated. Can we produce a mathematical equation that predicts which class an orange juice belongs to from its spectrum? There are several approaches for classification, which are introduced in Sections 31.5.2 – 31.5.4, but the common characteristic is that a model is developed between the spectra and the class each belongs to, which predicts whether a spectrum can be assigned to one class or another.

Once this is done, it is usual to determine how well the model predicts the groups. Table 31.1, often called a

Tab. 31.1 Confusion matrix for classification using a training set.

Predicted	Known			Overall
	Cyprus	Spain	Adulterated	
Cyprus	40	6	6	
Spain	7	44	8	
Adulterated	3	0	36	
Correct	40	44	36	120
%CC	80	88	72	80

confusion matrix or *contingency table*, illustrates a possible scenario. The higher the relative values down the diagonals, the better the class membership predicted. In this case, of the 150 spectra, 120 are correctly classified. The off-diagonal elements show which class is confused (or mistaken) for another Some classes are modeled better than others, 88% of the Spanish orange juices is correctly classified, but only 72% of the adulterated orange juices. A parameter percentage correctly classified (%CC) can be calculated and is 80% overall. There appears some risk of making a mistake, but the aim of a spectroscopic technique might be to perform screening, and there is a good chance that suspect orange juices (e.g., those adulterated) would be detected, which could then be subject to further detailed analysis. Chemometrics combined with spectroscopy acts like a "sniffer dog" in a customs checkpoint trying to detect drugs. The dog may miss some cases and may even get excited when there are no drugs, but there will be a good chance that the dog is correct. Proof, however, comes only when the suitcase is opened and usually spectroscopy is applied as a cheap and first method, which sifts samples, allowing a smaller number to be examined often by other methods in detail.

Sometimes the number of true positives (TPs), true negatives (TNs), false positives (FPs), and false negatives (FNs) can be computed as an alternative measure of the quality of a classification technique. However, this can be done only for what is called a *one class classifier*, that is, one class against the rest. However, we are interested in knowing whether an orange juice is adulterated or not. Table 31.2 illustrates the way in which such a contingency table can be

Tab. 31.2 Stages in developing a 2 × 2 contingency table for determining adulteration.

	Adulterated	Not adulterated
Adulterated	36	3
Not adulterated	14	97

	Adulterated	Not adulterated
Adulterated	72%	3%
Not adulterated	28%	97%

	Adulterated	Not adulterated
Adulterated	TP	FP
Not adulterated	FN	TN

developed. A "positive" is a spectrum arising from an adulterated orange juice that is known to be adulterated. The percentage of TPs is the percentage of orange juices that are actually positive and the mathematical model predicts to be positive. In the example, the %TP is 72% although the %TN is 97%; hence, the test is quite good at determining whether an orange juice is not adulterated but might miss some adulterated samples.

31.5.1.2 Test Sets, Cross Validation, and the Bootstrap

However, it is normal that the training set results in good predictions, but this does not necessarily mean that the method can safely be used in practice. A recommended second step is to test the quality of predictions often using a *test* set. This is a series of samples that has been left out of the original calculations, and is a bit like a "blind test." These samples are assumed to be unknowns at first. Usually the test set predictions are lower than those for the training set, and these are the best indicators as to how well the method can be used in the field. It is particularly important to understand that in spectroscopy it is often possible to obtain several hundreds or thousands of variables per sample. This is a unique property of chemometrics compared to conventional statistics; thus, for the spectroscopist often the sample-to-variable ratio is low, that is, there are many more wavelengths or frequencies than spectra in most typical datasets. In conventional applications of statistics, such as in biology, it is difficult to obtain many measurements per sample. Many classical multivariate methods first reported by biologists were first demonstrated by their application to data obtained by RA Fisher on the iris plant, involving only four variables (Fisher, 1936). However, it can be shown that if the number of variables strongly exceeds the number of samples, it is often possible to get high %CC even starting from completely random datasets (Brereton, 2006); thus, to determine the true effectiveness of the model, it is important to ensure that the model is carefully validated.

Using a test set to determine the quality of predictions is a form of *validation*. The test set could be obtained using samples that are left out of the original dataset. Alternatively, they could have been supplied by an independent laboratory. With modern computers, sometimes it is possible to iteratively generate test sets. For example, if 400 samples are recorded in the original dataset, 100 can be left out randomly at each iteration for testing, and the procedure is repeated many times: computing power increases every year; thus, methods described and implemented 10 years ago can often be improved significantly. What might have taken several hours in 1995 could take a few minutes in 2009 and thus can be repeated many times. With the growth of distributing, computing, and high-performance clusters, we are likely to see significant developments in this direction.

An alternative approach is *cross validation* (Wold, 1978). Only a single training set is required, but usually one (or a group) of the objects is removed at a time, and a model is determined on the remaining samples. Then the prediction on the object (or set of objects) left out is tested. The most common approach is leave one out (LOO) cross validation where one sample is left out at a time. This procedure is repeated until all objects have been left out and is discussed in Section 31.2.3 in the context of PCA. However, the aim in pattern recognition is to determine how well the samples are classified rather than the residual error in modeling the dataset.

Finally, mention should be made of the third alternative called the *bootstrap* (Efron and Tibshirani, 1993). This is a halfway house between cross validation and having a single independent test set and involves iteratively producing several internal test sets: not just removing samples once as in cross validation but also not just having a single test set. A set of samples may be removed, for example, 50 times, each time a different set (although the same samples will usually be part of several of these test sets). The prediction ability each time is calculated, and the overall predictive ability is the average of each iteration.

Naturally, using these it is also possible to calculate the FP or FN rate, and the criterion employed to judge whether a method is suitable or not depends very much on the perspective of the scientist.

If the model is not very satisfactory, there are a number of ways to improve it. The first is to use a different computational algorithm. The second is to modify the existing method – a common approach might involve, for example, instead of using an entire spectrum (many wavelengths are not very meaningful), selecting just the most diagnostic parts of the spectrum. Finally, if everything else fails, change the analytical technique.

It is important to remember that there are two different reasons for using the techniques described in this section. The first is to optimize a computational model. This means that different models can be checked and the one that gives the best prediction rate is retained. In this manner, the samples left out are actually used to improve the model. The second is as an independent test of how well the model performs on unknowns. This is a subtly different reason and sometimes both motivations are mixed up, which can lead to overoptimistic predictions of the quality of a model on unknowns; therefore, care must be taken as discussed by Brereton (2006).

31.5.1.3 Application

Once a satisfactory model is available, it can then be applied to unknown samples, using analytical data such as spectra or chromatograms, to make predictions. Usually by this stage, special software is required that is tailor made for a specific application and measurement technique. The software is expected to determine whether a new sample really fits into the training set or not. One major difficulty is the detection of outliers that belong to none of the previously studied groups, for example, if a Brazilian orange juice sample was to be measured when the training set consists just of Spanish and Cypriot orange juices. In areas such as clinical or forensic science, outlier detection can be quite important; otherwise, an incorrect conviction or inaccurate medical diagnosis would be obtained. There are specific protocols according to the application area and type of spectroscopy involved, but outliers and changing instrumental performance are factors that should always be taken into account.

Another problem is to ensure the stability of the method over time, for example, instruments tend to perform slightly differently every day. Sometimes this can have a serious influence on the classification ability of chemometrics algorithms. One way is to perform a small test on the instrument on a regular basis and only accept data if the performance of this test falls within certain limits. If a spectrometer's performance is demonstrably changing with time (as often happens in NIR), and if historic classification (and calibration) models are to be used for future samples, there are special methods called *calibration transfer* that allow adjustments.

31.5.2
Classical Discriminant Analysis

The majority of statistically based software packages contain substantial numbers of procedures, called by various names, such as discriminant analysis, canonical variates analysis, and so on.

31.5.2.1 Univariate Classification

The simplest form of classification is *univariate* where one measurement or variable is used to divide objects into groups. An example may be if a peak in an IR spectrum of a known amount of sample is above a certain predefined threshold. If it is, then the sample falls into one group, and if not, it falls into another group. This might be useful, for example, in process analysis, if this peak relates to an unacceptable impurity in the mixture. However, in most problems in the spectroscopy of mixtures, criteria for whether a sample falls into one group or another are more subtle.

31.5.2.2 Bivariate and Multivariate Discriminant Models

Often, several measurements are required to determine the group a sample belongs to. Consider performing two measurements, which may involve measuring the intensity of two peaks in a spectrum and producing a graph of the values of these measurements for two groups, as shown in Figure 31.13. The objects denoted by squares are clearly distinct from the objects denoted by circles, but neither of the two measurements, alone, can discriminate between these groups; therefore, both are essential for classification. It is, however, possible to draw a discriminating line between the two groups. Graphically this can be represented by *projecting* the objects onto a line at right angles to this line as demonstrated in Figure 31.14. The projection can now be converted to a position along a single line. Often these numbers are converted to a *class distance*, which is the distance of each object to the center of the classes. If the distance to the center of class A is greater than that to class B, the object is placed in class A and vice versa. It is not always possible to exactly divide the classes into two groups by this method, but the misclassified samples are far from the center of both classes, with the two class distances that are approximately equal.

More sophisticated approaches involve calculating the class centroids (which are the average measurements of all the

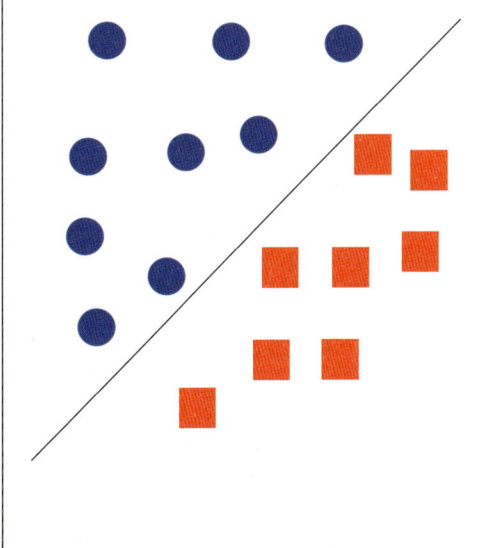

Fig. 31.13 Bivariate example where one measurement is insufficient to distinguish between classes.

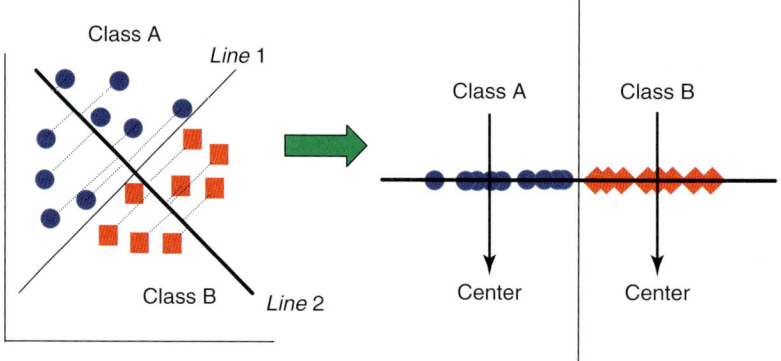

Fig. 31.14 Discriminant function.

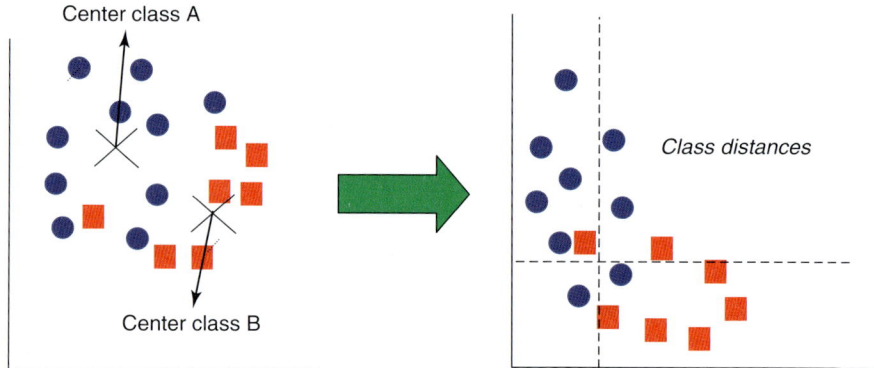

Fig. 31.15 Class distance plot using the Euclidean distance. Top left: almost certainly class A. Bottom left: unambiguous membership. Bottom right: almost certainly class B. Top right: unlikely to be a member of either class, sometimes called an *outlier*.

samples in one class) and the Euclidean distance to the centroids of each class defined as

$$d_A = \sqrt{(\mathbf{x} - \bar{\mathbf{x}}_A) \cdot (\mathbf{x} - \bar{\mathbf{x}}_A)'} \qquad (31.37)$$

where $\mathbf{x}$ is the vector representation of a specific spectrum, preprocessed if necessary, and $\bar{\mathbf{x}}_A$ is the centroid of class A. The distance to two or more class centroids can be calculated. The points no longer fall onto straight lines, but a class distance plot can be obtained as illustrated in Figure 31.15.

It is easy to extend the above-mentioned methods to multivariate situations where instead of two variables many (which can run to several hundreds in spectroscopy) are used to form the raw data. Normally, the class with the smallest distance is taken as the class an object belongs to, but there are several modifications for determining outliers or ambiguous samples or sometimes samples that genuinely belong to more than one class (e.g., a compound might contain both ketone and ester groups and thus exhibit spectroscopic properties of both types of compound and therefore belong to more than one class).

The Euclidean distance is not the only distance measure that can be used; a common enhancement is to employ the Mahalanobis distance defined as

$$d_A = \sqrt{(\mathbf{x} - \bar{\mathbf{x}}_A) \cdot \mathbf{S}^{-1} \cdot (\mathbf{x} - \bar{\mathbf{x}}_A)'} \quad (31.38)$$

where $\mathbf{S}$ is a variance–covariance matrix. There are two reasons for this modification to the Euclidean distance.

1. Different classes may be dispersed differently. One class may be quite scattered, whereas another might be very tight. Therefore, a larger Euclidean in from the centroid of the more scattered class has less significance than from the tighter class.
2. Variables may be correlated. This is quite common in spectroscopy, for example, several peaks may arise from a single feature; in NMR of mixtures, each compound will normally result in several features in the spectrum. However, other compounds that could be just as important for the discrimination may have many less spectral features.

A more subtle distinction is between linear discriminant analysis (LDA) and quadratic discriminant analysis (QDA). Both use the Mahalanobis distance, but the former takes the variance–covariance matrix of the entire training set ($\mathbf{S}$) and in fact forms linear boundaries between classes, whereas the latter involves using a separate variance–covariance matrix for each class in the training set ($\mathbf{S}_A$) and boundaries between classes can be represented as quadratic terms.

As with all classification models, once it is established on a training set, then it can be tested or applied to other samples as discussed in Section 31.5.1.

31.5.3
Discriminant Partial Least Squares

Another technique that can be employed for classification is discriminant PLS. We have already discussed PLS as a calibration method (Section 31.3.3). In that section we showed how PLS can be used to relate on block of data (e.g., "x"), such as a spectrum to another (e.g., "c"), such as the concentration of one or more components in a spectrum.

PLS can be used for discriminant analysis or classification (Dixon et al., 2007). The simplest approach is when there are only two classes. In such case, one sets up a "c" vector, which consists of a number according to which class a spectrum belongs to; there are two common variations, using "1" and "0" or "+1" and "−1" for classes A and B. We recommend the latter convention, although there is no real difference between either of the approaches. PLS can be used to form a model between the x block, for example, a set of spectroscopic measurements, and the c block, which may correspond to the origins of the spectra. The results of calibration are a number. A simple rule for interpreting this number may be to assign an object to class A if it is positive and to class B if it is negative. There are elaborations, for example, a large positive or negative number much greater than ±1 might indicate an outlier, and there may be a region around 0 where we feel that the classification is ambiguous; hence, it is possible to choose a different decision threshold. If the two classes are, for example, whether a patient is diseased or not, then the decision threshold can be adjusted to reduce the number of FPs or FNs according to the one that is most important for the specific application.

There are many advantages of using PLS. For example, one can choose the number of significant components using cross validation just as in normal calibration to see how well the data are predicted numerically, or one can determine the optimum number of components by looking at the percent correctly classified. There will be scores, loadings, and weights, and these can be used to provide graphical information on the class structure and about the variables (e.g., spectroscopic wavelengths) that are most diagnostic of each class.

When there are more than two classes involved, the method becomes quite tricky. Normally, there is no particular sequential relationship between the classes. A simple approach is to set up a "C" matrix of $N-1$ columns, where N is the number of classes. For example, if there are three classes, set up the first column with "+1" for class A and "−1" for either classes B or C, and the second column with "+1" for class B and "−1" for classes A or C. It is not necessary to have a third column. Obviously there are several (in this case three) ways of doing this. It can be a bit tricky to interpret the results of classification, but one can use an approach similar to that used in discriminant analysis when there are more than two classes. Alternatively, it is possible to use a "C" matrix where each column relates to membership of a single class; hence, for three classes we use three columns, with +1 for members and −1 for nonmembers. One should be careful while interpreting the result of calibration. For example, a negative number may be used to imply that an object does not belong to a specific class. What happens when all the three values from the calibration are negative? Do we say that the object is an outlier and belongs to no class? Or do we find the value closest to "+1" and assign an object to that specific class? There are a variety of ways of making choices and interpreting the results.

31.5.4
K-Nearest Neighbors

The methods discussed in Sections 31.5.2 and 31.5.3 involve producing statistical models. Nearest neighbor methods are conceptually much simpler and do not require elaborate statistical computations, and belong to the class of method often called *machine learning*, which had an influence on chemometric pattern recognition in its early days.

The K-nearest neighbor (KNN) method has been used by chemists for over 30 years (Kowalski and Bender, 1972). The algorithm starts with a number of objects assigned to each class.

The method is implemented as follows.

1. Assign a training set to known classes.
2. Calculate the distance of an unknown to all members of the training set. Usually the multivariate Euclidean distance is computed.
3. Rank these in order (1 = smallest distance and so on).
4. Pick the K smallest distances and see what classes the unknown in closest to.
5. Take the "majority vote" and use this for classification; usually K is an odd

number to prevent the problem of tied votes.
6. Sometimes it is useful to perform KNN analysis for a number of values of K, for example, 3, 5 and 7, and see if the classification changes. This can be used to spot anomalies.

This conceptually simple approach works well in many situations, but it is important to understand the limitations.

The first is that the numbers in each class of the training set should be approximately equal; otherwise, the "votes" will be biased toward the class with most representatives. The second is that for the simplest implementations, each variable is of equal significance. In spectroscopy, we may record hundreds of wavelengths, and some will either not be diagnostic or else be correlated. A way of getting around this is either to select the variables or to use another distance measure. The Mahalanobis distance is quite a useful alternative. The third problem is that ambiguous or outlying samples in the training set can cause major problems in the resultant classification. Fourth, the methods take no account of the spread or variance in a class. For example, if we were trying to determine whether a forensic sample is a forgery, it is likely that the class of forgeries has a much higher variance to the class of nonforged samples.

It is possible to follow procedures of validation (Section 31.5.1.2) just as in all other methods for supervised pattern recognition. There are quite a number of diagnostics that can be obtained using these methods.

However, KNN is a very simple approach that can be easily understood and programmed. Many spectroscopists like these approaches, while statisticians often prefer the more elaborate methods involving modeling the data. KNN makes very few assumptions, whereas methods based on modeling often inherently make assumptions such as normality of noise distributions that are not always experimentally justified, especially when statistical tests are employed to provide probabilities of class membership.

31.6
Multivariate Statistical Process Control

A major trend in chemometrics over the past decade has been in multivariate process control (MSPC) (Kresta, MacGregor and Marlin, 1991). This has its origins in industrial processes. Over the past decade, there has been a large development in online process spectroscopy, especially using miniaturized probes for NIR, MIR, UV/vis, and Raman spectroscopy. This allows real-time monitoring of processes and the ability to take spectra every few minutes if necessary. Some probes can be dipped into a reaction vessel, whereas others are pointed at the reaction mixture. In the pharmaceutical industry in particular, it is important to know whether a product is within accepted limits, and there are often legal requirements that impurities, often byproducts of a reaction and of unknown physiological properties, are below limits. If the composition of a product is outside these limits, then the process may have to be stopped or investigated. Chemometrics combined with online spectroscopy has a major role to play. MSPC can be used to decide whether a process is stable or whether there are unexpected changes. A major driving force has been the process analytical technology (PAT) initiative from the US Food and Drugs Administration (FDA) that has encouraged good practice in process monitoring (US Department of Health and Human Services, 2004).

MSPC has been quite slow to get off the ground and most process engineers still use simple univariate measures. A problem with the conventional univariate approach is that there may be an enormous number of parameters measured in time, at regular intervals; thus, in a typical day hundreds of thousands of pieces of information are available especially if spectroscopy is used for process monitoring. Many of the measurements are related; thus, it makes sense to form a multivariate model rather than several univariate models. Methods such as PCA and PLS come to our rescue. Why plot a single parameter when we could plot the PC (suitably scaled) of several parameters? And why not calibrate spectroscopic data to the concentration of a known additive, and use PLS predictions of concentrations as a variable that is monitored with time? In addition to following a number of physical (or process) measurements, the concentrations of specific compounds can be monitored with time, using multivariate calibration.

More sophisticated multivariate methods can be developed and used by chemical engineers who often first define regions of a reaction or batches of a product as normal operating conditions NOC samples; in the case of a reaction or a product, the samples arise from an acceptable process. These NOC samples represent a series of spectra that are considered normal or acceptable, and the aim of MSPC is to look for samples whose characteristics (normally spectroscopic) deviate significantly from the characteristics of these samples, by calculating a statistic for this deviation. There are two common methods.

1. The D-statistic is a measurement of distance a sample from the process is to the center of the NOC samples using a PC model and determines whether a specific sample has a systematic deviation from the steady state region. Usually the Mahalanobis distance from the NOC class of spectra is computed.
2. The Q-statistic calculates the residuals from the new samples when trying to fit them to a PC model, which is based on the NOC samples. The Q-statistic measures nonsystematic variations, which are not explained by the NOC model.

If both statistics are assumed to follow normal distributions, the control limit for these statistics can be computed; usually 95 and 99% limits are calculated. If samples appear to be exceeding these limits, it is probable that there is some difficulty in the process. It is possible to produce control charts in which the values of these statistics are plotted as a function of process time with the control limits indicated. Ninety-five percent limits are often called *warning limits* and allow the operator to investigate why the process appears to be behaving unusually, although one in 20 of the NOC samples will be expected to exceed these limits. The 99% limit is often called an *action limit*.

The use of chemometrics in online process control is highly developed theoretically and slowly being implemented in certain high-tech industries such as the pharmaceutical and food industries for front line products, but there is still a long way to go; the majority of processes are still monitored using fairly straightforward univariate control charts. Nevertheless, this is a major potential growth area, especially in terms of applications of chemometrics to spectroscopy.

31.7 Conclusions

There is a wide applicability of chemometrics to spectroscopy. With the rapid development of fast online computing power, it is possible to do much more compared to a couple of decades ago. What was feasible only off-line using powerful dedicated mainframes in the 1980s is feasible online and in real time now. Most modern chemometrics approaches had their origins several decades ago, but with user-friendly software, the ability to generate large quantities of data and powerful computing are now available to a wide range of users.

This chapter has reviewed some of the most common chemometrics techniques that have relevance to spectroscopy. There is a much wider ranger of methods, but most spectroscopists come across resolution, calibration, and pattern recognition in their basic work. Many approaches not discussed in this chapter have been developed by the machine learning and statistics community and are only slowly being realized by spectroscopists due to the lack of user-friendly software. PCA- and PLS- based approaches, in contrast, have been incorporated into readily available packages for many years and are now commonplace in some spectroscopies. The next few years are likely to see a rapid development in applications of chemometrics methods to a large variety of spectroscopies. However, most methods are based on a relatively small number of techniques such as MLR, PCA, and PLS, and understanding these basic building blocks will allow the spectroscopist to appreciate other approaches reported in the literature.

Glossary

Agglomerative Clustering: Forming relationships between samples by joining together those that are most similar.

ALS: Alternating least squares – a method of modeling usually spectra and concentration profiles that can incorporate constraints such as positivity.

Bootstrap: Alternative to cross-validation involving an iterative algorithm.

Confusion Matrix: See Contingency Table.

Contingency Table: A table whose rows and columns represent each group in a series of spectra, and containing information about how many samples are classified into each group, allowing misclassification rates to be determined.

Cross-Validation: Method for determining the number of PC or PLS components by leaving some samples out and predicting them.

Dendrogram: Graphic representation of cluster analysis.

Discriminant Analysis: Mathematical approaches for determining grouping in samples.

Dissimilarity Matrix: Like a similarity matrix, except that numerical values relate to dissimilarity.

D-statistic: Indicator used usually in MSPC to determine how close a sample is to the centre of an NOC model.

EFA: Evolving factor analysis – used to look at multivariate processes that evolve.

Eigenvalue: Size of PC.

Euclidean Distance: Difference between two spectra obtained by representing each as a point in multidimensional space.

Factor Analysis: Similar to PCA except that the components have an interpretable meaning, for example, as spectra and concentration profiles.

FN: False negative.

FP: False positive.

Inverse: Defined for most square matrices, for which the product is the unit matrix.

ITTFA: Iterative target transform factor analysis – a method similar to ALS.

KNN: K nearest neighbours – a method for classification.

LDA: Linear discriminant analysis – using Mahalanobis distance calculated on a training set consisting of several groups.

Loadings: Numerical values of PCs corresponding to variables.

Mahalanobis Distance: Difference between two spectra obtained by representing each as a point in multidimensional space and scaling each dimension according to variance.

Matrix: Representation of data in a two-dimensional array.

MLR: Multiple linear regression – an alternative to PLS that relates two sets of data, for example, a series of spectra, to concentrations of constituents.

MSPC: Multivariate statistical process control.

Multivariate: Measurements where there is more than one variable, for example, a spectrum at several wavelengths.

NIPALS: Nonlinear lterative partial least squares – a method for performing PCA.

NOC: Normal operating conditions – used in MSPC to indicate the set of spectra that correspond to samples that are of acceptable quality.

PC: Principal component – characterised by scores and loadings.

PCA: Principal components analysis – exploratory visualisation of data matrices, for example, a series of spectra.

PCR: Principal components regression – alternative to PLS and MLR.

PLS: Partial least squares – method for multivariate calibration of data matrices onto another block, calibrating a series of spectra onto concentrations of constituents.

PLS1: PLS algorithm for which there is a single c variable.

PLS2: PLS algorithm for which there is more than one c variable.

Pseudoinverse: Mathematical equivalent to an inverse for a rectangular matrix.

QDA: Quadratic discriminant analysis – using Mahalanobis distance calculated separately on each group in a training set.

Q-statistic: Indicator used usually in MSPC to determine how well a sample fits an NOC model.

Rectangular Matrix: Matrix for which the number of rows and columns are unequal.

RMSEC: Root mean square error of calibration – used for autopredictive models.

RMSECV: Root mean square error of cross validation – often used to determine how many PLS components are suitable for a model.

RMSEP: Root mean square error of prediction – used on an independent test set.

Rotation: Transforming a PC onto a target vector normally consisting of the concentrations of an analyte.

Sample: A row of a matrix usually representing a spectrum.

Scalar: Single number.

Scores: Numerical values of PCs corresponding to samples.

Similarity Matrix: A matrix whose rows and columns represent samples containing values that indicate how similar each sample is to every other sample.

Square Matrix: Matrix for which the number of rows and columns are equal.

Standardizing: Subtracting the mean and dividing by the standard deviation.

SVD: Singular value decomposition – a method for performing PCA.

Test Set: Samples used to determine how well a model performs on samples left out of the model building.

TN: True negative.

TP: True positive.

Training Set: Samples used to establish a model.

Transformation: See rotation.

Transpose: Swapping rows and columns of a matrix.

Univariate: Measurements where there is single variable.

Validation: Determining how effective a model is.

Variable: A column of a matrix usually representing a frequency or wavelength.

Vector: Representation of data in a one-dimensional array, for example, a column or row of a matrix.

References

Adams, M.J. (2004) *Chemometrics in Analytical Spectroscopy*, 2nd edn, Royal Society of Chemistry, Cambridge.

Beebe, K.R., Pell, R.J. and Seasholtz, M.B. (1998) *Chemometrics: A Practical Guide*, Wiley, New York.

Brereton, R.G. (ed.) (1992) *Multivariate Pattern Recognition in Chemometrics, Illustrated by Case Studies*, Elsevier, Amsterdam.

Brereton, R.G. (2000) Introduction to multivariate calibration in analytical chemistry. *Analyst*, **125**, 2125–2154.

Brereton, R.G. (2003) *Chemometrics: Data Analysis for the Laboratory and Chemical Plant*, Wiley, Chichester.

Brereton, R.G. (2006) Consequences of sample sizes, variable selection, model validation and optimisation for predicting classification ability from analytical data. *Trends Anal. Chem.*, **25**, 1103–1111.

Brereton, R.G. (2007) *Applied Chemometrics for Scientists*, John Wiley & Sons, Chichester.

Brereton, R.G., Scott, D.R., Massart, D.L., Dessy, R.E., Hopke, P.K., Spiegelman, C.H. and Wegscheider, W. (eds) (1992) *Chemometrics Tutorials II*, Elsevier, Amsterdam.

Bruno, P., Caselli, M., Curri, M.L., Genga, A., Striccoli, R. and Traini, A. (2000) Chemical characterisation of ancient pottery from south of Italy by inductively coupled plasma atomic emission spectroscopy (ICP-AES) statistical multivariate analysis of data. *Anal. Chim. Acta*, **410**, 193–202.

De Braekeleer, K. and Massart, D.L. (1997) Evaluation of the orthogonal projection approach (OPA) and the SIMPLISMA approach on the Windig standard spectral data sets. *Chemometr. Intell. Lab Syst.*, **39**, 127–141.

Dixon, S.J., Xu, Y., Brereton, R.G., Soini, A., Novotny, M.V., Oberzaucher, E., Grammer, K. and Penn, D.J. (2007) Pattern recognition of gas chromatography mass spectrometry of human volatiles in sweat to distinguish the sex of subjects and determine potential discriminatory marker peaks. *Chemometr. Intell. Lab. Systems*, **87**, 161–172.

Efron, B. and Tibshirani, R.J. (1993) *An Introduction to the Bootstrap*, Chapman and Hall, New York.

Eriksson, L., Johansson, E., Kettaneh-Wold, N. and Wold, S. (2001) *Multi- and Megavariate Data Analysis: Principles and Applications*, Umetrics, Umeå.

Esbensen, K.H. (2002) *Multivariate Data Analysis in Practice*, CAMO, Oslo.

Everitt, B.S., Landau, S. and Leese, M. (2001) *Cluster Analysis*, Arnold, London.

Fisher, R.A. (1936) The use of multiple measurements in taxonomic problems. *Ann. Eugen.*, **7**, 179–188.

Gampp, H., Maeder, M., Meyer, C.J. and Zuberbuhler, A.D. (1985) Calculation of equilibrium constants from multiwavelength spectroscopic data–III. Model-free analysis of spectrophotometric and ESR titrations. *Talanta*, **32**, 1133–1139.

Geladi, P. and Kowalski, B.R. (1986) Partial least-squares regression: a tutorial. *Anal. Chim. Acta*, **185**, 1–17.

Gemperline, P.J. (ed.) (2006) *Practical Guide to Chemometrics*, 2nd edn, CRC, Boca Raton.

Haswell, S.J. (ed.) (1992) *Practical Guide to Chemometrics*, CRC, Boca Raton.

Hoskuldsson, A. (1988) PLS regression methods. *J. Chemometrics*, **2**, 211–228.

Joliffe, I.T. (1987) *Principal Components Analysis*, Springer-Verlag, New York.

Kowalski, B.R. (ed.) (1984) *Chemometrics: Mathematics and Statistics in Chemistry*, Reidel, Dordrecht.

Kowalski, B.R. and Bender, C.F. (1972) K-nearest neighbor classification rule (pattern-recognition) applied to nuclear magnetic-resonance spectral interpretation. *Anal. Chem.*, **44**, 1405–1411.

Kramer, R. (1998) *Chemometrics Techniques for Quantitative Analysis*, Marcel Dekker, New York.

Kresta, J.V., MacGregor, J.F. and Marlin, T.E. (1991) Multivariate statistical monitoring of process operating performance. *Can. J. Chem. Eng.*, **69**, 35–47.

Malinowski, E.R. (2002) *Factor Analysis in Chemistry*, 3rd edn, Wiley, New York.

Malinowski, E.R. and Howery, D.R. (1980) *Factor Analysis in Chemistry.*, Wiley, New York.

Mark, H. and Workman, J. (2003) *Statistics in Spectroscopy (Hardcover)*, 2nd edn, Academic Press, New York.

Martens, H. and Martens, M. (2000) *Multivariate Analysis of Quality*, Wiley, Chichester.

Martens, H. and Næs, T. (1989) *Multivariate Calibration*, Wiley, Chichester.

Massart, D.L., Brereton, R.G., Dessy, R.E., Hopke, P.K., Spiegelman, C.H. and Wegscheider, W. (eds) (1990) *Chemometrics Tutorials*, Elsevier, Amsterdam.

Massart, D.L. and Kaufmann, L. (1983) *The Interpretation of Analytical Chemical Data by the Use of Cluster Analysis*, Wiley, New York.

Massart, D.L., Vandeginste, B.G.M., Buydens, L.M.C., De Jong, S., Lewi, P.J. and Smeyers-Verbeke, J. (1997) *Handbook of Chemometrics and Qualimetrics Part A*, Elsevier, Amsterdam.

Massart, D.L., Vandeginste, B.G.M., Deming, S.N., Michotte, Y. and Kaufman, L. (1988) *Chemometrics: A Textbook*, Elsevier, Amsterdam.

Meloun, M., Militky, J. and Forina, M. (1992) *Chemometrics for Analytical Chemistry*, Vol. 1, Ellis Horwood, Chichester.

Meloun, M., Militky, J. and Forina, M. (1994) *Chemometrics for Analytical Chemistry*, Vol. 2, Ellis Horwood, Chichester.

Naes, T., Isaksson, T., Fearn, T. and Davies, T. (2002) *A User Friendly Guide to Multivariate Calibration and Classification*, NIR Publications, Chichester.

Otto, M. (1998) *Chemometrics : Statistics and Computer Applications in Analytical Chemistry*, Wiley-VCH, Weinheim.

Sharaf, M.A., Illman, D.L. and Kowalski, B.R. (1986) *Chemometrics*, Wiley, New York.

Tauler, R., Kowalski, B. and Fleming, S. (1993) Multivariate curve resolution applied to spectral data from multiple runs of an industrial-process. *Anal. Chem.*, **65**, 2040–2047.

US Department of Health and Human Services, FDA (2004) *Guidance for Industry PAT – A Framework for Innovative Pharmaceutical Development, Manufacturing, and Quality Assurance*, September 2004 (http://www.fda.gov/cder/guidance/6419fnl.htm).

Vandeginste, B.G.M., Derks, W. and Kateman, G. (1985) Multicomponent self-modeling curve resolution in high-performance liquid-chromatography by iterative target transformation analysis. *Anal. Chim. Acta*, **173**, 253–264.

Vandeginste, B.G.M., Massart., D.L., Buydens, L.M.C., de Jong, S., Lewi, P.J. and Smeyers-Verbeke, J. (1998) *Handbook of Chemometrics and Qualimetrics Part B*, Elsevier, Amsterdam.

Windig, W., Heckler, C.E., Agblebor, F.A. and Evans, R.J. (1992) Self-modeling mixture analysis of categorized pyrolysis mass-spectral data with the SIMPLISMA approach. *Chemometr. Intell. Lab Syst.*, **14**, 195–207.

Wold, S. (1978) Cross-validatory estimation of number of components in factor and principal components models. *Technometrics*, **20**, 397–405.

Wold, S., Esbensen, K. and Geladi, P. (1987) Principal components analysis. *Chemometr. Intell. Lab Syst.*, **2**, 37.

Further Reading

Adams, M.J. (2004) *Chemometrics in Analytical Spectroscopy*, 2nd edn, Royal Society of Chemistry, Cambridge.

Beebe, K.R., Pell, R.J., and Seasholtz, M.B. (1998) *Chemometrics : a Practical Guide*, John Wiley & Sons, New York.

Brereton, R.G. (1990) *Chemometrics : Applications of Mathematics and Statistics to Laboratory Systems*, Ellis Horwood, Chichester.

Brereton, R.G. (ed) (1992) *Multivariate Pattern Recognition in Chemometrics, Illustrated by Case Studies*, Elsevier, Amsterdam.

Brereton, R.G. (2003) *Chemometrics : Data Analysis for the Laboratory and Chemical Plant*, John Wiley & Sons, Chichester.

Brereton, R.G. (2007) *Applied Chemometrics for Scientists*, John Wiley & Sons, Chichester.

Brereton, R.G., Scott, D.R., Massart, D.L., Dessy, R.E., Hopke, P.K., Spiegelman, C.H., and Wegscheider, W. (eds) (1992) *Chemometrics Tutorials II*, Elsevier, Amsterdam.

Coomans, D. and Broeckaert, I. (1986) *Potential Pattern Recognition in Chemical and Medical Decision Making*, Research Studies Press, Letchworth.

De Maesschalck, R., Jouan-Rimbaud, D., Massart, D.L. (2000) The Mahalanobis distance. *Chemometr. Intell. Lab Syst.*, **50**, 1–18.

Eriksson, L., Johansson, E., Kettaneh-Wold, N., and Wold, S. (2001) *Multi- and Megavariate Data Analysis: Principles and Applications*, Umetrics, Umeå.

Esbensen, K.H. (2002) *Multivariate Data Analysis in Practice*, CAMO, Oslo.

Gemperline, P.J. (ed) (2006) *Practical Guide to Chemometrics*, 2nd edn, CRC, Boca Raton, FL.

Jurs, P.C. and Isenhour, T.L. (1975) *Chemical Applications of Pattern Recognition*, John Wiley & Sons, New York.

Kowalski, B.R. (ed) (1984) *Chemometrics: Mathematics and Statistics in Chemistry*, Reidel, Dordrecht.

Kramer, R. (1998) *Chemometrics Techniques for Quantitative Analysis*, Marcel Dekker, New York.

Mark, H. and Workman, J. (2007) *Chemometrics in Spectroscopy*, Elsevier, London.

Massart, D.L., Brereton, R.G., Dessy, R.E., Hopke, P.K., Spiegelman, C.H., and Wegscheider, W. (eds) (1990) *Chemometrics Tutorials*, Elsevier, Amsterdam.

Massart, D.L. and Kaufmann, L. (1983) *The Interpretation of Analytical Chemical Data by the Use of Cluster Analysis*, John Wiley & Sons, New York.

Massart, D.L., Vandeginste, B.G.M., Buydens, L.M.C., De Jong, S., Lewi, P.J., and Smeyers-Verbeke, J. (1997) *Handbook of Chemometrics and Qualimetrics Part A*, Elsevier, Amsterdam.

Massart, D.L., Vandeginste, B.G.M., Deming, S.N., Michotte, Y., and Kaufman, L. (1988) *Chemometrics : A Textbook*, Elsevier, Amsterdam.

Meloun, M., Militky, J., and Forina, M. (1992, 1994) *Chemometrics for Analytical Chemistry Vols 1 and 2*, Ellis Horwood, Chichester.

Miller, J.N. and Miller, J.C. (2005) *Statistics and Chemometrics for Analytical Chemistry*, 5th edn, Pearson, Harlow.

NIST *Engineering Statistics Handbook*, http://www.itl.nist.gov/div898/handbook/. 20 May 2009.

Otto, M. (2007) *Chemometrics : Statistics and Computer Applications in Analytical Chemistry*, Wiley-VCH Verlag GmbH, Weinheim.

Sharaf, M.A., Illman, D.L., and Kowalski, B.R. (1986) *Chemometrics*, John Wiley & Sons, New York.

Strouf, O. (1986) *Chemical Pattern Recognition*, Research Studies Press, Letchworth.

Vandeginste, B.M.G., Massart, D.L., Buydens, L.M.C., de Jong, S., Lewi, P.J., and Smeyers-Verbeke, J. (1998) *Handbook of Chemometrics and Qualimetrics Part B*, Elsevier, Amsterdam.

Varmuza, K. (1980) *Pattern Recognition in Chemistry*, Springer, Berlin.

Varmuza, K. and Filmoser, P. (2009) *Introduction to Multivariate Statistical Analysis in Chemometrics*, CRC, Boca Raton, FL.

32
Fourier and Other Mathematical Transforms

Ronald N. Bracewell[†]

32.1	Introduction	1151
32.2	The Fourier Transform	1151
32.3	Continuous versus Discrete Transforms	1152
32.4	Some Common Transforms	1155
32.5	The Laplace Transform	1155
32.6	Convergence Conditions	1158
32.7	Why Transforms are Useful	1159
32.8	Fields of Application	1160
32.9	The Hartley Transform	1162
32.10	The Fast Fourier Transform	1163
32.11	The Fast Hartley Algorithm	1164
32.12	The Mellin Transform	1165
32.13	The Hilbert Transform	1165
32.14	Multidimensional Transforms	1166
32.15	The Hankel Transform	1167
32.16	The Abel Transform	1167
32.17	Tomography and the Radon Transform	1168
32.18	The Walsh Transform	1170
32.19	The Z Transform	1171
32.20	Convolution	1172
32.21	Summary	1173
	Glossary	1173
	References	1174
	Further Reading	1175

Encyclopedia of Applied Spectroscopy. Edited by David L. Andrews.
Copyright © 2009 WILEY-VCH Verlag GmbH & Co. KGaA, Weinheim
ISBN: 978-3-527-40773-6

32.1
Introduction

Fourier analysis, which gained prominence from the work of J. B. J. Fourier in the early 1800s, led immediately to applications in mechanics and heat conduction. It also contributed to the advance of pure mathematics as regards the basic notions of limit, convergence, and integrability; the impact on mathematics and applied physics has continued to this day. Applications of transform methods were developed in connection with differential and integral equations and became very powerful; more recently, numerical analysis, aided by electronic computing, has added an extra dimension to the applied relevance of mathematical transforms and especially of the Fourier transform. The analytical and computational aspects are dealt with first; among applied examples, heat conduction, Fourier-transform spectroscopy, diffraction, sampled data, and tomography are mentioned.

When one looks for antecedents from which Fourier analysis might have evolved, they are not hard to find. Euler had published trigonometric series, and the sum to infinity, in such statements as

$$\sin x - \tfrac{1}{2}\sin 2x + \tfrac{1}{3}\sin 3x + \cdots = \tfrac{1}{2}x \qquad (32.1)$$

Gauss analyzed motion in astronomical orbits into harmonics and indeed utilized the fast algorithm now favored for computing. Much earlier in Roman times, Claudius Ptolemy expressed motion in planetary orbits by the geometrical equivalent of trigonometric series and, according to Neugebauer (1983), the idea of epicycles has roots in Mesopotamian astronomy where the solar motion was matched by zigzag functions, rough approximations of the sinusoids to come.

32.2
The Fourier Transform

There are many transforms, each characterized by its own explicit operator, which we may call as **T**. The operand, or entity operated on, is a function such as $f(x)$, where x is a real variable ranging from $-\infty$ to ∞. The notation $\mathbf{T}\{f(x)\}$ signifies the outcome of applying the operator **T** to the function $f(x)$. To illustrate, the operation that converts a given function $f(x)$ to its Fourier transform, which is a different function $F(s)$, is as follows: "Multiply the function $f(x)$ by $\exp(-i2\pi s x)$ and integrate with

Encyclopedia of Applied Spectroscopy. Edited by David L. Andrews.
Copyright © 2009 WILEY-VCH Verlag GmbH & Co. KGaA, Weinheim
ISBN: 978-3-527-40773-6

respect to x from $-\infty$ to ∞". Applying this operation to $f(x) = \exp(-|x|)$, we find that $\mathbf{T}\{f(x)\} = F(s) = 2/[1 + (2\pi s)^2]$, which is the Fourier transform of $\exp(-|x|)$. The symbolic expression of the Fourier transform operation is

$$F(s) = \int_{-\infty}^{\infty} f(x) e^{-i 2\pi s x} dx \quad (32.2)$$

It is apparent that any particular value of $F(s)$ (for example, $F(2)$, which equals 0.0126) takes into account the whole range of x; that is, the value depends on the shape of $f()$ as a whole, not on any single point. Thus the Fourier operation is quite unlike the operation that converts $f(x) = \exp(-|x|)$ to $\sin[\exp(-|x|)]$; the outcome of this latter operation is referred to as a *function of a function*, and the resulting values each depend only on the single value of x. When the result depends on the shape of $f(x)$ over part or all of the range of x, an entity such as $F(s)$ is called a *functional* of $f()$. The variable s is called the *transform variable* and may have a physical meaning; if so, its units will be cycles per unit of x. A short list of Fourier transforms for illustration is shown in Table 32.1.

In this list, rect x is the unit rectangle function (equal to unity where $|x| < 0.5$, elsewhere zero) and sinc $x = (\sin \pi s)/\pi s$. The last five lines are representative theorems of the form, "If $f(x)$ has Fourier transform $F(s)$, then [modification of $f(x)$] has transform [modification of $F(s)$]". Extensive lists of such transform pairs and theorems are available from the reference texts; the short list given would cover a sizable fraction of the analytic forms encountered in the literature.

With some transforms – the Abel transform is an example – each transform value depends on only a part of, not all of, $f()$; and with other transforms, the transform variable does not necessarily have a different identity (as s is different from x) but may have the same identity (Hilbert transform). The integral from $-\infty$ to x is a transform with both of the above restrictive properties.

All the transforms dealt with here are linear transforms, which are the commonest type; they all obey the superposition rule that $\mathbf{T}\{f_1(x) + f_2(x)\} = \mathbf{T}\{f_1(x)\} + \mathbf{T}\{f_2(x)\}$ for any choice of the given functions $f_1(x)$ and $f_2(x)$. An example of a nonlinear transformation is provided by $\mathbf{T}\{f(x)\} = a + bf(x)$, as may be tested by reference to the superposition definition; clearly the term *linear* in "linear transform" does not have the same meaning as in Cartesian geometry.

Tab. 32.1 Selected Fourier transforms. The quantity a is a constant.

F(x)	F(s)
$e^{-|x|}$	$2/[1 + (2\pi s)^2]$
$\delta(x)$	1
$\cos(2\pi x/a)$	$\frac{1}{2}\delta(s + a^{-1}) + \frac{1}{2}\delta(s - a^{-1})$
rect x	sinc s
$e^{-\pi x^2}$	$e^{-\pi s^2}$
$e^{-\pi(x/a)^2}$	$|a| e^{-\pi(as)^2}$
$f(x/a)$	$|a| F(as)$
$f(x + a)$	$e^{i 2\pi as} F(s)$
$f'(x)$	$i 2\pi s F(s)$
Autocorrelation of $f(x)$	$|F(s)|^2$
$\int_{-\infty}^{\infty} f(x - u) g(u) du$	$F(s) G(s)$

32.3 Continuous versus Discrete Transforms

Before defining the main transforms succinctly by their operations $\mathbf{T}$, all of which involve integration over some range, it is worth commenting on a numerical aspect. One could take the point of view, as is customary with numerical integration, that the desired integral is an entity in its own right; that the integral may on occasion be subject to precise evaluation in analytic terms, as with $F(s) = 2/[1 + (2\pi s)^2]$; and that if numerical methods are required, a sum will be evaluated that is an approximation to the desired integral. One would then discuss the desired degree of approximation and how to reach it. Now this is quite unlike the customary way of thinking about the discrete Fourier transform. What we evaluate is indeed a sum, but we consider the sum as precise and not as an approximation to an integral. There are excellent reasons for this. Meanwhile, the important thing to realize is that there are both a Fourier transform and a discrete Fourier transform, each with its own definition. The discrete Fourier transform operation is

$$F(\nu) = \frac{1}{N} \sum_{\tau=0}^{N-1} f(\tau) e^{-i2\pi \nu \tau / N} \qquad (32.3)$$

The word "discrete" is used in antithesis to "continuous," and in the cases discussed here, it means that an independent variable assumes integer values. To understand the discrete Fourier transform, which is exclusively what we compute when in numerical mode, it is best to forget the Fourier integral and to start a fresh. Instead of starting with a complex function $f(x)$ that depends on the continuous real variable x, we start with N data (complex in general, but often real) indexed by an integer serial number τ (like time) that runs from 0 to $N - 1$. In the days when FORTRAN did not accept zero as a subscript, summation from $\tau = 0$ caused much schizophrenia, but the mathematical tradition of counting from zero prevailed and is now unanimous. In cases where $f(\ )$ is a wave form, as it often is, the quantity τ can be considered as time that is counted in units starting from time zero. Clearly, N samples can never fully represent $\exp(-|x|)$, for two reasons: the samples take no account of the function where x exceeds some finite value, and no account is taken of fine detail between the samples. Nevertheless, one may judge that, for a given particular purpose, 100 samples will suffice, and the confidence to judge may be bolstered by trying whether acquisition of 200 samples significantly affects the purpose in hand. Numerical intuition as developed by hand calculation has always been a feature of mathematical work but was regarded as weak compared with physical intuition. Nowadays, however, numerical intuition is so readily acquired that it has become a matter of choice whether to deal with questions about the size of N by traditional analytic approaches. A new mix of tools from analysis, finite mathematics, and numerical analysis is evolving.

The discrete transform variable ν reminds us of frequency. If τ is considered as time measured in integral numbers of seconds, then ν is measured in cycles per second, and is indeed like frequency (cycles per second or Hertz), but not exactly. It is ν/N that gives correct frequencies in Hertz, and then only for $\nu \leq N/2$. Where ν exceeds $N/2$, we encounter a domain where the discrete approach conflicts with the continuous. When the Fourier transform is evaluated as an integral, it is quite ordinary to contemplate negative values of s, and a graph of $F(s)$ will ordinarily have the vertical $s = 0$ axis in the middle, giving equal weight to positive and negative "frequencies." (The unit of s is always cycles per unit of x; if x is in meters, s will be a spatial frequency in cycles per meter; if x is in seconds, s will be a temporal frequency in cycles per second, or hertz.) However, the discrete Fourier transform, as conventionally defined, explicitly requires the transform variable ν to range from 0 to $N-1$, not exhibiting negative values at all. There is nothing wrong with that, but persons coming from continuous mathematics or from physics may like to know that, when ν is in the range from $N/2$ to $N-1$, the quantities $N-\nu$ correspond to the negative frequencies familiar to them as residing to the left of the origin on the frequency axis. This is because the discrete transform is periodic in ν, with period N.

In the familiar Fourier series

$$p(x) = a_0 + \sum_1^\infty (a_\nu \cos 2\pi \nu x + b_\nu \sin 2\pi \nu x) \quad (32.4)$$

for a periodic function $p(x)$ of period 2π, the first term a_0 represents the direct current, zero frequency, or mean value over one period as calculated from

$$a_0 = \frac{1}{2\pi} \int_0^{2\pi} p(x)\, dx \quad (32.5)$$

Therefore, the first term $F(0)$ of the discrete Fourier transform is the average of the N data values. This is the reason for the factor $1/N$ in front of the summation sign in Eq. (32.3), a factor that must be remembered when checking. In practical computing, it is efficient to combine the factor $1/N$ with other factors such as calibration factors and graphical scale factors that are applied later at the display stage. The remaining Fourier coefficients, given by

$$a_\nu = \frac{1}{\pi} \int_0^{2\pi} p(x) \cos 2\pi \nu x\, dx \quad (32.6)$$

$$b_\nu = \frac{1}{\pi} \int_0^{2\pi} p(x) \sin 2\pi \nu x\, dx \quad (32.7)$$

are related to the discrete Fourier transform by $a_\nu - ib_\nu = F(\nu)$. The minus sign arises from the negative exponent in the Fourier kernel $e^{-i2\pi sx}$. The reason for the choice of the negative exponent is to preserve the convention that d/dt be replaceable by $+i\omega$ in the solution of linear differential equations, as when the impedance of an inductance L to alternating voltage of angular frequency ω is written as $+i\omega L$ (more usually as $j\omega L$).

How to decide whether the discrete Fourier transform is an adequate approximation to the Fourier transform is a very interesting question. But the question itself is open to challenge. If I am studying cyclicity in animal populations, perhaps seasonal influence on bird migration, I may start with 365 reports of how many birds were seen each day of the year. In such a case, and in many other cases, discrete data mean that the integrals, even though convenient, are themselves the approximations; the discrete Fourier transform, given N equispaced data, is a valid entity in its own right. However, unexpected discrepancies may arise over the choice of N, which may be very large or very small. Slow computing (N very large), unwanted sensitivity to measurement error (N very small), and aliasing are its negative consequences. Aliasing is the word for the following phenomenon. Measurements are made of some time-varying phenomenon at regularly spaced time intervals, perhaps temperature is recorded twice a day or perhaps speech samples are taken at a 10-kHz rate. Such data can represent harmonic components with period longer than one day or longer than 2×10^{-4} s, but cannot faithfully follow faster harmonic variation. The samples will not ignore the presence of such high frequencies, because the high-frequency variations will indeed be sampled, but the samples will be consistent with, and indistinguishable from, a long-period sinusoidal component that is not actually present. The imperfectly sampled component emerges under the alias of a lower, counterfeit frequency.

32.4
Some Common Transforms

As a convenient reference source, definitions of several transforms (Laplace, Fourier, Hartley, Mellin, Hilbert, Abel, Hankel, Radon) are presented in Table 32.2. When one has the transform, there is a way of returning to the original function in all the cases chosen. In some cases, the inverse operation $\mathbf{T}^{-1}$ is the same as the defining operation $\mathbf{T}$ (e.g., Hartley and Hilbert, which are reciprocal transforms), but the majority differ, as shown. In addition, examples of each transform are presented. These will be found to convey various general properties at a glance and may be helpful for numerical checking.

32.5
The Laplace Transform

A long and diverse history (Deakin, 1985) characterizes the Laplace transform, which was in use long before Laplace, but became known to current generations mainly through its pertinence to the linear differential equations of transient behavior in electricity and heat conduction. Many tough technological problems of electric circuits that arose in connection with telegraphy, submarine cables, and wireless, and related industrial-process problems of thermal diffusion, were cracked around the turn of the century, sometimes by novel methods such as those of Heaviside (1970), which were to be justified subsequently (Nahin, 1987) to the satisfaction of academic mathematics by systematic application of the Laplace transform. Heaviside is remembered for stimulating the application of the Laplace transform to convergence of series and for Maxwell's equations, the delta function, the Heaviside layer, impedance, nonconvergent series that are useful for computing, fractional order derivatives and integrals, and operational calculus.

Table 32.2 gives, as an example, the Laplace transform of $f(x) = \exp(-x - 1.5)\mathbf{H}(x + 1.5)$. The Heaviside unit step function $\mathbf{H}(x)$ jumps, as x increases from 0 to 1, the jump being where $x = 0$; one of its uses is as a multiplying factor to allow algebraic expression of functions that switch on. The transform of $f(x)$, which is easy to verify by integration, is $(\exp 1.5s)/(1 + s)$; the transform variable s may be complex but must lie among those numbers whose real parts are greater than -1 (otherwise the integral does not exist). It is rather cumbersome to exhibit the complex transform graphically on the complex plane, and so an illustration is omitted. To invert the transform requires integration on the complex plane along a semicircular contour with indentations if necessary to circumvent points where the integrand goes to infinity (poles). The

Tab. 32.2 Transform definitions, inverses, and examples. (Ticks are at unit spacing.

Name of transform	Nature of function domain variable	Example of function	Nature of transform variable	Nature of transform	Example of transform	Defining formula and the inverse
Laplace	Continuous real	$f(x) = e^{-x-1.5}H(x+1.5)$	Continuous complex	Complex	$\dfrac{e^{1.5s}}{1+s}, \; -1 < \operatorname{Re} s$	$F_L(s) = \int_{-\infty}^{\infty} f(x)e^{-sx}dx$ $\dfrac{1}{2\pi i}\int_{c-i\infty}^{c+i\infty} F_L(s)e^{xs}ds$
Fourier	Continuous real	$f(x) = e^{-x-1.5}H(x+1.5)$	Continuous real	Complex	$F(s) = \dfrac{e^{i3\pi s}}{1+i2\pi s}$	$F(s) = \int_{-\infty}^{\infty} f(x)e^{i2\pi sx}dx$ $f(x) = \int_{-\infty}^{\infty} F(s)e^{i2\pi xs}ds$
Discrete Fourier	Discrete, real	$f(\tau) = e^{-\tau-1.5}H(\tau+1.5)$	Discrete, real	Complex		$F(v) = N^{-1}\sum_{\tau=0}^{N-1} f(\tau)e^{-i2\pi v\tau}$ $f(\tau) = \sum_{v=0}^{N-1} F(v)e^{i2\pi \tau v}$
Hartley[a]	Continuous real	$f(x) = e^{-x-1.5}H(x+1.5)$	Continuous real	Real	$H(s) = \dfrac{\operatorname{cas}(-3\pi s) + 2\pi s\operatorname{cas}3\pi s}{1+4\pi^2 s^2}$	$H(s) = \int_{-\infty}^{\infty} f(x)\operatorname{cas}2\pi sx\,dx$ $f(x) = \int_{-\infty}^{\infty} H(s)\operatorname{cas}2\pi xs\,ds$
Discrete Hartley	Discrete, real	$f(\tau) = e^{-\tau-1.5}H(\tau+1.5)$	Discrete, real	Real		$H(v) = N^{-1}\sum_0^{N-1} f(\tau)\operatorname{cas}2\pi v\tau$ $f(\tau) = \sum_{v=0}^{N-1} H(v)\operatorname{cas}2\pi v\tau$
Mellin	Continuous real	$f(x) = e^{-x}H(x)$	Discrete, real	Real		$F_M(s) = \int_0^{\infty} f(x)x^{s-1}dx$

32.5 The Laplace Transform

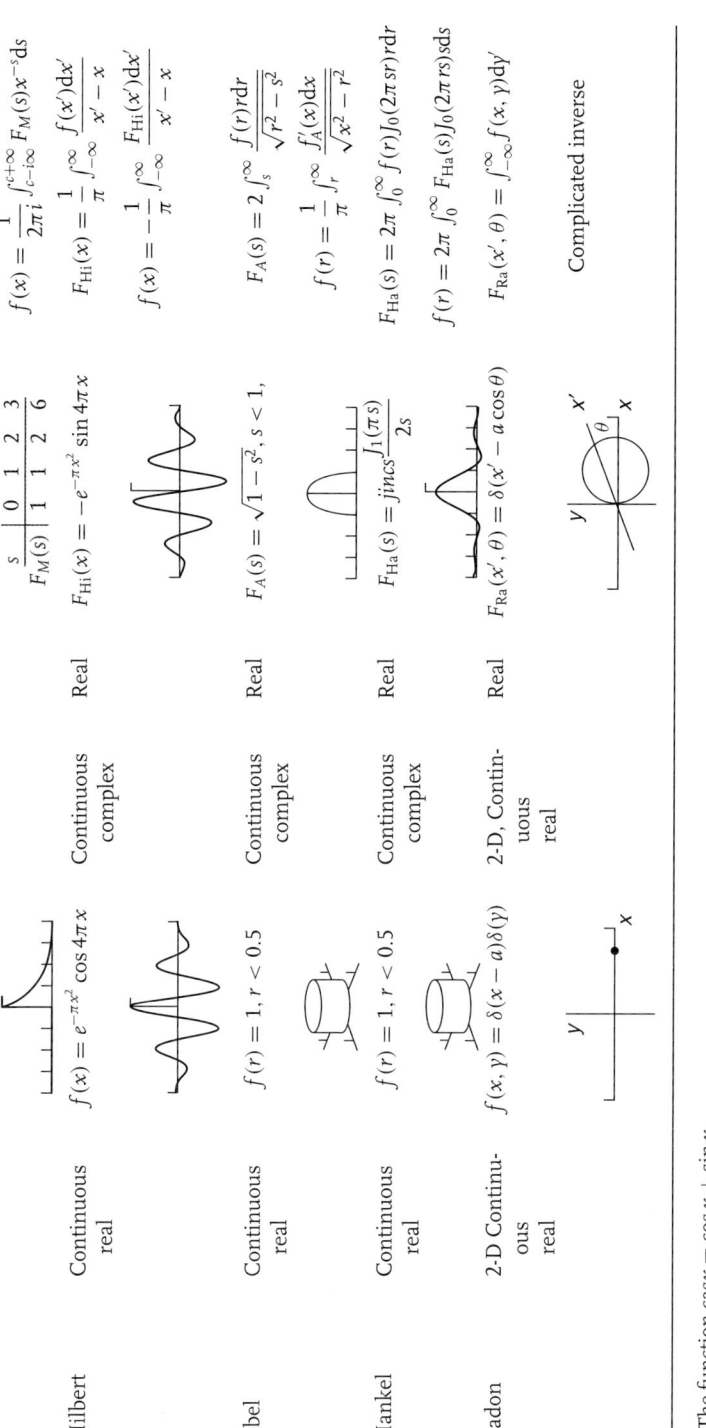

Hilbert	Continuous real	$f(x) = e^{-\pi x^2}\cos 4\pi x$	Continuous complex	$F_{\text{Hi}}(x) = -e^{-\pi x^2}\sin 4\pi x$
Abel	Continuous real	$f(r) = 1, r < 0.5$	Continuous complex	$F_A(s) = \sqrt{1-s^2}, s < 1$
Hankel	Continuous real	$f(r) = 1, r < 0.5$	Continuous complex	$F_{\text{Ha}}(s) = jincs\dfrac{J_1(\pi s)}{2s}$
Radon	2-D Continuous real	$f(x,y) = \delta(x-a)\delta(y)$	2-D, Continuous real	$F_{\text{Ra}}(x',\theta) = \delta(x' - a\cos\theta)$

Transforms (right column):

$$f(x) = \frac{1}{2\pi i}\int_{c-i\infty}^{c+\infty} F_M(s)x^{-s}\,ds$$

$$F_{\text{Hi}}(x) = \frac{1}{\pi}\int_{-\infty}^{\infty}\frac{f(x')\,dx'}{x'-x}$$

$$f(x) = -\frac{1}{\pi}\int_{-\infty}^{\infty}\frac{F_{\text{Hi}}(x')\,dx'}{x'-x}$$

$$F_A(s) = 2\int_s^\infty \frac{f(r)r\,dr}{\sqrt{r^2-s^2}}$$

$$f(r) = -\frac{1}{\pi}\int_r^\infty \frac{f_A'(x)\,dx}{\sqrt{x^2-r^2}}$$

$$F_{\text{Ha}}(s) = 2\pi\int_0^\infty f(r)J_0(2\pi sr)r\,dr$$

$$f(r) = 2\pi\int_0^\infty F_{\text{Ha}}(s)J_0(2\pi rs)s\,ds$$

$$F_{\text{Ra}}(x',\theta) = \int_{-\infty}^\infty f(x,y)\,dy$$

Complicated inverse

s	0	1	2	3
$F_M(s)$	1	1	2	6

[a] The function $\operatorname{cas} x = \cos x + \sin x$.

constant c in the inversion formula is to be chosen to the right of all poles.

To some extent, Laplace transforms were computed numerically, but more typically, development led to compilations of analytic transforms resembling the tables of integrals (Campbell and Foster, 1948; Erdélyi et al., 1954). Programs for deriving the Laplace transform of the impulse response from electrical networks given diagrammatically are also available. Consequently, it is hardly ever necessary to analytically derive Laplace transforms today. The analytic solution of transients in electric circuits, a subject traditionally used for sharpening the minds of electrical engineers, is obsolescent because impulse responses and transfer functions have been concisely published (McCollum and Brown, 1965). Furthermore, the advent of integrated circuits has meant that inductance is seldom included in new designs, and those circuits containing more than two or three elements have become less common. Mature programs are also available for step-by-step integration of circuit differential equations.

On the numerical side, the Laplace transform has also been largely eroded by use of the Fourier transform. This is because angular frequency ω is a mathematically real quantity, and it ought to be possible to compute the behavior of an electrical, acoustical, or mechanical system without reference to a complex frequency $\omega - i\sigma$. Certainly, the Laplace transform is computable over its strip of convergence from any single slice therein. Nevertheless, practitioners of control theory find it convenient to assume the complex plane of s in terms of poles and zeros that are off the real frequency axis, and theirs is one tradition that keeps the complex plane alive; the convenience stems from the fact that the Laplace transform is analytic, and thus specifiable by its poles and isolated zeroes. There are problems that used to be handled by the Laplace transform, paying strict attention to the strip of convergence, because the Fourier integral did not converge; but these situations are now universally handled by Fourier methodology with the aid of delta-function notation for impulses and their derivatives, and no longer call for special treatment. When it comes to discrete computing, the impulse, and its associated spectrum reaching to indefinitely large frequencies, may in any case be forgotten. Thus, it has been wondered (Körner, 1988) "whether the Laplace transform will keep its place in the standard mathematical methods course for very much longer," but it will never die out; a new balance between curricular segments will be struck.

32.6
Convergence Conditions

Much attention used to be given to the existence of the Fourier integral because of paradoxes with such wanted entities as $f(x) = 1, f(x) = \cos x$, or $f(x) = \delta(x)$, where $\delta(x)$ is the unit impulse at $x = 0$, none of which possessed a Fourier integral. Today, we reason as a physicist would, recognizing that a voltage waveform cannot have a value of 1 V forever, but must have turned on at some time in the past and will turn off at some time in the future. The finite-duration function does have a Fourier transform. We then consider a sequence of waveforms of longer and longer duration and the corresponding sequence of transforms, arriving at the concept of "transforms in the limit." This attitude has received mathematical respectability under the rubric of generalized functions (Lighthill, 1958) and is the basis for saying that the Fourier transform of $\delta(x)$ is 1 (while conversely the Fourier transform of 1 is $\delta(s)$). The elaborate conditions for the existence of a transform when generalized functions were excluded have thus lost interest. Even $\delta'(x)$ now has the indispensable transform $i2\pi s$; under the rules of analysis, $\delta'(x)$ was an unimaginable entity, certainly not qualifying as a function of x; it was considered as a commonplace dipole by physicists, and as a local load such as a moment applied at a point on a beam in mechanics.

The fact that the Laplace integral converged when the Fourier transform did not give the Laplace transform a certain prestige, even though convergence was achieved at the cost of tapering the given function by a real, exponentially decaying factor. In addition, the strip of convergence had to be specified for the complex transform variable s. The convenience of dealing with the real and physically intuitive frequency as the transform variable has shifted preference in favor of the Fourier and Hartley transforms. The only effective condition for the existence of a Fourier or Hartley transform today is that the given function should have a physical interpretation, or be representable by a sequence of physically interpretable functions whose individual transforms approach a limit. Consequently, it is no longer necessary to require that $f(x)$ be absolutely integrable ($\int_{-\infty}^{\infty} |f(x)| dx$ exists) or that any discontinuities be finite; on the contrary, the "shah function" $III(x) = \sum_{n=-\infty}^{n=\infty} \delta(x-n)$, which could be said to possess an infinite number of infinite discontinuities, now has a Fourier transform thanks to the theory of generalized functions (Bracewell, 1956). Interestingly, the Fourier transform of $III(x)$ is $III(s)$.

The function $\sin(x^{-1})$ raises a convergence question as a result of possessing an infinite number of maxima in any interval containing $x = 0$; this sort of behavior is without interest in the world of numerical computing but of considerable interest to the theory of integration. Possession of an infinite number of maxima does not in itself define the convergence condition because the Fourier integral may converge if the amplitude of the oscillation dies down so that the function exhibits bounded variation. Nor does bounded variation define the convergence condition because Lipschitz has demonstrated functions of unbounded variation whose Fourier integrals converge. However, the Lipschitz condition is not the ultimate convergence condition, as has been shown by Dini (Bracewell, 1986a). This style of analysis has lost practitioners as activity has moved in the direction of finite, or discrete, mathematics.

32.7
Why Transforms are Useful

Many problems can be posed in the form of a differential equation (or a difference equation, or an integral equation, or an integro-differential equation) that has to be solved for some wanted function subject to stated boundary conditions or initial conditions. Laplace's equation in three dimensions describes the potential distribution set up by an array of electric charges, and the diffusion equation describes the heat flow distribution set up by a given distribution of heat. By applying a transformation such as the Laplace or Fourier to each term of such an equation, we arrive at a new equation that describes the transform rather than the original wanted function. The interesting thing about this is that the new equation may be simpler, sometimes solvable just by algebra. We solve that equation for the transform of the solution and then invert. Not all differential equations are simplified in this way; those that do are characterized by linearity and coordinate invariance (such as time invariance), and the presence of these characteristics in nature is responsible for a good deal of the numerical activity with transforms. Transfer functions, such as the frequency response curves of amplifiers, are corresponding manifestations of these same characteristics. The passage of a speech waveform through an amplifier is described by a differential equation that may be hard to solve; but having used a Fourier transform to go to the frequency domain, we apply the transfer function, frequency by frequency, by complex multiplication to get the transform of the output. Then retransforming gives the output waveform.

There is also a differential equation, describing the bending of a beam under the influence of a load distribution that may be considered as a spatial input analogous to an input waveform, whereas the curve of deflection is analogous to the output waveform. Although Hooke's law, the first of the linear laws, may apply, we do not use transform methods. If we analyze the load distribution into spatially sinusoidal components and find the bending response to each component and linearly sum the responses, we will get the desired shape of the bent beam, but there is no transfer function to facilitate getting the individual responses by simple algebra. The reason is that we have linearity but not space invariance – if we shift the load, the response does not shift correspondingly without change of shape; a sinusoidal load does not produce sinusoidal deflection. If, on the contrary, we delay the input to an amplifier or a vibratory mechanical system, the response is correspondingly delayed but is unchanged as to shape; furthermore, a sinusoidal input produces a sinusoidal output.

32.8
Fields of Application

Fourier (Grattan-Guinness, 1972) originally thought of representing the temperature on a heat-conducting bar as the sum of sinusoids. To avoid a problem of integration, he considered the bar to be bent around on itself in a large circle, a distortion that is not harmful to the discussion of any given finite straight bar because the arc of interest can be made as straight as you wish by taking the circle large enough. Because the temperature distribution on the ring is of necessity now periodic in space, only a fundamental and harmonics need to be considered, plus the constant temperature a_0 representing the mean temperature. As time elapses, the temperature distribution varies as the heat flows under the influence of the temperature gradients, ultimately approaching the uniform value a_0 in the limit. Fourier found that the component sinusoids decay exponentially with a time constant proportional to the spatial period, or wavelength, the nodes of each sinusoid remaining fixed. By attenuating each component in accordance with the elapsed time and summing one gets the same result as if the spatially variable heat flow were followed in real time. This is an example of the duality of the function domain (space domain in this instance) and the transform domain (spatial frequency domain) that permeates Fourier applications.

Music can be considered in terms of the wave form of the wave that conveys the sound through the air (function domain), or in terms of the harmonic constituents (spectral domain) that are separately discernible by the ear and are treated separately by an amplifier. In crystallography, there is the arrangement of the atoms in space (crystal lattice domain) and the spatial Fourier components (reciprocal lattice domain) that under illumination by X-rays or neutron beams, evidence themselves by diffraction at defined angles. Image formation with cameras and radio telescopes can be conceived as operating on the object domain, or "sky plane," or we can think in terms of complex coherence measurements in the transform domain. All these dual modes of thought, under their respective terminologies, are fully equivalent; it helps to be familiar with both and to be able to translate from one domain to the other. In addition, it is most helpful to be able to translate between fields, converting a problem in one subject into the analogous problem in another subject where the solution may be intuitively obvious. As an example, persons who know very well that the diffraction from a pair of side-by-side pinholes is sinusoidal in space may not know that the spectrum of a pair of audible clicks in succession is sinusoidal in frequency. How much richer this knowledge becomes when they are able to translate from acoustics to optics and vice versa!

As a formal illustration of the methodology, let us calculate the response of an electric circuit consisting of an inductance L in series with a resistance R to which a voltage impulse of strength A is applied at $t = -1.5$. Equating the sum of the voltages in the circuit to zero, as taught by Kirchhoff, gives the differential equation

$$A\delta(t) = L\frac{di}{dt} + Ri \qquad (32.8)$$

where $i(t)$ is the current flow in response to the applied voltage. Taking the Fourier transforms term by term (Table 32.1), we find that

$$Ae^{i2\pi \times 1.5 s} = i2\pi s LI(s) + RI(s) \qquad (32.9)$$

where $I(s)$ is the transform of the wanted current. Solving this algebraic equation gives

$$I(s) = \frac{Ae^{i2\pi \times 1.5 s}}{R + i2\pi Ls} \qquad (32.10)$$

and taking the inverse Fourier transforms of both sides give

$$i(t) = \frac{A}{L}e^{-R(t+1.5)/L}\mathrm{H}(t+1.5) \qquad (32.11)$$

The transform involved is illustrated in Table 32.2 for the Fourier transform. The method for solving the same problem by the Laplace transform is similar but involves reference to convergence of the integral, a complication that is circumvented when generalized function theory is combined with the Fourier integral.

Newton showed how to split sunlight into its constituent colors with a prism, where we think in the spatial domain, but there is another way that we learned from Michelson that is explicable in the time domain. We split a beam of light, and then recombine the two beams on a photodetector, but not before a controlled delay is introduced into one of the beams, for example, by retroreflection from a movable plane mirror. The detector output reveals the autocorrelation of the light beam from which, by using the autocorrelation theorem (Table 32.1) and numerical Fourier transformation, we get the spectral distribution of power.

32.9
The Hartley Transform

Table 32.2 illustrates by example that the Fourier transform in general is a complex function of the real transform variable s; consequently, two transform curves must be drawn, one for the real part and one (broken) for the imaginary part. The example $f(\tau)$ for the discrete Fourier transform is based on samples of the previous $f(x)$. Imaginary values of the discrete transform $F(\nu)$ are shown as hollow circles. Three features may be noted: no matter how closely samples are spaced, some detail can be missed; no outlying parts beyond a finite range are represented; the indexing convention 0 to $N-1$ has the effect of cutting off the left side of $F(s)$, translating it to the right, and reconnecting it. To convey the nature of this third comment, the points for $\tau > N/2$ have been copied back on the left.

The Hartley transform differs from the Fourier transform in that the kernel is the real function cas $2\pi s x$ instead of $\exp(-i2\pi s x)$. The cas function, which was introduced by Hartley (1942), is defined by cas $x = \cos x + \sin x$ and

is simply a sinusoid of amplitude $\sqrt{2}$ shifted one-eight of a period to the left. The consequences of the change are that the Hartley transform is real rather than complex and that the transformation is identical to the inverse transformation. As may be obvious from the graphical example, the Hartley transform contains all the information that is in the Fourier transform and one may move freely from one to the other using the relations

$$H(s) = \text{Re } F(s) - \text{Im } F(s) \quad (32.12)$$

and

$$2F(s) = H(s) + H(N-s)$$
$$-iH(s) + iH(N-s) \quad (32.13)$$

The convenience that arises from familiarity with complex algebra when one is thinking about transforms loses its value in computing. What one thinks of compactly as one complex product still means four real multiplications to computer hardware, which must be instructed accordingly.

The Hartley transform is fully equivalent to the Fourier transform and can be used for any purpose for which the Fourier transform is used, such as spectral analysis. To get the power spectrum from the complex-valued Fourier transform, one forms $[\text{Re } f(s)]^2 + [\text{Im } F(s)]^2$; starting from the real-valued Hartley transform, one forms $[H(s)]^2 + [H(-s)]^2$. The phase is obtained from

$$\tan \phi(s) = \text{Im}(s)/\text{Re}(s)$$
$$= [H(-s)/H(s)] - \pi/4$$
$$(32.14)$$

We see that for purposes of spectral analysis by the Hartley transform, it is not necessary to work with complex quantities, because power spectrum is an intrinsic property independent of choice of kernel; the phase depends on the x origin which is locked to the peak of the cosine function in one case and the peak of the cas function in the other, hence the term $\pi/4$.

32.10 The Fast Fourier Transform

Around 1805, C.F. Gauss, who was then 28, was computing orbits by a technique of trigonometric sums equivalent to today's discrete Fourier synthesis. To get the coefficients from a set of a dozen regularly spaced data he could explicitly implement the formula that we recognize as the discrete Fourier transform. To do this, he would multiply the N data values $f(\tau)$ by the weighting factors $\exp(-i2\pi\nu\tau)$, sum the products, and repeat these N multiplications N times, once for each value of ν, making a total of N^2 multiplications. But he found that, in the case where N is a composite number with factors such that $N = n_1 n_2$, the number of multiplications was reduced when the data were partitioned into n_2 sets of n_1 terms. Where N was composed of three or more factors, a further advantage could be obtained. Gauss (1876) wrote, "illam vero methodum calculi mechanici taedium magis minuere, praxis tentatem docebit." He refers to diminishing the tedium of mechanical calculation, as practice will teach him who tries. This factoring procedure, usually into factors of 2, is the basis of the fast Fourier transform (FFT) algorithm, which is explained in many textbooks (Rabiner and Gold, 1975; IEEE, 1979; Elliott and Rao, 1982; Nussbaumer, 1982; Bracewell, 1986a; Press et al., 1986) and is available in software packages. The fast method (Cooley and Tukey, 1965) burst on the world of signal analysis in 1965 and was for a time known as the *Cooley-Tukey algorithm* (IEEE, 1967), but as the interesting history (Heideman et al., 1985) of prior usage in computing circles became known, the term FFT became universal.

Most FFT programs in use take advantage of factors by adopting a choice of N that is some power P of 2, that is, $N = 2^P$. The user may then design the data collection to gather, for example, $256 = 2^8$ readings. Alternatively, when such a choice does not offer, a user with 365 data points can simply append sufficient zeros to reach $512 = 2^9$ values. This might seem wasteful, but an attendant feature is the closer spacing of the resulting transform samples, which is advantageous for visual presentation. Perhaps one could do the job faster, say by factoring into 5×73. There are fast algorithms for 5 points and for many other small primes, but not for 73, as far as I know; it is simply not practical to store and select from lots of special programs for peculiar values of N. On the other hand, a significant speed advantage is gained if one elects more rigidity rather than more flexibility, tailors one's data collection to a total of 4^P values, and uses what is referred to as a *radix-4 program*. Because $1024 = 4^5$, the radix-4 approach is applicable to $N = 1024$ data samples (or to 256 for example), but not to 512 unless one appends 512 zeros. Packing with just as many zeros as there are data is commonly practiced because twice as many transform values result from the computation, and when the power spectrum is presented graphically as a polygon connecting the computed values, the appearance to the eye is much smoother.

Much practical technique is involved. If the sound level of an aircraft passing over a residential area is to be recorded as a set of measurements equispaced in time, the quantity under study begins and ends at zero value. But in other cases, such as a record of freeway noise, the noise is present when measurements begin and is still there when they cease; if the N values recorded are then packed with zeros, a discontinuity is introduced whose effects on the transform, such as overshoot and negative-going oscillation, may be undesirable. Packing with plausible (but unobserved) data can eliminate the undesired artifacts and is probably practiced in more cases than are admitted to. Authors often mitigate the effects of implied discontinuities in the data by multiplying by a tapering function, such as a set of binomial coefficients, that approaches zero at both the beginning and end of the data taken; they should then explain that they value freedom from negatives more than accuracy of amplitude values of spectral peaks or than resolution of adjacent peaks.

The FFT is carried out in P successive stages, each entailing N multiplications, for a total of NP. When NP is compared with N^2 (as for direct implementation of the defining formula), the savings are substantial for large N and make operations feasible, especially on large digital images, that would otherwise be unreasonably time consuming.

32.11
The Fast Hartley Algorithm

When data values are real, which is very commonly the case, the Fourier transform is nevertheless complex. The N transform values are also redundant (if you have the results for $0 \leq \nu \leq N/2$ you can deduce the rest). This inefficiency was originally dealt with by the introduction of a variety of efficient but unilateral algorithms that transformed in half of the time of the FFT, albeit in one direction only; now we have the Hartley transform, which for real data is itself real, is not redundant, and is bidirectional. The Hartley transform is elegant and simple and takes you to the other domain, regardless of which one you are in currently (Bracewell, 1986b; Buneman, 1989).

When a Hartley transform is obtained, there may be a further step required to get the more familiar complex Fourier transform. The time taken is always negligible, but even so the step is usually unnecessary. The reason is that although we are accustomed to thinking in terms of complex quantities for convenience, it is never obligatory to do so. As a common example, suppose we want the power spectrum, which is defined in terms of the real and imaginary parts of the Fourier transform by $P(\nu) = [\text{Re } F(\nu)]^2 + [\text{Im} F(\nu)]^2$. If we already have the Hartley transform $H(\nu)$, then it is not necessary to move first to the complex plane and then to get the power spectrum; the desired result is obtained directly as $\{[H(\nu)]^2 + [H(N-\nu)]^2\}/2$. Likewise phase $\varphi(\nu)$, which is required much less often than $P(\nu)$, is defined by $\tan [\varphi(\nu)] = \text{Im } F(\nu)/\text{Re } F(\nu)$; alternatively, one can get phase directly from $\tan [\varphi(\nu) + \pi/4] = H(N-\nu)/H(\nu)$, thus circumventing the further step that would be necessary to go via the well-beaten path of real and imaginary parts.

To illustrate the application to power spectra, take as a short example the data set {1 2 3 4 5 6 7 8}, whose discrete Hartley transform is

$$H(\nu) = \{4.5 - 1.707 - 1 - 0.707 \\ - 0.5 - 0.293 \; 0 \; 0.707\} \quad (32.15)$$

The first term, 4.5, is the mean value of the data set. The power spectrum for zero frequency is 4.5^2, for frequency $1/8$ ($\nu = 1$), $P(1) = (-1.707)^2 + (0.707)^2$, for frequency $2/8$ ($\nu = 2$), $P(2) = (-1)^2 + 0^2$. Similarly $P(3) = (-0.707)^2 + (-0.293)^2$ and $P(4) = (-0.5)^2 + (0.5)^2$. The highest frequency reached is $4/8$, corresponding to a period of 2, which is the shortest period countenanced by data at unit interval.

The encoding of phase by a real transform has added a physical dimension to the interest of the Hartley transform, which has been constructed in the laboratory with light and microwaves (Villasenor and Bracewell, 1987, 1988, 1990; Bracewell, 1989; Bracewell and Villasenor, 1990) and has suggested a new sort of hologram.

32.12
The Mellin Transform

The vast majority of transform calculations that are done every day fall into categories that have already been dealt with, and much of what has been said is applicable to the special transforms that remain to be mentioned. The Mellin transform has the property that $F_M(n+1)$ is the nth moment of $f(x)$ when n assumes a finite number of integer values 1, 2, 3, ... The special value $F_M(1)$ is the zeroth moment of, or area under, $f(x)$. But the transform variable does not have to be integral or even real, so one can think of the Mellin transform as a sort of interpolate passing through the moment values. When the scale of x is stretched or compressed, for example, when $f(x)$ is changed to $f(ax)$, the Mellin transform becomes $a^{-2} F_M(s)$, a modification that leaves the position of features on the s-axis unchanged and is useful in some pattern-recognition problems.

If we plot $f(x)$ on a logarithmic scale of x, a familiar type of distortion results, and we have a new function $f(e^{-x})$ whose Laplace transform is exactly the same as the Mellin transform of $f(x)$. An equally intimate relation exists with the Fourier transform. Consequently, the FFT may be applicable in numerical situations. Because of the intimate relationship with moments and with spectral analysis, Mellin transforms have very wide application. A specific example is given by the solution of the two-dimensional Laplace equation expressed in polar coordinates, namely $\partial^2 V/\partial r^2 + r^{-1} \partial V/\partial r + r^{-2} \partial^2 V/\partial \theta^2 = 0$. Multiply each term by r^{s-1} and integrate with respect to r from 0 to ∞. We get $d^2 F_M/d\theta^2 + s^2 F_M = 0$. Solve this for $F_M()$ and invert the transform to get the solution. In this example, a partial differential equation is converted to a simple differential equation by the transform technique.

32.13
The Hilbert Transform

As the example in Table 32.2 shows, the Hilbert transform, or quadrature function, of a cosinusoidal wave packet is a similar, but odd, waveform sharing the same envelope. But what do we mean by the envelope of an oscillation that only touches the intuitively conceived envelope at discrete points? The Hilbert transform provides an answer in the form $\sqrt{[f(x)]^2 + [f_{Hi}(x)]^2}$. Likewise, the original wave packet reveals its phase at its zero crossings. But what is the phase at intermediate points? The Hilbert transform supplies an instantaneous phase ϕ in the form $\tan \phi = f_{Hi}(x)/f(x)$. The operation **T** for the Hilbert transform is simply convolution with $-1/\pi x$. It is known that the Fourier transform of $-1/\pi x$ is i sgn s, where sgn s is 1 for $s > 0$ and -1 for $s < 0$. Therefore, by the convolution theorem (last line of Table 32.1), according to which the Fourier transform of a convolution is the product of the separate Fourier transforms, it would seem that a fast Hilbert transform of $f(x)$ could be calculated as follows. Take the FFT of $f(x)$, multiply by i for $0 < \nu < N/2$ and by $-i$ for $N/2 < \nu < N$, set $F(0)$ and $F(N/2)$ equal to zero, and invert the FFT to obtain the Hilbert transform. This sounds straightforward, but the procedure is fraught with peril, for two reasons. We propose to multiply a given function $f(x)$ by $-1/\pi[(x+ \text{const})]$ and to integrate from $-\infty$ to ∞, but we are only given N samples. The extremities of $-1/\pi x$ approach zero and have opposite signs, but there is infinite area under these tails no matter how far out we start. Consequently, we ask two oppositely signed large numbers to cancel acceptably. How can we expect satisfaction when the convolving function $-1/\pi x$ is not symmetrically situated about the extremes of the data range? The second reason is that we ask for similar cancellation in the vicinity of the pole of $1/x$. Experience shows that satisfactory envelopes and phases only result when $f(x)$ is a rather narrow-band function. Under other circumstances, an N-point discrete Hilbert transform can be defined and will give valid results free from worries about the infinities of analysis, but the outcome may not suit expectation.

An optical wave packet $\exp(-\pi t^2/T^2) \sin 2\pi \nu t$ of equivalent duration T easily meets the narrow-band condition when the duration T is much greater than the wave period $1/\nu$; it has a Hilbert transform $\exp(-\pi t^2/T^2) \cos 2\pi \nu t$. The square root of the sum of the squares yields $\exp(-\pi t^2/T^2)$ for the envelope, in full accord with expectation.

32.14
Multidimensional Transforms

The two-dimensional Fourier and Hartley transforms are defined respectively by

$$F(u, v) = \int_{-\infty}^{\infty} \int_{-\infty}^{\infty} f(x, y) \times e^{-i2\pi(ux+vy)} dxdy \quad (32.16)$$

$$F(u, v) = \int_{-\infty}^{\infty} \int_{-\infty}^{\infty} f(x, y) \times \text{cas}[2\pi(ux+vy)] dxdy \quad (32.17)$$

where the transform variables u and v mean spatial frequency components in the x and y directions. Work with images involves two dimensions, electrostatics and X-ray crystallography involve three, and fluid dynamics involves four. Multidimensional transforms can be handled numerically with a one-dimensional FFT subprogram or a fast Hartley as follows. Consider an $N \times N$ data array. Take the 1-D (one-dimensional) transform of each row and write the N transform values in over the data values. Now take the 1-D transform of each resulting column

(Bracewell, 1984; Bracewell et al., 1986). In three and four dimensions, the procedure is analogous (Hao and Bracewell, 1987; Buneman, 1989). Further simple steps lead to the Hartley transform and to the real and imaginary parts of the Fourier transform if they are required, but usually they are not; more often the quadratic content (power spectrum) suffices.

When a 2-D function has circular symmetry, as commonly arises with the response functions of optical instruments, not so much work is required, as explained below in connection with the Hankel transform. Cylindrical symmetry in 3-D is essentially the same, while spherical symmetry in 3-D is also referred to below.

32.15
The Hankel Transform

In two dimensions, where there is circular symmetry as expressed by a given function $f(r)$, the two-dimensional Fourier transform is also circularly symmetrical; call it $F_{Ha}(s)$. It can be arrived at by taking the full 2-D transform as described earlier, or it can be obtained from a single 1-D Hankel transform as defined in Table 32.2. The inverse transform is identical. There is apparently no opening for the Hartley transform because in the presence of circular symmetry the 2-D Fourier transform of real data contains no imaginary part. The kernel for the Hankel transform is a zero-order Bessel function, which is a complication that hampers the FFT factoring approach, but there is an elegant sidestep around this, which is explained below in connection with the Abel transform. Under spherical symmetry, the 3-D Fourier transform reduces to a different one-dimensional transform

$$4\pi \int_0^\infty f(r)\text{sinc}(2sr)r^2 dr \qquad (32.18)$$

The inverse transform is identical.

To illustrate by a well-known result from optical diffraction, we consider a telescope aperture $f(r)$ representable as $\text{rect}(r/D)$, a two-dimensional function that is equal to unity over a circle of diameter D. The Hankel transform is $D^2\text{jinc}Ds$, the familiar Fraunhofer diffraction field of a circular aperture. The jinc function [$\text{jinc}x = J_1(\pi x)/2x$], which is the Hankel transform of the unit rectangle function of unit height within a radius of 0.5, has the property that jinc $1.22 = 0$; this is the source of the constant in the expression $1.22\lambda/D$ for the angular resolution of a telescope.

32.16
The Abel Transform

Most commonly, although not always, the Abel transform arises when a 2-D function $g(x, y)$ has circular symmetry, as given by $f(r)$. The Abel transform (Table 32.2) then simplifies to $F_A(x) = \int_{-\infty}^{\infty} g(x, y) dy$. In other words, if the given $f(r)$ is represented by a square matrix of suitably spaced samples, then the Abel transform results when the columns are summed. There might not seem to be any future in trying to speed up such a basic operation, apart from the obvious step of summing only half way and doubling. However, when it is remembered that for each of $N^2/8$ matrix elements, we have to calculate $\sqrt{x^2 + y^2}$ to find r, and thence $f(r)$, it gives pause. The alternative is to proceed by equal steps in r rather than in y; then the oversampling near the x-axis is mitigated. But the variable radial spacing of elements stacked in a column needs correction by a factor $r/\sqrt{r^2 - s^2}$, which takes more time to compute than $\sqrt{x^2 + y^2}$. This is an excellent case for decision by using the millisecond timer found on personal computers. Of course, if many runs are to be made, the factors $r/\sqrt{r^2 - s^2}$ can be precomputed and the preparation time can be amortized over the successive runs.

Figure 32.1 shows a given function $g(x, y)$ and its one-dimensional projection (labeled P) which is derived by integrating along the y-axis in the (x, y) plane. Integrating along the y'-axis of a rotated coordinate system gives the projection P'. Now if $g(x, y)$ were circularly symmetrical, being a function $f(r)$ of r only, then the projections P and P' would be identical and equal to the Abel transform of $f(r)$. This is the graphical interpretation of the Abel transform.

Applications of the Abel transform arise wherever circular or spherical symmetry exists. As an example of the latter, consider a photograph of a globular cluster of stars in the outer reaches of the galaxy. The number of stars per unit area can be counted as a function of distance from the center of the cluster; this is the projected density. To find the true volume density as a function of radius requires taking the inverse Abel transform (Table 32.2) of the projected density.

With the Abel transform under control, we can now see a way of doing the Hankel transform without having to call up Bessel functions. The Abel, Fourier, and Hankel transforms form a cycle known as the *FHA cycle* (Bracewell, 1956), so that if we take the Abel transform and then take the FFT, we get the Hankel transform; the theorem is

$$\int_0^\infty dr J_0(2\pi \xi r) r \int_{-\infty}^\infty ds e^{i2\pi rs}$$

$$\int_s^\infty \frac{dx 2x f(x)}{\sqrt{x^2 - s^2}} = f(\xi) \qquad (32.19)$$

The FFT required will not be complex, except in the extraordinary case of complex 2-D data; consequently, it will in fact be appropriate to use the fast Hartley to get the Hankel transform. Because of symmetry, the result will also be exactly the same as obtained with the FFT, if after taking the FFT we pay no attention to the imaginary parts that have been computed, which all should be zero or close to zero.

The FHA cycle of transforms is a special case of the projection slice theorem, a theorem which refers to the more general situation where $g(x, y)$ is not circularly symmetrical. Circular symmetry characterizes instruments, especially optical instruments, which are artifacts. Lack of symmetry characterizes data; tomographic data will be taken as the illustration for the projection-slice theorem.

32.17
Tomography and the Radon Transform

Consider a set of rotated coordinates (x', y') centered on the (x, y) plane, but rotated through θ. The expression $\int_{-\infty}^\infty g(x, y) dy$ given for the Abel transform, representing a line integral in the y direction at a given value of x, would equal the line integral $\int_{-\infty}^\infty g(x, y) dy'$ in the rotated direction y' provided $g(x, y)$ had circular symmetry as specified for the Abel transform. But when $g(x, y)$

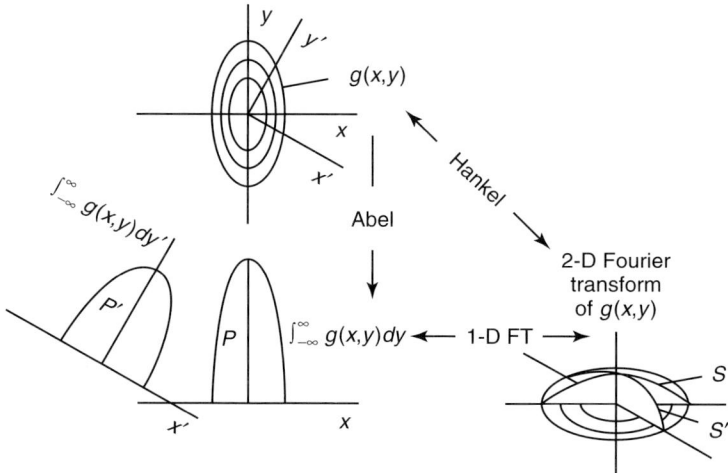

Fig. 32.1 Illustrating the projection-slice theorem, which states that if a distribution $g(x, y)$ has a projection P', in the y' direction, its 1-D Fourier transform is the slice S' through the 2-D Fourier transform of $g(x, y)$. The set of projections P' for all inclination angles of the (x', y') coordinates constitutes the Radon transform. In the presence of circular symmetry where $g(x, y) = f(r)$, the projection P in any direction is the Abel transform of $f(r)$. The 1-D Fourier transform of P is the slice S in any direction; this slice S is then the Hankel transform of $f(r)$. Thus the Abel, Fourier, and Hankel transforms form a cycle of transforms.

does not have symmetry, then the line-integral values depend both on x' and on the angle θ (Figure 32.1). The set of integrals with respect to dy' is the Radon transform of $g(x, y)$, named after Johann Radon (1917). Such integrals arise in computed X-ray tomography, where a needle beam of X-rays scans within a thin plane section of an organ such as the brain with a view to determining the distribution of absorption coefficient in that plane. If there are N^2 pixels for which values have to be determined, and because one scan will give N data, at least N different directions of scan spaced $180°/N$ apart will be needed to acquire enough data to solve for the N^2 unknowns. In practice, more than $2N$ directions are helpful to compensate for diminished sample density at the periphery. To compute a Radon transform is easy; the only tricky part is summing a given matrix along inclined directions. One approach is to rotate all the matrices and interpolate onto a rotated grid, for each direction of scan; but this may be too costly. At the other extreme one sums, without weighting, the matrix values lying within inclined strips that, independently of inclination, preserve unit width in the direction parallel to the nearer coordinate direction. How coarse the increment inclination angle may be dependent on acceptability as judged by the user in the presence of actual data.

The harder problem is to invert the line-integral data to retrieve the required absorption coefficient distribution. A solution was given by Radon

(1917). Later Cormack (1963, 1964, 1980), working in the context of X-ray scanning of a solid object, gave a solution in terms of sums of transcendental functions. Other solutions include the modified back-projection algorithm (Bracewell, 1956; Bracewell and Riddle, 1967) used in CAT scanners (Brooks and Di Chiro, 1976; Rosenfeld and Kac, 1982; Deans, 1983). The algorithm depends on the projection-slice theorem (see Figure 32.1). According to this theorem, (Bracewell, 1956) the 1-D Fourier transform of the projection P' (or scan) of $g(x, y)$ in any one direction is the corresponding central cross section or slice S' through the 2-D Fourier transform of the wanted distribution $g(x, y)$. The proof is as follows. Let the 2-D Fourier transform of $g(x, y)$ be $G(u, v)$ as defined by

$$G(u, v) = \int_{-\infty}^{\infty} \int_{-\infty}^{\infty} g(x, y) \times e^{-i2\pi(ux+vy)} dxdy \quad (32.20)$$

Setting $v = 0$, so as to have the representation $G(u,0)$ for the slice S, we get

$$G(u, 0) = \int_{-\infty}^{\infty} \int_{-\infty}^{\infty} g(x, y) e^{-i2\pi ux} dxdy$$

$$= \int_{-\infty}^{\infty} \left[\int_{-\infty}^{\infty} g(x, y) dy \right] \times e^{-i2\pi ux} dx$$

$$= \int_{-\infty}^{\infty} P(x) e^{-i2\pi ux} dx \quad (32.21)$$

where $P(x)$ is the projection of $g(x, y)$ onto the x-axis. Thus, the 1-D transform of the projection $P(x)$ is the slice $G(u, 0)$ through the 2-D transform of $g(x, y)$. If we rotate the coordinate axes to any other orientation (x', y'), we see that the same proof applies.

Because the density of polar coordinate samples is inversely proportional to the radius of the Fourier transform plane, a simple correction factor followed by an inverse 2-D Fourier transform will yield the solution. However, a solution was found (Bracewell and Riddle, 1967; Brooks and Di Chiro, 1976), based on this theoretical reasoning, to avoid numerical Fourier transforms entirely. An equivalent correction term, arrived at by convolving each projection P' with a few coefficients, can be directly applied to each P', after which the modified projections are accumulated on the (x, y) plane by back projection to reconstitute $g(x, y)$. Back projection means assigning the projected value at x' to all points of the (x, y) plane which, in the rotated coordinate system, have the abscissa x'. Accumulation means summing the back-projected distributions for all inclination angles.

32.18
The Walsh Transform

A function defined on the interval (0, 1) can be expressed as a sum of sines and cosines of frequency 1,2,3,..., but can also be expressed as a sum of many other sets of basis functions. Among the alternatives, Walsh functions (Walsh, 1923; Elliott and Rao, 1982; Hsu and Wu, 1987) are particularly interesting because they oscillate between values of +1, 0, and −1, a property that is most appropriate to digital circuits, telecommunications, and radar. Furthermore, multiplication by a Walsh function value takes much less time than multiplication by a trigonometric function. Walsh functions, not being periodic, are not to be confused with the periodic square cosine and sine functions $C(x) = \text{sgn}(\cos x)$ and $S(x) = \text{sgn}(\sin x)$; but on a finite support they do form a complete set from which any given function can be composed. They are also orthonormal (mutually orthogonal and with fixed quadratic content, as with Fourier components), which leads to simple relations for both analysis and synthesis. The Walsh (or Walsh-Hadamard) transform has found use in digital signal and image processing and for fast spectral analysis. Fast algorithms are available that use only addition and subtraction and have been implemented in hardware. A vast, enthusiastic literature sprang into existence in the 1970s, a guide to which can be found in the text by Elliott and Rao (1982).

32.19
The Z Transform

In control theory, in dealing with signals of the form

$$f(t) = \sum_{-\infty}^{\infty} a_n \delta(t-n) \quad (32.22)$$

and systems whose response to $\delta(t)$ is

$$h(t) = \sum_{0}^{\infty} h_n \delta(t-n) \quad (32.23)$$

the response $g(t)$ is the convolution integral

$$g(t) = \int_{-\infty}^{\infty} f(t')h(t-t')dt' \quad (32.24)$$

This response is a series of equispaced impulses whose strengths are given by $\Sigma_i a_i h_{n-i}$, an expression representable in asterisk notation for convolution by $\{g_n\} = \{a_n\} * \{h_n\}$ (in this notation the sequence $\{a_n\}$ sufficiently represents $f(t)$). For example, a signal $\{1\ 1\ 1\ 1\ 1\ 1\ \ldots\}$ applied to a system whose impulse response is $\{8\ 4\ 2\ 1\}$ produces a response

$$\{1\ 1\ 1\ 1\ 1\ 1 \ldots\} * \{8\ 4\ 2\ 1\}$$
$$= \{8\ 12\ 14\ 15\ 15\ 15 \ldots\} \quad (32.25)$$

This is the same rule as that that produces the coefficients of the polynomial that is the product of the two polynomials $\Sigma a_n z^n$ and $\Sigma h_n z^n$, as may be verified by multiplying $1 + z + z^2 + z^3 + z^4 + z^5 + \ldots$ by $8 + 4z + 2z^2 + z^3$. The z transform of the sequence $\{8\ 4\ 2\ 1\}$ is, by one definition, just the polynomial $8 + 4z + 2z^2 + z^3$; more often one sees $8 + 4z^{-1} + 2z^{-2} + z^{-3}$. If, conversely, we ask what applied signal would produce the response $\{8\ 12\ 14\ 15\ 15\ 15 \ldots\}$, we get the answer by long division:

$$(8 + 12z + 14z^2 + 15z^3 + 15z^4 + 15z^5$$
$$+ \cdots)/(8 + 4z + 2z^2 + z^3) \quad (32.26)$$

Occasionally, one of the polynomials may factor, or simplify, allowing cancellation of factors in the numerator and denominator. For example, the z transform of the infinite impulse response $\{8\ 4\ 2\ 1\ 0.5 \ldots\}$, where successive elements are halved, simplifies to $8/(1 - z/2)$. But with measured data, or measured system responses, or both, this never happens and the z notation for a polynomial quotient is then just a waste of ink compared with straightforward sequence notation such as $\{8\ 12\ 14\ 15\ 15\ 15 \ldots\} * \{8\ 4\ 2 \ldots\}^{-1}$. Whenever sampled data are operated on by a convolution operator (examples would be finite differences, finite sums, weighted running means, and finite-impulse-response filters), the z transform of the outcome is expressible as a product of z transforms. Thus to take the finite difference of a data sequence, one could multiply its z transform by $1 - z$ and the resulting polynomial would be the z transform of the desired answer; in a numerical environment one would simply convolve the data with $\{1 - 1\}$. In control theory and filter design, the complex plane of z is valued as a tool for studying the topology of the poles and zeroes of transfer functions.

32.20
Convolution

Sequences to be convolved may be handled directly with available subprograms for convolution and inverse convolution that operate by complex multiplication in the Fourier transform domain. When two real sequences are to be convolved, you can do it conveniently by calling the two Hartley transforms, multiplying term by term, and calling the same Hartley transform again to get the answer. Some subtleties are involved when the sequences are of unequal length or in the unusual event that neither of the factors has symmetry (even or odd) (Bracewell, 1986b). If one of the sequences is short, having less than about 32 elements, depending on the machine, then slow convolution by direct evaluation of the convolution sum may be faster, and a shorter program will suffice. When the Fourier transform is used, the multiplications are complex but half of them may be avoided because of Hermitian symmetry. Software packages such as CNVLV (Press et al., 1986) are available that handle these technicalities by calling two unilateral transforms, each faster than the FFT, or two equivalent subprograms; one fast Hartley transform, which is bilateral and, conveniently for the computer, real valued, now replaces such packages. Fast convolution using prime-factor algorithms is also available if general-purpose use is not a requisite.

As an example, suppose that $\{1\ 2\ 1\}$ is to be convolved with $\{1\ 4\ 6\ 4\ 1\}$, a simple situation where we know that the answer is the binomial sequence

$$\{1\ 6\ 15\ 20\ 15\ 6\ 1\} \qquad (32.27)$$

If we select $N = 8$ for the discrete transform calculation, the given factors become in effect

$$f_1(\tau) = \{1\ 2\ 1\ 0\ 0\ 0\ 0\ 0\} \qquad (32.28)$$

and

$$f_2(\tau) = \{1\ 4\ 6\ 4\ 1\ 0\ 0\ 0\} \qquad (32.29)$$

respectively, where the boldface emphasizes the zeroth elements $f_1(0)$ and $f_2(0)$. The sequence $\{1\ 2\ 1\}$ is commonly used to apply some smoothing to a data sequence, but because the center of symmetry at the element 2 is offset from the origin at $\tau = 0$, a shift will be introduced in addition to the smoothing. Therefore it makes sense to permute the sequence cyclically and use

$$f_1(\tau) = \{2\ 1\ 0\ 0\ 0\ 0\ 0\ 1\} \qquad (32.30)$$

To compute the convolution

$$\begin{aligned} f_3 &= f_1 * f_2 \\ &= \{2\ 1\ 0\ 0\ 0\ 0\ 0\ 1\} \\ &\quad * \{1\ 4\ 6\ 4\ 1\ 0\ 0\ 0\} \end{aligned} \qquad (32.31)$$

we take the two 8-element Hartley transforms to get the values H_1 and H_2 tabulated in Table 32.3. Multiply the corresponding values as shown under $H_1 H_2$ and take the Hartley transform again. The result is as expected; notice

Tab. 32.3 Performing convolution by multiplying Hartley transforms.

F_1	F_2	H_1	H_2	$H_1 H_2$	F_3
2	1	0.5	2	1	6
1	4	0.427	1.457	0.622	15.008
0	6	0.25	−0.5	−0.125	20
0	4	0.073	−0.043	−0.003	15.008
0	1	0	0	0	6
0	0	0.073	0.043	0.003	0.992
0	0	0.25	−0.5	−0.125	0
1	0	0.427	−1.457	−0.622	0.992

that the peak value 20 occurs where the peak value 6 of $f_2(\tau)$ occurs; this is a result of the precaution of centering {1 2 1} appropriately. The noninteger results are a consequence of rounding to three decimals for demonstration purposes, and these errors will be present, though smaller, if more decimals are retained.

32.21 Summary

A great analytic tradition of mathematical transform theory has gained far-ranging everyday importance by virtue of new numerical possibilities opened up by automatic computing machines.

Glossary

Alias: A sinusoid of low frequency spuriously introduced by insufficient sampling in the presence of a sinusoidal component of semiperiod shorter than the sampling interval.

Convolution of Two Functions: A third function composed by spreading each element of one given function out into the form of the second function and superimposing the spread components.

Discrete Transform: One suited to functions, such as those constituted by equispaced data samples, where the function values occur at discrete intervals and are usually finite in number.

Fast Fourier Transform (FFT): An algorithm for computing the discrete Fourier transform in less time than would be required to evaluate the sum of the products indicated in the defining formula.

Frequency, Negative: A convenient fiction arising from the representation of real sinusoids by complex quantities. The representation of the real function $\cos 2\pi ft$ in the form $\frac{1}{2}\exp[i2\pi ft] + \frac{1}{2}\exp[i2\pi(-f)t]$ involves clockwise rotation at frequency f and counterclockwise rotation at frequency $-f$.

Frequency, Spatial: The reciprocal of the period of a periodic function of space. Values of spatial frequency are expressed in cycles per meter, or in cycles per radian, according to whether the spatial variable is distance or angle.

Frequency, Temporal: The reciprocal of the period of a periodic function of time. Values are expressed in cycles per second, or hertz.

Heaviside Unit Step Function: A function $H(x)$ that is equal to zero to the left of the origin and equal to unity to the right. The value $H(0)$ at the origin has no effect on the value of integrals but may conventionally be taken as 0.5.

Inverse Transformation: An operation that, when applied to the transform of a function, effectively recovers the function.

Linear Transformation: A transformation with the property that the transform of the sum of any two functions is the sum of the separate transforms.

Tomography: Originally, a photographic technique for obtaining an X-ray image of a slice of tissue within the body; now applied in many fields to a technique of combining projections in many orientations to reconstruct an image.

Transform: A mathematical function, each value of which is derived from a set of values of a given function by an explicit operation.

References

Bracewell, R.N. (1956) *Aust. J. Phys.*, **9**, 198–217.
Bracewell, R.N. (1984) *Proc. IEEE*, **72**, 1010–1018.
Bracewell, R.N. (1986a) *The Fourier Transform and Its Applications*, 2nd edn rev., McGraw-Hill, New York.
Bracewell, R.N. (1986b) *The Hartley Transform*, Oxford University Press, New York.
Bracewell, R.N. (1989) *J. Atmos. Terrest. Phys.*, **51**, 791–795.
Bracewell, R.N., Buneman, O., Hao, H. and Villasenor, J. (1986) *Proc. IEEE*, **74**, 1283–1284.
Bracewell, R.N. and Riddle, A.C. (1967) *Astrophys. J.*, **50**, 427–434.
Brooks, R.A. and Di Chiro, G. (1976) *Phys. Med. Biol.*, **21**, 689–732.
Buneman, O. (1989) *IEEE Trans. ASSP*, **37**, 577–580, (Copyright held by the Board of Trustees of Leland Stanford Junior University.).
Campbell, G.A. and Foster, R.M. (1948) *Fourier Integrals for Practical Applications*, Van Nostrand, New York.
Cooley, J.W. and Tukey, J.W. (1965) *Math. Comput.*, **19**, 297–301.
Cormack, A.M. (1963) *J. Appl. Phys.*, **34**, 2722–2727.
Cormack, A.M. (1964) *J. Appl. Phys.*, **35**, 2908–2913.
Cormack, A.M. (1980) *Med. Phys.*, **7**, 277–282.
Deakin, M.A.B. (1985) *Math. Educ.*, **1**, 24–28.
Deans, S.R. (1983) *The Radon Transform and Some of its Applications*, Wiley, New York.
Elliott, D.F. and Rao, K.R. (1982) *Fast Transforms: Algorithms, Analyses, Applications*, Academic, New York.
Erdélyi, A., Oberhettinger, F., Magnus, W. and Tricomi, F.G. (1954) *Tables of Integral Transforms*, McGraw-Hill, New York.
Gauss, C.F. (1876) *Werke*, Vol. 3, Königliche Gesellschaft der Wissenschaffen, Göttingen.
Grattan-Guinness, I. (1972) *Joseph Fourier, 1768–1830*, The MIT Press, Cambridge, MA.
Hao, H. and Bracewell, R.N. (1987) *Proc. IEEE*, **75**, 264–266.
Hartley, R.V.L. (1942) *Proc. IRE*, **30**, 144–150.
Heaviside, O. (1970) *Electromagnetic Theory*, Vols 1–3, Chelsea, New York.
Heideman, M.T., Johnson, D.H. and Burrus, C.S. (1985) *Arch. Hist. Exact Sci.*, **34**, 265–277.
Hsu, C.-Y. and Wu, J.-L. (1987) *Electron. Lett.*, **23**, 466–468.
IEEE (1967) Special issue on Fast Fourier Transform, *IEEE Trans. Audio Electroacoustics* **AU-2**, 43–98.
IEEE (1979) Digital signal processing committee, in *Programs of Digital Signal Processing*, (eds IEEE ASSP Soc.), IEEE Press, New York.
Körner, T.W. (1988) *Fourier Analysis*, Cambridge University Press, Cambridge.
Lighthill, M.J. (1958) *An Introduction to Fourier Analysis and Generalized Functions*, Cambridge University Press, Cambridge.
McCollum, P.A. and Brown, B.F. (1965) *Laplace Transforms Tables and Theorems*, Holt, Rinehart and Winston, New York.
Nahin, P.J. (1987) *Oliver Heaviside, Sage in Solitude*, IEEE Press, New York.
Neugebauer, O. (1983) *Astronomy and History, Selected Essays*, Springer, New York.
Nussbaumer, H.J. (1982) *Fast Fourier Transform and Convolution Algorithms*, Springer, New York.
Press, W.H., Flannery, B.P., Teukolsky, S.A. and Vetterling, W.T. (1986) *Numerical Recipes*, Cambridge University Press, Cambridge.
Rabiner, L.R. and Gold, B.A. (1975) *Theory and Application of Digital Signal Processing*, Prentice Hall, Englewood Cliffs, NJ.
Radon, J. (1917) *Ber. Sächs. Akad. Wiss. Leipzig, Math.-Phys. Kl.*, **69**, 262–277.

Rosenfeld, A. and Kac, A.C. (1982) *Digital Picture Processing*, Vol. 1, Academic, New York.
Villasenor, J.D. and Bracewell, R.N. (1987) *Nature*, **330**, 735–737.
Villasenor, J.D. and Bracewell, R.N. (1988) *Nature*, **335**, 617–619.
Villasenor, J.D. and Bracewell, R.N. (1990) *J. Opt. Soc. Am.*, **7**, 21–26.
Walsh, J.L. (1923) *Am. J. Math.*, **45**, 5–24.

Further Reading

Texts treating the various transforms and computational methods are identifiable from their titles in the list of works cited. An indispensable source for locating recent material on any of the special branches mentioned in this article is *Mathematical Abstracts*.

Glossary

The following listing of acronyms and other terms with broader significance in applied spectroscopy supplements the topic-specific glossary terms given in individual chapters.

Aberration: The Gaussian approximation for a ray passing through a lens or reflected from a spherical mirror is that the angle of deviation θ is small enough for $\sin\theta$ to be replaced by θ. Any departure from this approximation is an *optical aberration*. For practical purposes, any departure from geometric optics that causes a point source to be imaged into a nonpoint image is an aberration.

Absorbance: $\log_{10}(I_0/I)$, where I_0 is the incident light intensity falling on an absorbing sample and I is the transmitted light intensity.

Absorption Coefficient: In a nonscattering medium the absorption coefficient μ_a is defined as the reciprocal of the distance d over which light of intensity $I(d=0) = I_0$ is attenuated (due to absorption) to $I(d) = I_0/e \approx 0.37 I_0$; the units are typically cm^{-1}; behind this definition is a fundamental process of photon absorption that is characterized by a photon absorption cross section.

Actinometry: Measurement of light intensities for quantum-yield determinations.

Agglomerative Clustering: Forming relationships between samples by joining together those that are most similar.

Alias: A sinusoid of low frequency spuriously introduced by insufficient sampling in the presence of a sinusoidal component of semiperiod shorter than the sampling interval.

ALS: Alternating least squares – a method of modeling spectra and concentration profiles that can incorporate constraints such as positivity.

Angular correlation: The angular distribution of two annihilation γ-photons about 180°, directly related to the momentum distribution of the annihilating pair (and hence, in condensed matter, essentially the electron momentum density).

Anisotropic Scattering: A scattering process characterized by a clearly apparent direction of photons.

Annihilation: The decay of a positron–electron pair, with the emission of energy in the form of γ-radiation; for free particles, two photons are most commonly emitted.

Encyclopedia of Applied Spectroscopy. Edited by David L. Andrews.
Copyright © 2009 WILEY-VCH Verlag GmbH & Co. KGaA, Weinheim
ISBN: 978-3-527-40773-6

ARPES: Angle-resolved photoelectron spectroscopy.

Auger Spectroscopy: Use of the Auger electrons emitted during the decay of core holes for elemental analysis at surfaces.

BIS: Bremsstrahlung isochromat spectroscopy. A form of inverse photoemission with constant photon energy and variable electron energy, originally used in the X-ray regime.

Blazed Gratings: Gratings for which the bar, groove, or line shape is structured so as to enhance reflection into a particular order.

Bond-Specific Photochemistry: Photochemistry having products determined by a specific bond or combination of bonds set into vibrational motion.

Bonded-Phase Chromatography: Liquid chromatography performed with a stationary phase that is chemically bonded to a particulate supporting material such as silica powder. See also Normal-Phase Chromatography, Reversed-Phase Chromatography.

Bootstrap: Alternative to cross-validation involving an iterative algorithm.

Bremsstrahlung: γ-Ray emitted by a high-energy decelerating electron, such as one that is interacting with the electric field of a nucleus or atomic electron.

Calorimeter: Device that measures the total energy it absorbs.

Capacity Factor: An old term that has fallen into abeyance. See Retention Factor.

Capillary Column: In gas or supercritical fluid chromatography, the small-diameter, open tubular column that contains the stationary phase. The inner wall of the tubing is coated with a solid or liquid film that functions as the stationary phase. The major type is a wall-coated open tubular (WCOT) column.

Carrier Gas: Mobile phase in gas chromatography.

Cascade Shower: The sequential energy-loss process in which γ-rays convert to electron–positron pairs, which then by a process of bremsstrahlung emit new γ-rays that in turn pair produce, and so on until all of the energy in the initial γ-ray is dissipated.

CCD: Charge-coupled device – a solid-state electronic device that serves as an imaging chip and is used in video cameras and fast spectrometers.

Cherenkov Radiation: Photons produced by a charged particle when its velocity in the medium exceeds that of the photons in the medium.

Chromatogram: The output from a chromatographic separation that shows the response of the detector as a function of time.

Chromatograph: The instrumentation used in chromatography (noun). To separate and analyze components in a mixture by chromatography (verb).

Chromophore: Light-absorbing chemical group.

Cis: A molecular structural configuration in which two like functional groups are adjacent to each other on the molecular skeleton.

Column: In chromatography, the glass, metal, or plastic tubing that contains the stationary phase.

Compton Scattering: Elastic scattering of a γ-ray off of an atomic electron.

Concerted Reaction: A reaction in which bond-breaking and -forming occur essentially simultaneously.

Conformation: The arrangement of atoms inside a molecule.

Conformers: Different, relatively stable, arrangements of atoms, each consisting of identical numbers and types of atoms. A dissociating or rearranging molecule may pass through different conformers. Conformers need not be capable of being physically or chemically isolated.

Confusion Matrix: See Contingency Table.

Conjugation: Alternating double and single bonds within a molecular chain or ring, resulting in bond stabilization via electron delocalization.

Contingency Table: A table whose rows and columns represent each group in a series of spectra, and containing information about how many samples are classified into each group, allowing misclassification rates to be determined.

Convolution of Two Functions: A function composed by spreading each element of one given function out into the form of the second function and superimposing the spread components.

Cross-Validation: Method for determining the number of PC or PLS components by leaving some samples out and predicting them.

CW: Continuous wave.

D-statistic: Indicator used usually in MSPC to determine how close a sample is to the centre of an NOC model.

Dark Current: Current from a photodetector, which is present even in the absence of illumination.

Dendrogram: Graphic representation of cluster analysis.

Diffraction: Modification of the intensity and/or phase of an electromagnetic wave by the presence of an object (such as a slit, hole, edge, etc.) in the path of the wave.

Diffusion Approximation (Diffusion Theory): The approximated diffusion-type solution of the *RTT*, which is accurate for describing of photon migration in infinite, homogeneous, highly scattering media.

Dipole Moment: A measure of the separation between centers of positive and negative charges in polar molecules.

Discrete Transform: One suited to functions, such as those constituted by equispaced data samples, where the function values occur at discrete intervals and are usually finite in number.

Discriminant Analysis: Mathematical approaches for determining grouping in samples.

Dispersive Spectrometer: A spectrometer that employs wavelength dispersion. Rays are dispersed (i.e., spatially separated) by the effects of diffraction. Dispersive spectrometers are sometimes called *wavelength-dispersive spectrometers*. The two classes of X-ray dispersive spectrometers are Bragg devices and diffraction gratings.

Dissimilarity Matrix: Like a similarity matrix, except that numerical values relate to dissimilarity.

Doppler broadening: The broadening due to nonzero momentum.

Doppler Drive: Device to change neutron velocity using Doppler effect.

EFA: Evolving factor analysis – used to look at multivariate processes that evolve.

Epi-Fluorescence: A microscopy technique in which the same objective is used for the illumination of a specimen and collection of the excited fluorescence.

ESCA: Electron spectroscopy for chemical analysis. Mapping of the core-level energies and their chemical shifts, typically using the K_α lines of Al or Mg at 1.49 keV and 1.25 keV, respectively. Synonym: XPS.

Escape Peak: For monochromatic incident X-ray photons with an energy E_x smaller than that of the absorption edge of detector atoms, the pulse height output is proportional to E_x. However, when the energy E_x of the monochromatic incident photons exceeds the absorption edge of the detector atoms, the output may contain two pulse height distributions. The additional or escape peak has a mean pulse height proportional to the difference between the energies of the incident photons E_x and the escaping (from active detector volume) photons $E_{K\alpha}$.

Euclidean Distance: Difference between two spectra obtained by representing each as a point in multidimensional space.

EXAFS: Extended X-ray absorption fine structure, determination of nearest-neighbor distances by interference effects in the core-level absorption spectrum.

Excimer: A complex formed from the excited state of a molecule with an identical molecule in the ground state.

Exciplex: A complex formed from an excited molecule and a different molecule in the ground state.

Exciton: In the context of solid-state physics, a bound state of an excited electron and a hole in the lattice of a semiconductor or insulator. Excitons can move around the crystal lattice of the material in which it is formed. In connection with molecular systems, the term is also used to identify an electronically excited unit.

Extinction Coefficient (Attenuation Coefficient): Specific absorbance of 1-cm pathlength of solution at 1 mole per litre concentration. The reciprocal of the distance over which light of intensity I is attenuated to $I/e \approx 0.37I$; attenuation or interaction coefficient, $\mu_t = \mu_a + \mu_s$, where μ_a is the *absorption coefficient* and μ_s is the *scattering coefficient*; the units are typically cm^{-1}.

Factor Analysis: Similar to PCA except that the components have an interpretable meaning, for example, as spectra and concentration profiles.

Fast Fourier Transform (FFT): An algorithm for computing the discrete Fourier transform in less time than would be required to evaluate the sum of the products indicated in the defining formula.

Fermi Chopper: Mechanical device to pulse and monochromatize a neutron beam.

Flash Photolysis: Experimental study of photochemical or photophysical processes initiated by a short pulse of light.

FN: False negative.

FP: False positive.

Frequency, Negative: A convenient fiction arising from the representation of real sinusoids by complex quantities. The representation of the real function $\cos 2\pi ft$ in the form $\frac{1}{2}\exp[i2\pi ft] + \frac{1}{2}\exp[i2\pi(-f)t]$ involves clockwise rotation at frequency f and counterclockwise rotation at frequency $-f$.

Frequency, Spatial: The reciprocal of the period of a periodic function of space. Values of spatial frequency are expressed in cycles per meter, or in cycles per radian, according to whether the spatial variable is distance or angle.

Frequency, Temporal: The reciprocal of the period of a periodic function of time. Values are expressed in cycles per second, or hertz.

FWHM: Full-width at half maximum.

Gas Chromatography (GC): Chromatography with a gaseous mobile phase.

Gas–Liquid Chromatography (GLC): Gas chromatography with a liquid stationary phase that is dispersed on or bonded to a particulate packing material or to the inner wall of the column.

Gas–Solid Chromatography (GSC): Gas chromatography with a solid stationary phase.

Gel Filtration Chromatography: Size-exclusion chromatography with an aqueous mobile phase.

Gel Permeation Chromatography: Size-exclusion chromatography with a non-aqueous mobile phase.

GKPF: Gegenbauer kernel phase function.

GMR: Giant magnetoresistance, a strong variation of the electrical resistance in a magnetic field induced in sandwiches of magnetic and nonmagnetic metal layers.

Gradient Elution: In liquid chromatography, a technique for improving resolution, which consists of continuously changing the composition of the mobile phase during the separation.

GW: Calculation method for quasiparticles, based on solving the Dyson equations with the full electron and photon propagators G and W (therefore the GW acronym), and a free particle approximation for the vertex. It goes beyond the LDA in determining the electronic structure for the excited state, which is measured in photoemission and inverse photoemission.

Heaviside Unit Step Function: A function $H(x)$ that is equal to zero to the left of the origin and equal to unity to the right. The value $H(0)$ at the origin has no effect on the value of integrals but may conventionally be taken as 0.5.

Height Equivalent of a Theoretical Plate: Originally, a quantity used to describe the efficiency of a particular separation scheme, but no longer used. See *plate height*.

Hermiticity: The degree to which a detector array is sensitive to all γ-rays, regardless of their directions or angles of emission from the source.

HGPF: Henyey–Greenstein phase function.

High-Performance Liquid Chromatography (HPLC): Liquid chromatography using mobile phases pumped at high pressures, stationary phases consisting of columns containing densely packed, small-diameter particles, and dynamic detection of the components eluting from the column.

Holographic Filter: An optical filter in which a three-dimensional diffraction pattern is formed in a photosensitive material. Depending on the pattern, the holographic filter can function as a wavelength sensitive mirror (also called a *notch filter*), diffraction grating or other type of filter.

HOMO: Highest occupied molecular orbital in molecules. Together with the LUMO, the HOMO is responsible for most chemical reactions of a molecule, in particular charge transfer reaction.

IAD: Inverse adding-doubling.

IMC: Inverse Monte Carlo.

Information Bandwidth: The speed with which information channels may be recorded. In a spectrometer, the information bandwidth may be increased either by increasing the efficiency of recording a single wavelength or by recording several wavelengths simultaneously.

Injector: In chromatography, the device used to introduce a sample onto the top of the column.

Internal Conversion: Energy loss within a given electronic excited or ground-state manifold.

Intersystem Crossing: Transition of an excited electron from a given multiplicity (such as the singlet) to a different one (such as the triplet).

Inverse Matrix: Defined for most (i.e. non-singular) square matrices, for which the product is the unit matrix.

Inverse Transformation: An operation that, when applied to the transform of a function, effectively recovers the function.

Ion Chromatography (IC): A group of liquid chromatography techniques used for the separation and analysis of ions.

Ion-Exchange Chromatography (IEC): Ion chromatography with an ion-exchange resin as the stationary phase.

Ion-Pair Chromatography: Ion chromatography with a reversed-phase or polymeric stationary phase, and a mobile phase containing chemical reagents that form neutral ion pairs with the sample ions.

Isomers: Different molecular conformations, consisting of identical atoms, that are stable, or metastable long enough to be physically or chemically separated.

Isotropic Scattering: An equality of scattering properties along all axes.

ITTFA: Iterative target transform factor analysis – a method similar to ALS.

KNN: K nearest neighbours – a method for classification.

LDA: Local density approximation, representing one of the most common methods for determining electronic states in solids and at surfaces.

LDA: Linear discriminant analysis – usually using Mahalanobis distance calculated on a training set consisting of several groups.

LEED: Low-energy electron diffraction, often used in conjunction with photoelectron spectroscopy to characterize the ordering at surfaces.

Lifetime: The mean life of a particle or excited state in a material: The time required for an excited species to decay to $1/e$ of its initial concentration.

Linear Transformation: A transformation with the property that the transform of the sum of any two functions is the sum of the separate transforms.

Liquid Chromatography (LC): Chromatography with a liquid mobile phase. See also *high-performance liquid chromatography*.

Liquid–Solid Chromatography (LSC): Liquid chromatography with a solid stationary phase.

Loadings: Numerical values of PCs corresponding to variables.

LUMO: Lowest unoccupied molecular orbital in molecules. Together with the HOMO, the LUMO is responsible for most chemical reactions of a molecule.

Mahalanobis Distance: Difference between two spectra obtained by representing each as a point in multidimensional space and scaling each dimension according to variance.

Many-Body Theory: Many-body theory stands for a theory dealing with an infinite number of interacting particles. This is mathematically simpler than dealing with 10^{20} electrons in a macroscopic solid.

Matrix: Representation of data in a two-dimensional array.

MCD: Magnetic circular dichroism, change of absorption between left- and right-handed, circularly polarized light, used for the determination of the magnetization at specific atomic sites.

MFP: Mean free path (of a photon).

MLR: Multiple linear regression – an alternative to PLS that relates two sets of data, for example, a series of spectra, to concentrations of constituents.

Mobile Phase (MP): In chromatography, the flowing gas, liquid, or supercritical fluid that carries the individual components of a mixture through the stationary phase.

Mode-Specific Photochemistry: Photochemistry having products determined by the vibrational mode set into motion.

Monochromator: An instrument used to isolate a single wavelength of light (or, more realistically, in a narrow range of wavelengths). The elements of dispersive spectrometers (crystals and gratings) can also be used as monochromators.

Monomers: Low-molecular-weight molecules or atoms that can join together to form a polymer.

MSPC: Multivariate statistical process control.

Multilayers: Synthetic crystals formed by sputtering or evaporating alternating layers of high-Z and low-Z materials onto a substrate. Multilayers can be used as X-ray spectrometers or monochromators and can be a part of an X-ray optics system to enhance the performance of the system over a narrow band and are also called *layered synthetic microstructures* or *LSMs*.

Multiphoton Absorption: Direct multiphoton absorption occurs when two or more photons are absorbed at once, mediated by higher order perturbation terms of the molecular Hamiltonian interaction with the radiation field. Sequential one-photon events are sometimes called by this name, but more properly may be called *multiple-photon absorption*.

Multivariate: Measurements where there is more than one variable, for example, a spectrum at several wavelengths.

Neutron Back-Scattering: Method to achieve very high energy resolution in neutron scattering using Bragg reflection with a Bragg angle of 90°.

Neutron Guide: Device to transport a neutron beam over long distances.

Neutron Spin Echo: Method to achieve very high energy resolution in neutron scattering using the Larmor precession of neutrons.

Neutron Three-Axis Spectrometer: Instrument consisting of three axes: monochromator, sample, and analyzer.

NEXAFS: Near Edge X-ray absorption fine structure, mapping of unoccupied states via transitions from core levels. Synonyms: XANES, XAS, Partial Yield Spectroscopy.

NIPALS: Nonlinear lterative partial least squares – a method for performing PCA.

NOC: Normal operating conditions – used in MSPC to indicate the set of spectra that correspond to samples that are of acceptable quality.

Nonadiabatic Transition: The transition between two adiabatic electronic states triggered by the motion of the nuclei.

Nondispersive Spectrometer: A spectrometer that does not employ wavelength dispersion. Some such devices operate by converting the photon energy into some other sort of particle or quasiparticle (e.g., electron–ion pairs or phonons); these are often called *energy-dispersive spectrometers*.

Nonpolar: A property of a material resulting from the distribution of electronic charge in the individual molecules. See also *polar*.

Nonradiative Energy Transfer: Loss of energy from an excited state, generally by transfer to another site or molecule, by any process that does not produce light.

Normal Mode: Classically, molecular vibration in which (nearly) all atoms move concertedly and periodically with the same frequency, but possibly not with the same phase. Quantum mechanically, an eigenfunction of the vibrational Hamiltonian.

Normal-Phase Chromatography (NPC): Liquid–solid chromatography or bonded-phase chromatography with a polar stationary phase and a nonpolar mobile phase.

Nuclear Spallation: Process in which a heavy nucleus fragments into lighter nuclei and neutrons, when hit by highly energetic particles like protons from an accelerator.

Number of Theoretical Plates: A measure of the efficiency of a particular separation scheme, now called *plate number*.

OCT: Optical coherence tomography.

Open Tubular Columns (OT): See *capillary column*.

Oscillator Strength: A measure of the area under an absorption band, or the degree to which a transition is allowed.

Packed Column: A chromatography column that contains the stationary phase as a packed powder or resin.

Pair Production: Transformation of a γ-ray interacting with the electric field of a nucleus or atomic electron into an electron–positron pair.

PC: Principal component – characterised by scores and loadings.

PCA: Principal components analysis – exploratory visualisation of data matrices, for example, a series of spectra.

PCR: Principal components regression – alternative to PLS and MLR.

Phase-Space Transformer: Device to transform a white collimated neutron beam into a monochromatic divergent beam using the Doppler effect.

PhD: Photoelectron diffraction, determination of the local atomic structure of surfaces.

Phonon: A vibrational mode of a crystal lattice.

Photodetachment: Ejection of an electron from a negative ion by the effect of light.

Photodissociation: Fragmentation of a molecule into two or more pieces by the effect of light.

Photoelectric Absorption: Absorption of light by an atomic electron.

Photomultiplier Tube: Device that generates a current pulse when struck by an optical photon, via a cascade multiplication of the secondary electron.

Photon Diffusion Coefficient: The proportionality coefficient between mean-square displacement of a photon within time interval τ: $\langle \Delta r^2 \rangle \sim D\tau$.

Plate Height (H): A measure of column efficiency equal to the column length (L) divided by the plate number (N).

Plate Number (N): A measure of column efficiency that is based on a comparison of an analyte's retention volume and its peak width.

PLS: Partial least squares – method for multivariate calibration of data matrices onto another block, calibrating a series of spectra onto concentrations of constituents.

PLS1: PLS algorithm for which there is a single c variable.

PLS2: PLS algorithm for which there is more than one c variable.

Polar: A property of a material resulting from the distribution of electronic charge in the individual molecules. A material is said to have polar character if any region of the individual molecules that compose the material possesses an excess or deficiency of electronic charge relative to the rest of the molecule.

Polarizability: The measure of atomic or molecular charge displacement by an externally applied electric field.

Polychromator: An instrument capable of isolating several monochromatic rays simultaneously from an incident beam.

Positron: The antiparticle of the electron.

Propagator: The propagator (or Green's function) describes the propagation of a particle from point A to point B in many-body theory. Mathematically, it is the expectation value of a particle creation operator at point A by an annihilation operator at point B.

Pseudoinverse: Mathematical equivalent to an inverse for a rectangular matrix.

Pump-Probe Technique: Technique in which a pump pulse of light initiates a process and a probe pulse measures the associated transient changes of observables.

Q-statistic: Indicator used usually in MSPC to determine how well a sample fits an NOC model.

QDA: Quadratic discriminant analysis – using Mahalanobis distance calculated separately on each group in a training set.

Quasiparticle: In many-body theory, an electron loses its simple particle character by dragging other particles along, such as phonons or other electrons in a solid. This whole particle cloud moves together and may be regarded as quasiparticle.

Radiative Energy Transfer: Loss of energy of an excited state by emission of light.

Rectangular Matrix: Matrix for which the number of rows and columns are unequal.

Reduced (Transport) Scattering Coefficient: A lumped property incorporating the *scattering coefficient* μ_s and the *scattering anisotropy parameter* g: $\mu_s' = \mu_s(1-g)$ (cm^{-1}); μ_s' describes the diffusion of photons in a random walk of step size of $1/\mu_s'$ (cm), where each step involves isotropic scattering; this is equivalent to description of photon movement using many small steps $1/\mu_s$ that each involve only a partial (anisotropic) deflection angle if there are many scattering events before an absorption event, that is, $\mu_a \ll \mu_s'$ (diffusion regime); μ_s' is useful in the diffusion regime, which is commonly encountered when treating how visible and near-infrared light propagates through tissues; for many *tissues*, the reduced scattering coefficient obeys a power law, $\mu_s' = q\lambda^{-h}$ (cm^{-1}, λ in μm).

Resolution (R_S): In chromatography, a quantity which describes the efficiency and degree of separation achieved in the chromatographic separation of two chemical compounds.

Resolving Power: The resolving power of a spectrometer (at an energy $E = hc/\lambda$) is usually defined as the ratio of the energy (or wavelength) of interest to the width of the response function of the spectrometer to a monochromatic line, that is, $R(E) = E/\Delta E = \lambda/\Delta\lambda$. The width of the line is usually chosen to be the FWHM of the spectral response function. Note that some authors invert this definition of resolving power so that it equals $\Delta E/E = \Delta\lambda/\lambda$ and often quote it as a percentage.

Response Function: The function that describes a material response to the applied external fields.

Retardation Factor for Column Chromatography (R): A measure of chromatographic retention behavior, equal to the average speed of a given analyte through the column divided by the average mobile phase velocity. Also equal to V_M/V_R.

Retardation Factor for Thin-Layer Chromatography (R_F): A measure of chromatographic retention behavior, equal to the distance traveled on a TLC plate by a given analyte divided by the distance traveled by the solvent (MP) front.

Retention Factor (k): A measure of chromatographic retention behavior equal to V'_R/V_M. Formerly called *capacity factor*, *capacity ratio*, and *partition ratio*, none of which should now be used.

Retention Time (t_R): In chromatography, the length of time a chemical compound spends in the column.

Reversed-Phase Chromatography (RPC): Bonded-phase chromatography using a nonpolar stationary phase and polar mobile phase.

RMSEC: Root mean square error of calibration – used for autopredictive models.

RMSECV: Root mean square error of cross validation – often used to determine how many PLS components are suitable for a model.

RMSEP: Root mean square error of prediction – used on an independent test set.

Rocking Curve: In crystal spectrometers, the rocking curve is a measure of the spectral resolving power of the crystal, independent of any geometric effects introduced by the geometry of the spectrometer. Specifically, it is the FWHM in degrees of the response of the crystal (as it is rotated or rocked) to a monochromatic X-ray beam.

Rotation: Transforming a PC onto a target vector normally consisting of the concentrations of an analyte.

Rowland Circle Configuration: This configuration is common in focusing X-ray spectrometers. The diverging X-ray source, the diffraction grating, and the detector all lie on a circle (the Rowland circle, after Henry Rowland) that has a diameter equal to the radius of curvature of the curved diffractor. Such a configuration gives one-dimensional focusing and disperses the spectrum along the circle. A variation, using a curved crystal rather than a grating, allows for imaging of small fields along the circle.

RTT: Radiation transfer theory.

Scalar: Single number with no directional attributes.

Scattering Anisotropy Parameter: A measure of the amount of forward direction retained after a single scattering event; if a photon is scattered by a particle so that its trajectory is deflected by a deflection angle θ, then the component of the new trajectory that is aligned in the forward direction is presented as $\cos\theta$; there is an average deflection angle and the mean value of $\langle\cos\theta\rangle$ is defined as the anisotropy, $g \equiv \langle\cos\theta\rangle$; the value of g varies in the range from -1 to 1: $g = 0$ corresponds to isotropic (Rayleigh) scattering, $g = 1$ to total forward scattering (Mie scattering at large particles), and -1 to total backward scattering.

Scattering Coefficient: A particle with a particular geometrical size redirects incident photons into new directions and so prevents the forward on-axis transmission of photons, this process constitutes scattering; the scattering coefficient μ_s (cm^{-1}) describes a medium containing many scattering particles at a concentration described as a volume density ρ (cm^3); the scattering coefficient is essentially the cross-sectional area σ_{sca} (cm^{-1}) per unit volume of medium: $\mu_s = \rho\sigma_{sca}$; a power law for dependence of the scattering coefficient on the wavelength is typical for many tissues: $\mu_s \propto \lambda^{-h}$, for different tissue structures parameter h is ranging from 1 to 2.

Scattering Length: Parameter defining the strength of a neutron sample interaction.

Scattering Phase Function: The function that describes the scattering properties of the medium and is, in fact, the probability density function for scattering in the direction $\vec{s}'$ of a photon traveling in the direction $\vec{s}$; it characterizes an elementary scattering act: if scattering is symmetric relative to the direction of the incident wave, then the phase function depends only on the scattering angle θ (angle between directions $\bar{s}$ and $\bar{s}'$).

Scores: Numerical values of PCs corresponding to samples.

Sensitizer: Compound that absorbs light at the irradiation wavelength, then transfers its excited-state energy to a substrate that cannot be directly excited.

Separation Factor (α): In chromatography, a measure of the ability of a separation scheme to separate two components. Also referred to as the *selectivity*.

SEXAFS: Surface extended X-ray absorption fine structure, determination of nearest-neighbor distances at surfaces by interference effects in the core-level absorption spectrum.

Similarity Matrix: A matrix whose rows and columns represent samples containing values that indicate how similar each sample is to every other sample.

Singlet: An electron configuration in which the sum of the electron-spin quantum numbers is 0, that is, all the electrons are paired with respect to their spins.

Size-Exclusion Chromatography (SEC): Liquid chromatography technique that separates chemical compounds on the basis of their physical size. See also *gel filtration chromatography, gel permeation chromatography*.

Specific trapping coefficient: The probability of (positron) trapping by a particular type of defect per second per defect site. The product of this coefficient and the defect concentration (per atom) gives the trapping rate.

Spectral Quality: The fraction of input radiation for which a measurement of its total energy is obtained.

Spectrograph: An instrument which produces a graphical record of an optical spectrum.

Spectrometer: Any instrument that measures the intensity of light as a function of wavelength.

Spectrophotometer: A spectrometer specialized for accurately recording the intensity of light.

Spin–orbit Coupling: Interaction of the orbital and spin angular momenta of an electron in an atom or polyatomic species.

Square Matrix: Matrix for which the number of rows and columns are equal.

Standardizing: Subtracting the mean and dividing by the standard deviation.

Stationary Phase (SP): In chromatography, the solid or liquid phase that resides in the column. The stationary phase is a powder or resin packed into the column or a film deposited on or bonded to the inner wall of the column.

Super-Mirror: Mirror for neutrons with an increased glancing angle using multilayer coating.

Supercritical Fluid: A material that is above its critical temperature and pressure.

Supercritical Fluid Chromatography (SFC): Chromatography with a supercritical fluid as the mobile phase.

Surface Plasmon: A surface electromagnetic wave that propagates parallel to the surface of a metal. Surface plasmons are excited at wavelengths that depend on the metal and the three-dimensional surface structure. Silver plasmons can be excited with visible radiation. Gold and copper plasmons can be excited with deep red and near-infrared radiation.

SVD: Singular value decomposition – a method for performing PCA.

Synchrotron Radiation: Versatile source of ultraviolet and X-ray light, emitted by high-energy electrons in a storage ring deflected by a magnetic field.

Temperature Programming (PTGC): In gas chromatography, a technique for improving resolution, which consists of continuously changing the column temperature during the separation.

Test Set: Samples used to determine how well a model performs on samples left out of the model building.

TEY: Total electron yield. One of the detection modes in XAS/NEXAFS/XANES spectroscopy where all electrons originating from the decay of the core hole are collected.

Thin-Layer Chromatography (TLC): Liquid chromatography with a stationary phase dispersed on a solid planar surface. Mobile-phase flow through the stationary phase occurs as a result of capillary forces.

Time-of-Flight Method: Determination of particle velocity by measuring its time of flight over a given distance.

TMFP: Transport mean free path of a photon.

TN: True negative.

Tomography: A three-dimensional imaging technique in which the image is constructed from a series of sections obtained at different points in a solid. Originally, a photographic technique for obtaining an X-ray image of a slice of tissue within the body; now applied in many fields to a technique of combining projections in many orientations to reconstruct an image.

TP: True positive.

Training Set: Samples used to establish a model.

Trans: A molecular structural configuration in which two like functional groups are opposite to each other on the molecular skeleton.

Transform: A mathematical function, each value of which is derived from a set of values of a given function by an explicit operation.

Transpose: Swapping rows and columns of a matrix, mirrored across the diagonal.

Trapping: The localization of positrons (and sometimes Ps) in defects sites; vacancy-type defects are deep traps and negatively charged impurities are shallow traps.

Triplet: An electron configuration in which the sum of electron-spin quantum numbers is 1, that is, there are two unpaired electrons of parallel spin.

$u \longrightarrow g$ **or** $g \longrightarrow u$ **Transitions:** Laporte-allowed transitions between wave functions that are even (gerade or g) or odd (ungerade or u) with respect to inversion about a center of symmetry.

Univariate: Measurements where there is single variable.

UPS: Ultraviolet photoelectron spectroscopy, mapping of the valence band structure of solids and surfaces with photon energies in the 5 to 50-eV range.

Validation: Determining how effective a model is.

Variable: A column of a matrix usually representing a frequency or wavelength.

Vector: Representation of data in a one-dimensional array, for example, a column or row of a matrix.

Velocity Selector: Mechanical device to monochromatize a neutron beam.

Vibrational Relaxation: Loss of internal vibrational energy into other modes of motion.

Wave Packet: A superposition of quantum mechanical states that does not correspond to an eigenstate of the Hamiltonian.

XANES: X-ray absorption near edge structure, mapping of unoccupied states via transitions from core levels. Synonyms: NEXAFS, Partial Yield Spectroscopy.

XAS: X-ray absorption spectroscopy. Related terms: NEXAFS, XANES, Partial yield spectroscopy.

XPS: X-ray photoelectron spectroscopy. Mapping of the core-level energies and their chemical shifts, typically using the K_α lines of Al or Mg at 1.49 keV and 1.25 keV, respectively. See also ESCA.

XSW: X-ray standing wave, determination of the local atomic structure of surfaces.

ZEKE: Zero kinetic energy spectroscopy, a high-resolution spectroscopy for studies of molecular ions.

Index

a

Abel transform, 1167–1168
Aberration-corrected holographic gratings, 342
Aberrations, of plane grating spectrometers, 334
Aberrations, of the imaging optics, 328
Absorption coefficients, 559–561, 570, 572–573, 575
Absorption of radiation, 424
Absorption spectroscopy, 298
– magnetic properties, 300
– in solids, 299–300
Absorption spectrum, 366
Absorption wavelength, for amino acid residues, 367
Accelerator mass spectrometry (AMS), 1010–1011
Achromatic colors, 384
Acoustic phonons, 120
Action spectroscopy, 377
Addition reactions, 737
Adiabatic-nuclei approximation, 263
Advanced photon source (APS), 60
Age-momentum correlation (AMOC), 131, 136
Air–hydrogen flame, 441
Alkali halide crystals, 370
Angiography, 524
Angle resolution, 33
Angular correlation of annihilation radiation (ACAR), 117, 125, 127–128
Angular differential cross sections, 215
Angular dispersion, of a grating, 323
Angular resolution, of a prism spectrometer, 326
Anharmonic wave, 20
Annihilation radiation
– Doppler shift, of, 129–131

Arc or spark excitation techniques, 430–432
Arcs, 430–432
Astigmatism, 329–330, 335–336
Asymmetry parameter, 68
Atmospheric pressure photoionization (APPI), 1036–1039
Atomic absorption spectrometry (AAS), 428
Atomic clusters, 793–795
Atomic fluorescence, 424
Atomic fluorescence spectrometry (AFS), 428, 460
Atomic spectra, 424–425
Atomic spectroscopy, 423
Atomizers
– in AAS, 440–446
– for AFS, 462
Atoms, 355–356
Attenuation coefficient a (E_g), 29
Auger electron spectroscopy (AES)
– detectors, 438–439
– role of the excitation source, 430–436
– spectrometers, 436–438
Autocorrelation, 752–753
Autocorrelator measurements, 751–752
Autoionization spectra, 229
Auxochromes, 366
Avalanche photodiode (APD), 349–350
Avoided level crossing (ALC-mSR), 177–180
Axially viewed plasma, 435

b

Background correction
– in HR-CS AAS, 455–459
– in LS AAS, 449
Background equivalent concentration (BEC), 429
Backscatter coefficients, 119
Back-scattering (BS) spectroscopy, 201–203

Encyclopedia of Applied Spectroscopy. Edited by David L. Andrews.
Copyright © 2009 WILEY-VCH Verlag GmbH & Co. KGaA, Weinheim
ISBN: 978-3-527-40773-6

Band mapping, 309–311
Barium sulfate (BaSO$_4$), 409, 411
Beam bunching, 134
Beam systems, 118
Beer–Lambert law, 355, 362 363, 428, 720, 897
Beer's law, 378
Bent-Crystal spectrometer, 95–98
Binary Encounter Bethe model, 258
Binary Encounter f-scaling method, 272
Biological tissues
– absorbing and scattering of tissues, 560–565
– index of refraction, 565–566
– light interactions, 559–561
– measuring techniques
– – human tissue, 575–613
– – inverse adding–doubling method, 571
– – inverse Monte Carlo method, 571–573
– – optical coherence tomography, 574
– – optical projection tomography (OPT), 616–617
– – refractivity measurements, 613–616
– – of the scattering phase function, 574–575
– – spatially resolved measurements, 573–574
– optical parameters of, 568–570
– – blood, 613
– – integrating sphere technique for measuring, 570
– – multiflux technique for measuring, 570–571
– short pulse interaction with biomaterials, 566–568
– sizes of cells and compartments, 558–559
– tissue optical models, 557–558
Birefringent crystal, 672
Bismuth germinate shield (BGO), 35
Bladder tumor detectors, 524–526
Blazed gratings, 101
M Bohr magnetons, magnetic moment of, 185
Boltzmann constant, 57
Boltzmann distribution, 426
Boltzmann's constant, 108
Born approximation, 251–252
Born–Oppenheimer approximation, 355, 358–361
Born–Oppenheimer separation
– of electronic and nuclear motions, 267, 773

Born series methods, 251–252
– distorted-wave (DW) methods, 252–253
– elastic scattering and excitation of atoms and ions at intermediate and high energies, 251
Bose–Einstein distribution, of phonons, 57
Bose statistics, 7
Bouguer-Beer-Lambert law, 561
BPP expression, 175
Bragg angle, 201
Bragg diffraction, 92
Bragg reflections, 60
– from crystals, 185
– electronic, 77
Bragg's law, 91–92
– angular dispersion (dl/dq), 93
Bragg spectrometer
– flat crystal, 93–95
– synthetic crystals (multilayers), 98–100
Brandeis positron reemission microscope, 146
Breit–Pauli R-matrix approach, 247
Breit–Rabi diagram, 176–177
Breit–Wigner formula, 62–63
Bremsstrahlung, 36–37, 286, 293, 471
Bremsstrahlung isochromat spectroscopy (BIS), 311
Brightness enhancement, 133
Brillouin scattering spectrometer, 196
BRISP, 196
B-splines, 242
BSR program, 242, 246
Bulk defect-free value P_b, 140

C
C_{60}, 177
C_{70}, 177
Calcite, 672
Calibration function, 428
Calibration standards, 428
Calorimetry technique, 39–42
Camera optics, 324
Capillary electrochromatography (CEC), 1096–1097
Cavity dumping, 680–681
Cavity-enhanced absorption spectroscopy (CEAS), 699
Cavity ring-down spectroscopy (CRDS), 378, 699
Cavity ring-down time, 378
CCC theory, for hydrogen-like systems, 234–235, 260
CdZnTe, 31

Characterization
- of objects, 388–389
- of observers, 391–393
Charge coupled devices (CCDs), 106, 193–194, 376
Charge parity (CP) invariance, 119
Charge-transfer-to-solvent (CTTS) transitions, 740
Charge-transfer transitions, 370, 739–740
CH_3COO, 313
Chemical sensing, 513–519
Chemical shift, 65
Chemical vapor generation (CVG), 444–446
Chemometrics, 362–363
- calibration, 1117–1126
- development of, 1106–1107
- exploratory data analysis, 1126–1132
- multivariate curve resolution, 1109–1117
- multivariate matrices, 1108
- multivariate process control (MSPC), 1141–1143
- notation, 1108–1109
- software and methodology, 1107–1108
- supervised pattern recognition, 1133–1141
Cherenkov materials, 40
Cherenkov radiation, 4, 40
Choppers, 192–193
Chroma, 384
Chromatic aberration, 330
Chromatic colors, 384
Chromatography
- bonded-phase, 1078
- chiral separations, 1095–1096
- classification, 1059
- definition, 1058
- gas, 1065–1069
- gas-liquid, 1069–1071
- gas-solid, 1071
- GC/GC, 1094
- GC/MS, 1091
- history of, 1057–1058
- ion-exchange, 1080–1082
- ion-pair, 1079–1080
- LC/GC, 1094
- LC/LC, 1094–1095
- LC/MS, 1091–1092
- liquid, 1072–1078
- mechanism of operation, 1059–1061
- preparative liquid, 1084–1085
- quantitative and qualitative analysis, 1064–1065

- reversed phase, 1078–1079
- separation mechanisms, 1093
- size-exclusion, 1082–1083
- spectrochemical detection, 1087–1090
- supercritical fluid, 1085–1087
- symbols, 1061–1064
- thin-layer, 1085
Chromophores, 365–367, 795
CIEDE 2000 color-difference equation, 403–405
CIE diffuse geometries, 408
CIE 1976 L*a*b* (CIELAB) color space, 396–397
CIELAB color-difference equation, 401–402, 411–412
CIE 1976 L*u*v* (CIELUV) color space, 397–398
CIELUV color-difference equation, 402–403
CIE standard illuminants, 386
Circle of least confusion, 328–329
Circular dichroism spectroscopy, 363–364
Circular polarization, 13
Clebsch–Gordan coefficients, 67
Close-coupling expansions, 238–243
- angle-differential cross section for a transition from a state, 239
- angle-integrated cross section illustration, 244–247
- asymptotic form of the wave function, 239
- Coulomb phase shift, 241
- dielectronic recombination, 250–251
- Dirac-Coulomb Hamiltonian, 244
- direct potential, 240–241
- discrete bound states, 240
- to intermediate and high energies, 242–243
- K-matrix asymptotic boundary conditions, 241
- Kohn variational method, 242
- long-range potentials, 241
- relativistic effects, 243–244
- resonance and quantum defect theory, 247–250
- role in dielectronic recombination, 250–251
- Schrödinger equation, for the $(N + 1)$-electron collision system, 239
- S-matrix, 241
- space and spin coordinates, 240
CMC(l : c) equation, 403
Cobalt-58, 125

Coherent sum, of two waves, 19
Coherent wave-packet dynamics, 790–792
Collimated (laser) beam, 561–565
Collimating optics, 324
Collision broadening, 427
Collision frame, 217
Color constancy, 384
– and metamerism, 384–385
Color-difference equation, 403
Color differences and tolerances, 398–400
Colorimeters, 410–411
Colorimetry
– characterizing objects, 388–390
– CIE system, 385
– color matching booths, 388
– light sources, 385–388
– standard observers, 390–395
Colorists, 384
Color-matching functions, 392
Color-matching light booth, 388
Color perception, 383–384
Color-tolerance equation, 403
Color-tolerance instruments, 406
Coma, 329
– of plane grating spectrometers, 334–335
Comatic flare
– of image, 329
Commission Internationale de l'Éclairage, 385
Complex-valued scattering amplitudes, 233–234
Compton scattering, 30–31
Concave grating, 322
Concave grating spectrometers
– non-Rowland Circle mounts, 341–342
– optical aberrations, 338–339
– Rowland Circle mounts, 339–341
Confocal optical microscopy, 638–640
Connes advantage, 877
Continuous (cw) photolysis, 729–730
Continuous wave (CW) lasers, 4, 697
Continuum distorted-wave eikonal initial-state (CDW-EIS) approach, 257
Continuum generation, of light, 684
Conventional polychromators, 437
Convergent close-coupling (CCC) method, 224, 240
Convergent conditions, 1158–1159
Conversion electron Mössbauer spectroscopy (CEMS), 62
Convolution, 1172–1173
Cooling cascade, of radiation, 33
Cooper minima, 291

Core-level spectroscopy
– of molecules adsorbed at surfaces, 303–305
– origin of core-level shifts, 301
– shake-up states and shake-off states, 302–303
– of solids, 305
Correlated color temperature, 386
Cotton effects, 364
Coulomb energy, 64
Coulomb phase shift, 241
Coulomb zone, 259
Counterfeit detection, using fluorescence, 528–529
Count-rate limitation, 33
Coupled integro-differential equations, 240
Coupling cases, 357
Cross-flow nebulizer, 434
Cryogenic detectors, 107–109
CRYRING, 269
– device, 232
Crystal-field splitting, 739
Crystal monochromators, 193
Cu nanoparticles, 142
Curvature of field, 330, 336
CuZnAl shape memory alloys, 141
3C wave function, 257
Cy3, 523
Cycloaddition reactions, 737
Czerny–Turner monochromators, 374
Czerny–Turner mounting, 436–437
Czerny–Turner spectrometer, 330

d

Darwin curve, 94
DBAR. see Doppler broadening spectroscopy (DBS)
Debye model, with a phonon density of states (DOS), 57
Debye temperature, 57
B-decay, 125
Decay of fluorescence, 630
Decay rate, 126
Degree of dissociation, 426
Depletion region, 103
Derivative spectroscopy, 363
Desorption electrospray/ionization (DESI), 1041
Desorption/ionization methods, field, 1019–1021
Detector, for HR-CS AAS, 454–455
Detectors, in LS AAS, 448
Detectors, in OES, 438

Deuterium lamp, 448
Dielectronic recombination, 250–251
Diffraction, 17–18
Diffraction gratings, 100–103
Diffraction orders, 373
Diffractometers, 194
Diffuse reflection, 406–407
DIN 6176, 405
DIN99 color-tolerance equation, 405
Dirac-Coulomb Hamiltonian, of Z atoms and ions, 244
Direct absorption, 698
Direct analysis in real time (DART), 1041–1042
Direct current plasma (DCP), 435
Direct geometry spectrometers, 195–196
Dissociation, of a diatomic molecule, 736
Distorted-Wave (DW) methods, 252–253
D50 (5000 K), 386
D65 (6500 K), 386
DNA photoprotection, 799–801
Doppler broadening, 427
– of annihilation radiation (DBAR), 124
Doppler broadening spectroscopy (DBS), 118, 129
Doppler drives, 193
Doppler shift
– of annihilation radiation, 129–131
– in photon energy, 129
– of the radiation, 33
D-orbital splitting patterns, 368
Double-beam spectrometers, 374
Double-crystal premonochromator Si(111), 60
Double-crystal spectrometer, 97
Double differential cross sections (DDCS), 221
Double electron ejection, 226
Double monochromators, 346–347
L-doubling, 357
DS3C function, 258
Duane–Hunt equation, 286
Dynamical beats (DBs), 78–79
Dynamic nuclear polarization (DNP), 959–960
Dynodes, 47, 438

e

Eagle mount, 330, 340–341
Echelle grating polychromator, 437
Echelles, 333
Effect pigments, 412
Efficiency, of the spectrometer, 323

EFG principal axes system, 70–71
E–H scattering, 242
Eikonal Born series (EBS), 252
Einstein coefficients for absorption and spontaneous emission, 630
Einstein formula, 298
Einstein model, of lattice vibrations, 56
Elastic scattering
– differential cross sections, 213–215
– and excitation of atoms and ions at intermediate and high energies, 251
Electic-field-induced SHG (EFISHG), 820
– measurements, 826–831
Electric quadrupole interactions
– Hamiltonian of, 69
– in a magnetically ordered solid, 71
– in Mössbauer spectroscopy, 68–73
– – combined with magnetic dipole, 71–72
– – saturation effect, 72–73
Electrocyclic ring-closure reactions, 736
Electromagnetic (EM) field, 3
Electromagnetic spectrum, 321
Electron cooling, 232
Electron cyclotron resonance (ECR), 293
Electron energy transfer, 734–735
Electronic photoabsorption spectroscopy, 732
Electron impact coherence parameters, 217
Electron-impact ionization process, 220
Electron-induced Auger electron spectroscopy (EAES), 138
Electron–ion interactions, 231–233
Electron–molecule scattering, 230–231
Electron momentum, 223–224
Electron scattering
– inclusion of nuclear motion, 265–270
– in laboratory frame of reference, 262–263
– in molecular frame representation, 263–265
– by molecules, 262
– – illustrative results, 270–274
Electron selection rule, 357–358
Electron spectrometer, 212–213
Electron spectroscopy for chemical analysis (ESCA), 287
Electron-spin correlation rules, 736
Electrothermal atomization (ETA), 442–444
Electrum, 117
Ellipsoidal highly oriented pyrolithic graphite (HOPG) crystals, 97
Emission spectrum, 733
Energy dispersive spectrometers (EDSs), 91

Energy-domain Mössbauer spectroscopy, 76
Energy partitioning, 725–729
Energy resolution, 33
Energy-versus-momentum band dispersions,
  of ferromagnetic nickel, 292
Entrance aperture, 323–324
Epifluorescence microscopy, 637
Escape cone, 298
Escape peaks, 106
E1 transitions, 358
European Spallation Source (ESS), 191
European Synchrotron Radiation Facility
  (ESRF), 60
ExB velocity filter, 132
Excimers, 495–497, 727
Exciplexes, 498, 727
Excitation energy transfer, 501–502
Excitation function, of the state, 215
Excitation interferences, 426
Excitation process, for an atomic target, 215
– angular and polarization correlations,
    216–220
– differential cross sections, 215
– integral cross sections, 215–216
Excitation source, in OES, 430
Excitation spectroscopy, 377
Excitation spectrum, 733
Excited state, 426
Excited-state deprotonation, 499–500
Exit aperture, 325
Extended X-ray absorption fine structure
  (EXAFS), 298, 312–313
Exterior complex scaling (ECS) method,
  255–256
Extinction coefficient, 924
Extinction methods from particles,
  924–926

f

Fabry–Perot interferometer, 324, 343–346
Fast Hartley algorithm, 1164–1165
Fastie–Ebert spectrograph, 373–374
Fellgett advantage, 877
^{57}Fe Mössbauer spectroscopy, 58
Femtochemistry, 733–734
Fermat's principle, 5, 7
Fermi chopper, 193
Fermi contact, 175
– field, 67
Fermi's golden rule
– for a differential cross section, 290
– of transition probability per unit time of the
    system, 188

Fermi surfaces, 142
Feshbach resonances, 249
Field effect transistor (FET), 830
F illuminants, 387
Finite-element discrete-variable
  representations (FEDVR), 256
First-order many-body theory (FOMBT),
  252
Fixed-nuclei approximation, 263–264
Flame atomizers, 440–442
Flame excitation, 433
Flame photometry, 433
Flare, 391
Flash photolysis, 730–732
Flippers, 193
Fluorescence, 479
Fluorescence anisotropy, 488–489
Fluorescence emission, characteristics of
– delayed fluorescence, 483–484
– emission and excitation spectra, 486–488
– internal conversion, 482
– intersystem crossing process, 483–484
– lifetimes and quantum yields, 484–486
– of photons, 482–483
– radiative and nonradiative states, 481–482
– Stokes shift, 488
– triplet–triplet annihilation, 484
Fluorescence excitation spectroscopy,
  376–377
Fluorescence LIDAR, 526–527
Fluorescence quenching, 490–492
Fluorescence yield, 467
Fluorescent materials, 412
Fluorescent molecular sensor, 513
Fluorescent molecular thermometers, 508
Fluorescent objects, 388
Fluorescent proteins, 522–523
Fluorescent sensor, 513
Fluorite prism, 322
Focusing devices, 193
Focusing or condensing optics, 324–325
Food systems, fluorescent substances in,
  527
Forensic science and fluorescent powders,
  527–528
Forman algorithm, 878
Förster cycle, associated with spectroscopic
  measurements, 500–501
Fourier analysis, 21
Fourier time, 204
Fourier transform, 1151–1152
– discrete vs continuous, 1152–1155
– fast, 1163–1164

– fields of application, 1160–1162
– use, 1159–1160
Four-wave mixing (FWM) processes, 783
Fovea, 391
Franck–Condon factor, 359
Franck–Condon principle, 355, 358–361
Fraunhofer diffractions, 17
Fraunhofer lines, 321
Free-electron laser, 691–692
Free induction decay (FID), 80
Frequency doubling, 685
Frequency modulation (FM) spectroscopy, 711–712
Frequency tripling technique, 687–689
Fresnel coils, 193
Fresnel diffraction, 17
Fresnel-Kirchhoff integral formula, 17
FRMII reactor, 205
Full width at half maximum (FWHM), 90, 94, 99, 104, 831
Fundamental emission anisotropy, 489

g
Gammasphere array, 35
Gammow–Teller decay strength, 36
Gas scintillation counter, 469
Gaussian derivative, of implantation profile, 120
GD-MS ion source, 1005
GEANT Monte Carlo program, 39
Ge detector, for DBAR measurements, 132
Germanium counters, 30, 33
G-factor, 66
Giant magnetoresistance (GMR), 309
Gibb's Principles of Mössbauer Spectroscopy, 65
Glow discharge optical emission spectrometry (GD OES), 431, 1006–1007
Glow discharges (GDs), 432
Goldanskii–Karyagin (G–K) effect, 57, 70
Gonioapparance, 413
Goniospectrophotometers, 413
Gratings, 100–103
Grating spectrometers
– angular dispersion, 330–331
– design and manufacture of gratings, 332
– echelles, 333
– holographic gratings, 333–334
– order-sorter for a, 345
– reflection gratings, 332–333
– resolution of, 331–332
– transmission gratings, 332

Grazing incidence, 341
Green's function G(A, B), 291
Green's operator, 251
GRETA, 35–36
GRETINA, 36
Ground state, 426

h
Half width at half maximum (HWHM), Lorentzian, 190
Hamiltonian eigenvalues, 69
Hankel transform, 1167
Harmonic time dependence, of the wave, 11
Harrison echelle, 345–346
Hartley transform, 1162–1163
He I, 293
Helicity, 135
Helium elastic scattering cross sections, 214
Helium–Neon laser, 8
Henyey–Greenstein function, 563
Herzberg–Teller effect, 369
Heterogeneous broadening, 707
HgI_2, 31
High-energy muons, 44
High-energy symmetric kinematics, 223
Higher order frequency mixing, 687
Higher order harmonic generation, 689–690
Highest energy occupied molecular orbitals (HOMOs), 298–299, 364, 735
High-intensity positron beams, 134
High-purity germanium (HPGe) detectors, 90
High-resolution continuum source atomic absorption
– spectroscopy (HR-CS AAS), 452–459
Hilbert transform, 1165–1166
Hollow cathode lamp (HCL), 446
Holographic gratings, 333–334
Hula-twist isomerization, 790
Human eye–brain evaluation, of color difference, 399
Hund's case, 357–358
Huygens's principle, 17
Hydrogen atom abstractions, 737
Hydrogen-bonded networks, 795
Hyperfine interaction, 175
Hyperfine interaction
– electron–muon, 176
– isotropic, 175
– magnetic, 72
– Mössbauer effect, 64

i

Ideal detector, 347–348
ILL reactor, 191
Illuminating and viewing geometries, 408–410
Imaging spectrometers, 347
IN4, 195–196
IN5, 195–196
IN6, 195–196
IN8, 200
IN11, 205
IN14, 200
IN15, 205
IN20, 200
Indirect (inverted) geometry spectrometers, 197–198
Induced emission, for transition, 424
Inductively coupled plasma (ICP), 433, 1007–1010
Inductively coupled plasma mass spectrometry (ICP-MS), 434
Inductively coupled plasma optical emission spectrometry (ICP OES), 434
Inelastic neutron scattering (INS), 187
Infrared (IR) spectroscopy
– asynchronous time-resolving system FT-IR spectrometers, 883–885
– dispersive, 870–872
– FT-IR spectrometers, 872–882
– phase correction, 877–878
– principles, 869–870
– step-scan interferometers, 882–883
Infrared microspectroscopy, 921–924
Infrared spectra, 926–931
Infrared vibrational spectroscopy
– classical and quantum-mechanical description of, 889–892
– detectors, 902
– gas-phase spectra and rotational fine structure, 896–897
– group frequencies and symmetry, 896
– measurements
– – of absorption spectra, 897
– – interferometry, 897–901
– – spectral processing, 897
– processing, 901–902
– reflection-based sampling methods, 905–921
– transmission sampling methods, 902–905
Inhomogeneities, of crystal, 43
Inner-shell ionization–Auger-electron decay, 230

In-plane photon detector, 217
Integrated cavity output spectroscopy (ICOS), 699
Interferences, 18, 425–426
Interferometric spectrometers, 342–345
Interferometry, 20
Intermediate-energy R-matrix (IERM) theory, 242
Intermolecular photophysical processes, on fluorescence, 489–490
Internal conversion, 479, 726
Intersystem crossing, 726
Intracavity laser absorption spectroscopy (ICLAS), 700
Intramolecular vibrational energy redistribution (IVR), 788
Intrinsic region, 103
Inverse Ps formation spectroscopy, 139
Ion formation, at atmospheric pressure, 1028–1030
Ionization process, 220, 426
– atmospheric, 1030–1032
– atmospheric pressure chemical, 1036–1039
– of atoms and ions
– – amplitude and cross section, 256–258
– – excitation–autoionization, 261–262
– at intermediate and high energies, 259–261
– threshold of the cross section, 258–259
– chemical, 1014–1016
– collisions involving ejection
– – of double electron, 226–227
– – of a single electron, 223–226
– double differential cross sections, 221–222
– electron, 1012–1014
– electrospray, 1032–1036
– excitation–autoionization process, 229
– excitation experiments, 227–229
– field, 1017–1019
– formation of negative ions, 1002–1003
– formation of positive ions, 1001–1002
– formation of positive radical ions, 999–1001
– formation of radical negative ions, 1002
– of gaseous molecules in vaccum, 1011–1012
– higher order differential cross sections, 222
– inner-shell, 230
– total cross sections, 220–221

Isomer shift, Mössbauer spectroscopy, 64–65
Isotropic jump-reorientation motion, 177

j

Jacquinot advantage, 877
Jahn–Teller coupling, 233, 270
Johann spectrometer, 97
Jones calculus, 15–16

k

K-fluorescence radiation, 82
K-matrix asymptotic boundary conditions, 241, 250
K-nearest neighbor (KNN) method, 1140–1141
Kramers–Kronig relation, 291
Kubelka-Munk model (KMM), 570–571

l

Laboratory-based beams, 132–134
Lagrangian principle, 6
Lamb–Mössbauer factor f_{LM}, 81
Langmuir–Blodgett soap films, 98
Laplace transform, 1155–1158
Laplacian electric field, 831
Laplacian law, 3
Laporte selection rule, 357–358
Larmor frequency, 175, 203
Laser-based methods, 376
Laser desorption/ionization (LDI), 1024–1025
Laser-excited atomic fluorescence spectrometry (LEAFS), 462
Laser-induced breakdown spectroscopy (LIBS), 436
Laser-induced fluorescence (LIF) spectroscopy, 733
– absorption spectrum, 704
– dispersed or resolved fluorescence, 705–706
– imaging, 705
– intermodulated fluorescence, 706
– light detection and ranging (LIDAR), 707–708
– quantum beat spectroscopy, 707
– saturation dip spectroscopy, 706–707
– single-molecule spectroscopy, 707
– stimulated emission pumping (SEP), 706
Laser magnetic resonance (LMR) spectroscopy, 700–701
Laser spectroscopy, 732–733

Layered synthetic microstructures (LSMs), 98
Leading edge, time resolution, 45
LEED, 313
Ligand-field (LF) transitions, 738–739
Ligand-to-metal (LMCT), 739–740
Light
– as an electromagnetic (EM) field, 3, 8–9
– – brightness, 13
– – diffraction, 17–18
– – energy density, 12
– – interference, 18
– – intensity, 12–13
– – longitudinal component, 16
– – Maxwell's equation for, 9–10
– – photons and particles, 24
– – polarization, 13–16
– – power, 13
– – superposition, 18–24
– – wave equations, 10–12
– Bose statistics, 7
– geometrical optics, 5–6
– half angle of emitted light, 4
– Huygen's views, 3
– and momentum conservation, 5
– Newton's views, 6
– physical optics, 6–7
– quantum, 7
– relationship between l and u, 4
– speed in a medium, 3
– speed in vaccum, 4
– visible region, 8
– wavelength spectrum of, 4, 7–8
– wave properties, 5
Light absorption, 355
Light-absorption principles, 720–721
Light-emitting diodes (LEDs), 371
Lightness, 384
Light-sensitive photodiode, 47
Limit of quantification (LOQ), 429
LINAC beam pulses, 134
Linear accelerator (LINAC), 125, 134
Linear spectroscopic techniques, probing with, 781–783
Line core, 427
Line tunable lasers, 701
Line wings, 427
Liquid crystals (LCs), 819
Liquid secondary-ion mass spectrometry (LSIMS), 1021–1024
Lithium-drifted germanium detectors [Ge(Li)], 105–106

Lithium-drifted silicon [Si(Li)] detectors, 104–105
Littrow configuration, 327
Littrow mount, 327
Local density approximation (LDA), 291
Lock–Crisp–West (LCW) procedure, 128
Longitudinal field component, 16
Longitudinal-field muon spectroscopy (LF-mSR), 172–173
Longitudinal-field muon spin relaxation, 173–177
Lorentz shift, 428
Low-energy electron diffraction (LEED), 138, 285
Low-energy positron diffraction (LEPD), 137
Lowest unoccupied molecular orbital (LUMO), 298, 364, 735
Low-spin systems, 369
Luminescence, 727
Luminosity, 322
– of a grating spectrometer, 331–332
– of a prism spectrometer, 326

m

M1, M2, and M3 transitions, 358
Magnetic angle changer (MAC), 212, 224
Magnetic dipole interactions, in Mössbauer spectroscopy, 65–68
Magnetic hyperfine interaction, 66–67
MAPS, 195–196
Massless photons, 7
Mass spectrometry (MS)
– concept, 991
– goal, 991–992
– mass spectral analysis methods, 996–998
– role of isotopes, 993–996
Matrix-assisted laser desorption/ionization (MALDI), 1026–1028
– atmospheric pressure, 1039–1041
Matrix effect, 426
Maupertu's principle, 6
Maxwell-displacement current (MDC) measurement, 829
Maxwellian distribution, 246
Maxwell's equations, 3
– in a medium, 10
McStas, 200
Mellin transform, 1165
Metal-to-ligand (MLCT), 739–740
Metamerism, 384, 391
MeV positron beams, 135–136
Michelson interferometer, 63, 324, 342–343

Michelson-Morley experiment, 9
Microchannel plates (MCPs), 375–376
Microfabricated systems, 519–521
Microspectroscopy, 300–301
Microwave-induced plasma (MIP), 435
Microwave-optical double resonance (MODR), 713
Mie scattering, 563
MINIS, 232
Mirrors, of neutron guides, 192
ML_6 octahedral complex, 368
Mn_{12}-acetate, 196
Mode locking, 681–683
Moderators, 192
Molecular 3DW (M3DW) method, 273
Molecular fluorescence, 479, 502–503
Molecular gas lasers
– CO_2 laser, 672–675
– dye laser, 677–680
– excimer laser, 676
– iodine laser, 675–676
– nitrogen laser, 675
Molecular R-matrix with pseudostates (MRMPS) method, 272–273
Molecules, 356–357
Molière radius, 37–38
Molière theory, of multiple scattering, 37–38
Momentum density $r(\mathbf{p})$, 124
Monochromanicity, 4
Monochromatic light, 4
Monochromator
– for HR-CS AAS, 454
– in LS AAS, 447–448
Monoenergetic signal output, 34
Mössbauer absorption spectrum, 63
Mössbauer effect, 53–58
– applications in magnetism, 68
– hyperfine interactions in, 64
Mössbauer isotopes, 58–60
Mössbauer nuclei, 61, 78
– elements, 68
Mössbauer spectrometer, 60–62
Mössbauer spectroscopy
– applications, 73–76
– electric quadrupole interactions, 68–71
– – combined with magnetic dipole, 71–72
– – saturation effect, 72–73
– hyperfine interactions, 64
– isomer shift, 64–65
– magnetic dipole interactions, 65–68
Mössbauer effect, 53–56
– overview, 53

- recoil-free fraction f,   56–58
- second-order Doppler effect,   73
- sources of radiation,   58–60
- spectral line shape and intensity,   62–63
- spectrometer,   60–62
- using SR radiation,   76–77

Mott detector,   236
Muller calculus,   15–16
Multichannel analyzer (MCA),   129
Multichannel quantum defect theory (MQDT),   250, 269–270
Multidimensional separation,   1093
Multidimensional transforms,   1166–1167
Multipass techniques,   698
Multiple hits, of nuclear reactions,   33
Munsell color space,   395–396
Muon
- beams,   157–158
- decayed beams,   158
- discovery,   155–157
- implanted in matter,   159
- – free radicals from,   166–168
- low-energy beams,   158–159
- range and range straggling within a sample,   159–161
- spin-flip transitions,   169
- surface beams,   158

Muon–electron system, with a schematic representation of the transitions,   170
Muonium,   163–166
Muon–proton flip-flop transition,   179
Muon spectroscopy,   155
- energies associated,   169
- theoretical aspects,   161–163
- variants of mSR,   169–170

n

NaI(Tl) scintillation counters,   61
Natural line width,   427
Nd : glass laser pulses,   95
Nd : YAG laser,   8
Near edge X-ray absorption fine structure (NEXAFS) measurements,   298–299
Nebulizer gas flows,   433
Neon-22,   126
Net planes,   91
Neutron
- Hamiltonian of the total system,   187

Neutron detectors,   193–194
Neutron guides,   192
Neutron reflectometry,   194
Neutrons, magnetic moment of,   185

Neutron scattering
- and eigenfunctions of the sample,   189
- instruments,   194
- intermediate scattering functions,   189–190
- magnetic,   190–191
- response functions,   190
- techniques
- – moderators,   192
- – neutron optics,   192–194
- – sources,   191–192
- theoretical aspects,   187–190
- types,   186

Neutron spectroscopy
- back-scattering,   201–203
- neutron spin-echo (NSE),   203–206
- time-of-flight (TOF),   194–201

Neutron spin-echo (NSE) spectroscopy,   185, 203–206
Nitrous oxide–acetylene flame,   440
Noble liquids,   41
4-n-octyl-4′-cyanobiphenyl (8CB) monolayers,   829
Nonadiabatic processes,   773
Nondispersive spectrometers,   91, 103–108
Nonequilibrium vacancy defects,   140
Nonlinear spectroscopic techniques
- experimental methods,   778
- generation of ultrashort laser pulses,   778–781
- nonperturbative methods,   778
- response function formalism,   776–778
Nonplanar waves,   12
Nonradiative energy transfers,   735
Nonresonance fluorescence,   461
Nonspectral interferences,   425
Nonthermalized positrons,   123
Nuclear exciton,   78
Nuclear fission reactors,   191
Nuclear magnetic resonance (NMR),   64
Nuclear magnetic resonance (NMR) spectrometry,   965
- chemical exchanges,   950–951
- chemical shift assignment,   969–970
- – in carbohydrates,   973–975
- – in nucleic acid,   973
- – in proteins,   970–973
- – structural determination from,   980–981
- cross polarization technique,   957
- dead time in wide-line,   955
- detection,   938–939

Nuclear magnetic resonance (NMR) spectrometry (contd.)
- equipment, 939–940
- experiments in liquids, 948–949
- general considerations, 936
- high-power decoupling technique, 956–957
- high-resolution solid-state, 956
- hydrogen bonds, 979
- imaging, 960–961
- isomeric and conformational analysis in solutions by, 953–954
- molecular modeling and structure calculation, 981–982
- nuclear overhauser effect (NOE), 975–977
- nuclear relaxation mechanism, 945–947
- nuclei in external magnetic field, 936–938
- paramagnetic centers in biological macromolecules, 980
- paramagnetic relaxation mechanism, 947
- polarization transfers, 949
- probes, 940
- relaxation mechanism in solids, 947–948, 955–956
- residual dipolar couplings (RDCs), 978–979
- sensitivity enhancement, 959–960
- solid state, 957–958
- – zero-field, 958–959
- spectra acquisition
- – nuclei, 966–967
- – pulse sequences and processing strategies, 968–969
- – sample preparation, 967–968
- spectral parameters, 940–942
- spectra of biological macromolecules, 965
- spin–spin coupling, 943–945
- torsion angles, 977–978
- translational molecular diffusion via, 951–953
- two-dimensional (2D), 950
- unpaired electron affects, 942–943
- wide-line, 954–955
Nuclear overhauser effect (NOE), 975–977

o

Oak Ridge National Laboratory (ORNL), 191
Obliquity factor, 17
Ochkur electron exchange amplitude, 252

Optical atomic spectroscopy
- principles, 428–429
Optical axis, 5
Optical density, 362
Optical electron spectroscopy. see Auger electron spectroscopy (AES)
Optical emission spectrometry (OES), 428
Optically pumped solid-state lasers
- argon laser, 671–672
- atomic and ionic gas lasers, 669
- color-center lasers, 662
- diode lasers, 664–666
- fiber lasers, 662–664
- helium–neon laser, 669–671
- krypton laser, 671–672
- neodymium lasers, 660–662
- quantum cascade laser, 667–668
- quantum dots, 667
- quantum microcavity lasers, 668–669
- ruby laser, 659–660
- vertical cavity surface emitter lasers (VCSEL), 666
- visible laser diodes, 666
- wavelengths available from commercial lasers, 658
- widely tunable lasers, 662
Optical–optical double resonance (OODR), 713
Optical parametric oscillator, 687
Optical path length (OPL), 5
Optical phonons, 120
Optical spectroscopy, 424
Optogalvanic spectroscopy, 710–711
Orbital angular momentum, of an electron, 355
Orbital field, 67
Ore Model, 124
Orientational order parameter, 819
Orientation averaged molecular orbital (OAMA) approach, 273
Osmetech/Roche OSPI critical care analyzer, 523–524
Overdamped excitations, 190

p

Pair production, 30
Paraxial approximation, 16
Paraxial rays, 5
Partial density of states (PDOS), 82
Partial ionization cross section, 221
Particle (or proton) induced X-ray emission (PIXE), 471

Paschen–Runge mount, 437
Paschen–Runge spectrograph, 340
Paul Scherrer Institute (PSI), 192
P-branch, 362
PECS theory, 259
Pellin–Broca mount, 330
Pellin–Broca prism, 328
Pellin–Broca spectrometer, 328
Penning-type trap, 136
Peptide bonds, 366
Peptide dynamics, 798–799
Perrin–Jablonski diagram, 481
Peterkop theorem, 257
Phase velocity, 11
PH dependent absorption and fluorescence spectra, 501
Phosphorescence, 479
Photoacoustic spectroscopy, 708
Photocathode, 438
Photochemical pathways, 728–729
Photochemical sigmatropic shift, 737
Photochemistry, 719, 789–790
– of inorganic and coordination compounds, 738–740
– of organic compounds, 735–738
Photodiodes, 349–351, 374–375
Photoelectric detector, 428
Photoelectric effect, 285
Photoelectric yield, 298
Photoelectron diffraction (PhD), 312–313
Photoelectron spectroscopy, 285, 708–709
Photoelectron spectroscopy
– detectors, 295–298
– light sources, 293–295
Photoemission, 285
– absorption coefficients, 298
– energy diagram for, 286
– inverse, 286, 311–312
– measurement in, 298
– and probing depth, 287
– quantum numbers, 289
– "single-step" calculations, 290
– theory, 288–293
– time-resolved, 311
– unoccupied orbitals, 298–299
Photoemission electron microscope (PEEM), 301
Photographic plates, 348
Photoinduced electron transfer (PET), 494–495, 792–793
Photoinduced proton transfer, 498–499
Photoisomerization, 736–737
Photolysis, 719

Photomultipliers, 194
Photomultiplier tube (PMT), 42, 47, 348–349, 375, 428
Photon, 6
Photon echo technique, 763–765
G-photons, 118–119
Photons and particles, 24
Photon wavefunction, 7
Photophysical basics, 719
– dye molecule, 634–635
– energy levels and transition states, 630–632
– energy transfers, 632–634
– signal-to-noise (SNR) ratio, 635–636
– wide-field fluorescence microscopy, 636–637
Photophysical pathways, 725–729
Photosensitization, 734
Photosynthesis, 796
Phthalocyanine (Pc), 820
Physical line width, 427
Picosecond continuum, 684
Planck constant, 57
Planck Law temperature, 371
Planck's constant, 720
Plane grating spectrometers
– aberrations, 334–336
– types
– – Czerny–Turner spectrometer, 337–338
– – Ebert–Fastie spectrometer, 336–337
– – Newtonian telescope, 338
Plane waves, 10–11
Plate factor, 323
P_N approximation, 563
Point spread function (PSF), of the microscope, 638
Polarization, 13–16
Polarization correlation measurements, 216–220
Polarization function S_P, 235
Polarization spectroscopy, 703
Polarized PAES, 139
Polarized positron beams, 135
Polarizers, 193
Polychromators, 436–437
Polycrystalline filters, 193
Polycrystalline solids, 289
Polyethylene glycol dimethacrylate, 143
Polytetrafluoroethylene (PTFE), 409
Portable (field) XRF instruments, 471
Position-sensitive detectors, 197
Positron
– annihilation of, 118–119

Positron (contd.)
– – observables, 123–124
– beams, 132
– diffusion, 120–121
– holograph, 139
– holography, 139
– implantation, 120
– lifetime, 126
– microbeams, 134–135
– nonthermalized, 123
– and positrinium chemistry, 146–148
– as a probes of open volume in polymers, 143–144
– reemission, 122–123, 132
– reemission microscopy, 146
– sources, 125–126
– spectroscopy, 126
– surface studies, 144–146
– trapping, 121
– – in solids, 118
– use of, 117
Positron-annihilation-induced Auger electron spectroscopy (PAES), 138
Positron annihilation lifetime spectroscopy (PALS), 124
Positron annihilation spectroscopy, 117
Positron backscattering, 119
Positron-beam-based spectroscopies, 136–137
Positron–electron enhancement factor e, 124
Positron-induced ion desorption, 139
Positronium, 117
– annihilation, 124–125
– emission spectroscopies, 138
– ground-state binding energy, 124–125
Positronium formation spectroscopy, 139
Positronium (Ps), 119
– diffraction, 139
– formation, 122, 148
– reflection, 138
Positron reemission microscopy (PRM), 134
B-positrons, 120
Positron scattering, 238
Positron-surface interactions, 121–123
Positron surface spectroscopies, 137–139
$^1P^o$ state, 216
POSTRAP, 136
Potential energy surfaces, 772–774
Precession coils, 193
Pressure dependence, of fluorescent compounds, 511–512

Principle of least action, 6
PRISMA, 198
Prism-coupled KBBF ($KBe_2BO_3F_2$) crystal, 295
Prism spectrometer
– aberrations, 328–330
– angular resolution, 326
– design, 326–328
– dispersion, 325–326
– luminosity, 326
Probe coherences, 783–784
Progression, 361
Protein biophysics
– folding property, 640
– unfolded state, 640–642
Pulse pair resolution, 46
Pulses, of light, 22–24
Pulse shaping, 784–785
Pulsing methods, 680
Pump-probe technique, 771
P-wave phaseshifts, 214

q

Q-switching, 681–683
Quadrupole splitting, 69–70
Quantum beats (QBs), 78, 80
Quantum dots, 523
Quantum efficiency, 47
Quantum-mechanical principles, of absorption, 721–725
Quantum-mechanical selection rules, 357
Quantum yield, 721
Quantum yield of fluorescence, 629
Quartz prism, 322
Quasi-continuous-wave (quasi-CW) mode, 295
Quasi-elastic backscattering, 119
Quasi-elastic neutron scattering (QENS), 187, 195, 198
Quenching collisions, 461

r

Radially viewed plasma, 435
G-radiation, annihilation of, 119
Radiation sources, for Mössbauer spectroscopy, 58–60
Radiative transitions, 424
Radiometric colorimeters, 416
Radon transform, 1168–1170
Raman spectroscopy, 701–703
– application in forensic field, 844–845, 850–851

- basic instrumentation, 838–839
- biological applications, 851–858
- characterization of inorganic materials using, 842–844
- characterization of semiconductor surfaces, 845–848
- data reduction, 841–842
- fiber optic probes, 839
- microprobes, 839–841
- in pharmaceutics, 848–850
- in polymer chemistry and physics, 850
- process control with, 848
- resonance enhancement and surface enhancement, 836–837
- spatially offset Raman spectroscopy (SORS), 837
- standard laser safety protocols, 837
- theory, 835–836
G -Ray detectors, 46
G -ray energy, 30
Rayleigh scattering, 563
G -Rays
- calorimeter designs, 42–46
- calorimetry techniques, 39–42
- detector systems
- - large solid-state systems, 32–34
- - modern arrays, 34–36
- - optical, 46–48
- - using semiconductor materials, 31–32
- energy deposition in materials, 29–30
- readout devices, 46–48
- shower development, 36–39
G-rays, 118
- 511-keV annihilation "death", 126
R-branch, 362
Reactive wave-packet dynamics, 786–787
Reactivity, of an excited species, 734
Readout devices, 48
Recoil-free fraction, 56–58, 70
Recoil-free g -photons, 56
Reemitted positron energy loss spectroscopy, 139
Reemitted positron spectroscopy, 139
Reflectance, 389, 410
Reflectance factor, 389
Reflected intensity (P), 94
Reflection grating, 332–333
Reflection high-energy positron diffraction (RHEPD), 138
Refraction, 5
Refraction, index of, 6

Refractive index, of air, 326
Regular transmittance, 415
Relative phase, of the wave, 11–12
Relativistic transformation, 127
Relaxation function, 174
Resolution
- of an instrument, 323
- of a diffraction grating, 331
Resolving power (R), 34, 323
Resonance, 423
Resonance energy transfer (RET), 535
- application to molecular biology
- - conformational changes, 546–547
- - intensity-based imaging, 547–548
- - lifetime-based imaging, 548–550
- - spectroscopic ruler, 545–546
- early developments of theory, 536
- Förster theory, 536–537
- history of, 535–536
- photophysics, 537
- - coupling of electronic transitions, 538
- - dexter transfer, 544–545
- - diffusion effects, 543–544
- - dissipative processes, 538–539
- - Förster equation, 539–540
- - Förster radius, 541–542
- - long-range transfer, 544
- - orientation dependence, 540–541
- - polarization features, 542–543
- - primary excitation process, 537–538
Resonance-enhanced multiphoton ionization (REMPI), 377, 709–710
Resonance lines, 427
Resonant excitation double autoionization (REDA), 229
Resonant two-photon ionization (R2PI), 377
RESTRAX, 200
Resultant "photopeak" spectra, 130
Reverse biasing, of diode junction, 31
Reverse bias photodiode, 374
Rhodopsins, 796–799
R-matrix with Pseudostates (RMPS) method, 243
Rocking curve, 94
Room-temperature devices, 106–107
Rotational fine structure, 361–362
Rowland circle, 97, 437
- radius of, 95
Russell–Saunders scheme, 357
Rydberg resonance, 250
Rydberg states, 355

S

Saint-Gobain (Bicron) plastic scintillators, 41
Sampling fluctuations, 43
S and W parameters, 130
Saturation effect, 72–73
Scalar harmonic waves, 11–12
Scanning transmission X-ray microscope (STXM), 301
Scattering lengths, 186, 189, 194
Schrödinger equation, 7, 772
Scintillation counter, 469
Scintillation crystal sodium iodide, 29
Second harmonic generation (SHG), 819
– measurements, 826–831
– spectroscopic considerations, 823–825
– sum frequency generation, 825–826
Selective volatilization, 431
Self-phase modulation, 684
Semiconductor-based detectors, 103–105
Semiconductor nanocrystals, 523
Semiconductors and ultrafast spectroscopy, 801–802
Se monolayer, 144
Sensitivity, 428–429
– of AAS instruments, 451
Sensitized fluorescence, 461
Sensitizer, 535
Sequential spectrometers, 436
Seya–Namioka mount, 330
Short wavelength coherent radiation, 684–687
Silicon, crystal lattice parameter of, 92
Silicon drift detectors (SDDs), 106
Si(Li) detector, 469
Simultaneous spectrometers, 436
Single-beam spectrometers, 374
Single-channel instruments, 468
Single-molecule fluorescence, on biological systems, 629
– enzymatic activity, 641
– – conformational changes and catalytic activity, 646–650
– – static and dynamic heterogeneity, 641–646
– observation of folding/unfolding transitions, 640–641
Single-number color scales, 400
Sisam spectrometer, 347
Si–SiO$_2$ region, 301
Small-angle neutron scattering (SANS) instruments, 194
Snell's law, 326

Sodium-22, 126
Solid-state detectors, 103
Soller collimators, 192
Solute-volatilization interferences, 425
Solvation, 788–789
Solvatochromism, 366
Solvent polarity, 366
Spallation, 191
Sparks, 431–432
Spark source mass spectrometry (SS-MS), 1004–1005
Spark-source OES instruments, 431
Spatial resolution, of the calorimeter, 45
Specimen's specular reflection geometry, 410
Spectral brightness, 13
Spectral interferences, 425
– in LS AAS, 448
Spectral line, 427
Spectral power distribution (SPD), 386
Spectral radiometers, 416
Spectrochemical buffer, 431
Spectrocolorimeters, 411
Spectrograph, 324
Spectrometer, 321, 325
– angular dispersion of a grating, 323
– efficiency of, 323
– elements of
– – collimating optics, 324
– – detector, 325
– – entrance aperture, 323–324
– – focusing or condensing optics, 324–325
– – wavelength selector, 324
– luminosity, 322
– resolution, 323
Spectrophotometers, 325, 390, 410, 415
G -spectroscopy, 31–32
Specular reflection, 406–407
Speed boundary, 4
"Speed" of a spectrometer, 331–332
Speed up, of initial decay, 79
Spin analyzers, 193
Spin angular momenta, of atom, 356
Spin–orbit interaction, 217
Spin orientation devices, 193
Spin polarization, of s-electrons, 67
Spin-polarized electron beams, 233, 237
Spin-polarized electron studies, 233–236
– apparatus for, 236–238
– exchange interaction, 234–235
– spin–orbit interaction, 235–236
Spontaneous emission, 424
Stabilized temperature platform furnace (STPF), 443

Stark spectroscopy, 713
Static exchange approximation, 240
Static quenching, 492–494
Steady-state emission anisotropy, 489
Steady-state spectroscopy, 772
Sternheimer antishielding factor, 70
Sternheimer shielding factor, 70
Stockholm ion storage ring, 232
Stokes direct-line fluorescence, 461
Stokes parameters, 217–219, 235–237
Stokes stepwise-line fluorescence, 461
Storage rings, 232
Strong-field ligands, 369
STU-parameter, 235
Sum-frequency generation (SFG), 819
Superconducting tunnel junctions (STJs), 107
Super Photon ring (SPring-8), 60
Superposition, 18–24
Surface barrier potential measurements, 139
Surface branching ratios, 123
Surface muon beams, 173
Surface optical second harmonic generation
– EIFSHG as a tool for probing excess surface charges, 823
– nonlinear polarization and, 820
– SHG as a tool for characterizing surface structure, 821–823
Surface trapping, 122
S-wavemodel, 256
S-wave phaseshifts, 214
Swiss Neutron Spallation Source (SINQ), 191–192
S-W plot, 137, 141
Synchronous pumping, 683
Synchrotron emission, from an electron, 372
Synchrotron Mössbauer spectroscopy, 82–83
Synchrotron radiation flux, 293
Synchrotron radiation (SR), 59–60, 294
– monochromatization, 77–78
– Mössbauer spectroscopy using, 76–77
– tunable, 303

t

Tabletop X-ray lasers, 692
Tandem mass spectrometry, 1042–1045
TAS spectrometer with polarization analysis (TASSE), 205
Taylor series, 20
Temperature dependence, of fluorescent compounds, 508–511

Thallium (Tl), 39
Thermal equilibrium $P(\infty)$, 174
Thermal ionization, 1003–1004
Thermal lensing spectroscopy, 712
Thermally-assisted fluorescence, 461
Thermally induced detrapping phenomenon, 121
Third-generation sources, 294
Three-axis spectroscopy (TAS), 198–201
Time-dependent close-coupling (TDCC), 224, 254–255, 260
Time-domain Mössbauer spectroscopy (TDMS), 78–81
– applications, 81–82
Time-of-flight (TOF) measurements, 185–186
Time-of-flight (TOF) spectroscopy, 194–198
Time-of-flight (TOF) techniques, 215
Time-resolved fluorescence, 489
Time-reversed LEED state, 290
Tint index, 400
TIRF microscopy, 637–638
Titanium sapphire lasers, 750
T-matrix, 243
T-matrix elements, 242
Toroidal crystals, 96
Total electron yield (TEY), 298
Total ionization cross section, 220–221
Total reflection XRF, 469–471
Tracers, 521–522
G-transition energy, 64
Transition-limited, 121
Transmission gratings, 101–102, 332
Transmittance measurements, 415–416
Transmitted radiant power, 428
Transport interferences, 425
Transport mean free path (TMFP), of a photon, 563
Transverse-field muon spectroscopy (TF-mSR), 170–172
Trap-based positron beams, 136
Triple differential cross sections (TDCS), 223–225, 228
Triple monochromators, 346–347
Triplet bottleneck, 631
Triplet state, 356
TRISP spectrometer, 205
Two-colour ionization, 377
Two-dimensional angular correlation of annihilation radiation (2D-ACAR), 124
Two-dimensional separation, 1093
Two-photon photoemission (2PPE), 804

u

Ultrafast lasers, 749–750
Ultrafast laser spectroscopy, 749
– experiments with
– – polarization spectroscopy, 756–759
– – transient absorption, 753–754
– – ultrafast fluorescence up-conversion, 754–756
– nonlinear optics and, 759–763
Ultrafast pulse compression, 683–684
Ultrafast spectroscopy, 771
– carrier envelope phase (CEP) and, 809–810
– in controlling the wavepacket dynamics, 805–807
– in the field of surface science, 804–805
– generation of high harmonics and attosecond pulses, 807–809
– photonic applications, 802–803
– time-resolved x-ray and electron diffraction, 810–811
Ultrafast supercontinuum, 684
Ultrafast time-resolved X-ray absorption spectroscopy, 109
Ultrashort pulse width measurements, 750–751
Ultraviolet photoelectron spectroscopy (UPS), 287
Ultraviolet spectrum, 321
Undulators, 294
US National Institute for Standards and Technology (NIST), 411
UV/visible absorption spectra, 355
UV/visible absorption spectrometer
– light sources, 371
– wavelength dispersive instruments, 372–373
UV/visible transitions
– in inorgnic molecules, 368–370
– in molecules, 364
– in solids, 370–371

v

Vacancy defects, 139–140
– nonequilibrium, 140
Vacuum fluctuations, 7
Valence electron spectroscopy
– band mapping, 309–311
– pump–probe measurements, 311
– surfaces and quantum wells, 308–309
– valence orbitals of free molecules, 307–308
– valence states, 305–306
Vapor-phase interferences, 426
Vclocity calibration, 63
Velocity modulation, using LIF technique, 712–713
Velocity selectors, 193
VEPAS techniques, 139
VEPFIT, 136
Vibrational relaxation, 725
Vibrational wave packets, 786
Vibronic transitions, 361
Viscosity, 503–508
Von Hamos-type spectrometer, 96

w

Wadsworth, 341–342
Walsh transform, 1170–1171
Wannier theory, 259
Water, dynamics of an excess proton, 795
Wave equations, 10–12
Wavelength dispersive instruments, 372–373
Wavelength-dispersive multichannel nstruments, 468
Wavelength-dispersive spectrometers (WDSs), 91
Wavelength-dispersive (WD) instruments, 467
Wavelength selector, in a spectrometer, 324
Wavelength-shifting (WLS) technique, 42
Wave packets, 774–776
– in small molecules, 786
Weak-field ligands, 369
Whiteness scales, 400–401
Wigglers, 294
Woodward–Hoffmann rules, 736

x

X-ray absorption fine structure (XAFS), 109
X-ray calorimeter, 108
X-ray fluorescence analysis, 110
X-ray fluorescence (XRF), 467
X-ray free-electron lasers (XFEL), 294
X-ray microcalorimeters, 108
X-ray photoelectron spectroscopy (XPS), 287
X-rays
– absorption of, 466
– emission of, 464–466

X-ray spectrometers, 89
– applications, 109–110
– dispersive, 91–103
– nondispersive, 103–108
– performance parameters of, 90
– resolving power, 90–91
X-ray spectroscopy (XRS), 463–464
X-ray transmission gratings, 101

y

Yellowness indices, 401
Yttrium aluminum garnet (YAG) phosphor, 372

z

Zeeman effect, 66
Zeeman energies, 174
Zeeman energy levels, 71, 173
Zeeman splitting, 69, 175
Zero field neutron spin echo (ZFNSE), 205
Zero kinetic energy spectroscopy (ZEKE), 296–297, 709
Zeroth-order absorption spectrum, 363
Z materials, 98
Z transform, 1171–1172